A TEXTBOOK FOR ENGINEERING STUDENTS

FLUID MECHANICS

INCLUDING HYDRAULIC MACHINES

Dr. A.K. Jain

Ph. D. (Roorkee University)

Formerly :

- *Dean, (Engineering Studies) Faculty of Engineering, VNS Group of Institutions, Bhopal*
- *Principal, Govt. Engineering College, Bhopal*
- *Addl. Director of Technical Education, (Retd.) Govt. of Madhya Pradesh, Bhopal*
- *Professor, HOD, Civil Engg. GEC, Raipur*

KHANNA PUBLISHERS®
Investing in Learning®

Operational Office :
4575/15, Onkar House, Opp. Happy School,
Ground Floor, Daryaganj, New Delhi 110 002
Phones : 011-45033819 • *Mob.* 09811541460
email : contactus@khannapublishers.in

***Published by*:**
Romesh Chander Khanna & Vineet Khanna
for **KHANNA PUBLISHERS**
2-B, Nath Market, Nai Sarak,
Delhi-110006 (India)

***Website* : www.khannapublishers.in**

Hologram & Description

To all readers of our books, to prevent yourself from being defrauded by piraters. Please make sure that there is an Hologram on the cover of our books with the below specifications. If you find any book without the Hologram or Description, please mail us at **contactus@khannapublishers.in**

Thanking you

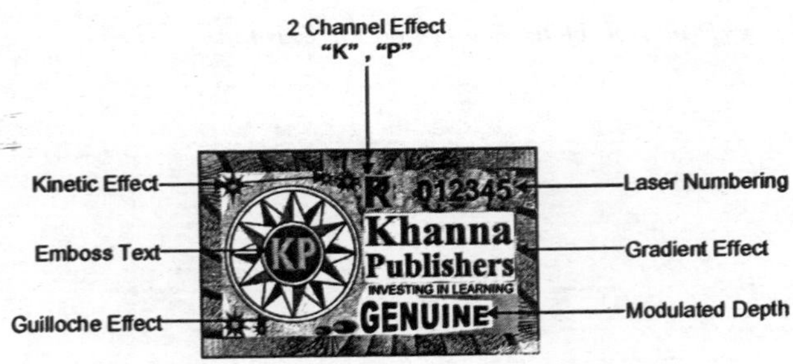

ISBN No. 978-81-7409-194-9

First Edition **: 1998**
Twelfth Edition **: 2014**
Sixth Reprint **: 2025**

Price : ` 999.00

DEDICATED TO
Wife ASHA,
and
Daughters ANITA and JUHI

ACKNOWLEDGEMENTS

The author is extremely grateful to the University of Roorkee, the Union Public Service Commission and the Institution of Engineers (India) for the permission to include typical questions from the respective question papers on Fluid Mechanics.

Thanks are also due to other reputed universities like Jawaharlal Nehru Technological University (JNTU), Hyderabad; Rajiv Gandhi Proudyogiki Vishwavidyalaya (RGPV), Bhopal for providing syllable and semester question papers. The author is grateful to all such universities that have included the book as a study material for their students.

Special thanks are also due to **M/s John Wiley and Sons, New York (U.S.A.), Publishers of "Elementary Mechanics of Fluids" by Hunter Rouse**, for permission to convert and use certain typical problems of the book.

Preface to the Ninth Edition

The purpose of writing of this book was prompted by a desire to meet the requirements of an average undergraduate student in engineering. It is the outcome of the author's association with the subject for the past four decades. While planning the course contents, the general requirements for civil, mechanical, aeronautical, chemical and electrical engineering students were kept in mind. In dealing with the various aspects of flow phenomena, efforts have been made to project a physical picture through description aided with illustrations. The main emphasis is thus on the physical aspect rather than rigorous mathematical treatment.

The author is of the opinion that the text alone is not generally adequate to provide full understanding and in order to achieve this objective, it should be supplemented by typical worked out examples exposing various facets of application of basic principles. All chapters therefore, contain many good examples to bring out clearly the application of fundamental concepts developed in the chapter. The historical developments in the Hydraulics and Fluid Mechanics are very interesting indeed and the author feels that the readers will be benefitted by a brief history included in the beginning of Chapter 1.

To apprise the readers with the type and quality of questions asked in the examinations conducted by institutions of national repute, many typical examples (both solved and unsolved from the examination question papers of the University of Roorkee, Engineering Services Examinations of the Union Public Service Commission (U.P.S.C.) and the Institution of Engineers Examination for Associate Membership have been included. Included also are some typical questions on the pattern of those asked in the examinations of Indian Institutes of Technology.

The book has been thoroughly revised. Valuable additions have been made in Chapters on Drag and Lift by adding a section on Drag on road vehicles and enlarging scope of Hydraulic Turbines by including photographic information on scroll casing, guide vanes and runners of different kinds, as also certain important hydroelectric installations of the country. Also included are sections on small hydropower plants and wind energy. A new chapter (no. 19) has been added on Hydraulic Systems and Devices, which also includes Air Lift Pump, Jet Pump, Rotary Pump, Gear Pump, Fluid Coupling and Torque Converters.

More worked out problems of typical nature drawn from reputed universities, UPSC Exams. and GATE Exams. have been added to enhance the usefulness of the book.

A special feature is the inclusion of several specially designed objective type questions given at the end of each chapter. Such questions are often asked in various competitive examinations. It is hoped that the book will now be more useful.

The book also contains outcome of some of the research with which the author was associated directly or otherwise at the University of Roorkee. It pertains to : (1) Model-Prototype Similarity law for Vortex Formation at Intakes ; (2) Accurate Explicit Equation for Friction Factor, Chapter IX ; (3) Modified Hazen-Williams Equation, Chapter IX ; (4) Explicit Equations for Pipe Flow Problems, Chapter XI ; (5) Direct Solution of Three-reservoir problem, Chapter XI ; and (6) Energy Loss due to Orificemeters, Chapter XIII ; (7) General Discharge Equation for Orificemeters, Chapter XIII.

In accomplishing this task the author has received encouragement, advice, suggestions and help from many of his teachers and colleagues. Thankful gratitude is expressed to

Dr. Bharat Singh, Professor, W.R.D.T.C., University of Roorkee (who subsequently became Vice Chancellor of the University) and Prof. V.V. Sarwate, Director of Techanical Education, Madhya Pradesh (formerly-Principal, G.C.E. and T. Raipur) for going through initial chapters and rendering valuable advice and suggestions. The author has been greatly benefitted by his association with Dr. K.G. Ranga Raju, Professor of Civil Engineering, University of Roorkee and wishes to record his gratefulness for useful discussions. Thanks are also due to Dr. R.N. Ingle, Professor and Head of the Civil Engineering Deptt. V.R.C.E., Nagpur and Prof. D.M. Kondap, College of Engineering, Amaravati (now at Poona) for their suggestions. Dr. N. Vittal, Reader in Civil Engineering, University of Roorkee and Dr. M.L. Agrawal, Professor in Mechanical Engineering G.E.C., Jabalpur, reviewed the chapters on Fluid Machines and Compressible Fluid Flow, respectively, and the author expresses his gratefulness to them. To Prof. S.V. Dandekar, Civil Engineering Deptt., G.C.E. and T. Raipur who helped in checking certain chapters of the typed script and Sarvashri S.S. Kulkarni and R.A. Mishra who prepared the initial tracings of the figures, the author wishes to record his sincere thanks.

Dr. A.K. JAIN

Preface to the Twelfth Edition

The author has thoroughly revised the book and in the process has shedded certain portions to make it leaner, and added useful items elsewhere to make it compact and more useful. Variety of typical questions drawn from reputed universities—like the RGPV—have been included to make it more versatile and useful. Aerodynamic drag on road vehicles, a brief history of their development, drag reduction measures and role of drag reduction on fuel efficiency have been added to the chapter 10 on Drag and Lift.

The objective type questions have been revisited and pruned to bring effectiveness and desired objectivity. The book in its present form aims to cater to the needs of B.E./ B.Tech. students of Civil, Mechanical, Aeronautical, Automobile, and Industrial & Production engineering branches of various universities.

Jan. 15, 2014 **Dr. A.K. JAIN**

Contents

CHAPTER 3

Kinematics of Fluid Motion **135—187**

Art.	Topic		Page

CHAPTER 10

Flow around Immersed Bodies Fluid-Dynamic Drag and Lift **466—522**

CHAPTER 11

Analysis of Pipe Flow **523—603**

Art.	Topic		Page

1

Introductory Notes and Fluid Properties

I.0. Fluid Mechanics

Fluid mechanics is a physical science concerned with the behaviour of fluid at rest and in motion. It combines the two separate approaches—the empirical hydraulics and the classical hydrodynamics developed by the hydraulicians and the mathematicians respectively. Hydraulics is mainly concerned with the motion of water. It is an applied science consisting of an enormous amount of experimental data which have been accumulated over a period of many centuries. The hydraulicians relied heavily on the field observations and laboratory tests. The data thus obtained are usually reduced to empirical formulas. Barring a few exceptions, these formulas are generally presented in a form such that they are not dimensionally homogeneous. Their applicability is limited to flow conditions similar to those for which these formulas were derived.

On the other hand, hydrodynamics is essentially a mathematical science dealing with flow analysis based on the concept of an ideal fluid—a fictitious fluid in which both fluid viscosity and fluid compressibility are assumed absent. The mathematical solutions of flow problems involving an ideal fluid thus have limited applicability to the motion of real fluids, even to those with small viscosity like water and air.

It is possible to experience applications of fluid mechanics in daily life. Some of the examples are :

(*i*) The flight of birds in the air and the motion of fish in the water are governed by the laws of fluid mechanics.

(*ii*) The cricket ball bowler depends upon circulation principle to provide the ball with desired spin and flight.

(*iii*) The dentated golf ball is designed to traverse longer distance with a minimum effort exerted by a golf player.

(*iv*) The circulation of blood in veins and arteries follows the law of fluid resistance.

(*v*) The human heart is a fine example of a pump delicately designed by nature to work continuously non-stop for many decades.

(*vi*) The designs of aeroplanes and ships are based on the theory of fluid mechanics.

(*vii*) The oil and gas pipelines, the water supply systems are designed on the principles of fluid mechanics.

We live in an environment of air and of water to such an extent that almost every thing we do is related in someway to the laws of fluid mechanics.

I.1. Historical Development of Fluid Mechanics

The following paragraphs briefly deal with certain important contributions made by various distinguished investigators since the dawn of history. The entire period of development has been divided into three parts depending upon the extent and the type of contribution made during the period. These are : Ancient and Medieval Period, Eighteenth and Nineteenth century period and Twentieth century period which marks the advent of fluid mechanics.

I.1.1. Ancient and Medieval Period Developments. The important contributions to hydraulics made upto the end of the seventeenth century have been covered here. The main contributors who influenced the course of future developments may be regarded as Archimedes, Leonardo da Vinci, Torricelli, Descartes, Mariotte, Pascal, Huygens. Isaac Newton and the Bernoulli brothers—Jakob and Johann. The major contributions of some of these are described briefly as under :

The recorded history of hydraulics[1] begins with **Archimedes** (287—212 B.C.). The most significant contribution of this Greek genius is his discovery of the principle of buoyancy and floatation. The development in the field of hydraulics was delayed until the experimental and observational approach gained a prominent place in the study of mechanics. No important contribution was made by any individual till the second half of the 15th century.

It was the Italian genius, **Leonardo da Vinci** (1452—1519), who first advocated the experimental approach to understand the flow behaviour. His basic premise was that, "*When dealing with water, we must begin with experiment and try through it to discover the reason*". As a result of his observations he was first to sketch and comment upon many hydraulic phenomena such as, profiles of free jet, formation of eddies at abrupt expansions and in wakes, velocity distribution in a vortex, hydraulic jump etc. He was also first to propose streamlining of bodies and the centrifugal pump. Credit also goes to Vinci for being the first to quantitatively state the principle of Continuity.

Evangelista Torricelli (1608—1647) was also an Italian who generalized the analysis of trajectories of projectiles. His major contribution is the principle of efflux, which is now commonly written as $V = \sqrt{2gh}$. The discovery of barometer is also attributed to him.

Isaac Newton (1642—1727) was the first Englishman to have earned an eminent place amongst the scientists of his time. His approach was that of a mathematician while his predecessors investigated nature from philosophical point of view. In the "Principia" (1687) he enunciated concisely the three basic laws of motion now named after him. It is these basic laws which form the basis of analysis of problems of mechanics. He carried out extensive research on fluid resistance, and was first to report on the fact that in viscous flow the shear is proportional to the relative velocity of the adjacent zone. The statement of this fact is now known as the law of viscosity and bears his name. Newton was also first to introduce the use of coefficient of contraction in problems of efflux.

The direct and indirect contributions of two mathematicians of Basel, Switzerland, who happened to be brothers were also significant to the advancement of physical science. The eldest, **Jakob Bernoulli** (1654—1705), was a professor of physics at the University of Basel, who trained his younger brother **Johann Bernoulli** (1667—1748). Johann then worked with a French mathematician, taught mathematics in Holland and succeeded his brother as professor.

The four essential steps emerged as a result of evolution of mathematical physics in the 17th century are :

(*i*) the use of plotted curve to describe a phenomenon ;

(*ii*) the expression of the curve in equation form ;

(*iii*) the determination of its area and slope ; and

(*iv*) the application of these procedures to practical problems.

It is not known as to who accomplished the first step, but it was definitely Descartes who accomplished the second step, Leibnitz (1846—1916, Germany) the third and the Bernoullis the fourth.

Of the contemporary mathematicians, to whom the credit for laying the foundation of hydrodynamics goes, two belonged to the Bernoulli School at Basel. The first one was Daniel Bernoulli (son of Johann) ; the second one was his close friend Leonardo Euler and the third one was d' Alembert of France.

---

1. History of hydraulics, by Hunter Rouse and Simon Ince, Dover publications, Inc., New York, 1963.

I.1.2. 18th and 19th Century Developments. Eighteenth and nineteenth centuries saw a tremendous advancement in the field of hydraulics and hydrodynamics. It was during this period that most of the experimental hydraulics was developed at the hands of the French, the Italian and the German engineers and hydraulicians. Noteworthy amongst them were Poleni, Pitot, Chezy, Borda, Venturi, Weber, Fourneyron, Belanger, Russel, Reech, Hagen, Poiseuille, Weisbach, Darcy, Bazin, Kutter, Manning and Froude. It was in the early 18th century when Daniel Bernoulli along with Leonhard Euler and d'Alembert founded the mathematical science of hydrodynamics. Others who significantly contributed to the development of hydrodynamics are Lagrange, Laplace, Navier, Stokes, Helmholtz, Kirchhoff, Boussinesq, Reynolds, Thomson (Lord Kelvin), Strutt (Lord Rayleigh) and Joukowsky. The important contributions of some of these distinguished investigators have been described in the following paragraphs.

Daniel Bernoulli (1700—1782) was a Swiss mathematician who published a treatise named "Hydrodynamica" dealing with various aspects of hydrostatics and hydraulics. He introduced for the first time the word 'Hydrodynamics' to encompass various topics of fluid statics and dynamics. He was first to use piezometer openings in the walls of conduits for pressure indication. His energy principle utilized only two terms, namely, the pressure and the velocity. Bernoulli's principle used for evaluating pressure seemed to indicate the constancy of pressure and velocity heads. He showed both experimentally and analytically that the pressure would become negative if the velocity increased sufficiently.

Leonhard Euler (1707—1783) was also a Swiss mathematician who worked mainly on hydrodynamics and hydraulic machinery. He was the first to explain the role of fluid pressure in fluid flow. Euler investigated the motion of a fluid under the action of an external force, rightly regarding the isotropic pressure as a function of space only ; formulated basic equations of motion (now known as Euler's equations of acceleration or motion). Assuming the fluid to be incompressible, flow to be steady and irrotational and utilizing the concept of a force potential he combined his three equations of acceleration to yield a single relationship involving pressure, velocity and elevation heads. This equation of Euler in its present familiar form is attributed to Bernoulli. Euler also contributed significantly to the hydrodynamics theory of centrifugal machinery. Analysing the performance of reaction turbines, he expressed for the first time the basic relationship by equating the torque to the change in the moment of momentum as the fluid is passed through the rotating part.

Jean le Rond d' Alembert (1717—1783), was a French mathematician first to introduce concepts of components of fluid velocity and acceleration and also the differential expression of continuity. Assuming similar condition in the rear of a body as in the front, the summation of the elementary pressures exerted on each part of the body surface led to the paradoxical result of a zero longitudinal force on the body. d'Alembert, however, left it on of future investigators to explain this anomaly between the theory and the reality. This is the paradox of zero resistance to steady non-uniform motion, known as *d' Alembert's paradox*.

Joseph Louis Lagrange (1736—1813) was a self-trained French mathematician who succeeded Euler as the world's leading mathematician. In the course of analysis of fluid motion, he introduced concepts of velocity potential and stream function which became of fundamental importance in describing the flow pattern. Lagrange was first to derive an equation for velocity of propagation of a wave of infinitesimal height in channel of finite depth ($V = \sqrt{gy}$).

Merquis Giovanni Poleni (1683—1761) is the only Italian who deserves a special mention for his contribution to the experimental hydraulics of the early 18th century. His major contributions were three-fold : (1) Based on measurements he obtained the coefficient of contraction for a sharp-edged orifice to be 0.62, a significant improvement over the value of $1/\sqrt{2}$ proposed by Newton. (2) He also experimented with short tubes (mouth-pieces) attached to the orifice and found that a maximum rate of discharge was obtained with an intermediate length of tube. (3) His last contribution was his treatment of the discharge through rectangular sharp- crested weirs. He considered the discharge as occurring in a series of horizontal elements, the velocity of each being assumed proportional to the square root of its depth below the liquid surface. The same approach was later used for deriving the discharge relationship for sharp-crested weirs, and hence the basic weir equation is often named after Poleni.

Henri de Pitot (1695—1771) was a French engineer whose principal contribution is the invention of a device (known after him as the Pitot-tube) for measuring fluid velocity. The original Pitot tube consisted of two parallel tubes mounted on a slender frame containing a scale and four petcocks ; one of the tubes being straight and the other bent through 90° at its lower end.

Antoine Chezy (1718—1798) was also a French engineer who contributed significantly to the understanding of the resistance in uniform open-channel flow. Credit goes to Chezy for not only presenting the first but also the most lasting resistance formula—later to be known by his name.

Jean Charles Borda (1733—1799) was a French military engineer who devoted himself to experiments in hydraulics and hydraulic machinery. His resistance studies verified the prevalent theory that the drag of an immersed body varied with the square of the relative velocity and showed that it would depend upon a still higher power if surface wave were produced. Borda was first to introduce the concept of elementary stream-tubes. He showed that not only the contraction of jet but also loss of energy must be taken into account in obtaining an expression for discharge. Making use of the momentum principle, he found that for the particular case of a re-entrant tube (*i.e.* the Borda mouth-piece) the coefficient of contraction has a value of 0.5.

Venturi's published work in Paris in 1797 reported his findings on various forms of mouthpieces fitted to the orifice. He demonstrated the effect of eddies formed at abrupt changes in section and, incidentally the change in discharge which would result from their elimination. Venturi observed that the replacement of the cylindrical tube with two conical sections essentially eliminated the eddies and increased the rate of flow, but it still produced the local reduction in pressure. This form of boundary is now used to measure flowrate in pipes and is rightly known as the venturimeter.

Giorgio Bidone (1781—1839) was an Italian hydraulician who is credited with having discovered the hydraulic jump. He was the first to study it systematically and to attempt its analysis.

Giuseppe Venturoli (1768—1816) also an Italian hydraulician, derived the elementary back water equation for rectangular channels. Through graphical integration, Venturoli succeeded in plotting various branches of the surface profile.

Claude Burdin (1790—1893), a French engineer, coined the word "turbine" and developed one with free efflux of water.

Benoit Fourneyron (1802—1867) improved upon Burdin's original device and developed a successful hydraulic turbine. More than hundred similar turbines were built by him for various parts of the world.

John Scott Russel (1808—1882), a Scottish engineer, was the first to study the problems of unsteady, non-uniform open-channel flow without discontinuity. He studied the effects of waves on the resistance of ships and proposed a reverse-curve form of bow which he believed would reduce the wave effect to a minimum.

A French contemporary of Russel, **Ferdinand Reech** (1805—1880) advocated the practicability of model tests and developed similitude principles based on Newton's laws of motion. Reech was first to express what is now known as the Froude criterion of similitude.

Gotthilf Heinrich Ludwig Hagen (1707—1884), a German hydraulic engineer made original contributions to the resistance of pipe flow. Based on extensive and accurate experiments on flow through small diameter tubes, he reasoned that the flow took place in series of cylindrical layers, the velocity of which varied linearly (an assumption which later proved wrong) from zero at the wall to a maximum at the centre. Hagen proposed an expression for the resistance to flow in small diameter tubes (*i.e.* laminar flow) based on the above assumption. He also carried out experiments on resistance of pipes in turbulent flow and correlated his measurements by means of a resistance equation.

Jean Louis Poiseuille (1799—1869), a French physician and not an engineer, was interested in experimental physiology. He conducted research on pumping power of the heart, the movement of blood in the veins and capillary vessels and the resistance to flow through tubes. He carried out accurate experiments on very small diameter tubes and presented an empirical relationship for the discharge in terms of the head loss, the tube diameter and the tube length. The resistance law for laminar flow was later named after Poiseuille rather than Hagen, and it still continues to be known by his name.

Julius Weisbach (1806—1871), a German hydraulician, wrote a treatise on hydraulics for engineering application which even now can be considered a textbook on hydraulics. Weisbach not only incorporated the best available experimental information, but in many cases supplemented it with the results of his own experiments. He advocated the use of the non-dimensional coefficient, and was first to express the resistance equation for pipes in the form $h = fLV^2/2gD$. Weisbach also modified the weir equation to include the velocity of approach and to eliminate the successive approximations involved in the determination of discharge he proposed instead and empirical equation of the type later adopted by Bazin.

Antoine Charles Bresse (1822—1883), a French engineer, accomplished integration of the equation of gradually varied open-channel flow and prepared tables of function, now known as Bresse's back water function, for general use. He also presented a correct formulation of the momentum characteristics of the hydraulic jump.

Henry Philibert Gaspard Darcy (1803—1858), also a French engineer, conducted studies on the flow of water in both pipes and permeable soils. His experiments included pipes of different sizes, materials and in various states of deterioration. His greatest contribution was his conclusive demonstration of the fact that the resistance depended upon the type and the condition of the boundary material. As a result of his studies on pipe flow, Darcy's name is commonly associated with that of Weisbach in designating the present-day resistance equation first formulated by Weisbach. On the basis of his filteration studies, Darcy concluded that the loss of head through a filter bed was proportional to the rate of flow rather than to its square root as was then generally believed.

Henri Emile Bazin (1829—1917), a French engineer and an associate of Darcy, conducted extensive experiments on open-channel resistance, propagation of waves and flow over weirs. Based upon his experiments in canals of various materials and shapes, Bazin proposed a formula of resistance in open-channel flow. He also carried out experiments for measurements of velocity distribution at various cross-sections with different linings. From these tests, he noted that the depth of point of maximum velocity varied with the relative width of cross-section, approaching zero as the width-depth ratio exceeded 5. Bazin's subsequent studies on discharge over vertical and inclined weirs included the precise determination of nappe profile and the distribution of velocity and pressure through the nappe. As a result of these studies he introduced a new dimensional term in the weir-discharge equation proposed by Weisbach.

Two Swiss engineers **Emile Oscar Ganguillet** (1818—1894) and **Wilhelm Rudolf Kutter** (1818—1888) are known for their contribution to the open-channel resistance. On the basis of several hundred experiments, they expressed the Chezy's coefficient C as a function of a roughness factor n, hydraulic radius R and the channel slope S.

Robert Manning (1816—1897), an Irish engineer, proposed in 1889 a relationship for open-channel flow of the form, $V = K R^{2/3} S^{1/2}$, which was in better agreement with the available data than the earlier ones. The present-day formula named after Manning was neither recommended nor ever devised in full by Manning himself. He also did not suggest use of Kutter's n—a coefficient now associated with the Manning's formula.

William Froude (1810—1879), an English engineer, developed and perfected the towing-tank techniques for testing of model ships. Froude believed that *"Experiments duly conducted on small scale model will give results truly indicating of the performance of the full size ships"*. He considered the total resistance to be made up of the skin friction resistance and the resistance due to other factors such as waves. It is interesting to note that Froude's name has been inseparably associated with a law of similarity and a non dimensional number, which were neither originated nor even used by him.

Lousi Marie Henry Navier (1785—1836), a French engineer, derived through a purely mathematical analysis equations of motion for a viscous flow. His name is, therefore, most frequently associated with the present day equations for viscous flow. He however, did not identify the fluid viscosity as a variable affecting the flow but instead considered molecular spacing.

George Gabriel Stokes (1819—1903), a British mathematician, was the one whose name becomes finally associated with that of Navier in designating the equations of motion for a viscous fluid. His paper "On the Theories of Internal Friction of Fluids in motion" published in 1845 contained the derivations of what are now known as the Navier-Stokes equations. The general coefficient ε appearing in Navier's equations was replaced by the dynamic viscosity μ. Stokes also derived an expression for the terminal velocity of fall of spheres, which is now known as the Stokes' law.

Osborne Reynolds (1842—1912), a British engineer, was : (1) the first to demonstrate the phenomenon of cavitation and attribute the accompanying noise to the collapse of vapour bubbles, just as in a kettle beginning to boil ; (2) the first to correlate the length and time scales in the study of distorted models ; and (3) the first to introduce the viscosity into a parameter now bearing his name, demarcating the limit between laminar and turbulent flows. Reynolds demonstrated experimentally in 1888 that the velocity at which eddy motion (*i.e.* turbulent flow) began did indeed vary with the tube diameter and fluid characteristics in such a manner as to yield a fairly definite value of this parameter (now known as the Reynolds number). The most important contribution of Reynolds was his application and extension of Navier-Stokes equations to turbulent flow. His lasting contribution, however, was the derivation of the equations for motion—now known as the Reynolds equations for turbulent flow.

I.1.3. 20th Century Developments—Advent of Fluid Mechanics. Until the early 20th century the two distinctly divergent approaches namely, the experimental hydraulics and the theoretical hydrodynamics had developed to such an extent that the apparent gap between theory and the practical reality was bothering the genius amongst the hydraulicians and the hydrodynamists. The emergence of aeronautics at a rapid pace also hastened the activities of engineering talent towards bridging the gap between theory and the fact. This was brought about by a new concept of boundary layer originated by Prandtl of Germany.

Ludwig Prandtl (1875—1953) is regarded as the founder of the present-day fluid mechanics. He realized the need for better correlation between theory and experiment in problems of fluid flow. Prandtl conducted his first experiments on the flow of air while he was employed as a engineer in a large machinery firm. Later he joined as a faculty member of Polytechnic Institute, Hannover and continued his research on flow of air. Within three years he presented a paper containing his findings before the Third International Congress of Mathematicians in 1904. In the eight-page paper he introduced the concept of boundary layer according to which the motion of fluids of low internal resistance (*i.e.* low viscosity) can be divided into two interdependent zones : (*i*) very close to the fixed boundaries, there is a transition layer in which the fluid velocity changes from zero at the boundary to practically the same value as in the free-stream at the edge of this layer. It is in this layer that effects of viscosity are predominant, and (*ii*) away from the boundary lies the zone of free-stream across which there is hardly any change in velocity and within which viscous effects are negligible. The flow in this part can be analysed on the basis of potential flow theory. The layer in the vicinity of the fixed boundary and to which the viscous effects are confined was given the name of '*boundary layer*'.

The emergence of the theory of boundary layer has come to play a vital role not only in aeronautics but also in hydraulics and other related fields. This has enabled a much rapid development of the science of motion of fluids known as the fluid mechanics.

As a result of his pioneering research on flow of low-viscosity fluids, Prandlt was invited as Professor and Director of Research Institute at the University of Göttingen. At Göttingen, he and his students further worked on the boundary layer theory, analyzed the phenomena of turbulence and drag and evolved dynamic principles of aerofoil behaviour. The first of his students to achieve recognition was **Paul Richard Heinrich Blasius** (1833—) of Berlin. He published in 1908 an analytical solution for velocity distribution and resistance of laminar boundary layer wherein Pandtl's qualitative theory was quantitatively verified by laboratory experiments.

Theoder von Karman (1881—1963) was perhaps the most illustrious alumnus of the Göttingen Institute where he worked under Prandtl. Karman was gifted with a combination of rare physical insight and mathematical ability. His primary contributions to fluid mechanics have been in the fields of form drag turbulence and surface resistance and the analogy between the sound and gravity waves. He investigated the problem of eddy formation behind circular cylinders and provided an analytical solution of what has since been known as the Karman vortex trail. Karman and Prandtl contributed to the analysis of velocity distribution and resistance to turbulent flow in pipes as well as long flat surfaces. The resulting logarithmic equations for resistance and velocity distributions are now known by their joint names.

Other notable investigators associated with the Göttingen Institute were : Walter Ludwig Christian **Schiller** (1882—) whose primary interest was in the field of pipe resistance. Walter Gustav Johannes **Tollmein** (1900—) distinguished himself by his analysis of flow stability and turbulent diffusion ; Hermann **Schlichting** (1907—) contributed greatly to the analysis of stability and boundary layer development. Carl **Wieselberger** (1887—1941) contributed significantly on the phenomena of drag ; Otty **Flachspart** (1898—1957) contributed on drag and particularly on wind pressures on buildings and Johann **Nikuradse** (1894—) contributed in the field of pipe resistance and is known for his famous experiments on artificially roughened pipes.

Credit goes to **Mortiz Weber** (1871—1951) of Berlin to put the general principles of similitude in their present form. It was he who specifically named the Frode and Reynolds numbers associated Cauchy's name with elastic similarity and introduced a capillarity parameter in his paper presented in 1919. The capillarity parameter **W** is subsequently named as the Weber number.

Geoffrey Ingram Taylor (1886—), a British physicist while employed as a meteorologist at the University of Cambridge, studied, eddy motion in the atmosphere. He published a series of papers dealing with the fundamental analysis of fluid turbulence by methods of statistics. Taylor presented his theory of diffusion by continuous movements and related the diffusive and dissipative characteristics of turbulent motion.

I.2. Systems of Measuring Physical Quantities

Any system of measurement is based on well defined units for certain basic quantities such as mass, length, time, temperature etc. The units for other quantities are derived from basis units by virtue of the relationship that exists among the quantities concerned. There is, however, an element of flexibility in making a choice for basic quantities. In the absolute system of units mass, length and time are considered the basic quantities and the corresponding units have been named as the basic or base units. The gravitational system of units, which has so far been used in engineering practice, considers force, length and time as basic quantities. While using the gravitational system of units, it is to be noted that the unit of force is not a universal constant, but varies from place to place owing to its dependence on the local acceleration due to gravity.

The metric (MKS) gravitational system of units was in engineering use in India for over past five decades. With growing acceptance of International System (SI) of units, the MKS units have been replaced by the SI units. In this book, SI units have been used alongwith the MKS to familiarize the reader with the new units.

I.2.1. Absolute (Coherent) and Gravitational Systems of Units. The basic or base units (mass or force, length and time) are inter-related to each other by Newton's second law of motion. This law states that a mass moving by virtue of an applied force will be accelerated and that the component of force in the direction of the acceleration is proportional to the product of the mass and the acceleration. Thus

$$F \propto m.a$$

$$= \frac{1}{C_n} m.a \qquad \text{...(1)}$$

where F = force in the direction of acceleration,

m = mass

a = acceleration, and

C_n = constant of proportionality known as Newton's constant.

If we select the units such that one unit of force acting on one unit of mass produces one unit of acceleration, the proportionality constant C_n will be unity. For this case we can write Newton's second law in the conventional form, $F = m.a$, expressing that the force equals mass multiplied by acceleration. The proportionality constant C_n will be unity only when the absolute system of units is used, that is when the system of measuring units is *coherent*.

A coherent system of units may be defined as the one in which the product or quotient of any two unit quantities involved in the phenomenon is the unit of the resultant quantity. As examples of this statement, in any coherent system the quotient of unit length and unit time gives unit velocity and similarly the product of unit mass and unit acceleration gives unit force. Any absolute system of units is essentially a coherent system.

I.2.2. SI Units[2]. In the International System (SI) of units the base quantities are mass, length, time and the thermodynamic temperature and the corresponding base units are kilogram (kg), metre (m), second (s) and Kelvin (K) respectively. From these base units the derived unit of force is newton (N) which is the force required to accelerate 1 kg mass at the rate of 1 m/s², and is obtained from Eq. (1).

$$1\text{ N} = \frac{1}{C_n} \times 1\text{ kg (mass)} \times 1\text{ m/s}^2$$

from which the Newton's constant is

$$C_n = \frac{1\text{ kg (mass) m}}{\text{N.s}^2} \qquad \text{...(2)}$$

I.2.3. MKS Gravitational Units. This system has the base quantities as force, length, time and thermodynamic temperature and the corresponding base units are kilogram (kg), metre (m), second (s) and Celsius (C). In the gravitational system, unit force is proportional to the product of unit mass and acceleration due to gravity. Unit force of one kilogram is defined as the force required to accelerate one kg mass at the rate of 9.81 m/s². Thus from Eq. (1), we obtain

$$1\text{ kg (force)} = \frac{1}{C_n} \times 1\text{ kg (mass)} \times 9.81\text{ m/s}^2.$$

$$C_n = 9.81 \frac{\text{kg (mass) . m}}{\text{kg (force) . s}^2} \qquad \text{...(3)}$$

Comparing Eqs. (2) and (3), 1 kg (force) = 9.81 newton (N) ...(4)

The derived unit of mass is called metric slug in analogy to slug—the British gravitational unit of mass. It is defined as the mass which will be accelerated at the rate of 1 m/s² when acted upon by a force of 1 kg. Equation (1) may be used to determine the Newton's constant.

---

2. "The International System of Units" Editors Chester H. Page and Paul Vigoureux, London, Her Majesty's Stationery Office 1973.

$$1 \text{ kg (force)} = \frac{1}{C_n} \times 1 \text{ metric slug} \times 1 \text{ m/s}^2.$$

$$C_n = \frac{\text{metric slug m}}{\text{kg (force) s}^2} \qquad ...(5)$$

Comparing Eqs. (3) and (5),

$$1 \text{ metric slug} = 9.81 \text{ kg (mass)} \qquad ...(6)$$

Table 1 shows SI and MKS (gravitational) units of various quantities used in fluid mechanics.

Table 1. SI and MKS (gravitational) units

Quantity	*System of units*			
	SI		*MKS (gravitational)*	
	Unit	*Symbol*	*Unit*	*Symbol*
Mass	kilogram	kg	metric-slug	m slug
Length	metre	m	metre	m
Time	second	s	second	s
Thermodynamic temperature	Kelvin	K	Celcius	C
Area	square metre	m^2	square metre	m^2
Volume	cubic metre	m^3	cubic metre	m^3
Velocity	metre per second	m/s	metre per second	m/s
Acceleration	metre per second square	m/s^2	metre per second squared	m/s^2
Force	newton	N	kilogram	kg
Moment of force	metre newton	Nm	metre kilogram	kgm
Pressure	newton per square metre (pascal)	N/m^2 (Pa)	kilogram per square metre	kg/m^2
Density	kilogram per cubic metre	kg/m^3	metric slug per cubic metre	m $slug/m^3$
Specific weight	newton per cubic metre	N/m^3	kilogram per cubic metre	kg/m^3
Dynamic viscosity	kilogram per metre second	kg/ms Pas	kilogram second per square metre	kgs/m^2
Surface tension	newton per metre	N/m	kilogram per metre	kg/m
Work, energy	joule	J = Nm	kilogram metre	kgm
Power	watt	W = J/s = Nm/s	kilogram per metre second	kg/ms

The conversion factors from MKS to SI, MKS to FPS and MKS to CGS units appear in Appendix E.

I.3. System

In the most general terms, a system may be defined as that region of space occupied by the quantity of fluid under consideration. The fluid contained within the system is separated from the surroundings by a boundary. The system may contain either a constant or a variable mass. Its boundaries may be fixed or deformable. It may be in motion or at rest with respect to a chosen co-ordinate system. The region outside the system's boundary is known as the surroundings. A system diagram shows the system's boundary and significant interactions between the system and its surroundings. If the surroundings of importance are forces then the system diagram reduces to the familiar '*free-body diagram*'.

I.3.1. Closed and Open Systems. A system is said to be a closed system if the same body of fluid remains within the system during a process. The process is then called a non-flow process. In such a process, work and heat may be transferred across the boundary but no fluid crosses the boundary. In Fig. 1 (*a*) dotted lines indicate the systems boundary and only the mass transfer has been considered.

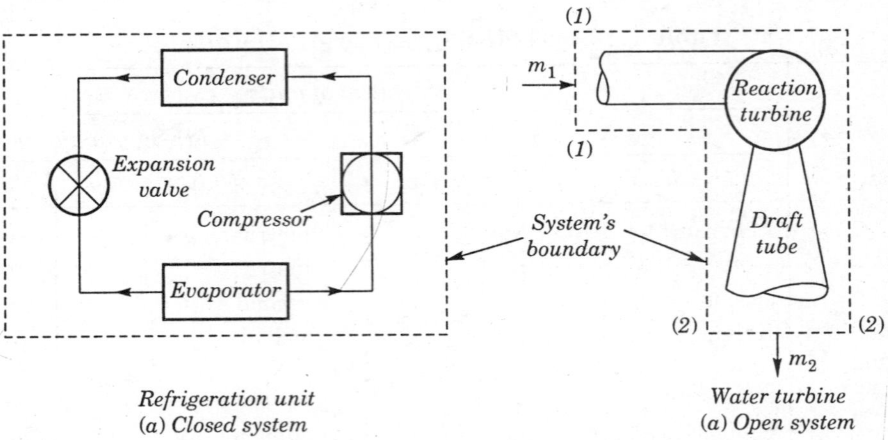

Fig. 1. Closed and open systems.

An open system is defined as one in which the fluid enters and/or leaves the system, and the process referred to as a flow process. In an open system fluid mass, momentum, energy and machinery all cross the boundary. Fig. 1 (*b*) shows such a system consisting of a water turbine. The mass rate of flow m_1 at (1)—(1) is the same as mass rate of flow m_2 at (2)—(2).

I.4. Control Volume

A control volume is an open-system which has its boundary fixed with respect to a fixed coordinate system. The control volume is thus, an arbitrary volume fixed in space and bounded by a closed surface which is known as the control surface. The fluid may enter and leave control volume by crossing the bounding surface (control surface) enabling transfer of mass, momentum and energy.

I.5. Free-body Diagram

As already pointed out, a free body diagram is a closed system in which the interactions between the system and its surroundings are the forces. The concept of a free-body diagram helps in cultivating a rational approach in respect of listing various forces which act on the system.

Consider the curved surface *AB* shown in Fig. 2 which closes an opening in the tank. The curved surface *AB* thus supports a liquid column contained in *ABCD*. The problem is pertaining to the statics of fluids, and the free-body of liquid *ABCD* must be in static equilibrium. The conditions of static equilibrium dictate that the algebraic sum of the force components in mutually perpendicular directions must be zero, and so also the algebraic sum of the moments of forces in the respective planes be zero, that is the forces must be co-planar.

The main problem is, therefore, of evaluating the forces acting on the free-body (magnitude, direction and their location). By the application of the above-stated principles of static equilibrium the unknown forces may be obtained. In Fig. 2, the forces acting on the free-body *ABCD* are :

(*i*) Weight *W*, of the liquid mass contained in *ABCD*.

(*ii*) Hydrostatic pressure force F_{H_2} exerted on the liquid by the tank wall.

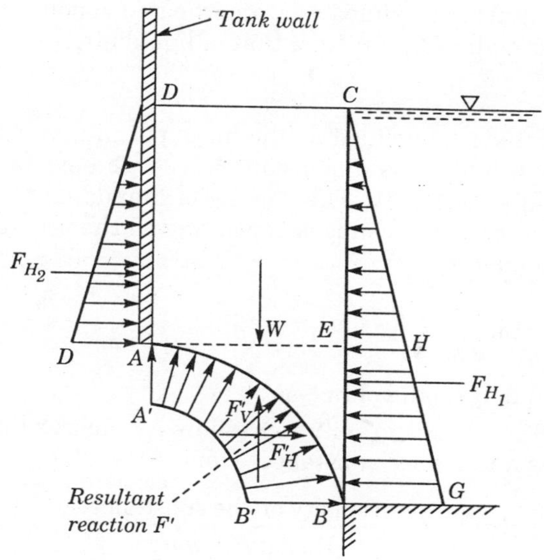

Fig. 2. Free-body diagram for *ABCD*.

(*iii*) Hydrostatic pressure F_{H_1} exerted on the free-body liquid by the surrounding liquid, and

(*iv*) The resultant reaction force F' (having components F_H' and F_V') exerted by the curved surface *AB* on the liquid.

The evaluation of hydrostatic forces is dealt with in Chapter II (refer articles 2.5 and 2.6).

PROPERTIES OF FLUID

1.0. Introduction

Matter can be distinguished by the physical form of its existence. These forms known as phases, are solid, liquid and gas.

Solid, Liquid and Gas

The liquid and gaseous phases are usually combined and given a common name of fluid, because of the common characteristics exhibited by liquids and gases. Solids differ from liquids and liquids from gases on account of their molecular structure (spacing of molecules and the ease with which they can move). The spacing of molecules is large in a gas, smaller in a liquid and extremely small in a solid. Very strong intermolecular attractive forces exist in solids which give them the property of rigidity. These forces are weaker in liquids and extremely small in gases.

Definition of a Fluid

The word fluid means a substance having particles which readily change their relative positions. A fluid may be defined as substance which deforms continuously under the action of shear stress, regardless of its magnitude.

Distinction between a Liquid and a Gas

A fluid may be either a liquid or a gas. The molecules of a liquid are very closely spaced as compared to those of a gas. While a liquid has a free surface and occupies a certain volume in a container, a gas does not possess a free surface and fills the entire space of the container regardless of its size. For all practical purposes, a liquid is incompressible while a gas is compressible and expands unless contained or enclosed in a container. In the words of Sir Oliver Lodge "*A solid has volume and shape, a liquid has volume but no shape, a gas has neither.*" A vapour is a gas whose temperature and pressure are such that it is very near to the liquid phase, and hence the steam is considered as a vapour. A gas may be defined as a superheated vapour. Air is regarded as a gas on account of its state being normally very far from that of liquid air.

1.2. Density

The density of a substance is defined as the mass per unit volume and is denoted by the symbol ρ (Greek letter rho). It has the dimension $[ML^{-3}]$. The expansion or contraction of the substance results in a change in density. The density of liquids may be considered as constant while that of gases will be subjected to changes depending on the pressure and temperature. The fluid density at a point is defined by

$$\rho = \lim_{\Delta V \to 0} \frac{\Delta m}{\Delta V}$$

where Δm = mass contained in a small volume ΔV.

Specific Gravity. The specific gravity represents a numerical ratio of two densities, and water is commonly taken as a reference substance. Thus

$$\text{Specific gravity of a substance} = \frac{\text{Density of the substance}}{\text{Density of water}}.$$

It is also called the relative density.

1.3. Specific Weight

It is the weight of a given substance per unit volume, and is commonly denoted by symbol γ (Greek letter gamma). The specific weight represents the force exerted by gravity on a unit volume of fluid and, therefore, must have the units of force per unit volume. It is related to the density by the following expression

$$\gamma = \rho g.$$

The specific weight of fresh water under normal conditions is 1000 kg/m^3 in MKS units and 9.81×10^3 N/m^3 in SI units.

Specific Volume. It is the volume occupied by a unit mass of fluid. It is commonly applied to gases. The specific volume is reciprocal of the density, *i.e.* $v = 1/\rho$.

Example 1.1. *Calculate the mass density, specific weight and weight of 1 litre of petrol, if its specific gravity is 0.72.*

Solution. Volume of petrol

$$= 1 \text{ litre} = 1000 \text{ cu. cm.}$$

$$= 1 \times 10^{-3} \text{ m}^3$$

$$\text{Specific gravity} = \frac{\text{Mass density of petrol}}{\text{Mass density of water}}$$

$$0.72 = \frac{\text{Mass density of petrol}}{1000 \text{ (kg/m}^3)}$$

$\therefore$ Mass density of petrol, ρ

$$\rho = 0.72 \times 1000$$

$$= 720 \text{ kg/m}^3$$

Specific weight of petrol, γ

$$= \text{weight of 1 m}^3 \text{ of petrol}$$

$$= \rho g$$

$$= 720 \times 9.81$$

$$= \mathbf{7063.2 \text{ N/m}^3}$$

Weight of 1 litre of petrol = γV

$= 7063.2\ (\text{N/m}^3) \times 10^{-3}\ (\text{m}^3)$

$= \mathbf{7.0632\ N.}$

Example 1.2. *Ten litres of a liquid of specific gravity 1.3 is mixed with 6 litres of a liquid of specific gravity 0.8. If the bulk of the liquid shrinks by 1.5% on mixing, calculate the specific gravity, density, volume and weight of the mixture.*

Solution. Weight of 10 litres of liquid of specific gravity 1.3

$= 10 \times 10^{-3} \times 9810 \times 1.3\ \text{N} = 127.53\ \text{N}$

Weight of 6 litres of liquid of sp. gr. 0.8

$= 6 \times 10^{-3} \times 9810 \times 0.8 = 47.1\ \text{N}$

Total volume of liquids before mixing

$= 10 + 6 = 16$ litres

Upon mixing the bulk shrinks by 1.5%

$\therefore$ New total volume $= 0.985 \times 16 = 15.76$ litres

Weight of equal volume of water $= 15.76 \times 10^{-3} \times 9810 = 154.6$ N

Weight of mixture $= 127.53 + 47.1 = 174.63$ N

(*i*) Specific gravity of mixture $= \dfrac{\text{Wt. of mixture}}{\text{Wt. of equal vol. of water}} = \dfrac{174.63}{154.6} = \mathbf{1.128.}$

(*ii*) Density of mixture $= \dfrac{\text{Mass}}{\text{Volume}} = \dfrac{174.63/9.81}{15.76 \times 10^{-3}} = \mathbf{1128\ kg/m^3.}$

(*iii*) Volume of mixture **= 15.76 litre.**

(*iv*) Weight of mixture **= 174.63 N.**

Example 1.3. *If 5 m³ of a certain oil weighs 40 kN, calculate the specific weight, mass density, specific volume and relative density of the oil.*

Solution. Specific weight $\gamma = \dfrac{\text{Weight}}{\text{Volume}} = \dfrac{40\ (\text{kN})}{5\ (\text{m}^3)} = 8\ \text{kN/m}^3$

Mass density $\rho = \dfrac{\text{Mass}}{\text{Volume}} = \dfrac{W/g}{V} = \dfrac{\gamma}{g} = \dfrac{8 \times 10^3\ (\text{N/m}^3)}{9.81\ (\text{m/s}^2)} = 815.49\ \text{kg/m}^3$

Specific volume, $v = \dfrac{1}{\rho} = \dfrac{1}{815.49} = 1.23 \times 10^{-3}\ \text{m}^3/\text{kg}$

Relative density $= \dfrac{\text{density of oil}}{\text{density of water}} = \dfrac{815.49}{1000.00} = \mathbf{0.81549.}$

1.4. Viscosity

Among all the fluid properties, viscosity is the most important and is recognised as the only single property which influences the fluid motion to a great extent. The viscosity is the property by virtue of which a fluid offers resistance to deformation under the influence of a shear force. For a given fluid, the rate of deformation is dependent upon the magnitude of shear force. The molecular friction or shear resistance within the fluid opposes such continuous deformation.

Let us consider a fluid contained in between two parallel plates as shown in Fig. 1.2, the bottom one being kept stationary while the top one moves at a constant speed *U* under influence of the applied shearing force *F*.

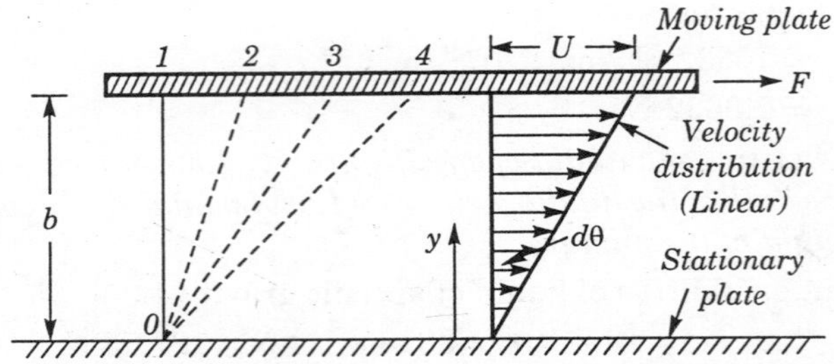

Fig. 1.2. Fluid motion between parallel plates, bottom plate stationary, top one moving.

Initially when the top plate is about to start moving all the fluid particles are at rest. The position of fluid particles lying along the vertical line 0—1 changes with time once the motion starts. The fluid particles sticking to the moving plate move with the same velocity U while those adhering to the bottom stationary plate are at rest. The velocity of the intermediate particles vary from 0 to U. If the gap separating the two plates is small, the velocity distribution will be linear (straight line) as shown, and the fluid particles originally lying on line 0—1 after a certain time, say $t = \Delta t$, will occupy the positions indicated by the line 0—2, and at times $t = 2\Delta t$ and $3\Delta t$ they will lie along the lines 0—3 and 0—4 respectively. The maximum deformation of fluid takes place at $y = b$, the magnitude of which in unit time is U, and the zero deformation occurs at $y = 0$, the bottom plate being stationary. The time rate of deformation is, therefore, equal to U/b. The rate of angular deformation $d\theta/dt$ is given by

$$b\,\frac{d\theta}{dt} = U \qquad \text{or} \qquad y\,\frac{d\theta}{dt} = u$$

$$\therefore \qquad \frac{d\theta}{dt} = \frac{U}{b} = \frac{u}{y}$$

where u is the velocity at a distance y from the stationary plate.

If A is the area of the moving surface and F is the force required to move the surface at a constant velocity, it has been established that the shear stress F/A is directly proportional to the time rate of deformation, thus,

$$\frac{F}{A} \propto \frac{d\theta}{dt}$$

or

$$\frac{F}{A} = \mu\,\frac{d\theta}{dt} = \mu\,\frac{U}{b} = \mu\,\frac{u}{y}$$

where μ is the proportionality constant and is called coefficient of viscosity. It is also known as the dynamic viscosity or simply viscosity. The shear stress which is usually denoted by the symbol τ (Greek letter tau) may be expressed as

$$\tau = \mu\,\frac{U}{b} = \mu\,\frac{d\theta}{dt} = \mu\,\frac{u}{y} \qquad \text{...(1.2)}$$

Equation (1.2) states that the rate of angular deformation is proportional to the shear stress.

Equation (1.2) in a differential form is expressed as

$$\tau = \mu\,\frac{du}{dy} \qquad \text{...(1.3)}$$

For a linear velocity distribution, du/dy, is a constant. But if the gap separating the parallel surfaces is large, the velocity distribution can no longer be assumed linear.

In case of a non-linear velocity distribution such as shown in Fig. 1.3, the rate of deformation is not constant, but changes from point to point. The shear stress, which depends on the rate of deformation will therefore not be constant throughout the fluid.

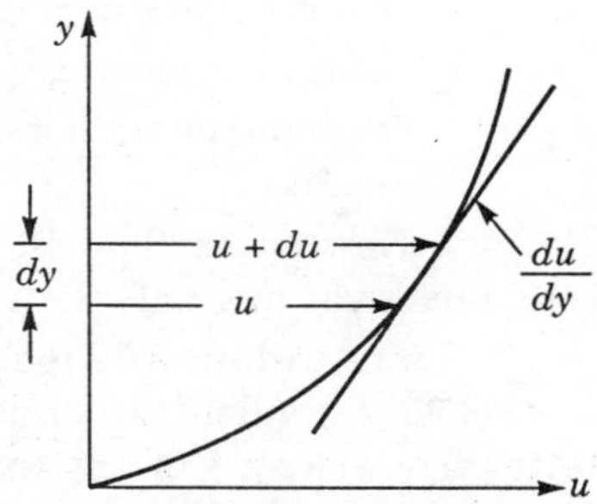

Fig. 1.3. Non-linear velocity distribution.

In Eq. (1.3), $\frac{du}{dy}$ represents the rate of shear deformation or rate of shear strain and is often called the velocity gradient. It relates the shear stress with the viscosity. Equation (1.3) is known as the *Newton's law of viscosity* or *Newton's law of fluid friction.* It may be noted here that when the fluid is at rest, no tangential forces exist and hence no fluid deformation takes place. In other words, as the velocity gradient, dy/dy is zero in a fluid at rest, no tangential or shearing force ever exists.

According to Newton's law of viscosity, for a given shear stress acting on a fluid element, the rate at which the fluid deforms is inversely proportional to the viscosity. This implies that for a constant shear stress the rate at which deformation takes place is larger for fluids of low viscosity. For solids, the resistance to shear deformation is due to modulus of elasticity whereas for fluids, the resistance to rate of shear deformation is on account of the viscosity. It is thus seen that the Hook's law for solids is analogous to the Newton's law of viscosity.

1.4.1. Newtonian and Non-Newtonian Fluids

A fluid which obeys Eq. (1.3) is known as a Newtonian fluid. Newtonian fluids have a certain constant viscosity, *i.e.*. the viscosity is independent of the shear stress. Many common fluids such as air, water, light oils and gasoline are Newtonian fluids under normal conditions. However, there are certain fluids which exhibit non-Newtonian characteristics—shear stress is not linearly dependent upon the velocity gradient. Non-Newtonian fluids, therefore, do not follow Newton's law of viscosity. Common examples of Non-Newtonian fluid are : human blood, lubricating oils, clay suspension in water, molten rubber, printer's ink, butter and sewage sludge. The following chart gives the classification of fluids :

Fluids

- *Newtonian*
- *Non-Newtonian*
  - Time-independent
    - (*i*) Dilatant
    - (*ii*) Bingham plastic
    - (*iii*) Pseudoplastic
  - Time-dependent
    - (*i*) Thixotropic
    - (*ii*) Rheopectic

A general relationship between shear stress and velocity gradient (rate of shear strain) for non-Newtonian fluid may be written as :

$$\tau = A\left(\frac{du}{dy}\right)^n + B \qquad ...(1.4)$$

where A and B are constants which depend upon the type of fluid and conditions imposed on the flow (shear stress).

The fluids which obey Eq. (1.4) are called *power-law fluids*. The additive constant B is zero for the fluids except Bingham plastic. Based on the value of power index n in Eq. (1.4), the non-Newtonian fluids are classified as :

(*a*) Dilatant, if $n > 1$, (example—quicksand, butter, printing inks)

(*b*) Bingham plastic, if $n = 1$, (*e.g.*, sewage sludge, drilling muds)

(*c*) Pseudoplastic, if $n < 1$, (*e.g.*, paper pulp, polymeric solutions such as rubbers, suspensions paints)

A Newtonian fluid is a special case of power law fluid having $n = 1$ and $B = 0$, and the constant A varying only with the type of fluid.

Time-independent fluids : In case of time-independent fluids, the rate of deformation or the velocity gradient depends only upon the shear stress, and is a single valued function of the latter. The viscosity of Newtonian fluids is independent of the shear stress, whereas in case of non-Newtonian fluids, the viscosity is a function of shear stress.

Time-dependent fluids : The rate of deformation and the viscosity depend upon both the shear stress and the duration of its application.

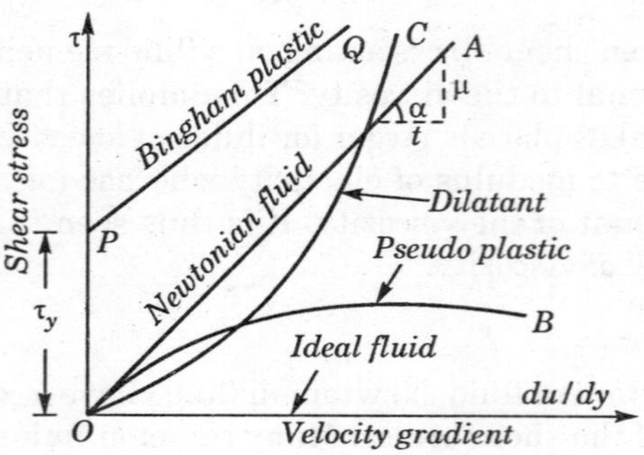

Fig. 1.4. Plot of τ *versus* $\dfrac{du}{dy}$.

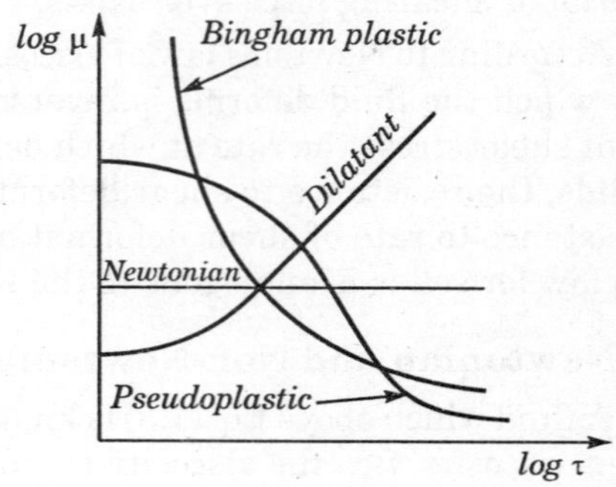

Fig. 1.5. Plot of log μ *versus* log τ.

Figs. 1.4 and 1.5 illustrate the shear or viscous characteristics of different fluids. Fig. 1.4 illustrates the shear stress-velocity gradient relationship for various type of fluids. The Newtonian fluids are characterised by linear relationship between the shear stress and the velocity gradient. They are represented by a straight line like *OA* passing through the origin and inclined at an angle α with the horizontal such that $\mu = \tan \alpha$.

The Bingham plastic fluids require a certain minimum shear stress τ_y known as the yield stress before they start flowing and exhibit a linear relationship between the shear stress and the velocity gradient as shown by the straight line *PQ*. The dilatant and pseudoplastic fluids are shown by curves marked *OC* and *OB* respectively.

Fig. 1.5 illustrates the viscosity μ and shear stress τ relationship for different fluids on log-log scale. For Newtonian fluids this relationship is a straight line parallel to τ-axis indicating that viscosity is independent of stress. For non-Newtonian fluids, the fact that the viscosity is a function of the shear stress can be noticed.

Thixotropic fluids are those which show an increase in apparent viscosity with time. Lipstic and certain paints and enamels exhibit thixotropic behaviour. The apparent viscosity may be defined as $\mu_{app} = \dfrac{\tau}{du/dy}$. Those fluids which show a decrease in the apparent viscosity with time are called *rheopectic*. Rheopectic fluids are much less common than thixotropic fluids. Gypsum suspensions in water and bentonite solutions are examples of rheopectic fluids. Thixotropy is an important property of paints and enamels. When subjected to high shear by the brush during application of paint, the apparent viscosity is reduced so that the paint covers the surface smoothly, and brush marks disappear subsequently.

1.4.2. Some Common Newtonian Fluids

The air and water both have relatively low viscosities, the viscosity of oils is much higher and the glycerine is the most viscous of the better known fluids (approximately one thousand times as viscous as water) having colourless appearance and is readily miscible with water, making it useful for laboratory purposes by providing glycerine-water solutions of any desired viscosity. At ordinary temperatures, a comparison of absolute viscosities of a few common fluids shows that water is about 50 times more viscous than air. As compared to water, castor oil is 1000 times more viscous, crude oil is 10 times more viscous and the gasoline is about $\frac{1}{3}$ times viscous than water.

Ideal Fluid. An ideal fluid is a conceptual fluid which is assumed non-viscous and incompressible. It is a concept that permits a fluid to possess nonexistential properties, like zero viscosity (inviscid fluid) and constant density implying zero compressibility. Such a concept was used by mathematicians for simplifying analysis of fluid motion. Truely speaking, there is no fluid that exists in nature that possesses zero viscosity and zero compressibility (*i.e.*, bulk modulus of elasticity having infinite value).

Real Fluid. All fluids that exist in nature are real fluids possessing properties like, viscosity, elasticity, surface tension and vapour pressure. Such fluids are viscous and compressible. Common examples are air, water, other gases and liquids.

1.4.3. Kinematic Viscosity

In the analysis of many fluid-flow problems, the dynamic viscosity divided by the density is commonly found to exist. This ratio of dynamic viscosity and mass density is called the kinematic viscosity and is denoted by the Greek letter ν (nu), thus

$$\text{Kinematic viscosity } \nu = \frac{\text{Dynamic viscosity } \mu}{\text{Mass density } \rho}$$

It is known as kinematic viscosity, because it can be defined dimensionally by only length and time dimensions, mass or force dimensions being not involved.

1.4.4. Dimensions of Dynamic and Kinematic Viscosities

The dimension of dynamic viscosity μ may be obtained by using Eq. (1.3)

$$\mu = \frac{\tau}{du / dy}$$

The fundamental dimensions are the mass (M), length (L) and time (T).

The dimensions of other quantities can be derived easily.

Writing the dimension of shear stress and velocity gradient, the dimension of the dynamic viscosity can be determined.

$$\text{Dimension of shear stress} = \frac{\text{Dimension of force}}{\text{Dimension of area}} = \left[\frac{MLT^{-2}}{L^2}\right] = ML^{-1}T^{-2}$$

Hence, $$\text{dimension of } [\mu] = \left[\frac{[MLT^{-2} / L^2]}{LT^{-1} / L}\right] = [ML^{-1}T^{-1}] \qquad \text{...(1/4 } a\text{)}$$

dimension of μ in F-L-T system will be

$$[\mu] = \left[\frac{F / L^2}{LT^{-1} / L}\right] = [FL^{-2}T] \qquad \text{...(1.4 } b\text{)}$$

the dimension of kinematic viscosity may be found as :

$$\text{Dimension of } [\upsilon] = \left[\frac{\text{Dimensions of } [\mu]}{\text{Dimensions of } [\rho]}\right] = \left[\frac{ML^{-1}\,T^{-1}}{ML^{-3}}\right] = [L^2\,T^{-1}] \qquad \text{...(1.5)}$$

The unit of measurement of viscosity μ, may be obtained using Eq. (1.4 b).

In MKS system, the viscosity is expressed as kg s/m^2 ; in CGS system, it is expressed by dyne-s/cm^2, and is called a poise ; and FPS it is measured in lb-sec/ft^2. In SI units it is expressed as N s/m^2. The unit of viscosity may be converted from one system to another by using the above definitions and basic conversion factors.

$$\text{(i) From FPS to CGS : } 1\frac{\text{lb}-s}{\text{ft}^2} = \frac{453.6\times 981}{(30.48)^2}\frac{\text{dyne}-\text{s}}{\text{cm}^2} = 479\ \frac{\text{dyne}-\text{s}}{\text{cm}^2} = 479 \text{ poise.}$$

$$\text{(ii) From FPS to MKS : } 1\frac{\text{lb}-\text{s}}{\text{ft}^2} = \frac{453.6/1000}{(0.3048)^2}\frac{\text{kg}-\text{s}}{\text{m}^2} = 4.87\ \frac{\text{kg}-\text{s}}{\text{m}^2}$$

$$\text{(iii) From MKS to SI : } 1\frac{\text{kg}-\text{s}}{\text{m}^2} = \frac{1000\times 981}{(100)^2}\frac{\text{dyne}-\text{s}}{\text{cm}^2} = 98.1 \text{ dyne-s/cm}^2 = 98.1 \text{ poise.}$$

$$1 \text{ kg-s/m}^2 = 9.81 \text{ N-s/m}^2 = 9.81 \text{ Pa.s}$$

$$98.1 \text{ poise} = 9.81 \text{ N-s/m}^2 \text{ ; } 1 \text{ poise} = \frac{1}{10} \text{ N-s/m}^2 = 0.1 \text{ N-s/m}^2 = 0.1 \text{ Pa.s}$$

The conversion factors of units are given in Appendix E.

Based on the dimensions of the kinematic viscosity $[L^2T^{-1}]$ the unit of ν in F.P.S. system is ft^2/sec, in C.G.S., it is cm^2/s and is called the stoke and in M.K.S. and SI units, it is expressed as m^2/s. These units can be converted from one system to the other using the same procedure as above :

$$1 \text{ ft}^2/\text{sec} = \frac{(30.48)^2}{1} \text{ cm}^2/\text{s} = (30.48)^2 \text{ stokes} = 930 \text{ stokes.}$$

$$1 \text{ stoke} = 1\ \frac{\text{cm}^2}{\text{sec}} = \left(\frac{1}{100}\right) \text{m}^2/\text{s} = 10^{-4} \text{ m}^2/\text{s.}$$

1.4.5. Variation of Viscosity with Temperature

The variation of viscosity with changes in temperature may be understood by knowing the factors contributing to viscosity. The viscosity of a fluid depends upon its intermolecular structure. In gases, the molecules are widely spaced resulting in a negligible intermolecular cohesion, while in liquids the molecules being very close to each other, the cohesion is much larger. The viscosity of a fluid is due to intermolecular cohesion and transfer of molecular momentum in a direction normal to the flow. In liquids, this momentum transfer is small as compared to the force of cohesion between the molecules and, therefore, the viscosity is primarily dependent upon the magnitude of intermolecular cohesive force. With the increase of temperature, the cohesive force decreases rapidly resulting in the decrease of viscosity. Thus, the viscosity of liquids decrease with the increase of temperature.

Poiseuille developed a formula for determining the kinematic viscosity of water at any temperature T,

$$\nu = \frac{0.0179}{1 + 0.0337\,T + 0.000221\,T^2}$$

in which ν is the kinematic viscosity in cm^2/s and T is the temperature in degree centigrade.

In case of gases, the viscosity is mainly due to transfer of molecular momentum in the transverse direction brought about by the molecular agitation. The contribution to the viscosity by the intermolecular cohesive force being negligible due to large spacing of molecules. As the molecular agitation increases with the rise of temperature the viscosity of gases also increases with temperature rise.

The viscosity of a fluid may thus be considered to be composed of two parts :

1. that due to intermolecular cohesion, and
2. that due to transfer of molecular momentum, and thus

Viscosity of a flowing fluid = contribution from intermolecular cohesion + contribution from transfer of molecular momentum.

Table 1.1 gives viscosity of air at different temperatures. For water and other liquids see Tables 1.5 and 1.6. Table 1.7 deals with properties of the standard atmosphere.

Table 1.1. Density and Viscosity of Air at Atmospheric Pressure

Temperature degree C	*Density* ρ kg/m^3	*Dyn. viscosity,* μ $10^5\mu$, $N\text{-}s/m^2$
0	1.29	1.71
20	1.202	1.81
40	1.125	1.90
60	1.059	2.00
100	0.945	2.18
150	0.833	2.39
200	0.745	2.58
300	0.615	2.95
400	0.524	3.28
500	0.456	3.58

The viscosity of gases, like that of liquids, changes with temperature but is practically unaffected by pressure. The kinematic viscosity, depending as it does on density, varies with both temperature and pressure. The following equation, given by Holman, may be used for determining the value of μ at different temperatures :

$$\mu = 1.7150 \times 10^{-4}\,(1 + 0.00275\,T - 3.4 \times 10^{-7}\,T^2) \quad \text{(poise)}$$

in which μ is in poise and T in degrees centigrade. The viscosity in kg s/m^2 and Ns/m^2 are given by :

$$\mu = 1.7500 \times 10^{-6}\,(1 + 0.00265\,T - 3.4 \times 10^{-7}\,T^2) \qquad \text{(kg-s/m}^2\text{)}$$

$$\mu = 1.715 \times 10^{-5}\,(1 + 0.00275\,T - 3.4 \times 10^{-7}\,T^2) \qquad \text{(N-s/m}^2\text{)}.$$

Example 1.4. *Calculate the velocity gradient at distances of 0, 100, 150 mm from the boundary if the velocity profile is a parabola with the vertex 150 mm from the boundary, where the velocity is 1 m/s. Also calculate the shear stresses at these points if the fluid has a viscosity of 0.804 N-s/m².*

Solution. Let the equation of the parabolic velocity profile be

$$u = Ay^2 + By + C \qquad \text{...(1)}$$

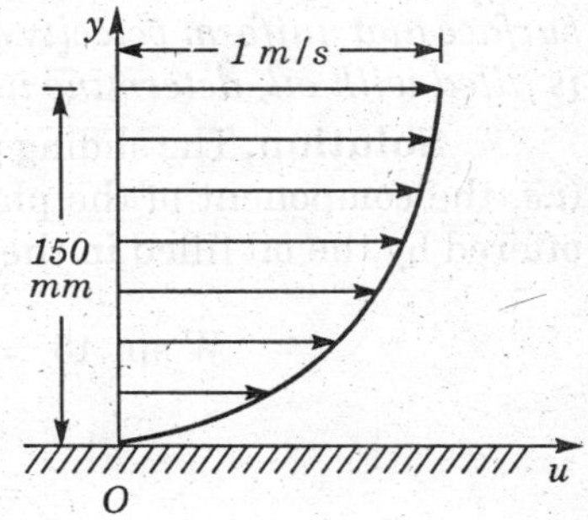

where *A*, *B* and *C* are constants to be determined from the following boundary conditions :

(*i*) $u = 0$ at $y = 0$

(*ii*) $u = 1$ m/s, at $y = 0.15$ m

(*iii*) $du/dy = 0$ at the vertex, *i.e.*, $y = 0.15$ m

Boundary condition (*i*) gives $C = 0$ and from (*ii*), we obtain

$$1 = A(0.15)^2 + B(0.15) \qquad \text{...(2)}$$

and from (*iii*), $\dfrac{du}{dy} = 2Ay + B$,

or $0 = 2A(0.15) + B$ ...(3)

Solving Eqs. (2) and (3), $A = -44.4$ and $B = 13.33$

Eq. (1) for the velocity profile now becomes

$$u = -44.4y^2 + 13.33y.$$

The velocity gradients and the shear stresses the desired points may be obtained as below :

(*a*) At $y = 0$ mm, $\frac{du}{dy} = -2 \times 44.4 \times 0 + 13.33 = 13.33 \text{ sec}^{-1}$

Shear stress $\tau = \mu \frac{du}{dy} = 0.804\ (13.33) = 10.8 \text{ N/m}^2$

(*b*) At $y = 100$ mm, $\frac{du}{dy} = -2 \times (44.4) \times (0.1) + 13.33 = 4.45 \text{ sec}^{-1}$

$$\tau = 0.804\ (4.45)$$
$$= 3.575 \text{ N/m}^2$$

(*c*) At $y = 150$ mm, $\frac{du}{dy} = -2 \times (44.4)(0.15) \times 15 + 13.33 = 0$

$$\tau = \mu \frac{du}{dy} = 0.$$

Example 1.5. *Two horizontal plates are placed 12.5 mm apart, the space between them being filled with oil of viscosity 14 poise. Calculate the shear stress in the oil if the upper plate moves with a velocity of 2.5 m/s.*

Solution. $1 \text{ poise} = 1 \text{ dyne-s/cm}^2 = \frac{1}{98.1} \text{ kg-s/m}^2 = 0.1 \text{ N-s/m}^2$

Shear stress $\tau = \mu \frac{du}{dy}$

Relative velocity between the plates, $du = 2.5$ m/s.

Distance between the plates, $dy = 1.25$ cm $= 0.0125$ m

Viscosity of the oil, $\mu = 14$ poise

$$= \frac{14}{98.1} = 0.143 \text{ kg-s/m}^2 = 1.4 \text{ N-s/m}^2$$

(since 1 Poise = 0.1 N-s/m²)

Substituting in the formula,

$$\tau = 0.143 \times \frac{2.5}{0.0125} = 28.55 \text{ kg/m}^2 = 280.0 \text{ N/m}^2.$$

Example 1.6. *A rectangular plate 1.2 m × 0.4 m, weighing 970 N slides down a 45° inclined surface at a uniform velocity of 2.25 m/s. If the 2 mm gap between the plate and the inclined surface is filled with oil, determine its viscosity.*

Solution. The sliding plate will attain the uniform velocity when the fore causing the motion (*i.e.* the component of the plates weight along the inclined surface) balances the fluid resistance offered by the oil filled in the gap.

$$W \sin 45° = \mu \frac{du}{dy} . A$$

$$= \mu \frac{V}{2/(10 \times 100)} \times 1.2 \times 0.4$$

or

$$970 \sin 45° = \mu \frac{2.25 \times 10^3}{2} \times 1.2 \times 0.4$$

$$\therefore \quad \mu = \frac{970 \times 1/\sqrt{2}}{0.54 \times 10^3} = 2.54 \text{ Ns/m}^2 = \textbf{25.4 Poise.}$$

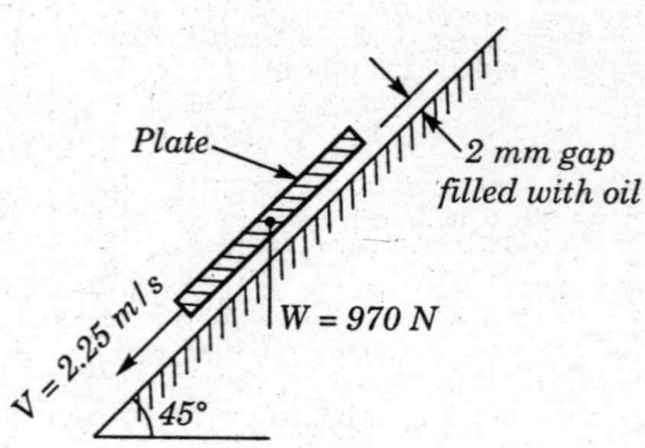

Example 1.7. *A piston 796 mm diameter and 200 mm long works in a cylinder of 800 mm diameter. If the annular space is filled with a lubricating oil of viscosity 5 cP, calculate the speed of descent of piston in vertical position. The weight of piston and the axial load are 9.8 N.*

Solution. Viscosity of oil

$$= 5 \text{ cP} = 5 \times 10^{-2} \text{ Poise}$$

$$= 0.5 \times 10^{-2} \text{ N-s/m}^2 = 5 \times 10^{-3} \text{ N-s/m}^2.$$

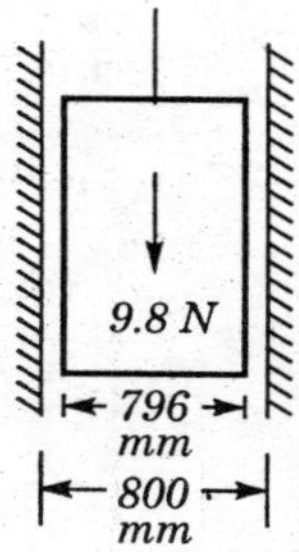

From Newton's law of viscosity, the shear stress

$$\tau = \mu \frac{du}{dy} \quad ...(i)$$

Shear stress developed by the vertical descent of the piston

$$= \frac{\text{Shear force (i.e. weight of piston etc.)}}{\text{Piston area in contact with oil}}$$

$$= \frac{9.8}{\pi \times 0.796 \times 0.20} = 19.61 \text{ N/m}^2.$$

Velocity gradient when the piston attains a constant velocity of V (m/s) in the annular gap of 2 mm

$$\frac{du}{dy} = \frac{V}{2 \times 10^{-3}}.$$

Substituting in the Newton's law, Eq. (*i*)

$$19.61 = 5.1 \times 10^{-3} \times \frac{V}{2 \times 10^{-3}}$$

from which, V = **7.841 m/s.**

Example 1.8. *A cylinder of 150 mm radius rotates concentrically inside a fixed cylinder of 155 mm radius. Both cylinders are 300 mm long. Determine the viscosity of the liquid which fills the space between the cylinders if a torque of 0.98 N-m is required to maintain an angular velocity of 60 r.p.m.*

Solution. The torque is transmitted through the fluid layers to the outer cylinder.

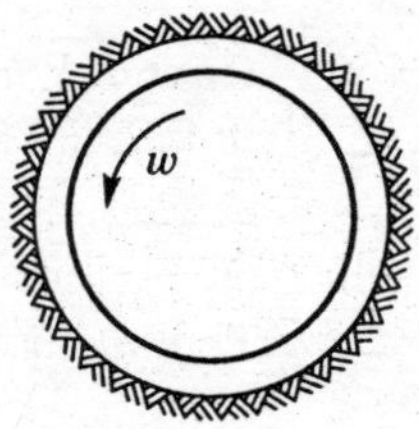

Tangential velocity of the inner cylinder = $r\omega$

$$= r \frac{2\pi n}{60}$$

$$= 0.15 \times \frac{2\pi \times 60}{60} = 0.943 \text{ m/s}.$$

For the small space between the cylinders, the velocity profile may be assumed to be a straight line, then

$$\frac{du}{dy} = \frac{0.943}{\frac{(15.5 - 15.0)}{100}} = 188.6 \text{ per sec.}$$

Torque applied = Torque resisted

$$0.98 = \tau \times \text{Area} \times \text{Lever arm} = \tau \times (2\pi \times 0.15 \times 0.30) \times 0.15$$

$$\tau = \frac{0.98}{2\pi \times 0.045 \times 0.15} = 23.15 \text{ N/m}^2$$

or

$$\mu = \frac{\rho}{du/dy} = \frac{23.15}{188.6} = 0.123 \text{ N-s/m}^2.$$

Example 1.9. *A circular disc of a diameter 'd' is slowly rotated in a liquid of large viscosity 'μ' at a small distance 'h' from a fixed surface. Derive an expression for torque 'T' necessary to maintain an angular velocity 'ω'.*

Solution. Consider an element of disc at a radius r and having a width dr.

Linear velocity at this radius = $r\omega$

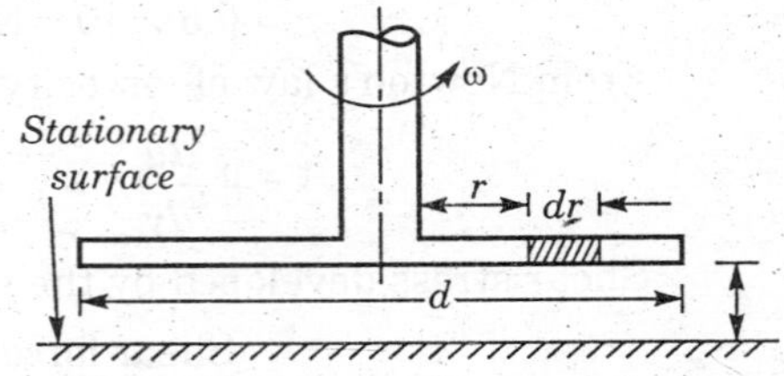

Shear stress $\tau = \mu \dfrac{du}{dy}$

Torque $= \text{Shear stress} \times \text{Area} \times r$

$= \tau \times 2\pi r\, dr \times r$

$= \mu \dfrac{du}{dy} . 2\pi r^2\, dr.$

Assuming the gap h to be small so that the velocity distribution may be assumed linear.

$$\frac{du}{dy} = \frac{r\omega}{h}$$

∴ Torque, dT, on the element

$$dT = \mu \frac{r\omega}{h} . 2\pi r^2\, dr = \frac{2\pi\mu\omega}{h} r^3 dr$$

Total torque,
$$T = \int_0^{d/2} \frac{2\pi\mu\omega}{h} r^3\, dr = \frac{2\pi\mu\omega}{h} \left| \frac{r^4}{4} \right|_0^{d/2} = \frac{\mu\pi d^4\omega}{32\, h}.$$

Example 1.10. *A space 25 mm wide between two large plane surfaces is filled with glycerine. What force is required to drag a very thin plate 0.75 sq metre in area between the surfaces at a speed of 0.5 m/s (i) if this plate remains equidistant from the two surfaces, (ii) if it is at a distance of 10 mm from one of the surfaces ? Take μ = 0.785 N-s/m².*

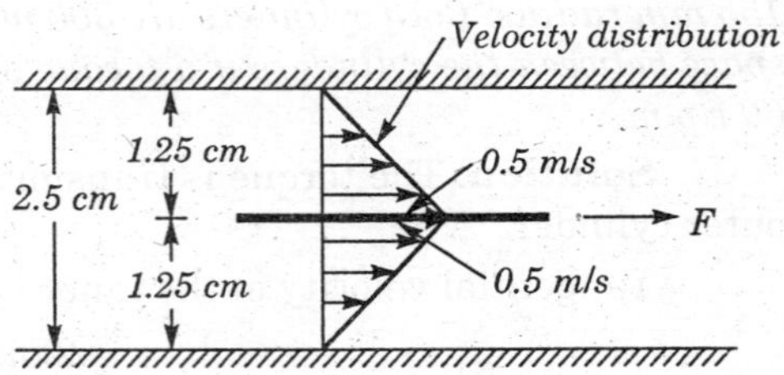

Solution. Total force required to drag the plate

= Sum of the forces on either side of the plate

∴ $F = F_1 + F_2$.

Case I. *When the plate is located midway between the surfaces.*

Since the space between the surfaces is small, the velocity distribution may be considered as straight line.

The shear force on the upper side of the plate

$$F_1 = \tau_1 \times \text{Area of the plate}$$

$$= \mu \left(\frac{du}{dy}\right)_1 \times 0.75 = 0.785 \times \left(\frac{0.5}{2.5/2 \times 100}\right) \times 0.75 = 23.5 \text{ N}$$

The force on the bottom side of the plate

$$F_2 = \mu \left(\frac{du}{dy}\right)_2 \times \text{Area of plate} = 0.785 \left(\frac{0.5}{2.5/2 \times 100}\right) \times 0.75 = 23.5 \text{ N}.$$

∴ The total resistance force (which is equal to the force required to drive it) experienced by the plate = $F_1 + F_2$ = 47.0 N.

Case II. *When the plate is located at a distance of 10 mm from one of the surfaces.*

Force on the upper side of the plate

$$F_1 = \mu \left(\frac{du}{dy}\right) \times 0.75 = 0.785 \left(\frac{0.5}{1.5/100}\right) \times 5 = 19.60 \text{ N}.$$

Force on the bottom side of the plate

$$F_2 = \mu \left(\frac{du}{dy}\right) \times 0.75 = 0.785 \left(\frac{0.5}{1/100}\right) \times 0.75 = 29.40 \text{ N}$$

Total force required $= F_1 + F_2 = 49.0$ N.

Example 1.11. *Lateral stability of a long shaft 150 mm in diameter is obtained by means of a 250 mm stationary bearing having an internal diameter of 150.25 mm. If the space between bearing and shaft is filled with a lubricant having a viscosity 0.245 N-s/m², what power will be required to overcome the viscous resistance when the shaft is rotated at a constant rate of 180 r.p.m. ?*

Solution. Circumferential velocity of the shaft

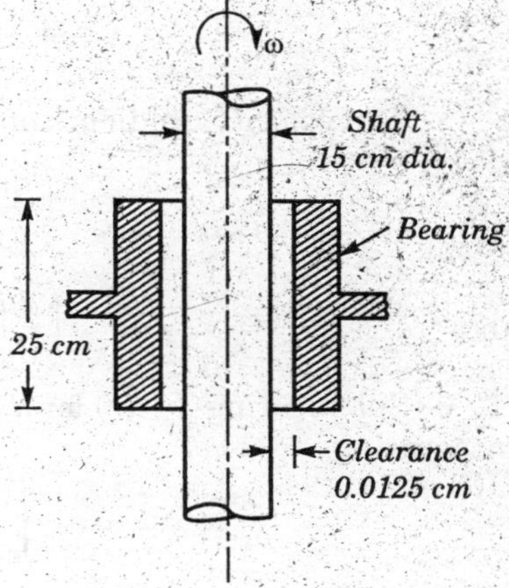

$$V = r\omega$$

$$= r \cdot \frac{2\pi n}{60} = \frac{15/2}{100} \times \frac{2\pi \times 180}{60} = 1.412 \text{ m/s}$$

Velocity gradient,

$$\left(\frac{du}{dy}\right) = \frac{\frac{1.412}{0.025}}{2 \times 100} = 1.13 \times 10^4 \text{ sec}^{-1}$$

Shear stress on the shaft

$$\tau = \mu \left(\frac{du}{dy}\right) = 0.245 \times 1.13 \times 10^4$$

$$= 2.77 \times 10^3 \text{ N/m}^2 = 2.77 \text{ kN/m}^2$$

Shear force on the shaft

$$F = \tau \,.\, 2\pi r l = 2.77 \times 10^3 \times 2 \times \pi \times \frac{15.2}{2 \times 100} \times \frac{25}{100} = 327.5 \text{ N}$$

Torque to be overcome by the shaft = Shear force × Radius of shaft

$$T = F \,.\, r$$

$$= 327.5 \times \frac{15}{2 \times 100} = 24.6 \text{ N-m.}$$

Power corresponding to this torque at a speed of $n = 180$ r.p.m.

$$P = \frac{2\pi n}{60} \times T = \frac{2\pi \times 180}{60} \times 24.6$$

$$= 463.0 \text{ N-m/s} = 463.0 \text{ W} = 0.463 \text{ kW.}$$

Example 1.12. *A cylinder 0.25 m in radius and 2 m length rotates coaxially inside a fixed cylinder of the same length and 0.30 m radius. Olive oil of viscosity 4.9 × 10⁻² Ns/m² fills the space between the cylinders. A torque 4.9 N-m is applied to the inner cylinder. After constant velocity is attained, calculate the velocity gradient at the cylinder walls, the resulting r.p.m., and the power dissipated by fluid resistance ignoring end effects.*

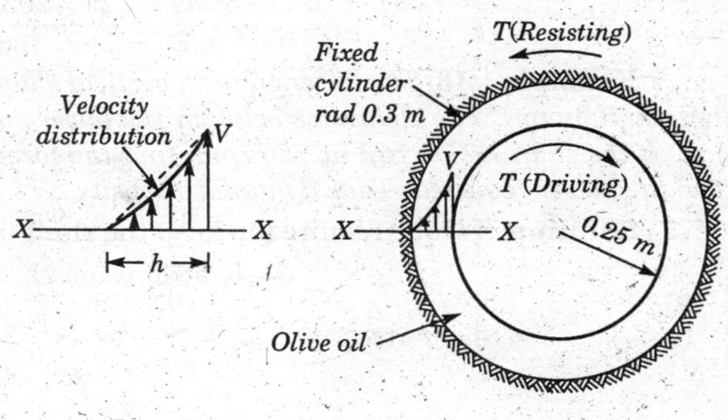

Solution. The surface area of the outer cylinder is larger than that of the inner one, since the former has a larger radius. Accordingly the shear force and the velocity gradient at the outer cylinder will be less than the respective quantities on the inner one. The velocity profile through the fluid will be non-linear as indicated in the figure, since the gap between the inner and outer cylinders is comparatively larger.

The torque of 4.9 Nm is transmitted from inner cylinder to the outer one through fluid friction (viscous effect). Let r be the radial distance of any fluid layer.

Then
$$4.9 = \tau \times (2\pi rl) \times r = \tau.2\pi r \times 2 \times r$$
$$= \tau.4\pi r^2 = \mu \frac{du}{dy}.\ 4\pi r^2$$
$$= 4.9 \times 10^{-2} \times 4\pi r^2 \frac{du}{dy}$$

$$\therefore \quad \frac{du}{dy} = \frac{100}{4\pi r^2} = \frac{7.95}{r^2}.$$

The velocity gradients at the inner at outer cylinders are :

$$\left(\frac{du}{dy}\right)_i = \frac{7.95}{(0.25)^2} = 127.2 \text{ sec}^{-1}$$

and
$$\left(\frac{du}{dy}\right)_0 = \frac{7.95}{(0.30)^2} = 88.3 \text{ sec}^{-1}.$$

Substituting $(-dr)$ for dy in the equation for du/dy since velocity decreases as r increases. Integrating,

$$\int_0^V du = -7.95 \int_{0.30}^{0.25} \frac{dr}{r^2}$$

$\therefore$ Velocity of inner cylinder,

$$V = 7.95 \left[\frac{1}{r}\right]_{0.30}^{0.25} = 5.30 \text{ m/s}$$

speed of inner cylinder,
$$\omega = \frac{V}{r} = \frac{5.3}{0.25} = 21.2 \text{ rad/sec.}$$
$$n = \frac{60\omega}{2\pi} = \frac{60 \times 21.2}{2\pi} = \mathbf{202.4 \text{ r.p.m.}}$$

Assuming the velocity profile to be linear for an approximate calculation
$$V = 127.1 \times 0.05 = 6.35 \text{ m/s}$$

and
$$n = 242.5 \text{ r.p.m.}$$

Since this result differs from the former by nearly 20%, the approximation is not satisfactory in this case.

The power dissipated in fluid friction

$$= \frac{2\pi nT}{60} = \frac{2\pi \times 202.4 \times 4.9}{60} = 104.0 \text{ Nm/s} = \mathbf{104.0 \text{ W.}}$$

Example 1.13. *The lower end of a vertical shaft of diameter 10 cm rests in a foot step bearing (length 100 mm). The clearance between the lower end of the shaft and the bearing surface is 0.5 mm. If the shaft has to run at 750 rpm, find the torque required to keep the shaft in motion. Find also the power required. Take dynamic viscosity as 1.5 poise.* *(RGPV, 2013 June)*

Solution. Torque required to keep the shaft in motion

$$= \text{shear force on shaft} \times \text{its radius}$$

Shear stress, $\tau = \mu \dfrac{d\mu}{dy}$

Referring to the figure, the foot-step bearing and the shaft is shown. The plan of the shaft and the bearing surface shows the velocity distribution in the annular clearance.

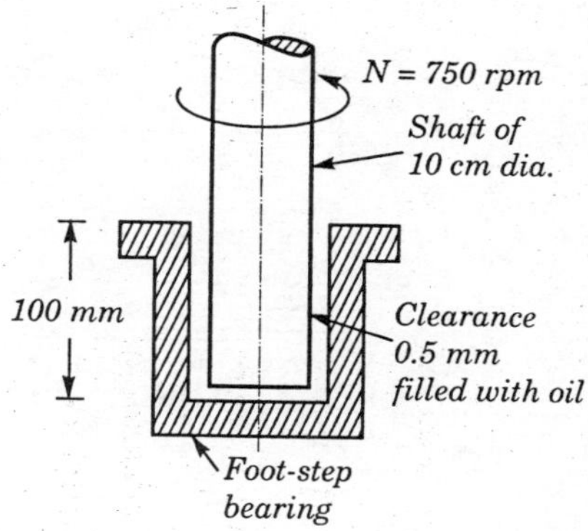

The shear stress on the shaft's outer surface is given by the Newton's law of viscosity, and will be calculated with the data given :

$$\mu = 1.5 \text{ poise} = 1.5 \times 0.1 \text{ (N.s/m}^2\text{)}$$

$$(10 \text{ poise} = 1 \text{ Ns/m}^2)$$

Speed of rotation of shaft, $N = 750$ rpm

Velocity of rotation, $V = R\omega$

$$= \frac{5}{100} \times \frac{2\pi \times 750}{60} \text{ (m/s)}$$

$$= 3.928 \text{ m/s}$$

The bearing surface being stationary, the velocity is zero, and

$$du = V - O \quad = 3.928 \text{ m/s}$$

$$dy = \text{clearance} = 0.5 \text{ mm} = 0.5 \times 10^{-3} \text{ m}$$

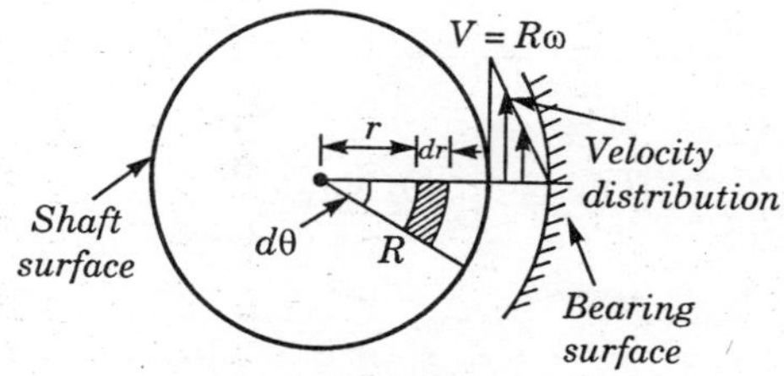

Shear stress, $\tau = 0.15 \times \dfrac{3.928}{0.5 \times 10^{-3}}$

$$= 1178.57 \text{ N/m}^2$$

Shear force on the shaft surface of area $2\pi Rl$

$$F = \tau.\ 2\pi Rl = 1178.57 \times 2\pi \times (10 \times 10^{-2}) \times (5 \times 10^{-2})$$

$$= 18.689 \text{ N}$$

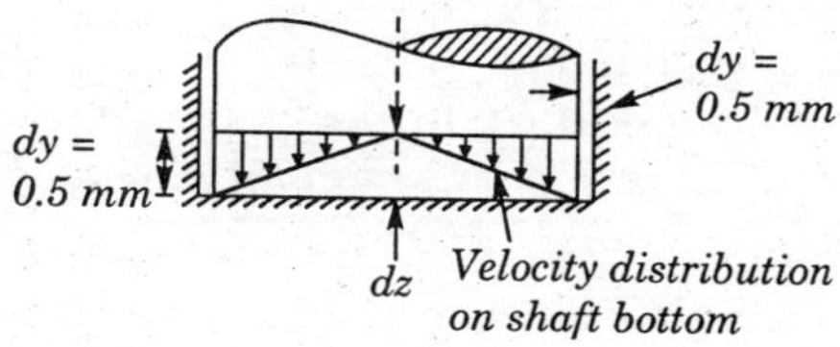

Torque exerted on the shaft

$$= F.R = 11.689 \times (5 \times 10^{-2})$$

$$= \mathbf{0.9344 \text{ Nm}}$$

$$\text{Power required} = \text{Torque} \times \omega = 0.9344 \times \frac{2\pi \times 750}{60}$$

$$= 73.35 \text{ Nm/s} = \mathbf{73.35 \text{ W}} \qquad ...(i)$$

To this, there is also the contribution of the shear force exerted on the shaft bottom.

To evaluate this, it is necessary to assume the velocity distribution at the bottom end of the shaft, as shown in the bottom-most sketch. The velocity at the shaft axis zero and it varies radially, there being a triangular velocity distribution with maximum velocity equal to $R\omega$. Consider the sectorial sketch in the plan, an element of shaft at a radius r, having thickness dr, the sector angle being $d\theta$.

Shear force on this element of area $rd\theta.dr$ is (dy being 0.5 mm)

$$dF = \tau \times \text{area of element} = \mu\frac{r\omega}{0.5 \times 10^{-3}} \times rd\theta\, dr$$

Torque exerted on this elementary area situated at radius r

$$dT = dF.r = \frac{\mu\omega r^2\, d\theta dr}{0.5 \times 10^{-3}}.r$$

$$= 2\mu\omega \times 10^3\, r^3\, d\theta dr$$

To get the total torque exerted on this shaft bottom, we integrate the expression for dT

$$T = \int_0^{2\pi}\int_0^R dT = 2\mu\omega \times 10^3 \int_0^{2\pi}\int_0^R r^3 dr\, d\theta$$

$$= 2 \times 10^3\ \mu\omega \int_0^{2\pi} \left| \frac{r^4}{4} \right|_0^R d\theta = 2 \times 10^3\ \mu\omega \int_0^{2\pi} \frac{R^4}{4}\, d\theta$$

$$= 2 \times 10^3\ \mu\omega |\ 0\ |_0^{2\pi} \frac{R^4}{4}$$

$$= \frac{2\mu\omega R^4}{4} . 2\pi = \pi R^4 \mu\omega$$

Power reqd. by the shaft bottom = $T.\omega$

$$= \pi R^4 \mu\omega^2$$

$$= \pi \times (5 \times 10^{-2})^4 \times 0.15 \times \left(\frac{2\pi \times 750}{60}\right)^2 \text{ Nm/s}$$

$$= \pi \times 625 \times 10^{-8} \times 0.15\ (25\ \pi)^2 \text{ Nm/s}$$

$$= 1962.5 \times 10^{-8} \times 0.15 \times 6162.25 \text{ Nm/s}$$

$$= 0.01814 \text{ Nm/s} = \mathbf{0.1814\ W} \quad ...(ii)$$

∴ Power required to keep the shaft in motion

$$= 73.35 + 0.1814 = \mathbf{73.5314\ W}$$

The contribution of the torque and therefore of power required by the shaft bottom to the total power

$$= \frac{0.01814}{73.5314} = 0.0002466 \text{ or } \mathbf{0.0247\%}$$

and is negligible.

Example 1.14. *A 90 mm diameter shaft rotates at 1200 rpm in a 100 mm long journal bearing of 90.5 mm internal diameter. The annular space in the bearing is filled with oil having a dynamic viscosity of 0.12 Pa.s. Estimate the power dissipated.*

Solution. Torque exerted on the rotating shaft due to viscous resistance offered to it by the journal bearing :

T = Shear force exerted on the shaft × radius of shaft

$$= \left(\mu \frac{du}{dy}\right) A \times r = \mu \frac{du}{dy} . \pi d\, L\, .\, r$$

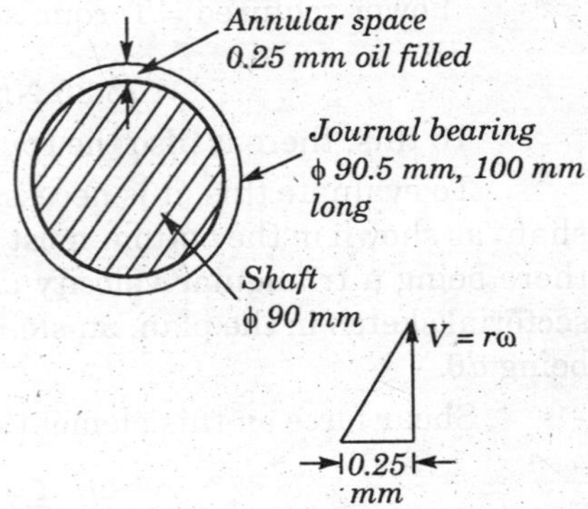

Assuming linear variation of velocity in the annular space, as shown, the velocity gradient ($du = v = rw$, dy = 0.25 mm)

$$\frac{du}{dy} = \frac{V}{0.25 \times 10^{-3}} = \frac{(2\pi r\, N/60)}{0.25 \times 10^{-3}}$$

$$= 2\pi \left(\frac{90 \times 10^{-3}}{2}\right) \times \frac{1200}{60} = 36\pi \times 10^2 \text{ m/s/m}$$

∴ $T = 0.12\ (\text{N.s/m}^2)\ (36\pi \times 10^2)\ (\pi \times 90 \times 10^{-3} \times 100 \times 10^{-3}) \times 45 \times 10^{-3}$

$= 1.728$ N.m

Power dissipated in overcoming viscous friction

$$P = \frac{2\pi\, NT}{60} = \frac{2\pi \times 1200 \times 1.728}{60}$$

$$= 217.06 \text{ N.m/s} = \mathbf{217.06\ W}.$$

Example 1.15. *A thin plate of large area is placed midway in a gap of height h filled with oil of viscosity* μ_0 *and the plate is pulled at a constant velocity V. If a lighter oil of viscosity* μ_1 *is then substituted in the gap, it is found that for the same velocity V, the drag force will be the same as before if the plate is located unsymmetrically in the gap but parallel to the walls. Find* μ_1 *in terms of* μ_0 *and the distance from the nearer wall to the plate.*

Solution. Case (*i*) *When the liquid of viscosity* μ_0 *fills the gap*. Since the plate is placed midway in the gap, the velocity profile on both the sides of the plate will be symmetrical, thus

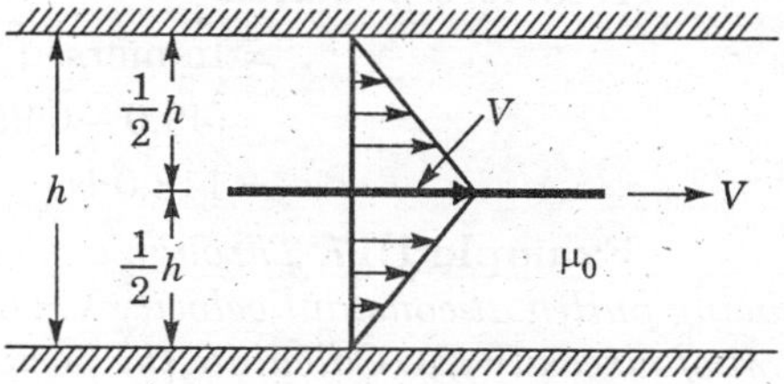

$$\frac{du}{dy} = \frac{V}{h/2} = \frac{2V}{h}.$$

The shear force on the upper and the bottom side of the plate will be same, and hence the drag force on the plate

$$F_1 = \left[\mu_0 \left(\frac{du}{dy}\right)_u + \mu_0 \left(\frac{du}{dy}\right)_b\right] A = \left[\mu_0 \frac{2V}{h} + \mu_0 \frac{2V}{h}\right] A = \frac{4V\mu_0 A}{h}$$

where A is the area of plate.

Case (*ii*) *When the liquid of viscosity* μ_1 *fills the gap*. Let the plate be placed at a distance of y from the bottom wall as shown.

The velocity gradients for the upper and bottom sides of the plate are :

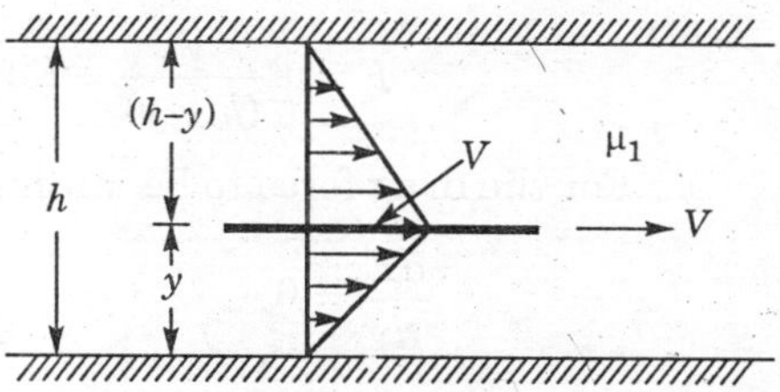

$$\left(\frac{du}{dy}\right)_u = \frac{V}{(h-y)}$$

and

$$\left(\frac{du}{dy}\right)_b = \frac{V}{y}.$$

Now, the drag force on the plate

$$F_2 = \left[\mu_1 \left(\frac{du}{dy}\right)_u + \mu_1 \left(\frac{du}{dy}\right)_b\right] A = \left[\mu_1 \frac{V}{(h-y)} + \mu_1 \frac{V}{y}\right] A = \frac{\mu_1 VhA}{y(h-y)}.$$

But since the drag forces F_1 and F_2 are equal, we have

$$\frac{4V\mu_0 A}{h} = \frac{\mu_1 VhA}{y(h-y)} \quad \text{or} \quad \mu_1 = 4\mu_0 \frac{y}{h}\left[1 - \frac{y}{h}\right]$$

Example 1.16. *A vertical gap 23.5 mm wide of infinite extent contains oil of specific gravity 0.95 and viscosity 2.45 N-s/m². A metal plate 1.5 m × 1.5 m × 1.5 mm weighing 49 N is to be lifted through the gap at a constant speed of 0.1 m/s. Estimate the force required.*

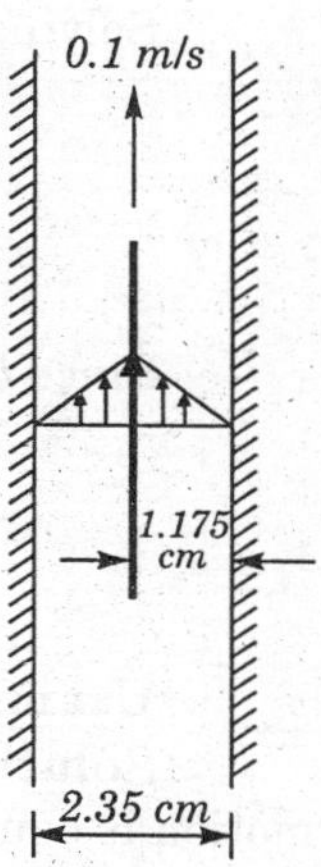

Solution. Let the plate be placed midway in the gap. The velocity gradient

$$\frac{du}{dy} = \frac{0.1}{1.10 \times 100} = 9.09 \text{ sec}^{-1}.$$

Viscous resistance to be overcome by the plate

= Shear force on the plate

= Sum of the shear force acting on each face

$$= 2 \cdot \mu \frac{du}{dy} \cdot A$$
$$= 2 \times 2.45 \times 9.09 \times (1.5 \times 1.5)$$
$$= 100.1 \text{ N}$$

Total force required

$$= \text{Immersed weight of the plate + Viscous resistance}$$
$$= (49.0 - 0.95 \times 9810 \times 1.5 \times 1.5 \times 0.0015) + 100.1$$
$$= 118.0 \text{ N.}$$

Example 1.17. *Through a very narrow gap of height h, a thin plate of a very large extent is being pulled at constant velocity V. On one side of the plate is oil of viscosity μ and on the other side oil of viscosity Kμ. Calculate the position of the plate so that drag force on it will be a minimum.*

Solution. Let the thin plate be placed at a distance y from one of the surfaces as shown. The drag force per unit area of the plate

F = Sum of shear forces per unit area on both the faces of plate

$$= \mu \left(\frac{du}{dy}\right)_u + K\mu \left(\frac{du}{dy}\right)_b$$

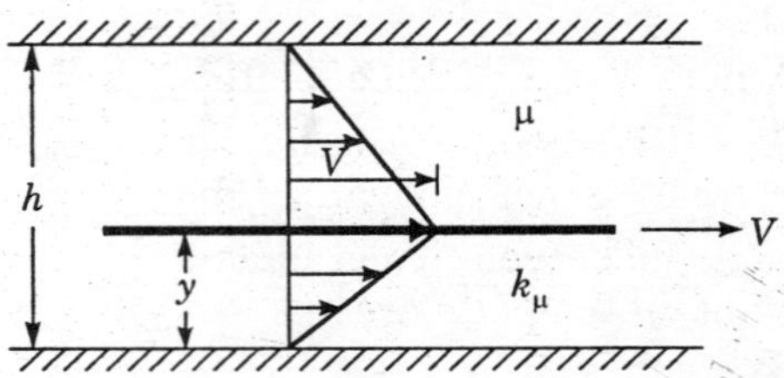

where the subscripts u and b refer to the upper and bottom sides of the plate, thus

$$F = \mu \frac{V}{(h - y)} + K\mu \frac{V}{y}$$

For the drag force to be minimum,

$$\frac{dF}{dy} = 0$$

$$\frac{dF}{dy} = \mu V \frac{1}{(h - y)^2} - K\mu \frac{V}{y^2} = 0$$

or $$\frac{\mu}{K\mu} = \frac{h^2 + y^2 - 2hy}{y^2} = \frac{h^2}{y^2} + 1 - 2\frac{h}{y}$$

or $$\frac{h^2}{y^2} - 2\frac{h}{y} + \left(1 - \frac{\mu}{K\mu}\right) = 0.$$

Solving the quadratic equation for h/y

$$\frac{h}{y} = \frac{2 \pm \sqrt{4 - 4\left(\frac{\mu}{K\mu}\right)}}{2} = 1 \pm \sqrt{\frac{\mu}{K\mu}}$$

Since h/y cannot be less than unity, using the plus sign

$$\frac{h}{y} = 1 + \sqrt{\frac{\mu}{K\mu}} \quad \text{or} \quad y = \frac{h}{1 + \sqrt{\frac{\mu}{K\mu}}}$$

Example 1.18. *Calculate the approximate viscosity of the oil for the following case :*

Solution. When the constant velocity of 0.5 m/s is attained, the viscous resistance to the motion is equal to the component of the weight of plate along the slope.

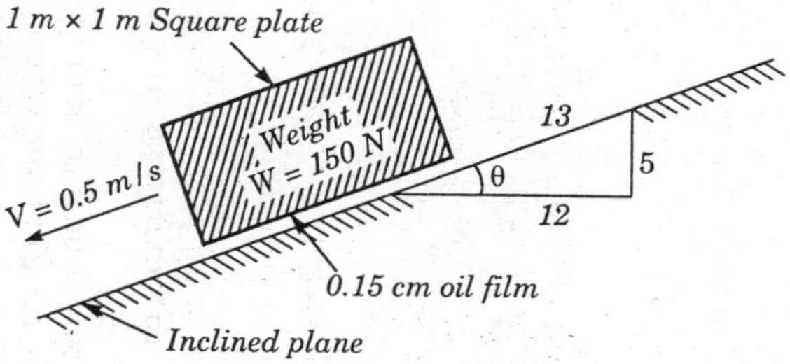

Component of weight along the slope

$$= W \sin\theta = 150 \times \frac{5}{13} = 56.6 \text{ N}$$

Viscous resistance $= \mu \cdot \dfrac{du}{dy} \cdot A = \mu \cdot \dfrac{0.5}{(0.15/100)} (1 \times 1) = 333.5\,\mu$

Hence $333.5\,\mu = 56.6$

$\therefore$ $\mu = \dfrac{56.6}{333.5} = 0.0173 \times 9.81 = 0.17 \text{ N-s/m}^2 = 0.17 \text{ Pa.s.}$

1.5. Surface Tension

The molecules of a liquid are held together by a force of attraction known as *cohesion*, the magnitude of which is very small, yet it enables the liquid to withstand a small tensile stress. The liquid molecules exert an attractive force upon all other molecules with which it comes into contact. The force of attraction between the molecules of two different liquids which do not mix or between the liquid molecules and molecules of solid boundary containing the liquid, is known as *adhesion*.

At surface of contact between a gas and a liquid (like air and water) or between two different immiscible (liquids that do not mix with each other) liquids, molecular attraction introduces a force which causes the interface (*i.e.* the contact surface) to behave like a membrane under tension. Within the body of a liquid a molecule is attracted equally in all directions by the other molecules surrounding it, but at the liquid-air interface or at the contact surface between two immiscible liquids the upward and downward attractions are unbalanced, giving rise to the phenomenon of surface tension. The liquid-air interface behaves as if it were an elastic membrane under tension. This surface tension is the same everywhere on the surface, and acts in the plane of the surface normal to any line in the surface. As shown in Fig. 1.6, a liquid molecule at the free surface will be exerted upon by a smaller force from the free surface side, giving rise to a resultant downward force acting at right angles to the free surface. This imbalance of molecular force gives rise to surface tension. It is a force which exists on the surface of a liquid when it is in contact with another fluid or a solid boundary. Its magnitude depends upon the relative magnitude of cohesive and adhesive forces. Surface tension is a force in the liquid surface and acts normal to a line of unit length drawn imaginarily on the surface. Thus, it is a line force. It represents surface energy per unit area (unit surface energy), and is denoted by Greek letter σ (sigma). It has dimensions $[FL^{-1}]$ or $[MT^{-2}]$ and is expressed in kg/m in MKS units and N/m in SI units.

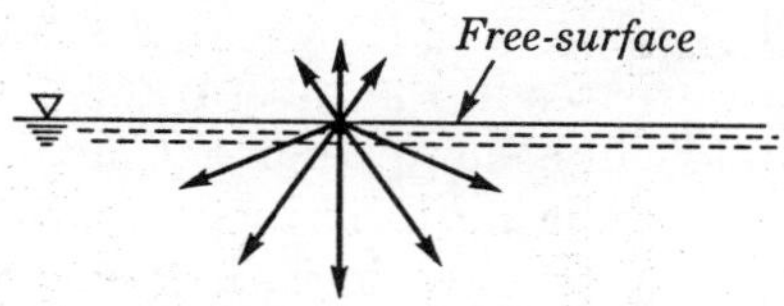

Fig. 1.6. Intermolecular force near a free surface.

A small needle gently placed on the liquid surface will not sink and will be supported by the surface tension. This is made possible by the localised curvature and depression caused by the needle.

1.5.1. Effect of temperature

As the surface tension depends directly upon the intermolecular cohesion, and since cohesion is known to decrease with temperature rise, the surface tension decreases with the rise of temperature. Its value for water-air contact (free-surface of water) reduces from 0.00745 kg/m (0.0731 N/m) at 17.8°C to 0.00596 kg/m (0.0585 N/m) at 100°C.

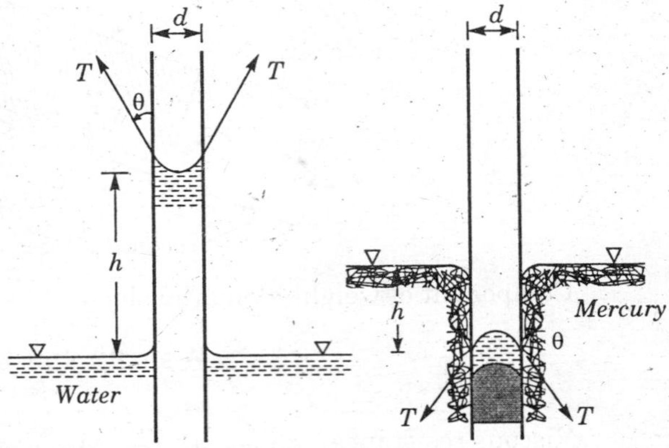

Fig. 1.7. Capillary action in a glass tube.

1.5.2. Capillary action

The molecules of a solid surface attract liquid molecules with a greater force than that which exist between the liquid molecules (except mercury). Because of this, most of the liquids completely wet the surface. The extent of wetting will depend upon the relative magnitude of the forces of adhesion and cohesion. If the adhesive force is greater than that of cohesion, the liquid tends to spread out and wet the surface. If the cohesive force is greater than the adhesive force, a small drop of the liquid placed on the solid surface will remain in the drop form *e.g.* a small mercury drop retains its almost spherical shape while resting on a solid surface. Mercury does not wet the surface because of its greater cohesion. If water drops are placed on a solid surface, they will completely spread out over the surface and will wet it. This is due to the fact that the adhesion between the molecules of water and the solid surface is greater than the cohesion between the water molecules.

The free surfaces of mercury and water will, therefore, behave differently at places where they come in contact with a solid surface. When a glass tube is dipped vertically into water, the water rises in the tube. If the glass tube is placed in mercury, the surface of mercury inside the tube will be lower than the outside level. Fig. 1.7 exhibits, what is known as the capillary action. The rise of water in the tube is called the capillary rise and the fall of mercury is termed as the capillary depression. The phenomenon of rise and fall of liquid in a capillary tube is known as *capillarity*. Its magnitude depends upon the diameter of tube, the specific weight of the liquid and its surface tension, and may be obtained by the following analysis. If the angle of contact between the liquid and the solid surface is θ, the water in the glass tube will continue to rise until the vertical component of the surface tension force ($T \cos \theta$) which acts over the wetted length (circumference of the tube) at the free surface equals the weight of the water column. Thus

$$T \cos \theta = \frac{\pi d^2}{4} hy$$

where $T = \sigma \pi d$. Substituting this value of T, the capillary rise h is given by

$$h = \frac{(4 \cos \theta)\sigma}{\gamma d} \quad \text{...(1.6)}$$

For pure water and clean glass surface, θ is almost equal to zero, but under actual conditions the water is neither pure nor the glass is clean. For water in contact with glass and air, Gibson has obtained the value of θ as 25° 32′, and that of σ = 0.0075 kg/m (0.0735 N/m). If h and d are expressed in mm. Eq. (1.6) reduces to

$$h = \frac{27.07}{d} \text{ mm} \quad \text{...(1.7)}$$

A similar analysis for glass tube placed in mercury shows that the mercury is depressed by an amount h given by Eq. (1.6). For mercury Gibson has obtained θ = 128° 52′ and the specific gravity and the surface tension may be taken as 13.55 and 0.53 kg/m (5.2 N/m) respectively. With h and d both expressed in mm, Eq. (1.6) for glass-air contact becomes

$$h = \frac{9.6}{d} \text{ mm} \qquad \text{...(1.8)}$$

Glass tubes are commonly used for measuring pressure of flow, and in order that they give correct pressure observations, it is necessary that the rise of water or any other liquid should not be influenced by the capillary action. To ensure this, the diameter of tube should be large enough so that the capillary rise is negligible.

Table 1.2 gives the values of h for glass tubes of different diameter for water and mercury as calculated from the above equation.

Table 1.2

	Capillary rise or depression, h in mm					
Tube diameter, d in mm	2	5	10	15	20	25
1. Water	13.4	5.4	2.7	1.8	1.3	1.1
2. Mercury	4.8	1.9	1.0	0.6	0.5	0.4

From this table it is obvious that if a smaller tube is used for pressure measurement, the height of liquid in the tube which indicates pressure of flow will not represent the correct pressure, as it includes the capillary rise. For this reason, the diameter of tube should never be less than 1 cm.

Liquid drop, jet and soap bubble :

In case of a liquid drop or inside a jet, the action of the surface tension is to increase the pressure inside, in relation to the outside pressure. In a liquid drop of diameter d, if Δp is the difference of pressure between the inside and outside of the drop, then using Fig. 1.8 (*a*).

$$\Delta p \,.\, \pi d^2/4 = \sigma \pi d$$

$$\Delta p = \frac{4\sigma}{d} \qquad \text{...(1.9)}$$

and in case of a liquid jet of diameter d and of unit length, we have from Fig. 1.8 (*b*).

$$2\sigma = \Delta p \,.\, d$$

$$\Delta p = \frac{2\sigma}{d} \qquad \text{...(1.10)}$$

A soap bubble in air has two surfaces in contact with air, one inside and the other outside, Fig. 1.8 (*c*). The forces that act on the hemispherical section are same as those for the drop, but the surface tension force is twice as great. The pressure difference is given by

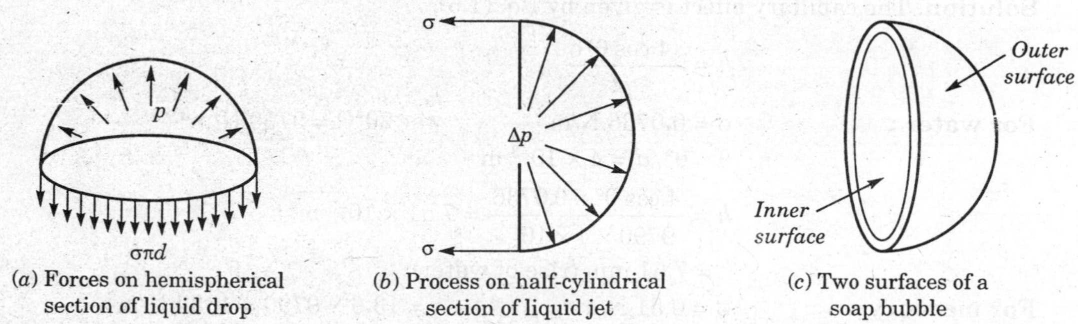

(*a*) Forces on hemispherical section of liquid drop (*b*) Process on half-cylindrical section of liquid jet (*c*) Two surfaces of a soap bubble

Fig. 1.8

$$\Delta p \, . \, \frac{\pi d^2}{4} = 2\pi d\sigma$$

or
$$\Delta p = \frac{8\sigma}{d} \quad ...(1.11)$$

A soap solution has a high value of σ, which causes a soap bubble to be larger in diameter for small pressure of blowing.

Example 1.19. *A soap bubble 25 mm in diameter has inside pressure of 20.0 N/m² above atmosphere. Calculate the tension in the soap film.*

Solution. Using Eq. (1.11), $\Delta p = \dfrac{8\sigma}{d}$

∴ Surface tension in soap film $\sigma = \dfrac{\Delta p \, . \, d}{8}$

$$= \frac{20}{8} (25/1000) = \mathbf{0.0625\ N/m.}$$

Example 1.20. *If the surface tension of water in contact with air is 0.075 N/m, what correction need be applied toward capillary rise in the manometric reading in tube of 3 mm diameter.*

Solution. Assuming the manometer tube made of glass, for water-air-glass contact gibson has determined the angle of contact θ as 25° 32′, and surface tension σ = 0.0075 kg/m. The capillary rise is given by Eq. (1.6) as

$$h = \frac{(4 \cos \theta)\, \sigma}{\gamma d}$$

$$= \frac{4 \cos 25.533° \times 0.075}{9810 \times (3/1000)} = 0.009198 \text{ m} = \mathbf{9.198\ mm}$$

However for pure water and clean glass, θ = 0°, and the capillary rise,

$$h = \frac{4\sigma}{\gamma d}$$

$$= \frac{4 \times 0.075}{9810 \times (3/1000)} = 0.01019 \text{ m} = 10.19 \text{ mm}$$

Note. From the above two values of h, it can be seen that the effect of impure water and unclean glass is reflected in decrease in the value of capillary rise.

Example 1.21. *Calculate the capillary effect in millimetres in a glass tube of 4 mm diameter, when immersed in (i) water and (ii) mercury. The temperature of the liquid is 20°C and the values of surface tension of water and mercury at 20°C in contact with air are 0.0736 N/m and 0.51 N/m respectively. The angle of contact for water is zero and that for mercury 130°.*

Solution. The capillary effect is given by Eq. (1.6),

$$h = \frac{(4 \cos \theta)\sigma}{\gamma d}$$

For water : σ = 0.0736 N/m γ at 20°C = 9790 N/m³

θ = 0°, $d = 4 \times 10^{-3}$ m

$$h = \frac{4 \cos 0° \times 0.0736}{9790 \times 4 \times 10^{-3}} = 7.51 \times 10^{-3} \text{ m}$$

= 7.51 mm (rise of water)

For mercury : σ = 0.51 N/m γ = 13.6 × 9790 = 133 kN/m³

θ = 130°, $d = 4 \times 10^{-3}$ m

$$\therefore \quad h = \frac{4 \times \cos 130° \times 4.51}{133 \times 10^3 \times 4 \times 10^{-3}} = -2.46 \times 10^{-3} \text{ m}$$

$$= -2.46 \text{ mm (depression).}$$

Example 1.22. *A U-tube is made of two capillaries of diameter 1.0 mm 1.5 mm respectively. The tube is kept vertically and partially filled with water of surface tension 0.0736 N/m and zero contact angle. Calculate the difference in the levels of miniscii caused by the capillarity.*

Solution. Capillary rise in a circular tube is given by Eq. (1.6),

$$h = \frac{4\sigma \cos \theta}{\gamma d}$$

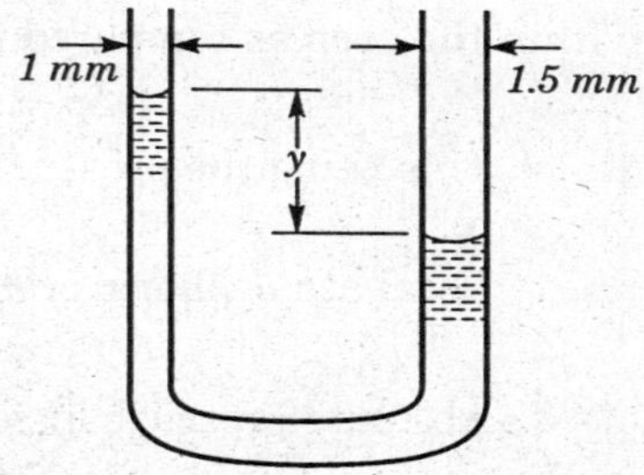

According to the data given

$$\theta = 0°, \sigma = 0.0736 \text{ N/m}$$

(*i*) Capillary rise in 1.0 mm tube

$$h_1 = \frac{4 \times 0.0736 \times 1}{9810 \times 1.0 \times 10^{-2}} = 0.030 \text{ m}$$

$$= 30 \text{ mm}$$

(*ii*) Capillary rise in 1.5 mm tube

$$h_2 = \frac{4 \times 0.0736 \times 1}{9810 \times 1.5 \times 10^{-2}} = 0.020 \text{ m}$$

$$= 20 \text{ mm.}$$

∴ The difference in levels of water in the two limbs caused by the surface tension effect

$$y = h_1 - h_2 = 30 - 20 = \mathbf{10 \text{ mm.}}$$

Example 1.23. *The diameters of the two glass limbs of a differential U-tube manometer were found to be 5 mm and 6 mm respectively. In an experiment the differential pressure readings of 50, 100, 250, 400 and 500 mm were indicated by the manometer. Determine the percentage error caused by the capillary effect. Take surface tension of water as 0.0736 N/m and the angle of contact as zero.*

Solution. Capillary rise in

(*i*) **5 mm tube**

$$h_1 = \frac{4\sigma \cos \theta}{\gamma d_1} = \frac{4 \times 0.0736 \times 1}{9810 \times 5 \times 10^{-3}} = 0.006$$

$$= 6 \text{ mm}$$

(*ii*) **6 mm tube**

$$h_2 = \frac{4\sigma \cos \theta}{\gamma d_2} = \frac{4 \times 0.0736 \times 1}{9810 \times 6 \times 10^{-3}}$$

$$= 0.005 = 5 \text{ mm}$$

∴ Capillary effect $= h_1 - h_2$

$$= 6 - 5 = 1 \text{ mm.}$$

Percentage Errors :

(*i*) When pressure difference is 50 mm, % error $= \pm \frac{1}{50} \times 100 = \pm 2\%$

(*ii*) When pressure difference is 100 mm, % error $= \pm \frac{1}{100} \times 100 = \pm 1\%$

(*iii*) When pressure difference is 250 mm, % error $= \pm \frac{1}{250} \times 100 = \pm 0.4\%$

(*iv*) When pressure difference is 400 mm, % error $= \pm \frac{1}{400} \times 100 = \pm 0.25\%$

(*v*) When pressure difference is 500 mm, % error $= \pm \frac{1}{500} \times 100 = \pm 0.20\%$.

Example 1.24. *Determine the absolute pressure and the gauge pressure that would exist within :*

(i) a spherical droplet of water 5 mm in diameter

(ii) a jet of water 5 mm in diameter.

Surface tension of water at the prevalent temperature is 0.0736 N/m and the barometer reading stands at 750 mm of mercury. Take specific gravity of mercury as 13.55 and specific weight of water as 9810 N/m³.

Solution. Case (*i*) For a spherical droplet of water, Eq. (1.9) gives differential pressure within the drop as compared to the outside atmosphere as

$$\Delta p = 4\sigma/d = 4 \times 0.0736/5 \times 10^{-3} = 58.86 \text{ N/m}^2$$

This being the pressure measured above the local atmospheric one, it represents the gauge pressure.

Local atmospheric pressure $= 13.55 \times 750$ mm water $= 9810 \times 13.55 \times 750 \times 10^{-3}$ N/m²

$$= 99.67 \text{ kN/m}^2 = 99.67 \text{ kPa.}$$

Absolute pressure inside the droplet

$$= \text{Gauge pressure} + \text{Local atmospheric pressure}$$

$$= 58.86 + 99.67 \times 10^3 = 99.73 \text{ kPa}$$

Case (*ii*) For the liquid jet, the differential pressure is

$$\Delta p = \frac{2\sigma}{d} = \frac{2 \times 0.0736}{5 \times 10^{-3}} = 29.43 \text{ N/m}^2 = 29.43 \text{ Pa}$$

Absolute pressure inside the liquid jet

$$= 29.43 + 99.67 \times 10^3 \text{ N/m}^2 = \mathbf{99.7 \text{ kN/m}^2}.$$

Example 1.25. *In measuring the unit surface energy of a mineral oil (sp. gr. = 0.85) by the bubble method, a tube having an internal diameter of 1.5 mm is immersed to a depth of 12.5 mm in the oil. Air is forced through the tube forming a bubble at the lower end. What magnitude of unit surface energy will be indicated by a maximum bubble pressure intensity of 147.15 Pa ?*

Solution. Specific weight of the mineral oil

$$\gamma = 0.85 \times 9810 = 8338.5 \text{ N/m}^3$$

Pressure at a depth of 12.5 mm $= \gamma h = 8338.5 \times \dfrac{12.5}{1000} = 104.2 \text{ N/m}^2 = 104.2 \text{ Pa}$

Pressure difference between inside and outside of the bubble

$$\Delta p = 147.15 - 104.2 = 42.95 \text{ Pa}$$

Using Eq. (1.9), the unit surface energy

$$\sigma = \frac{\Delta p \,.\, d}{4} = 4 \times \frac{1.5}{10^3} = 16.1 \text{ N/m.}$$

Example 1.26. *Air is introduced through a nozzle into a tank of water to form a stream of bubbles. If the bubbles are intended to have a diameter of 2 mm, calculate by how much the pressure of the air at the nozzle must exceed that of the surrounding water. Assume that surface tension is 71.6 mN/m.*

Solution. The pressure at the tip of the nozzle must be the same as the excess pressure inside the air bubble. The difference in pressure between inside and outside of the bubble is given by Eq. (1.9),

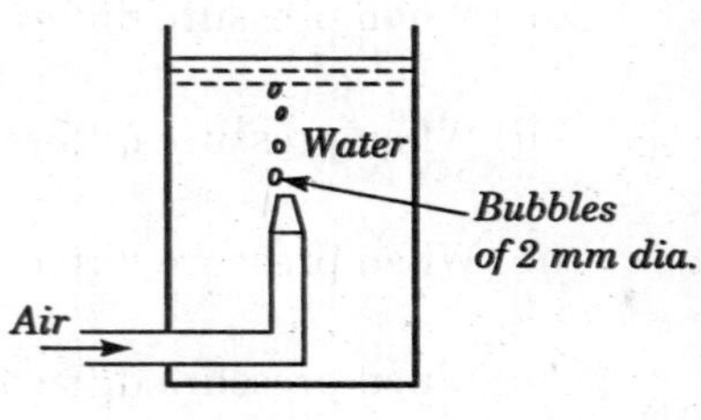

$$\Delta p = \frac{4\sigma}{d} = \frac{4 \times 71.6 \times 10^{-3}}{2 \times 10^{-3}}$$

$$= \mathbf{143.23 \text{ N/m}^2 = 143.23 \text{ Pa}.}$$

1.6. Vapour Pressure

All liquids possess a tendency to vaporise when exposed to air or gaseous atmosphere. The vaporisation takes place due to liquid molecules escaping from the free surface. The rate at which this vaporisation occurs depends upon the molecular energy of the liquid (which is dependent upon the nature of liquid and its temperature) and the condition of the atmosphere adjoining it. Consider a liquid contained in a sealed container, Fig. 1.9. Let a constant temperature be maintained within the container. Some of the liquid molecules have sufficient energy to break away from the liquid surface and enter the air space in the vapour state, see Fig. 1.9 (*i*). After a certain time, the air will contain enough liquid molecules to exert a partial pressure of air on some of the molecules forcing them to rejoin the liquid surface, Fig 1.9 (*ii*). Eventually, when equilibrium condition is established, the rate at which the molecules are leaving the liquid surface will be the same as the rate of return of molecules. In this condition the air above the liquid is saturated with liquid vapour molecules and the pressure exerted on the liquid surface is called the vapour pressure. A liquid with a high vapour pressure evaporates readily and is known as a volatile liquid. The boiling of a liquid is closely related to its vapour pressure. When the pressure impressed on the liquid surface is brought slightly below the vapour pressure limit, the liquid starts boiling. This means that the boiling can be achieved either by raising the temperature of the liquid so that its vapour pressure rises or by lowering the pressure of the overlying air below the vapour pressure of the liquid.

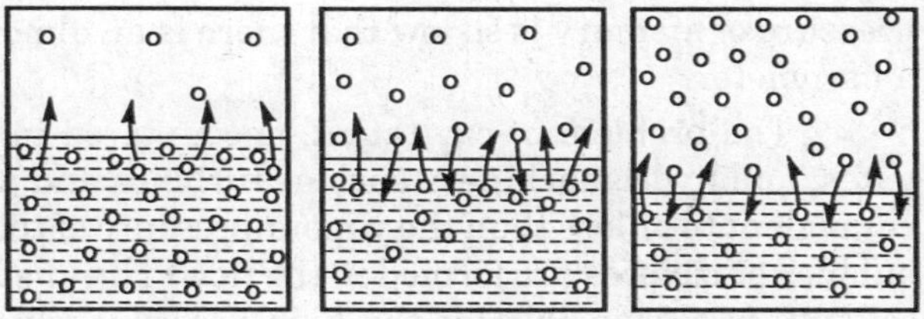

Fig. 1.9

The vapour pressure depends upon the molecular activity which is a function of temperature, as such it depends upon the temperature and increases with it. Table 1.3 gives the vapour pressure of some important liquids, while Table 1.4 shows the vapour pressure of water at different temperatures.

Table 1.3. Vapour Pressure at 20°C

Liquid	*N/m² (Pa)*
Water	2.345×10^3
Kerosine	3.310×10^3
Benzene	10.000×10^3
Petrol	30.400×10^3
Mercury	0.160

Table 1.4. Vapour Pressure of Water

T(°C)	*N/m²(Pa)*	*m of water*
0	0.610×10^3	0.063
10	1.230×10^3	0.125
15	1.62×10^3	0.165
20	2.345×10^3	0.239
30	4.27×10^3	0.437
40	7.400×10^3	0.762
50	12.36×10^3	1.275
60	20.0×10^3	2.075
80	47.5×10^3	4.960
100	101.5×10^3	10.790

The low vapour pressure of mercury, as evidenced from Table 1.3 (along with its high density) makes it very suitable for use in barometers and other pressure measuring devices. The vapour pressure of mercury is so low that there is an almost perfect vacuum above the mercury column in a barometer.

The problem of cavitation, encountered in hydraulic structures like spillways and sluice gates, and hydromachinery such as turbines and pumps, is a direct result of local pressures being equivalent to or less than the vapour pressure of the liquid. When such a situation develops, vapour bubbles or cavities are formed in the flow. The unsteady nature of these bubbles and their ultimate collapse is responsible for the high pressure which often leads to vibrations, noise, pitting and erosion of metal parts of machines and concrete surfaces of hydraulic structures.

Tables 1.5 and 1.6 show the properties of some common fluids and that of water respectively. Table 1.7 deals with the properties of standard atmosphere.

Table 1.5. Properties of some common fluids at 20°C and atmospheric pressure

	Density, ρ	*Specific wt., γ*	*Dyn. viscosity 10^5 μ*	*Kin. viscosity 10^7 υ*	*Surface tension* σ*	*Vapour pressure, p_v*	*Bulk Modulus of Elasticity, E_v*
	kg/m³	*N/m³*	*Ns/m²(Pa.s)*	*m²/s*	*N/m*	*N/m²(Pa)*	*N/m²(Pa)*
Air	1.235	12.12	1.853	150.0	—	—	—
Benzene	888.0	8650.0	65.3	7.43	0.0255	1.0×10^4	1.035×10^9
Carbon tetrachloride	159.40	15.64×10^3	96.5	6.04	0.0265	1.305×10^4	1.105×10^9
Castor oil	960.0	9410.0	8.04×10^4	1.0×10^4	0.0392	—	1.44×10^9
Ethyl alcohol	788.9	7730.0	119.7	15.0	0.0216	5.78×10^3	1.206×10^9
Glycerine	1270.0	12.45×10^3	9.04×10^4	6.3×10^3	0.0490	1.37×10^4	4.35×10^9
Kerosine	800.0	7850.0	188.0	22.9	0.0235	—	—
Mercury	13.53×10^3	13.28×10^4	155.0	1.16	0.5100	0.173	2.62×10^{10}
Water	998.0	8.9×10^3	100.0	10.0	0.0735	2.39×10^3	2.11×10^9

*In contact with air.

Table 1.6. Properties of water at different temperatures

Temperature T degree C	*Density, ρ*	*Specific wt., γ*	*Dyn. viscosity 10^5 μ*	*Kin. viscosity $10^7 υ$*	*Surface tension* $10^3σ$*	*Vapour pressure p_v*	*Bulk modulus of Elasticity $10^{-6} E_v$*
	kg/m³	*N/m³*	*N-s/m² (Pa.s)*	*m²/s*	*N/m*	*N/m² (abs) (Pa)*	*N/m² (Pa)*
0	1000.0	9810.0	179.3	17.93	75.7	611.0	2016.0
5	1000.0	9810.0	155.5	15.30	74.9	883.0	
10	1000.0	9810.0	131.2	13.20	74.2	1229.0	
15	999.0	9800.0	113.8	11.50	73.6	1618.0	
20	990.0	9780.0	100.1	10.00	72.9	2340.0	2110.0
25	997.0	9770.0	90.0	9.02	72.2	3217.0	
30	996.0	9760.0	79.9	8.09	61.2	4270.0	

35	994.0	9750.0	71.9	7.30	70.5	5585.0	
40	992.0	9730.0	65.1	6.62	69.6	7300.0	
45	990.0	9710.0	59.3	6.04	68.7	9510.0	
50	988.0	9690.0	55.0	5.53	68.0	12.36×10^3	2286.0
55	985.0	9660.0	50.7	5.21	67.1	16.03×10^3	
60	983.0	9640.0	46.6	4.82	66.0	20.45×10^9	
65	981.1	9620.0	43.6	4.50	65.3	24.90×10^3	
70	978.1	9600.0	40.7	4.17	64.4	31.05×10^4	
75	975.2	9560.0	38.0	3.90	63.6	38.55×10^3	
80	971.6	9520.0	35.6	3.68	62.7	47.60×10^3	
85	968.3	9500.0	33.5	3.46	61.8	57.30×10^3	
90	964.0	9450.0	31.5	3.28	60.9	69.20×10^3	
95	961.0	9425.0	29.9	3.01	59.9	84.75×10^3	2138.0
100	959.0	9400.0	28.4	2.90	59.0	101.30×10^3	

*In contact with air.

Table 1.7. Properties of standard atmosphere

Altitude Z *km*	Temperature T *degree C*	*Pressure absolute* *N/m² (Pa)* $10^{-3}\,p$	*Density* *kg/m³* ρ	*Kin. viscosity* *m²/s* $10^5\,\nu$	*Velocity of sound* *m/s* C
0	15	101.2	1.225	1.45	341.0
0.3	13.3	97.6	1.193	1.49	339.2
0.6	11.1	94.6	1.156	1.52	338.0
0.9	9.1	91.2	1.125	1.56	337.4
1.2	7.2	87.7	1.090	1.59	337.3
1.5	5.2	83.2	1.085	1.64	335.0
3.0	– 4.7	70.6	0.910	1.85	329.0
4.5	– 14.6	68.2	0.775	2.10	322.5
6.0	– 23.6	47.5	0.667	2.42	316.0
7.5	– 34.4	38.7	0.555	2.70	310.0
9.0	– 44.3	31.4	0.460	—	303.4
10.5	– 54.2	25.0	0.382	—	297.0
11.0	– 55.3	22.5	0.363	—	296.0
11.3	– 55.3	21.6	0.347	—	296.0

1.7. Properties of Gases

Gases are highly compressible fluids and are characterised by change in density. The change in density is achieved by both change in pressure and change in temperature. The absolute pressure p, the specific volume v and the absolute temperature T are related by the equation of state. For a perfect gas, the equation of state per unit weight is

$$pv = RT \qquad ...(1.12)$$

where R = a gas constant, the value of which depends upon the particular gas.

For air the value of the gas constant R is 287 N-m/kg-K or (J/kg-K) in SI. Equation (1.12) may also be written as

$$p = \rho RT \qquad ...(1.13)$$

which may be used to compute the density if the gas constant R is known. The absolute temperature (T) is measured in °K (Kelvin), and is related to temperature in Celcius by the equation

$$T = 273 + T_c, \text{ where } T_c = \text{Temperature in °C.}$$

Another fundamental equation for a perfect gas is

$$pv^n = p_1v_1^{\ n} = p_2v_2^{\ n} = \text{constant} \qquad ...(1.14)$$

in which n may have any value from zero to infinity depending upon the process to which the gas is subjected. By combining Eqs. (1.12), (1.13) and (1.14) the following useful relationship can be obtained :

$$\frac{T_2}{T_1} = \left(\frac{p_2}{p_1}\right)^{(n-1)/n} = \left(\frac{v_1}{v_2}\right)^{n-1} = \left(\frac{\rho_1}{\rho_2}\right)^{n-1} \qquad ...(1.15)$$

Example 1.27. *One cubic metre of air at 40°C and pressure 0.105×10^6 N/m^2 is compressed adiabatically to 0.5 m^3. What are the temperature and pressure of the gas ? If the process had been isothermal, what would be the temperature and pressure ?*

Solution. The gas constant for air, R = 287 N-m/kg °K

From the equation of state, $pv = RT$

Substituting the given data,

$$0.105 \times 10^6\, v_1 = 287(273 + 40), \text{ from which } v_1 = 1.17 \text{ m}^3/\text{kg}$$

For adiabatic process, $pv^k = \text{constant}$

$$p_2(0.5)^{1.4} = 0.105 \times 10^6(1.17)^{1.4}$$

$$p_2 = 0.277 \times 10^6 \text{ N/m}^2.$$

Also, $0.277 \times 10^6(1.17 \times 0.5) = 287(273 + T_2)$

$$T_2 = 292\text{°K}.$$

If the process is isothermal, T_2 = 40°C.

$$pv = \text{constant}$$

and

$$p = \frac{0.105 \times 10^6 \times 1}{0.5} = 52.5 \times 10^3 \text{ N/m}^2.$$

1.7.1. Isothermal Process

The compression and expansion of a gas may take place according to various laws of thermodynamics. If the temperature is kept constant, the process is known as *isothermal* and the value of n is unity. Eq. (1.13) is then written as : p/ρ = constant.

1.7.2. Isentropic Adiabatic Process

If the process is such that no heat is added to or withdrawn from the gas (the case of zero heat transfer), it is said to be an *adiabatic process*. An isentropic process is the one in which there is no friction and hence is a reversible process. An isentropic adiabatic process is accompanied by a decided change in temperature. The value of the exponent n in Eq. (1.14) is then denoted by k, the ratio of specific heats at constant pressure and constant volume. The ratio k is called the adiabatic constant for the gas. Eq. (1.14) then changes to :

$$p/\rho^k = \text{constant}.$$

For actual gases the exponent k is nearly constant over a wide range of states. If the process is adiabatic but not frictionless, it is described by an equation like Eq. (1.14) with an exponent slightly different from k. The exponent is smaller than k for expansion and is larger than k for compression.

1.8. Incompressible and compressible Fluids

Fluid mechanics deals with both incompressible and compressible fluids, that is, with fluids of constant and variable densities. Although there is no such thing in reality as an incompressible fluid, this term is applied where the change in density with pressure is so small as to be negligible. This is usually the case with all liquids. Gases, too, may be considered incompressible when the pressure variation is small compared with the absolute pressure.

In liquid pipelines at times of sudden or rapid valve closure, a high pressure wave is generated giving rise to the phenomenon of *water hammer*. In water hammer problems, it is, therefore, necessary to consider the compressibility of liquid.

The flow of air in a ventilating system is a case where a gas may be considered incompressible, because the pressure variation is so small that the change in density is of no importance. But in case of a gas or steam flowing through a pipe line at high velocity, the drop in pressure may be so great that the resulting change of density cannot be ignored. For an aeroplane flying at or less than 400 kmph, the air may be considered to be of constant density. But as an object moving through the air approaches the velocity of sound, which is around 1150 kmph, the pressure and density adjacent to the body become significantly different from those of the air at some distance away. Under such circumstances, the air must be treated as a compressible fluid.

1.8.1. Compressibility

The compressibility is the measure of change in volume (or density) when a substance is subjected to pressure. The reciprocal of coefficient of compressibility β is known as the *bulk modulus of elasticity*. Thus

Coefficient of compressibility = Percentage change in volume for a given change in pressure.

or
$$\beta = 1/E_v$$

A fluid may be compressed by the application of pressure, thereby reducing its volume and giving rise to a volumetric strain. Such a compressed fluid will expand to its original volume when the applied pressure is withdrawn. This property of compressibility of a fluid is expressed by the bulk modulus of elasticity. If by applying a pressure dp, the decrease in the fluid volume is dV, then the bulk modulus of elasticity is defined as

$$E_v = -\frac{dp}{dV/V} \qquad \text{...(1.16)}$$

where V = original fluid volume. The negative sign indicates a decrease in volume with the increase in pressure. Since most of the liquids have a comparatively high value of bulk modulus of elasticity, the compressibility is very close to zero and hence the liquids are considered practically incompressible under ordinary conditions. The bulk modulus of elasticity of fluid is not a constant but increases with increasing pressure. Tables 1.5 and 1.6 indicate E_v-values for some common fluids and water respectively.

As the density is equal to the mass divided by volume, we have

$$\rho = \frac{m}{V}$$

Since the mass m of a certain volume V is constant, differentiating ρ,

$$d\rho = d\left(\frac{m}{V}\right) = -\frac{mdV}{V^2} = -\rho\frac{dV}{V}$$

$$-\frac{dV}{V} = \frac{d\rho}{\rho} \qquad \text{...(1.17)}$$

From Eqs. (1.16) and (1.17),

$$E_v = \rho\frac{dp}{d\rho} \qquad \text{...(1.18)}$$

(*i*) For an isothermal process, p/ρ = constant.

$$\therefore \qquad \frac{dp}{d\rho} = \text{constant} = \frac{p}{\rho}$$

and the bulk modulus $E_v = p$ ...(1.19)

(*ii*) If the process is isentropic,

$$\frac{p}{\rho^k} = \text{constant.}$$

Differentiating, $dp = \text{constant.}\ k\rho^{k-1}\, d\rho = k\rho^{k-1}\, d\rho\, \frac{p}{\rho^k} = k\, \frac{d\rho}{\rho} p$

$$\therefore \qquad \frac{dp}{d\rho} = k\left(\frac{p}{\rho}\right)$$

and the bulk modulus, $E_v = \rho k \left(\frac{p}{\rho}\right) = kp$ ...(1.20)

The velocity of sound through a fluid medium is expressed by

$$C = \sqrt{\frac{dp}{d\rho}} \qquad \text{...(1.21)}$$

Small pressure disturbances travel through the fluid medium at a velocity which depends upon bulk modulus and the density of the fluid. Using Eqs. (1.18) and (1.21),

$$C = \sqrt{\frac{E_v}{\rho}} \qquad \text{...(1.22)}$$

in which C is the sonic velocity or the velocity of sound in the fluid medium.

The values of the bulk modulus E_v for air and water at standard atmospheric conditions are:

Fluid	*Value of* E_v (N/m²)
1. Air (isothermal process)	100×10^3
2. Air (isentropic process)	140×10^3
3. Water	2.11×10^9

The bulk modulus is also designated by E.

In the present age of high speed flight, the compressibility of fluid has assumed paramount importance. The Mach number M, which is the ratio of the velocity of flow V (or the velocity of a body) to the velocity of sound in the fluid medium ($M = V/C$) is a measure of the compressibility effects. If $M < 1$, the compressibility effects are of little importance, and when the Mach number exceeds unity ($M > 1$), the compressibility of the fluid affects the flow phenomena to an appreciable extent. Thus the Mach number offers a criterion to decide whether or not the fluid be assumed incompressible in the flow analysis. In case of an incompressible fluid, $E = \infty$ and so also is the sonic velocity $C = \infty$ resulting in a zero Mach number. It has been found through experience that if $M \leq 0.2$, the effects of compressibility are negligible. For air at room temperature, $M = 0.2$ corresponds to $V = 70$ m/s.

The Mach number of flow is also used to describe the flow as subsonic if $M < 1$, supersonic if $M > 1$, and hypersonic if $M > 5$.

Example 1.28. *The volume of a liquid is reduced by 1.2% by increasing the pressure from 0.40 MPa to 12.3 MPa. Estimate the modulus of elasticity of the liquid.*

Solution. The modulus of elasicity is given by

$$E_v = -\frac{dp}{dV/V} = -\frac{(12.3 - 0.40)}{\left(-\dfrac{1.2}{100} V\right)/V} = \mathbf{991.7\ MPa.}$$

Example 1.29. *A liquid with a volume of 0.2 m³ at 300 kPa is subjected to a pressure of 3000 kPa and its volume is found to decrease by 0.2%. Calculate the bulk modulus of eleasticity of the liquid.*

Solution. Bulk modulus of elasticity of a fluid is defined by Eq. (1.16),

$$E_v = -\frac{dp}{dV/V}$$

Pressure increase, $dp = 3000 - 300 = 2700$ kPa

Resulting dcrease in volume, $dV/V = 0.2\% = -\frac{0.2}{100} = \frac{(-2)}{1000}$

∴ Bulk modulus of the liquid

$$E_V = -\frac{2700}{(-2/1000)} = \mathbf{13.5 \times 10^5\ kPa.}$$

Example 1.30. *A cylinder contains 0.75 m³ of gas at 20°C and 2.5 bar pressure. After compression, the volume gets reduced to 0.15 m³. Determine final pressure and bulk modulus of compressed gas if compression takes place under :*

(i) Isothermal conditions

(ii) Adiabatic conditions (n = 1.4) *(RGPV, June 2013)*

Solution. (*i*) **Isothermal conditions :**

Equation of state, for a perfect gas Eq. (1.12) gives

$pv = RT$, which for isothermal conditions, T is constunt, hence = constant

or $p_1v_1 = p_2v_2$, where v_1 and v_2 are volumes per unit weight pressure after compression, *i.e.*,

$$p_2 = p_1\frac{v_1}{v_2} = 2.5 \times \frac{0.75}{0.15} = \mathbf{12.5\ bar}$$

∴ Final pressure is 12.5 bar.

Bulk modulus of compressed gas, E_v is given by Eq. (1.16) :

$$E_v = -\frac{dp}{dv/v} = -\frac{p_2 - p_1}{(v_2 - v_1)/v_1} = -\frac{12.5 - 0.75}{(0.15 - 0.75)/0.75} = 12.5 \text{ bar}$$

$$1 \text{ bar} = 10^5 \text{ N/m}^2 = 10^5 \text{ Pascal}$$

$$E_v = \mathbf{12.5 \times 10^5\ N/m^2}$$

(*ii*) **Adiabatic compression :**

For a perfect gas, Eq. (1.14) gives another relationship

$$pv^n = \text{constant}$$

or

$$p_1v_1^{\,n} = p_2v_2^{\,n}, \quad \therefore \quad p_2 = p_1\left(\frac{v_1}{v_2}\right)^n$$

$$= 2.5\left(\frac{0.75}{0.15}\right)^{1.4} = \mathbf{23.79\ bar}$$

∴ Final pressure after compression = 23.79 bar

Bulk modulus of compressed gas

$$E_v = -\frac{dp}{dv/v} = -\frac{23.79 - 2.5}{(0.15 - 0.75)/0.75} = 26.6125 \text{ bar}$$

$$= \mathbf{26.61 \times 10^5\ N/m^2}$$

Example 1.31. *Determine the velocity of sound at 20°C and 101.2 kN/m² in (i) air, (ii) water, and (iii) mercury. Use table 1.5 for ρ and E_v.*

Solution. The velocity of sound in a fluid medium is given by

$$C = \sqrt{E_v/\rho}\,.$$

(*i*) The thermodynamic process involving air may be considered as adiabatic and frictionless (isentropic) for which, $E_v = kp$. The velocity of sound then becomes

$$C = \sqrt{kp/\rho}$$

Using the equation of state $p = \rho RT$, the above equation may be written

$$C = \sqrt{kRT}$$

in which R has the dimensions $(L^2/T^2\theta)$.

For air, $k = 1.4$, $R = 287$ J/kg °K.

Making substitutions, $C = \sqrt{1.4 \times 287 \times (273 + 20)} = 343.0$ m/s.

(*ii*) From table 1.5, for water,

$$\rho = 998.0 \text{ kg/m}^3$$
$$E_v = 2.11 \times 10^9 \text{ N/m}^2$$
$$C = \sqrt{E_v/\rho} = 1470 \text{ m/s.}$$

(*iii*) From table 1.5, for, mercury,

$$\rho = 13530 \text{ kg/m}^3$$
$$E_v = 2.62 \times 10^{10} \text{ N/m}^2$$
$$C = \sqrt{E_v/\rho} = 1392 \text{ m/s.}$$

Example 1.32. *For determining the depth of sea at a place, a charge was exploded at 100 m below the sea water surface. The first reflected wave was recorded after 2.5 seconds at the surface. Calculate the depth assuming the sea has a flat bottom. Average value of bulk modulus of elasticity of sea water is 1.96×10^9 N/m^2 and its specific weight is 10×10^3 N/m^3.*

Solution. $E_v = 1.96 \times 10^9$ N/m^2, $\rho = 10^4/9.81 = 1020$ kg/m^3.

Velocity of sound in sea water $= \sqrt{E_v/\rho} = 1385$ m/s

Let the depth of sea below the sea water surface be d m. The distance travelled by the reflected sound wave

$$= (d - 100) + d = 2d - 100$$

Time taken by the first reflected wave to reach the surface is

$$2.5 = \frac{2d - 100}{1385}$$

from which d = **1781.25 m.**

Example 1.33. *At standard atmospheric conditions, determine the increase in pressure necessary to cause :*

(*i*) *1% reduction in the volume of water,*

(*ii*) *1% reduction in the volume of air subjected to isentropic compression,*

(*iii*) *1% reduction in the volume of air when subjected to isothermal compression.*

Take the volume of the bulk modulus of elasticity the same as given in Ex. 1.22.

Solution. From Eq. (1.16), $E_v = -dp/dV/V$

Reduction of volume by 1% results in, $-dV/V = \dfrac{1}{100} = 0.01$

(*i*) Increase in pressure of water to produce 1% reduction in its volume

$$dp = 1.96 \times 10^9 \times 0.01 = 1.96 \times 10^7 \text{ N/m}^2$$
$$= 19.6 \text{ MN/m}^2 = 19.6 \text{ MPa.}$$

(*ii*) For air subjected to isentropic process, the bulk modulus is given by Eq. (1.20) :

$$E_v = kp = 1.4 \times 101.3 \times 10^3 = 1.42 \times 10^5 \text{ N/m}^2 = 142 \text{ kPa.}$$

$\therefore$ Increase in pressure, $dp = 142 \times 10^3 \times 0.01 = 1.42 \times 10^3 \text{ N/m}^2 = 1.42 \text{ kPa.}$

(*iii*) For air under isothermal process, Eq. (1.19) gives the bulk modulus,

$$E_v = p = 101.3 \text{ kN/m}^2$$

$\therefore$ Increase in pressure, $dp = 101.3 \times 10^3 \times 0.01 = 1.01 \times 10^3 \text{ N/m}^2 = 1.01 \text{ kPa.}$

From the values, it is evident that the increase in pressure of water to cause 1% reduction in its volume is extremely large, being about 15,000 times greater, as compared to that required for air for similar reduction in volume. Liquids are, therefore, considered incompressible, unless subjected to sudden and large pressure changes.

Example 1.34. *To measure the depth of sea, a device employed takes the form of a solid steel vessel with a partition having a recess and a valve. The upper portion of the vessel is filled with water and the lower one with mercury. When the vessel is lowered, the seawater penetrates into the lower part of the vessel through a small orifice and forces the mercury up through the valve. The compressibility of water reduces its volume. Determine the depth of the sea if the upper part of the vessel contains 600 g of mercury when the device reaches the bottom. The volume of water in the upper part of the vessel is 1000 cm³. Assume the specific weight of water constant and equal to 1050 kg/m³. Take Bulk modules of elasticity of water E_v = 2.13 × 10⁴ kg/cm². Neglect the compressibility of mercury.*

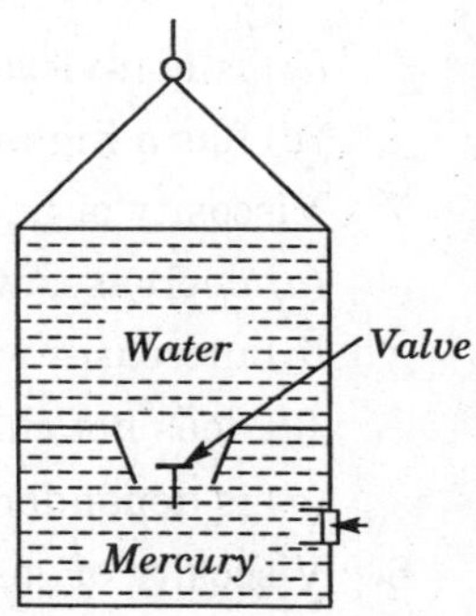

Solution. $$E_v = -\frac{\Delta p}{\Delta V/V} = \frac{\Delta p}{(-\Delta V/V)}$$

$$V = 1000 \text{ cm}^3 \text{ ; weight = volume} \times \text{density}$$

$$600 \text{ g} = \Delta V \times (13.6 \times 1 \text{ gm/cm}^3)$$

where ΔV = change in volume of water

or $$\Delta V = \frac{600}{13.6} = 44.12 \text{ cm}^3$$

$$\left(-\frac{\Delta V}{V}\right) = \frac{44.12}{1000} = 0.04412$$

Making substitutions,

$$2.13 \times 10^4 = \frac{\Delta p}{0.04412}$$

or $$\Delta p = 938.67 \text{ kg/cm}^2$$

From the pressure-depth relationship,

$$h = \frac{\Delta p}{\gamma_{sea}} = \frac{938.67}{(1/100)^2 \times 1050} = \mathbf{8939.74 \text{ m}}$$

$\therefore$ Depth of the sea = **8939.74 m**

OBJECTIVE TYPE QUESTIONS

Select the correct answer(s)

1. A fluid is defined as a substance which—

(*a*) takes the shape and volume of the container into which it is poured

(*b*) is highly compressible

(*c*) has a constant shear stress throughout

(*d*) deforms continuously under the action of a shear stress.

2. An ideal fluid is one which—
   (*i*) is compressible (*ii*) is non-viscous and incompressible
   (*iii*) has low density (*iv*) is elastic and viscous.
3. Fluid continuum is a concept in which—
   (*a*) fluid is non-homogeneous (*b*) fluid density is very low
   (*c*) fluid particles are very closely spaced
   (*d*) fluid particles are widely scattered in space.
4. Viscous deformations in fluid flow are—
   (*i*) inversely proportional to the dynamic viscosity
   (*ii*) directly proportional to the dynamic viscosity
   (*iii*) independent of shear stress and kinematic viscosity
   (*iv*) dependent on the pressure.
5. The Newton's law of viscosity is a relationship between—
   (*a*) shear stress and pressure (*b*) viscosity and temperature of fluid
   (*c*) shear stress and velocity gradient (*d*) pressure and viscosity
6. A Newtonian fluid is one which—
   (*i*) has a specific weight of 1 newton/m^3
   (*ii*) has a linear relationship between the shear stress and the resulting rate of deformation (*i.e.* velocity gradient)
   (*iii*) is non-viscous and incompressible
   (*iv*) has a high viscosity.
7. Viscosity of liquids—
   (*a*) decreases with decrease in fluid temperature
   (*b*) increases with decrease in fluid temperature
   (*c*) does not change with fluid temperature
   (*d*) is dependent of pressure.
8. Viscosity of gases—
   (*i*) decreases with decrease in fluid temperature
   (*ii*) increases with decrease in fluid temperature
   (*iii*) does not change with fluid temperature
   (*iv*) is dependent on pressure.
9. The dimensions of dynamic viscosity μ are—
   (*a*) $ML^{-1}T^{-2}$ (*b*) $ML^{-1}T^{-1}$
   (*c*) MLT^{-2} (*d*) ML^2T^{-1}
10. In case of solid mechanics, the law similar to Newton's law of viscosity is—
    (*i*) Hooke's law (*ii*) Newton's second law of motion
    (*iii*) Archemede's principle (*iv*) Newton's first law
11. The following numerical values represent the magnitude of kinematic viscosity (in m^2/s) of water at different temperatures—
    (*a*) 6.62×10^{-7} (*b*) 10.0×10^{-7} (*c*) 17.93×10^{-7}.
    If the temperatures involved are 0, 20 and 40 degree C, identify the above values with the corresponding temperature.

12. Match the kinematic viscosity of each liquid choosing the correct value from column *B*—

Col. (A) Name of liquid	*Col. (B)* Kinematic viscosity in m^2/s
(*i*) Water	(*a*) 150×10^{-7}
(*ii*) Glycerine	(*b*) 10×10^{-7}
(*iii*) Air	(*c*) 6.3×10^{-4}.

13. The following table shows the dynamic viscosity of air in N s/m^2 at atmospheric pressure at different temperatures—

(*i*) 2×10^{-5} (*ii*) 1.90×10^{-5}

(*iii*) 1.81×10^{-5} (*iv*) 1.715×10^{-5}

If the temperatures involved are 0, 20, 40 and 60 degree C, identify each viscosity value with its temperature.

14. MKS unit of 1 kg force is equal to how many newtons—

(*a*) 0.981 (*b*) 98.1 (*c*) 9.81 (*d*) 981.

15. To convert the MKS unit of dynamic viscosity (kg s/m^2) into poise, the multiplying factor is—

(*i*) 89.1 (*ii*) 981

(*iii*) 98.1 (*iv*) 9.81.

16. 1 poise is equal to

(*a*) 1 dyne s/cm^2 (*b*) 98.1 dyne s/cm^2

(*c*) 1 dyne s/m^2 (*d*) 1 kg s/m^2

(*e*) 1 kg s/cm^2.

17. To convert the MKS unit of viscosity (kg s/m^2) into its SI equivalent (N s/m^2) multiply by—

(*i*) 98.1 (*ii*) 981

(*iii*) 9.81 (*iv*) 0.981.

18. One stoke is equal to—

(*a*) 1 cm^2/s (*b*) 1 m^2/s

(*c*) 1 ft^2/s (*d*) 1 mm^2/s.

19. The FPS unit of kinematic viscosity, ft^2/s, is equal to how many stokes—

(*i*) 93 (*ii*) 930

(*iii*) 9.30 (*iv*) 9300.

20. Surface tension is a phenomenon due to—

(*a*) cohesion only (*b*) viscous force

(*c*) adhesion between liquid and solid molecules

(*d*) difference in magnitude between the forces due to adhesion and cohesion.

21. Weight of liquid that rises in a capillary tube is supported by—

(*i*) the friction between the tube wall and the liquid

(*ii*) the atmospheric pressure

(*iii*) the vertical component of force due to surface tension

(*iv*) the curvature of the miniscus.

22. The capillary depression in mercury is on account of—

(*a*) adhesion being greater than cohesion

(*b*) surface tension being larger than the viscosity

(*c*) cohesion being greater than the adhesion

(*d*) vapour pressure being small.

23. The capillary rise or depression in a small diameter tube is—

(*i*) directly proportional to the diameter

(*ii*) inversely proportional to the surface tension

(*iii*) directly proportional to the surface tension

(*iv*) inversely proportional to the diameter.

24. The following are the values of surface tension in N/m for certain liquids :

(*a*) 0.5100 (*b*) 0.0735 (*c*) 0.0235

If the liquids involved are water, kerosene and mercury, identify the above values with the corresponding liquid.

25. The pressure within a soap bubble is—

(*i*) the same as that of the surrounding atmosphere

(*ii*) greater than the external pressure

(*iii*) less than the external pressure

(*iv*) equal to the vapour pressure.

26. The following expressions give the pressure difference between inside and outside of a bubble, a liquid drop and a liquid jet (not in that order, necessarily) :

(*a*) $\frac{2\sigma}{d}$ (*b*) $\frac{8\sigma}{d}$ (*c*) $\frac{4\sigma}{d}$.

Identify each of thee above values with the corresponding item.

27. An incompressible flow is one in which—

(*a*) the temperature of fluid remains constant

(*b*) the density does not change with pressure

(*c*) the fluid is non-viscous

(*d*) the fluid compressibility is non-zero.

28. A measure of the effect of compressibility in fluid flow is the magnitude of a dimensionless parameter known as—

(*i*) Reynolds number (*ii*) Mach number

(*iii*) Weber number (*iv*) Froude number

(*v*) Strouhl number.

29. Gas-flows can be treated as incompressible when the Mach number is less than—

(*a*) 0.5 (*b*) 1.0

(*c*) 0.2 (*d*) 0.1

(*e*) 0.05.

30. For air flow at room temperature to be incompressible, the fluid velocity must not exceed—

(*i*) 100 m/s (*ii*) 70 m/s

(*iii*) 50 m/s (*iv*) 25 m/s.

31. Density, pressure and temperature in a gas flow are related by the—

(*i*) First law of thermodynamics (*ii*) Newton's law of viscosity

(*iii*) Equation of state (*iv*) Equation of motion.

32. A perfect gas is the one—

(*a*) which is incompressible and viscous

(*b*) which obeys the equations of state

(*c*) which follows the Newton's law of gravity

(*d*) which exists in isothermal flows only.

33. A fluid-flow process is isothermal only when—

(*i*) the fluid pressure does not change

(*ii*) the density change is small

(*iii*) the fluid temperature remains constant

(*iv*) there is no heat transfer.

34. An adiabatic fluid-flow is one in which—

(*a*) the fluid temperature does not vary

(*b*) the heat is neither added to nor withdrawn from the gas (the case of zero heat transfer)

(*c*) the pressure remains constant

(*d*) the heat transfer has a non-zero value.

35. An isentropic adiabatic process of fluid-flow is one in which—

(*i*) the heat-transfer is non-zero

(*ii*) there is no friction and the process is reversible

(*iii*) there is no change in the temperature

(*iv*) the process is irreversible.

36. The following table shows the vapour pressure of certain liquids at 20 degree C—

(*a*) 3100 (*b*) 0.0163

(*c*) 239 (*d*) 337.

If the liquids happen to be water, kerosene, mercury and petrol, identify these with their respective values.

37. Mercury is used in barometers on account of—

(*i*) its high density (*ii*) negligible capillarity effect

(*iii*) very low vapour pressure (*iv*) its low compressibility.

38. Spherical shape of droplets of mercury is due to—

(*a*) high density (*b*) high surface tension

(*c*) high adhesion (*d*) low vapour pressure.

39. Arrange the "evaporability" of the following liquids in the decreasing order of magnitude—

(*a*) Kerosene (*b*) Petrol

(*c*) Mercury (*d*) Water.

(**Note.** Evaporability is directly proportional to the vapour pressure).

40. Arrange the following fluids in the decreasing order of their compressibility—

(*a*) water (*b*) air at 1 atm. pressure

(*c*) a gas at 5 atm. pressure (*d*) air at 0.5 atm. pressure.

(**Note.** Compressibility is the reciprocal of the bulk modulus of elasticity).

41. Capillary rise of water in a glass tube depends primarily on its diameter and the angle of contact. Arrange the following sizes of glass tube in increasing order of capillary rise of water.

(*a*) 5 mm (*b*) 2 mm

(*c*) 10 mm (*d*) 1 mm

42. Glass tubes of the same diameter are dipped vertically in vessels containing different liquids. Arrange the following liquids such that the capillary effect in the respective tubes is in the decreasing order of magnitude—

(*a*) kerosene (*b*) water (*c*) mercury.

(**Note.** The capillary effect for the same tube but different liquids is proportional to σ/γ.)

43. Vapour pressure of a liquid is due do—
(*i*) the pressure of flow
(*ii*) the molecules of liquid which hang over the free-surface
(*iii*) the pressure of air above the free surface
(*iv*) the existence of free-surface.

44. The unit of dynamic viscosity of a fluid is
(*a*) m^2/s (*b*) $N.s/m^2$
(*c*) $Pa.s/m^2$ (*d*) $kg.s^2/m^2$.

ANSWERS TO OBJECTIVE TYPE QUESTIONS

1. (*e*)	**2.** (*ii*)	**3.** (*c*)	**4.** (*i*)	**5.** (*c*)
6. (*ii*)	**7.** (*b*)	**8.** (*i*)	**9.** (*b*)	**10.** (*a*)
11. (*a*) : 0°C, (*b*) : 20°C, (*c*) : 40°C		**12.** (*i*) (*b*), (*ii*) (*c*), (*iii*) (*a*),		
13. (*i*) 60°C, (*ii*) 40°C, (*iii*) 20°, (*iv*) 0°C			**14.** (*c*)	**15.** (*iii*)
16. (*i*)	**17.** (*iii*)	**18.** (*a*)	**19.** (*ii*)	**20.** (*d*)
21. (*iii*)	**22.** (*c*)	**23.** (*iii*), (*iv*)		
24. (*a*) mercury, (*b*) water, (*c*) kerosene			**25.** (*ii*)	
26. (*a*) jet, (*b*) bubble, (*c*) drop		**27.** (*b*)	**28.** (*ii*)	**29.** (*c*)
30. (*ii*)	**31.** (*iii*)	**32.** (*b*)	**33.** (*iii*)	**34.** (*b*)
35. (*ii*)	**36.** (*a*) petrol, (*b*) mercury, (*c*) water, (*d*) kerosene			
37. (*iii*), (*i*)	**38.** (*b*)	**39.** (*b*), (*a*), (*d*), (*c*),	**40.** (*d*), (*b*), (*c*), (*a*)	
41. (*c*), (*a*), (*b*), (*d*)	**42.** (*b*), (*c*), (*a*)	**43.** (*ii*)	**44.** (*b*).	

REVIEW QUESTIONS

1. Distinguish between solids, liquids and gases.
2. Is there any analogy of Hooke's law in fluids ? If so, state the parallel law in fluids.
3. Explain why the following statements are right or wrong :
   (*i*) Ideal fluid can sustain a shearing stress when in motion.
   (*ii*) Fluids cannot sustain shearing stress when at rest.
   (*iii*) Molecular viscosity may be expressed in gm sec^{-1} cm^{-1} or dyne sec^{-2}.
4. Enunciate Newton's law of viscosity and distinguish between Newtonian and non-Newtonian fluids.
5. Explain why petrol evaporates more readily than water at ordinary temperature ?
6. Explain how certain insects are able to walk on the surface of water ?
7. Under what conditions is the miniscus between two liquids in a class tube : (*i*) concave upwards and (*ii*) concave downwards ?
8. Comment on the role of capillary action in (*i*) an oil lamp and (*ii*) a fountain pen.
9. Why is it necessary in winter to use a *lighter* oil for automobiles than in summer ? To what property does the term *light* refer ?
10. Will the viscous resistance to the flow of honey be greater or lesser than the viscous resistance to the flow of water ?
11. Is the pressure intensity, (*a*) within a soap bubble (*b*) within a drop of liquid, probably greater than, equal to or less than that of the surrounding medium ?
12. Explain what do you mean by capillarity ?

13. Cite examples where surface tension effects play a prominent role.
14. Answer briefly the following questions : (*i*) define a fluid from mechanics point of view, (*ii*) can a fluid sustain (*a*) tension (*b*) volumetric compression (*c*) shear ? In each case state the relationship between the applied stress and the corresponding shear developed.
15. To what type of flow is the Newton's law of viscosity applicable ? Can it be used to determine shear stress in turbulent flow ?
16. Classify the fluids on the basis of existence of interface (Free-surface).
17. What are the characteristic fluid properties to which the following phenomena are attributable, *viz* :

    (*i*) rise of sap in a trees, (*ii*) spherical shape of a drop of liquid,
    (*iii*) cavitation, (*iv*) flow of a jet of oil in an unbroken stream,
    (*v*) water hammer ?

    Express the quantities involved in the metric system and also in terms of fundamental units.
18. Explain the property of fluids on the basis of molecular motion.
19. Define Newtonians fluids.

PROBLEMS

1.1. If the equation of velocity profile is $u = 3y^{2/3}$ (u in cm/s, y in cm), what is the velocity gradient at the boundary and at 10 cm from it ?

1.2. A plate 2.5×10^{-5} m distant from a fixed plate moves at 0.60 m/s and requires a force of 1.96 N/m² to maintain this speed. Determine the fluid viscosity of the substance that fills the space between the plates.

1.3. A piston 12 cm dia and 15 cm long moves down in a 12.04 cm dia cylinder. The oil filling the annular space has a viscosity of 8.0×10^{-2} Ns/m² and the weight of the piston is 9.81 N. Find the speed with which the piston slides down.

1.4. If two coaxial cylinders 10 cm and 9.75 cm in dia and 25 cm high have a certain liquid filled in between, find the viscosity of the liquid which produces a torque of 0.98 Nm upon the inner cylinder when the outer one rotates at the rate of 90 rpm.

1.5. If the velocity distribution over a plate is given by $u = \frac{2}{3}y - y^2$

in which u is the velocity in m/s at a distance y metres above the plate, determine the shear stress at $y = 0$ and $y = 0.15$ m. Take $\mu = 0.863$ Ns/m².

1.6. Derive the dimensions of :

(*i*) dynamic viscosity, and (*ii*) kinematic viscosity

and hence obtain their units in (*i*) MKS, (*iii*) FPS and (*iii*) SI systems.

1.7. A piece of pipe 0.5 m long weighing 9.81 N and having internal diameter of 5.25 cm is slipped over a vertical shaft 5.0 cm in diameter and allowed to fall. Calculate the approximate velocity attained by the pipe if a film of oil of viscosity 0.196 Ns/m² is maintained between pipe and shaft.

1.8. A piece of pipe of 5.25 cm internal diameter and 15 cm long slides down a vertical shaft of 5.0 cm diameter at a constant speed of 0.1 m/s. A vertical force 14.7 N is required to pull the pipe back up the shaft at the same constant speed. Calculate the approximate viscosity of oil which fills the small gap between the pipe and shaft.

1.9. A Newtonian fluid is filled in the clearance between a shaft and concentric sleeve. When a force of 490 N is applied to the sleeve parallel to the shaft the sleeve attains a speed of 70 cm/sec. If 2450 N force is applied, what speed will the sleeve attain ? The temperature of the sleeve remains constant.

1.10. A very large thin plate is centred in a gap of width 6 cm with different oils of unknown viscosities above and below, the viscosity of one being twice that of the other. When the plate is pulled at a velocity of 30 cm/sec, the resulting force on one square metre of plate due to viscous shear on both sides is 29.4 N. Assuming viscous flow and neglecting all end effects, calculate viscosities of the oils.

1.11. Two coaxial cylinders with the gap in between completely filled with a viscous fluid, one cylinder rotating while the other one remains stationary, show that the shear stress on the inner cylinder is always greater than that on the outer one. Indicate the velocity profiles for (*a*) inner cylinder rotating while outer one is stationary, and (*b*) outer cylinder rotating while the inner one is kept stationary.

1.12. A rotating viscometer consists of a disc pivoted above a stationary boundary, the fluid to be tested filling the very small space between the parallel surfaces. Means are available to measure the driving torque and the rotational speed of the disc. Determine by integration of the expression for shear stress over the lower surface of the disc., the torque-speed ratio which would be obtained for a liquid having a viscosity of 0.15 Ns/m^2 with a disc having 20 cm diameter and a boundary spacing of 0.19 cm.

1.13. (*a*) On a plot of shear stress *versus* velocity gradient represent a Newtonian and non-Newtonian fluid.

(*b*) Two large parallel flat plates are placed 1.25 cm apart. A 0.25 cm thick plate of 0.2 m^2 area is being towed in glycerine filled between the above plates with a constant force of 9.81 N. Calculate the towing speed of the plate when it is held equidistant from the two parallel plates. Take μ = 0.01 cm (mass) per cm sec.

1.14. A shaft of diameter 74.9 mm rotates in a bearing of diameter 75 mm and of length 75 mm. The annular space between the shaft and the bearing is filled with oil having a coefficient of viscosity of 0.2 stokes and the specific gravity 0.94. Determine the power in overcoming viscous resistance in this bearing at 1400 rpm.

1.15. Calculate the maximum capillary rise of water (20°C) to be expected in a vertical glass tube 1 mm in diameter. The surface tension at 20°C is 0.0718 N/m.

1.16. Derive an equation for theoretical capillary rise between vertical parallel plates.

1.17. Calculate the maximum rise of water (20°C) to be expected between two vertical, clean glass plates spaced 1 mm apart.

1.18. What force is necessary to lift a thin wire ring 2.5 cm in diameter from a water surface at 20°C ? Neglect the weight of ring.

1.19. A soap bubble 5 cm in diameter contains a pressure (in excess of atmosphere) of 20.07 N/m^2. Calculate the tension in the soap film.

1.20. Determine the velocity of sound in air at (*i*) 20°C and 101.2 kN/m^2, and (*ii*) 267°C and 706.0 kN/m^2 and find out the ratio of the two velocities.

1.21. Calculate the dynamic viscosity of standard air using table 1.7. Comment upon the effect of pressure on the dynamic viscosity.

1.22. Compute kinematic viscosity of air at atmospheric pressure using table 1.1. What conclusions can be drawn regarding the effect of temperature on the viscosity (both dynamic and kinematic) ?

1.23. A cubical block weighing 196.2 N and having a 20 cm edge is allowed to slide down on an inclined plane surface making an angle of 20° with the horizontal on which there is a thin film of oil having a viscosity of 2.16×10^{-3} Ns/m^2. What terminal velocity will be attained by the block, if the film thickness is estimated to be 0.025 mm ?

1.24. In a stream of glycerine in motion at a certain point the velocity gradient is 0.25 per second. The mass density of fluid is 1268 kg per cubic metre and the kinematic viscosity is 6.30 × 10^{-4} m^2/s. Calculate the shear stress at the point.

1.25. Calculate the capillary effect in millimeter in a glass tube of 4 mm diameter when immersed in (*i*) water and (*ii*) in mercury. The temperature of the liquids is 20°C, and the value of surface tension of water and mercury at 20°C in contact with air respectively 0.07357 N/m and 0.490 N/m. The contact angle for θ = 0°, and for mercury θ = 130° 24′.

ANSWERS

1.1. ∞, 0.9275 s^{-1}.

1.2. 81.8 × 10^{-6} Ns/m^2.

1.3. 0.434 m/s.

1.4. 0.7. Ns/m^2.

1.5. 0.574 N/m^2, 0.316 N/m^2.

1.6. (*i*) FTL^{-2}, kgs/m^2, lbs/ft^2, Ns/m^2. (*ii*) L^2T^{-1}, m^2/s, ft^2/s, m^2/s,

1.7. 1.515 m/s.

1.8. 3.72 Ns/m^2

1.9. 3.5 m/s.

1.10. 0.98 and 1.96 Ns/m^2

1.12. 2/3978.

1.13. 1.275 m/s.

1.14. 199.75 kW

1.15. 2.932 cm.

1.16. $h = \dfrac{2\sigma \cos \theta}{r_w d}$

1.17. 1.466 cm.

1.18. 11.28 × 10^{-2} N.

1.19. 0.1293 N/m.

1.20. 343 m/s, 465.5 m/s, 1.354.

1.21. 1.18 × 10^{-6} msl/m-s (1.775 × 10^{-5} kg/n-s), 1.50 × 10^{-6} msl/ms (1.50 × 10^{-6} kg/m-s).

1.22. 13.3 × 10^{-6} m^2/s, 7.85 × 10^{-6} m^2/s.

1.23. 19.4 m/s.

1.24. 19.94 × 10^{-2} N/m^2.

1.25. 7.5 mm in water, and 2.47 mm in mercury.

2

Fluid Statics

2.0. Introduction

A fluid at rest is characterised by absence of relative motion between adjacent or neighbouring fluid layers. Under such a condition, the velocity gradient, du/dy, has a zero value which results in a zero shear stress. Thus the viscosity of a fluid has no effect on fluids at rest.

There are two types of forces to be considered :

(i) due to gravity, and

(ii) due to static pressure.

There are no tangential or frictional forces as the shear stress is zero in the absence of any relative motion.

2.1. Variation of Static Pressure

Let us consider a fluid element of size $dx\ dz$ and of unit length as shown in the following figure. Let the static pressure at the centre of the element O having the co-ordinates (x, z), be p. The fluid element is in equilibrium under the influence the following five forces, which act in the horizontal and vertical directions x and z :

(i) the static pressure force on the left face of the element

$$= \left(p - \frac{dp}{dx}.\frac{dx}{2}\right)dz$$

(ii) the static pressure force on the right face

$$= \left(p + \frac{dp}{dx}.\frac{dx}{2}\right)dz$$

(iii) the static pressure force on the bottom face

$$= \left(p - \frac{dp}{dz}.\frac{dz}{2}\right)dx$$

(iv) the static pressure force on the top face

$$= \left(p + \frac{dp}{dz}.\frac{dz}{2}\right)dx,$$

and (v) the weight of the element,

$$dw = (dx.dz\ 1)\gamma$$

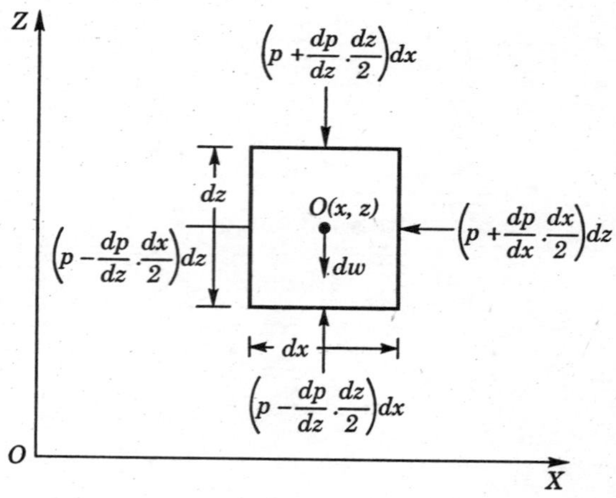

Fig. 2.1. Forces on a static fluid element.

The pressure is assumed to increase with increasing x and z, and, therefore, the pressure gradient in x and z directions, $\frac{dp}{dx}$ and $\frac{dp}{dz}$, are positive.

For static equilibrium, summation of forces in the x and z directions must be zero, thus in the x-direction, $F_x = 0$ (there being no acceleration) therefore :

$$\left(p - \frac{dp}{dx} \cdot \frac{dx}{2}\right)dz - \left(p + \frac{dp}{dx} \cdot \frac{dx}{2}\right)dz = 0$$

which simplifies to
$$\frac{dp}{dx} = 0 \qquad \text{...(2.1)}$$

This equation indicates that the pressure does not vary in the x-direction. In other words, the static pressure remains constant in the horizontal direction.

In the z-direction, $\Sigma F_z = 0$; therefore,

$$\left(p - \frac{dp}{dz} \cdot \frac{dz}{2}\right)dx - \left(p + \frac{dp}{dz} \cdot \frac{dz}{2}\right)dx - dx\, dz\, .\, \gamma = 0$$

which simplifies to
$$-\frac{dp}{dz} = \gamma. \qquad \text{...(2.2)}$$

From equations (2.1) and (2.2), it is evident that the static pressure p is a function of z only.

Integrating Eq. (2.2), for incompressible fluids (γ = constant)

$$\int_{p_1}^{p_2} dp = -\int_{z_1}^{z_2} \gamma dz$$

$$p_2 - p_1 = -\gamma(z_2 - z_1) \qquad \text{...(2.3)}$$

or
$$\frac{p_2}{\gamma} + z_2 = \frac{p_1}{\gamma} + z_1 \qquad \text{...(2.4)}$$

The expression $(p/\gamma + z)$ represents the piezometric head and Eq. (2.4) shows that for static fluid the piezometric head is constant throughout the fluid.

The result of integration of Eq. (2.2) may also be written as

$$p = -\gamma z + A \qquad \text{...(2.5)}$$

where A is a constant. At the free-surface of the liquid the pressure is atmospheric, and hence the static pressure is zero gauge.

Using Fig. 2.2, at $z = H$, $p = 0$.

Substituting this in Eq. (2.5), $A = \gamma H$, and the equation of pressure variation is

$$p = \gamma(H - z)$$

If the distances are measured from the free surface, the pressure at a distance h below the free surface will be

$$p = \gamma h \qquad \text{...(2.6)}$$

since $z = H - h$. Eq. (2.6) gives the variation of static pressure in the vertical direction, the distance h being measured from the free-surface. This equation describes the law of variation of static pressure, and is often termed as *hydrostatic law of pressure variation*.

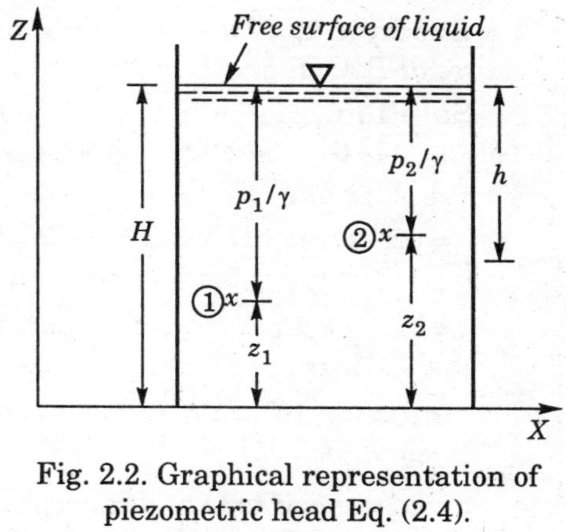

Fig. 2.2. Graphical representation of piezometric head Eq. (2.4).

Example 2.1. *The barometric pressure at sea level is 760 mm of mercury while that on a mountain top is 735 mm, if the specific weight of air is assumed constant at 11.77 N/m³, what is the elevation of the mountain top ?*

Solution. It is a known fact that as the elevation above the sea level increases, the atmospheric pressure decreases. Assuming the specific weight of the air as constant, one can write the following equation for pressure p at any elevation h above the sea level.

$$p = p_0 - \gamma_a \cdot h$$

where p_0 = standard atmospheric pressure, *i.e.* pressure at the mean sea level, and γ_a = specific weight of air.

The elevation of a place is given by

$$h = \frac{p_0 - p}{\gamma_a}$$

$$p_0 = \frac{760}{1000} \times 13.6 \times 9810 = 101.3 \times 10^2 \text{ N/m}^2 = 101.3 \text{ kN/m}^2$$

and $$p = \frac{735}{1000} \times 13.6 \times 9810 = 97.7 \times 10^3 \text{ N/m}^2 = 97.7 \text{ kN/m}^2$$

Therefore, $$h = \frac{(101.3 - 97.7)10^3}{11.77} = \mathbf{283.0 \text{ m.}}$$

Example 2.2. *The specific weight of the ocean water may be taken to vary according to the expression*

$$\gamma = \gamma_0 + \sqrt{C}\, h$$

where y_0 = the specific weight at the sea level, and h is the depth below the sea surface. Calculate the pressure as a function of the depth.

Solution. From Eq. (2.6), $dp/dh = \gamma = \gamma_0 + \sqrt{C}\, h$

Integrating w.r.t. h, $$p = \gamma_0 h + \frac{2}{3} C h^{3/2} + D$$

At the sea surface $h = 0, p = 0$

$\therefore$ $D = 0$

and the pressure-depth relationship is

$$p = \gamma_0 h + \frac{2}{3} C h^{3/2}$$

Example 2.3. *The container shown below holds kerosene and air as indicated. Compute the pressures at P, Q, R and S in kN/m². Take specific gravity of kerosene 0.8.*

Solution. The container at the left side is open to atmosphere. The pressure at the liquid surface at O is therefore atmospheric.

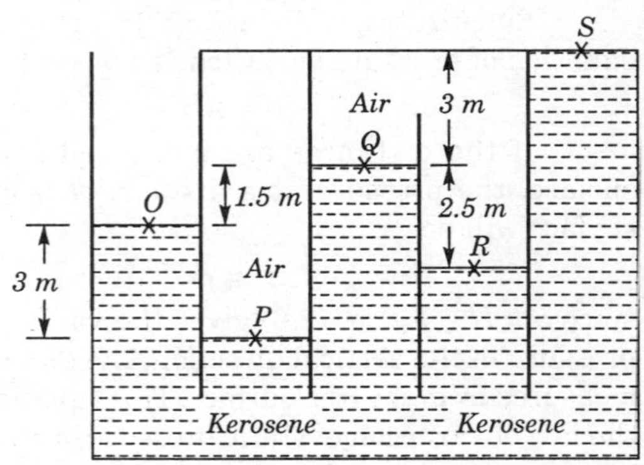

Pressure at P

= pressure at $O + 3 \times 0.8 \times 9810$ N/m²

= $101330 + 23544 = 124.9 \times 10^3$ N/m²

= 124.9 kN/m²

Pressure at Q

= $124.9 \times 10^3 - 4.5 \times 0.8 \times 9810$

= 89.6 kN/m²

Since the points Q and R are connected by air, pressure at Q and R are the same.

$\therefore$ Pressure at R = Pressure at Q = 89.6 kN/m²

Pressure at $S = 89.6 \times 10^3 - 5.5 \times 0.8 \times 9810 = 46.4$ kN/m².

2.1.1. Fluid with Variable Density

Equation (2.2) expresses a relationship between the pressure gradient and the specific weight ($\gamma = \rho g$) of a fluid, and is applicable to any fluid whether compressible or incompressible. For fluids with variable density, the specific weight is not constant, and therefore, Eq. (2.2) cannot be integrated to yield an expression for pressure p at any elevation z, where the fluid density is ρ, unless the relationship between p and γ, is known. Such problems of variable density are encountered in aeronautics, oceanography and meteorology. In oceanography, the pressure-specific weight relationship is worked out from elasticity considerations $\left(E = \dfrac{dp}{d\gamma/\gamma}\right)$. In aeronautics and meteorology, we are met with atmospheric air, the density, pressure and temperature of which change with elevation. For obtaining relationships among these variables, certain gas laws are used.

For gases (atmospheric air is treated as a perfect gas), the equation characterizing the polytropic process may be used.

$$\frac{p}{\rho^n} = A, \text{ in which } A \text{ is a constant.} \qquad ...(2.7)$$

Equation (2.7) is another form of Eq. (1.14) in which p the absolute pressure, ρ is the density and n a constant the value of which depends on the particular process involved. For a perfect gas under isothermal process, $n = 1$ and for adiabatic process $n = k$. From Eq. (2.2)

$$\frac{dp}{dz} = -\gamma = -\rho g. \qquad ...(2.2)$$

$$= -g\,(p/A)^{1/n}$$

Rearranging the terms $$\frac{dp}{\frac{1}{p^n}} = g\,\frac{dz}{\frac{1}{A^n}} \qquad ...(2.8)$$

Since the variation of g with z is very small (it decreases by about 0.033% per 1 km increase in elevation) it may be assumed constant. Integrating Eq. (2.8), the following may be obtained,

$$\frac{n}{n-1}\,p^{(n-1)/n} = -\frac{gz}{A^{1/n}} + B$$

The constant of integration B may be evaluated from known conditions of p and z at a place. At $z = 0$, let $p = p_0$, then

$$B = \frac{n}{n-1}\,p_0^{(n-1)/n}$$

Also from Eq. (2.7), $$\frac{p_0}{\rho_0^{\,n}} = A$$

Substituting for these constants,

$$\frac{n}{n-1}\,p_0^{(n-1)/n} = -\frac{g\,z\,\rho_0}{p_0^{1/n}} + \frac{n}{n-1}\,p_0^{(n-1)/n}$$

Rearranging terms, we obtain

$$\frac{n}{n-1}\left[p^{(n-1)/n} - p_0^{(n-1)/n}\right] = -\frac{g\,z\,\rho_0}{p_0^{1/n}}$$

$$\left(\frac{p}{p_0}\right)^{(n-1)/n} - 1 = -\frac{(n-1)}{n}\,\frac{gz\,\rho_0}{p_0^{(n-1)/n}\cdot p_0^{1/n}} = -\frac{(n-1)}{n}\,\frac{gz\,\rho_0}{p_0}$$

or $$\frac{p}{p_0} = \left[1 - \frac{(n-1)}{n}\,\frac{gz\,\rho}{p_0}\right]^{n/(n-1)}$$

but from the equation of state for a perfect gas.

$$p_0 = \rho_0\,RT_0 \qquad \text{or} \qquad \frac{p_0}{\rho_0} = RT_0.$$

Substituting for p_0/ρ_0, the following relationship between the pressure and the elevation may be written as

$$\frac{p}{p_0} = \left[1 - \frac{(n-1)}{n}\frac{gz}{RT_0}\right]^{n/(n-1)} \qquad ...(2.9)$$

In order to determine the pressure p at any elevation z, the binomial expression of Eq. (2.9) may be expanded, and the right hand side of the equation evaluated.

Temperature lapse-rate. The rate at which the temperature changes with the elevation is known as the *temperature lapse-rate*. The knowledge of temperature at any desired elevation is as important as the determination of pressure at that elevation, as the density of air is dependent on both the pressure as well as the temperature.

Making use of the equation of state $p = \rho RT$ and equation (2.7), one gets

$$A\rho^{n-1} = RT \quad \text{and} \quad A\rho_0^{n-1} = RT_0$$

which result in,

$$\left(\frac{\rho}{\rho_0}\right)^{n-1} = \frac{T}{T_0}$$

also

$$\frac{p}{p_0} = \frac{\rho}{\rho_0}\frac{T}{T_0} = \left(\frac{T}{T_0}\right)^{1/(n-1)}\frac{T}{T_0} = \left(\frac{T}{T_0}\right)^{n/(n-1)} \qquad ...(2.10)$$

Substituting for p/p_0 in terms of T/T_0 in Eq. (2.9)

$$\left(\frac{T}{T_0}\right)^{n/(n-1)} = \left[1 - \frac{n-1}{n}\frac{gz}{RT_0}\right]^{n/(n-1)}$$

or

$$\frac{T}{T_0} = \left[1 - \frac{n-1}{n}\frac{gz}{RT_0}\right] \qquad ...(2.11)$$

Equation (2.11) expresses the absolute temperature T as a function of z. Differentiating T with respect to z one obtains an expression for the lapse-rate.

$$\frac{dT}{dz} = -\frac{g}{R}\frac{(n-1)}{n} \qquad ...(2.12)$$

Solving Eq. (2.12) for n

$$n = \frac{1}{1 + \frac{R}{g}\left(\frac{dT}{dz}\right)} \qquad ...(2.13)$$

Since $n \geq 1$ and the elevation z is measured vertically upward, from Eq. (2.12), dT/dz is shown to have a negative value for $n > 1$, and is equal to zero if $n = 1$.

From the mean sea-level upto an elevation of about 11 km, the temperature of air decreases uniformly at the rate of 0.0065°C/m. From about 11 to 32 km, the temperature remains constant at – 56.5°C, on account of the lapse rate being zero. The atmospheric air is thus under isothermal conditions. The temperature rises again beyond an altitude of 32 km, as shown in Fig 2.3.

The *standard atmosphere* corresponds to values of n, p_0 and T_0 obtaining at 40° latitude. The atmosphere up to an altitude of 11 km is known as the *troposphere* and that beyond it and below 32 km is known as *stratosphere*. The value of n is 1.238 in the former, whereas it is 1 in case of the later.

The standard atmosphere is defined by the following conditions :

In MKS Units

p_0 = 10,330 kg/m^2, T_0 = 15°C, ρ_0 = 1.227 metric slug/m^3.

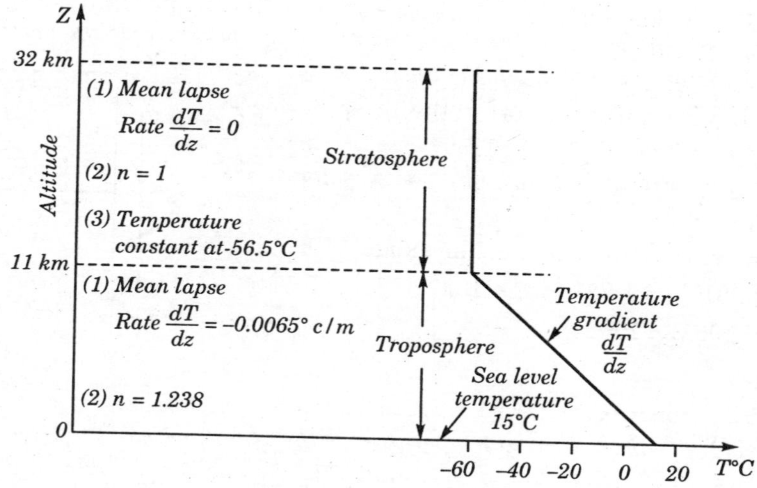

Fig. 2.3. Conditions at standard atmosphere.

In SI Units

$p_0 = 101{,}330$ N/m², $T_0 = 288$ K, $\rho_0 = 1.227$ kg (mass)/m³ and the temperature at any elevation z metres above the mean sea-level is given by

$$T = T_0 - 0.0065\, z$$

where T and T_0 can be either in Celcius or in Kelvin. The above temperature-altitude relationship is valid within the troposphere, *i.e.* $z \leq 11{,}000$ m. For the atmospheric air, temperature, pressure and density at different altitudes are tabulated as under :

Table 2.1

Altitude, z (m)	*Temp. T(C)*	*Pressure, p(N/m²)*	*Density, ρ, (kg/m³)*
1800	3.0	81,000	1.025
4600	− 14.7	57,000	0.77
8500 (Top of Mt. Everest)	− 40.0	33,000	0.493
14,000	− 56.5	15,000	0.237
21,000	− 56.5	4,500	0.0715
31,000	− 56.5	1,000	0.0171
46,000	45.0	144	0.0016
61,000	71.0	32	0.00032
76,000	− 22.0	55	0.000077

Example 2.4. *An open tank contains water for a depth of 2 m and above it oil for a depth of 1 m. If the oil has a specific gravity of 0.8, find the pressure intensity (a) at the interface of two liquids ; and (b) at the bottom of the tank.*

Solution. (*a*) Pressure intensity at the oil-water interface *A-B*

$$p = \gamma h$$
$$= (0.8 \times 9810)\,(1.0)$$
$$= \mathbf{7848\ N/m^2.}$$

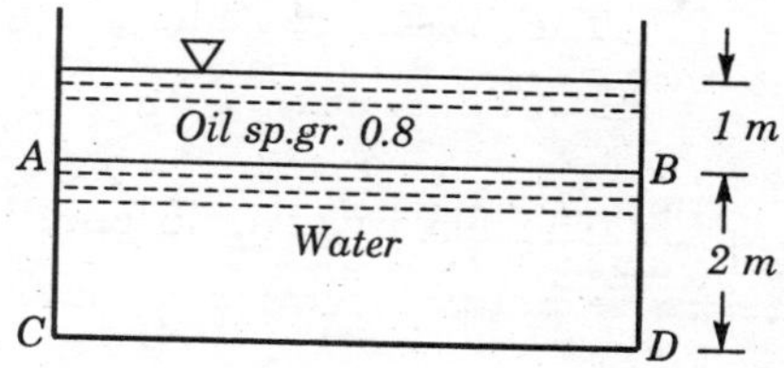

(*b*) Pressure intensity at the tank bottom *C-D*

$$p_{CD} = \gamma_0 h_0 + \gamma_w h_w$$
$$= (0.8 \times 9810)(1.0) + (9810)(2.0)$$
$$= 7848 + 19620 = \mathbf{27468.0\ N/m^2.}$$

Example 2.5. *A closed tank contains 0.5 m of mercury, 2 m of water, 3 m of oil of specific gravity 0.6 and there is air space above the oil. If the gauge pressure at the bottom of the tank is 196.2 kPa, what is the pressure of air at the top of the tank.*

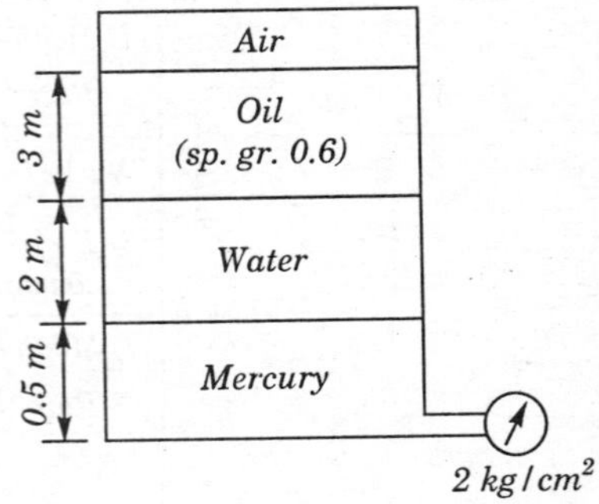

Solution. Let the gauge pressure in the air-space be p, then for the conditions given one can write,

p + pressure due to (3 m depth of oil + 2 m depth of water + 0.5 m depth of Hg)

= pressure indicated by the gauge

or $$p + 0.6 \times 9810 \times 3 + 9810 \times 2 + 13.6 \times 9810 \times 0.5 = 196.2 \times 10^3$$

from which,

$$p = \mathbf{92.2\ kN/m^2 = 92.2\ kPa.}$$

Example 2.6. *The pressure and temperature of the atmosphere at the ground level are 1 kgf/cm² and 26°C. Find the pressure and density of air at a height of 4,643 m. Assume an isothermal atmosphere.*

Solution. Treating atmospheric air as a perfect gas, for an isothermal process, the absolute pressure and the density are related by the following :

$$\frac{p}{\rho} = \frac{p_0}{\rho_0} = \frac{p_1}{\rho_1} = A \qquad ...(1)$$

It is a known fact that as the elevation (z) increases, atmospheric pressure decreases. One can therefore write

$$\frac{dp}{dz} = -\gamma = -\rho g = -\left(\frac{p}{A}\right)g \qquad ...(2)$$

or $$\frac{dp}{p} = -\left(\frac{g}{A}\right)dz$$

Integrating, $$\log p = -\frac{gz}{A} + B$$

At the ground level, let $z_0 = 0, p = p_0, B = \log p_0$

$$\log \frac{p}{p_0} = -\frac{gz}{A}, \text{ or } \frac{p}{p_0} = e^{-(gz/A)}$$

Using the equation of state, the deusity of atmospheric air at the ground level

$$\rho_0 = \frac{p_0}{RT_0}$$

Taking R = 274 Nm/kg°K, = 274 J/kg°K

$$\rho_0 = \frac{1 \times 10^4 \times 9.81}{274\,(273 + 26)} = 1.197 \text{ kg/m}^3$$

$$\frac{p_0}{\rho_0} = \frac{1 \times 10^4 \times 9.81}{1.197} = 81926, \therefore \quad A = 81926$$

The pressure at a height of 4643 m would be

$$p = p_0\, e^{-(9.81 \times 4643/81926)} = 1 \times 10^4 \times 9.81\, e^{-0.556}$$
$$= 56262.28 \text{ N/m}^2 = \mathbf{0.5735\ kgf/cm^2}$$

The density of atmospheric air at an altitude of 4643 m

$$\rho = \frac{p}{A} = \frac{56262.28}{81926} = \mathbf{0.687\ kg/m^3.}$$

Example 2.7. *The atmospheric pressure at the sea-level is 101.3 kN/m² and the temperature is 15°C, calculate the pressure 8 km above sea-level, assuming (i) air is incompressible, (ii) isothermal variation of pressure and density, and (iii) adiabatic variation of pressure and density.*

Assume sea-level density of air as 1.285 kg/m³. Neglect the variation of g with altitude.

Solution. (*i*) *Treating air as incompressible*

$$\frac{dp}{dz} = -\gamma.$$

Since γ is constant, integrating w.r.t. z

$$p = -\gamma z + A$$

At $z = 0, p = 101.33 \text{ kN/m}^2$

$$A = 101.33 \text{ kN/m}^2$$

$$p = 101.33 \times 10^3 - 1.285 \times 9.81z$$

At $z = 8 \text{ km}, p = 101.33 \times 10^3 - 12.61 \times 1800 = 450 \text{ N/m}^2$

(*ii*) *For isothermal conditions*

$$\frac{p}{\rho} = \text{constant } (A) = \frac{p_0}{\rho_0} = \frac{101330}{1.285} = 78{,}800.$$

Therefore, $\rho = \dfrac{p}{\text{constant}} = \dfrac{p}{A}$

We know that $\dfrac{dp}{dz} = -\rho g = -\dfrac{pg}{A}$ from which $\dfrac{dp}{p} = -\dfrac{g\,dz}{A}$

Integrating, $\log p = -\dfrac{gz}{A} + B$

When $z = 0, p = p_0, B = \log p_0$ and

$$\log \frac{p}{p_0} = -\frac{gz}{A}$$

$\therefore$ $\dfrac{p}{p_0} = e^{-(gz/A)}$

$$p = p_0 e^{-(gz/A)} = 101330e^{\frac{-9.81 \times 8 \times 1000}{78{,}800}} = 101330e^{-0.995}$$

or $= 37450 \text{ N/m}^2$

(*iii*) *Adiabatic conditions, n = k = 1.4*

From the equation of state, $p_0 = \rho_0 R T_0$

$\therefore$ $R = \dfrac{p_0}{\rho_0 T_0} = \dfrac{101330}{1.285 \times 288} = 274 \text{ Nm/kg °K}$

Pressure p is given by Eq. (2.9),

$$\frac{p_0}{p_0} = \left[1 - \frac{gz}{RT_0}\left(\frac{n-1}{n}\right)\right]^{n/(n-1)}$$

$$= \left[1 - \frac{9.81 \times 8 \times 1000}{247 \times (273 + 15)} \times \frac{(1.4-1)}{1.0}\right]^{1.4/0.4} = 0.311$$

$\therefore$ $p = 0.311 \times 101330 = 31514 \text{ N/m}^2.$

Example 2.8. *The pressure and temperature of air at sea-level are 101.3 kN/m² and 15°C respectively. Calculate the pressure and density at an altitude of 3,000 metres if the lapse rate is 7°C per 1000 m.*

Solution. Assuming sea level density = 1.285 kg/m³,

the gas constant $R = \dfrac{p}{\rho_0 T_0} = \dfrac{101330}{1.285 \times (273+15)} = 274 \text{ Nm/kg °K}, \quad \dfrac{dT}{dz} = -\dfrac{7}{1000}$

The power index, $n = \dfrac{1}{1+\dfrac{R}{g}\left(\dfrac{dT}{dz}\right)} = \dfrac{1}{1-\dfrac{247}{9.81}\left(\dfrac{7}{1000}\right)} = 1.243$

Pressure at 3000 m altitude from Eq. (2.9)

$$\frac{p}{p_0} = \left[1-\frac{gz}{RT_0}\left(\frac{n-1}{n}\right)\right]^{n/(n-1)} = \left[1-\frac{9.81\times 3000}{274\times(273+15)}\times\frac{0.243}{1.243}\right]^{\frac{1.243}{0.243}}$$

$$= 0.667$$

$$p = 101300 \times 0.667 = 68600 \text{ N/m}^2.$$

From the equation of state, $p = \rho RT$

Temperature at 3,000 m altitude $= 15 - \dfrac{7}{1{,}000} \times 3{,}000 = -6°\text{C}$.

$\therefore$ Density $\rho = \dfrac{p}{RT} = \dfrac{68600}{274\times(273-6)} = \dfrac{68600}{274\times 267} = 0.939 \text{ kg/m}^3.$

2.2. Absolute and Gauge Pressures

Pressures can be expressed in two different systems. In the first system, the pressure is measured above the absolute zero or complete vacuum, and is called the *absolute pressure*. In the second system, the pressure is measured and expressed as a difference between its absolute value and the local atmospheric pressure, and is called the *gauge pressure*. If the pressure is below the local atmospheric pressure, it is known as the negative or vacuum or suction pressure. The mechanical pressure gauges, such as the Bourdon gauge, record the gauge pressure whereas the Aneroid barometer registers the absolute pressure.

The gauge pressure, absolute pressure and the local atmospheric pressure are thus related by the following relationship :

Gauge pressure = Absolute pressure – Local atmospheric pressure ...(2.14 *a*)

and if the gauge pressure is negative, then

Vacuum or negative pressure = Local atmospheric pressure – Absolute pressure ...(2.14 *b*)

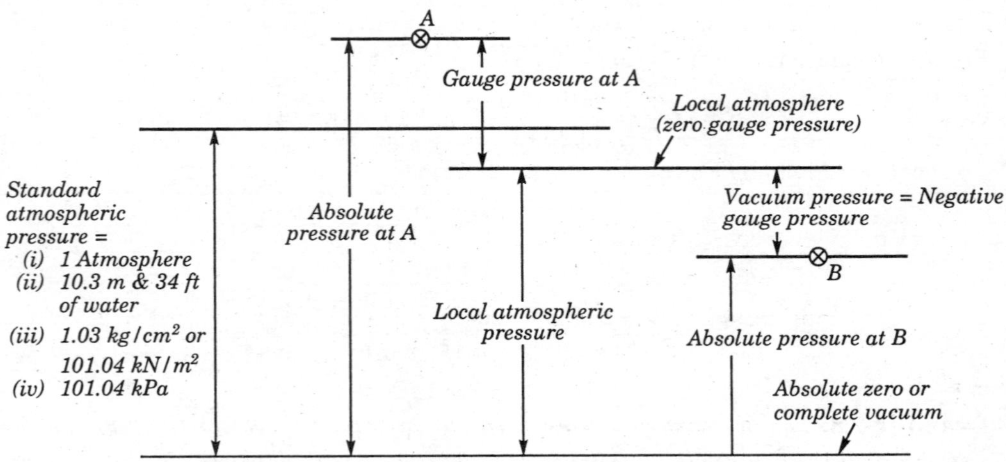

Fig. 2.4. Representation of absolute and gauge pressures.

These relations are schematically shown by Fig. 2.4. The *standard atmospheric pressure* is the mean pressure at the sea-level (at altitude of 40°), and is equal to 760 mm of mercury. The local atmospheric pressure is measured by means of a mercury barometer or an aneroid barometer. A mercury barometer consists of a glass tube closed at one end and is filled with mercury. It is inverted so that the open end is submerged in mercury. The space above the mercury contains mercury vapour exerting a negligible pressure (vapour pressure of mercury at ordinary atmospheric temperature is very small) which may be ignored safely. The pressure at p is the local atmospheric pressure and is related by the following equation :

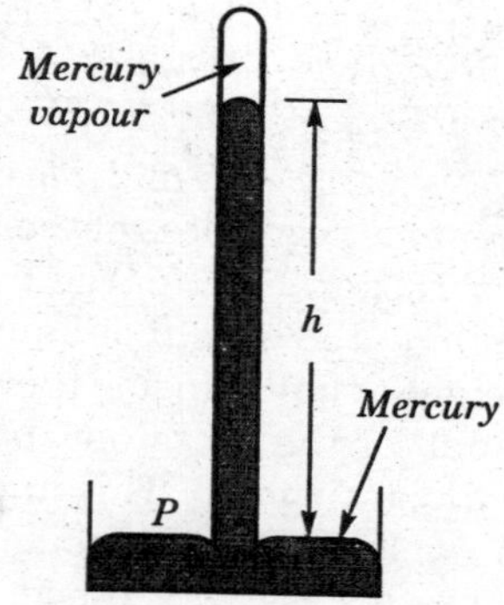

Fig. 2.5. Mercury barometer.

$$p_a - p_v = \gamma h \qquad \text{...(2.15 a)}$$

or

$$p_a/\gamma - p_v/\gamma = h$$

Here p_v is the vapour pressure of mercury and γ its specific weight. We know that mercury has a very low vapour pressure (at 20°C, p_v = 0.173 N/m^2), it can be safely dropped from Eq. (2.15 *a*), and the equation can be expressed as

$$p_a/\gamma = h \text{ metre of mercury column}$$
$$= 13.55\, h \text{ metre of water column} \qquad \text{...(2.15 b)}$$

(The height of mercury column h is generally measured and expressed in centimetres.)

The barometric pressure depends upon the geographical location of the place—its elevation above the sea-level and also the weather conditions.

In Fig. 2.4, pressure at A has been shown as absolute pressure and positive gauge pressure. At point B, where the pressure is below atmospheric, it has been expressed as negative gauge pressure and also in terms of absolute pressure. For example, if the pressure at B is 400 mm of mercury absolute, with barometer reading as 740 mm, the pressure may be expressed as – 340 mm of mercury gauge, 340 mm of mercury suction or 340 mm mercury vacuum. It is a usual practice to express pressure as gauge pressure and, therefore, all the pressures can be taken as gauge pressures, unless they are specified as absolute pressures.

2.2.1. Hydrostatic Paradox

Fig. 2.5 (*a*) shows five containers of different shapes having the same base area A, which are filled by the same liquid (of specific weight γ) upto the same weight h.

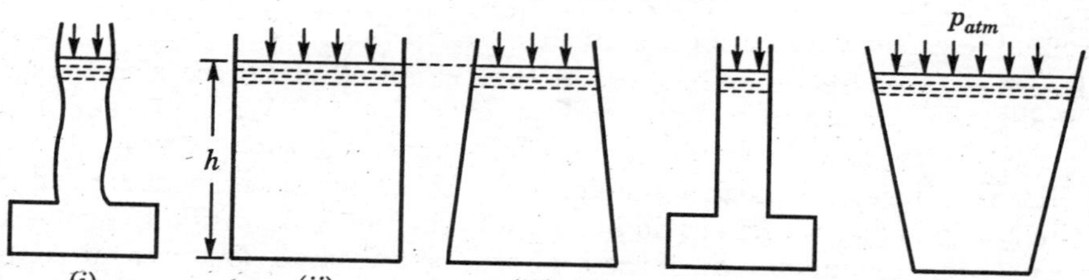

Fig. 2.5 (*a*) The hydrostatic paradox.

According to the hydrostatic law of pressure variation (Eq. 2.6), the pressure at the base of each container is

$$p = p_{atm} + \gamma h$$
$$(p - p_{atm}) = \gamma h \qquad \text{...(2.6 a)}$$

Here $(p - p_{atm})$ represents the gauge pressure, and

$$p = \text{absolute hydrostatic pressure}$$

p_{atm} = local atmospheric pressure acting over the liquid surface.

The hydrostatic force exerted on the base of each container

$$P = \text{Pressure} \times \text{Area} = (p - p_{atm}) \times A = \gamma h\, A \quad ...(2.6\ b)$$

Since A, h and γ are the same for all the containers, the hydrostatic force on the bottom is the same regardless of their shapes. The quantity represents the volume of the prism and therefore γAh is the weight of the liquid prism. The hydrostatic force on the base of each container is thus equal to the weight of liquid prism of base area A and height h. In containers (*i*), (*iii*) and (*iv*), the weight of liquid inside the container is less than the hydrostatic force, whereas these two forces are equal for the container (*ii*). For the container shown in (*v*), the weight of liquid is more than the hydrostatic force. This apparent contradiction in the hydrostatic force on the container-base and the weight of liquid inside the container is known as the *hydrostatic paradox*.

2.3. Pressure Measurement

Pressures are measured and quoted in two different systems : one relative (gauge) and the other absolute. In the later system, the arbitrary datum is the absolute zero, whereas in the first one it is the local atmospheric pressure. When a pressure is expressed as a difference between its value and a complete vacuum, it is called *absolute pressure*. When it is expressed as a difference between its value and local atmospheric pressure, it is called *gauge pressure*.

Mechanical pressure gauges—the Bourdon gauge being one such type—are devices used for measuring gauge pressures. Open piezometers are also used for indicating the gauge pressures. Local atmospheric pressure is measured by a mercury barometer or by an aneroid barometer.

2.3.1. Mechanical Pressure Gauges

The aneroid barometer and the Bourden gauge, shown in Fig. 2.5 (*b*), are the standard mechanical devices for measuring absolute and gauge pressures respectively. The aneroid barometer consists of a short cylinder (*C*), its one end being provided with an elastic diaphragm (*D*). The cylinder is evacuated (complete evacuation would result in the lowest possible pressure as there would be no fluid molecules to exert pressure) so that the inside pressure is very close to absolute zero. Pressures imposed on the other side of the diaphragm will cause it deflect inward. These deflections are a direct measure of the applied pressures, which may be transferred to a suitable scale (*S*) through appropriate linkage (*L*). The pressures indicated will be measured relative to the absolute zero and hence are absolute pressures.

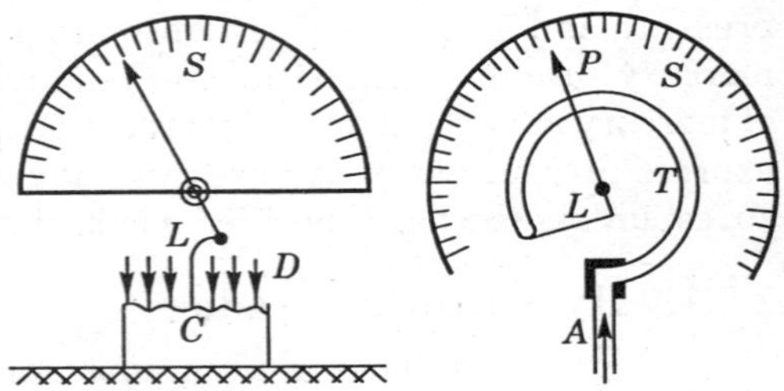

(*a*) *Aneroid* (*b*) *Bourdon*

Fig. 2.5 (*b*) Mechanical pressure gauges.

The Bourdon pressure gauge has a bent tube (*T*) of elliptical cross-section which is fixed rigidly at *A* and its free end is connected to a pointer (*P*) by link (*L*). When pressure is applied to the tube, its cross-section tends to become circular. This causes the tube to straighten which in turn causes the pointer to move over the graduated scale (*S*). The pointer rests at zero on the graduated scale when the gauge is disconnected ; no pressure is then exerted inside the tube. In this condition, the pressures inside and outside of the tube will be the same. The pressure outside the tube being usually atmospheric, it is obvious that such pressure gauges, are actuated by the difference between the pressures inside and outside of the tube. In the relative system of pressure measurement, the local atmospheric pressure becomes the zero of pressure-scale. For pressures greater than the local atmospheric the pointer will move to the right of zero, whereas for pressures less than that of the local atmosphere the pointer will move to the left. The gauge reading for the former case is positive and is called *gauge pressure* ; for the latter case the reading is negative and is designated as *vacuum*.

Ex. 2.7. *At the hill-station of Mussouri, the pressure of flow in a water supply pipeline was 245.25 kN/m^2 (gauge). The local atmospheric pressure recorded by a barometer was 80.93 kN/m^2. Calculate the absolute pressure of flow in the pipeline. At Trivandrum—a coastal place in Kerala—*

what would be the absolute pressure of flow in the pipeline, if the gauge pressure is the same and the atmospheric pressure stands at 101.53 kN/m².

Sol. At Mussouri $p_{abs} = p_{gauge} + p_{atm.} = 245.25 + 80.93 = 326.18$ kN.m².

At Trivandrum $p_{abs} = p_{gauge} + p_{atm.} = 245.25 + 101.53 = 346.78$ kN/m².

2.3.2. Pressure Measurement by Manometers

It has been shown that the pressure at a depth h below the free-surface in a liquid of specific weight γ is given by Eq. (2.6) as $p = \gamma h$ ...(2.6)

This hydrostatic law of variation of pressure is utilised in measuring pressures in fluid mechanics. The pressure measurement in a fluid, in its simplest form consists of installating glass tube at or near the point where the pressure is desired. This height, of the liquid that rises in the glass tube (commonly known as a *piezometer* or piezometer tube) measured above the point, is a measure of the pressure at that point. The pressure can then be obtained by knowing the specific weight of the liquid and using Eq. (2.6). It is a common practice to express the pressure in terms of height of liquid column that rises in a piezometer. The piezometers when employed for pressure measurement are called *manometers*. The manometers may be classified as :

(*i*) Simple manometers, (*ii*) Differential manometers, and (*iii*) Micro-manometers.

2.3.2.1. Simple manometer

The manometer already described above is called a simple manometer and is used for measuring pressure at a point. Fig. 2.6 shows simple manometers.

Fig. 2.6 (*a*) shows a simple manometer which can be used for measuring positive pressure at A. The height of liquid in the piezometer, h, above the point A, gives the pressure at that point, $p_A = \gamma h$, where γ is the specific weight of the liquid. This manometer can only be used for measuring small and moderate pressures because as the pressure gets higher the length of piezometer becomes correspondingly larger. It is also unsuitable for measurement of negative pressure at A, for then air will enter into the container through the piezometer. For negative pressures the piezometer may be shaped as shown in Fig. 2.6 (*b*) and the pressure at A will then be given by :

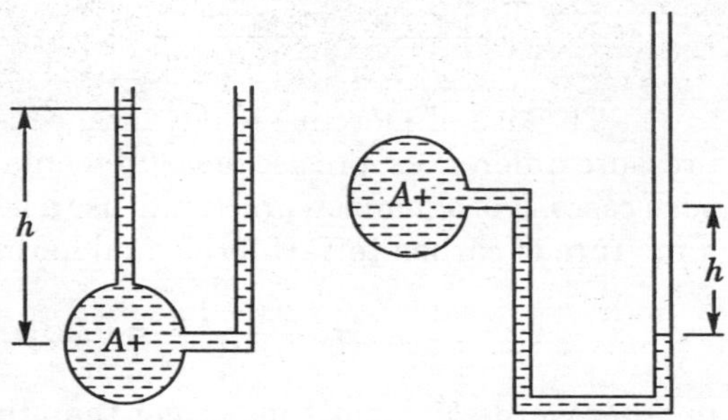

(*a*) For positive pressure at A. (*b*) For negative pressure at A.

Fig. 2.6. Simple Manometers.

$$p_A = -\gamma h.$$

For large positive and negative pressures, a second liquid of higher specific gravity is used in the manometer. It must not mix with the fluid whose pressure is being measured ; in other words it must be immisicible. The shape of the manometer will be the same as in Fig. 2.6 (*b*).

2.3.2.2. Differential Manometer

A differential manometer is used to measure the difference in pressure at two points. The manometer consists of a glass U-tube connected to the points between which the pressure difference is to be determined. A heavier liquid which is immisicible with the other two liquids, occupies the lower part of the U-tube. The petcocks at C and D permit the expulsion of air. Due to the difference in pressure a points A and B there exists a difference in level h in the manometer liquid.

It is obvious from Eq. (2.1) that in a continuous liquid, the pressure is same in the same horizontal plane. Therefore, equating pressures in the U-tube along X-X, refer Fig. 2.7,

$$p_A - S_1\gamma_w Z_1 = p_B - S_2\gamma_w Z_2 + S\gamma_w h,$$

in which γ_w = sp. wt. of water.

$$\therefore \quad p_A - p_B = S\gamma_w h + S_1\gamma_w Z_1 - S_2\gamma_w Z_2 \quad ...(2.6)$$

and in terms of head of water.

$$\frac{p_A - p_B}{\gamma_w} = Sh + S_1 Z_1 - S_2 Z_2 \quad ...(2.16\ a)$$

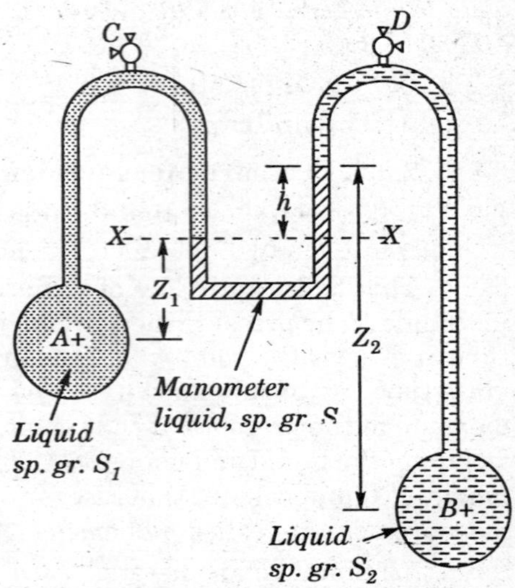

Fig. 2.7. A differential manometer.

If the manometer liquid is mercury, S = 13.6, and the liquid in both the containers is water, then $S_1 = S_2$ = 1.0, and further if both the containers are at the same elevation then $Z_2 - Z_1 = h$, Eq. (2.26 *a*) then becomes

$$\frac{p_A - p_B}{\gamma_w} = (S - 1)h = 12.6\ h \quad ...(2.17)$$

For the same liquid other than water, $S_1 = S_2$, and the pressure difference in terms of water head, Eq. (2.17) changes to

$$\frac{p_A - p_B}{\gamma_w} = (S - S_1)h \quad ...(2.17\ a)$$

In terms of head of liquid of specific gravity S_1

$$\frac{p_A - p_B}{\gamma} = \frac{(S - S_1)h}{S_1} = \left(\frac{S}{S_1} - 1\right)h \quad ...(2.17\ b)$$

The use of mercury is advantageous where the pressure difference is large, but for small pressure differences, precise measurements are difficult and, therefore, mercury is not suitable. For such cases, it is a common practice to use a liquid which is slightly heavier than water. If, the example, a mixture of carbon tetrachloride and kerosene having a specific gravity of 1.25 is used, then

$$\frac{p_A - p_B}{\gamma_w} = 0.25\ h$$

in other words h is four time larger than the actual difference in pressure heads.

In place of using a slightly heavier liquid than water, if we use a lighter liquid than water, the U-tube will then have to be used in an inverted position as shown in Fig. 2.8. For small pressure differences, large deflections of the manometer liquid in the inverted U-tube will be obtained. Equating pressures along *X-X* in Fig. 2.8 one gets

$$p_A - S_1\gamma_w Z_1 = p_B - S_2\gamma_w Z_2 - S\gamma_w h$$

giving the pressure difference

$$p_A - p_B = S_1\gamma_w Z_1 - S_2\gamma_w Z_2 - S\gamma_w h$$

and in terms of head of water

$$\frac{p_A - p_B}{\gamma_w} = S_1 Z_1 - S_2 Z_2 - Sh \quad ...(2.18)$$

If both the containers are at the same level, then $Z_1 - Z_2 = h$, and if both the liquids are water ($S_1 = S_2$ = 1.0), then

$$\frac{p_A - p_B}{\gamma_w} = h - Sh = h(1 - S) \quad ...(2.19)$$

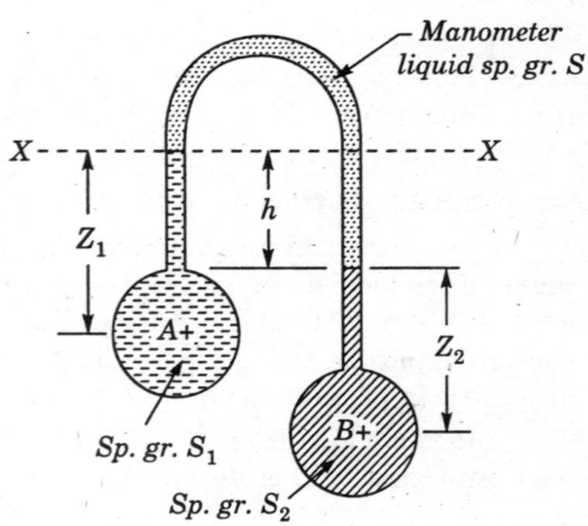

Fig. 2.8. Differential manometer with a lighter manometer liquid.

Using kerosene having a specific gravity 0.80, Eq. (2.19) reduces to

$$\frac{p_A - p_B}{\gamma_w} = 0.20\, h$$

and the deflection, h, is five times the difference in pressure heads at A and B.

If however the liquids are same but other than water, $S_1 = S_2$, and Eq. (2.19) may be written as

$$\frac{p_A - p_B}{\gamma} = h\left(1 - \frac{S}{S_1}\right) \qquad ...(2.19\ a)$$

It can be noted from Eqs. (2.17) and (2.19) that the deflection h of the manometer liquid is always dependent upon the difference in its specific gravity S and that of water. This deflection becomes larger as the two specific gravities approach each other in magnitude while it becomes smaller as they differ more widely.

2.3.2.3. Micromanometer

The micromanometers are used for measuring small difference of pressures. The micro-manometer shown in Fig. 2.9 utilises two manometer liquids which are immisicible with each other and also with the fluid whose pressure difference is to be measured. The heavier liquid fills the lower part of the U-tube upto O-O and then the lighter liquid is added on both sides, filling the tanks C and D upto the level X-X. The fluid (liquid or a gas) whose pressure difference is to be determined fills the space above X-X. When the pressure p_A is slightly greater than p_B, the liquid levels will be as shown in Fig. 2.9. The volume of the liquid displaced in each tank is equal to the volume of liquid displaced in the U-tube. If a is the area of cross-section of the U-tube and A that of the tank then

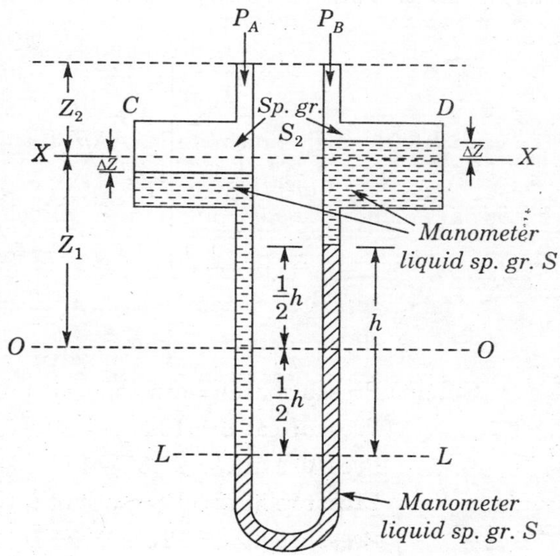

Fig. 2.9. Micromanometer with two manometer liquids.

$$A \,.\, \Delta z = \frac{h}{2} \,.\, a$$

An expression relating p_A and p_B may be obtained by equating pressures along L-L in the U-tube, thus

$$p_A + S_2\gamma_w(Z_2 + \Delta z) + S_1\gamma_w(Z_1 - \Delta z + h/2)$$

$$= p_B + S_2\gamma_w(Z_2 - \Delta z) + S_1\gamma_w\left(Z_1 - \frac{h}{2} + \Delta z\right) + S\gamma_w h$$

Substituting for $\Delta z = ha/2A$ and simplifying, we get the difference in the pressure head

$$\frac{p_A - p_B}{\gamma_w} = h\left[S - S_1\left(1 - \frac{a}{A}\right) - S_2\frac{a}{A}\right] \qquad ...(2.20)$$

The quantity within the large brackets is a constant for a given manometer and given liquids of sp. gr. S_1 and S_2 and known fluid and, therefore, the pressure difference is proportional to h,

thus $\dfrac{p_A - p_B}{\gamma_w} = Kh$

where K represents the quantity within the large brackets.

2.3.2.4. Inclined Manometer

An inclined manometer as shown in Fig. 2.10 is used for precise measurement of small pressures in low velocity gas flow. When the opening of the tank C is exposed to the atmosphere, the inclined glass tube has the liquid at level 0. But when a positive small pressure is applied to the tank liquid, the level of the liquid in the inclined tube rises and is measured by the inclined distance l. The change in the level of the tank is negligible if its cross-section is made very large as compared to that of the tube. The applied pressure will then be equal to

$$p = S\gamma_w l \sin\theta = \gamma l \sin\theta \qquad ...(2.21)$$

where γ is the specific weight of the manometer liquid.

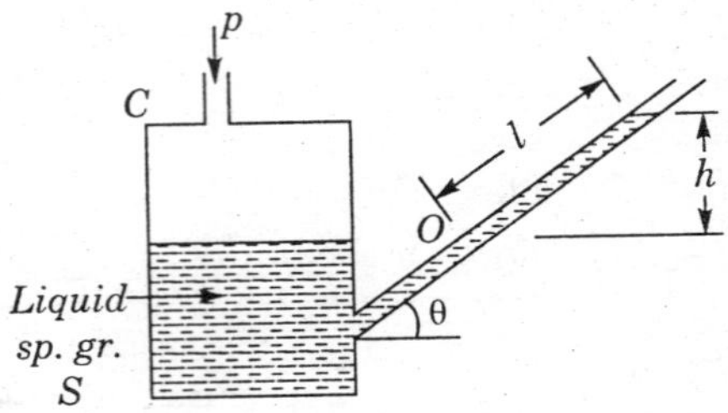

Fig. 2.10. Inclined manometer.

For a small pressure, the distance l can be observed accurately, whereas it is not possible to measure h with reasonable precision. The amplification of h is l/h which is equal to $1/\sin\theta$. This type of manometer, when calibrated to read directly in cm or inches or water, is called a draft gauge.

Ex. 2.9. *Determine h in meter of water for conditions shown in the following figure. If the parts filled with air were also full of water, what would be the new value of h ?*

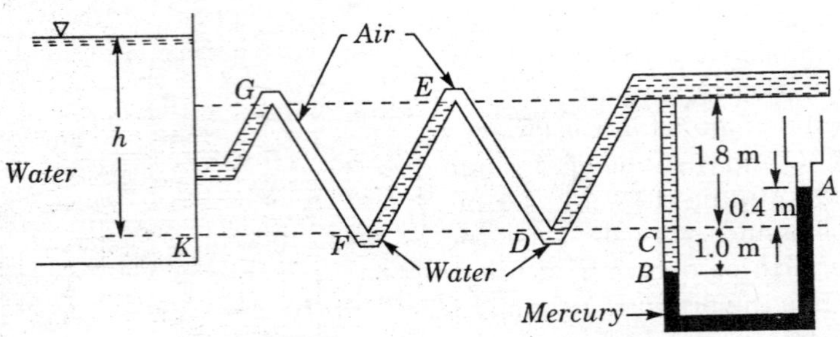

Sol. (*a*) Pressure at A = atmospheric pressure = 10.30 metres of water.

Pressure at B = 10.30 + (1.0 + 0.40) × 13.6 = 29.35 m of water.

Pressure at C = 29.35 – 1.0 = 28.35 m of water.

Pressure at D = Pressure at C (liquid being continuous and at the same level)

Pressure at E = Pressure at D (assuming no change in air pressure with elevation)

Pressure at F = 28.35 + 1.8 = 30.15 m of water

Pressure of G = Pressure at F

∴ Pressure at K = 30.15 + 1.8 = 31.95 m of water.

Since this pressure includes the atmospheric pressure, therefore,

$$h = 31.95 - 10.30 = 21.65 \text{ m of water.}$$

(*b*) **When the pipe is full of water**

Pressure at A = Atmospheric pressure = 10.30 m of water

Pressure at B = 10.30 + (0.40 + 1.00) × 13.6 = 29.35 m of water

Pressure at C = 29.35 – 1.00 = 28.35 m of water

Pressure at K = Pressure of F = Pressure at D = Pressure at C

= 28.35 m of water absolute

= 28.35 – 10.30 – 18.05 m of water (gauge)

∴ h = 18.05 m.

Ex. 2.10. *Two U-tube manometers are connected in series as shown in the figure. Determine difference of pressures between X and Y.*

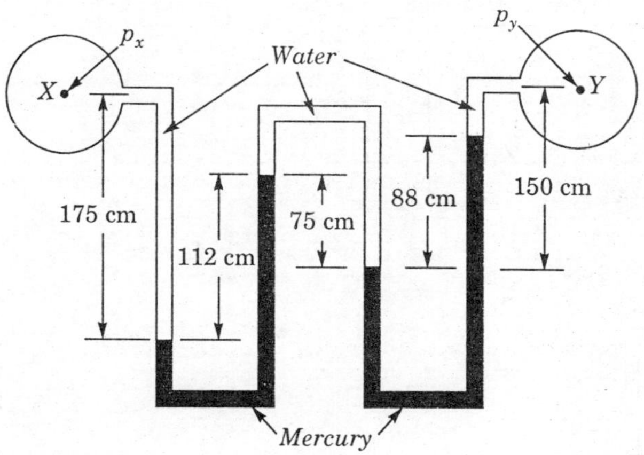

Sol. Starting from point X, the manometer equation may be written as

$$\frac{p_x}{\gamma} + 1.75 - 1.12 \times 13.6 + 0.75 - 0.88 \times 13.6 - (1.50 - 0.88) = \frac{p_y}{\gamma}$$

or
$$\frac{p_x - p_y}{\gamma} = 25.32 \text{ m of water.}$$

$$\therefore \qquad p_x - p_y = 25.32 \times \frac{9810}{10^4} = 248 \text{ kN/m}^2.$$

2.4. Hydraulic Press

A well-known application of the law of hydrostatics is the hydraulic press shown in Fig. 2.11. In order to exert a large force F on the piston of area A, it is necessary to create a pressure $p = F/A$. This can be done by applying a force $F(a/A)$ on the plunger, where a is the area of the plunger. Assuming very slow motion of the plunger, the friction and inertia forces may be neglected. The pressure variation due to the height may also be neglected. If the body below the piston, on which the work is being done, undergoes a reduction in its height by Δz, the work done by the piston is $F\Delta z = pA\Delta z$. This requires the transfer of liquid of volume $A\Delta z$ from the small cylinder to the large one. It follows, therefore, that the plunger must move down a distance

$$\Delta h = \frac{A\Delta z}{a}.$$

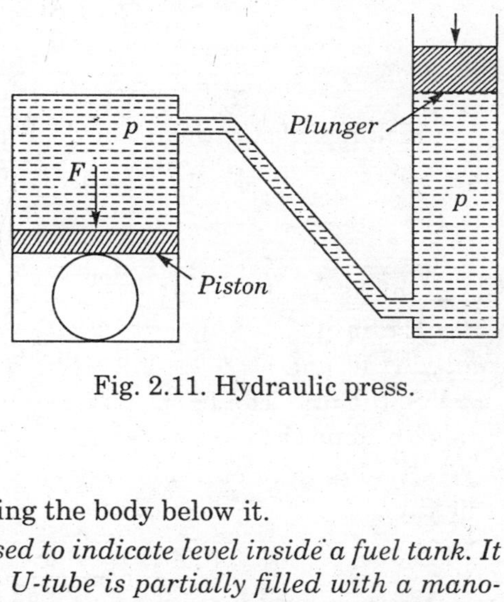

Fig. 2.11. Hydraulic press.

The work done by the plunger is $F . \dfrac{a}{A} . A \dfrac{\Delta z}{a} = F . \Delta z$,

which is equal to the work done by the piston in compressing the body below it.

Ex. 2.11. *In the following figure is shown a device used to indicate level inside a fuel tank. It is in the form of a U-tube with the lines crossing over. The U-tube is partially filled with a manometric fluid of specific weight γ_2 which is heavier than the fuel whose specific weight is γ_1. When the U-tube and the level indicating tube are vertical and the tank is full of fuel the reading on the scale is unity. Determine the relation between h_1, the distance through which the fuel level is lower in the tank and h_2, the distance through which the fuel level falls in the indicating tube.*

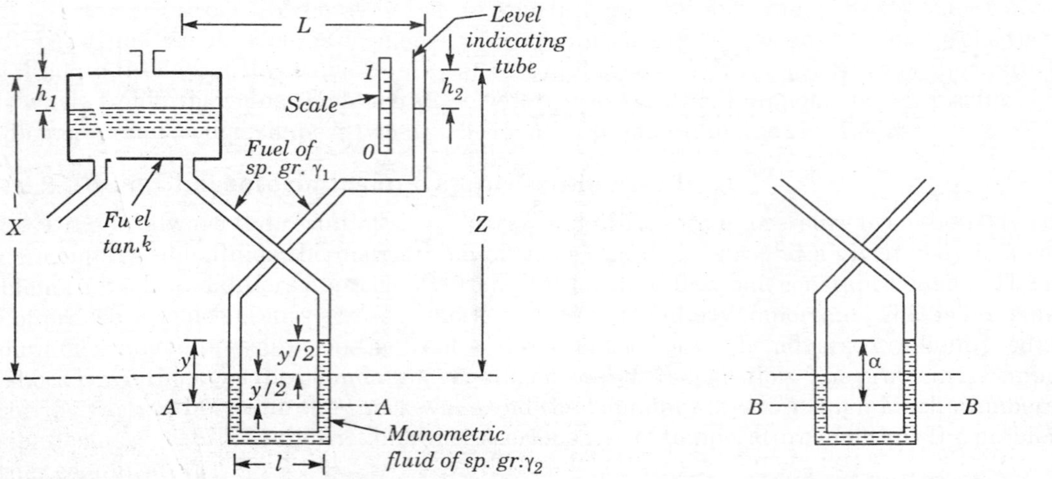

(a) When fuel tank is full. (b) When the fuel tank level falls.

Fig. Ex. 2.9. Fuel tank problem.

Solution. Let the deflection of the manometric liquid when the fuel tank is full be y, as shown in the figure. Equating pressures along A-A, working from the left limb first :

$$\gamma_1(Z + y/2) = \gamma_1(X - y/2) + \gamma_2 y$$

$$\gamma_1 Z + \gamma_1 y = \gamma_1 X - \gamma_2 y$$

or $$(\gamma_2 - \gamma_1)y = \gamma_1(Z - X) \quad ...(1)$$

(*b*) *When the level in fuel tank falls* :

Let the deflection of the manometric fluid be α, as shown in Fig. (*b*). Equating pressures along *B-B* :

$$\gamma_1(Z + \alpha/2 - h_2) = \gamma_1(X - h_1 - \alpha/2) + \gamma_2\alpha$$

$$\gamma_1 Z + \gamma_1\alpha - \gamma_1 h_2 = \gamma_1 X - \gamma_1 h_1 + \gamma_2\alpha$$

or $$\gamma_1(Z - X) + \gamma_1(h_1 - h_2) = \alpha(\gamma_2 - \gamma_1) \quad ...(2)$$

Making substitution from (1), Eq. (2) may be written as :

$$(\gamma_2 - \gamma_1)y + \gamma_1(h_1 - h_2) = \alpha(\gamma_2 - \gamma_1)$$

or $$\gamma_1 h_1 = \alpha(\gamma_2 - \gamma_1) - y(\gamma_2 - \gamma_1) + \gamma_1 h_2 = (\gamma_2 - \gamma_1)(\alpha - y) + \gamma_1 h_2$$

or $$h_1 = (\gamma_2/\gamma_1 - 1)(\alpha - y) + h_2.$$

This is the desired relationship between h_1 and h_2.

Example 2.12. *A hydraulic press has a ram 12.5 cm in dia and a plunger of 1.25 cm dia. What force would be required on the plunger to raise a weight of 10 kN on the ram ?*

Solution. Let F be the force required on the plunger to raise 1 tonne on the ram.

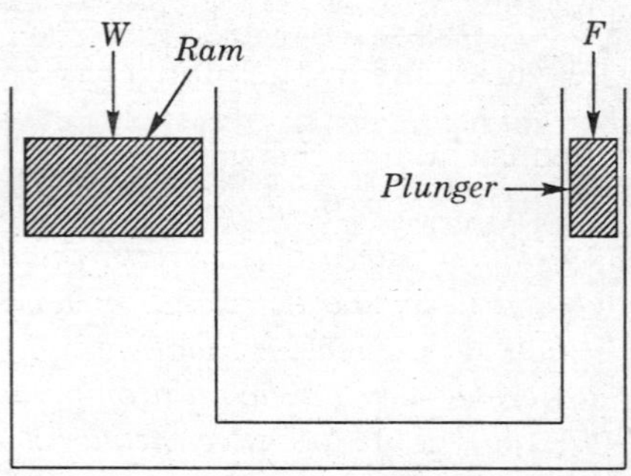

The pressure of fluid everywhere may be taken as same thus

$$\frac{10 \times 1000}{\frac{\pi}{4} \times \left(\frac{12.5}{100}\right)^2} = \frac{F}{\frac{\pi}{4}\left(\frac{1.25}{100}\right)^2}$$

$$F = 10^4 \times \left(\frac{1.25}{12.5}\right)^2 = 100 \text{ N}.$$

Example 2.13. *The cross-sectional area of one limb of a U-tube manometer (shown in the figure) is made 500 times larger than the other, so that the pressure difference between the two limbs can be determined by measuring 'h' on one limb of the manometer. Determine the percentage error involved.*

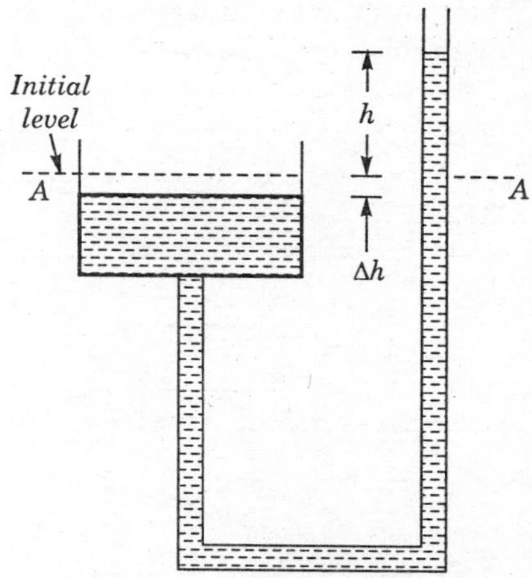

Solution. When both the limbs of the U-tube manometer are exposed to atmospheric pressure, the liquid level in the limbs stands at *A-A*. When a negative pressure is applied to the right, the left limb remaining exposed to atmospheric pressure, a pressure difference is created, and the manometer liquid rises in the right limb. The decrease in volume of liquid in the left limb must be equal to increase in its volume in the right limb. Hence,

Area of cross-section left limb × Δh = Area of cross-section of right limb × h

or $$\frac{\Delta h}{h} = \frac{\text{Cross section area of right limb}}{\text{Cross section area of left limb}} = \frac{1}{500}$$

Actual pressure difference in terms of head = $h + \Delta h$

Pressure difference that is being observed = h

Error $= \Delta h$

Percentage error $$= \frac{\Delta h}{h} \times 100$$

$$= \frac{1}{500} \times 100 = \mathbf{0.2\%}.$$

2.5. Forces on Plane Surfaces

The knowledge of magnitude, direction and location of the resultant hydrostatic force on surface submerged under a liquid is essential in the design of gates, locks and dams. The following analysis will develop relations enabling determination of these quantities.

In Fig. 2.12 a plane area is submerged in a liquid such that it makes an angle θ with the liquid surface. It has a shape as shown in the plane *XOY*. Let the centre of area be at a depth h_c below the liquid surface. Consider an elementary strip of area *dA* within the surface at a depth *h*. The hydrostatic force *dF* on the strip will be

$$dF = \gamma h\, dA = \gamma y dA \,.\, \sin\theta, \text{ since } h = y \sin\theta$$

The resultant hydrostatic force may be obtained by integrating the above expression

thus $$F = \int^A \gamma \sin\theta . y . dA = \gamma \sin\theta \int^A y . dA$$

in which the integral $\int^A y\, dA$, is the moment of the area about *O-X*, which is equal to $A \,.\, y_c$,

thus $$\int^A y . dA = A . y_c$$

and the resultant hydrostatic force *F* becomes

$$F = \gamma \sin\theta . A . y_c = \gamma A y_c \sin\theta = \gamma A h_c \qquad ...(2.22)$$

Equation (2.22) indicates that the resultant hydrostatic force on any submerged plane surface is equal to the product of the area *A* and the pressure at its centroid (centre of area) γh_c.

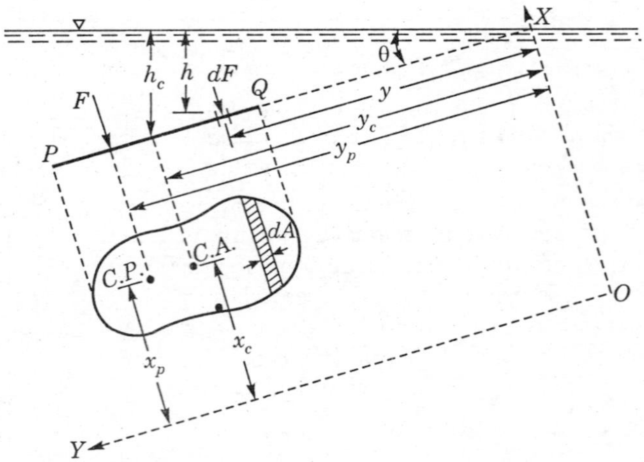

Fig. 2.12. Forces on a plane submerged area.

This resultant force F acts normal to the area, since in fluids at rest, the shear stresses are non-existent. The point through which the resultant force F acts is known as the *centre of pressure*. The location of the centre of pressure may be obtained by dividing the moment of the force about O-X by the magnitude of the force F.

Taking moment of force dF about O-X,

$$dM = dF \, . \, y = \gamma y \, dA \sin\theta \, . \, y = \gamma y^2 \, . \, dA \sin\theta.$$

The moment of the resultant force about O-X, is

$$M = \int^A \gamma y^2 dA \sin\theta = \gamma \sin\theta \int^A y^2 dA$$

in which $\int^A y^2 dA$ is the second moment of area about O-X, and may be written as I_{x-x},

thus $$M = \gamma I_{x-x} \sin\theta$$

and the location of the centre of pressure is

$$y_p = \frac{M}{F} = \frac{\gamma I_{x-x} \sin\theta}{\gamma A h_c} = \frac{I_{x-x} \sin\theta}{A y_c \sin\theta} = \frac{I_{x-x}}{A y_c} \quad ...(2.23)$$

Equation (2.33) may be simplified by using the theorem of parallel axes of moment of inertia, according to which $I_{x-x} = I_c + A y_c^2$

in which I_c is the second moment or moment of inertia of the area about an axis parallel to X-axis (O-X) and passing through the centre of area (centroid). Substituting for I_{x-x} in Eq. (2.23), one gets

$$y_p = \frac{I_c}{A y_c} + y_c \quad ...(2.24)$$

Equation (2.24) shows that the centre of pressure is always below the centre of area, since $y_p > y_c$, except when the plane surface is horizontal, in which case, the centre of pressure is at the centroid.

The x-coordinate of the centre of pressure may be obtained by taking the moments about the y-axis (O-Y), thus

$$x_p = \frac{1}{F}\int^A (\gamma y \sin\theta \, dA)\, x = \frac{1}{A\gamma y_c \sin\theta}\int^A \gamma x y \sin\theta \, dA = \frac{1}{A y_x}\int^A xy dA = \frac{I_{x-y}}{A y_c} \quad ...(2.25)$$

In Eq. (2.25), I_{x-y} is the product of inertia of the area.

When any of the centroidal axes, $x = x_c$ and $y = y_c$ is an axis of symmetry, $I_{x-y} = 0$, and the centre of pressure lies on $x = x_c$.

It is to be noted that y_c, y_p, x_c, x_p are the distances in the plane of the submerged surface. In general, we are concerned only with the vertical position of the centre of pressure and Eq. (2.25) is seldom used. From Eq. (2.23) it is clear that the position of the centre of pressure is independent of the angle θ. Hence we may vary θ by rotating the surface provided h_c remains unchanged.

2.5.1. Horizontal Surfaces

If the submerged plane surface is horizontal, the pressure at all points lying on it is same and the pressure distribution is rectangular with $p = \gamma h$ as the ordinate, where h is the depth of the surface below the liquid. The hydrostatic force on the surface is equal to the pressure multiplied by the area of the surface, thus

$$F = \gamma h \,.\, A$$

is obtained by using Eq. (2.22) in which the value of $h_c = h$. Fig. 2.13 shows the pressure distribution over a horizontal plane surface.

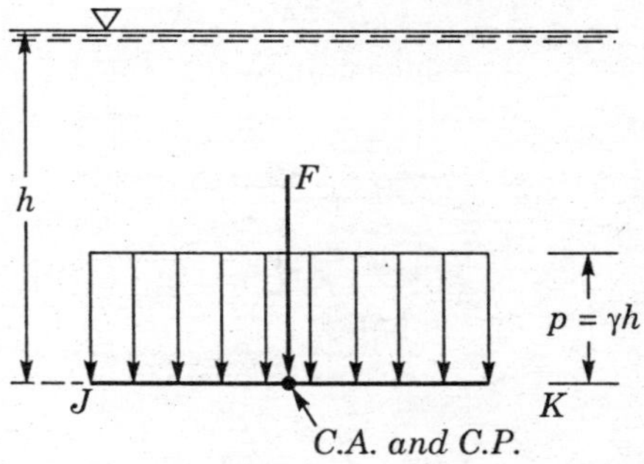

Fig. 2.13. Pressure distribution over a horizontal surface.

Owing to the uniform pressure distribution over a horizontal surface, the hydrostatic force F acts through the centre of area, and thus the centre of pressure coincides with the centre of area.

2.5.2. Vertical Surfaces

Unlike horizontal surfaces (where the pressure is uniform throughout), the pressure in case of a vertical surface changes in the vertical direction, its magnitude depending upon its location from the free surface. The following figures show the pressure variation for vertical surfaces.

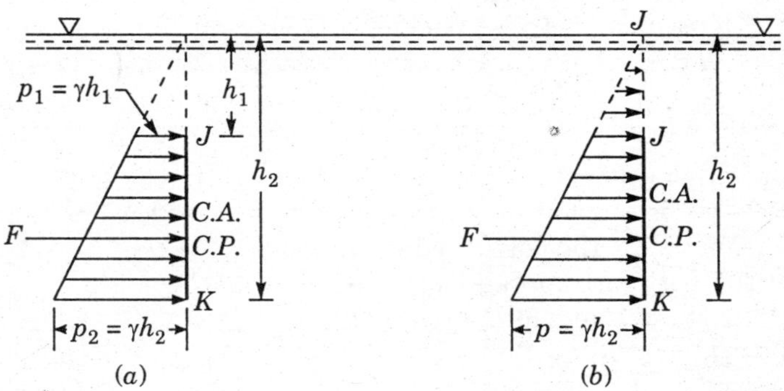

Fig. 2.14. Pressure distribution on vertical surfaces.

In Fig. 2.14 (*a*), the top edge of the surface *JK* is at a depth h_1 while the lower one is h_2 below the free surface. The pressure distribution is trapezoidal with pressure γh_1 at the top and γh_2 at the bottom and varying linearly in between. As shown in Fig. 2.14 (*b*), the submerged surface *JK* has its top edge *J* in the free-surface while the bottom one is at a depth h_2 from the free-surface. The pressure diagram is triangular, the pressure varying from zero at the top to γh_2 at the bottom.

From Figs. 2.13 and 2.14 it is evident that the hydrostatic force is equal to the volume of the pressure prism or pressure diagram over the surface, and is located at the centre of gravity of the pressure prism. Thus in case of Fig. 2.14 (*b*), the magnitude of hydrostatic force = $\frac{1}{2}\gamma h_2 A$, where A = area of the surface, and acts through a point $\frac{2}{3}h_2$ below the free surface.

Ex. 2.14. *A square surface 3 m × 3 m lies in a vertical plane. Determine the position of the centre of pressure and the total force on the square, when its upper edge is (a) in water surface and (b) 15 metres below the water surface.*

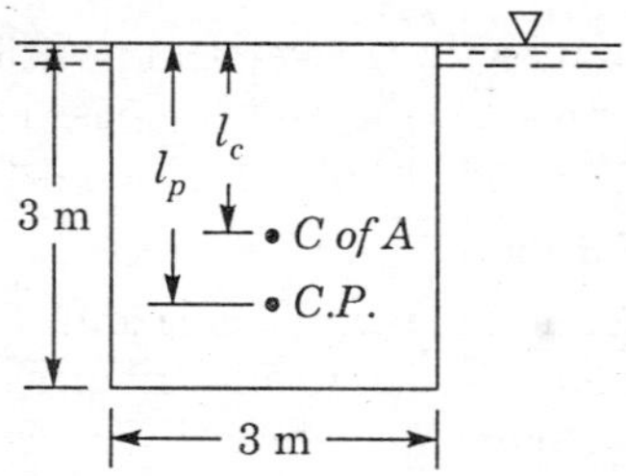

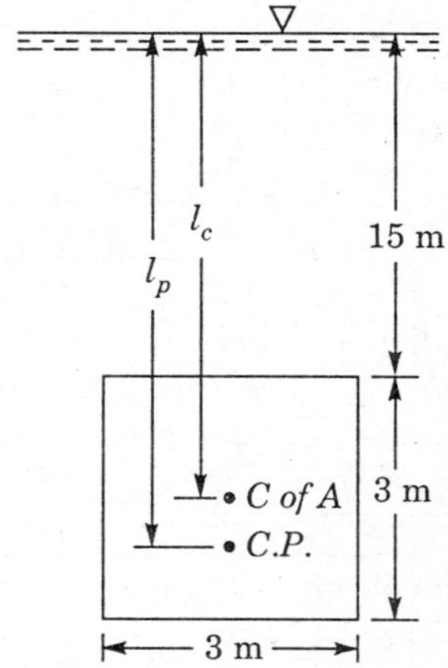

Sol. (*a*) Total force on the square

$$F = \gamma h_c \, . \, A = 9810 \times 1.5 \times 9 \text{ N}$$
$$= 132.3 \text{ kN}.$$
$$F = \gamma h_c \, . \, A$$

The location of the total force is given by

$$l_p = l_c + \frac{I_c}{A l_c}$$

where l_p and l_c are the distances of C.P. and centroid below the water surface.

$$= 1.5 + \frac{\frac{1}{12} \times 3 \times (3)^3}{9 \times 1.5} = 2.0 \text{ m}.$$

(*b*) Total force on the square,

$$F = \gamma h_c \, . \, A$$
$$= 9810 \times (1.5 + 15) \times 9 = 1457 \times 10^3 \text{ N} = 1457 \text{ kN}.$$

Distance between centre of pressure and centroid

$$= l_p - l_c = \frac{I_c}{A l_c} = \frac{\frac{1}{12} \times 3 \times (3)^3}{9 \times 16.5} = 0.0455 \text{ m} = 4.55 \text{ cm}.$$

Depth of centre of pressure = 16.5 + 0.0455 = 16.5455 m.

Ex. 2.15. *A vertical gate supports water on one side to a depth of 7.2 metres. Horizontal load on the gate is taken by three beams placed parallel to the water surface. Determine the position of the beams such that each beam carries one-third of the total load.*

Sol. Let each of three beams be located at depths z_1, z_2 and z_3 and support a force F, which is one-third of the total hydrostatic force. Let the top beam support the horizontal water load due to depth h_1, the middle beam that of the portion whose depth is from h_1 to h_2 and the bottom beam support the load coming on the vertical portion from depth h_2 to 7.2 m.

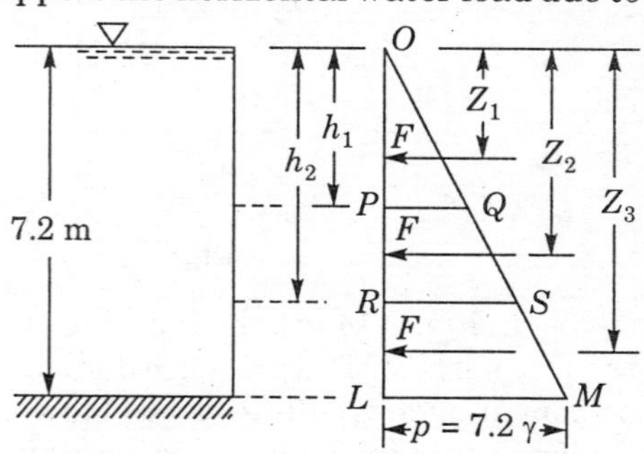

(*i*) **Location of the top beam**

$$\tfrac{1}{2} \gamma h_1^2 = \tfrac{1}{3} \times \tfrac{1}{2} \gamma \times (7.2)^2$$
$$h_1 = 4.15 \text{ m}.$$

Location of the top beam,

$$Z_1 = \tfrac{2}{3} h_1 = \tfrac{2}{3} \times 4.15$$
$$= 2.765 \text{ m}.$$

(*ii*) **Location of the middle beam**

$$\tfrac{1}{2} (\gamma h_1 + \gamma h_2)(h_2 - h_1) = \tfrac{1}{6} \gamma \times (7.2)^2$$
$$h_2 = 5.87 \text{ m}.$$

Let the middle beam be located at a depth Z_2 from the water surface. Taking moments of forces on portions *OR* and *OP* about *O*.

$$F \times Z_2 = 2F \times \tfrac{2}{3} h_2 - F \times \tfrac{2}{3} h_1$$
$$Z_2 = 5.055 \text{ m}.$$

(*iii*) **Location of the bottom beam**

$$F \times Z_3 = 3F \times \tfrac{2}{3} \times 7.2 - 2F \times \tfrac{2}{3}\, h_2$$

$$Z_3 = 6.575 \text{ m.}$$

The three beams should be located at depths 2.765, 5.055 and 6.575 m below the water surface.

Ex. 2.16. *A rectangular gate in the vertical side of a reservoir can turn freely about its upper edge which is horizontal and is fastened at its two lower corners. The gate is 1.0 m wide and 2.0 m high and its upper edge is 2.0 m below the water level. Determine the reactions at the lower corners assuming them to be equal.*

Sol. Consider the figure shown below. The gate *ABCD* is hinged along its upper edge to facilitate free turning. It is fastened at *B* and *D*.

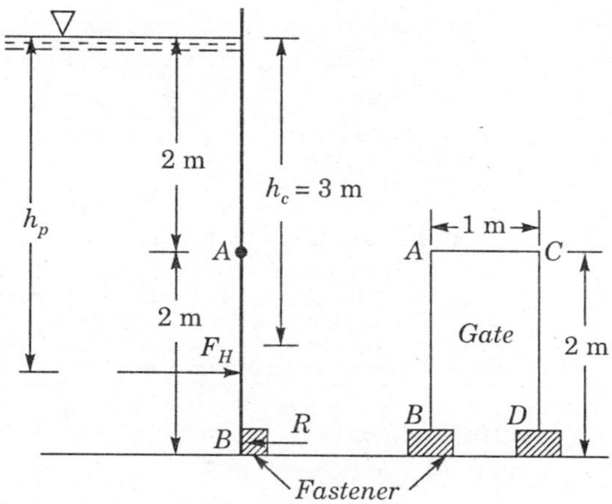

The resultant hydrostatic force acting on the gate is

$$F_H = \gamma h_c \times \text{Area of the gate} = 9810 \times 3 \times 1 \times 2 = 58750 \text{ N}$$

This force F_H acts at a distance h_p below the water surface

$$h_p = h_c + \frac{I_c}{h_c \times \text{Area of the gate}} = 3 + \frac{\left(\frac{1}{12}\right) \times 1 \times (2)^3}{3 \times (1 \times 2)} = 3.11 \text{ m.}$$

Let R be the reaction of each fastener. Taking moments of the forces about the top corner AC,

$$F_H \times (3.11 - 2.0) = 2R \times 2.0$$

which yields, $$R = \frac{58750 \times 1.11}{4} = 16340 \text{ N} = 16.34 \text{ kN.}$$

Ex. 2.17. *The gate shown in the figure is required to hold water on its left hand side. As the water rises, the gate will automatically topple about the hinge and allow water to flow over it. At what depth above the hinge will this occur. Neglect the weight of the gate.*

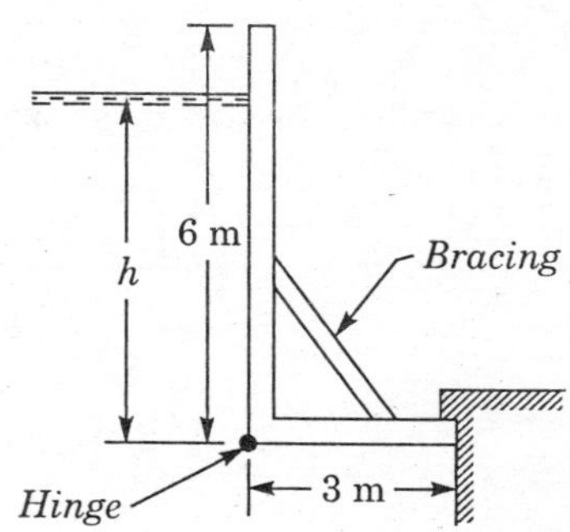

Sol. Let the limiting water level be h above the hinge, upto which the gate will be in equilibrium.

The adjoining figure shows the pressure distribution over the vertical and horizontal parts of the gate. Let F_V and F_H be the respective hydrostatic forces acting on each part.

For limiting equilibrium, the moments about the hinge must be equal.

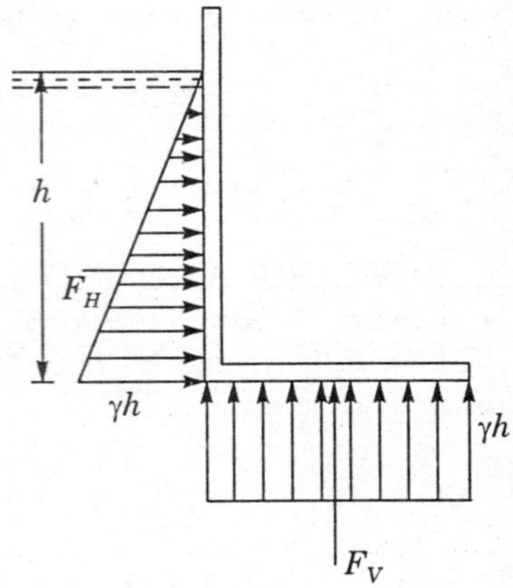

Pressure distribution

$$F_H \times \frac{h}{3} = F_V \times \tfrac{1}{2} \times 3 \qquad ...(1)$$

Considering unit length of the gate

$$F_H = \gamma h_c \,.\, A = \gamma \tfrac{1}{2}\, h \times (h \times 1) = \tfrac{1}{2}\, \gamma h^2$$

$$F_V = \gamma h \times (3 \times 1) = 3\gamma h.$$

Substituting for F_H and F_V in (1),

$$\tfrac{1}{2}\, \gamma h^2 \times \frac{h}{3} = 3\gamma h \times \tfrac{1}{2} \times 3$$

$$h^2 = 27$$

or $$h = \sqrt{27} = 5.20 \text{ metres.}$$

Ex. 2.18. *A sliding gate 3 metre wide and 1.5 metre high lies in a vertical plane and has a coefficient of friction of 0.20 between itself and guides. If the gate weighs 19.6 kN and if its upper edge is at a depth of 10 metre, determine the vertical force required to raise it ? Neglect buoyancy effects on the gate.*

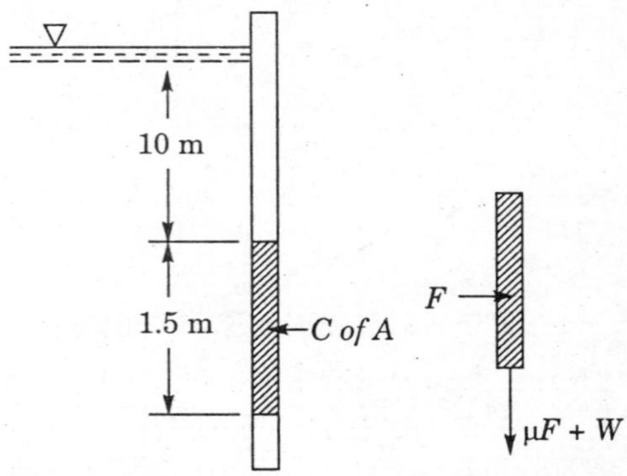

Sol. The hydrostatic force on the gate

$$F = \gamma h_c \,.\, A$$
$$= 9810 \times 10.75 \times 3 \times 1.5$$
$$= 474559 \text{ N} = 474.56 \text{ kN.}$$

Force required to raise the gate

$$= \mu F + \text{Weight of the gate}$$
$$= 0.20 \times 474.56 + 19.6$$
$$= 114.5 \text{ kN.}$$

Ex. 2.19. *The vertical side of a reservoir has a rectangular opening 2.75 m long × 1.2 m high. It is closed by a plate using 4 bolts placed at the corner of the opening. What would be the tension in the bolts if water stands to a height of 1.8 m above the top edge of the opening ?*

Sol. Resultant hydrostatic force exerted on the plate,

$$F_H = \gamma A h_c$$

$$= 9810\,(2.75 \times 1.2)\left(1.8 + \frac{1.2}{2}\right)$$

$$= 77.7 \times 10^3 \text{ N}$$

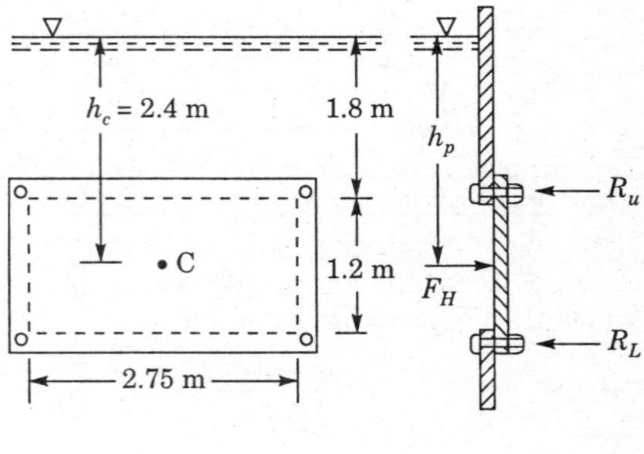

This force acts at a depth, h_p, below the water surface,

$$h_p = h_c + \frac{I_c}{A.h_c}$$

$$= 2.4 + \frac{1/12 d^3}{(bd)\, h_c} = 2.4 + \frac{(1.2)^2}{12 \times 2.4}$$

$$= 2.45 \text{ m.}$$

Let R_u and R_L be the part of the hydrostatic force transmitted to the upper and bottom bolts respectively

$$R_u + R_L = F_H = 77.7 \times 10^3 \text{ N}$$

Taking moment of forces about the bottom line of bolts.

$$R_u \times 1.2 = F_H (3.0 - h_p)$$
$$= 77.7 \times 10^3 (3.0 - 2.45) = 42.7 \times 10^3 \text{ N}$$

$\therefore$ $$R_u = 35610 \text{ N}$$

and $$R_L = 77.7 \times 10^3 - 35.6 \times 10^3 \text{ N}$$
$$= 42.1 \times 10^3 \text{ N}$$

Hence, (*i*) tension in each of the top bolt = $\dfrac{35610}{2}$ = **17805 N.**

(*ii*) tension in each of the bottom bolt = $\dfrac{42100}{2}$ = **21050 N.**

Ex. 2.20. *A tank, completely full of water, is 5 metres square at the top, 3 metres square at the bottom and 2 metres deep. The four sides are planes, each having the same trapezoidal shape. Find (a) the magnitude of the resultant pressure force on each side of the tank (b) the location of the above pressure force.*

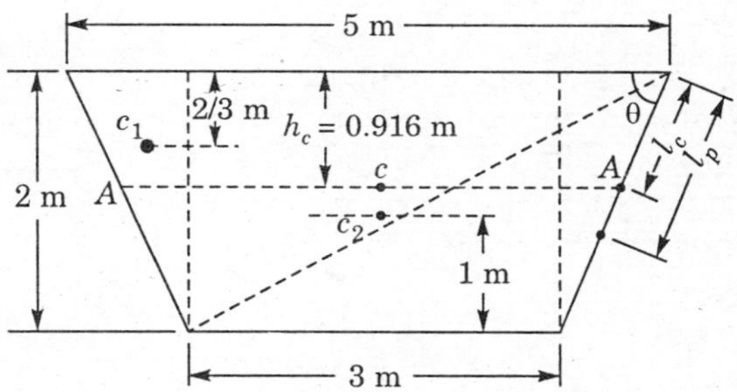

Sol. Let us first find out the centre of area of the trapezoid. Divide the trapezoid into two triangles as shown. Let the centre of area be at a distance h_c below the top, then

$$\tfrac{1}{2}(5 + 3) \times 2 \times h_c = \tfrac{1}{2} \times 5 \times 2 \times (\tfrac{1}{3} \times 2) + \tfrac{1}{2} \times 3 \times 2 (\tfrac{2}{3} \times 2)$$

or $$8h_c = \frac{10}{3} + 4 \qquad \text{or} \qquad h_c = 0.916 \text{ m.}$$

(*a*) Resultant pressure on each side = γh_c × Area of inclined surface

$$= 9810 \times 0.916 \times \tfrac{1}{2}(5 + 3) \times 2 \times \frac{\sqrt{5}}{2} = 80370 \text{ N}$$
$$= 80.37 \text{ kN.}$$

(*b*) Location of this force : This force is located at an inclined distance l_p (along the side) from the top.

Inclined length of the tank sides = $\sqrt{1+4} = \sqrt{5}$ m.

$\therefore$ $$\sin\theta = 2/\sqrt{5} \quad \text{and} \quad l_c \sin\theta = 0.916$$

or $$l_c = \frac{0.916\sqrt{5}}{2} = 1.015 \text{ m}$$

But, $$l_p = l_c + \frac{I_c}{l_c A}$$

I_c = Moment of inertia of area (trapezoidal) about *A-A*. Dividing the area into a rectangle and two triangles,

$$I_p = 2\left[\frac{1 \times (2)^3}{56} + \tfrac{1}{2} \times 1 \times 2 \times (0.916 - 0.666)^2\right] + \frac{1}{12} \times 3(2)^3 + (2 \times 3) \times (1 - 0.916)^2$$

$$= 0.569 + 2.042 = 2.611 \text{ m}^4$$

$$\therefore \quad l_p = 1.025 + \frac{2.611}{1.025 \times \frac{1}{2}(3+5) \times 2} = 1.343 \text{ m.}$$

Ex. 2.21. *Determine the centre of pressure for the vertical area of triangular shape shown in the following figure :*

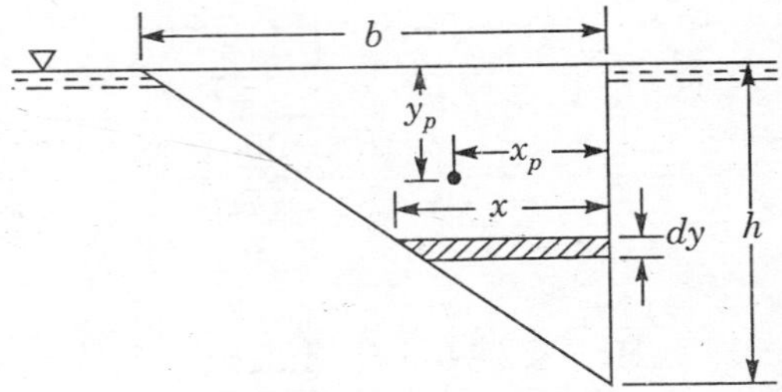

Sol. Consider a horizontal strip of length, x, and thickness, dy, situated at depth, y, below the water surface. The hydrostatic force over this strip is

$$dF_h = \gamma \, dA \, . \, h_c = \gamma(x \, . \, dy) \, . \, y = \gamma xy dy$$

From similar triangles, we have

$$\frac{b}{h} = \frac{x}{(h-y)} \quad \text{or} \quad x = \frac{b}{h}(h-y).$$

$$\therefore \quad dx = \frac{b}{h}(-dy) \quad \text{or} \quad dy = -\frac{h}{b} dx.$$

The centre of pressure in this case of small strip may be considered to be located at a distance $x/2$ from the vertical side and y below the water surface. Let the location of the centre of pressure for the given triangle be x_p and y_p as shown in figure, then

$$\gamma(\tfrac{1}{2} bh) \times \tfrac{1}{3} h \, x_p = \int_0^h \gamma xy \, dy \, . \frac{x}{2} = \int_b^0 \frac{1}{2}\gamma x^2 \left(h - x\frac{h}{b}\right)\left(-\frac{h}{b} dx\right)$$

$$= \int_0^b \tfrac{1}{2}\gamma x^2 \left(h - x\frac{h}{b}\right)\frac{h}{b} dx$$

or

$$\tfrac{1}{3} bh^2 x_p = \int_0^b x^2 \left(h - x\frac{h}{b}\right)\frac{h}{b} dx = \frac{h}{b}\left[\frac{x^3 h}{3} - \frac{h}{b}\frac{x^4}{4}\right]_0^b = \frac{h}{b}\left(\frac{b^3 h}{3} - \frac{hb^3}{4}\right)$$

$$x_p = \frac{h/b}{\frac{1}{3} bh^2}\left(\frac{4b^3 h - 3hb^3}{12}\right) = b/4$$

Similarly, we can obtain

$$\frac{1}{6} bh^2 \, . \, y_p = \int_0^h xy \, dy \, . \, y = \int_0^h \frac{b}{h}(h-y) \, y^2 \, dy = \frac{b}{h}\left[h\frac{y^3}{3} - \frac{y^4}{4}\right]_0^h = \frac{bh^3}{12}$$

or

$$y_p = \frac{h}{2}$$

$$\therefore \quad x_p = \frac{b}{4} \quad \text{and} \quad y_p = \frac{h}{2}.$$

Ex. 2.22. *An isosceles triangular plate of base 4 metres and altitude 4 metres is immersed vertically in fluid with a spacific gravity 0.9. The base of the triangle is touching the top of the surface of the fluid horizontally and rest of its portion is within the fluid. Determine the total pressure and centre of pressure of the plate.*

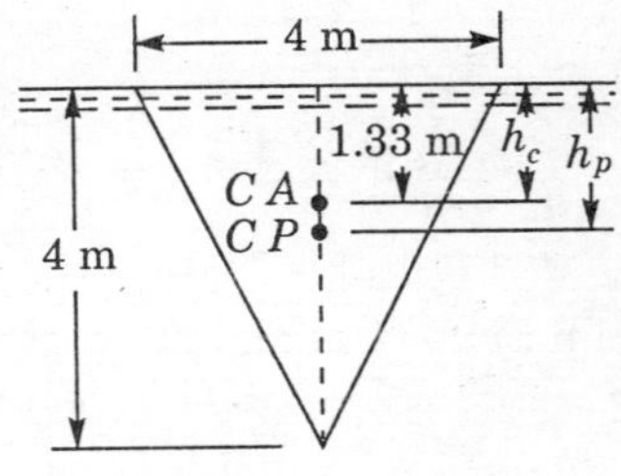

Sol. Total pressure on the triangle

$$F = p \, . \, A = \gamma h_c A$$

$$= (1000 \times 0.9)(1.33)(\tfrac{1}{2} \times 4 \times 4) = \mathbf{9600 \ kg}$$

$$= (9810 \times 0.9)(1.33)\left(\tfrac{1}{2} \times 4 \times 4\right)$$

$$= 94176 \text{ N} = \mathbf{9.18\ kN}$$

The centre of pressure (CP) lies on the vertical line of symmetry passing through the centre of area (CA), the location of which is given by Eq. (2.24). Noting that $y_p = h_p$,

$$h_p = \frac{I_c}{A\,h_c} + h_c = \frac{bh^3/36}{\left(\frac{bh}{2}\right)\left(\frac{h}{3}\right)} + \frac{h}{3} = \frac{h}{2} = \frac{4}{2} = \mathbf{2\ m.}$$

2.6. Forces on Curved Surfaces

In many engineering problems, it is necessary to know the magnitude, direction and location of the resultant hydrostatic force exerted by the liquid on the submerged curved surface. This resultant force may be obtained combining vectorially, its mutually perpendicular force components. These mutually perpendicular force components are generally taken as the horizontal and the vertical components. In Fig. 2.15, a curved surface *AB* is shown. The curved surface may be considered being acted upon by pressure which is always normal to the surface, the magnitude and the direction of which changes from point to point. The pressure force acting on any elementary area dA, inclined at an angle α to the horizontal is $dF = p.dA$ (Fig. 2.16). The horizontal component of this force dF is $dF \cos \alpha = p.dA \cos \alpha$ in which p is the pressure at depth h. But $dA \cos \alpha$ is the vertical projection of area dA. Therefore, the horizontal component of the force is equal to the force on the vertical projection of the area.

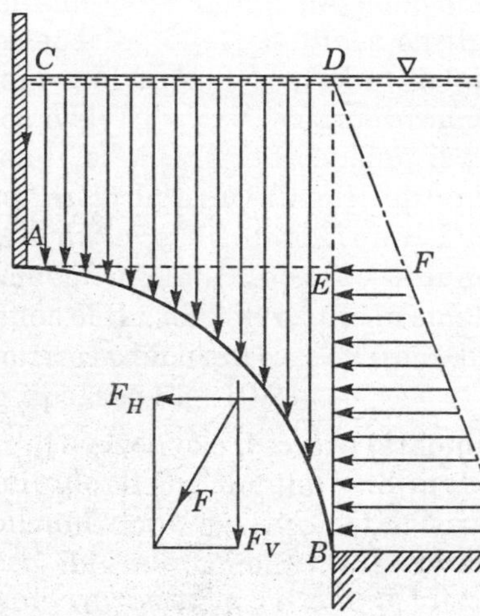

Fig. 2.15. Resultant force on a curved surface.

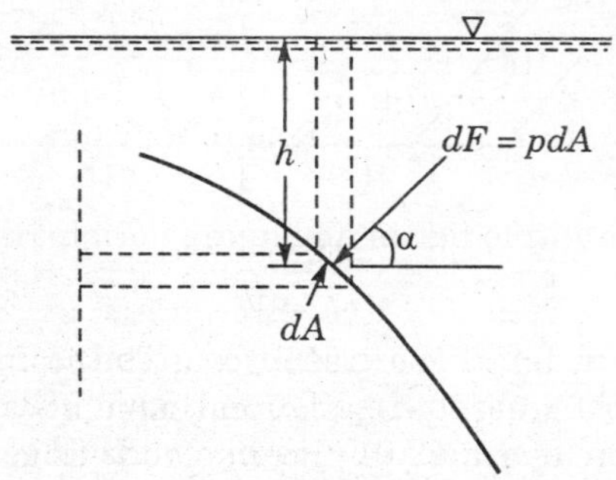

Fig. 2.16. Force on an elementary area.

The horizontal component F_H of the resultant hydrostatic force F may be computed by projecting the surface upon a vertical plane and multiplying the projected area by the pressure at its own centre of area (sometimes also called as centre of gravity of the area). The vertical projection of the curved surface *AB* is *EB* and *EFGB* is the pressure diagram, refer Fig. 2.15.

The vertical component of the hydrostatic force on the upper side of the surface *AB*, must be equal to the weight of the prism of liquid vertically above it. This vertical component of force F_V acts through the centre of gravity of the liquid prism represented by *ABEDC*. The determination of the weight (or volume) of the prism and its centre of gravity is simplified by dividing it into two convenient parts, whose centres of gravity are already known. The resultant hydrostatic force F can then be obtained as $\sqrt{F_{H^2} + F_{V^2}}$.

2.7. Buoyant Force (Archimedes' Principle)

The theory of floatation and fluid displacement due to submergence of a body in the fluid owes its origin to Archimedes (see page 2). When a body is submerged (partly or fully) in a fluid, it displaces a certain amount of fluid resulting in a vertically upward force, which is exerted by the fluid on the body. The vertically upward force is known as the *buoyant force*, the magnitude of which is equal to the weight of the fluid displaced. The Archimedes' principle states that a *submerged body is subjected to a buoyancy equal to the weight of the fluid displaced*.

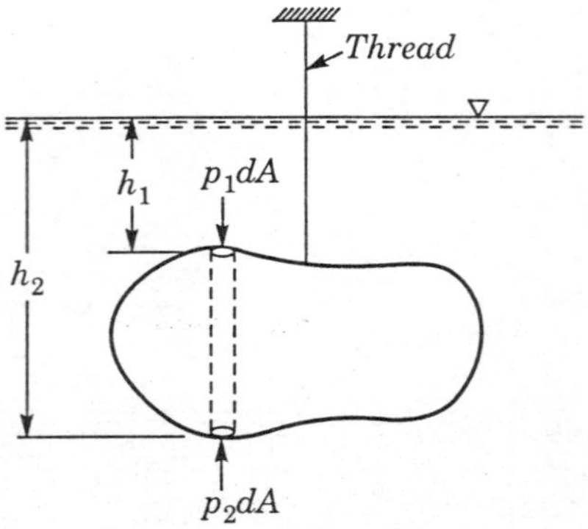

Fig. 2.17. Buoyant force on a submerged body.

Consider a solid body suspended by a thread below the surface of a liquid. Assume a vertical cylindrical element of cross-sectional area dA, as shown in Fig. 2.17. As dA is small, the pressure on the exposed ends of the cylinder may be taken as p_1 and p_2.

Since p_2 is greater than p_1, there will be an upward force $(p_2 - p_1)\, dA$ acting on the cylindrical element. Therefore, the buoyant force on the element is

$$dF_B = (p_2 - p_1)dA = \gamma(h_2 - h_1)dA = \gamma dV$$

in which dV is the volume of the cylindrical element. The entire submerged body may be considered to be made up of small cylindrical elements, and the buoyant force on the body as a whole is

$$F_B = \int^A (p_2 - p_1)dA = \int^A \gamma(h_2 - h_1)dA$$

$$= \gamma \int^A (h_2 - h_1)dA = \gamma V \qquad \text{...(2.26)}$$

in which V is the volume of the body, and γ is the specific weight of the liquid. If γ_b is the specific weight of the body, then weight of the body is

$$W = \gamma_b V$$

and the tension in the thread is

$$W - F_B = V(\gamma_b - \gamma) \qquad \text{...(2.27)}$$

Since the buoyant force is the sum of a number of elementary forces, each equal in magnitude to the weight of cylindrical element of fluid, it follows that the buoyant force must pass through the centre of gravity of the displaced fluid mass. The point through which the buoyant force acts is called the *centre of buoyancy* (C.B.) and is at the centre of gravity of the displaced fluid.

Ex. 2.23. *A metallic body floats at the interface of mercury (sp. gr. 13.6) and water in such a way that 40% of its volume is submerged in mercury and 60% in water. Find the density of the metallic body.*

Sol. Weight of the floating metallic body

= Buoyant force acting on the body

= Weight of water displaced + weight of mercury displaced

or $\rho g V = 9810 \times 0.6 \times V + 13.6 \times 9810 \times 0.4\, V$

where ρ is the density and V the volume of the metallic body. If S is the specific gravity of the body, then

$$S \times 9810 \times V = 9810 \times 0.6\, V + 13.6 \times 9810 \times 0.4\, V$$

or $\mathbf{S = 6.04}$

$$\rho = \frac{S \times 9810}{g} = \frac{6.04 \times 9810}{9.81} = \mathbf{6040\ kg/m^3}.$$

Example 2.24. *An odd shaped body when immersed in water weighs 1030 N and when immersed in kerosene (sp. gr. 0.8) weighs 1200 N. Determine the volume and weight of body in air.*

Solution. The odd shaped body is shown immersed in the desired liquid. The weight of the body (W_a) in air acts through its centre of gravity (marked by *CG*), which also happens to be the centre of buoyancy (marked by *CB*) through which the buoyant force (F_B) acts. It may be noted that the weight (W_a) is far in excess of the buoyant force (F_B). The submerged weight of the body in the fluid (W_f) may be written as,

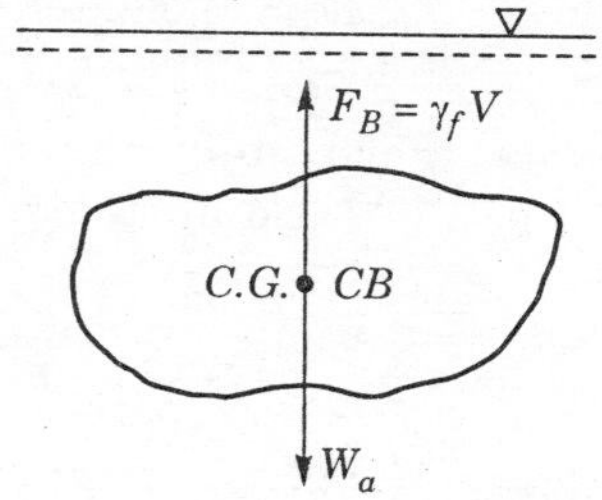

$$W_f = W_a - F_B$$
$$= W_a - \gamma_f V, \text{ in which } V = \text{volume of the body,}$$

and γ_f = sp. wt. of fluid.

Substituting the data given the above equation, we obtain

(*i*) In case of water $1030 = W_a - 9810V$ ...(1)

(*ii*) In case of kerosene $1200 = W_a - 0.8 \times 9810V$ ...(2)

Solving (1) and (2), the desired volume and weight are

$$V = 0.0866 \text{ m}^3, \text{ and}$$
$$W_a = \mathbf{1879.999\ N,\ say\ 1800\ N.}$$

Example 2.25. *A block of wood 5 m long × 2 m wide × 1 m thick is floating horizontally in water. The density of wood is 650 kg/m³. Find volume of water displaced and position of centre of buoyancy.*

Solution. Depth of immersion (*y*) of wooden black will be obtained by equating the weight of the block with the buoyant acting through the points *G* (representing the c.g.) and *B* (through which the buogant force acts) respectively, *i.e.*

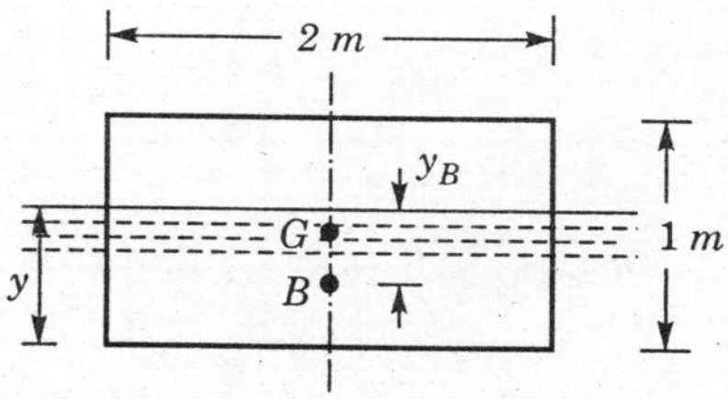

$$W = F_B \quad \text{...}(i)$$

Weight of block of wood,

$$W = 650 \times 9.81 \times 5 \times 2 \times 1 = 63765 \text{ N};$$

Buoyant force F_B = weight of water displaced

$$F_B = \gamma \,.\, V = 9810 \times 5 \times 2 \times y$$

From (*i*) $63765 = 98100\, y$

$\therefore$ $y = 0.65$ m

Location of centre of buogyancy is at the centroid (*i.e.* point *B*) of the volume of water displaced by the wooden block

$$y_B = \frac{y}{2} = \frac{0.65}{2} = \mathbf{0.325\ m}$$

$$\text{Volume of water displaced} = \frac{F_B}{\gamma} = \frac{63765}{9810} = \mathbf{6.5\ m^3}.$$

Example 2.26. *A light metal stick of square cross-section (15 cm × 15 cm × 4 m) weighs 686 N, and is shown in the figure below. Determine its angle of inclination when the water surface is 2 m above the pivot. What minimum depth of water will be required to bring the metal stick in a vertical position ?*

Solution. Let us first consider a general case in which θ is the inclination of the metal stick with the horizontal and *y* the depth of water above the pivot. The length of the stick below water surface = $y/\sin\theta$.

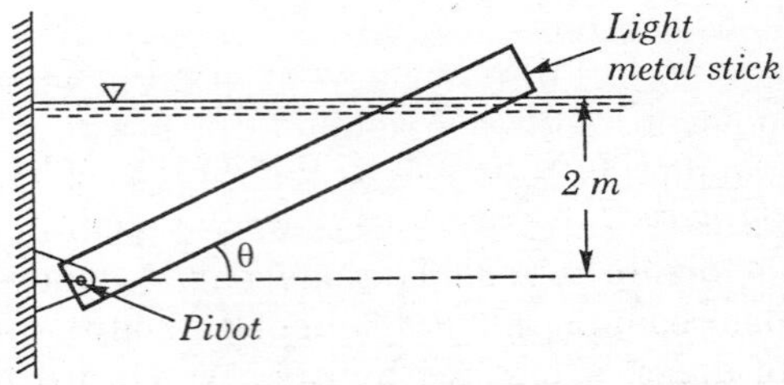

Volume of stick submerged in water

$$= 0.15 \times 0.15 \times \frac{y}{\sin\theta} = 0.0225y \operatorname{cosec}\theta$$

∴ Buoyant force,

F_B = Weight of liquid displaced

$$= 9810 \times 0.0225y \operatorname{cosec}\theta = 220.7y \operatorname{cosec}\theta$$

For equilibrium conditions, the moment of the weight of stick must be equal to the moment of the buoyant force about the pivot,

$$686 \times 2\cos\theta = 220.7y \operatorname{cosec}\theta \times \left(\frac{1}{2}\times\frac{y}{\sin\theta}\right) \times \cos\theta$$

which simplifies to $\frac{y}{\sin\theta} = \sqrt{\frac{280}{22.5}}$

(*i*) for $y = 2$ m

$\sin\theta = 0.567$ ∴ $\theta = 34.5°$

(*ii*) and when $\theta = 90°, y = 3.53$ m

For $y \geq 3.53$ m, the stick will stand vertically.

Example 2.27. *A two-dimensional gate having cross-section of a quarter cylinder is pivoted (hinged) at the bottom O and is in static equilibrium when subjected to differential fluids of specific weights γ_1 and γ_2. Assuming the gate weightless, calculate the relationship between γ_1 and γ_2.*

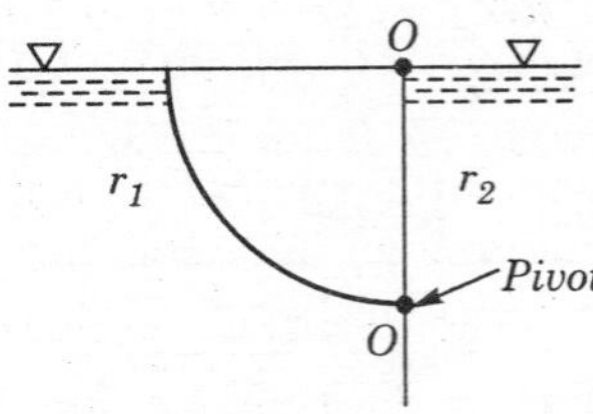

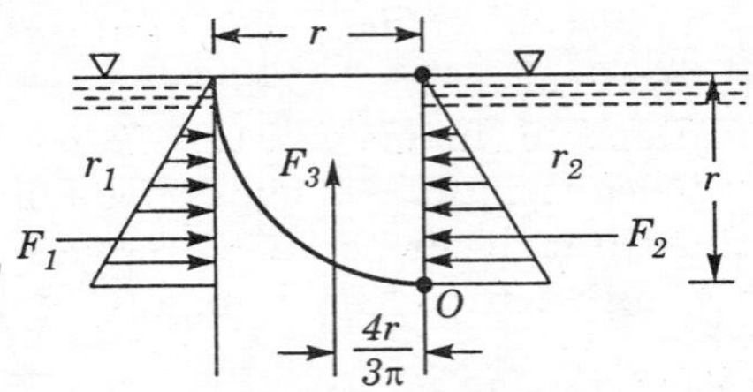

Solution. Forces acting on the quarter cylindrical gate per unit length are :

(*i*) Horizontal hydrostatic force $F_1 = \frac{1}{2}\gamma_1 r^2$, acting at a distance of $(r/3)$ above the pivot.

(*ii*) Horizontal hydrostatic force $F_2 = \frac{1}{2}\gamma_2 r^2$, acting at a distance $(r/3)$ above the pivot, and

(*iii*) The buoyant force over the weightless gate

$$F_B = \frac{\pi r^2}{4} \times 1.\gamma_1$$

and it acts at a distance of $4r/3\pi$ from O.

For static equilibrium, the clockwise and anti-clockwise moments of forces about the pivot must be equal :

$$(F_2 - F_1) \times \frac{r}{3} = F_B \times \frac{4r}{3\pi}$$

$$\left(\frac{1}{2}\gamma_2 r^2 - \frac{1}{2}\gamma_1 r^2\right)\frac{r}{3} = \frac{\pi r^2}{4}\,.\,\gamma_1 \times \frac{4\pi}{3\pi} \quad \text{or} \quad \frac{1}{2}(\gamma_2 - \gamma_1) = \gamma_1$$

$$\therefore \quad \gamma_1/\gamma_2 = \frac{1}{3}.$$

Ex. 2.28. *A cylindrical gate is 3 m long and has water on its both sides as shown. Determine the magnitude, location and direction of the resultant hydrostatic force exerted on the gate. Also calculate the minimum weight of the gate so that it will not float away from the floor.*

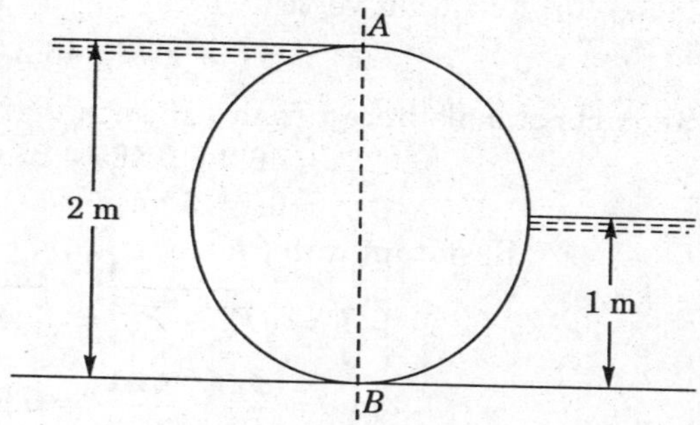

Sol. Let us break up the gate into two halves through the vertical plane AB.

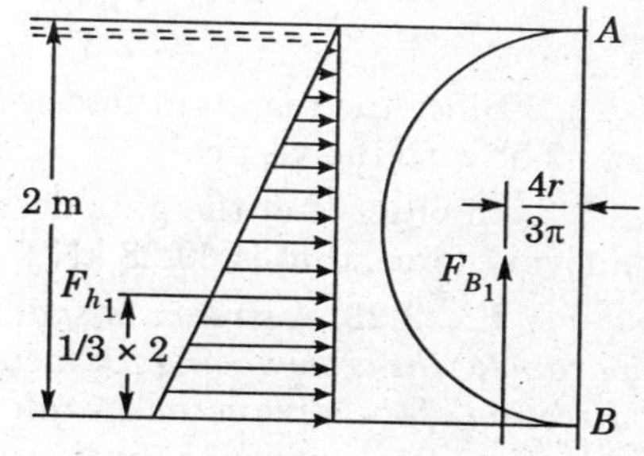

The forces acting on the left half are :

(*i*) Horizontal water force

$$F_{h_1} = \gamma h_c A,$$

where A = vertically projected area of the gate

$$F_{h_1} = 9810 \times 1.0 \times (2.0 \times 3.0)$$

$$= 58810 \text{ N} = 58.8 \text{ kN}$$

This force acts at a distance

$\frac{1}{3} \times 2 = 0.666$ m from the bottom.

(*ii*) Buoyant force F_{B_1} = Weight of the liquid displaced

$$= \gamma \frac{\pi r^2}{2} \times 3.0 = 9810 \times \frac{\pi \times (1.0)^2 \times 3}{2} = 46200 \text{ N} = 46.2 \text{ kN}$$

and acts at a distance of $\frac{4r}{3\pi} = \frac{4 \times 1.0}{3\pi} = 0.425$ m from AB.

Forces acting on the right half :

(*i*) Horizontal water force,

$$F_{h_2} = \gamma h_c \,.\, A$$

$$= 9810 \times 0.5 \times 1.0 \times 3.0 = 14700 \text{ N} = 14.7 \text{ kN}$$

and acts at a distance of $\frac{1}{3} \times 1.0 = 0.333$ m from the bottom.

(*ii*) Buoyant force F_{B_2}

$$= \gamma \times \frac{\pi r^2}{4} \times 3.0 = 9810 \times \frac{\pi \times (1.0)^2 \times 3.0}{4}$$

$$= 23080 \text{ N} = 23.08 \text{ kN}$$

and acts at a distance $\frac{4r}{3\pi} = 0.425$ m from AB.

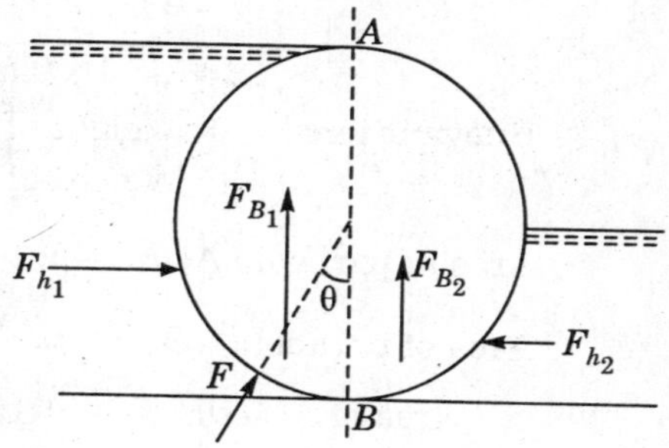

These forces have been shown on the cylindrical gate.

(*a*) Resultant horizontal force

$$F_h = F_{h_1} - F_{h_2} = 58.8 - 14.7 = 44.1 \text{ kN}$$

and acts at a distance y from the bottom, such that

$$44.1y = 58.8 \times \tfrac{2}{3} - 14.7 \times \tfrac{1}{3}$$

$$y = 0.777 \text{ m from the bottom.}$$

(*b*) Resultant vertical force,

$$F_B = F_{B_1} + F_{B_2} = 46.2 + 23.08 = 69.28 \text{ kN}$$

and acts at a distance x from AB, such that

$$69.28x = 46.2 \times 0.425 - 23.08 \times 0.425$$

$\therefore$ $x = 0.1415$ m.

$\therefore$ Resultant water force,

$$F = \sqrt{F_h^2 + F_B^2} = \sqrt{(44.1)^2 + (69.28)^2} = 82.1 \text{ kN}$$

$$\tan\theta = \frac{F_h}{F_B} = \frac{44.1}{69.28} = 0.637.$$

$\therefore$ $\theta = 32.5°$.

Since the resultant force acts normal to the surface, it is a radial force and makes an angle of 32.5° with the vertical.

In order that the gate does not float away, its weight must not be less than the resultant buoyant force, that is 69.28 kN.

Ex. 2.29. *A stream is spanned by a bridge which is a single masonry arch in the form of a parabola, the crown being 2.5 metre above the springings which are 9 metres apart. Measured in the direction of the stream the overall width of the bridge is 6 metres. During a flood the stream rises to a level 2 metres above the springings. Calculate the force tending to lift the bridge from its foundations if the arch remains water tight.*

Sol. Let the equation of the parabola be $x^2 = Ky$

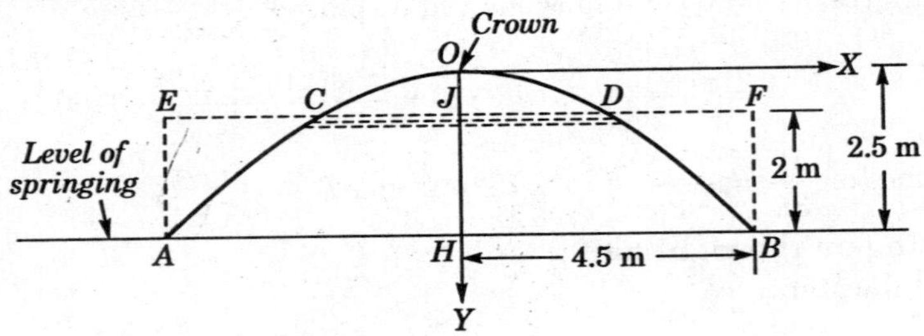

when $y = 2.5$ m, $x = 4.5$ m

$\therefore$ $K = \dfrac{(4.5)^2}{2.5} = 8.1$

at $y = 2.5 - 2.0 = 0.5$ m

$x = \sqrt{8.1 \times 0.5} = 2.01$ m

Buoyant force = Weight of water displaced by the arch
= Weight of volume ($ACE + BDF$)

Area of parabola AOB, $\frac{2}{3} \times \overline{AB} \times \overline{OH} = \frac{2}{3} \times 9 \times 2.5 = 15 \text{ m}^2$

Area of parabola COD $= \frac{2}{3} \times 4.02 \times 0.5 = 1.34 \text{ m}^2$

$\therefore$ Area of $ACJDB$ $= 15 - 1.34 = 13.66 \text{ m}^2$

$\therefore$ Area ($AEC + BDF$) $= 9 \times 2 - 13.66 = 4.34 \text{ m}^2$

$\therefore$ Volume of water displaced $= 4.34 \times 6 = 26.05 \text{ m}^3$

Force tending to lift the bridge $= 9810 \times 26.05 \text{ N} = 255500 \text{ N} = 255.5 \text{ kN}$.

Ex. 2.30. *A cylindrical gate as shown below consists of a circular cylinder and a plate, hinged at the weir. The gate is controlled in position by pumping water into or out of the cylinder. The centre* *of gravity of the empty gate is on the line of symmetry 1.40 m from the hinge. It is in equilibrium when empty in the position shown in the figure. Determine the volume of water that must be added per metre of cylinder length to hold the gate in its position when the water surface is raised by 1 m.*

Sol. Buoyant force, F_B, on the cylindrical gate

$$= \gamma \times \text{Volume of liquid displaced}$$

$$= 9810 \times \frac{\pi \times (1)^2}{2} \times 1 = 15400 \text{ N per metre length}$$

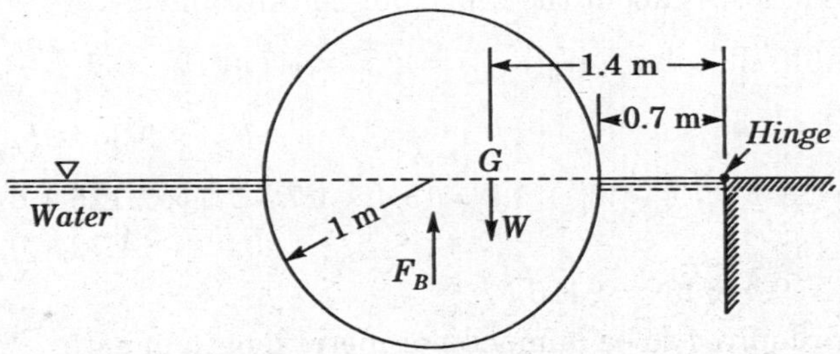

Let W be the weight of the cylindrical gate, taking moments about the hinge, we obtain

$$F_B \times 1.7 = W \times 1.4$$

$$\therefore \quad W = \frac{1.7}{1.4} \times 15400 = 18700 \text{ N} = 18.7 \text{ kN}.$$

When the water surface is raised by 1 m :

The system of forces that act on the cylinder are the horizontal forces F_{h_1}, F_{h_2}, the buoyant forces F_{B_1}, F_{B_2}, F_{B_3}, weight of the cylinder W, and the weight of water W_w required to maintain the gate in position. These forces are shown in the figure alongside, and are evaluated as follows.

$F_{h_1} = \gamma\, h_c A = \gamma \times 1 \times (2 \times 1) = 2\gamma$, which acts at 4/3 m from the water surface.

$F_{h_2} = \gamma h_c A = \gamma \times (1 + 0.5)\,(1 \times 1) = 1.5\,\gamma$, which acts at a distance of y below the water surface, which is given by

$$F_{h_2} \times y = \gamma \times 1 \times (2 \times 1) \times \frac{4}{3} - \gamma \times 0.5 \times (1 \times 1) \times 2/3$$

from which, $\quad y = \dfrac{(8/3 - 1/3)}{1.5} = 1.556$ m.

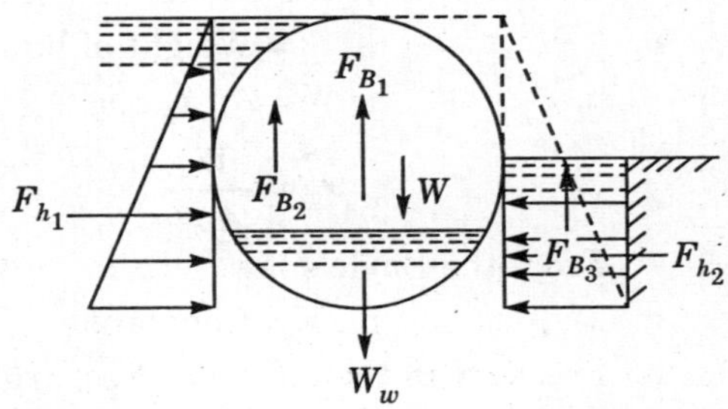

Total buoyant force = $F_{B_1} + F_{B_2} + F_{B_3}$

where F_{B_1} = Buoyand force on the bottom half cylinder

$$= \gamma V_1$$

$$= 9810 \times \frac{\pi r^2}{2}$$

$$= 9810 \times \frac{\pi \times (1)^2 \times 1}{2}$$

= 15400 N which acts at a distance of 1.7 m from the hinge.

F_{B_2} = Buoyant force on the quarter left top of the cylinder

$$= \gamma V_2 = 9810 \times \frac{\pi r^2}{4} \times 1 = 9810 \times \frac{\pi (1)^2 \times 1}{4}$$

$= 7700$ N, which acts at a distance of $\left(1.7 + \dfrac{4 \times 1}{3\pi}\right)$ from the hinge

F_{B_3} = Buoyant force on the plate = γV_3

$= 9810 \times 0.7 \times 1 \times 1 = 6867$ N, which acts at a distance of 0.35 m from the hinge.

Now, let W_w be the weight of water, (required to maintain the gate in position) pumped into the cylinder ; this weight acts at a distance of 1.7 m from the hinge.

Taking moments of forces about the hinge, for equilibrium

$$F_{h_1} \times (1.333 - 1.0) + W \times 1.4 + W_w \times 1.7 = F_{B_1} \times 1.7 + F_{B_2} \times (1.7 + 0.425)$$
$$+ F_{B_3} \times 0.35 + F_{h_2} \times (1.556 - 1.0)$$

or $$19.62 \times 0.333 + 18.7 \times 1.4 + W_w \times 1.7 = 15.4 \times 1.7 + 7.7 \times 2.125 + 0.687 \times 0.35 + 14.72 \times 0.556$$

from which $W_w = 9.416$ kN $= 9416$ N

Volume of water required to be pumped per metre length of gate

$$= \frac{9416}{9810} = 0.96 \text{ m}^3.$$

Ex. 2.31. *A cylindrical bucket with one end open is observed to be floating on a liquid of density ρ with the open end down. It is of weight W and is supported by air that is trapped inside it as shown below. The bucket floats with a height h above the liquid surface. If the air is assumed to follow isothermal law (pressure times volume is a constant), determine the force F necessary just to submerge the bucket. The internal area of cross-section of the bucket is A. The thickness of the wall is assumed to be negligible and the hydrostatic pressure due to the atmosphere may by neglected.*

Sol. Weight of the bucket

= Weight of liquid displaced

$\therefore \quad W = A \,.\, x_1 \,.\, \rho g \qquad$ ...(1)

$\therefore \quad x_1 = \dfrac{W}{A\rho g}$.

From the Boyle's law,

$p \,.\, V$ = Constant

Pressure in the air (p) = Pressure in the liquid at the liquid-air interface

$$= \rho g \,.\, x_1 = \rho g \,.\, \frac{W}{A\rho g} = \frac{W}{A}$$

Volume of air, $V = A(h + x_1) = A\left(h + \dfrac{W}{A\rho g}\right)$

$$\therefore \qquad p \,.\, V = \frac{W}{A} A\left(h + \frac{W}{A\rho g}\right) = W\left(h + \frac{W}{A\rho g}\right) \qquad ...(2)$$

Considering the final position of the bucket :

Total downward force

$= F + W$, must be equal to the Buoyant force on the bucket

Buoyant force on the bucket

= Weight of the liquid displaced $= A \,.\, x_2 \rho g$

Thus, $\quad F + W = A x_2 \rho g \qquad ...(3)$

From the Boyle's law $pV = p_1 V_1$,

but $\quad p_1 = \rho g \,.\, x_2$ and $V_1 = A \,.\, x_2$

$\therefore \quad p_1 V_1 = \rho g A x_2^2$

Making substitutions in the Boyle's law

$$A \rho g\, x_2^2 = W\left(h + \frac{W}{A\rho g}\right) \qquad \text{or} \qquad x_2 = \sqrt{\frac{W}{A\rho g}\left(h + \frac{W}{A\rho g}\right)}$$

Now from Eq. (3),

$$F + W = A \rho g \,.\, x_2$$

$$= \sqrt{A \rho g W\left(h + \frac{W}{A\rho g}\right)} = \sqrt{W^2 + A \rho g h W}$$

$$\therefore \qquad F = \left[\sqrt{W^2 + A h \rho g W} - W\right].$$

Ex. 2.32. *In the following figure is shown a float valve regulating the flow of oil of specific gravity 0.8 into a cistern. The spherical float is 15 cm in diameter. AOB is a weightless link carrying the float at one end and a valve at the other end which closes the pipe through which oil flows into the cistern. The link is mounted on a frictionless hinge at O and the angle AOB is 135°. The length OA is 20 cm and the distance between the centre of the float and the hinge is 50 cm. When the flow is stopped AO will be vertical. The valve is to be pressed on to the seat with a force of 10 N to completely stop the flow into into the cistern. It was observed that the flow of oil is stopped when the free surface of oil in the cirstern is 35 cm below the hinge. Determine the weight of the float.*

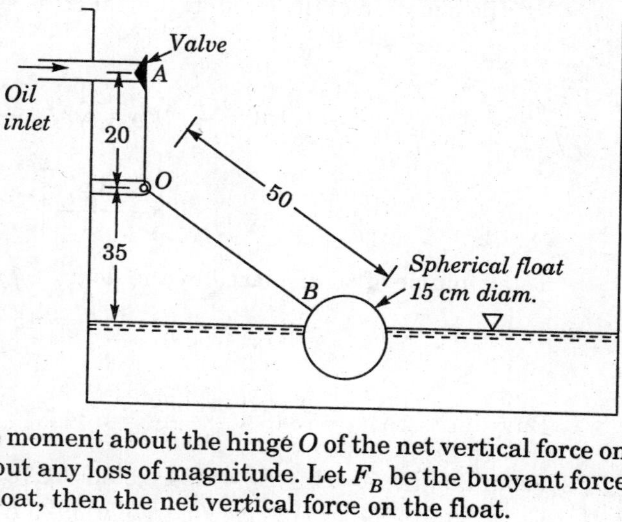

Sol. Since the hinge is frictionless, the moment about the hinge O of the net vertical force on the float is transferred to the valve at A without any loss of magnitude. Let F_B be the buoyant force acting on the float and W the weight of the float, then the net vertical force on the float.

$$F_V = F_B - W \qquad ...(i)$$

Taking moment of the forces about the hinge (see the following figure)

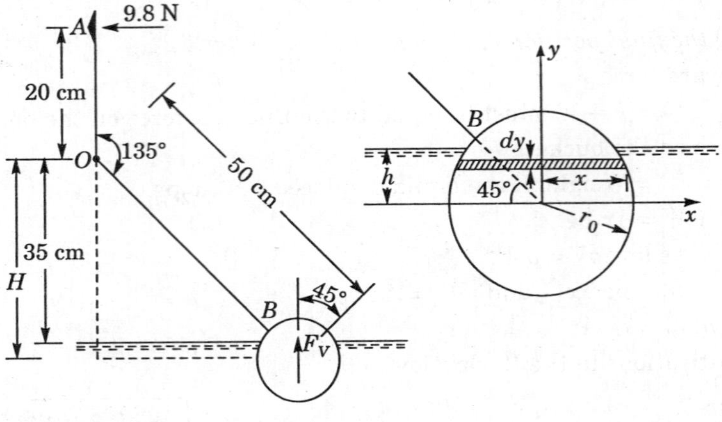

$F_V \cos 45° \times 50 = 10 \times 20 \quad \therefore \quad F_V = 5.56$ N.

Let H be the height of the hinge above the centre of the float, and h the height of the oil free-surface above the float centre, then

$$\sin 45° = \frac{H}{50}, H = 35.3 \text{ cm} \quad \text{and} \quad h = H - 35.0 = 0.3 \text{ cm}.$$

For finding out the volume of oil displaced by the float, consider a disc element of radius x, thickness dy and located at a distance y above the x-axis, see the right-hand figure. The volume of oil displaced by the portion lying above the x-axis is

$$V = \int_0^h \pi x^2 dy$$

Let the equation of the circular central cross-section of the float be ;

$$x^2 + y^2 = r_0^2$$

$$V = \int_0^h \pi(r_0^2 - y^2)dy = \pi\left[r_0^2 y - \frac{y^3}{3}\right]_0^h$$

$$= \pi\left[\left(\frac{15}{2}\right)^2 \times 0.3 - \frac{(0.3)^3}{3}\right] = 53.0 \text{ cm}^3.$$

Total volume of oil displaced by the float

$$= \frac{2}{3}\pi r_0^3 + 53.0 = \frac{2}{3}\pi \times \left(\frac{15}{2}\right)^2 + 53.0 = 935.5 \text{ cm}^3$$

Buoyant force $F_B = 7848 \times \frac{935.5}{(100)^3} = 7.34$ N

From Eq. (*i*), $W = F_B - F_V = 7.34 - 5.56 =$ **1.78 N.**

Ex. 2.33. *An iceberg of specific weight 8976 N/m³ extends above the surface of sea water of specific weight 10104 N/m³. What percentage of the total volume of the iceberg is visible to an observer ?*

Sol. Let V_a = Volume of iceberg in an air (above the sea surface)

V_w = Volume of iceberg inside the sea-water

V = total volume of iceberg

then $V = V_a + V_w$

From the principle of buoyancy and floatation

Weight of iceberg W = Buoyant force F_B

or $\gamma_i(V_a + V_w) = \gamma_w V_w$

from which, $\dfrac{V_a}{V_w} + 1 = \dfrac{\gamma_w}{\gamma_i}$

$V_a/V_w = \gamma_w/\gamma_i - 1$

$= \dfrac{10104}{8976} - 1 = \dfrac{1128}{8976}$

$\dfrac{V_w}{V_a} = \dfrac{8976}{1128}$

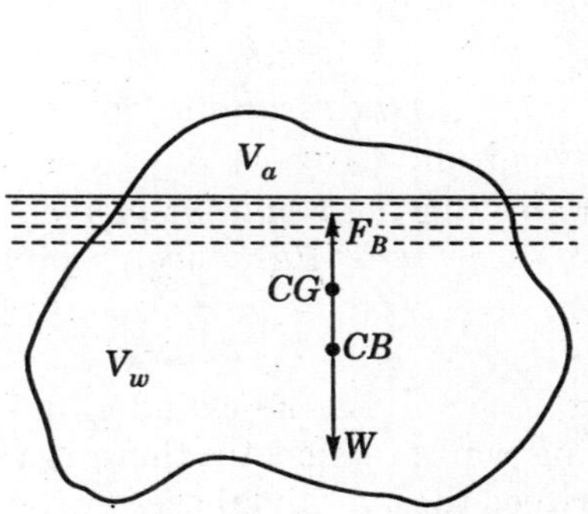

$\therefore \quad \dfrac{V_a + V_w}{V_a} = \dfrac{10104}{1128} = 8.957$

or $\quad \dfrac{V}{V_a} = 8.957$ or $V_a = 0.1116$ V.

Therefore, the percentage of *visible portion* of icebergy = **11.16%.**

Ex. 2.34. *A float valve of 'the ball-cock' type is required to close an opening of a supply pipe feeding a cisterm. The valve should close down when 60% of the volume of the spherical float is immersed in water. The valve has a diameter of 15 mm and the fulcrum of the operating lever is to be 150 mm from the valve and 0.50 m from the centre of the float. Determine the minimum diameter of the float if it is required to close the valve against a pressure of 147 kN/m² gauge. Neglect the weight of the lever and the float.*

Sol. With reference to the following figure :

The force required to close the valve

$$F = \frac{\pi d^2}{4} . p$$

$$= \frac{\pi(15 \times 10^{-3})^2}{4} \times 147 \times 10^3 \text{ N}$$

$= 26.0$ N.

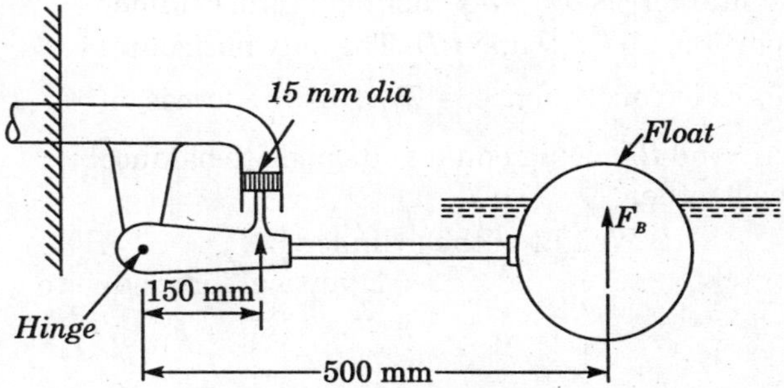

Float Valve

The corresponding buoyant force F_B required to be exerted by the float is obtained by taking moment of forces about the hinge,

$$F_B \times 500 = F \times 150$$

or $\quad F_B = 26.0 \times \dfrac{150}{500} = 7.8$ N.

Neglecting the weight of the lever and the float,

$$F_B = 0.60\gamma V$$

where V is the volume of spherical float. Substituting the values,

$$7.8 = 0.60 \times 9810 \times \frac{\pi d^3}{6}$$

from which, $\quad d$ = **13.6 cm** or **136 mm.**

To allow for the weight of the lever and float, and to ensure that the valve is firmly seated, a larger diameter of the float would be required.

Ex. 2.35. *A cylinder of radians 0.3 m is located in water as shown. The cylinder and the wall are smooth. For a 1.5 m length of cylinder, find*

(i) its weight

(ii) the resultant force exerted by the wall on the cylinder

(iii) the resultant moment around the centre of the cylinder due to water forces on the cylinder.

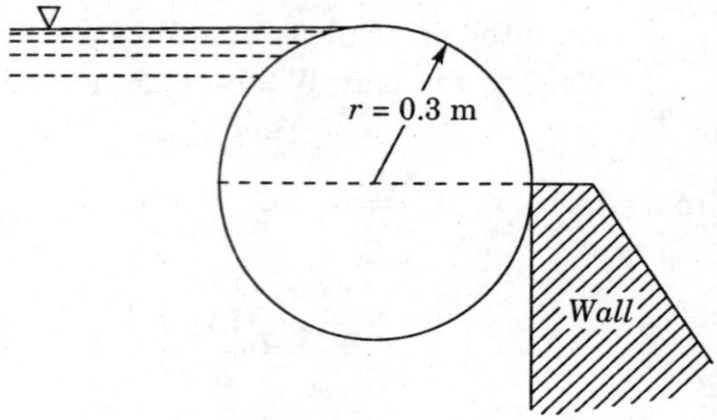

Sol. As the cylinder and the supporting wall are smooth, there is no friction between the two, and thesefore the weight of the cylinder must be equal to the net vertically upward hydrostatic force acting on the cylinder, considering the equilibrium of forces acting on the cylinder *ADBC* as shown. These forces are (*a*) the vertical forces $F_{V_1}, F_{V_2}, F_{V_3}$ of equal magnitude representing the weight of water contained in *APC*, *ASD* and *DTB*, and the hydrostatic force on *ST* of magnitude (γd), *d* being the diameter of cylinder. (*b*) The horizontal forces F_{H_1}, F_{H_2} acting on the cylinder portions *AC*, *AD* and *BD*. The only horizontal force of consequence is F_{H_1}, as the forces on *AD* and *BD* being equal and opposite balance each other.

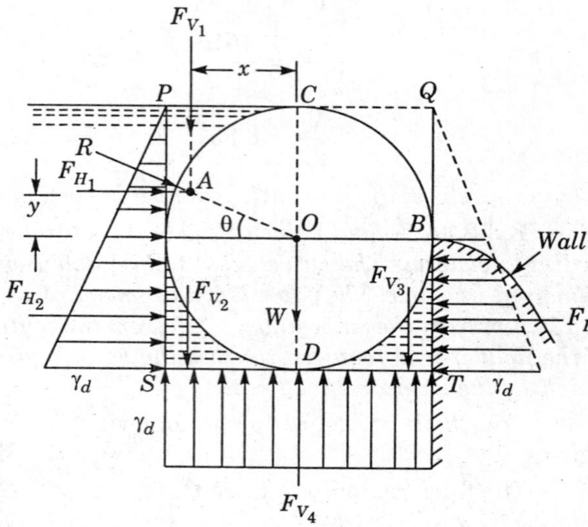

Hydrostatic Forces acting on cylinder

(*i*) Weight of the cylinder *W*

$$= \Sigma \text{ vertical forces due to water}$$

$$= F_{V_4} - (F_{V_1} + F_{V_2} + F_{V_3})$$

$$= \gamma d\, L - \left[3\gamma\left(r^2 - \frac{\pi r^2}{4}\right)L\right], \text{ since } F_{V_1} = F_{V_2} = F_{V_3}$$

$$= 1000 \times 0.6 \times 1.5 - \left[3 \times 1000\left(0.3^2 - \frac{\pi 0.3^2}{4}\right)1.5\right]$$

$$= 452.93 \text{ kg (or 4443.2 N)}$$

(*ii*) Resultant force *R* exerted by water on the cylinder is due to hydrostatic pressure on cylinder surface represented by *AC*, the effect of force acting on the bottom half of the cylinder balance each other.

Resultant hydrostatic force $R = \sqrt{F_{H1}^2 + F_{V1}^2}$

The horizontal force on *AC* F_H is given by

$$F_{H_1} = \frac{1}{2}(\gamma r)(r \times 1.5) = \frac{1}{2} \times 1000 \times (0.3)^2 \times 1.5 = 67.5 \text{ kg.}$$

This force acts at a distance of $y = \frac{r}{3} = \frac{0.3}{3} = 0.1$ m from *AB*.

Vertical component F_{V_1} acting on AC

$$= \text{Weight of water standing on curred surface } AC$$

$$= \gamma\left[r^2 - \frac{\pi r^2}{4}\right] L = 1000 \times (0.3)^2 \left[1 - \frac{\pi}{4}\right] \times 1.5 = 28.97 \text{ kg}$$

which acts at a distance x from OC. For evaluating x, taking moment of areas involved

$$r^2 \times \frac{r}{2} - \left(r^2 - \frac{\pi r^2}{4}\right)x = \frac{\pi r^2}{4}\frac{4r}{3\pi}$$

from which, $$x = \frac{2r}{3(4-\pi)}$$

$\therefore$ Resultant force $R = \sqrt{(28.97)^2 + (67.5)^2} = 73.45$ kg

acts radially on the surface AC as shown. The angle of inclination θ at which the resultant is inclined with AB

$$\tan\theta = \frac{F_{V_1}}{F_{H_1}} = \frac{28.97}{67.5} = 0.439, \Rightarrow \theta = \mathbf{23.23°}.$$

The resultant force exerted by the wall is equal to $F_{H_1} = 67.5$ kg, (which acts from B) as all vertical forces due to water are balanced by the weight of the cylinder and the horizontal forcs on the bottom portion ADB being equal and opposite balance each other.

(*iii*) Resultant moment around the centre of cylinder due to water forces on the cylinder

$$= F_{H_1} \cdot y - F_{V_1} \cdot x$$

$$= 67.50 - 28.97 \times 0.233 = 0.$$

Ex. 2.36. *The following figure illustrates the adjustment of gasolene (petrol) level in the float chamber of a carburettor of an automobile. Gasolene is fed into the chamber through a pipe of diameter d = 4 mm under a pressure of 0.3 atm. (gauge). The ball-lype float having a weight of 25 g and the needle shutting off the flow of gasolene are secured on a lever which can swing about a stationary axle O.*

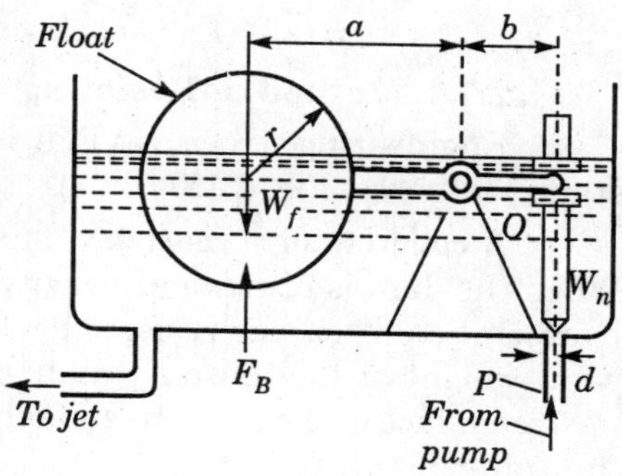

Calculate the size of the float (radius r) at which a constant level of gasolene is maintained in the chamber and the float is submerged by half at the moment the orifice opens. Weight of the needle in the gasolene is 12 g, and the specific weight of gasolene = 700 kg/m³. The distances :

a = 40 mm, and b = 15 mm. Neglect weight of the lever.

Sol. Let F_B be the buoyant force acting on the spherical ball (float) of radius 'r', which passes through the ball centre. Let W_f be the weight of the float and W_n that of the needle.

Taking moment of forces about the pivotal point of the axle O,

$$(F_B - W_f)\, a + W_n \cdot b - \frac{\pi d^2}{4}\, p \cdot b = 0$$

F_B = "Weight of gasolene displaced by half the float which remains submerged."

$$= \frac{1}{2}\left(\frac{4}{3}\pi r^3 \cdot \gamma_{gas}\right)$$

or $$\left(\frac{2}{3}\gamma_{gas}\, nr^3 . a - W_f . a + W_n . b - \frac{\pi d^2}{4} p . b\right) = 0$$

Simplifying, $$r = \sqrt[3]{\frac{\left(\frac{\pi d^2}{4} p - W_n\right) b + W_f . a}{\frac{2}{3}\pi\, \gamma_{gas} . a}}$$

from the given data,

$$d = 4 \text{ mm} = \frac{4}{10} \text{ cm} ; a = \frac{40}{10} \text{ cm} = 4 \text{ cm} ; \; b = \frac{15}{10} = 1.5 \text{ cm}$$

$p = 0.3$ atm (gauge) $\qquad \gamma_{gas} = 700$ kg/m³

$= 0.3 \times 10329$ kg/m² (gauge) $\qquad = 700 \times \frac{1000}{(10)^3}$ g/cm³

$= 3098.7$ kg/m² (gauge) $= 0.70$ g/cm³ $\qquad = 0.70$ g/cm³

$= 3098.7 \times \frac{1000}{(100)^2}$ g/cm² (gauge)

$= 309.87$ g/cm² (gauge)

Substituting in the expression for r,

$$r = \left[\frac{\left\{\frac{\pi}{4}\left(\frac{4}{10}\right)^2 \times 309.87 - 12\right\} 1.5 + 25 \times 4}{\frac{2}{3}\pi \times 0.7 \times 4}\right]^{1/3}$$

$= (23.94)^{1/3} =$ **2.88 cm or 2.9 cm**

2.7.1. Method of Resisting Hydrostaic Pressure

Engineering structures like dams, lock-gates, sluice-gates are required to withstand hydrostatic pressures. Fig. 2.17A illustrates these structures.

A concrete or a masonry dam is held in equilibrium by a system of forces and reactions. When the dam is of straight gravity type, Fig. 2.17A (*i*), the overturning moment exerted by the hydrostatic pressure is resisted by a stabilising moment of which the two elements are the downward weight of the masonry and the weight of the wedge-shaped water body supported by the upstream face of the dam. In an arch-dam built across a narrow-gorge, as shown in Fig. 2.17A (*ii*), the water pressure is mainly resisted by the reactions R_A and R_B of the rock walls of the gorge which act as abutments to the arch.

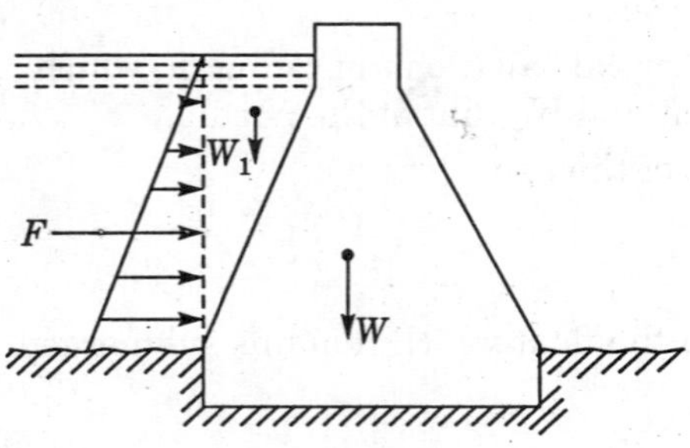

(*i*) Section of a gravity dam.

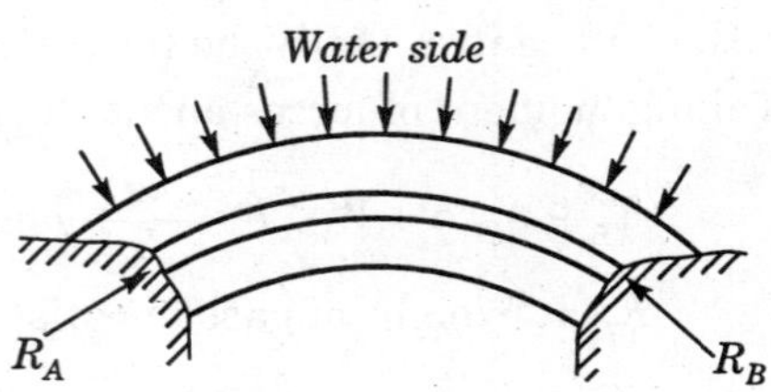

(*ii*) Plan of an arch-dam

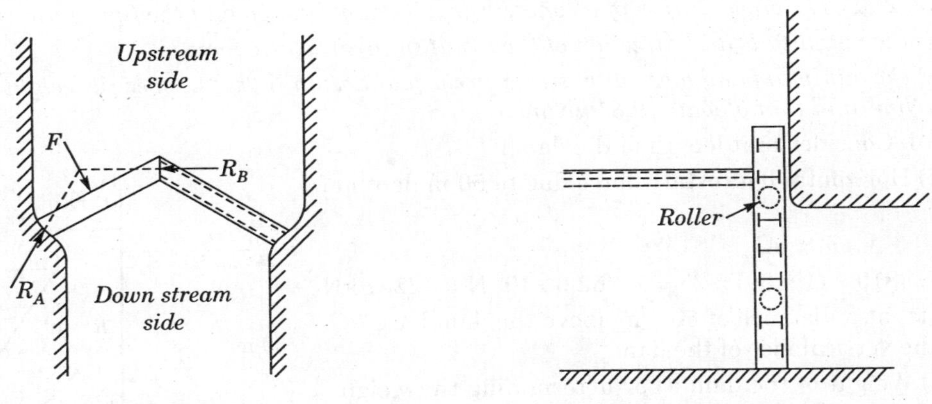

(*iii*) Plan of lock-gates (*iv*) A sluice-gate

Fig. 2.17. A-Engineering Structures to resist water pressure.

Conditions similar to arch-dam prevail in the conventional type of hinged lock-gates shown in Fig. 2.17 (*iii*) (Lock-gates are used in docks and at the tail end a river for navigation purposes). In this case, the hydrostatic thrust F on the leaf shown in full lines is resisted by the reaction R_A of the hinge and by the reaction R_B of the opposite gate shown dotted. In the sluice-gate (sluice-gates are used to regulate flow of water from reservoirs and canals) shown in Fig. 2.17A (*iv*), the horizontal beams supporting the skin-plates are closely spaced together near the bottom as compared to the top of the gate, in accordance with the hydrostatic law of pressure variation.

Ex. 2.37. *The following figure shows the cross-section of a dam. For the conditions indicated on the figure, determine the magnitude of the resultant force due to water, its inclination with the vertical and the distance from P its line of action cuts the horizontal base PQ.*

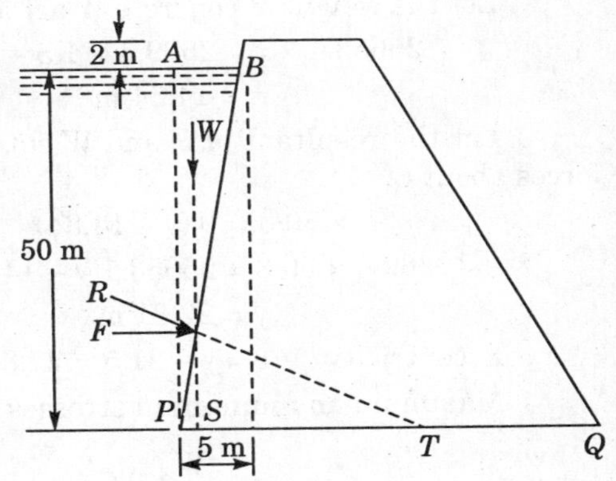

Sol. The hydrostatic horizontal force F acting per metre of dam length

$$F = \gamma A h_c = 9810 \times (1 \times 50) \times 25$$
$$= 12262 \times 10^3 \text{ N}$$
$$= 12262 \text{ kN per m length}$$

This force acts over AP at a distance of 50/3 m from P.

The weight of the triangular wedge ABP of water per metre of dam length

$$W = \left(\frac{1}{2} \times 50 \times \frac{5}{52} \times 50 \times 1\right) \times 9810 = 1197 \times 10^3 = 1197 \text{ kN per m length}$$

which acts at a distance of $\frac{1}{3} \times \frac{5 \times 50}{52} = 1.63$ m from AP

Resultant water force $R = \sqrt{F^2 + W^2} = \sqrt{(12262)^2 + (1197)^2} = 12331$ kN/m

Its inclination with the vertical

$$\tan\theta = \frac{F}{W} = \frac{12262}{1197} = 10.24, \ \theta = \tan^{-1}(10.24)$$

The distance PT where the resultant force R cuts the base

$$PT = PS + ST$$
$$= \frac{5 \times 50}{52 \times 3} + \frac{50}{3}\tan\theta = 1.603 + 170.6 = 172.203 \text{ m.}$$

Ex. 2.38. *The cross-section of a concrete gravity dam is shown in the figure. Determine :*

(i) the magnitude and direction of the resultant hydrostatic force

(ii) the minimum and maximum stress on the foundation. Take the specific weight of concrete as 24.5 kN/m³ and that of water 9.81 kN/m³.

Sol. Consider unit length of the dam :

(*i*) Horizontal hydrostatic force due to 50 m depth of water

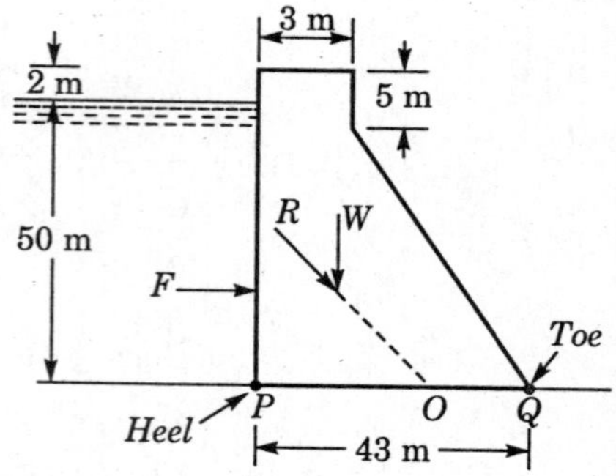

$$F = \gamma A h_c$$

$$= 9810 \times (1 \times 50) \times 25 = 12262.0 \times 10^3 \text{ N} = 12262 \text{ kN}$$

which acts, at a distance of 50/3 m above the dam base *PQ*, against the vertical face of the dam.

(*ii*) Weight of the dam : For determining the weight, dam section may be divided into a rectangle and a triangle.

(*a*) Weight of the rectangle

$$W_1 = (3 \times 52 \times 1) \times 24.5 \times 10^3 \text{ N} = 2599.6 \text{ kN},$$

which acts at a distance of 1.5 m from the vertical face of the dam

(*b*) Weight of the triangle $W_2 = (\frac{1}{2} \times 47 \times 40 \times 1) \times 24.5 = 23053.5$ kN

which acts at a distance of (3 + 40/3) m from the vertical dam face. The weight of the dam per unit length

$$W = W_1 + W_2 = 2599.6 + 23053.5 = 25653.1 \text{ kN}$$

Let the weight of concrete *W* act at a distance of *x* metres from *P* on *PQ*, then

$$25653.1 \times x = 2599.6 \times 1.5 + 23053.5(3 + 40/3)$$

$$x = 14.84 \text{ m}$$

Let the resultant of *F* and *W* cut the base *PQ* at *O*, such that *PO* = *y*. Taking moment of forces about *O*

$$F \times 50/3 = W(y - 14.84)$$

$$12262.0 \times 50/3 = 25653.1(y - 14.84)$$

$$y = 22.81 \text{ m}$$

Eccentricity, $e = y - 21.5 = 22.81 - 21.5 = 1.31$ m

Maximum and minimum stresses are given by

$$\sigma = \frac{W}{A}\left(1 \pm \frac{6e}{b}\right) = \frac{25653.1}{43 \times 1}\left(1 \pm \frac{6 \times 1.31}{43}\right) \text{ kN/m}^2$$

$$= 706.3 \text{ kN/m}^2 \text{ ; } 585.17 \text{ kN/m}^2.$$

Ex. 2.39. *The figure shows the parabolic section of a dam, the vertex of the parabola being at O. The axis of the parabola is vertical and is 15 m away from the dam face at the water level. Assume the centroid of the half-parabolic cross-section of water located at 5.5 m from the vertical through O.*

Determine the resultant water pressure per metre-run of dam length, its inclination with the vertical and the distance from O its line of action cuts the horizontal base OX.

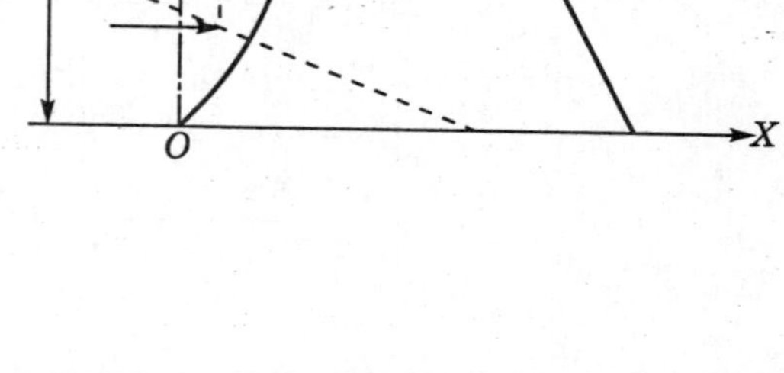

Sol. The horizontal water force on the parabolic face of the dam per metre length is :

$$F = \gamma A h_c = 9810 \times (1 \times 60) \times 30 = 17658 \times 10^3 \text{ N}$$
$$= 17658 \text{ kN}$$

and acts at a distance of 20 m above OX.

Let the equation of the parabola with the axes shown, be

$$y = cx^2$$

from the given data, for $y = 60$ m, $x = 15$ m which gives $c = 60/15 \times 15 = 4/15$. The equation of the parabola may be written as

$$x = \sqrt{\frac{15y}{4}}$$

The weight of the curved wedge OAB of water per metre of dam length is :

$$W = \int_0^{60} \gamma(1 \times x\, dy)$$

$$= \gamma \int_0^{60} \sqrt{\frac{15y}{4}}\, dy = 1.935\gamma \left[\frac{2}{3} y^{3/2}\right]_0^{60}$$

$$= 1.935 \times 9810 \times \tfrac{2}{3}\ (60)^{3/2} = 5886 \times 10^3 \text{ N} = 5886.0 \text{ kN}$$

The resultant water force $R = \sqrt{F^2 + W^2} = \sqrt{(17658)^2 + (5886)^2}$ kN

$$= 18609.57 \text{ kN}$$

and is inclined at an angle θ with the vertical

$$\tan\theta = \frac{F}{W} = \frac{17658}{5886} = 3.0$$

so that its line of action cuts OX at a distance of

$$= 5.5 + 20 \tan\theta = 5.5 + 20 \times 3 = 65.5 \text{ m from } O.$$

Ex. 2.40. *The following figure gives profile of a gravity dam with reservoir level as shown. If the coefficient of friction is 0.75, is the dam safe against sliding ? Take density of concrete 2.4 tonnes/m³.*

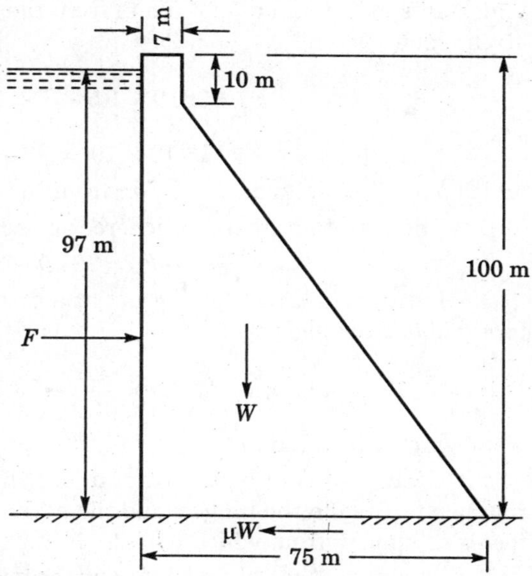

Sol. Consider one metre length of the dam for calculating various forces acting on the dam section. The force tending to cause sliding of the dam is the resultant hydrostatic force, F, and the force resisting this action is μW (which acts an the base of the dam) W being the weight of the dam. For the dam to be safe against sliding, the resisting force μW must be greater than the force causing sliding *i.e.* F.

The resultant hydrostatic force due to 97 m of water depth,

$$F = \gamma h\, A$$
$$= (1000 \times 97)(97 \times 1)$$
$$= 9409 \times 10^3 \text{ kg} = 9409 \text{ tonnes/m}$$

The weight of concrete gravity dam per metre length

$$W = 2.4 \left(10 \times 7 \times 1 + \frac{7 + 75}{2} \times 90 \times 1\right) = 9024 \text{ tonnes/m}$$

Force resisting the sliding $= \mu W$

$= 0.75 \times 9024$

$= 6768$ tonnes/m length of dam

Since $\mu W < F$, hence the dam is unsafe against sliding.

Ex. 2.41. *A pair of lock gates, each 3 m wide and 6 m high, when closed make an angle of 120° between them in plan. The depth of water is 4.5 m on one side and 1.5 m on the other side. Each gate is supported on two hinges situated at heights of 0.6 m and 5 m above the bottom. Determine the resultant water pressure on each gate and the reaction between the gates. Assuming that the reaction force acts at the same level as the resultant force due to water pressure, find the force on each hinge.*

Sol. Referring to the figures shown below, the hydrostatic force on the upstream side of one gate

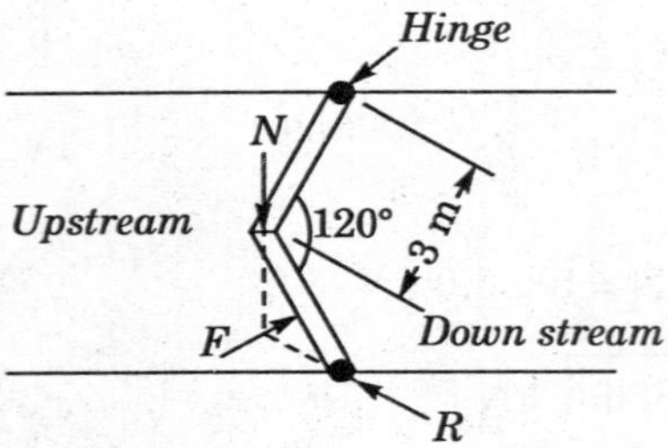

(*i*) Plan

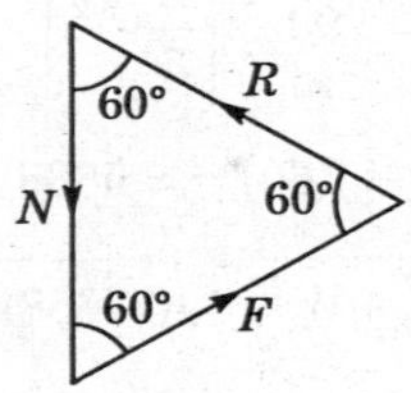

(*ii*) Triangle of forces forces on lock-gaters

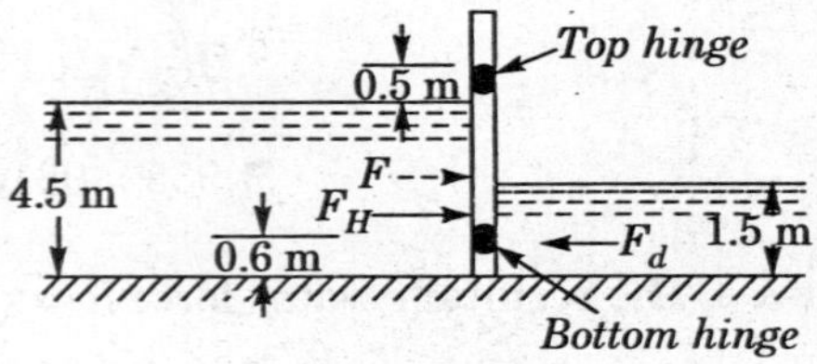

(*iii*) Elevation

$$F_u = \gamma A h_c = 9810 \times 4.5 \times 3 \times \frac{4.5}{2}$$

$$= 297.7 \times 10^3 \text{ N} = 297.7 \text{ kN}$$

which acts at a distance of 4.5/3 m from the bottom. Similarly, the hydrostatic force on the downstream side.

$$F_d = 9810 \times 1.5 \times 3 \times \frac{1.5}{2}$$

$$= 33.12 \times 10^3 = 33.12 \text{ kN}$$

which acts at a distance of 1.5/3 m from the bottom.

The resultant hydrostatic force on each gate acts normal to the gate as shown and is equal to

$$F = F_u - F_d = 297.7 - 33.12 = 264.62 \text{ kN}$$

Let the resultant force F act at a distance of x metres from the bottom. Taking moment of forces about the bottom,

$$264.62 \times x = 297.7 \times \frac{4.5}{3} - 33.12 \times \frac{1.5}{3}$$

from which, $x = 1.626$ m.

Each gate is in equilibrium under the action of the resultant water force F, the resultant reaction force R on the hinges, and the normal thrust N between the gates. These form a triangle of forces as shown, from which $N = R = F$.

Assuming that N and R act at the same level as F, the resultant hinge reaction R may be split into upper and lower hinge reactions, R_u and R_l. By taking moment of hinge reactions about the lower hinge :

$$264.62 \times (1.626 - 0.6) = R_u \times (5 - 0.6)$$

from which, $\mathbf{R_u = 61.7\ kN}$

$\mathbf{R_l} = R - R_u = 264.62 - 61.7 = \mathbf{202.9\ kN.}$

Ex. 2.42. *A tainter gate is installed on the crest of a spillway of a dam to control the water level in the reservoir. It is in the form of sector of a circle of radius of 8 m. The width of the gate between piers is 10 m. The gate-surface in contact with water (when the reservoir is full) makes an angle of 30 degrees at the centre of the circle where it is hinged. Determine the magnitude and direction of the resultant hydrostatic force on the gate. What is the vertical pull experienced by each hinge. The hinge is at the full reservoir level (F.R.L.).*

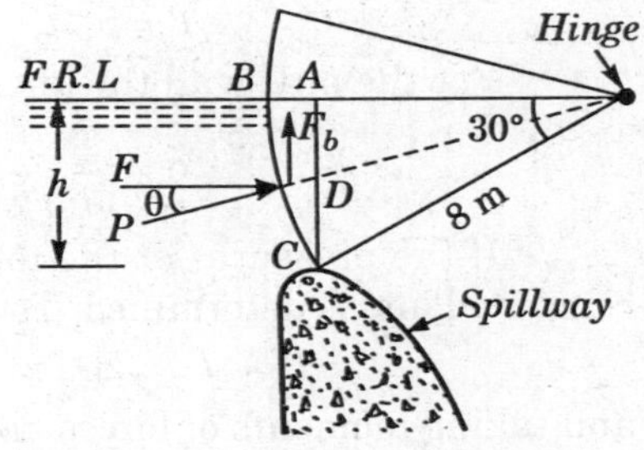

Tainter gate

Sol. Depth of water above the crest of spillway,

$$h = AC = 8 \sin 30° = 4 \text{ m}$$

The horizontal component F of the resultant hydrostatic force R is the force on the vertical projection of the gate surface BG.

$\therefore$

$$F = \gamma A h_c$$

$$F = 9810 \times (AC \times \text{width of gate}) \times h/2$$

$$= 9810 \times (4 \times 10)2$$

$$= 7848 \times 10^3 \text{ kg} = 784.8 \text{ kN}$$

The vertical component F_b of the resultant force is due to buoyancy and acts vertically upwards. Its magnitude is equal to the weight of water displaced by ABC.

$$F_b = \gamma \times \text{Area } ABC \times \text{width of gate}$$

$$= \gamma\left(\pi \times 8^2 \times \frac{30}{360} - \frac{1}{2} \times 4 \times 8 \cos 30°\right) \times 10$$

$$F_b = 9.81 \times 10^3 \times 2.89 \times 10 = 28.35 \times 10^4 \text{ N} = \mathbf{283.5\ kN}$$

Resultant hydrostatic force, $R = \sqrt{F^2 + F_b^{\,2}}$

$$= \sqrt{(784.8)^2 + (283.5)^2} = \mathbf{833.0\ kN}$$

The inclination of resultant with the horizontal

$$\tan\theta = F_b/F = \frac{283.5}{784.8} = 0.36$$

$$\theta = 19.75°$$

Since the surface of the gate is cylindrical (part of a circular cylinder), the resultant passes through the hinge.

Vertical pull on each hinge $= F_b/2 = 141.75$ kN.

2.7.2. Hoop Tension in a Pressure-Pipe

A circular pipe under the action of an internal pressure is in tension around its circumference. This hoop tension causes stretching of the pipe wall and the pipe material is subjected to tensile stresses. Consider Fig. 2.17 (*b*) showing longitudinal section and one-half cross-section of a circular pipe. Assuming that no longitudinal stress occurs, the walls are in tension as shown. Considering unit length of pipe and one-half of its circular section as a free-body, let the tensile force per unit pipe length at top and bottom be T_1 and T_2, respectively, as shown in the figure.

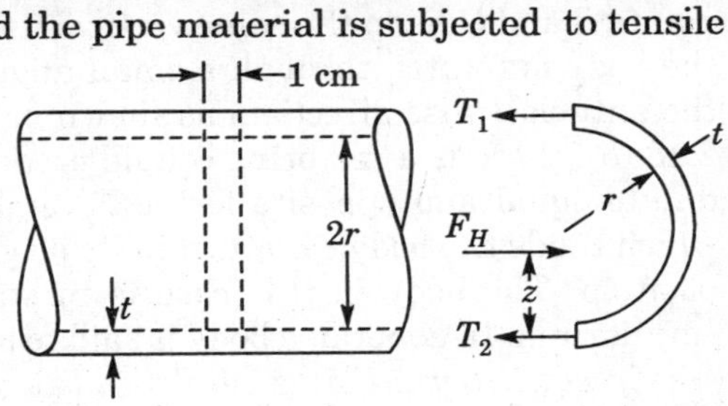

The horizontal hydrostatic force F_H acts through the centre of pressure of the projected area and has a magnitude $2rp$, in which p is the pressure at the pipe centre and r is the pipe radius. For penstock pipes which are subjected to high pressures, the centre of pressure may be taken at the pipe centre, and then

$$T_1 = T_2 = T$$

For equilibrium of forces $T_1 + T_2 = F_H = 2pr$

or $$2T = 2pr$$

from which, the hoop-tension force

$$T = p \,.\, r \qquad \text{...[2.17 (i)]}$$

For the pipe-wall thickness t, the tensile stress σ in the pipe-wall is

$$\sigma = \frac{T}{t} = \frac{p.r}{t} \qquad \text{...[2.27 (ii)]}$$

If the pressures at the top and bottom of pipe differ appreciably, the centre of pressure is computed and Z determined. To determine T_1 and T_2, the following equations may be written

$$T_1 + T_2 = 2p \,.\, r$$

and taking moments of forces about the pipe bottom

$$T_1 \,.\, 2r = F_H Z = 2\,pr \,.\, Z$$

Solving the above two equations

$$T_1 = p\,Z,\; T_2 = p\,(2r - Z) \qquad \text{...[2.27 (iii)]}$$

Ex. 2.43. *A 100 mm internal diameter steel pipe has 6 mm thickness of wall. Determine the maximum hydrostatic pressure of water and the corresponding water head which the pipe material can safely withstand, if the allowable tensile stress in the material is 78480 kN/m².*

Sol. $$\sigma = \frac{p.r}{t};\; p = \frac{\sigma.t}{r}$$

$$p = \frac{78480 \times 6 \times 10^{-3}}{50 \times 10^{-3}} = 9417.6 \text{ kN/m}^2$$

$$p = \gamma h$$

$$\therefore \quad h = p/\gamma = \frac{9417.6}{9.81} = \mathbf{960 \text{ m.}}$$

2.8. Stability of Submerged Bodies

A completely submerged body, such as a submarine, will remain in equilibrium at rest only if (*i*) the buoyant force F_B is equal to the weight W and (*ii*) the centre of buoyancy (B) lies directly above the centre of gravity (G). Submarines are required to use balancing tanks to make $F_B = W$ and trimming tanks to bring the centre of buoyancy above the centre of gravity. In order to bring the submarine on the free surface, the weight W must be decreased, by pumping out the water, such that $F_B > W$. In Fig. 2.18 (*a*), the submerged body is in equilibrium position and points B and G lie in the same vertical line. If the body is rotated through a small angle in the anti-clockwise direction, as shown in Fig. 2.18 (*b*), there is a restoring couple produced due to equal and opposite forces F_B and W, which tends the body to return to its original position. The body in this case is in stable equilibrium. In general, a body is said to be in *stable equilibrium*, if a slight rotational

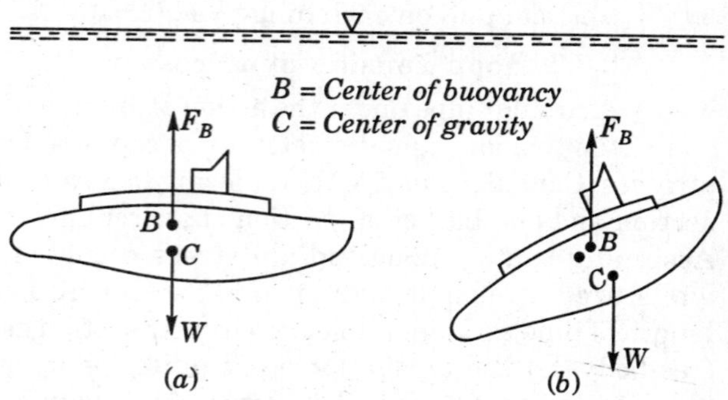

Fig. 2.18. Stability of a submerged body.

displacement generates forces which oppose the change of position and tend to bring the body to its original position.

A submerged body is said to be is *unstable equilibrium* if G lies above B, because the overturning couple produced due to a slight disturbance will cause the body to move away from its original position. If B and G coincide, the submerged body is said to be in *neutral equilibrium*.

In general, if the equilibrium is stable, the stability of the submerged body is said to be positive, if the equilibrium is unstable, the stability is negative and if the equilibrium is neutral, the body will have zero stability.

2.9. Stability of Floating Bodies

A body partly submerged in liquid is subjected to a buoyant force that is equal in magnitude to the weight of the liquid displaced by the body and acts through the centre of gravity of the displaced liquid.

A submerged body is in stable equilibrium, as already discussed, only when the centre of buoyancy (B) lies above the centre of gravity (G) of the body. On the contrary, a floating body may be in stable equilibrium against rolling, when B lies considerably below G. The criterion of stability of a floating body is the position of a fictitious point, known as *metacentre*, with respect to the centre of gravity. The metacentre, denoted by letter M, owes its origin to the French naval architect Bouguer, who discovered (in 18th century) that the initial stability of a floating body is inherent in the shape of the submerged portion of the hull. The concept of the metacentre has, since then, been utilised in determining the stability of floating bodies.

2.9.1. Metacentre and Metacentric Height

Fig. 2.19 shows the plan and vertical cross-sections of a floating vessel. The position of centre of gravity is represented by G and the centre of buoyancy of the displaced volume is at B. All sea going vessels are non-prismatic, as the vertical cross-section along the length of the vessel changes. Fig. 2.19 (*a*) shows the water line plan of a typical non-prismatic vessel. Let us consider an outside force applied causing the vessel to tilt or heel through a small angle θ, as shown in Fig. 2.19 (*c*). Due to change in the shape of displaced volume, the centre of buoyancy shifts to B'. The line of action of the buoyant force (acting through B') cuts the axis of symmetry *c-d* at a point M, forming a righting couple tending to restore the vessel to its original upright position. If, however, the point M lies in between G and B, an upsetting couple is produced which results in progressive titling of the vessel, (θ increases continuously) thus rendering it unstable.

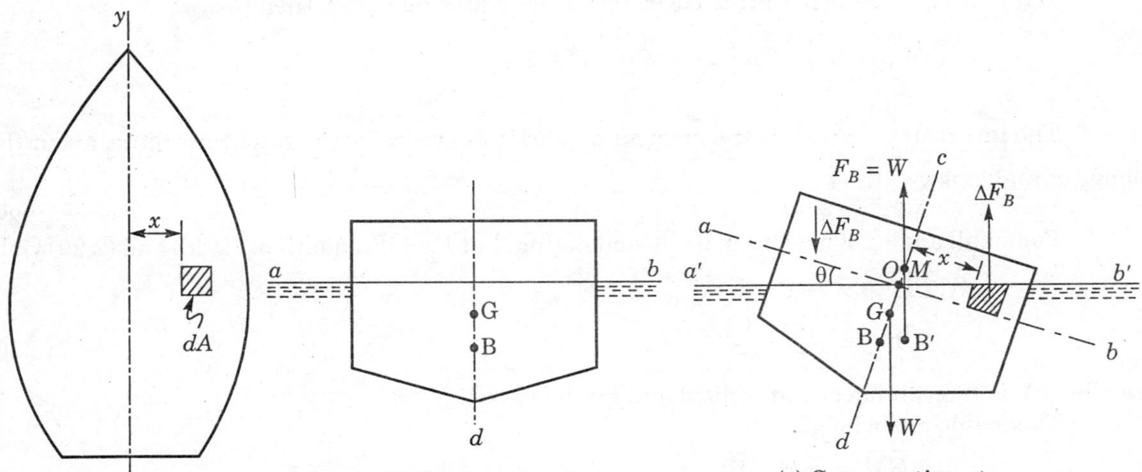

(*a*) Water-line plan (*b*) Cross-section in upright position (*c*) Cross-section at an angle of heel θ

Fig. 2.19. Stability of a non-prismatic floating vessel.

If the angle θ is increased, the centre of buoyancy shifts to the right of B', resulting in change in the position of M and increased $\overline{GM}$. On the other hand, if θ is decreased ; the point M draws closer to G, and it will be found that as θ approaches zero, the point M approaches a limiting position above G. This position of M is known as the *metacentre* and the distance between the centre of gravity G and the metacentre M is called the *metacentric height* $\overline{GM}$. The metacentric height will be positive when M lies above G, and represents the condition for stable equilibrium of the vessel. If, however, the metacentre M lies below G (in between B and G), the metacentric height is negative signifying unstable equilibrium.

2.9.2. Evaluation of Metacentric Height

An expression for the metacentric height $\overline{GM}$ may be obtained from the following analysis. Referring to Fig. 2.19 (*c*), it can be seen that a clockwise tilting of the vessel causes submergence of a wedge-shaped prism represented by *bob′* on the right side of *c-d*, while on its left side an identical wedge shaped prism *aoa′* emerges out of water, thus changing the shape of the displaced volume of water. These wedges represent a gain in the buoyant force on the right side and a corresponding loss in buoyancy on the left side of *c-d*. The gain may be represented by a vertically upward force, ΔF_B, acting through the centroid of the wedge-shaped prism *bob′*, while the loss may be indicated by an equal but opposite force (vertically downwards) ΔF_B, acting through the centroid of *aoa′*.

The buoyant force F_B acting through the new centre of buoyancy B' may be considered as the resultant of the original buoyant force through B, and the gain and loss in the buoyant force each equal to ΔF_B. Taking moment of these forces about B, we have

$$F_B \times \overline{BM} \sin\theta = \Delta F_B \times l \qquad \text{...(2.28)}$$

Consider an element of area dA in plan at a distance x from 0, refer Fig. 2.19 (*a*) and (*c*). The buoyant force acting on this element is $\gamma x\, \theta\, dA$. The gain in the buoyant force is

$$\Delta F_B = \int \gamma x\, \theta\, dA,$$

the integration to be performed is over half of the water line area. The moment of this force about axis *y-y* is

$$\Delta F_B \cdot \frac{l}{2} = \int \gamma\, \theta\, x^2\, dA$$

$$\Delta F_B\, l = 2 \int \gamma\, \theta\, x^2\, dA$$

If the integration is performed over the entire water line area, then

$$\Delta F_B \cdot l = \int^A \theta \gamma x^2\, dA = \gamma\theta \int^A x^2\, dA$$

The integral $\int^A x^2\, dA$, is the moment of inertia I of vessel area at the water-line, about the longitudinal axis *y-y*, thus

$$\Delta F_B . l = \gamma\, \theta\, I \qquad \text{...(2.29)}$$

For small angles of heel θ, $\sin\theta \approx \theta$, and noting that $F_B = W$, equations (2.28) and (2.29) yield

$$W . \overline{BM}\ \theta = \gamma\, \theta\, I$$

hence,

$$\overline{BM} = \frac{\gamma I}{\gamma V} = \frac{I}{V} \qquad \text{...(2.30)}$$

in which V is the volume of water displaced by the vessel.

The metacentric height

$$\overline{GM} = \overline{BM} - \overline{BG}$$

$$= \frac{I}{V} - \overline{BG} \qquad \text{...(2.31)}$$

It is to be noted for a vessel to float in stable equilibrium, $\overline{GM} > 0$ or $I/V > \overline{BG}$ ...(2.32)

If the centre of gravity G is below B, then the metacentric height is given by

$$\overline{GM} = \frac{I}{V} + \overline{BG}$$

The distance $BM = I/V$ is known as the *metacentric radius*. It depends upon the geometry of the vessel and also on the draft (the draft is the depth of submergence of the vessel). For a vessel having parallel sides, the moment of inertia I will remain constant, but the displaced volume V becomes less as load is removed from the vessel. The metacentric radius BM will therefore, increase.

The location of the C.G. influences the stability of a floating vessel to a great extent. The position of the C.G. changes as the cargo is loaded or is removed from the vessel. If the cargo is unloaded from somewhere near the bottom of the vessel, the C.G. will shift to a higher position, while a corresponding reduction in displaced volume of the liquid associated with the unloading of cargo will result in shifting of C.B. to a lower position. As a result, $\overline{BG}$ may increase to the extent that at light draft (the light draft refers of lighter loading conditions), the metacentric height may vanish, resulting in zero stability.

The metacentric height GM, may be considered as a measure of stability of a vessel. In order to achieve high stability, the vessel should be so shaped that I becomes large, and the cargo accommodation inside it is so arranged that it gives a lower C.G., making $\overline{BG}$ smaller.

In the foregoing analysis, the angle of roll, θ (heel or list, as it is also sometimes called) is assumed to be small. The stability measured in terms of the metacentric height represents the initial stability. The initial metacentric height for a passenger steamer varies from 30 to 60 cm and for ships and other naval vessels it varies from 90 to 120 cm. For large metacentric height indicating a greater stability, experience has shown that the period of rolling is shorter and the rolling is rather uncomfortable. It is for this reason that vessels meant for carrying passengers are designed to have a low metacentric height as compared to other vessels.

2.9.3. Period of Rolling

The period of rolling oscillation T, primarily depends upon the relative magnitudes of the righting moment $W\overline{GM}\,\theta$ and the moment of inertia of the body $(W/g)\,k^2$ where k is the radius of gyration with respect to the axis of roll. Neglecting the frictional damping of water and using the method of dimensional analysis, the time period T may be expressed as

$$T = \frac{k}{\sqrt{g.GM}}\, f(\theta) \qquad ...(2.33)$$

in which $f(\theta)$ is an unknown function and has been experimentally found to be constant for small angles of roll. From this equation, it is evident that large metacentric height GM will reduce period of rolling and *vice versa*. It may, therefore, be concluded that more the stable a vessel is, shorter will be its rolling period. The period of rolling is inversely proportional to the stability and directly proportional to the radius of gyration of the vessel.

2.9.4. Shape of Floating Body and its Stability

The shape of the submerged portion of a vessel plays an important role in giving it a desired stability. For a cylinder floating with its axis parallel to the liquid surface, the metacentre does not exist and thus metacentric height $\overline{GM}$ is zero. Such a cylinder has zero stability. A vessel with vertical sides is generally stable, but a vessel having sides diverging upwards has better stability while a floating body with sides converging upwards has a doubtful stability.

The provision of a reasonable initial stability in a floating vessel is not sufficient from safety point of view, if adequate freeboard is lacking. At times of high transverse winds or waves in the sea, a floating vessel with lesser free board may have its deck edge submerged during rolling, thus reducing the moment of inertia abruptly and endangering its stability. Hence, a provision of adequate free board is as important as metacentric height in providing safety and stability. Fig. 2.20 depicts different forms of floating bodies along with the stability associated with them.

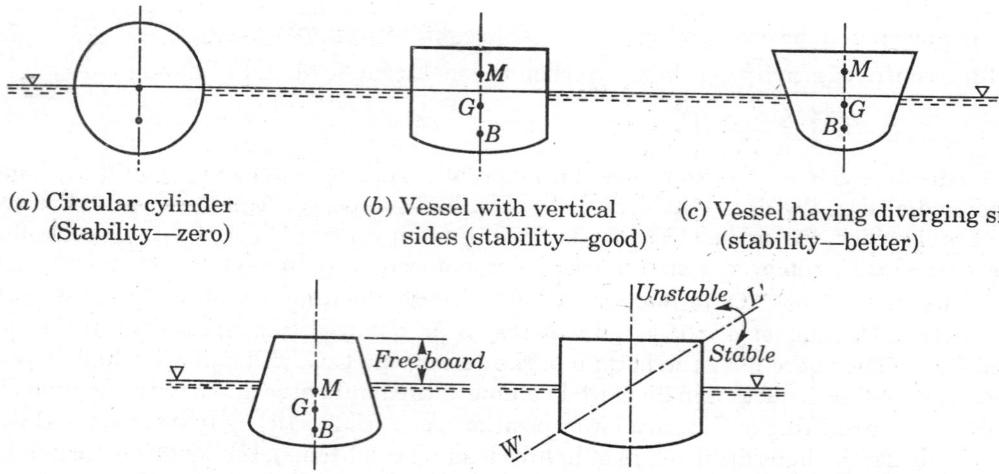

(*a*) Circular cylinder (Stability—zero)

(*b*) Vessel with vertical sides (stability—good)

(*c*) Vessel having diverging sides (stability—better)

(*d*) Vessel with converging sides (stability—doubtful)

(*e*) *W'L'* represents the limit of water surface during rolling upto which the vessel is stable.

Fig. 2.20. Stability of different forms of floating vessel.

2.9.5. Experimental Determination of Metacentric Height

The experimental determination of the metacentric height $\overline{GM}$ consists of moving a weight w across the deck through a certain distance l and observing the corresponding angle of hell or roll θ. The shifting of the weight w through distance l produces a moment $w.l$ which causes the vessel to tilt through an angle θ. This moment $w.l$ is balanced by the righting moment $W\overline{GM}\,\theta$, thus

$$w.l = W\overline{GM}\,\theta$$

giving,
$$\overline{GM} = \frac{w.l}{W\theta} \qquad ...(2.34)$$

It may be noted that the vessel, before the weight was moved, was in an upright (vertical) position, and W is the total weight of the vessel (including w). Fig. 2.21 shows the tilted position of the vessel.

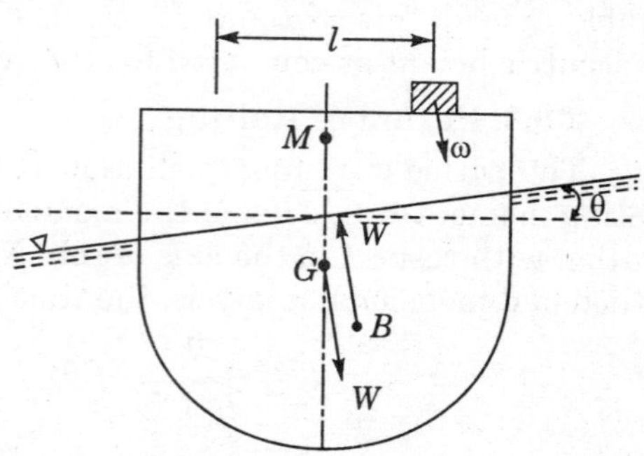

Fig. 2.21. Determination of metacentric height.

Ex. 2.44. *A empty tank (with all sides closed) is rectangular in plan, side elevation and end elevation with sides 12.5 m long, 0.7 m broad and 0.6 m high. If the sheet metal weighs 363 N/m² of the surface and the tank is allowed to float in fresh water with 0.6 m edges vertical, prove that the equilibrium is stable or otherwise. Derive the equation you may use. Weight of water is 9810 N/m³.*

Sol. Weight of tank

= Surface area of tank × weight of sheet per unit area

= (0.6 × 12.5 + 0.7 × 12.5 + 0.6 × 0.7) × 2 × 363

= 12101.4 N

∴ Weight of water displaced by the tank

= 12101.4 N

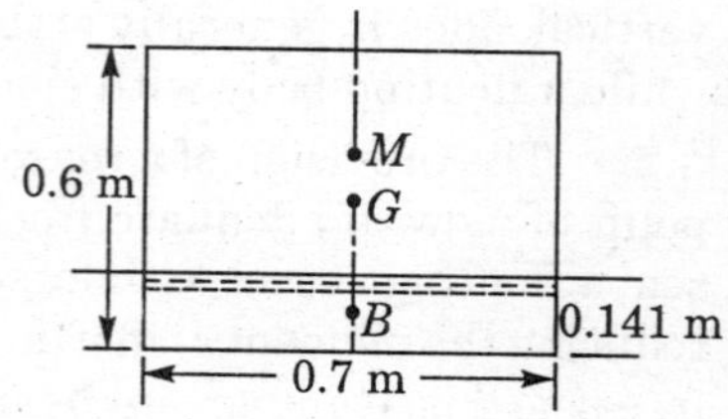

$$\text{Volume of water displaced} = \frac{12101.4}{9810} = 1.2336 \text{ m}^3$$

Depth of immersion of tank

$$= \frac{\text{Volume displaced}}{\text{bottom area}} = \frac{1.2336}{0.7 \times 12.5} = 0.141 \text{ m}$$

Distance between centre of gravity (G) and centre of buoyancy (B)

$$\overline{BG} = 0.3 - \frac{0.141}{2} = 0.2295 \text{ m}$$

Metacentric height $MG = \frac{I}{V} = - BG$

$$\overline{MG} = \frac{\frac{1}{12} bd^3}{bd \times 0.141} - 0.2295 = \frac{\frac{1}{12} d^2}{0.141} - 0.2295$$

$$= \frac{1}{12} \times \frac{(0.7)^2}{0.141} - 0.2295 = 0.06 \text{ m.}$$

Since the metacentric height is positive, the tank will float in stable equilibrium.

Ex. 2.45. *A rectangular barge is 20 m long, 7 m wide and 3 m deep. It has a draft of 2 m when fully loaded. The c.g. of barge is on the axis of symmetry at the water surface. Determine the stability condition of the barge and the metacentric height.*

Sol. A floating body remains in stable equilibrium so long as it has a positive metacentric height *i.e.*,

$$\overline{GM} = \overline{BM} - \overline{BG} = \frac{I}{V} - \overline{BG}$$

$$I = \frac{1}{12} bd^3 = \frac{1}{12} \times 20 \times (7)^3 ;$$

$$V = bd \times \text{depth of draft} ;$$

$$\overline{BG} = 2 - 1 = 1 \text{ m}$$

Making substitutions, the metacentric height is determined as

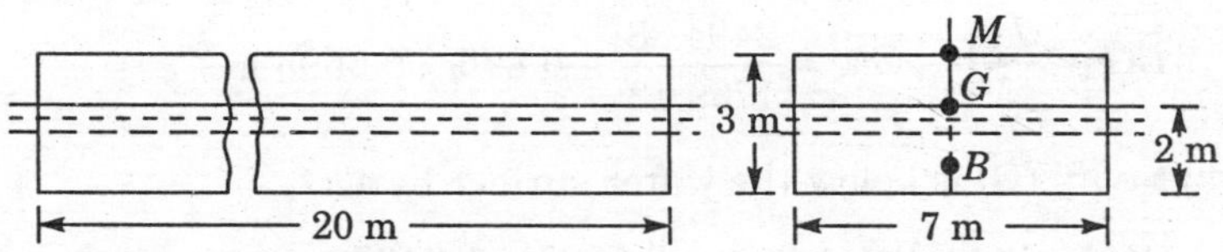

Side elevation *Cross-section*

$$\overline{GM} = \frac{\frac{1}{12} \times 20 \times (7)^3}{20 \times 7 \times 2} - 1 = 1.0417 \text{ m}$$

Under the given conditions, the barge is stable and its metacentric height is 1.0417 m.

Ex. 2.46. *A ship 65 m long and 8.5 m broad has a displacement of 1800 tonnes. A weight 30 tonnes is moved across the deck through a distance of 5.5 m. The ship is tilted through 6°. The moment of inertia of ship at water line about its longitudinal and transverse axis is 75% of the moment of inertia of the circumscribing rectangle. The centre of buoyancy is 2.2 m below the water-line. Determine the metacentric height and position of C.G. of ship. Specific weight of sea water is 1035 kg/m³.*

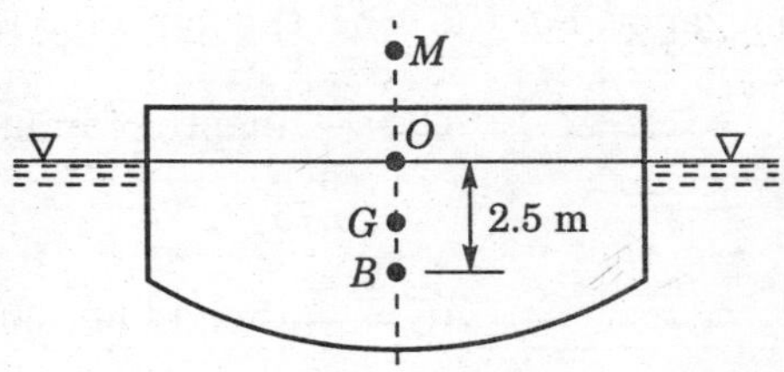

Fig. A. Upright position of ship

Sol. Given weight of ship

$$W = 1800 \text{ tonnes}$$

Moment of inertia about its longitudinal axis at water line

$I_{XX} = 0.75$ of I_{XX} of circumscribing rectangle.

The metacentric height

$$\overline{GM} = \overline{BM} - \overline{BG}$$

$$= \frac{I}{V} - \overline{BG}$$

Volume of displaced sea water

$$= \frac{W}{V} = \frac{1800 \times 1000 \text{ (kg)}}{1035 \text{ (kg/m}^3\text{)}}$$

$$= 1739.13 \text{ m}^3.$$

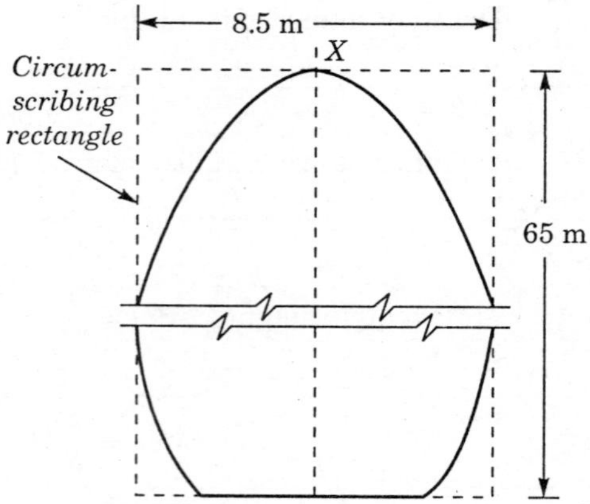

Fig. B. Waterline plan of ship.

Moving a weight weighing 30 tonnes through a distance of 5.5 m across the deck causes the ship to tilt through an angle of 6°, as shown in Fig. C.

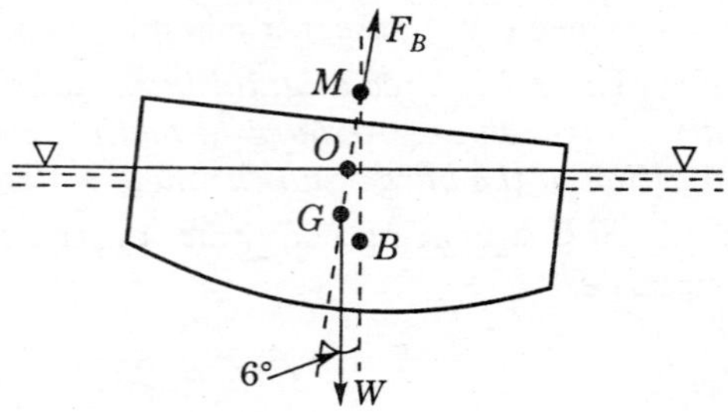

Hence, $\quad wl = W \,.\, \overline{GM} \sin 6°$

$$30 \times 5.5 = 1800 \,.\, \overline{GM} \,.\, \frac{\pi}{30}$$

$\therefore \quad \overline{GM} = 0.875 \text{ m}.$

Location of C.G. of ship :

$$\overline{GM} = \overline{BM} - \overline{BG} = \frac{I_{XX}}{V} - \overline{BG}$$

$180° = \pi$ radian
$6° = \frac{\pi}{180} \times 6 = \frac{\pi}{30}$ rad.

I_{XX} of ship $\quad = 0.75 \times \frac{1}{12} \times 65 \times (8.5)^3 = 2494.88 \text{ m}^4$

$\Rightarrow \quad \overline{BG} = \frac{I_{XX}}{V} - \overline{GM} = \frac{2494.88}{1739.13} - 0.875 = 0.5595 \text{ m}$

The centre of gravity (G) is below the water surface by $\overline{OG}$

$$\overline{OG} = \overline{BO} - \overline{BG} = 2.50 - 0.5595 = \mathbf{1.9405 \text{ m.}}$$

Ex. 2.47. *A solid cone floats in water with its apex downwards. Determine the least apex angle of cone for stable equilibrium. The specific gravity of the material of cone is 0.70.*

Sol. Consider a cone of diameter '*D*', height '*H*' having apex angle 2θ, floating in water with its apex downwards as shown. Let the cone be immersed in water upto a depth '*h*' so that it remains in stable equilibrium. Further, let *G* represent the centre of gravity of the cone, and *B* the centre of buoyancy. For the cone, the following relationships for the location of *B* and *G* can be written :

$$AB = \frac{3}{4} \text{ (depth of immersion)}$$

$$= 0.75y$$

$$AG = \frac{3}{4} \text{ (height of cone)}$$

$$= 0.75H$$

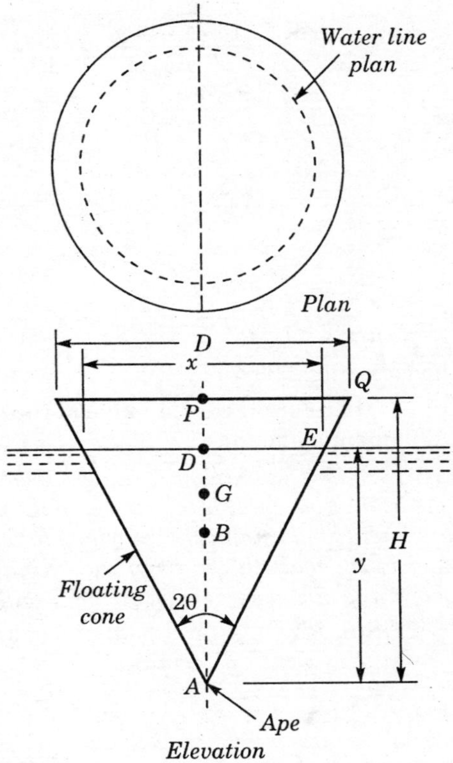

$$\therefore \qquad \overline{BG} = \overline{AG} - \overline{AB} = 0.75\,(H - y)$$

Volume of cone $= \dfrac{1}{3}$ (of the volume of circumscribing cylinder)

$$= \frac{1}{3}\left(\frac{\pi D^2}{4} H\right) = \frac{1}{12}\,\pi\, D^2 H$$

The density of cone-material (ρ)

$$= 0.70 \times 1000 = 700 \text{ kg/m}^3$$

Weight of cone $= (\rho g) = \left(\dfrac{1}{12}\pi D^2 H\right)$

$$= (700g)\left(\frac{1}{12}\pi D^2 H\right) \ (N)$$

Weight of water displaced by the cone

$$= (1000g)\left(\frac{1}{12}\pi x^2 y\right) \ (N)$$

Since the cone is floating freely, the two weight must balance, hence,

$$(700g)\left(\frac{1}{12}\pi D^2 H\right) = (1000g)\left(\frac{1}{12}\pi x^2 y\right)$$

or $$700 D^2 H = 1000 x^2 y \qquad \text{...(1)}$$

From ΔAPQ, $\dfrac{D}{2H} = \tan\theta$, or $D = 2H\tan\theta$

Similarly, $\dfrac{x}{2y} = \tan\theta$, or $x = 2y\tan\theta$

Using these relationships in (1),

$$700(4H^2 \tan^2\theta)\, H = 1000\,(4y^2 \tan^2\theta) y$$

which results in $$\frac{H^3}{y^3} = \frac{1000}{7000} \quad \text{or} \quad \frac{H}{y} = \left(\frac{1000}{700}\right)^{1/3} \qquad \text{...(2)}$$

For a floating body to remain in stable equilibrium, the metacentric height must be positive

or $$\overline{GM} > 0$$

$$\frac{I}{V} - \overline{BG} > 0$$

I = Moment of inertia of the cone in the plane of water surface $= \dfrac{\pi x^4}{64}$

V = Volume of water displaced by the cone $= \dfrac{1}{3}\left(\dfrac{\pi x^2}{4}\,.\,y\right)$

Hence, for stable equilibrium,

$$\frac{\pi x^4/64}{1/3\,(\pi x^2 y/4)} - \frac{3}{4}\,(H - y) > 0$$

or $$\frac{12}{64}\frac{x^2}{y} - \frac{3}{4}\,(H - y) > 0$$

or $$\frac{12(2y\tan\theta)^2}{64y} - \frac{3}{4}(H-y) > 0$$

or $$\frac{3}{4}y\tan^2\theta - \frac{3}{4}(H-y) > 0 \quad \text{or} \quad y(1+\tan^2\theta) - H > 0 \quad \text{or} \quad y\sec^2\theta > H$$

or $$\cos^2\theta > \frac{y}{H}, \text{ but from (2), } \frac{y}{H} = \left(\frac{700}{1000}\right)^{1/3}$$

$$\therefore \quad \cos^2\theta > \left(\frac{700}{1000}\right)^{1/3}$$

from which, $\cos\theta > 0.9423$, or $\theta > 19.56°$

the cone angle $2\theta >$ **39.12°**.

Therefore the least value of the apex angle of static equilibrium must be **39.12°.** The practical value may be take.

Ex. 2.48. *A hollow cylinder 30 cm in diameter and 50 cm high weighs 279.5 N and has a small hole in the bottom. It is lowered into the water so that its axis remains vertical. Calculate (1) the depth to which it will sink (2) the height to which water will rise in it, and (3) the air pressure inside it. Neglect the wall thickness of the cylinder and assume that compression of air follows Boyle's law. Determine also whether the cylinder will be stable in the vertical position when in equilibrium.*

Sol. Air pressure inside the cylinder = Pressure in water at the air-water interface.

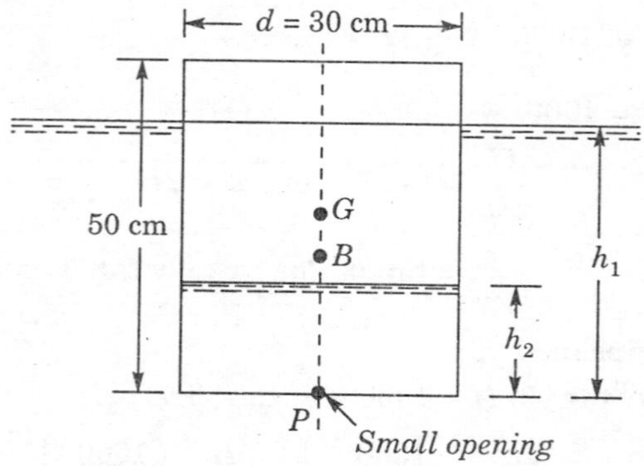

Let the atmospheric pressure be 101330 N/m²

$\therefore$ Air pressure (absolute)

$$p_1 = \gamma(h_1 - h_2) + 101330 = 9810\,(h_1 - h_2) + 101330 \text{ N/m}^2$$
$$= 9810\,(h_1 - h_2) + 101.3 \times 10^3 \text{ (N/m}^2)$$

Air volume inside the cylinder,

$$V_1 = \pi \times (0.15)^2 \times (0.50 - h_2)$$

where h_1 = submerged depth of the cylinder, and

h_2 = depth of water inside the cylinder.

For isothermal compression, $\quad p_0V_0 = p_1V_1 \quad$ ...(*i*)

in which the subscripts 0 and 1 refer to conditions before and after the compression.

Making substituting in Eq. (*i*),

$$101330 \times \pi \times (0.15)^2 \times 0.5 = [9810\,(h_1 - h_2) + 101.3 \times 10^3] \times \pi \times (0.15)^2\,(0.5 - h_2) \quad ...(ii)$$

Considering balance of forces acting on the cylinder :

Weight of (cylinder + water inside) = Buoyant force

$$279.5 = 9810 \times \pi \times (0.15)^2 \times (h_1 - h_2) \quad ...(iii)$$

From (*iii*) one gets : $h_1 - h_2 = 0.4032$ ...(*iv*)

From (*ii*) one may write : $[(h_1 - h_2) + 10.32] \times (0.5 - h_2) = 5.16$ ...(*v*)

Solving (*iv*) and (*v*), $h_2 = 0.0188 \text{ m} = 1.88 \text{ cm}$

$$h_1 = 0.4032 + 0.0188 = 0.422 \text{ m} = 42.2 \text{ cm}$$

Hence :

(*i*) Depth to which the cylinder will sink = h_1 = 42.2 cm

(*ii*) Height to which water will rise = h_2 = 1.88 cm

(*iii*) Air pressure inside the cylinder = 9810 (0.422 – 0.0188) = **3955 N/m² (gauge).**

Stability of the Cylinder. If G is the resultant centre of gravity, then taking moments about P, we can find its location.

Weight of water inside cylinder

$$= \frac{\pi}{4}(0.3)^2 \times \frac{1.88}{100} \times 9810 = 12.8 \text{ N}$$

Taking moments about P

$$(279.5 + 12.8) \times \overline{GP} \times 279.5 \times 25 + 12.8 \times \frac{1.85}{2}$$

or $\overline{GP} = 23.95$ cm.

$$\therefore \quad \overline{BG} = \overline{GP} - \overline{BP} = 23.95 - \frac{1}{2}(h_1 + h_2)$$

$$= 23.95 - \frac{45.85}{2} = 1.025 \text{ cm}$$

and metacentric radius,

$$\overline{BM} = \frac{I}{V} = \frac{\frac{\pi d^4}{64}}{\frac{\pi d^2}{4} \times (h_1 - h_2)} = \frac{d^2}{16\,(h_1 - h_2)} = \frac{(30)^2}{16 \times 40.32} \; 1.395 \text{ cm.}$$

$\therefore$ The metacentric height $\overline{MG}$

$$= \frac{I}{V} - \overline{BG} = 1.395 - 1.025 = +\,0.37 \text{ cm.}$$

since the metacentric height is positive, the cylinder is stable.

Ex. 2.49. *A raft is formed of three cylinders each 1 metre diameter, symmetrically placed with axes horizontal, the extreme breadth of the cylinders being 6 m. When so placed centre to centre spacing of cylinders is 2.5 m. The fore and aft axis is parallel to the axis of the cylinders. The raft floats with cylinders half immersed and its centre of gravity in this condition is 1.2 metre above the axis of the central cylinder. Find the metacentric height.*

Sol. The problem is schematically shown in the figure.

The metacentric height $\overline{GM}$ is given by

$$\overline{GM} = \overline{BM} - \overline{BG}$$

$\overline{BM} = I/V$, where I = water-line moment of inertia of the cylinder about the central axis.

and V = volume of water displaced.

The moment of inertia of the cylinders at the water-surface about Y-Y, per unit length of raft is

$$I = 2\left[\frac{1}{12} \times 1 \times (1)^3 + 1 \times 1 \times (2.5)^2\right] + \frac{1}{12} \times 1 \times (1)^3 = 12.7472 \text{ m}^4$$

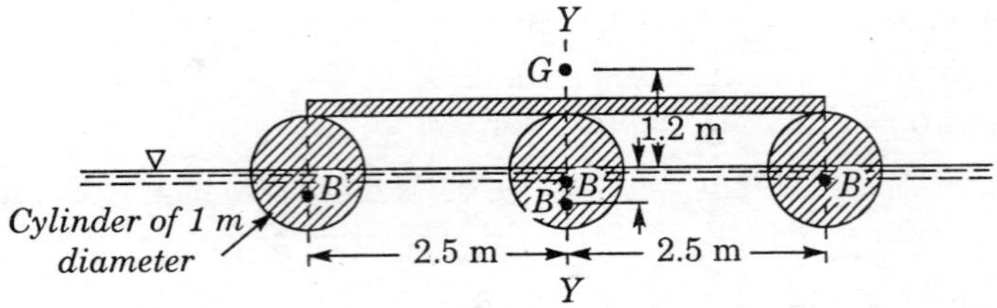

and $$V = 3 \times \frac{1}{2} \frac{\pi(1)^2}{4} \times 1 = 1.18 \text{ m}^3$$

$$\therefore \quad \overline{BM} = \frac{I}{V} = \frac{12.7422}{1.18} = 10.8 \text{ m.}$$

Let B be the centre of buoyancy of liquid displaced by each cylinder and B' be the resultant centre of buoyancy of the system of three cylinders. Let B' be located at a distance h below the water surface on the axis of symmetry Y-Y. Taking moments of displaced volumes about the water surface, we have

$$3 \times \left[\frac{1}{2} \times \frac{\pi(1)^2}{4} \times 1\right] \times \frac{4 \times \frac{1}{2}}{3\pi} = 3 \times \left[\frac{1}{2} \frac{\pi(1)^2}{4} \times 1\right] \times h$$

$$\therefore \quad h = \frac{2}{3\pi} \text{ m}$$

This shows that the point B' is located at B, and thus

$$\overline{BG} = \frac{2}{3\pi} + 1.2$$
$$= 1.412 \text{ m.}$$

and therefore the metacentric height (for rolling about Y-Y)

$$\overline{GM} = 10.80 - 1.412 = 8.388 \text{ m.}$$

Ex. 2.50. *A solid cylinder, 22.5 cm diameter consists of two parts made of different materials. The first part, at one end, is axially 5 cm long and made of a material having specific gravity of 1.5. The other part of cylinder is made of material having specific gravity 0.5. What should be the minimum length of this part in order that the cylinder may float freely in water ? If the overall length of the cylinder was 25 cm including the part 5 cm long of sp. gr. 1.5, find if the cylinder could float in stable equilibrium with its axis vertical.*

Sol. Weight of the composite cylinder

$$= \frac{\pi}{4} (0.225)^2 \times \frac{5}{100} \times 1.5 \times 9810 + \frac{\pi(0.225)^2}{4} \times \frac{h}{100} \times 0.5 \times 9810$$

$$= \frac{\pi}{4} (0.225)^2 \times 9810 \left(1.5 \times \frac{5}{100} + 0.5 \frac{h}{100}\right)$$

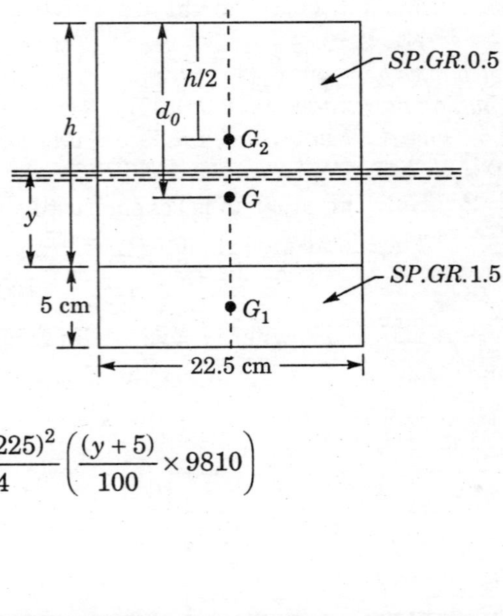

Let the depth of submergence of the composite cylinder be $(y + 5)$ cm, then

Weight of the cylinder = buoyant force = weight of liquid displaced

$$\frac{\pi}{4} (0.225)^2 \times 9810 \left(\frac{7.5}{100} + \frac{0.5h}{100}\right) = \frac{\pi(0.225)^2}{4} \left(\frac{(y+5)}{100} \times 9810\right)$$

or
$$\frac{7.5}{100}+\frac{0.5h}{100}=\frac{y+5}{100}$$

or
$$y = 0.5\,h + 2.5 \qquad ...(i)$$

Let the combined centre of gravity be G, located at distance d_0 below the top of the cylinder. Then

$$\frac{\pi}{4}\,(0.225)^2 \times \frac{5}{100} \times 1.5 \times 9810 \times \left(\frac{h+2.5}{100}\right)+\frac{\pi(0.225)^2}{4}\times\frac{h}{100} \times 0.5 \times 9810 \times \frac{h}{2\times 100}$$

$$= \frac{\pi}{4} \times (0.225)^3 \times 9810 d_0 \left(\frac{5\times 1.5}{100}+\frac{0.5y}{100}\right)$$

or
$$\frac{5\times 1.5}{100}\times\frac{(h+2.5)}{100}+\frac{h}{100} \times 0.5 \times \frac{h}{200}$$

$$= d_0 \times \left(\frac{7.5}{100}+\frac{0.5h}{100}\right)$$

$$\therefore \qquad d_0 = \frac{1\times 7.5(h+2.5)+0.25h^2}{100(7.5+0.5h)}\text{ m} \qquad ...(ii)$$

Centre of buoyancy, B, is at a depth $\left(h-y+\frac{y+5}{2}\right)$ cm.

$= (h - 0.5y + 2.5)$ cm below the cylinder top.

$$\therefore \qquad \overline{BG} = \frac{h-0.5y+2.5}{100}-\left[\frac{7.5\,(h+2.5)+0.25h^2}{100\,(7.5+0.5h)}\right]\text{ m.}$$

$$\overline{MG} = \overline{BM} - \overline{BG}$$

For floating stability, $\overline{MG} > 0$, or $\overline{BM} > \overline{BG}$

$$\overline{BM} = \frac{I}{V}=\frac{\dfrac{\pi d^4}{64}}{\dfrac{\pi d^2}{4}\,\dfrac{(y+5)}{100}}=\frac{100d^2}{16(y+5)}\text{ m}$$

or
$$\frac{100\times(0.225)^2}{16(y+5)}>\left[\frac{h-0.5y+2.5}{100}-\frac{\{7.5(h+2.5)+0.25h^2\}}{100(7.5+0.5h)}\right] \qquad ...(iii)$$

Eliminating y from (i), $y = 2.5 + 0.5h$

$$\frac{5.06}{16(7.5+0.5h)}>\left[\frac{h-\frac{1}{2}(2.5+0.5h)}{100}-\frac{(7.5h+18.75+0.25h^2)}{750+50h}\right]$$

which can be simplified to

$$h^2 + 6.67h - 109.3 > 0$$

or Solving which, $h = 7.65$ cm.

The minimum length of the upper part of the cylinder = 7.65 cm.

If $h = 25 - 5 = 20$ cm,

then $y = 2.5 + 0.5 \times 20 = 12.5$ cm.

$$\overline{BM} = \frac{100\times(0.225)^2}{16(12.5+5)}=\frac{5.06}{16\times 17.5}$$

$$= 0.0181 \text{ m or } 1.81 \text{ cm.}$$

or $$\overline{BG} = 20 - 0.5 \times 12.5 + 2.5 - \left[\frac{7.5(22.5) + 0.25 \times 400}{750 + 50 \times 20}\right] = 16.079 \text{ cm.}$$

Metacentric height

$$\overline{MG} = \overline{BM} - \overline{BG} = 1.81 - 16.097$$
$$= - 14.287 \text{ cm.}$$

Since the metacentric height is negative, the cylinder is unstable and, therefore, cannot float in a vertical position.

Ex. 2.51. *A cylindrical buoy is 2 m in diameter and 2.5 m long and weighs 21.6 kN. The density of sea water is 10055 N/m³. Show that the buoy does not float with its axis vertical.*

What minimum pull should be applied to a chain attached to the centre of the base to keep it vertical ?

Sol. (1) Check for Stability

Let h be the depth upto which the cylinder is submerged. By equating its weight with the buoyant force (noting that the cylinder is not anchored by the chain), one gets

$$21.6 \times 1000 = \frac{\pi \times (2)^2}{4} \times h \times 10055$$
$$h = 0.682 \text{ m}$$

The metacentric height,

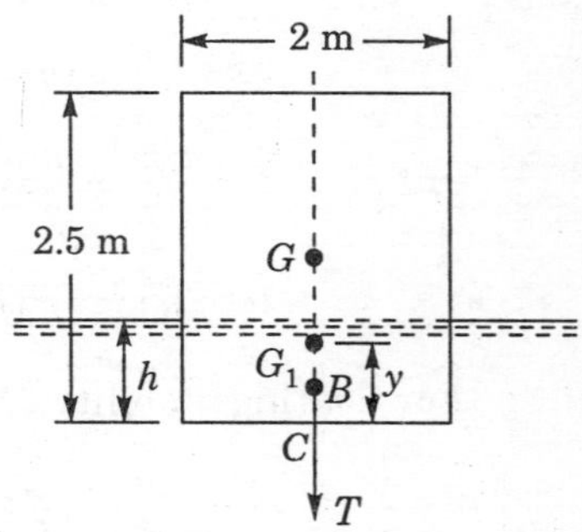

$$\overline{BG} = \frac{I}{V} - \overline{BG}$$
$$= \frac{\pi d^2 h/64}{\pi d^2/4} - \left(1.25 - \frac{h}{2}\right)$$
$$= \frac{(2)^2}{16 \times 0.682} - (1.25 - 0.341)$$
$$= - 0.545 \text{ m.}$$

Since the metacentric height is negative the cylinder is unstable and, therefore, cannot float with its axis vertical.

(2) Pull in the chain

In order to keep the cylinder vertical, let a chain be tied to the centre of its base C and a pull T be applied, as shown in the figure. With the increased downward force, it being $(W + T)$, where W represents the weight of the cylinder, the centre of gravity of the system changes. Let the new centre of gravity be located at G_1 at a height y above the cylinder base. Taking moments about C,

$$W \times 1.25 = (W + T) \times y \qquad ...(i)$$

Considering the equilibrium of the cylinder, the total downward force must equal the upward force due to buoyancy.

$$W + T = F_B \qquad ...(ii)$$

The depth of immersion of the cylinder will increase because of the increased buoyant force. Let this depth be h_1 and the new centre of buoyancy be at B_1, then

$$\overline{CB_1} = \frac{h_1}{2} \qquad ...(iii)$$

From Eq. (*ii*),

$$T = F_B - W = 10055 \times \frac{\pi d^2}{4} h_1 - 21.6 \times 10^3 \text{ (N)} \qquad ...(iv)$$

Let the new metacentre be at M_1, then

$$\overline{B_1M_1} = \frac{I}{V} = \frac{\pi \times (2)^4 \times 4}{64 \times \pi \times (2)^2 \times h_1} = \frac{1}{4h_1}$$

and

$$\overline{B_1G_1} = \overline{CG_1} - \overline{CB_1}$$

$$= y - \frac{h_1}{2} = \frac{1.25\,W}{F_B} - \frac{h_1}{2} = \frac{1.25 \times 2200 \times 4}{\pi \times (2)^2 \times h_1 \times 1025} - \frac{h_1}{2}$$

$$= \frac{0.853}{h_1} - 0.5h_1$$

To keep the cylinder vertical, $\overline{M_1G_1} > 0$, or , $\overline{B_1M_1} > \overline{B_1G_1}$.

Therefore, $$\frac{1}{4h_1} > \left(\frac{0.853}{h_1} - 0.5h_1\right)$$

which simplifies to,

$$h_1^2 - 1.206 > 0, \text{ or } h_1 > 1.1 \text{ m}$$

Using Eq. (*iv*),

$$T > 10055 \times \pi \times \frac{(2)^2}{4} \times 1.1 - 21.6 \times 10^3 (\text{N})$$

$$> 13.20 \text{ kN}$$

The minimum pull to be applied to the chain

= **13.20 kN.**

Ex. 2.52. *A spherical container 2 m in diameter is made up of two hemispheres, one resting on the other with interface horizontal. The sphere is completely filled with oil of sp. gr. 0.7, through a small hole on the top. What is the minimum weight of the upper hemisphere to prevent it from lifting ?*

Sol. Considering the equilibrium of upper hemisphere. Pressure at the horizontal interface is uniform

$$p = \gamma h$$

$$= 9810 \times (0.70) \times 1 = 6867 \text{ N/m}^2$$

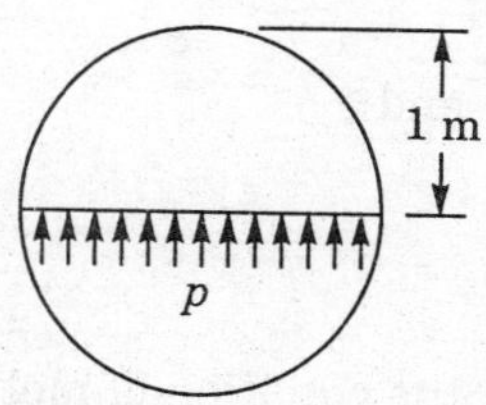

This pressure acts over an area of $\frac{\pi d^2}{4}$ therefore, vertically upward force

$$= \frac{\pi}{4} (2)^2 \times 6867 = 6867\,\pi \text{ (N)}$$

Vertically downward force at the plane of horizontal interface = Weight of oil in the upper hemisphere.

$$= 6867 \times \frac{\pi (2)^2}{12} = 6867 \times \frac{8}{12}\,\pi \text{ (N)}$$

Net vertically upward force

$$= 6867 \,.\, \pi \left(1 - \frac{8}{12}\right) = 7190 \text{ N}$$

∴ Minimum weight of the upper hemisphere

= 7190 N or **7.19 kN.**

2.10. Static Fluid Subjected to Uniform Acceleration

Fluid masses which do not have any relative motion between its particles are characterised by zero shear stress and are governed by the laws of statics of fluid. A fluid mass moving at uniform velocity in translation, is one such case in which the pressure variation is hydrostatic. However, in a fluid mass subjected to an acceleration, the conditions are not much different. Such problems fall essentially within the domain of fluid statics and are analysed by writing the Newton's second law of motion for a fluid element. When fluid masses move without relative motion, they may be considered as if they were solids, and are said to be in relative equilibrium.

Consider a fluid element of size $dx.dz$ and unit length, as shown in Fig. 2.22. It is subjected to acceleration a which has components a_x and a_z in x and z directions. Let p be the pressure at the centre of the element.

Applying Newton's second law of motion in the x and z directions, one obtains

$$\Sigma F_x = m.a_x \quad \text{and} \quad \Sigma F_z = m.a_z$$

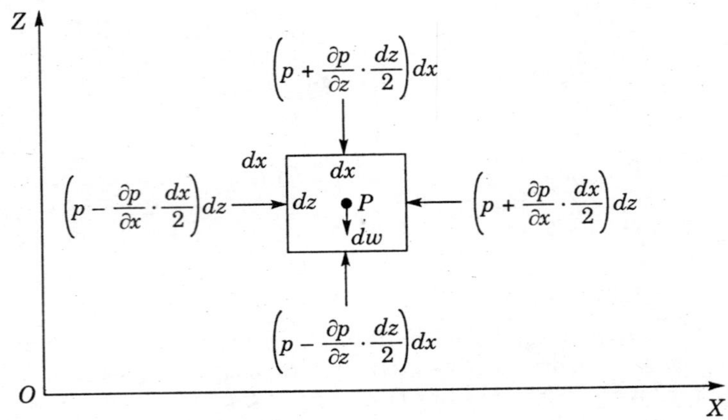

Fig. 2.22. Fluid mass subjected to acceleration.

but
$$\Sigma F_x = \left(p - \frac{\partial p}{\partial x}\frac{dx}{2}\right) dz - \left(p + \frac{\partial p}{\partial x} \cdot \frac{dx}{2}\right) dz = -\frac{\partial p}{dx}\, dx.dz$$

and
$$\Sigma F_z = \left(p - \frac{\partial p}{\partial z} \cdot \frac{dz}{2}\right) dx - \left(p + \frac{\partial p}{\partial z} \cdot \frac{dz}{2}\right) dx - \gamma \,.\, dx \,.\, dz$$

$$= -\frac{\partial p}{\partial z} \,.\, dx\, dz - \gamma\, dxdz$$

the equation of motion in x-direction gives $-\frac{\partial p}{\partial x}\, dx\, dz = \left(\frac{\gamma}{g}\, dx\, dz\right) a_x$

which simplifies to
$$-\frac{\partial p}{\partial x} = \frac{\gamma}{g}\, a_x \qquad \text{...(2.35)}$$

the equation of motion in z-direction yields :

$$-\frac{\partial p}{\partial z}\, dx\, dz - \gamma\, dx\, dz = \left(\frac{\gamma}{g}\, dx\, dz\right) a_z$$

resulting in
$$-\frac{\partial p}{\partial z} = \gamma\left(1 + \frac{a_z}{g}\right) \qquad \text{...(2.36)}$$

Equations (2.35) and (2.36) give the pressure gradients in x and z directions respectively. The pressure change dp may be expressed as

$$dp = \frac{\partial p}{\partial x}dx + \frac{\partial p}{\partial z}dz = -\frac{\gamma}{g}a_x\,dx - \gamma\left(1+\frac{a_z}{g}\right)dz.$$

At the surface of constant pressure, $dp = 0$, and accordingly,

$$-\frac{\gamma}{g}a_x\,dx - \gamma\left(1+\frac{a_z}{g}\right)dz = 0$$

therefore,
$$\frac{dz}{dx} = -\frac{a_x}{a_z + g} \qquad ...(2.37)$$

In Eq. (2.37), dz/dx is the slope of the surface of constant pressure. Since pressure variation follows hydrostatic law, the surfaces of constant pressure are parallel. Eq. (2.37) may be used to discuss some special cases as described below :

(*i*) **Acceleration in the vertical direction,** ($a_x = 0$). When $a_x = 0$, $\frac{dz}{dx} = 0$ and the surfaces of constant pressure are horizontal, and the pressure gradient is

$$\frac{dp}{dz} = -\gamma\left(1+\frac{a_x}{g}\right)$$

the negative sign indicates that the pressure decreases as z increases. If, however, the distance h is measured in the vertically downward direction from the free surface, the above equation takes the form

$$\frac{dp}{dh} = \gamma\left(1+\frac{a_z}{g}\right) \qquad ...(2.38)$$

which upon integration yields
$$p = \gamma\left(1+\frac{a_z}{g}\right)h + C$$

in which C is the constant of integration. At the free surface $h = 0$, $p = 0$, therefore, $C = 0$ and the pressure at a depth h is given by

$$p = \gamma\left(1+\frac{a_z}{g}\right)h \qquad ...(2.39)$$

(*a*) If the acceleration a_z is positive ($a_z > 0$), Eq. (2.39) gives pressure which is larger by an amount ($\gamma a_z h/g$) than the corresponding hydrostatic pressure γh.

(*b*) In case of negative acceleration ($a_z < 0$), the pressure in the liquid will be smaller than the corresponding hydrostatic pressure. The above observations have been shown graphically in Fig. 2.23.

If the fluid is acceleration such that $a_z = -g$ (freely falling fluid mass), Eq. (2.39) shows that the pressure throughout the fluid will zero (atmospheric), and this is confirmed from Eq. (2.38) which gives $\frac{dp}{dh} = 0$, indicating zero pressure gradient. If the fluid mass has a free surface, the pressure in the fluid will be atmospheric at all points.

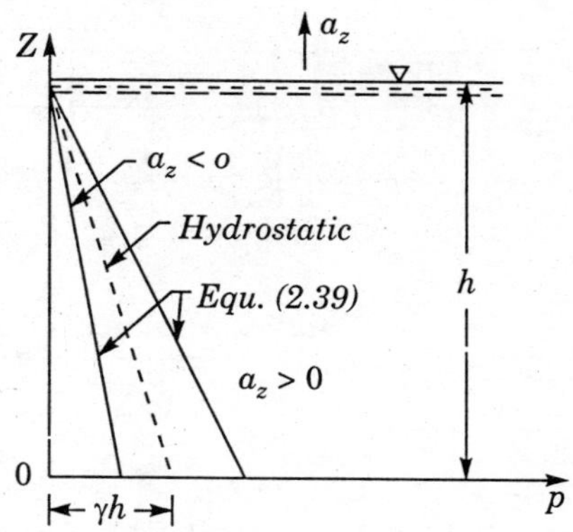

Fig. 2.23. Pressure distribution in the fluid subjected to vertical acceleration.

(*ii*) **Acceleration in the horizontal direction (a_z = 0).** From purely horizontal acceleration the slope of the constant pressure surface, from Eq. (2.37) is :

$$\frac{dz}{dx} = -\frac{a_x}{g} \qquad ...(2.40)$$

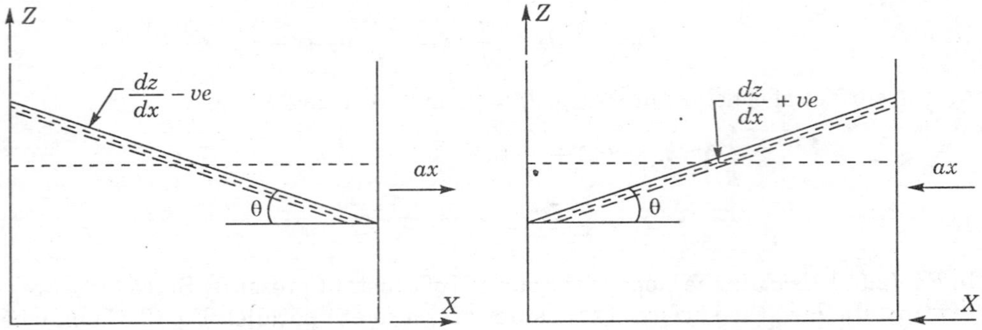

Fig. 2.24. Free surface during horizontal acceleration.

the negative sign shows that the surface is sloping downwards, if $a_x > 0$, and, if $a_x < 0$ (*i.e.* the acceleration in the negative x-direction), the surface will have a positive slope. These two conditions are shown in Fig. 2.24. If θ is the angle free surface makes with the horizontal then using Eq. (2.40).

$$\tan \theta = \frac{a_x}{g} \qquad \text{...(2.41)}$$

2.11. Fluid Rotation About a Vertical Axis

When a fluid is rotated about a vertical axis at constant speed, every particle of it has the same angular velocity and the fluid moves as if it were a solid. This type of motion is known as a *forced vortex*. The rotation of a bucket or a cylinder containing water will produce a forced vortex causing the water to rise near the sides and depressing it near the axis of rotation. A free vortex motion is charcterised by velocity which varies inversely with the distance from the axis and must be distinguished from a forced-vortex. An example of free-vortex motion is the flow from a wash-basin which has an opening in the bottom, the water having been initially filled in the basin keeping the opening plugged and then allowing water to flow through the opening. The motion in water thus produced is a *free vortex* motion.

The forced-vortex motion may be analysed by considering fluid inside a container which is rotated about a vertical axis at constant angular velocity. The fluid mass is subjected to radially inward acceleration (*i.e.* centripetal acceleration).

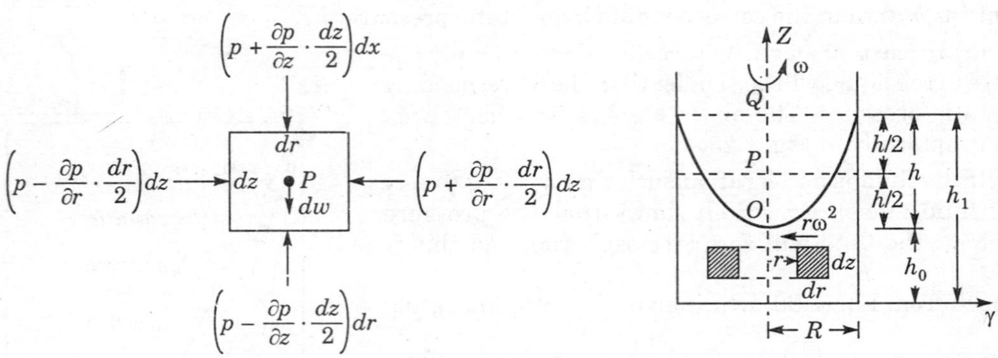

Fig. 2.25. Fluid mass subjected to uniform rotation.

Consider a circular ring element at a radius r and of cross-section $dz.dr$. Considering unit length of the element (circumferentially), the forces acting on this unit element are shown in Fig. 2.25. The fluid is subjected to an acceleration $- r\omega^2$, the negative sign indicating radially inwards direction with the co-ordinate axes z, and r chosen as shown. From the Newton's second law of motion in z-direction, we may write.

In z-direction. $\Sigma F_z = m.a_z$, and $a_z = 0$

therefore, $$\left(p - \frac{\partial p}{\partial z}.\frac{dz}{2}\right) dr - \left(p + \frac{\partial p}{\partial z}.\frac{dz}{2}\right) dr - \gamma dz.dr = 0$$

which simplifies to $$\frac{\partial p}{\partial z} = -\gamma \quad ...(2.42)$$

In r-direction. $\Sigma F_r = m.a_r$ and $a_r = -r\omega^2$,

therefore, $$\left(p - \frac{\partial p}{\partial r}.\frac{dr}{2}\right) dz - \left(p + \frac{\partial p}{\partial r}.\frac{dr}{2}\right) dz = \frac{\gamma}{g}\, dz.d\beta\, (-r\omega^2)$$

which reduces to $$\frac{\partial p}{\partial r} = \frac{\gamma}{g} r\omega^2 \quad ...(2.43)$$

The pressure change, $dp = \frac{\partial p}{\partial r} dr + \frac{\partial p}{\partial z} dz = \frac{\gamma}{g} r\omega^2\, dr + (-\gamma)\, dz.$

For surfaces of constant pressure, $dp = 0$, and we have

$$\frac{dz}{dr} = \frac{r\omega^2}{g} \quad ...(2.44)$$

which upon integration yields, $$z = \frac{r^2\omega^2}{2g} + \text{constant} \quad ...(2.45)$$

on the free-surface at $r = 0$, $z = h_0$ and the constant in Eq. (2.45) is equal to h_0, thus we have

$$z = \frac{r^2\omega^2}{2} + h_0 \quad ...(2.46)$$

at the sides of the container, $r = R$ and $z = h_1$, using Eq. (2.46)

$$h_1 = \frac{\omega^2 R^2}{2g} + h_0$$

from which $$h = h_1 - h_0$$

$$= \frac{\omega^2 R^2}{2g} \quad ...(2.47)$$

From Eq. (2.46), it is evident that the free surface and the surfaces of constant pressures are paraboloids of revolution. From the property of paraboloid of revolution, its volume is equal to half the volume of the circumscribing cylinder. From this it follows that the liquid is depressed in the centre by the same distance, as it rises at the sides. Thus in Fig. 2.25, $OP = PQ$.

It may be noted that the pressures in the vertical direction vary according to the law of hydrostatics as evidenced by Eq. (2.42), and the pressure along the bottom will vary as

$$p = \gamma z = \gamma(h_0 + \omega^2 r^2/2g) \quad ...(2.48)$$

Ex. 2.53. *A closed tank containing water to a depth of 1.5 m falls vertically with an acceleration of 2 m/s². What is the pressure on the bottom of the tank ? What is the acceleration required to have a vacuum pressure of 10 cm of mercury on the bottom of the tank ?*

Sol. Pressure variation with depth when the fluid mass is subject to vertical acceleration is given by Equation (2.39)

$$p = \gamma\left(1 + \frac{a_z}{g}\right) h \quad ...(i)$$

for $a_z = -2$ m/s², and $h = 1.5$ m,

$$p = 9810\left(1 - \frac{2}{9.81}\right) 1.5 = 11715.0 \text{ N/m}^2$$

In order to create a vacuum pressure of 10 cm of mercury on the tank bottom, the acceleration required can be obtained as follows :

Pressure at the tank bottom

$$= -\frac{10 \times 13.6}{100} \text{ m of water} = -9810\left(\frac{10 \times 13.6}{100}\right) \text{ N/m}^2$$

$$= -13341.6 \text{ N/m}^2 \quad \text{or} \quad -13.34 \text{ kN/m}^2$$

Substituting in Equation (*i*),

$$-13341.6 = 9810\left(1 + \frac{a_z}{9.81}\right)1.5$$

from which $a_z = \mathbf{-18.70\ m/s^2}$.

Ex. 2.54. *A circular cylinder of height h_0 and radius r_0 is open at the top and filled with a liquid. It is rotated about its vertical axis. Determine the speed of rotation so that half the area of the bottom gets exposed.*

Sol. Area of the bottom = πr_0^2.

Let r_1 be the radius of the exposed bottom area, then

$$\pi r_1^2 = \tfrac{1}{2}\pi r_0^2$$

or $$r_1 = r_0/\sqrt{2}.$$

The pressure head at r_0 is

$$\frac{p_{r_0}}{\gamma} = H = \frac{p_0}{\gamma} + \frac{\omega^2 r_0^2}{2g} \qquad ...(i)$$

where p_0 = pressure at the vertex *O* of the paraboloid of revolution.

Similarly, the pressure head at radius $r_1 = \frac{r_0}{\sqrt{2}}$ is

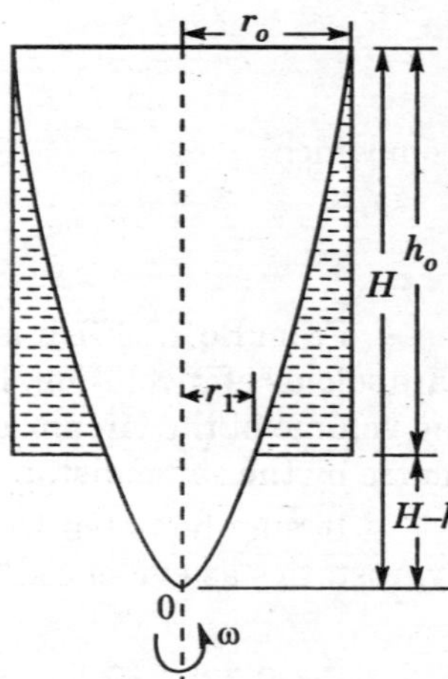

$$H - h_0 = \frac{p_{r_1}}{\gamma} = \frac{p_0}{\gamma} + \frac{\omega^2 r_1^2}{2g} \qquad ...(ii)$$

From Eqs. (*i*) and (*ii*), we have

$$h_0 = \frac{\omega(r_0^2 - r_1^2)}{2g} = \frac{\omega^2\left(r_0^2 - \frac{r_0^2}{2}\right)}{2g}$$

or $$\omega^2 = \frac{2gh_0}{r_0^2/2} = \frac{4gh_0}{r_0^2}$$

$\therefore$ $$\omega = \frac{2}{r_0}\sqrt{gh_0}.$$

Ex. 2.55. *A cylindrical tank 0.9 m in diameter and 2 m high open at the top is filled with water to a depth of 1.5 m. It is rotated at its vertical axis at N r.p.m. Determine the speed of rotation in r.p.m. which will raise the water level even with the brim.*

Sol. When the water reaches the edge of the tank and there is no spilling, the volume of water in the tank must remain the same, thus

$$\frac{\pi(0.9)^2}{4} \times (2 - h) + \frac{1}{2} \times \frac{\pi(0.9)^2}{4} \times h = \frac{\pi(0.9)^2}{4} \times 1.5$$

or $$2 - h + \frac{h}{2} = 1.5 \text{ or } h = 1.0 \text{ m}$$

but, $$h = \frac{r^2\omega^2}{2g}.$$

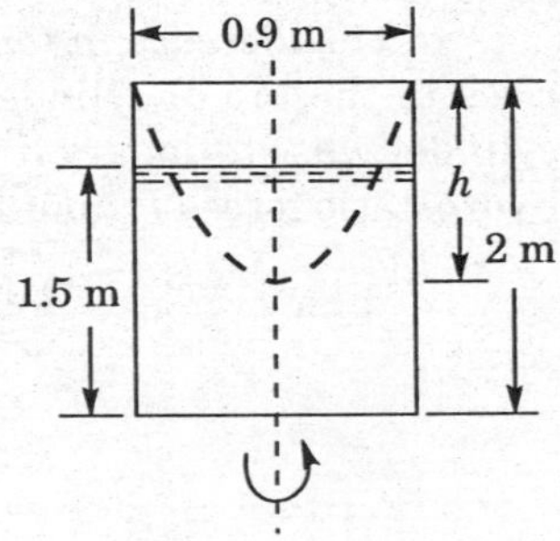

$$\therefore \quad \omega = \frac{\sqrt{2gh}}{r} = \frac{\sqrt{2 \times 9.81 \times 1.0}}{0.9/2}$$

$$= 9.84 \text{ rad/sec.}$$

Speed of rotation in r.p.m.

$$N = \frac{60\omega}{2\pi} = \frac{60 \times 9.84}{2\pi}$$

$$= \mathbf{94 \text{ r.p.m.}}$$

Ex. 2.56. *An open circular tank of 1 m height and 0.3 m diameter contains 588.6 N of water. If the cylinder is rotated about its axis, what is the maximum angular velocity so that no water spills. Find the difference in total pressure (i) at the bottom of the cylinder (ii) at the side of cylinder due to rotation.*

Sol. Let 'y' be the depth to accommodate 588.6 N of water in the tank,

$$9810 \times \frac{\pi(0.3)^2}{4} \times y = 588.6$$

$$\therefore \quad y = 0.849 \text{ m.}$$

As the tank is rotated, the water will rise at the sides upto the top, and will be depressed in the centre, the water surface forming a paraboloid of revolution. As there is to be no spilling, the total volume of water inside the tank must remain constant. Hence,

$$\frac{\pi(0.3)^2}{4}(1-h) + \left(\frac{1}{2}\right)\frac{\pi(0.3)^2}{4} h = \frac{\pi}{4}(0.3)^2(0.849)$$

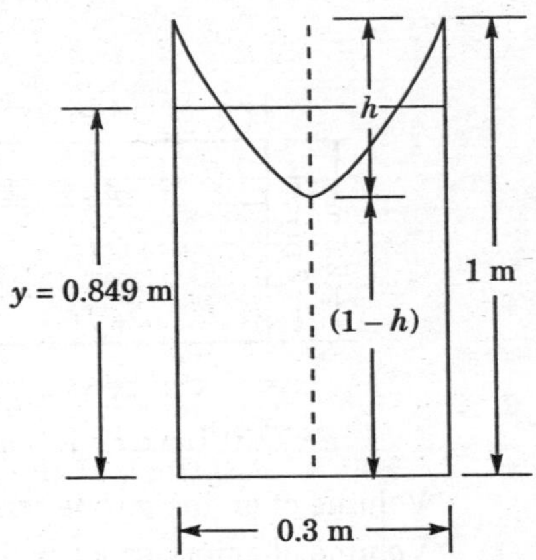

from which, $h = 0.302$ m.

Maximum angular velocity of rotation for no spilling to take place is given by

$$h = \frac{\omega^2 r^2}{2g},$$

or $$\omega^2 = \frac{2 \times 9.81 \times 0.302}{(0.15)^2} = 263.34$$

$$\omega = \mathbf{16.22 \text{ rad/sec.}}$$

Speed of rotation in r.p.m.

$$N = \frac{60\omega}{2\pi} = \frac{60 \times 16.22}{2\pi} = \mathbf{154.89 \text{ r.p.m.}}$$

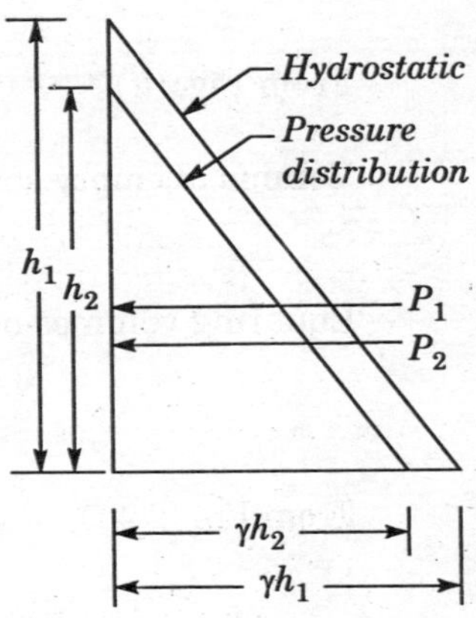

(*i*) *Difference in total pressure on the bottom of cylinder.* When the cylinder was stationary, the total pressure on the bottom was due to the weight of water resting on the bottom which was distributed uniformly at a depth of 0.849 m. But when the cylinder was rotated at 154.89 r.p.m., the water reached the edge of cylinder but did not spill. The pressure on the bottom varies with a minimum value at the centre of axis and increasing in the radial direction attaining a maximum value equivalent to the water depth at the edge. The total weight of water, however, remains the same as there is no spilling. Therefore the difference in the total pressure on the cylinder bottom is *zero*.

(ii) Difference in total pressure at the side of cylinder. Due to rotation of cylinder the water rises to the top of cylinder, the depth of water on the side wall is h_1 = 1 m, whereas when the cylinder was stationary it was h_2 = 0.849 m. The difference in total pressure is the difference in the hydrostatic forces P_1 and P_2.

$$P_1 - P_2 = \pi d\,[\tfrac{1}{2}\gamma h_1^2 - \tfrac{1}{2}\gamma h_1^2]$$

$$= \frac{1}{2} \times 9810 \times \pi \times 0.3\,[1^2 - 0.849^2]$$

$$= 1290.7 \text{ N.}$$

Ex. 2.57. *An oil tanker of 2.5 × 2.5 m square cross-section is 4 m long. Oil is filled up to a depth of 2 m. At what acceleration in the direction of its length the tanker be moved so that the corner A is exposed ? What is then the net horizontal force acting on the tanker sides ? Takes specific gravity of oil as 0.8.*

Sol. Corner *A* of the oil tanker will get exposed on account of uniform acceleration of the tanker in the direction indicated in the figure shown below. The oil surface which was initially horizontal (indicated by *BC*) now assumes a profile *ATS*. The angle of inclination θ of line *AT* (with the horizontal) will be determined from the condition that the quantity of oil inside the tanker remains the same.

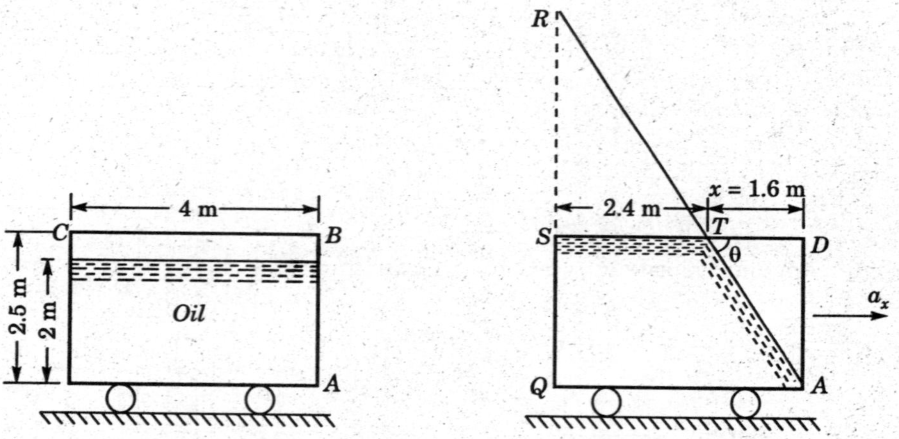

(*i*) Tanker stationary (*ii*) Tanker in motion

Volume of oil inside the tanker = 2 × 2.5 × 4 = 20.0 m²

Volume of empty space inside the tanker

$$= 0.5 \times 2.5 \times 4 = 5.0 \text{ m}^2$$

From Fig. (*ii*), $\tan\theta = \dfrac{AD}{TD} = \dfrac{2.5}{x}$ ...(*a*)

Volume of empty space $= \frac{1}{2} \times AD \times TD \times 2.5$

$$= \tfrac{1}{2} \times 2.5 \times x \times 2.5 = 3.125x$$

Equating volumes of empty space in the two cases, *i.e.* before and after the motion

$$3.125x = 5.0$$

$$x = 1.60 \text{ m.}$$

From Eq. (2.41), $\tan\theta = \dfrac{a_x}{g}$ ...(*b*)

From Eqs. (*a*) and (*b*), $\frac{a_x}{g} = \frac{2.5}{1.6}$, $a_x = \frac{2.5 \times 9.81}{1.6} = 15.32$ m/s^2.

From Δs *ARQ* and *RST*,

$$\frac{QA}{QR} = \frac{ST}{SR}$$

Let $SR = y$, making appropriate substitutions, one may write

$$\frac{4}{2.5 + y} = \frac{2.4}{y}$$

which results in $y = 3.74$ m.

Hydrostatic force on the left side face *QS*, Eq. (2.22) gives,

$$F_h = \gamma h_c A$$

$$= 7848 \times \left(3.74 + \frac{2.5}{2}\right) \times 2.5 \times 2.5 = 245{,}000 \text{ N} = 245 \text{ kN}.$$

Since the right side *AD* of the tank is exposed to the air, the net pressure force is equal to 245 kN and acts on the face *QS* (from right to left). The centre of pressure will be at the centroid of the trapezoidal pressure diagram *STAQ*.

Example 2.58. *An open rectangular tank 1.5 m wide, 2 m deep and 3 m long is half-filled with water. It is accelerated horizontally at 3.27 m/s² in the direction of its length. Determine the depth of water at each end of the tank. Also determine the total force on each end of the tank and check the difference between these forces from Newton's second law of motion.*

Solution. From Eq. (2.41),

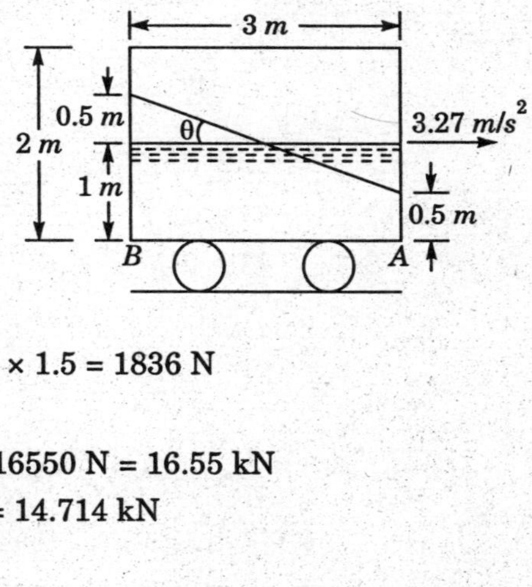

$$\tan\theta = \frac{3.27}{9.81} = \frac{1}{3}.$$

Depth at the corner *A*

$$= 1.0 - 1.5 \tan\theta = 0.5 \text{ m}$$

Depth at the corner *B*

$$= 1 + 1.5 \tan\theta = 1.5 \text{ m}.$$

Total force on the shallow end (corner *A*)

$$= \tfrac{1}{2}\gamma h^2 b = \tfrac{1}{2} \times 9810 \times (0.5)^2 \times 1.5 = 1836 \text{ N}$$

Total force on the deep end (corner *B*)

$$= \tfrac{1}{2} \times 9810 \times (1.5)^2 \times 1.5 = 16550 \text{ N} = 16.55 \text{ kN}$$

Net horizontal force $= 16550 - 1836 = 14714$ N $= 14.714$ kN

From Newton's second law,

$$F = m\,a_x = \frac{9810}{9.81} \times (1.5 \times 1 \times 3) \times 3.27 = 14700 \text{ N} = \mathbf{14.7\ kN.}$$

The net horizontal force determined from hydrostatic law tallies with that obtained from Newton's law.

Example 2.59. *A closed oil tanker 2 m deep × 1.8 m wide and 3.5 m long has been filled with an oil of specific gravity 0.8 upto a depth of 1.6 m. Calculate the acceleration which may be imparted to the tank in the direction of its length so that bottom front end of the tank is just exposed. Also calculate the net horizontal force acting on the tanker sides and show that this equals the force necessary to accelerate the liquid mass in the tanker. Take specific weight of water = 9807 N/m³.*

(*RGPV May 2013*)

Solution. Volume of oil in closed tanker $= 1.6 \times 1.8 \times 3.5 = 10.08$ m^3

Let the tanker move with acceleration a_x, with the axes X and Z as shown. As seen in the figure, the bottom front of tanker face BC is just exposed. In this condition, the volume of oil in the tanker when moving with acceleration a_x, must be the same as in the static condition, *i.e.*, 10.08 m³.

Let PC be the new exposed top having value x as shown, the oil surface PB making an angle θ with PC.

Equating volume of oil before and after the acceleration,

$$10.08 = \left[(3.5 - x) \times 2 + \frac{1}{2} \times x \times 2\right] \times 1.8$$

Solving, we obtain $x = 1.4$ m.

Let the axes X and Z be taken as horizontal and vertical directions, BP is the new free surface of oil, which is inclined at angle θ as shown. As x increases, Z decreases, from Eq. (2.40) which represents the case depicted in Fig. 2.24 that shows the free surface for horizontal acceleration, we obtain :

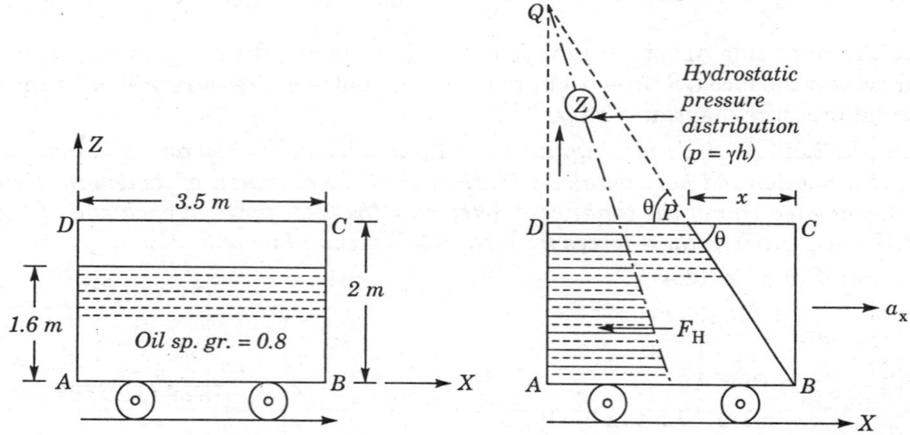

$$\frac{dZ}{dx} = -\frac{a_x}{g}$$

Considering Δs, PQD and BCP,

$$\tan\theta = \frac{DQ}{DP} = \frac{BC}{x} = \frac{2}{1.4}$$

$$\therefore \quad DQ = DP \tan\theta = (3.5 - 1.4) \times \frac{2}{1.4} = 3.0 \text{ m}$$

Horizontal hydrostatic force on tanker face AD, representing area ($AD \times 1.8$)

$$F_H = \text{Average pressure on the face} \times AD \times 1.8$$

$$= \frac{p_D + p_A}{2} \times 2 \times 1.8 = \frac{(3+5) \times 9807 \times 0.8}{2} \times 2 \times 1.8 \text{ N} = 112976.64 \text{ N}$$

Since the tanker face BC is exposed, there is no hydrostatic force acting on it. The net horizontal force acting on the tanker wall AD :

$$\mathbf{F_H = 112976.64\ N} \qquad \text{...(1)}$$

Force required to accelerate the liquid mass in the tanker with acceleration $a_x = g \tan\theta$

$$\text{Force} = \text{mass} \times \text{acceleration}$$

$$= \frac{\text{weight of oil}}{g} \times g \tan\theta = \frac{\gamma V}{g} \cdot g \tan\theta$$

$$= \frac{(0.8 \times 9807 \times 10.08)}{9.807} \times 9.807 \times \frac{2}{1.4}$$

$$= \mathbf{112976.64\ N} \qquad ...(2)$$

The two forces, (1) and (2) are equal, as was required to be proved.

Example 2.60. *An iron tyre for a wheel is cast in a mould which is rotated to impart a higher density to the iron. Determine the gauge pressure in the molten iron at the point K (shown in the following figure) in a stationary and a rotating mould if n = 300 rpm. The density of cast iron is 7000 kg/m^3, the diameter D_1 = 0.95 m, D_2 = 0.9 m, and h = 250 mm.*

Solution. (*i*) *When the mould is stationary*

The gauge pressure at *K* in the molten iron,

$$p = \gamma h = 7000 \times \frac{250}{1000}\ \text{Kg/m}^2 = \mathbf{1750\ kg/m^2\ (gauge)}$$

(*ii*) *When the mould rotates at 300 rpm.*

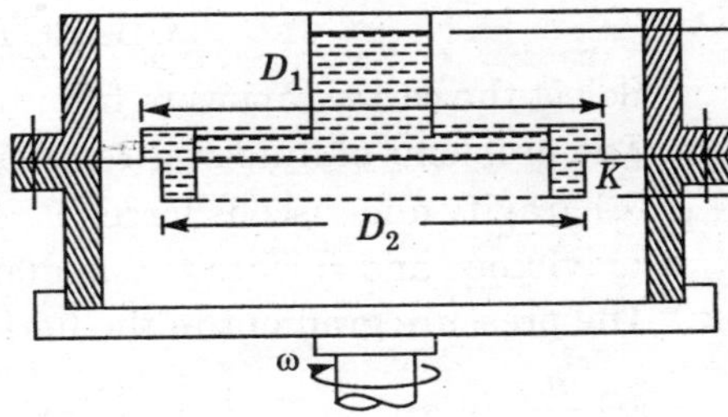

using Eq. (2.47), pressure at *K*

$$p = \gamma h = \gamma \frac{\omega^2 R^2}{2g}, \text{ where } R = \frac{D_2}{2} = \frac{0.95}{2}\ \text{m}$$

$$\text{Angular velocity } \omega = \frac{2\pi n}{60} = \frac{2\pi \times 300}{60}$$

$$= 10\ \pi \text{ rad/sec}$$

$$\therefore \qquad p = \frac{7000}{19.62}\left[(10\ \pi)^2 \left(\frac{0.95}{2}\right)^2\right] = \mathbf{79447.85\ kg/m^2}$$

Example 2.61. *A reservoir filled with a liquid to a height 'h' is rotated with a constant angular velocity 'ω' about an axis perpendicular to the plane of the drawing. The distance perpendicular to the surface of the liquid is 'R'. Determine the minimum number of revolutions at which the liquid will be held within the reservoir during rotation.*

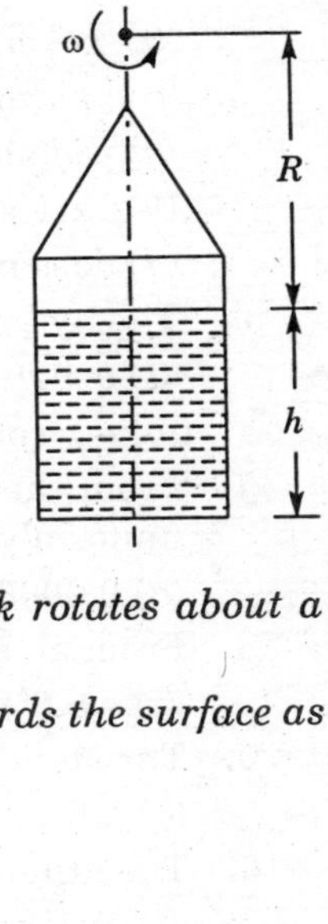

Solution. The minimum number of revolutions will be determined by the condition when the force due to centrifugal acceleration on the liquid mass is equal to the force due to gravity on the reservoir liquid. Hence,

$$mr\omega^2 = mg$$

$$\omega^2 = \frac{g}{r}, \text{ here } r = R, \text{ and } \omega = \frac{2\pi n}{60}, n \text{ is in rpm}$$

making substitutions, $\left(\frac{2\pi n}{60}\right)^2 = \frac{g}{R}$

from which, $n = \frac{30}{\pi}\sqrt{\frac{g}{R}}$ rpm

Example 2.62. *In a device used for milk analysis, a test tube with milk rotates about a vertical axis at a speed 'n' = 400 rpm at a distance 'r' = 250 mm from the axis.*

Determine the increase in force driving the particles of fat in the milk towards the surface as compared with the force on these particles at rest.

Solution.

The centrifugal force acting on milk particle of mass '*m*' = $m\ (r\omega^2)$

the gravitational force acting on the same particle = $m\ .\ g$

The angle of the test tube setting, α, is given by

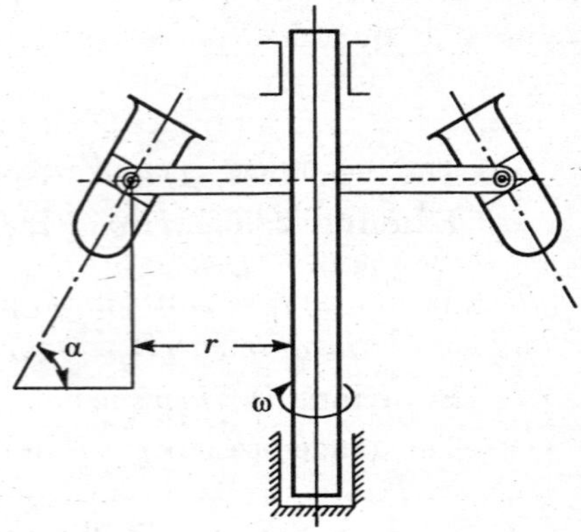

$$\tan \alpha = \frac{mg}{m(r\omega^2)} = \frac{g}{r\omega^2}$$

from the data given, $n = 400$ rpm, and $r = 250$ mm

$$\omega = \frac{2\pi n}{60} = \frac{2\pi \times 400}{60}$$

$$= 41.888 \text{ rad/sec}$$

$$\therefore \qquad \tan \alpha = \frac{9.81}{\left(\frac{250}{1000}\right)(41.888)^2} = \mathbf{0.02236}$$

The resultant force acting on mass 'm'

$$= \sqrt{m^2(r\omega)^2 + (mg)^2} = \boldsymbol{m\sqrt{g^2 + (r\omega^2)^2}}$$

OBJECTIVE TYPE QUESTIONS

Select the correct answers from the following :

1. The problems of fluid statics are influenced by the following forces :
(*a*) gravity and viscous forces (*b*) gravity and pressure forces
(*c*) viscous and surface tension forces (*d*) gravity and surface tension forces.

2. The pressure gradient in the horizontal direction in a static fluid is represented by :
(*i*) $\frac{\partial p}{\partial z} = -\gamma$ (*ii*) $\frac{\partial p}{\partial x} = \gamma$ (*iii*) $\frac{\partial p}{\partial x} = 0$ (*iv*) $\frac{\partial p}{\partial z} = 0$.
Here z-direction is vertically upwards, and x is along horizontal.

3. Indicate the variation of hydrostatic pressure with depth below the free surface :
(*a*) it decreases as the depth increases (*b*) there is no change with the depth
(*c*) it increases with the increase in depth
(*d*) after a certain depth there is no change in the pressure.

4. The piezometric head is expressed by :
(*i*) $\gamma z + p$ (*ii*) $z + p/\gamma$ (*iii*) $gz + p/\rho$ (*iv*) $\rho z + p/g$.

5. The piezometric head in a static liquid :
(*a*) remains constant in the horizontal direction, provided the liquid is continuous
(*b*) varies only in the vertical direction (*c*) remains constant throughout the fluid
(*d*) does not vary in the vertical direction.

6. The hydrostatic law of pressure variation is expressed by :
(*i*) $p = z/\gamma$ (*ii*) $p = \gamma z$ (*iii*) $p = -\gamma_z +$ constant
(*iv*) $p =$ constant

7. From the mean sea-level upto an altitude of about 11 km, temperature of air decreases uniformly at the rate of :
(*a*) 0.00065°C/m (*b*) 0.0065°C/m (*c*) 0.065°C/m (*d*) 0.65°C/m.

8. From an altitude of 11 km upto 32 km, the air temperature remains constant at
(*i*) – 5.6°C (*ii*) – 30°C (*iii*) 35.3°C (*iv*) – 56.5°C

9. The standard atmosphere relates to the air properties obtaining at a latitude of :
(*a*) 23° (*b*) 0° (*c*) 40° (*d*) 45°.

10. The atmosphere upto an altitude of 11 km is known as :
(*i*) stratosphere (*ii*) cosmosphere (*iii*) troposphere (*iv*) ionosphere.

11. Just as the Newton's law of gravitation is said to have been inspired by the fall of an apple from a tree, the principle of floatation and buoyancy was similarly discovered by (pick up the name from the following) ...... as a result of his absent-minded entry into a bath-tub full of water.

(*a*) Galileo (*b*) Archimedes (*c*) Newton (*d*) Pascal

12. Who is credited to have said "Eureka" (meaning I have found it !)

(*i*) Newton (*ii*) Torricelli (*iii*) Pascal (*iv*) Archimedes

13. The following figure shows four containers of different shapes but of the same base area 'A' which are filled by the same liquid (sp. wt. γ) upto the same height h :

Base area for all the containers = A

Liquid surface area, for (*i*) = A ; for (*ii*) = $A/2$; for (*iii*) = $A/4$; for (*iv*) = $2A$

The hydrostatic force on the base :

(*a*) is more for container (*iv*) and is equal to $2\gamma Ah$,

(*b*) has the minimum value for container (*iii*) and is equal to $\gamma Ah/4$,

(*c*) of container (*ii*) is equal to $\gamma Ah/2$

(*d*) of container (*i*) is the same as for the other remaining three and is equal to γAh.

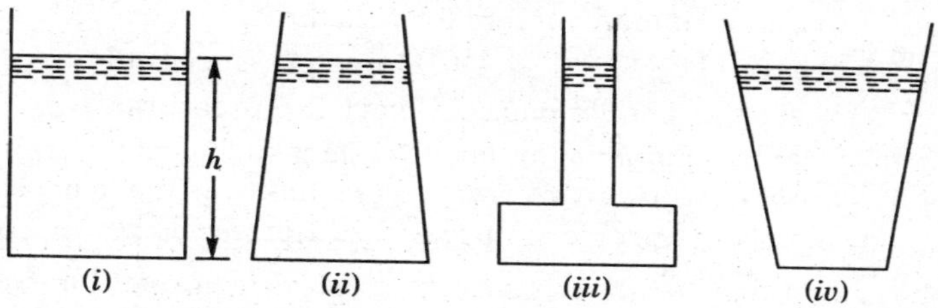

14. Absolute pressure in a flow-system :

(*i*) is always above local atmospheric pressure

(*ii*) is a vacuum pressure

(*iii*) may be above, below or equal to the local atmospheric pressure

(*iv*) is also called negative pressure.

15. Vacuum or negative pressure at a flow-section is

(*a*) a pressure measured always below the local atmospheric pressure

(*b*) a pressure measured by a Bourdon gauge

(*c*) a pressure indicated by a vacuum gauge

(*d*) always the pressure indicated by a differential manometer.

16. Local atmospheric pressure is measured by

(*i*) a mercury barometer (*ii*) a Bourdon gauge

(*iii*) a vacuum gauge (*iv*) a manometer.

17. Absolute pressure is measured by :

(*a*) a Bourdon gauge (*b*) an Aneroid barometer

(*c*) a differential manometer (*d*) a hook gauge.

18. Gauge pressures in flow systems are measured by :

(*i*) manometers (*ii*) Aneroid barometer

(*iii*) vacuum gauge (*iv*) Bourdon gauge.

19. Standard atmospheric pressure in terms of water column is :

(*a*) 9.81 m (*b*) 10.33 m (*c*) 8.75 m (*d*) 12.35 m.

20. The hydrostatic pressure on a plane surface of area A is given by :

(*i*) $0.5\ \gamma Ah_c$ (*ii*) γAh_c (*iii*) $2\gamma Ah_c$ (*iv*) $1.5\ \gamma Ah_c$.

where h_c = depth of centroid of the area below the liquid surface.

21. The point through which the resultant hydrostatic force acts is called :
(*a*) metacentre (*b*) centre of pressure
(*c*) centre of buoyancy (*d*) centre of gravity.

22. Location of centre of pressure is such that it is almost always :
(*i*) below the centroid of the plane surface
(*ii*) above the centroid of the plane surface
(*iii*) at the centre of buoyancy
(*iv*) below the metacentre.

23. The centre of pressure of a liquid on a plane surface immersed vertically in a static body of liquid, always lies below the centroid of the surface area, because :
(*a*) in liquids the pressure resting is same in all directions
(*b*) there is no shear stress in liquids at rest
(*c*) the liquid pressure is constant over depth
(*d*) the liquid pressure increases linearly with depth.

24. Mercury is generally used in manometers for measuring :
(*a*) low pressures accurately (*b*) large pressures only
(*c*) all pressures except the smaller ones (*d*) very low pressures.

25. In a mercury differential manometer used for measuring pressure differences across a venturimeter in a water pipeline, if an error of 2 mm has been made in observing a differential head of 10 mm, the percentage error in pressure difference is :
(*i*) 12.6 (*ii*) 25.2 (*iii*) 20 (*iv*) 2 (*v*) 10.

26. For precise measurement of small pressure differences in water systems, if a mixture of carbon tetrachloride and kerosene, having a sp. gr. 1.25 is used the magnification so obtained in the differential head is :
(*a*) 1.25 (*b*) 4 (*c*) 0.25 (*d*) 2.5.

27. The buoyant force on a floating/submerged body is a force :
(*i*) due to gravity and acts downwards
(*ii*) due to volume of liquid displaced by the body and acts vertically upwards
(*iii*) equal to the submerged weight of the body and acts vertically downwards
(*iv*) which acts horizontally on the vertical projection of the body.

28. A floating body displaces a volume of liquid equal to :
(*a*) its own volume (*b*) its own weight
(*c*) its submerged weight (*d*) none of the above.

29. Centre of buoyancy is :
(*i*) the point through which the submerged weight of the body acts
(*ii*) the point through which the buoyant force acts
(*iii*) the point through which the resultant hydrostatic force acts
(*iv*) the centre of gravity of the liquid displaced by the body.

30. In a water pipeline, the pressure at the highest elevation is a suction pressure of magnitude 4.9×10^2 N/m^2. A mercury barometer installed at the place recorded an atmospheric pressure of 98,100 N/m^2. The absolute pressure at the highest point of the pipeline is :
(*a*) 1019.37 N/m^2 (*b*) 509.7 N/m^2 (*c*) 5096.8 N/m^2 (*d*) 50.97 N/m^2.

31. The criterion for stability of a floating body depends on :
(*i*) the relative position of centre of buoyancy and the centre of gravity
(*ii*) the relative position of the metacentre and the centre of buoyancy
(*iii*) the relative position of the metacentre and the centre of gravity
(*iv*) none of these.

32. The metacentric height of a floating body depends :

(*a*) directly on the shape of its water-line area

(*b*) on the volume of liquid displaced by the body

(*c*) on the distance between the metacentre and the centre of gravity

(*d*) second-moment of water line area

33. One of the following statements is true with regard to bodies that float or are submerged in liquids :

(*a*) For a body wholly submerged in a liquid the stability is ensured if the centre of buoyancy is below the centre of gravity of the body.

(*b*) For a body floating in a liquid the stability is ensured if the centre of buoyancy is below the centre of gravity of the body.

(*c*) For a body floating in a liquid the stability is ensured if the metacentre is above both the centre of buoyancy and centre of gravity, regardless of the relative positions of the centres of buoyancy and gravity.

(*d*) For a body floating in a liquid the stability is ensured if the centre of buoyancy is below the centre of gravity and the metacentre is above both the centres of gravity and buoyancy.

34. A floating body will remain in stable equilibrium so long as :

(*i*) the metacentre M lies below the centre of gravity G

(*ii*) the metacentre M lies above the centre of gravity G

(*iii*) the metacentre and the centre of gravity remain at the same position

(*iv*) none of these.

35. Merchant ships, especially liners, are designed to have a relatively small metcentric height which lies within the range :

(*a*) 0.1—0.3 m (*b*) 0.3—0.6 m (*c*) 0.6—0.9 m (*d*) 0.9—1.2 m.

36. In case of warships, sea-kindlines is sacrified so that they have a larger reserve of stability. The metacentric height, therefore, falls within the range :

(*i*) 0.2—0.6 m (*ii*) 0.1—0.3 m (*iii*) 0.6—2.0 m (*iv*) 2.0—3.0 m.

37. Fluid masses subjected to uniform acceleration are analysed using :

(*a*) the Newton's law of viscosity (*b*) the Newton's second law of motion

(*c*) the Bernoulli's equation (*d*) the momentum equation

38. An oil tanker is partially filled and is then moved in translation at uniform acceleration. The free-surface of oil then :

(*i*) remains horizontal

(*ii*) is inclined to horizontal with smaller depth at the rear end

(*iii*) is inclined such that a larger depth occurs at the rear end

(*iv*) none of these.

39. A racing car has a fuel tank which is partially filled. The car gets into motion in the horizontal direction at a uniform acceleration equal to g. The free-surface of the liquid fuel in the tank will assume a slope (with the horizontal) of :

(*a*) 20° (*b*) 30° (*c*) 45° (*d*) 60°.

40. A space vehicle is launched by a powerful rocket which is propelled by a liquid fuel. The rocket moves in the vertically upward direction at a uniform acceleration of g. The pressure on the bottom of the fuel tank of depth h is :

(*i*) according to the hydrostatic law, its magnitude being equal to γh.

(*ii*) such that the pressure is larger than the corresponding hydrostatic value, it being equal to $2\gamma h$.

(*iii*) such that there is no variation of pressure throughout the liquid, the pressure remaining constant at all points

(*iv*) such that the pressure is less than the hydrostatic one with magnitude equal to 0.5 γh.

41. When a static liquid is subjected to uniform rotation in a container the free-surface assumes a shape of :

(*a*) a cone (*b*) a circular cylinder

(*c*) a paraboloid of revolution (*d*) an ellipsoid of revolution.

42. The free surface of a liquid partially filled in a cylindrical container when subjected to uniform rotation assumes a shape which can be represented by :

(*i*) $Z = \frac{r\omega^2}{2g} + h_0$ (*ii*) $Z = \frac{r^2\omega^2}{2g} + h_0$ (*iii*) $Z = \frac{r^2\omega}{2g} + h_0$ (*iv*) $Z = \frac{r^2\omega^2}{g} + h_0$

The coordinate axes (Z, r) have their origin at the centre of the bottom of container, h_0 being the central depth of liquid.

43. The vertical distance of the centre of pressure below the centroid of the plane area is :

(*a*) $\frac{I_c}{Ay_c}$ (*b*) $\frac{I_c \sin^2 \theta}{Ah_c}$ (*c*) $\frac{I_c \sin^2 \theta}{Ay_c}$ (*d*) $\frac{I_c \sin \theta}{Ay_c}$

44. A penstock pipeline of diameter d is carrying water at a high pressure p to a hydropower station. If f_s is the permissible tensile stress for the pipe material, the thickness of pipe wall from consideration of hoop tension is :

(*i*) $\frac{pd}{f_s}$ (*ii*) $\frac{pd}{2f_s}$ (*iii*) $\frac{2pd}{f_s}$ (*iv*) $\frac{3pd}{2f_s}$

45. A tank has in its side a very small horizontal cylinder fitted with a frictionless piston. The head of liquid above the piston is h and the piston area is a, the liquid having a specific weight γ. The force that must be exerted on the piston to hold it in position against the hydrostatic pressure is :

(*a*) $2\gamma ha$ (*b*) γha (*c*) $\frac{2\gamma ha}{3}$ (*d*) none of these.

46. Match the statements of column A with the corresponding statement of column B which pertain to resisting of hydrostatic force in various structures :

Column A	*Column B*
(*i*) gravity dam	(*a*) hydrostatic force is resisted by the abutment reactions
(*ii*) circular water tank	(*b*) it is resisted by the tensile strength of the cylindrical shell
(*iii*) rectangular water tank	(*c*) it is resisted by a system of tie rods or stays
(*iv*) lockgates	(*d*) it is mainly resisted by the weight of masonry
(*v*) arch dam	(*e*) it is resisted by the reaction of the hinge and the side reaction.

47. A circular plate 1 m in diameter is submerged vertically in water such that its upper edge is 8 m below the free surface of water. The total hydrostatic pressure force on one side of the plate is :

(*a*) 6.7 kN (*b*) 65.4 kN (*c*) 45.0 kN (*d*) 77.0 kN.

48. A container carrying water is moved in a horizontal direction with an acceleration of 2.45 m/s^2. The angle of inclination of free water surface to the horizontal is :

(*a*) 14.03° (*b*) 67.8° (*c*) 45.00° (*d*) 0°.

ANSWERS TO OBJECTIVE TYPE QUESTIONS

1. (*b*)	**2.** (*iii*)	**3.** (*c*)	**4.** (*ii*)	**5.** (*a*), (*b*)	**6.** (*iii*)	**7.** (*b*)	**8.** (*iv*)
9. (*c*)	**10.** (*iii*)	**11.** (*b*)	**12.** (*iv*)	**13.** (*d*)	**14.** (*iii*)	**15.** (*a*), (*c*)	**16.** (*i*)

17. (*b*)	**18.** (*i*), (*iv*)	**19.** (*b*)	**20.** (*ii*)	**21.** (*b*)	**22.** (*i*)	**23.** (*d*)	**24.** (*c*)
25. (*iii*)	**26.** (*b*)	**27.** (*ii*)	**28.** (*b*)	**29.** (*ii*), (*iv*)	**30.** (*b*)	**31.** (*iii*)	**32.** (*a*), (*b*)
33. (*d*)	**34.** (*ii*)	**35.** (*b*)	**36.** (*iii*)	**37.** (*b*)	**38.** (*iii*)	**39.** (*c*)	**40.** (*ii*)
41. (*c*)	**42.** (*ii*)	**43.** (*d*)	**44.** (*ii*)	**45.** (*b*)			
46. (*i*) (*d*), (*ii*) (*b*), (*iii*) (*c*), (*iv*) (*e*), (*v*) (*a*)				**47.** (*b*)	**48.** (*a*)		

REVIEW QUESTIONS

1. What forces act on a fluid element in static equilibrium ?

2. Which of the following statements are correct ?
(*i*) pressure at all points over a horizontal plane is same,
(*ii*) pressure at all points in a continuous fluid lying on a horizontal plane is same,
(*iii*) pressure in a continuous fluid does not vary in the horizontal direction.

3. On what factors does the pressure variation in the vertical direction depends ?

4. Express pressure in as many different ways as you can.

5. Does a (*i*) barometer, (*ii*) Bourdon gauge, (*iii*) piezometer tube measure absolute or gauge pressure ?

6. What is the relation between the vacuum (suction or negative) pressure and the absolute pressure ?

7. Study of pressure variation around objects in a wind tunnel requires measurement of very small pressures. What measuring fluid and what type of manometer would you suggest ?

8. If a tank filled with a certain liquid (sp. weight γ) upto a height h is allowed to drop freely under gravity how would the pressure vary in the vertical direction ?

9. If the tank of Q. 8 is moved vertically upward with constant acceleration g, what would be the variation of pressure in the vertical direction ?

10. A cylindrical tank half filled with a liquid is rotated about its vertical axis with a certain constant speed. If the liquid at the sides reaches at the top of the tank, what would be the level of water at the axis of rotation ?

11. Which of the following statements about *centre of pressure* are correct ?
(*i*) It is always above the centroid of the submerged area,
(*ii*) It is at the centroid of the submerged area,
(*iii*) It is the centroid of the pressure prism,
(*iv*) It is the point of intersection of the line of action of the resultant hydrostatic force with the submerged area.

12. The horizontal component of the resultant hydrostatic force on a curved surface is
(*i*) equal to the product of pressure at its centroid and area,
(*ii*) equal to the weight of liquid retained by the curved surface,
(*iii*) equal to the weight of the pressure prism over the vertical projection of the curved surface,
(*iv*) equal to the force on the vertical projection of the curved surface.

13. The vertical component of the resultant hydrostatic force on a curved surface is equal to
(*i*) the product of the pressure at its centroid and area,
(*ii*) the weight of a liquid vertically over the curved surface,
(*iii*) the buoyant force over the curved surface.

14. The centre of buoyancy is located at the
(*i*) centroid of the volume of the floating body,
(*ii*) centre of gravity of the floating body,
(*iii*) centroid of the volume displaced by the floating body.

15. A floating body is in stable equilibrium when

(*i*) the centre of gravity is above the centre of buoyancy,

(*ii*) the metacentre is below the centre of gravity but above centre of buoyancy,

(*iii*) when the metacentre is above the centre of gravity,

(*iv*) when $\left(\frac{I}{V} + \overline{BG}\right)$ is positive.

PROBLEMS

2.1. The pressure 3 metre below the free surface of a liquid is 13.72 kN/m^2. Determine its specific weight.

2.2. If the pressure at a point below the sea is 137.7 kN/m^2, what is the pressure 30 m below this point ? Specific weight of ocean water is 10.06 kN/m^2.

2.3. The specific weight of a liquid varies linearly with depth according to the relation

$$\gamma = \gamma_0 + Ch$$

obtain an expression for pressure as a function of depth.

2.4. At the top of a mountain the temperature is – 5°C. The pressure and the temperature at the sea-level are 101.3 kN/m^2 and 15°C. If the air density at the sea-level is 1.284 kg/m^3, calculate the height of the mountain above sea-level. Assume dry adiabatic conditions.

2.5. An oil of specific gravity 0.80 is under a pressure of 137.2 kN/m^2.

(*a*) What is the pressure head expressed in metres of water ?

(*b*) What is the pressure head expressed in metres of oil ?

2.6. How thick is the layer of liquid mud (sp. gr. 1.6) at the bottom of a river with water 8 m deep, if there is a pressure of 343 kN/m^2 at the bottom of the mud ? Treat the mud as a fluid.

2.7. Calculate the pressure difference between the pipes *A* and *B* for the inverted U-tube monometer reading shown in Fig. Prob. 2.07.

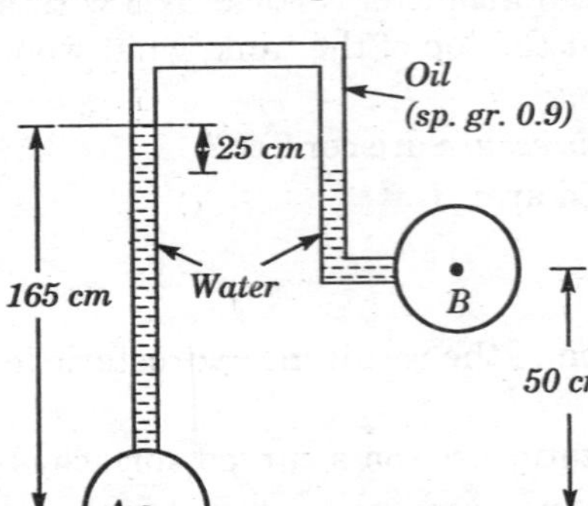

Fig. Prob. 2.07

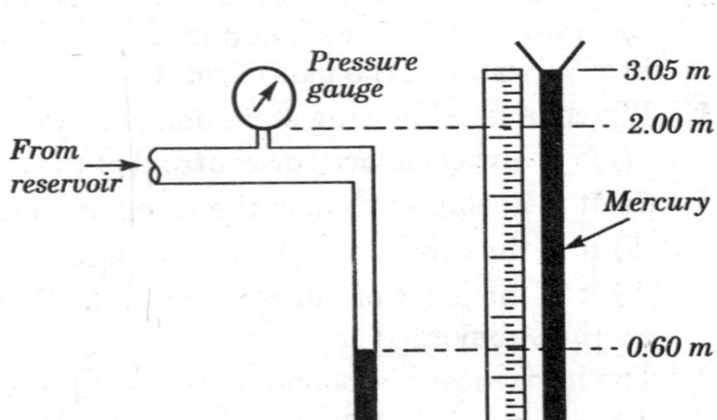

Fig. Prob. 2.08

2.8. A pressure gauge is to be calibrated using mercury column.

(*a*) For column readings shown in the Fig. Prob. 2.08, how high is the reservoir surface above the pressure gauge connection ?

(*b*) What is the correction to be applied to the gauge if it reads 304 kN/m^2 ?

2.9. The Fig. Prob. 2.09 shows water column with air trapped in the manometer bend. Compute the pressure head at *A* in metre of water.

2.10. If the barometric pressure is 90.2 kN/m^2, calculate the vapour pressure of the liquid and the reading of the pressure gauge *G*.

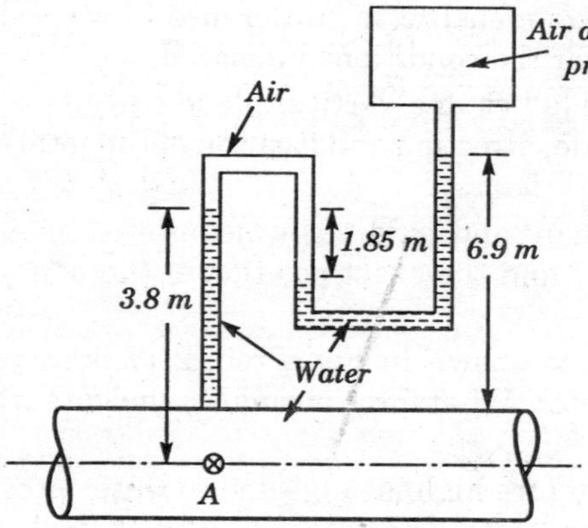

Fig. Prob. 2.09

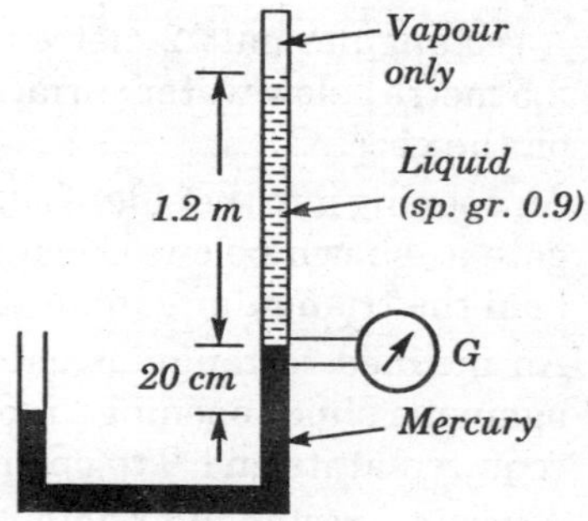

Fig. Prob. 2.10

2.11. With no weight placed on the pan, the reading in the manometer is as shown in Fig. 2.11. Compute the manometer reading after a weight 4.9 N is placed on the pan. Assume no leakage or friction between the piston and cylinder.

2.12. Calculate pressure gauge reading for the case shown in Fig. Prob. 2.12.

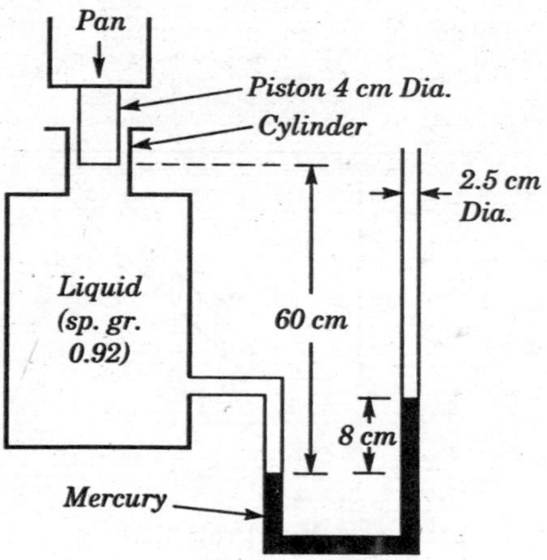

Fig. Prob. 2.11

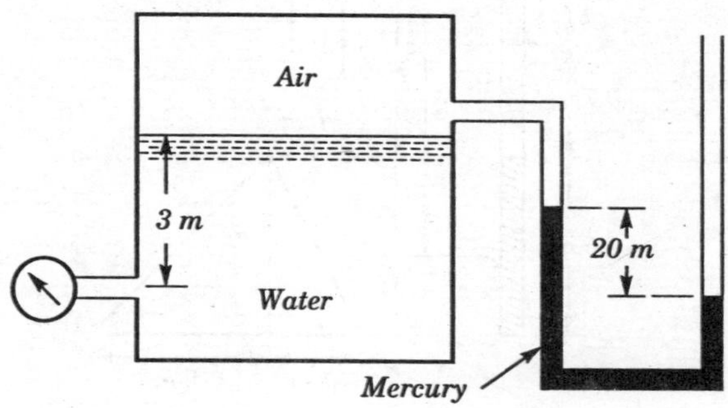

Fig. Prob. 2.12

2.13. Barometric pressure for Fig. Prob. 2.13 is 70 cm of mercury. The cock is opened and the air pumped out so that the gauge reads 50 cm of mercury vacuum. Determine the absolute pressure in the tank and the manometer reading. The change of water surface in the tank may be neglected.

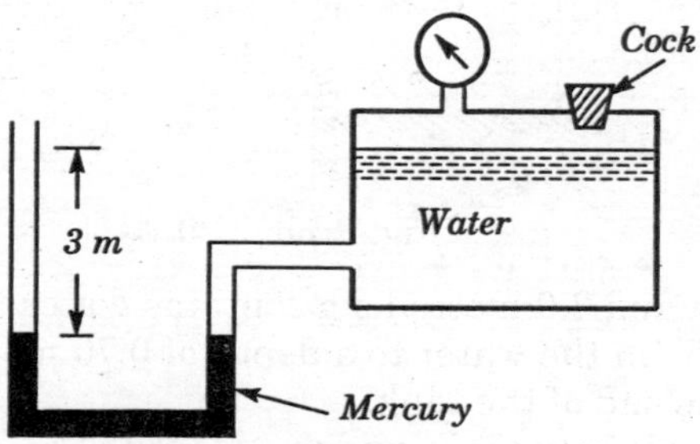

Fig. Prob. 2.13

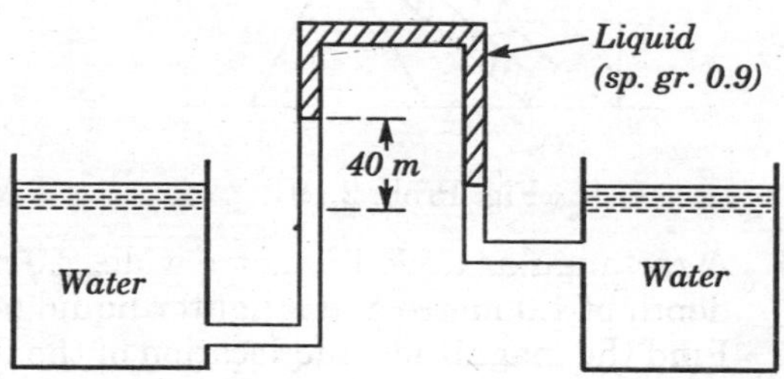

Fig. Prob. 2.14

2.14. The manometer shown in Fig. Prob. 2.14 is used to measure the difference in water level between the two tanks. Calculate this difference for the conditions indicated.

2.15. A rectangular gate 2 metre long and 1.5 metre wide lies in a vertical plane with its centre 2.5 metre below water surface. Calculate magnitude, direction and location of the total force on the gate.

2.16. An opening into the side of a tank is triangular in shape and is closed by means of a triangular gate as shown below. Calculate the moment $w \times l$ and the exact position of the pointer to hold the triangular gate in the position shown.

2.17. An inclined rectangular gate 1.22 m × 5 m size as shown in Fig. Prob. 2.17 is installed against a sluice opening to control discharge of water. What force normal to the gate will be required at its end *B* to open it, if *A* is hind ?

2.18. A vertical rectangular gate 2 metre wide and 2.5 metre high is subjected to water pressure on one side, the water surface being at the top of the gate. The gate is hinged at the bottom and is held by a horizontal chain at the top. Calculate the tension in the chain.

2.19. Predict the height *h* to which water must rise to tip (overturn) the rectangular gate shown in Fig. Prob. 2.19. Neglect the weight of the gate, which is inclined at an angle of 45°.

2.20. A vertical gate of triangular shape 3 m high and 2.25 m top width, which closes the end of triangular channel as shown in Fig. Prob. 2.20 is hinged at its upper edge. Determine how large must the weight be to keep the channel closed.

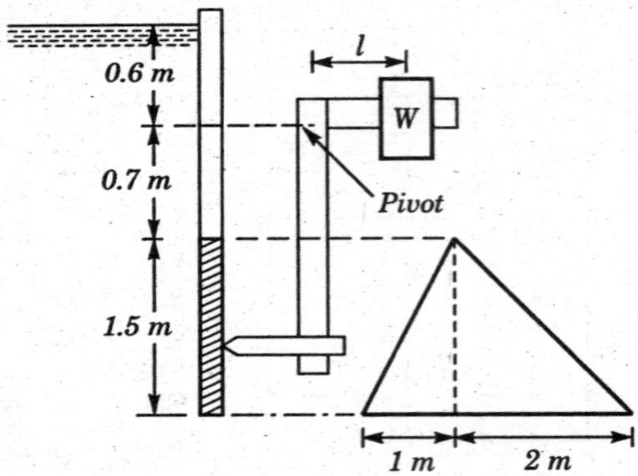

Fig. Prob. 2.16

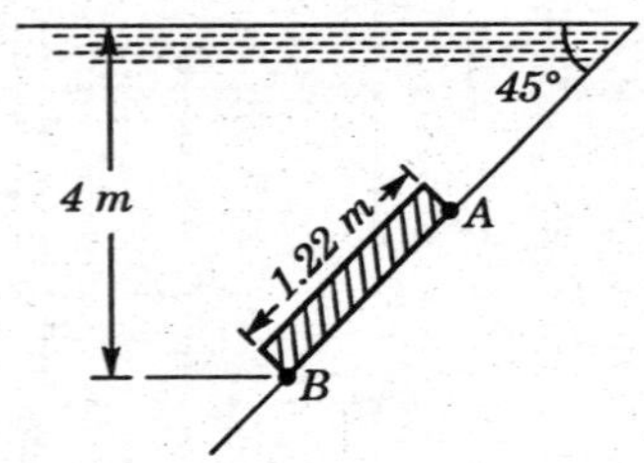

Fig. Prob. 2.17

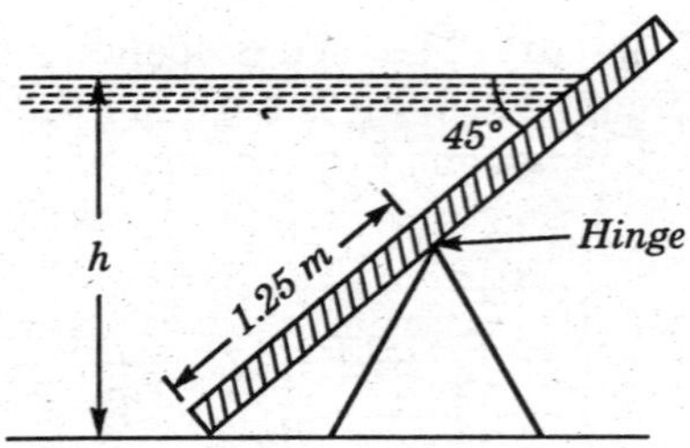

Fig. Prob. 2.19

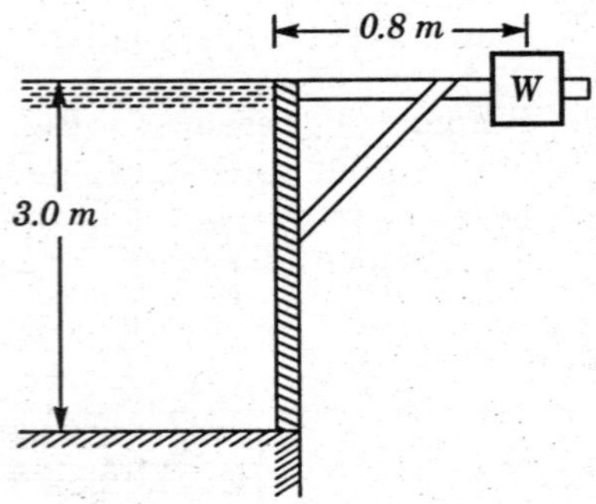

Fig. Prob. 2.20

2.21. A rectangular tank 1.5 metre wide, 2.0 metre high and 3.0 metre long contains water to a depth of 1.0 metre and a lighter liquid (sp. gr. 0.85) on the water to a depth of 0.70 metre. Find the magnitude and location of the force on one end of the tank.

2.22. From the principles of hydrostatics, determine the volume of water *ABCD* held by the circular gate *AD* shown in Fig. Prob. 2.22.

2.23. The gate in the form of a quarter cylinder *CD*, shown in Fig. Prob. 2.23 is 3 metre long. Determine magnitude, direction and location of resultant force of water on *CD*.

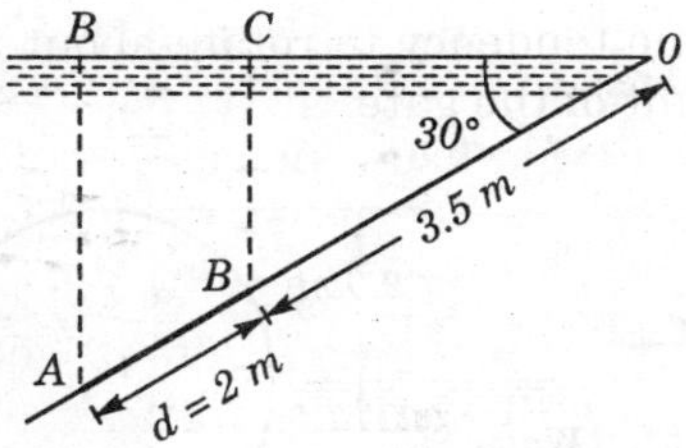

Fig. Prob. 2.22

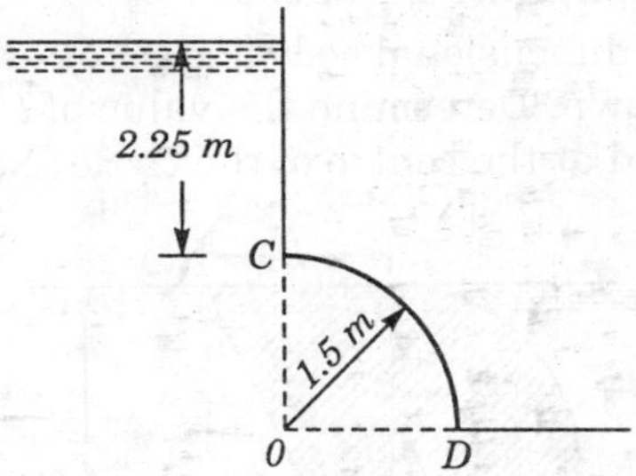

Fig. Prob. 2.23

2.24. The flash-boards on an overflow weir crest are 1.20 metre high and are supported on steel posts spaced 0.80 metre on centres. The posts are designed to fall under a bending moment of 8.33 kN- m. Determine the depth over the flash-boards which will cause the posts to fail ? Assume hydrostatic pressure distribution over the flash-boards.

2.25. A vertical gate in a tunnel 6.0 metre in diameter has water on the one side and air on the other. The water surface is 12.0 metre above the invert and the air pressure is 122.5 kN/m^2 ab. Determine the magnitude and location on the resultant force acting on the gate.

2.26. The cassion for closing the entrance to a dry dock is of trapezoidal form 15.50 metres at the top and 12 metres wide at the bottom and 7.60 metres deep. If the water on the outside is just level with the top and dock is empty find the total water pressure on it and the depth of the centre of pressure.

2.27. The profile of the water face of a dam is given by $10.85\,y = x^{2.5}$, where y is the height in metres above the floor level and x is horizontal distance of the face from the vertical line of reference in metres. The depth of water is 40 metres. Determine completely, the resultant force due to water pressure per metre length of the dam, giving its magnitude in MN, the inclination to the vertical and the point where this resultant force cuts the horizontal line at floor level.

2.28. A tank 1.8 m deep is 4.5 m square at the top and 3 m square at the bottom. The four sides are plane and each has the same trapezoidal shape. The tank is completely full of oil of specific gravity 0.9. Obtain working from first principles ; (*a*) the resultant pressure on each side, (*b*) the position of the point on each side through which the resultant pressure acts.

2.29. A vertical square area 1 m × 1 m is submerged in water with upper edge 50 cm below the surface. Locate the horizontal line on the surface of the square such that (*a*) force on the upper portion equals that on the lower portion ; (*b*) moment of force about the line due to the upper portion equals the moment due to lower portion.

2.30. Calculate the x and y co-ordinates of the centre of pressure of vertical area having the right-angled triangular shape as shown in Fig. Prob. 2.30.

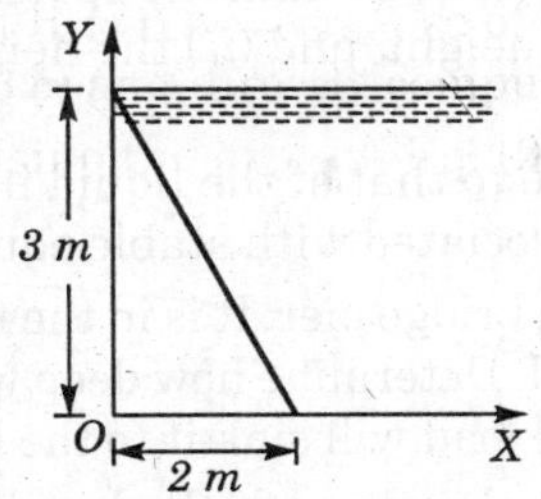

Fig. Prob. 2.30

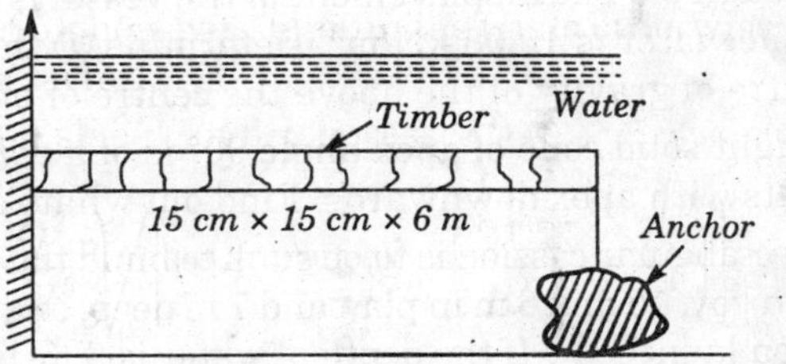

Fig. Prob. 2.31

2.31. A timber stick as shown in Fig. Prob. 2.31 has a specific weight of 5.34 kN/m^3 and is held in a horizontal position by the concrete anchor of sp. weight 23.5 kN/m^3. What is the minimum total weight of the anchor ?

2.32. A two dimensional body has a cross-section which is a combination of semi-circle and triangle as shown. Determine the value of l so that it has no tendency to rotate about the hinge O located at the centre of the circle. Neglect the weight of the gate.

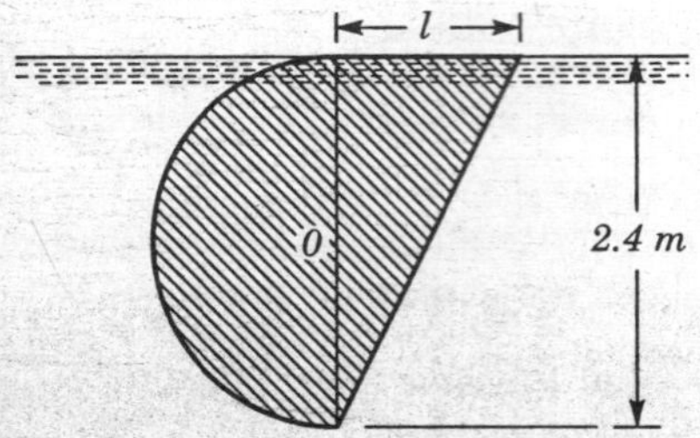

Fig. Prob. 2.32

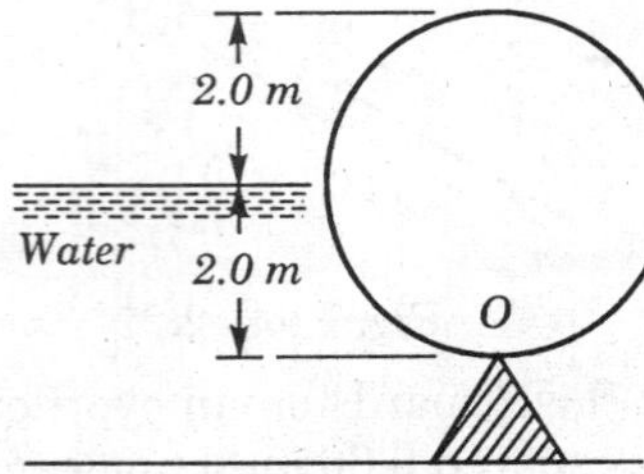

Fig. Prob. 2.33

2.33. The cylinder shown in Fig. Prob. 2.33 is 2.5 m long and is hinged at O. What is the torque necessary to hold the gate in position.

2.34. A cube 0.7 × 0.7 × 0.7 m is made up of two substances. Its upper half has a sp. gr. of 0.6 and the lower half of sp. gr. 1.4. The cube is submerged in a two-layered liquid, the upper layer being of sp. gr. 0.9 while the lower layer has sp. gr. 1.2. What portion of the cube is above the interface ?

2.35. A wooden beam of volume 0.2 m^3 and specific weight 6.37 kN/m^3 is required to be sunk in water. Calculate the weight of concrete (specific gravity 2.4) which must be attached to the beam to cause it to sink.

2.36. A hemispherical shell of diameter of d rests on a plane surface and is filled with a liquid of specific weight γ through a small tube at the top. Find the least weight of the shell that will prevent it from being lifted by the static pressure of the liquid.

2.37. A rigid dam of rectangular section is composed of material of specific weigh γ_m N/m^3 and the dam height is h metres. Calculate the minimum width b of the dam necessary to prevent it from overturning about the toe, when the water of sp. weight γ approaches the top of the dam. Assume that the maximum hydrostatic pressure acts over the bottom of the dam, uplift pressures are developed due to seepage flow under the dam.

2.38. A 60 cm long capillary tube closed at one end with the other one open to atmosphere is completely filled with water and rotated about the closed end at n revolutions per second. Assuming the water does not run out of the tube and that no voids are formed inside the liquid, compute the value of n required to reduce the pressure at the closed end to zero.

2.39. The shifting of a portion of cargo weighing 245 kN through a distance of 6 metres at right angle to the vertical plane containing the longitudinal axis of a vessel cause it to heel through 6°. The displacement of the vessel is 44.10 MN. I = 7,000 m^4 and the specific weight of the water is 10.05 kN/m^3. Determine (*i*) the metacentric height and (*ii*) the height of the centre of gravity of the above the centre of buoyancy.

2.40. A right solid cone of apex angle 90° is of a density S relative to that of the liquid in which it floats with apex downwards. Find out what range of S is associated with stable equilibrium.

2.41. A rectangular cassion is to be sunk to build the foundation of a bridge pier. It is in the form of an open boy, 15 m × 6 m in plan and 7 m deep, and weighs 785 kN. Determine how deep will it sink when launched ? If the depth of water is 6 m, what additional load will sink it to the bottom ?

2.42. A flat bottomed scow is built with vertical sides and straight sloping ends. Its length on deck is 26 m, on the bottom 20 m, its width is 6 m and its vertical depth 3.5 m. How much fresh water will it draw (find the draft) if it weighs 2550 kN ?

2.43. A stick of timber, weighing 6.86 kN/m^3 measures 15 cm × 30 cm × 6 cm. What load will it carry without sinking if placed in fresh water.

2.44. A ship with a total displacement of 16.68 MN rolls to one side through an angle of 2° when a deck load of 58.9 kN is moved laterally through a distance of 5 m. Calculate the metacentric height.

2.45. A rectangular pontoon is 25 m long, 12 m wide and 4 m deep. Its draft when launched is 1.5 and is increased to 3 m when fully loaded. Find the position of metacentre for both the drafts. Assume in each case the centre of gravity of the portion and load at the geometrical centre of the cross-section.

2.46. Calculate the metacentric height of the circular tube shown in Fig. Prob. 2.46.

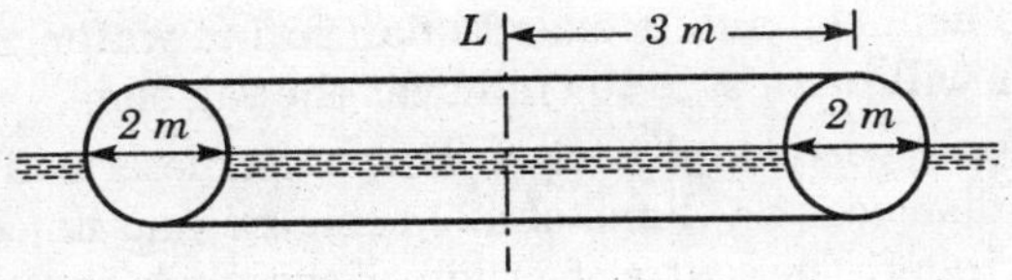

Fig. Prob. 2.46

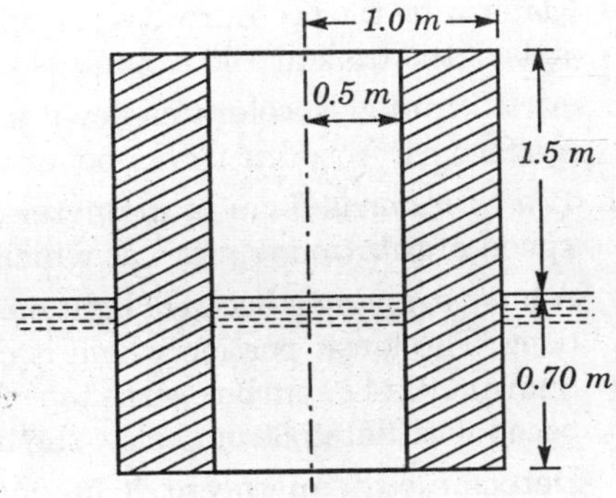

Fig. Prob. 2.47

2.47. A thick-walled cylinder is shown below. For the dimensions shown in Fig. Prob. 2.47, check its stability.

2.48. A cylindrical floating dock 12 m diameter floats in a harbour. It is submerged to a depth of 9 m and its centre of gravity is 3 m above the bottom. Find its metacentric height. If due to the displacement of weight inside the dock the dock tilts through an angle 15°, what is the righting couple ?

2.49. A right circular cylinder of radius r_0 and height h_0 has a specific gravity S. It is required to float in water with end faces horizontal. Determine the ratio of r_0/h_0 for stable equilibrium.

2.50. A circular tank 2 m depth and 12.0 m diameter is filled with liquid and accelerated uniformly in a horizontal direction. Determine the acceleration, if one-third of the liquid is spilled.

2.51. A tube, bent in the U-shape as shown below, is filled with liquid of sp. gr. 2.40. It is accelerated to the right at 2.45 m/s^2. Sketch the imaginary free surface and calculate the pressure at A. For P_A = 4.9 kN/m^2 vacuum, determine the horizontal acceleration.

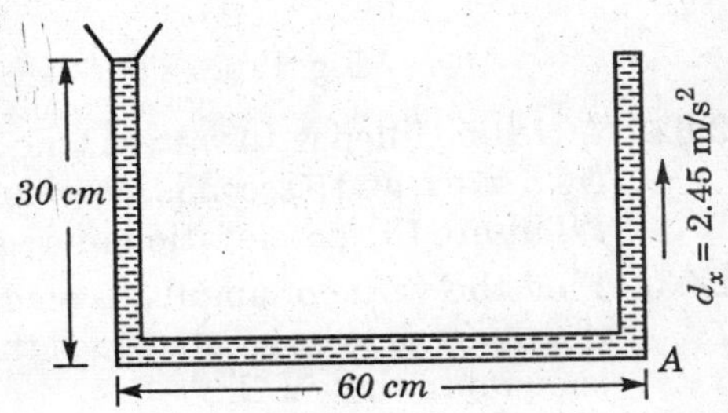

Fig. Prob. 2.51

2.52. A cubic box 1 × 1 × 1 m, open at the top and half filled with water, is placed on an inclined plane making a 30° angle with the horizontal. The empty box weighs 490 N and has a coefficient of friction of 0.30 with the inclined plane. Calculate the acceleration of the box and the angle made by the free water surface with the horizontal.

2.53. A circular cylinder containing liquid of sp. gr. 1.2 is rotated about a vertical axis. The pressure at a point 60 cm radially from the axis is the same as at another point 1.2 m from the axis and at an elevation 60 cm higher. Determine the speed of rotation of the cylinder.

2.54. A cube 0.5 × 0.5 × 0.5 m is filled with a liquid of sp. gr. 0.65 and is accelerated downward at 2.45 m/s^2. Find the resultant force on one side of the cube due to liquid pressure.

2.55. A cylinder 2 m in diameter and 6 m long is accelerated uniformally along its axis in a horizontal direction at 4.9 m/s^2. It is filled with liquid of specific weight 7.85 kN/m^3 and the pressure along its axis is 6.91 kN/m^2 gauge before acceleration commences. Find the net force exerted against the liquid in the cylinder.

2.56. A rectangular tank 2 m wide, 4 m long and 2.50 m deep contains water to a depth of 1.70 m. When it is accelerated horizontally at 3 m/s^2 in the direction of its length, determine the depth of water at each end of the tank and the total force on each end of the tank. Check the difference between these forces using Newton's second law.

2.57. A vertical cylindrical tank 2 m high and 1 m in diameter is filled with water to a depth of 1.70 m. The tank is then closed and the pressure in the space above the water raised to 6.86 kN/m^2. Determine the pressure at the corner of the tank bottom when the tank is rotated about vertical axis of the tank at 150 r.p.m. Also determine the resultant force on the bottom of the tank.

2.58. An oil tanker accelerates down a 30° inclined plane at 5 m/s^2. What is the slope of its free surface ? Determine the slope for the same acceleration when the tanker moves up the plane.

2.59. If a cylinder is 8 cm in diameter and 10 cm deep and is exactly half full of water, find the speed of rotation (r.p.m.) at which water will just begin to spill over the top edge.

2.60. A buoy carrying a becon of light has the upper portion cylindrical, 210 cm diameter and 120 cm deep. The lower portion which is curved displaces a volume of 420 cubic cm and its centre of buoyancy is 127 cm below the top of the cylinder. The centre of gravity of the whole buoy and the becon is situated 80 cm below the top of cylinder and the total displacement is 21.6 kN. Determine the metacentric height. (Refer Fig. Prob. 2.60).

2.61. The tank shown in Fig. Prob. 2.61 is filled with oil of sp. gr. 0.8 and accelerated at 4 m/s^2 as shown in the figure. Determine the pressure at *B*. Also determine the acceleration necessary to make the pressure at *B* zero.

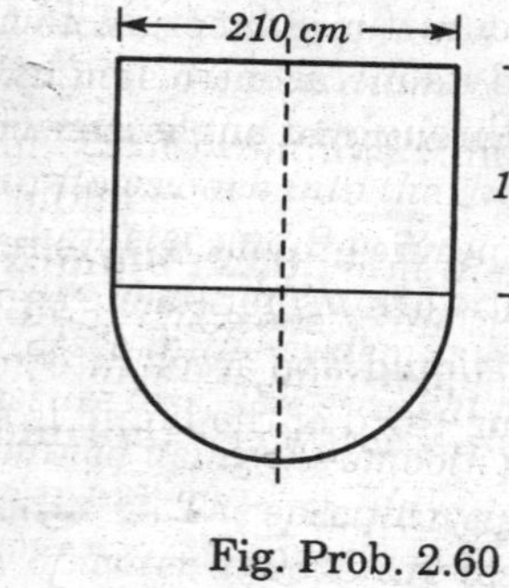

Fig. Prob. 2.60

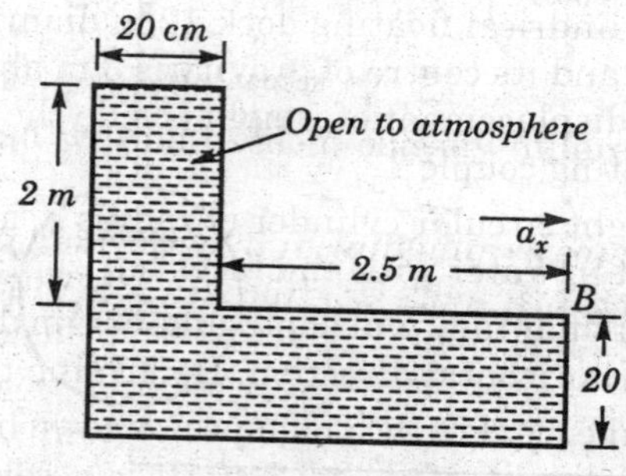

Fig. Prob. 2.61

2.62. A U-tube filled with water is mounted on a racing car. The car accelerates at a constant rate from start, and 5 seconds later the water in U-tube makes and angle of 20° with the horizontal. What are the acceleration and velocity of the car at that instant.

2.63. Find the value of angular speed at which the container shown in the figure is to be rotated such that the volume of fluid in portion *A* is equal to that in *B*.

2.64. Show that for fluid subjected to uniform rotation, the free surface assumes a paraboloid form. An open cylindrical tank 1.2 m in diameter and 0.2 m deep is filled with water and rotated at 60 r.p.m. Determine the volume of water spilled and depth of water along the axis.

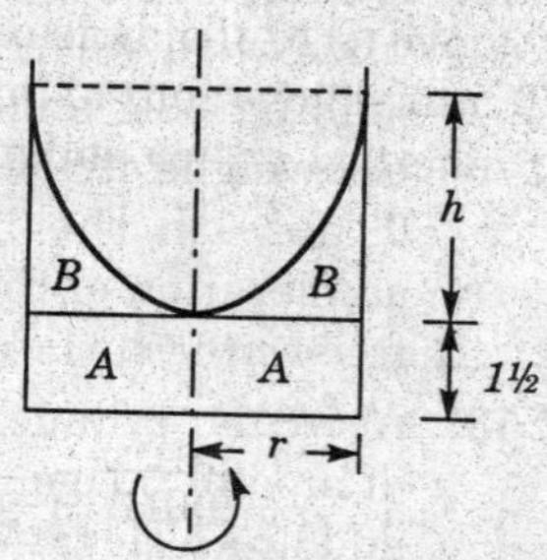

Fig. Prob. 2.63

2.65. (*a*) Explain how the magnitude and direction of the resultant of the curved surface *AB* can be determined in the two cases shown below.

(*b*) Show that for a plane surface immersed vertically in a liquid the moment of the hydrostatic thrust about the horizontal line passing through the c.g. of the surface is independent of the depth of immersion in the liquid.

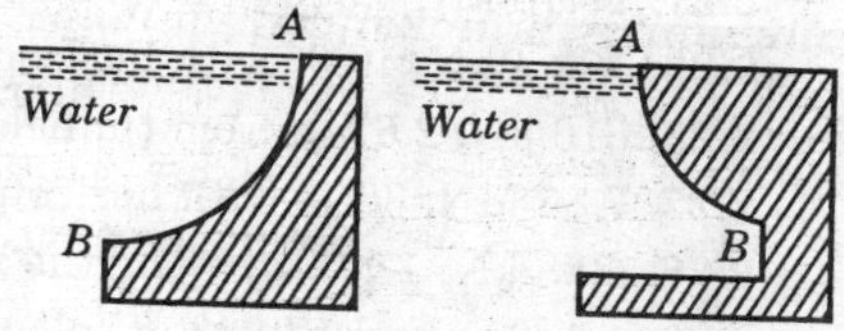

Fig. Prob. 2.65

2.66. A vertical cylinder 1.30 m in diameter and 1.60 m high is revolving on its vertical axis at 60 r.p.m. It contains a liquid which revolves with the cylinder. How deep will be the liquid at the centre ?

2.67. An open tank of water slides down an inclined plane. Show that the free surface will be :

(*a*) horizontal if velocity is constant ;

(*b*) parallel to the plane if the acceleration is equal to the component of *g* in the direction of motion, that is the tank slides without frictional resistance.

2.68. (*a*) Show that for a plane surface immersed in a liquid the 'centre of pressure' lies below the 'centre of gravity'. Is it always necessary that the centre of pressure be below the centre of gravity ?

(*b*) Differentiate between the stability of a 'floating body' and a 'submerged body'.

2.69. An open cylindrical bucket, 30 cm in diameter and 50 cm long whose wall thickness and weight can be considered negligible is forced open end first into water until its lower edge is 4 m below the surface. What force will be required to maintain this position assuming the trapped air to remain at constant temperature during the entire operation ?

[**Hint.** See solved example 2.21.]

2.70. An open tank 10 m long and 2 m deep is filled with 1.5 m of oil of specific gravity 0.82. The tank is accelerated uniformly from rest to speed of 90 m/s. What is the shortest time in which this speed may be attained without spilling any oil ?

2.71. A closed cylindrical vessel 65 cm diameter is half filled with water and then rotated with its axis horizontal at a speed of 600 r.p.m. If the water rotates as a forced vortex at the same angular velocity as the vessel and takes the form of a cylinder with hollow core what would be the total axial thrust tending to burst the flat ends ?

2.72. A U-tube differential manometer connects two pipes, carrying a liquid of specific gravity 2.14 at two points *L* and *M*. The liquid is flowing under a pressure of 147 kN/m^2 at *L* and 206 kN/m^2 at *M*. The point *L* is 3.0 m above the point *M*. Find the manometer reading if the manometric liquid is mercury.

2.73. An open cylindrical tank 2 m in diameter and 3 m deep is filled with water and rotated about its axis at 60 r.p.m. How much liquid is spilled and how deep is water at the axis while the tank is rotating ?

2.74. A cylindrical tank partly filled with liquid is rotated at a constant angular velocity ω about its vertical axis. What form will the surface assume ? How will the pressure vary along the bottom the tank ?

ANSWERS

2.1. 4.6 kN/m^3, 0.469. **2.2.** 440 kN/m^2. **2.3.** $p = \gamma_0 h + \frac{1}{2} Ch^2$.

2.4. 1954.86 m. **2.5.** (*a*) 14 m, (*b*) 17.5 m. **2.6.** 16.87 m.

2.7. 5.15 kN/m^2. **2.8.** (*a*) 31.9 m, (*b*) + 8.82 kN/m^2.

2.9. 13.9 m. **2.10.** 5.475 m of water. **2.11.** 11.0 cm of mercury.

2.12. 2.745 kN/m^2 **2.13.** 20 cm and 29 cm of mercury. **2.14.** 4 cm.

2.15. 73.5 kN, 2.575 m below w/s.

2.16. 89 kN m, pointer location 1.7544 m below pivot. **2.17.** 110.2 kN.

2.18. 20.45 kN. **2.19.** 2.65 m. **2.20.** 62.1 kN.

2.21. 19.18 kN, 0.555 m from tank bottom.
2.22. 6.12 m^3.
2.23. 174.4 kN, 40.6° with horizontal.
2.24. 1.075 m.
2.25. 610 kN, 2.67 m from tunnel bottom.
2.26. 3.72 MN, 4.87 m below w/s.
2.27. 7.85 MN, 36.7° with horizontal, 13.28 m from the floor.
2.28. 59.8 kN, 1.115 m along the side from top.
2.29. 1.118 m and 1.0883 m below w/s.
2.30. $x = 0.75$ m, $y = 0.75$ m.
2.31. 1.033 kN.
2.32. 1.696 m.
2.33. 98 kN-m anti-clockwise.
2.34. 47 cm.
2.35. 1.177 kN.
2.36. $\gamma\pi d^3/24$.
2.37. $b = h\sqrt{\gamma/3\,(\gamma_m - \gamma)}$.
2.38. 225.6 r.p.m.
2.39. 0.318 m, 1.277 m.
240. 1/8 to 1.
2.41. 4.51 MN.
2.42. 2.02 m.
2.43. 795 N.
2.44. 50.5 cm.
2.45. 6.75 m., 3.5 m.
2.46. 5.935 m.
2.47. Unstable.
4.48. 2.5 m, 6.46 MN-m.
4.49. $r_0/h_0\sqrt{2}\, s(1 - s)$.
2.50. 10.9 m/s^2.
2.51. 105 kN/m^2, 38.9 m/s^2.
2.52. 2.36 m/s, 13.3°.
2.53. 31.54. r.p.m.
2.54. 298.5 N.
2.55. 73.9 kN, (left to right).
2.56. 2.312 m, 1.088 m, 52.4 kN, 11.61 KN.
2.57. 10.273 m of water, 67.1 kN.
2.58. 30.7°, 19.4° (with horizontal).
2.59. 334 r.p.m.
2.60. 35.1 cm.
2.61. 7.69 kN/m^2, 7.85 m/s^2.
2.62. 3.565 m/s^2, 64.2 km/hr
2.63. $42.3\sqrt{h}/r$ r.p.m.
2.64. 1.275 m, 0.41 m^3.
2.65. 0.747 m.
2.69. 498 N.
2.70. 20.4 seconds.
2.71. 17.77 kN.
2.72. 3.66 cm.
2.73. 3.16 m^3
2.74. Paraboloid of revolution ; hydrostatic variation of pressure, Equation 2.48.

3

Kinematics of Fluid Motion

3.0. Introduction

The kinematics of fluid motion deals with space-time relationships (velocity and acceleration) without taking into account the forces associated with them. The fluid motion can be described completely by an expression describing the location of a fluid particle in space at different times, thus enabling determination of the magnitude and direction of velocity and acceleration in the flow field at any instant of time.

3.1. Methods of Describing Fluid Motion

Every particle of fluid in motion has an instantaneous value of velocity, pressure and density. As a fluid particle moves about its velocity, pressure and density may change with space and time. These flow characteristics are measured with respect to some co-ordinate system, fixed or moving. We may choose a co-ordinate system fixed in space and study the motion of fluid particles, for example—an observer sitting on the bank of a river and studying the movement of fluid particles or a floating body. This approach of flow-study is called *Eulerian method.* Alternatively, we might choose a co-ordinate system attached to a particular part of a moving object, and moving along with it. An observer sitting in a moving boat and studying the flow exemplifies a case in which the co-ordinates axes are moving. This way of studying the flow is known as *Lagrangian method.*

If the position (space co-ordinates x, y, z) of each and every fluid particle is known for every instant of time, we are in a position to obtain a complete picture of fluid-flow. The velocity and acceleration of a fluid particle can be found from changes in its position as the time passes. Once the velocity and acceleration are known, other flow parameters may then be determined from well established relationships connecting the unknown parameter with those already known.

For simplicity, we shall adopt the first system of co-ordinates (Eulerian method), and describe the flow characteristics with reference to a stationary co-ordinate system, thus

$$\left.\begin{aligned} v &= f(x, y, z, t) \\ p &= f(x, y, z, t) \\ \rho &= f(x, y, z, t) \end{aligned}\right\} \qquad \text{...(3)}$$

Eq. (3.1) states that the velocity v, pressure p and the density ρ are, in general, functions of space co-ordinates x, y, z, and time t.

3.2. Classification of Flow

The flow may be described as steady or unsteady, uniform or non-uniform, depending on the variation of flow and fluid characteristics with time and space. It may also be classified as one, two and three dimensional and laminar or turbulent flow

3.2.1. Steady and Unsteady Flow

If the fluid and flow characteristics such as density, velocity, pressure, acceleration etc., at a point do not change with time, the flow is said to be *steady,* thus for steady flow

$$\left(\frac{\partial \rho}{\partial t}\right)_{x_0, y_0, z_0} = 0, \left(\frac{\partial v}{\partial t}\right)_{x_0, y_0, z_0} = 0,$$

$$\left(\frac{\partial p}{\partial t}\right)_{x_0, y_0, z_0} = 0, \text{ and so on.}$$

In unsteady flow one or more of the above variables at a point may change with time. Mathematically speaking the above stated differential quantities will be non-zero. Thus, for unsteady flow

$$\left(\frac{\partial \rho}{\partial t}\right)_{x_0, y_0, z_0} \neq 0, \left(\frac{\partial v}{\partial t}\right)_{x_0, y_0, z_0} \neq 0,$$

$$\left(\frac{\partial p}{\partial t}\right)_{x_0, y_0, z_0} \neq 0$$

and so on. Here (x_0, y_0, z_0) refers to a fixed point in the fluid, where these variables are being measured with respect to time.

3.2.2. Uniform and Non-Uniform Flow

If the velocity vector at all points in the flow is same (in magnitude as well as in direction) at any instant of time, the flow is known as *uniform*. Thus, for uniform flow

$$\left(\frac{\partial v}{\partial s}\right)_{t_0} = 0$$

in which ∂v is the change in velocity corresponding to a displacement ∂s in any direction s. Conversely, the flow will be non-uniform if the velocity vector varies from point to point at any instant of time, thus

$$\left(\frac{\partial v}{\partial s}\right)_{t_0} \neq 0.$$

In real fluid flow, the velocity at a solid boundary is zero (unless the boundary itself in moving) and varies across a flow section. For such cases, the flow will be deemed uniform if the average velocity of flow does not vary from section to section. If the flow passage is straight and prismatic, the flow will be uniform (in uniform flow, the streamlines are straight and parallel).

3.2.3. Three, Two and One-Dimensional Flow

In general, the fluid flow is *three-dimensional* in the sense that the flow characteristics, such as the velocity, pressure, density etc. may vary in all the three mutually perpendicular directions. The velocity components, u along x-axis, v along y-axis and w along the z-axis in a three-dimensional flow are functions of the three space co-ordinates x, y, z and if the flow is unsteady, they are also dependent upon the time t. Thus for a three-dimensional unsteady flow

$$\left.\begin{aligned} u &= f_1(x, y, z, t) \\ v &= f_2(x, y, z, t) \\ w &= f_3(x, y, z, t) \end{aligned}\right\} \quad ...(3.1)$$

If the flow is steady, the flow parameters (velocity, pressure etc.) will be independent of time and the velocity components of Eq. (3.1) will be functions of the space co-ordinates only.

The flow is defined as *two-dimensional* when the flow parameters are functions of time and two rectangular space co-ordinates, say x and y. The variation of flow parameters in the third direction (*i.e.* z-direction) is assumed to be negligibly small and, therefore, the flow patterns, represented by streamlines at any instant, will be same in all planes parallel to x-y planes (*i.e.* planes perpendicular to the z-axis). The two dimensional unsteady flow may be represented by the following equations :

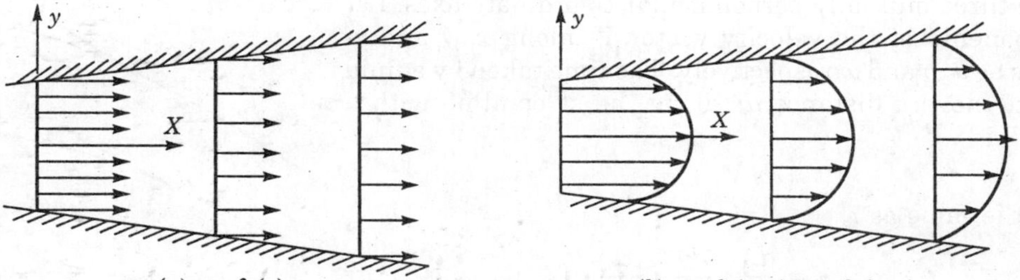

(a) $u = f_1(x)$
One dimensional steady flow

(b) $u = f_2(x, y), v = f_3(x, y)$
Two dimensional steady flow

Fig. 3.1. Velocity distributions in one and two-dimensional flow.

$$u = f_1(x, y, t), \qquad v = f_2(x, y, t), \qquad w = 0 \qquad \text{...(3.2)}$$

The *one-dimensional* flow may likewise be defined as that in which the flow parameters are functions of time and one space co-ordinate only, say x. This space co-ordinate is usually the distance measured along the flow direction. The variation of flow parameters in other two mutually perpendicular directions is assumed insignificant and hence negligible. It may be expressed by

$$u = f_1(x, t), \qquad v = 0, \qquad w = 0 \qquad \text{...(3.3)}$$

For example, the flow through a straight uniform diameter pipe is generally considered one-dimensional. The velocity, pressure etc., may vary along the pipe length, and any variation of these over the cross-section are assumed negligible. As a matter of fact, the flow is never truly one-dimensional because the viscosity causes the fluid velocity to decrease to zero at the boundaries. If, however, the non-uniformity of velocity is not appreciable, one-dimensional approach of flow analysis may be used considering average values of the flow parameters at any given section. The flow through a stream tube of small cross-section represents an example where flow is one-dimensional. Fig. 3.1 illustrates the velocity distributions for one and two-dimensional flow. It is to be noted that in one-dimensional approach of flow analysis, it is the average velocity across a section which is considered.

3.3. Laminar and Turbulent Flow

In laminar flow, the fluid particles move along smooth, regular paths which can be predicted well in advance. The particles thus move in laminas or layers gliding smoothy over the adjacent layers. The flow through a small diameter tube at low velocity is generally *laminar*. The streamlines are straight and, therefore, the laminar flow is also called as streamline flow. The *turbulent flow* is characterised by random and erratic movement of fluid particles resulting in the formation of eddies. This eddy formation is responsible for high energy loss as compared to loss in laminar flow. The additional loss in turbulent flow is attributed to the existence of turbulent shear stress, which is many times larger than the laminar shear stress given by Newton's law of viscosity.

The definition of steady flow needs a modification while dealing with turbulent flow. In turbulent flow, fluid particles move in an erratic and haphazard manner, causing fluctuations in the velocity, pressure etc., at a point. If the time average velocity (also known as temporal mean velocity) at a point does not change with time, the flow is termed as *steady*.

3.4. Streamline, Pathline and Streakline

3.4.1. Streamline

A streamline is an imaginary line drawn in a flow field such that a tangent drawn at any point on this line represents the direction of the velocity vector.

From the definition, it follows that there can be no flow across a streamline. Considering a particle moving along a streamline for a very short distance ds having its components dx, dy and dz

along the three mutually perpendicular co-ordinate axes. Let the components of the velocity vector V_s along x, y and z-directions be u, v and w respectively. The time taken by a fluid particle to move a distance ds along the streamline with a velocity V_s is

$$t = ds/V_s$$

which is the same as

$$t = \frac{dx}{u} = \frac{dy}{v} = \frac{dz}{w}$$

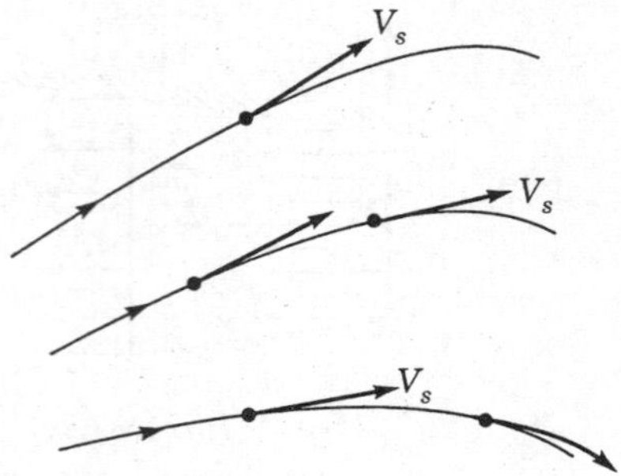

Fig. 3.2. Streamlines.

Hence the differential equation of the streamline may be written as :

$$\frac{dx}{u} = \frac{dy}{v} = \frac{dz}{w} \quad ...(3.4)$$

3.4.2. Pathline

A pathline is the locus of a fluid particle as it moves along. In other words, a pathline is a curve traced by a single fluid particle during its motion. Fig. 3.3 shows a streamline at time t_1 indicating the velocity vectors for particles A and B. At times t_2 and t_3, the particle A is shown to occupy the successive positions. The line connecting these various positions of A represents its *pathline*.

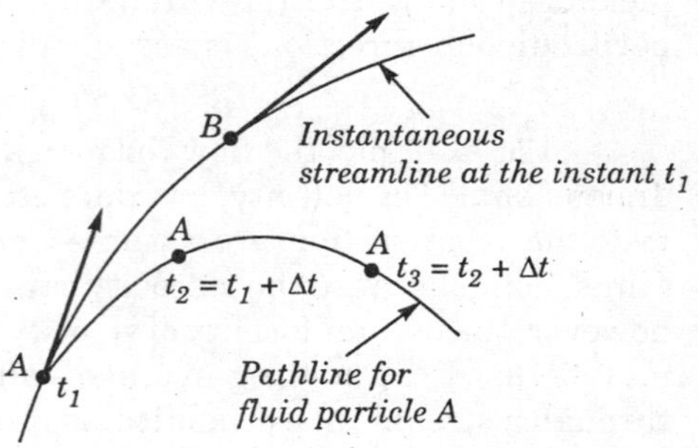

Fig. 3.3. Steamline and pathline for unsteady non-uniform flow.

3.4.3. Streakline

When a dye is injected in a liquid or smoke in a gas, so as to trace the subsequent motion of fluid particles passing a fixed point, the path followed by the dye or smoke is called the *streakline*. Thus, a streakline connects all particles passing through a given point.

In steady flow, the streamlines remain fixed with respect to the co-ordinate axes. Streamlines in steady flow also represent the pathlines and streaklines. In unsteady flow, a fluid particle will not, in general, remain on the same streamline (except for unsteady uniform flow). Hence streamlines and pathlines do not coincide in unsteady non-uniform flow, refer Fig. 3.3.

3.4.4. Instantaneous Streamline

In fluid motion which is independent of time, the position of streamline is fixed in space and a fluid particle following a streamline will continue to do so. In case of time-dependent flow, a fluid particle follows a streamline for only a short interval of time, before changing over to another streamline. The streamlines in such cases are not fixed in space, but change with time. The position of a streamline at a given instant of time is known as *instantaneous streamline*. For different instants of time, we shall have different instantaneous streamlines in the same space (See Fig. 3.4). The streamline, pathline and the streakline are one and the same, if the flow is steady.

3.4.5. Streamtube

If streamlines are drawn through a closed curve, they form a boundary surface across which fluid cannot penetrate. Such a surface bounded by streamlines is a sort of tube, and is known as a *streamtube*.

From the definition of streamline, it is evident that no fluid can cross the bounding surface of the streamtube. This implies that the quantity (mass) of fluid entering the streamtube at one end must be the same as the quantity leaving it at the other. The streamtube is generally assumed to be of a small cross-sectional area so that the velocity over it could be considered uniform.

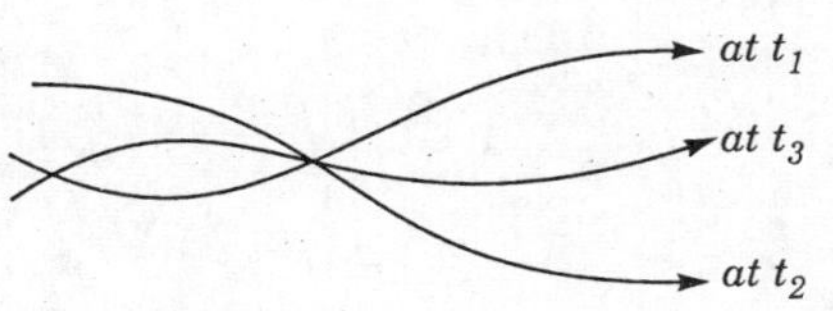

Fig. 3.4. Instantaneous streamline.

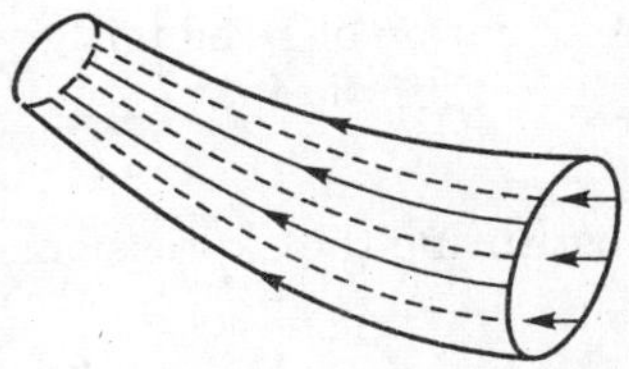

Fig. 3.5. Streamtube.

The concept of streamtube can be extended, and the entire flow region may be considered composed of innumerable streamtubes of small cross-section. The streamtubes may be of any shape regular or irregular. The solid boundaries of flow represent the surface containing the streamlines.

Example 3.1. *The velocity distribution for a three-dimensional flow is given by* $u = -x$, $v = 2y$ *and* $w = 3 - z$. *Find the equation of the streamline passing through (1, 1, 2).*

Solution. The differential equation of streamline is

$$\frac{dx}{u} = \frac{dy}{v} = \frac{dz}{w}$$

Substituting for u, v and w, we obtain

$$\frac{dx}{-x} = \frac{dy}{2y} = \frac{dz}{3-z}$$

The above expression represents two independent equations. Considering each separately, one gets

$$\frac{dx}{-x} = \frac{dy}{2y}$$

Integrating, $$-\log x = \frac{1}{2}\log y + A$$

Since the streamline passes through $x = 1, y = 1$, $\therefore$ $A = 0$ and we obtain :

$$(x)^{-1} = \sqrt{y} \qquad \text{...(1)}$$

Considering the second differential equation,

$$\frac{dx}{-x} = \frac{dz}{3-z}$$

Integrating, $$-\log(x) = -\log(3-z) + B$$

As the streamline has to pass through $x = 1$ and $z = 2$,

$\therefore$ $B = 0$, and we obtain

$$(x)^{-1} = (3-z)^{-1}$$

or $$x = (3-z) \qquad \text{...(2)}$$

Combining (1) and (2), we get

$$x = \frac{1}{\sqrt{y}} = (3-z).$$

Example 3.2. *The velocity vector in a flow field is expressed by*

$$\overline{V} = -xi + 2yj + (5-z)k$$

Derive the equation of streamline passing through point (2, 1, 1).

Solution. The velocity components u, v and w are :

$$u = -x, v = 2y, \text{ and } w = 5 - z$$

The equation of stream is given by Eq. (3.4), and is

$$\frac{dx}{u} = \frac{dy}{v} = \frac{dz}{w}$$

Substituting the expressions for u, v and w,

$$\frac{dx}{-x} = \frac{dy}{2y} = \frac{dz}{5-z}$$

From the above equation, one can write

$$\frac{dx}{-x} = \frac{1}{2}\frac{dy}{y}$$

Integrating, we obtain $\qquad -\log x = \frac{1}{2}\log y + \log c_1 \qquad$ ...(1)

Substituting $x = 2, y = 1$, c_1 can be evaluated :

$$-\log 2 = \frac{1}{2}\log 1 + \log c_1$$

or $\qquad \log (2)^{-1} = \log c_1 \qquad$ or $\qquad \log \frac{1}{2} = \log c_1$

substituting for $\log c_1$ in Eq. (1) above,

$$-\log x = \frac{1}{2}\log y + \log\frac{1}{2} = \log\sqrt{y} + \log\frac{1}{2} = \log\frac{\sqrt{y}}{2}$$

from which $\qquad \frac{1}{x} = \frac{\sqrt{y}}{2} \qquad$ ...(2)

Now considering, $\qquad -\frac{dx}{x} = \frac{dz}{5-z}$

Integrating, $\qquad -\log x = -\log(5-z) + \log c_2 \qquad$ ...(3)

For the streamline passing through $x = 2, z = 1$

$$-\log 2 = -\log(5-1) + \log c_2$$

or $\qquad \log c_2 = \log 4 - \log 2 = \log 2$

substituting $\log c_2$ in Eq. (3),

$$-\log x = -\log(5-z) + \log 2 = \log\frac{2}{5-z}$$

or $\qquad \frac{1}{x} = \frac{2}{5-z} \qquad$ ...(4)

The equation of streamline is ; $\qquad \frac{1}{x} = \frac{\sqrt{y}}{2} = \frac{2}{5-z}$.

Example 3.3. *The velocity distribution for a three-dimensional flow is given by :*

$$\overline{V} = axi + ayj - 2azk$$

Find the equation of streamline passing through the position vetocr $\overline{r} = 2i + 2j + 4k$.

Solution. The velocity components u, v and w can be written as :

$$u = ax,\ v = ay, \quad \text{and} \quad w = -2az$$

The streamline has to pass through the point (2, 2, 4).

The equation of streamline is obtained using Eq. (3.4) ; which is :

$$\frac{dx}{u} = \frac{dy}{v} = \frac{dz}{w}$$

or
$$\frac{dx}{ax} = \frac{dy}{ay} = \frac{dz}{-2az} \quad \text{or} \quad \frac{dx}{x} = \frac{dy}{y} = \frac{dz}{-2z} \qquad ...(1)$$

Integrating the first two terms leads to

$$\log x = \log y + \log A, \text{ where } A \text{ is a constant}$$

Since the streamline has to pass through $x = 2, y = 2$, we obtain,

$$\log A = \log 2 - \log 2 = 0,$$

Substituting this value of $\log A$,

$$\log x = \log y, \text{ which gives } x = y \qquad ...(2)$$

Considering the last two terms of eq. (1)

$$\frac{dy}{y} = \frac{dz}{-2z}$$

Integrating, we obtain : $\log y = -\frac{1}{2} \log z + \log B$

To get value of $\log B$, substitute $y = 2$ and $z = 4$

$$\log 2 + \frac{1}{2} \log 4 = \log B = \log 2 + \log (4)^{1/2}$$

or
$$\log B = \log 2 + \log 2 = 2 \log 2 = \log (2)^2 = \log 4$$

substituting for $\log B$,

$$\log y = -\frac{1}{2} \log z + \log 4 = -\log z^{1/2} + \log 4$$

$$= \log \frac{4}{\sqrt{z}}$$

Taking antilog,

$$y = \frac{4}{\sqrt{z}} \qquad ...(3)$$

Combining (2) and (3), we obtain the equation of streamline as :

$$\boldsymbol{x = y = \frac{4}{\sqrt{z}}}.$$

3.5. Equations for Acceleration

In the following analysis, expressions for acceleration in one, two and three-dimensional flow have been derived.

3.5.1. Acceleration in One-Dimensional Flow

In case of one-dimensional flow along a streamline, consider a fluid particle undergoing a small displacement ds in a short interval of time dt. The velocity V of this particle over the distance ds is

$$V = V_s = \frac{ds}{dt} \qquad ...(3.5)$$

in which V_s is the velocity tangential to the streamline. The velocity vector is tangential to the streamlines and, therefore, V and V_s are identical in magnitude.

Consider steady flow along a streamline shown in Fig. 3.6. The particle is assumed to be initially at the point P and moves a distance ds in time dt to anther point Q. The acceleration of the particle will be the change in its velocity at Q with respect to its velocity at P, divided by the time interval dt. In other words, the acceleration is the time rate of change of velocity,

$$a = \frac{dV}{dt} \quad ...(3.6)$$

Let the co-ordinate axes be along the tangential and normal directions to the streamline. The acceleration will then have components in the tangential (a_s) and normal directions (a_n), the normal acceleration will be directed towards the local centre of curvature O of the streamline.

Let V represent the initial velocity of the particle at P, which is tangential to the streamline at that point. When the particle reaches Q, the velocity is tangential to the streamline at that point, its direction having changed through an infinitesimal angle $d\theta = \frac{ds}{r}$. Since $d\theta$ is very small, $\cos(d\theta) \approx 1$ and $\sin(d\theta) \approx d\theta$.

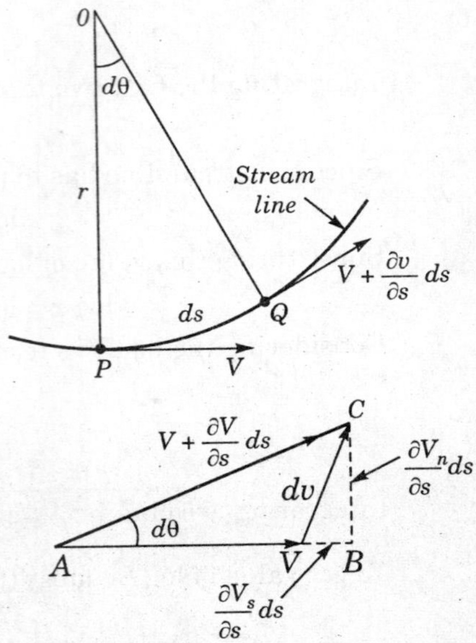

Fig. 3.6. Velocity variation along a streamline.

The magnitude of the velocity along the streamline has changed from V at P to $V + \frac{\partial V}{\partial s} ds$ at Q. The component of velocity at Q in the direction of tangent at P is

$$V_s = \left(V + \frac{\partial V}{\partial s}.ds\right)\cos(d\theta) = V + \frac{\partial V_s}{\partial s}.ds$$

using Eq. (3.6), the tangential acceleration

$$a_s = \frac{dV_s}{dt} = \frac{V + \frac{\partial V_s}{\partial s}ds - V}{dt} = \frac{\partial V_s}{\partial s}.\frac{ds}{dt}$$

making use of Eq. (3.5) and noting that

$$\frac{\partial V}{\partial s} = \frac{dV_s}{ds}$$

we obtain

$$a_s = V\frac{\partial V}{\partial s} \quad ...(3.7)$$

The velocity normal to the streamline at P is zero. At Q, the velocity in the normal direction to the streamline at P will be (refer Fig. 3.6)

$$V_n = \left(V + \frac{\partial V}{\partial s}ds\right)\sin(d\theta) = \left(V + \frac{\partial V}{\partial s}ds\right)\frac{ds}{r}$$

$$= V\frac{ds}{r} + \frac{1}{r}\frac{\partial V}{\partial s}(ds)^2 = V\frac{ds}{r}$$

since ds is very small, $(ds)^2$ is negligible. The acceleration in the normal direction

$$a_n = \frac{dV_n}{dt} = \frac{V\frac{ds}{r} - 0}{dt} = \frac{V}{r}.\frac{ds}{dt} = \frac{V^2}{r} \quad ...(3.8)$$

If the flow is unsteady, the total acceleration will be due to (*i*) change in velocity with space and (*ii*) change in velocity with time. The acceleration components in unsteady flow will, therefore, include a term which will account for the unsteadiness of flow, and Eqs. (3.7) and (3.8) will be correspondingly modified as below.

For tangential acceleration,

$$a_s = \frac{\partial V}{\partial t} + V\frac{\partial V}{\partial s} \qquad ...(3.9)$$

and for normal acceleration,

$$a_n = \frac{\partial V_n}{\partial t} + \frac{V^2}{r} \qquad ...(3.10)$$

The partial derivatives $\frac{\partial V}{\partial t}$ and $\frac{\partial V_n}{\partial t}$ express variation of tangential and normal velocities with time, and are known as *local accelerations* in the respective directions. The other partial derivative, $V\frac{\partial V}{\partial s}$ takes into account the change in velocity with space, and is termed as *convective acceleration*. Thus each component of total acceleration consists of two parts ; the first one is due to unsteadiness of flow and is known as local acceleration, and the second part is due to non-uniformity of flow and is called convective acceleration.

If the streamline spacing is constant, representing uniform velocity, the change in velocity with space $\partial V/\partial s$ is zero, and consequently the convective acceleration in turn will be zero. If the flow is steady, the local acceleration will also be zero. Thus in uniform steady flow, the tangential acceleration a_s is zero, and if the streamlines are straight ($r = \infty$) the normal acceleration will also be zero. In other words, when the streamlines are straight and parallel, and the flow in steady, both the tangential and normal accelerations are zero.

Example 3.4. *Along a straight streamline passing through the origin the velocity is given by* $V = 4\sqrt{(x^2 + y^2)}$, *determine the velocity and acceleration at the point (4, 3).*

Solution. Velocity $= 4\sqrt{(x^2 + y^2)} = 4s$

where $s = \sqrt{(x^2 + y^2)}$, and is the distance along the streamline.

Velocity at (4, 3)

$$= 4\sqrt{(4^2 + 3^2)} = 4 \times 5 = \mathbf{20\ m/s.}$$

Acceleration along the streamline

$$a_s = V\frac{dV}{ds}$$

$$= 4s(4) = 16s = 16 \times 5 = \mathbf{80\ m/s^2.}$$

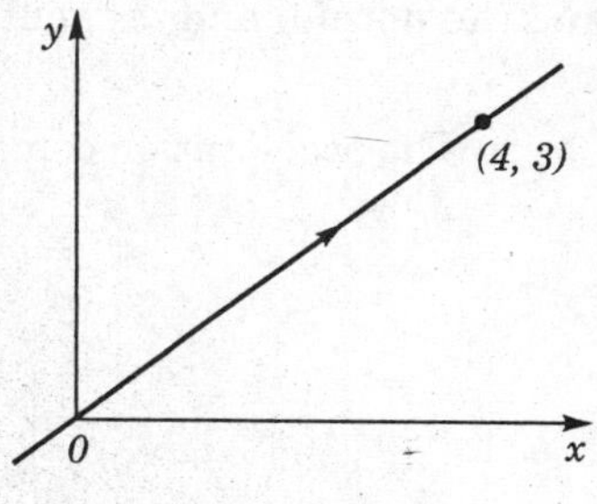

Fig. Ex. 3.2

Example 3.5. *The velocity along a circular streamline having its centre at the origin and radius of 1.5 m is constant at 2 m/s. Determine the radial and tangential components of acceleration at any point on the streamline. Calculate the horizontal, vertical, tangential and normal acceleration at (1.5, 30°) expressing these quantities in cartesian and polar co-ordinates.*

Solution. Since the velocity is constant, the tangential component of acceleration, $a_\theta = 0$, and since the radius of the streamline is 1.5 m and the velocity along it is 2 m/s, the radial acceleration

$$a_r = -\frac{v_\theta^2}{r} = -\frac{(2)^2}{1.5} = -2.67 \text{ m/s}^2$$

which is directed towards the centre of the circle.

The acceleration and velocity triangles (*PQR* and *PST* respectively) being similar, the following relationships can be written :

$$\sin \theta = \frac{y}{\sqrt{x^2 + y^2}} = -\frac{u}{v_\theta}$$

from which, $$u = -\frac{v_\theta\, y}{\sqrt{x^2 + y^2}} = -\frac{2y}{\sqrt{x^2 + y^2}}$$

The negative sign appears because of the negative direction of u-component.

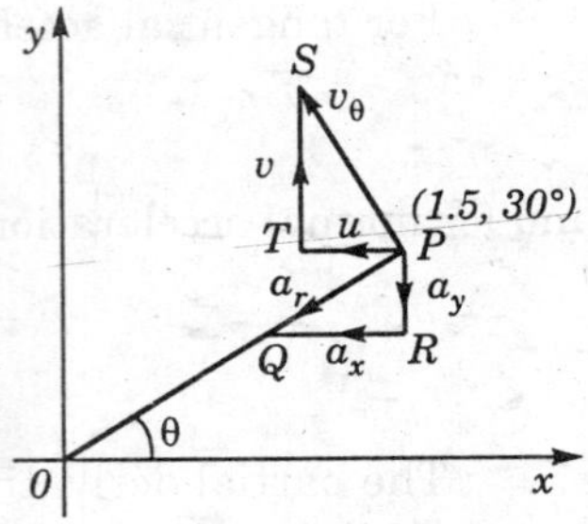

Fig. Ex. 3.3

$$\cos \theta = \frac{x}{\sqrt{x^2 + y^2}} = \frac{v}{v_\theta},$$

$$\therefore \qquad v = \frac{v_\theta\, x}{\sqrt{x^2 + y^2}} = \frac{2x}{\sqrt{x^2 + y^2}}$$

The given point P is defined by the polar co-ordinates (1.5, 30°), hence $r = 1.5$ m, and $\theta = 30°$. Also,

$$x = r \cos \theta = 1.5 \cos 30° = 1.299 \text{ m}$$

$$y = r \sin \theta = 1.5 \sin 30° = 0.75 \text{ m}$$

and $$r^2 = x^2 + y^2.$$

The horizontal and vertical components of velocity are :

$$u = -\frac{2y}{r} = -\frac{2 \times 0.75}{1.5} = \mathbf{-1.0 \text{ m/s}}$$

$$v = \frac{2x}{r} = \frac{2 \times 1.299}{1.5} = \mathbf{1.732 \text{ m/s}}$$

and the normal and tangential components are :

$$v_\theta = \mathbf{2 \text{ m/s}} \text{ and } v_r = \mathbf{0 \text{ m/s}}.$$

The acceleration components are given by Eq. (3.13) which for steady flow may be written as

$$a_x = u\,\frac{\partial u}{\partial x} + v\,\frac{\partial u}{\partial y}$$

$$= \left(-\frac{2y}{\sqrt{x^2 + y^2}}\right)\left[\frac{2xy}{(x^2 + y^2)^{3/2}}\right] + \left(\frac{2x}{\sqrt{x^2 + y^2}}\right)\left[\frac{2y^2}{(x^2 + y^2)^{3/2}} - \frac{2}{\sqrt{x^2 + y^2}}\right]$$

$$= -\frac{4x}{x^2 + y^2} = -\frac{4x}{(1.5)^2} = -1.78x$$

$$a_y = u\,\frac{\partial v}{\partial x} + v\,\frac{\partial v}{\partial y}$$

$$= \left(-\frac{2y}{\sqrt{x^2 + y^2}}\right)\frac{\partial}{\partial x}\left(\frac{2x}{\sqrt{x^2 + y^2}}\right) + \left(\frac{2x}{\sqrt{x^2 + y^2}}\right)\frac{\partial}{\partial y}\left(\frac{2x}{\sqrt{x^2 + y^2}}\right)$$

$$= -\frac{2y}{\sqrt{x^2 + y^2}}\left[\frac{2}{\sqrt{x^2 + y^2}} - \frac{2x^2}{(x^2 + y^2)^{3/2}}\right] + \frac{2x}{\sqrt{x^2 + y^2}}\left[-\frac{2xy}{(x^2 + y^2)^{3/2}}\right]$$

$$= -\frac{4y}{x^2 + y^2} = -\frac{4y}{(1.5)^2} = -1.78y$$

Substituting $x = 1.299$ m, $y = 0.75$ m

$$a_x = -1.78x = -1.78 \times 1.299 = -\mathbf{2.312\ m/s^2}$$

$$a_y = -1.78y = -1.78 \times 0.75 = -\mathbf{1.335\ m/s^2}.$$

Using equations for acceleration in polar co-ordinates, Eqs. (3.13*a*) and (3.13*b*) respectively :

$$a_r = v_r \frac{\partial v_r}{\partial r} + v_\theta \frac{\partial v_r}{r\partial\theta} - \frac{{v_\theta}^2}{r}$$

$$= 0 \frac{\partial}{\partial r}(0) + 2 \frac{\partial}{r\partial\theta}(0) - \frac{2\times 2}{1.5} - \mathbf{2.667\ m/s^2}$$

$$a_\theta = v_r \frac{\partial v_\theta}{\partial r} + v_\theta \frac{\partial v_\theta}{r\partial\theta} + \frac{v_r v_\theta}{r}$$

$$= 0 \frac{\partial}{\partial r}(2) + 2 \frac{\partial}{r\partial\theta}(2) + \frac{0\times 2}{1.50} = \mathbf{0\ m/s^2}.$$

Example 3.6. *A two-dimensional flow-field is described by*

$$v_\theta = r\omega \text{ and } v_r = 0$$

in which ω represents angular velocity which is a constant. Show that the flow is physically possible and sketch the flow-field (streamline pattern).

Solution. Since the radial component of velocity is everywhere zero, the streamlines for this flow must be concentric circles. This type of flow is called vortex flow. Substituting for v_r and v_θ in the continuity equation, Eq. (3.29), one obtains :

$$\frac{0}{r} + \frac{\partial}{\partial r}(0) + \frac{\partial}{\partial r\theta}(r\omega) = 0 + 0 + 0 = 0.$$

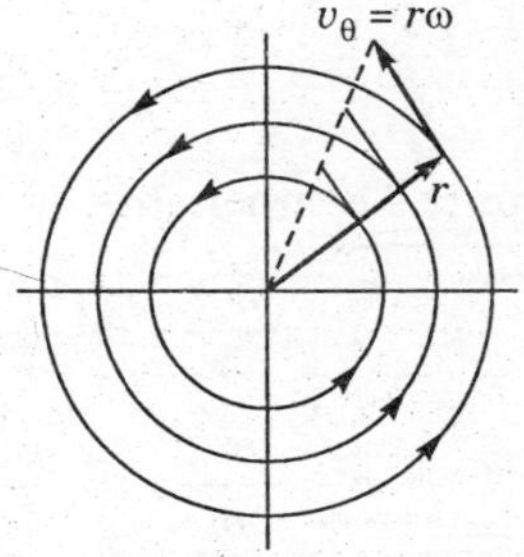

Fig. Ex. 3.4. Streamline Pattern.

It is thus seen that the flow-field satisfies the continuity equation and hence the flow is physically possible. The streamline pattern is shown in the Fig. Ex. 3.4.

3.5.2. Acceleration in Two-Dimensional Flow

Consider a two-dimensional flow, in which the flow pattern is similar in planes parallel to the *xy*-plane. Let the velocity of a fluid particle at *P*(*x*, *y*) be *V*, having components *u* and *v* in *x* and *y*-directions respectively. Let the particle have an acceleration *a* having *x* and *y* components as a_x and a_y respectively (Fig. 3.7 *a*). The velocity components in two-dimensional flow may be written as

$$u = f_1(x, y, t)$$

and $$v = f_2(x, y, t)$$

the velocities *u* and *v* may by written in terms of displacement and time as

$$u = \frac{dx}{dt}$$

and $$v = \frac{dv}{dt} \quad ...(3.11)$$

and the accelerations $$a_x = \frac{du}{dt}$$

and $$a_y = \frac{dv}{dt} \quad ...(3.12)$$

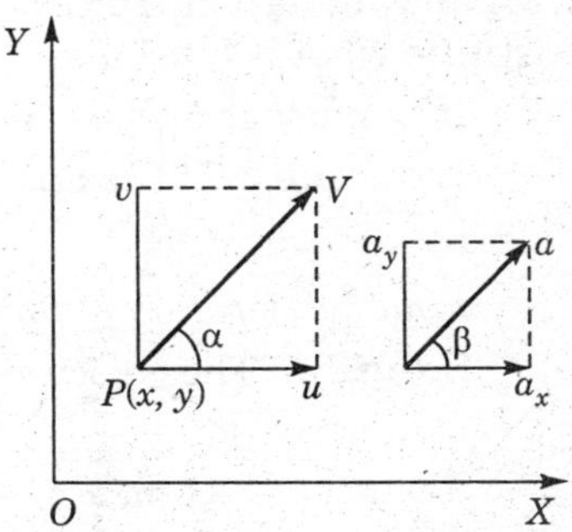

Fig. 3.7(*a*). Acceleration in two-dimensional flow.

the total derivative *du* may be written as

$$du = \frac{\partial u}{\partial t} . dt + \frac{\partial u}{\partial x} dx + \frac{\partial u}{\partial y} . dy$$

and, therefore, the acceleration a_x is

$$a_x = \frac{du}{dt} = \frac{\partial u}{\partial t} + \frac{\partial u}{\partial x} \cdot \frac{dx}{dt} + \frac{\partial u}{\partial y} \cdot \frac{\partial y}{\partial t}$$

Using Eq. (3.1),

$$a_x = \frac{\partial u}{\partial t} + u\frac{\partial u}{\partial x} + v\frac{\partial u}{\partial y}$$

and similarly

$$a_y = \frac{\partial v}{\partial t} + u\frac{\partial v}{\partial x} + v\frac{\partial v}{\partial y} \qquad ...(3.13)$$

Equations in Polar Co-ordinates

Velocity components in radial and tangential directions are :

$$v_r = \frac{dr}{dt}$$

$$v_\theta = r\frac{d\theta}{dt}$$

and the components of acceleration are :

$$a_r = \frac{\partial v_r}{\partial t} + vr\frac{\partial v_r}{\partial r} + \frac{v_\theta}{r}\frac{\partial v_r}{\partial \theta} - \frac{v_\theta^2}{r} \qquad ...(3.13\ a)$$

$$a_\theta = \frac{\partial v_\theta}{\partial t} + vr\frac{\partial v_\theta}{\partial r} + \frac{v_\theta}{r}\frac{\partial v_\theta}{\partial \theta} - \frac{v_r v_\theta}{r} \qquad ...(3.13\ b)$$

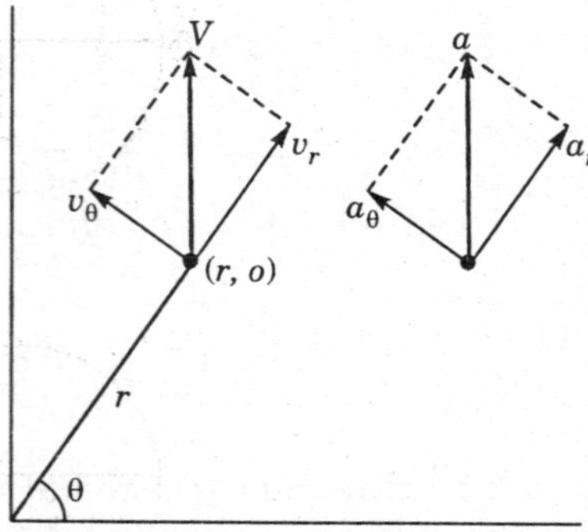

Fig. 3.7 (*b*). Notations for velocity and acceleration components in polar co-ordinates.

3.5.3. Acceleration in Three-dimensional Flow

In case of three-dimensional flow, the expressions for acceleration components a_x, a_y, a_z may be obtained in a similar manner, and expressed as

$$a_x = \frac{\partial u}{\partial t} + u\frac{\partial u}{\partial x} + v\frac{\partial u}{\partial y} + w\frac{\partial u}{\partial x}$$

and

$$a_y = \frac{\partial v}{\partial t} + u\frac{\partial v}{\partial x} + v\frac{\partial v}{\partial y} + w\frac{\partial v}{\partial z} \qquad ...(3.14)$$

$$a_z = \frac{\partial w}{\partial t} + u\frac{\partial w}{\partial x} + v\frac{\partial w}{\partial y} + w\frac{\partial w}{\partial z}$$

here w is the velocity component in the z-direction.

Equations (3.13) and (3.14) are general expressions applicable for all types for flow. If, however, the flow is steady the local acceleration term will become zero. The last two terms of Eq. (3.13) and the last three terms of Eq. (3.14) which take into account the non-uniformity of flow constitute what is known as the convective acceleration.

If the flow is uniform and steady, all the components of acceleration will be zero.

Example 3.7. *The velocity field in a fluid flow is given by*

$$\overline{V} = 10\,x^2yi + 15xyj + (25t - 3xy)\,k$$

Find acceleration of flow at (1, 2, – 1) m at time it = 0.5 sec.

Solution. The velocity components

$$u = 10\,x^2y,\ v = 15xy,\ w = 25t - 3xy$$

At the point (1 m, 2 m, – 1 m) and t = 0.5 sec, the velocity components are :

$$u = 10\,(1)^2 \times 2 = 20 \text{ m/s} ;\ v = 15 \times 1 \times 2 = 30 \text{ m/s}$$

$$w = 25 \times 0.5 - 3 \times 1 \times 2 = 6.5 \text{ m/s}$$

The acceleration components are given by Eq (3.14)

$$a_x = \frac{\partial u}{\partial t} + u\frac{\partial u}{\partial x} + v\frac{\partial u}{\partial y} + w\frac{\partial u}{\partial z}$$
$$= 0 + 20\,(20\,xy) + 30(10x^2) + 6.5\,(0)$$
$$= 20\,(20 \times 1 \times 2) + 30(10 \times 1^2) = 1100 \text{ m/s}^2$$

$$a_y = \frac{\partial v}{\partial t} + u\frac{\partial v}{\partial x} + v\frac{\partial v}{\partial y} + w\frac{\partial v}{\partial z}$$
$$= 0 + 20(15 \times 2) + 30(15 \times 1) + 6.5 \times 0$$
$$= 1050 \text{ m/s}^2$$

$$a_z = \frac{\partial w}{\partial t} + u\frac{\partial w}{\partial x} + v\frac{\partial w}{\partial y} + w\frac{\partial w}{\partial z}$$
$$= 25 + 20(-3y) + 30(-3x) + 6.5 \times 0$$
$$= 25 + 20(-3 \times 2) + 30\,(-3 \times 1) + 0 = -185 \text{ m/s}^2$$

The acceleration at (1, 2, – 1) is given by

$$a = \sqrt{a_x{}^2 + a_y{}^2 + a_z{}^2}$$
$$= \sqrt{(1100)^2 + (1050)^2 + (-185)^2} = 1531.9 \text{ m/s}^2.$$

Example 3.8. *The velocity vector in a fluid flow is given by :*

$$V = 2x^2 i + 4tk - 5x^2 yj$$

Find the velocity and acceleration of a fluid at (0, 2, 3) at time t = 1.

Solution. From the given velocity vector, the components are :

$$u = 2x^2,\ v = -5x^2y, \text{ and } w = 4t$$

The acceleration components are given be Eq. (3.14) as

$$a_x = \frac{\partial u}{\partial t} + u\frac{\partial u}{\partial x} + v\frac{\partial u}{\partial y} + w\frac{\partial u}{\partial z}$$

$$a_y = \frac{\partial v}{\partial t} + u\frac{\partial v}{\partial x} + v\frac{\partial v}{\partial y} + w\frac{\partial v}{\partial z}, \text{ and}$$

$$a_z = \frac{\partial w}{\partial t} + u\frac{\partial w}{\partial x} + v\frac{\partial w}{\partial y} + w\frac{\partial w}{\partial z}$$

The velocity and acceleration are given by

$$V = \sqrt{u^2 + v^2 + w^2} \quad \text{and} \quad a = \sqrt{a_x^2 + q_y^2 + a_z^2}$$

The derivatives of velocity components at point (0, 2, 3) and $t = 1$ are

$$\frac{\partial u}{\partial t} = 0,\ \frac{\partial u}{\partial x} = 4x = 4 \times 0 = 0,\ \frac{\partial u}{\partial y} = 0,\ \frac{\partial u}{\partial z} = 0$$

$$\frac{\partial v}{\partial t} = 0,\ \frac{\partial v}{\partial x} = -10xy = -10 \times 0 \times 2 = 0,\ \frac{\partial v}{\partial y} = -5x^2 = -5 \times (0)^2 = 0,$$

and $$\frac{\partial v}{\partial z} = 0.$$

$$\frac{\partial w}{\partial t} = 4,\ \frac{\partial w}{\partial x} = 0,\ \frac{\partial w}{\partial y} = 0,\ \frac{\partial w}{\partial z} = 0$$

and, $$u = 2x^2 = 2 \times (0)^2 = 0,\ v = -5x^2y = -5(0)^2 \times 2 = 0,\ w = 4 \text{ units}$$

Substituting the above values

$$V = \sqrt{0 + 0 + (4)^4} = \mathbf{4\ units}, \text{ and}$$

$$a = \sqrt{a_x^2 + a_y^2 + a_z^2}\,,\ a_x = 0 + 0 \times 0 + 0 \times 2 + 4 \times 0 = 0$$

$$a_y = 0 + 0 \times 0 + 0 \times 0 + 4 \times 0 = 0$$

$$a_z = 4 + 0 \times 0 + 0 \times 0 + 4 \times 0 = 4 \text{ units.}$$

$$\therefore \qquad a = \sqrt{0^2 + 0^2 + 4^2} = \mathbf{4\ units}$$

Example 3.9. *Water approaches the intake of a pump, the velocity varying inversely as the square of the radial distance from the intake. At a radial distance of 1.50 m, the velocity was found to be 0.58 m/s. What is the acceleration of flow at radial distances of 1.50 m, 1 m and 0.50 m from the intake ? The streamlines are radial.*

Solution. The velocity law is given to be

$$v = \frac{C}{r^2}$$

where r is the radial distance from the intake. Since the streamlines are radial, normal acceleration is zero.

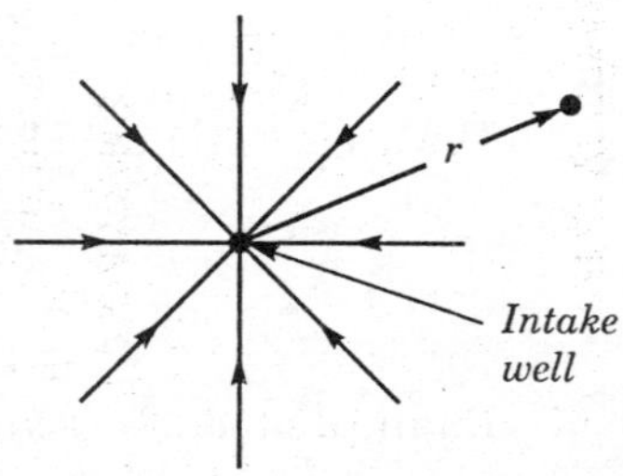

Radial flow towards a well.

$$\text{Acceleration} = \frac{\partial v}{\partial t} + v\frac{\partial v}{\partial s}.$$

Flow is steady since the velocity is dependent only on the radial distance from the intake, therefore, $\frac{\partial v}{\partial t} = 0$ and the acceleration

$$a = v\frac{\partial v}{\partial s}$$

but as $\qquad v = \frac{C}{r^2}$ and $\frac{\partial v}{\partial s} = \frac{\partial v}{\partial r}$

(since r is measured along the streamline)

$$\frac{\partial v}{\partial r} = -\frac{2C}{r^3}$$

$$\therefore \quad \text{Acceleration} \qquad a = v\frac{\partial v}{\partial s} = \frac{C}{r^2}\left(-\frac{2C}{r^2}\right) = -\frac{2C^2}{r^5}$$

At $\qquad r = 1.5$ m, $\qquad v = 0.68$ m/sec.

Substituting these values in the velocity law

$$v = C/r^2$$

$$\therefore \qquad C = vr^2 = 0.68 \times (1.5)^2 = 1.53 \text{ m}^3\text{/sec.}$$

Acceleration at the radial distance r, then becomes

$$a = -\frac{2 \times (1.53)^2}{r^5} = -\frac{4.68}{r^5}$$

(*i*) Acceleration at $r = 1.5$ m $= -\frac{4.68}{(1.5)^5} = -0.616$ m/sec^2.

(*ii*) Acceleration at $r = 1.0 = -4.68$ m/sec^2.

(*iii*) Acceleration at $r = 0.5$ m $= -\frac{4.68}{(0.5)^5} = -149.9$ m/sec^2.

The negative sign indicates the acceleration is directed towards the intake.

Example 3.10. *A pipe 15 cm in diameter carries an unsteady flow of an incompressible fluid. The pipe terminates in a nozzle 7.5 cm in diameter. Determine the velocity and acceleration of the parallel stream leaving the nozzle at the instant when the velocity and acceleration in the pipe are 2 m/s and 0.5 m/s².*

Solution. The acceleration along a streamline may be written as

$$\frac{du}{dt} = \frac{\partial u}{\partial t} + u\frac{\partial u}{\partial s}.$$

Since the streamlines are straight and parallel,

$$\frac{\partial u}{\partial s} = 0$$

and the acceleration $$\frac{\partial u}{\partial t} = \frac{\partial u}{\partial t}.$$

In the pipe $$\frac{\partial u}{\partial t} = 0.5 \text{ m/s}^2.$$

From the continuity equation, refer Equation 3.16

$$Q = A_1V_1 = A_2V_2.$$

∴ Velocity in the jet $$V_2 = \frac{A_1}{A_2}V_1 = \left(\frac{d_1}{d_2}\right)^2 V_1 = \left(\frac{15.0}{7.5}\right)^2 \times 2 = 8 \text{ m/s}.$$

Also from continuity, $A_1dV_1 = A_2dV_2$

where dV_2 and dV_1 are changes in velocities V_2 and V_1 in time dt, thus dividing by time dt.

$$A_1\left(\frac{dV_1}{dt}\right) = A_2\left(\frac{dV_2}{dt}\right)$$

$$\therefore \quad \frac{dV_2}{dt} = \left(\frac{A_1}{A_2}\right)\frac{dV_1}{dt} = \left(\frac{15}{7.5}\right)^2 \times 0.5 = 2 \text{ m/s}^2.$$

∴ Acceleration in the jet at the given instant = 2 m/s².

Example 3.11. *Given the velocity field :*

$$V = (6 + 2xy + t^2)\, i - (xy^2 + 10t)\, j + 25\, k.$$

What is the acceleration of a particle at (3, 0, 2) at time t = 1 ?

Solution. For the given velocity field, the x, y and z velocity components are :

$$u = 6 + 2xy + t^2$$
$$v = -(xy^2 + 10t)$$
$$w = 25.$$

The acceleration components in x, y and z directions are given by

$$a_x = \frac{du}{dt} = \frac{\partial u}{\partial t} + u\frac{\partial u}{\partial x} + v\frac{\partial u}{\partial y} + w\frac{\partial u}{\partial z}$$

$$a_y = \frac{dv}{dt} = \frac{\partial v}{\partial t} + u\frac{\partial v}{\partial x} + v\frac{\partial v}{\partial y} + w\frac{\partial v}{\partial z}$$

$$a_z = \frac{dw}{dt} = \frac{\partial w}{\partial t} + u\frac{\partial w}{\partial x} + v\frac{\partial w}{\partial y} + w\frac{\partial w}{\partial z}$$

since the acceleration is a vector quantity, it is equal to

$$a = \sqrt{a_x^2 + a_y^2 + a_z^2}.$$

From the velocity components,

$$\frac{\partial u}{\partial t} = 2t, \frac{\partial u}{\partial x} = 2y, \frac{\partial u}{\partial y} = 2x, \frac{\partial u}{\partial z} = 0$$

$$\frac{\partial v}{\partial t} = -10, \frac{\partial v}{\partial x} = -y^2, \frac{\partial v}{\partial y} = -2xy, \frac{\partial v}{\partial z} = 0$$

$$\frac{\partial w}{\partial t} = 0, \frac{\partial w}{\partial x} = \frac{\partial w}{\partial y} = \frac{\partial w}{\partial z} = 0.$$

Substituting these values, the acceleration components at (3, 0, 2) at $t = 1$

$$a_x = 2t + (6 + 2xy + t^2) . 2y + [-(xy^2 + 10t) . 2x] + 25 \times 0$$

$$= 2 + (6 + 0 + 1) 2 \times 0 + [(0 + 10) . 2 \times 3] = -58$$

$$a_y = 10 + (6 + 2xy + t^2)(-y^2) + [-(xy^2 + 10t)(-2xy)] + 0 = -10$$

$$a_z = 0$$

$$\therefore \quad a = \sqrt{(-58)^2 + (-10)^2 + 0} = +58.75 \text{ units.}$$

3.6. Continuity Equation

The continuity equation embodies the principle of conservation of mass according to which fluid matter can be neither created nor destroyed.

3.6.1. Continuity equation for one-dimensional flow

A stream tube as shown in Fig. 3.8, may be considered as bounded by the streamlines passing through the closed curves having areas dA_1 and dA_2.

If these areas are sufficiently small, the velocities at their mid-point will very closely represent the average velocities at those sections. As the fluid cannot cross the boundary surface of the stream tube, it follows that the mass rate of steady flow passing any normal section remains constant. If the fluid is incompressible (constant mass density), the volume rate of flow will be constant. The principle of conservation of fluid mass may be expressed mathematically as,

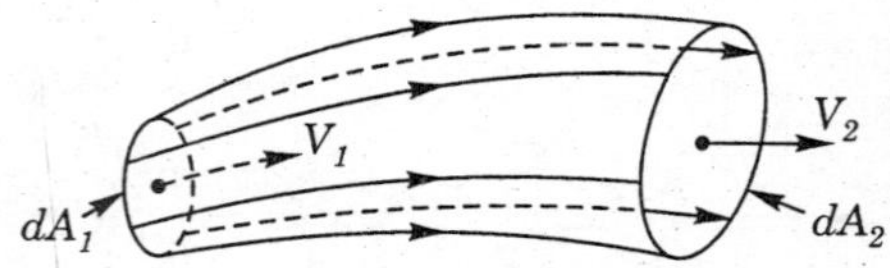

Fig. 3.8. Flow continuity, through stream tube.

$$\rho_1 \, dA_1 \, V_1 = \rho_2 \, dA_2 \, V_2 \qquad ...(3.15)$$

Equation (3.15) states that since no fluid mass can either be created or destroyed inside a stream tube, for steady flow the mass per unit time entering the tube must flow out at the same rate. For incompressible fluids ($\rho_1 = \rho_2$), the volume rate of flow past any normal section must be same, thus, if dQ represents the volume rate of flow, then

$$dQ = V_1.dA = V_2.dA_2.$$

If the normal cross-sectional area is A_1 and the average or mean velocity at this section is V_1, and the similar quantities at the other normal section are A_2 and V_2, then the volume rate of flow (commonly known as the discharge) is

$$Q = A_1V_1 = A_2V_2 \qquad ...(3.16)$$

Equation (3.16) relates the cross-sectional area and the normal mean velocity of flow and shows that if at any section the flow area decreases the velocity must increase accordingly. This equation is known as the *continuity equation*. For compressible fluids, however, the continuity equation will consider the mass flow rate, ρAV, and will be written as :

$$\rho_1A_1V_1 = \rho_2A_2V_2 \qquad ...(3.17)$$

Equation (3.17) expresses the fact that in steady flow the mass flow rate passing all sections of a stream tube is constant.

3.6.2. Continuity Equation for two-dimensional flow

When the velocity vector has its components in only two mutually perpendicular directions, the component in the third direction being everywhere zero, the flow is said to be two-dimensional. Under such circumstances, all streamlines must lie in parallel planes, all such planes having identical flow pattern. The intersection of a stream tube with one such plane will result into streamlines as shown in Fig. 3.9.

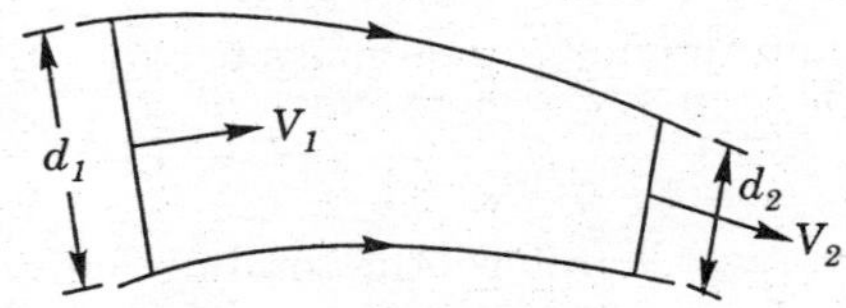

Fig. 3.9. Continuity relations in two-dimensional flow.

For incompressible fluids, from the principle of conservation of mass, the product of the velocity with the normal distance d between these streamlines will be the volume flowing per unit time per unit width, thus

$$dq = V.d \qquad \text{...(3.18)}$$

and the continuity equation for two-dimensional flow is

$$V_1.d_1 = V_2.d_2 \qquad \text{...(3.19)}$$

Total rate of flow Q. If the velocity distribution across a normal section is known, the rate of flow Q passing the section may be obtained by integrating Eq. (3.15), thus

$$Q = \int V.dA \qquad \text{...(3.20)}$$

If the velocity is expressible in the equation form, the integral $\int V.dA$ may be evaluated analytically. However, if the velocity distribution across a section is otherwise known the integration must be performed numerically.

3.6.3. Continuity equation in cartesian co-ordinates

Consider a three-dimensional fluid element (parallelopiped) of size $dx\ dy\ dz$, its centre O having co-ordinates x, y, z and velocity components, u, v and w being along x, y and z directions respectively (Fig. 3.10). Let ρ be the mass density of the fluid which may be a variable quantity.

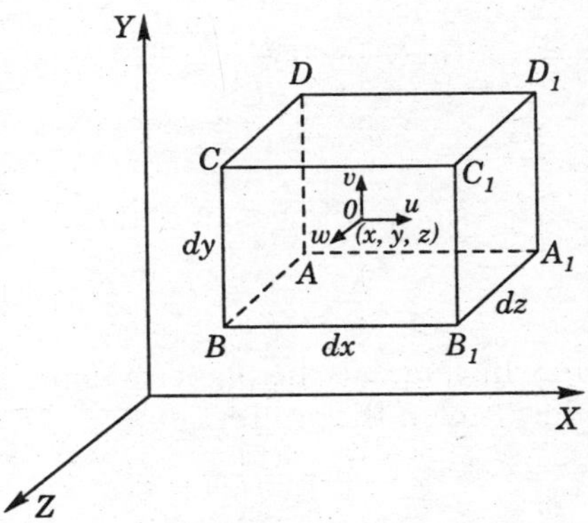

Fig. 3.10. Three-dimensional mass rate of flow.

The rate of mass flow passing the central surface containing O and parallel to $ABCD$, that is in the x-direction

$= \rho u\ dy\ dz$

Then the rate of flow of mass entering face $ABCD$ of the element

$$= \rho u dy dz - \frac{\partial(\rho u)}{\partial x}\frac{dx}{2}\, dy \,.\, dz = \left[\rho u - \frac{\partial(\rho u)}{\partial x}.\frac{dx}{2}\right] dy.dz$$

The mass rate of flow leaving the face $A_1B_1C_1D_1$

$$= \left[\rho u + \frac{\partial(\rho u)}{\partial x}.\frac{dx}{2}\right] dydz$$

The net mass rate of flow being retained in the element

= Mass rate entering – Mass rate leaving

$$= \left[\rho u - \frac{\partial(\rho u)}{\partial x}\frac{dx}{2}\right] dydz - \left[\rho u + \frac{\partial(\rho u)}{\partial x}.\frac{dx}{2}\right] dydz$$

$$= -\frac{\partial(\rho u)}{\partial x}\,.\, dx \,.\, dy \,.\, dz \qquad \text{...(3.21)}$$

Similar expressions may be obtained considering the faces normal to y- and z-axes. The net gain of mass per unit time in the element considering faces normal to y-axis will be

$$-\frac{\partial(\rho v)}{\partial y}\, dydxdz \qquad ...(3.22)$$

and the net gain of mass per unit time considering faces normal to z-axis will be

$$-\frac{\partial(\rho w)}{\partial z}\, dzdxdy$$

Therefore, the total net gain of mass per unit time within the element is

$$-\left\{\frac{\partial(\rho u)}{\partial x}+\frac{\partial(\rho v)}{\partial y}+\frac{\partial(\rho w)}{\partial z}\right\} dxdydz \qquad ...(3.24)$$

Considering the time rate of variation of density, the change in the mass of the element per unit time will be

$$\frac{\partial \rho}{\partial t}\,.\, dxdydz \qquad ...(3.25)$$

Equations (3.24) and (3.25) represent the same quantity, hence

$$\frac{\partial \rho}{\partial t}\, dxdydz = -\left[\frac{\partial(\rho u)}{\partial x}+\frac{\partial(\rho v)}{\partial y}+\frac{\partial(\rho w)}{\partial z}\right] dxdydz$$

dividing throughout by the volume of the element $dxdydz$ and taking the limit as $dxdydz$ approaches zero, the continuity equation at a point becomes :

$$\frac{\partial \rho}{\partial t}+\frac{\partial(\rho u)}{\partial x}+\frac{\partial(\rho v)}{\partial y}+\frac{\partial(\rho w)}{\partial z}=0 \qquad ...(3.26)$$

Equation (3.26) is the general equation of continuity in three-dimensions and is applicable to any type of flow and for any fluid whether compressible or incompressible.

However, for steady flow of incompressible fluids, the equation reduces to

$$\frac{\partial u}{\partial x}+\frac{\partial v}{\partial y}+\frac{\partial w}{\partial z}=0 \qquad ...(3.27)$$

In two-dimensional flow, Eq. (3.27) takes the form

$$\frac{\partial u}{\partial x}+\frac{\partial v}{\partial y}=0 \qquad ...(3.28)$$

in which $w = 0$, the planes of similar flow pattern being parallel to xy-plane. For one dimensional flow, say in the x-direction

$$\frac{\partial u}{\partial x}=0 \qquad ...(3.29)$$

or u = constant, $v = w = 0$ and if the area of flow is a, then the rate of flow is

$Q = a\,.\,u$ = constant (for steady flow)

which is same as the Eq. (3.17) and states that if the area of flow a is constant the velocity of flow u must also be constant.

3.6.4. Equation of Continuity in Polar Co-ordinates

For compressible fluids, the equation of continuity in polar co-ordinates may be written as :

$$\frac{1}{r}(\rho v_r)+\frac{\partial}{\partial r}(\rho v_r)+\frac{\partial}{r\partial\theta}(\rho v_\theta)=0 \qquad ...(3.29\ a)$$

which for incompressible fluids reduces to

$$\frac{v_r}{r}+\frac{\partial v_r}{\partial r}+\frac{1}{r}\cdot\frac{\partial v_\theta}{\partial\theta}=0 \qquad ...(3.29\ b)$$

Example 3.12. *A gas follows the law p = (constant) ρT and flows steadily in a horizontal pipe of constant diameter. If the flow is isothermal and the ratio of pressure at the two sections under consideration is* $\frac{p_1}{p_2} = \frac{8}{7}$. *Find the velocity ratio* $\frac{V_1}{V_2}$.

Solution. From the principle of conservation of mass, the mass rate of flow is constant, thus

$$\rho_1 A_1 V_1 = \rho_2 A_2 V_2$$

but since the pipe cross-section is uniform, $A_1 = A_2$.

$$\therefore \quad \rho_1 V_1 = \rho_2 V_2$$

and the velocity ratio

$$\frac{V_1}{V_2} = \frac{\rho_2}{\rho_1}$$

from the gas law, $\quad p = (\text{constant})\, \rho T.$

Since the flow is isothermal, the temperature remains same every where and law at the two sections yields

$$\frac{p_1}{p_2} = \frac{\rho_1}{\rho_2}$$

and, therefore, the velocity ratio

$$\frac{V_1}{V_2} = \frac{\rho_2}{\rho_1} = \frac{p_2}{p_1} \text{ but } \frac{p_2}{p_1} = \frac{8}{7} \qquad \therefore \qquad \frac{V_1}{V_2} = \frac{7}{8}.$$

Example 3.13. *A pipeline carries oil (sp. gr. 0.83), at a velocity of 2 m/s through a 20 cm pipe. At another section the diameter is 15 cm. Find the velocity at this section and the mass rate of flow.*

Solution. The continuity equation applied to the two sections under consideration yields :

$$Q = A_1 V_1 = A_2 V_2$$

$$V_2 = \frac{A_1 V_1}{A_2} = \frac{\frac{\pi}{4}(0.2)^2 \times 2}{\frac{\pi}{4}(0.15)^2} = 3.56 \text{ m/s.}$$

Volume rate of flow $= Q = A_1 V_1 = \frac{\pi}{4}(0.2)^2 \times 2 = 0.0629 \text{ m}^3\text{/s.}$

Mass rate of flow $= \rho Q = 1000 \times 0.83 \times 0.0629 = 52.2$ kg/s.

Example 3.14. *Water flows out through a long rectangular opening in the side of a tank. The sheet of water as it emerges out is horizontal and has a velocity of 7.5 m/s and its thickness is 3.0 cm. The velocity and the thickness of the jet change as it is deflected under the influence of gravity. Find the velocity and the thickness of the jet at a section where it has a slope of 60° to the horizontal.*

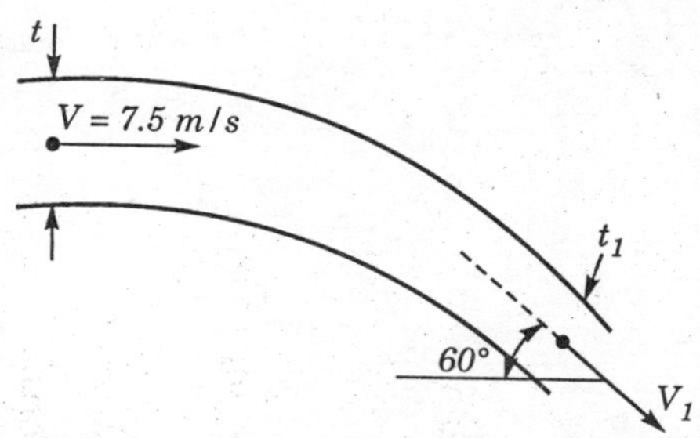

Solution. Since there is no acceleration in the horizontal direction, the horizontal component of velocity will remain unchanged, thus

$$V_1 \cos 60° = 7.5 \qquad \therefore \qquad V_1 = 15.0 \text{ m/s.}$$

Consider unit length of the rectangular opening and apply the continuity equation

$$q = V \,.\, t = V_1 t_1 \quad \therefore \quad t_1 = \frac{V}{V_1} t = \frac{7.5}{15.0} \times 3.0 = 1.5 \text{ cm.}$$

Example 3.15. *A pipeline 60 cm in diameter bifurcates at a Y-junction into two branches 40 cm and 30 cm in diameter. If the rate of flow in the main pipe is 1.5 m³/s, and the mean velocity of flow in the 30 cm pipe is 7.5 m/s, determine the rate of flow in the 40 cm pipe.*

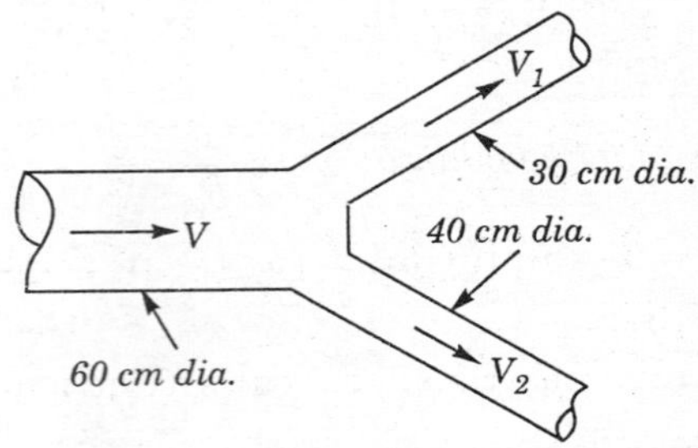

Solution. From the continuity relationship,

$$Q = Q_1 + Q_2 = A_1V_1 + A_2V_2.$$

Rate of flow in 30 cm pipe

$$= \frac{\pi}{4}(0.3)^2 \times 7.5 = 0.53 \text{ m}^3/\text{s}.$$

$$\therefore \quad Q_2 = Q - Q_1 = 1.5 - 0.53 = 0.97 \text{ m}^3/\text{s}.$$

Example 3.16. *Find the mean velocity of flow for the following two-dimensional flow velocity profiles in terms of the maximum velocity V_{max}.*

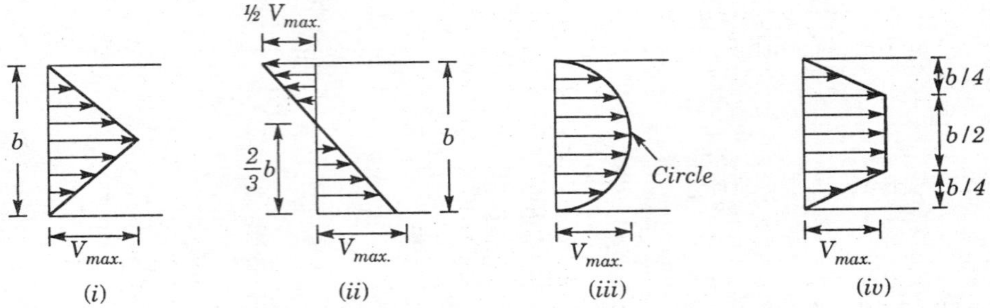

Solution. (*i*) Rate of flow per unit width,

$$q = 2\left(\frac{1}{2}V_{max} \times \frac{b}{2} \times 1\right) = \frac{V_{max}\, b}{2}$$

$$\therefore \quad \text{Average velocity} = \frac{q}{\text{Area per unit width}} = \frac{b\,V_{max}}{2 \times b} = \frac{V_{max}}{2}$$

(*ii*) Rate of flow per unit width

$$q = \left(\frac{2}{3}b \times 1\right) \times \frac{1}{2}V_{max} - \left(\frac{1}{3}b \times 1\right) \times \frac{V_{max}}{2 \times 2} = \frac{b\,V_{max}}{3} - \frac{b\,.\,V_{max}}{12} = \frac{b\,V_{max}}{4}$$

$\therefore$ Mean velocity of flow,

$$V = \frac{q}{b} = \frac{b\,V_{max}}{b \times 4} = \frac{V_{max}}{4}.$$

(*iii*) The equation of the circular velocity profile may be written as, with y measured from the bottom boundary :

$$u^2 + \left(y - \frac{b}{2}\right)^2 = \frac{b^2}{4}$$

at $\quad y = \frac{1}{2}b,\ u = V_{max} = b/2$

$$\therefore \quad b = 2V_{max} \quad \text{and} \quad u = \sqrt{\frac{b^2}{4} - \left(y - \frac{b}{2}\right)^2}$$

Rate of flow per unit width, $\quad q = 2\int_0^{b/2} u\,.\,dy = 2\int_0^{b/2} \sqrt{\frac{b^2}{4} - \left(y - \frac{b}{2}\right)^2}\, dy$

putting $\quad \left(y - \frac{b}{2}\right) = \frac{b}{2}\sin\theta$

$$dy = \frac{b}{2} \cos\theta \, d\theta$$

and
$$q = 2\int_{-\pi/2}^{0} \left(\sqrt{\frac{b^2}{4} - \frac{b^2}{4}\sin^2\theta}\right) \frac{b}{2} \cos\theta \, d\theta$$

$$q = 2\int_{-\pi/2}^{0} \frac{b^2}{4} \cos^2\theta \, d\theta = \frac{b^2}{2}\int_{-\pi/2}^{0} \left(\frac{1+\cos 2\theta}{2}\right) d\theta$$

$$= \frac{b^2}{2}\left(\frac{\theta + \frac{\sin 2\theta}{2}}{2}\right)_{-\pi/2}^{0} = \frac{\pi b^2}{8}$$

$\therefore$ Mean velocity of flow $= \frac{q}{b} = \frac{\pi b^2}{8b} = \frac{\pi b}{8}$

but $b = 2\,V_{max}$

$\therefore$ Mean velocity $= \frac{2\pi\, V_{max}}{8} = 0.786\, V_{max}$.

(*iv*) Rate of flow per unit width,

$$q = \left(\frac{b}{4} \times 1\right)\left(\frac{1}{2} V_{max}\right) + \left(\frac{b}{2} \times 1\right) V_{max} + \left(\frac{b}{4} \times 1\right)\left(\frac{1}{2} V_{max}\right)$$

$$= \frac{2b}{8} V_{max} + \frac{b\,V_{max}}{2} = \frac{3b\,V_{max}}{4}.$$

$\therefore$ Mean velocity of flow $= \frac{q}{b} = \frac{3b\,V_{max}}{4b} = \frac{3}{4} V_{max}$.

Example 3.17. *An incompressible fluid flows through the circular pipe as shown in the figure at the rate of Q m^3/s,*

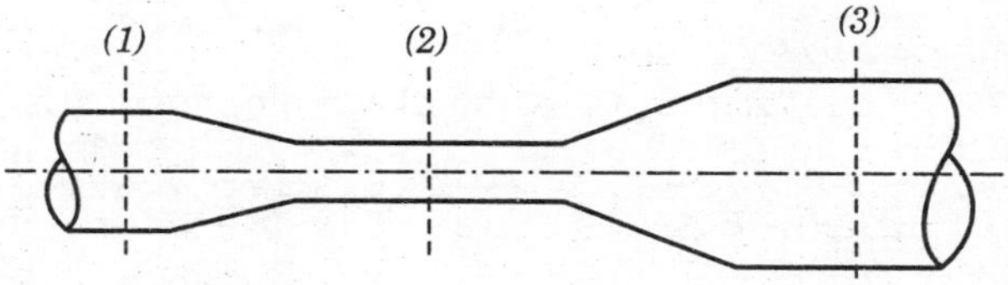

(i) If it is assumed that the velocities at sections (1), (2) and (3) are uniform, what are the velocities given that the diameters of the pipe at the three sections are d_1, d_2 and d_3 metre respectively.

(ii) Determine these velocities for,

$$Q = 0.3\ m^3/s,\ d_1 = 0.4\ m,\ d_2 = 0.2\ m \text{ and } d_3 = 0.75\ m.$$

(iii) Find the velocities when Q = 0.3 m^3/s at section (1) but the density changes in such a way that $\rho_2 = 0.5\,\rho_1$, $\rho_3 = 1.3\rho_1$. The flow is steady in all cases.

The diameters are same as given for (ii).

Solution. (*i*) From the continuity equation in one-dimensional flow

$$Q = AV.$$

$\therefore$ Velocity at section (1) $V_1 = \frac{Q}{\frac{\pi}{4}(d_1)^2} = \frac{4Q}{\pi d_1^2}$

Similarly, $V_2 = \frac{Q}{\frac{\pi}{4} d_2^2} = \frac{4Q}{\pi d_2^2}$ and $V_3 = \frac{Q}{\frac{\pi}{4} d_3^2} = \frac{4Q}{\pi d_3^2}$

(ii)
$$V_1 = \frac{0.3}{\frac{\pi}{4} \times (0.4)^2} = 2.39 \text{ m/s.}$$

$$V_2 = \frac{0.3}{\frac{\pi}{4} \times (0.2)^2} = 9.55 \text{ m/s. and } V_3 = \frac{0.3}{\frac{\pi}{4} \times (0.75)^2} = 0.691 \text{ m/s.}$$

(iii) From the principle of conservation of mass, mass rate of flow in a continuous fluid is the same, thus

$$\rho_1 V_1 \frac{\pi}{4} d_1^2 = \rho_2 V_2 \frac{\pi}{4} d_2^2 = \rho_3 V_3 \frac{\pi}{4} d_3^2$$

it is given that at section (1), $Q = 0.3 \text{ m}^3/\text{s}$, hence,

$$\rho_1 Q_1 = \rho_2 V_2 \frac{\pi}{4} d_2^2$$

$$\therefore \quad V_2 = \frac{\rho_1 \times 0.3}{0.5\, \rho_1 \times \frac{\pi}{4} (0.2)^2} = 19.10 \text{ m/s}$$

also
$$\rho_1 Q_1 = \rho_3 \frac{\pi}{4} d_3^2 \,.\, V_3$$

$$\therefore \quad V_3 = \frac{\rho_1 \times 0.03}{1.3\, \rho_1 \times \frac{\pi}{4} (0.75)^2} = 0.531 \text{ m/s and } \rho_1 Q_1 = \rho_1 \frac{\pi d^2}{4} V_1$$

$$V_1 = \frac{4Q_1}{\pi d_1^2} = \frac{4 \times 0.3}{\pi \times (0.4)^2} = 2.39 \text{ m/s.}$$

Example 3.18. *The velocity components in x and y directions are given as*

$$u = 2xy^3/3 - x^2y \text{ and } v = xy^2 - 2yx^3/3.$$

Indicate whether the given velocity distribution is :

(i) a possible field of flow *(ii) not a possible field of flow.*

Solution. A possible flow field must satisfy the continuity equation

$$\frac{\partial u}{\partial x} + \frac{\partial v}{\partial y} = 0$$

$$u = \frac{2xy^3}{3} - x^2y, \quad v = xy^2 - \frac{2yx^3}{3}$$

$$\frac{\partial u}{\partial x} = \frac{2}{3} y^2 - 2xy, \quad \frac{\partial v}{\partial y} = 2xy - \frac{2}{3} x^3$$

Making substitution,

$$\frac{\partial u}{\partial x} + \frac{\partial v}{\partial y} = \frac{2}{3} y^3 - 2xy + 2xy - \frac{2}{3} x^3 = \frac{2}{3} (y^3 - x^3)$$

Since the continuity equation is not satisfied, the given velocity components, therefore, do not represent a possible flow field.

3.7. Circulation and Vorticity

Consider a closed curve *ABC* within the fluid medium. The *circulation* around such a curve is defined as the summation of product of velocity component along the element (such as *PQ*) of the curve and the elemental length *ds*. Let the velocity at any arbitrary point be *V* and *ds* be the length of element along the curve. Let θ be the angle between the length element *ds* and the velocity vector *V*. The circulation denoted by Γ (Greek capital letter 'gamma') is given by

$$\Gamma = \oint V \cos\theta \, ds \qquad ...(3.30\ a)$$

Vectorially it is written as the dot product of the velocity vector and the length vector,

$$\Gamma = \oint V \cdot ds \qquad ...(3.30\ b)$$

$$= (ui + vj + wk) . (idx + jdy + kdz)$$

$$= \oint (udx + vdy + wdz) \qquad ...(3.30\ c)$$

in which the symbol $\oint$ denotes a line integral taken around the closed curve in an anti-clockwise direction. The concept of circulation in fluid mechanics is similar to the work done by a force in moving around a closed curve dealt with in mechanics of solids ($\oint F \cos\theta \, ds$) or vectorially $\oint F.ds$.

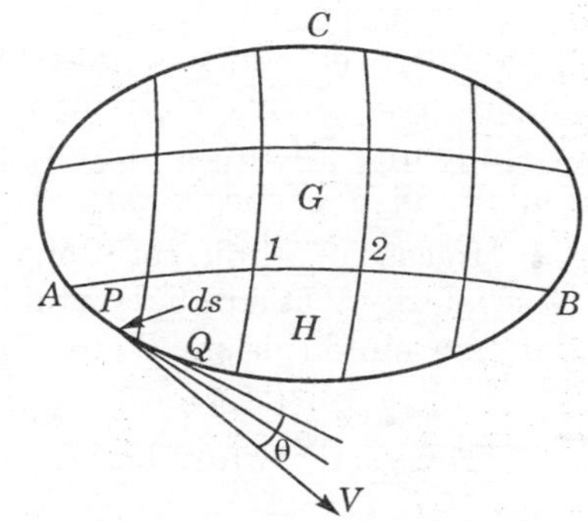

Fig. 3.11. Circulation around a closed curve.

The circulation around a large closed curve may be shown to be equal to the sum of circulations around small meshes into which the closed curve has been sub-divided. Two such typical meshes are represented by G and H, and their common boundary is marked by numerals 1-2 in Fig. 3.11. The line integral $\int_2^1 V \cdot ds$ along this common boundary has equal magnitude but opposing directions when the meshes G and H are considered. The contributions from all common boundaries to the circulation around the closed curve, therefore, cancel while summing up, and the final result consists of the circulation around the outer periphery of the closed curve.

Although the determination of circulation around any arbitrary closed curve is generally a tedious step-by-step integration, the principle may be applied easily to certain closed curves having regular shapes such as circles and rectangles.

Consider an elementary rectangle $ABCD$ of size $dxdy$ as shown in Fig. 3.12. The average velocities along each side of the rectangle are shown and their directions are inclined by the arrows. Proceeding from the corner A and remembering that circulation is considered positive in the anti-clockwise direction, its value around the rectangular element is

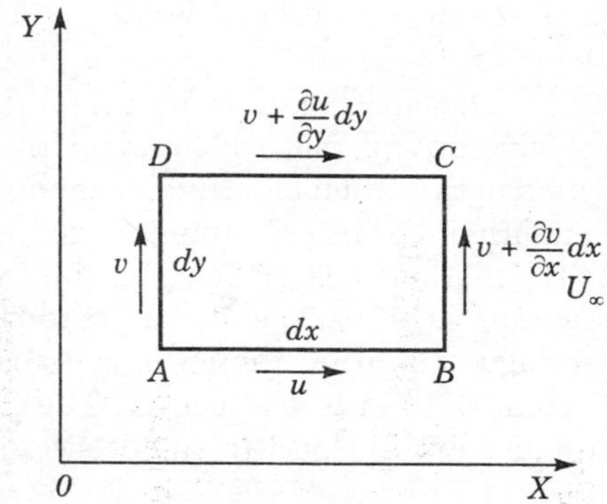

Fig. 3.12. Circulation around an elementary rectangle.

$$d\Gamma = udx + \left(v + \frac{\partial v}{\partial x} dx\right) dy - \left(u + \frac{\partial u}{\partial y} dy\right) dx - vdy$$

$$= \frac{\partial v}{\partial x} dxdy - \frac{\partial u}{\partial y} dxdy \qquad ...(3.31)$$

in which $d\Gamma$ is the differential circulation around the rectangular element of enclosed area $dxdy$.

The *vorticity* denoted by ξ (Greek letter xi) at a point is defined as the ratio of the differential circulation around an infinitesimal closed curve enclosing the point to the area of the closed curve. In other words, the vorticity is the circulation per unit area enclosed by the curve.

Thus from Eq. (3.31), the z-component of vorticity,

$$\xi_z = \frac{d\Gamma}{dxdy} = \frac{\partial v}{\partial x} - \frac{\partial u}{\partial y} \qquad ...(3.32)$$

The vorticity component in z-direction in polar co-ordinates is expressed as

$$\xi_z = \frac{v_\theta}{r} + \frac{\partial v_\theta}{\partial r} - \frac{1}{r}\frac{\partial v_r}{\partial \theta} \qquad ...(3.32\ a)$$

Equation (3.32) represents vorticity for two-dimensional flow. In case of three-dimensional flow, the above equation gives only a component of vorticity. Vorticity is a vector quantity having its direction perpendicular to the plane of the closed curve around which the circulation is measured.

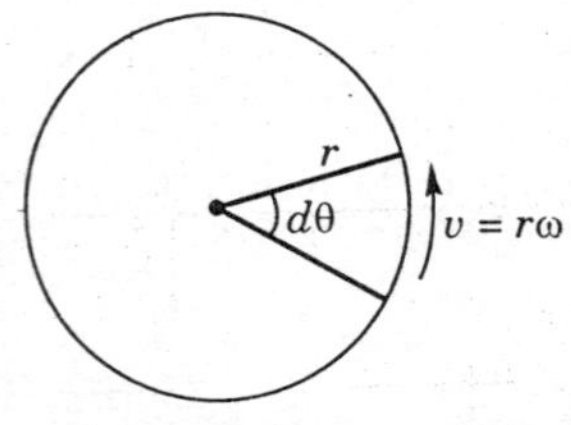

Fig. 3.13. Cirulation around a circular boundary.

In case of a fluid rotating at a certain constant angular velocity ω, the velocity V at a radial distance r is given by $r\omega$ (Fig. 3.13). The circulation around this small circular closed curve of radius r is given by

$$d\Gamma = \int v\, r d\theta = \int r^2 \omega d\theta = r^2\omega \int d\theta = r^2 \omega 2\pi.$$

The vorticity at a point,

$$\xi = \frac{d\Gamma}{\pi r^2} = 2\omega \qquad ...(3.33)$$

is two times the angular velocity of the particle at that point.

3.8. Irrotational and Rotational Flow

The concept of vorticity may be utilised for distinguishing a flow as either irrotational or rotational. If the vorticity is zero at all points in flow region (except at certain special points known as 'singular points', where the velocity or the acceleration is zero or infinite), the flow in that region is known as *irrotational*. Thus for irrotational flow, the vorticity must be zero. In such flow zones, where the vorticity is non-zero, the flow is said to be *rotational*. In actual flow problems, there may be zones possessing characteristics of both irrotational and rotational flows. As will be seen later, the flow outside the boundary layer has irrotational characteristics while that within the boundary layer is rotational.

The above definition of irrotational flow is purely based on the concept of vorticity and does not give any physical idea about the nature of flow. Irrotational flow is defined as that flow in which net rotation of a fluid element about its own mass centre is zero, as it moves from one place to another. An interesting and well-known example similar to irrotational flow is that presented by the motion of carriages of a Giant wheel—a most common feature of carnivals and village fairs. Each carriage is suspended at its top from the periphery of the wheel, and as the wheel revolves the carriages maintain their vertical position (as in case of stationary wheel) while moving along a circular path at low velocities. The carriage, therefore, does not rotate about its centre of gravity and also with respect to an observer standing nearby.

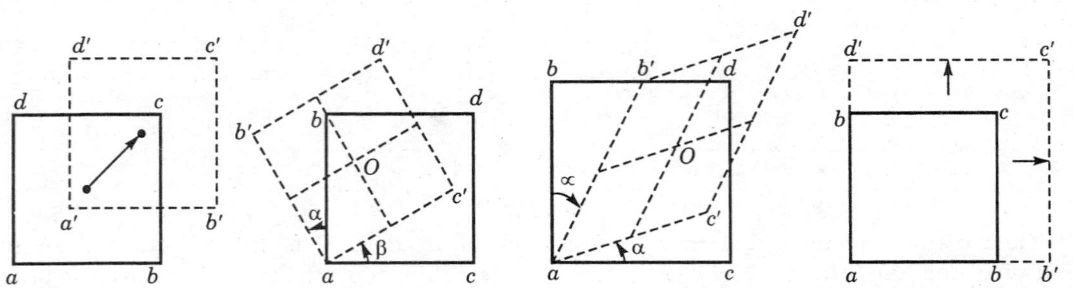

Fig. 3.14. Possible motions of a fluid element.

Let us now consider all possible motions of a cubical fluid element. A fluid element during its motion may undergo either translation, rotation, or distortion or any combination of these. Fig. 3.14 shows these various possibilities. Mere translation of fluid element and also its uniform rotation does not require the existence of shear stresses. Alternatively, if the shear stresses do not exist in any part of the flow, the fluid will either move in translation of uniform rotation. It then appears that due to the existence of shear stresses, the fluid element must suffer a distortion in its shape. The angular distortion results in an incompressible fluid, the volume of the fluid element remaining constant while the deformation takes place. This results in a shear-strain which causes a change in the angle between two adjacent sides of the element. The other type of deformation is volumetric distortion which might result in flow of a compressible fluid.

Irrotational flow exists when the net rotation of the fluid element about its mass centre is zero. In other words, if any part (a pair of parallel faces) of a continuous fluid element rotates in one direction, and the other part (pair of adjoining parallel faces) of the element rotates in the opposite direction such that the net rotation is zero, and there occurs only the distortion of the element. Fig. 3.14 (*b*) shows the net rotation of the fluid element about its mass centre. The faces *ab* and *cd* of the element undergo anti-clockwise rotation through an angle α with respect to its original position shown in full lines, while the faces *ac* and *bd* are rotated also in anti-clockwise direction through an angle β. The net rotation of the element about its mass centre is, therefore, $(\alpha + \beta)$, and the flow is rotational. Whereas in Fig. 3.14 (*c*), the faces *ab* and *cd* are rotated by angle α in the clockwise direction while the faces *ac* and *bd* suffer an anti-clockwise rotation through the same angle α. The net rotation of the element about its mass centre is, therefore, zero causing the flow to be irrotational.

It may again be emphasized that the rotation of fluid element can only be produced by shear stresses, caused by the viscous action, and if such stresses are of negligible magnitude, it follows that the flow may be considered irrotational. In case of an ideal fluid, since the shear stresses are non-existent due to the assumption of zero viscosity of fluid, the flow along a straight path is purely irrotational. It is for this reason that the ideal fluid flow is regarded as irrotational, except for flow along curved paths. The viscous action in a moving fluid generates shear stresses resulting in non-zero net rotation and vorticity. But since viscosity is always present in real fluids, we might think that in actual practice all fluid flows must be rotational. This presumption is, however, not totally correct. In most of the flow problems the effects of viscosity are confined to a rather thin region of flow near a solid boundary where the shear stresses are high. This thin region or layer of fluid, in the immediate neighbourhood of the solid boundary where the effects of viscosity are predominant is known as the *boundary layer*. Outside this boundary layer, the viscous effects are negligible, and the flow in this part which constitutes the main body of flow is generally treated as irrotational. Within the boundary layer, the viscous effects are of considerable importance resulting in very high shear stresses causing the flow to be rotational.

For flow between circular boundaries, such as in canals and pipe bends, the velocity varies inversely with the radius r, yielding a velocity relationship,

$$V.r = \text{constant}$$

causing higher velocities near the inner boundary as compared to the outer line. This velocity distribution is characteristic of irrotational flow. An example of irrotational flow is the 'drain-hole vortex' which develops above a drain hole (an outlet) in the bottom of a stationary tank. This type of vortex is also known as a *free vortex* or an *irrotational vortex*.

An example of rotational flow is the motion of liquid in a rotating tank, in which the velocity varies directly with the radial distance r measured from the axis of rotation, *i.e.*, $v = r\omega$. The vortex which develops within the tank is known as the *forced vortex*.

3.9. Examples of Various Types of Flow

1. **Steady flow.** Flow through a conduit at a constant flow rate Q m³/s is steady. It is characterized by velocity equation of the form $u = ax^2 + bx + c$, which is independent of time t.

2. **Unsteady flow.** If the rate of flow Q is either increasing or decreasing with time the flow will be termed as unsteady, which may be represented by a velocity equation, $u = ax^2 + bxt$. The flow in a pipe whose valve is being opened or closed gradually is an example of unsteady flow.

3. **Uniform flow.** Flow through a straight prismatic conduit is uniform. In other words, flow through a straight pipe of constant diameter will be uniform. A prismatic conduit has a constant size and shape.

4. **Non-uniform flow.** Flow through a non-prismatic conduit is non-uniform. Flow around a uniform diameter pipe-bend or a canal bend are examples of non-uniform flow.

5. **Laminar flow.** Flow in a conduit of small size at a low velocity is generally laminar *e.g.*, flow of blood in veins and arteries, flow through a capillary tube and ground water flow are typical examples of this type of flow.

6. **Turbulent flow.** High velocity flow in a conduit of large size will be turbulent. Nearly all fluid flow problems encountered in engineering practice have a turbulent character.

7. **Rotational flow.** Flow near the solid boundaries is rotational. Motion of liquid in a rotating tank is a typical example of this type of motion.

8. **Irrotational flow.** Flow of an ideal fluid is irrotational and in case of parallel boundaries, it is characterized by uniform velocity distribution (rectangular in shape). Flow above a drainhole of a stationary tank or a wash-basin is an example of irrotational flow. Flow outside the boundary layer is generally considered irrotational.

9. **One-dimensional flow.** Flow in a pipe is considered one-dimensional, where average flow-parameters are considered for analysis.

10. **Two-dimensional flow.** Flow between parallel plates of infinite extent is two dimensional. The flow in the main-stream of a wide river is generally regarded as two-dimensional.

11. **Three-dimensional flow.** The flow in a converging or diverging pipe or channel is three-dimensional. The flow in a prismatic open channel in which the width and the water depth are of the same order of magnitude is also an example of three-dimensional flow.

3.10. Velocity Potential

The velocity potential is represented by ϕ (Greek letter phi) and is defined by the following expression :

$$-\phi = \int V_s \,.\, ds \qquad ...(3.34)$$

in which V_s is the velocity along a small length element ds. From Eq. (3.34) we obtain

$$d\phi = -V_s\, ds \qquad \text{or} \qquad V_s = -(d\phi/ds).$$

The velocity potential is a scalar quantity dependent upon space and time. Its negative derivative with respect to any direction gives the velocity in that direction, that is

$$u = -\frac{\partial \phi}{\partial x}, \qquad v = -\frac{\partial \phi}{\partial y} \quad \text{and} \quad w = -\frac{\partial \phi}{\partial z} \qquad ...(3.35)$$

In polar co-ordinates (r, θ, z), the velocity components are

$$v_r = -\frac{\partial \phi}{\partial r}, \qquad v_\theta = -\frac{1}{r}\frac{\partial \phi}{\partial \theta}, \qquad v_z = -\frac{\partial \phi}{\partial z} \qquad ...(3.35\ a)$$

The velocity potential ϕ thus provides an alternative means of expressing velocity components. The minus sign in Eq. (3.35) appears because of the convention that the velocity potential decreases in the direction of flow just as the electrical potential decreases in the direction in which the current flows. The velocity potential is not a physical quantity which could be directly measured and, therefore, its zero position may be arbitrarily chosen.

The continuity equation for three-dimensional steady flow of incompressible fluids, Eq. (3.27), gives

$$\frac{\partial u}{\partial x} + \frac{\partial v}{\partial y} + \frac{\partial w}{\partial z} = 0$$

which may be written in terms of ϕ as

$$\frac{\partial^2\phi}{\partial x^2}+\frac{\partial^2\phi}{\partial y^2}+\frac{\partial^2\phi}{\partial z^2}=0 \qquad ...(3.36)$$

Using continuity equation in polar coodinates *i.e.* Eq. (3.29 *b*), one can write

$$\frac{1}{r}\frac{\partial\phi}{\partial r}+\frac{\partial^2\phi}{\partial r^2}+\frac{1}{r^2}\frac{\partial^2\phi}{\partial\theta^2} \qquad ...(3.36\ a)$$

Eq. (3.36) is three-dimensional form of the *Laplace's* equation.

The angular velocities of a fluid element about its mass centre are

$$\omega_x=\frac{1}{2}\left(\frac{\partial w}{\partial y}-\frac{\partial v}{\partial z}\right),\ \omega_y=\frac{1}{2}\left(\frac{\partial u}{\partial z}-\frac{\partial w}{\partial x}\right) \text{ and } \omega_z=\frac{1}{2}\left(\frac{\partial v}{\partial x}-\frac{\partial u}{\partial y}\right) \qquad ...(3.37)$$

From Eq. (3.33), it is obvious that the vorticity is two times as great as the angular velocity.

In case of a two-dimensional flow in the *xy*-plane, the angular velocity and the vorticity about the mass centre of an element are

$$\omega_z=\frac{1}{2}\left(\frac{\partial v}{\partial x}-\frac{\partial u}{\partial y}\right) \text{ and } \xi_z=\left(\frac{\partial v}{\partial x}-\frac{\partial u}{\partial y}\right),$$

the angular velocity components in *x* and *y* directions being zero, *i.e.*,

$$\omega_x=\omega_y=0$$

writing this in terms of ϕ,

$$\omega_z=\frac{1}{2}\left(-\frac{\partial^2\phi}{\partial x\partial y}+\frac{\partial^2\phi}{\partial y\partial x}\right) \qquad ...(3.38)$$

Since ϕ is a continuous function of *x* and *y*,

$$\frac{\partial^2\phi}{\partial x\partial y}=\frac{\partial^2\phi}{\partial y\partial x}$$

and Eq. (3.38) yields $\omega_z = 0$. The angular velocities ω_x and ω_y being already zero, the resultant angular velocity (rate of rotation)

$$\omega=\sqrt{\omega_x{}^2+\omega_y{}^2+\omega_z{}^2}=0$$

resulting in a zero vorticity. We know that if the vorticity or the rate of rotation ω of each fluid particle about its mass centre is zero, the flow must be irrotational. This sets up a condition for velocity potential to exit in a flow field. The existence of velocity potential in a flow field ensures that the flow must be irrotational. It is for this reason that an irrotational flow is often called as *potential flow*. Lines drawn in the flow field along which ϕ is constant are known as *equipotential lines*.

Equations (3.27) and (3.36) indicate that all flows in which ϕ exists and satisfies the continuity equation must also satisfy the Laplace's equation.

3.11. Stream Function

The concept of stream function ψ (Greek letter psi) is based on the principle of continuity and the properties of a stream line. It is applicable to two-dimensional flow cases. Consider the two points P_1 and P_2, P_1 lying on the streamline *D* while P_2 lies on another streamline *C* in Fig. 3.15. From the definition of a streamline, it is known that no flow crosses it and, therefore, the quantity of flow between any two streamlines must remain the same in accordance with the continuity equation. If the points P_1 and P_2 have stream function values of ψ and $\psi + d\psi$ respectively, then the flow across P_1P_2 is $d\psi$. Like the velocity potential, the stream

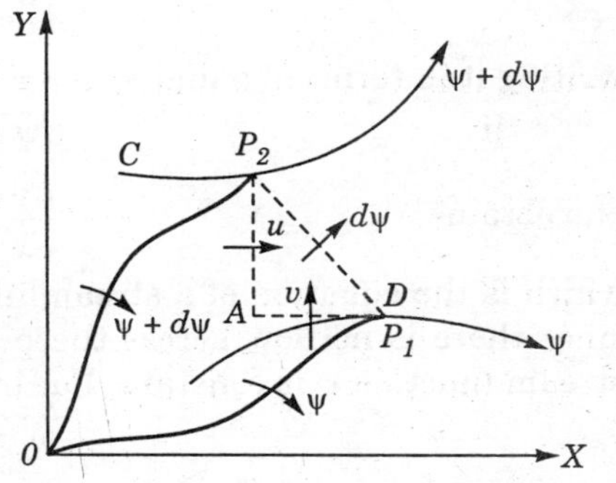

Fig. 3.15. Flow between two points and its relation to stream function.

function is also a scalar quantity, and is considered positive according to a sign convention in which an observer at O looking towards the point P_1 (that is in the direction OP_1) finds that the flow crosses OP_1 from left to right. In triangle AP_1P_2, the flow leaving the triangle (across P_1P_2) is $d\psi$ whereas the flow entering it is $udy - vdx$ in which dx and dy are the lengths of AP_1 and AP_2 respectively. Thus one may write,

$$d\psi = udy - vdx \qquad ...(3.39)$$

Now, if ψ is considered as a function x and y, that is

$$\psi = f(x, y)$$

$$d\psi = \frac{\partial \psi}{\partial x} dx + \frac{\partial \psi}{\partial y} dy \qquad ...(3.40)$$

Comparing Eqs. (3.39) and (3.40) we obtain expressions relating the velocity components u and v with the stream function, thus

$$u = \frac{\partial \psi}{\partial y} \text{ and } v = -\frac{\delta \psi}{\partial x} \qquad ...(3.41)$$

Eq. (3.41) states that the partial derivative of the stream function with respect to any direction gives the velocity component 90° clockwise with respect to that direction.

The mathematical expression for a streamline, for two-dimensional flow reduces to (using Eq. 3.4)

$$\frac{dx}{u} = \frac{dy}{v} \quad \text{or} \quad \frac{dy}{dx} = \frac{v}{u} \qquad ...(3.42)$$

Analogous relations in polar co-ordinates are

$$v_r = \frac{\partial \psi}{r\partial \theta} = v_\theta - \frac{\partial \psi}{\partial r} \qquad ...(3.42\ a)$$

in which v_r is positive radially outwards from the origin and v_θ and θ are positive anti-clockwise.

The continuity equation in two-dimensions (Eq. 3.28) takes the form

$$\frac{\partial u}{\partial x} + \frac{\partial v}{\partial y} = 0 \qquad ...(3.42\ b)$$

Substituting the values of u and v in terms of ψ

$$\frac{\partial^2 \psi}{\partial x \partial y} + \frac{\partial^2 \psi}{\partial y \partial x} = 0 \qquad ...(3.43)$$

Since ψ is continuous function of x and y, it may be observed that the stream function ψ satisfies the continuity equation. The total derivative of ψ is given by Eq. (3.40),

$$d\psi = \frac{\partial \psi}{\partial x} . dx + \frac{\partial \psi}{\partial y} . dy$$

writing it in terms of u and v, $d\psi = -vdx + udy$ ...(3.44)

If $\quad d\psi = 0,$

one obtains $\quad \dfrac{dy}{dx} = \dfrac{v}{u}$

which is the equation of a streamline, Eq. (3.42). Thus lines of constant ψ represent streamlines since there is no flow across these lines. Conversely, a streamline is that line along which the stream function ψ is constant. For irrotational flow,

$$\omega_z = 0$$

and, therefore, $\quad \dfrac{\partial v}{\partial x} = \dfrac{\partial u}{\partial y}$ ...(3.45) expressing u and v in terms of ψ, we get

$$-\frac{\partial^2 \psi}{\partial x^2} = \frac{\partial^2 \psi}{\partial y^2} \quad \text{or} \quad \frac{\partial^2 \psi}{\partial x^2} + \frac{\partial^2 \psi}{\partial y^2} = 0 \qquad ...(3.46)$$

Equation (3.46) shows that the stream function must also satisfy the Laplace's equation provided that the flow is irrotational. Conversely, fluid-flow problems which do not satisfy the Laplace equation in ψ are rotational flows.

Let us consider a two dimensional incompressible flow in an open channel. From the principle of continuity, the rate of flow per unit width of the channel is the same at all cross-sections, thus

$$q = \int_0^y u dy \qquad ...(3.47)$$

where q is the flow rate per unit per width, dy is the spacing between adjacent streamlines measured perpendicular to u, Fig. 3.16. As the flow is along the x-direction, the velocity component $v = 0$ and Eq. (3.44) reduces to $d\psi = u dy$, the integration of which between the limits 0 to y gives the difference in the stream function between the water surface and the channel bottom, thus

$$\psi_1 - \psi_5 = \int_0^y u dy \qquad ...(3.48)$$

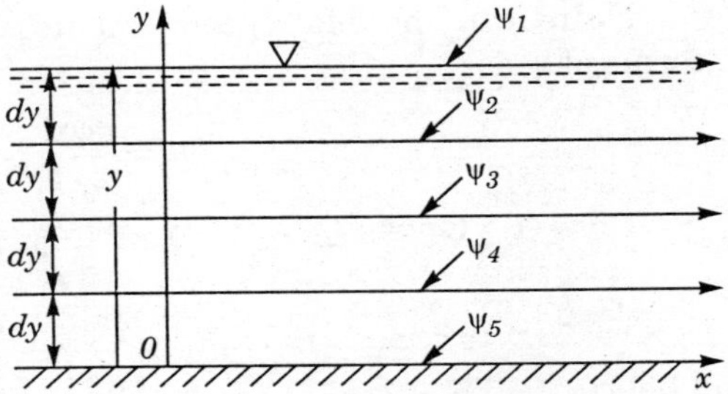

Fig. 3.16. Two-dimensional flow in an open channel.

The right hand sides of Eqs. (3.47) and (3.48) are equal and, therefore,

$$\psi_1 - \psi_5 = q \qquad ...(3.49)$$

Thus we may conclude that lines of constant stream function represent streamlines and that the numerical difference in ψ between two streamlines is equal to the flow rate per unit width passing between them.

If the two points P_1 and P_2 have stream function ψ_1 and ψ_2, then $\psi_1 - \psi_2$ represents the flow passing between them. If P_1 and P_2 are on the same streamline, then $\psi_1 - \psi_2 = 0$, as there is no flow across the streamline. Hence for all points lying on the same streamline ψ = constant.

It may be emphasized here that whereas the concept of velocity potential is, in general, applicable to three dimensional irrotational flow, the stream function finds its applications in two-dimensional flow.

Example 3.19. *Water flows through a pipe AB 1.2 m diameter at 3 m/s and then passes through a pipe BC 1.5 m diameter. At C, the pipe branches. Branch CD is 0.8 m in diameter and carries one-third of the flow in AB. The flow velocity in branch CE is 2.5 m/s. Find the volume rate of flow in AB, the velocity in BC, the velocity in CD and diameter of CE.*

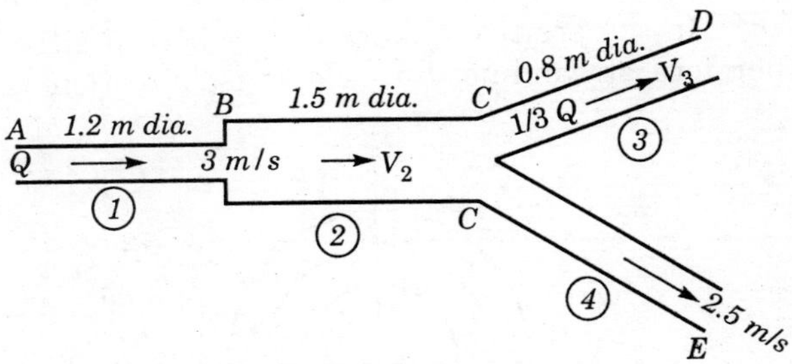

Solution. Discharge through AB, from continuity equation

$$Q = A_1 V_1 = \frac{\pi}{4}(1.2)^2 \times 3 = 3.393 \text{ m}^3/\text{s}$$

Velocity in pipe BC, $$V_2 = \frac{4Q}{\pi(1.5)^2} = \frac{4 \times 3.393}{\pi(1.5)^2} = 1.92 \text{ m/s.}$$

Velocity in CD, $$V_3 = \frac{4(Q/3}{\pi(0.8)^2} = \frac{4 \times 1.131}{2.0106} = 2.25 \text{ m/s.}$$

Discharge passing through pipe $CE = \frac{2}{3} Q = 2.262 \text{ m}^3/\text{s}$.

Area of pipe CE

$$= \frac{\left(\frac{2}{3}Q\right)}{V_4} = \frac{2.262}{2.5} = 0.9048 \ m^2$$

$$\frac{\pi D_4^2}{4} = 0.9048 \quad \therefore \quad D_4 = 1.073 \text{ m.}$$

Example 3.20. *Sketch the streamlines represented by $\psi = x^2 + y^2$ and $\psi = x^2 - y^2$ showing the direction of flow in each case.*

Also find out the velocity and its direction at (1, 1).

Solution. The velocity components u and v in two-dimensional steady flow may be expressed in terms of ψ as

$$u = \frac{\partial \psi}{\partial y} \quad \text{and} \quad v = -\frac{\partial \psi}{\partial x}$$

Case (I) : Given $\quad \psi = x^2 + y^2 \quad$ ...(1)

$$\therefore \quad \frac{\partial \psi}{\partial y} = 2y \quad \text{and} \quad \frac{\partial \psi}{\partial x} = 2x$$

and hence, $\quad u = 2y \quad$ and $\quad v = -2x \quad$ ...(2)

the resultant velocity, $\quad V = \sqrt{u^2 + v^2} = 2\sqrt{x^2 + y^2}$

the equation of streamline is $= \frac{dx}{u} = \frac{dy}{v}$

and, therefore, the slope of the velocity vector

$$\frac{dy}{dx} = \frac{v}{u} = -\frac{2x}{2y} = -\frac{x}{y} \quad \text{...(3)}$$

Integrating, $\frac{x^2}{2} + \frac{y^2}{2} = \text{constant.} \quad$ or $\quad x^2 + y^2 = \text{constant.}$

Therefore, streamlines are concentric circles.

For a streamline ψ = constant, and for non-zero values of ψ, equation (1) represents concentric circles with centre at the origin (0, 0) and radius $\sqrt{\psi}$.

In the positive quadrant (x, y both positive), $u = 2y$, is positive and $v = -2x$ is negative. Therefore, streamlines have clockwise direction as shown in the figure.

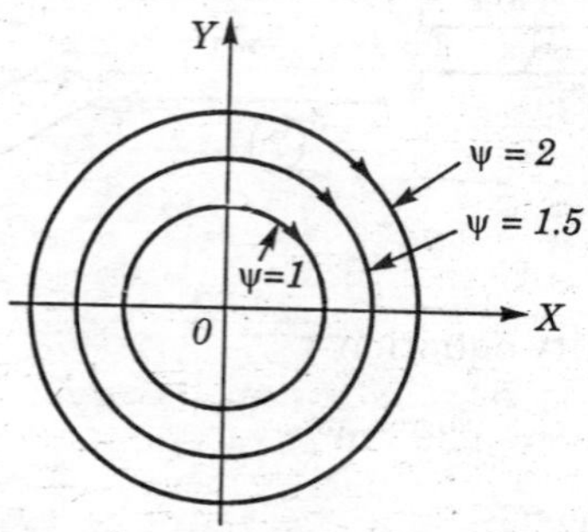

Streamline pattern (concentric circles) represented by $\psi = x^2 + y^2$

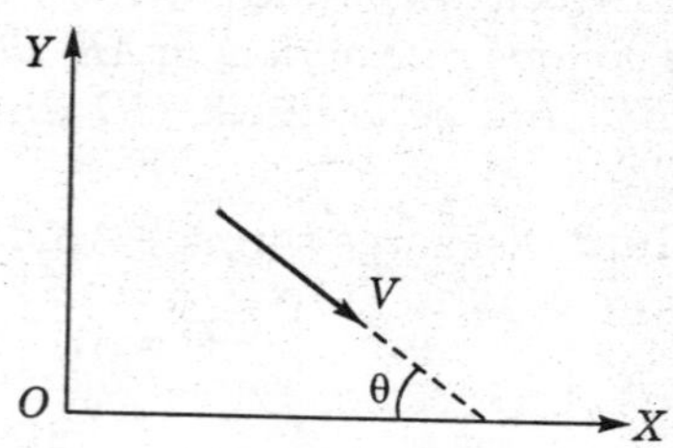

Velocity V at (1, 1) $= 2\sqrt{1^2 + 1^2} = 2\sqrt{2}$

and it makes an angle θ with the x-axis, as shown

$$\tan \theta = \frac{dy}{dx} = \frac{x}{y} = 1$$

$$\theta = 45°.$$

Case (II) : $$\psi = x^2 - y^2 \qquad ...(4)$$

then $$\frac{\partial \psi}{\partial y} = -2y \quad \text{and} \quad \frac{\partial \psi}{\partial y} = 2x$$

and, therefore, $u = -2y$ $\quad \therefore \quad V = \sqrt{u^2 + v^2} = 2\sqrt{x^2 + y^2}$.

and $v = -2x$.

The streamlines are lines of constant ψ, and for constant ψ, equation (4) represents hyperbola, which may be plotted for different values of ψ. The following table gives values of x and y for different values of ψ.

ψ	y	$x = \pm\sqrt{y^2 + \psi}$
1	0	± 1
	1	$\pm\sqrt{2}$
	2	$\pm\sqrt{5}$
	3	$\pm\sqrt{10}$
2	0	$\pm\sqrt{2}$
	1	$\pm\sqrt{3}$
	2	$\pm\sqrt{6}$
	3	$\pm\sqrt{11}$

For positive values of x and y the components of velocity u and v both are negative and, therefore, the streamlines have the direction as shown in the figure.

The velocity at (1, 1) is $V = 2\sqrt{2}$ and its direction has a slope

$$\frac{dy}{dx} = \frac{v}{u} = \frac{-2x}{-2y} = \frac{x}{y} = 1.$$

Therefore, the velocity vector is inclined at an angle of 45° to x-axis.

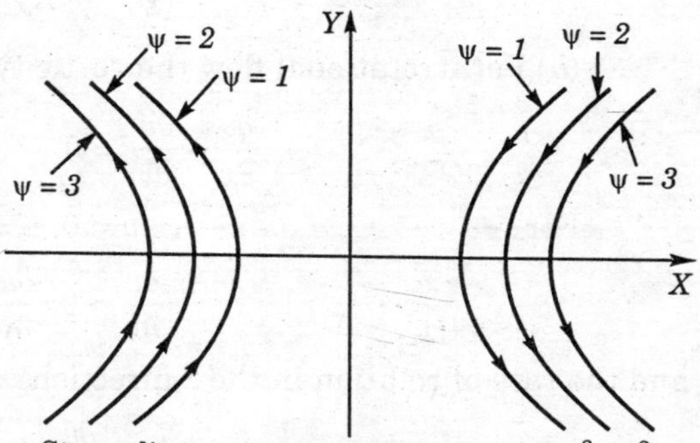

Streamline pattern represented by $\psi = x^2 - y^2$

Example 3.21. *For an incompressible flow represented by* $\psi = x^2 - y^2$*, calculate the total acceleration vector and show that it is proportional to the radius vector.*

Solution. The total acceleration

$$a = \sqrt{a_x^2 + a_y^2}$$

also, $$a_x = u\frac{\partial u}{\partial x} + v\frac{\partial u}{\partial y}; \qquad a_y = u\frac{\partial v}{\partial x} + v\frac{\partial v}{\partial y}$$

for the given stream function, $\psi = x^2 - y^2$

$$u = \frac{\partial \psi}{\partial y} = -2y \quad \text{and} \quad v = -\frac{\partial \psi}{\partial x} = -2x$$

$$\therefore \quad \frac{\partial u}{\partial x} = 0, \frac{\partial u}{\partial y} = 2 ; \frac{\partial v}{\partial x} = -2, \frac{\partial v}{\partial y} = 0.$$

$$\therefore \quad a_x = 0 + (-2x)(2) = -4x$$

and $$a_y = (2y)(-2) + 0 = -4y \quad \therefore \quad a = \sqrt{a_x^{\,2} + a_y^{\,2}} = 4\sqrt{x^2 + y^2} = 4r$$

Since acceleration $a = 4r$, it is proportional to the radius r.

Example 3.22. *A fluid flows along a flat surface parallel to the x-direction. The velocity u varies linearly with y, the distance from the flat surface, that is u = Ay. (a) Find the stream function of the flow and (b) Determine whether or not the flow is irrotational.*

Solution. Since the flow is parallel to the x-direction, and is given by $u = Ay$, therefore, $v = 0$. The stream function is related to the velocity components

$$u = \frac{\partial \psi}{\partial y} \quad \text{and} \quad v = -\frac{\partial \psi}{\partial x}.$$

The total differential, $$d\psi = \frac{\partial \psi}{\partial x} dx + \frac{\partial \psi}{\partial y} dy$$

From the given velocity components, $u = Ay$ and $v = 0$, one may write,

$$u = \frac{\partial \psi}{\partial y} = Ay \quad \text{and} \quad \frac{\partial \psi}{\partial x} = 0,$$

hence, $$d\psi = 0 + Aydy.$$

Integrating, $\psi = A\dfrac{y^2}{2}$ + constant of integration, since ψ is a function of y, the constant of integration is a pure number, thus

$$\psi = \frac{1}{2} Ay^2 + B$$

(*b*) For irrotational flow the vorticity (or the rate of rotation) is everywhere zero, *i.e.*

$$\frac{\partial v}{\partial x} - \frac{\partial u}{\partial y} = 0.$$

For $$v = 0, u = Ay$$

$$\frac{\partial v}{\partial x} = 0, \frac{\partial u}{\partial y} = A$$

and the rate of rotation in the z-direction,

$$\omega_z = \frac{\partial v}{\partial x} - \frac{\partial u}{\partial y} = 0 - A = -A$$

since the rate of rotation is not zero, the flow is rotational.

Example 3.23. *If $\phi = 3xy$, find x and y components of velocity at (1, 3) and (3, 3). Determine the discharge passing between streamlines passing through these points.*

Solution. The velocity components in terms of ϕ are given by

$$u = -\frac{\partial \phi}{\partial x} \quad \text{and} \quad v = -\frac{\partial \phi}{\partial y}$$

but $$\phi = 3xy$$

$$\therefore \quad \frac{\partial \phi}{\partial x} = 3y \quad \text{and} \quad \frac{\partial \phi}{\partial y} = 3x$$

hence the velocity components $u = -3y$ and $v = -3x$.

The velocity components at (1, 3) are $u = -9$

and $v = -3$ and at (3, 3), $u = -9$ and $v = -3$

The total derivative ψ may be written as

$$d\psi = \frac{\partial \psi}{\partial x}dx + \frac{\partial \psi}{\partial y}dy \quad \text{but} \quad u = \frac{\partial \psi}{\partial y} \quad \text{and} \quad v = -\frac{\partial \psi}{\partial x}$$

the total derivative $d\psi = vdx + udy = 3xdx + (-3y)\,dy$

Integrating $\psi = \frac{3}{2}x^2 - \frac{3}{2}y^2 + A.$

where A = constant of integration.

Discharge between the streamlines passing through (1, 3) and (3, 3)

$$= \psi_{(1,\,3)} - \psi_{(3,\,3)} = \frac{3}{2}(1-9) - \frac{3}{2}(9-9) = -12 \text{ units.}$$

Example 3.24. *If the expression for the stream function is described by $\psi = x^3 - 3xy^2$, determine whether the flow is*

(i) rotational *(ii) irrotational.*

If the flow is irrotational, then indicate the correct value of the velocity potential :

(a) $\phi = y^3 - 3x^2y$, *(b) $\phi = -3x^2y$.*

Solution. For a two-dimensional flow in $x-y$ plane to be irrotational, the vorticity vector in the z-direction must be zero, *i.e.*

$$\xi_z = \frac{\partial v}{\partial x} - \frac{\partial u}{\partial y} = 0$$

$$u = \frac{\partial \psi}{\partial y} \quad \text{and} \quad v = -\frac{\partial \psi}{\partial x}$$

Given $\psi = x^3 - 3xy^2$

$\therefore$ $u = -6xy$, and $v = -(3x^2 - 3y^2) = 3(x^2 - y^2)$

and $\frac{\partial u}{\partial y} = -6x, \; \frac{\partial v}{\partial x} = -6x$

Making appropriate substitutions

$$\xi_z = \frac{\partial v}{\partial x} - \frac{\partial u}{\partial y} = -6x - (-6x) = -6x + 6x = 0$$

Hence the flow is *irrotational*.

For an irrotational flow, the Laplace equation in ϕ must be satisfied, *i.e.*

$$\frac{\partial^2 \phi}{\partial x^2} + \frac{\partial^2 \phi}{\partial y^2} = 0.$$

Let us check the validity for each expression for ϕ

(a) $\phi = y^3 - 3x^2y$, $\frac{\partial^2 \phi}{\partial x^2} = -6y$ and $\frac{\partial^2 \phi}{\partial y^2} = 6y$

$\therefore$ $\partial^2\phi/\partial x^2 + \partial^2\phi/\partial y^2 = -6y + 6y = 0$

(b) $\phi = -3x^3y$, $\partial^2\phi/\partial x^2 = -6y^2$ and $\partial^2\phi/\partial y^2 = 0$, $\therefore \frac{\partial^2 \phi}{\partial x^2} + \frac{\partial^2 \phi}{\partial y^2} \neq 0$

Hence the correct value of $\phi = y^3 - 3x^2y$.

Example 3.25. *The stream function and velocity potential for a flow are given by* $\psi = 2xy$, $\phi = (x^2 - y^2)$. *Show that the conditions of continuity and irrotational flow are satisfied. Illustrate the type of flow represented by sketching a few streamlines.*

Solution. Conditions for the continuity and irrotationality of flow are :

$$\frac{\partial u}{\partial x} + \frac{\partial v}{\partial y} = 0$$

$$\frac{\partial v}{\partial x} - \frac{\partial u}{\partial y} = 0 \qquad \text{(vorticity being zero for irrotational flow)}$$

$$\psi = 2xy,\ \phi = x^2 - y^2$$

$$u = \frac{\partial \psi}{\partial y} = 2x \quad \text{and} \quad v = -\frac{\partial \psi}{\partial x} = -2y$$

$$\therefore \qquad \frac{\partial u}{\partial x} = 2,\ \frac{\partial u}{\partial y} = 0 \quad \text{and} \quad \frac{\partial v}{\partial y} = -2,\ \frac{\partial v}{\partial x} = 0.$$

Making appropriate substitutions in the above equations, it is seen that the continuity equation

$$\frac{\partial u}{\partial x} + \frac{\partial v}{\partial y} = 2 - 2 = 0$$

is satisfied. For irrotational flow, the vorticity component ξ_z must be zero

$$\xi_z = \frac{\partial v}{\partial x} - \frac{\partial x}{\partial y}\ ;\ \frac{\partial v}{\partial x} = 0,\ \frac{\partial x}{\partial y} = 0,\ \text{hence } \xi_z = 0 - 0 = 0$$

The flow is therefore irrotational.

Using the velocity potential ϕ and noting that $u = -\partial\phi/\partial x$ and $v = -\partial\phi/\partial y$, the above conditions can be verified.

For sketching the streamlines for arbitrarily assumed values of ψ, values of y can be computed for various assumed values of x as follows :

For $\psi = 1$

$$x = \tfrac{1}{4},\ \tfrac{1}{2},\ 1,\ 3/2,\ 2,\ 3,\ 4.$$

$$y = 2,\ 1,\ \tfrac{1}{2},\ \tfrac{1}{3},\ \tfrac{1}{4},\ \tfrac{1}{6},\ \tfrac{1}{8}$$

For $\psi = 2$

$$x = \tfrac{1}{4},\ \tfrac{1}{2},\ 1,\ 3/2,\ 2,\ 3,\ 4.$$

$$y = 4,\ 2,\ 1,\ \tfrac{2}{3},\ \tfrac{1}{2},\ \tfrac{1}{3},\ \tfrac{1}{4}.$$

The streamlines for $\psi = 1$ and $\psi = 2$ are shown plotted in the adjacent figure.

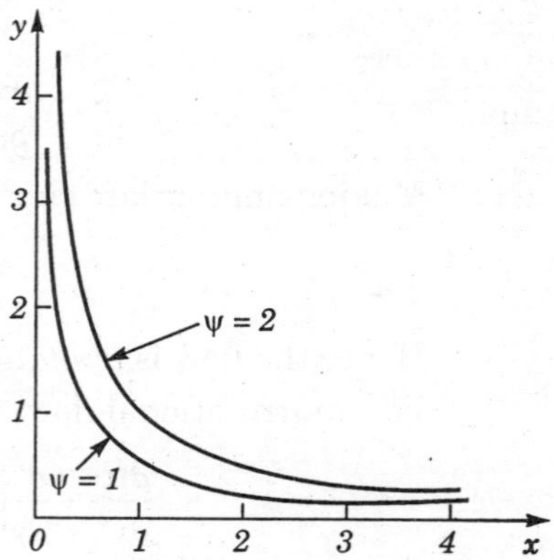

Fig. Ex. 3.21. Streemline pattern.

Determination of Flow Direction :

$$\psi = 2xy$$

For streamlines, ψ = constant

$$y = \frac{\psi}{2x}$$

$$\therefore \qquad dy = -\frac{\psi}{2x^2}\,dx \quad \text{or} \quad \frac{dy}{dx} = -\frac{\psi}{2x^2} = -\frac{\text{constant}}{2x^2}$$

Since dy/dx is negative, it indicates that as x increases, y must decrease. Hence the flow direction is as marked by the arrows.

Example 3.26. *A two-dimensional flow field is described by $\psi = x - y^2$. Calculate the horizontal and vertical components of velocity and obtain an expression for velocity of flow at any point in flow field. Determine the velocity of flow and state whether flow is rotational or irrotational. Sketch the streamlines for $\psi = 0$, 1 and 2.*

Solution. The horizontal and vertical components of velocity are given by

$$u = \frac{\partial \psi}{\partial y}, \; v = -\frac{\partial \psi}{\partial x}$$

but $\psi = x - y^2$, and hence $u = (-2y)$ m/s and $v = 1$ m/s.

From the equation for ψ it is seen that for ψ = constant, the streamlines are a family of parabolas symmetrical about x-axis. For $\psi = 0$, the streamlines pass through the origin. The directional arrows on the streamlines are decided from the expression for u and v determined above, and have been marked on the streamlines as shown. The resultant velocity

$$V = \sqrt{u^2 + v^2} = \sqrt{(-2y)^2 + 1^2} = \sqrt{1 + 4y^2}$$

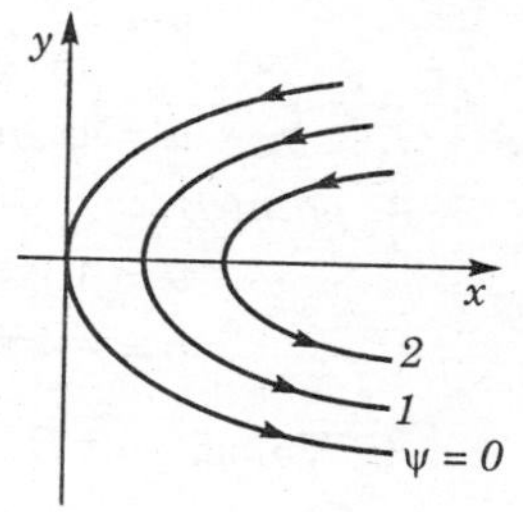

Fig. 3.21. Streamline pattern.

The vorticity is given by

$\xi = \partial v/\partial x - \partial u/\partial y = 0 - (-2) = 2 \text{ sec}^{-1}$ (anti-clockwise).

Since the vorticity is not zero, the flow is rotational.

Example 3.27. *What is the irrotational velocity field associated with the potential*

$$\phi = 3x^2 - 3x + 3y^2 + 16t^2 + 12zt.$$

Does the flow field satisfy the incompressible continuity equation ?

Solution. The velocity field is represented by

$$u = -\frac{\partial \phi}{\partial x}, \; v = -\frac{\partial \phi}{\partial y}$$

from the given velocity potential

$$\phi = 3x^2 - 3x + 3y^2 + 16t^2 + 12zt$$

$$u = -\frac{\partial \phi}{\partial x} = -6x + 3 \quad \text{and} \quad \frac{\partial u}{\partial x} = -6$$

$$v = -\frac{\partial \phi}{\partial y} = -6y \quad \text{and} \quad \frac{\partial v}{\partial y} = -6$$

The continuity equation for incompressible fluid is

$$\frac{\partial u}{\partial x} + \frac{\partial v}{\partial y} = 0$$

Substituting the values of the velocity gradients

$$\frac{\partial u}{\partial x} + \frac{\partial v}{\partial y} = -6 - 6 = -12.$$

This shows that the flow field does not satisfy the continuity equation.

Example 3.28. (*a*) *In two-dimensional incompressible flow, show that the flow rate per unit width between two streamlines is equal to the difference between the values of the stream function corresponding to these streamlines.*

(*b*) *The stream function $\psi = 4xy$, in which ψ is in cm²/second and x and y are in metres, describe the incompressible flow between the boundaries shown below :*

Calculate :

(*i*) *Velocity at B* (*ii*) *Convective acceleration at B*

(*iii*) *Flow rate per unit width across AB.*

Solution. $\psi = 4xy$, ψ is in cm²/s, and x and y in metres. Co-ordinates of A are (3, 0), and that of B are :

$$x = 3 \text{ m},$$

$$y = \frac{3}{x} = \frac{3}{3} = 1 \text{ m (obtained from } xy = 3).$$

Hence the co-ordinates of B are (3, 1).

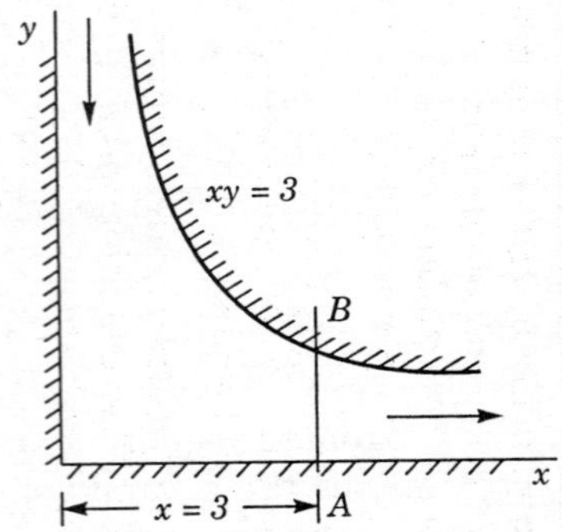

Fig. Ex. 3.24

Velocity components :

$$u = \frac{\partial \psi}{\partial y}, \; v = -\frac{\partial \psi}{\partial x}$$

$$\therefore \quad u = 4x, \text{ and } v = -4y$$

At point B

$$u = 4 \times 3 = 13 \text{ cm/s}$$

$$v = -4 \times 1 = -4 \text{ cm/s}$$

(*i*) Velocity at B, $V = \sqrt{u^2 + v^2}$

$$= \sqrt{(12)^2 + (-4)^2} = \mathbf{12.65 \text{ cm/s.}}$$

(*ii*) *Convective acceleration* :

$$a = \sqrt{a_x^2 + a_y^2}$$

$$a_x = u\frac{\partial u}{\partial x} + v\frac{\partial u}{\partial y} \quad \text{and} \quad a_y = u\frac{\partial v}{\partial x} + v\frac{\partial v}{\partial y}$$

$$u = 4x, \; v = -4y$$

$$\frac{\partial u}{\partial x} = 4, \; \frac{\partial u}{\partial y} = 0 \; ; \; \frac{\partial v}{\partial x} = 0, \; \frac{\partial v}{\partial y} = -4$$

$$\therefore \quad a_x = (4x)(4) + (-4y)(0) = 16x$$

$$a_y = (4x)(0) + (-4y)(-4) = 16y$$

and the convective accelerations at B (3, 1) is

$$a = \sqrt{(16x)^2 + (16y)^2}$$

$$= 16\sqrt{x^2 + y^2} = 16\sqrt{(3)^2 + (1)^2} = \mathbf{50.60 \text{ cm/s}^2.}$$

(*iii*) *Flowrate per unit width across AB*

$$q_{AB} = \psi_B - \psi_A = \psi_{(3,\,1)} - \psi_{(2,\,0)}$$

$$= 4 \times 3 \times 1 - 4 \times 3 \times 0 = \mathbf{12 \text{ cm}^3\text{/s/cm.}}$$

Example 3.29. *In a two-dimensional incompressible flow, the fluid velocity components are given by* $u = x - 4y$ *and* $v = -y - 4x$. *Show that velocity potential exists and determine its form as well as stream function.*

Solution. For velocity potential to exist, the flow must be irrotational. The vorticity component in the z-direction must therefore be zero.

$$\xi_z = \frac{\partial v}{\partial x} - \frac{\partial u}{\partial y}$$

$$u = x - 4y, \; \frac{\partial u}{\partial y} = -4$$

$$v = -y - 4x, \; \frac{\partial v}{\partial x} = -4 \quad \therefore \quad \frac{\partial v}{\partial x} - \frac{\partial u}{\partial y} = -4 - (4) = 0$$

Since the vorticity is zero, the flow is irrotational. Hence the velocity potential exists.

$$u = -\frac{\partial \phi}{\partial x} \quad \text{and} \quad v = -\frac{\partial \phi}{\partial y}$$

Total change in the velocity potential

$$d\phi = \frac{\partial \phi}{\partial x} dx + \frac{\partial \phi}{\partial y} dy$$

$$= -u\,dx - v\,dy = -(x - 4y)dx - (-y - 4x)dy$$

$$= -xdx + 4ydx - ydy + 4xdy$$

Integrating, $$\phi = -\frac{x^2}{2} + 4xy + \frac{y^2}{2} + 4xy = \tfrac{1}{2}(y^2 - x^2) + 8xy$$

For stream function, $$d\psi = \frac{\partial \psi}{\partial x} dx + \frac{\partial \psi}{\partial y} dy$$

We know that, $$u = \frac{\partial \psi}{\partial y} \quad \text{and} \quad v = -\frac{\partial \psi}{\partial x}$$

Therefore, $$d\psi = -v\,dx + u\,dy = (y + 4x)dx + (x - 4y)dy$$

Integrating $$\psi = xy + 4\,\frac{x^2}{2} + xy - 4\,\frac{y^2}{2} = 2(x^2 - y^2) + 2xy.$$

Example 3.30. *For a two-dimensional potential flow, the velocity potential is given by*

$$\phi = 4x\,(3y - 4)$$

Determine the velocity at point (2, 3). Determine also the stream function and its value at a point (2, 3).

Solution. Velocity components in terms of ϕ

$$u = -\frac{\partial \phi}{\partial x} = -(12y - 16)\,; \qquad v = -\frac{\partial \phi}{\partial y} = -(12x)$$

Velocity at (2, 3), $$V = \sqrt{u^2 + v^2} = \sqrt{(16 - 36)^2 + (-24)^2} = \mathbf{31.24\ units}$$

Velocity components in terms of the stream function ψ

$$u = \frac{\partial \psi}{\partial x}\,;\; v = -\frac{\partial \psi}{\partial x}$$

$$d\psi = -\frac{\partial \psi}{\partial x} dx + \frac{\partial \psi}{\partial x} dy = -vdx + udy = 12x\,dx + (16 - 12y)\,dy$$

Integrating, $$\psi = 12.\,\frac{x^2}{2} + 16y - 12.\,\frac{y^2}{2} = 6x^2 - 6y^2 + 16y$$

Its value at (2, 3), $\psi_{2,\,3} = 6(2)^2 - 6(3)^2 + 16(3) =$ **18 units.**

Example 3.31. *The stream function in a two-dimensional flow is* $\psi = 6x - 4y + 7xy$

Verify whether the flow is irrotational. Determine the direction of streamline at a point (1, – 1). Determine also the expression for the velocity potential.

Solution. For irrotational flow, the vorticity component must be zero.

$$\text{Vorticity } \xi_z = \left(\frac{\partial v}{\partial x} - \frac{\partial u}{\partial y}\right) \qquad \text{...(1)}$$

The velocity components in terms of the stream function are

$$u = \frac{\partial \psi}{\partial y} = -4 + 7x\,;\; \frac{\partial u}{\partial y} = 0$$

$$v = -\frac{\partial \psi}{\partial x} = -(6 + 7y)\,;\; \frac{\partial v}{\partial x} = 0$$

Substituting in the vorticity equation (1), we find that the vorticity is zero.

Hence the flow is irrotational.

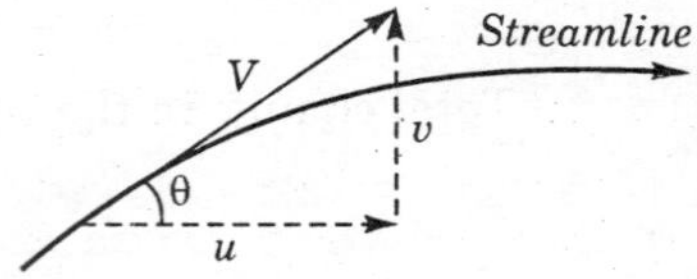

Direction of Streamline at (1, – 1)

$$\tan\theta = \frac{v}{u} = \frac{-(6+7y)}{-4+7x} = \frac{-(6-7)}{-4+7} = \frac{1}{3}$$

The steamline at (1, – 1) makes an angle θ with the x-axis such that $\tan\theta = \frac{1}{2}$.

Velocity Potential

$$d\phi = \frac{\partial\phi}{\partial x}\,dx + \frac{\partial\phi}{\partial y}\,dy = u\,dx - v\,dy = -(7x-4)dx + (6+7y)\,dy$$

Integrating, $$\phi = -\left(\frac{7x^2}{2} - 4x\right) + 6y + 7\,\frac{y^2}{2} = \mathbf{7(y^2 - x^2) + 8x + 12y.}$$

3.12. Relation Between Stream Function and Velocity Potential

Since each component of velocity is now expressible in terms of φ and ψ, the following relation between φ and ψ must exist. It is to be noted that φ exists only in irrotational flow whereas ψ exists in both rotational as well as irrotational flow.

$$u = -\frac{\partial\phi}{\partial x} = \frac{\partial\psi}{\partial y} \quad \text{and} \quad v = -\frac{\partial\phi}{\partial y} = \frac{\partial\psi}{\partial x}$$

therefore, $$\frac{\partial\psi}{\partial y} = -\frac{\partial\phi}{\partial x} \quad \text{and} \quad \frac{\partial\psi}{\partial x} = \frac{\partial\phi}{\partial y} \qquad ...(3.50)$$

Eq. (3.50) represents what are known as the Cauchy-Reimann conditions.

From Eq. (3.44) we know that for a streamline $d\psi = 0$, and

$$\frac{dy}{dx} = \frac{v}{u} \qquad ...(3.51)$$

The total derivative for the velocity potential φ may be written as

$$d\phi = \frac{\partial\phi}{\partial x}\,.\,dx + \frac{\partial\phi}{\partial y}\,.\,dy = -\,udx - vdy \qquad ...(3.52)$$

For an equipotential line, φ = constant and, therefore, $d\phi = 0$. From Eq. (3.52) we obtain for an equipotential line,

$$\frac{dy}{dx} = -\frac{u}{v} \qquad ...(3.53)$$

From Eqs. (3.51) and (3.53) it is seen that the equipotential lines (φ = constant) and streamlines (ψ = constant) intersect each other orthogonally. Thus the streamlines are normal to the equipotential lines. The streamlines and equipotential lines form a net of mutually perpendicular lines, which is called a *flownet*. The flownet, which utilise the concepts of stream function and velocity potential, is an important tool in analysing two-dimensional, irrotational, flow problems.

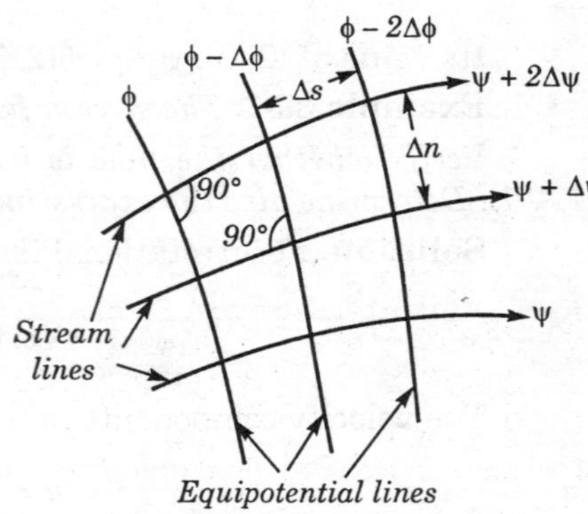

Fig. 3.17. (*a*) Flownet.

3.13. Flownet

The flownet is a graphical representation of two-dimensional irrotational flow and consists of a family of streamlines intersecting orthogonally a family of equipotential lines (they intersect at right angles) and in the process forming small curvilinear squares. In Fig. 3.17 (*a*), a family of streamlines represented by ψ, $\psi + \Delta\psi$ and $\psi + 2\Delta\psi$ are orthogonally intersected by a series of equipotential line ϕ, $\phi - \Delta\phi$ and $\phi - 2\Delta\phi$. It is to be noted that ϕ decreases in the direction of flow. If the distance between the streamlines at some small region of flow is Δn, and the distance between equipotential lines is Δs, then for a flownet $\Delta s = \Delta n$, and the smaller these dimensions, the better will be the representation of flow. This condition may be obtained using Eq. (3.50) expressing

$$\partial\phi/\partial n = \partial\psi/\partial s,$$

thus if

$$\Delta\phi = \Delta\psi,\ \Delta n = \Delta s.$$

3.13.1. Methods of Drawing the Flownet

1. **Mathematical Analysis.** The flownet can be drawn by finding the expressions for ϕ and ψ in terms of x and y and plotting them. The plot of ψ gives the streamlines while that of ϕ yields equipotential lines. This method can be applied to problems with simple and ideal boundary conditions.

2. **Hydraulic Models.** Streamlines can be traced by injecting a dye in a seepage model or Hele Shaw apparatus. The flownet is completed by drawing equipotential lines.

3. **Graphical Methods.** This method consists of drawing streamlines and equipotential lines such that they cut orthogonally and form curvilinear squares. This method, no doubt, consumes lot of time and requires lot of erasing in order to achieve the condition of orthogonal intersection combined with square shaped nets.

4. **Electrical Analogy.** If the characteristics of two or more systems, which are apparently different, can be expressed in identical mathematical forms, they are said to be analogous. To investigate the problem in one field, it is necessary to study similar problems in the analogous field.

The technique of electrical analogy is based on the similarity of flow of electric current through conductors in an electrical system on the one hand and, flow of water, for example, through a closed conduit (pipe-flow) in a hydraulic system on the other hand. The flow in the first system is governed by the Ohm's law whereas that in the latter case the governing equation is the Darcy-Weisbach equation. Hence problems of flow through pipe network can be analysed by studying flow of electric current through a similar network of homogeneous conductors.

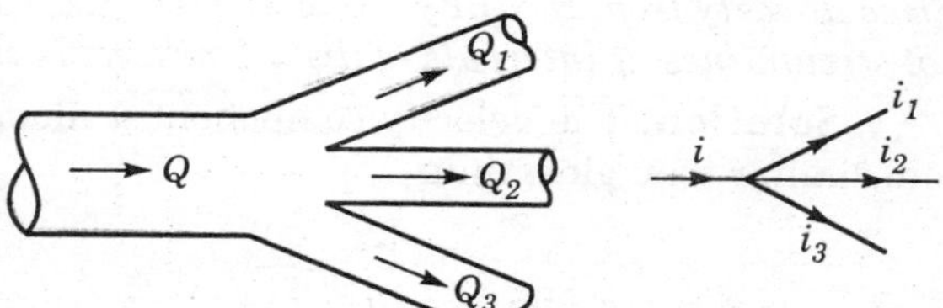

Fig. 3.17. (*b*) Similarity of flow through a pipe network with that of electric current through a similar network of conductors.

The similarity between the flow of water in a pipe network shown in Fig. 3.17 (*b*) and flow of electric current in a similar network of conductors may be established.

Flow of water in a pipe (pressure-flow)	*Flow of electric current*
1. Darcy-Weisbach equation $h_1 = f\dfrac{l}{d}\dfrac{V^2}{2g}$ governs the flow of water.	1. Ohm's law governs the flow of electric current in a conductor.

2. The piezometric head in hydraulic flow corresponds to voltage in flow of electric current.

3. Length of flow path in pipe corresponds to the length of the electrical conductor.

4. The discharge (rate of flow of water) approaching a junction point equals the discharge flowing out, thus $Q = Q_1 + Q_2 + Q_3$. In flow of electric current, the similar expression is $i = i_1 + i_2 + i_3$, where i is the current flowing.

The technique of electrical analogy can be applied with advantage to any system which satisfies the Laplace's equation in ϕ

$$\frac{\partial^2 \phi}{\partial x^2} + \frac{\partial^2 \phi}{\partial y^2} + \frac{\partial^2 \phi}{\partial z^2} = 0 \quad ...(3.54)$$

The above example establishes similarity between two analogous systems and identifies similar quantities. It does not show how the flownets are obtained from the study and measurement of analogous quantities in the electrical system. The ground water flow problems are usually analysed with the help of electrical analogy and involve the drawing of the flownets. If the two analogous systems satisfy the Laplace's equation expressed in terms of similar quantities such as the velocity potential and the electrical potential, the distribution of electrical potential in the electrical system may be used to plot the equipotential lines in the ground water flow field. The streamlines can then be drawn to satisfy the requirements of flownet (orthogonal intersection of streamlines and equipotential lines and idealized square shapes of the flownets). The properties of flownet may be used to determine any desired information such as the quantity of seepage flow underneath a dam or through the canal banks or through the body of an earthen dam.

3.13.2. Uses of flownet

The following are some of the important uses of flownet analysis :

1. For given boundaries of flow, the velocity and pressure distribution can be determined, if the velocity distribution and pressure at any reference section are known.
2. Loss of flow due to seepage in earth dams and unlined canals can be evaluated.
3. Uplift pressures on the underside (bottom) of the dam can be worked out.
4. Outlets can be designed for their streamlining.

Example 3.32. *In a parallel, two-dimensional flow in the positive x-direction the velocity varies linearly from zero at y = 0 to 32 metres/sec at y = 1 metre. Determine the expression for ψ and plot streamlines at intervals of $d\psi = 3$ m²/sec. Is the flow irrotational ?*

Solution. The velocity component u along the x-direction at distance y, may be obtained from similar triangles, thus

$$\frac{32}{u} = \frac{1}{y}$$

$\therefore$ $$u = 32\,y.$$

But $$u = \frac{\partial \psi}{\partial y}$$

and since there is no velocity in the y-direction

$$v = 0$$

and, therefore, we may write

$$u = \frac{d\psi}{dy}$$

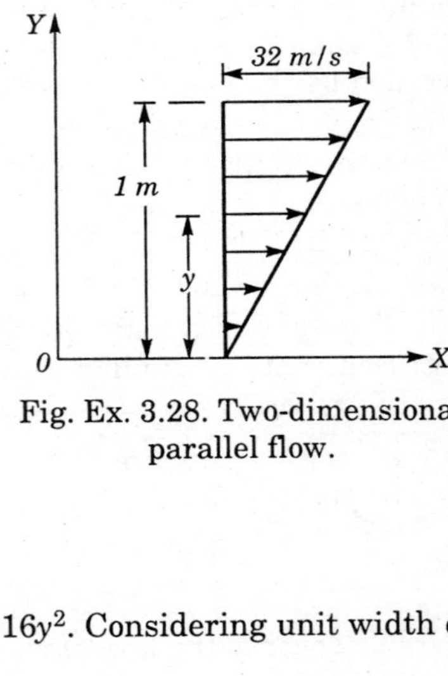

Fig. Ex. 3.28. Two-dimensional parallel flow.

since $$\psi = f(y)$$

or $$\psi = \int u\,dy = \int 32y\,dy = 16y^2 + A.$$

At $y = 0$, let $\psi = 0$, then $A = 0$, and the stream function $\psi = 16y^2$. Considering unit width of flow.

$$dq = d\psi.$$

Let $d\psi$ be the difference in ψ at $y = y$ and $y = 0$, then $d\psi = \psi_y - \psi_0 = 3 = (16y^2 - 0)$

$\therefore$ $$y = \sqrt{3/16}\text{ m} \quad \text{or} \quad y = 0.433\text{ m}.$$

The first streamline will be at $y = 0$ and parallel to x-axis, the second streamline will be a line at $y = 0.433$ m parallel to x-axis. The position of the third streamline may be obtained as

$$d\psi = \psi_y - \psi_{0.433} = 16y^2 - 16 \times (0.433)^2 = 16y^2 - 3$$

but since $d\psi = 3 \text{ m}^2/\text{sec}$

$\therefore$ $3 = 16y^2 - 3$

or $y = \sqrt{6/16} = 0.6125$ m.

The fourth streamline : $3 = 16y^2 - 6$

$\therefore$ $y = \sqrt{9/16} = 0.75$ m.

The fifth streamline : $3 = 16y^2 - 9$

$\therefore$ $y = \sqrt{12/16} = 0.866$ m.

The sixth streamline : $3 = 16y^2 - 12$

$\therefore$ $y = \sqrt{15/16} = 0.969$ m.

and so on.

The streamlines have been shown in the adjacent figure :

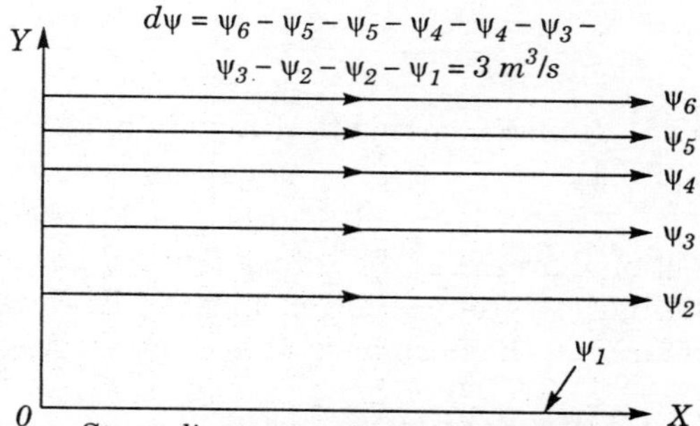

Streamline pattern in two-dimensional parallel flow, $d\psi = 3\ m^2/sec$

For irrotational flow :

$$\frac{\partial u}{\partial y} - \frac{\partial v}{\partial x} = 0 \quad \text{or} \quad \frac{\partial u}{\partial y} = \frac{\partial v}{\partial x}$$

The partial derivatives,

$$\frac{\partial u}{\partial y} = 32 \quad \text{and} \quad \frac{\partial v}{\partial x} = 0$$

and thus we find that

$$\frac{\partial u}{\partial y} \neq \frac{\partial v}{\partial x}$$

and hence the flow is rotational.

3.14. Vortex Flow

In a fluid motion, each small part of the fluid may be considered as a body in motion. Such a body may or may not be rotating. If it is rotating, it spins about an axis at a certain rate, and thus its spin (or rotation) may be represented by an arrow. The arrow, which points along the axis of rotation and has a length equal to twice the angular speed represents the vorticity vector (Fig. 3.18). In a fluid motion different particles may have different velocities, and the velocity of a given particle may change as the particle moves. On the surface of fluid, the vorticity may be observed by following a cork (upon which a cross is marked) placed on the free-surface. If the cork and hence arms forming the cross do not rotate the vorticity is zero ; if they do rotate presence of vorticity is indicated. Fluid motion in which the streamlines are concentric circles is known as the *vortex flow*. The fluid particles moving in these concentric circles may, or may not, rotate about their mass centres. For zero rotation of fluid particles about their respective mass centres, the flow is irrotational and the vortex is known as *irrotational* or *free-vortex*. It is also sometimes called as the potential vortex. If on the other hand, the circulatory flow is such that the fluid particles undergo rotation

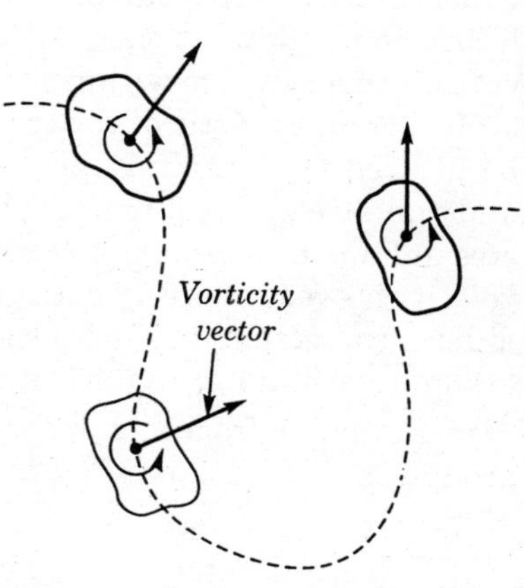

Fig. 3.18. Flow with vorticity.

about their mass centres, the flow is rotational and the resulting vortex is known as the *forced vortex*. The prefix forced used along with the word vortex is probably used to indicate application of some external torque essential for causing this type of vortex flow. Although the forced vortex has fewer engineering applications than the free-vortex, its study is of considerable interest for further understanding of the concept of vorticity. The free and the forced-vortices are discussed in detail in articles 3.14.2 and 3.14.3.

3.14.1. Vortex Lines and Vortex Tubes

The rate of rotation or angular velocity ω is a vector quantity represented by an arrow directed along the axis of rotation, and its distribution in space may be represented in a similar way as that of the velocity V. In other words, a series of vortex lines may be constructed in space such that the angular velocity (rate of rotation) ω is everywhere tangential. The relation between a vortex line and the angular velocity is the same as a streamline has with the velocity. Since the angular vorticity ω and the velocity are inter-related, it can be said that the vorticity vector is also tangential at every point of the vortex line. If ω_x, ω_y, ω_z are components of angular velocity ω along the three co-ordinate axes x, y and z, the differential equation of vortex line is

$$\frac{dx}{\omega_x} = \frac{dy}{\omega_y} = \frac{dz}{\omega_z} \qquad ...(3.55)$$

It may be noted that Eq. (3.55) is exactly similar to the differential equation of a streamline Eq. (3.4).

It may be recalled that a small group of streamlines form a stream tube of variable cross-sectional area dA ; the rate of flow dQ passing through all such sections is same at a given instant. Similarly, a group of vortex lines form a vortex tube of varying cross-sectional area dA, the circulation $d\Gamma$ around the periphery at all cross-sections remaining constant at any instant.

3.14.2. Free-Vortex

The free-vortex flow is described by concentric circular streamlines and the velocity distribution is such that the flow is irrotational. We shall, therefore, examine the conditions under which flow in a circular path is irrotational. For irrotational flow, it has been proved that the vorticity everywhere within the fluid is zero. This typical characteristic of the irrotational flow may be utilised for obtaining the velocity distribution and the stream function for the flow. Since the streamlines are circular the radial component of velocity is everywhere zero. Let us select a suitable differential element $PQRS$ as shown in Fig. 3.18 (a) for determining circulation around it.

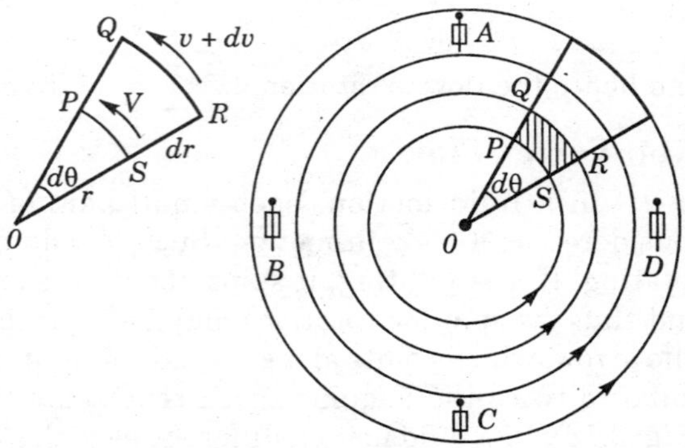

Fig. 3.18 (a). Flow pattern for free-vortex flow.

Starting from the corner R and proceeding in an anti-clockwise direction, the circulation $d\Gamma$ around the closed curve $PQRS$ may be written as

$$d\Gamma = (V + dV)(r + dr)d\theta + 0 - V.r \,.\, d\theta + 0.$$

Neglecting term of higher order of small magnitude, *i.e.* $dV \,.\, dr \,.\, d\theta$

$$\text{Vorticity } \xi = \frac{d\Gamma}{\text{Area}} = \frac{(Vdr + rdV)d\theta}{r \,.\, d\theta \,.\, dr} = \frac{V}{r} + \frac{dV}{dr}.$$

For irrotational flow, the vorticity $\xi = 0$, thus

$$\frac{V}{r} + \frac{dV}{dr} = 0,$$

which may be written as $Vdr + rdV = 0$,

dividing throughout by Vr, one obtains,

$$\frac{dr}{r} + \frac{dV}{V} = 0 \qquad \text{...(3.56)}$$

Integration of Eq. (3.56) results in

$$\log r + \log V = \log C$$

which may be written as $\qquad Vr = C \quad \text{or} \quad V = \dfrac{C}{r} \qquad$ ...(3.57)

in which C is the constant of integration.

Eq. (3.57) states that the velocity is inversely proportional to the radial distances and its variation with r is of hyperbolic nature. It shows that, as $r \to 0$, $V \to \infty$ indicating a point of singularity. In case of real fluids, as $r \to 0$, the viscous action becomes increasingly dominant, and the fluid in the central region (known as the *core* of vortex) tends to rotate as a solid body with velocity proportional to the radius. Eq. (3.57), therefore, is not applicable to the *core* region of the vortex where the flow has rotational characteristics. The vorticity within the core is non-zero and has an infinite value at the **axis** of the vortex (since the area of a closed curve at the axis approaches zero). The axis is, therefore, a vortex line and the core region represents a vortex tube.

The flow in a free-vortex is a combination of two types of flow : irrotational and rotational and is, therefore, known as the Rankine combined vortex. The flow within the core is rotational while that outside it is irrotational. Fig. 3.19 shows the characteristic of a combined vortex. The velocity characteristics within the core are similar to those of a tank of liquid rotating at a constant angular speed ω, *i.e.* $v = r\omega$. From the knowledge of energy relationship (Bernoulli's equation) discussed in Chapter 4, the nature of the free surface in the irrotational-vortex may be determined.

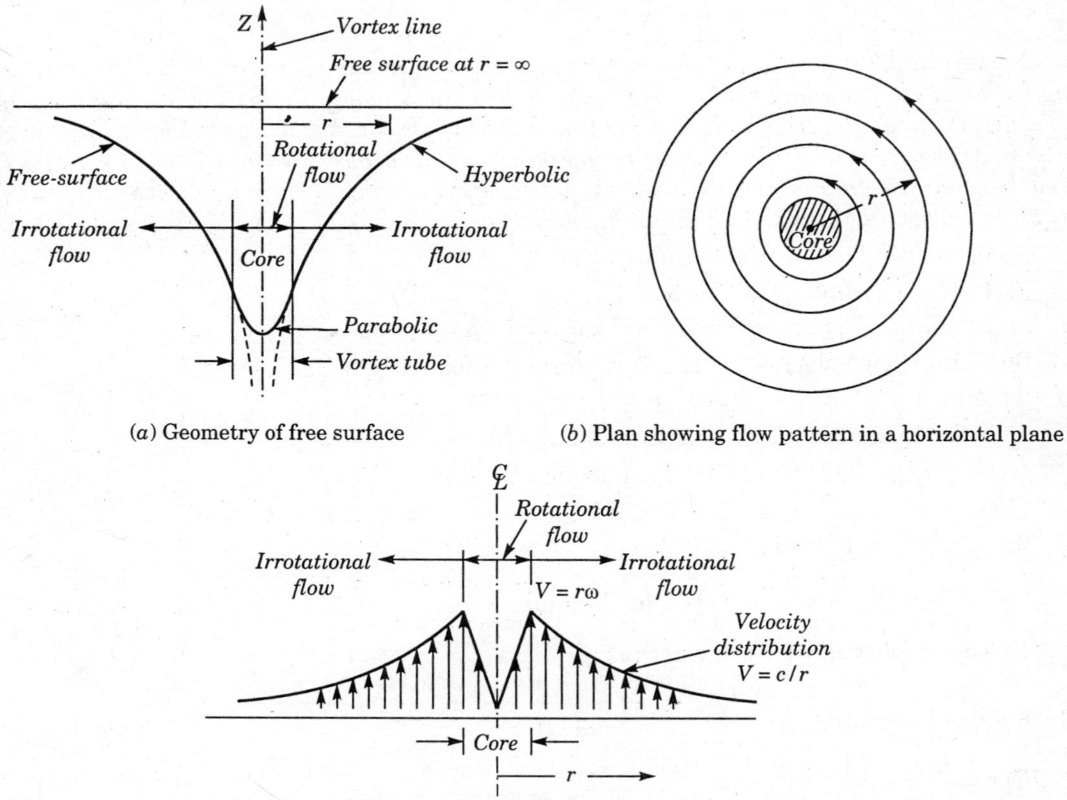

(*a*) Geometry of free surface

(*b*) Plan showing flow pattern in a horizontal plane

(*c*) Velocity distribution along r

Fig. 3.19. Characteristics of a combined vortex.

The irrotationality of a free vortex may be physically verified with the help of small pieces of light weight material such as cork placed floating on the free-surface. As these pieces move around it will be observed that they do not rotate above their mass centres—the fact may be visualised by marking a line on the floating piece which can be observed to remain parallel to itself during the motion. Four different positions of one such floating object are shown by A, B, C and D in Fig. 3.18 (a).

The circulation around a circular streamline is $\Gamma = 2\pi r.V = 2\pi\, C$ = constant, since $V = \dfrac{C}{r}$.

Here also, V is the tangential velocity.

As the circulation is independent of the radial distance, it shows that the circulation in a free-vortex is constant for all the streamlines. The stream function for the vortex may be obtained noting that the velocity is wholly tangential.

$$\psi = \int \frac{\partial\psi}{\partial r}\,dr + \int \frac{\partial\psi}{\partial\theta}\,\partial\theta \text{ but we known that } V = \frac{\Gamma}{2\pi r} = -\frac{\partial\psi}{\partial r} \quad \text{and} \quad V_r = \frac{\partial\psi}{r\partial\theta} = 0.$$

Making these substitutions,

$$\psi = \int\left(-\frac{\Gamma}{2\pi r}\right)dr + 0 = -\frac{\Gamma}{2\pi}\log r + A.$$

Let at $r = 1$, $\psi = 0$, then $A = 0$ and the stream function,

$$\psi = -\frac{\Gamma}{2\pi}\log r \qquad \text{...(3.58 } a\text{)}$$

The circulation Γ which is constant for a vortex is known as the *vortex strength*.

The stream function for a clockwise vortex may be written as

$$\psi = \frac{\Gamma}{2\pi}\log r \qquad \text{...(3.58 } b\text{)}$$

Example 3.33. *A cylindrical drum has an opening in the centre of its bottom. During its filling with water the opening is plugged so that no flow takes place through the opening. In order to draw water, the plug is opened and the water is allowed to flow out. As the water depth is progressively decreased, a spiral flow which is distinctly visible on the liquid surface is set up, the water from the sides approaching towards the centre follows a spiral path. This type of motion is designated as the free-spiral vortex. If the tangential velocity at a radial distance of 35 cm is found to be 25 cm/sec, determine :*

(*i*) *the velocity at a radial distance of 15 cm*

(*ii*) *the velocity-law for the vortex*

(*iii*) *the radial distance where the velocity is equal to 5 cm/sec.*

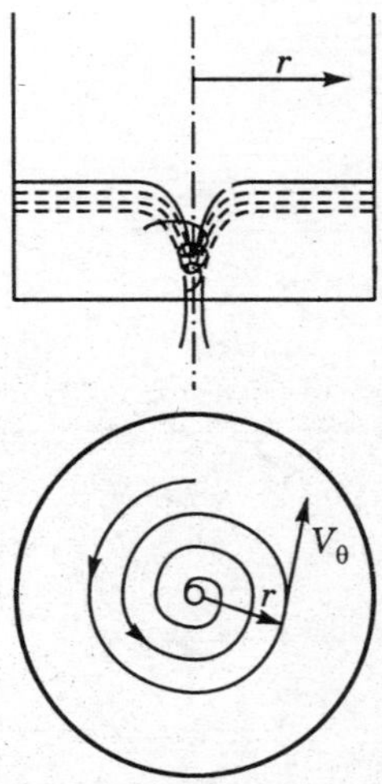

Free-spiral vortex

Solution. The velocity-law in a free-spiral vortex is written as

$$v_\theta \,.\, r = C$$

where v_θ is the tangential velocity at a radial distance r.

(*i*) Given v_θ = 25 cm/s at r = 25 cm, using this data,

$$C = 25 \times 35 = 875$$

Velocity at r = 15 cm is

$$v_\theta = \frac{C}{r} = \frac{875}{15} = \mathbf{58.33\ cm/s.}$$

(*ii*) The velocity law for the vortex is :

$$\mathbf{v_\theta r = 875}$$

in which v_θ is in cm/s and r in cm.

(*iii*) $r = \dfrac{C}{v_\theta} = \dfrac{875}{50}$ = 17.4 cm

At r = 17.4 cm, a velocity of 50 cm/s would be attained.

3.14.3. Forced-vortex

The fluid motion obtained when a tank with liquid inside is rotated about an axis at a constant angular speed is known as the *forced vortex* or a *rotational vortex*. The rotation of the liquid tank requires application of an external torque, thus justifying its naming as forced-vortex. The vorticity of flow everywhere within the fluid is non-zero and finite causing flow to be rotational. It is for this reason that this type of vortex is also called a rotational vortex. This type of flow has been analysed in Chapter 2 (Fluid Statics) while dealing with uniform rotation of static fluid. The streamlines are concentric circles and the velocity varies linearly with the radial distance, that is, $V = rw$. While dealing with circulation and vorticity, we have found that the vorticity for this type of flow is 2ω, which shows that it is directly related to the angular velocity of the fluid mass. Fig. 3.20 illustrates the distinguishing characteristics of a forced-vortex.

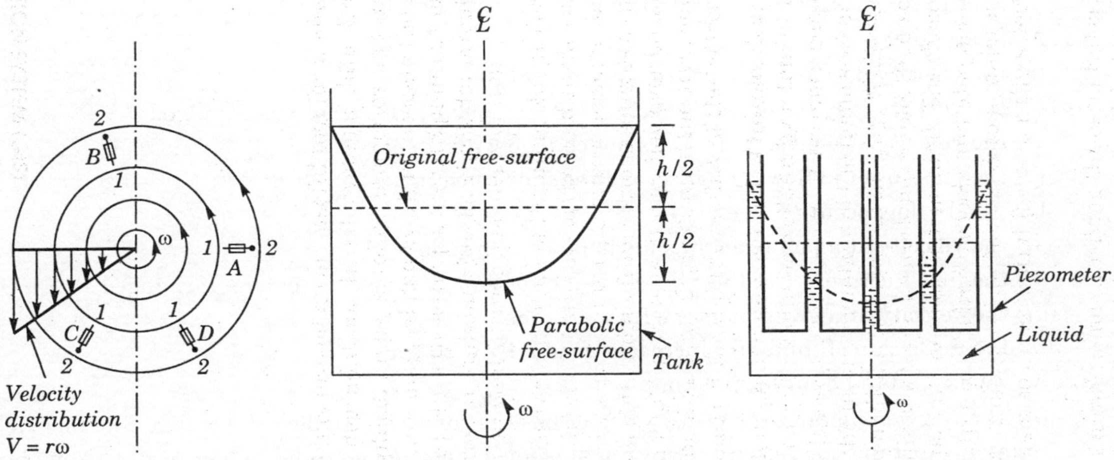

(*a*) Streamline pattern and velocity distribution in a horizontal plane

(*b*) Free-sruface inside a rotating tank

(*c*) Liquid under pressure between two circular rotating discs

Fig. 3.20. Characteristics of a forced vortex.

We know that since the streamlines are circular, the velocity vector being everywhere tangential has no radial component. An expression for the stream function may be obtained as below.

$$\psi = \int \frac{\partial \psi}{\partial r} dr + \int \frac{\partial \psi}{\partial \theta} d\theta$$

but
$$V = r\omega = \frac{\partial \psi}{\partial r}$$

and radial component
$$V_r = \frac{\partial \psi}{r\partial \theta} = 0.$$

Substituting these values and integrating, we obtain

$$\psi = -\frac{r^2\omega}{2} + \text{constant} \qquad \text{....(3.59)}$$

The negative sign indicates anti-clock wise direction of streamlines.

At $r = 0$, if $\psi = 0$, then the constant is equal to zero,

and
$$\psi = -\frac{r^2\omega}{2} \qquad \text{...(3.59 a)}$$

For clockwise streamlines

$$\psi = \frac{r^2\omega}{2} \quad \text{...(3.59 } b\text{)}$$

The rotational aspect of the forced-vortex may be physically visualised by means of small floating objects having a directional line marked on each of them. Various positions (indicated by, *A, B, C,* and *D*) of one such floating object as it moves along with the rotating liquid are shown in Fig. 3.20 (*a*). The directional line 1-2 changes its orientation from place to place indicating the existence of rotation of the object about its mass centre. This shows that the vorticity does exist at every point characterising rotational aspect of the flow.

OBJECTIVE TYPE QUESTIONS

Select the correct answer(s) from the following statements :

1. A steady flow is one in which :

(*a*) the velocity does not change from place to place

(*b*) the velocity may change its direction but the magnitude remains unchanged

(*c*) the velocity at a given point does not change with time

(*d*) the mass of fluid flowing per second remains constant

2. Unsteady flow occurs when :

(*i*) the velocity at a point remains unchanged with time

(*ii*) the flow fluctuates but the average values of flow parameters remain unchanged

(*iii*) the velocity at a given point varies with time

(*iv*) the mass rate of fluid flow increases gradually with time

3. A flow is said to be uniform when :

(*a*) at any given instant, the velocity vector has the same direction at all points in the flow field.

(*b*) at any instant, the velocity vector remains unchanged in magnitude and direction throughout the flow field

(*c*) at any instant, the velocity vector may change its direction but its magnitude does not change with space

(*d*) $\left(\frac{\partial V}{\partial s}\right) = 0$

4. A three-dimensional flow is one in which :

(*i*) the flow region has a width, depth and length

(*ii*) the velocity vector is inclined such that it can be resolved into three mutually perpendicular directions

(*iii*) the fluid moves in straight paths parallel to the main direction of flow

(*iv*) the velocity distribution over parallel planes is identical throughout the flow region

5. The flow in a river during the period of heavy rainfall is :

(*a*) steady, uniform, two-dimensional

(*b*) unsteady, uniform, three-dimensional

(*c*) unsteady, non-uniform and three-dimensional

(*d*) steady, non-uniform and three-dimensional

6. A water supply pipeline changes its alignment through a bend. When the flow in the pipeline is increased by operating a value, the flow in the bend is classified as :

(*i*) unsteady, uniform flow (*ii*) unsteady, non-uniform flow

(*iii*) steady, uniform flow (*iv*) steady, non-uniform flow

7. A streamline is a line drawn in the flow region such that :
   (*a*) it is always parallel to the main direction of flow
   (*b*) the velocity potential is constant on this line.
   (*c*) the tangent drawn at any point on the line indicates the direction of velocity
   (*d*) there is no flow across the line
8. A streamtube is one :
   (*i*) which is bounded by a closed surface containing the streamlines
   (*ii*) which has constant area throughout its length so that the velocity remains constant
   (*iii*) in which the flow cannot cross the bounding surface
   (*iv*) in which the stream function does not change
9. Total acceleration in fluid-flow consists of local and convective parts. Column A of the following table shows few situations of pipe flow. Identify the type of acceleration associated with each, choosing it from column B.

Column A	*Column B*
(*a*) flow at constant rate passing through a bend.	(*i*) zero acceleration
(*b*) gradually changing flow through a straight pipe	(*ii*) local and convective acceleration both
(*c*) gradually changing flow through a bend	(*iii*) convective acceleration
(*d*) flow at constant rate passing through a straight uniform diameter pipe	(*iv*) local acceleration
(*e*) flow at constant rate passing through a pipe of gradually changing diameter.	(*v*) convective acceleration.

10. Normal acceleration in fluid-flow situations exists only when
    (*i*) the streamlines are straight and parallel
    (*ii*) the flow is two-dimensional
    (*iii*) the streamlines are curved
    (*iv*) the flow is unsteady
11. The continuity equation in fluid mechanics is a mathematical statement embodying the principle of :
    (*a*) conservation of energy (*b*) conservation of mass
    (*c*) conservation of momentum (*d*) none of the above.
12. Irrotational flow is characterised as the one in which :
    (*i*) the fluid flows along a straight path
    (*ii*) the fluid does not rotate as it moves along
    (*iii*) the net rotation of fluid particles about their mass centres remains zero
    (*iv*) the streamlines of flow are curved and closely spaced
13. The existence of velocity potential in fluid-flow indicates that :
    (*a*) the flow must be irrotational
    (*b*) the flow is rotational and satisfies the continuity equation
    (*c*) the vorticity must be zero
    (*d*) the circulation around any closed curve must have a finite value
14. The concept of stream function which is based on the principle of continuity is applicable to :
    (*i*) three-dimensional flow (*ii*) two-dimensional flow only
    (*ii*) uniform flow cases only (*iv*) irrotational flow only

15. Streamlines in fluid-flow are :

(*a*) equipotential lines along which the velocity potential ϕ = constant

(*b*) lines along which the stream function ψ = constant

(*c*) lines along which the vorticity is zero

(*d*) lines which exist in irrotational flow only

16. A flownet is a graphical representation of streamlines and equipotential lines such that :

(*i*) these lines indicate the direction and magnitude of velocity vector

(*ii*) these lines intersect each other orthogonally forming curvilinear squares

(*iii*) these lines intersect each other at various different angles forming irregular shaped nets

(*iv*) the velocity potential ϕ increases in the direction of flow

17. In vortex motion, the streamlines are circular and the fluid appears to be rotating. Identify the type of vortex motion for the following cases choosing from the alternatives given :

(*a*) liquid rotating at constant speed in a vessel	(*i*) free-vortex
(*b*) flow in a pipe bend	(*ii*) forced-vortex
(*c*) flow in a wash-basin when it is being emptied through a central opening	(*iii*) free-vortex
(*d*) flow in a canal bend	(*iv*) forced-vortex
(*e*) flow in the impeller of a centrifugal fugal pump.	(*v*) free-vortex

18. The velocity potential function for a source varies with distance r as :

(*a*) $\dfrac{1}{r}$ (*b*) $\dfrac{1}{r^2}$

(*c*) e^r (*d*) $\ln r$.

Note. Use Eqs. (3.35*a*) and (3.36*a*).

19. A Newtonian fluid has the following velocity field :

$$\vec{V} = x^2y\,\hat{i} + 2\,xy^2z\,\hat{j} - yz^3\,\hat{k}$$

The rate of shear deformation ε_{yz} at the point $x = -2, y = -1$ and $z = 2$ for the given flow is :

(*a*) = – 6 (*b*) = – 2

(*c*) = – 12 (*d*) = – 4.

20. Two flow patterns are represented by their stream functions ψ_1 and ψ_2 as given below :

$$\psi_1 = x^2 + y^2$$
$$\psi_2 = 2xy$$

If these two patterns are superposed on one another, the resulting streamline pattern can be represented by one of the following :

(*a*) A family of parallel streamlines (*b*) A family of circles

(*c*) A family of parabolas (*d*) A family of hyperbolas.

21. A streamline and an equipotential line in a flow field :

(*a*) are parallel to each other (*b*) are perpendicular to each other

(*c*) intersect at acute angle (*d*) are identical.

ANSWERS TO OBJECTIVE TYPE QUESTIONS

1. (*c*), (*d*) **2.** (*iii*) **3.** (*b*), (*d*) **4.** (*ii*) **5.** (*c*) **6.** (*ii*) **7.** (*c*), (*d*) **8.** (*i*), (*iii*)
9. (*a*) (*iii*) ; (*b*) (*iv*) ; (*c*) (*ii*) ; (*d*), (*i*) ; (*e*) (*v*) **10.** (*iii*) **11.** (*b*) **12.** (*iii*) **13.** (*c*)

14. (*ii*) **15.** (*b*) **16.** (*ii*) **17.** (*a*), (*ii*) ; (*b*), (*i*) ; (*c*), (*iii*) ; (*d*), (*v*) ; (*e*) (*iv*) **18.** (*a*)
19. (*b*) **20.** (*d*) **21.** (*b*)

REVIEW QUESTIONS

1. Distinguish between streamlines, streaklines and pathlines.

2. Define steady, non-steady, uniform and non-uniform flow.

3. When is the flow in a pipe (*a*) unsteady and (*b*) non-uniform ?

4. What type of flow is represented by the following equations :

(*a*) $u = u_0\left(\dfrac{r_0^2 - r^2}{r_0^2}\right)$ (*b*) $u = kxyzt$

(*c*) $u = xy$ (*d*) $u =$ constant ?

5. Cite examples of unsteady flow. How can the unsteady flow be transformed to steady one ?

6. A lawn sprinkler issues jets at either end of an arm which rotates about a vertical axis. Is the flow from such a sprinkler steady or unsteady ?

7. Define average velocity of flow across an area.

8. What should be the acceleration in steady uniform flow ?

9. Sketch the velocity distribution for uniform irrotational flow.

10. What do you understand by vorticity ?

11. Explain the physical significance and use of the term 'stream function' in dealing with the problems of two-dimensional fluid motion.

12. Differentiate between a free and a forced-vortex.

13. What is vortex flow ?

14. What use can the velocity potential and the stream function be put to in analysing two-dimensional irrotational flow ?

15. Is the flownet analysis applicable to rotational flow ? If not, why ?

16. To what type of flow is the concept of (*a*) velocity potential (*b*) stream function applicable ?

17. If the points P_1 and P_2 have stream function ψ_1 and ψ_2, what will be the flow passing between them if flow is two-dimensional ?

18. If the points P_1 and P_2 of question 17 are lying on the same streamline, what is the relation between ψ_1 and ψ_2 ?

19. Does the existence of stream function in a flow problem imply the existence of velocity potential ?

20. What is a vortex line ?

21. What do you understand by (*a*) a potential vortex, and (*b*) a combined vortex ?

22. The velocity of flow in a problem changes from section to section along the flow direction. Is the type of flow : (*i*) unsteady (*ii*) turbulent (*iii*) non-uniform ?

23. Identify the type of flow for cases listed in column *A* choosing the flow type from column *B*.

A	*B*
(*i*) flow in a pipe bend	uniform,
(*ii*) $u = a + 2xt$	non-uniform,
(*iii*) flow in a river during a rainstorm	unsteady and non-uniform
(*iv*) flow in a constant diameter pipe	unsteady,
when a valve is being closed	unsteady, non-uniform and turbulent.

24. Show from the consideration of vorticity and rotation, that in case of ideal fluids the flow is irrotational.

25. Show that the streamlines and equipotential lines form a net of mutually perpendicular lines.

PROBLEMS

3.1. For the flow of an incompressible fluid, the velocity component in the x-direction is

$$u = ax^2 + by$$

and the velocity component in the z-direction is zero. Find the velocity component v in the y direction. To evaluate arbitrary functions which might appear in the integration assume that $v = 0$ at $y = 0$.

3.2. Incompressible fluid flows radially, outward in all directions from a point source. Assuming that the velocity V depends only upon r^-, the distance from the source, find the velocity V as a function of r.

3.3. Incompressible fluid flows radially outward from a line source. Assuming that the velocity V depends only upon r, the distance from the source, find the velocity V as a function r.

3.4. A 15 cm diameter horizontal pipe discharges water at the outlet. The jet strikes the ground at a horizontal distance of 3.2 m, the vertical drop being 1.4 m measured from the end of the pipe. Determine the rate of flow.

3.5. A nozzle with a base diameter of 8 cm and a 3 cm diameter tip discharges 25 litres per second. Find the velocity at the base at the tip of the nozzle.

3.6. A 5 m diameter pressure pipe has a velocity of 4.5 m/s. After passing through a reducing bend the flow enters into a 4.5 m diameter pipe. If the losses vary as the square of the velocity, how much greater are they through the 4.5 m pipe as compared to the 5 m pipe per 100 m length of pipe ?

3.7. Liquid flows through an orifice in the bottom of a cylindrical tank at a velocity of 4 m/s. If the jet diameter is 2.5 cm, what is the rate of flow in lit/sec. ? If the tank is 25 cm in diameter, at what rate is the liquid surface falling ?

3.8. A diffuser consists of two circular parallel plates 20 cm in diameter 0.5 cm apart, connected to a 3 cm pipe as shown in the figure. If the streamlines are assumed to be radial in the diffuser, what mean velocity in the pipe will correspond to an exit velocity of 0.5 m/s ?

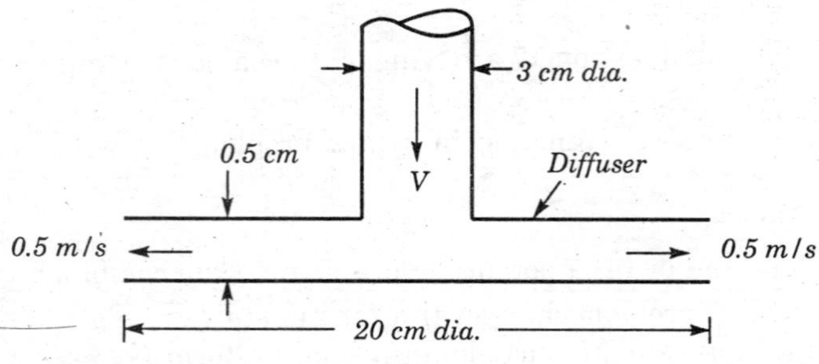

3.9. A nozzle is so shaped that the velocity of flow along the centre line changes linearly from 2 to 15 m/s in a distance of 0.5 m. Determine the convective acceleration and the force per unit volume acting on water at the beginning and end of the nozzle.

3.10. The rate of flow through a tappered pipe varies from 0 to 0.5 m^3/s in 15 seconds. Determine the acceleration at sections 30 cm, 22.5 and 15 cm in diameter.

3.11. A circular cylinder is rotated about its own axis at a constant rate of 1500 r.p.m. Determine the centripetal acceleration at a radial distance of 15 cm and find the ratio of this acceleration to the acceleration due to gravity.

3.12. An overflow section of a dam has a radius of 2 m at its topmost point. Determine the local velocity of flow at this section which will produce a normal acceleration equal to 50% of gravity acceleration.

3.13. The toe of a spillway is curved and has a radius of 5.75 m. Spillway is 25 m long and is designed for a flood discharge of 320 m^3/s. At this rate of flow, the thickness of the water sheet near the toe is 50 cm. What is the resulting centripetal acceleration ?

3.14. Water flows through a 15 cm diameter pipeline at a mean velocity of 1.35 m/s. Calculate the rate of flow in m^3/s, litre/sec, kg/s and N/s.

3.15. A 30 cm diameter horizontal pipeline leaves a reservoir through a circular hole in its vertical wall. The mean velocity of flow in the pipe has been determined as 5.24 m/s. Assuming that the water approaches the centre of the pipe entrance radially, determine velocity of water at distance 0.5, 1.0 and 1.5 m from the pipe entrance.

3.16. A channel, while negotiating a fall, has its bottom sloping at 30° to the horizontal. The vertical depth of flow at a section was observed to be 1.85 m and the velocity (parallel to the channel bottom) was estimated at 2.40 m/s. If the channel width is 10 m, what is the rate of flow in the channel ?

3.17. The depth of water at a place in a wide river was found to be 3.10 m. At the same place, the current meter observations yielded an average velocity of 0.78 m/s. Determine the discharge (rate of flow) per unit width of the river. At another section in the river where the depth of water is 2.65 m, what mean velocity of flow would exist ?

3.18. Show that the velocity components given by

$$u = \frac{2x}{t}, v = -\frac{y}{t} \text{ and } w = -\frac{z}{t}$$

represent a possible fluid flow case. Is the flow steady and irrotational ?

3.19. From the law of conservation of mass, show whether the flow field represented by

$$u = -3x + y^2 - \frac{1}{x} \quad \text{and} \quad v = x^2 + 3y + y \log x$$

is a possible velocity field for two-dimensional incompressible fluid flow.

3.20. A stream function is defined by $\psi = x^3 - y^3$ show that the flow cannot be a potential flow.

3.21. Show that in a compressible steady flow, the following equations suitably define the stream function

$$\rho u = \rho_0 \frac{\partial \psi}{\partial y}; \quad \rho v = \rho_0 \frac{\partial \psi}{\partial x}$$

where ρ_0 is a reference density.

3.22. Does the velocity distribution

$$u = -x, v = 2y \text{ and } w = 2 - z$$

which represents incompressible fluid flow, satisfies the continuity equation ?

3.23. The flow of a fluid has velocity components

$$u = 3x + y$$
$$v = 2x - 3y.$$

Determine whether or not the flow is irrotational.

3.24. A certain flow pattern has a velocity potential

$$\phi = 1/3x^2 - x^2 + xy^2 + y^2.$$

Determine the stream function.

3.25. If the velocity components at a point in a fluid flow are given by

$$u = 2xy$$

$$v = a^2 + x^2 - y^2.$$

Show that the flow is possible, and obtain the stream function ψ.

3.26. If the velocity components of a possible flow are

$$v = 4\,ay(3x^2 - y^2),\; u = -\,4ax(x^2 - 3y^2).$$

Determine whether or not the motion is possible, and obtain the stream function.

3.27. Show that

$$u = -\frac{2xyz}{(x^2 + y^2)^2},\quad v = \frac{(x^2 - y^2)z}{(x^2 + y^2)^2},\quad w = \frac{y}{x^2 + y^2}$$

are the velocity components of a possible fluid motion. Is this motion irrotational ?

3.28. A two-dimensional flow can be described by

$$u = -\frac{y}{b^2},\qquad v = \frac{x}{a^2}$$

Verify that this is the flow of an incompressible fluid and that the ellipse

$$\frac{a^2}{y^2} + \frac{y^2}{b^2} = 1 \text{ is a streamline.}$$

3.29. For the given three-dimensional flow field described by

$$V = (x + y)\,i + (y + z)j + (x^2 + y^2 + z^2)k$$

find the components of rotation at (1, 1, 1).

3.30. The x and y components in a three-dimensional flow are given by

$$u = x^2 + z^2,\; v = y^2 + z^2$$

Find the simplest z-component of velocity that satisfies the continuity equation.

3.31. Given a flow field $V = (16x^2 + y)i + 10j + yz^2k$, what is the angular velocity of fluid element at $r = 6i + 3j + 2k$?

3.32. What is the irrotational flow field associated with the function

$$\psi = 3x^2 - 3x + 3y^2 + 16z^2 + 12\,zt.$$

Does the flow field satisfy the incompressible continuity equation ?

3.33. Does the velocity distribution

$$q = i(4 - x^2 + y) + j(3 + 2y - z) + k\;2r\;(x - 1)$$

satisfy the continuity equation for incompressible flow in which q is the velocity vector ?

3.34. Does the velocity distribution

$$q = i\,\log x^2y^2 + j\left(\frac{2y}{g} - \log xt\right)$$

satisfy the continuity ?

3.35. A flow of an incompressible fluid is represented by the stream function $\psi = 2Vy$; (*i*) Plot the streamlines, (*ii*) find the quantity flowing between the streamlines $y = 2$ and $y = 3$ and (*iii*) find the x and y components of velocity at any point.

3.36. Given the stream function $\psi = Bxy$, in which B is a positive constant : (*a*) Sketch the streamlines, (*b*) find x and y velocity components at (1, 1) and (1, 4) and (*c*) find the flow crossing any line joining (1, 3) and (2, 5).

3.37. A two-dimensional flow is represented by $\psi = 3x^2y - y^3$. Plot the streamlines and determine the magnitude of the velocity at any point.

3.38. Which of the following velocity fields are possible fluid flow cases ? Which of these satisfy the conditions for irrotational flow ?

(*i*) $u = x + y + z,\; v = x - y + z,\; w = x + y + 3.$

(*ii*) $u = xy, y = yz, w = y^2 + z^2$.

(*iii*) $u = kxyzt, v = -kxyzt^2, w = \frac{1}{2} kz^2(xt^2 - yt)$.

3.39. Show that $u = -2xyz/(x^2 + y^2)^2, v = (x^2 - y^2) - z/(x^3 + y^2)^2, w = y/(x^3 + y^2)$ are velocity- components of a possible fluid motion.

Verify that the motion is irrotational and find the velocity potential.

3.40. A velocity field is given by $u = yz + t, v = zx > t, w = xy$. Is the flow steady ? Find the acceleration components of a fluid particle at (1, 1, 1) in terms of t.

3.41. If the velocity at the point $(x, y, z,)$ is (u, v, w) where $u = ax, v = ay, w = -2ax$, prove that the streamlines are the surfaces y^2z = constant are cut by the planes x/y = constant.

3.42. Determine the condition that $u = Ax + By, v = Cx + Dy$, may give the velocity components of a liquid. Show that the streamlines of this motion are conic sections in general, and rectangular hyperbolas when the motion is irrotational.

3.43. Show that $\psi = (x - t)(y - t)$ represents the velocity potential of an incompressible two-dimensional fluid. Show that the streamlines at time t are the curves $(x - t)^2 - (y - t)$ = constant.

3.44. If for two-dimensional flow the stream function is given by $\psi = 2xy$, calculate the velocity at the point (3, 6). Show that the velocity potential ϕ exists for this case and deduce it.

Draw the streamlines corresponding to $\psi = 100$ and $\psi = 300$; also equipotential lines corresponding to $\phi = 100$ and $\phi = 300$. About six points on each line to represent its trend would suffice.

3.45. Show how the linear velocity components may be expressed as a function of the partial derivatives of the stream function.

3.46. Define irrotational flow in the context of steady flow of an incompressible fluid. Explain the significance of this type of flow in the study of free vortex flow and in flownet applications.

ANSWERS

3.1. $v = -2axy$. **3.2** $V = K/r^2$. **3.3** $V = K/r$.

3.4. 0.106 m³/s. **3.5.** 4.97 m/s, 35.3 m/s. **3.6.** 52%.

3.7. 1.964 lit/s, 4 cm/s. **3.8.** 2.22 m/s.

3.9. 52 m/s², 390 m/s² ; 2 kN/m³, 391 kN/m³

3.10. 0.471 m/s², 0.83 m/s³, 1.89 m/s². **3.11.** 3700 m/s², 377.

3.12. 3.13 m/s. **3.13.** 114 m/s².

3.14. 2.38×10^{-2} m²/s, 23.8 lit/s, 238.5 kg/s, 233.6 N/s.

3.15. 0.236 m/s, 0.059 m/s, 0.262 m/s. **3.16.** 38.4 m³/s.

3.17. 2.42 m²/s/m, 0.912 m/s. **3.18.** Possible, unsteady, irrotational.

3.19. Not possible **3.22.** Satisfies. **3.23.** Rotational.

3.24. $\psi = 1/3y^3 - 2xy + 4xy$. **3.25.** $\psi = x(2y^2 - a^3 - 1/3x^2)$. **3.26.** Possible, $\psi = 8\,axy(y^3 - x^2)$.

3.27. Yes **3.28.** $\omega_x = 0.5\ s^{-1}, \omega_y = -1\ s^{-1}, \omega_z = -0.5\ s^{-1}$.

3.30. $\omega = -2z(x + y)$. **3.31.** $\omega = 2.06\ s^{-1}$.

3.32. $V = (-6x + 3)\,i - 6\,yj - 12\,tk$ and $a = (36x - 18)i + 36yj - 12k$, Not satisfied.

3.33. Satisfies. **3.34.** Not satisfied. **3.35.** $dq = 2\,V, u = 2\,V, v = 0$.

3.36. (*b*) $u = B, v = -B, u = B, v = -4B$, (*c*) $dq = 7B$ **3.37.** $V = 3(x^2 + y^2)$.

3.38. (*i*) Possible, $\phi = \frac{1}{2}(y^2 - x^2) - 3z - (yz + zy + xy) + f(t)$. (*ii*) Not possible.

(*iii*) Possible, Rotational motion. **3.39** $\phi = -yz(x^2 + y^2)$.

3.40. Unsteady, $a_x = 3 - t, a_y = 1 + t, a_2 = 2$. **3.44.** $2\sqrt{45}$, $\phi = y^2 + x^2$.

4

Dynamics of Fluid Flow

4.0. Introduction

In the previous chapter on Kinematics of Flow, we have dealt with velocities and accelerations without bothering about the forces associated with the flow phenomena. The present chapter on dynamics of fluid flow takes into considerations all such forces which affect the phenomena. The Newton's second law of motion, which relates the acceleration with the forces causing it, provides an important tool for analysing the dynamic behaviour of flow.

4.1. Transport Phenomena

In the study of dynamics of fluids, one is usually concerned with some aspect of the fluid transport phenomena. These aspects may be connected with the ability of fluids in motion to convey materials and properties from place to place and the mechanism by which these materials and properties are transmitted and diffused through the fluid medium. It is useful to classify the available methods of analysis in terms of various of types of transport process. The fundamental transport phenomena which are associated with the motion of a fluid are :

1. mass transport,
2. heat transport, and
3. momentum transport.

Each of these processes, in turn, is associated with a basic physical law. For example, the mass transport is governed by the principle of conservation of matter, the heat transport by the conservation of energy (first law of thermodynamics) and the momentum transport by the Newton's second law of motion.

4.1.1. Mass Transport

All fluid motions must satisfy the principle of conservation of matter. Fluids may be homogeneous or non-homogeneous. A homogeneous fluid is one which exists throughout the region of interest as a single specy. As an example, air may undergo changes in density, velocity and temperature, yet it may remain identifiable as the stable mixture of gases known as air. All liquids are treated as homogenous unless there is a change in phase. A non-homogeneous fluid is one which consists of two or more identificable species within the region of interest. Non homogeneous fluids are characterized by variation in the amount of one substance relative to another from point to point. The species or components may be of the same or different phases. For example a jet of carbon dioxide discharged into the atmosphere leads to variation in the concentration of carbon dioxide from point to point and the fluid (mixture of CO_2 and air) is non-homogeneous but of the same phase. A canal carrying sediment laden water in suspension is an example of non-homogeneous, two-phase flow. Similarly, a mixture of air bubbles and water (flow with air entertained) is another example of non-homogeneous two-phase flow.

In homogeneous fluids, the principle of conservation of matter leads to what is known as the continuity equation. In case of non-homogeneous fluids, the principle of conservation of matter must be satisfied for each component or species of the fluid mixture.

4.1.2. Heat Transport

In the dynamics of fluid motion, we are primarily concerned with transport of heat (energy) by fluid motion. By applying the principle of conservation of energy (also known as the first law of thermodynamics) to a flow process, an equation is obtained which provides a useful relation between pressure, velocity, elevation, density, temperature, mechanical work and heat input (or output). All kinds of fluid machines such as pumps, turbines and compressors involve energy transfer in flow process.

4.1.3. Momentum Transport

Newton's second law of motion provides the fundamental relation between the summation of forces acting on a particle and the time rate of change of its momentum. The resultant equations are known as equations of motion. Momentum transport phenomena are of interest in fluid mechanics since they deal with mechanics of fluid resistance, shear stresses and propulsion and forces on immersed bodies.

4.2. One-dimensional Method of Flow Analysis

Fluid flow problems, in general, are three dimensional in character, meaning thereby that both the velocity and the acceleration may have components along each of the three co-ordinate axes. The flow analysis of such problems is very complex. The one-dimensional method of flow analysis assumes that the velocity is same in magnitude and direction at all successive cross-sections. The average velocity, pressure and elevation are assumed to represent adequately the condition at a given section. In the following treatment, one dimensional method will be used to develop energy and momentum equations.

4.3. Laws Affecting Fluid Motion

The dynamics of fluid motion deals with kinetics which relates the kinematics with the forces responsible for causing the motion. This relationship of fluid motion is established by the use of laws of nature, namely :

(*i*) The principle of conservation of mass (*i.e.* the continuity relationship)

(*ii*) Newton's laws of motion

(*iii*) The first and second laws of thermodynamics.

In addition, the boundary conditions such as the 'no-slip' condition implying that a real fluid has zero velocity at the boundary relative to the boundary itself or that fluids cannot penetrate a boundary must also be satisfied. At times, the Newton's law of viscosity and the equation of state may also have to be used.

Newton's First Law of Motion

Every body remains in its state of either rest or of uniform motion unless its state is changed by some external forces impressed upon it. This law gives the conditions of equilibrium as :

$$\Sigma F = 0 \quad \text{and} \quad \Sigma M = 0$$

in which F stands for the force and M for the moment of the force. These conditions have been used in determining the hydrostatic force on submerged bodies and also in evaluating the stability of floating bodies.

Newton's Second Law of Motion

The rate of change of momentum (mass × velocity) is proportional to the impressed force, and takes place in the direction of that force.

$$F = \frac{d}{dt}(mV) = m\,\frac{dV}{dt} = m\,.\,a$$

The momentum may be linear or angular and the corresponding actions are force and moment (torque) respectively. This law of motion forms the backbone of dynamical analysis of fluid-flow situations.

Newton's Third Law of Motion

To every action there is always an equal and opposite reaction. This law finds application in determining the reaction force exerted by pipe bends, nozzle-jets, curved vanes and other objects the presence of which causes a change in the momentum of flow.

First Law of Thermodynamics

The change in total energy ΔE of a system is equal to the amount of heat δQ added to the system minus the work δW done by the system. Thus,

$$\Delta E = \delta Q - \delta W$$

Here the total energy E is a property of the system as it depends on the state of the system. The quantities Q and W represent the interactions of the system with its surroundings (and hence are not properties).

In problems of flow analysis, we are concerned with the rates of change rather than the absolute changes in the thermodynamic variables. The law may, therefore, be written as

$$\frac{DE}{Di} = \frac{dQ}{dt} - \frac{dW}{dt} = \dot{Q} - \dot{W}$$

where $\dot{Q}$ and $\dot{W}$ are the rate of heat transfer to and the rate of work done by the system respectively.

Second Law of Thermodynamics

The change of entropy equals or exceeds the heat exchange divided by absolute temperature. It equals for a reversible process and exceeds for an irreversible process.

$$\Delta s \geq \frac{\delta Q}{T}$$

The second law is utilized in dealing with the analysis of compressible processes.

4.3.1. Equations of Motion

The dynamic behaviour of fluid motion is governed by a set of equations, known as equations of motion. These equations are obtained by using the Newton's second law, which may be written as

$$F_x = m \cdot a_x \qquad ...(4.1)$$

where F_x is the net force acting in the x-direction upon a fluid element of mass m producing an acceleration a_x in the x-direction.

The forces which may be present in fluid flow problems are : gravity forces F_g, pressure force F_p, force due to viscosity F_v, force due to turbulence F_t, and force due to compressibility of fluid F_c. When volume changes are small, the force due to compressibility is negligible, and the general equation of motion in the x-direction using Eq. (4.1) may be written as

$$ma_x = (F_g)_x + (F_p)_x + (F_v)_x + (F_t)_x \qquad ...(4.2)$$

Similar expression for y and z-directions may also be written. When we substitute the expressions for various quantities involved in Eq. (4.2) and other similar equations, the resulting equations are known as the *Reynolds equations.*

For flow at low Reynolds number, the force due to turbulence is of no significance and, therefore, may be ignored from Eq. (4.2). The resulting equation of motion taking into account the gravity force, pressure force and the viscous force is

$$ma_x = (F_g)_x + (F_p)_x + (F_v)_x \qquad ...(4.3)$$

Eq. (4.3) together with similar expressions for y and z-directions are known as the *Navier-Stokes equations.*

If the flow is assumed to be ideal *i.e.* possessing no viscosity, the equation of motion in the x-direction may further be simplified and written as

$$ma_x = (F_g)_x + (F_p)_x \qquad ...(4.4)$$

The equations of motion, which take into account only the gravity and pressure forces are known as the *Euler's equations* of motion.

4.4. Euler's Equations of Motion

Let us consider a streamline with the flow taking place in s-direction (Fig. 4.1). Consider a cylindrical element of cross-sectional area dA and length ds. The pressure force on its upstream face $p.dA$ and that on the downstream face $\left(p+\frac{\partial p}{\partial s}ds\right)dA$; and act normal to these faces *i.e.* tangentially to the streamline. The weight of the element is $\gamma dA\, ds$. Let S and N be the body forces per unit mass in the direction of the streamline and normal to it respectively. According to Newton's law of motion, the equation of motion in the s-direction may be written as

$$(\rho\, dA\, ds)a_s = pdA - \left(p+\frac{\partial p}{\partial s}ds\right) dA + \rho d\, Ads\, S$$

$$= -\frac{\partial p}{\partial s} dA\, ds + \rho dA\, ds\, .\, S$$

dividing by the mass of the element $\rho\, dAds$, the acceleration

$$a_s = S - \frac{1}{\rho}\frac{\partial p}{\partial s}$$

Substituting from Eq. (3.9), we obtain

$$\frac{\partial V_s}{\partial t} + V_s \frac{\partial V_s}{\partial s} = S - \frac{1}{\rho}\frac{\partial p}{\partial s} \qquad ...(4.5)$$

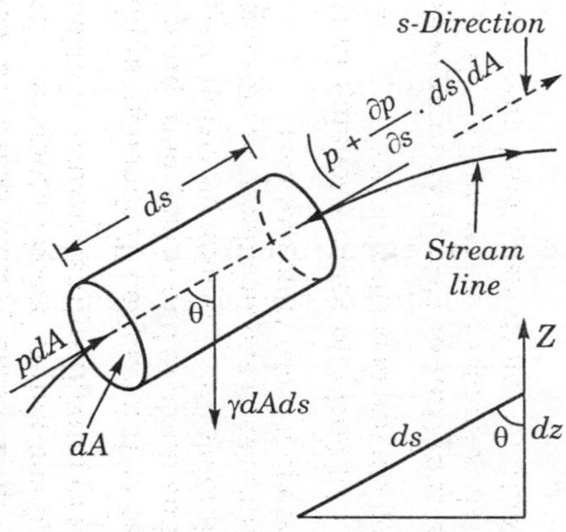

Fig. 4.1. Pressure and gravity forces on a cylindrical element along a streamline.

Equation (4.5) is known as the Euler's (pronounced as Oyler) equation of motion in the s-direction. Similarly, we may write Euler's equation in the normal direction.

$$a_n = \frac{\partial V_n}{\partial t} + V_s \frac{\partial V_n}{\partial s} = N - \frac{1}{\rho}\frac{\partial p}{\partial n} \qquad ...(4.6)$$

Equations (4.5) and (4.6) are also sometimes called as the Euler's equations of accelerations, since the left hand side of these equations represent the accelerations in the respective directions.

The Euler's equations of motion in the x, y and z-directions may be written as

$$\left.\begin{aligned} \frac{Du}{Dt} &= \frac{\partial u}{\partial t} + u\frac{\partial u}{\partial x} + v\frac{\partial u}{\partial y} + w\frac{\partial u}{\partial z} = X - \frac{1}{\rho}\frac{\partial p}{\partial x} \\ \frac{Dv}{Dt} &= \frac{\partial v}{\partial t} + u\frac{\partial v}{\partial x} + v\frac{\partial v}{\partial y} + w\frac{\partial v}{\partial z} = Y - \frac{1}{\rho}\frac{\partial p}{\partial y} \\ \text{and} \quad \frac{Dw}{Dt} &= \frac{\partial w}{\partial t} + u\frac{\partial w}{\partial x} + v\frac{\partial w}{\partial y} + w\frac{\partial w}{\partial z} = Z - \frac{1}{\rho}\frac{\partial v}{\partial z} \end{aligned}\right\} \qquad ...(4.6a)$$

where u, v and w are the velocity components and X, Y and Z are the body forces per unit mass along x, y and z directions respectively, and the differential operator D/Dt is known as the total deviative.

Example 4.1. *The rate at which water flows through a horizontal 25 cm pipe is increased linearly from 30 to 150 litres/sec in 3.5 seconds. What pressure gradient must exist to produce this acceleration ? What difference in pressure intensity will prevail between sections 8 m apart ? Take* $\rho = 1000\ kg/m^3$.

Solution. The Euler's equation of motion along the pipe axis may be written as

$$\frac{\partial v}{\partial t} + v\frac{\partial v}{\partial s} = S - \frac{1}{\rho}\frac{\partial p}{\partial s}$$

Since the pipe has a constant diameter,

$$\frac{\partial v}{\partial s} = 0$$

and since it is horizontal, the body force per unit mass S along the flow direction is also zero. The above equation of motion, therefore, reduces to :

$$\frac{\partial v}{\partial t} = -\frac{1}{\rho}\frac{\partial p}{\partial s}.$$

The change in velocity as the flow changes from 30 to 150 litres/sec is

$$\Delta v = \frac{150}{1000 \times \frac{\pi}{4} \times (0.25)^2} - \frac{30}{1000 \times \frac{\pi}{4} \times (0.25)^2} = 2.445 \text{ m/s}.$$

$$\therefore \quad \frac{\partial v}{\partial t} = \frac{2.445}{3.5} = 0.698 \text{ m/s}^2$$

and the pressure gradient $\dfrac{\partial p}{\partial s} = -\rho\dfrac{\partial v}{\partial t} = -1000 \times 0.698 = -698 \text{ N/m}^2\text{/m}$

Difference in pressure between sections 8 m apart

$$= \frac{\partial p}{\partial s} \times 8 = -698 \times 8 = -5.58 \text{ kN/m}^2.$$

4.5. Integration of Euler's Equation to Yield Bernoulli's Equation

Let Ω be the force potential such that

$$S = -\frac{\partial \Omega}{\partial s} \text{ and } N = -\frac{\partial \Omega}{\partial n}.$$

For steady flow Eq. (4.5) results in

$$V_s \frac{\partial V_s}{\partial s} = -\frac{\partial \Omega}{\partial s} - \frac{1}{\rho}\frac{\partial p}{\partial s}$$

or
$$\frac{1}{2}\frac{\partial}{\partial s}(V_s^2) = -\frac{\partial}{\partial s}\left(\frac{p}{\rho} + \Omega\right) \quad \text{or} \quad \frac{\partial}{\partial s}\left(\frac{V_s^2}{2} + \frac{p}{\rho} + \Omega\right) = 0$$

Integrating this expression with respect to s, taking limit $dA \to 0$, $ds \to 0$ and treating ρ as a constant, *i.e.* considering fluid as incompressible, we obtain

$$\frac{V_s^2}{2} + \frac{p}{\rho} + \Omega = \text{constant} \quad \text{...(4.7)}$$

From Fig. 4.1 noting that S is the body force per unit mass, and considering it due to gravity :

the body force component in the s-direction $= (\rho\, dAds)S$

and, the weight component in the same direction $= -\gamma dAds \cos\theta$

$$= -\gamma dAds \frac{dz}{ds}$$

Since the two force components are the same

$$(\rho\, dA\, ds)\, S = -\gamma\, dA\, ds \frac{dz}{ds}$$

$$\therefore \quad S = -g\frac{dz}{ds}$$

But from the concept of the force potential $S = -\dfrac{\partial \Omega}{\partial s}$.

$$\therefore \quad S = -\frac{\partial \Omega}{\partial s} = -g\frac{dz}{ds}$$

Integrating, $\Omega = gZ$ + constant. Substituting for Ω in Eq. (4.7), we obtain

$$\frac{V_s^2}{2} + \frac{p}{\rho} + gZ = \text{constant} \qquad \text{...(4.8)}$$

It may be noted that the right hand side of Eqs. (4.7) and (4.8) will have different values.

Eq. (4.8) is applicable to points on the same streamline. V_s represents the velocity of any point on the streamline, where the pressure is p and which is located at a distance Z above an arbitrarily chosen horizontal plane commonly known as the *datum plane*. It may be of interest to recall that Eq. (4.8) has to satisfy the following conditions for its applicability :

(*i*) flow is steady ;

(*ii*) fluid is incompressible ;

(*iii*) fluid is non-viscous since viscous force has been neglected ; and

(*iv*) it is applicable to points along a streamline.

If Eq. (4.8) is applied to points lying on different streamlines, the value of constant will be different, but if the points are situated on the same streamline, the constant will have the same value.

Considering the Euler's equation of motion in the normal direction, Eq. (4.6) gives

$$\frac{\partial V_n}{\partial t} + V_s\frac{\partial V_n}{\partial s} = N - \frac{1}{\rho}\frac{\partial p}{\partial n} = -\frac{\partial \Omega}{\partial n} - \frac{1}{\rho}\frac{\partial p}{\partial n}$$

$$= -\frac{\partial}{\partial n}\left(\frac{p}{\rho} + \Omega\right) = -\frac{\partial}{\partial n}\left(\frac{p}{\rho} + gZ\right) \qquad \text{...(4.9 a)}$$

adding and subtracting $V_s \dfrac{\partial V_s}{\partial n}$ from the left hand side,

$$\frac{\partial V_n}{\partial t} + V_s\frac{\partial V_n}{\partial s} - V_s\frac{\partial V_s}{\partial n} + V_s\frac{\partial V_s}{\partial n} = -\frac{\partial}{\partial n}\left(\frac{p}{\rho} + \Omega\right)$$

or

$$\frac{\partial V_n}{\partial t} + V_s\left(\frac{\partial V_n}{\partial s} - \frac{\partial V_s}{\partial n}\right) = -\frac{\partial}{\partial n}\left(\frac{V_s^2}{2} + \frac{p}{\rho} + \Omega\right)$$

For steady flow, we obtain

$$-\frac{\partial}{\partial n}\left(\frac{V_s^2}{2} + \frac{p}{\rho} + \Omega\right) = V_s\left(\frac{\partial V_n}{\partial s} - \frac{\partial V_s}{\partial n}\right) \qquad \text{...(4.9 b)}$$

In Eq. (4.9 *b*), the quantity $\left(\dfrac{\partial V_n}{\partial s} - \dfrac{\partial V_s}{\partial n}\right)$ is equal to twice the rate of rotation about an axis perpendicular to s and n-axes. If the flow is assumed to be irrotational,

$$\left(\frac{\partial V_n}{\partial s} - \frac{\partial V_s}{\partial n}\right) = 0$$

and Eq. (4.9 *b*) becomes $-\dfrac{\partial}{\partial n}\left(\dfrac{V_s^2}{2} + \dfrac{p}{\rho} + \Omega\right) = 0.$

Integration of which with respect to n, for an incompressible fluid yields,

$$\frac{V_s^2}{2} + \frac{p}{\rho} + \Omega = \text{constant} \qquad \text{...(4.10)}$$

The subscript s may now be dropped, noting that V is the velocity of flow along the streamline, substituting for

$$\Omega = gZ + \text{constant},$$

we obtain

$$\frac{V^2}{2} + \frac{p}{\rho} + gZ = \text{constant} \qquad \text{...(4.11)}$$

Eq. (4.11) shows that for irrotational flow, its application in the normal direction (that is, to points on different streamlines) will yields the same constant. Thus the validity of Eq. (4.11) is based on the following assumptions :

1. flow is steady,
2. flow is irrotational,
3. fluid is incompressible, and
4. fluid is non-viscous (ideal).

It may be noted here that if the flow is rotational

$$\frac{\partial V_n}{\partial s} \neq \frac{\partial V_s}{\partial n}, \text{ and}$$

Eq. (4.9 *b*) shows that the quantity,

$$\left(\frac{V^2}{2} + \frac{p}{\rho} + gZ\right), \text{ varies in the normal direction.}$$

Dividing Eq. (4.11) by g, we arrive at a very important expression, which, in fact, is one of the fundamental equations of fluid mechanics and has many applications in fluid flow problems. The equation expressed as,

$$\frac{V^2}{2g} + \frac{p}{\gamma} + Z = \text{constant} \qquad ...(4.12)$$

is known as the *Bernoulli's equation* in honour of Swiss mathematician, Daniel Bernoulli and the constant is called the Bernoulli constant.

Equation (4.12) is a suitable form for flow of liquids. For flow of gases, the following form of the Bernoulli's equation is more suitable.

$$\rho \frac{V^2}{2} + p = \text{constant} \qquad ...(4.13)$$

Since the specific weight of gases is very small, small changes in Z will not significantly affect the value of constant, and hence ommission of γZ from Eq. (4.13) is justified.

Ex. 4.2. *Water is flowing through a horizontal conical pipe whose diameter changes from 20 cm to 10 cm over length of 1 m. If the velocity of flow is 4 m / s at the bigger diameter, find the velocity heads at the two ends at the middle of the pipe. Calculate also the mass rate of flow.*

Sol. Applying the continuity equation between sections (1) and (2),

$$Q = A_1V_1 = A_2V_2$$

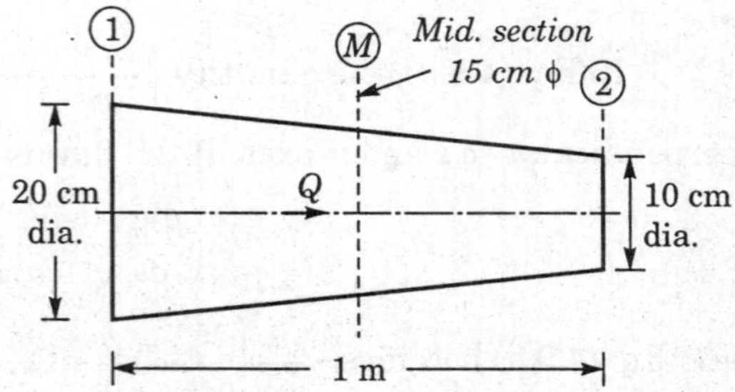

Area of flow,

$$A_1 = \frac{\pi}{4}(0.2)^2 = \frac{\pi \times 0.04}{4} \text{ (m}^2\text{)}$$

$$= 0.01\pi \text{ (m}^2\text{)}$$

Area of flow,

$$A_2 = \frac{\pi}{4}(0.1)^2 = \frac{0.01}{4}\pi \text{ (m}^2\text{)}$$

(*i*) Velocity at the smaller dia end, $V_2 = \frac{A_1}{A_2}V_1 = \frac{(0.2)^2}{(0.1)^2} \times 4 = \mathbf{16\ m/s.}$

(*ii*) Velocity at the mid-section where the pipe diameter is 15 cm

$$V_m = \left(\frac{0.20}{0.15}\right)^2 \times 4 = \mathbf{7.1\ m/s.}$$

(*iii*) Velocity heads :

$$(a)\ \frac{V_1^2}{2g} = \frac{4^2}{2\times 9.81} = 0.8155 \text{ m}$$

$$(b)\ \frac{V_m^2}{2g} = \frac{7.1^2}{2\times 9.81} = 2.569 \text{ m}$$

$$(c)\ \frac{V_2^2}{2g} = \frac{16^2}{2\times 9.81} = 13.048 \text{ m}$$

$$(iv)\ \text{Mass flow rate} = \rho Q = \frac{\gamma}{g} Q = \frac{1000}{9.81} \times 0.01\pi \times 4 = 12.815 \text{ metric slug/m}^3.$$

Example 4.3. (*a*) *What is irrotational flow ? Explain how a two-dimensional flow of an incompressible fluid along a curved boundary can be irrotational.*

The wind velocity in a cyclone may be assumed to vary according to free vortex law. If the velocity is 16 km. per hour 50 km. from the centre of the cyclone, what pressure gradient should obtain at this point ? What reduction in barometric pressure should occur over a radial distance of 10 km from this point towards the centre of the storm ? Take mass density of air as 1.208 kg per cubic metre.

Solution. In a cyclone the wind velocity is given to vary according to the free-vortex law. The velocity distribution in a free-vortex is given by Eq. (3.57), *i.e.*

$$V.r = C \qquad ...(3.57)$$

At a radical location defined by

$$r_1 = 50 \text{ km},\quad V_1 = 16 \text{ km/hr}.$$

Substituting this data in Eq. (3.57), the value of the constant

$$C = 16 \times 50 = 800 \text{ (km}^2\text{/hr)}$$

Velocity at a radial distance of (50 – 10), *i.e.* 40 km

$$V_2 = \frac{C}{r_2} = \frac{800}{40} = 20 \text{ km/hr}.$$

From the Bernoulli's equation expressed by Eq. (4.13), *i.e.*

$$p + \rho\frac{V^2}{2} = \text{constant}$$

one gets by differentiating w.r.t. *r*

$$\frac{dp}{dr} + \rho V\frac{dV}{dr} = 0$$

or

$$\partial p/dr = -\rho V\, dV/dr.$$

From Eq. (3.57) recalled above, one may write on differentiation,

$$r\frac{dV}{dr} + V = 0, \quad \text{or} \quad \frac{dV}{dr} = -\frac{V}{r}$$

and the pressure gradient

$$\frac{dp}{dr} = -\rho V\frac{dV}{dr} = \rho\frac{V^2}{r}.$$

The pressure gradient at a radial distance of 50 km where the velocity is 16 km/hr

$$\left(\frac{dp}{dr}\right)_{r=50\text{ km}} = 1.208 \times \frac{\left(\frac{16\times 1000}{3600}\right)^2}{50\times 1000} = 477 \times 10^{-6} \text{ N/m}^2 \text{ per metre}$$

$$= \mathbf{0.477\ N/m^2\ per\ km.}$$

Reduction in barometric pressure over a radial distance of 10 km from $r_1 = 50$ km to $r_2 = 40$ km

$$r_1 = 50 \text{ km}, V_1 = 16 \text{ km/hr} = 4.44 \text{ m/s}$$

$$r_2 = 40 \text{ km}, V_2 = 20 \text{ km/h} = 5.55 \text{ m/s.}$$

Using the Bernoulli's equation, the reduction the barometric pressure is obtained :

$$p_1 + \rho \frac{V_1^2}{2} = p_2 + \rho \frac{V_2^2}{2}$$

or $$p_1 - p_2 = \frac{1}{2} \rho(V_2^2 - V_1^2) = \frac{1.208}{2} (5.55^2 - 4.44^2) = \mathbf{6.73 \text{ N/m}^2}.$$

4.6. Significance of Each Term of Bernoulli's Equation

Eq. (4.12) states that the sum of the three quantities, *i.e.* $V^2/2g$, p/γ and Z is a constant. From the principle of dimensional homogeneity (refer Art. 5.2) it implies that each term of Eq. (4.12) must have the same dimensions and, therefore, the nature of quantities must be same. In order to have a clear picture of each term and to understand its significance, let us examine what each term of Eq. (4.12) stands for. The first term, $V^2/2g$, represents kinetic energy of flow per unit weight (per kg or per N) of fluid.

Kinetic energy = $\frac{1}{2}$ mass × V^2, and kinetic energy divided by the weight of the fluid

$$= \frac{\frac{1}{2}(\text{Mass})\, V^2}{\text{weight}} = \frac{\frac{1}{2}(\text{Mass})\, V^2}{(\text{Mass})g} = \frac{V^2}{2g}.$$

The second term p/γ represents ability of unit weight of fluid to do work by virtue of its pressure. Let a net pressure p act over a fluid element of cross-sectional area dA resulting in a pressure force pdA acting in the direction of flow. When the fluid element moves a distance ds, work is done on the element by the pressure force which is equal to $pdAds$. The work done by the pressure force per unit weight of fluid element is $pdAds/\gamma dAds = p/\gamma$. The term p/γ is, therefore, the work done by the pressure force per unit weight of fluid and it known as *flow work* or *flow energy*. It is sometimes misleadingly termed as pressure energy. When a pressure is applied to a fluid, it gets compressed to some extent and the elastic energy is imparted to the fluid. The flow energy p/γ has nothing to do with the elastic energy. The third term Z is the potential energy per unit weight of fluid and represents the energy given to a unit weight of fluid in raising it from datum level to a height Z above it. The potential energy of weight W of fluid in raising it to a height Z above the datum level is WZ, and, therefore, potential energy per unit weight of fluid is $WZ/W = Z$. The potential energy is also sometimes known as the gravitational energy.

Each term of the Bernoulli's equation,. Eq. (4.12), therefore, represents some form of energy per unit of weight of fluid and has the dimension

$$= \frac{(\text{Energy})}{(\text{Weight})} = \frac{[ML^2T]^{-2}}{[MLT^{-2}]} = [L]$$

Since each term has a length dimension, it can be represented graphically with reference to an arbitrarily chosen datum plane. The quantities $V^2/2g$, p/γ and Z are, therefore, known as *velocity head*, *pressure head* and *elevation* or *datum head* and their sum is called the *total head*.

Example 4.4. *An impeller, 300 mm diameter, rotating inside a closed cylindrical casing, 600 mm diameter, produces a forced-vortex inside the impeller and the free-vortex in the annular space between the impeller and the circular side of the casing. The velocity at the junction of the two vortices is the same and the pressure at the centre of impeller is 1.3 m of water. Find the pressure head at the impeller periphery and at the casing if the speed of rotation is 1200 rpm.*

Solution. Velocity distribution in the vortex regions is :

(*i*) Forced-vortex law $V = rw$ ...(*A*)

(*ii*) Free-vortex law, Eq. (2.57),

$$Vr = C \qquad ...(B)$$

Since the velocity at the junction of the two vortices is the same, *i.e.* at r = 150 mm, from Eqs. (*A*) and (*B*)

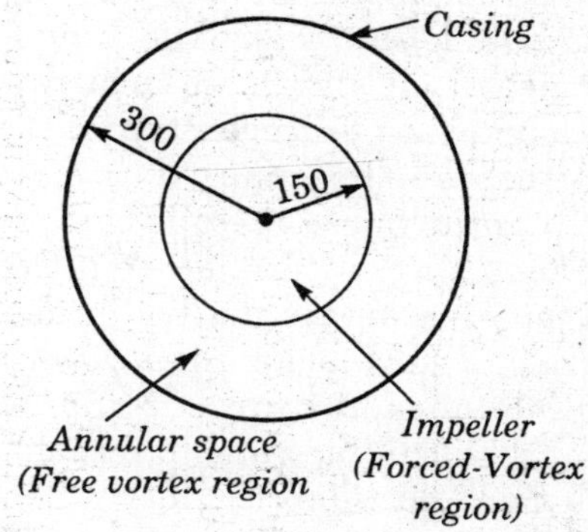

$$V = r\omega = \frac{C}{r} \qquad \therefore \quad C = r^2\omega$$

or
$$C = (0.15)^2 \times \frac{2\pi \times 1200}{60} = 2.827 \text{ m}^2/\text{s}$$

Velocity at r = 150 mm,

$$V_{150} = 0.15 \times \frac{2\pi \times 1200}{60} = 18.85 \text{ m/s.}$$

Velocity at r = 300 mm,

$$V_{300} = \frac{C}{r} = \frac{2.827}{0.30} = 9.423 \text{ m/s.}$$

When the impeller rotates, the flow of water follows the forced-vortex law and the water surface attains the shape of paraboloid of revolution. The height to which water will rise is

$$h = \frac{r^2\omega^2}{2g} = \frac{(0.15)^2 \times (125.664)^2}{9 \times 9.81} = 18.11 \text{ m}$$

Pressure at the centre of impeller (given) = 1.3 m of water

$\therefore$ Pressure at the periphery of impeller = 1.3 + 18.11 = **19.41 m of water.**

Applying Bernoulli's equation in the annular space, where the free-vortex law is applicable,

$$\frac{V_i^2}{2g} + \frac{p_i}{\gamma} = \frac{V_0^2}{2g} + \frac{p_0}{\gamma}$$

where the subscripts *i* and *o* indicate the inner and outer boundaries of the annular space. Substituting the numerical values,

$$\frac{(18.85)^2}{2 \times 9.81} + 19.14 = \frac{(9.423)^2}{2 \times 9.81} + \frac{p_0}{\gamma}$$

from which the pressure head at the casing is

$$p_o/\gamma = \mathbf{32.98 \text{ m of water.}}$$

4.7. Graphical Representation of Bernoulli's Equation

If the datum head, pressure head and the velocity head are plotted as vertical distances above an arbitrarily assumed horizontal datum, the line joining the points representing the total head

$$\frac{V^2}{2g} + \frac{p}{\gamma} + Z$$

is known as the energy grade line or simply the *energy line*. It is also sometimes known as the *total head line or total energy line* (*T.E.L.*).

In Fig. 4.2 is shown a portion of a pipeline terminating in a nozzle which discharges freely into the atmosphere. A gradually converging pipe *AB* leads the flow into a uniform diameter pipe *BC*, the downstream end of which is connected to nozzle *CD*. A horizontal plane located below the pipeline is arbitrarily chosen at any desired elevation as a reference or datum plane. For application of Bernoulli's equation (4.12), let us consider the points *A, B, C* and *D* lying on the centreline of the pipe system. The datum head or the elevation head is vertical distance between these points and the arbitrary datum. If manometer tube (also called the piezometer) is installed in the pipeline at A, and level of liquid risen in the tube measured above the point A, represents the pressure head p/γ at A. If, however, the liquid level is measured above the arbitrary datum plane, the piezometer tube is then said to measure the quantity $(p/\gamma + Z)$. This quantity $(p/\gamma + Z)$ is called the *piezometric head*. The line joining the points representing piezometric heads, *i.e.* locus of $(p/\gamma + Z)$, along the flow direction, is known as the *hydraulic grade line*. It is also sometime called piezometric head line. Addition of velocity head to the piezometric head gives the location of the point representing the total head. The line joining the points representing the total head, *i.e.*

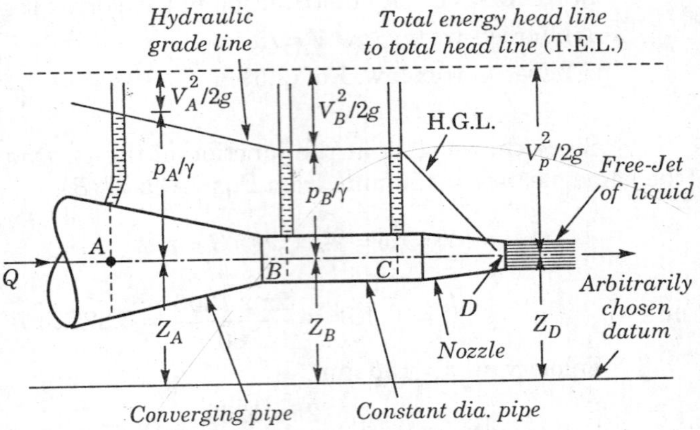

Fig. 4.2. Energy and hydraulic grade lines for flow of an ideal fluid.

$\left(\frac{V_2}{2g} + \frac{p}{\gamma} + Z\right)$, is the *total energy line* or *total head line*. In case of an ideal fluid, the energy line is horizontal as the total head remains constant along the flow. The energy line lies above the hydraulic grade line and is separated by a vertical distance equal to the velocity head $V^2/2g$. The Bernoulli's equation may be expressed as

$$\frac{V^2}{2g} + (p/\gamma + Z) = \text{constant}$$

or, Velocity head + Piezometric head = Constant ...(4.14)

For a converging pipe like AB, the continuity equation shows that the velocity increases in the downstream direction. According to Eq. (4.14), this results in the decrease in piezometric head and, therefore, the hydraulic grade line has a downward slope, signifying the lower piezometric head and a large velocity head. In the uniform pipe length BC, the velocity remains unchanged, and hence the energy line and the hydraulic grade lines are parallel. At the nozzle throat D, the jet of liquid is exposed to the atmosphere resulting in a zero pressure head. The total energy of flow at D accordingly consists of the kinetic energy ($V_{D^2}/2g$) and the potential energy (Z_D). The hydraulic grade line, therefore, joins the point D at the nozzle throat. The ordinate between the energy line and the nozzle throat measures the velocity head ($V_{D^2}/2g$).

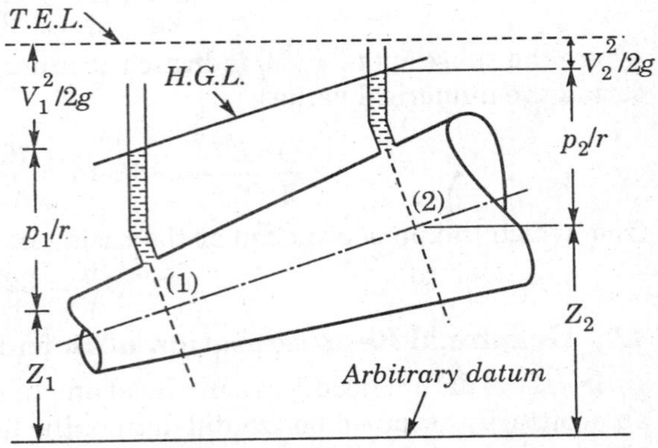

Fig. 4.3. Hydraulic grade line sloping upward in a diverging flow.

The hydraulic grade line slopes upward only when the piezometric head increases in the direction of flow. This may happen in case of a gradually diverging pipe as shown in Fig. 4.3. A decrease in the velocity causes an increase in the piezometric head as expressed by Eq. (4.14).

The concepts of total head and the piezometric head are applicable to the flow of liquids, on account of the fact that a liquid rises to a free-surface in a piezometer tube. These, therefore, have no physical significance for flow of gases.

Example 4.5. *At a certain section A of a pipeline carrying water, the diameter is 1 m. the pressure is 98.1 kN/m² and the velocity is 3 m/s. At another section B which is 2 m higher than A, the diameter is 0.7 m and the pressure is 59.2 kN/m². What is the direction of flow ?*

Solution. The flow in pipelines always takes place from higher energy to lower energy. Velocity at section B, from the continuity equation is

$$V_B = \frac{3 \times \frac{\pi}{4}(1)^2}{\frac{\pi}{4}(0.7)^2} = 6.12 \text{ m/s.}$$

Total energy per unit weight at section A, with the horizontal line through A as the datum

$$= \frac{V_A^2}{2g} + \frac{p_A}{\gamma} = \frac{(3)^2}{2 \times 9.81} + \frac{98.1 \times 10^3}{9810} = 10.459 \text{ m.}$$

Total energy per unit weight at section B, 2 m higher than section A

$$= \frac{V_B^2}{2g} + \frac{p_B}{\gamma} + Z_B = \frac{(6.32)^2}{2 \times 9.81} + \frac{59.2 \times 10^3}{9810} + 2 = 7.908 \text{ m.}$$

Since the total energy at A is greater than at B, the flow direction is from section A to section B.

Example 4.6. *Water is pumped at the rate of 150 litres/sec from a reservoir under the conditions indicated. If the pump supplies energy to the flow at the rate of 11 kW, determine the pressure intensities at points A and B.*

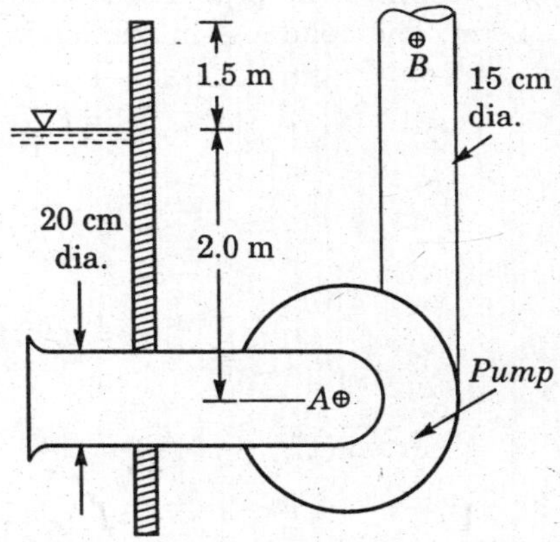

Solution. Energy per unit weight imparted to the fluid by the pump

$$= \gamma Q(\Delta H)$$

where ΔH = the head imparted to water by the pump.

$$\therefore \quad \Delta H = \frac{11 \times 10^3}{9810 \times \frac{150}{1000}} = 7.5 \text{ m of water.}$$

Velocity at A,

$$V_A = \frac{Q}{\text{Area}} = \frac{150/1000}{\frac{\pi}{4} \times (2.2)^2} = 4.77 \text{ m/s.}$$

Velocity at B,

$$V_B = \left(\frac{20}{15}\right)^2 V_A = 8.48 \text{ m/s.}$$

Applying the Bernoulli's equation to the free surface in the reservoir and the point A, taking the horizontal line through A as the datum.

$$2 + 0 + 0 = \frac{p_A}{\gamma} + \frac{(4.77)^2}{2 \times 9.81} = \frac{p_A}{\gamma} + 1.158$$

or

$$\frac{p_A}{\gamma} = 0.842 \text{ m of water.}$$

$$\therefore \quad p_A = 9810 \times 0.842 = 8260 \text{ N/m}^2.$$

Energy per unit weight of water at B is greater than at A by the energy per unit weight supplied by the pump. Applying Bernoulli's equation between A and B,

$$0 + \frac{V_A^2}{2g} + \frac{p_A}{\gamma} + \Delta H = 3.5 + \frac{V^2{}_b}{2g} + \frac{p_b}{\gamma}$$

or
$$9.5 = 3.5 + \frac{(8.48)^2}{2 \times 9.81} + \frac{p_b}{\gamma}$$

$$\frac{p_b}{\gamma} = 6.0 - 3.675 = 2.325 \text{ m of water.}$$

$\therefore$
$$p_B = 9810 \times 2.325 = 22.81 \text{ kN/m}^2.$$

Ex. 4.7. *A steel pipe of 15 cm diameter carries water at a rate of 30 litres / second from point A to B along the pipe, the point B being 20 m higher than point A and 600 m apart along the pipe. If the pressure at B is to be 2.8 kg / cm², what pressure must be maintained at A, if friction factor for the pipe is 0.024 ?*

What will be the capacity of the pipe after 15 years of service, whence the friction factor is tripled ? Assume that the same pressures are to be maintained at A and B. (kg = unit of force in metric system).

Sol. Given. Q = 30 l/s d = 15 cm

p_B = 2.8 kg/cm², f = 0.024

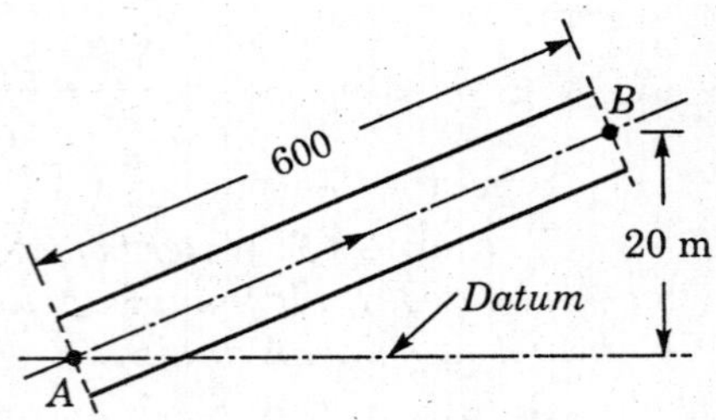

Applying Bernoulli's equation between A and B with horizontal line through A taken as the datum :

$$Z_A + \frac{p_A}{\gamma} + \frac{V_A^2}{2g} = Z_B + \frac{p_B}{\gamma} + \frac{V_B^2}{2g} + h_L \quad ...(1)$$

Since the pipe diameter is same $V_A = V_B$. $Z_B - Z_A$ = 20 m. The head loss in friction is given by Eq. (9.57) as

$$h_L = f\frac{L}{d}\frac{V^2}{2g} \quad ...(2)$$

$$V = \frac{4Q}{\pi d^2} = \frac{4 \times 0.03}{\pi (0.15)^2} = 1.6976 \text{ m/s}$$

$$h_L = 0.024 \times \frac{600}{0.15} \times \frac{(1.6976)^2}{2 \times 9.81} = 14.10 \text{ m}$$

From (1),
$$\frac{p_A}{\gamma} = \frac{p_B}{\gamma} + Z_B - Z_A + h_L = \frac{2.8 \times 10^4}{1000} + 20.0 + 14.10 = 62.10 \text{ m}$$

$\therefore$
$$p_A = \frac{62.10 \times 1000}{10^4} = \mathbf{6.21 \text{ kg/cm}^2}.$$

Since p_A and p_B are to be the same. From the Bernoulli's eq., the head loss

$$h_L = \frac{p_A - p_B}{\gamma} + Z_A - Z_B = 62.10 - 28.0 - 20 = 14.10 \text{ m}$$

Using Eq (2), the new velocity of pipe when $f = 3 \times 0.024 = 0.072$

$$14.10 = 0.072 \times \frac{600}{0.15} \times \frac{V^2}{2 \times 9.81}$$

from which V = 0.98 m/s, and the new discharge $Q = \frac{\pi d^2}{4} V$

$$Q = \frac{\pi (0.15)^2}{4} \times 0.98 = 0.01732 \text{ m}^3\text{/s} = \mathbf{17.32 \text{ lit/sec.}}$$

Ex. 4.8. *A pipe 250 m long has a slope of 1 in 100 and tapers from 1000 mm diameter at higher end to 500 mm at lower end. If 5000 litres of water is flowing through the pipe per minute and the pressure of water at higher end is 1 kg / cm² (98.1 kN / m²). Find the pressure at the lower end.*

Sol. Velocity at section (1) where pipe dia is 1000 mm

$$V_1 = \frac{Q}{A_1} = \frac{(5000/1000 \times 60)}{\pi(1)^2/4} = 0.106 \text{ m/s}$$

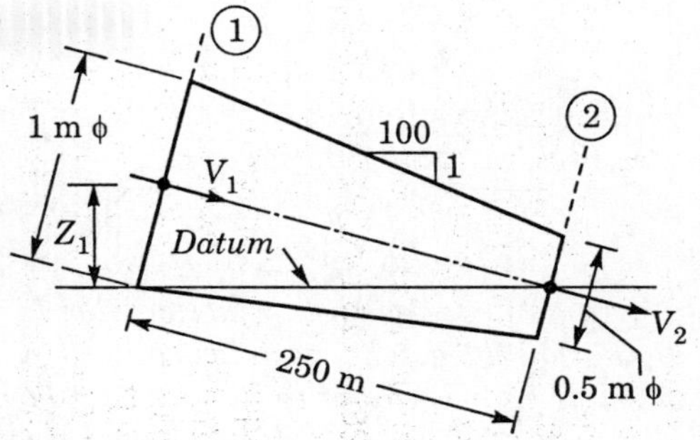

From continuity equation

$$Q = A_1V_1 = A_2V_2, \quad \therefore \quad V_2 = 4V_1 = 0.424 \text{ m/s}$$

With reference to the horizontal datum passing through the centre of section (2) as shown, the Bernoulli's equation applied to sections (1) and (2) results in (ignoring the energy loss due to friction)

$$Z_1 + \frac{p_1}{\gamma} + \frac{V_1^2}{2g} = Z_2 + \frac{p_2}{\gamma} + \frac{V_2^2}{2g}$$

With the given slope of 1 in 100, the fall in the datum level of section on (2) in pipe length of 250 m is 2.5 m, $\therefore Z_1 = 2.5$ m. Hence,

$$2.5 + \frac{1 \times 10^4 \text{ (kg/m}^2)}{1 \times 10^3 \text{ (kg/m}^3)} + \frac{(0.106)^2}{2 \times 9.81} = 0 + \frac{p^2}{1 \times 10^3 \text{ (kg/m}^3)} + \frac{(0.4242)^2}{2 \times 9.81}$$

from which $p_2/\gamma = 12.49$ m or $p_2 = \frac{1000 \times 12.49}{10^4} = \mathbf{1.249 \text{ kg/cm}^2}$.

$$= 9810 \times 12.49 \text{ (N/m}^2) = \mathbf{122.53 \text{ kN/m}^2}.$$

Example 4.9. *Two stationary, horizontal flat plates with an external diameter of 40 cm are placed 1 cm apart. A vertical pipe 5 cm in diameter delivers 0.005 m³/s of water to the centre of the plates. The water is discharged to the periphery of the plates at atmospheric pressure of 98 kN/m². Assuming radial flow and neglecting losses, determine the absolute pressure at the entrance of the flow.*

Solution. $Q = 0.005 \text{ m}^3/\text{s}$

Using the continuity equation, the velocity at the entrance to the angular space, $V_i = \frac{Q}{ndi \times t}$

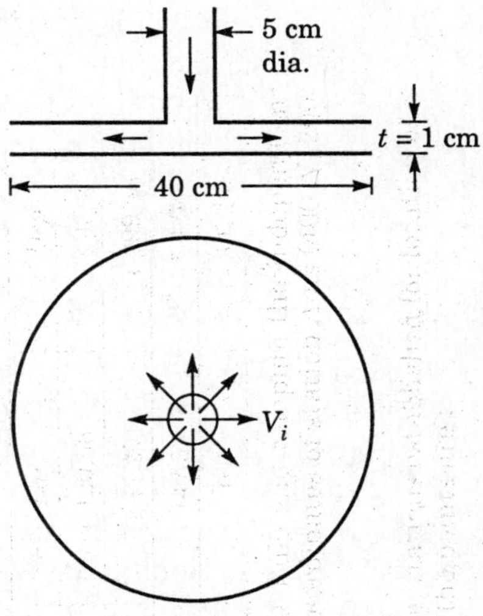

$$V_i = \frac{0.005}{\pi \times \frac{5}{100} \times \frac{1}{100}} = 3.18 \text{ m/s}$$

Velocity at the periphery or plates

$$V_o = \frac{Q}{\pi d_0 t}$$

$$= \frac{0.005}{\pi \times 0.40 \times \frac{1}{100}} = 0.398 \text{ m/s}$$

Applying the Bernoulli's equation between the inlet to and exit from the plates.

$$\frac{p_i}{\gamma} + \frac{V_i^2}{2g} = \frac{p_o}{\gamma} + \frac{V_o^2}{2g}$$

or

$$\frac{p_i}{\gamma} + \frac{(3.18)^2}{2 \times 9.81} = \frac{98 \times 10^3}{9800} + \frac{(0.398)^2}{2 \times 9.81}$$

from which $\frac{p_i}{\gamma}$ = **9.493 m of water absolute.**

$p_i = 9810 \times 9.493 = 93 \text{ kN/m}^2$.

Example 4.10. *Pressures on the suction side and the discharge side of a pump are read by the pressure gauge as*

(i) vacuum of 25 cm of mercury, and

(ii) 140 kN/m² (1.4 kgf/cm²), respectively.

The gauge on the suction side is placed 60 cm below the centreline of the pump whereas the gauge on the discharge side is placed 17 cm above the centreline. If the diameters of pipes on suction and discharge sides are 15 cm and 10 cm, respectively, and the quantity of water to be pumped is 30 litres per second, calculate the power required for the pump.

Solution. Pump discharge Q = 30 lit/sec = 0.03 m³/s

Velocity in the suction pipe, $V_s = \dfrac{0.03 \times 4}{\pi(0.15)^2} = 1.70$ m/s

Velocity in the delivery pipe, $V_d = \dfrac{0.03 \times 4}{\pi(0.01)^2} = 3.82$ m/s

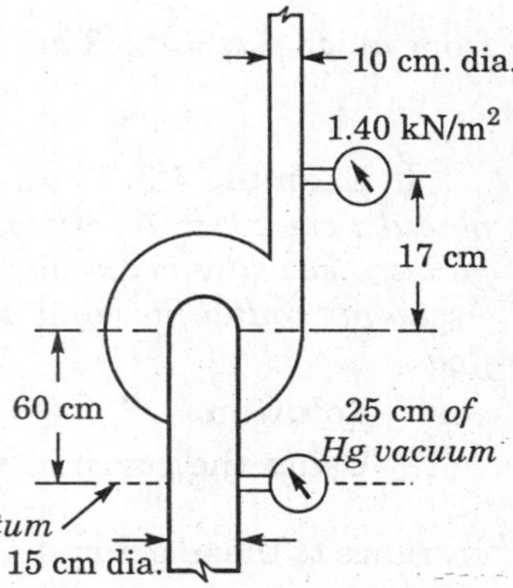

Applying Bernoulli's equation to the pressure guage locations at suction and delivery sides of the pipeline, with datum taken through the pressure gauge on the suction side,

$$0 + \left(\frac{-25 \times 13.6}{100}\right) + \frac{(1.7)^2}{2 \times 9.81} + \text{Head developed by the pump}$$

$$= (0.60 + 0.17) + \frac{14 \times 10^4}{9810} + \frac{(3.82)^2}{2 \times 9.81}$$

$\therefore$ Head developed by the pump

= 18.767 m of water

Power developed by the pump

= γQH = 9810 × 0.03 × 18.767 W = **5.51 kW.**

Assuming an efficiency of 65% for the pump, the horse power required

$$= \frac{5.51}{0.65} = \mathbf{8.48\ kW}.$$

4.8. Applications of Euler's Equations of Motion

4.8.1. Pressure Variation Across a Uniform Conduit

It is a known fact that for a real fluid in a uniform conduit straight in alignment (that is, constant diameter straight pipe), the velocity varies across any section normal to flow direction. Let us examine how the pressure changes across the section.

Euler's equation of motion in a direction normal to the streamline is given by Eq. (4.6).

$$a_n = \frac{\partial V_n}{\partial t} + V_s \frac{\partial V_n}{\partial s}$$

$$= N - \frac{1}{\rho}\frac{\partial p}{\partial n}$$

$$= -\frac{\partial}{\partial n}\left(\frac{p}{\rho} + gZ\right).$$

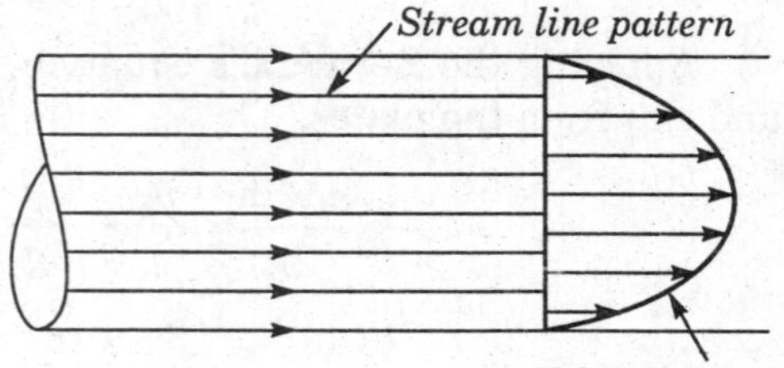

Fig. 4.4. Real fluid flow in a straight uniform pipe.

Since the conduit is uniform and straight the streamlines are parallel signifying that there could be no acceleration in the normal direction.

$$\therefore \qquad a_n = 0.$$

Hence
$$\frac{\partial}{\partial n}\left(\frac{p}{\rho} + gZ\right) = 0$$

or
$$\frac{p}{\rho} + gZ = \text{constant, and we find that in the normal direction}$$

$$\frac{p}{\gamma} + Z = \text{constant.}$$

Thus the piezometric head remains constant normal to the streamlines in case of a uniform diameter straight pipe.

4.8.2. Pressure Variation Across a Uniform Bend

Consider a uniform pipe bend shown in Fig. 4.5 (*a*). The equation governing the pressure distribution normal to the streamline is given by Eq. (4.9 *a*).

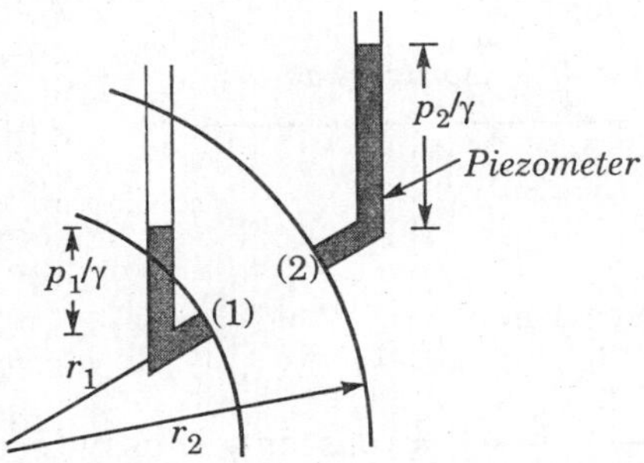

Fig. 4.5 (*a*) Flow through a bend.

$$-\frac{\partial}{\partial n}\left(\frac{p}{\rho} + gZ\right) = \frac{\partial V_n}{\partial t} + V_s\frac{\partial V_n}{\partial s}$$

$$= \frac{\partial V_n}{\partial t} + \frac{V^2}{r} \qquad \text{(see Eq. 3.10)}$$

where V is the velocity in the flow direction and r is the radius of the streamline. For steady flow,

$$\frac{\partial V_n}{\partial t} = 0$$

and the pressure variation in the radial direction is given by

$$-\frac{\partial}{\partial r}\left(\frac{p}{\rho} + gZ\right) = -\frac{V^2}{r}.$$

Since the acceleration V^2/r is directed towards the centre of curvature of streamlines, it is negative, and as such a negative sign appears in the above equation along with the acceleration. The above equation may be written as a gradient of $(p/\gamma + Z)$, the piezometric head,

$$\frac{\partial}{\partial r}\left(\frac{p}{\gamma} + Z\right) = \frac{V^2}{gr} \qquad \text{...(4.14 a)}$$

Equation (4.14 *a*) indicates that the gradient of the piezometric head, is positive in the r-direction. In other words, $(p/\gamma + Z)$ increases in a direction outward from the centre of the curvature. Thus, the piezometric head at the outer boundary of the pipe bend is greater than that at its inner boundary.

Flow in a pipe bend, considered to be irrotational, is characterised by the velocity law,

$$V.r = \text{constant}$$

or
$$V = \frac{C}{r}, \text{ where } C \text{ is a constant.}$$

This velocity-radius relationship shows that the velocity decreases with increasing radius. Hence the velocity at the outer boundary of the bend has a lower magnitude than that at the inner boundary.

Eq. (4.14 *a*) upon substituting for V,
$$\frac{\partial}{\partial r}(p/\gamma + Z) = \frac{C^2}{gr^3}$$

Integrating it with respect to r, $p/\gamma + Z = -\dfrac{C^2}{2gr^2}$ + constant.

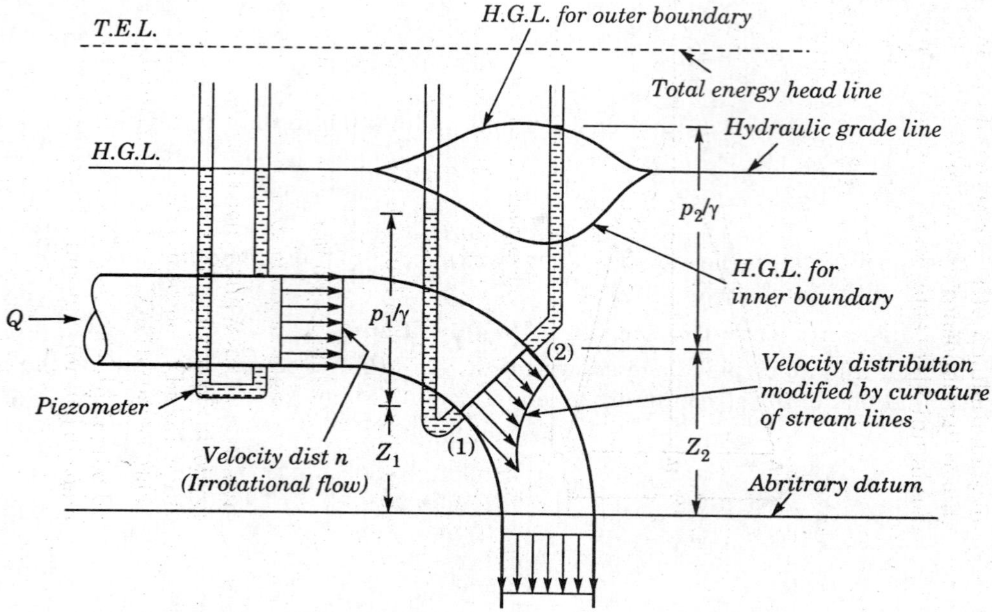

Fig. 4.5 (*b*). Pressure variation in a pipe bend.

This equation, by substituting V^2 for $\dfrac{C^2}{r^2}$ changes to $\dfrac{V^2}{2g} + \dfrac{p}{\gamma} + Z$ = constant. Thus we see that the constant of integration is the same as the total head or total energy per unit weight, which is a constant quantity for irrotational flow. Let us denote it by H. The expression relating pressure with the radial distance may be written as

$$p/\gamma + Z = H - \frac{C^2}{2gr^2}$$

or
$$p = \gamma H - \rho \frac{C^2}{2r^2} - \gamma Z \qquad ...(4.15\ a)$$

If $Z = 0$, the pressure distribution in the same horizontal plane as the datum is given by

$$p = \gamma H - \frac{\rho C^2}{2r^2} \qquad ...(4.15\ b)$$

Eq. (4.15 *b*) shows that the pressure at the outer boundary of the pipe bend is more than that at its inner boundary. Fig. 4.5 (*b*) shows the variation of piezometric head for the outer and inner boundaries of a pipe bend. The velocity distribution at the inlet to and exit from the bend is rectangular (for an ideal fluid), while at the central section of the bend, it is hyperbolic (Vr = constant).

Example 4.11. *A pipeline of a rectangular cross-section has a width b and height h. It has a circular bend of radius R measured upto the centreline of the bend. The velocity variation across radius may be approximated by a free-vortex distribution $v_\theta r$ = constant. Show that the pressure rise between the outside and inside of the bend is given by*

$$\frac{16\rho bRQ^2}{\left[h(4R^2 - b^2)^2 \log_e \dfrac{(2R+b)}{(2R-b)}\right]^2}$$ *where ρ and Q are the density and discharge respectively.*

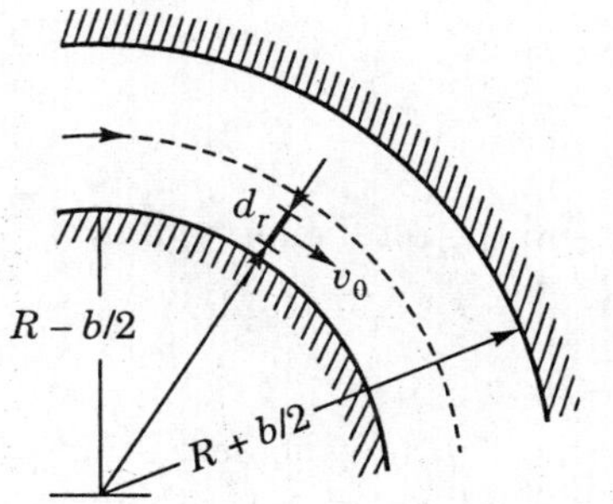

Solution. The discharge passing through an area element of length dr and height h is

$$dQ = hdr.\ v_\theta$$

$$\text{The total discharge, } Q = \int_{R-b/2}^{R+b/2} dQ$$

$$= \int_{R-b/2}^{R+b/2} hdr\ .\ \frac{C}{r}, \text{ since } (v_\theta r = C)$$

$$= C\,h\,\log_e \frac{2R+b}{2R-b} \quad ...(A)$$

Considering the bend to be in the horizontal plane and applying Bernoulli's equation between the outside and inside boundaries of the bend (noting that the Bernoulli constant in an irrotational flow is the same for all streamlines).

$$\frac{p_o}{\gamma} + \frac{V_o^2}{2g} = \frac{p_i}{\gamma} + \frac{V_i^2}{2g}$$

$$\frac{p_o - p_i}{\gamma} = \frac{V_i^2 - V_o^2}{2g} = \frac{C^2}{2g}\left[\frac{1}{\left(R-\frac{b}{2}\right)^2} - \frac{1}{\left(R+\frac{b}{2}\right)^2}\right]$$

Substituting for C from Eq. (A) and simplifying,

$$(p_o - p_i) = \frac{16\rho R b Q^2}{\left[h(4R^2 - b^2)\log_e \frac{2R+b}{2R-b}\right]^2}.$$

4.9. Alternative Forms of Bernoulli's Equation

The Bernoulli's equation expressed by **Eq.** (4.12) gives the total energy per unit weight of fluid. If each terms of the equation is multiplied by the specific weight of fluid γ, the Bernoulli's equation assumes the form :

$$\rho\,\frac{V^2}{2} + p + \gamma Z = \text{constant} \quad ...(4.16)$$

The first term of Eq. (4.16) represents kinetic energy per unit volume, which may be verified as follows. Kinetic energy = $\frac{1}{2}$ (ρ × Volume) V^2 and, therefore, K.E. per unit volume is $\frac{1}{2}$ (ρ × Volume) V^2/Volume = $\frac{1}{2}\rho V^2$. Hence each term of Eq. (4.16) denotes energy per unit volume, and in order to find out the total energy carried by the flow, we must multiply the left hand side of Eq. (4.16) by the volume rate of flow Q, thus

$$\text{Total energy of flow} = Q\left(\rho\frac{V^2}{2} + p + \gamma Z\right) \quad ...(4.17)$$

The total energy given by Eq. (4.17) will have N-m/s as unit of measurement.

In case of Eq. (4.12), which represents energy per unit of weight, the total energy of flow will be given by

$$\gamma Q\left(\frac{V^2}{2g} + \frac{p}{\gamma} + Z\right) \quad ...(4.18)$$

The Bernoulli's equation representing energy per unit of mass, may be expressed as

$$\frac{V^2}{2} + \frac{p}{\rho} + gZ = \text{constant}$$

and the total energy of flow, using Eq. (4.11) will be

$$\rho Q \left(\frac{V^2}{2} + \frac{p}{\rho} + gZ \right) \qquad \text{...(4.19)}$$

It may be noted that equations (4.17), (4.18) and (4.19) represent the same quantity but by using the concepts of energy per unit of volume, energy per unit of weight and energy per unit of mass respectively. The Bernoulli's equation may, therefore, be expressed in three different forms expressed by equations (4.11), (4.12) and (4.16). The value of the constant in each of these equations is evidently different and has different dimensions.

4.10. General Energy Equation

The principle of conservation of energy which is generally referred to as the first law of thermodynamics, may be stated thus : *energy can be neither created nor destroyed but can only be changed in form*. If a small quantity of heat is added to a homogeneous mass system, this thermal energy, by the first law of thermodynamics, can be transformed to other forms of energy, such as an increase in the internal energy of the system, the kinetic energy of the system proper, and the external mechanical work done by the system.

The internal energy comprises of molecular energy of the fluid, and in general is a function of temperature and density. For a perfect gas, the internal energy is a function of temperature only. The first law of thermodynamics, therefore, keeps an account of additions, subtractions and various transformations of energy which occur in a flow problem. In order to obtain an energy equation for steady flow of a fluid, let us consider a device shown in Fig. 4.6.

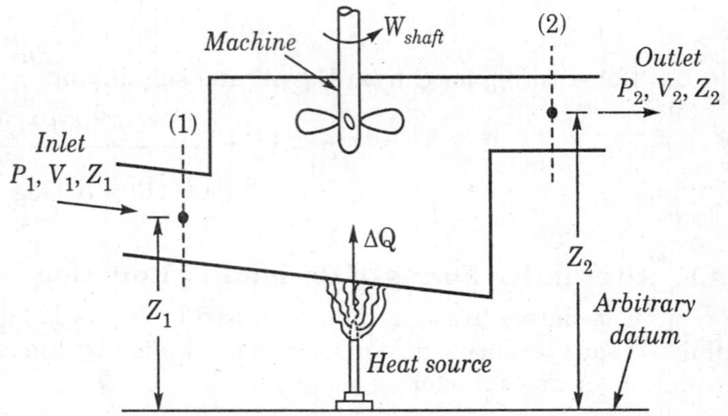

Fig. 4.6. Derivation of energy equation for steady flow.

The fluid enters the system from the left side and flows out from the right side of the system. Energy in the form of heat is supplied to the system at a steady rate and the external work is done by fluid on the machine. This work done on the machine is known as the shaft work, W_{shaft}. The machine may be, say, a turbine, a pump or a steam engine. The pressure, velocity and the elevation head at inlet and outlet are p_1, V_1, Z_1 and p_2, V_2, Z_2 respectively. We assume that all conditions of flow are steady.

Expressing u as the internal energy per unit mass, the sum of kinetic, potential and internal energies entering the system at section (1) is

$$\frac{V_1^2}{2} + gZ_1 + u_1$$

where u_1 is the internal energy per unit mass at section (1). Similarly the sum of the above three forms of energy at the outlet is

$$\frac{V_2^2}{2g} + gZ_2 + u_2$$

The flow work per second in forcing the fluid into the system is $p_1A_1V_1$ and the mass of fluid entering per second is $\rho_1A_1V_1$. The flow work per unit mass at inlet is

$$\frac{p_1A_1V_1}{\rho_1A_1V_1} = \frac{p_1}{\rho_1}.$$

Similarly, the flow work at outlet is p_2/ρ_1.

Let q be the heat supplied per unit mass and w_s the work done by the fluid on the machine per unit mass of fluid or the shaft work per unit mass. Then

$$q = \frac{Q}{\rho_1A_1V_1}$$

where Q is the heat per second supplied to the system

and $$w_s = \frac{W_{shaft}}{\rho_1A_1V_1}$$

According to the first law of thermodynamics :

Energy entering the system = Energy leaving it.

Taking into account various forms of energies, we may write,

$$\frac{V_1^2}{2} + gZ_1 + u_1 + \frac{p_1}{\rho_1} + q = \frac{V_2^2}{2} + gZ_2 + u_2 + \frac{p_2}{\rho_2} + w_s \quad ...(4.20)$$

Eq. (4.20) is the *steady flow energy equation*. In case of a pump installed within the system under consideration, the term w_s will be negative. Likewise, if the heat is transmitted by the fluid to the system surroundings, the heat transfer term q will be negative. Eq. (4.20) may be re-arranged and written as

$$q = \left(\frac{V_2^2}{2} + \frac{p_2}{\rho_2} + gZ_2\right) - \left(\frac{V_1^2}{2} + \frac{p_1}{\rho_1} + gZ_1\right) + u_2 - u_1 + w_s \quad ...(4.21)$$

If no heat is supplied to the fluid from outside the boundaries and that the temperature of the fluid and that of its surroundings are practically the same or if the boundaries are insulated so that there is no heat transfer, the term q may be taken as zero. If there is no machine between sections (1) and (2), w_s is also zero.

If the fluid is incompressible, *i.e.* $\rho_1 = \rho_2$ and also non-viscous (or frictionless), the temperature of fluid would, therefore, remain constant. The internal energy of the fluid is then also constant and Eq. (4.21) reduced to,

$$0 = \left(\frac{V_2^2}{2} + \frac{p_2}{\rho_2} + gZ_2\right) - \left(\frac{V_1^2}{2} + \frac{p_1}{\rho_1} + gZ_1\right)$$

dividing it by g, we obtain $\frac{V_1^2}{2g} + \frac{p_1}{\gamma} + Z_1 = \frac{V_2^2}{2g} + \frac{p_2}{\gamma} + Z_2 = \text{constant}.$

This may be identified as the Bernoulli's Eq. (4.12). The Bernoulli's equation is, therefore, a special form of energy equation, which is applicable to an ideal fluid and assumes that the transfer of heat into the fluid or out of it to the surroundings is zero and that the temperature of fluid remains constant so that the internal energy does not change. The effect of the presence of a machine is also ignored by the Bernoulli's equation.

4.11. Bernoulli's Equation for a Real Fluid

Application of Bernoulli's equation to the flow of real fluids requires a modification on account of viscosity. The assumption of zero-viscosity of the fluid results in constancy of total energy throughout the flow length, which is never true in case of flow of real fluids. A real fluid does possess viscosity and consequently it offers resistance to its flow. In order to overcome this viscous resistance and other resistances due to surface roughness and turbulence, a part of the total energy of flow is lost. This energy loss, in the strict sense of the word, is not a lost energy, as from the first law of thermodynamics, energy can neither be destroyed nor created ; but is converted into heat energy, thereby increasing the temperature of the fluid slightly. The increase of temperature of the fluid causes an increase in the internal energy. The increase in internal energy and the heat transfer from the fluid represent a loss of useful energy. The total loss of energy per unit mass of fluid is $(u_2 - u_1 - q)$, dividing it by g, we obtain the energy loss per weight in overcoming resistance. Therefore,

$$h_l = \frac{u_2 - u_1 - q}{g},$$

h_l is known as the head loss.

It is now obvious that the total energy of flow represented by the Bernoulli's Eq. (4.12) decreases in the flow direction, and consequently the energy line has a downward slope. The modified Bernoulli's equation for upstream section (1) and the downstream section (2) may now be written as

$$\frac{V_1^2}{2g} + \frac{p_1}{\gamma} + Z_1 = \frac{V_2^2}{2g} + \frac{p_2}{\gamma} + Z_2 + h_l \qquad ...(4.22)$$

where h_l is the head loss between the two sections.

Fig. 4.7 illustrates the graphical representation of modified Bernoulli's equation.

If a machine exists between the two sections under consideration, a term representing the energy per unit of weight utilised by the machine (a turbine utilises the fluid energy) or imparted to the fluid, as in case of a pump, is introduced. The Bernoulli's equation taking into account the machine energy may be expressed as,

$$\frac{V_1^2}{2g} + \frac{p_1}{\gamma} + Z_1 = \frac{V_2^2}{2g} + \frac{p_2}{\gamma} + Z_2 \pm H_m + h_l \qquad ...(4.23)$$

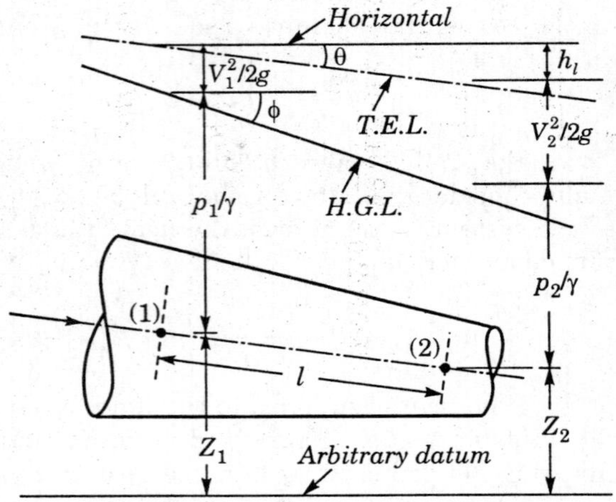

Fig. 4.7. Graphical representation of modified Bernoulli's equation.

The positive sign in Eq. (4.23) is used when the machine happens to be a turbine, and the negative sign if it is a pump. If the total energy at two sections is known, then the direction of flow can be ascertained. Provided that there is no machine in between the flow will take place from the higher energy to the lower energy.

The hydraulic gradient and the energy gradient are the slopes of the hydraulic grade line (H.G.L.) and the total energy line (T.E.L.) and are expressed as the sine of the angle these lines make with the horizontal. Thus, *energy gradient*

$$= \sin\theta = \frac{h_1}{l} = \frac{1}{l}\left[\left(\frac{V_1^2}{2g} + \frac{p_1}{\gamma} + Z_1\right) - \left(\frac{V_2^2}{2g} + \frac{p_2}{\gamma} + Z_2\right)\right]$$

and the *hydraulic gradient*

$$= \sin\phi = \frac{1}{l}\left(h_1 + \frac{V_2^2}{2g} - \frac{V_1^2}{2g}\right) = \frac{1}{l}\left[\left(\frac{p_1}{\gamma} + Z_1\right) - \left(\frac{p_2}{\gamma} + Z_2\right)\right]$$

For a uniform diameter pipe $V_1 = V_2$ and, therefore, $\sin\phi = \sin\theta$, the energy gradient and the hydraulic gradient will then be equal.

Example 4.12. *The following are the data given of a change in diameter effected in laying a water supply pipeline. The change in diameter is gradual from 20 cm at A to 50 cm at B. Pressures at A and B are 78.5 kN/m² and 58.9 kN/m² respectively with the end B being 3 m higher than A. If the flow in pipeline is 200 litres/sec, find :*

(i) direction of flow, and (ii) the head lost in friction between A and B.

Solution. Q = 200 litres/sec

$$= \frac{200}{1000} = 0.2 \text{ m}^3/\text{s}$$

Velocity of flow in the pipe at A,

$$V_A = \frac{4Q}{\pi d_A{}^2} = \frac{4 \times 0.2}{\pi \times (0.2)^2} = 6.366 \text{ m/s}$$

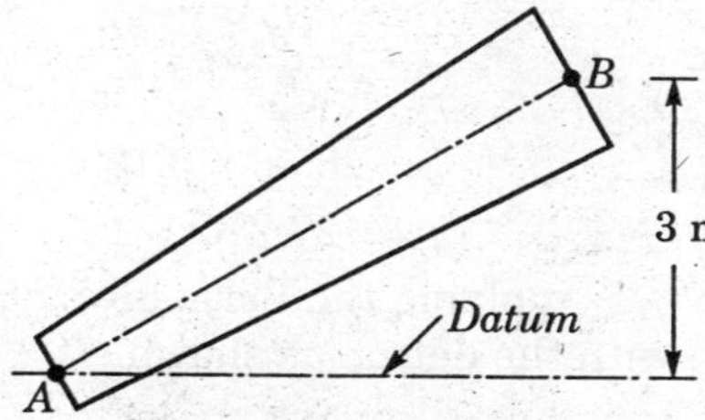

Velocity at B, $\quad V_B = \dfrac{0.2 \times 4}{\pi \times (0.5)^2} = 1.018$ m/s

Pressure at A, $\quad p_A = 78.5 \text{ kN/m}^2 = 78500 \text{ N/m}^2$

Pressure at B, $\quad p_B = 58.9 \text{ kN/m}^2 = 58900 \text{ N/m}^2$.

Applying the Bernoulli's equation to pipe sections at A and B, taking the horizontal line passing through point A as the datum, the total energy available in flow at A,

$$H_A = Z_A + \frac{p_A}{\gamma} + \frac{V_A{}^2}{2g} = 0 + \frac{78500}{9810} + \frac{(6.366)^2}{2 \times 9.81} = 10.065 \text{ m of water.}$$

Similarly, the total energy available at B

$$H_B = Z_B + \frac{p_b}{2g} = 3 + \frac{58900}{9810} + \frac{(1.018)^2}{2 \times 9.81} = 9.053 \text{ m of water.}$$

(*i*) Since the flow takes place from higher total energy to the lower one, and its value at A being higher than at B, **the direction of flow is from A to B.**

(*ii*) From Bernoulli's equation, one may write :

Total energy at A = Total energy at B + Head lost between A and B.

or $\quad 10.065 = 9.053 +$ Head lost

$\therefore$ Head lost in friction between A and B = 10.065 – 9.053 = **1.012 m of water.**

Ex. 4.13. *The centre line of a tapered pipe AB slopes upward from A to B at an angle of 30° to the horizontal. The distance AB is 5 m and the diameter increases uniformly from 10 cm at A to 15 cm at B. The pipe carries petrol (sp. gr. 0.74) and pressure gauges are installed at A and B. Find (a) the flow rate when the readings on the pressure gauges are equal (b) the pressure difference across AB for the same rate of flow when the direction of taper is reversed.*

Solution.

Case (a) Pressure at A (p_A) = pressure at B (p_B)

$$V_A = \frac{4Q}{\pi d_A{}^2} = \frac{4Q}{\pi (0.1)^2} = 127.27Q$$

$$V_B = \frac{4Q}{\pi d_B{}^2} = \frac{4Q}{\pi (0.15)^2} = 56.56Q$$

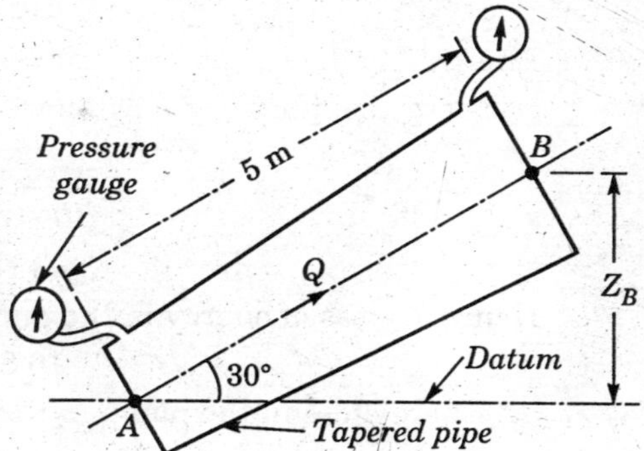

Applying Bernoulli's equation between A and B, flow taking place from A towards B, the datum line taken as a horizontal line passing the rough A, neglecting loss of energy :

$$0 + \frac{p_A}{\gamma} + \frac{V_A^2}{2g} = \frac{p_B}{\gamma} + \frac{V_B^2}{2g} + 5 \sin 30°$$

noting $p_A = p_B$, $$\frac{(127.27Q)^2}{2g} = \frac{(56.56Q)^2}{2g} + 8.66$$

from which, $Q = 0.1325 \text{ m}^3/\text{s} = \mathbf{132.5\ l/s.}$

Case (b) Direction of taper reversed, the same discharge 132.5 l/s flows.

$$V_A = \frac{4Q}{\pi d_A^2} = \frac{4Q}{\pi (0.15)^2} = 56.56Q$$

$$= 56.56 \times 0.1325 = 7.49 \text{ m/s}$$

$$V_B = 127.27Q = 127.27 \times 0.1325 = 16.86 \text{ m/s}$$

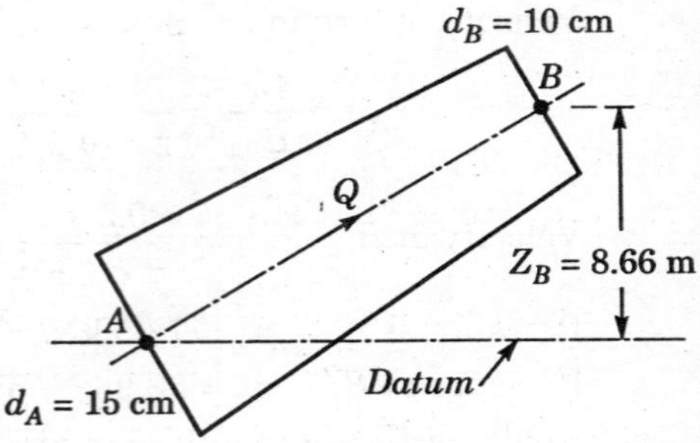

Applying the Bernoulli's equation between A and B, with the datum as shown.

$$0 + \frac{p_A}{\gamma} + \frac{(7.49)^2}{2 \times 9.81} = 8.66 + \frac{p_B}{\gamma} + \frac{(16.86)^2}{2 \times 9.81}$$

$$\therefore \quad \frac{p_A - p_B}{\gamma} = 20.29 \text{ m}$$

Pressure difference across $AB = (9810 \times 0.74) \times 20.29 = 147.3 \times 10^3 \text{ N/m}^2 = \mathbf{147.3\ kPa.}$

Example 4.14. *A pump is installed in a pipeline, 5 cm diameter, carrying oil of sp. gr. 0.83. It returns the oil to a 5 cm diameter pipe at the same elevation with a pressure increase of 13.7 kN/m². The quantity of oil flowing in the pipeline is 10 litres/sec. The motor driving the pump delivers 2.8 kW to the pump shaft. Calculate the loss of energy in the pump.*

Solution. Consider section (1) immediately upstream of pump and section (2) just downstream of the pump. The application of Bernoulli's equation gives

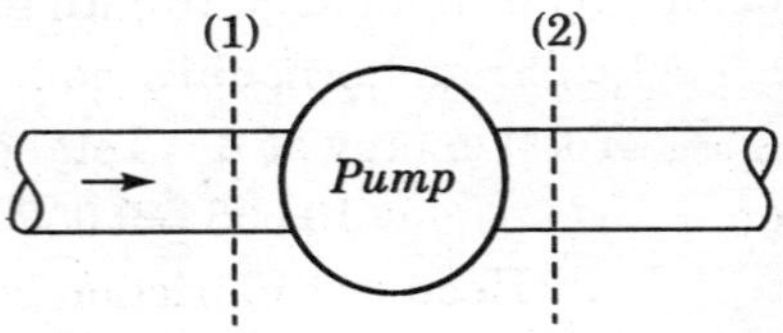

$$Z_1 + \frac{p_1}{\gamma} + \frac{V_1^2}{2g}$$

+ Energy supplied by pump/N weight of oil

$$= Z_2 + \frac{p_2}{\gamma} + \frac{V_2^2}{2g} + \text{Losses in the pump.}$$

But since $Z_1 = Z_2$ and $V_1 = V_2$ since diameter of the pipe is same, one obtains

$$\text{Losses in the pump} = \left(\frac{p_1 - p_2}{\gamma}\right) + \text{Energy supplied/N of oil.}$$

Since $p_2 > p_1$, $$\frac{p_2 - p_1}{\gamma} = \frac{13.7 \times 1000}{9810 \times 0.83} = 16.87 \text{ m of oil} = 16.78 \text{ Nm/N of oil.}$$

Energy supplied to pump shaft = 2.8 kW = 2800 W = 2800 Nm/s

∴ Energy supplied per newton weight of oil

$$= \frac{2800}{10 \times 10^{-3} \times 9810 \times 0.83} = 34.39 \text{ m of oil}$$

$$= 34.3 \text{ Nm/N of oil.}$$

Hence, (*i*) loss of energy in the pump

$$= -16.87 + 34.30 = 17.43 \text{ Nm/N of oil.}$$

(*ii*) Energy loss in the pump in watt = 17.43 × 81.4 = 1419.2 Nm/s = 1.93 W.

Example 4.15. *At a point in the pipeline where the diameter is 20 cm, the velocity of water is 4 m/s and the pressure is 343 kN/m². At a point 15 m downstream the diameter reduces to 10 cm. Calculate the pressure at this point, if the pipe is (a) horizontal (b) vertical with flow downward (c) vertical with flow upward.*

Solution. Velocity at the 10 cm diameter section

$$\frac{4 \times \frac{\pi}{4} \times (20)^2}{\frac{\pi}{4} \times (10)^2} = 16 \text{ m/s.}$$

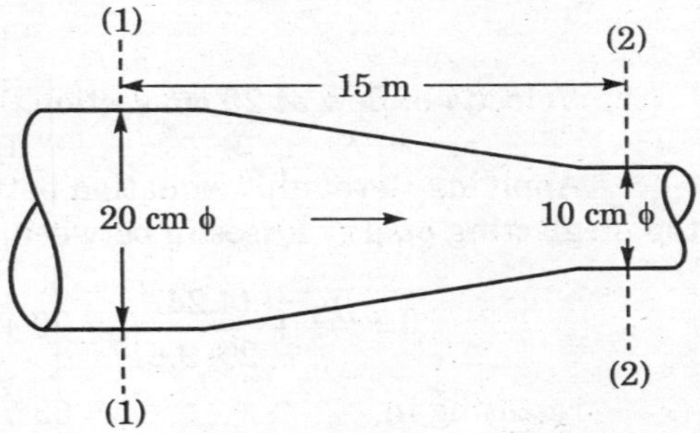

(*a*) **Pipeline horizontal.** Applying Bernoulli's equation to section (1) and (2) of the horizontal pipeline, taking the pipeline axis as the datum, $Z_1 = Z_2$, and neglecting losses.

$$\frac{V_1^2}{2g} + \frac{p_1}{\gamma} = \frac{V_2^2}{2g} + \frac{p_2}{\gamma}$$

or
$$\frac{(4)^2}{2 \times 9.81} + \frac{343 \times 10^3}{9.81 \times 10^3} = \frac{(16)^2}{2 \times 9.81} + \frac{p_2}{\gamma}$$

$$p_2 = (35.815 - 13.03) \times 9810 \text{ N/m}^2$$
$$= 223.5 \text{ kN/m}^2$$

(*b*) **Pipeline vertical (Flow downward).** Bernoulli's equation between section (1) and (2) gives

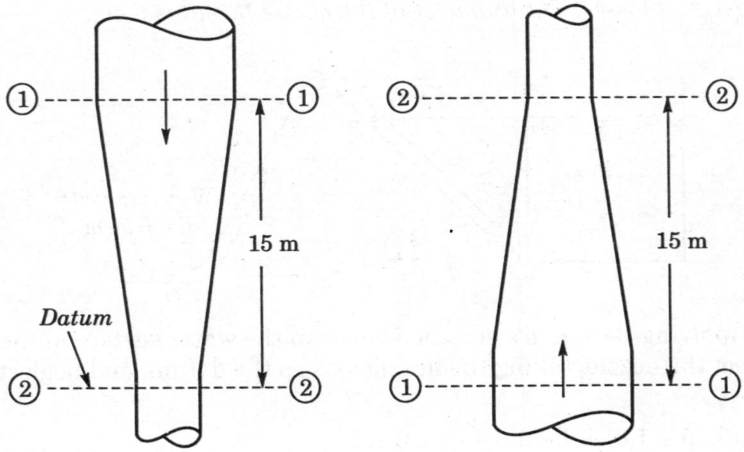

$$15 + \frac{(4)^2}{2 \times 9.81} + \frac{343 \times 10^3}{9.81(10)^3} = 0 + \frac{(16)^2}{2 \times 9.81} + \frac{p_2}{\gamma}$$

$\therefore$
$$p_2 = 370.6 \text{ kN/m}^2$$

(*c*) **Pipeline vertical (Flow upward).** Bernoulli's equation between sections (1) and (2) results in

$$0 + \frac{(4)^2}{2 \times 9.81} + \frac{343 \times 10^3}{9810} = 15 + \frac{(16)^2}{2 \times 9.81} + \frac{p_2}{\gamma}$$

From which,
$$p_2/\gamma = 7.785 \text{ m}$$
$$p_2 = 7.785 \times 9810 = 76370.85 \text{ N/m}^2 = \mathbf{76.37 \text{ kN/m}^2}.$$

Example 4.16. *Water is pumped at the rate of 300 litres / sec through a 30 cm pipe upto a hill top. On the hill top which has an elevation of 50 m, the diameter of pipeline reduces to 20 cm. If the pump maintains a pressure of 981 bar at the hill top, what is the pressure at the foot hills having zero elevation ? What is the power required to pump the water ?*

Solution. Discharge $Q = 300/1000 = 0.3$ m³/s.

Velocity of flow at 30 cm section, $V_1 = \dfrac{0.3}{\dfrac{\pi}{4}(0.3)^2} = 4.24$ m/s

Velocity of flow at 20 cm section, $V_2 = \dfrac{0.3 \times 4}{\pi(0.2)^2} = 9.55$ m/s.

Applying Bernoulli's equation between the pipeline sections at the foot hills and at the hill top, neglecting energy losses in between, noting that 1 bar = 100 N/m²

$$0 + \frac{p_1}{\gamma} + \frac{(4.24)^2}{2 \times 9.81} = 50 + \frac{981 \times 100}{9810} + \frac{(9.55)^2}{2 \times 9.81} = 64.65 \text{ m of water}$$

resulting in, $p_1 = 63.734 \times 9810$ N/m² = 625.2 kN/m² = **6252 bar.**

Pressure at the foot hills = 6252 bar.

Power required to pump water to the hill top

$= \gamma QH = 9810 \times 0.3 \times 64.65 = 190265$ Nm/s

$= 190265$ W = **190.265 kW.**

Example 4.17. *A pipe 20 cm in diameter is connected to a water tank as shown. A nozzle fitted at the end of the pipe discharges into the atmosphere. Calculate the flow rate and the pressure at A, B, C and D. Neglect losses. The diameter of the nozzle throat is 5 cm.*

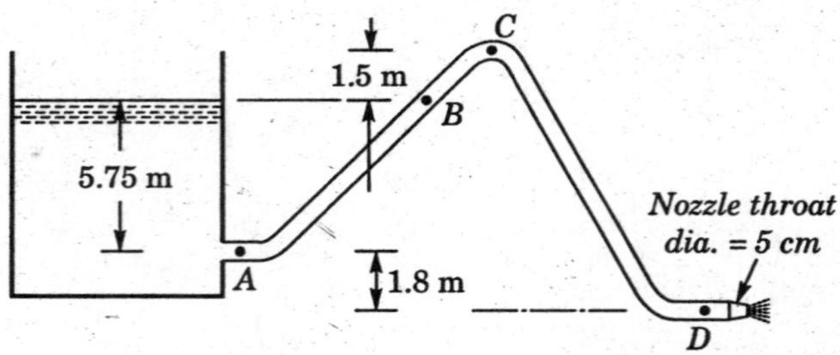

Solution. Applying Bernoulli's equation between the water surface in the tank and the jet of water issued from the nozzle, taking the nozzle axis as the datum, and neglecting losses,

$$(5.75 + 1.8) + 0 + 0 = \frac{V^2}{2g} + 0 + 0$$

$\therefore$ Velocity of jet, $V = \sqrt{2 \times 9.81 \times 7.55} = 12.17$ m/s.

The flow rate, $Q = \dfrac{\pi}{4}\left(\dfrac{5}{100}\right)^2 \times 12.17 = 23.9$ litres/sec.

Velocity in the 20 cm pipe

$$= \frac{0.0239 \times 4}{\pi \times (0.2)^2} = 0.76 \text{ m/s.}$$

The pressures at *A*, *B*, *C* and *D* may be obtained by applying Bernoulli's equation between the water surface in the tank and these points.

Pressure at A : $5.75 + 0 + 0 = \dfrac{(0.76)^2}{2 \times 9.81} + \dfrac{p_A}{\gamma}$

$\therefore \qquad \dfrac{p_A}{\gamma} = 5.7205$ m of water

$\therefore \qquad p_A = 5.72 \times 9810 \text{ N/m}^2 = 56.1 \text{ kN/m}^2 = 561 \text{ bar.} = 56.1 \text{ kPa.}$

Pressure at B : $\quad 0 + 0 + 0 = 0 + \dfrac{(0.76)^2}{2 \times 9.81} + \dfrac{p_B}{\gamma}$

$\therefore \qquad \dfrac{p_B}{\gamma} = -0.0295$ m of water

$\therefore \qquad p_B = -0.0295 \times 9810 = -289.4 \text{ N/m}^2 = -289.4 \text{ Pa.}$

Pressure at C : $\quad 0 + 0 + 0 = 1.5 + \dfrac{(0.76)^2}{2 \times 9.81} + \dfrac{p_c}{\gamma}.$

$\therefore \qquad \dfrac{p_c}{\gamma} = -1.5295$ m of water

$\therefore \qquad p_c = -1.5295 \times 9810 = -15.0 \text{ kN/m}^2 = -15.0 \text{ kPa.}$

Pressure at D :

$$5.75 + 1.8 + 0 + 0 = \frac{(0.76)^2}{2 \times 9.81} + \frac{p_d}{\gamma} + 0$$

$$\frac{p_d}{\gamma} = 7.5205 \text{ m of water}$$

$$= 7.5205 \times 9810 \text{ N/m}^2 = 73.8 \text{ kN/m}^2 = 73.8 \text{ kPa.}$$

Ex. 4.18. *A large tank partially filled with water, the air space above being under pressure. A 0.05 m hose connected to the tank discharges on the roof of a building 15 m above the level of the tank. The friction loss is 0.055 m. What is pressure must be maintained in the tank to deliver 0.001 m^3/s on the roof. ?*

Solution. With reference to the diagram shown alongside, let 'p' be the air pressure inside the tank. The velocity of water in the hose,

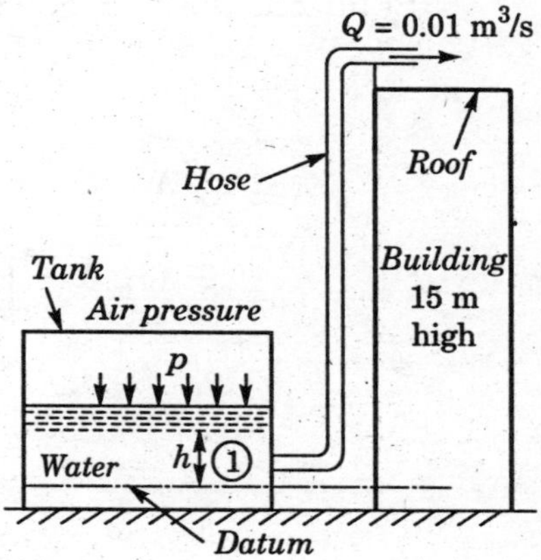

$$V = \frac{Q}{A} = \frac{0.001}{\pi(0.05)^2/4} = 0.509 \text{ m/s}$$

Applying the Bernoulli's equation to the inlet end (1) and the outlet end of the hose at 15 m height above the bottom level, taking the horizontal line passing through (1) as the datum.

$$p/\gamma + h = 15 + V^2/2g + 0.055$$

here p is the pressure of air in the tank and 'h' the water depth, γ being the specific weight of water. Assuming $h << 15$ m, neglecting its effect,

$$\frac{p}{\gamma} = 15 + \frac{(0.509)^2}{2 \times 9.81} + 0.055 = 15.068 \text{ m of water}$$

$\therefore$ Air pressure, $p = 15.068 \times (9810) \text{ N/m}^2 = 147819.10 \text{ N/m}^2 = 147.819 \text{ kN/m}^2.$ = **147.82 kPa.**

Example 4.19. *Water discharges from a two-dimensional rectangular opening into the air as indicated at A in the accompanying figure. At B water discharges from under a gate onto a floor.*

Calculate the discharges of A and B per metre length of the openings. Why do they differ ? Neglect all losses.

Solution. Since the jet through the two-dimensional slot discharges into air, it is an orifice flow. Neglecting the losses, the jet velocity

$$V = \sqrt{2gh} = \sqrt{2 \times 9.81(2.5 + 0.25)}$$
$$= 7.38 \text{ m/s.}$$

$\therefore$ Discharge per metre length through the opening A

$$= V \times 0.50 \times 1 = 7.38 \times 0.50 \times 1$$
$$= 3.69 \text{ m}^3\text{/s/m.}$$

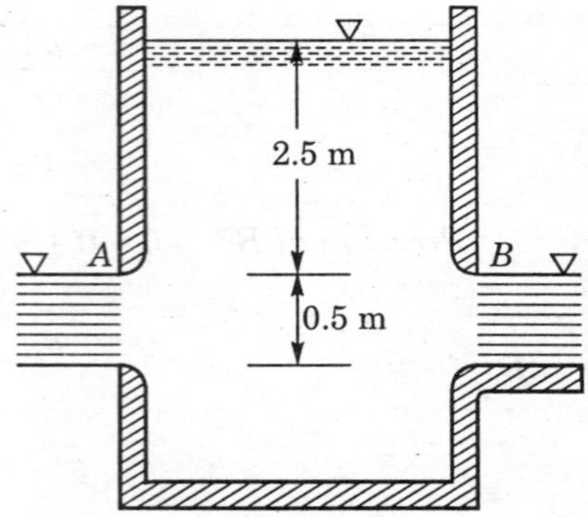

Fig. Ex. 4.19

Applying Bernoulli's equation to the water surface in the reservoir and water surface at B, just downstream of gate, where pressure is atmospheric.

$$2.5 \doteq \frac{V^2}{2g}$$

$$V = \sqrt{2g \times 2.5} = \sqrt{2 \times 9.81 \times 2.5} = 7.0 \text{ m/s.}$$

$\therefore$ Discharge through the gate opening B per metre length

$$= V \times 0.5 \times 1.0 = 7 \times 0.5 = 3.5 \text{ m}^3\text{/s/m.}$$

The difference in discharges is due to the difference in the pressure distribution in the jets at A and B. While at A the pressure in the jet remains atmospheric, it is hydrostatic at B. The flow through the opening B is the sluice flow, *i.e.* flow through the sluice gate.

Example 4.20. *The velocity distribution in a pipe is given by*

$$\frac{v}{v_{max}} = \left(\frac{y}{r_0}\right)^{1/n}$$

in which v is the velocity at a distance y from the pipe wall and r_0 the pipes radius. Determine the average velocity and the momentum correction factor.

Solution. The average velocity may be obtained as

$$V \,.\, A = \int_0^{r_0} 2\pi r \, dr \,.\, v$$

where $\qquad r = r_0 - y \quad$ and $\quad A = \pi r_0^2$

$$\pi r_0^2 V = \int_{r_0}^{0} 2\pi(r_0 - y)(-dy)\, v_{max} \left(\frac{y}{r_0}\right)^{1/n}$$

$$= \int_0^{r_0} 2\pi\, v_{max} \left\{ r_0 \left(\frac{y}{r_0}\right)^{1/n} - \frac{1}{r_0^{1/n}}\, y^{(1+n)/n} \right\} dy$$

$$= 2\pi\, v_{max} \left[r_0 \frac{n}{n+1} \cdot \frac{y^{(n+1)/n}}{r_0^{1/n}} - \frac{1}{r_0^{1/n}} \frac{n}{1+2n}\, y^{(1+n)/n} \right]_0^{r_0}$$

$$= 2\pi\, v_{max} \left[\frac{n}{n+1} r_0^2 - \frac{n}{1+2n} r_0^2 \right]$$

$$\therefore \qquad V = 2\, v_{max} \left[\frac{n(1+2n) - n(1+n)}{(1+n)(1+2n)} \right] = v_{max} \left[\frac{n^2}{(1+n)(1+2n)} \right]$$

The momentum correction factor β is defined as (see art. 4.20, page 232),

$$\left[\frac{\gamma}{g} AV\right] A \,.\, \beta = \int_A \frac{\gamma}{g}\, dA . v^2$$

or
$$\beta = \frac{1}{A}\int_A \left(\frac{v}{V}\right)^2 dA \qquad \text{where } V = \text{average velocity.}$$

Substituting for v and V

$$\beta = \frac{1}{\pi r_0^2}\int_0^{r_0} \left[\frac{(y/r_0)^{1/n}\, v_{max}\,(1+n)\,(1+2n)}{2\, v_{max}\, n^2}\right]^2 2\pi r dr$$

Substituting $\quad r = r_0 - y$

$$\therefore \qquad \beta = \frac{1}{\pi r_0^2}\int_{r_0}^{0} \left[\frac{(1+n)(1+2n)}{2n^2}\right]^2 \left(\frac{y}{r_0}\right)^{n/2} . \, 2\pi(r_0 - y)(-\,dy)$$

$$= \frac{1}{\pi r_0^2} . \, 2\left[\frac{(1+n)(1+2n)}{2n}\right]^2 \int_0^{r_0} \left(r_0 \frac{1}{r_0^{2/n}}\, y^{2/n} - \frac{1}{r_0^{2/n}}\, y^{(n+2)/n}\right) dy$$

$$= \frac{(1+n)^2 (1+2n)^2}{2 r_0^2 n^4}\left[\frac{r_0}{r_0^{2/n}}\,\frac{n}{n+2}\, y^{(2n+n)/n} - \frac{1}{r_0^{2/n}}\left(\frac{n}{2n+2}\right) y^{(2n+2)/n}\right]_0^{r_0}$$

which simplifies to,
$$= \frac{(1+n)\,(1+2n)^2}{4n^2\,(n+2)}.$$

Example 4.21. *The nozzle shown in the accompanying figure delivers a jet 2.5 cm in diameter. Calculate the discharge Q through the nozzle if p_1 = 1668 bar (gauge) and p_2 = 30 cm of mercury (gauge). What is the maximum height of the free jet above the nozzle ?*

Solution. Applying the Bernoulli's equation to the water surface in the left tank and the jet as it emerges from the nozzle, taking the centre line of nozzle as the datum,

$$3 + \frac{1668 \times 100}{9810} + 0 = 0 + \frac{V^2}{2g} + 0.3 \times 13.6$$

$$20 = \frac{V^2}{2g} + 4.08$$

yielding $\quad V = 17.66$ m/s.

$\therefore$ Discharge $Q = VA$

$$= \frac{\pi}{4}\left(\frac{2.5}{100}\right)^2 \times 17.66 \text{ m}^3\text{/s}$$

$= 0.00867$ m³/s

$= 0.00867 \times 1000$ litres/sec.

$= 8.67$ litres/sec.

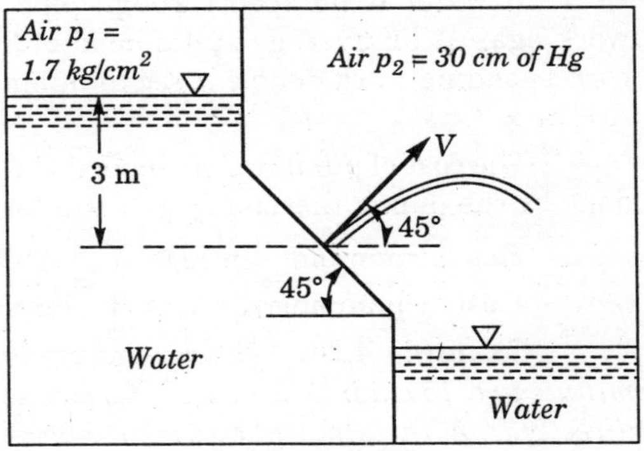

Fig. Example 4.21

Vertical component of jet velocity

$= V \sin 45° = 17.66 \sin 45° = 12.48$ m/s.

$\therefore$ Maximum height to which jet will rise $= \dfrac{(12.48)^2}{2 \times 9.81} = 7.93$ m.

Example 4.22. *Water moves steadily through the turbine shown in the figure at the rate of 0.23 m²/s. The pressures at (1) and (2) are 186.4 kN/m² and – 19.6 kN/m² respectively. Neglecting heat transfer, determine the horsepower delivered to the turbine from water.*

Solution. Velocity at the 20 cm section

$$= \frac{0.23 \times 4}{\pi(0.2)^2} = 7.32 \text{ m/s.}$$

Velocity at the 40 cm section

$$= \frac{0.23 \times 4}{\pi \times (0.4)^2} = 1.83 \text{ m/s.}$$

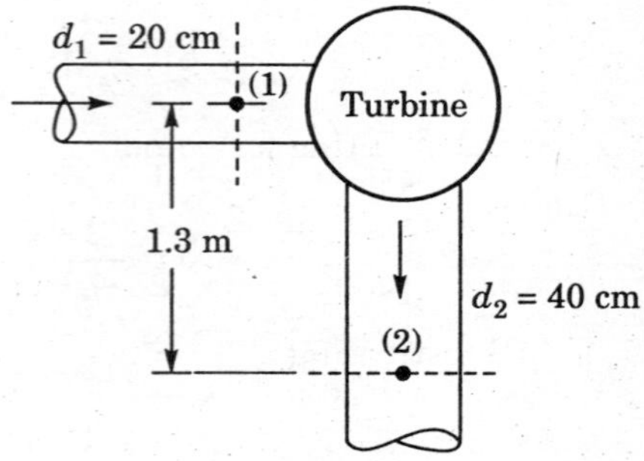

Fig. Example 4.22

Energy head available at (1) with reference to horizontal datum passing section (2)

$$= 1.30 + \frac{186.4 \times 10^2}{9810} + \frac{(7.32)^2}{2 \times 9.81} = 23.035 \text{ Nm/N}$$

Energy head in the flow at section (2)

$$= 0 - \frac{19.6 \times 10^3}{9810} + \frac{(1.83)^2}{2 \times 9.81}$$

$$= -2.0 + 0.171 = -1.829 \text{ Nm/N.}$$

Energy head utilised by the turbine

$$= 23.025 - (-1.829) = 24.864 \text{ Nm/N.}$$

∴ Power delivered to turbine

$$= \gamma Q \times 24.864 = 9810 \times 0.23 \times 24.864 = 56 \times 10^3 \text{ W} = 56 \text{ kW.}$$

Example 4.23. *The water level in a vertical cylindrical tank with top open, is to be raised by 6 m from a nearby reservoir. Compare the cost of pumping through pipes A and B.*

Solution. For flow through pipe B, the pump has to work against a static head of 6 m and the work done per second in pumping at the flow rate of Q will by $\gamma Q \times 6$.

If water is pumped through pipe A, it has to work against an average static head of 3 m, and the corresponding work done per sec in pumping through $A = \gamma Q \times 3$.

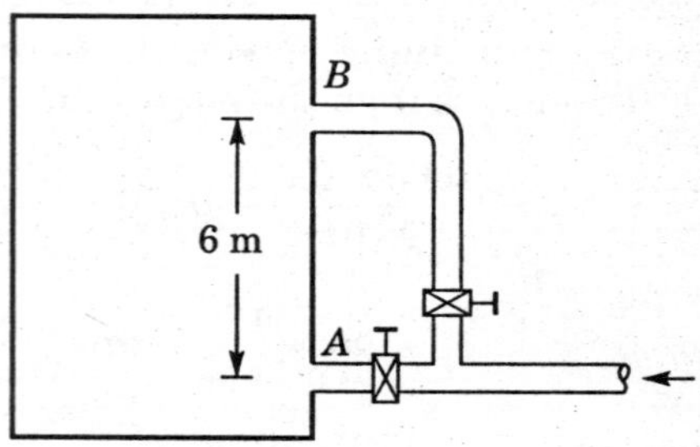

Fig. Example 4.23

The cost of pumping is proportional to the work done by the pump, therefore, the ratio

$$\frac{\text{Cost of pumping through } A}{\text{Cost of pumping through } B} = \frac{3\gamma Q}{6\gamma Q} = \frac{1}{2}.$$

Example 4.24. *A blower having an efficiency of 75% is to supply 14 cubic metre of air per minute to a 15 cm pipe under a pressure equivalent of 5 cm of water. If the 30 cm intake pipe draws directly from the atmosphere, what horsepower motor should be provided ? What will be the pressure intensity in the intake pipe ? ρ_{air} = 1.284 kg/m³.*

Solution. Velocity in 15 cm pipe

$$= \frac{14}{60 \times \frac{\pi}{4} (0.15)^2} = 13.2 \text{ m/s.}$$

Velocity in 30 cm pipe $= \frac{13.2}{4} = 3.3$ m/s.

Energy at section (2)

= Energy at section (1)

+ Energy supplied by the blower

$$0 + 0 = \frac{V_2^2}{2g} + \frac{p_2}{\gamma}$$

– Energy supplied by blower.

∴ Energy supplied by the blower per unit weight of fluid

$$= \frac{V_2^2}{2g} + \frac{p_2}{\gamma}$$

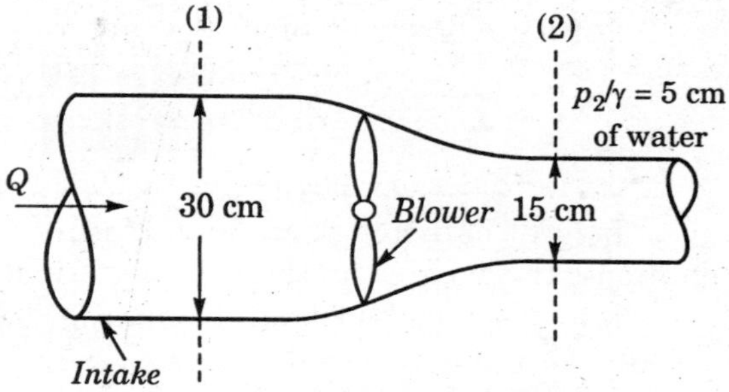

Fig. Example 4.24

$$p_2 = \frac{5}{100} \times 9810 \text{ N/m}^2 = 490 \text{ N/m}^2$$

$$\gamma = \rho g = 1.284 \times 9.81 = 12.6 \text{ N/m}^3$$

∴ Energy supplied by blower

$$= \frac{(13.2)^2}{2 \times 9.81} + \frac{490}{12.6} = 47.79 \text{ m/N of air} = 47.79 \text{ Nm/N of air}$$

Output of the blower $= 47.79\, Q\gamma = 47.79 \times 12.6 \times 14 = 138.4$ W

∴ Power of the motor $= \text{Output/Efficiency} = \frac{138.4}{0.75} = 186.7$ W.

Applying Bernoulli's equation to the intake and outside where pressure is atmospheric and total energy is zero, we obtain

$$0 = \frac{p_1}{\gamma} + \frac{V^2}{2g}$$

yielding $$\frac{p_1}{12.6} = \frac{(3.3)^2}{2 \times 9.81}$$

$$p_1 = -12.6 \times 0.555 = 7.0 \text{ N/m}^2 \text{ (i.e. 7.0 N/m}^2 \text{ below atmosphere).}$$

Example 4.25. *If a 920 W motor is required by a ventilating fan to produce a 60 cm diameter stream of air having a velocity of 15 m/s, what is the efficiency of the fan ? Assume that the intensity of pressure at inside and outside of the fan is same. Take* $\rho = 1.285$ *kg/m³.*

Solution. Kinetic energy of air stream $= \frac{1}{2}\rho Q V^2$

$$Q = \frac{\pi}{4} \times (0.6)^2 \times 15 \text{ m}^3/\text{s}$$

$$\text{K.E.} = \frac{1}{2} \times 1.285 \times \frac{\pi}{4} \times (0.6)^2 \times 15 \times (15)^2 = 612.14 \text{ Nm/s}$$

Output of the fan = 612.14 W

Input = 920 W

Efficiency of the fan $= \frac{\text{Output}}{\text{Input}} = \frac{612.14}{920} \times 100 = 66.6\%$.

Example 4.26. *An air-compressor draws air from atmosphere through a bellmouth entrance. The bellmouth is to be calibrated for measuring discharge passing through it in terms of height h of water column that rises in the single tube manometer as shown in the figure. If h = 0.20 m and the specific weight of the air is 10.8 N/m³, determine the discharge of air through the compressor. The diameter of the entrance duct where the manometer is installed is 15 cm.*

Solution. Pressure of air flow in terms of water column.

$$\frac{p}{\gamma_w} = 0.20 \text{ m}$$

$$\therefore \quad p = 9810 \times 0.20 = 1962 \text{ N/m}^2$$

It is to be noted that this pressure is negative and represents the average pressure at the section.

Applying the Bernoulli's equation between a point upstream of the entrance where the velocity is zero and the pressure atmospheric and the flow section where the manometer is installed :

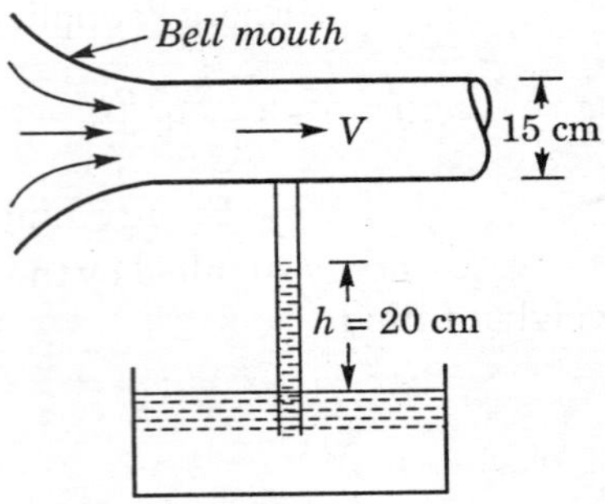

Fig. Ex. 4.26

$$0 + 0 + 0 = 0 - \frac{p}{\gamma_{air}} + \frac{V^2}{2g}$$

Substituting for p and γ_{air}, $\quad \frac{V^2}{2g} = \frac{1962}{10.8} = 181.7 \text{ m}$

From which, $\quad V = \sqrt{2 \times 9.81 \times 181.7} = 59.70 \text{ m/s}$

Air flow rate entering the bellmouth

$$Q = \frac{\pi d^2}{4} V = \frac{\pi}{4} (0.15)^2 \times 59.70 \text{ m}^3\text{/s}$$

$= \mathbf{1.054\ m^3/s}$ **or 1054 litre/second.**

Example 4.27. *The Hero's fountain could be designed by suitable arranging a system of three reservoirs as shown in the following figure. For the dimensions indicated, determine the velocity of water-jet issuing from the nozzle. If the nozzle tip diameter is kept as 2.5 cm, estimate the quantity of flow required. How high would the jet rise theoretically above the nozzle tip. Neglect pipe and air friction losses.*

Solution. Pressure at A on the liquid surface of tank III

$= \gamma h = 9810 \times 8 = 78480 \text{ N/m}^2$ (gauge)

Since the air space in tanks II and III are interconnected, the air pressure in the two must be the same.

Air pressure in tank III and II

$=$ pressure at $A = 78.48 \text{ kN/m}^2$.

Applying Bernoulli's equation between point B on the liquid surface of tank II and tip of the nozzle, where the jet velocity is V, taking water-surface in tank II as the datum, and neglecting pipe friction losses, one can write :

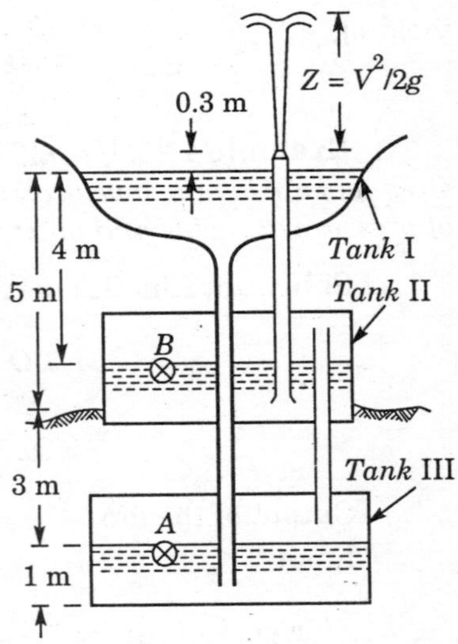

Fig. Ex. 4.27

$$0 + \frac{78480}{9810} + 0 = 4.3 + 0 + \frac{V^2}{2g}$$

from which, $V = \sqrt{2 \times 9.81 \times 3.7} = \mathbf{8.52\ m/s}$.

The water fountain will, therefore, have a velocity of 8.52 m/s at the nozzle-tip.

The quantity of water required per unit time

$$Q = \frac{\pi d^2}{4} V = \frac{\pi}{4}\left(\frac{2.5}{100}\right)^2 \times 8.52 \text{ m}^3\text{/s}$$

$= 41.77 \times 10^{-4} \text{ m}^3\text{/s} =$ **4.177 litre/second.**

Since the jet is directed vertically upward, the jet velocity V is in the vertical direction. Neglecting air resistance, the jet will theoretically rise upto a height

$$Z = \frac{V^2}{2g} = \frac{(8.52)^2}{2 \times 9.81} = \mathbf{3.70\ m.}$$

Example 4.28. *A jet of water 20 mm diameter nozzle leaves the nozzle tip with 15 m/s and is directed vertically upwards. If the jet remains circular, work out its diameter at a point 5 m above the nozzle tip. Neglect any loss of energy.* (*RGPV, June 2012*)

Solution. Applying the Bernoulli's equation to the jet at the nozzle tip marked *A-A*, and the jet at a height 5 m above the tip as shown, *i.e.* at *B-B*, neglecting loss of energy, we can write :

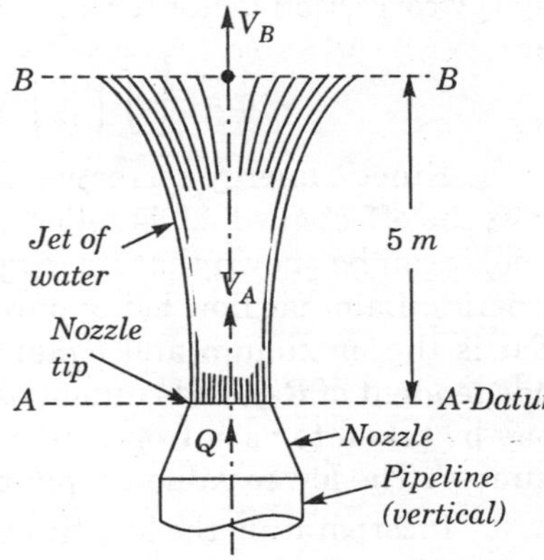

$$\frac{V_A^2}{2g} + \frac{p_A}{\gamma} + Z_A = \frac{V_B^2}{2g} + \frac{p_B}{\gamma} + Z_B$$

Taking the horizontal line *A-A* as the datum, and noting that the nozzle discharges into the atmosphere, we obtain :

$$Z_A = 0,\ Z_B = 5\ \text{m},\ p_A = p_B = 0\ \text{(gauge)}$$

given $V_A = 15$ m/s, and $d_A = 20$ mm $= 20 \times 10^{-3}$ m. Applying the continuity equation to the nozzle tip and at *B-B*,

$$Q = V_A \cdot \frac{\pi d_A^2}{4} = V_B \cdot \frac{\pi d_B^2}{4}$$

since V_A and d_A are known, $Q = 15 \times \dfrac{\pi (20 \times 10^{-3})^2}{4} = 0.471 \times 10^{-2}\ \text{m}^3/\text{s}$

Putting the data in the Bernoulli's equation,

$$\frac{(15)^2}{2 \times 9.81} + 0 + 0 = \frac{V_B^2}{19.62} + 0 + 5, \quad \therefore \quad V_B^2 = 19.62\,(11.468 - 5)$$

$$= 128.9, \quad \therefore \quad \mathbf{V_B = 11.265\ m/s}$$

From the continuity equation

$$0.471 \times 10^{-2} = V_B \times \frac{\pi d_B^2}{4}, \quad \therefore \quad d_B^2 = \frac{4 \times 0.471 \times 10^{-2}}{\pi \times 11.265} = 0.05326 \times 10^{-2}$$

or $$d_B = 0.2308 \times 10^{-1}\ \text{m} = \mathbf{23.08\ mm}$$

The diameter of the jet at 5 m height is **23.08 mm.**

4.12. Energy Correction Factor

The analysis of flow problems is usually based on the 'one-dimensional' approach. The entire flow is considered to be taking place through stream tube with average velocity V at each cross-section. The velocity distribution at any cross-section in real fluid flow is non-uniform, on account of viscous and boundary resistance and consequently the kinetic energy per unit of weight given by $V^2/2g$ does not represent the true kinetic energy across the section, *i.e.* average of $v^2/2g$ taken over the cross-section, v being the point velocity.

In order to compensate for the discrepancy, a coefficient known as *energy correction factor*, or Coriolis coefficient denoted by α is used. The multiplication of α with $V^2/2g$ yields the kinetic energy actually passing a section or the flux of kinetic energy.

Consider Fig. 4.8, which shows velocity distribution and geometry of flow section. The kinetic energy per unit time passing through an area element dA is $\frac{1}{2}(\rho dA\ v)v^2$, where v is the velocity at that level. Total kinetic energy passing the section is

$\int_A \frac{1}{2}\rho v^3 \, dA.$

But the actual kinetic energy based on average velocity V passing the section is equal to $\alpha \frac{1}{2}\rho V^3 A$.

Equating the two expressions representing the same quantity, we obtain an expression for the kinetic energy correction factor for an incompressible fluid,

$$\alpha = \frac{1}{A}\int \left(\frac{v}{V}\right)^3 dA \qquad \text{...(4.24)}$$

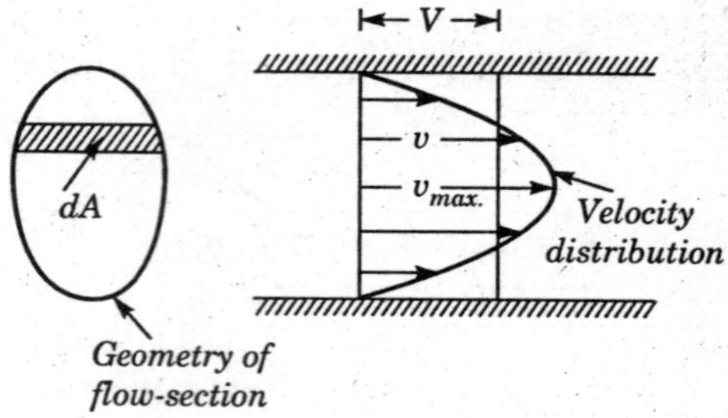

Fig. 4.8

Kinetic energy correction factor α is a measure of viscous resistance, generated in a given flow, the effect of which is reflected in the non-uniform nature of velocity distribution. For a given pipe, it can be shown that its magnitude is a function of the type of flow and its turbulent characteristics. Laminar flow being purely a viscous flow (and turbulence being non-existent), the value of α is the maximum and equal to 2.0. In case of fully developed turbulent flow in pipes, α is independent of Reynolds number and may be considered to have almost a constant value. Its value may be taken to vary from 1.01 to 1.15 depending upon the surface roughness and the Reynolds number. The lower value is applicable to relatively rough surface and high Reynolds number.

Incorporating the kinetic energy correction factor, the Bernoulli's equation may be written as

$$\alpha_1 \frac{V_1^2}{2g} + \frac{p_1}{\gamma} + Z_1 = \alpha_2 \frac{V_3^2}{2g} + \frac{p_2}{2\gamma} + Z_2 + h_l \qquad \text{...(4.25)}$$

In case of identical velocity distributions at the two sections under consideration $\alpha_1 = \alpha_2$.

Unless the accuracy of computation demand, it is usual to ignore the effects of the non-uniformity of velocity distribution by assuming $\alpha = 1.0$.

4.13. Pressure Distribution in Irrotational Flow

While dealing with the flownet in the chapter on Kinematics of Flow, it was pointed out that the flownet analysis is applicable only to irrotational flow. A flownet provides streamline picture of the flow problem, and enables us to study the velocity variation at a glance. From the Bernoulli's equation the pressure and the velocity are inter-related. Fig. 4.9 shows flownet for flow near the front of a bridge pier. Considering the flow in the horizontal plane, the Bernoulli's Eq. (4.16) applied between a point (1) in the undisturbed flow, where the velocity is V_0 and the pressure p_0 and a point (2), marked X, on the surface of bridge pier, where the velocity is V and the pressure p may be written as

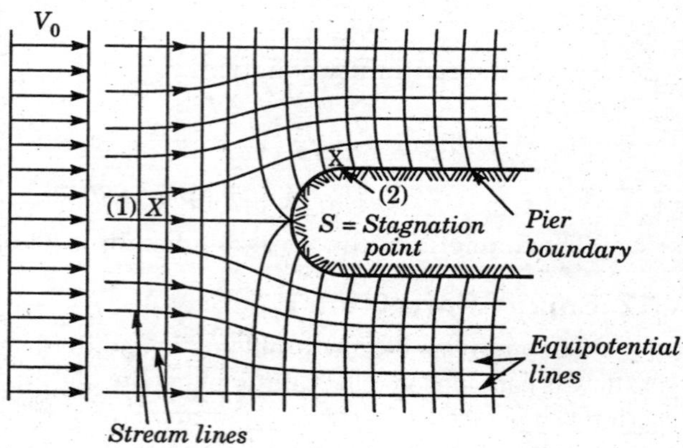

Fig. 4.9. Flownet at the front of a the bridge pier.

$$p_0 + \rho\frac{V_0^2}{2} = p + \rho\frac{V^2}{2}$$

or

$$\frac{p - p_0}{\rho\frac{V_0^2}{2}} = 1 - \left(\frac{V}{V_0}\right)^2$$

denoting the difference in pressure by Δp, one obtains,

$$\frac{\Delta p}{\rho \dfrac{V_0^2}{2}} = 1 - \left(\frac{V}{V_0}\right)^2 \qquad ...(4.26)$$

Eq. (4.26) is a dimensionless equation and $\Delta p/\rho \dfrac{V_0^2}{2}$ is known as the dimensionless pressure. The distribution of dimensionless pressure over the surface of the pier may be obtained simply by finding the ratio of V/V_0 which is readily available from the streamline pattern. It may be recalled that where the streamlines are closer, the velocities are higher and where they are widely spaced, the velocities are lower. This simple fact helps us in locating the zones of high and low velocities from the streamline pattern.

4.13.1. Stagnation Pressure

With reference to Fig. 4.9, it can be seen that the central streamline terminates at S, the head of the pier. This is on account of the inability of the streamline to take a sudden turn. The fluid flowing along the central streamline, therefore, stops moving as it reaches the point S. Hence the velocity of flow at this point is zero. This point is known as the *stagnation point*. Applying the Bernoulli's equation to point (1) in the undisturbed flow region and the stagnation point S, we obtain

$$p_s = p_0 + \rho \frac{V_0^2}{2} \qquad ...(4.27)$$

The pressure at the stagnation point is known as the *stagnation pressure* and is denoted by p_s. It is also sometimes called the total pressure, and is composed of the static pressure p_0 and the dynamic pressure $\rho \dfrac{V_0^2}{2}$.

Example 4.29. *Consider the saline drip bottle shown in the following figure. If ρ is the density of saline, find :*

(*a*) *pressure at A* (*b*) *the velocity of flow of saline through the tube.*

Neglect viscous losses in tube. Atmospheric pressure = p_{atm}.

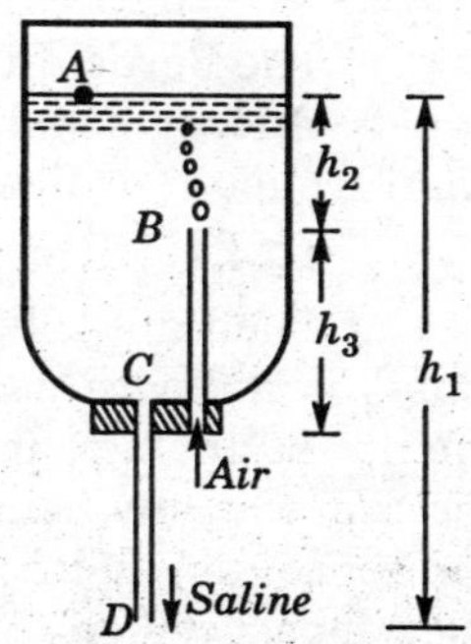

Solution. Pressure at B = Atmospheric pressure, p_{atm}.

Pressure at A = Pressure at $B - \rho g h_2$

$= p_{atm} - \rho g h_2$

To determine the velocity of flow of saline, apply Bernoulli's equation between points A and D (neglecting losses), taking horizontal line through D as the datum,

$$h_1 + \frac{(p_{atm} - \rho g h_2)}{\rho g} + 0 = 0 + \frac{p_{atm}}{\rho g} + \frac{V^2}{2g}$$

in which V is the velocity of flow at D where the pressure is atmospheric.

$$\frac{V^2}{2g} = h_1 - h_2 \quad \text{or} \quad V = \sqrt{2g(h_1 - h_2)}.$$

Example 4.30. *At a flow field, static pressure is 1.05 KPa and velocity is 20 m/s. What is the stagnation or total pressure if density is 1.2 kg/m³ ? Choose the correct answer from the following options.*

(*a*) *1.20 kPa* (*b*) *1.25 kPa*

(*c*) *1.29 kPa* (*d*) *1.32 kPa.*

Solution. Static pressure $p_0 = 1.05$ kPa, $V_0 = 20$ m/s, $\rho = 1.2$ kg/m^3

Stagnation or total pressure is given by Eq. (4.27)

$$p_s = p_0 + \rho \frac{V_0^2}{2}$$

$$= 1.05 \times 10^3 \text{ (N/m}^2) + 1.2 \frac{(20)^2}{2} \text{ (N/m}^2) = 1.29 \times 10^3 \text{ N/m}^2 = \mathbf{1.29 \text{ kPa.}}$$

Example 4.31. *When a body is moving horizontally through still water at a constant speed of 3 m/sec, the velocity at a point 1 m ahead of it is found to be 2 m/sec. The motion is irrotational. Determine the pressure intensity at this point relative to the initially undisturbed fluid.*

Solution. The motion of a body in an otherwise undisturbed fluid represents an unsteady flow case. This can be made steady by superimposing a velocity of 3 m/sec against the motion of the body. The body will thus become stationary. Stagnation point is created at the nose of the moving body. Applying the Bernoulli's equation between a point in the undisturbed flow and the nose of the body (stagnation point) where the velocity reduces to zero.

$$\frac{(3)^2}{2g} + \frac{p_0}{\gamma} = 0 + \frac{p_s}{\gamma}$$

here p_0 is the pressure in the undisturbed flow, and p_s the stagnation pressure.

Applying the Bernoulli's equation between a point 1 m ahead of the nose (where the velocity is given to be 2 m/s) the velocity after superimposition of 3 m/s velocity being 3 – 2 = 1 m/s at this point, and a point in the undisturbed flow where the velocity is 3 m/s and pressure p_0.

$$\frac{(3)^2}{2g} + \frac{p_0}{\gamma} = \frac{(1)^2}{2g} + \frac{p_2}{\gamma}$$

Here p_2 is the pressure at the point 1 m ahead of the nose. The difference in the pressure intensity

$$\frac{p_2 - p_0}{\gamma} = \frac{(3)^2 - (1)^2}{2 \times 9.81} = \frac{8}{19.62} = 0.4077 \text{ m of water}$$

or $$p_2 - p_0 = 0.4077 \times 9810 = \mathbf{4000 \text{ N/m}^2 = 4 \text{ kN/m}^2 = 4 \text{ kPa.}}$$

4.14. Practical Applications of Bernoulli's Equation

The Bernoulli's equation finds its application in every problem of incompressible fluid flow, where energy considerations are involved. Many measuring devices such as the Pitot tube, venturimeter, orificemeter etc. utilize the principle embodied in the Bernoulli's equation.

4.14.1. Pitot Tube

The Pitot-tube (pronounced as Pee-tou tube) is a device used for velocity measurement. It, in fact, measures the stagnation pressure at any point in the flow. The static pressure may be measured by a piezometer. The measurements of stagnation and static pressures are necessary and sufficient to evaluate the velocity at the point under consideration making use of Eq. (4.27).

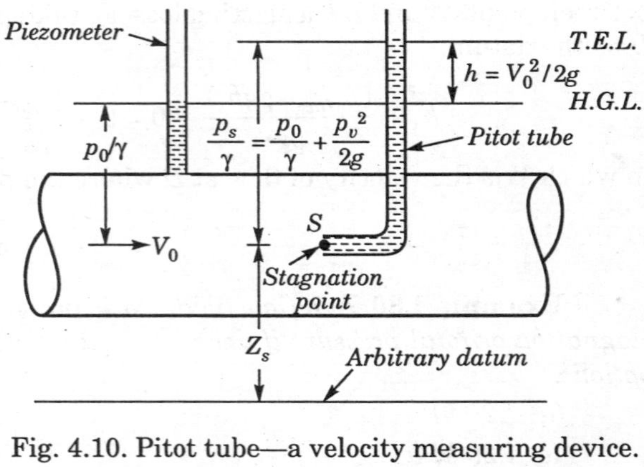

Fig. 4.10. Pitot tube—a velocity measuring device.

The pitot-tube consists of a tube having a 90° bend of shorter length, opened at its both ends. The bent leg is directed upstream, so that a stagnation point is created immediately ahead of its inlet (Fig. 4.10). On account of the stagnation pressure so caused, the liquid rises in the vertical limb. A piezometer installed on the pipe boundary as shown, indicates the static pressure p_0. The difference between the stagnation pressure and the static pressure is the dynamic pressure, Eq. (4.27) thus yields

$$V_0 = \sqrt{\frac{2(p_s - p_0)}{\rho}} = \sqrt{2g\left(\frac{p_s - p_0}{\gamma}\right)} = \sqrt{2gh} \qquad \text{...(4.28)}$$

where h represents the difference between the stagnation head p_s/γ and the static head p_0/γ.

Eq. (4.28) gives a relation for velocity of flow. The velocity distribution across the flow (in the normal direction) can be obtained by placing the Pitot-tube at desired points and obtaining the velocity head, as indicated by the difference in liquid levels between the Pitot-tube and the piezometer.

An inspection of Fig. 4.10 reveals that the liquid level in the Pitot-tube indicates the position of the energy line above any arbitrary datum. The Pitot-tube thus measures the total head and, therefore, it is also known as *total head tube*. The Pitot-tube and the piezometer can be used to indicate the positions of the energy line and hydraulic grade line, without actually knowing the magnitudes of pressure and velocity at different locations.

Having known what the Pitot-tube is and what for it can be used, we are in a position to utilise it for studying the variation of head across the flow in a rotational field. It is known that in rotational flow, the velocity varies normal to the flow, it being minimum at the boundaries and increasing towards the central zone of flow, as shown in Fig. 4.11.

In article 4.8, we have seen that the piezometric head in a uniform conduit remains constant in a direction normal to flow, even for a rotational flow. The total head may, therefore, be written as

$$\text{Total head} = \frac{V^2}{2g} + \left(\frac{p}{\gamma} + Z\right) = \frac{V^2}{2g} + \text{constant.}$$

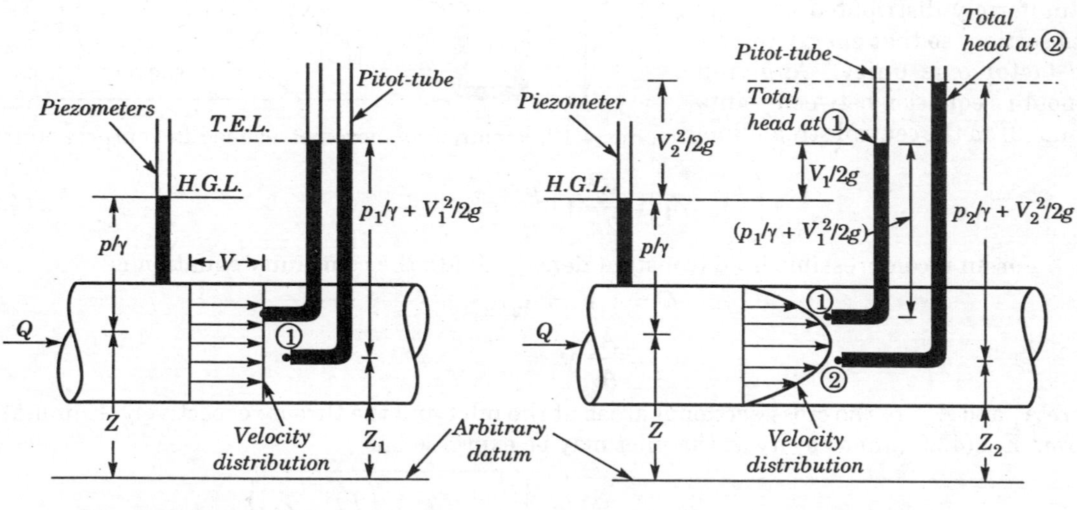

Fig. 4.11 (*a*) Total head in ideal fluid (irrotational) flow.

Fig. 4.11 (*b*) Total head in real fluid (rotational) flow.

This shows that the total head varies only with the velocity head. In real fluid flow (rotational flow), since the velocity varies throughout the normal section, it is obvious that the total head would also vary in a direction normal to the flow. Fig. 4.11 (*b*) shows that the total head at point (2) is greater than that at point (1). In case of an ideal fluid flow (irrotational flow), since the velocity at a normal section remains constant, the total head would also remain constant as shown in Fig. 4.11 (*a*).

4.14.2. Venturimeter

A venturimeter is a device for measuring rate of flow in a pipeline. It is based on the Bernoulli principle, that is, when the velocity head increases in an accelerated flow, there is a corresponding reduction in the piezometric head. It may be divided into *three component parts* :

(*i*) a converging entrance cone of angle of about 20°,

(*ii*) a cylindrical portion of short length, known as the *throat* and

(*iii*) a diverging part, known as diffuser, of cone angle 5° to 7°.

Entrance Cone

The accelerated flow is achieved in the converging entrance cone. With the gradual reduction in the area of flow in the converging portion, the velocity progressively increases in the flow direction. The velocity gradient dV/ds along the flow direction is positive and, therefore, the convective part of the acceleration $V\,(dV/ds)$ assumes a finite and a non-zero value. Thus even for steady flow the converging cone produces an accelerating flow. From continuity equation, the velocity at the throat, where the flow area is minimum, is the highest. The high velocity head at the throat of the venturimeter causes the hydraulic grade line to fall appreciably as shown in Fig. 4.12. The function of the converging cone is thus to convert the pressure energy into the kinetic energy.

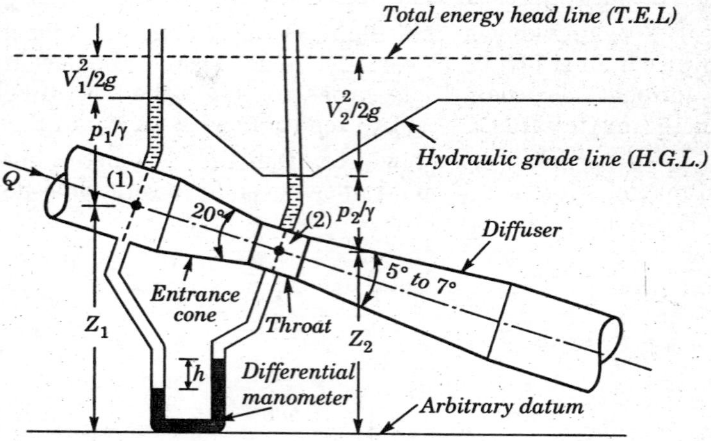

Fig. 4.12. Variation of piezometric head in venturimeter.

Assuming that the fluid is ideal (so that no energy is dissipated in overcoming frictional resistance) and that the velocities V_1 and V_2 at the inlet and the throat respectively, are uniformly distributed over the cross-section (so that energy correction factor α is unity). Applying Bernoulli's equation between points (1) and (2) on the central streamline,

$$\frac{V_1^2}{2g} + \left(\frac{p_1}{\gamma} + Z_1\right) = \frac{V_2^2}{2g} + \left(\frac{p_2}{\gamma} + Z_2\right) \qquad \text{...(4.29)}$$

For an incompressible fluid (constant density fluid), the continuity equation is

$$Q = A_1 V_1 = A_2 V_2$$

$$\therefore \qquad V_2 = \frac{A_1}{A_2} V_1$$

where A_1 and A_2 are the cross-sectional areas at the inlet and the throat respectively. Eliminating V_2 from Eq. (4.29), the velocity at the inlet may be expressed as

$$V_1 = \frac{A_2 \sqrt{2g\left[\left(\frac{p_1}{\gamma} + Z_1\right) - \left(\frac{p_2}{\gamma} + Z_2\right)\right]}}{\sqrt{A_1^2 - A_2^2}} \qquad \text{...(4.30)}$$

From Eq. (4.30) and the continuity equation $Q = A_1 V_1$, it can be seen that the difference in the piezometric heads at the inlet section (1) and the throat section (2) are related to the flow rate Q. The higher the difference the greater will be the flow rate.

If the inclination of the venturimeter is changed, that is, if it is rotated in the vertical plane, the difference in the piezometric heads between the inlet and the throat remains unchanged on account of the fixed positions of the T.E.L. and H.G.L. The discharge Q, therefore, does not depend upon the orientation of the venturimeter ; it may be kept horizontal ; inclined at any angle or even vertical without affecting the discharge.

The change in the piezometric head,

$$\left[\left(\frac{p_1}{\gamma}+Z_1\right)-\left(\frac{p_2}{\gamma}+Z_2\right)\right]$$

may be measured by a differential manometer. From the principles of manometry, it can be readil, proved that

$$\left(\frac{p_1}{\gamma}+Z_1\right)-\left(\frac{p_2}{\gamma}+Z_2\right)=\frac{(S_2-S_1)}{S_1}\,h=(S_2/S_1-1)h$$

in terms of flowing liquid, where S_2 is the specific gravity of manometer liquid, S_1 is the sp. gr. of the flowing fluid, and h is the difference of level in the differential manometer. The ideal flow rate Q may then be written as

$$Q_{ideal}=A_1V_1=\frac{A_1A_2}{\sqrt{A_1^2-A_2^2}}\sqrt{2g(S_2/S_1-1)\,h} \qquad ...(4.31)$$

Coefficient of Discharge

For flow of real fluids, loss of energy takes place between the sections under consideration. On this account, the actual velocity in the throat becomes less than the theoretical velocity. This causes a corresponding decrease in the actual discharge as compared to the theoretical one given by Eq. (4.31). Loss of energy between sections (1) and (2) also causes $(p_2/\gamma + Z_2)$ to be slightly less than its value for ideal fluid. This increases the difference in piezometric head at the two sections. In order to determine actual discharge, a coefficient C_d is introduced and used with Eq. (4.31). Thus, the actual discharge

$$Q=C_d\,.\,Q_{ideal}=\frac{C_d\,A_1A_2}{\sqrt{A_1^2-A_2^2}}\sqrt{2g(S_2/S_1-1)\,h} \qquad ...(4.31\,a)$$

The coefficient of discharge also takes into account the effects of non-uniformity of velocity distribution at the inlet and throat sections. C_d for a given venturimeter depends upon the Reynolds number of flow and the roughness of the surface. It varies from 0.95 to 0.99. The higher value is applicable for a highly turbulent flow indicated by a large Reynolds number.

Diffuser

The function of the diffuser (diverging part) is to convert the kinetic energy into pressure energy by reducing the velocity and to restore the pressure as nearly as possible to its original value at the inlet.

Flow through diverging boundaries has a tendency to separate from the boundaries resulting in the formation of eddies and consequent energy dissipation. The greater the angle of divergence the greater is the energy dissipation. A small angle would mean a large length of the diffuser. Cone angles of 5° to 7° have been found to give lower values of energy loss. About 85% of the pressure drop between inlet and throat is recovered in the diffuser.

Effect of Velocity Distribution

In certain special cases where the inner surface of the venturimeter is very smooth causing the loss of energy between the inlet and the throat to be negligibly small, the coefficient of discharge C_d may be slightly greater than unity. A possible explanation for this strange phenomenon is that in the rapidly converging flow, as is obtained between the inlet and the throat sections, the velocity distribution tends to become more uniform. Consequently, the K.E. correction factor at the throat section is almost equal to unity, while at the inlet section it is the slightly greater than unity owing to a relatively non-uniform distribution of velocity. With these values of α_1 and α_2, the application of Bernoulli's theorem between the inlet and throat yields

$$\alpha_1 \frac{V_1^2}{2g} + \frac{p_1}{\gamma} + Z_1 = \frac{V_2^2}{2g} + \frac{p_2}{\gamma} + Z_2$$

(since $\alpha_2 \approx 1.0$ and the energy loss is negligibly small)

Using the continuity equation

$$Q = A_1V_2 = A_2V_2,$$

the following equation may obtained

$$V_1 = \frac{A_2}{\sqrt{(A_1^2 - \alpha_1 A_2^2)}} \sqrt{2g\left[\left(\frac{p_1}{\gamma} + Z_1\right) - \left(\frac{p_2}{\gamma} + Z_2\right)\right]} \qquad \text{...(4.31 b)}$$

Since $\alpha_1 > 1.0$, Eq. (4.31 *b*) gives a slightly higher value of V_1 as compared to that given by Eq. (4.30). As we have already taken into account the effects of non-uniformity of velocity distribution and of friction (venturimeter inside surface being generally very smooth causing energy loss due to surface roughness tending to zero), Eq. (4.31 *b*) may be taken to predict the actual average velocity of flow. The discharge based on this velocity will then be the actual discharge. Comparison of this discharge which is slightly greater than the theoretical one, Eq. (4.31 *a*) would result in a C_d value which is somewhat greater than unity. (In cases where the manometer connections are faulty, the piezometric head difference may be so erroneous as to yield a C_d value slightly greater than unity. Proper manometer connections must, therefore, be ensured.)

Installation

In order to achieve accurate results it is desirable that the venturimeter should be preceded by a straight pipe of sufficient length, preferably 50 times the pipe diameter but not less than 30 times the diameter of straight uniform pipe free from fitting such as valves, so that the flow is free of large scale turbulence. For more accurate results straightening vanes may be provided upstream of the venturimeter to minimize the rotational tendencies and to reduce the scale of turbulent eddies.

The ratio of throat to inlet diameters d_2/d_1 may range between 0.75 and 0.25, but the most commonly used ratio is 0.50 ($A_2/A_1 = 0.25$ and $V_2/V_1 = 4.0$).

A smaller ratio gives a higher difference in the piezometric heads (between the inlet and the throat) which can be measured more accurately by a differential manometer. At the same time, a smaller throat (that is a lower d_2/d_1 ratio) will mean higher throat velocities which may cause pressures low enough to liberate dissolved gases thereby creating conditions for cavitation to set in (for cavitation refer Art. 4.16).

Factors Influencing the Coefficient of Discharge

We are now in a position to summarize the factors which influence the value of coefficient of discharge. In general the discharge coefficient C_d not only depends upon the inside roughness and the Reynolds number (which takes into account the effects of viscous forces) but also on other factors such as the diameter ratio d_2/d_1, placement of pipe fittings such as bends, valves, orifice plates upstream of the venturimeter. Consequently, it is necessary to calibrate a venturimeter, placing it in a position in which it is likely to be used, for obtaining accurate results. For a well designed venturimeter preceded by a straight uniform diameter pipe of sufficient length, free of any fitting, the discharge coefficient will vary slightly with Reynolds number provided it is greater than 2000 ; its variation may be within the range 0.95 to 0.99 for turbulent flow.

Example 4.32. *A pitot static tube is used to measure the velocity of an aeroplane. If a U-tube differential manometer connected to the pitot-tube registers a pressure difference of 90 mm of water, find the speed of plane, taking coefficient of tube as 0.98. Take specific weight of air = 12.02 N/m³. Neglect compressibility effects.*

Solution. The stagnation pressure and velocity are related by

$$V_o = \frac{\sqrt{2g(p_2 - p_0)}}{\gamma} = \sqrt{2gh}$$

Pressure difference, $h = \dfrac{p_2 - p_0}{\gamma} = 90 \times 10^{-3}$ m of water

$$p_s - p_o = 90 \times 10^{-3} \times 9810 = 882.9 \text{ N/m}^2$$

$$h = \frac{p_s - p_o}{\gamma_{air}} = \frac{882.9}{12.02} = 73.47 \text{ m of air}$$

Speed of aeroplane, $V_0 = \sqrt{2gh} = \sqrt{2 \times 9.81 \times 73.47} = \mathbf{37.97\ m/s}$

$$= \frac{37.97 \times 60 \times 60}{1000} = \text{km/hr} = \mathbf{136.7\ km/hr.}$$

Example 4.33. *When a Pitot-static tube is immersed in a water stream (density ρ_w) the differential head measured by a mercury (density ρ_m) manometer is h. What is the velocity of water stream ?*

Solution. Velocity of water stream,

$$V = \sqrt{2g \times \text{Differential head in terms of water column}}$$

Height of water column equivalent of mercury head h

$$= \left(\frac{\rho_m}{\rho_w} - 1\right) h$$

Hence, the velocity of water stream

$$V = \sqrt{2g\left(\frac{\rho_m}{\rho_w} - 1\right) h}$$

Example 4.34. *Water flows through a pipe of diameter 0.30 m. What would be the velocity V for the conditions shown in the figure below ?*

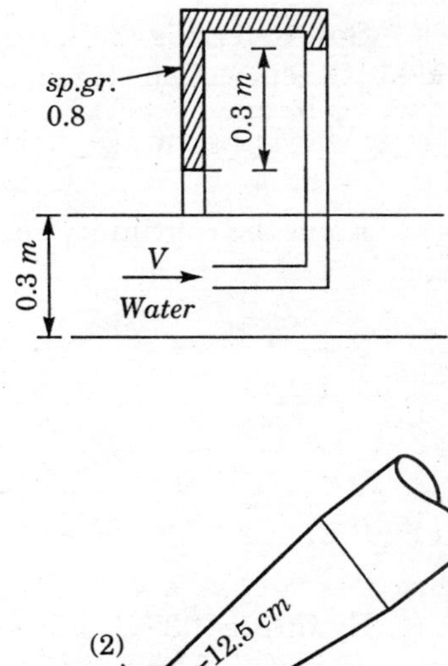

Solution.

$$V = \sqrt{2g \times \text{Differential head } h \text{ (water column)}}$$

Differential head h

$$= (1 - S) \times \text{Differential head of manometer}$$
$$= (1 - 0.8) \times 0.3 = 0.06 \text{ m of water}$$

$$V = \sqrt{2 \times 9.81 \times 0.06} = 1.085 \text{ m/s.}$$

Example 4.35. *A 20 cm water pipe has in it a venturimeter of throat diameter 12.5 cm as shown in the figure, which is connected to a mercury manometer showing a difference of 86.5 cm. Find the velocity in the throat and the discharge.*

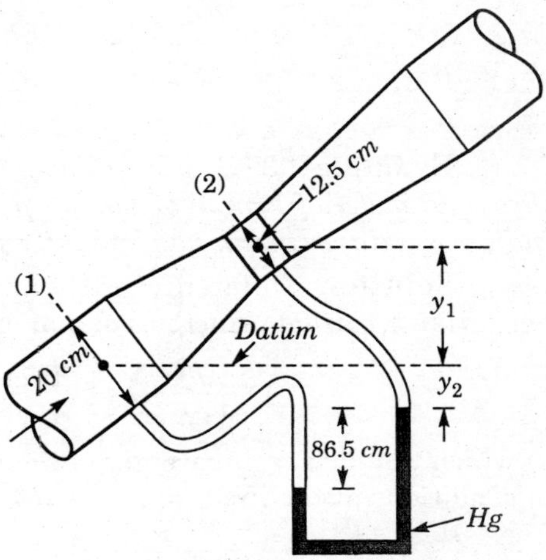

Fig. Ex. 4.34

Solution. Applying the Bernoulli's equation to the entrance section (1) and the throat section (2) of the venturimeter, taking the datum as shown,

$$\frac{V_1^2}{2g} + \frac{p_1}{\gamma} = \frac{V_2^2}{2g} + \frac{p_2}{\gamma} + y_1$$

From the principles of manometry,

$$\frac{p_1}{\gamma} + y_2 + 0.865 = \frac{p_2}{\gamma} + y_1 + y_2 + 13.6 \times 0.865$$

$$\therefore \quad \frac{p_1 - p_2}{\gamma} - y_1 = (13.6 - 1)0.865$$

$$= 12.6 \times 0.865 \text{ m of water.}$$

From the Bernoulli's equation,

$$\frac{V_2^2 - V_1^2}{2g} = \frac{p_1 - p_2}{\gamma} - y_1 = 12.6 \times 0.865 \text{ m}.$$

Using the continuity equation,

$$Q = \frac{\pi}{4} \times (0.2)^2 V_1 = \frac{\pi}{4} (0.125)^2 V_2$$

Substituting V_1 in terms of the throat velocity V_2.

$$V_1 = \left(\frac{0.125}{0.2}\right)^2 V_2 = 0.39 V_2$$

the Bernoulli's equation gives,

$$\frac{V_2^2 - 0.152 V_2^2}{2g} = 12.6 \times 0.865$$

resulting in $$V_2 = \sqrt{\frac{2 \times 9.81 \times 12.6 \times 0.865}{0.848}} = 15.86 \text{ m/s}.$$

Discharge $$Q = \frac{\pi}{4} (0.125)^2 \times 15.86 = 0.194 \text{ m}^3\text{/s}.$$

Example 4.36. *A venturimeter is used for the measurement of discharge of water in a horizontal pipeline. The upstream diameter is 300 mm, the throat is 150 mm diameter and the difference in pressure between the inlet and throat is 3 m head of water. If the loss of head through the converging section of the meter is 1/8 of the throat velocity head, calculate the discharge in the pipeline.*

Sol. Referring to figure 4.12, and treating the pipeline horizontal, Bernoulli's equation applied between sections (1) and (2) with the pipe axis taken as the datum, results in

$$0 + \frac{p_1}{r} + \frac{V_1^2}{2g} = 0 + \frac{p_2}{\gamma} + \frac{V_2^2}{2g} + \frac{1}{8}\left(\frac{V_2^2}{2g}\right)$$

From the continuity equation

$$Q = \frac{\pi d_1^2}{4} V_1 = \frac{\pi d_2^2}{4} V_2$$

$$V_1 = 4Q/\pi\, d_1^2, \text{ and } V_2 = 4Q/\pi d_2^2$$

$$(p_1 - p_2)/\gamma = \frac{9}{16g}\left[\frac{4Q}{\pi(0.15)^2}\right] - \frac{1}{2g}\left[\frac{4Q}{\pi(0.3)^2}\right]$$

But, $(p_1 - p_2)/\gamma = 3$ m of water head (given)

$\therefore$ $3 = 173.27Q^2$ or $Q = 0.131 \text{ m}^3\text{/s} =$ **131.0 l/s.**

Example 4.37. *A vertical venturimeter shown in the figure has an area ratio of 5. It has a throat diameter of 1 cm. When oil of sp. gr. 0.8 flows through it the mercury in the differential gauge indicates a difference in height of 12 cm. Find the discharge through the venturimeter.*

Solution. Difference in the piezometric heads between the inlet and the throat of the venturimeter, from principles of manometry

$$\left(\frac{p_1}{\gamma} + Z_1\right) - \left(\frac{p_2}{\gamma} + Z_2\right) = \left(\frac{13.6}{0.8} - 1\right) \times 0.12 \text{ m of oil}$$

in which γ is the specific weight of the flowing fluid (oil in this case). Here subscripts 1 and 2 refer to conditions at the inlet and the throat respectively.

Application of Bernoulli's equation to the inlet and throat of venturimeter gives

$$\frac{V_2^2 - V_1^2}{2g} = \left(\frac{p_1}{\gamma} + Z_1\right) - \left(\frac{p_2}{\gamma} + Z_2\right)$$

$$= (17 - 1) \times 0.12 = 1.92 \text{ m of oil.}$$

From the continuity equation,

$$Q = A_1V_1 = A_2V_2.$$

$$\therefore \quad V_1 = \frac{A_2}{A_1} V_2 = \frac{1}{5} V_2.$$

Substituting V_1 in terms of V_2

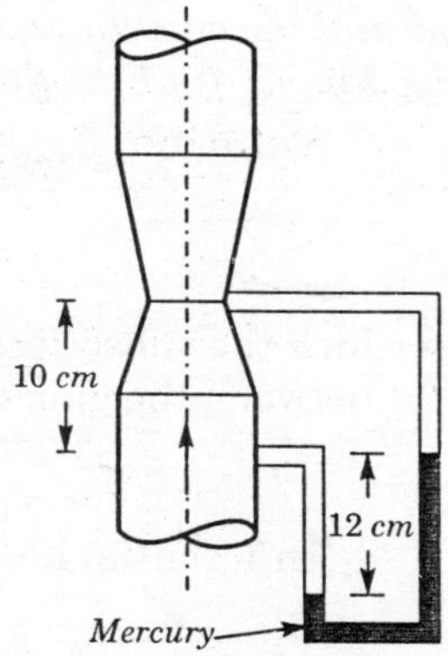

Fig. Ex. 4.36

$$\frac{V_2^2 - \frac{1}{25} V_2^2}{2g} = 1.92 \quad \text{or} \quad \frac{24V_2^2}{25 \times 2 \times 9.81} = 1.92$$

yielding, $\quad V_2 = 6.26$ m/s.

Discharge, $\quad Q = \frac{\pi}{4} \times \left(\frac{1}{100}\right)^2 \times 6.26 \times 1000$ lit/sec.

$$= 0.492 \text{ lit/sec.} = 29.5 \text{ lit/minute.}$$

Example 4.38. *A Venturimeter is to be fitted in a 150 mm dia. pipeline horizontally at a section where the pressure is 100 kN/m². If the maximum flow of water in the pipe is 150 litres/sec, find the diameter of the throat so that the pressure at the throat does not fall below 75 kN/m² (vacuum). Assume that 3% of the differential head is lost between the inlet and the throat.*

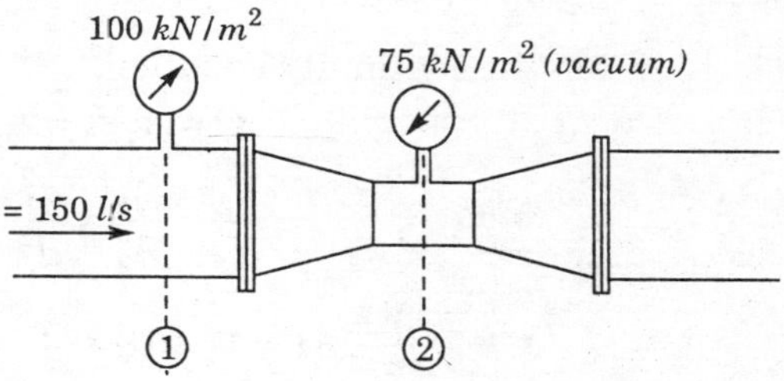

Solution. Data given :

$d_1 = 150$ mm, $\quad Q = 150$ lites/sec

$p_1 = 100$ kN/m²,

$p_2 = 75$ kN/m² (vacuum)

$h_L = 3\%$ of differential head

$$= \frac{3}{100} \times \left(\frac{p_1 - p_2}{\gamma}\right) = \frac{3}{100} \frac{[100 - (75)] \times 10^3}{9810} \text{ m} = 0.535 \text{ m}$$

From continuity eqn. $\quad Q = A_1V_1 = A_2V_2$

$$\therefore \quad V_1 = \frac{Q}{A_1} = \frac{4Q}{\pi d_1^2} = \frac{4 \times 150 \times 10^{-3}}{\pi (0.15)^2} = 8.485 \text{ m/s}$$

$$V_2 = \frac{4Q}{\pi d_2^2} = \frac{4 \times 0.15}{\pi d_2^2} = \frac{0.191}{d_2^2}$$

Applying Bernoulli's equation between sections (1) in the main pipe, and (2) at the throat, noting that the pipeline is horizontal,

$$\frac{V_1^2}{2g} + \frac{p_1}{\gamma} = \frac{V_2^2}{2g} + \frac{p_2}{\gamma} + h_L$$

$$\frac{(8.485)^2}{2 \times 9.81} + \frac{1000 \times 10^3}{9810} = \frac{\left(\frac{0.191}{d_2^2}\right)^2}{2 \times 9.81} + \frac{(-75 \times 10^3)}{9810} + 0.535$$

Simplifying, $\quad \frac{0.00186}{d_2^4} = 20.97 \quad \text{or} \quad d_2 = 0.097 \text{ m} = 97 \text{ mm}$

$\therefore$ Diameter of throat = 97 mm.

Example 4.39. *A venturimeter is installed in a pipeline 30 cm in diameter. The throat-pipe diameter ratio is 1/3. Water flows through the installation. The pressure in the pipeline is 137.7 kN/m² and the vacuum in the throat is 37.5 cm of mercury. If 4% of the differential head is lost between the gauges, find the flow in the pipeline.*

Solution. Applying Bernoulli's equation to the entrance and throat of the venturimeter

$$\frac{V_1^2}{2g} + \frac{p_1}{\gamma} + Z_1 = \frac{V_2^2}{2g} + \frac{p_2}{\gamma} + Z_2 + h_l$$

in which the subscripts 1 and 2 refer to the entrance and the throat respectively, h_l being the head lost between the gauges. Considering the venturimeter to be horizontal, $Z_1 = Z_2$.

$$p_2 = -37.5 \text{ cm of mercury} = -13.6 \times 37.5 = -5.1 \text{ m of water}$$

$$\text{Differential head} = \frac{p_1 - p_2}{\gamma} = 14.05 - (-5.1) = 19.15 \text{ m}$$

$$h_l = 4\% \text{ of differential head}$$

$$= \frac{4}{100} \times 19.15 = 0.766 \text{ m of water}$$

$$\frac{V_1^2 - V_2^2}{2g} = \left(-5.1 + 0.766 - \frac{137.7 \times 10^3}{9810}\right) = -18.384 \text{ m}$$

$$\frac{V_1^2}{2g}\left(\frac{V_2^2}{V_1^2} - 1\right) = 18.384$$

From the continuity equation,

$$Q = A_1V_1 = A_2V_2$$

$$\frac{V_2}{V_1} = \frac{A_1}{A_2} = \frac{d_1^2}{d_2^2} = (3)^2 = 9$$

$$\frac{V_1^2}{2g}(81 - 1) = 18.384$$

$$V_1 = 2.134 \text{ m/s.}$$

Hence the discharge, $Q = \frac{\pi}{4} \times (0.3)^2 \times 2.134 = 0.151 \text{ m}^3/\text{s}$ (151 lit/s).

Example 4.40. *A venturimeter with 30 cm diameter inlet and 15 cm throat is used for measuring flow of oil of sp. gr. 0.90. The oil-mercury differential gauge shows a reading of 33 cm. Assuming the coefficient of meter 0.98, calculate the discharge.*

Solution. From Eq. (4.31 *a*),

$$Q = C_d\left(\frac{A_1A_2}{\sqrt{A_1^2 - A_2^2}}\right)\sqrt{2g\left(\frac{S_2 - S_1}{S_1}\right)h} = C_d\frac{A_2}{\sqrt{1-(A_2/A_1)^2}}\sqrt{2g\left(\frac{S_2 - S_1}{S_1}\right)h}$$

$$= 0.98 \times \frac{\frac{\pi}{4}(15/100)^2}{\sqrt{1-(15/30)^2}} \times \sqrt{2 \times 9.81\left(\frac{13.6 - 0.9}{0.9}\right) \times 0.33} = 0.1911 \text{ m}^3/\text{s} = 191.1 \text{ l/s.}$$

Example 4.41. *A pressure gauge is installed in front of a valve on a water pipe of diameter D = 50 mm. When the valve is closed the gauge reads 6 atm. When the valve is opened the reading drops to 2 atm. Calculate the rate of discharge of water in the pipeline.*

Solution. When the valve is closed, the static pressure in the pipeline $= \frac{p \text{ static}}{\gamma}$ and there is no flow of water. When the valve is opened, the total energy is partly converted into kinetic energy and the rest remains as the pressure energy. Let the velocity of water be V_0, and the pressure p_0.

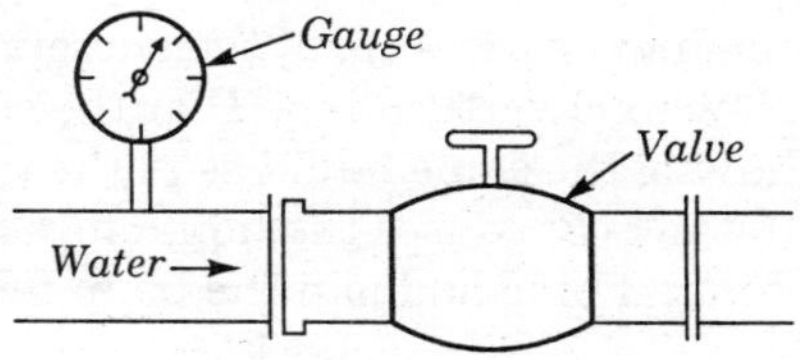

The Bernoulli's equation gives the following relationship :

$$0 + \frac{p_{static}}{\gamma} = \frac{V_0^2}{2g} + \frac{p_0}{\gamma} \quad \text{or,} \quad \frac{V_0^2}{2g} + \frac{p_{st} - p_0}{\gamma}$$

1 atm. pressure = 10329 kg/m^2

$$\text{The discharge, } Q = A_0 V_0 = \frac{\pi D^2}{4} V_0 = \frac{\pi D^2}{4}\sqrt{\frac{2g}{\gamma}(p_{st} - p_0)}$$

$$= \frac{\pi\left(\frac{50}{1000}\right)^2}{4}\sqrt{\frac{2 \times 9.81}{1000}(6-2) \times 10329} \text{ m}^3/\text{s}$$

$= 790.7 \times 10^{-4}$ m^3/s = **79.07 litre/sec.**

4.14.3. Suction Effect of a Flowing Fluid (*Principle of Jet Pump*)

Consider Fig. 4.12 (*a*) which shows flow of an incompressible liquid from the supply tank (*A*) through a pipe having a gradual contraction of cross-sectional area A_c at *C*. Let *A* be the flow area at the outlet located at a depth *H* below the level of liquid in the supply tank. Let the flow be steady. Ignoring the vertical pipe *D* connected to the minimum-flow-area section *C*, the Bernoulli's equation between the liquid surface and the contracted portion at *C* can be written as

$$\frac{V_1^2}{2g} + \frac{p_a}{\gamma} + h_c = \frac{V_c^2}{2g} + \frac{p_c}{\gamma}$$

in which V_1 = velocity of free surface in the tank (*A*), p_a = atmospheric pressure which exists over the liquid free-surface, h_c = depth of contracted portion *C* below the liquid surface, and V_c, p_c are the velocity and pressure at *C* and γ is the specific weight of the liquid.

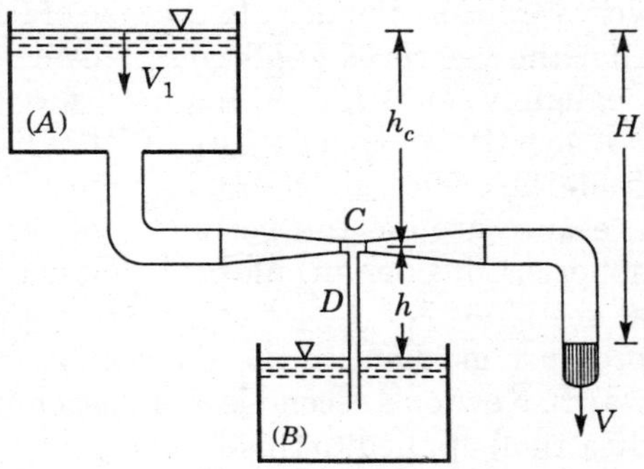

Fig. 4.12 (*a*) Principle of jet pump.

From the continuity equation

$$Q = A_1 V_1 = A_c V_c = AV$$

in which A_1 = area of supply tank normal to V_1, a is the outlet area where the velocity is *V*. Writing V_1 and V_c in terms of *A* and *V*.

$$V_1 = AV/A_1 \quad \text{and} \quad V_c = AV/A_c.$$

Substituting these quantities in the Bernoulli's equation, the pressure difference

$$\frac{p_a - p_c}{\gamma} = \frac{(V_c^2 - V_1^2)}{2g} - h_c = \frac{A^2V^2}{2g}\left(\frac{1}{A_c^2} - \frac{1}{A_1^2}\right) - h_c \quad \text{...(4.32)}$$

Since A_1 is very much greater than A_c, $1/A_1^2$ can be neglected in comparison to $1/A_c^2$. Neglecting all energy losses between the supply tank and the outlet, one can write from Torricelli's law (see Eq. 4.34), $V = \sqrt{2gH}$.

$$\text{The pressure difference simplifies to } \frac{p_a - p_c}{\gamma} = H\left[\frac{A}{A_c}\right]^2 - h_c.$$

If the pressure at *C* becomes negative, then $p_c < p_a$ and consequently $h_c < H\,(A/A_c)^2$. Under such a condition, if a hole is drilled at *C*, no liquid would flow out through this hole. On the contrary, the higher pressure of the surrounding atmosphere would force air into the pipe through the hole. Now fix a vertical pipe D at the contracted portion C with its lower end immersed in the tank (B) having the same liquid as in the tank (A). The atmospheric pressure p_a acting on the liquid surface in tank (B) causes the liquid to rise in the vertical pipe D. If $(p_a - p_c) > \gamma h$ the liquid will rise in the vertical pipe and join the main flow. This condition may be expressed as $(p_a - p_c) > \gamma h$

or
$$\left\{H\left(\frac{A}{A_c}\right)^2 - h_c\right\} > h$$

from which
$$\frac{A}{A_c} > \sqrt{\frac{h + h_c}{H}} \qquad \text{...(4.32 a)}$$

Equation (4.32 a) shows that for the liquid in tank (B) to be lifted up through the vertical pipe D and to join the main flow in the pipe, the ratio of the pipe area at the outlet to the contracted area must be greater than the square-root of the ratio $(h + h_c)/H$. This effect resulting from the negative or suction pressure created at C due to the flowing liquid forms the basis of the design of jet pumps. In practice, the accurate analysis will involve consideration of friction and mixing losses which occur in the system.

4.14.4. Flow Through a Reservoir Opening *(Orifice Flow)*

Consider an opening in the side of a reservoir as shown in Fig. 4.13. The reservoir is assumed to be very large as compared to the size of the opening, so that the velocities of all points in the reservoir are negligibly small. Let the liquid surface be at a height h above the centre of orifice. A jet of liquid is issued from the orifice under the influence of gravity. The pressure around the periphery of the jet is atmospheric. Consider an enlarged view of a fluid element at section (2) of the jet as shown. Let dp/dz be the pressure gradient in the Z-direction (vertically upward). Newton's second law of motion in the vertical direction, applied to this fluid element, gives

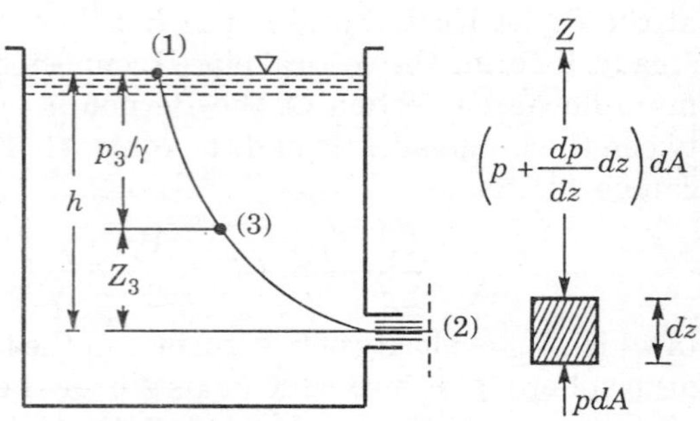

Fig. 4.13. Flow from a reservoir.

$$pdA - \left(p + \frac{dp}{dz}dz\right)dA - \gamma dAdz = -(\rho dAdz)\,g$$

which simplifies to $\dfrac{dp}{dz} = 0$, resulting in, p = constant.

This shows that the pressure gradient across the jet is zero and, therefore, the pressure remains constant throughout the jet section. Since the jet is surrounded by air at atmospheric pressure, the pressure at the centreline of the jet is also atmospheric and remain atmospheric everywhere.

Applying Bernoulli's equation to points (1) on the reservoir surface and (2) on the centreline of the emerging jet, considering horizontal plane through the centre of opening as the arbitrary datum, we obtain

$$0 + 0 + h = \frac{V^2}{2g} + 0 + 0 \qquad \text{...(4.33)}$$

In the above equation, velocity and pressure both at point (1) are zero and $Z_1 = h$. Likewise at point (2), the pressure and the datum (or elevation) head are zero. In other words, it can be said that the total energy at the reservoir surface, which exist in the potential form is converted into kinetic form as the liquid flows through the opening. The velocity of flow in the jet, from Eq. (4.33) is :

$$V = \sqrt{2gh} \qquad \text{...(4.34)}$$

Equation (4.34) is known as *Torricelli's theorem*. In 1643, Terricelli experimentally demonstrated that the velocity of a jet flowing out of a small opening (known as an orifice) is proportional to the square root of head of liquid above it.

Torricelli's theorem is now recognised as a special case of the Bernoulli's equation.

Instead of point (1) on the reservoir surface if any other point is chosen for the application of Bernoulli's equation, say, point (3) on the streamline, we find that $Z_3 + \frac{p_3}{\gamma} = h$, and since the reservoir area is large enough the velocity is negligibly small as compared to the velocity at (1). The Bernoulli's equation between points (3) and (2) yields the same result as expressed by Eq. (4.34).

4.14.5. Carburettors

Carburettors are used in reciprocating internal-combustion engines to meter, atomise and mix the fuel with air. The stream of air passes through a venturitube in which fuel jet is placed as shown in Fig. 4.14. The velocity of air increases in the venturi portion and the pressure decreases according to the Bernoulli's equation. The reduction in pressure at the venturi throat causes fuel to flow through the fuel jet into the air stream.

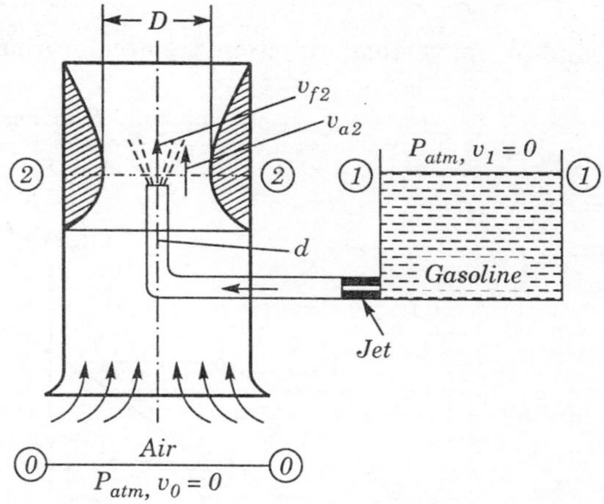

Fig. 4.14. Carburettor.

Let us designate various flow parameters, and diameters by the following notations :

W_f = weight rate of discharge of fuel,
W_a = weight rate of discharge of air,
D = diameter of venturi throat,
d = diameter of jet pipe
K_{La} = Loss coefficient of venturi throat
K_{Lf} = Loss coefficient of fuel jet
γ_a = specific weight of air
γ_f = specific weight of fuel (gasolene)
V_{f_2} = velocity of fuel jet ; and
V_{a_2} = Velocity of air a section (2) – (2)

Let us find the relationship between the weight rates of discharge of fuel (gasolene) and air, by writing the Bernoulli's equation for the air stream {sections at (0) – (0) and (2) – (2)} and for the fuel stream {sections (1) – (1) and (2) – (2)}, assuming $Z_1 = Z_2$ and kinetic energy correction factor $\alpha = 1$. The following equations can be written :

$$\frac{p_{atm}}{\gamma_a} = \frac{p_2}{\gamma_a} + \frac{V_{a2}^2}{2g} + K_{La}\frac{V_{a2}^2}{2g}, \text{ and } \frac{p_{atm}}{\gamma_f} = \frac{p_2}{\gamma_f} + \frac{V_{f2}^2}{2g} + K_{Lf}\frac{V_{f2}^2}{2g}$$

These equations lead to : $p_{atm} - p_2 = \gamma_a \frac{V_{a2}^2}{2g}(1 + K_{La})$, and $p_{atm} - p_2 = \gamma_f \frac{V_{f2}^2}{2g}(1 + K_{Lf})$

hence,
$$\gamma_a \frac{V_{a2}^2}{2g}(1 + K_{La}) = \gamma_f \frac{V_{f2}^2}{2g}(1 + K_{Lf})$$

The weight rate of discharge of fuel

$$W_f = \gamma_f \frac{\pi d^2}{4} V_{f2}, \text{ and the weight rate of discharge of air}$$

$$W_a = \gamma_a \frac{\pi D^2}{4} V_{a2}$$

The ratio of the two discharges : $\frac{W_f}{W_a} = \left(\frac{d}{D}\right)^2 \frac{\gamma_f}{\gamma_a}\frac{V_{f2}}{V_{a2}} = \left(\frac{d}{D}\right)^2 \cdot \frac{\gamma_f}{\gamma_a} \cdot \sqrt{\frac{\gamma_a}{\gamma_f}\frac{(1+K_{La})}{(1+K_{Lf})}}$

$$= \left(\frac{d}{D}\right)^2 \sqrt{\frac{\gamma_f(1+K_{La})}{\gamma_a(1+K_{Lf})}} \qquad ...(4.35)$$

4.15. Trajectory of Free-Jet

Consider the jet of liquid issued from a nozzle as shown in Fig. 4.14. The jet as it emerges from the nozzle has a velocity V_1 inclined at an angle θ to the horizontal. Under the influence of gravity, the jet follows a path indicated by its central streamline RPS. Neglecting frictional effect of the atmosphere, the Bernoulli's equation between points (1) and (2) gives

$$Z_1 + \frac{V_1^{\ 2}}{2g} = Z_2 + \frac{V_2^{\ 2}}{2g}$$

The pressure $p_1 = p_2 = p_a$, since the jet is surrounded by atmosphere throughout. If A_1 is the nozzle area at its exit, the continuity equation gives the flow rate $Q = A_1V_1$

There being no acceleration in the horizontal direction, $a_x = \dfrac{dv_x}{dt} = 0$

and in the vertical direction the acceleration is $(-g)$, hence $a_z = \dfrac{dv_z}{dt} = -g$.

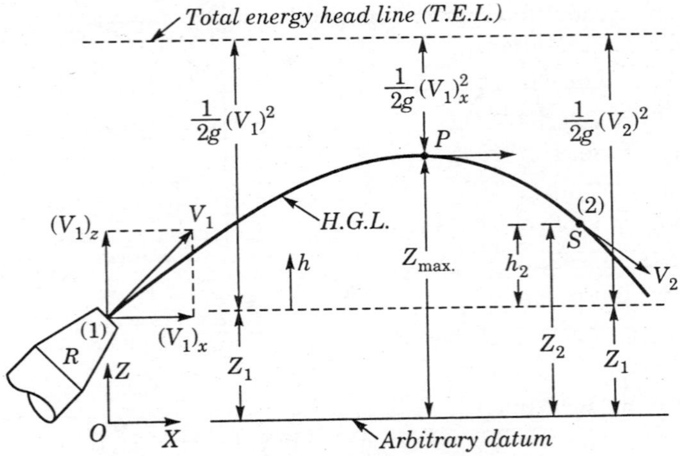

Fig. 4.14. Trajectory of a free-jet.

Integrating these expressions with respect to time, we obtain

$$V_x = \text{constant} = (V_1)_x \qquad ...(4.36)$$

and

$$V_z = -gt + \text{constant}.$$

At time $t = 0$, $V_z = (V_1)_z$ and the constant $= (V_1)$ thus,

$$V_z = (V_1)_z - gt \qquad ...(4.37)$$

The velocity $\quad V_x = \dfrac{dx}{dt} \quad$ and $\quad V_z = \dfrac{dz}{dt}$

Equations (4.36) and (4.37) can be written as

$$\frac{dz}{dt} = V_x = (V_1)_x$$

$\therefore$

$$x = (V_1)_x t + \text{constant}.$$

The constant of integration is zero, since at $t = 0$, $x = 0$. Hence

$$x = (V_1)_x t \qquad ...(4.38)$$

$$\frac{dz}{dt} = (V_1)_z = (V_1)_z - gt$$

$$\therefore \quad Z = (V_1)_z t - \frac{1}{2} g t^2 + \text{constant.}$$

At $t = 0$, $Z = Z_1$ and, therefore, the constant of integration is Z_1 and we obtain

$$Z - Z_1 = h = (V_1)_z t - \frac{1}{2} g t^2 \qquad \text{...(4.39)}$$

where h = height of jet above the nozzle tip.

Eliminating t between Eqs. (4.38) and (4.39)

$$h = \frac{(V_1)_z}{(V_1)_x} x - \frac{1}{2} g \frac{x^2}{(V_1)_x^2}.$$

But $\frac{(V_1)_z}{(V_1)_x} = \tan\theta$, and $(V_1)_x = V_1 \cos\theta$. Making these substitutions the equation of trajectory is

$$h = x \tan\theta - \frac{gx^2}{2V_1^2 \cos^2\theta} = x\tan\theta - \frac{g A_1^2 x^2}{2Q^2 \cos^2\theta} \qquad \text{...(4.40)}$$

Equation (4.40) shows that the trajectory has a parabolic shape. From Eq. (4.35), the velocity at point S is given by

$$V_2^2 = V_1^2 - 2gh_2$$

$$= V_1^2 - 2g\left(x\tan\theta - \frac{gx^2}{2V_1^2\cos^2\theta}\right) = \frac{Q^2}{A_1^2} - 2g\left(x\tan\theta - \frac{g A_1^2 x^2}{2Q^2\cos^2\theta}\right)$$

or

$$V_2 = \sqrt{\left(\frac{Q}{A_1}\right)^2 - 2g\left(x\tan\theta - \frac{gA_1^2 x^2}{2Q^2\cos^2\theta}\right)} \qquad \text{...(4.41)}$$

where V_2 is the velocity of jet at a horizontal distance x from the nozzle. At the highest point P of the trajectory the jet velocity becomes horizontal $V_p = (V_1)_x$ and the Bernoulli's equation gives the maximum height h_{max} above the nozzle exit,

$$\frac{V_1^2}{2g} = h_{max} + \frac{(V_1)_x^2}{2g}$$

or

$$h_{max} = \frac{V_1^2}{2g} - \frac{(V_1)_x^2}{2g} = \frac{(V_1)_x^2 + (V_1)_z^2}{2g} - \frac{(V_1)_x^2}{2g}$$

$$= \frac{(V_1)_z^2}{2g} \qquad \text{...(4.42)}$$

The central streamline of the jet is at a distance $(V_1)_x^2/2g$ below the total head line and is, therefore, the hydraulic grade line.

Example 4.41. *Water flows out through an opening, 20 cm in diameter, in the bottom of a constant level tank as shown in the figure. Obtain an equation for the profile of the jet expressing r in term of Z/H. The radius of the jet is r at a depth Z blow the tank bottom and H is the depth of water in the tank.*

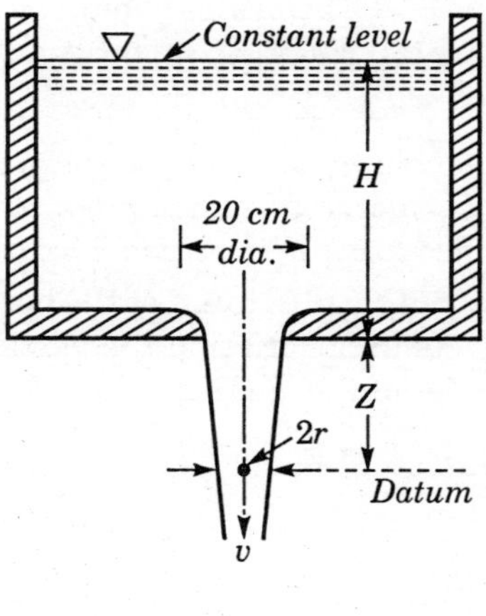

Solution. Applying Bernoulli's equation between the water surface in the tank and jet at a vertical distance Z below the tank bottom, as shown, taking the jet cross-section as the datum.

$$0 + 0 + (H + Z) = \frac{v^2}{2g} + 0 + 0$$

$$\therefore \quad v = \sqrt{2g(H+Z)}.$$

From the continuity equation, $Q = V \cdot \frac{\pi}{4}\left(\frac{20}{100}\right)^2 = v.\pi r^2.$

But the velocity at a depth H below the water surface is

$$V = \sqrt{2gH}$$

$$\therefore \qquad \sqrt{2gH}\,\frac{\pi}{4}\left(\frac{1}{5}\right)^2 = \sqrt{2g(H+Z)}\;\pi r^2$$

Hence, $$r^2 = \left(\frac{1}{100}\right)\frac{\sqrt{H}}{\sqrt{H+Z}}$$

$$\therefore \qquad r = \frac{1}{10\left(1+\dfrac{Z}{H}\right)^{1/4}}.$$

Example 4.42. *An orifice is to be placed in the side of a tank at such an elevation that the jet will attain a maximum horizontal distance from the tank at the level of its base. If the depth of the liquid in the tank is maintained at 1.5 m ; what is the proper distance from the orifice to the free surface ?*

Solution. See Fig. 4.13 ; using Eq. (4.34), the horizontal component of jet velocity is

$$(V_0)x = \sqrt{2gh}$$

and the vertical component $\qquad (V_0)z = 0,$

since the jet flows out is in the horizontal direction. The horizontal and the vertical distances covered in time t are,

$$x = \sqrt{2gh}\; t \qquad \text{and} \qquad z = -\tfrac{1}{2}\,gt^2.$$

The Z-direction being taken, as usual, vertically upwards.

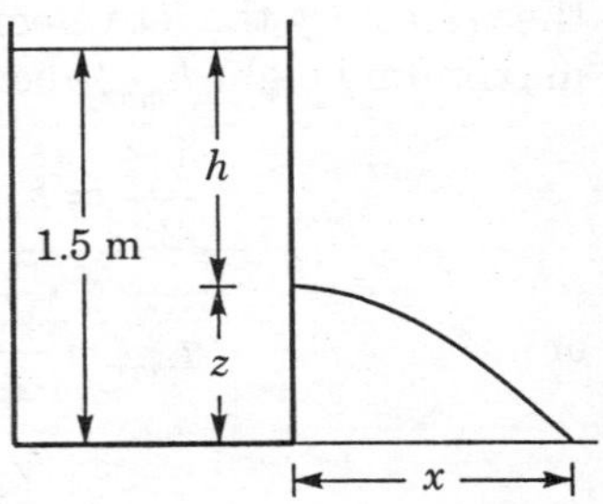

Elimination t, $$z = -\frac{1}{2}g\left(\frac{x}{\sqrt{2gh}}\right)^2 = -\frac{x^2}{4h}$$

$$z = 1.5 - h$$

$$\therefore \qquad x^2 = -4h(1.5-h) = 4h^2 - 6h.$$

For maximum x, $$\frac{dx}{dh} = 0$$

$$\therefore \qquad 2x\frac{dx}{dh} = 8h - 6 = 0$$

$$\therefore \qquad h = 0.75 \text{ m}.$$

The orifice should be located at a distance of 0.75 m below the free surface.

Example 4.43. *For the three orifices discharging as shown, show that*

$$h_1 y_1 = h_2 y_2 = h_3 y_3,$$

where h_1, h_2 and h_3 are the depths of the centres below the water surface, and y_1, y_2 and y_3 are the vertical distances travelled by the jets to cover the same horizontal distance.

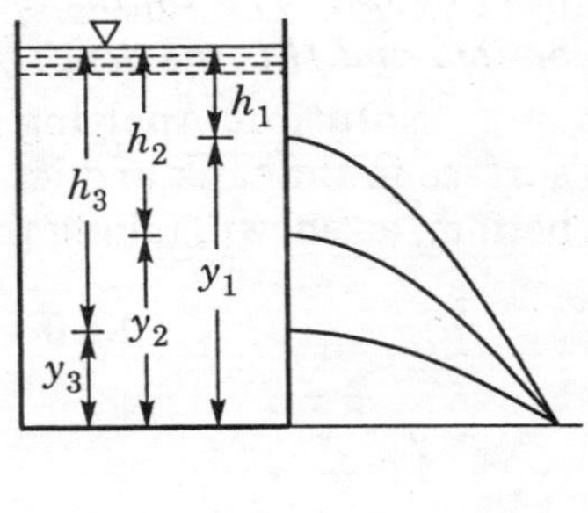

Solution. Let the horizontal distance covered by all the three jets in negotiating the vertical falls y_1, y_2 and y_3 be x. The trajectories of these are given by

$$x_1^2 = -4y_1h_1 \qquad \text{...(i)}$$

$$x_2^2 = -4y_2h_2 \qquad \text{...(ii)}$$

and $$x_3^2 = -4y_3h_3 \qquad \text{...(iii)}$$

From Eqs. (*i*), (*ii*) and (*iii*), since

$$x_1 = x_2 = x_3,$$

we have $$y_1h_1 = y_2h_2 = y_3h_3.$$

Note. For detailed exposition on orifices, article 13.4 and subsequent solved examples may be referred.

Example 4.44. *Air nozzle of an aircraft engine carburettor is shown in the accompanying figure. Determine the vacuum pressure in mm of mercury, if the air is sucked in directly from the atmosphere at zero altitude at a rate of Q = 0.15 m³/s. The diameter of the narrow section of the air nozzle D = 40 mm, the diameter of the gasolene atomiser d = 10 mm. The specific weight of air = 1.29 kg/m³.*

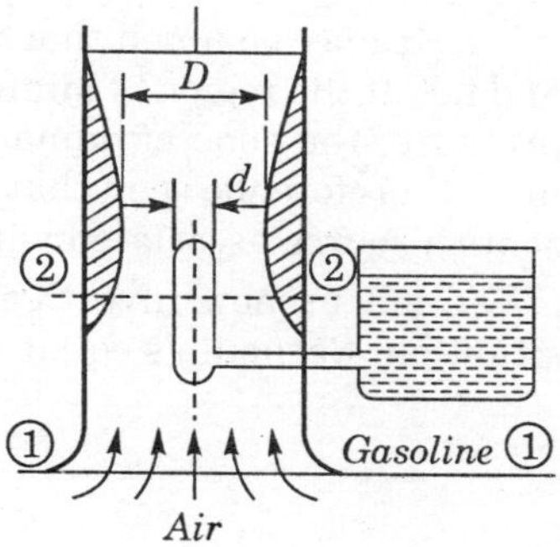

Solution. Applying the Bernoulli's equation between the atmosphere *i.e.* section (1)-(1) where the velocity will be negligibly small and can be ignored, and the narrow cross-section of the airnozzle *i.e.* (2)-(2) ;

$$\frac{p_{atm}}{\gamma_{air}} + \frac{V_1^2}{2g} = \frac{p_2}{\gamma_{air}} + \frac{V_2^2}{2g}$$

$$V_2 = \frac{Q}{A_2} = \frac{4Q}{\pi (D^2 - d^2)}$$

Given $D = 40$ mm $= 0.4$ m, $d = 10$ mm $= 0.01$ m, $\gamma_{air} = 1.29$ kg/m³, $p_{atm} = 0.0$ gauge, $V_1 = 0.0$

$$V_2 = \frac{4 \times 0.15}{\pi [(0.04)^2 - (0.01)^2]} = 127.3 \text{ m/s}$$

Making substitution, $$0 + 0 = \frac{p_2}{1.29} + \frac{(127.3)^2}{2 \times 9.81}$$

from which, $$p_2 = -1.29 \frac{(127.3)^2}{19.82} \text{ kg/m}^2 = -\frac{1.29}{13.6 \times 1000} \times \frac{(127.3)^2}{19.82} \text{ m of Hg}$$

$$= -7.834 \times 10^{-2} \text{ m} = \mathbf{-78.3 \text{ mm of Hg}}$$

4.16. Cavitation

In a liquid, if the pressure at any point reduces to its vapour pressure, the liquid boils and vaporization starts. If the liquid contains dissolved gases, they are liberated at such low pressures. Large number of small size bubbles of liquid vapour and liberated gas are thus formed. These are then carried by the flow. These bubbles upon reaching a zone of high pressure condense and collapse abruptly. The surrounding liquid rushes to fill the space or cavity created by the bubble collapse. The liquid rushing from all directions collides at the centre of cavity generating a pressure wave which travels almost at the velocity of sound thus giving rise to very high local pressures. If the collapse of bubbles take place at a solid boundary, it is subjected to these very high pressures. The boundary material may thus be stressed locally beyond its elastic limit, resulting ultimately in fatigue failure. Even if the bubble collapse and the consequent formation of cavities do not actually occur at the solid boundary, the intense pressures are propagated to the boundary through the pressure wave which originates from the cavity centre.

This sequence of formation of vapour bubbles, their transport to zones of high pressure and subsequent collapse repeats itself many times in a second, frequency depending directly on the velocity of flow. These very high pressures, acting repeatedly over the solid boundary as repeated hammer-blows at a high frequency, cause severe damage to the boundary surface. The surface becomes badly dented, and at some places its material appears to have been eroded by the liquid flow. This phenomenon is known as *pitting,* and the surface is said to have been pitted by the action of the flowing liquid. The material of the boundary surface ultimately fails by fatigue. The entire phenomenon characterised by the bubble formation and their subsequent collapse, is known as the *cavitation*. There may be considerable noise and vibrations associated with the cavitating flow. The efficiency of the system also decreases on account of reduced flow-area due to the space occupied by the bubbles. When cavitation occurs in a turbine or a pump, the noise may sound as if small pieces of stone and gravel are passing through it.

The cavitation may occur in high speed hydraulic machines such as turbines, pumps, marine propellers and in such hydraulic structures as spillways and sluice openings of high dams. It may also occur in certain portions of the pipeline and some devices used for flow measurement.

It may be noted that the damage to the solid boundary occurs at a point of bubble collapse and not at the region of minimum pressure. And since the collapse takes place in the region of high pressure, the zone of damage, therefore, lies downstream of the point of minimum pressure. It may, therefore, be seen that the formation of vapour bubbles, is not damaging in itself, but represents a source associated with severe damage of the boundary, considerable vibrations and noise.

Let us now investigate the factors which enhance the possibility of cavitation in a flow situation. Bernoulli's equation expressed as energy per unit volume is given by Eq. (4.19), *i.e.*

$$\rho \frac{V^2}{2} + p + \gamma Z = \text{a constant (say } E_v)$$

The pressure at a point is $p = E_v - \rho \dfrac{V^2}{2} - \gamma Z$ ...(4.43)

The total energy per unit volume E_v is a constant along a streamline in a rotational flow and at all points in an irrotational flow. The pressure p at a point, then, depends upon the velocity and the datum head, as is evident from Eq. (4.43). The pressure may get reduced either by increased velocity or on account of higher elevation head or due to a combination of both. At certain regions in the flow, these variables may be so combined as to reduce the pressure to its vapour pressure value. Once this pressure is attained, the cavitation is imminent. Engineers are, therefore, required to design and operate the hydraulic machines and structures in such a way that the possibility of occurrence of cavitation is ruled out. While designing, we must ensure that nowhere in the flow pressure would fall to such a low value as the vapour pressure of the liquid at the prevailing temperature. In case of water, the absolute pressure head should not be allowed to fall below 2.5 m in order to prevent cavitation.

4.17. Pressure distribution on Curved Boundaries (*Effects of Curvature of Streamlines*)

In order to understand the effect of curvature of streamlines in a better way, let us consider a convergent-divergent channel as shown in Fig. 4.15.

The upper boundary *BB* is curved, its curvature starts from point B_1 and the convexity of curvature changes at B_2 where the boundary assumes a reverse curvature. The portion of the boundary B_3B_5 again assumes curvature having convexity upwards. While dealing with the flow around a bend (Article 4.8.2) it was observed that the piezometric head increases on the outer boundary and *vice versa*. It may be deduced that the boundary portion B_1B_2 acts as an outer boundary, its centre of curvature lying below the bottom boundary. The bottom boundary becomes the inner boundary. Hence the piezometric head on the curved portion B_1B_2 is greater than that on the corresponding straight portion of the lower boundary. The portion B_2B_3 has a reverse curvature having its centre above the top boundary *BB*. The curved portion B_2B_3 thus becomes the inner boundary and the corresponding straight portion of the lower boundary acts as the outer boundary. Consequently, we have lower piezometric heads on the curved portion B_2B_3. The streamlines between B_2B_4 are in transition for changing their sign

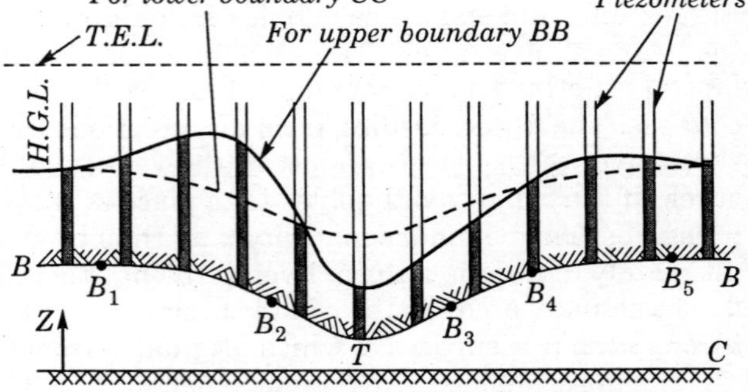

Fig. 4.15. Pressure distribution on curved (upper) and straight (lower) boundaries.

of curvature and, therefore, for most of the boundary length between B_3 and B_4, the piezometric head continues to be less than that on the bottom straight portion. The curved portion B_4B_5 having a large radius of curvature again becomes the outer boundary causing the piezometric head on its surface to be slightly greater than on the straight bottom of the channel. From the variation of piezometric head on both the boundaries the pressure head at any point of the boundary may be obtained ; it being the ordinate between the hydraulic grade line and the boundary itself. It may thus be seen that the hydraulic grade lines for the upper and lower boundaries, in fact, represent the pressure head distribution over these boundaries.

From Fig. 4.15, it may be observed that the minimum pressure anywhere in the channel occurs on the curved boundary at the section of minimum of area, *i.e.* point *T*. Thus, if cavitation at this section of least area is to occur, the fluid near the upper boundary is expected to contribute to the bubble formation due to vaporization of fluid, since pressures on the lower boundary are higher. The cavitation damage (*i.e.* pitting) will, however, be located downstream where the pressures become high, either on the lower or the upper boundary.

4.18. Momentum Principle

In many problems, we are required to determine the force produced on a solid body by the action of flowing fluid. The magnitude of such a force is determined by Newton's second law of motion, which may be expressed in a somewhat different and more useful from. The force acting on a fluid mass m is given by the Newton's law, $F = ma$, where a is the acceleration acting in the same direction as the force F. As the mass of the fluid is constant, we may express

$$F = m\frac{dv}{dt} = \frac{d(mv)}{dt} = \frac{d}{dt}(mv) \qquad ...(4.44)$$

Equation (4.44) embodies what is known as the *momentum principle* and expresses that the change in momentum of flow per unit time in any direction is equal to the net force acting on the fluid mass in that direction. It is usually known as the momentum equation.

Equation (4.44) when expressed as $F.\,dt = d(mv)$ is known as the *impulse momentum equation* and states that the impulse of a force F acting over a short interval of time dt is equal to the change in momentum in the direction of the force.

The force on a pipe-bend caused by the fluid flowing through it ; the force exerted by jet of fluid striking against a solid surface ; thrust on a propeller ; reaction of a jet etc. represent examples involving application of the momentum principle. All these forces are hydrodynamic forces arising from the change in momentum of fluid.

4.19. Steady Flow Momentum Equation

We shall now obtain momentum equation for steady flow which will be applicable to the entire fluid within a given space. According to the momentum principle the net force on fluid element is equal to the time rate of change of momentum of fluid, and acts in the same direction as that of the momentum-change. It may be noted that since the force and the momentum both are vector quantities, it is essential to specify the direction.

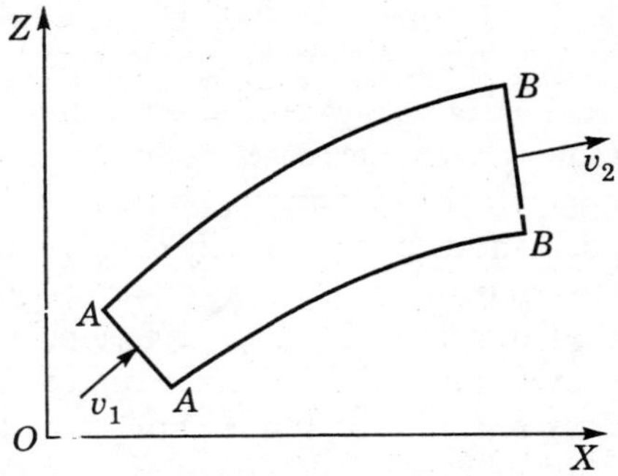

Fig. 4.16. Notations for momentum change in a streamtube.

The entire flow space may be considered to be made up of innumerable streamtubes. Let us consider one such streamtube, as shown in Fig. 4.16. The cross-section of the streamtube is sufficiently small so that the velocity may be considered uniformly distributed over the flow-section. The flow is steady and, therefore, the streamtube remains stationary with respect to the fixed co-ordinate axes. Let the velocity of flow at AA be v_1 and that at BB be v_2. Let the fluid flow at the volume rate of dQ per unit time. The momentum of fluid entering section AA per unit time in the x-direction is $\rho\ dQ(v_1)_x$, where $(v_1)_x$ is the component of velocity v_1 along x-axis. Similarly, the momentum leaving section BB per unit time in the same direction is $\rho\ dQ(v_2)_x$, where $(v_2)_x$ is likewise the velocity component along x-direction. The change in the momentum per unit time is equal to the momentum of fluid leaving the streamtube per unit time minus the momentum entering it in a unit time. Thus, from the momentum principle.

$$dF_x = \rho\ dQ[(v_2)_x - (v_1)_x]$$

where dF_x is the net force exerted on the fluid in the x-direction. If the quantity within the brackets is positive, it indicates that the momentum leaving is greater than the momentum entering the streamtube under consideration. This results in a positive force dF_x which acts on the fluid from left to right. Similar equation in the z-direction may be written as

$$dF_z = \rho dQ[(v_2)_z - (v_1)_z]$$

Considering the entire flow space consisting of a bundle of streamtubes (the space enclosed by all these streamtubes is known as *control volume*), the total force in the x-direction is given by

$$F_x = \int_A dF_x = \int_A \rho\ dQ\ (v_2)_x - \int_A \rho\ dQ\ (v_1)_x$$

$$= \int_A \rho v_2 (v_2)_x dA_2 - \int_A \rho v_1 (v_1)_x dA_1 \qquad \text{...(4.45 } a)$$

where dA_1 and dA_2 are the cross-sectional areas of the streamtube at sections AA and BB respectively. Similar equations for y-and z-directions may be written

$$F_y = \int_A \rho\ v_2\ (v_2)_y\ dA_2 - \int_A \rho\ v_1 (v_1)_y\ dA \qquad \text{...(4.45 } b)$$

and

$$F_z = \int_A \rho\ v_2 (v_2)_z\ dA_2 - \int_A \rho\ V_1 (v_1)_z\ dA_1 \qquad \text{...(4.45 } c)$$

It may be observed from Eq. (4.45) that the evaluation of force components F_x, F_y and F_z requires the knowledge of the flow conditions at inlet and exit of the control volume. The details of the flow behaviour between the inlet and the outlet sections are, therefore, not needed. Assuming that the fluid is incompressible and that the velocity and its various components do not vary significantly at the inlet and outlet sections the right hand side of Eq. (4.45) may be integrated to yield

$$F_x = \rho\ V_2 A_2 (V_2)_x - \rho\ V_1 A_1 (V_1)_x$$

$$= \rho\ Q[(V_2)_x - (V_1)_x] \qquad \text{...(4.46 } a)$$

and similarly

$$F_y = \rho\ Q[(V_2)_y - (V_1)_y] \qquad \text{...(4.46 } b)$$

and

$$F_z = \rho\ Q[(V_2)_z - (V_1)_z] \qquad \text{...(4.46 } c)$$

The force exerted by the fluid on its surroundings in the x, y and z directions will be, in accordance with the Newton's third law, equal in magnitude but opposite in direction to F_x, F_y and F_z of Eq. (4.46). In Eq. (4.46), V_1 and V_2 are the average velocities at the inlet and outlet of the control volume.

Let us now consider the determination of the net force, expressed by the left hand side of Eq. (4.46). Every force acting on a fluid element within the control volume must have an equal and opposite force acting upon its neighbouring elements, as per the Newton's third law of motion. All such internal forces will cancel each other, so we are left with those forces which are exerted externally. The evaluation of the net force, say F_x, involves all the external forces acting on the control volume in the x-direction. In general, the external forces are those due to the pressure, the gravity and the frictional resistance at the boundary. The force due to the boundary friction is not easy to evaluate and is generally not taken into account due to its relatively small magnitude. It may, however, affect the magnitude of quantities at the outlet section. In the application of momentum equation, the length along the boundary should, in general, be small so as to minimize the effect of boundary friction.

4.19.1. On Choosing a Control Volume

A brief introduction to control volume has been included in chapter 1. A control volume refers to a region in space and is useful in the analysis of situations where flow occurs into and out of space. The boundary of a control volume is its control surface. Its size and shape are entirely arbitrary, but frequently they are made to coincide with solid boundaries in parts, and in other parts they are drawn normal to the flow directions as a matter of simplification.

A greater degree of simplification in the flow analysis can be achieved by carefully selecting a suitable volume. The following salient points should kept in mind while demarcating the control volume:

1. The size and shape of control volume are entirely arbitrary.
2. The surface of control volume adjoining the solid boundary, is frequently made to coincide with the solid boundary.
3. Other parts of control volume are drawn normal to the flow direction. The control volume would thus enclose all the fluid between the inlet and the outlet.
4. The control surface should cut across the supports or fixtures to include external reactions on the control volume.

4.20. Momentum Correction Factor

The non-uniformity of velocity distributions at the inlet and outlet sections also affect the accuracy of momentum flux (the momentum per second passing through a section is known as the *momentum flux*) based on the average velocity of flow. To correct it, a coefficient β, known as *momentum correction factor* is used such that $\beta\rho QV$ represents the true momentum flux passing a section. Consider an element of area dA [see Fig. 4.16 (*a*)], where the velocity is v, the momentum of fluid passing this area element is $\rho v^2 dA$ and the momentum flux passing section (1) is

$$\int_A \rho v^2 \, dA.$$

Based on the average velocity V, the momentum flux is $\beta\rho AV^2$; equating the two expressions, one obtains,

$$\beta\rho AV^2 = \int_A \rho v^2 \, dA$$

the momentum correction factor, for an incompressible fluid

$$\beta = \frac{1}{A}\int_A \left(\frac{v}{V}\right)^2 dA \qquad \text{...(4.47)}$$

For laminar flow in a circular tube, β = 4/3. For turbulent flow in circular pipes β varies from 1.01 to 1.05, the lower value being applicable to highly turbulent flows. For any type of flow, it can be shown that β is always less than α, the energy correction factor.

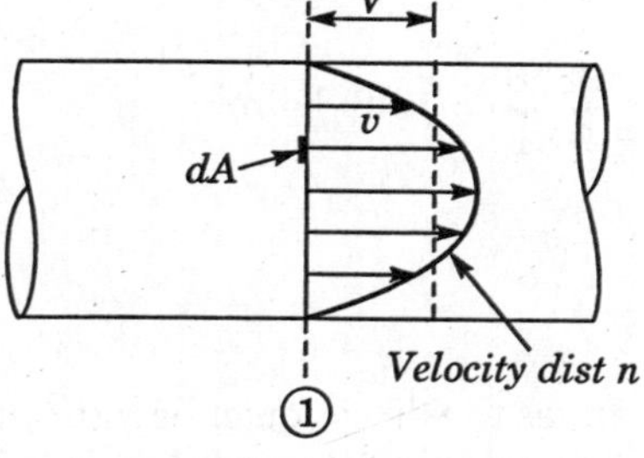

Fig. 4.16. (*a*) Notations for momentum correction factor.

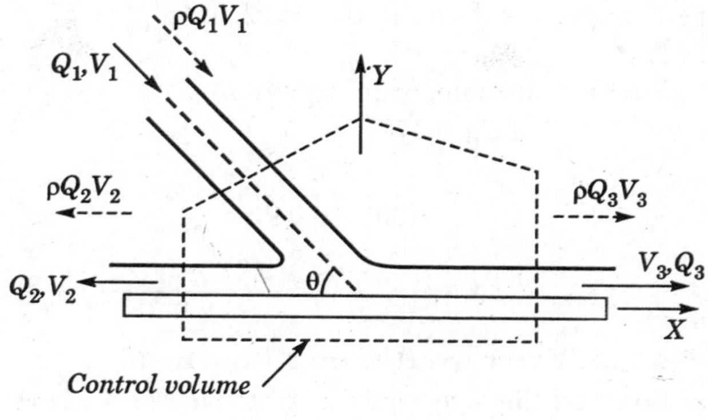

Fig. 4.17. Force on a flat surface due to momentum change.

4.21. Applications of Momentum Equation

4.21.1. Force caused by jet striking against a plane surface

Consider a jet moving at a velocity V_1 striking a plane surface as shown in Fig. 4.17. After striking the surface it gets split into two parts and flows along the surface as shown.

Assuming that the jet before and after the impact remains in the horizontal plane, the effects of gravity are eliminated. Let the x-direction be taken along the plane surface under consideration and the y-axis perpendicular to it. A suitable control volume indicated by the dotted lines may be selected.

Applying the momentum principle, the x and y components of the force acting on the fluid may be obtained. The rate at which x-momentum enters the control volume is $\rho Q_1 V_1 \cos\theta$ and the rate at which it leaves is $(\rho Q_3 V_3 - \rho Q_2 V_2)$. The net force acting on the fluid in the x-direction is

$$F_x = (\rho Q_3 V_3 - \rho Q_2 V_2) - \rho Q_1 V_1 \cos\theta \qquad ...(4.48)$$

This force, being parallel to the surface is a shear force exerted by the surface on the fluid. If the fluid is considered to be ideal, there would be no resistance offered to its flow and as such no shear force is possible. The force F_x for the ideal fluid would, therefore, be zero and the x-momentum starting and leaving the control volume would then be equal.

The rate at which y-momentum enters the control volume is $\rho Q_1 V_1 \sin\theta$. The momentum in y-direction leaving the control volume is obviously zero. The force component F_x exerted by the surface on the fluid to effect this change of momentum is given by :

$$F_y = 0 - (-\rho Q_1 V_1 \sin\theta) = \rho Q_1 V_1 \sin\theta \qquad ...(4.49)$$

The force components exerted by the fluid on the surface will be equal in magnitude but opposite in direction to F_x and F_y.

For flow over a curved surface, expressions similar to Eqs. (4.48) and (4.49) may be written to obtain the force exerted by the flowing fluid on the curved surface.

Example 4.45. *A vertical jet of water 7.5 cm in diameter leaving the nozzle, with a 9.2 m/s velocity strikes a horizontal and movable disc weighing 172 N (see the figure). The jet is then deflected horizontally. Determine the vertical distance y above the nozzle tip at which the disc will be held in equilibrium.*

Solution. Discharge through the nozzle,

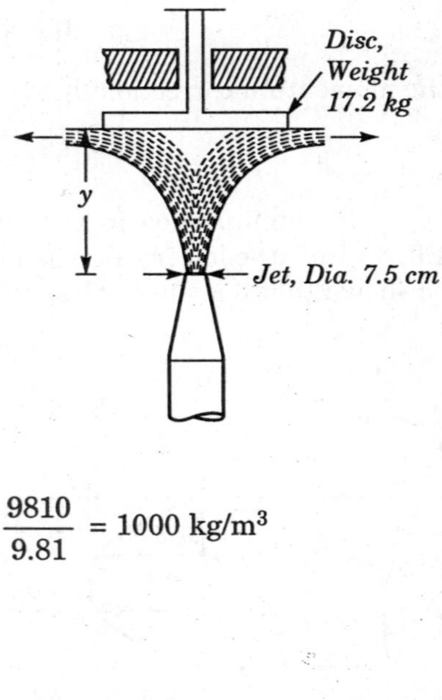

$$Q = \frac{\pi}{4}(0.075)^2 \times 9.2 = 0.0405 \text{ m}^3/\text{s}.$$

Let u be the jet velocity at an elevation y above the nozzle.

Applying the Bernoulli's theorem between the jet at the nozzle exit and the jet at an elevation y.

$$\frac{V^2}{2g} = \frac{u^2}{2g} + y,$$

where V is the jet velocity at the nozzle exit.

$$\therefore \qquad u^2 = V^2 - 2gy$$

Writing the momentum equation

$$\rho Q u = 172$$

$$u = \frac{172}{1000 \times 0.045} = 4.23 \text{ m/s} \qquad \because \quad \rho = \frac{9810}{9.81} = 1000 \text{ kg/m}^3$$

and
$$y = \frac{V^2 - u^2}{2g} = \frac{(9.2)^2 - (4.22)^2}{2 \times 9.81} = 3.402 \text{ m}.$$

4.21.2. Force exerted on a Pipe Bend

Consider the pipe bend illustrated in Fig. 4.18 (a) which changes the direction of flow through an angle θ. As a general case, a reducer bend has been selected which not only changes the direction of velocity, but also its magnitude on account of gradual reduction on in the cross-sectional area. It is assumed for simplicity that the axis of the bend is in the horizontal plane.

The control volume selected includes the inlet and outlet sections (1) and (2) of the pipe bend.

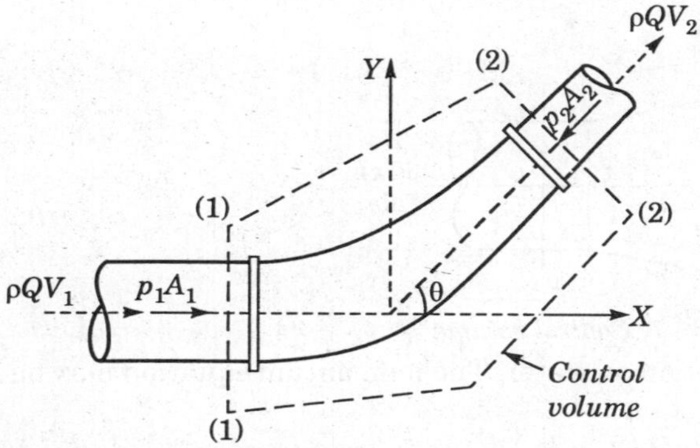

Fig. 4.18 (*a*) Momentum equation applied to the reducer bend.

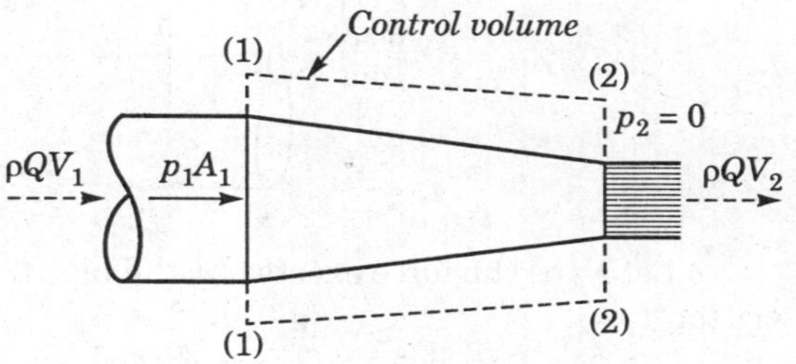

Fig. 4.18 (*b*) Notations for the force on the nozzle.

Let the average velocity, pressure and the area of flow at sections (1) and (2) be V_1, p_1, A_1 and V_2, p_2, A_2 respectively. Let F_x and F_y be the components of the force exerted on the fluid by the pipe bend in the x-and y-directions respectively. The other external forces acting over the fluid in the control volume area p_1A_1 acting over the section (1) and p_2A_2 over the section (2), these being the pressure forces. The momentum equation in the x-direction may now be written as

$$p_1A_1 - p_2A_2 \cos\theta + F_x = \rho Q(V_2 \cos\theta - V_1) \quad \text{...(4.50)}$$

Equation (4.50) together with Bernoulli's and continuity equations enable determination of F_x. Similarly F_y is determined from the momentum equation in the y-direction,

$$-p_2A_2 \sin\theta + F_y = \rho Q\,(V_2 \sin\theta - 0) \quad \text{...(4.51)}$$

With F_x and F_y known, the total force F exerted by the pipe-bend on the fluid can be readily determined. The force exerted by the fluid on the pipe-bend will be equal in magnitude but opposite in direction to F. It may be pointed out here that the pressures p_1 and p_2 are the gauge pressures, because the pipe bend being surrounded by the atmosphere throughout the net force due to atmospheric pressure acting over the bend surface is zero.

Flow through a nozzle which discharges into the atmosphere constitutes a special case in which there is a change only in magnitude of velocity, its direction remaining unchanged. The pressure p_2 at the exit being atmospheric is equal to zero. The force exerted by the nozzle in the direction a flow, Fig. 4.18 (*b*), is given by

$$p_1A_1 + F_x = \rho Q\,(V_2 - V_1) \quad \text{...(4.52)}$$

There is no force acting in the y-direction, as there is no momentum change in that direction.

Example 4.46. *Water flows at the rate of 0.71 m³/s through the pipe transition shown in the figure. If the pressure intensity at the centre line of the 90 cm section is 9810 N/m², what will be the centre line pressure in the 60 cm section ? What force will be required to produce the change in momentum of water as it passes through the transition.*

Solution. Velocity in the 90 cm section, $V_1 = \dfrac{0.71}{\dfrac{\pi}{4} \times (0.9)^2} = 1.115$ m/s.

Velocity in the 60 cm section $V_2 = \dfrac{0.71}{\dfrac{\pi}{4} \times (0.6)^2} = 2.51$ m/s.

Applying the Bernoulli's equation to 90 cm and 60 cm sections, noting that the pipe transition is horizontal

$$\frac{p_1}{\gamma} + \frac{V_1^2}{2g} = \frac{p_2}{\gamma} + \frac{V_2^2}{2g}$$

$$\therefore \quad \frac{p_2}{\gamma} = \frac{9810}{9810} + \frac{(.115)^2}{2 \times 9.81} - \frac{(2.51)^2}{2 \times 9.81} = 0.7425 \text{ m of water} = 7284 \text{ N/m}^2.$$

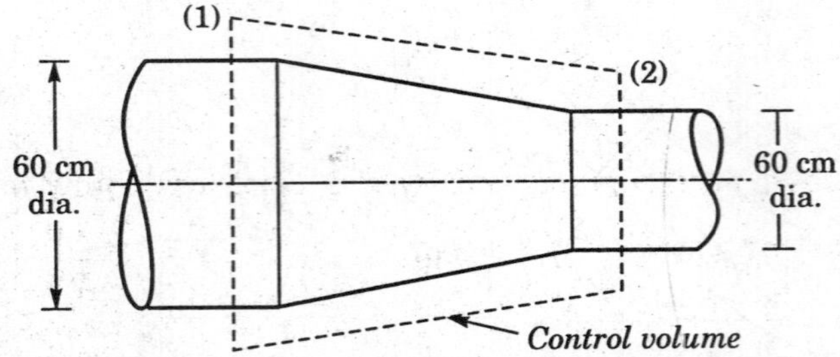

Let F_x be the force exerted by the pipe transition on water. The momentum equation may be written :

$$p_1A_1 + F_x - p_2A_2 = \rho Q\,(V_2 - V_1)$$

Making substitutions $9810 \times \frac{\pi}{4} \times (0.9)^2 + F_x - 7284 \times \frac{\pi}{4}\,(0.6)^2 = 1000 \times 0.71(2.51 - 1.15)$

$$F_x = -\,3198 \text{ N}.$$

The force required to produce the desired change in momentum is 3198 N and acts from right to left.

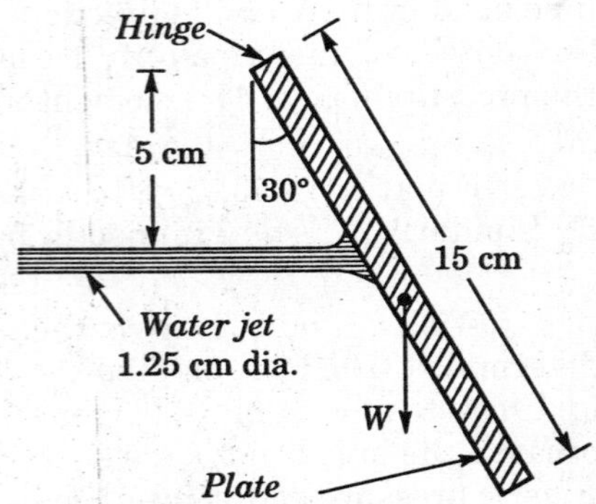

(15 cm × 15 cm × 0.64 cm)

Example 4.47. *A metal plate 0.64 cm thick and 15 cm square swings about a horizontal edge, see the figure. A horizontal jet of water 1.25 cm in diameter impinges with its axis perpendicular to and 5 cm below the edge of the hinge and keeps it steadily inclined at 20° to the vertical. Find the velocity of the jet if the metal plate has a sp. gr. of 8.*

Solution. The jet strikes the plate horizontally, and its momentum is completed destroyed by its impact upon the plate. The horizontal force due to impact of jet is

$$F = \rho QV = 102 \times \frac{\pi}{4}\left(\frac{1.25}{100}\right)^2 \times V^2$$

Weight of plate $W = 15 \times 15 \times 0.64 \times 10^{-6} \times 8 \times 9810 = 11.28$ N

Taking moment of these forces about the hinge

$$F \times \frac{5}{100} = W \times \frac{7.5}{100} \sin 30°$$

Substituting values of F and W

$$1000 \times \frac{\pi}{4}\left(\frac{1.25}{100}\right)^2 V^2 \times \frac{5}{100} = 11.28 \times \frac{7.5}{100} \times \frac{1}{2}$$

Resulting in, $V = 8.31$ m/s.

$\therefore$ The velocity of the jet = 8.31 m/s.

Example 4.48. *What is the force on the elbow-nozzle assembly shown in the figure. Water issues as a free-jet from the nozzle. The interior volume of the elbow-nozzle assembly is 0.1 cubic metre. The elbow-nozzle assembly is in a horizontal plane.*

Also determine the bending moment at section X-X due to the jet issuing from the nozzle by using the method of moment of momentum (indicate the control volume taken clearly).

Solution. From the continuity equation, the velocity of jet,

$$V_j = \frac{(\pi/4)\,(10)^2 \times 1.5}{(\pi/4)\,(2.5)^2} = 24.0 \text{ m/s}.$$

Let the control volume be as shown. Applying Bernoulli's equation, the pressure at the entrance to elbow may be obtained,

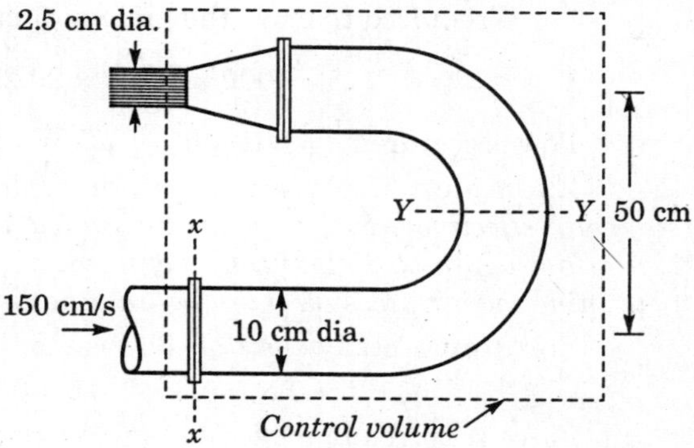

$$\frac{(1.5)^2}{2g} + \frac{p}{\gamma} = \frac{(24)^2}{2g}$$

yielding, $\frac{p}{\gamma} = 29.25$ m of water.

Momentum entering the control volume

$$= \rho Q V_1$$

$$= 1000 \times \left[\frac{\pi}{4} \times (0.10)^2 \times 1.5\right] \times 1.5$$

Momentum leaving the control volume

$$= \rho Q V_j = 1000 \times \left[\frac{\pi}{4} \times (0.1)^2 \times 1.5\right] \times 24$$

Let the force exerted by the elbow-nozzle assembly on water be F_x. Nothing that the pressure in the free-jet remains atmospheric, the momentum equation results in,

$$p.A + F_x = 1000 \left[\frac{\pi}{4} \times (0.1)^2 \times 1.5\right] (24.0 - 1.5) = 265.4 \text{ N}$$

or $$29.25 \times 9810 \times \frac{\pi}{4} \times (0.1)^2 + F_x = 265.4$$

$\therefore$ $$F_x = -1986.0 \text{ N}$$

The negative sign indicates that F_x acts from right to left.

The force exerted by water on the elbow-nozzle assembly is 1986.0 N and acts in a horizontal direction from left to right.

From the moment of momentum equation, the rate of change of moment of momentum is equal to a torque. The moment of momentum entering the control volume (about Y-Y)

$$= \rho Q V_1 \times 0.25$$

and the moment of momentum leaving the control volume is

$$= \rho Q V_j \times 0.25$$

Therefore, the torque $$= \rho Q (V_j - V_1) \times 0.25 = 1000 \left[\frac{\pi}{4} \times (0.1)^2 \times 1.5\right] (24.0 - 1.5) \times 0.25$$

$$= 66.3 \text{ Nm.}$$

The torque due to the jet exerted on the elbow-nozzle assembly is clockwise. Therefore, the bending moment about X-X is 66.3 Nm and is clockwise.

Example 4.49. *A jet of water 5 cm in diameter discharges 30 litres/sec. Calculate the force required to move a plate towards the jet with a velocity of 5 m/s. The jet strikes the plate perpendicularly. Determine the power required to move the plate.*

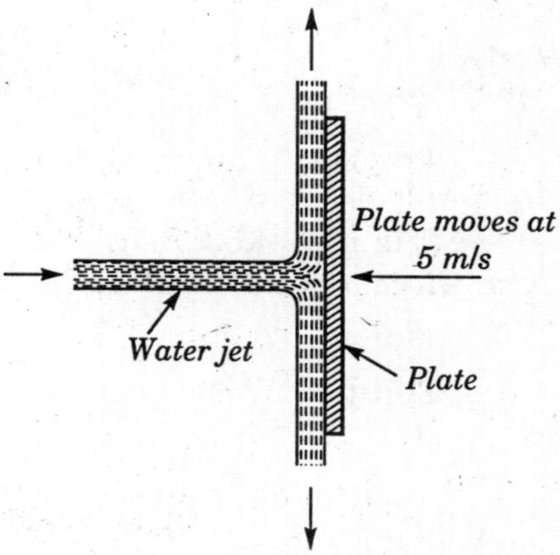

Solution. Absolute velocity of jet

$$= \frac{30 \times 4 \times 10^4}{1000 \times \pi \times 5^2} = 15.27 \text{ m/s.}$$

Velocity of jet relative to the moving plate

$$= 15.27 + 5 = 20.27 \text{ m/s.}$$

Rate of flow striking the plate

$$= \frac{\pi}{4} \times \left(\frac{5}{10}\right)^2 \times 20.27 \times 1000 = 39.8 \text{ lit/sec.}$$

Force required to move the plate at 5 m/s against the jet

$$= 1000 \times \frac{39.8}{1000} \times 20.27 \text{ N} = 805 \text{ N}.$$

Power required $= 805 \times 5 = 4025$ W = 4.025 kW.

Example 4.51. *A closed tank 1 m × 1 m × 4 m high and weighing 1.176 kN has a hole in the side with effective area 6.5 cm² and situated 15 cm above the tank bottom. The tank is filled with water to a depth of 3 m. If the co-efficient of friction between the ground and the wheels is 0.01, determine the air pressure p in the tank that is required to move it.*

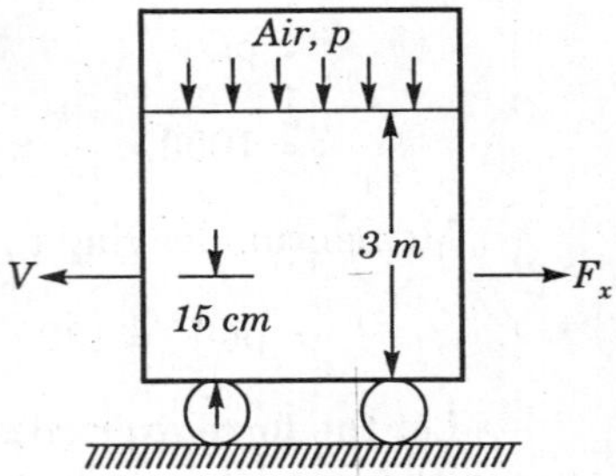

Solution. Total weight of the system,

$$W = 1176 + 1 \times 1 \times 3 \times 9810 = 30600 \text{ N} = 30.6 \text{ kN}$$

Force required to move the tank

$= \mu W$; μ being the coefficient of friction

$= 0.01 \times 30600 = 306$ N.

Force developed by change of momentum,

$$F_x = \rho Q(V - 0.0) \quad \text{or} \quad 306 = 1000 \times \frac{6.5}{100 \times 100} \times V^2$$

resulting in, $V = 21.6$ m/s.

Applying the Bernoulli's equation between the centre line of jet and the free surface of water where pressure is p,

$$\frac{V^2}{2g} = 2.85 + \frac{p}{\gamma}$$

$$\therefore \quad \frac{p}{\gamma} = \frac{(21.7)^2}{2 \times 9.81} - 2.85 = 21.10 \text{ metre of water.}$$

or $$p = 21.10 \times 9810 \text{ N/m}^2 = 207.0 \text{ kN/m}^2.$$

Example 4.52. *A large disc weighing 58.9 N is so mounted that it may move freely along a vertical axis, the plane of the disc thereby remaining horizontal. Directly below the disc is a nozzle delivering a vertical stream of water having an efflux velocity of 8 m/s and an initial jet diameter of 5 cm. Assuming that the disc deflects the jet horizontally, how high above the nozzle will the disc be held in equilibrium by the force of the jet ?*

Solution. Referring to Figure of Ex. 4.35 and its solution

$$u^2 = V^2 - 2gy = (8)^2 - 2 \times 9.81\, y$$

Rate of efflux, $$Q = \frac{\pi}{4}\left(\frac{5}{100}\right)^2 \times 8 \text{ m}^3/\text{s}$$

Writing the momentum equation (substituting $F = 58.9$ N),

$$58.9 = \rho Q u$$

$$58.9 = 1000 \times \frac{\pi}{4}\left(\frac{5}{100}\right)^2 \times 8 \times \sqrt{64 - 19.62y}$$

from which, $y =$ **2.54 m.**

Example 4.53. *Determine the forces exerted by the pipe boundary on fluid for the following case. Given $p_1 = 2.452$ bar (gauge), $V_1 = 4.0$ m/s, $D = 30$ cm, $\theta = 60°$, $D_2 = 15$ cm and $\rho = 1000$ kg/m³.*

Solution. Velocity at section (2), $$V_2 = \frac{\frac{\pi}{4}(0.3)^2 \times 4.0}{\frac{\pi}{4}(0.15)^2} = 16.0 \text{ m/s.}$$

Applying Bernoulli's equation between sections (1) and (2) assuming the pipe to be in the horizontal plane,

$$\frac{(4.0)^2}{2g} + \frac{2.452 \times 10^5}{9810} = \frac{(16.0)^2}{2g} + \frac{p_2}{\gamma}$$

$$\therefore \qquad \frac{p_2}{\gamma} = 12.766 \text{ m of water} = 125.2 \text{ kN/m}^2.$$

Let the force components exerted by the pipe boundary on the fluid be F_X and F_Y in the X and Y directions respectively. Applying the momentum equation to the control volume bounded by the pipe boundary and sections (1) and (2) in X and Y directions.

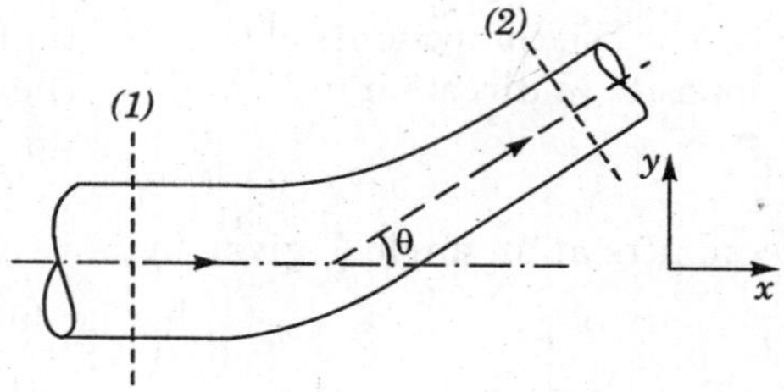

Momentum equation in the X-direction

$$p_1A_1 + F_X - p_2A_2 \cos\theta = \rho Q(V_2 \cos\theta - V_1).$$

Substituting the values,

$$2452 \times 100 \times \frac{\pi}{4}(0.3)^2 + F_X - 125.2 \times 10^3 \times \frac{\pi}{4}(0.15)^2 \cos 60°$$

$$= 1000 \times \frac{\pi}{4} \times (0.3)^2 \times 4(16 \cos 60° - 4.0)$$

$$\therefore \qquad F_X = -\ 15.10 \text{ kN}$$

Momentum equation in the Y-direction

$$F_Y - p_2A_2 \sin 60° = \rho Q\, V_2 \sin 60°$$

$$F_Y = 125.2 \times 10^3 \times \frac{\pi}{4} \times (0.15)^2 \times \frac{\sqrt{3}}{2} + 1000 \times \frac{\pi}{4}(0.3)^2 \times 4.0 \times \frac{16\sqrt{3}}{2}$$

$$= 5.83 \text{ kN}.$$

∴ The force components exerted by the pipe boundary on fluid in X and Y directions are 15.1 kN from right to left and 5.83 km acting vertically upward.

Example 4.54. *A pipeline, 60 cm diameter, conveying oil (sp. gr. = 0.85) at the flowrate of 1800 lit/sec. has a 90° bend in the horizontal plane. The pressure at the entrance to the bend is 1471 bar and the loss of head in the bend is 2.0 m of oil. Find the magnitude and direction of the force exerted by the oil on the bend and show the direction of the force on a sketch of the bend.*

Solution. Consider the 90° pipe-bend shown in the figure. The mean velocity in the bend

$$V_1 = V_2 = \frac{4Q}{\pi d^2} = \frac{4 \times 1.8}{\pi \times (0.6)^2} = 6.375 \text{ m/s}.$$

Applying the Bernoulli's equations between sections (1) and (2) including the loss of head, noting that 1 bar = 100 N/m²

$$\frac{p_1}{\gamma} + \frac{V_1^2}{2g} = \frac{p_2}{\gamma} + \frac{V_2^2}{2g} + 2$$

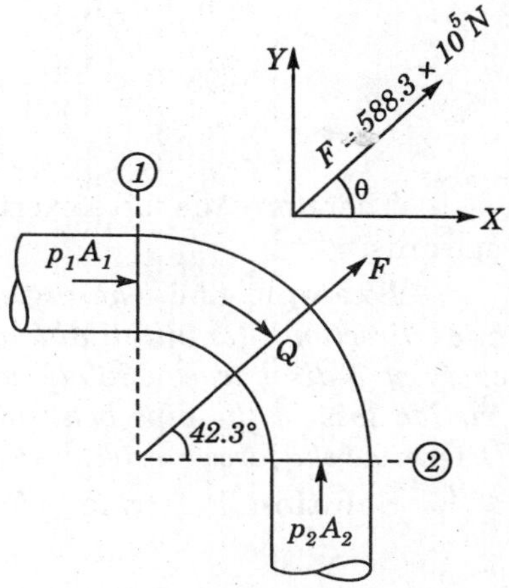

Since $\qquad V_1 = V_2, \dfrac{p_2}{\gamma} = \dfrac{p_1}{\gamma} - 2$

$$= \frac{1471 \times 100}{0.85 \times 9810} - 2 = 15.65 \text{ m of oil.}$$

$$p_2 = 8338 \times 15.65 = 130423 \text{ N/m}^2 = 130.42 \text{ kN/m}^2$$

Momentum equation in the X-direction :

$$p_1A_1 + F_x = \rho Q(0 - V_1)$$

$$1471 \times 100 \times \frac{\pi}{100}(0.6)^2 + F_x = \frac{8338}{9.81} \times 1.8(0 - 6.375)$$

which results in, $\qquad F_x = -\ 51335$ N

Momentum equation in the Y-direction

$$p_2A_2 + F_y = \rho Q(-V_2 - 0)$$

$$130423 \times \frac{\pi}{4} \times (0.6)^2 + F_y = \frac{8338}{9.81} \times 1.8(-6.375 - 0)$$

from which, $F_y = -46627$ N

The components of the resultant force exerted by oil on the bend are equal in magnitude but opposite in direction to F_x and F_y. The resultant force exerted by oil on the bend is

$$F = \sqrt{(51335)^2 + (46627)^2} = 69357 \text{ N}$$

and acts at an angle θ, given by,

$$\tan\theta = \frac{46627}{51335} = 0.91$$

$$\theta = 42.3°.$$

The resultant force F is indicated on the figure. It is radial force and passes through the centre of curvature of the bend.

Example 4.55. *0.91 m³/s of water flows up through a vertical conical enlargement, one metre high, 30 cm diameter at bottom and 60 cm diameter at top. The pressure at bottom is 981 bar. Calculate magnitude and direction of the force exerted by water on the enlargement.*

Solution. Velocities at the two ends of the enlargement are

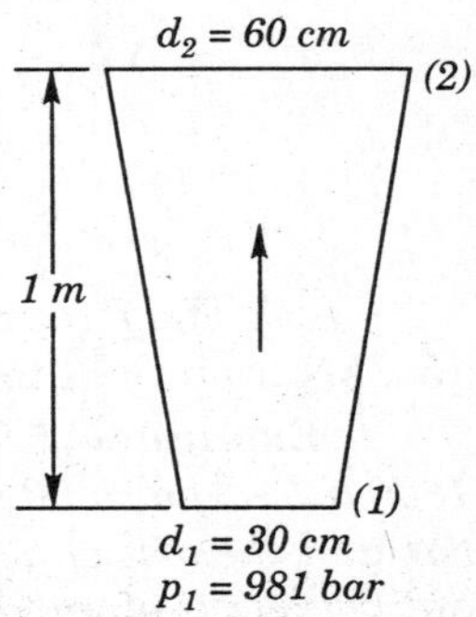

$$V_1 = \frac{4Q}{\pi d_1^2} = \frac{4 \times 0.28}{\pi\left(\frac{30}{100}\right)^2} = 3.96 \text{ m/s}$$

$$V_2 = \frac{3.96}{4} = 0.99 \text{ m/s.}$$

Applying the Bernoulli's equation between sections (1) and (2) to determine the pressure p_2, taking section (1) as the datum

$$0 + \frac{981 \times 100}{9810} + \frac{(3.96)^2}{2 \times 9.81} = 1 + \frac{p_2}{9810} + \frac{(0.99)^2}{2 \times 9.81}$$

which results in, $p_2 = 95648$ N/m²

Writing the momentum equation for the enlargement, the net force exerted on the fluid is equated to the change in momentum per second. If W represents the weight of water in the conical enlargement then

$$-W + p_1A_1 + F - p_2A = \rho Q(V_2 - V_1)$$

$$-9810 \times \frac{\pi}{4 \times 2}(0.3^2 + 0.6^2) \times 1 + 98100 \times 10^4 \times (\pi/4)(0.3)^2 + F - 95648(\pi/4)(0.6)^2$$

$$= 1000 \times 0.28(0.99 - 3.96), \quad \therefore \quad F = 23849 \text{ N}$$

Therefore, the force exerted by water on the enlargement = 23849 N, which acts vertically upwards.

Example 4.56. *The angle of a reducing bend is 60° (that is deviation from initial direction to final direction). Its initial diameter is 30 cm and final diameter 15 cm and is fitted in a pipeline, carrying a discharge of 360 lit/sec. The pressure at the commencement of the bend is 2943 bar. The friction loss in the pipe bend may be assumed as 10% of the kinetic energy at exit of the bend. Determine the force exerted by the reducing bend.*

Solution. Referring to Fig. 4.18 (*a*), the velocities at the inlet and exit of the bend are

$$V_1 = \frac{4Q}{\pi d_1^2} = \frac{4 \times 0.36}{\pi \times (0.3)^2} = 5.09 \text{ m/s}$$

$$V_2 = \frac{4Q}{\pi d_2^2} = \frac{4 \times 0.36}{\pi \times (0.15)^2} = 20.36 \text{ m/s}$$

Applying the Bernoulli's equation between sections (1) and (2) of the bend which lies in the horizontal plane

$$\frac{p_1}{\gamma} + \frac{V_1^2}{2g} = \frac{p_2}{\gamma} + \frac{V_2^2}{2g} + 0.1\frac{V_2^2}{2g}$$

$$\frac{p_2}{\gamma} = \frac{p_1}{\gamma} + \frac{V_1^2}{2g} - 1.1 \times \frac{V_2^2}{2g}$$

$$= \frac{294300}{9810} + \frac{(5.09)^2}{2 \times 9.81} - 1.1 \times \frac{(20.38)^2}{2 \times 9.81} = 8.052 \text{ m of water}$$

$$p_2 = 9810 \times 8.052 = 78990 \text{ N/m}^2$$

Momentum Equation in the X-direction [see Fig. 4.18 (*a*)] :

$$p_1A_1 - p_2A_2 \cos 60° + F_x = \rho Q(V_2 \cos 60° - V_1)$$

$$294300 \times \frac{\pi}{4} \times \left(\frac{30}{100}\right)^2 - 78990 \times \frac{\pi}{4}\left(\frac{15}{100}\right)^2 \times \frac{1}{2} + F_x = 1000 \times 0.36 \left(20.36 \times \frac{1}{3} - 5.90\right)$$

from which, $F_x = -18271$ N

it being the x-component of the force exerted by the bend on water.

Momentum Equation in the Y-direction :

$$F_y - p_2A_2 \sin 60° = \rho Q(V_2 \sin 60° - 0)$$

$$F_y - 78990 \times \frac{\pi}{4} \times \left(\frac{15}{100}\right)^2 \times 0.865$$

$$= 1000 \times 0.36(20.36 \times 0.865 - 0)$$

$$F_y = 7554 \text{ N}$$

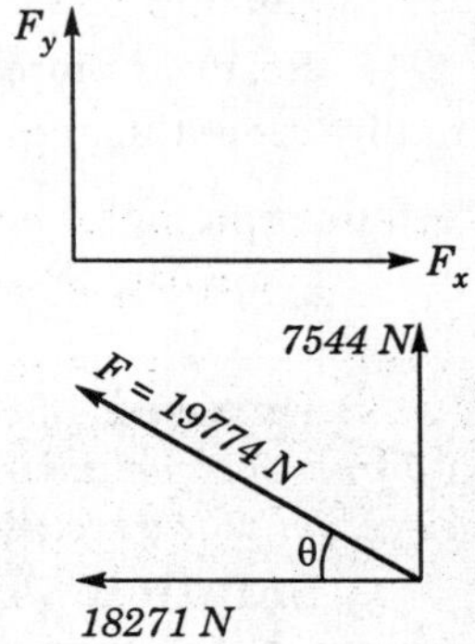

Resultant force exerted by the bend on water

$$F = \sqrt{(18271)^2 + (7554)^2} = 19774 \text{ N}$$

$$\tan\theta = 7754/18271 \; ; \; \theta = 22.45°$$

Therefore, the resultant hydrodynamic force exerted by water on the bend is 19774 N and acts in a direction opposite to that of F.

Example 4.57. *A pipe bend tapers from a diameter of 500 mm at inlet to a diameter of 250 mm at outlet and turn the flow through 45°. The pressures at inlet and outlet are 352 and 226 bar. If the pipe is conveying oil of specific gravity 0.85, calculate the magnitude and direction of the resultant force on the bend when the oil flow rate is 0.45 m³/s. The bend is in a horizontal plane.*

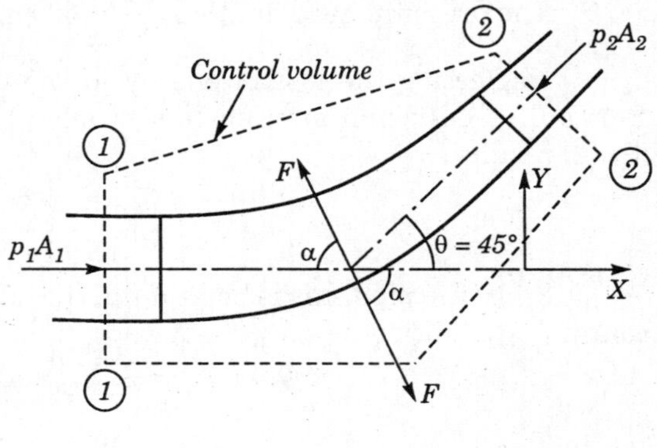

Solution. Data given :

Oil sp. gr. = 0.85

$Q = 0.45$ m^3/s

Bend in horizontal plane

$p_1 = 392$ bar $= 39200$ N/m^2

$p_2 = 226$ bar $= 22600$ N/m^2

$d_1 = 500$ mm, $d_2 = 250$ mm.

Let F be the force exerted by the pipe bend on the oil inside the bend, and let F_x and F_y be its components in the x and y-directions, respectively. Writing the momentum equation in X-direction, one gets

$$p_1A_1 - p_2A_2 \cos\theta + F_x = \rho Q(V_2 \cos\theta - V_1)$$

From the continuity equation,

$$Q = A_1V_1 = A_2V_2\,;$$

$$V_1 = \frac{4Q}{\pi d_1^2} = \frac{4 \times 0.45}{\pi(0.5)^2} = 2.292 \text{ m/s}$$

$$V_2 = 4V_1 = 9.167 \text{ m/s}$$

Making substitutions,

$$39200 \times \frac{\pi}{4}(0.5)^2 - 22600 \times \frac{\pi}{4}(0.25)^2 \cos 45° + F_x = \frac{8338 \times 0.45}{9.81}(9.167 \cos 45° - 2.292)$$

simplifying, $\quad F_x = \mathbf{-\ 5319.0\ N.}$

Momentum equation in Y-direction :

$$0 - p_2A_2 \sin\theta + F_y = \rho Q(V_2 \sin\theta - 0)$$

or

$$-\ 0.23 \times 10^4 \times \frac{\pi}{4}(0.25)^2 \sin 45° + F_y = \frac{8338 \times 0.45}{9.81}(9.167 \sin 45°)$$

from which, $\quad F_y = \mathbf{3262.0\ N.}$

Resultant force, F, exerted by the bend on the oil

$$F = \sqrt{F_x^2 + F_y^2} = \sqrt{(-\ 5319)^2 + (3262)^2} = \mathbf{6250.0\ N.}$$

Resultant force exerted by oil on the bend, $\overline{F}$, will be equal in magnitude to F and opposite in direction, the angle α it makes with the horizontal is given by

$$\tan\alpha = \frac{F_y}{F_x} = \frac{3262}{5319} = 0.6133$$

$$\alpha = 62°.$$

Example 4.58. *A 300 mm diameter pipe carries water at a head of 20 metres with a velocity 3.5 m/s. If the axis of the pipe turns through 45° in the horizontal plane, find the magnitude and direction of the resultant force on the bend.*

Solution. The sketch shown alongside illustrates the change in direction of flow through 45°. The diameter remains unchanged. Two sections marked (1)-(1) and (2)-(2) are chosen on either side of the bend for applying the momentum equation. The plane X-Y in which the pipe axis lies is horizonital.

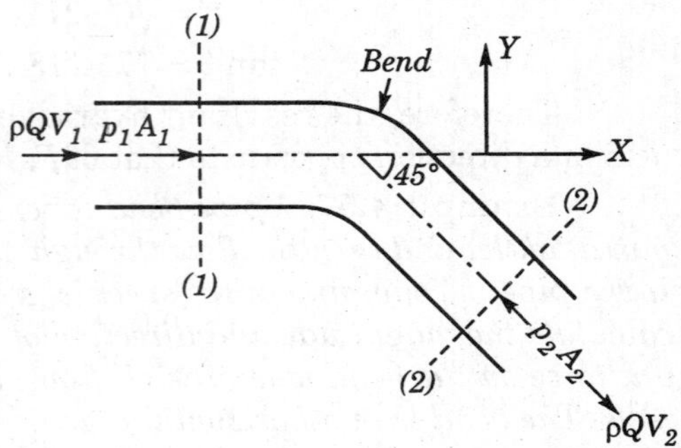

The pressure head at (1)-(1), $\dfrac{p_1}{\gamma} = 20$ m of water

Neglecting the loss of energy in the bend, the Bernoulli's equation applied to sections (1)-(1) and (2)-(2) results in :

$$Z_1 + \frac{p_1}{\gamma} + \frac{V_1^2}{2g} = Z_2 + \frac{p_2}{\gamma} + \frac{V_2^2}{2g} + h_L$$

Since the pipe axis is horizontal, the datum heads $Z_1 = Z_2$; also h_L is assumed zero ; from the continuity equation

$$Q = A_1V_1 = A_2V_2, \text{ and since pipe is uniform, } A_1 = A_2$$

$\therefore \qquad V_1 = V_2$ resulting in $p_1 = p_2$, or $\dfrac{p_1}{\gamma} = \dfrac{p_2}{\gamma} = 20$ m

(Given $V = 3.5 = 3.5$ m/s)

Applying momentum equation to the space enclosed between the two sections, and the pipe boundary, we obtain

In the X-direction :

$$p_1A_1 - p_2A_2 \cos 45° + F_x = \rho Q\,(V_2 \cos 45° - V_1) \qquad ...(1)$$

where F_x is the force exerted by the bend on the flow

$$Q = A.\,V = \frac{\pi}{4}\,(0.3)^2 \times 3.5 = 0.2473 \text{ m}^3\text{/s}$$

$$\gamma = 9810 \text{ N/m}^3, \qquad \therefore \quad \rho = \frac{\gamma}{g} = \frac{9810}{9.81} = 1000 \text{ kg/m}^3$$

$$p = \gamma h = 9810 \times 20 = 196200 \text{ N/m}^2$$

Substituting in Eq. (1), the equation results in

$$9810 \times 20\,(1 - \cos 45°) + F_x = 1000 \times 0.2473\,(\cos 45° - 1) \times 3.5$$

which gives, $F_x = -\,4315.78$ N

(The minus sign indicates that F_x acts in the negative X-direction).

Therefore, the force exerted by the flow on the bend in X-direction is equal to F_x in magnitude but opposite in direction, *i.e.* it in the positive X-direction, as shown below :

Momentum equation in Y-direction :

$$0 + p_2A_2 \sin 45° + F_y = \rho Q\,(V_2 \sin 45° - 0)$$

or $\qquad 9810 \times 20 \times 0.071 \sin 45° + F_y = 1000 \times 0.2473\,(3.5 \sin 45°)$

which gives, $F_y = -\,9339.74$ N (acts in negative Y-direction)

The force exerted by the bend on the fluid flow acts in the negative Y-direction. Hence, the components of the resultant force that act on the bend by water are as shown below :

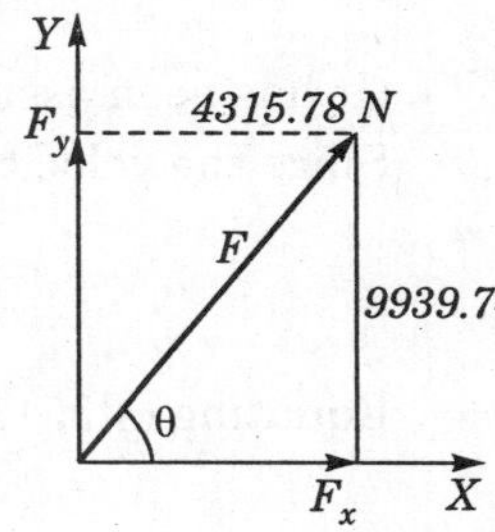

The resultant force on the bend

$$= \sqrt{(4315.78)^2 + (9939.74)^2}$$

$$F = \mathbf{10835.46\ N}$$

and the angle it makes with X-axis is

$$\tan\theta = \frac{9939.74}{4315.78} = 2.3031$$

$\therefore \qquad \theta = 66.5297° =$ **66° 31′ 47″.**

4.21.3. Propeller and Jet Propulsion

The thrust developed by a propeller and the reaction of a jet are used for propelling boats, ships and airplanes. The propeller changes the momentum of the fluid within which it is submerged and develops a thrust that is used for propulsion. Let us consider a propeller moving to the left with a velocity V, through the stationary fluid which is assumed to be incompressible and non-viscous (ideal fluid).

Fig. 4.19 illustrates the flow conditions up-stream and downstream of the propeller. The flow is undisturbed at the upstream section (1) and gets accelerated as it approaches the propeller, because of the reduced pressure at action (2) just upstream of the propeller. From the continuity and Bernoulli's equations it can be seen that the pressure p_2 at section (2) is smaller than p_1 and the pressure just downstream of the propeller is greater than p_4, this is obvious from the shape of the slipstream. As the fluid passes through the propeller, energy is added to it by the propeller resulting in increased pressure p_3. There is no change in the velocity of flow V, as the fluid passes through the propeller, that is between sections (2) and (3) on account of the same area of flow. The pressures p_1 and p_4 some distance upstream and downstream of the propeller are the same as that over the slipstream boundary.

Applying the momentum equation to the control volume enclosed by the slipstream boundary and sections (1) and (4), the force acting on the fluid is

$$F = \rho Q(V_4 - V_1) = \rho AV(V_4 - V_1) \quad ...(4.53)$$

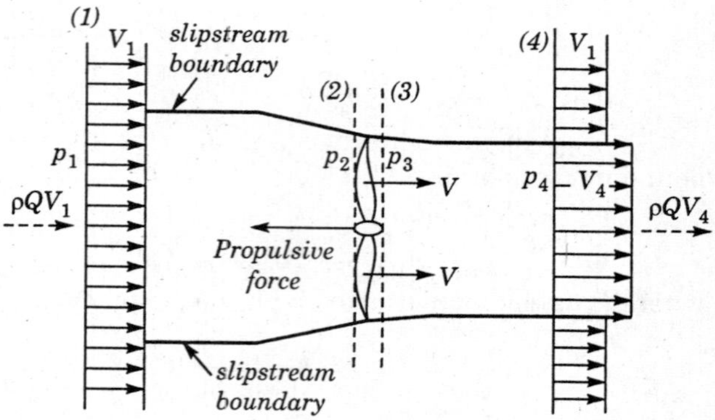

Fig. 4.19. Flow picture for moving propeller.

where A is the area swept by the propeller blades. The force F is also equal to the force due to the pressure difference between sections (2) and (3), thus

$$F = (p_3 - p_2) A$$

From equations (4.53) and (4.54),

$$(p_3 - p_2) A = \rho AV(V_4 - V_1) \quad ...(4.54)$$

or

$$(p_3 - p_2) = \rho V(V_4 - V_1) \quad ...(4.55)$$

Applying Bernoulli's equation between sections (1) and (2)

$$p_1 + \rho \frac{V_1^2}{2} = p_2 + \rho \frac{V_2^2}{2} \quad ...(4.56)$$

and between sections (3) and (4)

$$p_3 + \rho \frac{V_3^2}{2} = p_4 + \rho \frac{V_4^2}{2} \quad ...(4.57)$$

(It has been assumed that the density of air is practically the same throughout.)

Since the velocity $V_2 = V_3 = V$ and $p_1 = p_4$, Eqs. (4.56) and (4.57) yield

$$p_3 - p_2 = \rho\left(\frac{V_4^2 - V_1^2}{2}\right) \quad ....(4.58)$$

Equating (4.55) and (4.85), the velocity through the propeller,

$$V = \frac{V_1 + V_4}{2} \quad ...(4.59)$$

is the average of the velocities upstream and downstream of the propeller.

The work done per second by the propeller in moving through the still fluid at a velocity V_1 (m/s) is FV_1 and the horse power output

$$P_0 = \frac{FV_1}{75} = \frac{\rho Q(V_4 - V_1)V_1}{75} \text{ (metric h.p.)} \quad ...(4.60)$$

in which F is in kg. If F is expressed in Newton, then the power output

$$P_o = FV_1 \text{ (watts)} = \rho Q(V_4 - V_1)\, V_1 \text{ (watts)}$$

The power input to propeller is equal to the difference in the kinetic energy between sections (4) and (1), and is used to increase the velocity from V_1 to V_4. Thus power input,

$$P_i = \left[\frac{1}{2}\rho Q\,(V_4^2 - V_1^2)\right] \times \frac{1}{75} \text{ (metric h.p.)} = \left[\rho Q\,(V_4 - V_1)\left(\frac{V_4 + V_1}{2}\right)\right] \times \frac{1}{75}$$

$$= [\rho QV(V_4 - V_1)] \times \frac{1}{75} \quad ...(4.61)$$

The power input in watts, $P_i = \rho QV (V_4 - V_1)$

The theoretical efficiency of the propeller is

$$\eta = \frac{P_o}{P_i} = \frac{V_1}{V} = \frac{2V_1}{V_1 + V_4} \qquad \text{...(4.62)}$$

Since $V_1 < V$, the η is always less than 100%.

The force exerted by the fluid on the propeller is equal and opposite to the force F given by Eq. (4.53). The thrust F is called the propelling or propulsive force.

Example 4.59. *A boat requires a 1962 N thrust to keep it moving at 25 km/hr. At what rate water must be taken in at the bow and ejected through a 40 cm pipe at the stern (beneath the water surface) to maintain the motion ?*

What would be the required discharge if the water taken from a tank inside the boat and ejected from the stern through the same pipe ?

Solution.
$$V_1 = \frac{25 \times 1000}{60 \times 60} = 6.95 \text{ m/s.}$$

$$V_2 = \frac{4Q}{\pi \times (0.4)^2} = 7.95\ Q.$$

Force exerted by the boat on water,

$$F = \rho Q\ (V_2 - V_1)$$

or
$$1962 = 1000 \times Q(7.95\ Q - 6.95) = 811\ Q^2 - 709\ Q$$

or
$$Q^2 - 0.875\ Q - 0.247 = 0$$

$\therefore$
$$Q = 1.099 \text{ m}^3\text{/s.}$$

The discharge when the water is taken from the tank inside the boat, may be obtained noting that the initial momentum in direction of boat motion is zero,

$$1962 = 1000\ Q\ (7.95\ Q - 0.0) = 811\ Q^2.$$

$\therefore$
$$Q = 0.497 \text{ m}^3\text{/s.}$$

Example 4.60. *An airplane flies at 200 km/hr through still air of density $\rho = 1.283$ kg/m³. The propeller is 2.5 m in diameter, and its slipstream has a velocity of 300 km/hr relative to the air plane. Determine (1) the thrust on the propeller (2) power output (3) propeller efficiency (4) velocity through plane of the propeller and (5) the pressure difference across the propeller disc.*

Solution. Refer Fig. 4.19,

$$V_1 = \frac{200 \times 1000}{60 \times 60} = 55.55 \text{ m/s}, \; V_4 = \frac{300 \times 1000}{60 \times 60} = 83.3 \text{ m/s.}$$

Force exerted by the propeller on air,

$$F = \rho Q\ (V_4 - V_1)$$

$$V = \frac{V_1 + V_4}{2} = \frac{55.55 + 83.3}{2} = 69.42 \text{ m/s}$$

and
$$Q = \frac{\pi}{4} (2.5)^2 \times 69.42 = 340.7 \text{ m}^3\text{/s.}$$

$\therefore$
$$F = 1.283 \times 340.7\ (83.3 - 55.55) = 12150 \text{ N}$$

(1) Thrust on the propeller = 12150 N

(2) Power output using Eq. (4.60)

$$= 12150 \times 55.55 = 674932 \text{ N-m/s} = 674.93 \text{ kW}$$

(3) Propeller efficiency is given by Eq. (4.62)

$$= \frac{V_1}{V} = \frac{55.55}{69.42} = 0.80 \text{ or } 80\%.$$

(4) Velocity through the plane of the propeller, $V = 69.42$ m/s

(5) Pressure difference across the propeller disc, from Eq. (4.55) :

$$(p_3 - p_2)A = \rho AV(V_4 - V_1)$$

or
$$p_3 - p_2 = \rho V(V_4 - V_1)$$
$$= 1.283 \times 69.42\ (83.3 - 55.55) = 2471.6\ \text{N/m}^2 = 24.72\ \text{bar.}$$

4.21.4. Jet Propulsion

In jet propulsion of aeroplanes, the air which is initially at rest is taken into the jet engine consisting of a compressor, a combusting chamber, a turbine and a jet pipe forming a nozzle. The air is drawn in the upstream section and is passed through the compressor, where its pressure, temperature and density are raised. Fuel is added into the air and is burned just downstream of the compressor thus increasing the energy content of the air. The air and the combustion gases then pass through a gas turbine, which drives the compressor, on its way to the jet pipe (nozzle). The air gas mixture emerges out into the atmosphere at a high velocity and at a pressure greater or equal to that surrounding the jet pipe. Only a portion of the energy of the hot gases is utilised by the turbine, since higher kinetic energy is required at the downstream end to produce a high propulsive force.

Fig. 4.20 illustrates the entry of air at velocity V_1, pressure p_1 and the ejection of air-gas mixture at velocity V_2 and pressure p_2. Let A_2 be the area of the jet at the outlet end. A control volume of any vertical extent may be chosen ; section (1) is located some distance upstream and section (2) just at the outlet end of the jet pipe or nozzle.

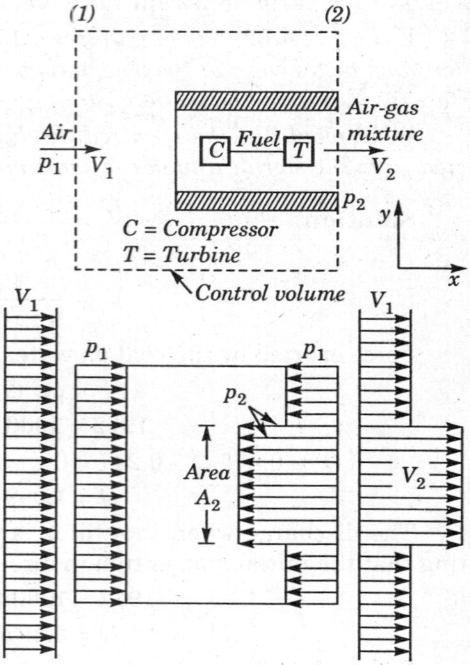

Fig. 4.20. Velocity and pressure distributions at sections (1) and (2) of the control volume.

Let F_x be the force exerted on the fluid by the jet engine, the application of momentum equation to the control volume permits evaluation of this force. The summation of force exerted on the fluid within the control volume results in,

$$\Sigma f_x = p_1 A_1 - p_1(A_1 - A_2) - p_2 A_2 + F_x$$
$$= (p_1 - p_2) A_2 + F_x \qquad \ldots(4.63)$$

If W_f is the weight of fuel injected per second and W_a is the weight rate of flow of air into the engine then, rate of change of momentum of fluid (fuel and air) between the sections (1) and (2) in the x-direction is equal to the net force in that direction,

$$\Sigma f_x = \frac{W_f}{g} V_2 + \frac{W_a}{g}(V_2 - V_1) \qquad \ldots(4.64)$$

From equations (4.63) and (4.64), one obtains,

$$(p_1 - p_2)A_2 + F_x = \frac{W_f}{g} V_2 + \frac{W_a}{g}(V_2 - V_1)$$

from which,
$$F_x = \frac{W_f}{g} V_2 + \frac{W_a}{g}(V_2 - V_1) + (p_2 - p_1)A_2$$

The force exerted by the fluid on the control volume is equal and opposite to F_x.

The kinetic energy per second of the discharged fluid is $\frac{1}{2}\frac{(W_f + W_a)}{g}(V_2 - V_1)^2$, where $(V_2 - V_1)$ is the absolute velocity at section (2), V_2 being the relative velocity at (2) and V_1 is the velocity of the plane. The work done per second (or the output) is $F_x V_1$, and the theoretical efficiency

$$\eta = \frac{\text{Output}}{\text{Output} + \text{K.E. of discharged fluid}}$$

$$= \frac{F_x V_1}{F_x V_1 + \frac{1}{2}\frac{(W_f + W_a)}{g}(V_2 - V_1)^2} \qquad ...(4.66)$$

It the weight rate of fuel flow is negligible, the propulsive force is given by

$$F_x = \frac{W_a}{g}(V_2 - V_1) + (p_2 - p_1)A = \rho Q(V_2 - V_1) + (p_2 - p_1)A \qquad ...(4.67)$$

Example 4.61. *The following figure shows a rocket of circular cross-section with 2 m as its maximum diameter. The rocket is moving at 200 m/s and the jet of gases leaves at 1000 m/s relative to the rocket. The outside air pressure is 98.1 kN/m² and that of the jet is 93.1 kN/m². The density of the outside air is 11.76 N/m³. The rocket breathes in air at the rate of 100 m³/s. Neglecting compressibility effects and assuming that the mass rate of flow of exhaust gases equals the mass flow rate of air breathed in, calculate :*

(i) Thrust developed by the rocket,

(ii) Energy supplied by the rocket to the air stream per unit weight of air flowing through the rocket,

(iii) Energy supplied per second by the rocket to the air stream, and

(iv) Power developed by the rocket.

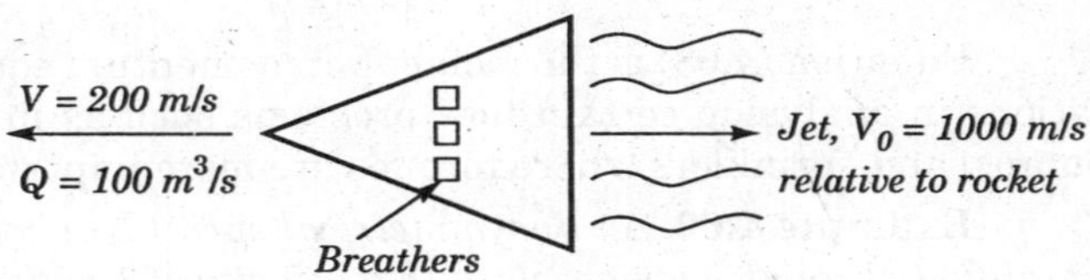

Solution. (*i*) Applying the momentum equation to the control volume, the force F exerted on the air by the rocket may be evaluated.

$$F + p_1A_1 - p_2A_2 = \rho Q(V_2 - V_1)$$

$$F = \rho Q(V_2 - V_1) + (p_2 - p_1)A_2.$$

(Since $A_1 = A_2$ and pressure p_1 is everywhere the same except in the jet.)

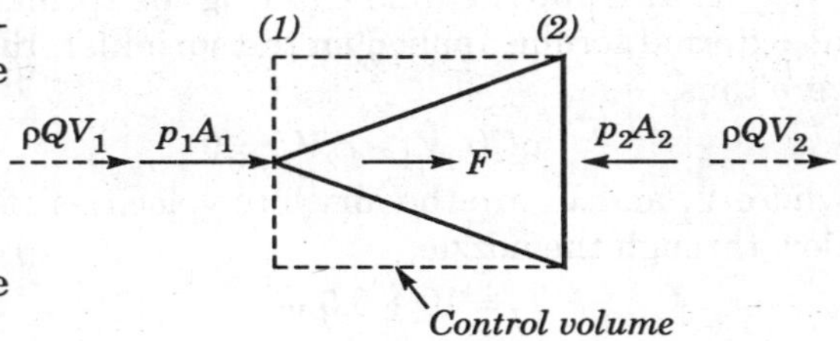

$$F = \frac{11.76}{9.81} \times 100\,(1000 - 200) + (93.1 - 98.1) \times 10^3 \times \frac{\pi}{4} \times 2^2$$

$$= 78.75 \text{ kN.}$$

Thrust on the rocket is equal and opposite to

$$F = 78.75 \text{ kN}$$

(*ii*) Absolute velocity of jet = 1000 – 200 = 800 m/s.

Energy per unit weight of air at section (2)

$$= \frac{V^2}{2g} + \frac{p}{\gamma} = \frac{(800)^2}{2 \times 9.81} + \frac{93.1 \times 10^3}{11.76}$$

$$= 4.055 \times 10^4 \text{ Nm/N of air.}$$

Energy per unit weight of air at section (1) (absolute velocity being zero)

$$= \frac{p}{\gamma} = \frac{98.1 \times 10^3}{11.76} = 0.8325 \times 10^4 \text{ Nm/N of air.}$$

$\therefore$ Energy supplied by the rocket per unit weight of air

$$= (4.055 - 0.8325) \times 10^4 \text{ Nm/N} = 32.225 \text{ kNm/N}.$$

(*iii*) Energy supplied per second

$$= \rho g Q \times 3.2225 \times 10^4 = 11.76 \times 100 \times 3.2225 \times 10^4 \text{ Nm}$$
$$= 37.9 \times 10^6 \text{ Nm}$$

(*iv*) Power developed by the rocket

$$= \text{Energy supplied per second} = 37.9 \times 10^6 \text{ Nm/s}$$
$$= \mathbf{37.9 \times 10^3 \ kW.}$$

4.22. Moment of Momentum Equation

When the moment of momentum of flow leaving a control volume is different from that entering it, the result is a torque acting over the control volume. This is yet another way of expressing the momentum principle. If the flow enters the control volume at a uniform velocity V_1 and at a steady rate Q, the momentum of flow entering the control volume is $\rho Q V_1$. If this flow has a curvature of radius r_1, the moment of momentum of flow entering the control volume is $\rho Q V_1 r_1$. Similarly, if V_2 and r_2 are the velocity and the radius of curvature at the outlet end of the control volume, the moment of momentum of flow leaving the control volume is $\rho Q V_2 r_2$. According to the moment of the momentum principle, the resulting torque is equal to the time rate of change in moment of momentum, thus,

$$T = \rho Q (V_2 r_2 - V_1 r_1) \qquad \text{...(4.68)}$$

Equation (4.68) is the moment of momentum equation for steady flow and has useful applications in analysing certain flow problems such as in turbomachinery (turbines and centrifugal pumps) and sprinklers where torques are more important than forces.

Example 4.62. *An unsymmetrical sprinkler shown in the figure has a frictionless shaft and equal flow through each nozzle with a velocity of 10 m/s relative to the nozzle. Determine its speed of rotation in r.p.m.*

Solution. The fluid entering the sprinkler has no moment of momentum and since there is no external torque applied on the sprinkler, the moment of momentum of fluid leaving it must be zero thus,

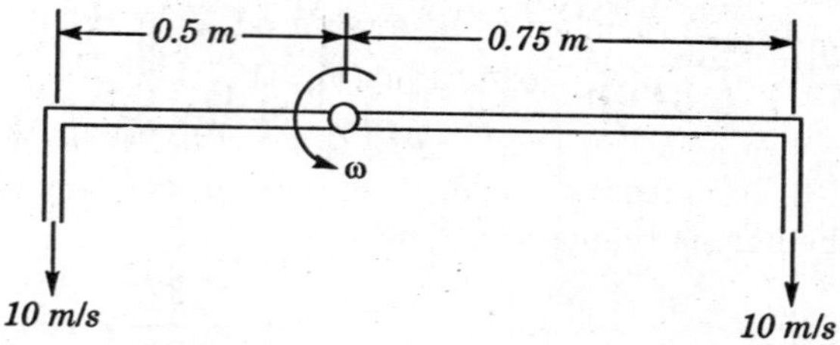

$$\rho Q (r_1 V_1 - r_2 V_2) = 0$$

where V_1 and V_2 are the absolute velocities of flow through the nozzle,

$$V_1 = 10 + 0.5\,\omega$$

and
$$V_2 = 10 - 0.75\,\omega$$

Substituting,

$$(10 + 0.5\,\omega) \times 0.5 - (10 - 0.75\,\omega) \times 0.75 = 0.$$

$$\therefore \qquad \omega = 3.08 \text{ rad/sec.}$$

and
$$N = \frac{60\omega}{2\pi} = \frac{60 \times 3.08}{2\pi} = 29.4 \text{ r.p.m.}$$

Example 4.63. *A lawn sprinkler shown in the figure has 1 cm diameter nozzle at the end of a rotating arm and discharges water at the rate of 8 m/s velocity. Determine the torque required to hold the rotating arm stationary. Neglecting friction, determine the constant speed of rotation of the arm.*

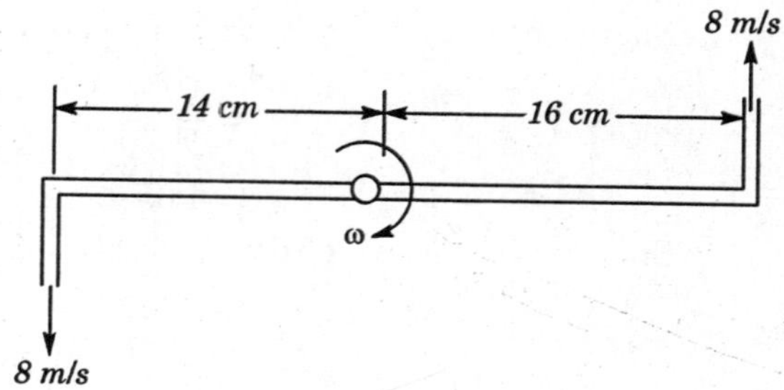

Solution. Time rate of change of moment of momentum is the torque required to hold the sprinkler stationary. It is to be noted that as water enters the sprinkler, it has no moment of momentum

$$T = \rho Q(V_1 r_1 + V_2 r_2) - 0$$

$$= 1000 \times \frac{\pi}{4}\left(\frac{1}{100}\right)^2 \times 8\,(8 \times 0.14 + 8 \times 0.16) = 1.50 \text{ Nm.}$$

If the sprinkler rotates at an angular velocity ω, the absolute velocities of flow through the nozzles are,

$$V_1 = 8 - 0.14\ \omega$$

and $$V_2 = 8 - 0.16\ \omega.$$

Since the moment of momentum of flow entering is zero and there is no friction, the moment of momentum leaving the sprinkler must also be zero, thus

$$\rho Q[(8 - 0.14\ \omega)\ 0.14 + (8 - 0.16\ \omega) \times 0.16] = 0$$

which simplifies to $$1.12 - 0.0196\ \omega + 1.28 - 00.256\ \omega = 0$$

or $$0.0452\ \omega = 2.4$$

$$\therefore \qquad \omega = \frac{2.4}{0.0452} = 3.1 \text{ rad/sec.}$$

and the speed in r.p.m. $$N = \frac{60\ \omega}{2\pi} = \frac{60 \times 53.1}{2\pi} = 507 \text{ r.p.m.}$$

OBJECTIVE TYPE QUESTIONS

Select the correct statement(s) :

1. One dimensional method of flow analysis is based on the assumption(s) that :

(*a*) the flow is steady

(*b*) the flow is along a straight line

(*c*) the velocity is same in magnitude and direction at all successive cross-sections

(*d*) the average velocity, pressure and elevation represent adequately the conditions at a given section

2. The laws of nature which govern fluid-flow phenomena are listed below in column A. Given in column B are certain flow situations where each of these laws finds application in flow analysis. Identify the law associated with each flow situation.

Column A	*Column B*
(*i*) Newton's first law of motion	(*a*) Analysis of laminar flow
(*ii*) Newton's second law of motion	(*b*) Analysis of compressible fluid flow
(*iii*) First law of thermodynamics	(*c*) Energy transformation in a flow system
(*iv*) Second law of thermodynamics	(*d*) Force exerted by fluid flow on a pipebend
(*v*) Newton's law of viscosity	(*e*) Hydrostatic force on submerged bodies and stability of floating bodies.

3. Fluid-flow problems in general are influenced by such forces as the gravity force F_g, viscous force F_v, pressure force F_p and the force due to turbulence F_t. Identify the combination of these forces which are considered in the following equations of motion :

(*a*) Navier-Stokes' equation	(*i*) F_g, F_p, F_v, F_t
(*b*) Reynolds' equation	(*ii*) F_g, F_p
(*c*) Euler's equation	(*iii*) F_g, F_p, F_v.

4. The integration of the Euler's equation results in the Bernoulli's equation. The Bernoulli constants for points lying on the same streamline and those which lie on other streamlines will have the same value if the flow is :

(*i*) incompressible (*ii*) steady
(*iii*) irrotational (*iv*) uniform

5. The derivation of Bernoulli's equation is based on certain assumptions. Identify these assumptions from those given in the following list :

(*a*) flow is uniform (*b*) flow is steady
(*c*) flow is irrotational (*d*) flow is turbulent
(*e*) flow is compressible (*f*) flow is non-viscous
(*g*) fluid is incompressible.

6. The Bernoulli's equation written in the conventional form

$$\frac{V^2}{2g} + \frac{p}{\gamma} + Z = \text{constant}$$

represents total energy per unit of certain quantity. Identify this quantity from the choice given below :

(*i*) energy per unit volume (*ii*) energy per unit mass
(*iii*) energy per unit weight (*iv*) energy per unit specific weight

7. The Bernoulli's equation can be written in the following different forms (see column A). Column B lists certain units each of which pertains to one of the possible forms of the equation. Identify the unit associated with each of the equations

Column A	*Column B*
(*a*) $V^2/2g + p/\gamma + Z$ = constant	(*i*) total energy per unit mass
(*b*) $\rho V^2/2 + p + \gamma Z$ = constant	(*ii*) total energy per unit weight
(*c*) $V^2/2 + p/\rho + gZ$ = constant	(*iii*) total energy per unit volume

8. If the average velocity, pressure and elevation head at a flow-section are V, p and Z respectively, and a fluid of specific weight γ (mass density ρ) is flowing at a discharge of Q, the total energy passing through the section is :

(*i*) $\rho Q \left(\frac{V^2}{2g} + \frac{p}{\gamma} + Z\right)$ (*ii*) $\gamma Q \left(\frac{V^2}{2g} + \frac{p}{\gamma} + Z\right)$

(*iii*) $Q \left(\frac{V^2}{2g} + \frac{p}{\gamma} + Z\right)$ (*iv*) $\frac{V^2}{2g} + \frac{p}{\gamma} + Z$

9. The piezometric head is the summation of :

(*a*) velocity head and elevation head (*b*) velocity head and pressure head
(*c*) pressure head and elevation head (*d*) none of these answers.

10. The hydraulic-grade-line is one which indicates the variation of :

(*i*) total energy of flow in the direction of flow
(*ii*) piezometric head in the direction of flow
(*iii*) velocity head in flow-direction
(*iv*) pressure head in the direction of flow

11. The total-energy-line is always higher than the hydraulic grade line, the vertical distance between the two represents :

(*a*) the datum head (*b*) the pressure head
(*c*) the velocity head (*d*) the piezometric head

12. The total-energy-line in pipe flow is a graphical representation of the Bernoulli's equation and represents the sum of velocity head, pressure head and the elevation head measured above :

(*i*) the centre line of pipe (*ii*) the top of the pipeline
(*iii*) the arbitrary horizontal datum (*iv*) the bottom of the pipe

13. The hydraulic grade line (HGL) normally slopes downwards along the direction of flow. However there are situations where it may have an upward slope. From the list of flow situations given below (in column A) identify the corresponding hydraulic grade line from the alternatives given in column B :

Column A	*Column B*
(*a*) steady flow in a straight uniform pipe	(*i*) HGL slopes downwards
(*b*) steady flow in a gradually enlarging pipe	(*ii*) HGL slopes downwards
(*c*) steady flow in a gradual contraction in pipe	(*iii*) HGL slopes upwards
(*d*) steady flow through a sudden enlargement	(*iv*) HGL slopes downwards
(*e*) steady flow through a sudden contraction.	(*v*) HGL slopes upwards.

14. The total energy represented by the Bernoulli's equation ($V^2/2g + p/\gamma + Z$) has the units :
(1) kg-m/m (N-m/m) (2) kg-m/kg (N-m/N)
(3) kg-m²/s (N-m²/s) (4) kg-m/s (N-m/s)
The units within the parenthesis are the SI equivalents.

15. The velocity head representing the kinetic energy per unit weight of fluid (with units kg-m/kg in MKS and N-m/N in SI) is denoted by :
(*a*) v^2/g (*b*) $v^2/2$ (*c*) $\sqrt{2gh}$ (*d*) $v^2/2g$
(*e*) none of these answers.

16. A liquid is capable of doing work by virtue of its sustained pressure. It has units of kg-m/kg in MKS and N-m/N in SI, and is represented by :
(*i*) p/ρ (*ii*) $\rho p/\gamma$ (*iii*) p/γ (*iv*) pz

17. The kinetic energy correction factor α is a measure of effect of non-uniform distribution of velocity which is caused on account of viscous and other resistances. It is expressed by :

(*a*) $\frac{1}{A}\int_A \left(\frac{v}{V}\right)^2 dA$ (*b*) $\frac{1}{A}\int \left(\frac{v}{V}\right) dA$

(*c*) $\frac{1}{A}\int_A \left(\frac{v}{V}\right)^3 dA$ (*d*) $\frac{1}{A}\int \left(\frac{V}{v}\right)^3 dA$

18. The magnitude of kinetic energy correction factor α depends on the type of flow, its turbulence characteristics, and the nature of pipe surface. For the following flow cases, select the proper value of α from the table given along side :

Flow situations	*Value of* α
(*i*) Laminar flow in pipes	(*a*) 1.01
(*ii*) Highly turbulent flow in a rough pipe	(*b*) 2.0
(*iii*) Highly turbulent flow in a smooth pipe	(*c*) 2.0
(*iv*) Laminar flow in a rough pipe.	(*d*) 1.15.

19. A stagnation point is a point :
(*a*) where the pressure is zero (*b*) where the total energy is zero

(*c*) where the velocity of flow reduces to zero
(*d*) where the total energy is maximum.

20. The stagnation pressure is the sum of :
(*i*) vacuum pressure and static pressure (*ii*) static pressure and dynamic pressure
(*iii*) dynamic pressure and vacuum pressure
(*iv*) absolute pressure and dynamic pressure

21. A pitot-tube is an instrument for measuring :
(*a*) pressure of flow (*b*) discharge of fluid
(*c*) velocity of flow (*d*) total energy

22. While using the Pitot-tube, it must be ensured that its alignment is such that :
(*i*) its horizontal leg is at right angles to the flow direction
(*ii*) its opening faces the downstream direction
(*iii*) its opening faces upstream and the horizontal leg is perfectly aligned with the direction of flow
(*iv*) the horizontal leg should be inclined at 45° in plan

23. A venturimeter is a device based on the Bernoulli principle and is used for measuring :
(*a*) piezometric head (*b*) velocity head
(*c*) flowrate (*d*) total energy

24. A venturimeter may be conveniently divided into three component parts indicated in column A. Identify the hydraulic function of each of the component from the list given in column B.

Column A	*Column B*
(*i*) converging entrance cone	(*a*) to obtain accelerated flow
(*ii*) cylindrical portion of short length called throat	(*b*) to convert kinetic energy into pressure energy
(*iii*) diverging portion called diffuser	(*c*) to obtain high velocity.

25. The coefficient of discharge C_d of a venturimeter takes which of the following items into account :
(*a*) roughness of the surface
(*b*) non-uniform velocity distributions at inlet and throat sections
(*c*) Reynolds number of flow (*d*) discharge
(*e*) length of throat (*f*) diameter of throat
(*g*) diameter ratio.
Select your answer from the following options :
(1) *a*, *b*, *d e* (2) *a*, *d*, *e*, *f* (3) *a*, *b*, *c*, *g* (4) *a*, *b*, *g*, *e*

26. The coefficient of discharge C_d of a venturimeter lies within the limits :
(*i*) 0.7 – 0.9 (*ii*) 0.6 – 0.8 (*iii*) 0.75 – 0.95 (*iv*) 0.95 – 0.99

27. In 1643, Torricelli experimentally demonstrated that the velocity of jet flowing out of a small opening (called orifice) depends, on the head of liquid measured above the centre of orifice. The Torricelli theorem gives the velocity of jet as :
(*a*) $\sqrt{gh}$ (*b*) $\sqrt{2gh}$ (*c*) $\sqrt{2gh/3}$ (*d*) $\sqrt{3gh^2/4}$

28. Cavitation in fluid flow occurs when :
(*i*) the total energy suddenly increases (*ii*) the total energy decreases suddenly
(*iii*) the pressure of flow decreases to a value close to its vapour pressure
(*iv*) the velocity head reduces to zero

29. When cavitation occurs in hydraulic structures and hydraulic machines, it is indicated by :

(*a*) formation of large number of air bubbles, their transport to zones of higher pressure where they suddenly collapse,

(*b*) increase in efficiency
(*c*) long life of the structure/machine
(*d*) noise and vibrations
(*e*) physical damage to the surface
(*f*) trouble free operation
(*g*) decrease in efficiency.

Select your answer from the following options :

(1) *a*, *b*, *c*, *d*. (2) *a*, *b*, *e*, *f*. (3) *a*, *d*, *e*, *g*. (4) *a*, *b*, *d*, *e*.

30. The derivation of momentum equation $F_x = \rho Q[(V_2)_x - (V_1)_x]$ is based on certain assumptions. Select the correct ones from the following :

(*i*) steady flow
(*ii*) uniform flow
(*iii*) velocity constant over the inlet and outlet cross-sections
(*iv*) irritation flow
(*v*) compressible fluid.

Select your answer from the following options :

1. (*i*), (*ii*) 2. (*i*), (*iv*) 3. (*i*), (*iii*) 4. (*iv*), (*v*)

31. While applying the momentum equation the net force that acts over the control volume is required to be evaluated. Which of the following forces are generally considered in determining the net force :

(*a*) gravity force
(*b*) viscous force
(*c*) pressure force
(*d*) boundary resistance
(*e*) elastic force
(*f*) surface tension force

Select your answer from the following options :

(1) *a*, *b*, *c* (2) *a*, *e*, *f* (3) *a*, *d*, *e* (4) *a*, *c*, *d*

32. The momentum correction factor β is used to account for :

(*i*) change in pressure
(*ii*) change in mass rate of flow
(*iii*) non-uniform distribution of velocities at inlet and outlet sections
(*iv*) change in total energy

33. The momentum correction factor β, being dependent on the velocity distribution, is greatly influenced by the type of flow. For the following category of flow, choose the corresponding value of β from the values given alongside :

(*a*) Laminar flow (*i*) 1.01
(*b*) Turbulent flow (*ii*) 1.05
(*c*) Highly turbulent flow (*iii*) 1.33

34. The analysis of the following cases of fluid dynamic problems, besides making use of the continuity equation certain other equations are used. Identify these equations for each of the flow cases :

Column A	*Column B*
(*I*) Flow through a venturimeter	(*a*) Momentum equation
(*II*) Force caused by a jet striking against a plane surface	(*b*) Bernoulli's equation
(*III*) Force exerted on a pipe bend	(*c*) Momentum principle and Bernoulli equation
(*IV*) Force exerted by the flow on a fire nozzle	(*d*) Moment of momentum principle
(*V*) Rotation of a lawn sprinkler	(*e*) Momentum equation.

35. The change in moment of momentum of fluid due to flow along a curved path results in :
(*i*) a dynamic force which passes through the centre of curvature
(*ii*) a torque
(*iii*) a change in pressure
(*iv*) a change in kinetic energy of jet

ANSWERS TO OBJECTIVE TYPE QUESTIONS

1. (*c*), (*d*). **2.** (*a*) (*v*) ; (*b*) (*iv*) ; (*c*) (*iii*) ; (*d*) (*ii*) ; (*e*) (*i*). **3.** (*a*) (*iii*) ; (*b*) (*i*); (*c*) (*ii*).
4. (*iii*). **5.** (*b*), (*c*), (*f*), (*g*). **6.** (*iii*). **7.** (*a*) (*ii*) ; (*b*) (*iii*) ; (*c*) (*i*).
8. (*ii*). **9.** (*c*). **10.** (*ii*). **11.** (*c*). **12.** (*iii*)
13. (*a*) (*ii*) ; (*b*) (*iii*) ; (*c*) (*i*) ; (*d*) (*v*) ; (*e*) (*iv*). **14.** (*ii*). **15.** (*d*). **16.** (*iii*).
17. (*c*). **18.** (*i*) (*c*) ; (*ii*) (*d*) (*iii*) (*a*) (*iv*) (*b*) **19.** (*c*) **20.** (*ii*)
21. (*c*) **22.** (*iii*) **23.** (*c*) **24.** (*i*) (*a*) (*ii*) (*c*) (*iii*) (*b*) **25.** (3).
26. (*iv*). **27.** (*b*). **28.** (*iii*). **29.** 3. **30.** 3. **31.** 4.
32. (*iii*). **33.** (*a*) (*iii*) ; (*b*) (*ii*) ; (*c*) (*i*).
34. (*I*) (*b*) ; (*II*) (*a*) ; (*III*) (*c*) ; (*IV*) (*e*) ; (*V*) (*d*). **35.** (*ii*).

REVIEW QUESTIONS

1. What simplifying assumptions are involved in the one-dimesnsional method of flow analysis ? Under what circumstances will each of these assumptions lead to appreciable error ?

2. What are the limitations of the Bernoulli's equation ?

3. How is the Bernoulli's equation related to the first law of thermodynamics ?

4. Enumerate various forms of energies which the Bernoulli's equation takes into account.

5. In determining the pressure distribution from the flow net, why is it necessary to distinguish between the steady and unsteady flow ?

6. "Bernoulli's equation is a special case of the generalised energy equation." Comment on the validity of this statement.

7. Which form of the Bernoulli's equation will be more suitable for application to flow of gases ?

8. A pipe carries a liquid of sp. wt. γ at a flow rate of Q m^3/s. If at a section, the datum head, the pressure and the velocity are Z, p and V respectively, what is the total energy passing that section ?

9. Why does the concept of fluid head no physical significance in gas flow ?

10. Does the total energy of flow increase or decrease as the liquid passes through a pump driven by a motor ? If the motor is suddenly stopped due to power failure before a valve on the delivery line could be closed, what would happen to the liquid ?

11. Does the steady flow through a converging pipe accelerate or decelerate ? Explain.

12. Why has the fluid weight no influence upon the flow pattern, if the fluid is confined by fixed boundaries.

13. For what type of flow will the stagnation opening of a pitot-tube record the same reading at all points at a flow-section ?

14. If a pitot-tube is directed against a flowing liquid, what quantity is represented by the liquid level that rises in the tube (measured above any given datum plane) ?

15. What factors are responsible for causing non-uniformity in the velocity distribution at a flow-section in case of real fluid flow ?

16. So for as energy considerations are concerned, how far are we justified in assuming the fluid as an ideal one ?

17. What do you understand by,
(*a*) stagnation pressure and (*b*) dynamic pressure.

18. To what maximum height, will a jet of liquid (sp. wt. γ) with a velocity V rise above nozzle base ? The inclination of nozzle can be changed at will to achieve the greater vertical rise.

19. What conditions lead to cavitation ? Is cavitation possible in a free-surface flow ?

20. Why does the zone of pitting (damaged region of the solid boundary due to cavitation) not coincide with section of minimum flow area ?

21. What precautions should normally be taken to obtain caviation-free-flow ?

22. Describe the nature of various forces normally included in the momentum equation.

23. A fireman holds a nozzle of fire-hose for extinguishing the fire. Will the nozzle be pushed toward him or away from him ? Explain.

24. Explain which type of flow problems are analysed with the help of moments of momentum equation.

25. If the flow direction in a venturimeter is reversed what objections would you have against its use as an efficient flow measuring device ?

26. Briefly comment on the conditions under which energy theorem and momentum theorem are applicable, and state what type of information can be obtained from each.

27. A taxi-driver when asked about the petrol consumed by the car in travelling a certain distance tells from his experience that when the car is driven faster it consumes more petrol as compared to covering the same distance at a slower speed. As a student of mechanics, how will you explain it in terms of power ?

28. Explain how the modern rockets are launched ? How is the tremendous thrust required to send it into a pre-determined trajectory achieved ?

29. What basic principles of fluid flow are essential to determine the power developed by a turbine or a centrifugal pump ?

30. Which of the following forces are accounted for in (*a*) Reynolds equation of motion (*b*) Navier Stokes equation and (*c*) Euler's equation

(*i*) gravity force (*ii*) pressure force
(*iii*) viscous force (*iv*) inertia force
(*v*) force due to turbulence (*vi*) force due to compressibility
(*vii*) surface tension force.

31. What quantities are measured by :

(*i*) a pitot-tube (*ii*) a venturimeter
(*iii*) a piezometer (*iv*) a manometer
(*v*) a combination of piezometer and pitot-tube.

32. Name the three forces considered in the derivation of Euler's equation of motion.

33. Euler's equation of motion is based on certain assumptions ? Identify these :

(*i*) motion along a streamline, incompressible fluid and unsteady flow ;
(*ii*) motion along a streamline, frictionless fluid and steady flow ;
(*iii*) frictionless, incompressible fluid and unsteady flow ;
(*iv*) frictionless, incompressible fluid and steady flow ;
(*v*) frictionless fluid only.

34. Select the correct statement :

(*i*) Bernoulli's equation relates various forces involved in a flow problem ;
(*ii*) Bernoulli's equation relates different forms of energy ;
(*iii*) Bernoulli's equation expresses the rate of change of momentum.

35. The following form of Bernoulli's equation expresses the total energy in either per unit of mass, weight or volume. Identify the proper units associated with each of the equation :

(*i*) $gZ + \frac{p}{\rho} + \frac{V^2}{2}$ = const. (*ii*) $Z + \frac{p}{\gamma} + \frac{V^2}{2g}$ = const.

(*iii*) $\gamma Z + p + \rho \frac{V^2}{2}$ = const.

36. Which of the following expression represents the total kinetic energy passing a flow section

(*i*) $\frac{V^2}{2g}$ (*ii*) $\rho \frac{V^2}{2}$ (*iii*) $\gamma Q \frac{V^2}{2g}$ (*iv*) $\frac{V^2}{2}$

in which Q is the flowrate and V the average velocity.

37. At a certain location in a flow system a hydraulic machine is installed. It is observed that the energy of flow is considerably decreased downstream of the machine. Identify the machine :

(*i*) pump (*ii*) turbine.

38. A pump with 15 cm suction pipe and 12 cm delivery pipe has an efficiency of 80% and is driven by a 11 kW motor. A differential manometer connected to suction and delivery sides show a maximum pressure head of 9.15 m of water. Predict the discharge of water through the pump.

39. Which of the following statement is correct in case of pipe flow :

(*i*) flow takes place from higher elevation to lower elevation ;

(*ii*) flow takes place from higher pressure to lower pressure ;

(*iii*) flow takes place from higher energy to lower energy ;

(*iv*) flow takes place from higher velocity to lower velocity.

PROBLEMS

4.1. A pipe carrying water tapers from cross-section of 0.3 m^2 at A to 0.14 m^2 at B. The average velocity at A is 1.8 m/s and the pressure 441 kN/m^2 (gauge). If frictional effects are negligible, determine the pressure at B which is 5.5 m above the level of A.

4.2. In an open rectangular channel the velocity varies linearly with depth, value at the free surface being twice that at the channel bottom. Show that the kinetic energy correction factor is 10/9.

4.3. A long bridge with piers 1.30 m. wide having a centre to centre spacing of 7 m spans a river. The depth of water upstream the bridge is 1.50 m and that between the piers under the bridge is 1.35 m. Calculate the flow passing through one span (*i.e.* between two adjacent piers), assuming that the river bed is horizontal and that frictional effects are negligible.

4.4. A drum of liquid having a density of 902 N per cubic metre is rotated about a vertical axis at a constant speed of 250 r.p.m. Determine the accelerative force per unit volume at a radial distance of 0 and 1 m from the axis.

4.5. What force would have to be exerted by a piston in a 25 cm pipe to accelerate a 15 m column of water at a rate of 1 m/s^2 ? (Take ρ = 1000 N/m^3).

4.6. A slender tube which is 50 cm long and closed at one end is half full of mercury. If the tube is rotated about the open end in a horizontal plane at the constant rate of 180 r.p.m., what pressure intensity will prevail at the closed end ?

4.7. A pitot-tube is mounted on a airplane to indicate the relative speed of the plane. What differential pressure intensity in N/m^2 will the instrument register when the plane is travelling at a speed of 200 km/hr in a wind blowing at 60 km/hr against the direction of motion of the plane. Take specific weight of air as 12.74 N/m^3.

4.8. A vertical tank 5 m in diameter and 20 m high is filled with water. How mush work done could be obtained from this water if run through a 100% efficient turbine that discharges into a reservoir situated 10 m below the bottom of the tank.

4.9. Water is flowing through a 50 cm diameter pipe at the rate of 1.5 m^3/sec. What is the kinetic energy of flow in N-m/s ?

4.10. Determine the height to which a vertical jet of water, with velocity 30 m/s will rise. Neglect the air resistance.

4.11. Water is discharged from a 2.4 cm horizontal nozzle held 1 m above the horizontal floor. It strikes the floor at a horizontal distance of 4 m from the end of the nozzle. Determine the rate of flow neglecting the resistance.

4.12. Determine the rate of flow passing through the orifice for the conditions shown in the following figure. The liquid level and pressure above it may be assumed constant. The atmospheric pressure may be taken as 100.7 kN/m^2.

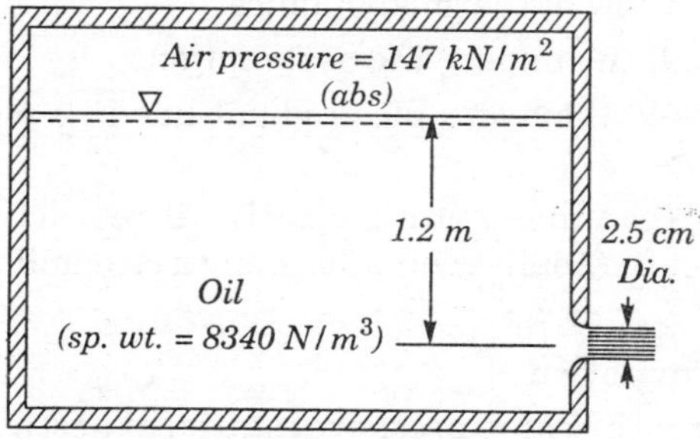

Fig. Prob. 4.12.

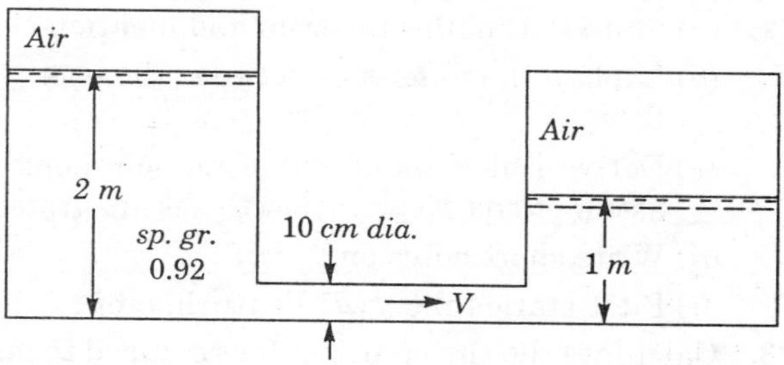

Fig. Prob. 4.13.

4.13. Determine the discharge of a liquid of sp. gr. 0.92 through the pipe 10 cm in diameter connecting tanks under conditions shown in the accompanying figure. Neglect the losses in the pipe and assuming the tank as having large cross-section area. The air pressure in the left tank is 490.5 bar and that in the right one is 1177 bar.

4.14. A tank having water filled upto 5 m height above an orifice opening of 10 cm diameter discharges a jet of water into the atmosphere. The pitot-tube placed in the jet just downstream of the orifice records a head of 4.80 m. Calculate the rate of flow, the loss of energy in N-m/N and also in watt.

4.15. An oil of sp. gr. 0.84 flows through a uniform diameter pipe at the rate of 375 lit/sec. The energy head losses are 25 N-m/N per km length of the pipe. Find the slope of the hydraulic grade and total energy lines and the power lost per km of pipe.

4.16. A horizontal 20 cm diameter pipe carries oil of sp. gr. 0.84 at the rate of 150 lit/sec and a pressure of 249 kN/m^2. At a section downstream its diameter reduces to 15 cm, calculate the pressure in the 15 cm section.

4.17. Water flows through the pipeline as shown. The velocity in the 10 cm section is 2.2 m/s, and the level of water in the open tube at B is 0.40 m. If the point B in the 6 cm section is 0.80 in above A, calculate the level of water in the open tube above point A.

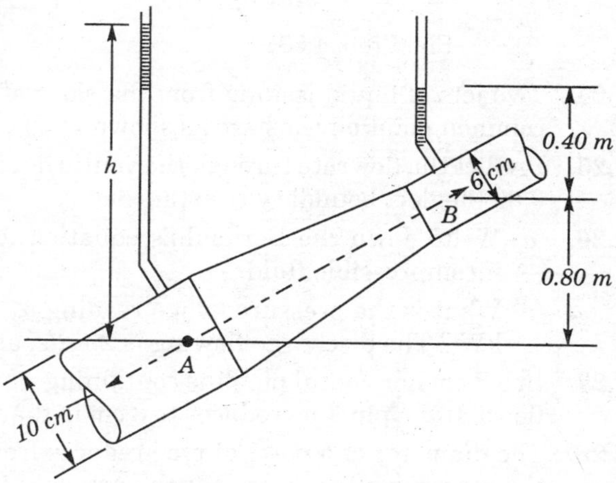

Fig. Prob. 4.17

4.18. A siphon consisting of 2 cm diameter flexible tube is used to drain water from a tank. The outlet of the tube is 2 m below the water surface and the bend in the flexible tube is 1 m above the water surface. Calculate the flow rate and the pressure in the bend.

4.19. A jet of water falling vertically has a velocity 25 m/s and a diameter of 5 cm at an elevation of 20 m. What is its velocity and jet diameter at elevation of 10 m ?

4.20. A jet of water has a velocity of 8.8 m/s and is inclined at an angle of 60° to the horizontal as it leaves a nozzle. Determine (*a*) initial radius of curvature of the jet (*b*) the maximum elevation which the jet will attain and (*c*) the radius of curvature of the jet at its highest point.

4.21. Write Bernoulli's equation in the form of energy per unit mass, energy per unit weight and energy per unit volume. Indicate the meaning of each term of the equation and discuss its limitations. Compare Bernoulli's equation with first law of thermodynamics.

4.22. (*a*) State Bernoulli's theorem and mention clearly all the underlying assumptions.

(*b*) Explain the difference between the energy line and the hydraulic gradient line in a pipe flow.

(*c*) Derive Euler's equation of motion along the streamline, stating clearly all restrictive assumptions. Explain how this is integrated to get Bernoulli's equation along a streamline.

(*d*) Write short notes on :

(*i*) Pitot-static tube ; (*ii*) Venturimeter ; (*iii*) Cavitation.

4.23. Calculate the throat diameter required to maintain the same pressure (static) at sections (1) and (2) of the vertical venturimeter shown in the figure. The velocity at section (1) is 7 m/s.

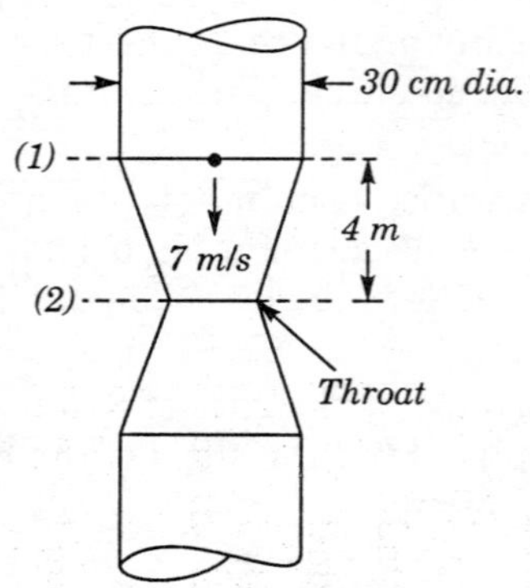

Fig. Prob. 4.23

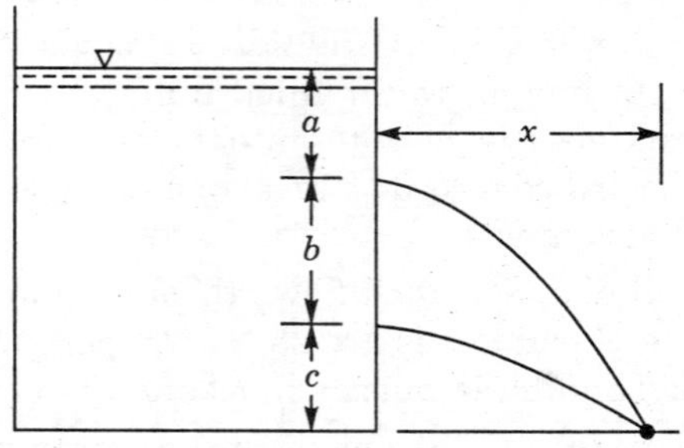

Fig. prob. 4.24

4.24. Two jets of liquid issuing from the sides of a tank under a constant free surface meet at a common point on the base as shown in the figure. Find the relationship between *a* and *c*.

4.25. Predict the flowrate through the venturimeter as shown, carrying gasoline of specific gravity 0.80 (neglect boundary resistance).

4.26. (*a*) Write down the Bernoulli's equation for steady and unsteady, irrotational flow of an incompressible fluid.

(*b*) What is the pressure gauge reading p_2, in the figure shown if the turbine develops 441 kW ? The discharge flowing is 2 m^3/s, and pressure at point (1) 785 bar.

4.27. In a 8 cm horizontal pipeline containing water at a pressure of 55.3 kN/m^2, 2 lit/sec of water flows. If the pipeline reduces to 3 cm in diameter, calculate the pressure in the 3 cm section.

4.28. The diameter of a vertical pipe reduces from 30 cm to *d* as shown in the figure. Determine the pipe diameter, *d*, so that the pressure in the two gauges is same.

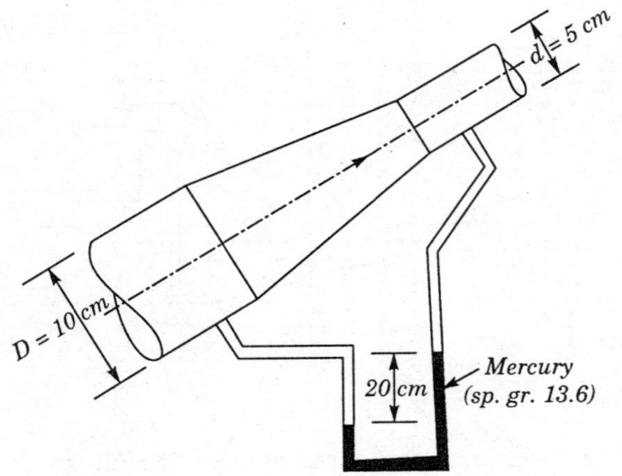

Fig. Prob. 4.25.

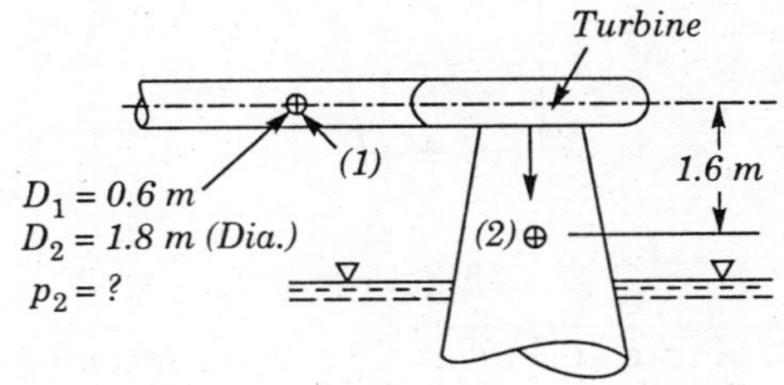

Fig. Prob. 4.26.

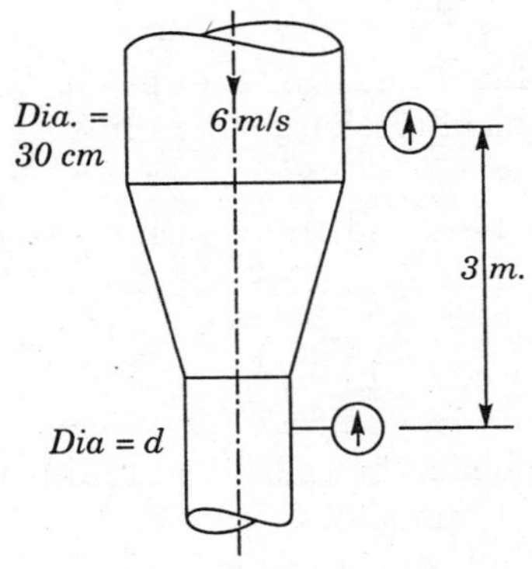

Fig. Prob. 4.28.

4.29. A conical pipe has diameters at the two ends of 0.5 m and 1.5 and is 15 metres long. It is vertical and the friction loss is 2.5 metres for flow of water in either direction, when the velocity at the smaller section is 10 m/s. If the smaller section is at the top and the pressure head there is 2 metre of water, find the pressure head at the lower end when the flow is (*a*) downwards (*b*) upwards.

4.30. Water flows through the turbine (shown in the accompanying figure) at the rate is 0.214 cu metre per sec and the pressures at *A* and *B* are 147 kN/m^2 and – 34.6 kN/m^2 respectively. Determine the power delivered to the turbine by the water.

4.31. Water is pumped from reservoir *A* at elevation 228.00 (metres) to reservoir *E* at elevation 244.00 (metres) through a 0.30 metre pipeline. The pressure in the 0.3 metre pipeline at point *D* at elevation 198.00 (metres) is 533.3 kN/m^2. The lost heads are :

A to pump section *B* = 0.70 metre
Through pump *C* to *D* = 38 $V^2/2g$
and *D* to *E* = 40 $V^2/2g$

Find the discharge *Q* and the power supplied by the pump *BC*.

4.32. The velocity of water in a pipe was measured with a pitot-tube consisting of a bent tube with orifice facing the direction of flow and the other orifice perpendicular to the first orifice. The difference of head at the centre of pipe was 10 cm of water. If the mean velocity of water is 2/3 of velocity at the centre, find the quantity of water flowing per minute. The pipe diameter is 25 cm.

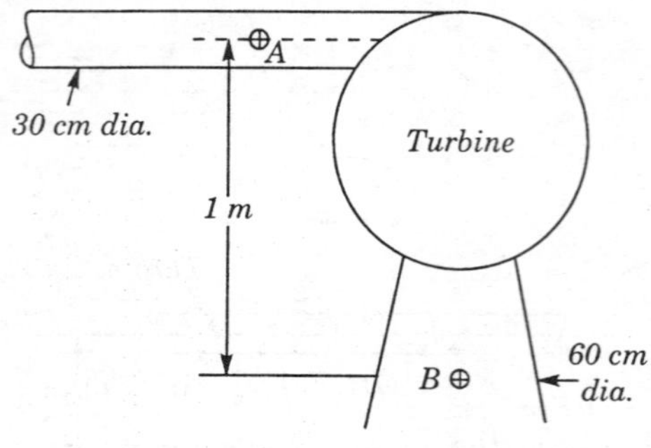

Fig. Prob. 4.30.

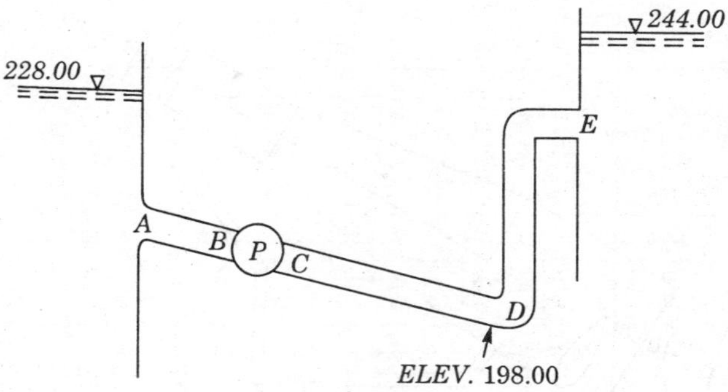

Fig. Prob. 4.31.

4.33. For a turbulent flow of an incompressible fluid through a circular pipe of radius r_0, the velocity distribution is given by

$$u = u_0 \left(1 - \frac{r}{r^0}\right)^{1/7}$$

where u is the velocity at radius r and u_0 is the velocity at the pipe axis. Calculate the total kinetic energy of flow in terms of discharge Q, radius r_0, specific weight γ and acceleration due to gravity g.

4.34. Oil flows through a circular pipe and the velocity distribution across the pipe section is given by

$$v = v_{max} \left(1 - \frac{r^2}{r_0^2}\right),$$

where r_0 is the pipe radius and v is the velocity at radius r from the pipe axis. Calculate the energy and momentum correction factors.

4.35. Water leaves the bucket of a high spillway at an angle of 45° and a velocity of 15 m/s. Determine the maximum height to which the water sheet will rise, and the location of the point at which it will again reach its initial elevation at the bucket level.

4.36. Liquid is discharged from a nozzle with an initial velocity of 10 m/s. At what two nozzle inclinations will the liquid strike an object situated at a horizontal distance of 8 m from the nozzle.

4.37. An air duct changes its diameter from 1.5 to 1.0 m. What rate of flow would be indicated by a head of 1 cm of water in the U-tube connected to the two sections ? Take ρ for air as 1.283 kg/m^3.

4.38. When the rate of flow of water through a pump is 80 lit/sec, a differential gauge connected to piezometer taps on either side of the pump indicates a 20 cm head of mercury. What power is delivered by the pump to the flow. The suction and the delivery pipes are 25 cm and 15 cm in diameter respectively. The difference in elevation between suction and delivery pipes where gauge is connected is 40 cm.

4.39. A blower delivers 350 cubic metre per minute of air having $\rho = 1.235$ kg/m^3, at an increase in pressure of 10 cm of water. It is 72% efficient. Determine the losses in the blower in N-m/N and also in watt, and determine the torque in the shaft if the blower turns at 1800 r.p.m.

4.40. When the momentum correction factor is unity, prove that the velocity must be uniform over the cross-section.

4.41. A water tank has an orifice in the bottom of the tank. The cross-sectional area of the jet where it leaves the orifice (at $x = 0$) is A_0. The height of water in the tank is h and this height is kept constant by continuously refilling the tank. The tank cross-section is much larger than the orifice area. Neglecting friction, surface tension etc., find the cross-sectional area A of the jet as a function of x. The distance x is measured vertically downward from the orifice.

4.42. A blunt symmetrical body is moving at a constant velocity V in the positive x-direction in an infinite frictionless incompressible fluid of density ρ, otherwise at rest. The fluid at the nose of the body must also move at velocity V in the same direction. What is the difference in pressure between the point on the nose and a point infinitely upstream where the velocity is zero.

4.43. In a pipe of 300 mm diameter, a venturimeter with throat diameter of 200 mm is installed to measure the flow of water. The pressure available at the inlet of the venturimeter is equivalent of 10 m of water above atmosphere. Determine the maximum allowable discharge through the venturimeter so that no negative pressure is created at the throat. Assume venturimeter coefficient as 0.98.

4.44. A pipe 12.5 cm in diameter is connected to a nozzle of 2.5 cm diameter by a flanged joint. If the nozzle discharges with a velocity of 43.2 m/s, find the magnitude and direction of the tension force in the flanged connections.

4.45. A suction device is arranged as shown. Find the flow rate through the main pipe at the instant at which the suction will begin for the following data :

$$A_2 = 4A_1, \quad A_1 = 4 \text{ cm}^2,$$
$$h = 92 \text{ cm}, \quad a = 61 \text{ cm},$$
$$g = 9.81 \text{ m/s}^2.$$

The fluid is incompressible and friction may be neglected. The external atmospheric pressure is uniform throughout the jet. Give the answer in litre/sec.

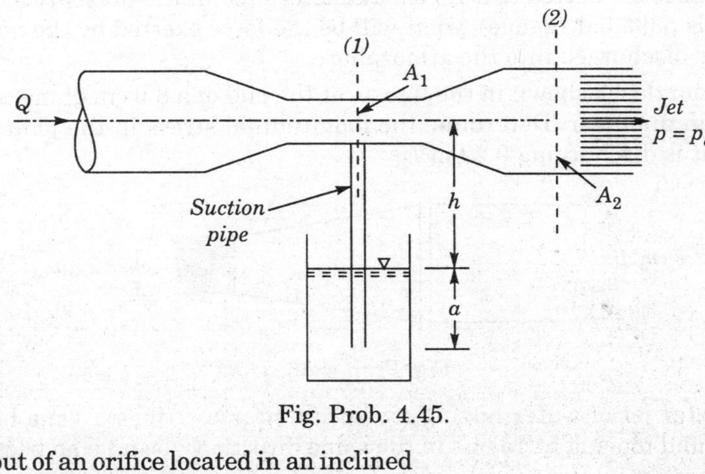

Fig. Prob. 4.45.

4.46. A jet flows out of an orifice located in an inclined plate as shown below. Calculate the maximum rise Y in terms of H and α. Neglect losses. Also find the corresponding horizontal distance.

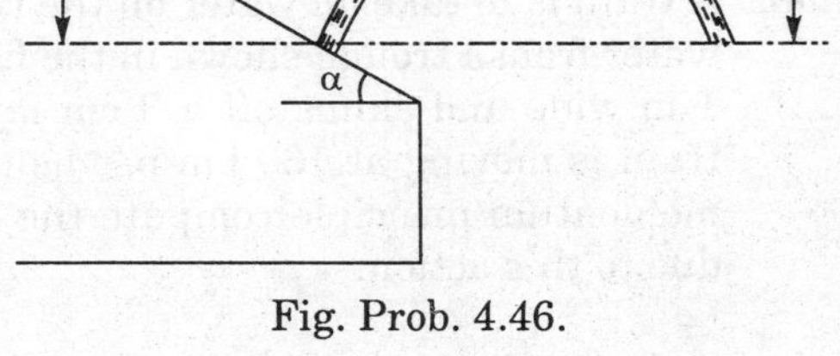

Fig. Prob. 4.46.

4.47. Show that the locus of the highest points of the jet of Problem 4.46 is given by

$$x^2 = 4Y(H - Y).$$

4.48. A jet of water 3 cm in diameter strikes a flat plate at an angle of 20° to the normal of the plate with a velocity of 40 m/s. Determine the velocity of the plate moving parallel to itself, if the normal force exerted by the jet 11.77 N.

4.49. A rectangular plate 10 cm wide 25 cm long hangs vertically hinges at its top edge. A jet of water 3 cm in diameter with a velocity 10 m/s strikes the plate at its centre. Determine the weight of the plate if the jet keeps it steadily inclined at 30° to the vertical.

4.50. (*a*) Define energy correction factor α and momentum correction factor β. What is their importance ? How important are they in turbulent flow ?

(*b*) In open channel flow where the velocity distribution is given by

$$\frac{u}{U} = \left(\frac{y}{D}\right)^{1/4}$$

Determine α and β.

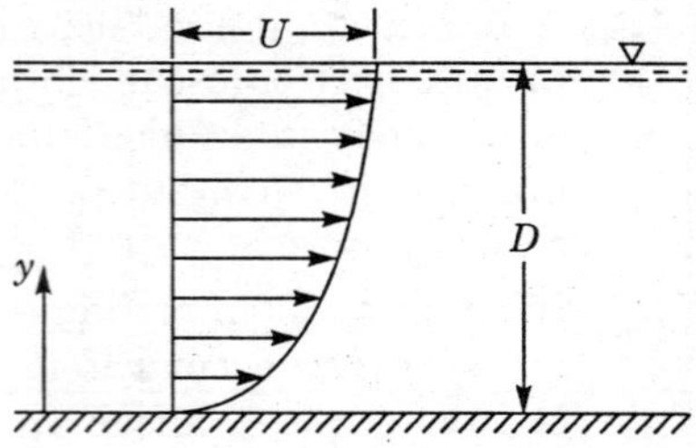

Fig. Prob. 4.50.

4.51. Water under a pressure of 415 kN/m^2 is flowing through a 23 cm diameter pipe at the rate of 0.30 cubic metre/sec. If the pipe is bent by 135° in the horizontal plane, find the magnitude and direction of the resultant force on the band.

4.52. If the sheet of liquid discharged at the rate q_0 from a long slot strikes plane boundary at an angle of 45° what will be the ratio q_1/q_2 for the divided flow. Assume no energy loss.

4.53. An 8 cm diameter jet of water having a velocity of 15 m/s is deflected through angle of 120° by a stationary curved plate. Determine the longitudinal and transverse components of force exerted upon the plate.

4.54. A pipe carrying 0.15 cu metre/sec of oil of specific gravity 0.80 contracts from 30 cm diameter to 15 cm diameter gradually. The pressure in the 30 cm section is 69.1 kN/m^2, determine the force acting on the contraction.

4.55. A 2.5 cm nozzle is connected to a 7.5 cm diameter pipe. If the pressure at the upstream end of the nozzle is 5395 bar (gauge), what will be the force exerted by the nozzle on the water, if the water is discharged into the atmosphere.

4.56. A horizontal nozzle, as shown in the figure, at the end of a 8.0 cm diameter hose produces a jet of 4.0 cm in diameter. Determine the longitudinal stress in the joint at the base of the nozzle when it is discharging 0.22 m^3/s.

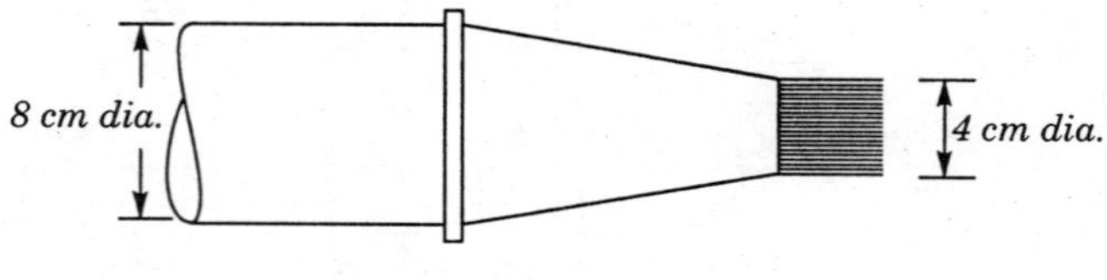

Fig. Prob. 4.56.

4.57. A 5 cm diameter jet of water having a velocity 40 m/s strikes a vane having a deflection angle of 150° and moving at 15 m/s in the same direction. Assume no friction and compute :

(*a*) the absolute velocity of jet when it leaves the vane,

(*b*) components of the force exerted by the fluid on the vane, and

(*c*) the power developed.

4.58. A train is to take on water on the run by scooping up water from a trough shown in the figure. The scoop is 1 m wide and skims off a 3 cm layer of water. The train is moving at 160 km per hour. Using impulse-momentum principle compute the drag on the train due to this action.

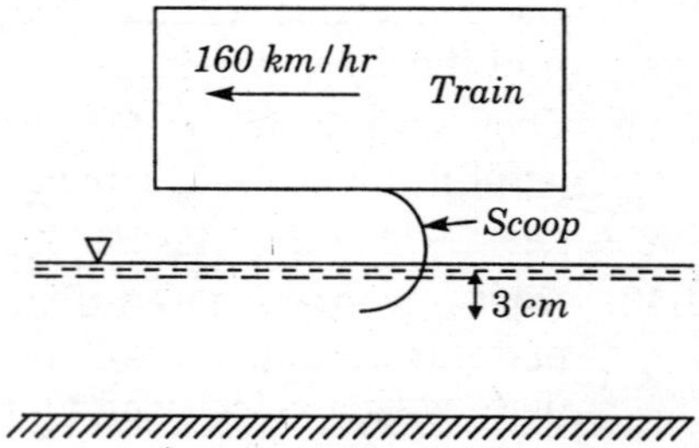

Fig. Prob. 4.58.

4.59. For turbulent flow of an incompressible fluid through a circular pipe of radius a, the velocity distribution is given $u = u_0 (1 - r/a)^{1/7}$, where u is the velocity at radius r and u_0 is the velocity on the axis. Determine the value of the K.E. correction factor α.

4.60. A boat travelling at 50 km/hr has a 60 cm propeller discharging 4.5 m³/s through its blades. Calculate (1) the thrust on the boat (2) the efficiency of the propulsion system, and (3) the power input to the propeller.

4.61. A jet propelled airplane travelling at 900 km/hr takes 78.5 N/s of air and discharges at 1800 m/s relative to the airplane. Neglecting the weight of the fuel, determine the thrust developed and the corresponding power.

4.62. A jet-propelled aeroplane travelling at 1000 km/hr takes in 73.6 N/s of air and uses 9.8 N of fuel for each 49 N of air. Determine the thrust developed when the exhaust gases have an absolute velocity of 1500 m/s, and what is the theoretical mechanical efficiency of the jet engine ?

4.63. A motor boat moves upstream in a river at a velocity of 10 m/s (absolute). The river flows at a velocity of 1.5 m/s. The boat is powered by a jet-propulsion unit which takes in water at the bow and discharges it at the stern (below the water surface). The jet velocity is observed to be 24 m/s. If the flowrate is 0.15 m³/s, determine the propulsive force and the power developed.

4.64. A propeller must produce 9.81 kN thrust to drive an aeroplane at 260 km/hr. An ideal propeller of what size must be provided if it is to operate at 85% efficiency, density of air is 1.284 kg/m³.

4.65. For a frictionless shaft of the symmetrical sprinkler shown in the figure, determine its speed in r.p.m. if the total flow is 1 lit/sec and nozzle tips are 0.5 cm in diameter.

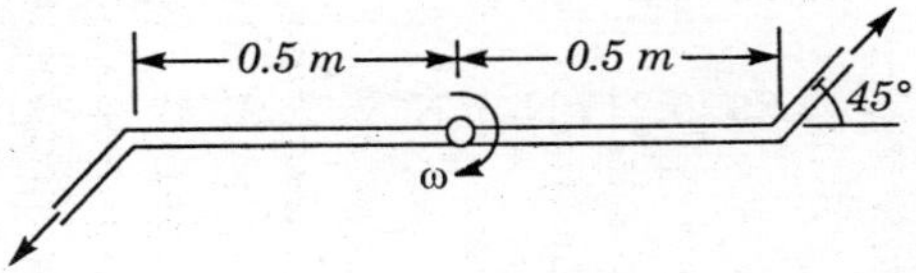

Fig. Prob. 4.65.

4.66. The sprinkler shown in figure has three nozzles as shown and the fluid passing through each of them is equal. For the relative velocities of jet shown, and a frictionless shaft, determine the speed of rotation in r.p.m.

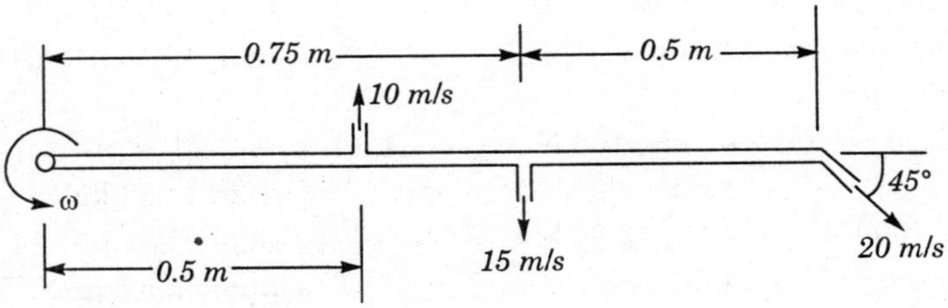

Fig. Prob. 4.66.

4.67. A water sprinkler consists of 1 cm jet at either end of a rotating arm as shown in the sketch. What torque must be applied to hold it stationary when the velocity of jet is 6 m/s.

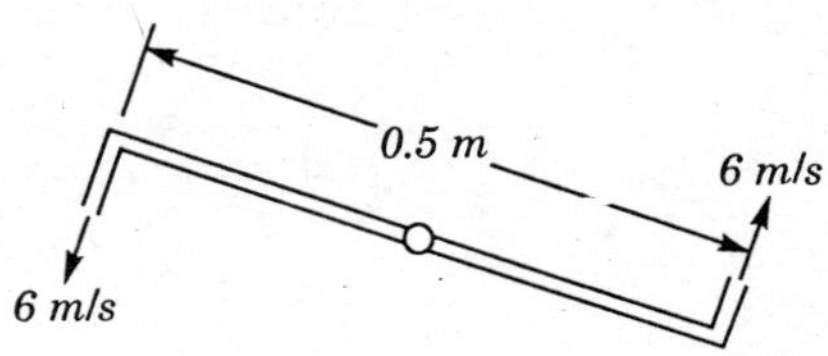

Fig. Prob. 4.67.

4.68. Compute (*i*) α-coefficient (*ii*) β-coefficient (*iii*) the depth at which average velocity occurs for a velocity distribution of

$$\frac{v}{V_m} = \left(\frac{y}{d}\right)^{2/5}$$

where d is the maximum depth and v_m is the velocity at that depth. Assume the flow takes place in a very wide rectangular channel.

4.69. Obtain an equation for flowrate for an orifice discharging into atmosphere from a closed vessel under pressure.

4.70. Determine the energy correction factor for a fully developed laminar flow through a circular pipe.

ANSWERS

4.1. 382 kN/m^2 **4.3.** 19.36 m^3/s **4.4.** 0.617 kN
4.5. 737 N **4.6.** – 901 kN/m^2. **4.7.** 12.12 N/m^2
4.8. 77 MN-m **4.9.** 42.9 kN-m/s **4.10.** 45.8 m
4.11. 4.35 lit/s **4.12.** 5.42 lit/s **4.13.** 89.25 lit/s
4.14. 76.2 lit/s, 0.2 N-m/N, 149.2 W **4.15.** 1 : 40, 77.15 kW
4.16. 228.5 kN/m^2 **4.17.** 2.853 m **4.18.** 1.967 lit/s, – 29.4 kN/m^2
4.20. 13.04 m, 2.446 m, 1.63 m **4.23.** 25.2 cm
4.24. $a = c$ **4.25.** 4.015 lit/s **4.26.** – 101.4 kN/m^2
4.27. 51 kN/m^2 **4.28.** 23.4 cm
4.29. (*a*) 19.537 m of water (*b*) 24.537 m of water **4.30.** 41.6 kW
4.31. 0.162 m^3/s, 59.7 kW **4.32.** 2.75 m^3/minute **4.33.** $\frac{0.53\gamma Q^3}{g\pi r_0^4}$
4.34. $\alpha = 2$, $\beta = 4/3$ **4.35.** 5.74 m, 22.95 m **4.36.** 25.85°, 64.15°
4.37. 12.43 m^3/s **4.38.** 2.69 kW
4.39. 0.317 N-m/N, 2.28 W, 42.6 × 10^{-3} N-m **4.41.** $A = A_0\sqrt{\frac{h}{h+x}}$
4.42. $\Delta p = \rho\,\frac{V^2}{2}$ **4.43.** 477.5 lit/s **4.44.** 877.5 N against flow
4.45. 1.76 lit/s **4.46.** $Y = H\cos^2\alpha$, $X = H\sin^2\alpha$
4.48. 5.85 m/s **4.49.** 141.7 N **4.50.** $\alpha = 125/112$, $\beta = 50/48$
4.51. 36 kN, 22.55° with horizontal **4.53.** 1.697 kN, 981 N
4.54. 3.38 kN acting towards right **4.55.** 1.815 kN acting towards left
4.56. 357.5 N acting towards right
4.57. (*a*) 17.87 m/s, (*b*) 2.09 kN, 165 N, (*c*) 31.3 kW
4.58. 59.25 kN **4.59.** 1.06 **4.60.** 14.32 kN, 89.6%, 202 kN
4.61. 12.4 kN, 3100 kW **4.62.** 13.86 kN, 27.6% **4.63.** 1275.3 N, 14.65 kW
4.64. 2.12 m **4.65.** 343 r.p.m. **4.66.** 96 r.p.m.
4.67. 0.237 N-m **4.68.** (*i*) $\alpha = 1.248$, (*ii*) $\beta = 1.09$, (*iii*) $y = 0.431\,d$
4.69. $Q = c_d \cdot a \cdot \sqrt{2g(H + \Delta p/\gamma)}$ **4.70.** 2.0.

5

Dimensional Analysis and Principles of Model Testing

5.0. Introduction

Dimensional analysis is a method which describes a natural phenomenon by a dimensionally correct equation among certain variables which affect the phenomenon. It reduces the number of variables and arranges them into dimensionless groups. Dimensional analysis finds its applications in nearly all fields of engineering, and is most extensively used in fluid mechanics and heat transfer. It has become an important tool for analysing fluid flow problems and is specially useful in presenting experimental results in a concise form. It is based on the principle of dimensional homogeneity and uses the dimensions of relevant variables affecting the phenomenon.

5.1. Fundamental Dimensions

The magnitudes of such quantities as mass, length, force, acceleration and the temperature are expressed differently in different units of measurement. The mass, length, time and the temperature are independent of each other and their units of measurement are prescribed by international standards. These four quantities may be regarded as the *fundamental quantities* and the units of all other quantities may be determined either from their definitions or from the physical laws describing them. In place of mass, the force is also sometimes considered as a fundamental quantity. The force and the mass are related by the Newton's law $F = m.a,$ and when the units of force, length and time are known, the unit of mass may be derived by this relationship.

The fundamental quantities are described by symbols such as (M) for mass, (L) for length, (T) for time, (F) for force and (θ) for the temperature. The symbols are known as the dimensions of the respective quantities. According to Langhaar, *"They (dimensions) are a code for telling us how the numerical value of a quantity changes when basic units of measurement are subjected to prescribed changes"*. When mass is taken as a fundamental quantity, the set of fundamental dimensions are (M), (L), (T) and (θ). When the mass dimension is replaced by the force dimension, the Newton's second law gives

$$\text{Force} = \text{Mass} \times \text{Acceleration}$$

Writing in terms of dimensions,

$$(F = (M)\,(LT^{-2}) \qquad \text{...(5.1)}$$

resulting in,

$$(M) = (FL^{-1}T^{2})$$

A quantity may either be expressed dimensionally in (M)-(L)-(T) system or (F)-(L)-(T) system. In thermodynamics and heat transfer problems, the temperature also comes into picture and we have the fourth fundamental dimension (θ) for the temperature. The following table gives the dimensions of various quantities in both the systems :

Table 5.1. Quantities used in Fluid Mechanics and Heat Transfer and their Dimensions

S. No.	*Quantity*	*Nature of quantity*	*Dimensions in*	
			(M)-(L)-(T)	*(F)-(L)-(T)*
1.	Length, breadth, height and depth	Geometric	(L)	(L)
2.	Area	,,	(L^2)	(L^2)
3.	Volume	,,	(L^3)	(L^3)
4.	Moment of inertia of an area	,,	(L^4)	(L^4)
5.	Velocity, V	Kinematic	(LT^{-1})	(LT^{-1})
6.	Velocity potential, ϕ	,,	(L^2T^{-1})	(L^2T^{-1})
7.	Angular velocity, ω ; rotational speed, N	,,	(T^{-1})	(T^{-1})
8.	Stream function, ψ ; circulation, Γ	,,	(L^2T^{-1})	(L^2T^{-1})
9.	Acceleration, a	,,	(LT^{-2})	(LT^{-2})
10.	Vorticity, ξ	,,	(T^{-1})	(T^{-1})
11.	Rate of flow, Q	,,	(L^3T^{-1})	(L^3T^{-1})
12.	Kinematic viscosity, ν	,,	(L^2T^{-1})	(L^2T^{-1})
13.	Density, ρ	Dynamic	(ML^{-3})	$(FL^{-4}T^2)$
14.	Force, F	,,	(MLT^{-2})	(F)
15.	Pressure, p ; shear stress, τ	,,	$(ML^{-1}T^{-2})$	(FL^{-2})
16.	Specific weight, γ	Dynamic	$(ML^{-2}T^{-2})$	(FL^{-3})
17.	Dynamic viscosity, μ	,,	$(ML^{-1}T^{-1})$	$(FL^{-2}T)$
18.	Surface tension, σ	,,	(MT^{-2})	(FL^{-1})
19.	Bulk modulus of elasticity, K	,,	$(ML^{-1}T^{-2})$	(FL^{-2})
20.	Momentum	,,	(MLT^{-1})	(FT)
21.	Angular momentum or moment of momentum	,,	(ML^2T^{-1})	(FLT)
22.	Energy, work	,,	(ML^2T^{-2})	(FL)
23.	Torque	,,	(ML^2T^{-2})	(FL)
24.	Power	,,	(ML^2T^{-3})	(FLT^{-1})
25.	Moment of inertia of a mass	,,	(ML^2)	(FLT^2)
26.	Temperature	Thermodynamic	(θ)	(θ)
27.	Thermal conductivity	,,	$(MLT^{-3}\theta^{-1})$	$(FT^{-1}\theta^{-1})$
28.	Entropy	,,	$(ML^2T^{-2}\theta^{-1})$	$(FL\theta^{-1})$
29.	Enthalpy per unit mass	,,	(L^2T^{-2})	(L^2T^{-2})
30.	Gas constant	,,	$(L^2T^{-2}\theta^{-1})$	$(L^2T^{-2}\theta^{-1})$
31.	Heat transfer	,,	(ML^2T^{-3})	(FLT^{-1})
32.	Internal energy per unit mass	,,	(L^2T^{-2})	(L^2T^{-2})

The use of dimensions of a quantity in converting its unit from one system of measurement to another may be understood from the following example. The average velocity of flow in an open channel is given by $V = 110\sqrt{RS}$, in which V is the velocity in ft/sec., R is the hydraulic radius in ft and S is the channel slope (dimensionless). It is desired to modify this formula so that it yields velocity in m/s and R in m.

The velocity equation is of the form,

$$V = C\sqrt{RS}.$$

The dimensions of $\quad C = \dfrac{(LT^{-1})}{(L^{1/2})} = (L^{1/2}\,T^{-1}).$

The given value of C is accordingly 110 $ft^{1/12}$/sec and that since 1 ft = 0.3048 m, it follows

$$C = 110 \times (0.348)\ m^{1/2}/sec = 60.5\ m^{1/2}/sec.$$

The velocity formula in the metric units is

$$V = 60.5\sqrt{RS}$$

5.2. Dimensional Homogeneity

The methods of dimensional analysis are based on Fourier's "principle of dimensional homogeneity" (1822). This principle has already been utilised in chapter I in obtaining the dimensions of dynamic and kinematic viscosities. The principle of dimensional homogeneity is greatly useful in ascertaining the forms of equations, required for solving problems involving physical phenomena, from the knowledge of relevant variables and their dimensions.

An equation is said to be *dimensionally homogeneous* if the form of the equation does not depend upon the units of measurement. As an example, the hydrostatic law of pressure variation, $p = \gamma h$, is valid whether the pressure p is measured in kg/m^2, or lb/ft^2, and the depth h in metre, or ft. Therefore, by definition, pressure-depth equation is dimensionally homogeneous. With a similar reasoning, the continuity equation $Q = A\,.V$, the Newton's second law $F = m.a$ and $v = u + ft$ are also dimensionally homogeneous. If however, $\gamma = 1000$ kg/m^3 is substituted in the equation for hydrostatic pressure, one obtains,

$$p = 1000\,h$$

this equation is valid only when p is expressed in kg/m^2 and h in metre. Thus it is no more dimensionally homogeneous. All empirical equations are valid only for one system of units of measurement and are, therefore, dimensionally non-homogeneous. Other examples of dimensionally non-homogeneous equations are,

$$V = 20\sqrt{RS}$$

$$Q = 3.3\,LH^{3/2}$$

$$v = u + 9.81\,t$$

$$t = 1.11\sqrt{l}$$

$$V = 5.2\,t, \text{ and so on.}$$

According to the principle of dimensional homogeneity, a mathematical equation expressing a physical phenomenon must be dimensionally homogeneous. In other words, the dimensions of each side of the equation must be same. If the equation

$$Q = A + B + C,$$

expresses a physical phenomenon it is dimensionally homogeneous, and the dimensions of the variables Q, A, B and C must be same. The empirical equations involving coefficients are dimensionally non-homogeneous only in one respect, and that is, they are valid only for one system of measurement units. But so long as they express physical phenomena, they are intrinsically dimensionally homogeneous. This fact enables us to determine the dimensions of the coefficient.

5.3. Dimensional Analysis

The dimensional analysis is a powerful tool in formulating problems of physical phenomena which defy analytical solution and must be solved experimentally. It helps in obtaining maximum information from a minimum of experiment. The dimensional analysis accomplishes this by formation of dimensionless groups containing relevant variables.

The application of dimensional analysis to any practical problem is based on the assumption that certain variables which affect the phenomenon are independent variables, and all variables other than these and the dependent variable, are irrelevant and have no bearing on the phenomenon. The total number of variables influencing the problem is equal to the number of independent variables plus one, one being the number of the dependent variable. The dimensional analysis is used to obtain a functional relationship between the dependent and independent variables. The first step in the dimensional analysis is to decide which variables enter the problem. The naming of these variables which affect the problem requires a thorough understanding of the phenomenon and is a matter of judgement and experience. The second step is the formation of dimensionless groups of the variables.

Of the two methods available for dimensional analysis, the Rayleigh's method (originated by Lord Rayleigh in 1899) gives a special form of relationship among the dimensionless groups, and has an inherent drawback that it does not provide any information regarding the number of dimensionless groups to be obtained as a results of dimensional analysis. It is for these reasons that it has become obsolete and is not favoured for use. The second method is the Buckingham method (developed by E. Buckingham in 1914) and is an improvement over the Rayleigh's method. Buckingham designated the dimensionless groups by the Greek capital letter π (Pi). It is, therefore, often called as, Buckingham-π method. It offers an advantage over Rayleigh's method in letting us know, in advance of the analysis, as to how many dimensionless groups are to be expected.

Example 5.1. *The drag (resistance force) D experienced by a ship having a characteristic length l and moving with a speed V depends upon the fluid mass-density* ρ*, viscosity* μ *and acceleration due to gravity g. Using Rayleigh's method of dimensional analysis, obtain an expression for the drag in terms of dimensionless parameters.*

Solution. The functional relationship between the dependent and independent variables can be expressed as

$$D = f(l, V, \rho, \mu, g)$$

Tabulating these variables and expressing them dimensionally, one obtains :

S. No.	*Variable*	*Symbol*	*Dimensions*
1.	Drag force	D	MLT^{-2}
2.	Characteristic length	l	L
3.	Velocity	V	LT^{-1}
4.	Mass density	ρ	ML^{-3}
5.	Viscosity	μ	$ML^{-1}T^{-1}$
6.	Gravitational acceleration	g	LT^{-2}

5.3.1. Use of Rayleigh's Method

From the principle of dimensional homogeneity it is apparent that the variables involved cannot be added or subtracted since their dimensions are different. This principle limits the equation to a combination of products of powers of the variables involved. Accordingly, from the dimensional considerations the following equation may be written :

$$D = C\, l^x\, V^y\, \rho^z\, \mu^r\, g^s \qquad \text{...(2)}$$

in which C is a dimensionless constant.

Expressing the above equation dimensionally

$$(MLT^{-2}) = (L^x)(LT^{-1})^y\, (ML^{-3})^z\, (ML^{-1}T^{-1})^r\, (LT^{-2})^s \qquad \text{...(3)}$$

Using the principle of dimensional homogeneity, the three equations for exponents of M, L and T are :

$$M : 1 = z + r$$

$$L : 1 = x + y - 3z - r + s$$

$$T : -2 = -y - r - 2s$$

Solving in terms of r and s, one obtains

$$z = 1 - r,\; y = 2 - r - 2s \text{ and } x = 2 - r + s$$

Substituting these values in Eq. (2)

$$D = C\, l^{(2-r+s)}\, V^{(2-r-2s)}\, \rho^{(1-r)} \mu^r\, g^s$$

grouping the terms, one may write

$$D = C\, \rho\, l^2\, V^2 \left(\frac{V^2}{lg}\right)^{-s} \left(\frac{\rho V l}{\mu}\right)^{-r} \qquad \text{...(4)}$$

Recognising the dimensionless groups as

$$\mathbf{F} = V/\sqrt{lg}, \quad \text{and} \quad \mathbf{R} = \frac{\rho V l}{\mu}$$

Eq. (4) may be written as

$$D = \rho\, l^2 V^2 f_1\,(\mathbf{F}, \mathbf{R}) \qquad \text{...(5)}$$

or

$$\frac{D}{\rho l^2 V^2} = f_1\,(\mathbf{F}, \mathbf{R})$$

which shows on the basis of dimensional analysis alone that $D/\rho l^2 V^2$ depends only upon the Froude and Reynolds numbers. This technique, originated by Lord Rayleigh, was improved upon by Buckingham with a broad generalization known as the π-theorem.

5.4. Buckingham-Pi Theorem

The Buckingham-Pi method is widely used in the dimensional analysis of a problem and expresses the resulting equation in terms of dimensionless groups (π-terms). According to this theorem, if a phenomenon involves n variables $a_1, a_2\, a_3, \ldots\, a_n$, such that a variable say a_1 depends upon independent variables $a_2, a_3, \ldots\, a_n$ and upon no other, the general functional relationship between the dependent variable a_1 and the independent variables may be expressed as,

$$a_1 = f\,(a_2, a_3, a_4, \ldots\ldots, a_n) \qquad \text{...(5.2)}$$

Equation (5.2) on account of the mathematical equilibrium between the dependent and independent variables may be written as

$$\phi(a_1, a_2, a_3, a_4, ... a_n) = 0 \quad ...(5.3)$$

The transformation of Eq. (5.2) into Eq. (5.3) may be understood like this,

$$x^2 = cy,$$

which may be written as, $x^2 = f(y)$ or $x^2 - cy = 0$

$\therefore$ $\phi(x, y) = 0$

where f and ϕ are unspecified functions.

The Buckingham-Pi theorem states that if all these n variables are described by m fundamental dimensions, they may be grouped into $(n - m)$ dimensionless π-terms. We then, have

$$f_1(\pi_1, \pi_2, \pi_3, ... \pi_{n-m}) = 0 \quad ...(5.4)$$

in which each π-term may contain maximum of $(m + 1)$ variables, only one of which is required to be changed from term to term.

As is evident from the foregoing discussion, the Pi-theorem not only arranges the variables into dimensionless groups, but also reduces the number of variables. In other words, the number of variables are reduced from n in Eq. (5.2) to $(n - m)$ in Eq. (5.4). If each of the fundamental dimensions (M, L, T) are involved, then $m = 3$ and the result will be an equation with $(n - 3)$ dimensionless terms.

5.4.1. Application of Π-Theorem

For applying the theorem, m repeating variables are selected from amongst the n variables, such that they in combination contain each of the m fundamental dimensions involved in the problem. These m repeating variables are then combined repeatedly with each of the other $(n - m)$ variables. Each combination yields a dimensionless group of a π-term and hence we obtain $(n - m)$ dimensionless groups, which are independent of each other. There is a considerable option in the selection of repeating variables, however, as already pointed out they must contain in themselves in combination all the fundamental dimensions involved. In problems in which all the three fundamental dimensions are involved, it is usually expedient to select repeating variables using the following guidelines :

(*i*) select the first repeating variable from those describing the geometry of flow,

(*ii*) select the second repeating variable from those representing in the fluid properties

and (*iii*) select the third repeating variable from those characterising the fluid motion.

The variables which can influence fluid phenomena, in general, can be divided into three categories :

(*a*) those variables which describe the geometry of the fluid motion : size and shape of the fluid passage or of the moving body such as diameter, length etc.

(*b*) those variables characterising the fluid : fluid properties such as density, viscosity, surface tension, elasticity and vapour pressure.

(*c*) those variables characterising the fluid motion : the kinematic and dynamic characteristics of flow such as velocity, acceleration, discharge, pressure, force, power.

It is usually conventional to select a length dimension, representing category (*a*), density from category (*b*), and velocity from category (*c*) as repeating variables. The length dimension may be the pipe diameter in case of pipe flow, or depth of flow or the hydraulic radius in open channel flow. It may, however, he emphasized that the nature or form of the dimensionless

π-terms would entirely depend upon the selection of repeating variables. If the dimensional analysis of a problem is attempted second time by selecting a completely different repeating variables, the result would be an entirely different set of dimensionless π-terms as compared to those obtained earlier.

Example 5.2. *The critical depth y_{cr} in a V-shaped channel depends upon the discharge Q, acceleration due to gravity g, and the vertex angle θ made by the two side-walls of the channel. Using the method of dimensional analysis derive an equation for the critical depth.*

Solution. The functional relationship for the critical depth may be written as

$$y_{cr} = f(Q, g, \theta)$$

Tabulating these variables and expressing them dimensionally, one obtains :

S. No.	*Variable*	*Symbol*	*Dimensions*
1.	Critical depth	y_{cr}	L
2.	Discharge	Q	L^3T^{-1}
3.	Acceleration due to gravity	g	LT^{-2}
4.	Vertex angle	θ	$M^\circ L^\circ T^\circ$

From the above table, one can easily notice that :

(*i*) total number of variables $n = 4$

(*ii*) number of fundamental dimensions $m = 2$.

Hence the number of dimensionless parameters

$$n - m = 4 - 2 = 2$$

The vertex angle θ can be identified as one of the dimensionless parameter. For obtaining the second one, let us select Q and g as the repeating variables, then

$$\pi_2 = Q^x g^y y_{cr}$$

Expressing it dimensionally,

$$M^0L^0T^0 = (L^3T^{-1})^x (LT^{-2})^y (L)$$

Using the principle of dimensional homogeneity

$$M : 0 = 0$$

$$L : 0 = 3x + y + 1$$

$$T : 0 = -x - 2y$$

Solving which, $\quad y = 1/5 \quad \text{and} \quad x = -2/5$

Substituting these values of the exponents

$$\pi_2 = (Q^2/g^{-1})^{-1/5} y_{cr}$$

$$f_1(\pi_1, \pi_2) = 0 \quad \text{or} \quad f_1[\theta, (Q^2/g)^{-1/5} y_{cr}] = 0$$

or $\quad y_{cr} = C(Q^2/g)^{1/5}$, where C is a function of θ.

Example 5.3. *The discharge Q over a rectangular weir depends on the head H over the weir, acceleration due to gravity g, length of weir crest L, height of the weir crest over the channel bottom Z and the kinematic viscosity ν of the liquid. By the method of dimensional analysis, obtain an expression for the discharge Q in terms of dimensionless parameters.*

Solution. The relationship between the dependent and independent variables may be expressed as

$$Q\, f(H, g, L, Z, \nu) \qquad \text{...(A)}$$

The following table gives the dimensions of the variables involved :

S. No.	*Variable*	*Symbol*	*Dimensions*
1.	Discharge	Q	(L^3T^{-1})
2.	Head	H	(L)
3.	Acceleration due to gravity	g	(LT^{-2})
4.	Weir crest length	L	(L)
5.	Weir height	Z	(L)
6.	Kinematic viscosity	ν	(L^2T^{-1})

From this table, the following observations may be made :

(*i*) total number of variables, $n = 6$

(*ii*) number of fundamental dimensions, $m = 2$.

Therefore, the number of dimensionless parameters, *i.e.*, π-terms are

$$n - m = 6 - 2 = 4.$$

Two repeating va~iables (equal to the number of fundamental dimensions m) are to be selected such that they in themselves contain m fundamental dimensions. The head over the weir is the most important independent variable, and the other variable which affects the flow in open channels is the acceleration due to gravity g. These two variables H and g satisfy the requirement of repeating variables. The dimensionless parameters may now be obtained by combining the repeating variables with each of the remaining variables.

(1) The first dimensionless parameter,

$$\pi_1 = g^{x1}\, H^{y1} Q$$

Writing in terms of dimensions, one obtains

$$(M^0L^0T^0) = (LT^{-2})^{x1}(L)^{y1}(L^3T^{-1})$$

Using the principle of dimensional homogeneity, that is equating the powers of M, L and T on each side of equation, we obtain

$$M : 0 = 0$$

$$L : 0 = x_1 + y_1 + 3 \quad \text{and} \quad T : 0 = -\, 2x_1 - 1$$

$$\therefore \qquad x_1 = \tfrac{1}{2} \quad \text{and} \quad y_1 = -\, 5/2$$

Therefore, the first dimensionless parameter

$$\pi_1 = g^{-1/2}H^{-5/3}\, Q = \frac{Q}{g^{1/2}(H)^{5/2}}$$

(2) The second dimensionless parameter

$$\pi_2 = g^{x2}\, H^{y2}\, L, \text{ writing in dimensional form}$$

$$(M^0L^0T^0) = (LT^{-2})^{x2}\,(L)^{y2}\,(L)$$

From the principle of dimensional homogeneity,

$$L : 0 = x_2 + y_2 + 1$$

$$T : 0 = -2x_2.$$

$$\therefore \quad x_2 = 0, \quad \text{and} \quad y_2 = -1$$

and the dimensionless parameter $\pi_2 = H^{-1} L = \dfrac{L}{H}$

(3) The third dimensionless parameter

$$\pi_3 = g^{x3} H^{y3} Z$$

by performing similar analysis, it may be easily shown that $\pi_3 = Z/H$.

(4) The fourth dimensionless parameter,

$$\pi_4 = g^{x4} H^{y4} \nu$$

writing it dimensionally$(M^0L^0T^0) = (LT^{-2})^{x4} (L)^{y4} (L^2T^{-1})$

Applying the principle of dimensional homogeneity

$$L : 0 = x_4 + y_4 + 2$$

$$T : 0 = 2x_4 - 1,$$

$$\therefore \quad x_4 = -\tfrac{1}{2} \quad \text{and} \quad y_4 = -3/2$$

and therefore, $\quad \pi_4 = \dfrac{\nu}{g^{1/2} H^{3/2}}$

The dimensionless parameters may now be written as,

$$f\left(\frac{Q}{g^{1/2}(H)^{5/2}}, \frac{L}{H}, \frac{Z}{H}, \frac{\nu}{g^{1/2}H^{3/2}}\right) = 0$$

or

$$\frac{Q}{g^{1/2}(H)^{5/2}} = f_1\left(\frac{\nu}{g^{1/2}H^{3/2}}, \frac{L}{H}, \frac{Z}{H}\right) \quad \text{...(B)}$$

From the equation of continuity $Q = AV$ and, therefore, $\dfrac{Q}{AV}$ is dimensionless. In case of a rectangular weir, one may write $A \propto LH$ and $V \propto g^{1/2} H^{1/2}$ and hence

$$\frac{Q}{LH g^{1/2} H^{1/2}} = \frac{Q}{L g^{1/2} H^{3/2}}$$

is a dimensionless quantity.

Thus the left hand side of Eq. (*B*) may be transformed and written as $\dfrac{Q}{Lg^{1/2}H^{3/2}}$ and Eq. (*B*) expressed as

$$\frac{Q}{Lg^{1/2}H^{3/2}} = f_1\left(\frac{\nu}{g^{1/2}H^{3/2}}, \frac{L}{H}, \frac{Z}{H}\right)$$

from which

$$Q = \sqrt{g}\, LH^{3/2} f_1\left(\frac{\nu}{g^{1/3}H^{3/2}}, \frac{L}{H}, \frac{Z}{H}\right) \quad \text{...(C)}$$

In Eq. (*C*), $\frac{\nu}{g^{1/2}H^{3/2}} = \frac{\nu}{H\sqrt{gH}}$ is the inverse of Reynolds number. Eq. (*C*) may be written as

$$Q = C\sqrt{g}\,LH^{3/2} \quad \text{...(D)}$$

where
$$C = f_2\left(\frac{H\sqrt{gH}}{\nu}, \frac{H}{L}, \frac{H}{Z}\right)$$

and since *g* is a constant for a place, Eq. (*D*) can be expressed as

$$Q = KLH^{3/2}.$$

Example 5.4. *The resisting force F of a plane during flight can be considered as dependent upon the length of aircraft l, velocity v, air viscosity μ, air density ρ and bulk modulus of air K. Express the functional relationship between these variables using dimensional analysis. Explain the physical meaning of the dimensionless groups.*

Solution. The functional relationship between the dependent variable *F* and the independent variables can be written as

$$F = f(l, v, \mu, \rho, K) \quad \text{...(i)}$$

Expressing these variables in terms of their dimensions in the tabular form as below :

S. No.	*Name of variables*	*Symbol*	*Dimensions*
1.	Resistance force	F	MLT^{-2}
2.	Length of aircraft	l	L
3.	Velocity	v	LT^{-1}
4.	Air viscosity	μ	$ML^{-1}T^{-1}$
5.	Air density	ρ	ML^{-3}
6.	Bulk modulus of air	K	$ML^{-1}T^{-2}$

From this table, it is obvious that

the total number of variables, $n = 6$

the number of fundamental dimensions, $m = 3$.

Therefore, the total number of dimensionless groups or π-terms to be obtained

$$= n - m = 6 - 3 = 3.$$

Selecting ρ, *v* and *l* as the repeating variables and making use of the Buckingham—Pi theorem, one can write the following dimensionless groups :

$$\pi_1 = \rho^{x_1}\, v^{y_1}\, l^{z_1}\, F \quad \text{...(ii)}$$

$$\pi_2 = \rho^{x_2}\, v^{y_2}\, l^{z_2}\, \mu \quad \text{...(iii)}$$

and
$$\pi_3 = \rho^{x_3}\, v^{y_3}\, l^{z_3}\, K \quad \text{...(iv)}$$

Expressing Eq. (*ii*) dimensionally

$$M^\circ L^\circ T^\circ = (ML^{-3})^{x1}\,(LT^{-1})^{y1}\,(L)^{z1}\,MLT^{-2}.$$

Using the principle of dimensional homogeneity, *i.e.* equating the exponents of *M, L* and *T* on both sides

$$M : 0 = x_1 + 1, \quad \therefore \quad x_1 = -1$$
$$L : 0 = -3x_1 + y_1 + z_1 + 1$$
$$T : 0 = -y_1 - 2 \quad \therefore \quad y_1 = -2 \text{ and } z_1 = -2$$

Making substitutions, $\pi_1 = \frac{F}{\rho l^2 v^2}$ ...(*v*)

Following similar steps as for π_1, one can obtain the other two dimensionless groups which will have the following form :

$$\pi_2 = \frac{\mu}{\pi\, vl} \quad ...(vi) \qquad \text{and} \qquad \pi_3 = \frac{K}{\rho v^2} \quad ...(vii)$$

From the formulation of these groups, one can easily recognise that π_1 is the ratio of resistance force F and a dynamic force $\rho l^2 v^2$. As will be seen subsequently while dealing with model testing, one can guess from Eq. (v) that the resistance force experienced by an aircraft is dependent on the length and the velocity. For any given aircraft, the resistance force will be proportional to the square of velocity.

Dimensionless group π_2 can be identified as the reciprocal of Reynolds number, and represents the role of viscous action on the resistance. Using Eq. (1.22), one can easily discern π_3 as the ratio of square of velocity in air to the square of aircraft velocity, and can be seen to be the reciprocal of square of the Mach number. It may be noted that π_3 takes into account the role of compressibility in influencing fluid resistance experienced by the aircraft.

Example 5.5. *The drag force F on a partially submerged body depends on the relative velocity V between the body and the fluid, characteristic linear dimension 'l', height of surface roughness k, fluid density* ρ, *the viscosity* μ *and the acceleration due to gravity g. Obtain an expression for the drag force, using the method of dimensional analysis.*

Solution. The functional relationship between the dependent and the independent variables may be written as

$$F = f(V, l, k, \rho, \mu, g)$$

A table may now be prepared showing the dimensions of all the variables involved.

S. No.	*Variables*	*Symbol*	*Dimensions*
1	Drag force	F	(MLT^{-2})
2	Velocity	V	(LT^{-1})
3	Characteristic length	l	(L)
4	Surface roughness height	k	(L)
5	Density	ρ	(ML^{-3})
6	Viscosity	μ	$(ML^{-1}T^{-1})$
7	Gravity	g	(LT^{-2})

From this table, we find that :

(i) total number of variables, $n = 7$, and

(ii) since all the three fundamental dimensions M, L, T are appearing in the last column of the table, $m = 3$ and, therefore, number of dimensionless parameters

$$= n - m = 7 - 3 = 4.$$

Selecting l, V and ρ as the repeating variables, the four dimensionless parameters π_1, π_2, π_3 and π_4 may be obtained from the method of dimensional analysis.

(1) $$\pi_1 = \rho^{x1} V^{y1} l^{z1}\, F$$

expressing it dimensionally

$$(M^0L^0T^0) = (ML^{-3})^{x1}(LT^{-1})^{y1}(L^{-z1})(MLT^{-2})$$

Using the principle of dimensional homogeneity, *i.e.* equating the power indices of M, L and T on both sides of the equation, we obtain

$$M : 0 = x_1 + 1$$

$$\therefore \quad x_1 = -1$$

$$L : 0 = -3x_1 + y_1 + z_1 + 1$$

$$T : 0 = -y_1 - 2,$$

$\therefore$ $y_1 = -2$ and $z_1 = -2.$

Therefore, $\pi_1 = \dfrac{F}{\rho l^2 V^2}$

(2) $\pi_2 = \rho^{x2} V^{y2} l^{z2} k.$

Transforming it in the dimensional form

$$(M^0L^0T^0) = (ML^{-3})^{x2}(LT^{-2})^{y2}(L)^{z2}(L)$$

Applying principle of dimensional homogeneity, we have

$$M : 0 = x_2$$
$$L : 0 = -3x_2 + y_2 + z_2 + 1$$
$$T : 0 = -y_2,$$
$$x_2 = y_2 = 0 \quad \text{and} \quad z_2 = -1 \quad \text{hence,} \quad \pi_2 = k/l.$$

(3) $\pi_3 = \rho^{x3} V^{y3} l^{z3} \mu$

it may be dimensionally written as

$$(M^0L^0T^0) = (ML^{-3})^{x_2} (LT^{-1})^{y_3} (L)^{z_3} (ML^{-1}T^{-1})$$

from the principle of dimensional homogeneity, we obtain

$$M : 0 = x_3 + 1,$$

$\therefore$ $x_3 = -1$

$$L : 0 = -3x_3 - y_3 + z_3 - 1$$
$$T : 0 = -y_1 - 1,$$

$\therefore$ $y_3 = -1,$

and $z_3 = -1.$

Substituting these values of x_3, y_3 and z_3, we have

$$\pi_3 = \frac{\mu}{\rho V l}$$

(4) $\pi_4 = \rho^{x4} V^{y4} l^{z4} g$

writing it dimensionally

$$(M^0L^0T^0) = (ML^{-3})^{x4}(LT^{-1})^{y4}(L)^{z4}(LT^{-2})$$

Principle of dimensional homogeneity yields

$$M : 0 = x_4$$
$$L : 0 = -3x_4 + y_4 + z_4 + 1$$
$$T : 0 = -y_4 - 2.$$

$\therefore$ $y_4 = -2$ and $z_4 = 1$

the dimensionless parameter π_4 now becomes,

$$\pi_4 = \frac{lg}{V^2}.$$

All the four dimensionless parameters may be expected as

$$f_1(\pi_1, \pi_2, \pi_3, \pi_4) = 0.$$

or $\pi_1 = \phi(\pi_2, \pi_3, \pi_4).$

Substituting the values of π_1, π_2, π_3 and π_4,

$$\frac{F}{\rho l^2 V^2} = \phi\left(\frac{k}{l}, \frac{\mu}{\rho V l}, \frac{lg}{V^2}\right)$$

$\therefore$ $F = \rho l^2 V^2 \phi\left(\dfrac{\mu}{\rho V l}, \dfrac{lg}{V^2}, \dfrac{k}{l}\right).$

The dimensionless parameter $\mu/\rho Vl$ is the reciprocal of the Reynolds number, the quantity lg/V^2 is the reciprocal of square of the Froude number, and k/l is known as the relative roughness and denoted by ε, thus

$$\frac{\mu}{\rho Vl} = \frac{1}{\mathbf{R}}, \quad \frac{lg}{V^2} = \frac{1}{\mathbf{F}^2} \quad \text{and} \quad \frac{k}{l} = \varepsilon$$

The above equation may, therefore, be written as

$$F = \rho l^2 V^2 \phi\,(\mathbf{R}, \mathbf{F}, \varepsilon).$$

Example 5.6. *The rate of flow Q of a liquid of density ρ and viscosity μ is to be measured by means of an orificemeter of diameter d installed in a pipeline of diameter D. The pressure drop across the orifice plate is Δp. Show by using the method of dimensional analysis that* $Q = Cd^2\sqrt{\frac{\Delta p}{\rho}}$ *in which C is a discharge coefficient dependent upon the parameters ρQ/μd and d/D.*

Solution. The functional relationship between the dependent and independent variables may be written as

$$Q = f\,(d, D\,\rho, \mu, \Delta p)$$

A table of variables involved in the problem alongwith their dimensions is given below :

S. No.	*Variables*	*Symbol*	*Dimensions*
1.	Rate of flow	Q	(L^3T^{-1})
2.	Orifice diameter	d	(L)
3.	Pipe diameter	D	(L)
4.	Density	ρ	(ML^{-3})
5.	Viscosity	μ	$(ML^{-1}T^{-1})$
6.	Pressure drop	Δp	$(ML^{-1}T^{-2})$

The total number of variables $n = 6$, and the number of fundamental dimensions $m = 3$.

Therefore, the number of dimensionless parameters

$$= n - m = 6 - 3 = 3$$

Selecting ρ, Q and d as repeating variables

(1) $\pi_1 = \rho^{x1}\, Q^{y1} d^{z1} D$

writing it dimensionally $(M^0L^0T^0) = (ML^{-3})^{x1}(L^3T^{-1})^{y1}(L)^{z1}(L)$

from the principle of dimensional homogeneity.

$$M : 0 = x_1$$
$$L : 0 = -\,3x_1 + 3y_1 + z_1 + 1$$
$$T : 0 = -\,y_1.$$

$\therefore$ $x_1 = 0,\, y_1 = 0,\, z_1 = -\,1.$

Therefore, $\pi_1 = \dfrac{D}{d}.$

(2) $\pi_2 = \rho^{x2} Q^{y2} d^{z2} \mu$

Substituting the dimensions of each term

$$(M^0L^0T^0) = (ML^{-3})^{x2}(L^3T^{-1})^{y2}(L)^{z2}(ML^{-1}T^{-1})$$

Using principle of dimensional homogeneity

$$M : 0 = x_2 + 1,$$

$\therefore$ $x_2 = -\,1$

$$L : 0 = -\,3x_2 + 3y_2 + z_2 - 1$$
$$T : 0 = -\,y_2 - 1,$$

$\therefore$ $y_2 = -\,1$ and $z_2 = 1.$

Therefore, $\pi_2 = \mu d/\rho Q$

(3) $\pi_3 = \rho^{x3} Q^{y3} d^{z3}\, \Delta p$

Writing it dimensionally

$$(M^0L^0T^0) = (ML^{-3})^{x3}(L^3T^{-1})^{y3}(L^z)^3(ML^{-1}T^{-2})$$

from the principle of dimensional homogeneity

$$\therefore \quad M : 0 = x_3 + 1$$

$$x_3 = -1$$

$$L : 0 = -3x_3 + 3y_3 + z_3 - 1$$

$$T : 0 = -y_3 - 2,$$

$$\therefore \quad y_3 = -2 \quad \text{and} \quad z_3 = 4$$

Therefore, $$\pi_3 = \frac{d^4 \Delta p}{\rho Q^2}$$

We may write

$$\phi(\pi_1, \pi_2, \pi_3) = 0$$

or $$\phi\left(\frac{D}{d}, \frac{\mu d}{\rho Q}, \frac{d^4 \Delta p}{Q^2 \rho}\right) = 0 \quad \text{or} \quad \phi\left(\frac{d}{D}, \frac{\rho Q}{\mu d}, \frac{Q^2 \rho}{d^4 \Delta p}\right) = 0$$

or $$\frac{Q^2 p}{d^4 \Delta p} = f_1\left(\frac{\rho Q}{\mu d}, \frac{d}{D}\right) \quad \text{or} \quad \frac{Q}{d^2\sqrt{\Delta p/\rho}} = f_2\left(\frac{\rho Q}{\mu d}, \frac{d}{D}\right)$$

$$Q = Cd^2\sqrt{\frac{\Delta p}{\rho}}$$

where C is dependent on $\dfrac{\rho Q}{\mu d}$ and $\dfrac{d}{D}$.

Example 5.7. *Fluid of density ρ and viscosity μ flows at an average velocity V through a circular pipe of diameter d. Show by dimensional analysis, that the shear stress of the pipe wall*

$$\tau_0 = \rho V^2 f\left(\frac{\rho V d}{\mu}\right)$$

Solution. The functional relationship between the dependent and independent variables is given by $\tau_0 = f(\rho, \mu, V, d)$.

A table showing relevant variables with their dimensions may be prepared.

S. No.	*Variables*	*Symbol*	*Dimensions*
1.	Shear stress at the pipe wall	τ_0	$(ML^{-1}T^{-1})$
2.	Density	ρ	(ML^{-3})
3.	Viscosity	μ	$(ML^{-1}T^{-1})$
4.	Velocity	V	(LT^{-1})
5.	Diameter	d	(L)

From the table, the total number of variables $n = 5$ and the number of fundamental dimensions $m = 3$, therefore, the number of dimensionsless parameters

$$= n - m = 5 - 3 = 2$$

Selecting ρ, V and d as the repeating variables.

(*i*) $\pi_1 = \rho^{x1} V^{y1} d^{z1} \tau_0$

Substituting the dimensions of each quantity

$$(M^0L^0T^0) = (ML^{-3})^{x1}(LT^{-1})^{y1}(L)^{z1}(ML^{-1}T^{-2})$$

from the principle of dimensional homogeneity

$$M : 0 = x_1 + 1$$

$\therefore$ $$x_1 = -1$$

$$L : 0 = -3x_1 + y_1 + z_1 - 1$$

$$T : 0 = -y_1 - 1$$

$\therefore$ $$y_1 = -2,$$

and $$z_1 = -3 + 2 + 1 = 0$$

$\therefore$ $$\pi_1 = \rho^{-1}V^{-2}\tau_0 = \frac{\tau_0}{\rho V^2}.$$

(*ii*) $$\pi_2 = \rho^{x2}V^{y2}d^{z2}\mu$$

Writing the dimensions of each quantity

$$(M^0L^0T^0) = (ML^{-3})^{x2}(LT^{-1})^{y2}(L)^{z2}(ML^{-1}T^{-1})$$

Using principle of dimensional homogeneity

$$M : 0 = x_2 + 1$$

$\therefore$ $$x_2 = -1$$

$$L : 0 = -3x_2 + y_2 + z_2 - 1$$

$$T : 0 = -y_2 - 1$$

$\therefore$ $$y_2 = -1 \quad \text{and} \quad z_2 = -3 + 1 + 1 = -1$$

Therefore, $$\pi_2 = \frac{\mu}{\rho V d} \quad \text{and} \quad \pi_1 = \phi(\pi_2)$$

or $$\frac{\tau}{\rho V^2} = \phi\left(\frac{\mu}{\rho V d}\right)$$

$\therefore$ $$\tau_0 = \rho V^2 f\left(\frac{\rho V d}{\mu}\right).$$

Example 5.8. *The power P developed by a water turbine depends on the rotational speed N, operating head H, gravity g, diameter D and breadth B of the runner, density* ρ *and* viscosity μ *of water. Show by dimensional analysis that*

$$P = \rho D^5 N^3 \,\phi\left[\frac{H}{D}, \frac{D}{B}, \frac{\rho D^2 N}{\mu}, \frac{ND}{\sqrt{gH}}\right]$$

Under what conditions it can be used to determine the characteristics of a similar machine ?

Solution. The functional relationship between the dependent and the independent variables may be written as $P = f(N, H, D, B, \rho, \mu, g)$

A table showing these variables, alongwith their dimensions may be prepared.

S. No.	*Variables*	*Symbol*	*Dimensions*
1.	Rotational speed	N	(T^{-1})
2.	Operating head	H	(L)
3.	Runner diameter	D	(L)
4.	Runner width	B	(L)
5.	Density	ρ	(ML^{-3})
6.	Viscosity	μ	$(ML^{-1}T^{-1})$
7.	Power	P	(ML^2T^{-3})
8.	Gravity	g	(LT^{-2})

From this table it may be observed that,

(*i*) the total number of variables : $n = 8$

(*ii*) the fundamental dimensions involved in the problem : $m = 3$,

and, therefore, the number of dimensionless parameters

$$= n - m = 8 - 3 = 5$$

Selecting ρ, N and D as the repeating variables, we may write

$$\pi_1 = \rho^{x1}N^{y1}D^{z1}H, \qquad \pi_2 = \rho^{x2}N^{y2}D^{z2}B,$$

$$\pi_3 = \rho^{x3}N^{y3}D^{z3}P, \qquad \pi_4 = \rho^{x4}N^{y4}D^{z4}\mu,$$

$$\pi_5 = \rho^{x5}N^{y5}D^{z}g$$

Determination of π-terms

$$\pi_1 = \rho^{x1}N^{y1}D^{z1}H$$

Writing π_1 dimensionally,

$$(M^0L^0T^0) = (ML^{-3})^{x1}(T^{-1})^{y1}(L)^{z1}(L)$$

From principle of dimensional homogeneity,

$$x_1 = 0, \qquad y_1 = 0, \qquad z_1 = -1$$

$$\therefore \qquad \pi_1 = \frac{H}{D}$$

Similarly, $$\pi_2 = \frac{B}{D}.$$

$$\pi_3 = \rho^{x3}N^{y3}D^{z3}P.$$

Writing in dimensional form, one gets

$$(M^0L^0T^0) = ML^{-3})^{x3}(T^{-1})^{y3}(L)^{z3}(ML^2T^{-3})$$

From the principle of dimensional homogeneity

$$M : 0 = x_3 + 1 \qquad \therefore \quad x_3 = -1$$

$$L : 0 = 3x_3 + z_3 + 2, \qquad \therefore \quad z_3 = -5$$

$$T : 0 = -y_3 - 3, \qquad \therefore \quad y_3 = -3$$

and the dimensionless parameter $\pi_3 = \dfrac{P}{\rho N^3 D^5}$

Writing π_4 dimensionally one gets

$$(M^0L^0T^0) = (ML^{-3})^{x4}(T^{-1})^{y4}(L)^{z4}(ML^{-1}T^{-1})$$

From the principle of dimensional homogeneity

$$M : 0 = x_4 + 1, \qquad \therefore \quad x_4 = -1$$

$$L : 0 = -3x_4 + z_4 - 1, \qquad \therefore \quad z_4 = -2$$

$$T : 0 = -y_4 - 1, \qquad \therefore \quad y_4 = -1$$

and the dimensionless parameter $\qquad \pi_4 = \dfrac{\mu}{\rho N D^2}$

The fifth dimensionless parameter

$$\pi_5 = \rho^{x5}N^{y5}D^{z5}g$$

Writing it in the dimensional form

$$(M^0L^0T^0) = (ML^{-3})^{x5}(T^{-1})^{y5}(L)^{z5}\,(LT^{-2})$$

from the principle of dimensional homogeneity

$$M : 0 = x_5, \qquad \therefore \quad x_5 = 0$$

$$L : 0 = -3x_5 + z_5 + 1, \qquad \therefore \quad z_5 = -1$$

$$T : 0 = -y_5 - 2, \qquad \therefore \quad y_5 = -2$$

and the dimensionless parameter $\pi_5 = \dfrac{g}{N^2D}$

Combining all the dimensionless parameters

$$f_1(\pi_1, \pi_2, \pi_3, \pi_4, \pi_5) = 0.$$

Substituting the value of each π-term, we obtain

$$\frac{P}{\rho N^3 D^5} = f_2\left(\frac{H}{D}, \frac{B}{D}, \frac{\mu}{\rho ND^2}, \frac{g}{N^2D}\right)$$

product of π_1 and π_5 yields a new π-term, gH/N^2D^2, and its square root $\sqrt{gH}/ND$ is also dimensionless. Thus the above relationship may be written as

$$\frac{P}{\rho N^3 D^5} = \phi\left(\frac{H}{D}, \frac{D}{B}, \frac{\rho ND^2}{\mu}, \frac{ND}{\sqrt{gH}}\right)$$

or

$$P = \rho N^3 D^5 \phi\left[\frac{H}{D}, \frac{D}{B}, \frac{\rho ND^2}{\mu}, \frac{ND}{\sqrt{gH}}\right]$$

If the two water turbines are dynamically similar, then all the dimensionless parameters must be the same in both the machines. In turbines, the viscous forces are less important as compared to the gravity forces and, therefore the parameter $\rho ND^2/\mu$ has little significance, thus for complete similarity between model and prototype of water turbines,

(*a*) $$\left(\frac{H}{D}\right)_p = \left(\frac{H}{D}\right)_m$$ and (*b*) $$\left(\frac{D}{B}\right)_p = \left(\frac{D}{B}\right)_m$$

give conditions for geometric similarity,

(*c*) $$\left(\frac{ND}{\sqrt{gH}}\right)_p = \left(\frac{ND}{\sqrt{gH}}\right)_m,$$

expresses a condition for kinematic similarity,

(*d*) $$\left(\frac{P}{\rho N^3 D^5}\right)_p = \left(\frac{P}{\rho N^3 D^5}\right)_m$$

states a condition for dynamic similarity.

The subscript p and m refer to the prototype and model respectively.

Example 5.9. *Vortices are frequently observed at reservoir intakes at low intake submergences. The critical submergence at which the vortices become strong enough to entrain air may be considered to have the following functional relationship for the given approach boundaries,*

$$S_c = f(Q, d, \Gamma, g, \rho, \mu, \sigma)$$

Using method of dimensional analysis, obtain the following non-dimensional relationship among the pertinent variables

$$\frac{S_c}{d} = f_i\left(\frac{\Gamma D}{Q}, \frac{\rho Vd}{\mu}, \frac{V}{\sqrt{gd}}, \frac{\rho V^2 d}{\sigma}\right)$$

in which Γ is the circulation in the approach flow at a radius r from the intake and is given by $\Gamma = 2\pi V_c$, where V_c is the tangential velocity at radius r and V is the average velocity in the intake. The other notations have the usual meaning.

Solution. A table showing the dimensions of the variables influencing the phenomenon may be prepared as under :

S. No.	*Variables*	*Symbol*	*Dimensions*
1	Critical Submergence	S_c	L
2	Diameter of intake	d	L
3	Discharge	Q	L^3T^{-1}
4	Circulation	Γ	L^2T^{-1}
5	Acceleration due to gravity	g	LT^{-2}
6	Mass density	ρ	ML^{-3}
7	Viscosity	μ	$ML^{-1}T^{-1}$
8	Surface tension	σ	MT^{-2}

From the above table, it is obvious that the total number of variables $n = 8$ and $m = 3$, and ML and T—all the three fundamental dimensions appear in the table. The number of dimensionless groups (π-terms) $= n - m = 8 - 3 = 5$.

Selecting ρ, Q and d as the repeating variables, the five π-terms may be determined as follows :

(1) $$\pi_1 = \rho^x Q^y d^z S_c.$$

Writing it dimensionally

$$M^0L^0T^0 = (ML^{-3})^x(L^3T^{-1})^y(L)^z L$$

Dimensional homogeneity results in

$$M : 0 = x$$
$$L : 0 = -3x + 3y + z + 1$$
$$T : 0 = -y$$

from which, $x = y = 0$, and $z = -1$ and therefore,

$$\pi_1 = S_c/d$$

(2) $$\pi_2 = \rho^{x1}Q^{y1}d^{z1}\Gamma,$$

Writing the dimensions of each variable

$$M^0L^0T^0 = (ML^{-3})^{x1}(L^3T^{-1})^{y1}(L)^{z1}(L^2T^{-1})$$

from the principle of dimensional homogeneity

$$M : 0 = x_1$$
$$L : 0 = -3x_1 + 3y_1 + z_1 + 2$$
$$T : 0 = -y_1 - 1$$
$$x_1 = 0, y_1 = -1, z_1 = 1$$

and $$\pi_2 = \frac{\Gamma d}{Q}.$$

(3) $$\pi_3 = \rho^{x2}Q^{y2}d^{z2}g,$$

expressing dimensionally,

$$M^0L^0T^0 = (ML^{-3})^{x2}(L^3T^{-1})^{y2}(L)^{z2}(LT^{-2})$$

from considerations of dimensional homogeneity

$$M : 0 = x_2$$
$$L : 0 = -3x_2 + 3y_2 + z_2 + 1$$
$$T : 0 = -y_2 - 2$$

which results in, $x_2 = 0, y_2 = -2, z_2 = 5$

$$\pi_3 = Q^{-2}d^5g = \frac{gd}{Q^2/d^4};$$

from the continuity principle, $Q = \frac{\pi}{4} d^2V$

$$\pi_3 \propto \frac{gd}{V^2}, \text{ and one may write}$$

$$\pi_3 = \frac{\sqrt{gd}}{V}$$

(4) $\pi_4 = \rho^{x3}\, Q^{y3}\, d^{z3}\, \mu$

writing in a dimensional form

$$M^0L^0T^0 = (ML^{-3})^{x3}(L^3T^{-1})^{y3}(L)^{z3}(ML^{-1}T^{-1})$$

from dimensional homogeneity

$$M : 0 = x_3 + 1$$
$$L : 0 = -\,3x_3 + 3y_3 + z_3 - 1$$
$$T : 0 = -\,y_3 - 1$$

which results in $x_3 = -\,1,\ y_3 = -\,1,\ z_3 = 1$, and hence

$$\pi_4 = \frac{\mu d}{\rho Q} \propto \frac{\mu}{\rho V d}, \qquad \text{(using the continuity relationship)}$$

one can, therefore, write $\pi_4 = \dfrac{\mu}{\rho V d}$.

(5) $\pi_5 = \rho^{x4}\, Q^{y4}\, d^{z4}\, \sigma$

writing the dimensions, $M^0L^0T^0 = (MT^{-3})^{x4}\,(L^3T^{-1})^{y4}\,(L)^{z4}\,(MT^{-2})$

dimensional homogeneity may be used to write

$$M : 0 = x_4 + 1$$
$$L : 0 = -\,3x_4 + 3y_4 + z_4$$
$$T : 0 = -\,y_4 - 2$$

from which $x_4 = -\,1,\ y_4 = -\,2,\ z_4 = 3$ and $\pi_5 = \dfrac{\sigma d^3}{\rho Q^2}$

Using the continuity principle, $\pi_5 \propto \dfrac{\sigma}{\rho V^2 d}$

which can also be written as, $\pi_5 = \dfrac{\sigma}{\rho V^2 d}$.

The functional relationship for the dependent π-term may be written as :

$$\pi_1 = f(\pi_2, \pi_3, \pi_4, \pi_5)$$

$$\frac{S_c}{d} = f\left(\frac{\Gamma d}{Q}, \frac{\sqrt{gd}}{V}, \frac{\mu}{\rho V d}, \frac{\sigma}{\rho V^2 d}\right)$$

$$= f_1\left(\frac{\Gamma d}{Q}, \frac{V}{\sqrt{gd}}, \frac{\rho V d}{\mu}, \frac{\rho V^2 d}{\sigma}\right).$$

Example 5.10. *The thrust T of a propeller depends on its diameter D, fluid density ρ, dynamic viscosity μ, the revolutions per unit time N, and the velocity of advance V with respect to the undisturbed fluid. By means of dymensional analysis, show that the appropriate dimensional parameters are :*

$$T = \rho V^2 D^2 f\left(\frac{\mu}{\rho V D}, \frac{ND}{V}\right). \qquad \textit{(RGPV, June 2013)}$$

Solution. The functional relationship between the dependent and independent variables can be expressed as

$$T = f(D, \rho, \mu, N, V)$$

Tabulating these variables and expressing each in terms of its dimensions, the following may be written.

S. No.	Variable	Symbol	Dimensions
1.	Thrust (a force)	T	MLT^{-2}
2.	Diameter	D	L
3.	Fluid density	ρ	ML^{-3}
4.	Dynamic viscosity	μ	$ML^{-1}T^{-1}$
5.	Revolution per unit time	N	T^{-1}
6.	Velocity of advance	V	LT^{-1}

From the table, we observe that the total number of variables affecting thrust T are 6. Hence $n = 6$. The number of dimensions (all three ; M, L and T are present) involved in totality are 3. Hence $m = 3$.

Therefore, the number of dimensionless parameters is $(n - m)$, that is $(6 - 3 = 3)$.3. Selecting three repeating variables, namely, ρ, V, and D, one can write each of the dimensions parameter *i.e.* π-terms as

$$\pi_1 = \rho^{x_1}V^{y_1}D^{z_1}T, \quad \pi_2 = \rho^{x_2}V^{y_2}D^{z_2}\mu, \text{ and } \pi_3 = \rho^{x_3}V^{y_3}D^{z_3}N$$

Taking π_1 first, and writing each of the terms by its dimensions, we can write :

$$M^0L^0T^0 = (ML^{-3})^{x_1}(LT^{-1})^{y_1}(L)^{z_1}MLT^{-2}$$

Applying the principle of dimensional homogeneity, we can evaluate the power indices x_1, y_1 and z_1 by equating power indices of each dimension M, L and T on both the sides. Thus, we obtain :

$$M : 0 = x_1 + 1 \quad \text{or} \quad x_1 = -1$$
$$L : 0 = -3x_1 + y_1 + z_1 + 1$$
$$T : 0 = -y_1 - 2, \quad \text{or} \quad y_1 = -2, \text{ and } -3(-1) + (-2) + z_1 + 1 = 0$$

which gives $z_1 = -2.$

The first, dimensionless parameter, $\pi_1 = \rho^{-1}V^{-2}D^{-2}T$

or
$$\pi_1 = \frac{T}{\rho V^2 D^2}$$

The second one, $\pi_2 = \rho^{x_2}V^{y_2}D^{z_2}\mu$

Expressing it dimensionally,

$$M^0L^0T^0 = (ML^{-3})^{x_2}(LT^{-1})^{y_2}(L)^{z_2}(ML^{-1}T^{-2})$$

Equating power indices of M, L and T on each side

$$M : 0 = x_2 + 1 \quad \text{or} \quad x_2 = -1$$
$$L : 0 = -3x_2 + y_2 + z_2 - 1$$
$$T : 0 = -y_2 - 1, \quad \text{or} \quad y_2 = -1$$

Substituting for $x_2 = -1$ and $y_2 = -1$, we obtain

$$L : 0 = -3(-1) + (-1) + z_2 - 1, \text{ from which, } z_2 = -1$$

The second dimensionless parameter

$$\pi_2 = \rho^{-1}V^{-1}D^{-1}\mu \quad \text{or} \quad \pi_2 = \frac{\mu}{\rho VD}$$

The third π-term, $\pi_3 = \rho^{x_3}V^{y_3}D^{z_3}N$

Writing it dimensionally,

$$M^0L^0T^0 = (ML^{-3})^{x_3}(LT^{-1})^{y_3}(L)^{z_3}(T^{-1})$$

Equating power indices of M, L, and T, we obtain

$$x_3 = 0, \quad y_3 = -1 \quad \text{and} \quad z_3 = 1,$$

and
$$\pi_3 = \rho^0V^{-1}D^{+1}N = \frac{ND}{V}$$

These π-terms can be expressed in terms of the dependent (π_1) and independent ones, (π_2 and π_3), as under

$$\pi_1 = f(\pi_2, \pi_3)$$

or $$\frac{T}{\rho V^2 D^2} = f\left(\frac{\mu}{\rho VD}, \frac{ND}{V}\right) \quad \text{or} \quad T = \rho V^2 D^2 f\left(\frac{\mu}{\rho VD}, \frac{ND}{V}\right).$$

5.5. Principles Governing Testing of Models (Dynamic Similitude)

The study of dynamically similar motions is based on the theory of models, and has applications in the design of experiments and the correlation of experimental data for solving many field problems. Exact analytical solutions are possible from the equations of motion only for a limited number of flow problems. However, one must rely on a large amount of experimental information to solve many problems in fluid mechanics. For such problems, the equations of motion and the boundary conditions are generally too complex to permit purely analytical prediction of the prototype behaviour.

In recent years hydraulic model studies have become increasingly popular and economically advantageous among designers and engineers in every field of fluid mechanics. The training of alluvial rivers, design of spillways, harbour works, pipe and open channel transitions etc., are studied in detail by working on small scale models ; modern aircrafts are manufactured after conducting exhaustive tests on small scale models in a wind tunnel ; the performance of fluid machines such as turbines, pumps, compressors, and fans are studied on small models ; the power requirement of a ship and its performance are studied by testing a small scale model of ship in a towing tank. The main objectives achieved through dimensional analysis and subsequent model testing may be :

(1) to obtain a relationship amongst variables influencing a problem in term of dimensionless parameters. The knowledge of this relationship is essential for planning and conducting the model tests and thereafter in presenting the test results ;

(2) to obtain experimental data from model tests for predicting the performance of its prototype under different conditions of flow it may be subjected to during its operation ; and

(3) to check the validity of assumptions made in the design of certain hydraulic structures. The design of many hydraulic structures is usually based on certain assumptions which may not be fully satisfied. It is always in the best interest, to find out whether the assumptions made are on the safer side and also to ascertain whether the structure so designed will serve the desired purpose.

Since it is not necessary to use the same fluid as will flow in the prototype, it is obvious that the model may be larger, smaller or even of the same size as the prototype. The prototype is the full-size machine, aircraft, ship or any device which is to be studied by model tests. The choice of the fluid and the geometrical scale will be limited only by the practical considerations. It has already been mentioned, that there exists a functional relationship among the several pertinent variables involved in the problem, and that these may be grouped together into a number of dimensionless groups of π-terms. If complete similarity is to exist between the flow in the prototype and the flow in its model, every dimensionless parameter referring to the conditions in the model must have the same numerical value as the corresponding parameter referring to the prototype. Since all geometric, kinematic and dynamic π-terms must then be the same in both the cases, it is evident that the model and the prototype must be completely similar :

(*i*) geometrically (*ii*) kinematically, and (*iii*) dynamically.

5.6. Types of Similarity

For complete similarity to exist between the model and its prototype, it is necessary that they must be geometrically, kinematically and dynamically similar. Thus there are three types of similarities.

5.6.1. Geometric Similarity

This type of similarity is obtained when the solid boundaries that control the flow of a fluid are geometrically similar, in addition to the model being geometrically similar to its prototype. The geometric similarity is thus the similarity of shape.

To satisfy this condition the ratios of the corresponding lengths in the model and its prototype must be same, thus

$$\frac{l_p}{l_m} = \frac{b_p}{b_m} = \frac{d_p}{d_m} = l_r$$

This ratio is also known as the scale factor. The subscripts p and m refer to the prototype and model respectively, and r stands for the ratio.

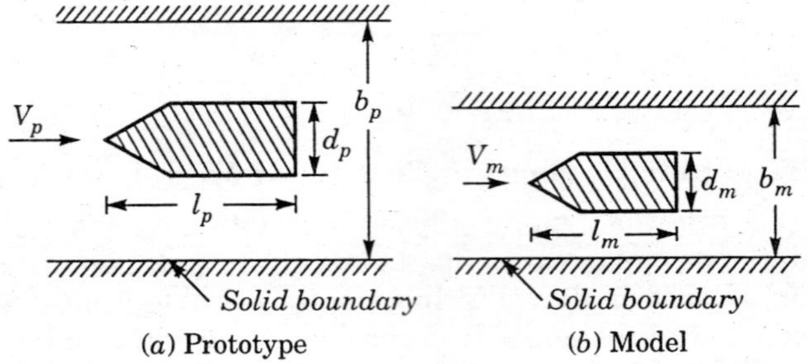

Fig. 5.1. Geometric similarity between model and prototype.

5.6.2. Kinematic Similarity

The kinematic similarity is the similarity of motion. For kinematic similarity to exist the streamline pattern in the model must be same as that in its prototype. It follows that the ratios of the kinematic quantities representing the flow characteristics such as the time, velocity, acceleration and discharge must be same at all corresponding points (the corresponding points are similarly located points in each system), thus for kinematic similarity

$$t_r = \frac{t_p}{t_m}, \qquad V_r = \frac{V_p}{V_m}, \qquad a_r = \frac{a_p}{a_m}, \qquad Q_r = \frac{Q_p}{Q_m},$$

where t, V, a and Q represent time, velocity, acceleration and discharge respectively.

The length, velocity and time are interrelated by

$$l = V \, . \, t$$

and the length ratio, velocity ratio, and time ratio may be expressed as

$$l_r = V_r . t_r, \qquad V_r = \frac{l_r}{t_r}$$

and acceleration and discharge ratios may be similarly obtained as

$$a = \frac{V}{t}$$

$$\therefore \quad a_r = \frac{V_r}{t_r} = \frac{l_r/t_r}{t_r} = \frac{l_r}{t_r^2} \quad \text{and} \quad Q = A.V = (l)^2 V.$$

$$\therefore \quad Q_r = l_r^2 . V_r = \frac{l_r^3}{t_r}.$$

Example 5.11. *If in a model study the length ratio is 1/40 and the velocity ratio is 1/6.34 ; find the discharge in the model required for a design discharge of 10,000 m³/s in the prototype, what duration of time in the model represents one day in the prototype ?*

Solution. Discharge, $Q = AV$

$$\therefore \quad \frac{Q_p}{Q_m} = \frac{A_p}{A_m}\frac{V_p}{V_m} = \left(\frac{l_p}{l_m}\right)^2 \frac{V_p}{V_m} = (40)^2 \times 6.34$$

$$\therefore \quad Q_m = \frac{Q_p}{1600 \times 6.34} = \frac{10{,}000}{1600 \times 6.34} = 0.985 \text{ m}^3\text{/s}.$$

Time ratio

$$t_r = \frac{l_r}{V_r} = \frac{1/40}{1/6.34} = \frac{6.34}{40}$$

$$t_r = \frac{t_m}{t_p}$$

$$t_m = t_p \times \frac{6.34}{40} = 0.1585 \, t_p = 0.1585 \times 24 \text{ hours} = 3.81 \text{ hours}.$$

5.6.3. Dynamic Similarity

In order to maintain the kinematic similarity, the ratio of forces acting on corresponding fluid masses in the two systems must have the same value throughout the flow. Thus, if the corresponding forces have the same ratio throughout the flow field, the flow in the model and its prototype will be *dynamically similar*. The dynamic similarity is thus, the similarity of forces. For each type of similarity (kinematic or dynamic) geometric similarity is a prerequisite. The requirement of a single ratio for corresponding forces shows that the polygons of forces on corresponding fluid elements in the two system are geometrically similar.

Thus, if the flow fields in the model and its prototype are dynamically similar, it is implied that they are necessarily, kinematically and geometrically similar. Alternatively, if the two flow fields are kinematically similar, they must also be dynamically and geometrically similar.

5.7. Forces Influencing Hydraulic Phenomena

The forces which may influence the flow characteristics of a problem are :

(*i*) *Inertia Force* . If a fluid mass m is acted upon by a resultant force F which produces an acceleration a in the same direction as that of the force, the Newton's second law gives

$$F = m \times a.$$

If an imaginary force of equal magnitude ($m \times a$) is applied to the fluid mass in a direction opposite to F, then the net force on the fluid element would be zero, the acceleration would also become zero. This imaginary force required to bring acceleration to zero is known as the *inertia force* and represents the reluctance of the fluid mass to be accelerated. Thus,

$$\text{Inertia force} = \text{Mass} \times \text{Acceleration}$$

$$= m.a = \rho\, l^3 \left(\frac{V^2}{l}\right) = \rho l^2 V^2 \qquad \text{...(5.5)}$$

(*ii*) *Viscous force* $\quad F_V = \tau \,.\, A = \mu \dfrac{du}{dy} \,.\, A$

$$= \mu \frac{V}{l} l^2 = \mu V l \qquad \text{...(5.6)}$$

(*iii*) *Gravity force* $\quad F_g = \text{Mass} \times \text{Acceleration due to the gravity} = M.g = \rho l^3.g$ ...(5.7)

(*iv*) *Pressure force* $\quad F_p = \Delta p \,.\, A = \Delta p. l^2$ ...(5.8)

(*v*) *Elastic force* $\quad F_e = E.A = El^2$ ...(5.9)

(*vi*) *Surface tension force* $F_s = \sigma.l$ ...(5.10)

In problems of fluid flow, the inertia force will always exist and hence it is customary to find out the force ratios with respect to the inertia force, thus

(*a*) the ratio of inertia to viscous forces is,

$$\frac{F_i}{F_v} = \frac{\rho l^2 V^2}{\mu V l} = \frac{\rho V l}{\mu} \qquad \text{...(5.11)}$$

since this is the ratio of two forces, it is dimensionless. The quantity $\rho Vl/\mu$ denoted by **R** is known as the Reynolds number.

$$\frac{F_i}{F_g} = \frac{\rho l^2 V^2}{\rho l^3 g} = \frac{V^2}{l.g} \qquad \text{...(5.12)}$$

and the dimensionless quantity $\dfrac{V}{\sqrt{lg}}$ is called the Froude (pronounced as Frood) number and is represented by **F**.

(*b*) The ratio of inertia to pressure forces is

$$\frac{F_i}{F_p} = \frac{\rho l^2 V^2}{\Delta p l^2} = \frac{\rho V^2}{\Delta p} \qquad \text{...(5.13)}$$

and the dimensionless quantity defined by

$$\mathbf{E} = \frac{V}{\sqrt{2\Delta p/\rho}}$$

is known as the Euler (pronounced as Oy-ler) number. Its slightly modified form $\Delta\rho/\frac{1}{2}\rho V^2$ which is also dimensionless is known as the *pressure coefficient.*

(*c*) The ratio of the inertia to elastic forces,

$$\frac{F_i}{F_e} = \frac{\rho l^2 V^2}{E l^2} = \frac{\rho V^2}{E} = \frac{V^2}{E/\rho} \quad \text{...(5.14)}$$

The parameter $\rho V^2/E$ is known as the Cauchy number. The dimensionless quantity,

$$\mathbf{M} = \frac{V}{\sqrt{E/\rho}}$$

is called the Mach number. Since $\sqrt{E/\rho}$ represents the velocity of sound in the fluid medium, the Mach number is generally regarded as the ratio of fluid velocity to the velocity of sound in the fluid medium.

(*d*) The ratio of inertia to surface tension forces.

$$\frac{F_i}{F_s} = \frac{\rho l^2 V^2}{\sigma l} = \frac{\rho V^2 l}{\sigma} \quad \text{...(5.15)}$$

The dimensionless number $\dfrac{\rho V^2 l}{\sigma} = \mathbf{W}$

is known as the Weber number.

The dynamic similarity between the model and the prototype involving all the above mentioned forces will be obtained when the relevant dimensionless numbers—Reynolds, Froude, Euler, Weber and Mach are the same in the two systems at the corresponding points. Thus, for dynamic similarity

$$(\mathbf{R})_p = (\mathbf{R})_m,\ (\mathbf{F})_p = (\mathbf{F})_m,\ (\mathbf{E})_p = (\mathbf{E})_m,\ (\mathbf{M})_p = (\mathbf{M})m \text{ and } (\mathbf{W})_p = (\mathbf{W})_m$$

These equations represent the equality of dimensionless force ratios for the prototype and its model.

In most of the engineering problems, some of the forces stated above may not exist or may be of negligible magnitude. A good understanding of the fluid phenomenon is, therefore, necessary to determine which forces are irrelevant or negligible.

5.8. Significance of Dimensionless Numbers

If the number of variables affecting a flow problem are known, these can be arranged into suitable dimensionless parameters by the method of dimensional analysis. From experience and judgment, less important parameters may be dropped out. Thus we are left with the most important parameters which have a far greater influence upon the phenomenon than the parameters dropped out. The dynamic similarity between the model and its prototype can now be attained by making these parameters same for the model and the prototype. Fortunately, in many problems the important dimensionless parameter can be reduced to one. But in certain problems there are two dimensionless parameters of the same order of magnitude, and obviously these parameters must be equal for model and prototype to obtain complete dynamic similarity.

5.8.1. Reynolds Number

It has been described as the ratio of inertia to viscous forces and, therefore, in problems where viscosity has an important role, this dimensionless parameter must have the same value for model and prototype to achieve dynamic similarity.

Flow in a pipe represents a case in which inertia and viscous forces are dominating. The pressure changes along the pipe are small and can be neglected. Surface tension effects are absent since pipe fluid is not in contact with any other fluid. Since the flow is in a closed conduit, gravity has no effect on the flow pattern, the weight of every fluid particle being counterbalanced by the static buoyant force, and if the flow is steady and the fluid incompressible, effects of compressibility are also absent. Thus out of total of five dimensionless parameters (Reynolds, Froude, Euler, Mach and Weber), only one has been found to have a dominating influence on the flow. Therefore, the equality of Reynolds number in model and its prototype will make the two systems dynamically similar.

Thus, $$\frac{\rho_p V_p l_p}{\mu_p} = \frac{\rho_m V_m l_m}{\mu_m}$$

For pipe flow, the characteristic length is the diameter, therefore, for dynamic similarity, one gets

$$\frac{\rho_p V_p d_p}{\mu_p} = \frac{\rho_m V_m d_m}{\mu_m}$$

if the same fluid is used in the model as in the prototype,

$$V_p d_p = V_m d_m$$

and accordingly the velocity in the model (a smaller diameter pipe) must be higher than the corresponding velocity in the prototype. If the model tests are impracticable at such higher velocities, it is permissible to use some other fluid of much lower viscosity for the model experiment.

It is of practical interest to observe that for experiments on a model aeroplane in air,

$$\frac{\rho_p}{\rho_m} = \frac{\mu_p}{\mu_m} = 1,$$

and $$V_p l_p = V_m l_m.$$

This means that a $2m$ model of a $20m$ wing expected to operate at 200 km per hour should be tested in a wind tunnel at $\frac{200 \times 20}{2}$ = 2000 km per hour,

if the viscous effects are to be truly represented. This is clearly impracticable, but if the wind tunnel air is compressed to 5 atmospheres, the test velocity is reduced to 400 km per hour.

Example 5.12. *A geometrically similar model of an air duct is built to 1/25 scale and tested with water which is 50 times more viscous and 800 times denser than air. When tested under dynamically similar conditions, the pressure drop is 200 kN/m^2 in the model.*

Find the corresponding pressure drop in the full scale prototype and express in cm of water.

Solution. For dynamic similarity, the Reynold numbers must be equal in the model and prototype, thus

$$\frac{\rho_p V_p l_p}{\mu_p} = \frac{\rho_m V_m l_m}{\mu_m}$$

$$\therefore \quad \frac{V_p}{V_m} = \frac{\mu_p}{\mu_m} \times \frac{\rho_m}{\rho_p} \times \frac{l_m}{l_p}$$

Given that $$\frac{\mu_p}{\mu_m} = \frac{1}{50}$$

$$\frac{\rho_p}{\rho_m} = \frac{1}{800} \quad \text{and} \quad \frac{l_p}{l_m} = 25$$

$$\therefore \quad \frac{V_p}{V_m} = \frac{1}{50} \times 800 \times \frac{1}{25} = \frac{16}{25}$$

The pressure $$p = \frac{F}{A} = \frac{\rho l^2 V^2}{l^2} = \rho V^2$$

$$\left(\frac{\text{Pressure drop in prototype}}{\text{Pressure drop in model}}\right) = \frac{\rho_p}{\rho_m}\frac{V_p^2}{V_m^2} = \frac{1}{800}\times\left(\frac{16}{25}\right)^2 = \frac{1}{1952}$$

∴ Pressure drop in the prototype

$$= \text{Pressure drop in the model} \times \frac{1}{1952} = 200 \times 10^3 \times \frac{1}{1952} \text{ N/m}^2$$

$$= 200 \times 10^3 \times \frac{1}{1952}\times\frac{1}{9810} \text{ m of water} = 200 \times 10^3 \times \frac{1}{1952}\times\frac{100}{9810} \text{ cm of water}$$

$$= 1.045 \text{ cm of water.}$$

Example 5.13. *Model tests have been conducted to study the energy loss in a pipeline of 1 m diameter required to transport kerosene of sp gr. 0.80 and dynamic viscosity 0.02 poise at the rate of 2000 litre/sec. Tests were conducted on a 10 cm diameter pipe using water at 20°C. What is the flow rate in the model ? If the energy head loss in 30 m length of the model is measured as 44.0 cm of water, what will be the corresponding head loss in the prototype ? What will be the friction factor for the prototype pipe.*

The absolute viscosity of water at 20°C = 1.0 × 1.0^{-2} *poise.*

Solution. For pipe flow, the dynamic similarity will be achieved if the Reynolds numbers in the model and the prototype are same, thus

$$\left(\frac{\rho VD}{\mu}\right)_m = \left(\frac{\rho VD}{\mu}\right)_p$$

$$\therefore \quad \frac{V_m}{V_p} = \frac{\rho_p}{\rho_m}\cdot\frac{\mu_m}{\mu_p}\cdot\frac{D_p}{D_m} = \frac{102\times 0.80}{102}\cdot\frac{1.0\times 10^{-2}}{0.02}\times\frac{100}{10} = 4.0$$

but
$$V_p = \frac{2000/1000}{\frac{\pi}{4}\times(1)^2} = 2.54 \text{ m/s.}$$

$$\therefore \quad V_m = 2.54 \times 4.0 = 10.16 \text{ m/s}$$

∴ Model discharge,
$$Q_m = A_m V_m = \frac{\pi}{4}\left(\frac{10}{100}\right)^2 \times 10.16 = 0.079 \text{ m}^3\text{/s} = 79.8 \text{ lit/sec.}$$

Resistance to flow,
$$R = \rho L^2 V^2 \phi\left(\frac{\rho VD}{\mu}\right)_p$$

As the Reynolds numbers of model and prototype are same, hence

$$\left(\frac{R}{\rho L^2 V_2}\right)_m = \left(\frac{R}{\rho L^2 V^2}\right)_p$$

But, being a force
$$R = \text{Pressure} \times \text{Cross- sectional area}$$
$$= \gamma h_f\left(\frac{\pi}{4}D^2\right) = \rho g h_f \frac{\pi}{4}D^2$$

Thus, we have
$$\left(\frac{\rho g h_f\cdot\frac{\pi}{4}D^2}{\rho L^2 V^2}\right)_m = \left(\frac{\rho g h_f\cdot\frac{\pi}{4}D^2}{\rho L^2 V^2}\right)$$

$$\frac{(h_f)_m}{(h_f)_p} = \frac{D_p^2}{D_m^2}\cdot\frac{L_m^2}{L_p^2}\cdot\frac{V_m^2}{V_p^2} = \left(\frac{100}{10}\right)^2\left(\frac{10}{100}\right)^2\times\left(\frac{10.16}{2.54}\right)^2 = 16.0$$

∴ Head loss in the prototype,

$$(h_f)_p = \frac{44.0}{16.0} = 2.75 \text{ cm.}$$

This head loss is in prototype length of 30 × 10 = 300 m.

The head loss due to friction in a pipe is given by,

$$h_f = f\frac{l}{d}\frac{V^2}{2g}$$

In case of prototype, $(h_f)_p = f_p\dfrac{l_p}{d_p}\cdot\dfrac{V_p^2}{2g}$

$$f_p = (h_f)_p \cdot \frac{d_p}{l_p}\cdot\frac{2g}{V_p^2} = \frac{2.75}{100}\times\frac{1.0}{300}\times\frac{2\times 9.81}{(2.54)^2} = 0.000279.$$

Since the friction factor depends upon the Reynolds number (for smooth pipes) and as the Reynolds numbers for model and prototype are the same, the friction factor for model and the prototype will also be same.

Example 5.14. *A model of submarine is scaled down 1/20 of the prototype and is to be tested in a wind tunnel. The speed of the prototype at which we are to estimate the drag is 8 m/s. What should be the free-stream velocity of the air ? What will be the ratio of the drag between the model and the prototype ? Explain why there would be no dynamic similarity if the submarine prototype is moved near the free surface.*

$\nu_{sea\ water} = 1.21 \times 10^{-2}\ cm^2/sec$; $\nu_{air} = 1.64 \times 10^{-4}\ cm^2/sec$

$\rho_{sea\ water} = 1027\ kg/m^3$; $\rho_{air} = 1.34\ kg/m^3$.

Solution. Since the submarine is to overcome the viscous resistance, the similarity of Reynolds number is essential for dynamic similarity between the model and the prototype.

$$\left(\frac{Vl}{\nu}\right)_p = \left(\frac{Vl}{\nu}\right)_m$$

$$\therefore \quad V_m = V_p\frac{l_p}{l_m}\cdot\frac{\nu_m}{\nu_p} = 8\times 20\times\frac{1.64\times 10^{-1}}{1.21\times 10^{-2}} = 2166 \text{ m/s}.$$

The drag force F = Mass × Acceleration = $\rho l^3 \dfrac{V}{l} = \rho l^2 V^2$

$$\therefore \text{ The ratio of drag force, } \frac{F_p}{F_m} = \frac{\rho_p}{\rho_m}\left(\frac{l_p}{l_m}\right)^2\times\left(\frac{V_p}{V_m}\right)^2$$

$$F_r = \frac{F_p}{F_n} = \frac{1027}{1.34}\times(20)^2\times\left(\frac{l_m}{l_p}\times\frac{v_p}{v_m}\right)^2 = \frac{1027}{1.34}\times 400\times\left(\frac{1}{20}\times\frac{1.21\times 10^{-2}}{1.64\times 10^{-1}}\right)^2 = 4.16.$$

If the submarine is moved near the free surface the gravity forces will also become important and the drag will consist of viscous resistance and the wave resistance. The model will then have to be tested in accordance with the similarity of Reynolds and Froude numbers.

Example 5.15. (*a*) *What are the various important dimensionless groups in fluid mechanics ? Under what circumstances is each of these groups important ?*

(*b*) *The drag of a small submarine hull is desired when it is moving far below the surface of water. A 1/10 scale model is to be tested. What dimensionless group should be duplicated between model and prototype and why ? If the drag of prototype at 1 knot is desired, at what speed should the model be moved to give the drag to be expected by the prototype ? Would this result still be true if the prototype were to move close to the surface ? Explain.*

Solution. When the submarine moves far below the surface of water, viscous forces predominate and the equality of Reynolds number satisfies the condition for dynamic similarity, *i.e.*

$$\left(\frac{VL}{\nu}\right)_{model} = \left(\frac{VL}{\nu}\right)_{proto.}$$

For the same fluid, $\nu_m = \nu_p.$

The Reynolds law then gives the velocity in the model as

$$V_m = V_p \times \frac{L_p}{L_m} = 1 \times 10 = \mathbf{10\ knots.}$$

The model must be moved at a speed of 10 knots in order to simulate dynamically similar flow conditions. When the submarine moves close to the surface, the waves offer additional resistance, and therefore, the gravity effect must also be considered. The above result would, therefore, not be valid then.

Example 5.16. *An orificemeter to carry water is calibrated with air in a geometrically similar model at 1/5 the prototype scale. Dynamically similar flow will be obtained when the discharge ratio (air to water) is*

(i) 0.4, *(ii) 2.5,* *(iii) 62.5.*

Assume the ratio of kinematic viscosity of air to water as 12.5.

Solution. In case of the orificemeter meant for carrying water but calibrated with air, the dynamically similar flow will be obtained when the Reynolds number in model and prototype are equal, thus

$$\frac{V_m L_m}{\nu_m} = \frac{V_p L_p}{\nu_p}$$

From the given data, $\frac{L_p}{L_m} = 5$ and $\frac{\nu_m}{\nu_p} = 12.5$

The velocity ratio, $V = V_p/V_m = \frac{L_m}{L_p} \times \frac{\nu_p}{\nu_m} = \frac{1}{5} \times \frac{1}{12.5} = \frac{1}{62.5}$

Discharge ratio (air to water) $\frac{Q_m}{Q_p} = \frac{L_m^2 V_m}{L_p^2 V_p} = \left(\frac{1}{5}\right)^2 \times 62.5 = \mathbf{2.5.}$

Example 5.17. *Resistance R, to the motion of a completely submerged body is given by $R = \rho l^2 V^2 \phi (Vl/\nu)$, where ρ and ν are density and kinematic viscosity of the fluid while l is the length of the body and V is the velocity of flow. If the resistance of a one-eight scale air-ship model when tested in water at 12 m/s is 216 N, what will be the resistance in air of the air-ship at the corresponding speed ? Kinematic viscosity of air is 13 times that of water and the density of water is 810 times that of air.*

Solution. From the resistance equation, it is obvious that flow in the model will be dynamically similar if the Reynolds number are equal in both the systems. Thus, if

$$\left(\frac{Vl}{\nu}\right)_m = \left(\frac{Vl}{\nu}\right)_p \quad ...(i)$$

then

$$\left(\frac{R}{\rho l^2 V^2}\right)_m = \left(\frac{R}{\rho l^2 V^2}\right)_p \quad ...(ii)$$

Thus the prototype-model similarity is obtained from Reynolds law, *i.e.* Eq. (*i*),

$$V_p = V_m \cdot \frac{l_m}{l_p} \cdot \frac{\nu_p}{\nu_m} = 12 \times \frac{1}{8} \times 13 = 19.5 \text{ m/s}$$

At this prototype velocity, the resistance of the airship is obtained from Eq. (*ii*),

$$R_p = R_m \frac{\rho_p}{\rho_m} \cdot \frac{l_p^2}{l_m^2} \cdot \frac{V_p^2}{V_m^2} = 216 \times \frac{1}{810} \times (8)^2 \times \left(\frac{19.5}{12}\right)^2 = \mathbf{45.06\ N.}$$

Example 5.18. (*a*) *Show by dimensional analysis that Darcy's friction factor is a function of the Reynolds number and relative roughness.*

(b) Water with kinematic viscosity of 0.0_ cm²/s flows in a 20 cm pipe with absolute roughness 0.004 cm with a mean velocity of 1 m/s and observed head loss is 60 cm in 100 m length of pipe. If a glycerine-water mixture of kinematic viscosity of 0.02 cm²/s flow in a 10 cm pipe of roughness 0.002 cm with a velocity of 4 m/s, how much head loss per 100 m of pipe should be expected ?

Solution. (*I*) Water-pipe line

$$\nu = 0.01 \text{ cm}^2/\text{s} \; ; \; d = 20 \text{ cm}, \; k_s = 0.004 \text{ cm}$$

$$V = 1 \text{ m/s}, \; h_L = \frac{0.60}{100} \; (0.6\%)$$

(*II*) Glycerine-water mixture

$$\nu = 0.02 \text{ cm}^2/\text{s}, \; d = 10 \text{ cm}, \; k_s = 0.002 \text{ cm}, \; V = 4 \text{ m/s}, \; h_L = ?$$

For dynamically similar flow conditions in both the pipe systems, the Reynolds number ; the relative roughness and the friction factor must be equal

$$\mathbf{R}_1 = \mathbf{R}_2 \; ; \; (k_s/d)_1 = (k_s/d)_2 \; ; \quad \text{and} \quad f_1 = f_2$$

upon substituting the given data, one can easily verify the above mentioned first two conditions.

Darcy-Weibach equation gives the head loss

$$h_{L_1} = f_1 \frac{L_1}{d_1} \frac{V_1^2}{2g}$$

Substituting the data for the water pipeline,

$$0.60 = f_1 \times \frac{100}{0.20} \times \frac{(1)^2}{2 \times 9.81}$$

$$f_1 = 0.0235, \text{ hence } f_2 = f_1 = 0.0235.$$

Head loss in flow of glycerine-water mixture

$$h_{L_2} = f_2 \frac{L_2}{d_s} \frac{V_2^2}{2g} = 0.0235 \times \frac{100}{0.10} \times \frac{(4)^2}{2 \times 9.81} = \mathbf{19.16 \text{ m.}}$$

Example 5.19. *A pipe of diameter 1.5 m is required to transport an oil of specific gravity 0.90 and viscosity 3 × 10⁻² poise at the rate of 3000 lit / sec. Tests were conducted on a 15 cm diameter pipe using water at 20°C. Find the velocity and rate of flow in the model. Viscosity of water at 20°C = 0.01 poise.*

(RGPV, Dec., 2005)

Solution. In pipe flow, the inertia and viscous forces are the dominant forces, it is the Reynolds number that will govern the dynamic similarity between the model and its prototype. Hence one can write

$$(\text{Reynolds number})_{\text{model}} = (\text{Reynolds number})_{\text{prototype}}$$

or,
$$\left(\frac{\rho V d}{\mu}\right)_m = \left(\frac{\rho V d}{\mu}\right)_p \qquad ...(i)$$

$$\text{sp. gr. of oil} = 0.90 = \frac{\text{density of oil, } \rho}{\text{density of water}} = \frac{\rho_{oil}}{1000} \; (\text{kg/m}^3)$$

or,
$$\rho_{\text{oil}} = 0.90 \times 1000 = 900 \text{ kg/m}^3,$$

In other words,
$$\rho_p = 900 \text{ kg/m}^3, \; \rho_m = 1000 \text{ kg/m}^3, \; d_p = 1.5 \text{ m}$$

$$\mu_p = 3 \times 10^{-2} \text{ poise}, \; \mu_m = 0.01 \text{ poise}, \; d_m = 15 \text{ cm} = 0.15 \text{ m}$$

Also given,
$$Q_p = 3000 \text{ lit/sec.}$$

$$\frac{\rho_p}{\rho_m} = \frac{900}{1000} = 0.9, \; \frac{\mu_p}{\mu_m} = \frac{3 \times 10^{-2}}{1 \times 10^{-2}} = 3, \; \frac{d_p}{d_m} = \frac{1.5}{0.15} = 10$$

Using Eq. (*i*), given above,

$$\frac{V_p}{V_m} = \frac{\rho_m}{\rho_p} \cdot \frac{\mu_p}{\mu_m} \cdot \frac{d_m}{d_p} = \frac{1}{0.9} \times 3 \times \frac{0.15}{1.5} = \frac{1}{3}$$

Velocity in the model, $V_m = 3V_p$

$$Q_m = A_m . V_m, \text{ and } Q_p = A_p . V_p, \text{ or } \frac{Q_m}{Q_p} = \frac{A_m}{A_p} \cdot \frac{V_m}{V_p}$$

or,

$$Q_m = Q_p . \frac{d_m^2}{d_p^2} \cdot \frac{V_m}{V_p} = Q_p \times \left(\frac{1}{10}\right)^2 \times 3 = \frac{3Q_p}{100} = \frac{3 \times 3000}{100} = \textbf{90 lit/sec.}$$

$\therefore$ Rate of flow in the model = 90 lit/sec.

Velocity in the prototype, $V_p = \frac{4Q_p}{\pi d_p^2} = \frac{4 \times 3000/1000 \text{ (m}^3\text{/s)}}{\pi \times (1.5)^2}$ = **1.697 m/s.**

5.8.2. Froude Number

In problems where a free surface is present or when there is an interface between two immiscible fluids the gravity forces are of importance. When surface waves are formed such as in the sea, the gravity force becomes far more important than the viscous force. In open hydraulic structures such as spillways, wiers, stilling basins, channel transitions, sluices etc, where change in the elevation of liquid surface are significant, the forces due to gravity are predominant. And, therefore, the equality of Froude number (which measures the ratio of inertia to gravity forces) for model and its prototype satisfies the condition for dynamic similarity. Thus

$$\frac{V_p}{\sqrt{l_p g_p}} = \frac{V_m}{\sqrt{l_m g_m}}$$

If the model experiments are conducted at the same place where the prototype is to operate, then $g_p = g_m$, and the following is obtained :

(*i*) $$V_p = V_m \sqrt{\frac{l_p}{l_m}}$$

the velocity ratio, $$V_r = \frac{V_p}{V_m} = \sqrt{\frac{l_p}{l_m}} = \sqrt{l_r} \quad \text{...(5.16)}$$

(*ii*) The ratio of corresponding times may be obtained as under :

$$t_m = \frac{l_m}{V_m} \text{ and } t_p = \frac{l_p}{V_p}$$

and hence the time ratio, $$t_r = \frac{t_p}{t_m} = \frac{l_p/V_p}{l_m/V_m} = \frac{l_p}{l_m} \cdot \frac{V_m}{V_p}$$

$$t_r = \frac{l_r}{\sqrt{l_r}} = \sqrt{l_r} \quad \text{...(5.17)}$$

(*iii*) The acceleration may be expressed as

$$a = \frac{V}{t}$$

and, therefore, the acceleration ratio

$$a_r = \frac{a_p}{a_m} = \frac{V_p/t_p}{V_m/t_m} = \frac{V_p}{V_m} \cdot \frac{t_m}{t_p} = \sqrt{l_r} \times \frac{1}{\sqrt{l_r}} = 1 \quad \text{...(5.18)}$$

The acceleration in the model is seen to be the same as that in the prototype.

(*iv*) The discharge, $$Q = A.V = l^2 \frac{l}{t} = \frac{l^3}{t}$$

$\therefore$ discharge ratio, $$Q_r = \frac{l_r^3}{t_r} = \frac{l_r^3}{\sqrt{l_r}} = l_r^{5/2} \quad \text{...(5.19)}$$

(*v*) The force may be expressed from the Newton's second law

$$F = m.a = \rho l^3 . \frac{V}{t} \rho l^2 V^2$$

and the force ratio, $F_r = \rho_r l_r^2 . V_r^2$
if the same fluid is used in the model then $\rho_r = 1$ and the force ratio

$$F_r = l_r^2 l_r = l_r^3 \quad \text{...(5.20)}$$

(*vi*) The pressure intensity, $p = \frac{F}{A} = \frac{\rho l^2 V^2}{l^2} = \rho V^2$

and, therefore, the pressure ratio $p_r = \rho_r V_r^2$
for the same fluid, $\rho_r = 1$ and then, $p_r = V_r^2 = l_r$ ...(5.21)

(*vii*) The energy or work done is given by the product of force and distance, thus

$$\text{Energy} = \text{Force} \times \text{Distance}$$

$$\therefore \quad E = F \times l$$

and the energy ratio $E_r = F_r . l_r = l_r^3 \times l_r = l_r^4$ ...(5.22)

(*viii*) The momentum or impulse

$$= mV = \rho l^3 V$$

and the momentum ratio $= \rho_r l_r^3 V_r = l_r^3 V_r$ (since $\rho_r = 1$)

$$M_r = l_r^3 \sqrt{lr} = l_r^{7/2} \quad \text{...(5.23)}$$

(*ix*) The torque is given by the product of force and its perpendicular distance from the centre of rotation,

$$T = F.l$$

and the torque ratio, $T_r = F_r . l_r = l_r^3 . l_r = l_r^4$ ...(5.24)

(*x*) The power being the time rate of doing work, is given by

$$p = \frac{F.l}{t}$$

and the power ratio is, $P_r = \frac{F_r . l_r}{t_r} = \frac{l_r^3 l_r}{\sqrt{l_r}} = l_r^{7/2}$.

All these ratios represent the ratio of the corresponding quantities in the model and its prototype for Froude's similarity and are useful in determining the prototype quantities from the known corresponding quantities in the model.

Example 5.20. *A model of spillway is made to test the flow. The discharge and the velocity of flow over the model were measured as 2.5 m³/s and 1.5 m/s respectively. Find the discharge and the velocity over the prototype which is 50 times larger than its model.*

Solution. Applying Froude number similarity

$$\frac{V_p}{\sqrt{l_p g_p}} = \frac{V_m}{\sqrt{l_m g_m}}$$

Assuming $g_p = g_m$

the ratio of velocities $\frac{V_p}{V_m} = \sqrt{\frac{l_p}{l_m}} = \sqrt{50}$

$$\therefore \quad V_p = 1.50\sqrt{50} = 10.62 \text{ m/s.}$$

Discharge $Q = AV = l^2 V$

$$\therefore \quad \frac{Q_p}{Q_m} \left(\frac{l_p}{l_m}\right)^2 \frac{V_p}{V_m} = (50)^2 \sqrt{50}$$

$$\therefore \quad Q_p = 2.5(50)^2 \sqrt{50} = 44{,}200 \text{ m}^3\text{/s.}$$

Example 5.21. *A ship model 1 m long with negligible friction is tested in a towing tank at a speed of 60 cm per second. To what ship velocity does this correspond if the ship is 60 metre long. A force of 4.9 N is required to low the model. What propulsive force does this represent in the prototype?*

Solution. Since friction (viscous force) is negligible, the Froude number must be same for model and prototype. For dynamic similarity, thus

$$\frac{V_p}{\sqrt{l_p g_p}} = \frac{V_m}{\sqrt{l_m g_m}}$$

If $$g_p = g_m$$

then $$\frac{V_p}{V_m} = \sqrt{\frac{l_p}{l_m}} = (l_r)^{1/2}$$

$\therefore$ The prototype velocity, $V_p = 0.6 \times (60)^{1/2} = 4.65$ m/s.

Force ratio $F_r = l_r^{\,3}$

$\therefore$ Prototype propulsive force, $F_p = F_m \,.\, l_r^{\,3} = 4.9\,(60)^3 = 1.06 \times 10^6$ N = 1.06 MN.

Example 5.22. *A model of rectangular pier 1.5 m wide and 4.5 m long in a river is built to a scale of 1/25. The average depth of water in the river is 3 m. The model was tested in a laboratory, where the velocity of flow was maintained constant at 0.65 m/s. It was observed that the force acting on the model was 3.92 N and the height of the standing wave was 3.5 cm.*

Determine for the prototype :

(a) the corresponding speed, *(b) the force acting,*

(c) the height of the standing wave at nose, and

(d) the coefficient of drag resistance (Make necessary assumptions).

Solution. Data given $l_r = \dfrac{l_p}{l_m} = 25$ $V_m = 0.65$ m/s

$F_m = 3.92$ N $d_m = 3.0$

$h_m = 3.5$ cm.

(*a*) The flow in a river being a free surface flow affected by gravity, the dynamic similarity between the model and its prototype will be achieved be equating the Froude number, thus

$$\frac{V_p}{\sqrt{l_p g_p}} = \frac{V_m}{\sqrt{l_m g_p}}$$

$$\therefore \qquad \frac{V_p}{V_m} = \sqrt{\frac{l_p}{l_m}}$$

$$\therefore \qquad V_p = V_m \sqrt{\frac{l_p}{l_m}} = 0.65 \times \sqrt{25} = 3.25 \text{ m/s.}$$

$\therefore$ Corresponding prototype speed = 3.25 m/s.

(*b*) Force = Mass × Acceleration $= \rho l^3 \dfrac{V^2}{l} = \rho l^2 V^2$

and the force ratio $$F_r = \frac{F_p}{F_m} \rho_r l_r^{\,2} V_r^2$$

For the same fluid in the model and the prototype, $\rho_r = 1.0$, and the corresponding prototype froce

$$F_p = F_m \,.\, l_r^{\,2} V_r^{\,2} = F_m l_r^{\,3} = 3.92 \times (25)^3 = 61.25 \text{ kN.}$$

(*c*) The height of the standing wave in the prototype $= h_m \times l_r$

$= 3.5 \times 25 = 87.4$ cm.

(*d*) The coefficient of drag resistance is defined by

$$F = C_D . A\rho \frac{V^2}{2}$$

$$\therefore \qquad C_D = \frac{F/A}{\rho V^2/2}$$

and the drag coefficient in the prototype (it will be same for the model)

$$(C_D)_p = \frac{F\rho/A_p}{\rho V_p^2/2} = \frac{61250/3 \times 1.5}{1000 \times \frac{(3.25)^2}{2}} = 2.57.$$

Example 5.23. *A model of a harbour has a length scale of 1/225. Storm waves of 20 metre amplitude and 10 m/s velocity strike against the break water of the prototype harbour. Determine for the model :*

(*a*) *the size and velocity of the waves, neglecting viscous effects.*

(*b*) *if the time between the tides in the prototype is 12 hours, what should be the tidal period in the model ?*

Solution. The similarity of Froude number will satisfy the condition for dynamic similarity between the model and the prototype, thus

$$\frac{V_p}{\sqrt{g_p l_p}} = \frac{V_m}{\sqrt{g_m l_m}}$$

assuming that $g_p = g_m$, one obtains,

$$\frac{V_p}{V_m} = \sqrt{\frac{l_p}{l_m}}.$$

(*a*) Velocity of waves in the model

$$V_m = V_p \sqrt{\frac{l_m}{l_p}} = 10\sqrt{\frac{1}{225}} = \frac{10}{15} = 0.66 \text{ m/s.}$$

and the size (amplitude of the model wave)

$$= \frac{20}{225} \text{m} = 8.89 \text{ cm.}$$

(*b*) $$\text{Time} = \frac{\text{Length}}{\text{Velocity}}. \quad \therefore \quad t_r = \frac{t_p}{t_m} = \frac{l_r}{V_r} = \sqrt{l_r}$$

$$\therefore \qquad t_m = t_p \cdot \frac{1}{\sqrt{l_r}} = 12 \times \frac{1}{\sqrt{225}} \text{ hours} = \frac{12}{15} = 0.8 \text{ hr.}$$

Example 5.24. *The force required to tow a 1 : 30 scale model of a motor boat in a lake at a speed of 2 m/s is 0.5 N. Assuming that the viscous resistance due to water and air is negligible in comparison with the wave resistance, calculate the corresponding velocity of the prototype of 50 m for dynamically similar conditions. What would be the force required to propel the prototype at that velocity in the same lake ?*

Solution. Data given : $L_r = L_p/L_m = 30$, $V_m = 2$ m/s, $F_m = 0.5$ N

The wave resistance being the dominant force in comparison with the viscous resistance, the dynamic similarity will be attained when the Froude numbers in model and prototype are equal,

$$(V/\sqrt{gL})_m = (V/\sqrt{gL})_p$$

from which $$V_p = V_m \sqrt{L_p/L_m}, \text{ since } g_m = g_p$$

$$= 2\sqrt{30} = 10.95 \text{ m/s.}$$

Force $F = \rho L^2V^2$, and Force ratio, $F_r = \dfrac{F_p}{F_m} = \dfrac{\rho_p L_p^2 V_p^2}{\rho_m L_m^2 V_m^2}$

or
$$\frac{F_p}{F_m} = \left(\frac{L_p}{L_m}\right)^2 \times \left(\frac{V_p}{V_m}\right)^2 = \left(\frac{L_p}{L_m}\right)^3 = (30)^3$$

Hence, the force required to propel the prototype,

$$F_p = F_m\,(30)^3 = 0.5 \times (30)^3 = \mathbf{13500\ N.}$$

Example 5.25. *The performance of a spillway of an irrigation project is to be studied by means of a model constructed to a scale of 1 : 9, neglecting the viscous and surface tension effects, determine :*

(i) rate of flow in the model for a prototype discharge of 1500 m³/s, and

(ii) the dissipation of energy in the prototype hydraulic jump, if the jump in the model dissipates 0.33 kW.

Solution. For dynamic similarity between the model and its prototype the Froude numbers must be equal, thus

$$\left(\frac{V}{\sqrt{gl}}\right)_p = \left(\frac{V}{\sqrt{gl}}\right)_m \qquad \therefore \qquad \frac{V_p}{V_m} = \sqrt{\frac{l_p}{l_m}}$$

the discharge ratio
$$Q_r = \frac{Q_p}{Q_m} = \frac{l_p^2 \,.\, V_p}{l_m^2 V_m} = \left(\frac{l_p}{l_m}\right)^{5/2}$$

$$\therefore \qquad Q_m = \frac{Q_p}{(l_p/l_m)^{5/2}} = \frac{1500}{(9)^{5/2}} = 6.17\ \text{m}^3/\text{s}.$$

Power,
$$P = \gamma QH$$

$\therefore$ Power ratio,
$$P_r = \frac{P_p}{P_m} = \frac{\gamma_p Q_p H_p}{\gamma_m Q_m H_m}\left(\frac{l_p}{l_m}\right)^{5/2} \frac{l_p}{l_m} = \left(\frac{l_p}{l_m}\right)^{7/2}$$

since
$$\frac{\gamma_p}{\gamma_m} = 1$$

$\therefore$ Power dissipation in the prototype

$$P_p = P_m \left(\frac{l_p}{l_m}\right)^{7/2} = 0.33 \times (9)^{7/2} = 721.71\ \text{kW}.$$

Example 5.26. *A model of spillway is constructed to a scale of 1 in 25. The discharge over the spillway is 8.4 m³/s per metre length of the spillway. If the model of the spillway is 0.6 m long what is the discharge required in the model ? If the head over the crest of the model is 0.12 m, find the corresponding head over the prototype spillway.*

Solution. Model testing of spillways being governed by the gravity force, the dynamic similarity of flow will be attained by ensuring equality of Froude numbers, *i.e.*

$$(F)_m = (F)_p \quad \text{or} \quad \frac{V_m}{\sqrt{gL_m}} = \frac{V_p}{\sqrt{gL_p}}$$

$$q = \frac{Q}{L} = \frac{A\,V}{L} = \frac{(L^2)\,V}{L} = L.V, \qquad \therefore \quad V = q/L$$

From Froude's similarity, $\dfrac{q_m}{L_m\sqrt{gL_m}} = \dfrac{qp}{L_p\sqrt{gL_p}}$, $\qquad \therefore \quad q_m = q_p\,\dfrac{L_m}{L_p}\sqrt{\dfrac{L_m}{L_p}}$

or
$$q_m = 8.4 \times \frac{1}{25} \times \sqrt{\frac{1}{25}} = 0.0672\ \text{m}^3/\text{s/m} = \mathbf{67.2\ lit/s/m}$$

$$H_p/H_m = 25, \quad \therefore \quad H_p = 25 \times 0.12 = \mathbf{3.0\ m.}$$

Example 5.27: *A 15 m long and 7.2 m ligh spillway dischanges 94 cunecs of water under a head of 2.0 m. If a 1 : 9 scale model of this spillway is to be construncted, determine model dimensions, head over the model and the dischange. If the model expoiences a force of 7500 N (764.53 kgf), determine the corresponding force on the proprhype.*

Solution. Flow over a spillway is a typical example of open channel flow where gravity force dominates and the conditions of dynamic similarity are satisfied by application of Froude's law :

$$\left(\frac{V}{\sqrt{gL}}\right)_m = \left(\frac{V}{\sqrt{gL}}\right)_p \text{, from which the velocity ratio,}$$

$$V_r = \frac{V_p}{V_m} = \sqrt{\frac{L_p}{L_m}} = \sqrt{L_r}, \qquad L_r = \frac{L_p}{L_m} = 9.$$

Model dimensions :

(*i*) Length of moded spillway $= \frac{15}{9} = 1.67$ m

(*ii*) Height of model spillway $= \frac{7.2}{9} = 0.80$ m

(*iii*) Head over the spillway $= \frac{2.0}{9} = 0.22$ m

(*iv*) Discharge $Q = A \,.\, V$

$$Q_r = \frac{Q_p}{Q_m} = \frac{A_p}{A_m}\frac{V_p}{V_m} = (L^r)^2\,(V_r) = (L_r)^{5/2}$$

$$\therefore \qquad Q_m = \frac{Q_p}{243} = \frac{94}{243} = 0.387 \text{ m}^3\text{/s.}$$

Force on the prototype

Force = mass × acceleration = $\rho\, L^3\,(V^2/L) = \rho\, L^2V^2$

$$\text{Force ratio} = \frac{F_p}{F_m} = \left(\frac{\rho_p}{\rho_m}\right)\left(\frac{L_p}{L_m}\right)^2\left(\frac{V_p}{V_m}\right)^2 = \rho_r\, L_r^{\,3}$$

$$= (1)\,(9)^2(9) = 729$$

$$\therefore \qquad F_p = 729\, F_m = 729 \times 7500 = \mathbf{5467.5\ kN}$$

$$= 557339.45 \text{ kg} = \mathbf{5.57 \times 10^5\ kg.}$$

Example. 5.28. *A 1.64 model is constmcted of an open channel in the concrete which has Manning's n = 0.014, find the value of 'n' for the model.*

Solution. For open channels, dynamic similoarily is governed by the Froude's law giving a relationship for the velocity ratios of follows

$$V_r = \sqrt{L_r}$$

Manning's formula is given by Eq. (12.8) and is reproduced below

$$V = \frac{1}{n} R^{2/3}\, S^{1/2}$$

$$\text{The velocity ratio :} \qquad V_r = \frac{V_p}{V_m} = \frac{\left(\frac{1}{n}R^{2/3}\,S^{1/2}\right)_p}{\left(\frac{1}{n}R^{2/3}\,S^{1/2}\right)_m} = \frac{n_m}{n_p}\left(\frac{L_p}{L_m}\right)^{2/3}$$

Assuming the slope ratio as unity or $S_p = S_m$. For Manning's n

$$n_m = n_p \left(\frac{L_m}{L_p}\right)^{2/3} \left(\frac{V_p}{V_m}\right) = n_p \left(\frac{1}{L_r}\right)^{1/6} = \frac{0.014}{(64)^{1/6}} = \mathbf{0.007.}$$

Example 5.29. *A model of a reservoir is drained in 6 minutes by opening the sluice gate. How long should it take to empty the prototype if the scale ratio is 1 : 256 ?*

Solution. Since the model pertains to open channel flow, the similarity is governed by Froude's law. The time ratio is given by Eq. (5.17),

$$t_r = \frac{t_p}{t_m} = \sqrt{\frac{l_p}{l_m}} = \sqrt{\frac{256}{1}} = 16$$

∴ Time to empty the prototye,

$$t_p = 16\, t_m = 16 \times 6 = 96 \text{ minutes} = \mathbf{1 \text{ hr } 36 \text{ minutes.}}$$

Example 5.30. *Wave motion inside a harbour is to be studied by means of a geometrically similar model constructed to a scale of 1 : 40. Neglecting viscous and surfaces tension effects, obtain the prototype to model scale ratios for (i) velocity (ii) time (iii) acceleration, and (iv) force.*

In the harbour, a 1.5 m high wave travels a certain distance in 30 seconds. What will be the corresponding wave height in the model and what time it will take to travel the corresponding distance ?

Solution. Since wave formation is governed by the gravity, the equality of Froude's number in model and prototype will satisfy the condition for dynamic similarity, thus

$$\left(\frac{V}{\sqrt{l.g}}\right)_p = \left(\frac{V}{\sqrt{l.g}}\right)_m$$

(*i*) Velocity ratio $\quad V_r = \dfrac{V_p}{V_m} = \sqrt{\dfrac{l_p}{l_m}} = \sqrt{40} = 6.325$

(*ii*) Time $\quad t = \dfrac{l}{V},$

therefore, time ratio, $\quad t_r = \dfrac{t_p}{t_m} = \dfrac{\dfrac{l_p}{V_p}}{\dfrac{l_m}{V_m}} = \dfrac{l_p}{l_m} \cdot \dfrac{V_m}{V_p} = 40 \times \dfrac{1}{\sqrt{40}} = 6.325$

(*iii*) Acceleration, $\quad a = \dfrac{V}{t} = \dfrac{V}{l/V} = \dfrac{V^2}{l}.$

∴ Acceleration ratio, $\quad a_r = \dfrac{a_p}{a_m} = \dfrac{(V_r)^2}{l_r} = \dfrac{40.0}{40.0} = 1.0.$

(*iv*) Force, $\quad F = \text{Mass} \times \text{Acceleration} = \rho l^3 \left(\dfrac{V}{t}\right) = \rho l^2 V^2$

∴ Force ratio, $\quad F_r = \dfrac{F_p}{F_m} = \dfrac{\rho_p l_p^2 V_p^2}{\rho_m l_m^2 V_m^2} = \left(\dfrac{l_p}{l_m}\right)^3 = (40)^3 = 64{,}000.$

Wave height in the model $= \dfrac{1}{40} \times 1.5 \times 100 \text{ cm} = 3.75 \text{ cm}.$

Time for the wave to travel the corresponding distance in the model.

$$\frac{t_p}{t_m} = 6.325$$

$$t_m = \frac{t_p}{6.325} = \frac{30}{6.325} = 4.75 \text{ sec.}$$

5.8.3 Euler Number

When pressure forces are more important in comparison to other forces, the dynamic similarity will be obtained by equating the Euler number in the model to that in the prototype. It may be recalled here that the Euler number represents the ratio of inertia and pressure forces.

In cavitations studies, the pressure force is important in addition to the viscous force, and the dynamic similarity will be obtained when the Reynolds and Euler numbers are kept the same for model and prototype.

5.8.4. Mach Number

The Mach Number can be regarded as the ratio of inertia and elastic forces. In problems where the compressibility of the fluid becomes important, the elastic forces must be considered. The high speed flow or the motion of objects like aeroplane and projectiles through air at supersonic speeds represent cases where the compressibility effects (elastic force) are of major importance. The Mach number is, therefore, a measure of the effects of compressibility.

Example 5.31. *High speed motion of missiles represents compressible fluid flow in which gravity and surface tension do not affect. Dynamic similarity will result from the actions of viscosity and elasticity characterised by the Reynolds and Mach numbers. Show that the prototype model length scale is given by* $\frac{l_p}{l_m} = \frac{\nu_p}{\nu_m}\frac{C_m}{C_p}$

in which ν *and* C *are the kinematic viscosity and the sonic velocity respectively.*

Solution. Dynamic similarity will result from equality of Reynolds and Mach numbers in the model and the prototype,

$$\left(\frac{Vl}{\nu}\right)_p = \left(\frac{Vl}{\nu}\right)_p \qquad \text{...(1)}$$

and

$$\left(\frac{V}{C}\right)_p = \left(\frac{V}{C}\right)_m \qquad \text{...(2)}$$

Eliminating V between (1) and (2)

$$\left(\frac{Cl}{\nu}\right)_p = \left(\frac{Cl}{\nu}\right)_m$$

from which the length scale $\frac{l_p}{l_m} = \frac{\nu_p}{\nu_m} \cdot \frac{C_m}{C_p}$.

5.8.5. Weber Number

If surface tension and inertia are the only important forces governing the flow, the dynamic similarity will be attained by keeping the ratio of these two forces—called the Weber number same for the model and prototype. In river and harbour models, reduction of scale often leads to appreciable viscous and capillary effects in the shallow regions of flow. The depth of flow in such cases should be sufficiently large so that capillary effects are negligible.

Example 5.32. *A model is to be built of a flow phenomenon which is dominated by the actions of gravity and surface tension. Derive an expression for the model scale in terms of physical properties of the fluid.*

Solution. For complete similarity between the model and the prototype, the equality of Froude number and the Weber number must be satisfied simultaneously, thus

$$\left(\frac{V}{\sqrt{gl}}\right)_p = \left(\frac{V}{\sqrt{gl}}\right)_m \qquad \therefore \quad \frac{V_p}{V_m} = \sqrt{\frac{l_p}{l_m}}$$

Equality of Weber number yields

$$\left(\frac{\rho V^2 l}{\sigma}\right)_p = \left(\frac{\rho V^2 l}{\sigma}\right)_m \quad \text{or} \quad \frac{\rho_p}{\rho_m} \cdot \frac{V_p^2}{V_m^2} \cdot \frac{l_p}{l_m} = \frac{\sigma_p}{\sigma_m}$$

from the Froude's similarity, $\frac{V_p}{V_m} = \sqrt{\frac{l_p}{l_m}}$, substituting this relationship,

$$\frac{\rho_p}{\rho_m}\left(\sqrt{\frac{l_p}{l_m}}\right)^2 \frac{l_p}{l_m} = \frac{\sigma_p}{\sigma_m}$$

The model scale $l_r = \frac{l_p}{l_m}$ is expressed by

$$\left(\frac{l_p}{l_m}\right)^2 = \frac{\sigma_p}{\sigma_m} \cdot \left(\frac{\rho_m}{\rho_p}\right) \quad \therefore \quad \left(\frac{l_p}{l_m}\right) = \sqrt{\frac{\sigma_p}{\sigma_m} / \frac{\rho_m}{\rho_m}}.$$

Example 5.33. *It is believed that the problem of air-entertainment in free surface vortex formation at intakes is influenced by forces of viscosity and surface tension. Show that for dynamic similarity between model and prototype then, the following relationship must be satisfied*

$$\left(\frac{\mu V}{\sigma}\right)_m = \left(\frac{\mu V}{\sigma}\right)_p$$

which for use of the same liquid results in the "equal-velocity" concept of model testing—a method empirically proposed by Denny.

Solution. When in an engineering problem, viscous and surface tension effects are of equal importance, the model will be dynamically similar to the prototype only when the Reynolds and Weber numbers in the two systems are identically similar. In other words, equality of these dimensionless parameters must be ensured, that is :

$$(\mathbf{R})_m = (\mathbf{R})_p \text{ ; and } (\mathbf{W})_m = (\mathbf{W})_p$$

Reynolds number $\quad \mathbf{R} = \frac{\rho V d}{\mu}$

Weber number $\quad \mathbf{W} = \frac{\rho V^2 d}{\sigma}$

Taking the ratio, $\quad \frac{\mathbf{W}}{\mathbf{R}} = \frac{\rho V^2 d/\sigma}{\rho V d/\mu} = \frac{\mu V}{\sigma}$

For dynamic similarity, therefore,

$$\left(\frac{\mu V}{\sigma}\right)_m = \left(\frac{\mu V}{\sigma}\right)_p$$

If the same liquid is used in the model as that in the prototype, $\mu_m = \mu_p$ and $\sigma_m = \sigma_p$. The above equation then becomes

$$(V)_m = (V)_p$$

according to which the velocity in model must be the same as the velocity in prototype. The "equal-velocity" method for vortex modelling of air-entertainment in pump sumps was suggested by Denny in 1956.

5.9. Model Testing of Ships

The drag (resistance to the motion) experienced by a ship consists of (*i*) the wave resistance (resistance offered by the waves generated on the sea surface) and (*ii*) the frictional or viscous resistance (which acts on the surface of contact between sea water and the ship). It is, therefore, evident that the phenomenon is influenced by the existence of three important forces—inertia, gravity and viscous. The gravity and the viscous forces are of the same order of magnitude and hence none can be ignored. The dimensional analysis among the relevant variables (F, V, l, ρ, μ, g) will yield the following equation for the drag force F :

$$F = \rho l^2 V^2 f\left(\frac{\rho V l}{\mu}, \frac{V}{\sqrt{gl}}\right) = \rho l^2 V^2 f(\mathbf{R}, \mathbf{F})$$

in which f stands for 'function of'.

The dynamic similarity will be achieved, if the Reynolds and Froude numbers both are same in the model and its prototype, thus if

$$(\mathbf{R})_p = (\mathbf{R})_m \text{ and } (\mathbf{F})_p = (\mathbf{F})_m, \text{ then}$$

$$\left(\frac{F}{\rho l^2 V^2}\right)_p = \left(\frac{F}{\rho l^2 V^2}\right)_m$$

From Reynolds and Froude similarities $\left(\frac{V_p l_p}{\nu_p}\right) = \left(\frac{V_m l_m}{\nu_m}\right)$ and $\frac{V_p}{\sqrt{l_p g_p}} = \frac{V_m}{\sqrt{l_m g_m}}$.

If $g_p = g_m$, dividing one equation by the other (eliminating the velocity), one gets

$$\frac{\nu_p}{\nu_m} = \left(\frac{l_p}{l_m}\right)^{3/2} = l_r^{\,3/2} \quad \text{...(5.26)}$$

This indicates that once the length ratio l_r is chosen, the ratio of kinematic viscosities of prototype and model fluids may be obtained by the above equation. If the same fluid is used for the model and the prototype, the model must be as large as the prototype itself. Since it may not be possible to find liquids of appropriate viscosities to satisfy Eq. (5.26) and testing of full size models is obviously not practicable, the condition of dynamic similarity, *i.e.* equality of Reynolds and Froude numbers for model and prototype cannot be satisfied simultaneously.

The testing of ship model has been made possible by the technique developed by William Froude in 1870, while conducting pioneering experiments with ship models. The model is operated such that $(\mathbf{F})_p = (\mathbf{F})_m$ and the total resistance of the ship model when towed in the towing tank is measured. The Reynolds number in the model is drastically reduced and the test results need to be corrected for the discrepancy caused by the inequality of Reynolds numbers. To do this, the frictional resistance is computed for the model and subtracted from the total resistance. The remainder is the wave resistance experienced by the model. The corresponding prototype wave resistance can be obtained from the force ratio given by Eq. (5.20). The frictional or viscous prototype resistance is then computed and added to the prototype wave resistance to yield the total resistance of the ship.

The above technique is applicable to any flow-field controlled by the combined action of gravity and viscous forces. Typical examples of such flow-fields are those obtaining in models of rivers, harbours, open channel flow problems, hydraulic structures and the like. Such models are, however, considerably more difficult to operate and interpret than ship models on account of ill-defined frictional resistance caused by variations of surface roughness and complex boundary geometry.

Example 5.34. *An 1 : 8 model of a boat is towed in water of kinematic viscosity 10^{-6} m²/s. What should be the speed of model to simulate a speed of 3.5 m/s if the resistance is due to* (*i*) *internal friction only and* (*ii*) *waves only ?*

Calculate the kinematic viscosity of the liquid in which model should be tested if the resistance due to internal friction and waves are to be considered.

Solution.

(*i*) **Considering internal friction only**

Reynolds number for prototype = Reynolds number for model

$$\frac{V_p l_p}{\nu_p} = \frac{V_m l_m}{\nu_m}$$

given that $\nu_p = \nu_m$

$$\therefore \quad V_m = V_p \times \frac{l_p}{l_m} = 3.5 \times 8 = 28.0 \text{ m/s.}$$

(*ii*) **Considering waves only**

Froude number for prototype = Froude number for model

$$\frac{V_v}{\sqrt{l_p g_p}} = \frac{V_m}{\sqrt{l_m g_m}}$$

assuming $g_p = g_m$

$$\therefore \quad V_m = V_p \sqrt{\frac{l_m}{l_p}} = 3.5 \times \sqrt{\frac{1}{8}} = 1.236 \text{ m/s.}$$

(*iii*) **Resistance due to internal friction and waves**

Similarity of Reynolds and Froude numbers to be considered. Reynolds similarity gives

$$\frac{\nu_p}{\nu_m} = \frac{V_p}{V_m} \cdot \frac{l_p}{l_m} \quad ...(1)$$

Froude similarity yields, $\frac{V_p}{V_m} = \sqrt{\frac{l_p}{l_m}}$. ...(2)

Eliminating $\frac{V_p}{V_m}$ between (1) and (2), $\frac{\nu_p}{\nu_m} = \sqrt{\frac{l_p}{l_m}} \cdot \frac{l_p}{l_m}$

$$\therefore \quad \nu_m = \nu_p \left(\frac{l_m}{l_p}\right)^{2/3} = 10^{-6} \left(\frac{1}{8}\right)^{3/2} = 4.42 \times 10^{-8} \text{ m}^2\text{/s.}$$

Example 5.35. *A proposed ocean going vessel is to have a length of 125 metres at the water line and wetted surface of 1600 sq. metres. Its cruising speed is to be 35 km/hour. Tests on the model of the vessel to a scale of 1 : 25 were made in a towing tank at a velocity corresponding to wave-making resistance. The total drag resistance of the model was 25.2 N. Calculate the total drag of the prototype. Adopt the following formula for skin-friction drag :*

$$F_s = \frac{1}{2} C_f A \rho V^2$$

where A = *area of wetted surface*

$$C_f = \textit{skin friction drag coefficient} = \frac{0.074}{(\mathbf{R})^{1/5}} \textit{ for } \mathbf{R} < 2 \times 10^7$$

$$C_f = \frac{0.01}{(\mathbf{R})^{1/5}} \text{ for } \mathbf{R} > 2 \times 10^7 \text{ ;}$$

R *being the Reynolds number. Total drag = Skin friction + Wave making drag*

$\rho_m = 1000 \ kg/m^3$; $\rho_p = 1027 \ kg/m^3$.

$\nu_m = 1.115 \times 10^{-2} \ cm^2/s$, *and* $\nu_p = 1.121 \times 10^{-2} \ cm^2/s$.

Solution. The model is to be based on the similarity of Froude's number, since the gravity forces are involved in the formation of waves, thus

$$\left(\frac{V}{\sqrt{gl}}\right)_p = \left(\frac{V}{\sqrt{gl}}\right)_m$$

Velocity in the prototype, $V_p = \dfrac{35 \times 1000}{60 \times 60} = 9.725$ m/s

the velocity in the model, $V_m = V_p \sqrt{\dfrac{l_m}{l_p}} = 9.725 \sqrt{\dfrac{1}{25}} = 1.944$ m/s

length corresponding to 125 m in the prototype,

$$l_m = \frac{125}{25} = 5 \text{ m.}$$

The Reynolds number in the model,

$$\mathbf{R}_m = \frac{V_m l_m}{\upsilon_m} = \frac{1.944 \times 5}{1.115 \times 10^{-3} \times 10^{-4}} = 8.72 \times 10^5$$

Since $\mathbf{R}_m < 2 \times 10^7$, the drag coefficient is given by,

$$C_f = \frac{0.074}{(\mathbf{R})^{1/5}}$$

$$\therefore \quad (C_f)_m = \frac{0.074}{(8.72) \times 10^6)^{1/5}} = 0.00303$$

The skin friction drag in the model

$$F_s = \frac{1}{2} C_f A \rho V^2$$

$$(F_s)_m = \frac{1}{2} \times 0.00303 \times \frac{1600}{(25)^2} \times 1000 \times (1.944)^2 = \mathbf{14.657\ N}$$

therefore, the wave-making drag in the model = 25.2 – 14.6 = 10.6 N

Force ratio for Froude's similarity

$$\frac{F_p}{F_m} = \frac{\rho_p l_p^2 V_p^2}{\rho_m l_m^2 V_m^2} = \frac{\rho_p}{\rho_m}\left(\frac{l_p}{l_m}\right)^3 = \frac{1027}{1000} \times (25)^3$$

$\therefore$ Wave-making resistance in the prototype

$$F_p = 10.6 \times \frac{1027}{1000} \times (25)^3 = 198.16 \text{ kN}$$

Reynolds number **in the** prototype

$$\mathbf{R}_p = \frac{V_p \cdot l_p}{\nu_p} = \frac{9.725 \times 125}{1.21 \times 10^{-2} \times 10^{-4}} = 10^9$$

drag coefficient for the prototype

$$(C_f)_p = \frac{0.01}{(\mathbf{R}_p)^{1/5}} = \frac{0.01}{(10^9)^{1/5}} = 0.0001586$$

Skin friction drag in the prototype,

$$F_s = \frac{1}{2} C_f A \rho V^2$$

$$(F_s)_p = \frac{1}{2} \times 0.0001586 \times 1600 \times 1027 \times (9.725)^2 = 12323.7 \text{ N.}$$

Hence, total prototype drag = Wave making drag + **Skin-friction** drag

= 198160 + 12323.7 = 210491 **N = 210.5** kN.

Example 5.36. (*a*) *Explain Buckingham's theorem. Derive* ***Froude*** *number.*

(*b*) *The resistance offered to the movement of 2 m long ship* ***model*** *in a towing tank full of water while moving with a speed of 2.5 m/s* was *471 N.*

(*i*) *If the prototype is 30 m in length what will be its corresponding velocity ?*

(*ii*) *What would be the force required to drive with a speed of 7.5 m/s prototype* ***40 m long*** *in sea- water of sp. gr. 1.025. Assume coefficient of drag was the same in both model and* ***prototype.***

Solution. The motion of ships is influenced by waves and viscous resistances. The dynamic similarity will, therefore, be obtained by simultaneously satisfying the Froude law as well as the Reynolds law. If however the wave-resistance is predominant and the viscous resistance could be neglected, the Froude number will govern the dynamic similarity between model and prototype :

$$\left(\frac{V}{\sqrt{Lg}}\right)_m = \left(\frac{V}{\sqrt{Lg}}\right)_p$$

from which $$V_p = V_m \sqrt{\frac{L_p}{Lg}} = 2.5\sqrt{\frac{30}{2}} = \mathbf{9.68\ m/s}$$

The drag force $$F = C_d \frac{1}{2} \rho A V^2$$

since the drag-coefficients are to be assumed having the same value, the force ratio

$$F_r = \rho_r A_r V_r^2 = \rho_r l_r^2 V_r^2$$

density ratio $$\rho_r = \frac{\rho_p}{\rho_m} = \frac{1.025}{1} = 1.025$$

length ratio $$l_r = l_p/l_m = 40/2 = 20$$

velocity ratio $$V_r = V_p/V_m = 7.5/2.5 = 3$$

$\therefore$ $$F_r = 1.025 \times (20)^2 \times (3)^2 = 3600$$

Hence, the force required to drive the prototype ship,

$$F_p = F_r \times F_m = 3690 \times 471 = 1738 \times 10^3 \text{ N} = 1738 \text{ kN}.$$

When the viscous resistance alone is important the dynamic similarity will be achieved by Reynolds number equality. Thus

$$\left(\frac{VL}{\nu}\right)_m = \left(\frac{VL}{\nu}\right)_p$$

Let the kinetic viscosities of water be same.

$\therefore$ $$V_p = V_m L_m/L_p = 2.5 \times \frac{2}{30} = \mathbf{0.167\ m/s}$$

Velocity ratio $$V_r = V_p/V_m = 2/15 = 0.133.$$

Force ratio, $$F_r = \rho_r l_r^2 V_r^2 = 1.025 \times (40/2)^2 \times (0.133)^2 = 7.2525$$

Force required to drive the prototype

$$F_p = F_m \times F_r = 471 \times 7.2525 = 3416 \text{ N}.$$

5.10. Distorted Models

The models in which it is not possible to maintain geometric similarity are known as *distorted models*. In river and harbour models, the horizontal and vertical scales are usually kept different for economic and physical reasons, resulting in departure from geometric similarity.

In a small size model, the forces which may have been entirely negligible in the prototype, may be of importance in the model. The effect of such forces must be taken into account or minimised if possible. For example, if a small model is made of a harbour, and water is used as a fluid in both the cases, surface tension forces which are completely negligible in the prototype, are of appreciable magnitude in the model. Therefore, steps should be taken to eliminate the effect of surface tension in the model if strict similarity is to be maintained. In other words, the Weber number should be kept at a high value so that in both the cases surface tension forces are insignificantly small as compared to other forces. In a harbour or river model, the ratio of model to prototype depth may usually be different from the ratio of horizontal distances to eliminate undesirable effects of surface tension.

In such models, it is difficult to reproduce the topography of the sea or river bed which may consist of rocks and sand grains of various sizes. In such situations, one must rely on experience to evaluate the usefulness of model results. However, we often find experiments with such imperfect models to be great value. The model study in such cases is intended to show at least the most important and salient features of flow behaviour. The discrepancy and deviation between the results of the imperfect model and the prototype is usually attributed to, what is know as the *scale effect*. To obtain some estimate of the scale effect, models of different size are constructed, and by comparing the behaviour of various models, conclusions may be drawn as to the relative importance of Reynolds Number, Froude Number and Weber Number.

The perfect geometric similarity is not always easy to attain. The geometric similarity not only implies similarity of size and shape of model to its prototype, but it also means similarity of roughness of the surfaces. For a small model, the surface roughness might not be reduced according to the length ratio. In the model experiments of river, a powder of high degree of fineness might be required to represent sand, in order to achieve perfect geometric similarity. The models in which complete geometric similarity cannot be achieved are also included under the category of distorted models. The scale effect is due to incomplete similarity ; it may be due to departure from complete dynamic similarity or due to imperfect geometric similarity or due to both of these imperfections.

Example 5.37. *In certain field problems of open channel flow such as the free-surface vortex formation at intakes, it has been established that the forces influencing the phenomenon are those due to gravity and viscosity. Show that for dynamic similarity, the ratio of kinematic viscosities must bear the following relationship in terms of the model scale.*

$$\frac{\nu_m}{\nu_p} = \left(\frac{l_m}{l_p}\right)^{3/2}$$

Solution. For dynamic similarity when the phenomenon is influenced by forces of gravity and viscosity, the Froude numbers and the Reynolds numbers in the model and prototype must be the same, hence

Froude number equality : $\left(\dfrac{V}{\sqrt{gl}}\right)_p = \left(\dfrac{V}{\sqrt{gl}}\right)_m$

and the equality of Reynolds numbers results in

$$\left(\frac{Vl}{\nu}\right)_p = \left(\frac{Vl}{\nu}\right)_m$$

Eliminating the velocity V, assuming that the acceleration due to gravity g does not vary, one obtains

$$\left(\frac{\nu}{l^{3/2}}\right)_p = \left(\frac{\nu}{l^{3/2}}\right)_m$$

$$\frac{\nu_m}{\nu_p} = \left(\frac{l_m}{l_p}\right)^{3/2}$$

Example 5.38. *Like model testing of ships, one of the recently advocated method for carrying out model studies on vortex formation at intakes is to conduct the tests at the same Froude number and then to correct the results for distortion in the Reynolds number. Show that when the model is tested at the same Froude number as that in the prototype, the Reynolds number in the model is reduced considerably according to the following relationship :*

$$\left(\frac{Vd}{\nu}\right)_m = \left(\frac{l_m}{l_p}\right)^{3/2} \left(\frac{Vd}{\nu}\right)_p .$$

Solution. Since the phenomenon of vortex formation at intakes is influenced by gravity and viscous forces, the dynamic similarity is obtained by equality of Froude and Reynolds numbers in the model and the prototype. From Froude similarity, one may write

$$\frac{V_m}{V_p} = \sqrt{\frac{l_m}{l_p}} \quad ...(1)$$

The ratio of Reynolds numbers

$$\frac{(V_m l_m/\nu_m)}{V_p l_p/\nu_p} = \frac{V_m l_m \nu_p}{V_p l_p \nu_m} \quad ...(2)$$

if the same fluid is used, then $\nu_p = \nu_m$.

From Eq. (1), $$\frac{\mathbf{R}_m}{\mathbf{R}_p} = \left(\frac{l_m}{l_p}\right)^{3/2}$$

and $$\left(\frac{Vd}{\nu}\right)_m = \left(\frac{l_m}{l_p}\right)^{3/2} \left(\frac{Vd}{\nu}\right)_p \quad ...(3)$$

From Eq. (3), it is evident that the Reynolds number in the model will be reduced by a factor of $m\sqrt{m}$, where $m = l_m/l_p$ and the model will consequently suffer a distortion on account of inequality of viscous effects. The model test data must, therefore, be corrected for this shortcoming.

5.11. Vortex Formation at Intakes

Intakes are widely used for drawing water and other liquids from rivers, reservoirs and sumps. Vortices are frequently formed at intakes under certain conditions of flow. At low submergences strong vortices are formed which may entrain air and cause noise, vibrations, reduction in discharge and fall in efficiency of hydraulic machines.

Prototype intake installations generally have irregular and complex approach boundaries which under various combinations of submergence and discharge would result in vortices of varying strengths. In order to predict formation of vortices at an intake, it is usually necessary to conduct a model study of the intake installation. From the results of model investigation it is possible to determine the prototype conditions by making use of valid similarity laws.

There are conflicting similarity conditions available for vortex formation studies at hydraulic intakes. While majority of researchers use Froude similarity, there are others who follow *equal-velocity law* and yet another school of thought advocate equality of Reynolds number in model and prototype. It can be rightly said that vortex formation at hydraulic intakes being a free-surface phenomenon, the gravity forces are greatly important. But at the same time, the circulation (or vorticity) present in the flow approaching the intake, which is solely responsible for causing formation of vortices, gets dissipated by viscous action of flowing fluid. This phenomenon, therefore, involves interaction of atleast gravity and viscous forces, and therefore one can conclude that the similarity laws currently in vogue do not take into account all the relevant forces. Based on the research work carried out by Jain *et . al*[1, 2, 3] relative roles of circulation, gravity and viscous forces in vortex formation at pipe intakes have been quantified, and a method has been proposed to transfer the model-test data to predict the corresponding hydraulic performanace of the intake.

5.11.1. Model-Prototype Similarity Law

The critical submergence of the intake at which air-entrainment becomes imminent is related to the circulation parameter N_Γ, the Reynolds number **R** and the Froude number **F** by the following expression proposed by Jain (1, 2) :

$$\frac{S_C}{d} = 0.19\ \mathbf{N}_\Gamma^{0.42}\ \mathbf{R}^{0.31} \cdot \mathbf{F}^{0.19} \quad ...(5.25)$$

where S_C = critical submergence of the intake of diameter d ;
N_Γ = a circulation parameter defined by $\Gamma S_C/Q$;
Γ = circulation ; Q = discharge through the intake ;
$\mathbf{R}$ = intake Reynolds number defined by $4Q/\pi d\nu$;
and $\mathbf{F}$ = Froude number $4Q/\pi \sqrt{gd^5}$.

Equation (5.25) is valid for $N_\upsilon \le 5 \times 10^4$, where $N_\upsilon = d \sqrt{gd/\nu}$.

For $N_\upsilon > 5 \times 10^4$, the vortices are free from viscous effects and the critical submergence is determined by

$$\frac{S_C}{d} = 5.6\, \mathbf{N}_\Gamma^{0.42}\, \mathbf{F}^{0.50} \qquad ...(5.26)$$

The circulation parameter N_Γ has been so formulated that for a geometrically similar model, this parameter has the same value in the model as in the prototype. Thus

$$(N_\Gamma)_m = (N_\Gamma)_p \qquad ...(5.27)$$

In the method proposed for model studies on vortex formation, the model is run according to the Froude law and the results are corrected for viscous distortion. Equation (5.25), however, takes into account this viscous effect. For appropriate conditions, Eqs. (5.25) and (5.26) can be directly used for determining the prototype critical submergence based on the corresponding model-test observation.

5.11.2. Minimum Submergence For Vortex-free Operation

In order to have a safe and trouble-free operation of the intake, adequate depth of water must always be provided above the intake opening. In case of intakes aligned horizontally or slightly inclined, the minimum value of safe submergence as suggested by Jain[4] may be expressed by

$$\frac{S_{min}}{d} = 1.5 + 2.0\, \mathbf{F}^{0.70} \qquad ...(5.28)$$

It is obvious that for safe and efficient operation, the submergence of intake must always be greater than the value obtained from Eq. (5.28).

OBJECTIVE TYPE QUESTIONS

1. The following table contains certain quantities used in fluid-flow problems. The table also presents dimensions of these quantities, but not in the order in which the quantities have been listed. Identify the dimensions of each quantity :

Quantity	**Dimensions**
(*a*) angular velocity, ω,	(*i*) L^2T^{-1}
(*b*) circulation, Γ	(*ii*) L^3T^{-1}
(*c*) rate of flow, Q	(*iii*) T^{-1}
(*d*) pressure, p	(*iv*) $ML^{-1}T^{-1}$
(*e*) dynamic viscosity, μ	(*v*) $ML^{-1}T^{-2}$
(*f*) energy	(*vi*) ML^2T^{-3}
(*g*) torque	(*vii*) ML^2T^{-2}
(*h*) power	(*viii*) ML^2T^{-2}.

(1) Jain, A.K., "Vortex Formation at Vertical Pipe Intakes", Ph.D. Thesis, University of Roorkee, April 1977.
(2) Jain, A.K., Ranga Raju, K.G. and Grade, R.J., "Vortex Formation at Vertical Pipe Intakes", Journal of Hydraulics Division, Proc. of American Society of Civil Engineers, Vol. 104, No. Hy. 10. Oct. 1978.
(3) Jain, A.K., "Physical Modelling of Vortices at Intakes," XIX Congress of the International Association for Hydraulic Research, New Delhi, India, Feb., 1981.
(4) Jain, A.K., "Vortex Problems at Horizontal Intakes and Their Prediction", Thriteenth National Conference on Fluid Mechanics and Fluid Power', Tiruchirapalli (Tamilnadu), Dec., 1984.

2. An equation is said to be "dimensionally homogeneous" if its form does not depend upon the units of measurement. The following equations describe certain physical phenomena. Check these equations for their dimensional homogeneity and state accordingly :

(*i*) $p = \gamma h$ (*ii*) $p = 1000h$ (*iii*) $F = ma$

(*iv*) $F = 25a$ (*v*) $V = C\sqrt{RS}$ (*vi*) $V = 100\sqrt{RS}$

(*vii*) $v = u + ft$ (*viii*) $v = 100 - 9.81t$ (*ix*) $Q = CLH^{3/2}$

(*x*) $Q = 3.3LH^{3/2}$.

3. The dimensional analysis is used to obtain a functional relationship between the dependent and independent variables. Identify the dependent variable from the following relationships describing certain physical phenomena :

(*i*) $f(Q, H, g, L, Z, \nu) = 0$ (*ii*) $F = f(V, l, k, \rho, \mu, g)$

(*iii*) $f(\tau_0, \rho, \mu, V, d) = 0$ (*iv*) $P = f(N, H, D, B, \rho, \mu, g)$.

4. The Buckingham-Pi theorem is widely used in the dimensional analysis and expresses the resulting equation in terms of :

(*a*) the dependent and independent variables (*b*) n dimensionless parameters

(*c*) $(n - m)$ dimensionless parameters (*d*) geometric, kinematic and dynamic variables

5. For applying the Pi-theorem, m repeating variables are selected from amongst the n variables influencing the phenomenon. The repeating variables are selected such that :

(*i*) they in combination contain each of the m fundamental dimensions involved in the problem

(*ii*) they belong to kinematic and dynamic category of variables

(*iii*) they represent the characteristic geometric, kinematic and dynamic variables

(*iv*) they must always contain the dependent variables

6. The geometric similarity between model and prototype is achieved by :

(*a*) using different model scales for horizontal and vertical dimensions

(*b*) having the solid boundaries that control the flow and the model itself as geometrically similar

(*c*) using the same model scale throughout

(*d*) having the same radius of curvature at identical places in model and prototype

7. Kinematic similarity between model and prototype is :

(*i*) the similarity of shape (*ii*) the similarity of streamline pattern

(*iii*) the similarity of forces influencing the fluid motion

(*iv*) the similarity of discharge

8. The flow in the model and prototype will be dynamically similar when :

(*a*) the forces in the two systems are the same (*b*) the two are geometrically similar

(*c*) the two are kinematically similar

(*d*) the forces at similar points in the two systems have the same ratio throughout the flow-field

9. Identify the correct combination of statements from the following :

(1) geometric similarity (*i*) is the similarity of forces

(2) kinematic similarity (*ii*) is the similarity of shapes

(3) dynamic similarity (*iii*) means the similarity of motions.

10. Various forces which may influence fluid-flow phenomena are listed below. The table also contains symbolic representation of these forces but their serial order is not in the same sequence. Identify each force with its corresponding formula containing the governing variables :

(*a*) inertia force (*i*) $E.l^2$

(*b*) viscous force (*ii*) $\sigma.l$

(*c*) gravity force (*iii*) $\Delta p.l^2$

(*d*) pressure force (*iv*) ρl^2V^2

(*e*) elastic force (*v*) μVl

(*f*) surface tension force (*vi*) $\rho l^2 g$.

11. In problems of fluid-flow, it is customary to express force ratios with respect to the inertia force. The various dimensionless parameters indicated below represent ratio of inertia to some other

force. Identify each dimensionless number with the corresponding characteristic dominating force :

(1) Reynolds number	(*a*) gravity force
(2) Froude number	(*b*) elastic force
(3) Mach number	(*c*) pressure force
(4) Weber number	(*d*) viscous force
(5) Euler number	(*e*) surface tension force.

12. In fluid-flow problems where a free-surface is present or when there is an interface between two immissible fluids, the gravity forces are important. The dynamic similarity in such cases is achieved by equality of Froude numbers. Identify from the following table, the model-prototype ratios with their corresponding relationship to the linear scale :

(*a*) velocity	(*i*) 1	(*f*) pressure ratio	(*vi*) $\sqrt{l_r}$
(*b*) acceleration ratio	(*ii*) $l_r^{5/2}$	(*g*) energy ratio	(*vii*) $l_r^{7/2}$
(*c*) time ratio	(*iii*) l_r^3	(*h*) momentum ratio	(*viii*) l_r^4
(*d*) discharge ratio	(*iv*) $\sqrt{l_r}$	(*i*) torque ratio	(*ix*) $l_r^{7/2}$
(*e*) force ratio	(*v*) l_r	(*j*) power ratio	(*x*) l_r^4.

13. The resistance experienced by a ship consists of wave resistance and frictional resistance. Identify from the list given below the forces which greatly influence the motion of a ship :

(1) inertia force	(2) surface tension force	(3) pressure force
(4) viscous force	(5) gravity force	(6) elastic force

14. Testing of a ship model has been made possible by the technique developed by William Froude in 1870. Which of the following statements are correct so far as this technique is concerned :

(*a*) the model is operated at equal Reynolds number

(*b*) the model is run at the same velocity

(*c*) the model is operated at equal Weber number

(*d*) the model is run at equal Froude number and the results are corrected for the discrepancy caused by the inequality of Reynolds number

15. The models in which it is not possible to maintain geometric similarity are known as the distorted models. River and harbour models belong to this category. In testing of such models :

(*i*) special type of liquid is used

(*ii*) horizontal and vertical scales are usually kept different

(*iii*) water-depth is kept small to bring in the surface tension effects

(*iv*) scale effects representing the discrepancy caused by the imperfect modelling are unavoidable.

16. In a model experiment with weir, if the dimensions of the model weir are reduced by a factor *k*, the flowrate through the model weir is the following fraction of the flowrate through the prototype

(*a*) $k^{5/2}$ (*b*) k^2 (*c*) 1 (*d*) k^{-2}.

ANSWERS

1. (*a*) (*iii*) ; (*b*) (*i*) ; (*c*) (*ii*) ; (*d*) (*v*) ; (*e*) (*iv*); (*f*) (*viii*) ; (*g*) (*vii*) ; (*h*) (*vi*)

2. (*i*), (*iii*), (*v*), (*vii*) and (*ix*)—dimensionally homogeneous ; others dimensionally non-homogeneous

3. (*i*) Q (*ii*) F (*iii*) τ_0 (*iv*) P.

4. (*c*). **5.** (*i*), (*iii*). **6.** (*b*), (*c*).

7. (*ii*) **8.** (*d*). **9.** (1) (*ii*) ; (2) (*iii*) ; (3) (*i*).

10. (*a*) (*iv*) ; (*b*) (*v*) ; (*c*) (*vi*) ; (*d*) (*iii*) ; (*e*) (*i*) ; (*f*) (*ii*).

11. (1) (*d*) ; (2) (*a*) ; (3) (*b*) ; (4) (*e*) ; (5) (*c*).

12. (*a*) (*iv*) ; (*b*) (*i*) ; (*c*) (*vi*) ; (*d*) (*ii*) ; (*e*) (*iii*) ; (*f*) (*v*) ; (*g*) (*viii*) ; (*h*) (*vii*) ; (*i*) (*x*) ; (*j*) (*ix*).

13. (1), (4), (5) **14.** (*d*) **15.** (*ii*), (*iv*). **16.** (*a*).

REVIEW QUESTIONS

1. Cite examples of dimensionally homogeneous and non-homogeneous equations.
2. How many π-terms are required to express the functional relationship.
$$V = f(l, t, a, v)\ ?$$
3. Which of the following quantities are dimensionless,
(*i*) Specific gravity (*ii*) Specific weight (*iii*) Mach number
(*iv*) Angular velocity (*v*) Dynamic viscosity.
Give the dimensions of the respective quantities.
4. Form dimensionless groups out of the following variables, (by inspection only) :
(*i*) V, t, a (*ii*) V, l, a (*iii*) ν, l, V (*iv*) ν, t, V.
Check their correctness by substituting the dimensions of respective quantities.
5. Mach number may be defined as the ratio of
(*i*) Viscous force to Elastic force (*ii*) Viscous force to Inertia force
(*iii*) Inertia force to Elastic force (*iv*) Inertia force to Gravity force
(*v*) Elastic force to Gravity force.
Select the correct answer.
6. If the two systems (model and prototype) are dynamically similar, is it implied that they are also kinematically and geometrically similar ?
7. If the streamline pattern in the two flow systems is geometrically similar will they be dynamically similar too ?
8. Is the scale effect due only to different geometrical scales (vertical and horizontal) used in the model ? Does it also include effects of other forces which exist only in the model ?
9. Discuss the circumstances which necessitate the use of distorted models.
10. Under what circumstances, will the Buckingham-π theorem yield incorrect number of dimensionless group.
11. If the number of fundamental dimensions appearing in the problem is $3(m = 3)$ in both the (M), (L), (T) and (F), (L), (T) systems, will the π-theorem then yield correct number of dimensionless π-terms ? (**Hint.** Refer dimensional matrix approach)
12. What forces influence the motion of
(*i*) a ship. (*ii*) a submarine.
(*iii*) an aeroplane flying at :
(*a*) subsonic speed (*b*) supersonic speed.
(*iv*) flow through a long capillary tube. (*v*) waves striking against a sea wall.

PROBLEMS

5.1. Show by dimensional analysis that the resistance R of a geometrically similar body moving at a velocity V under totally submerged condition, through a fluid having density ρ, viscosity ρ given by
$$R = C_D \rho\, l^2 V^2$$
in which l is a linear dimension of the body and C_D is a drag coefficient dependent only on the Reynolds number.

5.2. The torque T required to rotate a disc of diameter D depends on fluid density ρ and viscosity μ and the angular velocity ω at which the disc rotates.
Show that, $T = \rho\omega^2 D^5 f(Q)$ where Q is a dimensionless parameter depending on ρ, D, μ and ω.

5.3. Show that the pressure drop Δp in flow through pipes can be expressed by
$$\Delta p = \rho V^2 f\left(\frac{l}{d}, \frac{\rho V d}{\mu}\right)$$
where l = length of pipe, d = diameter of pipe,
V = average velocity, ρ = mass density, and μ = viscosity.

5.4. The viscous resistance F of a sphere of diameter d moving at a constant velocity V through a fluid of density ρ and viscosity μ may be expressed as $F = \frac{1}{\rho}\mu^2 f\left(\frac{\rho V d}{\mu}\right)$.

5.5. Show that the rate of flow Q of a fluid of density ρ and viscosity μ over a triangular notch may expressed by $Q = H^2 \sqrt{gH} f\left(\rho H \frac{\sqrt{gH}}{\mu}, \theta\right)$

in which H is the head over the apex and θ is the angle of the notch.

5.6. Show from dimensional analysis that the thrust F of a propeller having diameter D, moving forward with a velocity V in fluid of density ρ and viscosity μ and rotating at a speed N may be expressed by

$$F = \rho D^2 V^2 f\left(\frac{DN}{V}, \frac{\rho VD}{\mu}\right).$$

5.7. The losses per unit length of pipe $\frac{\Delta H}{l}$ in turbulent flow through a smooth pipe depend upon velocity V, diameter D, gravity g, viscosity μ and density ρ. By a process of dimensional analysis determine the various π-parameters, $f\left(\frac{\Delta h}{l}, V, D, \rho, \mu, g\right) = 0$.

5.8. State Buckingham's π-theorem. Explain the terms geometric, kinematic and dynamic similitude. Is it always possible to have complete similarity for model and prototype analysis ? If not, what best approximation is resorted to ?

5.9. The pressure loss (Δp) due to valve fixed in a pipe depends on the pipe diameter D, average velocity in the pipe V, mass density ρ, and viscosity μ, and characteristic length of orifice d. Determine general form of dimensionally homogeneous equation.

5.10. (*a*)Arrange the following groups into dimensionless parameters :

(*i*) $\Delta p, \rho, V$ (*ii*) ρ, g, V, F (*iii*) $\mu, F, \Delta p, t$.

(*b*) By inspection, arrange the following groups into dimensionless parameters :

(*i*) a, l, t (*ii*) ν, l, t (*iii*) A, Q, w.

(*c*) The Mach number M of flow of a perfect gas in a pipe depends upon the specific heat ratio K (dimensionless), the pressure p, the density ρ and the velocity V. Obtain by dimensional reasoning the form of the Mach number expression.

5.11. If $F = \phi(V, L, \rho, \mu, g, \sigma, E)$ expresses the functional relationship, determine the complete set of dimensionless parameters.

5.12. Find expressions for the ratio of prototype horse-power to model horse-power in terms of the geometric scale ratio $\left(\frac{l_m}{l_p}\right) = \lambda$

based on the,

(*i*) Reynolds number similarity, (*ii*) Froude number similarity, and

(*iii*) Mach number similarity.

5.13. Assuming that the drag force exerted by a flowing fluid is a function of the density, viscosity, velocity of the fluid and a characteristic length of the body, develop a general equation by dimensional analysis.

5.14. By dimensional analysis, obtain the following relationship for water turbines :

$$P = \rho D^5 N^3 \,\phi\left(\frac{\rho D^2 N}{\mu}, \frac{ND}{\sqrt{gH}}, \frac{D}{B}\right)$$

where P = power, B, D = Breadth and diameter of runner

N = speed in r.p.m., H = operating head,

μ, ρ = viscosity and density of water.

5.15. The power P required to drive a fan or blower depends upon the fluid density ρ, fluid viscosity μ, impeller diameter D, flow rate Q and rotational speed N. Express these variables in a dimensionless form. Show that $\frac{\rho ND^2}{\mu}$ is a form of Reynolds number.

5.16. (*a*) State and explain Buckingham-II theorem.

(*b*) Derive a general form of an equation for power lost when a liquid of mass density ρ, dynamic viscosity μ, flows through a conduit of diameter D, length L with velocity V.

5.17. (*a*) Drive by dimensional analysis the fundamental relationship between force of drag and the relevant basic parameters for motion in incompressible viscous fluid. Modify it in the case of very slow viscous motion to show that $\frac{F}{\mu DV}$ = constant

where F = drag force, D = diameter of body or some other pertinent linear dimension, V = uniform velocity of body and μ = dynamic viscosity.

(*b*) Find the viscosity of a liquid in which a steel ball of diameter 1.5 mm falls with uniform velocity through a distance of 22 cm in 10 sec. The specific gravity of liquid is 0.84 and that of steel 7.8. The constant in the above equation is 3π.

5.18. An aeroplane wing of 3 m chord length moves through a still air at 15°C and 100 kN/m² (absolute) at a speed of 230 kmph. A 1 : 20 scale model of this wing placed in a wind tunnel and dynamic similarity between model and prototype desired.

(*i*) What velocity is necessary in a tunnel where the air has the same pressure and temperature as that in flight ?

5.19. A spillway model is tested in a laboratory to predict the performance of the prototype during maximum discharge. What similarity law would govern the flow ? Why ? If q_p and q_m represent discharge per unit width in prototype and model respectively, L_p/L_m the model ratio, prove that $\frac{q_p}{q_m} = \left(\frac{L_m}{L_p}\right)^{3/2}$

5.20. (*a*) From first principles show that where gravity is predominant over all other forces, prototype ratio of discharge is equal to $l_r^{5/2}$, where l_r is the length ratio of model to prototype.

(*b*) In a prototype, both gravity and viscous forces predominate. A model study is conducted in an oil of viscosity 0.05 stokes. If the model-prototype ratio is 1 : 3, how many times more viscous should the prototype oil be ?

5.21. A hydraulic model of the spillway of Bhakhra dam was made to study the performance of the prototype. The length ratio was fixed as 1 : 200. Assume the flow in the model and prototype to be insensitive to changes in the Reynolds number.

(*a*) State under what conditions the model may be expected to predict adequately the behaviour of the prototype.

(*b*) The velocity at the bottom of the model spillway was measured and found to be 1.25 m/sec corresponding to a model flowrate of 30 litres per second. What are corresponding prototype quantities ?

5.22. The propeller of a ship operates in close proximity to the free surface. The propeller of an ordinary cargo ship 190 metres long is about 2.5 metres in diameter and turns at 100 r.m.p. If a 1 : 100 scale model of the ship and propeller is to be tested in a towing tank.

(*a*) What should be the speed of a model propeller ?

(*b*) What is the ratio of the power required by the model and that of the ship ?

5.23. At a temperature of 25°C and Reynolds number of 3.0×10^6 (based on length) the lift force on a model hydrofoil 20 cm in length was tested in a water tunnel and was found to be 1772 N. The prototype to be used in water at the same temperature, has a length of 80 cm. What will be the force on the prototype at the same Reynolds number ? What is the fluid velocity for the prototype ?

5.24. For small models of surface ships and overflow structures, the actions of gravity, viscosity and surface tension may be of equal importance. For dynamic similarity between model and prototype, what relation must exist between viscosity, surface tension and model scale ?

5.25. A ship 150 m long moves through fresh water at 20°C ($\mu = 10^{-3}$ N-s/m^2) at a speed of 40 km/hr. A model of this ship is to be tested in the towing tank containing a liquid having a specific weight 8947 N/m^3. The model is made to scale of 1 : 90. What viscosity must this model liquid have if the forces due to gravity and viscosity are equally dominant ? At what velocity must the model be towed ? What propulsive force on the ship corresponds to a towing force of 1 kg in the model ?

5.26. Model studies of a hydraulic jump are to conducted in a laboratory channel at a scale of 1 : 9. The prototype jump takes place from a depth of 0.5 to 1.5 m ; what rate of flow per unit width should be used in the model ? Determine the Froude number of flow before and after the jump.

5.27. An overflow structure 500 metres long is designed to pass a fluid flow of 34,000 m^3/s. A 1 : 20 model of the structure is built in a laboratory channel 1 metre wide. Calculate the required laboratory flowrate if the actions of viscosity and surface tension both are neglected. When the model is tested at this flowrate, the pressure at a point on the model is observed to be 5 cm of mercury vacuum. How should this be interpreted for the prototype ?

5.28. A seaplane is to take off at 140 km/hour. If the maximum speed available for conducting the model tests is 6 m/s, what is the largest model scale to which the model can be constructed ?

5.29. A ship model 1 m long with negligible skin friction is tested in a towing tank at a speed 0.7 m/s. What is the corresponding speed of the prototype ship if its length is 70 m ? What propulsive force in the prototype would be represented by a force of 4.9 N required to tow the model ?

5.30. An air-duct is to be modelled to a scale of 1 : 20 and tested with water which is 50 times viscous and 800 times more dense than air, when tested under dynamically similar conditions, the pressure drop between two sections in the model is 235 kN/m^2. What is the corresponding pressure drop in the prototype ?

5.31. Water at 5°C flow in a 9 cm horizontal pipe at mean velocity of 3 m/s. The pressure drop in 9 m of this pipe is 13.72 kN/m^2. With what velocity must benzene (20°C) flow in a geometrically similar 3 cm pipeline for the flows to be dynamically similar, and what pressure drop is to be expected in 3 m of the 3 cm pipe ?

μ water at 5°C $= 18.06 \times 10^{-4}$ N-s/m^2

μ benzene at 20°C $= 65.75 \times 10^{-5}$ N-s/m^2

and sp. gr. $= 0.88$.

5.32. A surface vessel of 120 m length is to be tested by a model 3 m long. If the vessel travels at 10 m/s, at what speed must model be towed for dynamic similarly between model and prototype ? If the drag of the model is 9.37 N, what prototype drag is to be expected ?

5.33. The discharge characteristics of a spillway 15 metres long are to be determined. It is designed to discharge 140 m^3/s at a head of 3 m. The supply available in the laboratory is 40 litres/sec. Calculate (*i*) prototype-model length ratio (*ii*) length of model-spillway and (*iii*) head over the model spillway.

5.34. A vessel 10 metres long is driven by a reaction jet at 120 km/hr. The viscous resistance is negligible as compared to the wave resistance. A similar model of the vessel, 0.5 m long was tested in the laboratory and found to require a force of 3.95 N to drive it over the water at the corresponding speed. At what speed must the model be tested and what is the thrust exerted on the prototype vessel by the jet. Neglect the air resistance.

5.35. (*a*) Explain Buckingham's theorem.

(*b*) The power required to drive a propeller is known to depend upon the diameter of the propeller the density, the viscosity and the stream velocity of the fluid, the velocity of sound in the fluid and the angular velocity of the propeller. Experiments were performed with various models of the propellers. What should be the dimensionless groups according to which the data should be plotted.

5.36. Explain what is meant by dynamic similarity between a flow system and its model.

If the dynamic behaviour of the flow system for an overflow spillway is governed by the Froude law of similarity, what would be the discharge ratio Q_r, when the (model) scale ratio is L_r ? Show that the same result would be obtained if the weir formula is used instead and it is assumed that the prototype and the model spillway have the same coefficient of discharge.

[Hint. See Article 5.8.2.]

ANSWERS

5.17. 0.381 N-s/m^2

5.18. (*i*) 88.9 m/s, (*ii*) 326 m/s (*iii*) 127 m/s

5.20. 5.2 times more viscous.

5.21. 17.67 m/s, 16,970 m^3/s

5.22. (*a*) 1000 r.p.m. (*b*) 10^{-7}.

5.23. 11.83 kN, 6.23 m/s.

5.24. $\frac{\sigma_p}{\sigma_m} = \frac{\mu_p}{\mu_m}\sqrt{\frac{l_p}{l_m}}$.

5.25. 10.75 × 10^{-7} N-s/m^2, 1.172 m/s, 7.88 MN.

5.26. 0.1 m^3/s, F_1 = 2.45, F_2 = 0.472.

5.27. 0.76 m^3/s, 100 cm of Hg vacuum.

5.28. 1 : 42.

5.29. 5.86 m/s, 1.69 MN

5.30. 189.3 N/m^2

5.31. 2.07 kN/m^2.

5.32. 1.58 m/s, 596.4 kN

5.33. (*i*) 1 : 26.5 (*ii*) 56.6 cm. (*iii*) 11.32 cm

5.34. 7.46 m/s, 31.6 kN.

6

Navier-Stokes Equations of Motion

(Equations of Motion for Viscous Fluids)

6.0. Introduction

In Chapter 4 it was pointed out that the equations of motion in which all important forces (such as those due to gravity, pressure, viscosity and turbulence) are taken into account are known as Reynold equations. The Reynolds equations are mathematically highly complicated because of the random and unpredictable phenomenon of turbulence on one hand, and the existence of variety of forces on the other—the force due to viscosity being the next single factor rendering the solution of the equation of motion impossible in most of the flow problems. In flow where the force due to turbulence is insignificant (flow at low Reynolds number, that is a low velocity flow of a highly viscous fluid through a passage of small dimension) it can be safely ignored, and the resulting equations which take into account the body force, pressure force and the force due to viscosity are known as the Navier-Stokes equations. The early research on equations of viscous fluid motion was done by Louis Navier (1785-1886) and George Stokes (1819-1903) and as a tribute to these poineer researchers, the equations of motion for a viscous fluid are aptly known as the Navier-Stokes equations. These are the fundamental equations used in the analyses of aerodynamics, fluid dynamics (low and medium Reynolds number flow) and gas dynamics problems, and are based on the Newton's second law of motion and the Newton's law of viscous friction.

6.1. Unknown Variables in Viscous Fluid Flow

For analysing the fluid flow problems, we are primarily interested in finding the velocity distribution as well as the condition of fluid over the entire space for any stipulated time of flow. There are six unknowns, namely the three components of velocity (u, v, w), density ρ, the pressure p and the temperature T in a three-dimensional flow of a compressible fluid. These variables are the functions of space co-ordinate and time. Determination of these unknowns require six equations connecting these variables. These equations are :

(1) Equation of state which relates the pressure, the density and the temperature of the fluid.

(2) Equation of continuity which is the statement of conservation of fluid mass.

(3) Equations of motion which are three in general and embody the principle of conservation of momentum. (It may here be emphasized that the momentum equation and the Newton's second law of motion are in fact the same).

(4) Energy equation which expresses the principle of conservation of energy.

We, therefore, have six equations as listed above connecting the six unknowns, the solutions of which yield the unknowns. For every flow problem, we must know the boundary conditions (initial and final) so that the constants appearing in the final equations representing the solutions are evaluated. For incompressible fluids and isothermal flows in a gravitational field, there are only four variables, u, v, w and p, which appear in the equations of motion.

6.2. The Navier-Stokes Equations of Motion

The derivation of the equations of motion for viscous fluids is substantially similar to the derivation of Euler's equation. However, beside the body forces and normal stresses acting on the fluid element, shear forces must also be considered.

In a viscous fluid flow the surface forces acting on a fluid element are more complicated. The resultant force on the surface of a fluid element is an inclined force producing a normal component as well as a tangential one. This results in the existence of (*i*) a normal stress similar to the pressure but may not be same in all directions and (*ii*) a tangential stress or a shear stress which acts tangential to the surface. In Fig. 6.1 is shown a surface area element of area dA, which has an outward pointing normal. The total force per unit area is shown inclined at an angle θ to the tangent to the surface. It produces a normal stress σ_n and shear stress τ. The notations followed in designating the stress system are the same as used in the theory of elasticity. The normal stress is denoted by σ and the shear stress by τ. Two subscripts are attached to each of the stress symbols : (*i*) the first symbol indicates the direction of the normal to the surface on which the stress acts and (*ii*) the second symbol represents the direction in which the stress acts. For example, τ_{xy} represents the shear stress acting in the y-direction on a surface whose normal points in the x-direction. The sign convention for the shear stresses is according to the right-handed co-ordinate system, in which the outwardly directed normal indicates the positive direction. In case of normal stresses which always act perpendicular to the surface, use of only one subscript, representing the direction of outward drawn normal, conveys the meaning and is hence adequate.

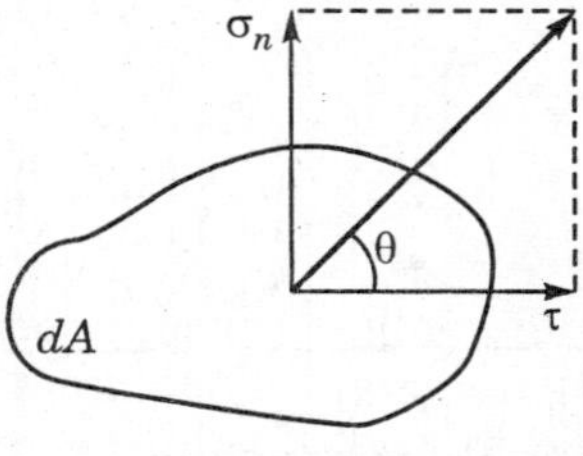

Fig. 6.1. Stresses on an element of surface.

Consider a fluid element of size $dx\, dy$ and of unit length in the z-direction in a two-dimensional flow ($\partial/\partial z = 0$). Let the stresses at the centre of this element be σ_x, σ_y, τ_{xy} and τ_{yx}. Referring to Fig. 6.2, the force acting on each surface are obtained by taking into account the variation of stresses with distance. The net force in the x-direction due to the stress shown in the figure is

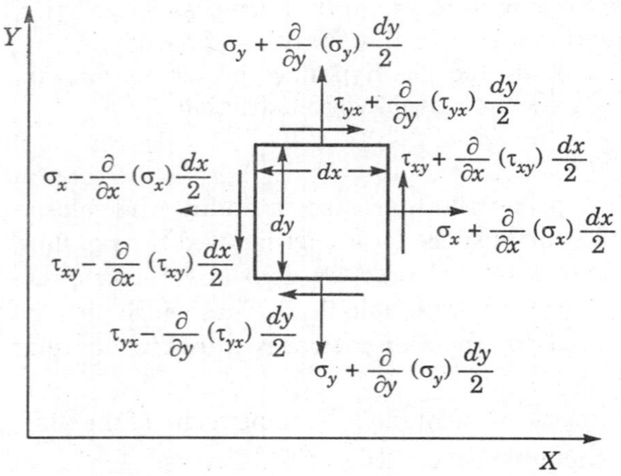

Fig. 6.2. Stresses on the surfaces of a fluid element.

$$\left\{\sigma_x + \frac{\partial}{\partial x}(\sigma_x)\frac{dx}{2}\right\}dy - \left\{\sigma_x - \frac{\partial}{\partial x}(\sigma_x)\frac{dx}{2}\right\}dy + \left\{\tau_{yx} + \frac{\partial}{\partial y}(\tau_{yx})\frac{dy}{2}\right\}dx - \left\{\tau_{yx} - \frac{\partial}{\partial y}(\tau_{yx})\frac{dy}{2}\right\}dx$$

$$= \frac{\partial}{\partial x}(\sigma_x)\, dy\, dx + \frac{\partial}{\partial y}(\tau_{yx})\, dy\, dx$$

If the flow were three dimensional and the z-dimension of the fluid element is taken as dz, then the net force in the x-direction would be

$$\frac{\partial}{\partial x}(\sigma_x) dy\, dx\, dz + \frac{\partial}{\partial y}(\tau_{yx})\, dy\, dx\, dz + \frac{\partial}{\partial z}(\tau_{zx})\, dy\, dx\, dz.$$

The role played by the pressure appearing in the Euler's equation of motion, Eq. (4.6*a*), is now replaced by these surface forces. Upon dividing the above equation by the volume $dx\, dy\, dz$ of the element, we obtain the equation of motion in the x-direction

$$\rho \frac{Du}{Dt} = \rho x + \frac{\partial}{\partial x}\sigma_x + \frac{\partial}{\partial y}\tau_{yx} + \frac{\partial}{\partial z}\tau_{zx} \qquad ...(6.1\, a)$$

Similar expressions in the y and z-directions are

$$\rho \frac{Dv}{Dt} = \rho Y + \frac{\partial}{\partial x} \tau_{xy} + \frac{\partial}{\partial y} \sigma_y + \frac{\partial}{\partial z} \tau_{zy} \qquad \text{...(6.1 } b\text{)}$$

and

$$\rho \frac{Dw}{Dt} = \rho Z + \frac{\partial}{\partial x} \tau_{xz} + \frac{\partial}{\partial y} \tau_{yz} + \frac{\partial}{\partial z} \sigma_z \qquad \text{...(6.1 } c\text{)}$$

in which the operator $D/Dt \equiv \partial/\partial t + u\, \partial/\partial x + v\, \partial/\partial y + w\, \partial/\partial z$.

It may be recalled here that X, Y and Z are the components of the body force per unit mass of fluid in the respective directions. In order to express equation (6.1) in a useful form, we require a relationship between the stress and the rate of strain (velocity gradient). Experimental investigations have shown that the stresses in Newtonian fluids are related linearly to the derivatives of velocities and that most fluids are *isotropic*—that is, the fluid properties are not dependent on the direction in space. In other words, an isotropic fluid has the same properties in all directions. The stresses do not explicitly depend upon the space co-ordinates and the velocity of the fluid. If the fluid element is considered to be very small in size, we may be justified in assuming that

$$\tau_{xy} = \tau_{yx},\ \tau_{yz} = \tau_{zy} \text{ and } \tau_{zx} = \tau_{xz} \qquad \text{...(6.2)}$$

The above mentioned assumptions lead to a unique form of relationship between the stresses and the velocity gradients, which may be expressed as

$$\begin{aligned} \sigma_x &= -p - \frac{2}{3} \mu\Theta + 2\mu \frac{\partial u}{\partial x} \\ \sigma_y &= -p - \frac{2}{3} \mu\Theta + 2\mu \frac{\partial v}{\partial y} \\ \sigma_z &= -p - \frac{2}{3} \mu\Theta + 2\mu \frac{\partial w}{\partial z} \end{aligned} \qquad \text{...(6.3 } a\text{)}$$

$$\begin{aligned} \tau_{xy} &= \tau_{yx} = \mu \left(\frac{\partial u}{\partial y} + \frac{\partial v}{\partial x} \right) \\ \tau_{xz} &= \tau_{zx} = \mu \left(\frac{\partial u}{\partial z} + \frac{\partial w}{\partial x} \right) \end{aligned} \qquad \text{...(6.3 } b\text{)}$$

and

$$\tau_{yz} = \tau_{zy} = \mu \left(\frac{\partial v}{\partial z} + \frac{\partial w}{\partial y} \right).$$

Expressions contained in Eq. (6.3 a) give the normal stresses while those in Eq. (6.3 b) are meant to yield the shear stresses. Equation (6.3 b) is known as the extended form of Newton's law of viscosity (or shear), and expresses the stress-strain relationship in general. For one dimensional flow, in which $v = w = 0$ and $\partial/\partial z = 0$, Eq. (6.3 b) yields

$$\tau_{xy} = \tau_{yx} = \mu \frac{du}{dy}$$

This expression is the famous Newton's law of viscosity (fluid friction) discussed in the chapter on fluid properties, see Eq. (1.3). In Eq. (1.3 a) Θ stands for $\frac{\partial}{\partial x} + \frac{\partial v}{\partial y} + \frac{\partial w}{\partial z}$. Substituting Eq. (6.3) in Eq. (6.1), the equations of motion take the form

$$\rho \frac{Du}{Dt} = \rho X - \frac{\partial p}{\partial x} - \frac{2}{3} \frac{\partial}{\partial x} (\mu\Theta) + 2 \frac{\partial}{\partial x} \left(\mu \frac{\partial u}{\partial x} \right) + \frac{\partial}{\partial y} \left\{ \mu \left(\frac{\partial u}{\partial y} + \frac{\partial v}{\partial x} \right) \right\} + \frac{\partial}{\partial z} \left\{ \mu \left(\frac{\partial u}{\partial z} + \frac{\partial w}{\partial x} \right) \right\},$$

$$\rho \frac{Dv}{Dt} = \rho Y - \frac{\partial p}{\partial y} - \frac{2}{3} \frac{\partial}{\partial y} (\mu\Theta) + \frac{\partial}{\partial x} \left\{ \mu \left(\frac{\partial u}{\partial y} + \frac{\partial v}{\partial x} \right) \right\} + 2 \frac{\partial}{\partial y} \left(\mu \frac{\partial v}{\partial y} \right) + \frac{\partial}{\partial z} \left\{ \mu \left(\frac{\partial v}{\partial z} + \frac{\partial w}{\partial y} \right) \right\}$$

...(6.4)

$$\rho \frac{Dw}{Dt} = \rho Z - \frac{\partial p}{\partial z} - \frac{2}{3}\frac{\partial}{\partial z}(\mu\Theta) + \frac{\partial}{\partial x}\left\{\mu\left(\frac{\partial u}{\partial z} + \frac{\partial w}{\partial x}\right)\right\} + \frac{\partial}{\partial y}\left\{\mu\left(\frac{\partial v}{\partial z} + \frac{\partial w}{\partial y}\right)\right\} + 2\frac{\partial}{\partial z}\left(\mu\frac{\partial w}{\partial z}\right)$$

Eq. (6.4) are known as the Navier-Stokes equations of motion for a Newtonian fluid of varying density and viscosity in a gravitational field. If the viscosity is assumed to be constant, these equations may be simplified and rearranged to yield,

$$\frac{Du}{Dt} = X - \frac{1}{\rho}\frac{\partial p}{\partial x} + \nu\left(\frac{\partial^2 u}{\partial x^2} + \frac{\partial^2 u}{\partial y^2} + \frac{\partial u^2}{\partial z^2}\right) + \frac{1}{3}\nu\frac{\partial \Theta}{\partial x}$$

$$\frac{Dv}{Dt} = Y - \frac{1}{\rho}\frac{\partial p}{\partial y} + \nu\left(\frac{\partial^2 v}{\partial x^2} + \frac{\partial^2 v}{\partial y^2} + \frac{\partial v^2}{\partial z^2}\right) + \frac{1}{3}\nu\frac{\partial \Theta}{\partial \psi}$$

and
$$\frac{Dw}{Dt} = Z - \frac{1}{\rho}\frac{\partial p}{\partial z} + \nu\left(\frac{\partial^2 w}{\partial x^2} + \frac{\partial^2 w}{\partial y^2} + \frac{\partial^2 w}{\partial z^2}\right) + \frac{1}{3}\nu\frac{\partial \Theta}{\partial z} \qquad ...(6.5)$$

Viscosity measurements indicate that μ is a function of temperature and very slightly a function of pressure. The effect of pressure is almost always negligible, and if temperature changes are not large, the assumption of constant viscosity corresponding to the mean temperature of fluid is justified.

If the density of the fluid is constant, Θ becomes zero, and Eqs. (6.5) reduce to the following simple form :

$$\frac{Du}{Dt} = X - \frac{1}{\rho}\frac{\partial p}{\partial z} + \nu\left(\frac{\partial^2 u}{\partial x^2} + \frac{\partial^2 u}{\partial y^2} + \frac{\partial^2 u}{\partial z^2}\right) = X - \frac{1}{\rho}\frac{\partial p}{\partial x} + \nu\,\nabla^2 u$$

$$\frac{Du}{Dt} = Y - \frac{1}{\rho}\frac{\partial p}{\partial y} + \nu\left(\frac{\partial^2 v}{\partial x^2} + \frac{\partial^2 v}{\partial y^2} + \frac{\partial^2 v}{\partial z^2}\right) = Y - \frac{1}{\rho}\frac{\partial p}{\partial y} + \nu\,\nabla^2 v \qquad ...(6.6)$$

$$\frac{Dw}{Dt} = Z - \frac{1}{\rho}\frac{\partial p}{\partial z} + \nu\left(\frac{\partial^2 w}{\partial x^2} + \frac{\partial^2 w}{\partial y^2} + \frac{\partial^2 w}{\partial z^2}\right) = Z - \frac{1}{\rho}\frac{\partial p}{\partial z} + \nu\,\nabla^2 w$$

where
$$\nabla^2 \equiv \frac{\partial^2}{\partial x^2} + \frac{\partial^2}{\partial y^2} + \frac{\partial^2}{\partial z^2}.$$

It may be noted here that Eqs. (6.4) are the cartesian form of the Navier-Stokes equations of the most general form applicable to any Newtonian compressible fluid of varying viscosity, whereas Eqs. (6.5) and (6.6) are simplified forms of the equations applicable to constant viscosity and constant viscosity-density fluids respectively.

6.2.1. The Navier-Stokes Equations in Cylindrical Co-ordinates

In many cases, it is convenient to work in a cylindrical co-ordinate system. The equations of motion and continuity may be suitably transformed into the desired co-ordinates. By co-ordinate transformation, the Navier-Stokes equation as well as the strain-stress relationships may be expressed in a cylindrical co-ordinate system. In the cylindrical system of co-ordinates (r, θ, z) the Navier-Stokes equations are given below in terms of V_r, V_θ and V_z, these being the velocity components in the respective directions.

(*i*) **In the r-direction**

$$\frac{\partial V_r}{\partial t} + V_r\frac{\partial V_r}{\partial r} + \frac{V_\theta}{r} + \frac{\partial V_r}{\partial \theta} + V_z\frac{\partial V_r}{\partial z} - \frac{V_\theta^2}{r}$$

$$= X_r - \frac{1}{\rho}\frac{\partial p}{\partial r} + \frac{1}{3}\nu\frac{\partial \Theta}{\partial r} + \nu\left\{\frac{1}{r}\frac{\partial}{\partial r}\left(r\frac{\partial V_r}{\partial r}\right) + \frac{1}{r^2}\frac{\partial^2 V_r}{\partial \theta^2} + \frac{\partial^2 V_r}{\partial z^2} - \frac{V_r}{r^2} - \frac{2}{r^2}\frac{\partial V_\theta}{\partial \theta}\right\}$$

(*ii*) **In the θ-direction**

$$\frac{\partial V_\theta}{\partial t} + V_r \frac{\partial V_\theta}{\partial r} + \frac{V_\theta}{r}\frac{\partial V_\theta}{\partial \theta} + V_z \frac{\partial V_\theta}{\partial z} + \frac{V_r V_\theta}{r}$$

$$= X_\theta - \frac{1}{\rho}\frac{\partial p}{r\partial \theta} + \frac{1}{3}\nu\frac{\partial \Theta}{r\partial \theta} + \nu\left\{\frac{1}{r}\frac{\partial}{\partial r}\left(r\frac{\partial V_\theta}{\partial r}\right) + \frac{1}{r^2}\frac{\partial^2 V_\theta}{\partial \theta^2} + \frac{\partial^2 V_\theta}{\partial z^2} - \frac{V_\theta}{r^2} + \frac{2}{r^2}\frac{\partial V_r}{\partial \theta}\right\}$$

(*iii*) **In the z-direction**

$$\frac{\partial V_z}{\partial t} + V_r \frac{\partial V_z}{\partial r} + \frac{V_\theta}{r}\frac{\partial V_z}{\partial \theta} + V_z \frac{\partial V_z}{\partial z}$$

$$= X_z - \frac{1}{\rho}\frac{\partial p}{\partial z} + \frac{1}{3}\nu\frac{\partial \Theta}{\partial z} + \nu\left\{\frac{1}{r}\frac{\partial}{\partial r}\left(r\frac{\partial V_\theta}{\partial r}\right) + \frac{1}{r^2}\frac{\partial^2 V_z}{\partial \theta^2} + \left\{\frac{\partial^2 V_z}{\partial z^2}\right\}\right\} \quad ...(6.7)$$

in which X_r, X_θ and X_z are the body forces per unit mass in the r, θ and z-directions respectively. Eq. (6.7) correspond to Eq. (6.5) of the cartesian system which are applicable for constant viscosity fluids. The stresses corresponding to Eq. (6.3) are given by

$$\sigma_r = -p - \frac{2}{3}\mu\Theta + 2\mu\frac{\partial V_r}{\partial r}$$

$$\sigma_\theta = -p - \frac{2}{3}\mu\Theta + 2\mu\left\{\frac{1}{r}\frac{\partial V_\theta}{\partial \theta} + \frac{V_\theta}{r}\right\}$$

$$\sigma_z = -p - \frac{2}{3}\mu\Theta + 2\mu\frac{\partial V_r}{\partial z}$$

$$\tau_{r\theta} = \mu\left\{r\frac{\partial}{\partial r}\left(\frac{V_\theta}{r}\right) + \frac{1}{r}\frac{\partial V_r}{\partial \theta}\right\} \quad ...(6.8)$$

$$\tau_{rz} = \mu\left\{\frac{\partial V_r}{\partial r} + \frac{\partial V_r}{\partial z}\right\}$$

$$\tau_{\theta z} = \mu\left(\frac{\partial V_\theta}{\partial z} + \frac{1}{r}\frac{\partial V_z}{\partial \theta}\right)$$

The equation of continuity in cylindrical co-ordinates is as follows :

$$\frac{\partial \rho}{\partial t} + \frac{\partial(\rho V_r)}{\partial r} + \frac{1}{r}\frac{\partial(\rho V_\theta)}{\partial \theta} + \frac{\partial(\rho V_z)}{\partial z} + \frac{\rho V_r}{r} = 0 \quad ...(6.9\ a)$$

In case of a constant density (incompressible) fluid, Eq. (6.9 *a*) becomes :

$$\frac{\partial V_r}{\partial r} + \frac{1}{r}\frac{\partial V_\theta}{\partial \theta} + \frac{\partial V_z}{\partial z} + \frac{V_r}{r} = 0 \quad ...(6.9\ b)$$

The equations of motion (6.7) are general equations applicable to fluids of variable density and constant viscosity. For liquids which normally are incompressible and also for gases where pressure changes are not considerable, Θ represented by left hand side of Eq. (6.9 *b*) is zero, and Eq. (6.7) can be written as,

***r*-direction :**

$$\frac{DV_r}{Dt} - \frac{V_\theta^2}{r} = X_r - \frac{1}{\rho}\frac{\partial p}{\partial r} + \nu\left(\nabla^2 V_r - \frac{V_r}{r^2} - \frac{2}{r^2}\frac{\partial V_\theta}{\partial \theta}\right)$$

θ-direction :

$$\frac{DV_\theta}{Dt} - \frac{V_r V_\theta}{r} = X_\theta - \frac{1}{r\rho}\frac{\partial p}{\partial \theta} + \nu\left(\nabla^2 V_\theta + \frac{2}{r^2}\frac{\partial V_\theta}{\partial \theta} - \frac{V_\theta}{r}\right) \quad ...(6.7\ a)$$

Z-direction :

$$\frac{DV_z}{Dt} = X_z - \frac{1}{\rho}\frac{\partial p}{\partial z} + \nu \nabla^2 V_z$$

in which the operators D/Dt and ∇^2 have the following meaning :

$$\frac{D}{Dt} \equiv \frac{\partial}{\partial t} + V_r \frac{\partial}{\partial r} + \frac{V_\theta}{r}\frac{\partial}{\partial \theta} + V_x \frac{\partial}{\partial z}$$

$$\nabla^2 \equiv \frac{\partial^2}{\partial r^2} + \frac{1}{r}\frac{\partial}{\partial r} + \frac{1}{r^2}\frac{\partial^2}{\partial \theta^2} + \frac{\partial^2}{\partial z^2}.$$

6.3. Significance of Body Force

The body forces due to gravity are important in flow problems in which free-liquid surface exists or when the fluid is non-homogeneous, that is its density changes from one point to another so that there exists a density gradient. In case of a homogeneous fluid-flow within closed boundaries, there is an equilibrium between the weight of a fluid element and the buoyant force acting on it. In such cases, the body force due to gravity does not influence the fluid motion, and are therefore omitted from the equations.

If the fluid is rotating about an axis, body force due to centrifugal action must be considered.

If the co-ordinate axes are so oriented that the z-axis and the vertical direction (measured positive upwards) coincide, then the body forces

$$X = Y = 0\ ; \qquad Z = -g \qquad \text{(Cartesian axes)}$$

also, $$X_r = X_\theta = 0\ ; \qquad X_z = -g \qquad \text{(Cylindrical axes)}$$

The negative sign appears because the gravitational acceleration acts in a direction opposite to z-axis.

6.4. Boundary Conditions

The fluid flow in most of the engineering problems is enclosed within solid boundaries. The solid boundary confining the fluid may be stationary, moving in translation or rotating about an axis. It is obvious that the fluid velocity component normal to the boundary surface must be equal to the corresponding component of the boundary itself, otherwise there would be flow across the boundary. Experimental evidences have shown that the liquids and gases (at ordinary temperatures and pressures) adhere to the solid boundary. The tangential velocity as well as the normal one at the solid boundary are thus equal to the corresponding velocity components of the boundary itself. In other words, there is no relative motion or *slip* between the fluid and the boundary and this behaviour of the fluid at the boundary is described by a *no-slip* condition. If the solid boundary is stationary, *i.e.* at rest, the fluid must also have zero velocity in the tangential as well as normal direction to the boundary.

6.5. Significance of the Viscous Term

Comparison of Eqs. (6.6) and (4.6 *a*) shows that the equations of motion for viscous fluids differ from those for an ideal fluid only in the viscous or friction terms, $\nu \nabla^2 u$, $\nu \nabla^2 v$ and $\nu \nabla^2 w$. Mathematically speaking, this point is very important, since the Euler's equations of motion contain only the first derivatives of velocity while the Navier-Stokes equations also contain their second derivatives in the viscous term. These equations are in other words, of higher order.

General solutions for the Navier-Stokes equations have not as yet been found because of the *non-linear, second order* nature of the partial differential equations. The non-linearity of these equations lies in the acceleration terms such as $u\,\partial u/\partial x$, $v\,\partial u/\partial y$, $w\,\partial u/dz$ and similar other terms. However, many particular solutions can be obtained by introducing various simplifications.

A justification for second-order differential equations, *i.e.* Eqs. (6.6) can be explained on physical basis by considering the boundary conditions which must be fulfilled at the plane of contact between the viscous fluid and a solid boundary. The 'no-slip' condition between the fluid and the boundary requires that the fluid velocity must be equal to that of the boundary (*i.e.* zero for a stationary boundary). In other words, both normal and tangential velocity components must be zero. In frictionless flow however, only the normal component of velocity is required to vanish. Two independent boundary conditions must, therefore, be fulfilled, which require a differential equation of second order. For the same reason, it is not permissible to ignore the viscous terms in the differential equations, even for very small values of ν, if the true behaviour of the viscous fluid is to be determined in the vicinity of the boundary.

The difficulties in the integration of Eq. (6.6) are particularly acute when the non-linear acceleration terms are of the same order of magnitude as the viscous terms. This occurs frequently in fluids of small kinematic viscosity or flows at large Reynolds numbers. But even if the viscous terms are very small, the difficulties remain since the boundary conditions can be fulfilled only when the viscous terms are taken into account. Some simplifications are however possible in 'creep' flow, *i.e.* flow at very small Reynolds number.

6.6. Limiting Cases of the Navier-Stokes Equations

The Navier-Stokes equations (abbreviated as N-S equations) of motion are exact mathematical statements of dynamic conditions within a viscous fluid and as such they are expected to apply correctly to each and every problem of viscous fluid flow. Owing to the mathematical complexity of the equations, it has already been emphasized that general solution of these equations is not possible. The following limiting cases are of importance and are, therefore, discussed briefly.

(*i*) Flow problems in which the viscous forces tend to zero, *i.e.* potential flow.

(*ii*) Flow problems wherein the viscous forces are of higher order of magnitude as compared to the inertial forces. These are the cases in which the flow is at very low Reynolds number. This type of flow is called *creep flow*.

(*iii*) Flow problems characterized by non-zero viscous forces but of smaller magnitude than the inertial forces. These are the cases in which the flow is at very high Reynolds number.

6.6.1. Potential Flow and N-S Equations

The first case represents flow of a non-viscous fluid, and if the fluid is also incompressible, the dynamics of such a fluid will be in accordance with the potential flow (ideal fluid flow) theory. In potential flow, we know that there exists a velocity potential and that the flow is irrotational.

For potential flow, from Eq. (3.35)

$$u = -\frac{\partial \phi}{\partial x},\ v = -\frac{\partial \phi}{\partial y},\ w = -\frac{\partial \phi}{\partial z}$$

the substitution of these in the continuity equation results in

$$\frac{\partial^2 \phi}{\partial x^2} + \frac{\partial^2 \phi}{\partial y^2} + \frac{\partial^2 \phi}{\partial z^2} = 0$$

expressing the viscous terms of the N-S equations in terms of the velocity potential

$$\nu\, \Delta^2 u = \nu \left(\frac{\partial^2 u}{\partial x^2} + \frac{\partial^2 u}{\partial y^2} + \frac{\partial^2 u}{\partial z^2} \right)$$

$$= \nu \left\{ \frac{\partial^2 u}{\partial x^2}\left(-\frac{\partial \phi}{\partial x}\right) + \frac{\partial^2}{\partial y^2}\left(-\frac{\partial \phi}{\partial x}\right) + \frac{\partial^2}{\partial z^2}\left(-\frac{\partial \phi}{\partial x}\right) \right\}$$

$$= -\nu \frac{\partial}{\partial x}\left(\frac{\partial^2 \phi}{\partial x^2} + \frac{\partial^2 \phi}{\partial y^2} + \frac{\partial^2 \phi}{\partial z^2} \right) = 0$$

Similarly, it can be shown that

$$\nu\, \Delta^2 v = v\left(\frac{\partial^2 v}{\partial x^2} + \frac{\partial^2}{\partial y^2} + \frac{\partial^2 v}{\partial z^2}\right) = 0,$$

and

$$\nu^2 w = \nu\left(\frac{\partial^2 w}{\partial x^2} + \frac{\partial^2 w}{\partial y^2} + \frac{\partial^2 w}{\partial z^2}\right) = 0 \qquad ...(6.10)$$

Equation (6.10) shows that in potential flow which is characterized by the existence of a velocity potential, the viscous terms in the N-S equations of motion, Eq. (6.6) are identically zero, and as a result these equations are reduced to the Euler's equations of motion.

Even though the solution of potential flow satisfies the N-S equations, it cannot satisfy all the boundary conditions. Owing to the assumption of zero-viscosity of the fluid in potential flow, the order of the differential equations is reduced by one. The resulting equations are, therefore, first order differential equations. The solution of the first order differential equations involves only one arbitrary constant, the evaluation of which requires only one boundary condition. As a consequence, we have to relax one of the boundary conditions of the original problem. In the flow of an actual fluid, ordinarily both the tangential and normal velocity components on the surface of a solid stationary boundary are zero owing to 'no-slip' condition. But in the potential flow problem, the motion is completely determined by the condition of zero normal velocity at the solid boundary, the tangential velocity being non-zero. There is no possibility, in the potential flow, of making the tangential velocity component zero simultaneously with the normal component. Hence the solutions of the N-S equations represent flow fields in general and are different from those of potential flow. The flow field of a viscous fluid is rotational. The dynamic behaviour of a viscous fluid differs from that of an ideal fluid only in the vicinity of the solid boundary, and the viscous effects are confined to a narrow region very close to the boundary. This narrcw region is known as the boundary layer. The concept of the boundary layer is discussed in detail in Chapter 7.

6.6.2. Flow at Very Small Reynolds Number (*Creep Flow*)

For very small Reynolds numbers (very small velocity, small linear dimensions of the body or of the flow passage and large viscosity of the fluid) the inertial forces are much smaller than the viscous forces. We know that for steady flow, the inertial force may be written as $\rho L^2 V^2$ while the viscous force can be expressed in a simplified as $\mu A \dfrac{du}{dx}$. It may be clearly seen from these expressions, that the inertial force contains velocity raised to power two, whereas the viscous force is proportional to the first power of velocity. Since, for very low Reynolds numbers the velocities are very small, we may neglect higher order terms of the inertial force, like $u \dfrac{\partial u}{\partial x}$, $v \dfrac{\partial u}{\partial r}$ etc., and express the N-S equation (6.6) as follows :

$$\begin{aligned} \frac{\partial u}{\partial r} &= X - \frac{1}{\rho}\frac{\partial p}{\partial x} + \nu\, \nabla^2 u \\ \frac{\partial v}{\partial t} &= Y - \frac{1}{\rho}\frac{\partial p}{\partial y} + \nu\, \nabla^2 v \\ \frac{\partial w}{\partial t} &= Z - \frac{1}{\rho}\frac{\partial p}{\partial z} + \nu\, \nabla^2 w \end{aligned} \qquad ...(6.11)$$

Equations (6.11) may be regarded as fundamental equations for very slow motion of viscous fluid known as *creep flow*. These are linear equations and are much easier to solve than the original non-linear equations. Creep flow analysis is of importance for laminar flow in pipes and open channels, for seepage flow of water and oil underground, for motion of very small bodies such as spheres in a highly viscous fluid and in the theory of lubrication.

6.6.3. Flow at Very Large Reynolds Number

Nearly all the problems in aerodynamics and fluid dynamics are cases involving large Reynolds number (large velocity, large linear dimensions of the body or of the flow passage and small viscosity of the fluid which is generally true for air and water). In such cases the viscous forces are much smaller than the inertial forces. The most obvious approach in solving such problems is to drop the viscous terms ($v\ \Delta^2 u$, etc.). The N-S equations are then reduced to the equations for the non-viscous fluids, and the order of the differential equations is lowered by one and one of the boundary conditions out of two cannot obviously be satisfied. Whereas the simplification of the problem by dropping the viscous terms gives good approximate results at places far away from the solid boundary, it is no longer a good approximation for the region near the boundary. Many important phenomena such as the existence of skinfriction and the resulting drag cannot be explained by treating the fluid as non-viscous. For large Reynolds numbers, the viscous effects are confined to a very thin boundary layer. The dynamics of flow in this part of the flow field will be dealt with in Chapter 7.

6.7. Simple Applications of the Navier-Stokes Equations

The Navier-Stokes equations for viscous fluid motion, as stated earlier are not mathematically amenable to solution owing to the non-linearity. It is only through certain valid assumptions that these equations are simplified and the main hurdle of non-linearity overcome. This is possible in case of a very few special problems wherein the boundary configuration is simple and the fluid characteristics such as the density and viscosity are almost constant. These assumptions of constant density and constant viscosity are valid for liquids and gases at low Mach number ($M < 1.0$). With these simplifying assumptions we can solve the N-S equations for the three velocity components and the pressure making use of the continuity equation.

To illustrate the applications of the N-S equations, we shall consider (*i*) laminar flow between two straight parallel boundaries (*ii*) laminar flow between concentric rotating cylinders and (*iii*) laminar flow (Hagen-Poiseuille) in a circular pipe.

6.7.1. Laminar Flow between Two Straight Parallel Boundaries

Consider a two-dimensional laminar flow of an incompressible fluid of constant viscosity between straight and parallel boundaries separated by a distance b apart as shown in Fig. 6.3. In laminar flow, the fluid moves in layers parallel to the solid boundaries. In order to maintain such a flow, the force due to pressure difference in the x-direction and the viscous (shear) resistance offered to the flow by the presence of boundaries must balance each other, that is their magnitudes must be equal and the directions in which they act must be opposite. For this type of flow, we can write,

$$u = f_1(x, y, t)$$
$$v = 0,\ w = 0$$
$$p = f_2(x, y, t),$$
$$\partial(\ )/\partial z = 0 \qquad \text{...(6.12)}$$

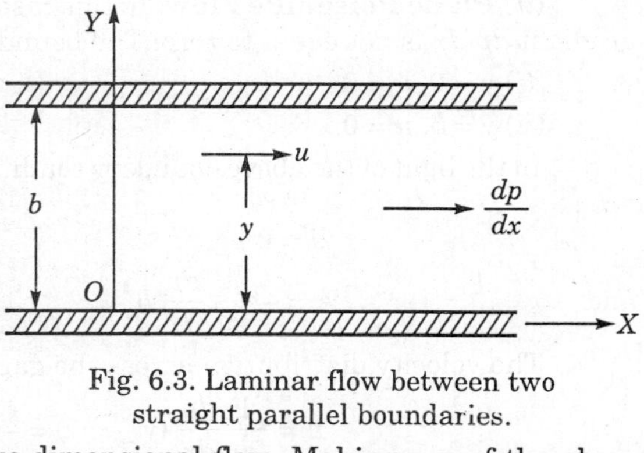

Fig. 6.3. Laminar flow between two straight parallel boundaries.

The last condition $\partial(\ )/\partial z = 0$, represents two-dimensional flow. Making use of the above conditions the equation of continuity reduces to

$$\frac{\partial u}{\partial x} = 0,$$

resulting in $\quad u = f_1(y, t) \qquad$ ...(6.13)

Assuming that the body forces are zero, using Eqs. (6.12) and (6.13), the N-S equations for an incompressible fluid expressed by Eq. (6.6) take the form

$$\frac{\partial u}{\partial t} = -\frac{1}{\rho}\frac{\partial p}{\partial x} + \nu\frac{\partial^2 u}{\partial y^2} \qquad ...(6.14)$$

$$0 = -\frac{1}{\rho}\frac{\partial p}{\partial y}$$

or $$p = f_2(x, t) \qquad ...(6.15)$$

Since u is not a function of x as is evident from Eq. (6.13), and p is not a function of y, it can be concluded from Eqs. (6.14) and (6.15) that dp/dx must be a constant or a function of time. Assuming steady flow ($\partial u/\partial t = 0$), integration of Eq. (6.14) with respect to y (treating dp/dx constant) yields

$$\frac{\partial u}{\partial y} = \frac{1}{\mu}\frac{dp}{dx}y + A$$

Integrating it once more, we obtain an equation for the velocity distribution

$$u = \frac{1}{2\mu}\frac{dp}{dx}y^2 + Ay + B \qquad ...(6.16)$$

where A and B are constants of integration to be determined using the boundary conditions.

We shall consider the following three cases embodying different boundary conditions for each case :

(*i*) **Simple Couette Flow.** In this type of flow, $dp/dx = 0$, one of the straight solid boundaries is stationary while the other one moves at a uniform speed U. The boundary conditions, referring to Fig. 6.3 are :

(*i*) $y = 0, u = 0,$

(*ii*) $y = b, u = U.$

An additional feature of this type of flow is that the gradient dp/dx does not exist. The above boundary conditions enable us to evaluate the arbitrary constants A and B of Eq. (6.16). We, therefore, have $B = 0$ and $A = U/b$. The equation of velocity distribution for this case becomes

$$u = \frac{y}{b}U \qquad ...(6.17)$$

Equation (6.17) indicates that the velocity distribution across the gap between the boundaries is linear (straight line variation).

(*ii*) **Plane Poiseuille Flow.** In this case both the boundaries are stationary and the pressure gradient dp/dx is not equal to zero. The boundary conditions are

(*i*) $y = 0, u = 0,$

(*ii*) $y = b, u = 0.$

In the light of the above boundary conditions, the arbitrary constants of the velocity equations are :

$$B = 0$$

and $$A = -\frac{1}{2\mu}\frac{dp}{dx}b.$$

The velocity distribution across the gap may be expressed as :

$$u = \frac{1}{2\mu}\frac{dp}{dx}(y^2 - by) = -\frac{1}{2\mu}\frac{dp}{dx}(by - y^2) \qquad ...(6.18)$$

Equation (6.18) shows that the velocity distribution in the y-direction is parabolic.

(*iii*) **Generalized Couette Flow.** This case is a combination of simple Couette and plane Poiseuille flow. The pressure gradient dp/dx is non-zero and one of the boundaries is in motion at a constant speed U. The boundary conditions are :

(*i*) $y = 0, u = 0,$

(*ii*) $y = b, u = U,$

which yield the constants $B = 0$

and $$A = \frac{U}{b} - \frac{1}{2\mu}\frac{dp}{dx}b$$

and the resulting velocity distribution is

$$u = \frac{U}{b}y - \frac{1}{2\mu}\frac{dp}{dx}(by - y^2) \quad ...(6.19)$$

As is evident from Eq. (6.19), the velocity distribution for a given flow problem depends upon U and dp/dx. The pressure gradient dp/dx may be negative or positive and U may also be negative. In case of plane Poiseulle flow, the pressure gradient is always negative. When dp/dx has a large positive value, some of fluid layers in the gap adjoining the stationary boundary will move in a direction opposite to that of the moving boundary. The critical pressure gradient is the limit of dp/dx which marks the beginning of backflow, and is obtained using the boundary conditions $y = 0$, $du/dy = 0$. From Eq. (6.19), the critical pressure gradient is

$$\left(\frac{dp}{dx}\right)_c = \frac{2U\mu}{b^2} \quad ...(6.20)$$

If the pressure gradient exceeds this critical value, some of the fluid near the stationary boundary will move in a direction opposite to the upper moving boundary, that is the back flow will take place near the lower stationary boundary.

6.7.2. Laminar Flow Between Concentric Rotating Cylinders (*Current Flow*)

Let us consider two concentric cylinders of radii r_1 and r_2 and length h rotating at uniform angular speeds ω_1 and ω_2 respectively. Let the annular gap between the two cylinders filled with a viscous liquid be small enough to produce laminar flow of the two-dimensional type (the cylinders length is assumed to be sufficiently large to make this possible). On account of uniform motion of cylinders, the laminar flow is steady, and in cylindrical co-ordinates (r, θ, z), one may write,

$$V_\theta = f_1(r),$$
$$V_r = V_z = 0$$

and $$p = f_2(r) \quad ...(6.21)$$

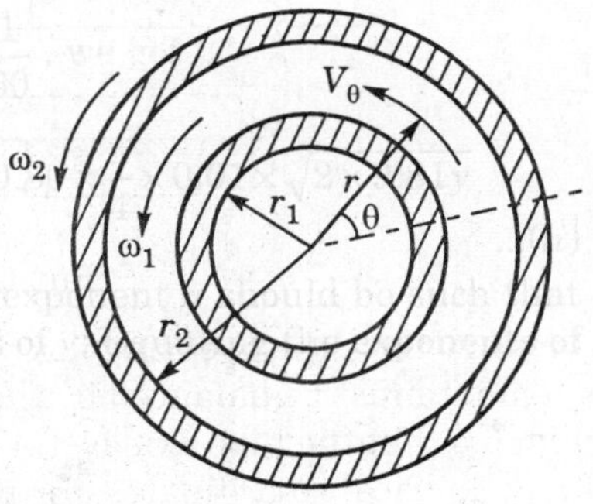

Fig. 6.4. Laminar flow between concentric rotating cylinders.

Substituting conditions expressed by Eq. (6.21) in the N-S equations of motion, neglecting body forces, we obtain :

(*i*) in *r*-direction $$\frac{V_\theta^2}{r} = \frac{1}{r} \times \frac{dp}{dr} \quad ...(6.22)$$

(*ii*) in θ-direction $$0 = \nu\left[\frac{1}{r}\frac{d}{dr}\left(\frac{rdV_\theta}{dr}\right) - \frac{V_\theta}{r^2}\right] = \frac{d^2V_\theta}{dr^2} - \frac{d}{dr}\left(\frac{V_\theta}{r}\right) \quad ...(6.23)$$

The boundary conditions for this case are

$$r = r_1, \quad V_\theta = r_1\omega_1;$$
$$r = r_2, \quad V_\theta = r_2\omega_2 \quad ...(6.24)$$

The solution of equation (6.22) using the boundary conditions of Eq. (6.24) are

$$V_\theta = \frac{1}{r_2^2 - r_1^2}\left[r(r_2^2\,\omega_2 - r_1^2\omega_1) - \frac{r_1^2 r_2^2}{r}(\omega_2 - \omega_1)\right] \quad ...(6.25)$$

and $$p = p_1 + \frac{p}{(r_2^2 - r_1^2)^2}\left[(r_2^2\omega_2 - r_1^2 w_1)^2\frac{r^2 - r_1^2}{2} - 2r_1^2 r_2^2(\omega_2 - \omega_1)(r_2^2\omega_2 - r_1^2\omega_1)\log\frac{r}{r_1} + r_1^4 r_2^4(\omega_2 - \omega_1)^2\left(\frac{1}{r^2} - \frac{1}{r_1^2}\right)\right] \quad ...(6.26)$$

in which p_1 is the pressure at $r = r_1$.

Couette was first to use this method of concentric rotating cylinders to determine the viscosity of liquid. He kept one of the cylinders stationary while the other one was rotated at a uniform speed. The determination of viscosity involved the measurement of torque required to maintain the motion of the rotating cylinder. This type of flow is, therefore, generally known as the *Couette flow.* For example let us consider the outer cylinder rotating and the inner one stationary. The torque exerted on the outer cylinder due to viscous shear may be expressed as below, using Eq. (6.8) for $\tau_{r\theta}$.

$$T = 2\pi r_2 h \tau_2 r_2 = 2\pi r_2^2 \, h\mu \left[r \frac{\partial (V_\theta / r)}{\partial r} \right]$$

$$= 4\pi \, \mu h \, \frac{r_1^2 r_2^2}{r_1^2 - r_2^2} \, \omega_2 \qquad \text{...(6.27)}$$

From Eq. (6.27), the viscosity μ can be determined.

6.7.3. Laminar Flow in a Circular Pipe (*Hagen-Poiseuille Flow*)

The most simple and practical case of laminar flow is the steady laminar flow through a long straight pipe of circular cross-section. Consider a pipe of radius r_0 and of straight length L through which a liquid of viscosity μ is flowing at a steady volumetric rate of Q. In order to maintain

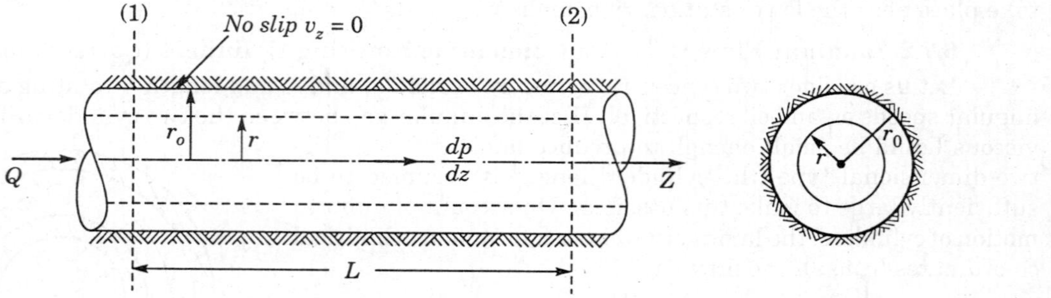

Fig. 6.5. Laminar flow through a circular pipe.

laminar flow, the shearing force (viscous resistance) on the surface of any cylindrical shape of fluid element must be balanced by the difference of the pressure force between the end faces (1) and (2), Fig. 6.5. Let the pipe axis be taken as the z-axis. The pressure p must, therefore, be a function of z. In cylindrical co-ordinates, for this particular case, we obtain

$$V_z = f_1(r), \quad V_\theta = V_r = 0$$

$$\partial(\ )/\delta\theta = 0, \qquad p = f_2(z) \qquad \text{...(6.28)}$$

In the light of Eq. (6.28), the N-S Equation of motion simplifies to the following form, ignoring the body components :

(*ii*) in r-direction : $$\frac{\partial p}{\partial r} = 0 \qquad \text{...(6.29)}$$

resulting in $p = f(z)$ for steady flow

(*iii*) in z-direction : $$-\frac{1}{\rho}\frac{\partial p}{\partial z} + \nu \left\{ \frac{1}{r}\frac{\partial}{\partial r}\left(r \frac{\partial Vz}{\partial r} \right) \right\} = 0$$

which simplifies to, $$\frac{dp}{dz} = \mu \left\{ \frac{d^2 V_z}{dr^2} + \frac{1}{r}\frac{dV_z}{dr} \right\} \qquad \text{...(6.30)}$$

From Eq. (6.29) since the pressure p is independent of r and is a function of z only, and V_z is a function of r only ; the pressure gradient dp/dz must be constant which can be reasoned out from Eq. (6.30). The first boundary condition is that on the pipe boundary at rest 'no-slip' condition must exist, that is, at $r = r_0$, $V_z = 0$, and the second one is from symmetry of flow : $r = 0$, $dV_z/dr = 0$. The solution of Eq. (6.30) yields,

$$V_z = \frac{1}{4\mu}\frac{dp}{dz}r^2 + A\log r + B \qquad ...(6.31\ a)$$

From the boundary conditions, $A = 0$,

$$B = -\frac{1}{4\mu}\frac{dp}{dz}r_0^2$$

and the velocity distribution becomes

$$V_z = -\frac{1}{4\mu}\frac{dp}{dz}(r_0^2 - r^2) \qquad ...(6.31\ b)$$

Eq. (6.31) shows that the velocity distribution across the pipe is parabolic similar to the plane Poiseuille flow. If p_1 and p_2 are the pressures at sections (1) and (2) distance L apart, the pressure gradient dp/dz is equal to $(p_1 - p_2)/L$. The maximum velocity occurs at the pipe and is equal to,

$$(V_z)_{max} = -\frac{1}{4\mu}\frac{(p_2 - p_1)}{L}r_0^2 = \frac{1}{4\mu}\frac{(p_1 - p_2)}{L}r_0^2 \qquad ...(6.32)$$

We know that in case of parabolic distribution, the average velocity V is half the maximum velocity. The volume of fluid Q flowing per unit time through the pipe is given by,

$$Q = \pi r_0^2 V = \pi r_0^2 \frac{1}{8\mu}\left(\frac{p_1 - p_2}{L}\right)r_0^2 = \frac{\pi r_0^4}{8\mu}\left(\frac{p_1 - p_1}{L}\right) \qquad ...(6.33)$$

The relationship expressed by Eq. (6.33) for the volumetric rate of flow, in terms of viscosity, pressure drop per unit length and the pipe radius was obtained experimentally by Hagen in 1839 and independently by Poiseuille in 1840. It is, therefore, jointly known as *Hagen-Poiseuille* equation and permits experimental determination of viscosity of the fluid.

The velocity distribution expressed by Eq. (6.31 *b*) will only be obtained when the flow reaches a fully developed state and remains laminar. The flow in pipe just after the entrance from a tank or reservoir is initially almost uniform (velocity distribution is similar to that in ideal fluid flow, that is, rectangular) across the pipe diameter. The velocity distribution is not fully transformed from an almost uniform profile to a parabolic one until the flow has travelled a distance of about 50 to 100 pipe-diameters from the entrance. Once the velocity profile becomes parabolic it remains unchanged in the direction of flow, and hence it is called *fully developed flow*. The distance (50 to 100 *d*) is known as the *entrance length* required for the establishment of fully developed laminar flow. (For further details see art. 11). So long as the Reynolds number,

$$R = \frac{V(2r_0)}{\nu}$$

(where V is the average velocity), has a value less than the critical Reynolds number, the laminar flow is usually possible to obtain. From experiments it has been shown that the critical Reynolds number is around 2000. As long as the Reynolds number is less than this value, the flow in the circular pipe will invariably be laminar.

The shear stress in Hagen-Poiseuille flow can be obtained using Eqs. (6.8) and (6.31 *b*),

$$\tau_{rz} = -\mu\frac{dV_z}{dr} \qquad ...(6.34)$$

The negative sign has been introduced because as r increases V_z decreases causing dV_z/dr negative, see Fig. 8.6. The velocity gradient from Eq. (6.31 *b*) is,

$$\frac{dV_z}{dr} = \frac{1}{2\mu}\frac{dp}{dz}r \qquad ...(6.35)$$

It may be noted that dp/dz is also negative (p decreases in the direction of flow).

If dp/dz is replaced by $(p_2 - p_1)/L$, the negative sign from Eq. (6.34) may be dropped. Using Eq. (6.35),

$$\tau_{rz} = -\mu\left\{\frac{1}{2\mu}\left(\frac{p_2 - p_1}{L}\right)r\right\} = \frac{(p_1 - p_2)r}{2L} \qquad ...(6.36)$$

Eq. (6.36) shows that the shear stress at any radius r is directly proportional to r and is independent of viscosity of the liquid. At $r = 0$, that is, at the pipe axis the shear stress is zero and is maximum at the pipe wall, where $r = r_0$. The variation of this stress is linear (straight line) across the pipe diameter in accordance with Eq. (6.36).

The following examples further illustrate the applications of fundamental equations embodying the basic principles of viscous fluid flow.

Example 6.1. *A viscous liquid of density ρ and viscosity μ flows down a wide inclined plate under the influence of gravity. The angle of inclination of the plate with horizontal is θ. For all practical purposes the flow is parallel to the plate and is fully developed. The velocity is not a function of the distance down the plate. The depth of liquid normal to the plate is h. The viscosity of the air in contact with the upper surface of the liquid may be neglected. For steady laminar two-dimensional flow of the incompressible liquid, determine :*

(*i*) *the velocity distribution normal to the flow,*

(*ii*) *the shear stress at the plate boundary,*

(*iii*) *the average velocity, and*

(*iv*) *the mass rate of flow per unit width of plate.*

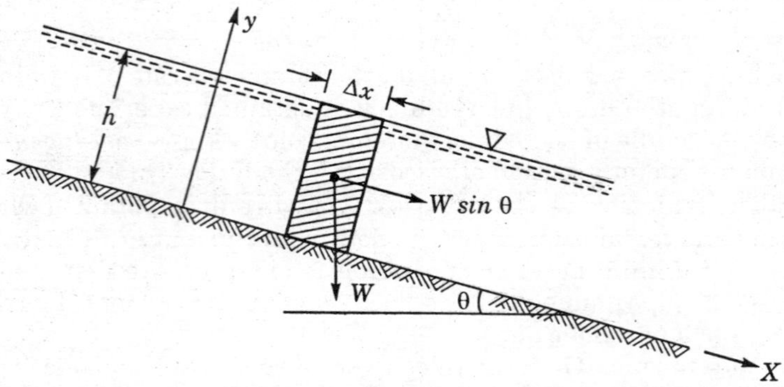

Solution. For steady two-dimensional flow of an incompressible viscous liquid, the Navier-Stokes equations of motion referring of Fig. (Ex. 6.1) are :

(1) in the x-direction

$$u\frac{\partial u}{\partial x} + v\frac{\partial u}{\partial y} + w\frac{\partial u}{\partial z} = X - \frac{1}{\rho}\frac{\partial p}{\partial x} + \nu\,\nabla^2 u \quad ...(a)$$

(2) in the y-direction

$$u\frac{\partial v}{\partial x} + v\frac{\partial v}{\partial y} + w\frac{\partial v}{\partial z} = Y - \frac{1}{\rho}\frac{\partial p}{\partial y} \quad \nu\,\nabla^2 v \quad ...(b)$$

Since the steady flow is in the x-direction and is fully developed, therefore, $u = f(y)$ and $v = 0$. On the liquid-air interface, the pressure is everywhere atmospheric and since the depth of flow is constant throughout, therefore, $\partial p/\partial x = 0$. Also since the flow is two-dimensional, $\partial(\ )/\partial z = 0$, $w = 0$. Under these conditions, the N-S equations are reduced to,

$$u\frac{\partial u}{\partial x} = \left(X + \nu\frac{\partial^2 u}{\partial y^2}\right) \quad ...(c)$$

and

$$0 = Y - \frac{1}{\rho}\frac{\partial p}{\partial y} \quad ...(d)$$

The continuity equation for steady three-dimensional flow is

$$\frac{\partial u}{\partial x} + \frac{\partial v}{\partial y} + \frac{\partial w}{\partial z} = 0.$$

Substituting the above conditions, we obtain $\partial u/\partial x = 0$ indicating that $u = f(y)$. Consider a length Δx and unit width of flow. The weight of the liquid contained in this volume is

$$W = \rho g\,(\Delta x.h)$$

the body force in the x-direction is $\rho g\,(\Delta x.h)\sin\theta$. The body force per unit mass of liquid is

$$X = \frac{\rho g h \Delta x \sin\theta}{\rho h \Delta x} = g\sin\theta \qquad ...(e)$$

Similarly the body force per unit mass in the y-direction :

$$Y = \frac{\rho g h \Delta x \cos\theta}{\rho h \Delta x} = g\cos\theta \qquad ...(f)$$

Equations (c) and (d) with the help of the continuity equation and Eqs. (e) and (f) take the following form :

$$0 = g\sin\theta + \frac{\mu}{\rho}\frac{d^2u}{dy^2} \qquad ...(g)$$

and

$$0 = g\cos\theta - \frac{1}{\rho}\frac{dp}{dy}$$

If angle θ is small, $\cos\theta \approx 1.0$, and we obtain

$$\frac{dp}{dy} = \rho g = \text{constant, for a given liquid} \qquad ...(h)$$

Equation (h) indicates the hydrostatic variation of pressure in the y-direction. The solution of Eq. (g) gives

$$u = -\frac{\rho g}{\mu}\sin\theta\,\frac{y^2}{2} + Ay + B \qquad ...(k)$$

The constants of integration are to be determined from the boundary conditions. One obvious boundary condition is that $u = 0$ for $y = 0$, which yields $B = 0$. The second boundary condition is not so obvious. The constant A is to be evaluated from the consideration of shear stress at $y = h$. Because of the absence of any adjacent liquid, it can be assumed that the shear stress at the free surface is zero. The shearing action of air may be small enough to be negligible. Thus at $y = h$, $\tau_{yx} = 0$, we know from Eq. (6.3 b)

$$\tau_{yx} = \mu\frac{du}{dy} \text{ which yields, } \frac{du}{dy} = 0 \text{ at } y = h.$$

From Eq. (k), $\dfrac{du}{dy} = -\dfrac{\rho g\sin\theta}{\mu}\,y + A$, resulting in $A = \dfrac{\rho g\sin\theta}{\mu}$

(i) The velocity distribution normal to the flow is thus given by,

$$u = \frac{\rho g\sin\theta}{2\mu}(2hy - y^2)$$

(ii) The shear stress at the plate, $(\tau_{yz})_{y=0} = \mu\,\dfrac{\rho g\sin\theta}{\mu} = \rho g h\sin\theta$

(iii) The average of velocity of flow, $V = \dfrac{1}{h}\displaystyle\int_0^h u\,dy\,\dfrac{\rho g h^2\sin\theta}{3\mu}$

(iv) Mass rate of flow per unit width of the plate $= \rho q = \rho h.V = \dfrac{\rho^2 g h^3\sin\theta}{3\mu}$.

This type of flow is possible as a result of continuous rainfall over a very vast plane land surface of gentle slope, and a known as *sheet flow.*

Example 6.2. *Obtain an expression for velocity distribution for laminar flow between two concentric cylinders. Also find expressions for the discharge through the annular space, the mean* *velocity and the shear stress. Find the value of r at which the shear stress will be zero. Examine the possibility of obtaining an expression for 'f' (Darcy-Weisbach coefficient) for this case.*

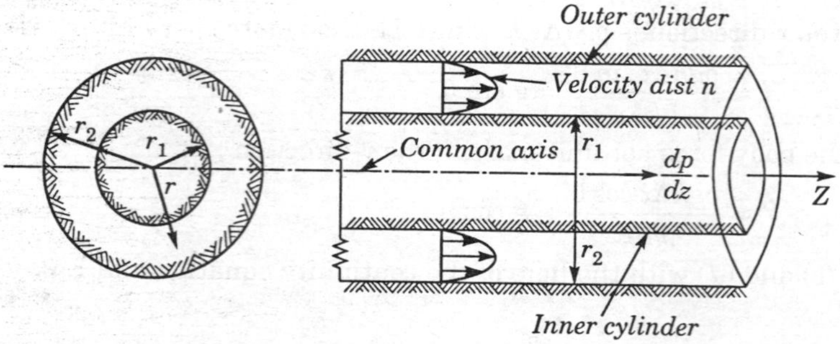

Solution. Using the Navier-Stokes equations in cylindrical co-ordinates (r, θ, z) and noting that for this type of problem

$$V_\theta = 0, \quad V_r = 0, \quad \partial/d\theta = 0 \qquad \text{...}(a)$$

In the light of Eq. (a) above, the N-S equations for steady flow reduce to

(*i*) in *r*-direction, $\frac{\partial p}{\partial r} = 0$, for steady flow this leads to $p = f(z)$

(*ii*) in *z*-direction *i.e.* along the common axis of the cylinders

$$-\frac{1}{\rho}\frac{dp}{dz} + \nu\left\{\frac{1}{r}\frac{\partial}{\partial r}\left(r\frac{\partial V_z}{\partial r}\right)\right\} = 0$$

which simplifies to,

$$\frac{dp}{dz} = \mu\left\{\frac{d^2V_z}{dr^2} + \frac{1}{r}\frac{dV_z}{dr}\right\} \qquad \text{...}(b)$$

Since p is independent of z and V_z is a function of r only, the pressure gradient must be a constant. Integration of Eq. (b) yields

$$V_z = \frac{1}{4\mu}\frac{dp}{dz}\, r^2 + A \log r + B \qquad \text{..}(c)$$

Owing to the existence of no-slip conditions at the pipe boundaries, we have the following boundary conditions to be satisfied

$$\text{at } r = r_1, \quad V_z = 0, \quad \text{and at} \quad r = r_2, \quad V_z = 0.$$

Using these boundary conditions, the constants of integration A and B are obtained as

$$A = \frac{1}{4\mu}\frac{dp/dz(r_2^2 - r_1^2)}{\log r_1/r_2}$$

and

$$B = -\frac{1}{4\mu}\frac{dp}{dz}\left[r_1^2 + \frac{r_2^2 - r_1^2}{\log r_1/r_2}\log r_1\right]$$

The substitution of which in Eq. (c) yields the velocity distribution law.

$$V_z = \frac{1}{4\mu}\frac{dp}{dz}\left[(r^2 - r_1^2) + \frac{r_2^2 - r_1^2}{\log r_1/r_2}\log r/r_1\right] \qquad \text{...}(d)$$

The discharge through the annular space is

$$Q = \int_{r_1}^{r_2} 2\pi r V_z dr$$

Upon substituting for V_z and integrating, we obtain

$$Q = \frac{\pi}{8\mu}\frac{dp}{dz}(r_2^2 - r_1^2)\left[\frac{r_2^2 - r_1^2}{\log r_2/r_1} - (r_2^2 + r_1^2)\right] \qquad \text{...}(e)$$

The mean velocity $V = \frac{Q}{\pi(r_2^2 - r_1^2)} = \frac{1}{8\mu}\frac{dp}{dz}\left[\frac{r_2^2 - r_1^2}{\log r_2/r_1} - (r_2^2 + r_1^2)\right]$ ...(f)

The shear stress $\tau = \mu\frac{dV_z}{dr} = \frac{1}{4}\frac{dp}{dz}\left[2r - \frac{1}{r}\frac{(r_2^2 - r_1^2)}{\log r_2/r_1}\right]$ ...(g)

Condition for zero shear stress gives

$$2r^2 = \frac{r_2^2 - r_1^2}{\log r_2/r_1} \quad \text{or} \quad r = \sqrt{\frac{1}{2}\frac{(r_2^2 - r_1^2)}{\log r_2/r_1}} \qquad ...(h)$$

Expression for friction factor f:

Under equilibrium conditions, the total force due to boundary resistance must be balanced by the force due to pressure gradient. Let τ_{o1} be the shearing stress at the inner cylinder boundary and τ_{o2} at the boundary of outer cylinder. Considering unit length of the annulus, we have

$$2\pi r_1\,\tau_{01} + 2\pi r_2\,\tau_{02} = \pi(r_2^2 - r_1^2)dp/dz$$

For simplicity of mathematical treatment, let us assume that

$$2\pi r_1\,\tau_{01} + 2\pi r_2\,\tau_{02} = 2\pi\,\frac{r_2 + r_2}{2}\,\tau_0$$

Using Eq. (f) for the average velocity V, we obtain

$$\tau_o = \frac{dp}{dz}\,(r_2 - r_1) = 8\mu\,V(r_2 - r_1)\Bigg/\left[\frac{r_2^2 - r_1^2}{\log r_2/r_1} - (r_2^2 + r_1^2)\right]$$

From the boundary shear stress-friction factor relationship, see Eq. (9.56), one may write :

$$\tau_0 = \frac{f}{8}\,\rho V^2$$

Eliminating τ_0,

$$f = \frac{64\mu/\rho\,\frac{(r_2 - r_1)}{V}}{\left[\frac{r_2^2 - r_1^2}{\log r_2/r_1} - (r_2^2 + r_1^2)\right]}$$

$$= \frac{64\,(1 - r_1/r_2)^2}{\left[\frac{1 - r_1^2/r_2^2}{\log r_2/r_1} - (1 + r_1^2/r_2^2)\right]\frac{\rho V(r_2 - r_1)}{\mu}} \qquad ...(i)$$

Selecting $(r_2 - r_1)$ as the characteristic length for defining the Reynolds number, we have $f = K.\,\frac{1}{\mathbf{R}}$, where, K = a constant depending purely on the radii ratio r_1/r_2.

$$f = f(\mathbf{R}) \qquad ...(j)$$

where $\mathbf{R}$ = Reynolds number = $\rho V(r_2 - r_1)/\mu$

Equation (i) gives a relationship for obtaining the friction factor in terms of fluid properties ρ and μ, the ratio of radii r_1/r_2 and the mean velocity V while equation (j) shows that it is a function of Reynolds number.

OBJECTIVE TYPE QUESTIONS

1. The Navier-Stokes equations of motion are the equations applicable for viscous fluids. Which of the following forces are taken into consideration in deriving these equations :

(a) inertia force (b) elastic force

(c) pressure force (d) viscous force

(e) gravity force (f) surface tension force.

2. For analysing viscous fluid flow problems in general, there are six variables namely u, v, w, ρ, p and T which influence a problem and are required to be determined. For isothermal flow of an incompressible fluid which of the following combination of the above variables would be applicable :

   (*i*) u, v, w, ρ (*ii*) u, v, w, p
   (*iii*) u, v, w, T (*iv*) u, v, p, T

3. In the derivation of the Navier-Stokes equation of motion, besides the body forces and normal stresses acting on the fluid element, shear forces must also be considered. Which of the following forces listed in column B fall into the three categories of column A :

Column A	*Column B*
(*a*) body forces	(*i*) pressure force
(*b*) normal forces	(*ii*) elastic force
(*c*) shear forces	(*iii*) inertia force
	(*iv*) gravity force
	(*v*) viscous force
	(*vi*) surface tension force.

4. The simplified form of the Navier-Stokes equation in x-direction may be written as :

$$\frac{Du}{Dt} = X - \frac{1}{\rho}\frac{\partial p}{\partial x} + v\left(\frac{\partial^2 u}{\partial x^2} + \frac{\partial_2 u}{\partial y^2} + \frac{\partial^2 u}{\partial z^2}\right)$$

   Its derivation is based on certain assumptions. Select the correct combination of assumptions from the following list for which the above equation is valid :

   (1) steady flow, (2) isotropic flow, (3) irrotational flow, (4) constant-density fluid, (5) constant viscosity, (6) Newton's law of viscosity is valid, (7) uniform flow

   (*a*) 1, 2, 3, 5 (*b*) 2, 4, 6, 7
   (*c*) 1, 3, 5, 7 (*d*) 2, 4, 5, 6

5. Make a choice from the following list as to which are the essential boundary conditions required to be satisfied by fluid flow at the solid boundary :

   (*i*) infinite velocity, (*ii*) full velocity,
   (*iii*) fluid velocity in the normal and tangential directions must be the same as the velocity of the boundary itself in the respective directions,
   (*iv*) viscosity is very large.

6. The general solution of the Navier-Stokes equations cannot be obtained because of the mathematical complexities introduced on account of their non linearity and second order nature. Identify from the following which terms of the equation contribute to non-linearity and which others are responsible for making these a second order differential equation :

(*a*) non-linear	(*i*) $\nu\nabla^2 u$
(*b*) second order	(*ii*) $u\ \partial u/\partial x$
	(*iii*) $\nu\nabla^2 v$
	(*iv*) $v\partial u/\partial y$
	(*v*) $\nu\nabla^2 w$
	(*vi*) $w\ \partial u/\partial z$.

7. Very slow motion of viscous-fluid is known as the "creep flow", the fundamental equations of motion for which are greatly simplified on account of which of the following :

   (*a*) the inertial force being very large
   (*b*) the viscous force being of negligible magnitude
   (*c*) the inertial force being extremely small as compared to the viscous force
   (*d*) the equations no longer remain non-linear

8. Creep flow analysis finds application in flow problems like :
(1) turbulent flow in pipes
(2) laminar flow in pipes and channels
(3) motion of submarines and ships
(4) seepage flow of water
(5) motion of very small bodies in a viscous fluid
(6) theory of lubrication
(7) flow of atmospheric air.
Choose your answer from the following alternatives :
(*a*) 1, 2, 3, 4 ; (*b*) 2, 4, 6, 7 ;
(*c*) 2, 3, 4, 7 ; (*d*) 2, 4, 5, 6 ;

ANSWERS

1. (*c*), (*d*), (*e*) **2.** (*ii*) **3.** (*a*)—(*iii*), (*iv*) ; (*b*)—(*i*), (*ii*) ; (*c*)—(*v*) **4.** (*d*)
5. (*iii*) **6.** (*a*)—(*ii*), (*iv*) (*vi*), (*b*)—(*i*), (*iii*), (*v*). **7.** (*c*), (*d*) **8.** (*d*).

REVIEW QUESTIONS

1. In what way are the Navier-Stokes equations of motion different from those of Euler's ?
2. What forces are accounted for by the N-S equations of motion ?
3. What are the fundamental equations (and the principles embodied in them) necessary for the complete solution of a problem involving the motion of a viscous compressible fluid ?
4. Comment on the validity of the application of N-S equations to the flow of rarified gases.
5. What is the extended form of the Newton's law of viscous shear ?
6. Write the N-S equations of motion for a viscous compressible fluid and explain the significance of each term.
7. Enumerate examples where the N-S equations in cylindrical co-ordinates find their applications.
8. Why are the exact solutions of the N-S equations not possible ? What valid approximations are usually made for analysing actual flow problems ?
9. Under what circumstances is the inclusion of body force in the force-field of the system necessary ?
10. What do you understand by 'no slip' condition ?
11. In case of a compressible fluid flow, the density

$\rho = f(x, y, z, t)$ and the total derivative of ρ

$$\frac{D\rho}{Dt} = \frac{\partial \rho}{\partial t} + u\frac{\partial \rho}{\partial x} + v\frac{\partial \rho}{\partial y} + w\frac{\partial \rho}{\partial z}$$

If in the above equation $\partial\rho/\partial t = 0$, could this equation be used for (*a*) steady laminar flow (*b*) incompressible laminar flow ?
12. Why it is necessary to relax one of the boundary conditions in case of solution of the N-S equations applied for an ideal fluid ? Identify the boundary condition.

PROBLEMS

6.1. A viscous liquid flows steadily in the annular space between two co-axial cylinders radii of r_1 and r_2. Using equations (6.31 *a*) show that the rate of volumetric flow is given by

$$Q = \frac{\pi r_1^4}{8\mu}\left(\frac{p_2 - p_1}{L}\right)\left[(n^4 - 1) - \frac{(n^2 - 1)^2}{\log n}\right]$$

where $(p_2 - p_1)/L$ represents the pressure drop per unit length and $n = r_2/r_1$.

6.2. Two viscous liquids share equally the gap between two parallel plates. The upper plate moves at a velocity U to the right and the lower plate is stationary. The pressure gradient is zero. The lower half of the gap is filled with a liquid of density ρ_1 and viscosity μ_1 and the upper half is filled with a liquid of density ρ_2 and viscosity μ_2. The vertical distance between the plates is b. Determine :

(*i*) the condition the shear stress must satisfy for $0 < y < b$, y being measured vertically from the stationary bottom plate.

(*ii*) the condition that must be satisfied by the velocity at the plate boundaries and at the interface of the two liquids.

(*iii*) the velocity profile in each of the two regions.

(*iv*) the shear stress at the lower plate.

6.3. A viscous liquid flows in the gap between two straight parallel plates. The lower plate is stationary while the upper one moves with a velocity u. Determine the relation between U and dp/dx for the following :

(*i*) the shear stress at the stationary plate is zero, and

(*ii*) the shear stress at the moving plate is zero.

6.4. Show that for plane Poiseuille flow the shearing stress is zero at the mid-point and that

$$\frac{\partial \tau_{yx}}{\partial y} = \frac{dp}{dx}.$$

6.5. Show that the average velocity for steady laminar flow of incompressible fluid in a circular pipe is given by

$$V = \frac{r_0^2}{8\mu}\left(-\frac{dp}{dz}\right) \quad \text{where } z \text{ is the direction of flow.}$$

6.6. In case of plane Poiseuille flow (flow between parallel stationary plates), show that the average viscous shear stress on the lower boundary expressed as a function of mean velocity may be written as $\tau_0 = \rho\mu V/b$. If the drag coefficient is defined as the ratio $\tau_0/\rho\,(V^2/2)$, express C_D as a function of the Reynolds number of flow and prove that $C_D = 12/R$.

6.7. For generalized couette flow, show that the discharge per unit width is given by :

$$q = U\frac{h}{2} - \frac{h^3}{12\mu}\frac{dp}{dx}.$$

6.8. Obtain an expression for the additional torque that will be experienced by the bottom surface of the inner cylinder in case of two concentric vertical cylinders separated by a small gap. The outer cylinder (radius r_2) rotates at an angular velocity of ω while the inner one (radius r_1) is held stationary.

6.9. For steady laminar flow between two concentric cylinders of radii R_1 and R_2 which are rotated at constant angular speeds ω_1 and ω_2, show that the tangential velocity at any radius r is given by

$$V_\theta = \frac{1}{R_2^2 - R_1^2}\left[(R_2^2\omega_2 - R_1^2\omega_1)r + \frac{R_1^2R_2^2\,(\omega_1 - \omega_2)}{r}\right].$$

ANSWERS

6.2. (*i*) τ = constant, (*ii*) $u = 0$ at $y = 0$, $u = U$ at $y = b$, u is continuous at $y = b/2$

(*iii*) for $y < b/2$, $u = \dfrac{2U/\mu_2}{b(\mu_1 + \mu_2)}\,y$; for $y > b/2$, $u = U - \dfrac{2U/\mu_1}{b(\mu_1 + \mu_2)}\,(b - y)$,

(*iv*) $2U\,\mu_1\mu_2/b(\mu_1 + \mu_2)$.

6.3. (*i*) $U = -\dfrac{1}{2\mu}\left(\dfrac{dp}{dx}\right)b^2$ (*ii*) $U = \dfrac{1}{2\mu}\left(-\dfrac{dp}{dx}\right)b^2$ **6.8.** $\pi\mu r_1^4\,\omega/2(r_2 - r_1)$.

7

Boundary Layer Flow

7.0. Introduction

The concept of boundary layer owes its existence to Ludwig Prandtl, a German aeronautic engineer, who in 1904 conceived and put forward the idea of a layer characteristic of fluids (even of very low viscosity) in the vicinity of a solid boundary. The flow of a viscous fluid, no matter how small is the viscosity, must satisfy the *no-slip* condition at the solid boundary. The *no-slip* condition implies that the velocity of fluid at a solid boundary must be the same as that of the boundary itself. For a boundary at rest, the fluid velocity must reduce to zero at the boundary surface. As soon as a flowing fluid comes in contact with a stationary solid body, its velocity reduces to zero in accordance with the no-slip condition. The fluid velocity increases in the direction normal to the boundary. The fluid layer in the neighbourhood of the solid boundary where the effects of fluid friction (viscous effects) are predominant is known as the *boundary layer*.

The theory of boundary layer flow constitutes the back-bone of modern fluid dynamics specially the flow of low viscosity fluids such as air and water which are encountered in everyday engineering problems. The development of the knowledge of boundary layer flow at the hands of Prandtl, Karman, Schlichting and many others has enabled the world to get such sophisticated flying machines as modern supersonic aircrafts, rockets, missiles and the like. With the aid of the boundary layer theory, the science of aeronautical engineering has made rapid progress in the last few decades. The other beneficiaries, as a result of development in the theory of boundary layer flow, are the fields of gas dynamics and aerofoil theory.

Prandtl succeeded in 1904 in showing that the flow of a low viscosity fluid (like water and air which are the most important fluids from their abundant availability in nature) is amenable to mathematical treatment. This was achieved by taking into account the viscous effects within the narrow region situated in the vicinity of the solid boundary. From the knowledge of classical hydrodynamics, the drag force experienced by a solid body immersed in a flowing fluid could not be theoretically explained until the emergence of the boundary layer concept.

The boundary layer exists in the flow of a real fluid past a solid body. For flow of water in an open channel ; in a pipe ; over a spillway ; over the land surface during and after the rains are some of the examples involving formation of the boundary layer. The flow of air over the earth surface and about large buildings ; the motion of aeroplane in the air mass ; in all these and such other cases where a flowing fluid comes in contact with a solid boundary, we have the existence of boundary layer.

The motion of fluids with "very little friction" (very low viscosity) amounts to fluid flow at very large Reynolds number. For such flows, Prandtl made the following observations :

(1) Viscous effects (effects of fluid friction) are confined to a very thin layer, called the boundary layer, which forms over the solid surface.

(2) The flow outside the boundary layer can be considered frictionless or ideal (fluid may be assumed inviscid or non-viscous) to a good approximation.

The assumption of zero-viscosity for fluids of very low viscosity enables us to use tools of the classical hydrodynamics—a science which deals with the flow of ideals fluids. This assumption leads to a very significant but physically impossible flow behaviour near the boundaries. The velocity distributions at and near a solid boundary for flow of both an ideal and a real fluids have been illustrated in Fig. 7.1.

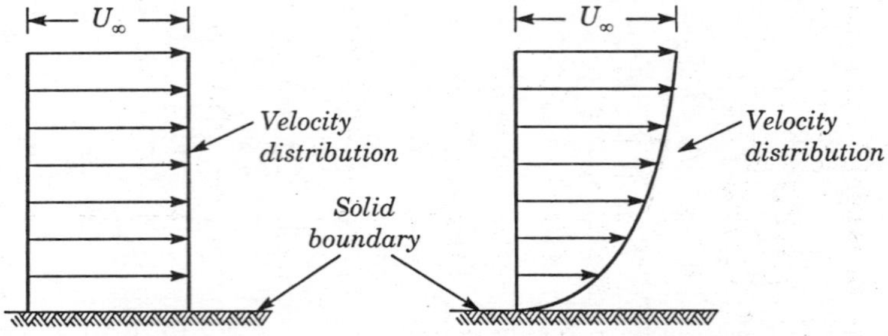

(*a*) Ideal fluid flow. (*b*) Flow of a low viscosity fluid.

Fig. 7.1. Velocity distributions for ideal and real fluids.

For an ideal fluid, the velocity distribution has a rectangular shape indicating that the velocity is same everywhere. This means that the velocity gradient has a zero value at all points irrespective of their positions. This leads to an important theoretical phenomenon that a fluid of negligible viscosity (like air) exerts no drag force on any object placed within the fluid stream. This is in contradiction to the physically observed fact that a body immersed in the flow of a fluid having a very low viscosity experiences a drag force. In accordance with the Newton's third law, the body also offers a certain resistance (equal and opposite to the drag) to the fluid flow.

The French mathematician D' Alembert, who developed a rational theory for calculating the resistance offered by a body immersed in a uniform stream of non-viscous fluid, showed that the drag force exerted on the body by the fluid is zero. His result was contradicted by engineers on the basis of physical evidence. Consequently, the theoretical non-existence of drag on a body placed in a uniform non-viscous fluid stream has since then been called *D' Alembert's paradox*.

7.1. Flow of Real Fluids

The real fluid flow differs from that of an ideal fluid only in respect of its being influenced by the fluid viscosity. However, in many cases the assumption of ideal fluid simplifies the equations of motion to large extent the enabling the mathematical treatment of the flow problem. It may be emphasized that the no-slip condition exists even in fluids of very low viscosity. In many problems, this condition of no-slip introduces large discrepancies in the dynamic behaviour of real and ideal fluids. As an example, the drag force exerted by the fluid on a body immersed in a fluid of low-viscosity has a finite and non-zero value, but if the same fluid is considered non-viscous, the ideal fluid theory results in a zero drag force. This discrepancy between the drag force caused by the flow of real fluid and that computed on the assumption of ideal fluid has its physical origin in the condition of no-slip at the solid boundary. The magnitude of viscosity of real fluids may vary between wide limits, and as such the fluid flow may be grouped on the basis of viscosity as described below.

7.1.1. Flow of fluid of very low-viscosity

The flow of such fluids is characterized by very high Reynolds numbers. It is this type of flow, the procedure for the analysis of which was outlined by Prandtl. The concept of boundary layer divides the field of high Reynolds number flow into two zones. In the first zone which is located very close to the solid boundary, the viscous forces are predominantly significant and far outweigh the inertial forces. The role of viscosity is very important in this region of flow field known as the boundary layer. The velocity at the stationary boundary is zero and increases towards the mainstream of flow. The shear stresses in this zone are very high owing to the existence of extremely high velocity gradients at and near the boundary. In the second zone (which lies in the main-body of flow a little distance away from the solid boundary—*i.e.* the region adjoining the boundary layer), the influence of viscosity is negligible, resulting in the inertial forces out-balancing the viscous forces. Thus the fluid in this region may be considered non-viscous, and the potential flow theory (often known as the theory of ideal fluid flow) may be used for analysing the flow. The governing dynamic equation in this zone is the Euler's equation of motion while those in the boundary layer region are the equations of Navier-Stokes.

7.1.2. Flow of Fluids of Moderate-Viscosity

In the flow of such fluids, the viscous and the inertial forces in the zone outside the boundary layer are of the same order of magnitude, and hence the approximation of negligible viscosity is not valid. The mathematical solution of the governing equations for the determination of flow characteristics in this region is not possible on account of the mathematical complexities involved. This type of flow has moderate Reynolds number and the boundary layer extends to a larger distance towards the mainstream as compared to the one that exists in the flow at very high Reynolds number.

7.1.3. Flow of Fluids of High-Viscosity

Flow of large viscosity fluids is characterized by very small Reynolds number. The viscous forces far outweigh the inertial forces, and the mathematical solution of the flow problems at very small Reynolds number becomes possible by ignoring the low magnitude inertial forces and thus simplifying the Navier-Stokes equations of motion. The boundary layer in this case extends far deeper into mainstream as compared to the above two cases.

7.2. Boundary Layer Growth Over a Flat Plate

Let us consider parallel flow of a low-density fluid such as air or water over a flat plate held at zero incidence (that is, the angle at which the flow direction makes with the plate is zero). The fact that at the solid boundary, the fluid adheres to it is in accordance with the condition of no-slip, and that the viscous for retards the fluid flow near the solid boundary. This results in formation of the boundary layer. Inside the boundary layer, the fluid velocity increases from zero at the boundary to its mainstream value which corresponds to the ideal fluid flow. There is, however, no sharp division between the boundary layer and the mainstream. The definition of the boundary layer thickness dealt with in article 7.3 is rather an arbitrary one.

Fig. 7.2 illustrates the velocity distribution in a two-dimensional laminar boundary layer, with the lateral dimensions considerably exaggerated.

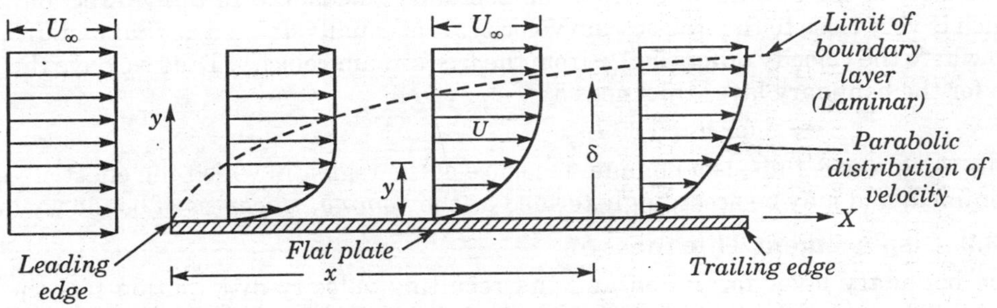

Fig. 7.2. Velocity distribution and growth of boundary layer in parallel flow over a flat-plate held at zero incidence.

The fluid approaches the plate with uniform velocity U_∞ known as an *undisturbed velocity* or the *free stream velocity* as shown by the extreme left velocity profile. As soon as the fluid comes in contact with the plate at its leading edge, its velocity is reduced to zero as the fluid particles adhere to the plate boundary in accordance with the no-slip condition. The velocity changes rapidly from zero at the boundary to its full value U_∞ within a very short distance causing very high velocity gradients to develop in the boundary layer. Even if the fluid viscosity is very small as in the case of air, extremely high velocity gradients generate excessively high shear stresses. As the fluid passes over the plate, the action of viscous shear retards more and more fluid in the lateral direction. The thickness of the boundary layer, therefore, increases along the plate in the downstream direction.

In the downstream direction from the leading edge the thickness, δ, of the retarded layer increases continuously. The viscous shear stresses ($\tau = \mu\, du/dy$) within the boundary layer, are very high owing to the existence of high velocity gradients across the flow, whereas outside the boundary layer they are very small. This suggests that the flow field in case of fluids of low viscosity can be divided into two regions for the sake of mathematical analysis. These are : (1) the thin boundary layer in the immediate vicinity of the solid boundary in which the viscous effects are predominant, (2) the region outside the boundary layer, where the viscous forces are small and may be neglected. In the region outside the boundary layer, the dominance of inertial effects over those of viscous ones permits the application of ideal fluid theory with satisfactory results.

In the boundary layer, the decelerated fluid particles do not, in all cases, remain in the narrow region of boundary layer which adheres to the solid boundary. Under certain conditions of flow and configuration of the boundary, the thickness of boundary layer increases considerably in the downstream direction so much so that the flow in the boundary layer is reversed. The decelerated fluid particles are forced away and the boundary layer is separated from the solid boundary. This phenomenon is known as the *boundary layer separation* (see art. 7.7). It occurs in case of flow past blunt bodies such as circular cylinders and spheres and is associated with the formation of vortices resulting in large energy losses. Just downstream of such a body there exists a region of highly decelerated flow. This region is known as the *wake zone* or simply the wake, and is characterized by pressure distribution different from that predicted by the ideal fluid theory.

7.3. Boundary Layer Thickness

The boundary layer thickness δ is an important parameter that must be determined in boundary layer problems. There are many definitions describing the boundary layer, but the displacement thickness δ^* is a more logical and mathematically useful definition.

7.3.1. Nominal Thickness δ

The definition of boundary layer thickness is somewhat arbitrary because the transition from zero velocity at the solid stationary boundary to the free-stream velocity U_∞ outside the boundary layer takes place asymptotically. The boundary layer thickness is defined in such a way that the velocity at a small distance from the boundary (measured in the y-direction) attains a value which is very close to the free-stream velocity. It is usually defined as that distance from the boundary where the velocity differs by 1% from the free stream velocity. Thus we have the following condition for the boundary layer thickness δ :

$$y = \delta \text{ for } u = 0.99\, U_\infty \qquad \text{...(7.1)}$$

This thickness δ, thus, is a nominal measure of the region in which the effect of viscosity is mostly confined, and may be accordingly termed as the *nominal thickness* of boundary layer.

7.3.2. Displacement Thickness δ^*

The boundary layer formation and the resulting velocity distribution has the effect of displacing the flowrate slightly outward from the solid boundary as compared with the ideal or potential flow that would exist in the absence of the boundary layer. Fig. 7.3 (*i*) shows the stream line pattern in the boundary layer. The deflecting stream lines and the widening gap in between indicate retardation of flow and a small vertical velocity component. The displacement thickness δ^* may be defined in any one of the following ways :

(*i*) it is the distance, measured perpendicular to the boundary, by which the free stream is displaced on account of formation of boundary layer ; or

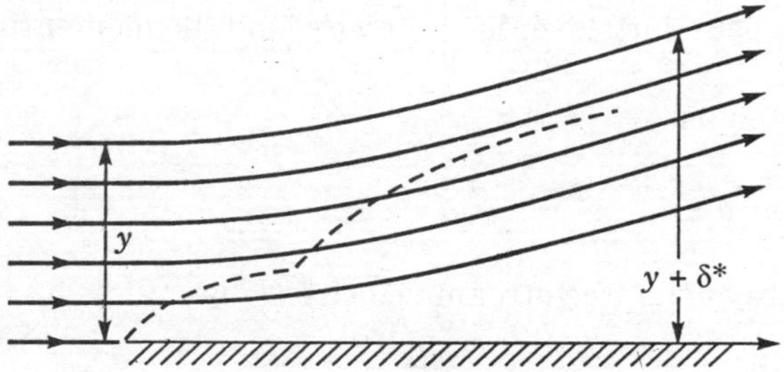

Fig. 7.3 (*i*) Stream line pattern in the boundary layer.

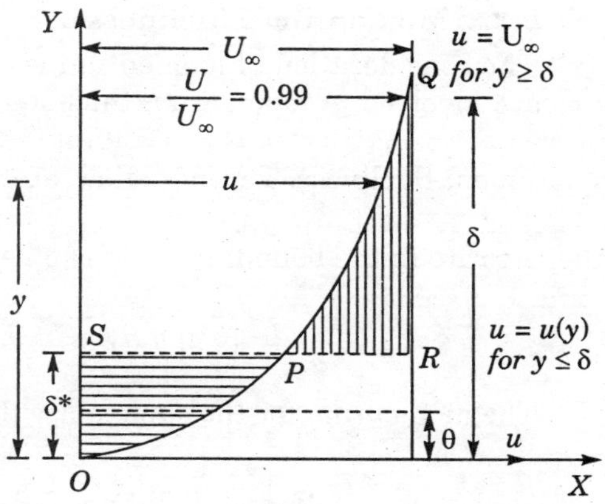

Fig. 7.3 (*ii*) Graphical representation of nominal, displacement and momentum thickness.

(*ii*) it can be defined as an additional "wall thickness" that would have to be added to compensate for the reduction in flowrate on account of boundary layer formation.

The displacement thickness δ^* for the two-dimensional boundary layer of Fig. 7.3 (*ii*) can be obtained as given below. Let b be the width of the flow section.

The total volume rate of flow past a section, considering influence of the boundary layer

$$Q = b\int_0^{\infty} u\,dy = b\int_0^{\delta} u\,dy + b\int_0^{\infty} U_\infty\,dy$$

The total volume rate of flow that would have passed had there been no boundary layer, *i.e.* if the flow is considered irrotational

$$Q_i = b\int_0^{\delta} U_\infty\,dy + b\int_{\delta}^{\infty} U_\infty\,dy$$

The reduction in flow rate caused by the boundary layer

$$Q_r = Q_i - Q$$

$$= b\int_0^{\delta} U_\infty\,dy - b\int_{\delta}^{\infty} u\,dy = b\int_0^{\delta} (U_\infty - u)\,dy$$

The displacement thickness δ^*, representing a shift in the free-stream, may be obtained by equating Q_r with frictionless flow passing through an area $(b\delta^*)$. (Shaded area *OPS* equals the area *PQR*). Thus,

$$b\delta^*\,U_\infty = b\int_0^{\delta} (U_\infty - u)\,dy$$

$$\delta^* = \frac{1}{U_\infty}\int_0^{\delta} (U_\infty - u)\,dy = \int_0^{\delta}\left(1 - \frac{u}{U_\infty}\right)dy \qquad ...(7.2)$$

Therefore, the same volume of fluid Q_r, with no boundary layer would have passed through the reduced cross-section $b\delta^*$. The quantity $(U_\infty - u)$ appears frequently in boundary layer work and is often called the *velocity defect*.

7.3.3. Momentum Thickness θ

The consideration of momentum is important in flow problems where pressure and shear forces are involved. A momentum thickness θ can be defined in a manner similar to the flowrate displacement thickness. The mass flowrate in the boundary layer is $\int_0^\delta \rho u\, dy$, and the momentum transport rate in the boundary layer can be written as

$$b \int_0^\delta (\rho u\, dy)\, u = b \int_0^\delta \rho u^2 dy$$

The momentum for the same mass flowrate at the free-stream velocity U_∞ is

$$b \int_0^\delta (\rho u\, dy)\, U_\infty = b \int_0^\delta \rho\, U_\infty\, u\, dy$$

The reduction in momentum or the momentum defect caused by the boundary layer is

$$b \int_0^\delta \rho(U_\infty\, u - u^2)\, dy = b \int_0^\delta \rho u\, (U_\infty - u)\, dy$$

If θ represents the momentum thickness, then the momentum transport through the flow area, $b\theta$, referred to the free-stream velocity U_∞ is $\rho\theta U_\infty^2$. Equating these momentum transport rates

$$\rho b\theta U_\infty^2 = b \int_0^\delta \rho u\, (U_\infty - u)\, dy$$

$$\theta = \int_0^\delta \frac{u}{U_\infty}\left(1 - \frac{u}{U_\infty}\right) dy \qquad \text{...(7.3)}$$

In the foregoing derivation the fluid density both within the boundary-layer and outside has been considered the same.

For parallel flow past a flat plate held at zero incidence, the approximate relationship among the three thicknesses of a boundary layer described above is

$$\delta = 3\, \delta^* = 7.5\, \theta$$

The ratio of displacement thickness to momentum thickness is called the shape factor and denoted by H. Therefore, $H = \delta^*/\theta$.

7.3.4. Energy Thickness δ_e

Following similar reasoning as for the momentum thickness, one can write the decrease in energy transport rate caused by the boundary friction as

$$\frac{1}{2} b \int_0^\delta \rho u\, (U_\infty^2 - u^2)\, dy$$

The energy thickness δ_e, is the thickness of a layer of fluid moving with a velocity U_∞ that represents the loss of energy transport rate. Therefore,

$$\frac{1}{2}\, \rho\, b\delta_e\, U_\infty^3 = \frac{1}{2} b \int_0^\delta \rho\, u\, (U_\infty^2 - u^2)\, dy$$

$$\delta_e = \int_0^\delta \frac{u}{U_\infty}\left(1 - \frac{u^2}{U_\infty^2}\right) dy \qquad \text{...(7.4)}$$

With the energy thickness known, the loss of energy per unit time can be determined from the relation

$$E_L = \frac{1}{2}\,\rho b \delta_e U_\infty^{\,3}.$$

Example 7.1. *For a linear distribution of velocity in the boundary layer on a flat plate, find the value of* δ^*/θ.

Solution. The velocity distribution in the boundary layer may be written as

$$\frac{u}{U_\infty} = \frac{y}{\delta}$$

The displacement thickness

$$\delta^* = \frac{1}{U_\infty}\int_0^\delta (U_\infty - u)dy = \int_0^\delta \left(1 - \frac{y}{\delta}\right)dy = \frac{\delta}{2}$$

The momentum thickness $\theta = \dfrac{1}{U_\infty^{\,2}}\displaystyle\int_0^\delta u(U_\infty - u)dy = \int_0^\delta \frac{y}{\delta}\left(1 - \frac{y}{\delta}\right)dy = \frac{\delta}{6}$

Therefore, $\dfrac{\delta^*}{\theta} = \dfrac{\delta/2}{\delta/6} = 3$, which is the value of the shape factor H.

Example 7.2. *The velocity distribution in a laminar boundary layer is given by*

$$\frac{V}{V_m} = 3\left(\frac{y}{\delta}\right) - 2\left(\frac{y}{\delta}\right)^2$$

in which V = velocity at distance y from the boundary ; V_m *= velocity at distance* δ*, the thickness of the boundary layer. Determine the value of the displacement thickness.*

Solution. Displacement thickness,

$$\delta^* = \int_0^\delta \left(1 - \frac{V}{V_m}\right)dy$$

$$= \int_0^\delta \left[1 - 3\left(\frac{y}{\delta}\right) + 2\left(\frac{y}{\delta}\right)^2\right]dy = \left[\delta - \frac{3}{2}\delta + \frac{2}{3}\delta\right] = \frac{\delta}{6}.$$

Example 7.3. *For steady Poiseuille flow in a pipe of radius R, obtain an expression for* δ^*/θ.

Solution. The velocity distribution for steady Poiseuille flow in a circular pipe is given by

$$u = \frac{1}{4\mu}\frac{dp}{dx}(R^2 - y^2)$$

y being measured from the centre of pipe of radius R.

The maximum velocity U occurs at $y = 0$.

$$U = -\frac{1}{4\mu}\frac{dp}{dx}R^2$$

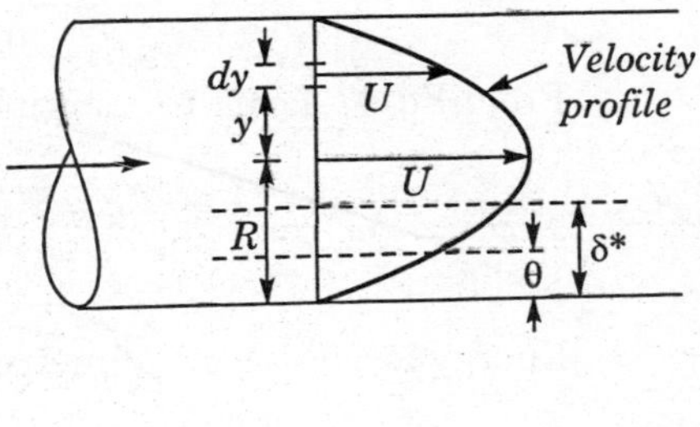

From the definition of displacement thickness

$$[\pi R^2 - \pi (R - \delta^*)^2]U = \int_0^R 2\pi y\,(U - u)\,dy$$

Since δ^* is very small, the term containing $(\delta^*)^2$ may be neglected. Simplifying, we obtain

$$2\,\pi R\delta^* U = \int_0^R 2\,\pi y(U-u)dy$$

which yields
$$\delta^* = \frac{1}{R}\int_0^R \left(1-\frac{u}{U}\right)y\,dy = \frac{1}{R}\int_0^R \frac{y^2}{R^2}y\,dy = \frac{R}{4}.$$

From the definition of momentum thickness, one may write,

$$[\rho\{(\pi R^2 - \pi(R-\theta)^2 U\}]\,U = \int_0^R \rho\,2\,\pi y\,dy\,u\,(U-u)$$

Simplification results in

$$2\,\pi\,R\,\theta\,U^2 = \int_0^R 2\,\pi\,yu\,(U-u)\,dy$$

which yields
$$\theta = \frac{1}{R}\int_0^R \frac{u}{U}\left(1-\frac{u}{U}\right)y\,dy = \frac{1}{R}\int_0^R \left(1-\frac{y^2}{R^2}\right)\frac{y^2}{R^2}y\,dy = \frac{R}{12}$$

Therefore,
$$\frac{\delta^*}{\theta} = \frac{R/4}{R/12} = 3.$$

Example 7.4. *The velocity distribution in the boundary-layer over the face of a high spillway was observed to have the following form :*

$$\frac{u}{U_\infty} = \left(\frac{y}{\delta}\right)^{0.22}$$

The free stream velocity U_∞ at a certain section was observed to be 20 m/s and a boundary layer thickness of 5 cm was estimated from the velocity distribution measured at the section. The discharge passing over the spillway was 5 m^3/s per metre length of spillway. Calculate the displacement thickness, the energy thickness, and the loss of energy upto the section under consideration.

Solution. The displacement thickness is given by Eq. (7.2)

$$\delta^* = \int_0^\delta \left(1-\frac{u}{U_\infty}\right)dy = \int_0^\delta \left\{1-\left(\frac{y}{\delta}\right)^{0.22}\right\}dy$$

$$= 0.18\,\delta = 0.18 \times 5 = 0.9 \text{ cm}.$$

For energy thickness, Eq. (7.4) gives

$$\delta_e = \int_0^\delta \frac{u}{U_\infty}\left(1-\frac{u^2}{U_\infty{}^2}\right)dy = \int_0^\delta \left(\frac{y}{\delta}\right)^{0.22}\left\{1-\left(\frac{y}{\delta}\right)^{0.44}\right\}dy$$

$$= 0.217\,\delta = 0.217 \times 5 = 1.085 \text{ cm}.$$

From Eq. (7.4 *a*), the energy loss per metre of spillway length is

$$E_L = \tfrac{1}{2}\rho\delta_e\,U_\infty{}^3 = \frac{1}{2} \times 1000 \times \frac{1.085}{100} \times (20)^3$$

$$= 43360 \text{ N-m/s} = \mathbf{43.36 \text{ kN-m/s}}$$

Energy loss in terms of metre of head,

$$= E_L/\gamma q = \frac{43{,}360}{9810 \times 5} = 0.885 \text{ m}.$$

7.4. Laminar and Turbulent Boundary Layers

7.4.1. Laminar boundary layer

It is an established fact that the flow in the initial stages of boundary layer development exhibits characteristics of laminar motion—that is the fluid particles downstream of the leading edge move in an orderly manner in laminas parallel to the solid boundary.

In a flow at very low Reynolds number through a straight pipe of uniform cross-section and smooth boundaries, every fluid particle moves at a certain constant velocity along the straight path parallel to the pipe axis. Viscous forces slow down the particles near the solid boundary as compared to the fluid motion in the central core of the pipe. The flow takes place in an orderly manner parallel to the pipe wall. The Reynolds number of flow in the boundary layer is expressed as $\mathbf{R}_x = U_\infty x/\nu$. From its composition it will be seen that a low value of $\mathbf{R}_x$ represents either a low free-stream velocity or a low value x (the distance from the leading edge) or a high kinematic viscosity or a judicious combination of these three variables. The velocity distribution in laminar flow, as already seen in chapter 6, is parabolic. The velocity in the upper reaches of the boundary, therefore, follows a parabolic law and is shown as such in Fig. 7.2, which depicts a two-dimensional laminar boundary layer.

As the boundary layer thickness increases with the distance from the leading edge, the velocity gradients which have a very high value just downstream of the leading edge decrease in magnitude along the length of the flat plate (in the x-direction) as indicated by flatter velocity profiles illustrated in Fig. 7.2. This results in very high viscous shear stresses just downstream of the leading edge thereby retarding more and more quantities of fluid and laterally extending the decelerating zone of the boundary layer ; the magnitude of these stresses decreases in the downstream direction.

The dimensionless boundary layer thickness depends on the free-stream velocity U_∞, the distance x measured from the leading edge and the kinematic viscosity ν of the fluid. For a two-dimensional laminar boundary layer it is given by

$$\frac{\delta}{x} = 5\sqrt{\frac{\nu}{U_\infty x}} = \frac{5}{\sqrt{\mathbf{R}_x}} \qquad \text{...(7.5 } a\text{)}$$

and the nominal thickness δ of the boundary layer is

$$\delta = 5\sqrt{\frac{\nu x}{U_\infty}} \qquad \text{...(7.5 } b\text{)}$$

From Eqs. (7.5 *a*) and (7.5 *b*), the following important conclusions can be drawn :

(1) At a given section along the plate length (that is, for x = constant) the boundary layer thickness δ decreases with the increasing Reynolds number

$$\mathbf{R}_x = U_\infty x/\nu$$

in accordance with Eq. (7.5 *a*). For a given fluid the Reynolds number $\mathbf{R}_x$ will depend on the free-stream velocity U_∞. Thus for a high velocity flow the boundary layer will be very thin. For the limiting case of frictionless flow, *i.e.* ideal fluid flow, with $\mathbf{R}_x \to \infty$, the boundary layer thickness vanishes.

(2) For a given fluid flowing at a certain free-stream velocity (that is ν and U_∞ both held constant), the boundary layer thickness increases with x according to Eq. (7.5 *b*).

(3) The boundary layer thickness increases with increasing kinematic viscosity of the fluid in accordance with Eq. (7.5 *b*).

With velocity-distribution known, it is possible to determine the shear stress at the boundary making use of the Newton's law of fluid friction, Eq. (1.3). This paves the way for finding an expression for the drag force exerted by the fluid on the flat plate. It may be noted that the boundary layer would form on both sides of the plate if the fluid motion takes place on the other side also. The total force of drag will then be two times the drag force on one of its side.

VELOCITY DISTRIBUTION LAWS FOR LAMINAR BOUNDARY LAYER

The velocity distribution laws generally followed for laminar boundary layers are

(i) $\dfrac{u}{U_\infty} = 2\left(\dfrac{y}{\delta}\right) - \left(\dfrac{y}{\delta}\right)^2$ (ii) $\dfrac{u}{U_\infty} = \dfrac{3}{2}\left(\dfrac{y}{\delta}\right) - \dfrac{1}{2}\left(\dfrac{y}{\delta}\right)^3$

(iii) $\dfrac{u}{U_\infty} = 2\left(\dfrac{y}{\delta}\right) - 2\left(\dfrac{y}{\delta}\right)^3 + \left(\dfrac{y}{\delta}\right)^4$

For any of the above stated laws, the boundary layer thickness δ, the boundary shear stress τ_0 and the drag coefficient C_D can be determined, and written as

$$\delta = (\text{coeff.}) \frac{x}{\sqrt{\mathbf{R}_x}}\ ;\ \tau_0 = (\text{coeff.}) \frac{\rho U_\infty^{\ 2}}{\sqrt{\mathbf{R}_x}}\ ;\ C_D = \frac{(\text{coeff.})}{\sqrt{\mathbf{R}_x}}$$

The coefficients for each of the above mentioned velocity laws are tabulated as under.

Velocity distribution	*Coefficients for* δ	τ_0	C_D
(i)	5.477	0.365	1.460
(ii)	4.641	0.323	1.292
(iii)	5.835	0.343	1.371
Values obtained by Blasius	4.910	0.332	1.328

Example 7.5. *The boundary layer thickness at a distance of 1 m from the leading edge of a flat plate kept at zero angle of incidence to the flow direction is 0.1 cm. The velocity outside the boundary layer is 25 m/s. The boundary layer thickness at a distance of 4 m is*

(i) 0.40 cm, *(ii) 0.20 cm,* *(iii) 0.10 cm.*

(Select the correct answer). Assume that the boundary layer is entirely laminar.

Solution. The thickness of laminar boundary layer is given by

$$\frac{\delta}{x} = \frac{5}{\sqrt{\mathbf{R}_x}}, \quad \text{or} \quad \delta = 5\sqrt{\frac{\nu x}{U_\infty}}$$

Given : $U_\infty = 25$ m/s, at $x_1 = 1$ m, $\delta_1 = 0.1$ cm.

For the same ν and U_∞, $x_2 = 4$ m

$$\frac{\delta_1}{\delta_2} = \sqrt{x_1/x_2}$$

or $$\frac{0.1}{\delta_2} = \sqrt{1/4}, \quad \therefore \quad \delta_2 = 0.1\sqrt{4} = 0.20 \text{ cm.}$$

Hence, the boundary layer thickness at a distance of 4 m is **0.20 cm.**

Example 7.6. *In case of laminar boundary layer over a flate plate, compare its growth for the following cases :*

(i) flat plate kept in a stream of air flowing at 2.5 m/s,

(ii) flat plate kept in a water stream flowing at 2.5 m/s, and

(iii) flat plate placed in a stream of air flowing at 10.0 m/s.

Take the specific weight of water and air as 9810 N/m^3 and 12.1 N/m^3 respectively and the viscosities of water and air respectively as 10^{-3} N-s/m^2 and 1.85×10^{-5} N-s/m^2.

Solution. The growth of laminar boundary layer is given by

$$\frac{\delta}{x} = \frac{5}{\sqrt{\mathbf{R}_x}}, \quad \text{or} \quad \delta = 5\sqrt{\frac{\nu x}{U_\infty}} = 5\sqrt{\frac{\mu}{\rho}}\sqrt{\frac{x}{U_\infty}}$$

Considering air :

$$\gamma = \rho g, \qquad \rho = \frac{12.1}{9.81} = 1.235 \text{ kg/m}^3.$$

$$\nu = \frac{1.85 \times 10^{-5}}{1.235} = 1.5 \times 10^{-5} \text{ m}^2/\text{s}$$

$$\sqrt{\nu_{air}} = \sqrt{1.5 \times 10^{-5}} = 0.0038$$

Considering water :

$$\rho = \frac{9810}{9.81} = 1000 \text{ kg/m}^3$$

$$\nu = \frac{\mu}{\rho} = \frac{10^{-3}}{1000} = 10^{-6} \text{ m}^2/\text{s}$$

$$\sqrt{\nu_{water}} = \sqrt{10^{-6}} = 0.001$$

For the same free-stream velocity U_∞, the thickness of laminar boundary layer at any given location x for flow of air and water, the following relationship can be written

$$\frac{\delta_{air}}{\delta_{water}} = \frac{\sqrt{\nu_{air}}}{\sqrt{\nu_{water}}} = \frac{0.0038}{0.001} = 3.8$$

Thus under similar conditions as above, the boundary layer thickness in air is 3.8 times as great as that in water.

In the same fluid, the thickness of laminar boundary layer at a given location is dependent on the free-stream velocity according to the following relation :

$$\frac{\delta_1}{\delta_2} = \sqrt{\frac{U_{\infty_2}}{U_{\infty_1}}} = \sqrt{\frac{10.0}{2.5}} = 2$$

or

$$\delta_2 = 0.5\,\delta_1$$

Thus in the same fluid, the boundary layer is shown to be thinner at higher velocities ; its thickness becomes half when the velocity is increased four times.

Example 7.7. *Velocity distribution in a boundary layer flow over a plate is given by*

$$\frac{u}{u_m} = 1.5\eta - 0.5\eta^3$$

where $\eta = y/\delta$; y is the distance measured normal to the plate ; δ is the boundary laye thickness ; and u_m is the maximum velocity at $y = \delta$. If the shear stress τ, acting on the plate i iven by

$$\tau = K(\mu\, u_m)/\delta$$

where, μ is the dynamic viscosity of the fluid find the value of K.

Solution. The shear stress is given by

$$\tau = \mu \frac{du}{dy} = \mu \frac{du}{d\eta} \cdot \frac{d\eta}{dy}$$

where $\eta = \frac{y}{\delta}$, $\therefore \; d\eta = \frac{1}{\delta} dy$ or $\frac{d\eta}{dy} = \frac{1}{\delta}$

From the velocity distribution

$$\frac{u}{u_m} = 1.5\,\eta - 0.5\,\eta^3$$

$$\frac{1}{u_m}\frac{du}{d\eta} = 1.5 - 0.5 \times 3\,\eta^2 = 1.5 - 1.5\left(\frac{y}{\delta}\right)^2$$

The shear stress acting on the plate *i.e.* at $y = 0$

$$\tau_0 = \mu \left(\frac{du}{d\eta}\right)_{y=0} \left(\frac{1}{\delta}\right) = \mu\,(1.5\,u_m)\,(1/\delta) = 1.5\mu\,u_m/\delta$$

which gives the value of **K = 1.5.**

7.4.2. Turbulent Boundary Layers

In many practical cases of real fluid flow, it has been established from experimental observation that in most of the engineering problems the flow is very often different from laminar flow. Such flow exhibits a typical characteristic which is known as turbulence. When the Reynolds number is increased the flow within the boundary layer undergoes a transition from the laminar to the turbulent regime. The orderly flow in the laminar boundary layer is not stable and ceases to exist at higher Reynolds number $\mathbf{R}_x$. At high Reynolds number, the flow changes from laminar to turbulent motion characterised by strong mixing of particles in a direction normal to the boundary. In turbulent flow, the main flow which has a direction parallel to the boundary is superimposed by a subsidiary motion at right angles to it. It is this subsidiary motion which is mainly responsible for transverse mixing of fluid particles. The pattern of stream lines at a fixed point becomes subjected to continuous fluctuations and the subsidiary motion causes an exchange of momentum in a transverse direction. As a consequence, the velocity distribution at a flow section is considerably more uniform in turbulent flow than in the laminar one.

7.4.3. Transition from Laminar to Turbulent flow

For flow of a real fluid at a fixed rate (U_∞ and ν both assumed constant), the Reynolds number $\mathbf{R}_x$ increases along the length of the solid boundary. If we consider a flat plate of sufficient length, at some point along the plate, the laminar flow in the boundary layer is expected to develop eddies (resulting from intense transverse mixing) which rapidly give rise to turbulence within the boundary layer. The critical value of Reynolds number $\mathbf{R}_x$ signifying the change of flow in the boundary layer from laminar to turbulent may range between 3×10^5 to 6×10^5, but may also vary considerable depending on such factors as the intensity and scale of turbulence of oncoming flow, the condition of the leading edge and the roughness of the boundary, boundary profile and its curvature and the pressure gradient. The boundary roughness near the leading edge hastens the transition from laminar to turbulent as does the presence of turbulence in the oncoming flow. For Reynolds number $\mathbf{R}_x$ less than 10^5, the laminar boundary layer is very stable, but as it increases and becomes greater than 2×10^5, it is difficult to prevent transition to turbulent state even when the boundary surface is smooth and the flow in the main stream is laminar. If the pressure gradient dp/dx is positive, that is if the flow is diverging (decelerating), the critical value of $\mathbf{R}_x$ at which transition occurs is lower ; on the other hand if dp/dx is negative the critical Reynolds number is higher. An average value of the critical Reynolds number based on the distance x from the leading edge may be taken as 5×10^5.

The presence of turbulence in the boundary layer not only considerably modifies the velocity distribution but also increases the rapidity with which the boundary layer expands laterally. The process of transverse mixing tends to make the velocity more uniform throughout except in the very immediate vicinity of the boundary where a very rapid change in velocity takes place. The velocity gradients in the turbulent boundary layer are, therefore, very high as compared to those in the laminar part. If the boundary or the wall is very smooth, the turbulent boundary layer has a very narrow zone very close to the boundary in which the flow is still laminar. This narrow region, exhibiting characteristics of laminar motion, is known as the *laminar sublayer*.

The reasons for the existence of laminar flow very close to the solid boundary even when the main flow is turbulent is that the transverse movement of particles must die out in this region of boundary layer.

TABLE SHOWING DATA PERTAINING TO LAMINAR BOUNDARY LAYER

S. No.	*Free-stream velocity, U_∞ (m/s)*	*Critical value of* $\mathbf{R}_x$	*Location of transition point x = L (cm)*	*Boundary Layer Thickness*		
				Nominal δ (cm)	*Displacement δ* (cm)*	*Momentum θ (cm)*
	For water at 20°C					
1.	0.10	5×10^5	500.0	3.530	1.210	0.469
2.	0.30	5×10^5	166.7	1.179	0.406	0.156
3.	1.00	5×10^5	50.0	0.353	0.121	0.047
4.	3.00	5×10^5	16.7	0.118	0.040	0.016
	For air at 20°C					
5.	1.00	5×10^5	94.5	0.67	0.230	0.089
6.	3.00	5×10^5	31.5	0.22	0.075	0.029
7.	6.00	5×10^5	15.7	0.11	0.038	0.015
8.	10.00	5×10^5	9.5	0.07	0.023	0.009

The effects of variation of free-stream velocity and viscosity of fluid on the boundary layer thickness is self-evident from the above table.

The transition from laminar to turbulent flow in the boundary layer becomes most clearly distinguishable by a sudden and large increase in the boundary layer thickness as well as the shear stress near the solid boundary. The thickness of a turbulent boundary layer is always larger than

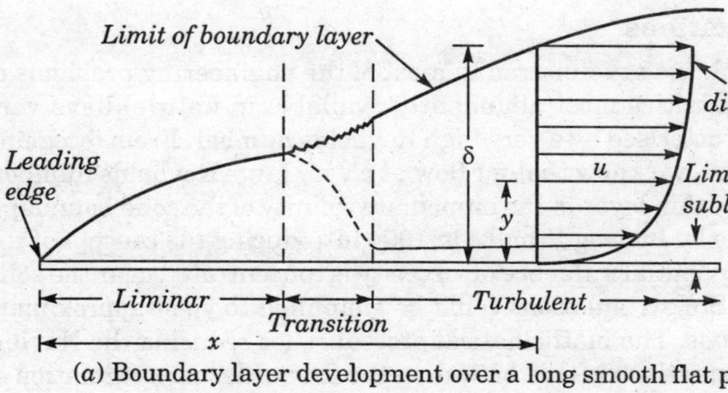

(*a*) Boundary layer development over a long smooth flat plate.

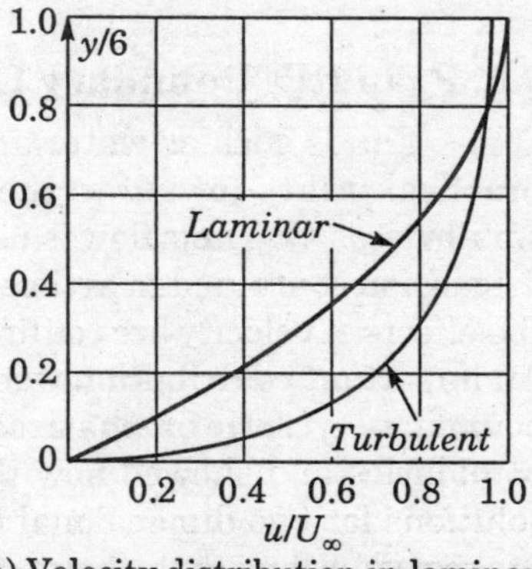

(*b*) Velocity distribution in laminar and turbulent boundary layers.

Fig. 7.4. Boundary layer characteristics on a smooth flat plate.

that of laminar boundary layer on account of greater shear resistance in the turbulent region. For a smooth flat plate at zero incidence, the turbulent boundary layer increases in thickness in the downstream direction from the leading edge in proportion to $x^{0.8}$, where x is the distance from the leading edge. It may be recalled here that in case of laminar boundary layer, its thickness varies according to square root of the distance from the leading edge, *i.e.* $x^{0.5}$ (Eq. 7.5 *b*). It can be shown that the variation of boundary layer thickness in turbulent flow is given by

$$\frac{\delta}{x} = 0.37\left(\frac{U_\infty x}{\nu}\right)^{-1/5} = \frac{0.37}{(\mathbf{R}_x)^{1/5}} \qquad \text{...(7.6)}$$

The velocity distribution in the turbulent boundary layer follows an almost logarithmic law whereas that in the laminar one is parabolic. Fig. 7.4 illustrates the development of boundary layer over a long flat plate having a smooth surface and shows a comparison of velocity distribution in laminar and turbulent parts of the boundary layer.

Example 7.8. *A smooth flat plate with a sharp leading edge is placed at zero incidence in a free-stream of water flowing at 3.5 m/s. Determine the distance from the leading edge where the transition from laminar to turbulent flow may commence. The viscosity of water is 1 centipoise. Calculate the boundary layer thickness at the transition point.*

Solution. Let the Reynolds number at which the transition from laminar to turbulent takes place be taken as 5×10^5, thus

$$\mathbf{R}_x = \frac{\rho U_\infty x}{\mu} = 5 \times 10^5$$

For water $\gamma = 9810 \text{ N/m}^3,\ \rho = 9810/9.81 = 1000 \text{ kg/m}^3$

$$98.1 \text{ poise} = 1 \text{ kg-s/m}^2,\ \therefore\ 1 \text{ centipoise} = \frac{1}{9810} \text{ kg-s/m}^2 = \frac{1}{1000} \text{ N-s/m}^2$$

Location of Transition point :

$$x = \frac{5 \times 10^5\, \mu}{\rho\, U_\infty} = \frac{5 \times 10^5 \times 1}{1000 \times 3.5 \times 1000} = 0.1428 \text{ m or } \mathbf{14.28\ cm.}$$

The thickness of laminar boundary layer is given by

$$\frac{\delta}{x} = \frac{5}{\sqrt{\mathbf{R}_x}}$$

$$\therefore \qquad \delta = \frac{x}{\sqrt{\mathbf{R}_x}} = \frac{5 \times 0.1428}{\sqrt{5 \times 10^5}} = 0.001 \text{ m} = \mathbf{0.10\ cm.}$$

7.5. Prandtl's Boundary Layer Equations

Fluids such as water and air which are encountered in most of the engineering problems of practical utility (owing to their vast quantities most abundantly available in nature) have very small viscosity. Their flow is usually characterised by a very high Reynolds number. From foregoing discussion contained in articles 7.0 and 7.1, we know that for flow at a very large Reynolds number, the effects of velocity are confined to a very thin layer in the immediate vicinity of the solid boundary. An important contribution was thus made by Ludwig Prandtl in 1904 introducing the concept of the boundary layer. He emphasized that the essential influence of viscosity is concentrated near the solid boundaries and showed how the Navier-Stokes equations could be simplified to yield approximate solutions for two dimensional flow problems. The mathematical procedure for reducing the Navier-Stokes equations to the boundary layer equations is that of finding the first order approximation of the former equations under flow at very large Reynolds number. It may be recalled here that in the main-stream of fluid (that part of fluid which is outside the boundary layer) the inertia forces predominate and the viscous forces tend to become vanishingly small. The deduction of boundary layer equations from the N-S equations with the underlying simplifications is outside the scope of this book, the readers interested in the detailed treatment are advised to refer to chapter 7 on "Boundary Layer Equations in Two-Dimensional Flow", of the book by H. Schlichting "Boundary Layer Theory". It is however, felt that the final simplified form of the Navier-Stokes equations along with necessary explanation be included here and is, therefore, incorporated.

The flow at a very high Reynolds number past a flat plate of infinite extent is essentially two dimensional flow. The flow, as suggested by Prandtl, may be divided into two regions, though the division between them is not very sharp :

(1) Region of boundary layer which lies in the immediate neighbourhood of the solid boundary and within which about 99% of the velocity change takes place. The transition from zero velocity at the plate surface to the full magnitude U_∞ at some distance from the wall takes place within a very short distance—the so called boundary layer. In this boundary layer the velocity gradient normal to the solid boundary, $\partial u/\partial y$, is very large with the result that even a low viscosity fluid exerts a very large shear stress $\tau = \mu\, \partial u/\partial y$. In this region the laws of viscous flow, *i.e.* the Navier-Stokes equations apply.

(2) The region outside the boundary layer, which constitutes the main body of flowing fluid, has velocities of the order of the free-stream velocity U_∞. The flow in this zone is governed by the potential flow (ideal fluid) theory : the viscous influence is negligibly small and the velocity gradients are also very small. The flow in this region is, therefore, assumed frictionless and the laws of potential flow are applicable.

The boundary conditions for analysing the flow in the boundary layer are : (*i*) existence of no-slip condition at the solid boundary, that is for $y = 0$, $u = v = 0$ and (*ii*) outside the boundary layer the velocity u tends to become equal to the free stream velocity U_∞, that is for $y \geq \delta$, $u = U_\infty$. The assumption that the dimensionless boundary layer thickness δ/x is very small as compared to unity ($\delta/x << 1$), is necessary for the derivation of boundary layer equations.

In the region outside the boundary layer $u \to U_\infty$, and since there are no large velocity gradients in this region, the viscous terms from the N-S equations vanish. Thus, ignoring the body force component, the Navier-Stokes equation in the x-direction for flow outside the boundary layer is reduced to

$$\frac{\partial U_\infty}{\partial t} + U_\infty \frac{\partial U_\infty}{\partial x} = -\frac{1}{\rho}\frac{\partial p}{\partial x} \quad \text{... (7.7 a)}$$

For steady flow the pressure p is function of x only, the partial derivative $\partial p/\partial x$ may, therefore, be replaced by the total derivative dp/dx, and Eq. (7.7 *a*) expressed in the following form :

$$U_\infty \frac{\partial U_\infty}{\partial x} = -\frac{1}{\rho}\frac{\partial p}{\partial x} \quad \text{...(7.7 b)}$$

Integrating,
$$p + \frac{1}{2}\rho U_\infty^2 = \text{Constant} \quad \text{...(7.8)}$$

The pressure distribution in a known potential flow (the free stream velocity U_∞ and its dependence on x and t must be known) may be determined by applying Eq. (7.7) which expresses the Bernoulli relationship in differential terms, along a streamline at the solid boundary.

The simplified Navier-Stokes equation may be written as :

$$\frac{\partial u}{\partial t} + u\frac{\partial u}{\partial x} + v\frac{\partial u}{\partial y} = -\frac{1}{\rho}\frac{\partial p}{\partial x} + \nu\frac{\partial^2 u}{\partial y^2} \quad \text{... (7.9 a)}$$

For steady flow using Eq. (7.7 *b*), Eq. (7.9 *a*) can be written as

$$u\frac{\partial u}{\partial x} + v\frac{\partial u}{\partial y} = U_\infty\frac{dU_\infty}{\partial x} + \nu\frac{\partial^2 u}{\partial y^2} \quad \text{...(7.9 b)}$$

(*ii*) The N-S equation in the *y*-direction, after ascertaining order-of-magnitude of various quantities involved, results in

$$\frac{\partial p}{\partial y} = 0 \qquad ...(7.10)$$

and the continuity equation for the two-dimensional steady flow is :

$$\frac{\partial u}{\partial x} + \frac{\partial v}{\partial y} = 0 \qquad ...(7.11)$$

Equations (7.9 *b*) and (7.11) are known as the *Prandtl's boundary layer equations* and are subject to the following boundary conditions (for stationary flat plate) :

(*i*) at $y = 0$, $u = v = 0$, and

(*ii*) at $y = \infty$, $u = U_\infty (x, t)$ ...(7.12)

Although the above equations are meant for rectilinear flow, they are applicable to curved flow as well provided that the radius of curvature is large enough as compared to the boundary layer thickness.

7.6. Momentum Integral Equation

The differential equation of the boundary layer and the equation of continuity. Eqs. (7.9) and (7.11) respectively together with the boundary conditions are basically sufficient to determine velocity components *u* and *v* within the boundary layer. The pressure *p* is taken as that in the free flow at the edge of the boundary layer and can be considered known. Integration of Eqs. (7.9) and (7.11) has been possible to date only for a few special cases, even then considerable mathematical difficulties arise.

This explains the necessity of devising approximate methods of solution. Approximate methods can be obtained if we do not insist on satisfying the boundary layer differential equation for every fluid particle. In other words the boundary layer equation is satisfied near the solid boundary and near the outer edge of the boundary layer making the transition to the external flow. For fluid within the boundary only a mean over the boundary layer differential equation is satisfied, the mean being taken over the whole thickness of the boundary layer. The Hungarian American engineer Theodore Von Karman (1881—1963) obtained an approximate method for boundary layer flow analysis based on the momentum principle. The momentum equation, which is the embodiment of momentum principle may be derived from the equation of motion by integration over the boundary layer thickness. This equation is known as the 'momentum integral equation' for boundary flow or may also be termed as Von-Karman integral equation.

7.6.1. Momentum Integral Equation Using Momentum Principle

We shall apply the momentum principle to the case of steady, two-dimensional, incompressible flow in a boundary layer over a flat plate over which there may be variation of pressure in the direction

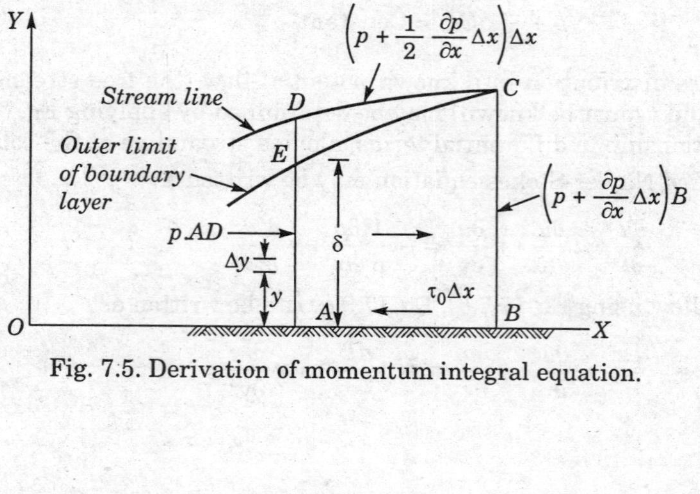

Fig. 7.5. Derivation of momentum integral equation.

of flow. Consider a flat plate of very large extent so that the 'side effects' are negligible and the flow is assumed wholly two dimensional. Fig. 7.5 shows a small length AB of the plate. Let $AB = \Delta x$ be the length of element of the boundary layer $ABCEA$ in the x-direction and δ be the thickness of the boundary layer at a distance of x from the leading edge of the plate. The outer edge of the boundary layer represented by EC is not a streamline because with increasing distance x more fluid continually enters the boundary layer. Let D be the point on the extended line AE such that it lies on the same streamline as C. As DC is a streamline, no fluid crosses it. For applying the momentum principle, let us select $ADCBA$ as a suitable control volume.

If the mean pressure over the face AD is p, the mean pressure over the face BC would be $\{p + (\partial p/\partial x)\Delta x\}$. Considering unit width of plate perpendicular to the plane of the paper, the net force in the x-direction due to difference of pressure is

$$pAD - \left(p + \frac{\partial p}{\partial x}\Delta x\right)BC + \left(p + \frac{1}{2}\frac{\partial p}{\partial x}\Delta x\right)(BC - AD) \quad \text{...(7.13)}$$

The last term of Eq. (7.13) contains $\left(p + \frac{1}{2}\frac{\partial p}{\partial x}\Delta x\right)$ which is the mean pressure over the streamline surface DC. In the limit as $\Delta x \to 0$, $BC \to AE$ and $\left(p + \frac{1}{2}\frac{\partial p}{\partial x}\Delta x\right) \to p$, Eq. (7.13) then reduces to $\left\{-\left(\frac{\partial p}{\partial x}\right)\Delta x\,(BC)\right\}$. The total force on the control volume in the x-direction, is equal to

$$-\tau_0 \Delta x - \frac{\partial p}{\partial x}\Delta x(BC) \quad \text{... (7.14)}$$

τ_0 in Eq. (7.14) is the shear stress at the solid boundary, *i.e.* at $y = 0$. According to the momentum principle this force is equal to the rate of change of momentum of fluid passing through the control volume, in the same direction as x. Consider an elementary strip of thickness Δy at a distance y from the boundary where the x-component of velocity is u. The mass rate of fluid flow through this strip per unit width is $\rho u\,\Delta y$, and corresponding rate at which momentum enters the control volume through this strip in the x-direction is $\rho u^2\,\Delta y$. Considering the entire boundary layer the x-component of rate of momentum entering the control volume through AE is

$$\int_0^\delta \rho u^2 dy \quad \text{... (7.15)}$$

The x-component of rate at which momentum leaves the control volume through the face BC is

$$\int_0^\delta \rho u^2 dy + \frac{\partial}{\partial x}\left(\int_0^\delta \rho u^2 dy\right)\Delta x \quad \text{...(7.16)}$$

Since AB represents a part of the solid boundary, no flow can enter the control volume through AB or leave it. And also since DC is part of the streamline, no fluid can cross it either way. We have yet to consider the rate at which momentum enters the control volume through ED, which may be seen to be equal to $[\rho U_\infty^2\,(ED)]$ where U_∞ is the velocity outside the boundary layer.

The net rate of change of x-momentum of fluid through the control volume may now be written as

= (x-momentum flux through BC) – (x-momentum flux through AE and ED)

$$= \left\{\int_0^\delta \rho u^2\, dy + \frac{\partial}{\partial x}\left(\int_0^\delta \rho u^2\, dy\right)\Delta x\right\} - \left\{\int_0^\delta \rho u^2\, dy + \rho U_\infty^2 (ED)\right\}$$

$$= \frac{\partial}{\partial x}\left(\int_0^\delta \rho u^2\, dy\right)\Delta x - \rho U_\infty^2 (ED) \qquad \text{... (7.17)}$$

The magnitude of ED may be evaluated in the following manner using the continuity principle :

Fluid mass entering through AD = fluid mass leaving through BC. In other words the mass of fluid entering through $(AE + ED)$ = mass of fluid leaving through BC.

Mass entering through $ED = \rho U_\infty (ED)$, which may be written as,

$$\rho U_\infty (ED) = \text{mass leaving } BC - \text{mass entering } AE$$

$$= \left\{\int_0^\delta \rho u\, dy + \frac{\partial}{\partial x}\left(\int_0^\delta \rho u\, dy\right)\Delta x\right\} - \int_0^\delta \rho u\, dy$$

$$= \frac{\partial}{\partial x}\left(\int_0^\delta \rho u\, dy\right)\Delta x$$

The last term of Eq. (7.17) may therefore be written as

$$U_\infty \frac{\partial}{\partial x}\left(\int_0^\delta \rho u\, dy\right)\Delta x$$

Applying the momentum principle to the control volume $ABCDA$ we obtain

$$-\tau_0 \Delta x - \frac{\partial p}{\partial x}\Delta x\,(BC) = \frac{\partial}{\partial x}\left(\int_0^\delta \rho u^2\, dy\right)\Delta x - U_\infty \frac{\partial}{\partial x}\left(\int_0^\delta \rho u\, dy\right)\Delta x$$

dividing throughout by $-\Delta x$,

$$\tau_0 + \frac{\partial p}{\partial x}(BC) = U_\infty \frac{\partial}{\partial x}\int_0^\delta \rho u\, dy - \frac{\partial}{\partial x}\int_0^\delta \rho u^2\, dy \qquad \text{...(7.18)}$$

For a flat plate, the variation of pressure p with y is negligible and the pressure within the boundary layer may be taken to be the same as that in the main-flow outside. The Bernoulli's relationship expressed by Eq. (7.8) or Eq. (7.7 *b*) may be applied to yield the pressure gradient within the boundary layer. The velocity U_∞ need not necessarily be equal to the velocity far upstream of the flat plate pertaining to the undisturbed flow. In flow problems where gravity forces are in existence the pressure p appearing in Eq. (7.18) and also in Bernoulli's Eq. (7.8) is required to be replaced by the piezometric pressure $(p + \gamma z)$.

Noting that as $\Delta x \to 0$, $BC \to AE = \delta$; substituting for $\frac{\partial p}{\partial x}$ from Eq. (7.7) into Eq. (7.18) and letting $BC = \delta = \int_0^\delta dy$, we get

$$\tau_0 - \rho U_\infty \frac{\partial U_\infty}{\partial x}\int_0^\delta dy = \rho U_\infty \frac{\partial}{\partial x}\int_0^\delta u dy - \rho \frac{\partial}{\partial x}\int_0^\delta u^2 dy \qquad \text{...(7.19)}$$

From the knowledge of differential calculus

$$\frac{\partial}{\partial x}(U_\infty u) = U_\infty \frac{\partial u}{\partial x} + u \frac{\partial U_\infty}{\partial x}$$

Using this relationship one may write,

$$\int \frac{\partial}{\partial x}(U_\infty u) dy = \int U_\infty \frac{\partial u}{\partial x} dy + \int u \frac{\partial U_\infty}{\partial x} dy$$

Equation (7.19) now becomes

$$\frac{\tau_0}{\rho} = \int_0^\delta \frac{\partial}{\partial x}(U_\infty u)\, dy - \int_0^\delta u \frac{\partial U_\infty}{\partial x} dy - \int_0^\delta \frac{\partial}{\partial x}(u^2) dy + \int_0^\delta U_\infty \frac{\partial U_\infty}{\partial x} dy$$

$$= \int_0^\delta \frac{\partial}{\partial x}[u(U_\infty - u) dy] + \frac{\partial U_\infty}{\partial x} \int_0^\delta (U_\infty - u) dy \qquad ...(7.20)$$

Since outside the boundary layer $(U_\infty - u) \to 0$, the upper limit of the above integral may be put as $y \to \infty$. From equations (7.2) and (7.3) which define the displacement and momentum thicknesses, Eq. (7.20) may be expressed in terms of δ^* and θ.

$$\frac{\tau_0}{\rho} = \frac{d}{dx}(U_\infty^2 \theta) + \delta^* U_\infty \frac{dU_\infty}{dx} \qquad ...(7.21)$$

Equation (7.21) expresses the modern form of the *momentum integral equation* for two-dimensional incompressible boundary layers. It is applicable to laminar as well as turbulent boundary layers as no assumption regarding the state of flow was made in its derivation. The right hand side of Eq. (7.21) can be expanded and the terms suitably grouped to yield an expression for the change of momentum thickness along the boundary, that is in the x-direction.

$$\frac{d\theta}{dx} = \frac{1}{\rho U_\infty^2}\left[\tau_0 + (2\theta + \delta^*)\frac{dp}{dx}\right] \qquad ...(7.22)$$

Equation (7.22) incorporates the influence of pressure gradient dp/dx, and the velocity distribution as reflected by the presence of θ and δ^*. For the case of zero pressure gradient which exists for flat plate at zero incidence the change of θ with x depends only on the boundary shear stress τ_0.

7.6.2. Momentum Integral Equation from Boundary Layer Equations

The momentum integral Eq. (7.21) may also be obtained by integrating the boundary layer equations, discussed in article 7.5. Equations (7.9) and (7.11) are reproduced here

$$\frac{\partial u}{\partial t} + u \frac{\partial u}{\partial x} + v \frac{\partial u}{\partial y} = -\frac{1}{\rho}\frac{\partial p}{\partial x} + \nu \frac{\partial^2 u}{\partial y^2} \qquad ...(7.9)$$

and

$$\frac{\partial u}{\partial x} + \frac{\partial v}{\partial y} = 0 \qquad ...(7.11)$$

Integrating Eq. (7.9) with respect to y, we obtain

$$\frac{\partial}{\partial t}\int_0^\delta u\, dy + \int_0^\delta u \frac{\partial u}{\partial x} dy + \int_0^\delta v \frac{\partial u}{\partial y} dy$$

$$= -\frac{1}{\rho}\int_0^\delta \frac{\partial p}{\partial x} dy + \nu \int_0^\delta \frac{\partial^2 u}{\partial y^2} dy \qquad ...(7.23)$$

from the knowledge of calculus, we may write

$$\int_0^\delta \frac{\partial (uv)}{\partial y} dy = \int_0^\delta u \frac{\partial v}{\partial y} dy + \int_0^\delta v \frac{\partial u}{\partial y} dy$$

from which $$\int_0^\delta v \frac{\partial u}{\partial y} dy = [uv]_0^\delta - \int_0^\delta u \frac{\partial u}{\partial y} dy \qquad ...(7.24)$$

Using the continuity Eq. (7.11), Eq. (7.24) becomes

$$\int_0^\delta v \frac{\partial u}{\partial y} dy = -U_\infty \int_0^\delta \frac{\partial u}{\partial x} dy + \int_0^\delta u \frac{\partial u}{\partial x} dy \qquad ...(7.25)$$

Substituting Eq. (7.25) into Eq. (7.23) and noting that $\tau = \mu \dfrac{du}{dy}$ vanishes at $y = \delta$,

we obtain

$$\frac{\partial}{\partial t}\int_0^\delta u dy + \int_0^\delta \frac{\partial}{\partial x}(u^2) dy - U_\infty \int_0^\delta \frac{\partial u}{\partial x} dy$$

$$= -\frac{1}{\rho}\frac{dp}{dx}\int_0^\delta dy - \frac{\tau_0}{\rho} \qquad ...(7.26)$$

Making use of the Leibnitz rule the second and the third terms of Eq. (7.26) may be expressed as

$$\int_0^\delta \frac{\partial}{\partial x}(u^2) dy = \frac{d}{dx}\int_0^\delta u^2 dy - U_\infty^2 \frac{d\delta}{dx}$$

and $$U_\infty \int_0^\delta \frac{\partial u}{\partial x} dy = U_\infty \frac{d}{dx}\int_0^\delta u dy - U_\infty^2 \frac{d\delta}{dx}$$

For steady incompressible two dimensional flow, Eq. (7.26) using Leibnitz rule and Eqs. (7.2), (7.3) and (7.7 *b*) may be simplified and expressed as

$$\frac{\tau_0}{\rho} = \frac{d}{dx}(U_\infty^2 \theta) + \delta^* U_\infty \frac{dU_\infty}{dx} \qquad ...(7.21)$$

Equation (7.21) can thus be obtained by either using the boundary layer equations or from the concept of momentum principle.

7.6.3. Pressure Distribution in the Boundary Layer

For two dimensional steady flow of an incompressible fluid, the application of Prandtl's boundary layer Eq. (7.7 *a*) outside the boundary layer (in the main flow) yields

$$U_\infty \frac{dU_\infty}{dx} = -\frac{1}{\rho}\frac{dp}{dx} \qquad ...(7.7\ b)$$

Eq. (7.7 *b*) is the differential form of the Bernoulli's equation (neglecting body forces). It indicates that the pressure distribution can be determined by the nature of flow outside of the boundary layer. Eq. (7.10) reveals that pressure does not vary in the y-direction, and therefore the same pressure as exists in the main flow is transmitted to the boundary layer. The pressure, therefore, is constant across the flow. This fact is expressed by saying that the external pressure is "impressed" on the boundary layer. The pressure gradient along the flow (x-direction) in the boundary layer will be the same as the pressure gradient dp/dx in the main flow.

7.7. Separation of Boundary Layer

Using the boundary layer Eq. (7.9 *a*) it is possible to make certain deductions even without its integration. Considering steady flow and applying the boundary conditions at the solid boundary where $y = 0$, $u = v = 0$, Eq. (7.9 *a*) reduces to

$$\left[\frac{\partial^2 u}{\partial y^2}\right]_{y=0} = \frac{1}{\mu}\frac{dp}{dx} \qquad ...(7.27)$$

Depending upon the curvature of the boundary (or the wall), there are three definite cases possible : (1) pressure drop in the direction of flow, that is, $dp/dx < 0$ (*i.e.*, the case of front surface of a circular cylinder, (Fig. 7.6 *a*) ; (2) no variation in pressure in the flow direction that is, $dp/dx = 0$ (flow along a flat plate) : and (3) pressure rise in the direction of flow, $dp/dx > 0$ (*i.e.*, the flow along the rear surface of a circular cylinder).

For case (1), the velocity profile must be convex towards the x-axis at the wall, whereas for case (3) it must be concave. For case (2), however, the curvature of the velocity profile is zero, so that the velocity gradient at the wall is constant.

It is of interest to have a physical interpretation of this behaviour. Fluid particles moving in the boundary-layer are decelerated because of viscous friction, and thus their kinetic energy is decreased. If there is a pressure drop in the direction of flow ($dp/dx < 0$), the pressure can overcome the deceleration of the particles so that they continue to move along the wall. A positive pressure gradient on the contrary, will tend to further decelerate particles in the boundary-layer. This added deceleration causes the thickness of the boundary-layer to increase and the particles slow down ; a little distance downstream they stop and a little further they may even reverse their direction of motion. A reverse-flow layer tends to push itself between the wall and the boundary-layer and in this way "separates" it from the wall. A plane of discontinuity across which a velocity jump takes place (known as a slip plane) is created between the reverse-flow layer and the boundary. The slip plane is unsteady and degenerates into vortices that are carried away by the flow. The energy required to produce the vortices cannot be recovered (and thus is a loss) and causes additional resistance between the wall (or the body) and the flouid. This phenomenon which separates the boundary-layer from the wall (or the body) is called "*separation*".

In order to get further acquainted with the phenomenon of boundary layer separation, let us consider the flow around a long circular cylinder representing a blunt body (Fig. 7.6 *a*).

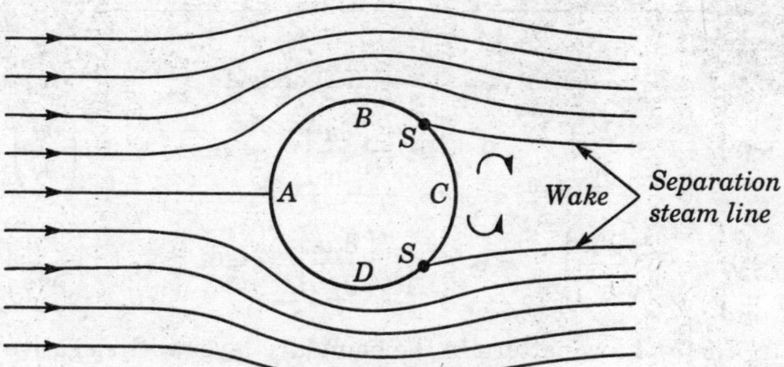

Fig. 7.6 (*a*) Streamline pattern about a circular cylinder as a result of boundary layer separation.

The flow past a long cylinder can be considered two-dimensional. The fluid particles are accelerated on the upstream side *AB* and *AD* indicated by the converging streamlines. The pressure, therefore, decreases from *A* to *B* (also from *A* to *D*), resulting in a negative pressure gradient ($dp/dx < 0$). The net pressure force on a fluid element in the boundary layer is in the direction of flow, and counteracts to some extent the "slowing effect" of the boundary on the fluid. The rate of growth of boundary-layer thickness, thus, is less than that for a flat plate with zero pressure gradient under similar conditions of flow (this statement is applicable to regions of the boundary where the pressure gradient is negative). The fluid particles negotiating the curve *BC* of the solid boundary are decelerated from *B* to *C* as indicated by diverging streamlines, thereby resulting in an increase of pressure. This sets up positive pressure gradients ($dp/dx > 0$) which tends to further retard the boundary layer fluid leading to its separation.

Fig. 7.6 (*b*) illustrates the pattern of flow in the boundary layer with negative and positive pressure gradients. The solid boundary *DASE* is a curved one. The portion *DA* of the boundary

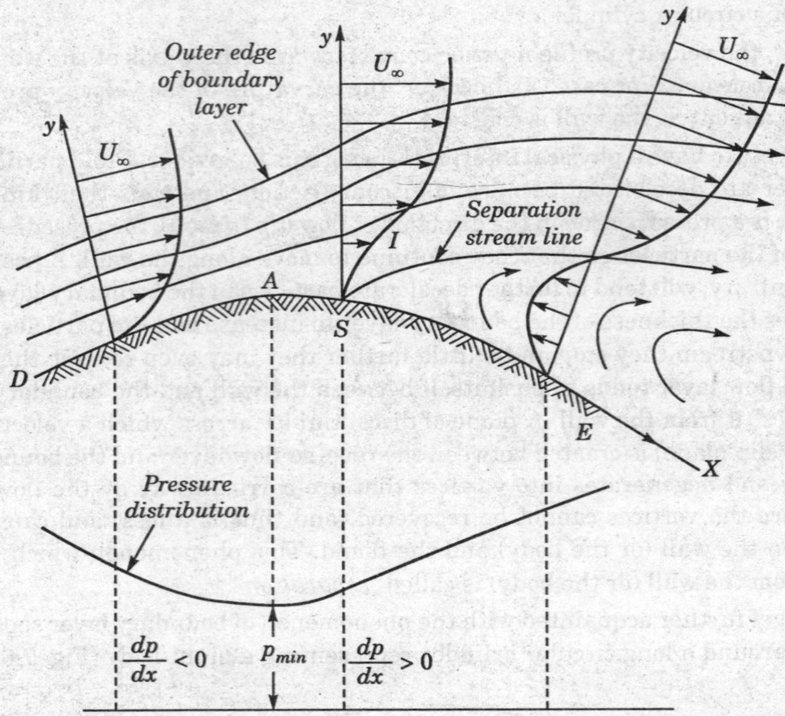

Conditions at section :

(*i*) *D*	(*ii*) *A*	(*iii*) near *A*	(*iv*) near *E*
$\left(\frac{\partial u}{\partial y}\right)_{y=0} > 0$	$\left(\frac{\partial u}{\partial y}\right)_{y=0} > 0$	$\left(\frac{\partial u}{\partial y}\right)_{y=0} = 0$	$\left(\frac{\partial u}{\partial y}\right)_{y=0} < 0$
$\left(\frac{\partial^2 u}{\partial y^2}\right)_{y=0} < 0$	$\left(\frac{\partial^2 u}{\partial y^2}\right)_{y=0} = 0$	$\left(\frac{\partial^2 u}{\partial y^2}\right)_{y=0} = 0$	$\left(\frac{\partial^2 u}{\partial^2 y}\right)_{y=0} > 0$

Fig. 7.6 (*b*). Flow pattern in the boundary-layer with negative and positive pressure gradients.

characterizes accelerated zone of fluid and therefore is the region of favourable (negative) pressures gradient ($dp/dx < 0$). Fig. 7.6 (*b*) also shows the velocity distribution in the boundary layer at various points along the boundary. The figure at the bottom indicates the variation of pressure in the *x*-direction. As long as $dp/dx < 0$, the entire boundary-layer moves forward and the velocity profile is convex towards the *x*-axis. Beyond the point of minimum pressure, marked *A*, the pressure rises and the pressure gradient is positive ($dp/dx > 0$). Although initially the velocity is still positive everywhere in the boundary layer, a reversal of curvature must exist somewhere within the boundary-layer.

In the immediate vicinity of the solid boundary the curvature of the velocity profile depends only on the pressure gradient, as evidenced from Eq. (7.27). In accelerated flow $dp/dx < 0$, which alongwith Eq. (7.27) yields $(\partial^2 u/\partial y^2)_{y=0} < 0$ and therefore for the entire thickness of the boundary layer $\partial^2 y/\partial y^2 < 0$. In the decelerated region $dp/dx > 0$, which results in $(\partial^2 u/\partial y^2)_{y=0} > 0$. Since, within the boundary layer at some distance from the boundary $\partial^2 u/\partial y^2 < 0$, and in the retarded flow $\partial^2 u/\partial y^2 > 0$ at and near the boundary, it follows that there must exit a point for which $\partial^2 u/\partial y^2 = 0$. This point is known as the point of inflexion indicated by I in the velocity profile at S (Fig. 7.6 *b*).

At some distance downstream from A along the boundary, point S is reached at which the velocity gradient is zero, *i.e.*, $(\partial u/\partial y)_{y=0} = 0$. Downstream of this point reverse-flow exists and point S can be considered a "separation point of the boundary-layer". Beyond S the velocity gradient is negative and back-flow occurs. The location of the separation point can be found from the condition.

$$(\partial u/\partial y)_{y=0} = 0 \qquad ...(7.28)$$

However this requires integration of the boundary-layer equations to obtain $u = u(x, y)$. The line of zero velocity dividing the forward flow and the back flow leaves the boundary surface at the separation point S, and is known as the *'separation streamline'* (which is a slipline). The region between the separation streamline and the boundary surface is known as the *'wake'*. The back-flow is the main cause of formation of wake which is characterized by large size irregular eddies (or vortices) resulting in a considerable energy loss. The vortices formed in the wake zone behind a blunt body are not shed in pairs but alternately from top and bottom and move downstream in the fluid. The formation of vortices and their subsequent movement downstream changes completely the flow-field in the wake. The pressure distribution in the wake suffers a radical change as compared to that in an ideal fluid flow.

At some distance downstream of the body it is possible to visualize a regular pattern of two rows of vortices which move alternately clockwise and anticlockwise. This system of vortices is known as a Karman vortex street and is shown in Fig. 7.6 (*c*).

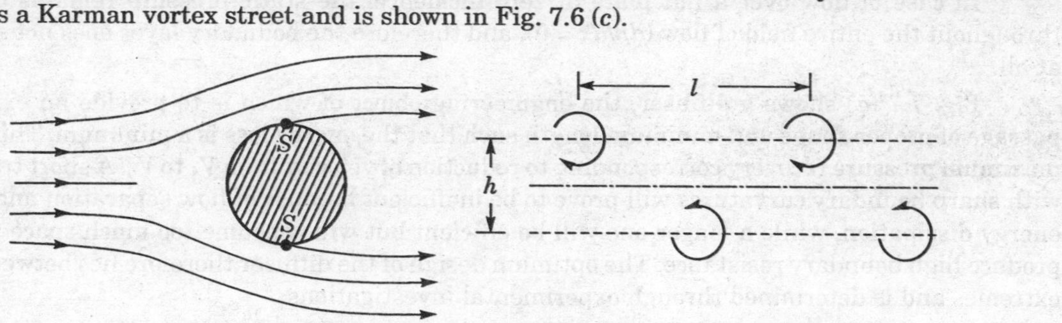

Fig. 7.6 (c) Karman vortex street behind a circular cylinder.

If h is the distance between the two lines of vortex centres and l the distance between the centres of consecutive vortices with the same sense of rotation (say clockwise), Karman has shown theoretically that the pattern of vortex formation is stable if the ratio $h/l = 0.28$. Measurements confirm this relationship in the early part of the wake, but at a larger distance the spacing h tends to increase. The vortex street moves with a velocity which is smaller than the flow velocity U_∞ in front of the body. The alternate breaking away of these vortices results in a periodic transverse force on the body. If the natural frequency of the body (cylinder in the case under consideration) in the transverse direction is in resonance with this forced transverse oscillation, severe vibrations of the body can take place. The vibration of telephone and transmission lines (the phenomenon of singing of suspended wires) at certain wind velocities and the aerodynamic instability of suspension bridges are familiar examples in sight.

From the foregoing discussion it can be easily understood that the separation of flow in the boundary layers occurs only in decelerated flow ($dp/dx > 0$). The separation may occur either in the

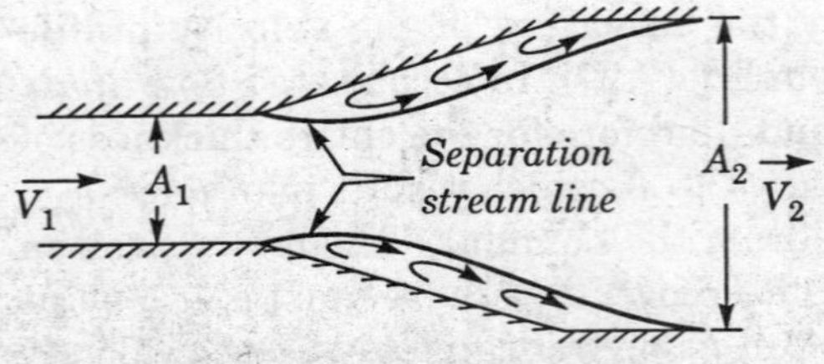

(*a*) Flow in a diverging-channel or a diffuser.

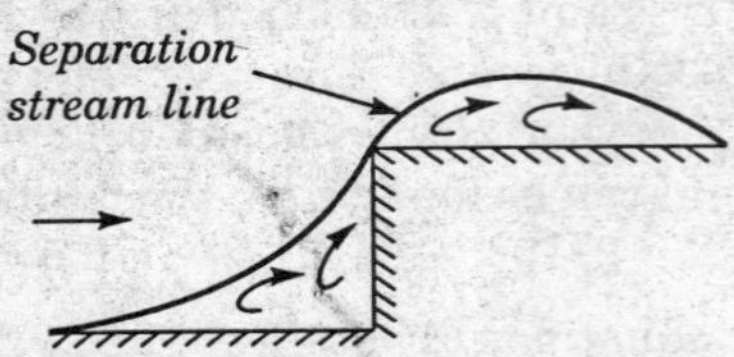

(*b*) Flow over sharp cornered broad crested weir.

Fig. 7.7. Examples of flow separation.

laminar portion of the boundary layer or in the turbulent part, but the laminar boundary layers are more susceptible to separation than turbulent ones. The reason being that in a laminar boundary layer the increase of velocity with distance *y* from the solid boundary is less rapid, and the existence of adverse pressure gradient can very easily stop the slow moving fluid in the vicinity of the boundary. A turbulent boundary layer can withstand the adverse pressure gradient for some distance before separating. In general, the greater the adverse pressure gradient, the sooner separation occurs. The boundary layer thickness increases rapidly in the region of positive pressure gradient, and the assumption that δ is small is no longer valid.

The separation phenomenon is not limited to blunt bodies or curved boundary surface. An adverse pressure gradient is obtained in a diffuser (diverging conduit). The boundary layer separation does also take place at places where the boundary surface suddenly changes its longitudinal profile providing sharp corners or edges (Fig. 7.7 *a*). The inability of streamlines to bend around the edges leads to the separation of flow. The flow over a sharp cornered broad crested weir or flow over a step is an example of flow separation due to sudden change in the profile of the boundary (Fig. 7.7 *b*). The flow around an aerofoil at a high angle of attack also separates on the upper side towards the downstream end.

In case of flow over a flat plate at zero incidence, the static pressure remains constant throughout the entire field of flow ($dp/dx = 0$), and therefore the boundary layer does not separate at all.

Fig. 7.7 (*a*) shows a diffuser, the engineering object of which is to provide an expanding passage of proper shape and minimum length such that the energy loss is a minimum. This means maximum pressure recovery corresponding to reduction of velocity from V_1 to V_2. A short transition with sharp boundary curvatures will prove to be inefficient because of flow separation and a large energy dissipation, while a longer one will be efficient but will consume too much space and will produce high boundary resistance. The optimum design of the diffuser therefore lies between these extremes and is determined through experimental investigations.

The prediction of separation may be simple in case of sharp-cornered obstructions, it becomes more complex a matter for streamlined (gently curved boundaries) objects or surfaces. In case of streamlined bodies, separation may not occur and the flow field is virtually identical with that of the ideal fluid except in the thin boundary layer region.

It is obvious that separation of the boundary-layer has far reaching consequences. In particular, the potential free flow-region does not recombine at the rear end of the body because of the existence of the vortex field. For this reason the pressure at the rear end of the body is considerably lower than that in the free-flow region. The boundary conditions applicable to potential flow are no longer true and the calculated pressure dp/dx obtained from Eq. (7.9 *a*) no longer applies at the boundary of the body. Beyond the point of separation the actual pressure drop can be determined from suitable experiments. In addition, the thickness of the boundary layer δ grows appreciably in the region of positive pressure gradient and the assumption of small δ is no longer valid.

Note. For more information on boundary-layers, refer to chapter X on "Drag and Lift".

Ex. 7.9. *For the velocity distribution given by* $\frac{u}{V} = -2\left(\frac{y}{\delta}\right) + \left(\frac{y}{\delta}\right)^2$ *determine whether the flow is separated or on the verge of separation or will attach with the surface.*

Solution. Conditions conducive for separation of boundary layer have been described in article 7.7 and illustrated in Fig. 7.6 (*b*). Accordingly :

(*a*) If the flow is separated :

$$\left(\frac{\partial u}{\partial y}\right)_{y=0} < 0, \text{ and } \left(\frac{\partial^2 u}{\partial y^2}\right)_{y=0} > 0$$

From the given velocity distribution, $\left(\frac{\partial u}{\partial y}\right)_{y=0} = -\frac{2}{\delta}$, and $\left(\frac{\partial^2 u}{\partial y^2}\right)_{y=0} = \frac{2}{\delta}$

The velocity distribution therefore, represents a case where the flow has been separated from the boundary.

OBJECTIVE TYPE QUESTIONS

1. Ludwig Prandtl is regarded as the father of modern fluid mechanics mainly on account of :
   (*a*) his fundamental research in the field of aircraft engineering
   (*b*) his keen interest and insight into the various phenomena of fluid dynamics.
   (*c*) his new interpretations on fluid resistance.
   (*d*) his poineering research on flow of low-viscosity fluid bringing forward a new concept of boundary layer.
2. The boundary-layer exists in :
   (*i*) flow of ideal fluids (*ii*) flow of real fluids
   (*iii*) only pipe-flow (*iv*) only flow over flat surfaces.
3. The existence of boundary-layer is on account of :
   (*a*) fluid density (*b*) gravitational effect
   (*c*) fluid viscosity (*d*) surface tension
4. The concept of boundary-layer divides the flow field into two zones. These zones are :
   (*i*) the region in which eddy-formation takes place
   (*ii*) the region in which the viscous effects are negligible and the flow is analysed by the potential flow theory
   (*iii*) the region where the turbulence and viscous effects both are present
   (*iv*) the region where the viscous effects are predominant.
5. As the fluid flows along the solid boundary, more and more fluid in the vicinity of the boundary gets retarded. This deceleration of fluid is on account of :
   (*a*) high velocity flow outside the boundary-layer
   (*b*) the assumption on fluid being ideal
   (*c*) high velocity gradients which exist at and near the boundary
   (*d*) high velocity of fluid.
6. The boundary-layer thickness is an important parameter in boundary-layer analysis. There are many definitions of which the displacement thickness is a more logical and mathematically useful one. It is defind as :
   (*i*) the thickness upto which the velocity approaches 99% of the free-stream velocity

(*ii*) the distance measured perpendicular to the boundary by which the free-stream is displaced on account of formation of boundary-layer

(*iii*) the layer which represents reduction in momentum caused by the boundary-layer

(*iv*) the layer in which the loss of energy is minimum.

7. The following table lists various kinds of boundary layer thicknesses and the corresponding equations are given in the second column, not necessarily in that order. Match each boundary-layer thickness with its mathematical-equivalent expression :

(*a*) nominal thickness (*i*) $\int_0^\delta \frac{u}{U_\infty}\left(1-\frac{u}{U_\infty}\right)dy$

(*b*) displacement thickness (*ii*) $y = \delta$ for $u = 0.09\, U_\infty$

(*c*) momentum thickness (*iii*) $\int_0^\delta \frac{u}{U_\infty}\left(1-\frac{u^2}{U_\infty^2}\right)dy$

(*d*) energy thickness (*iv*) $\int_0^\delta \left(1-\frac{u}{U_\infty}\right)dy$

8. The boundary-layer just downstream of the leading edge of a flat surface exhibit characteristics of laminar flow and is termed as laminar boundary-layer. The velocity distribution follows a parabolic law. The boundary-layer thickness increases in the direction of flow according to the following relationship :

(*i*) $\delta = 5\sqrt{\frac{vx}{U_\infty}}$ (*ii*) $\delta = 5\sqrt{\frac{v}{U_\infty}}\, x^{3/2}$

(*iii*) $\delta = 0.37\left(\frac{v}{U_\infty}\right)^{1/5} x^{4/5}$ (*iv*) $\delta = 3.7\left(\frac{v}{U_\infty}\right)^{4/5} x^{1/5}$

9. The laminar boundary-layer over a long flat plate becomes unstable and changes flow characteristics from laminar to turbulent when the plate Reynolds number approaches a value which lies in the range :

(*a*) $2 \times 10^6 - 5 \times 10^6$ (*b*) $3 \times 10^5 - 6 \times 10^5$

(*c*) $3 \times 10^4 - 5 \times 10^4$ (*d*) $5 \times 10^6 - 8 \times 10^6$.

10. The separation of boundary-layer is a phenomenon in which the boundary-layer gets detached (or separated) from the wall due to progressive deceleration of flow.

The boundary-layer separation occurs when :

(*i*) $\frac{dp}{dx} = 0$ (*ii*) $\left(\frac{\partial u}{\partial y}\right)_{y=0} = 0$

(*iii*) $\frac{dp}{dx} < 0$ (*iv*) $\left(\frac{du}{dy}\right)_{y=0} < 0$

ANSWERS

1. (*d*) **2.** (*ii*). **3.** (*c*) **4.** (*ii*), (*iv*). **5.** (*c*)

6. (*ii*) **7.** (*a*), (*ii*) ; (*b*), (*iv*) ; (*c*), (*i*) ; (*d*) (*iii*). **8.** (*i*) **9.** (*b*)

10. (*ii*).

REVIEW QUESTIONS

1. What is a boundary layer ?
2. Do you always need a wall or a solid boundary for a boundary layer to exist ?
3. Cite as many practical examples of boundary layer formation as you can.
4. What is D'Alembert's paradox ?
5. How was it possible for Prandtl to divide the flow field into zones of dominant viscous and inertial forces ? How is the flow analysis simplified this way ?
6. Sketch the boundary layer separation in a diverging channel by showing at least three velocity profiles.
7. Comment on the arbitrariness of boundary layer thickness.
8. Of the displacement and momentum thicknesses, which one is more representative ?
9. Place the momentum thickness, nominal thickness of the boundary layer and the displacement thickness in order of their relative magnitudes.
10. Differentiate between the characteristics of laminar and turbulent boundary layers.
11. What is a laminar sublayer ?
12. How are the Prandtl's boundary layer equations different from those Navier-Stokes ?
13. What are the difficulties associated with the analysis of boundary layer flow ?
14. Comment on the pressure distribution within the boundary layer ?
15. Can separation of boundary layer take place in case of :
    (*i*) negative pressure gradient. (*ii*) zero pressure gradient.
    Give reasons.
16. What are the causes which result in separation of boundary layer ?
17. What is "wake" and what are its characteristics ?
18. Cite examples where boundary layer separation may be readily predicted.
19. Why are boundary layer equations not applicable in the region downstream of point of separation.

PROBLEMS

7.1. Show that for the velccity distribution,

$$\frac{u}{U} = 2\eta - \eta^2$$

where $\eta = y/\delta$, that ratio $\delta/\delta^* = 3$.

7.2. Show that for linear distribution of velocity in the boundary layer, $\delta/\theta = 6$.

7.3. Show that the ratio of the boundary layer thickness to the displacement thickness (δ/δ^*) for the velocity distribution given by $u/U_\infty = \sin(\pi\delta/2)$ is 2.75.

7.4. Show that for the velocity distribution of problem (7.3), the ratio of boundary layer thickness (δ/θ) is 7.32.

7.5. For the velocity distribution in the laminar layer given by Blasius

$$\frac{u}{U_\infty} = 0.332\eta - 0.00229\delta^4 + 1.996 \times 10^{-6}\eta^7 ......$$

Show that $\delta/\delta^* = 1.199$ and $\delta/\theta = 7.74$, where $\delta = y/\delta$.

7.6. The velocity distribution in the boundary layer was found to fit the equation

$$\frac{u}{U_\infty} = \left(\frac{y}{\delta}\right)^n$$

Show that the displacement and the energy thicknesses can be expressed as

$$\delta^* = \delta - \frac{\delta}{n+1}; \quad \delta_e = \frac{\delta}{n+1} - \frac{\delta}{3n+1}.$$

7.7. The velocity distribution in the boundary layer over a high spillway face was found to have the following form

$$\frac{u}{U_\infty} = \left(\frac{y}{\delta}\right)^{0.22}$$

Prove that the displacement thickness, the momentum thickness and the energy thickness in term of δ—the boundary thickness—can be expressed as

$$\frac{\delta^*}{\delta} = 0.18\,; \quad \frac{\theta}{\delta} = 0.125\,; \quad \frac{\delta_e}{\delta} = 0.217.$$

7.8. For a velocity distribution within the boundary layer

$$\frac{u}{U_\infty} = \frac{3}{2}\left(\frac{y}{\delta}\right) - \frac{1}{2}\left(\frac{y}{\delta}\right)^3,$$

compute the ratio of displacement thickness of boundary layer to nominal boundary layer thickness.

8

Laminar Flow

8.1. Introduction

The flow of a real fluid differs from that of an ideal one in respect of viscous effects that take place in the vicinity of solid boundary. As an ideal fluid is non-viscous, the presence of a solid boundary does not affect the flow ; the fluid is assumed to slip over the boundary and as such the velocity distribution over the boundary is uniform throughout. In case of a real fluid, the flow is affected in the neighbourhood of the boundary. When a real fluid approaches a boundary surface (which is stationary) the fluid particles which come in contact with the stationary surface are brought to rest by viscous resistance of the fluid. Hence the fluid particles at the boundary have zero velocity relative to the boundary. This flow characteristic of a real fluid at the boundary is in accordance with the no-slip condition. The fluid particles away from the boundary attain higher velocity, and this change of velocity across the flow gives rise to a velocity gradient. The existence of velocity gradient is responsible for development of viscous shear resistance which opposes the motion.

In case of flow of an ideal fluid, the flow remains unaffected by the presence of a solid boundary on account of fluid being assumed non-viscous. This results in total absence of viscous resistance which causes velocities to remain uniform across the flow and hence the uniform velocity distribution. This uniform velocity distribution implies a zero velocity gradient-throughout the flow, resulting in, according to Newton's law of viscosity, a zero shear stress in the entire flow region. This zero shear is significant as it indicates that there is no resistance to flow, and consequently there would be no energy loss in the flow direction. This would mean that flow of an ideal fluid can be caused and maintained without expending power. In other words the power required to cause and maintain the flow is zero.

Actually, flow of a real fluid involves loss of energy in overcoming resistances that come into play, and contradicts the inference of zero resistance of ideal fluid flow.

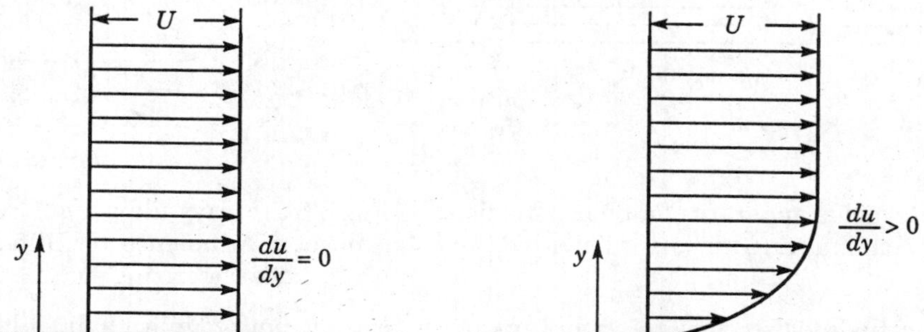

(*a*) Velocity distribution in ideal fluid flow ($du/dy = 0$) over a plane boundary

(*b*) Velocity distribution in real fluid flow over a plane boundary (Velocity in the vicinity of the boundary has been affected by viscous action, $du/dy > 0$)

Fig. 8.1. Comparison of ideal and real fluid flow.

The velocity distributions, Fig. 8.1, are characteristics of flow of ideal and real fluids.

It may be noted here that the concept of an ideal fluid is the creation of mathematicians to enable the flow phenomena amenable to mathematical analysis.

8.2. Types of Flow

It has long been known that two distinct kinds of flow occur. These are : the laminar motion and the turbulent flow. The simpler of these is the laminar flow which is characterised by motion of fluid in layers or laminas, parallel to the boundary surface. In case of straight boundaries, the path followed by any small mass of fluid is a straight line. Conditions favourable for laminar flow are high viscosity μ, low mass density ρ, low mean velocity V and small flow passage L. An appropriate measure of these conditions is the magnitude of a dimensionless parameter involving the above variables, $\rho VL/\mu$, which provides a criterion for ascertaining the type of flow.

Laminar flow through a circular tube was studied experimentally by Hagen who published his results in 1839, and independently by Poiseuille whose work was published between 1840 and 1846. Each deduced from his tests that the volume rate of flow is directly proportional to the pressure drop and to the fourth power of the radius and inversely proportional to the tube length. The equation relating these quantities to the viscosity is, thus, known as Hagen-Poiseuille equation.

A laminar flow becomes unstable and tends to change over to turbulent as the mean velocity is increased, other variables remaining unchanged. Osborne Reynolds was first to demonstrate in 1883, that the transition from laminar to turbulent flow depends not only on the mean velocity, but on the dimensionless quantity, $\rho VL/\mu$, known as the Reynolds number **R**. In case of circular tubes the characteristic length L is taken as the tube diameter. He showed that for pipe flow if **R** is less than about 2000, the flow is laminar while at higher values the flow tends to become turbulent.

8.2.1. Reynolds Experiment

Osborne Reynolds was first to show that the Reynolds number, Vd/ν is the criterion for determining the type of flow (laminar or turbulent) in a circular pipe. His apparatus is shown schematically in Fig. 8.2.

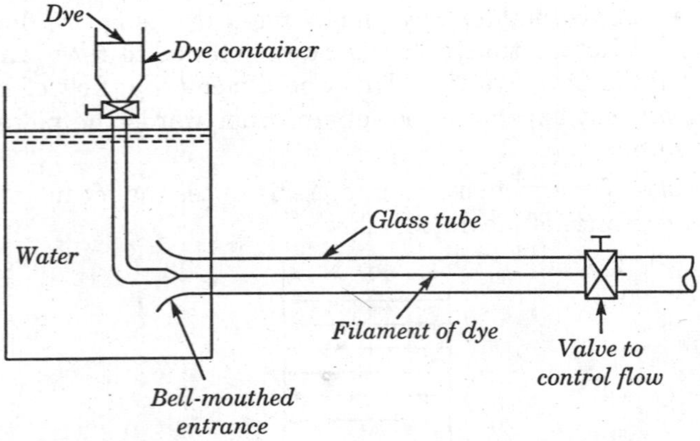

Fig. 8.2. Reynolds apparatus for demonstration of laminar and turbulent flow.

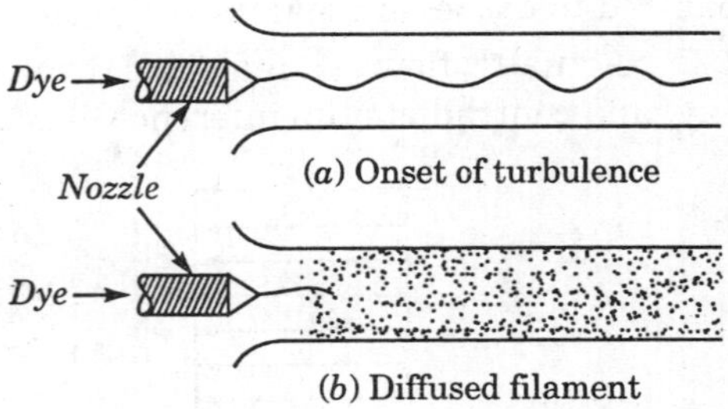

Fig. 8.3 (*a*) Wavy filament of dye indicating onset of turbulence. (*b*) Diffused filament characterizing turbulent flow.

The apparatus consisted of (*i*) a water tank (*ii*) an arrangement to inject a fine filament of dye into the bell-mouthed entrance of a glass tube through which water flows and (*iii*) a valve to control the flow through the tube as shown in Fig. 8.2. A fine filament of dye was introduced into the glass tube near the smoothly rounded, bell-mouthed (or trumpet-shaped) entrance. At low rates of flow the filament of dye appeared as a straight line parallel to the tube axis characterizing laminar flow. As the valve was further opened and greater velocities were attained, the dye filament became wavy leading to its breakup and finally diffusing into the flowing water, Fig. 8.3 (*a*) and (*b*) depict the flow conditions at increasing velocities characterizing the onset of turbulence and the flow completely changing over to turbulent.

A sudden breakdown of laminar motion occurs at a certain critical velocity, at which the water in the tube becomes evenly coloured by the mixing of the dye filament.

Critical Reynolds Number

By performing this experiment with tubes of different sizes and with water at various temperatures, Reynolds established the fact that turbulence sets in at a certain value of Reynolds number. This value he found to be very sensitive to the initial disturbances in the fluid that exist near the entrance. By allowing water to remain undisturbed for some hours he obtained a critical value of $\mathbf{R}$ = 13,000 upto which the flow was laminar. Ekman, working in 1910 with Reynolds original apparatus was able to obtain laminar flow upto a Reynolds number of 50,000 by quieting the water in the tank for many days before conducting the tests.

While there appears to be no definite upper limit of the critical value of $\mathbf{R}$ which characterizes full attainment of turbulence, all experimenters agree that a lower limit does exist and that its value is approximately 2,000. This lower critical Reynolds number is of greater engineering importance as it defines the limit below which all turbulence, no matter how severe, entering the flow from any source, will eventually be damped out by viscous action.

The upper limit of critical Reynolds number ($\mathbf{R}_C$) has been found to depend upon several factors such as (*i*) initial turbulence in the approach flow (*ii*) shape of the pipe entrance, and (*iii*) roughness of pipe. The upper limit of $\mathbf{R}_C$ was found by Reynolds to lie between $12{,}000 < \mathbf{R}_C < 14{,}000$. These high values of $\mathbf{R}_C$ are of little practical interest, and we may consider the upper limit of $\mathbf{R}_C$ to be defined by $2{,}700 < \mathbf{R}_C < 4{,}000$.

The concept of critical Reynolds number is a useful one in demarcating the regimes of laminar and turbulent flows. Applying this concept to circular pipes, we may predict that a flow will be laminar if $\mathbf{R} < 2{,}000$ and turbulent if $\mathbf{R} > 4{,}000$. It is to be noted that Reynolds number which in general is defined as $\rho VL/\mu$ is a function of the geometry of the flow area taken into account by the characteristic length L. This characteristic length is usually taken as diameter d in case of circular pipes whereas in case of flow through parallel plates it is the spacing of the plates b, in case of flow in wide open channels it is the depth of flow y, and for flow about a sphere it is the diameter of sphere. The lower critical Reynolds number for the above mentioned cases are $\mathbf{R}$ = 1000 (for parallel plates), $\mathbf{R}$ = 500 (for open channels) and $\mathbf{R}$ = 1 (for sphere).

8.3. Experimental Determination of Critical Velocity

The existence of two distinct types of flow, namely, laminar and turbulent, may be demonstrated indirectly by performing a simple experiment, refer Fig. 8.4 (*a*).

In this experiment, the energy loss in a certain length of straight circular pipe of constant diameter is determined by pressure drop measured by a differential manometer. The mean velocity of flow is calculated from the known volumetric rate of flow passing through the pipe. The energy loss h_L and the corresponding mean velocity V may be plotted to give a straight line variation ($h_L \propto V$) at smaller values of V, while at higher velocities a nearly parabolic relationship ($h_L \propto V^2$) will result. The straight line relationship between h_L and V indicates laminar flow while the parabolic one characterizes the turbulent flow. Between the two flow regimes lies a transition zone in which flow is in the process of changing over from laminar to turbulent. As V is increased the head loss h_L varies with the mean velocity V according to the law indicated by line *OPQRS* whereas for decreasing velocities it will be seen to follow the plot *SRPO* shown in Fig. 8.4 (*b*). From this experiment, it may be deduced that the points P and Q lying on the straight line *OP* define the lower and upper critical velocities respectively.

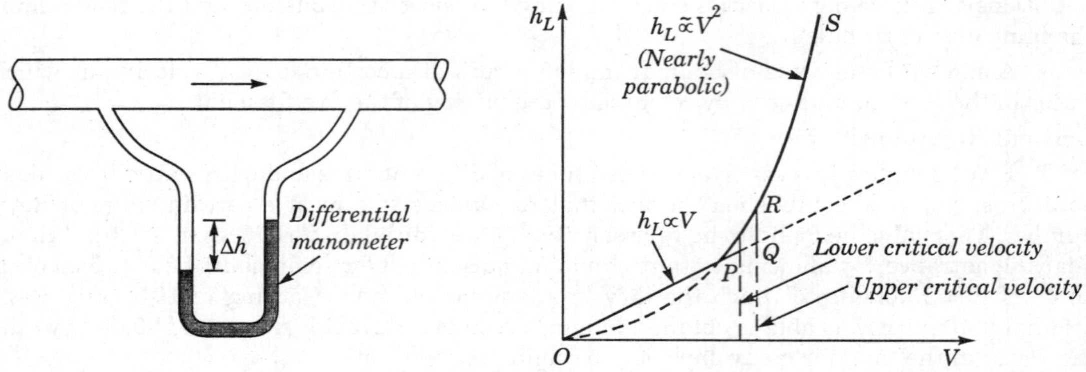

(*a*) Experimental set up. (*b*) Energy head loss variation with velocity.

Fig. 8.4

8.4. Steady Laminar Flow Through a Circular Tube

The laminar flow of an incompressible fluid flowing under steady conditions may be completely analysed using Newton's law of viscosity and Newton's second law of motion. Let us consider a horizontal tube of radius R_0, and a concentric cylindrical fluid element of radius r and length dx, as shown below in Fig. 8.5.

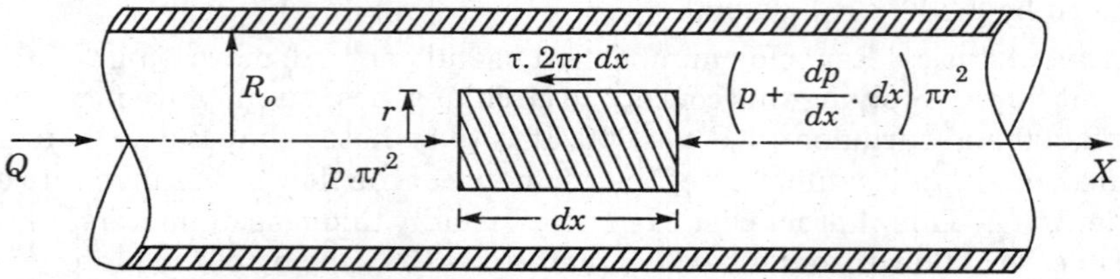

Fig. 8.5. Free body diagram for steady laminar flow through a horizontal tube.

The forces acting on the free body of the fluid element are :

1. The pressure, force ($p\pi r^2$), on the left face (acting from left to right).
2. The pressure force $\left(p + \frac{dp}{dx}\,dx\right)\pi r^2$ on the right face (acting from right to left).
3. The shear force ($\tau.2\pi r\,dx$) acting on the surface of the fluid element opposing the motion (acting from right to left).

As the flow is steady and the tube is of uniform size, *i.e.* the radius is same throughout the length under consideration, the total acceleration given by

$$\frac{du}{dt} = \frac{\partial u}{\partial t} + u\frac{\partial u}{\partial s} = 0.$$

Since the flow is steady, $\frac{\partial u}{\partial t} = 0$, and for uniform flow, $\frac{\partial u}{\partial s} = 0$. Newton's second law of motion in the direction of flow may be written as :

$$\Sigma F_x = m\ .\ a_x = 0$$

$$p\pi\, r^2 - \left(p + \frac{dp}{dx}\,dx\right)\pi r^2 - \tau 2\pi\, r dx = 0$$

which upon simplification yields

$$\tau = -\left(\frac{dp}{dx}\right).\frac{r}{2} \qquad \text{...(8.1)}$$

This equations gives the distribution of shear stress across a flow section. The pressure gradient, dp/dx, depends only on x and hence Eq. (8.1) provides a linear relationship between τ and r. It is evident from the equation that the shear will be maximum at the tube surface where $r = R_0$ and minimum at $r = 0$, where it will have a zero value. It is to be noted that Eq. (8.1) is applicable to any type of flow whether laminar or turbulent, as no assumption has so far been made regarding the type of flow. In case of a horizontal tube the pressure force is the only force available to overcome the resisting force, the term dp/dx is, therefore, always negative.

The shear stress is related to the velocity gradient by Newton's law of viscosity,

$$\tau = -\mu \frac{du}{dr}$$

It is to be noted that the velocity gradient, du/dr, is negative, *i.e.* as r increases u decreases. Making this substitution in Eq. (8.1) one obtains,

$$-\mu \frac{du}{dr} = -\left(\frac{dp}{dx}\right)\frac{r}{2}$$

$$\frac{du}{dr} = \frac{1}{\mu}\frac{dp}{dx}\left(\frac{r}{2}\right). \quad \text{Integrating, } u = \frac{1}{\mu}\frac{dp}{dx}\, r^2 + A$$

The velocity of a real fluid at a fixed boundary is always zero to satisfy the no-slip condition hence at $r = R_0$, $u = 0$, this boundary condition enables determination of the constant of integration A, thus

$$A = -\frac{1}{4\mu}\left(\frac{dp}{dx}\right)R_0^2$$

The velocity distribution is given by

$$u = \frac{1}{4\mu}\left(-\frac{dp}{dx}\right)(R_0^2 - r^2) \qquad \text{...(8.2)}$$

Evidently, the velocity distribution expressed by Eq. (8.2) is a parabolic one and the velocity distribution surface is a paraboiled of revolution. Equation (8.2) is identical with Eq. (6.31 *b*) which was derived for similar flow using Navier-Stokes equations.

For maximum velocity, $\frac{du}{dr} = 0$

or, $$\frac{1}{4\mu}\frac{dp}{dx}(-2r) = 0 \text{ giving } r = 0$$

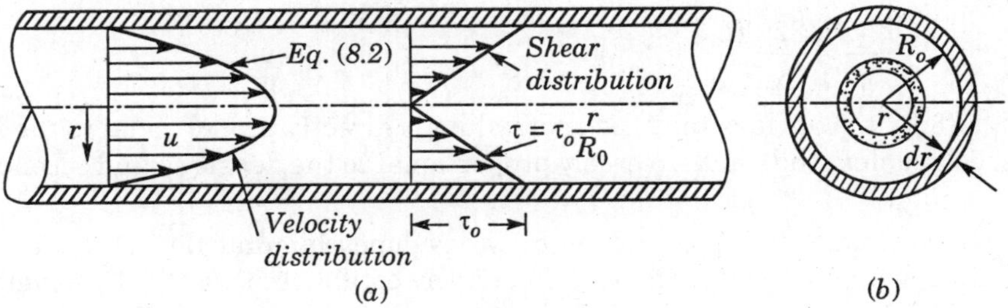

(*a*) Distribution of velocity and shear in laminar flow.
(*b*) Circular ring element for computation of rate of flow.

Fig. 8.6

and the maximum velocity $$u_{max} = -\frac{1}{4\mu}\left(\frac{dp}{dx}\right)R_0^2 \qquad \text{...(8.3)}$$

The quantity of fluid flowing per second through the tube may be obtained by considering the flow through a circular ring element of radius r and thickness dr, Fig. 8.6 (*b*). The discharge through this elementary ring

$dQ = 2\pi r.dr.u$, and discharge through the tube

$$Q = \int_0^{R_0} 2\pi r\, u.dr = \int_0^{R_0} 2\pi \frac{1}{4\mu}\left(-\frac{dp}{dx}\right)(R_0^2 - r^2)\, r\, dr$$

$$= \frac{1}{8\mu}\left(-\frac{dp}{dx}\right) \pi R_0^4 \quad ...(8.4)$$

The term $\left(-\frac{dp}{dx}\right)$ represents pressure drop per unit length of tube and may be written as $\frac{\Delta p}{L}$, consequently the discharge equation becomes

$$Q = \frac{\pi R_0^4 \Delta p}{8\mu L} \quad ...(8.5)$$

Replacing the radius R_0 by half the diameter $d/2$,

$$Q = \frac{\pi d^4\, \Delta p}{128\mu L} \quad ...(8.6)$$

This equation (8.6) is called *Hagen-Poiseuille equation*, after Hagen and Poiseuille who independently obtained it experimentally. The analytical proof was given by G.H. Wiedermann in 1856.

Shear stress at the pipe wall, from Eq. (8.1)

$$\tau_0 = (\Delta p/L)\, .\, r_0/2 \quad ...(8.7)$$

The average velocity through the tube may be expressed as

$$V = \frac{Q}{\pi R_0^2} = \frac{\Delta p R_0^2}{8\mu L} \quad ...(8.8)$$

and the pressure drop in length L, as

$$\Delta p = \frac{8\mu VL}{R_0^2} = \frac{32\mu VL}{d^2}$$

The energy loss per unit weight of fluid in laminar flow through a circular tube is

$$\frac{\Delta p}{\gamma} = \frac{32\mu VL}{\rho g d^2} \quad ...(8.9)$$

From Eq. (8.9) it is obvious that the energy loss is directly proportional to the first power of velocity V, viscosity μ, length L, and inversely proportional to the density ρ and second power of the diameter d.

The above equations are applicable to fully developed laminar flow and are not valid near the entrance of a tube. For detailed information on the establishment of fully developed laminar flow, see article 11.1 of the chapter on Analysis of Pipe Flow.

Eq. (8.9) may be written as

$$\frac{\Delta p}{\gamma} = \frac{32\mu VL}{\rho g d^2}\frac{2V}{2V} = \frac{64}{\rho V d/\mu} \cdot \frac{L}{d} \cdot \frac{V^2}{2g}$$

Comparing this equation with that of the Darcy-Weisbach, Eq. (9.57) which is reproduced below

$$h_L = f\,\frac{L}{d}\frac{V^2}{2g}$$

one obtains, $$f = \frac{64}{\rho V\, d/\mu} = \frac{64}{\mathbf{R}} \qquad ...(8.10)$$

Eq. (8.10) gives a relationship between friction factor f and the Reynolds number **R** for laminar flow through a circular tube.

In case of an inclined tube, an additional force, *i.e.* the gravity force component will act on the free body of the fluid element, and while applying Newton's Second Law of motion this force needs to be taken into account. Fig. 8.7 shows the free body diagram for an inclined tube.

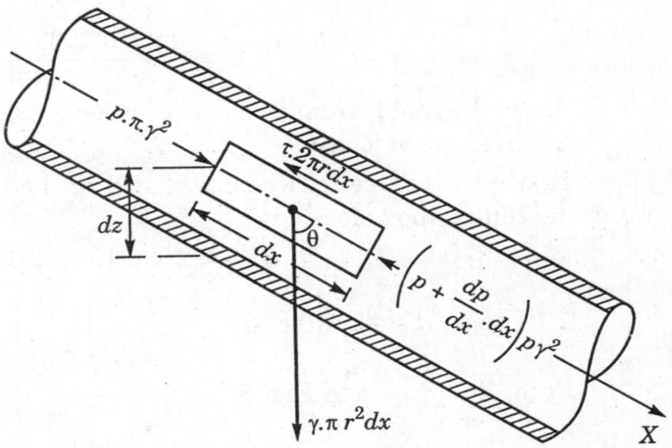

Fig. 8.7. Freebody diagram for steady laminar flow through an inclined tube.

While writing the Newton's law of motion along the direction of flow the component of the weight of fluid element $\gamma\pi r^2 dx \cos\theta$ will be an additional term. If z is measured vertically upwards, for an increase in distance x, z will decrease resulting in

$$\cos\theta = -\frac{dz}{dx}.$$

The expressions for the discharge, pressure drop, average velocity etc. may be obtained exactly in a similar manner as for a horizontal tube. The resulting equation will differ only in one respect, *i.e.* the pressure term p will be found to be replaced by $(p + \gamma z)$. The discharge equation will then be written as

$$Q = \frac{\pi R_0^4}{8\mu}\left\{-\frac{d(p+\gamma z)}{dx}\right\} \qquad ...(8.11)$$

The resisting force is overcome partly by the pressure force and partly by the gravity force. The requirement of energy for maintaining steady laminar flow is equal to the loss of energy given by Eq. (8.9).

Example 8.1. *A horizontal pipe of 5 cm diameter conveys oil of specific gravity 0.9 and dynamic viscosity 0.8 kg/ms. Measurements indicate a pressure drop of 20 kN/m² per metre of pipe length traversed. Make calculations for the :*

(i) flow rate of oil and centreline velocity.

(ii) wall shear stress and the frictional drag over 100 m length of pipe length.

(iii) power of pump required assuming an overall efficiency of 60%.

(iv) velocity and shear stress at 1 cm from the pipe surface. (*RGPV, June 2012*)

Solution. From Eq. (8.9), $\frac{\Delta p}{\gamma} = \frac{32\mu VL}{\rho g d^2}$ or $\frac{\Delta p}{L} = \frac{32\mu V}{d^2}$

Given : $\frac{\Delta p}{L} = 20$ kN/m²/m, $\mu = 0.8$ kg/ms $= 0.8 \times 9.81 = 7.85$ Ns/m²

Substituting the above data, $d = 5$ cm $= 5 \times 10^{-2}$ m.

$$V = \left(\frac{\Delta p}{L}\right)\frac{d^2}{32\,\mu} = \frac{20\times10^3}{1}\times\frac{(5\times10^{-2})^2}{32\times7.85} = \mathbf{0.1998\ m/s}$$

(*i*) Flow rate $Q = A.V = \frac{\pi}{4}d^2.V = \frac{\pi\times(5\times10^{-2})^2}{4}\times 0.1998 = \mathbf{3.92\times10^{-4}\ m^3/s}$

Since Eq. (8.9) is valid for laminar flow, let us confirm whether the flow is laminar or not, by determining the Reynolds number.

$$\mathbb{R} = \frac{\rho V d}{\mu}\ ;\ \rho = \text{sp. gr.} \times \text{density of water}$$

$$= 0.9 \times 1 = 0.9\ \text{kg/m}^3$$

$$\mathbb{R} = \frac{0.9 \times 0.1998 \times 5 \times 10^{-2}}{7.85} = 0.1145 \times 10^{-2}$$

As the Reynolds number is very low (< 2000), the flow is laminar. From Eqs. (8.3) and (8.8), $u_{max} = 2 \times$ average velocity V.

Hence the centreline velocity which in the laminar flow in a pipe happens to be the maximum, the centreline velocity is u_{max}.

$$u_{max} = 2 \times 0.1988 = \mathbf{0.4\ m/s}$$

(*ii*) The wall shear stress, $\tau_0 = \mu \left(\frac{du}{dy}\right)_{y=0}$, here y is measured from the boundary surface.

From Eq. (8.2), and Fig. 8.6

$$u = \frac{1}{4\mu}\left(\frac{\Delta p}{L}\right)(R_0^2 - r^2),\ y = R_0 - r,\ \text{at } y = 0,\ r = R_0$$

$$dy = -\,dr$$

$$\left(\frac{du}{dy}\right)_{y=0} = \left(-\frac{du}{dr}\right)_{r=R_0} = -\frac{1}{4\mu}\left(\frac{\Delta p}{L}\right)(-2R_0) = \frac{1}{4\mu}\left(\frac{\Delta p}{L}\right)2R_0$$

Wall shear stress, $\tau_0 = \mu \cdot \frac{1}{2\mu} \cdot \frac{\Delta p}{L} \cdot R_0 = \frac{\Delta p}{L} \cdot \frac{R_0}{2} = \frac{d}{4}\left(\frac{\Delta p}{L}\right)$

$$= (5 \times 10^{-2}/4)(20/1) = 25 \times 10^{-2}\ \text{kN/m}^2 = \mathbf{250\ N/m^2}$$

Friction drag over 100 m of pipe length

= Wall shear stress × circumferential area of pipe

$$= \tau_0 \times \pi d L = 250 \times \pi \times 5 \times 10^{-2} \times 100\ \text{N} = \mathbf{3928.57\ N}$$

(*iii*) Power of pump required at overall efficiency of 60%.

$$= \frac{Q(\Delta p/L) \times 100}{0.60} = \frac{3.92 \times 10^{-4} \times 20 \times 10^3 \times 100}{0.60}\ \text{Nm/s}$$

$$= 1306.7\ \text{Nm/s} = \mathbf{1306.7\ W}$$

(*iv*) Velocity and shear stress at 1 cm from pipe surface :

When $y = 1$ cm, $r = R_0 - y = 2.5 - 1 = 1.5$ cm

Velocity distribution by Eq. (8.2) : $u = \frac{1}{4\mu}\left(\frac{\Delta p}{L}\right)(R_0^2 - r^2)$

$$u_{r = 1.5\ \text{cm}} = \frac{1}{4 \times 7.86}\left(\frac{20 \times 10^{-3}}{1}\right)[(2.5 \times 10^{-2})^2 - (1.5 \times 10^{-2})^2] = 2.548\ \text{m/s}$$

Wall shear stress at $r = 1.5$ cm, $\tau_r = \mu \frac{du}{dr} = \mu \frac{\Delta p}{2\mu L} \cdot r = \frac{1}{2}\left(\frac{20 \times 10^3}{1}\right)(1.5 \times 10^{-2})$

$$= \mathbf{1500\ N/m^2}$$

Example 8.2. *A total of 12 litres per second of oil is pumped through two pipes in parallel, one 12 cm in diameter and the other 10 cm in diameter, both pipes being 1000 metres long. The specific gravity of the oil is 0.97 and the kinematic viscosity 9 cm² per second. Calculate the flowrate through each pipe and the power requirement of the pump.*

Solution. Since the pipes are in parallel, the total discharge Q is distributed in both the pipes. From the continuity principle

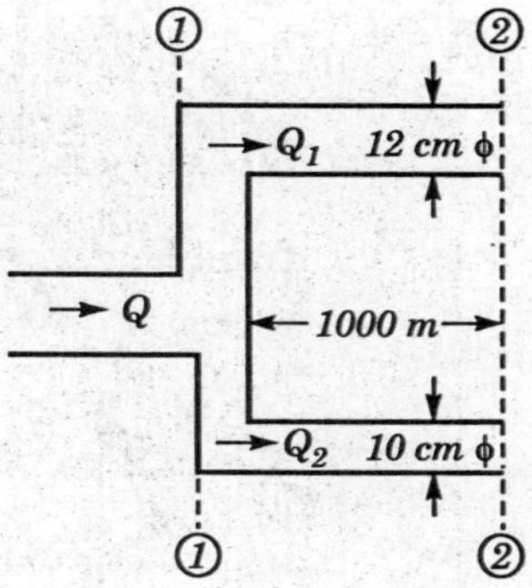

$$Q = Q_1 + Q_2$$
$$12 \times 1000 = Q_1 + Q_2 \qquad ...(1)$$

here Q_1 and Q_2 are in cm³/sec. Assuming laminar flow in both the pipes, the pressure difference between sections (1) and (2) is the same in both the pipes,

$$\Delta p = \frac{32\mu V_1 L}{d_1^2} = \frac{32\mu V_2 L}{d_2^2} \qquad ...(2)$$

From the continuity equation,

$$V_1 = \frac{4Q_1}{\pi d_1^2};\ V_2 = \frac{4Q_2}{\pi d_2^2}$$

From Eq. (2) $\quad \dfrac{V_1}{d_1^2} = \dfrac{V_2}{d_2^2}, \quad \text{or} \quad \dfrac{4Q_1}{\pi d_1^4} = \dfrac{4Q_2}{\pi d_2^4}$

From which, $\quad \dfrac{Q_1}{Q_2} = \left(\dfrac{d_1}{d_2}\right)^4 \quad \text{or} \quad Q_2 = Q_1\left(\dfrac{d_2}{d_1}\right)^4$

Substituting this in Eq. (1)

$$12 \times 1000 = Q_1 + Q_1(d_2/d_1)^4 = Q_1[1 + (10/12)^4] = 1.482\ Q_1$$

$\therefore \quad Q_1 = 8097$ cm³/s = **8.097 lit/sec**

$Q_2 = 12 - 8.097 =$ **3.903 lit/sec.**

Velocity of flow in 10 cm diameter pipe

$$V_2 = \frac{4Q_2}{\pi d_2^2} = \frac{4 \times 3.903 \times 1000}{\pi \times (10)^2} = 49.69 \text{ cm/s}$$

Velocity of flow in 12 cm diameter pipe,

$$V_1 = V_2(d_2/d_1)^2 = 49.69(10/12)^2 = 34.51 \text{ cm/s.}$$

Pressure drop in 1000 m length of 10 cm pipe carrying 3.903 lit/s,

$$\Delta p = \frac{32\mu V_2 L}{d_2^2}$$

Head loss $\quad h_L = \Delta p/\gamma = \dfrac{32\mu VL}{\gamma d_2^2} = \dfrac{32(\mu/p)V_2 L}{g d_2^2}$

$$= \frac{32 \times 9 \times 10^{-4} \times 0.4969 \times 1000}{9.81 \times (0.10)^2} = 145.88 \text{ m of oil}$$

Power lost in viscous friction $= \gamma Q h_L$

$= 9518 \times (12/1000) \times 145.88 = 16640.4$ W = **16.64 kW**

Assuming the overall efficiency of the pump as 65%, the power of the pump.

$$= \frac{16.64}{0.65} = \mathbf{25.6\ kW}$$

Reynolds number of flow in 10 cm diameter pipe

$$\mathbf{R}_2 = \frac{V_2 d_2}{\nu} = \frac{0.4969 \times 0.10}{9 \times 10^{-4}} = 55.2$$

Similarly, the Reynolds number of flow in 12 cm diameter pipe is $\mathbf{R}_1 = 46.01$.

Since the Reynold's numbers in both the pipes are less than 2000, the assumption of laminar flow in both the pipes is valid.

Example 8.3. *A pipe 10 cm in diameter and 1000 m long is used to pump oil of viscosity 8.5 poises and sp. gr. 0.92 at the rate of 1200 lit/min. The first 300 m of the pipe is laid along the ground* *sloping upwards at 10° to the horizontal and the remaining pipe is laid on the ground sloping upwards at 15° to the horizontal. State whether the flow is laminar or turbulent ? Determine the pressure required to be developed by the pump and the power of the driving motor if the pump efficiency is 60%. Assume suitable data for friction factor f, if required.*

Solution. $Q = 1200 \text{ lit/min} = \dfrac{1.2}{60} = 0.02 \text{ m}^3\text{/s}$

$1 \text{ poise} = 0.1 \text{ N-s/m}^2$

$\mu = 8.5 \text{ poise} = 0.85 \text{ N-s/m}^2$

$$\nu = \frac{\mu}{\rho} = \frac{\mu g}{\gamma} = \frac{0.850 \times 9.81}{0.92 \times 9810} = 9.27 \times 10^{-4} \text{ m}^2\text{/s}$$

Reynolds number of flow R

$$= \frac{Vd}{\nu} = \frac{4Q}{\pi d \nu} = \frac{4 \times (1.2/60)}{\pi \times 0.10 \times 9.27 \times 10^{-4}} = 274.7$$

Since the Reynolds number is less than 2000, the flow is laminar.

Velocity in the 10 cm diameter pipe

$$V = \frac{4Q}{\pi d^2} = \frac{4 \times 0.02}{\pi \times (0.10)^2} = 2.548 \text{ m/s}$$

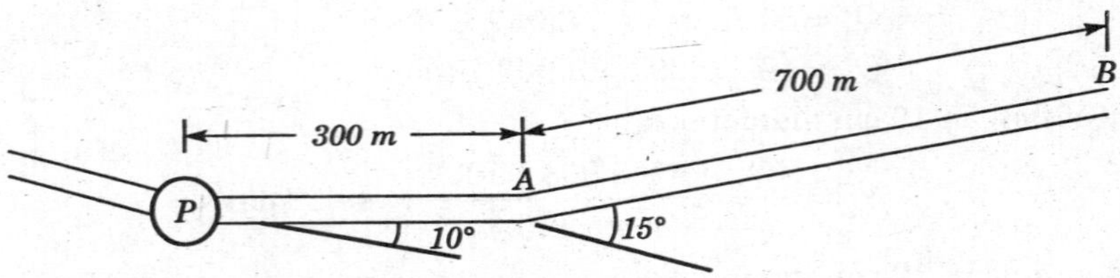

Height of the end point B of the pipeline above the pump centre P

$$= 300 \sin 10° + 700 \sin 15° = 233.26 \text{ m}$$

For laminar flow, the friction factor is given by

$$f = \frac{64}{R} = \frac{64}{274.7} = 0.23298$$

Head loss in friction in 1000 m length of pipeline

$$h_f = f \frac{L}{d} \frac{V^2}{2g} = 0.233 \times \frac{1000^2}{0.10} \times \frac{(2.548)^2}{2 \times 9.81} = 763.22 \text{ m of oil}$$

[Alternatively, the head loss can also be determined by computing the pressure drop $\Delta p = \dfrac{32\mu VL}{d^2}$]

Applying Bernoulli's equation between the pump outlet and the end of the pipeline (taking horizontal line through the pump centre as the datum) and assuming pressure in the pipeline at B as atmospheric, one obtains :

$$0 + \frac{p}{\gamma} + \frac{V^2}{2g} = 233.26 + 0 + \frac{V^2}{2g} + 763.22$$

in which p is the pressure just at the pump outlet.

∴ The pressure to be developed by the pump

$$= \frac{p}{\gamma} = \textbf{996.48 m of oil.}$$

or $p = 9025 \times 996.48 = 8993.4 \times 10^3 \text{ N/m}^2 = 8.99 \text{ kN/m}^2$

Power of the pump required to develop this pressure

$$= \gamma Q(p/\gamma) = Q \, . \, p = 0.02 \times 9025 \times 996.48 \text{ N-m/s (W)}$$

$$= 179.86 \times 10^3 \text{ W} = 179.86 \text{ kW}$$

Power of the driving motor $= \dfrac{179.86 \times 10^3}{0.60} = \mathbf{299.77\ kW.}$

Example 8.4. *A liquid of kinematic viscosity of 14 mm²/s is flowing down a tube inclined at 40° with the horizontal. When the Reynolds number for flow in the tube is 1500, the intensity of pressure remains constant at all the cross-sections of the tube. Determine the diameter of the tube and the discharge through it.*

Solution. Reynolds Number

$$\mathbf{R} = \frac{Vd}{\nu} = \frac{4Q}{\pi d \nu} \qquad ...(1)$$

For steady laminar flow through an inclined tube, [Refer Eq. 8.11 on Page 354]

$$Q = \frac{\pi R_0^4}{8\mu}\left[-\frac{d(p+\gamma z)}{dx}\right],$$ in which R_0 represents the pipe radius.

Since the intensity of pressure remains constant, $\dfrac{dp}{dx} = 0$

and

$$-\frac{dz}{dx} = \sin 40°$$

$$Q = \frac{\pi R_0^4}{8\mu}\left(-\gamma \frac{dz}{dx}\right) = \frac{g\pi R_0^4}{8\nu} \sin 40° \qquad ...(2)$$

Eliminating Q between (1) and (2)

$$\frac{\pi \nu d \mathbf{R}}{4} = \frac{g\pi R_0^4}{8\nu} \sin 40°$$

or

$$\frac{\pi \nu (2R_0) \mathbf{R}}{4} = \frac{g\pi R_0^4}{8\nu} \sin 40°$$

or,

$$r_0^3 = \frac{4\nu^2 \times 1500}{g \sin 40°} = \frac{4 \times (14 \times 10^{-6})^2 \times 1500}{9.81 \times 0.6428} = 0.1864 \times 10^{-6} \text{ m}^3$$

$$r_0 = 0.57 \times 10^{-2} \text{ m} = 5.7 \text{ mm}$$

Diameter of tube $= 2\,r_0 = 2 \times 5.7 = \mathbf{11.4\ mm.}$

Discharge through the tube

$$Q = \frac{g\pi r_0^4}{8\nu} \sin 40° = \frac{9.81\pi (0.0057)^4 \times 0.6428}{8 \times 14 \times 10^{-6}} = 179 \times 10^{-6} \text{ m}^3/\text{s}$$

$$= \mathbf{0.179\ lit/sec.}$$

Example 8.5. *Oil is pumped through a 7.5 cm diameter pipe, 1 km long. The dynamic viscosity of the oil is 0.1 N.s/m². A power input of 5.4 kW with an overall efficiency of 60% at the pump is noted. Determine the discharge through the pump. Also determine the Reynolds number of the flow if the specific gravity of the oil is 0.9.*

Solution. Pump efficiency (60%) $= \dfrac{\text{Power output}}{\text{Power input}} = \dfrac{\text{Power output}}{5.4 \text{ (kW)}}$

$\therefore$ Power output $= 5.4 \times 0.6 = 3.24$ kW ...(1)

But the power output of the pump in terms of Q and the head lost in friction H_L is $(\gamma Q H_L)$. Using Eq. (8.9) for the head lost

$$H_L = \frac{32\,\mu\,VL}{\gamma d^2} = \frac{32(0.1)\left(\dfrac{4Q}{\pi(7.5/100)^2}\right)(1\times 1000)}{(0.9\times 9810)(7.5/100)^2} = 1.4579 \times 10^4\,Q$$

Hence, the power output

$$= \gamma\,Q\,H_L = (0.9 \times 9810)Q\,(1.4579 \times 10^4\,Q) \qquad \text{...(2)}$$

Equating (1), and (2)

$$12871.8 \times 10^4\,Q^2 = 3.24 \times 10^3$$

$$Q = 9.03 \times 10^{-3}\ \text{m}^3/\text{s} = \textbf{9.03 litre/second.}$$

Example 8.6. *A pipe 100 mm in diameter is 1000 m long and used to pump oil of viscosity 3.5 poise and specific gravity 0.92 at the rate of 1200 lit/min. The first 300 m of the pipe is laid along the ground sloping up-wards at 10° to the horizontal and the remaining pipe is laid on the ground sloping upwards at 15° to the horizontal. State whether the flow is laminar or turbulent ? Determine the pressure required to be developed by the pump and the power of the driving motor if the pump efficiency is 60 percent. Assume suitable data for friction coefficient (factor), if required.*

(**Note.** *This problem is same as Ex. 8.3 except value of viscosity.*)

(*RGPV, May 2013*)

Solution. In order to determine the type of flow, Reynolds number of flow needs to be found.

Viscosity, 1 poise = 1 dyne-s/cm² = 0.1 N-s/m²

$$\mu = 3.5 \text{ poise} = 3.5 \times 0.1 = 0.35 \text{ N-s/m}^2$$

$$\gamma = \rho g = 0.92 \times 9810 \text{ N/m}^3, (0.92 \times 1000 \times 9.81)$$

$$= 9025.2 \text{ N/m}^3$$

$$\rho = \frac{\gamma}{g} = \frac{9025.2}{9.81} = 920 \text{ kg/m}^3$$

$$V = Q/A = \left(\frac{1200}{1000\times 60}\right)\times\frac{4}{\pi(0.1)^2} = 0.636 \text{ m/s}$$

$$\text{Reynolds No. } \mathbb{R} = \frac{\rho V d}{\mu} = \frac{920\times 0.636\times 0.1}{0.35} = 167.2$$

since $\mathbb{R} < 2000$, the **flow is laminar.**

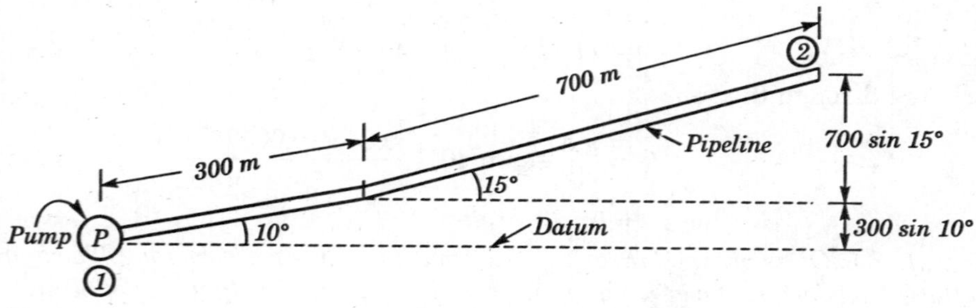

Head loss in friction, $h_f = f\,\dfrac{L}{d}\,\dfrac{V^2}{2g}$

For laminar flow, the friction is related to the Reynolds number by the relation :

$$f = \frac{64}{\mathbb{R}} \times \frac{64}{167.2} = 0.3828$$

and the head loss $= 0.3828 \times \dfrac{1000}{0.1} \times \dfrac{(0.636)^2}{2\times 9.81} = 78.91$ m of oil

Total head required to pump oil through 1000 m long pipeline, apply Bernoulli's equation between the pump delivery marked ① and pipeline end point marked ② in the sketch. Take the horizontal line passing through ① as the datum, we obtain :

$$\frac{p_1}{\gamma} + \frac{V_1^2}{2g} + Z_1 = \frac{p_2}{\gamma} + \frac{V_2^2}{2g} + Z_2 + h_f$$

since diameter of pipeline is constant, $V_1 = V_2$, $Z_1 = 0$, and $Z_2 = 300 \sin 10° + 700 \sin 15°$

Let H = The head required to be developed by the pump = $\dfrac{p_1 - p_2}{\gamma}$

or
$$H = \frac{p_1 - p_2}{\gamma} = 300 \sin 10° + 700 \sin 15° + 78.91$$
$$= 52.09 + 181.17 + 78.91 = 312.17 \text{ m of oil}$$

∴ Pressure required to be developed by the pump $= p_1 - p_2 = \gamma H = 9025.2 \times 312.17$
$$= 2817396.7 \text{ N/m}^2$$

Power required to develop 312.17 m of oil head by the pump, P_p :

$$P_p = \gamma QH = 9025.2 \times \left(\frac{1200}{1000 \times 60}\right) \times 312.17 \text{ N-m/s}$$
$$= 56347.93 \text{ N-m/s (or Watt)} = 56347.93 \text{ W}$$

Given the pump efficiency as 60%, the power of the driving motor, P_m :

$$P_m = \frac{P_p}{\eta} = \frac{56347.93}{0.6} = 93913.22 \text{ W} = \mathbf{93.913 \text{ kW}}$$

8.5. Flow Though an Annulus

The flow through the space between two concentric cylinders can be analysed and expressions may be obtained for velocity distribution, shear distribution and the discharge. Let us consider a horizontal annulus of outer radius R_1 and inner radius R_2 through which steady laminar flow of an incompressible fluid is taking place. The forces acting on an annular element of radius r, thickness dr and length dx are shown in Fig. 8.8.

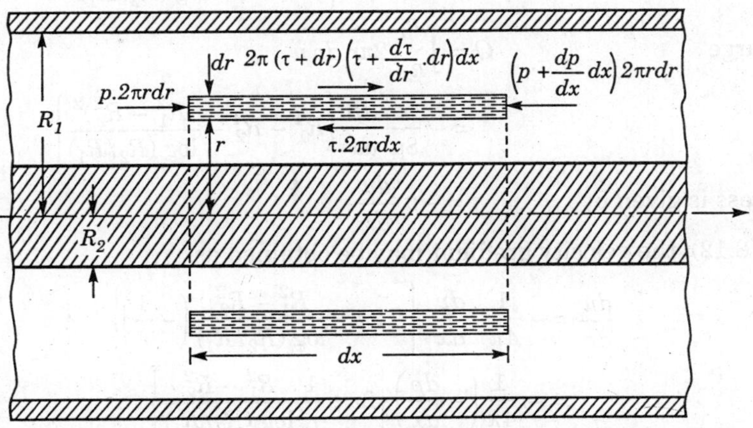

Fig. 8.8. Forces on an annular element of fluid in steady laminar flow through an annulus.

For steady uniform flow, the equation of motion may be written as

$$p.2\pi\, r\, dr - \left(p + \frac{dp}{dx}.\, dx\right) 2\pi\, r\, dr + 2\pi(r + dr)\left(\tau + \frac{d\tau}{dr}.\, dr\right) dx - \tau.2\pi\, r.\, dx = 0.$$

dividing by the volume of the element, $2\pi.rdr.dx$ and dropping the infinitesimal **term, we obtain**

$$\frac{dp}{dx} + \frac{d\tau}{dr} + \frac{\tau}{r} = 0$$

Since p is dependent on x and τ on r, the above equation may be expressed as

$$\frac{dp}{dx} + \frac{1}{r}\frac{d(\tau r)}{dr} = 0$$

Integrating w.r.t. r, we get

$$\frac{dp}{dx}\int r\,dr + \int \frac{d}{dr}(\tau\,.\,r)\,dr = A$$

$$\frac{r^2}{2}\frac{dp}{dx} + \tau\,.\,r = A$$

where A = constant of integration.

Substituting for τ from the Newton's law of viscosity

$$\tau = -\mu\frac{du}{dr}$$

the above equation may be written as :

$$\frac{dp}{dx}\frac{r^2}{2} - \mu\frac{du}{dr}r = A$$

or
$$\frac{dp}{dx}\cdot\frac{r}{2} - \mu\frac{du}{dr} = \frac{A}{r}$$

Integrating w.r.t. r, $\frac{dp}{dx}\cdot\frac{r^2}{4} - \mu u = A\log r + B$

where B = constant of integration.

The constants A and B can be evaluated from the known boundary conditions, *i.e.* at $r = R_1$, $u = 0$ and also at $r = R_2$, $u = 0$. After substituting these conditions and solving for A and B we obtain the velocity distribution as

$$u = -\frac{1}{4\mu}\frac{dp}{dx}\left\{R_1^2 + r^2 + \frac{R_1^2 - R_2^2}{\log(R_2/R_1)}\log\frac{R_1}{r}\right\} \quad ...(8.12)$$

The discharge
$$Q = \int_{R_2}^{R_1} 2\pi r\,dr\,u$$

$$= -\frac{\pi}{8\mu}\frac{dp}{dx}\left\{R_1^4 - R_2^4 - \frac{(R_1^2 - R_2^2)^2}{\log(R_2/R_1)}\right\} \quad ...(8.13)$$

and the shear stress is given by $\tau = -\mu\frac{du}{dr}$

From Eq. (8.12), the velocity gradient may be obtained as

$$\frac{du}{dr} = -\frac{1}{4\pi}\cdot\frac{dp}{dx}\left\{-2r + \frac{R_1^2 - R_2^2}{\log(R_2/R_1)}\left(-\frac{1}{r}\right)\right\}$$

$$= -\frac{1}{4\pi}\left(-\frac{dp}{dx}\right)\left\{2r + \frac{1}{r}\frac{R_1^2 - R_2^2}{\log(R_2/R_1)}\right\}$$

The shear stress distribution is given by

$$\tau = -\mu\frac{du}{dr}$$

$$= \frac{1}{4}\left(-\frac{dp}{dx}\right)\left\{2r + \frac{1}{r}\frac{R_1^2 - R_2^2}{\log R_2/R_1}\right\}. \quad ...(8.14)$$

Example 8.7. *A uniform circular tube of bore radius 'a' has a fixed coaxial cylindrical solid core of radius 'b'. An incompressible viscous fluid flows through the annular passage under a pressure gradient (– dp/dx). Determine the radius at which the shear stress in the stream is zero given that the flow is laminar under steady state condition.*

Solution. From Eq. (8.14), the shear stress distribution is given by

$$\tau = \frac{1}{4}\left(-\frac{dp}{dx}\right)\left[2r + \frac{1}{r}\frac{(a^2 - b^2)}{\log b/a}\right]$$

For zero shear stress, $\tau = 0$, and we obtain

$$2r + \frac{1}{r}\frac{(a^2 - b^2)}{\log b/a} = 0 \; ; \qquad 2r - \frac{1}{r}\frac{(a^2 - b^2)}{\log a/b} = 0$$

or

$$2r^2 = \frac{(a^2 - b^2)}{\log a/b}$$

$$r = \sqrt{\frac{1}{2}\frac{(a^2 - b^2)}{\log a/b}}.$$

Example 8.8. *A horizontal circular tube of radius "a" has a fixed coaxial cylindrical core of radius "b". τ_a and τ_b are the shear stresses along the tube and core surfaces when a viscous liquid is flowing through the annulus. The flow is laminar and the rate of variation of pressure along the length of the passage is – dp/dl. Show that*

$$a\tau_a - b\tau_b = \frac{1}{2}(a^2 - b^2)\frac{dp}{dl}.$$

Solution. Refer article 8.5 and Fig. 8.8. From the equation of motion, one may write the following equation :

$$\frac{dp}{dx} + \frac{d\tau}{dr} + \frac{\tau}{r} = 0$$

since p is dependent on x and τ on r, the above equation can be written as

$$\frac{dp}{dx} + \frac{1}{r}\frac{d}{dr}(\tau \,.\, r) = 0$$

integration of which w.r.t. r yields

$$\frac{r^2}{2}\frac{dp}{dx} + \tau \,.\, r = A$$

where A is a constant of integration.

At $\qquad r = a, \tau = \tau_a$ and at $r = b, \tau = \tau_b$.

Also $\qquad \dfrac{dp}{dx} = -\dfrac{dp}{dl}$

Substituting the above values, one obtains

$$\frac{a^2}{2}\left(-\frac{dp}{dl}\right) + a \,.\, \tau_a = A$$

and

$$\frac{b^2}{2}\left(-\frac{dp}{dl}\right) + b \,.\, \tau_b = A$$

Eliminating A,

$$\frac{1}{2}(a^2 - b^2)\left(-\frac{dp}{dl}\right) + a \,.\, \tau_a - b \,.\, \tau_b = 0$$

or

$$a \,.\, \tau_a - b \,.\, \tau_b = \frac{1}{2}(a^2 - b^2)\frac{dp}{dl}.$$

8.6. Flow Between Parallel Plates

Let us consider two parallel plates—bottom plate being stationary while the top one moving with a constant velocity U. The forces acting on an element of fluid of dimensions dx (length), dy (thickness) and unit width are as shown indicated in Fig. 8.9.

Applying the Newton's second law, the equation of motion for steady incompressible flow is

$$p \cdot dy - \left(p + \frac{dp}{dx} \cdot dx\right) dy + \left(\tau + \frac{d\tau}{dy} \cdot dy\right) dx - \tau dx = 0.$$

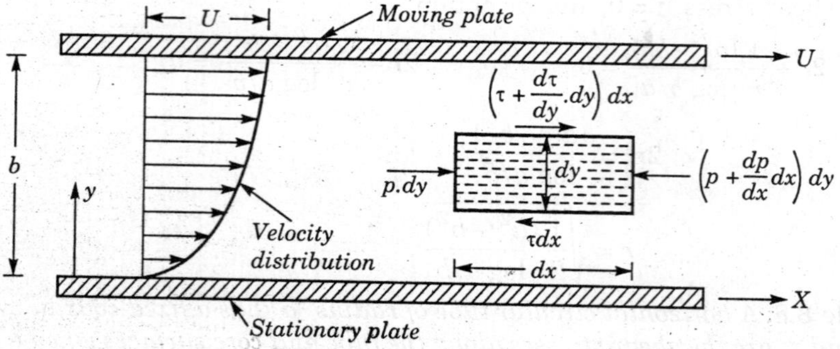

Fig. 8.9. Laminar flow between horizontal parallel plates.

Simplifying and dividing by the volume of the element $dx\ dy$, one obtains

$$\frac{dp}{dx} = \frac{d\tau}{dy} \quad \text{...(8.15)}$$

Eq. (8.15) shows the interdependence of shear and pressure gradient, and is applicable for laminar as well as turbulent flow. Accordingly the pressure gradient in the direction of flow is equal to the shear gradient across the flow.

In case of laminar flow, the shear stress is given by the Newton's law of viscosity

$$\tau = \mu du/dy$$

Substituting for τ in Eq. (8.15), we obtain

$$\mu \frac{d^2u}{dy^2} = \frac{dp}{dx}$$

Integrating twice,

$$u = \frac{1}{2\mu} \frac{dp}{dx} y^2 + Ay + B \quad \text{...(8.16)}$$

where A and B are the constants of integration to be evaluated from the known boundary conditions. In the present case the boundary conditions are

(*i*) at $y = 0, u = 0$

and (*ii*) at $y = b, u = U$

Substituting these conditions in Eq. (8.16), the constants are obtained as :

$$A = \frac{U}{b} - \frac{1}{2\mu} \frac{dp}{dx} b, \quad \text{and} \quad B = 0$$

Equation (8.16), for velocity distribution may be written as

$$u = \frac{U}{b} y - \frac{1}{2\mu} \frac{dp}{dx} (by - y^2) \quad \text{...(8.17)}$$

Eq. (8.17) represents the generalised Couette flow.

The discharge per unit width of plates may be obtained as

$$q = \int_0^b u\,dy = \int_0^b \left\{ \frac{U}{b} y - \frac{1}{2\mu} \frac{dp}{dx} (by - y^2) \right\} dy$$

$$= U \frac{b}{2} - \frac{b^3}{12\mu} \frac{dp}{dx} \qquad \text{...(8.18)}$$

The shear distribution may be computed using the Newton's law of viscosity

$$\tau = \mu \frac{du}{dy} = \mu \left\{ \frac{U}{b} y - \frac{1}{2\mu} \frac{dp}{dx} (b - 2y) \right\}$$

$$= \mu \frac{U}{2} - \frac{1}{2} \frac{dp}{dx} (b - 2y) \qquad \text{...(8.19)}$$

The type of flow discussed above, *i.e.* the flow between two parallel surfaces—one stationary and the other moving is known as the *generalised Couette flow*.

Both Plates at Rest :

If the upper plate is also kept stationary, then $U = 0$, and the equations for velocity, discharge q and the shear stress can be obtained from similar equations for generalised couette flow by putting $U = 0$. For flow between two stationary parallel boundaries, shown in Fig. 8.10, we have

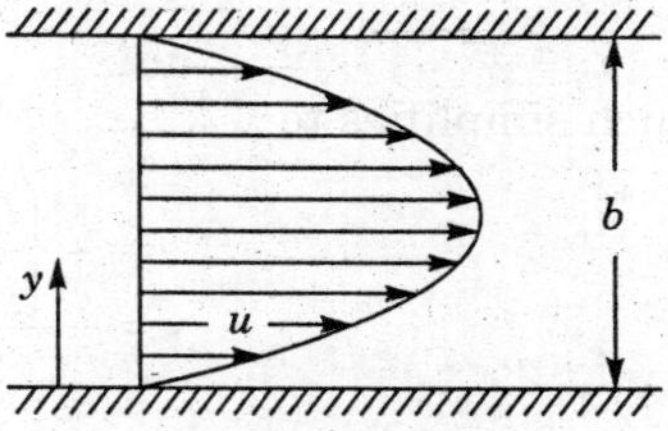

Fig. 8.10. Flow between stationary plates.

$$u = -\frac{1}{2\mu} \frac{dp}{dx} (by - y^2) \qquad \text{...(8.20)}$$

Eq. (8.20) represents the plane Poiseuille flow,

$$q = -\frac{b^3}{12\mu} \frac{dp}{dx} \qquad \text{...(8.21)}$$

$$\tau = -\frac{1}{2} \frac{dp}{dx} (b - 2y) \qquad \text{...(8.22)}$$

For zero pressure gradient, Eq. (8.17) reduces to the case of plane Couette flow, see article 6.7.1.

8.6.1. Both Plates Moving in Opposite Directions

The velocity distribution for flow between parallel plates as obtained in equation (8.16) is

$$u = \frac{1}{2\mu} \frac{dp}{dx} y^2 + Ay + B \qquad \text{...(8.16)}$$

The constants of integration A and B are to be evaluated for the boundary conditions of the problem, that is

(*i*) at $y = 0$, $u = -V$ (*ii*) at $y = b$, $u = U$.

Substituting these boundary conditions in Eq. (8.16), we obtain

$$-V = \frac{1}{2\mu} \frac{dp}{dx} \times 0 + A \times 0 + B,$$

from which,

$$B = -V$$

and

$$U = \frac{1}{2\mu} \frac{dp}{dx} b + Ab + (-V)$$

which yields,

$$A = \frac{U}{b} + \frac{V}{b} - \frac{1}{2\mu} \frac{dp}{dx} b$$

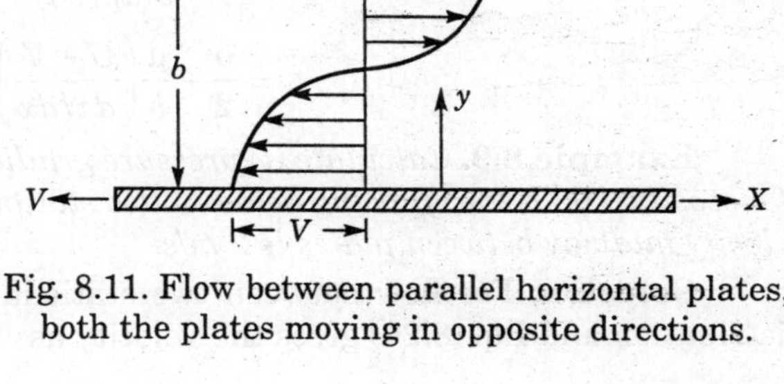

Fig. 8.11. Flow between parallel horizontal plates, both the plates moving in opposite directions.

and the resulting velocity distribution equation is

$$u = (U + V)\,\frac{y}{b} - \frac{1}{2\mu}\frac{dp}{dx}\,(by - y^2) - V \qquad \text{...(8.23)}$$

The distance y at which the velocity u is zero may by obtained as follows :

$$(U + V)\,\frac{y}{b} - \frac{1}{2\mu}\frac{dp}{dx}\,(by - y^2) - V = 0$$

or

$$\frac{1}{2\mu}\frac{dp}{dx}\,y^2 + \left(\frac{U+V}{b} - \frac{1}{2\mu}\frac{dp}{dx}\,b\right) y - V = 0$$

Solving this quadratic equation

$$y = \frac{-\left\{\frac{U+V}{b} - \frac{1}{2\mu}\frac{dp}{dx}\,b\right\} \pm \sqrt{\left(\frac{U+V}{b} - \frac{1}{2\mu}\frac{dp}{dx}\,d\right)^2 + 4\,.\,\frac{1}{2\mu}\frac{dp}{dx}\,V}}{2\,.\,\frac{1}{2\mu}\frac{dp}{dx}}$$

which simplifies to

$$y = \frac{-\left\{\frac{U+V}{b} - \frac{1}{2\mu}\frac{dp}{dx}\,b\right\} \pm \sqrt{\left(\frac{V+U}{b}\right)^2 + \frac{1}{4\mu^2}\left(\frac{dp}{dx}\right)^2 b^2 - \frac{(U-V)}{\mu}\frac{dp}{dx}}}{\frac{1}{\mu}\frac{dp}{dx}}$$

This equation gives two values of y of which only one, which is positive and less than b, will be physically meaningful and the other one may, therefore, be discarded.

The discharge per unit width of plates may be obtained as under

$$q = \int_0^b u\,dy = \int_0^b \left\{(U+V)\,\frac{y}{b} - \frac{1}{2\mu}\frac{dp}{dx}\,(by - y^2) - V\right\} dy$$

$$= (U - V)\,\frac{b}{2} - \frac{1}{12\mu}\frac{dp}{dx}\,b^3 \qquad \text{...(8.24)}$$

The shear stress distribution may be expressed using Newton's law of viscosity as

$$\tau = \mu\,\frac{du}{dy} = \mu\left\{\frac{U+V}{b} - \frac{1}{2}\frac{dp}{dx}\,(b - 2y)\right\} = \mu\,\frac{U+V}{b} - \frac{1}{2}\frac{dp}{dx}\,(b - 2y)$$

$$= (U + V)\,\frac{\mu}{b} - \frac{dp}{dx}\left(\frac{b}{2} - y\right) \qquad \text{...(8.25)}$$

The shear stress will be zero at a distance y, which may be obtained by setting Eq. (8.25) to zero.

$$0 = (U + V)\,\frac{\mu}{b} - \frac{dp}{dx}\,(b/2 - y)$$

$$\therefore \qquad y = \frac{b}{2} - \frac{(U+V)\mu}{b(dp/dx)}$$

$$= \frac{b}{2} - \frac{\mu}{b}\left(\frac{U+V}{dp/dx}\right) \qquad \text{...(8.26)}$$

Example 8.9. *Calculate (i) pressure gradient along flow, and (ii) average velocity for an oil of viscosity 0.02 Ns/m² flowing between two stationary plates 1 m wide maintained 1 cm apart. The velocity midway between plates is 2 m/s.* *(RGPV, Dec. 2005)*

Solution. For flow between two stationary parallel plates, Fig. 8.10 depicts the velocity distribution and Eq. (8.20) gives the velocity as

$$u = -\frac{1}{2\mu}\frac{dp}{dx}(by - y^2)$$

The maximum velocity, u_{max} occurs when $\frac{du}{dy} = 0$, for which we obtain:

$$\frac{du}{dy} = -\frac{1}{2\mu}\frac{dp}{dx}(b - 2y) = 0, \text{ resulting in } y = \frac{b}{2}$$

here 'b' is the distance between the stationary parallel plates. From this condition, it is clear that the maximum velocity occurs at the centre of the gap *i.e.* midway between the plates.

$$u_{max} = -\frac{1}{2\mu}\frac{dp}{dx}\left[b \cdot \frac{b}{2} - \left(\frac{b}{2}\right)^2\right] = -\frac{1}{2\mu}\frac{dp}{dx} \cdot \frac{b^2}{4}$$

$$= -\frac{b^2}{8\mu}\frac{dp}{dx}$$

Data given : $\mu = 0.02$ Ns/m², $b = 1$ cm, $u_{max} = 2$ m/s

$\therefore$ Pressure gradient along the flow, $\frac{dp}{dx} = -\frac{8\mu}{b^2}u_{max}$

or

$$\left(-\frac{dp}{dx}\right) = \frac{8 \times 0.02 \times 2}{(1 \times 10^{-2})^2} = \mathbf{32 \times 10^2\ N/m^2}$$

Discharge per unit width,

$$q = \int_0^b u\,dy = -\frac{1}{2\mu}\frac{dp}{dx}\left[b\frac{y^2}{2} - \frac{y^3}{3}\right]_0^b = -\frac{b^3}{12\mu}\frac{dp}{dx}$$

average velocity,

$$V = \frac{q}{b} = \frac{b^2}{12\mu}\left(-\frac{dp}{dx}\right) = \frac{(1 \times 10^{-2})^2}{12 \times 0.02}(32 \times 10^2) = \frac{32}{24} = \frac{4}{3} = \mathbf{1.33\ m/s}$$

8.7. Stokes' Law

The uniform motion of a sphere through a large expanse of viscous fluid at very small Reynolds numbers was first studied by Sir George G. Stokes (1819—1903). He based his solution on the assumption (*i*) that the inertia forces on the fluid particles are small and may be neglected in comparison to the viscous forces. This condition boils down to the motion of sphere at low Reynolds number. His other assumptions were, (*ii*) that the walls of the vessel containing the fluid are not so near as to affect the flow around the sphere. In other words, the fluid is infinite in extent and is also quiescent, (*iii*) that there is no slip between the fluid and the sphere, (*iv*) that the sphere is rigid, and (*v*) only a single sphere is considered.

The force resisting the motion of sphere (*i.e.* the drag on the sphere) was then obtained by Stokes as

$$F_D = 3\pi\mu\,Vd \qquad \text{...(8.27)}$$

where V = velocity of the sphere relative to the undisturbed fluid,
μ = dynamic viscosity of fluid, and
d = diameter of sphere.

It has been experimentally found that the Stokes' law expressed by Eq. (8.27) is valid only if the Reynolds number is less than 0.1. It is used to determine the fall of relatively small bodies through fluids of relatively high viscosity such as the fall of dust particles, mist droplets in the atmosphere and the settlement of silt in reservoirs.

A small solid particle falling through a fluid under its own weight will accelerate until the net downward force on it is zero. In other words, when the submerged weight of the particle is equal to force given by Eq. (8.27), it will have reached the steady state of its motion. No further acceleration is then possible and the particle is said to have reached its *terminal velocity*.

Thus, the submerged weight (weight of particle minus the buoyant force) = Resisting force

$$\frac{\pi d^3}{6}(\gamma_s - \gamma) = 3\pi\mu\, Vd \qquad \text{...(8.28)}$$

and the terminal velocity,
$$V = \frac{d^2}{18\mu}(\gamma_s - \gamma) \qquad \text{...(8.29)}$$

where γ_s and γ are the specific weights of the solid and the fluid respectively. Eq. (8.29) expresses the *Stokes' law*. It's validity lies within the Reynolds number $\rho Vd/\mu \le 0.1$.

8.8. Measurement of Viscosity

The devices which are used for viscosity measurements are based on the principle of existence of fully established laminar flow. It is, therefore, very essential that every care and precaution is taken to ensure that the laminar flow is fully developed, so that the observations and the corresponding results obtained from laminar flow theory are accurate. The devices used for the measurement of viscosity are known as "Viscosimeters" or "Viscometer", and may be classified as :

(1) Rotating Cylinder method (Rotary Viscometers)

(2) Falling Sphere method

(3) Capillary Tube method

(4) Industrial Viscometers.

8.8.1. Rotating Cylinder Method

As shown schematically in Fig 8.12 a rotating cylinder viscometer consists of two concentric cylinders separated by a small gap filled with a liquid whose viscosity is to be determined. The inner cylinder is suspended by a torsion wire or spring, the torque-deflection ratio of which is known from calibration.

The outer cylinder may be rotated at any desired speed. The torque T exerted on the inner cylinder by the deformed liquid may be shown to be equal to

$$T = \frac{2\pi R^2 \mu h V}{\Delta r} + \frac{\pi R^3 \mu V}{\Delta h} \qquad \text{...(8.30)}$$

the first term is due to viscous shear in the liquid contained between the cylinder walls, the second term is due to viscous shear between the bottom surfaces of the two cylinders.

The torque T can be obtained by observing the deflection of the torsion wire and reading the corresponding torque from calibration chart of the wire. The velocity $V = R\omega$ is known for angular velocity ω. All other quantities in the above equation such as R, h, Δr and Δh are easily measurable, and hence we can compute the viscosity μ of the liquid.

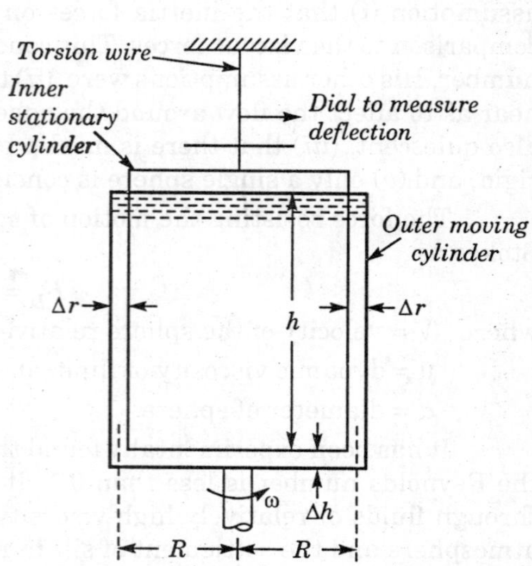

Fig. 8.12. Rotating cylinder viscometer.

8.8.2. Falling-Sphere Method

This method is based on the Stokes' law, according to which the drag (resisting force) on a small sphere moving under laminar conditions at constant velocity is given by Eq. (8.27)

$$F_D = 3\pi\mu Vd \qquad \text{...(8.27)}$$

where V = constant velocity of sphere, and

d = diameter of sphere.

When the sphere attains a constant velocity V, the drag F_D is the difference between the weight of sphere and the buoyant force acting on it. It is to be noted that the drag force, which always acts against the direction of motion, and the buoyant force both act vertically upward whereas the weight of the sphere acts vertically downward. The constant velocity V is easy to obtain by measuring time t required by the sphere to fall a certain length L,

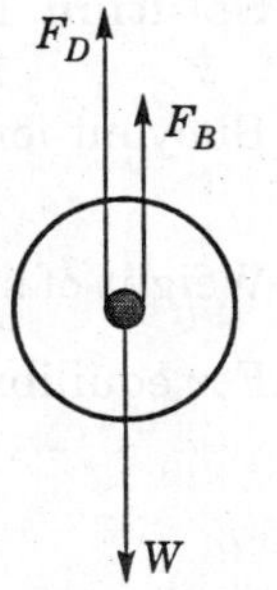

Fig. 8.13 (*a*) Forces acting on a falling sphere.

$$V = \frac{L}{t}$$

$$F_D = W - F_B$$

and $$F_D = \frac{\pi d^3}{6}\gamma_s - \frac{\pi d^3}{6}\gamma = \frac{\pi d^3}{6}(\gamma_s - \gamma)$$

where γ_s = specific weight of sphere,

and γ = specific weight of the liquid.

From Eq. (8.27),

$$\frac{\pi d^3}{6}(\gamma_s - \gamma) = 3\pi\mu V d$$

or, $$\mu = \frac{d^2(\gamma_s - \gamma)}{18V} \quad ...(8.31)$$

F_D = drag force.

F_B = buoyant force.

W = weight of sphere.

The experiment consists of gently placing a small sphere on the surface of the liquid (whose viscosity is to be determined) which has been filled in a cylindrical tank of diameter D, Fig. 8.13 (*b*). After the sphere has attained a constant velocity V, the time to fall a distance L can be observed. With the known diameter of sphere d, specific weights γ_s and γ, the viscosity μ of the liquid can be computed using Eq. (8.31).

It may be recalled here that the Stokes' law is derived for a liquid of infinite extent. But in this case, the extent of fluid is limited and the wall effect due to the proximity of the tank wall is considerably large. Wall effect depends on the ratio of the sphere diameter to the tank diameter and is expressed as a correction to the velocity by the following equation :

$$\frac{V}{V_0} = 1 + \left(\frac{9d}{4D}\right) + \left(\frac{9d}{4D}\right)^2$$

in which V_0 is the observed velocity of fall in the tank $\left(V_0 = \frac{L}{t}\right)$.

Applying this correction, the viscosity may be determined accurately by the following equation :

$$\mu = \frac{d^2(\gamma_s - \gamma)}{18V_0\left\{1 + \frac{9d}{4D} + \left(\frac{9d}{4D}\right)^2\right\}} \quad ...(8.32)$$

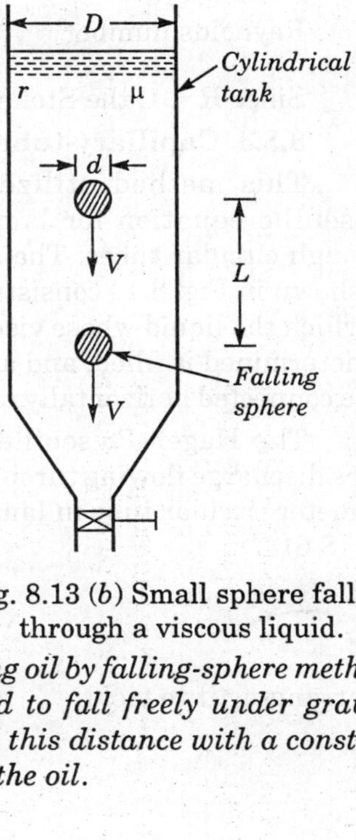

Fig. 8.13 (*b*) Small sphere falling through a viscous liquid.

Example 8.10. *In order to determine the viscosity of a lubricating oil by falling-sphere method, a steel spherical ball of sp. gr. 8.0 and diameter 10 mm is allowed to fall freely under gravity through a distance of 90 cm. The time taken by the steel ball to travel this distance with a constant velocity is 18 sec. The sp. gr. of oil is 0.90. Determine the viscosity of the oil.*

Solution. Drag force $F_D = 3\,\pi\mu Vd$

Buoyant force $F_B = \frac{\pi d^3}{6}\gamma = \frac{\pi d^3}{6} \times 0.9 \times 9810$ (N)

Weight of steel ball $W = \frac{\pi d^3}{6}\gamma_s = \frac{\pi d^3}{6} \times 8 \times 9810$ (N)

For equilibrium, one can write

$$F_D + F_B = W$$

or $$F_D = W - F_B$$

Making substituting, $3\pi\mu Vd = \frac{\pi d^3}{6} \times (78480 - 8829) = \frac{\pi d^3}{6} \times 69651$

from which, $$\mu = \frac{69651 d^2}{18\,V} = \frac{69651 \times (10/1000)^2}{18 \times (0.9/18)} = \mathbf{7.74\ Ns/m^2.}$$

Checking the Reynolds number, $\mathbf{R} = \rho\frac{Vd}{\mu} = \left(\frac{8829}{9.81} \times \frac{0.9}{18} \times \frac{1}{100}\right)\Big/7.74 = 0.55 < 1.$

Example 8.11. *An air bubble of 1.2 mm diameter rises in an oil medium with a uniform velocity of 9 mm/sec. Estimate the viscosity of oil.*

Solution. Neglecting the weight of the bubble, the buoyant force which causes the upward motion of the bubble = Drag force on the bubble

or, $$F_B = F_D$$

$$\frac{\pi d^3}{6}\gamma = 3\pi\mu Vd$$

$$\mu = \frac{\gamma d^2}{18V} \quad \text{or} \quad \nu = \frac{\mu}{\rho} = \frac{gd^2}{18V}$$

or $$\nu = \frac{9.81 \times (0.0012)^2}{18 \times 0.009} = \mathbf{8.72 \times 10^{-5}\ m^2/s}$$

Reynolds number $\mathbf{R} = Vd/\nu = \frac{0.009 \times 0.0012}{8.72 \times 10^{-5}} = 0.124$

Since $\mathbf{R} < 1$, the Stokes' law is valid.

8.8.3. Capillary-tube Method

This method utilizes Hagen-Poiseuille equation for laminar flow through circular tubes. The apparatus as shown in Fig. 8.14 consists of a tank in which the liquid whose viscosity is to be determined is filled, and a capillary-tube connected horizontally to the tank.

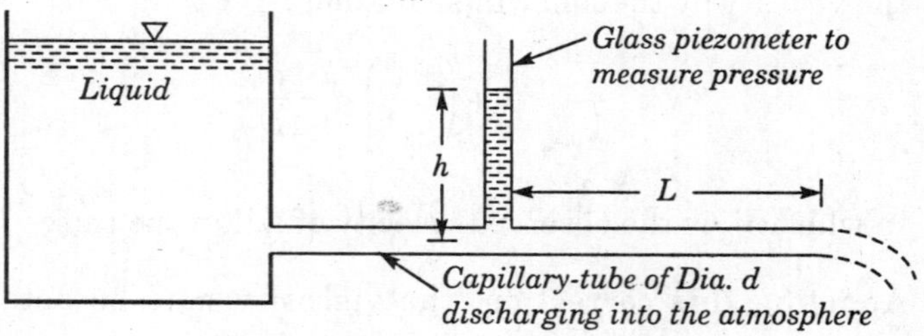

Fig. 8.14. Capillary-tube Viscometer.

The Hagen-Poiseuille equation gives discharge flowing through a small diameter circular tube in laminar flow, Eq. (8.6)

$$Q = \frac{\pi d^4 (p_1 - p_2)}{128\mu L}$$

re-arranging it to yield, $$\mu = \frac{\pi d^4 (p_1 - p_2)}{128 QL} \qquad \text{...(8.33)}$$

The discharge Q can be measured by collecting liquid in a measuring tank for a known interval of time. The pressure difference $p_1 - p_2$, is also known in terms of h, since $p_2 = 0$.

$$p_1 - p_2 = \gamma h$$

In order that the Hagen-Poiseuille equation be applicable it is necessary to measure the pressure difference between the two points both of which must lie in a region of fully developed flow (see art. 11.1). Hence the piezometer must be located at a reasonably large distance from the inlet. It is difficult to install manometer in a fine capillary tube capable of providing laminar flow. Further, it may be difficult to measure accurately the diameter of such a tube. A small error in measuring the diameter will cause an appreciable error in the viscosity since μ is proportional to fourth power of diameter. For these reasons this type of viscometer is not widely used.

Example 8.12. *To determine the viscosity of an oil (sp. gr. 0.93) in a laboratory set up, two pressure gauges placed 6 m apart in a 50 mm diameter horizontal pipe showed a pressure difference of 17 kN/m². During the test run of 3 minutes, oil collected in a tank was found to be 5150 N. Determine the viscosity of the oil.*

Solution. Assuming the flow to be laminar, the drop in pressure is given by,

$$\Delta p = \frac{32\,\mu\,VL}{d^2}$$

$$Q = \frac{5150}{3 \times 60} = 28.6 \text{ N/sec} = \frac{28.6}{9810 \times 0.93} = 0.00308 \text{ m}^3\text{/sec}$$

Velocity in the pipe, $V = \dfrac{4Q}{\pi d^2} = \dfrac{4 \times 0.00308}{\pi(50 \times 10^{-3})^2} = 1.57 \text{ m/s}$

The dynamic viscosity, $\mu = \dfrac{\Delta p \,.\, d^2}{32\,VL} = \dfrac{17 \times 10^3 \times (50 \times 10^{-3})^2}{32 \times 1.57 \times 6} = \mathbf{0.141\ N\text{-}s/m^2}$.

8.8.4. Industrial Viscometers

These viscometers require the measure-ment of time taken by a certain quantity of the liquid to flow through a short capillary tube. In many viscometers this capillary tube is so short that fully developed laminar flow is rarely achieved in the capillary tube. Thus Hagen-Poiseuille's law does not strictly holds good. The viscometer thus requires calibration with a liquid of known viscosity.

Viscometers of different designs but based on the same principle are in use in different countries. America (U.S.A.) favours the use of Saybolt viscometer while in the United Kingdom Redwood viscometer is more commonly used. The following description relates to Saybolt viscometer.

Saybolt Viscometer. The instrument shown in Fig. 8.15 consists of a reservoir in the bottom of which is a short capillary tube. These are surrounded by a constant temperature oil bath. The viscosity is determined by measuring the time required for 60 cu. cm. of liquid at a known temperature to flow out of the reservoir through the tube. The initial level of liquid in the reservoir is previously adjusted to a standard height. From the time measurement (Saybolt seconds), the kinematic viscosity of the liquid can be determined through the use of empirical formula or the calibration chart.

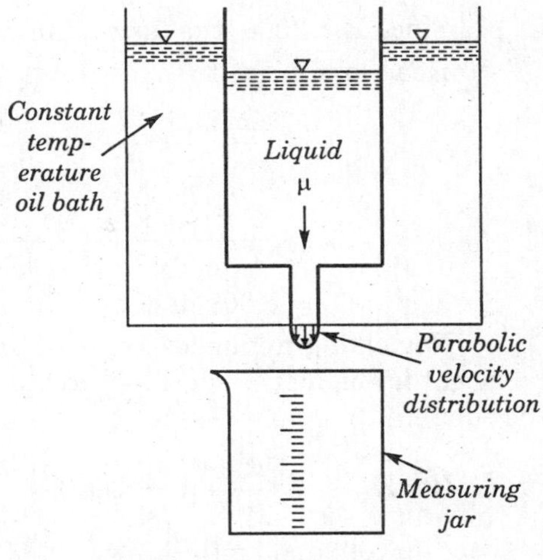

Fig. 8.15. Saybolt viscometer.

8.9. Application to Hydrodynamic Lubrication

Whenever two metallic surfaces are in contact with each other and are subjected to relative linear or rotational motion, there is a high degree of friction which results in loss of energy, mechanical wearing and generation of heat. These effects are greatly minimised by separating the two surfaces by a thin film of viscous fluid called lubricant. Laminar flow finds an important application to the field of lubrication of different types of bearings. In many cases high velocities may be involved, but as the thickness of the film of lubricant is so small that the Reynolds number is well within the laminar range. Hence laminar flow may safely be assumed in lubrication problems.

The most important function of lubricant is to separate the bearing surface so that there is no direct contact between them. The theory of hydrodynamic lubrication is utilised to determine (*i*) the clearances between the surfaces (*ii*) the viscosity and rate of flow of lubricant necessary to prevent the bearing surfaces coming into metal to metal contact and also to prevent excessive heating caused by dissipation of mechanical energy through viscous shear.

Example 8.13. *For lubrication and cooling of the guide bearing of the vertical shaft of a hydraulic turbine the arrangement shown in the following figure is used. A cylindrical reservoir of lubricant is mounted on the shaft and rotates with it. The lubricant is transferred into the cup above the bearing by a stationary tube which is bent at its lower end in the manner shown in the figure.*

Determine D_0, the diameter at which the inlet of the tube is to be located so that lubricant is supplied to the bearing at the rate of 0.15 lit/sec. when the turbine shaft is rotating at 200 rpm.

The free surface of the lubricant in the reservoir is at a diameter of 1 m in the horizontal plane containing the tube opening. The tube is 12 mm in diameter and 4 m long. The outlet of the tube is 3 m above its inlet. The kinematic viscosity of the fluid is 0.5×10^{-4} m²/sec. Neglect all minor losses.

Solution. Discharge $Q = 0.15$ lit/sec

Velocity in the tube,

$$V = \frac{4Q}{\pi d^2} = \frac{4 \times 0.15 \times 1000}{\pi (1.2)^2} \text{ cm/sec}$$

$$= 132.63 \text{ cm/sec} = 1.3263 \text{ m/s}$$

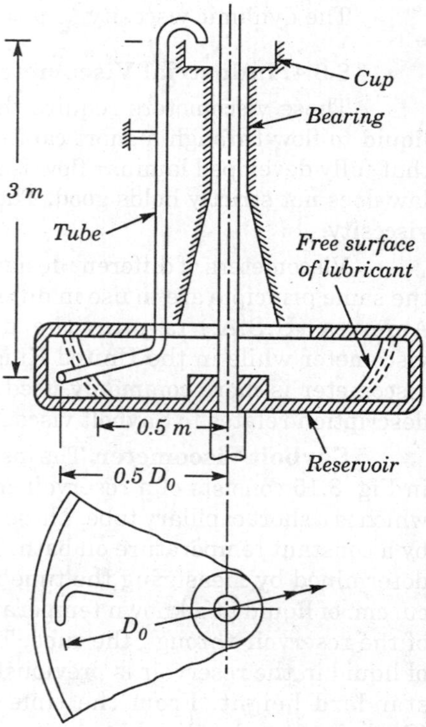

Reynolds number of flow in the tube

$$\mathbf{R} = \frac{Vd}{\upsilon} = \frac{1.3263 \times (1.2/100)}{0.5 \times 10^{-4}} = 318.3$$

Since **R** < 2000, the flow is laminar.

Head loss in the tube

$$h_L = \frac{\Delta p}{\gamma} = \frac{32 \mu V L}{\gamma d^2} = 32 \left(\frac{\mu}{P}\right) \frac{VL}{gd^2}$$

$$= \frac{32 \times (0.5 \times 10^{-4}) \times 1.3263 \times 4}{9.81 \times (1.2/100)^2}$$

$$= 6.009 \text{ m of oil.}$$

Neglecting minor losses, total energy required by the lubricant for being transferred into the cup (above the inlet of the tube)

$$H = Z + \frac{p}{\gamma} + \frac{V^2}{2g} + h_L = 3 + 0 + \frac{(1.3263)^2}{2 \times 9.81} + 6.009$$

$$= 9.098 \text{ m of oil.} \qquad \text{...(1)}$$

Head developed in the rotating reservoir of lubricant (in the plane of the tube inlet)

$$H = \frac{r_1^2 \omega^2}{2g} - \frac{r_2^2 \omega^2}{2g}$$

where $r_1 = 0.5\ D_0$ and $r_2 = 0.50$ m

$$\omega = \frac{2\pi N}{60} = \frac{2\pi \times 200}{60} = 20.94 \text{ per sec.}$$

$$\therefore \quad H = \frac{(20.94)^2}{2 \times 9.81}\,[(0.5\ D_0)^2 - (0.50)^2]$$

$$= 22.357 \times 0.25\,(D_0^2 - 1) \qquad \text{...(2)}$$

Equating Eqs. (1) and (2),

$$22.357 \times 0.25\,(D_0^2 - 1) = 9.098$$

$$D_0^2 = 2.6277 \quad \text{or} \quad D_0 = 1.621 \text{ m.}$$

8.10. Flow Through Porous Media—Darcy's Law

Steady flow of a fluid such as water or oil through porous soil mass (in particular fine sand) is quite similar to laminar flow in small-diameter pipes. In flow through porous media, it is to be noted that this is a case of laminar flow through small irregular passages instead of the flow through uniform-size conduit. The common examples of this type of flow include flow of water, oil and other fluids through filter beds, surface soil and porous rock. In general the porous-media flow is characterized by low velocity, high pressure drops, and very small pore diameters, so that the conditions for laminar flow (*i.e.*, flow at low Reynolds number) are fully satisfied. It is to be noted that in this type of flow since the velocities are very small, the inertia effects tend to be negligible and it is the viscous effects which are far more important, see article 6.6.2.

For laminar flow in pipes, pressure drop was found to be proportional to the average velocity and hence the volume flowrate, see Eq. (8.8). An analogous law holds for small Reynolds number flow in porous beds. This can be shown by the following experiment. Fig. 8.16 shows an inclined pipe filled with fine sand and fitted with two vertical piezometers at a distance L apart. If water is made to flow through this pipe in the direction indicated by the arrow, the difference in the water levels in piezometers (1) and (2), *i.e.* h, indicates the drop in the piezometric head in length L. In other words,

$$h = \left(\frac{p_1}{\gamma} + Z_1\right) - \left(\frac{p_2}{\gamma} + Z_2\right).$$

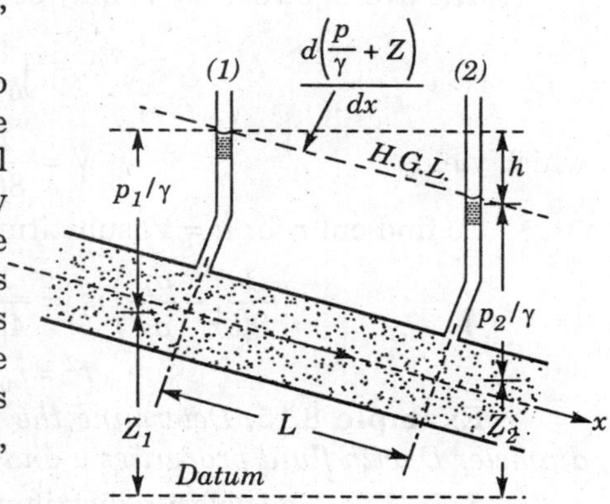

Fig. 8.16. Experimental set up for Darcy's law.

The change in the piezometric head per unit length is

$$\frac{d(p/\gamma + Z)}{dx} = \frac{1}{L}\left\{\left(\frac{p_2}{\gamma} + Z_2\right) - \left(\frac{p_1}{\gamma} + Z_1\right)\right\} = -\frac{h}{L} \qquad \text{...(8.34)}$$

In Porous-media flow, we are interested in the average discharge velocities rather than the local velocities in the pores. Let us define the *superficial or discharge velocity* as the ratio of the volume rate of flow through the pipe divided by the cross- sectional area of the pipe. It is obvious that this velocity is smaller than the actual velocity at which the water flows through the sand pores.

Experiments have shown that the superficial velocity u can be expressed as

$$u = k\,\frac{h}{L} = -k\,\frac{d(p/\gamma + Z)}{dx} \qquad \text{...(8.35)}$$

in which k is called the permeability. It has the dimensions of velocity and is equal to the superficial velocity if the piezometric gradient is unity. The magnitude of the permeability depends to a large extent on the grain size of the sand or soil, its porosity and the clay contents present.

Eq. (8.35) was first developed (based on experiments) by the French engineer, H. Darcy, in 1856 and has since been called *Darcy's law*. According to Ehrenberger, Eq. (8.35) is normally valid for velocities u less than 3 to 4.5 mm/s. The Darcy's law linear relationship between velocity and the gradient of $(p/\gamma + Z)$ used in seepage flow analysis holds only so long as the flow remains laminar and inertia effects remain unimportant. For sand beds a Reynolds number may be defined as

$$\mathbf{R} = \frac{ud_{50}}{\nu} \qquad \text{...(8.36)}$$

in which u is the superficial velocity, d_{50} is the grain size for which 50% of the sample by weight is finer and ν is the kinematic viscosity of the liquid. Experiments with sand beds give $\mathbf{R} \approx 10$ as an upper limit for the Darcy's law. This Reynolds number may be exceeded in rock aquifers and well casings. The sand-bed experiments have shown that for $\mathbf{R} > 10$, departure from Darcy's law occurs (while the flow is still laminar) due to fluid acceleration causing inertia effect. The transition from laminar into the fluctuating flow condition which is called turbulence is gradual, somewhere in the range $60 < \mathbf{R} < 600$. The flow resistance seems to become independent of Reynolds number at about $\mathbf{R} = 1000$.

Example 8.14. *At what distance 'r' from the centre of a tube of radius 'r_0' does the average velocity occur in laminar flow ?*

Solution. The velocity distribution in laminar flow is given by

$$u = -\frac{1}{4\mu}\left(\frac{dp}{dx}\right)(r_0^2 - r^2) \qquad \text{...(1)}$$

The average velocity V, may be obtained as

$$\pi r_0^2\, V = \int_0^{r_0} u.2\pi r\, dr = \int_0^{r_0} -\frac{1}{4\mu}\left(\frac{dp}{dx}\right)(r_0^2 - r^2)\, 2\pi r\, dr$$

which yields

$$V = \frac{1}{8\mu}\left(-\frac{dp}{dx}\right) r_0^2$$

To find out r for $u = V$, substituting $u = V$ in (1), we obtain

$$\frac{1}{8\mu}\left(-\frac{dp}{dx}\right) r_0^2 = \frac{1}{4\mu}\left(-\frac{dp}{dx}\right)(r_0^2 - r^2)$$

or

$$r^2 = r_0^2/2 \text{ resulting in } r = 0.707\, r_0.$$

Example 8.15. *Determine the maximum wall shear stress for laminar flow in a tube of diameter D with fluid properties* μ *and* ρ *given.*

Solution. The velocity distribution in laminar flow in a tube is given by,

$$u = \frac{1}{4\mu}\left(-\frac{dp}{dx}\right)(r_0^2 - r^2) \qquad \text{...(1)}$$

The shear stress distribution is given by

$$\tau = -\mu\,\frac{du}{dr}$$

From Eq. (1),

$$\frac{du}{dr} = \frac{1}{4\mu}\left(-\frac{dp}{dx}\right)(-2r) = \frac{1}{2\mu}\frac{dp}{dx}\, r$$

$\therefore$

$$\tau = -\mu \,.\, \frac{1}{2\mu}\left(\frac{dp}{dx}\right) . \, r = \frac{r}{2}\left(-\frac{dp}{dx}\right)$$

For shear stress at the pipe wall, $r = D/2$

$$\tau_0 = \frac{D}{4}\left(-\frac{dp}{dx}\right) = \frac{D}{4}\left(\frac{\Delta p}{L}\right) \qquad \text{...(2)}$$

The loss of energy in laminar flow in a circular tube is given by Eq. (8.8),

$$\Delta p = \frac{32\,\mu\, VL}{D^2}$$

or $$\frac{\Delta p}{L} = \frac{32\mu V}{D^2} \quad ...(3)$$

Wall shear stress will be maximum, when $\Delta p/L$ is maximum but $\Delta p/L$ will be maximum when V is maximum.

In laminar flow, maximum velocity will be attained when

$$\frac{\rho VD}{\mu} = 2000 \quad ...(4)$$

Eliminating V between Eqs. (3) and (4)

$$\frac{\Delta p}{L} = \frac{32\mu \times 2000\mu}{\rho D.D^2} = \frac{64{,}000\,\mu^2}{\rho D^3}$$

Substituting this value of $\Delta p/L$ in (2), we obtain the maximum wall shear stress as

$$\tau_0 = \frac{D}{4} \times \frac{64{,}000\,\mu^2}{\rho D^3} = \frac{16{,}000\mu^2}{\rho D^2}.$$

Example 8.16. *A thin film of very viscous fluid flows down a plane at an angle α to the horizontal. Assuming that the flow is laminar and uniform, show that*

$$\mathbf{F}^2 = \frac{\mathbf{R}}{3} \sin \alpha$$

where $$\mathbf{R} = \frac{V.d}{\nu} \text{ and } \mathbf{F} = \frac{V}{\sqrt{gd}}$$

V = mean velocity, d = film thickness,
ν = kinematic viscosity of fluid.

Solution. Considering fluid element of thickness y, length dx and unit width the forces acting on it are the shear resistance and the weight of the element. For steady and uniform flow down the inclined plane, we have

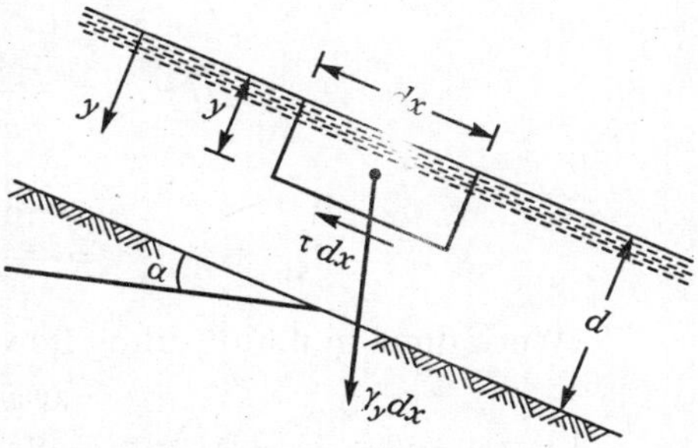

$$\gamma y\, dx \sin \alpha - \tau\, dx = 0 \quad \text{or} \quad \tau = \gamma y \sin \alpha$$

but since the flow is laminar, $\tau = -\mu \dfrac{du}{dy}$

the negative sign is introduced because as y increases u decreases.

$$\therefore \quad -\mu \frac{du}{dy} = \gamma y \sin \alpha$$

Integrating, $$u = -\frac{1}{\mu} \gamma \frac{y^2}{2} \sin \alpha + A$$

From the known boundary condition, *i.e.* at $y = d$, $u = 0$, the constant of integration is obtained as,

$$A = \frac{1}{2\mu} \gamma d^2 \sin \alpha$$

and the velocity distribution then becomes

$$u = \frac{1}{2\mu} \gamma \sin \alpha \,(d^2 - y^2)$$

The discharge per unit width

$$q = \int_0^d u\,dy = \int_0^d \frac{1}{2\mu}\,\gamma \sin\alpha\,(d^2 - y^2)dy = \gamma \sin\alpha \,.\, \frac{d^3}{3\mu}$$

but $\quad q = V \,.\, d$

$$\therefore \quad V = \frac{\gamma d^2 \sin\alpha}{3\mu}$$

$$V^2 = \frac{\gamma d^2 V \sin\alpha}{3\mu} \quad \text{or} \quad \frac{V^2}{gd} = \frac{\rho V d \sin\alpha}{3\mu} = \frac{\mathbf{R}}{3}\sin\alpha$$

$$\therefore \quad \mathbf{F}^2 = \mathbf{R}\,\frac{\sin\alpha}{3}.$$

Example 8.17. *Crude oil is pumped through a 15 cm smooth pipe which is subjected to seasonal changes in temperature. At the maximum temperature of 38°C, when the kinematic viscosity is 0.28 stokes, a power input of 2.2 kW per 300 metre is required to maintain a flow of 30 litres/sec. What power input would be required to maintain the same rate of flow at the minimum temperature of 0°C if the viscosity of the oil is then 10 times as great ? Assume a specific gravity of 0.90 at both temperatures.*

Solution. Reynolds number at 38°C $= \dfrac{Vd}{\nu} = \dfrac{\left\{\dfrac{0.03}{(\pi/4)(0.15)^2}\right\} \times 0.15}{0.28 \times 10^{-4}} = 9060$

Reynolds number at 0°C $= \dfrac{1}{10} \times 9060 = 906$

It is thus obvious that the flow at 38°C is turbulent, while at 0°C, it is laminar (since $R < 2000$).

Pressure drop in laminar flow per 300 m length, from Eq. (8.8).

$$p_1 - p_2 = \frac{32\mu VL}{d^2} = \frac{32\left(\dfrac{8829}{9.81} \times 0.28 \times 10^{-4} \times 10\right) VL}{d^2}$$

$$= \frac{32(0.245)\dfrac{0.03}{(\pi/4)(0.15)^2} \times 300}{(0.15)^2} = 182.5 \text{ kN/m}^2$$

Work done in maintaining flow,

$$= Q(p_1 - p_2) = 0.03 \times 182.5 \times 10^3 = 5474 \text{ Nm/s}$$

$\therefore$ Power required $\quad$ = Work done/sec = 5474 Nm/s = 5474 W = 5.474 kW.

Example 8.18. *Flow is in laminar motion between two parallel plates under the action of motion of one of the plates and also under the presence of a pressure gradient in such a way that the net forward discharge across any section is zero. Find out the point where minimum velocity occurs and its magnitude. Draw the velocity distribution graph across any section.*

Solution. The velocity distribution is given by Eq. (8.17).

$$u = \frac{U}{b}\,y - \frac{1}{2\mu}\frac{dp}{dx}\,(by - y^2)$$

and the discharge per unit width may be obtained from Eq. (8.18)

$$q = U\,\frac{b}{2} - \frac{b^3}{12\mu}\frac{dp}{dx}$$

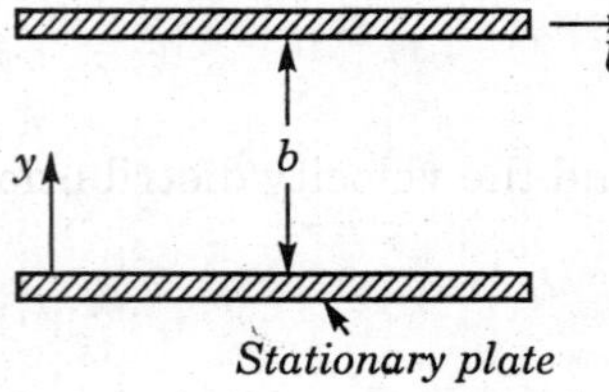

for net forward discharge, $q = 0$, thus yielding

$$\frac{dp}{dx} = \frac{6\mu}{b^3} bU = \frac{6\mu U}{b^2}$$

Minimum velocity occurs where $\frac{du}{dy} = 0$

thus, $$\frac{du}{dy} = \frac{U}{b} - \frac{1}{2\mu}\frac{dp}{dx}(b - 2y) = 0$$

yielding, $$y = \frac{b}{2} - \frac{\mu U}{b(dp/dx)} = \frac{b}{2} - \frac{\mu U b^2}{b(6\mu U)} = \frac{b}{3}$$

The magnitude of the minimum velocity is

$$u = \frac{U}{b}\left(\frac{b}{3}\right) - \frac{1}{2\mu}\frac{6\mu U}{b^2}\left(\frac{b^3}{3} - \frac{b^2}{9}\right) = -\frac{U}{3}.$$

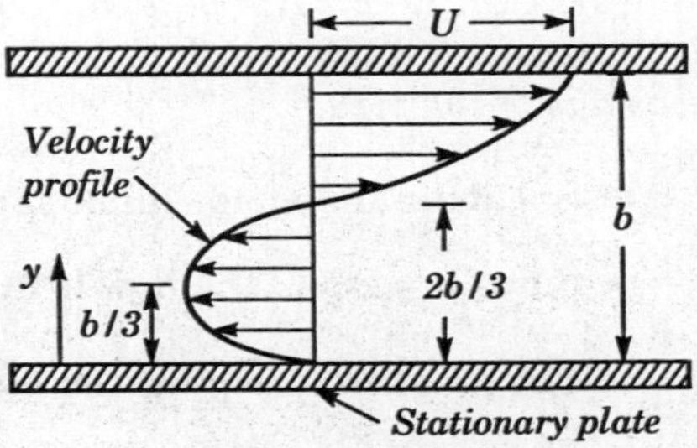

Velocity distribution

The velocity distribution graph may be obtained by substituting arbitrary values of y such as $0.1b$, $0.2b$ and so on in the equation

$$u = \frac{U}{b}y - \frac{3U}{b^2}(by - y^2).$$

and determining u in terms of U.

For $u = 0$, one can obtain $y = 0$ and $y = 2/3\, b$.

Example 8.19. *Prove that in the case of flow in a pipe line under viscous conditions, the velocity at the axis of the pipe is twice the mean velocity in the pipe.*

Oil flows through a pipe 15 cm in diameter and of 650 m length with a velocity of 0.5 m/s. If the kinematic viscosity of oil at the temperature is 18.7 stokes, find the loss of heat in friction. (Assume suitable values for data not given).

Solution. *Numerical part* : The pressure drop along the pipe length in viscous flow is given by Eq. (8.8).

$$(p_1 - p_2) = \frac{32\mu LV}{d^2}$$

The head loss in viscous friction is $\frac{p_1 - p_2}{\gamma}$, thus,

$$\frac{p_1 - p_2}{\gamma} = \frac{32\mu LV}{\rho g d^2} = \frac{32\nu LV}{g d^2} = \frac{32\,(18.7 \times 10^{-4}) \times 650 \times 0.5}{9.81 \times (0.15)^2} \text{ m of oil} = 88.1 \text{ m of oil.}$$

$$p_1 - p_2 = 88.1 \times 8829 \text{ N/m}^2 = 7.78 \times 10^5 \text{ N/m}^2$$

Assuming the sp. gr. of the oil as 0.9, the power lost in friction

$$= \gamma Q \times 88.1 = 8829 \times (\pi/4)\,(0.15)^2 \times 0.5 \times 88.1 = 6.87 \text{ kW}.$$

For obtaining the velocity at the pipe axis, use Eq. (8.2), put $r = 0$, and compare it with Eq. (8.8) to prove $u_{max} = 2V$.

Example 8.20. *An oil having a viscosity of 0.096 Ns/m² and a specific gravity of 1.59 flows through a horizontal pipe of 5 cm diameter with a pressure drop of 5.89 kN/m² per metre length of the pipe. Determine (i) the rate of flow in kg/minute (ii) the shear stress at the pipe wall (iii) the total drag for 100 m length of pipe, and (iv) the power required for 100 m length of the pipe to maintain the flow.*

Solution. In laminar flow, the discharge is related to the pressure drop by Eq. (8.6)

(*i*) $$Q = \frac{\pi d^4\, \Delta p}{128\mu L} = \pi\,\frac{(5/100)^4 \times 5.89 \times 10^3}{128 \times 0.096 \times 1} = 0.00937 \text{ m}^3\text{/s}$$

Weight flowrate $= (1.59 \times 9810) \times 0.00937 = 146.17$ N/s (*ii*) Shear stress at the pipe wall from Eq. (8.7)

$$\tau_0 = \frac{\Delta p}{L} \cdot \frac{r_0}{2} = \left(\frac{5.89 \times 10^3}{1}\right) \times \frac{1}{2}\left(\frac{2.5}{100}\right) = \mathbf{73.57\ N/m^2}$$

(*iii*) Total drag per 100 m length of pipe

$$= (\pi d L)\tau_0 = \pi \times \frac{5}{100} \times 100 \times 73.57 = \mathbf{1155.6\ N.}$$

(*iv*) Power required to maintain flow per 100 m length

$$= Q\left(\frac{\Delta p}{\Delta x} \times 100\right) = 0.00937 \times \frac{5.89 \times 10^3 \times 100}{1} = \mathbf{5.5\ kW.}$$

Example 8.21. *A pipe line 20 cm in diameter, 70 m long, conveys oil of specific gravity 0.95 and viscosity 0.23 Ns/m². If the velocity of oil is 1.38 m/s, find the difference in pressure between the two ends of the pipe.*

Solution. Reynolds number of flow $\mathbf{R} = \frac{\rho V d}{\mu} = \frac{(950) \times 1.38 \times 0.2}{0.23} = 1136$

Since **R** < 2000, the flow is laminar.

The pressure drop is given by $\Delta p = \frac{32 \mu V L}{d^2} = \frac{32 \times 0.23 \times 1.38 \times 70}{(20/100)^2} = 17854\ \text{N/m}^2$

Pressure drop in metre of oil $= \frac{\Delta p}{\gamma} = \frac{17854}{(0.95 \times 9810)} = \mathbf{1.915\ m.}$

Example 8.22. *Find the power required per 1000 m of pipe line to overcome the viscous resistance to flow of oil, density = 900 kg/m³, μ = 0.19 Ns/m², through a horizontal 8 cm diameter pipe at the rate of 19 m³/hr.*

Solution. Reynolds number of flow,

$$\mathbf{R} = \frac{\rho V d}{\mu} \qquad V = \frac{4Q}{\pi d^2} = \frac{4 \times 19/3600}{\pi (8/100)^2} = 1.05\ \text{m/s}$$

$$\mathbf{R} = \frac{900 \times 1.05 \times 8/100}{0.19} = 398$$

Since **R** < 2000, the flow is laminar,

$$\Delta p = \frac{32 \mu V L}{d^2} = \frac{32 \times 0.19 \times 1.05 \times 1600}{(8/100)^2} = 1599 \times 10^3\ \text{N/m}^2 = 1599\ \text{kN/m}^2$$

Power required $= Q \cdot \Delta p = (19/3600) \times 1599 \times 10^3 = 8452.5\ \text{W} = \mathbf{8.45\ kW.}$

Example 8.23. *An oil having specific gravity of 0.9 flows at the rate of 11.3 litres per second through a horizontal pipe 7.5 cm in diameter. The pressure drop over 300 m length of the pipe is 412 kN/m². Assuming laminar flow, find the viscosity of the oil. Also establish that the flow through the pipe is laminar.*

Solution. $\mu = \frac{\Delta p . d^2}{32 V L}$; $\quad V = \frac{4Q}{\pi d^2} = \frac{4 \times 11.3 \times 10^{-3}}{\pi (7.5)^2 \times 10^{-4}} = 2.55\ \text{m/s}$

$$= \frac{412 \times 10^3 \times (7.5 \times 10^{-2})^2}{32 \times 2.55 \times 300} = 9.47 \times 10^{-2}\ \text{Ns/m}^2 = \mathbf{94.7\ Pa\ s}$$

Reynolds number $\quad \mathbf{R} = \frac{\rho V d}{\mu} = \frac{900 \times 2.55 \times (7.5/100)}{9.47 \times 10^{-4}} = 1820$

Since **R** < 2000, the flow is laminar.

Example 8.24. *An oil having viscosity of 0.143 Ns/m² and specific gravity of 0.9 flows through a pipe 2.5 cm diameter and 300 cm long, at 1/10 of the critical velocity for which Reynolds number is 2500. Find : (a) the velocity of flow through the pipe, (b) the head in metres of oil across the pipe length required to maintain the flow, and (c) power of the flow.*

Solution. Critical Reynolds number

$$\mathbf{R}_{cr} = \frac{\rho V_{cr} d}{\mu} \; ; \; V_{cr} = \frac{\mathbf{R}_{cr}\mu}{\rho d}$$

$$V_{cr} = \frac{2500 \times 0.143}{900 \times (2.5/100)} = 15.87 \text{ m/s}$$

(*a*) Velocity of flow through the pipe

$$V = \frac{1}{10} V_{cr} = \mathbf{1.587 \text{ m/s.}}$$

(*b*) $$\frac{\Delta p}{\gamma} = \frac{32\mu VL}{\gamma d^2} = \frac{32 \times (1.43/9.81) \times 1.587 \times 3}{(0.9 \times 9810) \times (2.5/100)^2} = \mathbf{3.86 \text{ m of oil.}}$$

(*c*) Power required $= Q\Delta p = \dfrac{\pi d^2 V \,.\, \Delta p}{4} = \dfrac{\pi(2.5/100)^2 \times 1.587 \times (8829 \times 3.86)}{4}$ W $= \mathbf{0.265 \text{ kW.}}$

Example 8.25. *A 0.20 m diameter pipe 20 km long transports oil at a flowrate of 0.01 m^3/s. Calculate the power required to maintain the flow if the dynamic viscosity and density of oil are 0.08 Pas and 900 kg/m^3 respectively.*

Solution. Data given :

$d = 0.20$ m, $L = 20$ km, $Q = 0.01$ m^3/s, $\mu = 0.08$ Pas $= 0.08$ Ns/m^2

$\rho = 900$ kg/m^3 ($\gamma = \rho g = 900 \times 9.81$ N/m^3)

From the continuity equation, $V = \dfrac{4Q}{\pi d^2} = \dfrac{4 \times 0.01}{\pi(0.2)^2} = 0.318$ m/s

Reynolds number of flow, $R = \dfrac{\rho V d}{\mu} = \dfrac{900 \times 0.318 \times 0.2}{0.08} = 715.5$

Since $R < 2000$, the flow in pipe is laminar. The energy loss in laminar flow is given by

$$\frac{\Delta p}{\gamma} = \frac{32\mu VL}{\gamma d^2} = \frac{32 \times 0.08 \times 0.318 \times (20 \times 1000)}{9.81 \times 900 \times (0.2)^2} = 46.147 \text{ m of oil}$$

Power required to maintain flow

$$P = \gamma Q(\Delta p/\gamma) = 8829 \times 0.01 \times 46.147 = 4074.3 \text{ W} = \mathbf{4.074 \text{ kW.}}$$

Example 8.26. *Glycerine of viscosity 0.883 Ns/m^2 and specific gravity 1.26 is pumped along a horizontal pipe 65 m long of diameter 0.01 m at a flowrate of 180 litres per minute. Determine the flow Reynolds number and verify whether the flow is laminar or turbulent. Calculate the pressure loss in the pipe due to frictional effects and calculate the maximum flowrate for laminar flow conditions to prevail.*

Solution. Reynolds number, $\mathbf{R} = \dfrac{Vd}{\nu} = \dfrac{4Q}{\pi d \nu}$

Kinematic viscosity $\nu = \dfrac{\mu}{\rho} = \dfrac{\mu g}{\gamma} = \dfrac{0.883 \times 9.81}{12360} = 0.7 \times 10^{-3}$ m^2/s

$$\mathbf{R} = \frac{4 \times (180 \times 10^{-3}/60)}{\pi \times 0.01 \times 0.7 \times 10^{-3}} = 54.57$$

Since $\mathbf{R} < 2000$, flow is laminar.

Pressure loss $\Delta p = \dfrac{32\mu VL}{d^2} = \dfrac{128\mu QL}{\pi d^4} = \dfrac{128 \times 0.883 \times 3 \times 10^{-3} \times 65}{\pi \times (0.01)^4} = 7014 \times 10^5$ N/m^2

Maximum flowrate for laminar flow will be obtained for $\mathbf{R} = 2000$,

$$V = \frac{\nu \mathbf{R}}{d} = \frac{0.7 \times 10^{-3} \times 2000}{0.01} = 140 \text{ m/s}$$

$$Q_{max} = \frac{\pi d^2}{4} \times V = \frac{\pi(0.01)^2}{4} \times 140 = 0.01099 \text{ m}^3/\text{s} = 10.99 \text{ l/s} = \textbf{659.73 litres/minute.}$$

Example 8.27. *Laminar flow of a fluid of viscosity 0.883 Ns/m² and specific gravity 1.26 occurs between a pair of parallel plates of extensive width, inclined at 45° to the horizontal, the plates being 10 mm apart. The upper plate moves with a velocity of 1.5 m/s relative to the lower plate and in a direction opposite to the fluid flow. Pressure gauges mounted at two points 1 m vertically apart on the upper plate record pressures of 245 kN/m² and 78.5 kN/m² respectively. Determine the velocity and shear stress distribution between the plates, the maximum flow velocity and the shear stress on the upper plate.*

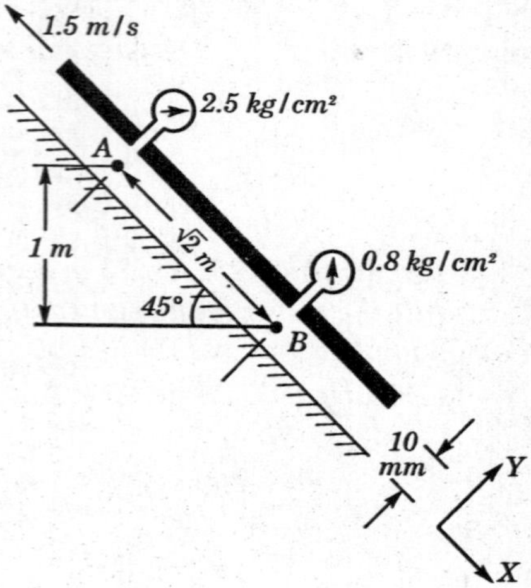

Solution. For inclined boundaries, Eq. (8.11) indicates that the pressure term p is to be replaced by $p + \gamma z$ in all expressions applicable for horizontal boundaries. Hence using Eq. (8.17), and noting that $U = -1.5$ m/s, one can write velocity distribution as

$$u = -1.5\frac{\gamma}{b} - \frac{1}{2\mu}\frac{d(p+\gamma z)}{dx}(by - y^2) \quad ...(i)$$

$$(p + \gamma z)_A = 245 \times 10^3 + 12360 \times 1 = 257 \times 10^3 \text{ N/m}^2$$

$$(p + \gamma z)_B = 78.5 \times 10^3 + 0 = 78.5 \times 10^3 \text{ N/m}^2$$

$$\frac{d(p+\gamma z)}{dx} = \frac{(p+\gamma z)_B - (p+\gamma z)_A}{\sqrt{2}}$$

$$= \frac{(78.5 - 257)10^3}{\sqrt{2}} \text{ N/m}^2/\text{m} = -12.65 \times 10^4 \text{ N/m}^2/\text{m}.$$

Making substitutions in Eq. (*i*) above, the velocity distribution is given by,

$$u = -1.5\frac{y}{10 \times 10^{-3}} - \frac{1}{2 \times 0.883}(-12.65 \times 10^4)(10 \times 10^{-3}\,y - y^2)$$

$$= -150\,y + 7.17 \times 10^4\,(0.01\,y - y^2) \quad ...(ii)$$

Using Eq. (8.19), with $U = -1.5$ m/s, the shear stress distribution can be expressed as

$$\tau = -\mu\frac{1.5}{b} - \frac{1}{2}\frac{d(p+\gamma z)}{dx}(b - 2y)$$

$$= -0.883 \times \frac{1.5}{10 \times 10^{-3}} - \frac{1}{2}(-12.65 \times 10^4)(0.01 - 2y)$$

$$= 500.6 - 126.6 \times 10^3 y \quad ...(iii)$$

For maximum velocity, $\frac{du}{dy} = 0$

or, $\quad -150 + 7.17 \times 10^4\,(0.01 - 2y) = 0$ from which $\quad y = 3.954 \times 10^{-3}$ m

and the maximum velocity

$$u_{max} = -150 \times 3.954 \times 10^{-3} + 7.17 \times 10^4\,(0.01 \times 3.954 \times 10^{-3} - 15.63 \times 10^{-6}) = \textbf{2.323 m/s.}$$

Shear stress on the upper plate :

$$\tau_{0.01} = 500.6 - 126.6 \times 10^3 \times 0.01 = \textbf{– 765.4 N/m}^2.$$

Example 8.28. *A water tank containing 4m depth of water has a horizontal crack of thickness 0.12 mm extending all the way through the 500 cm thickness of the wall at that location. The water seeps through this crack to atmosphere. Estimate the leakage rate per minute per metre width of the crack if the crack is located at 4 m depth. Take kinematic viscosity as 0.01 stoke.*

Solution. The flow through the horizontal crack can be approximated as laminar flow between parallel plates separated by a distance $b = 0.12$ mm. Using Eq. (8.20), the discharge through the crack per metre width,

$$q = \int_0^b u dy = \frac{1}{2\mu}\left(-\frac{dp}{dx}\right)\int_0^b (by - y^2)dy = \frac{1}{12\mu}\left(-\frac{dp}{dx}\right)b^3.$$

Average velocity, $V = \dfrac{q}{b} = \dfrac{1}{12\mu}\left(-\dfrac{dp}{dx}\right)b^2.$

If h_f is the head loss in length L, $\left(-\dfrac{dp}{dx}\right) = \dfrac{\gamma h_f}{L}$, and

$$V = \frac{1}{12\mu}\frac{\gamma h_f}{L} b^2, \quad \text{from which,} \quad h_f = \frac{12\mu VL}{\gamma b^2}.$$

Making substitution : $h_f = 4$ m, $b = 0.12 \times 10^{-3}$ m ; $L = 0.50$ m

$\mu = \rho\nu = 998 \times 0.01 \times 10^{-4}$ Ns/m², $\gamma = 9790$ N/m³

$$V = \frac{1 \times 9790 \times 4 \times (0.12 \times 10^{-3})^2}{12 \times 998 \times 0.01 \times 10^{-4} \times 0.50} = 0.0942 \text{ m/s}$$

Discharge per metre width, $q = V \,.\, b = 1.13 \times 10^{-5}$ m³/s

= 0.678 litre/min per metre width of crack.

Checking for laminar flow, the Reynolds number $\rho\dfrac{Vb}{\mu} = 11.3$, which is well within the limit for laminar flow.

Example 8.29. *For steady laminar flow, determine the expression for head loss in a conical contraction.*

Solution. Consider a conical contraction of length L having diameters d_1 and d_2 at its ends. The head loss in pipe flow can be expressed by Eq. (9.57) which in a differential form may be written as,

$$dh_f = f\frac{dx}{d}\frac{V^2}{2g}$$

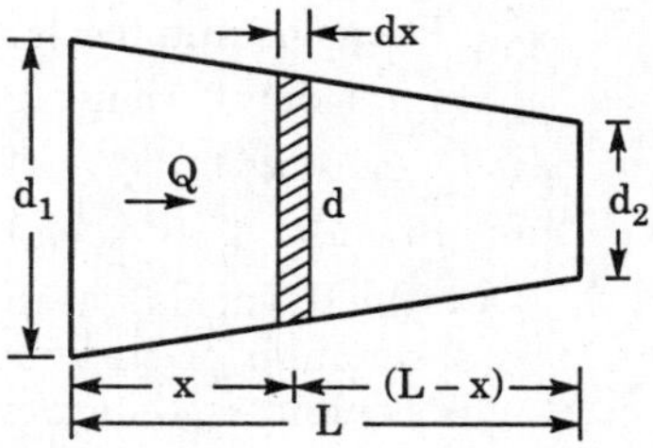

From the geometry of cone, one can write

$$\frac{d_1 - d_2}{L} = \frac{d_1 - d_2}{L - x}, \text{ and } d = d_2 + \frac{(d_1 - d_2)}{L}(L - x)$$

Differentiating, $d(d) = -\dfrac{(d_1 - d_2)}{L} \,.\, dx$

or
$$dx = -\frac{Ld(d)}{d_1 - d_2}$$

For laminar flow,
$$f = \frac{64\nu}{Vd} = \frac{64\nu \,.\, \pi d^2}{4Qd} = \frac{16\pi\nu d}{Q}$$

Making substitution, $dh_f = \dfrac{16\pi\nu d}{Q}\left(-\dfrac{Ld(d)}{d_1 - d_2}\right)\dfrac{16Q^2}{2g\pi^2 d^5} = \dfrac{128\nu LQ}{g\pi(d_1 - d_2)} \cdot \dfrac{d(d)}{d^4}$

Integrating,
$$h_f = \frac{128\nu LQ}{g\pi(d_1 - d_2)}\int_{d_1}^{d_2}\frac{d(d)}{d^4} = \frac{\mathbf{128\nu QL}}{\mathbf{3g\pi(d_1 - d_2)}}\left[\frac{\mathbf{1}}{\mathbf{d_2^3}} - \frac{\mathbf{1}}{\mathbf{d_1^3}}\right].$$

Example 8.30. *A steady-uniform laminar flow is taking place over an inclined flat plate with an average velocity, $\overline{u}$, equal to 100 mm/s. The flow depth normal to the plate, $T = 200$ mm. The kinematic viscosity of the fluid is 100 mm²/s. Determine the Darcy-Weisbach friction factor.*

Solution. For steady-laminar flow,

$$\overline{u} = \frac{\gamma(\sin\theta)T^2}{3\mu} = \frac{g(\sin\theta)T^2}{3\nu}$$

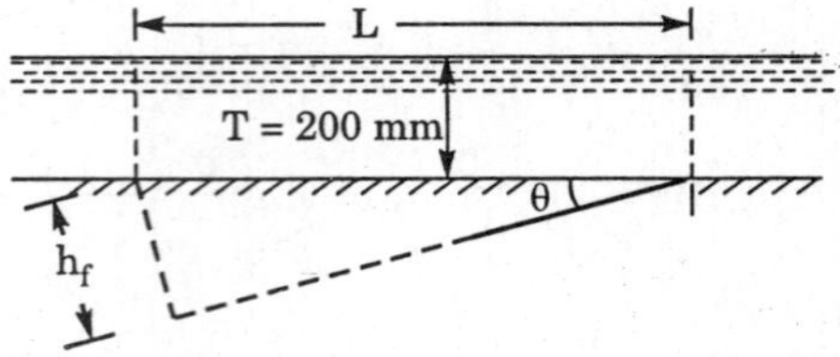

or $$\sin\theta = \frac{h_f}{L} = \frac{3\,\nu\bar{u}}{gT^2} \qquad ...(1)$$

but from Darcy-Weisbach equation $h_f = f\,\dfrac{L}{d}\,\dfrac{\bar{u}^2}{2g}$ ...(2)

Eliminating h_f between (1) and (2),

$$f = \frac{24\,\nu}{T\bar{u}} = \frac{24 \times 100\ (\text{mm}^2/\text{s})}{200\ (\text{mm}) \times 100\ (\text{mm/s})} = \mathbf{0.12.}$$

8.11. Fluidization

When a fluid is passed upwards at increasing velocities through a column of granular material of size d, a stage is reached when all the particles are entrained in the flow and the bed is in a fluid state. This is known as fluidization and the corresponding **superficial** velocity, V_f, as fluidization velocity (V_f = discharge/gross area) and is given by

$$V_f = \frac{(\gamma_s - \gamma)\, d^2 e^2}{18\mu} \qquad ...(8.37)$$

where e is the porosity of material. Eq. (8.37) is valid within the Stokes' range, and is often used in chemical engineering and in back washing of gravity filters far water purification.

OBJECTIVE TYPE QUESTIONS

1. The laminar flow is characterized by (select the correct statements) :
   (*i*) irregular motion of fluid particles
   (*ii*) fluid particles moving in layers parallel to the boundary surface
   (*iii*) high Reynolds number of flow (*iv*) existence of eddies
   (*v*) Reynolds number which is less than the critical value.
2. Laminar flow through a circular tube was studied experimentally by
   (*a*) Newton (*b*) Pascal (*c*) Hagen and Poiseuille
   (*d*) Prandtl (*e*) Mannings.
3. Osborne Reynolds was first to provide the criterion for determining the type of flow and classifying it as laminar or turbulent. The dimensionless parameter which quantifies the role of viscous effect and is used for flow classification is expressed as :
   (*i*) $V/\sqrt{gd}$ (*ii*) $\rho V d/\mu$ (*iii*) $\rho V^2 L/\sigma$ (*iv*) $V\sqrt{\rho/E}$.
4. The Reynolds experiments for determining the character of flow in pipes involved :
   (*a*) measurement of velocity by Pitot-tube (*b*) measurement of flow only
   (*c*) observation of the condition of dye-filament which become wavy or irregular at the onset of turbulence
   (*d*) measurement of intensity of turbulence.
5. The lower limit of the critical Reynolds number below which all disturbances (or sources of turbulence) in pipe flow are damped out by viscous action has a value approximately equal to :
   (*i*) 1 (*ii*) 500 (*iii*) 1000 (*iv*) 2000.
6. The upper limit of critical Reynolds number in pipe flow has been found to depend on many factors. Select the right combination involving these factors from the alternatives given below :
   (*a*) length of pipe (*b*) initial turbulence in the approach flow
   (*c*) shape of pipe entrance (*d*) surface tension of fluid
   (*e*) roughness of pipe surface.
   Alternatives : 1—a, b, c ; 2—b, c, d ; 3—c, d, e ; 4—b, c, e ; 5—a, d, e.
7. The shear-stress distribution in pipe flow does not depend on the type of flow, it being dependent upon the pressure gradient dp/dx and the radius, and is expressed by :

(*i*) $\tau = \frac{1}{r}(dp/dx)$, (*ii*) $\tau = -(dp/dx)\frac{r}{2}$ (*iii*) $\tau = -2r(dp/dx)$, (*iv*) $\tau = -\frac{r^2}{L}(dp/dx)$.

8. The velocity distribution in laminar flow through pipes follows a parabolic law attaining a maximum velocity at the pipe axis. The Hagen-Poiseuille equation which governs this type of flow may be expressed as :

(*a*) $Q = \frac{\pi d^4 \Delta p}{128 \mu L}$ (*b*) $Q = \frac{\pi d^2 \Delta p}{128 \mu L}$ (*c*) $Q = \frac{128 \mu L}{\pi d^2 \Delta p}$ (*d*) $Q = \frac{128 \mu L}{\pi d^4 \Delta p}$.

9. The pressure drop per unit length of pipe ($\Delta p/L$) in laminar flow is dependent on the velocity, viscosity and the diameter. It is equal to :

(*i*) $\frac{d^2}{32 \mu V}$ (*ii*) $\frac{32 \mu VL}{\gamma d^2}$ (*iii*) $\frac{32 \mu V}{d^2}$ (*iv*) $\frac{8 \mu V}{d^2}$.

10. In laminar flow through a circular tube, the Darcy-Weisbach friction factor depends only on the Reynolds number, and the two are related by :

(*a*) $f = 1/\mathbf{R}$ (*b*) $f = 16/\mathbf{R}$ (*c*) $f = \frac{0.316}{\mathbf{R}^{1/4}}$ (*d*) $f = 64/\mathbf{R}$.

11. Very slow motion of a sphere in a viscous fluid was first analysed by Sir George G. Stokes on the assumption that the inertia forces are negligibly small as compared to the viscous ones. The drag force resisting the motion of the sphere is given by :

(*i*) $\pi \mu V^2 d$ (*ii*) $3\pi \mu V d$ (*iii*) $3\pi \mu V d^2$ (*iv*) $3\pi \mu^2 V d$.

12. A small solid particle falling through a fluid mass under its own weight eventually attains a steady state and thereby a constant velocity known as the terminal velocity. This velocity is given by :

(*a*) $d^2 (\gamma_s - \gamma)/18\mu$ (*b*) $d (\gamma_s - \gamma)/18\mu$ (*c*) $d (\gamma_s - \gamma)/18\mu^2$ (*d*) $d^2 (\gamma_s - \gamma)/18\mu^2$.

13. The flow through a porous media is governed by the well-known Darcy's law which relates the velocity with the head loss and is usually expressed as :

(*i*) $u = kh/L$ (*ii*) $u = kL/h$ (*iii*) $u = k(h/L)^2$ (*iv*) $u = k(L/h)^2$.

14. According to Ehrenberger, the Darcy's law is valid for velocities u less than :

(*a*) 1 m/s (*b*) 10 cm/s (*c*) 3 to 4.5 mm/s (*d*) 5 to 10 mm/s.

15. For sand beds, the Darcy's law would be valid if the Reynolds number defined by $\mathbf{R} = ud_{50}/\nu$ is less than :

(*i*) 0.1 (*ii*) 1 (*iii*) 10 (*iv*) 1000.

16. The discharge in m^3/s for laminar flow through a pipe of diameter 0.04 m having a centreline velocity of 1.5 m/s is :

(*a*) $3\pi/59$ (*b*) $3\pi/2500$ (*c*) $3\pi/5000$ (*d*) $3\pi/10000$.

ANSWERS

1. (*ii*), (*v*)	**2.** (*c*)	**3.** (*ii*)	**4.** (*c*)	**5.** (*iv*)	**6.** 4
7. (*ii*)	**8.** (*a*)	**9.** (*iii*)	**10.** (*d*)	**11.** (*ii*)	**12.** (*a*)
13. (*i*)	**14.** (*c*)	**15.** (*iii*)	**16.** (*d*)		

REVIEW QUESTIONS

1. Enumerate distinguishing characteristics of laminar flow.

2. Sketch the Reynolds apparatus and explain how the laminar flow can be demonstrated with the help of this apparatus.

3. What is the physical significance of lower critical velocity ? What value of Reynolds number characterizes it for (*i*) pipe flow, (*ii*) flow between parallel plates and (*iii*) open channel flow ?

4. Sketch the velocity and shear stress distributions across the flow in a circular pipe (for fully developed laminar flow).

5. What factors account for the loss of energy in laminar flow ? How does the energy loss vary with velocity of flow ?

6. In what way the laminar flow equations are affected if in addition to pressure force, the gravitational force is also to be taken into account ?

7. Under what circumstance would the net flow rate be zero in case of laminar flow between parallel plates ? (both plates may be considered moving).

8. What are the limitations of Stokes' law ?

9. Compare the time of fall of two small spherical particles of the same material, one having its diameter two times as great as compared to the other. Both of them are falling freely through a glycerine tank. Assume Stokes' law to be applicable in both the cases and neglect the effect of proximity of walls. **[Ans.** 1 : 4]

10. What are the sources of error which might creep in while determining the viscosity of liquid using capillary tube method ?

PROBLEMS

8.1. Working from fundamentals, show that the flow per unit width past a fixed cross-section for laminar flow between the two parallel plates moving in opposite directions as shown in the figure is

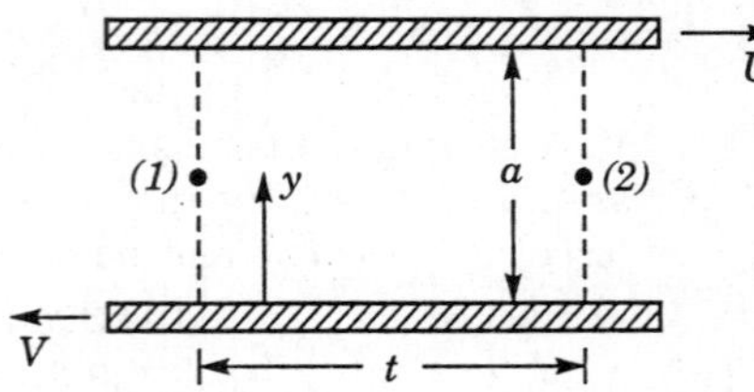

$$q = \frac{a}{2}(U - V) - \frac{1}{12\mu} a^3 \frac{dp}{dx}$$

8.2. Compute the kinetic energy and momentum correction factors for laminar flow between fixed parallel plates.

8.3. With a free body, as shown in figure of Ex. 8.10, page 383, for a uniform flow of a thin lamina of liquid down an inclined plane, show that the velocity distribution is

$$u = \frac{\gamma}{2\mu}(d^2 - y^2) \sin \theta$$

and the discharge per unit width is $q = \dfrac{\gamma}{3\mu} d^2 \sin \theta$.

8.4. An oil having a sp. gr. of 0.90 and dynamic viscosity of 0.98 Pas flows between parallel plates spaced 8 cm apart. Piezometers along the plate indicate that the pressure is decreasing in the direction of flow at the rate of 88.3 Pa per meter length. Determine :

1. velocity of flow midway between the plates
2. average velocity of flow, and
3. Reynolds number of flow.

8.5. Determine the power of a pump required to pump 250 litre/min. of linseed oil from a tank of free surface elevation 40 m to another tank having free surface at elevation 46 m. The pipe line 7.5 cm in diameter is 500 m long. If the oil is at a temperature of :

(*i*) 28°C and has a velocity 0.98 Pas (*ii*) 38°C having a viscocity of 0.078 Pas

Assume a sp. gr. of 0.90 at both the temperatures. Take pump efficiency as 75%.

8.6. Oil of sp. gr. 0.82 is pumped through a horizontal pipeline 15 cm in diameter and 3 km long, at the rate of 900 litres/minute. The pump has an efficiency of 68% and requires 7.35 kW, to pump the oil. If friction factor f is given by $f = \dfrac{64}{\mathbf{R}}$ where **R** is the Reynolds number, determine the viscosity of the oil. Verify that the flow is laminar.

8.7. In the figure shown below water is moving through a pipe. The velocity at some section is shown and is given mathematically as

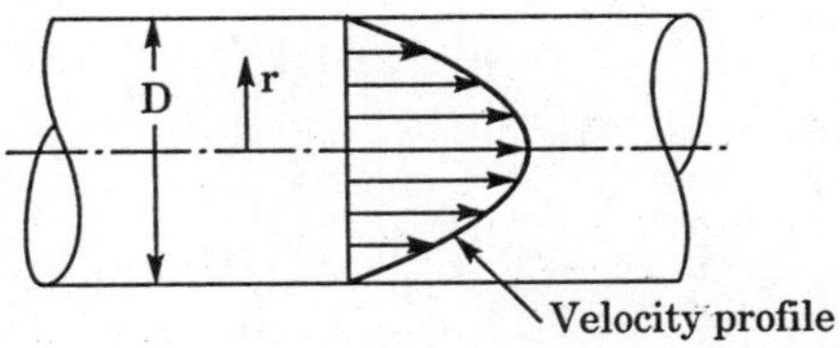

$$u = \left(\frac{\beta}{4\mu}\right)\left(\frac{D^2}{4} - r^2\right)$$

where β = A constant (non-dimensional)

r = Radial distance from centre line

u = Velocity at radius r.

What is the shear stress at the wall of the pipe ?

What is the shear stress at a position $r = D/4$?

If the above velocity profile persists for a distance L along the pipe, what drag is induced on the pipe surface by the flow of the water ?

8.8. The velocity of heavy oil flowing through a tube of radius r_0 is known to vary with distance r from the centre line according to the expression

$$u = 2V \frac{r_0^2 - r^2}{r_0^2}, \quad \text{where} \quad V = \text{Mean velocity}$$

Determine the maximum velocity if the tube is carrying 1 litre/sec and r_0 = 2.5 cm.

8.9. Oil of sp. gr. 0.92 flows through a pipeline 5 km long, having 30 cm diameter. The head lost in overcoming the viscous resistance has been found to be 20.0 m. The viscosity of the oil is constant at 0.08 Pas for the first 3 km of the pipeline, but in the last 2 km the viscosity increases to 0.98 Pas on account of cooling. Over this length, the viscosity changes linearly from the 3rd km to 5th km. Determine the rate of flow of oil in lit/min. Assume laminar flow throughout the pipe line.

8.10. Show that in laminar flow through a circular pipe, the total kinetic energy of fluid passing per second is twice the value obtained on the basis of average velocity.

8.11. Show that discharge per unit width between two parallel plates distance b apart, when one plate is moving at velocity U while the other one is held stationary, for the condition of zero shear stress at the fixed plate is $q = bU/3$.

8.12. What power will be required per one kilometre length of pipe line to overcome the viscous resistance to the flow of crude oil (μ = 0.19 Pas) through a horizontal 10 cm diameter pipe at the rate of 650 lit/hr.

8.13. Show from first principle that the loss of energy in terms of head in steady laminar flow along a horizontal tube of diameter d may be expressed as

$$h_l = \frac{64}{\mathbf{R}} \frac{L}{d} \frac{V^2}{2g} \quad \text{where} \quad \mathbf{R} = \frac{\rho V d}{\mu} \quad \text{is the Reynolds number of flow.}$$

8.14. Deduce an expression in terms of Reynolds number for friction co-efficient f in Darcy-Weisbach's equation for laminar flow in circular pipes.

8.15. (*a*) Enumerate important characteristic of laminar flow. Give examples where such a flow is encountered.

(*b*) Obtain an expression for the head loss in laminar flow in a circular pipe. Also write down the equation for head loss due to laminar flow between parallel plates and for flow down an inclined plane. Give the Reynolds numbers upto which these equations are valid.

8.16. (*a*) Derive an expression for pressure drop for a steady two-dimensional laminar flow between parallel plates.

(*b*) What power is required per kilometre of a line to overcome the viscous resistance to the flow of glycerine through a horizontal 10 cm diameter pipe at the rate of 10 litres per second ? Take μ = 0.78 Pas and $\nu = 6.3 \times 10^{-4}$ m²/s.

ANSWERS

8.2. $\alpha = 54/35$, $\beta = 6/5$ **8.4.** 0.72 m/s, 0.48 m/s, 352 **8.5.** 17.5 kW, 14.6 kW

8.6. 0.9 Pas **8.7.** $\beta D/4$, $\beta D/8$, $\pi D^2 \beta L/4$ **8.8.** 101.6 cm/s

8.9. 87 lit/sec. 4280, 3430 **8.12.** 0.0025 kW **8.16.** 21.9 kW.

9

Turbulent Flow in Pipes

9.0. Reynolds Observations on Pipe Flow

Osborne Reynolds in 1833 carried out systematic experiments to measure energy loss in lead pipes. He studied the nature of flow both at small and large velocities, and found great quantitative and qualitative variation. Visual observation of flow through glass tubes, with dye injected centrally into the liquid showed that : (1) at small velocities of flow the dye filament remains straight and flows parallel to the length of tube and, no mixing occurred between the dye and the liquid. This orderly flow in which the fluid appears to move in layers is the so-called laminar flow. (2) When the velocity of flow is gradually increased, a stage is reached when the dye filament begins to oscillate and finally breaks up diffusing itself completely with the liquid at a short distance from the tip of the dye injector (refer Fig. 8.3). Intense mixing occurred between the dye and the liquid and the condition of flow became irregular, disorderly and random. The flow is termed as the *turbulent flow*.

In case of laminar flow, the loss of pressure head was found to be proportional to the mean velocity (*i.e.* $h_L \propto V$). In the range of turbulent flow Reynolds found that the loss of pressure head was approximately proportional to the square of mean velocity (*i.e.* $h_L \propto V^n$, where n varies from 1.75 to 2.0). From numerous experiments on a variety of pipes, Reynolds concluded that the transition from laminar to turbulent flow was dependent not only on velocity but also on the pipe diameter and on the physical properties (*i.e.* viscosity and density) of the fluid. The dimensionless parameter involving the above mentioned variables which affect the nature of flow in pipes was derived by Reynolds, and has been named after him, as Reynolds number.

9.1. Transition form Laminar to Turbulent Flow

At low Reynolds number the flow through a straight pipe of uniform cross-section, every fluid particle moves with a constant velocity along a straight path. The flow is well-ordered and particles move in parallel streamlines or laminas. The velocities of the adjacent laminas are not the same. Experimentally it has been verified that this orderly pattern of flow (laminar flow) ceases to exist at higher Reynolds numbers and that strong mixing of fluid particles occurs. This phenomenon of mixing was first demonstrated by Osborne Reynolds in 1833 through the diffusion of dye filament. He showed that as soon as the flow becomes turbulent the dye filament diffuses into the fluid stream and the fluid appears uniformly coloured at a short distance down stream of the pipe entrance. The transverse mixing in turbulent flow is due to superimposition on the main flow having a direction along the pipe axis, of a subsidiary motion at right angles to it. The streamline pattern at a fixed point becomes subjected to continuous fluctuations and the subsidiary motion causes exchange of momentum in the transverse direction. On account of continuous mixing in the transverse direction, the velocity distribution over the cross-section is significantly more uniform in turbulent flow as compared to that in the laminar one, see Fig. 9.2.

Osborne Reynolds was also first to investigate in detail the circumstances leading to flow transition from laminar to turbulent. The Reynolds number, which is a non-dimensional parameter expressing the ratio of intertial force to viscous force, is regarded as a general criterion for ascertaining the type of flow. In case of pipe flow, the Reynolds number is defined as $\mathbf{R} = \rho Vd/\mu$, where $V = Q/A$ is the mean velocity of flow, d is the pipe diameter and ρ and μ are the density and viscosity of the fluid. The critical Reynolds number is that value of Reynolds number at which transition is expected to occur. Based on his experimental findings, Reynolds concluded that transition from laminar to turbulent flow in pipes always occurred at more or less the same Reynolds number. The approximate numerical value of the Reynolds number at which the laminar flow starts becoming unstable and breaks down was found to be of the order of 2,000, thus

$$\mathbf{R}_{cr} = \frac{\rho Vd}{\mu} = 2{,}000 \quad \text{...(9.1)}$$

Accordingly, for pipe flow if $\mathbf{R} < \mathbf{R}_{cr}$, the flow is laminar while for $\mathbf{R} > \mathbf{R}_{cr}$ the flow may change over to turbulent. Reynolds thought that the critical Reynolds number depends strongly on the initial disturbance prevailing in the flow and increases as the disturbances (in the fluid before its entry into the pipe) are decreased. V.W. Ekman, working on Reynolds apparatus succeeded in maintaining laminar flow upto a critical Reynolds number of 40,000 by providing a smooth entrance and creating conditions free from disturbance. The upper limit of the critical Reynolds number has no physical significance in so far as it depends on the initial disturbance existing in the incoming fluid. There however exists a lower limit of critical Reynolds number as verified through numerous experiments and is given by Eq. (9.1). Below this value of $\mathbf{R}_{cr}$ = 2,000, the flow remains laminar even in the presence of strong disturbances. Laminar flows are basically unstable at high Reynolds number because of fluid inertia.

The factors which affect the transition from laminar to turbulent flow may be listed as below :

(1) The free-stream turbulence (initial disturbances) prevailing in the in-coming fluid.

(2) The pressure gradient dp/dx influences the transition ; for accelerated flow characterized by $dp/dx < 0$, the critical Reynolds number signifying the change of flow state from laminar to turbulent increases, whereas for retarded flow the pressure gradient $dp/dx > 0$ causes the critical Reynolds number to decrease.

(3) The effect of roughness of the boundary is to decrease the critical Reynolds number. The presence of surface roughness is a source of additional disturbance.

(4) The curvature of boundary also affects the transition. The effect of curvature can be understood through the pressure gradient developed as a result of curvature of streamlines.

(5) Suction of the boundary layer—a method for preventing separation of boundary layer by sucking the fluid, has a very powerful influence on stabilizing the laminar flow ; but injection on fluid into the boundary layer will tend to decrease the critical Reynolds number.

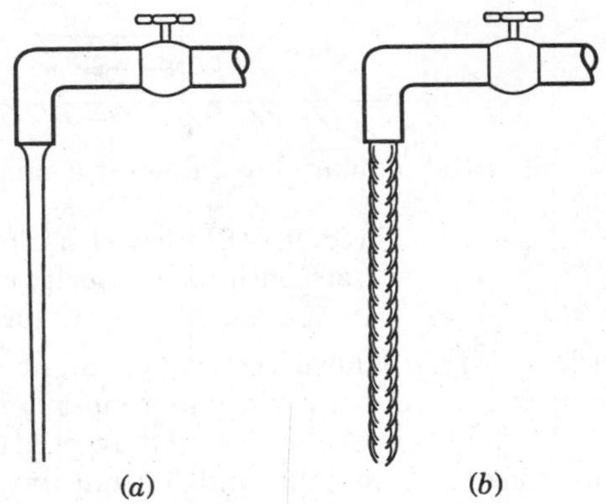

Flow from a pipe tap : (*a*) laminar flow at very small flow rates and (*b*) turbulent flow at higher flow rates.
Fig. 9.1 (A)

Transition from laminar to turbulent flow causes a considerable change in the law of resistance which influences the energy required to maintain flow. In laminar flow the longitudinal pressure gradient which maintains the flow is proportional to first power of average velocity, see Eq. (8.9).

On the contrary, in turbulent flow this pressure gradient becomes nearly proportional to second power of mean velocity (energy loss in turbulent flow is, therefore, nearly proportional to second power of velocity). The increase in resistance in turbulent flow as compared to that for laminar is due to turbulent mixing.

A rising column of smoke emerging out of a railway steam engine or from a factory chimney and its subsequent diffusion into the atmosphere ; the flow of water from a tap [Fig. 9.1 A (*b*)] and the smoke rising from a lighted cigarette are some of the common and familiar examples which illustrate the phenomenon of turbulence. The diffusion of a column of smoke and its subsequent disappearance into the atmospheric air is attributed to the turbulent mixing.

9.1.1. Causes of Turbulence

Turbulence is generated in any of the following ways :

(1) In a flow having abrupt discontinuous velocity distribution. The surface of discontinuity occurs whenever two fluid streams come together in such a way as to cause a sudden jump in velocity between adjacent layers, Fig. 9.1 (B) part (1), and (2).

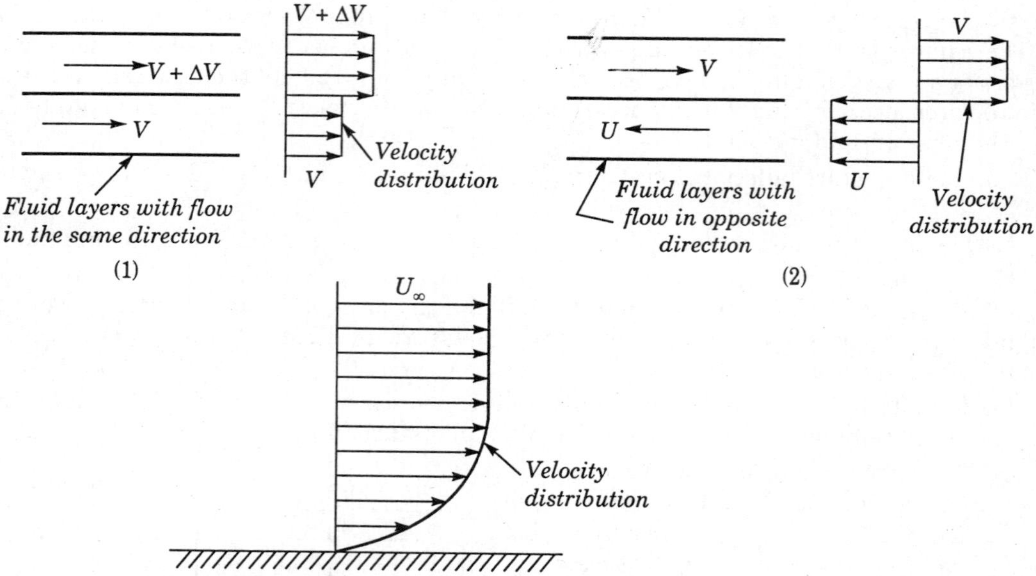

(3) Boundary layer flow—a common example of shear flow

Fig. 9.1 (B) Flow situation in (1) and (2) illustrate abrupt discontinuous velocity change resulting in turbulence ; (3) shows velocity profile in a typical shear flow.

(2) In a flow where velocity gradient occurs without abrupt discontinuity, *i.e.* shear flow (flow in which there is a variation of mean velocity in space is called a shear flow) see Fig. 9.1 (B) part (3). Reynolds experiment with a dyestreak in a glass tube, described in 8.2.1, is an example of shear flow becoming unstable and degenerating into turbulence. In ordinary shear flow, disturbances may be generated due to many reasons, and once they appear they might amplify or may disappear.

9.1.2. Instability of Disturbances

Of the several theories of instability of disturbances, the one which has a good experimental support is the small-perturbation theory (known as the Tollmein-Schlichting theory). It is based on the premise that the disturbances are composed of oscillations of a range of frequencies which can be selectively amplified by the hydrodynamic flow field. This theory leads to the conclusion that below a critical Reynolds number all disturbances will be suppressed by viscous damping. Above this critical Reynolds number certain frequencies will be amplified while others damped.

9.2. Characteristics of Turbulent Flow

The turbulent flow is characterized by random, irregular and haphazard movement of fluid particles, making it impossible to predict the motion of a fluid particle with respect to time and space. This type of motion gives rise to velocity fluctuations in all the directions. The velocity fluctuations in the transverse direction (normal to the flow) are responsible for transverse mixing of small masses of fluid. This mixing process of turbulent flow has been clearly evidenced in the Reynolds experiment through the diffusion of dye filament. In the process of turbulent mixing the fluid with high kinetic energy is carried by the transverse fluctuations towards the boundary, while the fluid moving at low kinetic energy near the boundary is carried away from the boundary into the region of high kinetic energy. This results in a more uniform velocity distribution in turbulent flow as compared to that in the laminar one. The transverse mixing of fluid tends to equalise the velocity except near the boundary where this mixing tends to vanish. Fig. 9.2 shows the velocity distribution in laminar and turbulent flow at the same Reynolds number.

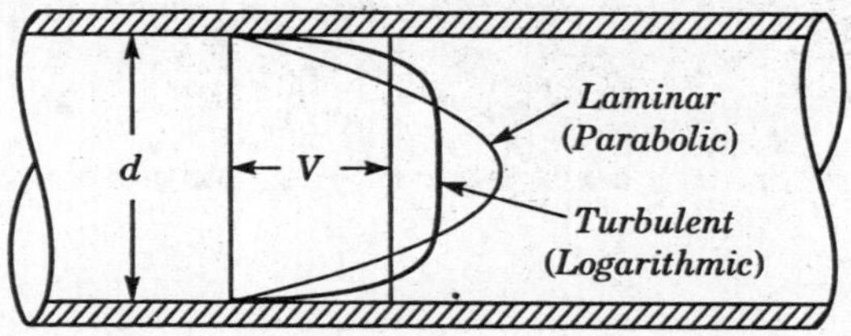

Fig. 9.2. Velocity distribution curves in laminar and turbulent flows in the same pipe at constant $R = \dfrac{\rho VD}{\mu}$.

The most essential feature of a turbulent flow is the fact that at a given point velocity and pressure are not constant with time but exhibit very irregular fluctuations of high frequency. Turbulent motion, once established, has a random, haphazard and irregular nature which makes exact description difficult. Because of this random nature, statistical methods are used to describe its characteristics. The statistical quantities which are useful include the root mean-square (r.m.s.) of the turbulent fluctuations, the degree of correlation defined by the correlation coefficient, and the spectrum of energy *versus* frequency.

H.L. Dryden measured the random velocity fluctuations using a hot wire anemometer. Fig. 9.3 shows nature of variation of instantaneous velocity of fluid measured at a point with respect to time.

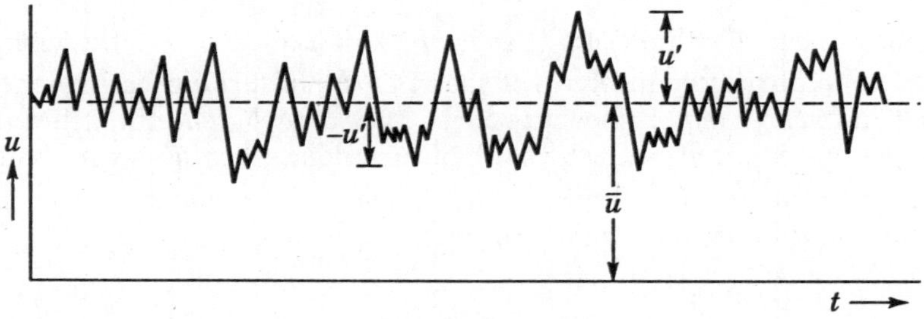

Fig. 9.3. Random velocity fluctuations at a point in turbulent flow.

The velocity in a turbulent flow may be considered as consisting of a time average portion ($\overline{u}$) and a fluctuating part (u'). The velocity could be written accordingly as

$$u = \overline{u} + u' \qquad \text{...(9.2)}$$

in which u = instantaneous velocity

$\overline{u}$ = time average or temporal mean velocity and

u' = velocity fluctuation.

By definition, the temporal mean of any quantity $a = f(x, y, z, t)$ may be denoted by $\overline{a}$ and expressed as

$$\overline{a} = \frac{1}{T}\int_0^T a\,dt$$

The temporal mean velocity components may therefore be expressed as

$$\overline{u} = \frac{1}{T}\int_0^T u\,dt$$

$$\overline{v} = \frac{1}{T}\int_0^T v\,dt$$

$$\overline{w} = \frac{1}{T}\int_0^T w\,dt \qquad ...(9.3)$$

and by the definition of u', the time average of u' or temporal mean of u' is

$$\overline{u}' = \frac{1}{T}\int_0^T u'\,dt = \frac{1}{T}\int_0^T (u - \overline{u})\,dt$$

$$= \frac{1}{T}\int_0^T u\,dt - \frac{1}{T}\int_0^T \overline{u}\,dt = \overline{u} - \overline{u} = 0$$

hence, $\overline{u}' = \overline{v}' = \overline{w}' = 0.$

The time T represents the interval over which the averaging is to be carried out. It is, therefore, the period of sampling which must be sufficiently large to include an adequate number of fluctuations.

In the light of above discussion, it would appear that a turbulent flow can never be considered steady since the velocity at a point changes with time. However as this change of velocity (that is the velocity fluctuation) is small as compared to the temporal mean velocity, the definition of steady of unsteady, uniform or non-uniform turbulent flow could be based on temporal mean velocity. Thus a turbulent flow will be steady only when the temporal mean velocity at a point remains constant with time.

The root-mean-square value of u', that is $\sqrt{\overline{u}'^2}$ is a measure of the violence of turbulent fluctuations and is a quantitative measure of the *intensity of turbulence*. The mean time interval between reversals in the sign of u' is measure of the *scale of turbulence*. In general, the intensity of turbulence increases with velocity and the scale of turbulence increases with boundary geometry.

The *isotropic turbulence* is defined as that turbulence in which the mean-square velocity fluctuation in three co-ordinate directions are equal to each other :

$$\overline{u}'^2 = \overline{v}'^2 = \overline{w}'^2$$

The isotropic turbulence may be obtained in a wind tunnel at a certain distance downstream of a fine-mesh screen. The level, degree or intensity of turbulence is defined by

$$\frac{1}{U_\infty}\sqrt{(\overline{u}'^2 + \overline{v}'^2 + \overline{w}'^2)/3} \qquad ...(9.4)$$

in which U_∞ is the free-stream velocity. The degree of turbulence in a wind tunnel with screens may be expressed as $\sqrt{\overline{u}'^2}/U_\infty$.

The average kinetic energy of turbulence per unit mass is defined as :

$$\frac{\text{Average K.E. of turbulence}}{\text{Mass}} = \frac{1}{2}(\overline{u}'^2 + \overline{v}'^2 + \overline{w}'^2) \qquad ...(9.5)$$

which is proportional to the sum of the squares of the intensities. The turbulent energy in a boundary-layer is maximum near the wall and decreases towards the free-stream.

As for any random process, the kinetic energy of turbulence can be decomposed into an energy spectrum of energy density *versus* frequency. The energy density $\Phi(f)$ at a point in the flow is defined as the limit of average kinetic energy per unit mass contained in a particular frequency bandwidth Δf divided by the bandwidth, or

$$\phi(f) = \lim_{\Delta f \to 0} \frac{\text{Average K.E./Mass contained in } \Delta f}{\Delta f}$$

in which f = frequency in cycles per second $\omega/2\pi$.

The integral of energies over the range of frequencies appearing in the turbulence gives the average value of kinetic energy, Eq. (9.5). Thus, we obtain

$$\frac{\text{Average K.E. of turbulence}}{\text{Mass}} = \int_0^\infty \phi(f)\, df = \frac{1}{2}(\overline{u'^2} + \overline{v'^2} + \overline{w'^2}) \qquad \text{...(9.6)}$$

9.3. Mechanism of Turbulence

It is known from the foregoing discussion that in turbulent flow small masses of fluid are actively participating in transverse mixing, *i.e.*, fluid mass moving with a lesser velocity such as at *A* moves transversely to a zone of higher velocity such as that represented by *B* (Fig. 9.4). In doing so, the momentum of the fluid mass undergoes change (momentum increases as the fluid mass is displaced from *A* to *B*), and exerts a drag on the fluid at *B*. Similarly if a fluid mass moves from a zone of higher velocity towards that of lower one (such as from *B* to *A*) it will tend to accelerate the fluid around its new position. These forces are the result of turbulent cross motion of small masses of fluid (say a group of few molecules) and are largely responsible for shear force within the fluid.

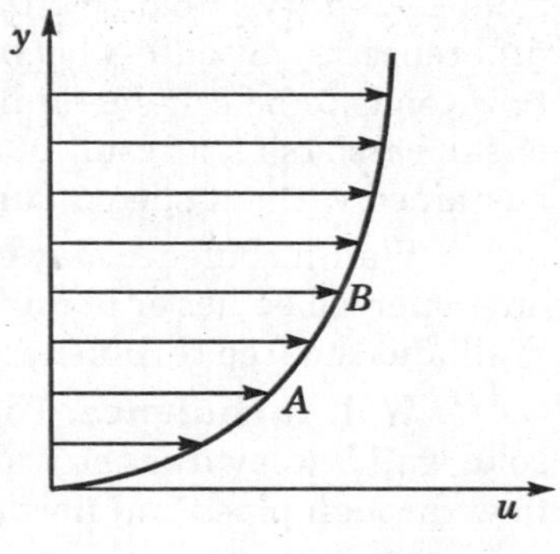

Fig. 9.4. Velocity distribution in turbulent flow.

Boussinesq introduced a coefficient of molar viscosity (as distinguished from molecular viscosity μ), by means of which the turbulent shear stress caused by momentum transport in the transverse direction might be expressed in terms of temporal mean velocity gradient. Thus for two dimensional parallel flow in which velocity varies perpendicular to flow, if the direction of flow is taken to be along the x-axis we have $\bar{u} = f(y)$, $\bar{v} = 0$, $\bar{w} = 0$ and the shear stress in turbulent flow reduces to

$$(\tau_{xy})_t = \eta\; d\bar{u}/dy \qquad \text{...(9.7)}$$

in which $(\tau_{xy})_t$ = turbulent shear stress,

$\frac{d\bar{u}}{dy}$ = velocity gradient based on temporal mean velocity and

η = a coefficient known as eddy viscosity.

This coefficient η (eta) is dimensionally the same as the dynamic viscosity μ (mu) and is known as the *eddy viscosity*. It may be considered as a dynamic coefficient of turbulence.

The total shear stress in turbulent flow, thus, consists of a laminar part and a turbulent part and may be expressed as

$$\tau = \tau_l + \tau_t$$

$$= \mu\frac{d\bar{u}}{dy} + \eta\frac{d\bar{u}}{dy} = (\mu + \eta)\frac{d\bar{u}}{dy} \qquad \text{...(9.8)}$$

Actually η varies widely in magnitude, not only with general conditions of flow but also from point to point in a given flow. While μ represents a fluid property independent of the type of flow, η depends upon the fluid density as well as upon the intensity of turbulent fluctuations.

In laminar flow there being no transverse fluctuation, η obviously has a zero magnitude. As the Reynolds number **R** increases beyond the critical value, the eddy viscosity becomes increasingly important, and as shown by experiments it evidently acquired such a high magnitude as to diminish the effect of dynamic viscosity completely.

The eddy viscosity may be recast and given added significance by dividing it by the mass density (ρ) of the fluid, thus

$$\eta/\rho = \varepsilon \text{ (epsilon)} \quad \text{...(9.9)}$$

Just as the ratio μ/ρ represents a quantity independent of mass dimension and results in its naming as the kinematic viscosity ν (nu), the ratio η/ρ is known as the *kinematic eddy viscosity* or the apparent kinematic viscosity and is denoted by ε. It is a direct measure of transporting capacity of the mixing process. Its magnitude in turbulent flow is much greater than the corresponding kinematic viscosity ν for laminar flow.

9.3.1. Classification of turbulence

Turbulence occurs most commonly in various forms of shear flow. Shear flows can be classified according to whether they occur along solid surface or at the interface between fluid zones having different velocities, such as the boundaries of a jet. Turbulence generated in the shear flow near a solid boundary is called *wall turbulence,* whereas, the one caused by velocity differential (*i.e.* difference in velocities) between fluid zones is called *free turbulence.* Turbulent motion in shear flow tends to be self-sustaining in the sense turbulence is generated as a result of shear and the shear persists as a result of turbulent fluctuations. Turbulent flow can neither be generated nor sustained without shear. Turbulent fluctuations can only decay in a shear-free field.

Wall turbulence may be produced in the shear flow with non-uniform boundary-layers found on immersed bodies or in uniform boundary layers which exist in fully developed uniform conduits. Wall and the free turbulence are defined as below :

Wall turbulence. The turbulence generated by the viscous effect due to the presence of a solid wall is known as the wall turbulence. The examples of this type of turbulence may be found in flow through pipes and flow past solid bodies.

Free turbulence. The turbulence generated by the fluid flowing at different velocities, in the absence of a solid wall is termed as free turbulence. Turbulent mixing of submerged jets and the turbulence in the wake regions are examples of this category of turbulence.

9.4. Shear Stress due to Turbulence (*Reynolds Stresses*)

Among the various investigators who studied the turbulence phenomena, Reynolds was first to show that the exchange of transverse momentum due to velocity fluctuations causes a tangential force between the adjacent layers. This can be better understood with the help of an example : consider two open railway wagons loaded with coal moving in the same direction on parallel tracks (which are close to each other) at slightly different speeds. Let the coal be exchanged between the two wagons simultaneously. The slowly moving wagon is continuously being attacked by pieces of coal which are also moving forward slightly faster than the wagon being hit by them. This striking of coal pieces against the slowly moving wagon results in an additional forward force on that wagon. Simultaneously, the coal pieces thrown from the slowly moving wagon strike the faster moving wagon. The two opposite forces thus generated on the respective wagons are equivalent to the frictional forces exerted by the wagons on each other as if the sides of the two wagons were rubbing together. This explains as to how the exchange of momentum in the transverse direction results in a frictional force or a shear stress. The movement of adjoining fluid layers at slightly different velocities is analogous to the motion of wagons and the exchange of coal pieces in the transverse direction is analogous to the velocity fluctuations resulting in the random movement of fluid masses.

Reynolds considered x-component of velocity fluctuation (u') as superimposed on the temporal mean velocity ($\overline{u}$) while the y-component (v') was considered as a cross-current or normal fluctuation (transverse to the flow direction). The function of v' is to cause an exchange of fluid particles from one layer to the other. Assuming that these transverse fluctuations v' are uniformly distributed over a surface of unit area (separating two adjacent layers), consider a control volume (Fig. 9.5) within a fluid and let us apply the momentum equation to this control volume. The average normal velocity component is zero, so that the average net mass flow through each side of the control volume is zero, but it is still possible to have an average net momentum flow through each side.

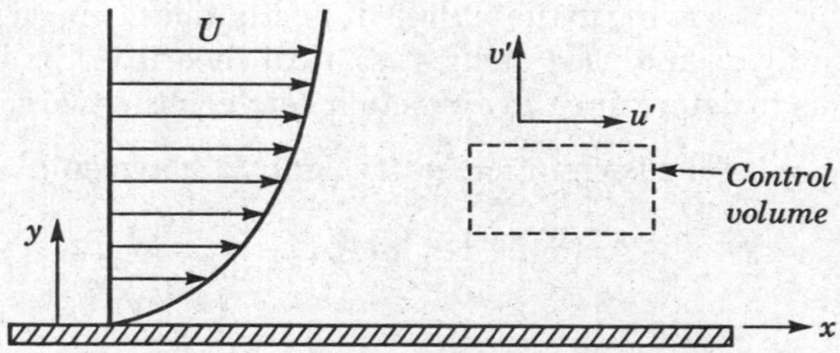

Fig. 9.5. Control volume near a wall to illustrate origin of turbulent shear stress.

The average net rate of momentum flow through the face parallel to flow per unit area is

$$\frac{1}{T}\int_0^T \rho u v'\,dt = \frac{1}{T}\int_0^T \rho(\overline{u}+u')\,v'\,dt = \frac{\rho\overline{u}}{T}\int_0^T v'\,dt$$

$$+\frac{\rho}{T}\int_0^T u'v'\,dt = \frac{\rho}{T}\int_0^T u'v'\,dt = \overline{\rho u'v'} \qquad ...(9.10)$$

From the momentum principle, Eq. (9.10) represents force per unit area and is, therefore, the *turbulent shear stress*. The turbulent stress is caused by the momentum transfer associated with the turbulent fluctuations.

Without going into the mathematical details, the equation of motion taking into account the effect of turbulence, expressed in terms of time average quantities in the x-direction may be written as

$$\rho\left(\overline{u}\frac{\partial\overline{u}}{\partial x}+\overline{v}\frac{\partial\overline{u}}{\partial y}+\overline{w}\frac{\partial\overline{u}}{\partial z}\right) = -\frac{\partial p}{\partial x}+\mu\left(\frac{\partial^2\overline{u}}{\partial x^2}+\frac{\partial^2\overline{u}}{\partial y^2}+\frac{\partial^2\overline{u}}{\partial z^2}\right)$$

$$-\left[\frac{\partial}{\partial x}(\overline{\rho u'^2})+\frac{\partial}{\partial y}(\overline{\rho u'v'})+\frac{\partial}{\partial z}(\overline{\rho u'w'})\right] \qquad ...(9.11)$$

Eq. (9.11) represents the *Reynolds equation* of motion in the x-direction ; the second terms on the right hand side of the equation represents the viscous stresses ($\overline{u}$ is the time-average or temporal mean velocity) while the third term within large brackets involves Reynolds stresses or the apparent stresses due to exchange of momentum resulting from turbulent mixing. The first stress $-\ (\overline{\rho u'^2})$ is a normal stress and represents the additional momentum leaving in the x-direction per unit area per unit time due to fluctuating velocity component u'. This is equivalent to additional dynamic pressure in the x-direction. The stresses $-(\overline{\rho u'v'})$ and $-(\overline{\rho u'w'})$ are the tangential stresses or shear stresses in the x-direction due to turbulent velocity fluctuations u', v' and w'. The Reynolds equations of motion in y- and z-directions can be similarly written from the Navier-Stokes equation by including the turbulent stresses on the pattern of Eq. (9.11).

It is an observed fact that a positive v' is associated with negative u' and *vice versa* with the result that the instantaneous rate of momentum transfer per unit area is positive and is equal to $-\rho u'v'$. We can now equate the mean rate of momentum transfer per unit area to a turbulent shearing stress,

$$(\tau_{yx})_t = -\overline{\rho u'v'} \qquad ...(9.12)$$

Owing to the turbulent fluctuations being a random phenomenon, the turbulent flow problems may be analysed using statistical theories. Application of statistical methods of analysis permits us to determine the correction between the two components of fluctuating velocities (u' and v') at a point. The correlation co-efficient $\overline{R}$ between u' and v' is defined as

$$\overline{R}\,(u, v) = \frac{-\overline{u'v'}}{\sqrt{\overline{u'^2}}\,\sqrt{\overline{v'^2}}} \qquad \text{...(9.13)}$$

For no correlation between u' and v', $\overline{u'v'} = 0$, leading to $\overline{R} = 0$, while for complete correction $\overline{u'v'} = \sqrt{\overline{u'^2}} = \sqrt{\overline{v'^2}}$ resulting $\overline{R} = 1$. In the turbulent boundary layer R has been experimentally found to have a value between 0.45 to 0.55.

It may be shown that the ratio of the turbulent shear stress τ_t and viscous shear stress τ_t has the form of a Reynolds number. The viscous shear stress (or laminar shear stress) may be written as

$$(\tau_{yx})_l = \mu u'/l \qquad \text{...(9.14)}$$

in which u' represents change in velocity in a distance l measured normal to the velocity. The ratio of shear stresses in turbulent and laminar flows may be expresses using Eqs. (9.10) and (9.14) as

$$(\tau_{yx})_t/(\tau_{yx})_l = \frac{\rho v' l}{\mu}$$

which has the form of a Reynolds number having a characteristic length l and characteristic velocity v'.

9.5. Semi-Empirical Theories of Turbulence

The most general type of fluid flow problems are three-dimensional in character involving forces due to turbulence in addition to those due to pressure, gravity and viscosity. The Reynolds equations including the continuity equation expressed in terms of time-average velocity components are the relevant equations of motion governing the turbulent flow. The turbulent flow being physically characterized by the disorderlines of fluid movement and random variation of various quantities with time and space. It is thus obvious that this type of flow is not exclusively amenable to mathematical analysis owing to the complexities imposed by the statistical phenomenon of turbulence. This is due to the Reynolds equations containing six unknown Reynolds stresses (three normal and three tangential or shear stresses) in addition to variables u, v, w and p. As the present knowledge of turbulence stands, the exact relationship between the Reynolds stresses and the time-average quantities is not yet known. The phenomenon of turbulence has, therefore, been studied statistically and the present theoretical knowledge is rather inadequate to make it applicable to problems of engineering interest. There are, however, several semi-empirical theories developed which have been found to yield useful and satisfactory results for many problems. Prandtl's mixing-length theory and Karman similarity hypothesis being well known semi-empirical theories have been described in brief in the following text.

9.5.1. Prandtl's mixing-length theory

Prandtl has developed a very useful theory of turbulence based on the concept of mixing-length using the analogy of mean-free-path of kinetic theory of gases. He introduced a linear dimension called the mixing length. In kinetic theory of gases the mean free path is defined as the average distance which a gas molecule travels before it strikes another molecule. Analogous to mean-free-path, the mixing length may be considered to represent the average distance perpendicular to the flow a small fluid mass will travel before its momentum is changed by the new environment. The turbulent shear stress expressed by Eq. (9.7) may be written in the form

$$(\tau_{xy})_t = \rho\varepsilon\; d\overline{u}/dy = -\rho u'\overline{v}' \qquad \text{...(9.15)}$$

From the analogy of kinetic theory of gases, it is reasonable to assume the following relation for turbulent flow :

$$\varepsilon = l u_* \quad ...(9.16)$$

where u_* is known as the *shear or friction velocity* ; and l represents a unique length characterizing the local intensity of turbulent mixing at any level. The magnitude of l may be much larger than the mean-free-path, and may also be a function of space (that is y), the mean velocity etc.

Prandtl's mixing-length theory is based on the assumption that there is complete correlation between the fluctuating velocity components u' and v' [R = 1.0, Eq. (9.13)]. From Eqs. (9.13) and (9.15) we obtain

$$u_*^2 = \frac{(\tau_{xy})_t}{\rho} = -\overline{u'v'} = \sqrt{\overline{u'^2}}\,\sqrt{\overline{v'^2}} \quad ...(9.17)$$

The turbulent shear stress may now written using Eqs. (9.15), (9.16) and (9.17)

$$(\tau_{xy})_t = \rho\, u_*^2 = \rho l u_* du/dy \quad ...(9.18)$$

which yields $$u_* = l\ d\bar{u}/dy \quad ...(9.19)$$

The Reynolds stress (*i.e.* the turbulent shear stress), substituting this value of u_* in Eq. (9.18) is obtained as

$$(\tau_{xy})_t = \rho l^2\,(d\bar{u}/dy)^2 \quad ...(9.20\ a)$$

As the Reynolds stress depends on the sign of the mean velocity gradient, Eq. (9.20 *a*) can be expressed as

$$(\tau_{xy})_t = \rho l^2\ |\ d\bar{u}/dy\ |\ d\bar{u}/dy \quad ...(9.20\ b)$$

in which $|\ d\bar{u}/dy\ |$ represents the magnitude or absolute value of the mean velocity gradient while $d\bar{u}/dy$ is used with its actual sign (negative or positive). Eq. (9.20 *b*) contains length l which is known as the Prandtl's *mixing length.* This equation is valid so long as $u' \sim v'$. Prandtl argued that the length l must be proportional to distance y as the turbulent fluctuations u' and v' must vanish at the solid boundary.

Expression for mixing length

In turbulent flow the total shear stress may be considered mainly due to momentum transport in the transverse direction ; the viscous shear stress being very small in comparison to the turbulent shear stress may be neglected. The turbulent flow may now be treated analytically by equating the expression for linear variation of shear stress caused by turbulence (Fig. 9.6).

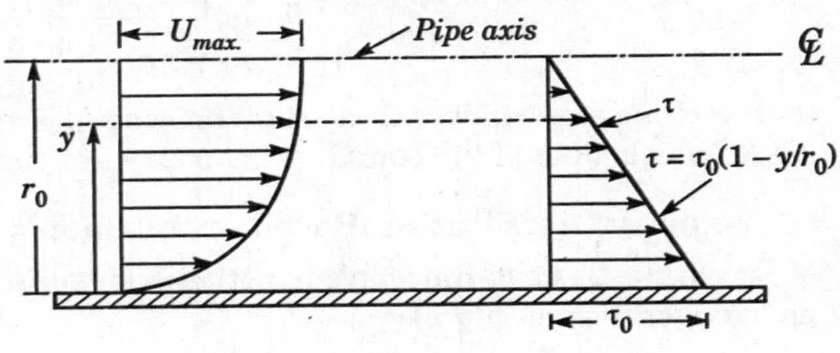

Fig. 9.6. Velocity and shear stress distribution in a pipe.

Thus

$$\tau = \tau_0(1 - y/r_0) = \rho l^2\,(d\bar{u}/dy)^2.$$

We can now drop the bar over the velocity, noting that the velocities are temporal mean velocities. An expression for mixing-length may now be written as :

$$l = \frac{\sqrt{\tau_0/\rho}}{du/dy}\sqrt{1 - y/r_0} \quad ...(9.21)$$

The mixing-length l varies across the flow (*i.e.* with y/r_0) as is evident from the Eq. (9.21), and its variation may be obtained from experimental observations. The measurement of velocity at different points in the y-direction by Pitot tube will yield a velocity profile and measurement of head loss will permit computation of τ_0 from, see Eq. (9.55),

$$\tau_0 = \frac{\gamma h_L r_0}{2L} \quad \text{...(9.22)}$$

in which γ = specific weight of the fluid,

r_0 = pipe radius, and

h_L = head loss in length L.

The values of du/dy can be obtained from the velocity profile, and the mixing-length computation at different values of y will yield distribution of mixing length across the flow. Fig. 9.7 shows the mixing-length distribution in a pipe at various Reynolds numbers.

It can be seen that mixing length trends to zero as the wall is approached and that all curves have a common tangent at the wall ($y = 0$), the value of the slope being 0.4.

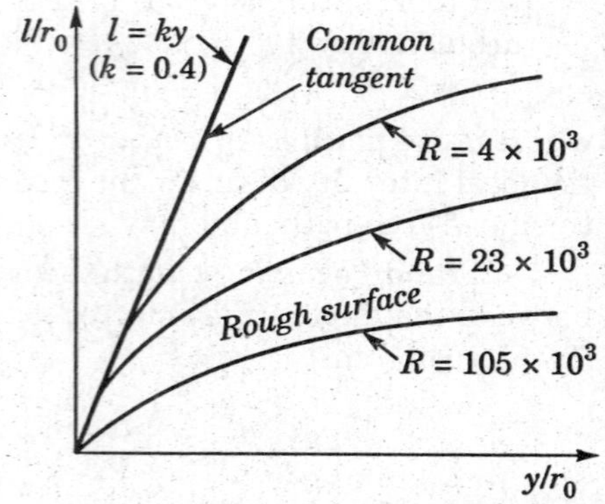

Fig. 9.7. Mixing-length disttribution in a pipe (not to scale).

9.5.2. Karman Similarity Hypothesis

Theodore von Karman developed an important relationship for the mixing-length l in terms of derivatives of mean velocity and based his hypothesis on similarity consideration. With reference to a two-dimensional mean motion parallel to x-axis, if a system of co-ordinates moving with the mean motion is selected, the motion becomes steady with respect to such a system. According to von Karman's basic hypothesis, the fluctuations at all points in such a system differ only in length and time scales. Expressed in other words, it amounts to saying that the mixing length l is a function of distribution of mean velocity $\bar{u}$ and is indirectly related to the distance y from a solid boundary. It is not proposed here to include the mathematical details, but only the relevant results are quoted below. An expression for the mixing-length suggested by von Karman is of the form

$$l = K \left| \frac{d\bar{u}/dy}{d^2\bar{u}/dy^2} \right| \quad \text{...(9.23)}$$

Eq. (9.23) indicates that the mixing length at a point is proportional to the ratio of first and second derivatives of mean velocity at that point. This equation enables study of the distribution of mixing-length perpendicular to the flow direction provided that the variation of mean velocity $\bar{u}$ with y is known. The turbulent shear stress using Eq. (9.21) is

$$(\tau_{xy})_t = \rho K^2 \frac{(d\bar{u}/dy)^4}{(d^2\bar{u}/dy^2)^2} \quad \text{...(9.24)}$$

where K (Kappa) is a universal constant known as Karman's constant (determined from experiments) and is independent of the boundary geometry, roughness and Reynolds number.

9.6. Velocity Distribution Based on Karman's Hypothesis

On the basis of Karman's hypothesis for turbulent flow in pipes, one may write (dropping the bar over the velocity u)

$$(\tau_{xy})_t = \tau_0(1 - y/r_0) = \rho K^2 \frac{(du/dy)^4}{(d^2u/dy^2)^2}$$

or

$$\sqrt{\frac{\tau_0}{\rho}} \sqrt{\left(1 - \frac{y}{r_0}\right)} = K \frac{(du/dy)^2}{d^2u/dy^2}.$$

The quantity $\sqrt{\tau_0/\rho}$ has the dimension of velocity and is, therefore, appropriately called *shear velocity* or friction velocity and is denoted by u_*. The above equation may now be written as

$$\frac{d^2u/dy^2}{(du/dy)^2} = \frac{K}{u_*\sqrt{1 - y/r_0}}$$

Since the second derivative d^2u/dy^2 is negative, introducing a negative sign in the above equation,

$$\frac{d^2u/dy^2}{(du/dy)^2} = -\frac{K}{u_*\sqrt{1 - y/r_0}}$$

putting $$\frac{du}{dy} = Z, \quad \frac{dZ/dy}{Z^2} = -\frac{K}{u_*\sqrt{1 - y/r_0}}$$

Integrating, $$\frac{1}{Z} = -\frac{2Kr_0}{u_*}\sqrt{1 - y/r_0} + A.$$

The value of the constant of integration A may be determined from the fact that as $y \to 0$, the velocity gradient du/dy becomes very large, so that $\frac{1}{du/dy}$ may be taken as zero. Using this condition,

$$0 = -\frac{2Kr_0}{u_*} + A,$$

$$\therefore \quad A = \frac{2Kr_0}{u_*}$$

$$\therefore \quad \frac{1}{du/dy} = -\frac{2Kr_0}{u_*}\left(\sqrt{1 - \frac{y}{r_0}} - 1\right)$$

or $$\frac{1}{du/dy} = \frac{2Kr_0}{u_*}\left(1 - \sqrt{1 - \frac{y}{r_0}}\right)$$

Let $1 - y/r_0 = x^2$

$\therefore \quad dy = -2r_0 x dx.$

Substituting and integrating

$$u = \frac{u_*}{2Kr_0}\int \frac{dy}{1 - \sqrt{1 - y/r_0}} = \frac{u_*}{2Kr_0}\int \frac{2r_0 x\, dx}{1 - x}$$

$$= -\frac{u_*}{K}\int \frac{xdx}{1 - x} = -\frac{u_*}{K}\,[x + \log_e(1 - x)] + B$$

Hence $$u = \frac{u_*}{K}\left[\sqrt{1 - \frac{y}{r_0}} + \log_e\left(1 - \sqrt{1 - \frac{y}{r_0}}\right)\right] + B$$

or $$\frac{u}{u_*} = \frac{1}{K}\left[\sqrt{1 - \frac{y}{r_0}} + \log_e\left(1 - \sqrt{1 - \frac{y}{r_0}}\right)\right] + B$$

The constant may be found by substituting $u = u_{max}$ at the centre of pipe, *i.e.* for $y = r_0$.

$$\therefore \quad \frac{u_{max}}{u_*} = B$$

The velocity distribution in a smooth pipe then becomes :

$$\frac{u}{u_*} = \frac{1}{K}\left[\sqrt{1-\frac{y}{r_0}} + \log_e\left(1-\sqrt{1-\frac{y}{r_0}}\right)\right] + \frac{u_{max}}{u_*}$$

from which,
$$\frac{u_{max}-u}{u_*} = -\frac{1}{K}\left[\sqrt{1-\frac{y}{r_0}} + \log_e\left(1-\sqrt{1-\frac{y}{r_0}}\right)\right] \qquad ...(9.25)$$

The quantity $(u_{max} - u)$ is called 'velocity deficiency' or the 'velocity defect' and has been found to be the function of y/r_0. Eq. (9.25), therefore, expresses what is known as the *velocity defect law.*

Nikusadse's experiments on smooth pipes ($5 \times 10^3 < \mathbf{R} < 3 \times 10^6$) and on pipes of uniform sand grain roughness (artificially roughened) showed that all velocity profiles could be represented by

$$\frac{u_{max}-u}{u_*} = -2.5 \log_e \frac{y}{r_0} \qquad ...(9.26)$$

Comparing Eqs. (9.25) and (9.26), it is found that the universal constant K has a value of 0.40.

The ratio of average velocity V and the maximum velocity u_{max} may be obtained by the principle of continuity. The average velocity V may be expressed as

$$V \,.\, \pi r_0^2 = \int_0^{r_0} 2\pi r u dr \qquad ...(9.27)$$

we may also write
$$u_{max}\pi r_0^2 = \int_0^{r_0} u_{max}\, 2\pi r\, dr \qquad ...(9.28)$$

Subtracting Eq. (9.27) from Eq. (9.28), we obtain

$$(u_{max} - V)\pi r_0^2 = \int_0^{r_0} (u_{max} - u)\, 2\pi r\, dr$$

substituting $(-2.5\, u_* \log_e y/r_0)$ for $(u_{max} - u)$ and $V\sqrt{f/8}$ for u_* [see Eq. (9.56)], the ratio of V/u_{max} may be written as

$$\frac{V}{u_{max}} = \frac{1}{1+4.7\sqrt{f/8}} = \frac{1}{1+1.33\sqrt{f}} \qquad ...(9.29)$$

The above equation which states that the velocity V/u_{max} decreases with increase in friction factor f (for f see Eq. 9.27) and *vice versa*, agrees well with the experimental observation. Eqs. (9.25) and (9.26) are the general equations for velocity distribution for turbulent flow in a circular pipe. As they have been developed from the turbulence characteristics without making any assumption as to the nature of the pipe wall, they are valid for smooth as well as rough pipes.

Experiments conducted by Fritsch on rectangular conduits show that for constant discharge the velocity profiles undergo substantial changes if the roughness of the pipe is varied (Fig. 9.8).

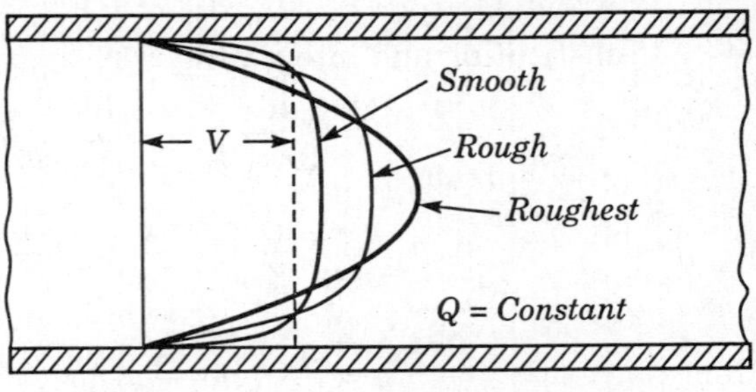

Fig. 9.8. Effect of conduit roughness on velocity distribution.

9.7. The Universal Velocity Distribution Law Near a Solid Boundary

Flow near a solid boundary is characterised by appreciable changes in velocity across the flow brought about by viscous action in the boundary layer. The region of flow affected by the viscous action is directly proportional to the kinematic viscosity ν. The velocity of flow u reduces towards the boundary due to the action of viscous shear and becomes equal to zero at the boundary itself, *i.e.* there being 'no-slip' between the fluid and the boundary.

For analysis of turbulent flow along a solid boundary it was suggested by Nikuradse, Karman and others that the flow should be divided into three zones. The first zone consists of a thin layer in the immediate neighbourhood of the boundary in which the action of viscous shear predominates while the Reynolds stress is negligible. This zone is the region of *laminar sub-layer*. The zone immediately above the laminar sub-layer is known as the 'buffer zone' or the laminar zone in which the viscous stress and the Reynolds stresses are of the same order of magnitude. The next zone adjoining the buffer zone is still farther away from the boundary in which the Reynolds stresses are much larger as compared to the viscous stresses, the viscous stresses may, therefore, be completely neglected in this zone. This zone is known as the *turbulent zone*. For convenience, however, the flow is usually divided into two zones. The first one adjoining the solid boundary accommodating the laminar sub-larger may be called the zone of laminar layer. In this laminar layer the viscous stresses are predominant over the Reynolds stresses (dynamic viscosity μ >> η the eddy viscosity). The second zone represents the remaining portion of flow in which the effect of turbulence is large enough to outweigh the viscous effects (η >> μ). This region of flow may be known as turbulent layer. Fig. 9.9 illustrates these zones diagramatically.

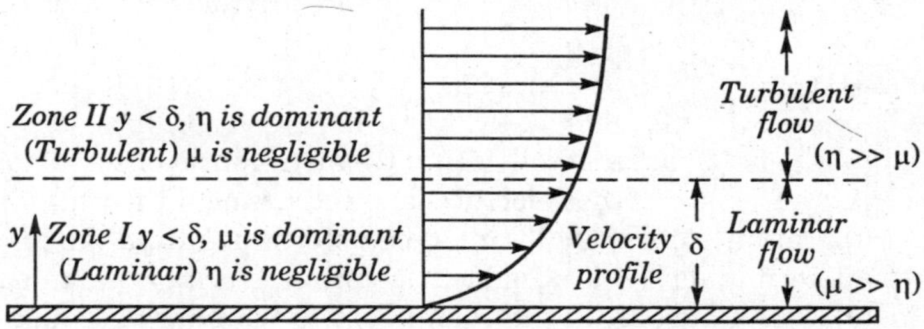

Fig. 9.9. Comparative role of dynamic viscosity μ and eddy viscosity η in affecting flow phenomena near a smooth boundary.

In between the two regions, there must also be a region in which μ and η are equally important. This is a region where the flow changes from laminar to turbulent and is, therefore, known as the transition region or the buffer zone as already described.

Velocity Distribution in Laminar region : (for $y \le \delta$)

The shear stress very close to the boundary may be taken as equal to the boundary shear stress (τ at $y = 0$). From Newton's law of viscosity,

$$\tau_0 = \mu\, du/dy = \mu\, u/y \left(\text{since } \frac{du}{dy} = \frac{u}{y} \text{ in laminar regin}\right) \text{dividing by } \rho,\ \frac{\tau_0}{\rho} = \frac{\mu}{\rho}\frac{u}{y}.$$

Substituting for $\tau/\rho = u_*^2$, we obtain

$$\frac{u}{u_*} = \frac{u_* y}{v} \qquad ...(9.30)$$

Eq. (9.30) shows a linear variation between u and y in the laminar film.

Velocity distribution in the turbulent region : (for $y > \delta$)

In the vicinity of the boundary the turbulent shear stress may be assumed, according to Prandtl, to be constant and equal to τ_0, thus in the turbulent region :

$$(\tau_{xy})_t = \tau_0 = \rho l^2 (du/dy)^2 \qquad ...(9.31)$$

We shall assume linear proportionality between the mixing-length l and the wall distance y, so that

$$l = Ky \qquad ...(9.32)$$

where K is the Karman's constant to be determined from experiments. This assumption is reasonable, because the turbulent shear stress at the solid boundary or wall is zero owing to the disappearance of fluctuations. Nikuradse's experimental data have shown that near the boundary, Eq. (9.32) is more or less satisfied supporting the assumption of linear dependence of mixing length on the distance from the boundary. Eliminating the mixing length from the Pradtl's Eq. (9.31) for the turbulent shear using Eq. (9.32)

$$\tau_0 = \rho K^2 y^2 (du/dy)^2$$

and rearranging, $$u_* = Ky\frac{du}{dy} \quad \text{or} \quad \frac{du}{u_*} = \frac{1}{K}\frac{dy}{y}$$

integrating, $$\frac{u}{u_*} = \frac{1}{K}\log_e y + A \qquad \text{...(9.33)}$$

Eq. (9.33) agrees well with the experimental observations and is useful even when τ depends on y, because most of the velocity change occurs near the wall where τ is almost constant. The constant of integration A of Eq. (9.33) must be determined from the condition at the wall. However, without even determining A it is possible to obtain a velocity law analogous to Eq. (9.25). Fully realizing that Eq. (9.33) is valid only in the vicinity of the boundary because of the assumption of τ = constant, we shall attempt to use it for the entire region of flow. At $y = r_0$, $u = u_{max}$, we then obtain

$$\frac{u_{max}}{u_*} = \frac{1}{K}\log\frac{r_0}{y} \qquad \text{...(9.34)}$$

Eq. (9.34) is one of the forms in which universal velocity distribution law due to Prandtl may be expressed.

The constant of integration A appearing in Eq. (9.33) should be determined from the condition that the turbulent velocity distribution must join the laminar velocity distribution in the immediate vicinity of the boundary where the laminar and the turbulent shear stresses are of the same order of magnitude. Determination of A involves the condition that $u = 0$ at a certain y_0 from the boundary. Eq. (9.33) thus yields

$$\frac{u}{u_*} = \frac{1}{K}\log(y/y_0) \qquad \text{...(9.35)}$$

The distance y_0 is of the order of the thickness of the laminar sub-layer. From dimensional considerations it may be seen that the distance y is proportional to the ratio of kinematic viscosity and the shear velocity, thus

$$y_0 = \beta\nu/u_*$$

where β is a non-dimensional constant. Substituting for y_0 into Eq. (9.35), we obtain

$$\frac{u}{u_*} = \frac{1}{K}\left(\log\frac{u_* y}{\nu} - \log\beta\right) \qquad \text{...(9.36)}$$

in which K and β are to be determined experimentally. Eq. (9.36) is the so-called *universal velocity profile* or the logarithmic velocity distribution law for turbulent flow. The values of K and β may be found by plotting u/u_* against $\log_s \frac{u_* y}{\nu}$. The experimental values of K and $\log\beta$ are :

For flat plates. K = 0.417 and $\log\beta = -5.84$, and the velocity distribution over a smooth flat plate may be expressed as

$$\frac{u}{u_*} = 5.52\log_{10}\frac{u_* y}{\nu} + 5.84 \qquad \text{...(9.37 a)}$$

For smooth pipes. Nikuradse's experiments on smooth pipes yielded a value of $K = 0.40$ and $\log\beta = -5.50$. Using these values the velocity distribution can be written as

$$\frac{u}{u_*} = 5.75\log_{10}\frac{u_* y}{\nu} + 5.50 \qquad \text{...(9.37 b)}$$

Equations (9.37 *a*) and (9.37 *b*) being meant for turbulent flow, are applicable only in regions where the laminar shear stress can be neglected in comparison to the turbulent shear stress. The above equations are known as the *Karman-Prandtl equation* for turbulent flow near smooth boundaries.

The experimental observations over a flat plate have shown that at low values of $\frac{u_* y}{\upsilon}\left(\frac{u_* y}{\nu} < 5\right)$, the velocity profile is well approximated by the linear relationship expressed by Eq. (9.30).

The measurement of velocity fluctuations as well as some heat transfer measurements support the hypothesis that the flow in the immediate neighbourhood of the wall is essentially laminar. This region in which

$$\mu \frac{du}{dy} >> \eta \frac{du}{dy}$$

is known as the *"laminar sub-layer"*. In the range $5 \frac{u_* y}{\nu} < 70$, contributions of the laminar and the turbulent frictions to the total resistance to fluid flow are of the same order of magnitude. This is the transition zone or the buffer zone for which Karman suggested a logarithmic equation for velocity distribution.

$$\frac{u}{u_*} = 11.5 \log_{10} \frac{u_* y}{\nu} + 3.05 \quad ...(9.38)$$

For $\frac{u_* y}{\nu} > 70$ the laminar contribution is negligible as compared to the turbulent friction and it is for this region that Eq. (9.37) is applicable. Thus we have :

(*i*) Purely laminar friction (laminar sub-layer for) $\frac{u_* y}{\nu} < 5$.

(*ii*) Laminar-turbulent friction (transition or buffer zone) for

$$5 < \frac{u_* y}{\nu} < 70.$$

(*iii*) Purely turbulent friction (turbulent zone) for $\frac{u_* y}{\nu} > 70$.

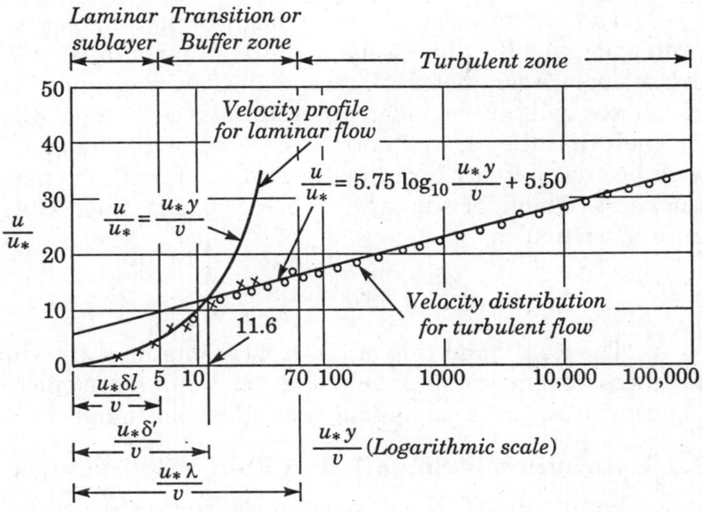

Fig. 9.10. Velocity distribution in a smooth pipe.

Fig. 9.10 shows the variation of u/u_* (ordinary scale) with $u_* y/\nu$ (log scale) and a graphical representation of Eqs. (9.30) and (9.37) and demarcates the limits for laminar, transition and turbulent zones. The condition $y = \delta_l$ defines the thickness of laminar sub-layer, whereas $y = \delta'$ is the thickness where the velocity profiles represented by Eqs. (9.30) and (9.37) intersect and is known as the *"nominal thickness"* of laminar sub-layer,

$$\frac{u_* \delta'}{\nu} = 11.6 \quad ...(9.39)$$

At $y = \lambda$, which is defined by $u_* \lambda/\nu = 70$, the flow becomes completely turbulent.

Instead of Eq. (9.37) an alternative relationship though approximate but valid upto Reynolds number 10^5 may be expressed as

$$\frac{u}{u_*} = 8.74 \left(\frac{u_* y}{\nu}\right)^{1/7} \qquad ...(9.40)$$

9.8. Flow Through Rough Pipes

Most of the pipes used in engineering practice such as cement pipes, cast iron pipes etc., cannot be regarded as hydraulically smooth especially at higher Reynolds numbers. In fact they actually behave as hydraulically rough. The resistance to fluid flow offered by rough boundaries is larger than that for smooth boundaries on account of formation of eddies behind roughness protrusions. The laws of resistance to fluid flow through rough pipes are, therefore, of great practical importance. The number of parameters describing the roughness of boundary surface is larger owing to great diversity in geometric forms. These parameters are the shape, the height, the manner in which the roughness protrusions are distributed over the surface and the density of distribution of such protrusions, that is, their number per unit area. L. Hopf investigated the problem of flow through rough pipes and open channels and found two types of roughnesses with a view to determine resistance laws for each type. The first type of roughness causes a resistance which is proportional to the square of velocity. The roughness elements are relatively coarse and very closely spaced and the coefficient of roughness is independent of the Reynolds number. In such cases, the nature of roughness projections can be characterized by a single non-dimensional parameter k/r_0 which is known as the *relative roughness*, where k is the height of a roughness protrusion and r_0 represents the pipe radius. For this type of roughness it has been found that the coefficient of resistance depends on the relative roughness only. The second type of roughness is characterized by gentle protrusions represented by a condition when a small number of roughness elements are distributed over a large area such as in wooden or commercial steel pipes. For flow through such pipes the resistance coefficient depends both on the Reynolds number as well as on the relative roughness.

The resistance to flow over rough boundaries seems to depend on the ratio of the height of protrusions to the thickness of the laminar sub-layer δ_l, so that k/δ_l becomes an important dimensionless parameter influencing the flow resistance. It needs no explanation that the boundary roughness will cause no increase in resistance in cases where the height of roughness elements is so small that they are all entirely covered by the laminar sub-layer. Thus if $k < \delta_l$ the boundary may be considered hydraulically smooth. From dimensional considerations, we can write the thickness of laminar sub-layer as δ_l = constant (ν/u_*). The non-dimensional roughness parameter may be written as,

$$\frac{k}{\delta_l} = \text{constant} \times \left(\frac{u_* k}{\nu}\right) \qquad ...(9.41)$$

The right hand side of Eq. (9.41) contains u_*k/ν which is a Reynolds number based on the roughness height and shear velocity u_*, and shows that effect of boundary roughness on the resistance to flow will be governed by roughness Reynolds number.

9.9. Hydraulically Smooth and Rough Boundaries

For laminar flow, all rough pipes irrespective of their roughness size and pattern offer the same resistance as that offered by a smooth pipe under similar conditions of flow (characterized by Reynolds number). As a matter of fact, there is no surface which may be regarded as perfectly smooth. The terms 'smooth' and 'rough' are relative. All boundaries have physically rough surfaces, though the degree of roughness of a new pipe may vary depending upon the process of manufacture and the finish given to the surface. It has been observed that in turbulent flow, there is a thin layer very close to the boundary in which the flow exhibits characteristics of laminar flow. This layer is the so-called '*laminar sub-layer*'. If the irregularities on any actual surface are such that the surface projections are completely covered by the laminar sub-layer, the surface is smooth from hydraulic point of view. But if the projections extend beyond the laminar sub-layer, the laminar layer gets disturbed and the surface then behaves as hydraulically rough.

Consequently, a surface will behave either as a smooth or a rough boundary depending upon the magnitude of the ratio of average height of roughness projections and the thickness of laminar sub-layer. The thickness of the laminar sub-layer may be taken as

$$\delta_l = \frac{5\nu}{u_*}$$

which may be expressed in terms of the Reynolds number by substituting $V\sqrt{\frac{f}{8}}$ for u_*.

Thus, we obtain
$$\frac{\delta_l}{d} = \frac{16.3\,\nu}{Vd\sqrt{f}} = \frac{16.3}{\mathbf{R}\sqrt{f}} \qquad \text{...(9.42 } a\text{)}$$

in which $\mathbf{R} = V\,d/\nu$ is the Reynolds number. A similar expression for the nominal thickness of the laminar sub-layer δ' using Eq. (9.39) may be obtained, and expressed as

$$\frac{\delta'}{d} = \frac{37.8}{\mathbf{R}\sqrt{f}} \qquad \text{...(9.42 } b\text{)}$$

From Eq. (9.42), it can be observed that for a given pipe diameter, the thickness of the laminar sub-layer decreases with increase in Reynolds number. (It is to be noted that the friction factor in turbulent flow decreases very slightly with increase in Reynolds number). As the thickness of the sub-layer in a given pipe decreases with an increase in Reynolds number, it can be concluded that the pipe will behave as hydraulically smooth at low Reynolds numbers while at high Reynolds number it will be hydraulically rough. Thus even a smooth pipe may behave as a rough one at high Reynolds numbers.

If k_s represents the mean height of roughness element, the roughness Reynolds number u_*k_s/ν may be taken as the criterion to characterize the type of boundary as hydraulically smooth, in transition or hydraulically rough.

1. **Hydraulically smooth boundary.** If the Reynolds number u_*k_s/ν is within the limit of 0 to 4, the boundary is hydraulically smooth and the friction factor f is independent of the relative roughness. Thus for hydraulically smooth boundaries.

$$0 \le \frac{u_*k_s}{\nu} \le 4, \text{ and } f = f(\mathbf{R}) \qquad \text{...(9.43 } a\text{)}$$

where $\mathbf{R}$ is the Reynolds number based on average velocity of flow, the characteristic length being the pipe diameter. The roughness elements are entirely contained within the laminar sub-layer.

2. **Boundary in transitional regime.** The roughness projections extend partly outside the laminar sub-layer into the main flow and the additional resistance is caused due to formation of eddies behind such individual roughness elements. The boundaries which are in transition from hydraulically smooth regime to hydraulically rough generally satisfy the following criterion :

$$4 \le \frac{u_*k_s}{\nu} \le 100,$$

and
$$f = f(k_s/r_0, \mathbf{R}) \qquad \text{...(9.43 } b\text{)}$$

The friction factor f is function of the relative roughness k_s/r_0 and Reynolds number $\mathbf{R}$.

3. **Completely rough (hydraulically) boundary.** When the thickness of laminar sub-layer becomes very small at higher Reynolds number as compared to the roughness height, the roughness elements get exposed to the main flow. The resistance to flow is then largely due to formation of eddies behind such roughness elements. For completely rough boundaries,

$$\frac{u_* k_s}{\nu} > 100,$$

and $$f = f(k_s/r_0) \qquad ...(9.43\ c)$$

The friction factor is thus dependent solely upon the relative roughness.

Experimental investigations have led to the conclusion that if $u_* k_s/\nu < 4$, the boundary may be considered hydraulically smooth. Since the thickness of laminar sub-layer is that pertaining to $u_* \delta_l/\nu = 5$, it is ensured that the mean roughness element would be submerged within the laminar sub-layer. For $4 < u_* k_s/\nu < 100$, the boundary is in the transitional range whereas for $u_*/k_s/\nu > 100$, the surface may be considered fully rough.

From Fig. 9.14, it can be noted that a pipe behaves as a smooth one for $\mathbf{R}(k_s/d)\sqrt{f} \le 8$, and becomes a completely rough pipe when $\mathbf{R}(k_s/d)\sqrt{f} \ge 200$. Using these values, the ratio of roughness height to thickness of laminar sub-layer may be worked out from (Eq. 9.42).

Thus, we have

(*i*) For hydraulically smooth pipe :

$$\frac{k_s}{\delta_l} < 0.5 \left(\text{or } \frac{k_s}{\delta'} < 0.25\right)$$

(*ii*) For hydraulically rough pipe :

$$\frac{k_s}{\delta_l} > 12 \left(\text{or } \frac{k_s}{\delta'} < 6\right) \qquad ...(9.44)$$

(*iii*) For $$0.5 < \frac{k_s}{\delta_l} < 12 \left(\text{or } 0.25 < \frac{k_s}{\delta'} < 6\right)$$

the pipe surface can neither be regarded as hydraulically smooth nor rough, but is in the transitional range.

The above criterion is generally used in ascertaining the hydraulic behaviour of the pipe surface under given conditions of flow. It may, thus, be concluded that the absolute size of roughness projections is not the measure of the effect of boundary roughness on resistance to flow. The flow resistance, in general, is governed by the ratio of mean height of roughness projections to the thickness of laminar sub-layer ; the latter being dependent upon the Reynolds number at large values of **R** the sub-layer practically vanishes and all the projections extend into the turbulent flow. Eddies are formed in the wake of each projection. These eddies are responsible for higher energy loss resulting in the quadratic resistance law ($h_L \propto V^2$), according to which the energy loss is proportional to the velocity square.

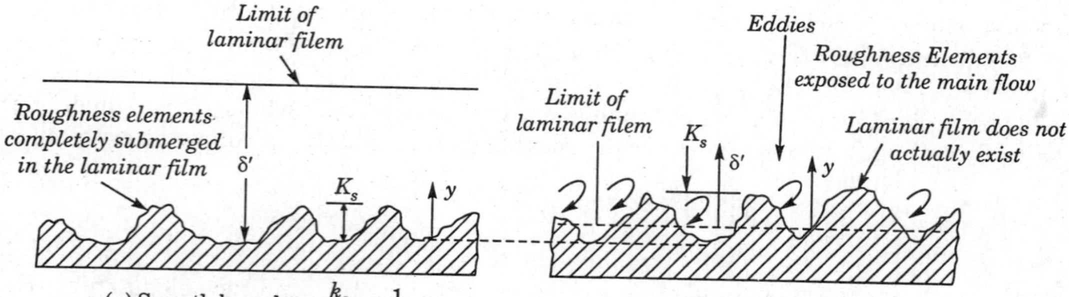

(*a*) Smooth boundary $\frac{k_s}{\delta'} < \frac{1}{4}$ (*b*) Rough boundary, $k_s/\delta' > 6$

Fig. 9.11. Hydraulically smooth and rough boundaries.

Fig. 9.11 shows the relative positions of the laminar film and the roughness projections for smooth and rough boundaries.

Example 9.1. *A test for determining the equivalent sand grain roughness of a certain pipe gave the following data :*

diameter of pipe = *30 cm*
discharge = *0.47* m^3/sec
head loss in 10 metres = *1.9 metres*
kinematic viscosity of fluid = 10^{-6} m^2/sec.

Determine the equivalent sand grain roughness of the pipe. What is the maximum roughness in order that the pipe may act as hydrodynamically smooth at the given discharge ?

Solution. The average velocity of flow,

$$V = \frac{Q}{A} = \frac{0.47}{\frac{\pi}{4} \times (0.3)^2} = \frac{0.47}{0.071} = 6.65 \text{ m/s}$$

the Reynolds number of flow, $\mathbf{R} = \frac{V.d}{\nu} = \frac{6.65 \times 0.3}{10^{-6}} = 1.995 \times 10^6$.

From the Darcy-Weisbach equation, the friction factor is obtained as

$$h_L = f\,\frac{L}{d}\,\frac{V^2}{2g}$$

$$\therefore \quad f = \frac{h_L.d.2g}{LV^2}$$

$$= \frac{1.9 \times 0.3 \times 2 \times 9.81}{10 \times (6.65)^2} = 0.0253.$$

From the resistance formula for rough pipes, Eq. (9.62)

$$\frac{1}{\sqrt{f}} = 2 \log_{10} \frac{r_0}{k_s} + 1.74$$

or $$\frac{1}{\sqrt{0.0253}} - 1.74 = 2 \log_{10} \frac{r_0}{k_s}$$

or $$2 \log_{10} (r_0/k_s) = \frac{1}{0.159} - 1.74 = 4.56$$

$$r_0/k_s = 190.5$$

or $$k_s = \frac{r_0}{190.5} = \frac{0.15}{190.5} \text{ m} = 0.785 \times 10^{-3} \text{ m}.$$

If δ' = nominal thickness of laminar sub-layer, then for hydrodynamically smooth pipe,

$$\frac{k_s}{\delta'} < \frac{1}{4}$$

The nominal thickness of laminar sub-layer is defined by

$$\frac{u_*\delta'}{\nu} = 11.6$$

where $$u_* = \sqrt{\frac{\tau_0}{\rho}} \text{ and } \tau_0 = \frac{f}{8}\,\rho V^2$$

$$\therefore \quad u_* = V\sqrt{\frac{f}{8}} \quad \text{or} \quad u_* = 0.65 \times \sqrt{\frac{0.02}{8}} = 0.37 \text{ m/s}$$

$$\therefore \quad \delta' = 11.6\,\nu/u_* = \frac{11.6 \times 10^{-6}}{0.37} = 31.35 \times 10^{-6} \text{ m}.$$

For the pipe to behave as hydrodynamically smooth at the given discharge, the maximum roughness,

$$k_s = \frac{1}{4}\delta' = \frac{31.35 \times 10^{-6}}{4} = 7.84 \times 10^{-6} \text{ m.}$$

Example 9.2. *A pipeline carrying water has surface protusions of average height 0.10 mm. If the shear stress developed is 7.85 N/m² determine whether the pipe surface acts as smooth, rough or in transition. For water take ρ = 1000 kg/m³ and kinematic viscosity ν = 0.93 × 10⁻² stokes.*

Solution. $k_s = 0.10$ mm, $\tau = 7.85$ N/m²

$$u_* = \sqrt{\frac{\tau_0}{\rho}} = \sqrt{\frac{7.85}{1000}} = 0.0877 \text{ m/s}$$

$$\frac{u_* k_s}{\nu} = \frac{0.0877 \times 0.10/1000}{0.93 \times 10^{-2} \times 10^{-4}} = 9.42$$

Since the roughness Reynolds number $\frac{u_* k_s}{\nu}$ lies between 4 and 100, the pipe surface acts is in transition.

9.10. Velocity Distribution in Rough Pipes

The velocity distribution law expressed by Eq. (9.35) for turbulent flow near a solid boundary may be used to determine the velocity distribution in rough pipes as well. In case of rough pipes, it may be assumed that as $u \to 0$, $y \to mk_s$, where m is an arbitrary constant and k_s is the mean roughness height. Eq. (9.33) yields

$$\frac{u}{u_*} = \frac{1}{K}(\log y/k_s - \log m) \quad ...(9.45)$$

The constant m appearing in Eq. (9.45) is dependent on the nature of roughness. Comparison of Eq. (9.45) with the Nikuradse's experimental data on pipes artificially roughened (by sand grains) has shown that it can be an equation of the form :

$$\frac{u}{u_*} = 5.75 \log_{10} y/k_s + B \quad ...(9.46)$$

in which the constant B has different values for the three types of surfaces (hydraulically smooth, transitional and rough) discussed in article 9.9. In case of completely rough pipes, $B = 8.5$ and the velocity distribution becomes

$$\frac{u}{u_*} = 5.75 \log_{10} \frac{y}{k_s} + 8.5 \quad ...(9.47)$$

In case of pipes of moderate roughness, that is $4 \le \frac{u_* k_s}{\nu} \le 100$, the value of B is no longer a constant but becomes a function of the roughness Reynold's number $u_* k_s/\nu$. Eq. (9.47) is also known as *Karman Prandtl equation for velocity distribution in rough pipes.*

Example 9.3. *The velocity of flow in a badly corroded 7.5 cm pipe is found to increase 20% as a pitot tube is moved from a point 1.0 cm from the wall to a point 2.0 cm from the wall. Estimate the height of roughness elements.*

Solution. The velocity distribution near rough boundaries in given by Eq. (9.47)

$$\frac{u}{u_*} = 5.75 \log_{10} \frac{y}{k_s} + 8.5$$

where k_s = average height of roughness elements. Let u be the velocity at a distance of 1 cm from the pipe wall, then

$$\frac{u}{u_*} = 5.75 \log_{10} \frac{1}{k_s} + 8.5 \quad ...(i)$$

At a distance of 2 cm from the wall, the velocity is found to be 1.2u, thus

$$\frac{1.2u}{u_*} = 5.75 \log_{10} \frac{2}{k_s} + 8.5 \quad \text{...}(ii)$$

Dividing (i) by (ii), $\dfrac{1}{1.2} = \dfrac{5.75 \log_{10} \dfrac{1}{k_s} + 8.5}{5.75 \log_{10} \dfrac{2}{k_s} + 8.5}$

or $\quad 5.75 \log_{10} \dfrac{2}{k_s} + 8.5 = 6.9 \log_{10} \dfrac{1}{k_s} + 10.2$

which simplifies to $\quad \log_{10} k_s = \dfrac{0.03}{1.15} = 0.0261$

$$k_s = 0.942 \text{ cm.}$$

9.11. Velocity Distribution in Terms of Average Velocity

In case of a smooth pipe, the average velocity may be obtained substituting $(r_0 - r)$ for y in Eq. (9.37 b) and integrating

$$Q = \pi r_0^2 V = \int_0^{r_0} 2\pi r u_* \left\{ 5.75 \log_{10} \frac{u_* (r_0 - r)}{\nu} + 5.5 \right\} dr$$

$$= \pi r_0^2 u_* \left(5.75 \log_{10} \frac{u_* r_0}{\nu} + 1.75 \right) \quad \text{...(9.48)}$$

therefore, the average velocity can be written as

$$\frac{V}{u_*} = 5.75 \log_{10} \frac{u_* r_0}{\nu} + 1.75 \quad \text{...(9.49)}$$

Subtracting Eq. (9.37 b) from Eq. (9.49) we obtain

$$\frac{V - u}{u_*} = 5.75 \log_{10} \frac{r_0}{y} - 3.75 \quad \text{...(9.50)}$$

For rough pipe, the average velocity can be obtained in a similar manner as above. Using Eq. (9.47) we can obtain an equation expressing the average velocity as

$$\frac{V}{u_*} = 5.75 \log_{10} \frac{r_0}{k} + 4.75 \quad \text{...(9.51)}$$

Subtracting Eq. (9.48) from Eq. (9.51)

$$\frac{V - u}{u_*} = 5.75 \log_{10} \frac{r_0}{y} - 3.75 \quad \text{...(9.52)}$$

Eqs. (9.50) and (9.52) are identical. It can thus be concluded that the velocity distributions, when referred to the mean velocity, become identical for smooth and rough pipes.

Example 9.4. *In the determination of mean velocity in wide rivers, it is a common practice to estimate the mean velocity in a vertical section by making a single measurement at a distance below the surface equal to six-tenths of the depth. From the equation of velocity distribution in very wide rivers, determine the actual distance below the surface at which u = V.*

Solution. The velocity distribution in very wide rivers may be represented by

$$\frac{V - u}{u_*} = 2.0 \log_{10} \frac{y}{y_0} + 0.88$$

in which y is the distance measured from the bottom of the river, and y_0 is the depth of water in the river. At $u = V$, the left hand side of the above equation becomes zero, thus

$$2 \log_{10} \frac{y}{y_0} + 0.88 = 0$$

$\therefore \qquad \log_{10} \dfrac{y}{y_0} = -0.44 = \bar{1}.56$

or $\qquad \dfrac{y}{y_0} = 0.363$

$\therefore$ The distance below the water surface at which $u = V$ is
$(1 - 0.363)\, y_0 = 0.637\, y_0$.

In other words, the average velocity occurs at distance of 0.637 times the depth of flow in the river measured below the water surface.

Example 9.5. *Show that for turbulent flow in pipes, the ratio of maximum velocity to average velocity is given by* $\dfrac{u_{max}}{V} = 1 + 1.33\sqrt{f}$.

Solution. From Eqs. (9.50) and (9.52), we obtain for rough and smooth pipes

$$\frac{V - u}{u_*} = 5.75 \log_{10} \frac{r_0}{y} - 3.75.$$

For $y = r_0$, $u = u_{max}$, the above equation reduces to

$$\frac{V - u_{max}}{u_*} = -3.75$$

Substituting $\sqrt{f/8}$ for u_* and simplifying, we get

$$\frac{u_{max}}{V} - 1 = 3.75\sqrt{f/8}$$

$$\frac{u_{max}}{V} = 1 + 1.33\sqrt{f}.$$

This is Eq. (9.29) obtained earlier.

9.12. Power Law for Velocity in Smooth Pipes

The velocity distribution formula developed in the foregoing treatment are logarithmic, and as such are not so convenient for use as are the exponential ones. Nikuradse carried out a very thorough experimental investigation to determine velocity distribution law in smooth pipes over a wide range of Reynolds numbers, $4 \times 10^3 \le \mathbf{R}\ 3.2 \times 10^6$. The dimensionless velocity distribution, based on his work may be represented by an empirical equation of the form

$$\frac{u}{u_{max}} = \left(\frac{y}{r_0}\right)^n \qquad \text{...(9.53)}$$

in which the exponent n is dependent on the Reynolds number **R**, and decreases with increasing **R**. For Reynolds number $\mathbf{R} = 4 \times 10^3$, $n = 1/6$ and reduces to 1/7 at $\mathbf{R} = 1.1 \times 10^5$ and to 1/10 at the Reynolds number $\mathbf{R} \ge 2 \times 10^6$.

For $n = 1/7$, the velocity distribution law becomes

$$\frac{u}{u_{max}} = \left(\frac{y}{r_0}\right)^{1/7} \quad \text{for } \mathbf{R} \le 10^5 \qquad \text{...(9.54)}$$

which agrees well with Nikuradse's experimental data upto $\mathbf{R} = 10^5$. This equation is known as the Blasius one-seventh power velocity distribution law. The logarithmic velocity distribution laws are all defective in the sense that they do not conform to no slip condition and also they do not yield a zero velocity gradient at the pipe centre. The exponential form of velocity distribution, Eq. (9.53) remedies the first defect (it satisfies the no slip condition), but remains faulty with respect to non-zero velocity gradient at the pipe axis.

Example 9.6. *For turbulent flow of an incompressible fluid through a circular pipe of radius r, the velocity distribution is given by*

$$u = u_0\left(1 - \frac{r}{r_0}\right)^{1/7}$$

where u is the velocity at radius r and u_0 is the velocity at the pipe axis. Calculate the total kinetic energy of flow in terms of discharge Q, radius r_0, specific weight γ and gravity g. Compare the kinetic energy with the kinetic energy when the flow in the pipe is completely laminar and the quantity of fluid flowing is constant at Q.

Solution. Kinetic energy of fluid per second = $\frac{1}{2}mu^2$

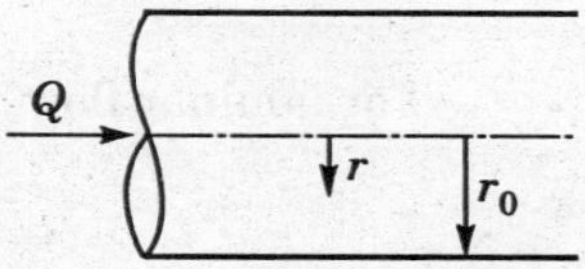

where m = mass of fluid flowing per sec, and

u = velocity of flow.

Considering a circular ring at radius r and of thickness dr, the rate of flow through this ring $dQ = 2\pi\, r.dr.u$.

and the mass of fluid flowing per second = $\frac{\gamma}{g}\, dQ$

and the total kinetic energy $= \int_0^{r_0} \frac{1}{2}\frac{\gamma}{g}\, dQ\, u^2$

$$= \int_0^{r_0} \frac{1}{2}\frac{\gamma}{g}\, 2\pi.r.dr.u.u^2 = \frac{\gamma\pi}{g}\int_0^{r_0} u^3 r\, dr$$

from the given velocity distribution

$$u = u_0\left(1 - \frac{r}{r_0}\right)^{1/7}$$

The total kinetic energy in turbulent flow

$$= \frac{\gamma\pi}{g}\int_0^{r_0} u_0^3\left(1 - \frac{r}{r_0}\right)^{1/7} r.dr$$

Let $1 - \frac{r}{r_0} = z, \;0 \quad \therefore \quad -\frac{1}{r_0}\, dr = dz$ and $r = r_0(1 - z)$

making these substitutions.

Total K.E. $= \frac{\gamma\pi}{g}\int_1^0 u_0^3 z^{3/7} r_0(1 - z)(-r_0 dz)$

$$= \frac{\gamma\pi\, u_0^3 r_0^2}{g}\int_0^1 (z^{3/7} - r^{10/7})dz = \frac{49}{170}\times\frac{\gamma\pi u_0^3 r_0^2}{g}$$

The discharge, $Q = \int_0^{r_0} 2\pi\, r\, dr\, u$

$$= \int_0^{r_0} 2\pi r\, u_0\left(1 - \frac{r}{r_0}\right)^{1/7} dr = \int_0^1 2\pi\, u_0 r_0 (1 - z)\, z.r_0\, dz$$

$$= 2\pi u_0 r_0^2 \int_0^1 (z^{1/7} - z^{3/7})dz = \frac{49}{60}\,\pi u_0 r_0^2$$

$$\therefore \qquad u_0 = \frac{60Q}{49r_0 2^2}.$$

Substituting the value of u_0 in the expression for kinetic energy, the total kinetic energy per sec in turbulent flow

$$= \frac{49}{170}\frac{\gamma\pi}{g}\left(\frac{60}{49}\frac{Q}{\pi r_0^2}\right)^3 r_0^2 = \frac{0.53\,\gamma\,Q^3}{g\pi^2\,r_0^4}$$

Kinetic energy in laminar flow

$$= \int_0^{r_0} \frac{1}{2}\frac{\gamma}{g}\, 2\pi r\, dr\, u.u^2 = \frac{1}{2}\frac{2\pi\gamma}{g}\int_0^{r_0} u^3\, r\, dr$$

For laminar flow, using Eqs. (8.2) and (8.3) the velocity given by

$$u = u_0\left(1 - \frac{r^2}{r_0^2}\right)$$

$$\text{Kinetic energy} \quad = \frac{\pi\gamma}{g}\int_0^{r_0} u_0^3\left(1 - \frac{r^2}{r_0^2}\right)^3 r\, dr = \frac{\pi\gamma\, u_0^3\, r_0^2}{dg}$$

the discharge,

$$Q = \int_0^{r_0} 2\pi r\, dr\, u = \int_0^{r_0} 2\pi r.u_0\left(1 - \frac{r^2}{r_0^2}\right)dr = \frac{1}{2}\pi r_0^2 u_0$$

or

$$u_0 = \frac{2Q}{\pi_0\, r_0^2}$$

Substituting for u_0, the kinetic energy in laminar flow

$$= \frac{\pi\gamma r_2^2\,(2Q/\pi r_0^2)^3}{8g} = \frac{\gamma Q^3}{\pi^2 g r_0^4}$$

Ratio of kinetic energy in laminar flow and turbulent flow

$$= \frac{\gamma Q^3/\pi^2 g r_0^4}{0.53\,\gamma Q^3/g\pi^2 r_0^4} = 1.888.$$

Example 9.7. *Sketch the velocity gradient du/dr along the diameter of a pipe through which a fluid is flowing (i) when the flow is laminar, (ii) when the flow is turbulent. Explain why such a wide difference in velocity distribution occurs with the change in the flow regime.*

Solution. (*i*) **When the flow is laminar :**

The velocity distribution in laminar flow may be expressed by

$$u = -\frac{1}{4\mu}\frac{dp}{dx}(R^2 - r^2) = u_{max}(1 - r^2/R^2) \qquad \text{...(1)}$$

in which u is the velocity at a radial distance r and R the radius of the pipe. Differentiating the velocity w.r.t. r one obtains the velocity gradient du/dr :

$$\frac{du}{dr} = \frac{u_{max}}{R^2}(-2r) = -\frac{2r}{R^2}u_{max} \qquad \text{...(2)}$$

The negative sign shows that as r increases, the velocity u decreases. From Eq. (2) it is evident that the velocity gradient varies linearly with the radial distance r. The velocity gradient attains a maximum value at the pipe wall where $r = R$, and is equal to

$$\left(\frac{du}{dr}\right)_R = -2\,\frac{u_{max}}{R}.$$

The minimum value of zero occurs at the pipe centre. The distribution of velocity gradient along the pipe diameter is shown in Fig. Ex. 9.7.

(*ii*) **When the flow is Turbulent :**

The velocity distribution for turbulent flow in pipes may be expressed by the following exponential equation :

$$u = u_{max}\left(1-\frac{r}{R}\right)^{1/7} \qquad ...(3)$$

$$\frac{du}{dr} = \frac{1}{7}\,u_{max}\left(1-\frac{r}{R}\right)^{-6/7}\left(-\frac{1}{R}\right) = -\frac{u_{max}}{7R\,(1-r/R)^{6/7}} \qquad ...(4)$$

The negative sign appearing in Eq. (4) indicates that the velocity u decreases as the radial distance r increases. For different assumed values of r, du/dr can be calculated and the values so obtained are given below :

Value of r	*Corresponding value of du / dr*	*Value of r*	*Corresponding value of du / dr*
0.00 R	0.143 u_{max}/R	0.75 R	0.468 u_{max}/R
0.25 R	0.183 u_{max}/R	0.90 R	1.027 u_{max}/R
0.33 R	0.201 u_{max}/R	1.00 R	∞
0.50 R	0.258 u_{max}/R		

Plotting the above values of velocity gradient along the pipe radius, one obtains the variation for turbulent flow. The distribution of velocity gradient for laminar flow using Eq. (2) has also been shown alongside for comparison, the figure being in non-dimensional form.

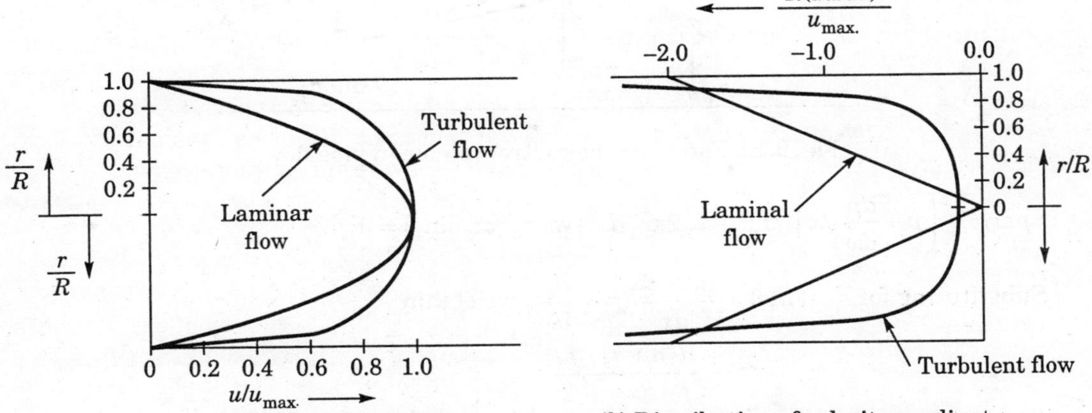

(*a*) Velocity distribution (*b*) Distribution of velocity gradient

Fig. Ex. 9.7. Distribution of velocity and its gradient in pipe flow expressed in non-dimensional form.

It is known that the nature of flow very close to the solid boundary exhibits characteristics of laminar flow even when the main flow is turbulent. The shear stress in the neighbourhood of the pipe boundary, therefore, is governed by the Newton's law of viscosity, namely,

$$\tau = \mu\,\frac{du}{dr} \qquad ...(5)$$

Since the velocity gradients in the vicinity of the pipe wall are very steep in turbulent flow (indicated by large values), causing the development of high shear stresses. The viscous effects in turbulent flow are thus confined to a thin layer near the solid boundary. This makes the velocity distribution more uniform as compared to the one obtaining in laminar flow.

The velocity distribution diagrams for both the regimes of flow can be obtained easily using Equations (1) and (3) above. The distributions of velocity and that of velocity gradients shown in the figure explain the role of the latter on the former with the help of Equation (5).

9.13. Resistance of Pipes Expressed in Terms of Friction-Factor

The fluid flow through pipes is subjected to resistances due to viscosity, fluid turbulence and roughness of the pipe surface. In order to overcome these resistances, the flow has to expend its energy and consequently, the available energy decreases in the direction of flow resulting in a downward sloping energy line.

Let us consider a sloping pipe through which a certain fluid (its viscosity μ and density ρ being known) is flowing. Consider a control volume of fluid bounded by the pipe surface and the two sections a distance dx apart. The following fluid exerts a shear stress τ_0 on the wall and the pipe wall in turn exerts an equal and opposite shear stress τ_0 on the fluid. The other forces acting on the control volume are the pressure forces on its end faces and the weight component acting in the direction of flow (Fig. 9.12). For steady uniform flow, the equation of motion in the flow direction yields,

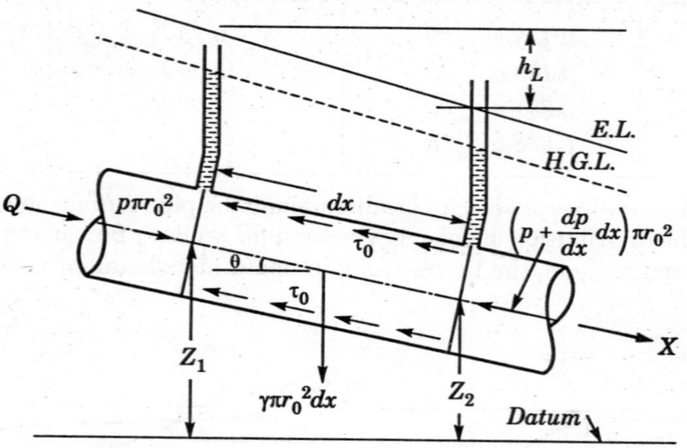

Fig. 9.12. Forces on the control volume in pipe flow.

$$p\pi r_0^2 - \left(p + \frac{dp}{dx}dx\right)\pi r_0^2 - \tau_0\, 2\pi r_0\, dx + \gamma\pi\, r_0^2\, dx \sin\theta = 0$$

Substituting for $\quad \sin\theta = \dfrac{z_1 - z_2}{dx} = -\dfrac{dz}{dx}$, we obtain

$$= -\frac{d(p+\gamma z)}{dx}\frac{r_0}{2}$$

In steady uniform flow, $\dfrac{d(p+\gamma z)}{dx}$, which represents the gradient of hydraulic grade line is also the gradient of energy line, thus

$$\frac{d(p/\gamma + z)}{dx} = -\frac{dh}{dx}, \qquad \text{where } h = \text{total energy head.}$$

or

$$\frac{d(p+\gamma z)}{dx} = \frac{\gamma h_L}{dx}$$

Hence, the wall shear stress τ_0 is equal to

$$\tau_0 = \frac{\gamma h_L r_0}{2dx} \quad ...(9.55)$$

but the shear strees τ_0 and friction factor f are related by

$$\tau_0 = \frac{f}{4}\rho\frac{V^2}{2} \quad ...(9.56)$$

Equating Eqs. (9.55) and (9.56), we obtain

$$\frac{\gamma h_L d}{4dx} = \frac{f}{4}\rho\frac{V^2}{2}$$

resulting in the head loss equation (for length dx)

$$h_L = f\frac{dx}{d}\frac{V^2}{2g} \quad ...(9.57)$$

The head loss (or energy head loss) in pipe length L, may be expressed as

$$h_L = f\frac{L}{d}\frac{V^2}{2g} \quad ...(9.57a)$$

Eq. (9.57) is known as the *Darcy-Weisbach* equation and shows that the energy loss is proportional to the length of the pipe, square of mean velocity V, and friction factor f which is a direct measure of resistance to flow and inversely proportional to the pipe diameter d. It is to be noted that f is a resistance coefficient known as friction factor and depends on the pipe Reynolds number and its surface roughness.

Example 9.8. *The population of a town is 4 lakhs. The town gets water from a reservoir through a pipe 20 km long. The per capita requirement of water is 150 litres/day and half of the total demand of the town is to be supplied in 8 hours. The flow is by gravity with a level difference of 18 m between the reservoir level and exit and of the pipe. If f = 0.025 for the pipe, find suitable diameter of the pipe.*

Solution. Total daily demand of water

$$= 4 \times 100{,}000 \times \frac{150}{1000}\ \text{m}^3 = 60{,}000\ \text{m}^3$$

Half of the total daily demand is to be supplied in 8 hours, hence, the pipe discharge,

$$Q = \frac{(60{,}000/2)}{8 \times 60 \times 60} = 1.0417\ \text{m}^3/\text{s}$$

From the Darcy-Weisbach equation (Eq. 9.57a), the head loss in friction, h_L

$$h_L = f\frac{L}{d}\frac{V^2}{2g}$$

The level difference of 18 m between the reservoir level and the exit end represents the head loss hence

$$18 = 0.025 \times \frac{20 \times 1000}{d}\frac{(4Q/\pi d^2)^2}{2 \times 9.81}$$

Substituting for Q, one gets

$$d^5 = 45.373 \qquad \text{or } d = \mathbf{2.1447\ m}$$

The practical size of the pipe may therefore be taken as **2.15 m.**

Example 9.9. *When the Reynolds number is less than about 10^5, the velocity profile in turbulent flow through a small pipe is well represented by*

$$\frac{u}{u_*} = 8.7\left(\frac{u_* y}{\nu}\right)^{1/7}$$

Using this formula show that the friction factor is

$$f = 0.31 \Big/ \left(\frac{V.d}{\nu}\right)^{0.25}$$

where V is the average velocity. Also show that the shear at the wall may be given by

$$\tau_0 = 0.0272 \ \rho u^2_{max} \Big/ \left(\frac{u_{max}\ r_0}{\nu}\right)^{0.25}$$

Solution. From the given velocity distribution, the discharge Q and the average velocity of flow V may be obtained

$$Q = \int_0^{r_0} 2\pi r.dr.u$$

in which, $$u = 8.7 \frac{u_*^{8/7}}{\nu^{1/7}} y^{1/17}.$$

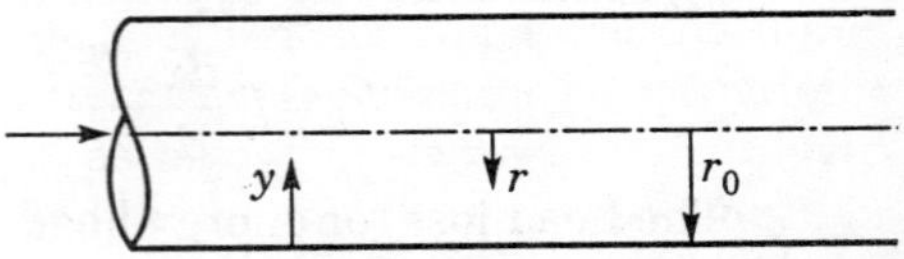

Fig. Ex. 9.8

Noting that, $y = r_0 - r$

$dy = -\ dr$

making these substitutions $$Q = \int_{r_0}^{0} 2\pi(r_0 - y)(-dy)\frac{8.7u_*^{8/7} y^{1/7}}{\nu^{1/7}}$$

$$= \int_{r_0}^{0} 2\pi \frac{8.7u_*^{8/7}}{\nu^{1/7}} (r_0 y^{1/7} - y^{8/7})dy$$

and the average velocity, $$V = \frac{Q}{\pi r_0^2} = \frac{1}{\pi r_0^2} \frac{17.4u_*^{8/7}\pi}{\nu^{1/7}}\left[r_0 \frac{7}{8} y^{8/7} - \frac{7}{15} y^{15/7}\right]_0^{r_0} = 7.1 \frac{u_*^{8/7} r_0^{1/7}}{\nu^{1/7}}$$

but $$\tau_0 = \frac{f}{8}\rho V^2 \quad \text{and} \quad u_*^2 = \frac{\tau_0}{\rho} = \frac{f}{8} V^2$$

from which, $$u_* = V\sqrt{f/8}$$

Substituting this value of u_*

$$V = \frac{7.1\,(V\sqrt{f/8})^{8/7}\ r_0^{1/7}}{\nu^{1/7}}$$

Rearranging and simplifying, we obtain

$$f = \frac{0.31}{(Vd/\nu)^{1/4}}.$$

The shear stress at the pipe wall is given by

$$\tau_0 = \frac{f}{8}\rho V^2 = \frac{0.31}{(Vd/\nu)^{1/4}} \rho \frac{V^2}{8}$$

From the velocity distribution

$$\frac{u_{max}}{u_*} = 8.7\left(\frac{ur_0}{\nu}\right)^{1/7}$$

The average velocity has been found to be

$$V = 7.1\ \frac{u_*^{8/7} r_0^{1/7}}{\nu^{1/7}} = 7.1\ u_* \left(\frac{u_* r_0}{\nu}\right)^{1/7}$$

$$= 7.1\ u_* \left(\frac{1}{8.7}\frac{u_{max}}{u_*}\right) = 0.816\ u_{max}$$

Substituting this value of V,

$$\tau_0 = \frac{0.31}{\left(\frac{0.816\,u_{max}\,d}{\nu}\right)^{1/4}}\frac{\rho}{8}\ (0.816\ u_{max})^2$$

$$= \frac{0.028}{(u_{max}\cdot r_0/\nu)^{1/4}}\ \rho u^2_{max}.$$

Example 9.10. *Water flows through a horizontal conical pipe, 2 m long and having a diameter of 20 cm at the inlet and 15 cm at the discharge end. A constant discharge of 40 litre per second flows through the pipe. Starting from first principles, determine the loss of head due to pipe friction. Take friction factor f = 0.04.* *(UPSC Exam, IES, 1992)*

Solution. The Darcy-Weisbach equation, Eq. (9.57), in a differential form can be written as

$$dh_f = f\ \frac{dx}{d}\frac{V^2}{2g}$$

in which dh_f is the head loss due to friction in a small length dx of the conical pipe. Let d_1 and d_2 be the diameters of the tapering pipe over a length L.

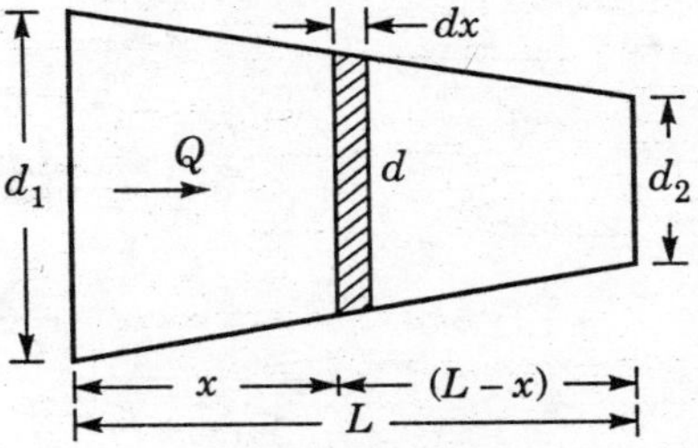

Fig. Ex. 9.9. Friction loss in a tapering pipe.

From the geometry of cone one can write,

$$\frac{d_1 - d_2}{L} = \frac{d - d_2}{L - x}$$

from which, $$d = d_2 + (d_1 - d_2)\ \frac{(L-x)}{L}$$

and, $$d(d) = -\ \frac{(d_1 - d_2)}{L}.\ dx.$$

$\therefore$ $$dh_f = f\ \frac{dx}{d}\left(\frac{16Q^2}{\pi^2 d^4}\right).\frac{1}{2g}.$$

Making substitution for dx

$$dh_f = -\ \frac{8Q^2 fL}{g\pi^2\,(d_1 - d_2)}\ \frac{d(d)}{d^5}$$

Integrating, $$h_f = -\ \frac{8Q^2 fL}{g\pi^2\,(d_1 - d_2)}\int_{d_1}^{d_2}\frac{d(d)}{d^5}$$

$$= \frac{8Q^2 fL}{g\pi^2\,(d_1 - d_2)}\left[\frac{1}{4d^4}\right]_{d_1}^{d_2}$$

$$= \frac{8Q^2 fL}{g\pi^2\,(d_1 - d_2)}\left[\frac{1}{d_2^{\,4}} - \frac{1}{d_1^{\,4}}\right].$$

The data given in the problem is : $L = 2$ m, $d_1 = 20$ cm, $d_2 = 15$ cm, $Q = 40$ lit./s, $f = 0.04$.

Making substitutions

$$h_L = \frac{2 \times 0.04 \times (40 \times 10^{-3})^2 \times 2}{9.81 \times \pi^2 \times \left(\frac{20}{100} - \frac{15}{100}\right)} \times \left[\frac{1}{\left(\frac{15}{100}\right)^4} - \frac{1}{\left(\frac{20}{100}\right)^4}\right]$$

$= 0.0715$ m $=$ **7.15 cm.**

Example 9.11. *A town of 2,00,000 population is to be supplied water from a source 2500 m way. The lowest water level in the source is 15 m below the water works of the town. The demand of water is estimated as 150 litres/capita/day. A pump of 220.5 kW is operated for 15 hours. If the maximum demand is 1.5 times the average demand, the velocity of flow through the rising main is 1.3 m/sec and the pump efficiency is 70%, determine*

(i) the hydraulic gradient, and

(ii) friction factor for the pipe.

Solution. Daily demand of water = 2,00,000 × (150/1000) m³.

Average discharge of the pump during 15 hours of pumping

$$= \frac{2{,}00{,}000 \times (150/1000)}{15 \times 60 \times 60} = 0.555 \text{ m}^3/\text{s}$$

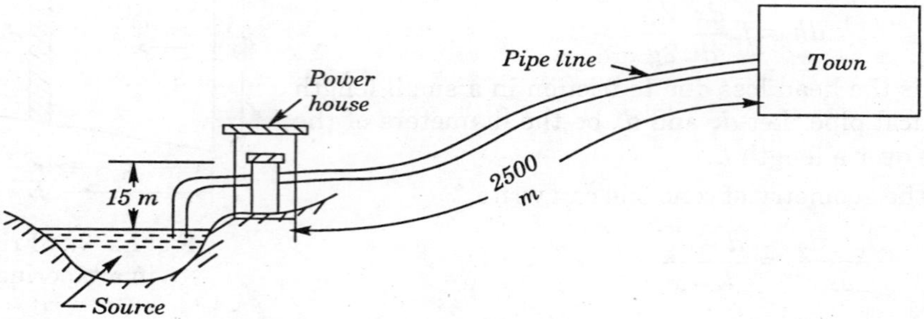

Maximum demand, $\quad Q = 1.5$ times average demand

$= 1.5 \times 0.555 = 0.8328 \text{ m}^3/\text{s}$

Pump output $= \dfrac{\gamma Q (h_L + 15)}{75}$, where h_L friction loss in pipe line

Input to pump $= 220.5 \times 1000 \text{ W} = 220.5 \times 1000 \text{ Nm/s}$

Overall pump efficiency $= \dfrac{\text{Pump output}}{\text{Input to pump}} = \dfrac{\gamma Q (h_L + 15)}{220.5 \times 1000}$

or
$$0.70 = \frac{9810 \times 0.8328 \times (h_L + 15)}{220.5 \times 1000}$$

from which
$$h_L = 3.91 \text{ m}$$

(i) Hydraulic gradient $= h_L / L = \dfrac{3.91}{2500} = \dfrac{\mathbf{1}}{\mathbf{639.4}}$.

(Head loss in friction etc./pipe length)

(ii) Head loss in friction is given by the Darcy-Weisbach eq. as

$$h_L = f \frac{L}{d} \frac{V^2}{2g}$$

Using the continuity equation, the pipe diameter can be found,

$$Q = \frac{\pi d^2}{4} V, \quad \therefore \quad d = \sqrt{\frac{4Q}{\pi V}} = \sqrt{\frac{4 \times 0.8328}{\pi \times 1.3}} = 0.903 \text{ m}$$

The friction factor $$f = \frac{2gdh_L}{LV^2} = \frac{2 \times 9.81 \times 0.903 \times 3.91}{2500 \times (1.3)^2} = \mathbf{0.0164.}$$

9.14. Nikuradse's Experiments of Artifically Roughened Pipes

On account of extreme complexity of naturally roughened surfaces, as those obtaining in commercial pipes used in engineering practice, most of the advances in understanding the parameters of pipe resistance have been developed around experiments on artificially roughened pipes.

J. Nikuradse, a German engineer, conducted a series of systematic, extensive and carefully planned experiments on pipes roughened artificially by glueing uniform sand grains as closely spaced as possible on the inner side of the pipe wall. By choosing pipe of varying diameters and by changing the size of sand grain, he was able to vary the relative roughness k_s/d, from about 1/1014 to 1/30. The results of tests conducted on smooth and artificially rough pipes (published in 1933) clearly demonstrate a relationship between friction factor f, Reynolds number **R** and the relative roughness defined by k_s/d, where k_s is the diameter of the sand grain. The concept of relative roughness introduced by Nikuradse has proved to be of great significance in determining the resistance of pipes (measured by the friction factor f) under different conditions of flow. Although the pattern of roughness in commercial pipes is entirely different, much more variable and much less definable than the artificial roughness of Nikuradse, his findings provide a basis for predicting the resistance of commercial pipes.

By performing dimensional analysis of the variables which affect the resistance to flow τ_0 (these variables are ρ, μ, d, V, k_s) for flow through artificially roughened pipes the following dimensionless parameters can be obtained :

$$\frac{\tau_0}{\rho V^2} = \phi\left(\frac{\rho V d}{\mu}, \frac{k_s}{d}\right) = \phi\,(\mathbf{R}, k_s/d)$$

and by comparing it with Eq. (9.55), we obtain

$$f = \phi_1\,(\mathbf{R}.\ k_s/d) \qquad \text{...(9.58)}$$

in which the symbol ϕ_1 stands for "function of".

Thus from Eq. (9.58), without having any prior knowledge of the experimental results, it has been shown that the friction factor f, depends upon the Reynolds number **R** and relative roughness k_s/d. Hence, the friction factors of two pipes will be same if their Reynolds numbers, roughness patterns (characterized in this case by uniform diameter closely spaced sand grains) and relative roughness are the same.

Nikuradse's test results plotted on logarithmic scale with friction factor as ordinate and Reynolds number as abscissae have been shown in Fig. 9.13 and are in complete agreement with eq. (9.58). This plot of friction factor veruse Reynolds number is called a "Stanton

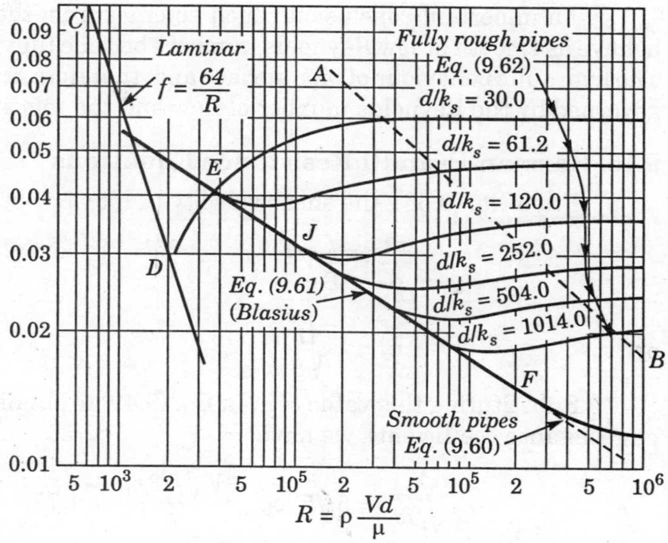

Fig. 9.13. Nikuradse's diagram for friction factor.

diagram". Blasius was first to correlate the smooth pipe experiments in turbulent flow and presented the results (in 1913) by an experimental formula, see Eq. (9.61).

A closer study of Fig. 9.13 reveals certain significant information pertaining to the resistance characteristics of laminar and turbulent flow in pipes. The following observations can be made.

1. Upto a Reynolds number of 2000, friction factor is seen to follow a linear relationship with Reynolds number (straight line *CD* representing the equation $f = 64/\mathbf{R}$). This represents the region of *laminar flow* in pipes and shows that the fricition factor is independent of the pipe roughness and is solely governed by the Reynolds number :

$$f = 64/\mathbf{R}.$$

2. For higher Reynolds numbers, pipes with very small relative roughness have their roughness elements completely covered by the laminar sub-layer. The friction factor is governed entirely by the Reynolds number. This is what is known as the *smooth turbulent flow* (turbulent flow in smooth pipes) represented by the line *EF*.

$$f = \phi(\mathbf{R})$$

3. For pipes with large relative roughness, it can be noted that f versus $\mathbf{R}$ curves depart from the line *EF*—friction factor first decreasing and subsequently increasing with the increase in Reynolds number. Between the lines *EF* and *AB*, the friction factor is dependent on both the Reynolds number as well as the relative roughness. The curves of different d/k_s start deviating from different points lying on *EF*—the point of departure being earlier for a rougher pipe than for a smoother pipe. This behaviour is on account of the fact that the laminar sub-layer gets destroyed earlier in rougher pipes while in smoother pipes it continues to envelop the roughness elements even at higher Reynolds numbers. The f *versus* $\mathbf{R}$ curves lying between *EF* and *AB* represent what is known as the transition from smooth turbulent to rough turbulent flow. In this transition region, the friction factor is influenced both by surface roughness as well as viscous shear.

$$f = \phi(\mathbf{R}, k_s/d).$$

4. Beyond the line *AB*, f versus $\mathbf{R}$ relationship becomes horizontal (or parallel to $\mathbf{R}$-axis) indicating that the friction factor no longer depends upon the Reynolds number. This region is identified as that of turbulent flow in fully rough pipes or simply as *rough turbulent flow*. The laminar sub-layer becomes so thin that it is completely destroyed by the protruding roughness elements. The friction factor is then solely governed by the relative roughness

$$f = \phi(k_s/d).$$

An important conclusion which emerges from the above discussion is that the pipes which behave as smooth at low Reynolds numbers become fully rough at high values of Reynolds number. The hydraulic behaviour of pipe surface and, therefore, its resistance characteristics are, in general, governed by the Reynolds number of flow and the relative roughness.

9.15. Karman-Prandtl Resistance Equations

From Eq. (9.56), the shear velocity u_* may be obtained as

$$\frac{\tau_0}{\rho} = \frac{f}{8}V^2$$

or

$$u_* = V\sqrt{\frac{f}{8}} \qquad \text{...(9.59)}$$

Substituting this value of u_* in Eq. (9.49) to obtain an equation for friction factor (also called as resistance coefficient), we have

$$\frac{V}{\sqrt{f/8}} = 5.75 \log_{10} \frac{V\sqrt{f/8r_0}}{v} + 1.75$$

which may be simplified to yield,

$$\frac{1}{\sqrt{f}} = 2.03 \log_{10} \frac{Vd}{\nu} \sqrt{f} - 0.91$$

In the light of experimental observations, the above equation necessitates a slight modification in the numerical values of the constants, and may be expressed as,

$$\frac{1}{\sqrt{f}} = 2 \log_{10} \mathbf{R} \sqrt{f} - 0.80 \qquad ...(9.60)$$

As is evident, evaluation of f at a given **R**, using Eq. (9.60) involves trial and error procedure. An alternative equation proposed by Blasius (applicable for $\mathbf{R} < 10^5$) may be used for direct determination of f. This equation is expressed as

$$f = \frac{0.3164}{\mathbf{R}^{0.25}} \qquad ...(9.61)$$

Eq. (9.60) is known as Karman-Prandtl resistance equation for turbulent flow in smooth pipes.

For rough pipes, the Karman-Prandtl resistance equation may be obtained using Eq. (9.51) and replacing u_* by $V\sqrt{f/8}$, thus

$$\frac{V}{V\sqrt{f/8}} = 5.05 \log_{10} \frac{r_0}{k_s} + 4.75$$

resulting in,

$$\frac{1}{\sqrt{f}} = 2.03 \log_{10} \frac{r_0}{k_s} + 1.68$$

and now adjusting the numerical constants to suit the experimental measurements, the above equation becomes

$$\frac{1}{\sqrt{f}} = 2 \log_{10} \frac{r_0}{k_s} + 1.74$$

or

$$\frac{1}{\sqrt{f}} = 2 \log_{10} \frac{d}{k_s} + 1.14 \qquad ...(9.62)$$

Eq. (9.62) is the Karman-Prandtl resistance equation for turbulent flow in rough pipes. This equation characterises the horizontal portions of the curves of Fig. 9.13 indicating that the loss of energy is due to roughness alone.

Example 9.12. *Compare the cost of pumping the same fluid at the same volumetric rate through 150 mm and 210 mm pipes having the same absolute roughness k_s = 0.3 mm. Assume that the Reynolds number is sufficiently high for viscous effects to be negligible.*

Solution. The friction factor for rough pipes is given by Eq. (9.62)

$$\frac{1}{\sqrt{f}} = 2 \log_{10} \frac{r_0}{k_s} + 1.74$$

For 15 cm pipe with absolute roughness

$$\frac{r_0}{k_s} = \frac{150/2}{0.3} = 250$$

and for 21 cm pipe with the same absolute roughness

$$\frac{r_0}{k_s} = \frac{210/2}{0.3} = 350$$

For 15 cm pipe : $\frac{1}{\sqrt{f}} = 2 \log_{10} 250 + 1.74$, resulting in $f = 0.0235$

For 21 cm pipe : $\frac{1}{\sqrt{f}} = 2 \log_{10} 350 + 1.74$, from which, $f = 0.0215$

Head loss, $$h_L = f\frac{L}{d}\frac{V^2}{2g} = f\frac{L}{d}\frac{\left(\frac{4Q}{\pi d^2}\right)^2}{2g} = f\frac{L}{d^5}\frac{\left(\frac{4Q}{\pi}\right)^2}{2g}$$

For the same discharge Q and same length L

$$\frac{h_{L_1}}{h_{L_2}} = \frac{f_1/d_1^{\,5}}{f_2/d_2^{\,2}} = \frac{f_1\, d_2^{\,5}}{f_2\, d_1^{\,5}}.$$

Cost of pumping will be proportional to the power required for pumping.

Power required = $\gamma Q h_L$, since γ and Q are the same for both the pipes, the ratio of

$$\frac{\text{Cost of pumping through 15 cm pipe}}{\text{Cost of pumping through 21 cm pipe}} = \frac{0.0235 \times (0.21)^5}{0.0215 \times (0.15)^5} = 5.86.$$

Example 9.13. *If the relative roughness (k_s/d) of a 150 mm diameter pipe, carrying water having kinematic viscosity of 0.009 cm²/s is 0.0005, what is the Reynolds number upto which the pipe behaves as smooth ?*

Solution. $$k_s/d = 0.0005$$
$$\nu = 0.009 \text{ cm}^2/\text{s}.$$

From Fig. 9.14, the pipe would behave as smooth when $\mathbf{R}(k_s/d)\sqrt{f} = 8$, thus

$$\mathbf{R}\sqrt{f} = \frac{8}{0.0005} = 1.6 \times 10^4$$

For smooth pipes, we have

$$\frac{1}{\sqrt{f}} = 2.0 \log_{10} \mathbf{R}\sqrt{f} - 0.8 = 2.0 \log_{10} 1.6 \times 10^4 - 0.8 = 7.609$$

which yields, $$f = 0.0173.$$

Therefore, the Reynolds number

$$\mathbf{R} = 1.6 \times 10^4/\sqrt{f} = 1.216 \times 10^5.$$

Example 9.14. *A rough plastic pipe 500 mm in diameter and 300 m in length is carrying water with a velocity of 3 m/s, and has an absolute roughness of 0.25 mm and a kinematic viscosity of 0.9 centistokes. Is the flow turbulent or laminar ? What is the head lost in friction ?*

For laminar flow $$f = \frac{64}{R}$$

For turbulent flow $$\frac{1}{\sqrt{f}} = 2 \log_{10} \frac{r_0}{k_s} + 1.74.$$

[*UPSC Civil Service (IAS) Exams, Civil Engg. Paper I, 1979*]

Solution. Velocity in the pipe $V = 3$ m/s

Kinematic viscosity $$\nu = 0.9 \text{ centistokes} = 0.9 \times 10^{-2} \text{ stokes}$$
$$= 0.9 \times 10^{-2} \text{ cm}^2/\text{s} = 0.9 \times 10^{-6} \text{ m}^2/\text{s}$$

Reynolds numbers of flow $$\mathbf{R} = \frac{Vd}{\upsilon} = \frac{3 \times 0.50}{0.9 \times 10^{-6}} = 1.667 \times 10^6$$

Since the Reynolds number of pipe of flow is greater than 2000, the flow is turbulent.

Friction factor for turbulent flow in rough pipes

$$\frac{1}{\sqrt{f}} = 2 \log_{10} \frac{0.25}{0.25 \times 10^{-3}} + 1.74 = 7.74$$

$$f = 0.0167$$

Head lost in friction is given by Darcy-Weisbach equation

$$h_L = f \frac{L}{d}\frac{V^2}{2g} = 0.0167 \times \frac{300}{0.50} \times \frac{(3)^2}{2 \times 9.81} = \mathbf{4.596\ m.}$$

9.16. Resistance of Commercial Pipes

The commercial pipes of such material as metal, wood and concrete vary greatly in roughness height, form and roughness pattern, that it is impossible to describe the roughness completely by any single length dimension, as was possible for pipes artificially roughened by uniform sand grains used by Nikuradse. It is because of these variations in the pattern, form and height of roughness projections that the Nikuradse's test results are not directly applicable to commercial pipes. However, the *equivalent sand grain roughness* k_s of any commercial pipe may by evaluated which would yield the same value of f in the same diameter pipe as for the flow under similar conditions in an artificially roughened pipe. This concept of equivalent sand grain roughness has been used in determining the resistance of commercial pipes.

Colebrook has shown that the Nikuradse's test results could be applied for obtaining the friction factor of commercial pipes. By subtracting 2 $\log_{10} d/k_s$ from both sides of Eqs. (9.60) and (9.62), we obtain

(*i*) for smooth pipes

$$\frac{1}{\sqrt{f}} - 2\log_{10}\frac{d}{k_s} = \log_{10}\mathbf{R}\left(\frac{k_s}{d}\right)\sqrt{f} - 0.80 \qquad \text{...(9.63 a)}$$

(*ii*) for fully rough pipes

$$\frac{1}{\sqrt{f}} - 2\log_{10}\frac{d}{k_s} = 1.14 \qquad \text{...(9.63 b)}$$

These equations may be plotted as shown in Fig (9.14) with $(1/\sqrt{f} - 2\log_{10} d/k_s)$ as ordinate and $\mathbf{R}(k_s/d)\sqrt{f}$ as abscissa. The two straight lines characterise resistance laws for smooth and rough pipe surface. A transition curve connects these lines asymptotically. The transition curve based on Nikuradse's data has also been shown on the figure. It may noted from the figure that the transition starts approximately at $\mathbf{R}(k_s/d)\sqrt{f} = 8$ and ends at $\mathbf{R}(k_s/d)\sqrt{f} \approx 200$. The object of presenting Nikuradse's tests results in this manner is to relate friction factor, Reynolds number and the relative roughness by a single equation by reducing the family of curve of Fig. 9.13 to a single curve.

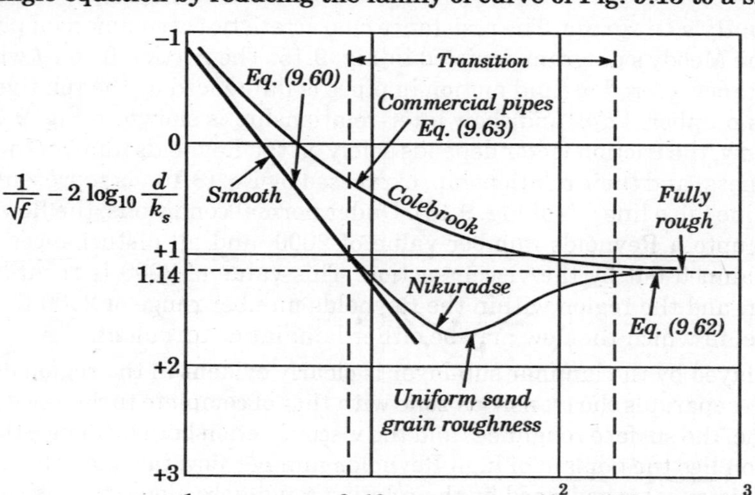

Fig. 9.14. Resistance diagram for commercial pipes compared with Nikuradse's artificially roughened pipes.

Colebrook found that at high Reynolds number, the friction factor of commercial pipe no longer varies with Reynolds number and is entirely dependent on relative roughness. This validity of Eq. (9.62) for commercial pipes enables determination of equivalent sand grain size, this would however, require experimentation at very high Reynolds numbers, Moody[1] has determined absolute roughness of commercial pipes by experiments in which f and $\mathbf{R}$ are computed and substituted in Colebrook equation (9.63). Table 9.1 shows the equivalent sand grain sizes, as obtained by Colebrook and Moody, for different pipe materials.

Table 9.1

Pipe material (Colebrook)	k_s *(mm)*	*Pipe material (moody)*	k_s *(mm)*
Uncoated cast iron	0.025	Rivetted steel	0.087 to 0.875
Galvanised iron	0.015	Concrete	0.030 to 0.30
Asphalted cast iron	0.012	Wood stave	0.017 to 0.085
Wrought iron	0.004		

Plotting his test results, Colebrook found that for the transition region an expression correlating friction factor, Reynolds number and relative roughness can be written as :

$$\frac{1}{\sqrt{f}} - 2\log_{10}\frac{d}{k_s} = 1.14 - 2\log_{10}\left[1 + \frac{9.35}{\mathbf{R}\left(\frac{k_s}{d}\right)\sqrt{f}}\right] \quad \text{...(9.63 } c\text{)}$$

which simplifies to,

$$\frac{1}{\sqrt{f}} = 1.14 - 2\log_{10}\left(\frac{k_s}{d} + \frac{9.35}{\mathbf{R}\sqrt{f}}\right) \quad \text{...(9.63 } d\text{)}$$

This equation is shown on Fig. 9.14 which shows $(1/\sqrt{f} - 2\log d/k_s)$ plotted against $\mathbf{R}\sqrt{f}\, k_s/d$. Although Eq. (9.63) summarise data on commercial pipes, it is not convenient for use because of the trial procedure involved in determination of friction factor. Eq. (9.63) has been plotted by Moody presenting friction factor at different Reynolds numbers and for various relative roughness (f *versus* $\mathbf{R}$ plotting is also known as Moody's diagram). Colebrook Eq. (9.63 *d*) reduces to Eq. (9.63 *a*) for smooth pipes for which $k_s \to 0$, and to Eq. (9.63 *b*) for rough pipes by letting $\mathbf{R} \to \infty$.

9.16.1. Moody's Diagram. The resistance characteristics of commercial pipes is graphically represented by the Moody's diagram depicted in Fig. 9.15. The friction factor f which is a measure of frictional resistance offered to fluid motion in pipes is influenced by the relative roughness, k_s/d, and the Reynolds number, Vd/ν, and their inter-relationship is shown in Fig. 9.15. It can be seen that in laminar flow, the friction factor depends solely on the Reynolds number (and is independent of the pipe roughness) and their relationship, expressed by Eq. (8.10), is represented by the steeply sloping straight line (also line *CD* of Fig. 9.13). Under normal conditions, the flow in pipes remains in laminar state upto a Reynolds number value of 2000, and all disturbances tending to cause turbulence are damped out by the viscous action. This value of 2000 is regarded as the critical Reynolds number, and the region within the Reynolds number range of 2000 to 4000 is known as the **critical zone**, in which the flow may be either laminar or turbulent.

The role played by the laminar sub-layer is clearly evident in the region demarcated by the dotted line, which separates the transition zone with that of complete turbulence in rough pipes. In the transition zone, the surface roughness and the viscous action both influence the pipe resistance. Outside this region lies the domain of high Reynolds number flow in which the friction factor (and hence the pipe resistance) is governed by the relative roughness alone, the viscosity ceases to play any role here on account of laminar sub-layer vanishing at very high Reynolds numbers. Viscosity,

1. Moody, L.F., "Friction Factors for Pipe Flow". Trans. A.S.M.E. 1944.

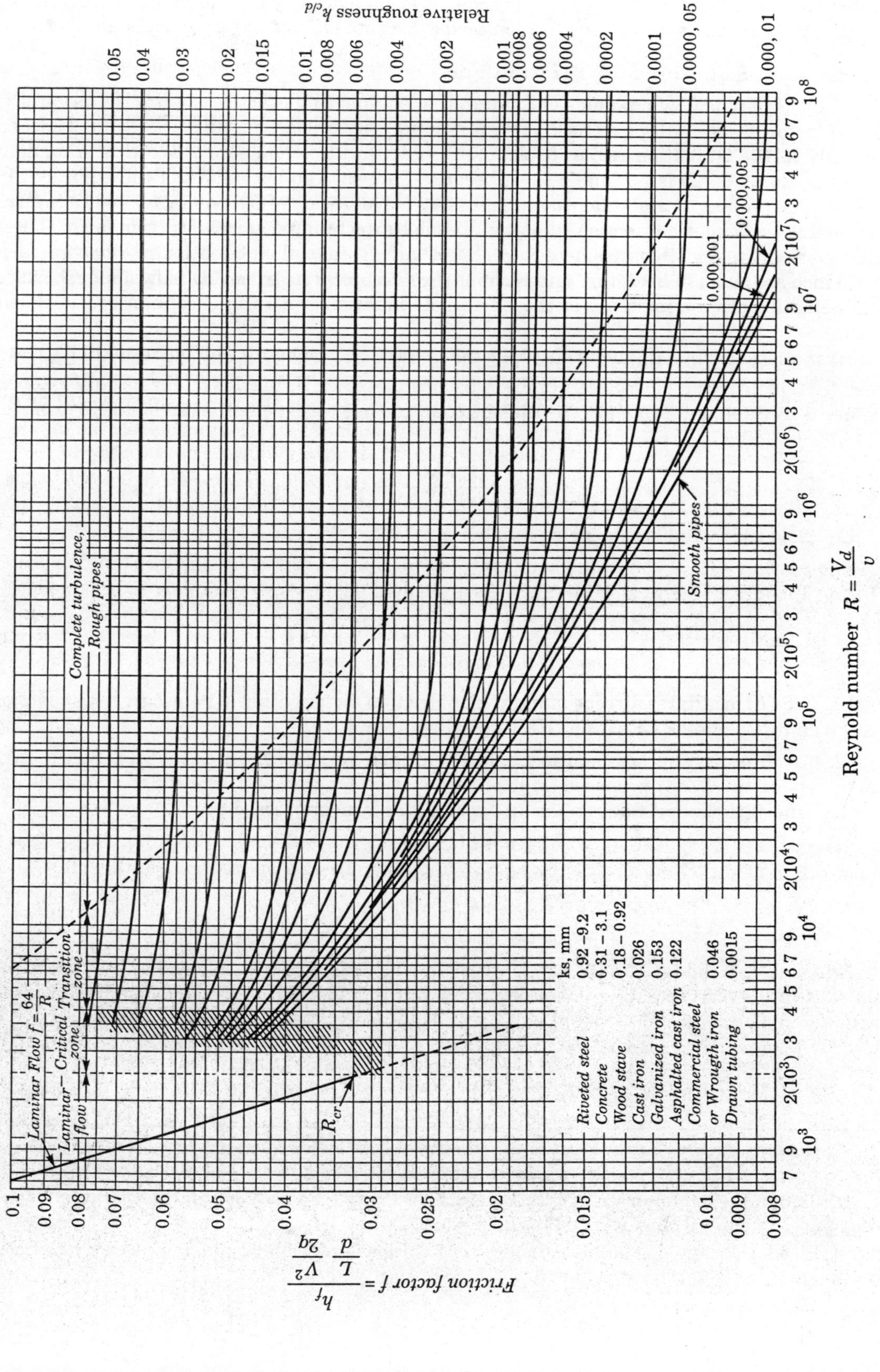

Fig. 9.15. Moody's Diagram.

therefore, does not affect energy loss in this zone, as is evidenced by the horizontal nature of $f-\mathbf{R}$ lines which indicate the fact that the friction factor does not change with Reynolds number.

9.16.2. Explicit Equations for Friction Factor

Analysis of pipe flow problems encountered in engineering practice requires determination of friction factor. Flow in all pipes used for water supply, hydroelectric generation and elsewhere, is generally in the transition region (Colebrook's observation lends support to this statement). Colebrook's[2] Eq. (9.63 *d*) is an implicit one because f appears on both sides of the equation. For given Reynolds number $\mathbf{R}$ and relative roughness k_s/d, evaluation of friction factor requires a trial procedure. Explicit equations, on the other hand, enable direct computation of friction factor thereby eliminating many trials otherwise necessary. Below are given explicit equations for friction factor some of which are latest in literature and perhaps the most accurate available so far. They, however, are mathematical approximations of the Colebrook equation, with the exception of the Blasius equation.

1. **For smooth-turbulent flow.** For hydraulically smooth pipes, one of the earliest relations between friction factor and Reynolds number was formulated by H. Blasius on the basis of extensive experiments by Saph and Schoder. The Blasius equation has an exponential form and is expressed as :

$$f=\frac{0.3164}{\mathbf{R}^{0.25}} \qquad \text{...(9.64 } a)$$

Eq. (9.64 *a*) is valid for Reynolds number $\mathbf{R} \leq 10^5$.

The author[3] (Jain) conducted studies to obtain an explicit equivalent of Colebrook's Eq. (9.63 *d*) for turbulent flow in smooth pipes. As a result the following equation was obtained :

$$\frac{1}{\sqrt{f}} = 1.8 \log_{10} \mathbf{R} - 1.5186 \qquad \text{...(9.64 } b)$$

Eq. (9.64 *b*) facilitates direct computation of friction factor within ± 1% accuracy as compared to that given by Colebrook's Eq. (9.63 *d*). It is valied upto $\mathbf{R} \leq 10^8$.

2. **For flow in the transition region (smooth-rough turbulent flow).** Based on the Colebrook equation, Wood[4] proposed an explicit equation for friction factor for the smooth-rough transition zone of turbulent flow. It has an exponential form expressed by

$$f = a + b\mathbf{R}^{-c} \qquad \text{...(9.65)}$$

in which $a = 0.094\,(k_s/d)^{0.225} + 0.53\, k_s/d$

$b = 88\,(k_s/d)^{0.44}$

and $c = 1.62\,(k_s/d)^{0.134}$

Eq. (9.65) is valid for $10^{-5} \leq k_s/d \leq 4 \times 10^{-2}$ and $5 \times 10^3 \leq \mathbf{R} \leq 10^8$ and computes friction factor within + 5% of the corresponding Colebrook's value. It is evident from the equation itself that it is not valid for smooth pipes.

---

2. Colebrook, C.F., "Turbulent Flow in pipes with particular reference to the Transition Region between Smooth and Rough Pipe Laws", Journal Inst. of Civil Engineers, Vol II, 1938-39, pp. 133-156.

3. Jain, A.K., "An Accurate Explicit Equation for Friction Factor". Journal of Hydraulic Division, American Society of Civil Engineers, May, 1976, pp. 674-77.

4. Wood, D.J., "An Explicit Friction Factor Relationship". Civil Engineering, American Society of Civil Engineers, Dec., 1966, pp. 60-61.

The author[3] (Jain) has obtained a simple and more accurate explicit equation for friction factor for flow in commercial pipes. The equation has a logarithmic form :

$$\frac{1}{\sqrt{f}} = 1.14 - 2 \log_{10} \left(\frac{k_s}{d} + \frac{21.25}{R^{0.9}} \right) \quad ...(9.66)$$

Collins[5] in his discussion on Eq. (9.64*b*) comments thus : "The author's (Jain's) proposed explicit expression for f is concluded to be an excellent approximation of the Colebrook-White equation in the range $5000 < \mathbf{R} < 10^8$. The ability to explicitly and accurately describe f over a large Reynolds number range can aid materially in efficient solution of pipe and pipe network problems."

Eq. (9.66) enables direct determination of friction factor for known values of k_s/d and **R**. The accuracy is well within ± 1% of the corresponding values obtained from the Colebrook equation. The equation reduces to Eq. (9.63 *a*) for flow in smooth pipes by putting $k_s/d \approx 0$, and to Eq. (9.63 *b*) for rough pipes by setting $\mathbf{R} \to \infty$. It is valid for the same range of k_s/d and **R** for which the Colebrook's equation is applicable, that is, $10^{-6} \le k_s/d \le 10^{-2}$, and $5 \times 10^3 \le \mathbf{R} \le 10^8$. Eq. (9.66) can, thus, be used for all categories of turbulent flow in pipes, namely, smooth, transitional and rough.

Example 9.15. *If a conduit of 2 m diameter is to carry a flow of 4.25 m^3/s of water at 20°C ($\nu = 1 \times 10^{-6}$ m^2/s) with minimum loss of energy, determine the permissible height of surface protrusions.*

Solution. Loss of energy will be minimum when the pipe surface is smooth.

Average velocity of flow

$$V = \frac{Q}{A} = \frac{4.25}{\pi/4\,(2)^2} = 1.35 \text{ m/s}.$$

Reynolds number of flow

$$\mathbf{R} = \frac{Vd}{\upsilon} = \frac{1.35 \times 2.0}{1 \times 10^{-6}} = 2.70 \times 10^6.$$

For flow through smooth pipes,

$$\frac{1}{\sqrt{f}} = 2.0 \log_{10} \mathbf{R} \sqrt{f} - 0.80.$$

For $\mathbf{R} = 2.7 \times 10^6$, solving the above equation for f by trial procedure we obtain $f = 0.01$. Use of explicit Eq. (9.64) also gives $f = 0.01$ in a single step.

Using resistance law for rough pipes Eq. (9.62), to determine permissible roughness k_s.

$$\frac{1}{\sqrt{f}} = 2.0 \log_{10} \frac{r_0}{k_s} + 1.74$$

for $\qquad r_0 = 1$ m and $f = 0.01$

$$\frac{1}{\sqrt{0.01}} = -\,2.0 \log_{10} k_s + 1.74$$

or $\qquad \log_{10} k_s = -\,4.13$

$$k_s = 7.41 \times 10^{-5} \text{ m}.$$

Example 9.16. *Water at 20°C flows in a 80 mm pipe at Reynolds number 80,000. The pipe is estimated to have an equivalent sand gain roughness of size 0.16 mm. Determine the head loss expected in 300 m length of the pipe. How much head loss would be expected if this pipe were smooth. Kinematic viscosity of water at 20°C is 10^{-6} m^2/s.*

5. Collins, Michael A., Discussion on (*i*) "Explicit Equation for Pipe Flow Problems" and (*ii*) "Accurate Explicit Equation for Friction Factor". Journal of Hydraulics Division, ASCE, Vol. 102, HY 11, Nov. 1976, pp. 1707-1709.

Solution. Using explicit equation (9.66) with **R** = 80,000 and $\frac{k_s}{d} = \frac{0.16}{80} = 0.002$

$$\frac{1}{\sqrt{f}} = 1.14 - 2\log_{10}\left(0.002 + \frac{21.25}{80{,}000^{0.9}}\right)$$

from which, $f = 0.0257$.

From the known Reynolds number, the mean velocity in the pipe

$$V = \frac{\upsilon \mathbf{R}}{d} = \frac{10^{-6} \times 80{,}000}{80/1000} = 1.0 \text{ m/s}$$

The head loss, $$h_L = f\frac{L}{d}\frac{V^2}{2g}$$

$$= 0.0257 \times \frac{300}{80/1000} \times \frac{(1)^2}{2 \times 9.81} = \mathbf{4.91\ m.}$$

For a smooth pipe, since, $\mathbf{R} < 10^5$, using Eq. (9.61) for fricition factor

$$f = \frac{0.3164}{\mathbf{R}^{0.25}} = \frac{0.3164}{(80{,}000)^{0.25}} = 0.0188$$

[**Note.** The explicit Eq. (9.65) could also have been used to determine f which results in almost the same value, see Ex. 9.13.]

For the same flow rate, diameter and length, the head loss is proportional to the friction factor. Hence if the pipe were smooth, the head loss

$$h_L = \frac{0.0188}{0.0257} \times 4.91 = \mathbf{3.587\ m.}$$

Example 9.17. *A rough pipe of diameter 450 mm and length 500 m carries water at the rate of 0.4 m^3/s. The wall roughness is 0.012 mm. Determine :*

(*i*) *the coefficient of friction*

(*ii*) *wall shear stress*

(*iii*) *centre-line velocity*

(*iv*) *velocity at a distance of 150 mm from the pipe wall.*

Solution. (*i*) The coefficient of friction or the friction factor in a rough pipe is given by :

$$\frac{1}{\sqrt{f}} = 2\log_{10}\frac{d}{k_s} + 1.14$$

$$= 2\log_{10}\frac{450}{0.012} + 1.14 = 2\log_{10} 37500 + 1.14$$

$$= 2 \times 4.5740 + 1.14 = 10.288$$

$$\therefore \quad f = 9.45 \times 10^{-3} = \mathbf{0.00945.}$$

(*ii*) The wall shear stress, τ_0 is expressed as

$$\frac{\tau_0}{\rho} = \frac{f}{8}V^2$$

$$V = \frac{Q}{A} = \frac{0.4}{\frac{\pi}{4}(0.45)^2} = 2.515 \text{ m/s}$$

$$\tau_0 = \rho\frac{f}{8}V^2 = \frac{1000}{9.81} \times \frac{0.00945}{8} \times (2.515)^2 = \mathbf{0.076\ kg/m^2.}$$

(*iii*) The velocity distribution in a rough pipe is given by :

$$\frac{V-u}{u^*} = 5.75 \log_{10} \frac{r_0}{y} - 3.75$$

in which u is the velocity at a distance y from the pipe wall.

$$u^* = \sqrt{\frac{\tau_0}{\rho}} = \sqrt{\frac{0.076 \times 9.81}{1000}} = \sqrt{0.0007456} = 0.0273 \text{ m/s}$$

The centre-line velocity is the value of u at $y = r_o$,

$$\frac{V-u_{cl}}{0.0273} = 5.75 \log_{10} 1 - 3.75 = 0 - 3.75$$

or $\frac{2.515-u_{cl}}{0.0273} = -3.75$, from which, $u_{cl} = 2.515 + 0.1024 =$ **2.6173 m/s.**

(*iv*) **Velocity at a distance of 15 from the pipe wall**

$$\frac{2.515-u}{0.0273} = 5.75 \log_{10} \frac{0.45/2}{0.15} - 3.75 = 1.0125 - 3.75 = -2.7374$$

or $2.515 - u = -0.0747$, from which

$u =$ **2.44 m/s.**

9.17. Ageing of Pipes

The values of equivalent sand grain roughness tabulated in Table 9.1 relate to the materials in new condition. Due to accumulation of dirt and corrosion of pipe wall, its roughness increases with age. This deterioration of pipe surface depends upon the nature of the material and also of the fluid. The accumulation of dirt not only increases the surface roughness but also reduces the effective diameter of pipe. The combined effect of these may lead to an extremely large increase in the friction factor after the pipe has been in service for a number of years. Colebrook and White have proposed that the roughness may be expected to increase with time according to the following expression :

$$k_s = k_0 + at \qquad \text{...(9.67)}$$

where k_0 = roughness of the new material, k_s = roughness at any time t, and α = time rate of increase of roughness.

For pipelines in service for a long period, say 20 years or more, the absolute roughness of old concrete or cement lined steel pipes may increase three times over that of new pipes ; for cast iron pipes the increase may be upto twenty times and for wrought iron and steel pipes it may reach even forty times.

9.18. Resistance of Non-circular Conduits : Use of Hydraulic Radius

Most of the conduits used in engineering practice happen to be of circular cross-section, however, there may be many situations where conduits of non-circular cross-section may have to be used. These could be : (*a*) a shell-and-tube exchanger, (*b*) square, and (*c*) rectangular air distribution ducts. The foregoing equations developed essentially for circular pipes may be used by applying the concept of hydraulic radius, which is used as the characteristic dimension 'l' in Eq. (5.11) to obtain the Reynolds number for closed non-circular cross-sections.

The hydraulic radius R_h, is defined as the ratio of flow area A to wetted perimeter P. In case of a circular pipe of diameter d, running full,

$$R_h = \frac{A}{P} = \frac{\pi d^2/4}{\pi d} = \frac{d}{4} \; ; d = 4R_h \qquad \text{...(9.68)}$$

this value of d may be substituted in the head loss equation (Darcy-Weisbach) and the Reynolds number. With the help of explicit Eq. (8.66) the friction factor can be directly obtained and the resulting head loss can then be calculated as below :

Reynolds number, $$R = \frac{\rho V(4R_h)}{\mu} \quad ...(9.69)$$

and the head loss, $$h_L = f\frac{L}{4R_h}\frac{V^2}{2g} \quad ...(9.70)$$

For a shell-and-tube heat exchanger, $A = \frac{\pi}{4}(D^2 - d^2)$, and $P = \pi(D + d)$, and $R_h = A/P$.

The calculation of head loss in non-circular conduits thus involves calculation of hydraulic radius and the use of friction factor obtained for an equivalent circular pipe of diameter $4R_h$. This method gives satisfactory results when flow in the conduit is turbulent, but if it is used for laminar flow, the results have been found to have large deviations. It may be noted that the relative roughness, k_s/d, becomes $k_s/4R_n$ for non-circular pipes.

Example 9.18. *Determine the loss of head and the pressure drop when air at 1.01 kN/m² and 20°C flows through a smooth rectangular duct of 450 mm × 300 mm cross-section and 600 m long with a mean velocity of 4 m/s. Take* $\nu = 1.5 \times 10^{-5}$ *m²/s,* $\gamma = 12.1$ *N/m³ air at 20°C.*

Solution. Using Eq. (9.68), the hydraulic radius

$$R_h = \frac{A}{P} = \frac{0.45 \times 0.30}{2 \times 0.45 + 2 \times 0.30} = 0.09 \text{ m}$$

Reynolds number, $$\mathbf{R} = \frac{V(4R_h)}{\upsilon} = \frac{4 \times (4 \times 0.09)}{1.5 \times 10^{-5}} = 96{,}000$$

From Eq. (9.64) the friction factor for a smooth conduit may be obtained,

$$\frac{1}{\sqrt{f}} = 1.8 \log_{10}(9.6 \times 10^4) - 1.5186$$

$$f = 0.018.$$

The head loss is given by Eq. (9.70)

$$h_L = f\frac{L}{(4R_h)}\frac{V^2}{2g}$$

$$= 0.018 \times \frac{600}{(4 \times 0.09)} \times \frac{(4)^2}{2 \times 9.81} = \mathbf{24.4 \text{ m of air.}}$$

Pressure difference, Δp, $\gamma h_L = 12.1 \times 24.4 =$ **295.6 N/m².**

9.19. Admissible Roughness

The height of roughness which is considered *admissible* in engineering practice is that height of individual roughness elements which causes no increase in drag over that for a smooth boundary. The practical importance of determining the admissible roughness for a given set of flow and boundary conditions lies in giving a particular type of finish to the surface. This is so because the admissible roughness determines the amount of labour required to be put in for the manufacture of boundary surface of the desired finish. For flat plates the admissible roughness may be expressed by

$$\frac{U_\infty K_{adm}}{\upsilon} \leq 100 \quad ...(9.71)$$

Eq. (9.71) is applicable for Reynolds number $\mathbf{R} < 10^6$, and indicates that the admissible height of roughness elements is solely determined by the velocity and the kinematic viscosity.

For engineering use, it is more convenient to express the admissible roughness, K_{adm}, in a non-dimensional form using the length L of the body under consideration (the body may be a ship, an aircraft wing, turbine or compressor blade). Eq. (9.71) may be rewritten as :

$$\frac{K_{adm}}{L} \leq \frac{100}{\mathbf{R}_L} \qquad ...(9.72)$$

where $\mathbf{R}_L = U_\infty L/\nu$ is the Reynolds number based on the length of the body L.

In case of ship's hulls, the admissible roughness is of the order of several hundredths of one millimetre. Such fine surfaces matching to this requirement of admissible roughness cannot be manufactured in practice and it is always necessary to allow for a large increase in drag due to roughness. For aircraft surfaces the range of admissible roughness lies between 0.01 to 0.1 mm ; with carefully executed manufacturing operations it is possible to obtain surfaces required for aircraft use. For steam turbine blades the admissible roughness range from 0.0002 to 0.002 mm ; the blade surface can hardly be smoothened to this degree of roughness. The height of roughness elements increases considerably due to corrosion and scale formation after the blade has been in operation for quite some time.

The height of a roughness element which causes transition in the laminar boundary layer is usually called *critical height or critical roughness*. The roughness affects the resistance offered by the boundary by shifting the point of transition upstream and depending upon the shape of the body, the drag may either be increased or decreased. The critical value of the roughness according to experimental work done by some Japanese researchers (J. Tani, R. Hama and S. Mituisi) is given by

$$\frac{u_* K_{crit}}{v} = 15 \qquad ...(9.73\ a)$$

More recent investigations carried out by K. Kraemer show that the critical roughness size which causes transition in the laminar boundary layer can be determined from

$$\frac{U_\infty K_{crit}}{v} \leq 900 \qquad ...(9.73\ b)$$

Eq. (9.73 *b*) is of the same form as Eq. (9.71). Both these equations represents Reynolds number based on admissible and critical roughness heights respectively.

9.20. Hazen-Williams Formula

Many empirical formulas are used for pipe flow analysis in engineering practice. These formulas have been obtained from experiments on flow of water under turbulent conditions. Use of such formulas for flow of other liquids is, therefore, not likely to give reliable results. The Hazen-Williams formula[6] is typical of such formulas and is probably the most widely used in water supply practice. In its original form, the formula is written as,

$$V = 1.318\, C R_h^{0.63} S^{0.54} \qquad ...(9.74)$$

$$= C' R_h^{0.63} S^{0.54} \qquad ...(9.74\ a)$$

in which V is the mean velocity in ft/sec, R_h the hydraulic radius in ft, and S the head loss per unit length of pipe, C is a roughness coefficient known as Hazen-Williams coefficient and $C' = 1.318\ C$. In deriving Eq. (9.74), Hazen and Williams used Chezy's equation and a considerable amount of experimental data on water flow through pipes of different materials and dimaters. The Chezy's formula is however expressed as,

$$V = C R_h^{0.50} S^{0.50} \qquad ...(9.75)$$

If b is taken as a multiplying factor to be used with Eq. (9.74 *a*) in order that Eqs. (9. 74 *a*) and (9.75) yield the same velocity for $C' = C$. We then obtain,

6. Williams, G.S., and Hazen A., "Hydraulic Tables" 3rd Ed. John Wiley and Sons, 1933, London, England 1966.

$$bR_h^{0.63}\,S^{0.54} = R_h^{0.50}\,S^{0.50}$$

which yields, $$b = R_h^{0.13}\,S^{-0.04} \qquad ...(9.76)$$

Hazen-Williams assumed R_h 1 ft. (0.3 m) and $S = 1/1{,}000$, to evaluate the multiplying factor b as 1.318, resulting in the commonly used form of the Hazen-Williams formula in FPS units, see Eq. (9.74). In MKS and SI units, Eq. (9.74) can be expressed as

$$V = 0.85\,CR_h^{0.63}\,S^{0.54} \qquad ...(9.77)$$

wherein V is in m/s and R_h in m. It is, therefore, obvious that the Hazen-Williams formula would give reliable results only when $R \approx 1$ ft (0.3) and $S \approx 1/1{,}000$; more the departure of these parameters, the greater will be the discrepancy in the results.

An important fact which merits attention is the basis of selection of the Hazen-Williams coefficient C. The coefficient, which is a measure of the resistance to flow of water, is selected solely on the basis of the pipe material and its condition. It, however, does neither considers the flow characteristics such as the Reynolds number nor the pipe dimater. Truely speaking, the friction factor in pipe flow (which, like the Hazen-Williams coefficient, is also a measure of resistance offered to fluid motion) is dependent on the relative roughness and the Reynolds number of flow. It is thus abundantly clear that the use of Hazen-Williams formula to problems of analysis and design of water supply pipelines would result in appreciable errors.

Despite the caution given by its authors (Hazen and Williams) against the universal usage of Eq. (9.74), design engineers seem to have no choice but to use the Hazen-Williams formula over wide ranges of diameters and friction slopes.

In order to ascertain the validity of Hazen-Williams formula, Jain, Mohan and Khanna[7] studied the accuracy aspects of the formula for its use under widely different conditions than those on which its derivation was based. The range of errors involved in the estimation of velocity and head loss due to the combined effect of changes in the values of the multiplying factor b and that of the coefficient C on account of changes in the hydraulic conditions and the pipe diameter are as follows :

Table 9.2. Showing Range of Errors Associated with the use of Hazen-Williams Formula

Nature of Error	*Range of Error in*	
	Velocity	*Head Loss*
Positive Error	4 to 52%	7 to 95%
Negative Error	10 to 63%	18 to 96%

The high percentage of error associated with the use of the Hazen-Williams formula, as evidenced by the above table, led Jain, Mohan and Khanna[7] to evolve a similar but accurate formula for analysis and design of water supply pipelines. The formula proposed by them has been called ***Modified Hazen-Williams Formula.*** The modified Hazen-Williams formula, which is based on the Colebrook-White equation for commercial pipes, may be expressed as :

$$V = \frac{3.83\,C_R d^{0.6575}\,(gS)^{0.5525}}{\upsilon^{0.105}} \qquad ...(9.78)$$

For water at temperature 20°C, $\nu = 10^{-6}$ m²/s and taking $g = 9.81$ m/s²,

$$V = 143.53\,C_R\,R_h^{0.6575}\,S^{0.5525} \qquad ...(9.79)$$

in which C_R is a coefficient of roughness. $C_R = 1$ for smooth pipes and for rough pipes $C_R < 1$. The coefficient of roughness depends on the relative roughness and a Reynolds number. Its value range from 0.57 to 1.00 for different conditions of flow and pipes of different materials and sizes (see Table 9.2).

7. Jain, A.K., Mohan, D.M. and Khanna, P., "Modified Hazen-Williams Formula". Journal of Environmental Engg. Division, American Society of Civil Engineers February 1978.

Table 9.3. Representative C_R Values For Commercial Pipes

Dia. in mm ↓ / Velocity in m/s →	*CI—New* $k_s = 0.00015\ m$			*CI—old* $k_s = 0.00024\ m$			*Spun concrete* $k_s = 0.00035\ m$			*A.C. New* $k_s = 0.000015\ m$		
	0.3	*0.9*	*1.2*	*0.3*	*0.9*	*1.2*	*0.3*	*0.9*	*1.2*	*0.3*	*0.9*	*1.2*
100	0.9524	0.9089	0.8922	0.6564	0.5886	0.5717			0.996			
125	0.9610	0.916	0.899	0.671	0.603	0.586						
150	0.9871	0.9216	0.9044	0.6840	0.6145	0.5958						
200	0.9760	0.9291	0.9117	0.7025	0.6331	0.6130						
250	0.9821	0.9342	0.9166	0.7158	0.6431	0.6329						
300	0.9865	0.9379	0.9202	0.7261	0.6523	0.6355						
350	0.9900	0.9406	0.9228	0.7343	0.6596	0.6407						
400	0.9927	0.9428	0.9249	0.7411	0.6657	0.6456						
450	0.9949	0.9445	0.9265	0.7468	0.6709	0.6515						
500	0.9967	0.9459	0.9279	0.7518	0.6753	0.6559						
550	0.9982	0.9471	0.9289	0.7561	0.6792	0.6597						
600	0.9945	0.9480	0.9299	0.7599	0.6826	0.6630			$C_R =$ 1.000	$C_R =$ 1.000	$C_R =$ 1.000	
650	1.000	0.9488	0.9306	0.7634	0.6857	0.6662						
700		0.9495	0.9313	0.7664	0.6884	0.6687	$C_R =$ 1.000	$C_R =$ 1.000				
750		0.9501	0.9318	0.7692	0.6909	0.6711						$C_R =$ 1.000
800		0.9506	0.9323	0.7718	0.6932	0.6733						
900		0.9514	0.9330	0.7762	0.6972	0.6772						
1000		0.9520	0.9336	0.7801	0.7007	0.6305						
1100		0.9524	0.9340	0.7834	0.7037	0.6834						
1200	$C_R =$ 1.000	0.9527	0.9342	0.7863	0.7063	0.6860						
1300		0.9529	0.9344	0.7889	0.7086	0.6882						
1400		0.9531	0.9346	0.7912	0.7107	0.6903						
1500		0.9532	0.9346	0.7933	0.7126	0.6921						
1600		0.9532	0.9346	0.7952	0.7143	0.6937			$C_R =$ 1.000			
1700		0.9532	0.9346	0.7970	0.7158	0.6953						
1800		0.9532	0.9346	0.7965	0.7176	0.6095						
1900		0.9532	0.9346	0.8000	0.7186	0.6979						
2000		0.9532	0.9346	0.8014	0.7198	0.6991						

Nomograph for ready use of water supply engineers has also been proposed. The formula (Eq. 9.78) is claimed to be as accurate and as representative as is the Colebrook-White equation. It can be used for any liquid and there are no limitations on its use. Table 9.2 gives representative values of C_R for commercial pipes. Slide rules based on Eq. (9.78) are also available for obtaining solutions of problems of design and analysis of flow in pipelines.

In order to use Hazen-Williams formula, however, one must know that value of the roughness coefficient C to be used. Table 9.3 gives typical values of the coefficient.

Table 9.3. Hazen-Williams Coefficient, C

Pipe material	*Hazen Williams Coefficient, C*	*Roughness height k_s metres*
Cast iron coated	130	0.0015
Cast iron (Old, Moderate corrosion, 30 years age	100	0.0024
Concrete (New)	130	0.000035
Steel (New)	140	0.00006

Example 9.19. *Water flows at 5 litre/second at 20°C ($\nu = 10^{-6}$ m²/s) through an 80 mm diameter extremely smooth and straight pipe. Determine the head loss in 300 m length of this pipe, using*

(1) Hazen-Williams formula,

(2) Modified Hazen-Williams formula of (Eq. 9.79) and

(3) Explicit equation for friction factor for smooth pipes.

Solution. Mean velocity through the pipe

$$V = \frac{4Q}{\pi d^2} = \frac{4 \times 5 \times 10^{-3}}{\pi (80/100)^2} = 1 \text{ m/s}$$

Hydraulic radius, $R_h = \frac{A}{P} = \frac{d}{4} = \frac{80/100}{4} = 2 \times 10^{-2}$ m.

(1) For a smooth pipe, Hazen-Williams coefficient $C = 140$. Using Eq. (9.77)

$$V = 0.85\, C R_h^{0.63}\, S^{0.54}$$

$$1 = 0.85 \times 140 \times (2 \times 10^{-2})^{0.63}\, (S)^{0.54}$$

from which, $S = 1/72.5 = h_L/300$; $h_L = 300/72.5 =$ **4.14 m.**

(2) From Eq. (9.79), $V = 143.53\, C_R\, R_h^{0.6675}\, S^{0.5525}$

for a smooth pipe $C_R = 1$, making appropriate substitutions

$$1 = 143.53 \times 1 \times (2 \times 10^{-2})^{0.6575}\, (S)^{0.5525}$$

which results in, $S = 1/75.85 = h_L/300$;

$$h_L = \frac{300}{75.85} = \mathbf{3.95\ m.}$$

(3) Reynolds number $\mathbf{R} = \frac{Vd}{\upsilon} = \frac{(1 \times 80 \times 10^{-2})}{10^{-6}} = 8 \times 10^4$

Using Eq. (9.64), $\frac{1}{\sqrt{f}} = 1.8 \log_{10} \mathbf{R} - 1.5186 = 1.8 \log_{10} 8 \times 10^4 - 1.5186$

from which $f = 0.01873$. The Darcy-Weisbach equation gives the head loss,

$$h_L = f\frac{L}{d}\frac{V^2}{2g} = 0.01873 \times \frac{300}{80 \times 10^{-3}} \times \frac{(1)^2}{2 \times 9.81} = \mathbf{3.58\ m.}$$

(Compare this with the head loss of Ex. 9.14).

Example 9.20. *A cast iron pipe 1.83 m in diameter and 1 km long carries a water discharge of 4.0 m^3/s at a temperature of 20°C($\nu = 10^{-6}\ m^2/s$). When new, the pipe surface is estimated to have an equivalent sand grain roughness of height k_s = 0.00015 m ; the corresponding Hazen-Williams roughness coefficient C may be taken as 130. Determine the head loss using :*

(1) the Hazen-Williams formula,

(2) the modified Hazen-Williams formula, Eq. (9.79), and

(3) the explicit equation for friction factor, Eq. (9.66).

After remaining in service for 30 years, it has been found that due to corrosion and accumulation of dirt, the pipe roughness k_s has increased to 0.0025 m, the corresponding Hazen-Williams coefficient is estimated at 100. Estimate the new value of the head loss using the above three equations.

Solution. Hydraulic radius

$$R_h = \frac{d}{4} = \frac{1.83}{4} = 0.457 \text{ m}$$

Mean velocity, $$V = \frac{4Q}{\pi d^2} = \frac{4 \times 4}{\pi (1.83)^2} = 1.52 \text{ m/s}$$

Reynolds number, $$\mathbf{R} = \frac{Vd}{\upsilon} = \frac{1.52 \times 1.83}{10^{-6}} = 2.78 \times 10^6$$

Case I : New Pipe

(1) Hazen-Williams formula :

$$C = 130$$

$$V = 0.85\, C\, R_h^{0.63}\, S^{0.54}$$

$$1.52 = 0.85 \times 130 \times (0.457)^{0.63}\, S^{0.54}$$

from which $$S = \frac{1}{1072} = \frac{h_L}{1000}$$

$$h_L = \frac{1000}{1072} = \mathbf{0.934\ m.}$$

(2) Modified Hazen-Williams formula, Eq. (9.79)

$$V = 143.53\, C_R\, R_h^{0.6575}\, S^{0.5525}$$

for d = 1830 mm and V = 1.52 m/s, C_R may be obtained from the table for cast iron pipe (new) given in Reference on page 441, (ASCE paper on Modified Hazen-Williams equation) ; C_R = 0.9205, making substitutions,

$$1.52 = 143.53 \times 0.9205 \times (0.457)^{0.6575}\, S^{0.5525}$$

which results in, $$S = \frac{1}{1319} = \frac{h_L}{1000}$$

$$h_L = \frac{1000}{1319} = \mathbf{0.758\ m.}$$

(3) Explicit equation for Friction factor, Eq. (9.66) :

$$\frac{1}{\sqrt{f}} = 1.14 - 2\log\left(\frac{k_s}{d} + \frac{21.25}{\mathbf{R}^{0.9}}\right)$$

$$\frac{k_s}{d} = \frac{0.00015}{1.83} = 0.000082\ ;$$
$$\mathbf{R} = 2.78 \times 10^6.$$

Substituting these values in the above formula, one obtains $f = 0.01226$. The head loss can now be determined,

$$h_L = f\frac{L}{d}\frac{V^2}{2g} = 0.01226 \times \frac{1000}{1.83} \times \frac{(1.52)^2}{2 \times 9.81} = \mathbf{0.788\ m.}$$

Case II : Old Pipe

(1) Hazen-Williams formula :

$$C = 100$$
$$V = 0.85\ C\ R_h^{0.63}\ S^{0.54}$$
$$1.52 = 0.85 \times 100 \times (0.457)^{0.63}\ S^{0.54}$$

which results in,
$$S = \frac{1}{660} = \frac{h_L}{1000}$$
$$h_L = \frac{1000}{660} = \mathbf{1.526\ m.}$$

(2) Modified Hazen-Williams formula, Eq. (9.79) :

$$V = 143.53\ C_R\ R_h^{0.6575}\ S^{0.5525}$$

for cast iron 30 year old pipe in service, one finds from the table given in reference on page 441 that for $V = 1.52$ m/s and $d = 1830$ mm, $C_R = 0.836$, hence,

$$1.52 = 143.53 \times 0.6836 \times (0.457)^{0.6575}\ S^{0.5525}$$

from which,
$$S = \frac{1}{741.5} = \frac{h_L}{1000}$$
$$h_L = \frac{1000}{741.5} = \mathbf{1.35\ m.}$$

(3) Explicit equation for friction factor, Eq. (9.66) :

$$\frac{k_s}{d} = \frac{0.0024}{1.83} = 0.00131\ ;\ \mathbf{R} = 2.78 \times 10^6.$$

Making substitutions in Eq. (9.66), one obtains

$$\frac{1}{\sqrt{f}} = 1.14 - 2\log\left[0.00131 + \frac{21.15}{(2.78 \times 10^6)^{0.9}}\right]$$

$f = 0.0211$; and the head loss from the Darcy-Weishach Eq.

$$h_L = f\frac{L}{d}\frac{V^2}{2g} = 0.0211 \times \frac{1000}{1.83} \times \frac{(1.52)^2}{2 \times 9.11} = \mathbf{1.356\ m.}$$

OBJECTIVE TYPE QUESTIONS

1. While carrying out systematic experiments to measure energy losses in lead pipes, Osborne Reynolds found that the nature of flow at small and large velocities showed great qualitative change. At higher velocities, the flow changes from laminar to turbulent, and the following statements characterise the changed conditions of flow. Select the correct statements pertaining to turbulent flow :

   (*i*) the dye filament remains straight and the flow is parallel to the length of the tube

(*ii*) the energy loss was found to be proportional to the mean velocity
(*iii*) intense mixing occurs between the dye and the liquid and the flow becomes irregular, disorderly and random
(*iv*) the loss of energy was very nearly proportional to the square of mean velocity
(*v*) the Reynolds number must be low.

2. The causes of turbulence in fluid-flow may be :
(*a*) high Reynolds number
(*b*) abrupt discontinuity in velocity distribution
(*c*) critical Reynolds number
(*d*) existence of velocity gradient without abrupt discontinuity (*i.e.* shear flow).

3. The most essential feature of a turbulent flow is :
(*i*) large discharge (*ii*) high velocity
(*iii*) velocity and pressure at a point exhibit irregular fluctuations of high frequency
(*iv*) velocity at a point remains constant with time.

4. A turbulent flow is considered steady when :
(*a*) the discharge remains constant
(*b*) the velocity at a point does not change with times
(*c*) temporal mean velocity at a point remains constant with time
(*d*) the algebraic sum of velocity fluctuations is zero.

5. Intensity of turbulence is :
(*i*) the frequency of turbulent fluctuations
(*ii*) the violence of turbulent fluctuations and is measured by the root mean square value of velocity fluctuations
(*iii*) the average kinetic energy of turbulence
(*iv*) the mean time interval between the reversals in the sign of velocity fluctuation.

6. In comparison to the kinematic viscosity, the kinetic eddy viscosity makes a much greater contribution to the total shear stress in a turbulent flow. The kinematic eddy viscosity is :
(*i*) a direct measure of transporting capacity of the mixing process
(*ii*) directly proportional to the viscous effects
(*iii*) the ratio of dynamic viscosity to mass density
(*iv*) the ratio of eddy viscosity to mass density.

Alternatives : *a*—(*i*), (*ii*) ; *b*—(*i*), (*iii*) ; *c*—(*ii*), (*iii*) ; *d*—(*i*), (*iv*) ; *e*—(*ii*), (*iv*).

7. Which of the following examples, listed in column 2, fall into the category of wall-turbulence and which into the other category of free-turbulence :

(*a*) wall-turbulence	(*i*) flow through pipes
(*b*) free-turbulence	(*ii*) turbulent mixing of submerged jets
	(*iii*) flow past solid bodies
	(*iv*) turbulence in the wake region
	(*v*) flow over spillway.

8. Which of the following terms represents the turbulent shear stresses :

(*i*) $-\rho \overline{u'^2}$ (*ii*) $-\rho \overline{u'v'}$

(*iii*) $-\rho \overline{v'^2}$ (*iv*) $-\rho \overline{u'w'}$

9. The velocity distribution in turbulent flow is a function of the distance y measured from the boundary surface and the friction velocity u_*, and follows a :

(*a*) parabolic law (*b*) logarithmic law

(*c*) hyperbolic law (*d*) linear law.

10. The turbulent flow over a smooth boundary may be divided into three zones depending on the relative role of laminar friction over that of turbulent one. Identify from the following list the corresponding criteria in terms of Reynolds number based on friction velocity :

Flow zones	*Possible criteria*
(*i*) where purely laminar friction dominates	(*a*) $\frac{u_* y}{\upsilon} > 70$
(*ii*) where laminar and turbulent frictions both dominate	(*b*) $\frac{u_* y}{\upsilon} < 5$
(*iii*) where purely turbulent friction dominates	(*c*) $5 < \frac{u_* y}{\upsilon} < 70$

11. Commercial steel and cast iron pipes carrying fluids under pressure are regarded as hydraulically smooth when :

(*a*) the boundary surface is relatively smooth

(*b*) the roughness projections are of low height

(*c*) the roughness elements are all completely covered by the laminar sub-layer

(*d*) the laminar layer is thin as compared to the average height of roughness elements

12. It is well known that with increasing Reynolds number the thickness of laminar sub-layer progressively decreases thereby changing the resistance characteristic of the pipe. A hydraulically smooth pipe at high Reynolds number would :

(*i*) still behave as hydraulically smooth

(*ii*) behave as a hydraulically rough pipe, if $k_s/\delta_l > 12$

(*iii*) be in the transitional range if $0.5 < k_s/\delta_l < 12$

(*iv*) be comparatively more smooth in its hydraulic behaviour.

13. The velocity distribution in rough pipes being dependent on the relative roughness alone may be expressed by :

(*a*) $\frac{u}{u_*} = 5.75 \log \frac{y}{k_s} + 8.5$ (*b*) $\frac{u}{u_*} = 8.74 \left(\frac{u_* y}{\upsilon}\right)^{1/7}$

(*c*) $\frac{u}{u_*} = 5.75 \log \frac{u_* y}{\upsilon} + 5.50$ (*d*) none of the above.

14. The Darcy-Weisbach friction factor f which is a direct measure of resistance to flow in pipes is dependent on :

(*i*) roughness height, diameter and velocity

(*ii*) relative roughness, diameter and viscosity

(*iii*) relative roughness, velocity and viscosity

(*iv*) roughness height, diameter, velocity and kinematic viscosity.

15. Nikuradse carried out extensive experiments on pipes roughened artificially with sand grains. The results plotted on a logarithmic scale between :

(*a*) friction factor and velocity with diameter as the third parameter

(*b*) friction factor and velocity, the roughness height being the third parameter

(*c*) friction factor and Reynolds number, the relative roughness being the third parameter.

(*d*) friction factor and diameter, the velocity being the third parameter.

16. An important observation from Nikuradse's friction diagram in respect of fully rough pipes is that the friction factor :

(*i*) becomes independent of Reynolds number and is dependent on the relative roughness only.

(*ii*) is solely dependent on the Reynolds number and is independent of the relative roughness

(*iii*) becomes constant and is independent of both Reynolds number and relative roughness

(*iv*) none of the above.

17. Identify the Karman-Prandtl resistance equations for turbulent flow in smooth and rough pipes from the alternatives given below :

(*a*) hydraulically smooth pipes	(1) $f = 0.3164/\mathbf{R}^{0.25}$
(*b*) hydraulically rough pipes	(2) $f = 64/\mathbf{R}$
	(3) $1/\sqrt{f} = 2 \log \mathbf{R}\sqrt{f} - 0.80$
	(4) $1/\sqrt{f} = 1.14 - 2 \log k_s/d$.

18. The resistance of commercial pipes was systematically studied by Colebrook utilising the concept of equivalent sand grain roughness. Select from the following the relevant conclusions drawn from the Colebrook resistance diagram :

(*i*) the variation of friction factor with Reynolds number was found to be identical with the trend obtained by Nikuradse

(*ii*) there was a good amount of divergence in the trends of curves of Nikuradse and Colebrook

(*iii*) it was found that at high Reynolds numbers, the friction factor no longer varies with Reynolds number and is entirely dependent on relative roughness

(*iv*) Colebrook observed that the commercial pipes are generally in the transition zone in which the surface roughness and the viscous effect both are important and contribute to flow resistance

(*v*) the friction factor is dependent only on the viscous effect represented by the Reynolds number.

19. Identify from the following table, which relationships are valid for the turbulent flow categorised below :

(*a*) smooth-turbulent flow	(*i*) $1/\sqrt{f} = 1.14 - 2 \log k_s/d$
(*b*) turbulent flow in the transition region	(*ii*) $f = 0.3164/\mathbf{R}^{0.25}$
(*c*) rough-turbulent flow	(*iii*) $1/\sqrt{f} = 1.8 \log \mathbf{R} - 1.5186$
	(*iv*) $1/\sqrt{f} = 1.14 - 2 \log \left(\frac{k_s}{d} + \frac{1.25}{\mathbf{R}^{0.9}}\right)$
	(*v*) $1/\sqrt{f} = 2 \log \mathbf{R}\sqrt{f} - 0.8$
	(*vi*) $1/\sqrt{f} = 1.14 - 2 \log \left(\frac{k_s}{d} + \frac{9.35}{\mathbf{R}\sqrt{f}}\right)$

20. The Hazen-Williams (H-W) formula for flow analysis in water pipelines was derived for conditions in the vicinity of hydraulic radius of 1 ft and energy slope of 1 in 1000. Greater the departure of actual conditions from the above values more will be the error in estimation of head loss etc. An accurate formula which is equally valid for all conditions and is similar to H-W formula is given by :

(*i*) $V = 1.318CR_h^{0.63} S^{0.54}$

(*ii*) $V = 143.53\, C_R R_h^{0.6575} S^{0.5525}$

(*iii*) $V = 3.85CR_h^{0.63} S^{0.54}$

(*iv*) $V = \dfrac{3.83\, C_R d^{0.6575} (gS)^{0.5525}}{\upsilon^{0.105}}$.

21. The parameters which determine the friction factor for turbulent flow in a rough pipe are :
(*a*) Froude number and relative roughness
(*b*) Froude number and Mach number
(*c*) Reynolds number and relative roughness
(*d*) Mach number and relative roughness.

22. In case of turbulent flow of a fluid through a circular tube (as compared to the case of laminar flow at the same flow rate) the maximum velocity is ...., shear stress at the wall is ......, and the pressure drop across a given length is .... The correct words for the blanks are, respectively :
(*a*) higher, higher, higher (*b*) higher, lower, lower
(*c*) lower, higher, higher (*d*) lower, higher, lower.

ANSWERS TO OBJECTIVE TYPE QUESTIONS

1. (*iii*), (*iv*) **2.** (*b*), (*d*) **3.** (*iii*) **4.** (*c*) **5.** (*ii*)
6. (*d*) **7.** (*a*)—(*i*), (*ii*), (*v*) ; (*b*)—(*ii*), (*iv*) **8.** (*ii*) **9.** (*b*)
10. (*i*)—(*b*) ; (*ii*)—*c* **11.** (*c*) **12.** (*ii*), (*iii*) **13.** (*a*) **14.** (*iv*)
15. (*c*) (*i*) **16.** (*i*) **17.** (*a*)—1, 3 ; (*b*)—4.
18. (*ii*), (*iii*), (*iv*) **19.** (*a*)—(*ii*), (*iii*), (*v*) ; (*b*)—(*iv*), (*vi*) ; (*c*)—(*i*) **20.** (*ii*), (*iv*)
21. (*c*). **22.** (*c*).

REVIEW QUESTIONS

1. What factors influence the stability of laminar flow ?
2. Describe the factor which affect the transition of flow from laminar to turbulent.
3. What is turbulence ? Give examples of common occurrence which characterize turbulent flow.
4. What are the characteristics of turbulent flow ? How does that head loss vary with mean velocity ?
5. What do you understand by (*i*) intensity of turbulence, (*ii*) scale of turbulence, (*iii*) isotropic turbulence ?
6. Explain what do you understand by wall turbulence and free turbulence ? Give examples where each type is generally encountered.
7. Discuss the mechanism of momentum transfer due to transverse movement of fluid particles, and show how this leads to a shearing stress.
8. Turbulent flow in a pipe may be visualized as the superimposition on the main flow directed along the pipe axis of a subsidiary motion having a direction at right angles to pipe axis. How does this cause transverse mixing and modifies the velocity distribution in turbulent flow ?
9. What are the semi-empirical theories of turbulence ? Explain the concept of mixing length introduced by Prandtl and state the relationship that exists between the turbulent shearing stress and the mixing length.
10. Describe Nikuradse's experiments on artificially roughened pipes and state his findings in respect of friction factor.
11. In what way does the flow through a rough pipe differ from that in a smooth pipe ?
12. Define eddy viscosity and molecular viscosity. How do they differ ?
13. (*a*) What is meant by a smooth boundary and a rough boundary ?
(*b*) Distinguish between effective and actual roughness of a conduit boundary.
14. On what factors does Darcy's friction factor f for pipes depend in (*i*) laminar flow, and (*ii*) turbulent flow ? Discuss Nikuradse's experiments on flow through pipes.
15. Derive the Karman-Prandtl velocity equations and discuss the role of laminar sub-layer and surface roughness on velocity distribution.
16. Why do pipes behave as a hydrodynamically smooth or rough when turbulent flow, takes place through then ?

17. Does the ratio of maximum to mean velocity in a pipe increase or decrease :
    (*i*) with increasing Reynolds number, and
    (*ii*) with increasing relative roughness.
18. In what ways does the resistance of commercial pipes differ from those used by Nikuradse ? Explain the concept of equivalent sand grain roughness.
19. Explain why the hydraulic loss in a pipe is influenced by the surface roughness only at higher Reynolds numbers.
20. What do you understand by 'aging of pipes' ?
21. What modifications in the flow equations etc., would be necessary for flow through non-circular conduits.
22. Explain what do you mean by the admissible roughness and the critical roughness ?

PROBLEMS

9.1. If in turbulent flow through smooth pipe, the velocity distribution is given by one-seventh power law, find the value of y/r_0 in a pipe, where the point velocity equals the average velocity.

9.2. Water is flowing in a 15 cm diameter smooth glass pipe at an average velocity of 15 metre/sec. Find the thickness of (1) the laminar sub-layer, and (2) the transition layer. The kinematic viscosity of water at the prevailing temperature is 0.009 cm²/sec.

9.3. When 0.30 m/s of water flow in a 30 cm pipeline 62.5 kW are lost in friction in 100 m length of pipe. Calculate the head lost, friction factor, shear velocity and shear stress at the pipe wall.

9.4. Show that the seventh-root law velocity profile gives velocity gradients equal to $u_{max}\ 1/7r_0$ at the centre of the pipe and infinite at the pipe wall.

9.5. A pump of what power is required to pump 600 litre/sec of water from a reservoir of surface elevation 100 m to another of surface elevation 150 m, if in the pump and the pipeline 12 metres of head are lost ?

9.6. If the friction factor for a 25 cm water pipeline is 0.030 and the shear stress in the water 5 cm from the pipe centre line is 13.7 Pa, what is the flow rate ?

9.7. When 290 litres/sec of water flow in a 30 cm pipeline, 60.3 kW are lost in friction in 300 metres length of pipe. Calculate head lost, friction factor, shear velocity and shear stress at the pipe wall.

9.8. For Reynolds number less than 10^5, the velocity distribution in a smooth pipe may be approximated by

$$\frac{u}{u_*} = 8.7\left(\frac{u_* y}{v}\right)^{1/7}$$

Show that the average velocity $V = 0.816\ u_{max}$ and the shear stress at the pipe wall.

$$\tau_0 = \frac{0.0272}{\left(\dfrac{u_{max}\ d}{v}\right)^{0.25}} \rho u_{max}^2$$

9.9. If the head loss in 150 metres of 7.5 cm smooth pipe is 21 metres when the flowrate is 8.5 litres/sec, is the flow laminar or turbulent ?

9.10. Measurements in a very wide river at a section where the depth is 3 metre yield a surface velocity of 2.5 m/sec when the rate of flow is 6 cubic metre/sec per metre width. What intensity of bed shear is indicated by these measurements ?

9.11. In order to determine the rate of flow of air through a ventilating duct of circular section, pitot-tubes were installed at the mid-point and at one quarter point of the 2 metre diameter duct. What rate of flow is indicated by simultaneous readings of 5.5 m/sec and 4.4 m/sec at the two points respectively ?

9.12. Oil having a kinematic viscosity of 4.64×10^{-5} m²/sec and a sp. gr. of 0.90 flows at the rate of 15 litres/sec through a smooth 10 cm pipe. What is the nominal thickness of laminar sub-layer (δ') ? What is the velocity at δ' from the pipe wall ?

9.13. Oil having viscosity of 0.049 Pas is being pumped at the rate of 1000 litre/min through a 10 cm diameter pipeline 1 km long. The oil has a sp. gr. of 0.92. If the friction f is given by $0.32/(\mathbf{R})^{1/4}$, where **R** is Reynolds number, find the head lost in overcoming the friction. The pump in addition to overcoming friction is required to lift the oil to a height of 18.5 m. If the pump has an efficiency of 70% determine the horse-power required by the pump.

9.14. Incompressible fluid is flowing through a circular pipe. If the flow is turbulent and the velocity distribution is given by $u = u_0 (y/r_0)^{1/7}$, where y = distance from the pipe wall, u_0 = velocity at the pipe axis and r_0 = pipe radius. Calculate the radius at which the velocity of flow will be equal to the mean velocity of flow.

9.15. Show that for turbulent flow in pipe ($\mathbf{R} > 10^5$), the energy and the power lost are proportional to $V^{7/4}$ and $V^{11/4}$ respectively.

[**Hint.** Workout shear stress in terms of pressure loss using Blasius equation.]

9.16. Obtain an expression for the mixing length distribution law using Karman's equation for mixing length.

9.17. Using Blasius law of friction, deduce the following expression :

$$\frac{u_{max}}{u_*} = 8.74 \left(\frac{u_* r_0}{v}\right)^{1/7}$$

For $\mathbf{R} < 10^5$, take (*i*) exponent $n = 7$ in the exponential velocity law $u/u_{max} = (y/r_0)^{1/n}$ and (*ii*) ratio of average to maximum velocity $V/u_{max} = 0.8$.

9.18. Plot velocity distributions for n = 1/5, 1/7 and 1/9 in the exponential velocity law $u/u_{max} = (1 - r/r_0)^n$ for pipe flow and compare them with the Poiseuille velocity profile (laminar flow).

9.19. A 50 cm diameter horizontal smooth pipeline if transporting water at the velocity of 1.5 m/s and at the Reynolds number 7.4×10^5, what power per kilometre would be required to maintain the flow ?

9.20. Oil having kinematic visocity of 4.65×10^{-5} m²/s is transported through a horizontal smooth pipe line 1.5 kilometre long at a rate of 6000 lit/min. Determine the power required per km to maintain the flow at a Reynolds number of 40,000.

9.21. Show that velocity distribution for turbulent flow in rough pipes can be expressed as

$$\frac{u_{max} - u}{V} = -2.03 \sqrt{f} \log_{10}\left(1 - \frac{r}{r_0}\right)$$

where V is the average velocity and r is the radial distance from the axis of pipe having a radius r_0.

9.22. Crude oil is pumped at the rate of 15 lit/sec through a 10 cm pipe sloping as shown. The specific gravity of the oil is 0.92, and the kinematic viscosity 4.65×10^{-5} m²/s. If on a very hot day the

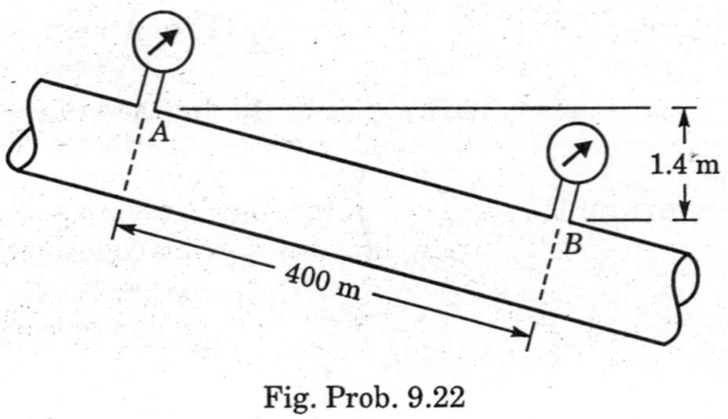

Fig. Prob. 9.22

increase in temperature caused the viscosity of the oil following in smooth pipe to decrease to one tenth of the value given, what would be the resulting difference in pressure intensity between point *A* and *B* for same rate of flow ?

9.23. Explain why resistance to flow through pipe is independent of the surface roughness at low Reynolds numbers whereas it depends only on the relative roughness at high Reynolds numbers.

9.24. Consider the steady flow of an incompressible fluid through a circular pipe. Sketch the velocity distribution along the diameter of the pipe for the two cases, namely, fully developed laminar flow and fully developed turbulent flow. Indicate the distinguishing feature of the two distributions. What are hydrodynamically smooth pipe flow and rough pipe flow ? What non-dimensional parameter is used to distinguish between these two types of flow ?

ANSWERS

9.1 $y/r_0 = 0.242$

9.2. 8.39×10^{-7} m, 11.73×10^{-6} m

9.3. 6.37 m, 0.069, 156.0 N/m^2 0.398 m/s

9.5. 364.6 kW

9.6. 148.4 lit/sec

9.7. 21.2, 0.0248, 0.228 m/s, 52.0 N/m^2

9.8. Laminar

9.10. 40.4 N/m^2

9.11. 10.53 m^3/sec

9.12. 4.02×10^{-3} m. 1.225 m/sec.

9.13. 92.6 m, 23.8 kN

9.14. 0.758 r_0

9.19. 8.14 kW

9.20. 11760 kW

9.22. 11.3 N/cm^2.

10

Flow Around Immersed Bodies Fluid-Dynamic Drag and Lift

10.1. Introduction

When an object is placed within a fluid stream, a force is exerted on the object that may be inclined to the flow direction. This force owes its existence to viscosity of fluid and the pressure distribution over the body surface. The component of this force parallel to the direction of flow is called the *drag* and the normal component is called the *lift*. Engineering problems requiring knowledge and understanding of drag and lift are varied and many. The determination of wind pressures on buildings and bridges, determination of efficient body shape of automobiles and trains which would result in minimum fluid resistance are some of the important problems which require a thorough understanding of phenomenon of drag. The principles of lift and drag are also used in the design of propellers, turbines and centrifugal pumps in addition to airfoils hydrofoils, aircrafts and fuel efficient cans.

The analysis of problems involving these phenomena, whether the object moves through the stationary fluid or the fluid passes around the stationary object is alike. It is, thus, possible to test airplane models in wind tunnels and submarine and torpedo models in water tunnels to predict satisfactorily the behaviour of their prototypes under different flow conditions.

10.2. Drag and Lift—Definitions and Fundamentals

An insight into the phenomena of drag and lift can be obtained by considering the flow pattern around an object. When flow occurs about an object which is either unsymmetrical or whose axis is not parallel to the flow direction, the flow field around it will be unsymmetrical causing local velocities and pressures on either side of the object to be different. The pressure difference over the surface causes a net force to act over the object. In addition to the force due to pressure there is another force which is everywhere tangential to the surface of object and is due to the action of viscous stress. The component of these forces parallel to the flow direction is the "drag" and that normal to it the "lift". Fig 10.1 (*a*) shows the drag and lift forces acting on a body placed in a fluid stream of uniform velocity U_∞.

An interesting example is that of an airfoil (or hydrofoil shown in Fig 10.1 (*b*)) which has an angle of attack α, the chord length c and the length of the foil perpendicular to the plane of the paper, termed as the span, is l. The resultant force F exerted by the fluid on the airfoil has two mutually perpendicular components : the lift F_L and the drag F_D. Consider a surface element dA of an immersed body (airfoil in this case) on which a pressure p and a shear stress τ_0 act. Let the tangent to the area element dA make an angle θ with the flow direction. The differential lift and drag on this area element are obtained by resolving the pressure and shear forces on the element perpendicular to and along the flow direction.

Considering Fig. 10.1 (*c*), one can write

$$dF_L = \tau_0 \, dA \sin\theta - p dA \cos\theta$$

and

$$dF_D = p dA \sin\theta + \tau_0 \, dA \cos\theta$$

which may be integrated to yield the lift and drag forces :

$$F_L = \int_A \tau_0 \, dA \sin\theta - \int_A pdA \cos\theta \qquad ...(10.1)$$

$$F_D = \int_A pdA \sin\theta + \int_A \tau_0 \, dA \cos\theta \qquad ...(10.2)$$

where the symbol $\int_A$ represents the integration over the entire body surface. The contribution of shear stresses to the lift may be neglected since such shear stresses are small as compared to the

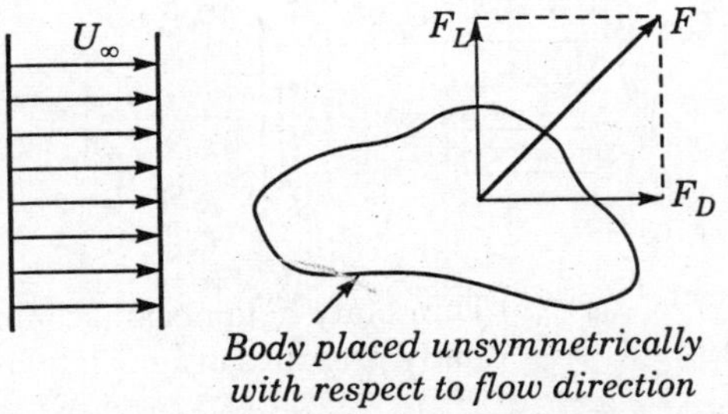

Fig. 10.1. (*a*) Definition sketch for lift and drag. (F_D = Drag force ; F_L = Lift force ; F = Resultant force)

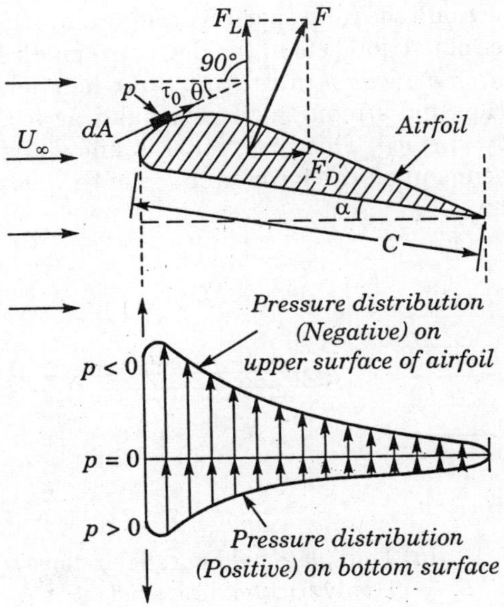

Fig. 10.1. (*b*) Pressure distribution over an airfoil

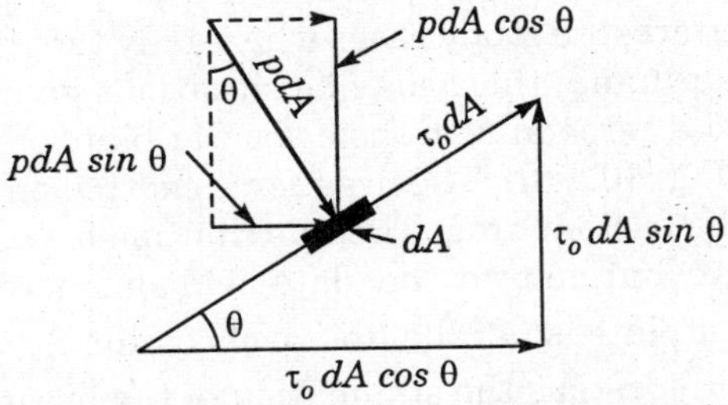

Fig. 10.1. (*c*) Normal stress (pressure *p*) and shear stress (τ_0) acting on area element *dA*.

pressure and act in a direction roughly perpendicular to F_L. The second integral in Eq. (10.1) for the lift may, therefore, be neglected and the resulting lift becomes

$$F_L = \int_A pdA \cos\theta \qquad ...(10.3)$$

Eq. (10.3) provides a mathematical expression for the graphical integration of pressure distribution of Fig. 10.1.

The terms in the drag Eq. (10.2) are of the same order of magnitude and are thus equally important. The first integral is called the "pressure drag F_{D_p}" which depends upon the form or shape of object and separation of flow while the second integral gives the "friction (or viscous) or skin-friction drag F_{D_f}" which depends upon the extent and character of the boundary layer.

10.3. Pressure Drag and Friction Drag

When flow occurs past a flat surface at zero incidence (flat surface held parallel to the flow direction), the fluid exerts a drag-force on the surface as consequence of viscous action. The resultant force of friction acts in the downstream direction and is usually known as the *friction drag* or *skin-friction drag* or *surface drag*. But when flow occurs past a surface which is not everywhere parallel to the fluid stream, an additional drag force results on account of difference of pressure over the body surface. This drag force is known as the *pressure drag*. Since the magnitude of this drag depends upon the form or shape of the body, it is also sometimes called as the *form drag*.

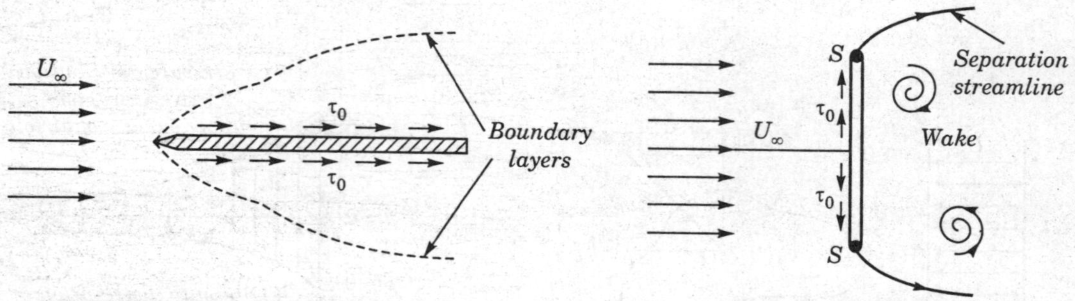

(*a*) Thin plate held parallel to flow (only friction drag exists)

(*b*) Thin plate held normal to flow (only pressure drag exists)

Fig. 10.2. Flow cases illustrating isolation of friction and pressure drags.

An accurate estimation of pressure drag involves prediction of separation points—a problem which is very complex dependent upon both the body shape and the boundary layer characteristics. It is, however, of interest to study cases where only the friction drag exists while the pressure drag is zero. On the other hand, there may be situations giving rise only to pressure drag, the friction drag having almost a zero value. Consider a flat plate of negligible thickness held in a fluid steam at zero incidence, Fig. 10.2 (*a*). The drag force exerted on the plate is due to viscous shear stresses resulting in only the friction drag. The contribution of pressure drag towards the total drag is zero as the pressure distribution over the plate held symmetrically to the flow is exactly similar.

Since the flat plate is parallel to the oncoming flow, $\theta = 0°$, $\sin \theta = 0$ and $\cos \theta = 1$. Eq. (10.2) which gives the total drag as the summation of the pressure drag F_{D_p} (given by the first integral) and the friction drag F_{D_f} (expressed by the second integral), for this case reduces to

$$F_D = F_{D_f} = \int_A \tau_0 \, dA$$

Fig. 10.2 (*b*) illustrates a flow situation in which a thin flat plate is held normal to the flow direction.

For this case, $\theta = 90°$, $\sin \theta = 1$ and $\cos \theta = 0$ and Eq. (10.2) results in,

$$F_{D_f} = 0$$

and therefore $$F_D = F_{D_p} = \int_A p dA$$

The boundary layer separates from the plate edges marked S. Just downstream of the plate large scale eddies are set up in the wake region. The formation of these eddies and their subsequent diffusion consumes a good amount of flow energy thereby reducing the pressure in the wake. The pressure difference upstream and downstream of the plate produces what is known as the pressure drag.

In almost all problems in which flow takes place around a solid body, the boundary layer separates from the surface towards the rear of the body leading to formation of eddies in the region of wake. The magnitude of pressure drag resulting from it depends very much the size of wake, and this in turn depends upon the position of the points of separation.

10.3.1. Streamlined body

If the shape of the body is such that separation occurs towards the rearmost part of the body, it results in a small wake. The drag is then mostly due to existence of a small wake which gives rise to a small pressure drag. The friction drag thus makes a major contribution to the total drag. Such a body is called a *streamlined body*. The friction drag for such bodies is many times greater than the pressure drag. Streamlining involves the easement of boundary curvature at the sides and rear for the purpose of reducing the form effect upon the drag.

10.3.2. Bluff body

If the body shape is such that the flow is separated much ahead of its rear end resulting in a large wake, the pressure drag is much greater than the friction drag. Body of such a shape in which the pressure drag is large as compared to the friction drag is called a *bluff body*.

10.4. Total Drag

It is usually the total drag (summation of pressure and friction drags) which is of interest as it gives the magnitude of resistance to flow. We are thus more interested in knowing as to how this total drag would vary by changing the body shape. Let us consider three bodies of different shapes but of same cross-sectional area normal to flow : these are (*i*) a circular disc, (*ii*) a sphere and (*iii*) a streamlined body shown in Fig. 10.3.

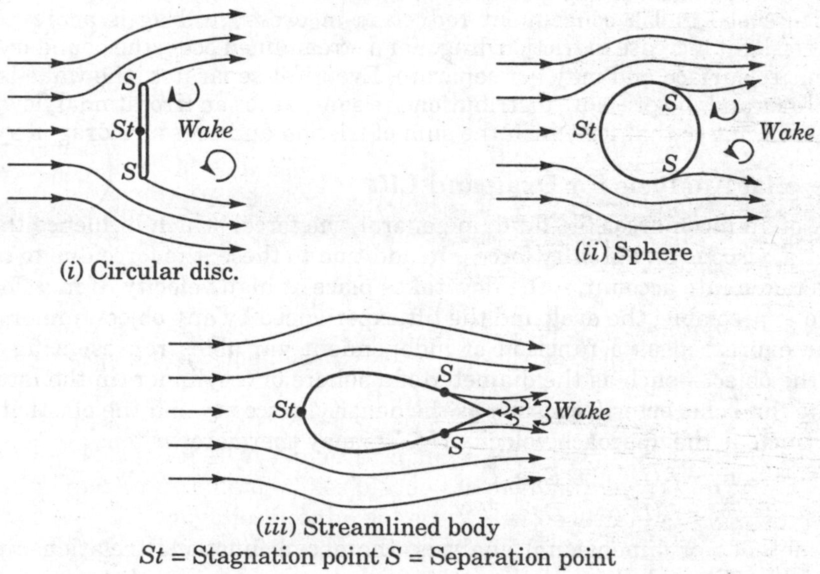

(*i*) Circular disc.

(*ii*) Sphere

(*iii*) Streamlined body

St = Stagnation point *S* = Separation point

Fig. 10.3. Comparison of size of wakes formed behind bodies of different shapes but of same cross-section.

The character of turbulent flow approaching these bodies is exactly similar. The flow in each case just near the front of the body is characterized by a stagnation point and maximum local pressure at that point.

In case of the disc placed normal to flow direction separation will occur at the edges marked by letter S representing the separation point. Near these edges occur high local velocity and consequently low pressure. This reduced pressure being adjacent to the wake will be transmitted into it causing the downstream pressure to be considerably smaller as compared to that on the upstream side of the disc. Since the shear stresses are at right angles to the main flow, the drag force is entirely caused by the difference in pressure over the two sides of the disc. This amounts to saying that there exists only the pressure drag while the friction drag has no contribution to the total drag.

For the sphere, the wake is smaller than that for the disc and contains somewhat higher pressure leading to a considerably smaller drag compared to that for the disc, see Fig. 10.3 (*i*) and (*ii*). Experiments have shown that the total drag on the sphere is about one-third of the drag for the disc. In this case, the friction drag is not zero since shear stresses acting on the sphere will have components along the direction of main stream. These shear stresses are however small resulting in a very small friction drag as compared to the pressure drag.

In case of a streamlined body, the wake may be very small depending on the extent of streamlining. The pressure in the wake is comparatively large since the gentle contour of the body permits deceleration of flow and consequent recovery of pressure without causing separation. Since the size of wake behind streamlined bodies is very small, the resulting pressure drag is a very small fraction of the drag for the disc. The friction drag for such bodies however, is considerably larger than that for the sphere simply because the streamlining increases the surface area in contact with the flow. For a well streamlined object, the friction drag is larger than the pressure drag, but both are so small that the total drag is about 1/40 of the drag for the disc.

The above discussion explains the necessity and importance of having the shape of bodies streamlined according to the streamline pattern. It may be noted that the viscosity of fluid is the root cause and is mainly responsible for producing drag. Viscosity affects the flow problem by causing (*i*) frictional or viscous effects (shear stresses) on the body surface and (*ii*) separation of boundary layer and creation of a low pressure wake giving rise to a pressure drag. By streamlining the body, the wake size is decreased and a consequent reduction in pressure drag is achieved though it is accompanied by a slight increase in friction drag. For a streamlined body, the boundary layer remains thin over the entire surface and may not separate. Even if it separates, the wake is rather small, and the streamlines and the pressure distribution are same as for an irrotational flow. The optimum amount of streamlining is that for which the sum of friction and pressure drag is a minimum.

10.5. Dimensional Analysis for Drag and Lift

For flow of an incompressible fluid, in general, the forces which influence the flow pattern are due to inertia, viscous and gravity forces. In addition to these, the force due to compressibility may have to be taken into account, if the flow takes place at high velocity. If in a flow problem, all these forces are appreciable, the drag and the lift experienced by any object immersed in the fluid stream may be expressed as a function of independent variables representing the geometric dimensions of the object—such as the diameter of a sphere or a cylinder (in the latter case length also affects), the fluid characteristics such as the density, viscosity and the elasticity and the flow characteristics such as the approach velocity U_∞. We may therefore write

$$F_D = f_1(L, \rho, \mu, E, g, U_\infty) \quad \text{...(10.4 } a)$$

and

$$F_L = f_2(L, \rho, \mu, E, g, U_\infty) \quad \text{...(10.4 } b)$$

By the methods of dimensional analysis, the above functional relationship between the dependent variable (F_D and F_L) and the independent variables may be expressed in terms of dimensionless parameters :

$$\frac{F_D}{\rho L^2 U_\infty^2} = f_3\left(\frac{\rho U_\infty L}{\mu}, \frac{U_\infty^2}{L_g}, \frac{U_\infty}{\sqrt{E/\rho}}\right) \quad \text{...(10.5 } a)$$

and
$$\frac{F_L}{\rho L^2 U_\infty^2} = f_4\left(\frac{\rho U_\infty L}{\mu}, \frac{U_\infty^2}{L_g}, \frac{U_\infty}{\sqrt{E/\rho}}\right) \qquad \text{...(10.5 b)}$$

It is customary to define the drag coefficient and the lift coefficient as

$$C_D = \frac{F_D}{\frac{1}{2}\rho U_\infty^2 A} \quad \text{and} \quad C_L \frac{F_L}{\frac{1}{2}\rho U_\infty^2 A} \qquad \text{...(10.6)}$$

where the symbol A stands for some area pertaining to the body in question. Except for airfoils, A usually represents the projected area of the body perpendicular to the oncoming flow, *i.e.* the frontal area of the body. For a thin plate at zero incidence the drag is exerted on its both the sides and hence area A represents both sides of the plate. In case of an airfoil A is equal to the product of the span and the mean chord. The area A used in the definition of C_D and C_L for a particular body should, therefore, be specified. The denominator of the expressions for the coefficients of drag and lift, Eq. (10.6), is the product of the dynamic pressure $\frac{1}{2}\rho U_\infty^2$ and the specified area A resulting in a dynamic pressure force. The drag and lift coefficients, thus, being ratios of two forces are dimensionless and for two dynamically similar flows, C_D and C_L values would be same.

The right hand side of Eq. (10.5) may be identified to contain Reynolds, Froude and Mach numbers. From Eqs. (10.5) and (10.6) one can obtain a functional relationship for C_D and C_L in terms of these dimensionless numbers. We may thus write

$$C_D = \phi_1(\mathbf{R}, \mathbf{F}, \mathbf{M}) \text{ and } C_L = \phi_2(\mathbf{R}, \mathbf{F}, \mathbf{M}) \qquad \text{...(10.7)}$$

In problems of engineering practice at least one of the three independent dimensionless numbers may be neglected on account of its low bearing on the phenomenon. If the flow velocities are smaller as compared to the velocity of sound in the fluid medium, the compressibility effects are generally insignificant and may, therefore, be ignored. Experiments have indicated that if the Mach number **M** of flow is less than 0.4, the compressibility effects may safely be neglected. When two immiscible fluids are involved such that $\Delta\rho/\rho \approx 1$, where $\Delta\rho$ is the difference in densities and ρ the density of the heavier fluid the gravity effects become important. In free-surface flow (which involves a liquid-air interface), the gravity force is in fact, the one which causes the flow. Thus in case of an incompressible fluid without a free surface, the coefficients of drag and lift are functions of the Reynolds number only.

Thus $$C_D = \phi_1(\mathbf{R}) \text{ and } C_L = \phi_2(\mathbf{R}) \qquad \text{...(10.8)}$$

Example 10.1. *Indicate the correct solutions to the following problem. Calculations if any, to justify the answer may be given briefly.*

Body A has twice the submerged weight, twice the projected area and twice the drag coefficient of body B. The terminal fall velocity of body A in air would be about

(i) 8, (ii) 2, (iii) $\sqrt{2}$, (iv) $1/\sqrt{2}$ that of B.

(UPSC, Combined Engg. Services Exam. 1977)

Solution. Under steady, uniform conditions of motion, weight of falling body,

$$W = \text{Drag force } F_D.$$

$$F_D = C_D(\tfrac{1}{2}\rho U_\infty^2 A), \quad \therefore \quad W = C_D(\tfrac{1}{2}\rho U_\infty^2 A), \qquad U_\infty = \sqrt{\frac{2W}{C_D \rho A}}$$

Terminal fall velocity of body A is terms of similar quantities of body B

$$= \sqrt{2(2W_B)/2(C_D)_B\,\rho(2A)_B} = \frac{1}{\sqrt{2}}\sqrt{\frac{2W_B}{(C_D)_B\,\rho\,A_B}}$$

Hence the correct solution is (*iv*).

Example 10.2. *Electrical transmission towers are stationed at 0.4 km intervals and a conducting cable 1.25 cm in diameter is stretched between them. If a wind at 100 km/hr blows transversely across the wire, compute the total drag force on one such wire between two towers. The specific weight of air is 13.4 N/m³ and the kinematic viscosity is 1.42 × 10⁻⁵ m²/s. The drag coefficient for Reynolds number greater than 10⁴ may be taken as 1.20.*

Solution. Reynolds number of flow,

$$R = \frac{U_\infty d}{\nu} = \frac{(100 \times 1000/3600)\dfrac{1.25}{100}}{1.42 \times 10^{-5}} = 2.33 \times 10^4$$

Since $R > 10^4$, $C_D = 1.2$.

Drag per metre length of wire

$$= C_D(1 \times d)\rho \frac{U_\infty^2}{2} = 7.89 \text{ N}$$

Total drag force on wire between two towers

$$= 7.89 \times 0.4 \times 1000 = 3.155 \text{ kN}.$$

Example 10.3. *An object has a projected area 1 m² normal to the direction of its motion. It has a drag coefficient of 0.4 for a Reynolds number of 10⁷ using a characteristic length of 1.5 m. For this Reynolds number, what is the drag on the object when moving through water at 20°C (μ = 9.81 × 10⁻⁴ N-s/m²).*

Solution. Velocity corresponding to Reynolds number of 10^7 is

$$U_\infty = \frac{\mathbf{R}\mu}{\rho l} = \frac{10^7 \times 10^{-4} \times 9.81}{1000 \times 1.5} = 6.53 \text{ m/s}.$$

$$\text{Drag on the object} = C_D A \rho \frac{U_\infty^2}{2} = 0.4 \times 1 \times 1000 \times \frac{(6.53)^2}{2} = 8530 \text{ N} = 8.53 \text{ kN}.$$

Example 10.4. *A kite weighing 2.45 N and having an effective area of 0.8 square metre assumes an angle of 15° to the horizontal and the chord attached to it makes an angle of 45° to the horizontal. The pull on the chord during a wind of 34 km/hr is 25.5 N. Determine the corresponding coefficients of lift and drag. The specific weight of air is 12.25 N/m³.*

Solution. Drag exerted on the kite

= Horizontal component of pull on chord

= 25.5 cos 45° = 18.0 N

Lift developed by the kite

= 25.5 sin 45° + 2.45 = 20.5 N.

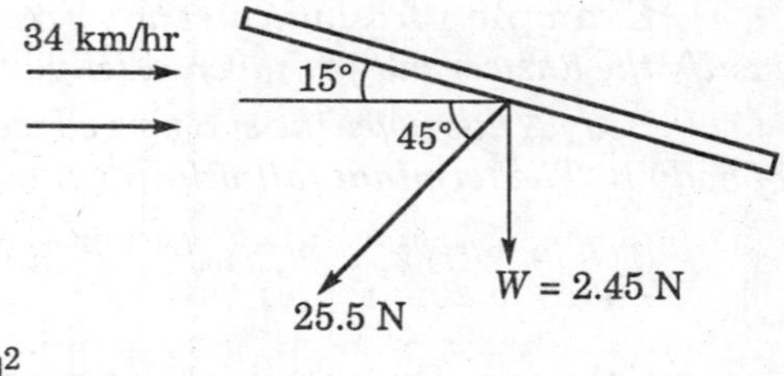

The drag $F_D = C_D A \dfrac{\rho U_\infty^2}{2}$

$$= C_D \times 0.8 \times \frac{12.25}{9.81} \times \frac{\left[\dfrac{34 \times 1000}{60 \times 60}\right]^2}{2} = 44.24\, C_D$$

$$\therefore \quad C_D = \frac{18.0}{44.24} = 0.4075$$

The lift $F_L = C_L A\rho \dfrac{U_\infty^2}{2} = C_L \times 0.8 \times \dfrac{12.25}{9.81} \times \dfrac{\left[\dfrac{1000 \times 24}{60 \times 60}\right]^2}{2} = 44.24\ C_L.$

$\therefore \quad C_L = \dfrac{20.5}{44.24} = 0.463.$

Example 10.5. *Experiments were conducted in a wind tunnel with a wind speed of 50 km/hr, on a flat plate of size 2 m long and 1 m wide. The specific weight of air is 11.28 N/m³. The plate is kept at such an angle that the coefficients of lift and drag are 0.75 and 0.15 respectively. Determine (i) lift force, (ii) drag force, (iii) resultant force, and (iv) power exerted by the air stream on the plate.*

Solution. $U_\infty = 50 \text{ km/hr} = \dfrac{50 \times 1000}{60 \times 60} = 13.9 \text{ m/s}$

(*i*) Lift force $F_L = C_L A\rho \dfrac{U_\infty^2}{2}$

$$= 0.75 \times (1.0 \times 2.0) \times \frac{11.28}{9.81} \times \frac{(13.9)^2}{2} = 166.8 \text{ N}$$

(*ii*) Drag force $F_D = C_D A\rho \dfrac{U_\infty^2}{2}$

$$= 0.15 \times (1.0 \times 2.0) \times \frac{11.28}{9.81} \times \frac{(13.9)^2}{2} = 33.36 \text{ N}$$

(*iii*) Resultant force,

$$F = \sqrt{F_L^2 + F_D^2} = \sqrt{(166.8)^2 + (33.36)^2} = 170.0 \text{ N}$$

inclined at an angle θ with the free-stream

$$\tan\theta = \frac{F_L}{F_D} = \frac{166.8}{33.36} = 5,$$

$$\therefore \quad \theta = 78^\circ\ 42'.$$

(*iv*) Power exerted by the air stream

$$= F_D \,.\, U_\infty = 33.36 \times 13.9 = 463.5 \text{ N m/s} = \mathbf{463.5\ W.}$$

10.6. Drag at Small Reynolds Numbers

We know that the Reynolds number **R** is a measure of the ratio of inertia and viscous forces. Consequently, for flow at very small Reynolds numbers (say **R** < 0.5), the inertia forces are negligible as compared to the viscous forces and the drag is almost proportional to the approach velocity U_∞. The flow at very small Reynolds number is known as the **creeping motion** and is largely laminar. Let us consider a sphere, the flow about which has been extensively studied. At low Reynolds number the flow closes behind the sphere and no wake forms. Sir George G. Stokes (1819—1903) analysed mathematically the drag for a sphere past which fluid of infinite extent flows under entirely laminar conditions (**R** < 1.0). Stokes has shown analytically that in laminar flow at very small Reynolds number where the inertia forces may be neglected, the drag experienced by the sphere referred to as the *deformation drag* is given by

$$F_D = 3\pi\mu U_\infty d \qquad \text{...(10.9)}$$

in which d is the diameter of the sphere, μ is the absolute viscosity and U_∞ is either the velocity of the undistributed fluid relative to the sphere or the terminal velocity of the sphere in motion. Of the total drag given by Eq. (10.9), two-thirds is contributed by skin-friction and the remaining one-third by pressure difference on the two-halves of the sphere. In other words the friction drag on a sphere is $\frac{2}{3}$ of the total drag and the pressure drag accounts for the remaining one-third. Utilizing the concept of drag and lift coefficients and using Eq. (10.6), the drag and lift forces may be expressed as

$$F_D = C_D A\rho \frac{U_\infty^2}{2} \qquad \text{...(10.10 a)}$$

and

$$F_L = C_L A\rho \frac{U_\infty^2}{2} \qquad \text{...(10.10 b)}$$

Comparing Eqs. (10.9) and (10.10 a), the drag coefficient C_D for the sphere under the conditions already stated may be found

$$F_D = C_D A\rho \frac{U_\infty^2}{2} = 3\pi\mu U_\infty d$$

Taking $A = \pi d^2/4$, the projected area of the sphere on a plane normal to flow direction and substituting, we obtain

$$C_D = \frac{24\mu}{\rho U_\infty d} = \frac{24}{\mathbf{R}} \qquad \text{...(10.11)}$$

Eq. (10.11) is in close agreement with experimental results only for $\mathbf{R} < 0.1$, because of the neglect of inertia terms. The experimental results begin to deviate from Eq. (10.11) as the inertial effects ingnored by Stokes begin to become appreciable. A Reynolds number slightly less than unity marks approximately the limit of deformation drag, beyond which Stokes' law, given by Eq. (10.9), is no longer applicable. Carl Wilhelm Oseen (1879—1944), a Swedish physicist, extended and improved upon the work of Stokes by accounting for the inertial effect. Oseen's solution, valid for $\mathbf{R} < 1.0$, is

$$C_D = \frac{24}{\mathbf{R}}\left(1 + \frac{3}{16}\mathbf{R}\right) \qquad \text{...(10.12)}$$

An approximate empirical relation which may be used upto $R = 100$ is :

$$C_D = \frac{24}{\mathbf{R}}\left(1 + \frac{3}{16}\mathbf{R}\right)^{1/3} \qquad \text{...(10.13)}$$

Eqs. (10.9), (10.10) and (10.12) are valid for extremely small sphere and velocities or for very viscous liquid satisfying the limits of Reynolds number for their applicability. Practical application of these equations are the sedimentation of dust particles in air and silt particles in water.

10.7. Drag at Large Reynolds Numbers

Flow at large Reynolds number is identified by inertia forces far exceeding the viscous forces. It has been found experimentally that fluid friction (*i.e.* viscous effects) at large values of Reynolds number can be neglected everywhere in the flow field except in the region of boundary layer. We know that in the boundary layer, the no slip condition must be satisfied causing the fluid velocity at the stationary boundary surface to reduce to zero, while the velocity at a short distance away has a magnitude equal to that of the undisturbed flow. Consequently, in the vicinity of the boundary surface the velocity change in the direction normal to the surface is very steep producing high velocity gradients. The shear stress $\tau = \mu\, du/dy$ is thus very large, even in a fluid of small viscosity like air and water. These high shear stresses exerted by the fluid on the body surface produce a friction drag.

Flow within the boundary-layer is rotational, the fluid particles are set into rotation by the shear forces. Outside the boundary-layer, the viscous forces are so small that the motion approximates closely to an irrotational flow.

In case of bluff bodies such as a thin plate or a disc held normal to flow, the flow separates at the edges provided that the Reynolds number is large enough to cause separation. The ratio of the projected areas of wake and the body normal to flow direction is, therefore, almost constant and so also is the coefficient of pressure drag $(C_D)_p$. The friction drag for bluff bodies is generally very small and, therefore, $(C_D)_p >> (C_D)_f$; the overall coefficient of drag C_D is thus nearly constant.

For a rounded body such as a sphere or a cylinder, however, the separation point depends upon the Reynolds number besides other factors such as the turbulence present in approaching flow and roughness of the body front. The boundary-layer separation and the consequent formation of wake result in a violent eddying motion just downstream of the body where the mean level of kinetic energy is higher and the mean level of pressure is less than that in the corresponding irrotational flow. As the pressure over the portion upstream of separation point approximates to that in irrotational flow, the pressure drag will be large if the ratio of projected areas of wake and the body is large.

For a streamlined body the boundary layer remains thin over the entire surface and does not separate, the wake is small, (Fig. 10.4), and the streamline pattern and the pressure distribution are nearly the same as for an irrotational flow. The pressure drag, consequently, will be very small for a streamlined body. Based on experiments it has been found that the friction drag is of the same order of magnitude as the pressure drag.

The pressure drag coefficient $(C_D)_p$ and the friction-drag coefficient $(C_D)_f$ may be defined by the following equations :

Pressure drag, $$F_{Dp} = (C_D)_p A\rho \frac{U_\infty^2}{2} \quad \text{...(10.14 a)}$$

and Friction drag, $$F_{Df} = (C_D)_f A\rho \frac{U_\infty^2}{2} \quad \text{...(10.14 b)}$$

Fig. 10.4 shows a streamlined body placed in a fluid stream of undistributed velocity U_∞. The boundary layer remains thin and the flow does not separate over the body surface. The figure shows that even without separation taking place at the rear of the body, the wake does form. This explains that for the formation of wake, the flow need not separate.

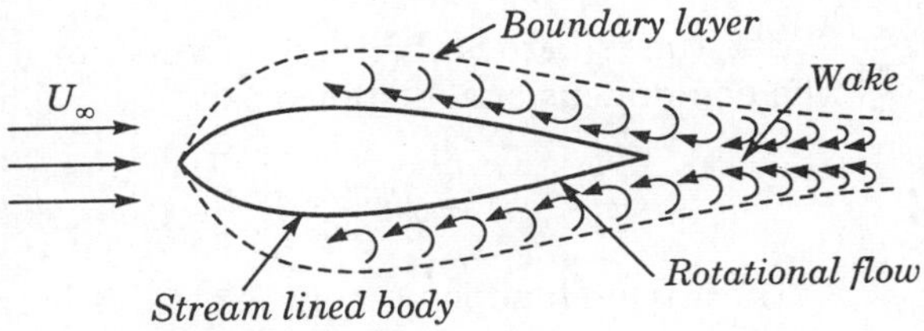

Fig. 10.4. Boundary layer growth and wake formation over a streamlined body in a non-separating flow.

10.8. Drag Over a Flat Plate

The drag exerted by the fluid over a thin flat plate held at zero incidence depends on whether the boundary layer formed over its surface is laminar or turbulent. We shall, therefore, consider drag characteristics of a flat plate for the cases : (*i*) when the boundary layer is completely laminar, (*ii*) when the boundary layer is fully turbulent, and (*iii*) when the flow in boundary layer changes from laminar to turbulent. It may be noted that the boundary layer flow will be laminar only when the plate length is short, whereas for a sufficiently long plate the flow will be laminar over a short length from the leading edge and afterwards it will change over to turbulent, and for the remaining length of plate the flow will continue to be turbulent. In order to evaluate the drag exerted by the fluid over the flat plate, we must know how the drag coefficient varies under different conditions of flow in the boundary layer.

10.8.1. Laminar boundary-layer over a flat plate

Consider the flow of an incompressible viscous fluid past a thin flat plate kept at zero incidence in a uniform fluid stream of velocity U_∞. Let the origin of the co-ordinates be at the leading edge of the plate, the x-axis being taken along the direction of uniform stream (and thus along the plate) and the y-axis along the normal to the plate (Fig. 10.5). Considering two-dimensional flow (*i.e.* the flow over a wide flat plate) the problem can be analysed by using the Prandtl's boundary layer equations.

The characteristic parameters involved in such problems are U_∞, ν, x and y. The velocity profile according to the law of similarity may be assumed as

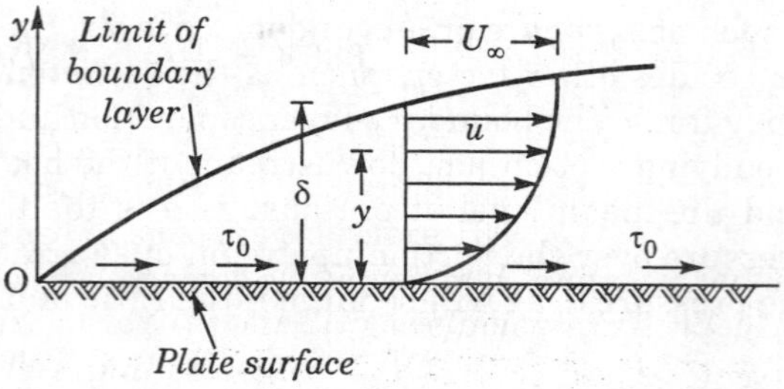

Fig. 10.5. Laminar boundary layer over one side of flat plate.

$$\frac{u}{U_\infty} = F(y, x, \nu, U_\infty) = F(\eta) \qquad ...(10.15)$$

in which $\eta = y/\delta$ and F stands for the function. From the knowledge of exact solution of unsteady motion of a flat plate (for which any standard text on boundary layer flow analysis may be referred to) we may write that the boundary layer thickness

$$\delta \sim \sqrt{\nu t} \sim \sqrt{\frac{\nu x}{U_\infty}} = \frac{x}{\sqrt{\mathbf{R}_x}} \qquad ...(10.16)$$

The time t here is the time required for a fluid particle to travel a distance x with velocity U_∞. The non-dimensional distance parameter η may be expressed as

$$\eta = y/\delta = y\sqrt{\frac{U_\infty}{\nu x}} \qquad ...(10.17)$$

The stream function ψ may be obtained in terms of the function of velocity using Eq. (10.15)

$$\psi = \int u\,dy = U_\infty\,\delta \int F(\eta)\,d\eta$$

$$= U_\infty \sqrt{\frac{\nu x}{U_\infty}} \int F(\eta)\,d\eta$$

or

$$\psi = \sqrt{U_\infty\,\nu x}\,f(\eta) \qquad ...(10.18)$$

where

$$f(\eta) = \int F(\eta)\,d\eta$$

The velocity u and the velocity gradient du/dy are given by

$$u = \frac{\partial \psi}{\partial y} = \frac{\partial \psi}{\partial \eta}\cdot\frac{\partial \eta}{\partial y} = \sqrt{U_\infty\,\nu x}\,f'(\eta)\,.\,1/\delta = U_\infty f'(\eta)$$

where $f'(\eta)$ is the first derivative of $f(\eta)$ with respect to η and is denoted by prime. The velocity gradient.

$$\frac{\partial u}{\partial y} = \frac{\partial u}{\partial \eta}\cdot\frac{\partial \eta}{\partial y} = U_\infty\,f''(\eta)\,\frac{1}{\delta} = U_\infty\sqrt{\frac{U_\infty}{\nu x}}\,f''(\eta) \qquad ...(10.19)$$

Shear stress on the surface of flat plate can be calculated from Newton's law of fluid friction and Eq. (10.19), thus

$$\tau_0 = \mu\left(\frac{\partial u}{\partial y}\right)_{y=0} = \mu U_\infty \sqrt{\frac{U_\infty}{\nu x}}\,f''(0).$$

From the Blasius solution of boundary layer flow analysis over a plate about $\eta = 0$, $f''(0) = 0.332$. The results for longitudinal and lateral velocities (*i.e.* u and v) show that velocity u approach its asymptotic value U_∞ very rapidly. Simultaneously the velocity gradient du/dy approaches zero. There is a finite v-component of the velocity at the outer edge of the boundary-layer, a fact we would expect from continuity principle [see article 7.3.2 and Fig. 7.3 (*i*)]. Making use of velocity profile obtained from the Blasius solution we can evaluate the boundary-layer thickness as

$$\delta = \frac{5x}{\sqrt{\mathbf{R}_x}} \quad \text{at} \quad \frac{u}{U_\infty} = 0.992 \qquad \text{...(10.20 a)}$$

It may be noted that this equation agrees with the form of Eq. (10.16). Use of Eqs. (7.2) and (7.3) gives the displacement and momentum thickness as

$$\delta_* = \frac{1.73\,x}{\sqrt{\mathbf{R}_x}} \qquad \text{...(10.20 b)}$$

$$\theta = \frac{0.664\,x}{\sqrt{\mathbf{R}_x}} \qquad \text{...(10.20 c)}$$

The local wall shear may be obtained from the velocity gradient at the wall,

$$\tau_0 = \mu\left(\frac{\partial u}{\partial y}\right)_{y=0} = 0.332\,\mu\,U_\infty\sqrt{\frac{\rho U_\infty}{\mu x}} = \frac{0.332}{\sqrt{\mathbf{R}_x}}\rho U_\infty^2 \qquad \text{...(10.20 d)}$$

where $\mathbf{R}_x = \rho U_\infty x/\mu$ is the local Reynolds number based on distance x from the leading edge. The local friction drag coefficient c_f is given by

$$c_f = \frac{\text{Friction drag}}{\text{Dynamic pressure force}} = \frac{\tau_0 A}{\frac{1}{2}\rho U_\infty^2 A} = \frac{\tau_0}{\frac{1}{2}\rho U_\infty^2}$$

$$= \frac{0.664}{\sqrt{\mathbf{R}_x}} \qquad \text{...(10.21)}$$

The friction drag over one side of the plate of length L per unit width is

$$F_{Df} = \int_0^L \tau_0\,dx = \int_0^L \frac{0.332}{\sqrt{U_\infty\,x/\nu}}\rho U_\infty^2\,dx = 0.664\,\rho U_\infty^2\sqrt{\frac{\nu L}{U_\infty}} \qquad \text{...(10.22)}$$

Eq. (10.22) shows the dependence of friction drag F_{Df} on fluid density ρ, 3/2 power of free-stream velocity U_∞, $\frac{1}{2}$ power of kinematic viscosity ν and $\frac{1}{2}$ power of the plate length L. The average coefficient of friction drag C_f may be expressed as

$$C_f = \frac{F_{Df}}{\frac{1}{2}\rho U_\infty^2 L} = \frac{0.664 U_\infty^2\sqrt{\dfrac{vL}{U_\infty}}}{\frac{1}{2}\rho U_\infty^2 L} = \frac{1.328}{\sqrt{\mathbf{R}_L}} \qquad \text{...(10.23)}$$

where R_L is the Reynolds number based on the plate length L, that is,

$$\mathbf{R}_L = \frac{U_\infty L}{\nu}.$$

Example 10.6. *Calculate the friction drag on a plate 15 cm wide and 45 cm long placed longitudinally in a stream of oil (specific gravity 0.925 and kinematic viscosity of 0.9 stokes) flowing with a free stream velocity of 6 m/s. Also find the thickness of the boundary layer and shear stress at the trailing edge.*

Solution. Plate Reynolds number,

$$\mathbf{R}_L = \frac{U_\infty L}{\nu} = \frac{6 \times 0.45}{0.9 \times 10^{-4}} = 3 \times 10^4$$

Since $\mathbf{R}_L < 5 \times 10^5$, the boundary-layer is laminar. The average friction drag coefficient for laminar boundary-layer is given by Eq. (10.23).

$$C_f = \frac{1.328}{\sqrt{\mathbf{R}_L}} = \frac{1.328}{\sqrt{3 \times 10^4}} = 0.766 \times 10^{-2}$$

Friction drag $F_{Df} = C_f \times \frac{1}{2} \rho U_\infty^2 \times$ Area of the plate

$$F_{Df} = 0.766 \times 10^{-2} \times \frac{1}{2} \times \frac{0.925 \times 9810}{9.81} \times (6)^2 \times 0.15 \times 0.45 = 8.6 \text{ N.}$$

Using Eq. 10.20 (*a*), the boundary-layer thickness at the trailing edge is

$$\delta_L = \frac{5L}{\sqrt{\mathbf{R}_L}} = \frac{5 \times 0.45}{\sqrt{3 \times 10^4}} = 1.3 \times 10^{-2} \text{ m} = 1.3 \text{ cm.}$$

The shear stress at the trailing edge, using Eq. 10.20 (*d*) is

$$\tau_0 = \frac{0.332}{\sqrt{\mathbf{R}_L}} \rho U_\infty^2$$

$$\tau_0 = \frac{0.332}{\sqrt{3 \times 10^4}} \times \left(\frac{0.925 \times 9810}{9.81}\right) \times (6)^2 = 63.8 \text{ N/m}^2.$$

Example 10.7. *Oil with a free stream velocity of 2 m/s flows over a thin plate 2 m wide and 2 m long. Calculate the boundary layer thickness and the shear stress at the trailing end point and determine the total surface resistance of the plate. Take specific gravity as 0.86 and kinematic viscosity as 10^{-5} m²/s.*

Solution. Reynolds number at the trailing edge

$$\mathbf{R}_L = \frac{UL}{\nu} = \frac{2 \times 2}{10^{-5}} = 4 \times 10^5$$

Taking the critical Reynolds number to be 5×10^5, the boundary layer flow can be assumed laminar. The thickness of laminar boundary layer is given by Eq. (10.20 *a*)

$$\frac{\delta}{x} = \frac{5}{\sqrt{\mathbf{R}_L}}$$

At the trailing edge, $x = L = 2$ m, $\mathbf{R}_x = \mathbf{R}_L = 4 \times 10^5$, and

$$\delta = \frac{5L}{\sqrt{\mathbf{R}_L}} = \frac{5 \times 2}{\sqrt{4 \times 10^5}} = 1.58 \times 10^{-2} \text{ m}$$

For laminar boundary layer, the local wall shear is given by Eq. (10.20 *d*)

$$\tau_0 = \frac{0.332}{\sqrt{\mathbf{R}_x}} \rho U^2$$

At the trailing edge, $$\tau_0 = \frac{0.332}{\sqrt{4 \times 10^5}} \times 860 \times (2)^2 = 1.805 \text{ N/m}^2$$

Total surface resistance can be obtained using Eq. (10.22)

$$F_D = C_f \frac{1}{2} (B \times L)\rho U^2$$

$$= \frac{1.328}{\sqrt{\mathbf{R}_L}} \times \frac{1}{2} (B \times L) \rho U^2 = \frac{0.664}{\sqrt{4 \times 10^5}} \times (2 \times 2) \times 860 \times (2)^2$$

$$= 1445.6 \text{ N.}$$

10.8.2. Turbulent boundary layer over a Smooth flat plate

In actual flow cases the boundary layer along a flat plate (or over any object) is not usually laminar over the entire plate, but it generally begins with laminar flow limited over a small length of the plate and changes to turbulent thereafter. The position of the transition point depends upon the intensity of turbulence in the approaching flow besides other factors mentioned in chapter 9 on turbulent flow, article 9.1. In the discussion that follows, we shall first determine the friction drag for a fully developed turbulent boundary-layer and subsequently the determination of total friction drag including the effect laminar of boundary layer over an initial short length of the plate will be taken up.

10.8.2. (*a*) Fully developed turbulent boundary layers

It will be assumed that the boundary-layer is turbulent right from the leading edge of the plate. For a turbulent boundary layer flow over a flat plate, the momentum integral Eq. (7.21) may be written in a slightly different form

$$\frac{d}{dx}\int_0^\delta u(U_\infty - u)\,dy = \frac{\tau_0}{\rho} \qquad \text{...(10.24)}$$

Shear stress over the plate surface, τ_0, is actually made up of viscous stress and the turbulent stress. The following analysis deals with fully developed turbulent flow in the boundary-layer where turbulent shear stresses are predominant.

Prandtl assumed that the velocity distribution within a turbulent boundary-layer over a flat plate is essentially the same as that in a large diameter pipe. Replacing (*i*) the maximum velocity which occurs at the pipe axis by the free-stream velocity U_∞ near the outer edge of boundary layer (*ii*) the radius r_0 by the outer edge of boundary layer thickness δ, Eq. (9.54) may be expressed as

$$\frac{u}{U_\infty} = \left(\frac{y}{\delta}\right)^{1/7} \qquad \text{...(10.25)}$$

Eq. (10.25) which expresses the Blasius one-seventh power law for velocity distribution in turbulent flow over a flat plate is valid for Reynolds numbers, $U_\infty L/\nu < 10^6$, where L is the length of the plate. For the turbulent shear stress τ_0 the following empirical formula for pipe flow may be used.

$$\tau_0 = 0.30955\,\rho V^2 \left(\frac{Vd}{\nu}\right)^{-1/4} \qquad \text{...(10.26)}$$

here d is the pipe diameter and V is the average velocity in the pipe. The shear stress in the turbulent boundary layer over a flat plate may by obtained using Eqs. (10.26) and (9.54) and noting that the average velocity :

$$V = 0.817\,U_{max} = 0.817\,U_\infty$$

The expression for wall shear stress in turbulent boundary layer may be expressed as

$$\tau_0 = 0.0233\,\rho U_\infty^2 \left(\frac{U_\infty \delta}{\nu}\right)^{-1/4} \qquad \text{...(10.27)}$$

Making use of Eqs. (10.24), (10.25) and (10.27), the dimensionless boundary layer thickness can be obtained as

$$\frac{\delta}{x} = \frac{0.379}{(\mathbf{R}_x)^{1/5}} \qquad \text{...(10.28)}$$

or

$$\delta = \frac{0.379x}{(U_\infty x/\nu)^{1/5}} \propto x^{4/5}$$

Eq. (10.28) indicates that the thickness of turbulent boundary layer over a flat plate increases with distance x according to $x^{4/5}$, whereas for the laminar boundary layer the thickness varies according to $x^{1/2}$, refer (7.5). Thus, a turbulent boundary layer thickens at a faster rate as compared to a laminar one. Eliminating δ between Eqs. (10.77) and (10.28), the boundary shear stress,

$$\tau_0 = 0.0295\, \rho U_\infty^2\, (\mathbf{R}_x)^{-1/5}$$

Local skin friction coefficient

$$C_f = \tau_0 / \tfrac{1}{2}\rho U_\infty^2 = \frac{0.0295 \times 2}{(\mathbf{R}_x)^{1/5}} = \frac{0.059}{(\mathbf{R}_x)^{1/5}} \qquad \text{...(10.29)}$$

The friction drag per unit width for one side of the plate of length L is given by

$$F_{Df} = \int_0^L \tau_0 dx = 0.0368\, L\rho U_\infty^2 (\mathbf{R}_L)^{-1/5} \qquad \text{...(10.30)}$$

The average coefficient of friction drag is,

$$C_f = \frac{F_{Df}}{\frac{1}{2}\rho U_\infty^2 L} = \frac{0.074}{(\mathbf{R}_L)^{1/5}} \qquad \text{...(10.31)}$$

Equation (10.31) is valid for a Reynolds number range of $5 \times 10^5 < \mathbf{R}_L < 10^7$. If the Reynolds number exceeds this limit the one-seventh power law for velocity distribution in the turbulent boundary layer has to be replaced by a logarithmic velocity distribution, Eq. (9.27). On making this change, Prandtl calculated the coefficient of friction drag as

$$C_f = \frac{0.455}{(\log_{10} \mathbf{R}_L)^{2.58}} \qquad \text{...(10.32)}$$

Eq. (10.32) is valid for the Reynolds number range from 10^6 to 10^9.

Example 10.8. *A thin plate 1 m wide and 2 m long is towed through water (kinematic coefficient of viscosity = 0.01 cm²/s) at a velocity of 2 m/s. Compute the drag on both sides of the plate assuming that the boundary layer is turbulent from the leading edge. The local skin friction is given as*

$$C_f = \frac{0.06}{\mathbf{R}_x^{1/5}} = \frac{\tau_0}{\frac{1}{2}\rho U^2}, \text{ where } \mathbf{R}_x = \frac{U_x}{\nu}$$

and U is the velocity outside the boundary layer, τ_0 = local boundary shear stress.

(U.P.S.C. Combined Engg. Services Exam. 1977)

Solution. The local boundary shear stress is expressible as

$$\tau_0 = C_f(\tfrac{1}{2}\rho U^2) = 0.06\left(\frac{U_x}{\nu}\right)^{-1/5} (\tfrac{1}{2}\rho U^2)$$

Drag on one side of plate of size 1 m × 2 m

$$= \text{Shear stress } \tau_0 \times \text{Area of one side of plate}$$

$$= \int_0^2 \tau_0 \times (1 \times dx)$$

Drag on both sides of the plate

$$F_D = 2\int_0^2 \tau_0\, dx = 2\int_0^2 0.06\left(\frac{U_x}{\nu}\right)^{-1/5} (\tfrac{1}{2}\rho U^2)\, dx$$

$$= 0.075\,\rho U^2 \left[\frac{x}{(Ux/\nu)^{1/5}}\right]_0^2 = 0.075\,\rho U^2 \left[\frac{x}{\mathbf{R}_x^{1/5}}\right]_0^2$$

$$= 0.075 \times 1000 \times (2)^2 \left[\frac{2}{(2 \times 2/0.01 \times 10^4)^{1/5}}\right] = \mathbf{28.7\ N.}$$

Example 10.9. *The drag coefficient due to boundary-layer for a thin plate 30 cm wide and 4.0 m long is given by :*

$$C_D = \frac{0.455}{(\log_{10} \mathbf{R}_L)^{2.58}}$$

where R_L is the Reynolds number based on length. If the plate is kept parallel to the flow of water with a free-stream velocity of 4.0 m/s, calculate the drag force on both sides of the plate. Assume kinematic viscosity of water = 0.01 stokes.

Solution. Reynolds number

$$R_L = \frac{U_\infty L}{\nu} = \frac{4 \times 4}{0.01 \times 10^{-5}} = 1.6 \times 10^7$$

average drag coefficient due to boundary-layer friction

$$C_D = \frac{0.455}{(\log_{10} R_L)^{2.58}} = \frac{0.455}{(\log_{10} 1.6 \times 10^7)^{2.58}} = 0.00274$$

Drag force on both sides of the plate

$$= C_D \times (\text{Area of both sides}) \times \rho\, \frac{U_\infty^2}{2}$$

$$= 0.00274 \times (2 \times 0.3 \times 4.0) \times 1000 \times \frac{(4)^2}{2} = 52.6\ \text{N.}$$

Example 10.10. *Calculate the drag and the power required to tow a smooth flat plate 1 m wide and 2.5 m long through still water at 1.2 m/s. Dynamic viscosity of water = 0.001 N-s/m².*

Solution. Reynolds no. of plate $= \rho\, \dfrac{VL}{\mu}$

$$(\mathbf{R}_L) = \frac{1000 \times 1.2 \times 2.5}{0.001} = 3.0 \times 10^6$$

Since $\mathbf{R}_L > 5 \times 10^5$, the boundary layer is turbulent, for which the average drag coefficient is given by Eq. (10.31)

$$C_f = \frac{0.074}{(\mathbf{R}_L)^{1/5}}$$

The friction drag per unit width for one side of the plate of length L is given by Eq. (10.30). Considering both the sides on which the friction drag will be acting while the plate is being towed in water, the total drag experienced by the plate will be two times. Thus, the drag force on plate of 1 m width is :

$$F_D = 2 \times 0.0368\, \rho\, L\, U_\infty^2\, (\mathbf{R}_L)^{-1/5}$$

$$= 0.0736 \times 1000 \times 2.5 \times (1.2)^2\, (3 \times 10^6)^{-1/5} = \mathbf{11.18\ N}$$

Power required to tow the plate at 1.2 m/s

$$= F_D \,.\, U_\infty = 11.18 \times 1.2 = 13.42\ \text{Nm/s} = \mathbf{13.42\ W.}$$

10.8.2. (*b*) Effect of Simultaneous Existence of Laminar and Turbulent Boundary Layers

If the plate length is sufficiently large, the boundary layer along a flat plate which begins as a laminar boundary layer, continues as such upto a certain initial length of the plate compatible with the stability of laminar flow, and subsequently changes to turbulent one downstream of this initial length. The existence of laminar boundary layer over a small initial plate length decreases the total friction drag exerted by the fluid over the plate. The comparison between the coefficient of friction drag in laminar and turbulent boundary layers may be obtained from Eqs. (10.23) and (10.31). The ratio of the two coefficients is

$$\frac{(C_f)_{lam}}{(C_f)_{turb}} = \frac{1.328\,(\mathbf{R}_L)^{1/5}}{0.074\,(\mathbf{R}_L)^{1/2}} = \frac{17.81}{(\mathbf{R}_L)^{3/10}} \quad ...(10.33)$$

It has been found experimentally that the boundary layer remains laminar upto a Reynolds number value of 5×10^5. For this Reynolds number, the ratio expressed by Eq. (10.33) is 0.35 and indicates that the friction drag in laminar boundary flow is 0.35 times of the drag in the corresponding turbulent boundary layer.

Let us consider Fig. 10.6 which shows a flat plate of length L. Let the flow over an initial small length x of the plate be laminar and beyond this length the boundary layer flow changes to

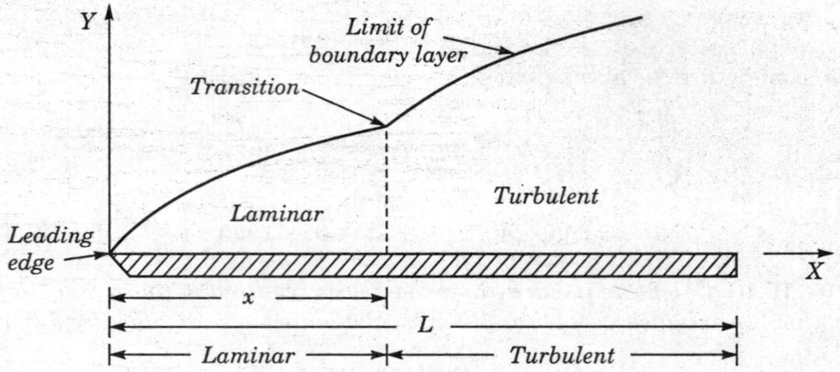

Fig. 10.6. Laminar and turbulent boundary layer over a flat plate.

turbulent. The friction drag over an initial length x over which the boundary layer remains laminar is obtained using Eq. (10.23) as

$$(F_{Df})_{lam} = \frac{1.328}{(\mathbf{R}_x)_{cr}^{1/2}} \left(\tfrac{1}{2}\,\rho U_\infty^2\, x\right) \quad ...(10.34)$$

where $(\mathbf{R}_x)_{cr}$ is the critical Reynolds number signifying the change of flow in the boundary layer from laminar to turbulent the value of which is generally taken as 5×10^5. The friction drag of turbulent boundary layer over the rear portion of the plate, *i.e.* from a distance x to L is given by

$$(F_{Df})_{turb} = \left[\frac{0.074L}{(\mathbf{R}_L)^{1/5}} - \frac{0.074x}{(\mathbf{R}_x)_{cr}^{1/5}}\right]\left[\frac{1}{2}\,\rho U_\infty^2\right] \quad ...(10.35)$$

The total friction drag over the flat plate is the sum of Eqs. (10.34) and (10.35) yielding

$$F_{Df} = (F_{Df})_{lam} + (F_{Df})_{turb}$$

$$= \left[\frac{1.328x}{(\mathbf{R}_x)_{cr}^{1/2}} + \frac{0.074L}{(\mathbf{R}_L)^{1/5}} - \frac{0.074x}{(\mathbf{R}_x)_{cr}^{1/5}}\right]\left(\frac{1}{2}\,\rho U_\infty^2\right)$$

Putting $(\mathbf{R}_x)_{cr} = 5 \times 10^5$ and $x = L\,\dfrac{(\mathbf{R}_x)_{cr}}{\mathbf{R}_L}$, the total friction drag becomes

$$F_{Df} = \left[\frac{0.074}{(\mathbf{R}_L)^{1/5}} - \frac{1700}{(\mathbf{R}_L)}\right]\left(\frac{1}{2}\rho U_\infty^2 L\right) \quad ...(10.36)$$

The average coefficient of friction drag which also accounts for the laminar boundary layer at the front portion of the plate can be expressed as

$$C_f = \frac{0.074}{(\mathbf{R}_L)^{1/5}} - \frac{1700}{(\mathbf{R}_L)} \quad ...(10.37)$$

Eq. (10.37) is valid for the Reynolds number ranging from 5×10^5 to 10^7. For Reynolds number, $\mathbf{R}_L$, exceeding 10^7, Eq. (10.32) gives more accurate results.

Fig. 10.7 illustrates the variation of coefficient of friction drag C_f with the plate Reynolds number $\mathbf{R}_L$.

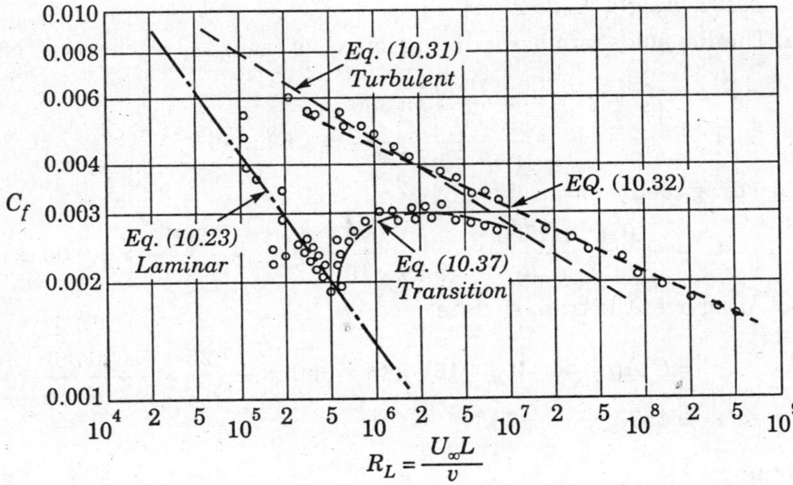

Fig. 10.7. Variation of average drag coefficient with Reynolds number.

Eqs. (10.23), (10.31), (10.32) and (10.37) expressing the coefficient of friction drag C_f as a function of the plate Reynolds number, R_L, for different regimes of flow, *i.e.* laminar, transitional and turbulent have been plotted in Fig. 10.7 and the experimental points marked by circles are intended to show the validity of each equation.

Example 10.11. *Find the power required to tow a plate of dimensions 1.25 m × 3.0 m in water at 20°C (kinematic viscosity = 9.3 × 10^{-3} stokes) with 3.0 m side in the flow direction, at 1.0 m/sec velocity. Make allowance for the fact that the boundary layer changes from laminar to turbulent on the plate. Find the length of the plate over which the boundary layer is laminar. Take critical Reynolds number as 5 × 10^5.* (*Roorkee 1970 ; Ravi Shankar Univ., 1978*)

Solution.

$$R_L = \frac{U_\infty L}{\nu} = \frac{1.0 \times 3.0}{9.3 \times 10^{-3} \times 10^{-4}} = 3.22 \times 10^6$$

Since $R_L > 5 \times 10^5$, the boundary layer is turbulent. The length of plate over which boundary layer in laminar is

$$x = \frac{R_{cr}\,\nu}{U_\infty} = \frac{5 \times 10^5 \times 9.3 \times 10^{-7}}{1.0} = 0.465 \text{ m.}$$

The average drag coefficient which takes into account the change of boundary layer flow from laminar to turbulent is

$$C_f = \frac{0.074}{(R_L)^{1/5}} - \frac{1700}{R_L} = \frac{0.074}{(3.22 \times 10^5)^{1/5}} - \frac{1700}{3.22 \times 10^4} = 0.003152$$

Total drag force F_D (considering both sides of plate)

$$= 2A\tau_0 = 2 \times (1.25 \times 3) C_f \rho \frac{U_\infty^2}{2}$$

$$= 2 \times 375 \times 0.00315 \times 1000 \times \frac{(1)^2}{2} = 11.8 \text{ N}$$

Power required to two the plate

$$= F_D U_\infty = 11.8 \times 1 = 11.8 \text{ W.}$$

Example 10.12. *A streamlined train is 300 m long with a typical cross-section having a perimeter of 8 m above the wheels. Evaluate the approximate surface drag (friction drag) of the train when running at 100 km/hr. The kinematic viscosity of air at the prevailing temperature is 1.49 × 10^{-5} m²/s and its specific weight 12.25 N/m³.*

Solution. The Reynolds number with the length of train as the characteristic length is

$$R_L = \frac{U_\infty L}{\nu} = \frac{\left(\frac{100 \times 1000}{60 \times 60}\right) \times 300}{1.49 \times 10^{-5}} = 5.59 \times 10^5$$

The average drag coefficient is given by

$$C_f = \frac{0.074}{(R_L)^{1/5}} - \frac{1700}{R_L} = \frac{0.074}{(5.59 \times 10^8)^{1/5}} - \frac{1700}{(5.59 \times 10^8)} = 0.001317$$

The friction drag over the train surface

$$= C_f A \rho \frac{U_\infty^2}{2} = 0.000137 \times (8 \times 300) \times \frac{12.25}{9.81} \times \frac{(1000/36)^2}{2}$$

$$= 1570 \text{ N} = 1.57 \text{ kN.}$$

10.9. Profile Drag

Although the total drag on any immersed object is always the sum of friction drag and pressure drag it will be found that in case of such objects as airfoils which experience a transverse force (lift force), this breakdown of the drag is not useful. The total drag in such cases is considered to be the sum of :

(*i*) the drag which would be developed if the airfoil is infinitely long causing the flow to be truly two dimensional, and

(*ii*) the drag produced by end effects due to finite length of the airfoil.

The first type of drag which depends only upon the shape or profile and orientation of the airfoil, is called the *profile drag*, whereas the second one which depends upon the airfoil plan form and is induced by the lift force is known as the *induced drag*. For objects which exhibit no lift force the induced drag will, obviously, be zero and the profile drag will then be equal to the total drag.

10.10. Drag Characteristics of Three Dimensional Bodies—(Sphere and Circular Disc)

The three-dimensional bodies are those around which the flow is of three-dimensional nature. Such bodies are primarily bodies of revolution, such as spheres, ellipsoidal objects etc., which have an axis of symmetry ; other objects having small length perpendicular to flow such as a disc and a circular cylinder of short length also fall under this category. In this article we shall consider the variation of drag coefficient C_D with the Reynolds number $\mathbf{R} = U_\infty d/\nu$ for a sphere and a circular disc.

Flow past a sphere at small Reynolds number has already been dealt with in article 10.6. At low Reynolds number, the flow will close behind the sphere and no wake will form, see Fig. 10.8 (*a*).

Under these conditions the total drag is entirely due to viscous effects, the contribution of pressure drag resulting from pressure variation over the body surface being zero. This is true for very low Reynolds numbers ($U_\infty d/\nu < 1$, in which d is the diameter of sphere) for which the flow about the sphere is completely viscous and the total drag termed as the *deformation drag* is given by the Stokes' law, Eq. (10.2). It has been indicated earlier that the drag coefficient ($C_D = 24/\mathbf{R}$) gives accurate values in close agreement with the experimental results for $\mathbf{R} < 0.1$, and shows that it depends only on the Reynolds number.

As the Reynolds number increases beyond unity (that is for $\mathbf{R} > 1$), the laminar boundary layer separates from the surface of the sphere, beginning first at the rear stagnation point where the adverse pressure gradient is the strongest. The C_D versus $\mathbf{R}$ curve starts deviating from an almost straight line law represented by $C_D = 24/\mathbf{R}$. The drag coefficient for a sphere continues to depend only upon magnitude of the Reynolds number and a plot of experimental results over a large range of Reynolds numbers for spheres of different sizes tested in different fluids gives a single curve shown in Fig. 10.9.

10.10.1. Separation and wake-formation

Separation of boundary layer begins at the downstream stagnation point and moves upstream as the Reynolds number increases. Approximately at $\mathbf{R} = 1000$, a stable separation position is reached at about 80°, Fig. 10.8 (*b*), from the flow direction, and at this Reynolds number the contribution of friction drag towards total drag is about 5% of the total drag. As the Reynolds number further increases, the contribution from the friction drag becomes smaller, thus rendering the drag problem at high Reynolds number a problem basically that of pressure drag.

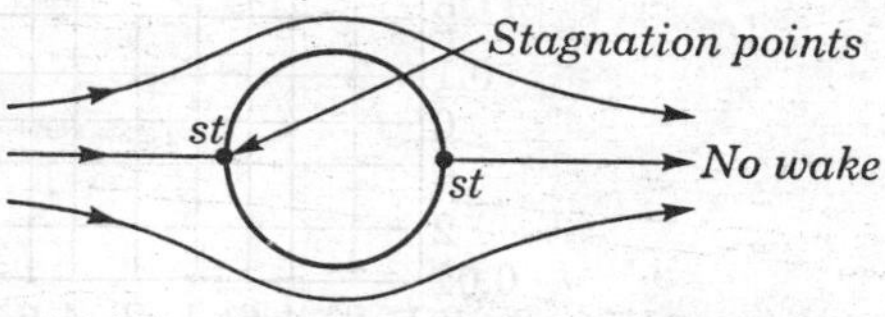

(*a*) Flow past a sphere at $\mathbf{R} < 0.1$

Fig. 10.8. (*a*)

At Reynolds number beyond 5000, the wake may be considered entirely turbulent, although the boundary-layer on the forward portion of the sphere remains laminar. From the shape of the curve shown in Fig. 10.9, three important regimes of flow are evident : (1) creeping motion $\mathbf{R} < 1$, (2) laminar boundary-layer and turbulent wake ($5 \times 10^3 \leq \mathbf{R} \leq 2.5 \times 10^5$) and (3) turbulent boundary-layer

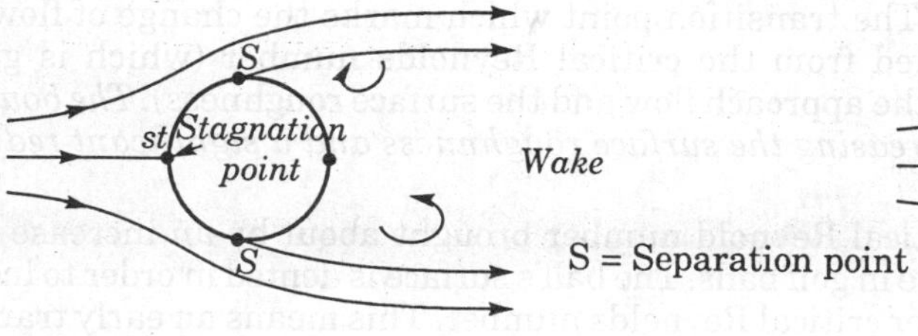

(*b*) $10^3 < \mathbf{R} < 2.5 \times 10^5$. Boundary layer before separation is laminar

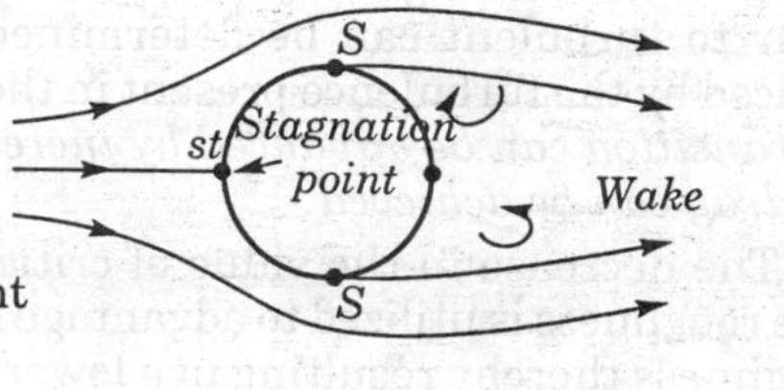

(*c*) $\mathbf{R} < 2.5 \times 10^5$. Boundary layer before separation is turbulent

Fig. 10.8. Flow past a sphere at different Reynolds numbers.

and turbulent wake ($\mathbf{R} > 2.5 \times 10^5$). The critical Reynolds number at which the boundary-layer transition (from laminar to turbulent) takes place is strongly influenced by two factors : the free-stream turbulence and the roughness of the sphere. An increase either in roughness or in free-stream turbulence has the effect of decreasing the critical Reynolds number.

The drag coefficient-Reynolds number variation of Fig. 10.9 shows a very rapid fall in the drag coefficient C_D with increasing Reynolds number : C_D undergoes a decrease from a value of about 100 at $\mathbf{R} \approx 0.2$ to $C_D = 1$ at $\mathbf{R} = 100$. The Stokes' law in seen valid for $\mathbf{R} < 1$. The drag coefficient may be considered practically constant at 0.4 in the Reynolds number range 10^3 to 10^4 and is thus independent of $\mathbf{R}$. For $10^4 < \mathbf{R} < 10^5$, C_D gradually increases to about 0.5 and then decreases to attain the earlier value of 0.4. At around $\mathbf{R} = 2.5 \times 10^5$, the boundary layer becomes turbulent before separating. At this stage the point of separation is moved further downstream, see Fig. 10.8 (*c*), resulting in a smaller wake. This results in a sharp drop in the drag coefficient by more than 50% of its value in the laminar boundary layer. The sudden drop in C_D is brought about by a smaller wake thereby resulting in a greatly reduced pressure drag. The total drag which is the combination of pressure and friction drags is, therefore, considerably reduced.

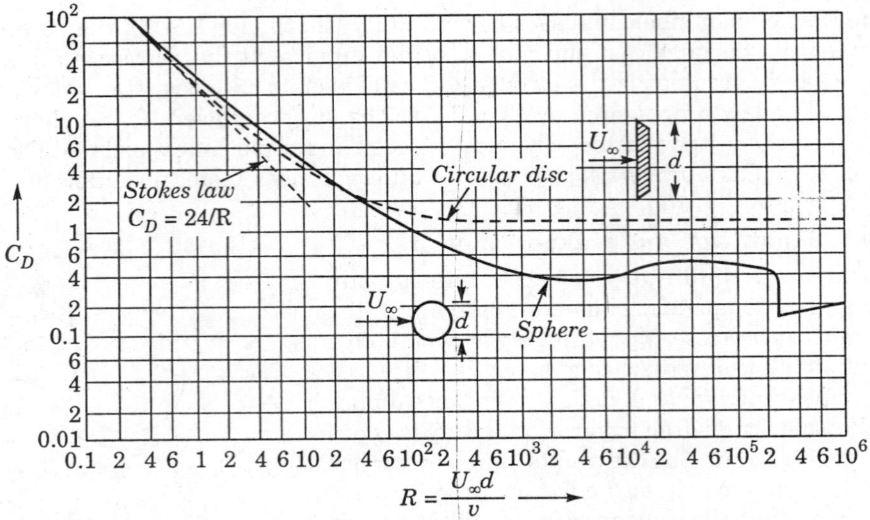

Fig. 10.9. Drag coefficient for sphere and disc (axially symmetrical bodies).

Comparison of Fig. 10.8 (*b*) and (*c*) reveals that so long as the boundary layer remains laminar, the points of separation S have been found from experiment located upstream of the midsection at about 80° from the front stagnation point resulting in a wide wake. Consequently, a larger pressure drag is obtained.

Once the character of boundary-layer flow changes from laminar to tubrulent, it is seen that the drag coefficient is grealty reduced. The transition point which marks the change of flow from laminar to turbulent can be determined from the critical Reynolds number (which is greatly influenced by the turbulence present in the approach flow and the surface roughness). *The boundary layer transition can be advanced by increasing the surface roughhness and a significant reduction in the drag can be achieved.*

The decrease in the value of critical Reynold number brought about by an increase in the surface roughness is utilized to advantage in golf balls. The ball's surface is dented in order to increase its roughness thereby resulting in a lower critical Reynolds number. This means an early transition and the boundary-layer flow becomes turbulent before separation of boundary-layer takes place. Consequently, the drag coefficient of the dented ball has a much lower value as compared to a smooth ball. The golf player can, thus, send the ball to a larger distance with the same driving power.

With a similar reasoning, it can be concluded that a cricket ball bowler can orient the seam in such a way as to reduce the drag coefficient (seam is the joint where the two-halves of the ball are united and its surface is rather rough). A proper utilization of the seam enables the bowler to send the ball to the batsman at a greater speed. Fig. 10.9. (*a*) shows three different positions of the seam,

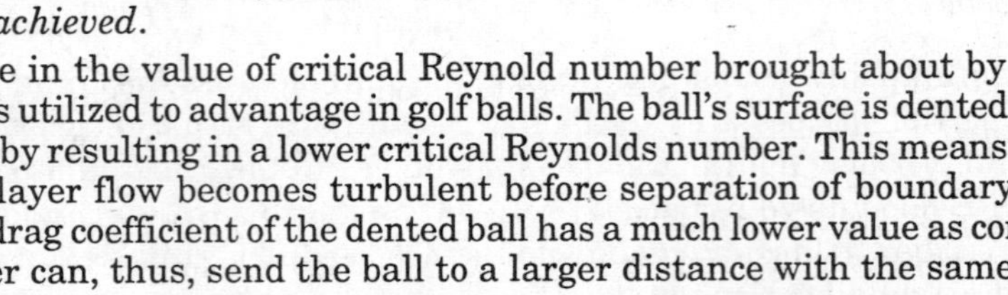

Fig. 10.9 (*a*) Reduction in the size of wake due to early boundary-layer transition (from laminar to turbulent flow).

the last one (with horizontal seam) results in the smallest wake which consequently produces the lowest drag coefficient.

For ideal fluids, the pressure distribution over the surface of a sphere is given by

$$\frac{p - p_\infty}{\frac{1}{2}\rho U_\infty^2} = 1 - \frac{9}{4}\sin^2\theta \qquad ...(10.37\ a)$$

This equation can also be written as

$$c_p = \frac{2\Delta p}{\rho U_\infty^2} \qquad ...(10.37\ b)$$

where $\Delta p = p - p_\infty$.

Here p_∞ and U_∞ are the pressure and velocity of the undisturbed fluid stream and θ the angle measured anti-clockwise from the downstream stagnation point refer Fig. 10.8 (*a*). Equation (10.37c) defines the pressure coefficient, c_p.

In case of thin circular disc placed normal to the flow the drag coefficient does not vary with Reynolds number, provided that **R** > 100. This is so because the separation point is fixed at the edge of the disc and cannot move either upstream or downstream regardless of the character of the boundary layer. Thus the width of the wake remains constant and so also does the drag coefficient.

The pressure distribution on the surface of an automobile is shown in Fig. 10.9 (*b*). In a modern automobile, the streamlining of its body-contour results in a smaller wake, which in turn is reflected in a low value of the drag coefficient. A small drag force thus obtained results in economy in consumption of fuel (petrol or diesel). The effect of streamlining of car profile on fuel efficiency and a brief history of development of cars follows.

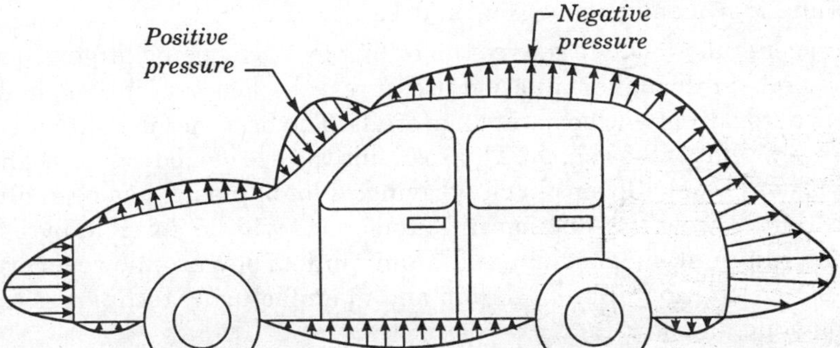

Fig. 10.9 (*b*) Pressure distribution over the surface of an automobile.

10.10.2. Aerodynamic Drag on Road Vehicles

Aerodynamic drag is the force that acts opposite to the direction of motion of a body moving through air, and causes reduction in its speed. The body may be an automobile or a truck moving on a road. The engine power—that is directly related to fuel consumption—needed to overcome aerodynamic drag makes the aerodynamic design of road vehicles a very important consideration.

The drag coefficient, C_D, is a measure of the vehicles's aerodynamic efficiency. From Eq. 10.6, one can write.

Aerodynamic drag, $\qquad F_D = (\rho/2)\, C_D\, A.V.^2 \qquad ...(10.37\ c)$

where ρ is air density, *A* the projected frontal area of the body, and V the velocity. Even though the aerodynamic drag is critically dependent on the velocity, it is the product of C_D and *A* that the designer can control. The reason that the aerodynamic losses are so important is that the power required to overcome these losses is a function of cube of the velocity.

Driving power, $P = F_D \cdot V = (\rho/2)\, C_D A.V.^3$ ...(10.37 *d*)

or, $P \propto (C_D A)V^3$

There are mainly three types of norms of speed for road vehicles as per the US classification. These are:

Type		Speed Limit
(*i*) Highway driving	......	> 90 kmph
(*ii*) Urban driving	......	30–90 kmph
(*iii*) City driving	......	< 30 kmph

At highway speeds, the aerodynamic drag is responsible for more than 50% of fuel consumption. In city driving, where stop-and-start driving below 30 kmph is more common, it is mainly the vehicle weight that contributes to fuel consumption. In the stage between city and highway driving (called urban driving cycle), overcoming aerodynamic drag accounts for nearly 25% of fuel consumption.

At speeds below 75 kmph, the rolling losses associated with the tyres and weight of the vehicle predominate the power required for a full-size car. At speeds of 75 kmph and above, it is the aerodynamic drag that consumes a larger portion of power. For an intermediate-size car, about 52% of the power goes towards overcoming aerodynamic drag on the highway at 90 kmph, and 24% during urban driving. Tyre consumes nearly 36% of the power in highway driving and 40% in the urban driving cycle (30-90 kmph); the remainder goes into overcoming mechanical losses, that is about 12% on highway and 36% on urban driving cycle.

For a subcompact car, 56% of the power is needed to overcome aerodynamic drag on the highway (as compared to 32% that accounts for tyre losses), and 27% for aerodynamic drag (and 38% for tyre losses) in urban driving. The remainder is consumed in overcoming mechanical losses, that is just 12% on highway and about 35% on urban driving.

Efforts are generally made to reduce the rolling losses by using larger diameter tyres, hard compounds in the rubber or higher inflation pressures; the last two options however, reduce the traction, and this puts a limit with existing materials. The best method of reducing, rolling losses seems to reduce the weight of the vehicle. This explains the use of modern-day light weight compact cars that are considered fuel efficient in city driving. A lot of plastic has been already in use, and any further substantial weight reduction in automobiles will be possible by using aircraft-type materials such as titanium, magnesium and aluminium. They are, however, very expensive as compared to steel and require different design and manufacturing techniques than are currently used in the automobile industry.

10.10.2.1. A Brief History of Development of Road Vehicles

In the early era of development of cars, aerodynamic drag was not of much consequence as the normal speeds were quite low. No wonder that during 1920s, the coefficients of drag (C_D) of cars were around 0.80. Research and development in the automobile industry gradually led to more advanced cars with lesser angular shape so that by 1940s, the C_D values were around 0.70. In the 1970s, the average drag coefficient was approximately 0.55. In the early 1980s the average C_D of cars dropped to 0.45, and presently auto designers and manufacturers are giving more attention to reducing drag. Almost all Japanese, Korean, European and U.S. automobile companies now have models with C_D values around 0.33, and some auto manufacturers claim to have developed new models with as low value of C_D as 0.25.

The streamlining of cars was first initiated by European manufacturers because the gasoline (petrol as it is known in India) prices were almost twice as high as compared to those prevailing in the United States. The following table shows C_D values for various contemporary car models starting from the Vintage 1932 Fiat Car.

Drag coefficients for different models of cars

Sr. No.	*Make and Model*	*Car profile*	C_D *value*
1.	Vintage 1932 Fiat (Balillo)	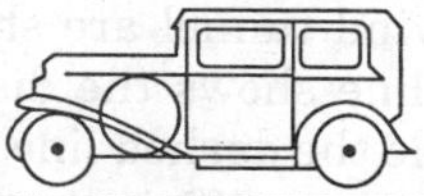	0.60
2.	Volkswagen (Bug)		0.46
3.	Volkswagen (Van)	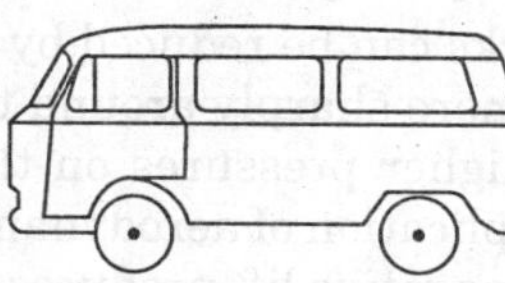	0.42
4.	Toyota Corolla (GT-S)	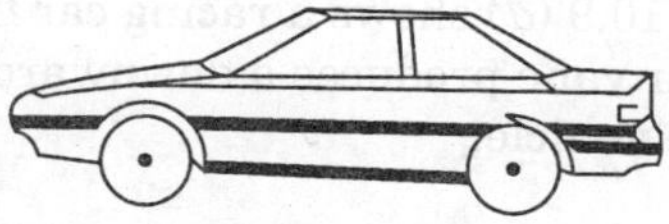	0.33
5.	Audi 90	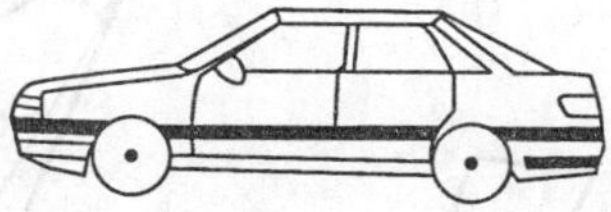	0.33
6.	Ford Probe	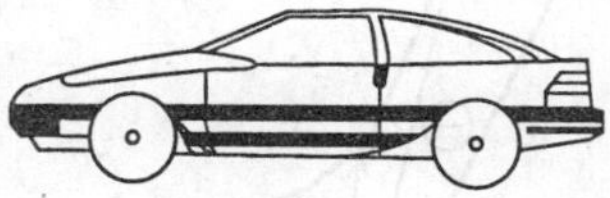	0.30
7.	Pontiac Grand Prix	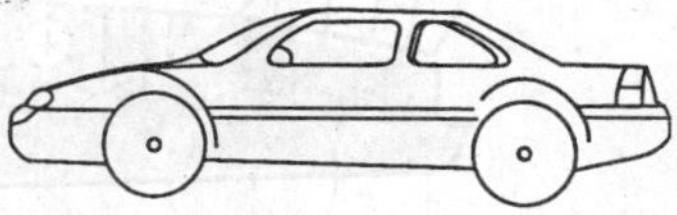	0.29
8.	Lincoln Machette (Concept car)	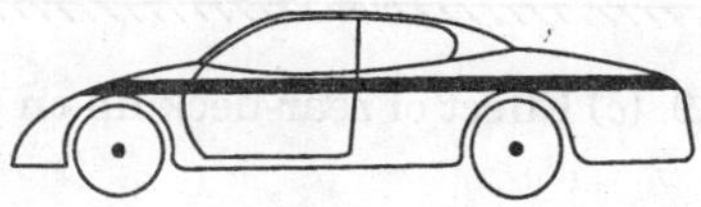	0.25

Rapid advances have been made in reducing the drag coefficients for passenger cars. One of the most streamlined cars was the "Bluebird", which in 1938 set a road-speed record in the world. It's coefficient of drag was 0.16. The minimum C_D for well streamlined cars is around 0.20. However, in case of passenger cars, significant progress in lowering the C_D will be very hard to achieve. Lowering its value below 0.30 will require exceptional design and workmanship. The stumbling block is the underside of most of the cars which is aerodynamically very rough due to axles, wheels, fuel tank, shock absorbers, muffler and so on. Other basic features which contribute to drag but are not easily amenable to drag-reduction modifications are interior air-flow systems for engine cooling, wheels, exterior features such as rear view mirrors and antennas, and other surface protusions.

A fuel efficient car will require the product of C_D and the frontal area A to be minimum, *i.e.*, C_DA directly affects the drag force for a given speed. Apart from a lower value of drag coefficient C_D, the frontal area A of the car needs to be minimised to cause reduction in the drag resistance. A reduced drag resistance will result in fuel economy.

10.10.3. Drag Reduction Measures

Wind tunnel tests are always helpful and are usually carried out for arriving at the most efficient aerodynamic design. Test results of experiments carried out by Schenkel on a conventional Sedan car model in a wind tunnel are shown in Fig. 10.9 (*c*) in the form of centreline pressure distribution. The solid line shows the distribution of pressure coefficient, Eq. (10.37 *b*), for the conventional Sedon while the dashed line shows the effect of its distribution for the car fitted with a rear-deck lip. The pressure coefficient, c_p, is plotted normal to the surface of the car. The presence of rear-deck lip causes the pressure on the rear of the car to increase (pressures are less negative) thereby reducing the drag on the car itself. It also decreases the lift thereby improving the traction and the road grip. The wind tunnel tests show that, the lip itself produces some drag, and the optimum height of the lip for greatest overall drag reduction is about 20 mm.

The drag of trucks can be reduced by installing vanes near the corners of the truck body to deflect the flow of air more sharply around the corner, thereby reducing the degree of separation. This in turn creates higher pressures on the rear surface of the truck, which reduces the drag coefficient. Another application of aerodynamics to road vehicles is the use of negative-lift vanes on racing cars, where the negative lift produces by the vanes improves the stability and traction of the racing car. Figure 10.9 (*d*) shows a racing car fitted with negative lift devices—a front vane and a rear vane. The rear vane produces a downward thrust (negative lift) which improves the stability and traction of the vehicle.

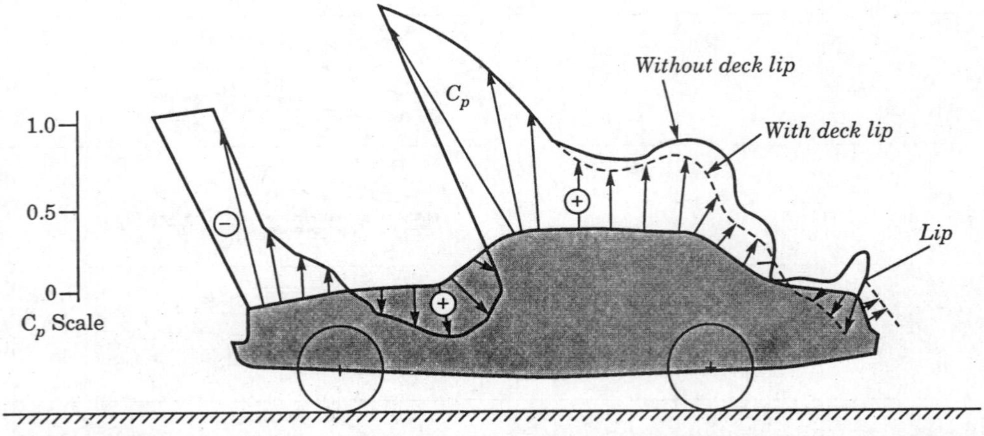

Fig. 10.9. (*c*) Effect of rear-deck lip on pressure distribution over the car surface.

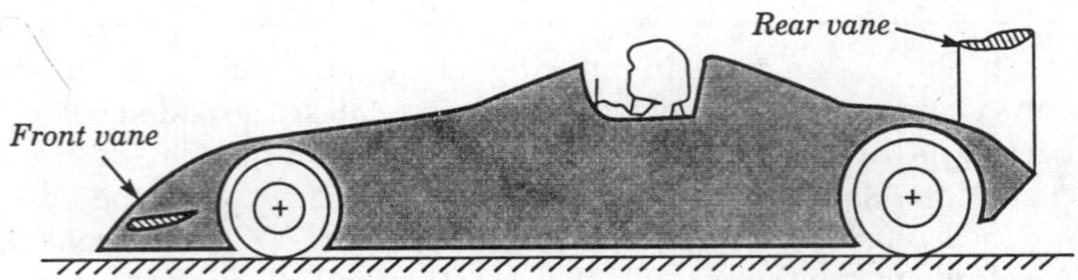

Fig. 10.9. (*d*) A racing car fitted with negative-lift devices.

10.10.4. Role of Drag Reduction on Fuel Efficiency

For minimising the drag force, it is not only the lower drag coefficient but also the frontal area of the vehicle on which the drag acts. In fact, it is the product of C_D and A that needs to be minimised to reduce the power required. The consumption of fuel is directly related to the power required to overcome various types of resistance, namely, aerodynamic drag, tyre friction, rolling resistance and other meachanical losses. In order to reduce drag, the underside of the vehicle requires to be made protrusion-free as far as possible, and also other external protrussions like rear view mirrors and antennas should be adjusted to cause minimum obstruction to the air flow.

To understand role of reduction in drag coefficient–for which wind tunnel tests are carried out on car model—it is to be noted that for no change in the frontal area, a 10% reduction in drag coefficient would result in 5% improvement in fuel economy on highway driving.

Exampl 10.13. *A Toyota car has a frontal projected area 1.65 m^2 and travels at a speed of 70 km/h. It has a drag coefficient of 0.33 based on the frontal area. Calculate :*

(i) the power required to overcome the air resistance by the car.

(ii) the speed of the car, if the drag coefficient is reduced to 0.30 by streamlining. The power expanded in overcoming air resistance remaining the same.

Take air density as 1.2 kg/m^3.

Solution. Speed of car, $V = 70 \text{ km/h} = \dfrac{70 \times 1000}{60 \times 60} = 19.44 \text{ m/s}$

(*i*) Drag force, $$F_D = C_D \rho \frac{AV^2}{2}$$

$$= 0.33 \times 1.2 \times 1.65 \times \frac{(19.44)^2}{2} = 123.46 \text{ N}$$

Power required to overcome air resistance

$$P = F_D \cdot V$$
$$= 123.46 \times 19.44 \text{ N.m} = 2400.15 \text{ Nm} = \mathbf{2.4\ kW}$$

(*ii*) For the same power, the new speed due to reduction in drag coefficient,

$$P = F_D \cdot V$$
$$= \rho C_D A \frac{V^2}{2} \cdot V = \rho C_D A \frac{V^3}{2}$$

Making substitutions,

$$2400.15 = 1.2 \times 0.30 \times 1.65 \times \frac{V^3}{2} = 0.297\ V^3$$

$\therefore$ $$V^3 = 8081.31$$
$$V = 20.07 \text{ m/s} = \mathbf{72.24\ km/h.}$$

Example 10.14. *Improved streamlining produces a 25% reduction in the drag coefficient of a torpedo. When travelling fully submerged what percentage increase in speed would this represent, assuming the driving power to remain the same ?*

Solution. Driving power $= \dfrac{F_D \times V}{75}$

But $$F_D = C_D A \rho \frac{V^2}{2}$$

for the same fluid and same cross-sectional area of the torpedo, the same driving power results in

$$F_{D_1} \times V_1 = F_{D_2} \times V_2$$

$$C_{D_1} A \frac{\rho V_1^2}{2} \times V_1 = C_{D_2} A \frac{\rho V_2^2}{2} \times V_2$$

or $$C_{D_1} V_1^3 = C_{D_2} V_2^3$$

given $$C_{D_2} = 0.75\ C_{D_1}$$

$$\therefore \qquad (V_2/V_1)^3 = \frac{C_{D_1}}{C_{D_2}} = 1.333$$

or $\qquad V_2/V_1 = 1.1$

$\therefore \qquad V_2 = 1.1\, V_1$

The percentage increase in speed is therefore 10%.

Example 10.15. *Two steel spheres of diameters d_1 and d_2 where $d_2 = 2d_1$, are dropped from rest at the free surface of a deep well. If their drag coefficients are independent of Reynolds number, determine the ratio of their terminal velocities V_1/V_2.*

Solution. When the falling spheres attain their terminal velocities, the drag force exerted on the spheres by the liquid will be equal to their submerged weights.

Drag force = Submerged weight

$$C_D \cdot \frac{\pi d^2}{4} \cdot \rho\, \frac{V^2}{2} = \frac{\pi d^3}{6}\,(s-1)\,\gamma$$

$$C_D V^2 = \frac{4}{3}\, d(s-1)\, g$$

Considering the spheres of diameters d_1 and d_2,

$$C_{D_1} V_1^2 = \frac{4}{3}\, d_1 (s-1)\, g \qquad ...(1)$$

and

$$C_{D_2} V_2^2 = \frac{4}{3}\, (2d_1)(s-1)\, g \qquad ...(2)$$

Since the drag coefficients are independent of Reynolds numbers

$$C_{D_1} = C_{D_2}$$

Dividing Eq. (1) by Eq. (2),

$$\frac{V_1^2}{V_2^2} = \frac{1}{2} \quad \text{or} \quad \frac{V_1}{V_2} = \frac{1}{\sqrt{2}}.$$

Example 10.16. *A spherical piece of boulder of small size rolls down the gentle slope of a hillock and falls into a deep lake. It weighs 52 N in air and 24.9 N in water. The temperature of water in the lake may be taken at 20°C at which its kinematic viscosity is 10^{-6} m²/s. The variation of drag coefficient C_D with Reynolds number **R** is as given below :*

R	10^3	4×10^3	10^4	4×10^4	10^5	2×10^5	3×10^5	10^6
C_D	0.40	0.35	0.40	0.50	0.45	0.40	0.15	0.20

Estimate the terminal velocity of fall of the boulder in water. Linear interpolation may be used for obtaining the drag coefficient at the desired Reynolds number.

Solution. Weight of spherical boulder, $W = 52$ N

Its weight in water = 24.9 N.

Force due to buoyancy, F_B, acting vertically upward on the boulder

= Loss of weight due to immersion

= Weight of water displaced

$\therefore \qquad F_B = 52 - 24.9 = 27.1$ N

Volume of water displaced by the boulder

$$= \frac{27.1}{9810}\ \text{m}^3 = 2.76 \times 10^{-3}\ \text{m}^3$$

Diameter, d, of the boulder is obtained as under :

$$\frac{4}{3}\pi\left(\frac{d}{2}\right)^3 = 2.76 \times 10^{-3}$$

$$d^3 = 5.27 \times 100^{-3}\ \text{m}^3$$

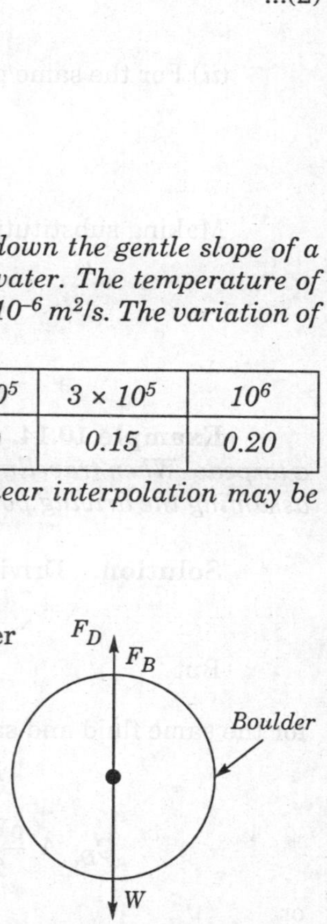

Force acting on the falling boulder.

$$= 5.27 \times 10^{-3} \times (10^2)^3 \text{ cm}^3$$
$$= 5.27 \times 10^3 \text{ cm}^3$$
$$\therefore \quad d = 17.40 \text{ cm.}$$

Assuming $C_D = 0.40$, the drag force, F_D, acting on the boulder is

$$F_D = C_D \times \frac{1}{2} \rho U_\infty^2 A$$
$$= 0.40 \times \frac{1}{2} \times 1000 \times (U_\infty^2) \times \frac{\pi}{4} (0.174)^2$$
$$= 4.76\, U_\infty^2.$$

Considering the equilibrium of forces

$$W = F_D + F_B \; 52 = 4.76\, U_\infty^2 + 27.1$$

from which, $\quad U_\infty = 2.29$ m/s.

Reynolds number $\mathbf{R} = \dfrac{U_\infty d}{\nu} = \dfrac{2.29 \times 0.174}{10^{-6}} = 3.98 \times 10^5$

Interpolating C_D corresponding to this Reynolds number

$$C_D = 0.15 + \frac{0.05}{7 \times 10^5} \times 0.98 \times 10^5 = 0.157$$

and the corresponding drag force

$$F_D = \frac{0.157}{0.40} \times 4.76\, U_\infty^2 = 1.86\, U_\infty^2$$

which must be equal to 24.9 N. Thus

$$1.86\, U_\infty^2 = 24.9 \; ; \; U_\infty = 3.65 \text{ m/s}$$

Corresponding to this velocity, the Reynolds number is

$$\mathbf{R} = \frac{3.65 \times 0.174}{10^{-6}} = 6.35 \times 10^5$$

At this $\mathbf{R}$, $\quad C_D = 0.15 + \dfrac{0.05}{7 \times 10^5} \times 3.35 \times 10^5 = 0.1739$

$$\therefore \quad F_D = \frac{0.1739}{0.40} \times 4.76\, U_\infty^2 = 2.07\, U_\infty^2$$

$$2.07\, U_\infty^2 = 24.9 \; ; \; U_\infty = 3.47 \text{ m/s}$$

$$\mathbf{R} = \frac{3.47 \times 0.174}{10^{-6}} = 6 \times 10^5 \text{ at which } C_D \cong 0.174.$$

Hence the terminal fall velocity is **3.47 m/s.**

Example 10.17. *Determine the weight of a 100 mm diameter spherical ball supported in equilibrium by a vertical stream of air of velocity 12 m/s. Kinematic viscosity of air is 15 × 10^{-6} m^2/s. Specific weight of air = 11.8 N/m^3. The values of the drag coefficient at Reynolds numbers 4 × 10^4 and 10^5 are respectively 0.50 and 0.45.*

Solution. Reynolds number $\mathbf{R} = \dfrac{U_\infty d}{\nu} = \dfrac{12 \times (100/1000)}{15 \times 10^{-6}} = 8 \times 10^4$

Drag coefficient at $\mathbf{R} = 8 \times 10^4$ is

$$C_D = 0.50 - \frac{0.05}{6 \times 10^4} \times 4 \times 10^4 = 0.467$$

Drag force on the ball, $F_D = C_D \times (\frac{1}{2} \rho U_\infty^2)A$

$$= 0.467 \times \frac{1}{2} \times \left(\frac{11.8}{9.81}\right) \times (12)^2 \times \frac{\pi}{4}\left(\frac{1}{10}\right)^2 = 0.317 \text{ N.}$$

Since the ball is in equilibrium,

Weight of ball = Drag force on ball = **0.317 N.**

10.11. Drag Characteristics of Two-Dimensional Bodies

(*Circular cylinder and long rectangular plate*)

The flow around a cylinder exhibits certain peculiar properties which are not ordinarily encountered in the three dimensional case of flow around a sphere. At very low values of Reynolds number (**R** < 0.5) inertia forces are negligible as compared to the viscous forces and the drag is almost proportional to U_∞. In other words, the drag coefficient C_D is approximately inversely proportional to the Reynolds number **R**, as indicated by the straight-line portion of the curve shown in Fig. 10.10. In this range of the Reynolds number, the friction drag accounts for a major part of the total drag. As the Reynolds number increases from 2 to about 30, the boundary layer separates symmetrically from the two sides and two weak, symmetrical standing vortices (or eddies) are formed. These vortices remain stably attached to the rear of the cylinder and do not get detached from it upto **R** ≤ 60. Under these circumstances when separation of boundary layer occurs, the pressure drag makes a proportionately larger contribution, and the slope of the C_D *versus* **R** curve decreases.

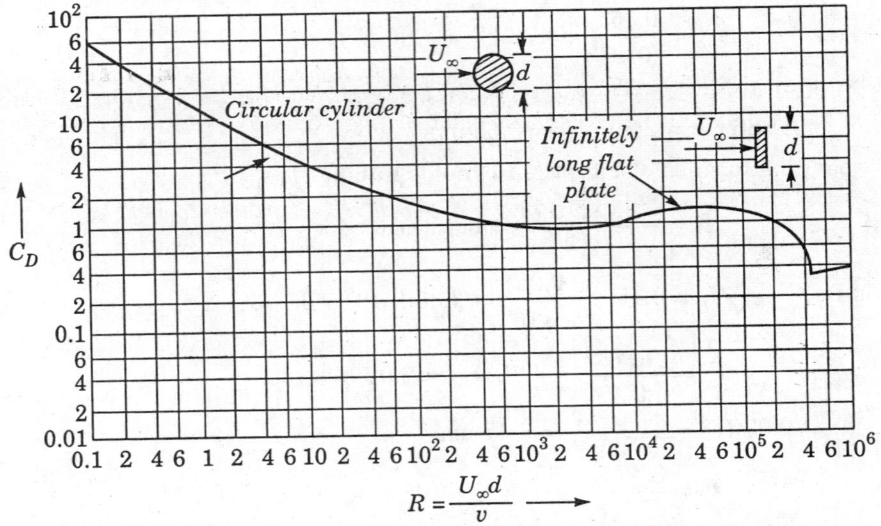

Fig. 10.10. Drag coefficients for circular cylinders and flat plates.

Observation of wake behind a two-dimensional body shows that at higher Reynolds numbers (60 < **R** < 5 × 10³), vortices form and leave the body surface alternatively from near the top and the bottom of the cylinder at a regular frequency and that they move downstream in two approximately parallel rows, Fig. 7.6 (*c*). The vortices in one row are staggered with respect to those in the other. Theodore von Karman studied the stability of these regular vortex patterns which are justly known as *Karman vortex street* or *vortex trail* (see article 7.7). Experiments have shown that the vortices are formed and then shed regularly and alternately from the opposite sides of the body (such as a cylinder) placed crosswise to a fluid flow. For Reynolds number range between 10^3 and 10^5, the vortices leave the body surface at a frequency of 0.21 U_∞/d, where d is the cylinder diameter. The formation of unsymmetrical vortices behind the cylinder and their shedding give rise to lateral thrust or lift with a frequency n. This leads to transverse oscillations which must be considered in the design of such elastic structures as suspension bridges, tall chimneys and the like which are exposed to wind. The worst condition will be obtained when the natural frequency of vibration of the structure coincides with the frequency of vortex shedding.

The peculiar characteristic of alternate shedding of vortices leads to what is known as the "singing" of telephone wires or electric power transmission lines, the phenomenon of fluttering of wires and poles and vibrations of chimneys and towers.

The frequency of vortex shedding is expressed in terms of a dimensionless parameter known as the **Strouhal number,**

$$\mathbf{S} = \frac{nd}{U_\infty} \quad \text{...(10.38)}$$

where n = frequency of shedding. As the Reynolds number increases, the frequency of shedding increases until at $\mathbf{R} = 10^3$, $\mathbf{S} = 0.21$. For $\mathbf{R} < 10^3$, the Strouhal number remains practically constant at 0.21. It varies from 0.12 to 0.21 for Reynolds number range of 50 to 10^3 respectively.

The variation of drag coefficient for a circular cylinder (placed across the flow as shown) with Reynolds number is shown in Fig. 10.10. The drag coefficient reaches a minimum value of about 0.95 when the Reynolds number is around 2000 and afterwards there is a rising trend resulting in a maximum $C_D \approx 1.2$ at about $\mathbf{R} = 3 \times 10^4$. The reason for this rise is the increased turbulence in the wake, and in addition the points of separation gradually advance upstream causing the pressure drag to increase correspondingly. At a Reynolds number of about 2×10^5 the boundary layer which was entirely laminar so far becomes turbulent before its separation. This results in the shifting of the separation points downstream of the mid-section of cylinder resulting in narrowing of the wake. Consequently, the drag coefficient falls appreciably. Experimental data for $\mathbf{R} > 5 \times 10^5$ are scarce, but it appears likely that after the sharp drop in C_D to about 0.3 there is a rise to about 0.7 within the Reynolds number range of 5×10^5 to 3×10^6.

The critical value of the Reynolds number at which the large drop in C_D occurs is smaller, both for increased roughness of the body surface upstream of separation points and for a greater of turbulence in the approach flow. If a small roughness element such as a wire is fixed to the cylinder surface upstream of separation points, the transition from laminar to turbulent boundary layer would occur at a smaller Reynolds number.

For bluff bodies, such as a thin flat plate held perpendicular to the flow, the friction drag is zero, since the shear stresses have no component in the flow direction. Except at low Reynolds numbers ($\mathbf{R} < 100$), the separation of boundary layer always occurs at the same place—the sharp edge of the plate, irrespective of the Reynolds number. The total drag is, therefore, proportional to the square of the velocity and C_D is independent of $\mathbf{R}$.

It is to be noted that the above results apply only to two-dimensional flow. This means that the cylinder and the flat plate should be long enough so that the 'end effects' are negligible. With reduced length of the body, the flow around the ends tends to reduce the pressure near the edges of the object. This reduction is substantial at the front face of the body. As a result the value of C_D decreases as the length of the body is reduced.

Pressure distribution for ideal fluid flow over the surface of a circular cylinder is given by Eq. (17.36) reprodcued here :

$$\frac{p - p_\infty}{\frac{1}{2}\rho U_\infty^2} = 1 - 4\sin^2\theta \quad \text{...(10.39)}$$

in which p_∞ and U_∞ are the pressure and the free stream velocity of undisturbed flow, and θ is the angle measured anti-clockwise with respect to the downstream stagnation point as shown in Fig. 10.11. As is evident from the figure, the actual pressure distribution 1 differs from that for ideal fluids. The two distributions are identical upto a point where the boundary layer remains attached to the surface, beyond which they differ widely.

From Fig. 10.11, it can be observed that :

(*i*) for ideal fluids, the pressure distribution is symmetrical about the flow direction as well as at right angles to it. There is, therefore, no lift and no drag developed.

(*ii*) for real fluids, the pressure distribution is symmetrical about the cylinder diameter parallel to the flow direction, but at right angles to it the distribution is unsymmetrical. There is thus the drag force only which results from the interaction between the fluid flow and the cylinder.

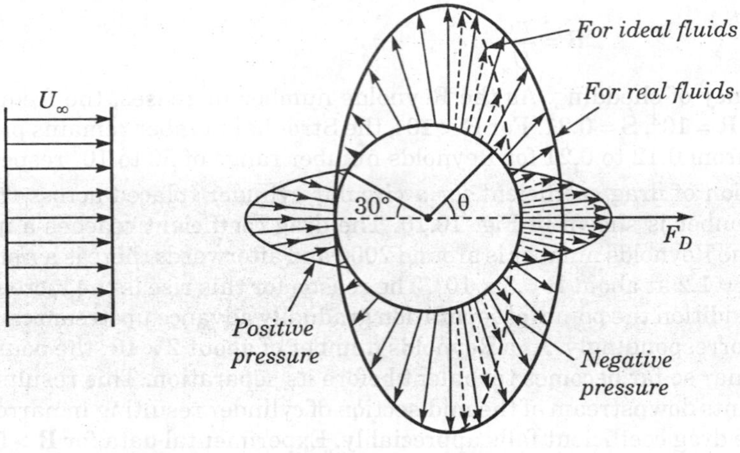

Fig. 10.11. Pressure distribution over a circular cylinder.

Example 10.18. *A cup anemometer, used in measuring wind velocity consists of two hollow hemispheres mounted in opposite directions at the ends of a horizontal rod which turns freely about a vertical axis as shown in the sketch below, what torque would be required to hold the rotating member stationary in a 50 km/hr wind ?*

For hemisphere hollow upstream, C_D = 1.33 while for the hemisphere hollow downstream C_D = 0.34, specific weight of air is 11.85 N/m³.

Solution. For the hemispheric hollow upstream the drag

$$F_{D_1} = C_D A\rho \frac{U_\infty^2}{2}$$

$$= 1.33 \times \frac{\pi}{4}\left(\frac{10}{100}\right)^2 \times \frac{11.85}{9.81} \times \frac{1}{2}$$

$$\times \left(\frac{500}{36}\right)^2 = 1.216 \text{ N}$$

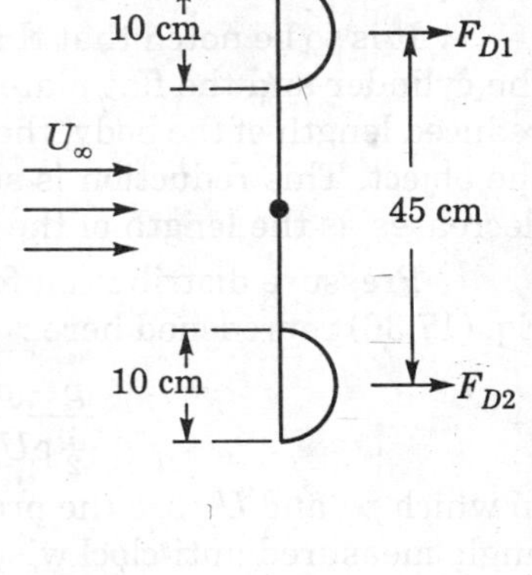

and for the hemisphere hollow downstream

$$= 0.34 \times \frac{\pi}{100}\left(\frac{10}{100}\right)^2 \times \frac{11.85}{9.11} \times \frac{1}{2} \times \left(\frac{500}{36}\right)^2$$

$$= 0.311 \text{ N}$$

Torque required to hold the rotating arm

$$= (F_{D_1} - F_{D_2})r = (F_{D_1} - F_{D_2}) \times \frac{45}{2 \times 100} = 0.204 \text{ N-m.}$$

Example 10.19. *A spherical balloon weighs 140 N and contains helium having density of 0.22 kg/m³. The surrounding air has a density of 1.22 kg/m³. Determine :*

(*i*) *the diameter of the balloon that permits a terminal rising velocity of 2.0 m/s ;*

(ii) the inclination of the cable with which if the balloon is tied to the ground when a wind of 15 km/h blows past the balloon.

Take kinematic viscosity of air at the prevailing temperature as 1.5×10^{-5} *m²/s ;* $C_D = 0.5$ *for Reynolds number range* 10^4 *to* 3×10^5*, and* $C_D = 0.2$ *for Reynolds number greater than* 3×10^5.

Solution. (*i*) **When the balloon is rising at a velocity of 2.0 m/s**

Buoyant force = Drag force + weight of balloon

$$(\rho_{air} - \rho_{He}) g \frac{\pi}{6} D^3 = C_D A \rho \frac{U^2}{2} + W$$

Assume $R > 3 \times 10^5$ so that $C_D = 0.2$, making substitutions,

$$(1.22 - 0.22) \times 9.81 \times \frac{\pi}{6} D^3 = 0.2 \left(\frac{\pi D^2}{4}\right) \frac{1.22 \times (2)^2}{2} + 140$$

$$5.1365 D^3 = 0.3833 D^2 + 140$$

Solving by triar and error, **D = 3.036 m.**

Reynolds number $= \dfrac{UD}{\nu} = \dfrac{2.0 \times 3.036}{1.5 \times 10^{-5}} = 4.048 \times 10^5$

Since the Reynolds number is greater than 3×10^5, $C_D = 0.2$ is correct.

(*ii*) **When the balloon is tied to the ground (i.e. tethered), see Fig. 10.16**

$$U_\infty = 15 \text{ km/h} = 4.17 \text{ m/s}$$

Let θ be the inclination of the cable with the horizontal, and T be the tension in the cable. Considering the forces acting on the balloon,

$$T \cos \theta = \text{Drag force } F_D = C_D A \rho \frac{U_\infty^2}{2}$$

$$T \sin \theta = \text{Lift force } F_L - \text{Weight of balloon } W$$

$$= (\rho_{air} - \rho_{He})\, g \frac{\pi D^3}{6} - W$$

$D = 3.036$ m, and $C_D = 0.2$ since Reynolds number $> 3 \times 10^5$

$$T \cos \theta = 0.2 \times \frac{\pi}{4} (3.036)^2 \times 1.22 \times \frac{(4.17)^2}{2} = 15.36$$

$$T \sin \theta = (1.22 - 0.22) \times 9.81 \times \frac{\pi (3.036)^3}{6} - 140 = 3.738$$

$$\tan \theta = \frac{3.738}{15.38} = 0.2434$$

$$\theta = 13.68° = \mathbf{13°40'48''}$$

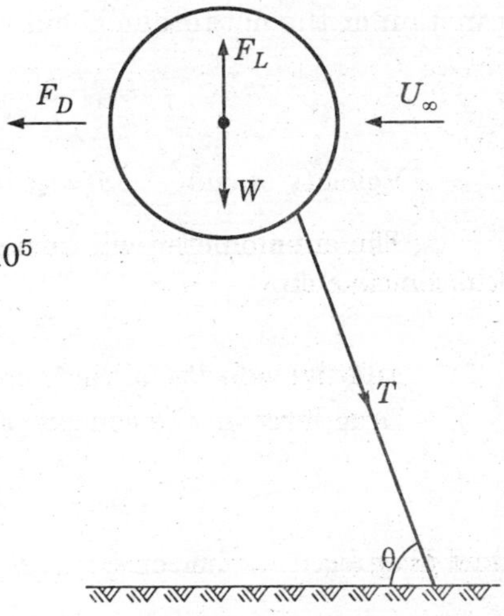

Fig. Ex. 10.19

Tension in the cable $T = \dfrac{3.738}{\sin 13.68°} = \mathbf{16.124\ N.}$

Example 10.20. *A high tension cable 5 cm in diameter is strung out between two towers. At a wind velocity of 100 km/h, determine the*

(i) drag force per metre length of the cable, and

(ii) frequency of vortex shedding.

Take $\rho_{air} = 1.22$ *kg/m³,* $\nu_{air} = 1.5 \times 10^{-5}$ *m²/s*

For a cylinder : $C_D = 1.20$ *for Reynolds number range* 10^4 *to* 5×10^5
$C_D = 0.35$ *for Reynolds number greater than* 5×10^5.

Solution. $U_\infty = 100$ km/h = 27.78 m/s

Reynolds number, $R = \dfrac{U_\infty D}{\nu} = \dfrac{27.78 \times 0.05}{1.5 \times 10^{-5}} = 9.26 \times 10^4$

Since $R < 5 \times 10^5$, $C_D = 1.20$

(*i*) *Drag force per metre length of cable* : ($A = LD$)

$$F_D = C_D A \rho \frac{U_\infty^2}{2} = 1.20 \times (1 \times 0.05) \times 1.22 \times \frac{(27.78)^2}{2}$$
$$= 28.25 \text{ N/m length of cable}$$

(*ii*) *Strouhal number* $S = 0.21$ for $R > 10^3$

$$\therefore \quad S = \frac{nD}{U_\infty}, \text{ frequency } n = \frac{U_\infty S}{D} = \frac{27.78 \times 0.21}{0.05} = \mathbf{116.68\ Hz}$$

Hence, frequency of vortex shedding is 116.68 Hz.

Example 10.21. *A cup anemometer required to measure wind velocity consists of two hemispherical cups of 10 cm diameter mounted in opposite directions at the ends of a horizontal rod which turns freely about a vertical axis at the middle. The centre to centre spacing of cups is 45 cms. Assuming zero friction at the bearings, calculate the rotional speed of the anemometer in a wind velocity of 50 km/hr. Take* $C_D = 1.33$ *and 0.34 depending on the frontal shape of the hemisphere, and specific weight of air = 11.85 N/m*^{3}.

Solution. Let ω be the angular velocity of the anemometer in a wind velocity of U_∞. For constant ω, the net torque about the axis of rotation is zero.

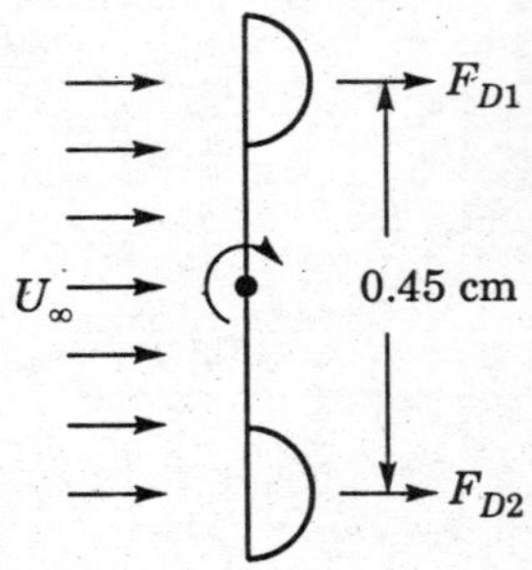

$$U_\infty = 50 \text{ km/hr} = \frac{50000}{3600} = 13.89 \text{ m/su}$$

Velocity of cup $\quad v = r\omega = \dfrac{0.45}{2}\,\omega = 0.225\,\omega$

The anemometer will turn clockwise, and the relative velocity of the concave cup

$$= U_\infty - r\omega, \text{ and}$$

relative velocity of the convex cup = $U_\infty + r\omega$

Drag force on the concave cup

$$F_{D_1} = 1.33 \frac{\pi d^2}{4} \frac{\rho}{2} (U_\infty - r\omega)^2,$$

and drag force on the convex cup

$$= 0.34 \frac{\pi d^2}{4} \frac{\rho}{2} (U_\infty + r\omega)^2$$

Torque $(F_{D_1} - F_{D_2}) \times r = 0, \quad \therefore \quad F_{D_1} = F_{D_2}$ (since $r \neq 0$)

or $\quad 1.33 \dfrac{\pi d^2 \rho}{8} (U_\infty - r\omega)^2 = 0.34 \dfrac{\pi d^2 \rho}{8} (U_\infty + r\omega)^2$

from which, $\quad \omega = 1.459\, U_\infty = 20.27$ rad/s

Rotational speed of the anemometer = $\dfrac{60\omega}{\pi(2r)} = \dfrac{60 \times 20.27}{\pi \times 0.45}$ = **860 RPM.**

10.12. Effect of Compressibility on Drag

Whenever the velocity of fluid approaches the velocity of sound in the fluid medium, the compressibility effects become important and a decided change takes place in the flow pattern. The forces affecting the drag phenomenon then are due to inertia, viscosity, and elasticity of fluid and the drag coefficient becomes, in general, a function of Mach (pronounced as Makh) number in addition to the Reynolds number. Thus we have

$$C_D = f(\mathbf{R}, \mathbf{M})$$

At very high Reynolds numbers, the viscous forces are of little importance as compared to the elastic forces and may thus be ignored. Under these conditions of flow, the drag co-efficient will be dependent only upon the Mach number.

10.12.1. Propagation of disturbance

In order to understand the effect of compressibility of fluid in a better ways, it is necessary to know about the propagation of elastic waves caused by the motion of objects such as rockets and aircrafts. Consider the motion of a small object moving through a fluid and generating waves of disturbances. If the fluid were truly incompressible, its modulus of elasticity would be infinite, the velocity of propagation of disturbance through the fluid would also be infinite resulting in a zero value of Mach number. The effects of disturbance would be felt instantaneously throughout the flow field and the fluid in front of the body would be aware of its presence and adjust itself accordingly as the body advances. Even in a real fluid when the flow velocities involved are small as compared to the velocity of sound in the fluid medium (which is the velocity of propagation of the disturbance), the propagation of disturbance is rapid enough to produce flow patterns affected by the compressibility of fluid.

An entirely different situation is developed when the source of disturbance (it may be an aircraft or a projectile) moves at a velocity appreciably greater than the velocity of sound in the fluid medium. The fluid upstream of the object is unaware of its presence and as the object encounters this fluid in the course of its motion the fluid must suddenly change its direction to adjust itself to sharp discontinuities known as the *shock waves*. The changes in flow properties (such as pressure, density, velocity and so on) across the surface of the shock wave may be considered abrupt and discontinuous.

In order to understand how the elastic wave of disturbance generated by the motion of an object travels in relation to the object itself, let us consider a point source of disturbance moving with velocity V through an otherwise still fluid. At the instance when this source passes through point B_0, the disturbance commences to radiate in all directions in the form of a spherical wave with velocity of sound C in the fluid medium. In successive instances the source passes through the points B_1, B_2 and B_3. While the source of disturbance has covered distance B_0B_3 with velocity V (say in time t) the sound wave travelling at a slower velocity C has traversed a radial distance B_0A_0 in the same time. Similar spherical waves emanating from B_1 and B_2 form conical envelope with vortex at B_3, where the source has just reached and the spherical pressure wave radiating from B_3 is about to start. This conical envelope is known as the *shock wave*. Referring to Fig. 10.12, the angle α is called the Mach angle and it is seen that

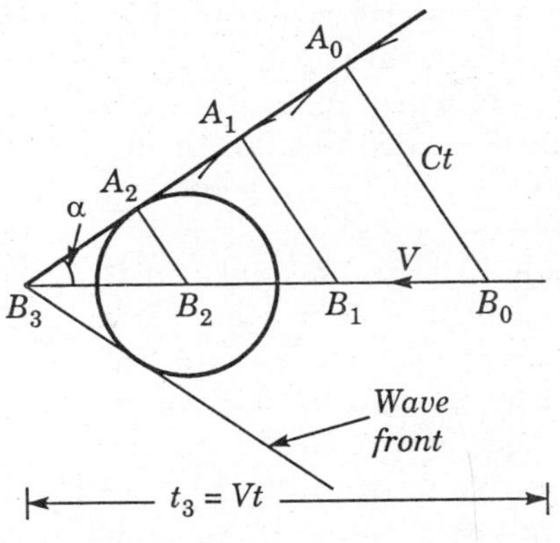

Fig. 10.12. Wave front of a shock wave.

$$\sin \alpha = \frac{B_0A_0}{B_0B_3} = \frac{C.t}{V.t} = \frac{C}{V}$$

It may be recalled that the dimensionless velocity ratio V/C is the Mach number.

As the waves of disturbance originate from different points occupied by the source at various instants of time, a wave front is developed which represents the line of advance of the elastic waves. The fluid to the left of this wave front is at rest and unaware of the presence of the disturbance. This wave front, known as the Mach wave will be conical (three-dimensional) if produced by the needle point and wedge-shaped (two-dimensional) if produced by the razor edge.

10.12.2. Drag Characteristics in Compressible Fluids

Drag is always the result of shear stress and difference of pressure over the body surface, but in compressible fluids the distribution of these quantities around a given body is a complex problem in itself and differs appreciably from that for fluid flow of a constant density. The abrupt rise of pressure which occurs across a shock wave is particularly important. This is not simply on account of changed pressures on the front and rear of the body. The adverse gradient produced by the shock wave thickens the boundary layer and encourages separation. The problem is complicated by the interaction between the shock wave and the boundary layer. At high Mach numbers, heat dissipation as a result of skin friction causes serious rise of temperature and thus the problem gets further complicated.

At low Mach numbers, the compressibility effects are unimportant whereas the corresponding Reynolds numbers are small enough for the viscous effects to be considered. When the velocity is high enough for compressibility effects to become important (*i.e.* at large Mach number), the Reynolds number becomes very large with the result that the viscous effects become negligibly small.

The drag coefficient of a given body rises abruptly as the Mach number of the oncoming flow approaches unity. For a blunt body for which the position of separation points is fixed by its shape, the friction drag is small and C_D continues to rise beyond $M = 1$, as a result of shock wave effects at or near the body front. These effects due to the shock wave contribute greatly to the total drag and thus for supersonic flow, the streamlining of the rear of the body has little effect on the total drag. The greatest reduction in drag is achieved by making the nose of the body a sharp point. The evidence of this fact lies in the pointed nose of modern supersonic aircrafts and high speed projectiles. For non-blunt bodies, the position of separation points is closely associated with the shock wave phenomenon. A shock wave first appears on the surface at the position of maximum velocity and separation occurs just behind it. There is thus a sharp rise in C_D, refer Fig. 10.13, which depicts C_D *versus* **M** variation for bodies of different shapes.

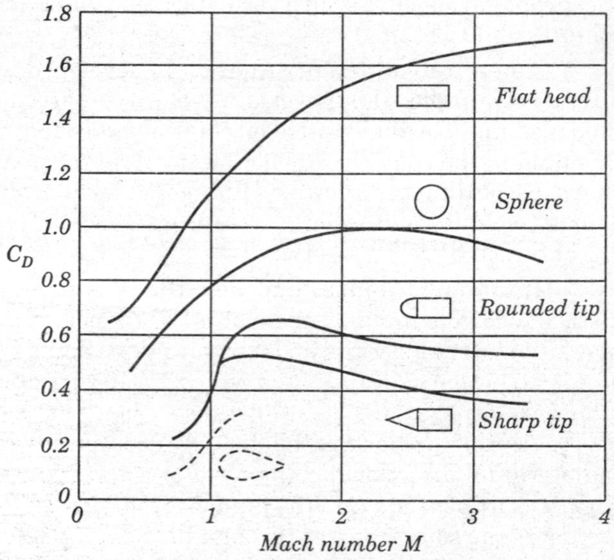

Fig. 10.13. Drag coefficient *versus* Mach number plot for different bodies of revolution.

For minimum drag in supersonic flow, the body should have a sharp conical nose and the shape of the rear is of secondary importance. This requirement is in contradiction to that which requires a body, at low Mach numbers (well below unity) to be rounded at the front and well tapered or streamlined at the rear. Table 10.1 given on page no. 495 shows approximate values of the drag co-efficients for objects of different shapes for given Reynolds number as indicated.

Example 10.22. *The drag co-efficient C_D of a 3224 N and 40 cm diameter bomb is a function of the Mach number M is tabulated below. Linear interpolation may be used between tabulated values :*

M	0.4	0.6	0.8	1.0	1.2
C_D	0.2	0.23	0.3	0.5	0.65

If the bomb is dropped from a high flying plane, estimate the terminal velocity with which it will strike the earth. Near the ground surface, assume atmospheric pressure as 103 kN/m² and temperature as 17°C. Assume the gas constant for air as 29.3 metres per degree Kelvin and adiabatic index k as 1.4.

Solution. The bomb falls under gravity due to downward force equal to its weight W. Initially the bomb, having started with zero velocity will experience acceleration till sufficient resistance (which is equal to the drag force F_D) is developed to match the force W, causing the motion. When these forces become equal, the velocity of descent (or the terminal velocity) attains a constant value. Hence

$$W = F_D = C_D\left(\frac{1}{2}\rho U_\infty^2 A\right)$$

Using the perfect-gas law

$$p = \rho RT$$

$$\therefore \quad \rho = \frac{p}{RT}$$

Atmospheric pressure $p = 103 \times 10^3$ N/m²

Absolute temperature $T = 273 + 17$
$= 290$°K

Gas constant $R = 29.3$ in/°K,

hence, $$\rho = \frac{103 \times 10^3}{29.3 \times 9.81 \times 290} = 1.243 \text{ kg/m}^3$$

Velocity of sound $= C = \sqrt{kRT}$

$$= \sqrt{1.4 \times (29.3 \times 9.81) \times 290} = \mathbf{341.35\ m/s.}$$

From the C_D *versus* Mach number data given in the problem, the velocity U_∞ and the drag force F_D were calculated and plots were drawn as shown in the figure. From the nature of curve, one can easily notice that linear interpolation will not give correct result. To solve this problem, trial and error method does not lead to a quicker solution. The following method has, however, been found to be accurate and easy. The steps involved are :

(*i*) determine the velocity U_∞ corresponding to the given Mach number values,

(*ii*) compute the corresponding drag force F_D,

(*iii*) plot variation of F_D *versus* U_∞,

(*iv*) read the value of U_∞ corresponding to $F_D = W = 3924$ N

(*v*) check the result by determining the Mach number **M** and the corresponding drag co-efficient C_D.

$$\text{Mach number } \mathbf{M} = \frac{\text{Terminal velocity } U_\infty}{\text{Velocity of sound } C} = \frac{U_\infty}{341.35}$$

$$\text{Drag force} \quad F_D = C_D\left[\frac{1}{2} \times 1.243 \times \frac{\pi}{4}(0.45)^2\right] U_\infty^{\ 2} = 0.0988\ C_D\ U_\infty^{\ 3}.$$

Using the above equation for F_D, the following table can be prepared :

M	0.4	0.6	0.8	0.9228	1.0	1.2
C_D	0.2	0.23	0.3	0.4	0.5	0.65
U_∞	136.54	204.81	273.08	315.0	341.35	409.62
F_D	368.0	953.4	2209.8	3922.7	5758.2	10779.2

Plotting F_D *versus* U_∞ (variation shown in the following figure), and reading U_∞ corresponding to F_D = 3924 N, one gets $\mathbf{U_\infty}$ **= 315 m/s.**

Check : $\mathbf{M} = \dfrac{U_\infty}{C} = \dfrac{315}{341.35} = 0.9228$

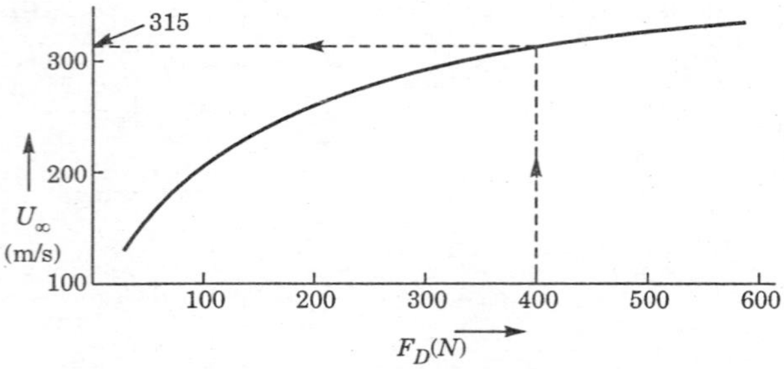

Fig. Ex. 10.22. Variation of F_D with U_∞.

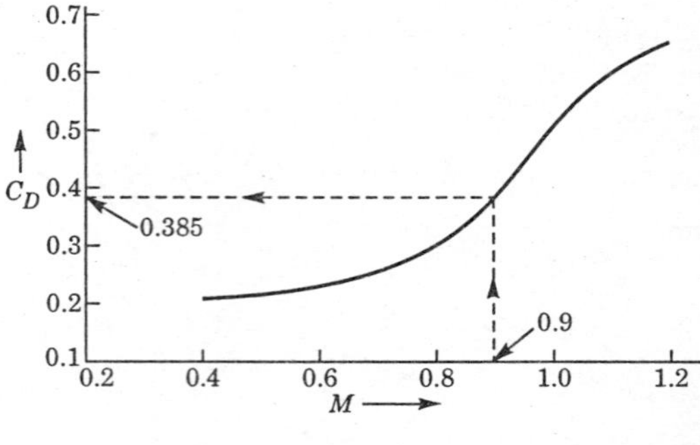

Fig. Ex. 10.22. Variation of C_D with Mach Number M.

From the C_D *versus* **M** plot, C_D = 0.4 and $F_D = 0.0988 \times 0.4 \times (315)^2 = 3922.7$ N hence the terminal fall velocity of 315 m/s is found to be correct. If however linear interpolation is used

$$C_D = 0.3 + \frac{0.5 - 0.3}{1.0 - 0.8} \times (0.9228 - 0.8) = 0.4228$$

and $$F_D = 0.0988 \times 0.4228 \times (315)^2 = 4146.3 \text{ N}$$

which confirms the graphical observation (C_D *versus* **M** curve) that linear interpolation does not yield correct result in this case.

10.13. Circulation, Lift and Magnus Effect

The concept of circulation was introduced in article 3.7 of Chapter 3 on kinematics of flow. If the product of velocity component along curved element and the length of element of closed curve is integrated around the curve, the magnitude of the integral gives circulation around the curve, thus

$$\Gamma = \oint v \cos \theta \, ds \qquad ...(3.30)$$

In order to understand the role played by circulation in causing lift, let us consider the flow field surrounding a free vortex. The free vortex motion, as already described earlier, is characterized by velocity equation $Vr = C$, where C is a constant. If we now choose a closed path around vortex centre to be a circular streamline, the computation of circulation is quite easy since the velocity is evidently constant and tangential to the streamline (causing $\cos \theta = 1$) and the line integral of ds is the circumference of the circle. Thus we have the circulation,

$$\Gamma = \oint V \, ds = V \oint ds = 2\pi r V \qquad ...(10.40)$$

but from the free vortex velocity equation $Vr = C$, hence

$$\Gamma = 2\pi C \qquad ...(10.41)$$

Eq. (10.41) demonstrates that the circulation around any curve enclosing vortex centre is a constant depending only upon the vortex strength C. It may be noted that the circulation is independent of the radius of closed curve.

In case of real fluid flow, the rotation of a cylinder produces local circulation through viscous shear. Let us now consider two separate cases of flow of an ideal fluid :

1. about a cylinder infinitely long (two-dimensional flow), and
2. circulatory flow about the cylinder caused by its spinning.

In the first case when a uniform fluid stream of velocity U_∞ flows past the cylinder, the velocity on the periphery of cylinder from the knowledge of classical hydrodynamics, is given by

$$v = 2U_\infty \sin \theta \qquad \text{...(10.42)}$$

The streamline pattern for this case is shown in Fig. 10.14 (*a*) which is completely symmetrical about the cylinder. Considering the second case of a circulatory flow about the cylinder, [Fig. 10.14 (*b*)] the circumferential velocity due to circulation is

$$v_c = \frac{\Gamma}{2\pi r_0} \qquad \text{...(10.43)}$$

where r_0 is the radius of the cylinder. Let us now combine the two flows superposing the circulatory flow over the uniform fluid motion past the cylinder to produce unsymmetrical flow of Fig. 10.14 (*c*).

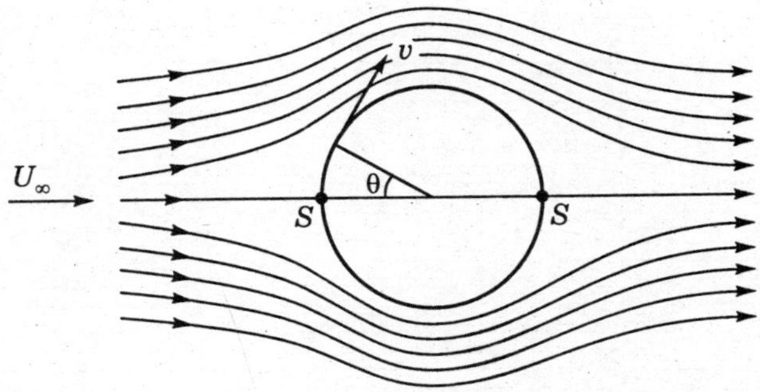

$v = 2U_\infty \sin \theta$

Fig. 10.14. (*a*) Irrotational flow (translatory) past a cylinder.

The velocity at the cylinder periphery, under these conditions of superpositions, is the sum of the two velocities expressed by Eqs. (10.42) and (10.43), thus we have

$$v = 2U_\infty \sin \theta + \frac{\Gamma}{2\pi r_0} \qquad \text{...(10.44)}$$

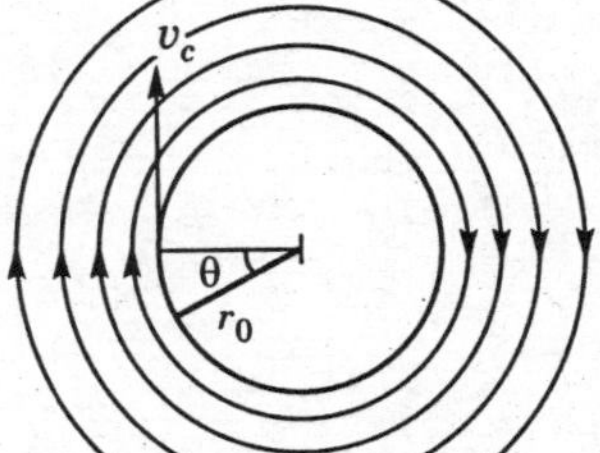

$v_c = \dfrac{\Gamma}{2\pi r_0}$

(*b*) Circulatory flow around a cylinder

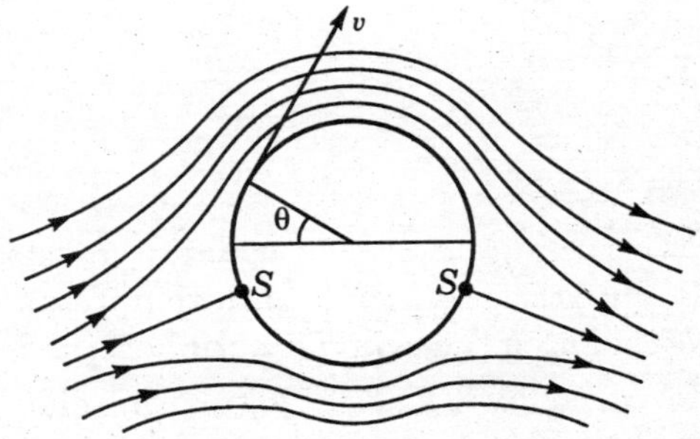

$v = 2U_\infty \sin \theta + \dfrac{\Gamma}{2\pi r_0}$

(*c*) Combined effect of translation and circulation

Fig. 10.14. Effect of circulation upon uniform irrotational flow past a cylinder (Magnus effect).

Table 10.1

S. No.	*Shape of object*	*L/d*	*Reynolds No.* **R**	C_D
1.	Circular disc (normal to flow)	—	$> 10^3$	1.12
2.	Rectangular flat Plate (normal to flow)	1 5 20 ∞	$> 10^3$	1.10 1.20 1.50 2.00
3.	Circular cylinder (axis parallel to flow)	0 1 2 4 7	$> 10^3$	1.12 0.91 0.85 0.87 0.99
4.	Circular cylinder (axis perpendicular to flow)	1 5 20 ∞ 5 ∞	10^5 $> 5 \times 10^5$	0.63 0.74 0.90 1.20 0.35 0.33
5.	Streamlined foil (1 : 3 airplane strut)		$> 4 \times 10^4$	0.07
6.	Hemisphere : (*i*) Hollow upstream (*ii*) Hollow downstream		$> 10^3$	1.33 0.34
7.	Sphere		10^5 $> 3 \times 10^5$	0.50 0.20
8.	Ellipsoid (1 : 2, major axis parallel to flow)		$> 2 \times 10^5$	0.07

In general, there will be two stagnation points to be determined from Eq. (10.44) by setting $v = 0$ and solving for the angle θ. The pressure at any point on the cylinder surface may be computed by writing the Bernoulli's equation between a point some distance upstream of the cylinder where the flow is almost undistributed and a point on the cylinder periphery.

We may thus write

$$p_\infty + \frac{1}{2}\rho U_\infty^2 = p + \frac{1}{2}\rho v^2 \qquad ...(10.45)$$

where p_∞ is the pressure upstream where the velocity is uniform and equal to U_∞ ; p and v are the pressure and velocity at any point on the cylinder periphery.

As indicated by Fig. 10.14 (*c*), the change in streamline spacing due to superposition of circulation over the flow pattern of Fig. 10.14 (*a*) results in velocity increase over the top half of the cylinder while on the bottom half the velocity decreases. This clearly indicates a rise in pressure intensity over the lower portion and a drop in pressure over the upper portion causing a transverse force perpendicular to the approaching flow. This transverse force, known as the *lift*, is produced by circulation. The lift force exerted upon the cylinder may be determined by

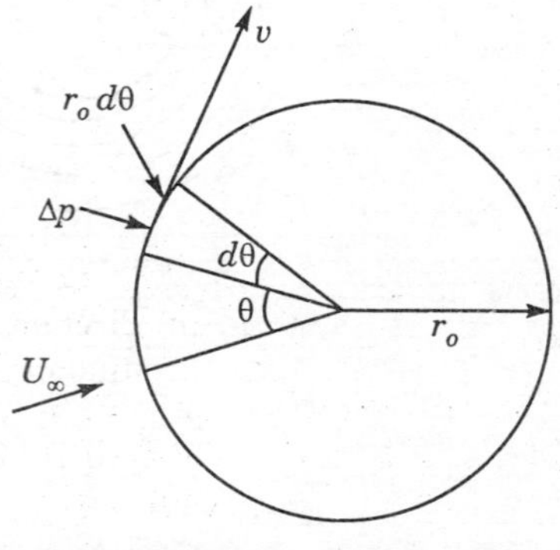

Fig. 10.15. Notations for derivation of lift exerted on a cylinder.

integrating over the surface of the cylinder the transverse component of force due to pressure upon an area element of cylinder surface. Using Eq. (10.45), the differential pressure acting over the cylinder is

$$\Delta p = p - p_{\infty} = \frac{\rho}{2}\,(U_{\infty}^{\,2} - v^2) \qquad \text{...(10.46)}$$

and with reference to Fig. 10.15, the lift developed upon a cylinder of length L will be,

$$F_L = L \int_0^{2\pi} (\Delta p\, r_0\, d\theta) \sin\theta$$

Substituting for Δp and v from Eqs. (10.46) and (10.44) the expression for lift force simplifies

$$F_L = \rho L U_{\infty} \Gamma \qquad \text{...(10.47)}$$

The transverse force given by Eq. (10.47) was first noted by a German physicist H.G. Magnus (1802-1870) who experimentally investigated it in 1852. The phenomenon of lift produced by imposing circulation over a uniform fluid stream is, therefore, known as the *Magnus effect*. The principle embodied in Eq. (10.47) may be used to explain the deviation of spinning tennis, cricket and golf balls from their normal trajectories. Similarly a rotating rifle bullet shot across the wind or spinning rod falling down under gravity have curved trajectories instead of straight ones.

Eq. (10.47) is known as the "Kutta-Joukowski Law" in honour of the German, M.W. Kutta (1867-1944) and the Russian, N.E. Joukowski (1847-1927) who independently showed that for a body of any shape in two dimensional flow, the transverse force per unit length is $\rho U_{\infty}\Gamma$ and is perpendicular to the direction of U_{∞}. The transverse force developed on a rotating cylinder has been used for ship propulsion. The rotor ship designed by A. Flettner (1885-1961) in 1924 had large vertical cylinders on the deck. These cylinders were rapidly rotated and were acted upon by prevailing winds. The combined effect was a transverse (propulsive) force normal to the direction of prevailing winds. Fig. 10.16 shows a Flettner rotor ship which utilizes the principle embodied in the Magnus effect.

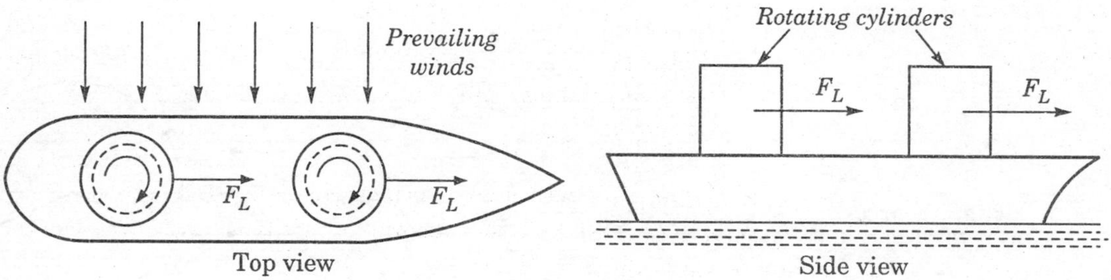

Fig. 10.16. (*a*) Propulsion of Flettner rotor ship.

The Magnus effect may be used with advantage in the games of tennis, cricket, table tennis and golf. Fig. 10.16 (*b*) indicates the type of spin given to the ball to achieve a rise, drop and curve in their trajectories, respectively.

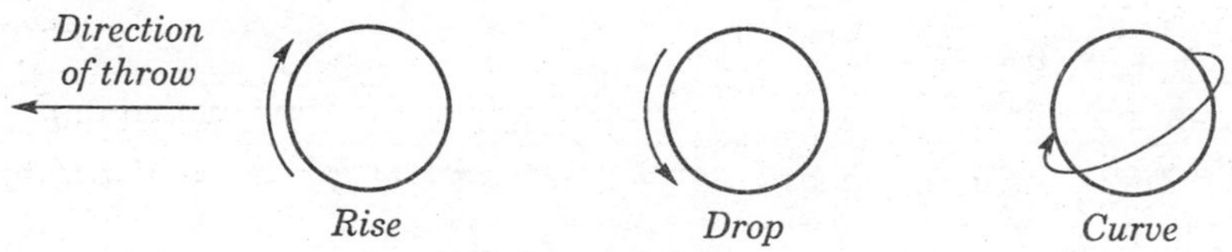

Fig. 10.16 (*b*)

For a circular cylinder the lift force may be expressed in terms of coefficient of lift as

$$F_L = C_L(L \cdot d)\rho \frac{U_\infty^2}{2} \quad \text{...(10.48)}$$

where L and d are the length and the diameter of cylinder respectively. The lift coefficient C_L must be obtained from experimental data as a function of v_c/U_∞, where v_c is the peripheral speed of the cylinder. Comparing Eqs. (10.47) and (10.48), we obtain

$$C_L = \frac{F_L/Ld}{\rho U_\infty^2/2} = \frac{2\Gamma}{U_\infty d} = 2\pi \frac{v_c}{U_\infty} \quad \text{...(10.49)}$$

Eq. (10.49) reveals that the rotation of a cylinder, which produces circulatory flow around it due to viscous effects, is essential for developing a lift force and shows that the coefficient of lift is a function of v_c/U_∞.

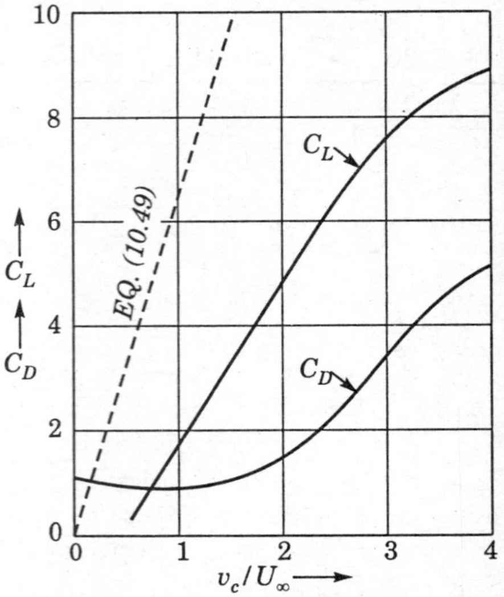

Fig. 10.17. Lift and drag characteristics of a rotating cylinder.

Fig. (10.17) shows variation of lift and drag co-efficients as a function of non-dimensional velocity v_c/U_∞. Eq. (10.49) is shown by the broken line. One can easily read from the figure that with increasing v_c/U_∞ ratio, the lift co-efficient C_L increases much rapidly than the drag coefficient C_D.

Example 10.23. *A cylinder 1.2 m in diameter and 8 m long rotates at 90 rpm with its axis perpendicular to an air stream with a wind velocity of 30 m/s. Assuming no slip condition between the cylinder and the circulatory flow, find*

(i) the magnitude of circulation, *(ii) the transverse or lift force, and*

(iii) the position of stagnation points.

Take specific weight of air as 12.25 N/m³.

Solution. (*i*) Tangential velocity

$$v_c = \frac{\pi dN}{60} = \frac{\pi \times 1.2 \times 90}{60} = 5.65 \text{ m/s.}$$

From Eq. (10.40) $\Gamma = 2\pi r_0 v_c = \pi \times 1.2 \times 5.65$

$= 21.3 \text{ m}^2\text{/s}$

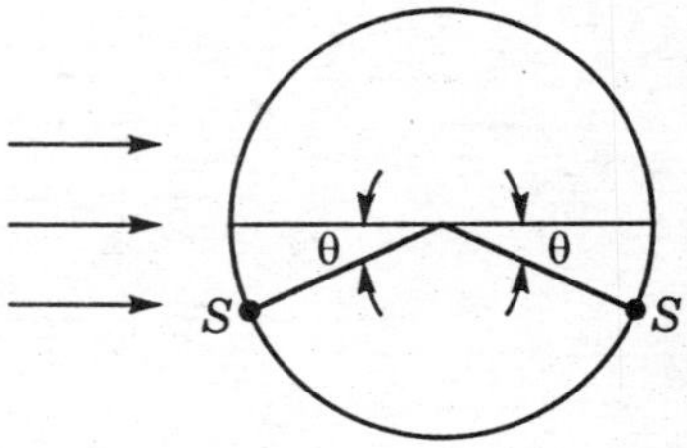

(*ii*) The lift force $F_L = \rho L U_\infty \Gamma = \frac{12.25}{9.81} \times 8 \times 30 \times 21.3$

$= 6383 \text{ N}$

(*iii*) The stagnation points are obtained by putting $v = 0$ in Eq. (10.44), we then have

$$\sin\theta = -\frac{\Gamma}{4\pi r_0 U_\infty} = -\frac{21.3}{4\pi \times \frac{1.2}{2} \times 30} = -0.0942$$

$\therefore$ $\theta = -5.4°$

The stagnation points will be located at an angle of – 5.4° as shown in the above figure.

Example 10.24. *As an application of the Magnus effect, a ship is built having two vertical rotors 10 m high and 3 m in diameter. The rotors are spun at 250 r.p.m. On a day when the air temperature is 20°C and the relative motion of the air to the ship results in 54 kmph wind, calculate the force emitted by the spinning rotors on the ship.* (*ρ of air 1.25 kg/m³*).

Solution. Circulation generated by the rotation of cylinder is given by Eq. (10.43),

$$\Gamma = 2\pi r_0 v_c$$

The circumferential velocity $v_c = \dfrac{2\pi r_0 N}{60} = \dfrac{2\pi \times 1.5 \times 250}{60} = 39.2$ m/s

$$\Gamma = 2\pi \times 1.5 \times 39.2 = 369.0 \text{ m}^2/\text{s}$$

Transverse force (perpendicular to the wind direction) developed by each rotating cylinder is given by Eq. (10.47)

$$F_L = \rho L U_\infty \Gamma$$

The wind velocity $U_\infty = \dfrac{54 \times 1000}{60 \times 60} = 15.0$ m/s

$$F_L = 1.25 \times 10 \times 15.0 \times 369.0 = 69187.5 \text{ N} = 69.2 \text{ kN}$$

Total force emitted by the spinning rotors = 2 × 69.2 = 138.4 kN.

10.14. Lift Characteristics of Airfoils

The most important application of the Kutta-Joukowski law Eq. (10.47) is in the theory of lift which finds applications in aircraft wings (also known as airfoils), blades of propellers, turbines, pumps and so on. The law states that, for a transverse force to be produced a circulation around the body is essential in addition to the velocity of translation. Thus, the circulation around the body is one of the essential requirements for lift force to be developed.

An airfoil, which has a section as shown in Fig. 10.18, is characterized by its chord length C and its dimension normal to the plane of figure known as the span L.

The boundary layer separation near the leading edge is prevented by rounding it and tapering the profile gradually towards the rear as shown. The angle a which the oncoming flow makes with the underside of airfoil or the chord length is called the *angle of attack*. In order to reduce the separation tendencies at moderately high values of α, the profile is given a slight camber.

Whereas in case of a cylinder placed in a real fluid stream, the circulation is caused by its rotation, in the case of an airfoil circulation is produced when a vortex known as the *starting*

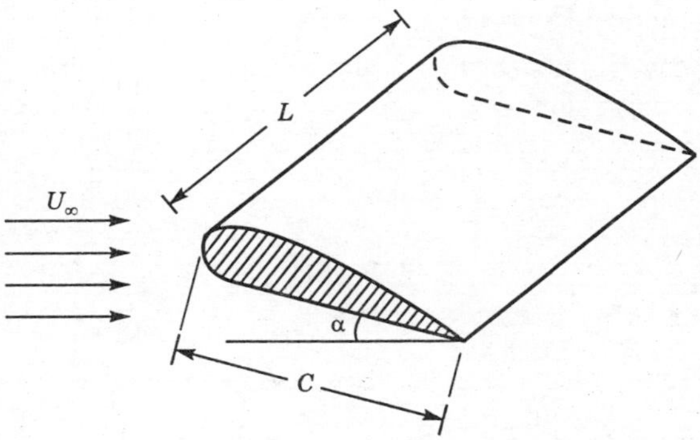

Fig. 10.18. Physical characteristics of an airfoil.

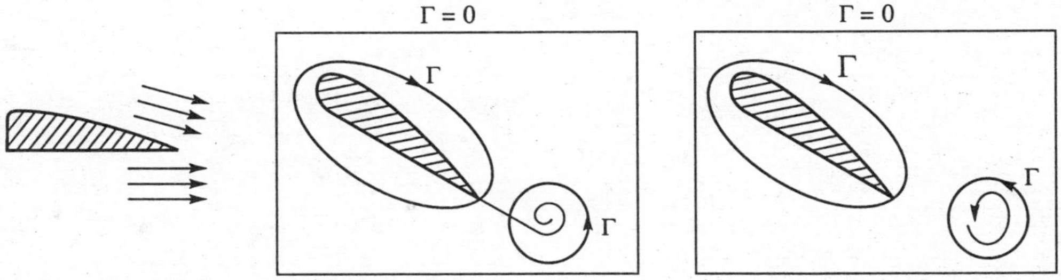

(*a*) Velocity discontinuity at the trailing edge

(*b*) Formation of starting votex

(*c*) Starting vortex breaks off

Fig. 10.19. Development of circulation around an airfoil.

vortex develops at the trailing edge of the airfoil. It is only then a lift force produced. Let us now investigate the conditions which lead to formation of a starting vortex. Before motion (either of the airfoil or of the fluid past it) begins, the circulation around it is obviously zero. As motion starts, the circulation around the airfoil tends to remain zero, but in a real-fluid the separation of flow near its sharp trailing edge cannot be prevented. As a matter of fact, the velocity of flow approaching the trailing edge from the upper side is much lower as compared to that from the bottom side of the airfoil. This results in velocity discontinuity at the trailing edge, see Fig. 10.19 (*a*). This discontinuity in velocity at the trailing edge gives rise to a vortex motion causing a starting vortex form as shown in Fig. 10.19 (*b*).

The formation of starting vortex develops an anticlockwise circulation at the trailing edge as shown. The Kelvin's theorem of constancy of circulation states that "the circulation around a closed curve in the fluid does not change with time". In order to satisfy this theorem, an equal and opposite circulation must be generated around the airfoil so that the net circulation becomes zero. After this circulation has been established, the starting vortex breaks off and is left behind as the airplane moves forward. The starting vortex keeps whirling around until it dies out due to viscous friction, see Fig. 10.19 (*c*).

The condition of flow at the trailing edge depends upon magnitude of circulation. The circulation required to produce a tangential streamline at the trailing edge is given by

$$\Gamma = \pi C U_\infty \sin\alpha \qquad \text{...(10.50)}$$

From Eqs. (10.47) and (10.50) the lift force

$$F_L = \rho L U_\infty (\pi C U_\infty \sin\alpha)$$

$$= 2\pi \sin\alpha\,(LU)\,\rho \frac{U_\infty^2}{2} \qquad \text{...(10.51)}$$

The lift and drag coefficients of an airfoil may be evaluated in terms of the coefficients C_L and C_D defined by the following expressions :

$$F_L = C_L (LC)\,\rho \frac{U_\infty^2}{2} \qquad \text{...(10.52)}$$

$$F_D = C_D (LC)\,\rho \frac{U_\infty^2}{2} \qquad \text{...(10.53)}$$

The coefficients of lift and drag are determined experimentally for any geometry of airfoil using Eqs. (10.52) and (10.53). Fig. 10.20 shows the experimental results for an airfoil of very large span.

From Eqs. (10.51) and (10.52), we obtain

$$C_L = 2\pi \sin\alpha \qquad \text{...(10.54)}$$

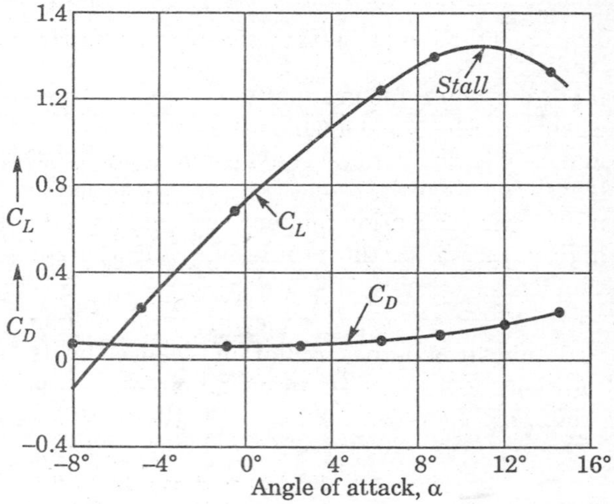

Fig. 10.20. Lift and drag characteristics of a typical airfoil.

The efficiency of an airfoil, which is defined by the ratio C_L/C_D, is much higher as compared to that of a rotating cylinder and as such an airfoil is a far superior lifting device. This explains the use of airfoil sections as airplane wings. At high angles of attack, even the best-shaped airfoil is subjected to separation near the leading edge, and when this happens it is followed by a rapid drop in the lift-drag ratio. This condition is known as the *stall* and is indicated by highest point on the C_L *versus* α curve.

Example 10.25. *A jet plane which weights 29.4 kN and has a wing area of 20 m² flies at a velocity of 950 km/hr when the engine delivers 7350 kW, 65% of the power is used to overcome the drag resistance of the wing. Calculate the coefficients of lift and drag for the wing. The specific weight of the atmospheric air is 11.85 N/m³.*

Solution. $U_\infty = 950 \text{ km/hr} = \dfrac{950 \times 1000}{60 \times 60} = 264 \text{ m/s}.$

Lift force $F_L = C_L A\rho \dfrac{U_\infty^2}{2}$

$\therefore$ $$C_L = \frac{29400}{20 \times \dfrac{11.85}{9.81} \times \dfrac{(264)^2}{2}} = 0.0349$$

Drag force $F_D = 0.65 \times (7350 \times 1000)/264 = 18{,}100 \text{ N} = 18.1 \text{ kN}$

The drag-lift ratio $= \dfrac{F_D}{F_L} = \dfrac{C_D}{C_L}$

or $$C_D = C_L \left(\frac{F_D}{F_L}\right) = 0.0349 \left(\frac{18100}{29400}\right) = \mathbf{0.0215.}$$

10.15. Airfoils of Finite Length—Induced Drag

The consideration of lift has so far been limited to strictly two- dimensional flow produced around an infinitely long airfoil. When the airfoil is of finite length (as is the case with almost all airplane wings), however, there are end conditions which affect both the lift and the drag. Since the pressure on the underside of the airfoil is greater than that on the upper side (on account of wider and closer spacing of streamlines on the bottom and topsides of the airfoil respectively), fluid will escape around the ends of the airfoil causing it to move inwards over airfoil top and outwards over the bottom. The movement of fluid upwards around the ends of the airfoil results in small tip vortices which are cast off from the wing tips, see Fig. 10.21.

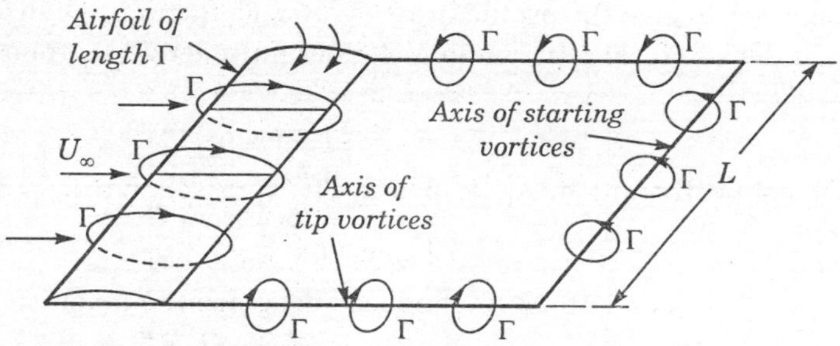

Fig. 10.21. Circulation around an airfoil of finite length.

The Kelvin's theorem of circulation still holds, since the tip vortices are of equal and opposite magnitude. If the circulation were to be computed about a hypothetical path passing through the airfoil and along the axes of the tip and starting vortices as shown in Fig. 10.21, it will still add up to zero. As a matter of fact, the circulation around the airfoil continues to exist, while the tip and starting vortices die out due to viscous resistance.

Since the pressure difference between the top and the bottom of an airfoil must reduce to zero at the tips, it is evident that the lift per unit length of span, being maximum at the centre reduces to zero at the tips. The variation of actual lift force per unit length over the span of an airfoil may be taken as elliptical, but for a rough estimate of lift produced on a wing of finite length the distribution may be assumed straight line, the lift per unit length of span being equal to $\rho U_\infty \Gamma$ everywhere over the span.

In case of an airfoil of finite length, circulation not only occurs around the foil itself but also about the tip, the vortices extending into the wake. The circulation about the tip induces a downward motion in the fluid and in so doing affect both the lift and drag by changing the effective angle of attack. As a result of this induced downward motion, an appreciable downwash is produced by the tip vortices, so that the relative velocity between the air has vertical as well as a horizontal component and hence differs from the absolute velocity of the foil. The vertical velocity is known as the downwash velocity V_i.

An airfoil of finite length is shown at an angle of attack α in the horizontal flow, see Fig. 10.22. The vertical velocity induced near the airfoil by the tip vortices decreases the angle of attack by a small angle α_i, making the effective angle of attack $\alpha - \alpha_i$.

Denoting this angle of attack by α_0

$$\alpha_0 = \alpha - \alpha_i \qquad ...(10.55)$$

Now, treating the airfoil as one of infinite span at an angle of attack α_0, the lift F_{L_0} exerted on such an airfoil is by definition normal to the direction of flow in which it is placed. The lift F_{L_0} is, therefore, normal to the effective velocity V_0 and at an angle α_i with the vertical. The lift F_L on the airfoil of finite length is normal to the approaching horizontal velocity U_∞ and is the vertical component of F_{L_0}. But F_{L_0} also has a component in the direction of velocity U_∞, which is a drag force F_{D_i}. This drag force is called the *induced drag* because its existence depends upon the downward velocity induced by the tip vortices. The induced drag may be represented in the standard form of a drag equation

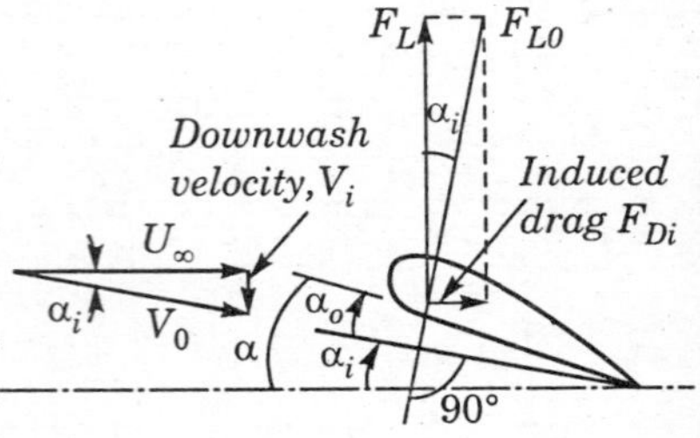

Fig. 10.22. Definition sketch for induced drag.

$$F_{D_i} = C_{D_i} A\rho \frac{U_\infty^2}{2} \qquad ...(10.56)$$

It is now necessary to distinguish between the two and three-dimensional cases of drag. The friction drag (skin friction) and pressure drag discussed earlier in this chapter can be grouped together to give the profile drag F_{D_0} which includes all drag forces acting on the profile of infinite length. The total drag on an airfoil of finite length is then the sum of the profile and the induced drags, thus

$$F_D = F_{D_o} + F_{D_i} \qquad ...(10.57)$$

which, by dividing by $A\rho U_\infty^2/2$, may be expressed in terms of dimensionless drag coefficient as

$$C_D = C_{D_o} + C_{D_i} \qquad ...(10.58)$$

From Fig. 10.22, the induced drag has the magnitude

$$F_{D_i} = F_{L_o} \sin \alpha_i \qquad ...(10.59)$$

Since angle α_i is small, $\cos \alpha_i \approx 1$ and $\sin \alpha_i \approx \alpha_i$ we may, therefore, write

$$F_{D_i} \approx F_{L_o} \alpha_i, F_L \approx F_{L_0},$$

$$U_\infty \approx V_o \cos \alpha_i \approx V_o$$

If the distribution of lift over an airfoil of finite span is taken to be semi-elliptical, it may be shown that

$$\alpha_i = \frac{C_L}{\pi(L/C)} \qquad ...(10.60)$$

in which, L/C, representing the ratio of span to chord, is known as the *aspect ratio*.

With the lift and the drag proportional to their respective coefficients, $F_{D_i} = F_{L_o} \alpha_i$ may be written as

$$\alpha_i = \frac{F_{D_i}}{F_{L_o}} = \frac{C_{D_i}}{C_L} \qquad ...(10.61)$$

Equating Eqs. (10.60) and (10.61), we obtain

$$C_{D_i} = \frac{C_L^2}{\pi(L/C)} \qquad ...(10.62)$$

Eq. (10.62) relates the lift and the induced drag through their dimensionless co-efficients and shows that the induced drag is inversely proportional to the aspect ratio, becoming zero at infinite aspect ratio (infinite span) and increasing as the aspect ratio and the span decrease.

10.16. Polar Diagram

Tests for determining the lift and drag characteristics of an airfoil having a particular aspect ratio are generally conducted in a wind tunnel. The results of such tests may be presented graphically as plots of the lift and drag coefficients *versus* the angle of attack. Since efficiency of airfoil is measured by the ratio of lift to drag, the value of C_L/C_D is also generally plotted. These three curves can be combined into a single curve (suggested by Prandtl) which is known as a *polar diagram*. The co-odinates of polar diagram are the lift and drag coefficients while the angles of attack are represented by different points along the curve. On this diagram (Fig. 10.23) Eq. (10.62) appears as

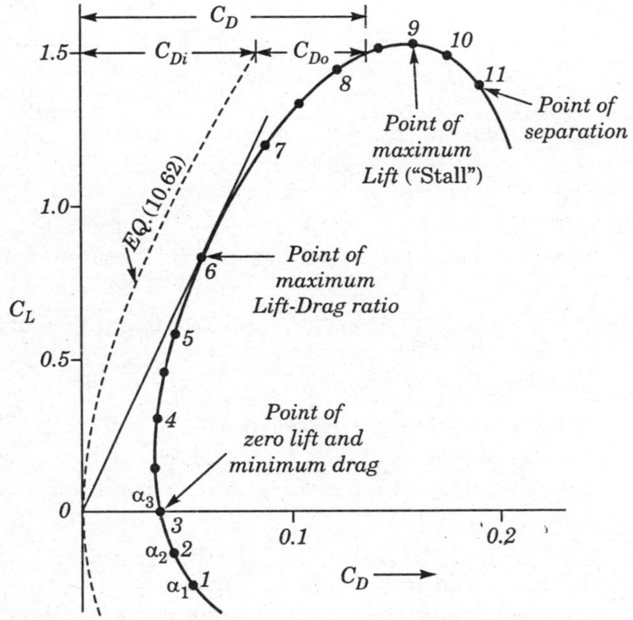

Point number	α
1	− 8°
2	− 6.7°
3	− 5.3°
4	1.5°
5	4.2°
6	6.9°
7	9.8°
8	12.9°
9	16.7°
10	17.3°
11	19.2°

Fig. 10.23. Polar diagram for a typical airfoil of aspect ratio 6.

a parabola passing through the origin and symmetrical about the C_D- axis, the position of the parabola depending upon the aspect ratio. The ratio of lift to drag is the slope of line from the origin to the point on the curve for which this ratio is to be found ; the maximum value of this ratio is the slope of a straight line tangent to the curve and passing through the origin. The diagram also shows the points of zero lift and minimum drag, and the point of maximum lift or *stall*, which determines the stalling angle above which lift no longer increases with angle of attack. From Fig. 10.23, the lift is seen to increase with the angle of attack (the value of α for different points on the curve are shown in the table alongside) upto the point of maximum lift or *stall* indicated by point 9. The end of the upper portion of the curve is point at which the flow separates completely from the upper side of the airfoil, forming a wide wake which increases the profile drag (and, therefore, the drag coefficient). This is accompanied by a large drop in lift and the lift coefficient because of the increased pressure on the upper side of the airfoil.

10.17. Aerodynamic Aspects of Formula One Race Cars

Historical Background

Motor car racing has been considered as one of the three real sports in the world, the other two being mountaineering and bullfighting. The real competition in motor car racing started when

Italian and German dictators Benito Mussolini and Adolf Hitler decided that their nations should be at the forefront of sports. For this reason, major financial back up was given to Italy's Alpha Romeo and Germany's Mercedes-Benz and Auto Union teams. In 1935, Mercedes-Benz developed a car W25 that produced 445 hp (331.5 kW) and was capable of speeds in excess of 280 kmph made possible on account of aerodynamics that made minimal drag possible. Its tyres were so skinny that they looked better suited to a bicycle. Their 1937 version, W125 was capable of producing 646 hp (481 kW) on account of its 5.7 litre, 8 cylinder engine that was fuelled by unrestricted mix of volatile chemical and petrol. The W125 could also reach top speeds in excess of 320 kmph and remained the most powerful Grand Prix car until turbocharged Formula One cars of 1980s.

Following the conclusion of world war in 1946, a new category of Grand Prix racing was created by the name Formula one. It referred to a 'Formula' of rules and regulations drafted to form the premier level of motor racing in the world. The cars remained front engined roadsters until the compact rear-engined version came in 1959. In the early days of Formula One, teams were solely interested in streamlining their cars to minimise the resistance to the flow of air, otherwise known as drag. The less drag a car has, the less power is required to push it through the air and the faster it goes. The advent of corporate sponsorship in 1968 set in motion a series of sweeping changes that gave us Formula One in its modern-day form.

Aerodynamics of Race Cars

Race car performance depends on elements such as the engine, tyres, suspension, road, aerodynamics and of course the skill of the driver. The aerodynamic design of modern race cars has two primary concerns :

(1) The creation of down force to help push the car tyres on to the track, and improve cornering force, and

(2) minimising the drag that gets caused by turbulence and acts to slow down of the car.

Modern Formula One (F1) car can withstand centrifugal forces upto 4g (g = acceleration due to gravity) without sliding off the track. The application of aerodynamics allows for higher cornering speeds (around the curves) than would be possible without down force, and so ensures a better performance besides more safety.

F1 race cars are provided with front and rear wings that are used to generate the down force. Race car wings operate on exactly the same principle as aircraft wings, but its profile is so designed as to develop a downward force. Air flows at different velocities over the two sides of the wings, and this creates difference in pressure in accordance with the Bernoulli's principle. Planes use their wings to create lift ; race cars use theirs to create the downforce.

The primary-wings mounted at front and rear are fitted with different profiles depending on the down force requirement of a particular track. The front wing of a F1 car creates about 25% of total cars downforce and about a third of car's total downforce can come from the rear wing. In modern F1 car, almost as much effort has been spent on reducing the drag as on increasing the downforce. This is mainly achieved by vertical end plates fitted to the wings to prevent vortices forming. The diffuser plates that are mounted low at the back help to re-equalise pressure of the faster-flowing air that passes under the car. If this is not done, it would create a low pressure 'balloon' at the rear end resulting in dragging effect. Illustration shown in Fig. 10.24 depicts various components of a F1 car.

New Modifications in Design of F1 car

In 2009, the International Automobile Federation (FIA) imposed a regulation effecting changes in design of wings to promote overtaking by making it easier for a car to closely follow another. The new rules prescribed lower and wider front wings, simultaneously making rear wings taller and narrower apart from change in specification of tyres. These changes are :

(*i*) **Front wings.** The front wing's height from the ground was reduced to 75 mm from 150 mm, and the width was increased by 400 mm.

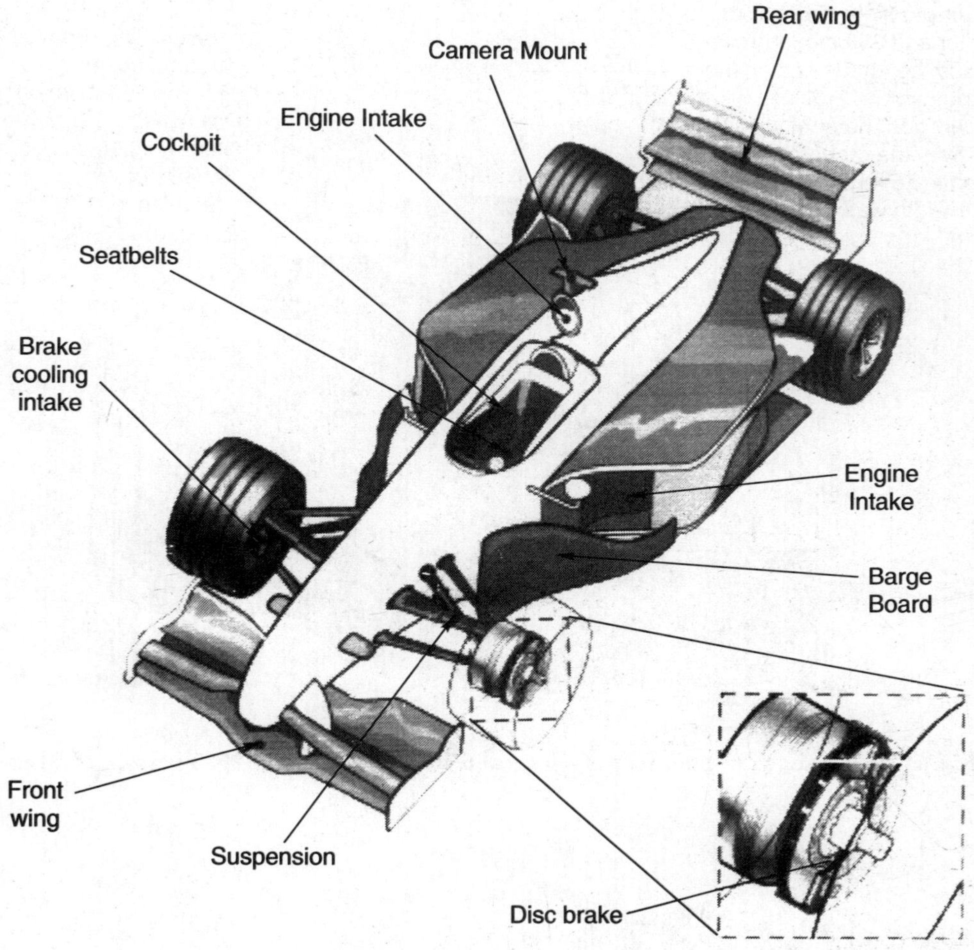

Fig. 10.24. Definition sketch of F1 car (2008 version)

(*ii*) **Rear wings.** These were made 150 mm taller and 250 mm narrower to reduce the effect of turbulence on the following car.

(*iii*) **Tyres.** The grooved tyres that had been in use since 1998 were done away with to give the cars more grip and negate the effects of wing turbulence.

The above stated changes have been schematically shown in Fig. 10.25, featuring the geometry of front and rear wings and grooved/ungrooved surface of tyres.

Perhaps the most interesting change, however, was the introduction of 'moveable aerodynamics' with driver able to make limited adjustments to the front wings from the cockpit during a race. Subsequently in 2011, a new system known as Drag Reduction System (DRS) involving the rear wing was introduced. This system allows a driver within a second of the driver ahead (who can't use it) to close within overtaking range. The system's availability is electronically governed and it can be used at any time in practice and qualifying, but during the race can only be activated when a driver is less than one second behind another car at predetermined points on the track. The system is deactivated once the driver applies the brakes. Figure 10.26 shows 2011 version of F1 car that used DRS.

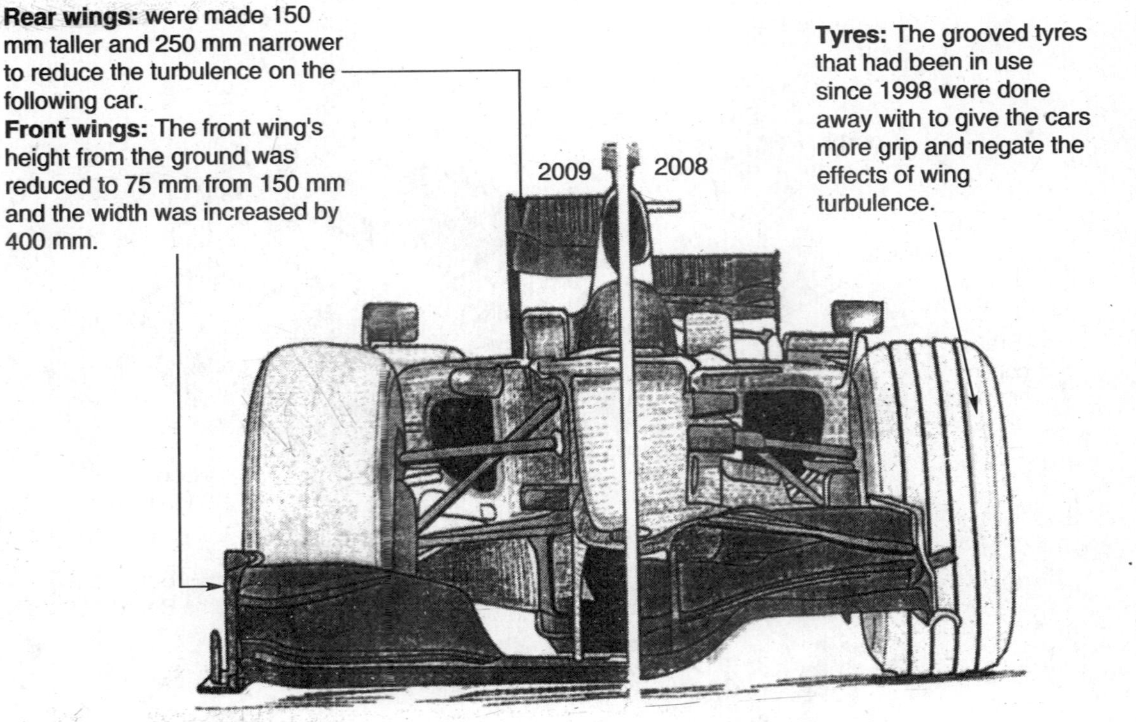

Fig. 10.25. Changes in geometry of wings and surface of tyres in 2009 version of F1 car.

Fig. 10.26. Formula one car (2011 version) with DRS

OBJECTIVE TYPE QUESTIONS

1. An object placed in a fluid stream may experience drag and lift forces. These forces are due to :

(*i*) pressure and gravity
(*ii*) pressure and viscosity
(*iii*) viscosity and turbulence
(*iv*) pressure and turbulence.

2. The drag force experienced by an object is :

(*a*) the horizontal component of the resultant fluid dynamic force acting on the object.
(*b*) the horizontal force due to pressure variation over the surface of object
(*c*) the component of resultant fluid dynamic force in the flow direction
(*d*) the resultant fluid dynamic force acting on the object.

3. The lift force that may act on an object is :

(*i*) the vertical component of the resultant fluid dynamic force acting on the body
(*ii*) the force due to shear stress that acts on the body surface
(*iii*) the buoyant force due to the fluid displaced by the body
(*iv*) the component of resultant fluid dynamic force in a direction normal to the general direction of flow.

4. In order to induce lift on an object it is necessary that :

(*a*) the object shall be symmetrical and the axis of symmetry be aligned parallel to the flow direction
(*b*) the object should be unsymmetrical and the axis should be parallel to the flow direction
(*c*) the object should be so shaped and suitably placed in the fluid stream that there are zones of high and low velocities resulting in a pressure difference between upperside and bottomside of the object
(*d*) the object-shape should be so designed that the pressure distribution over its surface is symmetrical.

5. Total drag experienced by an object consists of friction-drag and pressure-drag. The shape and curvature of its boundary surface affect these constituents. Given below are certain body shapes placed in a fluid stream. Identify the type of drag associated with each and the relative magnitude of the total drag :

(*i*) a circular disc held perpendicular to the flow direction
(*ii*) a sphere
(*iii*) a streamlined body

(*a*) friction-drag predominant, pressure-drag considerably small
(*b*) pressure and friction drags both contribute almost equally
(*c*) pressure drag very large on account of larger wake, friction drag is zero
(*d*) total drag is about 1/3 of that for the disc
(*e*) total drag is about 1/40 of that for the disc.

6. The pressure drag depends upon :

(*i*) the shear stresses generated on the body surface
(*ii*) the boundary layer formation
(*iii*) the separation of boundary layer and the size of wake
(*iv*) the characteristics of the oncoming flow.

7. The friction drag is primarily due to :

(*a*) pressure distribution over the body surface
(*b*) shear stresses generated due to viscous action

(c) weight component in the direction of flow
(d) separation of boundary layer.

8. A streamlined body is one which is so-shaped as to :
 (i) shift the boundary layer separation to the rearmost part thereby considerably reducing the wake-size
 (ii) decrease the friction drag considerably
 (iii) fix the separation points as mush ahead as possible
 (iv) make the streamline pattern symmetrical.

9. The drag coefficient is defined by :

(a) $\dfrac{F_D}{\rho U_\infty^2 A}$ (b) $\dfrac{F_D}{\frac{1}{2}\rho U_\infty^2 A}$

(c) $\dfrac{F_D}{\rho U_\infty^2 A^2}$ (d) $\dfrac{F_D}{\frac{1}{2}\rho U_\infty^2 A^2}$

10. Sedimentation of dust particles in air and silt particles in water represent examples of drag at small Reynolds number ($R < 1.0$). The drag is termed as deformation drag and is equal to :
 (i) $3\pi\mu U_\infty^2 d$ (ii) $\pi\mu U_\infty d$ (iii) $3\pi\mu U_\infty d$ (iv) $\pi\mu U_\infty d^2$

11. The deformation drag comprises friction and pressure drags, the latter having a smaller magnitude. Identify the magnitude of ratios of various combinations of these drags from the table given below :

(a) friction drag/total drag	(i) $\dfrac{1}{3}$
(b) pressure drag/total drag	(ii) 2
(c) friction drag/pressure drag	(iii) $\dfrac{2}{3}$.

12. Bluff bodies are those which have definite separation points fixed by their shapes. Which of the following statements are true for such objects :
 (i) friction drag is greater than the pressure drag
 (ii) total drag is considerably larger as compared to that for streamlined bodies
 (iii) total drag is much less as compared to that for streamlined bodies
 (iv) the friction drag is generally very small and the pressure drag almost entirely constitutes the total drag.

13. Which of the following statements are true for a perfectly streamlined body :
 (a) the pressure drag is larger than the friction drag
 (b) the boundary layer remains thin over the entire surface and the pressure drag is very small.
 (c) the flow separation takes place and a wake region is formed
 (d) the friction drag is small compared to the pressure drag.

14. The boundary layer flow over a sufficiently large flat plate is :
 (i) laminar throughout the entire length
 (ii) turbulent throughout
 (iii) laminar over a short initial length and, therefore, turns turbulent
 (iv) in transition from laminar to turbulent.

15. For the laminar boundary layer, its thickness is expressed by the relationship :

(a) $\delta = \dfrac{5x}{\sqrt{\mathbf{R}}}$ (b) $\delta = \dfrac{0.664x}{\mathbf{R}^{0.20}}$

(c) $\delta = \dfrac{0.664x}{\sqrt{\mathbf{R}_x}}$ (d) $\delta = \dfrac{1.75x}{\sqrt{\mathbf{R}_x}}$.

Where $\mathbf{R}_x = \dfrac{U_\infty x}{\upsilon}$ is the plate Reynolds number.

16. The average drag coefficient for a flat plate on which the boundary layer growth is laminar is given by :

(i) $C_f = \dfrac{5}{\sqrt{\mathbf{R}_L}}$ (ii) $C_f = \dfrac{1.328}{\sqrt{\mathbf{R}_L}}$

(iii) $C_f = \dfrac{1.328}{\sqrt{\mathbf{R}_x}}$ (iv) $C_f = \dfrac{5}{\sqrt{\mathbf{R}_x}}$.

17. The thickness of turbulent boundary layer over a flat plate increases with distance x along the plate more rapidly than in the laminar one. Other parameters remaining the same, it is proportional to :

(a) $x^{4/5}$ (b) $x^{1/2}$ (c) $x^{1/5}$ (d) x

18. The average coefficient of friction drag for turbulent boundary layer is expressed by :

(i) $C_f = \dfrac{1.328}{\sqrt{\mathbf{R}_L}}$ (ii) $C_f = \dfrac{0.074}{\mathbf{R}^{1/5}}$

(iii) $C_f = \dfrac{0.74}{\mathbf{R}_L{}^{4/5}}$ (iv) $C_f = \dfrac{0.379}{\mathbf{R}_L{}^{2/3}}$

19. The comparison between the average of coefficients of friction drag in laminar and turbulent boundary layers results in a ratio of the two coefficients which has the following value :

(a) $\dfrac{15}{\mathbf{R}_L{}^{1/3}}$ (b) $\dfrac{0.045}{\mathbf{R}_L{}^{4/5}}$

(c) $\dfrac{17.94}{\mathbf{R}_L{}^{3/10}}$ (d) $\dfrac{1794}{\mathbf{R}_L{}^{1/5}}$.

20. The average coefficient of friction drag which also accounts for the laminar boundary layer at the front portion of the flat plate is expressed as :

(i) $C_f = \dfrac{0.455}{(\log \mathbf{R}_L)^{2.58}}$ (ii) $C_f = \dfrac{0.074}{\mathbf{R}_L{}^{1/5}}$

(iii) $C_f = \dfrac{0.074}{\mathbf{R}_L{}^{4/5}}$ (iv) $C_f = \dfrac{0.074}{\mathbf{R}_L{}^{1/5}} - \dfrac{1700}{\mathbf{R}_L}$

21. In case of airfoils, the profile drag is one which is caused by :

(a) end effects due to finite length of airfoil
(b) the shape and orientation of airfoil
(c) the circulation induced around the airfoil
(d) the compressibility effects.

22. The drag on a spherical object is dependent on the character of boundary layer. The drag coefficient is greatly reduced when :

(i) the surface is highly smooth

(ii) The Reynolds number has a low magnitude

(iii) the surface roughness is increased as in case of golf balls

(iv) the separation points make an angle of 160° at the centre.

23. Flow around two-dimensional bodies like long circular cylinders in the range $60 < \mathbf{R} < 5 \times 10^3$ has peculiar properties of :

(a) formation of vortices near the top and bottom of cylinder and their alternate shedding

(b) vortices sticking to the surface

(c) vortices do not form at all

(d) the drag coefficient remaining practically constant.

24. Experiments have shown that for $10^3 < \mathbf{R} < 10^5$, vortices are formed and shed regularly and alternately from the opposite sides of a cylindrical body. The frequency of vortex-shedding is :

(i) $5\ U_\infty/d$ (ii) $0.21\ U_\infty/d$

(iii) $0.12\ U_\infty/d$ (iv) $0.15\ U_\infty/d$.

25. When an object travels at supersonic velocity, the fluid and flow properties, significantly change within the region :

(a) just upstream of the object (b) demarcated by the shock-wave front

(c) in the neighbourhood of the object (d) just at the rear of the object.

26. Drag characteristics of objects moving in compressible fluids are influenced by the interaction between the shock-wave and the boundary layer. The drag coefficient at supersonic velocities depends on :

(i) Reynolds and Mach numbers both (ii) Reynolds number only

(iii) Mach number only (iv) Strouhal number

27. Given below are six body-shapes and six values of drag coefficients. Match them correctly :

Object	*Drag coefficient*
(a) Parachute	(i) 0.15
(b) Submarine	(ii) 0.33
(c) Modern car	(iii) 1.33
(d) Tennis ball ($\mathbf{R} = 5 \times 10^5$)	(iv) 0.20

28. The phenomenon of lift produced on a circular cylinder by imposing circulation over a uniform fluid stream is known as the Magnus effect. The lift so produced is caused by :

(i) the shear stresses due to viscous action

(ii) the pressure over the top half being more than that on the bottom half of the cylinder

(iii) the pressure difference between the two-halves, the bottom half-being subjected to a higher pressure

(iv) the symmetrical streamline pattern.

29. The Magnus effect may be used with advantage in games of tennis, cricket, table-tennis and golf. In order to obtain a rise, drop and an oblique curve in the trajectory of the ball, it is to be given a typical spin. Match the trajectory to be obtained with its corresponding spin :

(a) rise in trajectory (i) oblique spin (not in vertical plane)

(b) drop in trajectory (ii) clockwise spin in vertical plane

(c) oblique trajectory (trajectory not in vertical plane) (iii) anti-clockwise spin in vertical plane.

30. In case of airfoils, the circulation around its periphery is necessary for lift to be produced. The circulation is generated when :
(*i*) like a rotating cylinder, airfoil is rotated
(*ii*) a vortex known as the starting vortex is developed at the trailing edge
(*iii*) a high velocity fluid stream flows past the airfoil
(*iv*) the airfoil is set at a suitable angle of attack.

31. An airfoil is considered a far superior lifting device on account of its much higher efficiency. The efficiency of an airfoil is measured by :
(*a*) C_L (*b*) $C_L C_D$
(*c*) circulation, Γ (*d*) angle of attack, α.

32. A streamlined body is defined as a body about which :
(*a*) the flow is laminar (*b*) the flow is along streamlines
(*c*) the flow separation is suppressed (*d*) the drag is zero.

ANSWERS

1. (*ii*) **2.** (*c*) **3.** (*iv*) **4.** (*c*) **5.** (*i*), (*c*) ; (*ii*) (*b*), (*d*) ; (*iii*) (*a*), (*e*) **6.** (*iii*) **7.** (*b*) **8.** (*i*) **9.** (*b*) **10.** (*iii*)
11. (*a*), (*iii*) ; (*b*), (*i*) ; (*c*), (*ii*) **12.** (*ii*), (*iv*) **13.** (*b*) **14.** (*iii*) **15.** (*a*)
16. (*ii*) **17.** (*a*) **18.** (*ii*) **19.** (*c*) **20.** (*iv*) **21.** (*b*)
22. (*iii*) **23.** (*a*) **24.** (*ii*) **25.** (*b*) **26.** (*iii*)
27. (*a*), (*iii*) ; (*b*), (*i*) ; (*c*), (*ii*), (*d*), (*iv*)
28. (*iii*) **29.** (*a*), (*ii*) ; (*b*), (*iii*) ; (*c*), (*i*) **30.** (*ii*), (*iv*) **31.** (*b*) **32.** (*c*).

REVIEW QUESTIONS

1. "No drag can be exerted upon a body by steady, irrotational flow" ; comment on the correctness of this statement. Why is it not true if the flow is unsteady ?

2. Distinguish between the friction drag, the pressure drag and the profile drag.

3. Which type of drag predominates in the motion of the following bodies ;
(*i*) a blunt body ; (*ii*) a streamlined body ;
(*iii*) a tennis ball ; (*iv*) an arrow ; and
(*v*) a parachute ?

4. Explain why the drag coefficient and hence the drag of a sphere in subsonic flow is suddenly decreased as the flow velocities are increased beyond a certain point.

5. A golf player seeks your advice in regard to the advantage of using a rough golf ball over a new one which is relatively smooth. How would you explain to him that the rougher a ball is the faster it goes ?

6. A sphere and an elongated streamlined object of the same cross-sectional area and mass are dropped in a very viscous liquid. Which one would fall more rapidly and why ?

7. Why is the drag of a hollow hemisphere greater when held with the concave side upstream than when held with the convex side upstream ?

8. Why is the streamlining of automobiles of less practical importance than that of subsonic airplanes ?

9. What is the cause of vibration of telephone wires in a strong wind blowing across the wire ? Under what circumstances could such vibration lead to rupture ?

10. How does the drag coefficient change with (*i*) surface roughness, (*ii*) turbulence level ? Explain why golf balls are designed to have dented surface ?

11. Under what conditions will roughening of the surface of a body (*a*) increase, (*b*) decrease, its drag coefficient ?

12. How does the boundary layer thickness for flow over a flat plate vary with distance from the leading edge for (*i*) laminar flow, (*ii*) turbulent flow ?
13. How does the compressibility of fluid affect the drag characteristics of an object ?
14. Comment on the shape of the body front and its rear for minimum drag in (*i*) subsonic flow, (*ii*) supersonic flow ?
15. What is Magnus effect ? Give examples of practical occurrence which illustrate the use of principle embodied in the Magnus effect.
16. In case of an ideal fluid, the rotation of cylinder will not generate circulation around it because of non-existence of shear stresses. Will then a lift force be developed if a rotating cylinder is placed in an ideal fluid stream ?
17. What are the practical disadvantages of the Flettner rotor ?
18. Explain the principle of kite.
19. Why do the airplanes take off and land against the wind ?
20. The efficiency of a wing may be measured in terms of maximum lift-drag ratio. At what point in an airplane's flight is a high drag (and hence a low efficiency) of great importance ?
21. An airplane is considered to be supported by a lifting force exerted upon the wing by the air. Using Newton's third law of motion (principle of action and reaction), can it be proved that the weight of the plane is actually transmitted to the surface of the earth ?
22. Explain what is induced drag, and what factors influence it ?
23. What is polar diagram and what useful information can be obtained from it ?
24. What is the "stalling angle" and what happens when the angle of attack is increased further?
25. In a game of tennis, a ball is served so that a rotation added to its motion of translation. If the rotation is about a horizontal axis, (perpendicular to the motion), discuss its effect on the flight of the ball.
26. A cricket ball has a seam around its diameter, which enables it to "swing" during its motion. Illustrate how the ball swings, when it is bowled with the seam held at an angle to the direction of bowling.
27. Explain how, under certain speed of hurling a rough sphere travels farther than a smoother one.
28. Describe with the help of a sketch, the variation of drag coefficient for a cylinder over a wide range of Reynolds number.
29. Describe with the help of sketches the variation of drag on a sphere over a wide range of Reynolds number.
30. How does circulation originate around an airfoil ?
31. Give reasons for the following observed phenomena :

    (*a*) Telephone wires produce a humming noise (singing) in wind.

    (*b*) A spinning table tennis ball curves in flight.

    (*c*) Air planes generally take off and land against wind.

    (*d*) Drag on a submerged body in an ideal fluid under steady flow conditions is zero whereas in real fluid flow the drag force is significant.

    (*e*) A pitted golf ball moves faster than a smooth ball of similar size and weight.

    (*f*) For the same given area, a square plate experiences a less drag than a rectangular one when placed normal to the flow.

    (*g*) High winds tend to lift roofs of buildings than to press them down.

    (*h*) When two fully submerged long cylinders are freely suspended in a plane normal to the flow and are separated by a small gap, the cylinders alternately come together and separate.

PROBLEMS

10.1. A thin plate 0.5 m wide and 1.0 m long is suspended and exposed parallel to air flowing with a velocity of 2.0 m/s. Calculate the drag force on the plate :

(*i*) when the 0.5 m side is parallel to the stream

(*ii*) when the 1.0 m side is parallel to the stream.

Assume laminar boundary layer and take $C_D = \dfrac{81.32}{\sqrt{R_L}}$. Account for the difference between the drags under the two conditions. The kinematic viscosity of air as 1.49×10^{-5} m²/sec and the specific weight of air as 11.95 N/m³.

10.2. The sails of a rotor ship consist of a pair of vertical cylinders 3 m in diameter and 20 m long rotating at a speed of 300 r.p.m. Determine the force on cylinders for a wind blowing crosswise at 50 km/hr relative to the cylinders. Take specific weight of air as 11.95 N/m³.

10.3. An automobile perfectly streamlined has as projected area of 2.4 m² and travels in air (20°C) at a maximum speed of 120 km/hr. What is the power required to overcome the air frictional resistance ? Take $C_D = 0.30$ and ρ for air = 1.235 kg/m³.

10.4. A thin plate 90 cm wide and 2.1 m long is dragged lengthwise through water at a velocity of 1.5 m/s. If the kinematic viscosity of water is 0.315 stokes, calculate the total drag on the plate.

10.5. A plate of size 1 m by 1.2 m moves at 12 m/s in still air at an angle of 12° with the horizontal. Using a coefficient of drag $C_D = 0.17$ and a coefficient of lift $C_L = 0.72$, determine :

(*i*) the resultant force exerted by air on the plate ;

(*ii*) the frictional force ; and

(*iii*) the power required to keep the plate moving. The specific weight of the air is 12.15 N/m³.

10.6. In investigating the possibility of using rotors in place of airplane wings, it is assumed that each of the two rotating cylinders would have a diameter of 1 m and length 5 m. If the weight of the entire plane is 44.1 kN determine the speed of rotation of rotors which will support drag. Take specific weight of air as 12.15 N/m³.

10.7. If a guide vane having the form of a symmetrical foil has a chord of 1 m, estimate from the equations of circulation theory the side thrust per metre length which would be exerted upon the vane by water travelling with a velocity of 4 m/s at an angle of 5° to the vane axis assume lift efficiency of 70%.

10.8. Estimate the maximum load which may be carried by a 5 m diameter parachute without attaining a velocity of descent of more than 6 m/s. Take $C_D = 1.33$ for a hemisphere hollow upstream. The specific gravity of air is 12.15 N/m³.

10.9. If a plate 1.6 m square is towed through water at a constant velocity of 1 m/s, it is found to attain a maximum drag of 1175 N when held at right angles to the direction of motion. What fraction of this drag would be encountered, when the plate is towed edgewise at the same velocity ? Take $\nu = 93 \times 10^{-3}$ stokes and $\rho = 1000$ kg/m³.

10.10. The screen of an open air theatre is 12 m long and 6 m high. What is the overturning moment about the base of the screen when a wind velocity of 100 km/hr strikes the screen normally. The drag coefficient under the prevailing flow conditions is 1.18 and the specific weight of air is 11.95 N/m³.

10.11. A parachute is in the form of a hemispherical shell. It is to be used for dropping a small object of weight 882 N so that it touches the earth at velocity not greater than 6 m/s. Calculate the diameter of the parachute, if the drag coefficient for a hemispherical shell with its concave side upstream is approximately 1.32 for $R > 10^3$. The specific weight of air is 11.96 N/m³.

10.12. A sign for advertising is mounted on the top of a delivery truck and is fixed near the front of the truck. The signboard is 1.5 m wide and 0.75 m high. If the drag coefficient for the signboard is 1.4, calculate the net force acting on it when the truck travel at 70 km/hr. Take specific weight of air as 11.85 N/m^2.

10.13. A steel ball of spherical shapes having diameter of 50 mm is dropped in a deep reservoir of water calculate its velocity of fall. The specific gravity of the steel ball is 8 and the drag coefficient may be taken as 0.40. Take kinematic viscosity of water as 10^{-6} m^2/s.
What will be the terminal velocity if the ball is dropped through the atmospheric air ? Kinematic viscosity of air may be taken is 15×10^{-6} m^2/s. The drag coefficient at Reynolds numbers of 3×10^5 and 10^6 may by taken as 0.10 and 0.20 respectively.

ANSWERS

10.1. 0.628×10^{-2} N, 0.432×10^{-2} N
10.2. 120 kN
10.3. 16.45 kW
10.4. 13.64 kN
10.5. 79.3 N at angle of 76.70° with horizontal.
10.6. 1082 r.p.m., 917 kW per rotor.
10.7. 3.06 kN
10.8. 593 N
10.9. 0.0063
10.10. 119.5 kN-m.
10.11. 6.23 m.
10.12. 397 N
10.13. 3.4 m/s, 210 m/s.

11

Analysis of Pipe Flow

11.0. Introduction

The knowledge and understanding of the principles of fluid flow discussed in the proceeding chapters find their applications in the analysis of pipe flow problems. Pipes of different shapes and sizes are used in industries for transporting fluids ; the air-conditioning (heating and ventilating) systems use pipes of various shapes ; the pipelines transporting oil and gas supply these fluids to consumers thousands kilometres away from the supply well. The municipal water supply involves extensive use of piping systems both large and small in size. These are some of the examples where pipes are used for hydraulic transmission of fluids.

A satisfactory analysis of pipe flow problems depends on the accuracy involved in estimating the dissipation of energy incurred in maintaining the flow. This requires the knowledge of boundary layer theory dealt with in Chapter 7, and of turbulence discussed in Chapter 9. The following treatment relates to homogeneous fluid of constant viscosity and constant density. The results are also applicable to gases provided that the density changes are small. Pipes in which the liquid does not occupy its entire cross-section (for example, sewers and culverts) have a free-surface and are in effect open channels. The open channel flow is treated separately in Chapter 12.

11.1. Establishment of Flow

The expressions for head loss considered in Chapters 8 and 9 are all applicable only to the *fully developed* flow obtaining some distance downstream of the pipe entrance or from the source of disturbance within the flow. The length of pipe from its entrance (along which the velocity profile varies from section to section) upto the point where the flow attains fully developed velocity profile and beyond which it remains unaltered for the remaining length downstream (provided that the pipe is straight and uniform) is known as the *entrance length*, see Fig. 11.1.

In case of a pipe leading from a reservoir, if the depth of liquid is large all particles entering it will start moving with practically the same velocity (as at *A* in Fig. 11.1). As soon as the fluid comes in contact with the pipe wall, the boundary layer begins to form. The retarding effect of the wall,

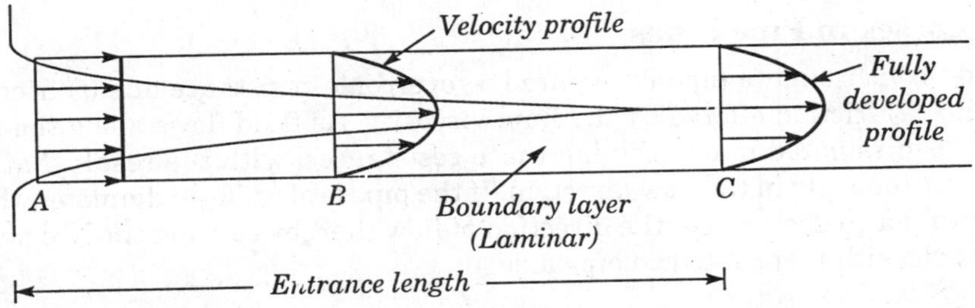

(*a*) Laminar flow

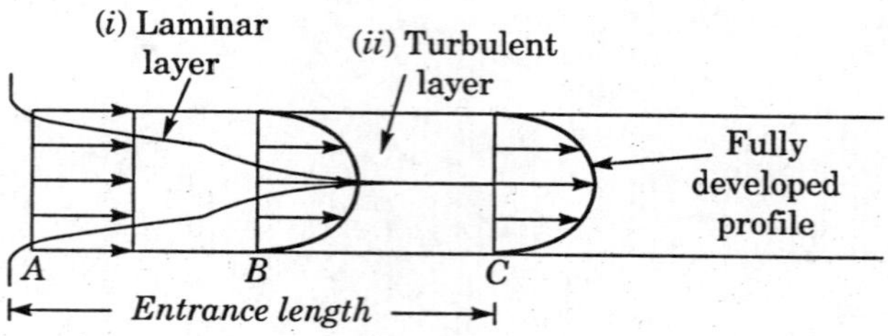

(b) Turbulent flow.

Fig. 11.1. Growth of boundary layer in a pipe.

where the velocity must be zero, at once operates so that the fluid near the wall is slowed down. In order to satisfy the continuity principle, the velocity of fluid near the pipe axis is increased to compensate for the reduction in velocity near the wall.

When the thickness of boundary layer becomes equal to pipe radius, the effect of retardation of flow reaches the central region of flow. Fig. 11.1 illustrates the growth of boundary layer in a pipe for laminar and turbulent flows. In case of laminar flow, the boundary layer starts as a laminar one and remains laminar throughout its length. The boundary layers from opposite sides meet at the pipe axis at a section downstream of which the flow is fully developed laminar flow, see Fig. 11.1 (a).

If the Reynolds number is high so that the flow in the pipe is turbulent, the boundary layer which starts as a laminar layer, a little distance downstream the transition occurs and the boundary layer becomes turbulent. This turbulent boundary layer increases in thickness much more rapidly with the result that the layers from opposite sides meet at the pipe axis much earlier. Downstream of this section the flow is fully developed turbulent flow. The entrance length for turbulent flow in the pipe is, therefore, less than that for laminar flow on account of rapid increase in the thickness of turbulent boundary layer.

The entrance length required to establish fully developed laminar flow, according to Langhaar, is given by

$$\frac{L_e}{d} = 0.057\ \mathbf{R} \qquad \text{...(11.1)}$$

where **R** is Reynold number Vd/ν and L_e the entrance length. For **R** = 2000, which may be taken as the maximum limit upto which laminar flow can exist in commercial pipes, the entrance length from Eq. (11.1) is 114 times the pipe diameter.

For turbulent flow, the fully developed state is reached earlier, and is given by

$$\frac{L_e}{d} = 0.7\ (\mathbf{R})^{1/4} \qquad \text{...(11.2)}$$

Experimental data have shown that the entrance length predicted by Eq. (11.2) is too short if part of the boundary layer is laminar. If, however, the entry is sharp edged or there are other factors producing turbulence at the inlet, the entrance length is reduced. For turbulent flow in smooth pipes, the entrance length is generally taken as 50 times the pipe diameter.

11.2. Energy Losses in Pipe Lines

The fluid flow through a pipe is required to overcome resistance due (i) friction offered by the pipe wall and (ii) friction offered by the fluid viscosity. As fluid flows along the pipe, a part of the fluid energy is dissipated in overcoming these resistances, with the result that the energy of fluid decreases continuously in the flow direction. If the pipe is of uniform diameter, the piezometric head $(Z + p/\gamma)$ would also decrease in the direction of flow thereby causing the hydraulic grade line to slope downwards with respect to the pipe axis.

With fluid flow and the pipe characteristics known (that is, ρ, μ, V, d and the pipe roughness k_s are known), the friction factor f of the Darcy-Weisbach equation., $h_L = f\frac{L}{d}\frac{V^2}{2g}$, can be estimated from f *versus* **R** relationship for known value of relative roughness k_s/d. The energy loss in terms of head can then be determined from the Darcy-Weisbach equation.

11.3. Minor Losses in Pipe Lines

The minor losses or local losses are those which are caused by change in pipe cross-section, presence of bends, valves and fittings of all kinds. Although in long pipelines, the losses due to the local disturbances caused by these fittings etc. are of minor importance and can often be neglected, they may however outweigh the friction losses, in a short pipe. These losses invariably, arise from sudden changes in velocity either in magnitude or in direction. The magnitude of these losses is proportional to the abruptness of the velocity change. The non-uniformity in flow thus caused generates large-scale turbulence in which energy is dissipated as heat.

The source of loss is usually confined to a very short length of pipe, but the turbulence produced may persist for a considerable distance downstream. For the analysis, it is assumed that the effects of pipe friction and of the additional large-scale turbulence can be separated and that the additional loss is concentrated at the device causing it. The total head lost in a pipe may then be calculated as the sum of the friction loss for the length of pipe under consideration and the additional minor losses.

A theoretical determination of the minor losses is seldom possible except for the loss due to sudden enlargement. Since the losses have been experimentally found to vary approximately as the square of the mean velocity, they are normally expressed in the form

$$h_L = K_L \frac{V^2}{2g} \qquad \text{...(11.3)}$$

in which K_L is known as the *loss coefficient*. For a given geometry, the value if K_L is practically constant at high Reynolds number. The magnitude of the loss coefficient is experimentally determined and is governed primarily by the shape of the obstruction or pipe fitting.

11.3.1. Loss Due to Sudden Enlargement

The loss of energy incurred when the flow encounters a sudden enlargement can be subjected to analysis. Fig. 11.2 illustrates two pipes of different sizes meeting at section (1) and causing sudden enlargement of pipe section from diameter d_1 to d_2. The fluid emerging from the smaller pipe being unable to take a sharp turn at the corners follows a path shown by the streamlines. Consequently, the pockets of turbulent eddies form in the corners and result in dissipation off energy in the form of heat. The lost hydraulic energy is thus converted into the thermal energy (raising slightly the fluid temperature) which is in turn lost to the surrounding medium.

The conditions at a section of sudden enlargement are shown in Fig. 11.2. There is a rise in pressure from sections (1) to (2) at the expense of velocity and in accordance with the Bernoulli principle, but this rise is not so large as it would have been had the loss of energy h_L not taken place. Excessive turbulence is generated between sections (1) and (2). Downstream of section (2), the outer streamlines merge with the pipe boundary. Section (2) is located downstream of section (1) at a distance of about 8 times the larger diameter. Let the pressure of flow just before entering the large pipe be p_1 and the average velocity V_1. Let the pressure in the annular space of area $\pi(d_2^2 - d_1^2)/4$ occupied by the eddying fluid at section (1) be p'. For simplicity the common axis of the pipes is assumed horizontal. If the pressure and average velocity at section (2) are p_2 and V_2 respectively, from the principle of continuity

$$Q = A_1V_1 = A_2V_2$$

$$= \frac{\pi}{4} d_1^2 V_1 = \frac{\pi}{4} d_2^2 V_2$$

or $$\frac{V_1}{V_2} = \left\{\frac{d_2}{d_1}\right\}^2 \qquad ...(11.4)$$

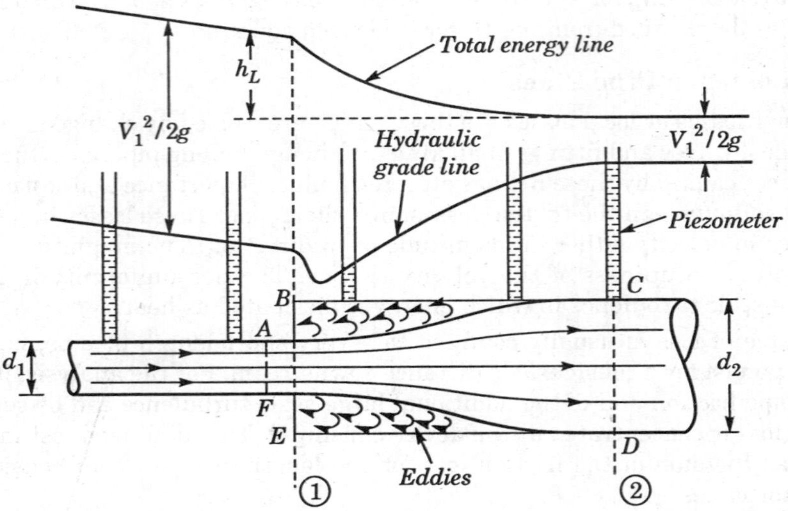

Fig. 11.2. Loss of energy due to sudden enlargement.

If the loss of energy in terms of head due to sudden enlargement is h_L, the Bernoulli's equation between sections (1) and (2) gives (assuming the datum heads to be the same) :

$$Z + \frac{p_1}{\gamma} + \frac{V_1^2}{2g} = Z + \frac{p_2}{\gamma} + \frac{V_2^2}{2g} + h_L \qquad ...(11.5)$$

An additional relationship between the pressures and the velocities may be obtained by applying momentum principle to the control volume *ABCDEF*. The shear force on pipe wall of short length between sections (1) and (2) being small has been neglected. The fact that the radial accelerations over the annular space *AB* are very small, we assume that the mean pressure p' of the eddying fluid is almost equal to the pressure p_1. The momentum equation applied to the control volume gives

$$\rho\, Q(V_2 - V_1) = p_1A_1 + p'(A_2 - A_1) - p_2A_2$$

$$= (p_1 - p_2)A_2 \qquad ...(11.6)$$

Therefore, $$p_1 - p_2 = \frac{\rho Q}{A_2}\,(V_2 - V_1)$$

$$= \rho V_2\,(V_2 - V_1)$$

Substituting this value of $(p_1 - p_2)$ into Eq. (11.5), the loss of energy head

$$h_L = \frac{p_1 - p_2}{\gamma} + \frac{V_1^2 - V_2^2}{2g}$$

$$= \frac{V_2(V_2 - V_1)}{g} + \frac{V_1^2 - V_2^2}{2g} = \frac{(V_1 - V_2)^2}{2g} \qquad ...(11.7)$$

From the continuity equation $V_1A_1 = V_2A_2$, Eq (11.7) may be alternatively written as

$$h_L = \frac{V_1^2}{2g}\left(1 - \frac{A_1}{A_2}\right)^2$$

$$= \frac{V_2^2}{2g}\left(\frac{A_2}{A_1} - 1\right)^2 \qquad ...(11.8)$$

Eq. (11.8) was first obtained by J.C. Borda (1733—99) and L. Carnot (1753—1823) and is frequently known as the Borda- Carnot head loss.

Exit loss. When a pipe discharges into a large reservoir as shown in Fig. 11.3, area $A_2 \to \infty$ and the loss at an abrupt enlargement tends to $V_1^2/2g$. The velocity head in the pipe ($V_1^2/2g$) which corresponds to the kinetic energy per unit weight is lost in turbulence of eddies in the reservoir. This loss is usually termed as the exit loss for the pipe.

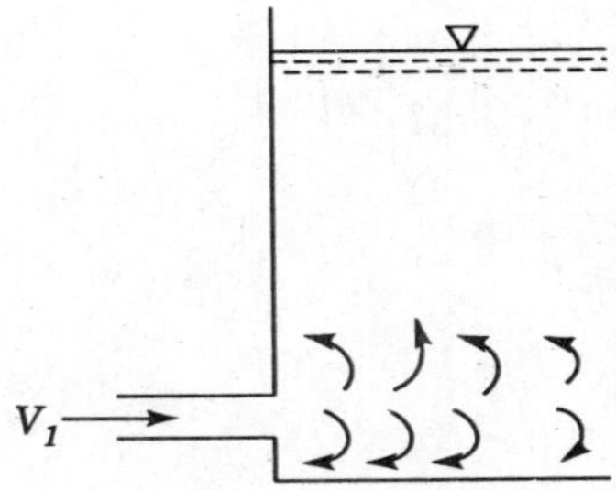

Fig. 11.3. Exit loss.

Example 11.1. *At a sudden enlargement of water line from a diameter of 240 mm to 480 mm, the hydraulic grade line rises by 10 mm. Estimate the quantity flowing in the line.*

Solution. Referring to Fig. 11.2 and applying the Bernoulli's equation to sections (1) and (2)

$$Z_1 + \frac{p_1}{\gamma} + \frac{V_1^2}{2g} = Z_2 + \frac{p_2}{\gamma} + \frac{V_2^2}{2g} + \frac{(V_1 - V_2)^2}{2g} \qquad ...(A)$$

in which the last term accounts for the head loss to sudden enlargement. The difference in the piezometric heads at sections (2) and (1) is 10 mm and, therefore,

$$\left(Z_2 + \frac{p_2}{\gamma}\right) - \left(Z_1 + \frac{p_1}{\gamma}\right) = \frac{1}{100} \text{ m}$$

From continuity relationship,

$$Q = A_1 V_1 = A_2 V_2$$

$$V_2 = \frac{d_1^2}{d_2^2} V_1 = \left(\frac{24}{48}\right)^2 V_1 = 0.25 V_1.$$

Substituting the above quantities in Eq. (*A*),

$$\frac{V_1^2}{2g} = \frac{1}{100} + \frac{(0.25V_1)^2}{2g} + \frac{(V_1 - 0.25V_1)^2}{2g}$$

From which, $V_1 = 0.723$ m/s

and the flowrate, $Q = \frac{\pi}{4} \times (0.24)^2 \times 0.723 = 0.0327 \text{ m}^3/\text{s} = 32.7$ lit/s.

Example 11.2. *For sudden expansions in pipe flow, what is the optimum ratio between the diameter of the pipe before expansion and the diameter of pipe after expansion so that the pressure rise may be maximum ?*

Solution. Referring to Fig. 11.2, and using Eqs. (11.5) and (11.7) one can write,

$$\frac{p_1}{\gamma} + \frac{V_1^2}{2g} = \frac{p_2}{\gamma} + \frac{V_2^2}{2g} + \frac{(V_1 - V_2)^2}{2g}$$

The pressure rise : $$\frac{(p_2 - p_1)}{\gamma} = \frac{V_1^2}{2g} - \frac{V_2^2}{2g} - \frac{(V_1 - V_2)^2}{2g}$$

From the continuity equation,

$$Q = A_1V_1 = A_2V_2$$

$$V_2 = \frac{A_1}{A_2} V_1$$

$$V_2 = \frac{d_1^2}{d_2^2} V_1$$

$$\frac{\Delta p}{\gamma} = \frac{p_2 - p_1}{\gamma} = \frac{V_1^2}{2g}\left[1 - \frac{d_1^4}{d_2^4} - \left(1 - \frac{d_1^2}{d_2^2}\right)^2\right]$$

$$= \left[2\left(\frac{d_1}{d_2}\right)^2 - 2\left(\frac{d_1}{d_2}\right)^4\right]\frac{V_1^2}{2g}$$

$$\Delta p = \rho\,\frac{V_1^2}{2}\left[2\left(\frac{d_1}{d_2}\right)^2 - 2\left(\frac{d_1}{d_2}\right)^4\right]$$

For maximum pressure rise,

$$\frac{d(\Delta p)}{d(d_1/d_2)} = 0$$

$$\frac{d(\Delta p)}{d\left(\frac{d_1}{d_2}\right)} = \left[4\left(\frac{d_1}{d_2}\right) - 8\left(\frac{d_1}{d_2}\right)^3\right] = 0$$

From which $\left(\frac{d_1}{d_2}\right)^2 = \frac{1}{2}$ and $\frac{d_1}{d_2} = \frac{1}{\sqrt{2}}$

The diameter ratio for the maximum pressure rise,

$$\frac{d_1}{d_2} = \frac{1}{\sqrt{2}}$$

Maximum pressure rise,

$$(\Delta p)_{max} = \rho\,\frac{V_1^2}{2}\left[2 \times \frac{1}{2} - 2 \times \left(\frac{1}{\sqrt{2}}\right)^4\right] = 0.5\,\rho\,\frac{V_1^2}{2}.$$

Example 11.3. *Oil of specific gravity 0.8 flows in a 80 mm diameter pipeline. A sudden expansion takes place into a second pipeline of such diameter that the maximum pressure rise is obtained. If the rate of flow through the pipeline is 12.5 lit/sec, find : (i) the loss of energy in the sudden expansion in centimetres of oil, and (ii) the differential gauge length h indicated on an oil-manometer connected between the two pipes.*

Solution. From Ex. 11.2, the diameter ratio for maximum pressure rise,

$$\frac{d_1}{d_2} = \frac{1}{\sqrt{2}} \quad \text{or} \quad d_2 = \sqrt{2}\,d_1$$

$$V_1 = \frac{4Q}{\pi d_1^2} = \frac{4 \times 12.5 \times 10^{-3}}{\pi \times (8 \times 10^{-2})^2} = 2.49 \text{ m/s}$$

$$V_2 = \frac{4Q}{\pi d_2^2} = \frac{4Q}{\pi\left(\sqrt{2}d_1\right)^2} = 1.245 \text{ m/s}$$

Head loss in sudden expansion,

$$h_L = \frac{(V_1 - V_2)^2}{2g} = \frac{(2.49 - 1.245)^2}{2 \times 9.81}$$

$$= 0.079 \text{ m of oil} = 79 \text{ mm of oil.}$$

From Eq. (11.5), the pressure rise is obtained as

$$\frac{p_2 - p_1}{\gamma} = \frac{V_1^2}{2g} - \frac{V_2^2}{2g} - h_L$$

$$= \frac{(2.49)^2}{2 \times 9.81} - \frac{(1.245)^2}{2 \times 9.81} - 0.079$$

$$= 0.158 \text{ m of oil} \qquad ...(i)$$

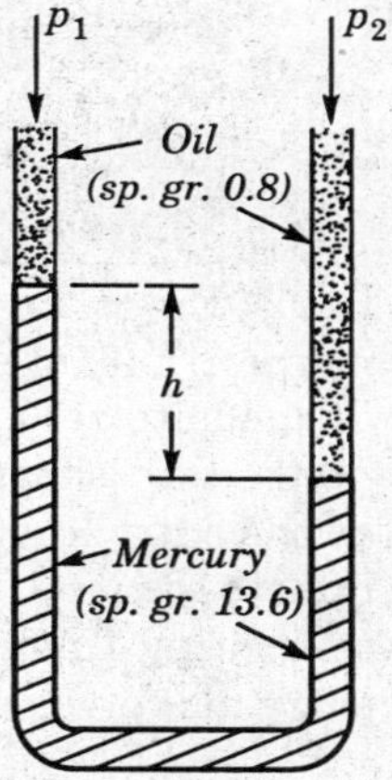

The manometric equation for oil-mercury differential manometer may be written as

$$\frac{p_1}{\gamma} + \frac{13.6}{0.8} h = \frac{p_2}{\gamma} + h$$

$$\frac{p_2 - p_1}{\gamma} = 16.0\, h \text{ m of oil} \qquad ...(ii)$$

Equating (*i*) and (*ii*)

$$16.0\, h = 0.158$$

$$\mathbf{h = 9.86 \text{ mm.}}$$

11.3.2 Loss Due to Sudden Contraction

If the flow in the pipe system of Fig. 11.2 is reversed, the flow from the large pipe would enter the pipe with smaller diameter causing the flow area to be suddenly reduced from A_1 to A_2, see Fig. 11.4. The momentum equation to the control volume bounded by the pipe wall and sections (1) and (2) cannot be applied because just upstream of the contraction the curvature of streamlines and the

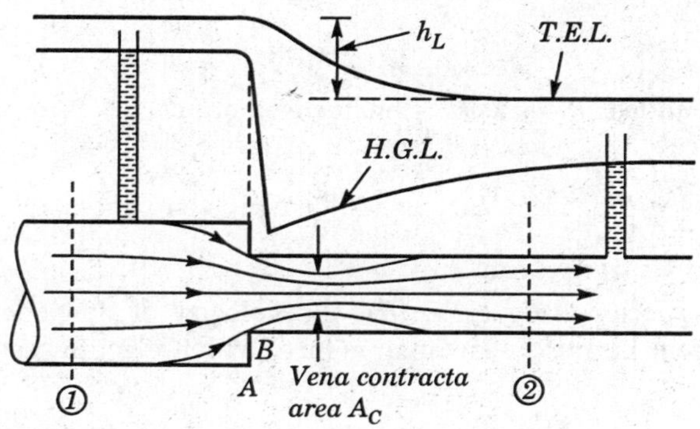

Fig. 11.4. Loss due to sudden contraction.

acceleration of fluid cause the pressure at the annular face AB to vary in an unknown manner ; immediately downstream of the junction a vena contracta is formed, after which the fluid stream widens to fill the pipe.

Eddies are formed between the pipe wall and the vena contracta, and it is these eddies which cause practically all the dissipation of energy. Between the vena contracta and the downstream section (2), where the flow has again become practically uniform, the flow pattern is similar to that which occurs at a sudden enlargement and so the head loss is accordingly given by Eq. (11.8).

$$h_L = \frac{V_2^2}{2g}\left(\frac{A_2}{A_c} - 1\right)^2 = \frac{V_2^2}{2g}\left(\frac{1}{C_c} - 1\right)^2 \qquad ...(11.9)$$

whereas A_c represents the area of cross-section at the vena contracta and the C_c is the coefficient of contraction defined by $C_c = A_c/A_2$.

Although the area A_1 is not explicitly involved in Eq. (11.9), the value of C_c is dependent upon the ratio A_2/A_1. For circular pipes and fairly high Reynolds numbers, Table 11.1 gives values of loss coefficient K_L defined by Eq. (11.3), for different values of diameter ratio d_2/d_1.

Table 11.1

d_2/d_1	0	0.2	0.4	0.6	0.8	1.0
K_L	0.5	0.45	0.38	0.28	0.14	0

Entry Loss. The flow from a large reservoir into a sharp-edged pipe (provided the pipe does not protrude into the reservoir) gives rise to a situation in which $A_1 \to \infty$ and the diameter ratio d_2/d_1 tends to zero. The loss coefficient K_L corresponding, to this condition is 0.5. Thus for a non-protruding, sharp-edged pipe entrance shown in Fig. 11.5 (*b*), the head loss of 0.5 $V_2^2/2g$ is known as the entry loss. A protruding pipe, as shown in Fig. 11.5 (*c*) causes a greater loss of head. If the inlet to the pipe is well rounded, as shown in Fig. 11.5 (*a*) the fluid can follow the boundary without separating from the pipe wall, and entry loss is considerably reduced. A gradually converging entry also gives a much lower loss than the sudden contraction.

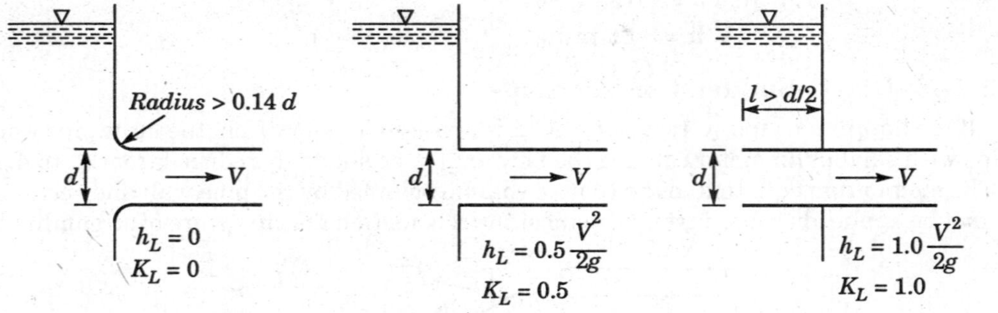

(*a*) Well rounded entrance. (*b*) Sharp-edged or square entrance. (*c*) Protruding or re-entrant entrance.

Fig. 11.5. Various kinds of pipe entrances.

11.3.2.1. Gradual Contraction

In a gradual contraction, [Fig. 11.5 (*d*)] the velocity of flow increases and the pressure decreases. The flow is from higher to lower pressure, which eliminates the causes of turbulence following formation of eddies. The loss of energy in a gradual contraction (also called a reducer) is due to friction mainly, and the resistance (energy loss) offered by a reducer is always less than that of a similar diffuser (refer article 11.3.3.1).

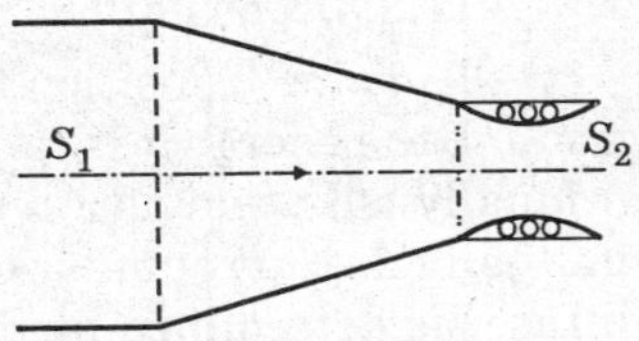

Fig. 11.5. (*d*) Gradual contraction of a pipe.

11.3.2.2. Nozzle

A slight turbulence, separation of the stream from the wall and contraction of the jet takes place at the reducer-outlet. To eliminate this and the resulting losses, the cone should be made to merge smoothly into the straight pipe that follows it, or a curved reducer can be used instead of a conic frustum to connect the two straight pipes, as illustrated in Fig. 11.5 (*e*).

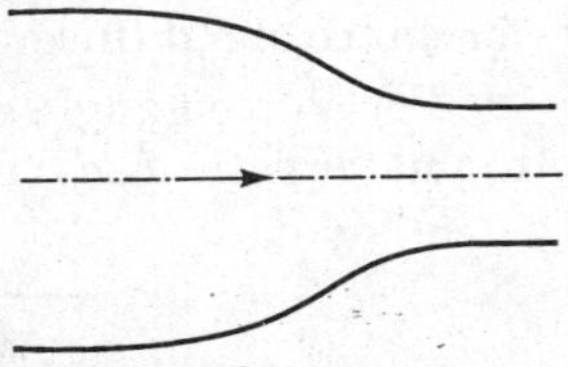

Fig. 11.5. (*e*) Nozzle.

In case of a curved contraction called a nozzle, the rate of contraction can be increased considerably, with a corresponding shortening of the reducer and reduction of energy losses. The loss coefficient, K_L, for such a curved contraction *i.e.* nozzle varies from 0.03 to 0.1, depending on the rate of contraction, geometry of curve and the Reynolds number (higher Reynolds number corresponds to lower value of the loss coefficient and vice versa).

Example 11.4. *A pipe 150 mm diameter, 15 metres long is connected to the bottom of a tank, 15 metres long by 12 metres wide. The original head over the open end of the pipe is 5 metres. Find the time of emptying the tank, assuming the entrance to the pipe is sharp-edged.*

Assume $f = 0.01$ *in* $h_L = \dfrac{flV^2}{2gd}$.

Solution. Writing the Bernoulli's equation between the liquid surface at height h and the lower end B of the pipe, considering the entrance and the friction losses, one gets

$$h + 15 = 0.5\frac{V^2}{2g} + \frac{V^2}{2g} + \frac{0.01 \times 15 \times V^2}{2g \times 0.15} = 2.5\frac{V^2}{2g}$$

$$V = \sqrt{\frac{2(h+15)}{2.5}}$$

Considering that the liquid surface falls a distance dh in time dt,

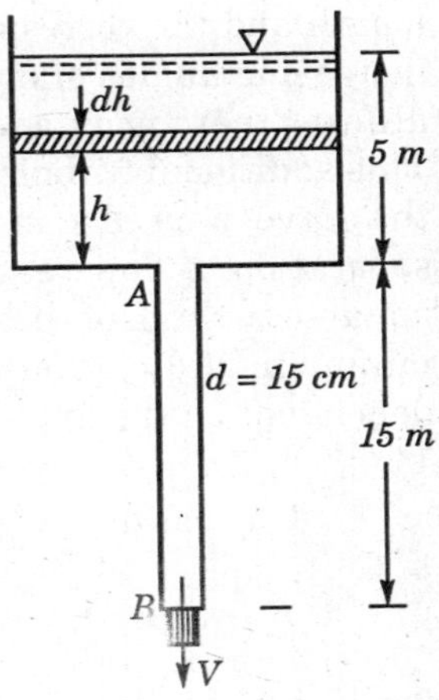

$$12 \times 15 \times (-dh) = Q.dt$$

$$= \frac{\pi}{4} \times (0.15)^2 \times \sqrt{\frac{2g(h+15)}{2.5}}\ .dt$$

$$= 0.0495\sqrt{(h+15)}\ dt$$

$$\int_0^T dt = -\frac{180}{0.0495}\int_5^0 \frac{dh}{\sqrt{h+15}}$$

Let $h + 15 = H$, then

$$T = -3635\int_{20}^{15} \frac{dh}{\sqrt{H}} = -3635 \times 2\left|\sqrt{H}\right|_{20}^{15}$$

$$= 4365 \text{ seconds} = 1.21 \text{ hours.}$$

11.3.3. Loss of Energy in Gradual Expansion

The pipe transition from cross-sectional area A_1 to A_2 instead of being abrupt, if it is made gradual, the energy loss can be reduced considerably. The device which reduces the velocity of fluid gradually is known as a *diffuser*. It is a tapering pipe the cross-sectional area of which increases gradually. It thus eliminates greatly the eddies responsible for dissipation of energy. The length of diffuser is determined by the angle of divergence and the ratio of the two areas. The loss of head which occurs in a diffuser depends on the angle of divergence θ and on the ratio of upstream and downstream areas. Because of the practical applications of diffuser in such cases at the draft tube of a reaction turbine, diffuser passage for a centrifugal pump, etc., the factors which influence the performance of a diffuser are described below :

For a given angle of divergence, the loss in a diffuser must increase as the area ratio A_2/A_1 (or diameter ratio d_2/d_1) increases owing to greater length of the diffuser.

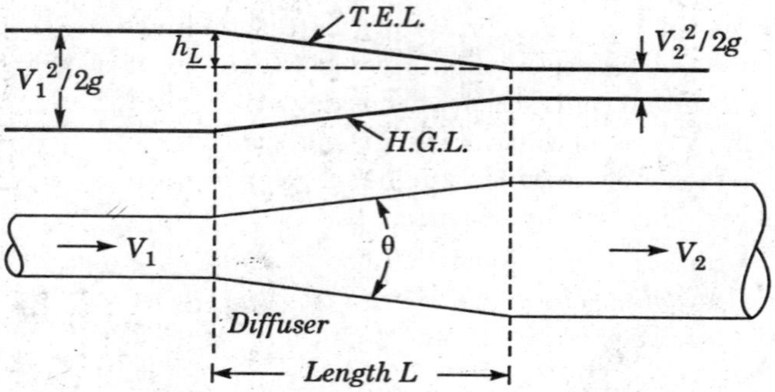

Fig. 11.6. Loss of energy in a gradual enlargement.

For a given ratio of A_2/A_1, the length of diffuser and head loss due to it will vary with the angle of divergence θ. The total loss of energy for flow through a diffuser can be considered as made up of two parts. One is the ordinary pipe friction loss, which may be expressed by

$$h_L = \int \frac{f}{d} \frac{V^2}{2g} \, dL \qquad \text{...(11.10)}$$

In order to integrate Eq. (11.10), it is necessary to express the variables *f*, *d* and *V* in terms of *L*. For the present purpose, it would suffice to note that the friction loss increases with the length of diffuser. Hence for a given ratio of areas, the larger the angle of divergence the lesser will be its length and the corresponding pipe friction will also be less. The decrease in the pipe friction for increasing angles of divergence is indicated by curve *F* in Fig. 11.7 (*a*). In case of flow through a diffuser there is an additional turbulence loss set up by the eddies formed in the separated flow. This additional turbulence loss will naturally increase with the angle of divergence, as is shown by the curve marked *T* in Fig. 11.7 (*a*) and if the angle of divergence θ is large enough, there may be separation of flow resulting in large scale eddy formation. The total loss in the diffuser is then equal to the sum of these two losses, represented by the curve marked *K*. The total loss is seen to be minimum at θ = 6° for a very smooth surface. The total loss coefficient, K_L, can also be seen to be dependent on the angle of divergence θ, and attains a minimum value of about 0.15 at θ = 6° [Fig. 11.7 (*a*)].

The loss due to gradual enlargement is expressed by

$$h_L = K_L \frac{(V_1 - V_2)^2}{2g} = K_L \left(1 - \frac{A_1}{A_2}\right)^2 \frac{V_1^2}{2g} \qquad \text{...(11.11)}$$

in which K_L is a loss coefficient dependent upon the shape of the enlargement. Based on experiments on conical enlargements the loss coefficient for different cone angles θ and area ratios A_2/A_1 are shown in Fig. 11.7 (*b*). The loss coefficient K_L is primarily dependent upon the cone angle θ but is also a

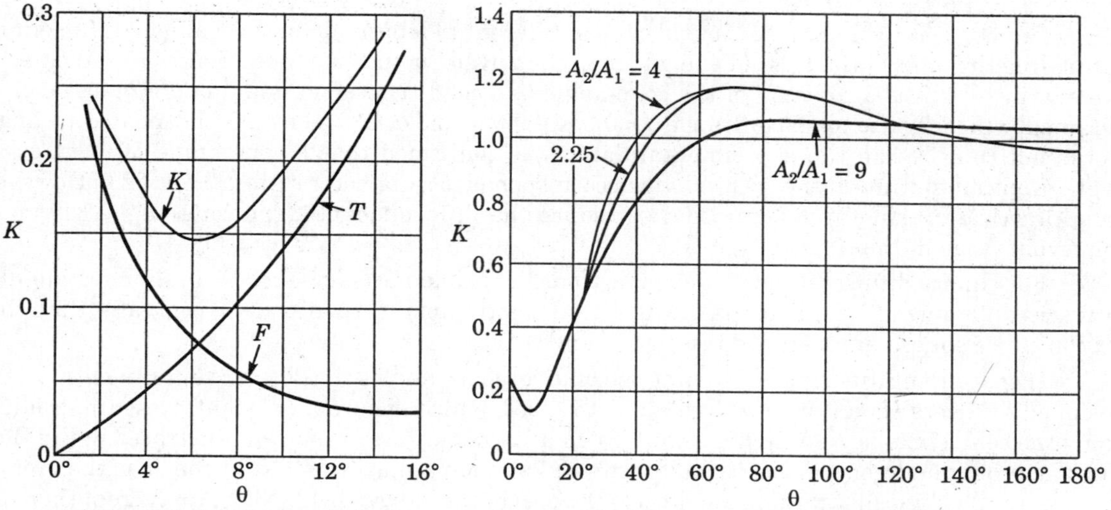

(*a*) Pipe friction loss, turbulence loss and total loss coefficients *versus* angle of divergence (cone angle),

(*b*) Total loss coefficient (K_L) *versus* angle of divergence (cone angle).

Fig. 11.7. Loss coefficient for conical enlargements.

function of the area ratio for $\theta > 20°$ as shown in figure. From Fig. 11.7 (*a*) and (*b*) it may be observed that (1) there is an optimum cone angle of about 6° where the combination of the effects of pipe friction and the additional turbulence due to eddies is a minimum, and (2) it is better to use a sudden enlargement in place of diffuser when cone angle is around 60°, since K_L is smaller for the former.

If the pipe surface is rougher, the value of the pipe friction F will be larger. This not only increases the value of coefficient K_L but also shifts the angle for minimum loss to around 8°. Thus the best angle of divergence increases with the roughness of the surface.

Gradual enlargements or diffusers of various forms are widely used in engineering practice for pressure recovery, that is for causing pressure rise in the direction of flow. The flow through a gradual contraction characterizing converging flow where pressure is being converted into kinetic energy represents a stable and an efficient process. Diverging flow, which occurs in a diffuser where most of the kinetic energy is transformed back into pressure is inherently unstable subject to eddies and separation resulting in excessive turbulence and is, therefore, less efficient.

11.3.3.1. Diffusers

A gradually expanding pipe is commonly known as a diffuser. In a diffuser, the velocity decreases gradually and the pressure increases in accordance with the Bernoulli's principle. The kinetic energy of fluid particles enables them to move against the growing pressure. The reduction in kinetic energy is along the diffuser and from the centre line to the boundary. The energy of fluid particles at the boundary becomes so small that the increased pressure may stop the flow or even reverse it. As a result, eddies form and the flow separates from the wall. Figure 11.7 (*A*) illustrates the phenomenon of eddy formation.

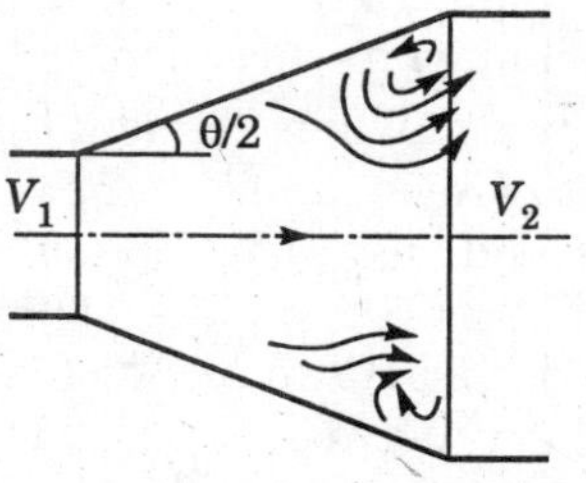

Fig. 11.7 (*A*). Eddy formation in a diffuser.

The greater the angle of a diffuser θ, the more pronounced is the formation of eddies, and the flow separates from the wall (boundary) resulting in greater loss of energy due to turbulence generated by eddies. Apart from this, there exist friction losses like those in pipes of uniform cross-section. The total loss of energy is therefore the sum of the above stated losses *i.e.* due to turbulence caused by eddies and the friction loss.

11.3.3.2. Industrial Diffusers

From theoretical analysis, considering a differential length along the wall of a diffuser and expressing the differential loss of head due to friction by using the basic equation (Darcy-Weibach), the energy, which is a function of the cone angle θ, a condition for minimum energy loss can be obtained, which for friction factor range, $f = 0.015 - 0.025$ and $d_2/d_1 = 1.4 - 2.0$, the mean cone angle works out to 6°. From the experimental data, it is confirmed that a cone angle of 6° gives an average optimum value of energy loss, which corresponds to a diameter ratio (d_2/d_1) of 2.0. However, from practical considerations, in order to reduce the diffuser length, the cone angle θ is made somewhat large in the range of 7° – 9°.

For square diffusers, the value of θ is commonly taken as same, *i.e.* 7 – 9°. For rectangular diffusers diverging in one plane (flat diffusers), the optimum divergence angle is greater than that for conical or square diffusers, and is in the range of 10 – 12°.

If space limitations do not permit use of divergence angles in the optimum range, specially designed diffusers should be used with $\theta > 15 - 25°$, which ensure a constant pressure gradient along the centreline, *i.e.* dp/dx = constant. Such a curved diffuser is shown illustrated in Fig. 11.7 (*B*). The reduction in energy loss in such diffusers as compared with straight cones, is the greater for larger cone angle θ, attaining a value of 40% at θ in the range of 40 – 60°. Apart from this, in a curved diffuser the flow is more stable.

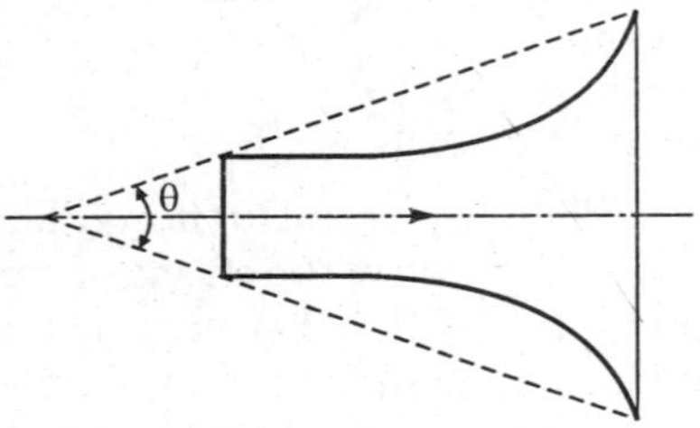

Fig. 11.7 (*B*). Diffuser with uniform pressure gradient

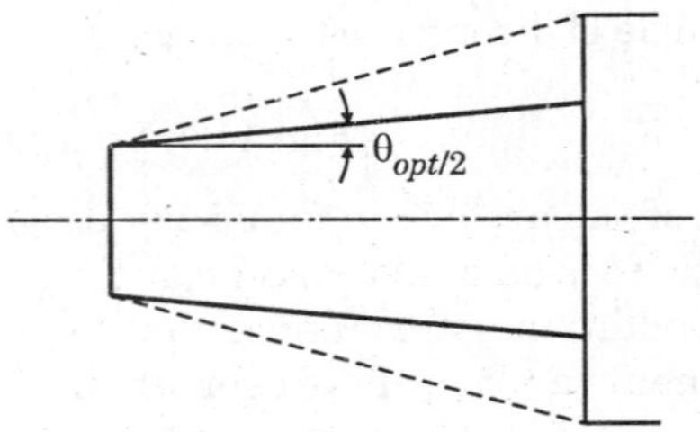

Fig. 11.7 (*C*). Stepped diffuser.

Very satisfactory results are also obtained by using stepped diffusers, consisting of an ordinary optimum angle diffuser followed by an abrupt expansion, as shown in Fig. 11.7 (*C*). The abrupt expansion does not cause high energy losses as the velocity drops considerably by the time the flow reaches the sudden enlargement. The total energy loss (resistance) incurred in such a diffuser is much less than that of an ordinary diffuser of the same length and rate of divergence.

11.3.4. Loss of Energy in Bends

When flow in a pipe encounters a bend, it is the direction of velocity which undergoes a change. The outer wall of the bend must exert an additional pressure against the flow in order to change its direction.

When a fluid flows in a curved path three must be a force acting radially inwards on the fluid to provide the inward acceleration. There is thus an increase of pressure near the outer wall of the bend, see Fig. 11.8 (*a*), starting from section (1) and rising to a maximum at 0. Simultaneously there is a reduction of pressure near the inner wall giving a minimum pressure at *I* and a subsequent rise from *I* to section (2). The adverse pressure gradient (that is, rise in pressure in the direction of flow) between section (1) and 0 in the outer wall and between *I* and section (2) in the inner wall leads to separation of flow from the boundary unless the radius of curvature of the bend is very large. The separation of flow results in loss of energy due to turbulence which mainly depends upon the relative radius of curvature, that is the ratio of the radius of bend to the pipe diameter.

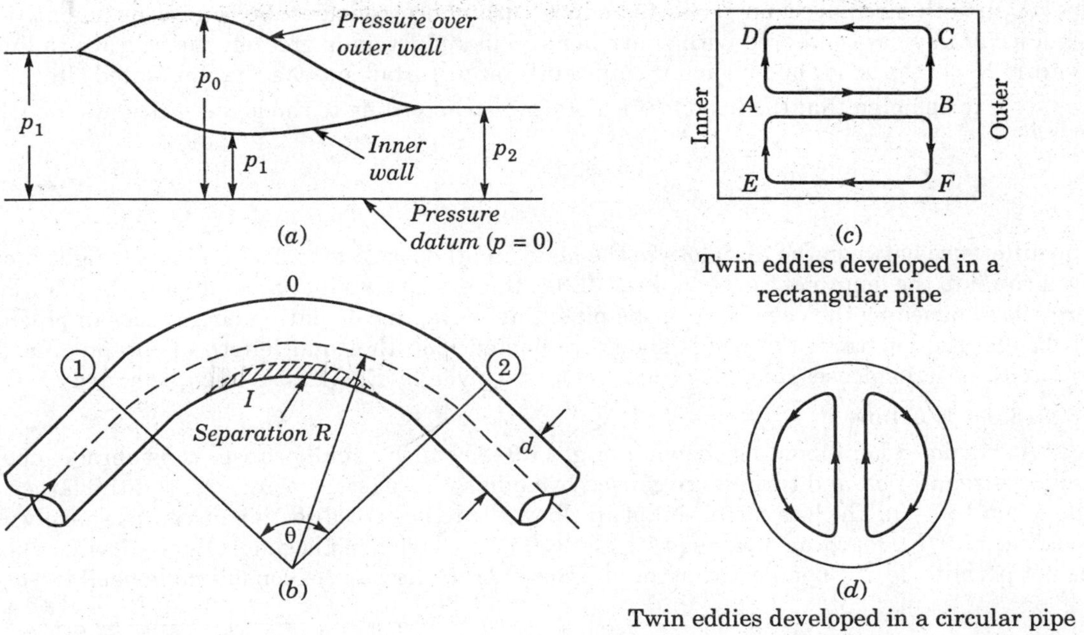

Fig. 11.8. Flow through a pipe bend.

There are also losses which arise from twin-eddy secondary flow. Considering flow in a rectangular pipe, Fig. 11.8 (*c*), the fluid adjacent to the upper and lower walls has its velocity reduced due to viscous action in the boundary layer. As a result the increase in pressure from inner to outer radius is less as compared to the pressure rise along the central line *AB*. Since the pressure at *B* is greater than at *C* and *F* and pressure at *A* is less than that at *D* and *E*, twin eddies are produced resulting in a secondary flow in the radial direction as indicated in the figure. Similar twin-eddy secondary flow is produced in circular pipe bends. This secondary flow combines with the main axial flow to produce double spiral flow which persists for some distance downstream (between 50 to 75 times the pipe diameter) until it is dissipated by the viscous friction. In fact more than half of the loss produced by a bend in the straight pipe following it is due to spiralling motion of fluid.

For rectangular pipes, the minimum resistance is obtained in a bend in which the ratio of the sides of the rectangle is of the order of two or slightly more, say 2.5 ; the larger side being parallel to the axis of curvature of the bend. Such an optimum cross-sectional shape used in a bend can reduce the energy loss to about 0.4 of an equivalent circular cross-section with identical values of R/d and θ. In some special cases, when losses must be reduced as much as possible, as in air ducts of some aircraft engines, such cross-sectional shapes are preferably used.

The loss of head in smooth pipe bends is thus caused by the combined effects of separation, wall friction and the twin-eddy secondary flow. For bends of large radius of curvature the last two effects will predominate whereas for small redius of curvature, separation and secondary flow will be more significant. The loss of energy in a pipe bend for turbulent flow depends upon the total angle of the bend θ and the relative radius of curvature R/d. The loss coefficient K_L is mainly a function of θ and R/d, it varying slightly with the Reynolds number usually encountered in pipe flow. The loss coefficient increases with the surface roughness.

A smooth circular bend tends to reduce turbulence considerably and consequently the resistance to flow, as compared with sharp bend. This reduction is greater, the relative radius of curvature R/d of the bend ; but when it is sufficiently large, turbulence may be eliminated altogether.

It is to be noted that the head loss is determined according to the loss coefficients, Eq. [11.3 (*A*)], *i.e.*,

$$h_L = (K_L)_{bend} \frac{V^2}{2g} \quad \text{...(11.3}A\text{)}$$

is the difference between the total loss in the bend and the loss due to friction in a straight pipe of length equal to the length of the bend. Eq. (11.3*A*) therefore takes into account only the additional energy loss caused by the curvature of the pipe. Therefore, in calculating energy loss in pipelines with bends, the length of the bends should be included in the total length of the pipe used to compute the friction losses, to which the bend losses given by Eq. (11.3*A*) must be added.

13.3.4.1. Elbow

A sharp bend or elbow, as shown in Fig. 11.9 (*A*) which usually causes considerable energy loss due to separation and turbulence caused by eddies, these losses being greater, the larger the angle α ($\alpha = 180 - \theta$). The loss coefficient of an elbow of circular cross-section increases sharply with α (see Fig. 11.9 (*b*), reaching a value of 1.1 at $\alpha = 90°$. A sharp bend is generally used, where space does not permit use of a bend of large radius. However, on account of considerable head losses in

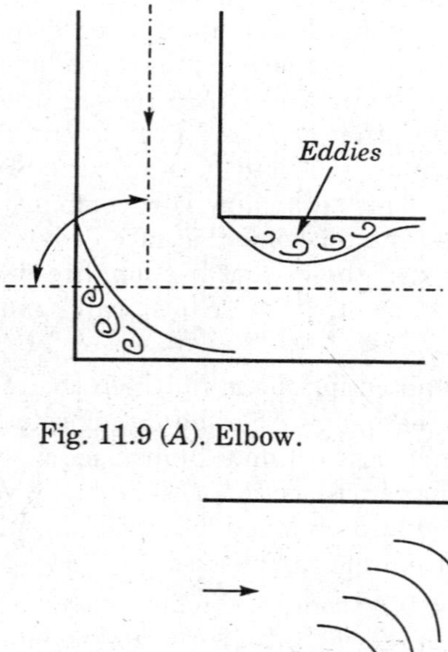

Fig. 11.9 (*A*). Elbow.

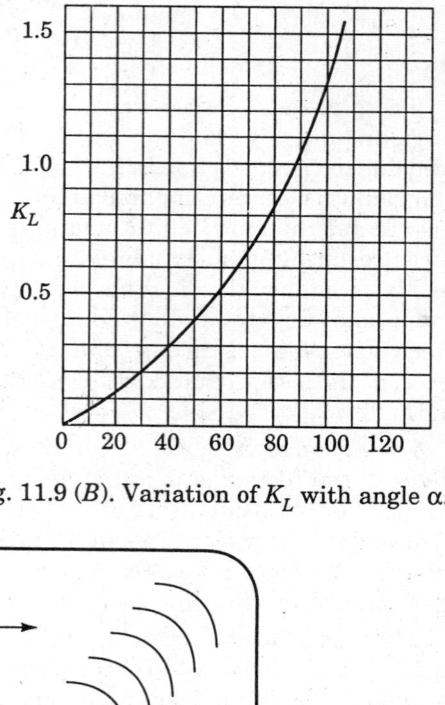

Fig. 11.9 (*B*). Variation of K_L with angle α.

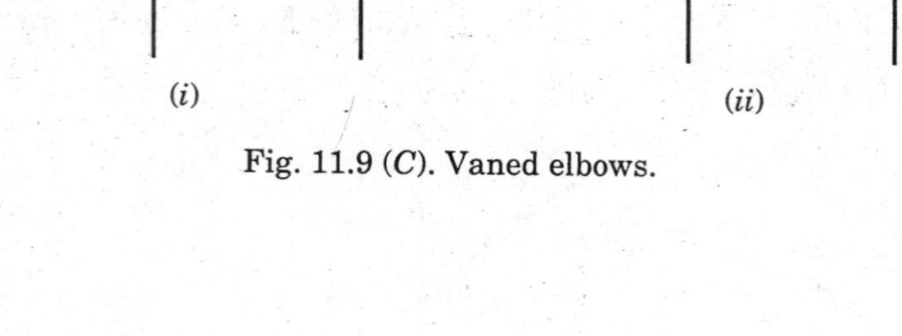

Fig. 11.9 (*C*). Vaned elbows.

unsounded elbows, they are not usually employed in pipelines. As a remedy, if a series of well designed curved guide vanes is installed (in large sized elbows), much of the separation and secondary flow, which are responsible for energy loss, is greatly reduced. An example of this can be found in a closed-circuit wind and water tunnels, to reduce the resistance of the large-sized elbow, a set of guide vanes is mounted in the bend, see Fig. 11.9 (*c*).

In the case of simple non-profiled vanes of circular elbow, see Fig. 11.9 (*c*)(*i*), the loss coefficient of the elbow drops to $K_L = 0.4$;

whereas with profiled vanes, Fig. 11.9(*c*)(*ii*), the loss coefficient can be reduced to a value of 0.25 or even 0.20.

11.3.5. Loss of Energy in Pipe Fittings

The loss of energy caused by commercial pipe fittings, such as valves, elbows, bends, etc., occur because of their rough and irregular interior surfaces which produce excessive large-scale turbulence. The loss of energy in commercial pipe fittings is generally expressed as already stated, by Eq. (11.3).

$$h_L = K_L \frac{V^2}{2g} \qquad ...(11.3)$$

in which V is the mean velocity in the pipe and K_L is the loss coefficient which is almost constant at high Reynolds numbers and the magnitude of which depends upon the shape of the fitting. Table 11.2 gives the values of K_L for various common pipe fittings.

Since the eddies generated by the fittings persists of some distance downstream, the total loss of head caused by various fittings installed close to each other is not necessarily the same as the sum of the losses each one would cause. Usually the total loss is less than the sum of the individual losses.

Table 11.2

S. No.	*Name of pipe fitting*	*Loss coefficient K_L*
1.	Globe valve, wide open	10.0
2.	Gate valve, wide open	0.2
	three-fourth open	1.15
	one-half open	5.6
	one-fourth open	24.0
3.	Pump's foot valve	1.5
4.	45° elbow	0.4
5.	90° elbow, short radius	0.9
	medium radius	0.75
	long radius	0.60
6.	Close return bend	2.2
7.	Angle valve, wide open	5.0
8.	Standard tee	1.8

Example 11.5. *The pressure in a large water main is maintained at a gauge pressure p_o which is independent of the flowrate. The velocity in the main can be neglected. A relatively small pipeline terminates in a nozzle of area A_n which has a velocity coefficient C_v and a contraction coefficient of unity. The loss coefficient for the entire pipeline between the main and the nozzle (including entrance effect at the main) is K based on the velocity in the pipe.*

(*i*) *Deduce an expression for the jet velocity,*

(*ii*) *Determine the power in the jet associated with the kinetic energy, and*

(*iii*) *Evaluate the nozzle area A_n for which this power is a maximum.*

Solution. Since the pressure in the main is maintained at p_0, the pressure of flow in the small pipeline at its entrance will also be the same.

(*i*) Applying the Bernoulli's equation to the entrance and end sections of branch the pipeline (the axes of the main and the branch pipelines are assumed to be in the same horizontal plane, thus the datum heads are same)

$$\frac{p_0}{\gamma} = \frac{p}{\gamma} + \frac{V_p^2}{2g} + K\frac{V_p^2}{2g} \quad ...(1)$$

here K is the loss coefficient between the two sections.

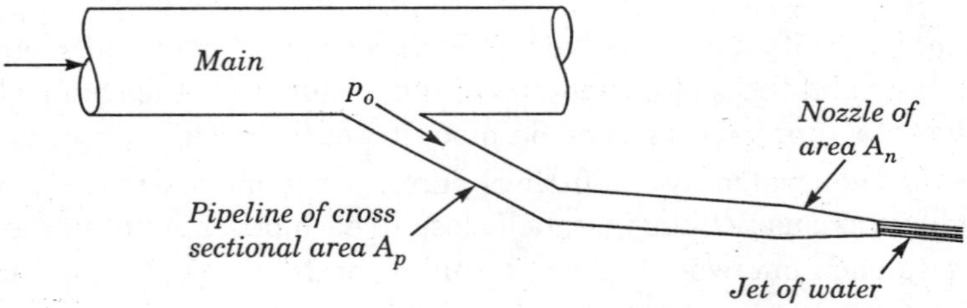

Now applying again the Bernoulli's equation to the base and the tip of nozzle, we obtain

$$\frac{p}{\gamma} + \frac{V_p^2}{2g} = 0 + \frac{V_j^2}{2g} + (h_L)_{nozzle} \quad ...(2)$$

From the continuity equation

$Q = A_p \,.\, V_p = A_j \,.\, V_j$, since the contraction coefficient is unity

$$A_j = A_n \,;\, V_p = V_j\frac{A_j}{A_p} = V_j\frac{A_n}{A_p} \quad ...(3)$$

From the definition of the coefficient of velocity

$$V_j = C_v\sqrt{2g\left(\frac{p}{\gamma} + \frac{V_p^2}{2g}\right)} \text{ or } \frac{p}{\gamma} + \frac{V_p^2}{2g} = \frac{1}{C_v^2}\cdot\frac{V_j^2}{2g}$$

Head loss in the nozzle, $(h_L)_{nozzle}$

$$= \frac{p}{\gamma} + \frac{V_p^2}{2g} - \frac{V_j^2}{2g} = \frac{1}{C_v^2}\cdot\frac{V_j^2}{2g} - \frac{V_j^2}{2g} = \frac{V_j^2}{2g}\left(\frac{1}{C_v^2 - 1}\right)$$

From (1), $$\frac{p_0}{\gamma} = \frac{1}{C_v^2}\frac{V_j^2}{2g} + K\frac{V_p^2}{2g}$$

but from (3), $$V_p = \frac{A_n \,.\, V_j}{A_p}$$

$$\therefore \qquad \frac{p_0}{\gamma} = \frac{1}{C_v^2}\frac{V_j^2}{2g} + K\frac{(A_n/A_p)^2 V_j^2}{2g} = \frac{V_j^2}{2g}\left[\frac{1}{C_v^2} K\left(\frac{A_n}{A_p}\right)^2\right]$$

from which, the jet velocity, $$V_j = \sqrt{\frac{2g\,p_0}{\gamma[1/C_v^2 + K(A_n/A_p)^2]}}$$

(*ii*) Kinetic energy of the jet = $\frac{1}{2}(\rho Q)\,V_j^2$

Power contained in the jet = Kinetic energy per second

$$P = \rho Q\,\frac{V_j^2}{2}$$

(*iii*) **Nozzle area for maximum power**

Substituting for Q in terms of V_j we obtain

$$P = \frac{\gamma A_n}{2g}\left[\frac{2g\,p_0}{\gamma\left\{[1/C_v^2] + K\,(A_n/A_p)^2\right\}}\right]^{3/2}$$

For maximum power, $\frac{dp}{dA_n} = 0$

$$\frac{dp}{dA_n} = \frac{\gamma}{2g}\left[\frac{2g\,p_0}{\gamma\left\{[1/C_v^2] + K\,(A_n/A_p)^2\right\}}\right]^{3/2}$$

$$+\frac{\gamma A_n}{2g}\left(\frac{2gp_0}{\gamma}\right)^{3/2}\left\{-\frac{3}{2}\left(\frac{1}{C_v^2} + K\frac{A_n^2}{A_p^2}\right)^{-5/2}\right\} \times 2K\frac{A_n}{A_p^2} = 0$$

The above condition simplifies to

$$\left(\frac{A_p}{A_n}\right)^2 = 2K\,C_v^2 \qquad \text{or} \qquad A_n = \frac{A_p}{\sqrt{2KC_v^2}}.$$

Example 11.6. *Oil is flowing in a pipeline at a rate of 0.6 cubic metre/sec. The energy loss in the pipe line is 9 metre per km. 367.5 kW pumps with an efficiency of 80% are available for the pumping station. At what intervals should the pumping stations be located. Assume that the density of the oil is the same as that of water.*

Solution. Output of the pumps = 367.5 × 0.8 = 294.0 kW

Power loss in friction = γQH

Head loss corresponding to 294.0 kW = 294 × 1000 W

$$H = \frac{294 \times 1000}{9810 \times 0.6} = 50 \text{ m}$$

Energy loss in the pipeline per km = 9 m

∴ Pumps are to be installed at $\frac{50}{9}$ = 5.55 km interval.

Example 11.7. *The tailrace of a turbine installation is h metre below the water level in a reservoir as shown in the following figure. The head loss due to friction in the pipeline etc., is $KV^2/2g$, where V is the velocity at the tailrace. The cross-sectional area A of the pipe is same everywhere and the water density ρ is constant. Determine :*

(i) head utilised by the turbine,

(ii) power developed by the turbine assuming that the turbine has an efficiency of 90%,

(iii) condition for maximum power output from the turbine, assuming constant h, constant K and constant turbine efficiency.

(iv) efficiency for maximum power output, if the overall efficiency of the installation is defined as the ratio of the head abstracted from the flow by the turbine to the elevation head h.

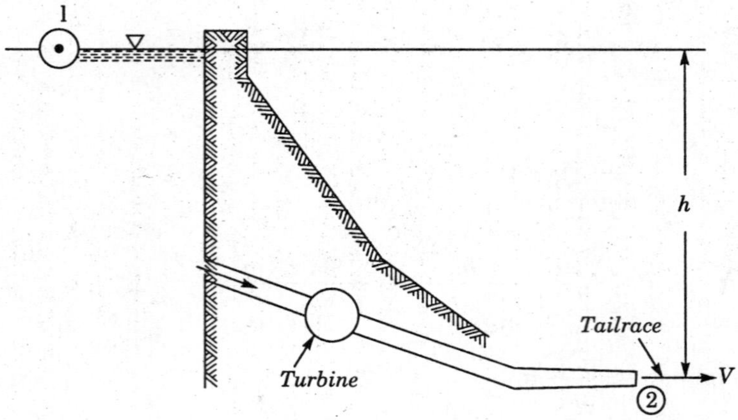

Solution. Applying the Bernoulli's equation between point (1) on the water surface in the reservoir and point (2) in the tailrace, we have

$$h = \frac{V^2}{2g} + K\frac{V^2}{2g} + \text{Energy head utilised by the turbine}$$

(*i*) Energy utilised by the turbine $H = h - (1 + K)\dfrac{V^2}{2g}$

(*ii*) Power developed by the turbine at 90% efficiency $P = \eta\, \gamma\, QH$.

$$P = 0.9\,(\rho g)\,(A.V)\,[h - (1 + K)\,V^2/2g]$$
$$= 0.9\,A[hV - (1 + K)\,V^3/2g]$$

(*iii*) Condition for maximum power is $\dfrac{dP}{dV} = 0$

$$h - 3(1 + K)V^2/2g = 0.$$

$$\therefore \qquad V^2 = \frac{2gh}{3(1+K)}$$

(*iv*) Overall efficiency for maximum power

$$= \frac{\text{Head utilised by the turbine}}{h}$$

$$= \frac{h - (1+K)\dfrac{V^2}{2g}}{h} = \frac{h - h/3}{h} = \frac{2}{3}.$$

Example 11.8. *A wind tunnel consists of a contracting nozzle, a test section where experiments are conducted, a closed circuit pipeline to recirculate the flow, and a fan to drive the air (as shown in the figure). The throat of the tunnel is 3 metre in diameter, and the upstream diameter is 6 metre. Assume that the nozzle velocity coefficient is 0.98, and that 0.2 of the velocity head in the throat is lost to friction in the remainder of the circuit. Determine :*

(*i*) *Pressure drop* Δp *across the nozzle if the velocity in throat is 70 m/s. The specific weight of the air is 11.77 N/m³. The air may be considered incompressible at these velocities.*

(*ii*) *If the fan efficiency is 85% what power is required to drive the fan ?*

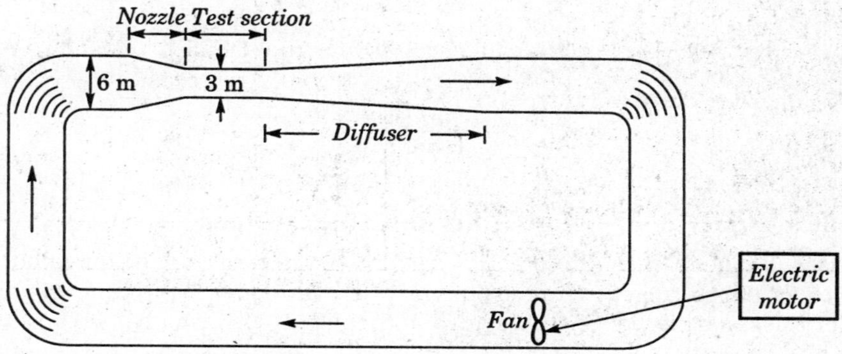

Solution. Velocity at the throat

$$V_2 = 70 \text{ m/s.}$$

Velocity at the base of the nozzle

$$V_1 = \frac{70}{4} = 17.5 \text{ m/s.}$$

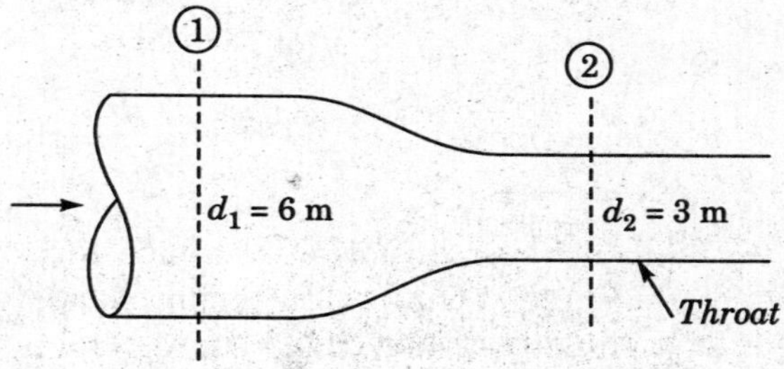

Applying the Bernoulli's equation between sections (1) and (2)

$$p_1/\gamma + V_1^2/2g = p_2/\gamma + V_2^2/2g + K\,V_2^2/2g$$

$$\therefore \quad \frac{p_1 - p_2}{\gamma} = \frac{V_2^2 - V_1^2}{2g} + K\frac{V_2^2}{2g}$$

$$= \frac{V_2^2 - V_1^2}{2g} + \left(\frac{1}{C_v^2} - 1\right)\frac{V_2^2}{2g}$$

since $$K = \frac{1}{C_v^2} - 1$$

or $$\frac{p_1 - p_2}{\gamma} = \frac{\Delta p}{\gamma}$$

$$= \frac{(70)^2 - (17.5)^2}{2 \times 9.81} + \left[\frac{1}{(0.98)^2} - 1\right]\frac{(70)^2}{2 \times 9.81} = 244 \text{ m of air}$$

$$\therefore \quad \Delta p = 11.77 \times 244 = 2870 \text{ N/m}^2$$

(*ii*) Head loss in the circuit

$$= K \frac{V_2^2}{2g} + 0.2 \frac{V_2^2}{2g}$$

$$= (1/C_v^2 - 1) \frac{V_2^2}{2g} + 0.2 \frac{V_2^2}{2g}$$

$$= 10 + 0.2 \times \frac{(70)^2}{2 \times 9.81} = 60.25 \text{ m of air}$$

$\therefore$ Power loss $= \gamma QH = 11.77 \times \frac{\pi}{4} (3)^2 \times 70 \times 60.25\ W = 350.88$ kW

Power required to drive fan if η of fan is 85%.

$$= \frac{350.88}{0.85} = \mathbf{412.8\ kW.}$$

11.4. Concept of Equivalent Length and Equivalent Pipes

The losses in pipe fittings are sometimes expressed in terms of an equivalent length which is the length of an unobstructed straight pipe in which an equal loss of energy (due to friction) would occur for the same discharge. The equivalent length L_{eq} is given by

$$K_L \frac{V^2}{2g} = f \frac{L_{eq}}{d} \frac{V^2}{2g}$$

which yields, $L_{eq} = K_L \frac{d}{f}$ [Here K_L is called loss coefficient] ...(11.12)

For a known value of f which depends upon the Reynolds number and the pipe roughness, the equivalent length for a given pipe fitting can be expressed in terms of pipe diameter. The effective length of pipe is obtained by adding to the actual length of pipe the equivalent lengths for different fittings installed in the piping system. This effective length can then be used in the Darcy-Weisbach equation for flow analysis.

In case of valves and fittings such as elbows, energy loss can be calculated using the equivalent length concept from the following equation

$$h_L = f \frac{L_{eq}}{d} \frac{V^2}{2g} \qquad \text{...(11.12 } A)$$

For many commercially available valves and fittings, the L_{eq}/d is approximately constant for a particular type of device regardless of size.

The loss coefficient K_L, is also used to calculate the energy loss in pipe fittings, which is expressed by the following formula

$$h_L = K_L (V^2/2g) \qquad \text{...(11.12 } B)$$

Eq. (11.12), which gives the equivalent length, L_{eq}, can also be written as

$$K_L = f\, L_{eq}/d \qquad \text{...(11.12 } C)$$

11.4.1. Non-Dimensional Equivalent Length

Resistance in values and fittings expressed as equivalent length in terms of pipe diameters, L_{eq}/d, is given below for different types of fittings.

S. No.	Type of fitting	Non-dimensional Equivalent length (Leq/d)	S. No.	Type of fitting	Non-dimensional Equivalent length (Leq/d)
	VALVES			**ELBOWS**	
1.	Globe valve		6.	90° Standard elbow	30
	—fully open	340	7.	90° Long radius elbow	20
2.	Angle valve		8.	90° Street elbow	50
	—fully open	145	9.	45° Standard elbow	16
3.	Gate valve		10.	45° street elbow	26
	—fully open	13		**BENDS AND TEES**	
	—3/4 open	35	11.	Close return bend	50
	—1/2 open	160	12.	Standard Tee	
	—1/4 open	900		—with flow through	
4.	Check valve			run	20
	—Swing type	135		—with flow through	
	—Ball type	150		branch	60
5.	Butterfly valve				
	—fully open	40			

11.4.2. Equivalent Pipes

A pipe of length L_1 diameter d_1 and friction factor f_1 will be equivalent to another pipe of length L_2, diameter d_2 and friction factor f_2, if

$$\frac{f_1 L_1}{d_1^5} = \frac{f_2 L_2}{d_2^5} \quad \text{... (11.12 A)}$$

This is the condition obtained for the same head loss in the two pipes for the same discharge, using Eq. (11.30).

If a set of pipes described by (L_1, d_1, f_1), (L_2, d_2, f_2), (L_3, d_3, f_3), etc. are connected in series, then an equivalent pipe (L_e, d_e, f_e) can be expressed using Eq. (11.28) as,

$$\frac{f_e L_e}{d_e^5} = \frac{f_1 L_1}{d_1^5} + \frac{f_2 L_2}{d_2^5} + \frac{f_3 L_3}{d_3^5} + \text{......} \quad \text{...(11.12 B)}$$

If the above set of pipes are connected in parallel between two points, then an equivalent pipe can be obtained using Eq.. (11.29) as,

$$\left(\frac{d_e^5}{f_e L_e}\right)^{1/2} = \left(\frac{d_1^5}{f_1 L_1}\right)^{1/2} + \left(\frac{d_2^5}{f_2 L_2}\right)^{1/2} + \left(\frac{d_3^5}{f_3 L_3}\right)^{1/2} + \text{......} \quad \text{...(11.12 C)}$$

which is based on the fact that the head loss for the parallel pipes is same. Using Eq. (11.30), one can write,

$$Q = \left(\frac{\pi^2 g h_L}{8}\right)^{1/2} \left(\frac{d^5}{fL}\right)^{1/2} = K\left(\frac{d^5}{fL}\right)^{1/2}$$

Making use of the continuity equation,

$$Q_e = Q_1 + Q_2 + Q_3 + \text{...}$$

Eq. (11.12 *C*) is obtained.

11.5. Problems in Pipe-Flow

All categories of problems of steady flow in pipes can be solved by using the equations of continuity, head loss (Darcy-Weisbach) and the Colebrook equation for friction or its explicit equivalent. There are five variables in any pipe flow problem : (1) the discharge Q, (2) the pipe diameter d, (3) the height of pipe wall roughness k_s, (4) the head loss per unit pipe length S_f (the friction slope), and (5) the kinematic viscosity of fluid ν. Three major problems commonly encountered in engineering practice are :

(*a*) determination of pipe diameter ;

(*b*) determination of discharge ; and

(*c*) determination of head loss.

Of the five variables listed above, four will always be known in a problem and the remainder required to be determined. Methods for determining the unknown quantity are briefly described below.

11.5.1. Determination of Pipe Diameter

The known quantities in this category of problems are Q, k_s, S_t and ν. The continuity equation, the head loss Eq. (9.57) and Colebrook Eq. (9.63 *c*) or its explicit equivalent, Eq. (9.66) provide the necessary tools for solving the pipe flow problems. Since the diameter itself is not known, the friction factor cannot be computed since the Reynolds number and the relative roughness are both not known. Therefore, a trial procedure has to be followed which is summarised below.

Procedure for Trial Solution. Adopt the following sequence of steps :

(1) Assume a trial value for the diameter.

(2) Calculate the Reynolds number, the relative roughness and determine the friction factor either by using Colebrook's Eq. (9.63 *c*) which will involve few trials or Moody's diagram or the explicit Eq. (9.66) which computes the friction factor directly.

(3) Use this value of friction factor in head loss Eq. (9.57) and evaluate the diameter.

(4) Check this diameter with the trial value assumed in step 1.

(5) Select a new suitable diameter for the next trial and repeat steps 1 to 4 till the diameter computed in step 3 matches with that in step 1.

Explicit Equation for Pipe Diameter. The author[6] was associated with a study on finding explicit equations for pipe flow problems. By suitably combining the continuity equation and Eqs. (9.57) and (9.63), grouping the variables to form dependent and independent non-dimensional parameters and using the curve fitting technique, the following explicit equation for pipe diameter was obtained :

$$D_* = 0.66\ [K_*^{1.25} + N_*]^{0.04} \qquad ...(11.13)$$

in which the non-dimensional parameters are defined as

$$D_* = d\ (gS_f/Q^2)^{0.2}$$

$$K_* = k_s\ (gS_f/Q^2)^{0.2}$$

and

$$N_* = \nu/(gS_f Q^3)^{0.2}.$$

The right-hand side of Eq. (11.13), which involves the known parameters K_* and N_*, enables direct determination of diameter.

In general, the diameter so obtained may not be a standard pipe size and the final selection of the diameter will have to be made from commercially manufactured sizes. Steel pipes of sizes larger than 1 metre in diameter are generally built to order and any size may be specified. While designing the pipelines for future, the reduction of flow on account of reduced cross-sectional area due to corrosion, silt and scale deposits and increase in the pipe roughness resulting in high energy loss in friction must also be taken into account.

11.5.2. Determination of Discharge

In problems requiring determination of discharge, the data known are d, k_s, S_f and ν. In the absence of discharge Q the Reynolds number is not known and, therefore, the friction factor and the head loss cannot be computed. The solution again involves a trial procedure, the steps for which are listed below.

Procedure for Trial Solution

1. Assume a trial value for the discharge.
2. Compute the Reynolds number and the friction factor as described in step 2 for determination of pipe diameter.
3. Use Darcy-Weisbach Eq. (9.57) for computing the friction slope S_f.
4. Check this computed value of S_f with the given value. If the computed S_f is less, the assumed discharge in step 1 was small, and *vice versa*. In the next trial assume a new value accordingly.
5. Select a new suitable value for the discharge and repeat steps 1 to 4 till the computed S_f and given S_f agree closely.

Explicit Equation for Discharge. Using the continuity equation, Eqs. (9.63 *c*) and (9.57), and suitably grouping the variables it is possible to arrive at the following explicit equation in non-dimensional form, which was first obtained by Swamee and Jain[6] :

$$\frac{Q}{d^2\sqrt{gdS_f}} = \frac{\pi}{2\sqrt{2}}\left[1.14 - 2\log_{10}\left(\frac{k_s}{d} + \frac{6.61\nu}{d\sqrt{gdS_f}}\right)\right] \qquad ...(11.14)$$

The right-hand side of Eq. (11.14) involves known quantities and permits direct computation of discharge.

11.5.3. Determination of Head Loss

For this category of problems, the data known are Q, d, k_s and ν. Using the Darcy-Weisbach Eq. (9.57), the friction slope can be written as

$$S_f = \frac{8}{\pi^2} f \frac{Q^2}{gd^5} \qquad ...(11.15\ a)$$

The right-hand side of Eq. (11.15 *a*) contains all known quantities except the friction factor *f*. Solution of Colebrook's Eq. (9.63 *c*) for *f* involves a trial procedure. The Moody's diagram, which is a graphical representation of Eq. (9.63 *c*), can be used to evaluate the friction factor for given values of **R** and *k*/*d*. The use of explicit Eq. (9.66), however, eliminates trials otherwise involved in Eq. (9.63 *c*) and directly computes the friction factor.

Using the explicit Eq. (9.66), the friction slope can be expressed as

$$S_f = \left(0.8115\frac{Q^2}{gd^5}\right)\Bigg/\left[1.14 - 2\log_{10}\left(\frac{21.25}{\mathbf{R}^{0.9}} + \frac{k_s}{d}\right)\right]^2 \qquad ...(11.15\ b)$$

6. Swamee, P.K., and Jain A.K., "Explicit Equation for Pipe Flow Problems", Journal of Hydraulics Division, Proceedings of American Society of Civil Engineers, May, 1976.

Eq. (11.15 *b*) computes the friction slope, from which the head loss $h_L = S_f.L$ can be readily obtained.

In the trial procedure for solving the pipe-flow problems as described in the foregoing, the trials can also be made, as an alternative, by assuming a trial friction factor instead of diameter and discharge. The various quantities to be computed in steps 2 to 4 will then accordingly change.

Example 11.9. *It is required to design the diameter of a pipeline 1.5 km long to carry water at 20°C under a 15 m head loss. A riveted steel pipe with smooth walls is to be used. It is assumed that the rivets cause a flow resistance equivalent to that produced by equivalent sand grain toughness $k_s = 10^{-3}$ m. Determine the diameter to convey a flow of 3 m³/s. The kinematic viscosity of water at 20°C is 10^{-6} m²/s.*

Solution. (1) **Using trial procedure.** The given discharge is a large one, the conveyance of which will require a large pipe diameter ranging between 1 to 2 m. The selection of the trial value for the diameter is a matter of judgment and experience, which to some degree can be acquired by working with similar problems. Following the procedure described in 11.5.1, select a trial diameter of 1 m.

Reynolds number $$\mathbf{R} = \frac{4Q}{\pi \nu d} = \frac{4 \times 3}{\pi \times 10^{-6} \times 1} = 3.82 \times 10^6.$$

Relative roughness $$\frac{k_s}{d} = \frac{10^{-3}}{1} = 10^{-3}.$$

Using explicit Eq. (9.66), the friction factor can be determined,

$$\frac{1}{\sqrt{f}} = 1.14 - 2\log_{10}\left\{\frac{k_s}{d} + \frac{21.25}{\mathbf{R}^{0.9}}\right\}$$

$$= 1.14 - 2\log_{10}\left\{10^{-3} + \frac{21.25}{(3.82 \times 10^6)^{0.9}}\right\}$$

which yields, $$f = 0.0196.$$

Making use of Eq. (11.15 *a*)

$$S_t = \frac{h_L}{L} = \frac{8}{\pi^2} f \frac{Q^2}{gd^5}$$

$$d^5 = \frac{8}{\pi^2} f \frac{Q^2}{gS_f} = \frac{8}{\pi^2} \times 0.0196 \times \frac{(3)^2}{9.81 \times 15/1500} = 1.442$$

$$d = 1.076 \text{ m or } 1076 \text{ mm}.$$

Since the computed diameter is in close agreement with the assumed trial diameter, subsequent trials are not needed.

(2) **Using Explicit (Eq. 11.13).** The independent dimensionless parameters appearing on the right hand side of Eq. (11.13) can be computed using the given data.

$$K_* = k_s\left(\frac{gS_f}{Q^2}\right)^{0.2} = 10^{-3}\left(\frac{9.81 \times \frac{15}{1500}}{9}\right)^{0.2} = 0.4051 \times 10^{-3}$$

$$N_* = \nu/(gS_f Q^3)^{0.2} = 10^{-6}\Big/\left(9.81 \times \frac{15}{1500} \times 27\right)^{0.2} = 0.824 \times 10^{-6}.$$

From Eq. (11.13),

$$D_* = 0.66[K_*^{2.25} + N_*]^{0.04}$$

$$= 0.66[(4.051 \times 10^{-4})^{1.25} + 8.24 \times 10^{-7}]^{00.4} = 0.41$$

but $$D_* = d\left(\frac{gS_f}{Q^2}\right)^{0.2} = 0.41$$

which yields $$d = 1.01 \text{ m or } 1010 \text{ mm.}$$

Example 11.10. *A horizontal commercial steel pipeline 100 mm in diameter and 400 m long carrying gasoline at 40°C has a pressure drop of 1.333 × 10⁵ N/m². Compute the flowrate when the specific gravity is 0.68. The kinematic viscosity of gasoline at 40°C is 4.2 × 10⁻⁷ m²/s. The equivalent sand grain roughness of pipe material is 4.92 × 10⁻⁵ m.*

Solution. Given :

$$d = 100 \text{ mm}, L = 400 \text{ m}, \Delta p = 1.333 \times 10^5 \text{ N/m}^2$$

$$\gamma = 0.68 \times 9810 = 6670 \text{ N/m}^3$$

$$k_s = 4.92 \times 10^{-5} \text{ m}, \nu_{40} = 4.2 \times 10^{-7} \text{ m}^2/\text{s}$$

$$h_L = \frac{\Delta p}{\gamma} = \frac{1.333 \times 10^5}{6670} = 20 \text{ m of gasoline}$$

$$\frac{k_s}{d} = \frac{4.92 \times 10^{-5}}{100/1000} = 4.92 \times 10^{-4}$$

Using the explicit Eq. (11.14),

$$\frac{Q}{d^2\sqrt{gS_f d}} = \frac{\pi}{2\sqrt{2}}\left[1.14 - 2\log_{10}\left(\frac{k_s}{d} + \frac{6.61\,\nu}{d\sqrt{gS_f d}}\right)\right]$$

$$d^2\sqrt{gS_f d} = \left(\frac{100}{1000}\right)^2\sqrt{9.81 \times \frac{20}{400} \times \frac{100}{1000}} = 2.215 \times 10^{-3}$$

$$\frac{d\sqrt{gS_f d}}{\nu} = \left(\frac{100}{1000}\right)^2\sqrt{\frac{9.81 \times \frac{20}{400} \times \frac{100}{1000}}{4.2 \times 10^{-7}}} = 5.275 \times 10^4$$

Making substitutions in the above formula

$$\frac{Q}{d^2\sqrt{gS_f d}} = \frac{\pi}{2\sqrt{2}}\left[1.14 - 2\log_{10}\left(4.29 \times 10^{-4} + \frac{6.61}{5.275 \times 10^4}\right)\right] = 8.40$$

$$Q = 8.40 \times 2.215 \times 10^{-3} \text{ m}^3/\text{s.}$$

$$= 8.40 \times 2.215 \text{ lit./s} = 18.60 \text{ lit./s.}$$

Example 11.11. *0.75 litre per minute of ethyl alcohol flows at 20°C through a 15 mm diameter 10 m long stainless-steel drawn tubing (k_s = 1.64 × 10⁻⁶ m). The kinematic viscosity of the fluid at 20° C is 1.77 × 10⁻⁶ m²/s. Determine the head loss and the power consumption. Specific gravity of ethyl alcohol is 0.79.*

Solution. Given = 15 mm, L = 10 m,

$$Q = 75 \text{ lit./min} = \frac{75}{60} = 1.25 \text{ lit./s}, k_s = 1.64 \times 10^{-6} \text{ m}$$

$$\nu = 1.77 \times 10^{-6} \text{ m}^2/\text{s}$$

$$\frac{k_s}{d} = \frac{1.64 \times 10^{-6}}{15/1000} = 1.093 \times 10^{-4}$$

$$\mathbf{R} = \frac{4Q}{\pi \nu d} = \frac{4 \times 1.25 \times 10^{-3}}{\pi \times 1.77 \times 10^{-6} \times 1.5 \times 10^{-6}} = 1.6 \times 10^5$$

$$\frac{Q^2}{gd^5} = \frac{(1.25 \times 10^{-3})^2}{9.81 \times (1.5 \times 10^{-2})^5} = 2.094 \times 10^2$$

Making use of the explicit Eq. (11.15 b)

$$S_f = \left(0.8115 \frac{Q^2}{gd^5}\right)\left[1.14 - 2 \log_{10}\left(\frac{k_s}{d} + \frac{21.25}{\mathbf{R}^{0.9}}\right)\right]^2$$

$$= \frac{0.8115 \times 2.094 \times 10^2}{\left[1.14 - 2 \log_{10}\left\{1.093 \times 10^{-4} + \frac{21.25}{(1.06 \times 10^5)^{0.9}}\right\}\right]^2} = 3.145.$$

Head loss in 10 m length, $h_L = S_f \cdot L = 3.145 \times 10$

$= 31.45$ m of alcohol.

Power consumption $= \gamma Q h_L = (0.79 \times 9810) \times 1.25 \times 10^{-3} \times 31.45$

$= 304.4$ Nm/s $= 304.4$ **W.**

Example 11.12. *A system for room heating which uses natural circulation is shown in the adjacent figure. Water is heated in the drum to a temperature of 95°C. It then flows up into the room heater which is at a height of 4 m above the drum through a pipe 5 cm in diameter and 8 m long. In the room heater the temperature of water falls to 50°C and flows back into the drum through a pipe 5 cm in diameter and 9 m long. The heat losses in the pipe are negligible and the pressure drop due to local losses in the system is 50% of the straight pipe pressure loss. The straight pipe head loss may be determined using the smooth pipe flow equation* $f = 0.316\ \mathbf{R}^{-0.25}$. *The specific weight and kinematic viscosity of water at 95°C are respectively 9437 N/m³ and 0.38 × 10⁻³ m²/s. The specific weight and kinematic viscosity of water at 50°C are 9614 N/m³ and 0.55 × 10⁻⁶ m²/s respectively. Determine the rate at which water circulates through the system.*

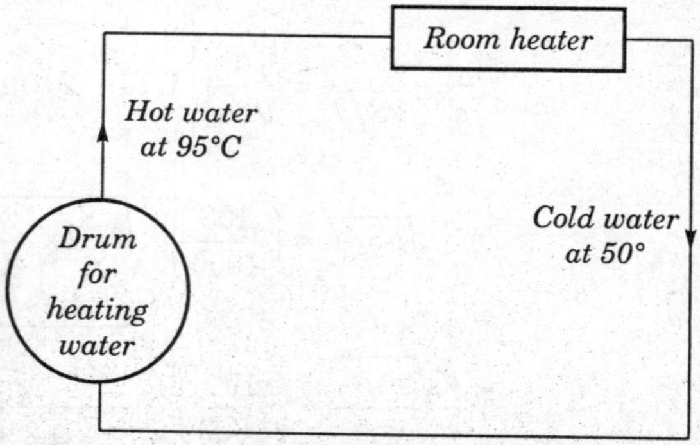

Solution. Let Q be the volume rate of flow at which water circulates through the system. The heating of water in the drum provides adequate energy to the flow to enable it to (i) overcome the friction and form drag of the system, and (ii) raise the hot water by 4 m in the room heater. The problem, therefore boils down to the evaluation of thermal energy supplied to the water in the drum and computation of energy losses in the system.

(1) The thermal energy supplied to the water in the drum

$$= m.s\,(T_1 - T_2) \qquad \ldots(a)$$

$= \rho_1 Q s_1 T_1 - \rho_2 Q s_2 T_2$

in which m = mass of water flowing per second,

s = specific heat of water (assuming $s_1 = s_2$),

T_1 = temperature of hot water leaving the drum,

T_2 = temperature cold water entering the drum,

ρ_1 = density of hot water, and

ρ_2 = density of cold water.

Using Eq. (*a*), the thermal energy supplied per second

$= [962 \times Q \times 1 \times (273.15 + 95) - 980 \times Q \times 1 \times (273.15 + 50)]$ kcals

$= 37.5 \times 10^3\ Q$ kcal/s $= 37.5 \times 10^3\ Q \times 4180$ Nm/s

$= 15.7 \times 10^7\ Q$ Nm/s or J/s ...(1)

(2) Loss of energy in overcoming various resistances consists of (*i*) the energy lost in friction in straight pipes and form resistance due to such non-uniformities as imposed by bends, valves, expansions and contractions, and (*ii*) the energy required to raise the hot water in the room heater.

(*i*) Energy lost in friction is $\gamma Q h_L$ where h_L is the loss of head given by

$$h_L = f \frac{L}{d} \frac{V^2}{2g}$$

$$f = \frac{0.316}{R^{0.25}} = \frac{0.316}{\left(\frac{4Q}{\pi d \nu}\right)^{0.25}}$$

$$h_L = 0.02456 \left(\frac{QL^{7/4}}{\nu^{1/4} d^{19/4}}\right) = 3.7275 \times 10^4\ LQ^{7/4}/\nu^{1/4}.$$

Head loss in straight pipe length upstream of the heater

$$= \frac{3.7275 \times 10^4 \times 8Q^{7/4}}{(0.38 \times 10^{-6})^{1/4}} = 1.196 \times 10^7\ Q^{7/4}$$

Local losses = 50% of straight pipe losses.

Therefore, the total head loss upstream of the heater

$H = 1.5 \times 1.196 \times 10^7\ Q^{7/4}$

$= 1.792 \times 10^7\ Q^{7/4}$

Energy loss upstream of the heater $= \gamma QH$

$= 962 \times 9.81 \times Q \times 1.792 \times 10^7 \times Q^{7/4}$ Nm/s

$= 1.69 \times 10^{11}\ Q^{11/4}$ Nm/s

Similarly, energy loss downstream of the heater

$= 1.773 \times 10^{11} \times Q^{11/4}$ Nm/s

Total energy loss $= (1.69 + 1.773) \times 10^{11} \times Q^{11/4}$ Nm/s

$= 3.463 \times 10^{11} \times Q^{11/4}$ Nm/s.

---

1 calorie = 1 gm (mass) × 1 deg C

1 kcal = 427 kg (force)-m = 427 × 9.81 Nm = 4180 Nm or joules.

(*ii*) Energy required to raise the hot water by 4 m to enable it to reach the room heater

$= \gamma Q \times 4 = 962 \times 9.81 \times Q \times 4$ Nm/s

$= 37.72 \times 10^3 Q$ Nm/s.

Total energy required to cause flow

= Energy loss in the system + Energy required to raise the water

$= 3.463 \times 10^{11} \times Q^{11/4} + 37.72 \times 10^3 Q$ ...(2)

Equating the thermal energy supplied and energy required, one obtains

$$15.7 \times 10^7 Q = 3.463 \times 10^{11} \times Q^{11/4} + 37.72 \times 10^3 Q$$

$$Q^{7/4} = 0.53 \times 10^{-4}$$

$$Q = 0.0123 \text{ m}^3/\text{s} = \mathbf{12.3\ lit/s.}$$

Example 11.13. *A smooth pipe carries 6.5 litres/sec of water at 20°C ($\nu = 10^{-6}$ m²/sec) with a head loss of 7.5 cm per 10 metre length. Determine the diameter of the pipe.*

Solution. From the Blasius equation $f = 0.316/(\mathbf{R})^{1/4}$

Reynolds number $\mathbf{R} = Vd/\nu = (4Q/\pi d^2)d/\nu = 4Q/\pi d\nu$

and from Darcy-Weisbach equation $h_L = f\dfrac{L}{d}\dfrac{V^2}{2g}$

The head loss equation may now be written as

$$h_L = \frac{0.316}{(4Q/\pi\, d\nu)^{1/4}} \times \frac{L}{d} \times \frac{(4Q/\pi d^2)^2}{2g}$$

Given that $h_L = \dfrac{7.5}{100}$ m, $L = 10$ m, $Q = \dfrac{6.5}{1000}$ m³/s, and $\nu = 10^{-6}$ m²/s.

Substituting these data in the head loss equation, and solving for d, we obtain

$$d = 9.75 \text{ cm.}$$

Example 11.14. *A good approximation to the Darcy friction factor over almost the whole turbulent range is the empirical equation*

$$f = 0.0055\left[1 + \left(20{,}000\frac{k_s}{d} + \frac{10^6}{\mathbf{R}}\right)^{0.33}\right]$$

in which k_s = equivalent sand grain roughness, d = diameter of pipe and $\mathbf{R}$ *= Reynolds number.*

Using this formula, design the diameter of a steel pipe to carry oil having a kinematic viscosity of 0.2 cm²/s at a velocity of 1.5 m/s. The head loss is to be limited to 40 cm per 100 m of pipe. The equivalent sand grain roughness for steel may be assumed as 0.0045 cm.

Solution. Following a trial procedure, assuming a diameter of 1 m, the Reynolds number

$$\mathbf{R} = \frac{Vd}{\nu} = \frac{1.5 \times 1}{0.2 \times 10^{-5}} = 7.5 \times 10^4$$

$$f = 0.0055\left[1 + \left\{20{,}000 \times \frac{0.0045}{10 \times 100} + \frac{10^6}{7.5 \times 10^4}\right\}^{0.33}\right] = 0.0187.$$

Darcy-Weisbach equation for head loss

$$h_L = f\frac{L}{d}\frac{V^2}{2g} = 0.0187 \times \frac{100}{1} \times \frac{(1.5)^2}{2 \times 9.81} = 0.2145 \text{ m} = 21.45 \text{ cm.}$$

This value of head loss is less than the value given, *i.e.* 40 cm. To increase it, the diameter must be decreased.

II Trial. Assuming $d = 0.7$ m,

$$\mathbf{R} = \frac{1.5 \times 0.7}{1.2 \times 10^{-4}} = 5.25 \times 10^4$$

$$f = 0.0055\left[1 + \left\{20{,}000 \times \frac{0.0045}{100 \times 0.7} + \frac{10^6}{5.25 \times 10^4}\right\}^{0.33}\right] = 0.01946.$$

$$h_L = 0.01946 \times \frac{100}{0.7} \times \frac{(1.5)^3}{2 \times 9.81} = 0.296 \text{ m} = 29.6 \text{ cm.}$$

Though this head loss is larger than the value obtained in the I trial, it is still less than 40 cm. The diameter, therefore, needs to be further reduced.

III Trial. Let $d = 0.5$ m,

$$\mathbf{R} = \frac{1.5 \times 0.5}{2.0 \times 10^{-4}} = 3.75 \times 10^4$$

$$f = 0.0054\left[1 + \left\{20{,}000 \times \frac{0.0045}{100 \times 0.5} + \frac{10^6}{3.75 \times 10^4}\right\}^{0.33}\right] = 0.0221$$

Head loss per 100 m length of pipe

$$h_L = 0.0221 \times \frac{100}{0.5} \times \frac{(1.5)^2}{2 \times 9.81}$$

$$= 0.509 \text{ m} = 50.9 \text{ cm.}$$

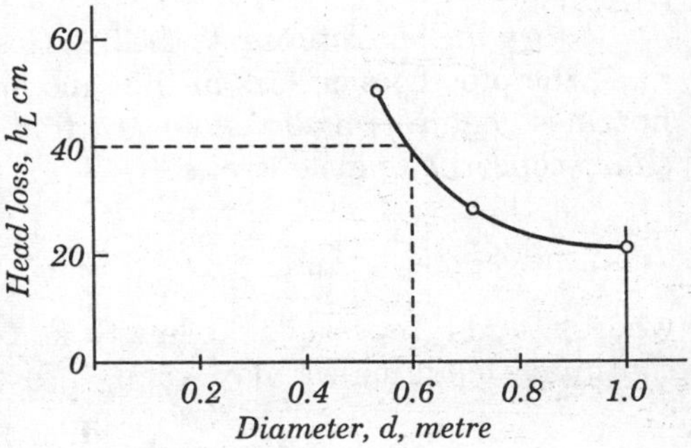

The computed head loss for $d = 0.5$ m is 50.9 cm, which is more than the given value of 40 cm. The pipe diameter must, therefore, lie between 0.5 m and 0.7 m. Plotting a graph of h_L versus d (see the figure), the diameter corresponding to $h_L = 40$ cm is seen to be equal to 0.57 m. The next trial would be made with this diameter.

IV Trial. Let $d = 0.57$ m.

$$\mathbf{R} = \frac{1.5 \times 0.57}{0.2 \times 10^{-4}} = 4.275 \times 10^4$$

$$f = 0.0055\left[1 + \left\{20{,}000 \times \frac{0.0045}{100 \times 0.57} + \frac{10^6}{4.275 \times 10^4}\right\}^{0.33}\right] = 0.0214$$

$$h_L = 0.0214 \times \frac{100}{0.57} \times \frac{(1.5)^2}{2 \times 9.81} = 0.431 \text{ m} = 43.1 \text{ cm.}$$

The head loss is slightly on the higher side, the diameter would be larger than 0.57 m.

V Trial. Let $d = 0.61$ m

$$\mathbf{R} = \frac{1.5 \times 0.61}{0.2 \times 10^{-4}} = 4.575 \times 10^4$$

$$f = 0.0055\left[1+\left\{20{,}000\times\frac{0.045}{100\times 0.61}+\frac{10^5}{4.575\times 10^4}\right\}^{0.33}\right] = 0.021$$

$$h_L = 0.021\times\frac{100}{0.61}\times\frac{(1.5)^2}{2\times 9.81} = 0.369 \text{ m} = \mathbf{39.6\ cm.}$$

Therefore, adopt a pipe diameter of 610 mm.

11.5.1. Flow From a Single Reservoir

Fig. 11.10 shows a pipeline emerging from a reservoir and terminating in a nozzle which discharges freely into the air. Applying Bernoulli's equation to any point on the reservoir free surface

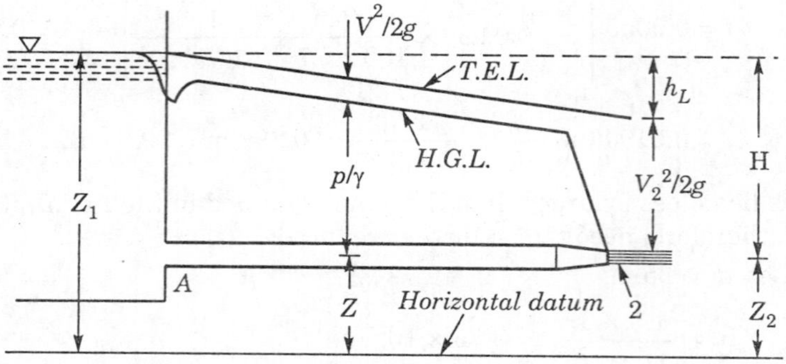

Fig. 11.10. Flow from a reservoir.

and to the jet of liquid issued from the nozzle, we obtain

$$Z_1 + 0 + 0 = Z_2 + \frac{V_2^2}{2g} + h_L \qquad \text{...(11.13)}$$

In which h_L represents the head lost from points (1) to (2) and includes the entry loss incurred at *A*, the pipe friction loss in the pipe length *AB* and the loss in the nozzle. The head loss in the nozzle is generally a very small fraction of the total loss and hence may be neglected, thus with a sharp-concerned pipe entrance,

$$h_L = 0.5\,\frac{V^2}{2g} + f\,\frac{L}{d}\,\frac{V^2}{2g} \qquad \text{...(11.14)}$$

where *L* and *d* represent the length and the diameter of pipe and *V* is the average velocity. From continuity equation we may write

$$Q = \frac{\pi}{4}\,d^2V = \frac{\pi}{4}\,D^2V_2 \qquad \text{...(11.15)}$$

$$V_2 = \left(\frac{d}{D}\right)^2 V$$

in which *D* is the diameter of the nozzle tip.

From Eq. (11.13) substituting for h_L and V_2, we obtain

$$Z_1 - Z_2 = \left(\frac{d}{D}\right)^2\frac{V^2}{2g} + 0.5\,\frac{V^2}{2g} + f\,\frac{L}{d}\,\frac{V^2}{2g}$$

but $Z_1 - Z_2 = H$ and, therefore,

$$H = \left[0.5 + f\,\frac{L}{d} + \left(\frac{d}{D}\right)^2\right]\frac{V^2}{2g} \qquad \text{...(11.16)}$$

Eq. (11.16) relates the head available *H* with the velocity in the pipe.

11.5.2. Flow Between Two Reservoirs

Fig. 11.11 illustrates the case of flow through a pipe connecting two reservoirs. The pipe diameter changes such that there is a sudden enlargement at B. All the junctions are supposed to be abrupt. If V_1 is the average velocity in pipe (1), there is a head loss of about 0.5 $V_1^2/2g$ at the inlet A and this is represented by a drop in the total energy line (T.E.L.) at that point. The head loss in friction in flow through the uniform dia-meter pipe is at a uniform rate and the T.E.L., therefore, has a constant downward slope.

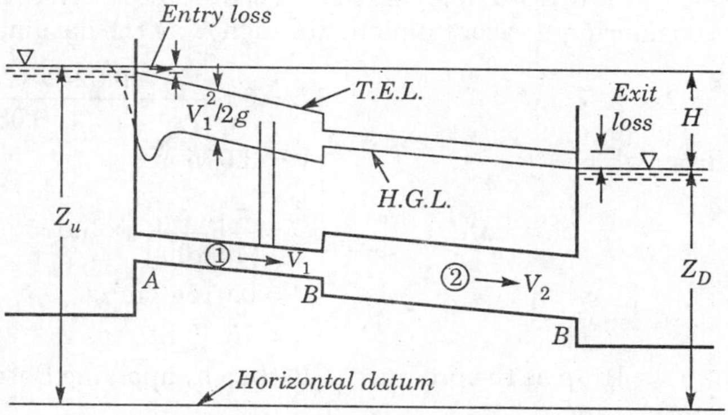

Fig. 11.11. Flow between two reservoirs.

At the sudden enlargement B, the head loss $(V_1 - V_2)^2/2g$ is represented by a sudden drop in the total energy line and at the end of the pipe line, the exit loss $V_2^2/2g$ is shown by a drop in the T.E.L. at C. The application of Bernoulli's equation, between any point on the upstream reservoir surface and a similar point on the free-surface of the downstream reservoir, taking into account the various head losses, yields

$$Z_u + 0 + 0 = 0.5\frac{V_1^2}{2g} + f_1\frac{L_1}{d_1}\frac{V_1^2}{2g} + \frac{(V_1 - V_2)^2}{2g} + f_2\frac{L_2}{d_2}\frac{V_2^2}{2g} + \frac{V_2^2}{2g} + Z_D$$

But $Z_u - Z_D = H$, and the above equation reduces to

$$H = \left(0.5 + f_1\frac{L_1}{d_1}\right)\frac{V_1^2}{2g} + \frac{(V_1 - V_2)^2}{2g} + \left(1 + f_2\frac{L_2}{d_2}\right)\frac{V_2^2}{2g} \qquad ...(11.17)$$

The two velocities and the pipe diameters are related by the equation of continuity, thus

$$Q = A_1V_1 = A_2V_2$$

$$= \frac{\pi}{4}d_1^2\,V_1 = \frac{\pi}{4}\,d_2^2\,V_2 \qquad ...(11.18)$$

For known pipeline characteristic, the friction factor and the head loss, Eqs. (11.17) and (11.18) can be used to compute the flow rate Q. Alternatively, if Q is given, the head loss H can be evaluated.

Example 11.15. *A pipe ABC connecting two reservoirs is 80 mm in diameter. From A to B the pipe is horizontal as shown in the following figure, and from B to C it falls by 3.5 metres. The length AB and BC are 25 and 15 m respectively. If the water surface in the reservoir at A is 4 m above the*

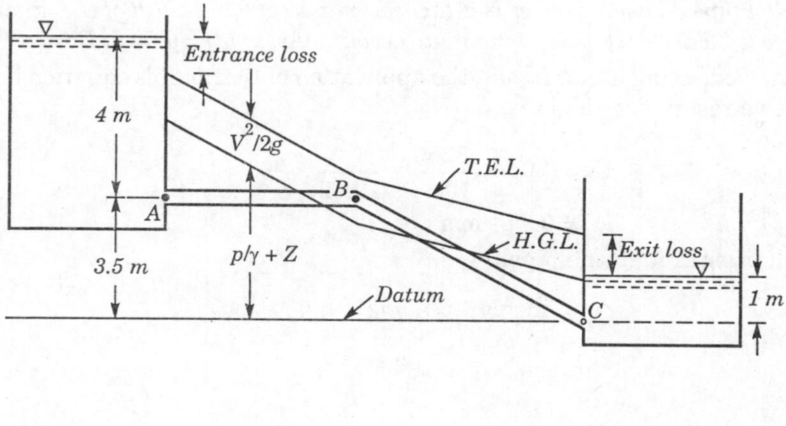

centre line of the pipe and at C 1 m above the centre line of the pipe, calculate the quantity of flow and also the pressure head in the pipe at B. Neglect the loss at the bend but consider all other losses.

Take Darcy f = 0.024 and $K_{entrance}$ *= 0.5. Draw the energy and hydraulic grade lines.*

Solution. Applying the Bernoulli's equation between the water surfaces in the two reservoirs, considering horizontal plane through C as the datum, we obtain

$$6.5 = 0.5\frac{V^2}{2g} + \frac{0.024 \times (25+15)}{0.08}\frac{V^2}{2g} + \frac{V^2}{2g} = 13.5\, V^2/2g$$

from which $V = 3.075$ m/s

$\therefore$ flow rate
$$Q = \frac{\pi}{4}\left[\frac{8}{100}\right]^2 \times 3.075$$
$$= 0.01546 \text{ m}^2\text{/s}$$
$$= 15.46 \text{ lit/sec.}$$

Let p be the pressure at B, then by applying Bernoulli's equation between upstream reservoir surface and B.

$$4 = 0.5\frac{V^2}{2g} + \frac{0.024 \times 25}{0.08} \times \frac{V^2}{2g} + \frac{p}{\gamma} + \frac{V^2}{2g}$$

Substituting $V = 3.075$ m/s

$$\frac{p}{\gamma} = 4 - 4.33$$
$$= -0.33 \text{ m of water below atmosphere.}$$

For plotting energy and hydraulic grade lines, the velocity head is same throughout and the two lines will be parallel and separated by a vertical distance equal to

$$\frac{V^2}{2g} = \frac{(3.075)^2}{19.62} = 0.48 \text{ m.}$$

Total energy at B with respect to horizontal datum through C

$$= 3.5 + \frac{p}{\gamma} + \frac{V^2}{2g}$$
$$= 3.5 - 0.33 + 0.48 = 3.65 \text{ m.}$$

The energy and the hydraulic grade lines are shown in the figure. It may be seen that near B the H.G.L. is below the pipeline thereby indicating negative pressures in that region.

Example 11.16. *An existing 300 mm diameter pipeline of 3200 m length connects two reservoir with 13 m difference in their water levels as shown in the following figure. Calculate the discharge* Q_1.

If a parallel pipe 30 cm diameter is attached to last 1600 m length of the existing pipeline, find the new discharge Q_2. *Take only wall friction into account. Assume f = 0.04 in Darcy-Weisbach formula.*

Solution. Neglecting minor losses, the application of Bernoulli's equation between the water surfaces of the two reservoirs yields

$$13 = 0.04 \times \frac{3200}{0.3} \times \frac{V^2}{19.62},$$

from which $V = 0.774$ m/s

The discharge through the pipeline

$$Q_1 = \frac{\pi}{4}(0.3)^2 \times 0.774 = 0.547 \text{ m}^2\text{/s.}$$

In case of the altered pipeline, since the two parallel pipelines have the same characteristics and have the same head loss, the discharge passing through each one is the same. If Q is the discharge in each of the parallel pipe, the combined flow passing through the first-half of the pipeline is $Q_2 = 2Q$. Bernoulli's equation between the two ends of the pipeline yields :

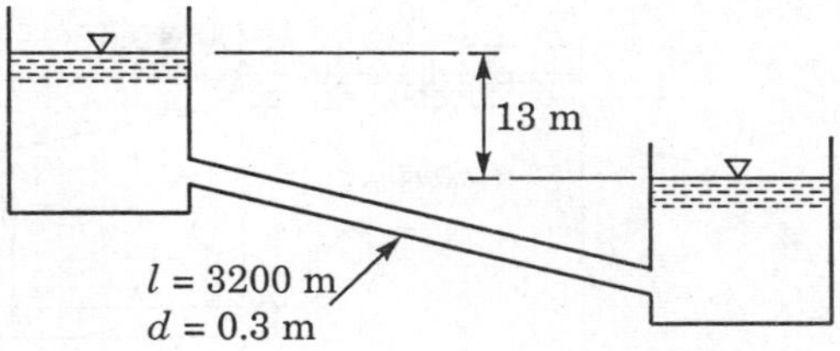

$$13 = 0.04 \times \frac{1600}{0.3}\frac{V_2^2}{2g} + \frac{0.04 \times 1600}{0.3}\frac{V^2}{2g}$$

From the continuity equation, $Q_2 = 2Q$

or $$\frac{\pi}{4}(0.3)^2 V_2 = \frac{2\pi}{4}(0.3)^2 V$$

$\therefore$ $$V_2 = 2V$$

Substituting $V_2 = 2V$, and simplifying, we obtain

$$V = 0.489 \text{ m/s}$$

and the increased flow rate

$$Q_2 = \frac{\pi}{4}(0.3)^2 \times 2 \times 0.489$$
$$= 0.698 \text{ m}^3\text{/s.}$$

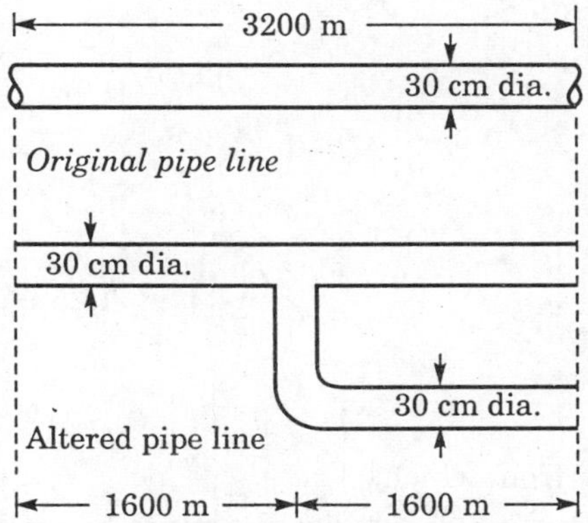

Example 11.17. *Two reservoirs A and C having a difference of level of 15.5 m are connected by a pipeline ABC the elevation of point B being 4.0 m below the level of water in reservoir A. The length AB of the pipeline is 250 m, the pipe being made of mild steel having a friction coefficient f_1, while the length BC is 450 m, the pipe having made of cast iron having a friction coefficient f_2. Both the lengths AB and BC have a diameter of 20 cm. A partially closed valve is located in the length BC at a distance of 150 m from reservoir C.*

If the flow through the pipeline is 3000 lit/min, the pressure head at B is 0.5 m and the head loss at the valve is 5.0 m, find the friction coefficients f_1 and f_2.

Draw the hydraulic grade line and indicate on the diagram head loss values at significant points.

Take into account head loss at entrance and exit points of the pipeline.

Solution. Velocity in the 20 m diameter pipeline

$$V = \frac{4Q}{\pi d^2} = \frac{4 \times \left[\dfrac{3000 \times 10^{-3}}{60}\right]}{\pi \times (0.2)^2} = 1.59 \text{ m/s}$$

1. Applying the Bernoulli's equation between the water surface in reservoir A and the pipe section at B, considering the entrance loss and the friction loss, taking the horizontal plane through the centre of pipe as datum,

$$4 = \frac{p_B}{\gamma} + \frac{V^2}{2g} + 0.5\frac{V^2}{2g} + f_1 \frac{L_{AB}}{d}\frac{V^2}{2g}$$
$$= 0.5 + \frac{(1.59)^2}{2 \times 9.81} \times 1.5 + f_1 \times \frac{250}{0.2} \times \frac{(1.59)^2}{2 \times 9.81}$$

which results in, $f_1 = \mathbf{0.0205}$.

2. Applying the Bernoulli's equation between the water surface in the reservoirs considering all losses in the pipeline ABC including the exit loss,

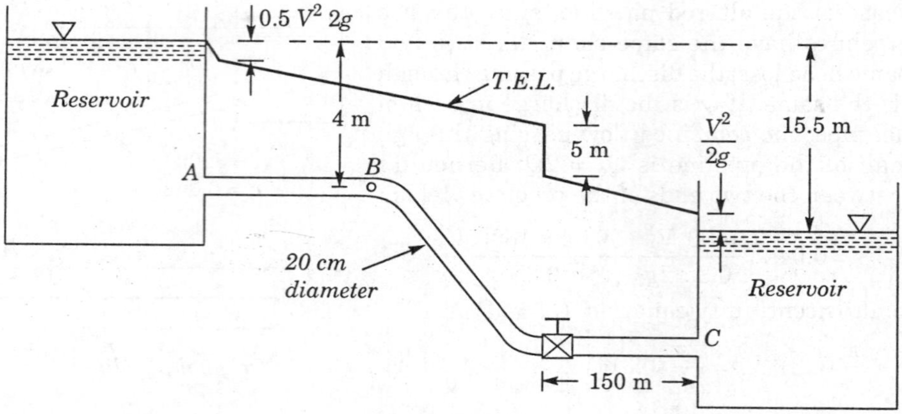

$$15.5 = 0.5\,\frac{V^2}{2g} + f_1 \frac{L_{AB}}{d}\frac{V^2}{2g} + f_2 \frac{L_{BC}}{2g}\frac{V^2}{2g} + 5.0 + \frac{V^2}{2g}$$

$$= 5.0 + 1.5 \times \frac{(1.59)^2}{2 \times 9.81} + 0.0205 \times \frac{250}{0.2} \times \frac{(1.59)^2}{2 \times 9.81} + f_2\,\frac{450}{0.2} \times \frac{(1.59)^2}{2 \times 9.81}$$

from which, $f_2 =$ **0.0241.**

The hydraulic grade line will be $V^2/2g$ below the total energy line (T.E.L.) indicated on the figure.

Example 11.18. *A 1800 m long, 300 mm diameter pipe connects two reservoirs with difference in water levels equal to 16 m. Neglecting minor losses, find the discharge. Take f = 0.018. To augment the discharge by 25% another pipe of same diameter and friction factor is connected to the original pipe midway and connected to the lower reservoir. Find the length of the new pipe.*

Solution. Neglecting minor losses, the difference in water levels *i.e.*, 16 m, of the two reservoirs represents the head loss in friction, hence

$$16 = f\,\frac{L}{D}\frac{V^2}{2g} = 0.018 \times \frac{1800}{0.3} \times \frac{\left(\frac{4Q}{\pi(0.3)^2}\right)^2}{2 \times 9.81}$$

from which, $Q = 0.12055\ \text{m}^3/\text{s}$

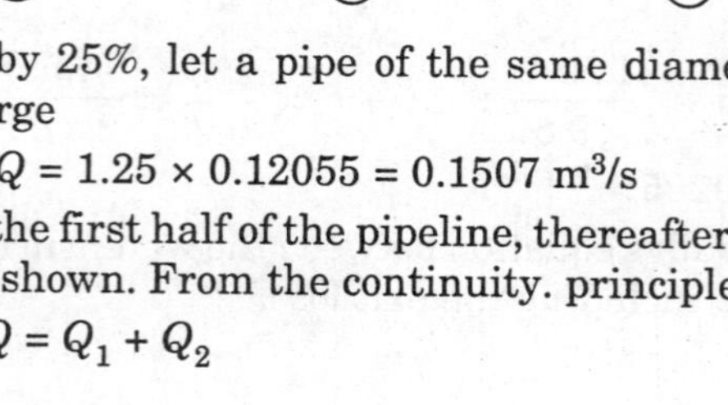

To augment discharge by 25%, let a pipe of the same diameter be connected midway as shown. The augmented discharge

$$Q = 1.25 \times 0.12055 = 0.1507\ \text{m}^3/\text{s}$$

This discharge flows in the first half of the pipeline, thereafter it gets divided into Q_1 and Q_2 which flow in the two pipes as shown. From the continuity. principle,

$$Q = Q_1 + Q_2$$

The head loss in the first half of the pipeline

$$(h_f)_{1-2} = f\frac{L_{12}}{d}\frac{V^2}{2g}\ ;\ V = \frac{Q}{A} = \frac{0.1507}{\left(\frac{\pi}{4}\right)(0.3)^2} = 2.1311 \text{ m/s}$$

$$= 0.018 \times \frac{900}{0.3} \times \frac{(2.1311)^2}{2 \times 9.81} = 12.50 \text{ m}$$

The remaining head (16 – 12.50 = 3.50 m) will be available for driving the flow Q_1 and Q_2 into the respective pipelines.

Thus, $(h_f)_{2-3} = 3.50$ m, given $L_{23} = 900$ m, $d = 0.3$ m.

$$f = \frac{L_{23}}{d}\frac{V_1^2}{2g} = 0.018 \times \frac{900}{0.3} \times \frac{\left(\frac{4Q}{\pi d^2}\right)^2}{2 \times 9.81} = 550.4\ Q_1^2$$

from which $Q_1 = 0.07974$ m³/s.

From the continuity relationship,

$$Q_2 = Q - Q_1 = 0.1507 - 0.07974 = 0.07095 \text{ m}^3\text{/s}$$

For this discharge to flow in the new pipeline connected in parallel to the original one, the length $L_{2\text{-}3}$ can be obtained as follows :

$$(h_f)_{2-3} = f\frac{L_{2-3}}{d}\frac{V_2^2}{2g} = f\frac{L_{2-3}}{d}\frac{(4Q_2/\pi d)^2}{2g}$$

Substituting the known quantities

$$3.50 = 0.018 \times \frac{L_{2-3}}{0.3} \times \frac{\left[\frac{4 \times 0.07095}{\pi \times (0.3)^2}\right]^2}{2 \times 9.81}$$

from which, $\mathbf{L_{2-3} = 1135.67}$ **m**

∴ The length of the new pipe = **1135.67 m.**

11.5.3. Pipeline with a Pump

A pump is used to lift liquid from a lower elevation to a higher one. The pump, therefore, imparts energy to flow, so that with the increased energy thus available, the fluid can rise to greater

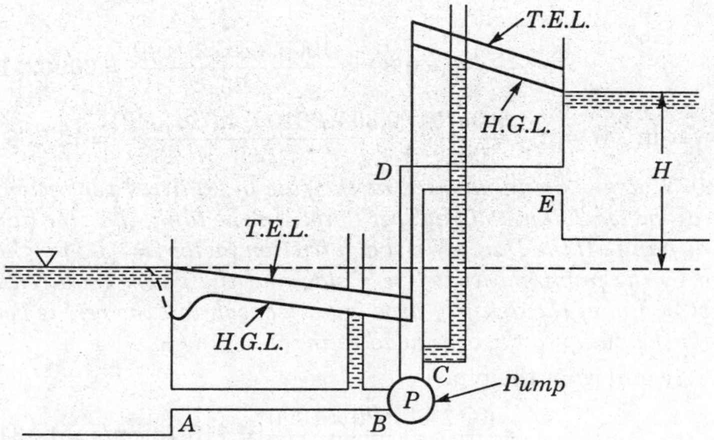

Fig. 11.12. Pipeline (with a pump) connecting two reservoirs.

heights. Fig. 11.12 shows a pipeline installed with a pump and discharging into a reservoir at a higher elevation.

The piezometers installed at C and B indicate the rise in the piezometric head due to increased flow energy supplied by the pump. The pump does work not only in lifting water through the height H, but it also helps in overcoming the head losses in the suction (AB) and discharge pipes (CDE). The total pumping head for this case is

$$h = H + h_L \quad \text{...(11.19)}$$

in which h_L includes all losses in the system from A to E. The power delivered to the liquid by the pump is $\gamma Q(H + h_L)$, where γ is the specific weight of liquid and Q is the flow rate. The power required to run the pump is greater than this, depending upon the efficiency of the pump. If the pump efficiency is η, power input to the pump is

$$\gamma Q(H + h_L)/\eta. \quad \text{...(11.20)}$$

Example 11.19. *Water has to be supplied to a town with one lakh population at the rate of 150 litre/capita/day from a river 2000 m away. The difference in elevation between the lowest water level in the sump and reservoir is 40 m. If the demand has to be supplied in 8 hours, determine the size of the main and brake horsepower of the pumps required. Assume maximum demand as 1.5 times the average demand. Assume f = 0.03, velocity in the pipe 2.4 m/s and efficiency of pump 80 percent.*

Solution. Average demand $= 100{,}000 \times \dfrac{150}{1000}$ m³/day = 15,000 m³ day

This average demand is to be supplied in 8 hours of pumping.

$\therefore$ Average discharge $= \dfrac{15000}{8 \times 60 \times 60} = 3.125$ m³/s

Maximum discharge $= 1.5 \times$ average discharge $= 1.5 \times 3.125 = 4.688$ m³/s

Using Darcy-Weisbach equation for head loss :

$$h_L = f\frac{L}{d}\frac{V^2}{2g} = f\frac{L}{d}\frac{(4Q/\pi d^2)^2}{2g} = \frac{8fLQ^2}{g\,\pi^2 d^5}$$

from which

$$d = \left(\frac{8fLQ^2}{g\,\pi^2 h_L}\right) = \left[\frac{8 \times 0.03 \times 2000 \times (4.688)^2}{9.81 \times \pi^2 \times 40}\right]^{1/5} = 1.222 \text{ m}$$

$\therefore$ Size of the main = 1.222 m

(The practical size of the rising main may be taken as 1.25 m) Brake horsepower of the pumps

$$= \eta_p \frac{rQh_L}{75} = 0.80 \times \frac{1000 \times 4.688 \times 40}{75} = \mathbf{2000.2\ hp.}$$

$$\text{Brake horsepower in kW} = \eta_p \cdot \frac{rQh_L}{1000} = \frac{0.80 \times 9810 \times 4.688 \times 40}{1000} = \mathbf{147.66\ kW.}$$

Example 11.20. *Water is being pumped at the rate of 20 litres per second to an overhead tank through a 15 cm diameter 300 m long delivery pipe. In the tank, the pipe discharges freely at height of 15 m above the pump. If the Darcy-Weisbach friction factor f = 0.03 for the pipe, determine the pressure developed by the pump on its delivery side and the power required to lift the water. Assuming that the first 285 m of the delivery pipe is horizontal and the rest is vertical, sketch the hydraulic gradient from the pump to the over head tank.*

Solution. Velocity of flow in the pipe

$$V = \frac{Q}{A} = \frac{4Q}{\pi d^2} = \frac{4 \times 20 \times 1000}{\pi \times (15)^2} = 113.13 \text{ cm/s} = 1.1313 \text{ m/s.}$$

Head loss due to friction, using Darcy-Weisbach equation,

$$h_L = f\frac{L}{d}\frac{V^2}{2g} = 0.03 \times \frac{300}{0.15} \times \frac{(1.1313)^2}{2 \times 9.81} = 3.914 \text{ m.}$$

Applying Bernoulli's equation to the section just on the delivery side of the pump, where the pressure (gauge) is p, and to the discharge end of the pipeline where the gauge pressure is zero, *i.e.* sections (1) and (2) respectively shown in the following figure. The central line of the horizontal portion of the pipeline has been taken as the datum.

$$0 + \frac{p}{\gamma} + \frac{(1.1313)^2}{2 \times 9.81} = 15 + 0 + \frac{(1.1313)^2}{2 \times 9.81} + 3.914$$

or $$p/\gamma = 18.914 \text{ m of water.}$$

Hence the pressure developed by the pump on the delivery side is 18.914 m of water.

Head loss in 15 m vertical length of pipeline

$$= \frac{15}{300} \times 3.914 = 0.1957 \text{ m.}$$

The hydraulic grade line (HGL) from the pump to the overhead tank plotted by marking the ordinates of piezometric head at (1), (2) and (3) and joining these by straight lines are shown in the figure below.

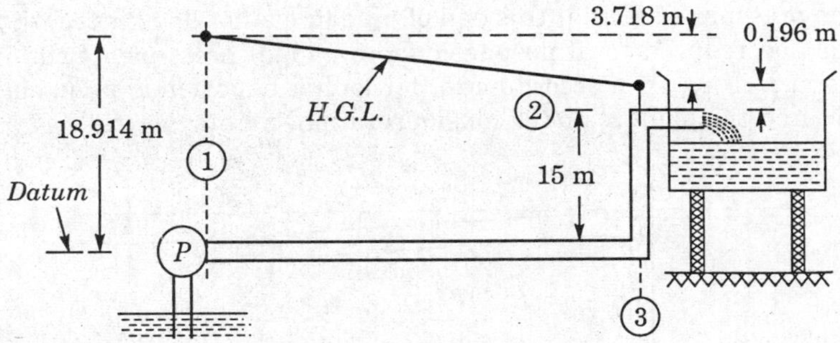

Piezometric head $(p/\gamma + z)$ at (1) $= 18.914$ m

Piezometric head at (3) $= 15.196$ m

Piezometric head at (2) $= 15.000$ m

Power required to lift the water $= \gamma Q/(p/\gamma)$

$= 9810 \times 0.02 \times 18.914 = 3710 \text{ W} = 3.710 \text{ kW}.$

11.5.4. Pipeline with a Turbine

The type of machine which is usually employed for converting the fluid energy into mechanical work is called a turbine. A turbine, thus consumes fluid energy and transforms it into mechanical energy, which is available in the form of rotation of turbine shaft. This is made possible by causing the fluid to pass over the curved blades of a reaction turbine or by the dynamic action of jet impinging on buckets of a Pelton turbine. It is the change in moment of momentum of fluid which produces a torque on the turbine shaft. This torque is responsible for causing the rotation of the turbine runner.

Fig. 11.13 shows a pipeline supplying water from a reservoir. At some suitable location, a turbine is installed. The energy available for conversion is spent in two ways : (*i*) a small part of it is used up in overcoming friction in the pipeline, (*ii*) the remainder is consumed by the turbine. Of that which is delivered to the turbine, a portion is lost in hydraulic friction and the rest is converted into mechanical work. If H is the head variables, the effective head under which the turbine operates is decreased by the friction loss in the pipe line. The head under which the turbine operates is

$$h = H - h_L \quad \text{...(11.22)}$$

where h_L is the head lost in pipe friction. The power delivered to the turbine is

$$P = \gamma Q(H - h_L) \quad \text{...(11.23)}$$

The power delivered by the turbine is less than the value given by Eq. (11.23) and depends upon both the hydraulic and mechanical losses in the turbine. If η is the overall efficiency of the turbine, then the power deli-vered by the turbine is

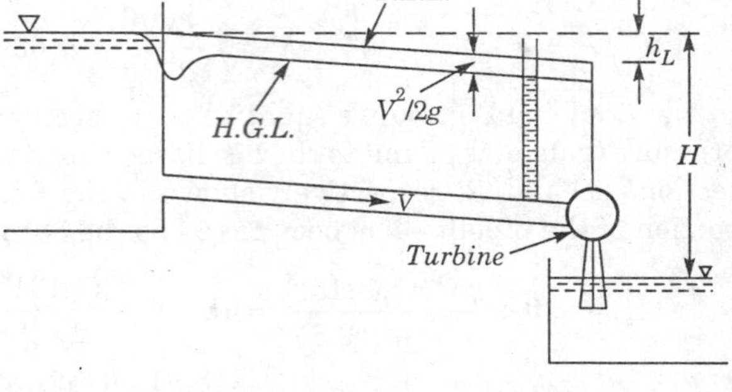

Fig. 11.13. Hydraulic energy utilised by a turbine.

$$P_0 = \eta\gamma Q(H - h_L) \quad \text{...(11.24)}$$

11.6. Hydraulic Power Transmission through a Pipe

Fig. 11.14 shows a pipeline taking supply from a reservoir and letting it out through a nozzle. Consider a point B which is located near the end of the pipeline. For no flow condition, the pressure at B is maximum, being equal to BX. But when flow occurs, the pressure at B drops to the value represented by BY. The greater the rate of flow, the steeper will be hydraulic grade line and the lesser will be the pressure at B. If at this end of pipe line a nozzle is fitted, the pressure of the issuing jet will then be reduced to atmospheric (zero gauge) pressure. If the head loss in the pipeline upto the point B is h_L, then according to Bernoulli's equation applied between a point on the reservoir surface and the point B under considereration, we may write

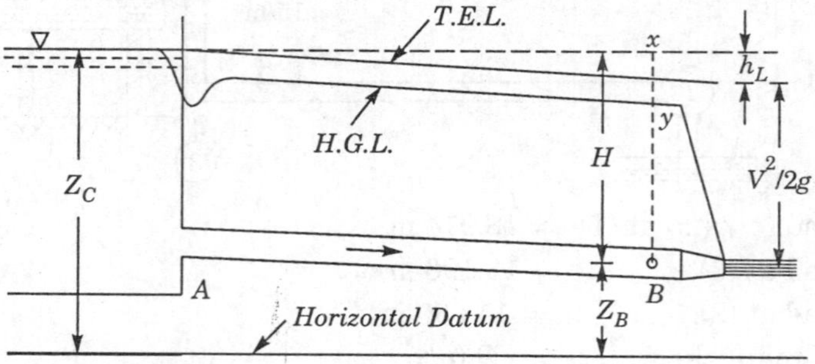

Fig. 11.14. Power delivered by a pipe.

$$Z_C + 0 + 0 = Z_B + \frac{p_B}{\gamma} + \frac{V_B^2}{2g} + h_L$$

the energy available with the flow at $B = Z_B + \dfrac{p_B}{\gamma} + \dfrac{V_B^2}{2g}$

but if the datum is changed to that passing through the point $B(Z_B = 0)$ the energy available at B is

$$\frac{p_B}{\gamma} + \frac{V_B^2}{2g} = H - h_L$$

in which h_L is the head loss upto location B.

If the rate of flow is Q, the available energy is equivalent to a power of $\gamma Q(H - h_L)$. Thus the power transmitted by the pipeline upto the point B is $\gamma Q(H - h_L)$. Let us now investigate the condition for maximum transmission of power. Since the head loss is proportional to the velocity square, we can write $h_L = K_L Q^2$, in which K_L is a constant for a given pipe. Thus the power

$$P = \gamma Q(H - h_L) = \gamma QH - \gamma K_L Q^3$$

$$\frac{dP}{dQ} = \gamma H - 3\gamma K_L Q^2 = \gamma(H - 3h_L)$$

For a maximum value of P, $\dfrac{dP}{dQ} = 0$, and we obtain

$$H - 3h_L = 0, \quad \text{or} \quad h_L = H/3 \qquad ...(11.25)$$

In other words, the power delivered by a given pipeline is maximum when the flow is such that one-third of the static head is consumed in pipe friction. The corresponding discharge

$$Q = \sqrt{H / 3K_L} \qquad ...(11.25\ a)$$

and the maximum power $\quad P = \gamma \sqrt{H / 3K_L} \cdot \dfrac{2H}{3} = \dfrac{2}{3\sqrt{3}} \dfrac{\gamma}{K_L} H^{3/2} \qquad ...(11.25\ b)$

The efficiency of a pipeline may by defined as the power delivered to the power supplied. But for a given discharge power is proportional to the head hence the efficiency is $(H - h_L)/H$. In the case of maximum transmission of power, one-third of H is consumed in friction hence the efficiency is only 66.66%.

Example 11.20.1. *A pipeline 10 km long delivers a power of 50 kW at its outlet end. The pressure at the inlet end is 5000 kN/m², and the pressure drop per km of pipeline is 50 kN/m². Find the size of the pipe and efficiency of transmission. Take friction factor f = 0.02.*

Solution. Pressure head at the inlet, $\left(\dfrac{p}{\gamma}\right) = \dfrac{5000\ (\text{kN/m}^2)}{1000 \times 9.81\ (\text{N/m}^3)} = \dfrac{5000 \times 1000}{1000 \times 9.81} = 509.68$ m

Pressure drop per km of pipeline = 50 kN/m²

∴ Head loss per km = $\Delta p/\gamma = 50 \times 1000/(1000 \times 9.81) = 5.0968$ m

This pressure drop is mainly due to friction loss in the pipeline.

Head loss per km, length, $h_f = f \dfrac{L}{d} \dfrac{V^2}{2g} = f \dfrac{L}{d} \dfrac{Q^2}{\left(\dfrac{\pi}{4} d^2\right)^2} \cdot \dfrac{1}{2g} = 8f \dfrac{L}{d^5} \dfrac{Q^2}{g\pi^2}$

Substituting for $h_f = 5.0968$ m and $L = 1000$ m

$$5.0968 = \frac{8 \times 0.02 \times 1000}{9.81 (3.1415)^2} \frac{Q^2}{d^5}$$

from which, $\quad \dfrac{Q^2}{d^5} = 3.06445$ or $\dfrac{Q}{d^{5/2}} = 1.750 \qquad ...(1)$

We know that, $\quad$ power, $P = \gamma QH$

∴ Power delivered at the outlet end *i.e.* 50 kW = γQ × Head available at the outlet

or, $\quad 50 \times 1000 = (9.81 \times 1000)\ Q\ (509.68 - 5.0968 \times 10)$ Nm/s

$$= 9810 \times 458.712\ Q$$

or, $\quad Q = 0.0111$ m³/s $\qquad ...(2)$

Substituting this value of Q in (1)

$$\frac{0.0111}{d^{5/2}} = 1.750$$

from which, $d = 0.132$ m, or 132 mm.

Commercially available size of pipe = **15 cm or 150 mm**

(*ii*) Efficiency of transmission = $\dfrac{\text{Power delivered at the pipe outlet}}{\text{Power available at the pipe inlet}}$

Power available at the pipe inlet = $\gamma Q\ (p/\gamma) = 9810 \times 0.0111 \times 509.68 = 55.49956$ kW.

∴ Efficiency of transmission = $\dfrac{50}{55.49956} \times 100 =$ **90.09%**

Example 11.21. *A fire engine supplies water to a hosepipe, 75 m long and 0.075 m in diameter, a pressure of 294.3 kN/m² (gauge). The discharge end of the hosepipe has a nozzle of diameter d fixed to it. Taking friction factor, f, as 0.032, determine the diameter d of the nozzle, so that the momentum of the issuing jet may be a maximum.*

Solution. Head loss in the hosepipe

$$h_L = f\frac{L}{d}\frac{V^2}{2g} = 0.032 \times \frac{75}{0.075} \times \frac{V^2}{2g} = 32\frac{V^2}{2g}$$

Applying the continuity equation to the pipe and the jet

$$Q = AV = A_jV_j; \qquad \therefore \quad V_j = \frac{A}{A_j}V = \left(\frac{0.075}{d}\right)^2 V$$

Bernoulli's equation applied to the hosepipe at the fire engine and to the nozzle jet, considering the hosepipe and the nozzle to be in the same horizontal plane

$$0 + \frac{294300}{9810} + \frac{V^2}{2g} = 0 + 0 + \frac{V_j^2}{2g} + 32\frac{V^2}{2g}$$

Here the energy loss in the nozzle being very small has not been considered. Substituting for V_j, the velocity V in the hosepipe can be obtained.

$$30 = \left(\frac{0.075}{d}\right)^4\frac{V^2}{2g} + 31\frac{V^2}{2g}\,;\; V = \sqrt{\frac{2\times30\times9.81}{31+\left(\frac{0.075}{d}\right)^4}}$$

Momentum of issuing jet,

$$M = \rho Q V_j = 102 \times \left\{\frac{\pi}{4}(0.075)^2 \times V\right\} V_j = 0.002532V^2/d^2$$

$$\text{Substituting for } V,\ M = \frac{0.002532}{d^2}\left[\frac{2\times30\times9.81}{31+(0.075/d)^4}\right] = \frac{1.4906}{d^2[31+(0.075/d)^4]}$$

For momentum to be maximum, $\dfrac{dM}{dd} = 0$

$$\frac{d}{dd}\left\{d^2\left[31+\left(\frac{0.075}{d}\right)^4\right]\right\} = 0\,; \qquad \text{or } 62d - \frac{(0.075)^4\times2}{d^3} = 0$$

or $\qquad d^4 = 2 \times (0.075)^4/62\,;\ d = 0.03178$ m.

Therefore, the diameter of the nozzle = 0.03178 m.

11.7. Negative Pressure in Pipeline

A pipe which rises above its hydraulic grade line has negative pressure and is known as a *siphon*. Theoretically the absolute pressure in a pipeline may fall to the vapour pressure of the liquid at which cavitation sets in. Even before such low pressures are attained, the dissolved gases will come out of solution and collect in the highest portion of the pipe in sufficient quantities to form an air lock. When this condition is reached the flow of liquid in a pipe would almost certainly stop. For these reasons the pressure of water in a pipeline should not be allowed to fall below about (– 7.6 kN/m²) gauge. This suggests that the pipeline should not be more than about 7.75 m (or 25.5 ft) higher than the hydraulic grade line. In case of siphons, somewhat higher difference between the pipe and the H.G.L. is possible because the emerging gas bubbles reduce the mean specific weight of the liquid.

11.8. Siphon

When water is lifted by a pipe to a greater height than the initial level in the supply reservoir as shown in Fig. 11.15, the pipe is called a siphon. Alternatively, a siphon is a device which carries liquid from higher level to a lower level through an intermediate high obstruction. A siphon is that portion of pipeline which lies above the hydraulic grade line (HGL).

It is necessary to exhaust the air in the pipe by some means in order to start the flow and if the flow is to continue the air collected at the summit (the highest point of the siphon) must be removed from time to time. The portion of the pipe which lies above the hydraulic grade line, marked by (1), (2), (3), has negative pressures. The highest point of the siphon (*i.e.* the summit) has the largest negative pressure, and since there is a limit upto which pressure can be lowered, the height of summit h_s above the liquid level in the supply reservoir, therefore, has a maximum value compatible with continued maintenance of flow and prevention of cavitation.

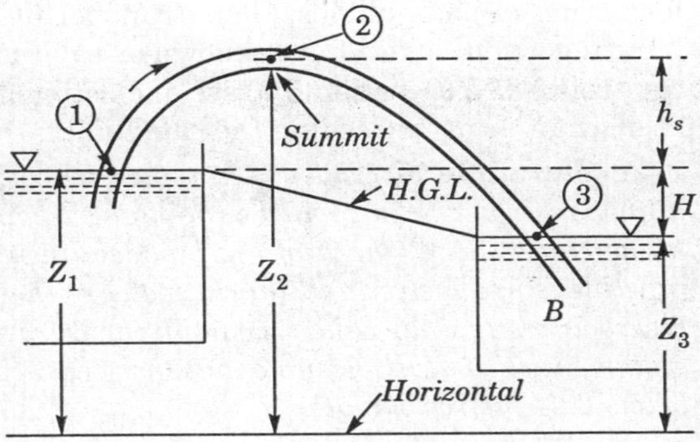

Fig. 11.15. Flow through a siphon.

The problem of flow through a siphon can be analysed using the equation of continuity and the Bernoulli's equation. Applying Bernoulli's equation between any point on the supply reservoir surface and point (3) in the pipe, we may write,

$$Z_1 + 0 + 0 = K\frac{V^2}{2g} + f\frac{L}{d}\frac{V^2}{2g} + \frac{V^2}{2g} + Z_3$$

or
$$Z_1 - Z_3 = H = \left(1 + K + f\frac{L}{d}\right)\frac{V^2}{2g} \qquad \text{...(11.26)}$$

It may be noted that the pressure at (3) is assumed atmospheric, though actually it will be somewhat less. In Eq. (11.26), K is the sum of all minor loss coefficient and L is the total length of pipe from (1) to (3), d is the diameter and f the friction factor. From Eq. (11.26) it is obvious that if H is large, the velocity of flow in the pipe will also be high. For a known discharge Q, if it is desired to determine H it is a matter of direct substitution of various known (directly or indirectly) quantities in Eq. (11.26). If however, H is given and it is required to estimate the rate of flow Q, the solution is accomplished by trial and error procedure, first assuming the velocity and calculating the friction factor and then checking H from Eq. (11.26).

To determine the pressure at the summit point situated at a height h_s or to determine the maximum height h_s for the permissible negative pressure at the summit, we are required to apply Bernoulli's equation between a point on the surface of supply reservoir and the summit marked by point (2). Thus we have

$$Z_1 + 0 + 0 = Z_2 + \frac{p_2}{\gamma} + \frac{V^2}{2g} + K_2\frac{V^2}{2g} + f\frac{L_2}{d}\frac{V^2}{2g}$$

from which the pressure p_2 may be expressed in terms of $h_s = Z_2 - Z_1$

$$\frac{p_2}{\gamma} = -h_s - \left(1 + K_2 + f\frac{L_2}{d}\right)\frac{V^2}{2g} \qquad \text{...(11.27)}$$

in which K_2 represents sum of all minor loss coefficients between points (1) and (2) and L_2 is the pipe length upto the summit point (2). From Eq. (11.27), it can be seen that the pressure at the summit is always negative ; the greater the summit height h_s, the lesser will be the pressure of flow. If however, the summit height is fixed, the summit pressure will vary with the rate of flow, it being lesser for higher rates of flow.

It may be emphasized here that there is a minimum limit for the pressure at the summit. It can never be lower than the vapour pressure of the liquid. It is known for certain that if the fluid pressure reaches in the vicinity of its vapour pressure, the liquid vaporizes and the gases dissolved in it are liberated. The liquid vapours and the liberated gases get deposited near the highest point of the pipe (summit) and when sufficient quantity of these is accumulated, they completely block the flow leading to either discontinuity of flow or its complete stoppage. The siphon then stops functioning. When the fluid pressure is reduced to its vapour pressure more serious problems arise due to the formation of gas bubbles, their transportation downstream by the flow and their subsequent bursting give rise to what is known as cavitation. A siphon should be so designed as to ensure its cavitation-free operation. In case of existing siphons, cavitation can be eliminated by operating it at discharges not exceeding the safe limit.

Example 11.22. *Two reservoirs are connected by a pipeline which rises above the level of the higher reservoir. What will be the highest point of the siphon above the water surface level in the higher tank if the length of the pipe leading upto this point is 450 metres, the total length of the pipeline is 915 metres and the diameter of the pipe is 30 cm ? The difference in the levels of the two reservoirs is 12.5 m. Take f = 0.04 in the Darcy-Weisbach equation for the head loss in the pipe. The siphon must run full. Assume that the separation will occur if the absolute pressure in the pipe falls below 2.44 metres of water.*

Solution. $L_{12} = 450$ m, $L_{13} = 915$ m, $d = 0.3$ m and $f = 0.04$.

Applying the Bernoulli's equation between points (1) and (3)

$$12.5 = 0.5\frac{V^2}{2g} + f\frac{L_{13}}{d}\frac{V^2}{2g} + \frac{V^2}{2g}$$

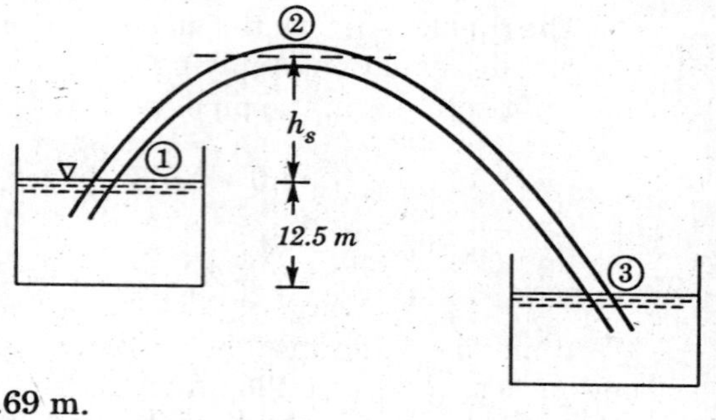

Substituting for the known quantities and solving for V, we obtain $V = 1.41$ m/s.

Bernoulli's equation between points (1) and (2) taking $p_1/\gamma = 10.35$ m of water absolute, we obtain

$$10.35 = 0.5\frac{V^2}{2g} + f\frac{L_{12}}{d}\frac{V^2}{2g} + \frac{V^2}{2g} + h_s + 2.44$$

Substituting for the known quantities, $h_s = 1.69$ m.

Example 11.23. *Water from a main canal is siphoned to a branch canal over an embankment by means of wrought iron pipes of 9 cm diameter. The length of pipeline upto the summit is 25 m and the total length 65 m. Entry loss may be assumed as one-half of the velocity head in the pipe. Assume friction factor, f = 0.03. Water surface elevation in the branch canal is 10 m below that of the main canal.*

(i) If the total quantity of water required to be conveyed is 60 lit/sec, how many pipe lines are needed.

(ii) What is the maximum permissible height of the summit above the water level in the main canal so that the water pressure at summit may not fall below 19.62 kN/m² absolute, the barometer reading being 10 m of water ?

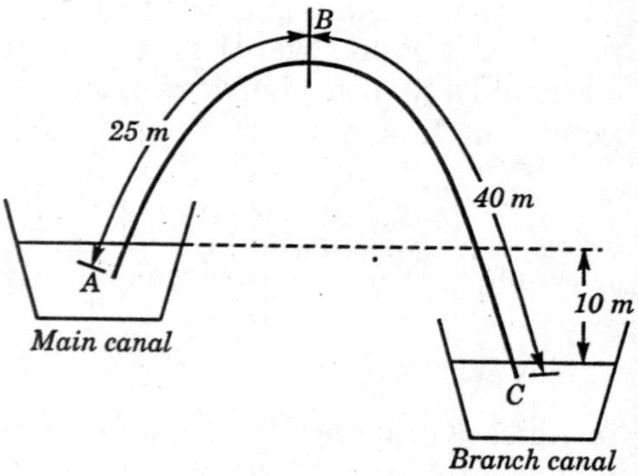

Solution. Applying Bernoulli's equation between the water surfaces of two canals

$$10 = 0.5\frac{V^2}{2g} + 0.03\frac{65}{9 \times 10^{-2}}\frac{V^2}{2g} + \frac{V^2}{2g}$$

which results in, $V = 2.91$ m/s.

Discharge through a 9 cm diameter pipe

$$Q = \frac{\pi}{4} \times \left(\frac{9}{100}\right)^2 \times 2.91 = 0.0185 \text{ m}^3/\text{s} = 18.5 \text{ lit/s}.$$

Number of 9 cm pipes to convey 60 lit./s = $\frac{60}{18.5}$ = 3.24, say 4.

Four number 9 cm diameter pipes will be required to convey a quantity of 60 lit./s.

(*ii*) Let h be the height of the summit above the water level in the main canal. Writing Bernoulli's equation between water surface in the canal and the summit point B (considering the absolute pressures and taking the water surface in the canal as the datum).

$$0+\frac{98100}{9810}+0=h+\frac{19620}{9810}+\frac{V^2}{2g}+0.5\frac{V^2}{2g}+0.03\times\frac{25}{9\times10^{-2}}\frac{V^2}{2g}$$

$$10=h+2+(1+0.5+8.33)\times\frac{(2.91)^2}{2\times9.81}$$

from which, $\quad h = \mathbf{3.76\ m.}$

11.9. Multiple Pipe Systems

Frequently, problems of divided flow in pipelines are encountered in engineering practice. These problems include looping pipes (pipe connected in parallel), branching pipes and pipe network ; and may be associated with additions to old systems or with the design of entirely new systems. In all such cases, the minor losses are, however, so small that these may be ignored. In such problems, apart from the minor losses, velocity heads and variation of friction factor f with Reynolds number **R** are usually neglected, and calculations are made on the basis of coincident energy and hydraulic gradelines.

11.9.1. Pipes in Series

If two or more pipes are connected in series, that is end to end as shown in Fig. 11.11, the same flow passes through each pipe. As the total energy line shows, the total head loss is the sum of the losses in each of the individual pipes and fittings. We may thus write

$$\left.\begin{aligned} Q &= Q_1 = Q_2 = Q_3 \\ \text{and} \quad h_L &= h_{L_1} + h_{L_2} + h_{L_3} \end{aligned}\right\} \quad ...(11.28)$$

Example 11.24. *A horizontal pipeline 40 m long is connected to a water tank at one end and discharges freely into the atmosphere at the other end. For the first 25 m of its length from the tank, the pipe is 15 cm diameter and then it is suddenly enlarged to 30 cm. The height of water level in the tank is 8 m above the centre of the pipe considering all losses of head which occur, determine the rate of flow. Take f = 0.01 for both sections of pipe.*

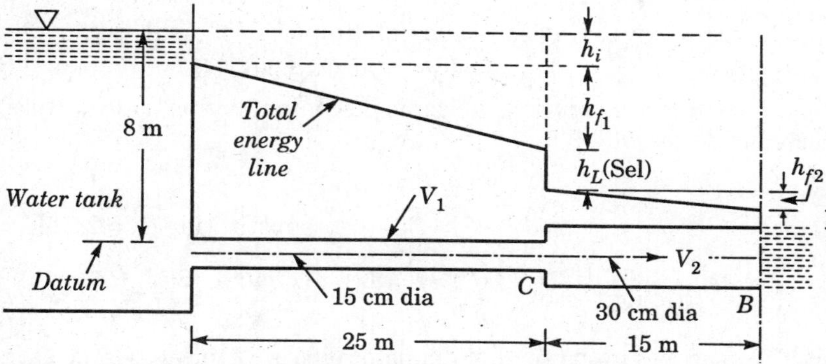

Solution. The horizontal pipeline AB taking off from the water tank at A and undergoing sudden enlargement at C is shown schematically in the figure. Various losses of energy (*i.e.* heads) which occur in pipe flow from A to B are as follows :

(*i*) Entrance loss at inlet A, $h_i = 0.50\,\dfrac{V_1^2}{2g}$

$$V_1=\frac{4Q}{\pi d_1^2}=\frac{4Q}{\pi(0.15)^2}=56.59\,Q\ ;\quad h_i=0.50\,\frac{(56.59)^2\,Q^2}{2\times9.81}=81.61\,Q^2$$

(*ii*) Friction loss in 15 cm dia 25 m long pipe

$$h_{f1}=f\,\frac{L_1}{d_1}\frac{V_1^2}{2g}=0.01\times\frac{25}{0.15}\times163.22\ ;\quad Q^2=272.03\,Q^2$$

(*iii*) Head loss in sudden enlargement at C from 15 cm dia to 30 cm dia

$$(h_L)_{SEL}=\frac{(V_1-V_2)^2}{2g}=\frac{16Q^2}{2g\,\pi^2}\left[\frac{1}{d_1^4}-\frac{1}{d_2^4}\right]=153.09\,Q^2$$

(*iv*) Friction loss in 30 cm dia 15 m long pipe

$$h_{f2} = f\frac{L_2}{d_2}\frac{V_2^2}{2g} = 0.01 \times \frac{15}{0.30} \times \frac{(4Q/\pi d_2^2)^2}{2 \times 9.81} = 5.1\,Q^2$$

(*v*) Exit loss at B, $h_e = \dfrac{V_2^2}{2g} = \dfrac{16\,Q^2}{2g\,\pi^2\,d_2^4} = 10.20\,Q^2$

Writing the Bernoulli's equation between tank water surface and the exit end, taking the pipe centreline which is horizontal, as the datum

$$8 + 0 + 0 = 0 + 0 + \frac{V_2^2}{2g}\,(= 10.2\,Q^2) + 81.61\,Q^2 + 272.03\,Q^2 + 153.09\,Q^2 + 5.1\,Q^2 = 522.03\,Q^2$$

from which $Q = \sqrt{8/522.03} = 0.1238$ m³/s = **123.8 litre/sec.**

Example 11.25. *Consider two pipes of the same length and having the same roughness coefficient but with the diameter of one pipe being half the other. Determine (i) the ratio of the discharges through these pipes if the friction loss for both of them is to be the same, and (ii) the ratio of friction losses, if the pipes carry the same discharge.*

Solution. Case (1) Friction loss in the two pipes is the same *i.e.* $h_{f1} = h_{f2}$; to find Q_1/Q_2. It is the case of pipes connected in parallel. (Refer article 11.9.2)

Given. $L_1 = L_2, f_1 = f_2$ and $d_1 = d_2/2$

Equating the friction loss $\quad f_1 = \dfrac{L_1}{d_1}\dfrac{V_1^2}{2g} = f_2\dfrac{L_2}{d_2}\dfrac{V_2^2}{2g}$

making the substitutions, $\quad \dfrac{V_1^2}{V_2^2} = \dfrac{d_1}{d_2} = \dfrac{1}{2}$ or $\dfrac{V_1}{V_2} = \dfrac{1}{\sqrt{2}}$

Ratio of discharges, $\quad \dfrac{Q_1}{Q_2} = \dfrac{V_1\,d_1^2}{V_2\,d_2^2} = \dfrac{1}{\sqrt{2}}\left(\dfrac{1}{2}\right)^2 = \dfrac{1}{4\sqrt{2}}$

The discharge in the larger diameter pipe *i.e.*, of dia d_2 is

$$Q_2 = 4\sqrt{2}\;Q_1 = \mathbf{5.657Q_1}$$

Case (2) Pipes carrying the same discharge, to find h_{f1}/h_{f2}. It represents the case of pipes connected in series.

$$Q_1 = Q_2$$

$$\frac{\pi\,d_1^2}{4}\,V_1 = \frac{\pi\,d_2^2}{4}\,V_2, \text{ substituting } d_2 = 2d_1.\ V_1 = 4V_2$$

Ratio of friction losses, $\quad \dfrac{h_{f1}}{h_{f2}} = \dfrac{f_1}{f_2}.\dfrac{L_1}{L_2}\,.\,\dfrac{d_2}{d_1}.\dfrac{V_1^2}{V_2^2} = \dfrac{2d_1}{d_1}\dfrac{16V_2^2}{V_2^2} = 32$

The head loss in the smaller diameter pipe is 32 times as large as in the larger diameter pipe.

Example 11.26. *Three pipes—300 m long, 30 cm in diameter ; 150 m long 20 cm diameter ; 200 m long, 25 cm diameter—are connected in series between two reservoirs. The friction factor values of for the three pipes are 0.018, 0.020 and 0.019 respectively. Determine the rate of flow if the difference in elevations of water levels between two reservoirs is 15 m. Use* $h_f = f\dfrac{L}{d}\dfrac{V^2}{2g}$ *for friction loss and account for all losses. Draw the hydraulic grade line for the flow. Contractions and expansions are sudden.*

Solution. Various types of losses which occur in the pipeline (1)—(2)—(3) are :

(*i*) Entrance loss $= 0.5\dfrac{V_1^2}{2g} = \dfrac{0.5}{2g}\left[\dfrac{4Q}{\pi d_1^2}\right]^2 = \dfrac{0.5 \times 16\,Q^2}{19.62 \times \pi^2 \times (0.3)^4} = 5.09\,Q^2$

(*ii*) Friction loss in pipe (1) of diameter 30 cm

$$= f_1\frac{L_1}{d_1}\frac{V_1^2}{2g} = \frac{f_1L_1}{2gd_1}\left[\frac{4Q}{\pi d_1^2}\right]^2 = \frac{0.018 \times 300}{19.62 \times 0.8}\left[\frac{4Q}{\pi \times (0.3)^2}\right] = 183.2\,Q^2.$$

(*iii*) Loss due to sudden contraction from diameter 30 to 20 cm $= K_L \dfrac{V_2^2}{2g}$

The loss coefficient K_L, for $d_2/d_1 = 20/30 = 0.67$, from table 11.1 has been obtained as 0.24. The loss of head is then

$$= \frac{0.24}{2g}\left[\frac{4Q}{\pi d_2^2}\right]^2 = \frac{0.24}{19.62}\left[\frac{4Q}{\pi\times(0.2)^2}\right]^2 = 12.37\,Q^2$$

(*iv*) Friction loss in pipe (2) of diameter 20 cm

$$= f_2\frac{L_2}{d_2}\frac{V_2^2}{2g} = \frac{8}{\pi^2} f_2 L_2 \frac{Q^2}{g d_2^5} = \frac{8}{\pi^2}\times 0.020\times 150\times\frac{Q^2}{9.81\times(0.2)^5} = 773.0\,Q^2.$$

(*v*) Loss due to sudden enlargement from diameter 20 to 30 cm

$$= \frac{(V_2 - V_3)^2}{2g} = \frac{8Q^2}{g\pi^2}\left[\frac{1}{d_2^4}-\frac{1}{d_3^4}\right] = \frac{Q^2}{9.81\pi^2}\left[\frac{1}{(0.2)^4}-\frac{1}{(0.25)^4}\right] = 30.45\,Q^2$$

(*vi*) Friction loss in pipe (3) of diameter 25 cm

$$= \frac{8}{\pi^2} f_3 L_3 \frac{Q^2}{g d_3^5} = \frac{8}{\pi^2}\times 0.019\times 200\times\frac{Q^2}{9.81\times(0.25)^5} = 321.0\,Q^2.$$

(*vii*) Exit loss $= \dfrac{V_3^2}{2g} = \dfrac{1}{2g}\left\{\dfrac{4Q}{\pi d_3^2}\right\}^2 \quad \dfrac{16Q^2}{19.62\times\pi^2\times(0.25)^4} = 21.1\,Q^2.$

Applying Bernoulli's equation between the water surfaces of the two reservoirs

$$15 + 0 + 0 = 0 + 0 + 0 + 5.09\,Q^2 + 183.2\,Q^2 + 12.37\,Q^2 + 773\,Q^2 + 30.45\,Q^2 + 321\,Q^2 + 21.1\,Q^2$$

$$15 = 1346.21\,Q^2\ ;\ Q = 0.1057\ \text{m}^3/\text{s} = \textbf{105.7 lit/sec.}$$

11.9.2. Pipes in Parallel (Looping Pipes)

In pipeline practice, laying a pipeline (2) parallel to an existing pipeline (1) and connected to it, as shown in Fig. 11.16, is a standard method of increasing the capacity (flow) of the pipeline. Application of the principle of continuity shows that the flow rate in the main pipeline is equal to the sum of the flow rates in pipes (1) and (2). Also the distribution of flow in the branches must be such that the same head loss occurs in each branch between the common points *A* and *B*. Thus the following relationships must be satisfied :

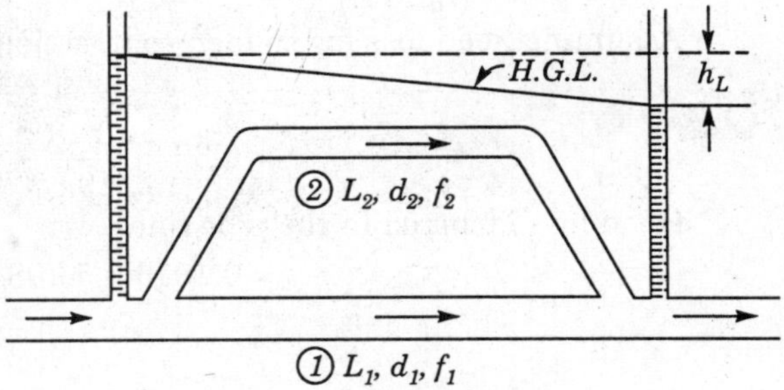

Fig. 11.16. Pipeline in parallel.

$$h_L = h_{L1} = h_{L2}$$

and $\quad Q = Q_1 + Q_2 \qquad$ ...(11.29)

The head loss may be expressed in terms of flow rate using Darcy-Weisbach equation

$$h_L = f\frac{L}{d}\frac{V^2}{2g} = \frac{fL}{2gd}\left(\frac{16Q^2}{\pi^2\times g d^4}\right) = \left(\frac{8fL}{\pi^2 g d^5}\right)Q^2 \qquad ...(11.30)$$

in which the quantity within brackets is a constant for a pipe of known characteristics. An empirical equation for pipe flow such as the Hazen-William formula (applicable for water only), see art. 9.20, which gives $h_L \alpha Q^{1.85}$, may also be used in place of the Darcy-Weisbach equation. In general, the head loss may be written in the form

$$h_L = KQ^n \qquad ...(11.31)$$

Substituting this in Eq. (11.29), we obtain, $K_1 Q_1^{\,n} = K_2 Q_2^{\,n}$

and $\quad Q = Q_1 + Q_2$

The solution of these simultaneous equations will yield the break up of the flow rate Q into Q_1 and Q_2 for known pipe characteristics. The increased flow rate obtainable by looping an existing pipeline can also be estimated by making use of the above principles.

Example 11.27. *Two pipes of diameter 40 cm and 20 cm are each 300 m long. When the pipes are connected in series and discharge through pipe line is 0.10 m³/sec, find the loss of head incurred. What would be the loss of head in the system to pass the same total discharge when the pipes are connected in parallel ? Take f = 0.0075 for each pipe.*

Solution. Pipe in Series. Velocity in 40 cm pipe,

$$V_1 = \frac{4Q}{\pi d_1^2} = \frac{4 \times 0.10}{\pi (0.4)^2} = 0.796 \text{ m/s}$$

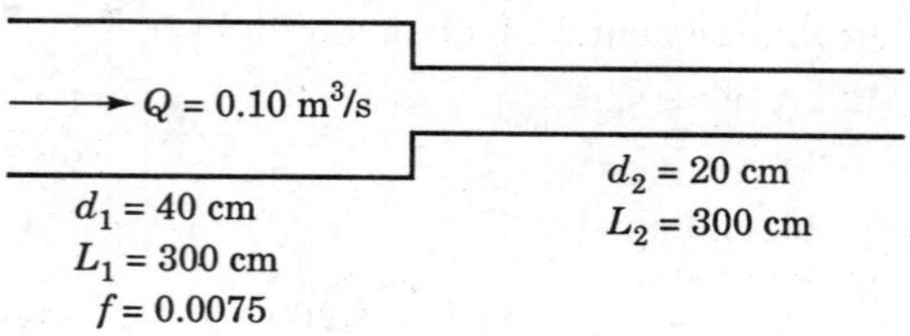

Velocity in 20 cm pipe,

$$V_2 = \frac{4Q_2}{\pi d_2^2} = \frac{4 \times 0.10}{\pi (0.2)^2} = 3.18 \text{ m/s}$$

$$\text{Head loss in 40 cm pipe} = 0.0075 \times \frac{300}{0.40} \times \frac{(0.796)^2}{19.62} = 0.1816 \text{ m}.$$

$$\text{Head loss in 20 cm pipe} = 0.0075 \times \frac{300}{0.20} \times \frac{(3.18)^2}{19.62} = 5.7984 \text{ m}$$

$$\text{Assuming head loss for sudden contraction } \left(\text{for } \frac{d_2}{d_1} = 0.5\right) = 0.33 \frac{V_2^2}{2g}$$

$$= 0.33 \times \frac{(3.18)^2}{19.62} = 0.17 \text{ m}$$

Head loss incurred in the pipe line

$$= 0.1816 + 5.7984 + 0.17 = \mathbf{6.15 \text{ m.}}$$

Pipes in Parallel

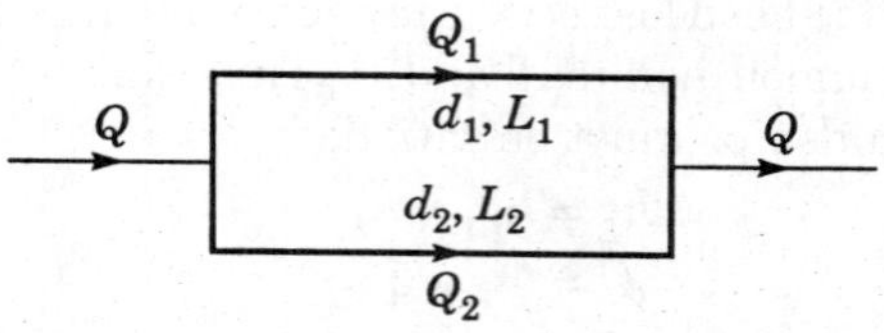

From the continuity relationship

$$Q = Q_1 + Q_2$$

or

$$0.10 = Q_1 + Q_2$$

$$= \frac{\pi}{4}(0.4)^2 V_1 + \frac{\pi}{4}(0.2)^2 V_2 = 0.1257 V_1 + 0.0314 V_2 \quad ...(1)$$

The pipes being connected in parallel, the head loss in each pipe will be the same

$$h_f = f_1 \frac{L_1}{d_1}\frac{V_1^2}{2g} = f_2 \frac{L_2}{d_2}\frac{V_2^2}{2g}$$

Since $f_1 = f_2$

$$\frac{L_1 V_1^2}{2gd_1} = \frac{L_2 V_2^2}{2gd_2}$$

$$\frac{300\,V_1^2}{0.40} = \frac{300\,V_2^2}{0.20}$$

or $\frac{V_1^2}{V_2^2} = 2$

$$V_1^2 = 2V_2^2 = 2\left(\frac{0.10 - 0.1257\,V_1}{0.0314}\right)^2 = 20.28 + 32\,V_1^2 - 50.96\,V_1$$

or $31\,V_1^2 - 50.96\,V_1 + 20.28 = 0$

$$V_1 = \frac{50.96 \pm \sqrt{2596.92 - 2514.72}}{62} = \frac{50.96 \pm 9.07}{62}$$

$$= 0.97 \text{ m/s} ;\ 0.6756 \text{ m/s}$$

Using $V_1 = 0.97$ m/s,

$$Q_1 = \frac{\pi}{4}(0.4)^2 \times 0.97 = 0.1216 \text{ m}^3\text{/s}$$

Since $Q_1 > Q$, $V_1 = 0.97$ m/s is not realistic.

Using the other value of $V_1 = 0.6756$ m/s

$$Q_1 = \frac{\pi}{4}(0.4)^2 \times 0.6756 = 0.0849 \text{ m}^3\text{/s}$$

$$Q_2 = 0.10 - 0.0849 = 0.0151 \text{ m}^3\text{/s}$$

Head loss $h_f = f\frac{L_1}{d_1}\frac{V_1^2}{2g} = 0.0075 \times \frac{300}{0.40} \times \frac{(0.6756)^2}{2 \times 9.81} = \mathbf{0.131\ m.}$

Example 11.28. *Two pipes of diameter D and d and of equal length L are considered. If the pipes are arranged in parallel, the loss of head for either pipe for a flow of Q is h. If the pipes are arranged is series and the same quantity Q flows through them, the loss of head is H. If d = 0.5 D, find the percentage of total flow through each pipe when placed in parallel and the ratio of H to h neglecting minor losses and assuming friction coefficients to be constants.*

Solution. The pipes connected in parallel and series are shown in the following figure :

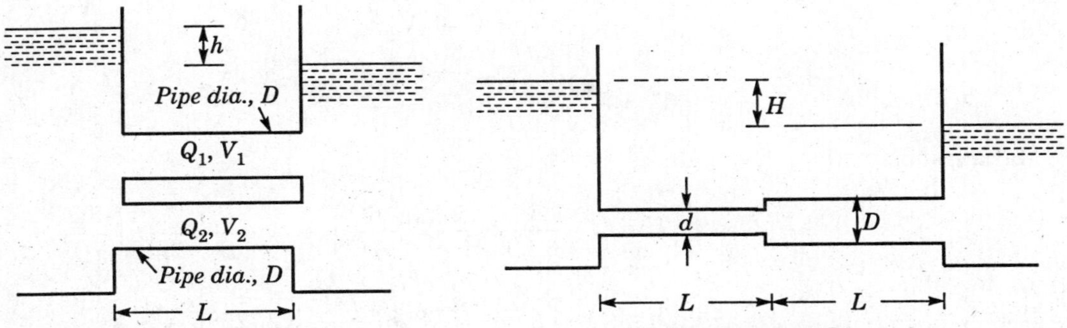

(*a*) Pipes in parallel (*b*) Pipes connected in series

Fig. Ex. 11.28

Considering Fig. Ex. 11.28 (*a*) in which pipes are shown connected in parallel, the head loss in both the pipes is the same,

$$h = f_1 \frac{L}{d}\frac{V_1^2}{2g} = f_2 \frac{L}{D}\frac{V_2^2}{2g}$$

Using the continuity equation $Q_1 = \frac{\pi}{4} d^2 V_1$ and $Q_2 = \frac{\pi}{4} D^2 V_2$

since $d = 0.5\, D$, one obtains $V_1 = \frac{4Q_1}{0.25\pi D^2} = \frac{5.091\, Q_1}{D^2}$ and $V_2 = \frac{4Q_2}{\pi D^2} = \frac{1.273\, Q_2}{D^2}$

Substituting these values, $h = f_1 \frac{L}{0.5D}\frac{1}{2g}\left(\frac{5.091\, Q_1}{D^2}\right)^2 = 2.642\, f_1 \frac{LQ_1^2}{D^5}$

also, $h = f_2 \frac{L}{d}\frac{1}{2g}\left(\frac{1.273 Q_2}{D^2}\right)^2 = 0.0818\, f_2 \frac{LQ_2^2}{D^5}$

from the above equations, one can write

$$Q_1 = \sqrt{\frac{hD^5}{2.642\, f_1 L}} \text{ and } Q_2 = \sqrt{\frac{hD^5}{0.0818\, f_2 L}}$$

The discharge ratio $\frac{Q_2}{Q_1} = \sqrt{\frac{2.642\, f_1}{0.0818\, f_2}} = 5.685\sqrt{\frac{f_1}{f_2}},\ \frac{Q_2 + Q_1}{Q_1} = 1 + 5.683\sqrt{\frac{f_1}{f_2}}$

Assuming $f_1 = f_2,\ \frac{Q_1 + Q_2}{Q_1} = 6.683$

Expressing the ratio in percentage

$$\frac{Q_1}{\text{Total flow } (Q_1 + Q_2)} = \frac{1}{6.683} \times 100 = 14.96\% \quad \textbf{say 15\%}$$

$$\frac{Q_2}{Q_1 + Q_2} = 100 - 14.96 = 85.04\% \quad \textbf{say 85\%}$$

Pipes in series

The head loss $H = f_1 \frac{L}{d}\frac{V_1^2}{2g} + f_2 \frac{L}{D}\frac{V_2^2}{2g}$

Since the same discharge flows

or $Q = \frac{\pi d^2}{4} V_1 = \frac{\pi D^2}{4} V_2$

or $V_1 = \frac{4Q}{\pi d^2} = \frac{16Q}{\pi D^2},\ V_2 = \frac{4Q}{\pi D^2}$

Making substitution, the head loss

$$H = f_1 \times \frac{L}{0.5D} \times \frac{1}{2g} \times \left(\frac{16Q}{\pi D^2}\right)^2 + f_2 \times \frac{L}{D} \times \frac{1}{2g} \times \left(\frac{4Q}{\pi D^2}\right)^2$$

and letting $f_1 = f_2 = f$, $H = 2.644\, f \frac{LQ^2}{D^5}\, 0.0818 f \frac{LQ^2}{D^2} = 2.7258 f \frac{LQ^2}{D^5}$ ...(1)

Considering the pipes arranged in parallel, and assuming $f_1 = f_2 = f$ as earlier, the flow in pipe of diameter D is $Q_2 = 0.85\ Q$. One can, therefore, write

$$h = f\frac{L}{D}\left(\frac{4\times 0.85\,Q}{\pi D^2}\right)^2 \times \frac{1}{2g} = 0.0591\frac{fLQ^2}{D^5} \qquad \text{...(2)}$$

Using Eqs. (1) and (2), the ratio of H and h,

$$\frac{H}{h} = \frac{2.7258}{0.0591} = 46.12$$

Check : Considering the pipe of diameter d in which the discharge is $Q_1 = 0.15\ Q$,

$$h = f\frac{L}{0.5D}\times\frac{1}{2g}\times\left(\frac{16\times 0.15Q}{\pi D^2}\right)^2 = 0.05949\frac{fLQ^2}{D^5} \qquad \text{...(3)}$$

Using Eqs. (1) and (3), $\dfrac{H}{h} = \dfrac{2.7258}{0.05949} = 45.82$

The two values of the ratio H/h can be seen to be almost equal. The slight difference is due to the approximate values of Q_1 and Q_2 used.

Example 11.29. *Two sharp ended pipes of diameters 50 mm and 100 mm respectively, each of length 100 m are connected in parallel between two reservoirs which have a difference of level of 10 m. If the co-efficient of friction for each pipe is (4f) = 0.32, calculate the rate of flow for each pipe and also the diameter of a single pipe 100 m long which would give the same discharge, if it were substituted for the original two pipes.*

Solution. From the Bernoulli's equation, one can write the following relationship. Difference in the level of the reservoirs

= Entrance loss + Friction loss + Exit loss

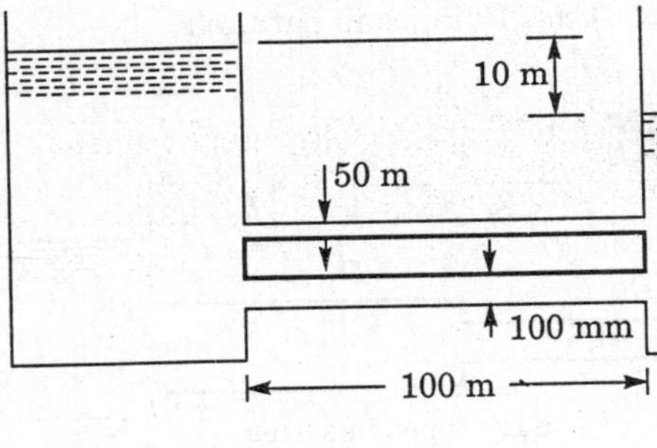

or $\quad 10 = 0.5\dfrac{V^2}{2g} + \dfrac{(4f)LV^2}{2gd} + \dfrac{V^2}{2g}$

$$= \frac{V^2}{2g}\left[0.5 + \frac{0.32\times 100}{d} + 1\right]$$

For 50 mm pipe

$$10 = \frac{V^2}{2g}\left[1.5 + \frac{32}{50\times 10^{-3}}\right] = 32.7\ V^2$$

or $\quad V = 0.553$ m/s

Rate of flow, $\quad Q = \dfrac{\pi}{4}d^2V = \dfrac{\pi}{4}(50\times 10^{-3})^2 \times 0.553 = 4.34\times 10^{-3}$ m^2/s $= 4.34$ lit/s

For 100 mm pipe $\quad 10 = \dfrac{V^2}{2g}\left[1.5 + \dfrac{32}{100\times 10^{-3}}\right] = 16.39\ V^2$

$V = 0.781$ m/s

Rate of flow, $\quad Q = \dfrac{\pi}{4}d^2V = \dfrac{\pi}{4}(100\times 10^{-3})^2 \times 0.781 = 6.134\times 10^{-3}$ m^3/s $= 6.134$ lit/s

Total rate of flow $\quad = 4.34 + 6.134 = 10.474$ lit/s

Diameter of single pipe for the same discharge

$Q = 10.474 \times 10^{-3}$ m^3/s

$$V = \frac{4Q}{\pi d^2} = \frac{4 \times 10.474 \times 10^{-3}}{\pi d^2} = \frac{0.0133}{d^2}$$

Assuming the same value of friction factor, one can write the following equation :

$$10 = \frac{V^2}{2g}\left[1.5 + \frac{0.32 \times 100}{d}\right] = \left(\frac{0.0133}{d^2}\right)^2 \times \frac{1}{2g}\left[1.5 + \frac{32}{d}\right]$$

Neglecting minor losses

$$10 = \left(\frac{0.0133}{d^2}\right)^2 \times \frac{1}{2 \times 9.81} \times \frac{32}{d} = \frac{2.885 \times 10^{-4}}{d^5}$$

from which, $d = 1.236 \times 10^{-1}$ m = 0.1236 m or 123.6 mm.

A pipe of nominal diameter 125 mm may be used to replace the two pipes.

Example 11.30. *Two pipes each of length L and diameters D_1 and D_2 are arranged in parallel ; the loss of head, when a total quantity of Q flows through them, being h_1. If the pipes are arranged in series and the same quantity of water, Q, flows through them, the loss of head is h_2. If $D_1 = 2D_2$ find the ratio of h_1 and h_2. Neglect minor loss and friction factor, f, to be constant.*

Solution. Given $D_1 = 2D_2$

f = constant

To find, h_1/h_2.

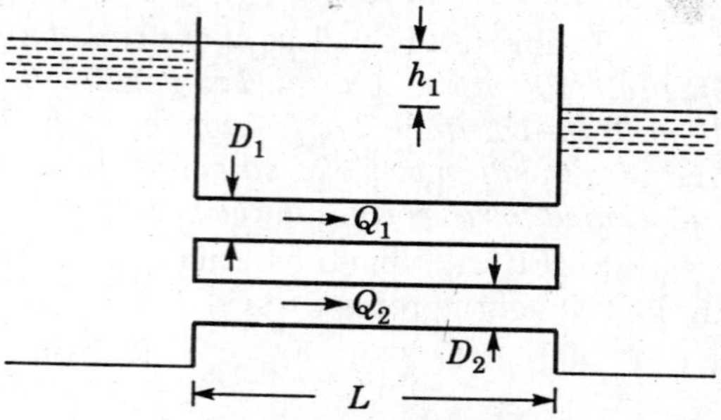

Case I : Pipes in parallel.

Case I. Pipes in parallel

From the continuity principle,

$$Q = Q_1 + Q_2 \qquad \text{...(1)}$$

Head loss,

$$h_1 = \frac{fL}{D_1}\frac{V_1^2}{2g} = \frac{fLV_2^2}{2gD_2}$$

from which $V_1 = \sqrt{\dfrac{2gD_1h_1}{fL}}$

and $V_2 = \sqrt{\dfrac{2gD_2h_1}{fL}}$

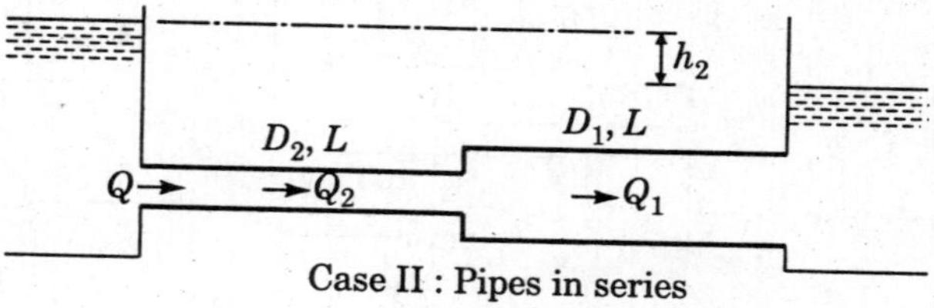

Case II : Pipes in series

$$Q_1 = \frac{\pi D_1^2}{4} V_1 = \frac{\pi D_1^2}{4}\sqrt{\frac{2g\,D_1h_1}{fL}}$$

$$= \frac{\pi}{4}\sqrt{\frac{2g}{fL}}\, D_1^{5/2}\, h_1^{1/2}$$

$$Q_2 = \frac{\pi D_2^2}{4} V_2 = \frac{\pi D_2^2}{4}\sqrt{\frac{2gD_2h_1}{fL}} = \frac{\pi}{4}\sqrt{\frac{2g}{fL}}\, D_2^{5/2}\, h_1^{1/2}$$

Substituting for Q_1 and Q_2 in (1)

$$Q = \frac{\pi}{4}\sqrt{\frac{2g}{fL}}\,[D_1^2\sqrt{D_1h_1} + D_2^2\sqrt{D_2h_1}\,] \qquad \text{...(2)}$$

Case II : Pipes in series.

$$Q = Q_1 = Q_2$$

Head loss, $$h_2 = \frac{fL}{D_2}\frac{V_2^2}{2g} + \frac{fL}{D_1}\frac{V_1^2}{2g} = \frac{fL}{2gD_2}\left(\frac{4Q}{\pi D_2}\right)^2 + \frac{fL}{2gD_1}\left(\frac{4Q}{\pi D_1^2}\right)^2$$

$$h_2 = \left(\frac{4Q}{\pi}\right)^2 \frac{fL}{2g}\left[\frac{1}{D_1^5} + \frac{1}{D_2^5}\right]$$

from which, $$Q = \frac{\pi}{4}\sqrt{\frac{2g}{fL}}\sqrt{\left(\frac{D_1^5\,D_2^5}{(D_1^5 + D_2^5)}\right)}\sqrt{h_2} \qquad ...(3)$$

Eliminating Q between (2) and (3),

$$[D_1{}^2\sqrt{D_1 h_1} + D_2{}^2\sqrt{D_2 h_1}] = \sqrt{\left(\frac{D_1^5\,D_2^5\,h_2}{(D_1^5 + D_2^5)}\right)}$$

or $$\sqrt{\frac{h_1}{h_2}} = \sqrt{\left(\frac{D_1{}^5\,D_2{}^5}{(D_1{}^5 + D_2{}^5)}\right)} \times \frac{1}{(D_1{}^{5/2} + D_2{}^{5/2})}$$

Substituting, $$D_1 = 2D_2$$

$$\sqrt{\frac{h_1}{h_2}} = \frac{1}{[(2D_2)^{5/2} + (D_2)^{5/2}]} \times \sqrt{\left[\frac{(2D_2)^5\,D_2{}^5}{(2D_2)^5 + D_2{}^5}\right]}$$

$$= \frac{2\sqrt{2}}{(2\sqrt{2}+1)\,(2^5+1)^{1/2}} = 0.1286$$

hence $h_1/h_2 =$ **0.01654.**

Example 11.31. *A pipeline of 0.6 m diameter is 1.5 km long. To augment the discharge, another pipeline of the same diameter is introduced parallel to the first in the second half of its length. Find the increase in discharge if f = 0.04 and head at the inlet is 30 m.*

Solution. Considering single pipeline 0.6 diameter 1.5 km long, and assuming that the 30 m head available at inlet is fully utilised by the flow in overcoming friction, one can write,

$$h_f = f\frac{L}{d}\frac{V^2}{2g}$$

or $$30 = 0.04 \times \frac{1.5 \times 1000}{0.6} \times \frac{V^2}{2 \times 9.81} = 5.097\,V^2$$

from which, $V = 2.426$ m/s

Discharge $$Q = \frac{\pi}{4}\,(0.6)^2 \times 2.426 = 0.6862 \text{ m}^3\text{/s}.$$

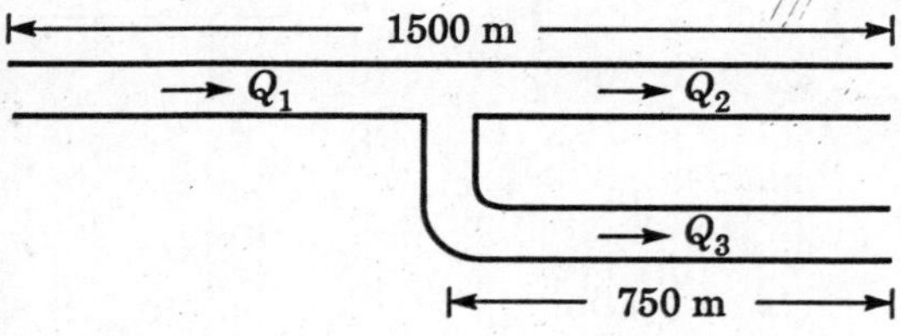

When a parallel pipe of the same diameter is introduced in the second-half of the pipeline, as shown, one can write the following equations :

$$Q_1 = Q_2 + Q_3 \qquad ...(1)$$

$$h_{f2} = h_{f3} \qquad ...(2)$$

$$30 = h_{f1} + h_{f2} \qquad ...(3)$$

from (2), $f\dfrac{(L/2)}{d}\dfrac{V_2^2}{2g} = f\dfrac{(L/2)}{d}\dfrac{V_3^2}{2g}$; hence $V_2 = V_3$

from (3), $30 = f\dfrac{(L/2)}{d}\dfrac{V_1^2}{2g} + f\dfrac{(L/2)}{d}\dfrac{V_2^2}{2g} = 0.04 \times \dfrac{750}{2 \times 9.81 \times 0.6}[V_1^2 + V_2^2]$ ...(4)

from (1), $\dfrac{\pi}{4}\, d^2V_1 = \dfrac{\pi}{4}\, d^2V_2 + \dfrac{\pi}{4}\, d^2V_3 = 2\,\dfrac{\pi d^2}{4}\, V_2$

$\therefore \quad V_1 = 2V_2$

Substituting for V_1 in (4), $30 = 12.75\, V_2^2$, hence $V_2 = 1.535$ m/s

Using (1), $Q_1 = 2Q_2 = 2 \times \pi(0.6)^2/4 \times 1.535 = 0.867\ \text{m}^3/\text{s}$

Hence, increase in discharge = 0.867 – 0.6862 = **0.1808 m³/s.**

Example 11.32. *Water is pumped from a low level reservoir to a high level reservoir through a main pipeline of 0.45 m diameter and 1400 m length. The pump is located at the low level reservoir. At a point along the main line at a distance of 450 m from the high level reservoir, a branch line of 0.3 m diameter and 360 m length takes off to discharge 180 litre / sec in the atmosphere.*

Level of water in the high level reservoir = + 30.00 m

Level of water in the open end of the 0.3 m diameter branch line = 25.50 m

Level of water in the low level reservoir = 18.00 m

Darcy's frictional coefficient for both pipes = 0.032.

Determine the flow rate into the high level reservoir and the theoretical power of the pump assuming the delivery valve at + 20.00 m.

Solution. Discharge through the branch line DC

= 180 lit./s = 0.8 m³/s.

Velocity through branch line DC

$$= \frac{Q}{A} = \frac{0.18}{\frac{\pi}{4}(0.3)^2} = 2.546 \text{ m/s.}$$

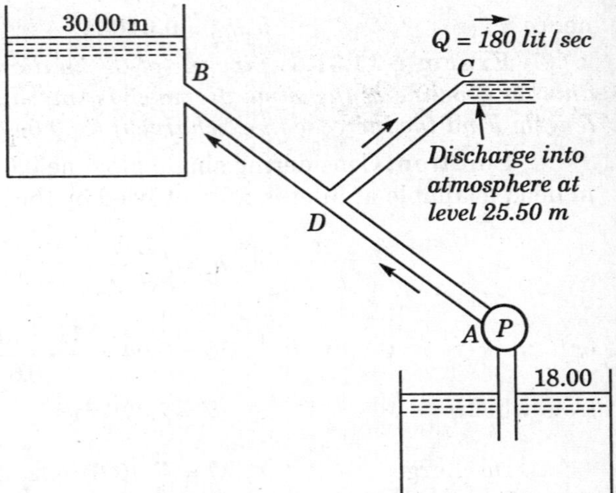

Head loss due to friction in branch line DC

$$= f\frac{L}{d}\frac{V^2}{2g} = (0.032)\,\frac{(360)}{(0.3)}\,\frac{(2.546)^2}{(2)(9.81)}$$

= 12.69 m.

Piezometric head at the junction D

= Piezometic head at the open end C

+ head loss in branch line DC

= 25.50 + 12.69

= 38.19 m.

Difference in head between junction D and high level reservoir

= 38.19 – 30.00

= 8.19 m.

Flow in the main pipe line from the junction D to high level reservoir is possible due to this head difference of 8.19 m required to overcome friction along DB. Writing the Darcy-Weisbach equation

$$8.19 = (0.032)\frac{(450)}{(0.45)}\frac{(V_{DB})^2}{19.62}$$

from which, $V_{DB} = 2.24$ m/s.

Discharge through pipe line DB,

$$Q_{DB} = \frac{\pi}{4}(0.45)^2 \times 2.24 = 0.36 \text{ m}^3\text{/s}$$

Discharge supplied by the pump,

$$Q_{AD} = Q_{DC} + Q_{DB} = 0.18 + 0.36 = 0.54 \text{ m}^3\text{/s.}$$

Velocity through pipe line AD,

$$V_{AD} = \frac{0.54}{\pi(0.45)^2/4} = 3.37 \text{ m/s.}$$

Velocity head at A, $$\frac{V_{AD}^2}{2g} = \frac{(3.37)^2}{2\times 9.81} = 0.58 \text{ m.}$$

Loss of head due to friction between A and D

$$= (0.032)\frac{(950)}{(0.45)}\frac{(3.37)^2}{19.62} = 39.10 \text{ m.}$$

The piezometric head at the pump delivery A will be higher than that at B by 39.10 m. Therefore the piezometric head at pump delivery

= 38.19 + 39.10 = 77.29 m.

Total head required at the pump delivery

= 77.29 + velocity head at A *i.e.* 0.58 = 77.87 m.

Applying Bernoulli's equation between water level in the lower level reservoir and the pump delivery where the total head required is 77.87, above the sump level (*i.e.*, 18.00 m),

18.00 + Head supplied by the pump = 77.87

∴ Head supplied by the pump = 59.87 m

Theoretical power of the pump = (9810) (0.54) (59.87) = **316.86 kW.**

11.9.3. Three-Reservoir Problem

Yet another example of practical importance involving pipe systems is the three-reservoir problem, which is met with sometimes in water supply systems. The three-reservoir problem has three interconnected pipelines meeting at a common junction and originating from three reservoirs having different water surface levels. The usual problem is to find the flowrates in the three pipes for given elevations of the reservoirs' water surface, the length, diameter and the material of the pipes (and hence the equivalent sand grain roughness k_s) being known. A correct solution of the problem requires consideration of the fact that friction factor f varies with Reynolds number and the relative roughness of the pipe surface. However, for the sake of simplicity, a constant value of f based on the pipe material and the likely range of Reynolds number may be assumed for each pipe. The conventional method is to assume a suitable elevation for the hydraulic grade line (H.G.L.) at the junction and then solve for the flowrates in the three pipes ; the correct solution being the one which satisfies the continuity requirement, that is the inflow into the junction equals the outflow from it.

Fig. 11.17 shows schematically the three-interconnected reservoirs. The reservoir A is the highest, B the intermediate and C the lowermost. The three pipes 1, 2 and 3 meet at the common junction J. Let Q_1, Q_2 and Q_3 be rates of flow in the respective pipes. Let the height of H.G.L. at a junction measured above the horizontal datum be Z_J as shown in the figure. It is obvious that the flow takes place from reservoir A (it being the highest) towards the junction and from the junction towards the lowermost reservoir C. But the direction of flow in the second pipe is yet uncertain.

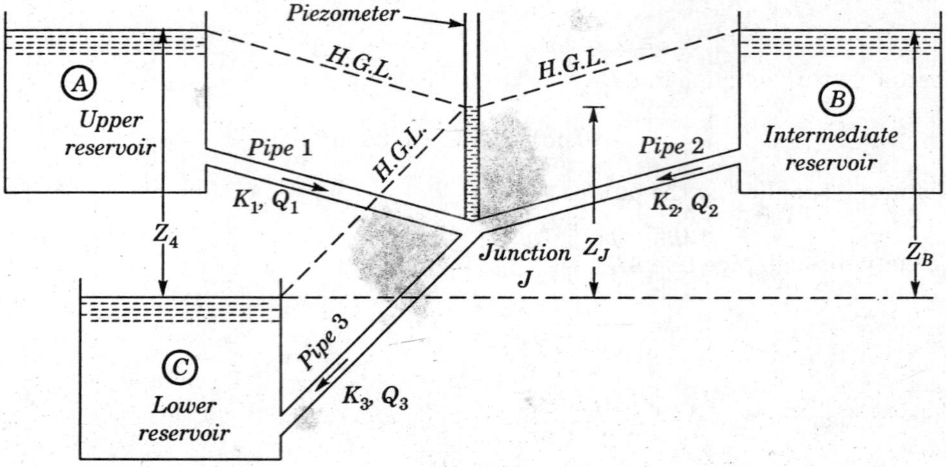

Fig. 11.17. Three-reservoir problem.

The solution of the three-reservoir problem hinges on determination of the direction of flow in pipe 2. It is in determining the flow direction that a trial procedure has to be followed in the conventional method of solving such problems. A method which directly determines the flow direction, and thereby eliminates the trials, is the one proposed by Ranga Raju and Sethruaman[7]. It assumes that the resistance law for each pipe can be expressed by

$$h_L = KQ^2 \qquad ...(11.32)$$

in which h_L is the head loss K, is the resistance parameter which is a constant for the given pipe and its value can be calculated if the pipe length L, diameter d are known and the friction factor f and the minor losses can be estimated. The procedure described below is based on this method.

Procedure for direct solution. Assume that $Z_J = Z_B$. Then obviously Q_2, the discharge in the second pipe is zero. From energy considerations (*i.e.* Bernoulli's equation) between reservoir A and the junction J, taking the water surface in the lower reservoir as the datum, one can write

$$Z_A - Z_J = Z_A - Z_B = H_i = K_1 Q_1^2 \qquad ...(11.32\ a)$$

Similarly for reservoir C and junction J, $Z_J = Z_B = H_b = K_3\, Q_3^2$ ...(11.32 b)

in which H_i and H_b are the differences in reservoir levels between the upper and the intermediate and the bottom reservoirs respectively, and K_1 and K_3 are the respective resistance parameters of pipes 1 and 3.

The assumed elevation for Z_J (*i.e.* $Z_J = Z_B$) would be correct only if

$$Q_1 = Q_3 \qquad ...(11.32\ c)$$

or

$$\sqrt{\frac{H_i}{K_1}} = \sqrt{\frac{H_b}{K_3}},\ \frac{H_i}{H_b} = \frac{K_1}{K_3}, \qquad ...(11.32\ d)$$

putting $H_i/H_b = H$ and $K_1/K_3 = R_1$, one obtains $H = R_1$

If Eq. (11.32 d) is satisfied, the Q_1 and Q_3 can be computed from Eqs. (11.32 b) and (11.32 c), Q_2 being zero.

7. Ranga Raju K.G. and Sethruaman, S. "Direct Solution of Three-Reservoir Problems". University of Roorkee Research Journal, Vol. XI, Paper No. C.E. 143, April 1969.

If, however, $Q_1 > Q_3$, the value of Z_J will have to be increased above its trial value to satisfy the continuity equation. This means that the flow takes place from the junction towards the reservoir B via pipe 2. The necessary condition for this case is given below :

$$\sqrt{\frac{H_i}{H_1}} > \sqrt{\frac{H_b}{H_3}} ; \sqrt{\frac{H_i}{H_b}} > \sqrt{\frac{K_1}{K_3}}$$

or $$H > R_1 \quad \text{...(11.32 e)}$$

Similarly, it can be shown that if, $H < R_1$ ...(11.32 *f*)

the flow takes place towards the junction from the reservoir B via pipe 2.

There are, therefore, the following three cases :

Case I. $H > R_1$

for which the continuity becomes $Q_1 = Q_2 + Q_3$ ...(11.32 *g*)

Case II. $H = R_1$

$$Q_1 = Q_2 = \sqrt{\frac{H_i}{K_1}} = \sqrt{\frac{H_b}{K_3}} ; Q_2' = 0 \quad \text{...(11.32 h)}$$

Case III. $H < R_1$, $Q_1 + Q_2 = Q_3$ ...(11.32 *i*)

A given problem belongs to which one of these cases can now be readily determined. The use of the corresponding continuity equation then leads to the distribution of flowrate in the pipelines. Let us consider case I in detail to understand the procedure for solving the problem. For this case the equation of continuity and head loss can be written as

$$Q_1 = Q_2 + Q_3 \quad \text{...(11.32 g)}$$

$$Z_A = K_1Q_1^2 + K_3Q_3^2 \quad \text{...(11.32 j)}$$

and $$Z_A - Z_B = K_1Q_1^2 + K_2Q_2^2 \quad \text{...(11.32 k)}$$

From Eqs. (11.32 *j*) and (11.32 *k*),

$$\frac{Z_A - Z_B}{Z_A} = \frac{K_1Q_1^2 + K_2Q_2^2}{K_1Q_1^2 + K_3Q_3^2}$$

Putting $\dfrac{Q_2}{Q_1} = x$, $\dfrac{Z_A}{Z_B} = h$, $\dfrac{K_1}{K_3} = R_1$ and $\dfrac{K_2}{K_3} = R_2$, the above equation simplifies to

$$(R_2h + 1 - h)x^2 + 2(h - 1)x + R_1 + 1 - h = 0 \quad \text{...(11.32 l)}$$

Solution of Eq. (11.32 *l*) for x yields

$$x = -\frac{h-1}{R_2h+1-h} \pm \sqrt{\left(\frac{h-1}{R_2h+1-h}\right)^2 + \frac{h+R_1-1}{R_2h+1-h}} \quad \text{...(11.32 m)}$$

Eq. (11.32 *m*) gives two values of x and it needs to be examined as to which of these corresponds to the physical case under consideration. With x known, using Eqs. (11.32 *g*) and (11.32 *j*), the rates of flow Q_1, Q_2 and Q_3 in the respective pipes can be determined.

Following a similar procedure problems falling under the category of cases II and III can be solved.

Example 11.33. *Three interconnected reservoirs have pipelines meeting at a common junction. The upper reservoir which has an elevation of 70.0 m is connected to the junction by means of a 900 mm diameter, 300 m long pipe having a friction factor f = 0.016. A pipe of 450 mm diameter and 600 m long (f = 0.024) takes off from the intermediate reservoir having water surface elevation of 50.0 m. The junction of these pipelines is connected to the lower reservoir, with water surface at 25.0 m, by means of a 600 mm diameter, 1200 m long pipe (f = 0.024). Determine the flowrate in each pipe and compute the piezometric head at the junction.*

Solution. Referring to Fig. 11.17, the following quantities can be determined,

$$Z_A = 70 - 25 = 45 \text{ m}, \qquad Z_B = 50 - 25 = 25 \text{ m}.$$

Neglecting the minor losses, the head loss in each pipe can be evaluated in terms of the discharge flowing in the pipe.

Using Eqs. (11.30) and (11.31), the resistance parameter

$$K = 8fL/\pi^2 gd^5 \qquad \text{...(A)}$$

The respective resistance parameters K_1, K_2 and K_3 for pipes 1, 2 and 3 can now be determined

$$K_1 = \frac{8}{\pi^2} \times 0.016 \times \frac{300}{9.81 \times (0.9)^5} = 0.672, \; K_2 = \frac{8}{\pi^2} \times 0.024 \times \frac{600}{9.81 \times (0.45)^5} = 64.4$$

$$K_3 = \frac{8}{\pi^2} \times 0.024 \times \frac{1200}{9.81 \times (0.6)^5} = 30.7$$

Using Eqs. (11.32 *a*) and (11.32 *b*)

$$H_i = Z_A - Z_B = 45 - 25 = 20 \text{ m}, \; H_b = Z_B = 25 \text{ m}$$

$$H = \frac{H_i}{H_b} = \frac{20}{25} = 0.8 \text{ and } R_1 = \frac{K_1}{K_3} = \frac{0.672}{30.7} = 0.0218$$

Since $H > R_1$, the problem belongs to Case I.

The flow takes place towards the reservoirs *B* and *C* from the junction, and the continuity equation is given by Eq. (11.32 *g*)

$$h = \frac{Z_A}{Z_B} = \frac{45}{25} = 1.8, \; R_2 = \frac{K_2}{K_3} = \frac{64.4}{30.7} = 2.1.$$

Making use of Eq. (11.32 *m*), the discharge ratio $Q_2/Q_1 = x$ is obtained as

$$x = -\frac{18 - 1.0}{2.1 \times 1.8 + 1 - 1.8} \pm \sqrt{\left[\frac{1.8 - 1.0}{2.1 \times 1.8 + 1 - 1.8}\right]^2 + \frac{1.8 + 0.0218 - 1}{2.1 \times 1.8 + 1 - 18}} = 0.311, -0.848.$$

Since the negative value of *x* has no physical significance to the problem, $x = 0.311$. Using Eqs. (11.32 *j*) and (11.32 *g*),

$$Q_1 = 1.704 \text{ m}^3/\text{s}, \; Q_2 = 0.53 \text{ m}^3/\text{s}, \text{ and } Q_3 = 1.174 \text{ m}^3/\text{s}.$$

Piezometric head at the junction, $Z_J = K_3 Q_3^2 = 30.7 \times (1.174)^2 =$ **52.2 m.**

Example 11.34. *The water levels in two reservoirs A and B are 104.5 m and 100 m, respectively above the datum. A pipe joins each to a common point D, where pressure is 98.1 kN/m² gauge and height is 83.5 m above datum. Another pipe connects D to another tank C. What will be the height of water level in C assuming the same value of 'f' for all pipes. Take f = 0.030.*

Pipe	*Length in metres*	*Diameter in mm*
AD	*240*	*300*
BD	*270*	*450*
CD	*300*	*600*

Solution. The pressure head at $D = \dfrac{98100}{9810} = 10$ m of water

$\therefore$ the piezometric head at $D = 83.5 + 10 = 93.5$ m

Head loss between *A* and $D = 104.5 - 93.5 = 11.0$ m

Head loss between *B* and $D = 100.0 - 93.5 = 6.5$ m

From Darcy-Weisbach equation, we obtain

$$11.0 = f\,\frac{240}{0.3}\,\frac{V_a^2}{2g} \qquad \text{...(1)}$$

$$6.5 = f\,\frac{270}{0.45}\frac{V_b^2}{2g} \qquad \text{...(2)}$$

Substituting $f = 0.032$, one obtains, $V_a = 3.0$ m/s

and $\qquad V_b = 2.66$ m/s

From the continuity equation $Q_a + Q_b = Q_c$

$$\therefore \qquad Q_c = \frac{\pi}{4}(0.3)^2 \times 3.0 + \frac{\pi}{4}(0.45)^2 \times 2.66 = 0.635 \text{ m}^3\text{/s}$$

and $$V_c = \frac{4Q_c}{\pi(0.60)^2} = 2.24 \text{ m/s}$$

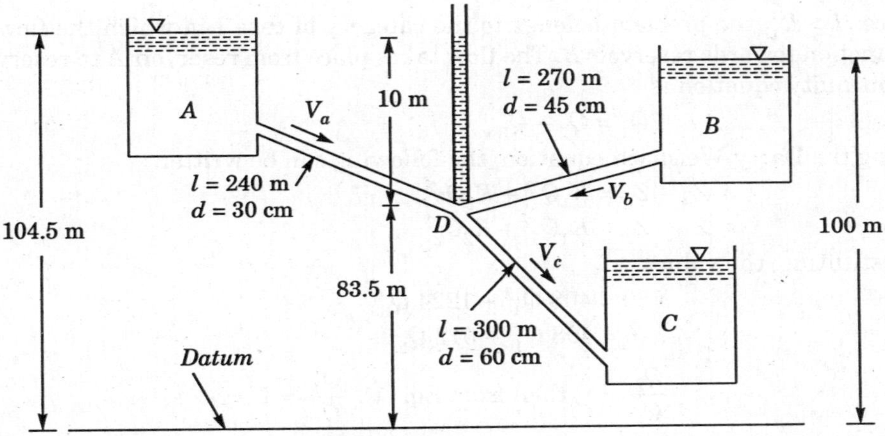

The head loss in pipe DC $= 0.3 \times \dfrac{300}{0.6} \times \dfrac{(2.24)^2}{2 \times 9.81} = 3.83$ m

$\therefore$ Water level in tank C $= 93.5 - 3.88 = 89.67$ m.

Example 11.35. *Three reservoirs A, B and C have elevations 40, 34 and 20 metres, respectively. A 30 cm diameter pipeline, 1500 m long, connects reservoirs A and B while another 30 cm diameter pipe, 2400 m long, connects reservoirs A and C. For a distance of 600 m from A both the pipes lie side by side upto a junction point J, where they are interconnected by a short branch pipe. Assuming that the pipe friction factor is 0.02, and neglecting losses at entry, exit and junction, find*

(a) the direction of flow in reservoir B, and

(b) the flows leaving or entering the reservoirs in lit/sec.

Solution. Heads loss in the pipe line

$$h_L = f\,\frac{L}{d}\frac{V^2}{2g}$$

$$= f\,\frac{L}{d}\frac{1}{2g}\left[\frac{4Q}{\pi d^2}\right]^2$$

$$= \left[\frac{8}{\pi^2} f \frac{L}{g d^5}\right] Q^2 = KQ^2$$

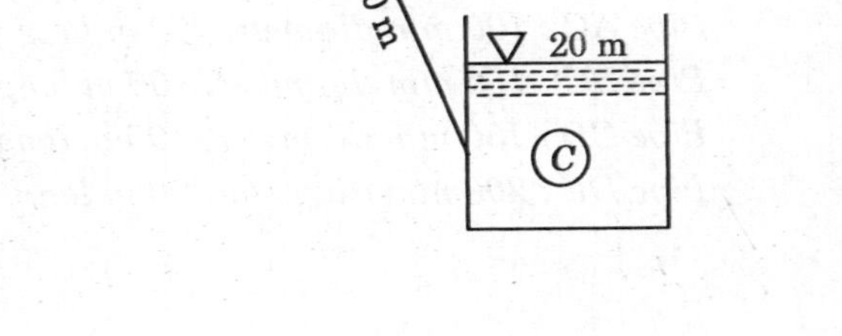

The resistance parameter K for each section of the pipeline, *i.e.* 600 m, 900 m and 1800 m lengths

$$K_1 = \frac{8}{\pi^2} \times 0.02 \times \frac{600}{9.81 \times (0.3)^5} = 408$$

Similarly, $K_2 = 612$ and $K_3 = 1224$

Following the procedure described in article 11.9.3

$$H_i = Z_A - Z_B = 40 - 34 = 6 \text{ m}$$

$$H_b = Z_B - Z_C = 34 - 20 = 14 \text{ m}$$

$$H = \frac{H_i}{H_b} = \frac{6}{14} = 0.4286$$

and
$$R_1 = \frac{K_1}{K_3} = \frac{408}{1224} = 0.333.$$

Since $H > R_1$, the problem belongs to the category of case I in which the flow takes place from the junction towards reservoir *B*. The flow takes place from reservoir *A* to reservoir *B* and *C*, and the continuity equation is

$$2Q_1 = Q_2 + Q_3 \qquad \text{...(1)}$$

Using the Darcy-Weisbach equation, the following can be written

$$Z_A - Z_C = K_1 Q_1^2 + K_3 Q_3^2,$$

and
$$Z_A - Z_B = K_1 Q_1^2 + K_2 Q_2^2.$$

Substituting, the values

$$20 = 408\, Q_1^2 + 1224\, Q_3^2 \qquad \text{...(2)}$$

$$6 = 408\, Q_1^2 + 612\, Q_2^2 \qquad \text{...(3)}$$

Let $\frac{Q_2}{Q_1} = x$, then from Eq. (1), $\frac{Q_3}{Q_1} = 2 - x$

From Eqs. (2) and (3), $\frac{20}{6} = \frac{1 + 3(2 - x)^2}{1 + 1.5x^2}$

$$6x^2 + 36x - 29 = 0$$

from which
$$x = 0.721$$

$$\frac{Q_3}{Q_1} = 2 - x = 2 - 0.721 = 1.279$$

Using Eq. (3),
$$6 = 408\, Q_1^2 + 612\, (Q_1 x)^2 = 726\, Q_1^2$$

$$Q_1 = 0.091 \text{ m}^3/\text{s} = 91 \text{ lit./s}$$

$$Q_2 = Q_1 x = 0.091 \times 0.721 = 0.0655 \text{ m}^3/\text{s} = 65.5 \text{ lit./s}$$

$$Q_3 = 1.279\, Q_1 = 1.279 \times 0.091$$

$$= 0.1164 \text{ m}^3/\text{s} = 116.4 \text{ lit./s.}$$

Flow leaving reservoir *A* $= 2Q_1 = 2 \times 0.091 =$ **0.182 m³/s.**

Flow entering reservoir *B* $= Q_2 =$ **0.0655 m³/s.**

Flow entering reservoir *C* $= Q_3 =$ **0.1252 m³/s.**

Example 11.36. *The reservoirs A and B with water levels of 20 m and 15 m, respectively, above datum are connected to a pipe system as shown in the figure. The pipe details are :*

Pipe AC : 100 mm diameter, 50 m long ;

Pipe BC : 150 mm diameter, 100 m long ;

Pipe CD : 150 mm diameter, 100 m long ;

Pipe DE : 200 mm diameter, 20 m long.

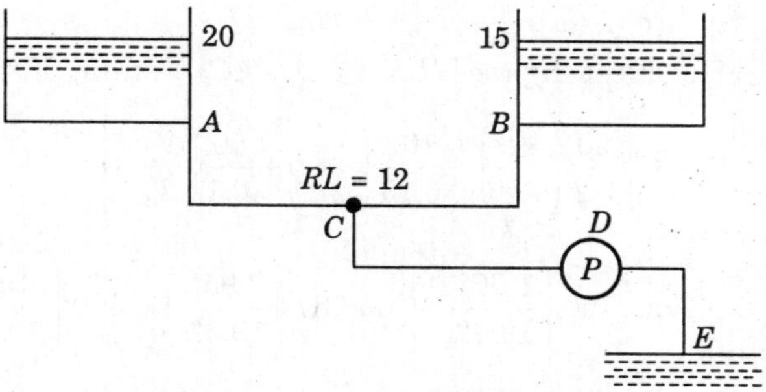

Coefficient of friction, f = 0.025, in Darcy-Weisbach head loss formula. If the flow to reservoir B is 0.04 m³/s, determine :

(i) pressure head at junction C if its RL is 12 m ;

(ii) total head developed by the pump.

Solution. If Q is the discharge of the pump, Q_A the discharge in pipe CA and Q_B in pipe CB, then from continuity principle

$$Q = Q_A + Q_B \quad ...(1)$$

Considering pipe AC. Application of Bernoulli's equation between end C of pipe AC and the water surface in reservoir A result in

$$12 + \frac{p_C}{\gamma} + \frac{V_A^2}{2g} = 20 + 0 + 0 + (0.025)\frac{(50)}{(0.1)}\frac{V_A^2}{2g} + \frac{V_A^2}{2g} \quad ...(2)$$

Considering pipe BC. Applying Bernoulli's equation between and C of pipe BC and the water surface in reservoir B,

$$12 + \frac{p_C}{\gamma} + \frac{V_B^2}{2g} = 15 + 0 + 0 + (0.025)\frac{(150)}{(0.15)}\frac{V_B^2}{2g} + \frac{V_B^2}{2g} \quad ...(3)$$

In Eqs. (2) and (3), p_C is the pressure at junction C

$$V_B = \frac{4Q_B}{\pi\,(0.15)^2} = \frac{4 \times 0.04}{\pi(0.15)^2} = 2.26 \text{ m/s.}$$

From Eq. (3), $$12 + \frac{p_C}{\gamma} = 15 + 25\,\frac{(2.26)^2}{2 \times 9.81} = 21.53$$

or $$\frac{p_C}{\gamma} = \mathbf{9.53\ m,}$$

∴ The pressure head at junction C = 9.53 m.

From Eq. (2), $12 + 9.53 = 20 + (0.025)\,(500\,V_A^2/2g)$

from which, $V_A = 1.55$ m/s

and $$Q_A = \frac{\pi}{4}(0.1)^2\,(1.55) = 0.0122 \text{ m}^3\text{/s.}$$

Pump discharge $Q = Q_A + Q_B = 0.0122 + 0.04 = 0.0522$ m³/s

Velocity in pipe ED, $$V_E = \frac{4Q}{\pi(0.2)^2} = \frac{4 \times 0.0522}{\pi(0.2)^2} = 1.66 \text{ m/s}$$

Velocity in pipe DC, $V_D = 4Q/\pi(0.15)^2$

$V_D = 2.95$ m/s.

Head developed by the pump

$$= \text{Head lost in pipes } (DE + CD + AC) + \text{static lift from } E \text{ to } A$$

$$= 0.025 \left[\frac{20}{0.2}\left(\frac{V_E^2}{2g}\right) + \frac{100}{0.15}\left(\frac{V_D^2}{2g}\right) + \frac{50}{0.1}\left(\frac{V_A^2}{2g}\right) \right] + 20$$

$$= 0.025 \left[100\left(\frac{1.66^2}{19.62}\right) + 666.67\left(\frac{2.95^2}{19.62}\right) + 500\left(\frac{1.55^2}{19.62}\right) \right] + 20$$

$$= 9.297 + 20 = \mathbf{29.297\ m.}$$

The total head developed by the pump = 29.297 m.

11.9.4. Distribution Mains

The network of distribution mains is always the most expensive item in a water supply project. In addition to the heavy capital investment, the cost of its unkeep generally represents a large proportion of the annual maintenance budget. The water supply engineer is, therefore, required to give considerable thought and care to the design of the most efficient distribution system. This entails an accurate prediction of flowrates and pressures in various pipe components.

The water demand depends upon the population, its per capita requirement in turn depending upon the local needs, people's habits and standard of living, the industrial or commercial importance of the city and similar factors. For towns where one-storeyed buildings are common and for supply to the ground level storage tanks in multi-storeyed buildings, the "Manual on Water Supply and Treatment (second edition), Ministry of Works and Housing, New Delhi, 1975" recommends the minimum residual pressure (at ferrrule point) of 7 m ; for direct supply where two-storeyed buildings are common it may be 12 m, and where three-storeyed buildings are prevalent, 16 m. The pressure required for fire fighting will have to be boosted by fire engines.

'Dead-end' mains are undesirable features since frequent flushing is usually necessary in order to prevent any tendency to stagnation. Distribution systems in urban areas are generally best laid out on a grid-iron or ring main pattern, mains being interconnected at distances not exceeding about 600 m. With this arrangement, water can be fed to any point in the system by more than one route, thus providing greater flexibility in meeting the demand and also enabling repairs to be carried out with the minimum disruption to supply.

In the design of a new supply network the topography of the land is very often a controlling factor, and due regard must be paid to the fact that frictional resistance causes a loss in pressure head. In urban areas where the ground contours vary by more than 60 m it may be advantageous to divide the supply network into pressure zones arranged in such a manner so us to avoid excessive pressures in the low-lying areas. Pressure relief valves and booster pumps are employed to obtain the desired pressure.

The standard pipe sizes for distribution mains range from 100 to 1800 mm in diameter. The frictional losses should be limited from 1 to 10 m per km, although higher losses are sometimes unavoidable with the smaller pipe sizes. Losses of this order are associated with velocities lying between 0.6 and 2 m/s, representing a range which is usually found to be satisfactory from both economic and operational points of view.

11.9.5. Working Pressure, Design Pressure and Test Pressure

The Indian Standard Specification, I.S. 468—1971, for concrete pipes (with and without reinforcement) gives the following definition for the working, design and test pressures.

Working pressure. It is the actual maximum pressure including abnormal conditions such as water hammer to which the pipe will be subjected during its operation.

Design pressure. It is the maximum pressure for which the pipe has been designed. This is equal to the product of the working pressure and a suitable multiplier factor to cover any abnormal increase in pressure in the pipeline due to unforeseen conditions.

Test pressure. It is the maximum pressure which the pipe can withstand without any leakage when tested for hydrostatic pressure in accordance with I.S. 458—1971 and I.S. 3597—1966 (method of test for concrete pipes).

11.9.6. Choice of Pipe Material

Pipes represent a large proportion of the capital invested in water supply undertakings and are, therefore, of particular importance. There are various types of pipes such as spun or cast iron (C.I.), steel, reinforced cement concrete (R.C.C.), prestressed reinforced concrete (P.S.C.), asbestos cement (A.C.) and polyethylene (low and high density) and polyvinyl chloride (rigid PVC).

The factors which affects the final choice of pipe material are internal pressure, hydraulic and operating conditions, maximum permissible diameters, external corrosion and special conditions of laying, if any.

Cast iron pipes are preferred for their strength and corrosion resistance. They are well suited for pressure mains and laterals where tappings are made for house connections. Steel pipes being light in weight are used for large diameters especially greater than 900 mm, where C.I. pipes become very heavy and costly. Prestressed concrete pipes are ideal for a pressure range of 490 to 1960 kN/m^2 (50 to 200 m of head), where C.I. and steel pipes are non-economical and R.C.C. not useable A.C. pipes are relatively more corrosion resistant than steel or C.I., light and easy to handle and these are used in lateral and minor distribution systems. These pipes, however, cannot take high internal pressures.

Unplasticised PVC and polyethylene pipes, being more flexible and more corrosion resistant compared to metallic and concrete pipes, can be extensively used for domestic service connections and plumbing systems. These pipes are well suited for use in mountainous terrain and undulating areas.

11.9.7. Classification of Pressure Pipes and their Diameters

(1) **Concrete pipes.** The pressure pipes are classified according to the test pressures and following is the description of pipes recommended by the I.S. Code. According to the specification, I.S. : 458—1971, for concrete pipes, the classification given below relates to the distribution mains.

Material	*Category of Pipes*	*Test Pressure*	*Situation where used*
Reinforced concrete	Clsss P 1	196 kN/m^2 (20 m head)	For use on gravity mains, design pressure not exceeding 2/3 of the test pressure.
"	Class P 2	392 kN/m^2 (40 m head)	For use on pumping main, the design pressure not exceeding 1/2 of the test pressure.
"	Class P 3	588 kN/m^2 (60 m head)	"

Internal diameter of the above category of pipes ranges from 80 to 1200 mm and the normal internal diameters available are : 80, 100, 150, 250, 300, 350, 400, 450, 500, 600, 700, 800, 900, 1000, 1100 and 1200 mm. The diameter range for the above mentioned categories is as under :

Class P 1 : 80—1200 mm

Class P 2 : 80—600 mm

Class P 3 : 80—400 mm.

(2) **Cast iron pipes.** The Indian Standard Specification, I.S. : 1536—1967, for centrifugally cast (spun) iron pressure pipes for water, gas and sewage (first revision) lays down recommendations

for test pressures and working pressures for different categories of pipes. The pipes are classified according to the wall thickness as class LA, class A and class B, the pipes of class B types being thicker than those of class A and class A pipes in turn thicker than class LA type. The following are the recommendation for the above categories of pipes in respect of the test pressures and working pressures.

Category of Pipes	*Test Pressure*	*Working Pressure*
Class LA	3435 kN/m^2	1175 kN/m^2
Class A	3435 kN/m^2	1765 kN/m^2
Class B	3435 kN/m^2	1960 kN/m^2

The normal internal diameters available are 80, 100, 125, 150, 200, 250, 300, 350, 400, 450, 500, 600, 700 and 750 mm.

(3) **Steel cylinder reinforced concrete pipes.** The Indian Standard Specification, I.S. : 1916—1963, for steel cylinder reinforced concrete pipes lays down requirements and methods of tests. The pipes have nominal internal diameters from 200 to 1800 mm for use in water mains, sewers, irrigation works and similar situations. The pipes shall be classified as under :

Category of Pipes	*Test Pressures*
Class 1	490 kN/m^2 (or 50 m head)
Class 2	980 kN/m^2 (or 100 m head)
Class 3	1470 kN/m^2 (or 150 m head)
Class 4	1960 kN/m^2 (or 200 m head)
Class 5	2450 kN/m^2 (or 250 m head)
Special class	Above 2450 kN/m^2 (or above 250 m head (the exact test pressure being specified by the purchaser.)

Note. As a general guide the corresponding working pressure for these pipes may be taken as 50% and 67% of the above values for pumping mains and gravity mains respectively.

(4) **Prestressed concrete pipes.** The Indian Standard Specification, I.S. 784—1959, covers prestressed concrete pipes in which permanent internal stresses are deliberately introduced by tensioned steel to counteract, to the desired degree, stresses caused in the pipe under service. These stresses are entirely independent of the stresses caused by external loads or internal pressures. This standard covers only fully prestressed concrete pipes in which no tensile stress is permitted in the concrete when the full internal test pressure is applied.

The internal diameter and length of pipes shall be as under :

Internal diameter of Pipes, mm,	*Length of Pipes, m*
(*i*) 80, 100, 125, 160, 250, 300, 350 and 400	2
(*ii*) 450, 500, 600, 700, 800, 900, 1000, 1100, 1200, 1400, 1600 and 1800	2.5

The pipes shall be subjected to a hydrostatic pressure test as described in Appendix B of I.S. : 3597—1966 ; the test pressure shall be twice that for which the pipe is designed. The pipe shall prove to be sound and watertight.

(5) **Polyethylene and PVC Pipes.** The unplasticized PVC and polyethylene pipes are now-a-days coming into increasing use for drinking water supply purposes. The Indian Standard

Specification, I.S. : 3076—1968, for low density polyethylene pipes for potable water supplies suggests on the basis of extensive investigations, that these pipes are safe and compare favourably with the conventionally used CI, GI and AC pipes in respect of bacterial quality of water, number of bacteria harboured in the internal surfaces and the problem of disinfection.

The pipes shall be classified by pressures ratings (working pressures) as follows :

245 kN/m^2, 392 kN/m^2, 588 kN/m^2 and 980 kN/m^2.

The outside diameters available are 10, 12, 16, 20, 25, 32, 40, 50, 63, 75, 90, 110, 125 and 140 mm. The pipes shall be in coils of nominal lengths of 25, 50, 100, 150 and 200 m unless otherwise agreed to between the purchaser and the supplier. The pipes shall conform to the hydraulic tests as specified in section 7.1 of the I.S. specification.

11.9.8. Hydraulic Analysis of Pipe Networks

In municipal water distribution systems, the pipelines are frequently branching and looping in a complex manner comprising what is known as a pipe network. Multiple pipe systems reach a limit of complexity in the problems of distribution of flow in pipe-networks. Fig. 11.18 shows a pipe network comprising multiple pipes connected to each other in a desired fashion. It is almost impossible to tell by inspection which way the flow travels, such as in pipe *CD*—whether the fluids flows from *C* to *D* or from *D* to *C* is not known. The most important single factor which makes the problem of flow analysis more complicated is the uncertainty about the flow direction in different pipes. The fluid flow in a pipe network must satisfy the basic relationships of continuity and energy as listed below :

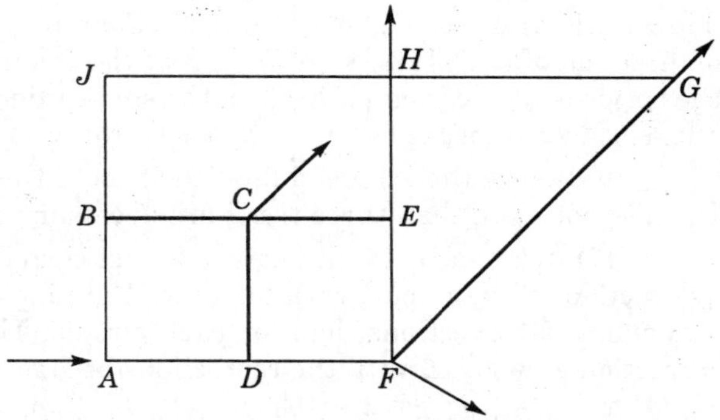

Fig. 11.18. Pipe network.

(1) The continuity principle demands that, the quantity of flow reaching any junction must equal the outflow from it.

(2) The flow in each pipe must satisfy the head loss equation (Darcy-Weisbach equation).

(3) The algebraic sum of the head losses around any closed circuit must be zero.

Accordingly, for each loop or circuit the head losses due to flow in clockwise direction must be equal to the head loss due to flow in anti-clockwise direction.

The pipe networks problems are generally solved by trial and error procedure. The most practical and widely used method of flow-analysis is that of successive approximations developed by **Hardy Cross.** It consists of the following sequence of computations :

(*i*) Assume the most reasonable flow in each pipe which satsififes the continuity equation, *i.e.* condition (1) listed above.

(*ii*) Compute the loss in each pipe by writing the head loss equation in the form expressed by Eq. (11.31)

$$h_L = KQ^n \qquad \text{...(11.31)}$$

in which K is a constant for each pipe which depends upon the fluid viscosity and the pipe size, length and roughness. In turbulent flow, when the pipe is rough $n = 2$, and when it is smooth $n < 2$ ($n = 1.85$ given by the empirical equation of Hazen-Williams). Eq. (11.31) is the statement of condition (2).

(*iii*) To investigate condition (3), compute the algebraic sum of the head losses around each elementary circuit, *i.e.* $\Sigma h_L = \Sigma KQ^n$. Consider head losses from clockwise flows as positive and from anti-clockwise flows as negative.

(*iv*) In order to satisfy condition (3) $\Sigma h_L = \Sigma KQ^n = 0$, adjustment of flow by applying a flow correction ΣQ may be necessary. The most important feature of this method (Hardy Cross) lies in the determination of ΔQ.

For any pipe we may write

$$Q = Q_0 - \Delta Q$$

in which Q_0 is the assumed flow and Q is the corrected flow. Then for each pipe

$$h_L = KQ^n = K(Q_0 - \Delta Q)^n$$
$$= K(Q_0{}^n - nQ_0{}^{n-1}\Delta Q \,...)$$

where the remaining terms of the series may be ignored if ΔQ is small as compared with Q_0. Now for a single circuit or loop, with ΔQ the same for all pipes

$$\Sigma h_L = \Sigma KQ^n = \Sigma KQ_0{}^n - \Delta Q \Sigma nKQ_0{}^{n-1} = 0$$

which results in $$\Delta Q = \frac{\Sigma KQ_0^n}{\Sigma nKQ_0^{n-1}} \quad ...(11.33)$$

It may be emphasized that the numerator of Eq. (11.33) is to be summed algebraically with due consideration of sign, while the denominator is summed arithmetically without any consideration of direction of flow. It may be noted that the assumption of considering only the first two terms of the series is not accurate if ΔQ is of the same order of magnitude as Q_0. The direction of ΔQ will be anti-clockwise if it is positive and *vice versa*.

(*v*) Revise the assumed flow and repeat the process until the desired accuracy is obtained. Usually not more than three trials are necessary except for very complex systems.

(*vi*) After each circuit is given a first correction, the losses will not still balance because of interaction of one circuit with another. The pipes which are common to two circuits receive two independent corrections, one for each circuit. The procedure is repeated arriving at a second correction and so on, until the corrections become negligible.

Example 11.37. *A water distribution network is an equivalent triangle ABC in shape. If the inflow at the junction A is 60 units and the outflow at the junctions B, C are 40 and 20 units respectively, find the discharge in each pipe. Take initial value of discharge from A to B as 15 units. Take values of 'K' in the expression $h_L = K.Q^2$ as AB = 4, BC = 1 and CA = 2. (Two trials are sufficient)*

Solution. Flow correction ΔQ is given by Eq. (11.33)

$$\Delta Q = \frac{\Sigma KQ_0^n}{\Sigma nKQ_0^{n-1}}$$

since $h_L = KQ^2$, the value of $n = 2$.

If ΔQ is positive, it is to be taken as anticlockwise and corrections applied accordingly *i.e.*, add ΔQ to anticlockwise flows.

Circuit ABC	
KQ^2	$2KQ$
(*i*) Head loss in AB is anticlockwise, hence taken as negative $-4 \times (15)^2 = -900$ (*ii*) $+1 \times (25)^2 = -625$ (*iii*) $+2 \times (45)^2 = 4050$	$2 \times 4 \times 15 = 120$ $2 \times 1 \times 25 = 50$ $2 \times 2 \times 45 = 180$
$\Sigma KQ^2 = 3775$	$\Sigma KQ = 350$

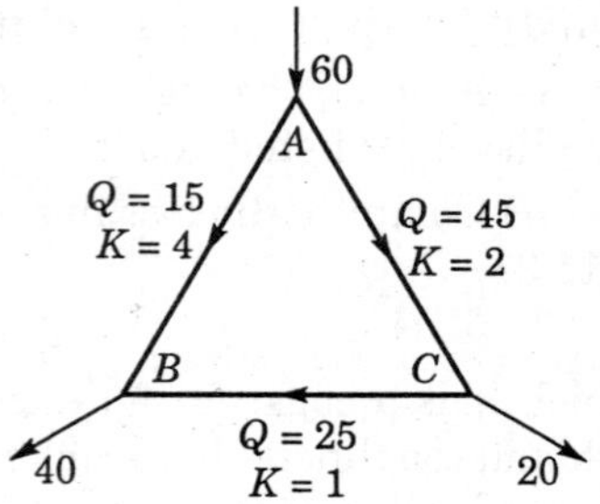

$$\therefore \quad \Delta Q = \frac{3775}{350} = 10.79 \text{ say } 11$$

The revised distribution of flow is shown below :

Second Trial :

For the revised distribution, the flow correction ΔQ is obtained as under :

Circuit ABC	
KQ^2	$2KQ$
$(-4)(26)^2 = (-) 2704$ $(1)(14)^2 = (+) 196$ $(2)(34)^2 = (+) 2312$	$2 \times 4 \times 26 = 208$ $2 \times 1 \times 14 = 28$ $2 \times 2 \times 34 = 136$
$\Sigma KQ^2 = -196$	$\Sigma 2KQ = 372$

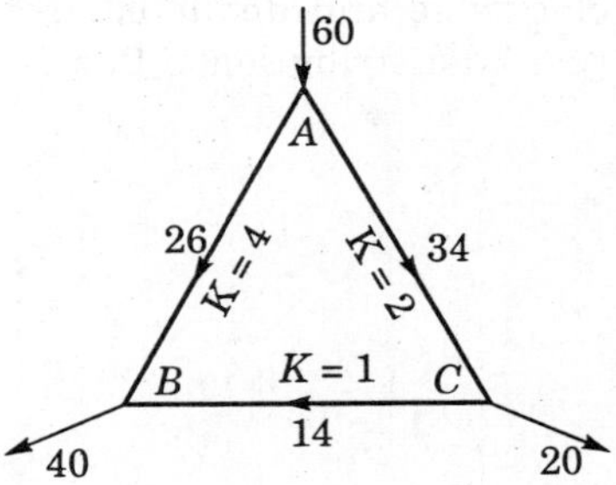

The flow correction

$$\Delta Q = \frac{\Sigma KQ^2}{\Sigma 2KQ} = -\frac{196}{372} = -0.52$$

In view of a small value for ΔQ, the distribution of flow in the network may be taken as shown.

Example 11.38. *Determine the distribution of flow in the pipe network shown in the following figure. The head loss $h_L = KQ^2$. The value of K for each pipe is indicated in the figure.*

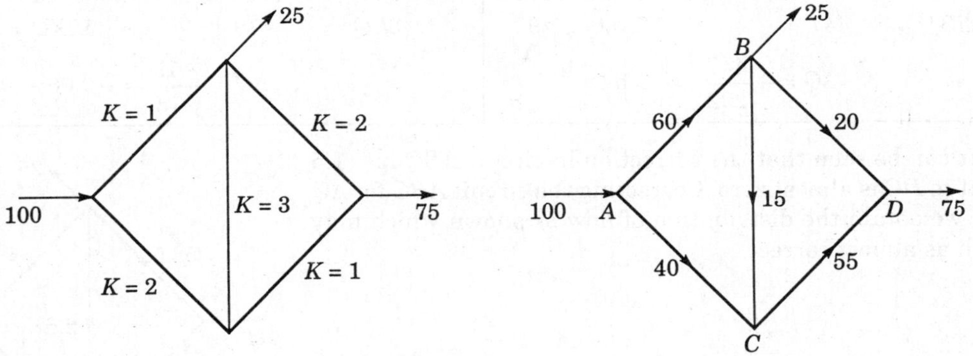

Solution.

Procedure. 1. Assume a suitable distribution of flow as shown above. Adopting the procedure suggested by Hardy Cross compute the correction ΔQ.

2. Compute the head loss $h_L = KQ^2$ for circuits ABC and BCD, considering clockwise flows as positive and anti-clockwise negative.

Circuit ABC		Circuit BCD	
KQ^2	$2KQ$	KQ^2	$2KQ$
$1 \times (60)^2 =$ 3600 $3 \times (15)^2 =$ 665 $2 \times (40)^2 = -3200$	$2 \times 1 \times 60 =$ 120 $2 \times 3 \times 15 =$ 90 $2 \times 2 \times 40 =$ 160	$2 \times (20)^2 =$ 800 $1 \times (55)^2 = -3025$ $3 \times (15)^2 = -665$	$2 \times 2 \times 20 =$ 80 $2 \times 1 \times 55 = 110$ $2 \times 3 \times 15 =$ 90
$\Sigma KQ^2 =$ 1065	$\Sigma 2QK =$ 370	$\Sigma KQ^2 = -2890$	$\Sigma 2KQ = 280$

for circuit ABC, $\quad \Delta Q = \frac{\Sigma KQ^2}{\Sigma 2KQ} = \frac{1065}{370} = 3$

for circuit BCD, $\quad \Delta Q = \frac{\Sigma KQ^2}{\Sigma 2KQ} = -\frac{2890}{280} = -11$

Positive ΔQ is to be added to anti-clockwise flow and deducted from clockwise and the negative ΔQ added to clockwise and deducted from anti-clockwise flow. The revised distribution of flow is shown in the figure.

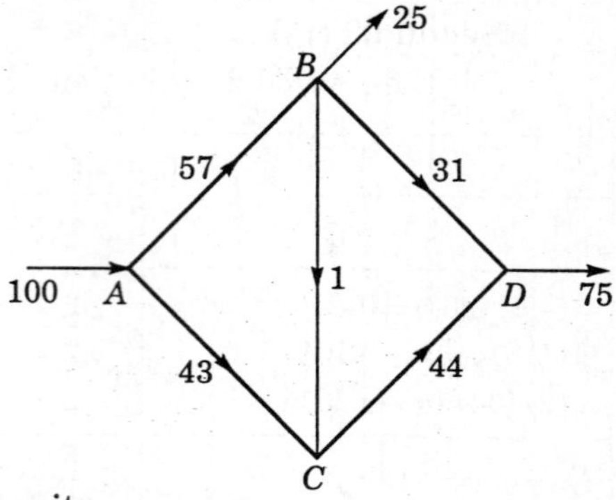

Repeating the same procedure as above for the two circuits.

Circuit ABC		*Circuit BCD*	
KQ^2	$2KQ$	KQ^2	$2KQ$
$1 \times (57)^2 = 3250$	$2 \times 1 \times 57 = 114$	$2 \times (31)^2 = 1920$	$2 \times 2 \times 31 = 124$
$3 \times (1)^2 = -3$	$2 \times 3 = 6$	$3 \times (1)^2 = -3$	$2 \times 1 \times 44 = 88$
$2 \times (43)^2 = -3700$	$2 \times 2 \times 43 = 172$	$1 \times (44)^2 = -1936$	$2 \times 3 \times 1 = 6$
$\Sigma KQ^2 = -447$	$\Sigma 2KQ = 292$	$\Sigma KQ^2 = -19$	$\Sigma 2KQ = 218$
$\therefore \ \Delta Q = -\dfrac{447}{292} = -1.5$		$\therefore \ \Delta Q = \dfrac{-19}{218} = -0.08$	

It can be seen that the correction in circuit *ABC* is – 1.5 and that in *DC* is almost zero. Correcting the circuit *ABC* for $\Delta Q = -1.5$, we obtain the distribution of flow as shown which may be taken as almost correct.

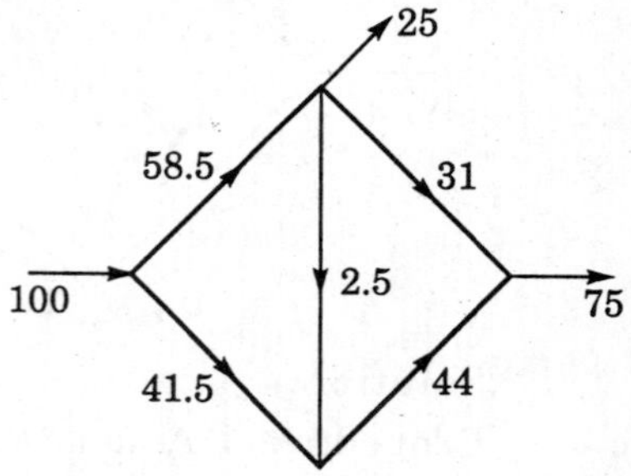

Example 11.39. *A pipe network is shown in figure in which Q and h_f refer to discharges and pressure drops respectively. Subscripts 1, 2, 3, 4 and 5 designate respective values in pipe lengths AC, BC, CD, DA and AC. Subscripts A, B, C and D designate discharges entering or leaving the junction points A, B, C and D respectively.*

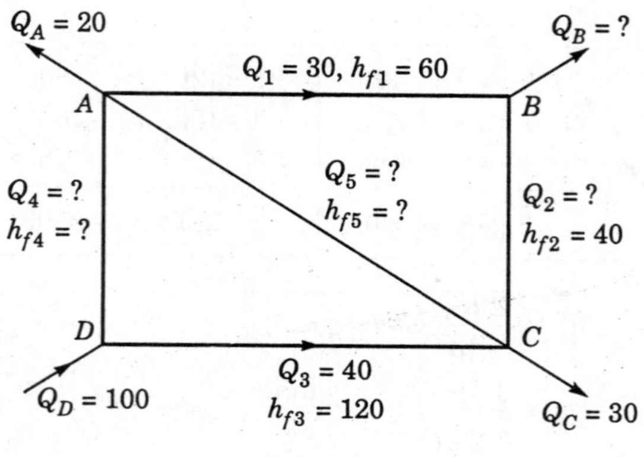

By sticking to the values given in figure find the following discharges Q_B, Q_2, Q_4 and Q_5; and pressure drops h_{f4} and h_{f5} and give these computed values at their respective places on a neat sketch of the network alongwith flow directions.

Solution. At junctions,

$\Sigma Q = 0$, *i.e.* Discharge entering junction = Discharge leaving it.

At D : $Q_D = Q_3 + Q_4$

$\therefore$ $Q_4 = 100 - 40 = 60$ (Leaving the junction)

At A : $Q_4 = Q_A + Q_1 + Q_5$

$\therefore$ $Q_5 = 60 - 20 - 30 = 10$ (Leaving the junction)

At C : $Q_3 + Q_5 + Q_2 = Q_C$

$40 + 10 + Q_2 = 30$

$\therefore$ $Q_2 = -20$ (Q_2 leaves the junction C)

At B : $Q_1 + Q_2 = Q_B$

$30 + 20 = Q_B = 50$ (Q_B leaves the junction B)

For each elementary circuit

$$\Sigma h_f = 0$$

Considering circuit ACB :

$$h_{f1} - h_{f2} - h_{f5} = 0$$

$$60 - 40 - h_{f5} = 0$$

$\therefore$ $h_{f5} = 20$

Considering circuit ACD :

$$h_{f4} + h_{f5} - h_{f3} = 0$$

$$h_{f4} + 20 - 120 = 0$$

$\therefore$ $h_{f4} = 100.$

The computed values and the flow directions are shown in the figure shown below :

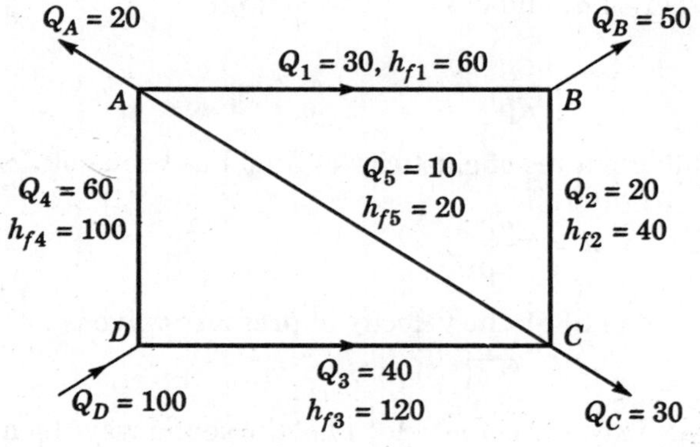

11.10. Velocity of Pressure Wave

When flow of a fluid in a pipe is stopped by the closure of a valve, the fluid just upstream of it is brought to rest. It is known that a change of velocity always gives rise to a corresponding change in pressure. As a result of elasticity of fluid, the changes of pressure do not take place instantaneously throughout the fluid but are propagated by pressure waves. The change of velocity of the fluid in the neighborhood of the valve brought about by the valve closure causes the pressures in the fluid to change so rapidly that a pressure wave is set up which travels upstream. The velocity of pressure wave may be determined by considering an elastic fluid at rest in a rigid pipe. Let A be the area of cross-section of the pipe. Let there be a piston at one end of the pipe which is suddenly moved with a velocity V for an interval of time dt. This will produce an increase in pressure which will travel through the fluid at velocity C. In time dt, the pressure wave will move a distance Cdt, while the piston moves a distance Vdt, see Fig. 11.19.

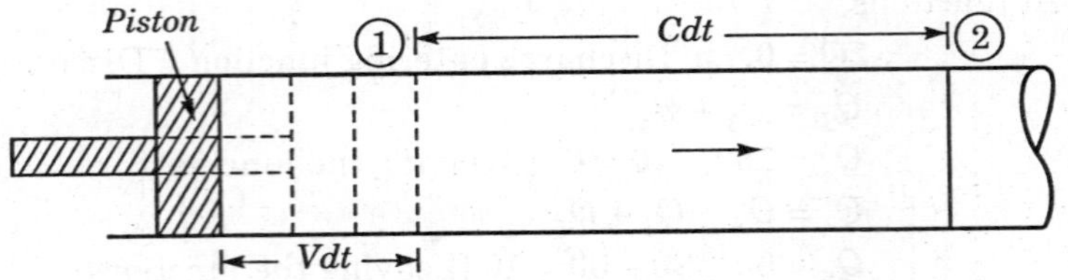

Fig. 11.19. Propagation of a pressure wave.

During this time the piston will displace a mass of fluid $\rho AVdt$, and during the same time the increase in pressure dp will increase the density of fluid between sections (1) and (2) by $d\rho$. Equating the mass displaced by the piston to the gain in mass between sections (1) and (2) due to increased density, we get

$$\rho AV\, dt = d\rho \,.\, ACdt$$

which yields,

$$C = V\frac{\rho}{d\rho} \qquad \text{...(11.34 a)}$$

From impulse momentum principle, the impulse of a force is equal to the resulting increase in momentum. The impulse of the force produced by the piston is $A\, dp\, dt$. The fluid mass $\rho ACdt$ was initially at rest, but as the pressure wave travels through it each element of it has its velocity increased to V, so that at the end of time dt the entire mass will have the velocity V. Hence the increase in momentum is $\rho ACdt \,.\, V$. Equating the impulse with the momentum change produced by it.

$$A\, dp\, dt = \rho AC\, dt \,.\, V$$

from which

$$C = \frac{dp}{\rho V} \qquad \text{...(11.34 b)}$$

Multiplying Eqs. (11.34 *a*) and (11.34 *b*), we obtain

$$C^2 = \frac{dp}{d\rho} \qquad \text{...(11.35)}$$

In art. 1.8, the bulk modulus of elasticity of fluid has been obtained vide Eq. (1.16) as

$$E = \rho\frac{dp}{d\rho} \qquad \text{...(1.16)}$$

From Eqs. (1.16) and (11.35), the velocity of pressure wave is

$$C = \sqrt{E/\rho} \qquad \text{...(11.36)}$$

The velocity C is also known as the velocity of the sound wave (sonic velocity) or the acoustic velocity.

In the preceding analysis for propagation of sound wave, we have considered the elasticity of fluid and the pipe has been taken as rigid. In fact the pipe material is elastic and the stretching of the pipe walls due to pressure wave makes the combined bulk modulus of elasticity less than that of the fluid alone. Let this new bulk modulus be expressed by K, and let the total volume change due to a pressure change dp be $dv = (dv)_1 + (dv)_2$, where $(dv)_1$ is due to compression of the fluid and $(dv)_2$ due to stretching of the pipe wall.

Then

$$K = -\frac{v\,dp}{(dv)_1 + (dv)_2} \qquad \text{...(11.37)}$$

in which v is the original volume. From Eq. (11.37), we have

$$\frac{1}{K} = -\frac{(dv)_1}{vdp} - \frac{(dv)_2}{vdp} \qquad ...(11.38)$$

The first term on the right hand side of Eq. (11.38) is $1/E$. If the pressure inside the pipe increases by dp, the increment of stress $dT = rdp/t$, (Fig. 11.20). If the circumference is stretched by an amount dl, the linear strain is $dl/2\pi r$. As $dl = 2\pi dr$, the strain becomes equal to dr/r, dr being increment in the radius. If the modulus of elasticity of the pipe material is E_m, then

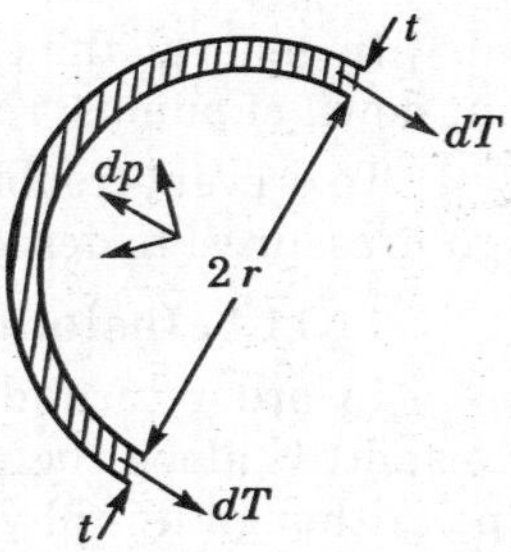

Fig. 11.20. Increment in circumferential stress.

$$E_m = \frac{dT}{dr/r} = \frac{r^2 dp}{t.dr}$$

from which, $$dp = E_m t\, dr/r^2 \qquad ...(11.39)$$

Considering unit length of the pipe, $v = \pi r^2$ and the increase in volume is equal to the increase in area so that $(dv)_2 = 2\pi r dr$. Substituting for dp, v and $(dv)_2$ in the second term of Eq. (11.38),

$$-\frac{(dv)_2}{vdp} = \frac{2r}{E_m t} = \frac{d}{E_m t}$$

and we obtain $$\frac{1}{K} = \frac{1}{E} + \frac{d}{E_m t}$$

from which $$K = \frac{E}{1 + (d/t)(E/E_m)} \qquad ...(11.40)$$

The velocity of pressure wave through an elastic fluid in an elastic pipe is, then

$$C = \sqrt{K/\rho} \qquad ...(11.41)$$

It is important to realize that the velocity of wave propagation is not a velocity with which fluid particles move. For this reason the word *celerity* is sometimes used for the velocity of wave propagation.

11.11. Water Hammer

The flow of a liquid in a pipeline has so far been considered as incompressible. But there are situations where the compressibility (or elasticity) of liquid as well as the elasticity of pipe material become important and must, therefore, be considered. If the velocity of liquid in the pipeline is rapidly decreased by operating a valve, large increase in pressure is caused which is propagated upstream throughout the fluid as a pressure wave. This phenomenon of abrupt increase of pressure caused by a rapid closure of valve is known as *water hammer*. The name water hammer is perhaps a misnomer because the phenomenon can occur in pipelines carrying liquids other than water or gases. The water hammer is a very important problem in case of hydroelectric plants, where the flow of water must be rapidly varied in accordance with the change of load on turbine.

It is a common experience that when a domestic water tap is closed very quickly a heavy knocking sound is heard and the entire pipe vibrates. The knocking sound and the vibrations are produced by the rise in pressure brought about by the rapid deceleration of water in the pipe when the tap is operated for closure. The fact that the knocking sound can be heard not only at the tap but elsewhere in the house clearly demonstrates that the disturbance caused by the sudden closing of the tap must travel along pipeline to other parts.

11.11.1. Definition of Water Hammer

Water-hammer in pipelines can be defined as the change in pressure above or below the normal working pressure caused by change in velocity of flow. Depending upon the rate and magnitude of the velocity change, this can lead to high pressure fluctuations in such pipelines as, for instance, supply line for drinking water and cooling water, discharge line for sewage water, oil-transmission line, penstock pipeline, etc.

11.11.2. Causes

In general, the changes in velocity are generated by valve operation, power failure, start or shut down of pumps, rupture of the line, etc.

To prevent trouble it is important to predict the transient pressures and in cases of excessively high pressures, to design a suitable protection.

11.11.3. Instantaneous Closure of Valve

In order to understand how the disturbance caused by sudden closing of the valve is transmitted along the pipe, let us consider the instantaneous closing of a valve at the end of the pipe. Although it is physically impossible to close a valve instantaneously, such a concept is useful as an introduction to the study of real cases. Fig. 11.21 shows a pipeline connected to a reservoir and having a valve near its downstream end *B*. Assume that the valve is closed instantaneously. Just before the closure, the pipe is full of fluid moving with a certain velocity towards the valve.

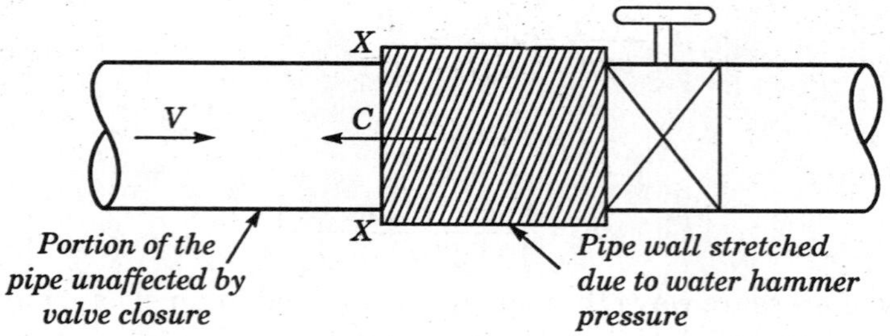

Fig. 11.21. Sequence of water hammer effects : fluid just u/s of the valve brought to rest, pressure increased, pipe wall stretched, pressure wave travels u/s and brings in more fluid to rest till the inlet end of the pipe is reached.

If the valve is suddenly closed, the fluid immediately upstream of the valve is stopped but the remaining fluid continues to move towards the valve as if nothing has happened. As the fluid in the immediate neighourhood of the valve has been brought to rest, it is slightly compressed by the fluid still flowing with its original velocity towards the valve. Not only will the fluid be compressed but the walls of the pipe surrounding this will also get stretched by the excess pressure produced. The next element of the fluid now finds an increased pressure with static conditions in front of it, therefore, it too comes to rest, is itself compressed and expands the pipe walls slightly. Each element of the fluid thus stops the element following it until all the fluid in the pipe has been brought to rest, Fig. 11.22. At any instant after closing the valve before all the fluid has stopped there is a discontinuity of conditions (such as pressure, velocity etc.) represented by the line xx in Fig. 11.21.

On the valve side of xx the fluid has stopped (shown shaded) and has been compressed and also the pipe is slightly expanded, see Fig. 11.22. On the left side of xx, however, the fluid is still moving with its initial velocity and initial pressure. As each successive element of the fluid is stopped, the discontinuity xx moves as pressure wave further away from the valve and towards the reservoir. When the pressure wave reaches the inlet at *A*, the entire fluid mass in pipe length *L* comes to rest and is subjected to an excess pressure throughout. There will be transient hydraulic grade line parallel to the original hydraulic grade line but at a height $\Delta p/\gamma$ above it, where Δp is the excess pressure.

It is impossible for a pressure which is greater than that due to the depth *DA* to exist at *A*. The pressure, therefore, instantly drops down to the value it would have for zero flow, that is the transient hydraulic grade line will join the water surface at *D*. But since the entire liquid in the

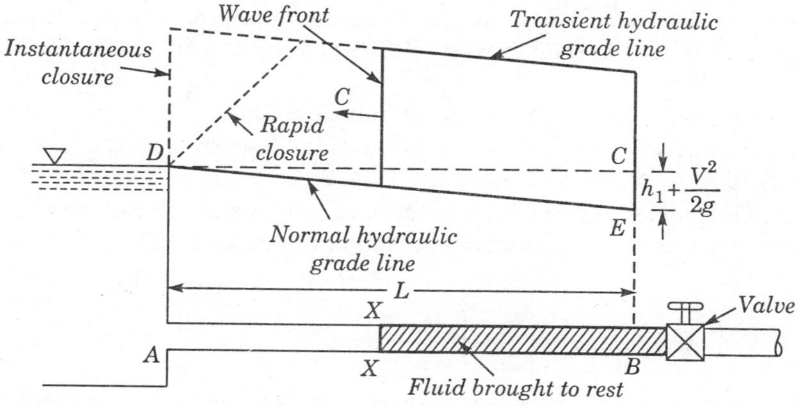

Fig. 11.22

pipe is under an excess pressure, it is slightly compressed and the pipe walls are stretched. As the liquid pressure everywhere in the pipe is higher than the pressure at A due to reservoir liquid depth DA, the pipe liquid starts to flow back into the reservoir and a wave of pressure unloading (or reflected wave) travels along the pipe from A to B. At the instant this pressure unloading wave reaches B, the entire liquid mass will be under the static pressure equivalent to the depth DA and indicated by the line DC. The liquid is still flowing back into the reservoir, and this reverse velocity must now be stopped instantaneously. This will produce a drop in pressure at B, which will be below the static pressure and the magnitude of which will be the same as it was an instant before, that is $(-\Delta p)$. This sets up a wave of rarefaction or a wave of negative pressure which travels towards the reservoir from B to A. As soon as this wave of rarefaction reaches the pipe inlet A, the velocity of the entire pipe liquid is brought to zero but the pressure everywhere in the pipe is negative Δp except at A where it must be the static pressure equivalent to depth DA. This pressure difference starts flow from the reservoir into the pipe and a wave of reflected negative pressure travels from A to B. As this wave reaches, B, the entire liquid in the pipe is moving with its original velocity towards the valve located at B, and the pressure in the pipe is brought to its initial value. This completes one full cycle of waves of positive and negative pressures generated as a result of instantaneous valve closure.

The time for a round trip of the pressure wave from B to A and back again is

$$T = \frac{2L}{C} \qquad \text{...(11.42)}$$

where L is the length of pipe and C is the velocity or celerity of the pressure wave. The mass of liquid in the pipe is ρAL, and the force required to change the velocity by an amount ΔV is ΔpA, where Δp is the excess pressure generated by the valve closure and A is area of pipe cross-section. From Newton's second law, the force and the change in momentum per second caused by it must be equal. Let Δt be the time required for the entire mass of liquid to reduce its velocity by ΔV, we then have

$$\Delta p \,.\, A = \rho AL \frac{\Delta V}{\Delta t} \qquad \text{...(11.43)}$$

but $\Delta t = L/C$,

substituting this value of Δt, we obtain

$$\Delta p \,.\, A = \rho AC\Delta V$$

from which $\Delta p = \rho C \Delta V$ ...(11.44)

If the valve closure is complete so that the velocity is reduced to zero, the maximum water hammer pressure is then given by

$$(\Delta p)_{max} = \rho CV \qquad \text{...(11.45)}$$

It will be seen from Eq. (11.44) that the water hammer pressure is independent of the pipe length and depends solely upon the acoustic velocity or velocity of sound in the fluid medium and change in the fluid velocity.

Example 11.40. *Water flows in a pipe of diameter 600 mm having a length of 2200 m at a velocity of 2 m/s. At the end of the pipe, flow is stopped in 20 sec by closing the valve. Determine the rise in pressure effected by the valve-closure.*

Solution. The momentum of flow is reduced to zero by the valve-closure, which generates a force causing increase in pressure. Newton's second law of motion relates these two by Eq. (11.42) as under :

$$\Delta p \, . \, \frac{\pi d^2}{4} = \rho \left(\frac{\pi d^2}{4}\right) L \frac{\Delta V}{\Delta t}$$

or

$$\Delta p = \rho L \frac{\Delta V}{\Delta t} = \frac{1000}{9.81} \times 2200 \times \frac{2}{20} \text{ (kg/m}^2\text{)}$$

$$= \frac{9810}{9.81} \times 2200 \times \frac{2}{20} \text{ (N/m}^2\text{)} = \mathbf{2.2 \times 10^5 \ N/m^2.}$$

11.11.4. Rapid Closure

It is physically impossible for a valve to be closed instantaneously and, therefore, we shall consider the real case where the valve is closed in a finite time t_c which is greater than zero but less than $2L/C$. From the actual pressure recordings, it has been observed that the maximum pressure rise for rapid closure is the same as for instantaneous closure of valve. Thus if $t_c \leq 2L/C$, the valve closure is rapid and the water hammer pressure is given by Eq. (11.45). The time $(2L/C)$ is called as 'critical time'. The time period of a complete cycle is $(4L/C)$.

11.11.5. Slow Closure

The slow closure is defined as the one wherein the time of valve closure is greater than $2L/C$. This will happen when the pipe length is not large or the valve closure is not rapid so that the pressure wave returns back to the valve while it is still being closed. In this case the maximum pressure rise will be less than that for a rapid valve closer because the wave of pressure unloading will reach the valve before the valve closure is completed and this will stop any further increase in pressure.

If the time of valve closure is many times greater than $2L/C$, the compressibility effects are no longer important and the pressure rise can be computed from Eq. (11.43). If the closure time is slightly greater than $2L/C$, the water hammer pressure Δp may be reduced in the proportion of $2L/Ct_c$ to yield

$$\frac{\Delta p}{(\Delta p)_{max}} = \frac{2L}{C} \frac{1}{t_c}$$

$$\Delta p = \rho VC \left(\frac{2L}{Ct_c}\right) = 2\rho VL/t_c \quad \text{...(11.46)}$$

Example 11.41. *A 50 cm diameter and 100 m long pipeline carrying 0.5 m³/s discharge is fitted with a valve at the downstream end. Calculate the rise in pressure caused within the pipe due to valve closure in (i) 1 second (ii) instantaneously. Take sonic velocity as 1430 m/sec.*

Solution. The critical time for valve closure is given by

$$t = \frac{2L}{C} = \frac{2 \times 100}{1430} = 0.1398 \text{ sec.}$$

Since the time taken for the valve closure (*i.e.* 1 second) is greater than t, it is the case of slow valve closure.

(*i*) The pressure rise can be computed using Eq. (11.46)

$$\Delta p = 2\rho VL/t_c = 2 \times \frac{1000}{9.81} \times \frac{4 \times 0.5}{\pi\,(0.5)^2} \times \frac{100}{1}$$

$$= 51948.1 \text{ kg/m}^2 = \mathbf{51.95 \times 10^3} \text{ kg/m}^2$$

(*ii*) For instantaneous valve closure, using Eq. (11.45)

$$(\Delta p)_{max} = \rho CV = \frac{1000}{9.81} \times 1430 \times \frac{4 \times 0.5}{\pi\,(0.5)^2}$$

$$= 371199.3 \text{ kg/m}^2 = \mathbf{371.2 \times 10^3 \text{ kg/m}^2}.$$

Example 11.42. *A 900 mm diameter steel pipe carries water at the rate of 1.5 m³/s. The pipe wall has a thickness of 1 cm. The elastic modulus of steel is 2 × 10¹¹ N/m² and the bulk modulus of water is 2.1 × 10⁹ N/m². Determine the increase in pressure if the valve at the end of 3.5 km long pipeline is closed in 3.5 seconds.*

Solution. The combined modulus of elasticity,

$$K = \frac{E}{1 + \dfrac{d}{t}\dfrac{E}{E_m}} = \frac{2.1 \times 10^9}{1 + \dfrac{90}{1} \times \dfrac{2.1 \times 10^9}{2 \times 10^{11}}}$$

$\therefore$ Velocity of pressure wave $C = \sqrt{\dfrac{K}{\rho}} = \sqrt{\dfrac{1.1 \times 10^9}{1000}} = 1049$ m/s

The time required for the pressure wave to travel from the valve to the inlet and back to the valve is

$$t = \frac{2L}{C} = \frac{2 \times 3500}{1049} = 6.67 \text{ seconds.}$$

Since the time of valve closure $t_c = 3.5$ seconds is less than 6.67 seconds, the valve closure is rapid.

Velocity of flow, $V = \dfrac{4Q}{\pi d^2} = \dfrac{4 \times 1.5}{\pi (0.9)^2} = 2.36$ m/s.

$\therefore$ The pressure rise $\Delta p = \rho CV = 1000 \times 1049 \times 2.36 = 2.48 \times 10^6$ N/m².

11.12. Surge Tank

The oscillation of flow in pipelines, when compressibility effects are not important, is referred to as "*surge*". As one of the means of eliminating water hammer, provision is made to permit liquid to surge into a tank, known as the *surge tank*. In hydro-electric power plants, the turbines must be supplied with water through a long pipe line or a tunnel cut into the rock. If the electric power demand is suddenly decreased, the turbines must accordingly adjust to this new situation by decreasing the water flow to the turbines. It is the consequent rapid deceleration of water in the pipeline which gives rise to water hammer. Since the water hammer may produce dangerously high pressure, it is desirable to minimise its effects. Fig. 11.23 shows a surge tank installed in the pipeline as close to the turbine as is practicable.

The simplest type of surge tank is an open vertical cylinder of large diameter as shown at *B*. The most desirable location for the surge tank is immediately upstream of the turbines, but it may not be practicable to install it there due to topography of the land. In such cases it should be located as close to turbines as possible. The upstream pipe line *AB* should be laid to a small slope so that the height of the surge tank is not large. The top of the surge tank must be higher than the water level in the reservoir. If the load on the turbines is suddenly reduced, the governing mechanism acts to decrease the flow accordingly. The rate of flow in the pipeline *AB* cannot at once drop to the required new value, and the temporary surplus of flow goes into the surge tank. A gradually rising hydraulic grade line, indicated by the line *DE* represents decrease in the load on the turbine thereby requiring a lesser flow of water. The rise of water in the surge tank provides a hydrostatic head which decelerates the flow in *AB*. It has been emphasized earlier that the length of pipe between the surge tank and the turbines should be as small as possible, because this pipe length is still subjected to water hammer effects and, therefore, it must be made strong enough to withstand the increased pressures.

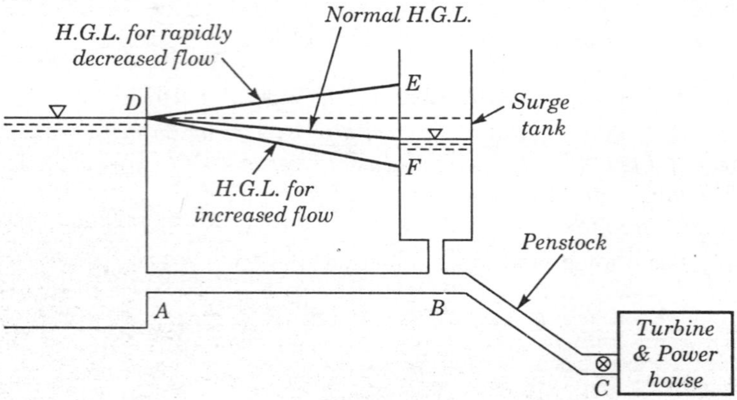

Fig. 11.23. A schematic diagram of a surge tank.

When the load on the turbines is increased suddenly, a large flow of water is required which is temporarily supplied from the water stored in the surge tank. In the absence of a surge tank the drop in pressure at the turbines could be excessive at the time of sudden demand. As the water level in the surge tank is lowered, the difference of head along *AB* is increased thereby accelerating the flow until the flow in *AB* equals that required by the turbines. This condition is indicated by the line *DF* in Fig. 11.23.

Surge tanks may be classified as simple, orifice and differential. The simple surge tank has an unrestricted opening into it and must be of adequate size so that is does not overflow. The orifice surge tank has a restricted opening or orifice between pipeline and the tank which allows more rapid pressure changes in the pipeline than the simple surge tank. The more rapid pressure change causes a more rapid adjustment of flow to the new conditions. The differential surge tank is a combination of an orifice surge tank and a simple surge tank of small cross-sectional area. In case of rapid valve opening a limited amount of water is directly available from the central riser and flow from the large tank supplements this flow. For sudden valve closures the central riser may be designed in such a way that it overflows into the outside tank. Fig. 11.24 illustrates all the three types of surge tanks discussed above.

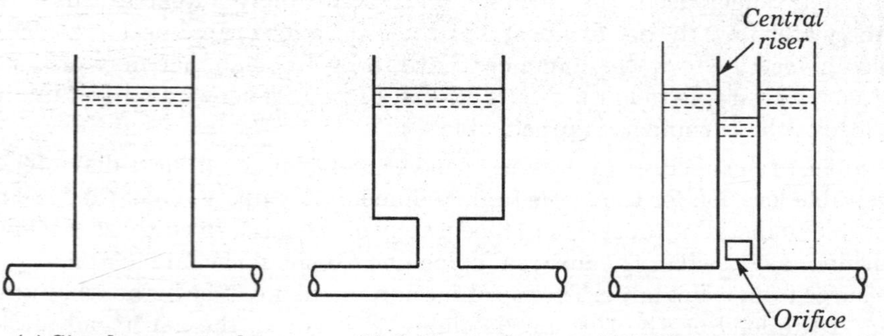

Fig. 11.24. Types of surge tank.

Example 11.44. *Water is flowing at 4 m/s in a pressure penstock 4500 m long. If the celerity of the pressure wave travelling in the pipe due to the sudden complete closure of a valve at the downstream end is given as 1500 m/s, what is the maximum pressure rise ? What is the period of oscillation ? Show how the pressure changes with time at the middle point of the penstock length. Ignore all frictional losses.*

Solution. Maximum pressure rise is given by Eq. (11.45)

$$(\Delta p)_{max} = \rho CV = 1000 \times 1500 \times 4 = 6 \times 10^6 \text{ N/m}^2$$

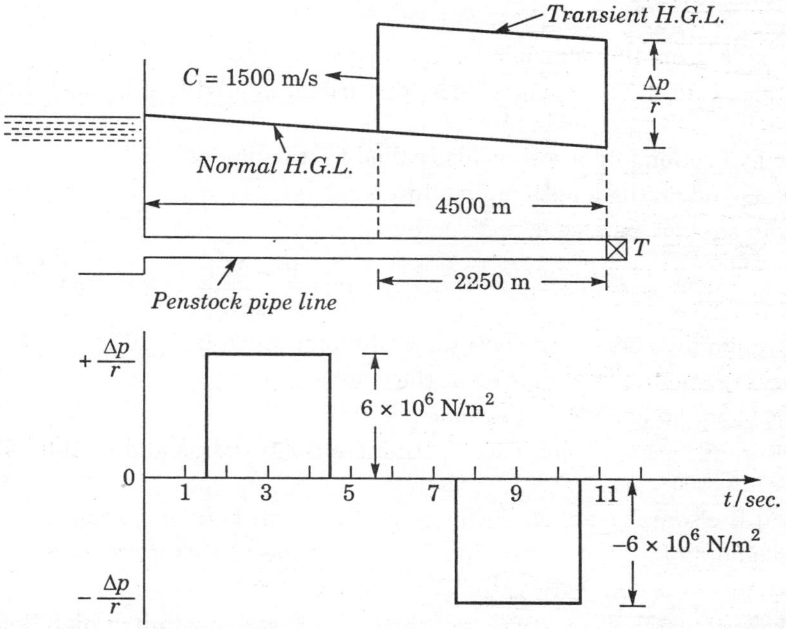

Fig. Ex. 11.36. Pressure variation at the middle point of penstock pipeline.

$$\text{Period of oscillation} = \frac{4L}{C} = \frac{4 \times 4500}{1500} = 12 \text{ seconds.}$$

At the middle point of the penstock length, the pressure wave reaches in a time of $\dfrac{4500}{2 \times 1500}$ = 1.5 seconds. The pressure at the middle point at this instant rises by 6×10^6 N/m^2 and remains so until the pressure wave returns as a wave of rarefaction (negative pressure). This happens at 4.5 seconds after the closure of the valve when the pressure in the penstock at the middle point is reduced by equal amount. The variation of pressure at the middle point with respect to time is shown in the figure. The next cycle will start at 13.5 seconds.

OBJECTIVE TYPE QUESTIONS

1. The entrance length or length of establishment of flow is :
   (*a*) the pipe length inside the reservoir
   (*b*) the initial length in which the flow develops fully such that the velocity profile does not change downstream
   (*c*) the length of pipe from its entrance in which the flow may be assumed irrotational
   (*d*) the length in which the boundary layer remains uniform.

2. For turbulent flow in smooth pipes, the entrance length is taken as :
   (*i*) 114 (*ii*) 75 (*iii*) 50 (*iv*) 100
   times the pipe diameter.
3. The energy loss in pipe lines is due to :
   (*a*) viscous action only (*b*) surface roughness only
   (*c*) friction offered by pipe wall as well as by the viscous action
   (*d*) turbulent shear stress alone.
4. Minor losses in pipe flow are those :
   (*i*) which have a small magnitude
   (*ii*) which are caused on account of local disturbances produced by such fittings as valves, bends etc.
   (*iii*) caused by friction and are thus also called friction losses
   (*iv*) which depend on the length of pipeline.
5. Head loss in sudden expansion is given by :
   (*a*) $\frac{V_1^2 - V_2^2}{2g}$ (*b*) $\frac{(V_1 - V_2)^3}{2g}$ (*c*) $\frac{(V_1 - V_2)^2}{2g}$ (*d*) $\frac{2(V_1^2 - V_2)^2}{g}$.
6. In flow through pipe bends, the pressures on inner and outer radii :
   (*i*) do not vary and are the same as at the centre of pipe
   (*ii*) vary, it being more on the inner one
   (*iii*) are different ; pressure increases with increase in radius and is, therefore, more at the outer radius
   (*iv*) stand at the same level increasing gradually towards the pipe centre.
7. The loss of energy in commercial pipe fittings is generally expressed by

$$h_L = K_L\ V^2/2g$$

   in which K_L is called the loss coefficient which is almost a constant at high Reynolds numbers. The following table lists certain pipe fittings alongwith possible K_L—values, given in a raondom order. Identify the K_L—value for each fitting :

(*a*) Globe valve, wide open	(*i*) 0.4
(*b*) Gate valve, half open	(*ii*) 0.9
(*c*) Gate valve, one-fourth open	(*iii*) 1.5
(*d*) Pump's foot valve	(*iv*) 24.0
(*e*) 90-degree elbow, short radius	(*v*) 5.6
(*f*) 45-degree elbow	(*vi*) 10.0

8. Listed below are three major pipe flow problems generally encountered in engineering practice. Also given alongside are certain groups of variables which are associated with these types of problems. Select the right group of variables for each type of problem :

I. determination of pipe diameter	(*a*) (Q, d, K_s, ν)
II. determination of discharge	(*b*) (Q, K_s, S_f, ν)
III. determination of head loss	(*c*) (d, K_s, S_f, ν)

9. Two reservoirs are joined together by means of two pipelines (connected in parallel) of different diameters but of same length. The water level in one reservoir is higher than that in the other one. Which of the following statements are correct :
   (*i*) the discharge in each pipeline is same
   (*ii*) the head loss for the smaller diameter pipe is more than that for the larger diameter one
   (*iii*) the head loss for both the pipes is the same
   (*iv*) the discharge in the larger diameter is more.

10. Two pipelines of equal length and of diameters 15 cm and 10 cm respectively connect two reservoirs such that the pipelines are in parallel. The difference in water levels in the reservoirs is 3 m. If the friction is assumed to be equal the discharge in the 15 cm diameter pipe is :

(*a*) 2.25 (*b*) 3.375 (*c*) 2.756 (*d*) 1.5

times the discharge in 10 cm diameter pipe.

11. The condition for maximum transmission of power through a pipeline is that one-third of the available head must be consumed in friction. The corresponding efficiency of the pipeline is :

(*i*) 66.67% (*ii*) 33.3% (*iii*) 100% (*iv*) $\sqrt{2/33}$.

12. If the absolute pressure in pipeline falls close to the vapour pressure of the fluid, cavitation may set in and the dissolved air and other gases upon liberation may get collected at the highest point in the pipeline thus forming an air lock. In such cases, the position of the hydraulic grade line with respect to the pipeline determines :

(*a*) the locations where the velocities are high

(*b*) the regions where the pressures are negative

(*c*) the portions of the pipeline where the potential energy is more

(*d*) none of the above.

13. In order to have a continuous flow through a siphon, no portion of the pipe be higher than :

(*i*) 10 m (*ii*) 10.33 m (*iii*) 5.5 m (*iv*) 7.75 m.

measured above the hydraulic grade line.

14. In the three-reservoir problem, the reservoirs are interconnected by means of three pipelines meeting at a common junction. The solution of this problem hinges on the determination of direction of flow in the pipe joining the intermediate reservoir. This involves :

(*a*) determination of flow in each pipe

(*b*) locating the position of hydraulic grade line at the junction

(*c*) knowing the height of the junction itself

(*d*) determining the head loss in each pipe.

15. The Hardy-Cross method of hydraulic analysis of pipe networks, besides satisfying the continuity and energy principles, must also satisfy the condition :

(*i*) that the flow into any junction must equal the outflow from it

(*ii*) that the flow in each pipe has head loss according to the Darcy-Weisbach equation

(*iii*) that the alogebraic sum of the head losses around any closed curve must be zero

(*iv*) that the momentum principle must be satisfied.

16. The most important feature of Hardy-Cross method lies in the determination of flow correction ΔQ which is expressed by :

(*i*) $\dfrac{\Sigma KQ_0^{n-1}}{\Sigma nKQ_0^{n}}$ (*ii*) $\dfrac{\Sigma KQ_0^{n}}{\Sigma nKQ^{n-1}}$ (*iii*) $\dfrac{\Sigma nKQ_0^{n-1}}{\Sigma KQ_0^{n}}$ (*iv*) $\dfrac{\Sigma nKQ_0^{n}}{\Sigma KQ_0^{n-1}}$.

17. The determination of flow correction ΔQ in the Hardy-Cross method is, based on the assumption that :

(*a*) the continuity equation must be satisfied

(*b*) the flow in each pipe must satisfy the head loss equation

(*c*) the correction ΔQ is small as compared to the assumed flow Q_0

(*d*) none of the above.

18. The pressure wave in a fluid medium travels as a sound wave, the velocity of which is given by :

(*i*) $C = \sqrt{\dfrac{E}{\rho}}$ (*ii*) $C = \sqrt{\rho K}$ (*iii*) $C = \sqrt{\dfrac{\rho}{E}}$ (*iv*) $C = \dfrac{E}{\rho}$.

19. The propagation of a pressure wave in pipes is influenced by the elasticity of pipe material which makes the combined bulk modulus of elasticity smaller than that for the fluid alone. The combined bulk modulus K is expressed by :

(*a*) $\dfrac{E}{\left[1+\left(\dfrac{t}{d}\right)\dfrac{E}{E_m}\right]}$ (*b*) $\dfrac{E}{1+\dfrac{d}{t}\dfrac{E}{E_m}}$ (*c*) $\dfrac{E}{1+\dfrac{d}{t}\dfrac{E_m}{E}}$ (*d*) $\dfrac{E}{1+\dfrac{t}{d}\dfrac{E_m}{E}}$.

20. Water hammer is a phenomenon which is caused by :

(*i*) sudden opening of a valve in a pipeline

(*ii*) sudden closure (partial or complete) of a valve in pipeflow

(*iii*) incompressibility of fluid

(*iv*) the pipe material being elastic.

21. The rapid closure of valve in a water pipeline will result in water hammer pressure of magnitude :

(*a*) $\rho C^2 V$ (*b*) $\rho C V^2$ (*c*) $\rho C V$ (*d*) $\rho C/V$.

22. The closure of valve is considered rapid when :

(*i*) the valve is closed almost instantaneously

(*ii*) the duration of valve-closure is less than $2L/C$

(*iii*) the duration of valve-closure is less than L/C

(*iv*) the duration of closure is greater than $2L/C$.

23. A surge tank is a device connected to the penstock pipeline and installed as close to the powerhouse as the topography of the area permits. The main function of the surge tank is to :

(*a*) restrict the water hammer effects to a small length of penstock

(*b*) provide a free-water surface near the turbines

(*c*) act as a reservoir

(*d*) protect the penstock pipeline from bursting.

ANSWERS TO OBJECTIVE TYPE QUESTIONS

1. (*b*) **2.** (*iii*) **3.** (*c*) **4.** (*ii*) **5.** (*c*)

6. (*iii*) **7.** (*a*) (*iv*) ; (*b*) (*v*) ; (*c*) (*iv*) ; (*d*) (*iii*) ; (*e*) (*ii*) ; (*f*) (*i*)

8. I, (*b*) ; II (*c*) ; III (*a*) **9.** (*iii*), (*iv*) **10.** (*c*) **11.** (*i*)

12. (*b*) **13.** (*iv*) **14.** (*b*) **15.** (*iii*) **16.** (*ii*)

17. (*c*) **18.** (*i*) **19.** (*b*) **20.** (*ii*) **21.** (*c*)

22. (*ii*) **23.** (*a*).

REVIEW QUESTIONS

1. What factors influence the establishment of fully developed flow ? What do you understand by entrance length ?

2. What are the minor losses ? Under what circumstances will they be negligible ?

3. What is the entrance loss ? Give the approximate values of loss coefficient for different types of pipe entrances ?

4. What is a diffuser and what function it is used for ?

5. What are the causes of loss of energy in pipe bends ?

6. One pipe system is said to be equivalent to another pipe system when the following two quantities are the same

(*i*) h_L, Q (*ii*) L, Q (*iii*) L, d (*iv*) f, d

(*v*) V, d.

7. What is a siphon and under what conditions would it stop functioning ?

8. The discharge of an existing pipeline is to be increased. You have additional pipe of any desired length and diameter at your disposal. Suggest how would you propose to augment the flow in the pipeline.

9. How are the sudden pressure changes transmitted in a pipeline ? Does the elasticity of the pipe material also affect the propagation of pressure wave ? Explain.

10. What do you understand by water hammer ? What provision is made in hydro-power plants to minimize the effects of water hammer ?

PROBLEMS

11.1. A pipeline leads water from one reservoir to another, which has its water surface at a lower elevation. For a discharge of 0.10 m^3/sec, determine the losses in Nm/N and also in Watts.

11.2. A hydroelectric power plant has a head (difference in elevation of headwater and tailwater) of 30 m and a flow of 3.3 m^3/s through turbines, which rotate at 180 r.p.m. The torque in the turbine shaft as measured is 43650 Nm and the power output of the generator is 779 kW. Determine the (*a*) loss of energy in terms of power ; (1) upto the turbine, (2) through the generator, and (*b*) efficiency of (*i*) the turbine and (*ii*) the generator.

11.3. In flow of liquid through a pipeline the losses are 1.47 kW for average velocity of 1.5 m/s and 2.94 kW for 2.25 m/s, what is the nature of flow ?

11.4. When tripling the flow in a pipe line causes the losses to increase by 7.64 times, how do the losses vary with velocity and what is the nature of flow ?

11.5. A short length of pipe is connected to a reservoir of height h. A nozzle of exit area A_j is connected to the pipe, the cross-sectional area being three times that of the nozzle exit. The velocity coefficient of the nozzle is C_v. The head loss in the pipe (including entrance loss) is given by $h_L = KV^2/2g$, where K is a known loss coefficient and V is the velocity in the pipe. Show that the jet velocity is given by

$$V_j = \sqrt{\frac{2gh}{\frac{1}{C_v^2} + \frac{K}{g}}}\,.$$

Find the numerical value for the jet velocity, if h = 15 m, C_v = 0.9 and K = 0.5.

11.6. A turbine is fed from a water reservoir whose level is 180 metres above the discharge pipe of the turbine. The pipe discharges into the atmosphere and the velocity at this point is 3 m/s. The friction loss in the entire pipeline (except for the losses in the turbine) is 15 metres, when the flow is 10 cu metre/sec. If the power output under these conditions is 13.97 MW, find the efficiency of the turbine.

11.7. A fire nozzle is to be used at an elevation of 9 metres above the level of a reservoir. The velocity of the water jet is to be 16 m/s. The ratio of the hose to nozzle area is 4. The loss coefficient between reservoir water and the inlet to the pump is 6 and the loss coefficient between the discharge side of the pump and the entrance to the nozzle is 7. The velocity coefficient of the nozzle 0.90, and the contraction coefficient is 1.0. The area of the inlet pipe is the same as that of the hose. (*a*) Find the net head to be supplied by the pump and (*b*) if the pump efficiency is 75%, find the power required to drive pump for a discharge of 4.9 kN/s.

11.8. A pump takes water from a reservoir and discharges it at an elevation of 15 metres higher (see the following sketch). The energy losses in the pipeline are given by $KV^2/2g$, where V is the velocity in the pipe and K is the loss coefficient which is considered constant. The contraction coefficient at the pipe discharge C_c = 1.0.

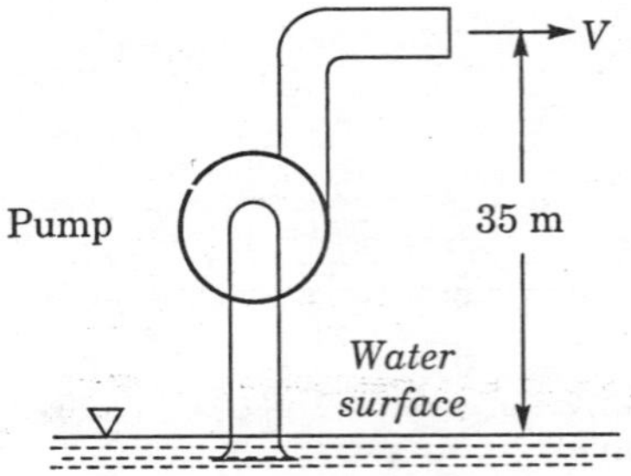

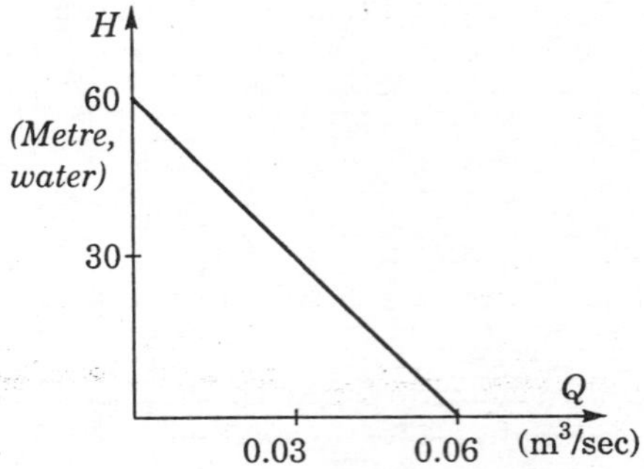

The relation between the total head rise across the pump (H) and the flowrate through the pump (Q) is given by the accompanying chart. If $K = 20$ and the pipe diameter is 10 cm

(*i*) what is the flow rate through the pump ?

(*ii*) what is the power required to drive the pump at 70% efficiency ?

11.9. A straight 25 cm pipeline 5 km long is laid between two reservoirs having a difference of levels of 40 metres. To increase the capacity of the system an additional 2.5 km long 2.5 cm pipe is laid parallel from the first reservoir to the mid-point of the original pipe. Assuming $f = 0.025$ for both pipes, find the increase in discharge due to installation of the new pipe.

11.10. A 60 cm pipe is supplied with water from a reservoir A and at a point D it is divided into branches of 45 cm and 30 cm diameter which discharge into reservoirs B and C respectively. Lengths of 60 cm, 45 cm and 30 cm pipes are 650 m, 1000 m and 500 m respectively. The surface levels in A, B and C and the piezometric head at D are 33.50 m, 21.30 m, 15.25 m and 24.40 m above datum respectively. Find the velocity of flow, in each pipe taking $f = 0.028$, check whether or not the given level at D, *i.e.* 24.40 m conforms to the principle of continuity of flow.

11.11. 162 kW power is to be transmitted through a pipeline, the pressure at the inlet of pipe being 9.81 MN/m^2. If the water pressure-drop per kilometre is to be 2.5 metres, find (*i*) the diameter of pipe (*ii*) efficiency of transmission in 20 km length of pipe. Take $f = 0.032$.

11.12. A pipeline of 0.6 m diameter is 1.5 km long. To augment the discharge another line of the same diameter is introduced parallel to the first in the second half of the length. Neglecting minor losses find the increase in discharge if $f = 0.04$ and the head above the outlet is 30 m.

11.13. What will be the discharge of water flowing through a 30 cm diameter rigid pipe, if sudden closure of the valve at the end of the pipe produces at pressure rise of 0.402 MN/m^2 ? Take E for water = 21.1×10^2 MN/m^2 and $\rho = 1000$ kg/m^2.

11.14. Determine the maximum time for rapid valve closure of a pipe 60 cm diameter, 450 m long made of steel ($E_m = 2.07 \times 10^{11}$ N/m^2) with a wall thickness of 1.25 cm. The pipe carries a liquid of specific gravity 0.88 and bulk modulus $E = 1.03 \times 10^9$ N/m^2 flowing at 1 m^3/s.

11.15. Flow takes place through the siphon as shown. The total length of the siphon is 6000 m and its diameter is 30 cm. The friction factor is 0.02. Determine (*i*) the discharge when the siphon is running full and (*ii*) the maximum length L_1 of the pipe upto the vertex if the minimum pressure allowed is 2 m water absolute.

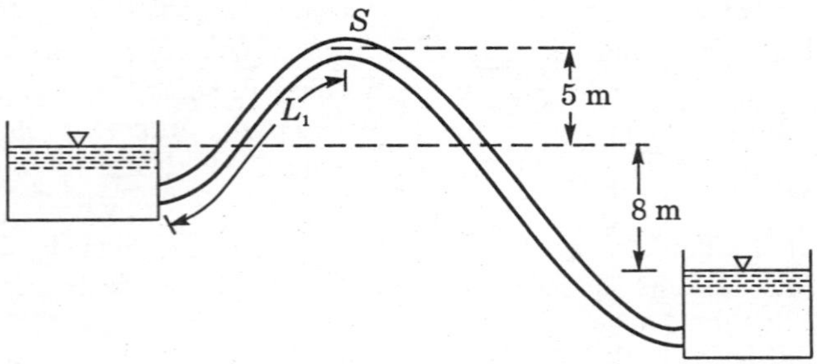

11.16. Determine the size of the clean wrought iron pipe (k_s = 0.000045 m) required to convey 250 litres/sec of fluid of kinematic viscosity $\nu = 9 \times 10^{-6}$ m²/s over a length of 3000 metres with a head loss of 22.5 metre.

11.17. An oil pipeline 60 cm in diameter and roughness height 0.00005 m carries a flow of 0.55 m³/s across the country. The pumping stations are located every 80 km. If the pump efficiency is 85%, determine the power input required at each station when the kinematic viscosity and the specific gravity of the oil are 2×10^{-6} m²/s and 0.88 respectively.

11.18. A pipeline 60 cm in diameter carrying water at the rate of 1.2 m³/s has its flow suddenly closed by a valve. If the velocity of pressure wave is 1200 m/s, what is the pressure rise at the valve ?

11.19. A 30 cm horizontal water pipeline enlarges to a 60 cm pipeline through a 20° conical enlargement for which the loss coefficient is 0.43. When 0.40 m³/s of water flow through the pipeline the pressure in the smaller pipe is 147 kN/m². Calculate pressure in the larger pipeline neglecting pipe friction.

11.20. At a sudden enlargement of a waterline from 200 mm to 400 mm diameter, the hydraulic gradient rises by 10 mm. Estimate the rate of flow.
[**Hint.** Similar to Solved Ex. 11.1]

11.21. Calculate the flow through each of the pipe of the network shown below. Take $h_L = KQ^2$.

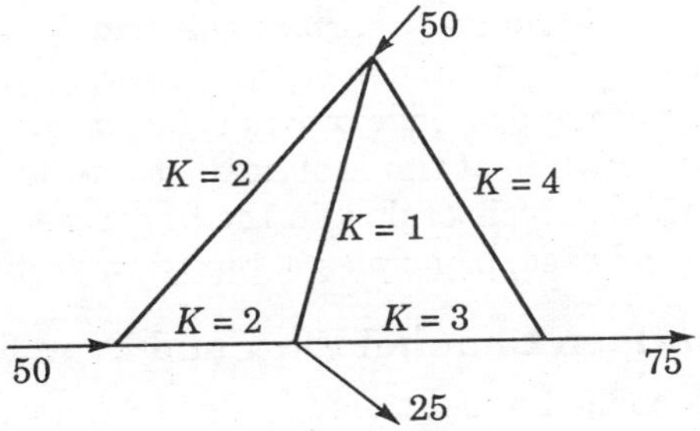

ANSWERS

11.1. 7 Nm/N, 6.86 kW
11.2. (*a*) 148.5 kW, 42.2 kW (*b*) 84.6%, 94.8%
11.3. Turbulent
11.4. $h = KV^{1.85}$, Turbulent
11.5. 15.1 m/s
11.6. 86.9%
11.7. 35.7 m, 233 kW
11.8. 0.02 m³/s, 11.2 kW
11.9. 9.5 lit/sec
11.10. 1.988 m/s, 0.99 m/s, 1.96 m/s, Continuity equation is not satisfied
11.11. 2.15 m, 95%
11.12. 0.177 m³/s or 25.6%
11.13. 194.4 lit/sec
11.14. 2.92 sec
11.15. 137.5 lit/sec, 235.5 m
11.16. 0.485 m
11.17. 1864 kW
11.18. 5.1 MN/m²
11.19. 158.3 kN/m²
11.20. 22.7 l/s
11.21.

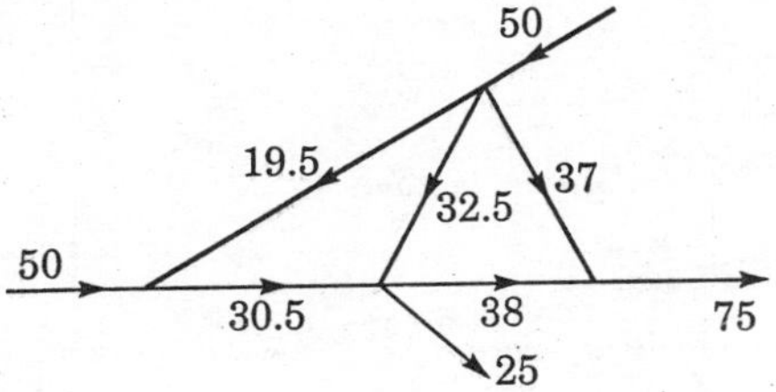

12

Flow Through Open Channels

PART A—UNIFORM FLOW

12.1. Introduction

The flow in open channels is characterized by the existence of a free-surface. The free-surface may be defined as the surface of contact between the liquid and the overlying gaseous fluid, (*i.e.*, liquid-gas interface), the interface being subjected to a constant pressure throughout its length and breadth. In most of the engineering problems of open channel flow, the gaseous fluid happens to be the air at local atmospheric pressure. In such cases, the free-surface is a liquid-air interface which is subjected to the local atmospheric pressure. Open-channel flow, however, in general, may have a free-surface arising out of a liquid-gas interface subjected to a constant pressure. The flow in a closed conduit may either be pipe flow or open channel flow, depending on whether it runs full under pressure, or has a free-surface. A drainage pipeline, *i.e.*, a sewer, under ordinary conditions of flow behaves as an open channel while at times of heavy rains, the sewer gets completely filled and runs full under pressure. The analysis of flow in the first case will involve principles of open channel flow whereas in the second case, principles of pipe flow will govern.

12.2. Comparison between Open Channel Flow and Pipe Flow

Ven Te Chow defines an open channel thus "*An open channel is a conduit in which water flows with a free-surface.*" This fact, that the flow in an open channel must have a free-surface, distinguishes it from pipe flow. The important points of differences between the two types of flow are given below.

(1) Since the open channel flow is distinguished by the presence of a free-surface exposed to the atmosphere, the pressure distribution in straight and continuous bottomed channels follows hydrostatic law, *i.e.* the pressure at a point situated at a vertical distance y below the free-surface will be γy. The flow in open channels is at the expense of the potential energy, and hence the flow is caused by gravity force provided by the sloping bottom, Fig. 12.1.

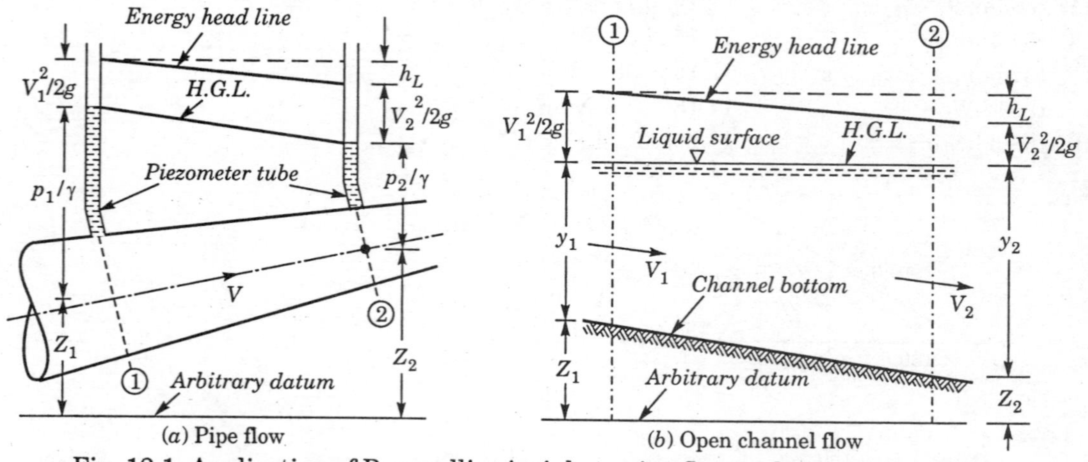

Fig. 12.1. Application of Bernoulli principle to pipe flow and open channel flow.

The pipe flow, on the other hand, does not have a free-surface. The pipe runs full and the flow, in general, takes place at the expense of hydraulic pressure. The pressure, therefore, continuously decreases in the direction of flow.

(2) The second point of difference lies in the geometry of the flow section. In pipes, which are generally round in cross-section the flowing liquid entirely fills the pipe section, and hence the cross-section of flow is fixed since it is completely defined by the geometry of the conduit. Open channels may have any shape : triangular, rectangular, trapezoidal, parabolic, circular or any irregular shape such as that attained by natural streams. In open channels, the position of the free surface is likely to change with space (along the direction of flow) and time. The free surface is subject to changes in elevation, resulting in change in depth of flow brought about by various factors affecting the flow. The boundaries of flow as defined by the channel perimeter are somewhat fixed, but the position of the free boundary (free surface) remains a factor to be predicted by the best available means. The determination of flow cross-section in open channels is, therefore, a matter of prediction and approximation.

(3) The surface roughness in open channels varies between wide limits. The channel surface varies from that of glass or polished metal used in the laboratory channels to that of rough and irregular river beds containing pebbles and boulders. It is believed with certainty, that the hydraulic roughness in open channels varies with the depth of flow also.

In case of pipes, the internal surface varies from that of new brass or steel pipe, having low roughness coefficient, to that of old and corroded iron or steel pipes, which have comparatively high roughness elements projecting out into the flow, resulting in a very high roughness coefficient.

(4) The application of the Bernoulli principle will show that the piezometric head in open channel flow is $(Z + y)$, where y is the depth of flow, as against $(Z + p/\gamma)$ in pipe flow. The hydraulic grade line, therefore, coincides with the water surface. Fig. 12.1 illustrates the diagrammatic representation of the Bernoulli principle.

(5) The velocity distribution in pipe flow and open channel flow differs considerably. The velocity distributions for both the types are shown in Fig. 12.2.

In pipe flow, the velocity distribution is symmetrical about the pipe axis, maximum velocity occurring at the pipe centre and the velocity at the pipe wall reducing to zero. In case of open channel flow, the maximum velocity occurs a little distance below the water surface. If the channel bed is mobile the velocity at the channel bottom may be non-zero. The shape of the velocity profile is also dependent on the channel roughness. It is understood that the channel roughness is likely to increase with increasing Froude number. It is believed that the presence of the free surface makes the channel rougher than the pipe.

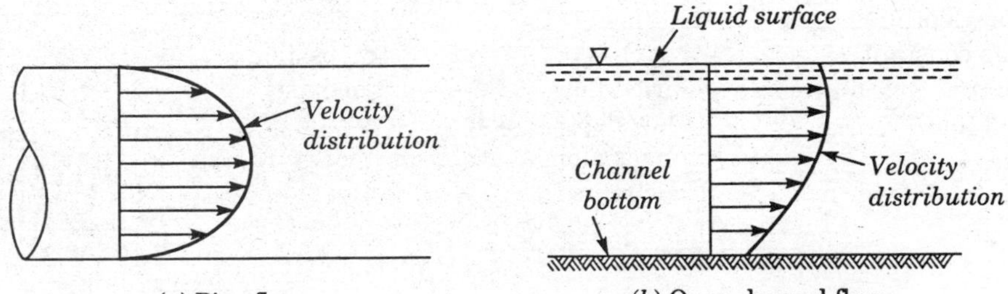

(a) Pipe flow (b) Open channel flow

Fig. 12.2. Velocity distribution in pipe flow and channel flow.

12.3. Classification of Flow

Open channel flow can be classified into many types. The following classification is due to Ven Te Chow and is based on the change in flow depth with respect to time and space.

(A) Classification with respect to Time

Steady Flow. Flow in an open channel is said to be steady if the depth of flow at a section does not change during the time interval under consideration.

Unsteady Flow. If the flow depth at a section changes with time, the flow is known as unsteady.

(B) Classification with respect to Space

Spatially Varied Flow. The steady flow in the main channel, which is joined by lateral channels at different points along its length is non-uniform. Such a flow where water flows into or out of the channel is known as spatially varied or discontinuous flow. The examples of this type of flow are : flow in road side gutters, flow in rivers having tributaries, flow in side channel spillways, flow in effluent channels around sewage-treatment tanks.

Uniform Flow. Flow in an open channel is said to be uniform when the depth of flow remains same at every section along the channel. Besides being uniform it may be steady or unsteady depending on whether or not depth remains constant with time. Uniform unsteady flow, however, may not be possible in practice.

Non-uniform or Varied Flow. Flow is said to be non-uniform or varied if the depth of flow changes along the length of the channel. It may be either steady or unsteady. In most of the natural open channels the flow is generally non-uniform as indicated by the varying depths across and along the channel.

The non-uniform flow may further be classified as either gradually varied or rapidly varied. If the changes in depth of flow are gradual, the flow is said to be gradually varied. On the contrary, if the depth changes abruptly within a short distance it is known as rapidly varied flow. It is called a local phenomenon, as it occupies a short length of the channel. Examples of this types are the hydraulic drop and the hydraulic jump. The following chart shows the above mentioned classification of flow.

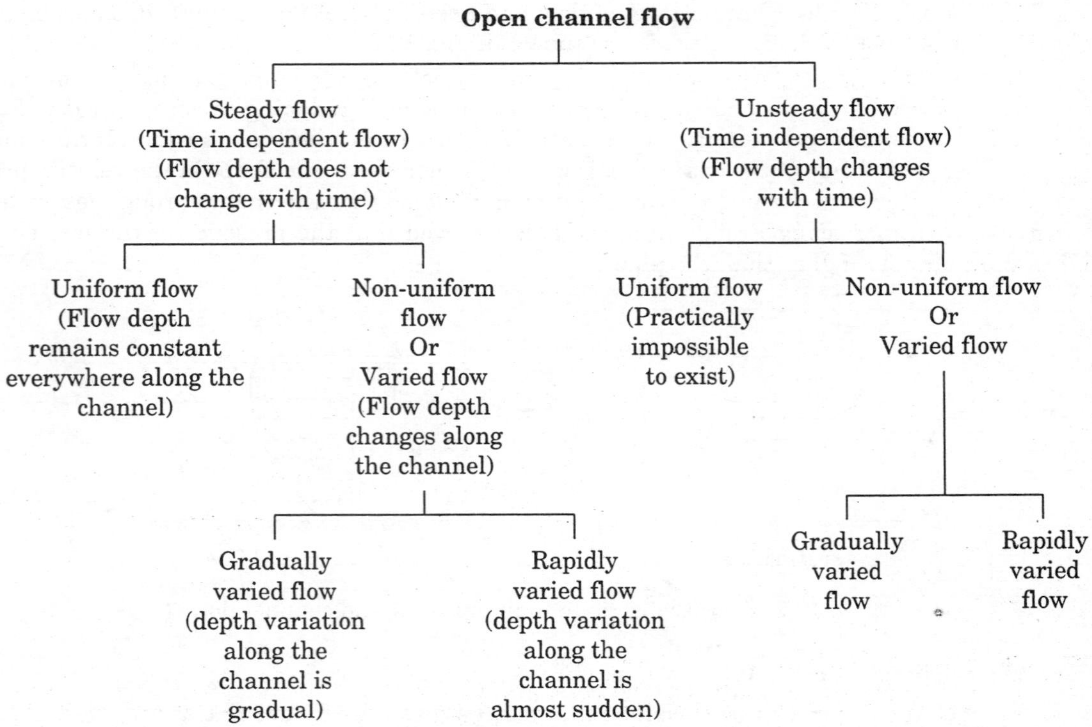

12.4. Effect of Gravity and Viscosity on Flow Behaviour

The open channel flow is greatly influenced by the effects of gravity and viscosity. The Froude number of flow is an index which takes into account the relative importance of inertial and gravity forces, and is expressed as the ratio of these forces as under :

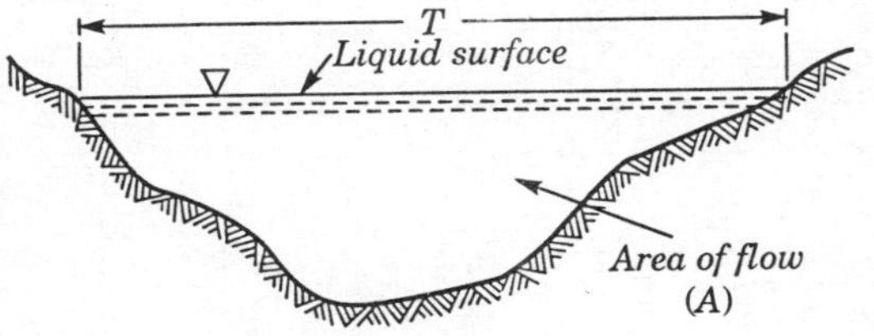

Fig. 12.3. Elements of Hydraulic depth.

$$\mathbf{F} = \frac{V}{\sqrt{gD}} \quad ...(12.1)$$

where **F** is the Froude number of flow, V the mean velocity of flow, g the acceleration due to gravity, D the hydraulic depth which is equal to A/T, the ratio of area of flow of the top width (width at the free surface). The magnitude of Froude number determines the type of flow carried by the channel.

(*i*) If $\mathbf{F} < 1$, the flow is said to be sub-critical or tranquil,

(*ii*) If $\mathbf{F} = 1$, the flow is known as critical, and

(*iii*) If $\mathbf{F} > 1$, the flow is called supercritical. It is also referred to as rapid or shooting flow.

The Reynolds number of flow, which is the ratio of inertial force to viscous force, is a measure of influence of viscosity as compared to inertia.

It may be expressed as

$$\mathbf{R} = \frac{\rho V L}{\mu} \quad ...(12.2)$$

where **R** is the Reynolds number of flow, ρ the mass density of fluid, μ the dynamic viscosity of fluid, V the mean velocity of flow and L is a characteristic length. In open channel flow, the characteristic length L is taken as the hydraulic radius (R) which is defined as the ratio of area of flow (A) to the wetted perimeter (P).

Based on the Reynolds number the flow in an open channel may be either laminar, turbulent or in transition from laminar to turbulent. The flow is considered laminar, if the Reynolds number based on the hydraulic radius is less than 500. If the Reynolds number is between 500 to 2,000 the flow is neither laminar nor turbulent, but is in the transitional state, *i.e.*, in the process of changing from laminar to turbulent. If, however, **R** is greater than 2,000 the flow may be taken as turbulent. It may be noted that limit of the transitional range, *i.e.*, 2,000 is arbitrary, and in certain cases under controlled conditions, the flow at even $\mathbf{R} = 2000$ may be laminar. But for all practical purposes, the upper limit is taken at 2,000 and the flow considered to be turbulent if $\mathbf{R} >$ 2,000.

In a wide rectangular channel of width b and depth of water y, the hydraulic radius R and the hydraulic depth D will be :

$$R = \frac{A}{P} = \frac{b\,.\,y}{b + 2y} \approx y, \text{ since } b >> y$$

and,

$$D = \frac{A}{T} = \frac{b\,.\,y}{b} = y$$

For water at 20°C, the kinematic viscosity $\nu = 10^{-6}\ \text{m}^2/\text{s}$ and taking $g = 9.81\ \text{m/s}^2$, the following table gives for various depths of flow y, the values of average velocity V below which the flow will be laminar and subscritical respectively.

Water depth y, (m) (1)	*Maximum limit of velocity V (m/s) for*	
	Laminar flow (2)	*Subcritical flow* (3)
0.10	5×10^{-3}	0.99
0.20	2.5×10^{-3}	1.36
0.30	1.67×10^{-3}	1.71
0.40	1.25×10^{-3}	1.98
0.50	1.0×10^{-3}	2.22
0.75	0.67×10^{-3}	2.71
1.00	0.5×10^{-3}	3.13

From the above table, it is obvious that the flow velocities in open channels are much higher than those indicated in column (2). The flow in open cannels is thus turbulent and lies mostly in the subcritical range.

12.5. Uniform Flow

When flow takes place in an open channel, two types of forces are developed. The first one is the force which causes the flow, and is due to the slope provided to the channel. The second one is the force of resistance which opposes the motion. The force causing the flow is equal to the component of weight of water (gravity force acting on the body of water) in the direction of flow. The uniform flow will be obtained when the resistance force equals the force causing the flow.

Based on extensive experimental investigations, many uniform-flow formulae have been developed. With the exceptions of Chezy's formula and Manning's formula, all other formulae have not achieved wide recognition and universal acceptability, perhaps for not taking into account all factors which affect the flow in open channels. The Chezy's and Manning's formulae are the best known to hydraulic engineers and are almost universally used.

12.5.1. Chezy's formula

It was Antoine Chezy, a French engineer who in 1770 developed the first uniform flow formula known to the engineering world. The Chezy's formula is based on the two assumptions listed below :

(1) The first assumption is in respect of flow resistance, according to which the force resisting the flow per unit of wetted area is proportional to the square of the velocity.

(2) The second assumption is the basic principle of uniform flow propounded by Brahms in 1754, according to which the force causing the flow must be equal to the force of resistance.

Referring to Fig. 12.4, considering the body of water contained in between sections (1) and (2) distance L apart, the resisting force in accordance with the first assumption is :

Resisting force per unit wetted area $\alpha\ V^2 = KV^2$

in which K is a constant.

Wetted area $= \text{Wetted perimeter } P \times \text{Length } L$

$= P.L$; where $P = b + 2s$

therefore, the resisting force $= K.LP.V^2$ ...(12.3)

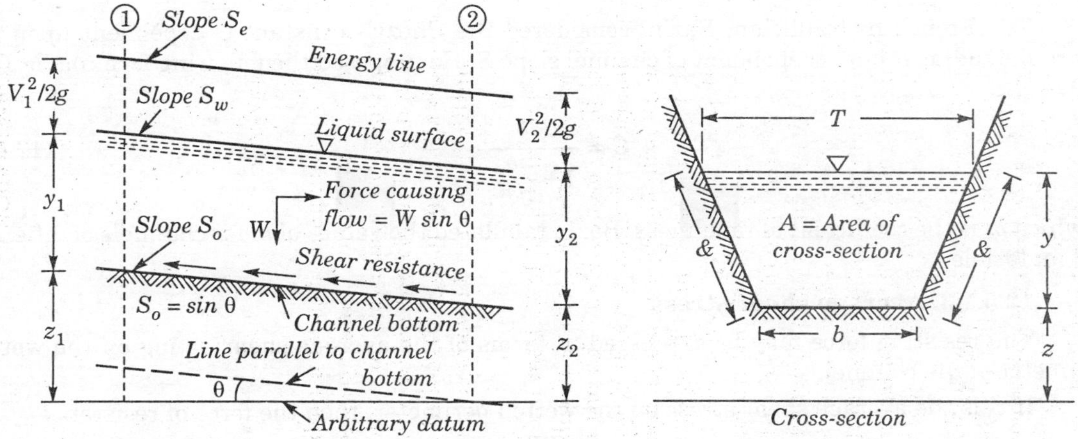

Fig. 12.4. Uniform flow in open channel ($S_o = S_w = S_e = S$).

The force causing flow is equal to the component of the weight of water W, in the direction of flow. If W = weight of water contained in between sections (1) and (2), it is equal to γAL and if θ = angle of inclination of channel bottom with the horizontal, then

Force causing flow $= W \sin\theta = \gamma AL \sin\theta = \gamma ALS$ ...(12.4)

It may be noted that in uniform flow, the channel slope, the water surface slope and the energy line slope are equal to each other, hence,

$$S_o = S_w = S_e = S = \sin\theta.$$

Using the second assumption, we may write

$$\gamma ALS = KPLV^2$$

which may be written as,

$$V = \sqrt{\left(\frac{\gamma}{K}\right)\left(\frac{A}{P}\right)S} = C\sqrt{RS} \quad \text{...(12.5)}$$

where $R = \dfrac{A}{P}$ is the hydraulic radius, and

$C = \sqrt{\dfrac{\gamma}{K}}$, the flow resistance factor known as Chezy's C.

Eq. (12.5) is known as the Chezy's equation. The Chezy's resistance factor C is believed to be dependent upon the channel slope S, the hydraulic radius R and the coefficient of roughness n.

12.5.1.1. Ganguillet and Kutter's formula

Based on extensive experimental studies on European rivers, two Swiss engineers—Ganguillet and Kutter—proposed in 1819 the following empirical formula for Chezy's constant.

$$C = \frac{23 + \dfrac{1}{n} + \dfrac{0.00155}{S}}{1 + \left(23 + \dfrac{0.00155}{S}\right)\dfrac{n}{\sqrt{R}}} \quad \text{...(12.6 a)}$$

in which n is a measure of roughness and is called Kutter's n and R is the hydraulic radius in metre. This formula is extensively used by European engineers and has been found to give satisfactory results.

12.5.1.2. Bazin's formula

The French hydraulician, Bazin, considered the Chezy's constant C dependent upon the hydraulic radius R but independent of channel slope S. He proposed the following formula for C in 1897 :

$$C = \frac{87}{1 + \dfrac{m}{\sqrt{R}}} \qquad \text{...(12.6 b)}$$

in which m is the coefficient of roughness. Bazin tabulated the values of m for channels of different bed materials.

12.5.1.3. Average shear stress

The resisting force may be expressed in terms of the average shear acting on the wetted perimeter of the channel.

If τ_0 is the average shear stress on the wetted perimeter, then the force of resistance

$$= \tau_0 PL \qquad \text{...(12.6 c)}$$

In case of uniform flow, the force of resistance, Eq. (12.6), and the force causing flow, Eq. (12.4) must be equal, hence

$$\tau_0 PL = \gamma ALS$$

or

$$\tau_0 = \gamma \frac{A}{P} S = \gamma RS \qquad \text{...(12.7)}$$

Eq. (12.7) gives the average shear stress τ_0 as a product of the specific weight of the flowing liquid, the hydraulic radius and the slope.

Example 12.1. *Water flows uniformly at a depth of 1.2 metres in a rectangular canal 3 metres wide laid on a slope of 1 metre per 1000 metres. What is the mean shear stress on the sides and bottom of the canal.*

Based on the Reynolds and Froude numbers, determine the type of flow in the canal. Take Chezy C as 70 and $\nu = 10^{-6}\ m^2/s$.

Solution. The mean shear stress on the sides and bottom is given by (Eq. 12.7) as

$$\tau_0 = \gamma RS$$

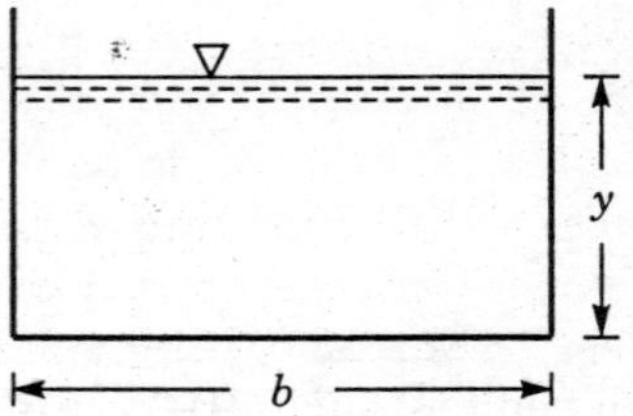

The hydraulic radius $R = \dfrac{A}{P} = \dfrac{by}{b + 2y} = \dfrac{2 \times 1.2}{3 + 2 \times 1.2} = 0.667$ m.

The mean shear stress $\tau_0 = 9810 \times 0.667 \times \dfrac{1}{1000} = 6.54\ \text{N/m}^2$

Average velocity, $V = C\sqrt{RS}$

$$= 70\sqrt{0.667 \times 1/1000} = 1.806 \text{ m/s}$$

Reynolds number of flow,

$$\mathbf{R} = \frac{VR}{\nu} = \frac{1.806 \times 0.667}{10^{-4}} = 1.203 \times 10^4$$

Since $\mathbf{R} > 500$, the flow is not in the laminar regime. Additionally, since the Reynolds number is very much higher than 2000, the flow is decidedly turbulent.

Froude number of flow, $\mathbf{F} = V/\sqrt{gD}$; $D = \dfrac{A}{T} = \dfrac{by}{b} = y$

$$\mathbf{F} = \frac{1.806}{\sqrt{9.81 \times 1.2}} = 0.527 \qquad \text{(since } \mathbf{F} < 1\text{, the flow is subcritical)}$$

The flow in the channel is thus in the subcritical state and turbulent regime.

Example 12.2. *A wide unlined channel carrying silt free water has a depth of 2.0 m. The maximum tractive stress permissibles on the bed to prevent scour is 1.96 N/m². What is maximum slope that can be given to the channel ?*

Solution. On a wide rectangular river bed, the tractive stress is

$$\tau_s = \gamma RS = \gamma y S$$

from which

$$S = \frac{\tau_s}{\gamma \cdot y} = \frac{1.96}{9810 \times 2} = \frac{1}{10^4}$$

The maximum slope that can be given to the channel is 1 in 10^4.

12.5.2. Manning's Formula

Robert Manning, an Irish engineer, presented in 1889, a formula which was later simplified and subsequently modified to the present form, expressed in metric units as :

$$V = \frac{1}{n} R^{2/3} S^{1/2} \qquad ...(12.8)$$

in which V is the mean velocity in metre/sec, R the hydraulic radius in metre, n the coefficient of roughness known as Manning's n (also known as the rugosity coefficient—a term generally used by the British engineers) and S the channel slope. Eq. (12.8) is known as the *Manning's formula*, and is widely used in open channel flow analysis on account of its simplicity in form and satisfactory results.

A relation between the Chezy's C and Manning's n may be obtained by comparing Eqs. (12.5) and (12.8) and expressed as :

$$C = \frac{1}{n} R^{1/6} \qquad ...(12.9)$$

The application of Manning's formula will yield satisfactory results only when a suitable value for Manning's roughness factor n is used. The selection of a representative value of n is a matter of experience and judgment on account of its dependence on many factors. In order of importance these factors are :

(*i*) *Surface roughness*. Size and shape of roughness particles forming the bed and banks of the channel affect the surface roughness. In general, fine grained material has a low value of n whereas coarse grained material results in high value of n.

(*ii*) *Vegetation*. The growth of vegetation not only increases the value of n, but it also decreases the carrying capacity of the channel. The main characteristics affecting this are its height, density and the type of vegetation.

(*iii*) *Channel irregularities and channel alignment*. Irregularities in the channel such as ridges and depressions, sand bars and ripples on the channel bed, and abrupt changes in the channel section result in an increase in the value of n. Curves in the channel alignments also affect the Manning's n and it increases with the sharpness in curvature.

(*iv*) *Silting and Scouring*. Deposition of silt in an irregular and rough channel makes it relatively uniform and smooth tending to decrease the Manning's n. Scouring, on the other hand, causes local depressions at places, thereby increasing the coefficient of roughness n.

(*v*) *Stage and Discharge*. With the increase in the stage (water surface elevation), the value of Manning's n decreases. Since discharge is directly proportional to the stage, the same statement is true for increasing discharge. However, at very high discharges the value of n may increase owing to the movement of bed material.

(*vi*) *Transport of suspended and bed material*. If the channel carries suspended and bed material requiring a part of the flow energy to sustain this movement, an additional head loss occurs on this account, thereby increasing the value of n.

Table 12.1 indicates representative values of Mannings's n for various types of regular channel surfaces and straight alignment.

Table 12.1. Manning's n for various types of channel surfaces
(for straight and regular channels)

Type of channel surface	*Manning's n*
Smooth Cement, Planed	0.010—0.014
Canvas, Unplaned wood	0.011—0.015
Finished concrete	0.011—0.013
Unfinished concrete	0.013—0.016
Brick masonry	0.012—0.020
Cast iron	0.013—0.017
Riveted steel	0.017—0.020
Rubble masonry	0.020—0.030
Earth	0.020—0.030
Gravel	0.022—0.035
Earth with vegetation	0.025—0.040

Example 12.3. *Determine the rate of flow through a 3 m wide concrete lined rectangular channel having a slope 1 in 10,000 when the depth of flow is 1 m.*

Take Manning's n = 0.012.

Solution. Area of flow,

$$A = b.y = 3 \times 1 = 3 \text{ m}^2$$

Wetted perimeter, $P = b + 2y = 3 + 2 \times 1 = 5 \text{ m}$

Hydraulic radius, $R = \dfrac{A}{P} = \dfrac{3}{5} = 0.6 \text{ m}$

Using Eq. (12.10), the flow rate

$$Q = \frac{1}{n} AR^{2/3} S^{1/2}$$

$$= \frac{1}{0.012} \times 3 \times (0.6)^{2/3} (10^{-4})^{1/2} = \mathbf{1.78 \ m^3/s.}$$

Example 12.4. *A rough timber flume (n = 0.012) in the form of an equilateral triangle (apex down) of 1.2 metres sides is laid on slope of 0.01. Calculate the uniform flow rate which occurs at a depth of 90 cm.*

Solution. Area of flow,

$$A = \frac{1}{2} \times 2 \times 0.9 \tan 30° \times 0.9 = 0.81 \times \frac{1}{\sqrt{3}} = 0.468 \text{ m}^2$$

Wetted perimeter,

$$P = 2\sqrt{(0.9)^2 + (0.9 \tan 30°)^2}$$

$$= 2.08 \text{ m}$$

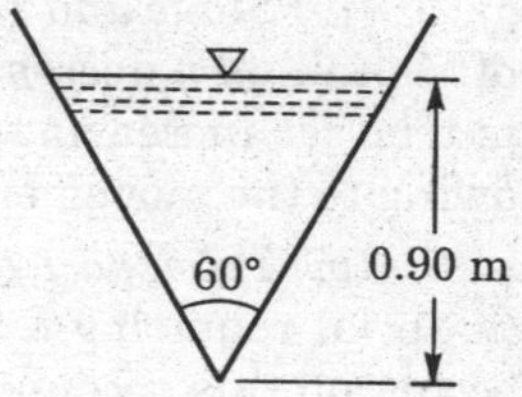

The hydraulic radius, $R = \dfrac{A}{P} = \dfrac{0.468}{2.08}$

$$= 0.225 \text{ m}$$

The mean velocity of flow,

$$V = \frac{1}{n} R^{2/3} . S^{1/2}$$

$$= \frac{1}{0.012} \times (0.225)^{2/3}(0.01)^{1/2} = 3.08 \text{ m/s}$$

∴ The flow-rate, $Q = AV = 0.468 \times 3.08 = 1.44 \text{ m}^3/\text{s}$.

Example 12.5. *Show that the friction factor of the Darcy-Weisbach formula, and the Manning's roughness factor n are inter-related by* $f = \frac{78.5n^2}{R^{1/3}}$ *in which R is the hydraulic radius.*

Solution. Darcy-Weisbach formula for, pipe flow is given by Eq. (9.57)

$$h_L = f \frac{L}{d} \frac{V^2}{2g}$$

In case of circular pipe, $R = \frac{A}{R} = \frac{\pi d^2/4}{\pi d} = \frac{d}{4}$

∴ $d = 4R$

In open channel flow, $S = \frac{h_L}{L}$, and the mean velocity by Manning's formula is

$$V = \frac{1}{n} R^{2/3} S^{1/2}$$

Making these substitutions in the Darcy-Weisbach formula, one gets

$$S = \frac{h_L}{L} = f \frac{\left(\frac{1}{n} R^{2/3} S^{1/2}\right)^2}{4R.2g}$$

$$S = \frac{fR^{4/3} S}{8gRn^2} \text{ which simplifies to } f = \frac{8gn^2}{R^{1/3}}$$

Substituting for $g = 9.81 \text{ m/s}^2$

$$f = \frac{8 \times 9.81\, n^2}{R^{1/3}} = 78.5 \frac{n^2}{R^{1/3}}.$$

Example 12.6. *A trapezoidal channel, shown in figure carries a water discharge of 10 m³/s to a depth of 1.5 m under uniform flow conditions. The longitudinal slope of the channel bed is 0.001. Compute the average shear stress in N/m² on the boundary. Also compute Manning's n value.*

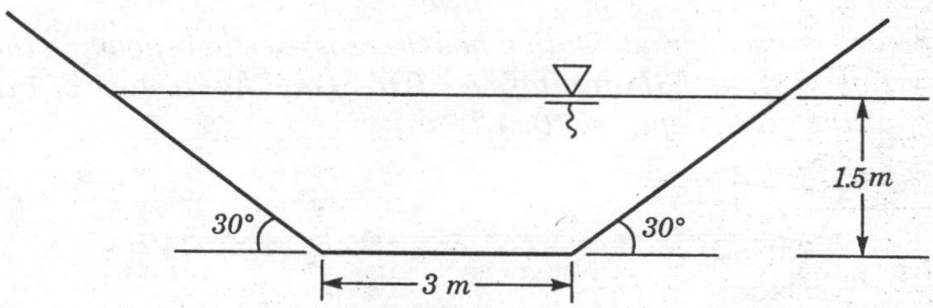

Solution. Average shear stress on the boundary

$$\tau_0 = \gamma RS$$

$$R = \frac{A}{P} = \frac{(b + yz)y}{b + 2y\sqrt{1 + z^2}} = \frac{(3 + 1.5 \times \sqrt{3})1.5}{3 + 2 \times 1.5\sqrt{1 + 3}} = 0.93 \text{ m}$$

$$\tau_0 = 9810 \times 0.93 \times 0.001 = \mathbf{9.156\ N/m^2}$$

$$V = \frac{Q}{A} = \frac{10}{5.6 \times 1.5} = 1.19 \text{ m/s}$$

From Manning's formula,

$$n = \frac{R^{2/3}\, S^{1/2}}{V} = \frac{(0.93)^{2/3}\, (0.001)^{1/2}}{1.19} = \mathbf{0.025.}$$

Example 12.7. *An open channel has U section semi circular at the bottom with vertical sides and is 1.2 m wide. It is laid at a bed slope of 0.375 m per km. Calculate the values of Chezy's C and Mannings 'n' if the depth of flow is 0.9 m while the discharge is 0.54 m^3/s.*

Solution. Given : $Q = 0.54$ m³/s, $d = 0.9$ m

$S = 0.375/1000$

From the Chezy's and Mannings equations :

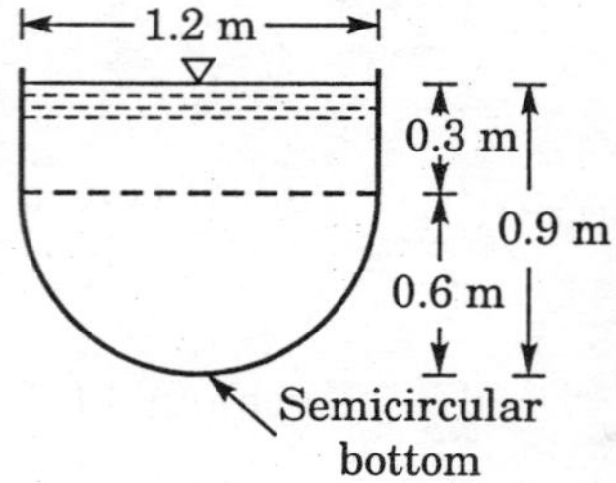

$$Q = AC\sqrt{RS} = \frac{1}{n} AR^{2/3}S^{1/2}$$

from which, $$C = \frac{Q}{A\sqrt{RS}}$$

and $$n = \frac{AR^{2/3}S^{1/2}}{Q}$$

Area of flow A $$= \frac{1}{2}\left(\frac{\pi \times 1.2^2}{4}\right) + 0.3 \times 1.2 = 0.9257 \text{ m}^2$$

Wetted perimeter $$P = \frac{\pi \times 1.2}{2} + 2 \times 0.3 = 2.4855 \text{ m}$$

Hydraulic radius $$R = \frac{A}{P} = \frac{0.9257}{2.4855} = 0.3724 \text{ m}$$

Making substitutions,

$$C = \frac{0.54}{0.9257\sqrt{0.3724 \times \left(\dfrac{0.375}{1000}\right)}} = 49.36$$

and $$n = \frac{0.9257 \times (0.3724)^{2/3}\left(\dfrac{0.375}{1000}\right)^{1/2}}{0.54} = \mathbf{0.01726.}$$

Example 12.8. *A river used for navigation has the cross-section shown in the following figure. If the channel cross-section is of earth (Manning's n = 0.025) and has a slope of 1 in 1000. What rate of flow should it carry at the given depth of 10 m ?*

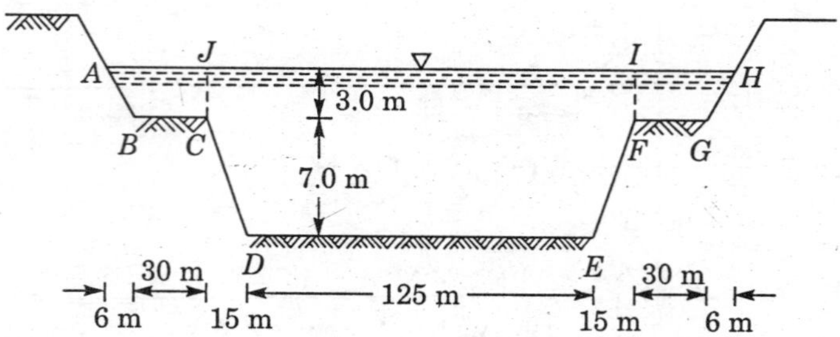

Solution. Dividing the channel section into three portions, *i.e., ABCJ*, *JCDEFI* and *FGHI* and finding the discharge for each portion.

1. Portion ABCJ : Area of flow,

$$A_1 = \frac{30 + 36}{2} \times 3 = 99 \text{ m}^2.$$

Wetted perimeter (*ABC*), $P_1 = 30 + \sqrt{(6)^3 + (3)^2} = 36.71$ m.

Hydraulic radius $R_1 = \frac{A_1}{P_1} = \frac{99}{36.71} = 2.7$ m

∴ Discharge through portion *ABCJ*,

$$Q_1 = \frac{1}{n} AR_1^{2/3} S_1^{1/2} = \frac{1}{0.025} \times 99 \times (2.7)^{2/3} \left(\frac{1}{1000}\right)^{1/2} = 243 \text{ m}^3\text{/s}$$

2. Portion JCDEFI : Area of flow,

$$A_2 = \frac{125 + 155}{2} \times 7 + 155 \times 3 = 1445 \text{ m}^2.$$

Wetted perimeter (*CDEF*), $P_2 = 125 + 2\sqrt{(15)^2 + (7)^2} = 158.08$ m.

∴ Hydraulic radius, $R_2 = \frac{A_2}{P_2} = \frac{1445}{158.08} = 9.15$ m.

Discharge through this portion,

$$Q_2 = \frac{1}{1.025} \times 1445 \times (9.15)^{2/3} \left(\frac{1}{1000}\right)^{1/2} = 8320 \text{ m}^3\text{/s}.$$

3. Portion FGHI. Since this portion is similar to *ABCJ*, the discharge through it will be 243 m³/s.

Therefore, discharge carried by the river,

$$Q = 2Q_1 + Q_2 = 2 \times 243 + 8320 = 8806 \text{ m}^3\text{/s}.$$

12.6. Prismatic and Non-prismatic Channels

A channel which has a constant shape and size and is laid to a constant slope is said to be a *prismatic channel*. An artificial channel will be prismatic so long as its cross-section (size and shape) and slope remain unchanged. Either a change in the size or shape or slope or any combination of these variables will make it *non-prismatic*. Such a channel, however, may again be prismatic in that reach in which its changed cross-section or changed slope or both, remain constant. The channel will be prismatic only for that much length in which its cross-section and slope remain unchanged.

All natural channels such as nallas and rivers are non-prismatic simply because they have variable cross-section and slope. An artificial channel having a gradually changing cross-section will also be a non-prismatic channel.

12.7. Hydraulically Efficient Channel Cross-section

The most efficient cross-section of an open channel, from hydraulic view-point, is the one which, for a given slope, roughness coefficient and the area of flow, carries the maximum flowrate.

From among all open channel cross-sections for a given area of flow, a semi-circular channel has the smallest wetted perimeter and is, therefore, the section of highest efficiency. There are practical limitations to using such cross-sections of greatest hydraulic efficiency. Dimensions of such cross-sections should, however, be known and adhered to as closely as conditions permit. From practical considerations, trapezoidal channels are widely used for conveying water for irrigation and hydroelectric power generation and other purposes.

For any given value of channel slope, surface roughness and cross-sectional area, the application of either Chezy's or Manning's formula gives the discharge as :

$$\left.\begin{aligned} Q &= AV = AC\sqrt{RS} \\ \text{Alternatively,} \quad Q &= \frac{1}{n} AR^{2/3} S^{1/2} \end{aligned}\right\} \quad \text{...(12.10)}$$

From Eq. (12.10), it is evident that for the given data (of channel slope, its roughness coefficient and area of flow) the discharge will be maximum, when the hydraulic radius R is maximum. From the definition of hydraulic radius, it may be seen that the hydraulic radius will be maximum, when the wetted perimeter is minimum (it is to be noted that the area of flow A is a constant). From hydraulic point of view, the most efficient channel is therefore one which has such a shape that provides minimum wetted perimeter for a given area of flow. Such a channel will not only be hydraulically most efficient, but will also be economical.

The hydraulically most efficient channel section will carry maximum discharge for a given area of flow. Conversely, for a given discharge such a channel will have a minimum area of flow. If the channel is to be excavated, a hydraulically most efficient section will also be the most economical since the quantity of the excavated material will then be minimum.

12.7.1. Hydraulically Efficient Rectangular Channel

For a given channel slope, surface roughness and area of flow, the hydraulically most efficient section is one which has the minimum wetted perimeter. In case of a rectangular channel, see Fig. 12.5, the following relationships can be written :

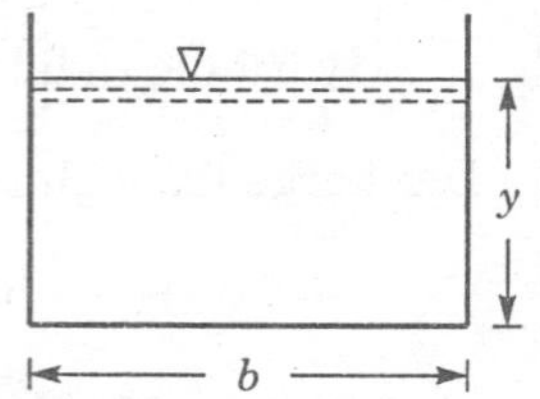

Fig. 12.5. Rectangular channel.

Area of flow $A = b.y$

and wetted perimeter

$$P = b + 2y = \frac{A}{y} + 2y$$

For P to be minimum, dP/dy should be zero ; differentiating P with respect to y, keeping A as constant,

$$\frac{dP}{dy} = -\frac{A}{y^2} + 2$$

letting, $\dfrac{dP}{dy} = 0$, we obtain, $A = 2y^2$

(The second derivative of P, *i.e.* $2A/y^3$ is positive and so the condition obtained is for minimum P)

but $A = b.y$ and, therefore, $b = 2y$

Also $R = \dfrac{A}{P} = \dfrac{2y^2}{4y} = \dfrac{y}{2}$

Thus, for hydraulically most efficient rectangular channel, the width should be two times the depth of flow, the hydraulic radius is then equal to half the depth of flow.

12.7.2. Hydraulically Efficient Trapezoidal Channel

Referring to Fig. 12.6, $A = (b + my)y$

$$P = b + 2y\sqrt{1+m^2} = \left(\frac{A}{y} - my\right) + 2y\sqrt{1+m^2}$$

differentiating P with respect to y, holding A and m constant

$$\frac{dP}{dy} = -\frac{A}{y^2} - m + 2\sqrt{1+m^2}.$$

(*i*) Condition for minimum P is that $\frac{dP}{dy} = 0$.

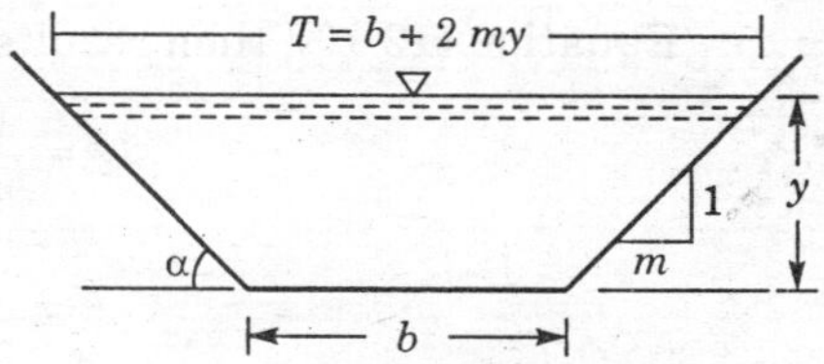

Fig. 12.6. Trapezoidal channel.

hence $$A = \left(2\sqrt{1+m^2} - m\right)y^2$$

but $$A = (b + my)y,$$

and, therefore, $$b = 2y\left(\sqrt{1+m^2} - m\right)$$

$$\frac{b+2my}{2} = y\sqrt{1+m^2}.$$

In other words, half the top width ($T/2$) is equal to the length of one sloping side of the cross-section.

(The second derivative of P with respect to y yields, $2A/y^3$, which is positive and the condition obtained is, therefore, for the minimum P.)

(*ii*) In the expression for P, there are two variables y and m, and therefore, a second condition for minimum P may be obtained by differentiating P with respect to m holding y constant.

$$\frac{dP}{dm} = -y + \frac{2my}{\sqrt{1+m^2}}$$

and letting $\frac{dP}{dm} = 0$, we obtain

$$\frac{2m}{\sqrt{1+m^2}} = 1 \quad \text{or} \quad m = \frac{1}{\sqrt{3}}$$

The angle α which the channel sides make with the horizontal is given by

$$\tan\alpha = \frac{1}{m} = \sqrt{3}, \text{ resulting in } \alpha = 60°.$$

Therefore, the wetted perimeter is minimum when $\alpha = 60°$. The first condition for minimum P may be used to determine the maximum value of the hydraulic radius.

$$R = \frac{A}{P} = \frac{(b+my)y}{b+2y\sqrt{1+m^2}} = \frac{(b+my)y}{b+b+2my} = \frac{y}{2}.$$

Thus, for maximum hydraulic efficiency, a trapezoidal channel should be so proportioned that (1) its hydraulic radius is equal to half the depth of flow, and (2) its sides make 60° angle with the horizontal. It may be shown that the length of each sloping side is the same as the bottom width. The above two conditions, therefore, show that the hydraulically most efficient trapezoidal section is half of a regular hexagon.

Example 12.9. *Determine the dimensions of the most economical trapezoidal earth lined channel (Manning's n = 0.020) to carry 14 m³/s at a slope of 4 in 10,000.*

Solution. For hydraulically efficient and economical trapezoidal channel, the hydraulic radius

$$R = \frac{A}{P} = \frac{y}{2}$$

Area of flow, $$A = (b + my)y$$

and $$b = 2y\left(\sqrt{1+m^2} - m\right)$$

but for an efficient section, $m = 1/\sqrt{3}$, which yields $b = 2y/\sqrt{3}$, and $A = \sqrt{3}y^2$.

Equation (12.10), then results

$$Q = \frac{1}{n} AR^{2/3} S_0^{1/2}$$

$$14 = \frac{1}{0.020} \times \sqrt{3}y^2 \left(\frac{y}{2}\right)^{2/3} \left(\frac{4}{10,000}\right)^{1/2} = 1.09\, y^{8/3}.$$

From which, $y = 2.58$ m.

and $$b = \frac{2 \times 2.58}{\sqrt{3}} = 2.93 \text{ m}.$$

Example 12.10. *A trapezoidal channel with side slopes of 1 : 1 has to be designed to convey 10 m³/s at a velocity of 2 m/s, so that the amount of concrete lining for the bed and sides is the minimum.*

(i) Calculate the area of lining required for one metre length of the canal.

(ii) If the rugosity coefficient n = 0.015, calculate the bed slope of the canal for uniform flow.

Solution. $Q = 10 \text{ m}^3/\text{s}, \quad V = 2 \text{ m/s}$

$$A = \frac{Q}{V} = \frac{10}{2} = 5 \text{ m}^2$$

For a trapezoidal channel, $A = (b + my)y$

$$P = b + 2y\sqrt{1+m^2}$$

(*i*) For minimum amount of concrete lining, the wetted perimeter must be minimum, hence $\frac{dP}{dy} = 0$, which gives,

$$A = (2\sqrt{1+m^2} - m)y^2$$

$$b = 2y\sqrt{1+m^2} - m)$$

substituting $m = 1$, $A = 5\ m^2$, we obtain

$$y = 1.652 \text{ m and } b = 1.37 \text{ m}.$$

Area of lining per metre length

$$= P \times 1$$

$$P = \left(\frac{A}{y} - my\right) + 2y\sqrt{1+m^2}$$

$$= \left(\frac{5}{1.652} - 1.652\right) + 2 \times 1.652\sqrt{1+1} = 6.043 \text{ m}$$

Hence the area $= 6.043 \times 1 = \mathbf{6.043\ m^2}$.

(*ii*) $$V = \frac{1}{n} R^{2/3} S_0^{1/2}$$

$$R = \frac{y}{2} = 0.826 \text{ m}$$

Making substitutions, $$2 = \frac{1}{0.015} \times (0.826)^{2/3} (S_0)^{1/2}$$

From which, $= S_0 = \mathbf{1.143 \times 10^{-3}}$ **or 1 in 875.**

Example 12.11. *A hydraulically efficient trapezoidal channel has side slopes of 1 : 1. It is required to discharge 14 m³/s with a gradient (channel slope) of 1 in 1000. If unlined, the value of*

Chezy's C is 45. If lined with concrete, the value is 70. If the cost per m^3 of excavation is three times the cost per m^2 of lining, will the lined or the unlined channel be cheaper ?

Solution. Hydraulically efficient trapezoidal channel must satisfy the following conditions :

(*i*) $b = 2y(\sqrt{1+m^2} - m)$, and (*ii*) $R = \frac{y}{2}$.

Unlined Channel.

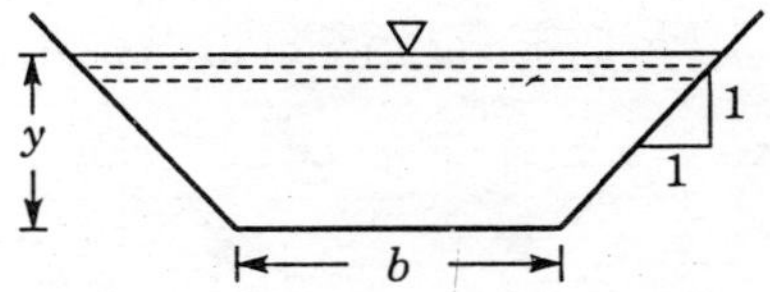

$$Q = AC\sqrt{RS}$$

$$= (b + my)yC\sqrt{\frac{y}{2}S}$$

for $m = 1, b = 2y(\sqrt{2} - 1) = 0.828\,y$.

Since $Q = 14$ m³/s, we obtain

$$14 = (0.828\,y + y)yC\sqrt{\frac{y}{2}S} = \frac{1.828y^2C}{\sqrt{2}}\sqrt{yS} = 1.29\,y^{5/2}\,C\sqrt{S}$$

Substituting, $S = 1/1000$ and $C = 45$, one gets

$$y = \left(\frac{14}{1.836}\right)^{2/5} = 2.25 \text{ m.}$$

Let K be the cost of lining per m² of the channel surface, then $3K$ will be the cost of excavation per m³. Considering one metre length of channel,

amount of excavation $= (b + my)y \times 1 = (0.828 \times 2.25 + 2.25)2.25$

$= 1.828 \times (2.25)^2$ m³.

Therefore, the cost of excavation

$$= 3K \times 1.828 \times (2.25)^2 = 27.75\,K.$$

Lined Channel $14 = 1.29y^{5/2}C\sqrt{S}$

Putting $C = 70$ for the lined channel, and $S = 1/1000$

$$14 = 1.29y^{5/2} \times 70\sqrt{\frac{1}{1000}}$$

$$y = \left(\frac{14}{2.86}\right)^{2/5} = 1.873 \text{ m.}$$

Considering one metre length of channel, cost of excavation

$$= (0.828 \times 1.873 + 1.873)\,1.873 \times 3K = 19.27\,K$$

Cost of lining $= (b + 2y\sqrt{1+m^2})\,K = (0.828 \times 1.873 + 2\sqrt{2} \times 1.873)\,K = 6.85\,K$

∴ Total cost of lining $= 19.27\,K + 6.85\,K = 26.12\,K$.

Since cost of excavation for the unlined channel (27.75 *K*) is greater than the total cost of lining (26.12 *K*), the lined channel will be cheaper.

Example 12.12. *Determine the rate of flow through a 3 m wide concrete lined rectangular channel having a slope of 1 in 10,000 when the depth of flow is 1 m. Take Manning's n = 0.012.*

Solution. Area flow, $A = b.y = 3 \times 1 = 3$ m³.

Wetted perimeter, $P = b + 2y = 3 + 2 \times 1 = 5$ m.

Hydraulic radius, $R = \frac{A}{P} = \frac{3}{5} = 0.6$ m.

Using Eq. (12.10), the flow rate

$$Q = \frac{1}{n} AR^{2/3} S_0^{1/2} = \frac{1}{0.012} \times 3 \times (0.6)^{2/3}(10^{-4})^{1/2} = 1.78 \text{ m}^3/\text{s}.$$

Example 12.13. *A certain stretch of lined trapezoidal channel has one vertical side wall and the other 45° sloping wall. If it is to deliver water at 30 m³/sec with a velocity of 1 m/sec, compute bed width and flow depth for minimum lining area.*

Solution. Discharge, $Q = 30 \text{ m}^3/\text{sec}$, Velocity = 1 m/sec

For minimum lining area, the wetted perimeter must be minimum.

Cross-sectional area of flow,

$$A = \frac{1}{2}(b + b + y)\,y = \frac{(2b + y)y}{2} \qquad \text{...(1)}$$

but, $$A = \frac{Q}{V} = \frac{30}{1} = 30 \text{ m}^2 \qquad \text{...(2)}$$

From (1) and (2),

$$\frac{(2b + y)y}{2} = 30\,; \quad \text{or} \quad b = \frac{60 - y^2}{2y} \qquad \text{...(3)}$$

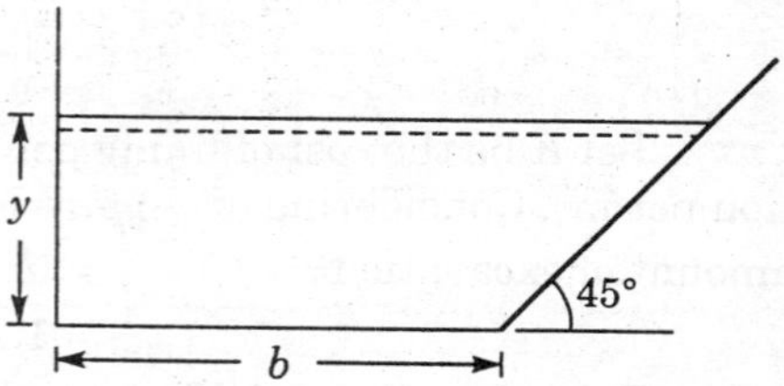

Wetted perimeter $P = b + y\sqrt{2y}$

$$= \frac{60 - y^2}{2y} + y(1 + \sqrt{2})$$

For minimum value of P, $dP/dy = 0$

$$\frac{dP}{dy} = -\frac{30}{y^2} - \frac{1}{2} + 1 + \sqrt{2} = 0$$

Solving which, $y = \mathbf{3.96\ m}$

and from (3), $$b = \frac{60 - (3.96)^2}{2 \times 3.96} = \mathbf{5.59\ m.}$$

Example 12.14. *Determine the dimensions of the most economical trapezoidal earth-lined channel (Manning's n = 0.020) to carry 14 m³/s at a slope of 4 in 10,000.*

Solution. For hydraulically efficient and economical trapezoidal channel, the hydraulic radius

$$R = \frac{A}{P} = \frac{y}{2}$$

$$m = \frac{1}{\sqrt{3}}$$

Area of flow, $A = (b + my)y$

and, bottom width, $b = 2y\left(\sqrt{1 + m^2} - m\right) = \frac{2y}{\sqrt{3}}$

Substituting for b, $A = \sqrt{3y^2}$

Eq. (12.10) then results in, $Q = \frac{1}{n} AR^{2/3} S_0^{1/2}$

or $$14 = \frac{1}{0.020} \times \sqrt{3y^2}\left(\frac{y}{2}\right)^{2/3}\left(\frac{4}{10{,}000}\right)^{1/2} = 1.09\,y^{8/3}$$

From which, $y = 2.58$ m,

and $$b = \frac{2 \times 2.58}{\sqrt{3}} = 2.98 \text{ m}.$$

Example 12.15. *Determine the dimensions of an economical trapezoidal section of an open channel with side slopes 2 H : 1 V laid at a slope of 1 in 1600 to carry a discharge of 36 m³/s assuming Chezy's coefficient C = 50.*

Solution. $Q = AV = AC\sqrt{RS}$

For most economical channel of trapezoidal section

(1) $R = y/2$

(2) $b = 2y\,(\sqrt{1+m^2} - m)$

here $m = 2.$

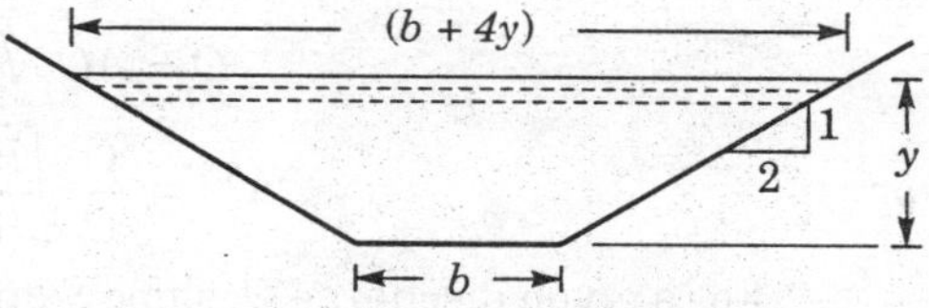

Area of flow, $A = (b + my)y$

$= (b + 2y)y$

$$b = 2y\,(\sqrt{1+4} - 2) = 0.472\,y$$

Substituting in the discharge equation

$$36 = (b + 2y)\, y \times 50\,\sqrt{\frac{y}{2} \times \frac{1}{1600}}$$

Putting the value of b and simplifying,

$$y^{3/2} = 16.4762, \quad \therefore \quad y = \mathbf{3.066\ m.}$$

and $$b = 0.472y = 0.472 \times 3.066 = \mathbf{1.447\ m.}$$

Example 12.16. *Design a concrete lined channel to carry a discharge of 500 cumecs at a slope of 1 in 4000. The side slopes of the channel may be taken as 1 : 1. The Manning's roughness coefficient for the lining is 0.014. Assume permissible velocity in the section as 2.5 m/s.*

Solution. $A = (b + y)y,\ P = b + 2\sqrt{2y^2}$

$$Q = A\,\frac{1}{n}\,R^{2/3}\,S^{1/2}$$

or $$500 = (b + y)y \times \frac{1}{0.014} \times \left[\frac{(b+y)y}{b + 2\sqrt{2y}}\right]^{2/3} \left(\frac{1}{4000}\right)^{1/2} \quad \text{...(1)}$$

Also, $$A = \frac{Q}{V}$$

or $$(b + y)y = \frac{500}{2.5} = 200 \quad \text{...(2)}$$

Simplifying (1), $$b + 2\sqrt{2}\,y = 60.7351 \quad \text{...(3)}$$

From (2), $$b + y = \frac{200}{y} \quad \text{or} \quad b = \frac{200}{y} - y$$

Substituting in (3), $$= \frac{200}{y} - y + 2\sqrt{2}\,y = 60.7351$$

or $$y^2 - 33.22y + 109.38 = 0$$

Solving which, **y = 3.705 m, and b = 50.276 m.**

∴ Width of the channel = 50.276 m.

depth of water in the channel = 3.705 m.

Example 12.17. *A trapezoidal channel has side slopes of 3 horizontal to 4 vertical and the slope of its bed is 1 in 2000. Determine the optimum dimensions of the channel, if it is to carry water at 0.5 m³/s. Take Chezy's constant as 80.*

Solution. Side slope 4 V to 3 H

$$= 1 \text{ V to } 0.75 \text{ H} \quad (\therefore \quad m = 0.75)$$

$$Q = 0.5 \text{ m}^3/\text{s}, C = 80, S = 1 \text{ in } 2000$$

Optimum dimensions of the canal are those which make it hydraulically efficient. For given discharge, slope and roughness coefficient, a hydraulically efficient channel will have a minimum area of flow.

$$A = (b + my)y$$

$$Q = AC\sqrt{RS} = AC\sqrt{(A/P).S} = A^{3/2}\, C\sqrt{S/P}$$

or

$$A^{3/2} = \frac{Q}{C}\sqrt{\frac{P}{S}}$$

For a given discharge Q, slope S and roughness coefficient C, area of flow A will be minimum when the wetted perimeter P attains a minimum value. Condition for minimum P is

$$b = 2y(\sqrt{1+m^2} - m) = 2y[\sqrt{1+(0.75)^2} - 0.75\} = y.$$

Wetted perimeter $\quad P = b + 2y\sqrt{1+m_2} = y + 2y \times 1.25 = 3.5\,y$

Substituting for b in the discharge equation,

$$Q = [(y + 0.75y)]^{3/2} \times 80 \times \sqrt{\frac{1}{2000 \times 3.5y}}$$

$$0.50 = 2.21\, y^{5/2}, \text{ from which, } y = 0.5516 \text{ m}$$

and since $b = y$, the bottom width of the channel is **0.5516 m.**

Example 12.18. *A most efficient trapezoidal section is required to give a maximum discharge of 21.5 m³/s of water. The slope of the channel bottom is 1 in 2500. Taking C = 70 m¹ᐟ²/s in Chezy's equation, determine the dimensions of the channel. Also determine the value of Manning's 'n', taking the velocity of flow as obtained for the channel by Chezy's equation.*

Solution. Referring to Fig. 12.6, for the most efficient trapezoidal channel, the following conditions must be satisfied

(*i*) $m = 1/\sqrt{3}$ or $\alpha = 60°$

(*ii*) $R = y/2$, and $\dfrac{b + 2my}{2} = y\sqrt{1+m^2}$ ...(1)

From continuity equation, $Q = A.V. = (b + my)y \,.\, C\sqrt{RS}$ ...(2)

Substituting for the known quantities, using (1)

$$\frac{b + 2y/\sqrt{3}}{2} = y\sqrt{1 + 1/3} \quad \text{or} \quad b = 2y\sqrt{3} \qquad ...(3)$$

From (2), $\quad 21.5 = (2y\sqrt{3} + y/\sqrt{3})y \,.\, 70\sqrt{\dfrac{y}{2 \times 2500}} = 1.715\, y^{5/2}$

from which, $\quad$ **y = 2.75 m**

and from (3), $\quad$ **b = 1.915 m**

From Manning's equation

$$V = \frac{1}{n} R^{2/3} S^{1/2} \text{ or } n = \frac{R^{2/3} S^{1/2}}{V} \qquad ...(4)$$

Area of flow, $A = (b + my)y = \left(1.915 + \frac{2.75}{\sqrt{3}}\right) 2.75 = 9.63 \text{ m}^2$

Average velocity of flow, $V = \frac{Q}{A} = \frac{21.5}{9.63} = 2.23$ m/s

Making substituting in (4)

$$n = \frac{\left(\frac{2.75}{2}\right)^{2/3} (1/2500)^{1/2}}{2.23} = \mathbf{0.011.}$$

Ex. 12.19. *A lined trapezoidal channel one side of which is vertical and other sloping 1 V to 2.0 H is to carry 20 m³/s at a mean velocity of 1 m/s. Find the bed width and depth of flow such that the bed slope is minimum. Manning's 'n' = 0.015. Also find the bed slope of the channel.*

Sol. $Q = 20 \text{ m}^3/\text{s}$ $V = 1$ m/s

$\therefore$ $A = Q/V = \frac{20}{1} = 20 \text{ m}^2$

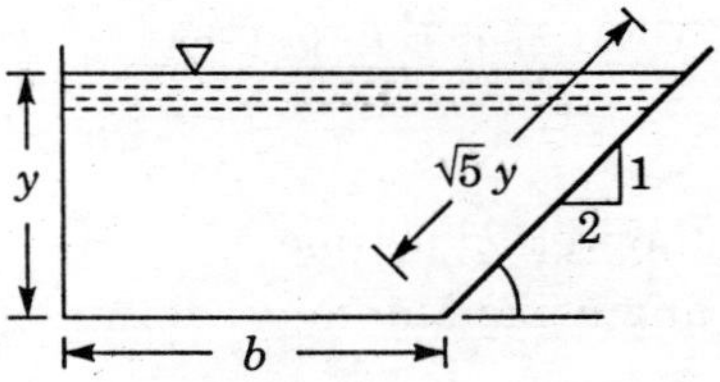

$n = 0.015$

Using Manning's formula

$$V = \frac{1}{n} R^{2/3} S_0^{1/2}$$

from which $S_0 = \frac{n^2V^2}{R^{4/3}}; R = \frac{A}{P}$

For minimum bed slope (S_0), since V is constant at 1 m/s, it is the hydraulic radius R which has to be maximum. As the area of flow (A) is fixed at 20 m², the condition of minimum bed slope boils down to minimum wetted perimeter.

Wetted perimeter, $P = y + b + \sqrt{5y}$

Area of flow, $A = by + \frac{1}{2} (y \times 2y) = by + y^2$

from which $b = (A - y^2)/y$

$\therefore$ $P = y + \frac{A}{y} - y + \sqrt{5}y = \frac{A}{y} + \sqrt{5}y$

For minimum P, $\frac{dP}{dy} = 0$, which results in

$$\frac{dP}{dy} = -\frac{A}{y^2} + \sqrt{5}, \therefore \ A = \sqrt{5}\,y^2$$

and $b = \frac{A - y^2}{y} = \frac{\sqrt{5}y^2 - y^2}{y} = 1.236\,y$

As $A = 20 \text{ m}^2, y = (20/\sqrt{5}\,)^{1/2} = \mathbf{2.991\ m}$

$b = 1.236 \times 2.991 = \mathbf{3.696\ m}$

Wetted perimeter $P = \frac{20}{2.991} + \sqrt{5} \times 2.991 = 13.375$ m

$\therefore$ $R = \frac{A}{P} = \frac{20}{13.375} = 1.495$ m

Minimum value of bed slope

$$S_0 = \frac{n^2V^2}{R^{4/3}} = \frac{(0.015)^2\,(1)^2}{(1.495)^{4/3}} = \frac{1}{6644}$$

∴ The channel bed slope is 1 in 6644.

12.7.3. Hydraulically Efficient Circular Channel

Closed conduits such as sewers and drains run partly full most of the times (except during floods when the drains run full), and therefore, have a free surface. Flow in such pipes is governed by the same principles as apply to open channel flow. For circular pipes the area of flow A corresponding to depth y is

$$A = \pi r_0^2 \times \frac{2\theta}{2\pi} - 2\left(\frac{1}{2} r_0 \sin\theta \times r_0 \cos\theta\right)$$

$$A = r_0^2\theta - 2\left(\frac{1}{2} r_0 \sin\theta\, r_0 \cos\theta\right) = r_0^2\left(\theta - \frac{1}{2}\sin 2\theta\right)$$

where r_0 is the radius of the conduit and θ defined by the relation,

$$\cos\theta = \frac{r-y}{r} = 1 - (y/r)$$

the wetted perimeter $P = 2r_0\theta$

and using Manning's formula, the discharge,

$$Q = A.\,\frac{1}{n}\,R^{2/3}\,S^{1/2}$$

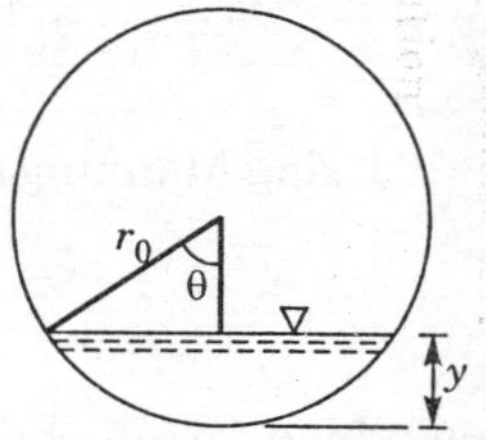

Fig. 12.7. Circular channel.

For a circular channel of given radius, slope and roughness flowrate can be expressed as

$$Q = KAR^{2/3}$$

in which K, known as the conveyance, is a constant the value of which depends upon roughness coefficient and the channel slope. For a given channel, K being a constant, the discharge Q is proportional to $AR^{2/3}$ which can be written in terms of r_0 and θ as

$$AR^{2/3} = A\left(\frac{A}{P}\right)^{2/3} = \frac{A^{5/2}}{P^{2/3}} = \frac{\left\{r_0^2\left(\theta - \frac{1}{2}\sin 2\theta\right)\right\}^{5/2}}{(2r_0\theta)^{2/3}}$$

For maximum discharge, $dQ/d\theta = 0$, which for a given channel reduces to

$$\frac{d(AR^{2/3})}{d\theta} = 0.$$

This condition determines $\theta = 151°\ 10'$ which enables conveyance of maximum discharge. The corresponding depth of flow is $y = 1.88\, r_0$. Similarly, the depth of flow for maximum velocity in the channel can be obtained. This depth is $y = 1.62\, r_0$. If we use Chezy's equation, $\theta = 154°$, and $y = 1.9\, r_0$.

12.7.4. Hydraulic Design of Open Channel

The economic design of an open channel of a given shape to convey a given flowrate under uniform flow conditions involves minimization of flow area and wetted perimeter. The problem is, therefore, of determining the economical dimensions from hydraulic considerations, for known flowrate, channel slope and roughness coefficient. The channel slope is governed by the topography of the land through which the channel has to pass and the roughness coefficient depends upon the nature of material forming the bed and banks of the channel (for fixed bed channels). For alluvial channels (which have cohesionless bed material), the roughness coefficient largely depends upon the ripples and dunes which are formed on the bed under certain conditions of flow.

The channels of trapezoidal cross-section are extensively used for the conveyance of water and other liquids. The side slope of the channel depends on the type of soil used in its construction. In case of rocks, the channel sides may be kept almost vertical, whereas the earthen channel banks will, ordinarily not be stable on slopes greater than 1 : $1\frac{1}{2}$ (horizontal). Those in sandy material may require side slopes as flat as 1 : 3.

Considering a trapezoidal section, Fig. 12.6, it is required to determine the economic geometric parameters of the channel section, *i.e.* the bed width, the side slope and the depth of flow, to convey a given flowrate Q, the channel slope and the roughness coefficient being S and n respectively. From Eq. (12.10), the area of flow can be written as

$$A = \frac{Qn}{\sqrt{S}} \cdot \frac{1}{R^{2/3}} \qquad \text{...(12.10}a\text{)}$$

For the channel section to be economical, for given Q, S and n, the area of flow A must be minimum. The area will be minimum when the hydraulic radius R is maximum. The following equations can now be written :

$$A = (b + my)y$$

$$P = b + 2y\sqrt{1+m^2}$$

$$R = \frac{A}{P} = \frac{(b+my)y}{b+2y\sqrt{1+m^2}} = \frac{by_*(1+y_*)}{m+2y_*\sqrt{1+m^2}} \qquad \text{...(12.10}b\text{)}$$

in which $y_* = my/b$. Eq. (12.10 a) can be expressed as

$$(1 + my)y = \frac{Qn}{\sqrt{S}} \cdot \frac{1}{R^{2/3}}$$

or

$$(1 + y_*)y_* = \frac{Qn}{\sqrt{S}} \cdot \frac{1}{R^{2/3}} \qquad \text{...(12.10 }c\text{)}$$

Eliminating y_* between Eqs. (12.10 b) and (12.10 c) one gets

$$\frac{\sqrt{S}}{Q_n} b^2(\cos\alpha - 2)R^{10/3} - 4\,\text{cosec}\,\alpha\, R^{8/3} + 2b(1-\cos\alpha)R^{5/3} + \frac{Qn}{\sqrt{S}}\cos\alpha = 0 \qquad \text{...(12.10 }d\text{)}$$

Eq. (12.10 b) shows that the hydraulic radius R is a function of bedwidth b, side slope m and depth of flow y. Eq. (12.10d) has however, been reduced to a form in which R is a function of b and m (or α) only. For R to be maximum,

$$\frac{\partial R}{\partial b} + \frac{\partial R}{\partial \alpha} = 0 \qquad \text{...(12.10 }e\text{)}$$

Using Eq. (12.10 d) and differentiating R with respect to b and putting $\frac{\partial R}{\partial b} = 0$, and then simplifying one gets

$$b = \frac{1-\cos\alpha}{2-\cos\alpha} \cdot \frac{Qn}{\sqrt{S}} \cdot \frac{1}{R^{5/2}} \qquad \text{...(12.10 }f\text{)}$$

Differentiating R now with respect to α and setting $\frac{\partial R}{\partial \alpha} = 0$ and simplifying the following can be written

$$b = \left\{1 + \frac{2R^{4/3}}{\sin\alpha}\sqrt{\frac{S^{1/2}}{Qn\tan\alpha}}\right\} \cdot \frac{Qn}{\sqrt{S}} \cdot \frac{1}{R^{5/3}} \qquad \text{...(12.10 }g\text{)}$$

Eliminating b from Eqs. (12.10 f) and (12.10 g), the following equation is obtained

$$\frac{Qn}{4\sqrt{S}\,R^{8/3}} = \cot\alpha\,(2\,\text{cosec}\,\alpha - \cot\alpha)^2 \qquad ...(12.10\ h)$$

Substituting for R from Eq. (12.10 h) into Eq. (12.10 g), b and R can be eliminated from Eq. (12.10 d) and one gets

$$\cos\alpha = \tfrac{1}{2} \quad \text{or} \quad \alpha = 60°$$

It may be recalled here that the same value of α was obtained in Article 12.7.2.

Using this value of α, the following dimensions of an economical trapezoidal section (which has sides sloping at $\alpha = 60°$) can be obtained.

$$b = \left(1.345\frac{Qn}{\sqrt{S}}\right)^{0.375}$$

and,

$$y = \left(0.9\frac{Qn}{\sqrt{S}}\right)^{0.375}$$

which results in $y/b = 0.669$. The minimum area of flow is

$$A_{min} = \left(1.904\frac{Qn}{\sqrt{S}}\right)^{0.75}$$

and the maximum average velocity

$$V_{max} = \frac{Q}{A_{min}} = \left(\frac{QS^{1.5}}{0.9\,n^2}\right)^{0.25}$$

It may be noted that the right hand sides of the above relationships are known for given values of Q, S and n, and hence the unknown parameters can be computed. For other channel shapes, a similar analysis can be made and the geometric parameters of the channel accordingly determined. For round bottom and rounded corner sections, however, the reference given in the footnote (given at the end of this page) can be consulted.

Example 12.20. *A sewer consists of a semicircular top and a semicircular bottom of radius 'r' separated by vertical sides of height 'r', the total height being 3 r. Determine the condition for maximum discharge and show that the angle subtended by the water surface at the centre of the upper semicircle is nearly 64°.*

If the water surface rises till it reaches the top of the sewer, calculate the percentage decrease in discharge. Use the Manning's formula.

Solution. Let the water surface subtend an angle of 2θ at the centre of the upper semicircle as shown in the figure.

Using the Manning's formula, the discharge can be expressed as

$$Q = A\,.\,\frac{1}{n}\,R^{2/3}\,S^{1/2} \qquad ...(1)$$

Here R represents the hydraulic radius of the flow reaction.

Area of flow $$A = \frac{\pi r^2}{2} + 2r^2 + \frac{\pi r^2}{2} - r^2\theta + r^2\sin\theta\cos\theta$$

$$= \pi r^2 + 2r^2 - r^2\theta + \frac{1}{2}\,r^2\sin 2\theta$$

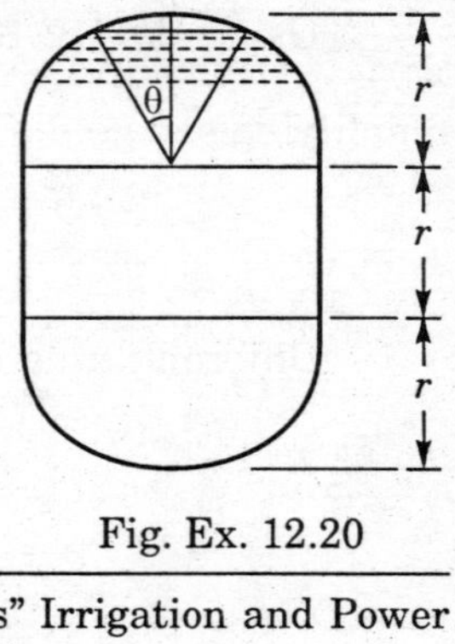

Fig. Ex. 12.20

8. Swamee, P.K. and Bhatia, K.G., "Economic Open Channel Sections" Irrigation and Power Journal, April 1972.

$$= \frac{1}{2} r^2 (2\pi + 4 - 2\theta + \sin 2\theta) \quad ...(2)$$

Wetted parameter $P = \pi r + 2r + \pi r - 2r\theta$

$$= 2r (2\pi + 1 - \theta) \quad ...(3)$$

Expressing Eq. (1) in terms A and P

$$Q = A \cdot \frac{1}{n} \left(\frac{A}{P}\right)^{2/3} S^{1/2}$$

$$= \frac{1}{n} \frac{A^{5/3}}{P^{2/3}} S^{1/2} = \frac{S^{1/2}}{n} \left(\frac{A^5}{P^2}\right)^{1/3} \quad ...(4)$$

For discharge to be maximum, from Eq. (4) it is seen that the quantity $\frac{A^5}{P^2}$ must be maximum.

Differentiating with respect to θ,

$$P^2(5A^4) \frac{dA}{d\theta} - A^2 (2p) \frac{dp}{d\theta} = 0$$

$$\frac{5}{2} p \frac{dA}{d\theta} = A \frac{dP}{d\theta} \quad ...(5)$$

Using Eqs. (2) and (3)

$$\frac{dA}{d\theta} = \frac{1}{2} r^2 (2 \cos 2\theta - 2) = r^2 (\cos 2\theta - 1)$$

and

$$\frac{dP}{d\theta} = 2R (-1) = -2R$$

Making substitutions in Eq. (5)

$$\frac{5}{2} \cdot 2r (\pi + 1 - \theta) r^2 (\cos 2\theta - 1) = \frac{1}{2} r^2 (2\pi + 4 - 2\theta + \sin 2\theta) (-2R)$$

Simplifying

$$5(1 - \cos 2\theta) = 2 + \frac{2 + \sin 2\theta}{\pi + 1 - \theta} \quad ...(6)$$

The above equation, which describes the condition for maximum discharge, is to be solved by trial and error method.

Substituting $\theta = 32°$, it is found that L.H.S. = R.H.S. = 2.81.

Thus for maximum discharge, the angle subtended by the water surface at the centre of the upper semicircle is $2 \times 32° = 64°$.

Using this condition, Eqs. (2) and (3) give

$$A = \frac{1}{2} r^2 \left(2\pi + 4 - 2 \times \frac{32\,\pi}{180} + \sin 64°\right) = 5.03\, r^2$$

$$P = 2r \left(\pi + 1 - \frac{32\,\pi}{180}\right) = 7.168\, r$$

Substituting these quantities in Eq. (4)

$$Q_{max} = \frac{1}{n} \left[\frac{(5.03 r^2)^5}{(7.168\, r)^2}\right]^{1/3} S^{1/2}$$

$$= \frac{3.98}{\pi} r^{8/3} S^{1/2} \qquad ...(7)$$

When the sewer runs full, the water surface rises to the top and $Q = 0°$. For this condition, Eqs. (2) and (3) result in

$$A = \frac{1}{2} r^2 (2\pi + 4) = 5.14r^2 \qquad ...(8)$$

$$P = 2r (\pi + 1) = 8.28\, r \qquad ...(9)$$

and the discharge from Eq. (4) becomes equal to

$$Q = \frac{1}{n}\left[\frac{(5.14\, r^2)^5}{(8.28\, r)^2}\right]^{1/3} S^{1/2} = \frac{3.74}{n} r^{8/3} S^{1/2} \qquad ...(10)$$

Comparing Eqs. (7) and (10), it is seen that the discharge decreases when the sewer runs full. The percentage decrease in discharge

$$= \frac{Q_{max} - Q}{Q_{max}} \times 100 = \frac{3.98 - 3.74}{3.98} \times 100 = \mathbf{6.03\%}$$

∴ The discharge decreases by 6.03%.

Example 12.21. *Water flows in a channel of the shape of isosceles triangle of bedwidth 'a' and sides making an angle of 45° with the bed. Determine the relation between depth of flow 'd' and the bedwidth 'a' for maximum velocity and for maximum discharge condition. Use Manning's formula and note that 'd' is less than 0.5 a.*

Solution. Manning's formula for velocity

$$V = \frac{1}{n} R^{2/3} S^{1/2}$$

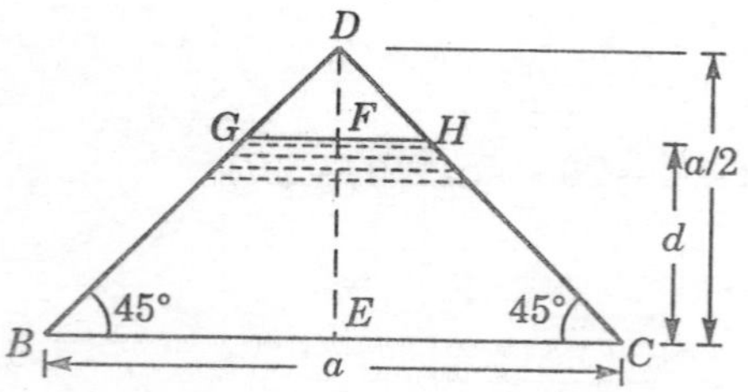

Area of flow

$$A = \tfrac{1}{2} (GH + BC) \times d$$

$$= \tfrac{1}{2} \{(a - 2d) + a\} \times d$$

$$= (a - d)d$$

Wetted perimeter $P = a + 2BG = a + 2\sqrt{2}\; d$

Hydraulic radius $\quad R = \dfrac{A}{P} = \dfrac{(a-d)d}{a + 2\sqrt{2}d}$

$$V = \frac{1}{n}\left\{\frac{(a-d)d}{a + 2\sqrt{2}d}\right\}^{2/3} S^{1/2}$$

For maximum velocity, $\dfrac{dV}{dd} = 0$

$$\frac{dV}{dd} = \frac{\sqrt{S}}{n}\left[\frac{(a + 2\sqrt{2}d)(a - 2d) - (a-d)d \times 2\sqrt{2}}{(a + 2\sqrt{2}\, d)^2}\right] \times \frac{2}{3}\left\{\frac{(a-d)d}{a + 2\sqrt{2}\, d}\right\}^{-1/3}$$

which upon simplification, results in for maximum velocity

$$a^2 - 2ad - 2\sqrt{2}\; d^2 = 0$$

from which, $\quad$ **d/a = 0.3382**

For maximum discharge, $\dfrac{dQ}{dd} = 0$

$$Q = A\,.V = (a-d)d \times \frac{1}{n}\left\{\frac{(a-d)d}{a+2\sqrt{2}d}\right\}^{2/3} S^{1/2}$$

$$\frac{dQ}{dd} = \frac{\sqrt{S}}{n} \times \left[\frac{(a+2\sqrt{2}d)^{2/3} \times \frac{5}{3}\{(a-d)d\}^{2/3} \times (a-2d) - \{(a-d)d\}^{5/3}\left\{\frac{2}{3}(a+2\sqrt{2}d)^{-1/3} \times 2\sqrt{2}\right\}}{(a+2\sqrt{2}d)^{4/3}}\right]$$

which for maximum discharge simplifies to

$$5a^2 - 1.5147\,ad - 22.6274\,d^2 = 0\ ;\ \text{from which, } \mathbf{d/a = 0.4378.}$$

12.8. Velocity Distribution in Open Channels

As in pipes, the velocities in open channels are retarded near the solid boundaries. An additional factor influencing the velocity profile in open channels is introduced on account of the free-surface. The surface tension of the liquid-air interface (*i.e.*, the free surface) produces a resistance to flow and causes maximum velocity to occur at some distance below the free-surface. A strong wing, when blowing in upstream or downstream direction, also affects the flow and accordingly modifies the velocity profile. In the vertical direction, the velocity is minimum at the channel bottom and it attains a maximum value at a distance from the free surface which varies from 0.05 to 0.25 times the flow depth, the higher value is applicable to section closer to the bank. At the free-surface, the velocity is somewhat less than the maximum value due to surface tension effects and possible air resistance. The sketches of Fig. 12.8 (*a*) show the pattern of velocity distributions in the horizontal and vertical directions for a rectangular channel. From the inspection of these velocity distribution curves the following observation can be made :

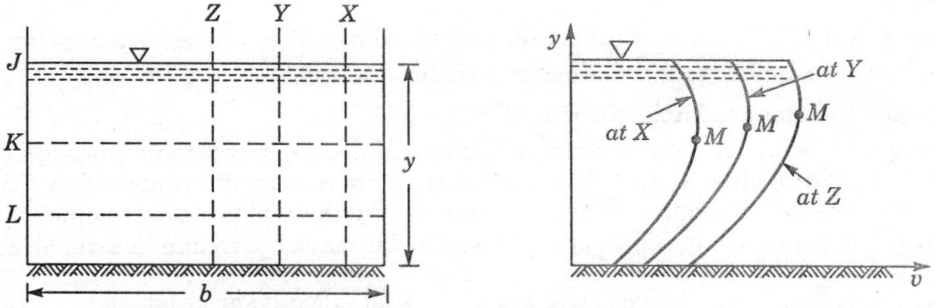

(*i*) Channel cross-section. (*ii*) Velocity distribution in the vertical plane.

Note. *M* represents the point of maximum velocity.

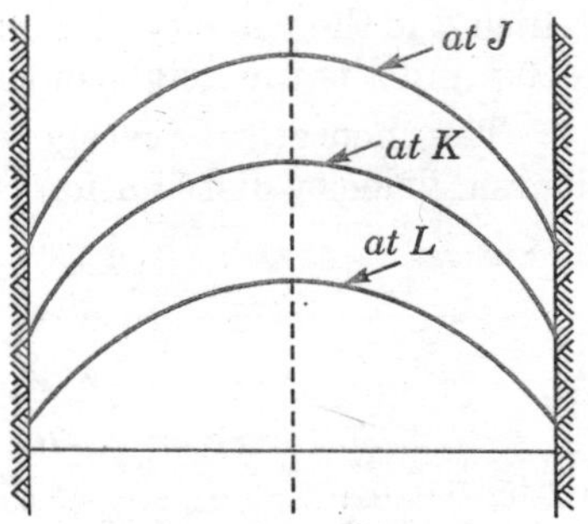

(*iii*) Velocity-distribution in the horizontal plane.

Fig. 12.8. (*a*) Velocity distribution in a rectagular channel.

(1) As we go away from the bank (evident from the locations of vertical sections *X*, *Y* and *Z*) the point of maximum velocity shifts upwards and at the central section *Z*, the maximum velocity occurs almost at the free surface.

The velocity distribution in channels depends upon the shape of the channel section, the channel roughness and the presence of bends in its alignment. The effect of roughness is to increase the curvature of the velocity distribution curve. Fig. 12.8 (*b*) shows the comparison between the velocity distribution curves for smooth and rough channels. The presence of bends also modifies the velocity distribution. If bend is a sharp one, the centrifugal acceleration which acts in the radially outward direction has a higher value, causing higher velocities near the outer bank of the channel.

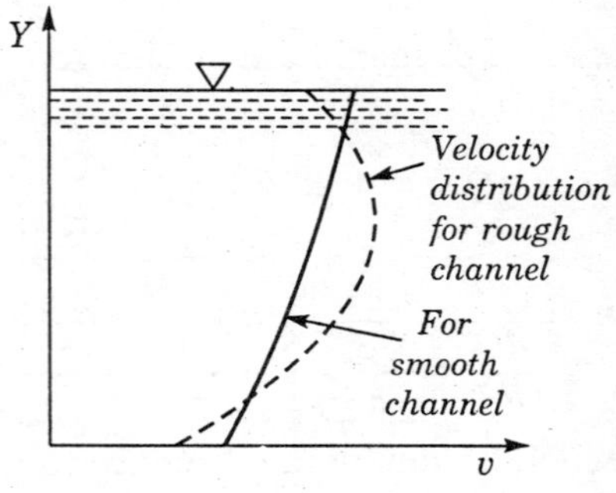

Fig. 12.8. (*b*) Velocity Distribution in rough and smooth channels.

From measurement of velocities in natural streams (*i.e.*, channels characterized by irregular cross-section) and the velocity distribution curves plotted from such observations, the following findings can be made :

(2) In general, the maximum velocity occurs somewhere between the liquid surface and one-third of the flow depth, the distance from the free-surface to the point of maximum velocity being proportionally greater for larger depths of liquid. For shallow streams, the maximum velocity is very near to the surface whereas for deep streams it may lie at about one-third of the depth. A strong wind blowing either upstream or downstream will affect distribution of velocity in the vertical plane.

(3) The mean velocity in the vertical plane is ordinarily at 0.55 to 0.65 of the flow depth. The velocity at 0.6 depth is usually within 5% of the mean velocity.

(4) The mean of velocities at 0.2 depth and 0.8 depth usually gives the mean velocity in the vertical within 2%.

(5) The mean velocity in the vertical is ordinarily 0.80 to 0.95 of the surface velocity. The smaller value applies to the shallower streams.

12.8.1. Equation for Velocity Distribution

The striking similarity between the observed and computed velocity distributions led Vanoni to demonstrate that the Prandtl-Karman universal logarithmic velocity distribution law for pipes is also applicable to two-dimensional open channel flow. Using the constants determined for pipe flow by Nikuradse, the following equations of velocity distribution in open channels can be written

$$u = 5.75\, u_* \log_{10} \frac{9yu_*}{\nu} \quad \text{...(for smooth surfaces)}$$

and

$$u = 5.75\, u_* \log_{10} \frac{30y}{k} \quad \text{...(for rough surface)} \qquad \text{...(12.11}a\text{)}$$

in which u is the velocity at a distance y from the channel bed, $u_* = \sqrt{gRS}$ is the shear or friction velocity and k is the height of surface roughness.

The theoretical average velocity analytically obtained by Keulegan, using Prandtl-Karman universal velocity distribution law and Nikuradse's and Bazin's data, is given below

$$\frac{V}{u_*} = 3.25 + 5.75 \log_{10} \frac{Ru_*}{\nu}, \text{ ... for smooth channels} \qquad \text{...(12.11}b\text{)}$$

and

$$\frac{V}{u_*} = 6.25 + 5.75 \log_{10} \frac{R}{k}, \text{ ... for rough channels} \qquad \text{...(12.11}c\text{)}$$

in which V = mean velocity of flow and R = hydraulic radius.

From Chezy's and Manning's formula

$$V = C\sqrt{RS} \text{ and } V = \frac{1}{n} R^{2/3} S^{1/2} = \frac{R^{1/6}\sqrt{RS}}{n}$$

and

$$u_* = \sqrt{\frac{\tau_0}{\rho}} = \sqrt{\frac{\gamma RS}{\rho}} = \sqrt{gRS}$$

The ratio $$\frac{V}{u_*} = \frac{C\sqrt{RS}}{\sqrt{gRS}} = \frac{C}{\sqrt{g}} \quad \text{...(12.11}d\text{)}$$

Using Manning's equation

$$\frac{V}{u_*} = \frac{\frac{1}{n}R^{2/3}S^{1/2}}{\sqrt{gRS}} = \frac{R^{1/6}}{n\sqrt{g}} \quad \text{...(12.11}e\text{)}$$

Eqs. (12.11 *d*) and (12.11 *e*) relate the Chezy's constant C with Manning's n and result in Eq. (12.9).

Example 12.21. *Uniform flow occurs in a rectangular channel 3 m wide at a depth of 2.0 m. If Chezy's coefficient is 64 $m^{1/2}$/sec, determine the approximate height of roughness projections. Take g = 10 m/s^3.*

Solution. $C = 64\ \text{m}^{1/2}/\text{s}$.

From Eq. (12.11 *c*), $$\frac{V}{u_*} = 6.25 + 5.75\log_{10}\left(\frac{R}{k}\right)$$

$$R = \frac{A}{P} = \frac{6}{7}$$

$$\frac{V}{u_*} = \frac{C\sqrt{RS}}{\sqrt{gRS}} = \frac{C}{\sqrt{g}} \quad \text{since } u_*^2 = \frac{\tau_0}{\rho} = gRS$$

$$\frac{C}{\sqrt{g}} = 6.25 + 5.75\log_{10}\frac{R}{k}$$

$$\frac{64}{\sqrt{10}} = 6.25 + 5.75\log_{10}\frac{R}{k}$$

$$\log\left(\frac{R}{k}\right) = 2.44$$

which yields $$\frac{R}{k} = 275, \quad \therefore \quad \frac{R}{275} = \frac{6}{7\times 275} = 0.00312\ \text{m} = 0.312\ \text{m}$$

12.8.2. Velocity Distribution Coefficients

The evaluation of energy flux at a flow section is usually based on the average velocity thereby ignoring the effect of non-uniformity of velocity distribution across the section. In reality, the velocity varies on account of viscous and boundary resistances ; it being smaller near the boundaries and increases towards the free-surface while considering the distribution in a vertical plane. For energy and momentum correction factors, attention is invited to articles 4.12 and 4.20 respectively.

It can be shown that the coefficients α and β individually are strongly dependent on the type of velocity distribution, but that the ratio $(\alpha - 1)/(\beta - 1)$ varies slightly in the range 2.7 to 2.8 as the flow changes from laminar to turbulent. The high value of α for laminar flow is of limited interest

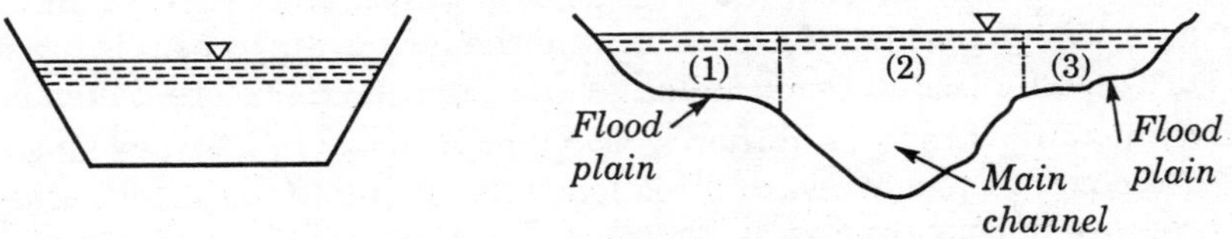

(*i*) Single (prismatic channel
$\alpha = 1.03$ to 1.36
$\beta = 1.01$ to 1.12

(*ii*) Natural river channel during flood
$\alpha = 2$, $\beta = 1.38$

Fig. 12.8. (*c*) Typical channel cross-sections ; (*i*) constructed channel, (*ii*) Naturally developed channel.

since laminar flow is very rarely encountered in practical problems. Experiments have shown that the kinetic energy correction factor α depends upon the ratio of maximum velocity to mean velocity and that for turbulent flow in straight prismatic channels, its value ranges from 1.03 to 1.36, Fig. 12.8 *c* (*i*). The α-value is generally higher for shallow and small sized channels and lower for large sized channels with large depth of flow. Natural rivers, on the other hand, frequently flow in channels which can be sub-divided into distinct regions, each with a different mean velocity. This is particularly true during the time of flood, when the river overflows onto its flood plains producing the channel section shown in Fig. 12.8 *c* (*ii*). In this case there are in effect three separate channels—portions marked (1) and (3) being on the plains while the main channel is represented by (2). Over the flood plains the mean velocity will be significantly less than that in the main channel, because of the higher resistance to flow due to smaller depth and rougher surface of the flood plains.

In case of natural channels, the irregularity of cross-section and the non-uniformity in channel roughness (specially during floods) have marked influence of the coefficients α and β. Under high flood conditions, α and β values may be even greater than 2.0 and 1.38 respectively. For straight prismatic channels β generally varies from 1.01 to 1.12 (variation of α being from 1.03 to 1.36).

Assuming logarithmic distribution of velocity, if can be shown that the velocity distribution coefficients can be expressed as

$$\alpha = 1+3\left(\frac{u_{max}}{V}-1\right)^2-2\left(\frac{u_{max}}{V}-1\right)^3 \;;\; \beta = 1+\left(\frac{u_{max}}{V}\right)^2 \qquad \text{...(12.11}f\text{)}$$

where u_{max} is the maximum velocity and V the average velocity. If the velocity distribution is assumed a linear one (*i.e.* straight line variation), the following relationships obtained by Rehbock can be used :

$$\alpha = 1+\left(\frac{u_{max}}{V}-1\right)^2 \;;\; \beta = 1+\frac{1}{3}\left(\frac{u_{max}}{V}-1\right)^3 \qquad \text{...(12.11}g\text{)}$$

12.9. Pressure Distribution in Open Channels

The pressure distribution at a flow section in open channels is governed by the pattern of streamlines in the vertical plane ; if they are straight and parallel, pressure follows the hydrostatic law. On the other hand, if the streamlines either have marked divergence or have appreciable curvature resulting in acceleration normal to flow direction, these factors alter the distribution of pressure. The flow in open channels may accordingly be described as

(*i*) Parallel flow, and (*ii*) Curvilinear flow.

In parallel flow, the streamlines are parallel and straight so that there is no acceleration normal to the flow. The pressure distribution at a flow section in parallel flow, is therefore, hydrostatic. The curvilinear flow has curved streamlines ; the effect of curvature of streamlines is to produce normal acceleration due to centrifugal effect. The departure of pressure distribution in curvilinear flow, from that of hydrostatic variation of pressure that would occur if the flow were parallel, is due to the curvature of streamlines. In gradually-varied flow the change in the depth along the flow direction is so small that the divergence or convergence of streamlines is insignificant to produce any appreciable normal acceleration, and the flow in such cases is practically a parallel flow. The pressure distribution in the gradually varied flow is, therefore, taken as hydrostatic.

The curvilinear flow may be either concave or convex. In concave flow, the centrifugal acceleration is directed in the downward direction thus adding to the acceleration due to gravity. The resulting pressure in concave flow is, therefore, greater than the hydrostatic pressure for the same depth of flow. In case of convex flow, the centrifugal acceleration has an upward direction opposing the gravitational acceleration, thus causing the pressure to be lower than the corresponding hydrostatic pressure. Similarly, if the streamlines have divergence enough to produce appreciable normal acceleration, the actual pressure distribution will be different from the hydrostatic distribution of pressure.

Consider a liquid element having a cross-sectional area of 1 m^2 and a depth h below the free surface, sec Fig. 12.9 a (*ii*). The centrifugal force acting vertically downward in case of concave flow over the element is $\frac{\gamma}{g} h \frac{v^2}{r}$ and the pressure acting over its base is

$$p_c = \frac{\gamma}{g} h \frac{v^2}{r} \qquad \text{...(12.12)}$$

Eq. (12.12) gives the pressure in excess of the hydrostatic pressure ($p_h = \gamma h$) due to the effect of curvature of streamlines. The pressure p_c at the channel bottom may be obtained from Eq. (12.12) by substituting the depth of flow for h, the radius of the channel bottom for r and the velocity v may be taken as the mean velocity of flow V. In case of convex flow, this pressure p_c has negative sign as it tends to decrease the pressure p at any point in the flow section. The pressure at a point situated at a depth h below the free surface in curvilinear flow, may be divided into two parts : (*i*) the hydrostatic pressure at the point, $p_h = \gamma h$, and (*ii*) the pressure p_c due to curvature effect, thus

$$p = p_h \pm p_c \qquad \text{...(12.12 } a\text{)}$$

The plus sign in Eq. (12.12 a) is applicable to concave flow while the negative sign is to be used for convex flow. The pressure head in terms of the height of the flowing liquid is obtained by dividing each term of Eq. (12.12 a) by the specific weight (γ) of the liquid, thus

$$\frac{p}{\gamma} = \frac{p_h}{\gamma} \pm \frac{p_c}{\gamma} \qquad \text{...(12.12 } b\text{)}$$

By using the concept of free-body diagram and Newton's second law of motion, the variation of pressure may be shown to follow a logarithmic law.

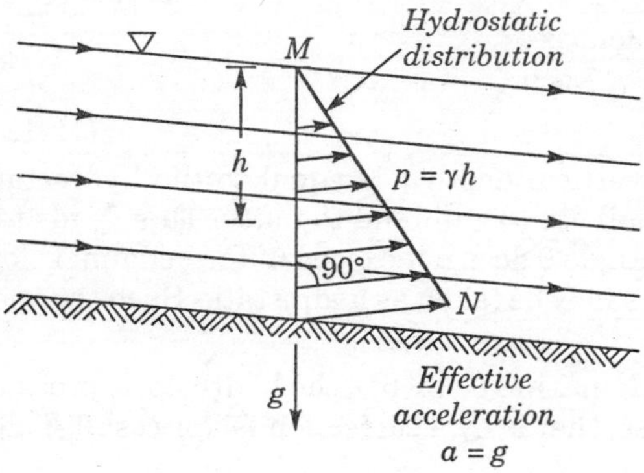

(*i*) Parallel flow.

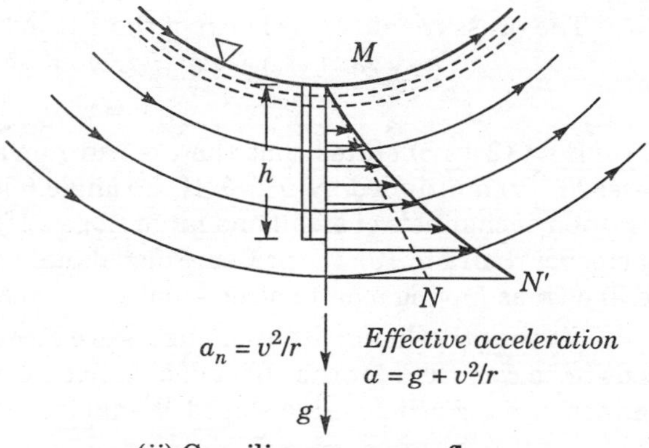

(*ii*) Curvilinear concave flow.

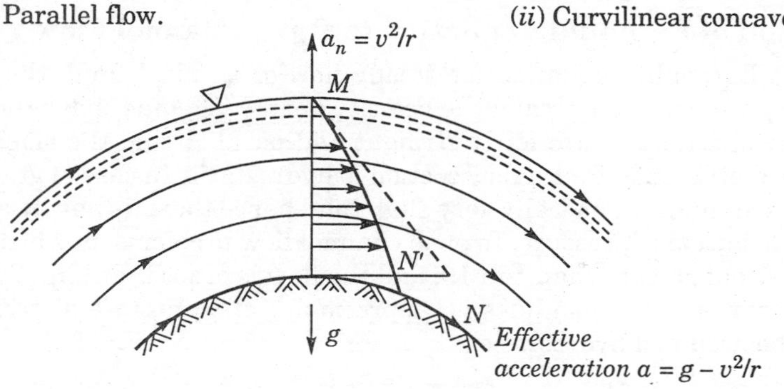

(*iii*) Curvilinear concave flow.

Fig. 12.9. (a) Pressure-distribution in curvilinear flow.

In parallel flow the pressure is hydrostatic, and the pressure head may be represented by the vertical depth of flow y. In curvilinear flow, the pressure head for simplicity may be expressed as $\alpha' y$ where α' is known as pressure distribution coefficient and represents a correction factor to account for the curvature effect.

12.9.1. Pressure Distribution in Channels of Large Slope (Parallel flow). Consider a liquid element of length dx in a channel of 1 m width having a large bottom slope of angle θ, shown in Fig. 12.9 (*b*). The weight of the shaded portion of the element is

$$\gamma dx(d) = \gamma y \cos\theta \, dx.$$

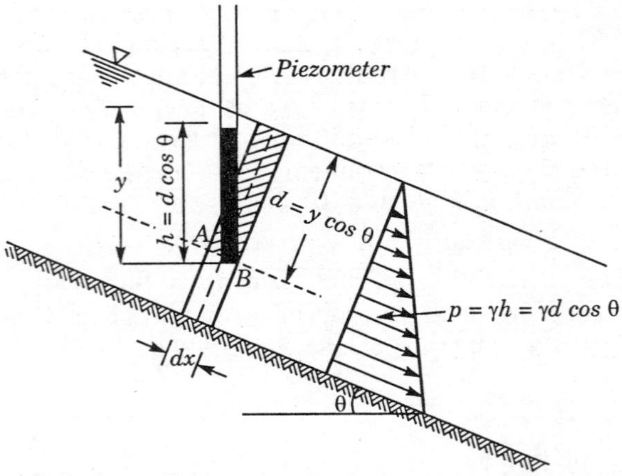

Fig. 12.9. (*b*) Pressure distribution in large-slope channels.

The pressure at the base AB of the shaded element is

$$\{(\gamma y \cos\theta \, dx) \cos\theta\}/dx = \gamma y \cos^2\theta = \gamma d \cos\theta$$

$$= \gamma h \qquad \text{...(12.12 } c)$$

Eq. (12.12 *c*) states that the pressure at any vertical depth y is equal to the hydrostatic pressure (γy) multiplied by $\cos^2\theta$. If the angle θ is small $\cos^2\theta$ will tend to unity. This leads to a definition of channels of small and large slopes. If the angle θ does not exceed 6° (*i.e.*, channel slope not steeper than 1 in 10), the pressure distribution may safely be taken as hydrostatic. Such channels are known as having small bottom-slope.

On the other hand, if the channels are steeper than 1 in 10 ($\theta > 6°$), the hydrostatic pressure needs to be corrected for the effect of channel steepness, that is by a correction factor $\cos^2\theta$. These channels are termed as large sloped channels.

12.10. Application of Bernoulli's equation to Open Channel Flow

According to Bernoulli's equation for steady flow of an ideal fluid, the summation of the datum head Z, the pressure head p/γ and the velocity head $V^2/2g$ must have the same value at all flow sections. If the streamlines are fairly straight and parallel, and the slope of the channel is small, the pressure distribution over a flow section is hydrostatic. In case of gradually varied flow, the curvature of streamlines is generally very slight and hence there is hardly any departure from the hydrostatic distribution of pressure. In open channel flow problems, in which the channel slope is not very large, say not steeper than 1 in 10, and the flow is other than rapidly varied (in rapidly varied flow, the curvature of streamlines is appreciable), the pressure distribution over a flow-section can safely be assumed hydrostatic.

Consider flow in a channel of small slope, Fig. 12.10. At point A, which is at a depth h below the free surface, the pressure according to hydrostatic law is $p = \gamma h$, and, therefore, for point A in the fluid stream the summation of pressure and datum heads, *i.e.*, piezometric head

$$= \frac{pA}{\gamma} + Z_A = h + Z_A.$$

It may be noted that $(h + Z_A)$ represents the height of the free-surface above the datum level. In other words, the free-surface represents the hydraulic grade line. In ideal fluid flow, the velocity does not very across a section, and hence the Bernoulli's equation at a flow section may written as

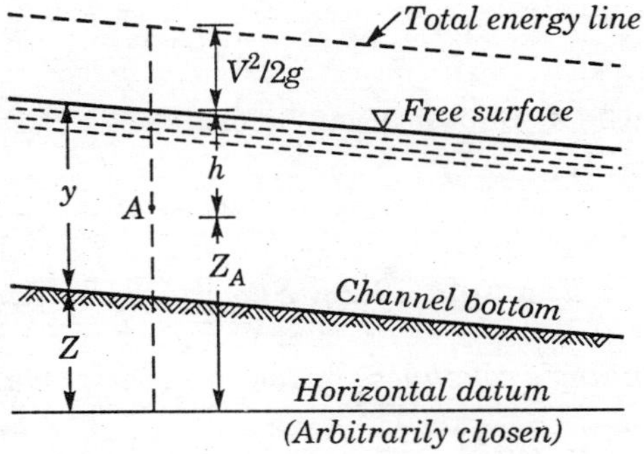

Fig. 12.10. Application of Bernoulli's equation.

$$Z_A + h + \frac{V^2}{2g} \qquad \text{...(12.13 a)}$$

If the fluid depth at the section under consideration is y, and the elevation head (datum head) of the channel bottom the above arbitrary horizontal datum is Z, then

$$Z_A + h = Z + y$$

and the Bernoulli's equation may now be expressed as

$$Z + y + \frac{V^2}{2g} \qquad \text{...(12.13 b)}$$

The assumption of ideal fluid flow results in a uniform distribution of velocity over a flow section. This is never achieved in practice. The Bernoulli's equation after having corrected for the non-uniform velocity distribution may be written as

$$Z + y + \alpha\frac{V^2}{2g} \qquad \text{...(12.13 c)}$$

where α is the kinetic energy correction factor.

Example 12.22. *Show that the theoretical discharge of the open channel flow may be expressed by*

$$Q = A_2\sqrt{\frac{2g(\Delta y - h_f)}{1-(A_2/A_1)^2}}$$

where A_1 and A_2 are the cross-sectional area of flow at sections (1) and (2) respectively, Δy is the drop in the water surface between the two sections and h_f is the energy head loss between the two sections.

Solution. Bernoulli's equation between sections (1) and (2) gives, see the following figure,

$$Z_1 + y_1 + \frac{V_1^2}{2g} = Z_2 + y_2 + \frac{V_2^2}{2g} + h_f$$

difference in the water surface elevation between (1) and (2)

$$\Delta y = (Z_1 + y_2) - (Z_2 + y_2)$$

Substituting this,

$$\Delta y + \frac{V_1^2}{2g} = \frac{V_2^2}{2g} + h_f$$

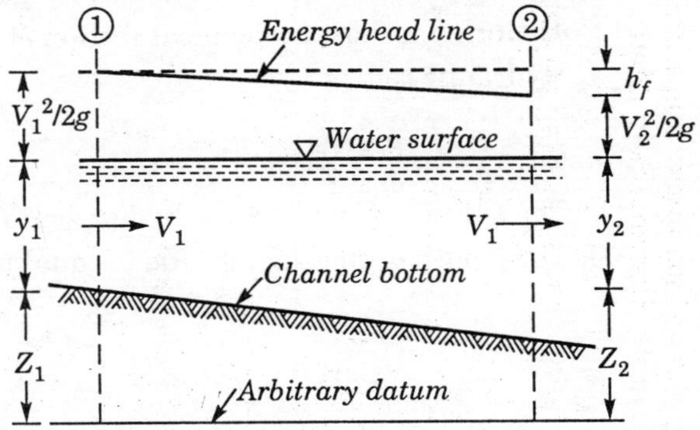

From the continuity equation, $Q = A_1V_1 = A_2V_2$, and we obtain

$$\Delta y + \frac{Q^2}{2gA_1^2} = \frac{Q^2}{2gA_2^2} + h_f$$

which on simplification yields

$$Q = A_2\sqrt{\frac{2g(\Delta y - h_f)}{1-(A_2/A_1)^2}}.$$

Example 12.23. *Show that the friction factor of the Darcy-Weisbach formula, and the Manning's roughness factor n are inter-related by* $f = \frac{78.5n^2}{R^{1/3}}$, *in which R is the hydraulic radius.*

Solution. Darcy-Weisbach formula for pipe flow

$$h_L = f\frac{L}{d}\frac{V^2}{2g}$$

In case of pipe flow, $R = \frac{A}{P} = \frac{\pi d^2/4}{\pi d} = \frac{d}{4}$

$\therefore \quad d = 4R$

In open channel flow, $S = \frac{h_L}{L}$, and the mean velocity by Manning's formula is

$$V = \frac{1}{n}R^{2/3}\,S^{1/2}$$

Making the substitutions in the Darcy-Weisbach formula, one gets

$$S = \frac{h_L}{L} = f\frac{([1/n]R^{2/3}S^{1/2})^2}{4R.2g} = \frac{fR^{4/3}S}{8gRn^2}$$

which simplifies to $\quad f = \frac{8gn^2}{R^{1/3}}$ Substituting for $g = 9.81$ m/s²

$$f = \frac{8\times 9.81n^2}{R^{1/3}} = 78.5\frac{n^2}{R^{1/3}}.$$

Example 12.24. *An open channel of most economical section, having the form of a half hexagon with horizontal bottom, is required to give a maximum discharge of 20.2 m³/s of water. The slope of the channel bottom is 1 in 2500. Taking C = 66 m$^{1/2}$/s in Chezy's equation, determine the dimensions of the cross-section.*

Taking the value of the velocity of flow in the channel as obtained by Chezy's equation, determine the value of n in the Manning's formula.

Solution. For most economical section of the trapezoidal channel,

1. Hydraulic radius

$$R = \frac{y}{2}.$$

2. The side slope angle with the horizontal is 60°, and

3. The length of the sloping side is equal to the bottom width.

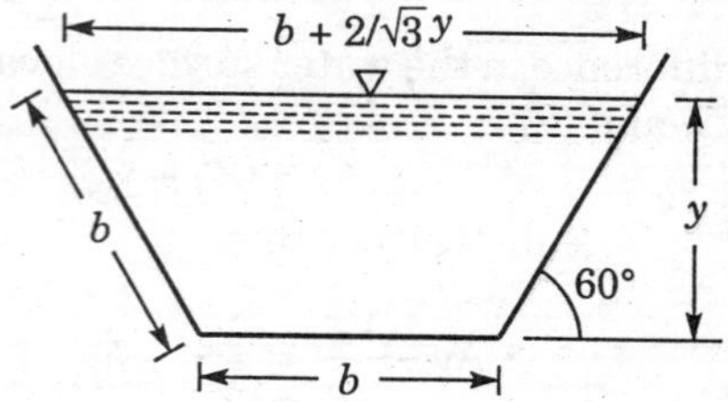

$$A = y\left(b + \frac{y}{\sqrt{3}}\right)$$

$$b = y/\sin 60° = \frac{2}{\sqrt{3}}\, y$$

$$A = y\left(\frac{2}{\sqrt{3}}y + \frac{1}{\sqrt{3}}y\right) = \sqrt{3}y^2.$$

Chezy's formula for discharge

$$Q = AC\sqrt{RS},$$

Making substitutions,

$$20 = \sqrt{3}y^2 \times 66 \times \sqrt{\frac{y}{2} \times \frac{1}{2500}} = 1.617y^{5/2}$$

$$y^{5/2} = \frac{20.2}{1.617} = 12.5$$

$\therefore$ $$y = (12.5)^{2/5} = 2.745 \text{ m}.$$

The depth of flow $y = 2.745$ m.

The bottom width $b = \frac{2}{\sqrt{3}}y = \frac{2}{\sqrt{3}} \times 2.745 = 3.17$ m.

Velocity from Chezy's formula

$$V = C\sqrt{RS} = 66\sqrt{\frac{2.745}{2} \times \frac{1}{2500}} = 1.543 \text{ m/s}.$$

Using Manning's formula

$$1.543 = \frac{1}{n}R^{2/3}S^{1/2}$$

$\therefore$ $$n = \frac{1}{1.543} \times \left(\frac{2.745}{2}\right)^{2/5} \times \frac{1}{50} = 0.016.$$

Example 12.25. *In a wide rectangular channel if the normal depth is increased by 20%, find the percentage increase in the discharge. Use Manning's equation.*

Solution. Manning's equation gives the velocity as

$$V = \frac{1}{n}R^{2/3}S^{1/2}$$

For a wide rectangular channel, $R \approx y$

The discharge $Q = A\,.\,V = (by)\,\frac{1}{n}(y)^{2/3}S^{1/2} = ky^{5/3}$

Differentiating, $dQ = \frac{5}{3}ky^{2/3}\,dy$

$$\frac{dQ}{Q} = \frac{5}{3}\frac{dy}{y} = \frac{5}{3} \times 20\% = 33.3\%$$

Percentage increase in discharge = **33.3%.**

OBJECTIVE TYPE QUESTIONS

1. Open channel flow is the one in which
   (*i*) a closed conduct is full of flowing liquid
   (*ii*) the liquid flowing in a closed conduit has a free-surface
   (*iii*) a covered channel flows full under pressure
   (*iv*) the solid boundaries confining the flow are open at the top.

2. The flow in a pipe takes place at the expense of pressure while that in an open channel is due to the gravity effect. Select from the following list, the flow situations which represent open channel flow

   (*a*) flow in a circular tunnel running full

   (*b*) flow in a circular tunnel which has a free-surface

   (*c*) flow in a sewerline during non-monsoon period

   (*d*) flow in a sewerline during heavy downpour when it runs full

   (*e*) flow in a covered canal which passes deep under a road embankment.

3. Which of the following statements are applicable to open channel flow

   (*i*) the hydraulic grade line (HGL) is higher than the liquid surface

   (*ii*) the HGL and the liquid surface coincide

   (*iii*) the pressure distribution is generally hydrostatic

   (*iv*) flow takes place at the expense of hydraulic pressure

   (*v*) the flow boundaries are fixed

   (*vi*) the velocity distribution is symmetrical about mid-depth

   (*vii*) the total energy line is above the liquid surface by a distance $V^2/2g$.

4. Uniform flow in open channels is characterised by

   (*a*) a changing depth of flow

   (*b*) a constant discharge passing down the channel

   (*c*) a constant depth of flow

   (*d*) no variation of velocity at a flow section with respect to time

5. Steady non-uniform flow in open channels occurs when

   (*i*) a constant discharge flows at the constant depth of flow

   (*ii*) a constant discharge flows in a channel laid at a fixed slope

   (*iii*) the discharge and the depth both vary along the channel length

   (*iv*) for a constant discharge the liquid depth in the channel varies along its length.

6. Listed below are certain flow situations pertaining to open channel flow. Given alongside are the possible flow classification. Identify for each flow situation, the corresponding category of flow classification

(*a*) constant discharge flowing at a uniform depth of flow	(*i*) unsteady non-uniform flow
(*b*) flow in a river during flood	(*ii*) steady non-uniform flow
(*c*) constant discharge flowing at varying depths	(*iii*) steady non-uniform flow
(*d*) flow at a constant rate in a channel bend	(*iv*) unsteady non-uniform flow
(*e*) flow with varying discharge in a lined channel.	(*v*) steady uniform flow.

7. The effects of viscosity and gravity are taken care of by the Reynolds and Froude numbers defined respectively by $\mathbf{R} = VR/\nu$ and $\mathbf{F} = V/\sqrt{gD}$. For the following values of these dimensionless parameters, select from the alternatives given, the type of flow that would exist

(1) $\mathbf{F} < 1, \mathbf{R} < 500$	(*a*) critical laminar flow
(2) $\mathbf{F} < 1, \mathbf{R} > 2000$	(*b*) supercritical turbulent flow
(3) $\mathbf{F} = 1, \mathbf{R} > 2000$	(*c*) subcritical laminar flow
(4) $\mathbf{F} = 1, \mathbf{R} < 500$	(*d*) critical turbulent flow
(5) $\mathbf{F} > 1, \mathbf{R} > 2000$	(*e*) subcritical turbulent flow.

8. Considering the two main forces which are developed in open channels, namely the force causing the flow and the resistance force, the uniform flow can be defined as one in which

(*i*) the force causing the flow is larger than the resistance force

(*ii*) the resistance force is larger than the force causing the force

(*iii*) the two forces are equal in magnitude

(*iv*) the resistance force always acts against the force causing the flow.

9. The Chezy's constant C which appears in the Chezy's equation for mean velocity in open channels

(*a*) is a dimensionless constant

(*b*) has a constant value for different type of channels

(*c*) had dimensions $L^{1/2}\,T^{-1}$

(*d*) does not depend on the quality of channel surface

10. The Manning's equation for flow in open channels is expressed as

(*i*) $V = C\sqrt{RS}$ (*ii*) $V = \dfrac{1}{n} R^{2/3} S^{1/2}$

(*iii*) $C = \dfrac{R^{1/6}}{n}$ (*iv*) $C = \dfrac{87}{\left(\dfrac{1+m}{\sqrt{R}}\right)}$

11. The following table shows the values of Manning's n for different types of channel surfaces. Identify its value characteristic of each channel surface

Type of channel surface	*Manning's n*
(1) Finished concrete	(*a*) 0.022—0.035
(2) Unfinished concrete	(*b*) 0.035—0.040
(3) Brick masonry	(*c*) 0.020—0.030
(4) Earth	(*d*) 0.012—0.020
(5) Earth with vegetation	(*e*) 0.013—0.016
(6) Rock cuts	(*f*) 0.011—0.013

12. Hydraulically efficient channel cross-section is one which

(*a*) has the minimum hydraulic resistance

(*b*) carries maximum discharge under given conditions of slope, roughness and flow area

(*c*) has the maximum possible slope consistent with the topography

(*d*) has the maximum wetted perimeter

13. The maximum velocity in open channels occurs

(*i*) at the mid depth (*ii*) at the free surface

(*iii*) a little below the free-surface (*iv*) near the channel bottom.

14. The mean velocity in the vertical occurs ordinarily at

(*a*) 0.2 (*b*) 0.6 (*c*) 0.8 (*d*) 0.3

times the depth of flow meaured from the free-surface.

15. The mean velocity in open channels can be estimated from the known velocity at the free-surface. It is approximately equal to

(*i*) 0.88 (*ii*) 0.75 (*iii*) 0.65 (*iv*) 1.10

time the surface velocity.

16. The pressure distribution in open channels follows hydrostatic law if the streamlines are straight and parallel. For curvilinear flows, the streamlines are curved and the resulting normal acceleration modifies the pressure distribution. For the following cases identify the pressure distribution :

(*a*) parallel flow	(*i*) more than hydrostatic
(*b*) concave flow as occurs near the foot of spillway crest	(*ii*) hydrostatic
(*c*) convex flow which occurs at the spillway crest	(*iii*) hydrostatic
(*d*) gradually varied flow	(*iv*) less than hydrostatic.

17. Pressure distribution in channels of large slopes (steeper than 1 in 10) deviates from the hydrostatic one. At the vertical depth in a channel sloping at an angle θ with the horizontal, the pressure is given by

(*i*) $\gamma y \cos \theta$ (*ii*) $\gamma y \cos^2 \theta$ (*iii*) $\gamma y \tan \theta$ (*iv*) $\gamma y \sin \theta$.

ANSWERS

1. (*ii*), (*iv*) **2.** (*b*), (*c*) **3.** (*ii*), (*iii*), (*vii*) **4.** (*c*)

5. (*iv*) **6.** (*a*), (*v*); (*b*) (*iv*); (*c*), (*iii*); (*d*) (*ii*); (*e*), (*i*)

7. (1), (*c*); (2), (*e*); (3), (*d*); (4), (*a*); (5), (*b*) **8.** (*iii*) **9.** (*c*)

10. (*ii*) **11.** (1), (*f*); (2), (*e*); (3), (*d*); (4), (*c*); (5), (*a*); (6), (*b*)

12. (*b*) **13.** (*iii*) **14.** (*b*) **15.** (*i*)

16. (*a*), (*ii*); (*b*) (*i*); (*c*), (*iv*); (*d*), (*iii*) **17.** (*ii*).

PROBLEMS

12.1. The cross-section of an open channel consists of semi-circular bottom 1.20 m in diameter and with vertical sides. If the depth of water is 1.20 m and the bed slope is 1 in 2500, calculate the discharge. Take Chezy's C as 65 $m^{1/2}$ per sec.

12.2. A sewer pipe is laid on a slope of 1 in 3000 and is to carry 2.5 m^3/s when the pipe flows full. What size pipe should be used if n in Manning formula is 0.015 ?

12.3. Prove that the discharge carried by a hydraulically efficient trapezoidal channel having sides slopping at 1 vertical to 2 horizontal is given by $Q = 1.15\, Cb^{5/2} \sqrt{S}$, where C is the Chezy's constant, b the bottom width and S the channel slope.

12.4. Show that the discharge in a hydraulically most efficient trapezoidal channel is given by

$$Q = 0.56\, Cb^{5/2} \sqrt{S},$$

in which C is the Chezy's constant, b the bottom width and S is the channel slope.

12.5. Prove that the flow rate in a hydraulically most efficient trapezoidal channel is given by

$$Q = \frac{0.742}{n} b^{8/3} \sqrt{S}$$

here n is the Manning's n, b the bottom width, S the channel slope.

12.6. A trapezoidal channel has an area of 2.5 m^2 and side slopes of 45°. Determine the most economical section, if the channel is to be concrete lined for which Manning's $n = 0.015$. Compute the discharge if the channel slope is 1 in 1000.

12.7. Design a trapezoidal channel of most economical section with side slope of 2 vertical to 1 horizontal to give a discharge of 6 m^3/s at a velocity of 1.5 m/s.

Find the saving in power per km length of channel in using the above section in comparison with a rectangular one 3 m wide and 1.2 m deep. Use Chezy's $C = 50$.

12.8. An open channel laid to a constant slope is to be designed to convey water at a constant velocity of 1.2 m/s for all depths of flow between a minimum flowrate of 1 m^3/s and a maximum of 2 m^3/s. Derive an expression for the profile of the channel cross-section. For the minimum discharge the channel is rectangular of most economical section. Use Chezy's formula.

12.9. A channel is of parabolic section defined by $T^2 = Ky$, T being the top width, K a constant and y the depth of flow. Show that the best hydraulic section is obtained when $K = 8$, resulting in a hydraulic radius of $\frac{1}{2}y$.

12.10. Show that the Chezy's C is related to the friction factor f of Darcy-Weisbach formula by $C = \frac{8.85}{\sqrt{f}}$.

12.11. Deduce the geometrical conditions governing the form of a channel of economical section (*i*) if the side slopes are fixed and (*ii*) if the slopes can be chosen.

12.12. An open channel of trapezoidal section, 2.5 m at the base and having sides inclined at 60° to the horizontal, has a bed slope of 1 in 500. It is found that when the flow is 1.5 m^3/s the depth of water in the channel is 0.5 m. Assuming the validity of the Manning's formula, calculate the flow when the depth is 0.7 m.

12.13. A lined irrigation canal with trapezoidal cross-section has 5 m bed width, 2.5 m depth and 1.5 (*H*) : 1 (*V*) side slope. Longitudinal bed slope of the canal is 1 in 1000 and Manning's $n = 0.016$. What is the maximum carrying capacity of the canal ?

ANSWERS

12.1. 0.402 m^3/s **12.2.** 2.628 m **12.6.** $b = 0.965$ m, $y = 1.17$ m, $Q = 3.69$ m^3/s

12.7. $b = 1.873$ m, $y = 1.516$ m, $S = 1$ in 842, 9.4 kW.

12.8. $y = 0.74 \log_{10} \{x - \sqrt{x^2 + 0.1045}\} - 0.06$

Top width = 6.32 m, depth = 1.178 m.

12.12. 3.56 m^3/s **12.13.** 58.186 m^3/s.

PART B—NON-UNIFORM FLOW

Discussed in this part are the concepts of specific energy, specific force, their significance and applications, the hydraulic jump (rapidly varied flow) and its various aspects, open channel waves and surges, detailed analysis of the gradually varied flow, back water and drawdown curves, determination of the length of water surface profiles, control sections and the location of the hydraulic jump.

12.11. Specific Energy

The concept of specific energy is very useful in the solution of many open channel flow problems. The total energy of flow per unit weight of liquid given by Eq. (12.13*b*).

$$Z + y + \frac{V^2}{2g}$$

where Z is the elevation of the channel bottom above the horizontal datum, y depth of flow and V the average velocity of flow. Now, if the channel bottom itself is taken as the datum, Fig. 12.11, then the total energy per unit weight of liquid,

$$E = y + \frac{V^2}{2g} \qquad \text{...(12.14)}$$

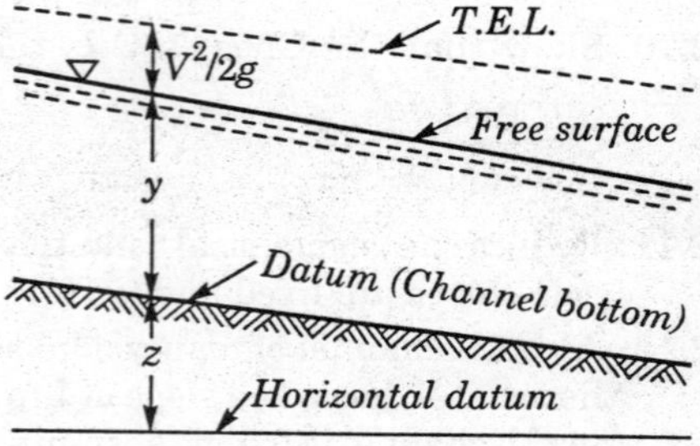

Fig. 12.11. Notation for specific energy.

This energy E is known as the *specific energy*. Thus the specific energy is the energy per unit weight of flowing liquid measured above the channel bottom and is equal to the depth of flow y plus the velocity head $V^2/2g$. It may be noted here that while the total energy decreases in the direction of flow on account of friction, the specific energy may decrease or increase or remain constant as in the case of uniform flow.

From Eq. (12.14), it is evident that the specific energy comprises of :

(1) the potential energy of flow y, and

(2) the kinetic energy of flow $\dfrac{V^2}{2g}$.

Using the continuity equation, $Q = AV$

$$E = y + \frac{Q^2}{2gA^2} \qquad \text{...(12.15)}$$

$$= f(y, Q) \quad \text{for a given channel.}$$

The area of flow A is dependent upon the channel geometry and the depth of flow, but for given channel section it will be solely dependent upon the depth of flow. The specific energy in a given channel is, therefore, a function of the depth of flow y and the rate of flow Q. The variation of the specific energy can be studied in two ways :

(1) keeping the discharge Q constant, the variation of specific energy E with depth of flow y can be obtained, and

(2) keeping the specific energy E constant, the variation of discharge Q with depth of flow y can be studied.

For the sake of simplicity, let us consider a channel of rectangular cross-section of width b. The area of flow $A = by$, and the discharge per unit width is Q/b which is normally denoted by q. Eq. (12.15) for a rectangular channel is

$$E = y + \frac{(q.b)^2}{2g(b.y)^2} = y + \frac{q^2}{2gy^2} = f(y, q) \qquad ...(12.16)$$

12.11.1. Specific Energy Curve

From Eq. (12.16) it is obvious that at low depths, the kinetic energy is very high and the potential energy low. Thus as $y \to 0$, $q^2/2gy^2 \to \infty$ and $E \to \infty$. On the other hand, for large depths the kinetic energy attains a low value and causes $E \to y$. It is, therefore, seen that as y varies from, a low value to a higher one, the specific energy first decreases attaining a minimum value at a certain depth, then increases with the depth till it is asymptotic to the potential energy line OP. This variation of E with y is depicted in Fig. 12.12 and is called *specific energy curve*. The depth of flow at which the specific energy attains a minimum value is called the *critical depth* and is denoted by subscript c, *i.e.*, y_c. The curve can be seen to be asymptotic to E-axis also.

If both y and E axes have identical scales, the potential energy line OP will make an angle of 45°. Fig. 12.12 shows specific energy curves for three different discharges Q_1, Q_2 and Q_3 such that $Q_3 > Q_2 > Q_1$. Points C_1, C_2 and C_3 represent points of minimum specific energy for the three curves respectively. As will be seen later (through use of Eq. 12.22), the line joining C_1, C_2, C_3 with the origin of axes O makes an angle α with the horizontal such that

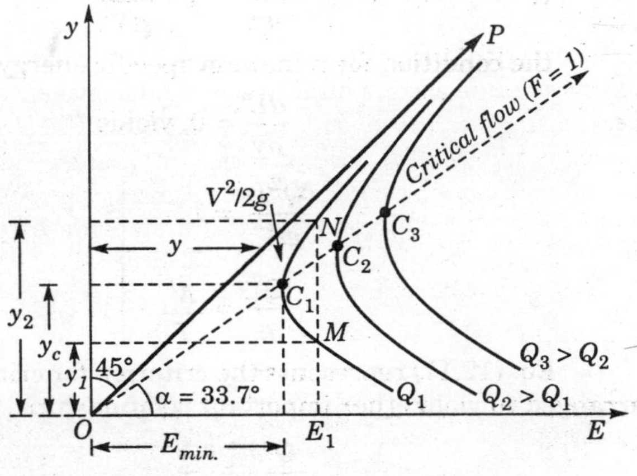

Fig. 12.12. Specific Energy Curve.

$$\tan \alpha = \frac{y_c}{E_{min}} = \frac{2}{3} ; \alpha = 33.7°C$$

In obtaining this result, the relationship between y_c and E_{min} expressed by Eq. (12.22) has been used.

The flow at the critical depth is called the *critical flow* and the velocity of flow known as the *critical velocity*. When the depth is less than the critical depth (*i.e.*, $y < y_c$) the velocity of flow is greater than the critical velocity and hence the flow is said to be *supercritical* or *rapid*. On the other hand, if the depth of flow is greater than the critical depth the velocity is less than the critical velocity and the flow is then called *subcritical* or *tranquil*.

If a vertical line MN, representing a given specific energy is drawn, it intersects the specific energy curve at two points M and N. These points represent two possible depths of flow which are known as *alternate depths*. From the specific energy curve it can be seen that for $E > E_{min}$, there are two possible depths :

(*i*) $y_1 < y_c$ (which corresponds to a point like M) represents super critical flow, and

(*ii*) $y_2 > y_c$ (corresponding to a point like N) representing subcritical flow.

Equation (12.16) is a cubic equation in y, the solution of which for a given specific energy will yield three roots-two positive and the third one being negative has no physical significance.

12.11.2. Critical Flow

The critical flow has been defined as that flow at which the specific energy is the minimum for a given discharge. Utilizing this definition, a general criterion may be arrived at as follows :

$$E = y + \frac{V^2}{2g} = y + \frac{Q^2}{2gA^2}.$$

For minimum specific energy, differentiating the expression for E with respect to y and equating $\frac{dE}{dy} = 0$.

$$\frac{dE}{dy} = 1 + \frac{Q^2}{2g}\left(-\frac{2}{A^3}\frac{dA}{dy}\right).$$

If T = top width of the channel, then

$$dA = T\,dy \text{ and } \frac{dA}{dy} = T.$$

$$\therefore \qquad \frac{dE}{dy} = 1 - \frac{Q^2T}{gA^3}$$

the condition for minimum specific energy,

i.e. $\frac{dE}{dy} = 0$, yields

$$\left.\begin{aligned} \frac{Q^2T}{gA^3} &= 1 \\ \frac{Q^2}{g} &= \frac{A^3}{T} \end{aligned}\right\} \qquad \text{...(12.17)}$$

Eq. (12.17) represents the criterion for critical flow in the channel. The equation may be re-arranged to yield other important relationships.

$$\frac{Q^2}{gA^2} = \frac{A}{T}$$

or $$\frac{V^2}{g} = D \qquad \left(D \text{ is the hydraulic depth} = \frac{A}{T}\right)$$

or $$\frac{V^2}{g} = \frac{D}{2} \qquad \text{...(12.18)}$$

from which $$\mathbf{F} = \frac{V}{\sqrt{gD}} = 1 \qquad \text{...(12.19)}$$

Eq. (12.18) states that in critical flow, the velocity head is equal to half the hydraulic depth while Eq. (12.19) shows that for critical flow the Froude number is unity.

Example 12.26. *Show that at the critical state of flow, the specific energy in a rectangular channel is equal to 1.5 times the depth of flow. Will the depth of flow be greater or less than $\frac{2}{3}$ × specific energy for the trapezoidal channel and why ?*

Solution. Rectangular channel

At critical flow, $$\frac{Q^2}{g} = \frac{A^3}{T}$$

$$\frac{V^2}{g} = \frac{A}{T} = \frac{by}{b} = y$$

The specific energy $$E = y + \frac{V^2}{2g} = y + \frac{y}{2} = \frac{3}{2}y.$$

Hence, specific energy = 1.5 × Depth of flow.

Trapezoidal channel :

$$A = y\,(b + my)\,;$$
$$T = b + 2my$$

At critical flow, $\dfrac{V^2}{g} = \dfrac{A}{T} = \dfrac{y(b+ym)}{b+2ym} < y$

$$\frac{V^2}{2g} < \frac{y}{2} \quad \text{and} \quad E = y + \frac{V^2}{2g} < \frac{3}{2}y$$

Therefore, depth of flow $y < \dfrac{2}{3} \times$ Specific energy.

Example 12.27. *A trapezoidal channel with a base of 6 m and side slopes of 2 horizontal to 1 vertical conveys water at 17 m³/sec with a depth of 1.5 m. Is the flow situation sub- or super-critical ?*

Solution. Froude Number $\mathbf{F} = \dfrac{V}{\sqrt{gD}}$

If **F** < 1, flow is subcritical, and if **F** > 1, it is supercritical.

$$A = (b + yz)\,y$$
$$= (6 + 1.5 \times 2) \times 1.5$$
$$= 13.5 \text{ m}^2$$
$$V = Q/A$$
$$= 17/13.5 = 1.259 \text{ m/sec}$$
$$T = b + 2yz$$
$$= 6 + 2 \times 1.5 \times 2 = 12 \text{ m}$$
$$D = A/T = 13.5/12 = 1.125 \text{ m}$$

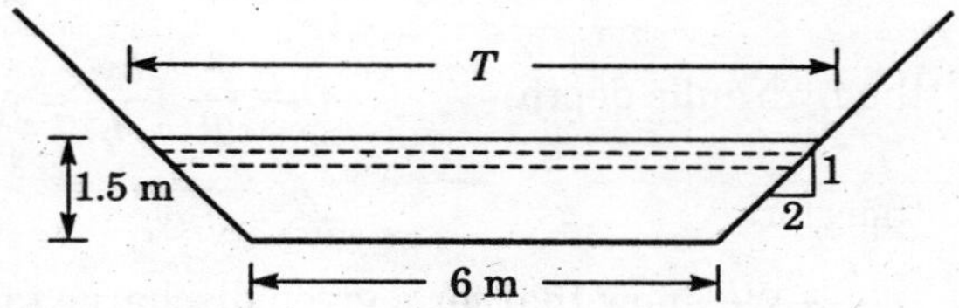

$\mathbf{F} = \dfrac{1.259}{\sqrt{9.81 \times 1.125}} = 0.379$; Since **F** < 1, **flow is subcritical.**

Example 12.28. *For a triangular channel having a vertex angle of 120°*, calculate the critical depth.

Solution. For critical flow in an open channel, the following condition must be satisfied,

$$\frac{Q^2}{g} = \frac{A^3}{T} \qquad \text{...(12.17)}$$

In a triangular channel of vertex angle 2θ,

Area of flow, $A = \dfrac{1}{2}T \,.\, y$ using the semi-vertex angle

$$\tan 60° = \frac{T/2}{y} \quad \text{or } T = 2\sqrt{3}y$$

$$A = \frac{1}{2}(2\sqrt{3}y)y = \sqrt{3}y^2$$

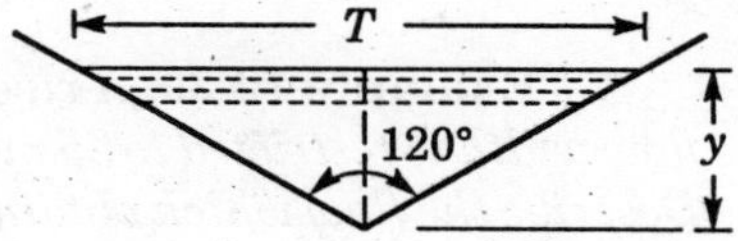

Substituting for A and T,

$$\frac{Q^2}{g} = \frac{(\sqrt{3}y^2)^3}{2\sqrt{3}y} = \frac{3}{2}y^5$$

From which, the critical depth, y_{cr}

$$\mathbf{y_{cr} = 0.584Q^{0.4}}.$$

12.11.3. Critical Flow in a Rectangular Channel

Eq. (12.17) may be used to obtain critical depth in any channel. In case of a rectangular channel of width b, carrying a discharge Q under critical conditions we obtain,

$$Q = q.b, A = by_c\ T = b$$

Eq. (12.17) gives :

$$\frac{Q^2}{g} = \frac{A^3}{T}$$

Substituting for Q, A and T,

$$\frac{(q.b)^2}{g} = \frac{(by_c)^3}{b}$$

which yields, $$y_c = \left(\frac{q^2}{g}\right)^{1/3} \quad ...(12.20)$$

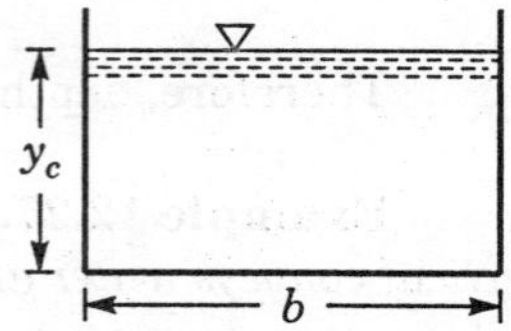

Fig. 12.13. Critical flow in a rectangular channel.

Eq. (12.20) relates the critical depth y_c and the discharge per unit width q and shows that the critical depth is independent of the channel slope.

From, Eq. (12.19), the critical velocity $V_c = \sqrt{gD}$,

the hydraulic depth $$D = \frac{A}{T} = \frac{by_c}{b} = y_c$$

therefore, $$V_c = \sqrt{gy_c} \quad ...(12.21)$$

We know that for a given discharge taking place at the critical depth, the specific energy is the minimum, thus

$$E_{min} = y_c + \frac{V_c^2}{2g} = y_c + \frac{y_c}{2} = \frac{3}{2}y_c \quad ...(12.22)$$

12.11.4. Critical Slope

Critical depth in a channel can be obtained from Eq. (12.17). The critical slope is the value of the slope (to be obtained either from Manning's or Chezy's equation) at which the uniform flow at specified rate Q occurs under critical flow conditions, *i.e.* the depth of flow becomes equal to the critical depth. Using suffix c to represent critical flow condition, the Manning's equation may be written as

$$Q = \frac{1}{n}(A_c R_c^{2/3}) S_c^{1/2} \quad ...(12.23)$$

where R_c = hydraulic radius at the critical depth y_c ; A_c is the area of flow at $y = y_c$ and S_c is the critical slope. Use of Chezy's equation would result in

$$Q = CA_c\sqrt{R_c S_c} \quad ...(12.24)$$

Example 12.29. *A river is 30 m wide and has a rectangular shape. At a bridge location the flow width is restricted to 25 m by the piers of the bridge and the river bed is approximately horizontal. Describe the flow which obtains underneath the bridge with minimum upstream depth when a flood of 450 m³/s flows in the river. Also determine the upstream depth.*

Solution. The upstream depth d will be minimum when the depth of flow underneath the bridge is critical (refer discharge curve). We know that

$$Q = 450 \text{ m}^3/\text{s}, b_1 = 30 \text{ m}, b_2 = 25 \text{ m}$$

discharge per unit width at section (2), see figure below

$$q_2 = \frac{Q}{b^2} = \frac{450}{25} = 18 \text{ m}^3/\text{s/m}$$

the critical depth at section (2),

$$y_c = \left(\frac{q_2^2}{g}\right)^{1/3} = \left(\frac{18 \times 18}{9.81}\right)^{1/3} = 3.21 \text{ m}$$

For horizontal river bed, assuming no energy loss,

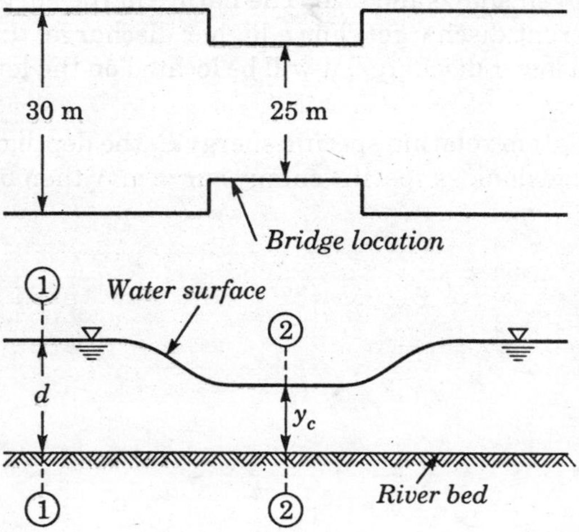

the upstream specific = energy specific energy at the contracted portion (2)

$$d + \frac{Q^2}{2gA^2} = 3.21 + \frac{V_c^2}{2 \times 9.81}$$

$$= 3.21 + \frac{3.21 \times 9.81}{2 \times 9.81}$$

or

$$d + \frac{(450)^2}{2 \times 9.81 \times (30d)^2} = 4.815.$$

Simplifying we obtain a cubic equation in d

$$d^3 - 4.815\, d^2 + 11.46 = 0.$$

Solution of which yields

$$d = 4.15 \text{ m}.$$

The flow underneath the bridge is critical.

12.12. Discharge Curve

For a constant specific energy E, if the discharge per unit width q is plotted against depth of flow y, we obtain what is known as the *discharge curve*. The discharge curve, Fig. 12.14, shows that q attains a maximum value at a particular value of y. The condition for maximum discharge q_{max} may be obtained using Eq. (12.16)

$$E = y + \frac{q^2}{2gy^2}$$

rearranging we obtain, $q^2 = 2gy^2 (E - y)$

Differentiating q with respect to y, noting that E is a constant,

$$2q \frac{dq}{dy} = 4gyE - 6gy^2$$

for maximum value of q, $\frac{dq}{dy} = 0$, and we obtain

$$E = \frac{3}{2} y \qquad \text{...(12.25)}$$

and the maximum discharge

$$q_{max}^{\ 2} = 2gy^2\left(\frac{3}{2}y - y\right)$$

$$= gy^3$$

or

$$y = \left(\frac{q_{max}^2}{g}\right)^{1/3} \qquad \text{...(12.26)}$$

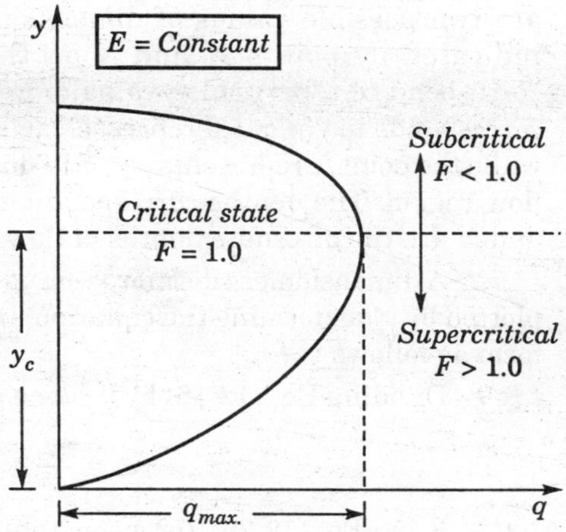

Fig. 12.14. Discharge curve.

Eqs. (12.25) and (12.26) are identical to Eqs. (12.22) and (12.20) respectively, and represent the condition for the critical flow. Hence it may be said that when the flow is critical, the discharge per unit width is the maximum for a given specific energy. It may be seen that q_{max} is equal to the critical discharge q_c.

12.13. Dimensionless Specific Energy and Discharge Curves

The specific energy curve is the graphical representation of the relationship between the specific energy E and the depth of flow y for a given discharge. For each discharge, therefore, there will be a fixed specific energy curve for a channel of given shape and size. The nature of the curve remaining same, its position will be different for different discharges. For a higher discharge, the specific energy curve will lie on the right side and for a lower discharge, it will be located on the left side of the curve as shown in Fig. 12.12.

For a rectangular channel, a dimensionless equation relating specific energy E, the depth of flow y and the critical depth y_c may be obtained. A dimensionless specific energy curve may then be plotted, which will be applicable to any value of q.

Eq. (12.16) gives $$E = y + \frac{q^2}{2gy^2}$$

dividing both sides by the critical depth y_c,

$$\frac{E}{y_c} = \frac{y}{y_c} + \frac{q^2}{2gy^2 y_c}$$

substituting for $\frac{q^2}{g} = y_c^3$, we obtain

$$\frac{E}{y_c} = \frac{y}{y_c} + \frac{1}{2}\left(\frac{y_c}{y}\right)^2 \quad \text{...(12.27)}$$

Eq. (12.27) is dimensionless, and is apparently independent of the discharge per unit width q though it is implicitly involved in y_c.

Figure 12.15 is the graphical representation of Eq. (12.27), and shows that for each value of dimensionless specific energy E/y_c greater than 1.5, there are two possible values of dimensionless depth y/y_c, indicated by points M and N on the curve (which correspond to a particular value of E/y_c). The point M corresponds to $y/y_c > 1.0$ representing subsrcitical flow, while the point N represents $y/y_c < 1.0$ in the supercritical flow region. The depths corresponding to the points M and N are the alternate depths of flow.

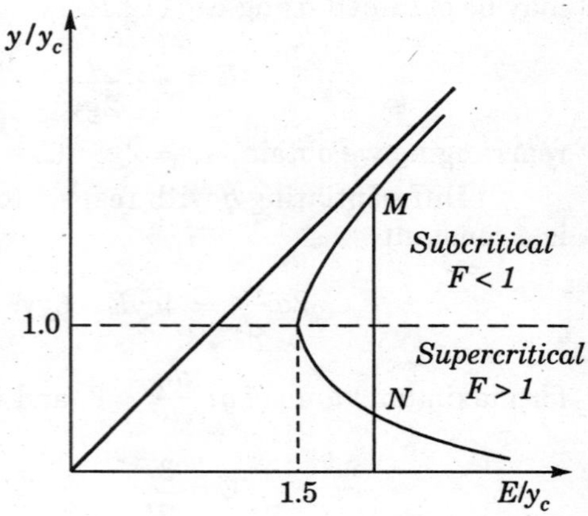

Fig. 12.15. Dimensionless specific energy curve.

A dimensionless discharge curve may be similarly plotted by transforming the equation in a dimensionless form as follows :

Dividing Eq. (12.16) by y_c^3, one gets

$$\frac{E}{y_c^3} = \frac{y}{y_c^3} + \frac{q^2}{2gy^2 y_c^3}$$

From Eq. (12.26), for a constant specific energy

$$y_c^3 = \frac{q^2_{max}}{g}$$

substituting for y_c^3 in the last term

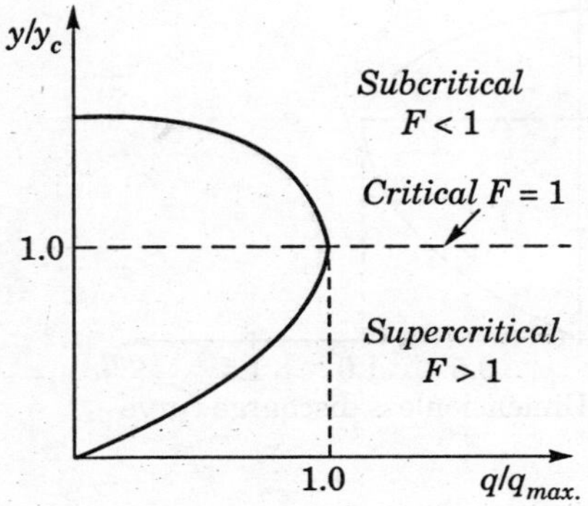

Fig. 12.16. Dimensionless discharge curve.

$$\frac{E}{y_c^3} = \frac{y}{y_c^3} + \frac{1}{2y^2}\left(\frac{q}{q_{max}}\right)^2$$

or
$$\frac{y^2}{y_c^3}E = \left(\frac{y}{y_c}\right)^3 + \frac{1}{2}\left(\frac{q}{q_{max}}\right)^2$$

For critical flow, $E = \frac{3}{2}\, y_c$, substituting for E,

$$\frac{3}{2}\left(\frac{y}{y_c}\right)^2 = \left(\frac{y}{y_c}\right)^3 + \frac{1}{2}\left(\frac{q}{q_{max}}\right)^2$$

or
$$\left(\frac{q}{q_{max}}\right)^2 = 3\left(\frac{y}{y_c}\right)^2 - 2\left(\frac{y}{y_c}\right)^3 \qquad ...(12.28)$$

Eq. (12.28) is a dimensionless equation relating the dimensionless discharge q/q_{max} with the dimensionless depth y/y_c. Figure 12.16 is a dimensionless plot of Eq. (12.28) and can be used for any value of specific energy.

Example 12.30. *Using concept of the specific energy, obtain for flow at a constant specific energy in a rectangular channel*

$$\frac{q}{q_c} = \sqrt{3\left(\frac{y}{y_c}\right)^2 - 2\left(\frac{y}{y_c}\right)^3}$$

where q_c is the critical discharge. Plot a dimensionless diagram with y/y_c as ordinate and q/q_c as abscissa.

Solution.
$$E = y + \frac{V^2}{2g} = y + \frac{Q^2}{2gA^2} = y + \frac{q^2}{2gy}$$

At constant E, for maximum q, $\frac{dq}{dy} = 0$

$$q^2 = 2gy^2\,(E - y)$$

differentiating,
$$2q\frac{dq}{dy} = 2g(2Ey - 3y^2)$$

setting,
$$\frac{dq}{dy} = 0, \quad E = \frac{3}{2}\,y$$

(which is the condition for critical flow).

But we know that when $y = y_c$ $q = q_c$,

$$E = \frac{3}{2}y_c = y + \frac{q^2}{2gy^2}$$

$$q_c = V_c y_c = \sqrt{gy_c}\,.\,y_c = y_c^{3/2}\sqrt{g}$$

$$\frac{3}{2}y_c = y + \frac{q^2 \times gy_c^3}{q_c^2 \times 2gy^2} = y + \frac{1}{2}\frac{y_c^3}{y^2}\cdot\frac{q^2}{q_c^2}$$

$$\frac{3}{2} = \frac{y}{y_c} + \frac{1}{2}\left(\frac{y_c}{y}\right)^2\frac{q^2}{q_c^2}$$

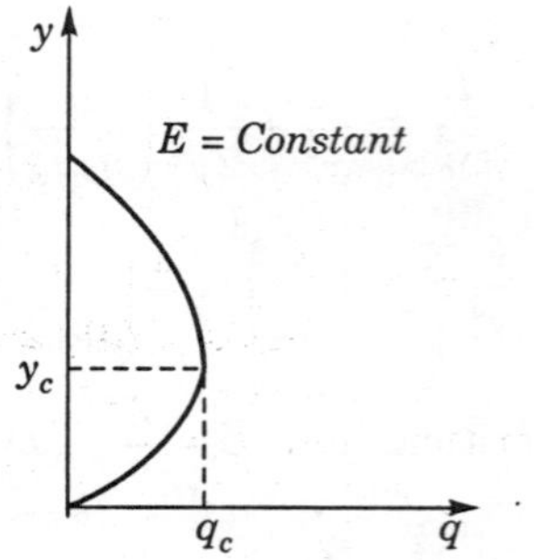

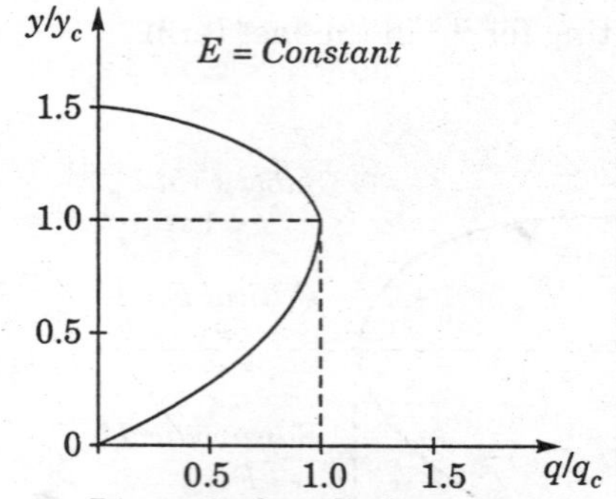

Dimensionless discharge curve

Multiplying throughout by $2\left(\frac{y}{y_c}\right)^2$

$$3\left(\frac{y}{y_c}\right)^2 = 2\left(\frac{y}{y_c}\right)^3 + \left(\frac{q}{q_c}\right)^2$$

from which
$$\frac{q}{q_c} = \sqrt{3\left(\frac{y}{y_c}\right)^2 - 2\left(\frac{y}{y_c}\right)^3}$$

It may be seen that this equation can be directly obtained by replacing q_{max} by q_c in Eq. (12.8). When, $q/q_c = 0$, the above equation has two positive roots,

(*i*) $\frac{y}{y_c} = 0$, and (*ii*) $3\left(\frac{y}{y_c}\right)^2 - 2\left(\frac{y}{y_c}\right)^3 = 0$, $\frac{y}{y_c} = \frac{3}{2} = 1.5$

A dimensionless curve q/q_c versus y/y_c may be plotted by assigning different values to y/y_c and obtaining corresponding q/q_c.

12.14. Applications of Specific Energy

The concept of specific energy finds its application for solving many open channel problems. Important among them are (*i*) analysis of flow through channel transition, (*ii*) flow over raised channel floor—broad crested weir, (*iii*) flow through sluice gate opening etc. To illustrate the use of specific energy, we shall first take up flow through a channel transition.

12.14.1. Flow through a Rectangular Channel Transition

When the width of a channel is to be reduced, it can be achieved either by suddenly contracting or by gradually reducing the width of the channel. The loss of energy in flow through a gradual contraction is much smaller than in sudden contraction. It is for this reason that in actual practice the reduction in width of the channel is made gradual. In case the size of the channel is to be enlarged, the increase in width of the channel is also obtained gradually. In either case, the portion of the channel in which the width changes gradually is known as the *transition*.

Let us consider a rectangular channel of width b_1 gradually reducing to width b_2, as shown in Fig. 12.17 (*a*). Let the flow approaching the transition be sub-critical. For a constant discharge Q, the discharge per unit width in the approach channel is $q_1 = Q/b_1$ and that in the throat portion is $q_2 = Q/b_2$. Now since $b_2 < b_1$, $q_2 > q_1$. From the discharge curve, Fig. 12.14, it can be observed that for sub-critical flow, when the discharge per unit width q increases, the depth of flow y decreases. Hence in the transition as well as in the throat the water surface will be lowered as indicated in Fig. 12.17 (*b*). The width at the throat can be reduced to such an extent (*i.e.* b_2') that results in the maximum discharge per unit width q_{max} for the given specific energy E_1, [see Fig. 12.17 (*c*) (*i*)]. In such a case, the depth of flow at the throat will be critical (y_{c2}) and the water surface profile will be

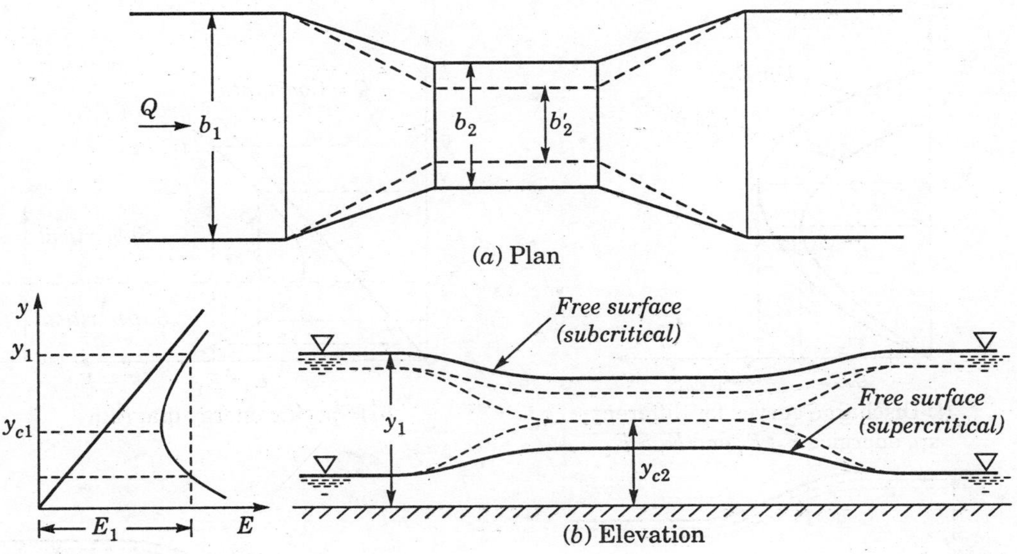

Fig. 12.17. Water Surface Profiles in Channel Transition.

as shown in Figs. 12.17 (*b*) and (*c*) (*iii*). The width b_2 is therefore the limiting width at which maximum unit discharge q_{max} passes through the throat under critical flow condition. If, however, the throat width is further reduced, that is $b_2 < b_2'$ resulting in $q_2 > q_{max}$, which is obviously not possible at the given value of specific energy E_1. The flow under these conditions necessitates an increase in the specific energy resulting in heading up of water upstream *i.e.*, the depth of flow y_1 will increase to $\bar{y}_1$ [see Fig. 12.17 (*c*) (*v*)], the increased specific energy being

$$E_1 = \bar{y}_1 + \left(\frac{Q}{b_2}\right)^2 \frac{1}{2g\bar{y}_1^2}.$$

In case the flow in the channel happens to be super-critical ($y_1 < y_{c1}$), the flow depth y_1 is the lower alternate depth corresponding to the specific energy $E_1 = y_1 + \dfrac{(Q/b_1)^2}{2gy_1^2}$. The discharge per unit width q_2 in the throat portion being greater than $q_1 = Q/b_1$, it is clearly seen from the discharge curve that for increasing q in the super-critical zone, the depth of flow also increases. Thus in the transition as well as in throat region, the water surface will rise as shown in Fig. 12.17 (*b*). For the limiting width b_2' at the throat the water surface will be shown by the dashed-line in Fig. 12.17 (*b*) and by full line in Fig. 12.17 (*c*) (*iv*), which joins the critical depth line at the throat. In case when the width at the throat $b_2 < b'_2$, the flow at the given specific energy E_1 is not possible, and the specific energy has to increase such that the discharge per unit width at throat, q_2 becomes the maximum discharge per unit width. From the specific energy curve, it can be observed that in the super-critical range of flow, for causing an increase in specific energy the depth of flow has to decrease. The water surface in the approach channel will, therefore, be lowered, Fig. 12.17 (*c*) (*vi*).

In the foregoing discussion, we have made use of the specific energy curve and the discharge diagram, and have considered three different widths at the throat : (1) b_2 (2) b'_2 and (3) width less than b'_2.

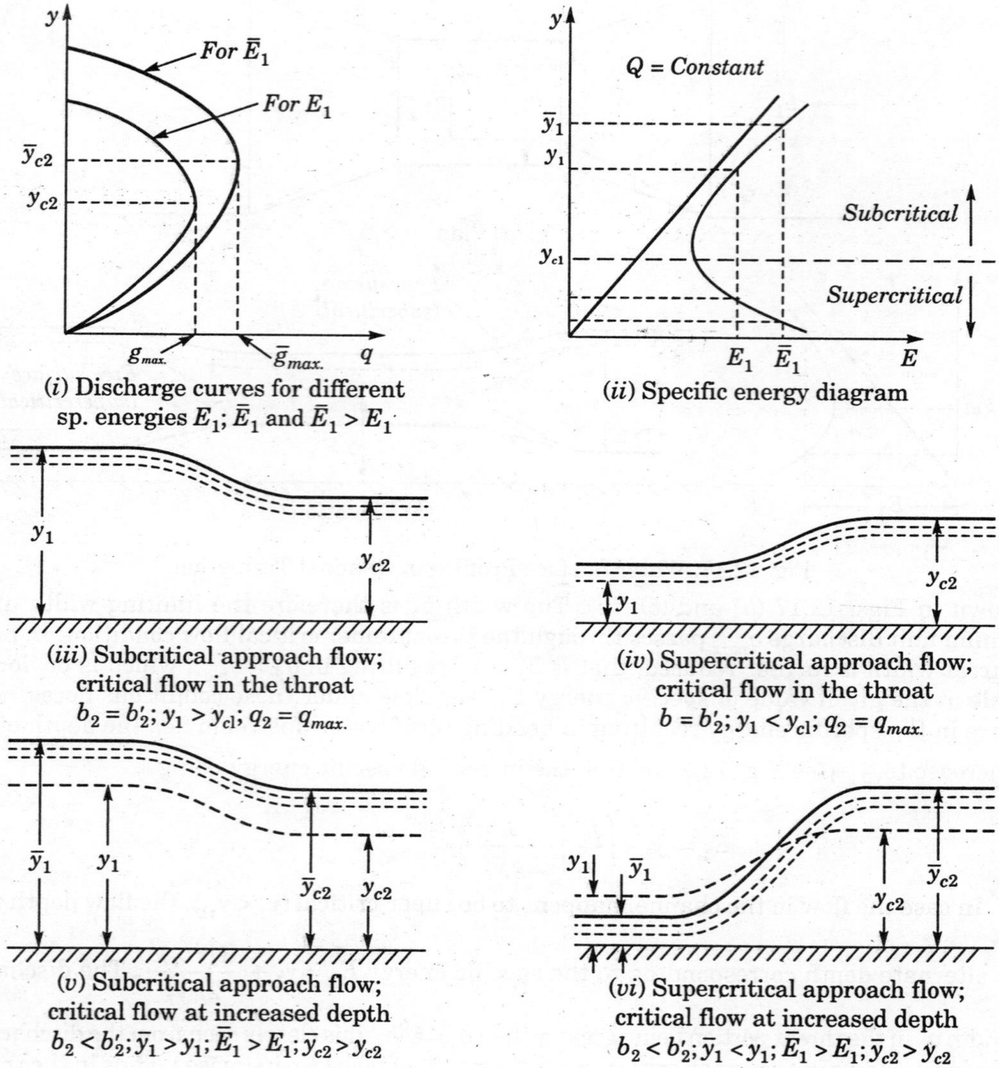

(i) Discharge curves for different sp. energies E_1; $\bar{E}_1$ and $\bar{E}_1 > E_1$

(ii) Specific energy diagram

(iii) Subcritical approach flow; critical flow in the throat $b_2 = b_2'$; $y_1 > y_{c1}$; $q_2 = q_{max.}$

(iv) Supercritical approach flow; critical flow in the throat $b = b_2'$; $y_1 < y_{c1}$; $q_2 = q_{max.}$

(v) Subcritical approach flow; critical flow at increased depth $b_2 < b_2'$; $\bar{y}_1 > y_1$; $\bar{E}_1 > E_1$; $\bar{y}_{c2} > y_{c2}$

(vi) Supercritical approach flow; critical flow at increased depth $b_2 < b_2'$; $\bar{y}_1 < y_1$; $\bar{E}_1 > E_1$; $\bar{y}_{c2} > y_{c2}$

Fig. 12.17. (c) Water surface profiles for different throat widths.

Example 12.31. *A rectangular channel 4 m wide is narrowed to 2 m width to cause critical flow in the contracted section. If the depth in this section is one metre, calculate the flow and depth in the 4 m section, neglecting energy losses in the transition. Sketch the energy line and water surface.*

Solution. At the 2 m wide section, where $y = y_c = 1$ m

$$1 = \left(\frac{q^2}{g}\right)^{1/3} \quad \text{from which,} \quad q = 3.135 \text{ m}^3/\text{s/m}$$

Flowrate $\quad Q = q.b = 3.135 \times 2 = \mathbf{6.27\ m^3/s}$

Equating the specific energies at the 4 m and 2 m wide sections,

$$y + \frac{V_1^2}{2g} = y_c + \frac{V_c^2}{2g} = y_c + \frac{y_c}{2} = \frac{3}{2}y_c = 1.5 \text{ m.}$$

From the continuity relationship

$$V_1 = \frac{Q}{b_1 y_1} = \frac{6.27}{4y_1} = \frac{1.567}{y_1}$$

Making substitution,

$$y_1 + \frac{(1.567)^2}{2\times 9.81 y_1^2} = 1.5$$

$$y_1 + \frac{0.125}{y_1^2} = 1.5$$

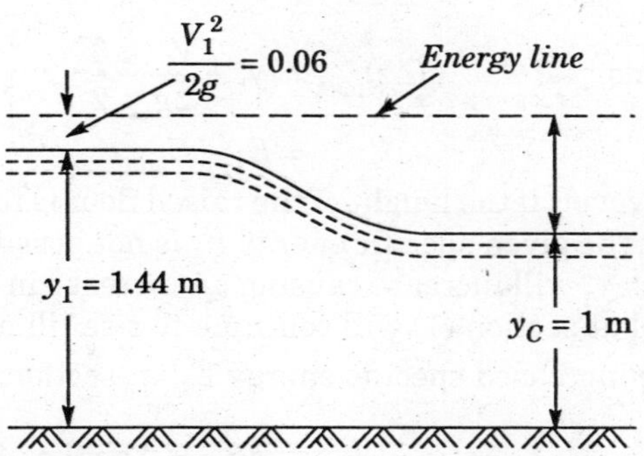

Solving by trial and error, $\mathbf{y_1 = 1.44}$ **m.**

The energy line and the water surface are shown in the figure.

12.14.2. Flow Over Local Rise in Channel Floor

Flow over a raised floor which is also sometimes called a hump is analysed by using the specific energy curve. Let us consider a rectangular channel having its bottom raised by an amount Δz, as shown in Fig. 12.18.

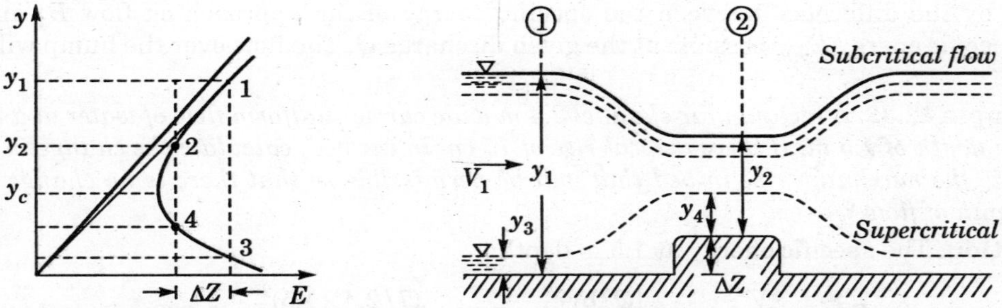

(*a*) Specific energy curve (*b*) Water surface profiles over a hump

Fig. 12.18

As the channel width remains constant, the critical depth will be same throughout. Let the channel be carrying a discharge Q under sub-critical flow conditions. The specific energy of flow approaching the raised floor is indicated by point 1 on the specific energy curve shown in Fig. 12.18 (*a*). Noting that the specific energy is measured above the channel floor, the specific energy at the raised floor is reduced by an amount Δz, where Δz is the height of the hump, and is shown by point 2 on the specific energy curve for which $y_2 < y_1$ where y_2 is the depth over the hump. The relation between energies at section (1) and (2) may be written as :

$$\left.\begin{aligned} E_1 &= E_2 + \Delta z \\ \text{or} \quad y_1 + \frac{V_1^2}{2g} &= y_2 + \frac{V_2^2}{2g} + \Delta z \end{aligned}\right\} \qquad \text{...(12.29)}$$

We have, thus, seen that if the approaching flow is sub-critical, the water surface over the raised floor or the hump is depressed. The maximum height of the hump, that is $(\Delta z)_{max}$, will be obtained when the point 2 on the specific energy curve coincides with the point representing the critical depth. The flow over the raised floor will then be critical. The expression for the $(\Delta z)_{max}$ may be written as :

$$y_1 + \frac{V_1^2}{2g} = y_c + \frac{V_c^2}{2g} + (\Delta z)_{max} = y_c + \frac{y_c}{2} + (\Delta z)_{max} = \frac{3}{2} y_c + (\Delta z)_{max}$$

which yields,
$$\left.\begin{aligned}(\Delta z)_{\max} &= y_1 + \frac{V_1^2}{2g} - \frac{3}{2}y_c \\ &= E_1 - E_{min}\end{aligned}\right\} \qquad ...(12.30)$$

However, if the height of the raised floor is further increased beyond $(\Delta z)_{max}$, *i.e.* $\Delta z' > (\Delta z)_{max}$, the flow at the given specific energy E_1 is not possible. This will result in heading up of water, that is the depth y_1 will increase causing an increase in the specific energy of the approaching flow. The water level at section (1) will continue to rise till at section (2) critical flow is obtained.

The increased specific energy E_1' at section (1) is given by

$$E'_1 = E_{\min} + \Delta' z = \frac{3}{2}y_c + \Delta z' \qquad ...(12.31)$$

where $\Delta z'$ is the height of the hump which is greater than $(\Delta z)_{max}$.

In case the flow in the channel is super-critical, the depth of flow and the specific energy of the approaching flow are represented by the point 3 on the specific curve of Fig. 12.18 (*a*). The specific energy of flow over the raised floor is decreased by Δz, the height of the hump and is shown by point 4 on the specific energy curve. It is thus seen that the depth of flow over the raised floor is greater than the depth of the approaching flow. The maximum height of hump $(\Delta z)_{max}$, will be determined by the difference between the specific energy of the approaching flow E_1 and the minimum specific energy E_{min} possible at the given discharge Q. The flow over the hump will then be critical.

Example 12.32. *A rectangular channel 2.4 m wide carries uniform flew of water at a rate of 7 m³/sec at a depth of 1.5 m. If there is local rise of 15 cm in the bed, calculate the change in water level. What is the maximum rise in bed that will be permissible so that there is no change in the upstream depth of flow ?*

Solution. The specific energy at 1.5 m depth

$$E = y + \frac{V^2}{2g} = y + \frac{(Q/by)^2}{2g} = 1.50 + \frac{(7/2.4\times1.5)^2}{2\times9.81} = 1.693 \text{ m}$$

When the channel bed is raised by 15 cm, the specific energy at the locally raised section (*i.e.* hump section) is increased as is evident from Fig. 12.18 (*a*), and is given by $E_1 = E + \Delta Z = 1.693 + 0.15 = 1.843$ m. This causes change in water depth to y_1

$$E_1 = y_1 + \frac{V_1^2}{2g} = y_1 + \frac{(Q/by_1)^2}{2g} = y_1 + \frac{(7/2.4y_1)^2}{2\times9.81} = y_1 + \frac{0.4336}{y_1^2}$$

Solving by trials, the new depth $y_1 = 1.69$ m. Therefore the change in water level

$= y_1 - y = 1.69 - 1.50 =$ **0.19 m or 19 cm.**

(*b*) **Maximum rise in bed without any change in upstream water level.**

The channel bed can be locally raised to such an extent that the flow over it becomes critical without affecting the upstream level. This means that the upstream specific energy should correspond to $y = 1.50$ m. With reference to Fig. 12.18 (*a*).

$$(\Delta Z)_{max} = E - E_{min}$$

but
$$E_{min} = \frac{3}{2}y_c\,;\ y_c = (q^2/g)^{3/2};\ q = \frac{Q}{b} = \frac{7}{24} = 2.92 \text{ m}^3/\text{s/m}.$$

$$y_c = \left[\frac{(2.92)^2}{9.81}\right]^{1/3} = 0.954 \text{ m}$$

$$E_{min} = 1.5 \times 0.954 = 1.43 \text{ m}$$

$\therefore$ $(\Delta Z)_{max} = 1.693 - 1.430 =$ **0.263 m**

The maximum rise in bed = **26.3 cm.**

Example 12.33. *A rectangular channel of 2.0 m width has a discharge of 0.4 m³/s. Find the height of the rectangular weir spanning the full width of the channel that can be used to pass this discharge while maintaining an upstream depth of 0.85 m.*

Solution. Consider Figs. (*a*) and (*b*) which fully illustrate the problem schematically.

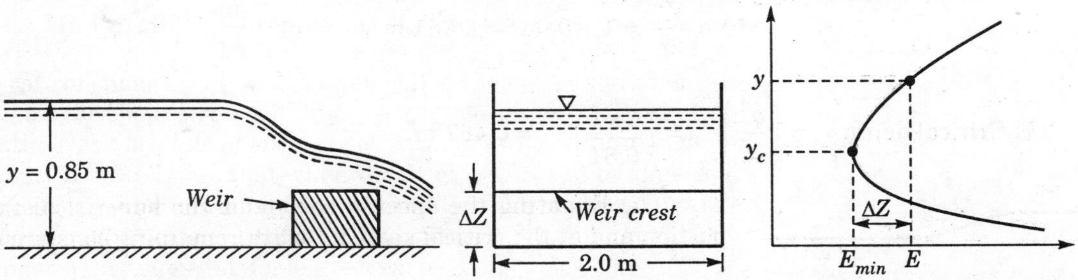

(*a*) Longitudinal section (*b*) Side view (*c*) Specific energy curve

Discharge per unit width of channel

$$q = \frac{Q}{b} = \frac{0.4}{0.2} = 0.2 \text{ m}^3\text{/s/m}$$

The critical depth, $$y_c = \left(\frac{q^2}{g}\right)^{1/3} = \left(\frac{0.2^2}{9.81}\right)^{1/3} = 0.16 \text{ m}$$

Minimum specific energy, $$E_{min} = y_c + \frac{V_c^2}{2g} = \frac{3}{2}y_c = 1.5 \times 0.16 = 0.24 \text{ m}$$

At $y = 0.85$ m, $$V = \frac{Q}{by} = \frac{0.4}{2 \times 0.85} = 0.235 \text{ m/s}$$

$$E_{0.85\text{ m}} = y + \frac{V^2}{2g} = 0.85 + \frac{(0.235)^2}{2 \times 9.81} = 0.8528 \text{ m}$$

Height of the rectangular weir ΔZ can be determined using the specific energy curve shown in Fig. (*c*).

$$\Delta Z = E_{0.85\ m} - E_{min} = 0.8528 - 0.24 = \mathbf{0.6128\ m.}$$

Example 12.34. *In a flume 1 m wide water flows at a depth of 1 m with a velocity 1 m/s. The flow is assumed to be free of losses. Find : (a) discharge in m³/s, (b) Froude number, (c) Is the flow sub-critical or super-critical, why ? (d) Reynolds number, ν = 1.01 × 10⁻⁶ m²/s, (e) Is the flow laminar or turbulent, why ? (f) Velocity head in metres, (g) Energy head from the flume bottom in metres, (h) Critical depth, (i) How far the bottom level has to be raised (Z in metre) so that the flow becomes critical (sketch it), (j) How far the width of the flume has to be reduced (symmetrically keeping the bottom line unchanged, horizontal and straight) so that the flow becomes critical, i.e. reduced width W′(k). Find a relation between Z and W so that the flow is critical.*

Solution. (*a*) Discharge $Q = AV = 1 \times 1 \times 1 = 1$ m³/s.

(*b*) Froude number $$\mathbf{F} = \frac{V}{\sqrt{gD}} = \frac{V}{\sqrt{gA/T}} = \frac{1}{\sqrt{9.81 \times 1}} = 0.3175.$$

(*c*) Flow is sub-critical since $\mathbf{F} < 1$.

Reynolds number $$\mathbf{R} = \frac{VR_h}{\nu} = \frac{V(A/P)}{\nu} = \frac{1 \times (1 \times 1)}{(1 + 2 \times 1) \times 1.01 \times 10^{-6}} = 3.3 \times 10^5$$

(*e*) Flow is turbulent since $\mathbf{R} > 500$.

(*f*) Velocity head in metre $= \dfrac{V^2}{2g} = \dfrac{1\times 1}{2\times 9.81} = 0.051$ m.

(*g*) Energy head (with flume bottom as datum)

$$= y + \frac{V^2}{2g} = 1 + 0.051 = 1.051 \text{ m.}$$

(*h*) Critical depth $y_c = \left(\dfrac{q^2}{g}\right)^{1/3} = \left(\dfrac{1\times 1}{9.81}\right)^{1/3} = 0.467$ m.

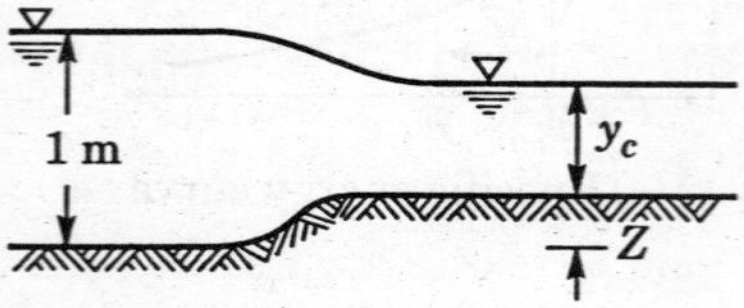

(*i*) Equating the specific energy at the subcritical flow portion and at the critical section (width remaining constant).

$$1 + 0.051 = \frac{3}{2}\times 0.467 + Z$$

from which $\qquad Z = 0.35$ m.

(*j*) Equating the specific energy at the sub-critical and critical sections (critical flow obtained by reducing the width). The critical depth in the contracted portion will be higher than the value determined in (*h*).

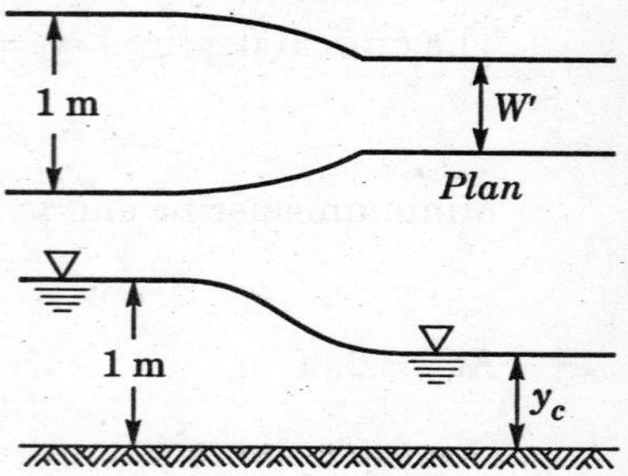

$$1.051 = y_c + \frac{V_c^2}{2g} = \frac{3}{2}y_c$$

$$\therefore \qquad y_c = \frac{2}{3}\times 1.051 = 0.7 \text{ m}$$

but in a rectangular channel $y_c = \left(\dfrac{q^2}{g}\right)^{1/3}$

$$q^2 = gy_c^{\,3} = 3.36$$

But $\qquad q = 1.832 \text{ m}^2/\text{s/m}$

$$Q = q\,.W'$$

$$W' = \frac{Q}{q} = \frac{1}{1.832} = 0.546 \text{ m.}$$

(*k*) Equating the specific energies at the sub-critical and critical sections :

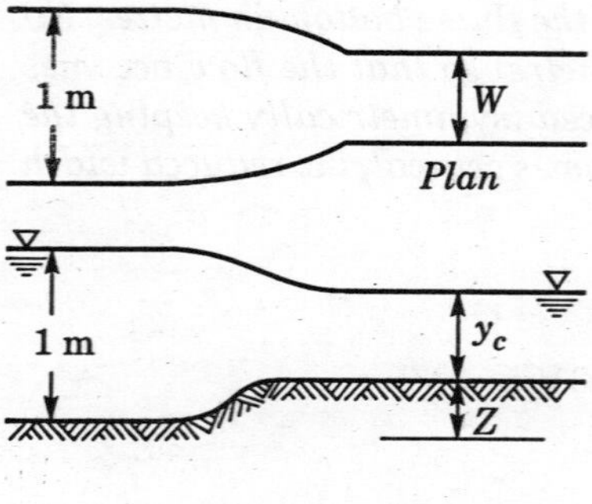

$$1 + 0.051 = Z + y_c + \frac{V_c^{\,2}}{2g} = Z + \frac{3}{2}y_c \qquad \ldots(1)$$

But $\qquad y_c = \left(\dfrac{q^2}{g}\right)^{1/3} = \left(\dfrac{Q^2}{W^2 g}\right)^{1/3}$

Substituting for y_c in (1)

$$1.051 - Z = \frac{3}{2}\left(\frac{Q^2}{W^2 g}\right)^{1/3} = \frac{3}{2}\left(\frac{1\times 1}{W^2\times 9.81}\right)^{1/3} = \frac{0.70}{(W)^{2/3}}$$

The required relation is : $\qquad 1.051 - Z = \dfrac{0.70}{(W)^{2/3}}$.

Example 12.35. *Uniform flow occurs at a depth of 1.5 m in a long rectangular channel 3 m wide and laid to a slope of 0.0009. If Manning's n = 0.015, calculate (a) maximum height of hump on the floor to produce critical depth (b) the width of contraction which will produce critical depth without increasing the upstream depth of flow.*

Solution. The flowrate $Q = A \,.\, \frac{1}{n} R^{2/3}\, S_0^{1/2}$

$$= 3 \times 1.5 \times \frac{1}{0.015} \times \left(\frac{4.5}{3+2\times1.5}\right)^{1/3} (0.0009)^{1/2} = 7.43 \text{ m}^3/\text{s}.$$

$$q = \frac{7.43}{3} = 2.477 \text{ m}^3/\text{s/m} \; ; y_c = \left(\frac{q^2}{g}\right)^{1/3} = \left\{\frac{(2.477)^2}{9.81}\right\}^{1/3} = 0.855 \text{ m}.$$

(*a*) Equating the specific energies upstream and at the hump

$$1.5 + \frac{(7.43/4.5)^2}{2\times9.81} = \Delta Z + y_c + \frac{V_c^2}{2g}$$

or $$1.5 + 0.139 = \Delta Z + 0.855 + \frac{0.855}{2} = \Delta Z + 1.2825$$

from which $$\Delta Z = 1.639 - 1.2825 = 0.3565 \text{ m}$$

The height of hump $= 0.3565$ m.

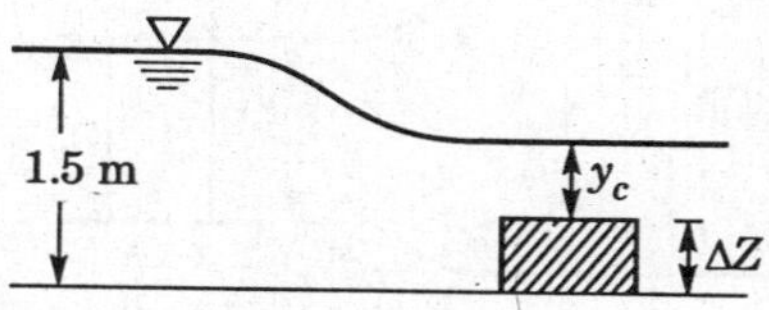

(*b*) Let *b* be the width at the contracted portion to produce critical depth.

Upstream specific energy = Specific energy at the contracted portion

$$1.639 = y_c + \frac{V_c^2}{2g} = \frac{3}{2} y_c = \frac{3}{2}\left(\frac{Q^2}{b^2\, g}\right)^{1/3} = \frac{3}{2}\left[\frac{(7.43)^2}{b^2 \times 9.81}\right]^{1/3}$$

$$\frac{(7.43)^2}{b^2 \times 9.81} = (1.092)^3$$

from which, $b = 1.98$ m.

Example 12.36. *A stream has a width of 30 m, depth of 3 m and a mean velocity of 1.25 m/sec. Find the height of a weir to be built on the stream floor to raise the water level by 1 m. Assume the discharge coefficient as 0.95.*

Solution. Discharge in the stream

$$Q = A\,.V = (30 \times 3) \times 1.25 = 112.5 \text{ m}^3/\text{s}$$

Velocity in the stream behind the weir where the depth is 4 m

$$V_1 = \frac{Q}{A_1} = \frac{112.5}{30\times4} = 0.9375 \text{ m/s}$$

Let the height of the weir be ΔZ.

The head over the weir (rectangular) to pass flowrate of 112.5 m³/s ;

$$Q = \frac{2}{3} C_d \sqrt{2g} B H^{3/2}$$

$$112.5 = \frac{2}{3} \times 0.95 \times \sqrt{2 \times 9.81} \times 20H^{3/2}$$

from which, $H = 1.2135$ m

Average velocity over the weir, $V = \dfrac{Q}{BH} = \dfrac{112.5}{30 \times 1.2135} = 3.09$ m/s

Equating the specific energy in the stream with the one at the wire section

$$y_1 + \frac{V_1^2}{2g} = \Delta Z + H + \frac{V^2}{2g}$$

or $$4 + \frac{(0.9375)^2}{2 \times 9.81} = \Delta Z + 1.2135 + \frac{(3.09)^2}{2 \times 9.81}$$

from which, ΔZ = **2.345 m.**

Therefore, the height of the weir to raise water level upstream by one metre is **2.345 m.**

Example 12.37. *A 3.6 m wide rectangular channel carries water to a depth of 1.8 m. In order to measure the discharge, the channel width is reduced to 2.4 m and a hump of 0.30 m height is provided in the bottom. Calculate the discharge if water surface in the contracted section drops by 0.15 m. Assume no losses.*

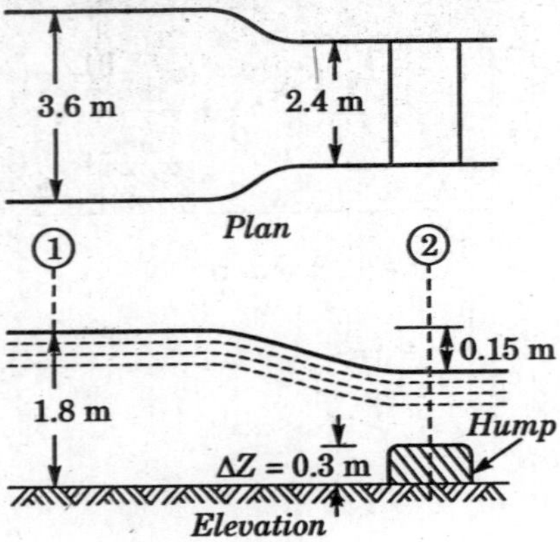

Solution. Applying continuity equation

$$Q = A_1V_1 = A_2V_2$$

$$\therefore \quad V_1 = \frac{Q}{A_1}; V_2 = \frac{Q}{A_2}$$

Equating the specific energies at the two sections

$$y_1 + \frac{V_1^2}{2g} = y_2 + \frac{V_2^2}{2g} + \Delta Z$$

or $$y_1 + \frac{Q^2}{2gA_1^2} = y_2 + \frac{Q^2}{2gA_2^2} + 0.30$$

$$A_1 = 3.6 \times 1.8 = 6.48 \text{ m}^2$$

$$y_2 = 1.8 - 0.3 - 0.15 = 1.35 \text{ m}$$

$$A_2 = 2.4 \times 0.35 = 3.24 \text{ m}^2$$

Substituting the values,

$$1.8 + \frac{Q^2}{2 \times 9.81 \times (6.48)^2} = 1.35 + \frac{Q^2}{2 \times 9.81 \times (3.24)^2} + 30$$

Simplifying and solving for Q, **Q = 6.418 m³/s.**

Example 12.38. *Water flows at a velocity of 1 m/s and a depth of 2 m in an open channel of rectangular cross-section, 3 m wide. At a certain section the width is reduced to 1.8 m and the bed is raised by 0.65 m. Will the upstream depth be affected ? If so, to what extent ?*

Solution. Specific energy at section (1)

$$E_1 = y_1 + \frac{V_1^2}{2g} = 2 + (1)^2/2 \times 9.81 = 2.051 \text{ m}$$

$$Q = A.V = b_1y_1V_1 = 3 \times 2 \times 1 = 6 \text{ m}^3/\text{s}$$

$$q_1 = \frac{Q}{b_1} = \frac{6}{3} = 2 \text{ m}^3/\text{s/m}$$

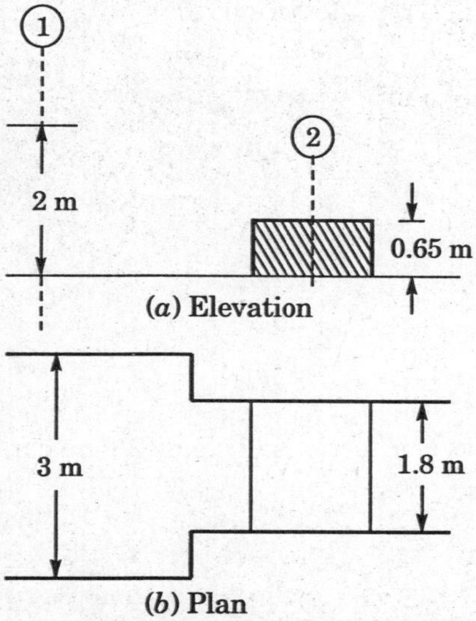

$$(y_c)_1 = \left(\frac{q_1^2}{g}\right)^{1/3} = \left(\frac{2 \times 2}{9.81}\right)^{1/3} = 0.7415 \text{ m.}$$

Since $y_1 > y_c$, the flow in the channel is sub-critical.

$$(E_{min})_1 = \frac{3}{2}\,(y_c)_1 = 1.5 \times 0.7415 = 1.1122 \text{ m.}$$

At the contracted and humped section

$$q_2 = \frac{Q}{b_2} = \frac{6}{1.8} = 3.333 \text{ m}^3\text{/s/m}$$

$$(y_c)_2 = \left(\frac{q_2^2}{g}\right)^{1/3} = \left(\frac{3.333 \times 3.333}{9.81}\right)^{1/3} = 1.0423 \text{ m}$$

$$(E_{min})_2 = \frac{3}{2}\,(y_c)_2 = 1.5 \times 1.0423 = 1.5634 \text{ m.}$$

Specific energy w.r.t. channel bed at section (2)

$$E_2 = (E_{min})_2 + \Delta Z = 1.5634 + 0.65 = 2.2134 \text{ m.}$$

Since $E_2 > E_1$, the upstream depth will be affected. The flow will be possible only when the upstream water level is increased such that

$$E_1 = E_2$$

or $$y_1 + \frac{V_1^2}{2g} = 2.2134 \quad \text{...(1)}$$

From the continuity equation

$$Q = b_1 y_1 V_1$$

or $$6 = 3\,y_1 V_1 \quad \therefore \quad V_1 y_1 = 2 \quad \text{...(2)}$$

From (1) and (2), $y_1 + \dfrac{(2/y_1)^2}{2g} = 2.2134$

or $$y_1 + \frac{0.204}{y_1^2} = 2.2134$$

Solving by trial and error, $y_1 = 2.17$ m. The water level on the upstream side will be headed up by **17 cm.**

Example 12.39. *A trapezoidal channel of base width 6 m and side slopes of 2 H to 1 V carries a flow of 60 m³/s at a depth of 2.5 m. There is a smooth transition to a rectangular section 6 m wide accompanied by a gradual lowering of the channel bed by 0.6 (i) Find the depth of water in the rectangular section and the change in water surface level. (ii) In case the drop in water surface level is to be restricted to 0.3 m, what is the amount the bed must be lowered ? (Assume no losses)*

Solution. The problem is illustrated by means of the following sketches :

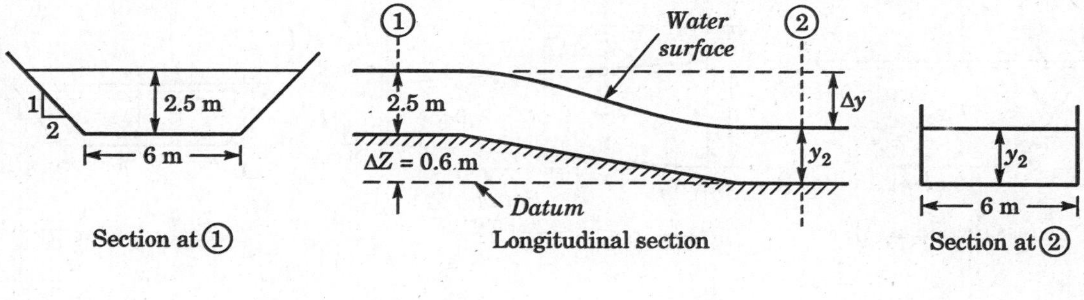

The specific energy in the trapezoidal section of the channel, *i.e.*, at section (1) is

$$E_1 = y_1 + V_1^2/2g$$

Area of flow, $A_1 = (b + my).y = (6 + 2 \times 2.5)2.5 = 27.5 \text{ m}^2$

Velocity of flow, $V_1 = \dfrac{Q}{A_1} = \dfrac{60}{27.5} = 2.18 \text{ m/s}$

Hence, $E_1 = 2.5 + (2.18)^2/(2 \times 9.81) = 2.7426 \text{ m}$

At section (2) : $A_2 = 6y_2 \; ; \; V_2 = \dfrac{Q}{A_2} = \dfrac{60}{6y_2} = \dfrac{10}{y_2}$

$$E_2 = y_2 + \frac{V_2^2}{2g} = y_2 + \frac{(10/y_2)^2}{2 \times 9.81} = y_2 + \frac{5.097}{y_2^2}$$

From energy considerations, neglecting loss of energy between sections (1) and (2) one can write

$$E_1 + \Delta z = E_2 \quad \text{...(1)}$$

$$2.7426 + 0.6 = y_2 + \frac{5.097}{y_2^2} \quad \text{or} \quad y_2^2 - 3.3426\, y_2 + 5.097 = 0$$

Solving by trials, $y_2 =$ **1.85 m.**

Change in water surface level

$$\Delta y = \Delta z + y_1 - y_2 = 0.6 + 2.5 - 1.85 = \mathbf{1.25\ m.}$$

(*ii*) *When the drop in water surface level is to be restricted to 0.3 m :*

$$\Delta y = 0.3, \text{ to find } \Delta z$$

$$\Delta y = \Delta z + y_1 - y_2$$

or

$$0.3 = \Delta z + 2.5 - y_2$$

$$\Delta z = y_2 - 2.2 \quad \text{...(2)}$$

Substituting for Δz in (1),

$$2.7426 + (y_2 - 2.2) = y_2 + \frac{5.097}{y_2^2}$$

$$y_2^2 = 9.393, \text{ resulting in } y_2 = 3.065 \text{ m}$$

Amount by which bed must be lowered, *i.e.*, $\Delta z = y_2 - 2.2 = 3.065 - 2.2 =$ **0.8649 m.**

Example 12.40. *A trapezoidal channel has a bed width of 10 m and side slopes of 2 horizontal to 1 vertical. At a section 200 m downstream the bed rises by 8 cm gradually, bed width increases to 15 m and side slopes become 3 horizontal to 1 vertical, rise in bed level, increase in bed width and flattening of side slopes are gradual. Flow rate in the channel is 200 m³/sec and the flow depth at the upstream section is 7 m. If the Manning n = 0.035, estimate the depth at the downstream section.*

Solution. The problem can be schematically represented by the following figure :

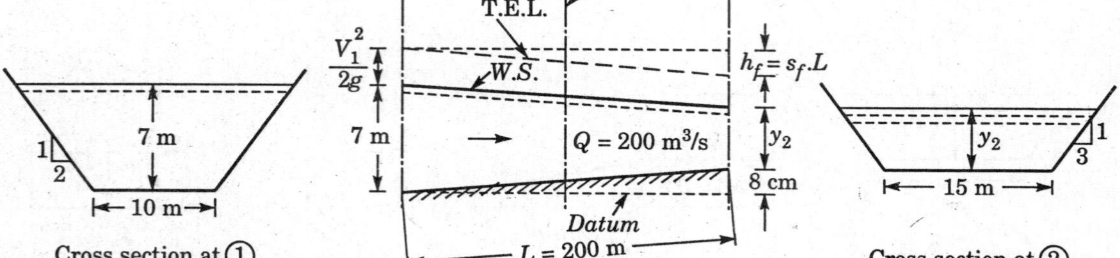

Cross section at ①

Cross section at ②

At the beginning of the channel, i.e. at section (1)

$$A_1 = \frac{1}{2}(10 + 38) \times 7 = 168 \text{ m}^2$$

$$V_1 = \frac{Q}{A_1} = \frac{200}{168} = 1.19 \text{ m/s}$$

$$\frac{V_1^2}{2g} = \frac{(1.19)^2}{2 \times 9.81} = 0.072 \text{ m} ; R_1 = \frac{A_1}{P_1} = \frac{168}{10 + 2\sqrt{7^2 + 14^2}} = 4.068 \text{ m}$$

Assuming $V_2 = V_1 = 1.19$ m/s at the 200 m *d/s* section *i.e.*, section (2).
Head loss in 200 m length of channel, assuming $R_2 = R_1 = 4.068$ m

$$V_2 = \frac{1}{n} R_2^{2/3} S_f^{1/2} \text{ or } 1.19 = \frac{1}{0.035} \times (4.068)^{2/3} S_f^{1/2}$$

or $$S_f = 0.000268 \text{ (or 1/3726)}$$

$$h_f = L \,.\, S_f = 200 \times \frac{1}{3726} = 0.0537 \text{ m}$$

Applying Bernoulli's equation to sections (1) and (2), with datum as indicated.

$$y_1 + \frac{V_1^2}{2g} = z_2 + \frac{V_2^2}{2g} + y_2 + h_f$$

$$7.0 + 0.072 = 0.08 + 0.072 + y_2 + 0.0537$$

As a first approximation, $y_2 = 6.87$ m.
With this depth at 200 m section, the depth at the mid-section, *i.e.*, at 100 m section,

$$y = \frac{y_1 + y_2}{2} = \frac{7.0 + 6.87}{2} = 6.93 \text{ m}$$

Width at the mid-section $= \dfrac{10 + 15}{2} = 12.5$ m

Side slopes at the mid-section $= \dfrac{2 + 3}{2} = 2.5$ horizontal to 1 V

Area at mid-section, $A = (12.5 + 47.15) \times 6.93 = 206.72 \text{ m}^2$

Velocity, $V = \dfrac{Q}{A} = \dfrac{200}{206.72} = 0.967 \text{ m/s}, \dfrac{V^2}{2g} = 0.0467 \text{ m}$

Hydraulic radius, $R = \dfrac{A}{P} = \dfrac{206.72}{12.5 + 2\sqrt{(6.93)^2 + (17.35)^2}} = 4.15 \text{ m}$

Friction slope, $S_f = \dfrac{V^2 n^2}{R^{4/3}} = \dfrac{(0.967)^2 (0.035)^2}{(4.15)^{4/3}} = 0.0001723$

Head loss in the channel, assuming the average friction slope

$$h_f = L \,.\, S_f = 200 \times 0.0001723 = 0.03446 \text{ m}$$

From the Bernoulli,s equation applied to sections (1) and (2),

$$y_1 + \frac{V_1^2}{2g} = z_2 + y_2 \frac{V_2^2}{2g} + h_f$$

$$7.0 + 0.072 = 0.08 + y_2 + 0.0467 + 0.03446$$

from which $$y_2 = \mathbf{6.91 \text{ m}},$$

Therefore, the estimated depth at the downstream section is **6.91 m.**

12.14.3. Flow Over a Broad Crested Weir

Consider Fig. 12.19 which shows a weir with broad horizontal crest extending to the full width of the channel. The height of the weir above the channel bottom is sufficiently large so that the area of approaching flow is much large as compared to the area of flow over the weir. The upstream edge of the weir is well rounded so as to prevent separation of flow and consequently to eliminate eddy formation near upstream edge thereby minimizing the loss of energy. Let the approaching flow be sub-critical ($y_1 > y_c$). As the flow passes over the weir, its surface is lowered to a depth y_2. If the height of the weir Z and the length of the crest (weir dimension in the direction of flow) are sufficiently large, and the friction is negligible, the water surface becomes parallel to the crest. If $Z \geq (E_1 - E_{min})$, the flow over the weir will be critical, that is $y_2 = y_c$. For a rectangular channel, we know that there exists a relationship between the critical depth y_c and the discharge per unit width, *i.e.*, Equation (12.20).

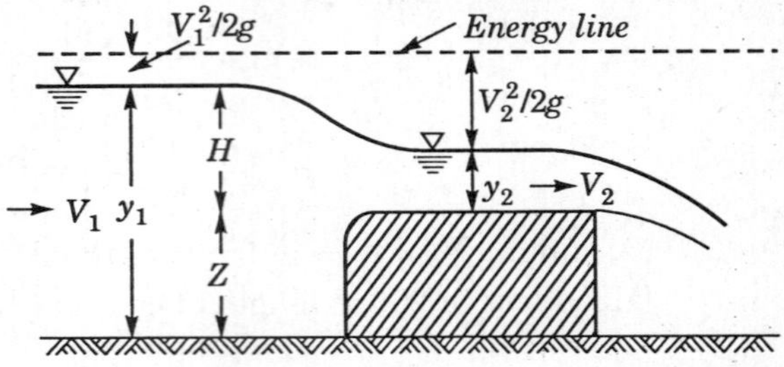

Fig. 12.19. Flow over a broad cast weir.

$$y_c = \left(\frac{q^2}{g}\right)^{1/3}$$

From this equation, the rate of flow can be determined by a single observation of critical depth y_c. Thus a broad crested weir is a useful device for measuring discharge in a rectangular channel.

For critical flow over the weir, the height of weir crest must satisfy the relationship

$$Z \geq (E_1 - E_{min}) \quad \text{...(12.32)}$$

From the concept of specific energy, we can write

$$E_1 = y_1 + \frac{V_1^2}{2g} = y_2 + \frac{V_2^2}{2g} + Z$$

from which, $$\frac{V_2^2}{2g} = E_1 - Z - y_2 = H - y_2,$$

since $$H = y_2 + \frac{V_2^2}{2g}$$

therefore, $$V_2 = \sqrt{2g(H - y_2)}$$

where H is the height of energy line above the weir crest. The discharge over the weir crest

$$Q = by_2V_2,$$

where b = width of the rectangular channel.

Substituting for V_2, $$Q = by_2\sqrt{2g(H - y_2)} \quad \text{...(12.33)}$$

Eq. (12.33) expresses the general discharge relationship and is applicable when Eq. (12.32) does not hold

$$q = \frac{Q}{b} = y_2\sqrt{2g(H - y_2)} = \sqrt{2g(Hy_2^2 - y_2^3)}$$

For a given discharge Q and a fixed height of the weir Z (H is fixed), the condition for maximum q is $dq/dy = 0$ differentiating the expression for q^2 with respect to y_2.

$$2q\frac{dq}{dy_2} = 2g(2Hy_2 - 3y_2^2),$$

and putting $dq/dy_2 = 0$

we obtain, $H = 3/2\, y_2$

Substituting for H, $q = \sqrt{2g\left(\frac{3}{2}y_2^3 - y_2^3\right)} = \sqrt{gy_2^3}$

which yields, $$y_2 = \left(\frac{q^2}{g}\right)^{1/3} \quad ...(12.34)$$

Eq. (12.34) shows that the depth of flow y_2 becomes the critical depth when maximum discharge passes over the weir. Substituting for y_2 in terms of H in Eq. (12.33), we get maximum discharge

$$Q = b\frac{2}{3}H\sqrt{2g\left(H - \frac{2}{3}H\right)} = \frac{2}{3\sqrt{3}}\,b\sqrt{2g}H^{3/2}$$

$$= 1.705\; b\; H^{3/2} \text{ (in M.K.S. and SI units)} \quad ...(12.35)$$

In Eq. (12.33), H is the energy of flow measured above the weir crest. If the velocity of approach $V_1^2/2g$ is negligible, then H becomes the depth of water to the weir crest. Eqs. (12.33) and (12.35) do not take into account the effects of friction, and curvature of streamlines, which reduce the rate of flow below the value given by the respective equations. The coefficient of discharge C_d is, therefore, used to account for these effects. Thus, the actual discharge may be expressed as

$$Q = 1.705\; C_d\; bH^{3/2} \quad ...(12.36)$$

The length of weir crest should be greater than three times the maximum head under which the weir is to operate, *i.e.*, $L \geq 3H_{max}$. This is to ensure : (1) the existence of streamlines parallel to the crest, and (2) that the flow is influenced not only by the curvature of streamlines but also by boundary resistance. Thus in a board crested weir, both accelerative and viscous effects are equally important in influencing the flow.

12.14.4. Flow in a Venturiflume

A venturiflume is essentially an artificial constriction in a channel which by producing a change in velocity and depth, facilitates the measurement of flowrate. It is usually constructed in concrete masonry, steel plate or timber, and the shape of its boundaries is most important. A venturiflume in its conventional form consists of a bell mouthed entry, parallel throat, and downstream diverging portion, Fig. 12.20.

The afflux or ponding back, h caused by the constriction is relatively small, and this is a great asset when the fall available is strictly limited as in case of irrigation canals in plains. A venturiflume provides a clear passage for the liquid to flow without causing deposition of the suspended material or blocking the bed material carried by the flow. Only a limited protection against scour is required as the regime of flow is not greatly disturbed. The venturiflume, which is also sometimes known as throated flume, is also well suited to measurement of flow in streams, small rivers and purification plants. The discharge capacity ranges from a few litres/second to about 25 m^3/s.

If the flow is sub-critical throughout as in Fig. 12.20 (*c*), the flume is said to be operating under submerged condition. This operating condition is obtained at exceptionally high discharges the measurement of which involves two liquid-level measurements (y_1 and y_2). Moreover, the surface at the throat tends to be unstable so that y_2 and the small difference $(y_1 - y_2)$ cannot be accurately determined, refer Eq. (12.43). For the flume to be a convenient and efficient flow measuring device a direct relationship between discharge Q and a single depth (y_1) is desirable. This is possible if the flow in some portion is super-critical or free, Fig. 12.20 (*b*), that is to say it is unaffected by conditions downstream, in which case it is only necessary to measure to upstream depth y_1. The is achieved by proportioning the throat, such that a hydraulic jump is formed just downstream of the throat, the channel bed remaining horizontal. A Venturiflume designed to operate under the latter conditions (that is when a hydraulic jump is ensured), is called a *standing-wave flame*.

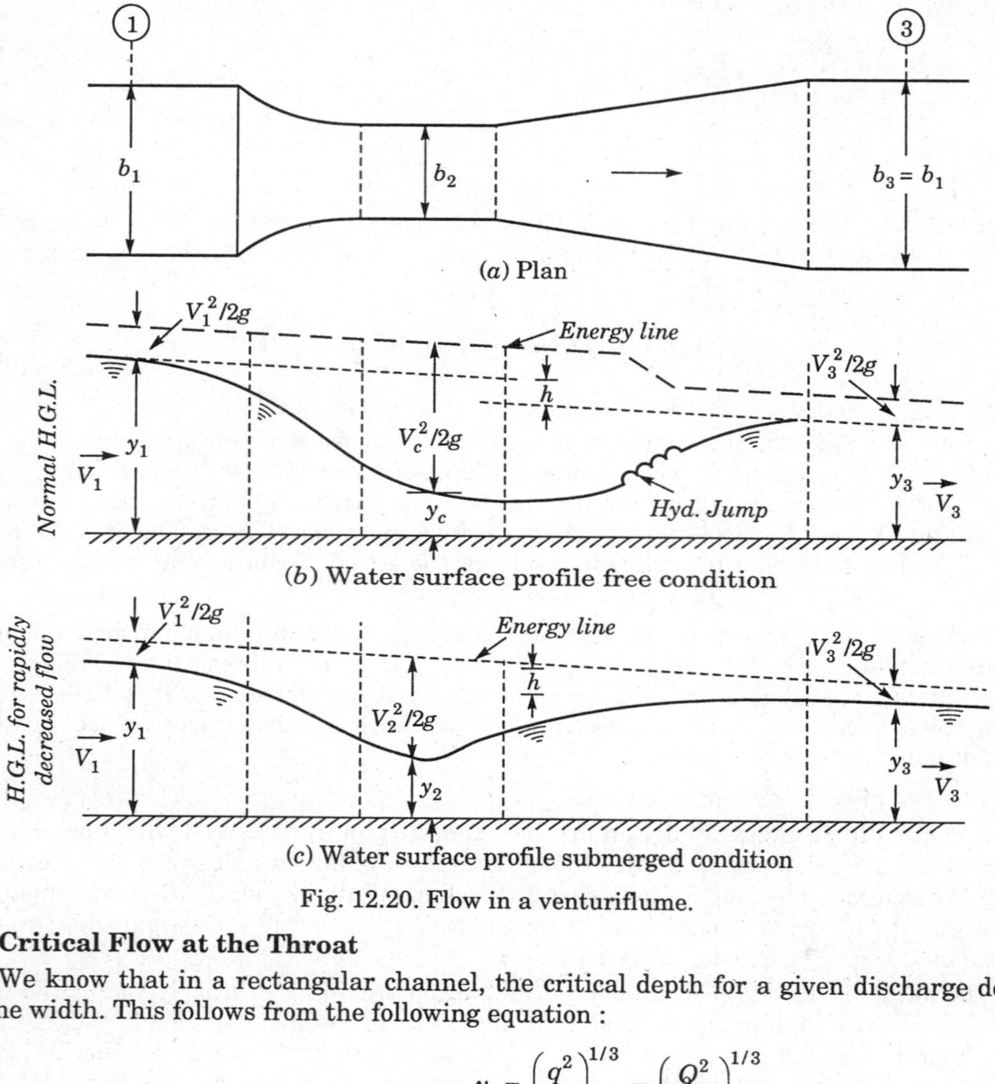

Fig. 12.20. Flow in a venturiflume.

Critical Flow at the Throat

We know that in a rectangular channel, the critical depth for a given discharge depends upon the width. This follows from the following equation :

$$y_c = \left(\frac{q^2}{g}\right)^{1/3} = \left(\frac{Q^2}{gb^2}\right)^{1/3}$$

At section (1) critical depth is given by $y_{c1} = \left(\frac{Q^2}{gb_1^2}\right)^{1/3}$

and at the throat section (2), it is $$y_{c2} = \left(\frac{Q^2}{gb_2^2}\right)^{1/3}$$

Now since $b_1 > b_2, y_{c2} > y_{c1}$, this indicates that the flow in the throat can be critical while the approaching flow remains sub-critical. The flow at the throat can be made critical by sufficiently reducing the width. It is thus seen that, a sub-critical flow in an open channel may become critical not only by passing it over a raised floor as in case of broad crested weir, but also by undergoing a reduction in width.

The flow at the throat is thus critical and the velocity is given by $V_2 = \sqrt{gy_2}$, and the discharge through the throat section is

$$Q_2 = b_2 y_2 V_2 = b_2 y_2 \sqrt{g y_2} \quad \text{...(12.37)}$$

Measurements of the throat width b_2 and the depth y_2 would be sufficient to determine the discharge Q. However, the exact position in the throat where critical conditions exist is not easy to determine and therefore, the exact measurement of y_2 may not be possible. Neglecting loss of energy in the upstream and converging part of the flume, we may write for the critical flow in the throat.

$$y_1 + \frac{V_1^2}{2g} = y_2 + \frac{V_2^2}{2g} = y_2 + \frac{y_2}{2} = \frac{3}{2} y_2 \quad \text{...(12.38)}$$

If the approach velocity V_1 is small, the velocity head $V_1^2/2g$ can be considered negligible. Eq. (12.38) then gives, $y_2 = 2/3\, y_1$ and substituting this in Eq. (12.37), we obtain

$$Q = b_2 \cdot \frac{2}{3} y_1 \sqrt{g \cdot \frac{2}{3} y_1} = \frac{2}{3\sqrt{3}} \sqrt{2g}\, b_2\, y_1^{3/2}$$

$$= 1.705 b_2 y_1^{3/2} \text{ ...(in Metric and SI units)} \quad \text{...(12.39)}$$

If however V_1 is appreciable, then

$$V_1 = \frac{Q}{b_1 y_1} = \frac{b_2 y_2 \sqrt{g y_2}}{b_1 y_1}$$

Substituting this expression for V_1 in Eq. (12.38)

$$y_1 + \frac{1}{2}\left(\frac{b_2}{b_1}\right)^2 \left(\frac{y_2}{y_1}\right)^2 y_2 = \frac{3}{2} y_2$$

or

$$\left(\frac{y_1}{y_2}\right)^3 + \frac{1}{2}\left(\frac{b_2}{b_1}\right)^2 = \frac{3}{2}\left(\frac{y_1}{y_2}\right)^2$$

$$2\left(\frac{y_1}{y_2}\right)^3 - 3\left(\frac{y_1}{y_2}\right)^2 + \left(\frac{b_2}{b_1}\right)^2 = 0 \quad \text{...(12.40)}$$

Eq. (12.40) is a cubic equation in y_1/y_2. Out of the three roots, only one has a physical significance, the second one is negative and the third root gives $y_1 < y_2$. For a given venturiflume since the ratio of widths (b_2/b_1) is fixed, it follows from the equation that the ratio y_1/y_2 is a constant and is independent of the discharge. Let this constant ratio $y_1/y_2 = k$, then $y_2 = y_1/k$ and Eq. (12.37) may be written as

$$Q = b_2 \sqrt{g} \left(\frac{y_1}{k}\right)^{3/2} = \frac{1}{k}\sqrt{\frac{g}{k}}\, b_2 y_1^{3/2} \quad \text{...(12.41)}$$

For a given width ratio, k can be determined from Eq. (12.40), and the discharge through the venturiflume computed by measurement of upstream depth y_1.

If b_1 is very large as compared to b_2 the last term of Eq. (12.40), *i.e.*, $(b_2/b_1)^2$ may be neglected, yielding $y_1/y_2 = 1.5$ and $k = 1.5$.

This value of k when substituted in Eq. (12.41) results in Eq. (12.39). Even for a ratio of $(b_1/b_2) = 1/3$, solution of Eq. (12.40) yields $y_1/y_2 = 1.474$ ($k = 1.474$). Eq. (12.41) then becomes :

$$Q = \frac{1}{1.474}\sqrt{\frac{9.81}{1.474}}\, b_2 y_1^{3/2} = 1.75 b_2 y_1^{3/2} \quad \text{...(12.42)}$$

If, however, Eq. (12.39) is used, the error involved in discharge due to neglect of the velocity of approach will be – 2.57%.

To account for the effect of friction between the inlet and the throat and for the curvature of streamlines, a coefficient of discharge C_d, the value of which varies from 0.94 to 0.92 is introduced in Eqs. (12.39), (12.41) and (12.42).

As a flow measuring device, the venturiflume has certain advantages over the broad crested weir. The first one is that the loss of energy in a venturiflume is much less than in a broad crested weir (due to elimination of eddies just upstream of the weir). And secondly there is no problem of silt deposition in the flume.

In case the venturiflume is not designed to produce critical flow at the throat, determination of discharge will then be possible by measuring the upstream depth y_1 and the depth at the throat y_2. Using Eq. (12.38)

$$y_1 + \frac{V_1^2}{2g} = y_2 + \frac{V_2^2}{2g}$$

and the continuity equation $Q_1 = b_1 y_1 V_1 = b_2 y_2 V_2$. Eliminating V_1, we obtain

$$V_2 = \frac{b_1 y_1}{\sqrt{b_1^2 y_1^2 - b_2^2 y_2^2}} \sqrt{2g(y_1 - y_2)}$$

and the discharge $$Q = b_2 y_2 V_2 = \frac{b_2 y_2 b_1 y_1}{\sqrt{b_1^2 y_1^2 - b_2^2 y_2^2}} \sqrt{2g(y_1 - y_2)}$$

The actual discharge can be expressed as

$$Q = C_d \frac{b_2 y_2 b_1 y_1}{\sqrt{b_1^2 y_1^2 - b_2^2 y_2^2}} \sqrt{2g(y_1 - y_2)} \qquad ...(12.43)$$

where C_d is coefficient of discharge,

Example 12.41. *A venturiflume is 1.30 m wide at entrance and 0.65 m in the throat. Neglecting hydraulic losses in the flume, calculate the flow if the depth at the entrance and throat is 0.65 m and 0.60 m. A hump is now installed at the throat, of height 20 cm, so that a standing wave (hydraulic) is formed beyond the throat. What is the increase in the upstream depth when the same flow as before passes through the flume ?*

Solution. From the continuity equation

$$Q = b_1 y_1 V_1 = b_2 y_2 V_2$$

$$Q = 1.3 \times 0.65\, V_1 = 0.65 \times 0.60\, V_2, \quad \therefore \quad V_2 = 2.17\, V_1$$

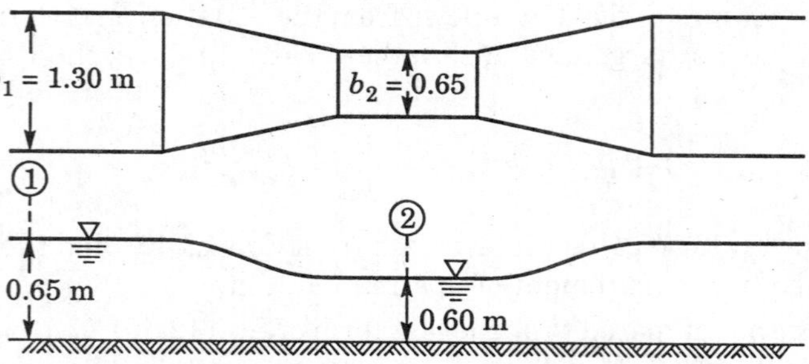

Neglecting losses, Specific energy at (1) = Specific energy at (2)

$$y_1 + \frac{V_1^2}{2g} = y_2 + \frac{V_2^2}{2g}$$

Making substitutions,

$$0.65 + \frac{V_1^2}{2g} = 0.65 + 4.7\frac{V_1^2}{2g} \quad \text{or} \quad 3.7\frac{V_1^2}{2g} = 0.05,$$

from which $$V_1 = \sqrt{\frac{0.05 \times 2 \times 9.81}{3.7}} = 0.515 \text{ m/s}.$$

The discharge $Q = b_1 y_1 V_1 = 1.3 \times 0.61 \times 0.515 = 0.435 \text{ m}^3\text{/s}.$

Critical depth in the contracted portion

$$y_c = \left(\frac{q^2}{g}\right)^{1/3}, q = \frac{0.435}{0.65} = 0.67 \text{ m}^3\text{/s/m}$$

$$y_c = \left[\frac{(0.67)^2}{9.81}\right]^{1/3} = 0.357 \text{ m}.$$

The new specific energy corresponding to critical flow at the throat when hump is installed :

$$= \Delta z + y_c + \frac{V_c^2}{2g} = \Delta z + \frac{3}{2}y_c = 0.20 + 1.5 \times 0.357 = 0.735 \text{ m}.$$

The upstream water surface will rise till the upstream specific energy equals 0.735 m.

$$0.735 = y_1 + \frac{V_1^2}{2g} = y_1 + \frac{Q^2}{2g(1.3 \times y_1^2)}$$

$$= y_1 + \frac{(0.485)^2}{33.2 y_1^2} = y_1 + \frac{0.0057}{y_1^2}$$

Solving by trial and error, $y_1 = 0.72$ m.

The increase in the upstream depth $= 0.72 - 0.65 = 0.07$ m $= 7$ cm.

12.15. Momentum Principle Applied to Open Channel Flow

The Newton's second law of motion when expressed in terms of momentum states that change of momentum per unit time in a given direction is equal to the resultant of all external forces acting on the body in the same direction. Consider a channel of large slope as shown in Fig. 12.21, and apply the momentum principle to the water contained between sections (1) and (2). The control volume is, therefore, the body of water bounded by sections (1) and (2), the wetted perimeter and the free surface.

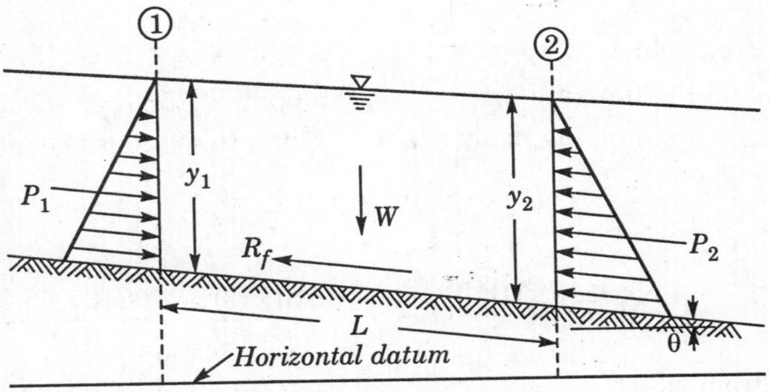

Fig. 12.21. Application of the momentum principle.

The subscripts 1 and 2 refer to the quantities at sections (1) and (2) respectively. The momentum of flow entering section (1) per second is $(\gamma Q/g)\, V_1$, where γ is the specific weight of the flowing liquid, and the momentum per second leaving section (2) is $(\gamma Q/\gamma)\, V_2$. The external forces acting on the body of water enclosed between sections (1) and (2) are :

(i) P_1, the hydrostatic pressure force at section 1

(ii) P_2, the hydrostatic pressure force at section 2

(*iii*) $W \sin \theta$, the component of the weight of water (enclosed between the sections) in the direction flow, and

(*iv*) R_f, the resistance force due to friction between channel surface and the liquid and acts along the contact surface.

Applying the momentum principle, we obtain

$$\gamma Q/g\,(V_2 - V_1) = P_1 - P_2 + W \sin \theta - R_f \quad ...(12.44)$$

Eq. (12.44) is known as the general *momentum equation* applicable to open channel flow.

Example 12.42. *A sluice across a channel 6 m wide discharges a stream 1 m deep. What is the flowrate when the depth upstream of the sluice is 7 m ? On the downstream side concrete blocks have been placed to create conditions for hydraulic jump to occur. Calculate the force on the blocks if the downstream depth is 3 m.*

Solution. From continuity equations between (1), (2) and (3)

$$Q = 7 \times 6 \times V_1 = 1 \times 6 \times V_2 = 3 \times 6 \times V_3.$$

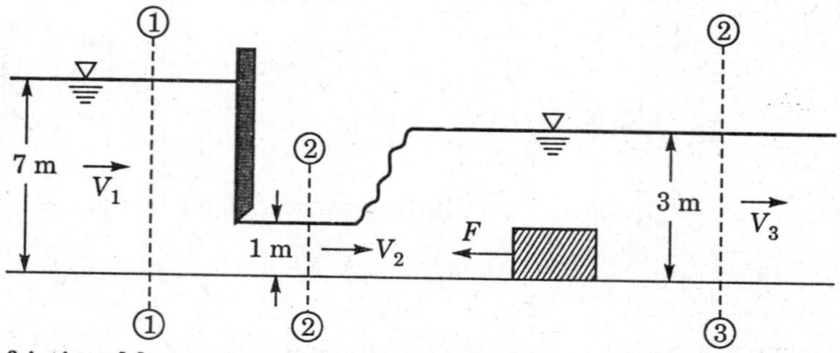

Neglecting frictional losses between sections (1) and (2) the specific energies at (1) and (2) are equal

$$7 + \frac{V_1^2}{2g} = 1 + \frac{V_2^2}{2g}$$

from continuity equation $V_2 = 7V_1$, and eliminating V_2 we obtain

$$7 + \frac{V_1^2}{2g} = 1 + 49\frac{V_1^2}{2g}$$

from which $V_1 = 1.565$ m/s

and the flowrate $Q = 6 \times 7 \times 1.565 = 65.7$ m³/s.

Applying momentum equation to sections (2) and (3) neglecting the boundary friction,

$$P_2 - F - P_3 = \frac{\gamma Q}{g}(V_3 - V_2)$$

$$\frac{6 \times 9810 \times (1)^2}{2} - F - \frac{6 \times 9810 \times (3)^2}{3} = \frac{9810 \times 65.7}{9.81}\left[\left(\frac{7}{3}V_1 - 7V_1\right)\right]$$

from which $F = 244 \times 10^3$ N = 244.0 kN

Therefore, the force on the concrete blocks = 244.0 kN which acts in a direction opposite to F.

12.16. Specific Force

In case of a horizontal channel, $\theta = 0$ and hence the weight component $W \sin \theta$ vanishes from the momentum equation. If, however, the length L involved between the sections is small, the resistance force R_f may be ignored and Eq. (12.44) then becomes,

$$(\gamma Q/g)\,(V_2 - V_1) = P_1 - P_2.$$

The hydrostatic forces P_1 and P_2 may be evaluated as $P_1 = \gamma\bar{z}_1 A_1$ and $P_2 = \gamma\bar{z}_2 A_2$, where $\bar{z}_1$ and $\bar{z}_2$ are the distances of centroids of flow areas A_1 and A_2 from their respective water surfaces. From continuity equation,

$$Q = V_1 A_1 = V_2 A_2.$$

Substituting for V_1 and V_2 in the momentum equation, one gets

$$\frac{\gamma Q^2}{g}\left(\frac{1}{A_2} - \frac{1}{A_1}\right) = \gamma\bar{z}_1 A_1 - \gamma\bar{z}_2 A_2 .$$

Re-arranging the terms, we get

$$\frac{Q^2}{gA_2} + \bar{z}_2 A_2 = \frac{Q^2}{gA_1} + \bar{z}_1 A_1 \qquad ...(12.45)$$

The two sides of Eq (12.45) are identical, and may be represented by F. Eq. (12.45) may then be expressed as

$$F = \frac{Q^2}{gA} + A\bar{z} \qquad ...(12.46)$$

The first term of Eq. (12.46), *i.e.*, Q^2/gA is the momentum of flow passing the channel section per unit time per specific weight γ of the liquid. The second term $A\bar{z}$ is the hydrostatic force per specific weight γ of the liquid. Both the terms being basically the forces per specific weight, their sum denoted by F is generally known as the *specific force*. Eq. (12.45) shows that for a short horizontal channel, the specific forces at sections (1) and (2) are equal.

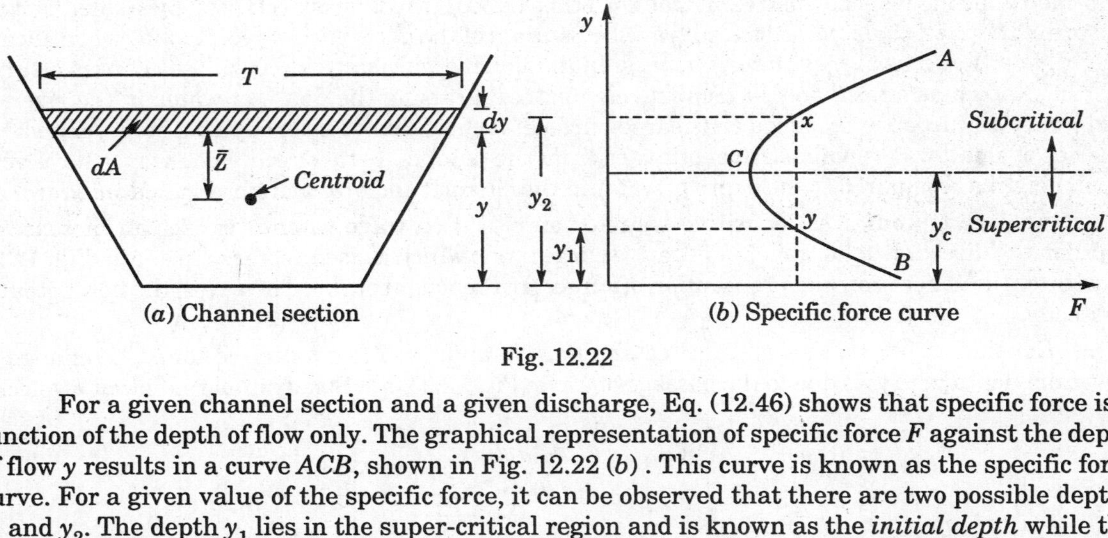

(*a*) Channel section (*b*) Specific force curve

Fig. 12.22

For a given channel section and a given discharge, Eq. (12.46) shows that specific force is a function of the depth of flow only. The graphical representation of specific force F against the depth of flow y results in a curve ACB, shown in Fig. 12.22 (*b*). This curve is known as the specific force curve. For a given value of the specific force, it can be observed that there are two possible depths y_1 and y_2. The depth y_1 lies in the super-critical region and is known as the *initial depth* while the second depth y_2 corresponds to sub-critical flow and is known as the *sequent depth*. These depths y_1 and y_2 are known as the *conjugate depths*. The point C corresponds to the minimum specific force, where the two depths merge into one to yield a single depth y_c. The depth y_c which corresponds to minimum specific force may be shown to be equal to the critical depth.

For a minimum value of specific force, $dF/dy = 0$

Differentiating Eq. (12.46) with respect to y,

$$\frac{dF}{dy} = -\frac{Q^2}{gA^2}\frac{dA}{dy} + \frac{d(A\bar{z})}{dy}$$

$d(A\bar{z})$ represents change in moment of flow area about the free surface brought about by a change dy in the flow depth, thus

$$d(A\bar{z}) = \left[A(\bar{z} + dy) + T\,dy\,.\frac{dy}{2}\right] - A\bar{z}$$

Neglecting the term $\dfrac{T(dy)^2}{2}$, being of small order of magnitude,

$$d(A\bar{z}) = Ady$$

and from Fig. 12.22 (*a*), $\quad dA = Tdy.$

Now we can write $\quad \dfrac{dF}{dy} = -\dfrac{Q^2T}{gA^2} + A$

Setting $\dfrac{dF}{dy} = 0$, we obtain the condition for the minimum specific force as

$$\frac{Q^2T}{gA^3} = 1.$$

From consideration of specific energy, it is known that the above condition is the criterion for critical flow, see Eq. (12.17). It may thus be stated that at the critical depth, like specific energy, the specific force also attains a minimum value.

12.17. Small Waves and Surges in Open Channels

If the flow in a channel is increased by sudden increase in the opening of a sluice gate, a wave is formed which travels downstream. One the other hand, if the flow is decreased by sudden partial closure of the gate, the wave so formed travels upstream of the gate. Such waves, having a significant height, are known as *surges* or *surge waves*. In a tidal river, a rising tide in the sea causes a surge to form, known as a *tidal bore*, which travels upstream raising the depth of water in the river. A temporary disturbance produced on the free-surface, such as caused by dropping a stone, generates a series of small waves which travel outward from the point of disturbance. Such waves have very small height as compared to the depth of water in the channel and are, therefore, called *small waves*.

A surge is known as a positive surge or an elevation wave when it causes an increase in depth in the direction of its travel. A negative surge is one which causes decrease in depth. Fig. 12.23 illustrates the generation and propagation of surges. In the neighbourhood of waves, the flow becomes unsteady.

Consider a positive surge in a rectangular channel with horizontal bottom. The change of flow from depth y_1 to y_2 is due to the passage of wave [Fig. 12.24 (*a*)]. Steady uniform flow at a velocity

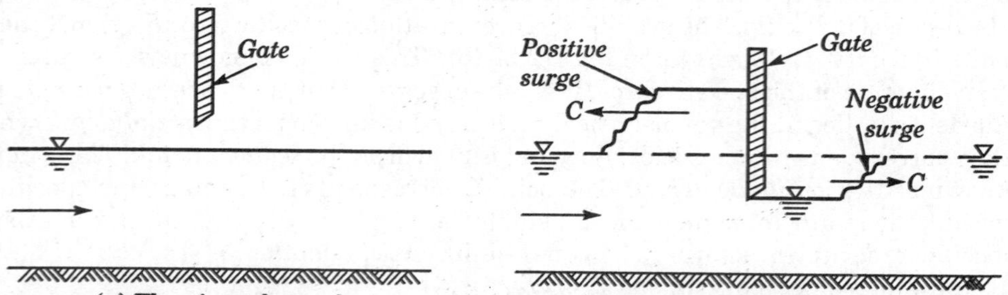

(*a*) Flow in a channel

(*b*) Flow affected by sudden lowering of gate. A positive surge travels upstream while a negative surge is transmitted down-stream of the gate

Fig. 12.23. Propagation of surges.

V_1 and depth y_1 is disturbed by partially closing a gate downstream which results in a positive, surge. The surge travels upstream with a constant velocity C relative to the channel bottom, producing the conditions of unsteady flow. The velocity of wave C is also known as *celerity* of the wave. A small distance downstream of the wave, the flow again becomes steady and uniform with velocity V_2 and depth y_2. For analysis the moving surge can be brought to rest by superposing an equal and opposite velocity C on the flow. After superposition, the velocity V_1 changes to $V_1 + C$ and V_2 to $V_2 + C$, Fig. 12.24 (*b*). The conditions of flow now are steady.

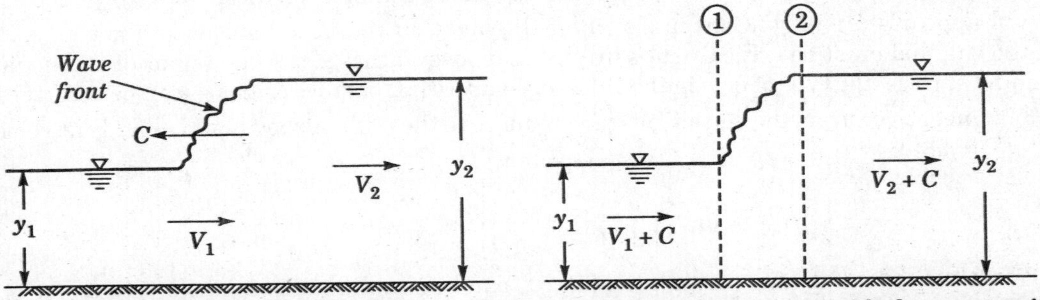

(*a*) Positive surge travelling upstream at a velocity C

(*b*) Surge becomes stationary with the superposition of velocity C over the flow. The flow now is steady

Fig. 12.24. Analysis of positive surge.

The momentum equation may be applied to yield a relationship among the depths y_1, y_2 and the velocities V_1 and C. As the channel bottom is horizontal, the component of the weight of liquid in the direction of flow is zero. The non-uniformity of flow exists over a short distance, that is between sections (1) and (2) and as such the resistance force due to friction, R_f, appearing in the momentum equation, Eq. (12.44), is negligible. The momentum equation applied to the liquid mass enclosed between sections (1) and (2) is

$$\frac{\gamma Q}{g}(V_2 - V_1) = P_1 - P_2$$

in which $(V_2 - V_1)$ is the change in velocity between the sections (1) and (2) [the velocity change is $V_2 + C - V_1 - C = (V_2 - V_1)$], Q = rate of flow $= by_1 (V_1 + C)$, and considering the pressure distribution at (1) and (2) to be hydrostatic, $P_1 = \frac{1}{2}\gamma b y_1^2$ and $P_2 = \frac{1}{2}\gamma b y_2^2$. Substituting these values, the momentum equation takes the from :

$$\frac{\gamma b y_1 (V_1 + C)}{g}(V_2 - V_1) = \frac{1}{2}\gamma b (y_1^2 - y_2^2) \quad \text{...(12.47)}$$

and the continuity equation between (1) and (2) is,

$$Q = by_1 (V_1 + C) = by_2 (V_2 + C) \quad \text{...(12.48)}$$

which yields,
$$V_2 = \frac{y_1}{y_2}(V_1 + C) - C.$$

Substituting for V_2 in Eq. (12.47) and simplifying,

$$V_1 + C = \sqrt{gy_2}\left[\frac{1}{2}\left(1 + \frac{y_2}{y_1}\right)\right]^{1/2}$$

$$= \sqrt{gy_1}\left[\frac{1}{2}\frac{y_2}{y_1}\left(1 + \frac{y_2}{y_1}\right)\right]^{1/2} \quad \text{...(12.49)}$$

For waves of small height, $y_1 \approx y_2 \approx y$, and Eq. (12.42) reduces to

$$V_1 + C = \sqrt{gy} \quad \text{...(12.50)}$$

Eq. (12.50) states that the velocity of a small wave relative to the undisturbed liquid moving at velocity V_1 is $\sqrt{gy}$. We know that for critical flow in a rectangular channel, the critical velocity is $\sqrt{gy}$. If the flow in the channel is subcritical ($V_1 < \sqrt{gy}$), the velocity of flow is less than the critical velocity and, therefore, small waves can travel upstream. However, if the flow is supercritical, the velocity of flow is greater then the critical velocity ($V_1 > \sqrt{gy}$) and, consequently, small waves on the liquid surface will be swept downstream. If the flow in the channel is critical, ($V_1 = \sqrt{gy}$) the small waves will appear stationary. Conversely, if small waves on the liquid surface do not move, the flow in the channel must be critical. Thus by physical observation of the behaviour of small waves on the liquid surface, the type of flow in the channel—subcritical or supercritical—can be ascertained.

For a negative surge, the velocity of wave relative to the undisturbed liquid moving at velocity V_1 is

$$C - V_1 = \sqrt{gy_1}\left[\frac{1}{2}\frac{y_2}{y_1}\left(1+\frac{y_2}{y_1}\right)\right]^{1/2} \qquad \text{...(12.51)}$$

Using the continuity Eq. (12.48), Eq. (12.51) can be simplified to yield the wave velocity

$$C = \frac{V_2y_2 - V_1y_1}{y_1 - y_2} \qquad \text{...(12.51 } a\text{)}$$

Example 12.43. *A tidal estuary is flowing at the rate of 6.5 km/hr and a depth of 2 m. Owing to the tide in the sea, the level rapidly rose and the resulting surge or 'bore' took one hour to reach a spot 22.5 km up the stream. Compute the height of the bore above the initial depth of flow. What speed and direction will the flow have after the bore has passed ?*

Solution. $V_1 = \dfrac{6.5\times1000}{60\times60} = 1.805$ m/s

$y_1 = 2$ m

$V_w = \dfrac{22.5\times1000}{60\times60} = 6.25$ m/s.

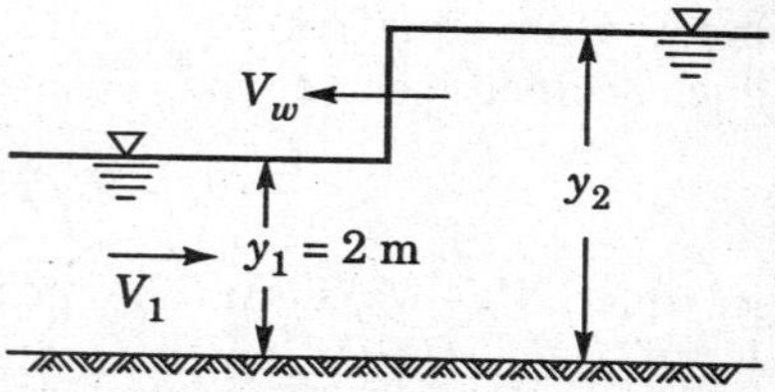

From Eq. (12.49) replacing C by V_w,

$$V_1 + V_w = \sqrt{\frac{gy_2}{2y_1}(y_1+y_2)}$$

making substitutions,

$$1.805 + 6.25 = \sqrt{\frac{9.81y_2(2+y_2)}{2\times2}}$$

$$(8.055)^2 = \frac{9.81y_2(2+y_2)}{2\times2}$$

or $\dfrac{54.8\times4}{9.81} = y_2(2+y_2)$ or $y_2^2 + 2y_2 - 26.47 = 0$

from which $y_2 = \dfrac{-2\pm\sqrt{4+105.88}}{2} = 4.24$ m.

Height of the bore $= y_2 - y_1 = 4.24 - 2.0 = 2.24$ m

Flow taken in by the bore $= (A_2 - A_1)V_w$

$= 2.24 \times 6.25$ m³/s/m width $= 14.4$ m³/s/m width.

Normal flow in the river $= 2.0 \times 1.805 = 3.61$ m³/s/m width

Net flow in the direction of bore

$= 14.0 - 3.61 = 10.39$ m³/s/m

Speed of flow after the bore has passed

$$= \frac{10.39}{y_2} = \frac{10.39}{4.24} = 2.475 \text{ m/s.}$$

The direction of flow is opposite to the normal direction of flow.

Example 12.44. *A gate is to be suddenly dropped into place closing a rectangular channel 2 m deep and 3 m wide in which 5.5 m³/s are flowing at a depth of 1.30 m. Will the channel overflow ? Assuming the sides of the channel to be sufficiently high to prevent it from overflowing, what will be the velocity of the surge produced ?*

Solution. Since gate is suddenly dropped

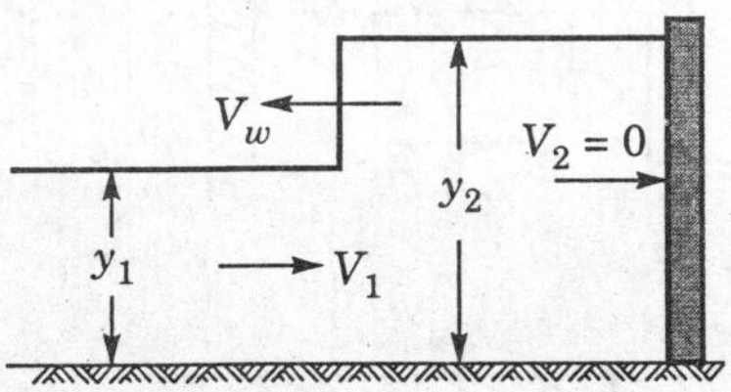

$$V_2 = 0$$
$$Q = 5.5 \text{ m}^3\text{/s},$$
$$y_1 = 1.3 \text{ m}, b = 3 \text{ m}$$
$$V_1 = \frac{Q}{A_1} = \frac{5.5}{3 \times 1.3} = 1.41 \text{ m/s}$$

Using Eq. (12.51 *a*), noting $V_w = C$

$$V_w = \frac{V_2 y_2 - V_1 y_1}{y_2 - y_1}$$
$$= \frac{0 - 1.41 \times 1.3}{y_2 - 1.3} = \frac{-1.833}{y_2 - 1.3} = \frac{1.833}{y_2 - 1.3}$$

$$V_w + V_1 = 1.41 + \frac{1.833}{y_2 - 1.3}$$

but $$V_1 + V_w = \sqrt{g \frac{y_2}{y_1} \left(\frac{y_1 + y_2}{2} \right)}$$

Making the substitutions,

$$1.41 + \frac{1.833}{y_2 - 1.3} = \sqrt{g \frac{y_2}{1.3} \left(\frac{1.3 + y_2}{2} \right)}$$

Solving for y_2 by trial and error, $y_2 = 1.85$ m which may be estimated from the following tabulation :

Let			
	$y_2 = 1.5$	LHS = 10.57	RHS = 3.98
	$y_2 = 2.5$	LHS = 2.937	RHS = 5.98
	$y_2 = 2.0$	LHS = 4.03	RHS = 4.98
	$y_2 = 1.9$	LHS = 4.465	RHS = 4.79
	$y_2 = 1.85$	LHS = 4.74	RHS = 4.68

Since $y_2 < 2.0$ m, hence there will be no overflow.

Velocity of the surge $V_w = \frac{1.833}{y_2 - 1.3} = \frac{1.833}{1.85 - 1.3} = 3.33$ m/s.

Example 12.45. *A horizontal rectangular channel of 3 m width and 2 m water depth conveys water at 18 m³/sec. If the flow rate is suddenly reduced to 2/3 of its original value, compute the magnitude and speed of the upstream ward surge.*

Assume that the front of the surge is rectangular and friction in the channel is neglected.

Solution. When the discharge in the channel is suddenly reduced by operating a gate, an upstream surge will be developed which will move upstream with a celerity *C*, as shown. An observer standing on the canal bank will notice the surge moving upstream. This unsteady flow case can be transformed into a steady one by superimposing flow with velocity C in the opposite direction. These two cases are shown in Fig. 12.24, and are depicted separately as below.

The continuity equation may be written as,

$$by_1 (V_1 + C) = by_2 (V_2 + C) \quad \text{or} \quad y_1 (V_1 + C) = y_2 (V_2 + C)$$

and the momentum equation,

$$\frac{b\gamma}{2}(y_2^2 - y_1^2) = \frac{b\gamma}{2} y_1(V_1 + C)(V_1 - V_2) \text{ or } (y_2^2 - y_1^2) = \frac{2y_1}{g}(V_1 + C)(V_1 - V_2)$$

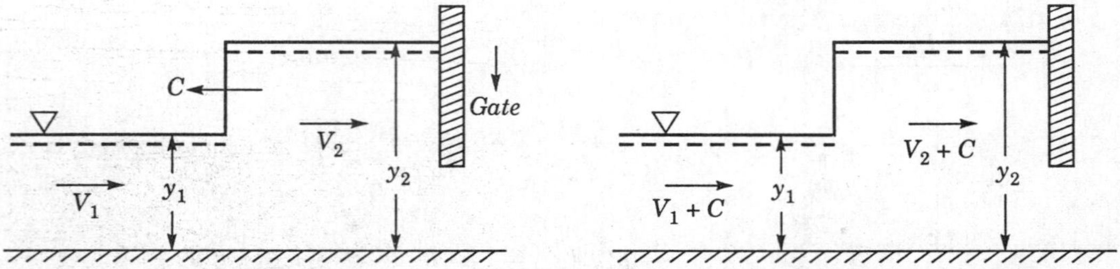

From the data given, $y_1 = 2$ m, $Q_1 = 18$ m³/s

$$V_1 = \frac{Q_1}{by_1} = \frac{18}{3 \times 2} = 3 \text{ m/s}$$

$$Q_2 = \frac{2}{3}Q_1 = \frac{2 \times 18}{3} = 12 \text{ m}^3\text{/s}$$

$$Q_2 = by_2 V_2, \qquad \therefore \quad V_2 y_2 = \frac{12}{3} = 4 \text{ m}^3\text{/s/m}$$

From the continuity equation,

$$V_1 y_1 = V_2 y_2 + C(y_2 - y_1)$$

$$3 \times 2 = 4\, C(y_2 - 2), \text{ or } C = \frac{2}{y_2 - 2}$$

Substituting in the momentum equation,

$$(y_2^2 - 4) = \frac{2 \times 2}{9.81}(3 + C)(3 - V_2)$$

or

$$y_2^2 - 4 = 0.41\left(3 + \frac{2}{y_2 - 2}\right)\left(3 - \frac{4}{y_2}\right)$$

Solving by trials, $\quad y_2 = 2.8$ m, and $V_2 = \dfrac{4}{2.8} = 1.4286$ m/s

Height of surge $\quad = y_2 - y_1 = 2.8 - 2.0 =$ **0.8 m.**

Celerity of surge, $\quad C = \dfrac{2}{2.8 - 2} =$ **2.5 m/s.**

12.18. Gradually Varied Flow

The flow in an open channel may be either uniform or non-uniform (also known as varied flow). In uniform flow, the depth of flow remains constant along the length of channel. This type of flow is possible in prismatic channels of sufficient length. The uniform flow is characterised by parallelism of channel bottom, liquid surface and the energy line. The non-uniform or varied flow is one in which the flow depth changes in the flow direction. The flow is said to be gradually varied when the depth changes gradually over a long distance, whereas in rapidly varied flow, the change in depth takes place in a short distance. The change in depth may sometimes be quite-abrupt, as in case of a hydraulic jump. Since this type of non-uniformity is limited to a short distance, a hydraulic jump and a hydraulic drop represent a local phenomena. The rapidly varied flow is, therefore, a local phenomenon.

Gradually varied flow may be caused as a result of one or more of the following factors :

(1) change in the shape and size of the channel cross-section ;

(2) change in the channel slope ;

(3) presence of obstruction, such as a weir ; and

(4) change in the frictional forces at the boundaries.

12.18.1. Equation of Gradually Varied Flow

In open channel flow computations, we are often called upon to predict depth of flow at a certain location or to estimate the distance over which backwater effects due to construction of a weir or a spillway would be transmitted upstream. If the depth of flow in a channel for a given discharge is known, the area and the mean velocity of flow can be determined. The position of liquid surface in relation to the channel bottom determines the depth of flow, which in turn is used to define the type of flow.

The following assumptions are necessary for analysing the gradually varied flow :

(1) that the flow is steady ;

(2) that the pressure distribution over the channel section is hydrostatic, *i.e.*, streamlines are practically straight and parallel ;

(3) that the head loss is same as for uniform flow ;

(4) that the channel slope is small, so that the depth measured vertically is the same as depth measured normal to the channel bottom ;

(5) that the channel is prismatic ;

(6) that the kinetic energy correction factor is very close to unity ; and

(7) that the channel roughness does not depend upon the depth of flow, and is constant along the channel length.

Referring to Fig. 12.25, let Z, y and $V^2/2g$ be the datum head, the depth of flow and the velocity head respectively at any section. Then according to the Bernoulli's equation, the total energy of flow per unit weight of liquid above the horizontal datum is

$$H = Z + y + \frac{V^2}{2g} \qquad ...(12.52)$$

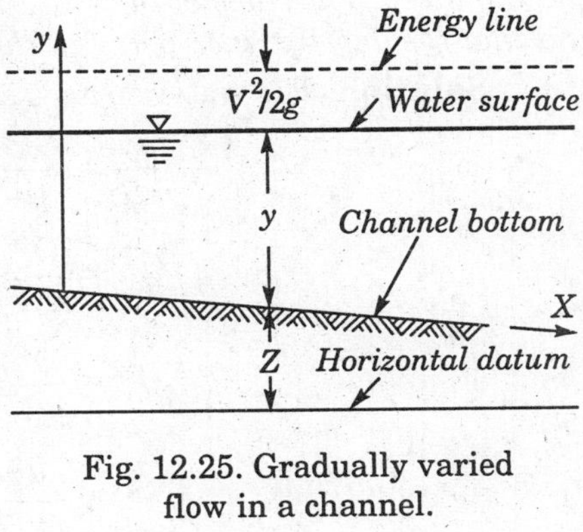

Fig. 12.25. Gradually varied flow in a channel.

Taking the bottom of the channel as the x-axis and the vertically upwards direction measured from the channel bottom, as the y-axis, differentiation of Eq. (12.52) with respect to x-yields

$$\frac{dH}{dx} = \frac{dZ}{dx} + \frac{dy}{dx} + \frac{d}{dx}\left(\frac{V^2}{2g}\right) \qquad ...(12.53)$$

Let S_e = slope of the energy line, and

S_o = slope of the channel bottom.

Now as the total energy H and the datum head Z both decrease in the direction of flow, the differential terms dH/dx and dZ/dx representing the energy line slope and the slope of the channel bottom both are negative. Hence Eq. (12.53) can be written

$$-S_e = -S_o + \frac{dy}{dx} + \frac{d}{dx}\left(\frac{V^2}{2g}\right) = -S_o + \frac{dy}{dx} + \frac{d}{dy}\left(\frac{Q^2}{2gA^2}\right)\frac{dy}{dx}$$

$$= -S_o + \frac{dy}{dx}\left(1 - \frac{Q^2}{gA^3}\frac{dA}{dy}\right)$$

from which the slope of the free surface

$$\frac{dy}{dx} = \frac{S_o - S_c}{1 - \dfrac{Q^2}{gA^3}\dfrac{dA}{dy}} \qquad \text{...(12.54)}$$

Eq. (12.54) is known as the equation of the gradually varied flow or the differential equation of gradually varied flow. It represents the change in the depth of flow y with respect to the channel bottom (x-axis). For channel of constant width, dA/dy of Eq. (12.54) can be replaced by channel width T at the free surface. We then obtain,

$$\frac{dy}{dx} = \frac{S_o - S_c}{1 - \dfrac{Q^2 T}{gA^3}} \qquad \text{...(12.55)}$$

We know that $$\frac{Q^2 T}{gA^3} = \frac{V^2}{g\left(\dfrac{A}{T}\right)} = \frac{V^2}{gD} = \mathbf{F}^2$$

where **F** is the Froude number of flow. Eq. (12.55) may now be expressed as

$$\frac{dy}{dx} = \frac{S_o - S_e}{1 - \dfrac{V^2}{gD}} = \frac{S_o - S_e}{1 - \mathbf{F}^2} \qquad \text{...(12.56)}$$

Eq. (12.56) contains valuable information, and will be used to classify different types of surfaces profiles.

Example 12.46. *Find the slope of the free water surface of a rectangular stream 15 m wide and 3 m deep. The slope of the stream bed is 1 in 5000. Total discharge is 29 m³/s. Assume C = 65 and that the depth is increasing in the direction of flow.*

Solution. The water surface slope is given by

$$\frac{dy}{dx} = \frac{S_o - S_e}{1 - \mathbf{F}^2}$$

$$V = \frac{Q}{A} = \frac{29}{15 \times 3} = 0.645 \text{ m/s}$$

$$\mathbf{F}^2 = \frac{V^2}{gy} = \frac{0.417}{9.81 \times 3} = 0.0142$$

$$S_e = \frac{V^2}{C^2 R} = \frac{(0.645)^2}{(65)^2 \times \dfrac{45}{21}} = \frac{1}{21{,}700}$$

$$\therefore \quad \frac{dy}{dx} = \frac{\dfrac{1}{5000} - \dfrac{1}{21{,}700}}{1 - 0.0142} = \frac{21{,}700 - 5000}{5000 \times 21{,}700 \times 0.9858} = \frac{1}{6400}.$$

Example 12.47. *Show that the gradually varied flow equation for flow in a rectangular channel of variable width b may be expressed as :*

$$\frac{dy}{dx} = \frac{S_o - S_e + \left(\frac{V^2}{gb}\right)\frac{db}{dx}}{1 - \mathbf{F}^2}.$$

Solution. Total energy $E = z + y + \frac{V^2}{2g}$

Differentiating w.r.t. x

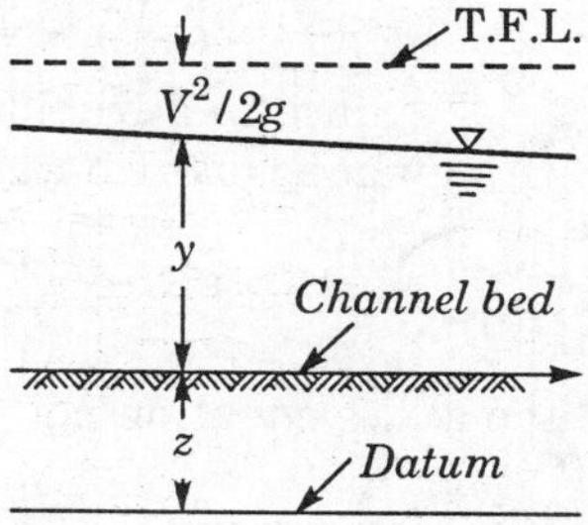

$$\frac{dE}{dx} = \frac{dz}{dx} + \frac{dy}{dx} + \frac{d}{dx}\left(\frac{V^2}{2g}\right)$$

or $$-S_e = -S_o + \frac{dy}{dx} + \frac{dy}{dx}\frac{d}{dy}\left(\frac{Q^2}{2gA^2}\right)$$

$$-S_e = -S_o + \frac{dy}{dx} - \frac{Q^2}{gA^3}\frac{dA}{dy}\frac{dy}{dx}$$

From continuity, $Q = byV$

Differentiating, $$\frac{dQ}{dx} = Vy\frac{db}{dx} + bV\frac{dy}{dx} + by\frac{dV}{dx}$$

Assuming $\frac{dQ}{dx} = 0$, dividing throughout by $by\,V$

$$0 = \frac{1}{b}\frac{db}{dx} + \frac{1}{y}\frac{dy}{dx} + \frac{1}{V}\frac{dV}{dx},$$

$$\frac{dV}{dx} = \frac{d}{dx}\left(\frac{Q}{A}\right) = -\frac{Q}{A^2}\frac{dy}{dx}\cdot\frac{dA}{dy}$$

Substituting for $\frac{dV}{dx}$, $$0 = \frac{1}{b}\frac{db}{dx} + \frac{1}{y}\frac{dy}{dx} - \frac{1}{A}\frac{dy}{dx}\frac{dA}{dy}$$

$$\frac{dA}{dy} = \frac{A}{b}\frac{db/dx}{dy/dx} + \frac{A}{y} = \frac{y(db/dx)}{dy/dx} + b, \text{ since } A = by$$

Substituting for $\frac{dA}{dy}$, $$-S_e = -S_o + \frac{dy}{dx} - \frac{Q^2}{gA^3}\left(\frac{dy}{dx}\right)\left(y\frac{db/dx}{dy/dx} + b\right)$$

$$S_o - S_e = \frac{dy}{dx} - \frac{Q^2 y}{gA^3}\cdot\frac{db}{dx} - \frac{Q^2 b}{gA^3}\frac{dy}{dx}$$

which yields, $$\frac{dy}{dx} = \frac{S_o - S_e + \frac{V^2}{gb}\cdot\frac{db}{dx}}{1 - \frac{Q^2 b}{gA^3}} = \frac{S_o - S_e + \left(\frac{V^2}{gb}\right)\frac{db}{dx}}{1 - \frac{V^2}{gy}}.$$

Example 12.48. *A frictionless liquid flows along a flat, open channel of variable breadth b subject to the influence of gravity g, obtain the expression*

$\frac{dy}{y} = -\frac{db/b}{1 - (gy/V^2)}$, *where y is the depth of the liquid and V is the mean velocity of flow.*

Solution. A flat channel here is used in the sence of having a horizontal bottom. Since the liquid is frictionless, no energy is lost in conveying the liquid. The flow in a horizontal channel is thus theoretically possible.

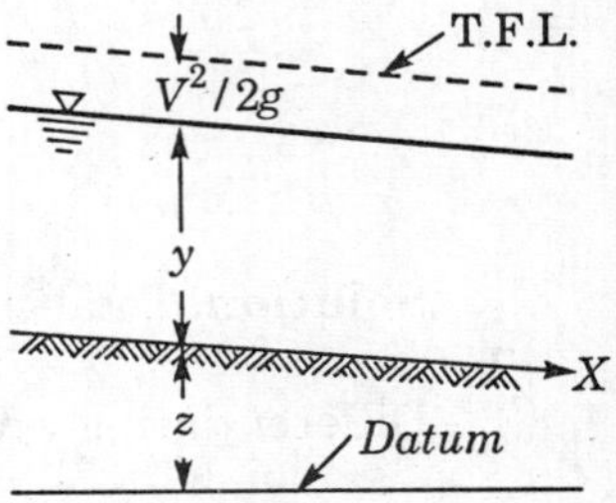

The total energy per unit of weight

$$E = z + y + \frac{V^2}{2g}.$$

Differentiating with respect to x

$$\frac{dE}{dx} = \frac{dz}{dx} + \frac{dy}{dx} + \frac{d}{dx}\left(\frac{V^2}{2g}\right).$$

Since in a frictionless liquid there is no energy loss, the total energy line is horizontal, thereby yielding

$$\frac{dE}{dx} = 0, \text{ and also } \frac{dz}{dx} = 0$$

the above equation now becomes, $0 = \frac{dy}{dx} + \left(\frac{V}{g}\right)\frac{dV}{dx}$ ...(1)

Assuming the channel to be rectangular, the continuity equation gives $Q = bVy$, and since the width b varies with x, differentiating the continuity equation with respect to x, we obtain

$$\frac{dQ}{dx} = bV\frac{dy}{dx} + by\frac{dV}{dx} + Vy\frac{db}{dx}$$

dividing throughout by bVy and noting that for constant discharge, $\frac{dQ}{dx} = 0$, thus

$$0 = \frac{1}{y}\frac{dy}{dx} + \frac{1}{V}\frac{dV}{dx} + \frac{1}{b}\frac{db}{dx} \quad ...(2)$$

From Eqs. (1) and (2), $\frac{dy}{dx} = -\frac{V}{g}\frac{dV}{dx} = \frac{V^2}{g}\left(\frac{1}{y}\frac{dy}{dx} + \frac{1}{b}\frac{db}{dx}\right)$

or $\frac{dy}{dx}\left(1 - \frac{V^2}{gy}\right) = \frac{V^2}{gy}\frac{db}{dx}$

which simplifies to, $\frac{dy}{y} = -\left(\frac{1}{1-(gy/V^2)}\right)\frac{db}{b}.$

Example 12.49. *Determine the approximate water surface slope (m per km) at a point in a rectangular channel in which the discharge is 0.84 cumec, the bed width is 3 m, the depth of flow is 0.75 m and the bed slope is 0.15 m per km. Assume C in Chezy's formula as 55 and take energy coefficient α, equal to unity.*

Solution. Water surface slope

$$\frac{dy}{dx} = \frac{S_o - S_e}{1 - \mathbf{F}^2}$$

$$R = \frac{A}{P} = \frac{3 \times 0.75}{3 + 2 \times 0.75} = 0.50 \text{ m} ; S_o = \frac{0.15}{1000}$$

$$V = C\sqrt{RS_e} \text{ or } S_e = \frac{V^2}{C^2R}$$

$$V = \frac{Q}{A} = \frac{0.84}{3 \times 0.75} = 0.3733 \text{ m/s}$$

$$D = \frac{A}{T} = \frac{by}{y} = y = 0.75 \text{ m}$$

Froude number $\quad \mathbf{F} = \frac{V}{\sqrt{gD}} = \frac{0.3733}{\sqrt{9.81 \times 0.75}} = 0.1378$

$$S_e = \frac{V^2}{C^2 R} = \frac{(0.3733)^2}{(55)^2 \times 0.50} = \frac{1}{10,854}$$

Substituting the values of various quantities

$$\frac{dy}{dx} = \frac{\frac{0.15}{1000} - \frac{1}{10.854}}{1-(0.1378)^2} = 0.000059 = \mathbf{0.059 \text{ m per km.}}$$

12.18.2. Derivation of Water Surface Slope S_w with Respect to Horizontal

Let dy/dx be the water surface slope with respect to the channel bottom (x-direction being along the channel bottom), S_o be the slope of channel bottom with respect to the horizontal and S_w be the water surface slope also with respect to the horizontal. From Fig. 12.26, the triangles OAC and OAB yield,

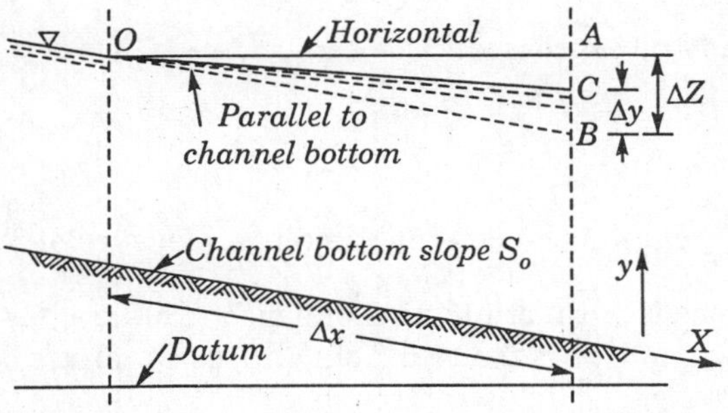

Fig. 12.26. Notation for derivation of S_w.

$$S_w = \frac{AC}{OC} = \frac{AB - BC}{OC} = \frac{\Delta Z - \Delta y}{\Delta x} = \frac{\Delta Z}{\Delta x} - \frac{\Delta y}{\Delta x}$$

writing in terms of differentials.

$$S_w = \frac{dz}{dx} - \frac{dy}{dx} = S_o - \frac{dy}{dx} \quad \text{...(12.56 } a\text{)}$$

Eq. (12.56 a) relates the channel slope S_o, the water surface slope dy/dx with respect to the channel bottom and S_w, the water surface slope with respect to horizontal.

12.18.3. Classification of Channel Slopes

When the discharge and the channel roughness are given, either Chezy's or Manning's formula can be used to determine the channel slope. If the flow in the channel is uniform, the channel is said to have a *normal slope* denoted by S_n. When the uniform flow occurs at the critical state, the channel has a *critical slope* S_c.

For a particular discharge and a given channel roughness, the critical slope may be used as a criterion for classifying the channel slope as mild, critical, steep, horizontal and adverse. The following definitions relate to a known discharge and a given channel roughness.

(*i*) *Mild slope*. If the slope of the channel bottom S_o is less than the critical slope S_c ($S_o < S_c$), the channel slope is known as *mild*. In channels of *mild slope*, the normal depth is always greater than the critical depth, *i.e.*, $y_n > y_c$.

(*ii*) *Critical slope*. When the channel slope S_o is equal to the critical slope S_c ($S_o = S_c$), the channel is said to have a *critical slope*. In such channels, the normal depth is equal to the critical depth, *i.e.*, $y_n = y_c$.

(*iii*) *Steep slope*. When the channel slope S_o exceeds the critical slope S_o ($S_o > S_c$), the channel slope is called *steep*. The critical depth for such channels is always greater than the normal depth, *i.e.*, $y_c > y_n$.

(*iv*) *Horizontal slope*. When the channel bottom is horizontal, *i.e.*, $S_o = 0$, the channel is said to have a *horizontal slope*. The normal depth tends to be infinite for such channels, *i.e.*, $y_n \to \infty$.

(*v*) *Adverse slope*. When the channel bottom rises in the direction of flow ($S_o < 0$), the channel slope is known as *adverse* or 'negative' slope. The normal depth in such channels dose not exist.

It may be noted that a given channel having a certain bottom slope S_o will behave as mild slope at lower discharges ; for the critical discharge, the channel slope will be equal to the critical slope and at still higher discharges, it will be a steep channel. This may be deduced from the fact that the critical depth in a given channel depends only on the discharge. For higher discharges, the critical depth will be large. From the Manning's formula the discharge

$$Q = \frac{1}{n} AR^{2/3} S_e^{1/2}$$

from which the energy line slope,

$$S_e = \frac{Q^2 n^2}{A^2 R^{4/3}} \qquad ...(12.57)$$

From Eq. (12.57), it can be seen that for each value of the depth, the channel must have a different energy line slope to maintain uniform flow at the given rate Q. For lower discharges, the critical depth has lower values for which the Manning's equation yields high values of critical slope such that the condition for a mild slope ($S_o < S_c$) is satisfied. Similarly a higher discharge will result in a lower critical slope such that $S_o > S_c$, the same channel thus now behaving as a steep channel. Thus it is seen that a channel laid to a certain slope will function as a mild, critical and steep channel under different discharge conditions.

Using Manning's formula, the discharge Q may be expressed as

$$Q = \frac{1}{n} AR^{2/3} S_e^{1/2} = K\sqrt{S} \qquad ...(12.58)$$

in which $K = \frac{1}{n} AR^{2/3}$. The term K is known is *conveyance* of the channel, and is a measure of its carrying capacity. For a given channel, the conveyance is a direct function of depth y. Making use of assumption 3 listed earlier, the energy line slope S_e may be computed from either Manning's or Chezy's formula. From Eq. (12.58), the energy line slope S_e may be written as

$$S_e = \frac{Q^2}{K^2} \qquad ...(12.59)$$

Since the conveyance K increases with increasing depth, Eq. (12.59) shows that the energy line slope decreases.

For given values of channel roughness n, discharge Q and the channel slope S_0, there in only one depth possible at which uniform flow occurs. It is known as the *normal depth* and is represented by y_n. The energy line slope is then equal to the bottom slope and $dy/dx = 0$. The channel slope, and the normal depth are related by the following equation :

$$S_o = \frac{Q^2 n^2}{A_r^2 R_n^{4/3}} = \frac{Q^2}{K_n^2} \qquad ...(12.60)$$

where A_n = area of flow at the normal depth y_n

R_n = hydraulic radius at the depth y_n, and

K_n = conveyance at the normal depth y_n.

When critical flow occurs, the slope is given by

$$S_c = \frac{Q^2 n^2}{A_c^2 R_c^2} = \frac{Q^2}{K_c^2} \qquad ...(12.60\ a)$$

For a mild slope $S_o < S_c$, which results in $K_n > K_c$ from Eqs. (12.60) and (12.60 *a*). Since the conveyance is a direct function of the depth y, $K_n > K_c$ yields $y_n > y_c$. Thus for a mild slope the normal depth is greater than the critical depth. In a similar way, it can be shown that for critical slope $y_n = y_c$ and for a steep slope $y_n < y_c$. The relative positions of the normal depth line (N.D.L.)

and the critical depth line (C.D.L.) above the channel bottom may, therefore, be used to distinguish and identity channel slopes. For horizontal slopes $S_o = 0$, and Eq. (12.60) yields $K_n = \infty$ resulting in $y_n = \infty$. The normal depth line, therefore, does not exist in horizontal channels. In case of adverse slopes, S_o is negative and since Q^2 cannot be negative, Eq. (12.60) indicates a negative value of K_n^2 resulting in an imaginary normal depth.

12.18.4. Classification of Surface Profiles

The surface profiles are primarily classified according to the channel slope. For example, surface profiles on the mild slope are designated by letter M. The space above the channel bottom is divided into three zones.

Zone 1— It includes the region which lies above the normal depth line and the critical depth line.

Zone 2— It represents the region which lies between the normal depth line and the critical depth line.

Zone 3— It is the region which lies below the normal depth line and critical depth line but above the channel bottom.

According to the magnitude of depth of flow y which may lie either in zone 1, 2 or 3, the profiles are further classified, and are accordingly assigned subscripts 1, 2 and 3 to the letters representing the channel slope ($M, C, S, H,$ and A). With these five channel slopes and three zones, it might appear that 15 different types of profiles can occur. As a matter of fact, zone 2 does not exist in critical slope as $y_c = y_n$; in horizontal channels $y_n = \infty$, and as such zone 1 is non-existent, and for adverse slopes y_n being imaginary, zone 1 is again out of picture. This way, it is not possible to have C_2, H_1 and A_1 surface profiles. Thus we are left with 12 surface profiles : $M_1, M_2, M_3, C_1, C_3, S_1, S_2, S_3, H_2, H_3, A_2$ and A_3.

Analysis of Eq. (12.56) to evaluate the sign of dy/dx will show that the numerator $(S_o - S_c)$ depends only upon the relative values of the normal depth and the depth of flow [this is evident from Eqs. (12.59) and (12.60)]. From Eq. (12.59), it can be seen that S_c decreases with increasing depth. For $y > y_n$, $S_e < S_o$ and $(S_o - S_e)$ is positive, and when $y < y_n$, $S_e > S_o$ and $(S_o - S_e)$ becomes negative. The sign of the denominator $(1 - \mathbf{F}^2)$ depends upon the relative magnitude of the depth of flow and the critical depth. Thus if $y > y_c$, $\mathbf{F} < 1$, the flow is subcritical and $(1 - \mathbf{F}^2)$ is positive. On the contrary, if $y < y_c$ so that $\mathbf{F} > 1$ representing supercritical flow, the denominator becomes negative. Having determined the signs of the numerator and the denominator under different conditions of flow, it is now easy to define various types of surface profiles.

12.18.5. Backwater and Drawdown Curves

In gradually varied flow, the flow properties vary along the channel length, thereby affecting the magnitude of quantities appearing in Eq. (12.56), causing the water surface slope dy/dx either positive or negative. When the depth of flow y increases in the direction of flow, dy/dx is positive and the water surface profile is known as a *backwater curve*. On the other hand, when the depth decreases along the flow direction, dy/dx becomes negative, and the surface profile is called a *drawdown curve*.

Backwater Curve. For a backwater curve, dy/dx is positive. For this condition Eq. (12.56) indicates two possible cases :

(1) $(S_o - S_e) > 0$ and $(1 - \mathbf{F}^2) > 0$, and

(2) $(S_o - S_e) < 0$ and $(1 - \mathbf{F}^2) < 0$.

Considering the first case $(S_o - S_e) > 0$ yields $y > y_n$ while the denominator $(1 - \mathbf{F}^2) > 0$ yields $y > y_c$. These two conditions combined together represent two surface profiles :

(*i*) $y > y_n > y_c$, since $y_n > y_c$ the channel has a mild slope and because $y > y_c$, the flow is subcritical and the surface profile is of M_1 type, and

(*ii*) $y > y_c > y_n$, we have subcritical flow $(y > y_c)$ in a steep channel $(y_c > y_n)$ and the surface profile is of S_1 type.

Similarly, analysis of the second case $(S_o - S_e) < 0$ and $(1 - \mathbf{F}^2) < 0$ yields $y < y_n$ and $y < y_c$. If $y < y_n < y_c$ it indicates supercritical flow in a steep channel and the surface profile is of S_3 type, and if $y < y_c < y_n$, it represents supercritical flow in a mild channel, with a M_3 type surface profile.

Drawdown Curve. For a drawdown curve, dy/dx is negative and Eq. (12.56) yields two possible cases :

(1) $(S_o - S_e) > 0$ and $(1 - \mathbf{F}^2) < 0$, and

(2) $(S_o - S_e) < 0$ and $(1 - \mathbf{F}^2) > 0$.

The analysis of the first case results in $y > y_n$ and $y < y_c$. If $y_c > y > y_n$, supercritical flow takes place in a steep channel and the profile is of S_2 type. Similarly the second case yields $y < y_n$ and $y > y_c$, indicating if $y_n > y > y_c$, it represents subcritical flow in a mild channel, and the profile is of M_2 type.

12.18.6. Examples of Various Types of Water Surface Profiles

When the water surface is parallel to channel bottom, thereby indicating no change in the depth of flow $dy/dx = 0$ and Eq. (12.56) yields $S_o = S_e$ resulting in $y = y_n$. This represents uniform flow. If $y = y_n = y_c$, it represents uniform critical flow. This flow is uniform subcritical, if $y = y_n > y_c$ and if $y_c > y_n = y$, the flow is then uniform supercritical.

For a horizontal slope, $S_o = 0$, and Eq. (12.60) yields $K_n = \infty$ which results in $y_n = \infty$, and Eq. (12.56) may be written as

$$\frac{dy}{dx} = \frac{-S_e}{1-\mathbf{F}^2}$$

For a backwater curve, $(1 - \mathbf{F}^2) < 0$ or $y < y_c$ which represents supercritical flow with a H_3 profile. The water surface profile will be a drawdown curve if $(1 - \mathbf{F}^2) > 0$ or $y > y_c$ indicating H_2 profile and subcritical flow.

In a channel of adverse slope, $S_o < 0$ and the normal depth is imaginary. The numerator of Eq. (12.56) becomes negative and there are two possible cases :

(1) $(1 - \mathbf{F}^2) > 0$ yielding $y > y_c$. This condition represents a drawdown curve of A_2 type with a subcritical flow.

(2) $(1 - \mathbf{F}^2) < 0$ resulting in $y < y_c$. This indicates a backwater curve of A_3 type having supercritical flow.

Inspection of Eq. (12.56) leads to the following important observations :

(*i*) as $y \to \infty$, $S_e \to 0$, $\mathbf{F} \to 0$ and $dy/dx = S_o$

This condition implies that for large depths, the water surface tends to become horizontal.

(*ii*) as $y \to y_n$, $S_e \to S_o$ and $dy/dx \to 0$.

It states that all surface profiles approach the normal depth line asymptotically. This is because uniform flow occurs at sections far away from obstructions such as weirs

(*iii*) as $y \to y_c$, $\mathbf{F} \to 1$ and $dy/dx \to \infty$,

this means that the surface profiles which cross the critical depth line do so perpendicularly. The assumption of straight and parallel streamlines is invalidated as y approaches y_c simply because the streamlines while crossing the critical depth assume sharp curvature and hence the gradually varied flow equation is not applicable in the neighbourhood of the critical depth. The water surface profiles are, therefore shown dotted in the vicinity of critical depth line.

Instances of physical occurrence of various types of surface profiles are illustrated in Figs. 12.27 to 12.31.

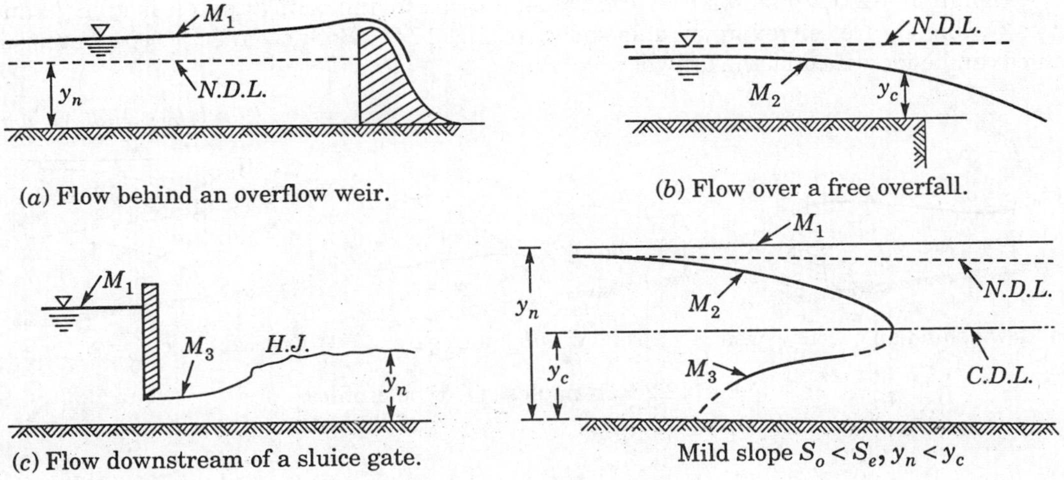

(a) Flow behind an overflow weir.

(b) Flow over a free overfall.

(c) Flow downstream of a sluice gate.

Mild slope $S_o < S_e, y_n < y_c$

Fig. 12.27. Examples of *M*-type surface profiles.

The surface profiles of the *C*-type will be obtained when the normal depth and the critical depth are equal. The water surface profiles will be classified as C_1 and C_3 depending on whether the depth of water is greater or less than the critical depth respectively.

For a wide rectangular channel, one can write the following expression for the water surface slope.

$$\frac{dy}{dx} = S_0 \frac{1-(y_n/y)^{10/3}}{1-(y_c/y)^3} \qquad \text{...(12.60 } a)$$

It may be noted that the normal depth line (N.D.L.) and the critical depth line (C.D.L.) merge into one and when the water surface approaches critical depth, *i.e.*, when $y \to y_c = y_n$, Eq. (12.60 *a*) shows that dy/dx attains an indeterminate form 0/0.

Let us find out the value of dy/dx in the limit,

$$\frac{y_n}{y} \to 1, \frac{y_c}{y} \to 1$$

Considering the numerator of Eq. (12.60 *a*),

Let $\quad \dfrac{y_n}{y} = 1 + h \quad$ as $\quad y_n/y \to 1, h \to 0$

$$\left(\frac{y_n}{y}\right)^{10/3} = (1+h)^{10/3} = 1 + \frac{10}{3}h + \frac{\frac{10}{3}\times\frac{7}{3}}{1\times 2}h^2 + \ldots$$

$$= 1 + \frac{10}{3}h \text{ (neglecting higher powers of } h)$$

or $$1-\left(\frac{y_n}{y}\right)^{10/3} = 1-\left(1+\frac{10}{3}h\right) = -\frac{10}{3}h$$

Similarly, by letting $(y_c/y) = 1 + h$, one can obtain

$$1-(y_c/y)^3 = 1-(1+3h) = -3h$$

Equation (12.60 *a*) may now be written as,

$$\frac{dy}{dx} = S_o \frac{-(10/3)h}{-3h} = \frac{10}{9}S_o \qquad \text{...(12.60}b)$$

Equation (12.60 *b*) shows that the value of *dy*/*dx* at the critical depth is 10 $S_o/9$ and the curves C_1 and C_3 are convex upward as shown in Fig. 12.28. Both C_1 and C_3 type water surface profiles can be identified as the drawdown curves.

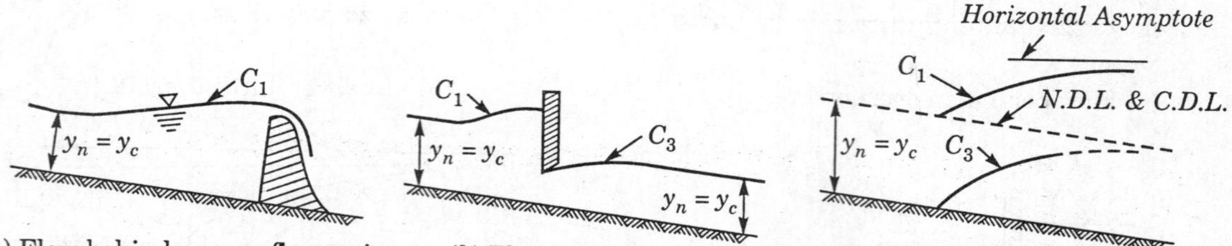

(*a*) Flow behind an overflow weir. (*b*) Flow at a sluice gate. Critical slope $S_o = S_c, y_n = y_c$

Fig. 12.28. Examples of *C*-type profiles.

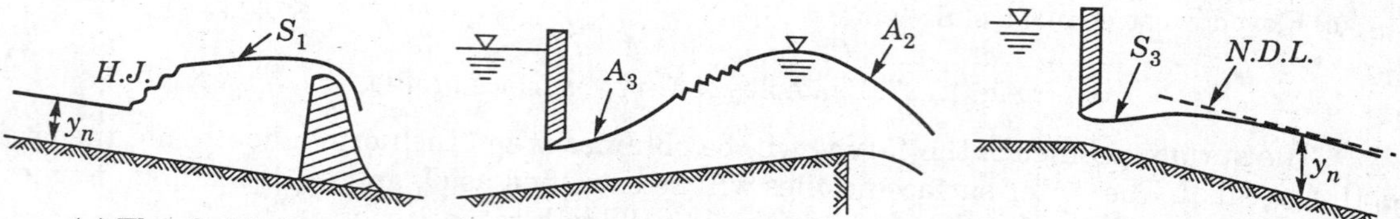

(*a*) Flow behind an overflow weir.

(*b*) Flow from steep to steeper channel.

(*c*) Flow downstream of a sluice gate

Steep slope
$S_o > S_c, y_c > y_o$

Fig. 12.29. Examples of *S*-type profiles.

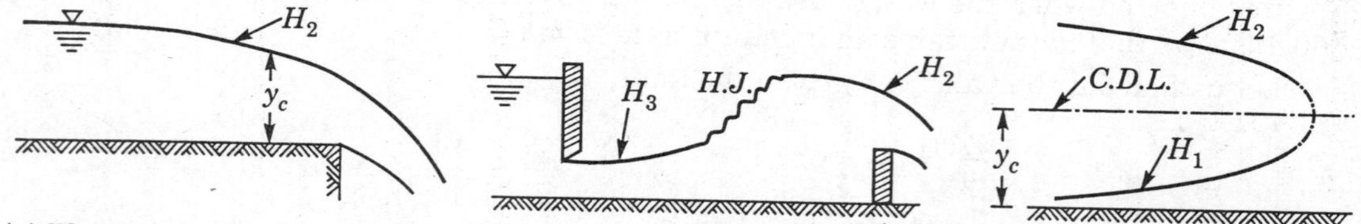

(*a*) Flow over a free overfall.

(*b*) Flow downstream of a sluice gate.

Horizontal slope
$S_o = 0, y_n = \infty$

Fig. 12.30. Examples of *H*-type profiles.

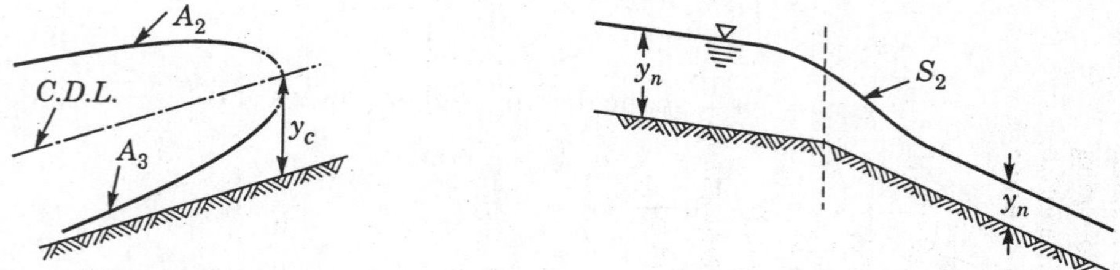

Adverse slope
$S_o < 0, y_n$ is imaginary.

(*a*) Flow downstream of a sluice and terminating in a free overall.

Fig. 12.31. Examples of *A*-type profiles.

Surface profiles of *S*, *H* and *A* types are illustrated in Figs. 12.29 to 12.31.

12.18.7. Length of the Surface Profile

In many of the gradually varied flow problems, we are required to find out the length of the curve upto which a particular depth is reached. The following methods may be considered.

12.18.7.1. The Numerical Integration Method. The differential equation of the gradually varied flow may be inverted and integrated to give the length of the surface profile between depths y_1 and y_2. The distance along the channel bottom using Eq. (12.56) may be written as

$$x = \int_{y_1}^{y_2} \frac{dx}{dy} \cdot dy = \int_{y_1}^{y_2} \frac{1 - \mathbf{F}^2}{S_o - S_e} dy \qquad \text{...(12.61)}$$

The integration of Eq. (12.61) is impossible for all types of channel sections, since the equation cannot be expressed explicitly as a function of y. Eq. (12.61) may be integrated graphically by plotting dx/dy against y, and computing the area under the curve enclosed by the **y-axis** and the ordinates of dx/dy corresponding to depths y_1 and y_2 (Fig. 12.32). Alternatively, the length of the curve may be computed by numerical integration of Eq. (12.61).

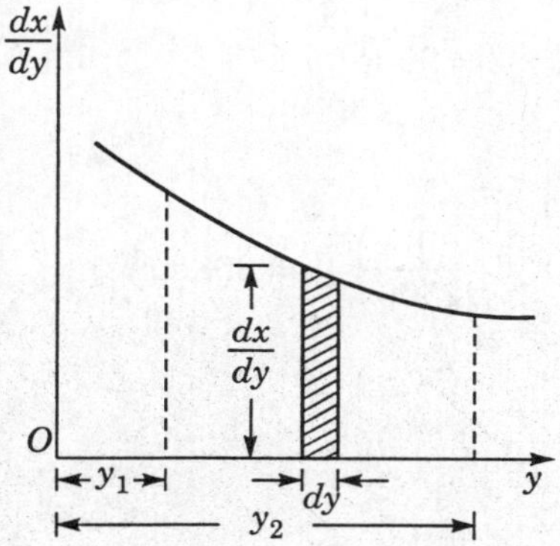

Fig. 12.32. Graphical integration method.

The numerical integration method involves division of channel length into number of short reaches and performing the calculation step by step from one end of the channel to the other.

The depths at the two ends of a short reach are known and, therefore, other flow parameters such as area, hydraulic depth, hydraulic radius, velocity (and hence the Froude number) and the energy line slopes can be determined at these sections. The average values of these quantities may then be obtained to represent the conditions of flow at the mean section. The inverted slope dx/dy using Eq. (12.56) and thereafter the length of the curve between the depths under consideration, can be determined by multiplying dx/dy with Δy, where Δy is the change in depth between the two ends of the reach. Such calculations are to be repeated for adjoining short reaches one after the other till the desired depth is reached. This method is simple and may be applied to prismatic as well as non-prismatic channels of any shape and slope.

12.18.7.2. The Direct Step Method. This method is also known as step by step method and is a simple method applicable to prismatic channels only. The channel is divided into short reaches and the computations are carried out from one end of the reach to the other, as in case of the numerical integration method. Fig. 12.33 (*a*) illustrates a short reach of the channel sections (1) and (2) situated at the ends of the reach, are separated by a distance Δx. With the horizontal datum as shown, application of the Bernoulli's equation to sections (1) and (2) yields,

$$S_o \Delta x + y_1 + \frac{V_1^2}{2g} = y_2 + \frac{V_2^2}{2g} + S_e \Delta x \qquad \text{...(12.62)}$$

Noting that the specific energy $E = y + V_2/2g$, Eq. (12.62) may be solved for Δx,

$$\Delta x = \frac{E_2 - E_1}{S_o - S_e} = \frac{\Delta E}{S_o - S_e} \qquad \text{...(12.63 a)}$$

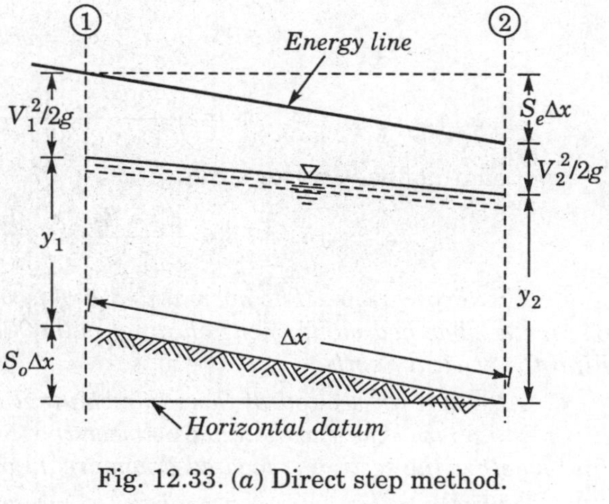

Fig. 12.33. (*a*) Direct step method.

The energy line slope S_e appearing in Eq. (12.63 *a*) is the average of its values at sections (1) and (2),

$$S_e = \frac{(S_e)_1 + (S_e)_2}{2}$$

In case of sub-critical flow in the channel, the computations should proceed in the upstream direction, and if it is super-critical they should be carried out in the downstream direction.

Example 12.51. *A river 90 m wide and 3 m deep has stable bed and vertical banks with a surface slope of 1 in 2500. Estimate the length of backwater curve produced by an afflux of 2 m. Assume Manning's n = 0.035.*

Solution. $b = 90$ m, $y = 3$ m

$$S_e = \frac{1}{2500}$$

$$A = 90 \times 3 = 270 \text{ m}^2 ; \qquad P = 90 + 2 \times 3 = 96 \text{ m}$$

$$R = \frac{270}{96} = 2.81 \text{ m.}$$

Discharge $\quad Q = \frac{1}{n} AR^{2/3} S_e^{1/2} = \frac{1}{0.035} \times 270 \times (2.81)^{2/3} \left(\frac{1}{2500}\right)^{1/2} = 308 \text{ m}^3/\text{s}.$

Specific energy at 3 m depth

$$E_1 = 3 + \frac{(308/90 \times 3)^2}{2 \times 9.81} = 3.0663 \text{ m.}$$

Specific energy at 3 + 2 = 5 m depth

$$E_2 = 5 + \frac{(308/90 \times 5)^2}{2 \times 9.81} = 5.0239 \text{ m.}$$

Energy line slope at the mean section where the depth is

$$\frac{3+5}{2} = 4 \text{ m}$$

$$S_e = \frac{Q^2 n^3}{R^2 A^{4/3}}, \qquad A = 90 \times 4 = 360 \text{ m}^2$$

$$P = 90 + 2 \times 4 = 98 \text{ m}, \qquad R = \frac{360}{98} = 3.67 \text{ m}$$

$$S_e = \frac{(308 \times 0.035)^2}{(360)^2 \, (3.67)^{4/3}} = 1.58 \times 10^{-4}.$$

Length of back water curve

$$\Delta x = \frac{E_2 - E_1}{S_o - S_e} = \frac{5.0239 - 3.0663}{4 \times 10^{-4} - 1.58 \times 10^{-4}} = \mathbf{8100 \text{ m.}}$$

Example 12.52. *The normal depth of flow of water, in a rectangular channel 1.5 m wide, is one metre. The bed slope of the channel is 0.0006 and Manning's roughness coefficient n = 0.012. Find the critical depth.*

At a certain section of the same channel the depth is 0.92 m while at a second section the depth is 0.86 m. Find the distance between the two sections (use one reach in the calculations). Also find whether the section is located downstream or upstream with respect to the first section.

Solution. $b = 1.5$ m, $y_n = 1$ m,

$$A = 1.5 \times 1 = 1.5 \text{ m}^2$$

$$P = 1.5 + 2 \times 1 = 3.5 \text{ m}$$

$$Q = \frac{1}{n} AR^{2/3} S_o^{1/2} = \frac{1}{0.012} \times 1.5 \times \left(\frac{1.5}{3.5}\right)^{2/3} \left(\frac{6}{10,000}\right)^{1/2} = 1.73 \text{ m}^3/\text{s}.$$

$$q = \frac{Q}{b} = \frac{1.73}{1.5} = 1.152 \text{ m}^3/\text{s/m}.$$

The critical depth, $y_c = \left(\frac{q^2}{g}\right)^{1/3} = \left[\frac{(1.152)}{9.81}\right]^{1/3}$ **= 0.512 m.**

Specific energy at 0.92 m depth

$$E_1 = 0.92 \times \frac{V_1^2}{2g}$$

$$V_1 = \frac{1.73}{1.5 \times 0.92} = 1.25 \text{ m/s}$$

$$E_1 = 0.92 + \frac{(1.25)^2}{2 \times 9.81} = 0.9995 \text{ m}.$$

Specific energy at 0.86 m depth, where the velocity

$$V_2 = \frac{1.74}{1.5 \times 0.86} = 1.34 \text{ m/s}$$

$$E_2 = 0.86 + \frac{(1.34)^2}{2 \times 9.81} = 0.9515 \text{ m}.$$

Energy line slope at the mean section where

$$y = \frac{1}{2}(y_1 + y_2) = 0.89 \text{ m}.$$

$$S_e = \frac{Q^2 n^2}{A^2 R^{4/3}} = \frac{(1.73 \times 0.012)^2}{(1.5 \times 0.89)^2 (0.407)^{4/3}} = 8.01 \times 10^{-4}.$$

Distance between the two sections

$$\Delta x = \frac{E_2 - E_1}{S_o - S_e} = \frac{0.9515 - 0.9995}{6 \times 10^{-4} - 8.01 \times 10^{-4}} = \mathbf{238.80 \text{ m}.}$$

Water surface slope $\frac{dy}{dx} = \frac{S_o - S_e}{1 - V^2/gy}$

Average depth of flow $y = 0.89$ m.

Velocity at the mean section,

$$V = \frac{1.73}{1.5 \times 0.89} = 1.28 \text{ m/s}$$

$$\frac{V^2}{gy} = \frac{(1.28)^2}{9.81 \times 0.89} = 0.188$$

$$\frac{dy}{dx} = \frac{6 \times 10^{-4} - 7.9 \times 10^{-4}}{1 - 0.188} = -2.34 \times 10^{-4}$$

Since dy/dx is negative, the second section is downstream.

Example 12.53. *A dam is built across a stream of rectangular cross-section which carries water at the rate of 5 m³/s. As a result the depth of flow just upstream of the dam is increased to 2.5 m. The stream is 3 m wide and has a bed slope of 1 in 4900. The channel is lined with concrete (Manning's n = 0.015). How far upstream is the depth within 10 cm of the normal depth ?*

Solution. The normal depth y_n may be determined by using Manning's formula

$$V = \frac{Q}{A} = \frac{1}{n} R^{2/3} S_e^{1/2}$$

$$\frac{5}{3y_n} = \frac{1}{0.015}\left(\frac{3y_n}{3+2y_n}\right)^{2/3}\left(\frac{1}{4900}\right)^{1/2}$$

Table. Computation for Ex. 12.45

Depth range y_2, y_1	Average $y = \frac{1}{2}(y_1+y_2)$	$A = 3y$	$P = 3 + 2y$	$R = \frac{A}{P}$	$V = \frac{Q}{A}$	$F^2 = \frac{V^2}{gy}$	$1 - F^2$	$S_o \times 10^4$	$S_e \times 10^4$	$(S_o - S_e) \cdot 10^4$	dx/dy	Δy	Δx	x
2.50—2.35	2.425	7.275	7.85	0.927	0.687	0.0199	0.9901	2.04	1.173	0.867	11,420	0.15	1712	1712
2.35—2.20	2.275	6.825	7.55	0.904	0.732	0.0240	0.9760	2.04	1.380	0.660	14,790	0.15	2220	3932
2.20—2.05	2.125	6.375	7.25	0.879	0.784	0.0295	0.9705	2.04	1.640	0.400	24,250	0.15	3640	7572

The distance between the dam face and the section, where y = 2.05 m is 7 km and 572 m.

Solving by trial and error, y_n = 1.95 m

The critical depth

$$y_c = \left(\frac{q^2}{g}\right)^{1/3} = \left(\frac{1.665 \times 1.665}{9.81}\right)^{1/3} = 1.415 \text{ m.}$$

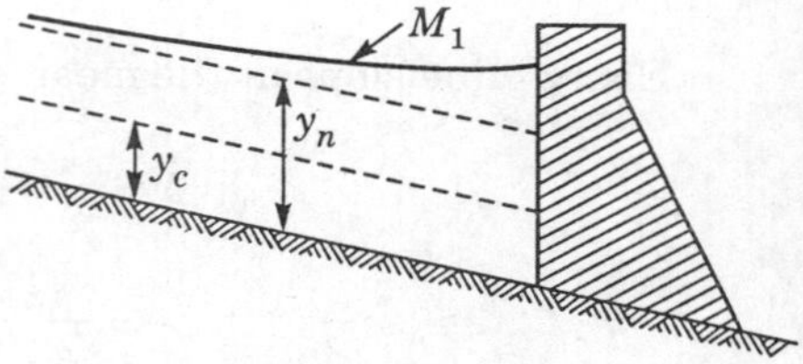

Since y (= 2.5 m) > y_n > y_c, the water surface curve is of M_1 type and the flow is sub-critical.

The length of the water surface curve is to be determined between the depths 2.5 and 1.95 + 0.10 = 2.05 m. Dividing this depth range 2.5 – 2.05 = 0.45 m into three equal parts of 0.15 m each and starting computation from the dam and proceeding upstream. The relevant computed quantities have been tabulated.

12.18.7.3. The circular Arc Method

The length of back-water curve, of the type shown in Fig. 12.33 (*b*) (*i.e.* M_1-type only) can be determined by the circular arc method. The afflux h created by construction of a weir will have its effect far upstream, as indicated by the back water curve QP. The curve joins the normal depth line asymptotically. Let the tangent at the water surface at point of afflux Q meet the normal depth line

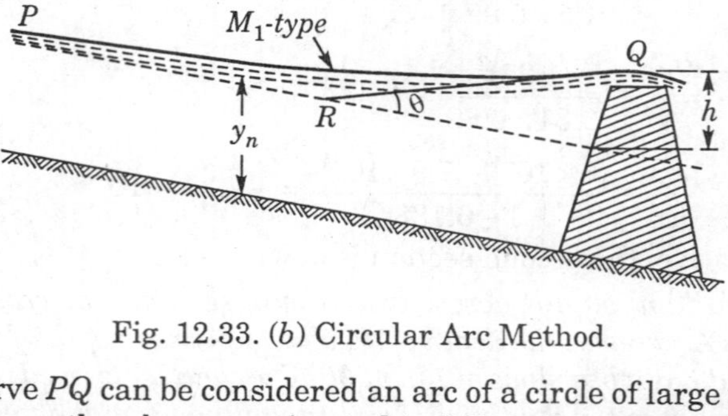

Fig. 12.33. (*b*) Circular Arc Method.

at R. The back-water curve PQ can be considered an arc of a circle of large radius. Let the length of the curve PQ be L ; QR may then be approximated to $L/2$, and assuming QR to be horizontal,

$$\tan\theta = \frac{h}{QR} = \frac{2h}{L} \quad \text{but} \quad \tan\theta = \frac{dy}{dx}$$

and hence, $$L = \frac{2h}{dy/dx} \qquad ...(12.63\ b)$$

in which dy/dx is the slope of water surface at the weir location, *i.e.* at Q, which can be calculated from Eqs. (12.56) and (12.57).

Example 12.54. *A river 45 m wide has a normal depth of flow of 3 m and an average bed slope of 1 in 10,000. A weir is built across the river raising the water surface level at the weir site to 5 m above the bottom of the river. Assuming that the back-water curve is an arc of a circle. Calculate the approximate length of the back-water curve. Consider that the river is prismatic. Take the value of n in Manning's formula as 0.025.*

Solution. $b = 45$ m, $y_n = 3$ m

$$S_0 = \frac{1}{10{,}000},\ y = 5 \text{ m},\ n = 0.025$$

Afflux $h = 5 - 3 = 2$ m.

Length of back-water curve, from Eq. (12.63 *b*)

$$L = \frac{2h}{dy/dx}$$

$$A = 45 \times 3 = 135 \text{ m}^3$$

$$R = \frac{45 \times 3}{45 + 2 \times 3} = 2.65 \text{ m},$$

$$Q = \frac{AR^{2/3}S_0^{1/2}}{n} = \frac{135 \times (2.65)^{2/3} \times (1/10{,}000)^{1/2}}{0.025} = 103.5 \text{ m}^3/\text{s}$$

At the weir site : $y = 5$ m, $V = \dfrac{Q}{45 \times 5} = \dfrac{103.5}{225} = 0.46$ m/s

$$R = \frac{225}{45 + 2 \times 5} = 4.09 \text{ m}$$

Water surface slope at the weir

$$\frac{dy}{dx} = \frac{S_0 - S_e}{1 - V^2/gy}$$

in which S_e is the slope of the total energy line at the weir ; V and y are the velocity and depth at the weir.

$$S_e = \frac{Q^2n^2}{A^2R^{4/3}} = \frac{V^2n^2}{R^{4/3}} = \frac{(0.46)^2(0.025)^2}{(4.09)^{4/3}} = 0.202 \times 10^{-4}$$

$$\frac{V^2}{gy} = \frac{(0.46)^2}{9.81 \times 5} = 0.0043,\ \frac{dy}{dx} = \frac{10^{-3} - 0.202 \times 10^{-4}}{1 - 0.0043} = 0.802 \times 10^{-4}$$

$$L = \frac{2 \times 2}{0.802 \times 10^{-4}} = 4.99 \times 10^4 \text{ m}$$

Length of back-water curve = **4.99 × 10⁴ m = 49.9 km.**

12.19. Control Section

A section where a definite relationship between depth of flow and the discharge exists is known as a *control section*. For critical flow, we already know that there is a definite depth discharge relationship and hence a critical depth section is a control section. A control section is, therefore, a suitable site for measuring flow in open channels. Sub-critical flows are affected by changes in downstream conditions. The control section for such flows is, therefore, located downstream. On the other hand, upstream conditions affect super-critical flow and hence the control section for such flows are located upstream.

Entrance to a channel (controlled by a sluice gate) and its exit under certain conditions become control sections. A gate located in the channel can be a control for both upstream and downstream reaches. A section where channel slope changes from mild to steep is also a control section. Figs. 12.34 to 12.37 illustrate control section, under different conditions of flow and channel slope.

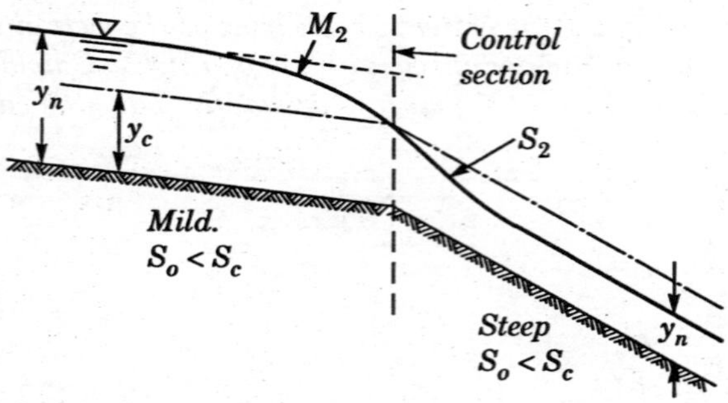

Fig. 12.34. Control section : Change of slope from mild to steep in a prismatic channel.

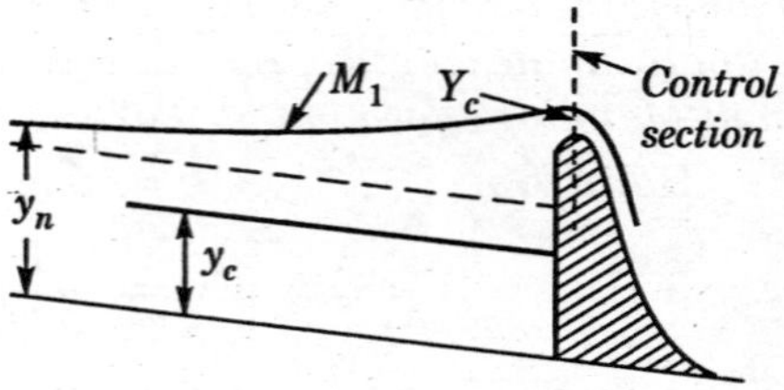

Fig. 12.35. Upstream flow controlled by an overflow spillway.

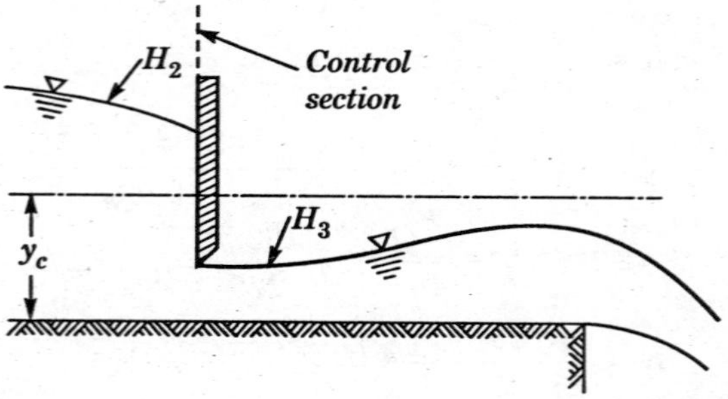

Fig. 12.36. Control section (a gate installed in a horizontal channel) influencing upstream and downstream conditions.

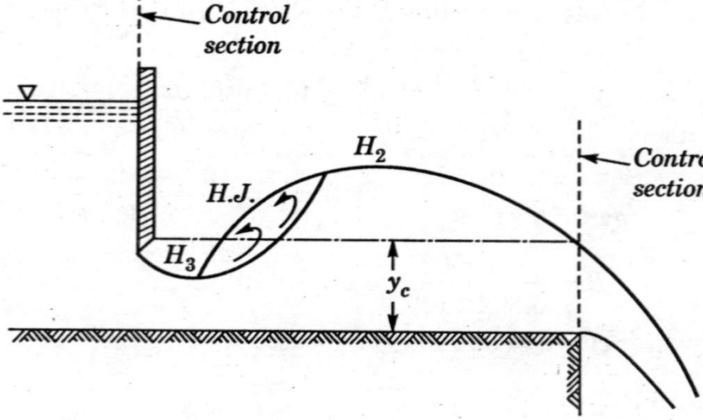

Fig. 12.37. Sluice gate and the free overfall constituting control sections with a hydraulic jump in between.

It is not necessary that a control section be also a critical section. A critical section, however, is a control section. When the channel is on a sub-critical slope a control section at the downstream end may be a critical section, such as that obtained on the top a spillway. A sluice gate or any other hydraulic structure may also be used to create a control section.

Critical flow may be achieved without change in the channel slope, by raising the channel bottom (as in broad crested weir) or by suitably contracting the channel width such as in a venturiflume.

Example 12.55. *A wide rectangular channel has a change in bed slope from a mild slope of 0.0004 to a steep slope of 0.01. If the discharge per metre width is 20 m³/s, find the depths on the mild slope and steep slope reaches. What is the depth at the point where the slope changes ? Sketch and classify the surface profiles. Take Mannings n = 0.01.*

Solution. From the continuity equation

$$q = Vy$$

or $$V = \frac{q}{y} = \frac{20}{y}, \text{ but from the Manning's equation}$$

$$V = \frac{1}{n} R^{2/3} S^{1/2} \quad \text{...(1)}$$

The depths attained by the flow will be the normal depths in the respective portions of the channel.

Normal depth in the Mild Channel

Using Eq. (1), and noting that for a wide rectangular channel $R \approx y$,

$$\frac{20}{y} = \frac{1}{0.01} \times (y)^{2/3} \times (0.0004)^{1/2}$$

simplifying $$y^{5/3} = 10, \qquad \text{or } y = 3.9810 \text{ m.}$$

This depth is the normal depth in the mild channel

$$\therefore \qquad y_n = 3.981 \text{ m.}$$

Normal depth in the Steep Channel

$$\frac{20}{y} = \frac{1}{0.01} \times (y)^{2/3} \times (0.01)^{1/2}$$

$$y^{5/3} = 2 \qquad \text{or} \qquad y = 1.5157 \text{ m.}$$

Hence the normal depth in the steep channel is $y_n = 1.5157$ m.

Critical Depth $$y_c = \left(\frac{q^2}{g}\right)^{1/3} = \left(\frac{20 \times 20}{9.81}\right)^{1/3} = 3.4346 \text{ m.}$$

Since the discharge per unit width q remains same in mild as well as steep reaches, the critical depth does not change.

In the mild reach $(y_n = 3.9810 \text{ m}) > (y_c = 3.4376 \text{ m})$

In the steep reach $(y_n = 1.5157 \text{ m}) < (y_c = 3.4376 \text{ m})$

At the break in the channel slope, *i.e.* point R marked on the figure shown below, the water surface passes through the point P (which is on the critical depth line) in order to attain the normal depth on the steep reach. Hence the depth at junction R of the slope is $y_c = 3.4376$ m.

The water surface profiles will be M_2 and S_2 in the mild and steep reaches respectively, and the same are shown in figure Ex. 12.55.

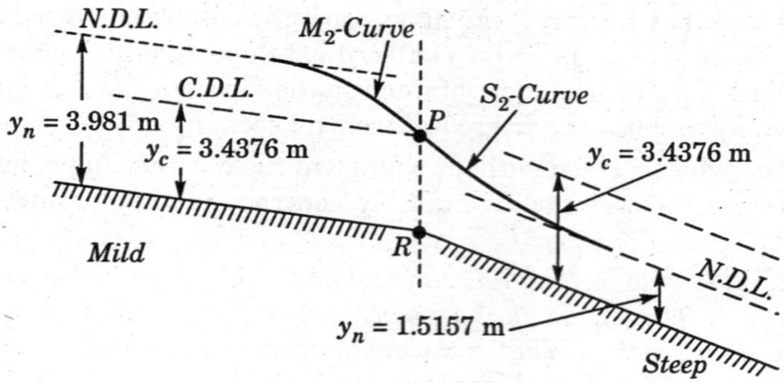

Fig. Ex. 12.55

Example 12.56. *Water flows from a lake into a steep channel 3 m wide. Determine the lake level above the channel bed at the outfall if the discharge is 40 m³/s.*

Solution. At the entrance to the steep channel, a control section exists *i.e.* the flow into the channel takes place at the critical depth, y_c. The water surface then joins the normal depth line as shown in the figure. The critical depth,

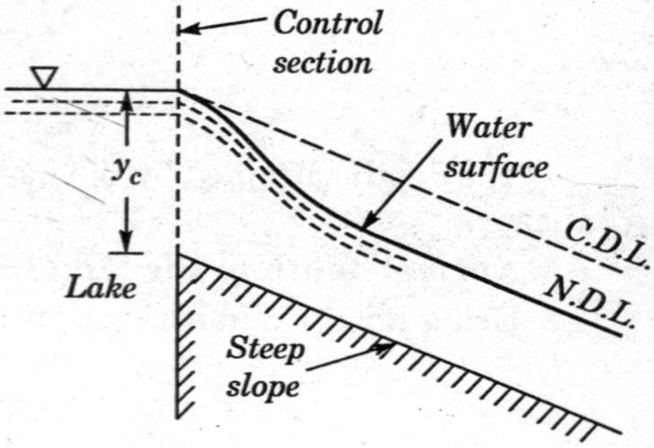

$$y_c = \left(\frac{q^2}{g}\right)^{1/3}$$

$$q = \frac{Q}{b} = \frac{40}{3} = 13.33 \text{ m}^3\text{/s/m}$$

$$y_c = \left(\frac{13.33^2}{9.81}\right)^{1/3} = 2.626 \text{ m}$$

$\therefore$ The water level in the lake above the channel bed at the outfall = **2.626 m.**

12.20. Hydraulic Jump

Super-critical flow in a channel of small slope is unstable, as the channel slope is unable to sustain it in the super-critical state. It must, therefore, change to sub-critical flow. We know that for super-critical flow, the depth of flow is less than the critical depth while that for the sub-critical flow is greater than the critical depth. Thus when flow changes from super-critical state to subcritical state, the depth of flow is increased, say from y_1 to y_2. This increase in depth occurs in a short length of the channel, and is known as the *hydraulic jump*. A hydraulic jump is, therefore, an abrupt rise in the depth of flow signifying a change of flow from super-critical state to sub-critical state. It represents a discontinuity in the liquid surface and involves (*i*) violent mixing due to high turbulence, and (*ii*) air entrainment.

In engineering practice hydraulic jump generally occurs at such places as the foot of spillway and downstream of a sluice gate where the velocities are very high. A high degree of turbulence, which is generated in the hydraulic jump, is responsible for high loss of energy. The hydraulic jump is, therefore, frequently employed as an effective energy dissipating device. The kinetic energy of flow after the jump is greatly reduced, which may sometimes be low enough to prevent erosion of the channel boundaries. It is also sometimes used as a mixing device to mix chemicals for water purification or sewage treatment. It may also be used to aerate water for municipal water supplies.

The hydraulic jump may be analysed by applying the continuity principle, momentum principle and the energy equation. For simplicity in mathematical treatment, we shall consider a horizontal channel of rectangular section. As hydraulic jump occupies a short length of the channel, resistance force due to friction (R_f) is negligible. The channel being horizontal, the component of weight of liquid, enclosed between the sections demarcating the jump, in the direction of flow ($W \sin\theta$) is zero.

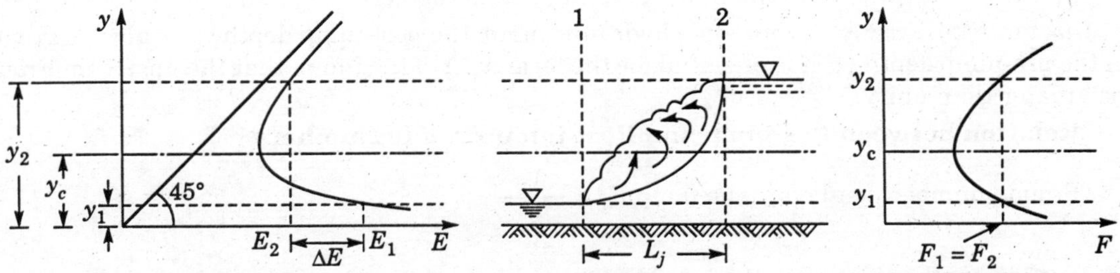

(a) Specific energy curve. (b) Hydraulic jump. (c) Specific force curve.

Fig. 12.38. Hydraulic jump in a horizontal rectangular channel.

The momentum equation applied to the body of water enclosed between sections (1) and (2) of the hydraulic jump, Fig. 12.38, may then be written as

$$\frac{\gamma\theta}{g}(V_2 - V_1) = P_1 - P_2 \qquad ...(12.64)$$

For a rectangular channel of width b, the hydrostatic force $P_1 = \gamma Z_1 A_1 = \gamma \times y_1/2 \times by_1 = \frac{1}{2}\gamma b y_1^2$ and similarly $P_2 = \frac{1}{2}\gamma b y_2^2$. Using the continuity equation

$$Q = b_1 y_1 V_1 = b_2 y_2 V_2$$

and introducing the discharge per unit with q, Eq. (12.64) becomes

$$\frac{\gamma(b\,.\,q)}{g}\left(\frac{q}{y_2} - \frac{q}{y_1}\right) = \frac{1}{2}\gamma b y_1^2 - \frac{1}{2}\gamma b y_2^2$$

which simplifies to
$$\frac{2q^2}{g} = y_1 y_2 (y_1 + y_2) \qquad ...(12.65)$$

Eq. (12.65) relates the conjugate (or sequent) depths y_1 and y_2 with the discharge per unit width q. Eq. (12.65) may be expressed as a quadratic equation in y_2,

$$y_2^2 + y_1 y_2 - \frac{2q^2}{g y_1} = 0$$

which for y_2 results in
$$y_2 = \frac{y_1}{2}\left[-1 \pm \sqrt{1 + \frac{8q^2}{g y_1^3}}\right] = \frac{y_1}{2}\left[-1 \pm \sqrt{1 + \frac{8V_1^2}{g y_1}}\right]$$

Disregarding the negative sign for obvious reasons,

$$\frac{y_2}{y_1} = \frac{1}{2}[\sqrt{1 + 8\mathbf{F}_1^2} - 1] \qquad ...(12.66)$$

where $\mathbf{F}_1 = V_1/\sqrt{g y_1}$, is the Froude number of approaching flow.

If the ratio y_1/y_2 is desired. Eq. (12.65) may be suitably re-arranged to yield,

$$y_1/y_2 = \frac{1}{2}(\sqrt{1 + 8\mathbf{F}_2^2} - 1)$$

In which $\mathbf{F}_2 = V_2/\sqrt{g y_2}$, is the Froude number downstream of the jump.

From Eq. (12.66), we see that the ratio of the conjugate depths y_2/y_1 is a function of the upstream Froude number $\mathbf{F}_1$ and that the larger the value of $\mathbf{F}_1$, the greater will be the depth ratio. Depending on the value of $\mathbf{F}_1$, Eq. (12.66) leads to the following three cases :

(*i*) for $\mathbf{F}_1 < 1, y_2/y_1 < 1$. This is the condition for a falling liquid surface (hydraulic drop in sub-critical flow) :

(*ii*) for $\mathbf{F}_1 = 1, y_2/y_1 = 1$. The flow in the channel is then critical ; and

(*iii*) for $\mathbf{F}_1 > 1, y_2/y_1 > 1$. Since in a hydraulic jump, the post-jump depth y_2, is always greater than the pre-jump depth y_1, it is essential for the formation of the jump, that the upstream Froude $\mathbf{F}_1$ be greater than unity.

Relation between Pre-jump and Post-jump Froude Numbers

Froude number just before the jump $\mathbf{F}_1 = \dfrac{V_1}{\sqrt{gy_1}}$

Froude number just after the jump $\mathbf{F}_2 = \dfrac{V_2}{\sqrt{gy_2}}$

From the continuity equation,

$$q = V_1 y_1 = V_2 y_2 = \mathbf{F}_1 \sqrt{gy_1} \cdot y_1 = \mathbf{F}_2 \sqrt{gy_2} \cdot y_2$$

or
$$\mathbf{F}_1 y_1^{3/2} = \mathbf{F}_2 y_2^{3/2}$$

from which,
$$\frac{y_2}{y_1} = \left(\frac{\mathbf{F}_1}{\mathbf{F}_2}\right)^{2/3}$$

Substituting for y_2/y_1 in Eq. (12.66), one obtains

$$\left(\frac{\mathbf{F}_1}{\mathbf{F}_2}\right)^{2/3} = \frac{1}{2}[\sqrt{1+8\mathbf{F}_1^2} - 1]$$

or
$$\mathbf{F}_2^2 = \frac{8\mathbf{F}_1^2}{[\sqrt{1+8\mathbf{F}_1^2} - 1]^3}$$

The post-jump Froude number $\mathbf{F}_2$ is thus dependent on the pre-jump Froude number $\mathbf{F}_1$ and may be determined as follows :

$$\mathbf{F}_2 = \frac{2\sqrt{2}\mathbf{F}_1}{[\sqrt{1+8\mathbf{F}_1^2} - 1]^{3/2}} \qquad \text{...(12.66 a)}$$

Example 12.57. *A control sluice spanning the entry of a 4.0 m wide rectangular channel having a mild slope admits 16.0 m³/s at a velocity of 3.0 m/s. Find whether a hydraulic jump is expected in the channel downstream from the sluice.*

Solution. $Q = 16 \text{ m}^3/\text{s}, b = 4 \text{ m}$

$$q = \frac{Q}{b} = 4 \text{ m}^3/\text{s/m}$$

$$y_c = (q^2/g)^{1/3} = (16/9.81)^{1/3} = 1.17 \text{ m.}$$

Sluice gate opening for $V = 3$ m/s

$$y = \frac{q}{V} = \frac{4}{3} = 1.333 \text{ m.}$$

Since $y > y_c$ **hydraulic jump is not expected.**

Example 12.58. *17 m³/s of water flow through a rectangular channel, 5 m wide. (a) Plot on a graph paper the specific energy curve for depths 0.5 to 3 m using the same scale for depth and specific energy. (b) Determine the conjugate depth corresponding to the depth of 0.70 m before the jump. (c) From the diagram, determine (i) the critical depth (ii) the minimum specific energy (iii) alternate depths for specific energy of 2.3 Nm/N and (iv) the loss of energy in the jump.*

Solution. $Q = 17 \text{ m}^3/\text{s}$

$$E = y + \frac{V^2}{2g} = y + \frac{Q^2}{2gA^2} = y + \frac{Q^2}{2gy^2} = y + \frac{17 \times 17}{2 \times 9.81 \times 25y^2} = y + \frac{0.59}{y^2}$$

Table Showing Co-ordinates of Specific Energy Curve

Depth y (m)	$\frac{0.59}{y^2}$ *(m)*	$E = y + \frac{0.59}{y^2}$ *(m)*
0.5	2.36	2.86
0.75	1.05	1.80
1.0	0.59	1.59
1.25	0.378	1.628
1.5	0.262	1.762
1.75	0.193	1.943
2.0	0.147	2.147
2.25	0.117	2.367
2.5	0.094	2.594
2.75	0.078	2.828
3.0	0.055	3.065

The following is a plot of specific energy for the given channel at $Q = 17 \text{ m}^3/\text{s}$.

(*b*) $\frac{y_2}{y_1} = \frac{1}{2}\left(-1 + \sqrt{1 + 8\mathbf{F}_1^2}\right)$

$\mathbf{F}_1 = \frac{V_1}{\sqrt{gy_1}}, \; y_1 = 0.7 \text{ m}$

$V_1 = \frac{17}{5 \times 0.7} = 4.85 \text{ m/s}$

$\mathbf{F}_1 = \frac{4.85}{\sqrt{9.81 \times 0.7}} = 1.85$

$\frac{y_2}{y_1} = \frac{1}{2}(-1 + 5.33) = 2.16$

Therefore, the conjugate depth corresponding to $y_1 = 0.7$ m is

$y_2 = 2.16\, y_1 = 2.16 \times 0.7 = 1.512$ m.

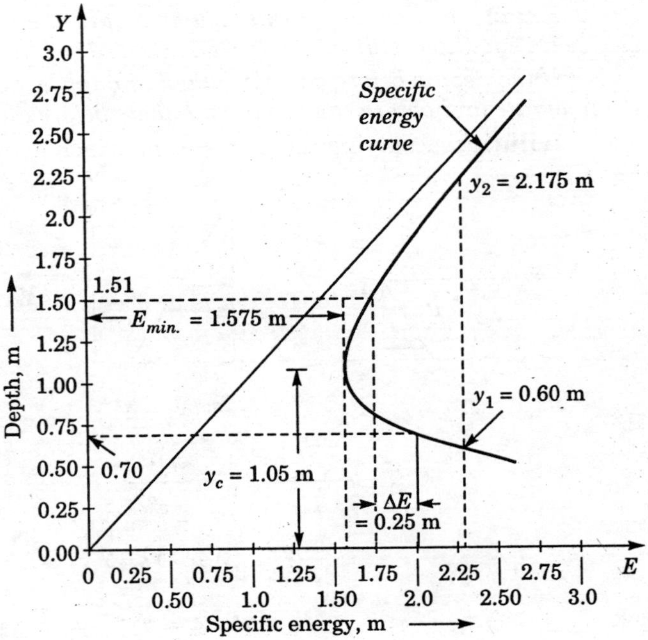

(*c*) From specific energy diagram :

(*i*) Critical depth $y_c = 1.05$ m,

(*ii*) $E_{min} = 1.575$ N-m/N

(*iii*) Alternate depths for $E = 2.3$ N-m/N are $y_1 = 0.6$ m, $y_2 = 2.175$ m

(*iv*) Energy loss in the jump = 0.25 N-m/N.

Example 12.59. *Water flows at the rate of 1.0 × 10⁶ cm³/s along a channel of rectangular section 1.75 metres in width. Calculate the critical depth. If a hydraulic jump formed at a point where the upstream depth is 25 cm, what would be the rise in water level and power lost in the jump ?*

Solution. $Q = 1.0 \times 10^6 \text{ cm}^3/\text{sec} = 1 \text{ m}^3/\text{sec}$

$$q = \frac{1}{1.75} = 0.5715 \text{ m}^3/\text{s/m}$$

$$\therefore \quad y_c = \left(\frac{q^2}{g}\right)^{1/3} = \left(\frac{0.327}{9.81}\right)^{1/3} = 0.322 \text{ m.}$$

The conjugate depths are related by

$$\frac{y_2}{y_1} = \frac{1}{2}(-1+\sqrt{1+8\mathbf{F}_1^2})$$

$$V_1 = \frac{Q}{by_1} = \frac{1}{1.75\times 0.25} = 2.285 \text{ m/s}$$

$$\mathbf{F}_1{}^2 = \frac{V_1^2}{gy_1} = \frac{5.23}{9.81\times 0.25} = 2.155$$

$$\frac{y_2}{y_1} = \frac{1}{2}(-1+4.27) = 1.635$$

$$y_2 = 1.635 \times 0.25 = 0.408 \text{ m.}$$

Rise in water level $= y_2 - y_1 = 0.408 - 0.25 = 0.158$ m.

Energy/N lost in the jump from Eq. (12.67)

$$\Delta E = \frac{(y_2 - y_1)^3}{4y_1y_2} = \frac{(0.158)^3}{4\times 0.025\times 0.408} = 0.00919 \text{ N-m/N}$$

Power lost in the jump (in watts) = $\gamma Q\Delta E = 9810 \times 1.0 \times 0.00919 = 90.15$ watts.

Example 12.60. *A hydraulic jump occurs in a V-shaped channel having sides sloping at 45°. Derive an equation relating the two depths and the flowrate.*

If the depths before and after the jump in the above channel are 0.50 m and 1.0 m, find out the flow rate and obtain the Froude numbers before and after the jump.

Solution. Applying the momentum principle

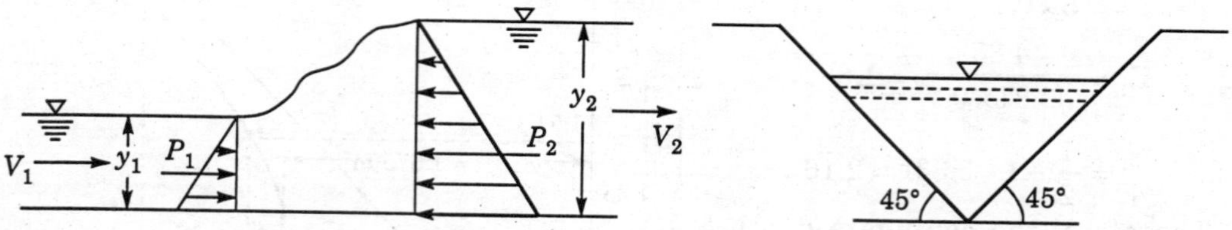

$$P_1 - P_2 = \frac{\gamma Q}{g}(V_2 - V_1)$$

or

$$P_1 + \frac{\gamma Q V_1}{g} = P_2 + \frac{\gamma Q V_2}{g}$$

$$A = \frac{1}{2}(2y \cdot y) = y^2, \quad \bar{Z} = \frac{1}{3}y$$

but

$$P_1 = \gamma A_1 \bar{Z}_1 = \gamma y_1^2 \frac{1}{3} y_1 = \frac{1}{3}\gamma y_1^3 \text{ and } P_2 = \gamma A_2 \bar{Z}_2 = \frac{1}{3}\gamma y_2^3$$

From the continuity equation,

$$Q = A_1V_1 = A_2V_2\,;\; V_1 = \frac{Q}{A_1} = \frac{Q}{y_1^2}$$

and

$$V_2 = \frac{Q}{A_2} = \frac{Q}{y_2^2}$$

Substituting these quantities in the momentum equation,

$$\frac{1}{3}\gamma y_1^3 + \frac{\gamma Q^2}{g y_1^2} = \frac{1}{3}\gamma y_2^3 + \frac{\gamma Q}{g y_2^2}$$

$$y_1^3 + \frac{3Q^2}{g y_1^2} = y_2^3 + \frac{3Q^2}{g y_2^2}\text{ ; from which } Q = y_1 y_2 \sqrt{\frac{g}{3}\left(\frac{y_2^3 - y_1^3}{y_2^2 - y_1^2}\right)}$$

This is the required relation between the depths and the discharge.

$$(0.5)^3 + \frac{3Q^2}{9.81 \times 0.25} = 1 + \frac{3Q^2}{9.81}$$

which yields,

$$Q^2 = 0.953$$
$$Q = 0.9775 \text{ m}^3/\text{s}.$$

Froude number, $\mathbf{F} = \dfrac{V}{\sqrt{gD}}$

$$V_1 = \frac{Q}{y_1^2} = \frac{0.9775}{0.25} = 3.91 \text{ m/s} \text{ ; } V_2 = \frac{Q}{y_2^2} = \frac{0.9775}{(1)^2} = 0.9775 \text{ m/s}$$

$$D_1 = \frac{A_1}{T_1} = \frac{y_1^2}{2y_1} = \frac{y_1}{2} = 0.25 \text{ m} \text{ ; } D_2 = \frac{A_2}{T_2} = \frac{y_2^2}{2y_2} = \frac{y_2}{2} = 0.5 \text{ m}$$

$$\mathbf{F}_1 = \frac{V_1}{\sqrt{gD_1}} = \frac{3.91}{\sqrt{9.81 \times 0.25}} \; 2.5 \text{ ; } \mathbf{F}_2 = \frac{V_2}{\sqrt{gD_2}} = \frac{0.9775}{\sqrt{0.81 \times 0.5}} = 0.441.$$

Example 12.61. *Show that for a submerged hydraulic jump just downstream of a sluice gate, in a horizontal rectangular channel*

$$\frac{y_s}{y_1} = \sqrt{2\mathbf{F}_1^2\left(\frac{y_1}{y_2} - 1\right) + \left(\frac{y_2}{y_1}\right)^2}$$

where y_1 is the depth of opening of the sluice gate, y_2 is the depth of flow downstream of the submerged hydraulic jump, y_s is the water depth on the downstream side of the sluice gate and $\mathbf{F}_1$ is the Froude number of flow through the sluice opening.

Solution. Applying momentum principle to control volume contained by sections (1) and (2) and the channel perimeter neglecting the force due to channel friction

$$\frac{\gamma g b}{g}\left(\frac{q}{y_2} - \frac{q}{y_1}\right) = P_1 - P_2$$

$$\frac{\gamma b q^2}{g}\left(\frac{y_1 - y_2}{y_2 y_1}\right) = \frac{1}{2} b\gamma y_s^2 - \frac{1}{2} b\gamma y_2^2$$

in which the velocities have been substituted using the continuity equation.

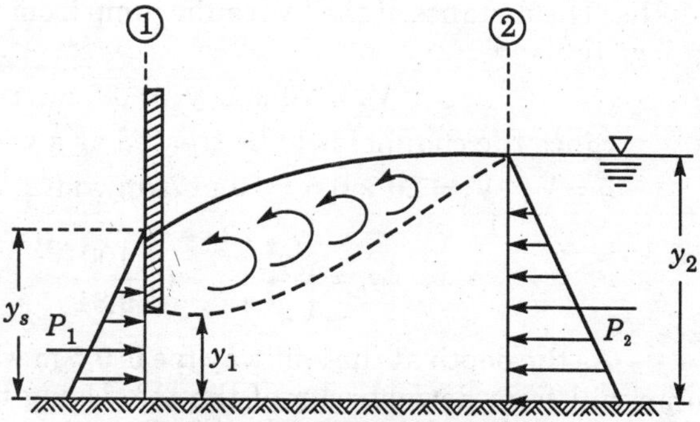

Submerged hydraulic jump

Simplifying, we obtain an equation governing the parameters influencing the submerged hydraulic jump,

$$\frac{2q^2}{g}\left(\frac{y_1 - y_2}{y_1 y_2}\right) = y_s^2 - y_2^2$$

dividing throughout by y_1^2,

$$\frac{q^2}{gy_1^3}\left(\frac{y_1}{y_2}-1\right) = \left(\frac{y_s}{y_1}\right)^2 - \left(\frac{y_2}{y_1}\right)^2$$

Substituting $\mathbf{F}_1^2$ for q^2/gy_1^3

$$2\mathbf{F}_1^2\left(\frac{y_1}{y_2}-1\right) = \left(\frac{y_s}{y_1}\right)^2 - \left(\frac{y_2}{y_1}\right)^2$$

from which
$$\frac{y_s}{y_1} = \sqrt{2\mathbf{F}_1^2\left(\frac{y_1}{y_2}-1\right)+\left(\frac{y_2}{y_1}\right)^2}$$

In place of y_1^2, if we divide the governing equation by y_2^2,

$$\frac{y_s}{y_2} = \sqrt{2\mathbf{F}_1^2\left(1-\frac{y_2}{y_1}\right)+1}\,.$$

Example 12.62. *Water emerges from a spillway with a velocity of 20 m / s and a depth of 0.5 m at the toe. Estimate the distance of the hydraulic jump from the spillway toe if no specific measures are taken for the formation of jump at the toe. Assume for the downstream channel, Manning's n = 0.020 and S_0 = 0.001. Calculations may be performed by direct step method, taking only one step. Assume channel to be a wide rectangular channel.*

Solution. The problem is illustrated in the figure shown alongside. The flow passing over the spillway emerges at its toe where the depth and velocity are 0.5 m and 20 m/s. Beyond the toe section marked by (*A*), the water-surface rises gradinally forming an M_3 curve which ends at section *B* where

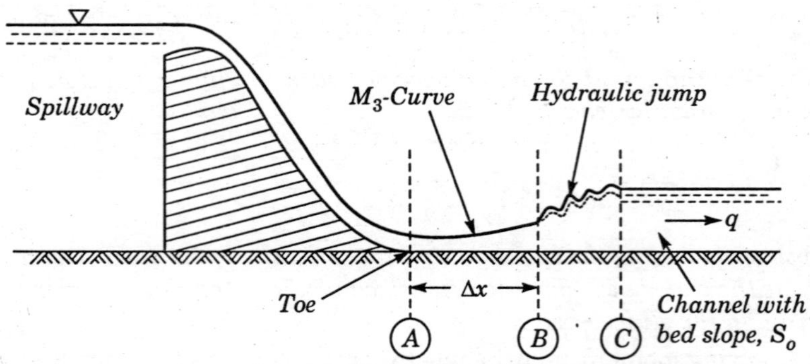

conditions are condusive for formation of hydraulic jump, which is shown occurring between sections (*B*) and (*C*). The flow in the downstream channel having a mild slope takes place at the normal depths. The distance of the hydraulic jump from the toe is shown to be Δx. From the data given, one can write :

$$V_A = 20 \text{ m/s}, y_A = 0.5 \text{ m}, n = 0.020, S_0 = 0.001$$

Since the channel is to be treated as a wide rectangular one,

$q = V_A \cdot y_A = 20 \times 0.5 = 10 \text{ m}^3\text{/s/m}$ width

$$y_c = \left(\frac{q^2}{g}\right)^{1/3} = \left(\frac{100}{9.81}\right)^{1/3} = 2.168 \text{ m}$$

As the depth at the spillway toe is 0.5 m which is less than the critical depth y_c, the hydraulic jump will from downstream of the toe. Using the Manning's equation, the normal depth in the downstream channel can be obtained as follows :

$$V_n = \frac{q}{y_n} = \frac{1}{n} R_n^{2/3} S_0^{1/2} = \frac{1}{n} y_n^{2/3} S_0^{1/2} \text{ (Since for a wide rectangular channel } R_n \approx y_n)$$

from which, $$y_n = \left(\frac{q \, . \, n}{S_0^{1/2}}\right)^{3/5} = \left[\frac{10 \times 0.020}{(0.001)^{1/2}}\right]^{3/5} = 3.024 \text{ m}$$

The depth of flow in the channel downstream of the hydranlic jump will be 3.024 m. The conjugate or sequent depths in a hydraulic jump are related by Eq. (12.65) reproduced below :

$$\frac{2q^2}{g} = y_1 y_2 (y_1 + y_2)$$

in which $y_1 = y_B$ and $y_2 = y_n = 3.024$ m. Substituting for the known quantities, and solving for y_1,

$$\frac{2 \times (10)^2}{9.81} = y_1 \times 3.024 \, (y_1 + 3.024)$$

from which $y_1 = 1.492$ m which represents the flow depth at section (B). Between sections (A) and (B), a water surface curve of M_3 type will be formed changing the depth of flow from 0.5 m to 1.492 m. A hydraulic jump with then be formed raising the depth from 1.492 m to 3.024 m. The length of water surface curve is given by direct step method (Eq. 12.63 a) as :

$$\Delta x = \frac{E_2 - E_1}{S_0 - S_e}$$

in which E_2 is the specific energy at the downstream end (B) of the curve where the depth is 1.492 m, and E_1 is its value at upstream section (A) having a depth of 0.50 m.

$$E_1 = y_A + \frac{V_A^2}{2g} = 0.50 + \frac{(20)^2}{2 \times 9.81} = 20.887 \text{ kg m/kg.}$$

$$E_2 = y_B + \frac{V_B^2}{2g}, \; V_B = \frac{q}{y_B} = \frac{10}{1.492} = 6.70 \text{ m/s}$$

$$E_2 = 1.492 + \frac{(6.70)^2}{2 \times 9.81} = 3.78 \text{ kg m/kg}$$

The energy line slope, S_e, at the mean section where $y = \frac{1}{2}(y_A + y_B)$ is obtainable using the Manning's equation.

$$y = \frac{1}{2}(0.50) + 1.492) = 0.996 \text{ m}$$

$$S_e = \frac{Q^2 n^2}{A^2 R^{4/3}} = \frac{(10 \times 1)^2 (0.020)^2}{(0.996 \times 1)^2 \times (0.996)^{4/3}} = 40.5 \times 10^{-3}$$

Therefore, $$\Delta x = \frac{3.780 - 20.887}{1 \times 10^{-3} - 40.5 \times 10^{-3}} = \mathbf{433.1 \text{ m}}$$

The hydranlic jump will form at a distance of 433.1 m from the spillway toe.

12.20.1. Loss of Energy in a Hydraulic Jump

Since the channel bottom is horizontal, the loss of energy in a hydraulic jump will be equal to the difference in the specific energies before and after the jump, thus

$$y_1 + \frac{V_1^2}{2g} = y_2 + \frac{V_2^2}{2g} + \Delta E$$

where ΔE is the loss of energy per unit weight of fluid, therefore,

$$\Delta E = \left(y_1 + \frac{V_1^2}{2g}\right) - \left(y_2 + \frac{V_2^2}{2g}\right) = (y_1 - y_2) + \frac{V_1^2 - V_2^2}{2g}.$$

Making use of the continuity equation (for eliminating V_1 and V_2)

$$q = y_1 V_1 = y_2 V_2$$

$$\Delta E = y_1 - y_2 + \frac{1}{2g}\left(\frac{q^2}{y_1^2} - \frac{q^2}{y_2^2}\right)$$

and using Eq. (12.65) for eliminating q, we obtain

$$\Delta E = \frac{(y_2 - y_1)^3}{4 y_1 y_2} \qquad ...(12.67)$$

This energy loss ΔE is represented on the specific energy curve, Fig. 12.38 (*a*), as the difference between E_1 and E_2.

Total energy loss in the jump $= \dfrac{\gamma Q (y_2 - y_1)^2}{4 y_1 y_2}$

where γQ = weight of the liquid flowing per second.

Power lost in the jump $= \dfrac{\gamma Q (y_2 - y_1)^3}{75 \times 4 y_1 y_2}$ (in metric h.p.)

in which γ is in kg/m³ (MKS units)

$= \dfrac{\gamma Q (y_2 - y_1)^3}{4 y_1 y_2}$ (in watts) where γ is in N/m³ (SI units).

Example 12.63. *In a rectangular channel, given the discharge intensity q = 1.62 cumecs/m, and the loss of energy in the jump h_L = 2.2 m, determine the two sequent depths. Take g = 10 m²/s.*

Solution. Given $h_L = 2.2$ m ; $q = 1.62$ m²/s/m.

We know that

$$\frac{y_2}{y_1} = \frac{1}{2}\left(-1 + \sqrt{1 + 8\mathbf{F}_1^2}\right)$$

$$h_L = \frac{(y_2 - y_1)^3}{4 y_1 y_2}$$

and

$$\mathbf{F}_1 = \frac{V_1}{\sqrt{g y_1}} = \frac{q}{y_1^{3/2}\sqrt{g}}.$$

Procedure. Solution for y_1 and y_2 is to be obtained by trial and error. First assume a suitable value of $\mathbf{F}_1$. Calculate y_1 and y_2/y_1 (hence y_2) and obtain h_L. If h_L thus computed does not tally with given value of h_L, choose a second trial value of $\mathbf{F}_1$ and repeat the above procedure till the given h_L is obtained.

Assume $\mathbf{F}_1 = 2$

then

$$y_1^{3/2} = \frac{q}{\mathbf{F}_1\sqrt{g}} = \frac{1.62}{2\sqrt{9.81}}$$

which yields $y_1 = 0.403$ m

$$\frac{y_2}{y_1} = \frac{1}{2}\left(-1 + \sqrt{1 + 8 \times 4}\right) = 2.37,$$

from which $y_2 = 0.956$ m

$$h_L = \frac{(0.956 - 0.403)^3}{4 \times 0.403 \times 0.956} = 0.11 \text{ m}.$$

Second trial. Let $\mathbf{F}_1 = 4$,

then the corresponding $y_1 = 0.247$ m,

$y_2/y_1 = 5.18$, and $y_2 = 1.282$ m

$$h_L = \frac{(1.282 - 0.247)^2}{4 \times 0.247 \times 1.282} = 0.85 \text{ m.}$$

Third trial. Let $\mathbf{F}_1 = 6$,

then $y_1 = 0.193$ m,

$y_2/y_1 = 8.15$, $y_2 = 1.573$ m

$$h_L = \frac{(1.573 - 0.193)^2}{4 \times 0.193 \times 1.573} = 2.17 \text{ m.}$$

The computed $h_L = 2.17$ m is very close to the given value of 2.2 m.

Therefore, the sequent depths are

$$y_1 = 0.193 \text{ m} \quad \text{and} \quad y_2 = 8.15 \times 0.193 = 1.573 \text{ m.}$$

Example 12.64. *State the conditions for the formation of a hydraulic jump in open channel flow. Establish a relationship between the depths of flow upstream and downstream of a hydraulic jump in terms of the upstream Froude number. A sluice gate discharges water into a horizontal rectangular channel with a velocity of 10 m/s and depth of flow of 1 metre. Determine the depth of flow after the jump and consequent loss in total head.*

Solution. The conjugate depths y_1 and y_2 are related by Eq. (12.66),

$$\frac{y_2}{y_1} = \frac{1}{2}[\sqrt{1 + 8\mathbf{F}_1^2} - 1]$$

Given $V_1 = 10$ m/s and $y_1 = 1$ m

$$\mathbf{F}_1 = \frac{V_1}{\sqrt{gy_1}} = \frac{10}{\sqrt{9.81 \times 1}} = 3.19$$

$$y_2 = \tfrac{1}{2} \times 1[\sqrt{1 + 8(3.19)^2} - 1] = \mathbf{4.039 \text{ m.}}$$

Loss of head in the hydraulic jump

$$h_L = \frac{(y_2 - y_1)^3}{4y_1y_2} = \frac{(4.039 - 1)^3}{4 \times 1 \times 4.039} = \mathbf{1.737 \text{ m.}}$$

Example 12.65. *A rectangular channel carrying a supercritical flow is to be provided with a hydraulic jump type of energy dissipater. Energy loss required in the hydraulic jump is 6.0 m and inlet Froude number is 10, determine the sequent depths.*

Solution. Ratio of sequent depths

$$\frac{y_2}{y_1} = \frac{1}{2}\left[\sqrt{1 + 8\mathbf{F}_1^2} - 1\right] = \frac{1}{2}\left[\sqrt{1 + 8(10)^2} - 1\right] = 13.65$$

Energy Loss
$$H_L = \frac{(y_2 - y_1)^3}{4y_1y_2}$$

or
$$\frac{H_L}{y_1} = \frac{\left(\frac{y_2}{y_1} - 1\right)^3}{4y_2/y_1} = \frac{(13.65 - 1)^3}{4 \times 13.65} = 37.07$$

But $H_L = 6.0$ m, $\therefore$ $y_1 = 6/37.07 = \mathbf{0.162 \text{ m}}$

and $y_2 = 13.65\, y_1 = 13.65 \times 0.162 = \mathbf{2.21 \text{ m.}}$

Example 12.65. *A hydraulic jump occurs in a rectangular channel and the depths of flow before and after the jump are 0.5 m and 2.0 m respectively. Calculate the critical depth of flow.*

Solution. The discharge per unit width and the sequent depths are related by the relationship as under

$$\frac{2q^2}{g} = y_1 y_2(y_1 + y_2) = 0.5 \times 2.0(0.5 + 2.0) = \mathbf{2.5}$$

or
$$q^2 = \frac{2.5 \times 9.81}{2} = 12.2625$$

The critical depth of flow,

$$y_c = \left(\frac{q^2}{g}\right)^{1/3} = \left(\frac{12.2625}{9.81}\right)^{1/3} = \mathbf{1.077\ m.}$$

Example 12.67. *A round crested spillway passes a design discharge of 1 m³/sec per metre length. The coefficient of discharge may be taken as C_d = 0.7. If the height of the crest above the downstream stilling basin floor level is 10 m, design the (i) depth and (ii) length of stilling basin. Depth of flow in the stream downstream of the spillway is 1 m at the design discharge of 1 m³/sec. Enquire if the stilling basin has to be depressed.*

Solution. Flow over the spillway can be written as

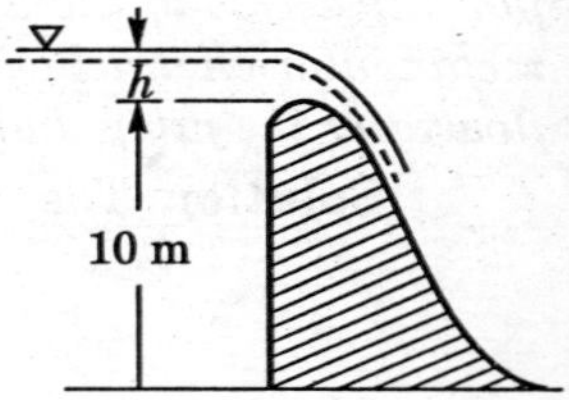

$$Q = \frac{2}{3} C_d L \sqrt{2g}\, h^{3/2}$$

Considering unit length of spillway,

$$1 = \frac{2}{3} \times 0.7 \times 1 \times \sqrt{2 \times 9.81}\, h^{3/2}$$

from which, $h = 0.62$ m.

Velocity of approach, $V_a = \dfrac{1}{10 + 0.62} = 0.094$ m/s

Velocity head, $V_a^2/2g$ = 0.00045 m, being very small can be neglected.

Average fall of water = 10 + 0.62/2 = 10.31 m

Theoretical velocity of flow at the foot of spillway

$$= \sqrt{2g \times 10.31} = 14.22 \text{ m/s}$$

If a coefficient of velocity of 0.9 is assumed, the actual velocity of flow

= 0.9 × 14.22 = 12.80 m/s

Depth of flow at the foot of spillway

$$= \frac{1}{12.80} = 0.078 \text{ m.}$$

Froude No. $\mathbf{F}_1 = \dfrac{V_1}{\sqrt{gy_1}} = \dfrac{1.80}{\sqrt{9.81 \times 0.078}} = 14.62$

Since $\mathbf{F}_1 > 1$, the flow is super-critical.

Flow on the downstream side where the depth is 1 m, and velocity = 1/1 = 1 m/s, the Froude number $= \dfrac{1}{\sqrt{9.81 \times 1}} = 0.319$. Since $\mathbf{F} < 1$, downstream flow is sub-critical. Hence formation of hydraulic jump is inevitable.

With 0.078 m depth before the jump, the depth after the jump

$$y_2 = \frac{y_1}{2}[\sqrt{1 + 8\mathbf{F}_1^2} - 1] = \frac{0.078}{2}[\sqrt{1 + 8(14.62)^2} - 1] = 1.574 \text{ m}$$

Depth available in the stream = 1.0 m.

The stilling basin has to be depressed by (1.574 – 1.0) = **0.574 m.**

Length of stilling basin is taken as 5 to 8 times the height of the jump.

Height of jump $= y_2 - y_1 = (1.574 - 0.078) = 1.496$ m

Length of stilling basin $= 6(y_2 - y_1) = 6 \times 1.496 =$ **8.976 m.**

Example 12.68. *A rectangular channel carrying super-critical flow is provided with a hydraulic jump type of energy dissipator. If it is expected to dissipate 5 m head of water in the formation of the jump, and if inlet Froude number is 8.5, find the sequent depths.*

Solution. Head loss in hydraulic jump is given by

$$H_L = \frac{(y_2 - y_1)^3}{4y_1y_2} = \frac{y_1\left(\frac{y_2}{y_1} - 1\right)^3}{4(y_2/y_1)}$$

The ratio of sequent depths,

$$\frac{y_2}{y_1} = \frac{1}{2}[-1 + \sqrt{1 + 8F_1^2}]$$

$$= \frac{1}{2}[-1 + \sqrt{1 + 8(8.5)^2}] = 11.53$$

Making substitution in expression for H_L,

$$5 = \frac{y_1(11.53 - 1)^3}{4 \times 11.53} = 25.32y_1$$

or $\quad y_1 = \mathbf{0.1975\ m}$

and $\quad y_2 = 11.53 \times 0.1975 = \mathbf{2.277\ m.}$

Example 12.69. *A 4 m wide rectangular channel has the Froude's number before the hydralic jump equal to 2.2. If the height of the jump is 2.0. m, find the depth of flow before and after the jump. Find also the loss energy due to the jump.*

Solution. Froude No. before the jump

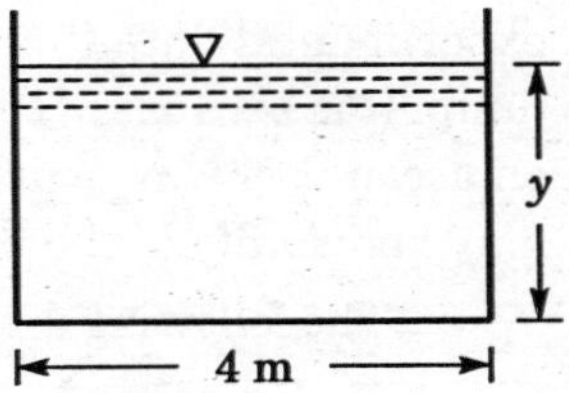

$$F_1 = V_1/\sqrt{gD_1} = 2.2$$

Hydranlic depth $\quad D_1 = A_1/T = \frac{4y_1}{4} = y_1$

Using Eq. (12.66), the ratio of conjugate depths

$$\frac{y_2}{y_1} = \frac{1}{2}[\sqrt{1 + 8F_1^2} - 1] = \frac{1}{2}[\sqrt{1 + 8(2.2)^2} - 1] = 2.65 \quad ...(1)$$

Height of the jump, $y_2 - y_1 = 2.0$ ...(2)

Solving for y_1 and y_2 using (1) and (2),

$$\mathbf{y_1 = 1.21\ m} \text{ and } y_2 = 2.65 \times 1.21 = \mathbf{3.21\ m}$$

Head loss in the jump, using Eq. (12.67)

$$\Delta E = H_L = (y_2 - y_1)^3/4y_1y_2 = \frac{(2)^3}{4 \times 1.21 \times 3.21} = \mathbf{0.515\ m.}$$

Example 12.70. *In a lined rectangular canal, the Fronde number of incoming flow is 3.0. A hydraulic jump forms when it meets the pool of water. The depth of flow after the jump formation is 1.51 m. Determine the Fronde number of flow after the jump.*

Solution. The relationship between the post-jump depth and the pre-jump depth in a hydraulic jump in terms of the incoming froude number is given by Eq. (12.66) :

$$\frac{y_2}{y_1} = \frac{1}{2}(\sqrt{1 + 8F_1^2} - 1), \text{ where } F_1^2 = \frac{q^2}{gy_1^3}$$

Making substitution of the given data, *i.e.* $y_2 = 1.51$ m, $F_1 = 3.0$

$$\frac{1.51}{y_1} = \frac{1}{2}(\sqrt{1+8(3)^2}-1) = 3.772$$

from which $$y_1 = 0.40 \text{ m}$$

The discharge per unit width,

$$q = F_1\sqrt{gy_1^3} = 3.0\sqrt{9.81(0.40)^3} = 2.377 \text{ m}^3/\text{s/m}$$

∴ The Froude number of flow after the jump,

$$F_2 = \frac{q}{g(y_2)^{3/2}} = \frac{2.377}{9.81\,(1.51)^{3/2}} = \mathbf{0.13.}$$

12.20.2. Elements and Characteristics of a Hydraulic Jump

The following quantities are generally known as the elements of the hydraulic jump :

(*i*) pre-jump depth y_1,

(*ii*) post-jump depth y_2,

(*iii*) height of the jump, $H_j = y_2 - y_1$,

(*iv*) length of the jump, $L_j = 5H_j$, approx.,

(*v*) specific energies before and after the jump (*i.e.* E_1 and E_2), and

(*vi*) the loss of energy, (ΔE), in the jump.

With the exception of the jump length L_j, all the remaining elements can be determined theoretically with the aid of the equations derived earlier. It can be noted that these elements are functions of the depths y_1 and y_2. It, therefore, appears that the conjugate depths constitute the most important elements of the jump. As regards the length of the jump, it has been observed from experimental data that there is a wide variation in the relationship between length and height of jump. It is because of the fact that the downstream end of the jump cannot be precisely demarcated on account of wavy surface which follows a jump. The length of the jump in rectangular channels may be usually of the order of five times its height.

The following dimensionless quantities are generally known as the characteristics of the jump :

1. Relative loss ($\Delta E/E_1$). It is defined as the ratio of energy loss and the specified energy before the jump.

$$\Delta E = \frac{(y_2 - y_1)^3}{4y_1y_2}$$

$$\frac{\Delta E}{y_1} = \frac{\left(\frac{y_2}{y_1}-1\right)^3}{4\frac{y_2}{y_1}} \qquad \text{...(12.68)}$$

and $$E_1 = y_1 + \frac{V_1^2}{2g}$$

or $$\frac{E_1}{y_1} = 1 + \frac{V_1^2}{2gy_1} = 1 + \frac{\mathbf{F}_1^2}{2} \qquad \text{...(12.69)}$$

From Eqs. (12.66), (12.68) and (12.69), we obtain

$$\frac{\Delta E}{E_1} = \frac{(\sqrt{1+8\mathbf{F}_1^2}-3)^3}{8(\mathbf{F}_1^2+2)(\sqrt{1+8\mathbf{F}_1^2}-1)} \qquad ...(12.70)$$

The relative loss being a ratio of energies is dimensionless.

2. Efficiency of the jump (E_2/E_1). The ratio of the specific energies after and before the jump is known as the efficiency of the jump. The efficiency may be expressed in terms of Froude number $\mathbf{F}_1$, thus

$$\frac{E_2}{E_1} = \frac{1+(8\mathbf{F}_1^2+1)^{3/2}-4\mathbf{F}_1^2}{8\mathbf{F}_1^2(2+\mathbf{F}_1^2)} \qquad ...(12.71)$$

Noting that $\frac{\Delta E}{E_1} = \frac{E_1 - E_2}{E_1} = 1 - \frac{E_2}{E_1}$, the above expression for efficiency $\frac{E_2}{E_1}$ may be obtained using Eq. (12.70).

3. Relative height of the jump (H_j/E_1). The height of the jump is defined as difference between the depths after and before the jump, thus

$$H_j = y_2 - y_1.$$

Since E_1 is the specific energy per unit weight of liquid, it has the dimensions of length, and hence jump height may be made dimensionless be dividing it by E_1

$$\frac{H_j}{E_1} = \frac{y_2 - y_1}{E_1} = \frac{\frac{y_2}{y_1}-1}{E_1/y_1}$$

Using Eqs. (12.66) and (12.69), we obtain

$$\frac{H_j}{E_1} = \frac{\sqrt{1+8\mathbf{F}_1^2}-3}{2+\mathbf{F}_1^2} \qquad ...(12.72)$$

12.20.3. Types of Jump

For jumps in which the ratio y_2/y_1 is not greater than 2.0, the liquid surface does not rise abruptly and has a number of undulations of gradually diminishing size. Such a jump is called as an *undular jump*. From Eq. (12.66) for $y_2/y_1 = 2.0$, the upstream Froude number $\mathbf{F}_1 = \sqrt{3}$. This fixes the upper limit of $\mathbf{F}_1$ for undular jumps. For higher values of y_2/y_1 ($y_2/y_1 > 2.0$ and consequently $\mathbf{F}_1 > \sqrt{3}$), the liquid surface rises fairly abruptly, and the hydraulic jump then is known as a *direct jump*.

The United States Bureau of Reclamation has classified the jump into the following five categories, depending upon the magnitude of the Froude number of the approaching flow F_1.

(1) **Undular jump.** The Froude number $\mathbf{F}_1$ ranges from 1 to 1.7 and the liquid surface shows undulations of gradually decreasing size.

(2) **Weak jump.** $\mathbf{F}_1$ ranges from 1.7 to 2.5, number of small rollers appear on the surface of the jump, and the downstream liquid surface remains smooth. The energy loss in the jump is low.

(3) **Oscillating jump.** For $\mathbf{F}_1$ ranging between 2.5 to 4.5, there is an oscillating jet which enters the jump bottom and oscillates to the surface. Each oscillation produces a large wave of irregular period and does extensive damage to the canal bed and banks while travelling miles downstream.

(4) **Steady jump.** This type of jump occurs in the Froude number range of 4.5 to 9.0. The fluctuations in the tail water depth have a very little effect on the position and the action of the jump. The energy dissipation may be in the range of 45 to 70%.

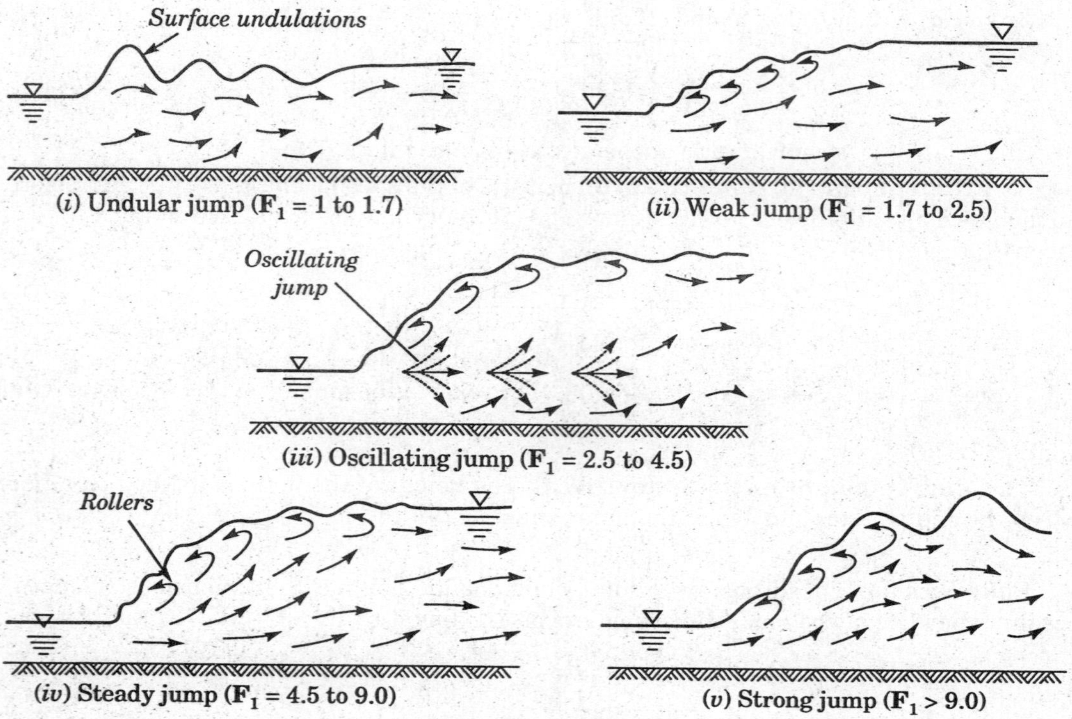

Fig. 12.39. Different types of jump.

(5) Strong jump. For Froude number greater than 9.0, the surface downstream of the jump is rough and the energy dissipation may be upto 85%. Fig. 12.39 illustrates the types of jump described above.

12.21. Location of Hydraulic Jump

The knowledge of location of the jump is essential for design of stilling basins. At what distance from the foot of spillway or from the sluice gate will jump from depends upon the initial depth, the sequent depth and the Froude number of approaching flow. The jump will form at a section where these three quantities satisfy Eq. (12.66). Eq. (12.66) is therefore, helpful in locating the position of a jump. However, for an accurate determination of the jump position, the length of the jump is taken into account. The following description deals with the approximate location of the jump position.

In Fig. 12.40, a hydraulic jump forms downstream of the sluice gate in a mild channel. The water surface curve *CD* is of M_3 type and that *AB* may be identified as of M_2 type. The sequent depths corresponding to initial depths represented by the M_3 curve *CD* are plotted and shown by *C′ D*. The intersection of the M_2 curve *AB* with the sequent depth curve *C′ D*, that is *Y*, gives the location

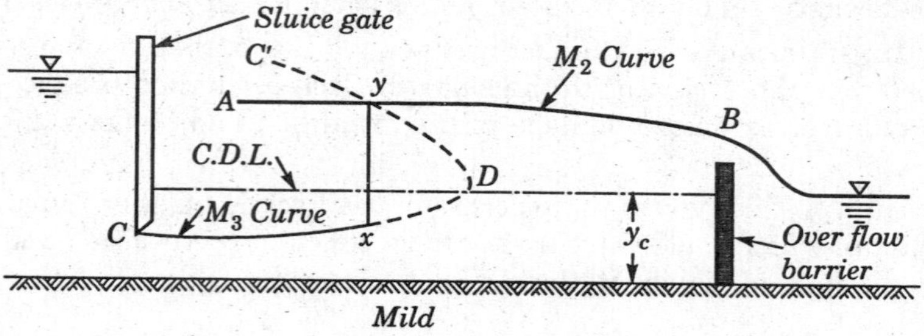

Fig. 12.40. Hydraulic jump downstream of a sluice gate.

of the jump. Thus XY represents the approximate position of the jump. For more accurate location, the jump length should be considered. The jump will then be shifted a little distance downstream. For flow downstream of the foot of the spillway, the same procedure will be applicable.

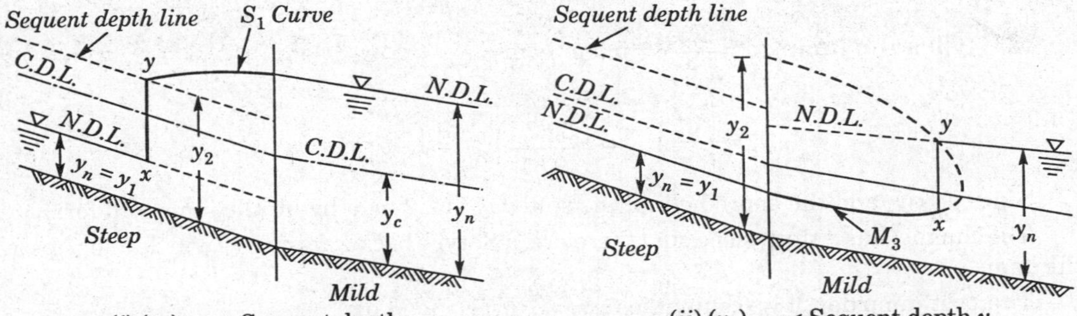

Fig. 12.41. Flow profile at break in slope from steep to mild.

Fig. 12.41 refers to the conditions when the channel slope changes from steep to mild. Except in the vicinity of break in channel slope, the flow may be assumed uniform. The jump may occur either in the steep channel or the mild channel depending on whether the downstream normal depth is greater or less than y_2, the depth sequent to upstream normal depth $y_n = y_1$. In Fig. 12.41 (*i*) the jump occurs in the steep channel, since $(y_n)_{mild}$ is greater than the sequent depth y_2. Downstream of the jump, the water surface assumes S_1 profile. If the downstream depth y_n is lowered to less than y_2, the jump will move into the mild channel. In this case, the jump may be located by using the procedure described for jump location downstream of the sluice gate.

Fig. 12.42 illustrates the jump formation behind an over-flow barrier installed in a steep channel. If the depth behind the barrier is greater than the sequent depth corresponding to the initial depth $(y_1 = y_n)$, the hydraulic jump will form. When the depth at the barrier is less than the sequent depth y_2, the flow will rise over the barrier forming a single wave crest as shown.

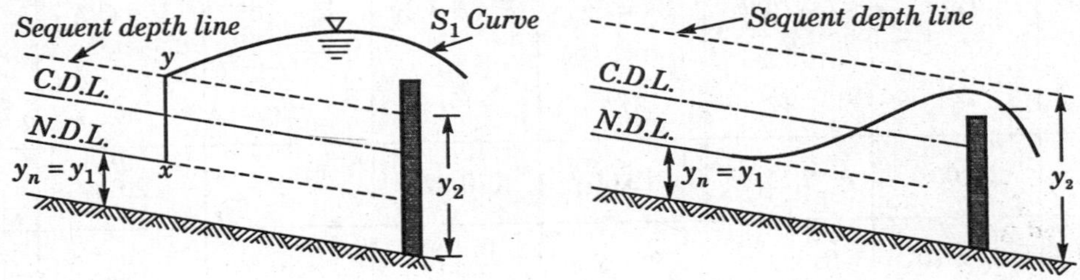

Fig. 12.42. Hydraulic jump behind an over-flow barrier in steep channel.

Example 12.70. *In a rectangular channel with bed width 6 m, n = 0.03 and bed slope 0.04, the depth of uniform flow is 1 m. A low overflow dam which keeps 2.2 m of water immediately behind it is built at the downstream end of the channel to create a hydraulic jump. Compute the following :*

(*i*) *the discharge over the spillway*

(*ii*) *the jump height*

(*iii*) *the distance of the jump from the dam treating the entire distance as one reach.*

Solution. Discharge in the channel, using Eq. (12.10),

$$Q = A \cdot \frac{1}{n} \cdot R^{2/3} S_0^{1/2}$$

$$= (6 \times 1) \times \frac{1}{0.03} \times \left(\frac{6 \times 1}{6 + 2 \times 1}\right)^{2/3} (0.04)^{1/2} = 33.0 \text{ m}^3/\text{s}$$

(1) Discharge over the spillway = 33.0 m³/s

(2) Hydraulic jump : $q = \frac{33.0}{6} = 5.5 \text{ m}^3/\text{s/m}$

$$y_c = \left(\frac{q^2}{g}\right)^{1/3} = \left(\frac{30.25}{9.81}\right)^{1/3} = 1.46 \text{ m.}$$

Since $y_n < y_c$ and the depth behind the spillway is 2.2 m, a hydraulic jump will form.

The channel has a steep slope, and therefore, a S_1 type back water curve will develop upstream of the dam.

The post jump depth y_2 (conjugate to $y_1 = 1$ m) is given by

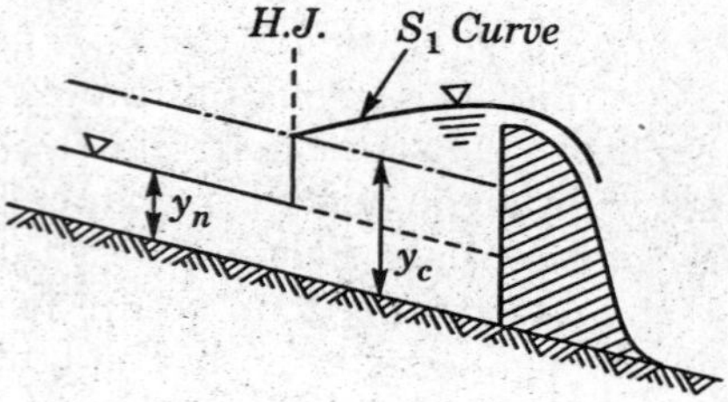

$$\frac{y_2}{y_1} = \frac{1}{2}(-1 + \sqrt{1 + 8\mathbf{F}_1^2})$$

$$\mathbf{F}_1 = \frac{V_1}{\sqrt{gy_1}} = \frac{(5.5/1)}{\sqrt{9.81 \times 1}} = 1.758$$

$$\frac{y_2}{y_1} = \frac{1}{2}(-1 + \sqrt{1 + 24.65}) = 2.03$$

$$y_2 = 2.03 \times 1 = 2.03 \text{ m}$$

Height of the jump $y_2 - y_1 = 2.03 - 1.0 = 1.03$ m.

(3) **Location of the jump.** The jump will form at a section where the water depth is 2.03 m. The following table shows the calculation for the jump location.

Length of Backwater Curve

y_2 (1)	y_1 (2)	Average y (3)	$A = 6y$ (4)	$P = 6 + 2y$ (5)	$R = A/P$ (6)	$V = Q/A$ (7)	$F^2 = \frac{V^2}{gy}$ (8)
2.2	2.03	2.11	12.66	10.22	1.24	2.61	0.329
$(1 - \mathbf{F}^2)$ (9)	$S_o \times 10^4$ (10)	$S_e \times 10^4$ Eq. (12.58) (11)	$(S_e - S_e) \times 10^4$ (12)	dx/dy (13)	Δy (14)	Δx (15)	x (16)
0.671	400	46	354	9.3	0.17	1.58	1.58

The jump is located at a distance of 1.58 m from the dam.

Example 12.71. *A rectangular flume 2 m wide carries discharge at the rate of 2 m³/s. The bed slope of the flume is 0.0004. At a certain section the depth of flow is 1 m. Calculate the distance of the section downstream where the depth of flow is 0.9 m. Solve by single step method. Assume rugosity coefficient as 0.014.*

Is the slope of the channel mild or steep ? How is this type of surface profile classified ?

Solution. Data given : $Q = 2$ m³/s, $b = 2$ m, $S_o = 0.0014$

$y_1 = 1$ m, $y_2 = 0.9$ m and $n = 0.014$

$$q = \frac{Q}{b} = \frac{2}{2} = 1 \text{ m}^3/\text{s/m}$$

Critical depth $$y_c = \left(\frac{q^2}{g}\right)^{1/3} = \left(\frac{1}{9.81}\right)^{1/3} = 0.467 \text{ m}$$

Using Manning's formula, the critical slope

$$S_c = \frac{Q^2 n^2}{A_c^2 R_c^{4/3}} = \frac{(2\times0.014)^2}{(2\times0.467)^2\left(\dfrac{2\times0.467}{2+0.934}\right)^{4/3}} = 0.00415$$

Since $S_0 < S_c$, the channel is Mild, and since the water surface is falling, it is an M_2 curve.

Energy line slope at the mean section where the water depth is $y = \frac{1}{2}(y_1 + y_2) = 0.95$ m.

$$S_e = \frac{Q^2 n^2}{A^2 R^{4/3}} = \frac{(2\times 0014)^2}{(2\times0.95)^2\left(\dfrac{2\times0.95}{2+1.9}\right)^{4/3}} = 5.66 \times 10^{-4}$$

The specific energies at the sections under consideration are

$$E_1 = y_1 + \frac{V_1^2}{2g} = 1.0 + \frac{1}{19.62} = 1.051 \text{ N-m/N}$$

$$E_2 = y_2 + \frac{V_2^2}{2g} = 0.9 + \frac{(1.11)^2}{10.62} = 0.963 \text{ N-m/N}$$

Using the step method

$$\Delta x = \frac{E_2 - E_1}{S_o - S_e} = \frac{0.963 - 1.051}{0.0004 - 0.000566} = 530 \text{ m.}$$

Example 12.72. *Water flows from under a sluice into a very wide rectangular channel. The channel has a bed slope of 1/1000. The sluice is regulated to discharge 6 m³/s/m width of channel, the depth at vena contracta being 0.5 m. Will a hydraulic jump form ? If so, determine its location. Use a single step for the computation with Manning's n = 0.015.*

Solution. Data given $q = 6$ m³/s/m, $S = 1/1000$,
$y_1 = 0.5$ m, $n = 0.015$.

Critical depth $$y_c = \left(\frac{q^2}{g}\right)^{1/3} = \left(\frac{36}{9.81}\right)^{1/3} = 1.542 \text{ m.}$$

Since the depth at vena contracta is less than the critical depth, the flow is supercritical, hence the hydraulic jump is likely to form.

Assuming the channel to be sufficiently long, the flow can be considered to have attained normal depth far downstream of the sluice. From the Manning's equation, the normal depth can be determined as follows :

$$V = \frac{1}{n} R_n^{2/3} S_o^{1/2}$$

For a wide rectangular channel, $R_n \approx y_n$, also

$$q = Vy_n = \frac{1}{n} y_n^{5/3} S_o^{1/2}$$

From which $$y_n = \left(\frac{qn}{\sqrt{S_o}}\right)^{3/5} = \left(\frac{6\times0.015}{\sqrt{1/1000}}\right)^{3/5} = 1.875 \text{ m.}$$

The sequent depth corresponding to post-jump depth of 1.875 m is given by

$$\frac{y_1}{y_2} = \frac{1}{2}\left[\sqrt{1+8F_2^2} - 1\right]$$

$$\mathbf{F}_2 = \frac{V_2}{\sqrt{gy_2}} = \frac{6/1.875}{\sqrt{9.81\times1.875}} = 0.745$$

$$\frac{y_1}{1.875} = \frac{1}{2}[\sqrt{1+8\times0.555} - 1] = 0.667$$

which yields $y_1 = 1.25$ m.

Since $y < y_c < y_n$, the channel is mild, and the super-critical flow will develop a M_3 type profile. Let the distance downstream of vena contracta where the jump is likely to form be Δx, then

$$\Delta x = \frac{E_2 - E_1}{S_o - S_e}$$

Energy line slope at vena contracta, where $y_2 = 0.5$ m.

$$S_{e1} = \frac{q^2n^2}{y_1^{10/3}} = \frac{(6\times0.015)^2}{(0.5)^{10/3}} = \frac{1}{1.225}$$

Energy line slope at $y_2 = 1.25$ m

$$S_{e2} = \frac{q^2n^2}{y_2^{10/3}} = \frac{(6\times0.015)^2}{(1.25)^{10/3}} = \frac{1}{25.9}$$

Average of the energy line slopes

$$\bar{S}_e = \frac{1}{2}(S_{c1} + S_{c2}) = \left(\frac{1}{1.225} + \frac{1}{25.9}\right) = \frac{1}{2.38}$$

$$V_2 = \frac{q}{y_2} = \frac{6}{1.25} = 4.8 \text{ m/s}$$

$$V_1 = \frac{q}{y_1} = \frac{6}{0.5} = 12.0 \text{ m/s}$$

$$E_2 = y_2 + \frac{V_2^2}{2g} = 1.25 + \frac{(4.8)^2}{19.62} = 2.472 \text{ N-m/N}$$

$$E_1 = y_1 + \frac{V_1^2}{2g} = 0.5 + \frac{(12.0)^2}{19.62} = 7.84 \text{ N-m/N}$$

$$\Delta x = \frac{2.472 - 7.84}{\frac{1}{1000} - \frac{1}{2.38}} = 12.8 \text{ m.}$$

12.22. Flow in Alluvial Channels

12.22.1. Rigid and Mobile Boundary Channels

Rigid boundary channels are those which have their geometry fixed (*i.e.* cross-sectional shape, size and plan form of the channel) and the liquid-flow is incapable of changing the cross-section, plan form and the boundary roughness of the channel. In case of channels with boundaries composed of loose non-cohesive granular material (*i.e.* sand which is normally found in natural streams) which can be transported by the flow, the interaction between the flow and the boundary tends to change the geometry of the channel and consequently determines its hydraulic roughness. Such channels are called *mobile boundary* channels or simply *alluvial* channels. While it is possible to predict with fair degree of accuracy the hydraulic roughness of rigid boundary channels, the mechanics of fluid flow in alluvial channels has however, not so far been understood to the extent as would permit an equally accurate estimate of its hydraulic roughness.

12.22.2. Initiation of Motion of Grains on Channel Bed

It is interesting to know the behaviour of an initially flat bed consisting of loose and cohesionless (mobile) solid particles when subjected to changing flow rates. To begin with, consider a very low discharge of fresh water flowing in an alluvial channel with plane bed. As soon as the liquid starts flowing, hydrodynamic forces are exerted upon the solid particles at the wetted perimeter. A further increase in the flowrate causes an increase in the magnitude of these forces (these being the drag and the lift). Thus, for a given flat stationary bed, a condition is eventually reached at which the bed particles are unable to resist the hydrodynamic forces. At the critical condition, the motive force is just adequate to overcome the resistance offered by the particle. Any further increase in discharge, beyond the critical one would cause the hydrodynamic forces to exceed the resistance, consequently, leading to movement of bed particles. The particles get first dislodged and then start to move downstream. The bed material, having a distributed size, has its finer particles initiated to motion earlier than the coarser ones. Ripples and dunes are known to form respectively at low and relatively high discharges. At still higher discharges and flow velocities dunes tend to change into a flat bed with particles in motion. There is thus a zone of transition from dune regime to flat bed. This flat bed when subjected to further high velocities tends to change into antidunes which have a *sinusoidal* geometry. While the ripples and dunes move downstream with velocity much less than that of the flowing liquid, the antidunes may move upstream, downstream or remain stationary (standing sand waves as they are sometimes called) depending upon properties of flow, fluid and sediment.

Shields' Analysis of Sediment Motion. The resultant force acting upon a typical sediment particle at the moment of its incipient motion was studied by Shields, who presented the most direct method of approach. He made certain assumptions in his analysis and then confirmed and supplemented his analysis experimentally. Shields assumed that the force exerted by the flow on the sediment particle could be expressed in terms of the usual drag coefficient, *i.e.*, $F = C_D A \, \rho V^2/2$. He further assumed that the resistance of the particle to motion should depend upon the form of the bed and the immersed weight of the particle. Introducing the concept of the critical tractive force to represent that value of the boundary shear at which the force exerted by the flow upon the particle just overcomes its resistance, a functional relationship for the tractive shear stress was obtained.

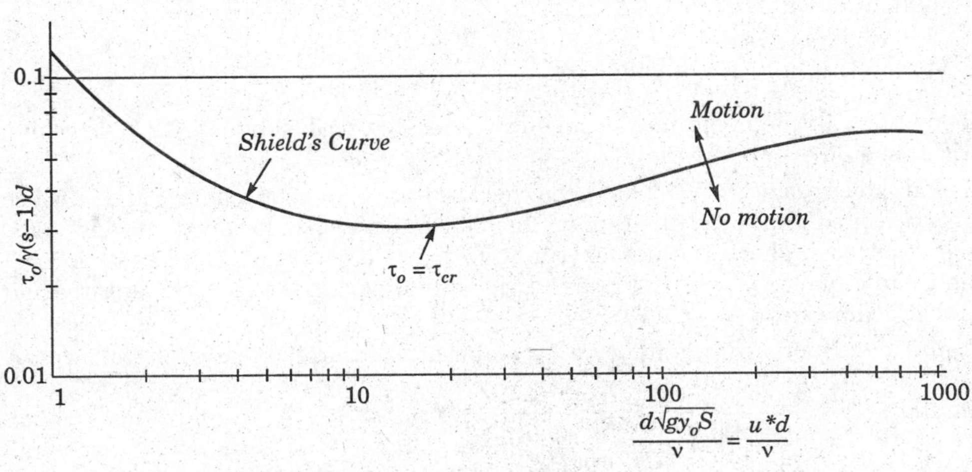

Fig. 12.43. Shields' diagram for incipient motion.

For the specific case of a level bed of uniform particle size investigated by Shields, the functional relationship is as follows :

$$\frac{\tau_{cr}}{\gamma(s-1)d} = f\left(\frac{d\sqrt{gy_o S}}{\nu}\right)$$

in which τ_{cr} is the critical tractive stress, s the specific gravity of the bed material, d the diameter of the particle, γ the specific weight of water, y_0 the water depth, S the channel slope and ν the kinematic viscosity of water at the prevailing temperature.

Fig. 12.43 shows form of the function as determined by Shields for a considerable range of each of the variables. Comparison of Figs. 12.43 and 9.14 shows striking similarity between flow through pipes artificially roughened by a coating of uniform sand grains (Nikuradse's experiment) and the result of Shields' experiments on level bed comprising uniform particle size. Since Shields' experiments, like those of many other investigators, were limited to particles of uniform size, it can be presumed that non-uniform bed material would yield a transition curve like that for commercial pipes—which primarily consists of non-uniform roughness—see Fig. 9.14, Colebrook transition. As is apparent from Shields' analysis, the configuration of the channel bed, which gradually develops after the particles have been initiated into motion, depends primarily upon the magnitude of the quantity, $\sqrt{gy_0 S/\nu}$.

There are three distinct zones noticeable in Fig. 12.43.

(1) Upto shear Reynolds number, $u_* d/\nu \approx 2$, the grains are covered by a laminar film (sub-layer), their movement is mainly due to viscous action and independent of turbulence.

(2) At intermediate Reynolds numbers there lies a zone of transition when the laminar film partially covers the grains and is partially interrupted by the existing bed roughness of the grains. This transition curve has a minimum value of 0.03 for the dimensionless shear stress at $u_* d/\nu \approx 10$. Shields' diagram thus indicates

$$\left[\frac{\tau_{cr}}{\gamma(s-1)d}\right]_{min} \approx 0.03 \qquad \text{...(12.73}a\text{)}$$

(3) At large shear Reynolds numbers, the laminar film gets fully interrupted by the existing bed roughness of grains. The rough boundary is the source of turbulence, and the dimensionless shear stress is independent of the shear Reynolds number. From Fig. 12.43, this condition is reached at $u_* d/\nu \geq 400$, for which

$$\frac{\tau_{cr}}{\gamma(s-1)d} = 0.06 \qquad \text{...(12.73 }b\text{)}$$

Below the value of non-dimensional shear stress prescribed by Eq. (12.73 a), non-motion of bed particles should ever occur.

Shields' diagram, Fig. 12.43, has been checked by many experimenters and is, consequently, now widely accepted. Since the diagram of Shields was developed for uniform sand material, it is reasonable to expect higher τ_{cr} values for non-uniform grain material as well as sticky or flocculant ones. The grain diameter d which appears in Shields' diagram is the 'representative' size and is taken as the median sieve size *i.e.* d_{50}.

For incipient motion of non-uniform bed material, Egiazaroff proposed the following equation :

$$\frac{\tau_{cr}}{\gamma(s-1)d} = \frac{0.1}{[\log 19\,(d/\bar{d})]^2} \qquad \text{...(12.73 }c\text{)}$$

in which d is the average diameter of grain for both gradation curves—for grains in motion, and for total bed material.

Example 12.73. *For large grain sizes, Shields' parameter is of the form*

$$\frac{\tau_c}{[\gamma(s-1)d]} = 0.06.$$

If the grain size d = 1 mm, the specific gravity s = 2.65, slope of the channel S = 0.0001, find the depth of the channel at which the sediment just begins to move.

Solution. From the Shields' parameter

$$\frac{\tau_c}{\gamma(s-1)d} = 0.06 \qquad ...(1)$$

The average boundary shear stress in the channel is given by

$$\tau_0 = \gamma RS$$

For a wide rectangular channel, $R \approx y$. When the boundary shear stress approaches the critical value, the sediment would begin to move. Hence

$$\tau_c = \gamma yS \qquad ...(2)$$

Equating (1) and (2),

$$0.06\,\gamma\,(s-1)\,d = \gamma yS$$

from which $$y = \frac{0.06\,(s-1)d}{S} = \frac{0.06\,(2.65-1)\times 1\times 10^{-3}}{0.0001} = 0.99 \text{ m}.$$

The sediment just begins to move when the depth of water becomes equal to 0.99 m.

12.22.3. Bedforms. *Ripples, Dunes and Antidunes*

Consider a flat and even, stationary bed consisting of loose, cohesionless, and movable solid particles. When water at constant flowrate is allowed to pass over such a bed, for established flow bedforms are generated from the bed material. The type of bed configuration and the bedforms—their shape and dimensions—are dependent upon various flow parameters and the characteristics of the bed materiel. The flow parameters, in turn, are influenced by the bedforms. For the purpose of classification of bedforms, flow regimes can be divided into three parts :

(1) a lower flow regime, **F** < 1 ;

(2) an upper flow regime, **F** > 1 ; and

(3) a transition in between ; **F** ≈ 1.

Here, $\mathbf{F} = V/\sqrt{gy_0}$ is the Froude number of flow ; y_0 is defined in Fig. 12.44.

In the lower flow regime are included both *ripples* and *dunes*, whereas the upper regime contains a *plane bed* with motion and *antidunes*. Between the lower and upper regimes there is a transition zone. For a given flow condition, the bed configuration is erratic ; it may range from developed dunes to a flat bed, or may consist of heterogeneous array of bedforms.

Table 12.2 gives details pertaining to bed configurations taken from ASCE Task Committee Report[1]. The table summarizes salient characteristics of various bedforms in the general order of their occurrence with increasing velocity and Froude number.

1. Task Committee Report on *Sediment Transportation Mechanics ; F. Hydraulic Relations for Alluvial Streams.* Journal of Hydraulics Division. ASCE. Vol. 97. No. Hy. 1, Jan. 1971.

Table 12.2. Different Types of Bed Forms

Bed form or configuration	*Dimensions*	*Shape*	*Behaviour and Occurrence*
Ripples	Wavelength less than approximately 30 cm (1 ft) ; height less than approximately 3 cm (0.1 ft).	Roughly triangular in profile, with gentle, slightly nearly equal to the angle of repose. Generally short crested and three-dimensional.	Move downstream with velocity much less than that of the flow. Generally do not occur in sediments coarser than about 0.6 mm.
Bars	Lengths comparable to the channel width. Height comparable to mean flow depth.	Profile similar to ripples. Plan form variable.	Four types of bars are distinguished (1) Point, (2) Alternating, (3) Transverse, (4) Tributary Ripples may occur on upstream slopes.
Dunes	Wavelengths and heights greater than ripples but less than bars.	Similar to ripples	Upstream slopes of dunes many be covered with ripples. Dunes migrate downstream in manner similar to ripples.
Transition	Vary widely.	Vary widely	A configuration consisting of a heterogeneous array for bed forms, primarily low amplitude ripples and dunes interspersed with flat regions.
Flat bed	—	—	A bed surface devoid of bed forms. May not occur for some ranges of depth and size.
Antidunes	Wavelength $= 2\pi \dfrac{V^2}{g}$ (approx.) (reported by Kennedy). Height depends on depth and velocity of flow.	Nearly sinusoidal in profile. Crest length comparable to wavelength.	In phase with and strongly interact with gravity water surface waves. May move upstream, downstream or remain stationary depending on properties of flow and sediment.

Fig. 12.44 shows typical sketches of flow in an alluvial channel with dunes and antidunes regimes occurring under appropriate conditions of flow. As a general rule, ripples form on the bed for flows at very low Froude number (**F** << 1), the ripple pattern being generally three dimensional in plan. At low values of Froude number dunes generally appear (**F** < 1) which may have ripples superimposed. For **F** > 1, antidunes are generated, the water surface becoming wavy and in phase with the antidunes, as shown in part (*b*) of Fig. 12.44. Earlier to this, at Froude number lower than unity dunes are washed away due to higher velocity, and this may lead to the transitional plane bed.

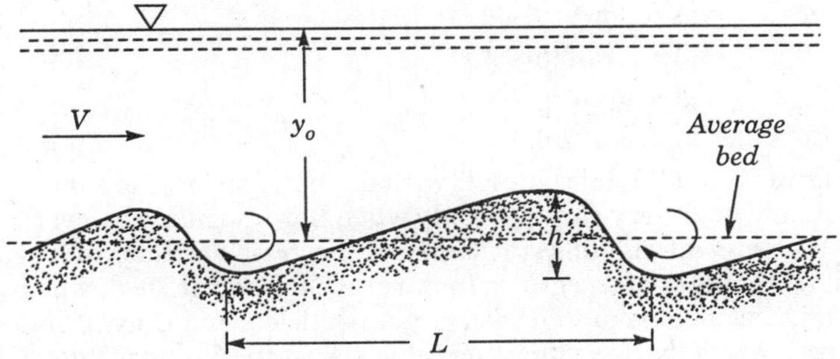

(*a*) A duned channel bed which moves slowly downstream (**F** < 1)

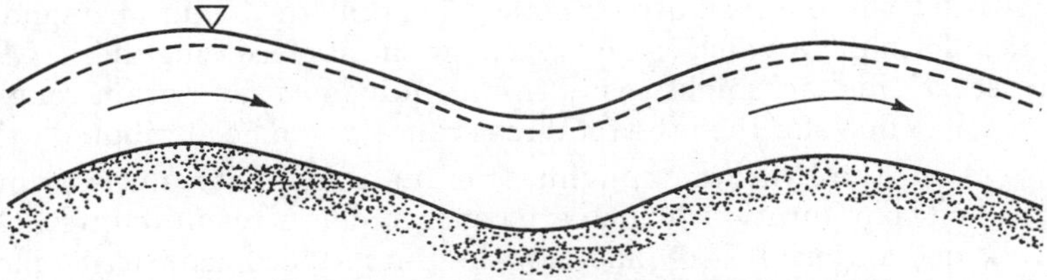

(*b*) Flow in a channel with antidunes and stationary wave-antidunes generally move upstream (**F** > 1)

Fig. 12.44. Flow in channels with dunes and antidunes.

12.22.4. Hydraulic Roughness of Channels

The resistance to flow in a channel of rigid or fixed boundary carrying clear water in steady uniform flow can be predicated quite accurately. But in channels composed of erodible boundaries, the problem of evolution of hydraulic resistance is a complex one because of the bed configuration which changes with the flow conditions. A brief account of resistance in rigid boundary channels is followed by that of alluvial channels.

Rigid Boundary Channels. The type of roughness found in rigid boundary open channel flow can be adequately represented by equivalent sand grain roughess of Nikuradse. Adopting a similar procedure as used for the development of the Karman-Prandtl resistance equations for pipe flow, *i.e.* Eqs. (9.60) and (9.62), the following resistance equations for two-dimensional open channel flow can be derived.

(*i*) For smooth boundary $\left(\text{defined by } \frac{u_* k_s}{\nu} < 4\right)$

$$\frac{1}{\sqrt{f}} = 2.03 \log (\mathbf{R}/\sqrt{f}) - 0.47 \qquad \text{...(12.74)}$$

in which $\mathbf{R} = 4RV/\nu$; R being the hydraulic radius.

(*ii*) For rough boundaries $\left(\text{for which } \frac{u_* k_s}{\nu} > 100\right)$

$$\frac{1}{\sqrt{f}} = 2.03 \log \frac{4R}{k_s} + 0.91 \qquad \text{...(12.75)}$$

which can be written, using the approximation $y_0 \approx R$ (which holds for a wide channel), as

$$\frac{1}{\sqrt{f}} = 2.03 \log \frac{y_0}{k_s} + 2.13 \qquad \text{...(12.76)}$$

Eqs. (12.75) and (12.76) can also be expressed in the following form :

$$\frac{1}{\sqrt{f}} = -C \log \left(\frac{k_s}{AR} + \frac{B}{\mathbf{R}\sqrt{f}}\right) \qquad \text{...(12.77)}$$

in which the A, B and C are as follows :

$$A = 11.55 \text{ to } 12.2$$
$$B = \text{upto } 3$$
$$C = 2 \text{ to } 2.03.$$

Alluvial channels. In alluvial channels with flat boundaries, the same concept of roughness height applies as for rigid boundary channels only when flows remains below the threshold of grain movement so that bed material remains stationary. If the grains are of uniform size, the roughness height will be equal to diameter of the grain. In natural channels its bed, which usually consists of non-uniform material, becomes armoured with larger particles. For alluvial channels, the hydraulic roughness, in general, depends upon the flow as well as on the bed material. In actual cases, natural streams seldom have flat bed. Under appropriate conditions, they develop ripples, dunes or antidunes. These bed forms have a much larger spacing than the Nikuradse sand grain roughness and have appreciable form drag which can overshadow the surface roughness (*i.e.* skin friction) of the bed form. Neglect of this fact might result in an equivalent sand grain roughness height, k_s, greater than the depth of flow, for then the total resistance would be attributed to the skin friction.

As already stated, the hydraulic roughness of an alluvial stream is primarily dependent upon the bed forms and configurations which are generated as a result of interaction between the flowing fluid and the bed material. Like pipe flow the Darcy-Weisbach friction factor is commonly used in channel flow and represents a measure of its hydraulic roughness. The friction factor for flow occurring in a given river-reach can vary by a factor of ten or more. (For example, during the passage of high flood its value may drop to as low a value as one-tenth of value at low flowrates).

In fact, the variability of alluvial-channel roughness is so great that it is possible for several discharges to occur at some depth-slope combinations.

Fig. 12.45 shows a typical friction factor *versus* mean velocity plot for two sections of Rio Grande, New Maxico (it is based on the plot prepared by Alam and Kennedy using data obtained by

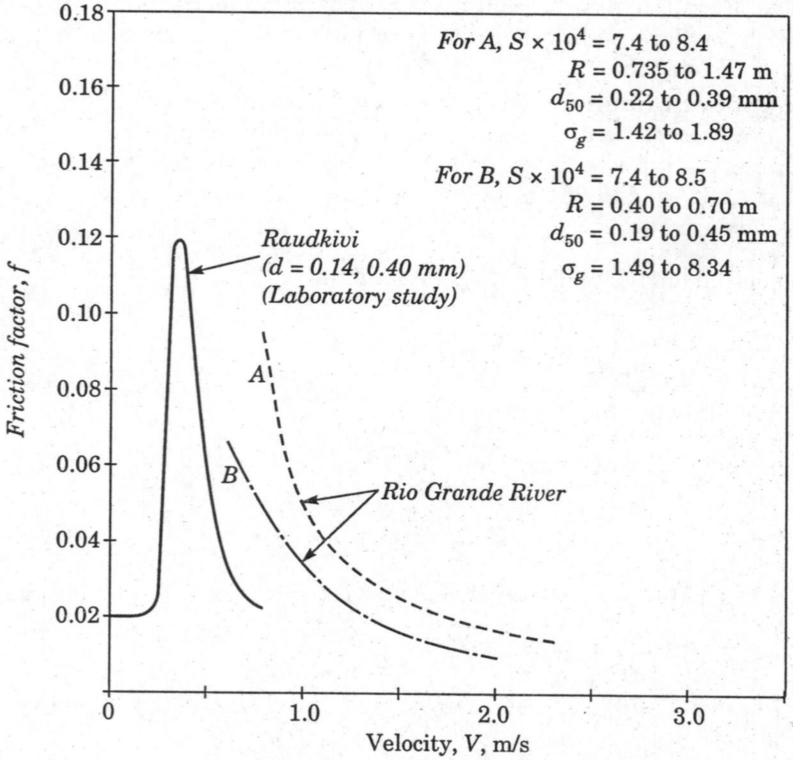

Fig. 12.45. Friction factor *versus* mean velocity.

Nordin). Also shown on the plot is the curve obtained by Raudkivi[1] based on his laboratory experiments. In case of Rio Grande river, the friction factors at the two reaches were found to decrease with the increasing velocity in the range of flow observation (the corresponding bed forms are dunes for velocity 0.5—0.8 m/s, and for higher velocities the dunes give place to the transition flat bed). According to the laboratory study conducted by Raudkivi it is seen that the friction factor remains almost constant for flows within the threshold condition and for velocities exceeding the critical limit different bed forms are generated. The friction factor may shoot upto a value as large as 6 to 10 times (as reported by Raudkivi and Alam and Kennedy respectively) the value at the threshold of particle movement. The friction factor then decreases as dunes give place to the transition (due to interaction among the flow, fluid and the moving bed particles) to a value as low as for the flat bed at the threshold condition.

12.22.5. Prediction of Stage-Discharge Relationship

The channel bed roughness, of which the friction factor is a measure, depends upon the depth of flow, mean velocity and the sediment transport rate. The discharge or the liquid flowrate in an alluvial channel can be predicted from the known slope and grain size distribution of the bed material. Of the many methods developed for predication of depth-discharge relations, only two methods—Engelund's and Ranga Raju's have been picked up for being included and are described as below.

12.22.5.1. Method Proposed by Engelund

Engelund[2], [3] developed a method for predicting water discharge in an alluvial stream for the known slope and grain-size distribution of bed material. He treated the energy line slope, S, to be consisting of two components

$$S = S' + S'' \qquad ...(12.78)$$

where S' = the grain-roughness slope and is equal to the slope that would be required for a given discharge to flow, at the specified hydraulic radius if the bed were flat non-erodible, and composed of non-cohesive material ; and S''' = the additional slope caused by the hydrodynamic drag on the bed forms, which Engelund considers it to be dependent on the height and wavelength of the bed forms and the water depth in the channel. He reasons that S'' is primarily due to the expansion losses in the separation zones downstream from the ripples and dunes and, therefore, utilizes the Carnots' formula for expansion losses in pipes to obtain an estimate for S''. He then introduces the dimensionless shear stress τ_* into his analysis, with its components τ_*' and τ_*''.

$$\tau_* = \tau_*' + \tau'' \qquad ...(12.79)$$

in which

$$\tau_* = \frac{\tau}{\gamma(s-1)d_s} = \frac{y_0 s}{(s-1)d_s} \qquad ...(12.80)$$

$$\tau_*' = \frac{y_0 s'}{(s-1)d_s} \qquad ...(12.81)$$

and

$$\tau_*'' = \frac{\mathbf{F}^2}{8}\frac{\alpha^2}{(s-1)d_s\lambda} \qquad ...(12.82)$$

Here γ is the specific weight of the fluid, s the specific gravity of the bed material, d_s a representative grain diameter (usually d_g or d_{50}, d_g being the geometric mean size), $\mathbf{F}$ the Froude number, $V/\sqrt{gy_0}$, α a geometric factor dependent upon h, L and y_0 (the height, wavelength and the water depth respectively).

---

(1) Raudkivi, A.J., "Study of Sediment Ripple Formation". Proc. ASCE, 89, No, Hy. 6, 15-33 Nov. 1961).

(2) Engelund, F., "Hydraulic Resistance of Alluvial Streams". Journal of the Hydraulics Division, ASCE, Vol. 92, No. Hy. 2 March 1966.

(3) Engelund F., Closure to "Hydraulic Resistance of Alluvial Stream", Journal of Hydraulics Division, ASCE, Vol. 93, No. Hy. 4 July 1967.

Engelund's analysis of flow in alluvial streams is based on similarity hypothesis which assumes that : (1) in two dynamically similar streams τ_*' takes on equal values ; and (2) in two dynamically similar streams the expansion loss (due to form roughness) is the same fraction of the total energy loss. To verify the hypothesis, Engelund used flume data reported by Guy, Simons and Richardson. The quantity τ_*' was calculated from

$$\tau_*' = \frac{\tau'}{\gamma(s-1)d_s} = \frac{y_0 S'}{(s-1)d_s} = \frac{y'S}{(s-1)d_s} \qquad \text{...(12.83)}$$

in which y' is the depth obtained from a logarithmic resistance formula Engelund obtained from flume experiments :

$$\frac{V}{\sqrt{gy'S}} = 6 + 2.5 \ln\left(\frac{y'}{2d_{65}}\right) \qquad \text{...(12.84)}$$

which can be, alternatively, written as

$$\frac{V}{\sqrt{gy'S}} = 5.76 \log\left(\frac{y'}{d_{65}}\right) + 4.26 \qquad \text{...(12.85)}$$

He plotted τ_* as a function of τ_*'. The resulting relationship between these parameters can be expressed by the following equation :

$$\tau_*' = 0.4\,\tau_*^2 + 0.06 \qquad \text{...(12.86)}$$

which is applicable to the lower flow regime in general and to the dunes in particular. The range of applicability of Eq. (12.86) is $0.01 \le \tau_*' \le 0.5$. For the transition regime, the relationship can be written as

$$\tau_*' = \tau_* \qquad \text{...(12.87)}$$

which holds in the range $0.5 < \tau_*' \le 1.0$.

Engelunds' similarity hypothesis is applicable only to rough-boundary channels. According to him, when the grain Reynolds number is larger than about 12, the channel surface will be hydraulically rough. If on the contrary, the Reynolds number is small, a viscous sub-layer exists and the bed configuration will then consist of ripples.

Procedure for depth-discharge computation

The procedure for determination of depth-discharge relationship, for a given channel of known slope and particle-size distribution of the bed material, based on Engelund's analysis, is as given below :

(1) Select an arbitrary value of average depth, y_o, and compute the corresponding value of τ_* from Eq. (12.80) (using $d_s = d_g$).

(2) Use Eq. (12.86) and compute the corresponding τ_*'.

(3) Calculate y' from Eq. (12.83) *i.e.*,

$$y' = \frac{\tau_*'(s-1)d_s}{S}$$

again taking $d_s = d_g$.

(4) Calculate the mean velocity, V, using Eq. (12.85).

(5) Calculate the cross-section area of flow and using the continuity relationship, the discharge through the channel.

Example 12.74. *The test section of a river has slope which can be represented by a trapezoidal cross-section having a bottom width of 91.5 m and banks sloping at 1 : 1. The river slope was found to be 2.5×10^{-4}. The grain size distribution of six samples, taken down to a depth of approximately 50 cm from representative location of the wetted perimeter yields the following results :*

$d_{65} = 0.367$ mm, $d_{50} = 0.321$ mm, $d_{35} = 0.287$ mm,

geometric mean size, $d_g = 0.294$ mm.

The specific gravity of the bed material was 2.65 and the kinematic viscosity of water at the prevailing temperature 10^{-6} m²/s. Determine the stage-discharge relationship and plot the data to yield the stage-discharge curve. Predict from the curve the discharge flowing at an average depth of 1.4 m.

Solution. The computations for the stage-discharge curve, based on Engelund's method, are given below.

Let the average depth be 0.5 m, *i.e.* $y_0 = 0.5$ m. Using Eq. (12.76) with $s = 2.65$, $s = 2.5 \times 10^{-4}$ and $d_s = d_g = 0.294$ mm.

$$\tau_* = \frac{y_0 S}{(s-1)d_s} = \frac{0.5 \times 2.5 \times 10^{-4}}{(2.65-1) \times 0.294 \times 10^{-2}} = 0.258$$

From Eq. (12.81), $\tau_*' = 0.4\tau_*^2 + 0.06 = 0.4(0.258)^2 + 0.06 = 0.0866$

Using Eq. (12.83), $y' = \dfrac{\tau_*'(s-1)d_s}{S}$; $y' = \dfrac{0.0866 \times 1.65 \times 0.294 \times 10^{-2}}{2.5 \times 10^{-4}} = 0.168$ m.

The average velocity from Eq. (12.80) is given by

$$\frac{V}{\sqrt{9.81 \times 0.168 \times 2.5 \times 10^{-4}}} = 5.76 \log \frac{0.168}{0.367 \times 10^{-3}} + 4.26$$

which yields, $V = 0.1624$ m/s

the cross-sectional area of flow at $y_0 = 0.5$ m is

$$A = \tfrac{1}{2}(91.5 + 92.5) \times 0.5 = 46.0 \text{ m}^2$$

and the corresponding discharge

$$Q = AV = 46.0 \times 0.1624 = 7.46 \text{ m}^3/\text{s}.$$

For various other depths (y_0 = 1.0, 1.5, 2.0 m), the corresponding discharge values, can be similarly determined. Table (Ex. 12.57) shows results of such computations.

Table : (Ex. 12.63). Depth-Discharge Relationship

Average depth yo (m)	*Non-dimensional total shear stress τ**	*Non-dimensional grain shear stress τ'**	*Average velocity V (m/s)*	*Discharge Q (m³/s)*
0.5	0.258	0.0866	0.1624	7.46
1.0	0.515	0.1658	0.596	55.2
1.5	0.773	0.298	0.855	119.2
2.0	1.03	0.484	1.145	214.0

The stage-discharge curve based on the foregoing depth-discharge values is plotted in the Fig. of Ex. 12.63.

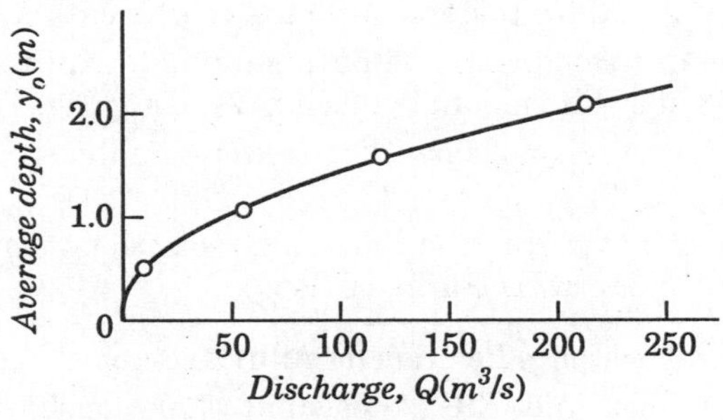

Fig. (Ex. 12.74). Stage-Discharge curve.

12.22.5.2. Method based on Ranga Raju's Analysis

Ranga Raju[4] utilised in his analysis a large amount of data on flumes, canals and natural streams collected by various investigators and presented a resistance relationship for alluvial channels. The graphical relationships evolved by him can be used to prepare the stage-discharge curve for any alluvial stream for which the sediment characteristics and the water surface slope are known.

The analysis of Ranga Raju [and the earlier one of Grade and Ranga Raju[5]] shows that the non-dimensional velocity, V_*, is a function of the non-dimensional slope S_*

$$V_* = f(S_*) \qquad ...(12.88)$$

in which

$$V_* = K_1\sqrt{\frac{R\Delta\gamma}{\rho_f}} \qquad ...(12.89)$$

$$S_* = K_2\left(\frac{R}{d}\right)^{1/3}\left(\frac{S}{\Delta\gamma_s/\gamma_f}\right) \qquad ...(12.90)$$

where V = mean velocity of flow ;

$\Delta\gamma_s = \gamma_s - \gamma_f$, γ_s and γ_f being the specific weights of the sediment and the fluid, respectively ;

ρ_f = mass density of the fluid ;

R = hydraulic radius ;

S = slope of the water surface ;

d = median size of the sediment ;

and K_1 and K_2 = the empirical coefficients which depend on the sediment size.

Fig. 12.46 is a plot of V_* *versus* S_* for all the data used in the analysis and is applicable for the ripple, dune, transition and antidune regimes. The coefficients K_1 and K_2 are unique functions of sediment size and their variation is shown in Fig. 12.47.

The range of variable covered by the data used in the analysis is given below :

Variable	*Range*
1. Sediment size, d	0.011 to 5.20 mm
2. Hydraulic radius, R	0.023 to 17.1 m
3. Slope, S	3.64×10^{-5} to 2.56×10^{-5}
4. Velocity, V	0.058 to 3.44 m/s
5. Specific gravity of sediment, s	1.052 to 4.22

Procedure for Depth-Discharge Relationship

The following steps are necessary for determining the discharge corresponding to a known depth (*i.e.*, stage) and sediment size and water surface slope :

1. For the given depth (or assumed depth in case of stage-discharge curve), calculate the hydraulic radius using the channel cross-section.
2. Use the given sediment size to determine the coefficients K_1 and K_2, from Fig. 12.47.
3. Calculate the non-dimensional slope parameter, S_*, expressed by Eq. (12.90) and read the corresponding non-dimensional velocity, V_* using Fig. 12.46.
4. Compute the mean velocity of flow, V, for known values of K_1 $\Delta\gamma_s$, ρ_f and R, from Eq. (12.89).
5. Determine the discharge using the continuity equation (that is, the flowrate equals the mean velocity multiplied by the area of flow).

---

4. Ranga Raju, K.G., "Resistance Relation for Alluvial Stream" La Houllie Blanche, No. 1, 1970.

5. Garde. R.J., and Ranga Raju, K.G., "Resistance Relationships for Alluvial Channel Flow", Journal of the Hydraulics Division, ASCE, July 1966.

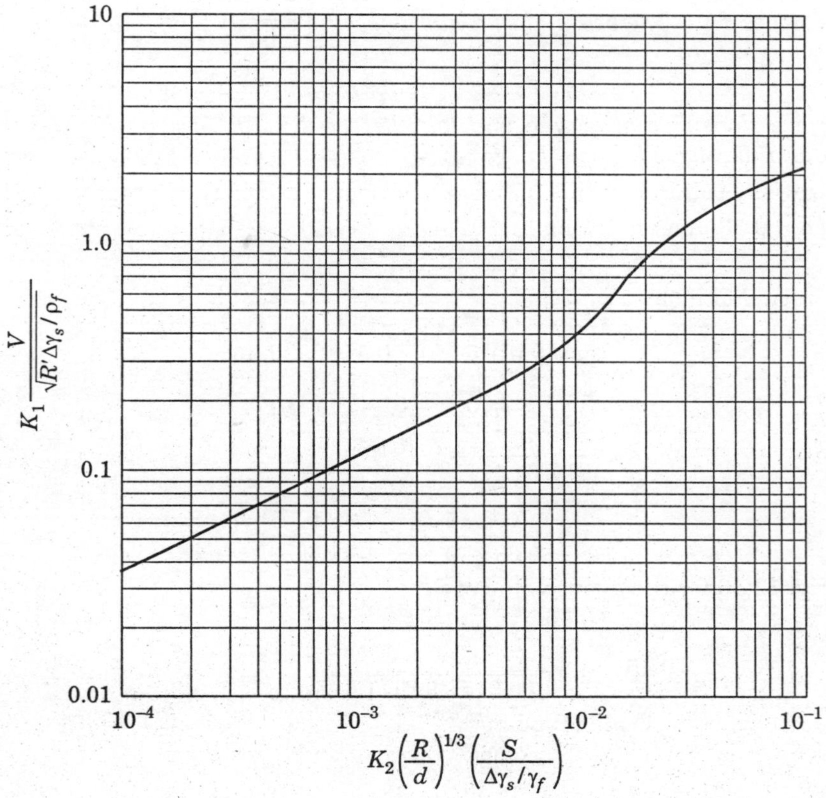

Fig. 12.46. Resistance law proposed by Ranga Raju.

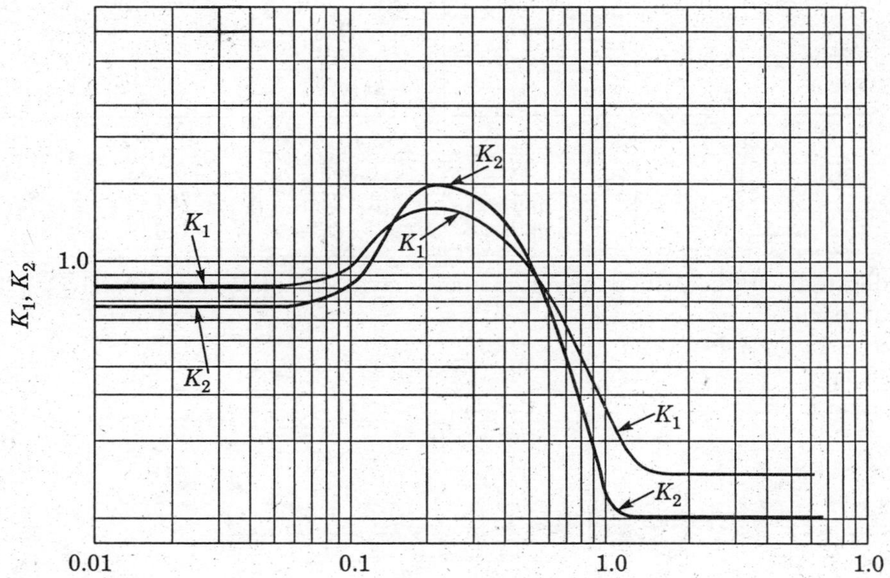

Fig. 12.47. Variation of K_1 and K_2 with sediment size, d mm.

Example 12.75. *A natural stream has an almost trapezoidal cross-section of bottom width 50 m and side slopes 1 : 1. The bed material has a specific gravity 2.68 and the grain size analysis yields the following results :*

$$d_{35} = 0.287 \text{ mm},\ d_{65} = 0.378 \text{ mm},\ d_g = 0.33 \text{ mm}.$$

The water surface slope was measured and found to be 2 × 10⁻⁴. Determine the depth-discharge relationship and prepare a stage-discharge curve.

Solution. The solution of this problem is based on the method proposed by Ranga Raju.

Taking $d = 0.33$ mm and using Fig. 12.47

$$K_1 = 1.3 \text{ and } K_2 = 1.6.$$

For average depth of 0.5 m

$$A = \frac{1}{2}(50.0 + 51.0) \times 0.5 = 25.25 \text{ m}^2.$$

$$P = 50.0 + 2\sqrt{0.5} = 51.41 \text{ m}$$

$$R = \frac{A}{P} = \frac{25.25}{51.41} = 0.4915 \text{ m}$$

$$S_* = 1.6\left(\frac{0.4915}{0.33\times10^{-2}}\right)^{1/3}\left(\frac{2\times10^{-4}}{1.68}\right) = 2.178 \times 10^{-3}$$

From Fig. 12.46 for the above value of S_*,

$$V_* = 0.15$$

$$0.15 = 1.3\frac{V}{\sqrt{0.4915\times9.81/1.68}}$$

from which $V = 0.1952$ m/s

$$Q = AV = 25.25 \times 0.1952 = 4.89 \text{ m}^3/\text{s}.$$

Table (Ex. **12.75**) shows the results of similar calculations **for** other depths.

Table (Ex. 12.75). Depth-Discharge and Other Related Parameters

Depth y_0 *(m)*	*Area* A *(m²)*	*Wetted Perimeter* P *(m)*	*Hydraulic radius* R *(m)*	*Velocity* V *(m/s)*	*Discharge* Q *(m³/s)*
0.5	25.25	51.41	0.4915	0.1982	4.89
1.0	51.0	52.83	0.965	0.308	15.70
1.5	77.25	54.24	1.425	0.422	32.60
2.0	104.0	55.67	1.867	0.558	58.00

The depth-discharge values tabulated above have been plotted in Fig. (Ex. 12.75) to yield a stage-discharge curve.

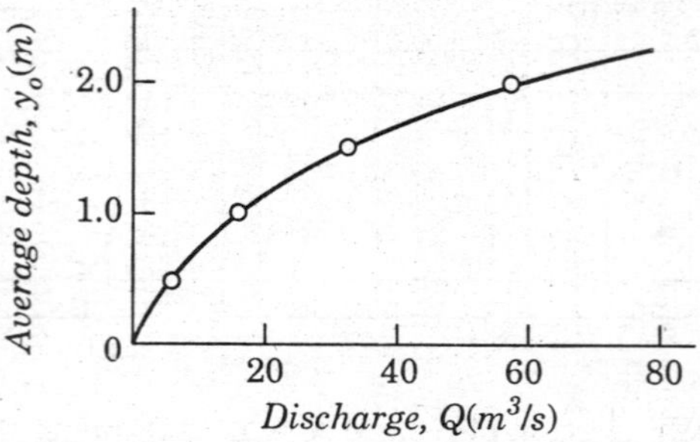

Fig. (Ex. 12.75). Stage-Discharge curve.

OBJECTIVE TYPE QUESTIONS

Select the correct answer :

18. The specific energy in open channels is :

(*a*) the total energy measured above a horizontal datum

(*b*) the total energy measured with respect to the channel bottom which is taken as the datum

(*c*) the kinetic energy plotted above the free-surface

(*d*) the total energy of a specified weight of liquid.

19. The specific energy in a given channel depends on :

(*i*) depth y and discharge Q (*ii*) roughness n and discharge Q

(*iii*) channel slope S_o and discharge Q (*iv*) depth y are roughness coefficient n.

20. The critical depth is the depth of flow at which :

(*a*) the specific energy is maximum (*b*) the unit discharge q is minimum

(*c*) the specific energy is minimum (*d*) the Froude number is greater than unity.

21. Which of the following conditions are applicable for critical flow :

(*i*) $\frac{QT^2}{gA^3} = 1$ (*ii*) $\frac{Q^2T}{gA^3} = 1$ (*iii*) $\frac{Q^2T}{gA^2} = 1$ (*iv*) $\frac{V}{\sqrt{gD}} = 1$.

22. Given below are certain flow situations which usually occur in open channels. Select from the alternatives given alongside the one which is most useful in the flow analysis of each situation :

(*a*) flow in a channel transition (*i*) specific force and specific energy diagrams

(*b*) flow over a hump (*ii*) discharge curve (y-q curve)

(*c*) flow in a hydraulic jump (*iii*) specific energy diagram

(*d*) flow in a venturi flume (*iv*) discharge curve (y-q curve).

23. The critical depth in a rectangular channel is given by :

(*i*) $(q^3/g)^{1/2}$ (*ii*) $(q^3/g)^{1/5}$ (*iii*) $(q^2/g)^{1/3}$ (*iv*) $(q^3/g)^{1/3}$.

24. In a subcritical flow, as the specific energy in a channel is decreased, the depth of flow :

(*a*) also decreases (*b*) increases (*c*) does not vary

(*d*) approaches to a minimum value which corresponds to the minimum specific energy.

25. In a channel transition, the width is gradually reduced till the desired width is obtained. For a given subcritical flow in the approach channel, the depth of flow in the contracted portion :

(*i*) increases resulting in an elevated water surface

(*ii*) is always equal to the critical depth

(*iii*) decreases resulting in a depressed water surface

(*iv*) may be less than the critical depth.

26. In order to create critical flow condition in a channel contraction, there is a limit to which the width can be reduced. If the width of the contracted portion is further reduced, then for subcritical approach flow, the flow in the contracted portion will :

(*a*) take place at the same critical depth

(*b*) be critical, but this will be possible only when sufficient heading up of water in the main channel has taken place to increase the specific energy

(*c*) not be critical (*d*) be supercritical.

27. If the flow in the approach channel happens to be supercritical, the width of contracted portion being large enough not to produce critical condition, the depth of flow in the transition :

(*i*) will be larger than that in the approach channel

(*ii*) will not undergo any change

(*iii*) will be smaller than that in the approach channel

(*iv*) will attain critical depth.

28. Alternate depths of flow in a channel are :

(*a*) the depths at which the total energies are same

(*b*) the depths which occur at the same specific energy

(*c*) the depths for the same specific force

(*d*) the conjugate depths.

29. In a rectangular channel of constant width, water flows at subcritical state. At a place where there is a local rise (hump) in the channel floor, the depth of water over the hump will be :

(*i*) the same as the upstream depth (*ii*) more than the upstream depth

(*iii*) less than the upstream depth (*iv*) less than the critical depth.

30. There is a maximum limit for the hump-height at which the flow over the hump is critical, the approach flow being subcritical. If the height of hump is further increased, the flow over the hump :

(*a*) will attain critical depth when the upstream specific energy has increased itself by heading up of water upstream

(*b*) will have a subcritical depth of flow

(*c*) will attain a supercritical depth

(*d*) will be such that there is no change in the flow depth.

31. For flow over a broad crested weir to be critical, the height Z of the weir should satisfy the following relationship :

(*i*) $Z < (E_1 - E_{min})$ (*ii*) $Z < (E_1 + E_{min})$

(*iii*) $Z \geq (E_1 - E_{min})$ (*iv*) $Z \geq (E_1 + E_{min})$.

32. The concept of specific force is useful in open channel flow analysis. It represents :

(*a*) the sum of all forces per unit volume acting in the flow direction

(*b*) the sum of momentum and hydrostatic force per unit specific weight γ considered in the flow direction

(*c*) the change in momentum of flow per unit weight in direction of flow

(*d*) the sum of momentum, frictional resistance and the hydrostatic force per unit volume acting in the flow action.

33. The specific force is expressed as :

(*i*) $\frac{\gamma Q_2}{gA} + \gamma A\bar{z}$ (*ii*) $\frac{Q_2}{gA} + A\bar{z}$ (*iii*) $\frac{Q_2 T}{gA^3} + \frac{A\bar{z}}{T^3}$ (*iv*) $\frac{Q^3}{A} + \gamma\bar{z}$.

34. Application of momentum principle to a horizontal channel of short length indicates constancy of specific force. The conjugate depths are :

(*a*) the depths at which flow at the same specific energy is possible

(*b*) the depths which correspond to a constant momentum flux

(*c*) possible in rectangular channels only

(*d*) the depths at which the magnitudes of specific force are same, the smaller one lying in the supercritical state while the one lies in the subcritical zone of flow.

35. Shall waves in open channels propagate with velocity equal to :

(*i*) $\sqrt{2gy}$ (*ii*) $\sqrt{gy}$ (*iii*) $\sqrt{gy + V}$ (*iv*) $\sqrt{y} + \sqrt{E/\rho}$.

36. The type of flow in open channels can be ascertained by observing the movement of ripples (small disturbances) on the water surface caused by dropping a stone. The flow will be subcritical when :

(*a*) the ripples are swept away downstream

(*b*) the ripples move in upstream and downstream directions

(*c*) no ripples are formed (*d*) the ripples travel sideways only.

37. Gradually varied flow in open channels is caused when :

(*i*) there is an equilibrium between the forces causing the flow and those opposing it

(*ii*) the pressure forces and the change of momentum are different from each other

(*iii*) when the force causing the flow is not equal to the resistance force

(*iv*) the channel slope is equal to the normal slope.

38. For the following open channel flow cases, how would the depth of flow change ?

(*a*) accelerating flow — (*i*) depth increases

(*b*) flow with zero acceleration — (*ii*) depth decreases

(*c*) decelerating flow — (*iii*) depth does not change.

39. The equation of gradually varied flow is expressed by

(*i*) $\frac{dy}{dx} = \frac{S_o - S_e}{1 + \mathbf{F}^2}$ (*ii*) $\frac{dy}{dx} = \frac{S_e - S_o}{1 + \mathbf{F}^2}$

(*iii*) $\frac{dy}{dx} = \frac{S_o - S_e}{1 - \mathbf{F}^2}$ (*iv*) $\frac{dy}{dx} = \frac{S_e - S_o}{1 - \mathbf{F}^2}$.

40. Given below are different types of channel slopes alongwith certain conditions of flow. Identify for each slope its relevant flow condition :

(*a*) normal slope — (*i*) channel bottom rises in the direction of flow

(*b*) critical slope — (*ii*) critical depth is less than the normal depth

(*c*) mild slope — (*iii*) flow in the channel is uniform subcritical

(*d*) steep slope — (*iv*) flow takes place under critical condition, the depth of flow being y_c

(*e*) adverse slope — (*v*) critical depth is more than the normal depth.

41. The conveyance of the channel defined by $K = \frac{1}{n} AR^{2/3}$ is a measure of its carrying capacity. In terms of K the discharge Q is equal to :

(*i*) $KC\sqrt{S_e}$ (*ii*) $K\sqrt{S_e}$ (*iii*) $KAS_e^{1/3}$ (*iv*) $K\sqrt{RS_e}$.

42. In case of uniform flow in the channel, the water surface slope dy/dx is equal to :

(*a*) 1 (*b*) ∞ (*c*) 0 (*d*) 1000.

43. For the energy line (S_e) and bed slope (S_o) as indicated below, interpret the conditions of flow choosing from the alternatives given :

(*a*) $S_e < S_o$ — (*i*) $y < y_n$

(*b*) $S_e > S_o$ — (*ii*) $y = y_n$

(*c*) $S_e = S_o$ — (*iii*) $y_n > 0$

(*d*) $S_e > 0$ — (*iv*) $y_n = \infty$

(*e*) $S_o = 0$ — (*v*) $y > y_n$.

44. Indicated below are certain relationships between depth of flow, normal depth and the critical depth. Shown alongside are possible types of flow in a specified channel. Match the relevant statement :

(*i*) $y > y_n > y_c$ — (*a*) supercritical flow in a steep channel

(*ii*) $y > y_c > y_n$ — (*b*) subcritical flow in a mid channel

(*iii*) $y < y_n < y_c$ — (*c*) supercritical flow in a mild channel

(*iv*) $y < y_c < y_n$ — (*d*) subcritical flow in a steep channel.

45. Identify conditions represented by the following, selecting the appropriate statement from column 2 :

(*a*) $S_o - S_e > 0$ (*i*) $y > y_e$
(*b*) $1 - \mathbf{F}^2 < 0$ (*ii*) $y > y_n$
(*c*) $S_o - S_e < 0$ (*iii*) $y = y_e$
(*d*) $1 - \mathbf{F}^2 > 0$ (*iv*) $y = y_n$
(*e*) $S_o - S_e = 0$ (*v*) $y < y_n$
(*f*) $1 - \mathbf{F}^2 = 0$ (*vi*) $y < y_c$.

46. A control section is one where a definite relationship between depth of flow and discharge exists. A critical depth section is a control section. Which of the following cases represent control sections :

(*i*) hydraulic jump (*ii*) flow under a sluice gate
(*iii*) flow over a small hump (*iv*) overflow spillway
(*v*) change of slope from mild to steep (*vi*) supercritical flow in a channel.

47. The most essential condition for a hydraulic jump to form is :

(*a*) the existence of subcritical flow before the jump
(*b*) the existence of critical flow before the jump
(*c*) the existence of supercritical flow before the jump
(*d*) the constancy of specific energy.

48. The pre-jump and post-jump depths are related to each other by the equation :

(*i*) $\dfrac{2q^3}{g} = y_1^2 y_2^2 (y_1 + y_2)$ (*ii*) $\dfrac{2q^2}{g} = y_1 y_2 (y_1 + y_2)$

(*iii*) $\dfrac{q^2}{2g} = y_1 y_2 (y_1 + y_2)$ (*iv*) $\dfrac{q^2}{g} = y_1 y_2 (y_1 + y_2)^2$.

49. The energy loss in a hydraulic jump is expressed as :

(*a*) $(y_2 - y_1)^2/4y_1 y_2$ (*b*) $(y_2 - y_1)^4/y_1^2 y_2^2$
(*c*) $(y_2 - y_1)^3/4y_1 y_2$ (*d*) $6(y_2 - y_1)^3/y_1 y_2$.

50. Certain characteristics of hydraulic jump are given below. Identify the relevant definition :

(1) Relative loss (*a*) E_2/E_1
(2) Efficiency of jump (*b*) H_j/E_1
(3) Relative height of jump (*c*) $\Delta E/E_1$.

51. In alluvial channels, the bed consists of cohesionless material which under certain flow conditions moves along the bed. The bed particles are initiated to motion when :

(*a*) the particles are of finer size
(*b*) the particles are loosely packed
(*c*) the bed shear stress is just sufficient to overcome the frictional resistance
(*d*) the Reynolds number has a low value.

52. Shield's analysis of sediment motion provides a means for predicting the motion of bed particles. The critical tractive stress is found to be a function of shear Reynolds number $u_* d/\nu$. The minimum value of the dimensionless critical tractive stress defined by $\tau_{cr}/\gamma(s-1)d$ is given by :

(*i*) 0.03 (*ii*) 0.06 (*iii*) 0.1 (*iv*) 0.15.

53. Under different conditions of flow, certain irregularities appear on an initially plane bed consisting of cohesionless materiel. Identify the flow conditions associated with each of the following bed forms :

(*a*) ripples	(*i*) **F** > 1 : water surface wavy in phase with the bed form
(*b*) dunes	(*ii*) at higher velocities ; **F** ≈ 1
(*c*) flat bed	(*iii*) **F** < 1 ; ripples are superimposed on the up stream slopes.
(*d*) antidunes	(*iv*) **F** << 1 ; do not occur in sediments coarser than 0.6 mm.

ANSWERS

18. (*b*) **19.** (*i*) **20.** (*c*) **21.** (*ii*), (*iv*) **22.** (*a*) (*ii*) ; (*b*) (*iii*) ; (*c*) (*i*) ; (*d*) (*iv*)
23. (*iii*) **24.** (*a*), (*d*) **25.** (*iii*) **26.** (*b*) **27.** (*i*) **28.** (*b*)
29. (*iii*) **30.** (*a*) **31.** (*iii*) **32.** (*b*) **33.** (*ii*) **34.** (*d*)
35. (*ii*) **36.** (*b*) **37.** (*iii*) **38.** (*a*) (*ii*) ; (*b*) (*iii*) ; (*c*) (*i*) **39.** (*iii*)
40. (*a*) (*iii*) ; (*b*) (*iv*) ; (*c*) (*ii*) ; (*d*) (*v*) ; (*e*) (*i*) **41.** (*ii*) **42.** (*c*)
43. (*a*) (*v*) ; (*b*) (*i*) ; (*c*) (*ii*) ; (*d*) (*iii*) ; (*e*) (*iv*) **44.** (*i*) (*b*) ; (*ii*) (*d*) ; (*iii*) (*a*) ; (*iv*) (*c*)
45. (*a*) (*ii*) ; (*b*) (*vi*) ; (*c*) ; (*v*) ; (*d*) (*i*) ; (*e*) (*iv*) ; (*f*) (*iii*) **46.** (*ii*), (*iv*), (*v*) **47.** (*c*)
48. (*ii*) **49.** (*c*) **50.** (1) (*c*) ; (2) (*a*) ; (3) (*b*) **51.** (*c*) **52.** (*i*)
53. (*a*) (*iv*) ; (*b*) (*iii*) ; (*c*) (*ii*) ; (*d*) (*i*).

REVIEW QUESTIONS

1. In what ways the open channel flow is different from the flow in closed conduits ?

2. What type of flow is represented by
(*i*) a river during floods (*ii*) a hydraulic jump
(*iii*) a channel in which its bottom and the water surface are parallel.

3. If water, honey and mercury are made to flow through identically similar channels having same bed slope and depth of flow, explain how the discharge of each of these liquids could be estimated, and what would be their ratio with that of water ?

4. What do you understand by the best hydraulic channel cross-section ? In how many ways can you express it ?

5. Discuss the possibility of uniform flow occurring in (*i*) a horizontal channel (*ii*) a fricitionless channel.

6. An open channel has the following boundary material :
(*i*) fine sand (*ii*) silt (*iii*) coarse sand and gravel
(*iv*) boulders (*v*) fine soil with acquatic weed growth.
Arrange these in order of increasing roughness indicating an approximate value of the Maning's roughness n.

7. Discuss the significance and utility of the concept of specific energy in the study of open channel problems.

8. For a given specific energy, show that the maximum discharge in a rectangular channel occurs at the critical depth.

9. What is Froude's number ? Show that in an open channel minimum specific energy at a given flow and maximum flow at a given specific energy occurs when Froude's number is unity.

10. If a rectangular channel is contracted, show that (*i*) the free-surface is depressed in the contracted portion if the approaching flow is sub-critical, and (*ii*) the water surface is elevated in the contracted portion if the approaching flow is supercritical.

11. If the Froude number of approaching flow in an open channel is less than unity, will (*i*) an increase in channel width and (*ii*) an increase in its bottom elevation cause the water surface to rise or fall ?

12. Which of the following statements are correct in relation to open channel flow, examine each one carefully and explain accordingly.
    (*a*) the hydraulic grade line is always parallel to the energy grade line
    (*b*) the energy grade line coincides with the free surface
    (*c*) the energy and the hydraulic grade lines coincide
    (*d*) the hydraulic grade line can never rise, and
    (*e*) the hydraulic grade line and the free-surface coincide.
13. Explain the difference between "alternate depth" and "sequent or conjugate depth" of flow in open channels.
14. Define critical depth in four different ways.
15. What conditions are essential for the formation of hydraulic jump ?
16. To what different uses can the hydraulic jump be put ?
17. What do you understand by a control section ? Explain the conditions under which a control may act as an upstream control and downstream control.
18. Explain what is a hydraulic bore ?
19. Discuss the circumstances which may lead to the formation of surge waves in open channels.
20. What is meant by "adverse slope" ? Discuss the various types of water surface curves which may be formed on adverse slopes. Illustrate your answer with sketches.
21. Under what conditions will a channel behave as (*i*) mild (*ii*) critical, and (*iii*) steep ?
22. What do you understand by the back water effect and afflux ?
23. Of what significance is the line of uniform depth in a horizontal channel.
24. Explain how the phenomena of hydraulic jump formation and travel of a translatory wave in an open channel are inter-related.
25. Cite examples which are generally encountered in practice wherein the water-surface profile is of (*i*) M_1, M_2 and M_2 types (*ii*) H_2 and H_3 types and (*iii*) A_2 and A_3 types.
26. What precautions should be undertaken to ensure the location of hydraulic jump within reasonable distance from the foot of a spillway ?
27. Represent a hydraulic jump and flow under a sluice on the force (specific force) and energy diagrams.
28. Using the generalised energy diagram for flow in open channels, sketch the changes of water level when an on coming fast flow meets narrow part of the channel.
29. Draw the hydraulic grade line and total energy line for gradually varied flow in open channel.
30. Prove that the Froude number is equal to unity in case of critical flow in open channels.
31. Prove that in uniform open channel the water surface is parallel to channel bed.
32. Describe conditions under which (*i*) water level rises, (*ii*) it falls with respect to the channel.
33. Describe the variations in pressure distributions for (*i*) parallel flow, (*ii*) convex flow and (*iii*) concave flow in a channel longitudinal section.
34. What are the characteristics of back water curve surface profiles in sustaining slope channels.
35. Write brief explanatory notes on :
    (*i*) Economical channel section (*ii*) Factors affecting Manning's *n*
    (*iii*) Control sections (*iv*) Celerity of a gravity wave
    (*v*) Practical applications of standing surge.
36. What is meant by critical depth in an open channel ? What is its relation with velocity of flow ?
37. Select and write correct statement. Critical depth in a non-rectangular channel is expressed by

    (*i*) $\frac{Q^2T}{gA^3} = 1,$ (*ii*) $\frac{QT^2}{gA^2} = 1,$

(*iii*) $\frac{Q^3A^2}{gT^2} = 1,$ (*iv*) $\frac{Q^2T}{gA^3} = 1,$

(*v*) $\frac{Q^2}{gA^3} = 1,$ (*vi*) None of these

38. Prove that for given discharge the specific energy of water flowing in an flowing in an open channel will be a minimum when the Froude's number is unity.

39. Show that for prismatic channel in general, when flow is critical

(*i*) $\frac{Q^2T}{gA^3} = 1,$ (*ii*) $D = y + \frac{A}{2T},$ (*iii*) $S = \frac{P}{C^2T}.$

40. Discuss the effects of :

(*i*) Curvature of flow on pressure distribution.

(*ii*) non-uniform velocity distribution on the energy and momentum equations in open channel flow.

41. What are the premises under which the specific force between two sections will remain the same ?

42. Explain the interaction between flow and alluvial boundaries of an open channel.

43. Discuss the variation of friction factor for alluvial channels.

44. Write about the variation of specific energy and friction slope along the flow direction in M_1 and M_2 profiles.

45. Starting from uniform subcritical flow in an open channel discuss about the conditions of flow up stream and over the hump when humps of increased heights are introduced on the bed. (Use specific energy diagram).

46. For a wide rectangular channel, derive the relation between (*i*) critical depth and discharge (*ii*) critical depth and minimum specific energy.

PROBLEMS

12.13. Show that at the critical state of flow, the specific energy in a rectangular channel is equal to 1.5 times the depth of flow. Will the depth be greater or less than 2/3 specific energy for the trapezoidal channel ?

12.14. A venturiflume of rectangular section has a horizontal base. It is 1.2 m wide at the inlet and 0.6 m wide at the throat. Neglecting frictional effects, determine the flowrate if the depths at the inlet and throat are 30 cm and 45 cm respectively.

12.15. Water flows at a depth of 1.25 m in a rectangular channel 3 m wide. Over a smooth hump 15 cm high on the channel bed, a drop of 20 cm in the water surface is observed. Determine the rate of flow.

12.16. In a rectangular channel 3 m wide, uniform flow occurs at a depth of 1 m. If Maning's n is 0.018 and the bed slope 1 in 1200, what is the minimum height of hump in the bed over which critical flow will be developed ?

12.17. Calculate the discharge through the venturiflume built in a rectangular channel 1 m wide and having its throat width 0.4 m. The upstream head $y_1 = 0.57$ m and the measured head at the throat was $y = 0.5$ m.

12.18. Fluid flows without friction over a horizontal bed with velocity V and depth d such that $V > \sqrt{gd}$. To what height must the channel bottom be raised to make the flow just critical ?

12.19. A rectangular channel 10 m wide carries 10 cumecs of water when flowing 1 m deep. Calculate (*i*) specific energy of flow (*ii*) specific force (*iii*) Is the flow subcritical or supercritical ?

12.20. From the concept of specific energy, obtain the following expression for a constant discharge.

$$E/y_c = \frac{1}{2}(y_c / y)^2 + y / y_c$$

in which all the ratios are dimensionless. Plot E/y_c as abscissas and y/y_c as ordinates to yield a dimensionless specific energy curve.

12.21. Show that the section of a channel in which the flow is critical at any stage takes the form expressed by

$$Q^2 = 80\, x^2 y^2$$

where x is the top width in metre and y is the distance of the water surface below the energy line in metres. Take g = 10 m/s^2.

12.22. Water flows at a depth of 1.5 m in a rectangular channel 3 m wide. A smooth hump 0.3 m high in the channel bed produces a drop of 15 cm in the water surface, what is the flowrate. Neglect losses.

12.23. Calculate the critical depth for a 4 cumecs in a V-shaped channel having sides sloped at 45°. Take g = 10 m/s^2.

12.24. Plot specific energy curve for a rectangular channel 5 m wide carrying 10 cumecs. The maximum depth of flow in the channel is to be 3 m. Determine the following from the curve :

(*i*) the critical depth of flow (*ii*) the minimum specific energy

(*iii*) the alternate depths corresponding to specific energy of 2.5 N-m/N.

12.25. Compute the critical depth for a discharge of 4 cumecs for the following cases :

(*i*) a rectangular channel 1 m wide

(*ii*) a triangular channel with slope 1 : 1

(*iii*) a trapezoidal channel with 2 m bottom width and side slopes 1 vertical to 2 horizontal.

12.26. Calculate the normal depth for a discharge of 4 cumecs flowing in the following channels laid to a slope of 1 in 1000, having Manning's n = 0.02.

(*i*) a rectangular channel 1 m wide

(*ii*) a triangular channel with slope 1 : 1

(*iii*) a trapezoidal channel 2 m wide with side slopes 1 vertical to 2 horizontal.

12.27. For a constant specific energy of 0.8 N-m/N, calculate the maximum discharge that may occur in a rectangular channel 5 m wide.

12.28. If a rectangular channel is contracted, show that (*i*) the free surface is depressed in the contracted portion if the approaching flow is subcritical. (*ii*) the water surface is elevated in the contracted portion when the approaching flow is supercritical.

12.29. For a given supercritical flow in an open channel, determine whether its sequent depth is greater, equal or less than its alternate depth.

12.30. Making use of the Chezy's Manning's and the Darcy-Weisbach's equations establish a relationship among the Chezy's coefficient, Manning's roughness coefficient and the friction factors.

12.31. A rectangular channel has a width of 6 m and n = 0.025, determine :

(*a*) normal slope at a normal depth of 1 m when discharge is 12 m^3/s

(*b*) the critical depth when the discharge is 12 m^3/s.

12.32. Water flows through a wide sluice opening into a rectangular channel 4 m wide at the rate of 6 m^3/s. A hydraulic jump is formed just downstream of a section, where the depth of water is 40 cm. Determine the depth downstream of the jump and the power dissipated in it.

12.33. The width of a stream is 15 m and the discharge is 50 m^3/sec. It passes through a sluice 70 cm deep and 15 m wide. Determine whether a hydraulic jump will be formed beyond the sluice and if so the sequent depth after the jump.

12.34. If y_1 and y_2 are alternate depths in a rectangular channel, show that $\dfrac{2y_1^2\, y_2^2}{y_1 + y_2} = {y_c}^3$

where y_c is the critical depth of flow. Using values of y_1/y_c as ordinate and y_2/y_c as abscissa, construct a dimensionless graph for the above equation and study its characteristics.

12.35. Prove that the loss of energy head in a horizontal hydraulic jump is equal to $(d_2 - d_1)^3/4d_1.d_2$, where d_1 and d_2 are conjugate depths.

12.36. Show that for a short horizontal channel of rectangular section, the Froude numbers before and after the jump are related by

$$(\sqrt{1 + 8\mathbf{F}_1^2} - 1)(\sqrt{1 + 8\mathbf{F}_2^2} - 1) = 4.$$

12.37. Show that for hydraulic jump in a horizontal rectangular channel

$$\mathbf{F}_2 = \mathbf{F}_1 \,(y_1/y_2)^{3/2}$$

where y_1 and y_2 are the conjugate depths.

12.38. What is meant by 'Hydraulic Bore' ?

A rectangular channel 3 m wide and 1.25 m deep discharging 10 cumecs of water suddenly has the discharge reduced to 7 cumecs at the downstream end. Determine the height and the velocity of the surge wave.

12.39. A steady flow at 3 m/s and 15 m depth is suddenly stopped at the downstream end of the 3600 m reach of a fricitionless channel of rectangular section. Determine the celerity of the surge produced and the time it takes to arrive at the upstream end of the reach.

12.40. A low overflows weir is installed at the end of a long rectangular channel. The channel is 3.25 m wide and its bed slope is 0.0001. If the depth just upstream of the weir is 2.50 m at a discharge 9.0 m^3/s, how far upstream the back water will cause a velocity reduction of 20%. Take n = 0.014.

12.41. A long rectangular channel of unfinished concrete, Manning's n = 0.014, has a width of 3.5 m and carries a flow of 7 m^3/s of water, the bottom slope changing abruptly from 0.01 to 0.0001 at an intermediate section. Show that the uniform depth of flow approaching the break in grade is less than the critical depth. A jump must evidently form in the vicinity of the transition. Will the jump occur on the steep or in the mild slope ?

12.42. Sketch to approximate scale the form of the free surface and the total energy line at an abrupt change in channel slope, assuming that the critical depth is 1.10 m and the uniform depth for the down-stream slope is 1.0 m.

12.43. Show that the flow profile in a wide horizontal channel may be expressed as

$$x = \frac{C^2}{g}\left(y - \frac{y^4}{4y_c^3}\right) + \text{Constant where } C = \text{Chezy's roughness coefficient.}$$

12.44. Show that for a rectangular horizontal channel the water surface slope is given by

$$\frac{dy}{dx} = \frac{f\,yQ^3}{8R(Q^2 - gb^2y^3)}$$

where f = the friction factor in the Darcy-Weisbach equation
R = the hydraulic radius
b = width of the channel, y = depth of flow, and
Q = discharge.

12.45. A trapezoidal canal having a bed width of 6 m, side slopes 1 horizontal to 2 vertical, bed slope S_0 = 0.0016 and Manning's n = 0.025 carries water at a normal depth of 0.7 m. A back water profile is created by a dam which backs up the flow to a depth of 1.5 m immediately upstream of the dam. At what distance from the dam will the depth of flow be 1.20 m, treating the entire distance as one reach.

12.46. A rectangular channel lined with concrete (Manning's n = 0.015) is 2 m wide and has a bottom slope of 1 in 1600. If ends in a free outfall. How far from the outfall is the depth 0.80 m when the flowrate is 2 m^3/s ?

12.47. A wide channel with C = 77.5 m$^{1/3}$/sec and S_0 = 1/4900 carries a discharge of 1.5 m^2/s/m width A weir is constructed across the channel at its downstream end and the depth of water behind the weir is 2.15 m. Find out at what distance (in km) from the weir the depth of water in the channel is 1.55 m.

12.48. A jet 100 cm deep with a total head of 10 metres is issued from a sluice gate 5 metres wide and then flows along a horizontal fricitionless channel to a narrow culvert. Determine the safe width for the culvert so that hydraulic jump is not formed upstream.

12.49. A river 50 m wide has a bed slope of 1 in 10,000. Compute the energy line slope between the two sections, where the water levels are 4 m and 7 m respectively. Take Manning's n as 0.03 for the section midway between the above sections.

12.50. A channel carries flow at the rate of Q m^3/s. The water surface slope $dy/dx > 0$ and the depth of how $y > y_c$. Is the specific energy increasing or decreasing, explain ?

12.51. Prove that for a gradually varied flow in a prismatic open channel, the distance x between two sections having depth y_1 and y_2 respectively, is given by the expression,

$$x = \int_{y_1}^{y_2} \frac{1 - \dfrac{Q^2 T}{gA^3}}{S_0 - \dfrac{n^2 Q^2}{A^2 (R)^{4/3}}} dy$$

where, Q is the discharge through the channel

T is the width of the cross-section at the water surface

A is the cross-sectional area

n is the roughness coefficient in the Manning's formula

R is the hydraulic radius.

12.52. The uniform depth in a very long channel of rectangular cross-section is 2.0 m, when the rate of flow is 4.0 cumecs per metre width. A sluice gate which controls the inflow at the upstream end produces under these conditions a sheet of water 30 cm deep, which gradually expands to yield a profile of the M_3 type. If the computed co-ordinates of this profile are as follows, how far down stream from the sluice gate will the depth be such that a stable hydraulic jump will form :

x	y
0 m	30.0 cm
30 m	33.5 cm
60 m	38.5 cm
90 m	44.5 cm
120 m	51.5 cm
150 m	60.0 cm
180 m	69.0 cm

where x is the distance in metres measured from the sluice gate along the channel and y is the depth of flow in cm.

12.53. Show that for gradually varied flow in rectangular channel, the water surface slope may be written as

$$\frac{dy}{dx} = S_0 \frac{1-(y_n/y)^{10/3}}{1-(y_c/y)^3}$$

where S_0 is the bed slope, y is the flow depth, y_n is the normal depth and y_c is the critical depth of flow.

12.54. For steady non-uniform flow in a rectangular channel, show that the slope of the water surface at any point is given by

$$\frac{dy}{dx} = S_0 \frac{1-(y_n/y)^2}{1-(y_c/y)^3}$$

12.55. What information may be obtained from the following :

(*i*) $\dfrac{dy}{dx} = 0$, (*ii*) $\dfrac{dy}{dx} = S_0$,

(*iii*) $\dfrac{dy}{dx} = \infty$, where $\dfrac{dy}{dx}$ is the slope of the water surface with respect to the channel bottom.

12.56. Derive the differential equation for gradually varied flow in prismatic channels, stating clearly the assumptions made in its derivation. Interpret this equation and obtain conditions for hydraulic jump to occur.

12.57. Illustrate the classification of all the water surface profiles in open channel flow. Discuss the characteristics of these profiles with the help of the dynamic equation of gradually varied flow.

12.58. What is the essential difference between gradually varied flow and rapidly varied flow ? Sketch the following types of back water curves and indicate the conditions and types of controls necessary to produce them ; (*i*) M_3 curve, (*ii*) S_1 curve, (*iii*) C_1 curve, (*iv*) H_2 curve and (*v*) A_2 curve.

12.59. If the discharge in a prismatic channel is suddenly increased from q_1 to q_2, a surge wave will develop and proceed downstream with a uniform velocity C. Show that the continuity and momentum equations can be written in the form :

$$(C - V_2)A_2 = (C - V_1)A_1$$

$$(\gamma A_2 \bar{Z}_2 - \gamma A_1 \bar{Z}_1) = \frac{\gamma}{g}(C - V_1)A_1(V_1 - V_2)$$

where V is the velocity, A is the area of flow, $\bar{Z}$ is the depth of C.G. of flow area below free surface, γ is specific weight of water and g is acceleration due to gravity.

12.60. Deduce an expression for computing the celerity of a gravity wave of moderate amplitude in shallow water.

12.61. Obtain the equation $(V_1 - V_2)^2 = (y_1 - y_2)^2 \dfrac{(y_1 + y_2)g}{2y_1y_2}$ which relates the initial and final depths of flow and velocities of the surge wave.

12.62. What are the general principles for locating the position of the hydraulic jump in the following cases : (*i*) downstream of a sluice, (*ii*) when the channel slope changes from steep to mild, and (*iii*) behind an overflow barrier.

12.63. Show that the pressure distribution coefficient α' may be expressed by

$$\alpha' = 1 \pm \frac{1}{\gamma} \int_0^A p_c v \, dA$$

and that its value is greater than 1.0 for concave flow, equal to 1.0 for parallel flow and less then 1.0 convex flow.

12.64. In fixed boundary channels, the velocity distribution obtained from boundary layer theory can be expressed by

$$u = 5.75 u_* \log \frac{32.2y}{k_s}$$

in which u = velocity at a distance y from the channel bottom ;

$u_* = \sqrt{g y_0 S}$, the shear velocity ;

k_s = Nikuradse's sand grain roughness ;

y_0 = depth of flow, and S = slope.

Integrating this equation, determine the mean velocity and the depth at which this velocity occurs. Compare the equation so obtained with that of Chezey's and plot the Chezy's coefficient C with the roughness parameter y_0/k_s. Using this plot calculate the value of y_0/k_s for which the Manning's formula is applicable.

12.65. A stream with nearly trapezoidal cross-section has a bed width of 9.4 m and the banks at 45° with the vertical. The slope is 3.5×10^{-4}. The bed material has a specific gravity 2.68 and its grain-size distribution yields.

$$d_{65} = 0.575 \text{ mm}, \quad d_{35} = 0.34 \text{ mm}.$$

Determine the discharge when the average depth is 1.0 m.

12.66. The river Pigeon Roost Creek near Byhalia, Missouri (USA) had a slopes of 9×10^{-4} at the time when discharge measurements were made. The bed material had the following characteristics :

$$\text{sp. gr.} = 2.68, \quad d_{65} = 0.45 \text{ mm}, \quad d_{35} = 0.35 \text{ mm}.$$

Predict the discharge per unit with at the average depth of 0.61 m and compare it with the corresponding observed flow of 0.52 m³/s/m.

12.67. The Colorado river at Tylors Ferry, Arizona (USA) was found to have a slope of 2.17×10^{-4}. The bed material has the following characteristics determined from grain-size analysis.

sp. gr. = 2.68 $d_{65} = 0.378$ mm, $d_{35} = 0.287$ mm.

Predict the depth-discharge per unit width relationship and plot the same on a plain graph paper.

12.68. Water flowing down a spillway at the rate of 30 m³/s/m width leaves the apron horizontally with a mean velocity of 15 m/s. Determine :

(*i*) the depth of tail water necessary to form a hydraulic jump ;

(*ii*) the specific energy of flow beyond the jump.

12.69. A wide river flows over a short horizontal stretch of very rough bed. The flow upstream of bed is 1 m deep and the average velocity of flow is 1 m/s. If the rough stone bed is estimated to experience a drag force of 2000 Newtons, determine the depth and velocity of the river downstream of the rough stretch.

12.70. A trapezoidal channel has a bottom width of 6.0 m and side slopes of 1 : 1. The depth of flow is 1.5 m at a discharge of 15 m³/s. Determine the specific energy.

12.71. Water discharging from an overfall spillway flows to a jump type basin. The discharge intensity is 30 m³/s/m and the water depth is 1.5 m. Find the conjugate depth for the jump to form.

12.72. Prove that the critical depth d_c in a V-shaped channel of central angle 2α is given by

$$d_c = \left(\frac{2Q^2}{g\tan^2\alpha}\right)^{1/5}$$

where Q is the discharge and g is acceleration due to gravity.

12.73. A rectangular channel 2.4 m wide carries uniform flow of 7 cumecs at a depth of 1.5 m. If there is a local rise of 0.15 m in the bed level, calculate the change of water surface elevation. What can be the maximum rise in the bed elevation such that the upstream depth is not affected ?

12.74. Show that hydraulically most efficient trapezoidal section is half of a regular hexagon and its hydraulic radius is equal to half the depth of flow. [**Hint.** Refer Art. 12.7.2]

ANSWERS

12.14. 0.3 m³/s. **12.15.** 6.165 m³/s **12.16.** 0.30 m. **12.17.** 0.25 m³/s.

12.18. $\Delta Z = d + \frac{V^2}{2g} - \frac{2(dgV)^{2/3}}{2g}$ **12.19.** (*i*) 1.051 N-m/N. (*ii*) 6.019 N/N/m³ (*iii*) Subcritical.

12.22. 7.55 m³/s. **12.23.** 1.26 m. **12.24.** (*i*) 0.74 m, (*ii*) 1.113 N-m/N. (*iii*) $y_1 = 0.03$ m, $y_2 = 2.45$ m. **12.25.** 1.177 m, 1.26 m. 0.61 m

12.26. 4.35 m, 1.61 m, 0.94 m. **12.27.** 20.6 m³/s. **12.30.** $C = \frac{y_0^{1/6}}{n} = \sqrt{\frac{g}{f}}$

12.31. (*a*) $S_n = 0.00367$ (*b*) $S_e = 0.00906$, $y_c = y_n = 0.742$ m (*c*) $S_{cn} = 0.009$, $Q = 18.78$ m³/s.

12.32. 0.891 m, 4.86 kW **12.33.** H.J. will be formed, $y_2 = 1.484$ m.

12.38. 1.8 m, 1.97 m/s. **12.39.** 11.53 m/s, 312 sec. **12.40.** 837 m.

12.41. $y_c = 0.742$ m, (*i*) $y_n = 0.54$ m (*ii*) $y_n = 2.9$ m jump will form on the steep slope.

12.45. 207.6 m **12.46.** 188.9 m. **12.47.** 4.116 km.

12.48. Less than 1.164 m (say 1.0 m). **12.49.** 1.421×10^{-4} **12.50.** Increasing.

12.52. 160 m. **12.64.** $V = 5.75\sqrt{gy_0S}\log\frac{12.1y_0}{k_s}$; $y = 0.368\, y_0$ for $u = V$, $y_0/k_s \approx 30$.

12.65. 7.6 m³/s. **12.68.** 8.629 m, 9.245 m. **12.69.** 0.042 m, 23.8 m/s

12.70. 1.5406 m. **12.71.** 10.335 m. **12.73.** 0.0764 m, 0.7391 m.

13

Flow Measurements

13.0. Introduction

Measurement of flow constitutes an essential part in all flow systems. Whether it is a fluid flow in a pipe line or liquid flow in an open channel, it is of primary importance to know the quantity of fluid passing through the system. Direct measurement of flow, either gravimetric or volumetric, is the most accurate method of flow measurement. However, the direct measurement of flow may not be practicable in large and continuous flow systems such as the main pipeline of a city water distribution system. This method can, nevertheless, be adopted for flow through small experimental pipelines and open channels generally used in laboratories. To measure the discharge by volume, it is necessary to have a calibrated tank or container into which the fluid can be discharged for a certain known period of time. The tank has to be calibrated so that the volume at various elevations is known. Knowing the volume of fluid collected in a known time, the volumetric rate of flow, commonly known as discharge, can be determined. Measurement of flow by gravimetric methods involves determination of weight of fluid collected in a weighing tank. Such a tank is either mounted on a platform scale or in case of large tanks special supporting system and balance must be designed for this purpose.

13.1. Measurement of Flow

The methods utilized for measuring flow in pipelines and channels are mostly indirect since the discharge for a known geometry of flow cross-section is determined by multiplying the area of cross-section with the mean velocity of flow. The problem of flow determination thus boils down to finding the average velocity of flow which enables estimation of flow by use of the continuity principle. The continuity principle expresses the flowrate as the product of cross-sectional area and the average velocity. For pipeline flow measurement, any of the following devices can be made use of : venturimeter, nozzle meter, orificemeter and bend meter. The measurement of flow in open channels is generally made by means of weirs and gates. In large open channels and rivers the flow is estimated by dividing the flow section into number of smaller sections and determining the average velocity for each by means of current meter.

13.2. Measurement of Flow in Pipes

13.2.1. Venturimeter. The rate of flow can be determined quite accurately with very little loss of energy by means of a venturimeter. Reference is invited to Article 4.14.2 which deals with the venturimeter in detail. If h is the difference in the piezometric heads (in terms of the flowing fluid) between the inlet and the throat, the discharge Eq. (4.32) may be written as

$$Q = \frac{C_d A_1 A_2}{\sqrt{A_1^2 - A_2^2}} \sqrt{2gh} \qquad \text{...(13.1 } a)$$

For a venturimeter of known geometric dimensions the areas A_1 and A_2 are known. The discharge coefficient C_d varies with the Reynolds number, the variation is very small and if an average value is assumed, the discharge equation for a particular meter reduces to

$$Q = K\sqrt{h} \qquad \text{...(13.1 } b)$$

in which K is a constant and is called the venturimeter constant. The value of K can be determined from the calibration of venturimeter. It is evident from Eq. (13.1 b) that there exists a distinctive relationship between the discharge Q and the differential head h. The overall pressure loss in the venturimeter ranges from 10 to 16% of the differential head h for diffuser cone angles of 5° to 15° respectively, the throat to pipe diameter ratio being 0.5. The pressure loss decreases with increasing diameter ratios and *vice versa*. For a venturitube with 15° recovery cone (*i.e.*, diffuser cone angle of

15°), the overall pressure loss ranges from about 30% to 10% of the differential head h as the diameter ratio increases from 0.2 to 0.8.

Example 13.1. *Show that for incompressible flow the loss per unit weight of fluid between the upstream section and throat of a venturimeter is* $K_L V_2^2/2g$ *where* $K_L = (1/C_v^2 - 1)\,[1 - D_2/D_1)^4]$ *and* V_2 *and* D_2 *are the velocity and the diameter at throat respectively.*

Solution. For simplicity in the analysis, let the venturimeter and the pipe axis be horizontal. Applying Bernoulli's equation to the upstream section (1) and the throat section (2), neglecting losses,

$$p_1/\gamma + V_1^2/2g = p_2/\gamma + V_2^2/2g \quad \text{...(1)}$$

From the continuity equation, we have $Q = A_1V_1 = A_2V_2$

or $$D_1^2V_1 = D_2^2V_2\,;\ V_1 = (D_2/D_1)^2/V_2.$$

Substituting for V_1 in Eq. (1) and solving for V_2, we obtain

$$V_2 = \frac{1}{\sqrt{1-(D_2/D_1)^4}}\sqrt{2g\left(\frac{p_1-p_2}{\gamma}\right)} \quad \text{...(2)}$$

Since the losses were neglected, Eq. (2) represents the theoretical velocity. Using a coefficient of velocity C_v, the actual velocity

$$V_2 = C_v\frac{1}{\sqrt{1-(D_2/D_1)^4}}\sqrt{2g\left(\frac{p_1-p_2}{\gamma}\right)} \quad \text{...(3)}$$

Applying Bernoulli's equation between sections (1) and (2), taking losses into consideration

$$p_1/\gamma + V_1^2/2g = p_2/\gamma + V_2^2/2g + h_{:L}$$

From which $$h_L = \frac{p_1-p_2}{\gamma} + \frac{V_1^2 - V_2^2}{2g}$$

Substituting for $\frac{p_1-p_2}{\gamma}$ from Eq. (3), we obtain

$$h_L = \frac{V_2^2}{C_v^2 2g}[1-(D_2/D_1)^4] + [(D_2/D_1)^4 - 1]\frac{V_2^2}{2g}$$

$$= \left[\frac{1}{C_v^2} - 1\right][1-(D_2/D_1)^4]\,V_2^2/2g = K_L\frac{V_2^2}{2g}$$

where $$K_L = \left[\frac{1}{C_v^2} - 1\right][1 - D_2/D_1)^4].$$

Note. *For more solved examples on venturimeter, see Ex. 4.34 to Ex. 4.39.*

13.2.2. Nozzle meter or Flow nozzle. If the diverging cone of the venturimeter be omitted, a flow nozzle of the type shown in Fig. 13.1 would be obtained. The flow nozzle is thus a truncated form of the venturimeter. It is simply a constriction with well rounded entrance placed in the pipeline. The flow nozzle is simpler than the venturimeter and can be installed between the flanges of a pipeline. It serves the same purpose, though at the expense of an increased energy loss due to sudden expansion of flow downstream of the nozzle. The discharge and other equations for the flow nozzle are the

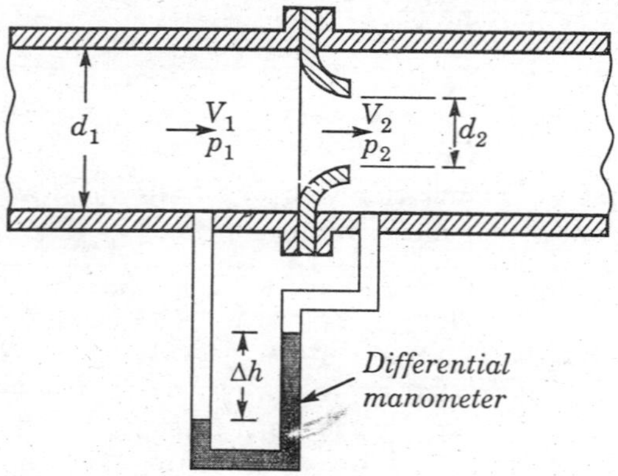

Fig. 13.1. Nozzle meter (or flow nozzle).

same as for the venturimeter and can be derived in the same manner. The coefficient of contraction $C_C = 1$ for all discharges through the nozzle. As is envisaged for the venturimeter, the flow nozzle should similarly be preceded by at least 10 diameters of straight pipe for accurate measurements. Also for accurate results, the flow nozzle, like the venturimeter, must be calibrated in place. The overall pressure loss for flow nozzle ranges from about 30 to 90% of the inlet to throat pressure drop as the throat to pipe diameter ratio decreases from 0.8 to 0.2.

13.2.3. Sharp-Edged Concentric Orifice meter

Orificemeter is an extensively used flow measuring device in pipes. It is possibly one of the oldest devices for measuring or regulating the flow of fluids and historical records show that the Romans of Caeser's time used it to measure water to householders.

In its simplest and most familiar form, the orifice is merely a circular hole in a thin flat plate which is clamped between the flanges at a joint in the pipe line so that its plane is perpendicular to the axis of the pipe and the hole is concentric with the pipe. A thicker plate is sometimes used. It is then chamfered around the hole on the outlet side so as to leave only a thin edge (hence the prefix—sharp edged), while the inlet face remains flat with a sharp 90 degree corner at the edge of hole. Pressure tappings, for being connected to a differential gauge, are made in the pipe wall on both sides of the plate (one pipe diameter upstream and half pipe diameter downstream of orifice plate is commonly used) so that the difference in the fluid pressure on both sides of the orifice can be measured. This difference of pressure or differential head is utilized to determine the flow.

Certain special orifice forms—eccentric and segmental—are used for measuring sediment laden liquids or liquids carrying extraneous matter. The specific location of the orifice permits complete drainage of the extraneous matter. A circular eccentric orifice is constructed and installed so that its centre is not on the axis of the pipe. Generally, the eccentricity is such that one side of the circular opening is almost flush with the inside wall of the pipe. In case of the segmental orifice, the shape of the restricting area is a segment of circle of approximately the same diameter as that of the pipe. Sometimes with very large pipes, only that part of the orifice plate forming the segment is used as the opening.

Theoretical analysis of flow through the orifice plate utilizes the fundamental relationships of fluid flow, namely, the continuity and energy equations. Fig. 13.2 illustrates the flow pattern at and near the orifice plate. The fluid flow approaching the orifice gets accelerated and the boundary streamlines confining the flow assume the shape as shown in the figure. The radially inward component of fluid acceleration causes reduction in the flow area of the jet issued from the orifice. There is, therefore, a vena contracta which is located within a distance of one pipe diameter. By applying the equation of continuity to sections (1) and (2) we obtain

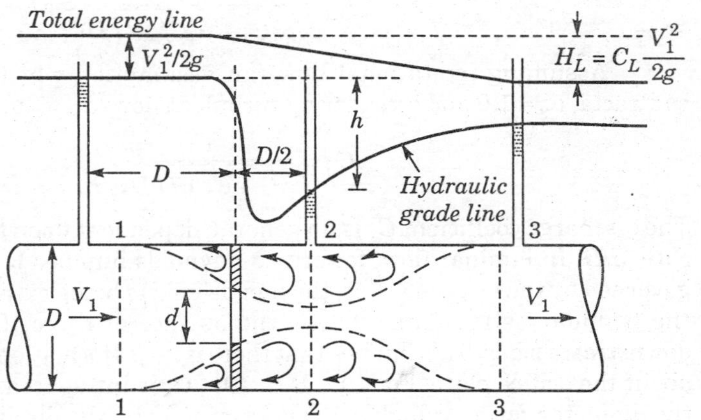

Fig. 13.2. Flow through orifice meter.

$$Q = \frac{\pi}{4} D^2 V_1 = C_c \frac{\pi}{4} d^2 V_2$$

$$D^2 V_1 = C_c d^2 V_2$$

in which Q is the volume flowrate, V_1 the velocity in the pipe of diameter D, V_2 the velocity at section (2) which is the section of vena contracta, d the diameter of orifice, and C_c the coefficient of jet contraction.

Applying the Bernoulli's equation between sections (1) and (2), considering the pipe axis horizontal

$$\frac{p_1}{\gamma} + \alpha_1 \frac{V_1^2}{2g} = \frac{p_2}{\gamma} + \alpha_2 \frac{V_2^2}{2g} \qquad \text{...(13.2 } a\text{)}$$

Since the flow between these sections is accelerating, the energy loss will be extremely small justifying its neglect. Here p_1, p_2 and α_1, α_2 are the pressures and kinetic energy correction factors at the respective sections, γ is the specific weight of the fluid and g the gravitational acceleration.

Eq. (13.2 *a*) on combination with the continuity equation results in,

$$\frac{p_1 - p_2}{\gamma} = h = \frac{V_1^2}{2g}\left[\frac{\alpha_2 D^4}{C_c^2\, d^4} - \alpha_1\right] \qquad \text{...(13.2 } b\text{)}$$

and the velocity can be written as

$$V_1 = \frac{C_c d^2}{\sqrt{\alpha^2 D^4 - \alpha_1 C_c^2 d^4}} \sqrt{2gh}$$

in which h is the differential head across the orifice and is equal to $(p_1 - p_2)/\gamma$, the flowrate is then,

$$Q = \frac{\pi}{4} D^2 V_1 = \frac{\pi}{4} D^2 \frac{C_c d^2}{\sqrt{\alpha^2 D^4 - \alpha_1 C_c^2 d^4}} \sqrt{2gh}$$

which one can write as

$$Q = C_d \frac{\pi}{4} d^2 \sqrt{2gh} \qquad \text{...(13.3 } a\text{)}$$

Here C_d is the coefficient of discharge given by

$$C_d = \frac{C_c}{\sqrt{\alpha_2 - \alpha_1 C_c^2 (d/D)^4}} \qquad \text{...(13.3 } b\text{)}$$

Assuming irrotational flow (an assumption which seems reasonable) in the jet at vena contracta, $\alpha_2 \approx 1.0$ and considering turbulent flow, $\alpha_1 \approx 1.0$. Eq. (13.3 *b*) for turbulent flow reduces to

$$C_d = \frac{C_c}{\sqrt{1 - C_c^2 (d/D)^4}} \qquad \text{...(13.3 } c\text{)}$$

The discharge coefficient C_d is, in general, dependent upon the Reynolds number **R** and the diameter ratio *d*/*D*. In laminar flow, for orifice Reynolds number $\mathbf{R}_d \le 10$, the discharge coefficient is solely governed by the viscous forces, the inertia effects being extremely small. Under such a circumstance, the boundary streamlines follow the surface of the orifice plate on both upstream as well as downstream sides. This implies that there is no jet formation and the flow pattern will be symmetrical about the orifice plate. For $10 < \mathbf{R}_d \le 10^4$, C_d is influenced by both the Reynolds number as well as the diameter ratio. At higher Reynolds numbers, the discharge coefficient becomes independent of viscous effects and is solely governed by the flow geometry, *i.e.*, the diameter ratio. We can now summarize the foregoing by writing

$$C_d = f(\mathbf{R}_d), \qquad \text{for } \mathbf{R}_d \le 10.$$

$$[C_d = 0.1 \text{ to } 0.5 \text{ for } \mathbf{R}_d \text{ ranging from } 0.4 \text{ to } 10]$$

$$C_d = f(\mathbf{R}_d, d/D), \qquad \text{for } 10 < \mathbf{R}_d \le 10^4$$

and

$$C_d = f(d/D), \qquad \text{for } \mathbf{R}_d > 10^4.$$

In the turbulent range of flow, the discharge coefficient ranges from 0.8837 to 0.5933 for *d*/*D* ratio varying from 0.8 to 0.2, the range of pipe Reynolds number, $\mathbf{R}_D$, being 5×10^3 to 10^6, respectively.

The discharge coefficient decreases with decreasing d/D and increasing Reynolds number. The pipe Reynolds number and the orifice Reynolds number are related by the diameter ratio.

$$\mathbf{R}_D = \frac{4Q}{\pi D \nu}, \quad \mathbf{R}_d = \frac{4Q}{\pi d \nu}$$

$$\mathbf{R}_D = \mathbf{R}_d\,(d/D).$$

13.2.3.1. Energy Loss in Orifice meter. Let section (3) of Fig. 13.2 represent the section where the flow regains uniformity and, consequently, the flow area becomes equal to the area of pipe cross-section. The energy loss between sections (1) and (3) can be expressed by

$$H_L = C_L\, V_1^2/2g \qquad \text{...(13.4)}$$

where H_L is the head loss, C_L the loss coefficient, and V_1 the velocity in the pipe.

As a result of an analytical-cum-experimental study[8] on "Energy Loss due to Sharp-edged Orificemeters", it is possible to predict the energy loss accurately. The theory has been excellently supported by experimental data for laminar and turbulent flows. The loss coefficients in turbulent flow, predicted by the theory, are tabulated in Table 13.1 for various diameter ratios.

Table 13.1. Loss Coefficients For Orificemeter in Turbulent Flow

Diameter ratio d / D	*Loss Coefficient C_L, Eq. (13.4)*
0.2	1670.0
0.3	306.0
0.4	86.0
0.5	29.5
0.6	11.0
0.7	4.4
0.8	1.3

In case the velocity is not known, head loss can be obtained approximately from the following relation

$$\frac{H_L}{h} = 2 - \left(\frac{d}{D}\right)^2$$

in which h is the differential head cross the orifice plate in terms of the fluid.

13.2.3.2. Use of Uncalibrated Orificemeter. The orificemeter should be calibrated in position for accurate measurement of discharge. The orifice plate should be preceded and followed by straight pipe of at least 100D on the upstream and 25D on the downstream side (ASME Report of Research Committee on Fluid Metres, 1959). If this much length of straight pipe is not possible, then straightness of flow can be ensured by placing straightening vanes in the pipe at relatively short distances.

An equation[9] for predicting discharge through an orificemeter provided with pressure tappings at 1D upstream and 0.5D downstream and the plate being preceded and followed by straight pipes of about 100D and 25D respectively, is as follows :

$$Q = 0.616\left(\frac{d}{D}\right)^{0.02} \frac{\pi}{4} d^2 \sqrt{2gh} \quad \text{for } 0.1 \le d/D \le 0.5. \qquad \text{...(13.5 a)}$$

and

$$Q = 0.889\left(\frac{d}{D}\right)^{0.52} \frac{\pi}{4} d^2 \sqrt{2gh} \quad \text{for } 0.5 \le d/D \le 0.8. \qquad \text{...(13.5 b)}$$

(8) Ranga Raju, K.G. and Jain A.K. "Energy Loss due to Sharp-Edged Orifice Meters". Journal of Irrigation and Power, July 1978.

(9) Jain, A.K. and Asawa, G.L. "Discharge Equation for Uncalibrated Sharp-crested Concentric Orifice Meters". Journal of Irrigation and Power, April 1977.

Eq. (13.5) predicts the discharge through the uncalibrated sharp-edged concentric circular orificemeter within ± 2%.

Ex. 13.2A. *An orifice meter with orifice diameter 15 cm is inserted in a pipe of 30 cm diameter. The water pressure on the upstream and downstream sides of the meter are 14.7 N/cm² and 9.81 N/cm² respectively. Find the discharge, if the discharge coefficient, $C_d = 0.6$. Also find the coefficient of contraction of the jet.*

Solution. Referring to Fig. 13.2, the pressure drop across the orifice meter in terms of head of water is,

$$h = \frac{p_1 - p_2}{\gamma} = \frac{(14.7 - 9.81)(100)^2 (\text{N/m}^2)}{1000 \times 9.81\ (\text{N/m}^3)} = 4.985 \text{ m}$$

Discharge passing through the orifice meter is given by Eq. (13.3*a*),

$$Q = C_d \frac{\pi}{4} . d^2 \sqrt{2gh} = 0.6 \frac{\pi}{4} \left(\frac{15}{100}\right)^2 \sqrt{2 \times 9.81 \times 4.985} = \mathbf{0.10483\ m^3/s}$$

The coefficient of discharge is related to contraction coefficient by Eq. (13.3*c*),

$$C_d = \frac{C_c}{\sqrt{1 - C_c^2 (d/D)^4}}$$

squaring both sides, $$(C_d)^2 = \frac{C_c^2}{\sqrt{1 - C_c^2 (d/D)^4}}$$

from which, $$C_c = \frac{C_d}{\sqrt{1 - C_d^2 (d/D)^4}} = \frac{0.6}{\sqrt{1 + (0.6)^2 (15/30)^4}} = \mathbf{0.575}$$

13.2.4. Bend meter

The bend meter or elbow meter utilizes the fact that in curved flow pressure increases with radius (see article 4.8.2), and consequently a pressure difference exists between the outer and inner walls of a pipe bend. Use of a suitable manometer allows measurement of this pressure difference which is a function of the rate of flow. The relationship expressing discharge through the bend may be written as

$$Q = C_d A \sqrt{2gh} \quad \text{...(13.6)}$$

in which C_d is a dimensionless discharge coefficient, A is the cross-sectional area of the pipe and h is the difference in the piezometric head in terms of the flowing fluid given by

$$h = (p_0 - p_i)/\gamma.$$

As the bend itself is a part of the piping system, no additional fitting is required for this meter. Accurate results can be obtained from the bend meter if it is calibrated in place. The discharge for large Reynolds numbers can, however, be measured within an accuracy of 10 per cent by determining the discharge coefficient as follows,

$$C_d = \sqrt{\frac{R}{2d}} \quad \text{...(13.7)}$$

provided there is atleast 30 diameters of straight pipe upstream of bend. In Eq. (13.7), R is the centreline radius of the bend and d is the pipe diameter. In general, C_d can vary from 0.55 to 1.2. Any bend fitted in a pipeline can be used as a bendmeter by providing two pressure tappings along a radial line as shown in Fig. 13.3.

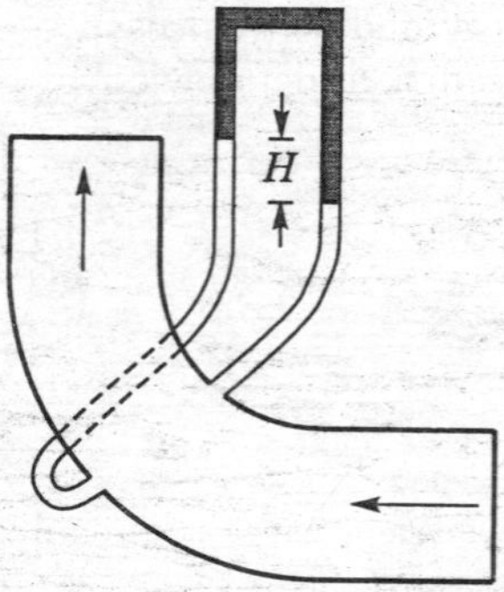

Fig. 13.3. Bendmeter.

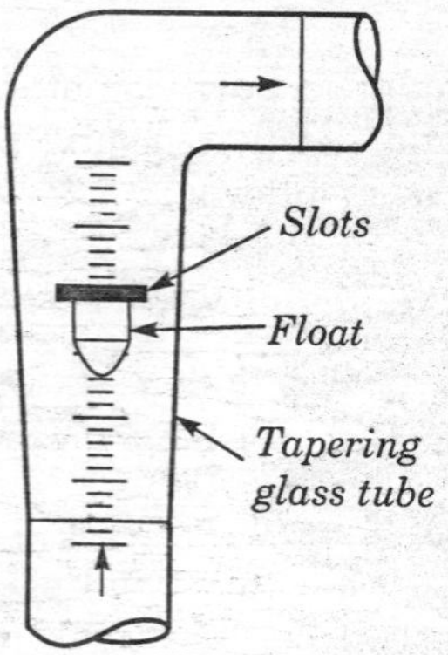

Fig. 13.4. Rotameter.

13.2.5. Rotameter

It is a device used for measuring discharge in pipelines generally used in chemical industries where high degree of accuracy is not required and the flow varies within a limited range. It is a variable area meter installed in a vertical segment of a pipeline. A rotameter consists of an accurately ground glass tube diverging upwards and the flow enters from the bottom. It contains a float made of a denser material than the fluid flowing through the tube. The float adjusts its position inside the tapering glass tube on account of the dynamic action of the flowing fluid. The location of the float inside the tube is decided by the drag force exerted on the float by the flowing fluid. The float thus moves up when the flowrate is increased. The position of the float is calibrated to indicate the flowrate corresponding to its different locations. For a given discharge, the position of the float is determined by a balance between drag force and buoyancy both acting vertically upward and the weight of float acting vertically downward. The float is provided with a conical end and at the top rim slanting slots are cut to set it rotating for stability against tilting. The float is thus kept in a vertical position.

13.3. Measurement of Velocity

Since the determination of velocity at a number of points in a cross-section permits the evaluation of discharge, the measurement of velocity is an important item which precedes accurate estimation of flow. Apart from this, in boundary layer experiments accurate measurement of velocity forms a basis for determining such parameters as δ, δ^*, θ, c_f and C_f.

In this article we shall discuss the Pitot-tube, hot-wire anemometer and current meter.

13.3.1. Pitot tube

Pitot tube is one of the most accurate devices used for measuring velocity. Simple Pitot tube has been discussed in article 4.14.1. For measuring velocity in open channels, a single tube (total head or stagnation tube) with a right angled bend may be used. In case of closed conduits (pipes), measurement of velocity by pitot tube involves determination of static and stagnation pressures. Whereas a Pitot tube measures the stagnation pressure, the static pressure is measured by a piezometer installed in a hole drilled perpendicular to the pipe surface. A tube which has arrangements for measuring both the static as well as the stagnation pressure is known as the *Pitot-static tube*. A Pitot-static tube consists of two circular concentric tubes one inside the other with an annular space in between, as shown in Fig. 13.5. The static pressure is measured through two or more holes drilled through the outer tube into an annular space and it is assumed that the fluid flows along the outside of the tube in such a way that the true static pressure is obtained.

For round-nosed body of revolution with its axis parallel to the flow, the stagnation pressure is caused at the tip (marked by A in the figure). When the Pitot tube is placed in the fluid stream, the flow along its outer surface gets accelerated and causes the static pressure to decrease. But the effect of the stem which is at right angles to the stream is to produce an excess pressure head which diminishes upstream from the stem. If the piezometer hole in the side of the tube is located where the excess pressure produced by the stem equals the decrease in pressure caused by the flow around the nose and along the tube, the true static pressure will be

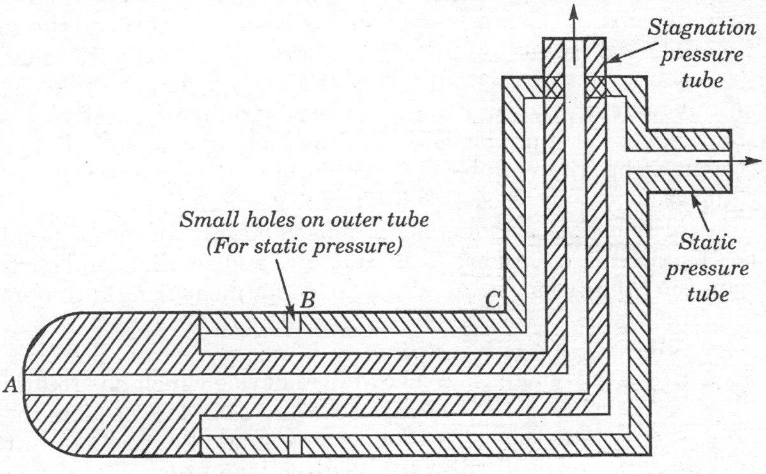

Fig. 13.5. Pitot-static tube.

obtained. If h is the difference in the stagnation pressure head, and the static pressure head, the velocity of flow is given by

$$V = C\sqrt{2gh} \quad ...(13.8)$$

where C is a corrective coefficient which takes into account the effect of the nose and the stem and is called the Pitot tube coefficient.

A particular form of the Pitot-static tube which has a slightly rounded nose is shown in Fig. 13.5. The tube conforming to the following dimensions in terms of the outer diameter of the tube is called the *Prandtl type Pitot tube* or simply *Prandtl tube*.

(1) Diameter of the nose and the pitot tube = d
(2) Diameter of the stagnation pressure hole at $A = 0.3\ d$
(3) Front length of the tube from A to $B = 3\ d$
(4) Rear length of the tube from B to $C = 9\ b$

where b is the width of the tube, which in case of a rounded tube will be the same as the diameter.

Prandtl type tube has been so designed that the effects of the nose and the stem cancel resulting in $C = 1$ in Eq. (13.8). The great advantage of the Prandtl tube is that it always has a unity coefficient thus yielding

$$V = \sqrt{2gh}\,.$$

Example 13.2. *A Pitot-static tube having a coefficient of 0.98 is used to measure the velocity of water in a pipe. The stagnation pressure recorded is 3 m and the static pressure 2 m. What velocity does this indicate ?*

Solution. From Eq. (13.8), the velocity is given by

$$V = C\sqrt{2gh}$$

in which C = Pitot-tube coefficient and $h = \dfrac{p_s - p_o}{\gamma}$; p_s and p_o being the stagnation and static pressure respectively.

$$h = \frac{p_s}{\gamma} - \frac{p_o}{\gamma} = 3 - 2 = 1 \text{ m}$$

$$V = 0.98\sqrt{2 \times 9.81 \times 1} = 4.35 \text{ m/s}.$$

13.3.2. Hot-wire Anemometer

Hot-wire anemometer is an instrument which is used for measuring air and gas velocities very accurately. It has a very quick response to changes in velocity and may be used where the velocity gradients are large, for example in boundary layers. The underlying principle of this instrument is that the electrical resistance of a wire is a function of its temperature. A short length of fine platinum wire is held between the ends of two pointed prongs, and is heated by an electric current. As the resistance to the flow of electricity through the wire is a function of its temperature, the continual passage of fluid stream around the hot wire cools it thereby changing its resistance.

By keeping constant either the voltage across the wire or current through the wire by a suitable circuit, the change in current or voltage respectively becomes a measure of velocity past the hot wire. It must be calibrated by placing it in the fluid stream of known velocity. The heat transfer from the wire to its surroundings depends also upon the density of the fluid and so the calibration should be carried out in the same fluid as the one whose velocity is to be measured.

The hot-wire anemometer is used mainly for measuring the velocity of gases. It has proved less successful in liquids because bubbles and smaller solid particles tend to collect on the wire and spoil the calibration. It can also be used for measuring turbulence characteristics of flow using the proper probe.

13.3.3. Current meter

A current meter is a mechanical device which has rotation elements the speed of rotation of which is a function of the velocity of flow. It is used to measure the velocity of liquid in open channels. The current meter consists of hollow hemispheres or cones mounted on spokes so as to cause rotation about a shaft perpendicular to the direction of flow. The drag on a hollow hemisphere or cone is greater when its open side faces the fluid stream, and so there is a net torque on the assembly when flow comes from any direction in the plane of rotation. The magnitude of the fluid velocity determines the speed of rotation. With an electric circuit and headphones, an audible signal marked by a ticking sound is detected. The number of signals in a given time period is a function of velocity. The current meter is calibrated by towing it through liquid in a towing tank at known speeds.

Air velocities are measured by either cup-type or vane (propeller) type anemometers. The number of revolutions made are usually indicated by a mechanical counter.

13.4. Flow Through Orifices

An orifice is an opening, usually round., through which the fluid flows. It may be used for measuring flow from a reservoir or through a pipe as in case of an orificemeter. An orifice in a tank or reservoir may be located in the side wall or in the bottom. The main feature of orifice flow is that most of the potential energy of fluid is converted into the kinetic energy of the free jet issued through the orifice. The orifice flow has already been introduced in article 4.14.4, followed by solved examples 4.40, 4.41 and 4.42.

Consider a large tank filled with a liquid with a small orifice located in the wall at a fairly large depth H below the free surface Fig. 13.6. The liquid flows out through the orifice into the atmosphere. If the circular orifice is considered located in a thin wall (or if the wall is thick the orifice is made sharp-edged), the out flowing liquid touches the orifice wall along a line as shown in the figure. As the liquid approaches the orifice it tends to contract due to inability of streamlines to take a sharp turn and results in contraction of jet. The fluid jet therefore tends to contract owing to inability of the streamlines to take a sharp turn at the opening. The contraction of the jet is limited to a distance of about one-half to one diameter downstream from the opening.

Fig. 13.6 (*b*) shows the curvature of streamlines as they approach the orifice and explain why should the jet contract as it passes through the opening. The cross-section where the contraction is

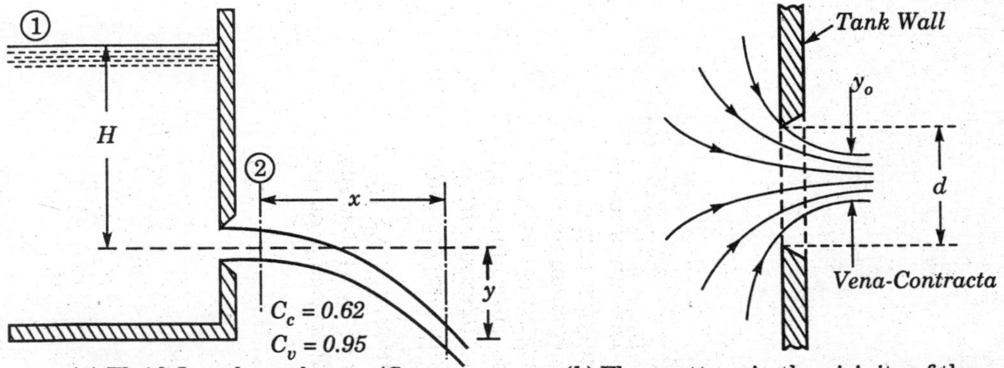

(*a*) Fluid flow through an orifice (*b*) Flow pattern in the vicinity of the orifice

Fig. 13.6. Orifice flow from a tank.

greatest or where the area of the jet is minimum is called the *vena contracta*. The streamlines at this section are parallel throughout the jet and the pressure is atmospheric. The degree of contraction is characterized by coefficient of contraction, C_c, defined by

$$C_c = \frac{\text{Area of jet at vena contracta}}{\text{Area of the orifice}}.$$

The head on the orifice, H, is measured from the centre of the orifice to the free surface. Assuming the head on the orifice to be held constant, the Bernoulli's equation applied between point (1) on the free surface and the centre of the vena-contracta, neglecting losses, may be written as (with elevation datum taken as the horizontal plane passing through the orifice centre) :

$$H + 0 + 0 = 0 + 0 + \frac{V^2}{2g}$$

or

$$V = \sqrt{2gH} \quad ...(13.9)$$

Eq. (13.9) which is the same as Eq. (4.34), gives the theoretical velocity because the losses between the two points were neglected. The ratio between the actual velocity at the vena contracta and the theoretical velocity expressed by Eq. (13.9) is known as the coefficient of velocity, C_v, thus

$$C_v = \frac{\text{Actual velocity } V_a}{\text{Theoretical velocity } V_t}$$

hence

$$V_a = C_v\sqrt{2gH} \quad ...(13.10)$$

Because of loss of energy in friction, the actual velocity is always less than the theoretical one and hence $C_v < 1.0$.

If A is the area of the orifice and A_o is the area of jet at the vena-contracta, then from the definition of coefficient of contraction

$$C_c = A_o/A \text{ or } A_o = AC_c.$$

The actual discharge passing through the orifice is the product of the actual velocity at the vena-contracta and the area of the jet, thus

$$Q = (AC_c)(C_v\sqrt{2gH}) = C_c\,C_v A\sqrt{2gH} \quad ...(13.11)$$

It is customary to combine the two co-efficients into one and to call it a discharge coefficient, C_d, thus

$$C_d = C_c C_v \quad ...(13.12)$$

and the actual discharge

$$Q = C_d A\sqrt{2gH} \quad ...(13.13)$$

The coefficient of discharge C_d is defined as the ratio of the actual discharge to the theoretical discharge and is expressed by

$$C_d = \frac{Q}{A\sqrt{2gH}} \quad ...(13.14)$$

By measuring the orifice area A, the head H and the discharge Q (either by gravimetric or volumetric means) C_d can be evaluated using Eq. (13.14). The coefficient of velocity must also be determined experimentally. Its value varies from 0.95 to 0.99 for sharp-edged and well-rounded orifices. There are several methods for determining one or more of the above coefficients which are called the hydraulic coefficients. The methods of determining C_v and C_d are described below.

13.4.1. Determination of Hydraulic Coefficients of Orifice

(1) ***Trajectory method.*** By measuring the position of a point on the trajectory of the free jet [*i.e.*, by observing x and y co-ordinates of any point such as shown in Fig. 13.4 (a)] from the vena contracta (vena-contracta taken as origin) the actual velocity V_a can be determined if air resistance is neglected. If t is the time a fluid particle takes in travelling from vena contracta to the point marked by the dot, one can write,

$$x = V_a \cdot t \quad ...(13.15)$$

In the same time, the fluid particle has travelled vertically a distance y under the influence of gravity,

$$y = \frac{1}{2}gt^2 \quad ...(13.16)$$

Eliminating t between the two equations, we obtain

$$V_a = \frac{x}{\sqrt{2y/g}} \quad ...(13.17)$$

The coefficient of velocity may now be expressed in terms of x, y and H

$$C_v = \frac{V_a}{V_t} = \frac{V_a}{\sqrt{2gH}} = \frac{x}{\sqrt{2y/g}\,\sqrt{2gH}} \qquad ...(13.18)$$

or

$$C_v = \frac{x}{\sqrt{4yH}}.$$

Example 13.3. *A tank of square cross-section, each side 30.5 cm, is open at the top and is fixed in the upright position. A 6 mm diameter orifice is situated in one of the vertical sides near the bottom. Water flows into the tank at the top at a constant rate of 0.283 m³/hr. At a particular instant the jet strikes the floor at a point 0.64 m from the vena contracta measured horizontally and 0.535 m down below the centreline measured vertically. Determine whether the water surface in the tank is rising or falling at the instant. Also find the height of the surface above the orifice and the rate of change of height.* $C_v = 0.97$; $C_d = 0.62$.

Solution. The coefficient of velocity is given by

$$C_v = \sqrt{\frac{x^2}{4yH}}, \text{ from which } H = \frac{x^2}{4yC_v^2}$$

Substituting the given data, *i.e.*, $x = 0.64$ m, $y = 0.535$ m and $C_v = 0.97$ m, we obtain $H = 0.204$ m.

The discharge through the orifice, $Q = C_d A \sqrt{2gH}$

or

$$Q = 0.64 \frac{\pi}{4}\left(\frac{0.6}{100}\right)^2 \sqrt{2 \times 9.81 \times 0.204} \text{ m}^3\text{/s} = 0.213 \text{ m}^3\text{/hr}.$$

but the flow into the tank = 0.283 m³/hr. Now, since the inflow is greater than the outflow, the water surface in the tank is rising.

Height of water surface above the orifice = 0.204 m.

The rate at which the water surface rises is

$$\frac{dH}{dt} = \frac{Q_{in} - Q_{out}}{\text{Area of tank}} = \frac{0.283 - 0.213}{(0.305)^2} \text{ m/hr} = 1.645 \text{ m/hr}.$$

(2) ***Pitot-tube method.*** The pitot-tube may be placed at the vena-contracta and the actual velocity obtaining there determined. The direct measurement of velocity this way, enables the determination of the velocity coefficient. Determination of jet velocity may not be accurate as the insertion of the Pitot-tube may alter the velocity considerably.

(3) ***Callipers method.*** With outside callipers, the diameter of the jet at the vena-contracta may be approximately determined. This permits the estimation of coefficient of contraction, but is not a precise method and is less satisfactory than other methods.

Example 13.4. *Oil of specific gravity 0.85 issues from a 5 cm diameter orifice under a pressure of 117.72 kN/m² (gauge). The diameter of the jet at the vena contracta is 4.0 cm and the discharge is 1.2 m³/minute. What is the coefficient of velocity ?*

Solution. Head under which orifice is discharging

$$H = \frac{117.72 \times 10^3}{0.85 \times 9810} = 14.12 \text{ m of oil.}$$

Theoretical velocity, $V_t = \sqrt{2gH}$

$$= \sqrt{2 \times 9.81 \times 14.12} = 16.63 \text{ m/s}$$

Actual velocity at vena-contracta

$$= \frac{Q}{\text{Area at vena-contracta}}$$

or $$V_a = \frac{(1.2/60)}{\frac{\pi}{4}\left(\frac{4}{100}\right)^2} = 15.92 \text{ m/s.}$$

$\therefore$ Coefficient of velocity

$$C_v = \frac{V_a}{V_t} = \frac{15.92}{16.63} = 0.957.$$

(4) ***Momentum method.*** A small tank suspended on knife-edges as shown in Fig. 13.7 can be used to determine the coefficient of velocity using the momentum equation. With the orifice opening closed the tank is levelled by adding or removing weights. When the orifice is discharging, the liquid jet exerts a force F on the tank wall (as shown) due to change of momentum. By putting additional weight W, the tank is again levelled.

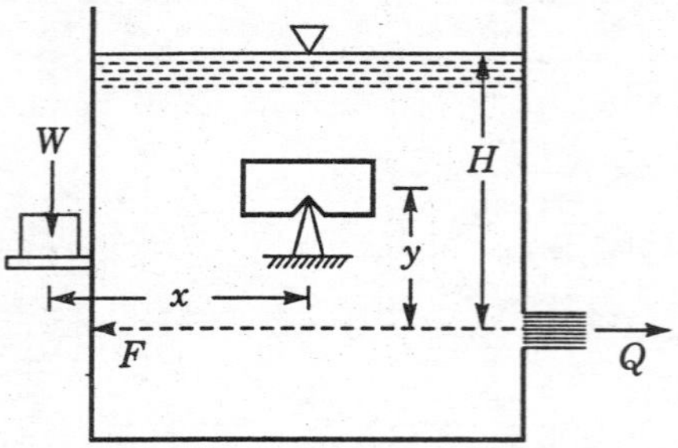

Fig. 13.7. Momentum method.

From the momentum equation, we obtain

$$F = \frac{\gamma Q}{g} V_a \qquad \text{...(13.19)}$$

From Fig. 13.7, by taking moments about the knife-edge.

$$F \,.\, y = W \,.\, x \quad \therefore \quad F = W\frac{x}{y} \qquad \text{...(13.20)}$$

From Eqs. (13.19) and (13.20)

$$W\frac{x}{y} = \frac{\gamma Q}{g} V_a \qquad \text{...(13.21)}$$

In Eq. (13.21) all quantities except V_a are known from the experiment. Now with known V_a, the velocity coefficient can be determined using Eq. (13.10).

Example 13.5. *A 10 cm diameter orifice discharges 45 litres/sec of water under a head of 2.75 metre. A flat plate held normal to the jet just downstream from the vena-contracta requires a force of 310 N to resist the impact of jet. Find C_c, C_v and C_d.*

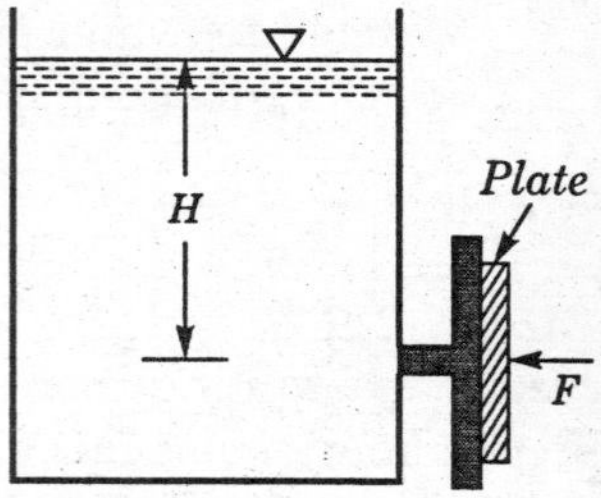

Solution. Applying momentum equation, the force required to hold the plate in position is

$$F = \frac{\gamma Q}{g} V$$

where V = Velocity at the vena-contracta

or $$V = \frac{Fg}{\gamma Q} = \frac{310 \times 9.81}{9810 \times 45 \times 10^{-3}} = 6.8 \text{ m/s.}$$

Theoretical velocity

$$V_t = \sqrt{2gH} = \sqrt{2 \times 9.81 \times 2.75} = 7.38 \text{ m/s.}$$

Coefficient of velocity, $$C_v = \frac{V}{V_t} = \frac{6.8}{7.33} = 0.927$$

Theoretical discharge, $$Q_t = \frac{\pi}{4} d^2 \sqrt{2gH} = \frac{\pi}{4}\left(\frac{10}{100}\right)^2 \times 7.38 = \frac{\pi \times 7.33}{4 \times 100} \text{ m}^3\text{/s}$$

$$= \frac{\pi \times 7.33}{4 \times 100} \times 1000 \text{ litres/sec} = 57.7 \text{ litres/sec.}$$

Coefficient of discharge $C_d = \frac{45}{57.7} = 0.78$

and coefficient of contraction

$$C_c = \frac{C_d}{C_v} = \frac{0.78}{0.927} = 0.84.$$

Loss of energy in orifice flow. The loss of energy in flow through an orifice is determined by applying Bernoulli's equation between point (1) and the vena-contracta considering loss of energy, thus

$$H = \frac{V_a^2}{2g} + H_L$$

where H_L is the head lost between points (1) and (2), see Fig. (13.6) (*a*).

$$H_L = H - \frac{V_a^2}{2g} = H(1 - C_v^2) \qquad ...(13.22)$$

$$= \left(\frac{1}{C_v^2} - 1\right)\frac{V_a^2}{2g}$$

the loss can, therefore, be expressed in terms of H and C_v or V_a and C_v.

Example 13.6. *Explain briefly how the coefficient of velocity of a jet issuing through an orifice can be experimentally determined.*

Find an expression for head loss in an orifice flow in terms of coefficient of velocity and jet velocity.

The head loss in flow through a 5 cm diameter orifice under a certain head is 16 cm of water and the velocity of water in the jet is 7.0 m/s. If the coefficient of discharge be 0.61, determine, (a) head on the orifice causing the flow, (b) the diameter of the jet, and (c) the coefficient of velocity.

Solution. Bernoulli's equation between the reservoir surface and the vena-contracta, (see Fig. 13.6), yields

$$H = \frac{V^2}{2g} + h_L = \frac{(7)^2}{2 \times 9.81} + 0.16 = 2.66 \text{ m.}$$

$$V = C_v \sqrt{2gH}$$

$$C_v = \frac{V}{\sqrt{2gH}} = \frac{7.0}{\sqrt{2 \times 9.81 \times 2.66}} = 0.97$$

$$C_c = \frac{C_d}{C_v} = \frac{0.61}{0.97} = 0.63.$$

From the definition of the coefficient of contraction

$$C_c = \frac{\pi d_j^2/4}{\pi d^2/4} = \frac{d_j^2}{d^2}$$

diameter of the jet $d_j = d\sqrt{C_c} = 5\sqrt{0.63}$ **= 3.96 cm.**

Example 13.7. *A tank has two identical orifices, each 5 cm in diameter, in one of its vertical sides and are situated one above the other. The upper orifice is 4 m below the water surface and the lower one is 6 m below the water surface. The coefficients of contraction and velocity are 0.64 and 0.98 for both the orifices. Find :*

(i) the combined rate of discharge from the two orifices, and

(ii) the distance of point of intersection of the two jets from the vertical side.

Solution. Velocity of jet from upper orifice

$$= C_V\sqrt{2gh} = 0.98\sqrt{2 \times 9.81 \times 4} = 8.68 \text{ m/s}$$

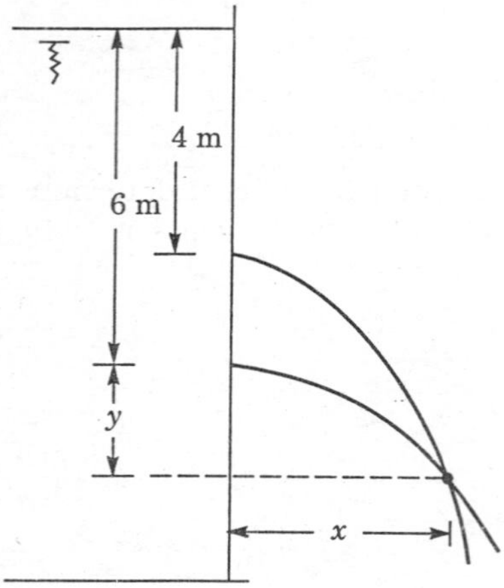

Velocity of jet from lower orifice

$$= 0.98\sqrt{2\times 9.81\times 6} = 10.63 \text{ m/s}$$

Discharge from upper orifice

= area of jet at vena-contracta × vel. of jet

$$= C_C \frac{\pi d^2}{4} C_V \sqrt{2gh} = 0.64 \times \frac{\pi\left(\frac{5}{100}\right)^2}{4} \times 8.68$$

$$= 109.1 \times 10^{-4} \text{ m}^3\text{/s} = 10.91 \text{ lit/sec}$$

Discharge from lower orifice

$$= 0.64 \times \frac{\pi}{4}\left(\frac{5}{100}\right)^2 \times 10.63$$

$$= 133.5 \times 10^{-4} \text{ m}^3\text{/s} = 13.35 \text{ lit/sec.}$$

Combined discharged from the two orifices

= 10.91 + 13.36 = **24.27 lit/sec.**

Distance of point of intersection

Let the lower jet travel a horizontal distance x where it intersects the upper jet, the vertical distance travelled being y. Let the time taken by the jet to reach the point of intersection be t, then

$$y = \frac{1}{2}gt^2 \,; \quad t = \sqrt{\frac{2y}{g}}$$

$$x = V\,.\,t = V\sqrt{\frac{2y}{g}} = 10.63\sqrt{\frac{2y}{g}} \qquad \text{...(1)}$$

Considering the upper jet, it is evident that it travels a vertical distance of $(2 + y)$ for covering the same horizontal distance x, hence

$$2 + y = \frac{1}{2}gt^2 \,; \quad t = \sqrt{\frac{2(2+y)}{g}}$$

$$x = V\,.\,t = 8.68\sqrt{\frac{2(2+y)}{g}} \qquad \text{...(2)}$$

Equating (1) and (2),

$$10.63\sqrt{\frac{2y}{g}} = 8.66\sqrt{\frac{2(2+y)}{g}} \quad \text{From which } y = 4 \text{ m}$$

Substituting in (1), $\quad x = 10.63\sqrt{\frac{2\times 4}{9.81}} = 9.60 \text{ m}$

13.4.2. Time of emptying a tank

Consider a tank of area A fitted with an orifice of area a. Let there be no inflow into the tank so that with the orifice functioning, the liquid level in the tank falls down. Let at any instant of time the liquid surface be at a distance h above the orifice and in time dt the level falls by dh (Fig. 13.8).

The discharge through the orifice,

$$Q = C_d a\sqrt{2gh}\,,$$

provided that the drop in liquid surface is slow enough for Beroulli's equation to be applicable without appreciable error.

During time dt the volume of liquid discharged by the orifice = $Q.\,dt$ and in the same period the volume of liquid in the tank is reduced by Adh. We can, therefore, write

$$-Adh = Qdt \quad ...(13.23)$$

The negative sign is introduced because with the increase of time the head h decreases. Substituting for Q in Eq. (13.23)

$$-Adh = C_d\, a\sqrt{2gh}\; dt\,.$$

$$\therefore \quad dt = -\frac{A}{C_d a}\frac{dh}{\sqrt{2gh}}$$

The time t required to lower the liquid level from H to h is

$$t = -\int_H^h \frac{A}{C_d a}\frac{dh}{\sqrt{2gh}} = \frac{2A}{C_d a\sqrt{2g}}(\sqrt{H}-\sqrt{h}) \quad ...(13.24)$$

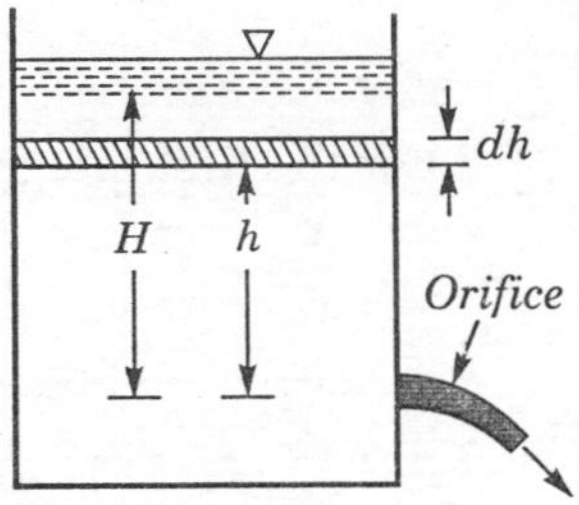

Fig. 13.8. Emptying of tank through orifice.

Example 13.8. *A vertical circular tank of 50 cm diameter and 4 m high is full of water. If has two orifices each of 10 square cm area, one at the bottom and other on mid height of side of the tank above bottom. Determine the time required to empty the tank, take coefficient discharge for both the orifices as 0.7.*

Sol. Both the orifices work letting out water till the water-level reaches 2 m height when the middle orifice stops functioning.

Let the water surface fall by 'dh' in time 'dt' as shown in the figure. Equating the volume of water depleted from the tank with the volume of water let out through the orifices at the rate Q in time dt :

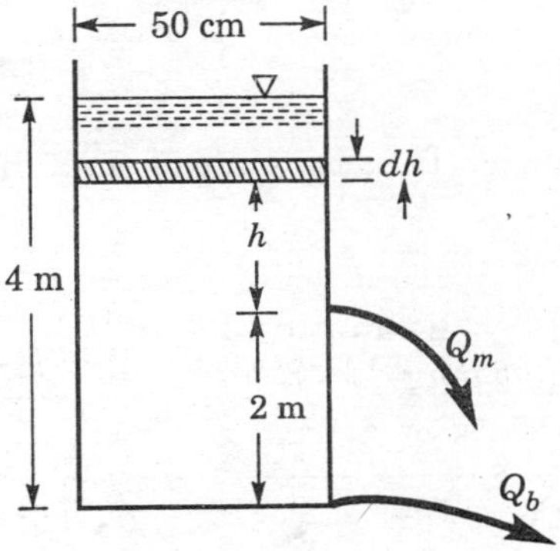

$$-A\,.\,dh = Q\,.\,dt \quad ...(a)$$

Here, A is the surface area of the tank, and Q the total discharge.

Total discharge from the tank at the instant when the water surface is at height 'h' above the middle orifice

$$Q = Q_m + Q_b$$

$$= c_d\,.\,a\,.\,\sqrt{2gh} + C_d\,.\,a\,.\sqrt{2g(h+2)}$$

$$= C_d\,.\,a\,.\sqrt{2g}\,[\sqrt{h}+\sqrt{(h+2)}] \quad ...(b)$$

Time to empty the tank

$$T = T_1 + T_2$$

in which T_1 = time taken when both the orifices work *i.e.* the water surface e is lowered from 4 m to 2 m, and

T_2 = time taken by the bottom orifice to lower the surface from 2 m to completely empty the tank.

Using Eqs. (a) and (b),

$$T_1 = \frac{A}{C_d\,.\,a\,.\sqrt{2g}}\int_0^2 \frac{dh}{[\sqrt{h}+\sqrt{(h+2)}]}$$

Integrating, $$\int_0^2 \frac{dh}{[\sqrt{h}+\sqrt{(h+2)}]} \times \frac{\sqrt{h+2}-\sqrt{h}}{\sqrt{h+2}-\sqrt{h}}$$

$$= \int_0^2 \frac{(\sqrt{h+2}-\sqrt{h})\,dh}{(h+2-h)} = \frac{1}{2}\left[\int_0^2 \sqrt{h+2}\,dh - \int_0^2 \sqrt{h}\,dh\right]$$

For the first integral, let $h + 2 = y$, $dh = dy$

$$\int_0^2 (\sqrt{h+2})\,dh = \int_2^4 \sqrt{y}\,dy = \left[\frac{2}{3}y^{3/2}\right]_2^4 = \frac{2}{3}[4\sqrt{4}-2\sqrt{2}] = 3.448$$

$$\int_0^2 \sqrt{h}\,dh = \left[\frac{2}{3}h^{3/2}\right]_0^2 = 1.885$$

Therefore, $$T_1 = \frac{A}{c_d \cdot a\sqrt{2g}} \times \frac{1}{2}\left[\int_0^2 \sqrt{h+2}\,dh - \int_0^2 \sqrt{h}\,dh\right]$$

$$= \frac{\frac{\pi}{4}\times(0.5)^2}{0.7\times10\times10^{-4}\times\sqrt{2\times9.81}}\times\frac{1}{2}\times[3.448-1.885] = \mathbf{49.475\ seconds.}$$

When the water level falls from 4 m to 2 m, the middle orifice stops discharging and the bottom one continues letting out water. The time of emptying from 2 m to zero.

$$T_1 = \frac{A}{c_d \cdot a \cdot \sqrt{2g}} \int_0^2 \frac{dh}{\sqrt{h}} = \frac{2\sqrt{2}A}{c_d \cdot a \cdot \sqrt{2g}} = \mathbf{89.55\ seconds}$$

Thus the time of emptying the tank,

$$T = T_1 + T_2 = 49.475 + 89.55 = 139.025 \text{ seconds}$$
$$= 2 \text{ min } 19.025 \text{ seconds.}$$

Example 13.9. *In the following figure is shown a truncated cone having vertex angle $\theta = 60°$. How long does it take to draw the liquid surface from $y = 3$ m to $y = 1$ m ? Take $C_d = 0.85$.*

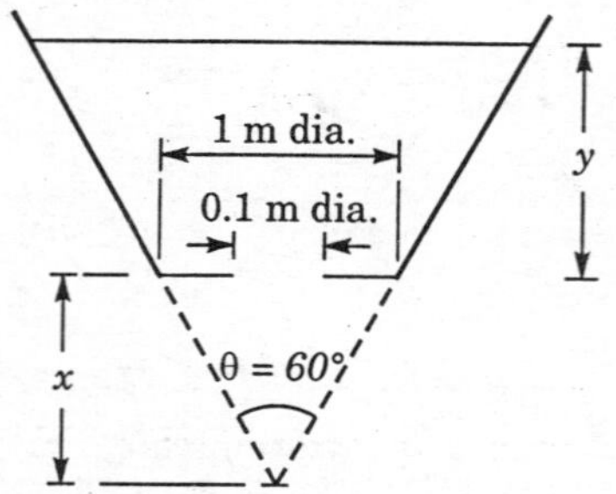

Solution. From the geometry of the truncated cone

$$\tan 30° = \frac{0.5}{x},$$

$$x = 0.5\sqrt{3} = 0.865 \text{ m}$$

radius of truncated cone at a distance $x + y$ from the vertex is obtained as follows :

$$\frac{r}{x+y} = \tan 30°$$

$$r = (0.865 + y) \times \frac{1}{\sqrt{3}}$$

Let in time dt, the liquid surface fall a distance dy, then from the continuity principle

$$\pi r^2 (-dy) = Q.dt$$

or $$-\pi(0.865+y)^2 \times \frac{1}{3}dy = C_d \frac{\pi}{4}(0.1)^2 \times \sqrt{2gy}\,dt$$

The time to empty the cone from $y = 3$ m to $y = 1$ m.

$$t = -\int_3^1 \frac{4}{3}\frac{(0.75 + y^2 + 1.730y)}{C_d \times 0.01 \times \sqrt{2gy}}\, dy$$

$$t = \int_1^3 \frac{4}{0.01 \times 3 \times 0.85 \times \sqrt{2 \times 9.81}} \times \left[\frac{0.75}{\sqrt{y}} + y^{3/2} + 1.73y^{1/2}\right] dy$$

$= 458$ sec. $= 7$ min 38 sec.

Example 13.10. *Select the reservoir of such size and shape that the liquid surface falls at the rate of 1 metre per minute over a 3 metre distance for flow through a 10 cm diameter orifice. Take $C_d = 0.80$.*

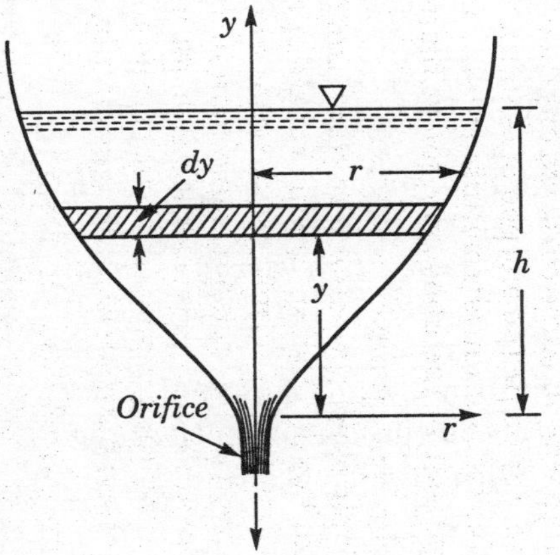

Solution. Let the reservoir shape be defined as

$$r = Ky^n$$

where r and y are variables as shown in the figure.

Let at any instant, the reservoir surface fall through a distance dy in time dt, then from the continuity relation, we have

$$\pi r^2\left(-\frac{dy}{dt}\right) = C_d \frac{\pi}{4}(0.1)^2 \sqrt{2gy} \quad ...(i)$$

Substituting for r and

$$\left(-\frac{dy}{dt}\right) = \frac{1}{60}, \text{ we get}$$

$$\pi K^2 y^{2n} \times \frac{1}{60} = 0.80 \times \frac{\pi}{4} \times 0.01 \times \sqrt{2 \times 9.81 y} \quad ...(ii)$$

Since Eq. (*ii*) contains only one variable *i.e.* y, the value of exponent n should be such that the left and right sides of this equation remain equal for all values of y. Equating the exponents of y on each side of the equation.

$$2n = \frac{1}{2}, \qquad \therefore \quad n = \frac{1}{4}$$

for obtaining K, equate the constants on each side, thus

$$K^2 = \frac{0.80 \times 0.01 \times 60\sqrt{2 \times 9.81}}{4} = 0.53$$

$$K = 0.728$$

$\therefore$ The reservoir shape is given by $r = 0.728\, y^{1/4}$

in which r is the radius in meter at a vertical distance y also measured in metres. This equation defines both the size and the shape of the reservoir.

Example 13.11. *A prismatic vessel has compartments A and B, communicating by a standard orifice 15 cm square, its centre being 1 m above the bottom of the vessel. The horizontal cross-section of A is 10 sq. metre and that of B is 20 sq. metre. At a certain time the water stands 4 m deep in A and 2 m deep in B. How soon thereafter will surface reach a common level ? Assume $C_d = 0.62$.*

Solution. Let y be the depth of water in A at any instant and dy be the change in depth during any interval of time dt.

The rise in B's level during the same time will be $10/20\ dy$ and the net change in head will be

$$dh = dy - (-\ 10/20\ dy) = 3/2\ dy$$

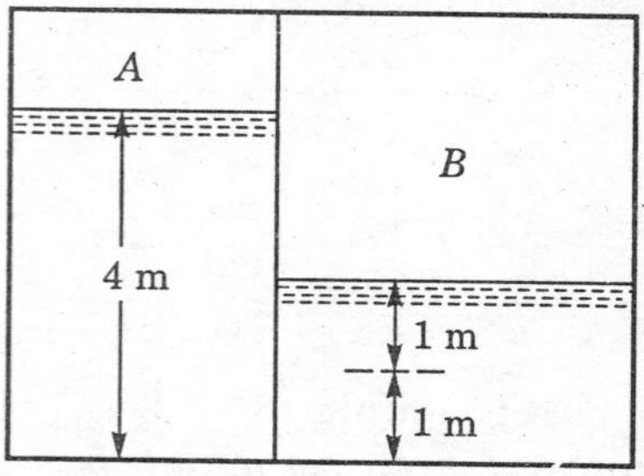

Discharge through orifice in time dt

$$= Qdt = C_d\, a\, \sqrt{2gh}\ dt$$

which must be equal to 10 dy, the decrease in volume of water in time dt

$$10\ dy = 10 \times 2/3\ dh = 20/3dh$$

$$\therefore \qquad -\ 20/3dh = C_d a\ \sqrt{2gh}\ dt$$

$$= 0.62 \times \frac{15 \times 15}{100 \times 100} \times \sqrt{2 \times 9.81}\ \sqrt{h}\ dt$$

or
$$t = -\ 108 \int_2^0 \frac{dh}{\sqrt{h}} = 108\left[2\sqrt{h}\right]_0^2 = 305.5 \text{ seconds.}$$

Example 13.12. *A cylindrical tank 3 m in diameter and 6 m high has an orifice 15 cm in diameter at the bottom centre of the tank. A constant discharge of 85 litres of water per second is fed into the tank. At the same time water is being discharged through the orifice. Determine the time taken to lower the water surface level in the tank from 5 m to 2.5 m above the centre of orifice. Take $C_d = 0.72$. The top of the tank is open to the atmosphere.*

Solution. Let A be the area of tank, a the area of orifice, q the inflow and Q the outflow (*i.e.*, the discharge through the orifice). Under the conditions of simultaneous inflow and outflow, let dh be the fall in the water surface in the tank in time dt. From the continuity principle

$$-\,Adh = (Q - q)\ dt = (C_d\, a\ \sqrt{2gh}\ - q)\ dt$$

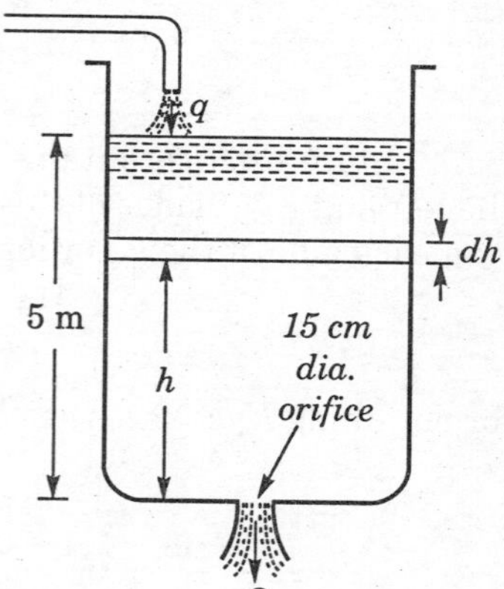

$$dt = -\frac{Adh}{(C_d a\sqrt{2gh} - q)}.$$

Let $\qquad Q = K\sqrt{h}$

where $\qquad K = C_d\, a\, \sqrt{2g}$

and $\qquad K\sqrt{h} - q = x,$

then $\qquad dx = \dfrac{Kdh}{2\sqrt{h}}$

Time taken to lower the water surface from a height H_1 to H_2

$$T = -\int_{H_1}^{H_2} \frac{Adh}{(K\sqrt{h} - q)} = -\frac{2A}{K^2}\int \frac{(x+q)\,dx}{x}$$

$$= -\frac{2A}{K^2}\int\left(1 + \frac{q}{x}\right)dx = \frac{2A}{K^2}\left[x + q \log x\right]$$

$$= \frac{2A}{K^2}\left[(K\sqrt{h} - q) + q \log\,(K\sqrt{h} - q)\right]_{H_2}^{H_1}$$

$$= \frac{2A}{K^2}\left[K(\sqrt{H_1} - \sqrt{H_2}) + q \log_e \frac{(K\sqrt{H_1} - q)}{(K\sqrt{H_2} - q)}\right]$$

Substituting the given data

$$K = C_d \,.\, a \,.\, \sqrt{2g}$$

$$= 0.72 \times \frac{\pi}{4}\left(\frac{15}{100}\right)^2 \times \sqrt{2 \times 9.81} = 5.64 \times 10^{-2}$$

$$A = \frac{\pi}{4}(3)^2 = \frac{9\pi}{4} \text{ m}^3$$

$$T = \frac{2 \times 9\pi/4}{(5.64 \times 10^{-2})^2}\Big[5.64 \times 10^{-2}\,(\sqrt{5}) - \sqrt{(2.5)} + 85 \times 10^{-3} \times 2.303 \log_{10} \frac{5.64 \times 10^{-2}\sqrt{5} - 85 \times 10^{-3}}{5.64 \times 10^{-2}\sqrt{2.5} - 85 \times 10^{-3}}\Big]$$

$= 1028$ seconds $= 17$ minutes 8 seconds.

Ex. 13.13. *A mouthpiece is fitted at the bottom of a boiler drum for the purpose of emptying it. The drum is horizontal and full of water. It is 10 m long and 2 m diameter. Find the time required to empty the boiler if the diameter of the mouthpiece is 5 cm. Assume coefficient of discharge 0.80.*

Sketch the boiler drum and derive the equation used.

Solution. Consider that at any instant the water surface in the boiler drum be at a height h above the mouthpiece exit. Head over the mouthpiece

$$h = R + R \cos \theta \qquad \ldots(1)$$

here R is the radius of the drum.

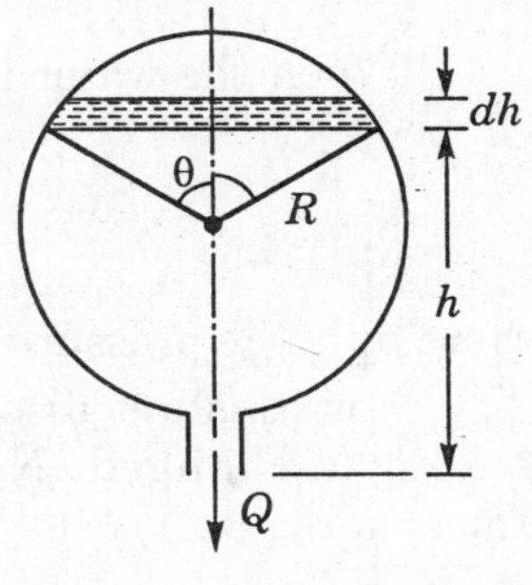

Fig. Ex. 13.13. Emptying of a Boiler drum through a mouthpiece.

Area of water surface in the drum

$$A = 2R \sin \theta \times \text{Length of drum}$$
$$= 2RL \sin \theta \qquad \ldots(2)$$

From Eq. (1), $dh = -R \sin \theta \, d\theta$

$$Adh = 2RL \sin \theta \,(-R \sin \theta \, d\theta)$$
$$= -2R^2L \sin^2 \theta \, d\theta.$$

Let in time dt, the water surface be lowered by dh, then from the continuity principle

$$-Adh = Q\,dt$$

Q being the discharge through the mouthpiece which is a function of time. Making substitutions

$$2R^2L \sin^2 \theta \, d\theta = (C_d \,.\, a \,.\, \sqrt{2gh}\,)\, dt$$

$$dt = \frac{2RL \sin^2 \theta \, d\theta}{C_d \,.\, a \,.\, \sqrt{2g\,(1 + \cos \theta)}}$$

here $\quad a =$ Area of mouthpiece.

Time of emptying the boiler drum completely

$$T = \frac{2RL}{C_d \,.\, a \,.\, \sqrt{2g}} \int_0^{\pi} \frac{\sin^2 \theta}{\sqrt{1 + \cos \theta}} \, d\theta$$

Let $1 + \cos \theta = x^2$, then one can write

$$T = \frac{2RL}{C_d \,.\, a \,.\, \sqrt{2g}} \int_0^{\sqrt{2}} x^{3/2} \, dx = \frac{3.8RL}{C_d \,.\, a \,.\, \sqrt{2g}}$$

In the problem, $R = 1$ m, $L = 10$ m, $C_d = 0.8$ and

$$a = \frac{\pi}{4}\left(\frac{5}{100}\right)^2$$

Making these substitutions, the time of emptying the boiler drum

$$T = \frac{3.8 \times 1 \times 4 \times 10^5}{0.8 \times \pi \times (5)^2 \times \sqrt{2 \times 9.81}} \text{ seconds}$$

$$= 5.46 \times 10^3 \text{ seconds} = 1.516 \text{ hours.}$$

Example 13.14. *The tank shown in the following sketch is used to give a constant discharge from the orifice O so long as the water level in the tank is above the bottom of the air inlet B. Determine the water levels in the piezometers C and D under the conditions shown in the sketch. What is the pressure intensity in the tank above the water level and at level B ? What happens to the air pressure in the tank and the discharge through the orifice when the level of water goes down below the air inlet ?*

Solution. Let the pressure of air inside the tank be p. For the conditions shown in the figure, the head over the orifice

$$H_1 = 0.75 + 0.25 + \frac{p}{\gamma} \quad ...(1)$$

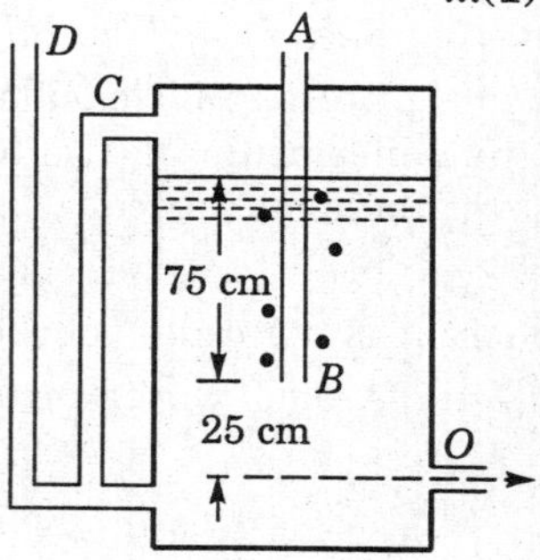

When the water level falls by a distance h, the orifice head will be

$$H_2 = (0.75 - h) + 0.25 + \frac{p_1}{\gamma} \quad ...(2)$$

where p_1 is the pressure in air above the water surface.

For constant discharge through the orifice, the head must remain unchanged. Noting that the pressure at B remains atmospheric

$$p/\gamma = -0.75 \text{ m of water (gauge)}$$

$$p_1/\gamma = -(0.75 - h) \text{ m of water (gauge)}$$

Equating the orifice heads H_1 and H_2 :

$$H_1 = 1 + \frac{p}{\gamma} = (0.75 - h) + 0.25 - (0.75 - h) = 0.25 \quad ...(3)$$

Equation (3) shows that the orifice works under a head of 0.25 m of water for all conditions of water levels above the air inlet B.

Water levels in different piezometers under the conditions shown in the figure will be as follows :

(1) at B, the water level will be 25 cm above the centre of orifice.

(2) at C, it will be 100 cm above the orifice centre.

(3) since piezometer at D is exposed to atmosphere, the water level will stand 25 cm above the centre of orifice.

When the water level goes down below the air inlet B, the air pressure becomes equal to atmospheric and the discharge decreases, it being proportional to the square root of head.

13.4.3. Bell-mouthed orifice

It is an opening constructed to conform to the shape of a jet which issues from a sharp-edged orifice and is called a bell-mouthed orifice. It has practically no contraction, hence $C_c = 1.0$. Fig. 13.9 shows such an orifice.

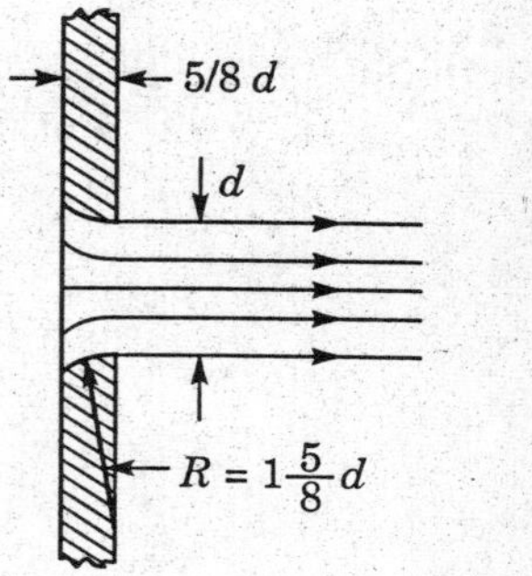

Fig. 13.9. Bell-mouthed orifice ($C_c = 1$).

13.4.4. Large orifice

When an orifice is large or the head over the orifice is small such that the mean velocity over the orifice area is no longer equivalent to the velocity at the centre of the orifice, it becomes necessary to take into account the variation of velocity over the orifice.

Consider a rectangular orifice of width B and height D, discharging under a head H_1 above its top as shown in Fig. 13.10.

The flow through an elementary strip of area Bdh is

$$dQ = C_d \, Bdh \sqrt{2gh}$$

integration of which between the limits H_1 and H_2 gives

$$Q = C_d B \sqrt{2gh} \int_{H_1}^{H_2} \sqrt{h}\, dh = \frac{2}{3} C_d B \sqrt{2g}\, (H_2^{3/2} - H_1^{3/2}) \quad ...(13.25)$$

When the velocity of approach V_0 is also taken into consideration, Eq. (13.25) taken the form

$$Q = \frac{2}{3} C_d B \sqrt{2g} \left[\left(H_2 + \frac{V_0^2}{2g} \right)^{3/2} - \left(H_1 + \frac{V_0^2}{2g} \right)^{3/2} \right] \quad ...(13.26)$$

Here V_0 is equal to the discharge Q divided by the area of flow normal to V_0.

From the above discussion it may be concluded that an orifice may begin to function as a large orifice when the head over it is reduced below a certain limit. When $H_1 < 1.5\, D$, the orifice may be regarded as a large one and the variation of velocity across it must be considered.

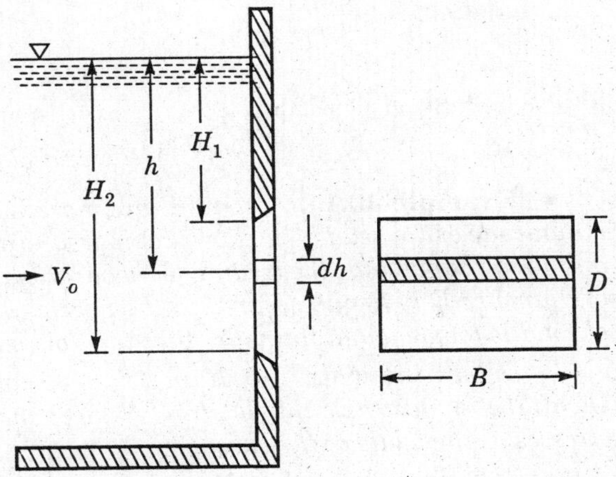

Fig. 13.10. Flow through a large orifice.

13.4.5. Standard orifice

A standard orifice is one with sharpedges as shown in Fig. 13.6, so that there is only a line contact with the fluid. The orifice is characterized by the fact that the thickness of the wall or plate is very small in comparison to the size of the opening. An orifice in a wall of small thickness with square-edges may also behave as a standard orifice provided that the fluid makes only a line contact with it. The coefficient of contraction for such an orifice is 0.62.

13.5. Mouthpieces or Short Tubes

If a short pipe equal to the size of the orifice is connected to the orifice and extended on the downstream side upto a length usually not more than 2.5 times the diameter, such an arrangement is known as a *mouthpiece* or a short tube. The mouthpieces may be cylindrical, converging or diverging. Cylindrical or parallel mouthpieces with sharp-edged entrance have considerable friction loss not because of initial contraction of the stream entering, but because of its subsequent re-expansion to fill the mouthpiece. Rounding the entrance as shown in Fig. 13.11 (*b*) greatly reduces the losses. For diverging mouthpieces, the coefficient of discharge varies with the angle of divergence and the length of the mouthpiece.

Experiments conducted by Venturi have shown that for maximum discharge to pass through a diverging mouthpiece its length should be equal to 9 times its diameter at the inlet and that the angle of divergence be equal to 5°. If the divergence is too great, the mouthpiece will not flow full. For converging mouthpieces also, the coefficient of discharge depends upon the angle of convergence.

13.5.1. Standard mouthpiece

A cylindrical mouthpiece having a length of 2 to 3 diameters with the inner end flush with the flat wall so as to form a sharp cornered entrance is called a standard mouthpiece or a standard short tube, see Fig. 13.11 (*a*). In such tubes the issuing jet first contracts, then expands filling the tube. Since the tube is full at its outlet, the coefficient of contraction is considered unity ($C_c = 1$) and the mean value of C_d is taken as 0.82.

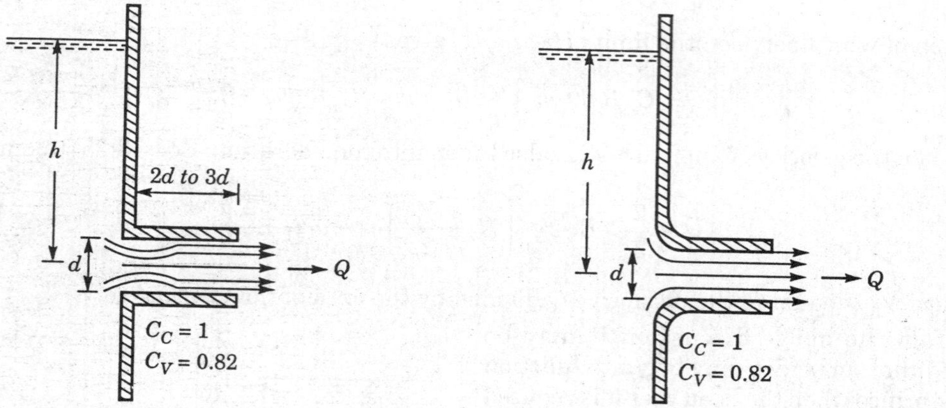

(*a*) Standard mouthpiece (*b*) Mouthpiece with round entrance

Fig. 13.11. Flow through cylindrical mouthpiece.

Example 13.15. *A streamlined nozzle of diameter d is supplied at a constant head, the magnitude of the head being large compared to d. The nozzle discharges directly into the atmosphere and is so shaped that the issuing jet is parallel at the nozzle exit. To increase the flow rate a shroud of diameter D is firmly secured to the nozzle as shown. The jet expands to fill the shroud and the shroud is long enough to ensure that the flow leaving is steady and parallel.*

Determine what the diameter of shroud should be to maximise the flowrate. Neglect shear stresses at the walls of the shroud.

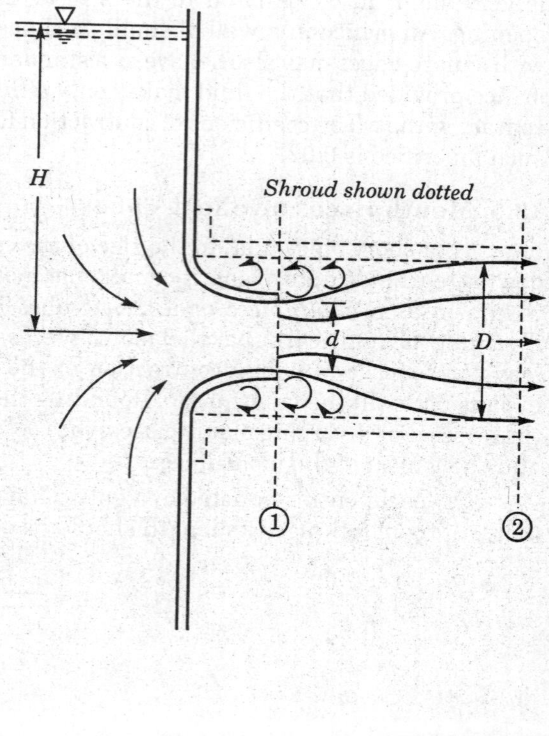

Solution. Given that $H > d$.

Writing the Bernoulli's equation between sections (1) and (2), including the loss due to sudden expansion,

$$\frac{p_1}{\gamma} + \frac{V_1^2}{2g} = 0 + \frac{V_2^2}{2g} + \frac{(V_1 - V_2)^2}{2g} \quad ...(1)$$

and between the water-surface in the tank and section (1)

$$H = \frac{p_1}{\gamma} + \frac{V_1^2}{2g} \quad ...(2)$$

From the continuity equation,

$$Q = \frac{\pi}{4} d^2 V_1 = \frac{\pi}{4} D^2 V_2 \quad ...(3)$$

Eliminating p_1 between Eqs. (1) and (2) and using (3), one can write

$$H = \frac{V_2^2}{2g} + \frac{(V_1 - V_2)^2}{2g}$$

$$= \frac{1}{2g}\left(\frac{4Q}{\pi D^2}\right)^2 + \frac{1}{2g}\left(\frac{4Q}{\pi d^2} - \frac{4Q}{\pi D^2}\right)^2$$

$$= \frac{8Q^2}{\pi^2 g}\left[\frac{2}{D^4} + \frac{1}{d^4} - \frac{2}{D^2 d^2}\right]$$

from which $$Q = \frac{\sqrt{\pi^2 gH}}{8} \frac{D^2 d^2}{\sqrt{d^4 + (D^2 - d^2)^2}}.$$

For maximum discharge, the shroud diameter can be determined from the condition $dQ/dD = 0$. Differentiating the discharge equation,

$$\frac{dQ}{dD} = \frac{1}{\sqrt{d^4 + (D^2 - d^2)^2}} [2Dd^2] + D^2 d^2 \left[-\frac{1}{2}\{d^4 + (D^2 - d^2)^2\}^{-3/2}\right] 2(D^2 - d^2).2D$$

equating it to zero and simplifying, one gets

$$D = \sqrt{2}d$$

The maximum discharge,

$$Q_{max} = \sqrt{\frac{\pi^2 g}{8}} \frac{2d^2 . d^2}{\sqrt{d^4 + (2d^2 - d^2)^2}} = \frac{\pi d^2}{2\sqrt{2}}\sqrt{2gH}.$$

Percentage increase in discharge

$$= \frac{Q_{max} - Q}{Q} \times 100 = \frac{\frac{\pi d^2}{2\sqrt{2}}\sqrt{2gH} - \frac{\pi d^2}{4}\sqrt{2gH}}{\frac{\pi d^2}{4}\sqrt{2gH}} \times 100$$

$$= (\sqrt{2} - 1) \times 100 = \mathbf{41.4\%}.$$

Example 13.16. *An external mouthpiece converges from inlet upto the vena-contracta to the shape of the jet and then diverges gradually. The diameter at the vena-contracta is 2 cm and the head over the centre of the mouthpiece is 1.44 m. The head loss in the contraction may be taken as 1% and that in the divergent portion 5% of the total energy head before the inlet. What is the maximum discharge that can be drawn through the outlet and what should be the corresponding diameter at the outlet ?*

Assuming that the pressure in the system may be permitted to fall upto 8 m below atmosphere, the liquid conveyed being water.

Solution. Bernoulli's equation between the reservoir surface and the thoat of the mouthpiece, *i.e.*, section (1), see figure below :

$$0 + 0 + 1.44 = 0 - 8.0 + \frac{V_1^2}{2g} + \frac{1}{100} \times 1.44$$

from which, $$V_1 = 13.6 \text{ m/s}$$

maximum discharge, $$Q = \frac{\pi \times (2)^2}{4 \times 10^4} \times 13.6 \text{ m}^3\text{/s} = 4.26 \text{ lits/ec.}$$

Bernoulli's equation between the reservoir surface and the exit end of the mouthpiece, yields

$$0 + 0 + 1.44 = 0 + \mathbf{0} + \frac{V_2^2}{2g} + \frac{6}{100} \times 1.44 \text{ ; } V_2 = 5.075 \text{ m/s.}$$

Using the continuity equation

$$Q = \frac{\pi D^2}{4} V_2,$$

$$D = \sqrt{\frac{4Q}{\pi V_2}}$$

$$= \sqrt{\frac{4 \times 4.261 \times 10^3}{\pi \times 5.075 \times 10^2}}$$

$$= \mathbf{3.27\ cm.}$$

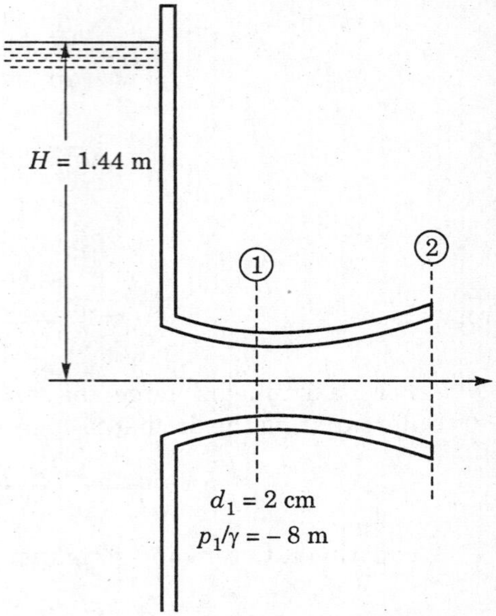

Example 13.17. *A convergent divergent mouthpiece is fitted into the vertical side of a tank containing water. Assuming that there are on losses in the convergent part of the mouthpiece, and that the losses in the divergent part are equivalent to 0.2 times the velocity head at exit and that the maximum absolute pressure head at the throat is 2.44 m of water for a barometric pressure of 760 mm of mercury, determine the throat and exit diameters of the mouthpiece when the discharge is 4.25 lit/sec for a head of 1.52 metres.* *(U.P.S.C. Exams. 1967)*

Solution. Referring to the figure of Ex. 13.15, for the present problem,

$$H = 1.52 \text{ m},$$

$$\frac{p_1}{\gamma} = 2.44 \text{ m of water absolute.}$$

Writing the Bernoulli's equation between the water surface in the tank and the throat section (1),

$$1.52 + \frac{760 \times 13.35}{1000} = 2.44 + \frac{V_1^2}{2g}$$

from which $V_1 = \sqrt{19.62 \times 9.38} = 13.54$ m/s

diameter at the throat is given by the continuity equation,

$$d_1 = \sqrt{\frac{4Q}{\pi V_1}} = \sqrt{\frac{4 \times 4.25 \times 1000}{\pi \times 13.54}} = 2 \text{ cm.}$$

The Bernoulli's equation between the reservoir and the exit section (2) gives

$$1.52 + \frac{760 \times 13.55}{1000} = \frac{760 \times 13.55}{1000} + \frac{V_2^2}{2g} + 0.2\frac{V_2^2}{2g}$$

$$1.2\frac{V_2^2}{2g} = 1.52\ ; \quad V_2 = 4.96 \text{ m/s}$$

from the continuity equation the exit diameter

$$d_2 = \sqrt{\frac{4Q}{\pi V_2}} = \sqrt{\frac{4 \times 4.25 \times 1000}{\pi \times 4.96}} = 3.3 \text{ cm.}$$

13.5.2. Re-entrant mouthpieces

Re-entrant mouthpieces are those which project from the wall into the tank as shown in Fig. 13.12. For a cylindrical re-entrant mouthpiece, if the length is between 2 to 3 times its diameter with flow full at the outlet, $C_c = 1$, $C_v = 0.75$ and the head lost is $0.78\ V^2/2g$.

Borda's mouthpiece is a short cylindrical re-entrant tube which projects into the tank Fig. 12.12 (*b*). Its length is about one diameter and has sharp edges to ensure perfect contraction so that the jet does not touch the sides of the tube. The following are the average values of the coefficients :

$$C_c = 0.52,\ C_v = 0.98,\ C_d = 0.51.$$

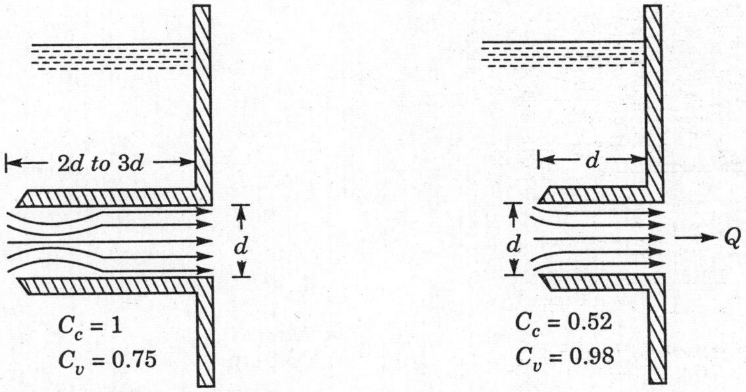

(*a*) Re-entrant mouthpiece (*b*) Borda's mouthpiece

Fig. 13.12. Re-entrant mouthpieces.

In case of Borda's mouthpiece, on account of its shorter length and sharp curvature of streamlines, the coefficient of contraction is lower than for any other type. If the mouthpiece runs full, $C_c = 1.0$, but because of surface friction, the velocity-coefficient is reduced. But if the jet springs clears as shown in Fig. 13.12 (*b*), the coefficient of velocity is as high as for a sharp edged orifice.

The coefficient of contraction can be calculated for Borda's mouthpiece by applying the momentum principle. The force causing flow through the mouthpiece is due to the hydrostatic pressure acting over the opening, and is equal to the change in momentum of flow. If A is the area of the opening, A_j the area of the jet and V the velocity of flow in the jet, then

Force causing change in momentum = γhA ...(13.27)

where h is height of liquid above the opening.

Change of momentum per second = $\gamma QV/g$ ...(13.28)

From the momentum principle, we obtain

$$\gamma hA = \frac{\gamma QV}{g} = \frac{\gamma A_j V^2}{g}$$

Substituting, $V = C_v\sqrt{2gh}$ and $A_j/A = C_c$

$$h = C_c C_v^2 \frac{2gh}{g}$$

$$2C_c C_v^2 = 1$$

or $$2C_d C_v = 1 \quad \text{...(13.29)}$$

If the fluid is considered frictionless $C_v = 1.0$,

$$V = \sqrt{2gh} \text{ and } C_d = C_v$$

From Eq. (13.29), then $C_c = 0.5$ ...(13.30)

Eq. (13.30) shows that for frictionless fluid, the contraction coefficient for Borda's mouthpiece is 0.5.

Example 13.18. *A borda mouthpiece 5 cm in diameter has a discharge coefficient of 0.51, what is the diameter of the issuing jet ?*

Solution. From Eqs. (13.27) and (13.28),

$$A\gamma h = \frac{\gamma Q}{g} V \quad ...(1)$$

but,

$$V = C_v \sqrt{2gH}$$

and

$$Q = C_d A \sqrt{2gH}$$

Substituting these in (1),

$$A\gamma h = \gamma \frac{C_d A \sqrt{2gh}}{g} C_v \sqrt{2gh}$$

$$\therefore \quad 2C_d C_v = 1, \; C_v = \frac{1}{2C_d} = \frac{1}{2 \times 0.51} = 0.98$$

and

$$C_c = \frac{C_d}{C_v} = \frac{0.51}{0.98} = 0.52$$

Area of jet $= C_c . A$

Diameter of jet $= \sqrt{C_c d^2} = 5 \times \sqrt{0.52} = 3.54$ cm.

13.6. Measurement of Flow in Open Channels

The flow in small channel is generally measured by means of weirs and in large channels by using the current meter. Sometimes the gates which are installed to regulate the flow in channels are also used for flow measurement. We shall discuss various types of weirs which are generally used for flow measurement in open channels, and also the flow under a sluice gate.

13.7. Weir

The weir has been a standard device for the measurement of flow in open channels. It is an obstruction in the channel that causes the liquid to rise behind the weir and then flows over it. By measuring the height of upstream liquid surface, the rate of flow is determined. Weirs may be constructed from a sheet of metal, masonry or from other material. If the jet of liquid, known as the *nappe,* springs free as it leaves the upstream face, the weir is known as the sharp-crested weir. The broad crested weirs are those which support the falling nappe over its crest, in the longitudinal direction.

The weirs may be of different shapes such as the rectangular, triangular, trapezoidal, parabolic etc. They may discharge under free conditions or may function as submerged weirs under appropriate downstream conditions. When the liquid level downstream of weir is above its crest level, the weir is said to be a *submerged weir*.

13.7.1. Sharp-Crested Rectangular Weir

It may be considered as large orifice of the rectangular shape placed in the channel in such a way that the head on its upper edge is zero. Consequently the upper edge is eliminated leaving only the lower edge known as the crest. The rate of flow is determined by measuring the head H over the weir crest, at a distance upstream at least four times the maximum head to be used. As the liquid approaches the weir, its surface assumes a curvature with convexity upwards, Fig. 13.13. The drop

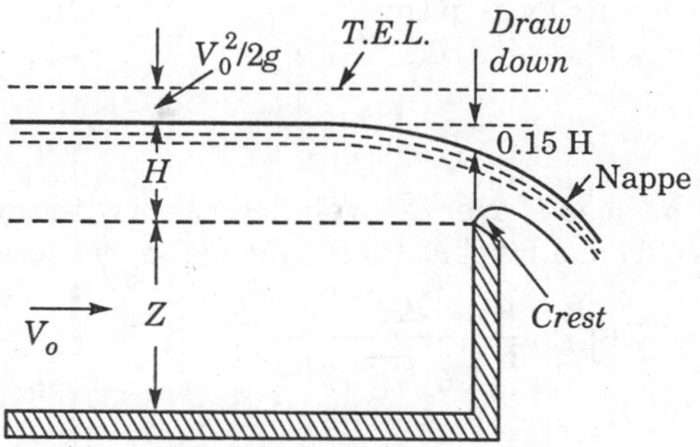

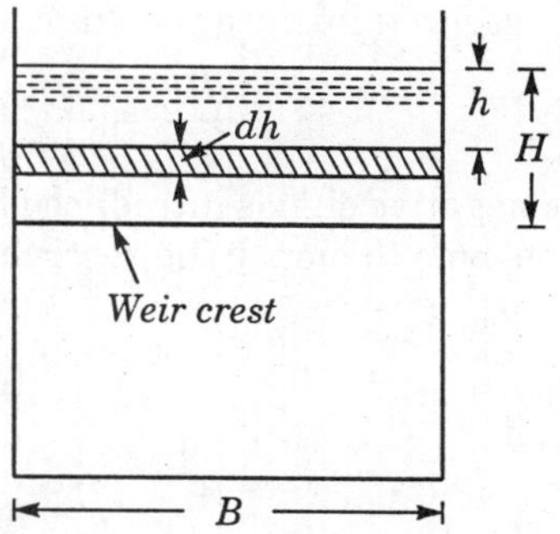

Fig. 13.13. Flow over a sharp-crested rectangular weir.

in liquid surface in the plane of the weir is known as the *drawdown*. The drawdown at the weir crest is about 0.15 H. Consider an elementary strip of area Bdh at a depth h below the free surface.

Theoretical flow through the strip

$$dQ = Bdh\sqrt{2gh}\ .$$

The theoretical flow passing over the weir is

$$Q = \int_0^H B\sqrt{2gh}\ dh = \frac{2}{3}B\sqrt{2g}\ H^{3/2} \quad ...(13.31)$$

This ideal flow will be decreased slightly by fluid friction and decreased much more due to following reasons :

(1) the flow area in the plane of the crest in not BH, but less on account of the drawdown.

(2) the velocity at $h = 0$, is not zero, and

(3) the streamlines in the plane of the crest are curved and not normal to the area of the flow.

To correct for these discrepancies, it is necessary to introduce an experimental coefficient of discharge. This then gives the basic weir formulas :

$$Q = C_d\left(\frac{2}{3}B\sqrt{2g}H^{3/2}\right) = \frac{2}{3}C_dB\sqrt{2g}\ H^{3/2} \quad ...(13.32)$$

If the velocity of approach V_0 at the section, where H was measured is appreciable, the limits of integration should be $(H + V_0/2g)$ and $(V_0^2/2g)$ and the resulting discharge equation will be

$$Q = \tfrac{2}{3}C_dB\sqrt{2g}\left[\left(H + \frac{V_0^2}{2g}\right)^{3/2} - \left(\frac{V_0^2}{2g}\right)^{3/2}\right] \quad ...(13.33)$$

The following empirical formula may be use to determine the discharge coefficient

$$C_d = 0.611 + 0.075\ H/Z$$

in which Z is the height of the weir crest above the channel floor.

Example 13.19. *Show that the time required to reduce the water level from H_1 to H_2 by means of a rectangular weir is given by*

$$t = \frac{3A}{C_dL\sqrt{2g}}\left(\frac{1}{\sqrt{H_2}} - \frac{1}{\sqrt{H_1}}\right)$$

in which A is the area of reservoir, C_d is the discharge coefficient and L is the length of the weir.

Solution. The discharge over the rectangular weir length L and a head H is

$$Q = \tfrac{2}{3}C_dL\sqrt{2g}\ H^{3/2}.$$

As the liquid is being discharged by the weir, its level in the reservoir is lowered. The rate at which the reservoir surface is falling is given by

$$-\frac{dH}{dt} = \frac{Q}{A}.$$

The negative sign is introduced because the head H over the weir decreases with the increase of time. The time to empty the reservoir surface from a level H_1 to H_2 above the weir crest is

$$t = -A\int_{H_1}^{H_2}\frac{dH}{Q} = -A\int_{H_1}^{H_2}\frac{dH}{\frac{2}{3}C_d L\sqrt{2g}\,H^{3/2}}$$

$$= \frac{3A}{2C_d L\sqrt{2g}}\int_{H_1}^{H_2}\frac{dH}{H^{3/2}} = \frac{3A}{C_d L\sqrt{2g}} = \left(\frac{1}{\sqrt{H_2}} - \frac{1}{\sqrt{H_1}}\right).$$

13.7.1.1. Suppressed Rectangular Weir. When the length of crest of the weir is the same as the width of the channel, the weir is said to be a *suppressed weir*. The width of the nappe for suppressed weir is equal to the length of the crest. Thus in this case, the effect of sides or ends of the weir on contraction of the nappe is eliminated or suppressed.

Thus for suppressed weirs,

Length of weir crest = Width of the channel.

The discharge equations developed above are applicable to suppressed weirs.

13.7.1.2. Contracted Weirs. When the crest length B of a rectangular weir is less than the width of the channel, there will be a lateral contraction of the nappe so that its length is less than B. Such type of weirs have end contractions and are known as contracted weirs or weirs with end contractions. The effective length of the nappe is thus reduced by the amount of contraction at each end.

If W is the width of the channel and B is the length of the weir crest and if $B < W$, the weir will be a contracted weir. Francis concluded that the effect of each side contraction is to reduce the length of nappe by 0.1 H. Thus if there are n end contractions, the length of the nappe will be reduced by 0.1 nH and will be equal to (B – 0.1 nH) and the discharge equation (13.32), will then be written as

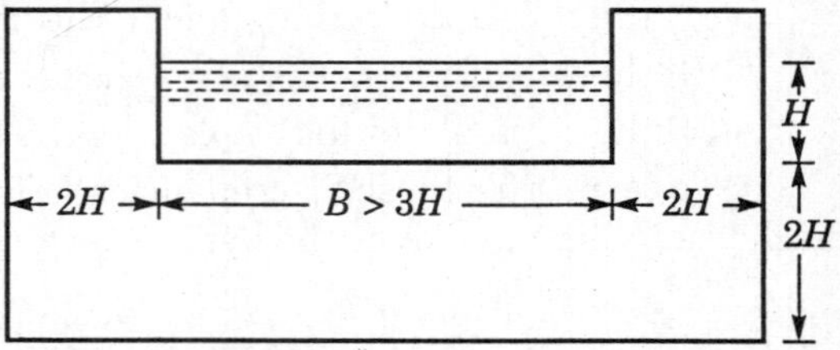

Fig. 13.24. Minimum proportions of standard contracted weir.

$$Q = \tfrac{2}{3}C_d(B - 0.1\,nH)\sqrt{2g}\,H^{3/2} \qquad ...(13.34)$$

From Eq. (13.34), it would appear that at higher heads a condition may be reached when this formula would indicate zero flow, and at still higher heads would compute negative flow. If Eq. (13.34) is to be used the minimum proportions as indicated in Fig. 13.24 are desirable.

Example 13.20. *Water flows over a rectangular sharp-crested weir 1 m long, the head over the sill of the weir being 0.66 m. The approach channel is 1.4 m wide and the depth of flow in the channel is 1.2 m. Starting from first principles, determine the rate of discharge over the weir. Consider also the velocity of approach and the effect of end contractions. Take the coefficient of discharge for the weir as 0.60.*

Solution. The discharge over a rectangular sharp-crested contracted weir is given by Eq. (13.34),

$$Q = \tfrac{2}{3}C_d(B - 0.1\,nH)\sqrt{2g}\;H^{3/2}$$

$$= \frac{2}{3} \times 0.60 \times (1 - 0.1 \times 2 \times 0.56)\sqrt{2 \times 9.81} \times (0.66)^{3/2} = 0.845 \text{ m}^3/\text{s}.$$

Velocity of approach, $V_0 = \frac{Q}{A} = \frac{0.845}{1.4 \times 1.2} = 0.503$ m/s

$$\frac{V_0^2}{2g} = \frac{(0.503)^2}{2 \times 9.81} = 0.0129 \text{ m}$$

$$H + \frac{V_0^2}{2g} = 0.66 + 0.0129 = 0.6729 \text{ m}.$$

Discharge considering the velocity of approach

$$Q = \frac{2}{3} C_d (B - 0.1\, nH)\sqrt{2g} \left[\left(H + \frac{V_0^2}{2g} \right)^{3/2} - \left(\frac{V_0^2}{2g} \right)^{3/2} \right]$$

$$= \frac{2}{3} \times 0.60 \times 0.868 \times \sqrt{2 \times 9.81}\ [(0.6729)^{3/2} - (0.0129)^{3/2}] = 0.847 \text{ m}^3/\text{s}.$$

13.7.2. Francis Weir Formula

On the basis of a large number of experiments on sharp-crested rectangular weirs, J.B. Francis found that the flow varied as $H^{1.47}$, but for grater convenience he adopted $H^{1.5}$. He then selected a constant average value of $C_d = 0.622$, so that for suppressed weir with a negligible velocity of approach

$$Q = \frac{2}{3} C_d B \sqrt{2g} H^{3/2}$$

$$= \frac{2}{3} \times 0.622 \sqrt{2 \times 9.81}\, BH^{3/2} = 1.837\, BH^{3/2} \text{ (MKS and SI units)} \quad ...(13.35)$$

In F.P.S. units, substituting $g = 32.2$ ft/s^2,

the discharge equation becomes $Q = 3.33\, BH^{3/2}$ (FPS units) ...(13.36)

Example 13.21. *Water flows over a 4.5 m wide rectangular notch at a head of 0.6 m. The notch has two end contractions. The approach channel is 6 m wide and the depth of flow in the channel is 0.9 m. Using Francis formula and considering the end contractions and the velocity of approach, determine the rate of flow in the channel.*

Solution. Using the Francis formula, the discharge through the rectangular notch

$$Q = 1.837 (B - 0.1\, nH) \left[\left(H + \frac{V_0^2}{2g} \right)^{3/2} - \left(\frac{V_0^2}{2g} \right)^{3/2} \right]$$

Initially neglecting the velocity of approach,

$$Q = 1.837(4.5 - 0.1 \times 2 \times 0.6)(0.6)^{3/2} = 3.739 \text{ m}^3/\text{s}$$

Velocity of approach, $V_0 = \frac{Q}{6 \times 0.9} = \frac{3.739}{5.4} = 0.692$ m/s

Velocity head, $V_0^2/2g = 0.0244$ m

Considering the velocity of approach,

$$Q = .837\ (4.5 - 0.1 \times 2 \times 0.6)\ [(0.6 + 0.0244)^{3/2} - (0.0244)^{3/2}]$$

$$= \mathbf{3.94\ m^3/s.}$$

Example 13.22. *A rectangular notch 50 cm. long is used for measuring a discharge of 30 litres / sec. An error of 2 mm, was made in measuring the head over the notch. Calculate the percentage error in the discharge. Take $C_d = 0.60$.*

Solution. Discharge through a rectangular notch

$$Q = \frac{2}{3}C_d\sqrt{2g}\,LH^{3/2} \qquad ...(1)$$

$$\frac{30}{1000} = \frac{2}{3}\times 0.60\sqrt{2\times 9.81}\left(\frac{50}{100}\right)H^{3/2}$$

from which $H = 0.1019$ m or 10.19 cm

Differentiating (1) w.r.t. H

$$dQ = \frac{2}{3}C_d\sqrt{2g}L\left(\frac{3}{2}H^{1/2}dH\right) \qquad ...(2)$$

Dividing (2) by (1),

$$\frac{dQ}{Q} = \frac{3}{2}\frac{dH}{H}$$

from which the percentage error in discharge

$$\frac{dQ}{Q} = \frac{3}{2}\left(\frac{2}{101.9}\right) = 2.94\times 10^{-2}, \quad \therefore \quad \text{Percentage error in } Q = \mathbf{2.94\%}.$$

Example 13.23. *A weir 36 m long is divided into 12 equal bays by vertical posts each 60 cm wide. Determine the discharge over the weir if the head over the crest is 1.2 m and the velocity of approach is 2 m/s.*

Solution. There will be 11 posts equally spaced to form 12 bays.

Net length of the weir = 36 – 0.6 × 11 = 29.4

Number of contractions, $n = 11\times 2 + 2$ (due to abutments) = 24

Head due to velocity of approach, $h_a = \frac{(V_a)^2}{2g} = \frac{(2)^2}{2\times 9.81} = 0.20$ m

Length of nappe, $B = 29.4 - 0.1\,nH = 29.4 - 0.1\times 24\,(1.2 + 0.2) = 26.04$ m

Using the Francis formula, the discharge over the rectangular weir is,

$$Q = 1.837\,BH^{3/2} = 1.837\,B\,[(h + h_a)^{3/2} - h_a^{\,3/2}]$$

$$= 1.837\times 26.04\,[1.4^{3/2} - 0.2^{3/2}] = \mathbf{36.66\ m^3/s}.$$

Example 13.24. *While conducting flow measurement using a rectangular notch, an error of 2% in head over the notch and an error of – 3% in length of the notch occurred. Determine the percentage error in discharge.*

Solution. Discharge passing over a rectangular notch

$$Q = \tfrac{2}{3}C_d\sqrt{2g}\,LH^{3/2} = KLH^{3/2}$$

$$\frac{\partial Q}{\partial L} = KH^{3/2}; \quad \frac{\partial Q}{\partial H} = \frac{3}{2}KLH^{1/2}$$

Total change in discharge

$$dQ = \frac{\partial Q}{\partial L}dL + \frac{\partial Q}{\partial H}dH = KH^{3/2}dL + \frac{3}{2}KLH^{1/2}\,dH$$

$$= \frac{Q}{L}dL + \frac{3}{2}\frac{Q}{H}dH = Q\left(\frac{dL}{L} + \frac{3}{2}\frac{dH}{H}\right)$$

or

$$\frac{dQ}{Q} = \frac{dL}{L} + \frac{3}{2}\frac{dH}{H}$$

Since $\frac{dL}{L} = -3\%$ and $\frac{dH}{H} = 2\%$

$$\frac{dQ}{Q} = -3 + \frac{3}{2} \times 2 = 0$$

The percentage error in discharge is **zero.**

Example 13.25. *Determine the height of a rectangular weir of length 6 m to be built across a rectangular channel. The maximum depth of water on the upstream side of the weir is 1.8 m and the discharge is 2000 litres per second. Take $C_d = 0.6$ and neglect side contraction.*

(RGPV, 2009 June)

Solution. The discharge over a rectangular weir is given by Eq. (13.32) :

$$Q = \frac{2}{3} C_d B\sqrt{2g}\, H^{3/2}$$

Given $Q = 2000 \text{ l/s} = \frac{2000}{1000} \text{ m}^3\text{/s} = 2 \text{ m}^3\text{/s}$

$C_d = 0.6, B = 6$ m

Substituting the data in the equation

$$2 = \frac{2}{3} \times 0.6 \times 6\sqrt{2 \times 9.81}\, H^{3/2}$$

$H^{3/2} = 0.188$, which gives $\mathbf{H = 0.3283}$ **m**

But the maximum depth of water u/s of weir is limited to 1.8 m, hence the height of weir = $1.8 - H = 1.8 - 0.3283 =$ **1.4717 m**

13.7.3. Triangular Weir

A triangular weir is particularly useful where the discharge is to vary over a large range and the same accuracy is desired for both small and large discharges. For measuring small discharges accurately and to avoid surface tension effects which are associated with flow at low heads and to obtain higher measurable heads, a triangular weir is preferable over the rectangular one. A triangular weir is also known as a V-notched weir. Fig. 13.15 shows a triangular weir with a vertex angle 2θ. The rate of flow through an elementary strip of area $dA = 2xdh$ is $dQ = 2xdh\sqrt{(2dh)}$, where h is the depth of strip below the liquid surface and dh is its thickness. From triangle OAB, we have

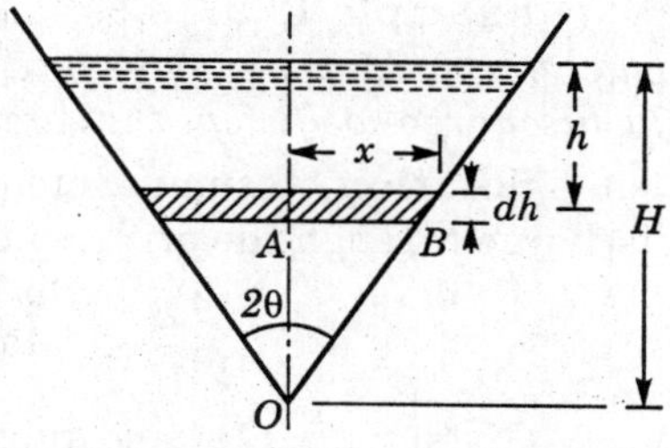

Fig. 13.15. Triangular weir.

$$\frac{x}{H-h} = \tan\theta$$

$$x = (H-h)\tan\theta.$$

The discharge flowing through the V-notched weir may be obtained by integrating dQ between the limits $h = 0$ and $h = H$

$$Q = \int_0^H 2xdh\sqrt{2gh} = \int_0^H 2(H-h)\tan\theta\,\sqrt{2gh}\,dh = \frac{8}{15}\sqrt{2g}\tan\theta\, H^{5/2} \quad ...(13.37)$$

Eq. (13.37) gives the theoretical discharge. The actual discharge is less than the theoretical one due factors already discussed while dealing with the rectangular weirs. The actual discharge is obtained by multiplying Eq. (13.37) with the coefficient of discharge C_d thus

$$Q = \left(\frac{8}{15} C_d \sqrt{2g}\tan\theta\right) H^{5/2} \quad ...(13.38)$$

For a given notch, the discharge varies as $H^{2.5}$, and the discharge head relationship may be expressed as

$$Q = KH^{5/2} \qquad ...(13.39)$$

in which K is almost a constant for the weir and is expressed by the bracketed quantity in Eq. (13.38).

Example 13.26. *When water flows through a right angled V-notch, show that the discharge is given by $Q = KH^{5/2}$ in which K is a constant and H is the height of the surface of water above the bottom of the notch. If H is measured in cm and Q in litres/sec, and the coefficient of discharge is 0.6, what is the value of K?*

Solution. The discharge through a triangular notch is given by

$$Q = \frac{8}{15} C_d \sqrt{2g} \tan \theta \, H^{5/2} = KH^{5/2} \qquad ...(i)$$

For a right angled V-notch, this equation reduces to

$$Q = \frac{8}{15} C_d \sqrt{2g} \, H^{5/2} \qquad ...(ii)$$

This equation is dimensionally homogeneous, *i.e.* the dimension of both sides of the equation are same. Therefore, if H is measured in cm, Q must also be expressed in cm³/sec. But since Q is to be expressed in litre/sec, using Eq. (*ii*)

$$Q \times 1000 = \frac{8}{15} \times 0.6 \times \sqrt{2 \times 9.81} \, H^{5/2}$$

$$Q = \frac{8/15 \times 0.6 \times \sqrt{1962}}{1000} H^{5/2} = 14.2 \times 10^{-3} H^{5/2}$$

$$\therefore \qquad K = 14.2 \times 10^{-3}.$$

Example 13.27. *A 90° triangular weir discharges water at a head of 0.15 metre into a tank which has a 7.5 cm sharp-edged orifice in the bottom. Predict the depth of water in the tank. Take the discharge coefficient the same for both.*

Solution. Assuming steady flow conditions, the discharge over the triangular weir and the discharge through the orifice must equal, thus

$$\frac{8}{15} C_{d1} \tan \theta \sqrt{2g} \, H_1^{5/2} = C_{d2} \frac{\pi}{4} d^2 \sqrt{2g \, H_2}$$

Given $C_{d1} = C_{d2}$, and substituting the given data

$$\frac{8}{15} \tan 45° \sqrt{2 \times 9.81} \, (0.15)^{5/2} = \frac{\pi}{4}\left(\frac{7.5}{100}\right)^2 \sqrt{2 \times 9.81 \, H_2}$$

From which $H_2 = 1.07$ m

$\therefore$ The depth of water in the tank = 1.07 m.

Example 13.28. *Water flows over a rectangular weir 1 m wide at a depth of 15 cm and afterwards passes through a triangular right angled weir. Find the depth of water through the triangular weir. Discharge coefficients for the rectangular and triangular weirs are 0.62 and 0.59 respectively.*

Solution. Discharge over the rectangular weir

$$Q_1 = \frac{2}{3} C_d \sqrt{2g} \, BH_1^{3/2} = \frac{2}{3} \times 0.62 \times \sqrt{2 \times 9.81} \times 1 \times (0.15)^{3/2} = 0.1062 \text{ m}^3/\text{s}$$

Since the same discharge passes over the right angled triangular weir,

$$Q_2 = \frac{8}{15} C_d \sqrt{2g} \, H_2^{5/2}$$

or $$0.1062 = \frac{8}{15} \times 0.59 \times \sqrt{2 \times 9.81}\, H_2^{5/2}$$

from which, $H_2 =$ **0.357 m** or **35.7 cm.**

Example 13.29. *Find the head over a right angled V-notch if the discharge over it is 0.05 m³/s. Take $C_d = 0.6$. If 2 mm error was made in measuring this head, what is the corresponding error in the discharge.*

Solution. Discharge through a 90° V-notch is given by, $Q = \frac{8}{15} C_d \sqrt{2g}\, H^{5/2}$

from which the head over the notch, H can be obtained.

$$H^{5/2} = \frac{15Q}{8C_d\sqrt{2g}} = \frac{15 \times 0.05}{8 \times 0.6 \times \sqrt{19.62}} = 0.0353$$

$\therefore$ $$H = 0.2625 \text{ m}$$

Error in Q due to 2 mm error in measuring H :

$$dQ = \frac{8}{15} C_d \sqrt{2g} \left(\frac{5}{2}\right) H^{3/2}\, dH = \frac{5}{2} Q \frac{dH}{H}$$

or $$\frac{dQ}{Q} = \frac{5}{2} \frac{dH}{H}$$

$\therefore$ $$dQ = Q \left(\frac{5}{2}\right)\left(\frac{dH}{H}\right) = 0.05 \times \frac{5}{2} \times \frac{2\,(\text{mm})}{262.5\,(\text{mm})}$$

$$= 9.52 \times 10^{-4} \text{ m}^3/\text{s} = \mathbf{0.952 \text{ litre/sec.}}$$

The error in measuring *discharge* is 0.952 litre/sec.

Example 13.30. *In an experiment on a 90° V-notch the flow is collected in a 0.9 m diameter vertical cylindrical tank. It is found that the depth of water increases by 0.685 m in 16.8 s when the head over the notch is 0.2 m. Determine the coefficient of discharge of notch.*

Solution. Discharge through a 90° V-notch (θ = 45°) is given by

$$Q = \frac{8}{15} C_d \sqrt{2g} \tan\theta\, H^{5/2} = \frac{8}{15} C_d \sqrt{2g}\, H^{5/2}$$

Quantity of water collected in the cylindrical tank in 16.8 seconds

$$V = \frac{\pi (0.9)^2}{4} \times 0.685 = 0.4559 \text{ m}^3$$

$\therefore$ $$\text{Discharge } Q = \frac{V}{t} = \frac{0.4359}{16.8} = 0.02595 \text{ m}^3/\text{s} = 25.95 \text{ l/s}$$

and coefficient of discharge

$$C_d = \frac{15\,Q}{8\sqrt{2g}H^{5/2}} = \frac{15 \times 0.02595}{8\sqrt{2 \times 9.81}\,(0.2)^{5/2}} = \mathbf{0.614.}$$

Example 13.31. *If there is an error of 5% in observing the head over (i) a rectangular weir, and (ii) a triangular weir, determine the error in the discharge resulting from it. Assume identical conditions for both the weirs.*

Solution. (*i*) **Rectangular weir**

Since the discharge $Q = KH^{3/2}$ ...(*i*)

$$\therefore \quad \frac{dQ}{dH} = \frac{3}{2} KH^{1/2}$$

or $$dQ = 1.5\, K\sqrt{H}\, dH \quad ...(ii)$$

Dividing (*ii*) and (*i*)

$$\therefore \quad \frac{dQ}{Q} = 1.5\, \frac{dH}{H}.$$

If the error in reading H is 5%, *i.e.* dH/H = 5%, then the corresponding error in Q will be

$$\frac{dQ}{Q} = 1.5 \times 5 = 7.5\%.$$

(*ii*) **Triangular weir**

$$Q = KH^{5/2}$$

$$\therefore \quad \frac{dQ}{dH} = \frac{5}{2} KH^{3/2}$$

then $$\frac{dH}{Q} = 2.5\, \frac{dH}{H}$$

an error of 5% in H will mean an error of 12.5% in the discharge. This shows that head observations for the triangular weir must be made very accurately.

Example 13.32. *A triangular notch is used to measure flow in a channel under a head of 20 cm. If the discharge is to be measured within 3% accuracy, what is the maximum velocity of approach that can be neglected ?*

Solution. When the velocity of approach is considered, the discharge through the triangular notch, can be written as :

$$Q = K\left[\left(H + \frac{V_0^2}{2g}\right)^{5/2} - \left(\frac{V_0^2}{2g}\right)^{5/2}\right]$$

Neglecting the effect of velocity of approach, the approximate discharge is given by

$$Q_0 = KH^{5/2}$$

and the error in discharge so caused is

$$dQ = Q - Q_0$$

$$\frac{dQ}{Q} = 1 - \frac{Q_0}{Q} = 1 - \frac{H^{5/2}}{\left[H + \left(\frac{V_0^2}{2g}\right)\right]^{5/2} - \left[\frac{V_0^2}{2g}\right]^{5/2}}$$

$$= 1 - \frac{H^{5/2}}{H^{5/2}\left(1 + \frac{V_0^2}{2gH}\right)^{5/2} - \left(\frac{V_0^2}{2g}\right)^{5/2}}$$

Neglecting the term $(V_0^2/2g)^{5/2}$, being very small

$$\frac{dQ}{Q} = 1 - \frac{1}{\left(1 + \frac{V_0^2}{2gH}\right)^{5/2}}$$

$$= 1 - \frac{1}{\left(1 + \frac{5}{2} \times \frac{V_0^2}{2gH}\right)} = 1 - \frac{1}{1 + \frac{5}{2} \times \frac{V_0^2}{2 \times 9.81 \times 0.2}}$$

Since $dQ/Q = 3\%$, one obtains $\frac{3}{100} = 1 - \frac{1}{1 + 0.637\, V_0^2}$

from which $V_0 = 0.2202$ m/sec = **22.02 cm/sec.**

Example 13.33. *The flow of water in a canal varies from 425 lit/sec. to 680 lit/sec. It is desired to discharge not less than 340 lit/sec of water per second and not more than 425 lit/sec over a 90 degree V-notch weir into one channel, while the remainder goes over a sharp-crested weir. Find the length of the rectangular weir and the maximum head on each weir. Take $C_d = 0.58$ for both weirs.*

Solution. Maximum discharge to pass over the 90° V-notch = 425 lit/s

Minimum discharge to pass over the 90° V-notch = 340 lit/s

Maximum discharge to pass over the rectangular weir = 680 – 340 = 340 lit/s

Minimum discharge to pass over the rectangular weir = 425 – 340 = 85 lit/s.

90° V-notch weir

$$Q = \frac{8}{15} C_d \sqrt{2g}\, H^{5/2}$$

Head required to pass 340 lit/s $\frac{340}{1000} = \frac{8}{15} \times 0.58 \times \sqrt{2 \times 9.81}\, (H_1)^{5/2}$

$H_1 = 57.5$ cm.

Similarly, the head required to pass 425 lit/s, $H_2 = 64.55$ cm.

Sharp-crested rectangular weir $Q = \frac{2}{3} C_d B \sqrt{2g}\, H^{3/2}$.

Let the head over the rectangular weir to pass the remainder flow, *i.e.* 425 – 340 = 85 lit/s be h cm, then

$$85 \times 1000 = \frac{2}{3} \times 0.58 \times B \times \sqrt{2g} \times h^{3/2} \qquad ...(1)$$

Taking the vertex of the V-notch as the zero level, the crest level of rectangular weir

$= (57.5 - h)$ cm.

Head over the rectangular weir to pass 340 lit/s (water surface level at the V-notch = 64.55 cm)

$= 64.55 - (57.5 - h) = (7.05 + h)$ cm

Hence, $$340 \times 1000 = \frac{1}{3} \times 0.58 \times B \times \sqrt{2g}\, (7.05 + h)^{3/2} \qquad ...(2)$$

From Eqs. (1) and (2) $\frac{340}{85} = \frac{(7.05 + h)^{3/2}}{h^{3/2}}$

solving which, $h = 4.63$ cm.

Crest length may now be obtained from Eq. (1) above

$$85 \times 1000 = \frac{2}{3} \times 0.58 \times B \times \sqrt{2 \times 9.81}\ (4.63)^{3/2}$$

$$B = 368 \text{ cm.}$$

∴ (1) Length of rectangular weir crest = 368 cm

(2) Maximum head on rectangular weir = 7.05 + h = 11.68 cm

(3) Maximum head on V-notch weir = 64.55 cm.

13.7.4. Trapezoidal Weir

A trapezoidal weir has an opening of the trapezoidal shape and may be regarded as a combination of a rectangular and a triangular weir. Fig. 13.16 shows a trapezoidal weir.

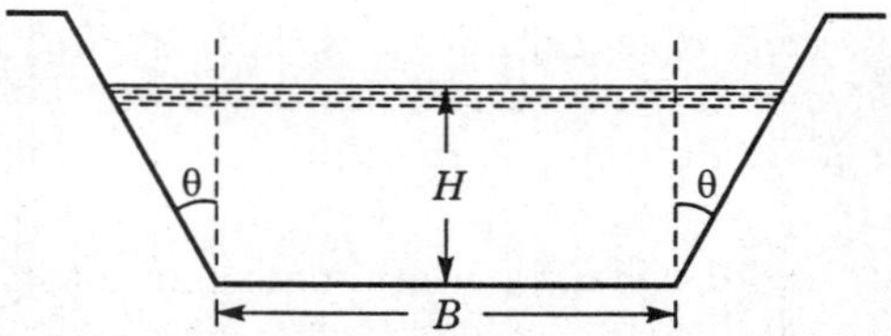

Fig. 13.16. Trapezoidal weir.

The trapezoidal weir section can be broken into a rectangular portion of width B and a triangular portion of vertex angle 2θ, for the purpose of computing discharge. Discharge through the trapezoidal weir is equal to the sum of the discharge through the rectangular and triangular portions, thus

$$Q = \frac{2}{3} C_d \sqrt{2g}\ BH^{3/2} + \frac{8}{15} C_d \sqrt{2g}\ \tan\theta\ H^{5/2} \qquad ...(13.40)$$

The coefficients of discharge involved in Eq. (13.40) for the two portions of the weir may be slightly different.

Example 13.34. *A sharp edged weir is in the form of a symmetrical trapezium. The horizontal base is 10 cm wide, the top is 50 cm wide and the depth is 30 cm. Develop a formula relating the discharge to the upstream water level and estimate the rate of flow when the upstream water surface is 50 cm above the weir crest. Assume that the velocity of approach is negligible and that coefficient of discharge C_d = 0.6.*

Solution. Consider a strip parallel to weir crest at a depth h and of width b and depth dh. From the geometry of figure,

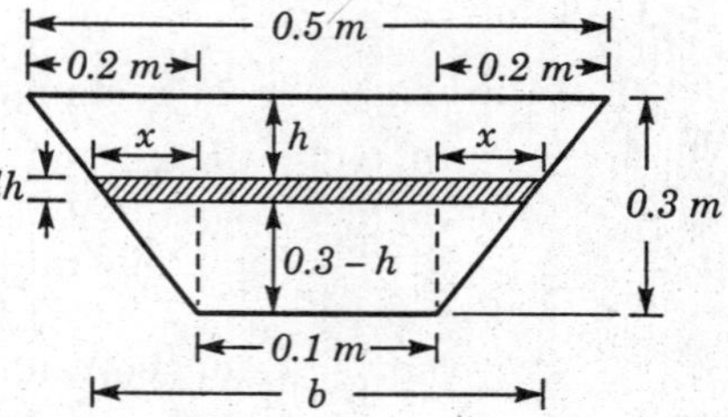

$$\frac{0.2}{0.3} = \frac{x}{0.3 - h}$$

∴ $$x = \frac{2}{3}(0.3 - h)$$

and $$b = 0.1 + 2x.$$

The flow through the elementary strip of area bdh is $dQ = bdh\sqrt{2gh}$.

If the head over the crest is H, the total discharge passing over the weir

$$Q = \int_0^H (0.1 + 2x)\, dh\sqrt{2gh} = \int_0^H \left[0.1 + \frac{4}{3}(0.3 - h)\right]\sqrt{2gh}\, dh$$

$$= \sqrt{2g}\int_0^H \left(0.5\sqrt{h} - \frac{4}{3}h^{3/2}\right) dh = \sqrt{2g}\left(\frac{1}{3}H^{3/2} - \frac{8}{15}H^{5/2}\right)$$

The actual discharge, $Q = C_d \sqrt{2g}\left(\frac{1}{3}H^{3/2} - \frac{8}{15}H^{5/2}\right)$

For $H = 30$ cm, the discharge

$$Q = 0.6\ \sqrt{2 \times 9.81}\left[\frac{1}{3} \times (0.3)^{3/2} - \frac{8}{15}(0.3)^{5/2}\right]$$

$$= 0.6 \times 4.43 \times 0.0285 = 0.0758 \text{ m}^3/\text{s}.$$

Example 13.35. *The depth of water in a 300 m wide rectangular river is 2 m and the velocity of flow is 0.6 m/s. What should be the height of a weir to be built across the river if the depth of water in the river on the upstream side of the weir is not to exceed 3.5 m ? Neglect velocity of approach and take coefficient of discharge for the weir equal to 0.62.* *(Nagpur Univ. 1987)*

Solution. Discharge flowing in the rectangular river,

$$Q = A\ .\ V = (300 \times 2)\ 0.6 = 360 \text{ m}^3/\text{s}$$

If a weir of height Z is constructed across the river, the same discharge will pass over it with a head h such that,

$$Z + h = 3.5$$

Discharge over the weir

$$Q = \frac{2}{3}\ C_d\ .\ L\ .\ \sqrt{2g}\ h^{3/2}$$

or

$$360 = \frac{2}{3} \times 0.62 \times 300 \times \sqrt{2 \times 9.81}\ h^{3/2}$$

from which $h = 0.7544$ m

and $Z = 3.5 - h =$ **2.7455 m.**

13.7.5. Cipolletti Weir

This is a type of trapezoidal weir in which the sides have a slope of 1 horizontal to 4 vertical. This weir is claimed to have an advantage over a rectangular weir and that is : the decrease in discharge due to end contraction is balanced by the discharge through the triangular portion and that a rectangular wire formula can then be used to compute the discharge. These conclusions were based on the experiments by Cipolletti on such weirs and, therefore, such a weir known after him as Cipolletti weir.

The reduction in the discharge of a rectangular weir due to two end contractions

$$= 2/3\ C_d \sqrt{2g}\ \ (0.2\ H)\ H^{3/2}$$

The discharge through the triangular weir of vertex angle 2θ

$$= \frac{8}{15} C_d \sqrt{2g}\ \ \tan\theta\ H^{5/2}$$

Since in a Cipolletti weir these two discharges must balance, thus,

$$\frac{2}{15} C_d \sqrt{2g}\ \ H^{5/2} = \frac{8}{15} C_d \sqrt{2g}\ \ \tan\theta\ H^{5/2}$$

Solving for θ, $\tan\theta = \frac{1}{4}$.

This shows that the side slopes of a trapezoidal weir in which the reduction in discharge due to end contractions is balanced by the discharge through the triangular portion should be 1 horizontal to 4 vertical. Cipolletti gave an equation for discharge through such a weir as

$$Q = 3.367\ BH^{2/2} \text{ (in F.P.S. units)} \quad ...(13.41)$$

the metric equivalent of which is

$$Q = 1.86\ BH^{3/2} \quad ...(13.42)$$

in which B and H are measured in metre and Q is in m^3/s. For determining discharge in irrigation canals, where this type of weir may be found in common use, we need only the observation of head H over the weir crest, as is evident from Eqs. (13.41) or (13.42).

13.7.6. Submerged Weir

When the water on the downstream side of a weir rises above the level of the crest, the weir is said to be a sumerged weir. Fig. 13.17 shows a submerged weir in which the upstream water is at a height H_1 above the weir crest and H_2 is the similar height on the downstream side.

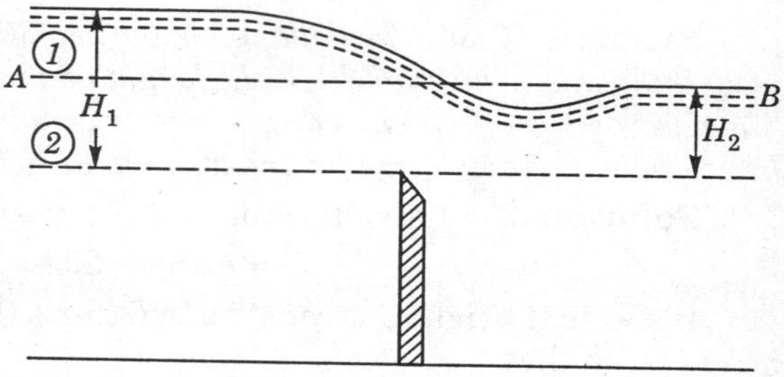

Fig. 13.17. Flow over a weir under submerged condition.

The flow over the submerged weir may be considered by dividing the flow into two portions :

(1) flow over the upper part of the section (lying above the line AB) of depth $(H_1 - H_2)$ may be considered as a free discharge into the air ; it is a weir flow and

(2) flow through the remaining lower part of depth H_2 (lying below the line AB) may be considered as a discharge through a submerged orifice.

Thus for the upper portion, the weir formula gives the flowrate

$$Q_1 = \frac{2}{3} C_{d1} \sqrt{2g}\, B(H_1 - H_2)^{3/2}$$

and for the lower portion, the orifice formula yields,

$$Q_2 = C_{d2} B H_2 \sqrt{2g\,(H_1 - H_2)}\,.$$

If we assume $C_{d1} = C_{d2} = C_d$, then the total discharge

$$Q = Q_1 + Q_2$$

$$= C_d B\, \sqrt{2g\,(H_1 - H_2)} \left[\frac{2}{3} H_1 - \frac{1}{3} H_2\right] \qquad \text{...(13.43)}$$

The discharge coefficient C_d varies with the submergence ratio, H_2/H_1, the variation of which for a given weir may be obtained from experiments.

J.R. Villemonte studied the discharge characteristics of a number of sharp-crested submerged weirs. He conducted tests on rectangular, triangular, parabolic and proportional weirs and showed that the result for all types could be represented by a single equation as given below :

$$Q = Q_1 \left[1 - \left(\frac{H_2}{H_1}\right)^n\right]^{0.385} \qquad \text{...(13.44)}$$

where Q = discharge for submerged conditions

Q_1 = free discharge given by $Q_1 = K H_1^{\,n}$

n = exponent in the free discharge equation.

Eq. (13.44) shows that the discharge of the weir under submerged conditions depends upon the submergence ratio H_2/H_1 and that the weir will yield zero discharge when $H_2/H_1 = 1$, that is when the upstream and the downstream water surfaces are at the same level above the weir crest.

13.7.7. Broad-Crested Weir

Weirs in which the sheet of flowing liquid is supported by the surface of the crest are called broad crested weirs. The theory of broad-crested weirs has been dealt with in Art. 12.14.3. If the length of weir crest is less than about 2/3 H, the nappe will ordinarily spring clear and the weir is in fact a sharp crested one. As the length of crest increases the form of the nappe changes, as shown in Fig. 13.18, and the sharp-crested weir coefficients no longer apply.

The discharge over a broad-crested weir may be computed by the following formula :

$$Q = C\,BH^{3/2} \quad \text{...(13.45)}$$

If the velocity of approach is significant, the above formula would change to

$$Q = CB\left[\left(H + \frac{V_0^2}{2g}\right)^{3/2} - \left(\frac{V_0^2}{2g}\right)^{3/2}\right] \quad \text{...(13.46)}$$

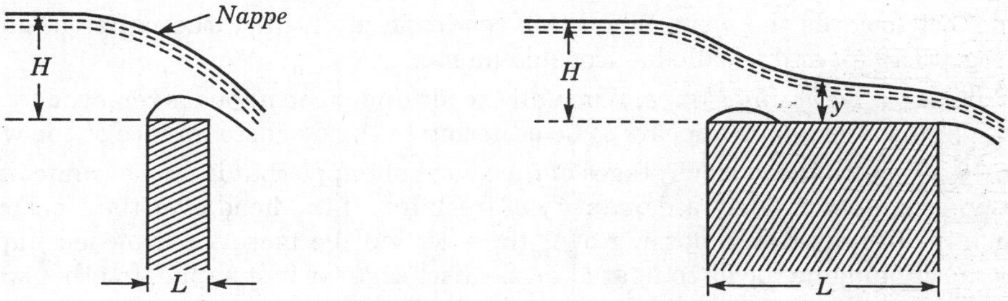

(a) $L < \frac{2}{3}H$
Nappe springing clear, weir functioning as sharp crested

(b) Broad-crested weir with square upstream corner

Fig. 13.13. Criteria for weir to be broad-crested.

For flat crested weirs with crest length ranging from 0.3 to 3 m (1 to 10 ft) and square upstream corner, for L/H values from 2 to 5, King found a value of the coefficient C of about 2.7 in FPS units (in metric units, $C = 1.49$). As L/H decreases below 2, C increases reaching a value of 3.3 (its metric equivalent being 1.825) which corresponds to the sharp crested condition and the nappe clears the crest.

The separation of nappe from the crest which occurs at a sharp cornered entrance can be avoided by adequate rounding of the entrance. The rounding reduces the amount of contraction and increases the coefficient of discharge.

13.8. Aeration on Nappe

It is an established fact that in case of the suppressed weir, the free access of air to the underside of nappe is prevented. The air entrapped between downstream face of the weir and the underside of the nappe is gradually evacuated in due course of time by dynamic action of the liquid. This eventually leads to subatmospheric pressure in the so evacuated region and this causes the discharge to increase. Bazin has thoroughly investigated the effect of restricting the free passage of air below the nappe upon the discharge and the form of nappe itself. He found that when the flow was sufficient to prevent the air from getting under the nappe, it may assume one of the three distinct forms as shown in Fig. 13.19, and the discharge for the unventilated nappe may be 28 per cent greater than when the air is freely admitted or the nappe is 'free'. Which of these forms the nappe assumes and the amount by which the discharge is greater

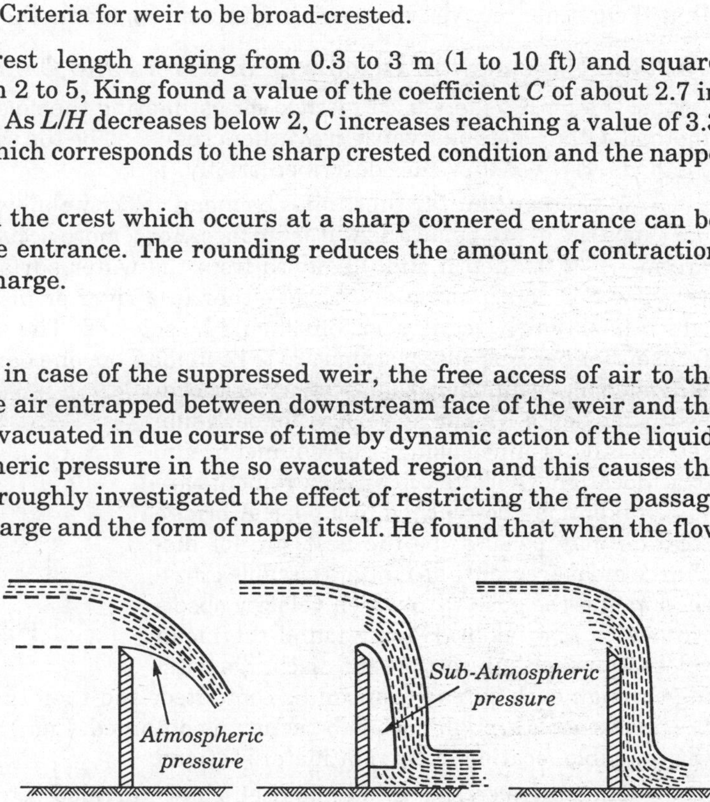

(a) Free nappe (b) Drowned nappe (c) Clinging nappe

Fig. 13.19. Ventilated and unventilated nappe.

than that for the 'free nappe', depends largely upon the head over the weir, the height of the weir and the water level in the downstream channel.

1. *Free Nappe*. The pressure under the falling nappe is maintained at atmospheric level by proper ventilation of the weir. This type of weir in which air below the nappe exists at atmospheric pressure is known as a ventilated weir or an aerated weir.

2. *Drowned Nappe*. When no ventilation is provided to the underside of the nappe, the air enclosed between the nappe and the downstream face of the weir gets gradually mixed up and carried away by the flowing liquid. This causes the pressure under the nappe to fall below the atmospheric level. The pressure difference on the two sides of the nappe—the upper side exposed to atmospheric pressure while the bottom one has pressure less than atmospheric—causes the nappe to deflect towards the weir. When this condition is reached, the nappe assumes a shape shown in Fig. 13.19 (*b*) and is called a drowned nappe.

3. *Clinging or Adhering Nappe*. When all the air under the nappe is removed by the dynamic action of the liquid, the nappe appears to be adhering to the downsteam face of the weir.

Fig. 13.19 shows the various stages of the shape of nappe that it may assume in the absence of ventilation or supply of adequate quantity of fresh air. If the head over the weir crest is maintained constant, the discharge passing over the weir will be more for drowned nappe and still greater for the clinging nappe as compared to the discharge for free nappe. In other words, a given discharge will pass over the weir at a decreased head for drowned nappe, and a further decreased head for a clinging nappe.

Thus when a weir is used as a measuring device, care must be taken to ensure that the nappe is fully aerated, *i.e.*, the pressure beneath the nappe must be maintained atmospheric so that the discharge given by the weir formula represents the correct discharge.

13.9. Measurement of Discharge in Rivers and Canals

The most widely used method for estimating the river and canal flow is the current meter method. Other methods which are rather crude may be the surface float, rod float and surface slope methods. Each of these are described briefly.

1. **Current meter method.** The mean velocity of flow in open channels is known to occur at 0.6 times the depth below the water surface. For a more accurate value, the velocity measurements are made at 0.2 depth and 0.8 depth from the water surface and their average gives the mean velocity. For determining discharge either in a river or in a big canal, a straight uniform reach which has a fairly regular bottom should be selected. The site should be such as to be accessible from either bank as observations are to be made from one bank to the other. The river cross-section is divided into a number of sub-sections of known width and depth. The mean velocity for each sub-section is determined using a current meter. The exact location where velocity measurements are to be taken may be reached in a boat. A bridge location may perhaps be the best site for discharge measurements as it offers a stable cross-section and the ease with which velocity observations can be taken. Having found the mean velocity for each sub-section, the discharge passing through each sub-section can be computed, and then the total discharge, which is the sum of the discharges through each sub-section, is obtained. Fig. 13.20 shows the river cross-section and the probable location of current meter stations.

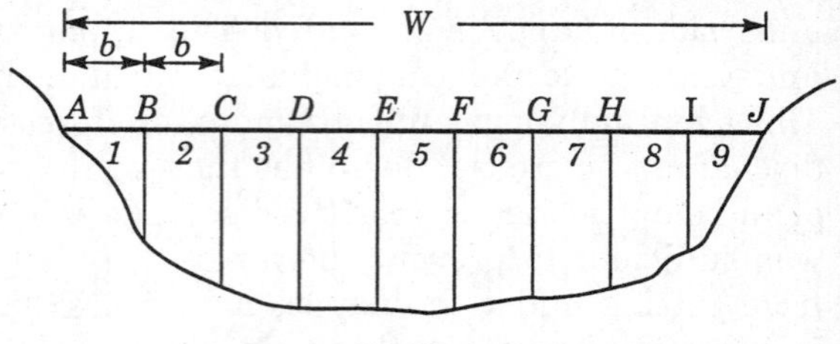

Fig. 13.20. Discharge measurement by current meter.

The entire cross-section has been shown divided into 9 sub-sections of equal width b (this number depends on the width and the accuracy desired). The current meter stations where the velocities are to be observed have been marked by capital letters *A, B, C* and so on. If d_1, d_2 etc.

represent the mean depth for each sub-section marked 1, 2 respectively, then the discharge through the sub-section will be

$$Q_1 = bd_1 \left(\frac{0 + V_B}{2}\right)$$

where d_1 is the mean of depths at A and B,

$$d_1 = \frac{d_A + d_B}{2}, \quad Q_2 = bd_2 \left(\frac{V_B + V_C}{2}\right), d_2 = \frac{d_B + d_C}{2}$$

and so on. The total discharge

$$Q = Q_1 + Q_2 + Q_3 + \ldots\ldots + Q_9.$$

2. **Surface Float method.** A surface float, which may be a corked bottle loaded inside so that its top floats just above the water surface may be used. The time taken by the float in travelling about 100 metres of straight length will permit determination of the surface velocity. The mean velocity of flow may be 0.80 to 0.95 of the surface velocity depending upon the depth and the boundary roughness. The mean velocity may thus be estimated from which the discharge $Q = AV$ can be determined. The cross-sectional area is to be determined from the depth observations.

3. **Rod Float method.** A rod float consists of a telescopic rod loaded at the bottom so that it floats vertically at about 0.6 depth from the water surface. The time taken by the rod to move in the downstream direction for 100 metres will enable determination of the mean velocity and the knowledge of the mean channel cross-section will permit computation of discharge.

4. **Surface Slope method.** This method which is also known as the slope-area method involves the use of either Manning's equation or Chezy's equation for determination of mean velocity

$$V = \frac{1}{n} R^{2/3} S^{1/2}.$$

The slope S is determined by observing the fall h in the water surface level in a distance L, thus yielding $S = h/L$. The hydraulic radius R can be estimated from the known channel cross-section. The only problem is to select a proper value of roughness n, which, of course, is a matter of experience and judgement.

This method is particularly useful for estimating flood flow in natural streams. During the periods of high floods it may not be possible to measure discharge by a more accurate method such as the current meter method thereby necessitating use of an indirect method which might utilize the history of the flood by referring to the impressions or high water marks left by it. The water surface slope and the cross-sectional area are determined by the high water marks along each of the banks. A straight reach of channel of considerable length must be selected if the slope and the area measurements are to be truly representative. The estimation of Manning's roughness n may pose some difficulty and is a source of considerable error in such measurements.

13.10. Flow Under a Sluice Gate

The flow under a sluice gate differs from that through an orifice on account of the fact that the emerging jet is not exposed to the atmospheric pressure throughout, but is guided by a horizontal floor. Consequently, the pressure in the jet is not atmospheric but is distributed hydrostatically. Fig. 13.21 shows a sluice gate installed in a channel of horizontal floor. The figure also shows distribution of pressure on the gate as well as on the channel floor. Note-worthy here is the influence of curvature of streamlines as the flow passes underneath the gate. The upward convexity of streamlines on the upstream side of the gate is responsible for decreasing the pressure on the floor, below the hydrostatic value. With a similar reasoning, the pressure on the floor downstream of the gate are larger than hydrostatic. When the streamlines regain their original character—straight and parallel to the floor—the pressures then are distributed hydrostatically.

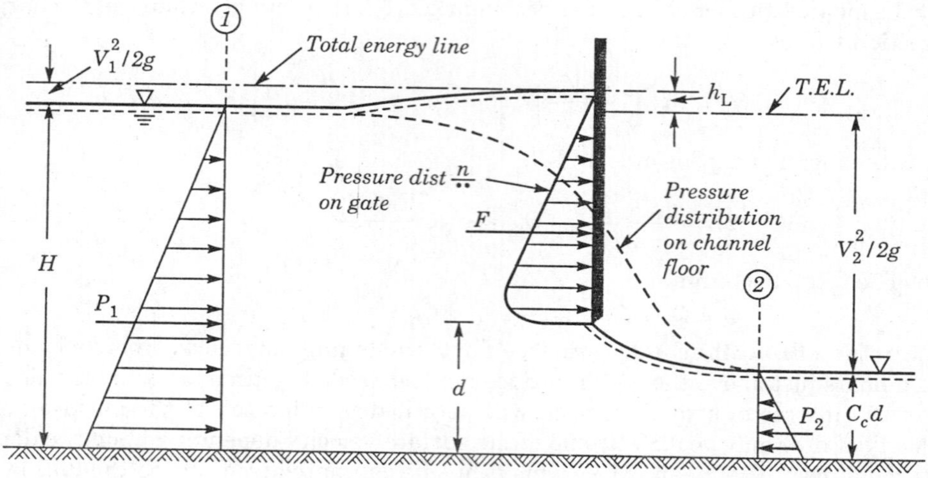

Fig. 13.21. Flow under a sluice gate.

The continuity equation between sections (1) and (2) yields

$$q = V_1 H = V_2 C_c d \quad ...(13.47)$$

in which q is the discharge per unit length of the gate, V_1 and V_2 are the respective velocities at the sections under consideration, H is the water depth upstream of the gate, d is the gate opening, and C_c is the coefficient of contraction of the jet.

Writing the Bernoulli's equation between these sections, neglecting the energy loss in between

$$\frac{V_1^2}{2g} + H = \frac{V_2^2}{2g} + C_c d \quad ...(13.48)$$

From Eqs. (13.47) and (13.48), one can write

$$V_2 = \sqrt{\frac{2gH}{1 + C_c d/H}} \quad ...(13.49)$$

and using Eq. (13.47),

$$q = \frac{C_c}{\sqrt{1 + C_c d/H}} d\sqrt{2gH} \quad ...(13.50)$$

Introducing a coefficient of discharge,

$$q = d\, C_d \sqrt{2gH} \quad ...(13.51)$$

where

$$C_d = C_c / \sqrt{1 + C_c d/H} \quad ...(13.52)$$

Fig. 13.22 shows the variation of C_d and C_c with the non-dimensional gate opening, d/H. Also shown on the figure is the variation of C_d with $H/(H + Z)$ for a sharp crested rectangular weir. The opposite trend exhibited by the C_d curves (for the weir and the sluice gate) is attributable to different pressure conditions within the contracted jet.

The degree of contraction for flow under a sluice gate remains practically same over a considerable range of d/H, as shown in Fig. 13.22, thereby yielding an almost horizontal curve for C_c : d/H. At low values of d/H, H and $V_2^2/2g$ differ by a narrow margin, see Fig. 13.21. But as d/H increases the disparity between the two increases resulting in decreased C_d values as in evident from Eq. (13.52). This explains for the downward trend of C_d for the sluice gate.

The continuity, energy (*i.e.* Bernoulli's) and momentum equations can be applied to evaluate the energy loss and the force exerted by the liquid upon the sluice gate. Applying the energy equation between sections (1) and (2) of Fig. 13.21.

$$H + \frac{V_1^2}{2g} = C_c d + \frac{V_2^2}{2g} + h_L \qquad ...(13.53)$$

where h_L is the energy loss between the sections. From Eq. (13.47)

$$V_1 = \frac{q}{H},\ V_2 = \frac{q}{C_c d}$$

using these, the energy loss

$$h_L = H - C_c d + \frac{q^2}{2gH^2}\left(1 - \frac{H^2}{C_c^2 d^2}\right) \qquad ...(13.54)$$

Expressing it in a non-dimensional form

$$\frac{h_L}{H} = 1 - C_c\frac{d}{H} + \frac{q^2}{2gH^3}\left(1 - \frac{H^2}{C_c^2 d^2}\right) \qquad ...(13.55)$$

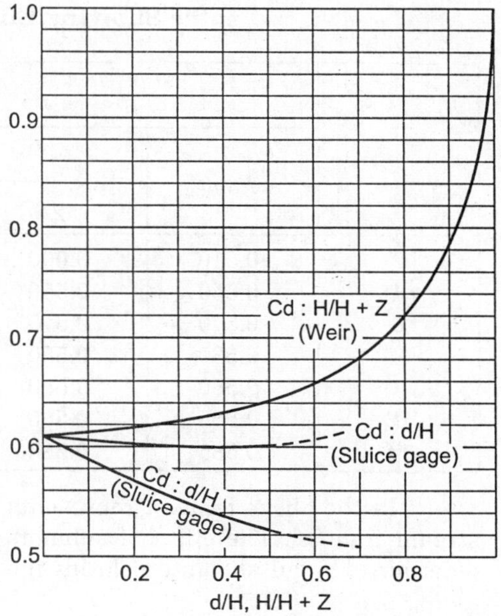

Fig. 13.22. Variation of hydraulic coefficients with flow geometry.

The momentum equation, applied to the control volume enclosed by sections (1) and (2), the gate and the channel floor yields

$$P_1 - P_2 + F = \frac{\gamma Q}{g}(V_2 - V_1) \qquad ...(13.56)$$

where P_1 and P_2 are the hydrostatic forces at the respective sections, F is the force exerted by the gate upon the control volume. Here the frictional force on the channel floor has been considered negligible. Substituting for P_1 and P_2, and using the continuity relationship, one gets

$$\frac{1}{2}\gamma BH^2 - \frac{1}{2}\gamma B(C_c d)^2 + F = \frac{\gamma Bq}{g}\left(\frac{q}{C_c d} - \frac{q}{H}\right)$$

which yields,

$$F = \frac{\gamma Bq^2}{gC_c d}\left(1 - C_c\frac{d}{H}\right) - \frac{\gamma_B}{2}(H^2 - C_c^2 d^2) \qquad ...(13.57)$$

Here B is the length of the sluice gate. Writing Eq. (13.57) non-dimensionally

$$\frac{F}{\gamma BH^2} = \frac{q^2}{gH^2 C_c d}\left(1 - C_c\frac{d}{H}\right) - \frac{1}{2}\left(1 - C_c^2\frac{d^2}{H^2}\right) \qquad ...(13.58)$$

For given conditions of flowrate, water depth behind the gate and the opening, one can determine h_L and F using Eqs. (13.55) and (13.58).

13.10.1. Sluice Gate Operating Under Submerged Conditions

When the sluice gate operates under submerged conditions, under certain downstream depths (in relation to the upstream ones) the discharge is considerably decreased as compared to the corresponding free-discharge. This is reflected in the lower values of the coefficients of discharge, the representative values of which are given in the following table :

Table Showing Effect of Submergence on Discharge

$\frac{H}{d}$	Discharge coefficient, C_d					
	For Free Flow	h/d				
		2	4	6	8	10
2	0.510	0.000	0.00	0.00	0.00	0.00
4	0.550	0.550	0.00	0.00	0.00	0.00
6	0.570	0.570	0.40	0.00	0.00	0.00
8	0.580	0.580	0.50	0.34	0.00	0.00
10	0.580	0.580	0.58	0.42	0.29	0.00
12	0.585	0.585	0585	0.47	0.37	0.25
14	0.585	0.585	0.585	0.51	0.43	0.33

In the above table, h represents the depth of submergence, *i.e.* water depth in the downstream channel. The effect of submergence on the hydraulic performance of sluice gate can be determined by suitably interpolating the value of discharge coefficient and making use of Eq. (13.51).

OBJECTIVE TYPE QUESTIONS

1. Which of the following devices are usually used in measuring pipe flow :

(*i*) mouthpiece (*ii*) cipolletti weir (*iii*) venturimeter

(*iv*) pitot tube (*v*) orifice meter.

2. Indicate the range of discharge coefficient for the following measuring devices in the turbulent range of flow :

(*a*) venturimeter (*i*) 0.65—0.75

(*b*) orificemeter (*ii*) 0.95—0.99

(*iii*) 0.59—0.88

(*iv*) 0.85—0.95

3. In a venturimeter, the lesser the diameter of throat, the greater will be the differential head which can be measured accurately. But the throat diameter cannot be made too small for then :

(*i*) the loss of energy will be large

(*ii*) the cone angle will be large and a greater length of diffuser will be required

(*iii*) the pressure at the throat will come down to such a low value as to cause cavitation

(*iv*) separation of flow will take place.

4. Arrange the following flow measuring devices in the decreasing order of head loss caused by them :

(*a*) nozzlemeter (*b*) venturimeter (*c*) orificemeter.

5. The head loss in orificemeter may be expressed by $h_L = C_L\, V^2/2g$. The loss coefficients for turbulent flow depends only on the orifice to pipe diameter ratio. Identify the loss coefficient for each ratio :

Diameter ratio	Loss coefficient
(*a*) 0.3	(*i*) 4.4
(*b*) 0.4	(*ii*) 11.0
(*c*) 0.5	(*iii*) 86.0
(*d*) 0.6	(*iv*) 29.5
(*e*) 0.7	(*v*) 306.0

6. The Prandtl type pitot tube is used to measure :

(*i*) stagation head (*ii*) velocity head

(*iii*) static pressure head (*iv*) discharge.

7. In orifice flow, the vena contracta represents :

(*a*) the section where the jet has the maximum flow area

(*b*) the section where the pressure is above atmospheric

(*c*) the location where the jet area is minimum, the streamlines parallel and the pressure atmospheric.

(*d*) the opening of the orifice itself.

8. Arrange the hydraulic coefficients of orifice in increasing order of magnitude :

(*i*) C_v (*ii*) C_c (*iii*) C_d.

9. A mouthpiece and an orifice, both of the same diameter *d*, are discharging under the same head *H*. The discharge through the mouthpiece will be :

(*a*) the same as that throuth the orifice

(*b*) more than the discharge through the orifice

(*c*) less than that through the orifice.

10. The discharge through a sharp-crested rectangular weir is given by :

(*i*) $C_d B \sqrt{2gH}$ (*ii*) $\frac{2}{3} C_d B \sqrt{2gH}$

(*iii*) $\frac{2}{3} C_d B \sqrt{2g}\ H^{3/2}$ (*iv*) $C_d B \sqrt{2g}\ H^{3/2}$.

11. As compared to a rectangular weir, a triangular weir measures low discharges more accurately, the discharge through the latter being equal to :

(*a*) $\frac{8}{15} C_d \sqrt{2g}\ \tan \theta\ H^{5/2}$ (*b*) $\frac{8}{15} C_d \sqrt{2g}\ \tan \theta\ H^{3/2}$

(*c*) $\frac{8}{15} C_d \tan \theta \sqrt{2gH}$ (*d*) $\frac{8}{15} C_d B \tan \theta\ H^{5/2}$.

12. A cipolletti weir is :

(*i*) a rectiangular weir with sharp edges

(*ii*) a high triangular notch

(*iii*) a trapezoidal notch with sides sloping at 45°

(*iv*) a trapezoidal notch with sides inclined at 1 *H* : 4 *V*.

13. A submerged weir is one in which :

(*a*) the water on the downstream side is below the crest level

(*b*) the water level downstream just touches the crest

(*c*) the water level downstream is above the crest level

(*d*) the water level downstream stands at the same level as that on the upstream side.

14. A weir is said to be a broad crested one when :

(*i*) the crest has a sharp edge

(*ii*) the width of the crest is greater than (2/3) *H* so that the sheet of water is supported by the crest

(*iii*) the crest width is less than (2/3) *H*

(*iv*) the crest width is equal to (2/3) *H*.

15. The head over a 90° *V*-notch weir increases from 0.15 to 0.3 m. The ratio of the new discharge to the original discharge is :

(*a*) 1.414 (*b*) 2.000 (*c*) 4.000 (*d*) 5.657.

16. The time taken for a tank (filled to a height h above its flat base) to empty through an orifice in the base varies as the following power of h :

(a) 1 (b) $\frac{1}{2}$ (c) $-\frac{1}{2}$ (d) -1.

ANSWERS

1. (iii) 2. (a) (ii) ; (b) (iii) 3. (iii) 4. (ii), (i), (iii)
5. (a) (v) ; (b) (iii) ;(c) (iv) ; (d) (ii) ; (e) (i) 6. (ii) 7. (c)
8. (iii), (ii), (i) 9. (b) 10. (iii) 11. (a) 12. (iv) 13. (c)
14. (ii) 15. (d) 16. (b).

REVIEW QUESTIONS

1. What are the various methods for measuring flow in pipelines ? Discuss their advantages and disadvantages.
2. Distinguish between a (i) Pitot tube (ii) Pitot-Static tube. What is the advantage of using Prandtl tube over the other types ?
3. Which type of Pitot-tube would you think will be suitable for velocity measurement in open channels ?
4. If a Pitot tube is directed (i) upstream (ii) downstream (iii) sideways in a fluid stream, indicate the liquid level that will rise in the tube in each case.
5. Explain the working of an orificemeter for measuring pipe flow. What factors mainly govern its coefficient of discharge. Do you think it is advisable to calibrate it in place. If so, why ?
6. Explain the advantages and disadvantages of each of the following :
   (i) Venturimeter, (ii) Nozzle meter, and (iii) Orificemeter. What factors account for a higher coefficient of discharge in a venturimeter as compared to the other two meters ?
7. Under what circumstances will an orifice function as a large orifice ?
8. What is a standard orifice? Under what conditions will an opening made in a thick wall behave as a sharp crested orifice ?
9. What arrangements would you suggest to increase the discharge without increasing the size of the orifice for the same head ?
10. Why should the coefficient of velocity be less for mouthpieces as compared to orifices ?
11. Under what conditions will a re-entrant mouthpiece flow full ? What is Borda's mouthpiece ?
12. What are the various methods for measuring flow in open channels ? Discuss the limitations of each.
13. Differentiate between a sharp-crested and a broad-crested weir. Under what conditions of flow may a broad-crested weir function as a sharp-crested one ?
14. What factors are taken care of by the coefficient of discharge ? How does it vary with the weir height and head over it ?
15. What do you understand by (i) a suppressed weir, and (ii) a contracted weir ? Why is the ventilation of suppressed weirs necessary ?
16. What are the advantages of a triangular weir over a rectangular one ? Which one is better suited for a wide range of discharge variation ?
17. What is a cipolletti weir and what is its chief advantage ?
18. What do you mean by (i) free flow, and (ii) submerged flow ? Express the submerged flow as a function of free flow over the weir and submergence ratio.
19. Briefly describe how discharge measurement could be made in a rectangular channel.
20. Briefly describe how you would use the surface slope method (slope-area method), if you were assigned the job of determining the magnitude of a recent but unmeasured flood.

21. You are required to determine flow in a wide river during a monsoon day. Outline a procedure that will be followed using a current meter.

22. Differentiate between orifice flow and sluice flow.

23. Mention the names of the various head meters available for measuring rate of flow through a pipe. State the circumstances under which each device can be advantageously used.

PROBLEMS

13.1. An orifice of area 30 cm^2 located in a vertical plate has a head of 1.1 metre of oil of sp. gr. 0.92. It discharges 6023 kg of oil in 74.8 sec. Trajectory measurememts yield x = 2.15 m, y = 1.24 m. Determine C_c, C_v and C_d.

13.2. Select the size of orifice that permits a tank of horizontal cross-section 1.5 m^2 to have the liquid surface drawn at the rate of 18 cm/sec for 3 m head on the orifice. Take $C_d = 0.63$.

13.3. A 10 cm orifice in the side of a 1.8 m diameter tank draws the surface down from 2.25 to 1.2 m above the orifice in 82.5 sec. Calculate the discharge coefficient.

13.4. Determine the dimensions of a tank such that the surface velocity varies inversely as the distance from the centreline of an orifice draining the tank. When the head is 0.30 m, the velocity of fall of the surface is 2 m/minute and the orifice diameter is 1.5 cm. Take $C_d = 0.65$.

13.5. How long does it take for the water surface in the tank shown in the accompanying sketch to rise by 2 metre ? The left hand surface represents a large reservoir of constant water-surface elevation.

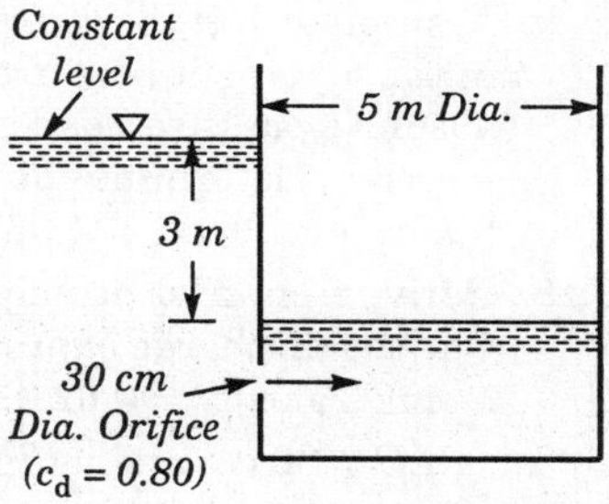

Fig. Prob. 13.5

13.6. A large tank is filled with a liquid of mass density ρ to a height h above a small opening known as orifice. The area of the orifice is A, the coefficient of contraction is C_c, and the coefficient of velocity C_v. Find the force on the tank due to the discharging jet.

13.7. Water discharges through a well rounded frictionless orifice under a head H. The jet impinges on a large vertical plate which covers the orifice of a second tank in which the water stands to a heights h above the orifice. Both the orifices have the same area and are at the same elevation. Find the relation between the two heads, h/H, such that the force on the plate due to the jet is just sufficient to balance the hydrostatic force on the plate. The water level in the first tank may be considered constant.

13.8. In a given waterfall the quantity of water was measured by a rectangular notch above the fall. The notch had a length of 4 m and mean head over the notch was measured as 50 cm. The velocity of approach at a point, where the head was measured was 3 km/hr, compute discharge in litres/minute. Take $C_d = 0.6$.

13.9. The dimensions of a swimming pool with a horizontal floor are : length 20 m, breadth 10 m and depth 1.80 m. The pool is full. Find the time taken to lower its level by 0.60 m and also the time to empty the pool through a rectangular orifice at the bottom of the pool having an area of 0.25 sq. metre ; coefficient of discharge is 0.6.

13.10. A trapezoidal channel with a bottom width of 1.5 m and side slopes of 1 horizontal to 2 vertical has a 90° V-notched weir at the outlet end, the vertex of the notch being 45 cm above the bottom of the channel. The water upstream has a depth of 1.20 m. Taking $C_d = 0.58$ for the weir, estimate the percentage error in the discharge when the velocity of approach is neglected.

13.11. The centre of an orifice is situated 15 cm above the bottom of a vessel containing water of depth 75 cm. Assuming a coefficient of velocity of 0.98, estimate how far the water in the jet issuing from the orifice in the side of the vessel will fall in moving horizontally a distance 60 cm from the vena contracta. Neglect air resistance.

13.12. Water flows through a circular orifice 2.5 cm in diameter in the side of a tank. The constant head of water above the centre of the orifice is 80 cm. The co-ordinates of the centreline of jet are 30 cm horizontally from the vena contracta and 3 cm vertically below the centreline of the orifice. The discharge from the orifice is 72 litres/minute. Find C_v, C_c and C_d.

13.13. The daily average rainfall in a particular catchment area is 0.20 million cubic metre. If 80% of this rain water reaches a storage reservoir and then passes over the overflow section of the dam 100 metres high and 3.0 metres long. Find

(*i*) the height of water above the crest ;

(*ii*) the theoretical horsepower available for power generation

(Assume $C_d = 0.65$)

(Assume rectangular weir formula to be applicable.)

13.14. A rectangular sharp-crested weir is 1.2 m long. How high should it be placed in a channel to maintain an upstream depth of 1.50 m for a discharge of 0.46 m^3/s ? Take $C_d = 0.64$.

13.15. A submarine submerged in sea water (specific weight 10055 N/m^3) travels at 20 km/hr. Calculate the pressure at the front stagnation point situated 20 m below the sea surface.

13.16. The head upstream of a rectangular weir 2 m long is 10 cm. If $C_d = 0.6$ and the cross-sectional area of the upstream flow is 0.3 m^2, estimate to a first approximation the discharge allowing for the velocity of approach.

13.17. A convergent-divergent mouthpiece is fitted into the vertical side of a tank containing water. Assuming that there are no losses in the convergent part of the mouthpiece, and that the losses in the divergent part are equivalent to 0.2 times the velocity head at exit and that the minimum absolute pressire head at the throat is 2.44 m of water for a barometric pressume of 760 mm of mercury, determine the throat and exit diameters of the mouthpiece when the discharge is 4.25 lit/sec for a head of 1.52 metre.

13.18. Derive from first principles a discharge formula for a 90° V notch in the form $Q = KH^n$ in which Q is the discharge expressed in lit/sec and H is the head over the vertex of the V-notch expressed in cm. Take $C_d = 0.62$. State the assumptions in deriving the formula.

If $H = 36$ cm and a probable error of 2 mm is made in reading H, find the percentage accuracy of the value of discharge which may be obtained by the above formula.

13.19. A reservoir has an area of 90,000 m^2 and is provided with a Francis weir, 5 m long with end contractions.

(*i*) If the initial reservoir level is 1 m above the sill of the weir how long will it take for the water-level to fall to 50 cm above the sill ?

(*ii*) What would be the time required for the water level to reach the sill level ?

ANSWERS

13.1. $C_c = 0.695$, $C_v = 0.922$, $C_d = 0.641$

13.2. 26.7 cm

13.3. $C_d = 0.72$

13.4. $r = 0.1273\ y^{3/4}$

13.5. 114.7 sec.

13.6. $\rho C_c C_v^2 A\ 2gh$

13.7. $h/H = 2$

13.8. 16.35×10^7 lit/min.

13.9. 147.3 sec, 807 sec.

13.10. 1.33%

13.11. 15.8 cm

13.12. $C_c = 0.637$, $C_v = 0.913$, $C_d = 0.582$

13.13. 0.47 m, 2474 h.p.

13.14. 1.16 m

13.15. 216.8 kN/m^2 (gauge)

13.16. 0.1242 m^3/s

13.17. 2 cm, 3.3 cm

13.18. $Q = 0.0147\ H^{5/2}$, error 1.39%, accuracy 98.61%.

13.19. 2.265 hr, ∞

14

Compressible Fluid Flow

14.0. Introduction

It is an established fact that liquids do not change appreciably their volume when subjected to changes in pressure and temperature and that they can be treated incompressible (or constant density fluids) for all practical purposes. On the other hand, the density of gases and vapours varies with both pressure and temperature and hence they are compressible. However, there are many examples of flow of gases in which density does not change appreciably, and the theory relating to constant density fluids adequately describes the flow phenomena for such cases.

When the pressure changes produced by flow are large, the assumption of fluid being incompressible is no longer valid and the corresponding changes in density must be considered. It is known through experience that the compression and expansion of a gas involve work done on and by the gas respectively, and this results in changes in the temperature of the gas. This introduces thermodynamic effects in fluid flow problems, and hence the laws of fluid mechanics alone are insufficient for complete solution of compressible flow problems. In addition, the principles of thermodynamics are necessary to completely determine the quantities involved in a flow problem. The study of the so-called 'compressible fluid flow' is thus a lot more complex than that of constant-density fluid flow. In this chapter an attempt has been made to introduce certain basic elements of compressible flow.

There is no sharp dividing line between flows in which the density changes are important and those in which they are not so important. Appreciable changes in density of a gas may be expected : (1) if the velocity (either of the gas itself or of a body moving through it) approaches or exceeds the velocity of sound through the gas, (2) if the gas is subjected to sudden acceleration or (3) if there are large changes in elevation. The last condition is rarely encountered except in meteorology and will not be considered in this chapter.

14.1. Thermodynamic Concepts

14.1.1. Equation of State

A perfect gas is defined as the fluid that has constant specific heats and follows the law given by Eq. 1.12

$$p = \rho RT \qquad ...(1.12)$$

in which p and T are absolute pressure and the absolute temperature respectively, ρ is the density and R the gas constant. Eq. (1.12) is known as the *equation of state* for a perfect gas, and the gas which obeys it is called a perfect gas.

14.1.2. Internal Energy and Enthalpy

The molecular energy of a compressible fluid is caused by the activity of the molecules, which increases as the temperature of the fluid increases. In a gas, this molecular activity also creates the pressure which is that part of molecular energy usually converted into mechanical work. In thermodynamics, the molecular energy is known as the enthalpy

$$h = u + pv = u + p/\rho \qquad ...(14.1)$$

in which h is the enthalpy or molecular energy per unit mass, v is the specific volume, u is the internal energy per unit mass which is that part of the molecular energy other than the pressure energy per unit mass p/ρ. The internal energy is the energy associated with the kinetic energy of molecules and the forces between them and depends upon the temperature ; high and low temperatures implying high or low internal energies respectively.

14.1.3. Specific Heats

The specific heat C is defined as the quantity of heat needed to raise a unit mass of fluid by a unit temperature,

$$C = \frac{dq}{dT} \qquad ...(14.2)$$

where dq is the heat added to a unit mass of the system and dT is the rise in temperature associated with it.

The value of the specific heat depends upon the process by which the heat is added. We are particularly interested in specific heats determined at constant volume C_v and at constant pressure C_p. Thus,

$$C_v = \left(\frac{dq}{dT}\right)_{constant\ volume} \qquad ... (14.2\ a)$$

$$C_p = \left(\frac{dq}{dT}\right)_{constant\ pressure} \qquad ...(14.2\ b)$$

14.1.4. Thermodynamic Processes of Perfect Gas

(1) *Isothermal Process.* The compression and expansion of a gas may take place according to various laws of thermodynamics. If the temperature is held constant, the process is known as 'isothermal' and the pressure-density relationship is given by Boyle's law, Eq. (1.13)

$$\frac{p}{\rho} = \text{constant} \qquad ...(1.13)$$

(2) *Adiabatic Process.* If the process is such that no heat is added to or withdrawn from the gas (heat transfer is zero), it is said to be an 'adiabatic' process.

If the process is reversible (or frictionless) adiabatic, it is called *isentropic* since it is accompanied by no change in entropy, and the following relationship between pressure and density, given by Eq. (1.14), exists

$$\frac{p}{\rho^k} = \text{constant} \qquad ...(1.14)$$

where $k = C_p/C_v$ is the ratio of specific heats.

(3) *Polytropic Process.* In general we can express a relation between the pressure and the density for each of the preceding process by

$$\frac{p}{\rho^n} = \text{constant} \qquad ...(14.3)$$

where n is a given constant for each process. Thus

(*i*) if $n = 0$, p = constant, the process is isobaric

(*ii*) if $n = 1$, T = constant, the process is isothermal

(*iii*) if $n = k$, Entropy = constant, the process is isentropic.

14.1.5. Reversible and Irreversible Processes

When the physical properties of a gas (*e.g.* pressure, density and temperature) are changed, it is said to undergo a process. The process is said to be reversible if the gas and its surroundings could subsequently be completely restored to their initial conditions by adding to (or extracting from) the gas exactly the same amount of heat and work taken from (or added to) it during the process.

A reversible process is an ideal one which is never achieved in practice. Viscous effects and friction dissipate mechanical energy as heat which cannot be converted back to mechanical energy without further changes occurring. Also heat passes by conduction from hotter to cooler parts of the system considered, and heat flow in the reverse direction is not possible. In practice, therefore, all processes are irreversible.

14.1.6. First Law of Thermodynamics

The principle of conservation of energy, which is generally referred to as the first law of thermodynamics, states that energy can neither be created nor destroyed but can only be changed in form. Experiments have demonstrated that heat is a form of energy which can be expressed in terms of mechanical units through the mechanical equivalent of heat and *vice versa*.

If a small quantity of heat is added to a simple homogeneous system (a system composed of a single gas or liquid whose thermodynamic properties are uniform throughout in all respects), this energy by the first law of thermodynamics can be converted into other forms of energy, such as an increase in the kinetic energy of molecules (the internal energy), the kinetic energy of the system proper, and external mechanical work done by the system. It is assumed that the system is stationary or nearly stationary, so that change in kinetic energy of the system need not be considered. Under these conditions the first law of thermodynamics states that a small quantity of heat added to any simple system is equal to the change in kinetic energy of molecules plus the internal mechanical work done by the system. Thus

$$dq = du + dw \qquad ...(14.4)$$

where dq is the amount of heat added to a unit mass of the system, du is the increase in the internal energy per unit mass of the system, and dw is the external work done per unit mass by the system. If p is the pressure and v the volume per unit mass, then Eq. (14.4) may be written in the form

$$dq = du + p\,dv = du + pd\left(\frac{1}{\rho}\right) \qquad ...(14.4\ a)$$

The relations among the specific heats, C_p, C_v, their ratio k, and the gas constant R are given by

$$C_p - C_v = R \qquad ...(14.5)$$

$$C_p = \frac{k}{k-1}R \qquad ...(14.6)$$

$$C_v = \frac{k}{k-1}R \qquad ...(14.7)$$

14.1.7. Energy

Entropy of a gas may be defined as the measure of availability of heat energy for transformation into mechanical work. If dq is the heat given to the fluid per unit mass, and s is the entropy per unit mass of fluid, then the change in entropy per unit mass due to heat absorbed by the fluid is given by

$$ds = \frac{dq}{T} \qquad ...(14.8)$$

where T is the absolute temperature of fluid and ds is the change in entropy per unit mass of fluid.

If dq_e is the heat added externally to the fluid per unit mass and dq_i is the heat developed internally per unit mass because of dissipation of mechanical energy into heat due to friction, then the total heat received by the fluid is

$$dq = dq_e + dq_i \qquad ...(14.9)$$

From Eqs. (14.8) and (14.9), we obtain

$$ds = \frac{dq_e}{T} + \frac{dq_i}{T} \qquad ...(14.10)$$

Constant entropy ($ds = 0$) requires $dq = 0$, a condition which is achieved if no heat passes between the fluid and its surroundings, *i.e.* $dq_e = 0$, and no mechanical energy is converted into heat energy by friction, *i.e.* $dq_i = 0$. In a frictionless process, which is also called a reversible process, $dq_i = 0$. The process, in which the fluid does not absorb heat from, or give heat to its surroundings, is said to be *adiabatic*. Hence for an adiabatic process $dq_e = 0$. A process in which the entropy does not change (*i.e.* $ds = 0$) is known as *isentropic*. A frictionless adiabatic process is, therefore, isentropic. In practice such a process is closely approximated if there is little friction and the changes occur rapidly enough for little heat transfer to take place across the boundaries.

If the heat added from external sources is equal to zero ($dq_e = 0$), the process is adiabatic, and from Eq. (14.10) we obtain

$$ds = \frac{dq_i}{T} > 0 \qquad ...(14.11)$$

Eq. (14.11) shows that an isentropic process is not equivalent to an adiabatic process.

Since the loss of mechanical energy into heat always corresponds to a positive heat of dissipation ($dq_i > 0$), it is evident that entropy always increases in an adiabatic process ($dq_e = 0$).

14.1.8. Second Law of Thermodynamics

The second law of thermodynamics which, like the first law, is the result of observation and experiment may be stated in the following manner :

1. Heat cannot be transferred from a body of lower temperature to one of higher temperature without other simultaneous changes occurring in the two systems or their environments.

2. Heat from a single source cannot be converted into mechanical work without simultaneous changes in the system or the environment.

3. The transfer of mechanical work into heat by friction is irreversible.

4. In an isolated system (no heat transfer), the entropy cannot decrease. This can be easily seen from Eq. (14.11). On the other hand, the entropy is always increased if the process is irreversible.

14.2. Fundamental Equations Governing Compressible Fluid Flow

In the study of incompressible fluid flow we need only to find the velocity and pressure at every point in space. In case of flow of a compressible fluid we are interested in determining the velocity, pressure, density and temperature of fluid. In order to determine these quantities, one vector and three scalars, we must have one vector equation and three scalar equations. These equations are provided by :

1. Equation of motion,
2. Equation of continuity,
3. Equation of state for a perfect gas, and
4. Energy equation.

14.2.1. Equation of motion

The equation of motion along a streamline for compressible fluid flow are the same as developed for incompressible fluid in chapter 4 article 4.4.

14.2.2. Equation of continuity

The equation of continuity in cartesian co-ordinates for fluid flow in general is derived in chapter 3 article 3.7, which for compressible fluids may be expressed as

$$\frac{\partial \rho}{\partial t} + \frac{\partial(\rho u)}{\partial x} + \frac{\partial(\rho v)}{\partial y} + \frac{\partial(\rho w)}{\partial z} = 0 \qquad ...(3.26)$$

The mass rate of flow along a narrow stream tube may be expressed by Eq. (3.15) as

$$\rho AV = \text{constant} \quad \text{...(3.15)}$$

differentiating Eq. (3.15) and dividing throughout by ρAV, we obtain

$$\frac{d\rho}{\rho} + \frac{dA}{A} + \frac{dV}{V} = 0 \quad \text{...(14.12)}$$

The equation of state for a perfect gas is expressed by Eq. (1.12).

14.2.3. Energy Equation for Compressible Fluids

Although the momentum equation is completely independent of compressibility effects, the energy equation is very much dependent upon variation in density. The general energy equation for the steady flow of any fluid was presented in chapter 4 article 4.10. This equation may be applied to any two points along a streamline. If no heat is added to (or extracted from) the fluid between these two points, and no mechanical work is done, we may put $q = 0$, and $w_s = 0$ in Eq. (4.21) and obtain

$$0 = \left(\frac{V_2^2}{2} + \frac{p_2}{\rho} + gZ_2\right) - \left(\frac{V_1^2}{2} + \frac{p_1}{\rho} + gZ_1\right) + u_2 - u_1 \quad \text{... (14.13)}$$

Since enthalpy per unit mass is given by Eq. (14.1) as $h = u + p/\rho$ and as points 1 and 2 were arbitrarily chosen, Eq. (14.13) may be expressed as

$$h + \frac{V^2}{2} + gZ = \text{constant} \quad \text{...(14.14)}$$

Eq. (14.14) expresses the general form of the energy equation for steady, adiabatic flow in which the fluid (gas, liquid or vapour) neither does work on its surroundings nor has work done on itself. If, however, the fluid is a perfect gas, then the enthalpy $h = C_pT$ and we obtain

$$C_pT + \frac{V^2}{2} + gZ = \text{constant} \quad \text{...(14.15)}$$

From the equation of state for a perfect gas,

$$p = \rho RT \quad \text{or} \quad T = \frac{p}{\rho R} = \frac{p}{\rho(C_p - C_v)}$$

Substituting for T in Eq. (14.15)

$$\frac{C_p}{C_p - C_v}\frac{p}{\rho} + \frac{V^2}{2} + gZ = \text{constant} \quad \text{...(14.16)}$$

Using $C_p/C_v = k$, the above equation takes the form

$$\frac{k}{k-1}\frac{p}{\rho} + \frac{V^2}{2} + gZ = \text{constant} \quad \text{...(14.17)}$$

The potential energy in a compressible fluid exists because of the elevation of the fluid. If the fluid is a gas, the potential energy term will be negligible as compared to other forms of energy because of very low specific weight of the gas. The potential energy per unit mass gZ may, therefore, be omitted from Eq. (14.17), and the resulting energy equation is,

$$\frac{k}{k-1}\frac{p}{\rho} + \frac{V^2}{2} = \text{constant} \quad \text{...(14.18 } a\text{)}$$

and Eq. (14.14) becomes,

$$h + \frac{V^2}{2} = \text{constant} \quad \text{...(14.18 } b\text{)}$$

Eq. (14.18 a) is more useful when expressed in terms of temperature ; using $p = \rho RT$, one gets

$$\frac{k}{k-1}RT + \frac{V^2}{2} = \text{constant} \quad \text{...(14.19)}$$

It may be noted that the above equations are applicable to steady adiabatic flow in which the gas neither does work on its surroundings nor has work done on itself.

14.3. Controlling Parameters in Compressible Flow

There are four parameters which control the flow phenomena of viscous compressible fluids. They are : (1) the Mach number, (2) the ratio of specific heats, (3) the Reynolds number, and (4) the Prandtl number.

The Mach number is a measure of the effect of fluid compressibility and is defined as the ratio of the free-stream velocity (or the velocity of the body moving through the fluid) and the corresponding velocity of sound in the fluid medium and is expressed as

$$\mathbf{M} = \frac{V}{C} \qquad \text{...(14.20)}$$

in which V is the velocity of fluid (or of the moving body) and C is the velocity of sound in the fluid medium. For isentropic flow from Eqs. (1.18) and (1.19)

$$C = \sqrt{kp/\rho} = \sqrt{kRT} \qquad \text{...(14.21)}$$

The second parameter is the ratio of specific heats of the fluid at constant pressure to that at constant volume, and is expressed by

$$k = \frac{C_p}{C_v} \qquad \text{...(14.22)}$$

which is a measure of the relative internal complexity of the fluid molecules.

The third parameter is the familiar Reynolds number which is a measure of the viscous effects of the fluid. The fourth parameter is the ratio of the kinematic viscosity to the thermal diffiusivity of the fluid and is known as the Prandtl number

$$\mathbf{P} = \frac{\mu/\rho}{K/\rho C_p} \qquad \text{...(14.23)}$$

in which **P** is the Prandtl number K is the thermal conductivity. The Prandtl number is a measure of the relative importance of heat conduction and viscosity of a fluid.

In order to have compressible flows around two bodies completely similar not only must the bodies be geometrically similar, but also the four non-dimensional parameters listed above namely the Mach number, ratio of specific heats, Reynolds number and the Prandtl number in the flow systems be equal. For flow of an inviscid (non-viscous) compressible fluid, the Reynolds number and the Prandtl number are out of consideration and we are thus left with the remaining two controlling parameters, namely the Mach number and the ratio of the specific heats.

14.4. Compressible Flow Regimes

As already stated, one of the most important non-dimensional flow parameters occurring in the analysis of compressible flow is the Mach number **M**. Based on the value of the Mach number, five flow regimes are generally defined as follows :

1. ***Incompressible flow regime.*** The Mach number is small as compared to unity ($0 < \mathbf{M} < 0.3$) for a perfect gas. In this range the effects of compressibility are generally considered to be negligible.
2. ***Subsonic flow regime.*** The Mach number is less than unity but of sufficient magnitude to be outside the incompressible flow regime ($0.3 < \mathbf{M} < 1.0$).
3. ***Transonic flow regime.*** The Mach number is in the immediate neighbourhood of unity, that is, it ranges from values slightly less than one to slightly greater than one ($0.9 < \mathbf{M} < 1.1$).
4. ***Supersonic flow regime.*** The Mach number is larger than one ($\mathbf{M} > 1$).

5. ***Hypersonic flow regime.*** The Mach number is very much larger than one (**M** >> 1). The dividing line between the supersonic and hypersonic regimes may be taken as approximately 5.

14.5. Mach Cone, Mach Line and Shock Wave

In chapter 11, article 11.10, we have discussed that a small pressure disturbance (or pressure impulse) in a stationary compressible fluid propagates with the speed of sound uniformly in all directions. Hence it may be stated that the pressure disturbances are propagated as spherical wave surfaces.

Consider a tiny object (such as a small projectile) moving from right to left in a stationary fluid at a velocity smaller than the velocity of sound ($0 < V < C$). The motion of the object creates pressure disturbances which are propagated spherically outward from the object with the speed of sound C. If the object was not in motion, the wave fronts would spread out spherically and would have the positions shown in Fig. 14.1 (*a*) for successive time intervals $dt = (t_2 - t_1) = (t_3 - t_2)$.

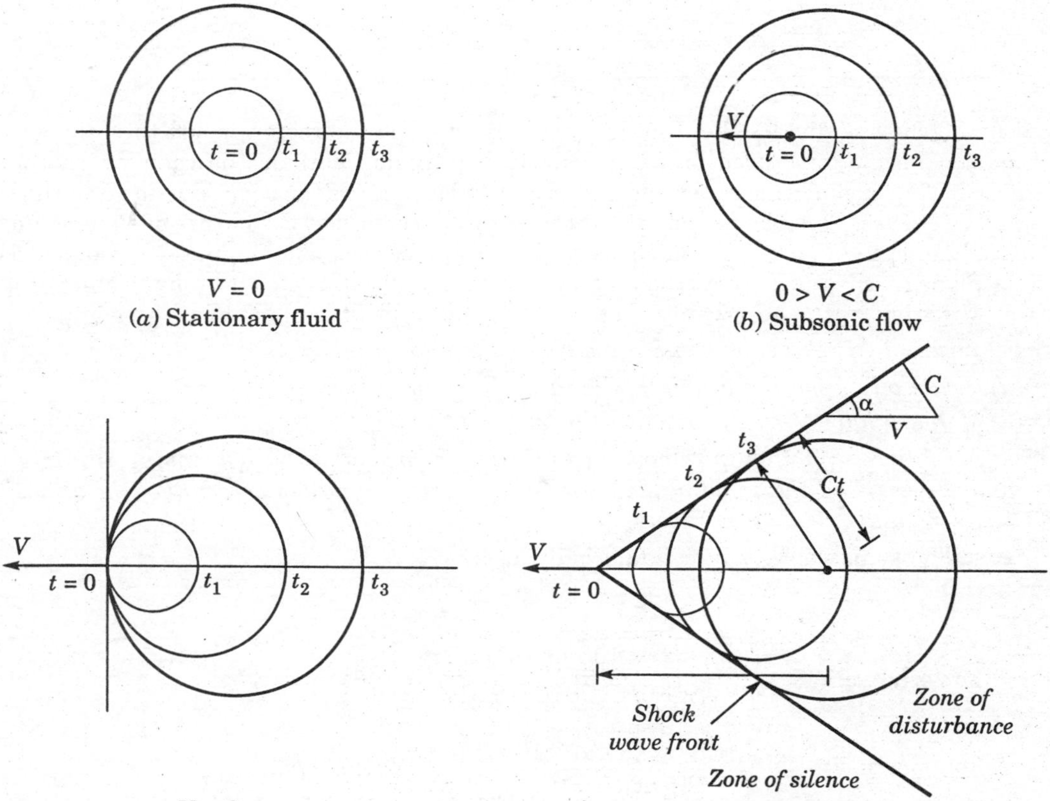

$V = 0$
(*a*) Stationary fluid

$0 > V < C$
(*b*) Subsonic flow

$V = C$
(*c*) Transonic flow

$V > C$
(*d*) Supersonic flow

Fig. 14.1. Wave patterns for a tiny object moving with different velocities.

The position of wave fronts for $0 < V < C$ is shown in Fig. 14.1 (*b*). The portion of the wave in front of the object progresses more slowly than for Fig. 14.1 (*a*), the velocity now being $(C - V)$. If the velocity of object is increased to the speed of sound C, so that $V = C$, see Fig. 14.1 (*c*), the waves do not move ahead of the object but appear to remain stationary. The wave fronts merge to form a plane wave front which is tangential to the upstream side shown, and the downstream wave moves at a speed of $(V + C)$. In this case, the pressure wave is unable to move upstream against the oncoming flow and the fluid ahead of this plane wave front is not influenced by the motion of body. Finally, when the velocity of the object is increased beyond the speed of sound so that $V > C$ and $\mathbf{M} > 1$, the individual pressure waves now combine to form a conical wave front. This conical pattern of wave fronts is known as the *Mach cone*. Fluid ahead of the cone is undisturbed but suddenly undergoes changes in pressure, temperature and density, as it passes through the Mach cone. An actual wave through which the flow experiences sudden changes in density, pressure and temperature is called a *shock wave*. We have thus seen that the effect of pressure disturbance is restricted to the interior of the envelope of the spherical pressure waves which is a cone whose vertex coincides with the source of disturbance (the assumed small projectile) and whose semi-vertex angle known as the Mach angle, is given by Eq. (10.39).

$$\sin \alpha = \frac{V}{C} \qquad \text{...(10.39)}$$

In two dimensional flow, the Mach cone reduces to a pair of intersecting lines each of which is called *Mach line* or *Mach wave*. From Eq. (10.39), it is evident that the Mach number,

$$\mathbf{M} = \frac{V}{C}$$

$$= \frac{1}{\sin \alpha} \text{ coses } \alpha \qquad \text{...(14.24)}$$

The Mach cone divides the fluid into two regions : (1) the region inside the cone which is disturbed by the motion of the projectile and (2) the region outside the cone which remains undisturbed. The former region was referred to by von Karman as the *zone of action* and the latter as the *zone of silence*. This phenomenon of shock wave characterizes the nature of supersonic flow. If an aeroplane is flying at a subsonic speed, the noise created by its engines can be heard long before the plane reaches directly overhead ; however, the situation created by a supersonic aeroplane is entirely different as the observer cannot hear the noise until the plane has crossed him.

Example 14.1. *Find the sonic velocity for the following fluids :*

(i) Crude oil of specific gravity 0.8 and bulk modulus of 1530 MN/m².

(ii) Mercury having a bulk modulus of 26487 MN/m².

(b) A schlieren photograph showing the wave formed by a bullet moving in air gave a Mach angle of 40°. Find the speed of the bullet if the pressure and temperature of the atmosphere are 88.3 KN/m² and – 2°C respectively.

Ans. The sonic velocity in a liquid is given by

$$C = \sqrt{k/\rho}$$

(*i*) For crude oil, $$C = \sqrt{\frac{1530 \times 10^6}{[(0.8 \times 9810)/9.81)]}} = \mathbf{1383\ m/s}$$

(*ii*) For mercury, $$C = \sqrt{\frac{26487 \times 10^6}{[(13.6 \times 9810/9.81)]}} = \mathbf{1395.5\ m/s.}$$

(*b*) The sonic velocity in air under the given conditions is :

$$C = \sqrt{kRT} = \sqrt{1.4 \times 287 \times (273 - 2)} = 329.98 \text{ m/s}$$

Mach angle represents the half angle of the Mach cone, which is related to the Mach number by Eq. (14.24)

$$\mathbf{M} = \frac{V}{C} = \frac{1}{\sin \alpha} \qquad \therefore \quad V = \frac{C}{\sin \alpha} = \frac{329.98}{\sin 40°} = \mathbf{513.35\ m/s}$$

The speed of the bullet is 513.35 m/s.

Example 14.2. *A supersonic aircraft flies at an altitude of 2000 m where the air temperature is 3°C. Determine the speed of the aircraft if its sound is heard 4.5 seconds after its passage over the head of an observer.*

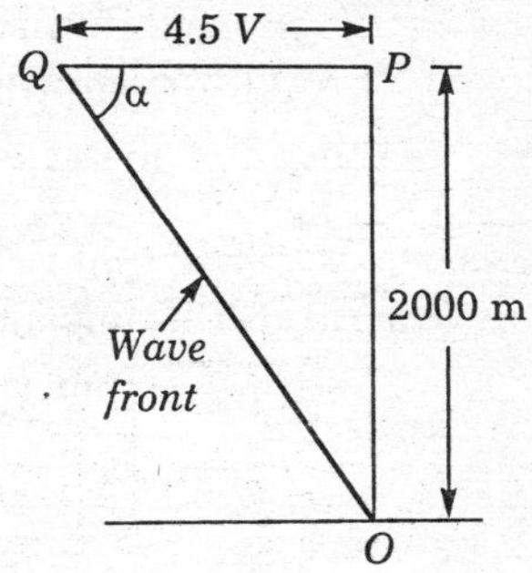

Solution. With reference to the figure, let O represent the observer and P the position of aircraft just vertically over the observer. After 4.5 seconds, the aircraft reaches a position represented by Q. Line OQ represents the wave front and α the Mach angle.

The velocity of sound in air

$$C = \sqrt{kRT} = \sqrt{1.4 \times 287 \times (273+3)} = 333 \text{ m/s}$$

From the figure,

$$\tan \alpha = \frac{2000}{4.5V} \quad \text{...(1)}$$

but

$$\mathbf{M} = \frac{V}{C} = \frac{1}{\sin \alpha} \quad \text{or} \quad V = \frac{C}{\sin \alpha}$$

Inserting the value of V in (1),

$$\tan \alpha = \frac{2000 \sin \alpha}{4.5 \times 333} \quad \text{or} \quad \cos \alpha = 0.7492$$

$$\sin \alpha = \sqrt{1-\cos^2 \alpha} = 0.6623$$

$$\therefore \quad V = \frac{333}{0.6623} = \mathbf{502.79 \text{ m/s.}}$$

The speed of aircraft is 502.79 m/s.

Example 14.3. *A cruise missile under test is moving horizontally at Mach number 2.0 in the atmosphere at an elevation of 305 m above the earth's surface. How long does it take for an observer on the ground to hear disturbance from the instant it is directly over head ? Suppose an observer in a plane is moving in the same direction as the missile at a speed of 1/2 speed of sound at an elevation of 305 m. Above the missile. What is the time elasped between the instant when the missible is directly below and the instant when the observer hears the sound. Neglect changes of speed of sand from 305 m to 610 m elevation.*

Solution. Assuming the air temperature at the ground level as 20°C, the velocity of sound in air

$$C = \sqrt{kRT} = \sqrt{1.4 \times 287 \times (273+20)} = 343.11 \text{ m/s}$$

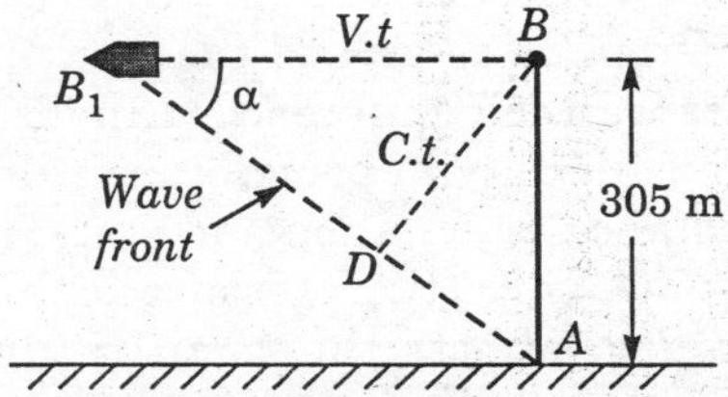

Fig. (a) Observer stationary at A

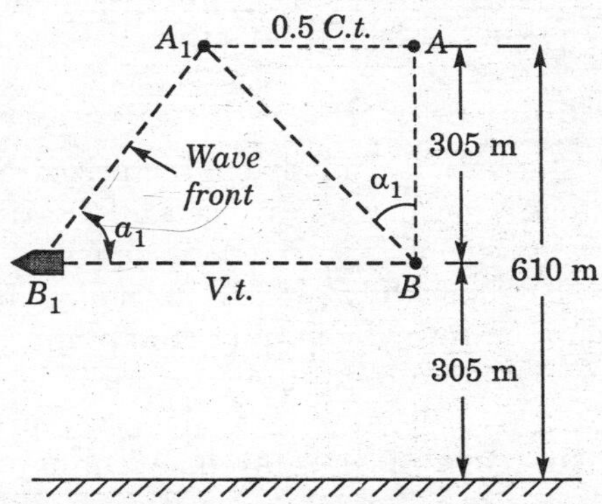

Fig. (b) Observer moving in a plane at 610 m elevation

Mach No., $(\mathbf{M}) = \dfrac{\text{Speed of missible } (V)}{\text{Velocity of sound } (C)}$

$\therefore$ Speed of missile $V = 2.0 \times 343.11 = 686.23$ m/s.

(*a*) **When the observer is stationary at the ground surface :**

Let the time taken by the observer to hear the disturbance (*i.e.* the noise created by the missile) be t. The missile moves from B to B_1 in this time, during which the sound wave with velocity c travels and the wave-front reaches the observer at A. It is then the observer hears the sound of the missile. Using ΔDBB_1

$$\sin \alpha = \frac{DB}{BB_1} = \frac{C.t}{V.t} = \frac{C}{V} = \frac{343.11}{680.23}\frac{1}{2} = 0.5$$

$$\therefore \quad \angle\alpha = 30°,\ \angle DAB = 60°,\ \sin 60° = \frac{C.t}{305}$$

from which, $$t = \frac{305 \sin 60°}{343.11} = \mathbf{0.77\ seconds.}$$

(*b*) **The observer is moving in plane in the same direction as the missile.**

Figure (*b*) shows the position of the observer A travelling in a plane in the same direction, the missile represented by B happens to be vertically below the observer at the given instant. In time t the wave front reaches the observer as both move to their new positions A_1 and B_1. The Mach angle of the wave-front generated by the missile can be obtained as :

Using ΔBA_1B_1, $$\sin \alpha_1 = \frac{BA_1}{BB_1} = \frac{c.t}{V.t} = \frac{C}{V} = \frac{343.11}{686.23} = 0.50$$

$$\therefore \quad \alpha_1 = 30°,\ \angle A_1BA = \alpha_1 = 30°$$

From ΔABA_1, $$\tan \alpha_1 = \frac{0.5\,Ct}{305}, \quad \text{or } t = \frac{305 \tan 30°}{0.5 \times 343.11} = \mathbf{1.026\ seconds}$$

Therefore, the desired time elapsed = 1.026 seconds.

14.6. Pressure Equation or Bernoulli's Equation

Eq. (14.18 *a*) is called the pressure equation for steady irrotational flow. If p_0 and ρ_0 are the pressure and density respectively at a point where the fluid is at rest, Eq. (14.18 *a*) may be expressed as

$$\frac{k}{k-1}\frac{p}{\rho} + \frac{V^2}{2} = \frac{k}{k-1}\frac{p_0}{\rho_0} \qquad \text{...(14.25)}$$

Eq. (14.25) is generally referred to as the pressure equation or the Bernoulli's equation for compressible adiabatic flow.

14.7. Stagnation Point and its Properties

A stagnation point is characterized by existence of zero velocity at that point ; we know that a stagnation point is caused at the tip of Pitot-tube. When a Pitot-static tube is used to measure the velocity of a constant-density fluid, the stagnation pressure and the static pressure need not be separately measured, it is sufficient to measure their difference. A high velocity gas stream, however, may undergo an appreciable change of density in being brought to rest at the front of the Pitot-static tube, and under these circumstances stagnation and static pressure must be separately measured. In a compressible fluid flow, since the pressure, density and temperature are inter-related, a change in pressure also, affects the temperature. At a stagnation point where the velocity is zero, the pressure is known the *stagnation pressure* and the temperature there is called the *stagnation temperature*.

Using the relation between the enthalpy and velocity expressed by Eq. (14.18 *b*), and substituting for enthalpy in terms of temperature ($h = C_pT$), Eq. (14.18 *b*) becomes

$$C_pT + \frac{V^2}{2} = \text{constant} \qquad ...(14.26)$$

At a stagnation point where the velocity is zero, and the pressure, density and temperature are p_0, ρ_0, T_0 respectively. Eq. (14.26) may be written as

$$C_pT + \frac{V^2}{2} = C_pT_0 \qquad ...(14.27)$$

in which T_0 is the stagnation temperature. Dividing Eq. (14.27) by C_pT, we obtain

$$1 + \frac{V^2}{2C_pT} = \frac{T_0}{T} \qquad ...(14.28)$$

we know from Eq. (14.21) that for isentropic flow of a perfect gas $C^2 = kRT$ and from Eq. (14.6) $C_p = kR/(k - 1)$. Eq. (14.28) now takes the form

$$1 + \frac{k-1}{2}\frac{V^2}{C^2} = \frac{T_0}{T} \qquad ...(14.29)$$

and from the definition of Mach number we may write

$$\frac{T_0}{T} = 1 + \frac{k-1}{2}\mathbf{M}^2 \qquad ...(14.30)$$

Thus the ratio of the stagnation temperature to the temperature of the undisturbed stream (which is called the static temperature) is a direct function of the Mach number and k.

For an isentropic flow of a perfect gas we have the following relations :

$$\frac{p}{\rho^k} = \text{constant, and } p = \rho\, RT.$$

eliminating ρ between the two equations, we may write

$$\frac{T_2}{T_1} = \left\{\frac{p_2}{p_1}\right\}^{k-1/k} \qquad ...(14.31)$$

Using Eq. (14.31), the ratio of stagnation pressure p_0 to the undisturbed stream pressure p (known as the static pressure) can be related directly to the temperature ratio T_0/T of Eq. (14.30) :

$$\frac{p_0}{p} = \left(\frac{T_0}{T}\right)^{k/k-1} = \left(1+\frac{k-1}{2}\mathbf{M}^2\right)^{k/k-1} \qquad ...(14.32)$$

Eq. (14.32) shows that the pressure ratio p_0/p is also a function of **M** and k.

Finally, we may find the ratio of the stagnation density ρ_0 and the static density ρ for isentropic flow

$$\frac{\rho_0}{\rho} = \left(\frac{p_0}{p}\right)^{1/k} = \left(1+\frac{k-1}{2}\mathbf{M}^2\right)^{1/k-1} \qquad ...(14.33)$$

Like temperature ratio T_0/T and pressure ratio p_0/p, the density ratio ρ_0/ρ is also a function of the Mach number and the ratio of specific heats. Eqs. (14.30), (14.32) and (14.33) are not valid for supersonic flow because of the shock wave forming at the tip of the Pitot tube and as such the fluid is not brought to rest isentropically.

If the right hand side of Eq. (14.32) is expanded by the binomial theorem*, we obtain

$$\frac{p_0}{p} = 1 + \frac{k}{2}\mathbf{M}^2 + \frac{k}{8}\mathbf{M}^4 + \frac{k(2-k)}{48}\mathbf{M}^6 + ... = 1 + \frac{k\mathbf{M}^2}{2}\left[1+\frac{\mathbf{M}^2}{4}+\frac{2-k}{24}\mathbf{M}^4 + ...\right]$$

---

* Binomial Theorem : $(1+x)^n = 1 + nx + \frac{n(n-1)}{\underline{|2}}x^2 + \frac{n(n-1)(n-2)}{\underline{|3}}x^2 + ...$

or
$$p_0 = p + \frac{pk\mathbf{M}^2}{2}\left[1+\frac{\mathbf{M}^2}{4}+\frac{2-k}{24}\mathbf{M}^4+...\right]$$

Using Eqs. (14.6), (14.28), (14.30) and (1.12) [*i.e.* the equation of state], one can write :

$$p_0 = p + \rho\,\frac{V^2}{2}\left[1+\frac{\mathbf{M}^2}{4}+\frac{2-k}{24}\mathbf{M}^4+...\right] \quad ...(14.34)$$

For subsonic flow, higher order terms in **M** will be of negligible values.

Comparison of Eq. (14.34) with the stagnation pressure $p_0 = p + \frac{\rho V^2}{2}$ given by Eq. (4.27) for a constant-density fluid shows that the effects of compressibility have been isolated in the bracketed quantity and that these effects depend only upon the Mach number. The bracketed quantity may thus be considered a *compressibility correction factor*. It is seen that for $\mathbf{M} < 0.2$, the compressibility affects the pressure difference $(p_0 - p)$ by less than 1% and the simple formula for flow at constant density is then sufficiently accurate. For larger values of **M** as the terms of the binomial expansion become significant, the compressibility effect must be taken ino account. The Pitot-static tube used for measuring aircraft speed needs to be corrected to take into account the effects of compressibility when the Mach number exceeds a value of about 0.3. The correction factor to be multiplied to the velocity recorded by the Pitot-tube calibrated at the ground level may be called the velocity correction factor. The true velocity of aircraft can be determined from the velocity as recorded by the pilot-tube calibrated at the ground level (where the fluid is considered incompressible). The velocity recorded by the Pitot-tube with normal calibration can be expressed by $V^2 = \frac{2(p_0 - p)}{\rho}$, and the true velocity by Eq. (14.34). Using these equations, one can write

$$V_{aircraft} = \frac{V_{pilot-tube}}{1+\frac{M^2}{4}+\frac{2-k}{24}\mathbf{M}^4} = VCF \times V_{pitot\text{-}tube} \quad ...(14.35)$$

where V.C.F. is the velocity correction factor

$$VCF = \frac{1}{\sqrt{1+\frac{\mathbf{M}^2}{4}+\frac{2-k}{24}\mathbf{M}^4}} = \frac{1}{\sqrt{CCF}} \quad ...(14.35A)$$

Example 14.4. *Air has a velocity of 1000 km/hr at a pressure of 9.81 kN/m² vacuum and a temperature of 47°C. Compute its stagnation properties, and the local Mach number. Take atmospheric pressure = 9.81 kN/m², R = 287 Nm/kg °K*

What would be the compressibility correction factor for a pitot-static tube to measure the velocity at a Mach number of 0.8 ?

Solution. Absolute temperature of air = 273 + 47 = 320°K

Sonic velocity $C = \sqrt{kRT} = \sqrt{1.4\times 287\times 320} = 358.53$ m/s

Velocity of air $V = \frac{1000\times 1000}{60\times 60} = 277.78$ m/s

Mach number $M = \frac{V}{C} = \frac{277.78}{358.53} = 0.7747$

The stagnation pressure is given by Eq. (14.32),

$$\frac{p_0}{p} = \left(1+\frac{k-1}{2}\mathbf{M}^2\right)^{k/k-1} = \left(1+\frac{1.4-1}{2}\times 0.7746^2\right)^{1.4/0.4} = 1.486$$

$\therefore \quad p_0 = 1.486\,p = 1.486 \times (98.1 - 9.81) = \mathbf{131.2\ kN/m^2}$

The stagnation temperature

$$\frac{T_0}{T} = 1 + \frac{k-1}{2}\mathbf{M}^2 = 1 + \frac{1.4-1}{2} \times 0.7747^2 = 1.1198$$

$\therefore$ $T_0 = 1.1198 \times 320 = \mathbf{358.33°K}$

From the equation of state,

$$\rho_0 = \frac{p_0}{RT_0} = \frac{131.2 \times 10^3}{287 \times 358.33} = \mathbf{1.275\ kg/m^3}$$

Compressibility correction factor

$$= 1 + \frac{M^2}{4} + \frac{2-k}{24}\mathbf{M}^4 + ... = 1 + \frac{0.7747^2}{4} + \frac{2-1.4}{24} \times 0.7747^4 = \mathbf{1.159.}$$

Example 14.5. *Determine the speed which would be indicated on a pitot-static tube of normal calibration (without taking compressibility effect into account) and calculate the percentage error when the tube is indicating the speed of an aircraft which is actually flying at 280 m/s. The velocity of sound in air at the aircraft altitude may be taken as 340 m/s.*

Solution. From Eq. (14.34), neglecting higher power terms of **M**

$$\frac{p_0 - p}{\frac{1}{2}\rho V^2} = 1 + \frac{\mathbf{M}^2}{4} + \frac{2-k}{23}\mathbf{M}^4,$$

in which V is the velocity affected by the compressibility of fluid, hence the actual speed of aircraft can be expressed as

$$V^2_{actual} = \frac{2(p_0 - p)}{\rho(1 + \mathbf{M}^2/4) + \frac{2-k}{24}\mathbf{M}^4} \quad ...(1)$$

The speed recorded by the pitot-static tube with normal calibration

$$V^2 = \frac{2(p_0 - p)}{\rho} \quad ...(2)$$

From Eqs. (1) and (2), $\dfrac{V}{V_{actual}} = \sqrt{1 + \dfrac{\mathbf{M}^2}{4} + \dfrac{2-k}{24}\mathbf{M}^4}$

$$\sqrt{CCF} = \frac{\text{Velocity from Pitot tube}}{\text{Actual Velicity of Aircraft}} = \sqrt{1 + \frac{\mathbf{M}^2}{4} + \frac{2-k}{24}\mathbf{M}^4}$$

$$= \sqrt{1 + \frac{(0.8235)^2}{4} + \frac{2-1.4}{24}(0.8235)^4} = \sqrt{1 + 0.1695 + 0.01149} = \mathbf{1.0867}$$

$$\therefore \quad V_{actual} = \frac{V_{pitot-tube}}{\sqrt{1 + \frac{\mathbf{M}^2}{4} + \frac{2-k}{24}\mathbf{M}^4}} = V_{pitot\text{-}tube} \times \text{Velocity correction factor.}$$

When the aircraft flies at 280 m/s, $\mathbf{M} = \dfrac{280}{340} = 0.8235$, and

$$\frac{V}{V_{actual}} = 1.0867 \qquad \therefore \quad V = 304.28 \text{ m/s.}$$

The pitot-static tube records a speed of 304.28 m/s as against the actual speed of 280 m/s. The percentage error

$$= \frac{V - V_{actual}}{V_{actual}} \times 100 = \frac{304.28 - 280}{280} \times 100 = 8.67\%.$$

Example 14.6. *At a certain section of a duct in which air is flowing at a temperature of 32°C and pressure of 79.95 kN/m² with a velocity of 365 m/s. Assuming isentropic flow (reversible adiabatic) determine : (i) the velocity and temperature at a section where the pressure is* ***122.63*** *kN/m² and (ii) the Mach number at both the sections. [Take R = 287 Nm/kg °K and k for air =* ***1.4.****]*

Solution. From Eq. (14.18 *a*), $\dfrac{V^2}{2} + \dfrac{k}{k-1}\dfrac{p}{\rho}$ = constant

which may be written as

$$\frac{V_1^2 - V_2^2}{2} = \frac{k}{k-1}\left(\frac{p_2}{\rho_2} - \frac{p_1}{\rho_1}\right) = \frac{1.4}{0.4}\left(\frac{p_2}{\rho_2} - \frac{p_1}{\rho_1}\right) = 3.5\left(\frac{p_2}{\rho_2} - \frac{p_1}{\rho_1}\right)$$

for isentropic flow, $\dfrac{p}{\rho^k}$ = constant, or $\dfrac{p_2}{p_1} = \left(\dfrac{\rho_2}{\rho_1}\right)$

$$\therefore \quad \frac{V_2^2 - V_1^2}{2} = 3.5\,\frac{p_1}{\rho_1}\left(1 - \frac{p_2}{p_1}\frac{\rho_1}{\rho_2}\right) = 3.5\,\frac{p_1}{\rho_1}\left\{1 - \left(\frac{p_2}{p_1}\right)^{(k-1)/k}\right\}$$

or

$$V_2^2 - V_1^2 = 7RT_1\left\{\left(1 - \frac{p_2}{p_1}\right)^{(k-1)/k}\right\}$$

inserting the numerical values :

$$V_1 = 365 \text{ m/s}, \; R = 287 \text{ Nm/kg°K}$$

$$\frac{p_2}{p_1} = \frac{122.63}{79.95} = 1.534, \; T_1 = 273 + 32 = 305 \text{ °K}$$

$$V_2^2 = (365)^2 + 7 \times 287 \times 305 \{(1 - 1.534)^{1/3.5}\}$$

or

$$V_2 = \mathbf{231.46 \text{ m/s}}$$

From Eq. (14.31),

$$\frac{T_2}{T_1} = \left(\frac{p_2}{p_1}\right)^{(k-1)/k} = 1.13$$

$$\therefore \quad T_2 = 1.13 \times 305 = \mathbf{344.65°K} \text{ or } 71.65°\text{C}$$

The Mach numbers, $\mathbf{M}_1 = \dfrac{V_1}{C_1} = \dfrac{365}{\sqrt{kRT_1}} = \dfrac{365}{\sqrt{1.4 \times 286.9 \times 305}} = \mathbf{1.043}$

$$\mathbf{M}_2 = \frac{V_2}{C_2} = \frac{231.46}{\sqrt{1.4 \times 286.9 \times 344.65}} = \mathbf{0.622.}$$

Example 14.7. *The pressure leads from a Pitot-static tube mounted on an aircraft are connected to a pressure gauge in the cockpit. The dial of the pressure gauge is calibrated to read the aircraft speed in metres/sec. The calibration is done on the ground by applying a known pressure across the gauge and calculating the equivalent velocity using incompressible Bernoulli equation and assuming that the density is 1.224 kg/m³.*

The gauge having been calibrated in this way the aircraft is flown at 9200 m, where the density is 0.454 kg/m³ and the ambient pressure is 30,000 N/m². The gauge indicates a velocity of 152 m/s. What is the true speed of the aircraft ?

Solution. The incompressible Bernoulli equation may be written as,

$$\rho\,\frac{V^2}{2} + p = \text{constant}$$

and the stagnation pressure created at the Pitot-static tube :

$$p_0 = p + \rho\,\frac{V^2}{2} \text{ or } V = \sqrt{\frac{2(p_0 - p)}{\rho}}$$

inserting the numerical values

$$152 = \sqrt{\frac{2(p_0 - 30{,}000)}{1.224}} \qquad \therefore \quad p_0 = 44139.6 \text{ N/m}^2$$

Neglecting compressibility effect, the speed of aircraft when related to $\rho = 0.454$ kg/m^3 is

$$V = \sqrt{\frac{2(p_0 - p)}{\rho}} = \sqrt{\frac{2(44139.6 - 30{,}000)}{0.454}} = 249.57 \text{ m/s.}$$

Sonic velocity $\quad C = \sqrt{kRT} = \sqrt{k\frac{p}{\rho}} = \sqrt{1.4 \times \frac{30{,}000}{0.454}} = 304.15$ m/s

Mach number $\quad \mathbf{M} = \frac{V}{C} = \frac{249.57}{304.15} = 0.818$

Compressibility correction factor $= \left(1 + \frac{\mathbf{M}^2}{4} + \frac{2-k}{24}\mathbf{M}^4\right) = \left(1 + \frac{(0.818)^2}{4} + \frac{2-1.4}{24}(0.818)^4\right)$

$= 1.178$

$\therefore$ True speed of aircraft (from Eq.14.34) $= \frac{249.57}{\sqrt{1.178}} =$ **229.92 m/s.**

Example 14.8. *A jet airplane is propelled at a velocity of 1000 km/hr through air in which the atmospheric pressure is 88.3 kN/m². The temperature is – 5°C and the wind velocity is negligible. Calculate the pressure intensity, the temperature, the density of air at the stagnation point on the nose of the jet, and determine its Mach number. Take R = 290 Nm/kg °K and k = 1.4.*

Solution. Velocity of sound in the atmospheric air

$$C = \sqrt{kRT} = \sqrt{1.4 \times 290 \times (273 - 5)} = 330.5 \text{ m/s.}$$

From the equation of state

$$\rho = \frac{p}{RT} = \frac{88.3 \times 10^3}{290 \times (273 - 5)} = 1.138 \text{ kg/m}^3$$

Applying Eq. (14.34), the stagnation pressure

$$p_0 = p + \rho\frac{V^2}{2}\left(1 + \frac{\mathbf{M}^2}{4} + \frac{2-k}{24}\mathbf{M}^4\right)$$

$$\mathbf{M} = \frac{V}{C} = \frac{1000 \times 1000/60 \times 60}{330.5} = 0.84$$

$$\therefore \qquad p_0 = 88.3 \times 10^3 + \frac{1.138 \times (278)^2}{2}\left[1 + \frac{(0.84)^2}{4} + \frac{2-1.4}{24}(0.84)^4\right] = 163.84 \text{ kN/m}^2$$

$\therefore$ Stagnation pressure $= 163.84$ kN/m^2

Applying Eq. (14.29)

$$\frac{k}{k-1}RT + \frac{V^2}{2} = \frac{k}{k-1}RT_0$$

From which $\frac{V^2}{2} = \frac{k}{k-1}R\,(T_0 - T)$

$$T_0 = \frac{(k-1)V^2}{2kR} + T = \frac{(1.4-1)\,(278)^2}{2 \times 1.4 \times 290} + (273 - 5) = 307.4°\text{K absolute.}$$

$$T_0 = 34.4°\text{C}$$

Density at the stagnation point,

$$\rho_0 = \frac{p_0}{RT_0} = \frac{140 \times 10^3}{290 \times 307.4} = 1.57 \text{ kg/m}^3.$$

Example 14.9. *Measuring the Mach angles α is an accurate means of measuring supersonic velocities for sharp nosed objects such as projectiles. Estimate the velocity and the Mach number of a projectile which reflects a shock wave of angle α = 30° in air at a pressure of 93.2 kN/m² and at a temperature of 3°C.*

Solution. The Mach angle is defined as, $\sin \alpha = \frac{C}{V} = \frac{1}{\mathbf{M}}$ assuming a frictionless process

$$C = \sqrt{kRT} = \sqrt{1.4 \times 290 \times (273 + 3)} = 344 \text{ m/s}$$

$$\sin 30° = \frac{344}{V} \quad \therefore V = 344 \times 2 = 688 \text{ m/s.}$$

Mach number $\mathbf{M} = \frac{V}{C} = 2.$

Example 14.10. *An aircraft is flying at a height at which the pressure and temperature are 0.5 × 10⁵ N/m² and – 17°C. A window of area 0.1 m² is situated towards the nose of the aircraft. Calculate the force acting on the window if the aircraft interior is pressurized at 1 × 10⁵ N/m², and determine the air speed and Mach number at the window when*

(i) **M** *= 0.2 for the aircraft, C_p = – 0.8 for the window :*

(ii) **M** *= 0.7 for the aircraft, C_p = – 0.5 for the window.*

Solution. The aircraft-nose can be taken as the stagnation point where the pressure and temperature are expressed by Eqs. (14.32) and (14.30) as :

$$\frac{p_0}{p} = \left(1 + \frac{k-1}{2} M^2\right)^{k/k-1} \qquad ...(14.32)$$

The velocity of sound, $C = \sqrt{kRT}$

Given : $p = 0.5 \times 10^5$ N/m², $T = -17$°C, Let $k = 1.4$, $R = 287$ J/kg°K

$$C = \sqrt{1.4 \times 287 \times (273 - 17)} = 320.72 \text{ m/s}$$

(i) **M = 0.2, $\mathbf{C_p}$ = – 0.8 :**

$$\text{Mach No, } M = \frac{\text{Vel. of aircraft, } V}{\text{Vel. of sound, } C}; \; 0.2 = \frac{V}{320.72}, \quad \therefore \quad \mathbf{V = 64.14 \text{ m/s.}}$$

$$\text{Stagnation pressure, } p_0 = 0.5 \times 10^5 \left(1 + \frac{1.4-1}{2} \times 0.2^2\right)^{\frac{1.4}{1.4-1}} = 0.514 \times 10^5 \text{ N/m}^2$$

As the window is located near the nose of the aircraft which represents the stagnation point, the pressure at the window can be considered having the same magnitude. The negative value of the pressure coefficient indicates the existence of negative pressures outside the aircraft, whereas the inside pressure is always positive. The combined effect of both the pressures is additive. Hence, the force acting on the window is

$$F = C_p A p_0 + A p_i = 0.8 \times 0.1 \times 0.514 \times 10^5 + 0.1 \times 1 \times 10^5$$
$$= 0.141 \times 10^5 \text{ N} = \mathbf{14.1 \text{ kN}} \text{ (acting outwardly)}$$

(ii) **M = 0.7, $\mathbf{C_p}$ = – 0.5 :**

Velocity of air craft, V = C.M. = 320.72 × 0.7 = **224.504** m/s

$$\text{Stagnation pressure, } p_0 = 0.5 \times 10^5 \left(1 + \frac{1.4-1}{2} \times 0.7^2\right)^{\frac{1.4}{0.4}} = 1.591 \times 10^5 \text{ N/m}^2$$

Force acting on the window

$$F = C_p\, A\, p_0 + A\, p_i = 0.5 \times 0.1 \times 1.591 \times 10^5 + 0.1 \times 1 \times 10^5$$
$$= \textbf{17.955 kN} \text{ (acting outwards)}$$

Due to the existence of "no-slip" condition at the aircraft-air contact surface, the air velocity at the window would be the same as that of the aircraft. The Mach number world have to be calculated by considering the orientation of the window with respect to the direction of motion of the aircraft, and the air velocity being taken in the same direction.

14.7.1. Rise in Skin Temperature

At a stagnation point where the velocity is reduced to zero, the stagnation temperature is given by Eq. (14.27)

$$C_p T_0 = C_p T + \frac{V^2}{2} \qquad \text{...(14.27)}$$

The stagnation temperature T_0 remains constant along the flow. At a stagnation point the dynamic temperature rise ΔT above the static temperature T is

$$\Delta T = T_0 - T = \frac{V^2}{2C_p}.$$

For air the formula may be reduced to a simple expression if V is in km per hour and ΔT in degree C, then

$$\Delta T = \left[\frac{V}{161}\right]^2$$

If an attempt is made to measure a flowing gas by placing a thermometer in a stream, the temperature recorded will be greater than T. Although the stagnation temperature would be reached at the stagnation point on a thermometer bulb, the temperature would rise less at other points on it and so the mean temperature recorded would be somewhat less than the stagnation temperature. The nose cone of a rocket traveling through air at 5100 km/hour must be able to withstand a temperature rise of about 1000 degree C.

Example 14.11. *A jet propelled aircraft is flying at 1000 km/hr at sea level, air temperature T = 25°C. Calculate the Mach number and rise in skin temperature.*

Solution. Velocity of sound in air at 25°C

$$C = \sqrt{kRT} = \sqrt{1.4 \times 290 \times (273 + 25)} = 348 \text{ m/s}$$

Mach number $$\mathbf{M} = \frac{V}{C} = \frac{1000 \times 1000/(60 \times 60)}{348} = 0.798$$

Rise in skin temperature

$$\Delta T = \left(\frac{V}{161}\right)^2 = \left(\frac{1000}{161}\right)^2 = 38.5°\text{C}.$$

14.8. One Dimensional Flow With Negligible Friction

From article 4.2 we recall that from the concept of one-dimensional flow all relevant quantities, such as the velocity, pressure, density and temperature, are considered constant over any cross-section of the conduit. Thus the flow can be described in terms of only one co-ordinate, that is the distance along the conduit axis, and time. The pressure is assumed constant over the flow section and the absence of friction (isentropic flow) permits no variation of velocity. The flow of a real fluid is never strictly one-dimensional because of the presence of boundary layer. But in many problems in which the boundary layer is not very thick, the assumption of one-dimensional flow provides satisfactory solutions. With pressure and velocity assumed constant throughout the flow cross-section, constancy of density and temperature follows from Eqs. (14.31) to (14.33).

For a cross-section of area A over which velocity V and density ρ are constant, the continuity equation for steady flow is

$$\rho AV = \text{constant} \quad \text{...(3.15)}$$

differentiating and dividing throughout by ρAV, we obtain Eq. (14.12)

$$\frac{d\rho}{\rho} + \frac{dA}{A} + \frac{dV}{V} = 0 \quad \text{...(14.12)}$$

If significant changes in cross-section area and, therefore, in V and ρ occur over only a short length of conduit, such as in a nozzle, frictional effects may be neglected. Euler's equation of motion for steady, frictionless flow neglecting the gravity and body forces from Eq. (4.5) may be written as

$$VdV + \frac{dp}{\rho} = 0 \quad \text{...(14.35)}$$

Multiplying the second term of the Eq. (14.35) by $dp/d\rho$, and noting that the velocity of sound is given by Eq. (11.35) as $C^2 = dp/d\rho$, Eq. (14.35) becomes

$$\frac{d\rho}{\rho} + \frac{VdV}{C^2} = 0.$$

Substituting for $d\rho/\rho$, from Eq. (14.12)

$$\frac{dA}{A} + \frac{dV}{V} = \frac{VdV}{C^2} \quad \text{or} \quad \frac{dA}{A} = \frac{dV}{V}\left(\frac{V^2}{C^2} - 1\right)$$

which may be expressed in terms of Mach number

$$\frac{dA}{A} = \frac{dV}{V}\,(\mathbf{M}^2 - 1) \quad \text{...(14.36)}$$

An equation similar to Eq. (14.36) for the pressure change dp may be determined using Eqs. (14.12) and (14.35). From Eq. (14.35),

$$\frac{dV}{V} = -\frac{dp}{V^2}$$

Substituting this value of dV/V in Eq. (14.12)

$$\frac{d\rho}{\rho} - \frac{dp}{\rho V^2} + \frac{dA}{A} = 0$$

from which
$$\frac{dA}{A} = \frac{dp}{\rho V^2} - \frac{d\rho}{\rho} = \frac{dp}{\rho V^2}\left(1 - V^2\frac{d\rho}{dp}\right)$$

and since $\dfrac{dp}{d\rho} = C^2$ we obtain, $\dfrac{dA}{A} = \dfrac{dp}{\rho V^2}\,(1 - \mathbf{M}^2)$

Solving for dp,
$$dp = \frac{\rho V^2}{A}\,\frac{1}{1-\mathbf{M}^2}\,dA \quad \text{...(14.37)}$$

Using Eqs. (14.36) and (14.37), we can determine as to how the velocity and pressure vary with the change in flow area under subsonic flow conditions. From the equations some far reaching and somewhat surprising conclusions may be deduced as below in article 14.8.1.

14.8.1. Flow through Ducts of Varying Area

(1) *Subsonic Flow* ($\mathbf{M} < 1$). Eq. (14.36) shows that for subsonic flow since ($\mathbf{M}^2 - 1$) is negative, dA and dV must be of opposite sign, that is an increase of cross-sectional area causes a reduction of velocity and *vice versa*. This result is similar to that obtained for a constant-density fluid.

Because the term $\rho V^2/A$ of Eq. (14.37) is positive, for subsonic flow, dp and dA are of the same sign. That is an increase in cross-sectional area causes an increase in pressure in accordance with Eq. (14.37). We may, therefore, conclude that in subsonic isentropic flow, an increase in flow area is accompanied by an increase in pressure and a decrease in velocity.

(2) *Supersonic flow* ($\mathbf{M} > 1$). For supersonic flow since $\mathbf{M} > 1$, Eq. (14.36) shows that dA and dV are of the same sign. An increase in the flow area is accompanied by an increase in the flow velocity and *vice versa*. Eq. (14.37) for supersonic flow shows that $(1 - \mathbf{M}^2)$ being negative, dp and dA are of opposite sign. Thus if the cross-sectional area increases, Eq. (14.37) shows that the pressure must decrease and *vice versa*.

In conclusion, for a supersonic isentropic flow, an increase in flow area is accompanied by an increase in flow velocity and a decrease in pressure, whereas a decrease in flow area is accompanied by a decrease in velocity and an increase in pressure.

(3) *Sonic flow* ($\mathbf{M} = 1$). When $V = C$, $\mathbf{M} = 1$, therefore dA must be zero and (since the second derivative is positive) A must be minimum. If the velocity of flow equals the sonic velocity anywhere, it must, therefore, do so where the cross-section is of minimum area.

Referring to Eq. (14.37), the value of $(1 - \mathbf{M}^2)$ is now zero and therefore either dp becomes infinite as $\mathbf{M}$ tends to unity or that dA is also zero simultaneously. The expression for dp would then be indeterminate. It can be shown that $\mathbf{M} = 1$ only when $dA = 0$. This condition could occur if a duct changes from a converging duct to a diverging duct of *vice versa*.

14.9. Variation of Density and Temperature With Area

From the expression for velocity of sound, we know

$$C^2 = \frac{dp}{d\rho}$$

from which, $$d\rho = \frac{1}{C^2}\, dp \qquad \text{...(14.38)}$$

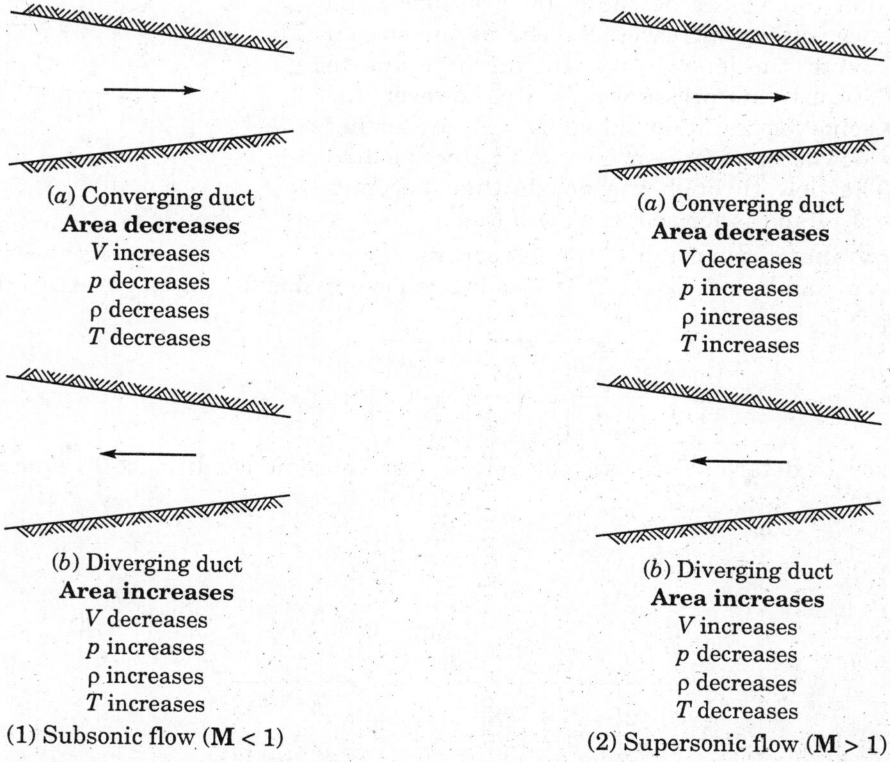

Fig. 14.2. Variation of Flow variables with Area for Subsonic and Supersonic Flow Conditions.

Eq. (14.38) shows that the density varies in the same manner as the pressure.

In the above analysis we have not made use of the equation of state. Therefore, the result from Eq. (14.36) onwards are applicable to the isentropic flow of any compressible fluid. Confining our attention now to the isentropic flow of an ideal gas, using the isentropic relation p = (constant) ρ^k and the equation of state $p = \rho RT$, we can obtain

$$T = (\text{constant})\ p^{(k-1)/k}$$

Differentiating it and eliminating the constant, we get

$$\frac{dT}{T} = \frac{k-1}{k}\frac{dp}{p} \qquad ...(14.39)$$

From Eq. (14.39) it is evident that the temperature varies in the same manner as the pressure. Combining Eqs. (14.38) and (14.39), we must conclude that the density and temperature vary in the same manner as does the pressure.

From the foregoing analysis, we can study the variation of velocity, pressure, density and temperature with changes in flow area for subsonic and supersonic flow conditions. Fig. 14.2 shows variation of flow variables with area for isentropic flow of an ideal gas.

In a subsonic flow, a diffuser is a diverging duct whereas for a supersonic flow the diffuser has a converging passage, Fig. 14.2 (*a*) and (*b*).

14.10. Flow through a Convergent Nozzle

Consider a convergent nozzle through which a gas is flowing, see Fig. 14.3. The fluid jet is set into motion by the difference of pressure acting on it. It is obvious that when p_2 is close to p_1, the flow will be subsonic throughout. The gas stream would then emerge in a similar way as would a liquid jet except that it would possess considerable compressibility as well as inertia. As the downstream pressure p_2 is reduced, the jet velocity increases. So long as the stream remains subsonic, the jet velocity will increase due to reduction of downstream pressure p_2. Once, however, the fluid reaches sonic velocity at the throat ($M = 1$), any further reduction in downstream pressure cannot be propagated upstream. The flow through the nozzle then becomes independent of p_2 and is termed as *choked flow*.

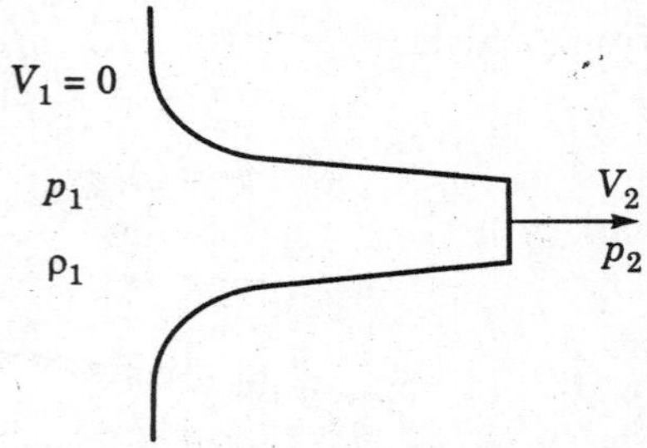

Fig. 14.3. Flow of a gas through a convergent nozzle.

For adiabatic reversible flow from reservoir conditions p_1, ρ_1 and $V_1 = 0$, Eq. (14.18 *a*) leads to an expression for velocity V at a section where the pressure is p :

$$V = \sqrt{\left[\frac{2k}{k-1}\frac{p_1}{\rho_1}\left\{1-\left(\frac{p}{\rho_1}\right)^{(k-1)/k}\right\}\right]} \qquad ...(14.40)$$

It can be used to determine the throat velocity for subsonic conditions. The corresponding mass flow rate

$$m = \rho AV$$

For reversible adiabatic flow,

$$\rho = \rho_1\left(\frac{p}{p_1}\right)^{1/k} \quad \text{and hence,}\ m = \rho_1\left(\frac{p}{p_1}\right)^{1/k} AV$$

$$= A\sqrt{\left[\frac{2k}{k-1}p_1\rho_1\left(\frac{p}{p_1}\right)^{2/k}\left\{1-\left(\frac{p}{p_1}\right)^{(k-1)/k}\right\}\right]} \qquad ...(14.41)$$

In the above expressions, the pressure at the throat is $p = p_2$ so long as the conditions are subcritical. As soon as the Mach number at the throat reaches unity, the throat velocity, pressure and density assume their 'critical' values V_c, p_c, ρ_c for choked flow. Thus

$$V_c = \sqrt{\left[\frac{2k}{k-1}\frac{p_c}{p_1}\left\{1-\left(\frac{p_c}{p_1}\right)^{(k-1)/k}\right\}\right]}$$

and
$$m_c = A\sqrt{\left[\frac{2k}{k-1}p_1\rho_1\left(\frac{p_c}{p_1}\right)^{2/k}\left\{1-\left(\frac{p_c}{p_1}\right)^{(k-1)/k}\right\}\right]} \quad ...(14.42)$$

Since for the choked flow, the local Mach number reaches unity, and $p_2 = p_c$, Eq. (14.32) when used for these conditions results in

$$\frac{p_c}{p_1} = \left(\frac{2}{k+1}\right)^{k/(k-1)} \quad ...(14.43)$$

For air with $k = 1.4$, the critical ratio

$$\frac{p_c}{p_1} = \left(\frac{2}{1.4+1}\right)^{3.5} = 0.528$$

Thus the mass flowrate through a convergent nozzle is given by Eq. (14.41) so long as $p_2 > 0.528\, p_1$. and Eq. (14.42) when $p_2 < 0.528\, p_1$. Fig. 14.4 shows variation of mass flow rate with pressure ratio p_2/p_1.

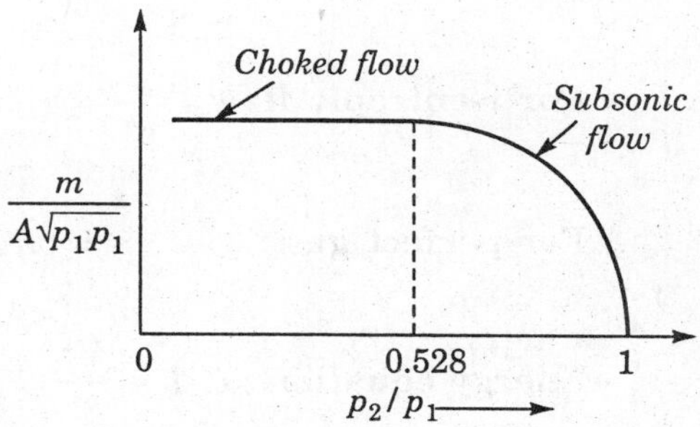

Fig. 14.4. Mass flowrate through a convergent nozzle.

Example 14.12. *A larger container holds carbon dioxide at a temperature of 15°C and pressure of 6 atmosphere. The gas is discharged through an orifice 10 mm in diameter. Determine the velocity of jet and the mass flowrate. Take k = 1.30 for carbon dioxide and atmospheric pressure = 101.24 kN/m², R = 189 Nm/kg °K.*

Solution. Density of CO_2 in the container

$$\rho_1 = \frac{p_1}{RT_1} = \frac{6\times 101.24\times 10^3}{189\times 288} = 11.16 \text{ kg/m}^3$$

Critical pressure ratio for choked flow

$$\frac{p_c}{p_1} = \left(\frac{2}{k+1}\right)^{k/(k-1)} = \left(\frac{2}{1.3+1}\right)^{1.3/0.3} = 0.545$$

$$\frac{p_2}{p_1} = \frac{1\times 101.24}{6\times 101.24} = \frac{1}{6} = 0.167$$

Since $p_2/p_1 < p_c/p_1$, the flow is a choked one, and the velocity of jet is equal to the sonic velocity in CO_2 at temperature T_c

$$\frac{T_c}{T_1} = \left(\frac{p_c}{p_1}\right)^{(k-1)/k} = \frac{2}{2.3}$$

or
$$T_c = \frac{2}{2.3}\times 288 = 250.43°K$$

and the sonic velocity
$$V_c = C = \sqrt{kRT_c} = \sqrt{1.3\times 189\times 250.43} = 248 \text{ m/s}$$

∴ The velocty of jet is 248 m/s.

Density of CO_2 at the orifice,

$$\rho_2 = \frac{p_c}{RT_c} = \frac{0.545 \times 6 \times 101.24 \times 10^3}{189 \times 250.43} = 6.99 \text{ kg/m}^3$$

Hence, the mass rate of flow

$$m = \rho_c A V_c = 6.99 = \frac{\pi}{4} \times \left(\frac{10}{1000}\right)^2 \times 248$$

$$= 0.1361 \text{ kg/sec} = \mathbf{8.17 \text{ kg/min.}}$$

14.11. Flow Variables in Terms of Mach Number

It is desirable to express changes in velocity, pressure, temperature and density in terms of the Mach number. To obtain these relationships use is made of the continuity, perfect gas, isentropic flow and the energy equations.

Continuity equation

$$\rho AV = \text{constant}$$

differentiating and dividing throughout by ρAV

$$\frac{d\rho}{\rho} + \frac{dA}{A} + \frac{dV}{V} = 0.$$

For isentropic flow : $\frac{p}{\rho^k}$ = constant or $\frac{dp}{p} = k\frac{d\rho}{\rho}$

For perfect gas : $p = \rho RT$ or $\frac{dp}{p} = \frac{d\rho}{\rho} + \frac{dT}{T}$

Energy equation : $C_pT + \frac{V^2}{2}$ = constant ; $C_p = \frac{k}{k-1}R$ $C_p dT + VdV = 0$

or $$\frac{kR}{k-1}dT + VdV = 0 \quad \text{or} \quad \frac{kR}{k-1}\frac{dT}{V^2} + \frac{dV}{V} = 0$$

$$C = \sqrt{kRT} \qquad \therefore \quad kR = \frac{C^2}{T}$$

or $$\frac{1}{(k-1)\mathbf{M}^2}\frac{dT}{T} + \frac{dV}{V} = 0 \qquad ...(14.44)$$

From the Mach number relationship

$$\mathbf{M} = V/\sqrt{kRT}$$

$$\frac{d\mathbf{M}}{\mathbf{M}} = \frac{dV}{V} - \frac{1}{2}\frac{dT}{T} \qquad ...(14.45)$$

Substituting for dT/T from Eq. (14.44)

$$\frac{d\mathbf{M}}{\mathbf{M}} = \frac{dV}{V}\left[1 + \frac{(k-1)}{2}\mathbf{M}^2\right]$$

or
$$\frac{dV}{V} = \frac{1}{\left[1 + \frac{(k-1)}{2}\mathbf{M}^2\right]} \frac{d\mathbf{M}}{\mathbf{M}} \qquad ...(14.46)$$

since the quantity within the brackets is always positive, the trend of variation of velocity and Mach number is similar. For temperature variation, one can write

$$\frac{dT}{T} = \left[\frac{-(k-1)\,\mathbf{M}^2}{1 + \frac{k-1}{2}\mathbf{M}^2}\right] \frac{d\mathbf{M}}{\mathbf{M}} \qquad ...(14.47)$$

since the right hand side is negative, the temperature changes follow an opposite trend to that of Mach number. Similar trends are obtained for pressure and density,

$$\frac{dp}{p} = \left[\frac{-k\,.\,\mathbf{M}^2}{1 + \frac{k-1}{2}\mathbf{M}^2}\right] \frac{d\mathbf{M}}{\mathbf{M}} \qquad ...(14.48)$$

and
$$\frac{d\rho}{\rho} = \left[\frac{-\mathbf{M}^2}{1 + \frac{k-1}{2}\mathbf{M}^2}\right] \frac{d\mathbf{M}}{\mathbf{M}} \qquad ...(14.49)$$

In case of area changes,
$$\frac{dA}{A} = \left[\frac{-(1-\mathbf{M}^2)}{1 + \frac{k-1}{2}\mathbf{M}^2}\right] \frac{d\mathbf{M}}{\mathbf{M}} \qquad ...(14.50)$$

The quantity within the brackets may be positive or negative depending upon the magnitude of Mach number. Equation (14.50) can be integrated to obtain a relationship between the critical throat area A_c, where the Mach number is unity and the area A at any section where $\mathbf{M} \lesssim 1$

$$\frac{A}{A_c} = \frac{1}{\mathbf{M}}\left[\frac{2 + (k-1)\,\mathbf{M}^2}{k+1}\right]^{\frac{k+1}{2(k-1)}} \qquad ...(14.51)$$

Equation (14.51) is unique in the sense that the Mach number is determined by the area ratio and k only.

Example 14.13. (*a*) *In case of isentropic flow of a compressible fluid through a variable area duct, show that*

$$\frac{A}{A^*} = \frac{1}{\mathbf{M}}\left[\frac{1 + \frac{1}{2}(k-1)\,\mathbf{M}^2}{\frac{1}{2}(k+1)}\right]^{\frac{k+1}{2(k-1)}}$$

where k is the ratio of specific heats, **M** *is the Mach number at a section whose area is A and A^* is the critical area of flow.*

(*b*) *A supersonic nozzle is to be designed for air flow with Mach number 3 at the exit section which is 20 cm in diameter. The pressure and temperature of air at the nozzle exit are to be 7.85 kN/m² and 200°K respectively. Determine the reservoir pressure and temperature and the throat area. Take k = 1.4.*

Solution. Since the velocity in the reservoir is zero, the temperature and pressure there correspond to the stagnation condition.

From Eq. (14.32),

$$p_0 = p\left[1+\frac{k-1}{2}\mathbf{M}^2\right]^{\frac{k}{(k-1)}} = 7.85\left[1+\frac{1.4-1}{2}3^2\right]^{3.5} = 287.5 \text{ kN/m}^2$$

using Eq. (14.30),

$$T_0 = T\left(1+\frac{k-1}{2}\mathbf{M}^2\right) = 200\left[1+\frac{0.4}{2}(3)^2\right] = 560°\text{K}$$

From Eq. (14.51),

$$\frac{A}{A_c} = \frac{1}{3}\left[\frac{2+(1.4-1)(3)^2}{1.4+1}\right]^{\frac{2.4}{2\times 0.4}} = 4.23$$

or

$$A_c = \frac{A}{4.23} = \frac{\frac{\pi}{4}(0.2)^2}{4.23} = \mathbf{0.00742\ m^2}.$$

14.12. Normal Shock

An abrupt change from supersonic conditions to the subsonic ones takes place through a shock wave (similar to hydraulic jump in open channel flow). Whereas the Mach number and velocity drop across a shock, there occurs a sudden rise in pressure, density, temperature and entropy. There may be two types of shocks :

(*i*) normal shocks which are almost perpendicular to the flow.

(*ii*) oblique shocks which are inclined to the flow direction.

Normal and oblique shocks may mutually interact to form a shock pattern. A shock may be attached or detached with respect to the body over which it may occur. A normal shock may occur in pipes, diverging section of a nozzle, diffuser of a supersonic wind tunnel or in front of bodies having blunt noses. For analysing a normal shock wave, use will be made of the continuity, momentum, perfect gas and energy equations.

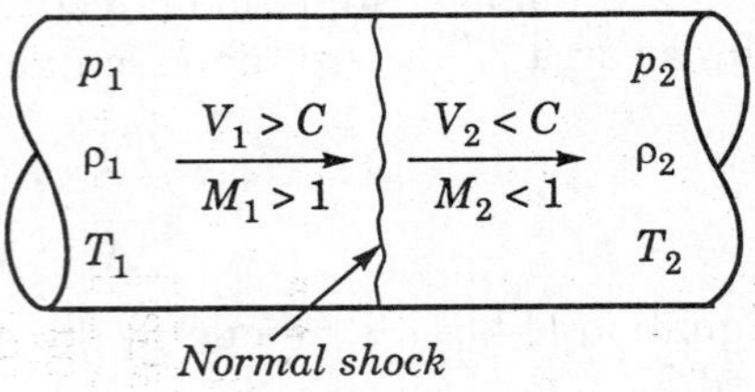

Fig. 14.5. Normal Shock Wave.

The upstream flow is supersonic ($\mathbf{M}_1 > 1$) with velocity V_1, pressure p_1, density ρ_1 and temperature T_1. On passing through the shock the flow becomes subsonic ($\mathbf{M}_2 < 1$) with velocity V_2, pressure p_2, density ρ_2 and temperature T_2. A shock wave involves dissipation of energy and as such cannot be considered isentropic. For a perfect gas undergoing adiabiatic flow in a constant area duct $A_1 = A_2 = A$. Considering Fig. 14.5 for a normal shock, the following equations can be written :

(*i*) **Continuity equation :**

$$\rho_1 A_1 V_1 = \rho_2 A_2 V_2$$

since $A_1 = A_2$, one gets

$$\rho_1 V_1 = \rho_2 V_2 \qquad \text{...(14.52)}$$

using the equation of state for a perfect gas and substituting the velocities in terms of the Mach number

$V = C\mathbf{M} = \mathbf{M}\sqrt{kRT}$, one may write Eq. (14.52) as

$$\frac{p_1\mathbf{M}_1\sqrt{kRT_1}}{RT_1} = \frac{p_2\mathbf{M}_2\sqrt{kRT_2}}{RT_2}$$

or

$$\frac{p_1\mathbf{M}_1}{\sqrt{T_1}} = \frac{p_2\mathbf{M}_2}{\sqrt{T_2}} \qquad \text{...(14.53)}$$

(*ii*) **Momentum equation :**

Neglecting the effect of boundary friction, the momentum equation gives

$$\text{Pressure force} = \text{Mass flowrate} \times \text{Change of velocity}$$

$$(p_1 - p_2) A = \rho_1 A V_1 (V_2 - V_1)$$

or

$$(p_1 - p_2) = \rho_2 V_2^2 - \rho_2 V_1^2$$

or

$$p_1 + \rho_1 V_1^2 = p_2 + \rho_2 V_2^2 \quad \text{or} \quad p_1 + \frac{p_1 V_1^2}{RT_1} = p_2 + \frac{p_2 V_2^2}{RT_2}$$

Substituting $V = \mathbf{M}C$, where $C = \sqrt{kRT}$, one gets

$$p_1 + k\,\mathbf{M}_1^2 p_1 = p_2 + k\,\mathbf{M}_2^2 p_2$$

from which

$$\frac{p_2}{p_1} = \frac{1 + k\mathbf{M}_1^2}{1 + k\mathbf{M}_2^2} \qquad ...(14.54)$$

Equation (14.54) represents static pressure ratio across a normal shock. Since $\mathbf{M}_1 > 1$ and $\mathbf{M}_2 < 1$, it is evident that $p_2 > p_1$, meaning thereby that the static pressure increases across a shock wave.

(*iii*) **Energy equation :**

For adiabatic condition one may write,

$$C_p T_1 + \frac{V_1^2}{2} = C_p T_2 + \frac{V_2^2}{2} \ ; \ \text{and } T_{o1} = T_{o2}$$

indicating that the stagnation temperatures are the same on both sides of a normal shock. From Eq. (14.30),

$$\frac{T_0}{T} = 1 + \frac{k-1}{2}\mathbf{M}^2$$

Equating the stagnation temperatures,

$$T_1\left(1 + \frac{k-1}{2}\mathbf{M}_1^2\right) = T_2\left(1 + \frac{k-1}{2}\mathbf{M}_2^2\right)$$

or

$$\frac{T_2}{T_1} = \frac{1 + \frac{k-1}{2}\mathbf{M}_1^2}{1 + \frac{k-1}{2}\mathbf{M}_2^2} \qquad ...(14.55)$$

since $\mathbf{M}_1 > 1$ and $\mathbf{M}_2 < 1$, Eq. (14.55) indicates that $T_2 > T_1$.

From Eq. (14.53)

$$\frac{p_2}{p_1} = \sqrt{\frac{T_2}{T_1}}\frac{\mathbf{M}_1}{\mathbf{M}_2}$$

$$= \frac{\mathbf{M}_1}{\mathbf{M}_2}\left[\frac{1 + \frac{k-1}{2}\mathbf{M}_1^2}{1 + \frac{k-1}{2}\mathbf{M}_2^2}\right]^{0.5} \qquad ...(14.56)$$

Equating Eqs. (14.54) and (14.56), one obtains a quadratic relation between $\mathbf{M}_1$ and $\mathbf{M}_2$. Neglecting the obvious trivial solution $\mathbf{M}_1 = \mathbf{M}_2$ for no-shock, one can write

$$\mathbf{M}_2^2 = \frac{2 + (k-1)\,\mathbf{M}_1^2}{2k\mathbf{M}_1^2 - (k-1)} \qquad ...(14.57)$$

Eq. (14.57) indicates that with $\mathbf{M}_1$ increasing, the denominator becomes larger and larger resulting in a progressively decreasing value of $\mathbf{M}_2$. By substituting the value of $\mathbf{M}_2$, the following equations are obtained :

$$\frac{p_2}{p_1} = \frac{2k\mathbf{M}_1^2 - (k-1)}{k+1} \qquad ...(14.58)$$

$$\frac{T_2}{T_1} = \frac{[(k-1)\,\mathbf{M}_1^2 + 2]\,[2k\mathbf{M}_1^2 - (k-1)}{(k+1)^2\,\mathbf{M}_1^2} \qquad ...(14.59)$$

$$\frac{\rho_2}{\rho_1} = \frac{p_2 / p_1}{T_2 / T_1} = \frac{(k+1)\,\mathbf{M}_1^2}{(k+1)\,\mathbf{M}_1^2 + 2} \qquad ...(14.60)$$

14.12.1. Shock Strength

The strength of shock is defined as the ratio of pressure rise across the shock to the upstream pressure, *i.e.*,

$$\text{Strength of shock} = \frac{p_2 - p_1}{p_1} = \frac{p_2}{p_1} - 1 = \frac{2k}{k+1}\,(\mathbf{M}_1^2 - 1) \qquad ...(14.61)$$

Example 14.14. *A normal shock wave occurs in a duct in which air is flowing at a Mach number of 1.50. The static pressure and temperature upstream of the shock wave are 1.5 bar and 27°C. Determine the pressure, temperature and the Mach number downstream of the shock. Also calculate the strength of shock. Take k = 1.4.*

Solution. $\frac{p_2}{p_1} = \frac{2k\mathbf{M}_1^2 - (k-1)}{k+1} = \frac{2 \times 1.4 \times (1.5)^2 - (1.4-1)}{1.4+1} = 2.458$

$\therefore \quad p_2 = 2.458 \times 1.5 = 3.687$ bar

$$\frac{T_2}{T_1} = \frac{[(k-1)\,\mathbf{M}_1^2 + 2\,[2k\mathbf{M}_1^2 - (k-1)]}{(k-1)^2\,\mathrm{M}_1^2}$$

$$= \frac{(0.4 \times 1.5^2 + 2)\,(2 \times 1.4 \times 1.5^2 - 0.4)}{2.4^2 \times 1.5^2} = 1.32$$

$\therefore \quad T_2 = 1.32(273 + 27) = \mathbf{396.06\ °K}$ or 123 °C

$$\mathbf{M}_2^2 = \frac{2 + (k-1)\,\mathbf{M}_1^2}{2k\mathbf{M}_1^2 - (k-1)} = \frac{2 + 0.4 \times 1.5^2}{2 \times 1.4 \times 1.5^2 - 0.4} = 0.49 \quad \text{or} \quad \mathbf{M_2 = 0.70}$$

Strength of shock $= \frac{p_2}{p_1} - 1 = 2.458 - 1 = \mathbf{1.458.}$

Example 14.15. *Establish an analogy between shock waves and open-channel waves. Show that the critical velocity in open-channel is equivalent to the sonic velocity in compressible flow and the liquid depth in the channel to the fluid density.*

Solution. The speed of an elementary surface wave in an open channel is

$$V = \sqrt{gy} \qquad ...(1)$$

where y is the depth of liquid in the channel. From the continuity equation, for a channel of constant width

$$V.y = \text{constant} \qquad ...(2)$$

and the continuity equation for compressible flow in a tube of constant cross-section is

$$V.\rho = \text{constant} \qquad ...(3)$$

Comparison of the two continuity Eqs. (2) and (3) reveals an analogy between the open-channel depth y and the compressible fluid density ρ.

The analogy between the critical velocity in open channel and the sonic velocity of compressible flow can be established from the consideration of energy equations. For open channel flow, the energy equation gives

$$\frac{V^2}{2g} + y = \text{constant}$$

differentiation of which yields

$$VdV + gdy = 0.$$

When the velocity in the channel is such that $V = V_c = \sqrt{gy}$, where V_c = critical velocity, elimination of g yields

$$VdV + V_c^2 \frac{dy}{y} = 0 \qquad \text{....(4)}$$

For a compressible fluid flow, we may write the energy equation, from Eq. (14.35), as

$$VdV + \frac{dp}{\rho} = 0$$

but the sonic velocity $C^2 = \dfrac{dp}{d\rho}$, eliminating dp, we obtain

$$VdV + C^2 \frac{d\rho}{\rho} = 0 \qquad \text{...(5)}$$

Comparing the two energy equations, Eqs. (4) and (5), we find that

(1) the flow depth y and the density ρ are analogous and

(2) the critical velocity V_c in the open-channel is analogous to the sonic velocity C.

14.13. Flow through a Convergent-Divergent Nozzle

Consider a convergent-divergent nozzle as shown in Fig. 14.6 with T_0, p_0 and ρ_0 as the stagnation properties and T, p, ρ, the static properties at the exit. When the back pressure equals the inlet pressure there will be no flow and the pressure will remain constant throughout the nozzle. The flow starts as the back pressure is reduced. Depending upon the value of the back pressure, the following flow conditions may be obtained :

(1) When the back pressure is reduced to a value such that the pressure at the throat is above the critical pressure, the flow is subsoinc throughout the nozzle. Flow in the nozzle is similar to that in a venturimeter (Curve B).

(2) Further lowering of back pressure may cause the sonic velocity at the throat, and the flow in the diverging portion returns to subsonic conditions (Curve C).

(3) With the sonic velocity attained at the throat, further reduction of pressure does not affect the flow in the convergent portion. Low exit pressure such as that represented by D causes the flow to be supersonic upto a certain distance from the throat, and then a shock wave occurring

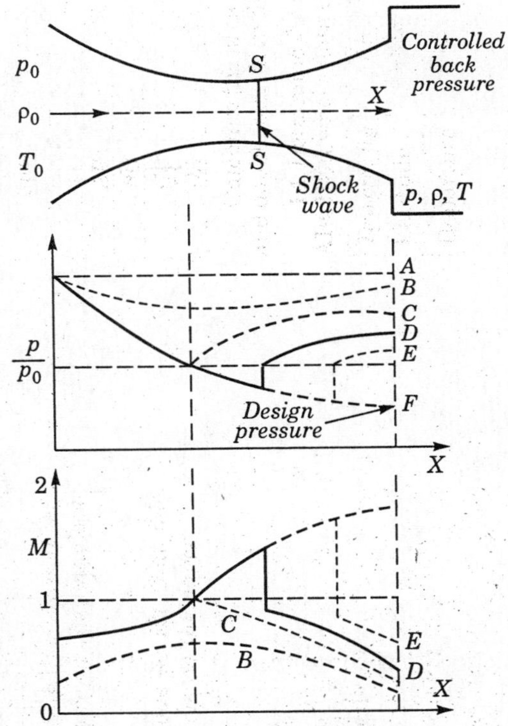

Fig. 14.6. Flow conditions in a Convergent-Divergent nozzle.

within the diverging portion causes subsonic conditions to exist in the remaining portion.

(4) When the exit pressure equals the nozzle design outlet pressure, there is a steady acceleration in the entire passage. The flow being subsonic in the convergent portion, sonic at the throat and supersonic in the diverging portion (Curve F).

(5) Still further reduction of back pressure below the design value, the expansion occurs outside the nozzle.

A nozzle operating below the design pressure is said to be under expanding and the one operating above the design value is called over expanding.

14.14. Compressible Flow Through a Venturimeter

The energy equation for a compressible fluid flow is given by Eq. (14.17), reproduced as follows :

$$\frac{k}{k-1}\frac{p}{\rho}+\frac{V^2}{2}+gZ = \text{constant} \qquad ...(14.17)$$

In case of compressible fluids, the term representing potential energy per unit mass, (gZ), has a very low value as compared to other forms of energy, on account of low value of mass density. The term (gZ) can therefore be neglected, and the energy equation reduces to,

$$\frac{k}{k-1}\frac{p}{\rho}+\frac{V^2}{2} = \text{constant} \qquad ...(14.18\ a)$$

For a detailed information, refer article 4.14.2. Referring to Fig. 4.12 for notations used for flow through a venturimeter, point 1 represents the entrance portion while point 2 refers to the throat section. Applying the energy equation, Eq. (14.18a) to these sections, one can write :

$$\frac{k}{k-1}\frac{p_1}{\rho_1}+\frac{V_1^2}{2}=\frac{k}{k-1}\frac{P_2}{\rho_2}+\frac{V_2^2}{2}$$

Rearranging the terms,

$$\frac{1}{2}(V_2^2-V_1^2)=\frac{k}{k-1}\left(\frac{p_1}{\rho_1}-\frac{p_2}{\rho_2}\right)$$

or,

$$\frac{V_2^2}{2}\left(1-\frac{V_1^2}{V_2^2}\right)=\frac{k}{k-1}\frac{p_1}{\rho_1}\left(1-\frac{p_2}{p_1}\frac{\rho_1}{\rho_2}\right) \qquad ...(i)$$

For adiabatic flow, the pressure and density are related by Eq. (1.14), *i.e.*

$$\frac{p}{\rho^k} = \text{constant},$$

and hence,

$$\left(\frac{\rho_1}{\rho_2}\right)=\frac{p_1}{p_2}, \quad \text{or} \quad \frac{\rho_1}{\rho_2}=\left(\frac{p_1}{p_2}\right)^{1/k}$$

and we obtain,

$$\frac{V_2^2}{2}\left(1-\frac{V_1^2}{V_2^2}\right)=\frac{k}{k-1}\frac{p_1}{\rho_1}\left[1-\frac{p_2}{p_1}\left(\frac{p_1}{p_2}\right)^{1/k}\right]$$

$$=\frac{k}{k-1}\frac{p_1}{\rho_1}\left[1-\left(\frac{p_2}{p_1}\right)^{1-\frac{1}{k}}\right] \qquad ...(ii)$$

From the continuity equation, the mass flow rate,

$$m = \rho_1 A_1 V_1 = \rho_2 A_2 V_2$$

from which,

$$\frac{V_1}{V_2}=\frac{\rho_2 A_2}{\rho_1 A}$$

Substituting in Eq. (*ii*) above, and rearranging the terms,

$$\frac{V_2^2}{2}\left[1-\left(\frac{\rho_2}{\rho_1}\right)^2\left(\frac{A_2}{A_1}\right)^2\right]=\frac{kp_1}{(k-1)\,\rho_1}\left[1-\left(\frac{p_2}{p_1}\right)^{\frac{k-1}{k}}\right]$$

or,

$$\frac{V_2^2}{2}\left[1-\left(\frac{p_2}{p_1}\right)^{2/k}\left(\frac{A_2}{A_1}\right)^2\right]=\frac{kp_1}{(k-1)\,\rho_1}\left[1-\left(\frac{p_2}{p_1}\right)^{\frac{k-1}{k}}\right]$$

or,

$$V_2=\sqrt{\frac{\frac{2k}{k-1}\cdot\frac{p_1}{\rho_1}\left[1-\left(\frac{p_2}{p_1}\right)^{(k-1)/k}\right]}{1-\left(\frac{p_2}{p_1}\right)^{2/k}\left(\frac{A_2}{A_1}\right)^2}} \quad \text{...(14.62)}$$

The mass rate of flow through the venturimeter,

$$m=\rho_2A_2V_2$$

$$=\rho_2A_2\sqrt{\frac{\frac{2k}{k-1}\cdot\frac{p_1}{\rho_1}\left[1-\left(\frac{p_2}{p_1}\right)^{(k-1)/k}\right]}{1-\left(\frac{p_2}{p_1}\right)^{2/k}\left(\frac{A_2}{A_1}\right)^2}}$$

$$=\rho_2A_2\sqrt{\frac{\frac{2k}{k-1}\cdot\frac{p_1}{\rho_1}\left[1-\left(\frac{p_2}{p_1}\right)^{(k-1)/k}\right]}{1-\left(\frac{p_2}{p_1}\right)^{2/k}\left(\frac{d_2}{d_1}\right)^4}} \quad \text{...(14.63)}$$

where d_2 and d_1 are the diameters at the throat and the entrance sections. Eq. (14.63) is applicable for subsonic flow, for which the pressure ratio p_2/p_1 must be greater than the critical value 0.528.

Ex. 14.15. *Determine the mass flowrate of air through a venturimeter installed in a pipeline having internal diameter of 300 mm. The venturimeter has a throat diameter of 150 mm. The pressure at the inlet of venturimeter is 150 kN/m² absolute and the temperature 15°C. The pressure at the throat was found to be 140 kN/m² absolute. Given R = 287 J/kg.k and k = 1.4.*

Solution. The data given :

Absolute Temperature (T_1) = 273 + 15 = 288°K

Pressure at the inlet p_1 = 150 kN/m², diameter d_1 = 300 mm

Pressure at the inlet p_2 = 140 kN/m², diameter d_2 = 150 mm

The mass density at the inlet ;

$$\rho_1=\frac{p_1}{RT_1}=\frac{150\times10^3}{287\times288}=1.845\text{ kg/m}^3$$

For adiabatic flow, the mass density and pressure are related :

$$\frac{\rho_1}{\rho_2}=\left(\frac{p_1}{p_2}\right)^{1/k}, \text{ from which}$$

$$\rho_2 = \rho_1 \left(\frac{p_2}{p_1}\right)^{1/k} = 1.845 \left(\frac{140}{150}\right)^{1/1.4} = 1.845(0.933)^{0.71428}$$

$$= 1.845 \times 0.90747 = 1.674 \text{ kg/m}^3$$

The mass rate of flow, using Eq. (14.63) :

$$m_2 = \rho_2 A_2 \sqrt{\frac{\frac{2k}{k-1} \cdot \frac{p_1}{\rho_1}\left[1 - \left(\frac{p_2}{p_1}\right)^{(k-1)/k}\right]}{1-(p_2/p_1)^{2/k}(d_2/d_1)^4}}$$

$$= 1.674 \times \frac{\pi}{4}\left(\frac{150}{1000}\right)^2 \sqrt{\frac{\frac{2\times 1.4}{1.4-1} \cdot \frac{150\times 10^3}{1.845}\left[1-\left(\frac{140}{150}\right)^{(1.4-1)/1.4}\right]}{1-\left(\frac{140}{150}\right)^{2/1.4}\left(\frac{150}{300}\right)^4}}$$

$$= 1.674 \times 0.0177 \sqrt{\frac{569.11\times 10^3 \times (1-0.9805)}{1-0.9061\times 0.0625}}$$

$$= 0.02963 \sqrt{\frac{569.11\times 10^3 \times 0.0195}{1-0.0566}}$$

$$= 0.02963 \sqrt{\frac{11097.6}{0.9434}} = \mathbf{3.214 \text{ kg/s.}}$$

14.15. Pitot Tube Placed in Compressible Flow

Article 4.14.1 describes the pitot tube which is used to measure velocity of flowing fluid, which is incompressible, and is given by Eq. 4.28 reproduce below :

$$V_0 = \sqrt{\frac{2g\,(p_s - p_0)}{\gamma}} = \sqrt{2gh} \quad \text{...(4.28)}$$

Furthermore, article 13.3.1 describes the details of a Pitot-static tube, which directly measures the difference in the stagnation pressure and static pressure in terms of head of liquid, that is related to the velocity of flow by Eq. (13.8).

But in case of compressible fluids, its compressibility affects the measurement of velocity, using Eq. (14.32), the stagnation pressure at the nose of the pitot tube can be written as :

$$\frac{p_s}{p_0} = \left[1 + \frac{k-1}{2} M^2\right]^{k/k-1}$$

where p_0 is the static pressure in the fluid.

For subsonic flow, the Mach number, $M < 1$, and the above equation can be expressed by Eq. (14.34)

$$p_s = p_0 + \rho \frac{V_0^2}{2}\left[1 + \frac{M^2}{4} + \frac{2-k}{24} M^4 + ...\right] \quad \text{...(14.64)}$$

Here V_0 represents the velocity of flowing compressible fluid.

In case of incompressible fluids, Eq. (4.28) can be recast as follows :

$$p_s = p_0 + \frac{\gamma V_0^2}{2g} = p_0 + \frac{\rho g V_0^2}{2g}$$

$$= p_0 + \rho \frac{V_0^2}{2} \quad \text{...(14.65)}$$

Comparing Eqs. (14.64) and (14.65), one can notice that the compressibility effect is confined in the bracketed quantity, *i.e.*

$$\left[1+\frac{M^2}{4}+\frac{2-k}{24}M^4+....\right] \qquad ...(14.66)$$

This equation represents the compressibility effect and is known as the "*compressibility correction factor*".

Example 14.16. *For measuring speed of air craft (aeroplane), a Pilot-tube is installed at a suitable location on the aircraft. The pressure leads from the tube are connected to a pressure gauge in the cockpit. The dial of the pressure gauge is calibrated to read the aircraft speed in metre/sec. The calibration is done on the ground by applying a known pressure across the gauge and calculating the equivalent velocity using the Bernoulli equation treating air as incompressible, and adopting the value of air density at the ground level, ρ = 1.225 kg/m³ (sea level density for the standard atomosphere).*

As the aircraft rises in altitude and gains the desired level, the air density reduces considerably, its value being 0.4134 kg/m³ at an altitude of 10 km, where the absolute ambient pressure reduces to 26.50 kN/m². Flying at this altitude, the pilot observes the air craft speed on the pressure gauge as 170 m/s (612 km/hr). Determine the true speed of the aircraft. Take k = 1.4.

Solution. Using the Bernoulli's equation for an incompressible fluid, the stagnation pressure developed at the nose of the Pitot-static tube can be expressed as,

$$p_0 = p + \rho\,\frac{V^2}{2} \quad \text{or} \quad V = \sqrt{\frac{2\,(p_0 - p)}{\rho}}$$

in which p_0 is the stagnation pressure, and p and ρ are the ambient pressure and density.

The velocity of sound or the sonic velocity,

$$C = \sqrt{kRT} = \sqrt{k\,p/\rho} = \sqrt{1.4\times 26.5\times 10^3/0.4134} = 299.57 \text{ m/s}$$

The Mach Number, $M = \dfrac{V}{C} = \dfrac{170}{299.57} = 0.567$

The compressibility correction factor (CCF),

$$CCF = \left[1+\frac{M^2}{4}+\frac{2-k}{24}M^4\right]$$

$$= \left[1+\frac{(0.567)^2}{4}+\frac{2-1.4}{24}(0.567)^4\right] = 1.081$$

The velocity correction factor (VCF) $= \dfrac{1}{\sqrt{CCF}} = \dfrac{1}{\sqrt{1.081}} = 0.962$

The true speed of the aircraft,

$$V_{true} = VCF \times \text{velocity recorded by the Pitot tube}$$

$$= 0.962 \times 170 = 163.5 \text{ m/s}$$

$$= \frac{163.54\times 3600}{1000} \text{ km/hr} = 588.74 \text{ km/hr.}$$

It is therefore, evident that the correct aircraft speed is 588.74 km/hr, whereas the one recorded by the pressure gauge *i.e.* 612 km/hr (calibrated at the ground level considering air as incompressible) is on the higher side. This is mainly due to the fact that at higher attitudes the ambient pressure and the air density both get considerably reduced. The percentage error involved in the speed recorded by the pressure gauge (connected to the pitot-tube) is

$$\% \text{ Error} = \frac{V_{press-gauge} - V_{true}}{V_{true}} \times 100 = \frac{612 - 588.74}{588.74} \times 100 = 3.95\%$$

Therefore, the % correction to be applied to the pressure gauge reading is (– 3.95%).

OBJECTIVE TYPE QUESTIONS

1. Mach number is a measure of compressibility effects in fluid flow. Indicate upto what Mach number can a fluid flow be considered incompressible :

(*i*) 1 (*ii*) 0.1 (*iii*) 0.3 (*iv*) 0.8 (*v*) 1.1.

2. Pressure and density relation in a fluid flow can be represented by

$$p/\rho^n = \text{constant}$$

in which the exponent n has different values for different types of processes. Match the value for each of the following processes :

(*a*) isobaric (*i*) $n = k$

(*b*) isothermal (*ii*) $n = 0$

(*c*) isentropic (*iii*) $n = 1$.

3. An isentropic flow is one which is :

(*i*) adiabatic (*ii*) isothermal

(*iii*) adiabatic and reversible (*iv*) adiabatic and irreversible

4. A shock wave which occurs in a supersonic flow represents a region in which :

(*a*) there is no change in pressure, temperature and density

(*b*) there is a sudden change in pressure, temperature and density

(*c*) a zone of silence exists (*d*) velocity is zero.

5. A stagnation point is one at which the fluid velocity attains a zero value. The stagnation pressure p_0 and temperature T_0 are :

(*i*) more as compared to their ambient counterparts

(*ii*) the same as in the ambient flow

(*iii*) less than their ambient counterparts

(*iv*) different ; whereas the pressure p_0 increases, the temperature T_0 decreases.

6. A diffuser in a subsonic flow is a passage of gradually increasing area. In supersonic flow, a diffuser is a conduit having :

(*a*) constant area throughout its length

(*b*) gradually increasing area as in a subsonic diffuser

(*c*) gradually decreasing area (*d*) converging-diverging passage.

7. The sonic velocity in a fluid medium is directly proportional to :

(*i*) pressure (*ii*) square root of temperature

(*iii*) density (*iv*) Mach number.

8. A normal shock may occur when a flow is supersonic and the conditions down-stream require it to be subsonic. Across a normal shock :

(*a*) pressure and temperature decrease (*b*) there is no energy loss

(*c*) pressure and density decrease (*d*) temperature and density decrease

(*e*) pressure and temperature increase.

ANSWERS

1. (*iii*) **2.** (*a*) (*ii*) ; (*b*) (*iii*) ; (*c*) (*i*) **3.** (*iii*) **4.** (*b*) **5.** (*i*)

6. (*c*) **7.** (*ii*) **8.** (*e*).

REVIEW QUESTIONS

1. Comment on the correctness of the statement : "In a compressible flow of an ideal gas the sum of the thermal and kinetic energies is constant".
2. A model of an aeroplane is put in high-speed wind tunnel, which parts of the aeroplane you expect to become the hottest ? Explain with reasons.
3. An aeroplane climbs from a low altitude over a hot desert to a high altitude where the temperature is significantly lower. If the airspeed is constant, what change, if any will occur in the flight Mach number ?
4. A man observes a lightening flash and counts 4 seconds before he hears the thunder. He then says, "The lighting was about 6.5 km away". He turns to you as an engineer and asks,"How does this technique of counting seconds give an estimate of the distance to the flash" ? Give him a technical answer and tell him whether the technique is correct.
5. Are principles of fluid mechanics alone sufficient to analyse problems of compressible fluid flow ? If not what other concepts are necessary ? Explain.
6. How is the first law of thermodynamics different from the Bernoulli's theorem ? Why is the Bernoulli's theorem inadequate for compressible flow problems ?
7. For compressible adiabatic flow, will the entropy increase, decrease or remain constant ? Explain.
8. State the fundamental equations which govern the compressible fluid flow and what are the controlling non-dimensional parameters governing such a flow ?
9. Explain what do you mean by
   (*i*) Transonic flow (*ii*) Hypersonic flow (*iii*) Shock wave (*iv*) Mach angle.
10. Prove that the temperature at the stagnation point is greater than the temperature in the free-stream.
11. Upto which limit of Mach number can we neglect the compressibility effects and treat the fluid as incompressible ?
12. How does the velocity and pressure vary with area for
    (*i*) Subsonic flow, and (*ii*) Supersonic flow.

PROBLEMS

14.1. Show that the Reynolds number for isothermal compressible gas flow in a uniform conduit remains constant.

14.2. Show that the velocity of sound in an adiabatic gas flow is given by $C = \sqrt{kRT}$.

14.3. Show that for an isothermal atmosphere the pressure variation in the vertical direction is given by

$$p = p_0 \, e^{-z/p_0/\rho_0}$$

where z is the distance measured vertically upward and p_0, ρ_0 are the pressure and density at $z = 0$. Assuming gravitational acceleration to be constant, compare the pressure given by this equation to the hydrostatic pressure distribution $p = p_0 - \gamma z$. Show this comparison graphically.

14.4. Show that the velocity of sound in the atmosphere under adiabatic conditions varies with the temperature and hence find out whether the sonic velocity will decrease or increase in magnitude with increasing elevation above the earth.

14.5. Show that for subsonic flow in a pipe the velocity must increase in the downstream direction. Take into account the losses in the pipe.

14.6. Determine the velocity of sound in air at 20°C, 50°C, and 100°C. Take gas constant $R = 290$ N-m/kg °K, $k = 1.4$.

14.7. Air is flowing in a duct at atmospheric pressure with a velocity of 500 m/s. If the air temperature is 70°C, what is the Mach number ?

ANSWERS

14.6. 345 m/s, 356.5 m/s, 389 m/s. **14.7.** 1.34.

15

Fluid Machines

15.0. Introduction

Energy is the primary input to every activity of sustenance or development. Electrical energy is the most convenient form. The per capita electrical energy consumed is considered to be an index of development and progress of a nation. We shall consider the fluid machines in which the fluid is in continuous motion and constitutes a medium for energy conversion. A fluid machine is a device either for converting the energy held by a fluid into mechanical energy or *vice versa*. Fluid machines such as turbines, centrifugal pumps, compressors, fans, hydraulic couplings and torque converters fall under the category of rotodynamic machines. The rotodynamic machines are also called turbomachines and have a rotating element which plays an important role in causing the energy transfer between the fluid and the rotating element. The energy transfer may take place from the fluid to the machine as in case of turbines or it may just be the reverse as in case of centrifugal pumps, blowers, fans and compressors.

Turbines convert hydraulic energy of water into mechanical energy of rotating element which in turn is converted into electrical energy through generator and associated system. On the other hand, pumps, blowers, fans and compressors use electrical energy to carry out the designed function. Apart from turbomachines, there is a second category of fluid machines known as the positive displacement machines. In such machines, the fluid is displaced by some moving mechanical element such as a gear system rotating in a closed casing as in case of a gear pump or a piston moving in a cylinder as in case of a reciprocating pump. In positive displacement machines, the static pressure is developed by a displacement action rather than by a velocity or kinetic energy change. The present chapter deals with the hydraulic turbines and the centrifugal pumps because of their importance and wide applications in the engineering practice. The theory of all turbomachines is the same and, therefore, the equations developed in this chapter will be applicable to turbines, centrifugal pumps and blowers as well.

15.0.1. High and Medium Head Hydropower

The ever increasing shortage of electrical power in a developing economy with the attendant problems of rising population and abnormal price hike of fossil fuels (coal and crude oil), harnessing of hydropower becomes a dependable and attractive alternative.

In early days of hydropower development, only the high and medium head sites were chosen, large capacity turbines were used to result in low cost energy. However, high and medium head hydroelectric projects require huge investments and have long gestation period.

Hydropower represents renewable, non-consumptive and non-polluting energy development as large quantities of surface water is available on a recurring basis. It represents highest prime moving efficiency and spectacular operational flexibility. A turbine acts as a prime mover for generating electricity as the shaft of the turbine is suitably coupled to the generator.

The hydraulic turbines utilize the energy of water and convert it into the mechanical energy of the rotating shaft. According to the principle of moment of momentum, if the moment of momentum of water is changed as it flows through the rotating element, there results a torque which rotates the turbine shaft. The hydraulic energy is thus converted into the mechanical energy. The hydraulic

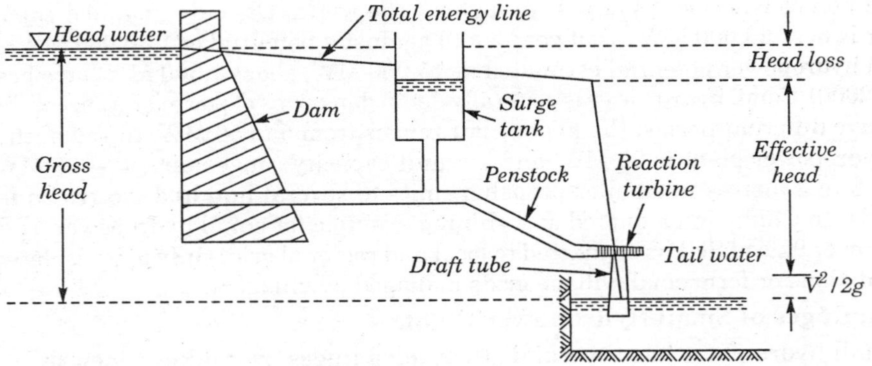

Fig. 15.1. Typical installation of a reaction turbine.

turbines constitute an important and essential item of a hydro-electric power plant. The primary function of a hydraulic turbine is to rotate the electric generator the rotation of which produces electrical power. Since the turbine main shaft is coupled to the generator shaft, the rotation of the turbine shaft ensures the rotation of the generator. The water from the storage reservoir is allowed to flow through a pressure pipe known as penstock, to the turbine (Fig. 15.1).

The total energy of flow at turbine entrance is considerably high, and as it flows through the turbine and comes out at the turbine exit, its moment of momentum is considerably utilized and the energy of flow at the exit is appreciably reduced. This change in the moment of momentum of flow produces a torque on the turbine runner causing it to rotate. The hydraulic energy of water is thus converted into the mechanical energy represented by the rotation of the turbine shaft.

The function of a pump is to increase the energy content of the flowing liquid. The increase in fluid energy is made possible by the rotation of the impeller (the rotating element of a pump is called an impeller) and represents the mechanical energy imparted to the pump. This mechanical energy is converted into the hydraulic energy by causing the liquid to flow through the impeller. The pumps are employed when energy of a flowing liquid needs to be increased. The situations where the use of a pump is necessary are :

1. When the liquid is to be pumped over long distances necessitating the fluid to possess sufficient flow energy to overcome the resistance to its flow.

2. When the liquid is to be pumped to higher elevations. Some of the common examples where pumps are used may be found in irrigation from ground water, water supply system, oil transportation and sewage disposal.

The functions of the turbine and the pump are thus opposite to each other ; while a turbine extracts energy from the flowing water and converts it into the useful mechanical energy, a pump imparts hydraulic energy through its impeller (by means of a shaft driven by a motor) to the liquid and thus increases its energy content.

The blowers and compressors are pumps dealing with air or gas. The primary function of a compressor is to increase pressure of air, and since high pressure changes are involved density changes must be considered. On the other hand, a fan or blower is primarily used for causing the movement of air. In this case, the change in pressure is quite small-usually sufficient to overcome the resistance to its flow and so the variation of density is negligible and the fluid may be regarded as incompressible.

15.0.2. Small Hydropower Plants (SHP)

Small hydropower plants are recognised as affordable, environmentally sound and clean source of energy most suited for rural areas. They are non-polluting, renewable and efficient energy source for electric power. Small hydropower represents the first place in generation of electricity from amongst all non-conventional sources throughout the world. Globally installed capacity of small hydropower is over 50,000 MW (as of year 2000) against a potential of 1,80,000 MW. In India, out of the small hydropower potential estimated at 15,000 MW, about 1,500 MW have been developed so far (year 2000). Small hydropower is generally defined in terms of power station capacity. Different countries have different norms, the upper limit ranges from 5 to 50 MW. In India, the upper limit for small hydropower plant is 25 MW, and the unit capacity ranges upto 10—25 MW. Heads can range from 2 to 3 metres in smaller propeller units to several hundred metres as in Pelton and Turgo wheels. In China, often quoted as a shining example for small hydropower development for rural areas, over 92% of SHP is connected to local and regional grid. Only a few isolated mini-micro plants stand alone or form small village grids managed by villagers.

Advantages of Small Hydropower Plants

A small hydro plant has several distinct advantages over other renewable alternatives, such as

1. Large untapped potential at abundant sites available (15,000 MW at over 5,000 sites).
2. Low initial investment cost (Rs. 1.80—5 crores/MW).
3. Short gestation period of less than two years and quick returns.
4. Cheapest amongst new renewable sources of energy (Rs. 1.00-2.25/unit) (as of year 2000).
5. Minimum transmission and distribution losses.
6. High efficiency (85%—95%).
7. Long project life.
8. Amenable to standardisation.

The only disadvantage is higher cost of SHP as compared to conventional hydropower, because cost per megawatt increases as capacity reduces and cost of energy per unit increases with reduction in plant load factor.

Sources for small hydropower

Small river systems, for example, Narmada, Tapti, Mahanadi, Godavari, Chambal, Betwa etc. take birth in Madhya Pradesh and adjoining Chhattisgarh State which hold several river basins. Small river planning for SHP oriented to cascading, trans-basin and multipurpose development includes storage, conveyance, generation and distribution. Storage may be eliminated in case of SHP on perennial streams as in Himalayan and north-eastern regions of the country but not in other states where rains and flows are seasonal.

Another source of small hydropower is in canal falls that are located on irrigation canals criss-crossing the rural India. Madhya Pradesh has planned 23 major irrigation projects on Narmada basin alone and can provide a large number of drops. The estimated SHP potential is 5000 MW. The canal flows are regulated, predictable and free from debris and the heads are constant. The falls are also standardised and can lead to mass production of standardised SHP units that may lead to further economy. Outlets from reservoirs are another similar source.

15.0.3. Wind Energy

Among the renewable alternate energy sources, wind energy has become cost effective and economical, next only to small hydropower. Technologically exploitable wind energy worldwide is placed at 20,000 billion kWh/yr at par with hydro. Out of the gross potential of 20,000 MW, and technically exploitable potential of 8950 MW in India, the share of Madhya Pradesh is estimated at 775 MW, of which only 19 MW (as of 2001) have been developed. Madhya Pradesh has low and medium wind velocity sites. More potential sites are located along sea coast and adjacent hill ranges. Wind energy is also site specific to be taken as and when and where available. It is most conveniently pumped into the grid, though capacitors could be used to store energy produced in low demand periods. The power factor of wind genereators is low at 12% to 15% as against world figure of 20-30%. The wind turbine generators require an area of about 8 hectares per MW. There are world renowned companies who have developed expertise in manufacturing, installation and commissioning, operation and maintenance of windmills. Some of these companies have extensive Research & Development wings that have made it possible to achieve a high level of transmission and efficient conversion of energy by continuously upgrading systems to withstand adverse climatic conditions prevailing in India, *e.g.*, high ambient temperature, erratic grid conditions etc. The entire electronic control circuits have been tropicalised with higher thermal tolerances because of high ambient temperature. Electrical and electronic systems have been upgraded to withstand surges and fluctuation in voltages.

Considerations influencing windmill sites

More thorough and scientific measurements and analysis of atmospheric boundary layer at wind energy sites can lead to optimum tower heights and economical design of wind turbines. Wind velocity measurements at three to five metre heights will yield more accurate polynomial boundary layer velocity distribution. Selection of suitable hill slopes preceding a wind farm can beneficially accelerate the atmospheric boundary layer flow. Similarly, water fronts reduce resistance to wind flow, reduce boundary layer thickness and yield higher wind velocities at lower heights, rendering dam tops potential sites for wind turbines, which can conveniently supply to the grid if it is a hydropower dam. Fluming effect of the river valley is also highly beneficial. Energy banking, as prevalent in Karnataka State, can encourage expeditious development of wind energy in other states.

Preliminary considerations for windmill design

A wind mill is a regulated machine having a three-bladed rotor which is coupled to the generator through flanges. The blades are carefully designed to give high performance and are made of glass reinforced plastic material. The length of a blade may vary between 10 to 15 metres. The wind mill is mounted on a sturdy lattice tower. The tallest tower installed in India is of 50 metres height, and the largest capacity of windmills is 350 kW with rotor diameter of 33.4 m, cut-in wind speed of 3.5 m/s, cut-off wind speed of 25 m/s and survival wind speed of 67 m/s. The rotor speed may range between 24 to 32 r.p.m.

Wind power generation as a technology is relatively young in the Indian subcontinent and involves many crucial activities apart from manufacturing of windmills. World renowned companies have started their operations in India with Total Power Solution on a turn-key basis, customised for needs of clients such as :

1. Site and wind data assessment and micrositing.
2. Wind farm designing and engineering.
3. Economic feasibility analysis and finance services.
4. Procurement of land and site development.
5. Civil and electrical works associated with erection of windmills.
6. Complete infrastructure facilities for development of large wind farms.
7. Establishment of independent substations.
8. Operation and maintenance.

Fig. 15.1 (A) shows an individual windmill installation with wind turbine mounted on a lattice tower while another.

Fig. 15.1 (B) shows a wind farm with a number of windmills.

Fig. 15.1. (A) A windmill installation showing rotor blades and the lattice tower.

Fig. 15.1. (B) A wind farm showing number of windmills.

15.1. Absolute and Relative Velocities

When the fluid enters a turbomachine and comes in contact with the rotating element, it becomes necessary to deal with both the absolute as well as the relative velocities of flow. The absolute velocity of a body is its velocity relative to earth.

The relative velocity of a body is its velocity relative to a second body which may, in turn, be in motion relative to the earth. *The absolute velocity of the first body is the vector sum of its velocity relative to the second body and the absolute velocity of the latter.*

A better understand of the relationship between the absolute and relative velocities may be obtained by considering the motion of a truck with a uniform velocity u. A man on the truck walks over from A to B at uniform speed w relative to the truck. If the time taken by him in travelling from A to B is Δt, the truck moves a distance $u\Delta t$ in the meantime. Fig. 15.2 illustrates the situation at time Δt. By the time he reaches the point B, the point B itself would have moved a distance $u\Delta t$ relative to earth in the direction in which the truck is moving (indicated by C). Thus the path of the man relative to the truck is AB, but relative to earth it is AC. The velocity V is the absolute velocity of the man and w is his velocity relative to the truck. The velocity diagram relating the three velocities namely u, V and w will be a triangle similar to triangle ABC.

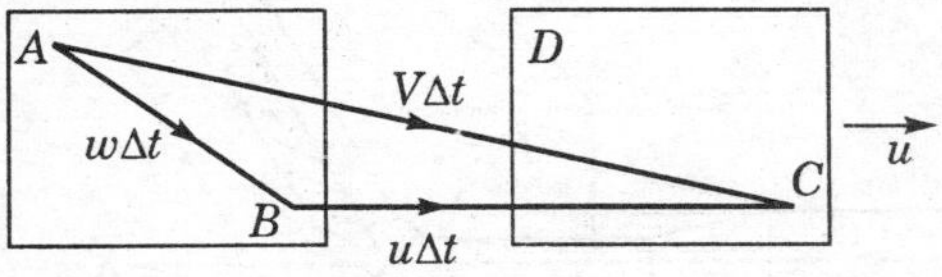

Fig. 15.2. Relation between absolute and relative velocities.

The angle which the absolute velocity vector V makes with the direction of motion of the body, that is with the velocity vector u, is generally denoted by α and the angle made by the relative velocity vector w with the velocity vector u by β. There are three different velocity diagrams possible depending upon the magnitude of u and magnitude and direction of V. These are shown in Fig. 15.3.

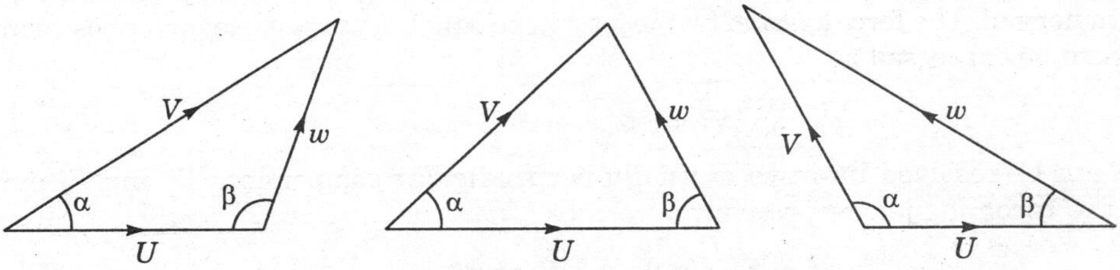

Fig. 15.3. Velocity diagrams for different values of α and β.

The following relationships amongst the three velocities and the two angles may be written.

$$\left.\begin{aligned} V\cos\alpha &= u - w\cos\beta \\ V\sin\alpha &= w\sin\beta \\ \text{and}\qquad \frac{V}{\sin\beta} &= \frac{w}{\sin\alpha} = \frac{u}{\sin(180^\circ - \alpha - \beta)} \end{aligned}\right\} \qquad \text{...(15.1)}$$

15.2. Impact of Free Jets

When momentum of a flowing fluid is changed by a body, a dynamic force is exerted upon it. The evaluation of this force is easy if the body is stationary. The momentum principle is employed to determine this force. The dynamic force exerted by the fluid jet is equal to the rate of change of momentum in the desired direction. If W is the weight of fluid striking the body per second and ΔV_x is the change in velocity of fluid in the x-direction caused by the body, then the dynamic force F_x in the x-direction is given by the momentum equation

$$F_x = \frac{W}{g}\,\Delta V_x \qquad \text{...(15.2)}$$

If the body is in motion, either translatory or rotary, the evaluation of the dynamic force becomes complicated on two accounts : (*i*) determination of the weight of fluid W striking the body, and (*ii*) determination of change in velocity. Evaluation of these quantities involves the consideration of both the absolute velocity and the relative velocity thereby complicating the process of determination of dynamic force.

15.2.1. Jet Striking a Stationary Inclined Plate

Consider a stationary smooth plate held at an angle θ to the direction of flow, as shown in Fig. 15.4. Let a fluid jet of area A strike the plate with a velocity V. After the impact, the jet is divided into two parts as shown. It is assumed that since the plate is smooth, the frictional resistance offered is negligible. The velocity of jet after the impact can, therefore, be assumed unchanged.

The mass of fluid striking the plate per second

$$= \rho AV.$$

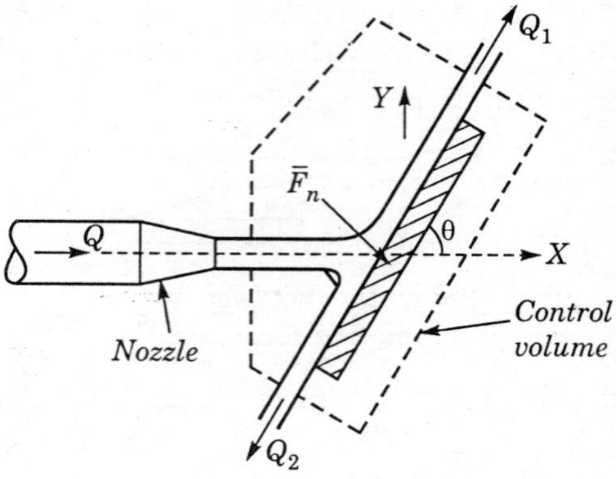

Fig. 15.4. Jet striking a stationary inclined plate.

Momentum equation in the normal direction

Momentum per second of jet before impact in a direction normal to the plate = $(\rho AV)\, V \sin \theta$.

Since the jet after the impact is deflected tangentially, the momentum of jet in the normal direction is zero. According to the momentum principle, the force exerted on the liquid by the plate in the normal direction

$$F_n = \rho AV\,(0 - V \sin \theta)$$
$$= -\,\rho AV^2 \sin \theta.$$

The negative sign indicates that the force exerted by the plate on the jet is in the direction of the outward normal. The force exerted by the jet will be equal in magnitude but opposite in direction to F_n, and can be expressed as

$$\overline{F}_n = \rho AV^2 \sin \theta \qquad \text{...(15.3)}$$

This force can be resolved into two mutually perpendicular components F_x and F_y parallel and normal to the oncoming jet, respectively :

$$F_x = \overline{\mathbf{F}}_{\mathbf{n}} \sin \theta = \rho AV^2 \sin^2 \theta$$

$$F_y = \overline{\mathbf{F}}_{\mathbf{n}} \cos \theta = \rho AV^2 \sin \theta \cos \theta$$

Momentum equation in the tangential direction

Since the plate surface is smooth, there can be no force exerted by the plate on the fluid jet in the tangential direction. One can thus write :

$$F_t = \rho Q_1 V - \rho Q_2 V - \rho QV \cos \theta = 0$$

or

$$Q_1 - Q_2 - Q \cos \theta = 0 \qquad \text{...(15.4)}$$

From the continuity equation

$$Q = Q_1 + Q_2 \qquad \text{...(15.5)}$$

Solving Eqs. (15.4) and (15.5), one gets

$$Q_1 = \frac{Q}{2}\,(1 + \cos \theta) \qquad \text{...(15.6)}$$

$$Q_2 = \frac{Q}{2}\,(1 - \cos \theta) \qquad \text{...(15.7)}$$

In case of a vertical plate, θ = 90°, and the corresponding expression can be obtained by making appropriate substitutions.

Example 15.1. *A 5 cm jet of water strikes a stationary inclined flat plate with a velocity of 20 m/s. The plate makes an angle of 30° with the axis of the jet. Determine the force exerted on the plate and the distribution of flow after the impact of jet on the plate.*

Solution. The normal force exerted by the jet is given by Eq. (15.3) as

$$\overline{F}_n = \rho A V^2 \sin \theta$$

$$= \frac{9810}{9.81} \times \frac{\pi}{4} (5 \times 10^{-2})^2 \times 20^2 \sin 30° = 392.7 \text{ N}.$$

After the impact, the jet splits into two parts Q_1 and Q_2 as shown in Fig. 15.4. The ratio of discharge, using Eqs. (15.6) and (15.7) is

$$\frac{Q_1}{Q_2} = \frac{1 + \cos \theta}{1 - \cos \theta} = \frac{1 + \cos 30°}{1 - \cos 30°} = \frac{1.866}{0.134} = \mathbf{13.9}.$$

15.2.2. Jet Striking a Moving Inclined Plate

Consider a smooth flat plate moving in translation with a constant velocity u in the direction of the jet. Let a fluid jet area A and velocity V strike this moving plate. The velocity with which the jet strikes the plate is $(V - u)$.

The fluid mass striking the moving plate per second

$$= \rho A (V - u)$$

Momentum equation in the normal direction

Momentum of flow entering the control volume per second in a direction normal to the plate

$$= \rho A (V - u)(V - u) \sin \theta = \rho A (V - u)^2 \sin \theta$$

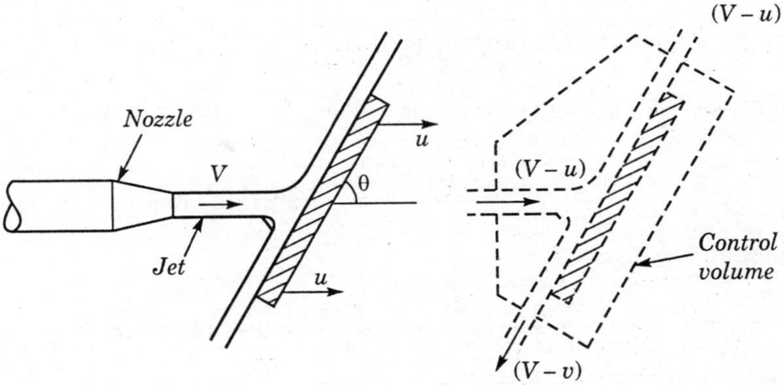

Fig. 15.5. Fluid jet striking a moving inclined plate.

Similar momentum leaving the control volume has a zero magnitude on account of the flow being tangential to the plate. Application of the momentum equation in the normal direction results in the force exerted on the fluid by the plate :

$$F_n = \rho A (V - u)^2 (0 - \sin \theta) = - \rho A (V - u)^2 \sin \theta.$$

The force due to impact of jet on the plate is equal in magnitude but opposite in direction to F_n, *i.e.*

$$\overline{F}_n = \rho A (V - u)^2 \sin \theta \quad \text{...(15.8)}$$

The component of this force in the direction of the jet

$$F_x = \overline{F}_n \sin \theta = \rho A (V - u)^2 \sin^2 \theta$$

The work done by this force in the direction of the jet

$$= F_x \,.\, u = \rho A (V - u)^2 \sin \theta \,.\, u \quad \text{...(15.9)}$$

If the plate is held perpendicular to the direction of the jet, then $\theta = 90°$. For this case, various quantities can be determined by substituting this value of θ.

Example 15.2. *A 15 m/s velocity jet of water 5 cm in diameter strikes perpendicularly a flat smooth plate. Determine the force exerted by the jet on the plate, if (i) the plate is at rest (ii) it moves in the direction of jet with a velocity of 5 m/s. Also determine the work done in each case and the efficiency of jet in the second case.*

Solution. (*i*) Normal force exerted by the jet on a stationary flat plate is given by

$$F = \rho A V^2 = \frac{\gamma Q V}{g} = \frac{W}{g} V$$

Weight of water striking the plate per second

$$W = \gamma Q = \gamma \frac{\pi d^2}{4} V = 9810 \times \frac{\pi (5 \times 10^{-2})^2}{4} \times 15 = 288.9 \text{ N}$$

$$F = \frac{288.9}{9.81} \times 15 = 441.7 \text{ N}$$

since the plate is stationary, $u = 0$, and the work done per second $= F \cdot u$ is zero.

(*ii*) When the plate is moving, the force exerted by the jet is given by

$$F = \rho A (V - u)^2$$

$$= \frac{9810}{9.81} \times \frac{\pi (5 \times 10^{-2})^2}{4} \times (15 - 5)^2 = 196.3 \text{ N.}$$

Work done per second by the jet $= F \cdot u$

$$= 196.3 \times 5 = 981.5 \text{ N-m/s}$$

Kinetic energy per second of oncoming jet $= \frac{1}{2} \rho A V^2$

$$= \frac{1}{2} \times \frac{9810}{9.81} \times \frac{\pi (5 \times 10^{-2})^2}{4} \times 15^2 = 2169.7 \text{ N-m/s}$$

$\therefore$ Efficiency of jet

$$\eta = \frac{\text{Work done by jet}}{\text{Kinetic energy of jet}} = \frac{981.5}{2169.7} = 0.4443 = 44.43\%.$$

Example 15.3. *A rectangular metal plate weighing 49 N is suspended vertically by a hinge and can swing freely about the upper horizontal edge. A horizontal jet of water 2.5 cm in diameter with its axis 21 cm below the hinge impinges perpendicularly on the plate with a velocity of 8 m/s. The centre of gravity of plate is 14 cm from the hinge. Determine :*

(i) the horizontal force required to be applied at the centre of gravity to hold the plate in a vertical position.

(ii) the inclination of the plate with vertical if the horizontal force is withdrawn.

Solution. (*i*) Consider the following figure which shows case (*i*). The force F exerted by the jet on the plate

$$F = \rho A V^2$$

$$= \frac{9810}{9.81} \times \frac{\pi}{4} (2.5 \times 10^{-2})^2 \times 8^2 = \mathbf{31.4 \text{ N.}}$$

Let R be the desired horizontal force. Taking moment of forces about the hinge from which

$$R = F \times \frac{21}{14} = 31.4 \times \frac{21}{14} = \mathbf{47.1 \text{ N.}}$$

(*ii*) The horizontal force F exerted by the jet on the inclined plate is located 21 cm below the hinge.

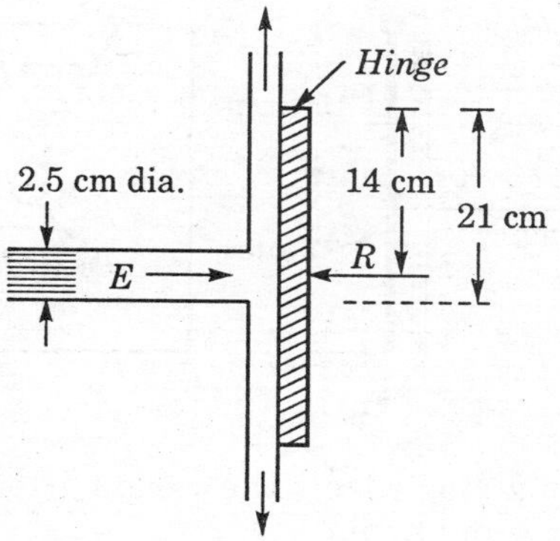

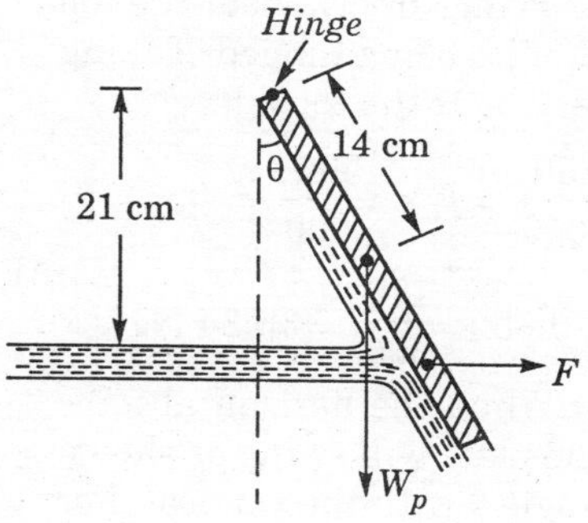

Weight of the plate $W_p = 49$ N

Taking moments about the hinge :

$$W_p \times 14 \sin \theta = F \times 21 = 31.4 \times 21$$

$$\sin \theta = \frac{31.4 \times 21}{49 \times 14} = 0.96$$

or $\theta = \mathbf{73.74°}$.

Example 15.4. *A jet of water issues, from a nozzle with a velocity 20 m/s and it impinges normally on a flat plate moving away from it at 10 m/s. The cross-sectional area of the jet is 0.01 m², and the density of water is 100 kg/m³. Determine the force developed on the plate.*

Solution. Mass of fluid striking the moving plate

$= \rho A_j (V_j - u)$

$= 1000 \times 0.01 \times (20 - 10)$

$= 100$ kg/m³

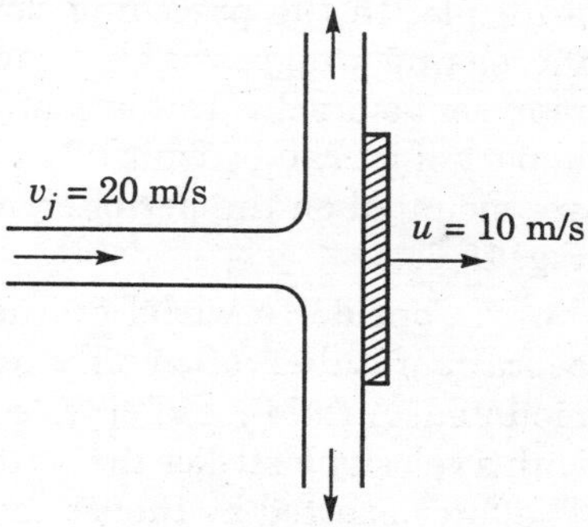

Force developed on the plate

= mass striking per sec

× vel. of jet relative to the plate

$= 100 (20 - 10)$

$= 1000$ kg (mass) m/s²

= **1000 N.**

Example 15.5. *A square plate of mass 12.7 kg, uniform thickness and 300 mm length of edge, is hung so that it can swing freely about its upper horizontal edge. A horizontal jet 20 mm in diameter strikes the plate with a velocity of 15 m/s. When the plate is vertical the jet strikes the plate normally at the centre. (a) what force must be applied at the lower edge of the plate to keep it vertical ? (b) what inclination to the vertical the plate will assume under the action of the jet if allowed to swing freely.*

Solution.

Case (a) Jet strikes the freely swinging plate in the centre, and the plate is held vertical by application of an external force F at the lower edge as shown.

Force exerted on the plate by the jet

$$F_j = \left(\frac{\gamma}{g} \frac{\pi d^2}{4} V\right) V = \frac{9810}{9.81} \times \frac{\pi}{4}\left(\frac{20}{1000}\right)^2 \times (15)^2 = 70.68 \text{ N}$$

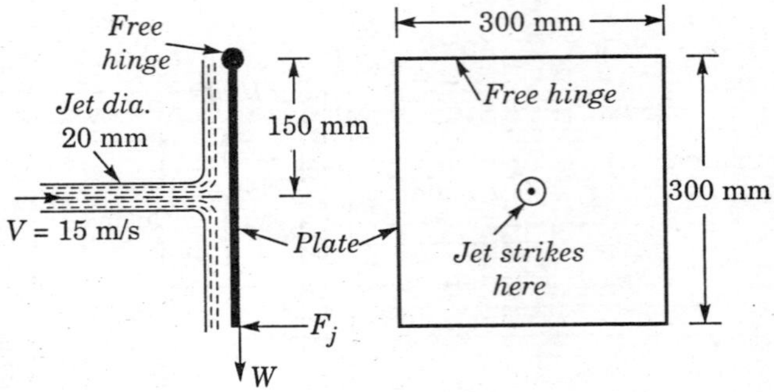

The plate swings freely about the top edge on account of its being hinged. Taking moment of forces about the top edge,

$$F_j \times \frac{150}{1000} = F \times \frac{300}{1000}.$$

$$\therefore \quad F = 70.68 \times \frac{170}{300} = \mathbf{35.34\ N.}$$

Case (b) When the horizontal force is withdrawn, the plate will swing as shown inclined at an angle θ with the vertical. For equilibrium, taking moment of forces about the hinge (top edge). The weight W of the uniformly thick plate acts through its centre of gravity, and is equal to

$$W = \text{mass} \times g = 12.7 \times 9.81 = 124.587 \text{ N}$$

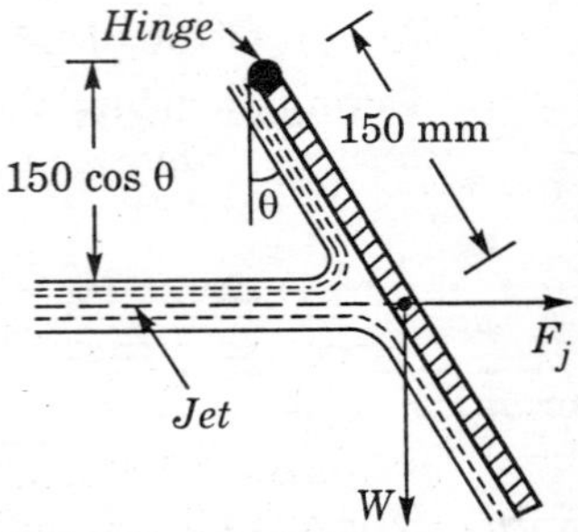

$$W \times \frac{150}{1000} \sin\theta = F_j \times \frac{150}{1000} \cos\theta$$

from which $\tan\theta = \frac{F_j}{W} = \frac{70.68}{124.587} = 0.5673$

and $\theta = \mathbf{29.57°}.$

15.2.3. Flat Plates Mounted on a Wheel

The foregoing cases of dynamic action of jet amply demonstrate the application of momentum principle. In the preceding flow-situation, the distance between the jet and the plate goes on increasing progressively, and, therefore, does not represent a practical situation. The force exerted by the impact of jet can be fruitfully utilised if the flat plates are mounted on the periphery of a wheel as shown in Fig. 15.6.

Consider a wheel on the periphery of which a certain number of evenly spaced flat plates are mounted. A fluid-jet of cross-sectional area A moving with a velocity V strikes the bottom most plate as shown. The force exerted by the jet causes the rotation of the wheel. The flat plates thus occupy the bottom-most position according to their turn. The number and location of plates is so arranged that no portion of jet goes waste without doing work on the plate. Thus the entire fluid mass issuing from the nozzle is considered to strike the plates.

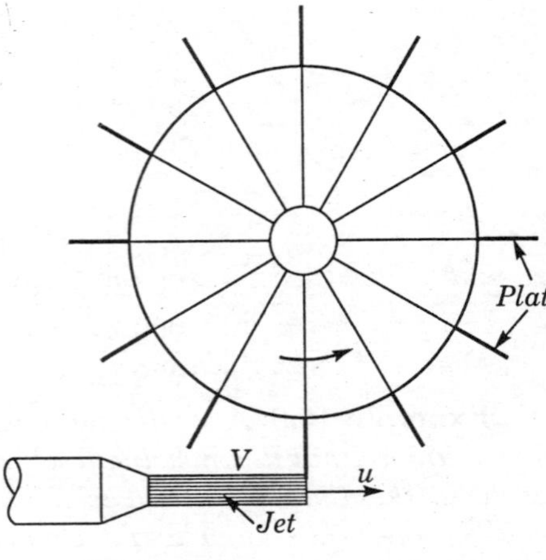

Fig. 15.6. Flat plates mounted on a wheel.

Fluid mass striking the plate per second

$$= \rho AV$$

Velocity with which the jet strikes the plates moving with a tangential velocity $u = V - u$

Force exerted on the plate by the jet

$$= (\rho AV)(V - u).$$

Work done per second

$$= \text{Force} \times \text{Distance moved per second}$$

$$= \rho AV(V - u) \,.\, u \qquad ...(15.10)$$

Kinetic energy of jet per second $= \frac{1}{2}\ (\rho AV)\ V^2$

$$\text{Efficiency of jet} = \frac{\text{Work done on the plate}}{\text{Kinetic energy on jet}}$$

or
$$\eta = \frac{\rho AV\ (V-u) \,.\, u}{\rho AV^3/2} = \frac{2(V-u)\,u}{V^2} \qquad ...(15.11)$$

For a given jet-velocity V, the efficiency η will be maximum when $d\eta/du = 0$, hence

$$\frac{d\eta}{du} = V - 2u = 0\ ; \quad u = V/2$$

The maximum efficiency occurs when the wheel moves with a peripheral velocity u equal to half the jet-velocity V. The maximum efficiency is

$$\eta_{max} = \frac{2\,(V - V/2)\,V/2}{V^2} = \frac{1}{2} = 50\%.$$

Example 15.6. *A 25 m/s velocity jet having a cross-sectional area of 40 cm² strikes a series of flat plates mounted on the periphery of a wheel such that each plate appears successively before the jet. Determine :*

(*i*) *the total force exerted by the jet on the plate,*

(*ii*) *the torque on the wheel,*

(*iii*) *the work done per second on the plate, and*

(*iv*) *the efficiency of jet.*

The jet strikes the plates at a radial distance of 1 m from the wheel axis, the wheel rotating at a speed of 30 r.p.m.

Solution. Velocity with which the plate moves

$$u = \frac{2\pi r\,N}{60} = \frac{2\pi \times 1 \times 30}{60} = \pi \text{ m/s.}$$

(*i*) Force exerted on the plate $F = \rho\,AV\,(V - u)$

$$= \frac{9810}{9.81} \times \frac{40}{(100)^2} \times 25\,(25 - \pi) = 2185.9 \text{ N}$$

(*ii*) Torque on the wheel = $F.r$ = 2185.9 N = 2185.9 N-m

(*iii*) Work done per second = $F.u$ = 2185.9 × π = 6867 N-m/s.

(*iv*) Efficiency of jet, from Eq. (15.11), is given by

$$\eta = \frac{2\,(V-u)u}{V^2} = \frac{2(25-\pi)\pi}{25^2} = 0.2197 = \mathbf{21.97\%}.$$

15.2.4. Force Exerted by a Jet on a Curved Vane

Consider a symmetrical curved vane having smooth surface. Let a fluid jet of cross-sectional area A and velocity V strike the vane at its centre as shown in Fig. 15.7. After the impact the jet is deflected equally along the vane surface. Since the vane surface is considered smooth, the velocity of the jet at exit may be assumed unchanged. The momentum of jet entering the control volume per second in the direction of oncoming jet is :

$$M_{in} = \rho AV\ V = \rho\ AV^2$$

momentum leaving the control volume per second in the direction of oncoming jet

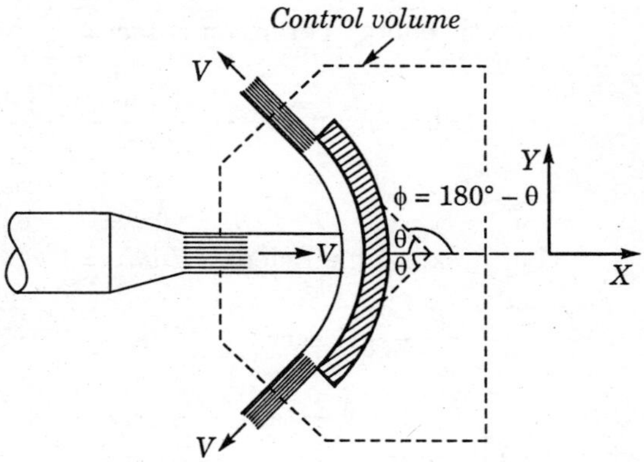

Fig. 15.7. Jet striking a curved vane at its centre.

$$M_{out} = -\frac{\rho AV}{2} \,.\, V\cos\theta - \frac{\rho AV}{2} \,.\, V\cos\theta = -\rho AV^2\cos\theta.$$

Force exerted by the curved vane on the jet in the direction of oncoming jet

$$F_x = M_{out} - M_{in}$$
$$= -\rho AV^2\cos\theta - \rho AV^2 = -\rho AV^2(1+\cos\theta).$$

The negative sign indicates that the force acts in the negative x-direction. The force exerted by the jet on the curved vane, $\overline{F_x}$, will be equal in magnitude but opposite in direction to F_x, thus,

$$\overline{F_x} = \rho AV^2(1+\cos\theta) \qquad ...(15.12)$$

Since the vane is symmetrical, the net momentum of jet in the y-direction will be zero.

Limiting cases

(*i*) For $\theta = 90°$, the vane becomes a flat plate held at right angles to the direction of jet, and $\overline{F_x} = \rho AV^2$. Comparing this with Eq. (15.12), we find that the force exerted by the jet on a curved vane is more than that for a flat plate.

(*ii*) For $\theta = 0°$, the vane becomes semi-circular, and the force exerted on the vane, $\overline{F_x} = 2\rho AV^2$ which is twice as great as on a flat plate.

15.2.4.1. Curved Vane Moving in Translation

If the vane is considered moving in translation with a velocity u, the mass of fluid that strikes the vane per second is $\rho A(V-u)$.

Momentum entering the control volume per second in the direction of the oncoming jet

$$M_{in} = \rho A(V-u)\,.\,(V-u) = \rho A(V-u)^2.$$

Momentum leaving the control volume in the same direction

$$M_{out} = -\rho A(V-u)^2\cos\theta.$$

Force exerted on the jet by the vane in the direction of oncoming jet $F_x = M_{out} - M_{in}$

$$= -\rho A(V-u)^2(1+\cos\theta) \qquad ...(15.13)$$

The force exerted by the jet on the vane in the same direction being equal in magnitude and opposite in direction to F_x is

$$\overline{F_x} = \rho A(V-u)^2(1+\cos\theta) \qquad ...(15.14)$$

Work done by the jet per second $= \overline{F_x} \,.\, u$

$$= \rho A(V-u)^2(1+\cos\theta)\,.\,u$$

Kinetic energy per second of jet $= \frac{1}{2}\rho AV\,.\,V^2 = \frac{1}{2}\rho AV^3$

$$\text{Efficiency } \eta = \frac{\text{Work done by jet}}{\text{Kinetic energy of jet}} = \frac{\rho A(V-u)^2(1+\cos\theta)}{\frac{1}{2}\rho AV^3}$$

$$= 2(1+\cos\theta)(V-u)^2 u/V^3 \qquad ...(15.15)$$

For maximum efficiency, $d\eta/du = 0$

or
$$\frac{2}{V^3}(1+\cos\theta)(V^2+3u^2-4Vu) = 0$$

since $1+\cos\theta \neq 0$, $\therefore$ $V^2+3u^2-4Vu = 0$ or $(V-u)(V-3u) = 0$

which results in $u = V$ and $u = V/3$.

When $u = V$, the work done reduces to zero. For maximum efficiency, therefore, the condition is $u = V/3$, and the maximum efficiency is

$$\eta_{max} = \frac{2\,(1+\cos\theta)\,(V - V/3)^2 \,.\, V/3}{V^3}$$

$$= \frac{8}{27}\,(1 + \cos\theta) \qquad ...(15.16)$$

For a semi-circular vane, $\theta = 0°$, and $\eta_{max} = 16/27$.

Example 15.7. *A water jet of 5 cm diameter moves with a constant velocity of 20 m/sec and strikes a stationary and symmetrical curved plate at the centre. If the jet is deflected through an angle of 120° at the outlet end of the curved plate, find the force exerted by the water jet in the direction of the jet.*

Solution. Referring to Fig. 15.7, $\theta = 60°$, $V = 20$ m/s, $\phi = 120°$.

Force exerted by the water jet on the curved vane is given by Eq. (15.12)

$$F_x = \rho\, AV^2\,(1 + \cos\theta)$$

$$= \frac{9810}{9.81} \times \frac{\pi}{4}\left(\frac{5}{100}\right)^2 (20)^2\,(1 + \cos 60°) = 1178.10 \text{ N} = \mathbf{1.178\ kN.}$$

15.2.4.2. Curved Vanes Mounted on a Wheel

The case of a single moving curved vane just discussed above does not present a practically feasible proposition for performing useful work as the distance between the vane and the jet will progressively increase. The work done by the jet can be fruitfully utilized if the vanes are suitably mounted on the periphery of a wheel. Let the wheel rotate such that the tangential velocity is u and a jet moving with a velocity V strike it similarly as shown in Fig. 15.7. The fluid mass striking the vane per second is ρAV.

The fluid momentum per second before impact in the direction of jet

$$M_{in} = \rho AV\,(V - u)$$

and its value after the impact in the same direction

$$M_{out} = -\,\rho AV\,(V - u)\cos\theta.$$

Force exerted by the jet on the vane in the direction of jet is

$$F = M_{in} - M_{out}$$

$$= \rho AV\,(V - u)\,(1 + \cos\theta) \qquad ...(15.17)$$

Work done per second on the vane $= F \,.\, u = \rho AVu\,(V - u)\,(1 + \cos\theta)$ ...(15.18)

Kinetic energy per second of jet $= \frac{1}{2}\,\rho\, AV^2$

$$\text{Efficiency of the wheel, } \eta = \frac{\rho AVu\,(V - u)\,(1 + \cos\theta)}{\frac{1}{2}\,\rho AV^3}$$

$$= \frac{2u\,(V - u)\,(1 + \cos\theta)}{V^2} \qquad ...(15.19)$$

For a given jet velocity V, the efficiency will be maximum when $d\eta/du = \theta$

or $\qquad V - 2u = 0\,; \qquad u = V/2$

and the maximum efficiency $\eta_{max} = 2 \,.\, \dfrac{V}{2}\left(V - \dfrac{V}{2}\right)(1 + \cos\theta)/V^2 = \dfrac{1}{2}\,(1 + \cos\theta).$

For a semi-circular vane $\theta = 0°$, and $\eta_{max} = 1$ or 100%. A Pelton turbine works on a similar principle. The vanes of this turbine are called buckets and have a double hemispherical shape. The jet of water strikes at the centre of each bucket and is deflected through an angle 165° to 170° depending on the design of bucket.

Example 15.8. *A 5 cm jet of water having a velocity 15 m/s strikes at the centre of a hemispherical vane. Determine the force exerted on the vane if it moves with a linear velocity of 5 m/s.*

If a series of such vanes are mounted on the periphery of a wheel and the jet strikes each vane in succession, what would be the force exerted by the jet ?

Solution. Relative velocity with which the jet approaches the moving vane

$$= V - u = 15 - 5 = 10 \text{ m/s}$$

Fluid mass striking the vane per second

$$= \rho A (V - u) = \frac{9810}{9.81} \times \frac{\pi}{4} (5 \times 10^{-2})^2 \times 10 = 19.62 \text{ kg/s}$$

Change in velocity $= (V - u) - [-(V - u) \cos \theta]$

$$= (V - u)(1 + \cos \theta)$$

for a hemispherical vane, $\theta = 0°$; the force exerted by the jet on the vane

$$F = \text{Mass/sec} \times \text{Change in velocity}$$

$$= 19.62 \times (15 - 5)(1 + \cos 0°) = \mathbf{392.4 \text{ N.}}$$

This result could have also been obtained by making substitution in the formula itself.

Jet striking series of vanes

Fluid mass striking the vane $= \rho AV$

$$= \frac{9810}{9.81} \times \frac{\pi}{4} (5 \times 10^{-2})^2 \times 15 = 29.43 \text{ kg/s}$$

Change in velocity remains the same hence the force exerted on the vane by the jet

$$F = 29.43 \times (15 - 5)(1 + \cos 0°) = \mathbf{588.6 \text{ N.}}$$

15.2.5. Dynamic Force on a Curved Blade Moving in Translation

Consider Fig. 15.8 showing a curved blade moving in translation at a uniform velocity u from left to right. A jet of fluid moving with a velocity V_1 and cross-sectional area a_1 strikes the blade as shown. Consider the extreme case when the blade moves faster than the jet, that is $u > V_1$. Since the blade is moving faster than the jet, there will always be a certain gap between the two and the jet will not be able to reach and strike the blade. This situation will continue till $u = V_1$. It is only when $u < V_1$, that the jet can strike the blade. The weight of fluid striking it will be proportional to $(V_1 - u)$, the jet velocity relative to the blade. If W' is the weight of fluid striking the blade per unit time, then

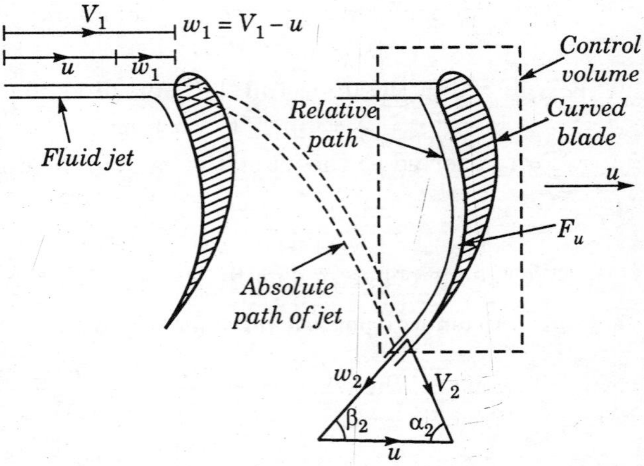

Fig. 15.8. Fluid jet striking a blade in translation.

$$W' = \frac{\gamma}{g} a_1 (V_1 - u) = \frac{\gamma a_1 w_1}{g} \quad \text{...(15.19 a)}$$

The momentum per second of fluid entering the control volume in the u-direction is $(W'V_1)/g$, and the momentum per second leaving the control volume in the same direction is $(W'V_2 \cos \alpha_2)/g$. The dynamic force exerted on the fluid by the blade is

$$F_u = \frac{W'}{g} (V_2 \cos \alpha_2 - V_1) \quad \text{...(15.20)}$$

From Newton's third law which states that the action and its reaction are equal in magnitude but opposite in direction, the force F_u' exerted by the fluid on the blade will be equal to F_u, but its direction will be opposite to F_u and is given by Eq. (15.20). The dynamic force exerted on the blade may be expressed as

$$F_u' = \frac{W'}{g}(V_1 - V_2 \cos \alpha_2) \quad ...(15.21)$$

Comparing Eqs. (15.19 *a*) and (15.20), we find that in computing the weight of fluid striking the blade per second the relative velocity is to be used but while determining the velocity change it is the absolute velocity which must be considered.

15.2.5.1. Jet Striking a moving Vane Tangentially at one Tip

In order to obtain a higher efficiency, it is always desirable that the relative velocity vector w_1 strike the vane tangentially at the inlet.

Consider a curved vane moving with a speed u and a fluid jet striking it tangentially with an absolute velocity V_1 at its inlet tip. Vectorial subtraction of vane speed u from the absolute jet velocity V_1 results in the relative velocity w_1. It is with this relative velocity that the jet strikes the vane tangentially at the inlet tip. The inlet velocity diagram *ABC*, see Fig. 15.9, shows the vane velocity vector u_1 represented by *AB*, the absolute velocity V_1 by *AC* and the relative velocity w_1 by *BC*. The angle β_1 which the relative velocity vector makes with the direction of motion of the vane at the inlet tip, *i.e.*, *AB*, is called the vane or the blade angle at inlet. For fluid entry without shock and minimum energy loss, the relative velocity vector w_1 must be tangential to the blade tip at inlet.

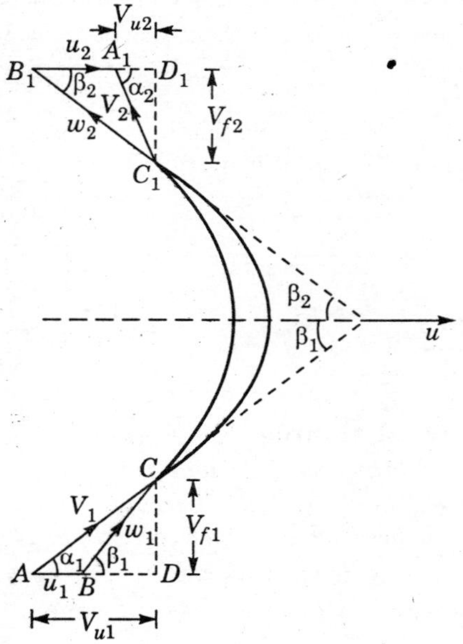

Fig. 15.9. Velocity diagrams for a fluid jet striking tangentially a moving vane.

The direction of the fluid jet changes as it flows over the smooth surface of the vane, and at the outlet tip it emerges with a relative velocity w_2 inclined at angle β_2 as shown in the figure. The angle β_2 is known as the blade or vane angle at the outlet. If the vane surface is considered very smooth, the relative velocities w_1 and w_2 may be assumed equal ($w_1 = w_2$). But in practice, however smooth may be the blade surface, there is always some friction which reduces the magnitude of w_2, thus $w_2 < w_1$. The effect of blade friction is expressed by the ratio

$$K = w_2/w_1$$

where K is called the blade velocity coefficient. The velocity triangle $A_1B_1C_1$ represents flow conditions at the outlet. From the inlet and outlet velocity diagrams, it is evident that the absolute velocity vectors V_1 and V_2 have two mutually perpendicular components :

(*i*) the component in the direction of blade motion (*i.e.* along u) which is denoted by V_u and known as the velocity of whirl.

(*ii*) the component at right angles to the direction of blade motion, denoted by V_f is called velocity of flow.

It is the change in the whirl velocity V_u between inlet and outlet sections that produces the force on the vane and is responsible for doing work. From the momentum principle, the force on the vane in the direction of vane motion is

$$F = \frac{W}{g}[V_{u_1} - (-V_{u_2})] = \frac{W}{g}(V_{u_1} + V_{u_2})$$

here W is the weight of fluid striking the vane per second. The above expression is valid when $\alpha_2 < 90°$. If $\alpha_2 > 90°$, F will be given by

$$F = \frac{W}{g}(V_{u_1} - V_{u_2})$$

Thus the general expression for F can be written as $F = \frac{W}{g}(V_{u_1} \pm V_{u_2})$ ...(15.22)

The rate at which work is done on the vane is equal to the product of the driving force and the vane velocity.

∴ Work done per second = $F \cdot u$

$$= \frac{W}{g}(V_{u_1} \pm V_{u_2}) \cdot u$$

Vane efficiency is defined as the ratio of work done to energy supplied

$$\eta = \frac{W(V_{u_1} \pm V_{u_2}) \cdot u/g}{WV_1^2/2g} = \frac{2(V_{u_1} \pm V_{u_2}) \cdot u}{V_1^2} \quad ...(15.24)$$

If the blade friction is negligible, then the work done by the jet can be taken as equal to the change in kinetic energy. The vane efficiency may then also be written as

$$\eta = \frac{\text{Change in kinetic energy}}{\text{Kinetic energy of jet}}$$

$$= \frac{W}{2g}(V_1^2 - V_2^2) \Big/ \frac{WV_1^2}{2g} = 1 - \left(\frac{V_2}{V_1}\right)^2$$

The efficiency would be maximum when V_2 becomes minimum.

Example 15.9. *A jet of water moving at 12 m/s impinges on a concave shaped vane to deflect the jet through 120° when stationary. If the vane is moving at 5 m/s, find the angle of jet so that there is no shock at inlet. Also compute the absolute velocity of jet at exit both in magnitude and direction, and the work done per second per N of water. Assume that the vane is smooth.*

Solution. Assume that the vane is symmetrical. For no shock, the jet strikes the vane inlet tangentially.

$$\beta_1 = \beta_2 = 30°$$

Given : $V_1 = 12$ m/s, $u_1 = u_2 = 5$ m/s

From the velocity triangle at inlet, one can write

$$\frac{V_1}{\sin(180° - \beta_1)} = \frac{u_1}{(\sin \beta_1 - \alpha_1)} = \frac{w_1}{\sin \alpha_1}$$

or
$$\frac{12}{\sin 150°} = \frac{5}{\sin(30° - \alpha_1)}$$

$$\sin(30° - \alpha_1) = \frac{5}{12}\sin 150° = 0.20833$$

∴ $$\alpha_1 = \mathbf{18°} \text{ and } w_1 = 12 \times \frac{\sin 18°}{\sin 150°}$$

$$= 12 \times \frac{0.3090}{0.500} = 7.416 \text{ m/s.}$$

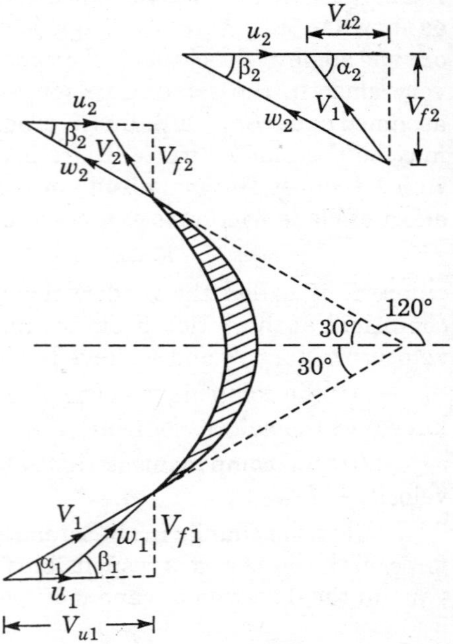

Since the vane is smooth, the blade friction can be neglected and $w_1 = w_2$. From the outlet velocity triangle,

$$V_{f_2} = w_2 \sin \beta_2 = 7.416 \sin 30° = 3.708 \text{ m/s}$$

$$w_2 \cos \beta_2 = 7.416 \cos 30° = 6.422 \text{ m/s}$$

$$V_{u_2} = w_2 \cos \beta_2 - u = 1.422 \text{ m/s}$$

$$\tan \alpha_2 = \frac{V_{f_2}}{w_2 \cos \beta_2 - u} = \frac{3.708}{6.422 - 5} = 2.607 \quad \text{or} \quad \alpha_2 = \mathbf{69°}.$$

The absolute velocity at the vane outlet V_2 can be obtained as

$$V_{f_2} = V_2 \sin \alpha_2,$$

$$\therefore \quad V_2 = \frac{V_{f_1}}{\sin \alpha_2} = \frac{3.708}{\sin 69°} = 3.972 \text{ m/s}$$

Since $\alpha_2 < 90°$, positive sign will be applicable in Eq. (15.22). The work done per second per N of water

$$= \frac{(V_{u_1} + V_{u_2})\,u}{g}$$

$$V_{u_1} = V_1 \cos \alpha_1 = 12 \cos 18° = 12 \times 0.951 = 11.412 \text{ m/s}$$

$$V_{u_2} = 1.422 \text{ m/s}, \quad u = 5 \text{ m/s, making these substitutions,}$$

$$\text{Work done/sec/N of water} = \frac{(11.412 + 1.422) \times 5}{9.81} = \mathbf{6.54 \text{ N-m/s/N.}}$$

Example 15.10. *A 40 m/s velocity jet of water strikes without shock a series of vanes moving at 10 m/s. The jet is inclined at an angle of 20° to the direction of motion of vanes. The relative velocity of jet at outlet is 0.9 times of the value at inlet and the absolute velocity of water at exit is to be normal to the motion of vanes. Determine :*

(i) vane angles at entrance and exit,

(ii) work done on vanes per second per N of water supplied by the jet, and

(iii) the hydraulic efficiency.

Solution. (*i*) With reference to the velocity triangles shown, the data given is :

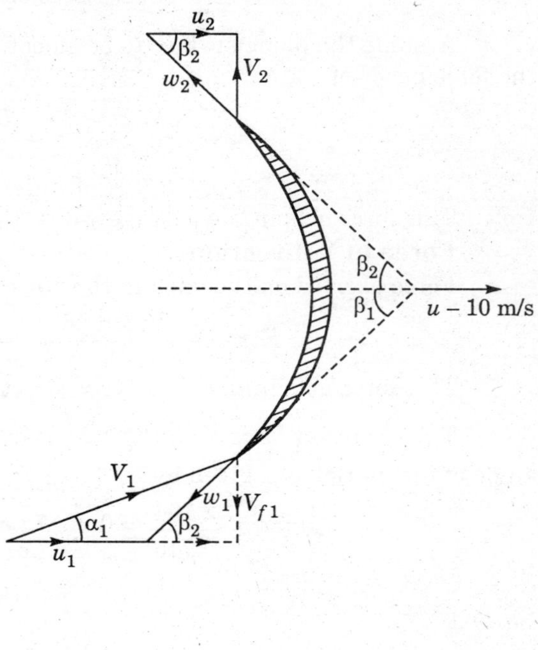

$V_1 = 40$ m/s, $u = u_1 = u_2 = 10$ m/s

$\alpha_1 = 20°$, $w_2 = 0.9\, w_1$

$\alpha_2 = 90°$

From the inlet velocity triangle,

$$\frac{V_1}{\sin (180° - \beta_1)} = \frac{w_1}{\sin \alpha_1}$$

$$V_{f_1} = V_1 \sin \alpha_1 = 40 \sin 20° = 13.69 \text{ m/s}$$

$$V_{u_1} = V_1 \cos \alpha_1 = 40 \cos 20° = 37.59 \text{ m/s}$$

$$\tan \beta_1 = \frac{V_{f_1}}{V_{u_1} - u_1} = \frac{13.68}{37.59 - 10} = 0.4958\,;$$

$$\beta_1 = \mathbf{26.3375°}$$

$$w_1 = V_1 \frac{\sin \alpha_1}{\sin \beta_1} = 40 \times \frac{\sin 20°}{\sin 26.3375°} = 30.838 \text{ m/s}$$

$$w_2 = 0.2\, w_1 = 0.9 \times 30.838 = 27.754 \text{ m/s}$$

From the outlet velocity diagram

$$\cos \beta_2 = \frac{u_2}{w_2} = \frac{10}{27.754} = 0.360 \; ; \; \beta_2 = \mathbf{68.9}$$

The blade angles at entrance and exit are 26.3375° and 68.9° respectively.

(*ii*) Work done per second per N of water

$$= \frac{V_{u_1} . u}{g} \qquad \text{(since } V_{u_2} = 0 \text{ on account of } \alpha_2 \text{ being 90°)}$$

$$= \frac{37.59 \times 10}{9.81} = \mathbf{38.318 \text{ N-m/s/N}}$$

(*iii*) Hydraulic efficiency $\eta = \dfrac{\text{Work done per N per sec}}{\text{Kinetic energy supplied per N per sec}}$

$$= \frac{38.318}{V_1^2/2g} = \frac{2 \times 9.81 \times 38.318}{(40)^2} = 0.46987 \quad \text{or} \quad \mathbf{46.987\%.}$$

Example 15.11. *A stationary vane having an inlet angle of zero degree and an outlet angle of 25° receives water at a velocity of 50 m/s. Determine the components of force acting on it in the direction of jet velocity and normal to it. Also find the resultant force in magnitude and direction.*

(a) If the vane stated above is moving with a velocity of 20 m/s in the direction of jet, calculate the force components in the direction of the vane velocity and across it, also the resultant force in magnitude and direction. Calculate the work done and horse power developed per N of the flow.

Solution.

Force in X-direction :

Force exerted on the vane in the direction of jet

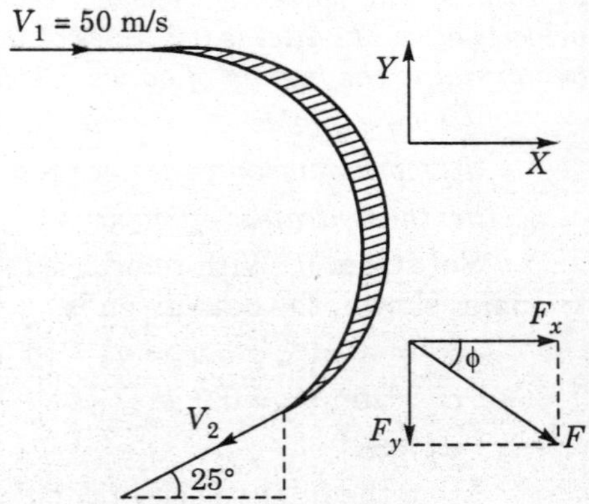

$$F_x = \frac{W}{g} [V_1 - (-V_2 \cos 25°)]$$

$$= \frac{W}{g} (V_1 + V_2 \cos 25°)$$

Assume the blade surface to be smooth, $V_1 = V_2$, and the force per N of water

$$= \frac{V_1 + V_2 \cos 25°}{9.81} = \frac{50 + 50 \cos 25°}{g}$$

$= 9.716$ N/N of water.

This force acts in the positive x-direction as shown.

Force in Y-direction

Force exerted on the vane in the normal direction to the jet per N of water

$$F_y = \frac{(0 - V_2 \sin 25°)}{9.81} = -\frac{50 \times 0.4226}{9.81} = -2.15 \text{ N/N of water.}$$

This force acts in the negative y-direction as shown.

The resultant force $F = \sqrt{F_x^2 + F_y^2} = \sqrt{(9.716)^2 + (-2.154)^2} = 9.95$ N/N of water

angle of inclination ϕ is given by

$$\tan \phi = \frac{2.154}{9.716} = 0.2217 \; ; \; \phi = 12.5° \text{ with the horizontal.}$$

(*b*) **Vane moving in the direction of jet :**

$$V_{u_2} = V_1 = 50 \text{ m/s}$$

$$w_1 = V_1 - u_1 = 50 - 20 = 30 \text{ m/s.}$$

Assuming that the blade surface is smooth, $w_1 = w_2 = 30$ m/s

$$V_{u_2} = w_2 \cos \beta_2 - u_2 = 30 \cos 25° - 20 = 7.189 \text{ m/s.}$$

Force in X-direction

Force generated on the vane per N of water in the direction of jet

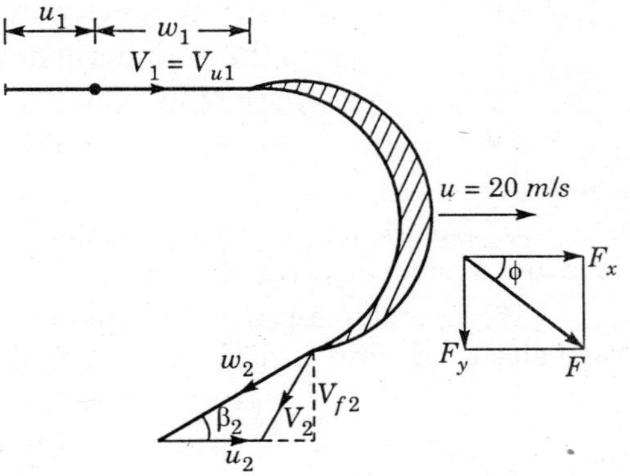

$$F_x = \frac{V_{u_1} + V_{u_2}}{g}$$

$$= \frac{50 + 7.189}{9.91}$$

$$= 5.829 \text{ N/N of water } (\longrightarrow)$$

Force in Y-direction

$$F_y = \frac{V_{f_2} - V_{f_2}}{g}$$

$$= \frac{0 - w_2 \sin \beta_2}{g} = \frac{30 \sin 25°}{9.81} = -1.292 \text{ N/N of water } (\downarrow)$$

This force acts in the negative y-direction.

Resultant force $F = \sqrt{F_x^2 + F_y^2} = \sqrt{(5.829)^2 + (-1.292)^2} = 5.97$ N/N of water

$$\tan \phi = \frac{1.292}{5.829} = 0.22165 \; ; \; \phi = 12.5°$$

Work done per sec per N of water = $F.u$

$$= 5.829 \times 20 = 116.58 \text{ N m/sec/N of water}$$

Power developed per N of water = 116.58 W/N of water.

15.2.5.2. Curved Vanes Mounted on a Wheel

Practical application of the case discussed in 15.2.5.1 is the one in which a series of such vanes is fixed radially to the periphery of a wheel. For rotation of the wheel at constant angular speed ω, the blade tip velocities will be

$$u_1 = r_1 \omega \text{ and } u_2 = r_2 \omega$$

r_1 and r_2 being the radii at the inlet and outlet respectively. This case is discussed more in detail in the succeeding article 15.3.

15.3. Theory of Turbomachines

The turbomachines (the Latin word *turbo* means a circular motion) consists of two main parts.

(1) The rotating element which is generally known as the runner if it is a part of a turbine and the impeller if it is a part of pump, and

(2) the casing which encloses the rotating element.

The runner or the impeller is rigidly mounted on the shaft. The rotating element (runner or impeller) has a number of vanes or curved blades mounted on it, and there takes place a transfer of energy between the fluid and the rotating element. If the fluid does work on the rotating element, the machine is a turbine and when the rotating element does work on the fluid the machine is called a pump. Turbines are thus turbomachines which extract energy from the flowing fluid and convert it into useful mechanical work. On the contrary, centrifugal pumps, compressors and blowers are turbomachines which add energy to the flowing fluid.

When a fluid flows through the runner of a turbomachine its radius usually varies along its path. Hence it is desirable to compute torque rather than the force. Consider flow through the runner of a turbomachine. Let the flow enter the runner blade with an absolute velocity V_1 at an angle α_1 and leave it with absolute velocity V_2 at an angle α_2. Fig. 15.10 shows the velocity diagrams at inlet and outlet. It is preferable to draw these diagrams on the blade tips as shown. Using the subscript 1 for conditions at the inlet and subscript 2 for the conditions at the blade outlet, the components of the velocity diagram may be defined.

(1) V represents the absolute velocity of fluid.

(2) u represents the peripheral or tangential velocity of blade.

(3) w represents the fluid velocity relative to the blade.

(4) α is the angle which the absolute velocity vector V makes with the tangential blade velocity u.

(5) β is the angle which the relative velocity vector w makes with tangential blade velocity u. It is also known as the blade angle.

(6) V_u is the tangential component of the absolute velocity V, that is along u. It is also called the velocity of whirl. From the velocity diagram of Fig. 15.10.

$$V_{u_1} = V_1 \cos \alpha_1$$

and

$$V_{u_2} = V_2 \cos \alpha_2.$$

(7) V_f is the radial component of the absolute velocity and is known as the velocity of flow, thus

$$V_{f_1} = V_1 \sin \alpha_1$$

$$V_{f_2} = V_2 \sin \alpha_2.$$

Consider the control volume contained between the dotted circular lines enclosing the runner, as shown in Fig. 15.10. The rate of flow passing through the control volume depends upon only the radial component of the absolute velocity V_1 or V_2.

$$Q = 2\pi r_1 b_1 V_1 \sin \alpha_1 = 2\pi r_1 b_1 V_{f_1}$$

and

$$Q = 2\pi r_2 b_2 V_2 \sin \alpha_2 = 2\pi r_2 b_2 V_{f_2} \qquad \text{...(15.25)}$$

where r_1 and r_2 are the radii, b_1 and b_2 the widths of the flow passages within the runner at inlet and outlet respectively. From the principle of moment of momentum, the torque acting on the fluid within the control volume must be equal to the moment of momentum leaving minus the moment of momentum entering the control volume per unit time. Torque exerted by fluid on the runner is equal and opposite to the torque acting on the fluid within the control volume. The torque exerted on the axis of rotation depends upon only the tangential components of the absolute velocities V_1 and V_2, and is equal to the moment of momentum in the tangential direction entering the runner minus the moment of momentum leaving it in the same direction per unit time :

$$T = \rho Q (V_1 \cos \alpha_1 r_1 - V_2 \cos \alpha_2 r_2)$$

$$= \rho Q (V_{u_1} r_1 - V_{u_2} r_2) \qquad \text{...(15.26)}$$

Eq. (15.26) gives the torque exerted by the fluid upon the runner of the turbomachine. This equation is based on the assumption that conditions at entry and exit from the runner are uniform, that is, all fluid particles at the radius considered have the same velocity. Eq. (15.26) was first given by Leonhard Euler (1707-83) about two hundred years ago and is, therefore, known as Euler's equation for turbomachines.

The torque available from the shaft of a turbine is somewhat less than that given by Eq. (15.26) because of friction in bearings and friction between the fluid and the runner. The power delivered to the runner or work done per unit time is obtained by multiplying the torque with the angular speed ω.

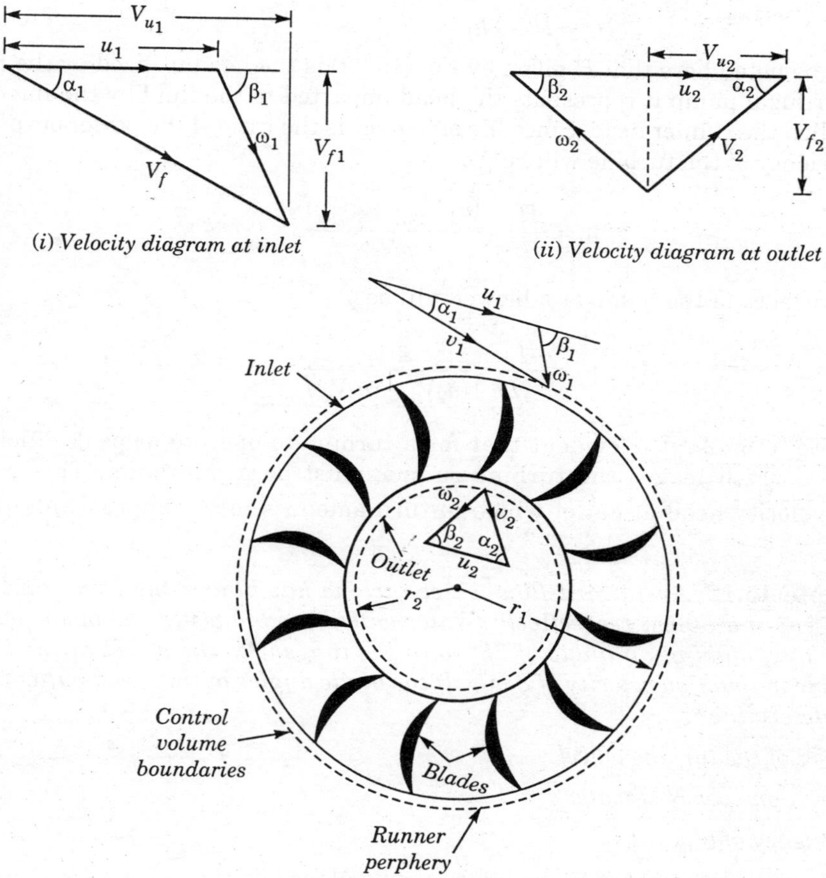

Fig. 15.10. Velocity diagrams at inlet and outlet.

Power delivered to the runner = $T\omega$

or
$$P = \rho Q \omega \, (V_{u_1} r_1 - V_{u_2} r_2)$$

Since,
$$u = r\omega$$

$$P = \rho Q \, (u_1 \, V_{u_1} - u_2 \, V_{u_2}) \qquad \text{...(15.27)}$$

Neglecting the loss of energy, if H is the head utilized by the turbine, then the power delivered to the turbine

$$P = \gamma Q H \qquad \text{...(15.28)}$$

From Eqs. (15.27) and (15.28)

$$H = \frac{u_1 V_{u_1} - u_2 V_{u_2}}{g} \qquad \text{...(15.29)}$$

Taking into account the energy losses which take place on account of surface friction, possible separation from the blades and such other losses as occur in scroll or spiral casing, guide vanes and draft tube, and representing these losses by Σh_l, the total head under which the turbine operates may be written as the sum of the useful work done on the runner given by Eq. (15.29) and the total head loss. Thus, for the turbine the operating head is

$$H_t = H + \Sigma h_l \qquad \text{...(15.30)}$$

and for a centrifugal pump, the head produced will be

$$H_p = H - \Sigma h_l \qquad ...(15.31)$$

It may be noted here that H given by Eq. (15.29) is the head utilized by the runner, and in case of a centrifugal pump it represents the head imparted to the fluid by the impeller. It is also sometimes called the runner head. Since the efficiency is the ratio of the power output to the power input the efficiency of the turbine will be

$$\eta_t = \frac{H}{H_t} = \frac{(u_1 V_{u_1} - u_2 V_{u_2})}{gH_t} \qquad ...(15.32)$$

The efficiency of the pump or a blower will be

$$\eta_p = \frac{H_p}{H} = \frac{gH_p}{(u_2 V_{u_2} - u_1 V_{u_1})} \qquad ...(15.33)$$

From Eq. (15.32), it is evident that for a turbine to operate at peak efficiency the total energy of flow which leaves the turbine runner must be a minimum. This condition leads to minimum velocity head at outlet which in turn amounts to V_{u_2} approaching to a value very close to zero.

Example 15.12. *An inward flow water turbine has blades the inner and outer radii of which are 30 cm and 50 cm respectively. Water enters the blades at the outer periphery with a velocity of 45 m/s, making an angle of 25° with the tangent to the wheel at the inlet tip. Water leaves the blade with a flowvelocity of 8 m/s. If the blade angles at inlet and outlet are 35° and 25° respectively, determine :*

(*i*) *speed of the turbine wheel*

(*ii*) *work done per N of water*

(*iii*) *efficiency of blading.*

Solution. With reference to the figure shown alongside

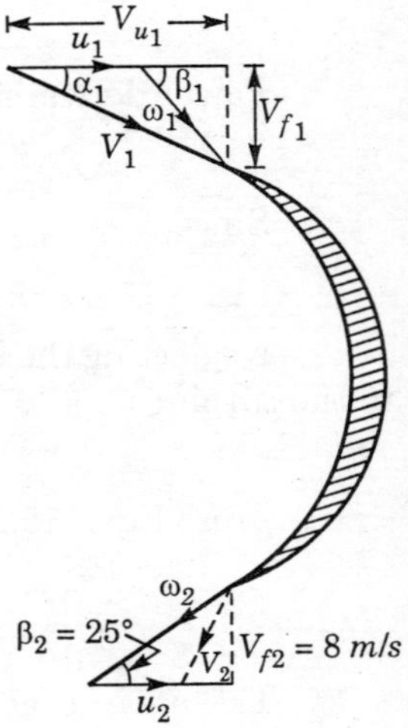

$$u_1 = r_1\omega = \frac{2\pi r_1 N}{60}$$

$$u_2 = r_2\omega = \frac{2\pi r_2 N}{60}$$

$$V_1 = 45 \text{ m/s}, \beta_1 = 35°, \alpha_1 = 25°$$

$$V_{f_1} = V_1 \sin\alpha_1 = 45° \sin 25° = 19.02 \text{ m/s}$$

$$V_{u_1} = V_1 \cos\alpha_1 = 45 \cos 25° = 40.78 \text{ m/s}$$

$$\tan\beta_1 = \frac{V_{f_1}}{V_{u_1} - u_1} ; \tan 35° = \frac{19.02}{40.78 - u_1}$$

from which $u_1 = 13.54$ m/s

$$\therefore \quad N = \frac{60 u_1}{2\pi r_1} = \frac{60 \times 13.54}{2\pi \times 0.50} = \mathbf{129.3\ RPM.}$$

From the outlet velocity triangle

$$\tan 25° = \frac{8}{u_2 + V_{u_2}};$$

$$u_2 = \frac{2\pi r_2 N}{60} = \frac{2\pi \times 0.3 \times 129.3}{60} = 4.06 \text{ m/s}$$

$$V_{u_2} = \frac{8}{\tan 25°} - u_2 = 13.09 \text{ m/s}$$

(*ii*) Work done per second per N of water

$$= (V_{u_1} u_1 + V_{u_2} u_2)/g$$

$$= \frac{40.78 \times 13.54 + 13.09 \times 4.06}{9.81} = \textbf{61.70 N-m/s/N of water.}$$

(*iii*) Efficiency of blading

$$\eta = \frac{\text{Work done per sec per } N}{\text{Kinetic energy per sec per } N} = \frac{61.70}{V_1^2/2g}$$

$$= \frac{61.70 \times 2 \times 9.81}{(45)^2} = \textbf{0.5978 or 59.78\%.}$$

15.4. Hydraulic Turbines

The hydraulic turbine is a prime mover that uses the energy of flowing water and converts it into the mechanical energy in the form of rotation of the runner. (A Prime mover is a machine which uses the raw energy of a substance and converts it into the mechanical energy). Since the fluid medium is water, these turbines are also known as the 'water turbines'. Hydraulic turbines coupled with hydro-generators form the so called 'hydrounits' which are widely used now-a-days for generating electrical power.

Hydropower generation constitutes the largest and the cheapest source of renewable supply of energy. There is a large utilizable hydroelectric potential in India still remaining to be developed. According to the assessment made by the Central Water and Power Commission, the economically utilizable hydropower potential of the country is approximately 25.26 million kW continuous. The hydroelectric installations in operation at present (as of 2007) in the country utilize only 23% of the total estimated potential of the country. The hydropower schemes presently under construction would raise the level of its development to approximately 28% of the estimated potential. Some advanced countries of the world have utilized their water-power potential to a large extent. Sweden and Norway lead the world in exploiting their hydro power potential to 98% , this percentage for European countries is 50% , for North America it is about 30% and for Russia 20%. India has to go a long way in harnessing its hydro-power resources comparable to those of advanced countries in Europe.

15.5. Classification of Hydraulic Turbines

The hydraulic turbines are classified mainly on the basis of process of conversion of hydraulic energy into the mechanical one. Turbines may be placed in one of the two general categories : (*a*) reaction, and (*b*) impulse. In both types the fluid passes through a runner having blades. The momentum of fluid in the tangential direction is changed so that a tangential force on the runner is produced. The runner thus rotates under the influence of a torque produced by the change in moment of momentum and performs useful work, while the fluid leaves it with reduced energy. In the reaction turbine, the static pressure decreases as the fluid passes through the runner. On the other hand, the important feature of the impulse turbine is that there is no change of static pressure across the runner. The impulse turbine has one or more fixed nozzles in each of which the available fluid energy is converted into the kinetic energy of jet. The fluid jet then impinges on the moving blades of the runner where it losses practically all its kinetic energy.

According to the Bernoull's equation, the energy per unit weight of fluid at the entrance of a runner is

$$H_1 = Z_1 + \frac{p_1}{\gamma} + \frac{V_1^2}{2g}$$

and at the exit of the runner, it is

$$H_2 = Z_2 + \frac{p_2}{\gamma} + \frac{V_2^2}{2g}.$$

The energy of flow converted into the mechanical energy represented by the rotation of the turbine shaft (ignoring the energy losses) is

$$H = H_1 - H_2 = Z_1 - Z_2 + \frac{p_1 - p_2}{\gamma} + \frac{V_1^2 - V_2^2}{2g} \qquad ...(15.34)$$

In case of reaction turbines the following relation holds good :

$$\frac{p_1 - p_2}{\gamma} + Z_1 - Z_2 > 0$$

Reaction turbines utilize both potential and kinetic parts of total hydraulic energy. Using Eq. (15.34), the coefficient of reaction (C_r) is defined as

$$1 = \frac{Z_1 - Z_2}{H} + \frac{p_1 - p_2}{\gamma H} + \frac{V_1^2 - V_2^2}{2gH} \approx C_r + \frac{V_1^2 - V_2^2}{2gH}$$

where $$C_r = \frac{(p_1 - p_2)}{\gamma H} \qquad ...(15.35)$$

Greater the value of C_r, greater is the part of potential energy used in reaction turbine. A reaction turbine is characterised by complete admission of water to the runner, *i.e.* water flows from the periphery of the runner and the pressure of flow within the turbine space changes gradually.

15.6. Reaction Turbines

The main types of reaction turbines are the mixed flow and axial flow which differ mainly in the design of runner and its position relative to the turbine shaft. They both have certain common features which are described subsequently.

The main distinguishing features of a reaction turbine are that only a part of the total energy is converted into the kinetic energy before the runner is reached, and the working fluid completely fills the passages in the runner. The pressure of the flowing water changes gradually as it passes through the runner and, therefore, the runner must be enclosed within a water tight casing. Such turbines may also be called pressure turbines. Water is led from the storage reservoir through a penstock (pressure pipe) and enters the runner through the spiral casing (also called a volute or scroll casing) and the distributor with a system of stationary guide vanes. The water after doing work on the runner leaves it through the draft tube and joins the tail race. (Tail race is the channel which carries the used water away and discharges into a natural stream). The turbine space of a reaction turbine is formed by the following hydraulic elements : spiral casing, stay ring, guide wheel (distributor), runner and the draft tube.

15.6.1. Spiral Casing

It is a closed passage of gradually reducing cross-sectional area enclosing the runner. Its function is to distribute the flow over the periphery of the runner such that the entrance velocity is same all around it.

The spiral casing is of logarithmic form and designed to maintain optimum flow conditions. It is manufactured in parts, depending on transport limitations. For larger sizes, spiral casing

Fig. 15.11. (*a*) Spiral casing erection at a power house.

Fig. 15.11. (*b*) Spiral casing installed at a power house.

segments are welded at site. The welding seams are tested by radiography and the complete casing is tested under hydraulic pressure. Pictorial view of a spiral casing during erection at a power house is shown in Fig. 15.11 (*a*). Figure 15.11 (*b*) depicts the completed form of the spiral casing.

15.6.2. Stay Ring

Large turbines may also have stay ring of fixed vanes outside the ring of guide vanes. The main function of the stay vanes is to act as columns helping to support the weight of the non-rotating parts of the turbine and transmit it to the power house foundations. They also direct the water from

Fig. 15.11. (*c*) Stay ring being installed to the spiral casing at the site.

Fig. 15.11. (*d*) Stay ring at the manufacturing unit.

the casing to the turbine guide wheel. They are shaped like airfoils to conform to the streamlines of flow approaching the guide vanes.

A view of the stayring being installed to the inner periphery of the spiral casing is shown in Fig. 15.11 (*c*), while Fig. 15.11 (*d*) shows a stay ring being given final touches at the manufacturing unit. The stay rings upto diameters of 12.5 m have been manufactured, and they are welded to the spiral casing at the power house site.

15.6.3. Guide Wheel (Distributor)

From the spiral casing the water flows through the radial passages provided by the stay ring and reaches the guide wheel. The guide wheel consists of evenly situated movable vanes mounted around its periphery. The vanes which have appropriate shape are fixed in the guide gate between upper and lower rings. Turning the wheel by means of special kinematic mechanism and the servomotors, the position of vanes is simultaneously changed through the same angle. As a consequence, the value of the angular momentum and the flow pattern before it reaches the runner vary resulting in a change of discharge through the turbine. Guide wheel, therefore serves to change the turbine discharge and to stop the delivery of water to the turbine in case of emergency. By regulating the water supply it is possible to develop the required turbine output to meet the load on the hydro-unit.

15.6.4. Runner

The runner constitutes the most important hydraulic component of a turbine. It consists of a number of curved blades fixed on its periphery. When water is allowed to flow through the runner, its angular momentum is reduced thereby producing a torque which rotates the runner. The runner is keyed to the main shaft. While passing, the water is guided by the runner blades and runner bend. From the runner the flow is led directly into the top of the draft tube.

Fig. 15.11 shows the plan and sectional view of a Francis type reaction turbine.

15.6.5. Draft Tube

A draft tube is a gradually expanding closed passage which connects the runner with the tail race. In the reaction turbine, as the flow leaves the runner it possesses a considerably large amount of kinetic energy. It is with this kinetic energy that the flow enters the draft tube. The function of the draft tube is to transform a large part of this high kinetic energy of flow into the pressure energy. The energy finally rejected into the tail race is greatly reduced by the draft tube. The lower end of the draft tube, under all conditions of operation, is kept submerged below the level of water in the tail race. The use of draft tube results in a higher efficiency of the turbine, and enables the turbine to be installed above the tail race level without loss of additional hydraulic energy.

The draft tube being a gradually diverging conduit, it causes the velocity to progressively decrease in such a way that at its exit end it is considerably reduced. Application of Bernoull's equation between the two ends of the draft tube shows that the action of the tube is to reduce the pressure at its entrance to less than the atmospheric value. This enables utilization of the height between the turbine and the tail race for increasing the effective head across the runner. The increased effective head would be equal to the difference in elevation between the head race and tail race levels minus the head losses.

Referring to Fig. 15.11 (*c*), application of Bernoull's equation between the draft tube entrance (2) and the exit (3) yields, the tail water level being taken as the datum.

$$Z_2 + \frac{p_2}{\gamma} + \frac{V_2^2}{2g} = -Z_3 + Z_3 + \frac{V_3^2}{2g} + \text{Losses}$$

Note that point (2) also corresponds to the runner exit. The pressure at the runner exit and the entrance to the draft tube is obtained as

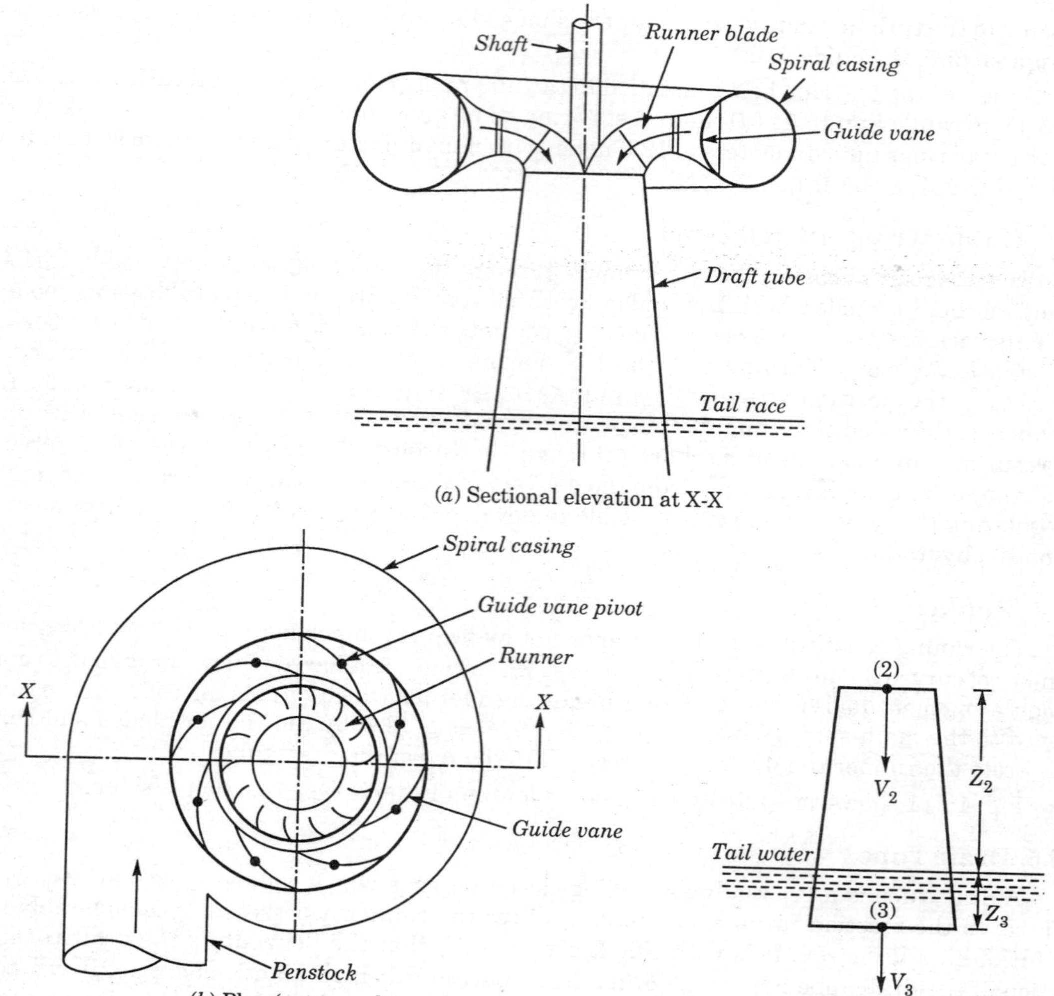

(*a*) Sectional elevation at X-X

(*b*) Plan (not to scale)

(*c*) Notations for draft tube

Fig. 15.11. Hydraulic elements of a Francis type reaction turbine.

$$\frac{p_2}{\gamma} = -Z_2 - \left(\frac{V_2^2 - V_3^2}{2g}\right) + \text{Losses} \qquad ...(15.36)$$

It may be noted here that p_2 is the gauge pressure.

Since the losses in the draft tube and exit velocity V_3 are usually quite small, Eq. (15.36) shows that considerable vacuum is produced at section (2), which effectively increases the head across the runner.

Expressing the loss of energy in the draft tube as

$$h_f = k\,\frac{(V_2^2 - V_3^2)}{2g}$$

one can write

$$\frac{p_2}{\gamma} = -\left[Z_2 + (1-k)\,\frac{(V_2^2 - V_3^2)}{2g}\right] \qquad ...(15.37)$$

Consider a case in which the draft tube is made cylindrical so that $V_3 = V_2$ and assume that the head loss is negligible. Equation (15.36) shows that the pressure head at the runner exit would be less than the atmospheric pressure by an amount Z_2. It is thus evident that with a cylindrical draft tube, a turbine would not lose the head Z_2 because of equal reduction in pressure head at the runner exit. In case of more common draft tubes which have gradually increasing cross-sectional area, Eq. (15.38) shows that the value of (p_2/γ) is further decreased by an amount $(1 - k)\ (V_2^2 - V_3^2)/2g$. In other words, in a draft tube a large portion of kinetic energy is converted into pressure energy.

∴ The net gain in pressure head = $(1 - k)\ \dfrac{(V_2^2 - V_3^2)}{2g}$ and, the efficiency of draft tube

$$\left.\begin{aligned}\eta_d &= \frac{\text{Net gain in pressure head}}{\text{Velocity head at entrance of draft tube}}\\ &= \frac{\dfrac{(V_2^2 - V_3^2)}{2g} - h_f}{V_2^2/2g}\\ &= 1 - \left(\frac{V_3}{V_2}\right)^2 - \frac{2gh_f}{V_2^2}\\ &= (1 - k)(1 - V_3^2/V_2^2)\end{aligned}\right\} \quad ...(15.39)$$

Example 15.13. *A turbine runner discharges water into a draft tube at a velocity of 6 m/s and the velocity at the tube exit is 1.25 m/s. The tail water is 5 m below the entrance of the draft tube and the losses in the draft tube are 0.10 m. Determine the pressure head at the draft tube entrance, and the draft tube efficiency.*

Solution. Using Eq. (15.37) with Z_2 = 5 m, V_2 = 6 m/s, V_3 = 1.25 m/s and losses = 0.10 m.

$$\frac{p_2}{\gamma} = -5 - \frac{[6^2 - (1.25^2)]}{2 \times 9.81} + 0.10 = -6.655 \text{ m.}$$

Pressure head at the draft tube entrance

= 6.655 m below atmospheric pressure.

$$\eta_d = \frac{6^2 - (1.25)^2 - 2 \times 9.81 \times 0.1}{6^2} = 0.902 \quad \text{or} \quad 90.2\%.$$

Comparison between Reaction and Impulse Turbines

The following table shows a comparison between the reaction and impulse turbines.

S. No.	*Reaction turbine*	*Impulse turbine*
1.	Only a fraction of the available hydraulic energy is converted into kinetic energy before the fluid enters the runner.	All the available hydraulic energy is converted into kinetic energy by a nozzle and it is the jet so produced which strikes the runner blades.
2.	Both pressure and velocity change as the fluid passes through the runner. Pressure at inlet is much higher than at the outlet.	It is the velcoity of jet which changes, the pressure throughout remaining atmospheric.
3.	The runner must be enclosed within a watertight casing (scroll casing).	Water-tight casing is not necessary. Casing has no hydraulic function to perform. It only serves to prevent splashing and guide water to the tail race.
4.	Water is admitted over the entire circumference of the runner.	Water is admitted only in the form of jets. There may be one or more jets striking equal number of buckets simultaneously.
5.	Water completely fills at the passages between the blades and while flowing between inlet and outlet sections does work on the blades.	The turbine does not run full and air has a free access to the buckets.

6.	The turbine is connected to the tail race through a draft tube which is a gradually expanding passage. It may be installed above or below the tail race.	The turbine is always installed above the tail race and there is no draft tube used.
7.	The flow regulation is carried out by means of a guide-vane assembly. Other component parts are scroll casing, stay ring, runner and the draft turbine.	Flow regulation is done by means of a needle valve fitted into the nozzle.

15.6.6. Radial-flow, Axial-flow and Mixed-flow Turbines

The reaction turbines are, in turn, sub-divided into several types depending upon the direction of flow within the runner with respect to the turbine axis.

1. ***Radial flow turbines.*** In a radial flow turbine, the water flows in a radial direction, the velocity vector always remaining in plane perpendicular to the turbine axis. The turbine may be radially inward flow type or a radially outward flow type depending upon whether the flow is from the outer periphery towards the inner one or *vice versa*.
2. ***Axial flow turbines.*** The flow within an axial flow turbine is parallel to the turbine axis. The propeller type turbines (with fixed blade), bulb turbines and the Kaplan turbines are essentially axis-flow machines. These turbines are used for low heads ranging from 3 to 80 metres.
3. ***Mixed flow turbines.*** In mixed-flow turbines the flow changes its direction from radial to axial within the runner. The Francis turbine is a mixed-flow turbine.

The position of blades with respect to the turbine axis, their shape and number influence the runner speed of rotation. Axial runners have the least number of blades (3 to 10) and the greatest specific speed.

15.6.7. Requirement of Head and Flow

The modern reaction turbines are used for heads ranging from 3 to 600 metres. In the past two decades great advances in regard to inventing and developing more efficient hydraulic turbines have been made. The Kaplan turbines, which until recently were used for heads upto 40 or 50 m, are now being designed for heads upto 80 m. The axial-flow turbines are basically used for low heads and high flowrates. Having adjustable blades, the Kaplan turbines which are very suitable for operating under varying load and head conditions are replacing Francis turbines within the head range $H = 40$ to 80 m.

The Francis turbines are used for medium heads, say $H = 80$ to 600 m and medium flowrates. The new developments in the design of Francis turbines have replaced the Pelton turbines in the range of 400 to 600 m.

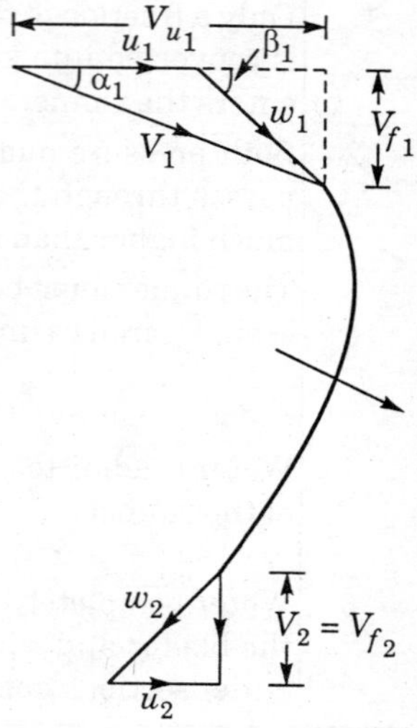

Example 15.14. *In a vertical shaft inward flow reaction turbine the sum of the pressure and kinetic heads at entrance to the spiral casing is 120 m and the vertical distance between this section and the tail race level is 3 m. The peripheral velocity of the runner at entry is 30 m/sec, the radial component of velocity of water (velocity of flow) is constant at 10 m/sec and the discharge from the runner is without whirl, i.e. radial discharge. The hydraulic losses are : (1) between turbine entrance and discharge from guide vanes, 4.5 m, (2) in the runner, 8 m, (3) in the draft tube, 0.8 m, (4) kinetic energy rejected to the tail race, 0.5 m. Determine : (a) the guide vane angle and the runner blade angle at inlet, and (b) the pressure heads at entry to and discharge from the runner.*

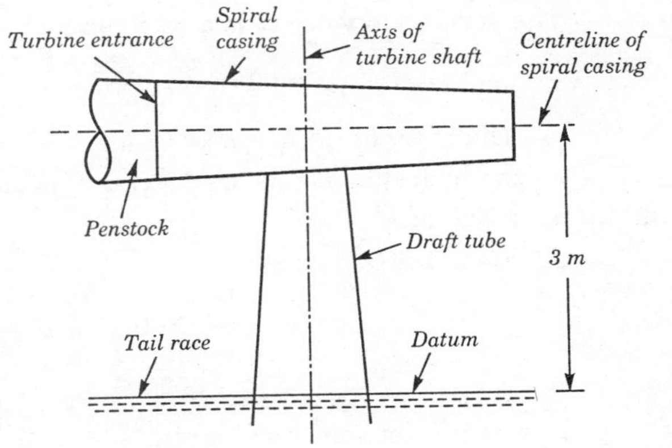

Fig. Prob. 15.14

Solution. Data given

$V_{f_1} = V_{f_2} = 10$ m/s, $u_1 = 30$ m/s, $V_{u_2} = 0$.

Head utilized by the runner

$H = 120 + 3 - 4.5 - 8 - 0.8 - 0.5 = 109.2$ m.

Head utilized by the runner is given by

$$H = \frac{u_1 V_{u_1} - u_2 V_{u_2}}{g}$$

$$109.2 = \frac{30 \times V_{u_1}}{9.81},$$

since $V_{u_2} = 0$ as the discharge from the runner is without whirl.

$$\therefore \quad V_{u_1} = 35.75 \text{ m/s.}$$

From the inlet velocity diagram $\tan \alpha_1 = \dfrac{V_{f_1}}{V_{u_1}} = \dfrac{10}{35.75} = 0.28$

The guide vane angle $\alpha_1 = 15.64°$.

$$\tan \beta_1 = \frac{V_{f_1}}{V_{u_1} - u_1} = \frac{10}{35.75 - 30} = 1.742.$$

The blade angle at inlet

$$\beta_1 = 60.15°. \qquad V_{f_1} = V_1 \sin \alpha_1.$$

$$\therefore \quad V_1 = \frac{10}{\sin 15.64°} = 37.1 \text{ m/s} \quad \text{and} \quad V_2 = V_{f_2} = 10 \text{ m/s.}$$

It is to be noted that the turbine entrance is taken as the entrance to the spiral casing.

Applying Bernoulli's equation to the entrance to spiral casing and to the runner inlet taking tail race level as datum, we obtain (refer the figure)

$$120 + 3 = \frac{V_1^2}{2g} + \frac{p_1}{\gamma} + Z_1 + 4.5$$

substituting $Z_1 = 3$ m and $V_1 = 37.1$ m/s,

the pressure head at inlet, $p_1/\gamma = 45.2$ m.

To obtain pressure head at the runner outlet, applying Bernoull's equation to the turbine entrance and the runner outlet,

$$120 + 3 = \frac{V_2^2}{2g} + \frac{p_2}{\gamma} + Z_2 + H + 4.5 + 0.8 \text{ ; here } \frac{p}{\gamma} + \frac{V^2}{2g} = 120 \text{ m at the entrance to the turbine}$$

Substituting, $Z_2 = 3$ m, $V_2 = 10$ m/s and $H = 109.2$ m,

the pressure head at the runner outlet,

$$p_2/\gamma = -6.8 \text{ m.}$$

Example 15.15. *An inward flow reaction turbine having an overall efficiency of 80% is required to deliver 136 kW. The head H is 16 m and the peripheral velocity is $3.3\sqrt{H}$. The radial velocity of flow at inlet is $1.1\sqrt{H}$. The runner rotates at 120 rpm. The hydraulic losses in the turbine are 15% of the available energy. Determine (a) the diameter of the runner, (b) guide vane angle, (c) the runner blade angle at inlet and (d) the discharge through the turbine.*

The discharge from the runner is radial.

Solution. Peripheral velocity at inlet, $u_1 = 3.3\sqrt{16} = 13.2$ m/s.

Radial velocity at inlet, $V_{f_1} = 1.1\sqrt{16} = 4.4$ m/s.

The peripheral velocity is also given by

$$u_1 = 13.2 = \frac{\pi d \times 120}{60}$$

$\therefore$ Diameter of runner, $d = 2.1$ m.

As the hydraulic losses are 15%, the head utilized by the runner is 0.85 × 16 m. Since the discharge is radial $V_{u_2} = 0$.

$\therefore$ Head utilized by runner

$$= \frac{u_1 V_{u_1}}{g} = 0.85 \times 16$$

or $$V_{u_1} = 10.1 \text{ m/s.}$$

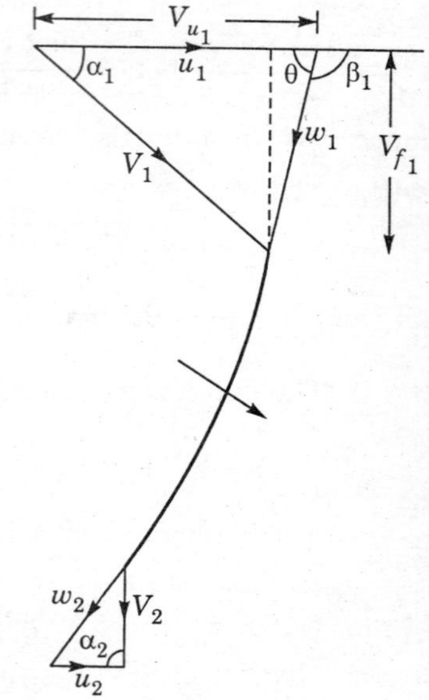

From the inlet velocity diagram

$$\tan \alpha_1 = \frac{V_{f_1}}{V_{u_1}} = \frac{4.4}{10.1} = 0.435$$

$\therefore$ Guide vane angle $\alpha_1 = 23.5°$.

The blade angle at inlet is given by

$$\tan \theta = \frac{V_{f_1}}{u_1 - V_{u_1}} = \frac{4.4}{13.2 - 10.1}$$

$\therefore$ $$\theta = 54.9°$$

The blade angle β_1 is $(180° - 54.9°) = 125.1°$.

$$\text{Overall efficiency } \eta = \frac{\text{Output}}{\gamma Q H} = \frac{136 \times 1000}{9810 \times Q \times 16}$$

$\therefore$ The required discharge, Q

$$= \frac{136000}{9810 \times 0.80 \times 16} = 1.083 \text{ m}^3/\text{s.}$$

Example 15.16. *An inward flow reaction turbine of inlet diameter 1.2 m operates under a head of 150 m and requires a discharge of 6 m³/s at a rotational speed of 400 r.p.m. The guide vane angle is 20° and the water leaves the runner blade radially. If the runner is 10 cm wide at the inlet, calculate (i) the torque and power applied to the shaft and (ii) the turbine efficiency.*

Solution. (*i*) Neglecting the thickness of blades, discharge

$$Q = \pi d_1 b_1 V_{f_1}$$

$\therefore$ $$V_{f_1} = \frac{6}{\pi \times 1.2 \times 0.1} = 15.9 \text{ m/s.}$$

Blade velocity at inlet

$$u_1 = \frac{\pi d_1 n}{60} = \frac{\pi \times 1.2 \times 400}{60} = 25.1 \text{ m/s.}$$

The velocity of whirl at inlet, refer Fig. 15.10. $V_{u_1} = V_1 \cos \alpha_1 = V_{f_1} \cot \alpha_1$

The power applied to shaft $P = \rho Q (u_1 V_{u_1} - u_2 V_{u_2})$

Since water leaves the runner radially, $\alpha_2 = 90°$ and $\therefore$ $V_{u_2} = 0$.

$$\text{Power } P = \frac{9810}{9.81} \times 6(25.1 \times 15.9 \times \cot 20°) = 6533 \text{ kN-m/s} = 6533 \text{ kW}$$

$$\text{Torque} = \frac{P}{\omega} = \frac{P}{u_1/r_1} = \frac{6533 \times 10^3}{21.1/0.6} = 156 \text{ kN-m}$$

(*ii*) Turbine efficiency $$\eta = \frac{P}{\gamma Q H} = \frac{6533 \times 10^3}{9810 \times 6 \times 150} = 0.74 \text{ or } 74\%.$$

Example 15.17. *A vertical shaft inward flow reaction turbine has as overall efficiency of 90% and runs at 420 r.p.m. and requires a discharge of 15.5 m³/s. The velocity at the inlet of the spiral casing is 9 m/s and the pressure head at this point is 250 m. The centre-line of the spiral casing inlet is 3 m above the tail water level. The diameter of runner at inlet is 2.3 m and the width at inlet is 30 cm. The flow is radial at the outlet from the runner, and the hydraulic efficiency is 94%. Calculate (i) the power developed by the turbine (ii) the specific speed (iii) the guide vane angle (iv) the runner blade angle at inlet and (v) the percentage of net head which is kinetic at entry to the runner. Neglect the thickness of blades.*

Solution. Head at inlet to the spiral casing $H = 250 + \dfrac{(9)^2}{2 \times 9.81} + 3 = 257.125$ m.

Overall efficiency of turbine, $\eta = \dfrac{\text{Power delivered by turbine}}{\gamma QH}$

$\therefore$ Power delivered by the turbine $= 0.9 \times 9810 \times 15.5 \times 257.125 = 35 \times 10^6$ N-m/s

$= 3.5 \times 10^4$ kW.

Specific speed $n_s = \dfrac{n\sqrt{P}}{H^{5/4}} = 420 \dfrac{\sqrt{3.5 \times 10^4}}{(257.125)^{5/4}} = 76.3.$ (Refer Eq. 15.69)

Peripheral velocity of blade at inlet, $u_1 = \dfrac{\pi \times 2.3 \times 420}{60} = 50.6$ m/s.

Velocity of flow at inlet

$$V_{f_1} = \frac{Q}{\pi d_1 b_1} = \frac{15.5}{\pi \times 2.3 \times 0.3} = 7.15 \text{ m/s.}$$

Since the exit discharge is radial, $V_{u_2} = 0$ and the hydraulic efficiency is

$$\eta_h = \frac{u_1 V_{u_1}}{gH} \quad \text{or} \quad V_{u_1} = \frac{0.94 \times 9.81 \times 257.125}{50.6} = 46.85 \text{ m/s.}$$

From the inlet velocity diagram (see Fig. of Ex. 15.15)

$$\tan \alpha_1 = \frac{V_{f_1}}{V_{u_1}} = \frac{7.15}{46.85} = 0.1525$$

the guide vane angle $\alpha_1 = 8.67°$

$$\tan \theta = \frac{V_{f_1}}{u_1 - V_{u_1}} = \frac{7.15}{50.6 - 46.85} = 1.907, \text{ where } \theta = 180° - \beta_1$$

$$\therefore \quad \theta = 62.35° \quad \text{or} \quad \beta_1 = 180° - 62.35° = 117.65°.$$

The blade angle at inlet β_1 is 117.65°.

Absolute velocity at inlet,

$$V_1 = \frac{V_{u_1}}{\cos \alpha_1} = \frac{46.85}{\cos 8.67°} = 47.4 \text{ m/s.}$$

Velocity head at inlet, $\dfrac{V_1^2}{2g} = \dfrac{(47.4)^2}{2 \times 9.81} = 114.5$ m

Kinetic head expressed as percentage of net head $= \dfrac{114.5}{257.125} \times 100 = 44.55\%.$

Example 15.18. *An inward flow reaction turbine develops 735 kW at 750 r.p.m. under a net head of 100 m. The guide vanes make an angle of 15° with the tangent at inlet. The axial length of the blade at inlet is 0.1 times the outer diameter. The radial velocity of flow through the wheel is constant* *and the discharge from the wheel is radial. The blade thickness blocks 5 per cent of the area of flow at inlet. The hydraulic efficiency of the wheel is 88 per cent and the overall efficiency is 84 per cent. Determine (a) the wheel diameter, (b) the wheel width and (c) the blade angle at inlet.*

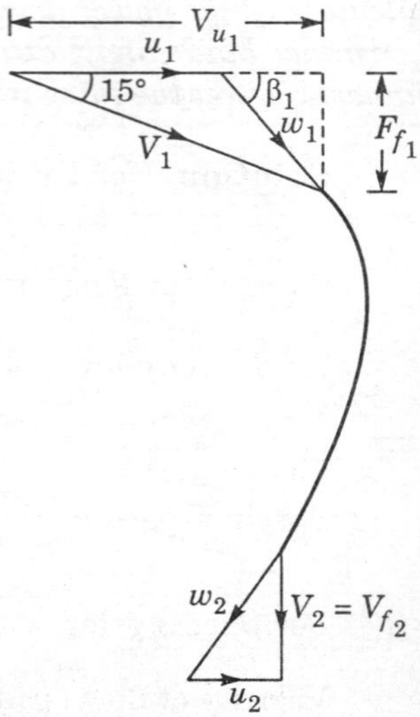

Solution. $\eta = \dfrac{\text{Power output}}{\gamma Q H}$

from which, $Q = \dfrac{735 \times 10^3}{9810 \times 0.84 \times 100} = 0.893 \text{ m}^3/\text{s}$

From inlet velocity diagram

$$\tan 15° = \frac{V_{f_1}}{V_{u_1}}$$

$$\therefore \quad V_{u_1} = 3.735\ V_{f_1}$$

$$u_1 = \frac{\pi d_1 \times 750}{60} = 39.3\ d_1$$

$$Q = 0.95\pi\ d_1\ (0.1\ d_1)\ V_{f_1} \quad \therefore \quad V_{f_1} = \frac{2.99}{d_1^2}$$

The hydraulic efficiency, $\eta_h = 0.88$

$$\eta_h = \frac{u_1 V_{u_1}}{gH} = \frac{39.3 d_1 \times 3.735 \times 2.99}{9.81 \times 100\ d_1^2} = 0.88$$

which yields, $d_1 = 0.57$ m.

Runner width $= 0.1\ d_1 = 5.74$ cm

Blade angle at inlet, β_1, is given by

$$\tan \beta_1 = \frac{V_{f_1}}{V_{u_1} - u_1} = \frac{5.05}{33.85 - 22.55} = 0.8\ \beta_1 = 38.7°.$$

Example 15.19. *A Francis turbine has an inlet diameter of 2.5 m and an outlet diameter of 1.5 m. The breadth of the blade is constant at 0.20 m. The runner rotates at a speed of 300 r.p.m with a discharge of 10.00 m³/s. The vanes are radial at the inlet and discharge is radially outwards at the outlet. Calculate the angle of the guide vane to be set at the inlet and the blade angle at the outlet.*

Solution. Since the vanes are radial at inlet,

$$\beta_1 = 90°,\ w_1 = V_{f_1} \text{ and } V_{u_1} = u_1$$

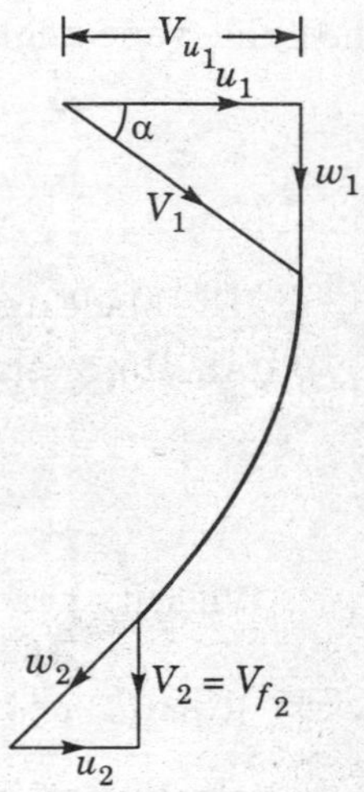

The guide vane angle at inlet is given by,

$$\tan \alpha_1 = \frac{V_{f_1}}{u_1}$$

The discharge $Q = \pi d_1 b_1 V_{f_1} = \pi\ d_2 b_2 V_{f_2}$

Data given : $d_1 = 2.5$ m, $d_2 = 1.5$ m, $b_1 = b_2 = 0.20$ m

$n = 300$ r.p.m, $Q = 10.00 \text{ m}^3/\text{s}$

Making substitutions,

$$V_{f_1} = \frac{10.00}{\pi \times 2.5 \times 0.20} = 6.365 \text{ m/s}$$

$$V_{f_2} = \frac{10.00}{\pi \times 1.5 \times 0.20} = 10.61 \text{ m/s}$$

The peripheral velocities of blade at inlet and outlet are given by

$$u_1 = \pi d_1 n/60 \text{ and } u_2 = \pi\, d_2\, n/60$$

$$= \frac{\pi \times 2.5 \times 300}{60} = 39.27 \text{ m/s}$$

and,
$$u_2 = \frac{\pi \times 1.5 \times 300}{60} = 23.56 \text{ m/s.}$$

Since discharge is radially outwards, $\alpha_2 = 90°$, and $V_{u_2} = 0$

hence,
$$\tan \beta_2 = \frac{V_{f_2}}{u_2}$$

$$\therefore \quad \tan \alpha_1 = \frac{6.365}{39.27} = 0.1621, \qquad \therefore \quad \alpha_1 = \mathbf{9.21°}$$

and
$$\tan \beta_2 = \frac{10.61}{23.56} = 0.4503, \qquad \therefore \quad \beta_2 = \mathbf{24.24°}$$

The guide vane angle to be set at the inlet, $\alpha_1 = 9.21°$, and the blade angle at the outlet $= 24.24°$.

Example 15.20. *For an inward flow reaction turbine, the blade angle at inlet is 90° and velocity of flow is constant. The discharge from the runner wheel is radial. Show that the hydraulic efficiency is expressed by* $\frac{2}{2 + tan^2\, \alpha}$ *where α is the guide blade angle.*

Solution. For radial discharge, the hydraulic efficiency is given by $\eta_h = \frac{u_1 V_{u_1}}{gH}$.

The net head,

$$H = \text{Head utilized by the runner} + \frac{V_2^2}{2g} + \text{Losses in the turbine.}$$

Neglecting losses, $H = \frac{u_1 V_{u_1}}{g} + \frac{V_2^2}{2g}$.

From the velocity diagrams, $V_{u_1} = u_2$, $V_2 = V_{f_2}$

Since velocity of the flow is constant, $V_{f_1} = V_{f_2}$

$$\tan \alpha = \frac{V_{f_1}}{u_1}, \quad \therefore \quad V_{f_1} = u_1 \tan \alpha$$

Making these substitutions,

$$\eta_h = \frac{u_1^2 / g}{\frac{u_1^2}{g} + \frac{u_1^2 \tan^2 \alpha}{2g}} = \frac{1}{1 + \frac{\tan^2 \alpha}{2}} = \frac{2}{2 + \tan^2 \alpha}.$$

Example 15.21. *An inward flow reaction turbine has radial tips at the inlet (outer periphery) while at the exit (inner periphery) the blades make an angle of 30° with the forward tangent. The* *radius of the outer periphery is two times the radius of inner one, and the turbine operates under a total head of 21 m. Assuming a constant radial velocity of flow and that the blade friction accounts for a dissipation of energy equivalent of 12% of kinetic energy at the outlet, determine the runner velocity at the rim, and the hydraulic efficiency of the turbine if the turbine discharges radially.*

Solution. Referring to Fig. 15.10, it can be seen that

$\beta_1 = 90°$ (blade tips at inlet being radial)

$V_{u_2} = 0,$

$\alpha_2 = 90°$ (discharge being radial)

$\beta_2 = 30°,$

$V_{f_1} = V_{f_2} = V_2$ (radial velocity of flow being constant)

$V_{u_1} = u_1.$

The velocity triangles at inlet and outlet both are thus right angled.

Work done per kg of water

$$= \frac{V_{u_1} u_1}{g}$$

From energy considerations :

Energy supplied = Energy transferred to the runner + Energy lost in blade friction + Kinetic energy at outlet

or
$$H = \frac{V_{u_1} u_1}{g} + 0.12\frac{V_2^2}{2g} + \frac{V_2^2}{2g} = \frac{u_1^2}{g} + 1.12\,\frac{V_2^2}{2g}$$

From the outlet velocity triangle,

$$V_{f_2} = V_2 = u_2 \tan 30°$$

$$H = \frac{u_1^2}{g} + 1.12\,\frac{(u_2 \tan 30°)^2}{2g}$$

Given that $\quad u_1 = r_1\,\omega \quad$ and $\quad u_2 = 0.5\,r_1\,\omega$

$\therefore \quad u_1/u_2 = 2$

$$H = \frac{8u_2^2}{2g} + 1.12\,\frac{(0.577\,u_2)^2}{2g} \qquad \text{or} \qquad H = 8.372\,\frac{u_2^2}{2g}.$$

Substituting $\quad H = 21$ m,

$$u_2^2 = \frac{2 \times 9.81 \times 21}{8.372} = 49.2$$

$$u_2 = 7.02 \text{ m/s.}$$

Runner velocity at the rim, *i.e.* at the inlet

$$u_1 = 2u_2 = 2 \times 7.02 = 14.04 \text{ m/s}$$

$$\text{Hydraulic efficiency} = \frac{V_{u_1} u_1}{gH} = \frac{u_1^2}{gH} = \frac{(14.04)^2}{9.81 \times 21} = 0.958 \quad \text{or} \quad 95.8\%.$$

15.7. Francis Turbine

It is a reaction turbine designed and developed by an American engineer, James B. Francis, in 1849. The modern Francis turbine is a mixed-flow type in which the water enters the runner radially at its outer periphery and leaves it in the axial direction. It is to be noted that the earlier

Francis turbine had a purely radial flow runner, the inlet and outlet velocities remaining in the same plane normal to the axis of the runner.

Runner and Blades

Runner of Francis turbine is a rotor which has passages formed between crown and shroud on the one side and any two consecutive blades on the other. Water flows through these passages

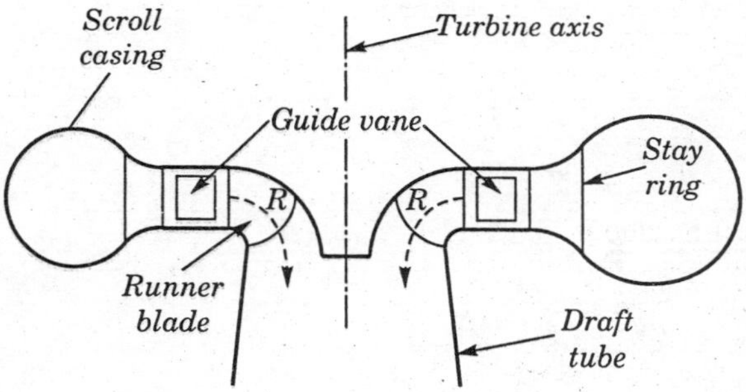

Fig. 15.12. Hydraulic Elements of a Francis turbine.

Fig. 15.12. (*a*) Francis Runner.

entering at the outer periphery in the radially inward direction and leaving it in a direction parallel to the turbine axis, see Fig. 15.12 which shows hydraulic elements of the turbine.

The number of blades usually varies between 16 and 24. For small units the runner is made of cast iron while the bigger units are made of stainless steel or a non-ferrous metal like bronze when the available water is likely to be chemically impure and there is danger of corrosion. The runner is keyed to the main shaft which is generally kept vertical.

A Francis runner being lowered at a power house site is shown in Fig. 15.12 (*a*).

15.7.1. Working Proportions of Francis Runner

(*a*) Ratio of wheel width B to runner diameter D is denoted by n, *i.e.*

$$n = B/D$$

and its value ranges from 0.10 to 0.45.

(*b*) Speed ratio ϕ represents the ratio of peripheral velocity u to the spouting velocity $\sqrt{2gH}$.

$$\phi = \frac{u}{\sqrt{2gH}}$$

The value of ϕ ranges from 0.6 to 0.9.

(*c*) Flow ratio ψ is the ratio of the velocity of flow at inlet, V_{f_1}, to the spouting velocity, $\sqrt{2gH}$, *i.e.*

$$\psi = \frac{V_{f_1}}{\sqrt{2gh}} \text{ its value varies from 0.15 to 0.30.}$$

15.7.2. Design Parameters of Francis Runner

To generate a given amount of power P, the turbine running at N r.p.m under a head of H, the design of Francis runner involves determination of size and the blade angles. The procedure consists of assuming probable values of hydraulic efficiency η_h, overall efficiency η_o, n and ψ and working out the following steps :

(*i*) Determine the discharge Q from the relation for shaft power

$$P = \eta_0 \frac{\gamma QH}{75}.$$

(*ii*) Determine the velocity of flow from the known Q and the flow area

$$A = (\pi D - Zt) B = K\pi DB.$$

where D, B and t are the diameter, width and the blade thickness of the runner, Z being the number of blades. Here K is a factor which is known as the vane thickness coefficient, its value is obviously less than unity and is of the order of 0.95 or so. The velocity of flow is given by

$$V_f = \frac{Q}{K\pi DB} = \frac{Q}{Kn\pi D^2}.$$

(*iii*) The flow velocity using the flow ratio given by

$$V_f = \psi \sqrt{2gH} = \frac{Q}{Kn\pi D^2}$$

from which,
$$D = \left[\frac{Q}{\psi n K\pi \sqrt{2gH}}\right]^{1/2}.$$

(*iv*) Determine the peripheral velocity (rim velocity) u_1 and the velocity of whirl V_{u_1} from the relations :

$$u_1 = \frac{\pi D_1 n}{60}; \qquad \eta_h = \frac{V_{u_1} . u_1}{gH}.$$

(*v*) Use the inlet velocity triangle to determine the guide vane angle α_1 and the blade angle β_1

$$\tan \alpha_1 = \frac{V_{f_1}}{V_{u_1}}; \tan \beta_1 = \frac{V_f}{V_{u_1} - u_1}.$$

(*vi*) Assume runner diameter at outlet D_2 to be approximately equal to one-half the diameter at inlet :

$$D_2 \approx \frac{1}{2} D_1; \qquad u_2 = 0.5 u_1$$

(*vii*) From the continuity equation

$$Q = K_1 \pi D_1 B_1 V_{f_1} = K_2 \pi D_2 B_2 V_{f_2}$$

$$\frac{V_{f_1}}{V_{f_2}} = \frac{K_2 D_2 B_2}{K_1 D_1 B_1}$$

It is generally assumed that $V_{f_1} = V_{f_2}$, $K_1 = K_2$, one then gets

$$B_2 = 2B_1.$$

(*viii*) Assume radial discharge at runner exit ($\alpha_2 = 90°$) and obtain the runner blade angle β_2 from the outlet velocity triangle as

$$\tan \beta_2 = \frac{V_{f_2}}{u_2}.$$

15.7.3. Speed for Shock-free Entry and Radial Discharge

From the velocity triangle at inlet, Fig. 15.10, by the use of the sine law, one obtains :

$$\frac{u_1}{\sin (\beta_1 - \alpha_1)} = \frac{V_1}{\sin (180° - \beta_1)} \quad \text{or} \quad \frac{u_1}{V_1} = \frac{\sin (\beta_1 - \alpha_1)}{\sin \beta_1} \qquad ...(15.40)$$

For radial discharge, $\alpha_2 = 90°$ and $V_{u_2} = 0$, the head utilized by the turbine, using Eq. (15.30) is

$$h = \frac{u_1 V_{u_1}}{g} = \frac{u_1 V_1 \cos \alpha_1}{g} \qquad ...(15.41)$$

If H is the operating head of the turbine, the hydraulic efficiency is

$$\eta_h = \frac{h}{H} \quad \text{or} \quad h = \eta_h \, . \, H \qquad ...(15.42)$$

From Eqs. (15.41) and (15.42),

$$V_1 = \frac{g \, \eta h \, . H}{u_1 \cos \alpha_1} \qquad ...(15.43)$$

Making substitution in Eq. (15.40)

$$\frac{u_1^2 \cos \alpha_1}{gH \, \eta h} = \frac{\sin (\beta_1 - \alpha_1)}{\sin \beta_1}$$

From which, $u_1 = \sqrt{\dfrac{\eta_h \, . \sin (\beta_1 - \alpha_1)}{2 \sin \beta_1 \cos \alpha_1}} \sqrt{2gH}$ ...(15.44)

Equation (15.44) is an important relationship which shows that :

(*i*) the best speed depends upon the values of α_1 and β_1 and, therefore, can be altered by changing these values ;

(*ii*) every runner has a best speed that varies with the square-root of the head under which it runs ;

(*iii*) the best speed may be expressed by

$$u_1 = \phi \sqrt{2gH} \, .$$

The value of ϕ varies : (*a*) from about 0.58 to 1.0 or more for the mixed-flow turbine, (*b*) from about 1.0 to 2.0 or more for axial or propeller type turbine.

15.7.4. Influence of Speed and Discharge on Runner Design

The head and the available discharge vary widely among the hydropower installations. Generally, large discharges are associated with low heads and relatively small rates of flow are available at high heads. The speed under a high head is inherently high since $u_1 = \phi \sqrt{2gH}$, and ϕ is limited within a certain range of values. Under a low head the speed is likewise low, and if an operative speed is to be maintained, a runner with a low value of ϕ must be used for high heads and a runner with a high value of ϕ for low heads. Runners are classified as low, high or medium speed according to their value of ϕ.

The result of various conditions of head, speed and discharge is that runners of low speed and discharge are generally demanded for high heads, while runners of high speed and discharge are necessary for low heads. The speed requirements can be met by alterations in design, giving α_1 and β_1 such values as will yield the desired value of ϕ. The discharge requirement is similarly met by altering the design to produce large or small passages through the guides and runner.

Fig. 15.13 shows four types of runners which for ease of comparison have the same diameter D and operate under approximately the same head. Fig. 15.13 (*a*) shows a runner having restricted flow areas and such values for α_1 and β_1 which result in a low value of ϕ. It is essentially a high head runner of relatively low speed and discharge. In Fig. 15.13 (*b*) is shown a runner having a slightly large discharge, and under the same head it would develop a larger power. Its value of ϕ is also considerably larger. It represents a turbine of medium speed and discharge. Runner (*c*) is an extreme

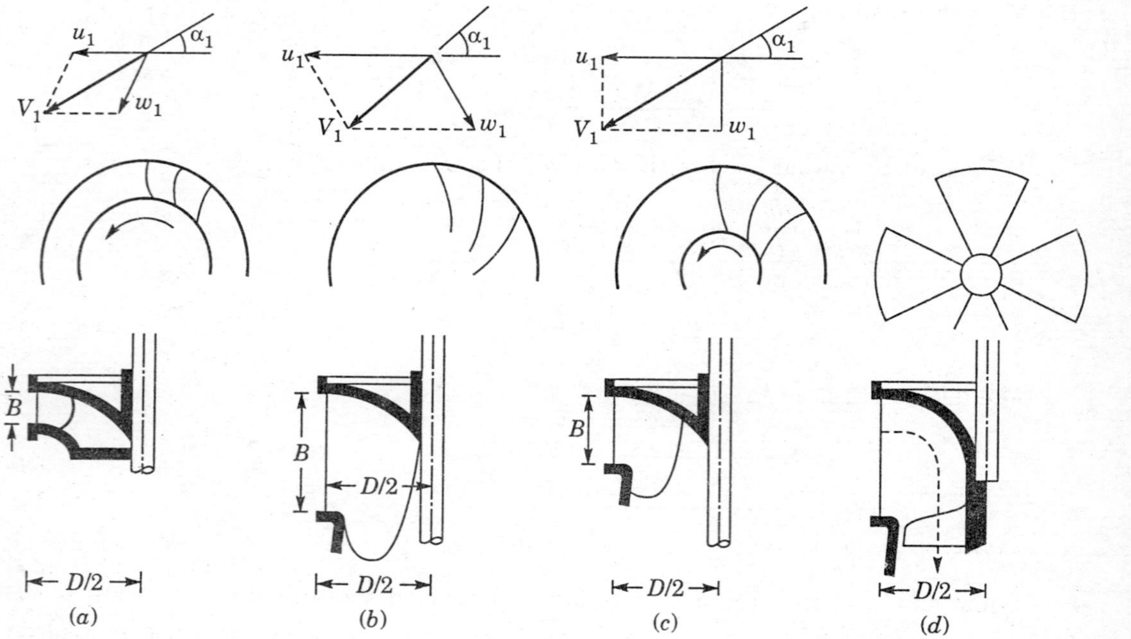

Fig. 15.13. Reaction Turbine runners. *a, b, c*—mixed flow runners for increasing discharge *d*—axial flow runner.

type meant for large discharge and a high value of ϕ. It is specially adapted to low heads where large power at high speeds can only be obtained, by large discharges and keeping u_1 (hence ϕ) as high as possible. Runner (*d*) is of the propeller type and is the most extreme development of high-speed, high discharge turbine.

The significant difference in (*a*), (*b*) and (*c*) types of runners is the vertical height *B*, of the guide passages which increases as the discharge increases. The blades also extend nearer to the shaft in the successive runners and the water undergoes a greater change in direction while in the runner. Such a design is necessary for the water to be guided without excessive turbulence at the inlet to the draft tube.

Example 15.22. *An inward flow reaction turbine is required to develop 294 kW at 220 r.p.m. The effective head on the turbine is 20 m. Determine the inside and outside diameters, inlet and exit angles for the vanes and the guide blade angle. Assume inlet diameter to be twice the outlet diameter, hydraulic efficiency as 82 per cent and constant radial velocity of flow 3.5 m/s. Assume ratio of width of wheel at inlet to its diameter as 0.1 and that 5 per cent of the area of flow in the runner is blocked by the blades. Assume radial discharge.*

Solution. Given : $D_1 = 2D_2$, $V_{f_1} = V_{f_2} = 3.5$ m/s, $B_1/D_1 = 0.1$

Power developed = 294 kW, N = 220 r.p.m., H = 20 m.

The hydraulic efficiency

$$\eta_h = \frac{(V_{u_1} . u_1 - V_{u_2} . u_2)}{gH}$$

For radial discharge, $V_{u_2} = 0$, hence

$$u_1 \, V_{u_1} = \eta_h \, . \, gH = 0.82 \times 9.81 \times 20 = 160.9$$

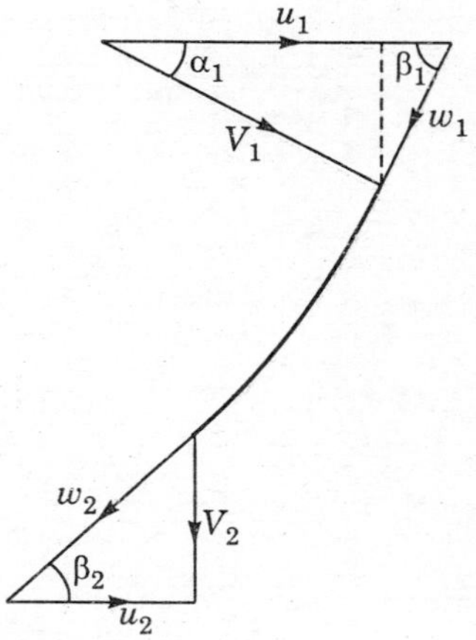

Assuming an overall efficiency of 80%

$$\eta_0 = \frac{\text{power developed in N-m/s}}{\gamma QH}$$

or $$0.80 = \frac{294 \times 10^3}{9810 \times Q \times 20} \text{ or } Q = 1.875 \text{ m}^3/\text{s}$$

But $$Q = 0.95\, \pi\, D_1\, B_1\, V_{f_1} = 0.95\, \pi \times 0.1\, D_1^2 \times 3.5$$

$\therefore$ $$D_1^2 = \frac{Q}{0.095\pi \times 3.5} = \frac{1.875}{0.095\pi \times 3.5} = 1.795$$

or $$\mathbf{D_1 = 1.339 \text{ m}} \text{ and } \mathbf{D_2 = 0.6695 \text{ m}}.$$

$$u_1 = \frac{\pi D_1 N}{60} = \frac{\pi \times 1.339 \times 220}{60} = 15.42 \text{ m/s}$$

$$u_2 = \frac{\pi D_2 N}{60} = 7.71 \text{ m/s}\ ; \quad V_{u_1} = \frac{160.9}{u_1} = \frac{160.9}{15.42} = 10.42 \text{ m/s}$$

$$\tan \alpha_1 = \frac{V_{f_1}}{V_{u_1}} = \frac{3.5}{10.42} = 0.3358,$$

$$\alpha_1 = \mathbf{18° \, 33'}$$

$$\tan \beta_1 = \frac{V_{f_1}}{u_1 - V_{u_1}} = \frac{3.5}{15.42 - 10.42} = 0.7, \quad \therefore \quad \beta_1 = \mathbf{34° \, 59'}$$

$$\tan \beta_2 = \frac{V_{f_1}}{u_2} = \frac{3.5}{7.71} = 0.4539, \quad \therefore \quad \beta_2 = \mathbf{24° \, 25'}.$$

Example 15.23. *A Francis turbine has to be designed to develop 367.5 kW under a head of H = 70 m while running at N = 750 r.p.m. Ratio of width of runner to diameter of runner, n = 0.1, inner diameter is half the outer diameter. Flow ratio ϕ = 0.15, hydraulic efficiency = 95 per cent, mechanical efficiency = 84 per cent. Four per cent of the circumferential area of runner to be occupied by the thickness of vanes, velocity of flow is constant and discharge is radial at exit. Calculate*

(i) the diameter of the wheel,

(ii) the quantity of water supplied, and

(iii) the guide vane angle at inlet and runner vane angles at inlet and exit.

Solution. Data given :

$P = 367.5$ kW $\qquad H = 70$ m,

$N = 750$ r.p.m, $\qquad n = B/D = 0.1$,

$D_2 = 0.5\, D_1$ $\qquad \psi = 0.15$,

$\eta_h = 0.95$, $\qquad \eta_m = 0.84$,

Blockage $= 4\%$. $\qquad V_f$ = constant, and discharge radial

$$\eta_0 = \eta_m \cdot \eta_h = \text{Power developed in N-m/s (or Watts)}/\gamma QH$$

$$Q = \frac{367.5 \times 10^3}{9810 \times 0.95 \times 0.84 \times 70} = 0.671 \text{ m}^3/\text{s}$$

$$V_{f_1} = V_{f_2} = \psi \sqrt{2gH}$$

$$= 0.15 \sqrt{2 \times 9.81 \times 70} = 5.559 \text{ m/s}$$

$$Q = 0.96\, \pi\, D_1\, B_1\, V_{f_1} = 0.96\, \pi\, D_1 \times 0.1\, D_1 \times 5.559 = 1.676\, D_1^2$$

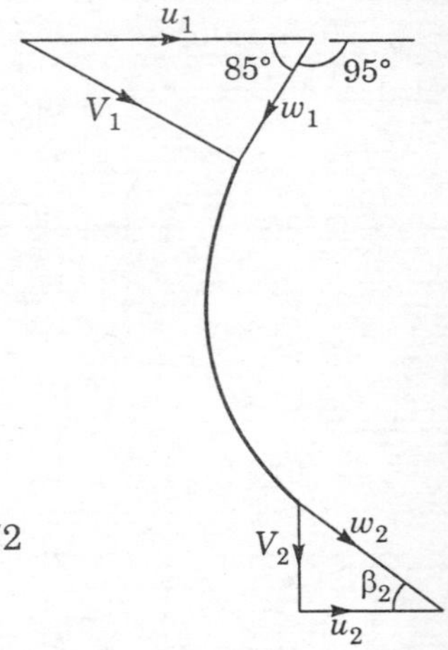

Equating this to the numerical value of Q

$$D_1^2 = \frac{6.671}{1.676}$$

$$D_1 = \mathbf{0.633\ m}.$$

(*i*) The diameter of the wheel is **0.633 m.**

(*ii*) Quality of water supplied = **0.671 m³/s.**

(*iii*) From the outlet velocity triangle,

$$u_2 = \frac{\pi D_2 N}{60} = \frac{\pi(0.633/2)\times 750}{60} = 12.429 \text{ m/s}$$

$$\tan \beta_2 = \frac{V_{f_2}}{u_2} = \frac{\psi\sqrt{2gH}}{u_2} = \frac{0.15\sqrt{2\times 9.81\times 70}}{12.429} = 0.4472$$

$\therefore$ $\beta_2 = \mathbf{24°\ 5'}.$

From the inlet velocity triangle,

$$\eta_h = \frac{u_1 . V_{u_1}}{gH},$$

$$V_{u_1} = \frac{\eta h . gH}{u_1} = \frac{0.95\times 9.81\times 70}{24.858} = 26.24 \text{ m/s}$$

$$\tan \alpha_1 = \frac{V_{f_1}}{V_{u_1}} = \frac{5.559}{26.24} = 0.2118,\ \alpha_1 = \mathbf{11°\ 58'}.$$

$$\tan \beta_1 = \frac{V_{f_1}}{V_{u_1} - u_1} = \frac{5.559}{26.24 - 24.858} = 4.022 \qquad \therefore \quad \beta_1 = \mathbf{76°\ 50'}.$$

Example 15.24. *An inward flow reaction turbine of Francis type has a runner 60 cm diameter and 5 cm width at the outer rim, the inner diameter being 39 cm. The wheel blade angles at the inlet and outlet are 95° (inside angle of the velocity triangle = 85°) and 14° respectively. The velocity of flow is uniform throughout the wheel and 8 per cent of the circumferential area of the runner is blocked by the blade thickness. The head on the turbine is 54 m, the hydraulic efficiency is 88 per cent, the overall efficiency 81 per cent and the discharge at the outlet is radial. Determine (a) the speed in r.p.m. and (b) the shaft power of the turbine.*

Solution. Data given :

$D_1 = 0.60$ m, $\quad D_2 = 0.39$ m,

$B_1 = 5$ cm ; $\quad \beta_1 = 85°$,

$\beta_2 = 14°$, $\quad V_{f_1} = V_{f_2}$,

Blockage = 8%, $\quad H = 54$ m

$\eta = 88\%$, $\quad \eta_0 = 80\%$, discharge radial

$$u_1 = \frac{\pi D_1 N}{60} = \frac{\pi\times 0.60\times N}{60} = 0.0314\, N$$

$$u_2 = \frac{\pi\times 0.39\, N}{60} = 0.020\, N$$

$$\tan \beta_1 = \frac{V_{f_1}}{u_1 - V_{u_1}} \quad \text{and} \quad \tan \beta_2 = \frac{V_{f_2}}{u_2}$$

dividing, $$\frac{\tan \beta_1}{\tan \beta_2} = \frac{u_2}{u_1 - V_{u_2}}$$

$$\eta_h = \frac{V_{u_1} . u_1}{gH} = \frac{\pi D_1 N V_{u_1}}{60\, gH} \quad \text{or} \quad V_{u_1} = \frac{60\, gH\, \eta h}{\pi D_1 N}$$

$$= \frac{60 \times 9.81 \times 54 \times 0.88}{\pi \times 0.60 \times N} = \frac{14838.7}{N}$$

Substituting, $$\frac{\tan \beta_1}{\tan \beta_2} = \frac{\tan 85°}{\tan 14°} = \frac{0.020\, N}{0.0314\, N - \dfrac{14{,}838.7}{N}}$$

or $$1.429\, N^2 = 680{,}251.4 \qquad \text{or} \quad N = \mathbf{692.38\ r.p.m.}$$

$$u_2 = 0.02\, N = 0.02 \times 692.38 = 13.847 \text{ m/s}$$

$$V_{f_1} = V_{f_2} = u_2 \tan \beta_2 = 13.847 \times 0.2493 = 3.45 \text{ m/s}$$

$$Q = 0.92\, \pi D_1 B_1 V_{f_1}$$

$$= 0.92 \times \pi \times 0.60 \times \frac{5}{100} \times 3.45 = 0.299 \text{ m}^3\text{/s}$$

$$\eta_0 = \frac{\text{Power developed}}{\gamma QH}$$

$\therefore$ $$P = \eta_0 \gamma QH = 0.81 \times 9810 \times 0.299 \times 54 \text{ N-m/s (or W)} = \mathbf{128.16\ kW.}$$

Example 15.25. *A Francis turbine supplied through a 6 m diameter penstock has the following particulars :*

Output of installation	*63,500 kW*
Flow	*117 m³/s*
Speed	*150 r.p.m*
Hydraulic efficiency	*92%*
Mean diameter of turbine at entry	*4 m*
Mean blade height at entry	*1 m*
Entry diameter of draft tube	*4.2 m*
Velocity in tail race	*2.4 m/s.*

The static pressure head in the penstock measured just before entry to the runner is 57.4 m. The point of measurement is 3 m above the level of the tail race. The loss in the draft tube is equivalent to 30% of the velocity head at entry to it. The exit plane of the runner is 2 m above the tail race and the flow leaves the runner without swirl. Determine

(a) the overall efficiency

(b) the direction of flow relative to the runner at inlet

(c) the pressure head at entry to the draft tube.

Solution. Velocity in the penstock

$$V = \frac{117}{\frac{\pi}{4}(6)^2} = 4.138 \text{ m/s}$$

Velocity at entry to the draft tube

$$V = \frac{4Q}{\pi (4.2)^2} = \frac{4 \times 117}{\pi (4.2)^2} = 8.44 \text{ m/s}$$

Head just before entry to the runner

$$= \frac{p}{\gamma} + z + \frac{V^2}{2g} = 57.4 + 3 + \frac{(4.138)^2}{2 \times 9.81} = \mathbf{61.27\ m}$$

Effective head = Head at entry to runner – Kinetic energy in tail race

$$\therefore \quad H = 61.27 - \frac{(2.4)^2}{2 \times 9.81} = \mathbf{60.97 \ m.}$$

Output power = 63,500 kW

Input power = γQH = 9810 × 117 × 60.97 = 70 × 10^6 W = 70 × 10^3 kW

$$\therefore \text{ Overall efficiency } = \frac{\text{Output}}{\text{Input}} = \frac{63500}{70000} = 0.90 = \mathbf{90\%.}$$

Direction of flow relative to the runner

Since the flow leaves the runner without swirl, the hydraulic efficiency is

$$\eta_h = \frac{u_1 \,.\, V_{u_1}}{gH}$$

from which, $$V_{u_1} = \frac{\eta h \,.\, gH}{u_1}$$

$$u_1 = \frac{\pi D_1 n}{60} = \frac{\pi \times 4 \times 150}{60} = 31.41 \text{ m/s.}$$

$$V_{u_1} = \frac{0.92 \times 9.81 \times 60.97}{31.41} = 17.52 \text{ m/s}$$

From the inlet velocity diagram,

$$\tan \beta_1 = \frac{V_{f_1}}{V_{u_1} - u_1}$$

$$V_{f_1} = \frac{Q}{\pi DB} = \frac{117}{\pi \times 4 \times 1} = 9.31 \text{ m/s}$$

$$\therefore \quad \tan \beta_1 = \frac{9.31}{17.52 - 31.41} = -0.67 \quad \text{and} \quad \beta_1 = \mathbf{180° - 33.82° = 146.18°}.$$

Applying Bernoulli's equation between the entrance to the draft tube and the tail race

$$\frac{p_2}{\gamma} + \frac{V_2^2}{2g} + z_2 = 0 + \frac{V_e^2}{2g} + 0.3 \frac{V_2^2}{2g}$$

$$\therefore \quad \frac{p_2}{\gamma} = z_2 - 0.7 \frac{V_2^2}{2g} + \frac{V_e^2}{2g} = -2 - 0.7 \times \frac{(8.44)^2}{2 \times 9.81} + \frac{(2.4)^2}{2 \times 9.81}$$

$$= -4.26 \text{ m.}$$

Example 15.26. *An inward flow reaction turbine works under a head of 30 m and discharge of 10 cum/sec. The speed of the runner is 300 r.p.m. The inlet tip of runner vane, the peripheral velocity of wheel is 0.9 $\sqrt{2gH}$ and the radial velocity of flow is 0.3 $\sqrt{2gH}$, where H is the head on the turbine. If the overall efficiency and the hydraulic efficiency of the turbine are 80% and 90% respectively, determine :*

(i) the power developed in kW *(ii) diameter and width of runner at inlet*

(iii) guide blade at inlet *(iv) inlet angle at runner vane*

(v) diameter of runner at outlet.

Assume that the discharge at outlet is radial.

Solution. Referring to Fig. 15.10, and the data given as under

$$H = 30 \text{ m}, \; Q = 10 \text{ m}^3\text{/s}, \; N = 300 \text{ r.p.m.}$$

$$u_1 = 0.9\sqrt{2gH}, \qquad V_{f_1} = 0.3\sqrt{2gH}, \; \eta_h = 90\%, \; \eta_0 = 80\%$$

discharge at outlet being radial, $\alpha_2 = 90°$, $V_f = V_2$

(*i*) **Power developed :**

Overall efficiency $\eta_0 = \dfrac{\text{Power developed } (P_0)}{\gamma\, QH}$

$\therefore$ Power developed $(P_0) = \eta_0\, \gamma QH$

$= 0.80 \times 9810 \times 10 \times 30$ W

$= 2354400$ W $=$ **2354.4 kW**

(*ii*) **Diameter and width of runner :**

Peripheral velocity of wheel at inlet, $u_1 = \dfrac{\pi D_1 N}{60}$

$$\therefore \quad D_1 = \frac{60 u_1}{\pi N} = \frac{60 \times 0.9\sqrt{2gH}}{\pi N} = \frac{60 \times 0.9\sqrt{2 \times 9.81 \times 30}}{\pi \times 300} = \mathbf{1.39\ m.}$$

The discharge $Q = \pi D_1 b_1 V_{f_1}$

$= \pi D_1 b_1 \times 0.3\sqrt{2gH}$

$10 = \pi \times 1.39\, b_1 \times 0.3\sqrt{2 \times 9.81 \times 30}, \quad \therefore \quad b_1 = \mathbf{0.315\ m}$

(*iii*) **Guide blade angle at inlet :**

Using the velocity diagram at inlet, $\tan \alpha_1 = \dfrac{V_{f_1}}{V_{u_1}}$

Hydraulic efficiency, $\eta_h = \dfrac{u_1 V_{u_1}}{gH}$

$$V_{u_1} = \frac{\eta_h\, gH}{u_1} = \frac{0.9 \times 9.81 \times 30}{0.9\sqrt{2 \times 9.81 \times 30}} = 12.13 \text{ m/s}$$

$$\therefore \quad \tan \alpha_1 = \frac{0.3\sqrt{2gH}}{12.13} = \frac{0.3\sqrt{2 \times 9.81 \times 30}}{12.13} = 0.60, \text{ from which } \alpha_1 = \mathbf{30.96^\circ}$$

(*iv*) **Inlet angle at runner vane :**

From velocity diagram at inlet, $\tan \beta_1 = \dfrac{V_{f_1}}{V_{u_1} - u_1}$

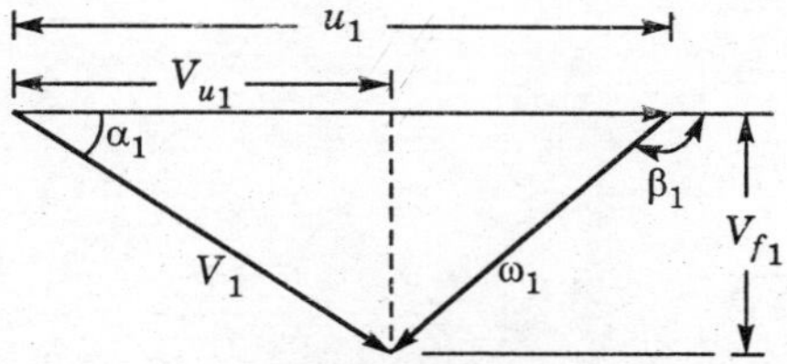

Velocity diagram at inlet

Making substitutions,

$$\tan \beta_1 = \frac{0.3\sqrt{2gH}}{12.13 - 0.9\sqrt{2gH}}$$

$$= \frac{0.3\sqrt{2 \times 9.81 \times 30}}{12.13 - 0.9\sqrt{2 \times 9.81 \times 30}} = -0.75$$

$\therefore \quad \beta_1 = 180^\circ - 36.88^\circ = \mathbf{143.12^\circ.}$

(*v*) **Diameter of runner at outlet :**

$$\sin(180^\circ - \beta_1) = \frac{V_{f_1}}{w_1}, \quad \therefore \quad w_1 = \frac{V_{f_1}}{\sin 36.88^\circ} = \frac{0.3\sqrt{2 \times 9.81 \times 30}}{0.06} = 12.13 \text{ m/s}$$

Assuming $V_{f_1} = V_{f_2}$, $V_2 = V_{f_2}$; also assuming no blade friction $w_1 = w_2$

$$\sin \beta_2 = \frac{V_2}{w_2} = \frac{7.278}{12.13} = 0.60, \quad \therefore \quad \beta_2 = 36.87^\circ$$

$$\cos \beta_2 = \frac{u_2}{w_2} \qquad \therefore \quad u_2 = w_2 \cos \beta_2 = 12.13 \cos 36.87° = 9.70 \text{ m/s}$$

But
$$u_2 = \frac{\pi D_2 N}{60} \qquad \therefore \quad D_2 = \frac{60 u_2}{\pi N} = \frac{60 \times 9.70}{\pi \times 300} = \mathbf{0.618 \ m.}$$

15.8. Propeller and Kaplan Turbines

A propeller turbine is an axial flow reaction turbine suitable for low head installations (upto 30 m) where high discharges are available. The blades of a propeller turbine remain fixed in position under all conditions of operation. Propeller turbines operate at high efficiency under full load conditions, but their efficiency is considerably reduced when they are subject to changing loads. Credit goes to Kaplan of Czechoslovakia for conceiving a propeller turbine with adjustable blades whose position may be varied with load on the turbine and hence with the discharge. As the blades are adjusted for a shock-free entry of flow, a high efficiency is maintained over a wide range of turbine output. A Kaplan runner, designed for a given head and discharge also, therefore works efficiently under much lower heads and at reduced discharge.

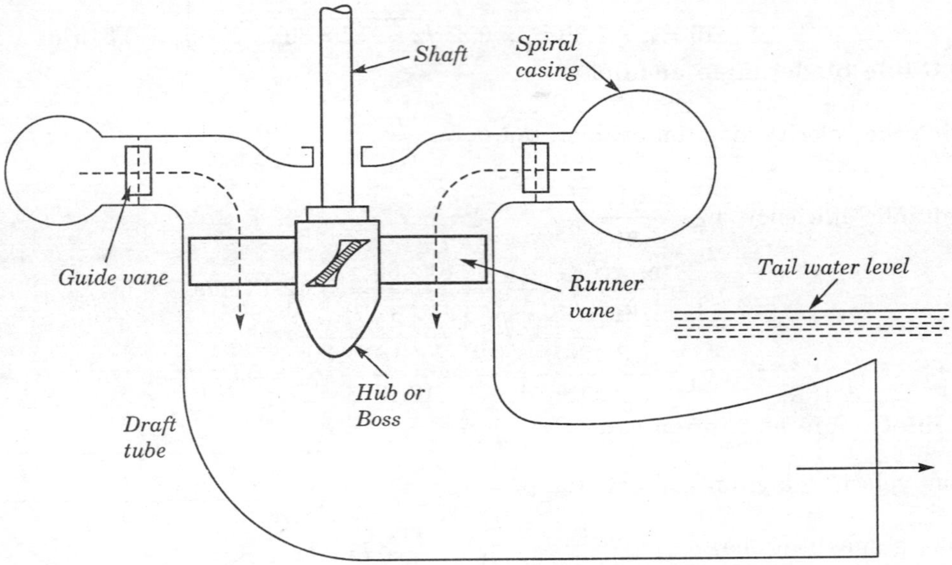

Fig. 15.14. A typical Kaplan turbine installation.

Except for the runner itself, the main component parts of a Kaplan turbine are spiral casing, guide vanes and draft tube which are the same as for a Francis turbine. Between the guide vanes and the runner, the water turns through a right-angle and flows parallel to the turbine shaft. A typical Kaplan runner is shown in Fig. 15.14.

The number of vanes of a Kaplan runner varies from 3 to 6 which are made of stainless steel and mounted on the hub. Kaplan turbines are also capable of taking overloads upto 15 to 20% and give high efficiency at all gate openings.

A Kaplan runner under installation at a powerhouse is shown in Fig. 15.14 (*a*).

Fig. 15.14. (*a*) Kaplan runner.

Fig. 15.15 shows the velocity diagrams for a Kaplan turbine at full and part-load conditions (speed and head remaining constant). As the discharge changes consequent upon a change in the turbine output, the blade angles get simultaneously adjusted in such a way that the flow enters the runner blades without shock. Consequently the eddy losses, which are inevitable in Francis and fixed blade propeller turbines, are eliminated in the Kaplan turbine.

Expressions for work-done, efficiency and power developed for propeller and Kaplan turbines are obtained in a similar manner as for a Francis runner.

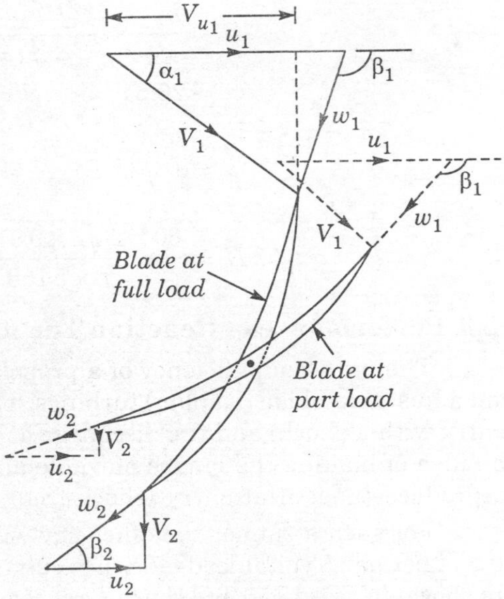

Fig. 15.15. Velocity diagrams for a Kaplan runner at full and part loads.

15.8.1. Working Proportions for Kaplan Turbine

The working proportions are obtained in a similar manner as for a Francis turbine, however, the following changes are to be noted :

(*i*) Ratio n of the hub diameter d to the runner outside diameter D ; $n = d/D$. From the continuity operation

$$Q = \frac{\pi}{4} (D^2 - d^2)V_f$$

$$Q = \frac{\pi}{4} (D^2 - d^2)\psi \sqrt{2gH}$$

$$= \frac{\pi}{4} D^2 (1 - n^2) \psi \sqrt{2gH} \qquad ...(15.45)$$

Generally, the diameter ratio n varies from 0.35 to 0.60. The flow ratio ψ has a value close to 0.70.

(*ii*) The blade speed u depends on the radius of the point under consideration and thus varies along the length of the blade. The blade angles thus vary continuously from the hub to the rim. The runner blades of a Kaplan turbine are thus warped or twisted.

Example 15.27. *A Kaplan turbine operating under a net head of 20 m develops 36750 kW with an overall efficiency of 86 per cent. The speed ratio is 2.0 and the flow ratio is 0.69. The hub diameter of the wheel is 0.35 times the outside diameter of the wheel. Find the diameter and speed of the turbine.*

Solution. Given : $H = 20$ m, $P = 36750$ kW

$\eta_0 = 86\%$, $\phi = 2.0$

$\psi = 0.60$, $n = 0.35$

$\therefore$ $d = 0.35\ D$

Discharge $Q = \frac{\pi D^2}{4} (1 - n^2)\ V_f$; $V_f = \psi \sqrt{2gH}$

Overall efficiency, $\eta_0 = \frac{P}{\gamma Q H}$

Substituting the data, $Q = \frac{36750 \times 10^3}{9810 \times 20 \times 0.86} = 218.025$ m³/s

$$D^2 = \frac{4Q}{\pi(1-n^2)\,\phi\,\sqrt{2gH}} = \frac{4 \times 218.025}{\pi[1-(0.35)^2] \times 0.60 \times \sqrt{2 \times 9.81 \times 20}}$$

$$= 26.61 \quad \text{or} \quad D = 5.158 \text{ m}$$

$$u = \phi \sqrt{2gH}\ ;\ N = \frac{60\,u}{\pi D} = \frac{60 \times \phi \sqrt{2gH}}{\pi D}$$

$$N = \frac{60 \times 2 \sqrt{2 \times 9.81 \times 20}}{\pi \times 5.158} = \textbf{146.67 r.p.m.}$$

15.9. Performance of Reaction Turbines : (Load-Efficiency Curves)

The part-load efficiency of a propeller (fixed blade) turbine is less than those for a Francis and adjustable-blade (Kaplan) turbines. This is mainly due to turbulence produced at the entrance (entry with a shock) and the discharge at the runner exit being far from radial. Kaplan conceived the idea of making the blades movable about a pivot, there by making it possible to adjust them to produce tangential entry (shock-free) for a wide range of gate movement.

For such a runner the efficiency will decrease but slightly as the load decreases from 40 to 30% of normal. Typical load-efficiency curves for propeller turbines (both fixed and adjustable blades) are shown in Fig. 15.16 and the difference in performance at part-loads is noticeable. For comparison, a similar curve for a Francis turbine is also shown. The superiority of a Kaplan runner over those of fixed blade propeller and Francis turbines for part-load conditions is thus established.

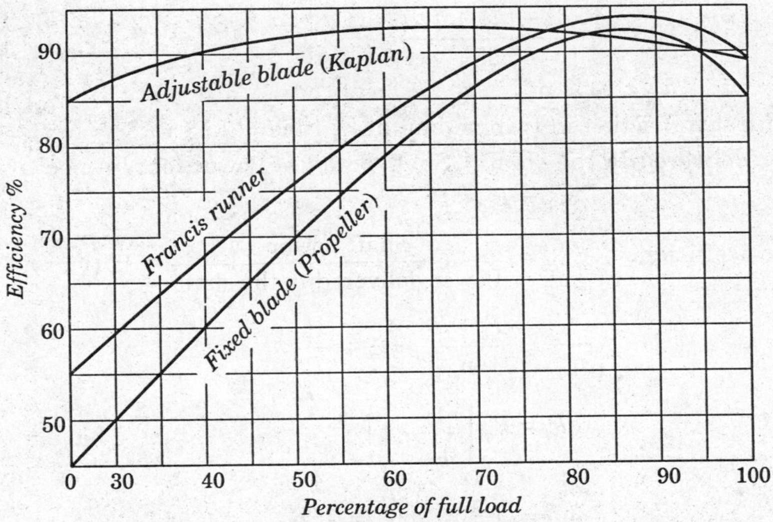

Fig. 15.16. Load-Efficiency Curves for Reaction Turbines.

Example 15.28. *A reaction turbine and its draft tube have a vertical axis. The pressure head in the spiral casing at inlet is 40 m above atmospheric pressure and the velocity of water is 5 m/s. The water flow through the tube is 1.75 m³/s, and the hydraulic and overall efficiencies are 0.30 and 0.80 respectively. The top of the draft tube is 1 m below the center line of the spiral casing while the tail race is 0.25 m below the top of the draft tube. The diameter of draft tube at inlet is 0.75 m and that at the tail race level 1.05 m. Determine :*

(a) the total head across the turbine *(b) the shaft power*

(c) the head lost in friction in turbine and draft tube

(d) the power lost in mechanical friction.

Solution. Total head across the turbine (above the tail race level)

= Total head in the spiral casing measured above the tail race.

or

$$H = \frac{p}{\gamma} + \frac{V_2}{2g} + (1 + 3.25)$$

$$= 40 + \frac{(5)^2}{2 \times 9.81} + 4.25 = \mathbf{45.52\ m.}$$

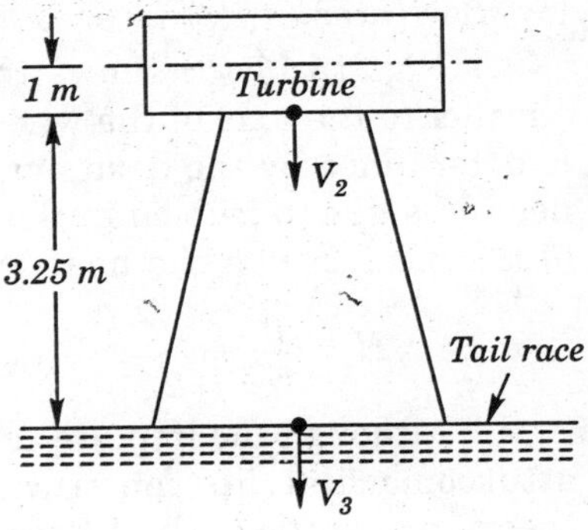

(b) Shaft power $P = \eta_0\ \gamma QH$

$= 0.80 \times 9810 \times 1.75 \times 45.52$

$= 624500$ N-m/s $=$ **624.5 kW.**

$$V_3 = \frac{4Q}{\pi d^2} = \frac{4 \times 1.75}{\pi\,(1.05)^2} = 2.02 \text{ m/s.}$$

(c) Head utilized by the turbine $= H - h_{ft} - h_{fd} - \dfrac{V_3^2}{2g}$

$$= 45.52 - (h_{ft} + h_{fd}) - \frac{(2.02)^2}{2 \times 9.81} = 45.31 - (h_{ft} + h_{fd})$$

Hydraulic efficiency

$$\eta_h = \frac{\text{Head utilized by the turbine}}{\text{Head supplied of the turbine}}$$

$$0.83 = \frac{45.31 - (h_{ft} + h_{fd})}{45.52}$$

or $$h_{ft} + h_{fd} = \mathbf{7.53\ m.}$$

∴ Head lost in friction in turbine and draft tube = **7.53 m.**

(*d*) Power delivered to the shaft = Shaft power + Power lost in mechanical friction

$$= P_s + P_f$$

Mechanical efficiency, $$\eta_m = \frac{\text{Shaft power}}{\text{Power delivered to the shaft}} = \frac{P_s}{P_s + P_f}$$

or $$\frac{1}{\eta_m} = \frac{P_s + P_f}{P_s} = 1 + \frac{P_f}{P_s}$$

or $$P_f = P_s \left(\frac{1}{\eta_m} - 1\right);$$

$$\eta_m = \frac{\eta_0}{\eta_h} = \frac{0.80}{0.83} = 0.96$$

$$P_f = 624.5\left(\frac{1}{0.98} - 1\right) = \mathbf{26.02\ kW.}$$

∴ The power lost in mechanical friction = 26.02 kW.

15.10. Impulse Turbines

In the impulse turbines, the total flow energy is converted into the kinetic energy of jet before the water strikes the runner. The conversion of the potential energy into the kinetic energy is made possible by passing the flow through a nozzle. As pressure around the jet or jets of water striking the runner is the same before and after the impact, the impulse turbines are also called constant pressure turbines. The pressure of flow at the entrance and exit of the runner is thus equal and is usually atmospheric, *i.e.* $p_1 = p_2 = p_{atm}$.

The impulse turbines are used for very high heads ranging between 600 to 2000 m and low flowrates.

Fig. 15.17 (*b*) shows the sectional plan of a bucket such that the plane of section is parallel to the axis of the wheel and contains the axis of the jet. The absolute velocity of the jet is determined by the head available at the base of the nozzle, that is, the gross head minus the head loss due to friction in the pipe line. If p and V are the pressure and the velocity at the base of the nozzle, then the head available at the nozzle base is

$$H = \frac{p}{\gamma} + \frac{V^2}{2g} \qquad \text{...(15.45)}$$

Although the impulse turbine is a turbomachine, its construction and theory are sufficiently different from those of centrifugal pumps and reaction turbines to justify special treatment.

Pelton turbine, named after Lester A. Pelton, an American engineer who contributed much to its development around 1880, is essentially an impulse turbine which is most commonly used and is

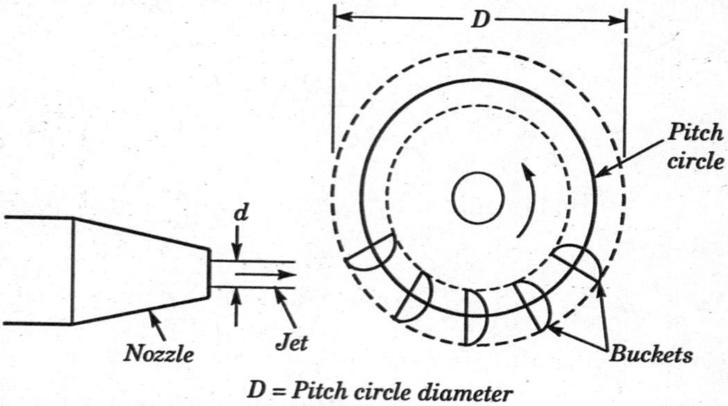

D = Pitch circle diameter
d = Jet diameter
(*a*) Elevation (not to scale)

particularly well suited to high heads. The runner consists of a circular disc with a number (not less than 15 but usually ranging between 18 to 22) of spoon-shaped buckets evenly spaced around its periphery, see Fig. 15.17 (*a*). One or more nozzles are mounted so that each directs a jet along a tangent to the circle through the centres of the buckets (known as the pitch circle). The nozzle which is attached to penstock is a specially designed needle valve which controls, by means of a governor, the discharge to the runner in accordance with power demand. In the centre of each bucket is a splitter ridge which divides the oncoming jet into two equal portions. The bucket is so shaped that after flowing round its smooth inner surface, the water leaves it with a relative velocity almost opposite in direction to the original jet.

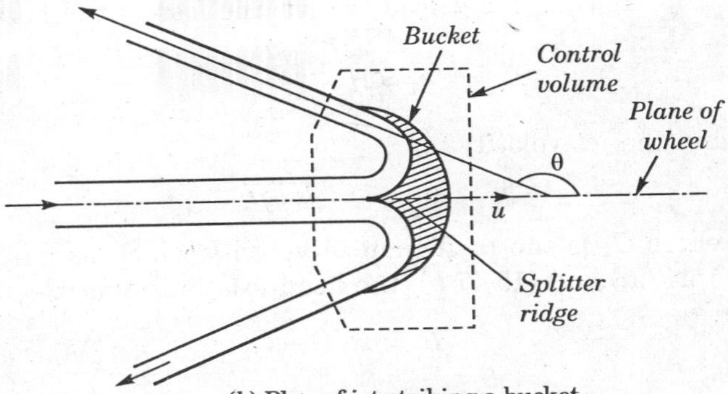

(*b*) Plan of jet striking a bucket

Fig. 15.17. Single jet Pelton wheel.

The buckets are cast from iron (for low heads), steel or bronze (for high heads) and have their surfaces polished and the splitter ground to an edge in order to minimise fractional resistance. They are generally fastened by two or more bolts to the runner disc. There is a notch (cut) in the outer rim of each bucket to make the jet face the bucket only when it has come almost normal to the jet.

The transfer of energy from the water jet to the buckets takes place according to the momentum principle. The dynamic force exerted by water on the buckets in a direction tangential to the pitch circle is the force which produces a torque and causes the rotation of the runner. We are, thus, mainly interested in finding the change of momentum in a direction tangential to the periphery of the runner. Fig. 15.17 (*b*) shows a Pelton wheel with a jet striking against a bucket.

A Pelton wheel runner installed at a power house is shown connected to the shaft, see Fig. 15.17 (*c*). The notches cut into the bucket and the splitter edge are clearly depicted in the photograph.

Fig. 15.17. (*c*) A Pelton wheel runner.

If V_1 is the velocity of jet, the loss in the nozzle is

$$h_l = H - \frac{V_1^2}{2g} \qquad \text{...(15.46 } a\text{)}$$

and the jet velocity is

$$V_1 = C_v \sqrt{2gH} \qquad \text{...(15.46 } b\text{)}$$

where C_v is the coefficient of velocity which usually ranges between 0.97 to 0.99. From Eqs. (15.47 *a*) and (15.47 *b*), the head loss in the nozzle

$$h_l = H\ (1 - C_v^2) \qquad \text{...(15.47)}$$

Fig. 15.18. Inlet and outlet velocity diagrams.

The relative velocity w_1 at the moment when water strikes the bucket is given by

$$w_1 = V_1 - u \qquad \text{...(15.48)}$$

The relative velocity w_2 with which the fluid leaves the bucket is somewhat less than the initial relative velocity w_1. There are two reasons for this : (1) although the inner surfaces of the buckets are polished so as to minimize the frictional losses as water flow over them, such losses cannot be entirely eliminated. (2) some additional loss is inevitable as water strikes the splitter ridge, because the ridge cannot have zero thickness. These losses reduce the relative velocity between water and bucket. We can, therefore, write

$$w_2 = Kw_1 \qquad \text{...(15.49)}$$

where K is blade friction coefficient which has a value slightly less than unity. Since the inlet and outled tips of the bucket are at the same radial distance from the centre of shaft,

$$u = u_1 = u_2 \qquad \text{...(15.50)}$$

There are three outlet velocity diagrams possible depending upon whether the runner is slow, medium or fast (that is, upon the magnitude of bucket speed u). For slow runner $\alpha < 90°$, for medium one $\alpha = 90°$ and for fast runner $\alpha > 90°$ (Fig. 15.18). Efforts are made to make the outlet absolute velocity $V_2^2/2g$ as small as possible since it represents the energy rejected to the tail race.

From the outlet velocity diagram, taking the direction of u as positive, the component of absolute velocity V_2 in the tangential direction is

$$V_{u_2} = u - w_2 \cos \beta$$

from inlet velocity diagram $V_{u_\tau} = V_1$.

Selecting a control volume as indicated by the dotted lines in Fig. 15.17 (*b*), and applying the momentum equation the force exerted by the bucket upon the fluid is

$$F' = \rho Q\ (V_{u_2} - V_{u_1})$$

and the dynamic force F exerted by the fluid upon the bucket is equal and opposite to the force F' thus

$$F = \rho Q\ (V_{u_1} - V_{u_2}) = \rho Q\ (V_1 - u + w_2 \cos \beta)$$

$$= \rho Q w_1\ (1 + K \cos \beta) \qquad \text{...(15.51)}$$

Eq. (15.52) gives the driving force causing the rotation of the Pelton wheel. The torque on the wheel is, therefore, $F.\,r$, where r is the radius to the centre of the bucket, and the power delivered to the wheel is

$$P = Fr\omega = F\,.\,u = \rho Q w_1\,(1 + K\cos\beta)\,u$$
$$= \rho Q\,(V_1 - u)\,(1 + K\cos\beta)\,u \qquad ...(15.52)$$

The energy of flow approaching the bucket is in the form of kinetic energy of jet and is given by $\frac{1}{2}\rho Q V_1^2$ per unit time. Hence the hydraulic efficiency of the Pelton wheel is

$$\eta_h = \frac{\rho Q(V_1 - u)\,(1 + K\cos\beta)u}{\frac{1}{2}\rho Q V_1^2} = \frac{2u(V_1 - u)\,(1 + K\cos\beta)}{V_1^2} \qquad ...(15.53)$$

If the blade surface is made perfectly smooth *i.e.* frictionless, then

$$K = 1 \quad \text{and} \quad \eta_h = \frac{2u\,(V_1 - u)}{V_1^2}\,(1 + \cos\beta)$$

Considering the power input to the nozzle for defining the hydraulic efficiency.

$$\eta_h = \frac{\rho Q(V_1 - u)\,(1 + K\cos\beta)u}{\gamma QH} = \frac{u(V_1 - u)\,(1 + K\cos\beta)}{gH} \qquad ...(15.54)$$

For a given jet velocity V_1 and the bucket tip angle β at the outlet, the condition for maximum efficiency may be obtained by differentiating Eq. (15.54) with respect to u and equating to zero.

$$\frac{d\eta_h}{du} = \frac{2(V_1 - 2u)\,(1 + K\cos\beta)}{V_1^2} = 0$$

which yields $\quad u = \dfrac{V_1}{2}.$

This states the best runner speed for maximum efficiency is equal to one-half the jet velocity. The maximum hydraulic efficiency is given by

$$(\eta_h)_{max} = \frac{1 + K\cos\beta}{2} \qquad ...(15.55)$$

The hydraulic efficiency represents the effectiveness of the wheel in converting the kinetic energy of the jet into mechanical energy of rotation. Not all this energy of rotation is available at the output shaft of the machine, because some energy is consumed in overcoming friction in bearing and some overcoming the 'windage' that is the friction between the wheel and the atmosphere in which it rotates. The overall efficiency is, therefore, less than the hydraulic efficiency. Even so, and overall efficiency of 85 to 90% may usually be achieved in large machines. As the losses due to bearing friction and windage increase rapidly with speed, the maximum overall efficiency occurs when the ratio u/V_1 is slightly less than the theoretical value of 0.5 ; its value ranges between 0.43 to 0.48. For maximum efficiency, it is evident from Eq. (15.54) that the bucket outlet tip angle β should be equal to zero or the angle θ should be 180°. This is, however, not possible in practice, because if the jet is deflected through 180°, the outgoing jet will strike the back of the backet just following it and will exert a retarding force on it. For the outgoing jet to keep clear of the bucket following it, the bucket tip angle β is kept in the range 10° to 20°.

Example 15.29. *The buckets of a Pelton impulse turbine deflect the jet through a total angle of 165° and owing to the surface friction the relative velocity of the water leaving the bucket is 0.85 times that at entry. Draw the velocity vector diagrams at entry and exit and find the ratio of bucket velocity to jet velocity in order that the water shall leave the buckets without whirl. In such a turbine the available head at the nozzle is 650 m, the coefficient of velocity for the nozzle is 0.97, the jet diameter 10 cm and the mean bucket circle diameter 1.20 m. Using the conditions referred to above determine :*

(a) best running speed in r.p.m.

(*b*) *the impulsive force of the buckets at this speed*
(*c*) *the power developed by the buckets*
(*d*) *the efficiency of buckets.*

Solution. (*a*) $V_1 = 0.97\sqrt{2g \times 650} = 109.6$ m/s

Refering to Fig. 15.18, $w_2 = 0.85(V_1 - u) = 0.85(109.6 - u)$

From the outlet velocity diagram, since water leaves without whirl, $\alpha = 90°$

$$w_2 \cos 15° = u$$

or $$0.15\,(109.6 - u) \times 0.9656 = u \quad \text{and} \quad u = 49.4 \text{ m/s}$$

Speed in r.p.m., $$u = \frac{\pi \times 1.2 \times n}{60}$$

$\therefore$ $$n = \frac{49.4 \times 60}{\pi \times 1.2} = 786 \text{ r.p.m.}$$

Discharge $$Q = \frac{\pi}{4}\left(\frac{10}{100}\right)^2 \times 109.6 = 0.86 \text{ m}^3\text{/s.}$$

(*b*) Impulsive force on the buckets from Eq. (15.22)

$$F = \rho Q(V_1 - u)\,(1 + K \cos \beta)$$

$$= \frac{9810 \times 0.86 \times (109.6 - 49.4) \times (1 + 0.85 \times 0.9659)}{9.81} = 94470 \text{ N}$$

(*c*) Power developed $= Fu = 94470 \times 49.4 = 4660 \times 10^3$ N-m/s $= 4660$ kW

(*d*) Efficiency of buckets $= \dfrac{\text{Power developed}}{\text{Power input to buckets}}$

Power input to buckets $= \gamma Q\,(V_1^2/2g)$

$$= 9810 \times 0.86 \times \frac{(109.6)^2}{2 \times 9.81} = 5145 \times 10^3 \text{ N-m/s} = 5145 \text{ kW}$$

$\therefore$ Efficiency $= \dfrac{4660}{5145} = 0.905$ or 90.5%.

Example 15.30. *A Pelton wheel has a mean bucket speed of 12 m/s and is supplied with water at the rate of 750 lit/sec under a head of 35 m. If the buckets deflect the jet through an angle of 160°, find the power and the efficiency of the turbine. Take the coefficient of velocity as 0.98.*

Solution. Velocity of jet

$$V_1 = 0.98\sqrt{2 \times 9.81 \times 35} = 25.68 \text{ m/s.}$$

$$w_1 = (V_1 - u) = (25.68 - 12) = 13.68 \text{ m/s.}$$

Neglecting friction, $w_1 = w_2$.

Power developed by the turbine

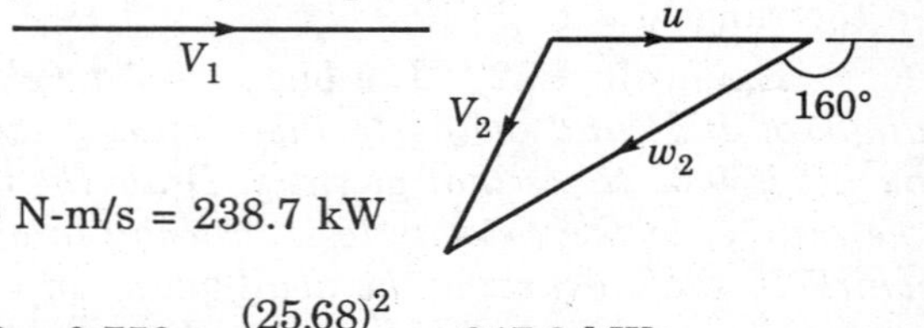

$$P = \frac{\gamma Q}{g}\,(V_1 - u)u(1 + \cos \beta)$$

$$= \frac{9810 \times 0.750}{9.81} \times 13.68 \times 12(1 + \cos 20°) \text{ N-m/s} = 238.7 \text{ kW}$$

Power input $$= \frac{\gamma Q}{75}\left(\frac{V_1^2}{2g}\right) = 9810 \times 0.750 \times \frac{(25.68)^2}{2 \times 9.81} = 247.3 \text{ kW}$$

$\therefore$ Efficiency of turbine $= \dfrac{238.7}{247.3} \times 100 = 96.6\%$.

15.10.1. Speed ratio, jet ratio and Working Proportions

(*i*) The ideal velocity of jet expressed by $\sqrt{2gH}$, where H is the effective head, is known as the spouting velocity. The ratio of the bucket speed u and the theoretical velocity of jet $\sqrt{2gH}$ is known as the speed ratio, and is denoted by ϕ, *i.e.*

$$u = \phi \sqrt{2gH} \quad ...(15.56)$$

where ϕ is known as the speed ratio. The value of ϕ ranges from 0.43 to 0.48.

(*ii*) The actual velocity of jet is slightly less than $\sqrt{2gH}$ on account of energy loss in the nozzle. It is expressed as

$$V = C_v \sqrt{2gH}$$

where C_v is a velocity coefficient which ranges from 0.98 to 0.99.

(*iii*) The angle through which water gets deflected by the bucket is normally 165°.

(*iv*) The ratio of the pitch circle diameter D and the jet diameter d is known as the jet ratio and is generally denoted by m, thus

$$m = D/d \quad ...(15.57)$$

For maximum efficiency, the value of m as determined from experiments ranges from 10 to 14.

(*v*) The main dimensions of the bucket are :

Axial width $B = 3d$ to $4d$; radial length $L = 2d$ to $3d$; depth $D = 0.8\ d$ to $1.2\ d$.

The jet diameter d is governed by the output power and the available head.

15.10.2. Number of Buckets

The number of buckets for a Pelton wheel is decided from two considerations : (1) the number of buckets should be as few as possible so as to reduce the frictional losses and (2) the jet must be fully utilized and no portion of it should leave the runner without doing work on it. These two principles, as is evident, are contradictory in nature and cannot be satisfied simultaneously. An empirical formula given by Taygun expresses the number of buckets as

$$Z = \frac{D}{2d} + 15 = 0.5\ m + 15 \quad ...(15.58)$$

This equation holds good for all values of m ranging from 6 to 35.

Example 15.31. *The water jet in a Pelton wheel is 8 cm in diameter and has a velocity of 93 m/s. The rotational speed of the wheel is 600 r.p.m and the deflection angle of the jet is 170°. If the speed ratio is 0.47, determine the diameter of wheel and the power developed.*

Solution. With reference to Fig. 15.18, $\theta = 170°$, $\beta = 180° - \theta = 10°$

$$V_1 = 93 \text{ m/s, speed ratio } \phi = u/\sqrt{2gH} = 0.47$$

$$\text{Discharge } Q = \frac{\pi}{4}\left(\frac{8}{100}\right)^2 \times 93 = 0.4676 \text{ m}^3\text{/s}$$

Assuming jet velocity coefficient $C_v = 0.98$

$$V_1 = C_v\sqrt{2gH}, \text{ from which } \sqrt{2gH} = V_1/C_v = 93/0.98 = 94.898$$

$$u = \phi\sqrt{2gH} = 0.47 \times 94.898 = 44.60 \text{ m/s}$$

Also, $u = \dfrac{\pi DN}{60}$, $\quad \therefore \quad D = \dfrac{60\,u}{\pi N} = \dfrac{60 \times 44.60}{\pi \times 600} = \mathbf{1.419\ m}$

Power $\quad P = \rho Q\,(V_1 - u)\,(1 + K\cos\beta)\,u$

assuming $\quad k = 1$,

Power developed $P = (1000) \times 0.4676\ (93 - 44.6)\ (1 + \cos 10°) \times 44.6 = \mathbf{2016.04\ kW.}$

Example 15.32. *Develop on expression for power input to the shaft of a Pelton wheel and hence prove that the hydraulic efficiency*

$$\eta_h = 2\phi\,(C_v - \phi)\,(1 - K \cos \beta)$$

where ϕ = *speed ratio,* C_v = *coefficient of velocity of jet*
K = *smoothness factor,* β = *bucket angle.*

Solution. From Eq. (15.55), the hydraulic efficiency

$$\eta_h = \frac{u(V_1 - u)(1 + K \cos \beta)}{gH}$$

Making use of Fig. 15.18, in which the blade angle is 180° – β, one can write the following :

$$\eta_h = \frac{\phi \sqrt{2gH}\,(C_v \sqrt{2gH} - \phi \sqrt{2gH})\,[1 + K \cos (180° - \beta)]}{gH}$$

$$= 2\phi\,(C_v - \phi)\,(1 - K \cos \beta).$$

Example 15.33. *A Pelton wheel, 2.45 m in diameter, operates under the following conditions :*
Net head = 370 m ; $C_v = 0.98$; $\phi = 0.47$; $K = 0.90$; $\beta = 160°$
Diameter of jet = 88 cm. Determine —
(i) the input power to the shaft, and *(ii) the r.p.m. of the wheel.*

Solution. (*i*) Input power to the shaft

$$= \frac{1}{2}\frac{\gamma Q}{g} V_1^2 \quad V_1 = C_v \sqrt{2gH} = 0.98 \sqrt{2 \times 9.81 \times 370} = 83.498 \text{ m/s}$$

$$Q = \frac{\pi}{4} d^2 V_1 = \frac{\pi}{4}\left(\frac{18}{100}\right)^2 \times 83.498 = 2.1247 \text{ m}^3\text{/s}$$

$$\text{Input power} = \frac{1}{2} \times \frac{9810 \times 2.1247}{9.81} \times (83.498)^2$$

$$= 7549.8 \times 10^3 \text{ N-m/s} = \mathbf{7549.8\ kW.}$$

(*ii*) Speed of wheel in r.p.m.

$$u = \frac{\pi D n}{60}, \quad \therefore \quad n = \frac{60u}{\pi D}$$

Substituting the data, $n = \dfrac{60 \times \phi \sqrt{2gH}}{\pi D} = \dfrac{60 \times 0.47 \sqrt{2 \times 9.81 \times 370}}{\pi \times 2.45}$ = **312.15 r.p.m.**

Example 15.34. *A Pelton wheel has to develop 13230 kW under a net head of 800 m while running at a speed of 600 r.p.m. If the coefficient of the jet* $C_v = 0.97$, *speed ratio* $\phi = 0.46$ *and the ratio of the jet diameter is 1/16 of wheel diameter, calculate the number of jets required for the Pelton wheel. Calculate also :*

(i) the diameter of the jets ; *(ii) the diameter of the pitch circle, and*
(iii) the quantity of water supplied to the wheel.

Assume overall efficiency as 85%.

Solution. The overall efficiency,

$$\eta_0 = \frac{bhp}{\gamma QH} = \frac{13230 \times 10^3}{9810 \times Q \times 800}$$

$$\therefore \quad Q = \frac{1350}{0.85 \times 800} = 1.985 \text{ m}^3\text{/s}$$

$$u = \phi \sqrt{2gH} = 0.46 \sqrt{2 \times 9.81 \times 800} = 57.63 \text{ m/s}$$

$$V_1 = C_v \sqrt{2gH} = 0.97 \sqrt{2 \times 9.81 \times 800} = 121.52 \text{ m/s}$$

Also $u = \dfrac{\pi D n}{60} \quad D = \dfrac{60u}{\pi n} = \dfrac{60 \times 57.63}{\pi \times 600} = 1.834 \text{ m}$

Diameter of jet $d = \dfrac{1}{16} D = \dfrac{1.834}{16} = 0.1146 \text{ m}$

Discharge through each nozzle $= \dfrac{\pi d^2}{4} \, . \, V_1$

$$= \frac{\pi (0.1146)^2}{4} \times 121.52 = 1.254 \text{ m}^3/\text{s}$$

Number of jets required $= \dfrac{\text{Total discharge}}{\text{Discharge perjet}} = \dfrac{1.985}{1.254} = 1.58$, say 2.

Hence two jets will be required.

Let the total discharge be equally divided between the two jets

$$= \frac{Q}{2} = \frac{1.985}{2} = 0.9925 \text{ m}^2/\text{s}$$

the diameter of jet is given by

$$\frac{\pi d^2}{4} V_1 = Q_j. \qquad \therefore \; d = \sqrt{\frac{4Q_j}{\pi V_1}} = \sqrt{\frac{4 \times 0.9925}{\pi \times 121.52}}$$

$$= 0.101197 \text{ m} \quad \text{or} \quad \mathbf{10.2 \text{ cm.}}$$

(*i*) Pitch circle diameter $D = 1.834$ m

(*ii*) Quantity of water supplied to the wheel = 1.985 m^3/s.

Example 15.35. *In a Pelton wheel the buckets deflect the jet by 170° and the relative velocity is reduced by 12% due to bucket friction. For a speed ratio of 0.47, calculate from first principles the hydraulic efficiency of the wheel.*

The bucket circle diameter of the wheel is 90 cm and there is one jet for which $C_v = 0.98$. The actual efficiency of the wheel is 0.9 times its theoretical efficiency. The wheel develops 1700 kW under a head of 550 m. Calculate

(a) The speed of the wheel in r.p.m. and (b) The diameter of the nozzle.

Solution. Data given : $\beta = 10°$ ($\theta = 170°$), $w_2 = 0.88\, w_1$, $\phi = 0.47$

Force exerted by the jet on the bucket,

$$F = \rho Q (V_{u_1} - V_{u_2}) \quad \text{or} \quad F = \rho Q (V_1 - u)(1 + K \cos \beta)$$

Torque exerted by this force $= F \, . \, r$

Power developed $= F.r.\omega = F \, . \, u = \rho Q u (V_1 - u)(1 + K \cos \beta)$

Power input to the wheel = Kinetic energy of the jet $= \frac{1}{2} \rho Q V_1^2$

$$\therefore \; \text{Hydraulic efficiency, } \eta_h = \frac{\rho Q (V_1 - u) u (1 + K \cos \beta)}{\frac{1}{2} \rho Q V_1^2} = \frac{2u (V_1 - u)(1 + K \cos \beta)}{V_1^2}$$

$$u = \phi \sqrt{2gH} = 0.47 \sqrt{550 \times 2 \times 9.81} = 48.82 \text{ m/s}$$

$$V_1 = C_v \sqrt{2gH} = 0.98 \sqrt{2 \times 9.81 \times 550} = 101.80 \text{ m/s}$$

$$K = 0.88, \; \beta = 10°$$

$\therefore \qquad \eta_h = \dfrac{2 \times 48.82 \times (101.80 - 48.82)(1 + 0.88 \cos 10°)}{(101.8)^2} = 0.99$

Over efficiency, $\eta_0 = 0.9 \times 0.99 = \mathbf{0.891}$ $P = 1700$ kW

Speed of wheel, $= \dfrac{\pi D n}{60}$ or $n = \dfrac{60\,u}{\pi D}$

$$n = \frac{6 \times 48.82}{\pi \times 0.9} = \mathbf{1035.6 \text{ r.p.m.}}$$

$$\eta_0 = \frac{P}{\gamma Q H} \text{ or } Q = \frac{1700 \times 10^3}{\eta_0 \gamma H} = \frac{1700 \times 10^3}{0.891 \times 9810 \times 550}$$
$$= 0.3539 \text{ m}^3\text{/s}$$

Diameter of the nozzle, $d = \sqrt{\dfrac{4Q}{\pi V_1}} = \sqrt{\dfrac{4 \times 0.3539}{\pi \times 101.8}} = \mathbf{0.066 \text{ m}}$ or **6.6 cm.**

Example 15.36. *The following are the design particulars of a large Pelton turbine :*

Head at distributor	*= 630 m*
Discharge	*= 12.5 m³/s*
Power	*= 65 megawatts*
Speed of rotation	*= 500 r.p.m*
Runner diameter	*= 1.96 m*
Number of jets	*= 4*
Jet diameter	*= 0.192 m*
Angle through which the jet is deflected by the bucket	*= 165°*
Mechanical efficiency of the turbine	*= 96%*

Determine the hydraulic power losses in the distributor nozzle assembly and the buckets.

Solution. Mechanical efficiency

$$\eta_m = \frac{\text{Power developed}}{\text{Power developed by the runner, } P}$$

Power developed by one jet $= \dfrac{65{,}000}{4} = 16{,}250$ kW

Discharge per jet $= \dfrac{12.5}{4} = 3.125$ m³/s

Work done by the runner,

$$P = \frac{\gamma Q}{g}(V_1 - u)(1 + K \cos \beta)\, u$$

Power developed by the runner,

$$P = (16{,}250)/0.96 = 16927 \text{ kW}$$

$$V_1 = \frac{4Q}{\pi d^2} = \frac{4 \times 3.125}{\pi \times (0.192)^2} = 107.93 \text{ m/s}$$

$$u = \frac{\pi D n}{60} = \frac{\pi \times 1.96 \times 500}{60} = 51.31 \text{ m/s}$$

$\beta = 15°$, substituting these data, one can write,

$$16927 \times 10^3 = \frac{9810 \times 3.125}{9.81}(107.93 - 51.31)(1 + 0.9659\,K) \times 51.31$$

from which, $K = 0.8965, \quad w_2 = K w_1 = K(V_1 - u)$

$$V_1 = C_v \sqrt{2gH} \quad \text{or} \quad C_v = \frac{V_1}{\sqrt{2gH}} = \frac{107.93}{\sqrt{2 \times 9.81 \times 6.30}} = 0.97$$

Head loss in the nozzle, $h_L = (1 - C_v^2)\ H = [1 - 0.97^2] \times 630 = 37.23$ m

$$\text{Head loss in the buckets} = \frac{w_1^2 - w_2^2}{2g} = \frac{w_1^2}{2g}(1 - K^2)$$

$$= \frac{(V_1 - u)^2}{2g}(1 - K^2) = \frac{(56.62)^2}{2 \times 9.8}[1 - 0.8965^2] = 32.04 \text{ m}$$

The head loss in the nozzle and the buckets

$= 32.04 + 37.23 = 69.27$ m

and the corresponding power loss $= \gamma Q H_L = 9810 \times 12.5 \times 69.27 = 8485.4 \times 10^3$ N-m/s

$=$ **8485.4 kW.**

Example 15.37. *A jet water 5 cm diameter impinges on the buckets of a Pelton wheel deflecting it through an angle of 170°, the blade friction coefficient being 0.95. The head at the nozzle is 30 m and the coefficient of velocity is 0.97. Determine the force exerted by the jet on a series of such buckets revolving at a velocity of 0.45 times that of the jet. Find the power developed by the buckets. Assuming that the windage and mechanical friction loss amount to 4% of the water power supplied, find the shaft power and the overall efficiency of the turbine.*

Solution.

$$V_1 = C_v \sqrt{2gH} = 0.97\sqrt{2 \times 9.81 \times 30} = 23.53 \text{ m/s}$$

$$u_1 = 0.45\ V_1 = 0.45 \times 23.53 = 10.59 \text{ m/s}$$

$$w_1 = V_1 - u = 23.53 - 10.59 = 12.94 \text{ m/s}$$

$$w_2 = Kw_1 = 0.95 \times 12.94 = 12.25 \text{ m/s}$$

Weight of water striking the bucket per second

$$W = \gamma A V_1 = 9810 \times \frac{\pi}{4}\left(\frac{5}{100}\right)^2 \times 23.53 = 453.2 \text{ N}$$

$$\text{Force exerted on the bucket} = \frac{W}{g}(w_1 + w_2 \cos 10°)$$

$$= \frac{453.2}{9.81}(12.94 + 12.25 \cos 10°) = 1155.4 \text{ N}$$

Force developed by water $= F \,.\, u$

$1155.4 \times 10.59 = 12236.0$ N-m/s

$= 12236$ W $= 12.236$ kW

Water power contained in the jet

$$= \frac{1}{2}\frac{W}{g}V_1^2$$

$$= \frac{453.2}{2 \times 9.81} \times (23.53^2) = 12789.3 \text{ N-m/s} = 12789 \text{ W} = 12.79 \text{ kW}$$

Losses in windage and mechanical friction

$= 0.04 \times 12.79 = 0.51$ kW

$\therefore$ Shaft power $=$ Power developed by water $-$ Losses

$= 12.236 - 0.51 = 11.726$ kW

Over efficiency, $\eta_0 = \dfrac{11.726}{12.79} =$ **0.9168 or 91.68%.**

15.11. Definitions

15.11.1. Gross Head

Gross head at the power plant is the difference between the water level in the reservoir (head water level) and the tail race (tail water level). See Fig. 15.1.

15.11.2. Net or Effective Head

The net or effective head is defined as the difference between the total energy per unit weight at entrance and exit of the turbine (exit of turbine means the outlet of the draft tube as the draft tube is an integral part of a reaction turbine). In other words, the net head is equal to the head available at the turbine entrance minus the velocity head at the exit of the draft tube. Thus,

Net or Effective head = Gross head – Head loss in penstock – Velocity head at the exit of draft tube
= Head available at turbine entrance – Velocity head at the exit of draft tube.

15.11.3. Rated Quantities

The rated quantities are those for which the turbine is designed. The rated head is the net head under which the turbine is to deliver the rated output power at the rated synchronous speed. The rated discharge is the turbine discharge at the rated head producing rated power at rated speed.

15.11.4. Energy Losses and Efficiencies

Conversion of hydraulic energy of flow into mechanical energy of revolving turbine rotor is inevitably associated with energy losses. The performance of hydraulic turbines is characterised by its overall efficiency,

$$\eta_0 = \frac{\text{Shaft power, } P_s}{\text{Water horse power, } \gamma Q H}$$

The magnitude of energy losses depends on the type of the turbine, its dimensions and regime of operation. Modern powerful reaction turbines have high values of overall efficiency at the design-point regime, say η_0 = 92 to 95%, and at the rated regime its magnitude may be between 86 to 90%. The impulse turbines have a smaller magnitude of overall efficiency such as $\eta_0 \leq 91\%$.

The following efficiencies are generally used in relation to hydraulic turbines. The different types of efficiencies are due to various kinds of losses, such as hydraulic, volumetric and mechanical that take place within a turbine.

Hydraulic efficiency. The hydraulic efficiency of the turbine is defined as the ratio of the hydraulic power utilized by the runner and the net power supplied by the water at the entrance to the turbine. For reaction turbines, it may be expressed as

$$\eta_h = \frac{\rho Q(u_1 V_{u_1} - u_2 V_{u_2})}{\gamma Q H}$$

$$= \frac{u_1 V_{u_1} - u_2 V_{u_2}}{gH} \quad ...(15.59)$$

in which H is the net or effective head. The hydraulic losses within the turbine take place mostly on account of blade friction, eddy formation and friction in draft tube. If these losses amount to ΔH, then the hydraulic efficiency becomes

$$\eta_n = \frac{H - \Delta H}{H} \quad ...(15.60)$$

Volumetric Efficiency. As the fluid passes through the turbine, a certain fraction of its discharge escapes through the gap between the guide apparatus and the runner without doing work on the runner. This way some hydraulic energy goes waste there by decreasing the turbine efficiency. If ΔQ is that part of the discharge which flows out through the gap without doing work on the runner, and Q is the total discharge entering the turbine, then the volumetric efficiency is

$$\eta_v = \frac{Q - \Delta Q}{Q} \qquad ...(15.61)$$

Mechanical Efficiency. The power developed by the runner is decreased by the mechanical losses caused by friction between the rotating parts like the shaft and the runner and stationary parts like guide and thrust bearings and sealings. On account of these losses, the power available at the turbine shaft is less than the power delivered to it by the runner. The mechanical efficiency is, therefore, equal to

$$\eta_m = \frac{\text{Power available at the turbine shaft}}{\text{Power delivered to the shaft by the runner}} \qquad ...(15.62)$$

Overall Efficiency of the turbine. It is defined as the ratio of the power available at the turbine shaft to the power input to the turbine. The overall efficiency is, therefore, equal to

$$\eta_o = \frac{\text{Power available at the turbine shaft}}{\gamma QH}$$

$$= \eta_h . \eta_v . \eta_m \qquad ...(15.63)$$

If η_g is the efficiency of the generator, then the power output of the hydraulic (turbine and the hydro-generator combined) is equal to $(\gamma QH)\eta_o\eta_g$. The product of η_o and η_g is known as the 'plant efficiency', η_p.

Example 15.38. *A Pelton wheel produces 44 kW at the shaft when available water head across the turbine is 40 m. The energy transferred from water to the runner is 35.0 N-m/N of water. If the mechanical efficiency is 92%, determine the overall efficiency and the discharge through the turbine.*

Solution. Hydraulic efficiency

$$\eta_h = \frac{\text{Energy head transferred to runner}}{\text{Water head}} = \frac{35.0}{40.0} = 0.875$$

Overall efficiency $\eta_o = \eta_m \times \eta_h = 0.92 \times 0.875 = 0.805$ or 80.5%

Shaft horsepower $P = \eta_o \gamma QH$

$$\therefore \quad Q = \frac{44 \times 10^3}{0.805 \times 40 \times 9810} = \mathbf{0.14\ m^3/s.}$$

15.11.5. Spherical and butterfly valves

For regulating the flow into the turbine, the main inlet valve of either spherical or butterfly type are used depending on the design head. The valves are custom-built and their sizes range from 500 mm to 5000 mm diameter. Main inlet valves and penstock valves of butterfly type are used for low head applications. For high head applications, spherical valves are used. These types of valves have been used in many hydropower installations. Salient features of some of these installations in various Indian states are shown in the following section. Butterfly valves (BV) and spherical valves (SV) alongwith their nominal diameters for different power stations have also been shown.

15.11.6. Typical hydropower installations of India

Salient features of some of the typical hydropower installations of our country equipped with Francis, Kaplan and Pelton runners are tabulated on next page.

15.12. Similarity Laws

Applying the principles of dynamic similarity it is possible to predict the performance of turbomachines from the results of a test on a geometrically similar machine. It also becomes possible to predict the performance of the same machine under conditions different from the test conditions. It may be recalled that geometric similarity which means similarity of size and shape is a prerequisite of dynamic similarity. Dynamic similarity also requires kinematic similarity, that is the corresponding velocities should have a constant ratio. The kinematic similarity amounts to geometric similarity of velocity diagrams at inlet and outlet for geometrically similar machines. Two geometrically similar turbomachines having similar velocity diagrams at inlet and outlet are said to be *homologous*. They will also have geometrically similar streamlines.

Considering the fluid to be incompressible, the power P transferred between the fluid and the runner depends upon the mass density ρ, viscosity μ, runner diameter D, discharge Q, rotational speed n and the potential energy per unit mass (gH). The functional relationship between the dependent variable and the independent variables may be expressed as

$$P = f(\rho, \mu, D, Q, n, gH) \qquad \text{...(15.64)}$$

Choosing ρ, n and D as the repeating variables and using the methods of dimensional analysis, the following four dimensionless π-terms may be obtained :

$$\pi_1 = \frac{Q}{nD^3}, \quad \pi_2 = \frac{gH}{n^2D^2}$$

$$\pi_3 = \frac{\rho n D^2}{\mu}, \quad \pi_4 = \frac{P}{\rho n^3 D^5} \qquad \text{...(15.65)}$$

Let us examine the significance of these π-terms. All lengths of the machine are proportional to D and thus all areas to D^2. The ratio Q/D^2 may be regarded as the typical velocity of fluid. The peripheral velocity of the runner (and also the blade velocity u) is proportional to nD. Thus π_1 represents the ratio of the fluid velocity and the blade velocity. Now if π_1 is same for two geometrically similar machines, it will mean similarity of velocity diagrams. In other words, kinematic similarity of two geometrically similar machines is achieved if Q/nD^3 is the same for both. The dimensionless term π_2 may also be visualised as the ratio of the square of fluid velocity and the square of the blade velocity, since gH is proportional to the square of fluid velocity.

The term π_3 may be shown to be proportional to the Reynolds number. The product of π_1 and π_3 results in $(Q/D^2)(\rho D/\mu)$, which represents the Reynolds number. If π_1 is held constant to obtain kinematic similarity, then π_3 is proportional to the Reynolds number. The flow in the turbomachines is highly turbulent in the normal range of speed and size and, therefore, the vicosity has a very little effect on its performance. The dimensionless term π_3 may, therefore, be ignored.

For dynamic similarity between two geometrically similar machines, the dimensionless term $\pi_4 = P/\rho n^3 D^5$ must be equal. Combining π_4 with π_1 and π_2, we obtain the ratio

$$\frac{\pi_4}{\pi_1 \pi_2} = \frac{P}{\gamma Q H} \qquad \text{...(15.66)}$$

For a turbine, Eq. (15.66) represents hydraulic efficiency, since γQH is the energy given up by the fluid in unit time in passing through the turbine, and P represents the energy transferred between the fluid and the runner. For a pump, the hydraulic efficiency is given by $\gamma QH/P$. If the volumetric and mechanical efficiencies are assumed constant, the shaft power may be used in place of P and the overall efficiency in place of hydraulic efficiency. Eq. (15.65 A) may now be expressed in dimensionless terms as

$$\frac{P}{\rho n^3 D^5} = f_1\left(\frac{Q}{nD^3}, \frac{gH}{n^2D^2}\right) \qquad \text{...(15.67}a\text{)}$$

and the efficiency may be expressed as

$$\eta = f\left(\frac{gH}{n^2D^2}, \frac{P}{\rho n^3 D^5}\right) \qquad \text{...(15.67}b\text{)}$$

which using Eq. (15.67 a), can also be written as

$$\eta = f\left(\frac{Q}{nD^3}, \frac{gH}{n^2D^2}\right) \qquad \text{...(15.67}c\text{)}$$

SOME TYPICAL HYDROPOWER INSTALLATIONS

S. No.	Name of Power station	Owned by	Year of commis-sioning	No. of units and capacity (MW)	Turbine Parameters			Main Inlet Valve	
					Rating (kW)	Design Head (m)	Speed (R.P.M)	Type	Nominal diameter (mm)
(1)	(2)	(3)	(4)	(5)	(6)	(7)	(8)	(9)	(10)
				TURBINE RUNNER : FRANCIS					
1.	Khadakwasla	Govt. of Maharashtra	1990	2 × 8	8,330	45.7	300.0	BV	2,000
2.	Sumbal Sindh	Govt. of Jammu and Kashmir	1973	2 × 11.3	11,800	149.0	500.0	SV	1,150
3.	Kundah-V	Tamil Nadu Electricity Board	1988	1 × 21.6	22,380	259.1	750.0	SV	1,000
4.	Kakkad	Kerala State Electricity Board	1991	2 × 25	25,860	123.5	428.6	BV	2,000
5.	Doyang	North Eastern Electric Power Corporation	1993	3 × 25	26,000	67.0	250.0	BV	2,900
6.	Khodri	U.P. State Electricity Board	1983	4 × 27.9	32,000	57.9	200.0	BV	3,480
7.	Giri Bata	Himachal Pradesh State Electricity Board	1978	2 × 30	31,500	147.5	428.6	SV	1,800
8.	Loktak	National Hydro-Electric Power Corporation Ltd.	1983	3 × 35	36,100	298.0	500.0	SV	1,300
9.	Kopili	North Eastern Electric Power Corporation	1988	2 × 50	51,500	326.5	600.0	SV	1,400
10.	Vaitarna	Govt. of Maharashtra	1976	1 × 60	67,000	277.6	375.0	SV	1,800
11.	Pong	Beas Construction Board	1977	6 × 60	63,400	65.5	166.7	BV	4,500
12.	Chibro	U.P. State Electricity Board	1975	4 × 56.7	62,000	110.0	250.0	BV	3,810
13.	Lower Periyar	Kerala State Electricity Board	1991	3 × 60	67,400	184.0	333.3	BV	2,200
14.	Subernarekha	Bihar State Electricity Board	1977	2 × 65	73,600	133.0	300.0	BV	2,600
15.	Ramganga	U.P. State Electricity Board	1975	3 × 66	68,000	97.2	187.5	BV	4,115
16.	Koyna Stage-III	Govt. of Maharashtra	1975	4 × 80	82,400	109.7	214.0	BV	4,115

S. No.	Name of Power station	Owned by	Year of commis-sioning	No. of units and capacity (MW)	Turbine Parameters			Main Inlet Valve	
					Rating (kW)	Design Head (m)	Speed (R.P.M)	Type	Nominal diameter (mm)
(1)	(2)	(3)	(4)	(5)	(6)	(7)	(8)	(9)	(10)
17.	Pench	M.P. Electricity Board	1986	2 × 80	82,400	150.0	231.0	BV	3,300
18.	Upper Kolab	Govt. of Orissa	1987	4 × 80	82,050	241.7	428.6	SV	2,150
19.	Tons	M.P. Electricity Board	1990	2 × 105	107,000	175.0	250.0	BV	2,800
20.	Lower Silieru	Andhra Pradesh State Electricity Board	1977	2 × 115	119,000	193.7	300.0	SV	2,600
21.	Kalinadi Stage-I (Nagjhari)	Karnataka Power Corporation Ltd.	1979	6 × 135	139,100	347.0	375.0	SV	2,150
22.	Ranjit Sagar Dam	Punjab State Electricity Board	1995	4 × 150	154,000	100.0	166.7	BV	5,000
23.	Dehar	Beas Construction Board	1977	6 × 165	169,300	282.0	300.0	SV	2,600
				TURBINE RUNNER : KAPLAN					
24.	Balimela Dam	Andhra Pradesh State Electricity Board	1995	2 × 30	31,000	35.8	187.5	BV	5,000
25.	Sutlej-Yamuna Link P.H. II	Punjab State Electricity Board	1991	2 × 7.43	7,740	13.5	166.7	—	—
26.	Kallada	Kerala State Electricity Board	1990	2 × 7.5	7,810	43.0	428.6	BV	2,000
27.	Warna	Govt. of Maharashtra	1991	2 × 8	8,350	29.5	300.0	BV	2,000
28.	Pawna	Govt. of Maharashtra	1988	1 × 10	10,450	25.0	250.0	BV	3,600
29.	Dudhganga	Govt. of Maharashtra	1992	2 × 12	12,495	42.0	300.0	BV	2,600
30.	Bansagar P.H. II	Madhya Pradesh Electricity Board	1993	2 × 15	15,530	18.0	166.7	—	—
31.	Ghatprabha	Karnataka Power Corporation Ltd.	1990	2 × 16	18,000	32.0	214.3	BV	4,000
32.	OBRA	Uttar Pradesh State Electricity Board	1970	3 × 33	33,860	20.4	115.4	—	—
33.	Sardar Sarovar CHPH	Sardar Sarovar Narmada Nigam Ltd.	1994	5 × 50	51,265	36.0	136.4	—	—

S. No.	Name of Power station	Owned by	Year of commissioning	No. of units and capacity (MW)	Turbine Parameters			Main Inlet Valve	
					Rating (kW)	Design Head (m)	Speed (R.P.M)	Type	Nominal diameter (mm)
(1)	(2)	(3)	(4)	(5)	(6)	(7)	(8)	(9)	(10)
				TURBINE RUNNER : PELTON					
34.	Chennai	Govt. of Jammu & Kashmir	1975	2 × 4.6	4,900	365.8	600.0	SV	600
35.	Lower Lagyap	Govt. of Sikkim	1979	2 × 6	6,200	530.7	750.5	SV	600
36.	Rammam	West Bengal State Electricity Board	1993	4 × 12.5	12,820	492.3	600.0	SV	600
37.	Bassi	Himachal Pradesh State Electricity Board	1970	4 × 15	15,460	335.7	500.0	SV	838
38.	Tata Bhira	Tata Electric Company	1974	5 × 26	27,000	440.0	375.0	SV	760
39.	Sanjay (Bhabha)	Himachal Pradesh State Electricity Board	1989	3 × 40	46,340	887.2	500.0	SV	750
40.	Shanan	Punjab State Electricity Board	1981	1 × 50	56,100	487.7	375.0	SV	1,400
41.	Tillari	Govt. of Maharashtra	1986	1 × 60	62,600	628.8	500.0	SV	1,200
42.	Sharavathy	Karnataka Power Corporation Ltd.	1975	2 × 89	104,500	439.5	300.0	SV	1,600
43.	Varahi	Karnataka Power Corporation Ltd.	1989	2 × 115	134,600	460.0	250.0	SV	1,800
44.	Andhra	Himachal Pradesh State Electricity Board	1987	3 × 5.65	5,860	311.3	500.0	SV	500

Nomenclature : BV = Butterfly Valve
SV = Spherical Valve

Source : Information Brochure titled "Hydro-electric Installations", brought out by BHEL, India.

Data obtained from the tests on a model turbine are plotted so as to show variation of non-dimensional parameters Q/nD^3, gH/n^2D^2, $P/\rho n^3D^5$ and η with one another, the graphs so obtained are applicable to any turbine in the same homologous series.

For a given incompressible fluid and a given turbine working under conditions of constant speed and constant head, ρ, D, n and H are constant, the dimensionless term gH/n^2D^2 is also a constant, and from Eq. (15.67 *b*) η is a function of P.

Example 15.39. *A Kaplan turbine is to develop 17640 kW when running at 240 r.p.m. under a net head of 49 m. In order to predict its performance a 1 : 5 scale model is tested under a net head of 25 m. At what speed should the model run and what power would it develop ?*

Determine the discharge in the model and in the full scale turbine, if the overall efficiency of the model at the best operating point was 85%.

Solution. When operating at the best point, the relevant dimensionless parameters of Eq. (15.64), for the full scale turbine must have the same values as for the model, thus

$$\left(\frac{gH}{n^2D^2}\right)_p = \left(\frac{gH}{n^2D^2}\right)_m \qquad \text{....(1)}$$

and

$$\left(\frac{Q}{nD^3}\right)_p = \left(\frac{Q}{nD^3}\right)_m \qquad \text{...(2)}$$

For the model : $H_m = 25$ m, $D_m = D_p/5$, $n_m = ?$

For the full scale turbine,

$$H_p = 49 \text{ m}, \; n_p = 240 \text{ r.p.m., and } P_p = 17640 \text{ kW}$$

From (1),

$$n_m = \left(\frac{H_m}{H_p}\right)^{1/2} \frac{D_p}{D_m} n_p = \left(\frac{25}{49}\right)^{1/2} \times \frac{5}{1} \times 240 = 857 \text{ r.p.m.}$$

$$\left(\frac{P}{\rho n^3D^5}\right)_p = \left(\frac{P}{\rho n^3D^5}\right)_m$$

$$\frac{P_p}{P_m} = \frac{\rho_p n_p^3 D_p^5}{\rho_m n_m^3 D_m^5}$$

for water

$$\rho_p = \rho_m$$

$$\frac{P_p}{P_m} = \left(\frac{n_p}{n_m}\right)^3 \times \left(\frac{D_p}{D_m}\right)^5 = \left(\frac{240}{857}\right)^3 (5)^5 = 68.5.$$

Power developed by the model,

$$P_m = \frac{P_p}{68.5} = \frac{17640}{68.5} = 257.5 \text{ kW}$$

Overall efficiency

$$\eta_0 = \frac{\text{Power developed}}{\gamma QH}$$

Discharge required in the model

$$Q_m = \frac{P_m}{\gamma \eta_0 H_m} = \frac{257.5 \times 10^3}{9810 \times 0.85 \times 25} = 1.23 \text{ m}^3/\text{s}.$$

From Eq. (2), discharge in the full scale turbine

$$Q_p = Q_m \frac{n_p}{n_m}\left(\frac{D_p}{D_m}\right)^3 = 1.23 \times \frac{240}{857} \times (5)^3 = 43.06 \text{ m}^3/\text{s}.$$

Example 15.40. *A model turbine constructed to a scale of 1/10 when tested under a head of 8 m at 400 r.p.m. gave an efficiency of 77 per cent. Determine the r.p.m. of the prototype and the ratio of powers developed by the model and prototype if the prototype works under a head of 100 m. What will be the efficiency of the prototype if scale effect is considered ?*

Solution.

Data given : $\dfrac{D_p}{D_m} = 10,\ H_m = 8 \text{ m},\ n_p = 400 \text{ r.p.m.},\ \eta_m = 77\%$

$H_p = 100$ m.

For similarity between model and prototype

$$\left(\frac{gH}{n^2D^2}\right)_m = \left(\frac{gH}{n^2D^2}\right)_p$$

$$n_p = n_m \times \left(\frac{D_m}{D_p}\right)\left(\frac{H_p}{H_m}\right)^{1/2} = 400 \times \frac{1}{10} \times \sqrt{\frac{100}{8}} = \mathbf{141.42\ rpm.}$$

Ratio of powers developed

$$\frac{P_p}{P_m} = \frac{n_p^3}{n_m^3} \cdot \frac{D_p^5}{D_m^5} = \left(\frac{141.42}{400}\right)^3 (10)^5 = \mathbf{4419.3}$$

The efficiencies are related by

$$\frac{n_p}{n_m} = \frac{D_m}{D_p}\sqrt{\frac{H_p}{H_m}}\sqrt{\frac{\eta_p}{\eta_m}}$$

from which $$\frac{\eta_p}{\eta_m} = \left(\frac{n_p}{n_m}\right)^2\left(\frac{D_p}{D_m}\right)^2\left(\frac{H_m}{H_p}\right) = \left(\frac{141.42}{400}\right)^2 (10)^2 \times \frac{8}{100} = 0.999$$

$\therefore$ $\eta_p = 0.999 \times 0.77 = \mathbf{0.769\ or\ 76.9\%}.$

15.12.1.1. Specific Speed of Turbine

For a turbine using a given fluid the operating conditions are expressed by n, P and H. An important parameter involving these variables is obtained by dividing $(\pi_4)^{1/2}$ by $(\pi_2)^{5/4}$

$$K_n = \frac{(\pi_4)^{1/2}}{(\pi_2)^{5/4}} = \frac{n\sqrt{P}}{\sqrt{\rho}(gH)^{5/4}} \quad ...(15.68)$$

Eq. (15.67) represents a dimensionless quantity. It is customary to use values of n, P and H which correspond to maximum efficiency. Since ρ and g are practically constant, it is customary to omit constant ρ and g terms from Eq. (15.67). The resulting quantity is expressed as

$$n_s = \frac{n\sqrt{P}}{H^{5/4}} \quad ...(15.69)$$

and is known as the *specific speed* of a turbine. The operating conditions expressed by power output P, effective head H and speed n are also called the "duty variables".

The specific speed of a turbine is defined as the speed of a geometrical similar turbine of such a size as to generate one unit of power (*e.g.*, 1 kW) under one unit of head (*e.g.*, 1 m) at maximum efficiency. This definition may be easily derived from Eq. (15.68). The specific speed is not dimensionless, since ρ and g terms have been dropped, nor is it a speed. As it is not dimensionless, its numerical value depends on the units in which the magnitude of its constituents are expressed. In metric system, the power P representing shaft power or brake horsepower is expressed in metric horsepower, n in revolutions per minute and H in metres. (The corresponding units in FPS system are hp, rpm and ft). The dimensionless ratio K_n is often known as the dimensionless specific speed. [n_s is metric units = 4.45 n_s, in British (FPS) units].

In SI units, power P is expressed in kW, n in r.p.m., and H in metre. The conversion from metric units into SI units is as follows :

$$(n_s) \text{ in SI units} = 0.857\ (n_s) \text{ in metric units}$$

From Eq. (15.68) it can be seen that for high heads, the turbine will have a lower specific speed, and that for smaller heads it will be higher. The concept of specific speed is very useful in the selection of turbine suitable for a given set of conditions. When the site of installation and the output required from a turbine are known, the specific speed n_s can be calculated and the type of machine best suited to these conditions selected. For the common types of turbines, experience has shown the following values as most suitable.

Table 15.1. Specific Speed for Turbine

S. No.	*Type of turbine*	*Approximate range of specific speed n_s*	
		Metric units	*SI units*
1. (*a*)	Pelton (Single jet)	10–35	8.5–30
1. (*b*)	Pelton (Multiple jets)	35–60	30–50
2.	Francis and mixed-flow (for heads not greater than about 370 m)	60–400	50–340
3.	Kaplan and axial flow (for heads not greater than about 60 m)	300–1200	250–1025

The Pelton wheel gives efficient performance at very high heads and required a low flowrate. The efficiency in most of the Pelton wheel installations range from 82 to 90 per cent. The efficiency of the Francis turbine is maximum when it works at the design load but at part load its efficiency drops rapidly. Kaplan turbines maintain a fairly high efficiency at part loads because of adjustable blades. Kaplan turbines are suitable for low heads and high flowrates.

The specific speed depends on the operating head. Creager and Justin recommend :

(*i*) for Francis turbine,

$$n_s = \frac{22{,}200}{3.28H + 32} + 83.5$$

(*ii*) for Kaplan turbine, $n_s = \dfrac{30{,}800}{3.28H + 32} + 154.$

Based on the study of a wide range of turbines of different makes, the following expressions give the average trend :

for Francis, $n_s = 1750\ H^{-0.51}$

for Kaplan, $n_s = 1377\ H^{-0.33}$

Specific Speed of Francis Runner

The discharge through a Francis turbine

$$Q = K\pi DBV_f\ ;\ V_f = \psi\sqrt{2gH}$$

or

$$Q = K\pi DB\psi\sqrt{2gH}$$

Power developed at the shaft

$$P = \eta_0\,\frac{\gamma QH}{75} = \eta_0\,\frac{\gamma . K\pi DB\psi\sqrt{2gH}\,.H}{75} \quad \text{(metric h.p.)}$$

$$= \frac{\pi \times 1000 \times \sqrt{2 \times 9.81}}{75} \eta_0 DB\psi KH^{3/2} = 185.5\ \eta_0\ DB\psi KH^{3/2} \text{ (metric h.p.)}$$

$$= 136.34\ \eta_0\ DB\ \psi\ KH^{3/2} \text{ (kW)}$$

Speed of rotation $n = \dfrac{60u}{\pi D} = \dfrac{60\phi\sqrt{2gH}}{\pi D}$

Substituting these quantities in the expression for the specific speed

$$n_s = n\frac{\sqrt{P}}{H^{5/4}} = \frac{60\phi\sqrt{2gH}}{\pi D}\frac{\sqrt{185.5\eta_0 DB\psi KH^{3/2}}}{H^{5/4}}$$

$$= 1152.2\ \phi\ \sqrt{\eta_0\psi K}\ \sqrt{\frac{B}{D}}$$

When B is expressed as a fraction of D, *i.e.* $B = \alpha D$, α being less than unity, the specific speed

$$n_s = 1152.2\ \phi\ \sqrt{\alpha\eta_0\psi K}$$

For the usual values of ϕ, η_0, ψ, α as follows :

$$\phi = 0.75,\ \eta_0 = 0.85,\ \psi = 0.25,\ \alpha = 0.15$$

and neglecting the vane thickness (assuming $k = 1$), one gets :

$$n_s = 1152.2 \times 0.75\ \sqrt{0.15 \times 0.85 \times 0.25 \times 1} = 154.28 \text{ (metric units)}$$

$$= 132.26 \text{ (SI units)}$$

Each of the above factors which influence the specific speed has a definite range of variation for which there would be a corresponding range of specific speed.

Specific Speed of Kalpan and Propeller Runners

$$Q = \frac{\pi}{4}(D^2 - d^2)\,V_f = \frac{\pi}{4}(D^2 - d^2)\,\psi\sqrt{2gH}$$

$$n = \frac{60u}{\pi D} = \frac{60\phi\sqrt{2gH}}{\pi D}$$

$$P = \eta_0\frac{\gamma QH}{75} = \eta_0\gamma\frac{\pi}{4}(D^2 - d^2)\,\psi\sqrt{2gH}\,.\,H/75 \text{ (metric units)}$$

Specific speed $n_s = n\dfrac{\sqrt{P}}{H^{5/4}}$

$$= \frac{60\phi\sqrt{2gH}}{\pi D}\cdot\sqrt{\frac{\eta_0\gamma\pi(D^2 - d^2)\psi\sqrt{2gH}\,.H/300}{H^{5/4}}}$$

$$= 576.1\ \phi\sqrt{\eta_0\psi(1 - d^2/D^2)}$$

For a Kaplan turbine with $\eta_0 = 0.90$, $d/D = 0.35$, $\psi = 0.70$,

$$n_s = 576.1\ \phi\sqrt{0.9 \times 0.7[1 - (0.35)^2]} = 428.3\phi \text{ (metric units)}$$

$$= 367.19\ \phi \text{ (SI units)}$$

Specific Speed of a Pelton Runner

The following relation can be written for a Pelton runner.

$$Q = \text{Area of jet} \times \text{Velocity of jet} = \frac{\pi d^2}{4}\,.\,C_v\sqrt{2gH}$$

Power output is given by

$$P = \eta_0 \frac{\gamma QH}{75} \eta_0 \gamma \frac{\pi d^2}{4} C_v \sqrt{2gH} \;. H/75 \text{ (metric h.p.)}$$

and,
$$u = \phi \sqrt{2gH}, \quad \text{also } u = \frac{\pi D.n}{60}$$

$\therefore$
$$n = \frac{60\phi\sqrt{2gH}}{\pi D}.$$

Substituting these values in the expression for specific speed,

$$n_s = n = \frac{\sqrt{P}}{H^{5/4}} = \frac{60\phi\sqrt{2gH}}{\pi D} \cdot \frac{\sqrt{\gamma \eta_0 \pi d^2 C_v H \sqrt{2gH/300}}}{H^{5/4}}$$

$$= 575.9 \, \phi \frac{d}{D} \sqrt{\eta_0 C_v} \text{ (metric units)}$$

If
$$\phi = 0.45, C_v = 0.97, \eta_0 = 0.89$$

$$n_s = 575.9 \times 0.45 \times \sqrt{0.89 \times 0.97} \cdot \frac{d}{D} = 240.8 \frac{d}{D} \text{ (metric units)}$$

$$= 206.44 \frac{d}{D} \text{ (SI units)}$$

and if
$$\phi = 0.46, C_v = 0.98, \eta_0 = 0.88$$

$$n_s = 246.0 \frac{d}{D} \text{ (metric units)} = 210.90 \frac{d}{D} \text{ (SI units)}$$

Evidently, the specific speed of a Pelton wheel is inversely proportional to the jet ratio m. The above expressions are for a single jet. Multiple jets are used when more power is to be developed. The power developed by N jets will be N times the power produced by a single jet. Hence

$$n_s = n \frac{\sqrt{NP}}{H^{5/4}} = \sqrt{N} \left(n \frac{\sqrt{P}}{H^{5/4}} \right)$$

$= \sqrt{N}$ times the specific speed for a single jet.

Example 15.41. *A hydraulic turbine develops 800 KW under a head of 12 m at a speed of 90 r.p.m. Its efficiency is 0.93. Calculate the discharge and the specific speed of the turbine.*

Solution. Turbine efficiency, $\eta = \dfrac{\text{Output}}{\text{Input}} = \dfrac{800 \times 1000}{9810 \times Q \times 12}$

since $\eta = 0.93$,
$$Q = \frac{800 \times 1000}{0.93 \times Q \times 12} = \mathbf{7.307 \ m^3/s}$$

Specific speed of turbine,
$$n_s = n \frac{\sqrt{P}}{H^{5/4}} = 90 \frac{\sqrt{800}}{(12)^{5/4}} = \mathbf{114.}$$

Example 15.42. *In a small hydro development, a Kaplan turbine runs under a head of 2.1 m. If has a runner of 3.5 m diameter and develops 600 KW at 800 r.p.m. Assuming an overall efficiency of 80%, determine the discharge and specific speed of the machine. If a homogeneous turbine is to be tested at a head of 3.0 m what are the rotational speed, discharge and power of the unit ?*

Solution. Power output $= \eta_0(\gamma QH)$

$$600 \times 1000 = 0.80 \times 9810 \times Q \times 2.1, \quad \therefore \quad Q = \mathbf{36.41 \ m^3/s}$$

Specific speed $\quad n_s = n \dfrac{\sqrt{p}}{H^{5/4}} = 80 \dfrac{\sqrt{600}}{(2.1)^{5/4}} = \mathbf{775.15}$

For homogeneous turbines, using Eq. (15.67 *a*), the equality of dimensionless parameters π_1, π_2 and π_4 yields,

(*i*) Rotational speed : $\eta_m = n_p (D_p/D_m) \sqrt{H_m/H_p} = 80 \times \dfrac{3.5}{1.5} \times \sqrt{\dfrac{3.0}{2.1}} = \mathbf{223.11\ r.p.m.}$

(*ii*) Discharge : $Q_m = Q_p (n_m/n_p)(D_m/D_p)^3 = 36.41\ (223.11/80)(1.5/3.5)^3 = \mathbf{7.99\ m^3/s.}$

(*iii*) Power : $\quad P_m = P_p (n_m/n_p)^3 (D_m/D_p)^5 = 600 \left(\dfrac{223.11}{80}\right)^3 \left(\dfrac{1.5}{3.5}\right)^5 = \mathbf{188.06\ kW.}$

Example 15.43. *A double-jet Pelton turbine is required to develop 5512.5 kW at a specific speed of 17. The water is supplied through a pipeline 1 km long from a reservoir the level of which is 350 m above the nozzles. Assume coefficient of velocity for the nozzles as 0.97, ϕ = 0.46, overall η = 85% and Darcy's friction factor = 0.024. Allowing for a friction loss in the penstock pipeline to be 5% of the gross head, determine :*

(*i*) *turbine speed in r.p.m* (*ii*) *diameter of the jets*

(*iii*) *mean diameter of bucket circle* (*iv*) *diameter of penstock pipeline.*

Solution.

(*i*) Specific speed, $\quad n_s = n \dfrac{\sqrt{P}}{H^{5/4}}$

$$H = \text{Gross head} - \text{Head lost in penstock}$$

$$= 350 - \frac{5}{100} \times 350 = 332.5 \text{ m}$$

Speed of rotation, $\quad n = n_s \dfrac{H^{5/4}}{\sqrt{P}} = 17 \dfrac{(332.5)^{5/4}}{\sqrt{5512.5}} = \mathbf{328\ r.p.m.}$

(*ii*) Velocity of jets, $\quad V = C_v \sqrt{2gH} = 0.97\sqrt{2 \times 9.81 \times 332.5} = 78.35$ m/s

Velocity of buckets, $\quad u = 0.46\ V = 0.46 \times 78.35 = 36.04$ m/s

also, $\quad u = \dfrac{\pi D n}{60}$

$\therefore$ Mean bucket diameter

$$D = \frac{60u}{\pi n} = \frac{60 \times 36.04}{\pi \times 328} = \mathbf{2.1\ m.}$$

(*iii*) Power developed through each jet

$$= \frac{5512.5}{2} = 2756.25 \text{ kW}$$

Power supplied by each jet $= \dfrac{\text{Power developed}}{\text{Overall } \eta} = \dfrac{2756.25}{0.85} = 3242.6$ kW

Power contained in each jet $= \dfrac{1}{2}\dfrac{W}{g} V^2 = \dfrac{1}{2} \dfrac{\gamma \frac{\pi}{4} d^2 V}{g} V^2$

$$= \frac{1}{2g}\frac{\gamma \pi d^2}{4} V^3 = \frac{1}{2 \times 9.81} \times 9810 \times \frac{\pi}{4} d^2 \times (78.35)^3$$

Equating the two representing the same quantity

$$d^2 = \frac{3242.6 \times 10^3 \times 75 \times 2 \times 9.81 \times 4}{9810 \times \pi \times (78.35)^3} = 0.01718 \text{ m}^2$$

or $\quad d = 0.131$ m or **13.1 cm.**

$\therefore$ Diameter of each jet = 13.1 cm.

(*iv*) Discharge requirement, $Q = 2 \times \frac{\pi}{4} (0.131)^2 \times 78.35 = 2.11 \text{ m}^3/\text{s}$

Head lost in penstock, $\quad h_f = \frac{5}{100} \times 350 = 17.5$ m

but
$$h_f = f \frac{L}{D} \frac{V^2}{2g} = \frac{8}{g\pi^2} fL \frac{Q^2}{D^5}$$

or
$$17.5 = \frac{8}{9.81\,\pi^2} \times \frac{0.024 \times 1000 \times (2.11)^2}{D^5}$$

$\therefore \quad D^5 = 0.5045 \quad$ or $\quad D = 0.974$ m

$\therefore$ The diameter of penstock = **0.974 m.**

Example 15.47. *A hydraulic turbine develops 882 kW under a head of 10 m at a speed of 90 r.p.m. and gives an efficiency of 92%. Find the water consumption and the specifie speed. If a model 1/10 full size is constructed to operate under a head of 8 m, what must be its speed, power developed and water consumption to run under the conditions similar to the prototype ?*

Solution. Efficiency, $\eta = \frac{\text{Power developed}}{\gamma QH}$

$$Q = \frac{882 \times 10^3}{0.92 \times 9810 \times 10} = 9.77 \text{ m}^3/\text{s}.$$

Specific speed, $\quad n_s = n \frac{\sqrt{P}}{H^{5/4}} = 90 \frac{\sqrt{882}}{(10)^{5/4}} = 150.3$

For similar turbines, each of the dimensionless terms,

$$\pi_1 = \frac{Q}{nD^3} \quad \text{and} \quad \pi_2 = \frac{gH}{n^2D^2}$$

must be same for both model and the prototype.

Omitting g, we obtain $\left(\frac{H}{n^2D^2}\right)_m = \left(\frac{Q}{m^2D^2}\right)_p$

model speed
$$n_m = n_p \frac{D_p}{D_m} \sqrt{\frac{H_m}{H_p}} = 90 \times 10 \sqrt{\frac{8}{10}} = 804 \text{ r.p.m.}$$

To obtain the discharge in the model, we have

$$\left(\frac{Q}{nD^3}\right)_m = \left(\frac{Q}{nD^3}\right)_p$$

$$Q_m = \frac{n_m}{n_p} \left(\frac{D_m}{D^p}\right)^3 Q_p = \frac{804}{90} \left(\frac{1}{10}\right)^3 \times 9.77$$

$$= 0.0872 \text{ m}^3/\text{s} = 87.2 \text{ lit/sec.}$$

For similar turbines, the specific speed must be same. Thus,

$$\left(n\frac{\sqrt{P}}{H^{5/4}}\right)_m = \left(n\frac{\sqrt{P}}{H^{5/4}}\right)_p$$

$\therefore$ Power developed by the model turbine

$$P_m = P_p\left(\frac{n_p}{n_m}\right)^2\left(\frac{H_m}{H_p}\right)^{5/2} = 882\left(\frac{90}{804}\right)^2\left(\frac{8}{10}\right)^{5/2} = 6.35\text{ kW}.$$

15.12.2. Unit Quantities

There are five important quantities pertaining to any turbomachine. They are (*i*) Head H, (*ii*) Discharge Q, (*iii*) Power P, (*iv*) Speed n and (*v*) Efficiency η. In order to predict the performance of a machine, it is necessary to know the nature of variation of these quantities with respect to one another. It is however customary to study the nature of variation of unit quantities instead of actual quantities involved. The unit quantities, such as unit speed n_u, unit discharge Q_u and the unit power P_u, are defined as the speed, discharge and power when the head on the machine is unity.

The dimensionless terms expressed in Eq. (15.35) may be used to predict the performance of the same machine under different conditions. Thus for the same machine, equating

$$\frac{gH}{n^2D^2} = \frac{gH_1}{n_1^2D^2}$$

which yields,

$$n_1 = n\sqrt{\frac{H_1}{H}} \qquad \text{...(15.70)}$$

where n_1 and H_1 represent the speed and head under changed conditions of operation. If H_1 is unity then the unit speed n_u is given by

$$n_u = \frac{n}{\sqrt{H}} \qquad \text{...(15.71)}$$

An expression for unit discharge may be obtained by dividing π_1 by $\sqrt{\pi_2}$, thus

$$\frac{\pi_1}{\sqrt{\pi_2}} = \frac{Q}{n\,D^3} \times \frac{nD}{\sqrt{gH}} = \frac{Q}{D^2\sqrt{gH}}$$

for the same machine operating under different conditions of Q and H, we may write

$$\frac{Q}{D^2\sqrt{gH}} = \frac{Q_1}{D^2\sqrt{gH_1}}$$

or

$$Q_1 = Q\sqrt{\frac{H_1}{H}} \qquad \text{...(15.72)}$$

Since unit discharge Q_u is the discharge for unit head ($H_1 = 1$), we obtain

$$Q_u = \frac{Q}{\sqrt{H}} \qquad \text{...(15.73)}$$

For obtaining an expression for unit power we require a relation between the power and the head. Dividing π_4 by $(\pi_2)^{3/2}$, we get

$$\frac{\pi_4}{(\pi_2)^{3/2}} = \frac{P}{\rho D^2(gH)^{3/2}}$$

For the same machine operating under different conditions of power and head, we may write

$$\frac{P}{\rho D^2(gH)^{3/2}} = \frac{P}{\rho D^2(gH_1)^{3/2}}$$

which gives,

$$P_1 = P\left(\frac{H_1}{H}\right)^{3/2} \qquad \text{...(15.74)}$$

and introducing the unit power P_u as the power produced under unit head,

$$P_u = \frac{P}{H^{3/2}} \qquad ...(15.75)$$

The specific speed can be written as $n_s = n_u \sqrt{P_u}$.

Example 15.48. *Define 'unit speed' and 'unit power' of a hydraulic turbine. Explain how a (graph) plot of unit power and efficiency against unit speed at different gate openings can be used to predict the performance of a turbine under any working condition.*

A reaction turbine, 50 cm diameter, when running at 608 r.p.m. develops 191 kW the flow through the turbine being 0.75 m³/s. Under the above operating condition the pressure head at the entrance to the turbine is 28 m the elevation of the turbine above the tail water level is 1.3 m and the velocity of flow at entrance to the turbine runner is 3.7 m/s. Calculate the effective head on the turbine and the efficiency.

If the same turbine is made to run under an effective head of 67.5 m, find the speed, the brake horse power and the discharge of the turbine for homologous operating condition.

Solution. Neglecting the velocity head rejected to the tail race, the effective head under which the turbine works,

$$H = \frac{p}{\gamma} + \frac{V^2}{2g} + z = 28 + \frac{(3.7)^2}{2 \times 9.81} + 1.3 = \mathbf{29.998\ m.}$$

$$\eta = \frac{P}{\gamma QH} = \frac{191 \times 10^3}{9810 \times 0.75 \times 29.998} = \mathbf{0.8667} \quad \text{or} \quad \mathbf{86.67\%.}$$

If the same machine is made to run under an effective head $H_1 = 67.5$ m,

$$n_1 = n\sqrt{\frac{H_1}{H}} = 600\sqrt{\frac{67.5}{29.998}} = \mathbf{896.7\ r.p.m.}$$

$$Q_1 = Q\sqrt{\frac{H_1}{H}} = 0.75\sqrt{\frac{67.5}{29.998}} = \mathbf{1.1208\ m^3/s}$$

$$P_1 = P\left(\frac{H_1}{H}\right)^{3/2} = 191\left(\frac{67.5}{29.998}\right)^{3/2} = \mathbf{645\ kW.}$$

Example 15.49. *Tests were conducted on a Francis turbine of 0.8 m diameter under a head of 9 m. The turbine running at 240 r.p.m. developed 84.5 kW and the water consumption was 1.2 m³/s. If the same turbine is operated under a head of 16 m, predict its new speed, discharge and power developed.*

Solution. For the same turbine, we have the following relations obtained from the dimensionless terms π_1, π_2 and π_3 of Eq. (15.65).

$$\frac{H_1}{n_1^2} = \frac{H_2}{n_2^2}, \frac{Q_1}{n_1} = \frac{Q_2}{n_2} \quad \text{and} \quad \frac{P_1}{H_1^{3/2}} = \frac{P_2}{H_2^{3/2}}$$

Quantities for $H_2 = 16$ m are

$$n_2 = n_1\sqrt{\frac{H_2}{H_1}} = 240\sqrt{\frac{16}{9}} = 320 \text{ r.p.m.}$$

$$Q_2 = Q_1\frac{n_2}{n_1} = 1.2 \times \frac{320}{240} = 1.6 \text{ m}^3/\text{s}$$

$$P_2 = P_1\left(\frac{H_2}{H_1}\right)^{3/2} = 84.5\left(\frac{16}{9}\right)^{3/2} = 200.28 \text{ kW.}$$

15.13. Performance Curves

The turbines are usually designed for particular values of H, Q, P, n and η which are known as the design conditions. Although desirable, it is not always possible for a turbine to run at design conditions. In actual practice of operations, the turbines may often be required to work at conditions different from those for which they have been designed. Hence it is necessary to have knowledge how these quantities (n, Q, H, P and η) will vary with respect to one another. Relation between any two of the above quantities is called a characteristic of the machine. The speed n, discharge Q and the head H are the physical independent variables. It is, therefore customary to study the variation of all other quantities with respect to variation of any one of the independent quantities. To determine the behaviour of the machine under varying conditions, tests are conducted either on the actual machine or its small scale model. The tests results are graphically plotted between the unit quantities and the resulting curves are known as the performance curves or characteristic curves. The characteristic curves are of the following three types :

1. ***Constant Head Characteristics.*** These are the curves plotted between unit discharge-unit speed, unit power-unit speed and efficiency-unit speed for different gate openings under constant head operation of the turbine.

2. ***Constant Speed Characteristics.*** Most turbines are required to run at constant speed so that the electrical generators to which they are coupled provide a fixed frequency and voltage. These characteristics are also known as the operating characteristics and are plotted between the efficiency and the percentage of full load or percentage of rated power. These characteristics show how efficient is the performance of the machine at part loads.

3. ***Constant Efficiency Curves.*** On the unit power-unit speed curve points of equal efficiency are projected using efficiency-unit speed curve for each gate opening. Isoefficiency curves are then plotted and the best performance curve is drawn. From this curve, for a given head and power, the speed and the gate opening can be determined for the best performance.

15.14. Cavitation in Reaction Turbines (Refer article 4.16 also)

When a liquid flows into a region where pressure is reduced to its vapour pressure at the prevailing temperature, the liquid boils and small bubbles or cavities form in large numbers. The appearance of cavities at a place depends on the difference between the absolute pressure p and the vapour pressure p_v. However, dissolved air usually separates out of water before such a low pressure as p_v is reached and the performance of the turbine begins to deteriorate. The cavitation occurs in those regions in the runner blades where the absolute pressure attains a lower value than the absolute pressure at the entrance to the draft tube.

If the pressure anywhere reduces to such a low value that liquid boils—for example on the convex side of curved blades, then vapour pockets form and remain as such until the liquid has moved to a region of higher pressure where the vapour bubbles collapse suddenly and cause noise, vibration and pitting of the blades and may ultimately cause failure of metal. Cavitation thus causes three undesirable effects :

(*i*) efficiency of the machine is lowered, (*ii*) damage to blades due the pitting, and

(*iii*) noise and vibrations.

Growth of bubbles and their collapse take place at a high speed ranging between 10,000 to 20,000 per second and it is likely that the damage of the metal is largely due to fatigue. The collapse of the cavities in the flow and/or on the metal surface and subsequent impact of water causes the formation of micro-cracks on the metal surface where the strength of the metal is the least. These microscopic cracks are the subsequent places of severe cavitation damage and the damaged metal surface looks like sponge.

Separation of 'air or air cavitation' usually occurs before 'vapour cavitation', the latter being detected by an increase in noise and vibrations and both by an appreciable reduction in thrust and efficiency of the machine.

If a turbine is run with a draft tube head (Z_2), see Fig. 15.11 (*c*), which is gradually increased from zero to a maximum value, it is found that the performance of the turbine begins to deteriorate.

This indicates that setting of a turbine above the tail race level greatly influences the turbine performance.

Since cavitation begins when the pressure reaches too low a value, it is likely to occur at points where the velocity or the elevation is high, and particularly at such points where high velocity and high elevation are combined. In a reaction turbine, the point of minimum pressure is usually at the outlet end of a blade on its convex side. For flow between such a point and the water level in the tail race, the Bernoulli's equation may be written as

$$\frac{p_{min}}{\gamma} + \frac{V^2}{2g} + Z = \frac{p_a}{\gamma} \qquad ...(15.76)$$

where Z is the elevation of the runner blade outlet (or the height of draft tube) above the tail race, p_a is the atmospheric pressure and V is the velocity at the point under condition. Rearranging Eq. (15.76), we obtain

$$\sigma = \frac{V^2}{2gH} = \frac{p_a - p_{min} - \gamma Z}{\gamma H} \qquad ...(15.77)$$

where σ is the Thoma's cavitation number and H is the net or effective head on which the turbine operates. For cavitation not to occur, p_{min} must be greater than the vapour pressure of the liquid p_v.

If either the height of the turbine runner above the tail water level, *i.e.* Z or H is increased, σ is decreased. To determine the maximum elevation Z_{max} of the turbine above the tail water level for cavitation-free operation, the following equation may be used

$$Z_{max} = \frac{p_a}{\gamma} - \frac{p_v}{\gamma} - \sigma_c H \qquad ...(15.78)$$

where σ_c is the critical cavitation parameter which is usually about 0.1. It is the value of σ corresponding to decrease in efficiency and appearance of noise and vibrations. Eq. (15.78) shows that greater the net head H on the turbine, the lower it must be placed in relation to tail water level.

The cavitation coefficient of turbine installation at the hydropower plant, σ_{inst}, which is called Thoma's coefficient, is given by

$$\sigma_{inst} H = \frac{p_{atm} - p_v}{\gamma} - H_s$$

or

$$\sigma_{inst} = \frac{H_{atm} - H_v - H_s}{H} \qquad ...(15.79)$$

Here H_{atm} is the atmospheric (barometric) pressure in m of water column, H_v is the vapour pressure at the prevailing temperature in m of water column and H_s is the suction head of the turbine, *i.e.* height of runner setting above the tail water level and H is the turbine head.

It follows from Eq. (15.79) that the regime of operation (discharge Q and turbine head H) and the suction head H_s considerably influence and determine the magnitude of minimum absolute pressure within the runner. If H_s is varied, the magnitude of vacuum pressure and its location within the runner change. At a certain value of H_s the magnitude of σ_{inst} will be small enough to cause developed cavitation within the runner which causes the turbine characteristics to deteriorate. Fig. 15.19 shows the variation of σ_{inst} with efficiency. With increase in H_s, a corresponding decrease in σ_{inst} is obtained and this has no effect on the turbine performance as is seen from the constant efficiency trend. A stage is however reached

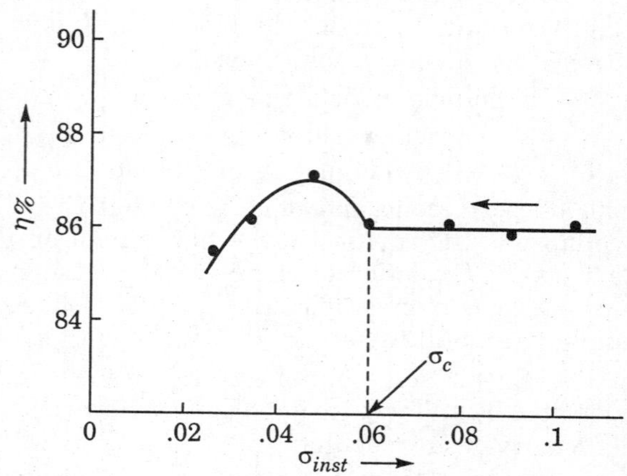

Fig. 15.19. Variation of σ_{inst} with efficiency η.

when any further increase in H_s deteriorates the turbine performance and the turbine efficiency falls. The critical value of cavitation coefficient, σ_c, is determined by this point marking the change in the trend of η.

For safe operation (cavitation-free) of turbine, it is evident that

$$\sigma_{inst} > \sigma_c.$$

The critical cavitation coefficient, σ_c, for the different types of turbines may be determined as follows :

For Francis, $\sigma_c = 0.0317\ (n_s/100)^2$

For Propeller, $\sigma_c = 0.3 + 0.0024\ (n_s/100)^{2.73}$

For Kaplan, σ_c = 1.1 times σ_c for propeller.

The modern trend in hydraulic turbines is to increase their specific speed, and for same discharge and power to have lesser dimension and weight by realising greater values of unit discharge and unit speed. This is possible at higher values of flow velocities within the turbine space. This growth of flow velocities is associated with the decrease in the absolute pressure in the turbine which may lead to phenomenon of cavitation. Higher the turbine specific speed, the larger is the pressure reduction in the absolute flow. The increase of specific speed, therefore, leads to further decrease of the mean pressure on the suction side of the blade.

Example 15.50. *A vertical shaft Kaplan turbine operating under a head of 9.8 m has a runner diameter of 9.3 m. At the optimum regime the turbine runs at 51.7 r.p.m. and develops 45,000 kW power discharging 535 m³/s of water. Determine the values of unit speed, unit discharge, unit power and specific speed of the turbine*

If the critical value of cavitation coefficient for the runner is 0.95, determine the location of the runner with respect to the tail water level. The atmospheric and saturated vapour pressures may be taken as 96 kN/m² and 24.5 kN/m² respectively.

Solution. Unit speed $n_u = \dfrac{n}{\sqrt{H}} = \dfrac{51.7}{\sqrt{9.8}} = \mathbf{16.515}$

Unit discharge $Q_u = \dfrac{Q}{\sqrt{H}} = \dfrac{535}{\sqrt{9.8}} = \mathbf{170.9}$

Unit power $P_u = \dfrac{P}{H^{3/2}} = \dfrac{45{,}000}{(9.8)^{3/2}} = \mathbf{1466.8}$

Specific speed $n_s = n\dfrac{\sqrt{P}}{H^{3/2}} = 51.7\dfrac{\sqrt{45{,}000}}{(9.8)^{5/4}} = \mathbf{632.5}$

Location of runner with respect to tail water level :

$$Z_{max} = \frac{P_a - P_v}{\gamma} - \sigma_c H = \frac{96 \times 10^3}{9810} - \frac{24.5 \times 10^3}{9810} - 0.95 \times 9.8 = -\mathbf{2.01\ m.}$$

The runner should be located at a depth not less than 2.01 m below the tail water level.

Example 15.51. *In a hydroelectric station, water is available at the rate of 175 m³/s under a head of 18 m. The turbines run at a speed of 150 r.p.m, with overall efficiency of 82%. Find the number of turbines required, if they have the maximum specific speed of 460.*

Solution. The specific speed of a turbine is given by, Eq. (15.69)

$$n_s = n\frac{\sqrt{P}}{H^{5/4}}$$

Data given, $n_s = 460$, $n = 150$ r.p.m, $H = 18$ m, P = Power output of turbine in KW

P_0 = Power output of hydroelectric station = $\eta_o\ (\gamma QH) = 0.82 \times 9810 \times 175 \times 18$ (Nm/s or W)

$= 25339.23$ kW

Power out of each turbine, $P = \frac{n_s^2 H^{5/2}}{n^2} = \frac{(460)^2(18)^{5/2}}{(150)^2} = 12927.4 \text{ kW}$

No. of turbines required $= \frac{\text{Total power output of the plant}}{\text{Power output per turbine}} = \frac{25339.23}{12927.4} = 1.96$

∴ No. of turbines required = **2.**

Example 15.52. *The model of a water turbine tested in the laboratory at a head of 20 m just-began to cavitate when the suction head was raised to 7m. Its optimum hydraulic efficiency was 82%. The prototype turbine is required to work at a head of 50 m. Determine the maximum safe setting for the turbine above the tail race. Atmospheric conditions in the laboratory and site are given below.*

	Atmospheric pressure	*Vapour pressure corresponding to water temperature*
At site	*1 × 10^5 N/m^2*	*0.035 × 10^5 N/m^2*
In laboratory	*1.012 × 10^5 N/m^2*	*0.033 × 10^5 N/m^2*

Solution. The critical cavitation coefficient is given by

$$\sigma_c = \frac{H_{atm} - H_v - H_s}{H}$$

From similarity considerations,

$$(\sigma_c)_{model} = (\sigma_c)_{proto}$$

or
$$\left[\frac{\left(\frac{1.012\times10^5}{98.10}\right) - \left(\frac{0.033\times10^5}{9810}\right) - 7}{20}\right] = \left[\frac{\left(\frac{10^5}{9810}\right) - \left(\frac{0.035\times10^5}{9810}\right) - H_s}{50}\right]$$

Solving which, one obtains

$$\mathbf{H_s = 2.387 \text{ m.}}$$

The maximum safe setting of the turbine above the tail race is 2.387 m.

15.15. Governing of Hydraulic Turbines

A hydraulic turbine is required to run at the same constant speed as that of the electric generator to which it is directly coupled. This synchronous speed must be maintained for all conditions of load on the generator. But since the demand for power fluctuates with time, the hydrounit must also adjust itself to such variations in load in such a way that the turbine speed remains unaffected. Whenever the load on the turbine decreases, if the discharge entering the turbine is not adjusted accordingly, its speed will increase. On the other hand, increase in load will tend to reduce the turbine speed unless the turbine discharge is increased in accordance with the now load. Such regulation of discharge with a view to maintain a synchronous speed of the turbine runner is called governing of turbine and is done automatically by means of a governor.

Centrifugal governors are usually used with modern hydraulic turbines, and have the following component parts :

(1) Pendulum or actuator which is belt or gear driven by the turbine main shaft.

(2) Servomotor.

(3) Relay valve, which is also known as control valve.

(4) Oil pump driven by belt connected to turbine main shaft.

(5) Oil supply pipes connecting the oil sump with the relay valve, and relay valve with servomotor.

15.15.1. Governing of Impulse Turbines

As already stated, the governing of turbines is done by regulating the flow of water to suit the load conditions on the turbines. In impulse turbines, the discharge is varied by adjusting the annular area of flow of the nozzle. The nozzle fitted at the end of the penstock is provided with a spear having a streamlined head which is fixed to one end of a rod. It is the movement of the spear which controls the annular area of flow and helps in changing the discharge. It is to be noted that the velocity of jet issuing from the nozzle remains unchanged. When the discharge is to be reduced or completely stopped quickly, a deflector which is provided near the nozzle is operated to deflected a part of the entire jet downward. The deflector is a movable steel plate attached to the tip of the nozzle. When there is a sudden drop in the load, the deflector is brought in between the nozzle and the buckets, thus deflecting a part or whole of the jet and preventing it from striking the buckets. In the mean time the spear is moved forward slowly by the governing mechanism. When the spear has adjusted the discharge according to the reduced load on the turbine, the deflector is brought back to its initial position and a clear jet again strikes the buckets. The spear and the deflector both are operated automatically by means of a centrifugal governor.

Let us consider a situation in which the load on the turbine decreases resulting in an increase in its speed. The pendulum or actuator being connected to the turbine main shaft, the increased speed causes the flyballs to move further apart due to centrifugal action, Fig. 15.20. This results in the upward movement of the sleeve. As the sleeve moves up, the main lever at the left end is raised which causes the bell crank lever to move downward. Alongwith this, the position of the control valve is pushed down in its cylinder. The downward motion of bell crank lever brings the deflector in front of the jet thereby diverting a portion of the jet downwards into the tail race. With the downward movement of the piston, oil which is under pressure finds entry into the pipe from the control valve cylinder and is transmitted to the left side of the servomotor. This causes the piston of the servomotor to move to the right which in turn causes the spear to move forward. The forward motion of the spear reduces the annular opening of the nozzle thereby reducing the discharge of the jet. The speed of the turbine is thus restored to its normal value. As the cam is connected to the spear rod, its movement towards the right causes the crank lever to move up. The upward movement of the bell crank lever brings the deflector back to its initial position. The direction of arrow shown on Fig. 15.20 indicates the direction of movement.

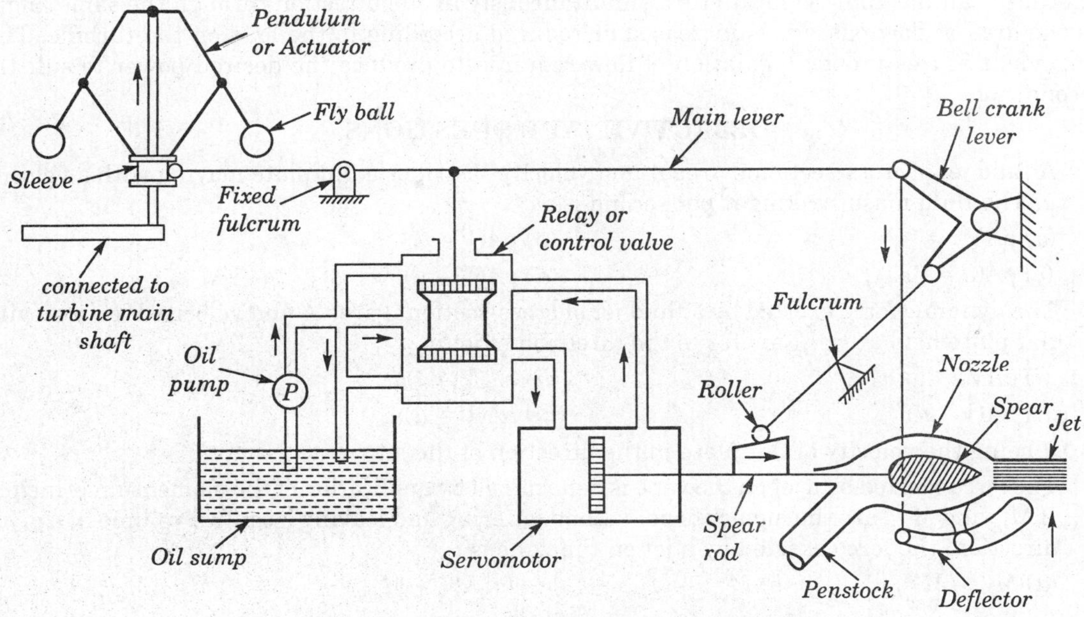

Fig. 15.20. Governing mechanism for Pelton Turbine.

In the event when the load on the turbine increases, the turbine speed would necessarily tend to decrease. This will also decrease the centrifugal force on the flyballs which in turn will cause a downward movement of the sleeve. The main lever at its left end will be lowered pulling the piston in the control valve up in the cylinder. This upward movement of the piston allows oil under pressure to flow into the pipe, and through it to the right side of the servomotor cylinder. The piston of the servomotor then moves to the left thereby moving the spear equally in the same direction. This causes an increase in the annular area of flow which results in the increased discharge. It is due to this increased discharge which is in accordance with the new load on turbine that the normal speed is restored.

15.15.2. Governing of Reaction Turbines

The discharge in reaction turbines is regulated by changing the gate opening of guide vanes which controls the area of flow passage. The governing mechanism is designed to change the position of guide vanes uniformly, which controls the area of flow passage. The mechanism changes the position of guide vanes to suit the load on the turbine. The system consists of a centrifugal governing mechanism described earlier, linkages, servomotor with its oil pressure governor and the guide wheel. The guide vanes are pivoted and connected by levers and links to the regulating ring. Two regulating rods are attached to the regulating ring at one end and to the regulating lever at the other, Fig. 15.21. The regulating lever is connected to a regulating shaft which is operated by the servomotor piston of the oil pressure governor. The motion of the servomotor piston is transmitted to the regulating ring which causes all the guide vanes to turn simultaneously in one direction through the same angle. Thus the area of flow passage is increased or reduced according to the load on the turbine. This arrangement esures proper regulation of flow required to produce the desired power to suit the load condition.

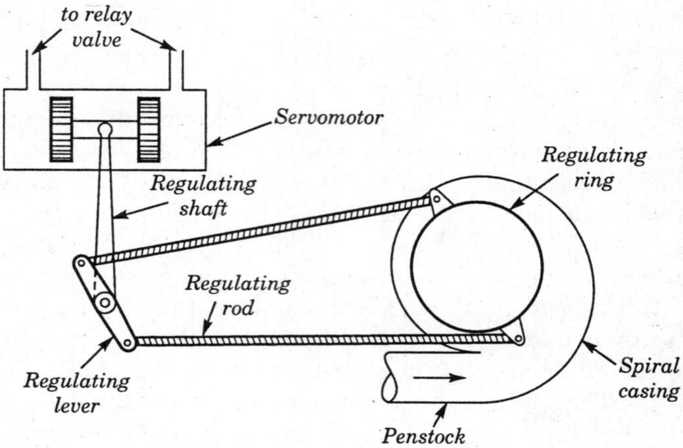

Fig. 15.21. Governing mechanism for Reaction Turbine.

OBJECTIVE TYPE QUESTIONS

1. A fluid jet of cross-sectional area A and velocity V strikes a flat plate moving with a velocity u. The fluid mass striking it per second is :

(*a*) ρAV (*b*) $\rho A(V - u)$

(*c*) $\rho A(u - V)$ (*d*) $\rho A(V + u)$.

2. The dynamic force exerted by a fluid jet of cross-sectional area A and velocity V on a moving flat plate held at right angles to the direction of jet is :

(*i*) ρAVu (*ii*) $\rho A(V - u)u$

(*iii*) $\rho A(V - u)^2$ (*iv*) $\rho A(V + u)u$.

u being the velocity of flat plate in the direction of the jet.

3. The force exerted by a jet on the vane is determined by application of the momentum principle. If M_{in} and M_{out} are the moments per second entering and leaving a control volume in a given direction, the force exerted by a jet on the vane is :

(*a*) $M_{in} - M_{out}$ (*b*) $M_{out} - M_{in}$ (*c*) $M_{in} + M_{out}$ (*d*) $M_{in} \cdot M_{out}$.

4. When a fluid jet strikes a curved moving vane, the velocity triangles represent the flow conditions at the entrance and exit. The force on the vane in the direction of its motion is :

(*i*) $\frac{W}{g}(w_1 - w_2)$ (*ii*) $\frac{W}{g}(V_{f_1} - V_{f_2})$ (*iii*) $\frac{W}{g}(V_{u_1} - V_{u_2})$ (*iv*) $\frac{W}{g}(V_{u_1} - V_{u_2})\, u$.

Here W is the weight of water striking the vane per second, and w, V_f and V_u are the relative velocity, velocity of flow and the velocity of whirl respectively. Suffix 1 and 2 represent inlet and outlet points of the vane.

5. Hydraulic turbines are classified as impulse and reaction depending on the manner in which the flow strikes the vanes. Which of the following statements pertain to each of the category :

(*a*) impulse turbine
(*b*) reaction turbine

(*i*) all the available energy is converted into kinetic energy.
(*ii*) pressure and velocity both change as flow passes through the turbine.
(*iv*) pressure throughout remains atmospheric and the velocity alone changes.
(*iv*) water is admitted over the entire circumference of the runner.
(*v*) water is admitted only in the form of jets.
(*vi*) turbine is connected to the tail race through a draft tube.

6. The head utilised by the runner of a reaction turbine is :

(*i*) $H = \dfrac{u_2 V_{u_2} - u_1 V_{u_1}}{g}$ (*ii*) $H = \dfrac{u_1 w_1 - u_2 w_2}{g}$

(*iii*) $H = \dfrac{u_1 V_{f_1} - u_2 V_{f_2}}{g}$ (*iv*) $H = \dfrac{u_1 V_{u_1} - u_2 V_{u_2}}{g}$.

7. Which of the following component parts pertain to reaction and impulse turbines :

(*a*) reaction turbines
(*b*) impulse turbines

(*i*) nozzle and spear
(*ii*) spiral casing
(*iii*) buckets
(*iv*) guide vanes
(*v*) draft tube
(*vi*) stay ring.

8. Identify the range of head for which the following types of turbines are generally used :

(*a*) Pelton turbine
(*b*) Francis turbine
(*c*) Axial flow turbine

(*i*) 3—80 m
(*ii*) 80—600 m
(*iii*) 600—2000 m.

9. By shock-free entry is meant :

(*i*) that the flow should enter the runner smoothly.
(*ii*) that the relative velocity vector should have the same direction as that of the tangent at the inlet tip of the blade.
(*iii*) that the absolute velocity vector should be tangential to the inlet tip of the blade.
(*iv*) that the velocity of flow should be natural.

10. Listed below in column *A* are the components of a reaction turbine, the function of each being given in column *B*, but not necessary in the same order. Identify the function of each component.

Column A	*Column B*
(*a*) Spiral casing	(*i*) transforms the hydraulic enery into the mechanical one.
(*b*) Guide vanes	(*ii*) transforms the major part of the exit kinetic energy into the pressure energy and thereby improves the turbine efficiency.
(*c*) Runner	(*iii*) are used to regulate flow into the runner and to direct the water on the runner at an appropriate angle.
(*d*) Draft tube	(*iv*) distributes the flow over the periphery of the runner such that the entrance velocity is same at the inlet tips.

11. The overall efficiency of a turbine is dependent on the hydraulic losses, volumetric losses and the mechanical losses represented by the respective efficiencies. It is related to these by the following relationship :

(*i*) $(\eta_h + \eta_v + \eta_m)/3$ (*ii*) $(\eta_h\, \eta_v\, \eta_m)^{1/3}$ (*iii*) $\eta_h \cdot \eta_v \cdot \eta_m$ (*iv*) $\sqrt{\eta_h \cdot \eta_v \cdot \eta_m}$.

12. Select the number of blades for each of the following category of turbine :

(*a*) Pelton turbine	(*i*) 16—21
(*b*) Francis turbine	(*ii*) 3—6
(*c*) Kaplan turbine	(*iii*) 18—33.

13. Turbines give best performance (*i.e.* work at peak efficiency) when they are operated at full or design load. The performance of many turbines deteriorates considerably at part-loads. Which of the following turbines is best suited for operation at part loads :

(*i*) Pelton turbine (*ii*) Francis turbine (*iii*) Propeller turbine (*iv*) Kaplan turbine.

14. The net or effective head at the turbine is :

(*a*) The sum of gross head plus the head loss in the penstock.

(*b*) The sum of gross head plus head loss in penstock and the velocity head at the turbine exit.

(*c*) The difference between gross head minus the head loss in penstock.

(*d*) The difference between the gross head minus head loss in penstock and the velocity head at the turbine exit.

15. The specific speed of a turbine is expressed as

(*i*) $n\,\dfrac{\sqrt{P}}{H^{5/4}}$ (*ii*) $n\,\dfrac{H^{5/4}}{\sqrt{P}}$ (*iii*) $n\,\dfrac{\sqrt{P}}{\rho H^{5/4}}$ (*iv*) $n\,\dfrac{\sqrt{P}}{(gH)^{5/4}}$.

16. The power which appears in the expression for the specific speed is :

(*a*) the water horse power

(*b*) the horse power developed by the runner

(*c*) the shaft horse power (brake horse power)

(*d*) the power input to the turbine.

17. The specific speed, which takes into account all the three operating variables into consideration, offers a criterion for selection of a suitable type of turbine. Select the range of specific speed associated with each type :

(*a*) Pelton (single jet)	(*i*) 50—340
(*b*) Pelton (multiple jet)	(*ii*) 250—1025
(*c*) Francis	(*iii*) 30—50
(*d*) Kaplan	(*iv*) 8.5—30.

18. The unit quantities are the quantities reduced to a head of unity. Identify from the following expressions for :

(*a*) unit speed	(*i*) $Q\sqrt{\dfrac{H_1}{H}}$	(*ii*) $\dfrac{Q}{\sqrt{H}}$
(*b*) unit discharge	(*iii*) $P(H_1/H)^{3/2}$	(*iv*) $P/H^{3/2}$
(*c*) unit power	(*v*) $n/\sqrt{H}$	(*vi*) $n\sqrt{H_1/H}$.

19. The power obtainable from an impulse turbine is proportional to the number of nozzles used. A Pelton turbine with six nozzles has a specific speed of 8.1. The specific speed per nozzle is :
(*i*) 1.35 (*ii*) 3.3 (*iii*) 2 (*iv*) 8.1.

20. The difference between the power obtained from the turbine shaft and the power supplied by water at its entry to the turbine is equal to :
(*a*) the hydraulic losses (*b*) mechanical losses
(*c*) hydraulic and mechanical losses (*d*) mechanical and volumetric losses

21. The hydraulic efficiency of a turbine is defined as :
(*i*) the ratio of power utilised by runner to that supplied by the water at entry to the turbine.
(*ii*) the ratio of power supplied by water at entry to the power utilised by runner.
(*iii*) the ratio of power of available at the turbine shaft to that supplied to it by the runner.
(*iv*) the ratio of the power supplied by the runner to the power available at the shaft.

22. Cavitation, being related to occurrence of low pressures in flow, would definitely occur if :
(*a*) the pressure anywhere in the flow approaches the atmospheric value.
(*b*) the pressure anywhere in the flow falls to a value very close to the vapour pressure of the liquid at the prevailing temperature.
(*c*) the pressure is lower than the vapour pressure.
(*d*) none of the above.

23. Cavitation damage in turbine runner occurs :
(*i*) near the inlet on the concave side of blades.
(*ii*) near the outlet on the convex side of blades.
(*iii*) near the inlet on the convex side of blades.
(*iv*) near the outlet on the concave side of blades.

24. For cavitation-free operation of turbines, the runner must be suitably installed with respect to the tail race. The Thoma's coefficient $\sigma_{inst} = (H_{atm} - H_v - H_s)/H$ for safe operation must satisfy the following relationship with the critical coefficient σ_c
(*a*) $\sigma_c > \sigma_{inst}$ (*b*) $\sigma_c = \sigma_{inst}$ (*c*) $\sigma_{inst} > \sigma_c$ (*d*) $\sigma_c + \sigma_{inst} = 1$.

25. Cavitating conditions in turbines cause many undesirable effects. Select from the following the serious problems which arise from cavitation :
(*i*) reduction in discharge (*ii*) fall in efficiency
(*iii*) eddying motion (*iv*) noise and vibrations
(*v*) damage to blade surface (*vi*) damage to penstock
(*vii*) damage to spiral casing.

26. The main components of a centrifugal pump are the impeller and its casing. Identify the hydraulic function of each :

(*a*) impeller	(*i*) it converts the hydraulic energy into the mechanical one.
(*b*) casing	(*ii*) it converts the pressure energy into the hydraulic one.

(*iii*) it converts the mechanical energy into hydraulic energy.

(*iv*) it transforms most of the kinetic energy into the pressure energy.

27. The head developed by a centrifugal pump may be expressed as :

(*i*) $H = \dfrac{u_1 V_{u_1} - u_2 V_{u_2}}{g}$ (*ii*) $H = \dfrac{u_2 V_{u_2}}{g}$ (*iii*) $H = \dfrac{u_1 V_{u_1}}{g}$ (*iv*) $H = \dfrac{u_2 V_{u_2} + u_1 V_{u_1}}{g}$

28. The performance of a centrifugal pump is greatly dependent upon the outlet angle of the impeller blades. Identify the limit of the blade angle for the following :

(*a*) slow runner (*i*) $\beta_2 < 90°$

(*b*) medium runner (*ii*) $\beta_2 > 90°$

(*c*) fast runner (*iii*) $\beta_2 = 90°$.

29. Two centrifugal pumps can individually deliver liquid at the rate of Q_1 and Q_2 against heads H_1 and H_2 respectively. If the liquid at the rate $(Q_1 + Q_2)$ is to be delivered, indicate how the pumps are to be connected :

(*i*) the pumps are to be connected in series such that the discharge from the pump is fed into the other.

(*ii*) the pump must be connected in parallel so that the discharge from each is fed into the main pipe.

(*iii*) none of the above.

30. Which of the following statements pertaining to centrifugal pump installation are correct :

(*i*) the discharge control valve is fitted in the suction pipe.

(*ii*) the suction pipe has larger diameter as compared to the discharge pipe.

(*iii*) the suction pipe is provided with a foot valve and a strainer.

(*iv*) the delivery pipe is provided with a foot valve and a strainer.

(*v*) the discharge control valve is fitted in the delivery pipe.

31. The manometric head is :

(*a*) the difference in elevation between the water surface in the high level reservoir and the water level in the sump.

(*b*) the height to which water is lifted by the pump measured above the pump centre-line.

(*c*) the difference in the piezometric heads between the points on the delivery and suction pipes as close to the pump as possible.

(*d*) the head developed by the pump.

32. The centrifugal pump should be so installed above the water level in the sump such that :

(*i*) its height is not more than 10.3 m at ordinary temperature of liquid.

(*ii*) its height is not allowed to exceed 6.7 m.

(*iii*) the negative pressures do not reach as low a value as the vapour pressure.

(*iv*) the negative pressures are not allowed to develop in the impeller.

33. In order to prevent cavitation, the suction lift must be considerably lower than the maximum limit expressed by :

(*i*) $\dfrac{p_a - p_v}{\gamma} - \dfrac{V_s^2}{2g} - h_{fs}$ (*ii*) $\dfrac{p_a - p_v}{g} - \dfrac{V_s^2}{2g} + h_{fs}$

(*iii*) $-\dfrac{p_a - p_v}{\gamma} + \dfrac{V_s^2}{2g} + h_{fs}$ (*iv*) $\dfrac{p_a - p_v}{\gamma} + \dfrac{V_s^2}{2g} - h_{fs}$.

34. The net positive suction head (NPSH) which represents the suction head at the impeller eye is given by

(*a*) $\frac{p_a - p_v}{\gamma} + h_s - h_{f_s}$ (*b*) $\frac{p_a - p_v}{\gamma} - h_s - h_{f_s}$

(*c*) $\frac{p_a - p_v}{\gamma} - h_s + h_{f_s}$ (*d*) $\frac{p_a - p_v}{\gamma} + h_s + h_{f_s}$

35. The minimum head across the impeller for the delivery to commence is :

(*i*) $h + (h_{f_s} + h_{f_d}) - \frac{V_s^2}{2g} - \frac{V_d^2}{2g}$ (*ii*) $h + (h_{f_d} + h_{f_s}) + \frac{V_s^2}{2g} + \frac{V_d^2}{2g}$

(*iii*) $h + (h_{f_s} + h_{f_d}) + \frac{V_s^2}{2g} - \frac{V_d^2}{2g}$ (*iv*) $h - (h_{f_s} + h_{f_d}) - \frac{V_s^2}{2g} - \frac{V_d^2}{2g}$

ANSWERS

1. (*b*) **2.** (*iii*). **3.** (*a*) **4.** (*iii*) **5.** (*a*) ; (*i*), (*iii*), (*v*) ; (*b*) (*ii*), (*iv*), (*vi*) **6.** (*iv*) **7.** (*a*) ; (*ii*), (*iv*), (*v*), (*vi*) ; (*b*) : (*i*), (*iii*)
8. (*a*), (*iii*) ; (*b*), (*ii*), (*c*), (*i*) **9.** (*ii*) **10.** (*a*), (*iv*) ; (*b*), (*iii*) ; (*c*), (*i*) ; (*d*), (*ii*)
11. (*iii*) **12.** (*a*), (*iii*) ; (*b*), (*i*) ; (*c*), (*ii*) **13.** (*iv*) **14.** (*d*) **15.** (*i*)
16. (*c*) **17.** (*a*), (*iv*) ; (*b*), (*iii*) ; (*c*), (*i*) ; (*d*), (*ii*) **18.** (*a*), (*v*) ; (*b*), (*ii*) ; (*c*), (*iv*)
19. (*ii*) **20.** (*c*) **21.** (*i*) **22.** (*b*) **23.** (*ii*) **24.** (*c*)
25. (*ii*), (*iv*), (*v*) **26.** (*a*), (*iii*) ; (*b*), (*iv*) **27.** (*ii*) **28.** (*a*), (*ii*) ; (*b*), (*iii*) ; (*c*), (*i*)
29. (*ii*) **30.** (*ii*), (*iii*), (*v*) **31.** (*c*) **32.** (*ii*), (*iii*) **33.** (*i*) **34.** (*b*)
35. (*iii*).

REVIEW QUESTIONS

1. State the principle on which turbomachines are based.
2. Under what conditions of flow and head would you use axial flow, radial flow and impulse turbines ?
3. What are the hydraulic functions of spiral casing, guide vanes and the draft tube ?
4. Which of the following turbines require complete initial conversion of total energy into the kinetic energy ? (*i*) Kaplan (*ii*) Francis (*iii*) Pelton.
5. Which type of turbine is best suited to conditions of changing loads on the turbine, and why ?
6. For maximum conversion of hydraulic power into mechanical power, what should be the shape of velocity diagram at the outlet in case of a reaction turbine ?
7. Why should it be necessary for the relative velocity vector at inlet to be tangential to the blade tip ?
8. Comment on the actual range of efficiency of hydraulic turbines. What factors should be considered for improving it ?
9. How does the specific speed of turbine give us valuable information regarding the selection of turbine under given set of conditions ?
10. What factors affect occurrence of cavitation in turbines ? What measures are taken to prevent it ? If cavitation were to occur, at what locations the cavitation damage is likely to take place ?
11. Under what conditions would you suggest use of double-suction pump and a multistage pump ?
12. Why centrifugal pumps are less efficient as compared to turbines ?
13. How would you compute the specific speed for a (*i*) double-suction pump and (*ii*) multi-stage sump ?
14. What are the limitations for installing a centrifugal pump above the liquid surface in the pump ? How is liquid temperature related to the height of the installation of the pump ?
15. What precautions are to be taken while starting and closing the centrifugal pump ?
16. If the power supply for a centrifugal pump fails, the flow sometimes changes direction before valve could be closed. The pump then behaves, in effect, like a turbine. Would its efficiency as such be high or low ? Why ?
17. Under what conditions would a reaction turbine work as a pump ?

PROBLEMS

15.1. Water enters an inward flow reaction turbine at an angle of 22° to the tangent to the outer rim and leaves the turbine radially. If the speed of the wheel is 300 r.p.m. and the velocity of flow is constant at 4 m/s, find the inlet and outlet angles of the blade when the inner and outer diameters of the runner are 30 cm and 60 cm respectively. If the width of the wheel at inlet is 15 cm, calculate the power developed. Neglect the thickness of blades.

15.2. An inward flow reaction turbine with a supply of 20 cumecs under a head of 20 m develops 330 kW at 375 r.p.m. The inner and outer diameters of the runner are 50 and 75 cm respectively. The velocity of flow at the exit is 3 m/s assuming that discharge is radial and that the width of the wheel is constant determine (*i*) the overall efficiency, (*ii*) hydraulic efficiency, and (*iii*) inlet angle of the guide and wheel vanes.

15.3. An inward flow reaction turbine has a wheel 60 cm in diameter and 5 cm wide at the outer rim the inner diameter being 0.65 times the outer diameter. The blade angles at inlet and outlet are 75° and 14° respectively. The velocity of flow is uniform throughout the wheel. 8% of the circumferential area of the runner is blocked by the blade thickness. If the head on the turbine is 55 m, the hydraulic efficiency 88% and overall efficiency 81%, determine the speed and output of the turbine and quantity of water if would require. Assume that the flow is radial at outlet.

(*UPSC 1960*)

15.4. A hydraulic turbine develops 7350 kW under a head of 16 m at a speed of 90 r.p.m. gives an efficiency of 90%, calculate the water consumption and the specific speed.

15.5. An inward flow turbine rotates at 100 r.p.m. and discharges 1 m^3/s. The diameter of runner at inlet is 50 cm and at outlet it is 25 cm. The guide vane angle is 15° and the blade angle at outlet is 45°. The width of the runner is 5 cm. Calculate (*i*) the inlet blade angle (*ii*) the torque developed by the turbine (*iii*) the power output from the turbine and (*iv*) head absorbed by the turbine.

15.6. The guide vane angle of an inward flow reaction turbine makes an angle α with the tangent to the runner wheel. The runner vane angle at inlet to the wheel is 90° and the radial velocity of flow of water at exit is c times that at entry. The discharge is radial and the wheel operates under a head of h. Develop an experssion for the peripheral velocity of the wheel in terms of h, c and α.

15.7. An inward flow reaction turbine discharges radially and the velocity of flow is constant, show that the hydraulic efficiency can be expressed by

$$\eta = \frac{1}{1 + \dfrac{0.5 \tan^2 \alpha}{1 - \left(\dfrac{\tan \alpha}{\tan \theta}\right)}}$$

where α and θ are the guide and vane angles at inlet.

15.8. A Pelton wheel is required to run at 600 r.p.m. The water jet is 7.5 cm in diameter and has a velocity of 100 m/s. The bucket deflects the jet through 170° and the ratio of the bucket speed of initial jet speed is 0.47. Neglecting losses, determine (*a*) the diameter of wheel to centreline of buckets, (*b*) power developed, and (*c*) kinetic energy per N remaining in the fluid.

15.9. An impulse turbine has a diameter of 1.25 m and runs at 360 r.p.m. The bucket angle is 160%, the coefficient of velocity is 0.98 and the speed ratio is 0.45. The relative velocity of water is decreased by 10% due to bucket surface friction. The diameter of the jet is 15 cm. Determine the power developed by the turbine.

15.10. The water jet in a Pelton wheel is 8 cm in diameter and has a velocity of 93 m/s. The wheel makes 660 r.p.m. and the jet is deflected through an angle of 170°. If the speed ratio $\phi = 0.47$, determine (*a*) the diameter of wheel to centreline of buckets, (*b*) the power developed, and (*c*) the kinetic energy per N remaining in the fluid.

15.11. An inward flow reaction turbine has external and internal diameters as 1.0 m and 0.5 m respectively. The water enters the wheel with a velocity of 30 m/s at an angle of 10°. The width of the wheel at inlet and outlet is 15 and 30 cm respectively. The angle is 90° at inlet and 25° at outlet. Determine (*i*) the tangental velocity of runner at inlet and (*ii*) absolute velocity of water at outlet.

15.12. An impulse turbine develops 7350 kW running at 550 r.p.m. The head on the turbine is 290 m. The jet ratio is 10 and the overall efficiency 85%. Determine (*a*) the discharge (*b*) the wheel diameter and (*c*) the number of jets. Assume the coefficient for the nozzle as 0.98 and speed ratio 0.48.

15.13. The following data were obtained from test on a pelton wheel :

Diameter of jet	= 12.5 cm
Discharge through nozzle	= 280 lit/sec
Head at the nozzle base	= 27 metres
Power developed	= 63 kW
Power absorbed in mechanical resistance	= 3.675 kW

Find (*i*) power lost in the nozzle, (*ii*) power lost due to hydraulic resistance in the runner.

15.14. At a hydropower plant, the turbine operates under a head of 100 m and consumes water at a rate of 12 m³/s. If the turbine runs at 240 r.p.m. and gives an efficiency of 90%, what is the specific speed of the turbine ?

15.15. Under a head of 16 m, the maximum feasible specific speed is 445. If under this head, an installation of 14700 kW is required, and if the speed is to be 160 r.p.m., how many turbines should be used ?

15.16. A turbines is required to work under a head of 30 m and to run at 420 r.p.m. Its quarter-scale model is tested under head of 14 m. At what speed must the model be run ? If it developes 95.5 kW and uses water at the rate or 1.2 m³/s at this speed, what power will be obtained from the full-scale turbine, if its efficiency is 40% better than that of the model. What is the specific speed of the full-scale turbine ?

15.17. A water turbine develops 128.6 kW at 240 r.p.m. under a head of 16 m. Find the scale ratio and speed of similar machine which will generate 661.5 kW at 25 m head otherwise under the same conditions.

15.18. In an inward flow reaction turbine the tangential velocity of the wheel at inlet is given by $u = \phi \sqrt{2gH}$ and the radial velocity of flow $V_f = \psi \sqrt{2gH}$. The breadth of the wheel at inlet is n times the diameter of wheel to inlet. If the turbine efficiency is 85% and the area blocked by the blades is 5% of the peripheral area, prove that the specific speed of the turbine is $1035\, \phi \sqrt{\psi n}$ in metric units.

15.19. At a proposed hydroelectric station the available head is 76 m and it is estimated that 35,700 litres of water per second will be available throughout the year for operating the water wheels. The turbines are to run at 500 r.p.m. and the overall efficiency of the wheels is 82%. The specific speed of the turbines is nearly 184 in SI units. Determine the maximum available power and the number of wheels is required.

ANSWERS

15.1. 105.2 kW. **15.2.** (*i*) 84.4%, (*ii*) 87.5%, (*iii*) 8.75°, 131°.

15.3. 698 r.p.m., 134.6 kW. **15.4.** 52.1 m³/s, 281.5.

15.6. (*i*) 15.8°, (*ii*) 14950 N-m, (*iii*) 156.5 kW, (*iv*) 15.97 m.

15.7. $h = \dfrac{u_1^2}{g}\left(1 + \dfrac{c^2}{2}\tan^2\alpha\right)$ **15.8.** (*a*) 1.496 m, (*b*) 2187 kW (*c*) 5.69 N-m/N

15.9. 1090 kW. **15.10.** (*a*) 1.265 m. (*b*) 2003 kW (*c*) 4.96 N-m/N

15.11. (*a*) 29.54 m/s, (*ii*) 6.32 m/s at 124.7°.

15.12. (*a*) 2.585 m/s (*b*) 1.218 m (*c*) 3.

15.13. (*i*) 1.43 kW (*ii*) 5.78 kW **15.14.** 91.1.

15.15. 3. **15.16.** 970 r.p.m., 8335 kW **15.17.** 1/1.62

15.19. 21.793 kW

16

Reciprocating Pumps

16.1. Introduction

On account of many advantages offered by centrifugal pumps such as non-pulsating flow, low initial cost, light weight, compact, easy installation etc., reciprocating pumps have now-a-days become generally obsolete. In the following pages, however, a brief description of reciprocating pumps has been given.

In all pumps, the pumping action is based on the same general principle. If the pump is to lift water from a well, it is first necessary to expel the air from its working chamber. This will create a partial vacuum within the pump. The pressure of the atmosphere on the surface of water in the sump well will push it up in the evacuated chamber.

16.2. Working of Reciprocating Pump

Fig. 16.1 shows a single-acting reciprocating pump consisting of a ram or plunger (plunger may be replaced by a piston) which moves forwards and backwards (in a reciprocating manner) in a

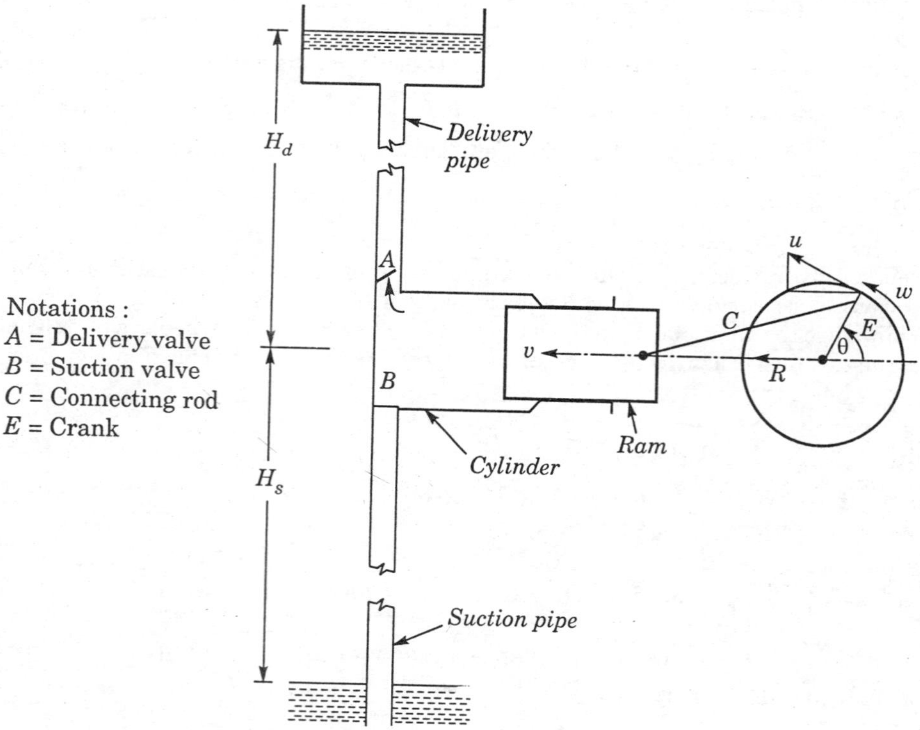

Fig. 16.1. Line diagram of a single-acting Reciprocating pump.

closely fitting cylinder. The reciprocating motion is provided by a crank-connecting rod arrangement as shown in Fig. 16.1. The self-acting lift valves are provided on the suction and delivery sides. Suction and delivery pipes are connected to the cylinder as indicated.

Before starting the reciprocating pump for the first time, the air is forced out of the cylinder by filling it with water from an outside source. This operation is called priming, and it is not usually required in case of a reciprocating pump after the initial start. This is because the cylinder will remain almost full of water each time the pump is stopped and will, therefore, remain primed and ready for use. For starting the pump, the crank is rotated at uniform speed by the driving engine or motor. As a result, the cylindrical ram or plunger moves to and fro in the cylinder. On the backward stroke (the ram moving from left to right), a partial vacuum is created inside the cylinder. This is the suction stroke. During the suction stroke, the suction valve B is lifted by the water entering to fill the vacuum, but is clapped shut as soon as the plunger reverses its direction. As soon as the discharge or delivery stroke starts (the ram moving from right to left), the pressure imparted to the water will lift the delivery valve A to allow the water to flow out in the delivery pipe.

If H_s is the suction head and H_d the delivery head, then $(H_s + H_d)$ is called the static head.

Let
R = crank radius
N = speed of crank shaft in r.p.m.
D = diameter of ram or plunger
Q = theoretical discharge
Q_a = actual discharge.

The volume swept per stroke $= \dfrac{\pi D^2}{4} . 2R$

Under ideal conditions, the average discharge per second is

$$Q = \frac{\pi D^2}{4} . \frac{2RN}{60} \qquad \text{...(16.1)}$$

On account of leakage through the valves and plunger, and due to lag in the closure of valves, the actual discharge Q_a is nearly always less than Q. The ratio of Q_a/Q is known as the volumetric efficiency and the percentage slip is defined as

$$= \frac{Q - Q_a}{Q} \times 100$$

The actual discharge in terms of the slip

$$Q_a = \frac{\pi D^2}{4} . \frac{2RN}{60} \left(\frac{100 - \text{Slip}}{100} \right) \qquad \text{...(16.2)}$$

For pumps maintained in good condition, the percentage slip is of the order of 2%. But when the delivery pressure is high, above 6867 kN/m^2, the compressibility of the liquid tends to increase the slip, especially if the clearance volume of the cylinder is excessive.

Force on the ram/plunger during suction stroke $= \gamma H_s \dfrac{\pi D^2}{4}$ ...(16.3)

Force on the ram/plunger during delivery stroke $= \gamma H_d = \dfrac{\pi D^2}{4}$ ...(16.4)

Neglecting losses, the power delivered by the pump $= \gamma Q \, (H_s + H_d)$ ...(16.5)

16.3. Double Acting Pump

In a double acting pump, shown in Fig. 16.2, each stroke of the piston carries out both the functions namely, suction as well as delivery. In other words, when piston moves from left to right as shown, the water moves into the cylindrical chamber while the other suction valve S_2 is kept shut. The suction pressure developed in front of the piston closes the delivery valve D_1 while the positive pressure on the rear side causes the delivery valve D_2 to open and the water is forced into the delivery pipe. The average discharge is given by

$$Q = \frac{\pi D^2}{4} \cdot \frac{2RN}{60} + \frac{\pi}{4}(D^2 - d_0^2)\frac{2NR}{60} \quad ...(16.6\ a)$$

in which d_0 = diameter of the piston rod and D the diameter of the piston itself. If $D >> d_0$, then

$$Q \approx 2\frac{\pi D^2}{4} \cdot \frac{2RN}{60} \quad ...(16.6\ b)$$

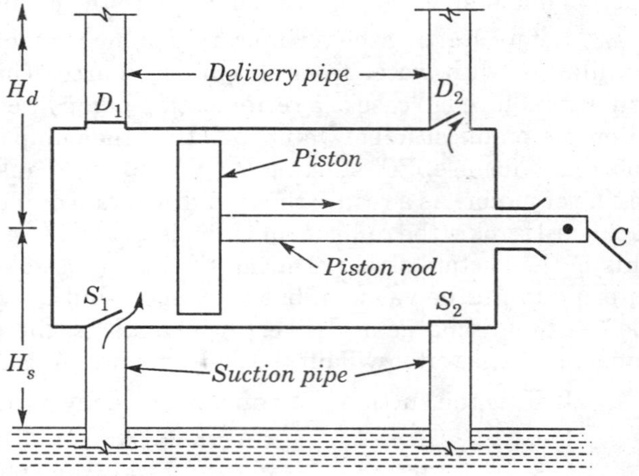

Notations:
D_1 = Delivery valve (closed)
D_2 = Delivery valve (being opened)
S_1 = Suction valve (being opened)
S_2 = Suction valve (closed)
C = Connecting rod

Fig. 16.2. Double Acting Piston Pump—a line diagram.

16.4. Instantaneous Rate of Discharge

The rate of discharge in the delivery pipe of a single acting reciprocation pump at any instant will vary quite widely, having a zero value during the entire suction stroke and rising to maximum during the delivery stroke.

If Q_i = instantaneous discharge in the delivery pipe at any instant,
V = instantaneous velocity in the delivery pipe at that instant,
v = velocity of the ram at that instant,
θ = angle through which the crank has turned from the outer dead centre at that instant, ωt
u = uniform peripheral speed of the crank-pin, $R\omega$
d = diameter of the delivery pipe.

From Fig. 16.1, it is evident that the velocity of the ram at any instant is identical with the horizontal component of the crank-pin velocity,

$$v = u \sin\theta = u \sin \omega t = R\omega \sin \omega t \quad ...(16.7)$$

In other words, the ram moves in simple harmonic motion under ideal conditions, the quantity of water displaced by the ram in a short interval of time is equal to the quantity of water raised in the delivery pipe in that time. The instantaneous rate of discharge

$$Q_i = \frac{\pi}{4} D^2 v = \frac{\pi}{4} d^2 V \quad ...(16.8\ a)$$

From Eqs. (16.7) and (16.8 a), one can write

$$V = \frac{D^2}{d^2} v = \frac{D^2}{d^2} u \sin\theta = \frac{D^2}{d^2} \cdot \frac{2\pi RN}{60} \sin\theta \quad ...(16.9)$$

$$Q_i = \frac{\pi}{4} D^2 \frac{2\pi RN}{60} \sin\theta \qquad ...(16.9)$$

Equation (16.9) expresses the relationship between the instantaneous discharge Q_i and the crank angle θ, see Fig. 16.3. The maximum instantaneous discharge in the pipe

$$Q_{max} = \frac{\pi}{4} D^2 \frac{2\pi RN}{60} \qquad ...(16.10)$$

Comparing Eq. (16.1) with Eq. (16.10), it is seen that the maximum instantaneous discharge is π times the mean discharge.

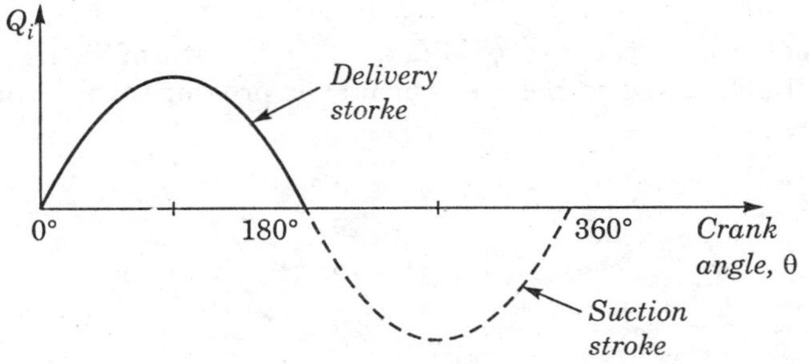

Fig. 16.3. Variation of discharge in a single acting reciproacting pump.

16.5. Acceleration of Piston and Water

The fluctuations in the discharge and hence in the pipe velocity give rise to inertia pressures due to effect of acceleration. These pressures can be readily determined by finding at any instant the acceleration or deceleration dV/dt of water in the pipe, and substituting it in the modified form of Eq. (11.43),

$$\Delta p_i = \rho L \frac{dV}{dt} \qquad ...(16.11)$$

in which L is the length of pipe. The acceleration

$$\frac{dV}{dt} = \frac{dV}{d\theta} \cdot \frac{d\theta}{dt}, \quad \text{where } \theta = \omega t$$

since $\quad v = u \sin\theta = R\omega \sin\omega t,$

acceleration of piston $\quad = dv/dt = R\omega^2 \cos\omega t$

and $\quad V = v \frac{D^2}{d^2}$

hence $\quad V = u \sin\theta \cdot \frac{D^2}{d^2}. \quad \therefore \quad \frac{dV}{d\theta} = u \cos\theta \cdot \frac{D^2}{d^2},$

and $\quad \frac{d\theta}{dt} = \omega = \frac{2\pi N}{60}$

Substituting for $\quad u = \frac{2\pi RN}{60}$, one obtains

$$\frac{dV}{dt} = \frac{2\pi NR}{60} \cos\theta \cdot \frac{D^2}{d^2} \cdot \frac{2\pi N}{60} = \left(\frac{2\pi N}{60}\right)^2 R \cos\theta \cdot \frac{D^2}{d^2}$$

The inertia pressure, from Eq. (16.11), is

$$\Delta p_i = \rho L \left(\frac{2\pi N}{60}\right)^2 R \cos\theta \,.\, \frac{D^2}{d^2} = \rho L\, \omega^2 R \cos\theta \,.\, \frac{D^2}{d^2} \qquad \text{...(16.12 } a)$$

$$\text{Acceleration head of water} = \Delta p_i / r = \frac{L}{g} (D/d)^2 \omega^2 R \cos\theta \qquad \text{...(16.12 } b)$$

The maximum and minimum values of Δp_i are attained at the dead centers, where $\cos\theta = \pm 1$.

Equation (16.12) would determine inertia pressure in the delivery pipe when the length L and the diameter d of the pipe are used in the formula. Similarly, the inertia pressure in the suction pipe could be obtained if appropriate values of L and d are used.

The true values of inertia pressure are influenced by the ratio of the length of the connecting rod to crank radius. Let this ratio be n, then the true inertia pressure at the inner dead centre (ram fully inside the cylinder) is

$$\Delta p_i \left(1 + \frac{1}{n}\right) \qquad \text{...(16.13)}$$

whereas its value at the outer dead centre is $\Delta p_i (1 - 1/n)$; Δp_i being the value calculated from Eq. (16.12) for appropriate value of θ.

If p_a = atmospheric pressure

p_v = vapour pressure of the liquid

p_{sv} = pressure required to force suction valve open

Δp_i = the value calculated from Eq. (16.12) for $\theta = 0°$ or $180°$:

Then the maximum permissible suction lift

$$(H_s)_{max} = \frac{p_a - p_v - p_{sv} - \Delta p_i (1 + 1/n)}{\gamma} \qquad \text{...(16.14)}$$

Example 16.1. *A single acting reciprocating pump has the following data :*

(i) piston diameter = 125 mm, *(ii) stroke length = 300 mm,*

(iii) suction pipe diameter = 75 mm, *(iv) suction pipe length = 5.0 m,*

(v) speed = 40 r.p.m., *(vi) min. permissible pressure head = 2.4 m atm.*

Determine the maximum suction head which will not cause separation at the beginning of suction stroke. Take H_{atm} = 9.8 m of water (absolute).

Solution. Inertia head at the beginning of suction stroke

$$\frac{(\Delta p_i)_{max}}{\gamma} = \frac{L_s}{g} \omega^2 R \left(\frac{D}{d_s}\right)^2 ; \ \omega = \frac{2\pi N}{60} = \frac{2\pi \times 40}{60} = \frac{4\pi}{3} \text{ rad/s}$$

$$= \frac{5}{9.81} \left(\frac{4\pi}{3}\right)^2 \frac{150}{1000} \left(\frac{125}{75}\right)^2 = 3.729 \text{ m}$$

Maximum suction head = Permissible suction head

$$H_s = H_a - H_v - (\Delta p_i)_{max} / \gamma$$

Taking vapour pressure (at 20°C) of water = 0.239 (Table 1.4 p. 33)

Maximum suction head = 9.8 – 0.239 – 3.727 = **5.832 m.**

Example 16.2. *A single-acting reciprocating pump is to be used for lifting water at a temperature of 50°C. The pump is to run at 40 r.p.m. at a place which is at an elevation of 900 m above mean sea level. The pump has the following particulars :*

Bore of the cylinder	= 12 cm,	*Length of stroke*	= 30 cm
Length of connecting rod	= 65 cm,	*Diameter of suction pipe*	= 10 cm
Length of suction pipe	= 10 cm.		

If no air-vessel is fitted on the suction side, what would be the maximum permissible suction lift ? The vapour pressure of water at 50°C = 12360 N/m² : specific weight of water at 50°C = 9790 N/m² ; barometric pressure at 900 m altitude = 91.2 × 10² N/m² ; pressure required to force the suction valve open = 2943 N/m².

Solution. The maximum permissible suction lift is given by Eq. (16.14),

$$(H_s)_{max} = \frac{p_a - p_v - p_{sv} - \Delta p_i\,(1 + 1/n)}{\gamma}$$

Inertia pressure at the end of stroke, from, Eq. (16.12)

$$\Delta p_i = \rho L \omega^2 R \frac{D^2}{d^2}$$

$$= \frac{9790}{9.81} \times 10 \times \left(\frac{2\pi \times 40}{60}\right)^2 \times \frac{15}{100} \times \left(\frac{12}{10}\right)^2 = 37670 \text{ N/m}^2$$

Considering the effect of length of connecting rod, corrected value of inertia pressure

$$= \Delta p_i \left(1 + \frac{1}{n}\right) 37670 \left(1 + \frac{15}{65}\right) = 46401 \text{ N/m}^2$$

$$(H_s)_{max} = \frac{91.2 \times 10^3 - 12.36 \times 10^3 - 2943 - 46.4 \times 10^3}{9790} = \mathbf{3.015\ m.}$$

Example 16.3. *A single acting reciprocating pump has vertical suction pipe 2.5 m long having 5 cm diameter. The stroke of the plunger is 30 cm and its diameter 15 cm. At what maximum speed should the pump be run to avoid cavitation at the suction valve when pumping water. The barometric pressure head is 10.3 m of water. Pressure required to force suction valve open is 2943 N/m² and cavitation occurs at the beginning of the stroke at 1.8 m of water absolute.*

Solution. Neglecting the effect of connecting rod, the maximum possible suction lift,

$$(H_s)_{max} = \frac{p_a - p_v - p_{sv} - (\Delta_{pi})_{max}}{\gamma}$$

or
$$\left(\frac{\Delta p_i}{\gamma}\right)_{max} = \left(\frac{p_a - p_v - p_{sv}}{\gamma}\right) = (H_s)_{max}$$

$$= \frac{p_a}{\gamma} - \frac{p_v}{\gamma} - \frac{p_{sv}}{\gamma} - (H_s)_{max} = 10.3 - 1.8 - \frac{2943}{9810} - 2.5 = 5.6 \text{ m}$$

but
$$(\Delta p_i)_{max} = \rho L \omega^2 R \left(\frac{D}{d}\right)^2 \quad \text{or} \quad \left(\frac{\Delta p_i}{\gamma}\right)_{max} = \frac{L\omega^2 R}{g}\left(\frac{D}{d}\right)^2$$

or
$$5.6 = \frac{2.5 \times \omega^2 \times 12}{9.81 \times 100} \times \left(\frac{15}{5}\right)^2$$

$$\omega^2 = 16.25, \qquad \therefore \quad \omega = 4.03 \text{ rad/sec.}$$

$$N = \frac{60\omega}{2\pi} = \frac{60 \times 4.03}{2\pi} = \mathbf{38.5\ r.p.m.}$$

16.6. Effects of Friction and Inertia Pressure

The head loss in friction in flow through pipes is given by Darcy-Weisbach Eq. (9.57) reproduced here

$$h_L = f \frac{L}{d} \frac{V^2}{2g} \qquad \text{...(9.57)}$$

in which d is the diameter, L the length and V the velocity through the pipe.

Let L_s = length of suction pipe,
d_s = diameter of suction pipe,
V_s = velocity in the suction pipe.

The velocity in the suction pipe is proportional to that of the plunger or piston by the relation,

$$\frac{\pi}{4} d_s^2 \, V_s = \frac{4}{\pi} D^2 v$$

Using Eq. (16.7),

$$V_s = u \sin\theta \,.\, \frac{D^2}{d_s^2} = \frac{\omega R D^2}{d_s^2} \sin\theta \qquad \text{...(16.15)}$$

where ω = angular velocity of the crank pin.

From Eq. (16.15), it is evident that the velocity in the suction pipe attains a maximum value when $\theta = 90°$ and the minimum value of zero is obtained when $\theta = 0°$ and $180°$. The loss of head in friction, being proportional to the velocity-square, also accordingly varies from zero value at the end each stroke to a maximum value at the mid-stroke. The head loss in friction in suction pipe is

$$h_{LS} = f \frac{L_s}{2gd_s} \left(\frac{\omega R D^2 \sin\theta}{d_s^2} \right)^2 \qquad \text{...(16.16 } a\text{)}$$

From which the maximum frictional head loss, $(h_{LS})_{max}$, is given by

$$(h_{LS})_{max} = f \frac{L_s}{2gd_s} \left(\frac{\omega R D^2}{d_s^2} \right)^2 \qquad \text{...(16.16 } b\text{)}$$

Since the variation of frictional head loss with stroke length (also crank angle, θ) is parabolic, the average frictional head loss in suction pipe is

$$(h_{LS})_{average} = \frac{2}{3} (h_{LS})_{max} = f \frac{L_s}{3gd_s} \left(\frac{\omega R D^2}{d_s^2} \right)^2 \qquad \text{...(16.16 } c\text{)}$$

A similar expression may be written for maximum head loss in the delivery pipe.

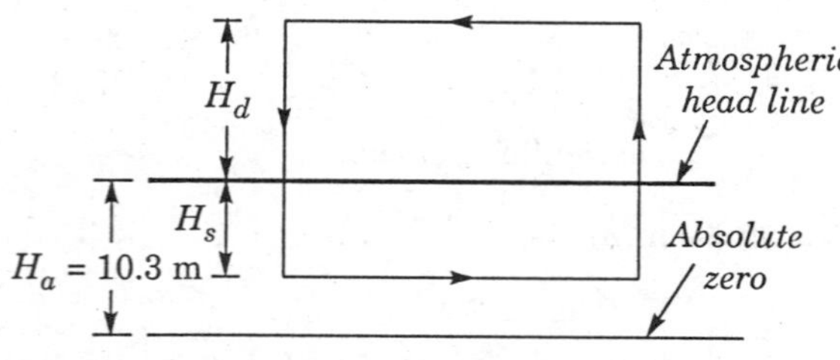

Fig. 16.4. (*a*) Hypothetical Indicator Diagram.

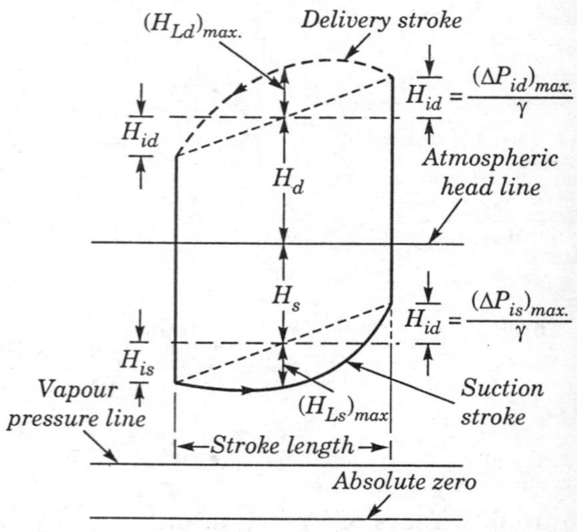

Fig. 16.4. (*b*) Indicator diagram showing head in cylinder during Suction and Delivery strokes.

The effects of inertia pressure and friction on the performance of the pump can conveniently be shown by plotting the indicator diagram. If these effects are eliminated or neglected, the indicator diagram would have a rectangular shape as shown in Fig. 16.4 (*a*). But when the effects of inertia pressure and friction are taken into account, the indicator diagram takes the shape of Fig. 16.4 (*b*).

It may be noted here that when the effect of inertia pressure alone is considered, the area of indicator diagram does not change and the power required to run the pump remains the same. Noteworthy is also the fact that Δp_i and $(h_{LS})_{max}$ both increase with increasing rotational speed. The speed of the crank shaft, therefore, governs the negative pressure in the cylinder, as the gap between the lower part of the indicator diagram during the suction stroke and the vapour pressure head line progressively reduces with increase in the rotational speed. At higher speeds, the negative pressure in the cylinder may reach such a low value as to cause separation of dissolved gases and the pump may cavitate. Reciprocating pumps are, therefore, designed to run at slower speeds.

Example 16.4. *The piston of a double-acting reciprocating pump has a diameter of 20 cm and a stroke of 60 cm. The pump runs at 20 r.p.m. and discharges through a 15 cm main 76 m long (friction factor f = 0.03), the vertical lift being 46 m. If no air-vessel is used, determine the heads in the cylinder at the ends and middle of the stroke.*

Solution. Data given : $D = 20$ cm, $2R = 60$ cm,
$N = 20$ r.p.m., $L_d = 76$ m,
$H_d = 46$ m, $d_d = 15$ cm,
$f = 0.03$.

Inertia pressure $\Delta p_i = \rho L_d\, \omega^2 R \cos\theta \dfrac{D^2}{d^2}$

(1) **At the start of delivery stroke, θ = 0°**

$$(\Delta p_i)_{max} = \frac{9810}{9.81} \times 76 \times \left(\frac{2\pi \times 20}{60}\right)^2 \times \frac{30}{100} \times \left(\frac{20}{15}\right)^2 \text{ N/m}^2 = 178.5 \text{ kN/m}^2$$

$$\frac{(\Delta p_i)_{max}}{\gamma} = 18.2 \text{ m.}$$

Head lost in friction at the beginning of stroke = 0.

∴ Head at the beginning of delivery stroke

$$= H_d + \frac{(\Delta p_i)_{max}}{\gamma} = 46 + 18.2 = 64.2 \text{ m.}$$

(2) **Head at the mid-stroke, θ = 90°**

Inertia head, $(\Delta p_i)/\gamma = 0$

Head lost in friction is the maximum

$$(H_{Ld})_{max} = f \frac{L_d}{2gd_d}\left(\frac{\omega R D^2}{d_d^{\,2}}\right)^2$$

$$= 0.02 \times \frac{76 \times 100}{2 \times 9.81 \times 15}\left[\left(\frac{2\pi \times 20}{60} \times \frac{30}{100}\right)\left(\frac{20}{15}\right)^2\right]^2 = 0.97 \text{ m}$$

∴ Head at the mid-stroke

$$= H_d + (H_{Ld})_{max} = 46 + 0.97 = 46.97 \text{ m.}$$

(3) **Head at the end of delivery stroke, θ = 180°**

Inertia head, $\dfrac{(\Delta p_i)_{max}}{\gamma} = -\ 18.2$ m.

Head lost in friction is zero.

∴ Head at the end of the stroke

$$= H_d + \frac{(\Delta p_i)_{max}}{\gamma} = 46 - 18.2 = 27.8 \text{ m above atmosphere.}$$

Example 16.5. *A double acting reciprocating pump has a stroke of 25 cm and a piston diameter of 12.5 cm. The centre of the pump is 4 metres above the level of water in the sump and 30 metres below*

below delivery water-level. The lengths of suction and delivery pipes are 6 metres and 35 metres respectively and their diameters are both 9 cm. Friction factor for the pipes f = 0.04.

If the pump is working at 30 r.p.m., find the pressure head in metres of water on the piston at-

(i) the beginning of suction stroke, *(ii) the middle of suction stroke, and*

(iii) the end of suction stroke.

If the mechanical efficiency is 75%, what is the power required to drive the pump ? Take atmospheric pressure as 10.3 metres of water.

Solution. Data given :

$D = 12.5$ m, $2R = 25$ cm,
$H_s = 4$ m, $H_d = 30$ m,
$L_s = 6$ m, $L_d = 35$ m,
$d = 6$ cm, $f = 0.04$,
$N = 30$ r.p.m., $\eta = 75\%$.

Considering the Suction Side

(*i*) **Pressure head at the beginning of suction stroke, $\theta = 180°$.**

Inertia head, $\dfrac{(\Delta p_i)_{max}}{\gamma} = -\dfrac{L_s}{g}\omega^2 R\left(\dfrac{D}{d}\right)^2$

$$= -\frac{6}{9.81} \times \left(\frac{2\pi \times 30}{60}\right)^2 \times \frac{12.5}{100} \times \left(\frac{12.5}{6}\right)^2$$

$= -3.28$ m of water.

Friction head being zero, the pressure head on the piston

$$= H_s + \frac{(\Delta p_i)_{max}}{\gamma} = -4 - 3.28$$

$= -7.28$ m of water $= 7.28$ m of water below atmosphere.

(*ii*) **Pressure head at middle of suction stroke, $\theta = 270°$.**

Inertia head = 0

Friction head with no air-vessel

$$(H_{LS})_{max} = f\frac{L_s}{2gd}\left(\frac{\omega R D^2}{d^2}\right)^2$$

$$= 0.04 \times \frac{6 \times 100}{2 \times 9.81 \times 6}\left[\left(\frac{2\pi \times 30}{60} \times \frac{12.5}{100}\right)\left(\frac{12.5}{6}\right)^2\right]^2$$

$= 0.6$ m of water.

∴ Pressure head on the piston at the middle of suction stroke

$= -4 - 0.6 = -4.6$ m of water $= 4.6$ m of water below atmosphere.

(*iii*) **Pressure head at the end of suction stroke, $\theta = 360°$**

Inertia head, $\dfrac{(\Delta p_{is})_{max}}{\gamma} = +3.28$ m of water

Friction head $= 0$

∴ The pressure at the end of suction stroke

$= -4 + 3.28 = 0.72$ m of water $= 0.72$ m below atmosphere.

Neglecting the effect of diameter of piston rod, the discharge Q

$$= 2 . \frac{\pi D^2}{4} . \frac{2RN}{60} = 2 \times \frac{\pi \times (0.125)^2}{4} \times \frac{25 \times 30}{100 \times 60}$$

$$= 3.06 \times 10^{-3} \text{ m}^3/\text{s} = 3.06 \text{ litre/sec.}$$

Considering the Delivery Side

$$\frac{(\Delta p_i)_{max}}{\gamma} = \frac{L_d}{g} \omega^2 R \left(\frac{D}{d}\right)^2 = \frac{35}{9.81} \times \left(\frac{2\pi \times 30}{60}\right)^2 \times \frac{12.5}{60} \times \left(\frac{12.5}{6}\right)^2$$

$$= 19.14 \text{ m of water.}$$

Friction head at mid-stroke

$$(H_{Ld})_{max} = \frac{L_d}{L_s}(H_{Ls})_{max} = \frac{35}{6} \times 0.6 = 3.5 \text{ m of water.}$$

Maximum head against which the pump has to work is larger of :

(*i*) $H_s + (H_{Ls})_{max} + H_d + (H_{Ld})_{max} = 4 + 0.6 + 30 + 3.5 = 38.1$ m

(*ii*) $H_s + \left(\frac{\Delta p_{is}}{\gamma}\right)_{max} + H_d + \left(\frac{\Delta p_{id}}{\gamma}\right)_{max} = 4 + 3.28 + 30 + 19.14 = 56.42$ m.

$\therefore \quad H = 56.42$ m.

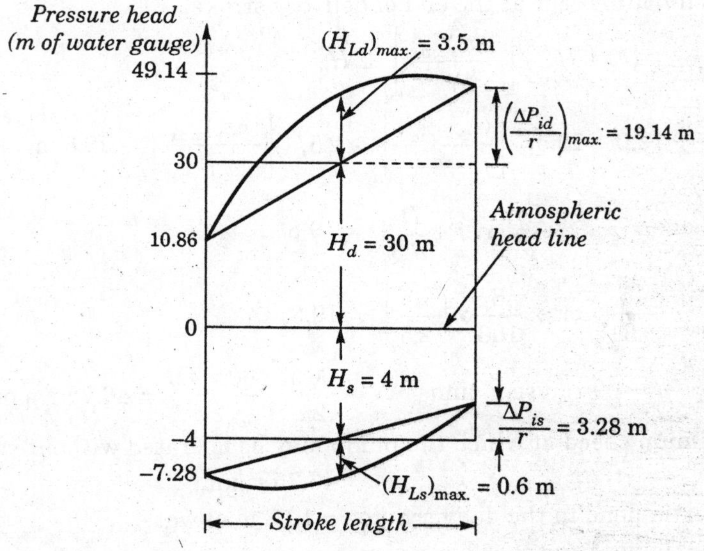

Fig. Ex. 16.5

Power delivered by the pump = $\gamma QH = 9810 \times 3.06 \times 10^{-3} \times 56.42$ N-m/s = 1.69 kW

Power required to drive the pump $= \frac{1.69}{0.75}$ = **2.25 kW.**

Example 16.6. *A single-acting reciprocating pump (with no air-vessel) has a plunger 7.5 cm diameter and a stroke of 12 cm. It draws water from a sump 3 metres below the pump through a suction pipe 3.5 cm diameter and 5 metres long. It delivers water to a tank 12 metres above the pump through a delivery pipe 2.5 cm diameter and 15 metres long. If separation occurs at a pressure of 76.5 kN/m² below atmospheric pressure, find the maximum speed at which the pump may be operated* *without separation. Assume that the plunger moves with simple harmonic motion. Sketch the indicator diagram of the pump for the above speed indicating values of pressure at the corner points.*

Solution. Data given : $D = 7.5$ cm, $2R = 12$ cm,

$H_s = 3$ m, $d_s = 3.5$ cm,

$L_s = 5$ m, $H_d = 12$ m,

$d_d = 2.8$ cm, $L_d = 15$ m,

$p_v = 76.5$ kN/m² below atmospheric pressure

$= 101.0 - 76.5 = 24.5$ kN/m² absolute

Inertia head at the beginning of suction stroke.

$$\frac{(\Delta p_i)_{max}}{\gamma} = \frac{L_s}{g}\omega^3 R\left(\frac{D}{d_s}\right)^2$$

Permissible suction head, $H_s = H_a - H_v - \dfrac{(\Delta p_i)_{max}}{\gamma}$

or
$$\frac{(\Delta p_i)_{max}}{\gamma} = H_a - H_v - H_s = 11.3 - \frac{24.5 \times 10^3}{9810} - 3 = 4.8 \text{ m}$$

$$\frac{L_s}{\gamma}\omega^2 R\left(\frac{D}{d_s}\right)^2 = 4.8 \qquad \text{or} \qquad \frac{5}{9.81} \times \omega^2 \times \frac{6}{100} \times \left(\frac{7.5}{3.5}\right)^2 = 4.8$$

$$\omega^2 = 34.2 \qquad \text{or} \qquad \omega = 5.85 \text{ rad/sec}$$

$$N = \frac{60\omega}{2\pi} \times \frac{60 \times 5.85}{2\pi} = 55.8 \text{ r.p.m.}$$

For separation not to occur at the end of delivery stroke,

$$H_d + H_a - \left[\frac{(\Delta p_i)_{max}}{\gamma}\right]_d \geq H_d$$

or
$$12.0 + 10.3 - \left[\frac{(\Delta p_i)_{max}}{\gamma}\right]_d \geq 2.5; \quad \left[\frac{(\Delta p_i)_{max}}{\gamma}\right]_d \leq 19.8 \text{ m}$$

for maximum speed ω,
$$\frac{L_d}{g}\omega^2 R\left(\frac{D}{d_d}\right)^2 = 19.8$$

$$\frac{15}{9.81} \times \omega^2 \times \frac{6}{100} \times \left(\frac{7.5}{2.5}\right)^2 = 19.8,$$

from which
$$\omega = 4.9 \text{ rad/sec} \quad \text{and} \quad N = \frac{60\omega}{2\pi} = \frac{60 \times 4.9}{2\pi} = 46.75 \text{ r.p.m.}$$

∴ The maximum speed at which the pump may be operated without separation

$= 46.75$ r.p.m.

Maximum inertia head in the delivery pipe = 19.8 m of water.

Maximum inertia head in the suction pipe

$$\left[\frac{(\Delta p_i)_{max}}{\gamma}\right]_s = \frac{L_s}{g}\omega^2 R\left(\frac{D}{d_s}\right)^2 = \frac{5}{9.81} \times (4.9)^2 \times \frac{6}{100} \times \left(\frac{7.5}{3.5}\right)^2$$

$= 3.48$ m of water.

The various pressure heads are as follows :

(*i*) at the beginning of suction stroke

$$= H_a - E_s - \left[\frac{(\Delta p_i)_{max}}{\gamma}\right]_s = 10.3 - 3.0 - 3.48$$

$= 3.82$ m of water absolute.

(*ii*) at the end of suction stroke

$$= H_a - H_s + \left[\frac{(\Delta p_1)_{max}}{\gamma}\right]_s$$

$$= 10.3 - 3.0 + 3.48$$

$= 10.78$ m of water absolute.

(*iii*) at the beginning of delivery stroke

$$= H_a + H_d + \left[\frac{(\Delta p_i)_{max}}{\gamma}\right]_d$$

$$= 10.3 + 12.0 + 19.8$$

$= 42.1$ m of water absolute.

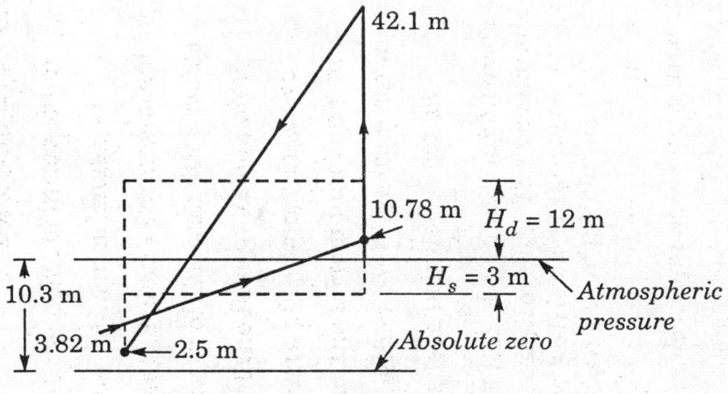

Fig. Ex. 16.6. Indicator diagram.

(*iv*) at the end of delivery stroke

$$= H_a + H_d - \left[\frac{(\Delta p_i)_{max}}{\gamma}\right]_d = 10.3 + 12.0 - 19.8$$

$= 2.5$ m of water absolute

The indicator diagram is shown in the figure.

Example 16.7. *A double-acting single cylinder reciprocating pump of 20 cm bore and 40 cm stroke runs at 35 r.p.m. The piston rod is 7.5 cm in diameter. The suction and delivery lifts are respectively 1 m and 30 m. The suction pipe is 2.5 m long and the delivery pipe is 40 m long and both of them are 10 cm in diameter. No air vessel is provided either on the suction or delivery pipe. The local losses can be neglected and for both pipes the friction factor can be taken as 0.03. The motion of the piston can be assumed to be simple harmonic motion.*

Determine the net force due to fluid pressure on the piston when it has moved through a distance of 10 cm from the inner dead centere.

Solution. In a double-acting reciprocating pump, every stroke is a suction-cum-delivery stroke. For any intermediate position of the piston, the net force on it will be the sum of force due to negative pressure (on suction side) and positive pressure (on delivery side) acting on either faces of the piston.

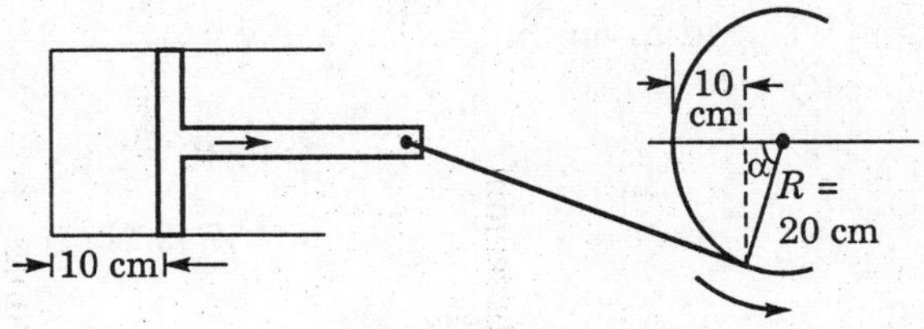

Fig. Ex. 16.7

With reference to the above figure and Fig. 16.1,

$$20 - 20 \cos \alpha = 10,$$

$$\therefore \qquad \alpha = 60°, \quad \theta = 180° + 60°.$$

Considering the suction side

The various heads for the position of the piston as indicated in the figure are as follows :

Inertia head, $$\frac{\Delta p_i}{\gamma} = \frac{L_s}{\gamma} \omega^2 R \left(\frac{D}{d_s}\right)^2 \cos \theta$$

$$= \frac{2.5}{9.18} \times \left(\frac{2\pi \times 35}{60}\right)^2 \times \frac{20}{100} \left(\frac{20}{10}\right)^2 \times \cos(180° + 60°)$$

$= -1.365$ m of water (gauge)

$= 1.365$ m of water below atmosphere

Friction head, $$h_{Ls} = \frac{fL_s}{2gd_s}\left(\frac{\omega RD^2 \sin\theta}{d_s^{\,2}}\right)^2$$

$$= 0.03 \times \frac{2.5}{2 \times 9.81 \times 0.1}\left[\left(\frac{2\pi \times 35}{60} \times 0.2\right) - \left(\frac{20}{10}\right)^2 \sin(180° + 60°)\right]^2$$

$$= 0.245 \text{ m of water.}$$

Total suction head on the piston

$$h_s = H_s + (\Delta p_i/\gamma) + h_{Ls}$$

$$= 1 + 1.365 + 0.245 = 2.16 \text{ m below atmosphere}$$

Considering the delivery side

Inertia head,

$$\left(\frac{\Delta p_i}{\gamma}\right)_{delivery} = \frac{L_d}{g}\omega^2 R\left(\frac{D}{d_d}\right)^2 \cos\theta = \frac{L_d}{L_s}\left(\frac{\Delta p_i}{\gamma}\right)_{suction} \qquad [\because\ d_d = d_s]$$

$$= \frac{40}{2.5} \times 1.365 = 21.82 \text{ m of water.}$$

Friction head, $$H_{Ld} = \frac{fL_d}{2gd_d}\left(\omega R\frac{D^2}{d_d^{\,2}}\sin\theta\right)^2 = \frac{L_d}{L_s} \times H_{Ls}$$

$$= \frac{40}{2.5} \times 0.245 = 3.92 \text{ m}$$

Total delivery head, $$h_d = H_d + \left(\frac{\Delta p_i}{\gamma}\right)_d + H_{Ld} = 30 + 21.82 + 3.92$$

$$= 55.74 \text{ m of water above atmosphere.}$$

Net force on the piston $$= \gamma h_s \frac{\pi D^2}{4} + \gamma h_d \frac{\pi(D^2 - d^2)}{4}$$

$$= 9810 \times 2.61 \times \frac{\pi \times (0.2)^2}{4} + 9810 \times 55.74 \times \frac{\pi\,[(0.2)^2\ (0.075)^2]}{4}$$

$$= \mathbf{15539\ N.}$$

16.7. Use of Air Vessels

An air vessel is a closed chamber connected to the pipe (separate vessels for a suction and delivery sides) at a point as close to pump as possible. By the use of air-vessels the following objectives are fulfilled:

(1) almost constant discharge is maintained in the suction and delivery pipes,

(2) fluid acceleration in the pipes is minimized,

(3) friction losses are considerably reduced as they now correspond to the mean velocity in the suction and delivery pipes.

Another advantage of fitting the air vessel on the suction side of pipeline is that the pump can run at a higher speed without danger of separation, and that it can be placed at a greater elevation above the water level in the sump.

Fig. 16.5 (*a*) shows a single-cylinder single-acting pump with air vessels on suction and delivery sides. The air vessels in its top portion contain air which for the vessel on the suction side must have sub-atmospheric pressure, whereas the one provided on the delivery side will have compressed air. When the level of water in the air vessel rises, the air in the top portion gets compressed and subsequently when pressure in the pipeline falls, the compressed air forces out the water into the delivery pipe. Fig. 16.5 (*b*) and (*c*) depict the variation of discharge with crank angle (which for a constant angular velocity represents time) without and with air vessel fitted in the delivery pipe. For most part of the delivery stroke, the piston is forcing water into the delivery pipe at a rate greater than the mean flowrate, the excess water thus flowing into the air vessel. At the end of the stroke when the flowrate is less than the mean value, the water stored in the air vessel at a higher pressure will be released to make up the deficiency. During the ensuing suction stroke, the pump discharge is zero and the flow in the delivery pipe is maintained by the stored water given out by the air vessel. The shaded portion of the discharge diagram represents excess discharge from the pump which must be stored in the air-vessel during the delivery stroke and the same must be given out during the suction stroke in order to equalize the discharge. Some change in the air pressure will naturally take place as the water surface in the vessel rises and falls, but by making the capacity of air vessel large enough, these periodical changes can be brought within

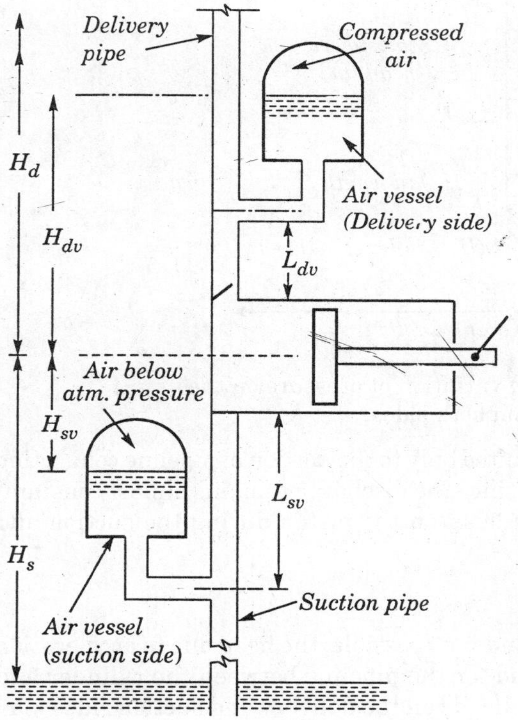

Fig. 16.5 (*a*) Air vessels on Suction and Delivery sides.

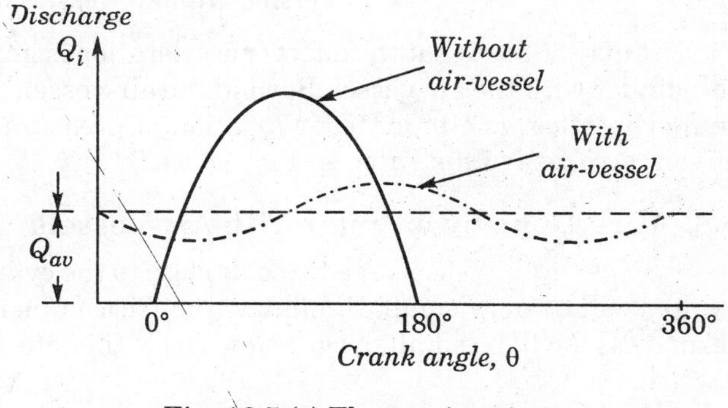

Fig. 16.5 (*c*) Fluctuations in discharge in delivery pipe.

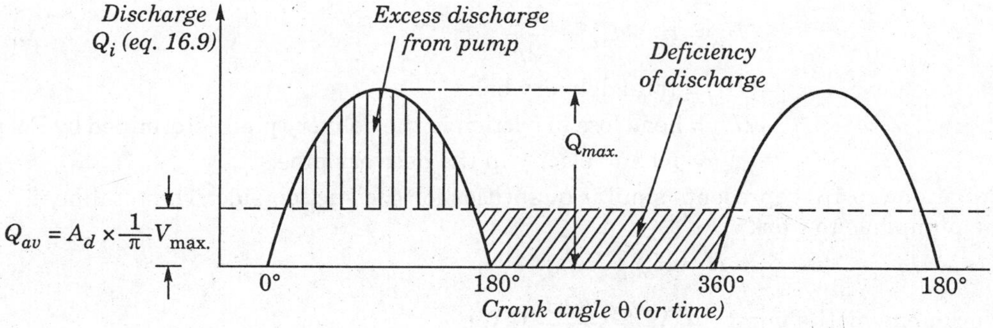

Fig. 16.5 (*b*) Discharge diagram for single-acting pump (without air-vessels).

prescribed limits. For a single-cylinder single-acting pump, it is usual to keep the volume of the air vessel six to nine times the pump displacement volume. The role played by the air-vessel in effecting uniformity in discharge is evidenced by broken line (dash-dot line) in Fig. 16.5 (*a*). The role played by the air-vessel in effecting uniformity in discharge is evidenced by broken line (dash-dot line) in Fig. 16.5 (*c*).

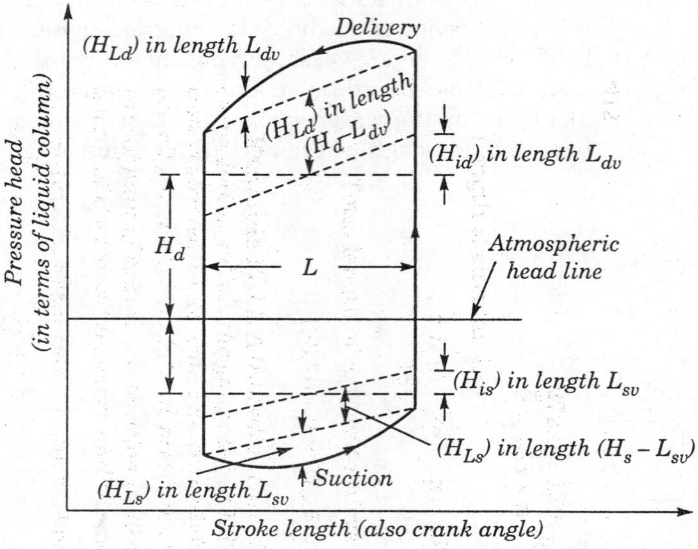

Fig. 16.5 (*d*) Indicator diagram showing variation of pressure on the piston with air-vessels on both sides.

It may be seen that the effect of acceleration is limited only to the portion of pipeline connecting the cylinder with the air-vessel. Beyond the air-vessel, since the discharge remains almost constant, the inertia effect is eliminated. Variation of pressure head on the piston during the suction and delivery strokes is illustrated in Fig. 16.5 (*d*).

16.8. Work Done by a Pump with Air Vessels

Since the air vessels are fitted as close to the cylinder as possible, the head due to acceleration of water will be very small. Similarly, the friction head for the pipeline between the cylinder and the air-vessel will be small. Neglecting these, the total heads on the delivery and suction sides may be written as

$$H_{td} = H_d + H_{Ld} + \frac{V_d^2}{2g} \qquad ...(16.17)$$

and

$$H_{ts} = H_s + H_{Ls} + \frac{V_s^2}{2g} \qquad ...(16.18)$$

where H_{td} = total delivery head

H_{Ld} = head loss in friction in the delivery pipe determined by Eq. (9.57).

V_d = mean velocity in the delivery pipe.

Equation (16.18) represents similar quantities for the suction side. The total head against which the pump has to work

$$H = H_{td} + H_{ts} \qquad ...(16.19)$$

Work done by the pump $= (W) = \dfrac{\gamma QH}{1000}$ (kW).

Example 16.8. *The plunger diameter of a single-acting reciprocating pump is 15 cm and its stroke 30 cm. The pump runs at 50 r.p.m. and lifts water through a height of 25 m. Find the theoretical discharge and the theoretical power required to run the pump. If the actual discharge is 4.2 litre/ sec., find the percentage slip.*

The delivery pipe 22 m long has a diameter of 10 cm, determine the acceleration head at the beginning of the delivery stroke. If a large air vessel is fitted very close to the cylinder on the delivery side, determine the head in the cylinder necessary to overcome friction in the delivery pipe. The friction factor in the Darcy-Weisbach equation has a value of 0.03.

Solution. Data given :

$$2R = 30 \text{ cm}, D = 15 \text{ cm}, N = 50 \text{ r.p.m.}, H_s + H_d = 25 \text{ m}.$$

Theoretical discharge Q = from Eq. (16.1) is

$$Q = \frac{\pi D^2}{4} \cdot \frac{2RN}{60} = \frac{\pi(0.15)^2}{4} \times \frac{30 \times 50}{100 \times 60} \text{ m}^3\text{/s}$$

$$= 4.41 \times 10^{-3} \text{ m}^3\text{/s} = 4.41 \text{ lit/s.}$$

$$\text{Percentage slip} = \frac{4.41 - 4.2}{4.41} \times 100 = \mathbf{4.76\%}.$$

Theoretical power required to run the pump

$$= \gamma QH = 9810 \times 4.41 \times 10^{-3} \times 25 = 1080 \text{ W} = \mathbf{1.08 \text{ kW}}.$$

Acceleration head at the beginning of delivery stroke

$$= \frac{L}{g} \cdot \omega^2 R \left(\frac{D}{d}\right)^2 = \frac{22}{9.81} \times \left(\frac{2\pi \times 50}{60}\right)^2 \times 0.15 \times \left(\frac{10}{15}\right)^2 = 16.7 \text{ m}.$$

With air-vessels fitted, the velocity in the delivery pipe

$$V_d = \frac{4Q}{\pi d^2} = \frac{4 \times 4.2 \times 10^{-3}}{\pi \times (0.1)^2} = 0.535 \text{ m/s}.$$

Friction head required to overcome friction

$$= f \frac{L}{d} \frac{V_d^2}{2g} = 0.03 \times \frac{22 \times (0.535)^2}{2 \times 9.81 \times 0.1} = \mathbf{0.096 \text{ m}.}$$

16.9. Power Saved by Air Vessels in Overcoming Pipe Friction

From Eqs. (9.57) and (16.15), it can be seen that the head loss in friction has a parabolic relation with the crank angle ; it being zero at the beginning and end of each stroke and attaining a maximum value given by Eq. (16.16), see Fig. 16.4 (*b*). Since the area under a parabola is 2/3 of the enclosing rectangle, the average friction head lost is

$$\frac{2}{3} (h_L)_{max} = \frac{2}{3} f \frac{L}{2gd} \left(\frac{\omega R D^2}{d^2}\right)^2$$

Power expended in overcoming friction when no air-vessels are fitted,

$$P_1 = \gamma Q \frac{2}{3} (h_L)_{max}$$

However, with air vessel fitted, the discharge being almost constant, the friction loss will also be constant. The power spent in overcoming friction will be

$$P_2 = \gamma Q h_L$$

where $\quad h_L = f \dfrac{L}{d} \dfrac{V^2}{2g} \quad$ and $V = \dfrac{4Q}{\pi d^2}$.

From Eq. (16.1), $\quad Q = \dfrac{\pi D^2}{4} \cdot \dfrac{R\omega}{\pi} \quad$ hence, $V = \dfrac{D^2}{d^2} \dfrac{R\omega}{\pi}$.

Making substitutions, $P_2 = \gamma Q\left[f\frac{L}{2gd}\left(\frac{D^2}{d^2}\frac{R\omega}{\pi^2}\right)^2\right]$

$$= \gamma Q\left[f\frac{L}{2gd}\frac{1}{\pi^2}\left(\frac{\omega RD^2}{d^2}\right)^2\right] \quad ...(16.21)$$

∴ % Power saved in pipe friction by fitting air vessel

$$= \frac{P_1 - P_2}{P_1}\times 100 = \left\{\frac{\frac{2}{3}-\frac{1}{\pi^2}}{2/3}\right\} = 84.8\%.$$

Similarly, for a double-acting pump for which the average velocity in the pipe $V = 2D^2 R\omega/d^2\pi$, the percentage saving in power will be 39.2%.

Example 16.9. *A single-cylinder single-acting reciprocating pump has a plunger of 20 cm diameter and a stroke of 40 cm. The delivery pipe is 10 cm diameter and 45 m long. Determine the power saved in overcoming friction in the delivery pipe by fitting a large air vessel when the pump runs at a speed of 55 r.p.m. Take f = 0.04 in the Darcy-Weisbach formula.*

Solution. Without air vessel, the maximum head loss in the delivery pipe, using Eq. (16.16)

$$(h_{Ld})_{max} = f\frac{L_d}{2gd_d}\left(\frac{\omega RD^2}{d_d^2}\right)^2$$

$$= 0.04\times\frac{45}{2\times 9.81\times 0.10}\left[\left(\frac{2\pi\times 55}{60}\times 0.20\right)\left(\frac{20}{10}\right)^2\right]^2 = 19.5 \text{ m.}$$

Since the head loss in friction has a parabolic variation, the average value over the delivery stroke

$$(H_{Ld})_{average} = \frac{2}{3}\times 19.5 = 13.0 \text{ m.}$$

With a large air vessel, water may be assumed to be discharged at a steady rate into the delivery pipe. The velocity of water in the delivery pipe is obtained using Eq. (16.1).

$$V_d = \frac{D^2}{d_d^2}\cdot\frac{2RN}{60} = \left(\frac{20}{10}\right)^2\cdot\frac{40\times 55}{100\times 60} = 1.47 \text{ m/s.}$$

Head loss in friction with air vessel fitted on the delivery side,

$$H_{Ld} = f\frac{L_s}{2gd_d}V_d^2$$

$$= 0.04\times\frac{45}{2\times 9.81\times 0.10}\times(1.47)^2 = 1.97 \text{ m.}$$

Volume of water pumped without slip, from Eq. (16.1),

$$Q = \frac{\pi D^2}{4}\cdot\frac{2RN}{60} = \frac{\pi}{4}(0.2)^2\times\frac{40}{100}\times\frac{55}{60}$$

$$= 1.15\times 15^{-2} \text{ m}^3/\text{s.}$$

Power saved by fitting a large air-vessel on the delivery side

$$= \gamma Q[(H_{Ld})_{average} - H_{Ld}]$$

$$= 9810\times 1.15\times 10^{-2}[13.0 - 1.97] \text{ W} = \mathbf{1.24\ kW}.$$

Example 16.10. *A single-acting reciprocating pump has a plunger diameter of 12 cm and stroke of 20 cm. The suction pipe has a diameter of 9 cm and is 4.3 m long. The friction factor may* *be taken as 0.04. If separation takes place at an absolute pressure of 1.5 m, determine the maximum speed at which the pump could be run without separation taking place. The barometric pressure head is 10.5 m of water, and the water level in the sump is 3 m below the pump cylinder axis. Also determine the power expended in overcoming friction when running at this speed.*

Solution. Data given :

$$D = 12 \text{ cm}, \qquad 2R = 24 \text{ cm},$$
$$d_s = 9 \text{ cm}, \qquad L_s = 4.3 \text{ m},$$
$$f = 0.04, \qquad \frac{p_v}{\gamma} = 1.5 \text{ m},$$
$$H_a = 10.5 \text{ m}, \qquad H_s = 3 \text{ m}.$$

Neglecting p_{sv} and the effect of length of connecting rod

$$(H_s)_{max} = \frac{pd - p_v - (\Delta p_i)_{max}}{\gamma}$$

$$\frac{(\Delta p_i)_{max}}{\gamma} = H_a - H_v - (H_s)_{max} = 10.5 - 1.5 - 3.0 = 6.0 \text{ m}$$

but $$(\Delta p_i)_{max} = \rho L_s \omega^2 R \frac{D^2}{d_s^2}$$

or $$\frac{(\Delta p_i)_{max}}{\gamma} = \frac{L_s \omega^2 R D^2 / d_s^2}{g}$$

or $$6 = \frac{4.3}{9.81} \times \omega^2 \times \frac{12}{100}\left(\frac{12}{9}\right)^2$$

from which $$\omega = 8 \text{ rad/sec},$$

and $$N = \frac{60 \times 8}{2\pi} = \mathbf{76.5\ r.p.m.}$$

$$(H_{Ls})_{max} = f \frac{L_s}{2gd_s}\left(\omega R \frac{D^2}{d_s^2}\right)^2$$

$$= 0.04 \times \frac{4.3 \times 100}{2 \times 9.81 \times 9}\left[8 \times \frac{12}{100}\left(\frac{12}{9}\right)^2\right]^2 = 0.283 \text{ m}.$$

Average head loss over the stroke

$$= \frac{2}{3} \times 0.283 = 0.189 \text{ m}.$$

Volume of water pumped per second,

$$Q = \frac{\pi}{4} D^2 \frac{2RN}{60} = \frac{\pi}{4} \times \left(\frac{12}{100}\right)^2 \times \frac{24 \times 76.5}{100 \times 60}$$
$$= 3.46 \times 10^{-3} \text{ m}^3\text{/s}.$$

Power expended in overcoming friction

$$= \gamma Q (H_{Ls})_{average} = 9810 \times 3.46 \times 10^{-3} \times 0.189 \text{ W} = \mathbf{6.4\ W}.$$

Example 16.11. *A single cylinder double acting reciprocating pump of 300 mm bore and 400 mm stroke runs at 35 r.p.m. It discharges 31 litres / sec of water under a total head of 15 m. What will be the volumetric efficiency, work done per second and power required if the overall efficiency of the pump is 72% ? Also work out the mechanical efficiency.*

Solution. Using Eq. (16.6 *b*), the average theoretical discharge of a double acting pump

$$Q = 2\left(\frac{\pi D^2}{4} . \frac{2RN}{60}\right) = 2\left(\frac{\pi \times 0.3^2}{4} . \frac{0.4 \times 35}{60}\right)$$

$$= 0.03297 \text{ m}^3/\text{s} = 32.97 \text{ lit/sec}$$

Volumetric efficiency $= \dfrac{\text{Actual discharge}}{\text{Theoretical discharge}} = \dfrac{31.00}{32.97} = \mathbf{0.94}$ or **94%**

Mechanical efficiency $= \dfrac{\text{Overall efficiency}}{\text{Volumetric efficiency}} = \dfrac{72}{94} = \mathbf{0.766}$ or **76.6%**

Work done per second $= \dfrac{\gamma QH}{1000} \text{ (kW)} = \dfrac{9810 \times 0.031 \times 15}{1000} = \mathbf{4.56\ kW}$

Power required $= \dfrac{4.56}{0.72} = \mathbf{6.336\ kW}$.

Example 16.12. *A single cylinder single acting pump of 200 mm bore and 400 mm stroke runs at 30 r.p.m. The suction and delivery lifts are respectively 3.0 m and 20 m. The suction pipe of 100 mm diameter is 5.0 m long while the delivery pipe of the same size is 20 m long. If the atmospheric pressure is equal to 10.3 m of water, compute the pressures (absolute) at the beginning, middle and end of the delivery stroke. Also determine, the work done during the suction and delivery strokes, and the power required. Assume Darcy-Weisbach friction factor of 0.03 for both the pipes.*

Solution. The following data is given :

$D = 200$ mm ; $2R = 400$ mm ; $N = 30$ r.p.m.

$d_s = 100$ mm ; $L_s = 5.0$ m ; $H_s = 3.0$ m

$d_d = 100$ mm ; $L_d = 20.0$ m ; $H_d = 20.0$ m

$H_a = 10.3$ m of water ; $f = 0.03$ for both the pipes.

Angular velocity of the crank,

$$\omega = \frac{2\pi N}{60} = \frac{2\pi \times 30}{60} = 3.142 \text{ rad./sec}$$

Using Eq. (16.12 *b*), the maximum acceleration head during the delivery stroke

$$H_{id} = \left(\frac{\Delta p_{is}}{\gamma}\right)_{max} = \frac{L_d}{g}\left(\frac{D}{d_d}\right)^2 \omega^2 R$$

$$= \frac{20}{9.81}\left(\frac{200}{100}\right)^2 \times (3.142)^2 \times 0.20 = 16.108 \text{ m}$$

Maximum frictional head loss in the delivery pipe is obtainable from Eq. (16.16 *a*)

$$(h_{Ld})_{max} = f\frac{L_d}{2gd_d}\left(\frac{\omega R D^2}{d_d^2}\right)^2$$

$$= \frac{0.03 \times 20}{2 \times 9.81 \times 0.100}\left(3.142 \times 0.20 \times \frac{200^2}{100^2}\right)^2 = 1.934 \text{ m}$$

$$(h_{Ld})_{av} = \text{Average frictional head loss} = \left(\frac{2}{3} 1.934\right) = 1.289 \text{ m}$$

(*i*) Pressure head at the beginning of the delivery stroke

$$= H_a + H_d + H_{id}$$

$= 10.30 + 20.00 + 16.108 =$ **46.41 m of water** (abs.)

(*ii*) Pressure head at the middle of the delivery stroke

$$= H_a + H_d + (h_{Ld})_{max}$$

$$= 10.30 + 20.00 + 1.934 = \mathbf{32.234\ m\ of\ water}\ \text{(abs.)}$$

(*iii*) Pressure at the end of the delivery stroke

$$= H_a + H_d - H_{id}$$

$$= 10.30 + 20.00 - 16.108 = \mathbf{14.192\ m\ of\ water}\ \text{(abs.)}$$

Maximum head loss in the suction pipe

$$(h_{Ls})_{max} = f\frac{L_s}{2gd_s}\left(\frac{\omega RD^2}{d_s^2}\right) = \frac{0.03 \times 5.0}{2 \times 9.81 \times 0.10}\left(3.142 \times 0.20 \times \frac{200^2}{100^2}\right)^2$$

$$= 0.484 \text{ m of water}$$

Average frictional head loss in the suction pipe

$$(h_{Ls})_{av} = \frac{2}{3} \times 0.484 = 0.323 \text{ m.}$$

(*iv*) Work done during suction stroke

$$= \gamma \frac{\pi D^2}{4} (2R)\,[H_d + (h_{Ls})_{av}]$$

$$= 9810 \times \frac{\pi \times 0.20^2}{4} \times (0.40)\,[3.0 + 0.323] = 408.9 \text{ Nm}$$

(*v*) Work done during delivery stroke

$$= \gamma \frac{\pi D^2}{4} (2R)\,[H_d + (h_{Ld})_{av}]$$

$$= 9810 \times \frac{\pi \times 0.20^2}{4}\ (0.40)\ [20 + 1.289] = 2619.3 \text{ Nm}$$

(*vi*) Work done during the delivery stroke being larger, the power required by the pump

$$= \frac{\text{Work done during the stroke}}{\text{Time taken in sec}}$$

$$= \frac{2619.3}{1} = 2619.3 \text{ W} = \mathbf{2.619\ kW.}$$

PROBLEMS

16.1. For a single acting reciprocating pump, piston diameter is 150 mm, stroke length is 300 mm, rotational speed is 50 r.p.m. The pump is required to lift water to the height of 18 m. Determine the theoretical discharge. If the actual discharge is 4.0 lit/sec, and the mechanical efficiency 80%, determine the volumetric efficiency, slip, theoretical power and the actual power required.

16.2. A single acting reciprocating pump has a piston diameter of 125 mm and stroke of 300 mm. The length of suction pipe is 10 m and diameter 75 mm. Find acceleration head at the beginning, middle and end of suction stroke. If the suction head is 3.0 m, determine the pressure head in the cylinder at the beginning, middle, and end of stroke when the pump runs at 30 r.p.m. Assume atmospheric pressure head of 10.23 m.

16.3. A single acting reciprocating has a stroke of length 300 mm and a piston of 125 mm diameter. A 75 mm diameter suction pipe of length 5.0 m is used against a suction head of 3.0 m. What is the maximum speed at which it can be run without causing separation during suction stroke ? Take safe minimum pressure head of 2.0 m (abs.) and the atmospheric pressure head as 10.23 m (abs.).

16.4. The following particulars of a single cylinder single acting reciprocating pump are known :

$R = 250$ mm ; $D = 125$ mm ; $H_{atm} = 10.2$ m

$L_s = 5.0$ m ; $d_s = 100$ mm ; $H_s = 3.0$ m

$L_d = 15.0$ m ; $d = 100$ mm ; $H_d = 10.0$ m

Safe minimum pressure head = 2.4 m.

Darcy-Weisbach friction factor for suction and delivery pipes may be taken as 0.02. Neglecting slip, determine :

(*i*) minimum permissible speed, and

(*ii*) energy required to drive the pump, if an air vessel is provided on the delivery side very close to the cylinder.

16.5. A single cylinder double acting reciprocating pump has the following details :

$R = 200$ mm ; $D = 200$ mm

$N = 120$ r.p.m. ; $d_s = 150$ mm

If an air vessel is fitted on the suction side, determine the crank angle at which there is no flow to the air vessel or from the air vessel.

16.6. A single acting reciprocating pump has the following particulars :

$D = 150$ mm ; $R = 100$ mm (stroke length = 200 mm) ; $N = 60$ r.p.m.

$d_s = 75$ mm ; $L_s = 10$ m ; $H_s = 3.0$ m

$d = 50$ mm ; $L_d = 40$ m ; $H_d = 30$ m

Air vessels have been provided at 3.0 m and 5.0 m on the suction and delivery sides. Determine the power required if the pump efficiency is 85%. Assume a friction factor of 0.03 for both pipes.

Answers

16.1. 4.416 l/s ; 90.58% ; 9.42% ; 0.778 kW ; 0.881 kW

16.2. 4.195 m, 0 m—4.195 m ; 3.035 m abs., 7.23 m abs., 11.425 m abs.

16.3. 47.38 r.p.m.

16.4. (*i*) 46.86 r.p.m. (*ii*) 0.621 kW

16.5. 39.47°

16.6. 1.619 kW.

17

Centrifugal Pumps

17.0. Rotodynamic Pumps

A rotodynamic pump is essentially a reaction turbine "in reverse"; mechanical energy is transferred from the rotor to the fluid. Like turbines, pumps are classified according to the direction of fluid path through them. They are thus radial flow (or centrifugal), axial flow (propeller type) and combination of the two called mixed-flow. For high heads and low discharges centrifugal pump with two or more stages (two more impellers in series) is best suited. For large discharges and small heads, the axial-flow pump is more suitable. The mixed-flow pump is used for medium head and medium discharge.

The equations developed in article on turbomachines apply just as well to pumps and blowers as to turbines.

17.1. Introduction

The principal parts of a centrifugal pump are the impeller mounted on a shaft and the casing which surrounds it. A centrifugal pump thus consists of an impeller rotating inside a spiral or volute casing. Liquid is admitted to the impeller in an axial direction, through a central opening in its side called the 'eye'. It then flows radially outward and is discharged around the entire cirumference into a casing. As the liquid flows through the rotating impeller energy is imparted to the fluid which results in increase in both the pressure and the kinetic energies. Since a large part of the energy of liquid leaving the impeller is kinetic, it is necessary to reduce the absolute velocity of flow and transform a large portion of this velocity head, into pressure head.

The name of the pump '*centrifugal*' is derived from the fact that the discharge of the liquid from the rotating impeller is due to the centrifugal head created in it. In chapter 2, art. 2.11, which deals with fluid rotation about a vertical axis, it was shown that when a liquid mass is rotated in a vessel, there results a pressure rise throughout the mass, the rise at any point being proportional to the square of angular velocity and square of the distance of the point from the axis of rotation. This high pressure near the impeller outlet forces the liquid out and causes it to rise in the delivery pipe. At the eye of the impeller a partial vacuum is created, and the atmospheric pressure in the sump forces the liquid through the suction pipe to replace the liquid that is being discharged from the impeller.

17.1.1. Classification

Centrifugal pumps are divided mainly into two categories according to the type of casing provided :

1. Volute pumps, and
2. Diffuser or turbine pumps.

In volute type of centrifugal pumps the impeller is surrounded by a spiral casing, the outer boundary of which is a curve called volute. The absolute velocity of the fluid leaving the impeller is

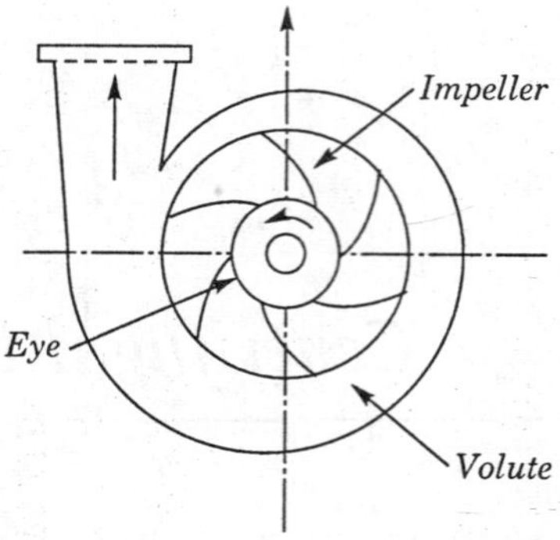

(a) Volute type.

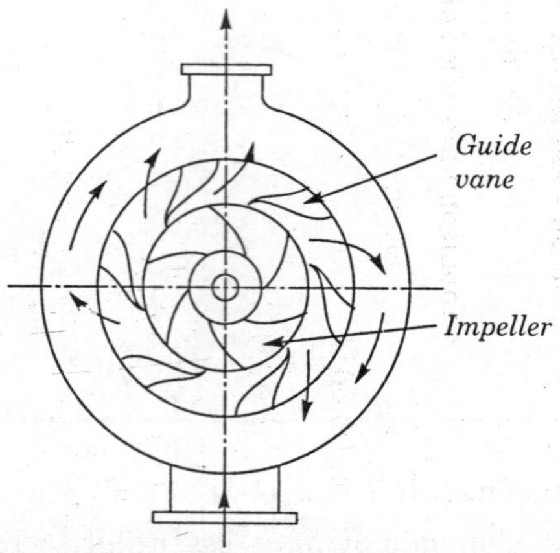

(b) Diffuser (or turbine) type.

Fig. 17.1. Centrifugal pumps.

reduced in the volute casing. By passing the fluid through the gradually increasing cross-sectional area of the volute casing, a part of its kinetic energy gets converted into the pressure energy. Fig. 17.1 (*a*) shows a volute type centrifugal pump.

In the diffuser pump, shown in Fig. 17.1 (*b*), the impeller is surrounded by a series of stationary guide vanes mounted on a diffuser ring. The guide vanes also known as diffuser vanes provide gradually enlarging passages so as to result in gradual reduction in velocity. Because of the superficial resemblance to a reaction turbine, this type of pump is often called a turbine pump. A diffuser pump is considered more efficient because of gradual reduction of velocity through the guide vanes that results in less energy loss.

Yet there is another type of pump which uses a vortex or whirlpool chamber. In this type an annular space is provided between the volute and impeller. This arrangement arrests the formation of eddies and gives an improved performance.

A well designed diffuser is capable of converting as much as 75% of the kinetic energy at the impeller outlet into pressure energy.

The efficiency of a pump is generally less than that of a turbine. The flow passages in a centrifugal pump are diverging (flow is decelerating) whereas those of a turbine are coverging (flow is accelerating). The losses in a pump are more because of possibility of flow separation and formation of eddies in a diverging flow.

Pumps are often further classified as 'single-suction' or 'double-suction' pumps according to whether the impeller receives water on one side or on both sides. For handling large quantities of liquid against low heads, it is more economical to use a double-suction pump, drawing the liquid from a common suction pipe. Impeller of a double-suction pump, may be considered as equivalent of two single-suction impellers placed back to back. The head developed by it is practically the same as that developed by a single-suction impeller of the same diameter and running at the same rotational speed, but its discharge is the combined discharges of both sides. It has the advantage that the side thrusts on each side, resulting from the change in direction of flow are hydraulically balanced.

17.1.2. Single and Multi-stage Pumps

In a single-stage pump a single impeller is used to develop pressure necessary to lift required discharge against the given head. The practical limiting head against which a single impeller may be used is about 38 metres. For higher head it is not economical to use a single impeller because then either a high rotational speed or an impeller of large diameter must be used. In either case, high mechanical stresses result and poor efficiency is obtained because of disc friction and leakage losses. Because of the clearance space between the impeller and the surrounding casing, liquid which has passed through the impeller may flow in limited quantity back to the suction side, thereby decreasing the efficiency of the pump. This is known as the leakage loss. Under very high heads, the leakage loss may become considerably high due to great difference in pressure between the discharge and suction sides of the impeller.

For high heads it is customary to use two or more impellers in series, so placed and connected that the water discharged with increased pressure from one impeller flows to the suction opening of the second and so on. The quantity pumped is the quantity passing a single impeller, but the total pressure head created by the pump is the sum of the pressure heads developed by each impeller. Such a pump is known as a multi-stage pump, the several impellers and their connecting passages being contained in a single casing. Pumps using upto ten stages have been built, but it is better from mechanical construction point of view to limit the number of stages to five and use two such pumps in series, if required.

17.1.3. Pumps in Series and Parallel

If a single pump is not capable of providing the required discharge against given head, then two pumps connected in series may be used. When more quantity of liquid is to be handled, two

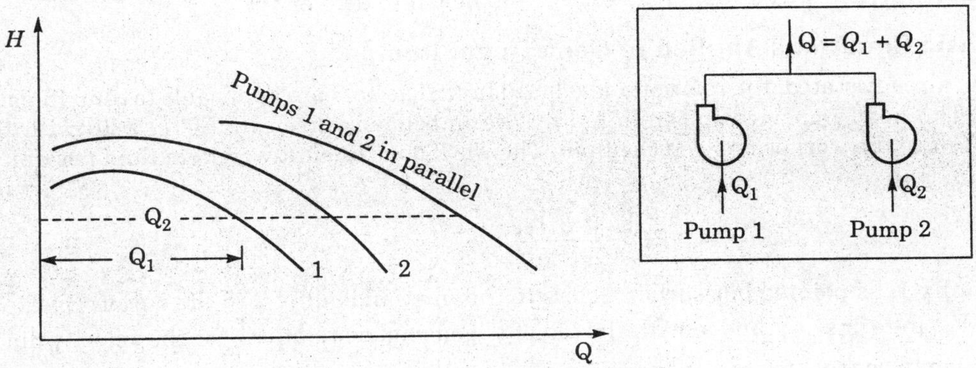

Fig. 17.2. Operation of pumps in parallel.

centrifugal pumps are to be connected in parallel in which case the discharge of each pump will be fed into a common main as shown in Fig. 17.2. The head-discharge characteristic of each pump alongwith their operation in parallel are shown in a separate figure. For parallel operation

$$Q_1 + Q_2 = Q$$

and, $$H_1 = H_2 = H$$

When the liquid is to be lifted higher up beyond the capacity of a single pump, the two pumps are required to be connected in series, as shown in Fig. 17.3. The discharge from the first pump will be fed into the other. The energy supplied by each pump is thus added to result in an increased head, thus for operation of

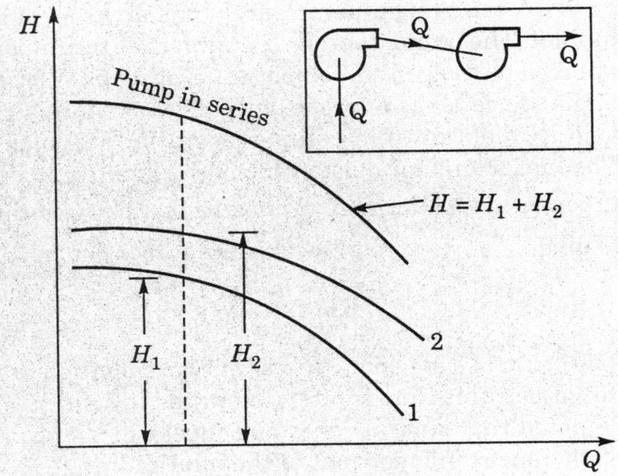

Fig. 17.3. Operation of pumps in series.

pumps in series :

$$H_1 + H_2 = H \quad \text{and} \quad Q_1 = Q_2 = Q.$$

17.1.4. Suction and Delivery Pipes

The pipe which connects the sump (sump is a low level reservoir from where the liquid is required to be pumped) and the inlet of the pump is called the suction pipe simply because of the existence of negative pressures in it. In order to avoid cavitation, the negative pressure at the inlet to the pump is kept within permissible limits. This restriction on the inlet pressure limits the height of suction pipe above the liquid level in the sump and requires that the losses in the suction pipe be as small as possible. For this purpose, suction pipe is quite often kept larger in diameter than the delivery pipe, and bends should be avoided as far as possible. The velocity in the suction pipe being thus small would in turn keep the friction losses on the lower side. The lower end of the suction pipe is provided with a foot valve which is immersed in the liquid and serves to prevent the liquid from flowing back into the sump when the pump is not working. A foot valve is a non-return valve which opens only in the upward direction. It is also provided with a strainer to prevent large size impurities and debris to reach impeller of the pump and cause damage to it.

The delivery pipe is connected at its lower end to the outlet of the pump and delivers the liquid to the required height. A delivery valve is provided on the delivery pipe near the outlet of the pump. The function of this valve is to control the flow into delivery pipe.

17.2. Basic Equations Applied to Centrifugal Pumps

As already stated, the relations developed in Article 15.3 are applicable to centrifugal pumps as well. The head imparted to the liquid by the impeller is given by Eq. (15.30) with the sign reversed since the work is done on the fluid. The work done on unit weight of fluid per unit time is thus

$$H = \frac{u_2 V_{u_2} - u_1 V_{u_1}}{g} \qquad \text{...(17.1)}$$

Suffix 1, as pointed out earlier, refers to the inlet and suffix 2 to the outlet. By expressing V_{u_2} and V_{u_1} in terms of V and w with the aid of trigonometric relations for the velocity diagrams., Eq. (17.1) may be transformed to the following form :

$$H = \frac{1}{2g}\left[(V_2^2 - V_1^2) + (u_2^2 - u_1^2) + (w_2^2 - w_1^2)\right] \qquad \text{...(17.2)}$$

A centrifugal pump rarely has any sort of guide vanes at inlet. The liquid, therefore, approaches the impeller without appreciable tangential component of absolute velocity. The inlet angle of the blade is so designed as to produce a right-angled velocity diagram at inlet under operation at design conditions. The initial angular momentum of the liquid may, therefore, be taken as zero, and we may substitute $V_{u_1} = 0$ in the Eq. (17.1) to give

$$H = \frac{u_2 V_{u_2}}{g} \qquad \text{...(17.3)}$$

Eq. (17.3) which represents work done on unit weight of liquid per unit time is often known as the *Euler head*. It may be noted that since u_1 does not appear in the expression, the work done is independent of the inlet radius.

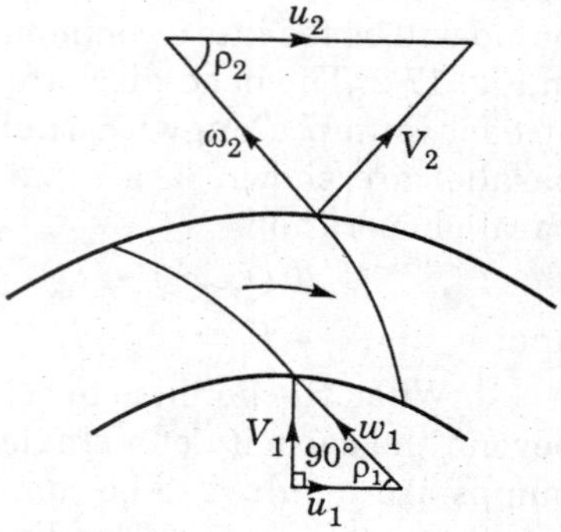

Fig. 17.4. Inlet and outlet velocity diagrams.

17.3. Velocity Diagrams at Outlet

The performance of a pump, among other things, depends upon the outlet angle of the impeller blades. Three typical cases arise according to whether the blade angle at outlet β_2 is equal to, greater or less than 90°. Fig. 17.5 shows these three cases.

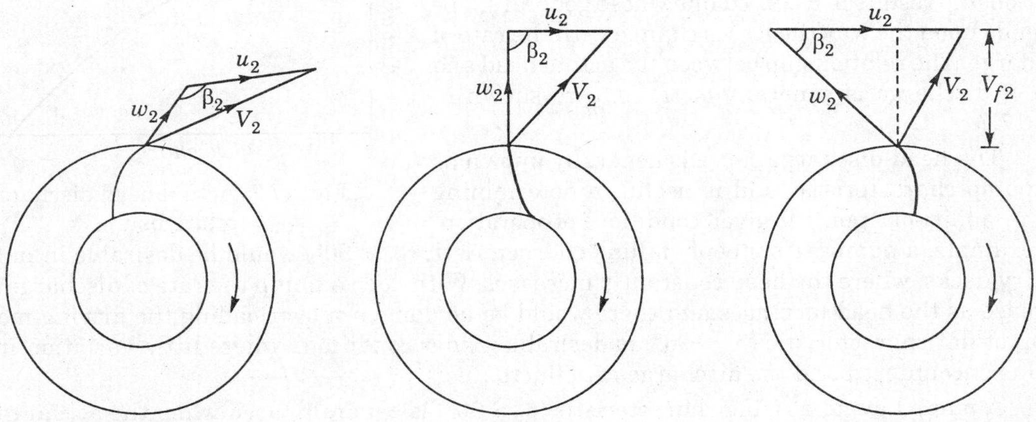

$\beta_2 > 90°$ (a) Forward-facing blades (Slow runner)

$\beta_2 = 90°$ (b) Radial blades (Medium runner)

$\beta_2 < 90°$ (c) Backward facing blades (Fast runner)

Fig. 17.5. Velocity diagrams at outlet.

Fig. 17.5 (*a*) shows a forward-facing blade ; the blade outlet faces the direction of rotation such that $\beta_2 > 90°$. The impeller having such blades is called a slow runner. In Fig. 17.5 (*b*) is shown a radial blade for which the blade angle at outlet $\beta = 90°$. The liquid leaves the impeller with relative velocity in a radial direction. Such blades are suitable for medium speed runners. In case the outlet tip of the blade curves in a direction opposite to the direction of rotation, as shown in Fig. 17.5 (*c*), the blade is called a backward facing blade for which $\beta_2 < 90°$. The impeller using backward-facing blades is known as a fast runner. The pumps run by electric motor usually have high speed and, therefore, represent the case shown in Fig. 17.5 (*c*).

If the fluid is assumed frictionless, the difference of total heads across the pump would be $u_2 V_{u_2}/g$. Using the velocity diagram of Fig. 17.5 (*c*), we may write

$$V_{u_2} = u_2 - V_{f_2} \cot \beta_2 .$$

If Q represents the rate of discharge through the pump, and A_2 the area of flow perpendicular to V_{f_2}, then $V_{f_2} = Q/A_2$.

The ideal head then is,

$$H = \frac{u_2 V_{u_2}}{g} = \frac{u_2}{g}\left(u_2 - \frac{Q}{A_2} \cot \beta_2\right) \quad ...(17.3)$$

The blade velocity u_2 is proportional to the rotational speed n and so the ideal head equals

$$H = An^2 - BnQ \quad ...(17.4)$$

where A and B are constants. Thus for a fixed speed n the variation of H with Q is linear as shown in Fig. 17.6. Three typical cases arise according to whether β_2 is equal to, greater or less than 90°. For $\beta_2 = 90°$, H is constant regardless of the rate of discharge. For $\beta_2 < 90°$, H increases with Q whereas for $\beta_2 > 90°$, H decreases with increase in Q.

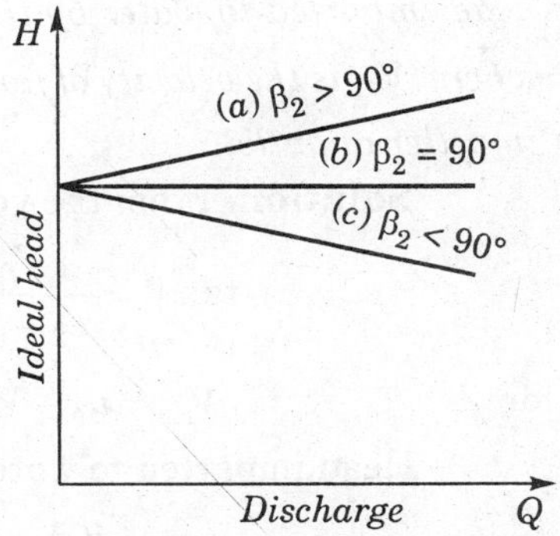

Fig. 17.6. Head discharge relationship.

It is however of greater interest and importance to know the relationship between Q and the head actually developed by the pump. This relationship in addition to angle β will be affected by the hydraulic losses which must be considered. The graphical plotting of this relationship results in a curved line whose general shape depends upon the way the losses change with the rate of discharge. The relationship between the actual head and rate of discharge is generally of the nature shown in Fig. 17.7.

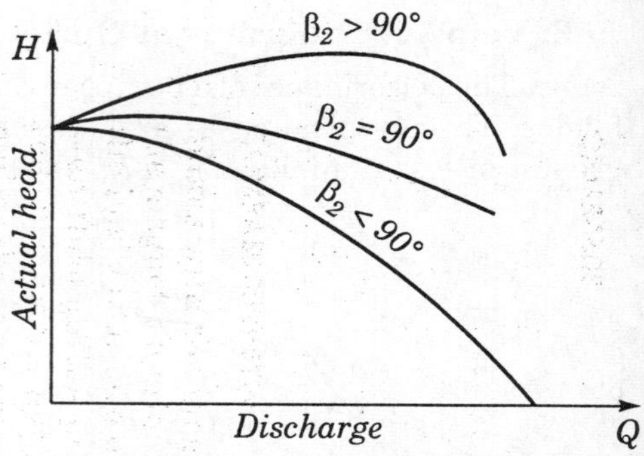

Fig. 17.7. Actual head discharge relationship.

The head-discharge curve is generally known as the pump characteristic, and is useful in determining the suitability of a pump to given conditions of operation. For example, a pump with steeply falling characteristic ($\beta_2 < 90°$) would be desirable in pumping out dry-docks, where the head constantly increases. With such a pump the rate of discharge would decrease as the head increases and there would be no danger of overloading the driving motor. A pump of flat characteristic ($\beta_2 = 90°$) is desirable under conditions where little variation in total head is encountered and the discharge rate fluctuates.

A pump having a rising characteristic ($\beta_2 > 90°$) is generally used when the actual lift is of small and constant amount, and the friction head large and variable with the rate of discharge. This would be the case of a sump required to supply a long pipe system where the demand for water varies but a constant pressure is desired.

Example 17.1. *A centrifugal pump delivers water at the rate of 40 litres / sec for a total head of 30 m. Find the output of the pump. If the overall efficiency of the set consisting of the pump and the electric motor is 75%, find the power required to drive the motor.*

Solution. (*a*) Pump output

$$= \gamma QH = 9810 \times \frac{40}{1000} \times 30 \ (W)$$
$$= 11.77 \times 10^3 \ W = \mathbf{11.77 \ kW.}$$

(*b*) Power required to drive the motor

$$= \frac{\text{Pump output}}{\text{Overall efficiency}} = \frac{11.77}{0.75} = \mathbf{15.70 \ kW.}$$

Example 17.2. *The impeller of a centrifugal pump has blades which are curved backwards and make an acute angle β with the tangent at the outlet periphery. The velocity of flow is constant and the absolute velocity of water at inlet is radial. Neglecting frictional losses show that the ratio of rise in pressure head to the head imparted to water by the impeller is $\frac{1}{2}\,[1 + V_{f_2}(\cot\beta)/u_2]$, where V_{f_2} is the velocity of flow and u_2 is the peripheral velocity of impeller at outlet.*

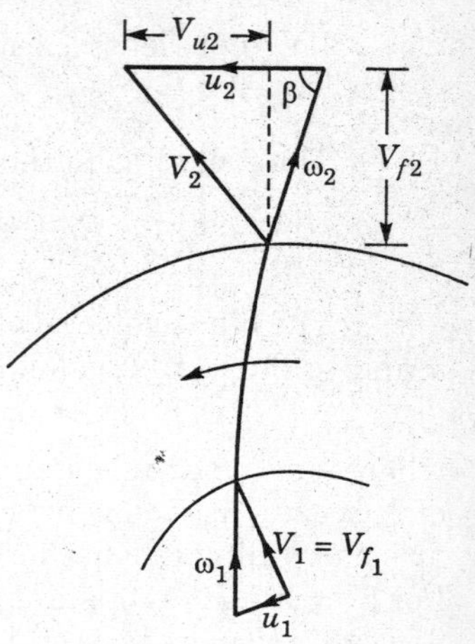

Solution. From the velocity diagram at outlet,

$$\tan\beta = \frac{V_{f_2}}{u_2 - V_{u_2}}$$

or $$V_{u_2} = u_2 - V_{f_2}\cot\beta\,; \quad V_2^{\,2} = V_{f_2}^2 + V_{u_2}^2$$

Head imparted to water by the impeller for radial entry

$$H = \frac{u_2 V_{u_2}}{g} = \frac{u_2(u_2 - V_{f_2}\cot\beta)}{g}$$

Applying Bernoulli's equation to the inlet and outlet of impeller, neglecting losses

$$\frac{p_1}{\gamma} + \frac{V_1^2}{2g} + \frac{u_2 V_{u_2}}{g} = \frac{p_2}{\gamma} + \frac{V_2^2}{2g}$$

Rise in pressure head,

$$\frac{p_2 - p_1}{\gamma} = \frac{u_2 V_{u_2}}{g} + \frac{V_1^2 - V_2^2}{2g}$$

Since velocity of flow is constant

$$V_{f_1} V_{f_2}$$

Absolute velocity at inlet being radial,

$$V_1 = V_{f_1} = V_{f_2}$$

$$V_2^2 = V_{f_2}{}^2 + (u_2 - V_{f_2} \cot \beta)^2$$

Substituting values for V_1 and V_2 the rise in pressure head,

$$\frac{p_2 - p_1}{\gamma} = \frac{u_2 V_{u_2}}{g} + \frac{V_{f_2}{}^2 - V_{f_2}{}^2 - (u_2 - V_{f_2} \cot \beta)^2}{2g}$$

$$= \frac{1}{2g}(u_2{}^2 - V_{f_2}{}^2 \cot^2 \beta)$$

$\therefore$ Ratio of rise in pressure head to head imparted by impeller

$$\frac{p_2 - p_1}{\gamma H} = \frac{\frac{1}{2g}(u_2{}^2 - V_{f_1}{}^2 \cot^2 \beta)}{\frac{u_2(u_2 - V_{f_2} \cot \beta)}{g}} = \frac{1}{2}\left(1 + \frac{V_{f_2} \cot \beta}{u_2}\right)$$

Example 17.3. *A centrifugal pump rotates at 1000 r.p.m. and delivers 300 lit/sec of water. The impeller has a diameter of 35 cm and a width of 5 cm at the periphery. The blades tip-angles are inclined backwards 60° from the radius (the blade angle = 30°). Determine the velocity and direction of water as it leaves the impeller.*

Solution. Referring to Fig. 17.5, it can be seen that this problem represents category shown in figure (*c*) for which $\beta_2 < 90°$.

$$u_2 = r_2 \omega = \frac{35}{2 \times 100} \times \frac{2\pi \times 1000}{60} = 18.33 \text{ m/s.}$$

$$V_{f_2} = \frac{Q}{2\pi r_2 b} = \frac{0.3 \times 2 \times 100 \times 100}{2\pi \times 35 \times 5} = 5.45 \text{ m/s}$$

$$V_{u_2} = u_2 - V_{f_2} \cot \beta_2 = 18.33 - 5.45 \cot 30°$$

$$= 8.88 \text{ m/s}$$

$$V_2 = \sqrt{V_{u_2}{}^2 + V_{f_2}{}^2} = \sqrt{(8.88)^2 + (5.45)^2} = \mathbf{10.42 \text{ m/s.}}$$

Direction of the absolute velocity

$$\tan \alpha_2 = \frac{V_{f_2}}{V_{u_2}} = \frac{5.45}{8.88} = 0.613$$

$$\alpha_2 = \mathbf{31.5°.}$$

Example 17.4. *A constant speed test of a centrifugal pump resulted in the following relationship :*

$$H = 43.8 + 251\, Q - 3768\, Q^2$$

where H is the total head in metre and Q the discharge in m³/s. The pump is to be used to deliver water through a pipe line 1 km long and 35 cm diameter, the static lift being 25.8 m. Calculate the operating head and the pump discharge taking the friction factor as 0.03 and neglecting the velocity heads. For the particular H versus Q point at which the pump operates ; determine the power required to drive the pump if the overall efficiency is 72%.

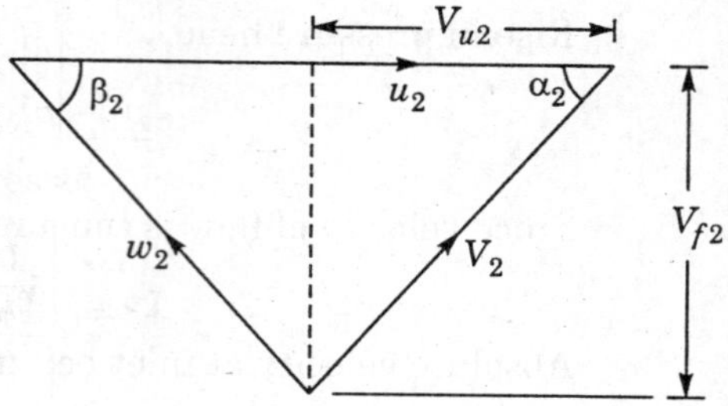

Solution. Head developed by pump (H)

$$= \text{Static lift + Head lost in pipe friction} = 25.8 + \frac{fL}{2gd}\left(\frac{4Q}{\pi d^2}\right)^2$$

$$= 25.8 + \frac{0.03 \times 1000}{2 \times 9.81 \times 0.35}\left(\frac{4Q}{\pi \times 0.35 \times 0.35}\right)^2 = 25.8 + 475\, Q^2$$

Also $\quad H = 43.8 + 251\, Q - 3768\, Q^2$

From the two expressions for head, we obtain

or $\quad 4235\, Q^2 - 251\, Q - 18.0 = 0$

From which the pump discharge

$$Q = 0.097 \text{ m}^3/\text{s} = 97 \text{ lit/sec.}$$

Operating head $\quad H = 25.8 + 475\,(0.097)^2 = 30.265$ m

Power output of the pump

$$P_0 = \gamma QH = 9810 \times 0.097 \times 30.265 = 28738.5 \text{ W}$$
$$= 28.74 \text{ kW}$$

Power required to drive the pump

$$= \frac{P_0}{\eta_0} = \frac{28.74}{0.72} = \mathbf{39.9\ kW.}$$

17.4. Loss of Head Due to Changed Discharge

A centrifugal pump attains its peak efficiency when it operates at the designed speed. It is only when the pump runs at this speed and delivers the normal discharge that the flow enters the impeller without shock. But when the pump delivers either a reduced flow or an increased one with respect to the normal discharge, the flow does not enter the impeller tangentially at the inlet tip. In other words, it enters with a shock and results in an additional loss of energy. The loss of head so caused at the inlet due to changed discharge decreases the pump efficiency.

Fig. 17.8 shows the inlet velocity triangles for reduced and increased discharges.

Considering the figure, *abc* represents the velocity triangle for normal-discharge condition when the flow enters tangentially at the inlet tip (shockless entry). But when the discharge is reduced, the radical velocity of flow reduces to V_f shown by *bd*, speed of pump remaining the same. The new velocity triangle at inlet is *abd* and the new relative velocity is *ad* which will obviously not be tangential to the blade tip and a shock will be experienced. But since the blade angle is β_1, the new velocity triangle will be *ebd*. Comparison of the new velocity triangle with the one corresponding to

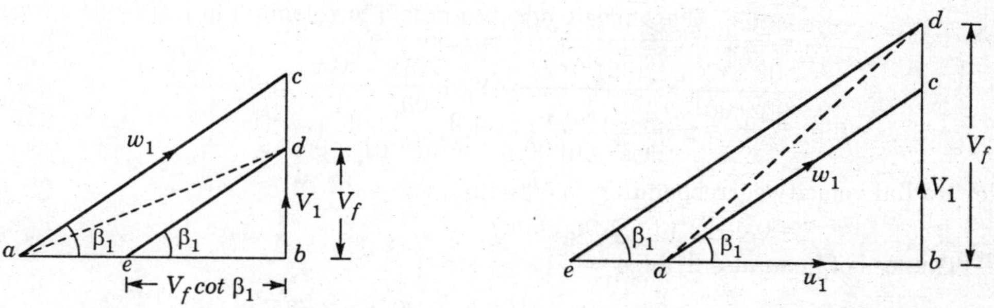

(*i*) For reduced discharge (*ii*) For increased discharge

Fig. 17.8. Velocity triangles at inlet for changed flow conditions.

the normal discharge, it can be seen that the tangential velocity suddenly changes from u_1 to $V_f \cot \beta_1$ resulting in a shock.

.ange in tangential velocity $= u_1 - V_f \cot \beta_1$

$$\text{Loss of head at inlet} = \frac{(\text{Change in velocity})^2}{2g} = \frac{(u_1 - V_f \cot \beta_1)^2}{2g} \qquad ...(17.5)$$

For the case when the discharge is increased above the normal value, the second figure shows the inlet velocity diagram in which the change in velocity will be $(V_f \cot \beta_1 - u_1)$.

Example 17.5. *A centrifugal pump has an impeller of 50 cm outer diameter and 25 cm inner diameter. When running at 6000 r.p.m. the pump discharges 130 litres of water per second against a gross head of 12 m. At this discharge, water enters the impeller without shock. The vane angle at outlet is 45° to the tangent at outlet, and the area of flow which is constant from inlet to outlet of the impeller is 0.61 m². Determine :*

(*i*) *the manometric efficiency of the pump and*

(*ii*) *the loss of head at inlet to the impeller when the discharge is reduced by 50%, the speed of rotation being unchanged.*

Solution. (*i*) Manometric efficiency

$$\eta_m = gH_m / u_2 \,.\, V_{u_2}$$

$$u_2 = \frac{\pi D_2 n}{60} = \frac{\pi \times 0.5 \times 600}{60} = 15.70 \text{ m/s}$$

$$V_{f_2} = \frac{Q}{A_{f_2}} = \frac{0.13}{0.061} = 2.13 \text{ m/s}$$

$$V_{u_2} = u_2 - V_{f_2} \cot \beta_2 = 15.7 - \frac{2.13}{\tan 45°} = 13.57 \text{ m/s}$$

$$\therefore \qquad \eta_m = \frac{9.81 \times 12}{15.7 \times 13.57} = \mathbf{0.5525} \quad \textbf{or} \quad \mathbf{55.25\%.}$$

(*ii*) Loss of head at entrance due to reduced flow

$$= (u_1 - V_f \cot \beta_1)^2 / 2g$$

When the discharge is reduced by 50%, there will be shock losses.

$$Q = V_{f_1} \,.\, A_{f_1} = V_{f_2} \,.\, A_{f_2}$$

$$V_f = V_{f_1} = V_{f_2} = 2.13 \text{ m/s}$$

$$u_1 = \frac{\pi D_1 n}{60} = \frac{\pi \times 0.25 \times 600}{60} = 7.85 \text{ m/s}$$

$$\tan \beta_1 = \frac{V_{f_1}}{u_1} = \frac{2.13}{7.85} = 0.2713 \text{ ; } \cot \beta_1 = 3.685$$

New radial velocity corresponding to 50% flow

$$= 0.5 \times 2.13 = 1.065 \text{ m/s}$$

∴ The loss of head at entrance

$$= (u_1 - V_f \cot \beta_1)^2/2g = \frac{(7.85 - 1.065 \times 3.685)^2}{2 \times 9.81} = \mathbf{0.785 \text{ m.}}$$

Example 17.6. *A centrifugal pump lifts water against a static head of 32.067 m of which 3.054 m is suction lift. The suction and delivery pipes are both 12.7 cm in diameter. The loss of head in suction pipe is 1.07 m of water and in delivery pipe is 5.955 m of water. The impeller is 30.54 cm in diameter and 2.54 cm wide at the outer diameter. It revolves at 1450 r.p.m. and the blade angle at exit is 35°. The manometric efficiency of the pump is 80% and its overall efficiency is 68%. Determine :*

(a) The discharge of the pump

(b) The power required to drive the pump, and

(c) The pressures at the two branches of the pipe.

Neglect the effect of vane thickness on the area of the flow.

Solution.

(*i*) Manometric efficiency, $\eta_m = \dfrac{gH}{u_2 V_{u_2}}$

Overall efficiency, $\eta_0 = \dfrac{\gamma QH}{\text{Power required to drive the pump}}$

$$u_2 = \frac{\pi Dn}{60} = \frac{\pi \times 0.3054 \times 1450}{60} = 23.18 \text{ m/s}$$

Neglecting the velocity head in the delivery pipe

$$H = (h_s + h_d) + h_{fs} + h_{fd}$$

$$= 32.067 + 1.07 + 5.955 = 39.09 \text{ m}$$

$$V_{u2} = \frac{gH}{u_2 \eta_m} = \frac{9.81 \times 39.09}{23.18 \times 0.08} = 20.68 \text{ m/s}$$

But $V_{u_2} = u_2 - V_{f_2} \cot \beta_2, \quad \therefore \quad V_{f_2} = (u_2 - V_{u_2}) \tan \beta_2$

or $V_{f_2} = (23.18 - 20.68) \tan 35° = 1.75 \text{ m/s}$

∴ Discharge $Q = \pi D b V_{f2} = \pi \times 0.3054 \times \dfrac{2.54}{100} \times 1.75$

$$= \mathbf{0.0426 \text{ m}^2\text{/s} = 42.6 \text{ lit/sec.}}$$

(*ii*) Power required to drive the pump

$$= \frac{\gamma QH}{\eta_0} = \frac{9810 \times 0.0426 \times 39.09}{0.68}$$

$$= 24023.4 \text{ N-m/s} = 24.02 \text{ kW.}$$

(*iii*) Velocity in suction and delivery pipes

$$V = \frac{4Q}{\pi d^2} = \frac{4 \times 0.0426}{\pi \times (12.7/100)^2} = 3.36 \text{ m/s}$$

Pressure in the suction pipe just before impeller inlet

$$\frac{P_s}{\gamma} = -h_s - h_{f_s} - \frac{V_s^2}{2g} = -3.054 - 1.07 - \frac{(3.36)^2}{2 \times 9.81}$$

$$= \mathbf{-4.7 \text{ m of water.}}$$

Pressure in the delivery pipe must be sufficient to raise the water through the delivery head

$$\frac{P_d}{\gamma} = h_d + h_{f_d} + \frac{V_d^2}{2g}$$

$$= (32.067 - 3.054) + 5.955 + \frac{(3.36)^2}{2 \times 9.81} = \mathbf{35.54 \text{ m of water.}}$$

Example 17.7. *A pump impeller is 37.5 cm diameter and discharges water with velocity components of 2 and 12 m/s in the radial and tangential directions respectively. The impeller is surrounded by a concentric cylindrical chamber with parallel sides, the outer diameter being 45 cm. If the flow in this chamber is a free-spiral vortex, find the component velocities of water on leaving and the increase in pressure if there is no loss.*

Solution. From the free-vortex law

$$V_{u_1} \cdot r_1 = V_{u_2} \cdot r_2 = V_u \cdot r$$

According to the data given

$$V_{f_2} = 2 \text{ m/s}, V_{u_2} = 12 \text{ m/s}, D_2 = 37.5 \text{ cm}, D = 45 \text{ cm}.$$

At the periphery of the shroud (vortex chamber), let the tangential velocity be V_u and the radial velocity V_f. Then

$$V_u = \frac{V_{u_2} \cdot r_2}{r} = \frac{12 \times 37.5/2}{45/2} = 10 \text{ m/s}.$$

$$Q = \pi D_2 b_2 V_{f_2} = \pi D b V_f$$

Since the shroud is cylindrical, $b = b_2$

$$\therefore \quad V_f = V_{f_2} \cdot \frac{D_2}{D} = 2 \times \frac{37.5}{45} = 1.67 \text{ m/s}$$

To determine increase in pressure within the shroud, apply Bernoulli's equation between the outlet of impeller and the periphery of shroud (assuming no loss)

$$\frac{p}{\gamma} + \frac{V^2}{2g} = \frac{p_2}{\gamma} + \frac{V_2^2}{2g}$$

$$V = \sqrt{V_f^2 + V_u^2} = \sqrt{(1.67)^2 + (10)^2} = 10.14 \text{ m/s}$$

$$V_2 = \sqrt{V_{f_2}^2 + V_{u_2}^2} = \sqrt{(2)^2 + (12)^2} = 12.165 \text{ m/s}$$

$$\therefore \text{ The pressure rise } = \frac{p - p_2}{\gamma} = \frac{V_2^2 - V^2}{2g} = \frac{(12.165)^2 - (10.14)^2}{2 \times 9.81} = \mathbf{2.30 \text{ m of water.}}$$

17.5. Definitions

17.5.1. Static Head

The difference between the liquid levels in sump and high level reservoir is called the static head or static lift and is represented by h in Fig. 17.9. It can be divided into two parts : (1) suction head h_s, which is the height of the centre line of the pump above the liquid level in the sump, and (2) delivery head h_d is the height of liquid level in the high level reservoir measured from the centre line of the pump.

Thus $h = h_s + h_d$.

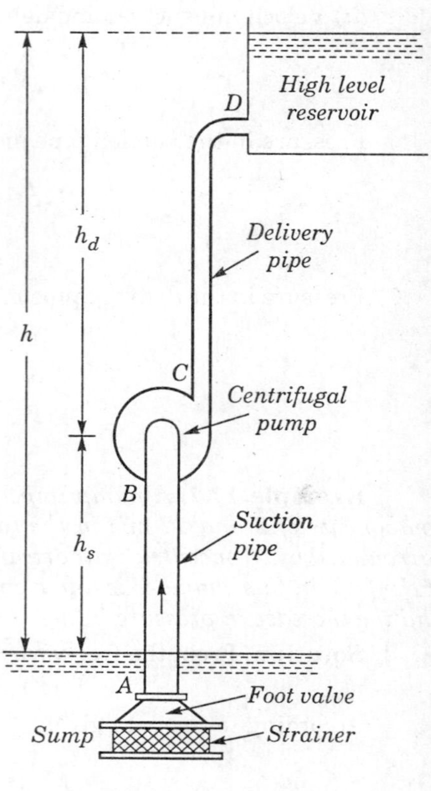

Fig. 17.9. Typical centrifugal pump installation.

17.5.2. Total Head

The head which a pump delivers must equal the static head (or lift) plus all losses in suction pipe, impeller and delivery pipe, in addition to the velocity head in the delivery pipe. The loss term must include the energy loss in the strainer, valves and bends in the pipes. Let h_{f_s} and h_{f_d} denote the head loss in the suction and delivery pipes respectively, the total head or total lift is

$$H = h + h_{f_s} + h_{f_d} + \frac{V_d^2}{2g}$$

$$= h + h_f + V_d^2/2g \qquad ...(17.6)$$

where V_d is the velocity in the delivery pipe and h_f is the total head loss in the pipe system.

17.5.3. Manometric Head

The total head across the pump is less than $u_2 V_{u_2}/g$ because of the energy dissipated in eddies and in friction in the impeller. The difference in the piezometric heads between a point on the discharge side and a point on the suction side of the impeller, both as close to the pump as possible, is called the '*manometric head*' H_m. It is the difference of head which would be recorded by a manometer connected between the inlet and outlet flanges of the pump. The manometric head can also be determined by installing two pressure gauges at inlet and outlet points.

17.5.4. Manometric Efficiency

The manometric efficiency is defined as the ratio of the manometric head developed by the pump to the head delivered to the liquid by the impeller. Thus

$$\eta_{mano} = \frac{gH_m}{u_2 V_{u_2}} \qquad ...(17.7)$$

The manometric efficiency is also called hydraulic efficiency.

Example 17.8. *Calculate the total power in kW to be supplied under following conditions :*

(*i*) *Suction lift = 3.00 m* (*ii*) *Loss of head in suction pipe = 1.00 m*

(*iii*) *Delivery head = 20 m* (*iv*) *Loss of head in delivery pipe = 10.00 m*

(*v*) *Discharge = 30 lit/sec* (*vi*) *Efficiency of pump = 60%*

(*vii*) *Efficiency of motor = 90%.*

Solution. Data given : $Q = 30$ 1/s, $H_d = 20$ m, $H_s = 3$ m

$H_{Ld} = 10$ m, $H_{Ls} = 1$ m,

$\eta_p = 60\%$ $\eta_m = 90\%$

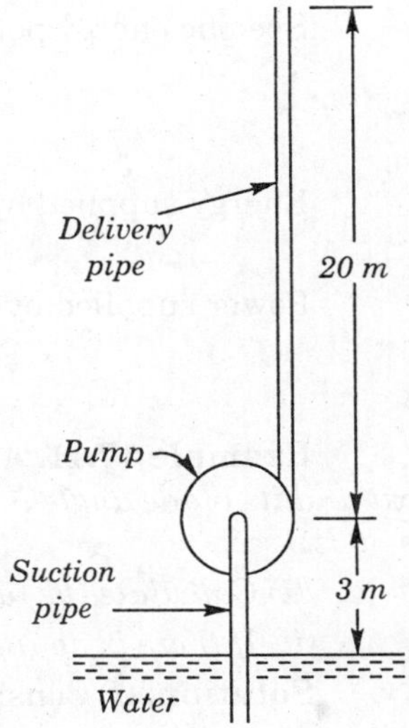

Power output of pump

$$= \gamma Q(H_s + H_d + H_{Ls} + H_{Ld})$$

$$= 9810 \times \frac{30}{1000}(3 + 20 + 1 + 10) \text{ Nm/s}$$

$$= 10006.2 \text{ Nm/s} = 10006.2 \text{ W} = 10.01 \text{ KW}$$

Input power required by the pump = Output of motor

$$= \frac{\text{Pump output}}{\text{Pump efficiency}} = \frac{10.01}{0.6} \text{ (kW)}$$

Total power required by the motor-pump system

$$= \frac{\text{Output power of motor}}{\text{Efficiency of motor}}$$

$$= \frac{10.01}{0.60 \times 0.90} = \mathbf{18.53 \ (kW).}$$

Example 17.9. *Static pressure rise across a water pump is 200 mm of water column. The suction and delivery pipes are respectively 100 mm and 80 mm. Flowrate is 19.6 litres / sec. Assuming losses are negligible, calculate power delivered by the pump to the water.*

Solution. Assuming that the points on suction and delivery pipes connected to the pressure measuring device are at the same level. Taking the connecting horizontal line to be the datum, the Bernoulli's equation applied to these points can be written as

$$0 + \frac{p_s}{\gamma} + \frac{V_s^2}{2g} = 0 + \frac{Pd}{\gamma} + \frac{V_d^2}{2g} - H_{pw}$$

in which H_{pw} is the head delivered by the pump to water. Subscripts s and d represent the quantitites in suction and delivery pipes.

$$\frac{p_d - p_s}{\gamma} = 200 \text{ mm of water column}$$

$$V_s = \frac{4Q}{\pi d_s^2} = \frac{4 \times (19.6/1000)}{\pi (100/1000)^2} = 2.4955 \text{ m/s}$$

$$V_d = \frac{4 \times 0.0196}{\pi (0.08)^2} = 3.899 \text{ m/s}$$

Head delivered by the pump to water

$$H_{pw} = \frac{p_d - p_s}{\gamma} + \frac{V_d^2 - V_s^2}{2g} = \frac{200}{1000} + \frac{(3.899)^2 - (2.4955)^2}{2 \times 9.81} = 0.6575 \text{ m}$$

Power delivered $= \gamma Q H_{pw} = 9810 \times 0.0196 \times 0.6575$

$$= 126.42 \text{ Nm/s} = \mathbf{126.42 \ W.}$$

Example 17.10. *Specific energy at the inlet of a pump is 110 m²/s², while that at the exit is 520 m²/s². If the energy loss is neglected, what is the power supplied in kW if the mass flow is 10 kg/s ?*

Solution. Assuming the inlet to and exit from the pump to be at the same level and neglecting the energy loss,

$$(E_s)_{inlet} + \text{Energy supplied by the pump per unit mass of fluid flow, } (E_s)_p = (E_s)_{exit}$$

$$\therefore \quad (E_s)_p = (E_s)_{exit} - (E)_{inlet} = 520 - 110 = 410 \text{ m}^2/\text{s}^2$$

Specific energy per unit fluid mass flow

$$= \frac{V^2}{2} + p_\rho$$

Energy supplied by the pump per unit fluid mass (E_s)

$$= 410 \text{ m}^2/\text{s}^2$$

Power supplied by the pump

$$= \text{mass flowrate} \times (E_s)_p = 10 \text{ (kg/s)} \times 410 \text{ (m}^2/\text{s}^2)$$

$$= 4100 \text{ N-m/s} = 4100 \text{ W} = \mathbf{4.1 \text{ kW}}.$$

Example 17.11. *A centrifugal pump impeller is 40 cm in outer diameter and 2.5 cm wide at exit and its blade angle is 150°. When run at a speed of 2100 r.p.m., the flowrate through the pump is 80 lit / sec.*

(*i*) *Calculate the* (*a*) *radial,* (*b*) *relative and* (*c*) *absolute fluid velocities at the impeller exit.*

(*ii*) *If there is no inlet whirl, what would be the head added to the water by the impeller ?*

Solution. (*i*) Consider velocity diagram shown in Fig. 17.5 (*a*). The rate of flow at the pump exit can be written as

$$Q = \pi \, d_2 \, b_2 \, V_{f2}$$

from which (*a*) the radial velocity at exit, V_{f_2} is

$$V_f = \frac{Q}{\pi \, d_2 \, b_2} = \frac{0.08}{\pi \times 0.40 \times (2.5/100)} = \mathbf{2.546 \text{ m/s}}$$

The peripheral blade velocity at exit

$$u_2 = \frac{\pi \, d_2 N}{60} = \frac{\pi \times 0.40 \times 2100}{60} = 43.98 \text{ m/s}$$

(*b*) From the outlet velocity diagram, the relative velocity w_2:

$$V_{f_2} = w_2 \sin \beta_2 \text{ or } w_2 \frac{V_{f_2}}{\sin \beta_2} = \frac{2.546}{\sin 30°} = \mathbf{5.093 \text{ m/s.}}$$

(here $\beta_2 = 180° - 150° = 30°$)

$$V_{u_2} = u_2 - w_2 \cos \beta_2 = 43.98 - 5.092 \cos 30° = 39.57 \text{ m/s.}$$

(*c*) The absolute velocity V_2 is given by

$$V_2 = \sqrt{V_{f_2}^2 + V_{u_2}^2} = \sqrt{(2.546)^2 + (39.57)^2} = \mathbf{39.65 \text{ m/s.}}$$

(*ii*) For zero-whirl at inlet, the head imparted to wated by the impeller

$$H = \frac{u_2 \, V_{u_2}}{g} = \frac{43.98 \times 39.57}{9.81} = \mathbf{177.40 \text{ m.}}$$

Example 17.12. *A centrifugal water pump has an impeller of outer diameter of 60 cm and inner diameter 20 cm. It is 2 cm wide at outlet and 5 cm wide at inlet. The blade angles at inlet and outlet are 20° and 10° respectively. The impeller rotates at 1800 r.p.m. Neglecting losses and vane thickness, determine :*

(*a*) *the discharge for shockless radial entry*

(*b*) *the theoretical head*

(*c*) *the power required, and*

(*d*) *the pressure rise through the impeller.*

Solution. Given $\beta_1 = 20°$, $\beta_2 = 10°$

$\alpha_1 = 90°$ (for radial entry)

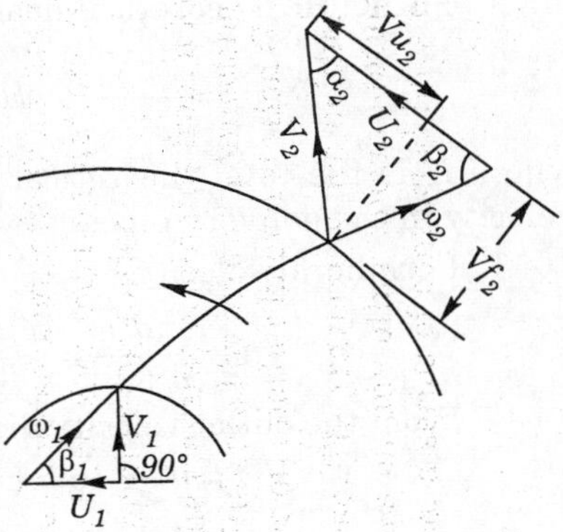

The peripheral speeds are

$$u_1 = \frac{\pi d_1 n}{60} = \frac{\pi \times 0.2 \times 1800}{60}$$

$$= 18.84 \text{ m/s}$$

$$u_2 = \frac{\pi d_2 n}{60} = \frac{\pi \times 0.6 \times 1800}{60} = 56.5 \text{ m/s.}$$

From the velocity diagram at entrance

$$V_1 = V_{f_1} = u_1 \tan 20° = 18.84 \times 0.3640 = 6.86 \text{ m/s.}$$

(*a*) Discharge $Q = \pi d_1 b_1 V_{f_1} = 0.2 \times \frac{5}{100} \times 6.86 = 0.2155 \text{ m}^3/\text{s}$.

At exit, the radial velocity of flow

$$V_{f_2} = \frac{Q}{\pi d_2 b_2} = \frac{0.2155}{\pi \times 0.6 \times \frac{2}{100}} = 5.71 \text{ m/s.}$$

$$V_{u_2} = u_2 - V_{f_2} \cot \beta_2 = 56.5 - 5.71 \cot 10° = 24.15 \text{ m/s.}$$

(*b*) Theoretical head developed

$$H = \frac{u_2 V_{u_2}}{g} = \frac{56.6 \times 24.15}{9.81} = 139.2 \text{ m.}$$

(*c*) **Power developed**

$$= \gamma QH = 9810 \times 0.2155 \times 139.2 = 293853 \text{ W} = 293.85 \text{ kW}$$

(*d*) Applying Bernoulli's equation between the impeller inlet and outlet assuming these at the same level, and neglecting losses

$$\frac{p_1}{\gamma} + \frac{V_1^2}{2g} + H = \frac{p_2}{\gamma} + \frac{V_2^2}{2g}, \; V_2 = V_{u_2} \cos \alpha_2 = 24.15 \text{ m/s}$$

$$\tan \alpha_2 = \frac{V_{f_2}}{V_{u_2}} = \frac{5.71}{24.15} = 0.2363$$

or $$\alpha_2 = 13.3°$$

$$V_2 = \frac{V_{u_2}}{\cos \alpha_2} = \frac{24.15}{0.9732} = 24.8 \text{ m/s.}$$

Pressure rise through the impeller

$$\frac{p_2 - p_1}{\gamma} = H + \frac{V_1^2 - V_2^2}{2g} = 139.2 + \frac{(6.86)^2 - (24.8)^2}{2 \times 9.81} = 110 \text{ m of water.}$$

or $$p_2 - p_1 = 9810 \times 110 \text{ N/m}^2 = 1079 \text{ kN/m}^2.$$

Example 17.13. *A centrifugal pump having an impeller 30 cm outside diameter rotates at 1050 r.p.m. The vanes are radial at exit and are 7.5 cm wide. The velocity of radial flow through the impeller is 3 m/s. The velocity in the suction and delivery pipes are 0.5 m/s and 1.5 m/s respectively. Neglecting frictional losses, determine the height through which the pump lifts, and the horse power of the pump. Assume radial entry.*

Solution. Theoretical head imparted to the pump

$$= \frac{u_2 V_{u_2}}{g} = H + \frac{V_d^2}{2g}$$

where H is the total theoretical lift of pump or the design head and V_d is the velocity in the delivery pipe.

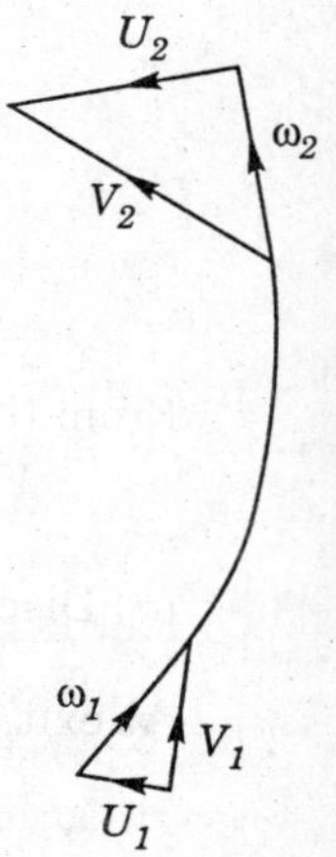

Peripheral velocity

$$u_2 = \frac{\pi d_2 n}{60} = \frac{\pi \times 0.3 \times 1050}{60} = 16.5 \text{ m/s.}$$

From the outlet velocity diagram

$$V_{u_2} = u_2 = 16.5 \text{ m/s}, V_{f_2} = w_2 = 3 \text{ m/s}$$

$$\therefore \quad H = \frac{u_2 V_{u_2}}{g} - \frac{V_d^2}{2g} = \frac{16.5 \times 16.5}{9.81} - \frac{(1.5)^2}{2 \times 9.81} = 26.68 \text{ m.}$$

$$Q = \pi d_2 b_2 V_{f_2} = \pi \times 0.3 \times \frac{7.5}{100} \times 3 = 0.212 \text{ m}^3\text{/s.}$$

Power of pump $= \gamma Q(H + V_d^2/2g) = 9810 \times 0.212 \times 26.8$ N-m/s $= 55.64$ kW.

Example 17.14. *If the centrifugal pump of Example 17.3 is designed to admit water radially, determine the torque exerted by the impeller on the water. If mechanical and hydraulic efficiencies are 88 and 75% respectively, calculate the shaft power and the lift developed by the pump.*

Solution. Torque developed by a centrifugal pump

$$T = \rho Q(V_{u_2} r_2 - V_{u_1} r_1) = \rho Q \, V_{u_2} r_2,$$

since for radial entry $V_{u_1} = 0$. From Ex. 17.3, $Q = 0.3$ m³/s, $V_{u_2} = 8.88$ m/s, $r_2 = \frac{35}{2 \times 100}$ m

$$T = 1000 \times 0.3 \times 8.88 \times \frac{35}{2 \times 100} = 465.975 \text{ N-m.}$$

Power exerted on water by the impeller

$$P_i = T\omega = 465.975 \times \frac{2\pi \times 1000}{60} = 48755.7 \text{ N-m/s} = 48.756 \text{ kW.}$$

Shaft horse power $= \frac{P_i}{\eta_m} = \frac{48.756}{0.88} = 55.4$ kW

Hydraulic efficiency $\eta_h = \frac{\gamma QH}{P_i}$

Lift developed by the pump

$$H = \frac{\eta_h P_i}{\gamma Q} = \frac{0.75 \times 48756}{9810 \times 0.3} = \mathbf{12.42 \text{ m.}}$$

Example 17.15. *A centrifugal pump is to discharge 0.118 m³/s at a speed of 1450 r.p.m. against a head of 25 m. The impeller diameter is 25 cm, its width at outlet is 5 cm and the manometric efficiency is 75%. Determine the vane angle at the outer periphery of the impeller.*

Solution. Theoretical head,

$$H = \frac{u_2 V_{u_2}}{g} = \frac{u_2(u_2 - V_{f_2} \cot \beta)}{g}$$

Manometric efficiency, $\eta_m = \frac{gH}{u_2 V_{u_2}} = \frac{gH}{u_2(u_2 - V_{f_2} \cot \beta)}$ ...(*i*)

$$u_2 = \frac{\pi D n}{60} = \frac{\pi \times 0.25 \times 1450}{60} = 18.98 \text{ m/s}$$

$$Q = \pi D B V_{f_2}$$

$$\therefore \quad V_{f_2} = \frac{0.118}{\pi \times 0.25 \times 0.05} = 3.005 \text{ m/s}$$

Substituting values in equation (*i*) above,

$$0.75 = \frac{9.81 \times 25}{18.98\,(18.98 - 3.005 \cot \beta)}$$

solving which, one obtains, $\beta =$ **59.77°**.

17.6. Minimum Head and Speed to Commence Flow

With reference to Fig. 17.9 assuming the pump to be working, the Bernoulli's equation between the water surface in the sump where the pressure is atmospheric and the eye of the pump impeller may be written as :

$$0 = \frac{p_s}{\gamma} + \frac{V_s^2}{2g} + h_s + h_{f_s}$$

The suction pressure at the inlet

$$\frac{p_s}{\gamma} = -\left(\frac{V_s^2}{2g} + h_s + h_{f_s}\right)$$

Similarly, the Bernoulli's equation between the delivery pipe at the pump exit and the water surface in the overhead tank

$$\frac{p_d}{\gamma} + \frac{V_d^2}{2g} = h_d + h_{f_d}$$

or

$$\frac{p_d}{\gamma} = h_d + h_{f_d} - \frac{V_d^2}{2g}$$

Hence,

$$\left.\begin{aligned} \frac{p_d - p_s}{\gamma} &= h_d + h_s + h_{f_s} + h_{f_d} + \frac{V_s^2}{2g} - \frac{V_d^2}{2g} \\ &= h + (h_{f_s} + h_{f_d}) + \frac{V_s^2}{2g} - \frac{V_d^2}{2g} \end{aligned}\right\} \quad \text{...(17.8)}$$

Before delivery commences, minimum pressure difference $(p_d - p_s) = \gamma h$ must be created across the impeller by the centrifugal action.

Referring to Article 2.11 on "Fluid rotation about a vertical axis" and making use of Eqs. (2.46) and (2.47), one can write

$$\frac{p_d - p_s}{\gamma} = \frac{\omega^2 (r_2^2 - r_1^2)}{2g} = \frac{u_2^2 - u_1^2}{2g}$$

Delivery will not commence until

$$\frac{p_d - p_s}{\gamma} = h = \frac{\omega^2 (r_2^2 - r_1^2)}{2g} \quad \text{...(17.9)}$$

Example 17.16. *The impeller tips of a centrifugal pump have radii of 10 cm and 30 cm. Determine the minimum speed at which delivery will commence against a static head of 15 m.*

Solution.

$$h = \frac{\omega^2 (r_2^2 - r_1^2)}{2g}$$

Substituting the given data

$$15 = \frac{\omega^2}{2g}\,[(0.30)^2 - (0.10)^2]$$

from which, $$\omega = \sqrt{\frac{2 \times 9.81 \times 15}{[(0.3)^2 - (0.1)^2]}} = 60.65 \text{ rad/s}$$

But $$\omega = \frac{2\pi n}{60}$$

$$\therefore \quad n = \frac{60\omega}{2\pi} = \frac{60 \times 60.65}{2\pi} = \mathbf{579.18 \text{ r.p.m.}}$$

The minimum speed is, therefore, **580 r.p.m.**

Example 17.17. *The impeller of a centrifugal pump has an outer diameter of 25 cm and an effective outlet area of 170 cm². The blades are bent backwards so that the direction of the outlet relative velocity makes an angle of 148° with the tangent drawn in the direction of impeller rotation. The diameters of the suction and delivery pipes are 15 cm and 10 cm respectively. The pump delivers 1860 lit/min at 1450 r.p.m. The gauge points on the suction and delivery pipes close to the pump show heads of 4.6 m below and 18.0 m above atmosphere respectively. The head losses in the suction and delivery pipes are 2.0 m and 2.9 m respectively. The motor driving the pump supplies 9.85 kW.*

Find the manometric efficiency assuming that water enters the pump without shock and whirl. Also find the overall efficiency of the pump.

Solution. Velocity in the suction pipe

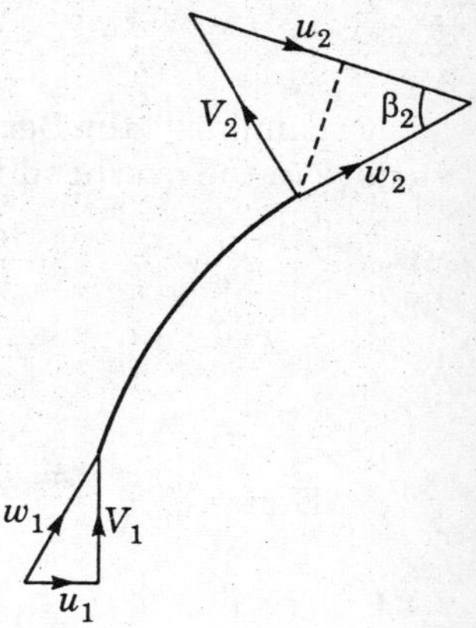

$$V_s = \frac{4Q}{\pi d_s^2} = \frac{4 \times 1.860/60}{\pi (0.15)^2} = 1.754 \text{ m/s}$$

Applying Bernoulli's equation between the water surface in the sump and the inlet of pump where pressure gauge is fitted

$$0 = h_s + h_{f_s} + \frac{p_s}{\gamma} + \frac{V_s^2}{2g}$$

$$= h_s + 2 - 4.6 + \frac{(1.754)^2}{2 \times 9.81}$$

or $$h_s = 2.44 \text{ m.}$$

Bernoulli's equation between the impeller outlet where the delivery pressure gauge is fitted and the outlet of delivery pipe gives

$$\frac{p_d}{\gamma} + \frac{V_d^2}{2g} = h_d + h_{f_d} + \frac{V_d^2}{2g}$$

or $$h_d = \frac{p_d}{\gamma} - h_{f_d} = 18.0 - 2.9 = 15.1 \text{ m.}$$

From the outlet velocity triangle

$$\tan 32° = \frac{V_{f_2}}{u_2 - V_{u_2}} \quad \text{or} \quad V_{u_2} = u_2 - V_{f_2} \cot 32°$$

$$u_2 = \frac{\pi D_2 n}{60} = \frac{\pi \times 0.25 \times 1450}{60} = 18.98 \text{ m/s}$$

$$V_{f_2} = \frac{Q}{a_2} = \frac{1.86/60}{170 \times 10^{-4}} = 1.82 \text{ m/s}$$

$$\therefore \quad V_{u_2} = 18.98 - 1.82 \cot 32° = 16.06 \text{ m/s}$$

$$V_d = \frac{4Q}{\pi d^2} = 3.946 \text{ m/s}, \frac{V_d^2}{2g} = 0.794 \text{ m}$$

$$H = h_s + h_d + h_{f_s} + h_{f_d} + \frac{V_d^2}{2g}$$

$$= 2.44 + 15.1 + 2 + 2.9 + 0.784 = 25.674 \text{ m.}$$

Manometric efficiency, $\eta_m = \frac{gH}{u_2 V_{u_2}} = \frac{9.81 \times 25.674}{18.98 \times 16.06} = 0.8263$ or **82.863**

Overall efficiency $\eta_0 = \frac{\gamma QH}{\text{Input to pump}} = \frac{9810 \times 1.860 \times 25.674}{60 \times 9.85 \times 10^3} = 0.7926$ or **79.26%.**

17.7. Specific Speed of Pump

For a pump the important variables are the speed n, head H and the rate of flow Q and therefore, the specific speed is expressed in terms of these variables. Using dimensionless π-terms expressed in Eq. (15.35), the ratio of $(\pi_1)^{1/2}$ to $(\pi_2)^{3/4}$ gives

$$C_n = \frac{(\pi_1)^{1/2}}{(\pi_2)^{3/4}} = \frac{n\sqrt{Q}}{(g)H^{3/4}} \quad \text{...(17.10)}$$

Eq. (17.10) is a dimensionless expression relating n, Q and H. Omitting g since it is a constant for a given geographical location, the conventional expression for the specific speed of pump becomes

$$n_s = \frac{n\sqrt{Q}}{H^{3/4}} \quad \text{...(17.11)}$$

It is obvious that the specific speed given by Eq. (17.11) is not a dimensionless quantity. It is conventional to express n in rev/minute, Q in lit/min and H in metres. In FPS system, Q is in gallons per minute and H in feet. The specific speed n_s in metric system is equal to 0.67 times n_s in FPS system, thus in an equation form

$$n_s \text{ in metric units} = 0.67\, n_s \text{ FPS units} \quad \text{...(17.12)}$$

The physical concept of n_s may be obtained by noting that it is the speed of a geometrically similar pump which would discharge at the rate of 1 lit/min against a head of 1 metre. In metric units, Eq. (17.11) may be expressed

$$n_s = \frac{n(\text{r.p.m.})\sqrt{Q(\text{lit/min})}}{[H(\text{metre})]^{3/4}} \quad \text{...(17.13)}$$

The value of n_s depends upon the type of impeller and ranges from about 300 to 5000. The values of Q and H correspond to maximum efficiency of the pump working at its normal speed. For pumps discharging large quantities at low heads such as axial flow pumps, the specific speed has a large value. Pumps lifting liquid to greater heights at lower flow rates will have smaller values of specific speed.

The value of n_s for a number of impellers on a single shaft (multi-stage pumps) is to be computed by using the head and discharge developed by a single impeller. The head per stage or that developed by a single impeller will be equal to the total head developed by the multistage pump divided by the number of stages. For a double-suction pump, the discharge Q to be used for computing the specific speed n_s is equal to half the discharge delivered by the pump.

Example 17.18. *A single stage centrifugal pump has an impeller of 30 cm diameter which rotates at 1780 r.p.m. and discharges 4000 lit/min to a height of 25 m at an efficiency of 70%. Obtain the number of stages of a similar multistage pump to lift 5000 lit/min to 175 m at 1500 r.p.m.*

Solution. Head developed by the single stage pump

$$= \frac{25}{0.7} = 35.7 \text{ m}$$

Specific speed of single-stage pump

$$n_s = n\frac{\sqrt{Q}}{H^{3/4}} = 1780 \times \frac{\sqrt{4000}}{(35.7)^{3/4}} = 7715.$$

If h is the actual lift of multi-stage pump, then the specific speed of multi-stage pump

$$n_s = 1500\frac{\sqrt{5000}}{(h/0.7)^{3/4}}$$

Equating the two specific speeds

$$7715 = 1500\frac{\sqrt{5000}}{(h/0.7)^{3/4}}.$$

or $$h = 23.15 \text{ m}$$

$$\therefore \quad \text{Number of stages} = \frac{175}{23.15} = 7.56 \quad \text{say} \quad 8.$$

Example 17.19. *A three-stage centrifugal pump has impeller 40 cm in diameter and 2 cm wide. The blade angle at outlet is 45° and the area occupied by the thickness of the vanes may be assumed 8% of the outlet area. If the pump delivers 3.6 cubic metres of water per minute when running at 920 r.p.m., calculate :*

(i) power of the pump, *(ii) manometric head, and* *(iii) specific speed.*

Assume mechanical efficiency as 88% and manometric efficiency as 77%.

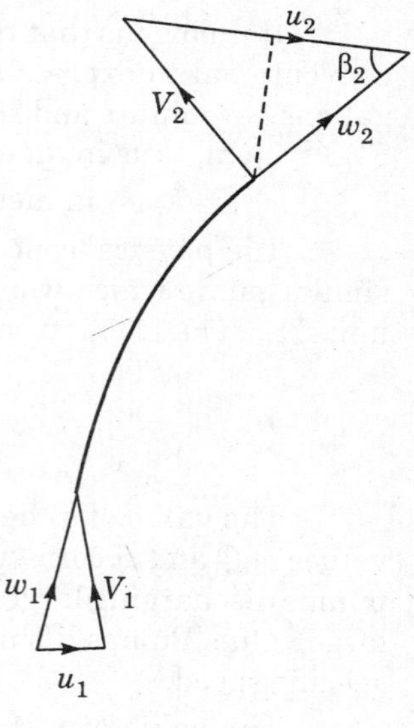

Solution. $$u_2 = \frac{\pi D_2 n}{60} = \frac{\pi \times 0.4 \times 920}{60} = 19.27 \text{ m/s}$$

Outlet area $= \pi Db$

Allowing for blade thickness,

Net outlet area $= 0.92\,\pi\, Db$

$$V_{f_2} = \frac{Q}{\pi Db \times 0.92} = \frac{3.6/60}{\pi \times 0.4 \times \frac{2}{100} \times 0.92} = 2.59 \text{ m/s.}$$

From the outlet velocity triangle

$$\tan \beta_2 = \frac{V_{f_2}}{u_2 - V_{u_2}}$$

$$\tan 45° = \frac{2.59}{19.27 - V_{u_2}}$$

$$\therefore \quad V_{u_2} = 19.27 - 2.59 \cot 45°.$$

Assuming radial, entry, head developed by each impeller

$$H_m = \eta_{mano} \times \frac{u_2 V_{u_2}}{g} = 0.77 \times 19.27\,\frac{(19.27 - 2.59 \cos 45°)}{9.81} = 25.23 \text{ m.}$$

Head developed by three impellers (manometric head).

$$H = 3 \times 25.23 = \mathbf{75.69\ m.}$$

Overall efficiency $= \eta_m \times \eta_{mano} = 0.88 \times 0.77 = 0.6776$

$$\eta_0 = \frac{\gamma QH}{\text{Power to pump}}$$

$\therefore$ Power of pump $= \dfrac{\gamma QH}{\eta_0} = \dfrac{9810 \times (3.6/60) \times 75.69}{0.6776}$ N-m/s (or W) = **65.68 kW.**

Specific speed $n_s = n\dfrac{\sqrt{Q}}{H_m^{3/4}}$

Head per impeller $H_m = 25.23$ m.

$Q = 3.6$ m^3/min = 3.6 × 1000 lit/min = 3600 lit/min.

$$\therefore \quad n_s = 920\,\frac{\sqrt{3600}}{(25.23)^{3/4}} = 2388.$$

Example 17.20. *A multistage pump is required to feed preheated water to a boiler. The quantity of water to be handled is 35 lit/sec against a pressure difference of 3202 kN/m^2, the speed of rotation of impeller being 2900 r.p.m. The specific weight of the preheated water is 9417 N/m^3. All impellers are identical and the specific speed per stage is not to be less than 5375. Determine :*

(i) the minimum number of stages and the head h per stage

(ii) the diameter of the impeller assuming a peripheral velocity $0.95\sqrt{2gh}$

(iii) the shaft power required to drive the pump, if the overall efficiency is 75%.

Solution. Total head created by the pump

$$= \frac{p_d - p_s}{\gamma} = \frac{3202 \times 10^3}{9417} = 348.96 \text{ m}$$

$Q = 35$ lit/sec = 35 × 60 = 2100 lit/min.

(*i*) If there are N stages, the head per stage

$h = (348.96/N)$

Specific speed, $n_s = n\dfrac{\sqrt{Q}}{h^{3/4}}$

or $$5375 = 2900\,\frac{\sqrt{2100}}{(348.96/N)^{3/4}}$$

from which $N = 4.865$, say 5

Therefore, the number of stages = 5,

and $$h = \frac{348.96}{5} = 69.8 \text{ m}$$

(*ii*) Peripheral speed of impeller

$$u = 0.95\sqrt{2gh} = 0.95\sqrt{2 \times 9.81 \times 69.8} = 37.0 \text{ m/s}$$

since $$u = \frac{\pi D n}{60}$$

$$\therefore \quad D = \frac{60u}{\pi n} = \frac{50 \times 37.0}{\pi \times 2900} = \mathbf{0.2436\ m.}$$

(*iii*) Shaft power $= \dfrac{\text{Power output}}{\eta_0} = \dfrac{\gamma QH}{\eta_0} = \dfrac{9417 \times 0.035 \times 348.96}{0.75}$ = **153.2 kW.**

17.8. Installation and Operation

Where pumps are set above the level of the sump, the pressure at the inlet will be less than atmospheric and may be determined by writing the Bernoulli's equation between the liquid level in the sump (where the pressure is atmospheric) and the pump inlet

$$\frac{p}{\gamma} + \frac{V^2}{2g} + Z + h_f = 0$$

or
$$\frac{p}{\gamma} = -\left(\frac{V^2}{2g} + Z + h_f\right) \quad ...(17.14)$$

where V is the velocity in the suction pipe at inlet, Z is the height of inlet above the sump level, and h_f is the head lost in the strainer, foot valve and the suction pipe.

Assuming that $V^2/2g$ and h_f each has value of 0.5 m, the pressure at inlet would be 0.935 kg/cm^2 or 91.73 N/m^2 below atmospheric pressure if Z were made equal to about 8.35 m. No pump could operate under such conditions because of rapid vaporization of water at this low pressure. For cavitation-free operation of the pump, it is very necessary that the pressure near the pump inlet should never be allowed to approach the vapour pressure of the liquid. It is commonly agreed that Z should never exceed a value of about 6.7 metres, and experience has shown that for dependable operation it should not exceed about 4.5 metres with water temperatures between 10°C to 20°C. For higher temperatures, Z must be decreased with increase in temperature, it becoming zero at about 65°C. When water is hotter than 65°C, it is necessary to provide positive pressures at the pump inlet.

17.8.1. Priming

A centrifugal pump must be primed before putting it into operation. Priming is carried out by filling the pump with the liquid to be pumped so that air is driven out, and the pump impeller is able to create the desired suction. Priming requires that valves must be provided before and after the pump. A foot valve is normally fitted at the lower end of the suction pipe for keeping the water in the pump. When the pump is started it opens to permit the liquid to enter the suction pipe, but closes as soon as the flow stops when the pump is switched off. The closure of the foot valve does not permit draining of the liquid into the sump. This enable the suction pipe and the pump to remain full of liquid and ready for use again without priming.

Over a period of time, due to frequent closures during periodic operations, the foot valve may develop leak and the suction pipe may not remain full of liquid during idle periods, thus requiring maintenance or even replacement. To eliminate this, an inlet valve may be installed before the pump which should be closed before the pump is switched off, so that the liquid remains inside the pump.

The delivery valve must be closed while starting the pump, and not opened until the speed has attained, its normal value. The delivery valve is then opened gradually. Similarly before stopping the pump, the delivery valve should be closed slowly with the pump running at normal speed. The closed valve does not harm the pump.

17.8.2. Problems in pump operation

The problems in pump operation may relate to 'no flow' or inadequate flow ; insufficient pressure of supply, and vibration and noise. The causes for these problems could be many and are described as under.

(*a*) No flow or Inadequate flow :

The reasons for such a condition may be :

(*i*) Pump is not primed.

(*ii*) Speed is too low ; wiring should be thoroughly checked.

(*iii*) Delivery head is too high.

(*iv*) Suction head is higher than required.

(*v*) Impeller rotates in wrong direction.

(*vi*) Suction pipe is clogged.

(*vii*) Impeller is plugged.

(*viii*) Air leak in suction pipe.

(*ix*) Foot valve opening is too small.

(*x*) Suction pipe end not properly submerged ; low submergence causing air-entrainment in suction pipe.

(*b*) Insufficient Pressure of Supply

The possible causes for such a supply could be :

(*i*) Presence of air through leaks in suction pipe.

(*ii*) Impeller diameter is too small.

(*iii*) Speed is too low.

(*iv*) Valve setting is inadequate/improper.

(*v*) Impeller is damaged

(*vi*) Packing in casing is defective.

(*c*) Vibrations and Noise

Occurrence of vibrations and noise may be due to :

(*i*) Cavitation

(*ii*) Motor not properly balanced

(*iii*) Bearings damaged

(*iv*) Water hammer in the pipeline

(*v*) Suction pipe picks up air due to vortex formation in the sump

(*vi*) Impeller out of balance.

17.8.3. Pump and System Characteristics

A centrifugal pump has a definite head-discharge relationship known as the pump characteristic. When it is installed in a system it is required to develop sufficient head to provide for the static lift and the losses in the system ; this is called the system characterstics. The point of intersection between the pump characteristics and the system characteristics determines the point of operation P of the pump as shown in Fig. 17.10.

The system characteristics between the head to be developed and the discharge to be delivered may be written as

$$H = h + KQ^2 \qquad \text{...(17.15)}$$

where h is the static lift and K is a coefficient which takes into account the effect of head loss in the system. As is evident from Eq. (17.15), the system characteristic rises with discharge whereas that of the pump has a falling trend.

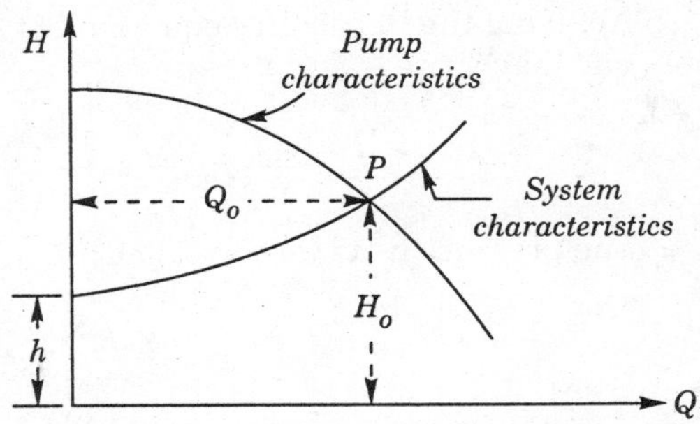

Fig. 17.10. Operation of a centrifugal pump (H_0 = Operating head ; Q_0 = Operating discharge)

17.9. Net Positive Suction Head

The net positive suction head (NPSH) is defined as :

$$\text{NPSH} = \begin{pmatrix}\text{absolute pressure}\\ \text{head at inlet to}\\ \text{pump}\end{pmatrix} - \begin{pmatrix}\text{vapour pressure}\\ \text{head of liquid}\\ \text{being pumped}\end{pmatrix} + \begin{pmatrix}\text{velocity head}\\ \text{in the suction}\\ \text{pipe}\end{pmatrix}$$

$$= \frac{p_s}{\gamma} - \frac{p_v}{\gamma} + \frac{V_s^2}{2g}$$

$$= \left(\frac{p_{atm}}{\gamma} - h_s - h_{fs} - \frac{V_s^2}{2g}\right) - \frac{p_v}{\gamma} + \frac{V_s^2}{2g}$$

$$= \frac{p_{atm} - p_v}{\gamma} - h_s - h_{fs} \qquad \text{...(15.94)}$$

where p_{atm} is the atmospheric pressure on the liquid surface in the sump.

It can be seen that NPSH represents the suction head at the impeller eye. This much suction head must be created by the impeller so that the liquid is lifted up from the sump into the impeller through the suction pipe.

Example 17.21. *A pump can lift water at a discharge of 0.15 m^3/s to a head of 25 m. The critical cavitation number (σ_c) for the pump is found to be 0.144. The pump is to be installed at a location where the barometric pressure is 9.8 m of water and the vapour pressure of water is 0.30 m of water. The intake pipe friction loss is 0.40 m. Using the minimum value of NPSH (Net positive suction head), find the maximum allowable elevation above the sump water surface at which the pump can be located.*

Solution. Thoma's cavitation factor is given by Eq. (17.19)

$$\sigma = \frac{(p_{atm} - p_v)/\gamma - (V_s^2/2g + hs + h_{fs})}{H_m}$$

Data given : $p_{atm}/\gamma = 9.80$ m ; $p_v/\gamma = 0.30$ m ; $Q = 0.15$ m^3/s

$\sigma = \sigma_c = 0.144$; $H_m = 25$ m ; $h_{fs} = 0.40$ m

Ignoring the effect of velocity head, $V_s^2/2g$, the maximum allowable elevation of the pump location,

$$h_s = \frac{p_{atm} - p_v}{\gamma} - \sigma_c H_m - h_{fs}$$

$$= 9.8 - 0.30 - 0.144 \times 25 - 0.40 = \mathbf{5.50\ m.}$$

17.10. Cavitation in Pumps

Applying the Bernoulli's equation between the water level in the sump and the inlet to the pump, one may write

$$\frac{p_{atm}}{\gamma} = \frac{p_s}{\gamma} + \frac{V_s^2}{2g} + h_s + h_{fs}$$

The pressure head (absolute) at the inlet to the pump is

$$\frac{p_s}{\gamma} = \frac{p_{atm}}{\gamma} - \left(h_s + h_{fs} + \frac{V_s^2}{2g}\right) \qquad \text{...(17.17)}$$

here h_{fs} includes the head loss in the foot valve, strainer and the suction pipe. Since the lowest limit of p_s is the vapour pressure (p_v) of the liquid at the prevailing temperature, the limiting case is obtained when $p_s = p_v$. The suction head for this case is

$$h_s = \frac{p_{atm} - p_v}{\gamma} - \frac{V_s^2}{2g} - h_{fs} \qquad \text{...(17.18)}$$

Equation (17.18) expresses the maximum limit (theoretical) of the suction lift. As the suction lift approaches the above limit, rapid vaporisation of liquid will take place giving rise to the phenomenon of cavitation. The suction lift h_s, must, therefore, always be less than the maximum limit expressed by Eq. (17.18).

For pumps, Thoma's cavitation factor is defined as

$$\sigma = \frac{(p_{atm} - p_v)/\gamma - (V_s^2/2g + hs + h_{fs})}{H_m} \qquad \text{...(17.19)}$$

here H_m represents the manometric head. Cavitation will occur it $\sigma < \sigma_c$. The critical valve σ_c represents that value at which cavitation begins to deteriorate pump's performance marked by a drop in the efficiency. The critical cavitation factor σ_c depends on the design of the pump and the specific operating condition. For pumps of normal design,

$$\sigma_c = 0.103\,(n_s/1000)^{4/3} \qquad ...(17.20)$$

where n_s is the specific speed of the pump.

From Eq. (17.19) it is obvious that σ can be kept within safe limits ($\sigma > \sigma_c$) by a proper choice of the suction lift h_s ; σ increases as h_s decreases. Sometimes the pump may have to be installed (when hot liquids are to be handled) either at the liquid surface or even below it.

In centrifugal pumps, the pressure is lowest on the underside of the vane at inlet and it is here that vapour bubbles tend to form. These bubbles are carried by flow to regions of higher pressure near the vane outlet where they collapse causing pitting, noise, vibrations and severe damage to the metal surface. The vane tips (exit points) are the most vulnerable spots for cavitation attack.

Example 17.22. *The minimum value of NPSH recommended by the pump manufacturer is 6.5 m. Water is being pumped from a reservoir at a rate of 0.30 m³/s. The water level in the reservoir is 1.25 m below the pump. Atmospheric pressure is 98.70 kN/m² and the water temperature is 20°C. If the total head loss in the suction pipe is 1.20 m, is the pump safe from cavitation effects ? The specific weight of water at 20°C is 9.78 kN/m³. Take vapour pressure of water as 2.34 kN/m².*

Solution. NPSH is given by Eq. (17.16). The data given is as follows :

$$p_{atm} = 98.70 \text{ kN/m}^2,\ p_v = 2.34 \text{ kN/m}^2,\ \gamma = 9.78 \text{ kN/m}^3$$

$$h_s = 1.25 \text{ m},\ h_{fs} = 1.20 \text{ m},\ (\text{NPSH})_{min} = 6.50 \text{ m}$$

Making substitutions in the equation for NPSH

$$\text{NPSH} = \frac{p_{atm} - p_v}{\gamma} - h_s - h_{fs}$$

$$= \frac{(98.70 - 2.34)\times 10^3}{9.78\times 10^3} - 1.25 - 1.20 = 9.85 - 2.45 = 7.40 \text{ m}$$

Since the NPSH value of 7.40 m is greater than the minimum recommended value of 6.50, the pump is safe from cavitation effects.

17.11. Shut-off Speed

If during the running of the pump, a valve on the discharge line is gradually closed, the discharge and the power output fall to zero. The pump motor then work only to overcome the bearing and water friction (disc friction). The head developed which now exists in the form of pressure is equal to $u_2^2/2g$.

$$\text{Therefore, the shut off head} = \frac{u_2^2}{2g} \qquad ...(17.21)$$

The shut-off head must always exceed the static lift against which the liquid is to be pumped, otherwise pumping cannot begin when the discharge valve is opened. The shut off speed can be determined from u_2.

17.12. Operating Characteristics of Centrifugal Pumps

For optimum performance (*i.e.* operation at maximum efficiency) a centrifugal pump is required to run at its design-speed, which happens to be the speed of the driving motor. At this speed, the head-discharge, efficiency-discharge, and the shaft power-discharge curves are called the operating characteristics. A typical set of these characteristics is shown in Fig. 17.11.

The head and the discharge corresponding to the maximum efficiency η are known as the normal (or design) head and normal (or design) discharge. From these characteristics it is possible to ascertain whether or not the pump will be able handle a given quantity of liquid against the desired head. From the P versus Q curve, the size of the motor can also be determined.

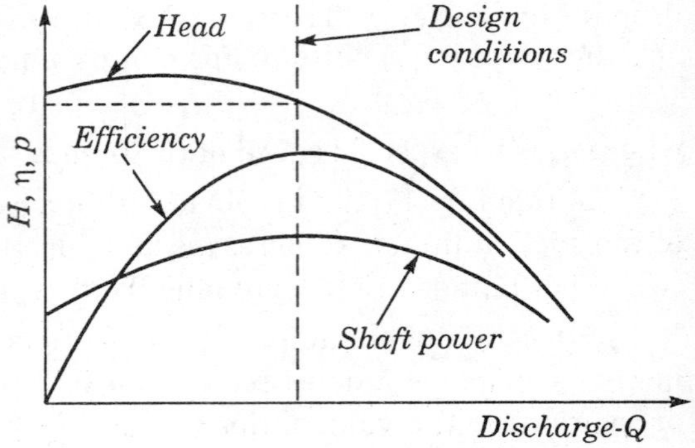

Fig. 17.11. Operating characteristics.

Example 17.23. *The head-discharge characteristics of a centrifugal pump is given below :*

Discharge (lit/sec)	*0*	*10*	*20*	*30*	*40*	*50*
Head (metres)	*25.3*	*25.5*	*24.5*	*22.2*	*18.7*	*12.0*

The pump delivers fresh water through a 500 m long × 15 cm diameter pipeline having a friction coefficient of f = 0.025. The static lift is 15 m. Neglecting minor losses in the pipe flow, find (a) the discharge of the pump under the above conditions (b) driving power of the pump motor. Assume a pump efficiency of 72%.

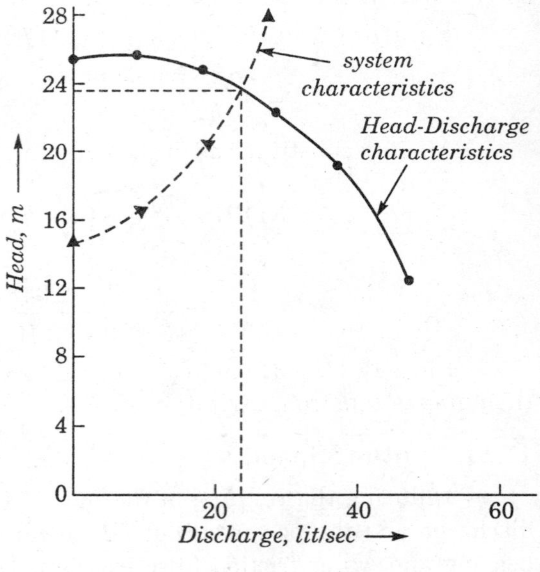

Solution. Head loss in the pipe line

$$h_f = f\frac{L}{d}\frac{V^2}{2g} = \frac{fl}{2gd}\left(\frac{4Q}{\pi d^2}\right)^2$$

$$= \frac{0.025 \times 500 \times 16}{\pi^2 \times (0.15)^5 \times 2 \times 9.81} Q^2$$

$$= 13{,}601.1\ Q^2$$

The system-characteristics can be written as $H = 15 + 13{,}601\ Q^2$

When Q is expressed in lit/sec

$$H = 15 + 13{,}601\left(\frac{Q}{1000}\right)^2 = 15 + 0.0136\ Q^2$$

The head H computed from the above equation for different values of Q is tabulated as below :

Q, (lit/sec)	0	10	15	30	40	50
H, (m)	15	16.36	20.44	27.24	36.76	49

The above tabulated data representing the system-characteristics is shown plotted by a dotted line in the graph. The point of intersection of this with the head-discharge characteristics (also shown on the graph on previous page) gives

$$Q = 25 \text{ lit/sec}$$

and the corresponding head is $H = 23.5$ m.

(*ii*) The driving power of the pump motor

$$= \frac{\gamma QH}{\eta} = \frac{9810 \times 25 \times 10^{-3} \times 23.5}{0.72} = \mathbf{7.99\ kW.}$$

Example 17.24. *The head-discharge relationship of a pump is given by $h = 38 - 650\,Q^2$ in which h = head in metres, and Q = discharge in cubic metres per second. It is fitted in a 800 m long, 200 mm diameter pipe having Hazen-Williams coefficient of 120. If the static lift is 15 m, find :*

(*i*) *the head and discharge at the operating point, and*

(*ii*) *the power of the pump if its efficiency is 72%.*

Solution. Hazen-Williams formula is given by Eq. (9.77)

$$V = 0.85\, C\, R_h^{0.63}\, S^{0.54}$$

$$\text{Hydraulic radius } R_h = \frac{A}{P} = \frac{\pi d^2/4}{\pi d} = \frac{d}{4}$$

$$Q = A.\,V = \frac{\pi d^2}{4}\,(0.85\,C)\left(\frac{d}{4}\right)^{0.63}\left(\frac{h_f}{L}\right)^{0.54}$$

$$= \frac{\pi (0.2)^2}{4}\,(0.85 \times 120)\left(\frac{0.2}{4}\right)^{0.63}\left(\frac{h_f}{800}\right)^{0.54} = \frac{0.486}{36.955}\, h_f^{\,0.54}$$

The head loss, $h_f = 3043\, Q^{1.852}$

Head developed by the pump

$$= \text{Static lift + head loss}$$

$$= 15 + 3043\, Q^{1.852} \quad ...(1)$$

The head-discharge relationship is

$$h = 38 - 650\, Q^2 \quad ...(2)$$

At the operating point, the head given by Eq. (2), must be equal to the head developed, Eq. (1) hence

$$38 - 650\, Q^2 = 15 + 3043\, Q^{1.852}$$

$$650\, Q^2 + 3043\, Q^{1.852} - 23 = 0$$

Solving this equation by trial and error method, one obtains

$$Q = 0.067 \text{ m}^3/\text{s}$$

(*i*) Hence the pump discharge at the operating point = **0.067 m³/s.**

The corresponding head developed

$$h = 15 + 3043\, Q^{1.852} = 15 + 20.379 = \mathbf{35.379\ m}$$

(*ii*) Power of the pump at 72% efficiency

$$= \frac{\gamma Qh}{\eta} = \frac{9810 \times 0.067 \times 35.379}{0.72} = \mathbf{32.27\ kW.}$$

Example 17.25. *Two identical pumps are connected to a pipe system as shown in the figure to pump water into the high level reservoir. The head-discharge characteristic of the pumps is given by the equation*

$$H = 40 + 0.14\, Q - 0.0012\, Q^2$$

where Q is the pump discharge in lit/sec and H is the head developed by the pump in m.

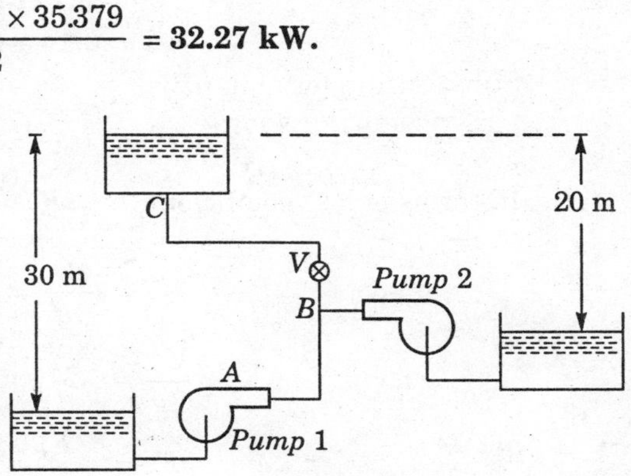

The pipe AB is 100 m long and is of diameter 30 cm. The pipe BC is 45 cm in diameter and 200 m

long. For both the pipes the friction factor can be taken as 0.03 "V". is a regulating valve on the pipeline BC. The static lifts against which the pumps are working are 20 m and 30 m.

For a given position of the valve "V" the pump operating against a static lift of 30 m develops a discharge of 140 lit/sec. Determine the discharge of the second pump and the loss coefficient of the valve, neglecting all other local losses.

Solution. Let the head loss in pipe *AB* be h_{f_1} and that in *BC* is h_{f_2}. From the continuity relationship,

$$Q_1 + Q_2 = Q \quad \text{or} \quad 0.140 + Q_2 = Q$$

Head loss in pipe *AB*, $h_{f_1} = f\frac{L}{d}\frac{V^2}{2g} = \frac{8}{\pi^2} f \frac{L}{d^5}\frac{Q^2}{2g}$

$$= \frac{8}{\pi^2} \times 0.03 \times \frac{100}{(0.3)^5} \times \frac{(0.14)^2}{2 \times 9.81} = 2.0 \text{ m}$$

Head loss in *BC*, $h_f = \frac{8}{\pi^2} \times 0.03 \times \frac{200}{(0.45)^2} \times \frac{Q}{2 \times 9.81} = 26.87\, Q^2$

Head developed by pump 1

$$H_1 = 40 + 0.14\, Q - 0.0012\, Q^2$$

$$= 40 + 0.14 \times (140) - 0.0012\, (140)^2 = 36.0 \text{ m} \qquad ...(i)$$

Head utilised $= \text{Static lift} + \text{Head losses}$

$$H_1 = 30 + h_{fAB} + h_{fBC} + h_v = 30 + 2.0 + 26.87\, Q^2 + h_v \qquad ...(ii)$$

Equating (*i*) and (*ii*) $\quad 26.87\, Q^2 + h_v = 4 \qquad ...(iii)$

Head developed by pump 2

$$H_2 = 40 + 0.14\, (1000\, Q_2) - 0.0012\, (1000\, Q_2)^2$$

$$= 40 + 140\, Q_2 - 1200\, Q_2^2 \qquad ...(iv)$$

Head utilised by pump 2

$$H_2 = 20 + h_{fBC} + h_v = 20 + 26.87\, Q^2 + h_v \qquad ...(v)$$

Equating (*iv*) and (*v*)

$$40 + 140\, Q_2 - 1200\, Q_2^2 = 20 + 26.87\, (0.14 + Q_2)^2 + h_v$$

or $\quad 20 + 140\, Q_2 - 1200\, Q_2^2 = 26.87\, (0.14 + Q)_2 + h_v$

from (*iii*), the RHS is equal to 4, hence

$$16 + 140\, Q_2 - 1200\, Q_2^2 = 0$$

or $\quad 1200\, Q_2^2 - 140\, Q_2 - 16 = 0$

Solving which, one obtains

$$Q_2 = \mathbf{0.1875\ m^3/s}$$

$$Q = 0.14 + 0.1875 = 0.3275 \text{ m}^3\text{/s}$$

Substituting for *Q* in (*iii*)

$$26.87\, (0.3275)^2 + h_v = 4, \quad \text{or} \quad h_v = 1.12 \text{ m}$$

In terms of loss coefficient, $h_v = K_L \frac{V^2}{2g}$

$$V = \frac{4Q}{\pi d^2} = \frac{4 \times 0.3275}{\pi (0.45)^2} = 2.06 \text{ m/s}$$

$$\frac{V^2}{2g} = 0.215 \text{ m}$$

$$\therefore \quad K_L = \frac{h_v}{V^2/2g} = \frac{1.12}{0.216} = \mathbf{5.18.}$$

17.13. Non-dimensional Performance Curves

Experimental data obtained from tests on a model pump are plotted to show variation of the non-dimensional parameters : efficiency η, gH/n^2D^2 and $P/\rho n^3D^5$ with respect to Q/nD^3. The performance curves shown in Fig. 17.12, are known as 'unit' curves and are applicable to any pump in the homologous series. These curves are very useful to the designer in the study of head and discharge characteristics of centrifugal pumps operating at any speed. Unit head multiplied by n^2D^2 gives the actual head and similarly the actual discharge is obtained by multiplying the unit discharge by nD^3 (Also refer article 15.12 on Similarly Laws).

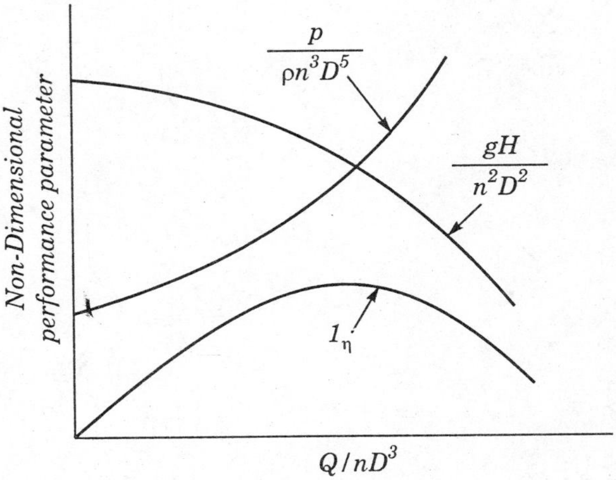

Fig. 17.12. Performance curves of centrifugal pumps.

Example 17.26. *Two geometrically similar pumps are to run at the same speed of 1200 r.p.m. If one of the pumps which has an impeller diameter of 0.30 m lifts water at the rate of 20 lit/sec through a head of 16 m, determine the head and size of the other pump to deliver half the quantity of water.*

Solution. Specific speed of pump from Eq. (15.60)

$$n_s = n\frac{\sqrt{Q}}{H^{3/4}} = \frac{1200\sqrt{20 \times 60}}{(16)^{3/4}} = 672.$$

Discharge of the other pump

$$= \frac{1}{2} \times 20 = 10 \text{ lit/sec.}$$

For the other pump, $672 = \dfrac{1200\sqrt{10 \times 60}}{H^{3/4}}$

from which $H = 154.2$ m,

From Eq. (15.35) $\left(\dfrac{Q}{nD^3}\right)_p = \left(\dfrac{Q}{nD^3}\right)_m$

For the same speed, $\dfrac{D_p}{D_m} = \left(\dfrac{Q_p}{Q_m}\right)^{1/3} = \left(\dfrac{20}{10}\right)^{1/3} = 1.26.$

Impeller diameter of the other pump

$$D_m = \frac{D_p}{1.26} = \frac{0.30}{1.26} = \mathbf{0.238\ m.}$$

Example 17.27. *A centrifugal pump is discharging 4500 lit/min. and operates at 1800 r.p.m. against a head of 35 m. The power input to the pump is 3 kW. If this pump is required to operate at 1200 r.p.m. assuming its efficiency remains constant, estimate (a) its discharge in lit/min, (b) the theoretical head it imparts to the liquid and (c) the power input to the pump.*

Solution. For the same turbomachine, the dimensionless parameters yield (keeping D constant) (*a*) $$\frac{Q_1}{n_1} = \frac{Q_2}{n_2}$$

or $$Q_2 = Q_1 \frac{n_2}{n_1} \times 4500 \times \frac{1200}{1800} = 3000 \text{ lit/min.}$$

(*b*) $$\frac{H_1}{n_1^2} = \frac{H_2}{n_2^2}$$

or $$H_2 = H_1 \left(\frac{n_2}{n_2}\right)^2 = 35 \left(\frac{1200}{1800}\right)^2 = 15.56 \text{ m.}$$

(*c*) $$\frac{P_1}{n_1^3} = \frac{P_2}{n_2^3},$$

or $$P_2 = P_1 \left(\frac{n_2}{n_1}\right)^3 = 3 \left(\frac{1200}{1800}\right)^3 = 0.89 \text{ kW.}$$

17.14. Centrifugal Fans

The principle and design method of a centrifugal fan are practically the same as that of a centrifugal pump. Basically, a pump is used to lift a liquid against a pressure head whereas a fan handles gases, more commonly air the density of which does not vary from the inlet to the outlet. The maximum pressure rise is of the order of 250 mm of water gauge. For higher pressure differentials, a blower or a compressor is used.

17.14.1. Outlet Blade Angle

For centrifugal pump impellers, three velocity diagrams at the outlet are possible which are sketched in Fig. 17.5. These are influenced by the magnitude of the outlet blade angle β_2. Though a pump impeller is usually backward facing ($\beta_2 < 90°$), all the three varieties are prevalent in the fan industry. The single most important parameter influencing the performance of a centrifugal impeller is the blade angle at outlet (β_2). Considering Fig. 17.5, it is noteworthy that :

(*i*) When $\beta_2 < 90°$, the blade is backward curved, and $V_{u2} < u_2$;

(*ii*) When $\beta_2 = 90°$, it is radial, and $V_{u_2} = u_2$;

(*iii*) When $\beta_2 > 90°$, the blade is forward curved, and $V_{u_2} < u_2$.

Noting that β_2 is measured clockwise from the positive u_2 direction to the relative velocity vector w_2, the Euler head for a centrifugal impeller is given by Eq. (17.3)

$$H = \frac{\Delta p}{\rho} = \frac{\Delta p}{\rho g} = \frac{u_2 V_{u_2}}{g}$$

from which, $$\frac{\Delta p}{\rho} = u_2 V_{u_2} = u_2 (u_2 - V_{f_2} \cot \beta_2)$$

The volumetric flowrate Q

$$Q = \pi d_1 b_1 V_{f_1} = \pi d_2 b_2 V_{f_2}$$

making substitution for V_{f_2},

$$\frac{\Delta p}{\rho} = u_2 \left(u_2 - \frac{Q}{\pi d_2 b_2} \cot \beta_2 \right) \quad \text{...(17.22)}$$

Defining the following coefficients :

Pressure coefficient $$\psi = \frac{\Delta p}{\left(\frac{1}{2} \rho u_2^2\right)} \quad \text{...(17.23)}$$

Volume coefficient $\phi = \frac{4Q}{(\pi d_2^2 u_2)}$ ...(17.24)

and, power coefficient, $\lambda = \phi\psi$ ...(17.25)

Eq. (17.22) can be written as

$$\frac{\Delta p}{\frac{1}{2}\rho u_2^2} = 2\left[1 - \frac{Q}{\pi d_2 b_2 u_2}\cot \beta_2\right]$$

or $$\psi = 2\left[1 - \frac{\phi d_2}{4b_2}\cot \beta_2\right] \quad ...(17.26)$$

and, $$\lambda = 2\phi\left[1 - \frac{\phi d_2}{4b_2}\cot \beta_2\right] \quad ...(17.27)$$

Variation of pressure coefficient and power coefficient with the volume coefficient for each category of β_2 is shown in Fig. 17.13, a perusal of which helps in drawing the following conclusions :

1. Fans with $\beta_2 > 90°$, (*i.e.* forward curved blades) will develop the highest pressure for a given impeller diameter and speed (*i.e.* for ϕ = constant).

2. For small increase in flowrates (*i.e.* ϕ) the power requirement (λ) increases steeply for forward curved fans.

3. The pressure developed decreases fast, as in reflected by the falling trend of ψ, with increasing flowrate (ϕ) in a backward curved fan.

17.14.2. Salient Characteristics

Backward curved fans ($\beta_2 < 90°$) are the most efficient centrifugal fans and the falling power characteristic makes them suitable for changing operating conditions.

Forward curved fans ($\beta_2 > 90°$) are used for large flowrates but suffer from higher power requirement. Losses in fluid machines being proportional to Q^2, the forward curved fans are suitable only for fixed duty operation such as in a ventilating system. They are unsuitable for off-design operating conditions.

Fans with radial blades are less efficient compared to backward curved and are preferred for handling dust laden fluids so as to prevent solid particles from sticking to the curved portion of the blades.

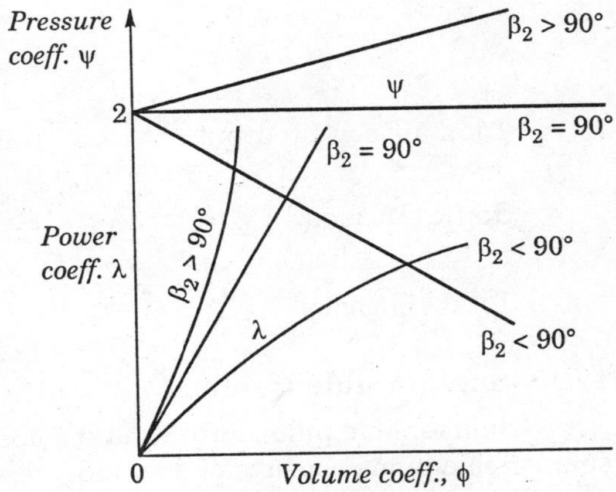

Fig. 17.13. Theoretical performance curves.

Quantitative information on various parameters pertaining to backward curved and forward curved fans is tabulated below :

Centrifugal Fan

		Backward curved ($\beta_2 < 90°$)	*Forward curved* ($\beta_2 > 90°$)
(*i*)	No. of blades	12—16	60
(*ii*)	Blade angle β_2	40°	145°
(*iii*)	Pressure coeff., ψ	0.80	2.15
(*iv*)	Volume coef., ϕ	0.18	0.50
(*v*)	Hydraulic efficiency	80%	70%

Example 17.28. *A centrifugal fan is required to deliver air at the rate of 380 m_3/min at a static pressure of 35 mm water gauge when operating at 450 r.p.m. The power input to the fan is 4.5 kW. The diameter of the outlet duct is 600 mm. Determine the static and total efficiencies of the fan. The pressure and temperature of air are 1.013 bar (1 atm) and 20 C for which the air density is 1.2 kg/m^3.*

Solution. Area of the outlet duct $= \frac{\pi}{4}\left(\frac{600}{1000}\right)^2 = 0.2827 \text{ m}^2$

Air velocity through the outlet duct

$$V_d = \frac{380}{60 \times 0.2827} = 22.40 \text{ m/s}$$

Velocity head

$$= \frac{V_d^2}{2g} \text{ (m of air)} = \frac{V_d^2}{2g}\frac{\rho_{air}}{\rho_{water}} \text{ (m of water)}$$

$$= \frac{(22.40)^2}{2 \times 9.81} \times \frac{1.2}{1000} = 0.03069 \text{ m of water}$$

$$= 30.69 \text{ mm of water gauge}$$

Total head $= 35 + 30.69 = 65.69$ mm of water

Static air power output $= Q\,(\rho g)$ (static pressure head)

$$= \frac{380}{60}(1000 \times 9.81)\, 35 \times 10^{-3} \text{ (N-m/s)}$$

$= 2174.55\ W =$ **2.175 kW**

Total air power output $= 2.175 \times \frac{65.69}{35} = 4.082$ kW

Static efficiency $= \frac{\text{Static power output}}{\text{Input}} = \frac{2.175}{4.5} = 0.483$ or 48.3%

Total efficiency $= \frac{4.082}{4.5} = 0.907$ or **90.7%.**

17.15. Submersible Pumps

Submersible pumps are used for deep well/bore well applications requiring high head. A submersible pump consists of a suitably designed impeller made of special material (Noryl). It is installed with a well designed diffuser to enable the pump to discharge water efficiently to the desired height. Modern pumps of this category use vacuum impregnated and dynamically balanced rotor and a thrust bearing made of a special alloy-nickel tin bronze for least friction and long life. A ferrobesto lined rotating disc for best bearing performance and a stainless steel ball mounted thrust bearing pivot for self alignment are used. High grade zinc free bushing is used to suit water as lubricant. A highly rigid coupling is used for ensuring proper alignment of pump and rotor shaft. A non-return valve made of special material (Noryl) is fitted for giving best performance.

Submersible pumps are designed to have corrosion and abrasion free internals, dynamically balanced rotor to achieve high efficiency and consume lower power during operation. These pumps find applications in domestic water supply, industries, farms, drip sprinkler irrigation, high rise buildings, fire fighting and similar other purposes. An idea of the performance of 100 mm (4 inch) submersible pumps for deep wells can be obtained from the following table.

Performance of Submersible Pumps*

Sr. No.	*Outlet size mm*	*Power consumption*		*No. of stages*	*Head in metres attained for different discharge, lph/lpm*							
		HP	*kW*		*720*	*1440*	*2160*	*2880*	*3600*	*5040*	*5760*	*7200*
					12	*24*	*36*	*48*	*60*	*84*	*96*	*120*
1.	38	0.50	0.37	8	40	33	15	—	—	—	—	—
2.	38	0.75	0.55	10	50	41	19	—	—	—	—	—
3.	38	1.00	0.75	13	—	62	51	32	—	—	—	—
4.	38	1.50	1.10	15	—	72	75	70	63	40	—	—
5.	38	2.00	1.50	10	—	—	—	—	—	54	50	40

In the above table, the discharges are given in litres per hour as well as in litres per minute *i.e.* lph, lpm, as indicated in first and second rows respectively.

PROBLEMS

17.1. The centre of impeller of a centrifugal pump is placed at a height of 4 m above the water in the sump. The suction pipe is 15 cm in diameter and the discharge is 1800 lit/min. The total head through which water is lifted is 25 m. The blades are curved back and make an angle of 30° with the tangent at the outlet tip of the wheel. The radial velocity of flow at exit is 3 m/s and the pump efficiency is 70%. Determine (*a*) the pump speed if the impeller diameter is 50 cm ; (*b*) the pressure at inlet and outlet of the wheel. Assume that the energy loss takes place after the water leaves the wheel, and that the entry is radial.

17.2. A centrifugal pump lifts water against a static head of 40 m of which 4 m is suction. The suction and delivery pipes are both 15 cm in diameter. The head lost in the suction pipe is 2 m and in the delivery pipe 7.5 m. The impeller is 40 cm in diameter and 2.5 cm wide and runs at 1250 r.p.m. The blade angle at exit is 30°. If the monometric efficiency of the pump is 80% and the mechanical efficiency 70%, determine the power required to drive the pump and the discharge in lit/min.

17.3. A centrifugal pump of 1.2 m diameter runs at 200 r.p.m. and pumps 1.8 m³/s, the average lift being 5.5 m. The angle which the blades make at exit with the tangent to the impeller is 30°, and the radial velocity of flow is 2.5 m/s. Determine the useful power and the efficiency. Find also the lowest speed to start pumping against a head of 5.5 m, the inner diameter of the impeller being 0.6 m.

17.4. The axis of a centrifugal pump is 3 m above the water level in the sump and the static lift from the pump centre is 35 m. The friction losses in the suction and the delivery pipes are 1 m and 8 m respectively. The suction and delivery pipes are each 15 cm in diameter. The impeller is 30 cm in diameter and 1.5 cm wide at outlet and its speed is 1700 r.p.m. The water at inlet has radial flow and the blade angle at outlet is 30° to the tangent to the periphery. Calculate the power to be supplied and the discharge in lit/min assuming a manometric efficiency of 77% and mechanical efficiency of 72%.

17.5. Show that the rise in pressure head through the impeller of a centrifugal pump is given by

$$\frac{1}{2g}(V_{f_1}^2 + u_2^2 - V_{f_2}^2 \text{ coses}^2 \beta)$$

where V_{f_1} and V_{f_2} are the velocities of flow at inlet and outlet respectively, u_2 is the peripheral velocity of the impeller at exit and β is the blade angle at exit. The flow enters radially.

*Source : Submersible Pumps, ISO 9002, Sarvottam Pumps Ltd., Faridabad, India.

17.6. A four stage pump is required to pump 20.5 lit/sec of water against a head of 125 m at 2400 r.p.m. Find the specific speed of the pump. If, however, single stage, double-suction pump is used, what will be the speed. Which one will you select for installation ?

17.7. Estimate the main dimensions and blade angles for an inward flow reaction turbine to suit following conditions :

Net head = 70 m ; speed = 720 r.p.m. ; output = 410 h.p. Assume hydraulic efficiency = 92% ; mechanical efficiency = 84% ; flow ratio = 0.15 ; ratio of wheel width to diameter = 0.1 ; inner diameter of runner = 0.5 outer diameter.

ANSWERS

17.1. 822 r.p.m., 6.2 m of water absolute and 28.03 m of water absolute.

17.2. 3072 lit/min, 35.8 kW **17.3.** 52.1%, 97 kW, 191 r.p.m.

17.4. 2013 lit/min, 21.5 kW **17.6.** 9770.

17.7. $B_1 = 0.057$ m, $D_1 = 0.57$ m, $D_2 = 0.285$ m, blade angles $\beta_1 = 35° \, 06'$, $\beta_2 = 28° \, 55'$.

18

Ideal Fluid Flow

The knowledge of higher mathematics is being increasingly used in the solution of engineering problems. In flow problems where simplifying assumptions can be used, the flow is generally treated as one-dimensional, *i.e.* flow in which the average velocity at each cross-section is used and variations across the section are neglected. But flow fields in which such assumptions are not justified, one dimensional methods of flow analysis cannot be applied and the use of methods of higher mathematics become necessary. Many problems in fluid flow require more exact knowledge of velocity and pressure distributions such as in flow over curved boundaries of an airplane wing or over the crest of spillway. Treating the fluid as non-viscous and incompressible, many real fluid flow situations can be analysed to provide theoretical solutions from which greater insight and understanding of the fluid-flow phenomenon can be obtained. The treatment in this chapter is confined to the steady flow of incompressible and non-viscous (*i.e.* ideal) fluids and two-dimensional flow fields mostly in the horizontal plane.

In Chapter 3 on "Kinematics of Fluid Motion", circulation and vorticity, velocity potential, stream function, relation between stream function and velocity potential, flownet, vortex flow, free-vortex and forced vortex have been dealt with in articles 3.7 to 3.15.

18.1. Rectilinear (Uniform) Flow

In analysing the flow about solid bodies immersed in a fluid stream, the approaching flow is frequently of infinite extent and possesses straight and parallel streamlines and uniform velocity distribution. Such a flow with velocity U in horizontal direction (along X-axis) is shown in Fig. 18.1.

For this type of flow $u = U, \qquad v = 0$

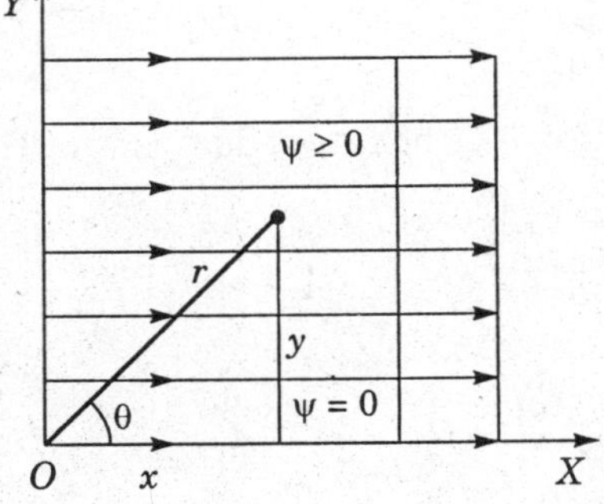

(*a*) Notations for horizontal uniform flow.

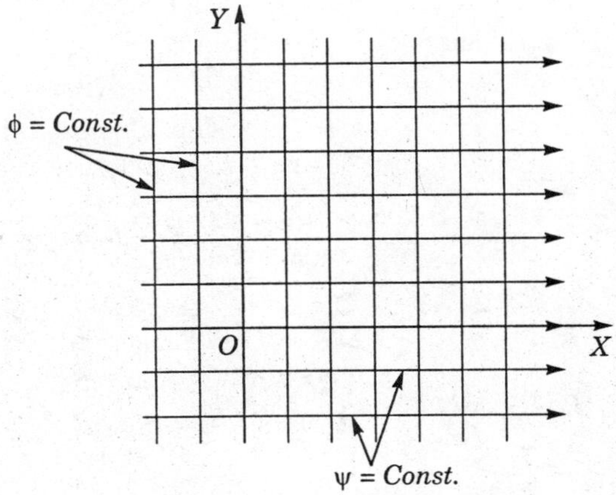

(*b*) Streamlines and equipotential lines for uniform flow.

Fig. 18.1

and the stream function $\psi = \int \frac{\partial \psi}{\partial y} . dy + \int \left(-\frac{\partial \psi}{\partial x}\right) dx + C = \int U\, dy + \int (0)\, dx + C = U.y + C$

If the streamline at $y = 0$ is termed as $\psi = 0$, the constant C vanishes and the stream function for the flow field becomes

$$\psi = U . y \qquad \text{...(18.1)}$$

From Eq. (18.1) it is evident that the streamlines for which ψ = constant are lines parallel to the X-axis. In polar co-ordinates

$$\psi = U . r \sin \theta \qquad \text{...(18.2)}$$

The velocity potential for this flow field may be obtained as

$$u = -\frac{\partial \phi}{\partial x} = U, \; v = -\frac{\partial \phi}{\partial y} = 0$$

$$\phi = \int \frac{\partial \phi}{\partial x} dx + \int \frac{\partial \phi}{\partial y} dy + D$$

$$= \int (-U) dx + \int (0)\; dx + D = -Ux + D$$

when $x = 0$, let $\phi = 0$, $\therefore \; D = 0$

and $\phi = -U . x = -U . r \cos \theta$ ...(18.3)

It may be recalled that ϕ = constant lines are the equipotential lines, which from Eq. (18.3) can be seen to be parallel to the Y-axis. Fig. 18.1 (*b*) shows a flownet representation consisting of streamlines intersected orthogonally by a set of equipotential lines. As can be seen the flownet is made up of perfect squares which maintain the same size throughout. Both the velocity as well as the pressure thus remain constant throughout the flow field.

18.2. Source and Sink

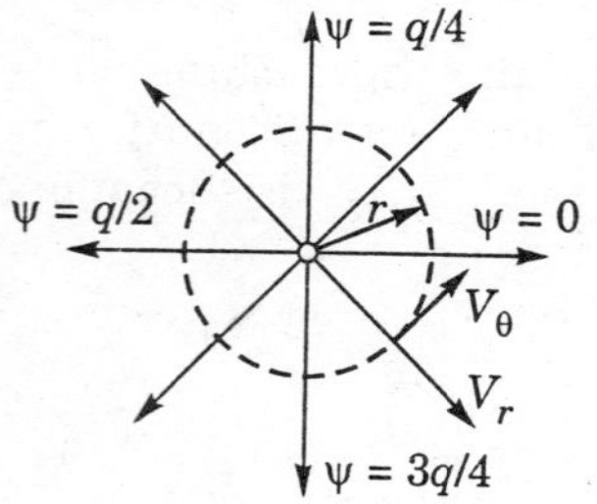

Fig. 18.2. Source flow.

Consider a symmetrical flow field consisting of radial streamlines directed outward from a common point (Fig. 18.3). Continuity principle shows that the velocities will diminish as the streamlines spread and the symmetry will require that all velocities will be the same at the same radial distance from the origin. Across all circles, the same discharge will pass, and thus the velocity at any point r in the flow field may be determined thus :

$$V_r = \frac{q}{2\pi r} \quad \text{and} \quad V_\theta = 0$$

The "*strength*" of the source q is the rate of flow passing through any circle enclosing the source.

$$\psi = \int \frac{\partial \psi}{r \partial \theta} . r d\theta + \int \frac{\partial \psi}{\partial r} . dr + C$$

$$= \int \left(\frac{q}{2\pi r}\right) r d\theta + \int (0)\, dr + C = \frac{q\theta}{2\pi} + C$$

Let for $\theta = 0°$, $\psi = 0$ $\therefore \; C = 0$

$$\psi = \frac{q}{2\pi} . \theta \qquad \text{...(18.4)}$$

and the velocity potential $\phi = \int \left(\frac{\partial \phi}{r\partial \theta}\right) r d\theta + \int \left(\frac{\partial \phi}{\partial r}\right) dr + C$

$$= \int (-V_\theta) r d\theta + \int \left(-\frac{q}{2\pi r}\right) dr + C = C - \frac{q}{2\pi} \log r$$

Since $V_\theta = -\frac{\partial \phi}{r\partial \theta} = 0$

and $V_r = -\frac{\partial \phi}{\partial r} = \frac{q}{2\pi r}$

Let when $r = 1, \phi = 0,$ $\therefore$ $C = 0$ and

$$\phi = -\frac{q}{2\pi} \log r \quad \text{...(18.5)}$$

The exact opposite of the outward flow from a source is the inward flow to a *sink* whose stream function may be written as

$$\psi = -\frac{q\,\theta}{2\pi} \quad \text{...(18.6 } a\text{)}$$

and the velocity potential for sink may be expressed as

$$\phi = \frac{q}{2\pi} \log r \quad \text{...(18.6 } b\text{)}$$

A *sink* is thus the reverse of a source. It is a point into which fluid flows radially from all directions. Algebraically, a sink is represented as the negative of a source.

18.3. Combining Flow Fields by Superposition

Flow fields of considerable engineering importance may be developed by suitably combining uniform rectilinear, source, sink and vortex flows. Consider two flow fields represented by stream functions ψ_A and ψ_B. Let at a point P in a flow field ψ_A, the velocity tangential to the streamline passing through P be V_A ; at the same point in a flow filed ψ_B, the similar velocity be V_B. When these flows are superposed the resulting velocity at the point must be resultant (vector sum) of the two velocities V_A and V_B. At other points the same vector addition of velocities occurs and a combined flow field ψ results.

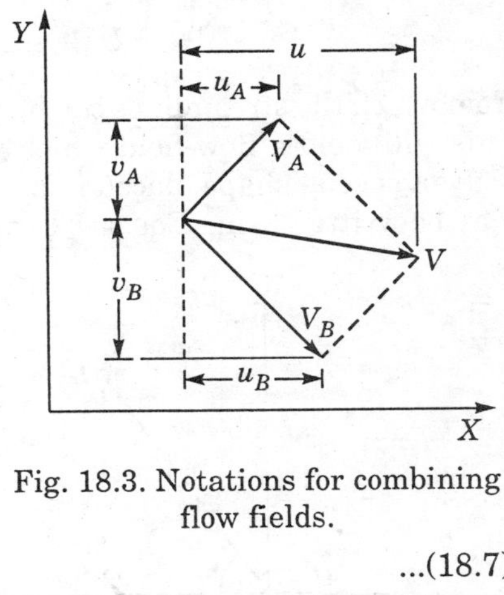

Fig. 18.3. Notations for combining flow fields.

From Fig. 18.3. one may write

$$u = u_A + u_B$$

and using the relationship between velocity component and stream function the above equation may be written as

$$\frac{\partial \psi}{\partial y} = \frac{\partial \psi_A}{\partial y} + \frac{\partial \psi_B}{\partial y} = \frac{\partial}{\partial y}(\psi_A + \psi_B)$$

from which $\psi = \psi_A + \psi_B$ ...(18.7)

It may thus be seen that the stream-function of resultant (combined) flow field is obtained by the algebraic addition of the stream functions of the individual flow fields. When the stream functions of a flowfield are not known as a function of x and y, the graphical method is the only alternative which may be used to combine the flow-fields. The graphical approach to such problems uses the definition of the stream-function and considers the flowrates between streamline and the origin for both the individual and combined flow fields. Let us consider the two flow fields represented by the stream functions ψ_A and ψ_B. Fig. 18.4 shows the stream lines for both the flows.

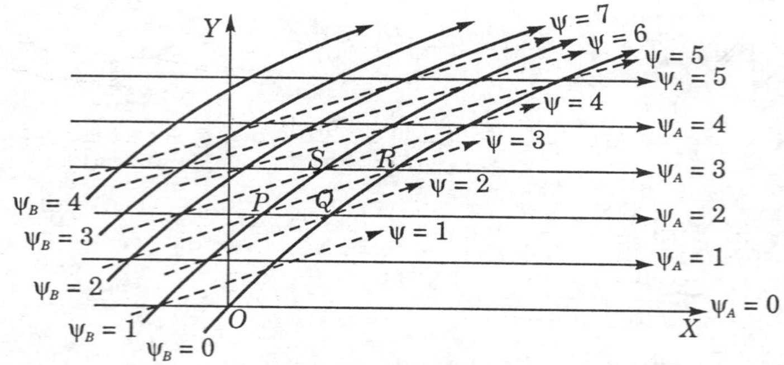

Fig. 18.4. Graphical method of determining streamlines of combined flow.

Across any line drawn between O and P will pass a flowrate of 2 cumecs/m in flow field A and 1 cumecs/m in flow field B. Therefore, for the combined flow field the flow rate will be 3 cumecs/m. Thus for the combined flow field the streamline passing through P has a value $\psi = 3$. The same streamline will pass through all other points at which $\psi_A = \psi_B = 3$. The streamlines for the combined flow can thus be plotted by joining the appropriate diagonally opposite corners (like PR) of parallelograms. One such parallelogram is $PQRS$; joining corners P and R we get points on $\psi = 3$. This method of determining ψ and constructing streamlines is called *"Rankine's method"*.

Obtaining ψ for the combined flow this way, the velocity components at any point in the flow field may be obtained from $\partial\psi/dx$ and $\partial\psi/dy$ and from these the resultant velocity vector at that point. The Bernoulli's equation may then be used to determine the pressure at the point.

18.4. Some Combined Flow Fields of Engineering Importance

18.4.1. Source in a Rectilinear Flow

Considering the source at the origin of co-ordinates, its superposition with the uniform flow will yield the stream function for the combined flow

$$\psi = U \, . \, r \sin \theta + \frac{q\,\theta}{2\pi} = U \, . \, y + \frac{q\,\theta}{2\pi} \quad ....(18.8)$$

from which the streamlines can be easily drawn using Rankine's method described in the foregoing. This will yield a flow-field which will represent the streamline picture about the nose of a solid body of special shape placed in a free stream (Fig. 18.5 *b*). The velocity potential for the flow-field may be written using Eqs. (18.3) and (18.5)

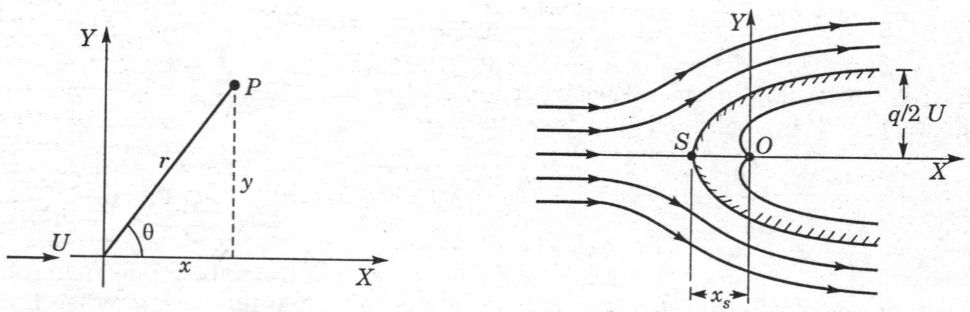

(*a*) Notations for co-ordinates systems.

(*b*) Flow pattern due to source placed in a rectilinear flow.

Fig. 18.5

$$\phi = -\,U \, . \, x - \frac{q}{2\pi} \log r = -\,U \, . \, x - \frac{q}{2\pi} \log \sqrt{x^2 + y^2}$$

$$= -U \, . \, r \cos\theta - \frac{q}{2\pi} \log r \qquad \text{...(18.9)}$$

and the velocity components in (x, y) and (r, θ) directions are :

$$u = \frac{\partial \phi}{\partial x} = U + \frac{\partial}{\partial x}\left(\frac{q}{2\pi} \log \sqrt{x^2 + y^2}\right)$$

$$= U + \frac{qx}{2\pi r^2} = U + \frac{q \cos\theta}{2\pi r} \qquad \text{...(18.10)}$$

$$v = -\frac{\partial \phi}{\partial y} = \frac{qy}{2\pi r^2} = \frac{q}{2\pi r} \sin\theta \qquad \text{...(18.11)}$$

$$v_r = -\frac{\partial \phi}{\partial r} = U \cos\theta + \frac{q}{2\pi r} \qquad \text{...(18.12)}$$

$$v_\theta = -\frac{\partial \phi}{r \partial \theta} = -\frac{1}{r}\,[U \, . \, r \sin\theta] = -U \sin\theta \qquad \text{...(18.13)}$$

Velocity components of flow due to the source located at O vary inversely with the radial distance and near the source they are very high. Accordingly, the effect of the source will be zero at a large distance from O and the free stream velocity at such a location will not be affected by the presence of the source. In the negative X-direction, however, at some distance from O, the velocity U of the free stream must equal the velocity due to source. This condition determines the stagnation point S, whose distance x, from the origin is given by

$$U = \frac{q}{2\pi x_s} \qquad \text{...(18.14)}$$

which shows x_s to be dependent upon the source strength q and the free-stream velocity U. Fluid particles leaving the source and moving to S cannot proceed further to the left, so they must move up or down from S and are carried to the right along a streamline which separates the source flow from that of the free stream. This stream line which passes through the stagnation point S [shown hatched in Fig. 18.5 (b)] may be considered to be the contour of a solid body placed in the original free-stream flow of velocity U. The equation of the body-contour may be completely determined by knowning the value of ψ for the streamline. Using the co-ordinates of stagnation point and Eq. (18.8), the value of ψ for the streamline passing through the stagnation point is

$$\psi_s = U \, . \left(\frac{q}{2\pi U}\right) \sin\pi + \frac{q\pi}{2\pi} = \frac{q}{2} \qquad \text{...(18.15)}$$

and the equation of the stream line passing through S, $\psi = \psi_s$ and hence that of the body contour is given by

$$Ur \sin\theta + \frac{q\theta}{2\pi} = \frac{q}{2}$$

or

$$y = r \sin\theta = \frac{q}{2U}\left(1 - \frac{\theta}{\pi}\right) \qquad \text{...(18.16)}$$

Equation (18.16) represents the streamline passing through the stagnation point S and divides the flow into two regions. Hence, this dividing streamline can be replaced by a solid wall. The maximum value of the height of the wall occurs at $\theta = 0$ ($x \to \infty$) and is given by Eq. (18.16) as

$$y = \frac{q}{2U} \qquad \text{...(18.17)}$$

Equation (18.17) shows that the body contour of the solid wall approaches this line asymptotically.

The body contour formed by a source in a rectangular flow is unique in the sense that it has a head or nose but no tail and is known classically as a "half-body". Half-bodies may be used in studying the velocity and pressure distribution near upstream of symmetrical objects such as bridge piers, airplane stabilizers, struts or torpedo and aircraft fuselages for the axisymmetric three-dimensional case.

18.4.2. Doublet

A doublet is a limiting combination of a source and a sink of equal strength, which are allowed to approach each other in such a manner that the product of their strength and the distance between them remains a constant. The axis of the doublet is from the sink towards the source.

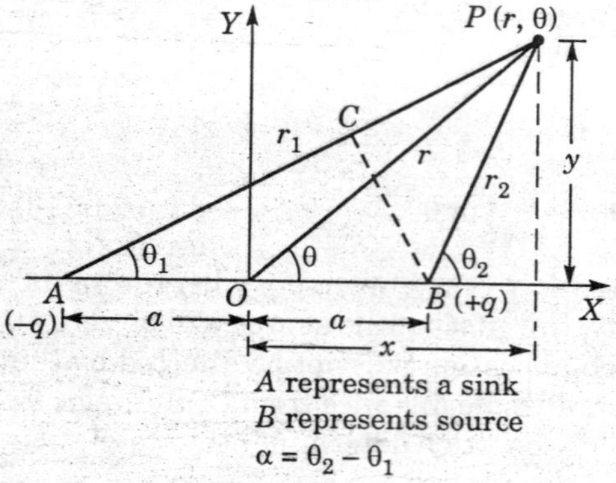

Fig. 18.6. Notations for a two dimensional doublet.

In Fig. 18.6, a source of strength q is located at $(a, 0)$ and a sink of equal strength at $(-a, 0)$. The stream function of the combined flow field will be

$$\psi = \frac{q}{2\pi}(\theta_2 - \theta_1)$$

$$= \frac{q}{2\pi} \qquad \text{...(18.18)}$$

Considering the common side BC of triangles ABC and BCP

$$AB = r_2 \sin \alpha = 2a \sin \theta_1 \qquad \text{...(18.19)}$$

As $a \to 0$, the following quantities assume their limiting values as follows

$$\alpha \to 0, \sin \alpha \to \alpha, r_2 \to r_1 \to r$$

and $\quad \theta_2 \to \theta_1 \to \theta.$

For this limiting case, Eq. (18.19) reduces to

$$r\,\alpha = 2a \sin \theta \qquad \text{...(18.20)}$$

and Eq. (18.18) upon making substitution from Eq. (18.20) changes to

$$\psi = \frac{q}{2\pi}\left(\frac{2a \sin \theta}{r}\right) \qquad \text{...(18.21)}$$

Considering now a condition under which as $a \to 0$, q is increased such that $2qa$ remains constant equal to m,

$$m = 2qa$$

here m is known as the "*strength of the doublet*". The stream function of the doublet may now be written as

$$\psi = \frac{m \sin \theta}{2\pi r} \qquad \text{...(18.22)}$$

The velocity potential for the doublet

$$\psi = \frac{q}{2\pi}(\log r_1 - \log r_2)$$

To take the limit as a approaches zero and q increased in such a manner that $2qa = m$, the form of the above expression for ϕ must be modified as under :

$$r_2^2 = r^2 + a^2 - 2ar \cos \theta = r^2\left[1 + \left(\frac{a}{r}\right)^2 - 2\left(\frac{a}{r}\right)\cos \theta\right]$$

$$r_1^2 = r^2 + a^2 + 2ar \cos \theta = r^2\left[1 + \left(\frac{a}{r}\right)^2 + 2\left(\frac{a}{r}\right)\cos \theta\right]$$

Rewriting the expression for ϕ as follows :

$$\phi = \frac{q}{4\pi}(\log r_1^2 - \log r_2^2)$$

$$= \frac{q}{4\pi}\left[\log r^2 + \log\left\{1 + \left(\frac{a}{r}\right)^2 + 2\left(\frac{a}{r}\right)\cos\theta\right\} - \log r^2 - \log\left\{1 + \left(\frac{a}{r}\right)^2 - 2\left(\frac{a}{r}\right)\cos\theta\right\}\right]$$

Using the following expression :

$$\log(1 + x) = x - \frac{x^2}{2} + \frac{x^3}{3} - \frac{x^4}{4} + \ldots\ldots$$

$$\phi = \frac{q}{4\pi}\left[\left(\frac{a}{r}\right)^2 + 2\left(\frac{a}{r}\right)\cos\theta - \frac{1}{2}\left\{\left(\frac{a}{r}\right)^2 + 2\left(\frac{a}{r}\right)\cos\theta\right\}^2 + \frac{1}{3}\left\{\left(\frac{a}{r}\right)^2 + 2\left(\frac{a}{r}\right)\cos\theta\right\}^3 - \left\{\left(\frac{a}{r}\right)^2 - 2\left(\frac{a}{r}\right)\cos\theta\right\} + \frac{1}{2}\left\{\left(\frac{a}{r}\right)^2 - 2\left(\frac{a}{r}\right)\cos\theta\right\}^2 - \frac{1}{3}\left\{\left(\frac{a}{r}\right)^2 + 2\left(\frac{a}{r}\right)\cos\theta\right\}^3\right]$$

$$= \frac{aq}{\pi}\left[\frac{\cos\theta}{r} - \left(\frac{a}{r}\right)^2\frac{\cos\theta}{r} + \left(\frac{a}{r}\right)^4\frac{\cos\theta}{r} + \ldots\ldots\right]$$

Using the definition of strength of doublet, *i.e.* $2aq = m$, and taking in the limit as $a \to 0$

$$\phi = \frac{m}{2\pi}\frac{\cos\theta}{r} \qquad \text{...(18.23)}$$

If the positions of sink and source are interchanged so that source is placed upstream, the strength of the doublet becomes $(-m)$ and the stream function

$$\psi = -\frac{m\sin\theta}{2\pi r} \qquad \text{...(18.24 a)}$$

and for the velocity potential

$$\phi = -\frac{m\cos\theta}{2\pi r} \qquad \text{...(18.24 b)}$$

For any constant value of ψ, Eq. (18.22) can be written as

$$r = r_0 \sin\theta \qquad \text{...(18.25)}$$

where $r_0 = m/2\pi\,\psi_c$ and ψ_c is an arbitrary constant for ψ. Equation (18.25) represents a circle which passes through the origin with centre on the y-axis and a diameter of r_0. Similarly for any arbitrary constant value of ϕ, Eq. (18.23) can be expressed as :

$$r = R_0 \cos\theta \qquad \text{...(18.26)}$$

where $R_0 = m/2\pi\,\phi_c$; ϕ_c being an arbitrary constant for ϕ. The above equation represents a circle which passes through the origin with centre on the x-axis and a diameter of R_0. The streamlines and equipotential lines for the doublet flow are shown in Fig. 18.7 (a) and 18.7 (b) respectively.

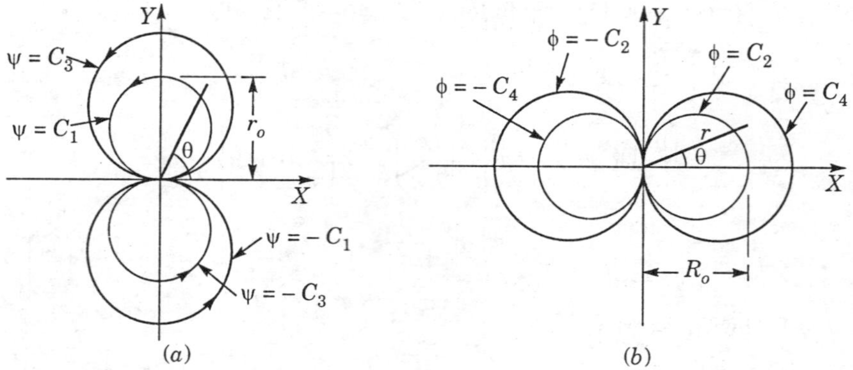

Fig. 18.7. Streamlines for doublet flow. Equipotential lines for doublet flow.

18.4.3. Doublet in Rectilinear Flow

If a doublet is placed in a uniform rectilinear flow of magnitude U in the positive direction of x-axis (see Fig. 18.8), a very useful flow field is obtained. The stream function for the combined flow is determined, using Eqs. (18.1) and (18.24)

$$\psi = Ur \sin\theta - \frac{m \sin\theta}{2\pi r} \quad ...(18.27)$$

For streamline $\psi = 0$, the above equation reduces to

$$\left(U - \frac{m}{2\pi r^2}\right) \sin\theta = 0 \quad ...(18.28)$$

the solution of which results in

$$\left.\begin{aligned} &\theta = 0 \text{ or } \pi, \quad \text{and} \\ &r = \left(\frac{m}{2\pi U}\right)^{1/2} = R \end{aligned}\right\} \quad ...(18.29)$$

Hence $\psi = 0$ represents the streamline which is a circle of radius R joined by x-axis on both sides. A cylinder of radius R may be taken as a solid boundary. The stream function and the velocity potential for uniform flow around a circular cylinder of radius R, by substituting for m from Eq. (18.29) are :

$$\psi = U\left(r - \frac{R^2}{r}\right) \sin\theta \quad ...(18.30)$$

$$\phi = -U\left(r + \frac{R^2}{r}\right) \cos\theta \quad ...(18.31)$$

The tangential and radial components of velocity are

$$v_\theta = -\frac{\partial\phi}{r\partial\theta} = -U\left(1 + \frac{R^2}{r^2}\right) \sin\theta \quad ...(18.32)$$

$$v_r = -\frac{\partial\phi}{\partial r} = U\left(1 - \frac{R^2}{r^2}\right) \cos\theta \quad ...(18.33)$$

The validity of the following boundary conditions can be easily checked

(*i*) at $r = R$, $v_r = 0$; this is obvious from Eq. (18.33),

(*ii*) at $r \to \infty$, $u = U$ and $v = 0$; this can be verified by writing the following equation from Fig. 18.8.

$$v = v_r \sin\theta + v_\theta \cos\theta$$
$$= U(\sin\theta\cos\theta - \cos\theta\sin\theta) = 0$$
$$u = v_r \cos\theta - v_\theta \sin\theta\, U\cos^2\theta + U\sin^2\theta = U.$$

From Eq. (18.32), the tangential velocity component on the surface of the cylinder ($r = R$) is

$$v_\theta = -2U\sin\theta \qquad \text{...(18.34)}$$

As shown in Fig. 18.8, a positive sign for v_θ indicates an anti-clockwise direction. The negative sign appearing in Eq. (18.34) indicates that v_θ is in clockwise direction. The equation also reveals that the stagnation points are located at $\theta = 0$ and π. At $\theta = \pm\pi/2$, the tangential velocity v_θ attains a maximum equal to $2U$.

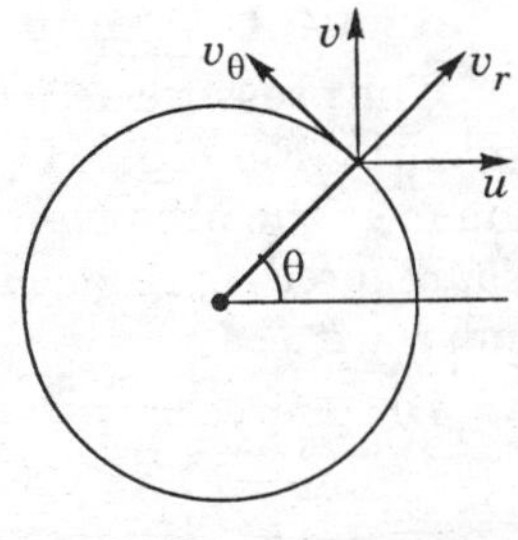

Fig. 18.8. Velocity components on the surface of a cylinder.

$$|v_\theta|_{max} = 2U \qquad \text{...(18.35)}$$

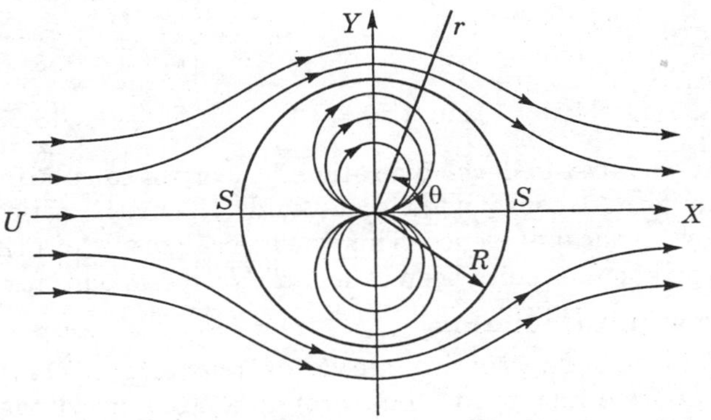

Fig. 18.9. Streamline pattern in a combined flow due to a doublet placed in a rectilinear flow.

The pressure at any point on the surface of the cylinder can be determined by using the Bernoulli's equation between a point far away in the free stream (where pressure is p_0 and velocity U) and a point on the cylinder where pressure is p and velocity v_θ. One may, therefore, write :

$$p + \rho\frac{v_\theta^2}{2} = p_0 + \rho\frac{U^2}{2}$$

By substituting for v_θ from Eq. (18.34).

$$p - p_0 = \frac{\rho}{2}(U^2 - v_\theta^2)$$

$$= \rho\frac{U^2}{2}(1 - 4\sin^2\theta) \qquad \text{...(18.36)}$$

The above equation gives the values and locations of maximum, minimum and zero pressure as follows :

(*i*) maximum pressures : $\dfrac{p - p_0}{\dfrac{\rho U^2}{2}} = 1$ at $\theta = 0$ and π ...(18.37 *a*)

(*ii*) minimum pressures : $\dfrac{p - p_0}{\dfrac{\rho U^2}{2}} = -3$ at $\theta = \dfrac{\pi}{2}$ and $\dfrac{3\pi}{2}$ ...(18.37 *b*)

(*iii*) zero pressure : $\dfrac{p - p_0}{\dfrac{\rho U^2}{2}} = 0$ at $\theta = \pm\dfrac{\pi}{6}$ and $\pm\dfrac{5\pi}{6}$ ...(18.37 *c*)

From symmetry of the pressure distribution, it is easy to show that there are no drag and lift forces acting on the cylinder in a rectilinear flow. Streamline pattern for such a flow is shown in Fig. 18.9.

18.4.4. Drag and Lift Forces on Cylinder

The pressure acting on an area element $Rd\theta$ of the cylinder surface in given by Eq. (18.36) and the differential pressure $(p - p_0)$ is shown acting over the area element, Fig. 18.10. The x and y components of the pressure force representing the drag and lift forces per unit length of the cylinder are :

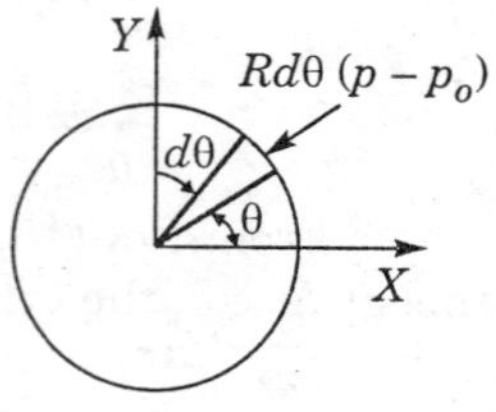

Fig. 18.10

Drag force
$$D = F_x$$
$$= - \int_0^{2\pi} (p - p_0)\, Rd\theta \cos\theta$$
$$= -\rho \frac{U^2}{2} \int_0^{2\pi} (1 - 4\sin^2\theta)\cos\theta\, d\theta = 0 \qquad ...(18.38)$$

and lift force
$$L = F_y$$
$$= - \int_0^{2\pi} (p - p_0)\, R\, d\theta \sin\theta$$
$$= -\rho \frac{U^2}{2} \int_0^{2\pi} (1 - 4\sin^2\theta)\sin\theta\, d\theta = 0 \qquad ...(18.39)$$

The French mathematician d'Alembert was first to develop a rational theory of calculating the resistance acting on a body immersed in a uniform stream of non-viscous fluids. His mathematical proof that drag was zero was contradicted by physical evidence. This theoretical result of non-existence of drag on body placed in a uniform stream has since been called *"d' Alemberts paradox"*.

18.4.5. Doublet in Rectilinear Flow with Circulation

Another very useful flow field may be obtained by superposing free vortex (refer article 3.14.2), doublet and rectilinear flow fields. The combination results in a flow around a circular cylinder with circulation (clockwise direction for Γ is assumed). The stream function for clockwise vortex is

$$\psi = U\left(r - \frac{R^2}{r}\right)\sin\theta + \frac{\Gamma}{2\pi}\log r \qquad ...(18.40)$$

The first term of Eq. (18.40), representing a doublet in the rectilinear flowfield, implies that the body contour (a circular boundary of radius R) is unchanged by the superposition of the vortex flow. This is due to the fact that the velocities induced by the vortex are wholly tangential and therefore possess no component normal to the body.

The velocity potential for the combined flow field using Eqs. (3.58 c) and (18.31) is given by

$$\phi = -U\left(r + \frac{R^2}{r}\right)\cos\theta + \frac{\Gamma\theta}{2\pi} \qquad ...(18.41)$$

The velocity components anywhere in the flow field are :

$$v_\theta = -\frac{\partial\psi}{\partial r} = -U\left(1 - \frac{R^2}{r^2}\right)\sin\theta - \frac{\Gamma}{2\pi r} \qquad ...(18.42)$$

$$v_r = -\frac{\partial\psi}{r\partial\theta} = U\left(1 - \frac{R^2}{r^2}\right)\cos\theta \qquad ...(18.43)$$

From Eqs. (18.42) and (18.43), it is evident that the radial velocity v_r remains unchanged from its value for the case without circulation, and the tangential velocity v_θ does not alter the condition at $r \to \infty$ as the second term of Eq. (18.42) tends to zero.

On the body contour (*i.e.* on the surface of the cylinder) $r = R$, the velocity components are :

$$v_\theta = -2U \sin\theta - \frac{\Gamma}{2\pi R} \quad \text{...(18.44)}$$

$$v_r = 0,$$

from which the stagnation points may be determined. At stagnation points, where v_θ must vanish, let $\theta = -\alpha$

$$0 = -2U \sin(-\alpha) - \frac{\Gamma}{2\pi R}$$

$$\sin\alpha = \frac{\Gamma}{4\pi RU} \quad \text{...(18.45)}$$

When the flowfield is plotted, it will be found that for $\Gamma < 4\pi RU$, there are two symmetrically located stagnation points as shown in Fig. 18.11. Increasing the strength of circulation, the value of α is seen to attain a maximum value of $\pi/2$ where the two stagnation points merge into one.

Hence for $\quad \alpha = \frac{\pi}{2}$,

$$\sin\alpha = 1$$

and $\quad \Gamma = 4\pi UR$

The flow field for this condition is shown in Fig. 18.12. For $\Gamma > 4\pi UR$, the stagnation point will be found to be located at some distance below the circle, on the vertical axis of symmetry as shown in Fig. 18.13.

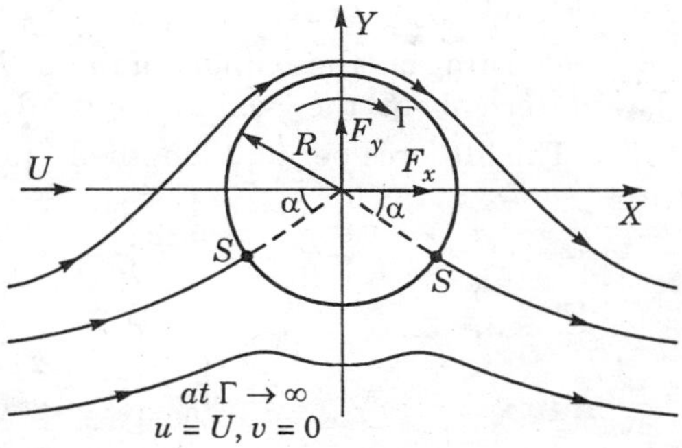

Fig. 18.11. Flow field in combined flow represented by a Doublet, free vortex and horizontal rectilinear flow when $\Gamma < 4\pi UR$.

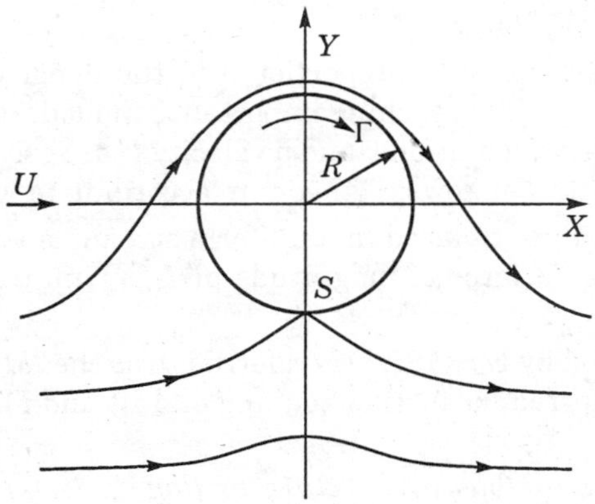

Fig. 18.12. Flow pattern for $\Gamma = 4\pi UR$.

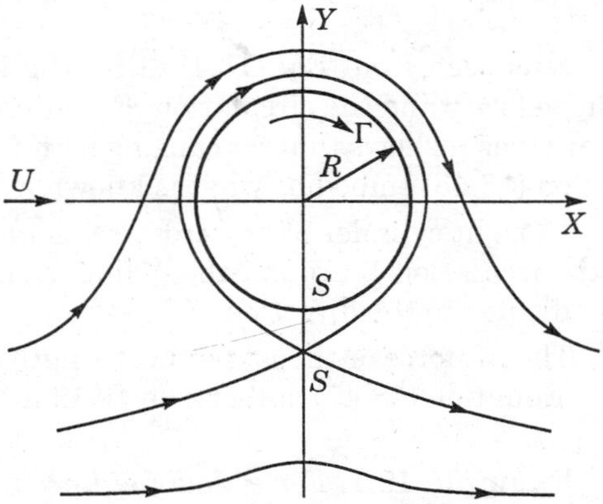

Fig. 18.13. Flow pattern for $\Gamma > 4\pi UR$.

The pressure at any point in the flow field may be found by using Bernoulli's equation. Applying this equation between a point in the free stream (where the velocity is U and pressure p_0) and another point on the surface of cylinder where pressure is p and velocity is

$$v_\theta = -2U \sin\theta - \frac{\Gamma}{2\pi R},$$

one may write

$$p + \frac{\rho}{2}\left(2U \sin\theta + \frac{\Gamma}{2\pi R}\right)^2 = p_0 + \frac{\rho}{2} U^2$$

or

$$p - p_0 = \frac{1}{2}\rho U^2 - \frac{1}{2}\rho\left(2U \sin\theta + \frac{\Gamma}{2\pi R}\right)^2 \quad ...(18.46)$$

With reference to Fig. 18.10, the x and y-components of pressure force representing the drag and lift forces are

Drag force

$$D = F_x$$

$$= -\int_0^{2\pi} (p - p_0)\, Rd\,\theta \cos\theta.$$

The drag on the cylinder of radius R is again zero [refer Eq. (18.38)] because of symmetry of flow pattern about the y-axis.

The lift force per unit length of the cylinder, however, is

$$L = F_y = -\int_0^{2\pi} (p - p_0)\, Rd\theta \sin\theta\, (p - p_0)\, Rd\theta \sin\theta$$

$$= -\int_0^{2\pi}\left[\frac{1}{2}\rho U^2 - \frac{1}{2}\rho\left(2U\sin\theta + \frac{\Gamma}{2\pi R}\right)^2\right] R \sin\theta\, d\theta \quad ...(18.47)$$

since $\int_0^{2\pi} \sin\theta\, d\theta = \int_0^{2\pi} \sin^3\theta\, d\theta = 0$

Equation (18.47) simplifies to

$$L = \rho\,\frac{U\Gamma}{\pi}\int_0^{2\pi} \sin^2\theta\, d\theta = \rho\,\frac{U\Gamma}{\pi}\int_0^{2\pi} \frac{1}{2}(1 - \cos 2\theta)\, d\theta = \rho U\Gamma \quad ..(18.48)$$

It is seen from Eq. (18.48) that the lift force is directly proportional to the density, free stream velocity and the circulation. Equation (18.48) is of far-reaching importance in modern fluid mechanics as it shows that for causing a lift force both a circulation Γ and a velocity U are essential. Equation (18.48) embodies what is known as the 'Kutta-Joukowski" theorem, according to which :

"A right cylinder of any cross-sectional shape at rest placed in a uniform stream of velocity U with circulation Γ around the cylinder experiences a force of magnitude $\rho U\Gamma$ per unit length perpendicular to the direction of flow."

The lift force or thrust per unit length expressed by Eq. (18.48) is referred to as the "*Magnus effect*" [Equation (18.47) and article 10.13 may be referred for further reading on Lift and Magnus effect.]

Example 18.1. *For a flow field produced by a doublet and rectilinear flow, determine the velocity potential and sketch the equipotential lines.*

Solution. The radial and tangential components of velocity are :

$$v_r = U\left(1 - \frac{R^2}{r^2}\right)\cos\theta$$

$$v_\theta = -U\left(1 + \frac{R^2}{r^2}\right)\sin\theta$$

which are related to the velocity potential as under

$$v_r = -\frac{\partial\phi}{\partial r} \quad \text{and} \quad v_\theta = -\frac{\partial\phi}{r\partial\theta}$$

and,
$$\phi = \int \frac{\partial\phi}{\partial r}\,dr + \int \frac{\partial\phi}{r\partial\theta}\,r d\theta + B.$$

Here B is the constant of integration. Making appropriate substitutions.

$$\phi = \int\left\{-U\left(1-\frac{R^2}{r^2}\right)\cos\theta\,dr + \int U\left(1+\frac{R^2}{r^2}\right)\sin\theta\,rd\theta + B\right.$$

$$= -U\left[\int(\cos\theta)\,dr - \int(r\sin\theta)\,d\theta\right] + UR^2\left[\int\left(\frac{\cos\theta}{r^2}\right)dr + \int\left(\frac{\sin\theta}{r}\right)d\theta\right] + B$$

The quantities within the brackets may be replaced by the left hand side of the following equations :

$$\int d(r\cos\theta) = \left[\int(\cos\theta)\,dr - \int(r\sin\theta)d\theta\right]$$

and,
$$\int d\left(\frac{-\cos\theta}{r}\right) = \left[\int\left(\frac{\cos\theta}{r^2}\right)dr + \int\left(\frac{\sin\theta}{r}\right)d\theta\right]$$

Hence,
$$\phi = -U\int d\,(r\cos\theta) + UR^2\int d\left(\frac{-\cos\theta}{r}\right) + B$$

$$= -U\,r\cos\theta - \frac{UR^2\cos\theta}{r} + B \quad = -U\left(r+\frac{R^2}{r}\right)\cos\theta + B$$

assuming $B = 0$, the following may be obtained :

(*i*) For $\phi = 0$, $\theta = \pi/2$ or $3\pi/2$, thus the equipotential line $\phi = 0$ coincides with the y-axis.

(*ii*) For $\theta \to \pi$ and $r \to \infty$, $\phi \to +\infty$, and for $\theta = 0$ and $r \to \infty$, $\phi \to -\infty$.

(*iii*) For $\theta = \pi/3$ and $r = R$, $\phi = -U$

(*iv*) For $\theta = \pi/4$ and $r = R$, $\phi = -\sqrt{2}\,U$

(*v*) For $\theta = \pi/6$ and $r = R$, $\phi = -\sqrt{3}\,U$.

The equipotential lines are sketched and shown in Fig. Ex. 18.1.

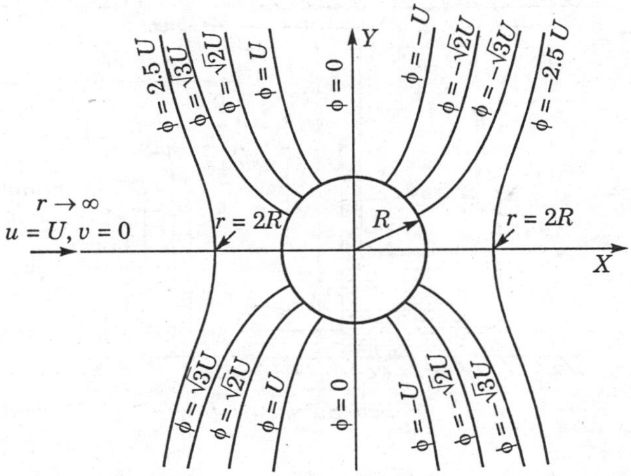

Fig. Ex. 18.1. Equipotential Lines.

19

Hydraulic Systems and Devices

A hydraulic system or a device is an arrangement consisting of items like cylinders, ram/piston, check valves and connecting passages through which liquid pressure is transmitted to carry out the assigned job. The principle of all hydraulic systems is based on Pascal's law which states that *"Pressure exerted on a confined liquid is transmitted undiminished in all directions"*.

A hydraulic transmission may be defined as a system of energy transmission by means of a fluid such as oil. The working element of a hydraulic transmission is basically a cylinder with a movable piston and rod. Hydraulic systems are widely used in aircrafts for operating retractible undercarriages, flaps etc. Such systems are powered by pumps like gear pump used with automatic relief valves which bypass the hydraulic fluid when peak pressure is reached.

Hydraulic energy is utilised in many practical devices such as hydraulic press, hydraulic lift, hydraulic accumulator, hydraulic intensifier, hydraulic coupling, hydraulic torque converter.

An accumulator is commonly incorporated in hydraulic systems to cater for short-period peak demands above pump capacity. This enables a relatively by capacity pump to be used. Additionally, in systems with automatic bypass an accumulator compensates for leakage from high pressure to the low pressure side when the pump is idling.

19.1. Hydraulic Press/Jack

In Fig. 19.1 is shown a hydraulic device which can operate either as a press or a jack. When working as a press, the object to be pressed is placed between the top rigid support attached to the foundation by vertical columns and the ram of diameter D. In case of its working as a jack, the load to be lifted will be placed on the movable ram. The hand-driven pump, with a plunger of diameter d having suction and discharge valves as shown, is used to supply pressure into the cylinder, and

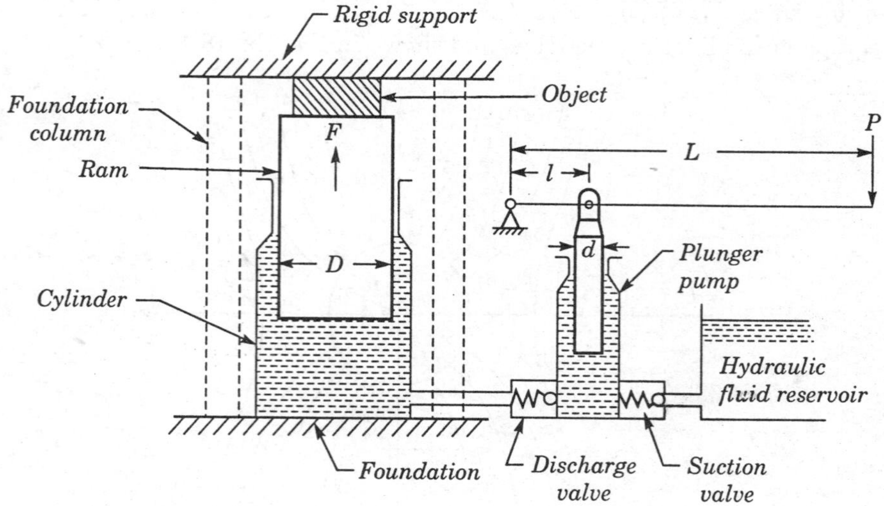

Fig. 19.1. Hydraulic Press-a schematic diagram.

helps in developing the desired thrust F, by application of effort P on the operating lever at the other end.

Various parts of a hydraulic press are shown in Fig. 19.1. The rigid support is attached to the foundation by means of vertical columns shown by broken lines. The object to be pressed is placed in between the support and the ram as shown. The hand-driven plunger pump is operated by a lever arrangement through which an effort P is applied to move the plunger. When the plunger moves up, the suction valve is opened while the delivery valve remains shut. The hydraulic fluid from the reservoir is sucked into the pump. During the downward movement of the plunger, the suction valve is closed due to the pressure developed and the delivery valve gets opened. The hydraulic fluid is thus pumped into the cylinder. The hydraulic pressure so developed causes an upward thrust F and the ram.

When an effort of magnitude P is applied on the operating lever, the force transmitted on the plunger is

$$P = \left(\frac{L}{l}\right)$$

Here L = total length of the operating lever as shown, and

l = distance between the hinge and the plunger as shown.

The pressure developed in the hydraulic fluid of the pump

$$P = \frac{4PL}{\pi d^2 l} \quad ...(19.1)$$

is transmitted into the cylinder due to which the ram experiences an upward force F

$$F = p\,.\frac{\pi D^2}{4} = P\left(\frac{L}{l}\right)\left(\frac{D}{d}\right)^2 \quad ...(19.2)$$

19.1.1. Mechanical Advantage. It is defined as the ratio of the weight lifted to the force applied on the lever.

$$\text{Mechanical Advantage} = \frac{F}{P} \quad ...(19.3)$$

The ratio $\left(\frac{L}{l}\right)$ is called the leverage of the hydraulic press.

Example 19.1. *A hydraulic press has a ram diameter of 250 mm and a plunger diameter of 30 mm. For a force of 40 N applied on the plunger, find the force exerted on the object by the hydraulic press.*

Solution. Pressure developed in the fluid by the plunger,

$$p = \frac{\text{Force applied}}{\text{Area of plunger}} = \frac{40}{\frac{\pi}{4}\left(\frac{30}{1000}\right)^2} \text{ N/m}^2$$

Following Pascal's law, the same fluid pressure is transmitted to the ram. Hence the force exerted on the object by the ram

$$F = p \times \text{Area of ram} = \frac{40}{\left(\frac{30}{1000}\right)^2} \times \left(\frac{250}{1000}\right)^2 = \mathbf{2777.8\ N.}$$

Example 19.2. *A hydraulic press has a plunger diameter 20 mm and the ram diameter 160 mm. The stroke of the plunger is 200 mm, and the weight lifted by the press is 10 kN. The distance travelled by the weight is 600 mm in 10 minutes. Determine :*

(*i*) *Force applied to the plunger,*

(*ii*) *power required to move the plunger, and*

(*iii*) *the number of strokes of the plunger*

Solution. Referring to Fig. 19.1, the pressure developed in the fluid in the ram cylinder

$$p = \frac{10 \times 1000}{\frac{\pi}{4} \times \left(\frac{160}{10 \times 100}\right)^2} \text{ N/m}^2 = 497.265 \text{ kN/m}^2$$

On account of continuity of fluid, the same pressure is transmitted in the plunger pump, hence

(*i*) *the force applied to the plunger* = p × area of plunger

$$= 497.265 \times \frac{\pi}{4}\left(\frac{20}{10 \times 100}\right)^2 = 0.1562 \text{ kN} = \mathbf{156.2 \text{ N}}$$

(*ii*) *the power required to move the plunger* = $\dfrac{\text{Work done}}{\text{Time in sec.}}$

$$= \frac{\text{Force} \times \text{stroke of plunger}}{\text{Time (sec.)}} = \frac{156.2 \times 200/(10 \times 100)}{10 \times 60} \text{ Nm/s (W)}$$

$$= \mathbf{0.0521 \text{ Nm/s}} \text{ (Watt)}$$

(*iii*) *the number of strokes of the plunger :*

In order to raise the weight, the hydraulic fluid is to be pumped by the plunger into the ram cylinder. Let the number of strokes be n, then

Volume of fluid transferred from the plunger into the ram × n

= Volume of liquid required to raise the weight by 600 mm

or, $$\frac{\pi}{4}(20)^2 \times 200 \times n = \frac{\pi}{4}(160)^2 \times 600$$

∴ $$n = \left(\frac{160}{20}\right)^2 \times \frac{600}{200} = \mathbf{192}$$

No. of strokes of the plunger = 192

Example 19.3. *In a hydraulic jack, a force of 200 N is applied on the lever having a leverage of 10. If the diameter of ram is 300 mm and that of the plunger 30 mm, determine the weight of the object that can be lifted.*

Solution. Given P = 200 N, Leverage, $\frac{L}{l} = 10$, $\frac{D}{d} = \frac{300}{30} = 10$

From Eq. (19.2), the weight which can be lifted is :

$$F = P\left(\frac{L}{l}\right)\left(\frac{D}{d}\right)^2 = 200 \times 10 \times (10)^2 = 200 \times 10^3 \text{ N} = \mathbf{200 \text{ kN.}}$$

Example 19.4. *A hydraulic jack with a plunger of 30 mm diameter has a stroke of 100 mm, and a ram of 300 mm diameter. A lever arrangement with a leverage of 10 is used to drive the plunger. If a weight of 5 kN is to be lifted, determine the force to be applied on the lever. Calculate the height to which the weight will be raised in 50 strokes, and the work done involved.*

Solution. The weight to be lifted and the effort to be applied on the operating lever are related by Eq. (19.2), hence

$$F = P\left(\frac{L}{l}\right)\left(\frac{D}{d}\right)^2$$

Making substitutions

$$5 \times 1000 = P\,(10)\left(\frac{300}{30}\right)^2 = 100\,P$$

∴ $$P = \mathbf{5 \text{ N}}$$

The force to be applied on the lever = 5 N

Volume of hydraulic fluid transferred to the cylinder in 50 strokes of the plunger

$$= 50 \times \frac{\pi}{4}\left(\frac{30}{1000}\right)^2 \times \frac{100}{1000} \text{ m}^3$$

The height to which the weight will be lifted by the ram

$$= \frac{\text{Volume of fluid}}{\text{Area of ram}}$$

$$= 50 \times \frac{(30/1000)^2}{(300/1000)^2} \times \frac{100}{1000} \times 1000 \text{ mm} = \mathbf{50\ mm}$$

The worked done involved = weight × height lifted

$$= 5 \times 1000 \times \frac{50}{1000} \text{ Nm} = \mathbf{250\ Nm.}$$

Example 19.5. *A hydraulic jack has a ram of 250 mm diameter and a plunger of 30 mm diameter. The stroke of the plunger is 150 mm. If a weight of 2000 N is to be lifted through 1.0 m in 5 minutes, determine :*

(*i*) *the force to be exerted on the plunger*

(*ii*) *the effort to be applied on the lever, if the leverage is 10,*

(*iii*) *the power required to drive the plunger, and*

(*iv*) *the number of strokes required to lift the weight.*

Solution.

(*i*) Force on the plunger $= \frac{4W}{\pi D^2} \times \frac{\pi d^2}{4} = W\left(\frac{d}{D}\right) = 2000\left(\frac{30}{250}\right)^2 = \mathbf{28.80\ N}$

(*ii*) Effort to be applied on the lever

$$= \frac{28.80}{\text{Leverage}} = \frac{28.80}{10} = \mathbf{2.88\ N}$$

(*iii*) Work done in lifting the weight = Weight lifted × lift height = 2000 × 1.0 Nm

Power required to drive the pump $= \frac{2000 \times 1.0}{5 \times 60}$ N-m/s = 6.67 N-m/s = **6.67 W**

(*iv*) Number of strokes $= \frac{\text{Total volume of liquid displaced}}{\text{Volume displaced per stroke}}$

$$= \frac{\frac{\pi D^2}{4} \times 1.0}{\frac{\pi d^2}{4} \times \left(\frac{150}{1000}\right)^4} = \left(\frac{250}{30}\right)^2 \times \frac{1000}{150} = 462.986 \text{ say } \mathbf{463\ strokes.}$$

19.2. Hydraulic Accumulator

A hydraulic accumulator is a device used for temporarily storing pressure energy of a pumped liquid when it is not needed by the machine. Many hydraulic machines like lifts or cranes do a huge amount of work in a small interval or time. This happens when the lift is moving up or the crane moves a load from one place to another. The energy required by these machines is supplied by pumps. But when the lift moves down, no large external energy is required during that period, but as the pump works continuously, the high pressure flow is diverted to the accumulator where it is temporarily stored.

The hydraulic accumulator consists of a fixed hollow cylinder inside which a sliding ram moves. A heavy weight is placed on the ram. High pressure water delivered by the pump is passed into the hollow cylinder when not needed by the machine. This arrangement is shown in Fig. 19.2.

Initially the sliding ram is at the lowermost position and the pump is supplying water to the machine, which is doing work while lifting/moving the weight (as is done by a lift or a crane). When the machine does not require high pressure water (as is the case when the lift moves down), it is diverted to the accumulator for being stored in the cylinder. The high pressure of liquid inside the cylinder causes the loaded ram to move upwards. When the ram attains its uppermost position, the cylinder is filled with water having high pressure and the accumulator is said to have stored the maximum pressure energy. When the machine (lift or crane) requires a large amount of energy, the energy stored in the accumulator is released by opening the outlet valve, and in doing so the ram moves in the downward direction.

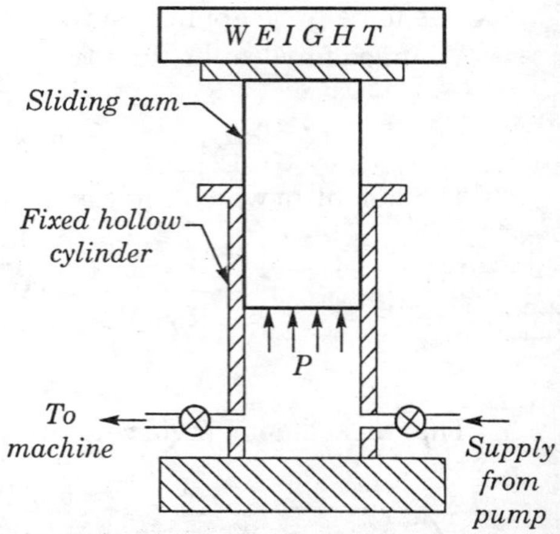

Hydraulic Accumulator—a line diagram.
Fig. 19.2

19.2.1. Capacity of Hydraulic Accumulator. The hydraulic energy stored in the accumulator represents the capacity of the hydraulic accumulator.

If p = pressure of water supplied by the pump
A = area of the sliding ram
L = total lift of the ram or the stroke

then the total weight of the ram (including the weight placed on it)

$$W = p \, . \, A$$

The work done in raising the ram by a height L

$$= W \, . \, L = p \, . \, A \, . \, L$$
$$= p \times \text{volume of accumulator}$$

The work done in lifting the ram is also the energy stored in the accumulator.

$\therefore$ The energy stored in the accumulator $= p \, . \, A \, . \, L$

and, the capacity of accumulator = Energy stored

$$= p \, . \, A \, . \, L = p \times \text{volume of accumulator}$$

The volume of accumulator is also called its displacement and is expressed in volumetric units.

Example 19.6. *A pump supplies water at a pressure of 150 kN/m² to a hydraulic accumulator having a ram of diameter 1.35 m and lift of 6.5 m. Determine :*

(i) the displacement
(ii) the capacity of accumulator
(iii) the total weight placed on the ram, including the external weight.

Solution. (*i*) Displacement of accumulator :

Volume of accumulator $= A \, . \, L = \frac{\pi}{4} (1.35)^2 \times 6.5 = 9.3040 \text{ m}^3$

$\therefore$ Displacement of accumulator = 9.3040 m³ = **9304.0 litres**

(*ii*) Capacity of accumulator = $p \times$ volume of accumulator

$$= 150 \times 10^3 \times 9.3040 \text{ Nm} = \textbf{1395.6 kN-m}$$

(*iii*) Total weight placed on the ram

$$W = p \, . \, A = 150 \times 10^3 \times \frac{\pi}{4} (1.35)^2 \text{ N} = \textbf{214.71 kN.}$$

Example 19.7. *The ram of an accumulator has a diameter of 350 mm and weights 50 kN. Determine the additional weight required to be placed over the ram to develop a pressure of 5000 kN.*

Solution. Let W_1 be the additional weight to be placed on the ram. The total weight, W, required to develop the desired pressure is

$$W = p \,.\, A = 5000 \times \frac{\pi}{4}\left(\frac{350}{1000}\right)^2 = 481.06 \text{ kN}$$

But the total weight in kN is

$$W = W_1 + 50$$

$$\therefore \quad W_1 = 481.06 - 50 = \mathbf{431.06 \text{ kN.}}$$

Example 19.8. *A hydraulic accumulator has a ram of 750 mm diameter. The total weight (including the self weight of the ram) placed on the ram is 500 kN. If the frictional resistance offerred by the packing against the movement of ram is 4% of the total weight, determine :*

(a) the pressure developed when the ram slides down at a uniform rate,

(b) the pressure developed when the ram moves up steadily.

Solution. Total weight of the moving ram = 500 kN

Frictional resistance = 0.04 × 500 kN = 20 kN

(*a*) When the ram slides down at uniform rate (*i.e.*, steadily), the force due to frictional resistance acts upwards, and the effective total weight of the ram

= 500 – 20 = 480 kN

The pressure developed to support this weight

$$p = \frac{480 \times 10^3}{\frac{\pi}{4}\left(\frac{750}{1000}\right)^2} = 1086.54 \times 10^3 \text{ N/m}^2 = \mathbf{1086.54 \text{ kN/m}^2.}$$

(*b*) When the ram moves up steadily, the frictional force acts downwards.

The effective total weight = 500 + 20 = 520 kN

and the pressure required to raise the ram is

$$p = \frac{520 \times 10^3}{\frac{\pi}{4}(0.75)^2} = 1170.08 \times 10^3 \text{ N/m}^2 = \mathbf{1170.80 \text{ kN/m}^2.}$$

Example 19.9. *A hydraulic accumulator has a sliding ram of diameter 750 mm and a stroke of 10 m. The total weight (including the self weight of ram) placed on the ram is 500 kN. The ram falls through the full height steadily in 150 seconds while supplying the stored energy to the hydraulic machine. The pump connected to the accumulator supplies water at the rate of 100 litres / second. If the packing resistance to the movement of ram is 5% of the total weight, determine :*

(i) the work done by the accumulator per unit time,

(ii) the work done by the pump per unit time, and

(iii) the power delivered to the hydraulic machine.

Solution. When the energy stored in the accumulator is to be utilised, the ram is made to move downwards by opening the outlet valve. The packing resistance therefore acts upwards causing a decrease in the net total weight.

The net total weight of ram (W) = 500 – 0.05 × 500 kN = 475 kN

(*i*) Work done per unit time by the accumulator

$$= \frac{WL}{150} = \frac{475 \times 10}{150} \text{ kNm/s} = 31.67 \text{ kNm/s}$$

Power delivered by the accumulator = 31.67 kNm/s = **31.67 kW** (*ii*) The pressure developed in the cylinder due to downward movement of ram

$$p = \frac{475}{\frac{\pi}{4}(0.75)^2} = 1075.22 \text{ kN/m}^2$$

The corresponding pressure head $H = \frac{p}{\gamma} = \frac{1075.22 \times 10^3 \text{ (N/m}^2)}{9810 \text{ (N/m}^3)} = 109.60$ m

The work done by the pump per unit time $= \gamma QH = Qp$

$$= \left(\frac{100}{1000}\right) \times 1075.22 \times 10^3 \text{ Nm/s}$$

$$= 107.522 \times 10^3 \text{ Nm/s} = \mathbf{107.522 \text{ kW}}$$

∴ Power delivered by the pump = 107.522 kW

(*iii*) Power delivered to the hydraulic machine

= Power delivered by the accumulator + Power delivered by the pump

= 31.67 + 107.522 = **139.192 kW.**

Example 19.10. *A 100 mm diameter pipe 1.5 km long is used to transmit 30 kW of power from an accumulator. Allowing 3% power loss in friction, determine the diameter of the ram which is loaded with 1250 kN. The pipe friction factor may be taken as 0.04.*

Solution. Darcy-Weisbach equation gives the head loss in friction :

$$h_f = f\frac{L}{d}\frac{V^2}{2g} = 0.04 \times \frac{1.5 \times 1000}{(100/1000)} \times \frac{V^2}{2 \times 9.81} = 30.58\, V^2$$

Weight of water flowing per second

$$= \gamma Q = \gamma \frac{\pi}{4} d^2 . V = 9810 \times \frac{\pi}{4}(0.1)^2 . V = 77.04\, V$$

Power lost in friction $= \gamma Q h_f = 77.05\, V \times 30.58\, V^2$

$= 2356.19\, V^3$ (W) $= 2.356 V^3$ (kW)

But the power loss in friction is given to be 3% of power to be transmitted, thus

$$2.356\, V^3 = 0.03 \times 30$$

from which, $V = 0.7256$ m/s

Flowrate in the pipeline, $Q = A \,.\, V = \frac{\pi}{4} d^2 \,.\, V$

$$= \frac{\pi}{4}(0.1)^2 \times 0.7256 = 0.005698 \text{ m}^2\text{/s} = 0.0057 \text{ m}^2\text{/s}$$

Power developed by the accumulator

$$= \frac{\gamma QH}{1000} \text{ (kW)}$$

where H = pressure head developed in the accumulator $= p/\gamma$

The pressure, p, required in the accumulator to transmit 30 kW of power is given by

$$30 = \frac{\gamma Q (p/\gamma)}{1000} = \frac{9810 \times 0.0057}{1000}\left(\frac{p}{9810}\right)$$

∴ $$p = \frac{30 \times 1000}{0.0057} = 5263.16 \text{ kN/m}^2$$

The diameter of ram required to develop the requisite pressure in order to support the given load (1250 kN) on the ram is given by

$$p \cdot \frac{\pi D^2}{4} = 1250$$

$$5263.16 \times \frac{\pi}{4} D^2 = 1250$$

from which $D = 1.0526 \text{ m} = $ **1052.6 mm.**

19.2.2. Differential Hydraulic Accumulator. It is a device in which high pressure in the accumulator is developed by placing a smaller external load on the sliding cylinder, as compared to heavy weight used in a simple accumulator. The device consists of a fixed vertical cylinder having a central vertical cavity of small diameter, as shown in Fig. 19.3. The fixed cylinder is encased by a brass bush which in turn is encased by an inverted sliding cylinder (or ram) provided with a collar to facilitate its loading.

Water from pump mains is passed into the fixed cylinder from its bottom through the vertical cavity. The water moves up and exerts an upward pressure on the annular roof of the sliding cylinder which is loaded as shown. The annular space is the same as the cross-sectional area of the bush.

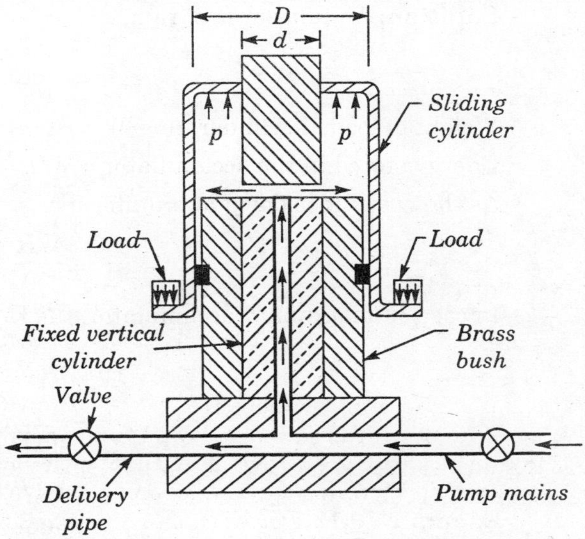

Fig. 19.3. Line diagram of a Differential Hydraulic Accumulator.

Let, W = total weight of the loaded sliding cylinder,

p = pressure intensity of water coming from pump mains

H = stroke or total lift of the accumulator

D = internal diameter of the sliding cylinder

d = external diameter of the fixed cylinder

a = annular area of the brass bush

$$= \frac{\pi}{4}(D^2 - d^2)$$

From consideration of static equilibrium,

$$p = \frac{W}{a} \quad \text{...(19.4)}$$

Total energy stored in the accumulator when the sliding cylinder rises to the upper end of stroke (*i.e.*, full height)

$$= WH \quad \text{...(19.5)}$$

As the sliding cylinder (or ram) moves down, the high pressure water flows out through the delivery pipe. It may be noted that proper valves at the inlet (pump mains) and outlet (delivery pipe) will have to be installed to regulate the flow as desired.

As can be seen, a high pressure intensity of water can be attained by making the annular area 'a' small keeping the weight 'W' at a convenient valve.

Example 19.11. *The two parts of the ram of a differential hydraulic accumulator have diameters of 200 mm and 170 mm. The stroke of the sliding cylinder is 1.25 m. The pressure of water* *is 9800 kN/m² when the total load is in equilibrium position at the upper end of the stroke. Calculate the weight of the loaded cylinder and the energy stored by the accumulator.*

Determine the power supplied by the accumulator if the ram completes the stroke at a uniform rate in 90 seconds.

Find also the diameter of ram of an ordinary accumulator to displace the same load at the given water pressure.

Solution. Annular area of ram, $a = (D^2 - d^2)$

$$= \frac{\pi}{4}(0.2^2 - 0.17^2) = 0.00872 \text{ m}^2$$

Weight of loaded cylinder, $W = p \cdot a = 9800 \times 0.00872 = \mathbf{85.456\ kN}$

Energy stored in the accumulator = $WH = 85.456 \times 1.25 = \mathbf{106.82\ kN\text{-}m}$

As the ram slides down at a uniform rate, the power delivered by the accumulator

$$= \frac{WH}{90} = \frac{106.82}{90} = 1.187 \text{ kNm/s} = \mathbf{1.187\ kW}$$

In case of an ordinary accumulator, the ram diameter D is given by

$$W = p \cdot \frac{\pi D^2}{4}$$

or

$$D = \sqrt{\frac{4W}{\pi p}} = \sqrt{\frac{4 \times 85.456}{\pi \times 9800}} = \mathbf{0.1053\ m.}$$

Example 19.12. *Power is to be transmitted hydraulically to a distance of 8 km by means of a number of 10 cm pipes laid horizontally, for which the coefficient of friction may be taken as 0.03. The pressure at the accumulator is maintained constant at 66.5 kg/cm². Determine the number of pipes required to ensure an efficiency of at least 92 percent, when the power delivered is 220 h.p.*

Solution. To ensure an efficiency of atleast 92%, the maximum less of power is limited to 8% of power delivered.

∴ Power loss = 0.08 × 220 hp = 17.6 hp

Since the pipes are laid horizontally, the power will be the same in each pipe. Let the velocity in the pipe be V m/s, then,

$$\text{Power loss} = \frac{\gamma Q H_f}{75} = 1000 \times \frac{\pi}{4}\left(\frac{10}{100}\right)^2 V \cdot H_f \qquad ...(1)$$

Let there be 'n' number of pipes, then the power transmitted by each pipe

$$= \frac{220}{n} \text{ hp.}$$

1 metric hp = 735 W = 735 Nm/s

$$\therefore \text{ power transmitted by each pipe} = \frac{220 \times 735}{n} \text{ Nm/s} \qquad ...(2)$$

The pressure in the pipe at the starting point of its 8 km long journey (*i.e.* at the accumulator) = 66.5 kg/cm² = 66.5 × 9.81 N/cm².

The power loss due to frictional resistance in the pipe = 17.6 hp = 17.6 × 735 Nm/s

This power loss is reflected in the corresponding pressure loss, which is 8% of the pressure at the accumulator.

∴ Pressure loss in frictional resistance = 0.08 × 66.5 kg/cm² = 5.32 kg/cm² and the corresponding head loss in friction $= \dfrac{\Delta p}{\gamma} = H_f$

$$= \frac{5.32 \times (100)^2 \text{ (kg/m}^2)}{1000 \text{ (kg/m}^3)} = 53.2 \text{ m} \qquad \ldots(3)$$

The head loss by Darcy-Weisbach equation is given by ;

$$H_f = f\frac{L}{d}\frac{V^2}{2g} = 0.03 \times \frac{8 \times 1000}{(10/100)} \frac{V^2}{(2 \times 9.81)} \qquad \ldots(4)$$

Equating (3) and (4), we obtain, $V = 0.6595$ m/s.

Power transmitted by each pipe,

$$P = \gamma Q (H - H_f)$$

$$= 1000 \times \frac{\pi}{4}\left(\frac{10}{100}\right)^2 \times 0.6595 \times \frac{(66.5 - 5.32) \times (100)^2}{1000} \text{ kg m/s}$$

$$= 3168.9 \text{ kg m/s}$$

Since, $\quad 1 \text{ kg} = 9.81 \text{ N}$

$$P = 3168.9 \times 9.81 \text{ Nm/s} \qquad \ldots(5)$$

Equating (2) and (5)

$$\frac{220 \times 735}{n} = 3168.9 \times 9.81$$

$$\therefore \quad n = 5.2$$

In order to maintain an efficiency of at least 92%, the number of 10 cm pipes required = 6.

Example 19.13. *The accumulator of an aircraft hydraulic system takes the form of a hollow sphere with an inner diameter $D = 300$ mm, composed of two halves held together by bolts. The inner space of the accumulator is separated into two parts by a rubber diaphragm clasped between the flanges. The upper part contains liquid of specific weight $\gamma = 1020$ kg/m^3, and the lower part, compressed air. Determine the force that tends to detach the upper hemisphere from the lower one, if the pressure of liquid in the parting plane is 100 kg/cm^2.*

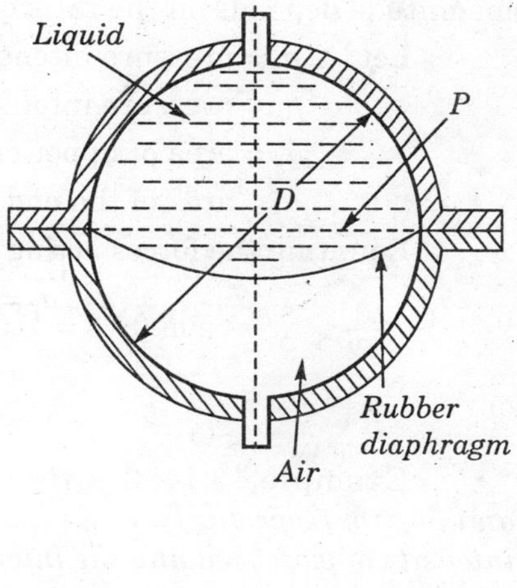

Solution. Let the force required to detach the upper hemisphere from the lower one be F.

$$F = \text{Area of the parting surface} \times \text{pressure}$$

$$= \frac{\pi D^2}{4} \times \text{pressure on the surface}$$

$$= \frac{\pi (0.3)^2}{4} \times 100\,(100)^2 \text{ kg} = \mathbf{70685 \text{ kg}}$$

19.3. Hydraulic Intensifier

This is a device meant for increasing the intensity of pressure of water by utilizing the hydraulic energy of larger quantity of water available at a lower pressure. Such a device is used when the pressure of water supplied to a hydraulic machine is not high enough as required.

The hydraulic intensifier consists of a fixed ram conveying high pressure water to the hydraulic machine. A hollow sliding ram which contains high pressure water is mounted over the fixed ram. The sliding ram is surrounded by a fixed cylinder containing water at low pressure from the main supply. Hydrostatic force due to low water pressure acts on the upper end of the sliding ram which forces it downwards on to the fixed ram. In this process, the water in the sliding ram is compressed and its pressure is increased. This pressurised water is then supplied through the fixed ram to the hydraulic machine. The working of a hydraulic intensifier is shown in Fig. 19.4.

As the high pressure water is drawn out, the sliding ram moves down till it reaches the bottom of its stroke. At this stage, the valve allowing high pressure water to the machine is closed, and the water collected in the fixed cylinder is released to exhaust. The low water pressure is then admitted into the sliding ram through the fixed ram which causes the sliding ram to rise till it reaches the upper limit of the stroke. At this stage, the supply of low pressure water into the sliding ram is stopped by closing the inlet valve, and the valve admitting high pressure water (*i.e.*, outlet valve) to the machine is opened. Simultaneously, the exhaust valve of the fixed cylinder is closed to admit low water pressure into the fixed cylinder. The hydraulic force exerted on sliding ram causes it to move downwards compressing the water inside it. The high pressure water so available is then supplied to the machine.

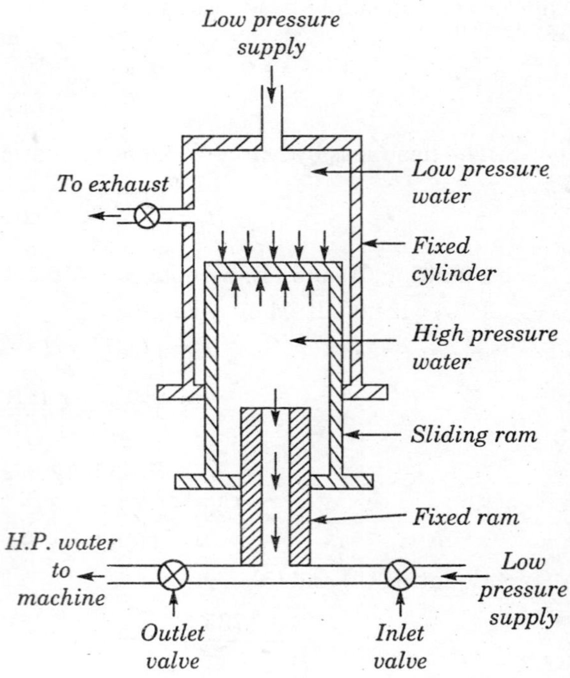

Fig. 19.4. Hydraulic Intensifier—a schematic diagram.

The high pressure is developed in the intensifier during the downward stroke only, and its magnitude depends on the ratio of area of sliding ram and the fixed ram, as determined below.

Let, p_f = pressure intensity of low pressure water in the fixed cylinder,

p_s = pressure intensity of high pressure water in the sliding ram,

a_s = area of upper end of sliding ram,

a_f = area of the end of fixed ram

Equating the forces acting upward and downward,

$$p_s \,.\, a_f = p_f \,.\, a_s$$

or

$$p_s = p_f \cdot \left(\frac{a_s}{a_f}\right) = p_f\left(\frac{d_s}{d_f}\right)^2 \qquad ...(19.6)$$

Example 19.14. *A hydraulic intensifier has sliding and fixed rams of diameters 100 mm and 50 mm respectively, and is supplied with water at a pressure of 180 kN/m². Find the pressure intensity of water leaving the intensifier.*

Solution. Given : d_f = 50 mm, d_s = 100 mm, p_f = 180 kN/m²

Intensity of high pressure water is given by Eq. (19.6), as :

$$p_s = p_f\left(\frac{d_s}{d_f}\right)^2 = 180\left(\frac{100}{50}\right)^2 = \textbf{720 kN/m}^2.$$

Example 19.15. *A hydraulic intensifier is used to increase the pressure of water from 5 MN/m² to 22 MN/m². The stroke of the intensifier is 1.75 m and the capacity is 35 litres. Find the diameters of the sliding ram and the fixed ram of the intensifier.*

Solution. $p_f = 5 \text{ MN/m}^2 = 5000 \text{ kN/m}^2$

$p_s = 22 \text{ MN/m}^2 = 22000 \text{ kN/m}^2$

Stroke = 1.75 m, Capacity = 35 litres = 0.035 m^3

Capacity of intensifier = Volume swept by the sliding ram = Area of ram × stroke

$$= \frac{\pi}{4} (d_f)^2 \times 1.75$$

or

$$0.035 = \frac{\pi}{4} (d_f)^2 \times 1.75$$

from which $d_f = 0.1595 \text{ m} = \mathbf{159.5 \text{ mm}}$

From Eq. (19.6), equating the upward and downward forces acting on the sliding ram, one obtains,

$$p_s = p_f \left(\frac{d_s}{d_f}\right)^2 \quad \text{or} \quad 22000 = 5000 \left(\frac{d_s}{159.5}\right)^2$$

from which, $d_s = \mathbf{334.57 \text{ mm}}$

∴ The diameter of the sliding ram = 334.75 mm,
and, the diameter of the fixed ram = **159.5 mm.**

19.4. Hydraulic Lift

A hydraulic lift is a device used for carrying goods or passengers from one floor to another in a multi-storeyed building. It can be either a direct acting type or a suspended type as described below.

19.4.1. Direct Acting Hydraulic Lift. It consists of a platform or cage supported by a vertical ram which slides in a fixed vertical cylinder as shown in Fig. 19.5. Water under pressure is pumped into the fixed cylinder which exerts a force on the sliding ram causing it to move vertically up. The cage is thus raised to the desired height. Its downward movement is ensured by releasing water from the fixed cylinder.

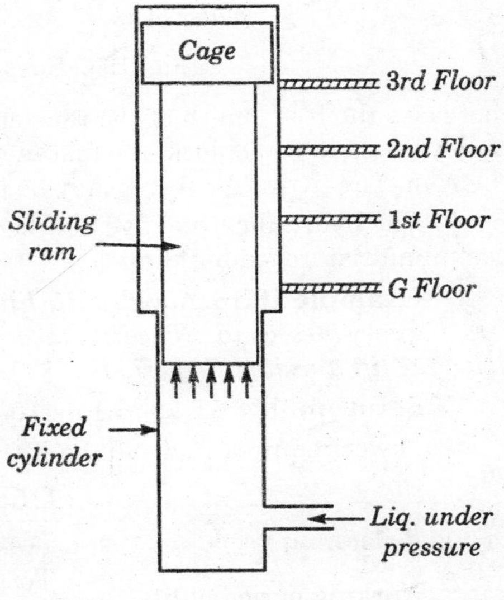

Fig. 19.5. A schematic view of a Direct Acting Hydraulic Lift.

19.4.2. Suspended Hydraulic Lift. In a suspended type of lift, a jigger arrangement is used to multiply the stroke of the sliding ram. The hydraulic jigger consists of a fixed cylinder, a sliding ram and two sets of pulley blocks as shown in Fig. 19.6. One set of pulley block is movable while the other one is fixed. The movable pulley block is mounted on the sliding ram whereas the fixed pulley block is installed at the head of the fixed cylinder. Each set of pulley block has the same number of pulleys.

A wire rope, one end of which is fixed to the body of the cylinder and the other end is taken around all pulleys of the movable and fixed blocks and finally passes over the guide pulleys to sypport the cage as shown in the figure.

The raising and lowering of the cage of the lift is done by the hydraulic jigger. When a fluid under pressure is admitted into the cylinder, the sliding ram moves out increasing the distance between the fixed and movable pulley blocks. The wire rope connected to the cage is pulled and the cage is lifted. For lowering the cage, the fluid inside the cylinder is taken out gradually causing the sliding ram to move inwards shortening the distance between the movable and fixed pulleys. This

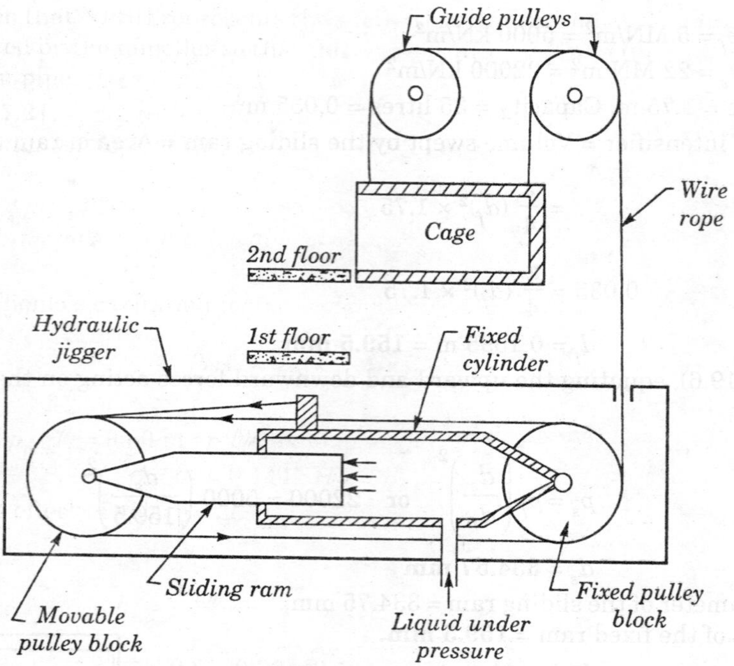

Fig. 19.6. Suspended Hydraulic Lift-a line diagram.

increases the free length of the rope enabling the cage to come downwards. If 'n' is the number of pulleys used in each block, the distance moved by the free end of the rope is 'n' times the stroke of the ram. The hydraulic jigger may be placed horizontally or vertically.

The hydraulic lifts have now become obsolete and have been superseded by electric lifts for use in multistoreyed buildings.

Example 19.16. *A hydraulic lift is required to raise a load of 15 kN through a height of 12 m once in every 90 seconds. The lift moves with a velocity of 0.75 m/s. Determine the power required to drive the lift and its idle period.*

Solution. $W = 15 \text{ kN} = 15 \times 1000 \text{ N}$, $H = 12$ m, Time = 90 seconds

Power required to drive the lift = Work done per second in lifting the load

$$= \frac{(15 \times 1000) \times 12}{90} = 2000 \text{ Nm/s} = 2000 \text{ W} = \mathbf{2\ kW}$$

$$\text{Working period of lift} = \frac{\text{Total lift}}{\text{Velocity}} = \frac{12.0}{0.75} = 16 \text{ seconds}$$

$\therefore$ Idle period of lift = Total time – Working period = 90 – 16 = **74 seconds.**

Example 19.17. *A hydraulic lift raises a load of 75 kN through a height of 15 m once in every 120 seconds. The lift moves with an average speed of 0.5 m/s. During the working stroke, water from accumulator and the pump is supplied to the lift at a pressure of 3700 kN/m^2. If the pump works with an efficiency of 78% and the lift operates at 72% efficiency, determine the power required to drive the pump, and the minimum capacity of the accumulator. Neglect frictional losses.*

Solution. Noting that the water is supplied to the lift from accumulator as well as from the pump.

Work done per second in raising the lift

= Load lifted × Height traversed by the lift per second

= (75 × 1000) × 0.5 Nm/s = 37500 Nm/s

$$\text{Power output of lift} = \frac{37500}{1000} = 37.5 \text{ kW}$$

$$\text{Power input to the lift} = \frac{\text{Power output}}{\text{Lift efficiency}} = \frac{37.5}{0.72} = 52.08 \text{ kW}$$

This power has been supplied jointly by the accumulator and the pump.

Let P be the output of the pump in kW, then

Output of the accumulator = $52.08 - P$ ...(I)

$$\text{Working period of the lift} = \frac{\text{Height of lift}}{\text{Speed of lift}} = \frac{15.0}{0.5} = 30 \text{ seconds}$$

Idle period of lift = Total time – Working period = 120 – 30 = 90 seconds

During this idle period, the energy supplied by the pump will be stored in the accumulator and during the working period the stored energy will be supplied by the accumulator to the lift.

Energy stored in accumulator during the idle period

= Output of pump × Idle period

= $(P \times 1000) \times 90$

This energy will be supplied by the accumulator in 30 seconds.

$$\therefore \quad \text{Output of accumulator} = \frac{(P \times 1000) \times 90}{30} \text{ Nm/s}$$

$= 3P$ kW ...(II)

Equating (I) and (II) as both represent the same quantity,

$$52.08 - P = 3P$$

or

$$P = \frac{53.08}{4} = 13.02 \text{ kW}$$

$\therefore$ Output of the pump = 13.02 kW

$$\text{Input to the pump} = \frac{\text{Output}}{\text{Pump efficiency}} = \frac{13.02}{0.78} = 16.69 \text{ kW}$$

$\therefore$ Power required to drive the pump = **16.69 kW**

Minimum Capacity of the Accumulator

= Energy stored in the accumulator = $(P \times 1000) \times 90$

= $13.02 \times 1000 \times 90$ Nm = **1171.8 kNm.**

19.5. Hydraulic Crane

A hydraulic crane is a device meant for raising, lowering or transfering heavy loads. It is used in workshops, warehouses and docks for loading and unloading of ships.

It consists of mainly two parts, namely, the jigger and the crane. The jigger, which consists of a cylinder and a sliding ram, is attached to the mast of the crane. The crane consists of a vertical mast, a jib and a tie. The jib and tie are suitably connected to the mast and their other end joins a guide pulley as shown in Fig. 19.7. The mast has its own pedestal and foundation. The mast alongwith the jib, tie, guide pulley and the load can be rotated about its vertical axis. A set of 2 to 6 pulleys is provided to the upper end of the sliding ram. Another set of equal number of pulleys is provided at the bottom end of the fixed cylinder. It is only the top set of pulleys carried by the ram that move vertically with the ram, while the lower set of pulleys fixed with the cylinder does not have any vertical movement.

A wire rope, one end of which is fixed to the movable pulley, is passed over all the pulleys of both the sets and is then taken over the guide pulley. The other end of the rope is provided with a suitable hook to suspend the load. The number of pulleys in a block determines the velocity ratio.

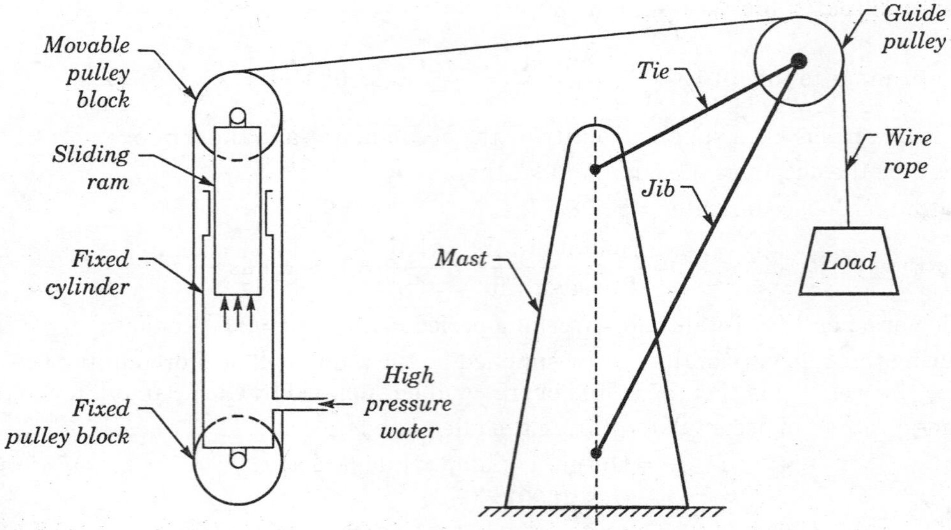

Fig. 19.7. Schematic view of a Hydraulic crane.

Normally, a six-sheaf pulley block system is used giving a velocity ratio of six. It means that the displacement of the suspended load is six times the displacement of the ram.

For lifting the load, water under pressure is admitted into the cylinder which causes the sliding ram to move vertically up, which in turn causes the movable pulley block to move up vertically. This action increases the distance between the two pulley blocks causing the free end of the wire rope to be pulled up. Thus the load attached to the wire rope is raised.

Example 19.18. *A hydraulic crane lifts a load of 150 kN through a height of 9.50 m utilizing 375 litres of water under a pressure of 7.50 N/mm². Determine its efficiency.*

Solution. Load = 150 kN = 150 × 1000 N, Pressure = 7.50 N/mm^2 = 7.50 × 10^6 N/m^2

Output of the crane = Work done by the crane in lifting the load
= Weight lifted × displacement of the load
= (150 × 1000) × 9.50 Nm = 1425.0 × 10^3 Nm

Input to the crane = Energy supplied to the crane
= Weight of water supplied × pressure head

$$= (\gamma \times \text{volume supplied}) \times \frac{p}{\gamma}$$

= Volume supplied × pressure

$$= \frac{375}{1000} \times 7.50 \times 10^6 \text{ Nm} = 2812.5 \times 10^3 \text{ Nm}$$

$$\therefore \quad \text{Efficiency of crane} = \frac{\text{Output}}{\text{Input}} = \frac{1425.0 \times 10^3}{2812.5 \times 10^3} = 0.507 = \mathbf{50.7\%}.$$

Example 19.19. *Water at a pressure of 5 N/mm² is supplied to a hydraulic crane which has a ram diameter of 200 mm. The crane has a velocity ratio of 6 and works at an efficiency of 55%. Determine the load lifted and the quantity of water required by the crane to lift the load to a height of 10.5 m.*

Solution. Pressure on ram = 5 N/mm^2 = 5 × 10^6 N/m^2

$$\text{Area of ram} = \frac{\pi}{4}\left(\frac{200}{1000}\right)^2 = 0.0314 \text{ m}^2$$

Pressure force on the ram = Pressure × Area

$= 5 \times 10^6 \times 0.0314$ N = 157079 N

Velocity ratio $= \dfrac{\text{Displacement of load}}{\text{Stroke of ram}} = \dfrac{10.5}{6} = 1.75$ m

Output of crane = Weight lifted × displacement of load

$= W \times 10.5$ Nm

Input to the crane = Pressure force × stroke of ram

$= 157079 \times 1.75$ Nm = 274888 Nm

Efficiency $= \dfrac{\text{Output}}{\text{Input}}$ or $0.55 = \dfrac{W \times 10.5}{274888}$

∴ $W = 14398$ N = 14.398 kN

Load lifted = **14398 N**

Quantity of water required = Area of ram × Stroke

$= 0.0314 \times 1.75 = 0.05495$ m^3 = **54.95 litres.**

Example 19.20. *A hydraulic crane fitted with a jiggar has a ram of 250 mm diameter and the velocity ratio of 6. It is required to lift a load of 18 kN through a height of 12 m at 1.20 m/s speed every two minutes. The crane, which operates with an efficiency of 75%, is fed from an accumulator alongwith a pump at a steady rate at a pressure of 5 N/mm². Determine :*

(a) the necessary pressure in the cylinder during lifting,

(b) the duty of the pump in litres, and

(c) the minimum capacity of the accumulator.

Solution. $W = 18$ kN $= 18 \times 1000$ N, $H = 12$ m, $p = 5$ N/mm^2 $= 5 \times 10^6$ N/m^2

Stroke of ram $= \dfrac{\text{Lift Height}}{\text{Velocity ratio}} = \dfrac{12}{6} = 2.0$ m, $D = 250$ mm $= 0.25$ m

$\eta = 0.75$, Time = 2 minutes = $2 \times 60 = 120$ seconds

(a) Pressure in the cylinder

Efficiency $= \dfrac{\text{Output}}{\text{Input}}$

Output of crane $= W \times H = (18 \times 1000) \times 12$ Nm

Input to crane $= p \times \dfrac{\pi}{4} (D)^2 \times$ stroke of ram

$= p \times \dfrac{\pi}{4} (0.25)^2 \times 2$

Making substitutions,

$$0.75 = \frac{18 \times 1000 \times 12}{p \times \frac{\pi}{4} (0.25)^2 \times 2}$$

from which, $p = 2934.78 \times 10^3$ N/m^2 = **2.935 N/mm²**

∴ Necessary pressure in the cylinder during lifting = **2.935 N/mm²**

(b) Duty of the Pump

The stroke of the ram = 2 m

The ram is displaced by 2 m in every 120 seconds

∴ Volume swept by the ram $= \dfrac{\pi}{4} (0.25)^2 \times 2$ m^3 $= 0.09817$ m^3

$$\text{Volume swept per second} = \frac{0.09817}{120} = 0.000818 \text{ m}^3/\text{s} = 0.818 \text{ litre/s}$$

$\therefore$ Pump duty = **0.818 litre/second**

(c) *Minimum Capacity of the Accumulator*

The load of 18 kN is moved through 12 m height at a speed of 1.2 m/s

$$\therefore \quad \text{The time of operation} = \frac{12}{1.2} = 10 \text{ seconds}$$

Volume swept by the ram during this interval of time = 0.09817 m^3

During this period of operation, the quantity supplied by the pump

$$= \text{pump duty} \times \text{time of operation}$$
$$= 0.000818 \times 10 = 0.00818 \text{ m}^3$$

The balance quantity (0.09817 – 0.00818 = 0.08999 m^3) must be supplied by the accumulator.

$\therefore$ Minimum capacity of the accumulator = 0.08999 m^3 = **89.99 litres.**

19.6. Hydraulic Ram

The hydraulic ram is a device used for raising a small quantity of water to a considerable height by utilising a large quantity of water available at a small height. It utilizes the principle of water hammer according to which high pressures are created when flowing water is suddenly stopped.

Such a device is useful in hilly areas, where large quantity of water at small heights is available in rivulets, for irrigating fields at higher elevations.

A hydraulic ram consists of a chamber fitted with valves and an air vessel. It is connected to the source of water by means of a pipeline at one end of the chamber. The delivery pipe is connected to the air vessel as shown in Fig. 19.8 and supplies water to the high level tank. The chamber has two valves, one for allowing water to go waste and the other for letting the water enter into the air vessel. Both the valves open and close automatically. A valve is also fitted in the supply pipe for regulating the flow from the low level water source as shown in the figure.

Water from the low level supply tank enters the supply pipe as soon as the regulating valve is opened. The water enters the chamber under a small head h. The waste valve which is already open due to its own weight allows water to go waste. As the water flows down the supply pipe, water flowing out exerts a dynamic force on the waste valve which starts closing. This force continues to increase till the waste valve gets closed. When this happens, the entire flowing water is suddenly brought to rest, causing instant increase in pressure. This sudden large increase in pressure causes the delivery valve to open and the water is forced to enter into the air vessel compressing the air inside it. The high air pressure forces the water to rise into the delivery pipe and get stored in the high level delivery tank. In this process, the pressure in the chamber is gradually decreased which causes the delivery valve to close

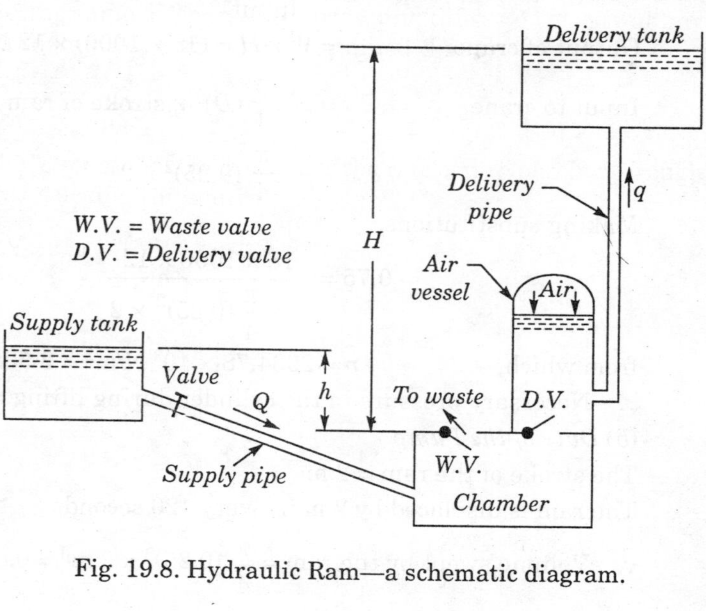

Fig. 19.8. Hydraulic Ram—a schematic diagram.

automatically and the waste valve opens allowing the water to flow out to waste. The flow is again caused in the supply pipe from the low level tank and the above mentioned cycle is repeated. The function of the air vessel is to act as a storage of water and to regulate its flow into the delivery pipe even when the delivery valve is closed.

Let Q = Flow rate entering the chamber through supply pipe

q = Flow rate entering the delivery tank

H = Height of water raised above the chamber

h = Height of water in the supply tank above the chamber.

Energy supplied to the ram per second (Input to ram)

= Weight of water supplied per second to the chamber × height of fall

$= \gamma Q h$

Energy supplied by the ram per second (Output of ram)

= Weight of water supplied per second by the ram × height raised

$= \gamma q H$

Efficiency of ram $= \dfrac{\text{Output}}{\text{Input}} = \dfrac{\gamma q H}{\gamma Q h}$

$$\eta = \frac{qH}{Qh} \qquad \text{...(19.7)}$$

This is known as D' Aubuisson's efficiency. The top of the chamber is taken as the datum. However when the water level in the supply tank is taken as the datum, the actual height through which water is lifted by the ram becomes $(H - h)$. According to Rankine,

the energy supplied to the ram = $\gamma(Q - q)h$

the energy delivered by the ram = $\gamma q(H - h)$

then the efficiency

$$\eta = \frac{\gamma q(H-h)}{\gamma(Q-q)h} = \frac{q(H-h)}{(Q-q)h} \qquad \text{...(19.8)}$$

This efficiency is called the Rankine's efficiency, and its value is always less than the D' Aubuisson's efficiency.

Example 19.21. *A hydraulic ram uses water at the rate of 2000 lit / minute which is supplied from a height of 3.2 m, and raises 250 lit / minute to a height of 20 m above the ram. The delivery pipe has a diameter of 70 mm and is 100 m long. The friction factor of the pipe is 0.03. Determine the efficiency of the ram.*

Solution. $Q = 2000 \text{ lit/min} = \dfrac{2000}{60} = 33.33 \text{ lit/s}$

$$q = 250 \text{ lit/min} = \frac{250}{60} = 4.167 \text{ lit/s}$$

$$H = 20 \text{ m},\ h = 3.2 \text{ m},\ d = 70 \text{ mm} = 0.07 \text{ m},\ L = 100 \text{ m}$$

Velocity of flow in the delivery pipe

$$V = \frac{4q}{\pi d^2} = \frac{4 \times (4.167/100)}{\pi\,(70/1000)^2} = 1.081 \text{ m/s}$$

Head loss in friction in the delivery pipe

$$h_f = f\frac{L}{d}\frac{V^2}{2g} = 0.03 \times \frac{100}{0.07} \times \frac{(1.081)^2}{2 \times 9.81} = 2.55 \text{ m}$$

Effective head developed by the ram = $H + h_f$

$= 20 + 2.55 = 22.55$ m

$$\text{D'Aubuisson's efficiency} = \frac{q \times \text{Effective head}}{Q \times h} = \frac{4.167 \times 22.55}{33.33 \times 3.2} = 0.881 \quad \text{or} \quad \mathbf{88.1\%}$$

$$\text{Rankine's efficiency} = \frac{q\,(\text{Effective head} - h)}{(Q - q)\,h} = \frac{4.167 \times (22.55 - 3.20)}{(33.33 - 4.167) \times 3.2} = 0.862 \quad \text{or} \quad \mathbf{86.2\%.}$$

19.7. Air Lift Pump

An air lift pump is a device used for lifting water from a well by making use of compressed air. It consists of an air pipe fitted with a nozzle at one end. The nozzle is surrounded by the delivery pipe near its bottom end as shown in Fig. 19.9. This device is based on the principle that when air bubbles are mixed with water, the density of the mixture is reduced as compared to that of water. Hence a small height of water column can balance a greater height of water-air mixture.

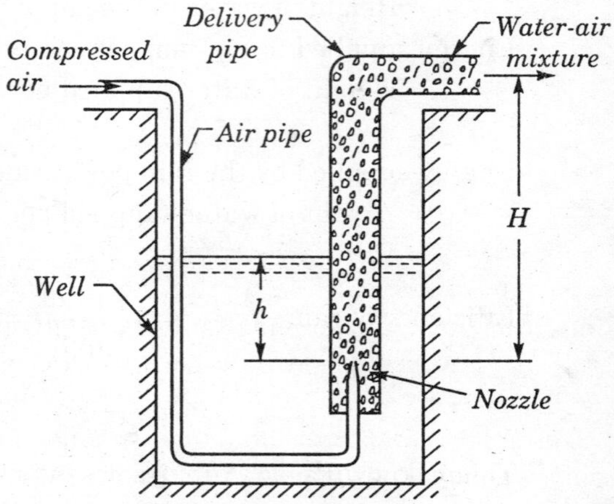

Fig. 19.9. Schematic diagram of an air lift pump.

Compressed air is introduced into the delivery pipe through one or more nozzles fitted at the lower end of the rising delivery pipe as shown in the figure. When water present in the delivery pipe is mixed with air bubbles, a mixture of water and air formed which has a lower density. This causes the mixture to rise to a greater height (H) as compared to the small height of water column (h) available in the well. Consequently the flow takes place and the water-air mixture is discharged out of the delivery pipe. The flow will continue as long as the flow of compressed air is available.

The figure shows the submergence depth (h) of the nozzle which represents the depth at which the nozzle is located below the water level in the well, and (H) is the height to which the water-air mixture is lifted above the tip of the nozzle. ($H - h$) is known as the useful lift. Satisfactory results are obtained when the useful lift ($H - h$) is less than the submergence depth (h) of the nozzle.

Obviously the ratio $\dfrac{h}{H-h} > 1$

This ratio generally varies between 4 and 1. For $h = 30$ m, the ratio has a value of about 4, but when $h = 90$ m, the ratio reduces to a value close to 1. For intermediate values of h, the value of the ratio can be interpolated.

The air-lift pump has no moving parts and hence there are no chances of suspended solid particles damaging the pump. This type of pump is found to lift more quantity of water through a bore hole of given diameter than any other pump. However, the efficiency of the pump is low and lies between 20%–40%.

19.8. Jet Pump

A jet pump is used to raise the suction limit of a pump beyond the normal limit. It consists of a stream of high pressure water which is passed through a nozzle to convert its pressure energy into kinetic energy, a part of which is imparted to the water surrounding the jet. The nozzle is enclosed in a constricted passage in which the high velocity jet discharged by the nozzle causes local pressure drop in the throat area. On account of the low pressure in the throat, water enters the suction pipe from the well (or sump) and is raised to the suction reach of the pump. There is a considerable loss of energy which takes place in the process of mixing of jet with the surrounding water. Consequently, the efficiency of this type of pump is around 25%. Fig. 19.10 shows the general arrangement of a jet pump.

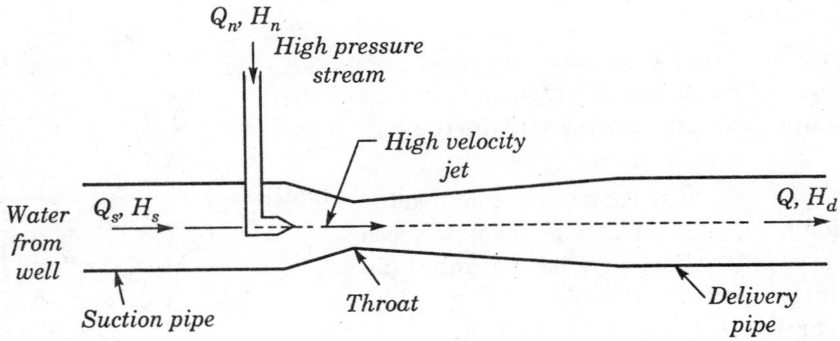

Fig. 19.10. Jet pump-a line sketch.

Various quantities associated with the performance of a jet pump are defined as under.

Q_s = flowrate in the suction pipe
Q_n = flowrate through the nozzle
Q = flowrate in the delivery pipe
H_n = supply head to the nozzle
H_s = suction head
H_d = delivery head

One can now write, $$Q = Q_s + Q_n$$

and the efficiency of the pump $$\eta = \frac{Q_s(H_s + H_d)}{Q_n(H_n - H_d)} \quad \text{...(19.9)}$$

19.9. Rotary Pumps

These are positive displacement pumps which deliver liquid by a pure rotation of the pumping elements. Any rotary pump consists mainly of a stationary housing in which is nested a power driven rotor carrying the pumping elements. The operation of a rotary pump consists of the following steps :

(*i*) the hydraulic fluid entering the inlet port is trapped by the pumping elements,
(*ii*) the pumping elements carry it around to the outlet port, and
(*iii*) from the outlet port it is expelled into the delivery pipeline.

A rotary pump needs no inlet or outlet valves. Moreover, a rotary pump does not require a crankshaft mechanism enabling it to be operated at a very high speed upto 5000 r.p.m. as compared to reciprocating pumps which operate in the range of 250–500 r.p.m. The theoretical delivery or discharge of a rotary pump is given by

$$Q = \frac{vzN}{60} \quad \text{...(19.10)}$$

where, v = displacement of each pumping element (characteristic volume delivered per revolution of pump shaft),
z = number of pumping elements (varies between 4 to 12),
N = revolution per minute of pump shaft.

Because of overlapping characteristic volumes in rotary pumps, a greater uniformity of delivery is achieved as compared to the reciprocating pumps.

On account of the above stated advantages, rotary pumps have found numerous application in many fields like aviation. In aircraft systems, rotary pumps are used in engine fuel and lubricating systems and in hydraulic power transmissions.

Rotary pumps are mainly of three types : Gears pumps, Vane pumps and Piston pumps. Of these, the gear pump will be described in the following paragraph.

19.10. Gear Pumps

These may be of spur-gear or screw type. Spur-gear pumps deliver the hydraulic fluid at right angles to the axis of rotation whereas the screw pumps deliver the fluid along the rotational axis.

Gear pumps are most commonly constructed with a pair of identical spur-gears with involute teeth revolving in a closely fitted housing. The hydraulic fluid coming at the intake port fills the spaces between the teeth and is carried around the periphery of the revolving gears to the discharge side. The teeth are meshing between the two gears and thus prevent the return of the hydraulic fluid from the discharge side to the suction side. However, as the spaces between the teeth are larger than the teeth, some hydraulic fluid gets back to the suction side. Fig. 19.11 shows the schematic diagram of the gear pump.

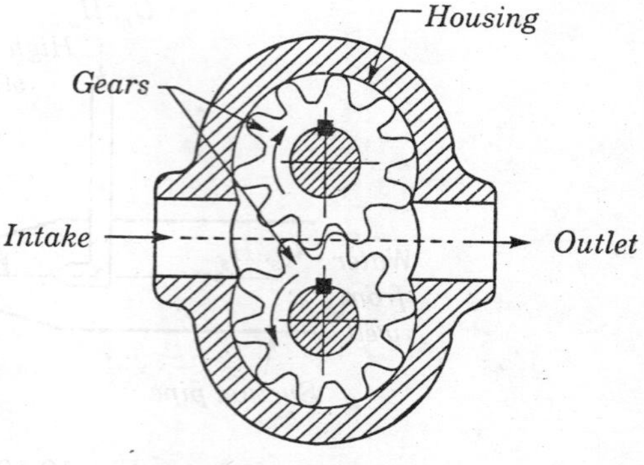

Fig. 19.11. Gear pump—a schematic diagram.

The characteristic volume, therefore, is not the volume of a space but is the volume of a tooth. The number of such volumes delivered per revolution is equal to the number of teeth of the two gears.

Let v_t = Volume of a tooth

Z = number of teeth in a gear

N = number of revolutions per minute of gear shaft

then the total displacement of a gear pump

$$Q = \frac{v(2Z)N}{60} \quad ...(19.11)$$

Eq. (19.11) represents the discharging capacity of the pump.

19.11. Fluid Couplings and Torque Converters

These devices are employed for the transmission of energy in the form of mechanical work between pairs of rotating shafts. They have been used in the transmission of automotive vehicles and are very advantageous in many other situations where transmission of power is required.

19.11.1. Fluid or Hydraulic Coupling.

A fluid or hydraulic coupling is a device used for transmitting power between rotating shafts. The power is transmitted from the driving shaft to the driven shaft with the help of hydraulic fluid (generally oil). A fluid coupling is a combination of an impeller of a rotodynamic pump and a runner of a radial hydraulic turbine enclosed within a single casing completely filled with a hydraulic oil. Fig. 19.12 shows the general arrangement of a fluid coupling.

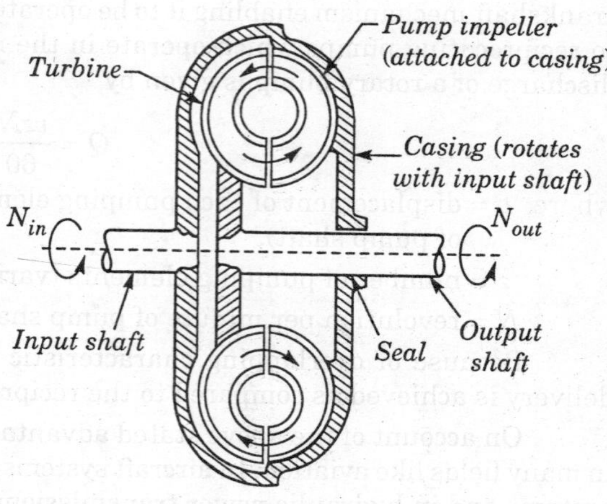

Fig. 19.12. General arrangement of a fluid coupling.

The pump impeller is connected to the input shaft (driving shaft) and the turbine runner to the output shaft (driven shaft). There is no stator or set of stationary guide vanes. When the input shaft is rotated the hydraulic fluid starts moving from the inner radius of the impeller to its outer radius, thereby increasing the angular momentum of the fluid. The fluid moves in the direction as indicated by the arrow. The fluid with increased angular momentum enters the runner of the reaction turbine and flows towards the inner radius, and in the process its angular momentum decreases. This decrease in angular momentum exerts a torque on the turbine runner and causes the output shaft to rotate.

The hydraulic fluid then flows back into the pump impeller as indicated by the arrows, and the same cycle of fluid circulation is repeated. It is therefore obvious that the angular momentum of fluid is changed by both the pump impeller as well as the turbine runner. Circulation of fluid between pump impeller and turbine runner is shown by arrows.

There being no direct mechanical connection between the input and output shafts, a fluid coupling transmits only a small fraction of torsional shock loads and vibrations between the two shafts. The speed of the output shaft is always slightly less than that of the input shaft. The efficiency of a well designed coupling is around 97%.

$$\text{Efficiency of a fluid coupling} = \frac{\text{Power transmitted to the output shaft}}{\text{Power available at the input shaft}}$$

$$\text{But shaft power} = \frac{2\pi NT}{6000} \propto NT$$

If,
N_{in} = Speed of input shaft
N_{in} = Torque at the input shaft
N_{out} = Speed of output shaft
T_{out} = Torque at the output shaft

then, power transmitted to the output shaft

$$P_{out} \propto N_{out} \times T_{out}$$

and power available at the input shaft

$$P_{in} \propto N_{in} \times T_{in}$$

Hence the efficiency of the fluid coupling

$$\eta = \frac{P_{out}}{P_{in}} = \frac{N_{out} \times T_{out}}{N_{in} \times T_{in}}$$

The torque transmitted between the shafts can be taken as unchanged, thus

$$T_{out} = T_{in}$$

and therefore

$$\eta = \frac{N_{out}}{N_{in}} \qquad \text{...(19.12)}$$

The slip of the fluid coupling is defined as the ratio of the difference in the speeds of the output and input shafts to the speed of the input shaft, that is

$$\text{Slip} = \frac{N_{in} - N_{out}}{N_{in}} = 1 - \frac{N_{out}}{N_{in}} = 1 - \eta \qquad \text{...(19.13)}$$

19.11.2. Torque Converter. A torque converter differs from a fluid coupling only in respect of having a stator or a set of stationary guide vanes through which the fluid flow between the pump impeller and the turbine runner. Fig. 19.13. shows the general configuration of a single-stage torque converter.

Torque converters are normally designed with blade angles selected in such a manner that the torque transmitted to the output shaft when it is stalled is many times the torque of the input shaft. When a torque converter is operated at a constant input speed, the torque necessary to

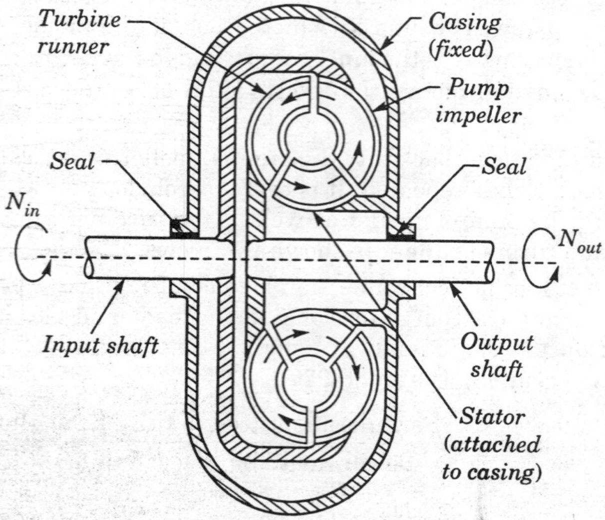

Fig. 19.13. General arrangement for a single-stage torque converter.

maintain this input speed is called the "primary torque". Torque converters can be made with several stages and with other refinements such as variable-angle blades.

The speed of the output shaft is always less than that of the input shaft, whereas the torque transmitted to the output shaft is always greater than the torque of the input shaft. The performance of the torque converter depends mainly on the output-input shafts speed ratio. The maximum efficiency may be around 80–82%.

20

Aircraft Hydraulic Systems

20.1. Introduction—An Overview

This chapter deals mainly with applications of fluid mechanics to hydraulic systems used in operation of aircrafts while taking off from the runway, getting airborne and during various stages of high altitude flight. Many aircraft hydraulic systems include a gas-loaded accumulator for energy storage. A component of aircraft hydraulic systems is an automatic relief valve connected to a pump, which charges a hydraulic accumulator when the fluid in the system is idling. In aircraft hydraulic system, cavitation may occur because of a reduced barometric pressure with rise in altitude during a flight. In such a case, the cavitation zone extends over a considerable portion of the low-pressure pipelines (suction pipes) and sometimes even along the whole length of pipeline. When this happens, the liquid stream in the pipe divides into a liquid and a vapour phase.

Local features in aircraft hydraulic systems and transmissions include filters, taps, valves, bends, other fittings and units of diverse geometrical shapes. Flow through such features may be either laminar or turbulent, depending on the velocity and the temperature of the fluid which affects its viscosity.

Aeronautical engineers usually have to deal with discharge from closed vessels (tanks) connected with the atmosphere by an orifice or a tube of small diameter. In such a case, the liquid efflux causes a vacuum in the tank and atmospheric air flows in through the orifice (or the tube). Consequently, the discharge is retared. Another example of discharge with varying head is found in the operation of under carriage shock-absorber.

Injectors are used in gas-turbine aircraft engines as also in liquid-propellant rocket engines. An injector is a specially designed tube or nozzle which atomises the discharged liquid *i.e.* jet into the atmosphere (or a space filled gas under pressure) is broken up into a fine spray.

USE of PUMPS

In aircraft hydraulic systems, pipelines with pumps are used to supply fuel to desired locations. Each pump has a suction pipeline which is subjected to negative or vacuum pressure. The minimum permissible absolute pressure at pump intake is of special importance in aircraft hydraulic systems. The maximum permissible attitude of flight for an aircraft is determined from the consideration of minimum permissible atmospheric pressure which is determined from a standard atmospheric table. The pressure above the atmospheric one is produced by supercharging the space which may contain an inert gas.

An increase in pressure in the suction (or supply) tank air space increases the pressure in the suction pipeline and helps attaining greater altitude range for the aircraft. Therefore, greater altitude range is usually attained by installing an additional pump at the beginning of the suction line to produce higher pressure in the suction pipeline, thereby preventing cavitation at the intake of the main pump.

There are certain special features of centrifugal pumps that are used in aeronautical and rocket engineering. The basic criteria governing the selection of pump design are Q, H and n. These quantities enable determination of the specific speed n_s, which suggests the type of rotodynamic pump to be preferred for the given conditions. In aircrafts with gas-turbine engines, centrifugal pumps are used mainly as booster pumps in fuel systems. They are usually installed inside the fuel tanks or close to them and are meant to increase the pressure in the entire suction pipeline, thereby improving the operating conditions for the main fuel pump and minimising the possibility of cavitation.

In rocket motors, centrifugal pumps are widely used to deliver fuel and oxidiser from the tanks to the combustion chamber. They must generate high head enough to overcome resistance of pipelines and the pressure in the combustion chamber and to ensure required pressure difference in the injectors. The required head is several hundreds of metres and the pressure tens of atmospheres. Displacement pumps are commonly used for such high pressures, but if a large discharge is also required (as is the case in rocket motors), and there is a possibility of providing a very high speed drive (*e.g.* a gas turbine), centrifugal pumps are then more suitable.

With impeller rotation at high speed and appropriately designed low specific speed centrifugal pumps can deliver liquids at high pressure. The principal feature of centrifugal pumps used in liquid-propellant motors is, therefore, their low specific speed.

Hydraulic Transmissions

Hydraulic transmission is a system of energy transmission by means of a fluid. Any hydraulic transmission consists of two basic elements : (*i*) a pump to convert basic mechanical energy into hydraulic energy, and (*ii*) a hydraulic motor of reciprocating or rotating type to reconvert fluid energy into mechanical work. Hydraulic transmissions are widely used in aviation engineering for operating the aircraft retractable undercarriages, flaps, ailerons etc.

Power controls are employed in all modern high-speed airplanes and helicopters as the force required to actuate the flight controls is much greater than the pilot's strength.

A detailed description of some of the hydraulic systems used in aircraft's operations is given in the following sections.

20.2. Gas-Loaded Accumulator

Many aircraft hydraulic systems include a gas-loaded accumulator for energy storage. (For additional information, refer chapter 18, articles 18.2 and 18.2.1 for hydraulic accumulator and differential hydraulic accumulator.) An accumulator basically consists of a cylinder with a piston on one side of which there is air under pressure p, and on the other side the hydraulic fluid, which is injected by pump (Fig. 20.1). Another type is a spherical accumulator with an elastic diaphragm separating the fluid from the compressed air (Fig. 20.2). The energy is stored by the fluid pushing the piston to the left and compressing the air volume V. The stored energy is delivered by the expanding air.

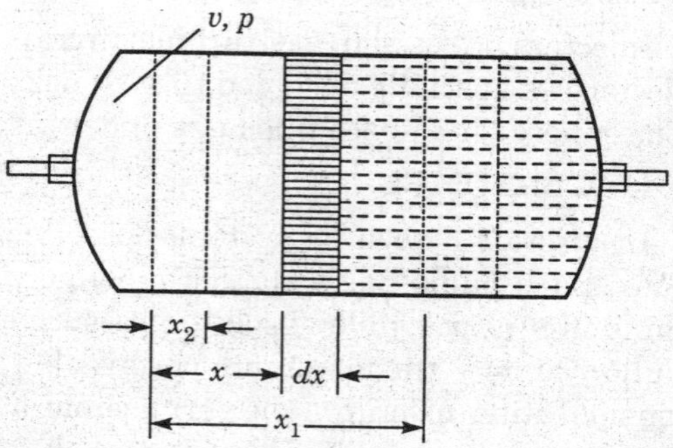

Fig. 20.1. Gas-loaded accumulator.

Example 20.1. *Determine the energy stored in an energy accumulator with a pressure load of 150 kg/cm², if it can be unloaded to a pressure of 75 kg/cm², which corresponds to an air volume of 3 litres. Assume that the compression and expansion of the air follows an isothermal process, that the air volume is proportional to the displacement of the piston, and that the piston moves without friction.*

Solution. Referring to Fig. 20.1, when the piston is displaced by dx, the elementary work done by the accumulator is

$$dW = p \,.\, A \,.\, dx \qquad ...(20.1)$$

where A is the area of the piston, and p the air pressure. As the air follows the isothermal law,

$$p = \frac{p_1 V_1}{V}, \text{ and } V = V_1 \frac{x}{x_1}$$

Here x and x_1 are the linear distances corresponding to the respective volumes. Making substitutions in Eq. (20.1)

$$dW = p_1 x_1 A \frac{dx}{x} = p_1 V_1 \frac{dx}{x}$$

The stored energy is obtained by integration :

$$W = p_1 V_1 \int_{x_2}^{x_1} \frac{dx}{x}$$

$$= p_1 V_1 \log \frac{x_1}{x_2} = p_1 V_1 \log \frac{p_2}{p_1} \qquad ...(20.2)$$

Substituting the numerical values,

$$W = 75 \times 10^4 \times 3 \times 10^{-3} \times \log \frac{150}{75} = \mathbf{1550\ kg.m}$$

20.3. Automatic Relief Valve

A component of aircraft hydraulic systems is an automatic relief valve as illustrated in Fig. 20.2. It works in the following manner. When the fluid in the system is idling, the pump charges a spherical hydraulic accumulator until a pressure p_{max} is reached at which piston 'd'

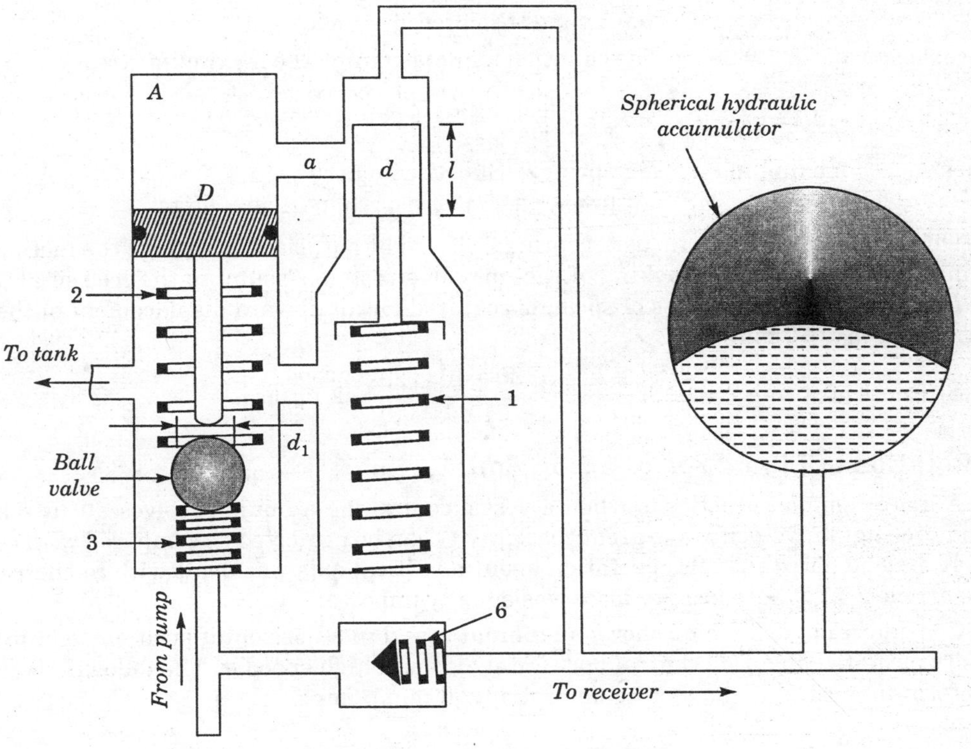

Fig. 20.2. Automatic relief valve

more down, letting the fluid into space 'A' via pipe 'a'. The pressure p_{max} drives piston 'D' down ; the piston rod opens a ball valve connecting the pressure pipe of the pump with the tank. The pump delivers the fluid back into the tank as long as the pressure in the tank remains above p_{min}, at which the piston 'd' moves up and the fluid in space A drains into the tank. When this happens, spring 2 pushes piston 'D' up, the ball valve closes and the pump supplies pressure to the receiver.

Example 20.2. *For the automatic relief valve used in aircraft hydraulic system shown in Fig. 20.2, calculate the characteristics of springs 1 and 2 if p_{max} = 140 atm, p_{min} = 80 atm, diameter of small piston d = 8 mm, stroke length l = 16 mm, diameter of the large system D = 10 mm, pipe diameter a = 2 mm and the opening at the ball valve d_1 = 6 mm. Assume the frictional force of large piston packing and other resistances to be 0.1 times the pressure force and the thrust of spring 3 = 30 kg.*

Solution. The maximum and minimum loads on spring 1 are determined as :

$$P_{max} = p_{max} \frac{\pi d^2}{4} = 140 \times \frac{\pi (0.8)^2}{4} \text{ kg} = 70.4 \text{ kg}$$

(as 1 atm. pressure is approximately equal to 1 kg/cm²)

$$P_{min} = p_{min} \frac{\pi d^2}{4} = 80 \times \frac{\pi (0.8)^2}{4} = 40.4 \text{ kg.}$$

The deflection of the spring is $h_1 = l - a = 16 - 2 = 14$ mm

$$\text{Stiffness of spring 1} = \frac{P_{max} - P_{min}}{h_1} = \frac{70.4 - 40.2}{14} = 2.16 \text{ kg/mm}$$

For opening the ball valve, the thrust on the piston should be sufficient to overcome the resistance of spring 2, the friction of the piston packing, the pressure acting on the ball valve and the thrust of spring 3.

It is given that the combined frictional resistance

= 0.1 times the pressure force, and

the thrust of spring 3, T = 30 kg. The equation for determining the maximum stress can be written :

$$(1 - 0.1)\, p_{max} \frac{\pi D^2}{4} = F_{max} + p_{max} \frac{\pi d_1^2}{4} + T$$

in which F_{max} = maximum stress of spring 2. Substituting the values,

$$F_{max} = 140 \times 0.25\, \pi\, (1^2 \times 0.9 - 0.6^2) - 30 = 30 \text{ kg}$$

On the other hand, the thrust of spring 2 should be sufficient to reverse the motion of piston D against the friction of the packing of the unloaded piston. Assuming the frictional force to be 10 kg, F_{min} = 10 kg. The stiffness of spring 2 required for an upward displacement of the piston D by 10 mm (*i.e.* h_2 = 10 mm) is

$$\text{Stiffness of spring 2} = \frac{F_{max} - F_{min}}{h_2} = \frac{30 - 10}{10} = \frac{20}{10} = \mathbf{2 \text{ kg/mm}}$$

20.4. Rotation of Liquid About a Horizontal Axis

A more common practice is when a vessel containing a liquid revolves about a horizontal axis and the angular velocity is so great that gravity can be neglected as compared with centrifugal forces. It may be noted that fluid rotation about a vertical axis was dealt with in chapter 2-Fluid statics, article 2.11, where gravity force was equally important.

Consider Fig. 20.3 which shows an annual vessel in a horizontal position containing liquid of density ρ rotated about its central horizontal axis at a high speed ω. The following analysis will help determining axial thrust acting on centrifugal pump impeller.

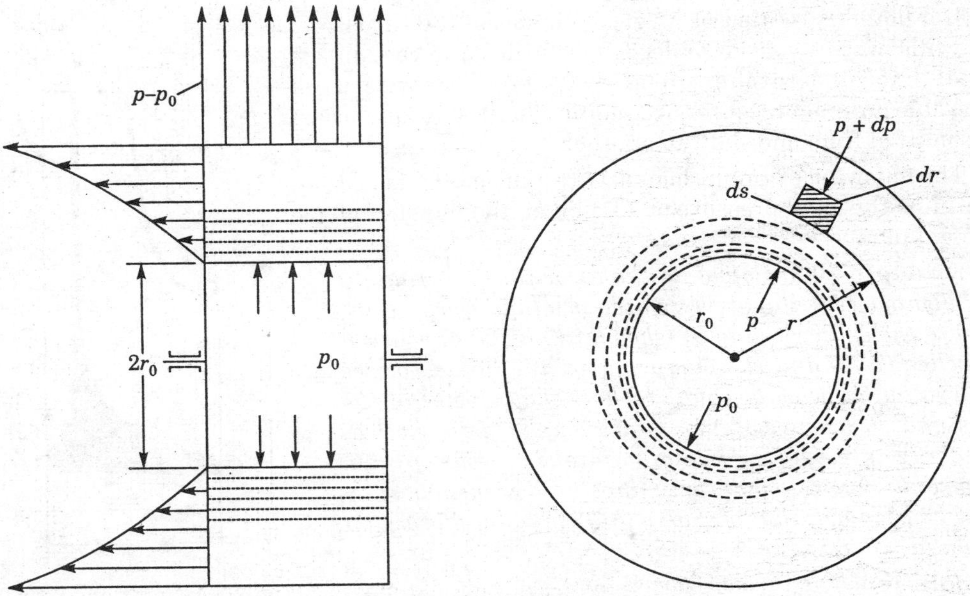

Fig. 20.3. Rotation of a vessel with liquid about its horizontal axis.

The pressure gradient in the liquid can be easily obtained by considering the equilibrium of forces for an elementary volume of base area dA and height dr taken along the radius. It is subjected to pressure and centrifugal forces. Let p be the pressure at the centre of area dA, r the distance of dA from the axis of rotation, $p + dp$ the pressure on the face opposite to dA, and $r + dr$ the distance of this face from the axis of rotation.

Considering the equilibrium of forces acting on the elementary volume in the radial direction, one can write

$$pdA - (p + dp)\, dA + \rho r \omega^2\, drdA = 0$$

from which,

$$dp = \rho\, \omega^2\, rdrdA$$

upon integration,

$$p = \rho\, \omega^2\, \frac{r^2}{2} + C$$

The constant C is found from the condition that, at $r = r_0$, $p = p_0$, hence,

$$C = p_0 - \rho\, \omega^2\, \frac{r_0^2}{2}.$$

The above equations result in a relation between p and r in the following form :

$$p = p_0 + \rho \frac{\omega^2}{2} (r^2 - r_0^2) \qquad \text{...(20.3)}$$

The surfaces of equal pressure are obviously circular cylinders whose centre lines are on the axis of rotation of the liquid *i.e.* they are concentric with the axis of rotation. If the vessel is only partially filled, the free surface as one of the surfaces of equal pressure, is a cylinder of radius r_0 and the pressure is p_0.

It is often necessary to determine the thrust of a liquid rotating with a vessel exerted on the wall normal to the axis of rotation. For this the thrust on an elementary ring surface of radius r and width dr can be determined from Eq. (20.3) as follows :

$$dp = p \,.\, dA = \left[p_0 + \rho \frac{\omega^2}{2} (r^2 - r_0^2) \right] 2\pi r dr \qquad \text{...(20.4)}$$

Integration of this equation gives the **desired** result.

If a liquid is rotated at a very high speed, the thrust on the walls may be considerable. This is utilized in certain types of friction clutches in aircraft engines where considerable normal pressures are required for the torque to be transmitted from one shaft to another.

The following example demonstrates the method used to determine the axial pressure of a liquid on the impeller of centrifugal pumps.

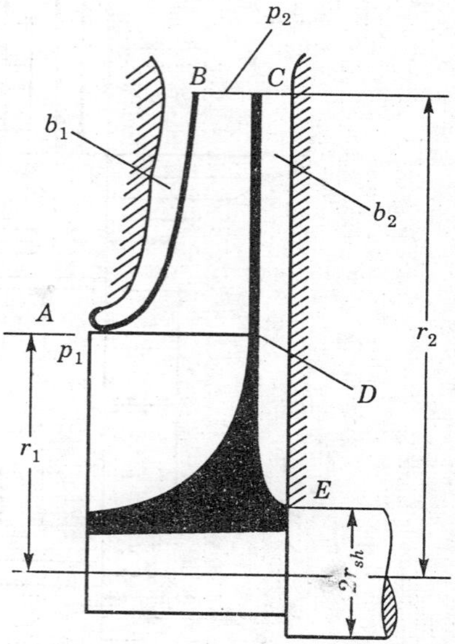

Fig. 20.4. Notations for calculating axial thrust acting on centrifugal pump impeller.

Example 20.3. *Calculate the axial thrust acting on the impeller of a centrifugal pump used in a liquid-propellant rocket if in the spaces b_1 and b_2 (shown in Fig. 20.4) between the impeller vanes and the pump casing, the liquid rotates with an angular velocity equal to half the velocity of the impeller and the leakage at A is small enough to be neglected. The discharge pressure $p_2 = 38$ atm, intake pressure $p_1 = 0$, speed of rotation $n = 16{,}500$ rpm ; impeller dimensions : $r_2 = 50$ mm, $r_1 = 25$ mm, radius of shaft $r_{sh} = 12$ mm, specific weight of the liquid = 918 kg/m³.*

Solution. The thrust exerted by the liquid on almost identical surfaces AB and CD is balanced. The only unbalanced force is the axial pressure on surface DE, *i.e.*, a ring area bounded by circles of radii r_1 and r_{sh}. The required force is directed towards the left and is equal to

$$F = 2\pi \int_{r_{sh}}^{r_1} prdr$$

substituting for p from Eq. (20.3), and substituting p_2 and r_2 for p_0 and r_0, respectively, we obtain

$$p = p_2 - \gamma \frac{\omega_f^2}{2g}(r_2^2 - r^2)$$

in which ω_f = angular velocity of the fluid. Therefore, the axial thrust is

$$F = 2\pi \int_{r_{sh}}^{r_1} \left[p_2 - \gamma \frac{\omega_f^2}{2g}(r_2^2 - r^2) \right] rdr$$

$$= 2\pi \left[p_2 \frac{r^2}{2} - \frac{\gamma}{2g} \omega f^2 \left(r_2^2 \frac{r^2}{2} - \frac{r^4}{4} \right) \right]_{r_{sh}}^{r_1}$$

$$= \pi (r_1^2 - r_{sh}^2) \left[p_2 - \gamma \frac{\omega_f^2}{2g} \left(r_2^2 - \frac{r_1^2 + r_{sh}^2}{2} \right) \right]$$

Substituting the numerical values, noting that $\omega_f = \dfrac{\omega}{2} = \dfrac{2\pi n}{2(60)} = \dfrac{\pi n}{60} = \dfrac{\pi \times 16{,}500}{60}$ rad/sec

$$F = \pi (2.5^2 - 1.2^2) \left[38 - \frac{981}{(100)^3} \left(\frac{\pi \times 16{,}500}{60} \right)^2 \times \frac{1}{2 \times 981} \times \left(5^2 - \frac{2.5^2 + 1.2^2}{2} \right) \right]$$

= **460 kg.**

∴ The axial thrust acting on the impeller of the pump = 460 kg.

20.5. Tank Pressurisation

Tank pressurisation is widely employed in aircraft to increase the pressure in fuel and other tanks. If the tank is opened to atmosphere through a small opening, the pressure over the fuel surface will be the atmospheric pressure. But if an arrangement like the one shown in Fig. 20.5 is used, the atmospheric pressure p_0, and the air velocity v_0 (which is the same as the flight velocity), can be used to obtain the increased pressure inside the tank where the air velocity is almost reduced to zero. The Bernoulli's equation gives us the pressure above the liquid (fuel) surface as

$$p_1 = p_0 + \rho \frac{v_0^2}{2} \quad \text{...(20.5)}$$

Eq. (20.5) demonstrates that at low flight velocities, the gauge pressure in the tank is approximately equal to the dynamic pressure $\left(i.e.\ \rho \frac{v_0^2}{2}\right)$ as calculated from the velocity of flight and air density.

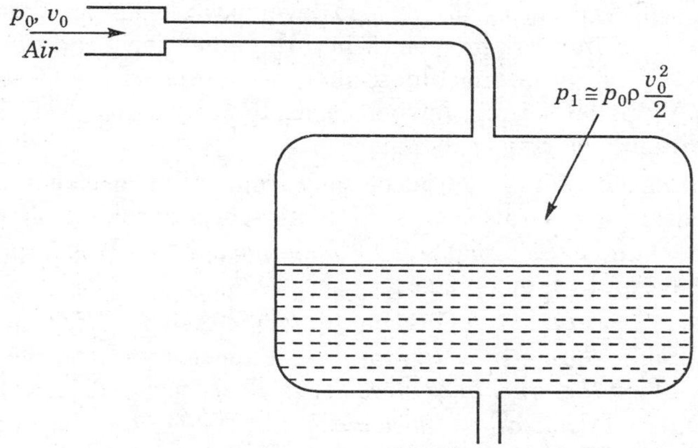

Fig. 20.5. Tank pressurisation

20.6. Problems Associated with Cavitation

In certain cases of fluid motion in closed conduits, a peculiar phenomenon takes place which is due to a change in the physical state of liquid. Vapourisation and evolution of gases dissolved in the liquid takes place.

When a liquid flows through a converging part in a pipeline, its velocity increases and pressure intensity decreases. If the absolute pressure drops to the vapour pressure of the liquid at the prevailing temperature, evaporation commences and gases are liberated. The liquid simply begins to boil locally. When the fluid stream diverges after passing the narrowing portion, the velocity decreases and the pressure increases. The boiling stops and the vapours condense partially or completely and the gases redissolve. This local boiling of flowing liquid is known as cavitation.

The phenomenon of cavitation is described in detail in chapter 4 article 4.16. A simple experimental device enables the phenomenon to be observed visually. Consider Fig. 20.6 where a transparent venturitube is installed in a pipeline. Water or some other liquid is brought under pressure to a valve through which it flows into a glass tube with a venturi contraction.

When the valve is opened slightly is discharge is low, the velocity of the stream is small, the pressure drop at the throat of the tube is also small and the stream is transparent, *i.e.* no cavitation takes place. The wider the valve opening, the faster is the velocity of stream and the lower pressure occurs in the narrowing portion. A stage comes when the pressure intensity reduces to the saturation vapour pressure. This is the stage when a distinct cavitation zone appears, which increases in size as the valve is further opened.

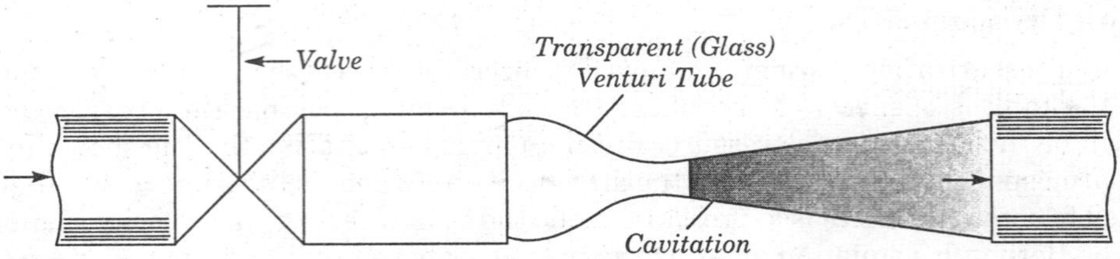

Fig. 20.6. A device for demonstrating cavitation.

Cavitation announces itself by characteristic noise and vibrations, as was described in chapter 4, article 4.16. Prolonged cavitation has an erosive effect on metal walls. The pitting of walls occurs not at the point where the vapour pockets appear, but at the locations where they collapse. Cavitation thus has an adverse effect on pipelines and hydraulic systems. When cavitation develops, the resistance of pipes increases sharply, with a corresponding reduction in discharge.

Cavitation may develop in any local converging portions followed by expansions, such as faucets, valves, gates, orifices and jets. It may also develop when a narrowing portion is not followed by a diverging section, and in straight pipes when the elevation head or energy losses increase. Cavitation may occur in hydraulic machines, such as pumps or turbines, and on the blades of rapidly revolving ship propeller screws. In such cases, the result is a sharp decline in the efficiency of a machine and a gradual wearing of its parts.

In aircraft hydraulic systems, cavitation may occur because of reduced barometric pressure with height. In this case, the cavitation zone extends over a considerable portion of low-pressure pipelines (suction pipes) and even along the whole length. When this happens, the stream in the pipe divides into a liquid and vapour phase.

Example 20.4. *Find out the flow conditions of a given hydraulic fluid in a pipe of an aircraft hydraulic system if the pipe diameter is 12 mm, the flowrate 0.25 litre / sec and the temperature is zero degree calcius at which the kinematic viscosity is 42 cst (centistoke). At what temperature will the flow conditions change ? The value of kinematic viscosity of the fluid at temperatures of 50°C, 20°C, 0°C and – 20°C are 10, 16, 42 and 130 centistokes.*

Solution. The kinematic viscosity of the fluid $\nu = 42$ centistokes $= \dfrac{42}{100}$ cm²/sec = 0.42 cm²/sec

The Reynolds number of flow, $\mathbf{R} = \dfrac{Vd}{\nu} = \dfrac{4Q}{\pi d\nu} = \dfrac{4 \times 250}{\pi \times 1.2 \times 0.42} = 632$

(*i*) Since the Reynolds number is less than 2000, the flow is Laminar.

(*ii*) The flow regime will change when the flow changes from the laminar state to the turbulent one, for which we may consider the critical Reynolds number as equal to 2000. The kinematic viscosity corresponding to this Reynolds number,

$$\nu = \frac{4Q}{\pi d\, R_{cr}} = \frac{4 \times 250}{\pi \times 1.2 \times 2000} = 0.132 \text{ cm}^2/\text{sec} = 13.2 \text{ centistokes}$$

(*iii*) Plotting a graph between the kinematic viscosity against the temperature as per the given data, we find that at a temperature of 41°C the desired value of kinematic viscosity will be obtained.

Example 20.5. *Determine the absolute pressure at pump intake in mm Hg for an aircraft lubricating system using a given hydraulic oil, having kinematic viscosity $\nu = 0.11$ stokes (1 stoke = 1 cm² / sec) and specific with $\gamma_{oil} = 900$ kg / m³, aircraft operating in horizontal flight at an altitude of 16,000 m (where atmospheric pressure is p_{atm} / γ_{Hg} = 77.1 Hg). Figure 20.7 shows a sketch of the aircraft oil supply system. The length of the pipeline is 2 m, the diameter of pipe 18 mm ; the elevation of the oil surface in the tank above the pump is 0.7 m ; the pressure in the oil tank is atmospheric. The required flowrate to ensure the necessary heat transfer into the oil at maximum*

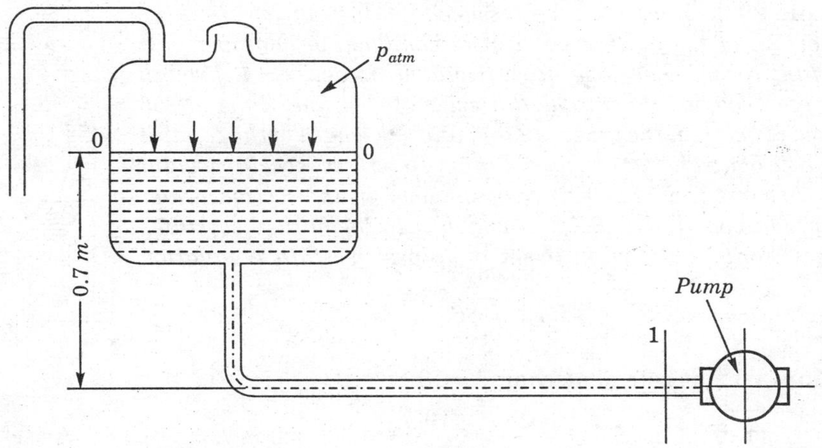

Fig. 20.7. Schematic drawing of aircraft oil supply system.

performance of the engine is 16 litre/minute ; Kinematic viscosity of the lubricating oil is 0.11 stokes, its specific weight 900 kg/m³. Neglect local losses.

Solution. Velocity of oil in the pipeline

$$V = \frac{4Q}{\pi d^2} = \frac{4\,(16 \times 10^{-3}/60)}{\pi\,(18 \times 10^{-3})^2} = 1.05 \text{ m/s}$$

The Reynolds number of flow,

$$\mathbf{R} = \frac{Vd}{\nu} = \frac{1.05 \times (18 \times 10^{-3})}{0.11 \times (10^{-2})^2} = 1718$$

Since the Reynolds number is less than 2000, the flow is laminar.

Pressure at the pump intake can obtained by applying Bernonlli's equation between sections 0-0 and 1-1.

$$Z + \frac{p_{atm}}{\gamma_{oil}} = \frac{p_1}{\gamma_{oil}} + \alpha \frac{V^2}{2g} + h_f$$

Friction less h_f in the intake pipe, $h_f = f \dfrac{l}{d} \dfrac{V^2}{2g}$

For laminar flow, $f = \dfrac{64}{\mathbf{R}} = \dfrac{64}{1718} = 0.03725,$

and kinetic energy correction factor, $\alpha = 2.0$

Making substitutions of values, one obtains

$$\frac{p_1}{\gamma_{oil}} = Z + \frac{p_{atm}}{\gamma_{oil}} - \alpha \frac{V^2}{2g} - h_f = 0.7 + \frac{77.1 \times 10^{-3} \times 13.6}{0.9} - \frac{2(1.05)^2}{2 \times 9.81} - h_f$$

$$h_f = 0.03725 \times \frac{2}{18 \times 10^{-3}} \times \frac{(1.05)^2}{2 \times 9.81} = 0.232 \text{ m}$$

substituting the value of h_f,

$$\frac{p_1}{\gamma_{oil}} = 1.52 \text{ m of oil}$$

$$= 1.52 \times \frac{0.9}{13.6} = 0.10 \text{ m of Hg} = \mathbf{100 \text{ mm of Hg}}$$

Example 20.6. *Calculate the pressure loss due to friction in the cylindrical portion of the cooling conduit of combustion chamber of a liquid-propellant rocket. The coolant (nitric acid) is delivered through the space between coaxial tubes. The diameter of the inner tube is 155 mm, the clearance betweern the tubes is 2 mm and the length of the conduit 500 mm. The rate of discharge of the coolant is 10 kg/sec, specific weight 1,510 kg/m^3. Assume the temperature of the acid in the section under consideration to be constant at 80°C at which the kinematic viscosity is 0.25 centistokes. For turbulent flow in smooth pipes, use Jain's* equation for friction factor :* $\frac{1}{\sqrt{f}} = 1.8 \log_{10} \mathbf{R} - 1.5186.$

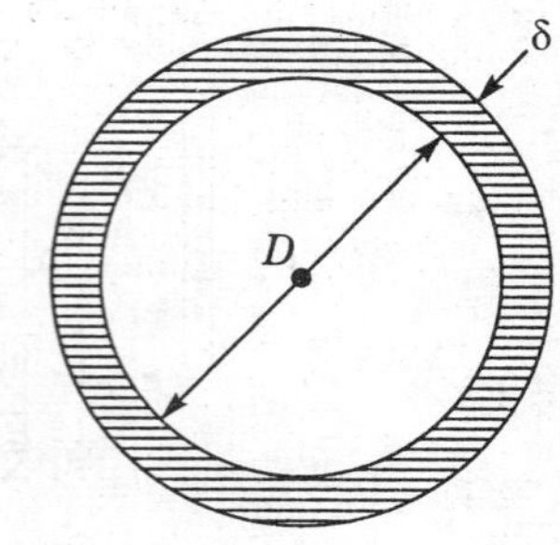

Fig. Ex. 20.6.

Solution. Weight rate of discharge = 10 kg/sec

Volumetric rate of flow, $Q = \frac{10}{\gamma_{acid}} = \frac{10}{1510} = 0.006623 \text{ m}^3\text{/s} = 6623 \text{ cm}^3\text{/sec}$

Velocity of flow through the clearance $= \frac{Q}{(\pi D\delta)} = \frac{10}{(1510 \times \pi \times 0.155 \times 0.002)} = 6.8$ m/s

where D is the diameter of the inner tube and δ the clearance between the tubes.

The hydraulic radius of the conduit

$$R_h = \frac{A}{P} = \frac{\pi[(D+2\delta)^2 - D^2]}{4\pi[(D+2\delta)+D]} = \frac{\delta}{2} = \frac{2}{2} = 1 \text{ mm}$$

The Reynolds number

$$\mathbf{R} = \frac{4R_nV}{\nu} = \frac{4 \times 0.1 \times 680}{0.25 \times 10^{-2}} = 110{,}000$$

$$\left(\text{for a circular pipe, } R_h = \frac{\pi d^2}{4\pi d} = \frac{d}{4}, \quad \therefore \quad d = 4R_h, \text{ and } \mathbf{R} = \frac{Vd}{\nu} = \frac{4R_hV}{\nu}\right)$$

At such a high Reynolds number, the flow is highly turbulent. For turbulent flow in smooth pipes, Jain's equation provides direct determination of friction factor (see Eq. 9.64 *b*) :

$$\frac{1}{\sqrt{f}} = 1.8 \log_{10} \mathbf{R} - 1.5186$$

The pressure loss due to friction in terms of head $= \frac{p_f}{\gamma}$

$$h_f = \frac{p_f}{\gamma} = f\,\frac{l}{d}\,\frac{V^2}{2g}$$

and the pressure loss, $p_f = \gamma_f \frac{l}{d}\frac{V^2}{2g}$, here $d = 4R_h$

$$\therefore \qquad p_f = 1510 \frac{1}{(1.8 \log_{10} 1{,}10{,}000 - 1.5186)^2} \times \frac{500}{4 \times 1} \times \frac{(6.8)^2}{2 \times 9.81} \text{ kg/m}$$

$= 8000 \text{ kg/cm}^2 = 8000 \times 10^{-4} \text{ kg/cm}^2 = 0.8 \text{ kg/cm}^2 =$ **0.8 atm**

20.7. Local Features in Aircraft Hydraulic Systems

Local features in aircraft hydraulic systems and transmissions include filters, taps, valves bends and other fittings and units of diverse geometrical shapes. Flow through such features may be either laminar or turbulent, depending upon the velocity and temperature (which determines

*Jain, A.K., "An Accurate Explicit Equation for Friction Factor," Jnl of Hydraulics Division, American Society of Civil engineers, May 1976, pp 674-77.

the viscosity) of the fluid. The Reynolds number varies within a fairly wide limits, and the loss coefficient, K_L, for these features must be considered as a function of Reynolds number, **R**.

The variation of loss coefficient, K_L, with **R** for some typical local features in aircraft hydraulic systems is plotted on log-log scale in Fig. 20.8. It is to be noted that the dependence of K_L on **R** in a felt filter, is a linear function upto **R** = 5000. The reason is that in laminar flow through the filter pores, friction is very great and practically no turbulence takes place. In the plots for a tap, valve and elbow the linear portions are much shorter ; they are followed by transient portions, and finally the K_L is no longer dependent on **R**, the plot becoming almost horizontal indicating K_L = constant.

The steeper slope of the non-return valve curve is due to the fact that with increasing R, the opening of the valve increases on account of faster flow through it.

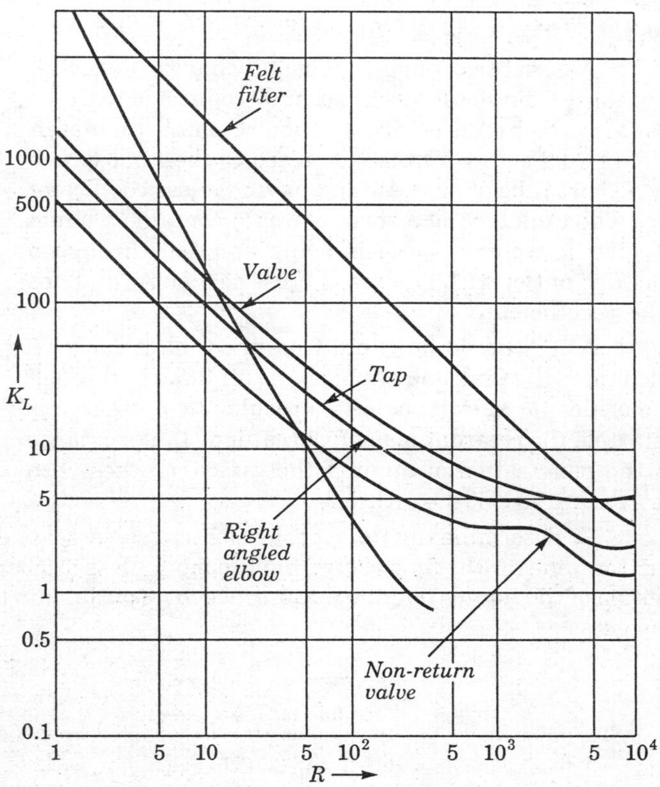

Fig. 20.8. Variation of K_L versus **R**.

In aircraft fuel systems, the Reynolds number is usually much higher than in hydraulic systems, and it can be assumed without much error, that the local loss coefficients of fuel pipes do not depend on **R**.

The following table presents loss coefficients of some common units and local features of aircraft fuel systems. The values of K_L are referred to the velocity head at the intake pipe of the unit or feature

Local feature of fuel system	K_L
Flexible pipe connection	0.3
Standard 90° elbow (bored body)	1.2-1.3
Tee	3.5
Fuel tap	1-2.5
Nonreturn valve	2.0
Gauze filter	1.5-2.5
Flowmeter sensor with impeller rotating	7.0
with impeller braked	11-12
Pipe entrance (tank outlet)	0.5-1.0
Pipe outlet (tank inlet)	1.0

20.8. Emptying of a Closed Tank

Aeronautical engineers usually have to deal with discharge from closed vessels (tanks) connected with the atmosphere by an orifice or tube of small diameter. Figure 20.9 shows a tank connected to the atmosphere by a short tube of area A_1, and having a loss coefficient K_{L1}. The tank having a surface area A contains a liquid of specific weight γ, and is being emptied through a rounded ontlet with an area of cross section A and has a loss coefficient K_{L2}.

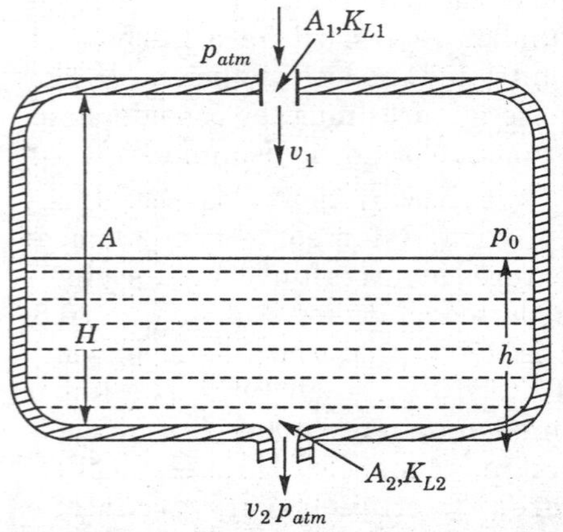

Fig. 20.9. Emptying of a closed tank.

As the liquid flows out from the tank it causes a vacuum in the tank and atmospheric air flows in through the orifice (or short tube). As a result, the discharge is retarted, the retardation being greater as the resistance to the penetration of air into the tank increases with decreasing size of the orifice.

To determine the time it takes for such a vessel to empty, two Bernonlli's equations must be written : one for the flow of air from the motionless atmosphere into the vessel, and another for the motion of the liquid from the free surface to its outlet into the atmosphere. Writing these equations, one gets

$$\frac{p_{atm}}{\gamma_{air}} = \frac{p_0}{\gamma_{air}} + (1 + K_{L1}) \frac{V_1^2}{2g}$$

and
$$h + \frac{p_0}{\gamma} = \frac{p_{atm}}{\gamma} + (1 + K_{L2}) \frac{V_2^2}{2g}$$

Neglecting the compressibility of air, the volume rates of discharge of air and the liquid can be equated :

$$Q = A_1 V_1 = A_2 V_2$$

(If the jet contracts, the coefficient of contraction C_c should be introduced)

The above equations can be rewritten as

$$p_{atm} = p_0 + (1 + K_{L1}) \frac{Q^2}{2gA_1^2} \gamma_{air}$$

$$\gamma h + p_0 = p_{atm} + (1 + K_{L2}) \frac{Q^2}{2gA_2^2} \gamma.$$

Solving these equations for Q,

$$Q = \sqrt{\frac{2gh\gamma}{\left(\frac{1+K_{L1}}{A_1^2}\gamma_{air}\right) + \left(\frac{1+K_{L2}}{A_2^2}\gamma\right)}} \quad ...(20.6)$$

When the tank is emptying, let the free surface of the liquid fall by dh in a differential time interval dt. The volume equation can be written as

$$A\,dh = -Q\,dt$$

Substituting Eq. (20.6) into the volume equation, solving for dt and integrating from $h = H$ to $h = 0$, one obtains the time of emptying as :

$$t = \frac{2AH}{\sqrt{2g\gamma H}} \sqrt{\frac{1+K_{L1}}{A_1^2}\gamma_{air} + \frac{1+K_{L2}}{A_2^2}\gamma} \quad ...(20.7)$$

This formula is valid for the case of air intake and liquid out flow through pipes. Introducing the following expression for the respective loss coefficients

$$K_L = f\frac{l}{d} + \Sigma K_L$$

where l and d are the length and diameter of the pipe respectively and ΣK_L the sum of the local loss coefficients in the pipe, necessary computations for the time of emptying using Eq. (20.7) can be made.

20.9. Hydro-Pneumatic Shock-Absorber

An example of discharge with varying head is found in the operation of an undercarriage shock-absorber. Figure 20.10 shows a schematic diagram of a hydropneumatic shock-absorber, which contains compressed air at the upper end the pressure of which during flight is p_0. When the airplane wheels touch the ground, the hydraulic fluid flows through an orifice of area A_0, its discharge coefficient being C_d, and the air compressed to p_1. The height of the air column is reduced from h_0 to h_1.

To simplify the actual process of shock absorption, let us assume that a force F is suddenly applied to the shock-absorber, and that the air is compressed according isothermal law. Let us determine the duration of the shock-absorption process. In other words, the time it takes for the liquid to flow through the orifice until the equilibrium is restored.

Consider that in a differential time dt, the piston of area A moves up through a distance dh, pushing the liquid through the orifice (area A_0) at a discharge rate Q. The volume equation can be written as

$$A\,dh = -Q\,dt$$

The rate of discharge through the orifice is given by

$$Q = C_d A_0 \sqrt{\frac{2g\,(p_1 - p_0)}{\gamma}}$$

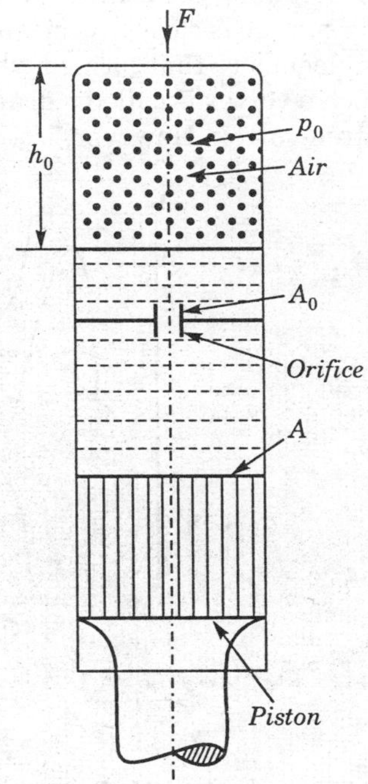

Fig. 20.10. Shock-absorber.

where $p_1 = F/A$ is the constant pressure over the piston, and

p = the varying air pressure given by the isothermal law, pV = constant.

or $$(p + p_{atm})\,h = (p_0 + p_{atm})\,h_0$$

from which we obtain,

$$p = (p_0 + p_{atm})\frac{h_0}{h} - p_{atm}$$

Substituting the expressions for p, p_1 and Q in the volume equation, one obtains

$$dt = -\frac{A\sqrt{\gamma}}{C_d \cdot A_0\sqrt{2g}\,(p_1 + p_{atm})}\sqrt{\frac{h}{h - h_1}}\,dh \qquad \text{...(20.7)}$$

The required time t can be determined after integrating Eq. (20.7) with h varying from h_0 to h_1.

20.10. Injectors

An injector is a specially designed tube or nozzle which atomises the discharged liquid, *i.e.* the jet into the atmosphere (or a space filled with gas under pressure) is broken up into a fine spray.

20.10.1. Swirl Injector

The most common type of injector used in gas-turbine aircraft and liquid-propellant rocket engines is the so called swirl injector for spraying fuel into the combustion chamber. In such an injector, as shown in Fig. 20.11, whirl is imparted to the stream by forcing it to follow a spiral path, which is then contracted. The angular momentum created by the tangential approach of the liquid remains approximately constant during flow through the injector. Consequently, as the stream converges, the whirl velocity component increases considerably and large centrifugal forces develop which throw the liquid to the walls, forming a thin film which issues from the injector in a fine spray. Along the centreline of the injector, there is formed air (gas) vortex in which the pressure at the surface is close to atmospheric (if the discharge is into the atmosphere). This vortex is completely analogous to that formed when liquid flows out of a hole in the bottom of vessel (or a wash basin). Such a vortex is called a drain-hole vortex, as depicted in Fig. 20.12. This type of vortex is known as a free-vortex. However, in the injector the vortex is much more intensive.

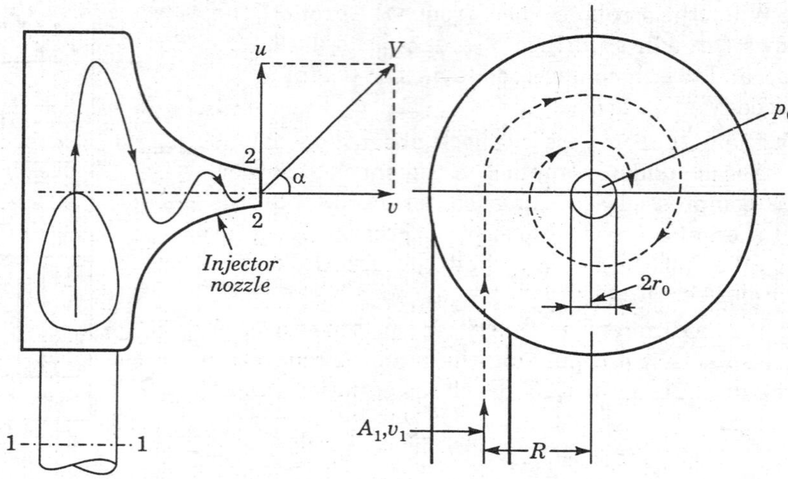

Fig. 20.11. Swirl injector.

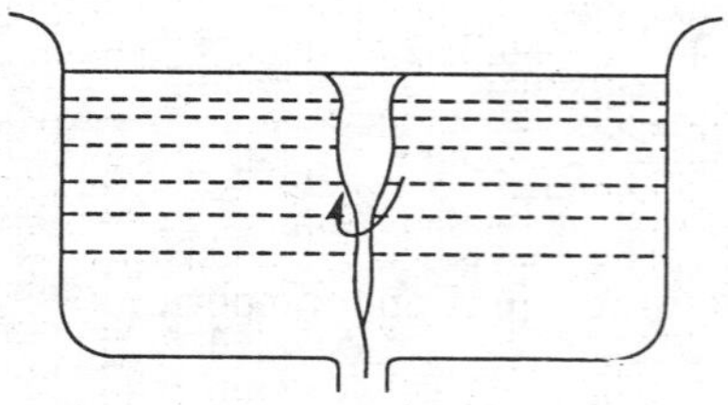

Fig. 20.12. Free vortex.

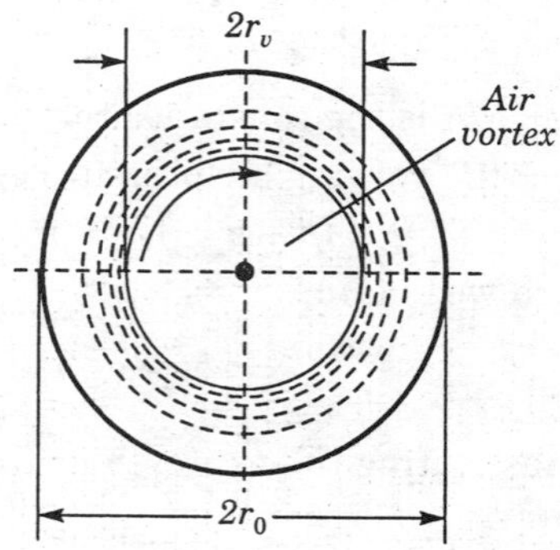

Fig. 20.13. Cross-section of a vortex in a swirl injector.

Thus, the jet issuing from a swirl injector does not fill the whole cross-section of the injector nozzle ; it is annualar, with an air vortex along the centre, as is illustrated in Fig. 20.13. Therefore, the coefficient of contraction, C_c, of a swirl injector is usually much less than unity. Because of this, and also because the resultant velocity of discharge V is directed at an angle to the cross-section of the injector nozzle (as shown in Fig. 20.11), the tangent of which is equal to the ratio of the whirl component u to the axial component v, the coefficient of discharge is always much less than unity, varying within wide limits depending on the geometry and relative dimensions of the injector.

For determing the rate of discharge of an injector, the basic equation gives

$$Q = C_d \cdot A_0 \sqrt{p/\gamma} \qquad \text{...(20.8)}$$

in which the precise value of the coefficient of discharge C_d must be known.

20.10.2. Injector with Helical Nozzle

In many liquid-propellant rocket injectors, whirl is imparted to the fluid by means of a two- or three-step helix built into the nozzle. The injector with helical nozzle is shown in Fig. 20.14, in which r_m represents mean radius of helical groove.

In modern gas-turbine engines, adjustable swirl injectors are more commonly used. In such injectors, the coefficient of discharge (or the area of the outlet) varies automatically with changing fuel pressure. Such nozzles enable the range of fuel supply to be increased for a given pressure range while maintaining the required quality of spraying.

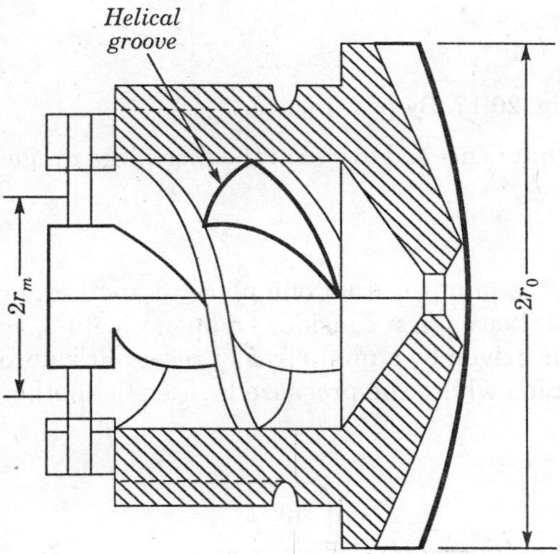

Fig. 20.14. Injector with helical nozzle.

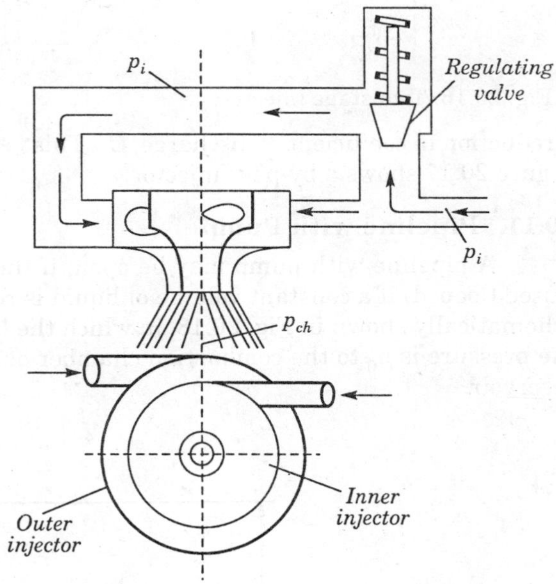

Fig. 20.15. Twin-nozzle Injector.

20.10.3. Twin-Nozzle Injector

Twin-nozzle injectors, two-stage injectors and injectors with by-pass arrangement are widely employed. Common to them all is a valve device which opens (or closes) an auxiliary conduit when the pressure rises, thereby increasing the coefficient of discharge or the outlet area.

The twin-nozzle actually comprises of two injectors, one inside the other, as shown in Fig. 20.15. At low pressures, the valve is closed and the inner injector operates ; when the pressure rises the valve opens and the second injector become operative. The supply of fuel increases sharply.

20.10.4. Two-Stage Injector

The two-stage injector has one nozzle and one swirl chamber with two inlet pipes, as shown in Fig. 20.16. At low pressures, the fuel is supplied through one pipe, at high pressures, through both. As a result the coefficient of discharge, C_d, increases.

20.10.5. By-Pass Injector

The by-pass injector is provided with a valve controlled drain. The lower the pressure, the wider is the opening of the valve ; at peak pressure the valve is closed completely. Thus, at low pressure the intake velocity is higher, which is equivalent to a smaller inlet area, and this means

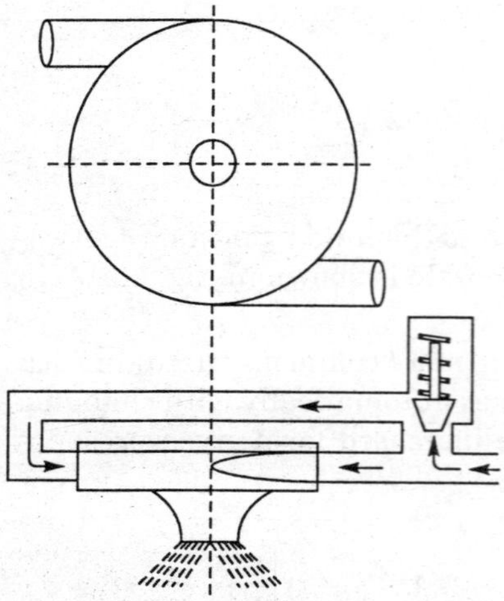

Fig. 20.16. Two-stage injector.

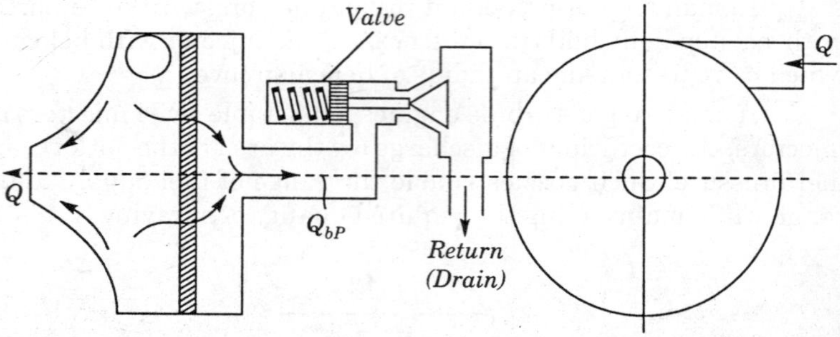

Fig. 20.17. By-pass injector.

a reduction in coefficient of discharge, C_d. This is just what is needed to widen the discharge range. Figure 20.17 shows a by-pass injector.

20.11. Pipeline with Pump

A pipeline with pump may be open, if the liquid is pumped from one place to another, or closed (looped) if a constant volume of liquid is recycled. Let us first consider an open pipeline, as schematically shown in Fig. 20.18, in which the liquid is driven by a pump from a reservoir where the pressure is p_0 to the combustion chamber of an engine where the pressure is p_3 or to another reservoir.

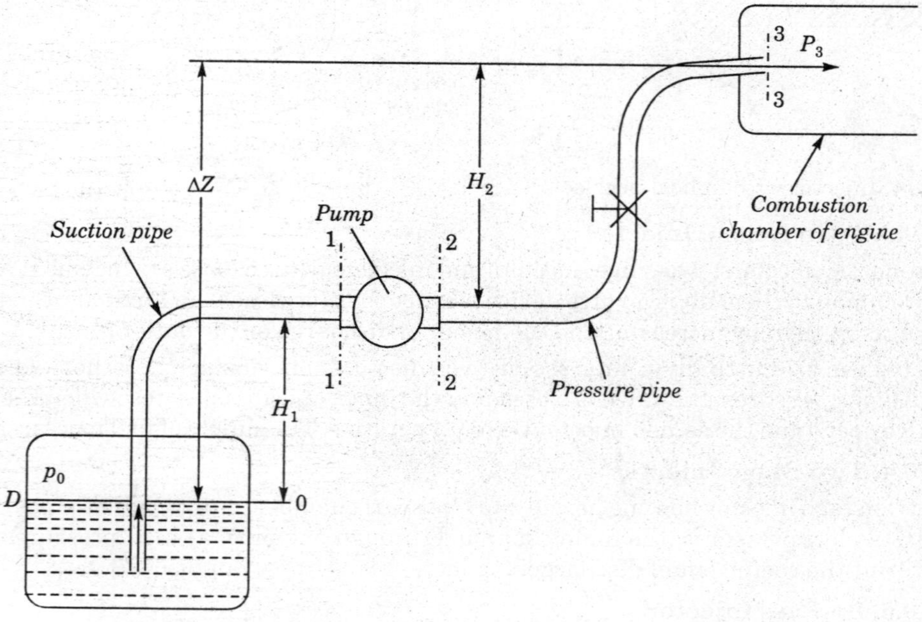

Fig. 20.18. Schematic representation of pipeline with pump.

The elevation of the pump shaft above the lower level H_1 is the suction head in the vertical plane, the pipe along which the liquid flows to the pump being the suction or the intake pipe. The elevation of the end section of the pipeline, or the upper level of the liquid is the vertically delivered head and the pipe through which the liquid travels from the pump is called the pressure or discharge or the outlet pipeline. Writing the Bernonlli's equation for flow of liquid in the suction pipeline between sections 0-1 and 1-1,

$$\frac{p_0}{\gamma} = H_1 + \frac{p_1}{\gamma} + \alpha \frac{V_1^2}{2g} + \Sigma h_{L0-1} \quad \text{...(20.9)}$$

This equation shows that the pump develops suction (lifts the liquid to height H_1), imparts kinetic energy to the liquid, $\alpha V_1^2/2g$, and overcomes hydraulic resistances (represented by the term Σh_{L0-1}) by expending the pressure p_0. As this pressure is very limited, it should be expended in such a way as to continuously maintain some pressure reserve p_1 at the pump intake, which is essential for the pump to work without cavitation. It is therefore necessary that the suction pipeline should be designed hydraulically with special care, which may require following solutions to be worked out :

1. For the given dimensions, and flowrate, to determine absolute pressure at pump intake. The absolute pressure p_1 found from Eq. (20.9) is compared with the minimum permissible pressure.
2. For the given minimum permissible absolute pressure $p_{1\text{-min}}$ at pump intake, to determine one of the following permissible limiting quantities :

$$H_{1\,\max},\ Q_{\max},\ d_{\min} \text{ or } p_{0\text{-min}}$$

Determination of the last quantity is of special importance in aircraft hydraulic systems, in which $p_0 = p_{atm} + \Delta p$. Here Δp is the pressure above the atmospheric pressure produced by supercharging or pressure of an inert gas. The minimum permissible atmospheric pressure $(p_{\text{atm}})_{\min}$ is determined from $(p_0)_{\min}$, then the maximum permissible altitude of flight for an aircraft equipped with such a system is determined from a standard atmosphere table.

An increase in the pressure p_0 increases the pressure in the whole of the suction pipeline, which provides for greater altitude range of the system. On the other hand, high pressure in the supply tank means stronger walls, and consequently, greater weight of the tank. Therefore, greater altitude range of aircraft hydraulic systems is usually attained by installing an additional pump at the beginning of the suction pipeline to produce a higher pressure in the suction line, thereby preventing cavitation at the intake of the main pump.

Application of Bernoulli's equation between section 2-2 and 3-3 of the pressure pipe results in the following relationship :

$$\frac{p_2}{\gamma} + \alpha_2 \frac{V_2^2}{2g} = H_2 + \frac{p_3}{\gamma} + \alpha_3 \frac{V_3^2}{2g} + \Sigma h_{L2-3} \quad \text{...(20.10)}$$

If the pressure pipe discharges into a reservoir, there is no velocity head in the right-hand side of Eq. (20.10), but the loss of head due to enlargement must be considered.

From Eq. (20.9), one can write :

$$\frac{p_1}{\gamma} + \alpha_1 \frac{V_1^2}{2g} = \frac{p_0}{\gamma} - H_1 - \Sigma h_{L0-1}$$

Let us determine the energy obtained by every kilogram of the liquid in passing through the pump. This energy is imparted by the pump, and is called the pumping head, H_{pump}. This is obtained as :

$$H_{pump} = \left(\frac{p_2}{\gamma} + \alpha_2 \frac{V_2^2}{2g}\right) - \left(\frac{p_1}{\gamma} + \alpha_1 \frac{V_1^2}{2g}\right)$$

$$= H_1 + H_2 + \frac{p_3 - p_0}{\gamma} + \alpha_3 \frac{V_3^2}{2g} + \Sigma h_{L0-1} + \Sigma H_{L2-3}$$

or
$$H_{pump} = \Delta Z + \frac{p_3 - p_0}{\gamma} + CQ^2 + KQ^n \quad ...(20.11)$$

where ΔZ = total static elevation head of liquid (see Fig. 20.18)

$CQ = \alpha_3 V_3^2 / 2g$;

KQ^n = total hydraulic losses in suction and pressure pipelines.

Adding the pressure head difference $(p_3 - p_0)/\gamma$ to the actual change in elevation ΔZ, we obtain an equivalent added lift

$$\Delta Z' = \Delta Z + \frac{p_3 - p_0}{\gamma}$$

and Eq. (20.11) can be rewritten as

$$H_{pump} = \Delta Z' + CQ^2 + kQ^n \quad ...(20.12)$$

The required head, H_{req}, is obtained by using the pipe characteristics as :

$$H_{req} = \Delta Z + CQ^2 + kQ^n \quad ...(20.13)$$

Comparing Eqs. (20.13) and (20.12), it is evident that

$$H_{pump} = H_{req} \quad ...(20.14)$$

This equality can be extended to all cases of steady operation of a pump in a pipeline. Equation (20.14) is the basis for a method of calculating relevant parameters in pipeline with pumps which consists in plotting two curves to the same scale on one diagram, the characteristic curve of the pipeline (also called system characteristics) $H_{req} = f_1(Q)$ and the characteristic curve of the pump, $H_{pump} = f_2(Q)$ and finding their point of intersection, as shown in Fig. 16.31 for article 16.24.3-Pump and system characteristics.

Example 20.7. *Compute the required head at the outlet of a booster pump in an aircraft fuel system to supply kerosene fuel at a rate of 1200 kg/hr from the service tank to the engine fuel pump if the length of the duralumin piping is 5 m, its diameter 15 mm, required pressure at the fuel-pump*

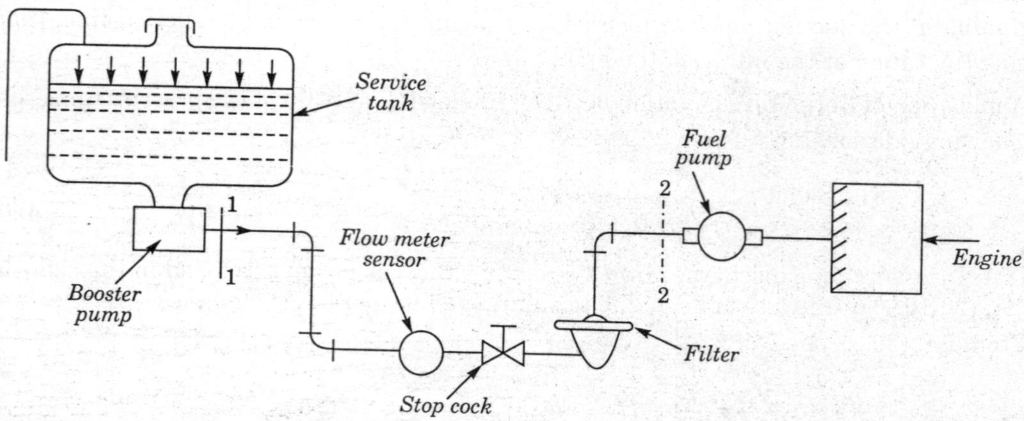

Fig. Ex. 20.7

intake p_2 *= 1.3 kg/cm*^{2}*. Local disturbance (i.e. fittings) as shown in the line sketch Fig. ex. 20.7. The physical properties of the fuel are : (i) kinematic viscosity of kerosene* ν *= 0.045 cm*^{2}*/sec, and its specific weight 820 kg/m*^{3}*. Neglect the height of liquid in the tank. Assume friction factor f = 0.0328. Take loss coefficients of the filter, stopcock, flowmeter sensor, and elbows as 2.0, 1.5, 7.0 and 1.2 respectively.*

Solution.

(*i*) Volumetric rate of flow $= \dfrac{\text{Weight/hr}}{\text{Sp. weight}} = \dfrac{1200}{820}\ \text{m}^3\text{/hr} = \dfrac{1200/3600}{820}\ \text{m}^3\text{/sec}$

$$= \frac{1}{3 \times 820} \times (100)^3\ \text{cm}^3\text{/s} = \frac{10^6}{2460}\ \text{cm}^3\text{/s}$$

$$= 406.5\ \text{cm}^3\text{/sec} = 0.4065\ \text{litre/sec}$$

Velocity of flow in the pipeline, $V = \dfrac{4Q}{\pi d^2} = \dfrac{4 \times 406.5}{\pi (1.5)^2} = 240$ cm/sec

(*ii*) Reyolds number of flow, $\mathbf{R} = \dfrac{Vd}{\nu} = \dfrac{240 \times 1.5}{0.045} = 8000$

The liquid flow in the pipeline is turbulent since **R** > 2000.

The pipe material being duralumin, the pipe interior surface is smooth, and for turbulent flow in a pipeline without incrustation, the Moody's diagram gives a value of 0.032 corresponding to **R** = 8000. Alternatively, any explicit equation like Blasius or the author's (Jain's) own explicit formula, Eq. (9.64 *b*) can be used to obtain the value of the friction factor. However, using the given value of $f = 0.0328$, and the loss coefficients $(K_L)_{\text{filter}} = 2.0$, $(K_L)_{\text{stop cock}} = 1.5$, $(K_L)_{\text{flowmeter sensor}} = 7.0$ and $(K_L)_{elbow} = 1.2$, one can write, noting that there are three elbows used in the piping :

(*iii*) pressure drop in the pipeline from booster pump to fuel pump :

$$\Sigma p_{drop} = \gamma_k \left(f\frac{l}{d} + 3K_{lelbow} + K_{Lsensor} + K_{Lcock} + K_{Lfilter} \right) \frac{V^2}{2g}$$

$$= 820 \left(0.0328 \times \frac{500}{1.5} + 3 \times 1.2 + 7.0 + 1.5 + 2.0 \right) \times \frac{(240)^2}{2 \times 9.81}$$

$$= 0.6\ \text{kg/m}^2$$

(*iv*) Hence the pressure required at the booster pump outlet :

$$p_1 = p_2 + \Sigma p_{drop} = 1.3 + 0.6 = \mathbf{1.9\ kg/m^2}$$

Example 20.8. *Hydraulic fuel (blended Kerosene) is delivered from two underwing tanks to the service tank, as shown in Fig. Ex. 20.8 by a pipeline through higher pressure in the former tanks, the pressure being higher by 0.2 kg/cm*^{2}*. Determine the diameter of the piping if the two under-wing tanks are to be emptied simultaneously and the total rate of discharge of the fuel is 1500 kg/hr. The volume of each under-wing tank is 450 litre. The piping is of duralumin of length 7 m. The kinematic viscosity of kerosene* ν *= 0.045 cm*^{2}*/sec and its specific weight is 830 kg / m*^{3}*. Neglect the height of liquid columns in the under-wing tanks.*

The loss coefficients the local features be taken as

$$(K_L)_{entrance} = 0.5,\ (K_L)_{valve} = 2.0,\ (K_L)_{elbow} = 1.2,\ (K_L)_{tee} = 3.5,\ (K_L)_{outlet} = 1.0.$$

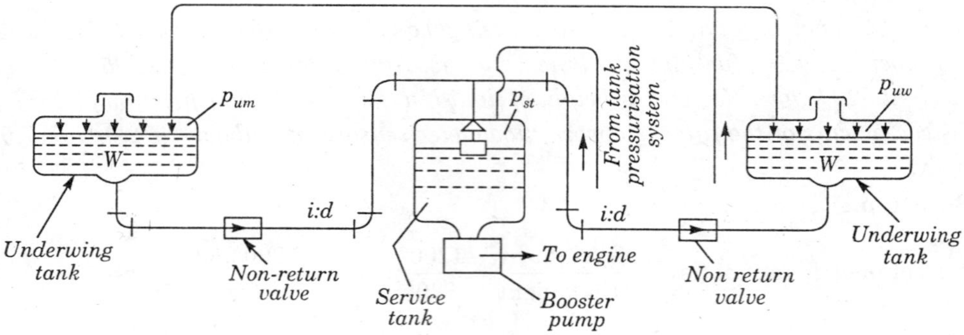

Fig. Ex. 20.8.

The friction factor can be obtained using Jain's equation for turbulent flow in smooth pipe i.e. $\frac{1}{\sqrt{f}} = 1.8 \log_{10} \mathbf{R} - 1.5186.$

Solution. Time of engine operation until under-wing tanks are emptied :

$$T = \frac{2 \times \text{volume of } u - w \text{ tank} \times \text{specific weight}}{\text{weight rate of discharge}}$$

$$= \frac{2 \times 450 \text{ (lit.)} \times 0.830 \text{ (kg/lit.)}}{1500 \text{ (kg/hr)}} = 0.5 \text{ hr.}$$

Rate of discharge from each tank

$$Q = \frac{\text{Vol.}}{\text{Time}} = \frac{450}{0.5} = 900 \text{ lit/hr} = 0.25 \text{ lit/sec}$$

Head available on account of higher pressure, $H_{avail} = \frac{\Delta p}{\gamma} = \frac{0.2 \times (100)^2}{830} = 2.41$ m (in the under-wing tanks)

Head required to overcome various types of resistances due to local features and pipe friction, noting that there are three elbows used on each side :

$$H_{req.} = \left[(K_L)_{entrance} + (K_L)_{valve} + 3(K_L)_{elbow} + (K_L)_{tee} + (K_L)_{outlet} + f\frac{l}{d}\right]\frac{V^2}{2g}$$

$$= \left[0.5 + 2.0 + 3 \times 1.2 + 3.5 + 1.0 + f\frac{7.0}{d}\right]\frac{16Q^2}{2g\,\pi^2 d^4}$$

This equation needs to be solved by trial method, assigning different values to the pipe diameter and finding the Reynolds number and thereafter the friction factor. Let us assume d = 12 mm ; 14 mm ; 16 mm ; and 18 mm.

(*i*) Velocity of flow in the pipe of diameter 12 mm

$$V = \frac{4Q}{\pi d^2} = \frac{4 \times 0.25 \times 1000 \text{ (cm}^3\text{/s)}}{\pi (1.2)^2 \text{ (cm}^2\text{)}} = 221.05 \text{ cm/s}$$

(*ii*) Reynolds number of flow

$$\mathbf{R} = \frac{Vd}{\nu} = \frac{221.05 \times 1.2}{0.045} = 5495$$

(*iii*) Using the given formula for the friction factor

$$\frac{1}{\sqrt{f}} = 1.8 \log_{10} 5495 - 1.5186 = 1.8 \times 3.73996 - 1.5186$$

$$\therefore \quad f = 0.0368$$

(*iv*) Substituting the values of f, Q and d, the head required

$$H_{req} = \left[10.6 + 0.0368 \times \frac{7}{1.2 \times 10^{-2}}\right] \frac{16 \times (0.25 \times 10^{-3})^2}{2 \times 9.81 \times \pi^2 (1.2 \times 10^{-2})^4}$$

$$= [10.6 + 21.47]\ \frac{16 \times 0.0625 \times 10^{-6}}{19.62\pi^2 \times 1.44 \times 10^{-8}} = 7.985 \text{ m}$$

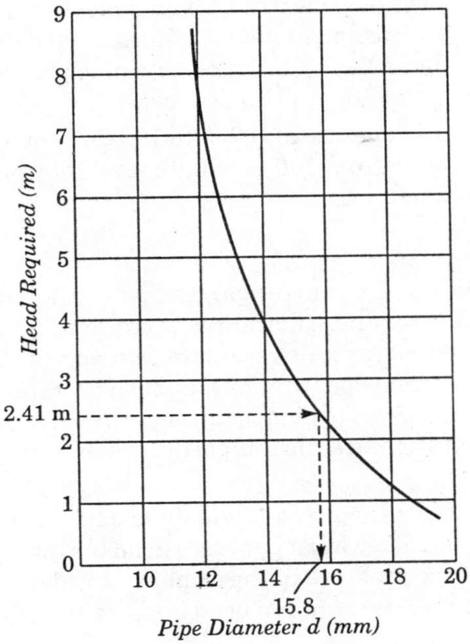

Fig. Ex. 20.8 (A)

For other assumed values of pipe diameter, various parameters as calculated in steps (*i*) to (*iv*) can be obtained, and a plot $H_{req.}$ and diameter d can be plotted, as shown in figure Ex. (20.8).

For the given pressure difference, Δp, the head available ($\Delta p/\gamma$) is 2.41 m. For this available head, the required pipe diameter can be estimated as 15.8 mm.

Hence the commercial pipe diameter to ensure the required flowrate for simultaneous delivery of fuel from under-wing tanks is 16 mm.

20.12. Selection of Pump Type for Aeronautical & Rocket Engineering

As a rule, the basic criteria governing the selection of pump design are the discharge rate Q, Head H and speed n. These quantities enable determination of the specific speed n_s, suggesting the type of rotodynamic pump to be preferred for given conditions.

(i) When specific speed is very high :

If the calculations give a very high specific speed-more than 1200-then it means that two or more pumps joined in parallel should be used.

(ii) When specific speed is too low :

On the other hand, if the specific speed turns out to be too low, and the speed of rotation cannot be increased, a multi-stage pump must be used with N, number of impellers mounted on a common shaft. The head developed by each impeller in this case is less by a factor N, and the specific speed greater by a factor of $N^{3/4}$, than a single-stage pump.

If such a design is undesirable, displacement pumps are used. In this case, however, such fluid properties as viscosity and chemical activity must be taken into account.

20.12.1. Aircraft with Gas-Turbine Engines

In aircrafts with gas-turbine engines, centrifugal pumps are used mainly as booster pumps in fuel systems. They are usually installed inside the fuel tanks or close to them and are meant to increase the pressure in the whole of the suction pipeline, thereby improving operating conditions for the main fuel pump and minimising the possibility of cavitation.

Booster Pump

When a booster pump is placed inside a tank, a propeller is commonly mounted at the impeller eye (on the same shaft) in order to improve its operating conditions and separate vapour and gas from the liquid. The shaft propeller must be designed for a much higher delivery than the centrifugal pump as some of the liquid delivered by it is hurled away.

Booster fuel pumps usually operate at 6,000 to 10,000 rpm and develop a pressure of 0.6 to 1.2 atm. Their specific speed varies from 100 to 200, that is, they belong to the class of normal- or even high speed centrifugal pumps.

20.12.2. Recent Trends

A recent tendency is to employ centrifugal pumps as the main fuel pumps of gas-turbine engines. The main reason is that very high discharge is required and less viscous and more volatile fuels are being used. Centrifugal pumps, furthermore, can operate at very high rotational speeds (several tens of thousands revolutions per minute). In such conditions, centrifugal pumps of much smaller size and weight than other types of pumps can ensure the required fuel delivery at sufficiently high pressures. These consideration are also important in selecting pumps for liquid-propellant rocket motors.

In rocket motors, centrifugal pumps are widely used to deliver fuel and oxidiser from the tanks to the combustion chamber. They must generate head (*i.e.* pressure) high enough to overcome the resistance of pipelines and the pressure in the combustion chamber, and to ensure the required pressure difference in the injectors. The required head is several hundreds of meters and the pressure tens of atmospheres.

Displacement pumps are used for such high pressures, but if large discharge is also required (as is the case in rocket motors) and there is a possibility of providing a high speed drive (for example, a gas turbine), centrifugal pumps are normally to be preferred. With a high impeller rotational speed and appropriate design, centrifugal pumps can deliver liquids at high pressures. The principal feature of centrifugal pumps used in liquid-propellant motors is low specific speed.

20.13. Hydraulic Transmissions

A hydraulic transmission may be defined as a system of energy transmission by means of a fluid. Any hydraulic transmission consists of two basic elements : (*i*) a pump to convert mechanical work into hydraulic energy, and (*ii*) a hydraulic motor of reciprocating or rotary type to reconvert fluid energy into mechanical work.

There are two main classifications of hydraulic transmissions :

1. Systems with intermittently operating (translational or reciprocating) motors, and
2. Systems with rotary motors.

Both types of systems are used in aviation engineering.

20.13.1. Translational Hydraulic Actuator

The working element of a translational hydraulic transmission is essentially a cylinder with movable piston and rod. A typical hydraulic actuator used in aircraft design is shown diagrammatically in Fig. 20.19. The dark lines denote the pipes along which oil flows during working operation, when the piston is moving up or down. The dashed lines represent the auxiliary piping.

Oil from a tank is pumped to high pressure and directed to a control valve which admits it to the desired side of the actuator cylinder (the lower end in Fig. 20.19). The oil pushes the piston up

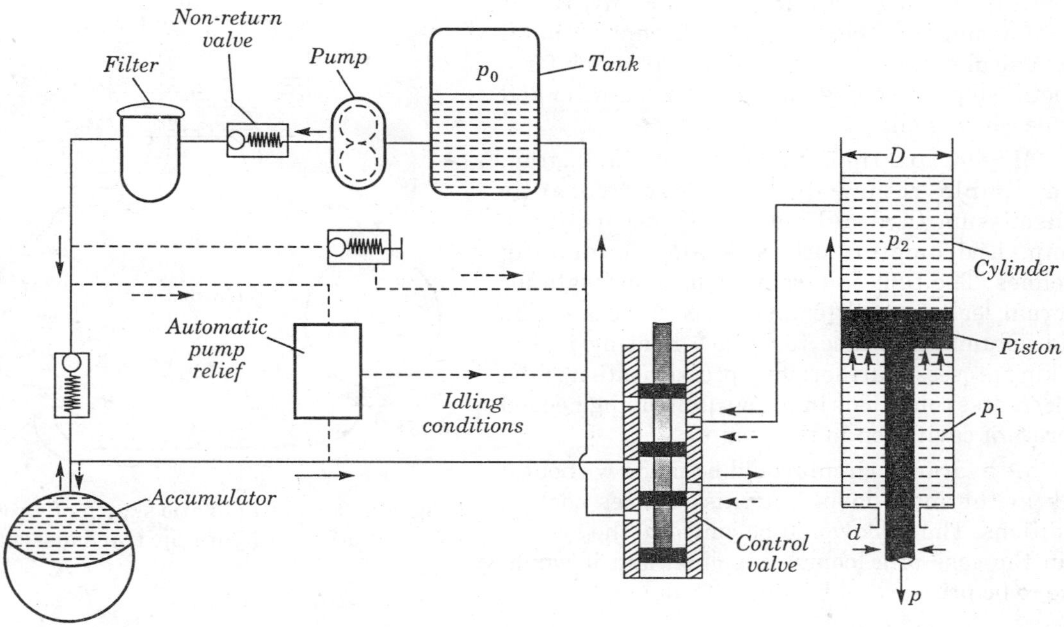

Fig. 20.19. Schematic sketch of a translational hydraulic actuator.

against load P *acting on the rod. Simultaneously, on the low pressure side oil is emptied from the cylinder and returned through the respective ports of the spool valve to the tank.*

Such hydraulic systems are widely used in aircraft for operating retractable under carriages, flaps, ailerons, etc. These systems are powered by gear, radial piston or swashplate pumps. Gear pumps are used with automatic relief valves which by-pass the hydraulic fluid when peak pressure is reached. Swashplate pumps are usually of the variable-displacement type which automatically adjust the delivery to maintain the required pressure in the hydraulic system.

An accumulator is commonly incorporated in the hydraulic system to cater for short-period peak demands above maximum pump capacity. This enables a relatively low-capacity pump to be used. Furthermore, in systems with automatic bypass an accumulator compensates for leakage from the high-pressure to the low-pressure side when the pump is idling.

In most of the aircraft systems, a blended hydraulic fluid consisting of heavy kerosene to which a special thickening fluid is added to increase viscosity. At low temperatures, the flow of such a fluid through pipes is laminar. However, the relatively low viscosity of the fluid sometimes results in turbulence developing in a hydraulic system. Thus, in an under carriage retracting system, if it is cold the flow regime in the low-pressure piping is laminar, while in the pressure line, where the fluid is warmer, the flow may be turbulent.

20.13.2. Rotary Hydraulic Transmissions

As per the accepted classification of pumps into rotodynamic and positive displacement machines, rotary hydraulic transmissions may also be classified into rotodynamic or hydrokinematic, and positive displacement or rotary positive-displacement transmissions.

Rotary displacement hydraulic transmissions represent a combination of a pump and a rotary displacement hydraulic motor. The motor is basically a pump working as an actuator. All rotary displacement pumps possess this quality of reversibility and can be used either as a pump or as a motor. This means that if hydraulic fluid is supplied to a rotary pump under sufficient pressure, its rotor will revolve and do work.

The main advantage of rotary hydraulic transmissions, as compared with conventional mechanical systems, is their ability to allow for a smooth, stepless change in transmission ratio and torque conversion.

Figure 20.20 represents a schematic diagram of a simple rotary displacement hydraulic transmission (hydraulic drive) incorporating a pump, hydromotor, oil tank and connecting pipelines. The tank is essential to compensate for external leakage and temperature changes of the fluid volume, and also to reduce heating in the working process. Furthermore, pressurisation of the tank creates conditions in the suction piping needed to prevent cavitation in the pump.

A hydraulic circuit could be made without a tank in the main loop, but this reduces cooling conditions. The tank would be joined to the system as in the case of a looping pipeline, and it would have to be pressurised by a booster pump.

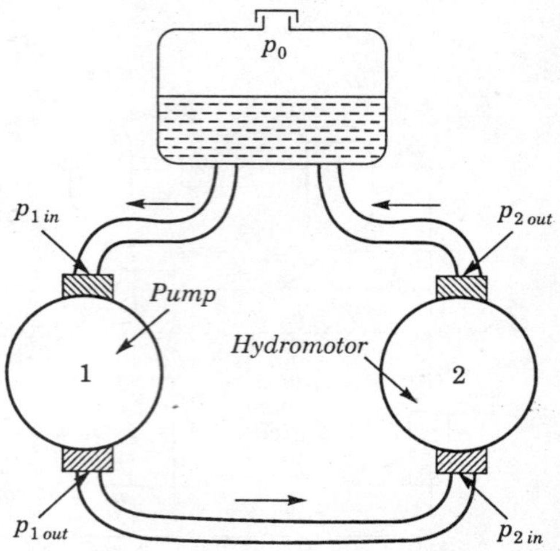

Fig. 20.20. Diagrammatic sketch of a rotary displacement hydraulic transmission.

21

Fluid Power and Pneumatics

21.0. Introduction

Fluid power is concerned with the use of pressurised fluids in a confined system to accomplish a given task. Liquids and gases are both used in fluid power systems. Hydraulic systems circulate liquids, usually petroleum oils, but often synthetic oils and water-base fluids are used for safety reasons. Pneumatic systems use air which is exhausted to the atmosphere after it has accomplished the assigned task.

The fluid used in hydraulic system exhibits the characteristics of a solid when confined, and provides a very rigid medium to transfer power through the system. Conversely, the air used in pneumatic systems exhibits spongy characteristics, and provision must be made in its control to provide smooth operation of actuators. Air power is, however, relatively less expensive to produce and use than hydraulic power.

Fluid power is used in variety of ways to accomplish two main objectives :

(1) It provides a substantial fluid force to move actuators in locations away from the power source where the connection between the two is obtained using one or two conductors. This means that a power source - a gasolene (petrol) or diesel engine-coupled to a hydraulic pump, can be housed in one area to power a small cylinder or rotary hydraulic motor hundreds of metres away. This is a significant advantage over systems using a mechanical drive train, particularly as the distance increases.

(2) Fluid power systems accomplish with relative ease highly accurate and precise movement of the actuator. This is particularly important in such applications as the machine tool industry where tolerances are often specified to one-hundredth of a mm to be repeatable over several million cycles.

21.1. Applications of Fluid Power

In the early twentieth century, the marine industry has applied fluid power to cargo handling and winch systems, controllable-pitch propellers, submarine control systems, shipboard aircraft elevators, aircraft and missile launch systems.

Hydraulics and pneumatics have almost unlimited applications in the production of goods and services in nearly all sectors of economy. Several industries are dependent on the capabilities that fluid power affords. Among these are agriculture, aerospace and aviation, construction, defence, manufacturing and machine tools, marine, materials handling, mining, transportation, undersea technology and public utilities like fire fighting.

Many developed countries including our own-India-have been in the forefront in applying hydraulics to solve variety of problems as in agriculture production for which many equipments and machines have been developed. Figure 21.1 illustrates wheel tractors that features extensive use of hydraulic power.

Fig. 21.1. Two views of a Wheel Tractor used in Agriculture operations.

It uses a three-pump system to power remote control valves, implement hitch, power take off control, power brake control, power steering, multiple disc clutch, and cooling and filtering transmission fluid.

Figure 21.2 illustrates a few of many applications of hydraulic power to aviation industry.

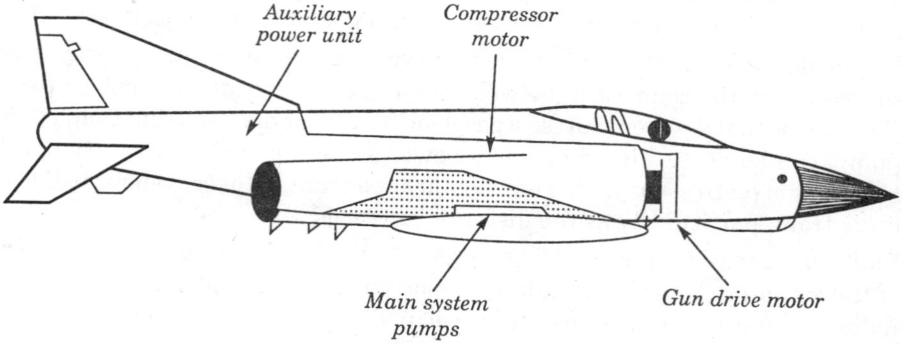

Fig. 21.2. Hydraulic Units in an Aircraft.

Modern construction industry has been greatly benefitted by the power of hydraulics and pneumatics. Crawler tractors, road graders, bucket loaders, trenchers, hydraulic shovels and pan scrapers are just a few of the many applications. Figure 21.3 shows a typical Loader and Hydraulic Pavement Breaker.

An innovation that uses pneumatics is the inflammable aircraft landing system. The elastic doughnut attached to the bottom of the fuselage inflates and allows air to escape through holes in the bottom to support the plane and reduce friction when landing on any reasonably smooth surface such as water and snow. During flight, it contracts to a sheltered aerodynamic fit against the aircrafts underside.

The manufacturing and machine-tool industry is greatly dependent on hydraulic power to provide power and close tolerance necessary in controlled production. A high-capacity forging press (say, 2000 ton) utilizes hydraulic fluid supplied by a number of pumps (as high as 20), each having a capacity of 135 litres per minute. At an operating pressure of 140 kg/cm^2, the press consumes about 600 kW. The press is driven by fluid supplied from central hydraulic pumping system consisting of ten double-end electric motors driving 20 hydraulic pumps.

Fig. 21.3. A Typical Loader and Hydraulic Pavement Breaker.

Material handling systems convey products from the point of production to the user end. Several systems are in use and nearly all of them depend on hydraulics and pneumatics in transport of goods. These include garbage trucks, conveyors, hydraulic levelling dock systems, hoists, and even industrial robots.

Transportation systems provide examples of most varied uses of fluid power. Road and off-the-road vehicles are typically suspended by pneumatic tyres. Air assisted brakes (both high pressure and vacuum assisted) are standard equipment. Power steering systems relieve the driver (or the operator) of the fatigue commonly associated with physical exertion over long period of time. Hydrostatic transmissions are very common on all types of vehicles. Suspension systems are dampened with hydraulic shock absorbers, and some combine with pneumatics by using the compressible nature of gases as the basis for air-oil suspension systems.

Fluid power has wide applications in continuous coal mining operations. Machines equipped with several hydraulic cylinders, jacks and motors dig and load coal at the rate of about two tons per minute. Conventional cutting and blasting are not required. Other mining applications utilizing fluid power pneumatics include rock drills, shuttle cars, roof bolting machines, and conventional hydraulic jacks.

Utilities and communications are two industries where use of fluid power is of vital importance. Utility vehicles are now a must to support personnel working above the ground. Similar vehicles are used in applications in developed countries throughout industry ranging from fruit tree maintence and picking, food processing to jumbo jet servicing. Almost unlimited movement capability is available to the operator with these systems.

21.2. Working Principle of Fluid Power

Fluid power works in accordance with the laws governing the behaviour of the fluid itself. The two most common fluids used are air in pneumatics and oil in hydraulic systems. The Pascal's law (year 1650) which states the pressure exerted by a confined fluid acts undiminished in all directions at right angles to the inside wall of the container. The Pascal's law can be extended to include transmission and multiplication of force, as illustrated in Fig. 21.4.

Fluid power works by applying a force against a movable surface. In a typical fluid power system, a fluid power compressor or pump delivers fluid through the control valve to system

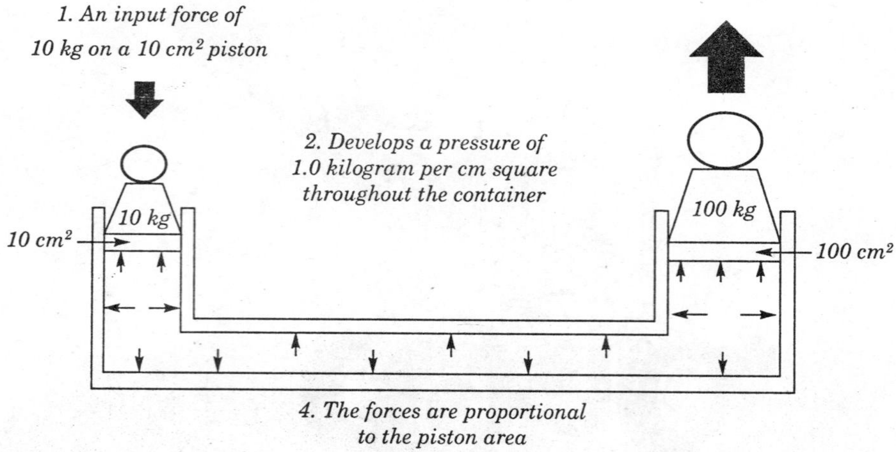

Fig. 21.4. Transmission and Multiplication of Fluid Power Force.

actuators, such as a cylinder, through tubes at high pressure. Unused fluid in hydraulic systems is returned to a reservoir at low pressure for cooling, storage and later use. Figure 21.5 shows a simulated hydraulic system used for lowering and raising the nose wheel of an aircraft landing gear. Movement of the control valve to lower the wheel causes fluid to be delivered at high pressure from the pump to the blank end of the cylinder. To raise the wheel, the control valve directs the fluid to the rod end of the cylinder. In both cases, fluid power works by applying a force against the area of the each side of the piston in the cylinder.

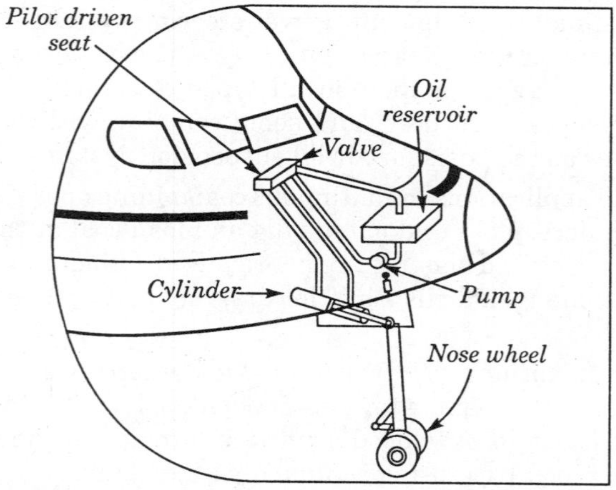

Fig. 21.5. Working of Nose-wheel of Aircraft Landing Gear.

21.3. Single-Acting Linear Systems

Many fluid power systems are single-acting. Simply speaking, they transfer power in one direction only. Simple application of such system is shown in Fig. 21.6, in which a single-acting piston pump transfers fluid to a single-acting hydraulic cylinder which acts against the load. In operation, the pump transfers fluid from the reservoir (a tank) to the hydraulic cylinder causing it to lengthen in space. When the cylinder is made to return to its earlier position, the fluid is returned directly to the reservoir. Simple hydraulic jacks, the cut-section of which is shown in Fig. 21.7,

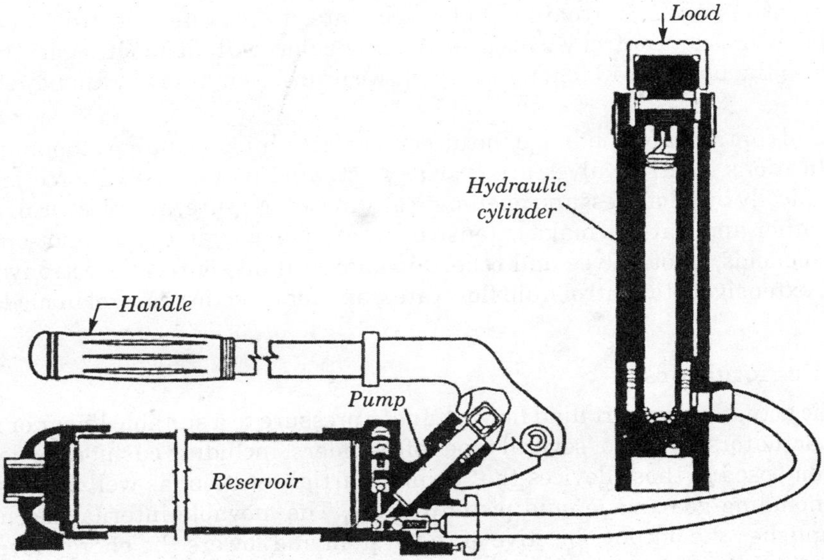

Fig. 21.6. A single-acting hand operated hydraulic system.

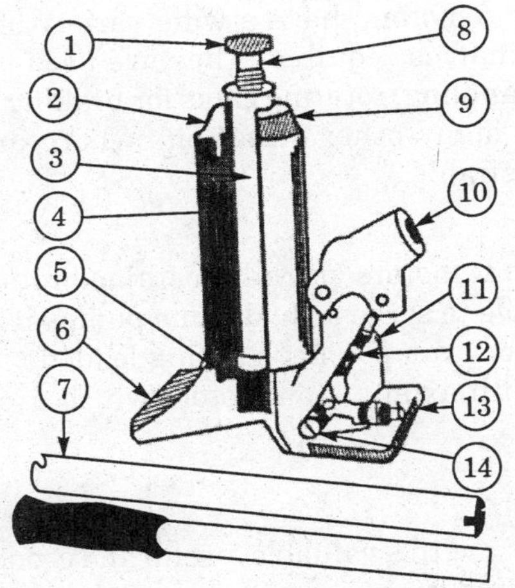

1. Screw plate
2. Ram seal
3. Ram
4. Ram guide
5. Reservoir
6. Jack base
7. Handle spacing
8. Screw adjustment
9. Gland nut
10. Handle holder
11. Pump seal
12. Pump piston
13. Bleed valve
14. Chock valve cover screw

Fig. 21.7. A single-acting hand operated jack.

hand pumps, car hoists, barber chairs, truck bed lifts, grease guns are only a few examples of applications using single-acting fluid power systems.

Other examples include a high force single-acting ram cylinder which has many applications for pull and push devices such as auto-body straighteners, gear pullers, and portable presses. Another example of a single-acting system is a single-acting cylinder used to manoeuver a concrete chute. This simple labour saving device raises, holds and lowers the chute at the command of the operator.

If the system is of the return type, as is the case with hydraulic jacks, some metal shears and worklifters, the energy required to return the fluid to the reservoir (tank) and the actuator (hydraulic cylinder) to its original position-is supplied by the load resistance.

21.4. Hydraulic Valves

Hydraulic valves are used to control the pressure, flowrate, and direction of fluid in the system in accordance with the basic principles of fluid flow. Pressure control valves protect the hydraulic system from damage resulting from peak pressures. They also ensure the safety of the operator. Flow control valves control the fluid output rate of the system, *i.e.* the rate at which the work is accomplished. Flow control valves also divide the flow of fluid on a priority or proportional basis to maximise the use of available flow to one, two or more circuits working alternately in the system.

Directional control valves manage the direction of the fluid, including stopping and starting. In mobile applications, control valves are typically actuated by a physical force provided by an operator hydraulically by pilot pressure, pneumatically by air pressure, and electrically by solenoids. Industrial and other applications make extensive use of solenoid valves and pilot-operated valves controlled by solenoids, pilot valves and other air and oil fluid sources. Electrohydraulic servo-valves are used extensively to control fluid flow rates, and for direction control in modern hydraulic systems.

21.5. Hydraulic Actuators

Hydraulic actuators convert fluid flowing under pressure to a suitable linear or rotary output motion. Linear actuators consist of several types of cylinders, including cushioning devices. Rotary actuators are composed of those devices that exhibit partial rotation, as well as those that rotate continuously, including gear, vane, and piston motors. The movable internal mechanism of the actuator accomplishes the output objective of the system and lowers the energy level of the fluid which is returned to the reservoir at low pressure.

21.5.1. Limited Rotation Actuators

Limited rotation actuators, also called torque motors, have a wide variety of applications where a specific degree of rotation at the output shaft is required. They are used extensively in industry for actuating clamping devices, material handling, rotating cams for braking mechanisms, tumbling and dumping, bending operations and many other situations where an economical application of fluid power for limited rotation is desirable.

21.5.2. Continuous Rotation Actuators

Continuous rotation actuators, called hydraulic motors, provide sustained rotation in either direction. Some hydraulic motors are also convertible to serve as hydraulic pumps if a mechanical drive is applied to the output shaft. Vane motors, for example, have spring loaded vanes whereas vane pumps usually do not. This is not the case with axial piston motors which are regarded as combination pump-motors.

21.6. Introduction to Pneumatics

Pneumatics provides various sectors of industry with a valuable, useful and economic medium of power for production and control. Unlike the relatively incompressible nature of hydraulic fluid, air used in pneumatics is compressible, and practically obeys the laws for a perfect gas. Pneumatics has a variety of uses in industry. Air at 5.6 – 8.8 kg/cm^2 is universally provided at construction sites and plants. In construction, it is an indispensable source of power for such tools as air drills, hammers, wrenches and even air-cushion supported structures, as also many vehicles using air suspension, braking and pneumatic tyres.

In manufacturing, air is used to power high speed clamping, drilling, grinding and assembly using pneumatic wrenches and riveting machines. Air is also used to power hoists and cushion supports to transport loads throughout the plant.

Operational characteristics make air a desirable medium for energy transfer. This because air can be inducted and exhausted directly into the atmosphere, a return line is not necessary as it is essential in hydraulic systems to return the fluid to the reservoir. Air is also safe in hazardous environments from ignition and burning where leaks occur around sparks or open flames.

As the air contains about 20% oxygen and is not sufficient alone to provide adequate lubrication to the moving parts and seals, oil is usually introduced into the air stream near the actuator to provide this lubrication preventing excessive wear and oxidation.

21.7. Gas Laws Applied to Pneumatic Systems

Air has mass (is compressible) and exerts a pressure on the surface of the earth. A barometer, consisting of an inverted glass tube closed at the top, will support a column of mercury (Hg) at exactly 760 mm at sea level when measuring at standard conditions at 0°C (32°F). Since mercury has a specific gravity of 13.6, this is equivalent of (760 × 13.6/1000) 10.336 metre of water or 1.0336 kg/cm^2 pressure, which is also equal to 1.0139 × 10^5 N/m^2. Pressures above the atmospheric one (*i.e.* 1.0336 kg/cm^2) are positive, whereas pressure below it cause a vacuum to be formed. Both positive pressures and vacuum pressures have useful purposes in pneumatics. Vacuum measurement is usually given in mm. of Hg and then converted into a holding force for such devices as suction pads and cylinders having specific diameters.

Pressure in gases is considered to result from molecular activity. As temperature decreases, molecular activity also decreases until at absolute zero – 273°C (zero degree Kelvin), it is thought to decrease altogether. For calculations involving gases, absolute temperature can be defined as

Absolute temperature °K = °C + 273

The compressible nature of air is governed by the laws for a perfect gas. Strictly speaking, air-which is composed of 80% nitrogen and 20% oxygen by volume is not a perfect gas, but the deviations are not sufficient to appreciably affect computed results. Boyle's law relates the absolute pressure (p) and volume (v) of a given quantity of gas held at constant temperature by the following relationship :

$$pv = \text{Constant} \qquad ...(21.1)$$

or

$$p_1 v_1 = p_2 v_2$$

which is the same as the pressure-density relationship expressed by Eq. (1.13).

Charle's law relates the absolute temperature (T) and volume (v) of a given quantity of gas held at constant pressure as follows :

$$\frac{T_1}{T_2} = \frac{v_1}{v_2}$$

or

$$T_1 v_2 = T_2 v_1 \qquad ...(21.2)$$

where T_2 and v_2 represent the desired quantities.

Gay Lussac's law relates the absolute pressure and absolute temperature of a given quantity of gas held at constant volume such that

$$\frac{p_1}{p_2} = \frac{T_1}{T_2}$$

or,

$$p_1 T_2 = p_2 T_1 \qquad ...(21.3)$$

Boyle's, Charle's and Gay Lussac's laws are combined in the general gas law and expressed as

$$\frac{p_1 v_1}{T_1} = \frac{p_2 v_2}{T_2} \qquad ...(21.4)$$

which may be solved for p_2, v_2 or T_2 if other quantities are known.

In all the above stated equations, v represents the volume of a given mass of gas and is called the specific volume. For a perfect gas, the equation of state is given by Eq. (1.12) reproduced below

$$pv = RT \qquad ...(1.12)$$

in which R represents gas constant, which has different valves for different gases. Its value for air is 287 Nm/kg K. Eq. (1.12) can also be written alternatively as

$$p = \rho RT \qquad ...(1.13)$$

For any gas, the gas constant (R) is the amount of work required to raise one kg mass by one degree Kelvin.

The specific heat at constant pressure (isothermal) and at constant volume (adiabatic) are used to determine the coefficients of expansion for Boyle's law.

For isothermal compression, the Boyle's law gives :

$$p_1 v_1 = p_2 v_2$$

or
$$\frac{p_1}{\rho_1} = \frac{p_2}{\rho_2} \qquad ...(21.1)$$

and for adiabatic compression, the Boyle's law becomes

$$(p_1 v_1)^{1.4} = (p_2 v_2)^{1.4} \qquad ...(21.5)$$

21.8. Compressors

Compressors may be classified as positive or non-positive displacement, single or multiple-stage, reciprocating or rotary, by delivery, and by pressure level or by rated power. Some of the these are Y-type two-stage water cooled industrial compressors rated at 150 kW or so. Also available are screw-type compressors in sizes upto 370 kW and delivery upto 70 cubic metre per minute (70,000 liter per minute). Compressor units may be stationary for use at large plants and located with other similar equipment in a pumping room, or made mobile for transport to job-sites above ground or below ground level in mines.

Unit air compressors, sometimes called automotive compressors because of their wide spread use in service stations and garages, are available in variety of types and sizes from less than 30 lpm for the hobbyist to 700 lpm or more for the commercial users. A typical unit air compressor is shown in Fig. 21.8. They are characterised by their completeness and usually include the compressor, pressure controls, receiver, and all instrumentation for immediate hook-up and use.

Fig. 21.8. Typical Unit Air Compressor.

Positive displacement compressors include piston, sliding or rotary vane, and lobe impeller types. Most large stationary industrial compressors are of the reciprocating piston type that uses the piston in single or double-acting fashion to compress the air.

There is a practical limit to the pressure that can be developed in single-stage piston type air compressors. This is reached at about 10.5 kg/cm^2 (10.355 × 10^5 N/m^2 or 10.335 × 10^5 Pa) where the compression ratio (10.335 × 10^5/1.0133 × 10^5 = 10.2 : 1), compressing chamber size and heat of compression impede efficient pumping action.

Staging has the effect of increasing pumping efficiency by dividing the total pressure between two or more stages and reducing the heat associated with compression by inter-cooling between the stages. Multistage piston-type compressors pipe the exhaust of each succeeding cylinder into the next, thereby stepping up the pressure. Cylinder size decreases in the progression and interceding removes the heat of compression, thus increasing the air density and volumetric efficiency of the compressor. Typical pressure ranges with the degree of staging are indicated below :

Staging of Piston-type compressors	*Pressure range* (kg/cm^2)
Single-stage	5.5 – 10.5
Two-stage	7.0 – 17.5
Three-stage	28.0 – 175.0
Four-stage	140.0 – 350.0

21.9. Pneumatic Systems

Pneumatic systems consist of appropriately-sized compressors, receivers, and plumbing to supply clean, dry air in adequate volumetric rates at relatively constant pressure to single or manifold connections near work stations. It is important to have adequate awareness of the air consumption rate at actuators, pressure drop in transmission lines, air conditioning equipment, and guidelines of sizing the compressor itself.

As the pressure is relatively constant at the actuator, for example through the entire stroke of a pneumatic cylinder, air consumption can be computed in the same way as fluid consumption in hydraulic systems. Air requirements for various types of pneumatic tools are shown in the following table. Allowance is made for the compressible nature of air after consumption at the specified pressure is determined using Boyle's law for approximate values, or the general gas law if conversion to standard conditions is specified.

Table 21.1. Air Requirement for various pneumatic tools

Name of Tool	*Air Requirement (liter per minute, lpm)*
Rotary drills, 1.5 – 16 mm	115 – 200 lpm
Rotary drills, 6.5 mm	500 – 550 lmp
Rotary drills, 9.5 mm	550 – 1100 lpm
Rotary drills, 12.5 – 19.5 mm	2000 lpm
Rotary drills, 22 – 25 mm	2250 lpm
Piston drills, 12.5 – 30 mm	1275 lpm
Piston drills, 22 – 25 mm	2125 – 2250 lpm
Piston drills, 38 – 50 mm	2500 lpm
Piston drills, 50 – 75 mm	2800 – 3100 lpm
Air hoist (cylinder)	30 – 120 lpm
Air hoist (motor)	55 – 120 lpm
Air hammer	450 lpm
1 kW Air motor	385 lpm
2 kW Air motor	580 lpm
3 kW Air motor	770 lpm

While plumbing air systems, it is necessary to minimise losses between the receiver located near the compressor and the point of use. Loss of 10% pressure is allowable under normal operating conditions with less than half the loss being attributed to the main transmission line.

21.10. Control Valves

Pneumatic control valves are used to regulate the pressure, air velocity and flow rate, and to manage the direction of the fluid. Pressure relief valves protect the compressor and components by setting the maximum system pressure within safe working limits. Pressure control valves, sometimes also called air line regulators, are installed at or near the pressure drop from the main transmission line and establish the working pressure of the branch circuit. Flow or speed control valves regulate the volume of air flow at a specified working pressure to the component and thus regulate the speed of operation. Directional control valves are used to stop, start, divert and otherwise manage the direction of fluid (air) in two or more ways to effect the desired direction of component movement.

21.11. Cylinders and Rotary Actuators

Linear and rotary actuators put air to work by a variety of methods. Some of these differ very little from hydraulic applications, but because air instead of hydraulic oil is the medium, lower pressures, lighter construction and higher cycle rates are used. Air is usually exhausted directly into the atmosphere through a muffler connected directly to rotary motors for single rotation, or to the control valve for dual rotation. Both single and double-acting cylinders often exhaust directly into the atmosphere through remote controlled valves placed at cylinder ports, thus eliminating further pressure drop through the return line to the main control valve. Lubrication for cylinder seals and bearing supports is normally introduced with the air stream.

21.12. Fluid Conditioners

Fluid conditioners function to make air a more acceptable medium to the system and operating personnel. They consist of aftercoolers, moisture-separators, air dryers, air receivers, filters, regulators, lubricators and mufflers. Improperly conditioned fluid, for example, air used in a system containing excessive moisture and contaminants, can cause oxidation of plumbing and component seizing resulting in erratic operation, failure, downtime, and costly repairs. Other environmental hazards affecting the human beings are numerous. For example, airborne particles and high exhaust noise level associated with exhausting directly into the atmosphere can endanger operating personnel with possible impairment to vision and hearing.

Air from the atmosphere contains moisture to varying degrees depending on atmospheric conditions. On humid days, more moisture is present. The amount of moisture present in air is measured in terms of the relative humidity, which is the ratio of the amount of water actually present in the air to the amount of water the air contains when it is saturated. The point of saturation is called the dew point, and is defined as that temperature at which moisture begins to condense out of the air. The lower the dew point, the lower is the temperature necessary before free moisture will condense and separate out of the air stream, causing damage to transmission lines, hoses, and system components.

When atmospheric air is compressed, say to 7.0 kg/cm^2, about 8 times the amount of moisture is pumped through the compressor and into the system. This can be understood as under :

1 atmospheric pressure $= 1.034$ kg/cm^2

The amount of moisture when air is compressed to 7.0 kg/cm^2

$$= \frac{7.0 + 1.034}{1.034} = 7.77 \approx 8$$

The amount of water present in saturated air at various temperatures is dependent on the pressure air is subjected to. For example, air at temperature of 38°C subjected to 7.0 kg/cm^2 pressure will contain water weighing 0.875 kg per 100 cubic metre of saturated free air. At the same temperature, however, 1.293 kg of moisture will be present in the saturated free air at zero gauge pressure.

21.12.1. Aftercoolers

Aftercoolers are placed immediately downstream of the compressor and function to remove most of the moisture, usually about 85% drawn in and pumped through the compressor. They are a necessity for plant air systems handling large air volumes. The shell and tube-type water coolers are widely used because of low operating cost and low maintenance cost as well. Automatic moisture traps are sometimes located between the aftercooler and the receiver or in transmission lines near drops to separate and discharge this moisture from the system before it enters the receiver.

If dry air is required by a pneumatic system, that is, air that does not release moisture at the temperature and pressure encountered by the system, dryers are installed downstream of the aftercooler and moisture trap. To remove moisture, the air may be refrigerated, heated in a furnace to remove water vapours.

21.12.2. Receivers

Receivers are designed to consider system output pressure and flowrate requirements, air input capability from the compressor, and duty of operation, that is, time related parameters. An over load factor and future expansion of production facilities also should be incorporated in the design to justify the initial investment in this unit as it is a one-time-buy item. To properly size a receiver, its function must be understood. The receiver is an air reservoir. It dampens pressure pulses from the compressor at the inlet and supplies air at substantially constant and steady pressure at the outlet. Pulsation generated in the outlet line from valve shifting and component operation is transmitted back to the receiver and damped. During times of excessive demand, the receiver supplies an output flow in excess of the compressor input delivery capability. Receivers with appreciable volumes also reduce the frequency of compressor operation, thus reducing operating costs and wear associated with excessive starting and stopping of the unit. Moisture that accumulates in the compressor is condensed and drained off through the petcock or automatic drain provided at the bottom of the receiver.

21.12.3. Airfilters

Airfilters are located in the compressor intake, in the pressure line before components, and immediately preceding power tools. They protect against catastrophic failure from large particles as well as gradual failure from micron particles that promote wear by lapping action using dirt and other particulate matter as an abrasive agent. Due care must be exercised so that the filter does not become clogged, causing higher than normal suction at the inlet. This would starve the compressor and result in reduced volumetric and pumping efficiency.

21.12.4. Mufflers

Mufflers are mounted to the exhaust ports of air valves, cylinders, and air motors to reduce noise and eliminate the possibility that the air stream or airborne particles will cause injury to personnel. To prevent the necessity of requiring employees to wear ear drum protective devices or operation of pneumatic systems from acoustical conditioned chambers, noise should be engineered out at the source. Mufflers are designed to reduce noise without creating back pressure sufficient to reduce the operating efficiency of the system.

Appendix A

Density and Specific Weight of Common Liquids at Standard Atomospheric Pressure

Liquid	*Temperature* °C	*Density* ρ kg/m^3	*Specific weight* γ kN/m^3
Mercury	0	13.6×10^3	133.4
	10	13.6 } $\times 10^3$	133.4
	15	13.6	133.4
	20	13.6	133.4
	25	13.5	132.4
	40	13.5	132.4
	100	13.37×10^3	131.16
Medium fuel oil	15	0.856×10^3	8.40
Glycerine	20	1.258×10^3	12.34
Benzene	20	0.878×10^3	8.60
Gasoline (Petrol)	15	0.722×10^3	7.08
Lubricating oil	20	0.897×10^3	8.80

Appendix B

Absolute and Kinematic Viscosities of Water at Standard Atomospheric Pressure

Temperature °C	μ, Ns/m^2 (*Pas*)	ν, m^2/s
0	1.79×10^{-3}	17.95×10^{-7}
5	1.52 } $\times 10^{-3}$	15.2 } $\times 10^{-7}$
10	1.31	13.1
15	1.14	11.4
20	0.995	9.76
25	0.879	8.85
30	0.788	7.96
35	0.716	7.24
45	0.597	6.07
65	0.435	4.45
85	0.336	3.48
100	0.283	2.97

Appendix C

Density and Specific Weight of Air (at atmospheric pressure of 760 mm of Hg)

Temperature, °C	ρ, *kg/m*^{3}	γ, *N/m*^{3}
– 15	1.37	13.46
– 10	1.34	13.18
– 5	1.31	12.90
0	1.29	12.70
5	1.27	12.49
10	1.24	12.21
15	1.22	11.93
20	1.20	11.73
25	1.17	11.45
30	1.15	11.25
35	1.14	11.15
40	1.13	11.15
50	1.10	10.77
60	1.06	10.40
90	0.96	9.43
150	0.84	8.24
200	0.74	7.31
260	0.67	6.54

Appendix D

Absolute and Kinematic Viscosities of Air at atmospheric Pressure of 760 mm of Hg

Temperature, °C	μ, *Ns/m*^{2} *(Pas)*	ν *m*^{2}*/s*	
– 15	16.05×10^{-6}	11.97×10^{-6}	
– 10	16.58	12.4	
– 5	16.84	12.85	
0	17.16	13.3	
5	17.31	13.68	$\times 10^{-6}$
10	17.56	14.13	
15	17.85	14.7	
20	18.15	15.15	
25	18.34	15.6	
30	18.59	16.0	$\times 10^{-6}$
40	19.01	17.58	
50	19.42	17.7	
60	19.81	18.7	
70	20.06	19.6	
80	20.65	20.7	
90	21.19	21.9	$\times 10^{-6}$
120	23.15	26.0	
150	23.79	28.5	
200	25.70	33.0	
260	27.96	42.4	

Note. See Appendix F for conversion factors.

Appendix E

CONVERSION FACTORS

Important. Do not use fullstop, dot, dash or plurals while writing units, For example
(*i*) Write Nm and not N.m, N-m
(*ii*) Write 65 Nm and not 65 Nms.

(A) MKS to SI units :

1 metric slug = 9.81 kg (mass) ;
1 kgm = 9.81 J ;
1 metric h.p. = 735 W.

1 poise = 0.1 Ns/m^2 = 0.1 Pa s
1 kg/cm^2 = 9.81 N/m^2
= 98.1 kN/m^2
= 98.1 kPa
= 0.98 bar

1 kg = 9.81 N
1 kg m/s = 9.81 W
1 Nm/s = 1 W
1 hp = 0.735 kW
1 British hp = 550 lb-ft/s
= 0.745 kW
= 1.014 metric hp

(B) MKS to FPS units :

1 metric slug = 0.672 slug
1 m = 3.28 ft ;
1 m^3 = 35.3 ft^3 ;
1 kg = 2.205 lb ;
1 lb/in^2 = 702 kg/m^2 ;
1 kg/m^2 = 62.5 × 10^{-3} lb/ft^3 ;
1 metric slug/m^3 = 0.0191 slug/ft^3 ;
1 m^3/s = 35.3 ft^3/s ;
1 kg m/s = 7.23 lb ft/s ;
1 kcal/m^2 = 0.3687 BTU/ft^2 ;
1 calorie = 1 gm (mass) × 1 deg C ;

1 m^2 = 10.76 ft^2 ;
1 kg/m^2 = 0.205 lb/ft^2 ;
1 lb/in^2 = 6890 N/m^2 ;
1 kg m = 7.23 lb ft
1 kg m = 9.29 × 10^{-3} BTU
1 kcal = 427 kg-m
= 427 × 9.81 Nm or joule = 4187 joule = 4.187 kJ

1 lb = 0.454 kg ;
1 litre = 0.22 British gallon ;
1 litre = 0.26 US gallon
1 gallon (UK) = 4.55 litre
1 gallon (US) = 3.78 litre

(C) MKS to CGS units :

1 metric slug = 9810 gm ; 1 m = 100 cm ;
1 m^3 = 1000 litre 1 kg = 9.81 × 10^5 dyne
1 kg/m^2 = 98.1 × dyne/cm^2 ;
1 kg/m^3 = 0.981 × dyne/cm^3 ;
1 kg-m = 9.81 × 10^{-7} dyne cm = 9.81 × 10^{-7} erg.

(D) Some important conversion factors :

1 Pa = 1 pascal = 1 N/m^2 ;
1 bar = 10^5 N/m^2 = 10^5 Pa ;
1 atm = 101330 N/m^2
= 101.33 kN/m^2 = 101.33 kPa ;
10^3 N = 1 kN ;
10^6 N = 1 MN ;
10^9 N = 1 GN ;
1 psi = 0.0703 kg/cm^2 ;
1 atm = 14.7 pound per sq. inch (psi)

1 inch = 2.54 cm
1 foot = 0.3048 m
1 mile = 1.6093 km
1 gallon = 4.546 × 10^{-3} m^3
1 pint = 0.5683 × 10^{-3} m^3
1 litre = 1.0 × 10^{-3} m^3
1 Joule = 1 Nm
1 Watt = 1 Nm/s = 1 J/s
1 atm = 760 mm of Hg at 0°C

Note. Pascal (Pa) and bar are used as units for pressure.
°C is the temperature in Celcius
°F is the temperature in Fahranheit

°C × 9/5 = °F – 32

4 qt = 1 gal ; 550 ft lb/s = 1 h.p. ; 1 pound = 0.4536 kg
4.448 N = 1 lb ; 778 ft lb = 1 Btu ; 1 pound/sq in = 6.895 kPa
7.48 gal = 1 ft^3 ; 200 lb = 1 ton ; 1 gal/min = 7 × 10^{-3} m^3/s = 4.55 lit/min
14.59 kg = 1 slug ; 1 cusec = 28.3 × 10^{-3} m^3/s = 28.3 lit/sec

Appendix F

Properties of Plane Geometrical Areas

Shape description	*Figure*	*Area*	*Location of centroid, hc*	*I or I_{gg} (About the dotted line)*
Rectangle	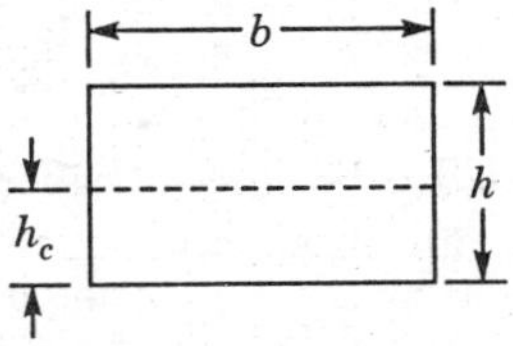	bh	$\frac{h}{2}$	$I_{gg} = \frac{bh^3}{12}$
Trapezium	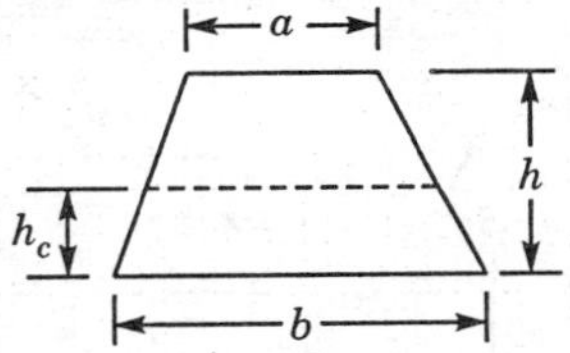	$\left(\frac{a+b}{2}\right)h$	$\frac{(2a+b)h}{3(a+b)}$	$I_{gg} = \frac{h^3}{24}(3b-a)$
Triangle	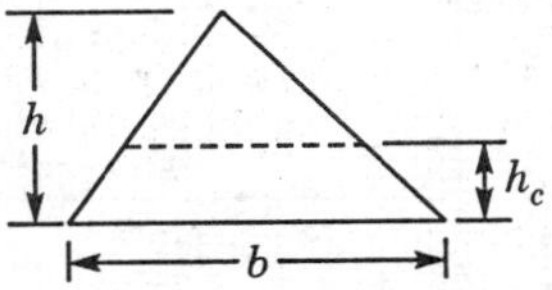	$\frac{bh}{2}$	$\frac{h}{3}$	$I_{gg} = \frac{bh^3}{36}$
Circle	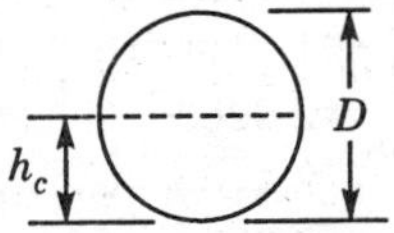	$\frac{\pi D^2}{4}$	$\frac{D}{2}$	$I_{gg} = \frac{\pi D^4}{64}$
Semicircle	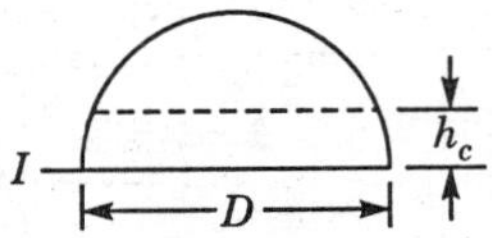	$\frac{\pi D^2}{8}$	$\left(\frac{4D/2}{3\pi}\right) = \frac{2D}{3\pi}$	$I = \frac{\pi D^4}{128}$
Parabola	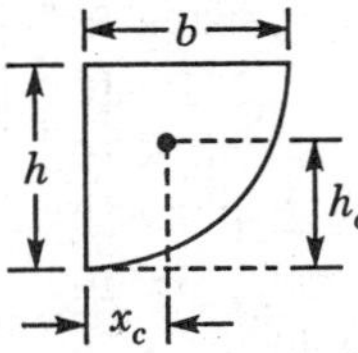	$\frac{2bh}{3}$	$x_c = \frac{3b}{8}$, $h_c = \frac{3h}{5}$	$I = \frac{2bh^3}{7}$

Cont...

Shape description	*Figure*	*Area*	*Location of centroid, hc*	*I or I_{gg} (About the dotted line)*
Ellipse	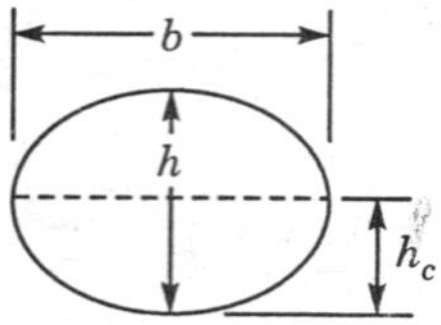	$\frac{\pi bh}{4}$	$\frac{h}{2}$	$I_{gg} = \frac{ibh^3}{64}$
Semiellipse	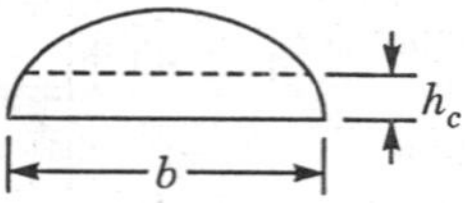	$\frac{\pi bh}{4}$	$\frac{4h}{3\pi}$	$I_{gg} = \frac{\pi bh^3}{16}$

The relation between I and I_{gg} is $I = I_{gg} + Ah^2$ where h is the distance of centroid from the axis about which I is being determined. I_{gg} is moment of inertia about an axis passing through centroid.

Appendix G

INTERNATIONAL STANDARD ATMOSPHERE

Absolute above sea level (m)	*Absolute pressure (kN/m^2)*	*Absolute temperature (K)*	*Velocity of sound (m/s)*	*Mass density, ρ (kg/m^3)*	*Kinematic viscosity, ν (m^2/s)*	*Dynamic viscosity, μ ($N\ s/m^2$)*
0	101.325	288.15	340.3	1.2250	1.461×10^{-5}	17.90×10^{-6}
1000	89.88	281.7	336.4	1.1117	1.581×10^{-5}	17.58×10^{-6}
2000	79.50	275.2	332.5	1.0066	1.715×10^{-5}	17.26×10^{-6}
4000	61.66	262.2	324.6	0.8194	2.028×10^{-5}	16.26×10^{-6}
6000	47.22	249.2	316.5	0.6602	2.416×10^{-5}	15.95×10^{-6}
8000	35.65	236.2	308.1	0.5258	2.904×10^{-5}	15.27×10^{-6}
10 000	26.50	223.3	299.5	0.4134	3.525×10^{-5}	14.57×10^{-6}
11 500	20.98	216.7	295.1	0.3375	4.213×10^{-5}	14.22×10^{-6}
14 000	14.17	216.7	295.1	0.2279	6.239×10^{-5}	14.22×10^{-6}
16 000	10.35	216.7	295.1	0.1665	8.540×10^{-5}	14.22×10^{-6}
18 000	7.565	216.7	295.1	0.1216	11.686×10^{-5}	14.22×10^{-6}
20 000	5.529	216.7	295.1	0.088 92	15.989×10^{-5}	14.22×10^{-6}
22 000	4.047	218.6	296.4	0.064 51	22.201×10^{-5}	14.32×10^{-6}
24 000	2.972	220.6	297.7	0.046 94	30.743×10^{-5}	14.43×10^{-6}
26 000	2.188	222.5	299.1	0.034 26	42.439×10^{-5}	14.54×10^{-6}
28 000	1.616	224.5	300.4	0.025 08	58.405×10^{-5}	14.65×10^{-6}
30 000	1.197	226.5	301.7	0.018 48	80.134×10^{-5}	14.75×10^{-6}
32 000	0.889	228.5	303.0	0.013 56	109.620×10^{-5}	14.86×10^{-6}

STP CONDITIONS: **Altitude** 0 **Pressure** 101.325 kN/m^2 **Temperature** 288.15 K **Density** 1.225 kg/m^3

Appendix – H

Basic Physical Properties of Some Liquids Used in Aircraft and Rocket Systems

Liquid	*Specific gravity,* γ	*Kinematic viscosity* ν *at temperature T, centistokes*						*Vapour pressure p*ν*, mm Hg*			*Volume modulus K, kg/cm²*	
		+ 70°C	+ 50°C	+ 20°C	+ 0°C	– 20°C	– 50°C	+ 60°C	+ 40°C	+ 20°C		
Aviation gasoline	0.750	—	0.54	0.73	0.93	1.26	2.60	—	195	90	13,300	
Kerosene (Heavy)	0.800 0.850	1.2	1.5	2.5	4.0	8.0	25	59	27	11.5	13,000	
Kerosene (Lighter)	0.755	—	—	1.05	2.0	—	5.5	—	100	—	13,000	
Lubricating oil (Heavy)	0.895	65	155	1,100	10,100	congeals						
Lubricating oil (Lighter)	0.895	—	8.3	30	—	498	—					
Hydraulic fluid (Heavy Kerosene mixed with (Vinipol)	0.850	7.5	10	16	42	130	1.250				13,300	
Nitric acid (98%)	1.510	0.30	0.38	0.58	0.70	0.83	1.72	355	156	60		
Ethyl alcohol	0.790	—	—	1.52	—	—	6.5	352	135	44		
Hydrogen peroxide (80%)	1.34	—	—	0.95	1.42	—	—	61.4	56.7	46.4		
Liquid oxygen	1.15	T	– 175°C	– 184°C	– 190°C	– 200°C	– 204°C	t	– 140°C	– 160°C	– 190°C	– 200°C
	1.25	ν, cst	0.125	0.170	0.193	0.257	0.297	p_v, atm	21	7	0.45	0.11

Index

A

B

C

I

J

K

L

M

N

O

P

T

U

V